Chemistry of the Upper and Lower Atmosphere

Chemistry of the Upper and Lower Atmosphere

Theory, Experiments, and Applications

Barbara J. Finlayson-Pitts
Department of Chemistry
School of Physical Sciences
University of California, Irvine
Irvine, California

James N. Pitts, Jr.
Department of Chemistry
School of Physical Sciences
University of California, Irvine
Irvine, California

ACADEMIC PRESS
An Imprint of Elsevier

San Diego   San Francisco   New York   Boston   London   Sydney   Tokyo

Front cover photograph: Polar stratospheric clouds in the upper atmosphere seen from the NASA DC-8 aircraft at nearly 39,000 feet, north of Stavanger, Norway; the authors are grateful to NASA and O. B. Toon for providing this photograph. Photograph of breaking waves at "the Wedge," Newport Beach, California, and of multiple light beam reflections in aerosol chamber by J. N. Pitts, Jr.

Back cover photograph: Image © 1999 Photodisc, Inc.

Dedication quote (facing page): AS TIME GOES BY, by Herman Hupfeld, © 1931 (Renewed) Warner Bros. Inc. All rights reserved. Used by permission. WARNER BROS. PUBLICATIONS U.S. INC., Miami, FL 33014. All rights relating to the interest of Herman Hupfeld in Canada and the reversionary territories are controlled by BIENSTOCK PUBLISHING COMPANY on behalf of REDWOOD MUSIC LTD. Used by permission. All rights reserved.

Permissions may be sought directly from Elsevier's Science and Technology Rights Department in Oxford, UK. Phone: (44) 1865 843830, Fax: (44) 1865 853333, e-mail: permissions@elsevier.co.uk. You may also complete your request on-line via the Elsevier homepage: http://www.elsevier.com by selecting "Customer Support" and then "Obtaining Permissions".

This book is printed on acid-free paper. ∞

Copyright © 2000 by ACADEMIC PRESS

All Rights Reserved.
No part of this publication may be reproduced or transmitted in any form or by any means, electronic or mechanical, including photocopy, recording, or any information storage and retrieval system, without permission in writing from the publisher.

Academic Press
An Imprint of Elsevier
525 B Street, Suite 1900, San Diego, California 92101-4495, USA
http://www.apnet.com

Academic Press
24-28 Oval Road, London NW1 7DX, UK
http://www.hbuk.co.uk/ap/

Library of Congress Catalog Card Number: 99-63218

International Standard Book Number: 0-12-257060-x

PRINTED IN THE UNITED STATES OF AMERICA
05 06 07 08 MM 9 8 7 6 5 4 3 2

*To our parents, Colin and Jean Finlayson and James and Esther Pitts,
who, across gaps of a continent and time,
imbued us with common interests in the environment,
values reflected in this book, and with the philosophy,*

*"The fundamental things apply
as time goes by."*

(From the song "As Time Goes By,"
Herman Hupfeld, 1931.)

Contents

Preface xvii
About the Authors xix
Acknowledgments xxi

1
Overview of the Chemistry of Polluted and Remote Atmospheres

A. REGIONS AND CHARACTERISTICS OF THE ATMOSPHERE 2

B. AIR POLLUTION AND THE CHEMISTRY OF OUR TROPOSPHERE 3

1. Historical Perspectives: Ancient and Medieval Times 3
2. "London" Smog: Sulfur Dioxide, Acidic Aerosols, and Soot 3
3. "Los Angeles" Smog: Ozone and Photochemical Oxidants 4
   a. Historical 4
   b. Photochemical Air Pollution 5
   c. Nighttime Chemistry of NO_2 7
4. Acid Deposition 8
   a. Historical 8
   b. Overview of Acidic Rain and Fogs 9

C. CHEMISTRY OF THE NATURAL TROPOSPHERE: REMOTE ATMOSPHERES 9

D. CHEMISTRY OF THE STRATOSPHERE 10

E. GLOBAL CLIMATE CHANGE 11

F. INDOOR AIR POLLUTION 13

G. DISCUSSION TOPIC AND OZIPR MODEL 13

1. Discussion Topic: "Background Ozone" 13
   Some Points for Discussion 13
2. OZIPR Model 13
   REFERENCES 13

2
The Atmospheric System

A. EMISSIONS 15
1. Oxides of Nitrogen 17
2. Volatile Organic Compounds (VOC) 18
3. Carbon Monoxide 20
4. Sulfur Compounds 20
5. Total Suspended Particles (TSP), PM10, and PM2.5 21
6. Lead 25

B. METEOROLOGY 26
1. Lapse Rate: Temperature and Altitude 26
2. Potential Temperature 28
3. Temperature Inversions 28

C. REMOVAL FROM THE ATMOSPHERE: WET AND DRY DEPOSITION 30

D. TYPICAL AMBIENT CONCENTRATIONS AND AIR QUALITY STANDARDS 33
1. Units of Concentrations and Conversions 33
   a. Parts per Million, Parts per Hundred Million, Parts per Billion, and Parts per Trillion 33
   b. Number per Cubic Centimeter 34
   c. Micrograms per Cubic Meter 34
2. Criteria and Noncriteria Pollutants and Air Quality Standards 35

E. EFFECTS ON VISIBILITY AND MATERIALS 37

F. ECONOMICS 38

G. ATMOSPHERIC CHEMISTRY: RISK ASSESSMENTS AND PUBLIC POLICIES FOR AIR POLLUTION CONTROL 38

H. PROBLEMS 39

REFERENCES 39

3
Spectroscopy and Photochemistry: Fundamentals

A. BASIC PRINCIPLES 43

1. Molecular Energy Levels and Absorption and Emission Spectroscopy 43
   a. Diatomic Molecules 43
   b. Polyatomic Molecules 49
2. Fates of Electronically Excited Molecules 50
   a. Photophysical Processes 50
   b. Photochemical Processes 51
   c. Quantum Yields 51

B. ABSORPTION OF LIGHT 52

1. Basic Relationships 52
2. The Beer–Lambert Law 53

C. ATMOSPHERIC PHOTOCHEMISTRY 55

1. Solar Radiation and Its Transmission through the Atmosphere 55
   a. The Sun and Its Relationship to the Earth: Some Important Definitions for Atmospheric Chemistry 55
   b. Solar Spectral Distribution and Intensity in the Troposphere 57
2. Calculating Photolysis Rates in the Atmosphere 61
   a. Photolysis Rate Constant $[k_p]$, Radiance $[L(\lambda)]$, Actinic Flux $[F(\lambda)]$, and Irradiance $[E(\lambda)]$ 61
   b. Estimates of the Actinic Flux, $F(\lambda)$, at the Earth's Surface 64
   c. Effects of Latitude, Season, and Time of Day on $F(\lambda)$ 65
   d. Effect of Surface Elevation on $F(\lambda)$ 65
   e. Effect of Height above Earth's Surface on $F(\lambda)$ 67
   f. Sensitivity of Calculated Actinic Fluxes to Input Values for Surface Albedo and Ozone and Particle Concentrations 67
   g. Effects of Clouds on $F(\lambda)$ 72
   h. Comparison of Calculated Actinic Fluxes to Experimentally Measured Values 75
   i. Actinic Fluxes in the Stratosphere 76
3. Procedure for Calculating Photolysis Rates 76
4. Example: Photolysis of Acetaldehyde at the Earth's Surface 81

D. PROBLEMS 83

REFERENCES 84

4
Photochemistry of Important Atmospheric Species

A. MOLECULAR OXYGEN 86
1. Absorption Spectra 86
2. Photochemistry 89

B. OZONE 90
1. Absorption Spectra 90
2. Photochemistry 91

C. NITROGEN DIOXIDE 95
1. Absorption Spectra 95
2. Photochemistry 96

D. NITRIC ACID 98

E. NITROUS ACID 99

F. PEROXYNITRIC ACID 100

G. NITRATE RADICAL 100

H. DINITROGEN PENTOXIDE 101

I. NITROUS OXIDE 101

J. ORGANIC NITRATES AND PEROXYACETYL NITRATE 102
1. Organic Nitrates 102
2. Peroxyacetyl Nitrate 103

K. SULFUR DIOXIDE AND SULFUR TRIOXIDE 103
1. SO_2 103
2. SO_3 105

L. HYDROGEN PEROXIDE AND ORGANIC HYDROPEROXIDES 107

M. ALDEHYDES AND KETONES 107

N. CHLORINE NITRATE ($ClONO_2$) AND BROMINE NITRATE ($BrONO_2$) 111

O. HCl AND HBr 113

P. THE HALOGENS 114

Q. ClO, BrO, AND IO 114

R. ClOOCl 114

S. OClO 115

T. HOCl, HOBr, AND HOI 115

U. NITROSYL CHLORIDE (ClNO) AND NITRYL CHLORIDE ($ClNO_2$) 117

V. HALOGENATED METHANES AND ETHANES   117

W. PROBLEMS   117

REFERENCES   126

5
Kinetics and Atmospheric Chemistry

A. FUNDAMENTAL PRINCIPLES OF GAS-PHASE KINETICS   130

1. Definitions   130
    a. Elementary vs Overall Reactions   130
    b. Rate Laws, Reaction Order, and the Rate Constant   131
    c. Half-Lives and Lifetimes   132
2. Termolecular Reactions and Pressure Dependence of Rate Constants   133
3. Temperature Dependence of Rate Constants   138
    a. Arrhenius Expression   138
    b. Predictions of Collision Theory and Transition State Theory   139
    c. Example of Importance of Temperature Dependence: PAN Decomposition   141

B. LABORATORY TECHNIQUES FOR DETERMINING ABSOLUTE RATE CONSTANTS FOR GAS-PHASE REACTIONS   141

1. Kinetic Analysis   142
2. Fast-Flow Systems   142
3. Flash Photolysis Systems   145
4. Pulse Radiolysis   146
5. Cavity Ring Down Method   147
6. Static Techniques   148

C. LABORATORY TECHNIQUES FOR DETERMINING RELATIVE RATE CONSTANTS FOR GAS-PHASE REACTIONS   149

D. REACTIONS IN SOLUTION   151

1. Interactions of Gaseous Air Pollutants with Atmospheric Aqueous Solutions   151
2. Diffusion-Controlled Reactions of Uncharged Nonpolar Species in Solution   152
3. Reactions of Charged Species in Solution   153
4. Experimental Techniques Used for Studying Solution Reactions   155

E. LABORATORY TECHNIQUES FOR STUDYING HETEROGENEOUS REACTIONS   156

1. Analysis of Systems with Gas- and Liquid-Phase Diffusion, Mass Accommodation, and Reactions in the Liquid Phase or at the Interface   158
2. Knudsen Cells   165
3. Flow Tube Studies   167
4. Falling-Droplet Apparatus   167
5. Bubble Apparatus   168
6. Aerosol Chambers   168
7. Liquid Jet Apparatus   169
8. DRIFTS   171
9. Surface Science Techniques   171
10. Other Methods   172

F. COMPILATIONS OF KINETIC DATA FOR ATMOSPHERIC REACTIONS   172

G. PROBLEMS   174

REFERENCES   175

6
Rates and Mechanisms of Gas-Phase Reactions in Irradiated Organic – NO_x – Air Mixtures

A. SOURCES OF OXIDANTS IN THE TROPOSPHERE: OH, O_3, NO_3, HO_2, AND Cl   179

1. OH   179
2. O_3 180
3. NO_3 180
4. HO_2 180
5. Cl   180

B. LIFETIMES OF TYPICAL ORGANICS IN THE TROPOSPHERE   181

C. REACTIONS OF ALKANES   182

1. Hydroxyl Radical (OH)   182
2. Nitrate Radical (NO_3)   184
3. Chlorine Atoms (Cl)   184

D. REACTIONS OF ALKYL (R), ALKYLPEROXY (RO_2), AND ALKOXY (RO) RADICALS IN AIR   185

1. Alkyl Radicals (R) 185
2. Alkylperoxy Radicals (RO_2) 185
   a. Reactions with NO 185
   b. Reactions with HO_2 and RO_2 186
   c. Reactions with NO_3 187
   d. Reactions with NO_2 187
   e. Fate of RO_2 under Typical Tropospheric Conditions 188
3. Alkoxy Radicals (RO) 188
4. Summary of R, RO_2, and RO Radical Reactions in the Troposphere 191

E. REACTIONS OF ALKENES (INCLUDING BIOGENICS) 191
1. Hydroxyl Radical (OH) 191
2. Ozone (O_3) 196
3. Nitrate Radical (NO_3) 201
4. Chlorine Atoms (Cl) 205
5. Nitrogen Dioxide (NO_2) 206

F. REACTIONS OF ALKYNES 206
1. Hydroxyl Radical (OH) 206

G. REACTIONS OF SIMPLE AROMATIC HYDROCARBONS 207
1. Hydroxyl Radical (OH) 207
2. Nitrate Radical (NO_3) 212
3. Chlorine Atoms (Cl) 212

H. REACTIONS OF OXYGEN-CONTAINING ORGANICS 213
1. Reactions of OH, NO_3, and Cl 213
2. Hydroperoxyl Radical (HO_2) 216

I. REACTIONS OF NITROGENOUS ORGANICS 217
1. Peroxyacetyl Nitrate and Its Homologs 217
2. Alkyl Nitrates and Nitrites 220
   a. Alkyl Nitrates 220
   b. Alkyl Nitrites 221
3. Amines, Nitrosamines, and Hydrazines 221
   a. Amines 221
   b. Nitrosamines 223
   c. Hydrazines 223

J. CHEMISTRY OF REMOTE REGIONS 225
1. Emissions of Biogenic Organics 225
2. Chemistry 231
   a. Biogenic Hydrocarbons 231
   b. General Remote Boundary Layer Chemistry 234
3. Upper Troposphere 239
4. Arctic 241

K. ATMOSPHERIC CHEMISTRY AND BIOMASS BURNING 244

L. PROBLEMS 247

REFERENCES 248

7
Chemistry of Inorganic Nitrogen Compounds

A. OXIDATION OF NO TO NO_2 AND THE LEIGHTON RELATIONSHIP 265

B. OXIDATION OF NO_2 266
1. Daytime Gas-Phase Reaction with OH 266
2. Nighttime Reactions to Form NO_3 and N_2O_5 267
3. Reactions of NO and NO_2 with Water and Alcohols 268
   a. Uptake into and Reaction with Liquid Water 268
   b. "Heterogeneous" Reaction of NO_2 with Water Vapor 269
   c. Reaction with Alcohols 272
4. Other Reactions of NO_2 272
   a. Organics 272
   b. Sea Salt Particles and Mineral Oxides 272

C. ATMOSPHERIC CHEMISTRY OF HONO 273
1. Formation of HONO 273
2. Atmospheric Fates of HONO 274

D. REACTIONS OF NO_3 AND N_2O_5 276
1. Reactions of NO_3 276
   a. Reactions with NO_2 and NO 276
   b. Reactions with Organics 276
   c. Thermal Decomposition 276
   d. Reaction with Water 277
2. Reactions of N_2O_5 279
   a. Hydrolysis 279
   b. Other Reactions 280

E. ATMOSPHERIC CHEMISTRY OF HNO_3 281
1. Formation 281
2. Tropospheric Fates 281

F. "MISSING" NO_y 286

G. AMMONIA (NH_3) 286

H. PROBLEMS 287

REFERENCES 288

8
Acid Deposition: Formation and Fates of Inorganic and Organic Acids in the Troposphere

A. CONTRIBUTION OF H_2SO_4, HNO_3, HONO, AND ORGANIC ACIDS 294

B. SOLUBILITY OF GASES IN RAIN, FOGS, AND CLOUDS: HENRY'S LAW AND AQUEOUS-PHASE EQUILIBRIA 295

C. OXIDATION OF SO_2 296

1. Field Studies 296
2. Oxidation in the Gas Phase 298
   a. *Hydroxyl Radical* 298
   b. *Criegee Biradical* 299
   c. *Other Gas-Phase Reactants* 300
   d. *Computing Oxidation Rates in the Atmosphere* 301
3. Oxidation in the Aqueous Phase 301
   a. *S(IV) Aqueous Equilibria* 301
   b. *Physical and Chemical Steps in Aqueous-Phase Oxidation* 306
   c. *Oxidation by O_2 (Catalyzed and Uncatalyzed)* 308
   d. *Oxidation by O_3* 311
   e. *Oxidation by H_2O_2 and Organic Peroxides* 313
   f. *Oxidation by Oxides of Nitrogen* 314
   g. *Free Radical Reactions in Clouds and Fogs* 315
   h. *Effect of Droplet Size on S(IV) Oxidation* 322
   i. *Fog–Smog–Fog Cycles* 323
4. Oxidation on Surfaces 324
5. Relative Importance of Various Oxidation Pathways for SO_2 325

D. ORGANIC ACIDS 326

E. OXIDATION OF SULFUR COMPOUNDS OTHER THAN SO_2 328

1. Reactions of Dimethyl Sulfide (CH_3SCH_3) 329
   a. *Reaction with OH* 329
   b. *Reaction with the Nitrate Radical (NO_3)* 332
   c. *Oxidation by Chlorine Atoms* 332
   d. *Oxidation by Halogen Oxides: IO, BrO, and ClO* 333
   e. *Oxidation by Ozone in the Aqueous Phase* 334
2. Dimethyl Disulfide (CH_3SSCH_3) 334
3. Methyl Mercaptan (CH_3SH) 334
4. Hydrogen Sulfide (H_2S) 335
5. Carbon Disulfide (CS_2) 335
6. Carbonyl Sulfide (COS) 336

F. PROBLEMS 336

REFERENCES 337

9
Particles in the Troposphere

A. PHYSICAL PROPERTIES 349

1. Some Definitions 349
2. Size Distributions 351
   a. *Number, Mass, Surface, and Volume Distributions* 351
   b. *Atmospheric Aerosols and Log-Normal Distributions* 358
3. Particle Motion 362
   a. *Gravitational Settling* 362
   b. *Brownian Diffusion* 363
4. Light Scattering and Absorption and Their Relationship to Visibility Reduction 365
   a. *Light Scattering and Absorption* 365
   b. *Relationship of Light Scattering and Absorption to Visibility Reduction* 368

B. REACTIONS INVOLVED IN PARTICLE FORMATION AND GROWTH 375

1. Nucleation, Condensation, and Coagulation 375
2. Reactions of Gases at Particle Surfaces 379
3. Reactions in the Aqueous Phase 380
4. Relative Importance of Various Aerosol Growth Mechanisms 380

C. CHEMICAL COMPOSITION OF TROPOSPHERIC AEROSOLS 380

1. Inorganic Species 381
   a. *Size Distribution* 381
   b. *Trace Elements as Tracers* 386
   c. *Source Apportionment Models* 386
   d. *Forms of Inorganics in Tropospheric Particles* 388
   e. *Unique Sources of Particles* 391
2. Organics 393
   a. *Biogenically Derived Organics in Aerosol Particles* 393
   b. *Organics in Anthropogenically Influenced and Aged Aerosol Particles* 396

D. GAS–PARTICLE DISTRIBUTION OF SEMIVOLATILE ORGANICS 412

1. Adsorption on Solid Particles 413
2. Absorption into Liquids 417
3. Octanol–Air Partitioning Coefficients 420

E. PROBLEMS 423

REFERENCES 423

10
Airborne Polycyclic Aromatic Hydrocarbons and Their Derivatives: Atmospheric Chemistry and Toxicological Implications

A. NOMENCLATURE AND SELECTED PHYSICAL AND SPECTROSCOPIC PROPERTIES OF POLYCYCLIC AROMATIC HYDROCARBONS (PAHs) AND POLYCYCLIC AROMATIC COMPOUNDS (PACs) 440

1. Combustion-Generated PAHs and PACs 440
2. Structures and IUPAC Rules for Nomenclature 440
   a. PAHs 440
   b. PACs 444
3. Solubilities and Vapor Pressures 451
4. Gas–Particle Partitioning, Sampling Techniques, and Ambient Levels of Selected PAHs and PACs 453
   a. PAHs 454
   b. PACs 461
5. Absorption and Emission Spectra of Selected PAHs and PACs 461

B. BIOLOGICAL PROPERTIES OF PAHs AND PACs. I: CARCINOGENICITY 466

1. Historical Perspective: Benzo[*a*]pyrene, the "Classic Chemical Carcinogen" 466
2. Carcinogenicity of PAHs, Cancer Potencies, and Potency Equivalence Factors 467
3. Carcinogenicity of Nitroarenes and Other Nitro-PACs 473

C. BIOLOGICAL PROPERTIES OF PAHs AND PACs. II: MUTAGENICITY 475

1. Short-Term Tests for Genetic and Related Effects 475
2. The Ames *Salmonella typhimurium* Reversion Assay 475
   a. Principle of Method: Direct vs Activatable Bacterial Mutagens 475
   b. Assay Procedure 478
   c. The "Microsuspension Modification" 478
   d. The Salmonella typhimurium Reversion Assay for Gas-Phase Mutagens 479
   e. Accuracy and Precision 480
   f. Some Mutagenic Potencies in the Salmonella typhimurium Assay 480
   g. Microbiology and Analytical Chemistry: Bioassay-Directed Fractionation and Chemical Analysis of Complex Mixtures of POM 482
3. The *Salmonella* TM677 "Forward Mutation" Assay 483
4. Human Cell Mutagenicities of PAHs and PACs 484

D. BACTERIAL AND HUMAN CELL MUTAGENICITIES OF POLLUTED AMBIENT AIR 486

1. Bacterial Mutagenicity of Urban Air: A Worldwide Phenomenon 486
   a. Background 486
   b. Particle Size Distribution of PAHs and Mutagenicity 487
   c. Variables Influencing Mutagenicity Levels 488
2. Sources, Ambient Levels, Transport, and Transformation: Some Case Studies 491
   a. Bacterial Mutagenicity 491
   b. Human Cell Mutagenicity 497
3. Bioassay-Directed Chemical Analysis for PAHs and PACs in Fine Ambient Aerosols Using a Human Cell Assay 497
4. Bioassay-Directed Chemical Analysis for Vapor-Phase and Particle-Phase PAHs and PACs in Ambient Air Using Bacterial Assays 502

E. ATMOSPHERIC FATES OF PARTICLE-ASSOCIATED PAHs: HETEROGENEOUS REACTIONS 504

1. Background 504
2. Theoretical and Experimental Structure–Reactivity Relationships 505
3. Field Studies of Atmospheric Reactions: Transport and Transformation 507
   a. Decay of Particle-Associated PAHs in Ambient Air 507
   b. Time–Concentration Profiles of Particle-Phase PAHs during Transport across an Air Basin 508
4. Photochemical Reactions of Particle-Associated PAHs 510
   a. Photooxidation in Solution and "Liquid-like" Surfaces of Organic Aerosols 510
   b. Photooxidations on Inorganic Solid–Air Surfaces 512
5. Gas–Particle Reactions 513
   a. Ozonolysis 513
   b. Nitration 515
6. Atmospheric Fates of Particle-Associated Nitroarenes 518
   a. Gas–Particle Reactions 518
   b. Photochemical Reactions 518

F. REACTIONS OF GAS-PHASE PAHs: ATMOSPHERIC FORMATION OF MUTAGENIC NITROARENES 519

1. Combustion-Generated Primary Emissions of Nitroarenes   519
2. Atmospheric Formation of Nitro-PAHs and Nitro-PACs   520

REFERENCES   527

11
Analytical Methods and Typical Atmospheric Concentrations for Gases and Particles

A. GASES   548

1. Optical Spectroscopic Techniques   548
   a. Chemiluminescence   548
   b. Fluorescence   548
   c. Infrared Spectroscopy (IR)   549
   d. DOAS (UV–Visible Absorption Spectroscopy)   556

2. Mass Spectrometry   561
   a. Sample Introduction   561
   b. Ionization   562
   c. Mass Filters   564
   d. Detectors   566

3. Filters, Denuders, Transition Flow Reactors, Mist Chambers, and Scrubbers   567
   a. Filters   567
   b. Denuders   567
   c. Transition Flow Reactors (TFRs)   568
   d. Mist Chambers and Scrubbers   568

4. Methods for, and Tropospheric Levels of, Specific Gases   569
   a. NO, NO_2, NO_x, and NO_y 569
   b. O_3 583
   c. CO   583
   d. SO_2 584
   e. NMHC and VOC   585
   f. Aldehydes, Ketones, Alcohols, and Carboxylic Acids   589
   g. PAN, Other Peroxynitrates, and Alkyl Nitrates   594
   h. H_2O_2 and Organic Peroxides   595
   i. HO_x Free Radicals   598

5. Generation of Standard Gas Mixtures   607

B. PARTICLES   608

1. Sampling and Collection of Particles   608
   a. Filters   608
   b. Impactors   610
   c. Electrostatic Precipitators   611
   d. Sedimentation Collectors   611

2. Measurement of Physical Characteristics: Mass and Size   612
   a. Mass   612
   b. Size   613
   c. Typical Particle Concentrations in the Atmosphere   618

3. Measurement of Chemical Composition   619
   a. Inorganic Elements   619
   b. Inorganic Ions   622
   c. Total Carbon: Organic versus Graphitic (Elemental)   623
   d. Speciation of Organics   625
   e. Artifacts   626

4. Real-Time Monitoring Techniques for Particles   626
   a. Single-Particle Laser Ionization Techniques   627
   b. Alternate Potential Mass Spectrometric Methods for Sizing and Chemical Composition   630
   c. Depth Profiling of Particle Composition   631

5. Generation of Calibration Aerosols   632
   a. Atomizers and Nebulizers   633
   b. Vibrating-Orifice Generator   634
   c. Spinning-Disk Generator   634
   d. Dry Powder Dispersion   634
   e. Tube Furnaces   634
   f. Condensation   635

C. PROBLEMS   635

REFERENCES   636

12
Homogeneous and Heterogeneous Chemistry in the Stratosphere

A. CHEMISTRY OF THE UNPERTURBED STRATOSPHERE   657

1. Stratosphere–Troposphere Exchange (STE)   658
2. Chapman Cycle and NO_x Chemistry   660

B. HIGH-SPEED CIVIL TRANSPORT (HSCT), ROCKETS, AND THE SPACE SHUTTLE   662

1. HSCT   662
2. Space Shuttle and Solid Rocket Motors   667

C. CHLOROFLUOROCARBONS   669

1. Types, Nomenclature, and Uses   669
2. Lifetimes and Atmospheric Fates of CFCs and Halons   670
3. Gas-Phase Chemistry in the Stratosphere   673
4. Antarctic "Ozone Hole"   675
5. Polar Stratospheric Clouds (PSCs) and Aerosols   680
   a. Nature of Aerosols and PSCs   680
   b. Uptake of HCl into PSCs   686
   c. Heterogeneous Chemistry on PSCs and Aerosols   688
6. Effects of Volcanic Eruptions   690

7. Ozone Depletion in the Arctic 696
8. Ozone Destruction in the Midlatitudes 700

D. CONTRIBUTION OF BROMINATED ORGANICS 701
1. Sources and Sinks of Brominated Organics 701
2. Bromine Chemistry in the Stratosphere 702

E. CONTRIBUTION OF IODINE-CONTAINING ORGANICS 706

F. SUMMARY 707

G. PROBLEMS 707

REFERENCES 708

13
Scientific Basis for Control of Halogenated Organics

A. INTERNATIONAL AGREEMENTS ON PHASEOUT OF HALOGENATED ORGANICS 727

B. OZONE DEPLETION POTENTIALS (ODP) 730

C. TRENDS IN CFCs, THEIR REPLACEMENTS, STRATOSPHERIC O_3, AND SURFACE UV 733
1. Trends in CFCs and Their Replacements 733
2. Trends in Stratospheric O_3 736
3. Trends in Surface Ultraviolet Radiation 741

D. TROPOSPHERIC CHEMISTRY OF ALTERNATE CFCs 744
1. Kinetics of OH Reactions 744
2. Tropospheric Chemistry 746
   a. *Chemistry of HFC-134a (CH_2FCF_3)* 746
   b. *Chemistry HCFC-125 (CHF_2CF_3)* 748
   c. *Chemistry of HCFC-123 ($CHCl_2CF_3$)* 749
   d. *Chemistry of HCFC-141b (CH_3CFCl_2)* 750
   e. *Tropospheric Fates of Halogenated Products of HCFC Oxidation* 750

E. SUMMARY 753

F. PROBLEMS 753

REFERENCES 753

14
Global Tropospheric Chemistry and Climate Change

A. RADIATION BALANCE OF THE ATMOSPHERE: THE GREENHOUSE EFFECT 763

1. Global Absorption and Emission of Radiation 763
2. Radiative Transfer Processes in the Atmosphere 766
   a. *Macroscopic View* 766
   b. *Molecular-Level View* 767
   c. *From Molecules to the Global Atmosphere* 768
3. Dependence of Net Infrared Absorption on Atmospheric Concentrations 769

B. CONTRIBUTION OF TRACE GASES TO THE GREENHOUSE EFFECT 770
1. Infrared Absorption by Trace Gases 770
2. Trends in Trace Gas Concentrations 773
   a. CO_2 773
   b. CH_4 777
   c. N_2O 779
   d. O_3 780
   e. *CFCs, HCFCs, and HFCs* 783
   f. *Other Gases* 783
3. Radiative Forcing by Greenhouse Gases and Global Warming Potentials 783
   a. *Instantaneous and Adjusted Radiative Forcing* 783
   b. *Absolute and Relative Global Warming Potentials* 784

C. AEROSOL PARTICLES, ATMOSPHERIC RADIATION, AND CLIMATE CHANGE 788
1. Direct Effects 789
   a. *Scattering of Solar Radiation* 789
   b. *Absorption of Solar Radiation* 796
   c. *Absorption of Long-Wavelength Infrared* 798
2. Indirect Effects of Aerosol Particles 799
   a. *Clouds* 799
   b. *Heterogeneous Chemistry Involving Climate Species* 814

D. SOME OTHER FACTORS AFFECTING GLOBAL CLIMATE 814
1. Absorption of Solar Radiation by Clouds 814
2. Feedbacks: Water Vapor, Clouds, and the "Supergreenhouse Effect" 819
3. Solar Variability 821
4. Volcanic Eruptions 822
5. Oceans 822

E. OBSERVATIONS OF CLIMATE CHANGES 823
1. Observed Temperature Trends 823
   a. *Trends over the Past Century* 823
   b. *Temperatures and Other Proxies for Climate Change over the Past $\sim 10^5$ Years* 825
2. Other Climate Changes 828

F. THE FUTURE 828

G. PROBLEMS 829

REFERENCES 829

15

Indoor Air Pollution: Sources, Levels, Chemistry, and Fates

A. RADON 844

B. OXIDES OF NITROGEN 846

1. Levels of NO_x 846
2. HONO and HNO_3 847

C. CO AND SO_2 849

D. VOLATILE ORGANIC COMPOUNDS 850

E. OZONE 859

F. INDOOR VOC–NO_x–O_3 CHEMISTRY 859

G. PARTICLES 861

H. PROBLEMS 865

REFERENCES 865

16

Applications of Atmospheric Chemistry: Air Pollution Control Strategies and Risk Assessments for Tropospheric Ozone and Associated Photochemical Oxidants, Acids, Particles, and Hazardous Air Pollutants

A. TROPOSPHERIC OZONE AND ASSOCIATED PHOTOCHEMICAL OXIDANTS 871

1. Environmental Chambers 872
    a. *Types of Chambers* 872
    b. *Preparation of Reactants, Including "Clean Air"* 876
    c. *Light Sources* 876
    d. *Typical Time–Concentration Profiles of Irradiated VOC–NO_x–Air Mixtures* 878
    e. *Advantages and Limitations of Environmental Chambers* 880
2. Isopleths for Ozone and Other Photochemically Derived Species 882
3. Models 886
    a. *Simple Models* 886
    b. *Mathematical Models* 887
    c. *Simple Mathematical Models* 892
    d. *Grid Models: Urban to Regional Scales* 893
    e. *Models Incorporating Particles* 907

B. REACTIVITY OF VOC 907

1. Typical Reactivity Scales 907
2. Application to Control of Mobile Source Emissions 909

C. FIELD OBSERVATIONS OF VOC, NO_x, AND O_3 913

D. ALTERNATE FUELS 918

1. Reformulated Gasolines 918
2. Compressed Natural Gas (CNG) 919
3. Liquefied Petroleum Gas (LPG) 920
4. Alcohol Fuels and Blends with Gasoline 920
5. Hydrogen 921
6. Electric Vehicles 921

E. CONTROL OF ACIDS 921

F. CONTROL OF PARTICLES 923

G. ATMOSPHERIC CHEMISTRY AND RISK ASSESSMENTS OF HAZARDOUS AIR POLLUTANTS 925

H. PROBLEMS 930

REFERENCES 932

Appendix I: Enthalpies of Formation of Some Gaseous Molecules, Atoms, and Free Radicals at 298 K 943
Appendix II: Bond Dissociation Energies 945
Appendix III: Running the OZIPR Model 947
Appendix IV: Some Relevant Web Sites 949
Appendix V: Pressures and Temperatures for Standard Atmosphere 951
Appendix VI: Answers to Selected Problems 952
Subject Index 957

Preface

What is written without effort is in general read without pleasure.

Samuel Johnson (1709–1784)
In William Seward *Biographia*

Given this admonition, written long ago by English poet and critic Samuel Johnson, we trust that this book may be read with some pleasure by a "spectrum" of readers.

The field of atmospheric chemistry has undergone dramatic changes since our first book on this subject was published in 1986. Since then, a number of new, exciting, and highly relevant research areas have emerged. We treat these here, along with the fundamentals of spectroscopy, photochemistry, and reaction kinetics and mechanisms of atmospheric systems. For example, the discovery of the Antarctic ozone hole has left no doubt that chlorofluorocarbons (CFCs) have led to depletion of stratospheric ozone and has highlighted the importance of heterogeneous chemistry on polar stratospheric clouds. Atmospheric measurements of the CFCs have documented changes in their global concentrations in response to control measures. The importance of emissions and chemistry for climate changes on a global scale has become an area of intense popular interest and scientific research. Furthermore, the formation, chemistry, and fates of airborne particulate matter, particularly that less than 2.5 μm in diameter (i.e., PM2.5), are now recognized not only as important to global climate issues but also as of concern for their toxicological effects. The similarity in much of the chemistry of indoor air pollutants and those outdoors is now evident, and the importance of understanding the chemistry of airborne toxic chemicals (also known as "hazardous air pollutants," HAPs) for the development of sound risk assessments is clear. A great deal more is known about the role of emissions and atmospheric reactions of polycyclic aromatic hydrocarbons in the mutagenicity and carcinogenicity of urban atmospheres. In addition to these major developments in the field, there has been a significant increase in our understanding of the gas-phase atmospheric chemistry and photochemistry of organics, oxides of nitrogen, and SO_2 and rapidly increasing evidence for the importance of a wide range of heterogeneous reactions in the troposphere.

We have attempted in this new book to present the current understanding of the chemistry of the natural and polluted upper and lower atmosphere in such a way that it will be useful to a range of atmospheric chemists as well as to atmospheric scientists and engineers working in this field. However, the fundamentals (e.g., theories, rates, and mechanisms of homogeneous and heterogeneous reactions and of spectroscopic and photochemical processes) are also emphasized. We believe that this approach is useful in providing the necessary background and tools for graduate students as well as for scientists and engineers in related fields who wish to enter this exciting and dynamic area. Problems are provided at the end of most chapters (and answers to selected problems are in Appendix VI) to enhance the book's use in teaching.

The literature is assessed through the end of 1998 and, in some cases, into 1999. We have cited only papers in the peer-reviewed literature or, in a few instances, government agency reports that are readily available. Our approach has been to consider primarily typical examples of major papers in the refereed literature in the relevant areas. Because of the enormous breadth of the field today, we have not been able to reference all papers in all relevant areas, which will unavoidably lead to some omissions. We apologize in advance to our colleagues whose work might not have been cited in the relevant area.

We are deeply indebted to our many colleagues in the field whose outstanding work, generous sharing of results, and helpful discussions have made this work possible. It is our hope that this book does justice to the current state of this exciting, rapidly maturing, and scientifically and societally relevant discipline.

Barbara J. Finlayson-Pitts
James N. Pitts, Jr.

Fawnskin, California
August 19, 1999

About the Authors

Barbara J. Finlayson-Pitts is Professor of Chemistry at the University of California, Irvine. Her research program focuses on laboratory studies of the kinetics and mechanisms of reactions in the atmosphere, especially those involving gases with liquids or solids of relevance in the troposphere. Reactions of sea salt particles to produce photochemically active halogen compounds and the subsequent fates of halogen atoms in the troposphere are particular areas of interest, as are reactions of oxides of nitrogen at aqueous and solid interfaces. Her research is currently supported by the National Science Foundation, the Department of Energy, the California Air Resources Board, the Dreyfus Foundation, and NATO. She has authored or coauthored more than 80 publications in this area, as well as a previous book, *Atmospheric Chemistry: Fundamentals and Experimental Techniques*.

At UCI, she teaches graduate-level courses in atmospheric chemistry on a regular basis. In addition, she teaches such classes as undergraduate instrumental analysis, in which she is developing a new laboratory curriculum centered around the analysis of complex environmental mixtures. This work has been supported by the Dreyfus Foundation and UCI.

She received her undergraduate B.Sc. (Hons) in 1970 from Trent University in Peterborough, Ontario, Canada, where her interest in atmospheric chemistry was first sparked by discussions of the spectroscopy of auroras in a physical chemistry class taught by Professor R. E. March. She obtained her master's degree (1971) and Ph.D. (1973) from the University of California, Riverside. After a year's postdoctoral work at UCR, she joined the faculty of California State University, Fullerton, where she taught and carried out research in physical and atmospheric chemistry. In 1994, she joined the faculty at UCI.

Dr. Finlayson-Pitts is a member of a number of professional societies, including the American Chemical Society, the American Geophysical Union, the American Association for the Advancement of Science, and Iota Sigma Pi. She is the recipient of a number of awards, including the Governor General's Medal at Trent University, a Woodrow Wilson Fellowship, a National Research Council of Canada Science Scholarship, Golden Key National Honor Society, and a Japan Society for the Promotion of Science Fellowship. She has been elected a Fellow of the American Association for the Advancement of Science and has several awards for undergraduate teaching.

James N. Pitts, Jr., is a Research Chemist at the University of California, Irvine, and Professor Emeritus from the University of California, Riverside. He was Professor of Chemistry (1954–1988) and cofounder (1961) and Director of the Statewide Air Pollution Research Center (1970–1988) at the University of California, Riverside. His research has focused on the spectroscopy, kinetics, mechanisms, and photochemistry of species involved in a variety of homogeneous and heterogeneous atmospheric reactions, including those associated with the formation and fate of mutagenic and carcinogenic polycyclic aromatic compounds. He is the author or coauthor of more than 300 research publications and three books: *Atmospheric Chemistry: Fundamentals and Experimental Techniques, Graduate School in the Sciences—Entrance, Survival and Careers,* and *Photochemistry*. He has been coeditor of two series, *Advances in Environmental Science and Technology* and *Advances in Photochemistry*. He served on a number of panels in California, the United States, and internationally. These included several National Academy of Science panels and service as Chair of the State of California's Scientific Review Panel for Toxic Air Contaminants and as a member of the Scientific Advisory Committee on Acid Deposition.

He received his B.S. (1945) and Ph.D. (1949) from the University of California, Los Angeles; his research advisor was Professor Francis E. Blacet, who first identified the photolysis of NO_2 as the anthropogenic source of ozone in photochemical air pollution. From 1942 to 1945, he participated in laboratory and field studies in chemical warfare. He was on the faculty at Northwestern University from 1949 to 1954, leaving to join the faculty at the new University of California, Riverside, campus. He was a Guggenheim Fellow at University College, Oxford, in 1961 and a Research Fellow of Merton College, Oxford, in 1965.

Dr. Pitts is a member of a number of professional societies, including the American Chemical Society, the American Geophysical Union, the American Association for the Advancement of Science, and the American Physical Society. He has received a number of awards for his research, including the Clean Air Award of the California Lung Association (1979), the Frank A. Chambers Award for "Outstanding Achievement in the Science and Art of Air Pollution Control" from the Air Pollution Control Association (1982), the Richard C. Tolman Medal (1983), the UCR Faculty Research Lectureship (1965), the F. J. Zimmerman Award in Environmental Science, and the Clean Air Award (1992) from the South Coast Air Management District. He is an elected Fellow of the American Association for the Advancement of Science. He has also received numerous commendations from local, state, and federal legislators for his application of fundamental atmospheric chemistry to air pollution problems.

Acknowledgments

Many individuals and organizations contributed to making this book a reality. First, our assistant, Mae Minnich, devoted innumerable long hours to the manuscript; her organization and professional talents were indispensable. Without her outstanding skills in word processing and manuscript preparation, combined with her patience, wonderful sense of humor, and tireless enthusiasm, this undertaking would not have been possible. Kelly Donovan took almost indecipherable drafts of figures and turned them into clear and attractive drawings, some of which reflect major creativity on her part. The cheerful dedication, professional skills, and willingness to "go the extra mile" on the part of both of these individuals helped the authors through some long days.

The assistance of a number of undergraduate and graduate students, postdoctoral fellows, and colleagues was critical in this effort. Christopher Elliott and Ellen Fleyshman assembled and organized the thousands of references and, along with Lisa Wingen, Alisa Ezell, and John Elliott, also provided computer and technical assistance. Their unflagging assistance in the midst of the final fray and the moral and concrete support of others, including Stacie Tibbets and Mike and Connie Ezell, reenergized us during the final "countdown." The assistance of Bill Barney, Mike and Alisa Ezell, Krishna Foster, Michael Gebel, Matt Lakin, Lisa Wingen, and Weihong Wong with figures and final page proofing was very helpful, as was the able library assistance of Jean Miller.

Sasha Madronich generously not only reviewed the section on atmospheric radiation, but provided his unpublished calculations of actinic fluxes at different altitudes in a form useful to the atmospheric chemistry community for estimates of photolysis rates from the troposphere through the stratosphere. A number of colleagues reviewed chapters or portions of chapters, and their insightful comments and suggestions are greatly appreciated. They include Janet Arey, Roger Atkinson, Thorsten Benter, Theo Brauers, Carl Berkowitz, Don Blake, Chris Doran, Anders Feilberg, Mario Molina, Ole John Nielsen, Torben Nielsen, Joyce Penner, F. Sherwood Rowland, David Rusch, Stephen Schwartz, Chet Spicer, Jochen Stutz, Darin Toohey, Douglas Worsnop, Y. L. Yung, and Paul Ziemann.

We are indebted to many colleagues who provided figures, data, and stimulating discussions, especially Ed Baum, John Barker, Thorsten Benter, Bill Carter, Glenn Cass, Ralph Cicerone, Bart Croes, David Crosely, Donald Dabdub, Paul Davidovits, Bill De More, Leon Dolislager, Alisa Ezell, Michael Ezell, Jerome Fast, Jeff Gaffney, David Golden, Les Grant, Bill Harger, John Holmes, John Jayne, John Johnson, Jake Hales, John Hemminger, Wolfgang Junkermann, George Kirk, Charles Kolb, William Lockett, Alan Lloyd, Nancy Marley, Mike Nicovich, Randy Pasek, Shankar Prasad, Michael Prather, Ralph Propper, Scott Samuelsen, Rolf Sander, Ken Schere, Jim Seiber, Brian Toon, Ernie Tuazon, Charles Weschler, Marvin Wesely, Hal Westberg, Arthur Winer, Paul Wine, Ed Yotter, Mark Zahniser, and Barbara Zielinska.

Michael Gery, who developed the OZIPR model, graciously provided advice on its use as well as electronic copies of the documentation. This model, which contains the two major chemical mechanism schemes for gas-phase, VOC-NO_x chemistry in use in atmospheric chemistry, is available on the Academic Press Web site (http://www.academicpress.com/pecs/download). A number of problems using this model are included in the book, and it is a valuable teaching tool for assessing the effects of various model input parameters on predicted concentrations of a wide variety of gas-phase species. His assistance and that of Marcia Dodge of the U.S. EPA in making it available are appreciated.

The authors have been very fortunate over the years to have a number of accomplished and creative mentors who generously shared their knowledge of, and enthusiasm for, the fields of fundamental photochemistry, spectroscopy, and kinetics; these include F. E. Blacet, E. J. Bowen, P. A. Leighton, W. A. Noyes, Jr., E. W. R. Steacie, and R. E. March. We hope that the central importance of these fundamentals to understanding atmospheric chemistry is evident in this book. We also appreciate the tremendously talented and enthusiastic researchers, too numerous to mention here, who have spent time in our research groups over the years. Many of them have gone on to careers in atmospheric chemistry; it has given us a great deal of satisfaction and delight to see the "academic lineage" being passed on from our mentors to them.

The authors have also very much appreciated the scientific interactions with, and financial support of, a number of public and private agencies and foundations, which have allowed us to conduct research in various areas in atmospheric chemistry. Over several decades, key individuals at these organizations have been helpful in many ways, and we are grateful to them. Agencies and individuals include Jarvis Moyers, Anne-Marie Schmoltner, and the late Richard Carrigan at the National Science Foundation; David Ballantine, George Stapleton, Ari Patrinos, Michael Riches, Michelle Broido, Rickey Petty, and Peter Lunn at the Department of Energy; John Holmes, Bart Croes, Ralph Propper, Jack Suder, Randy Pasek, and Eileen McCauley at the California Air Resources Board; Ron Patterson and Marcia Dodge at the Environmental Protection Agency; Brian Andreen and the late Hal Ramsey at The Research Corporation; and Robert Lichter at the Dreyfus Foundation.

We have been impressed indeed by the professional skills, patience, thoughtful and imaginative ideas, and dedication of the staff at Academic Press. Special thanks and appreciation go to Executive Editor David Packer, as well as his colleagues at AP, including Cheryl Uppling (Senior Production Editor), Mike Early (Book Production Manager), Liz Novelozo and Kim Schettig (Marketing), Linda Klinger (Editorial Coordinator), design artist Amy Stirnkorb, and David Phanco.

On a more personal note, the support of our colleagues at the University of California, Irvine, particularly the Chairs in the Department of Chemistry, John Hemminger and Richard Chamberlin, has been essential to the completion of this project. The hospitality and encouragement of the Science Education Project staff, especially Ann and George Miller, Lynne Davanzo, and Frank Potter, are appreciated. We are also grateful to Edward J. McIntyre for his professional skills and wise and effective counsel and to Bill and Maura Dickerson for their timely advice during an interesting period in the genesis of this book.

Successful completion of this project was only possible through the professional and personal support of our friends and neighbors, particularly in University Hills at UCI as well as in our Fawnskin mountain "hamlet" (population 360 and elevation 6500 ft), especially Doris Layne and Linda Neuman. Regular visits to our home in Fawnskin by a variety of wildlife of the nonhuman type, including eagles, coyotes, bear, quail, and wild turkeys, provided much sanity and perspective during the writing of this book.

Finally, we are very grateful to all of our colleagues, families, and friends, as well as to our golden retriever contingent, Babe, Maj, and BR, who have been extremely patient with our being "missing in action" for too many years while finishing this book. They (and we, at times) wondered if it would ever end... but in large part through all of their patience, support, and encouragement, it finally has!

CHAPTER 1

Overview of the Chemistry of Polluted and Remote Atmospheres

Atmospheric chemistry is an exciting, relatively new field. It encompasses the chemistry of the globe, from polluted to "clean," remote regions and from the region closest to the earth's surface, the *troposphere* ($\lesssim$ 10–15 km), through the *tropopause* ($\sim$ 10–15 km) into the upper atmosphere—which, for the purposes of this book, we restrict to the *stratosphere* ($\sim$ 10–50 km). Chemical and physical processes occurring at the earth's surface—emissions, transport, lifetimes, and fates of certain anthropogenic and biogenic/geogenic chemicals—can impact the stratosphere—and vice versa. Thus, even though in some early studies the tropopause was perceived as being a "barrier" between the lower and upper atmospheres, it has become increasingly clear that the troposphere and stratosphere are intimately connected. Witness the vertical transport of long-lived ozone-destroying anthropogenic emissions of chlorofluorocarbons (CFCs) and conversely the downward transport of stratospheric ozone into the troposphere. Hence, even though we devote separate chapters to the chemistry of the troposphere and stratosphere, the emphasis throughout is on *one integrated "system" of global atmospheric chemistry*.

In this regard, there are several topics that fall outside the scope of this book, including the evolution of the earth's atmosphere. For reviews, the reader is referred to articles by Kasting (1993) and Allègre and Schneider (1994). Although we point out throughout this book the interconnectedness of the lower and upper atmospheres, practicality and length preclude extension to the obvious interrelationships and feedbacks with other components of the earth system, including the controversial *Gaia Hypothesis* (named after the Greek goddess of the earth). The reader is referred to articles by Lovelock (1989), Kirchner (1989), Schneider (1990), and Lenton (1998) for discussions of the latter.

Although atmospheric chemistry is sometimes viewed as an "applied" science, its foundations rest on *fundamental* research in diverse areas of chemistry. These include theoretical and experimental aspects of spectroscopy, photochemistry, and the kinetics and mechanisms of homogeneous and heterogeneous organic and inorganic reactions. We believe it is useful for today's students, researchers, and educators to be aware that major resources for Leighton's masterful 1961 treatment of the newly emerging field of atmospheric chemistry, *Photochemistry of Air Pollution*, were, in fact, pioneering, basic research monographs published years earlier. They include *Photochemistry of Gases* by W. A. Noyes, Jr., and P. A. Leighton (1939), two editions of *Atomic and Free Radical Reactions* by E. W. R. Steacie (1946, 1954), and G. Herzberg's classics *Atomic Spectra and Atomic Structure* (1944); *Molecular Spectra and Molecular Structure I: Spectra of Diatomic Molecules* (1950); and *Infrared and Raman Spectra of Polyatomic Molecules* (1945).

Throughout the body of this book, we address the basic chemistry driving key atmospheric processes in the natural and polluted troposphere and stratosphere and illustrate their critical interactions on local, regional, and global scales. In so doing, our treatment overall reflects the message that Sam sings so eloquently to Bogart and Bergman in the classic movie *Casablanca*,..."The fundamental things apply, as time goes by."

In this chapter we provide an overview of the chemistry of the lower and upper atmospheres. In Chapter 2, we illustrate how this chemistry plays a critical role in the concept of an *integrated* "atmospheric chemistry system"—a loop that starts with emissions (anthropogenic and natural) and ultimately closes with scientific health and environmental risk assessments and associated risk management decisions for the control of air pollutants.

Chapters 3–12 present a detailed examination and explanation of how one applies the theoretical and experimental *fundamentals* of photochemistry, spectroscopy, and kinetics and mechanisms (structure and reactivity) to the most important homogeneous and heterogeneous processes that take place in our natural and polluted atmosphere.

We conclude by illustrating how our understanding of these chemical processes in our clean and polluted troposphere and stratosphere plays a crucial role in generating the *"exposure"* portions of scientific health risk assessments. Such assessments provide the foundation for sound, health-protective and cost-effective strategies for the control of tropospheric ozone, particles, acids, and a spectrum of "hazardous air pollutants" (including carcinogens and pesticides)—as well as for the mitigation of stratospheric ozone depletion.

A. REGIONS AND CHARACTERISTICS OF THE ATMOSPHERE

Figure 1.1 shows the different regions of the atmosphere. (See also Appendix V for typical pressures and temperatures as a function of altitude.) In the troposphere, the temperature generally falls with increasing altitude (except in the presence of inversions; see Chapter 2.B.3). This is due to the strong heating effect at the surface from the absorption of radiation. Because hot air rises, this causes strong vertical mixing so that species emitted at the earth's surface can rise to the *tropopause*, the region separating the troposphere from the stratosphere, in a few days or less, depending on the meteorological conditions. Essentially all of the water vapor, clouds, and precipitation are found in the troposphere, which provides an important mechanism for scavenging pollutants from the atmosphere.

However, at the tropopause the temperature profile changes, increasing with altitude throughout the stratosphere. The reason for this increase is a critical series of photochemical reactions involving ozone and molecular oxygen. The "Chapman cycle," reactions (1)–(4), hypothesized in the 1930's by Sir Sydney Chapman,

$$O_2 + h\nu \rightarrow 2O, \qquad (1)$$

$$O + O_2 \xrightarrow{M} O_3, \qquad (2)$$

$$O + O_3 \rightarrow 2O_2, \qquad (3)$$

$$O_3 + h\nu \rightarrow O + O_2, \qquad (4)$$

is responsible for generating a steady-state concentration of O_3 in the stratosphere.

Stratospheric ozone is essential for life on earth as we know it, because it strongly absorbs light of $\lambda < 290$ nm. As a result, sunlight reaching the troposphere, commonly referred to as *actinic radiation*, has wavelengths longer than 290 nm. This short-wavelength cutoff sets limits on tropospheric photochemistry; thus only those molecules that absorb radiation at wavelengths longer than 290 nm can undergo photodissociation and other primary photochemical processes.

Ozone absorbs light strongly between approximately 200 and 310 nm and weakly up into the visible. Dissociation to electronically excited O_2 ($^1\Delta_g$) and $O(^1D)$ requires light energetically equivalent to 310 nm. Therefore, the excess energy available after absorption of light up to this threshold value is released as heat; energy is also released from the $O + O_2$ reaction (2). Both give rise to the increase in temperature in the stratosphere. Relatively little vertical mixing occurs in the stratosphere, and no precipitation scavenging occurs in this region. As a result, massive injections of particles, for example, from volcanic eruptions such as the Mt. Pinatubo eruption, often produce layers of particles in the stratosphere that persist for long periods of time (see Chapter 12).

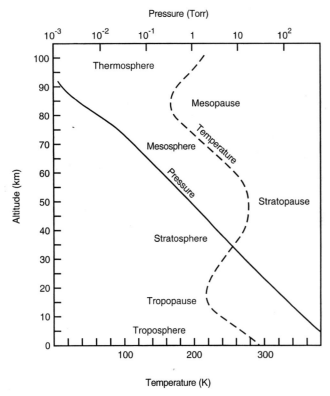

FIGURE 1.1 Typical variation of temperature with altitude at mid-latitudes as a basis for the divisions of the atmosphere into various regions. Also shown is the variation of total pressure (in Torr) with altitude (top scale, base 10 logarithms) where 1 standard atmosphere = 760 Torr.

In the *mesosphere*, from ~50 to ~85 km, the temperature again falls with altitude and vertical mixing within the region occurs. This temperature trend is due to the decrease in the O_3 concentration with altitude. At about 85 km the temperature starts to rise again because of increased absorption of solar radiation of wavelengths < 200 nm by O_2 and N_2 as well as by atomic species. This region is known as the *thermosphere*.

The transition zones between the various regions of the atmosphere are known as the *tropopause*, *stratopause*, and *mesopause*, respectively. Their locations, of course, are not fixed, but vary with latitude, season, and year. Thus Fig. 1.1 represents an average profile for mid-latitudes. Specific temperatures, pressures, densities, winds, and the concentrations of some atmospheric constituents as a function of altitude, geographic position, and time are incorporated into a NASA model, the *G*lobal *R*eference *A*tmosphere *M*odel (GRAM); information on obtaining this model and data is included in Appendix IV.

We shall see throughout this book that different chemical and physical processes occur in the troposphere and stratosphere, and we shall frequently refer to different regions in Fig. 1.1. However, it is important to put the atmosphere in perspective with respect to the size of the earth itself. The earth's average diameter is 12,742 km, yet the average distance from the earth's surface to the top of the stratosphere is only ~50 km, less than 0.4% of the earth's diameter! The space shuttle orbits outside the atmosphere, but at an altitude of only several hundred miles, which is less than the distance from Los Angeles to San Francisco. Clearly, the atmosphere is a very thin, and as we shall see, fragile shield upon which life as we know it on earth depends.

B. AIR POLLUTION AND THE CHEMISTRY OF OUR TROPOSPHERE

1. Historical Perspectives: Ancient and Medieval Times

Concern over air pollution has been well documented (Brimblecombe, 1978). The impacts of atmospheric chemistry on human health and the environment can be traced back many centuries, indeed some two thousand years. For example, the Mishnah Laws in Israel in the first and second centuries A.D. required that, because of the odors emitted, tanneries be located at least 30 m away from the town and only on the east side, due to prevailing westerly winds (Mamane, 1987).

In the twelfth century, the Hebrew philosopher, scientist, and jurist Moses Maimonides (1135–1204) wrote (Goodhill, 1971)

> "Comparing the air of cities to the air of deserts and arid lands is like comparing waters that are befouled and turbid to waters that are fine and pure. In the city, because of the height of its buildings, the narrowness of its streets, and all that pours forth from its inhabitants and their superfluities...the air becomes stagnant, turbid, thick, misty, and foggy.... If there is no choice in this matter, for we have grown up in the cities and have become accustomed to them, you should...select from the cities one of open horizons...endeavor at least to dwell at the outskirts of the city....

> "If the air is altered ever so slightly, the state of the Psychic Spirit will be altered perceptibly. Therefore you find many men in whom you can notice defects in the actions of the psyche with the spoilage of the air, namely, that they develop dullness of understanding, failure of intelligence and defect of memory...."

To this day, many of us can relate to his view of the health and psychological impacts of heavy smog episodes—whether they be of the London or Los Angeles variety.

2. "London" Smog: Sulfur Dioxide, Acidic Aerosols, and Soot

In the seventeenth century, John Evelyn published a major treatise on air pollution in London, caused by the widespread domestic use of high-sulfur coal. In it, he noted effects not only on materials:

> "It is this horrid Smoake which obscures our Church and makes our Palaces look old, which fouls our Cloth and corrupts the Waters, so as the very Rain, and refreshing Dews which fall in the several Seasons, precipitate to impure vapour, which, with its black and tenacious quality, spots and contaminates whatever is exposed to it."

but also on health:

> "But, without the use of Calculations it is evident to every one who looks on the yearly Bill of Mortality, that near half the children that are born and bred in London die under two years of age.[a] Some have attributed this amazing destruction to luxury and the abuse of Spirituous Liquors: These, no doubt, are powerful assistants; but the constant and unremitting Poison is communicated by the foul Air, which, as the Town still grows larger, has made regular and steady advances in its fatal influence."

[a]"A child born in a Country Village has an even chance of living near 40 years..."

Evelyn's air pollution classic, and an article by Barr, *The Doom of London*, are reprinted in the book *The Smoake of London. Two Prophecies* (Lodge, 1969). They make interesting and useful reading and help place our present problems in perspective.

TABLE 1.1 Some Incidents of Excess Deaths Associated with Smog[a]

Year	Place	Number of excess deaths
1930	Meuse Valley, Belgium	63
1948	Donora, Pennsylvania	20
1952	London	4000
1962	London	700

[a] From Firket (1936), Wilkins (1954), Rouaché (1965), and Cochran *et al.* (1992).

In more recent times, a number of air pollution episodes have been quite dramatic. Table 1.1 lists some of the most severe in which excess deaths (i.e., deaths beyond what is expected for that location and time of year based on past statistics) have been attributed to air pollution. During these episodes, there tended to be heavy fogs and low inversion levels that concentrated the pollutants in a relatively small volume.

A particularly interesting prediction was made by Firket (1936). It was based on a devastating smog episode that occurred December 1–5, 1930, along the Meuse Valley in Belgium (Table 1.1). On examining the combination of emissions and meteorological conditions that led to 63 excess deaths, as well as several hundred people with severe respiratory problems, he commented that:

> "This apprehension was quite justified, when we think that, proportionally the public services of London, e.g., might be faced with the responsibility of 3200 sudden deaths if such a phenomenon occurred there."

His prediction turned out to be remarkably accurate and prophetic, when some 16 years later, there were 4000 excess deaths in the 1952 London episode.

The actual pollutants or combination of pollutants responsible for the excess deaths in London have not been identified, although in all cases there were greatly increased levels of SO_2 and particulate matter in the presence of dense fog and very low, strong meteorological inversions. Figure 1.2 shows the concentrations of SO_2 and "smoke," i.e., particles, during the 1952 episode (Wilkins, 1954). Clearly, the death rate tracks these two pollutants. It is this combination of smoke and fog that led to the now commonly used term, "smog."

Subsequent to the 1952 London episode, Britain passed a Clean Air Act to reduce emissions. Although meteorological conditions similar to those in December 1952 occurred in 1962, as seen in Table 1.1, the number of excess deaths that occurred declined dramatically.

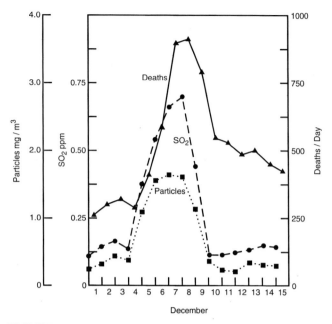

FIGURE 1.2 Concentrations of SO_2 and "smoke" as well as the death rate during the 1952 smog episode (adapted from Wilkins, 1954).

3. "Los Angeles" Smog: Ozone and Photochemical Oxidants

a. Historical

In the late 1940's, a remarkable air pollution phenomenon began to impact the Los Angeles area. In sharp contrast to "London" smog, the ambient air contained strongly oxidizing, eye-watering and plant-killing pollutants—and occurred on *hot* days with *bright* sunshine. Plant pathologists at the University of California, Riverside, observed a unique type of damage to agricultural crops in areas of the Los Angeles basin impacted by this "plague" and reported it to be an entirely new form of air pollution—Los Angeles smog (Middleton *et al.*, 1950).

Shortly thereafter, in a classic series of papers in the early 1950's, Arie Haagen-Smit and co-workers reported that these plant damage symptoms observed outdoors in ambient air could be duplicated in the laboratory by irradiating plants with sunlight and concurrently exposing them to synthetic polluted air containing alkenes and nitrogen dioxide:

Organics + NO_x + Sunlight →

$$O_3 + \text{"Other products."} \quad (5)$$

Similar effects were observed with sunlight-irradiated, diluted auto exhaust, which contains NO_x (NO_x = $NO + NO_2$) and a variety of hydrocarbons (HC)

(Haagen-Smit, 1952; Haagen-Smit et al., 1952, 1953; Haagen-Smit and Fox, 1954, 1955, 1956).

Since then, high ozone levels have also been measured throughout the world, e.g., in Athens, Greece, and in regions downwind from Sydney, Australia. In Mexico City, ozone levels over 400 ppb have been measured. Thus, although photochemical air pollution was first recognized in the Los Angeles area, it is now recognized to be a world-wide problem in areas where volatile organic compounds (VOC) and NO_x emissions from major mobile and stationary sources are "trapped" by thermal inversions and irradiated by sunlight during transport to downwind regions. Encouragingly, in southern California, ozone peaks have declined significantly since around 1980 due to increasingly tight controls on VOC and NO_x.

Table 1.2 summarizes some aspects of London and photochemical air pollution that have traditionally been considered to differentiate the two. However, as we shall see, it has become clear since the mid-1970's that these two, along with such phenomena as the fate of airborne toxic chemicals, are inextricably linked through their atmospheric chemistry. It is this common linkage that forms the central core of this book.

b. Photochemical Air Pollution

Today in many major urban areas around the world, air pollution is characterized more by the formation of ozone and other oxidants rather than by SO_2, particles, and sulfuric acid. In these regions, the primary pollutants are NO_x (mainly NO) and volatile organic compounds (VOC), which undergo photochemical reactions in sunlight to form a host of secondary pollutants, the most prominent of which is O_3. Some of these are criteria pollutants for which air quality standards have been set, such as O_3, SO_2, CO, NO_2, PM_{10}, and $PM_{2.5}$ (ambient particulate matter less than 10 or 2.5 μm in diameter, respectively). Others are so-called "trace" noncriteria pollutants, e.g., gaseous peroxyacetyl nitrate ($CH_3C(O)OONO_2$, PAN), nitric acid (HNO_3 or $HONO_2$), formaldehyde (HCHO), and formic acid (HCOOH). The overall reaction is now written as

$$VOC + NO_x + h\nu \rightarrow$$
$$O_3 + PAN + HNO_3 \ldots + \text{Particles, etc.} \quad (6)$$

Certain reproducible features of time–concentration profiles for pollutants are observed in "smoggy" ambient air. Figure 1.3, a classic example of historical interest, shows such profiles for NO, NO_2, and total oxidant (mainly O_3) in Pasadena, California, during a severe photochemical air pollution episode in July 1973.

Reproducible features include the following:

- In the early morning, the concentration of NO rises and reaches a maximum at a time that approximately coincides with the maximum emissions of NO, in this case, peak automobile traffic;
- Subsequently, NO_2 rises to a maximum;
- Oxidant (e.g., O_3) levels, which are relatively low in the early morning, increase significantly about noon when the NO concentration drops to a low value.

In this instance, near an urban center, the O_3 reaches a maximum after NO_2 peaks. Downwind from urban

TABLE 1.2 Historical Aspects of Sulfurous (London) and Photochemical (Los Angeles) Air Pollution

Characteristics	Sulfurous (London)	Photochemical (Los Angeles)
First recognized	Centuries ago	Mid-1940s
Primary pollutants	SO_2, soot particles	VOC, NO_x
Secondary pollutants	H_2SO_4, sulfate aerosols, etc.	O_3, PAN, HNO_3, aldehydes, particulate nitrate and sulfate, etc.
Temperature	Cool (< 35°C)	Hot (> 75°F)
Relative humidity	High, usually foggy	Low, usually hot and dry
Type of inversion	Radiation (ground)	Subsidence (overhead)
Time air pollution peaks	Early morning	Noon to evening

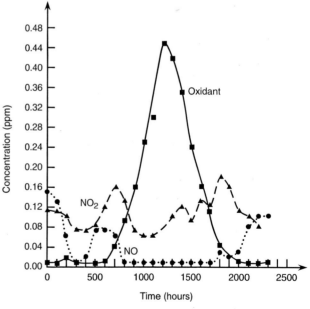

FIGURE 1.3 Diurnal variation of NO, NO_2, and total oxidant in Pasadena, California, on July 25, 1973 (adapted from Finlayson-Pitts and Pitts, 1977).

centers, the profiles are shifted and O_3 may peak in the afternoon, or even after dark, depending on emissions and airshed transport phenomena. Thus, although O_3 is no longer formed after sunset, a dirty, urban air mass containing O_3 and other secondary pollutants formed during the day can be transported many kilometers downwind to an otherwise relatively clean rural site.

In the early 1950's, soon after the new phenomenon of photochemical air pollution had been reported, the fundamental chemistry responsible for many of these general features began to be established. Thus, as first suggested by F. E. Blacet in 1952, photodissociation of NO_2 in air was shown to form O_3 (Blacet, 1952):

$$NO_2 + h\nu(\lambda < 430 \text{ nm}) \rightarrow NO + O, \quad (7)$$

$$O + O_2 \xrightarrow{M} O_3. \quad (2)$$

Reaction (2) still remains the sole significant source of anthropogenically produced ozone.

The nitric oxide formed in reaction (7) was also shown to react relatively rapidly with O_3, re-forming NO_2:

$$NO + O_3 \rightarrow NO_2 + O_2. \quad (8)$$

Because of reaction (8), significant concentrations of O_3 and NO cannot co-exist, and the delay in the oxidant (O_3) peak until NO has fallen to low concentrations, shown in Fig. 1.3, is explained.

Three major questions on the overall atmospheric chemistry of photochemical "smog," not readily answered in the early studies, are:

- How is NO oxidized to NO_2?
- What is the role played by organics?
- What reactions are responsible for the rapid loss of organics?

It was first suggested in the 1950's that NO was thermally oxidized by O_2:

$$2NO + O_2 \rightarrow 2NO_2. \quad (9)$$

Indeed, in the laboratory, at Torr concentrations, the clear, colorless NO is oxidized in air virtually instantaneously to dark, red-brown NO_2.

However, the rate of this reaction is second order in NO; that is, the speed of oxidation increases as the square of the NO concentration. Thus when one lowers the NO from high (Torr) concentrations to ambient part per trillion (ppt) or part per billion (ppb) levels (ppt = parts in 10^{12}; ppb = parts in 10^9), the speed of the oxidation drops to the point where the rate is very small. For example, at 100 Torr NO ($\sim 1.3 \times 10^5$ ppm),

about 85% of the NO is oxidized in ~ 15 s. However, at 100 ppb NO, approximately 226 days would be required to achieve the same net oxidation! As a result, the so-called thermal (i.e., nonphotochemical) oxidation of NO by reaction (9) is generally too slow to be of importance in the atmosphere.

One exception to this generalization is the case where high concentrations of NO (e.g., several thousand ppm) may be emitted from sources such as uncontrolled power plants. In the initial seconds as the plume enters the atmosphere before it has had a chance to become completely diluted with the surrounding air, the NO may be sufficiently concentrated that the oxidation (9) by O_2 is significant. For example, at 2000 ppm NO, 90% of the reactant would be oxidized to NO_2 within 30 min if this high concentration were to be maintained for that long. The plume integrity is generally not maintained for this period of time; however, under some meteorological conditions, the plume can be sufficiently stable that a significant fraction of the NO can undergo thermal oxidation by O_2, and NO_2 can be directly formed many kilometers from the stack.

In summary, it soon became evident that in ambient photochemical smog, the thermal oxidation of NO could not explain the relatively rapid conversion of NO to NO_2.

With respect to the role of the organics, it was suggested about 1969–1970 that the hydroxyl radical drives the daytime chemistry of both polluted and clean atmospheres (Heicklen et al., 1969; Weinstock, 1969; Stedman et al., 1970; Levy, 1971). Thus, OH initiates chain reactions by attack on VOC or CO. These chains are then propagated through reactions such as those in Fig. 1.4. In this cycle, the organic is oxidized to a ketone, two molecules of NO are converted to NO_2, and OH is regenerated. Of course, the ketone can then photodissociate into free radicals or itself be attacked by OH, and a similar cycle occurs, leading to further NO oxidation.

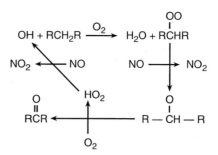

FIGURE 1.4 Typical sequence of elementary reactions in which OH initiates the oxidation of an alkane in the troposphere.

The chain reactions are eventually terminated by such reactions as

$$OH + NO_2 \xrightarrow{M} HNO_3, \quad (10)$$

$$HO_2 + HO_2 \xrightarrow{M, H_2O} H_2O_2 + O_2, \quad (11)$$

$$RO_2 + HO_2 \rightarrow ROOH + O_2. \quad (12)$$

A major source of OH in both clean and polluted air is the photodissociation of O_3 by actinic UV radiation in sunlight to produce an electronically excited oxygen atom, $O(^1D)$,

$$O_3 + h\nu(\lambda < 340 \text{ nm}) \rightarrow O(^1D) + O_2, \quad (13)$$

followed by a very rapid reaction, in competition with deactivation, of the excited oxygen atom with water vapor, which is always present in the atmosphere:

$$O(^1D) + H_2O \rightarrow 2OH. \quad (14)$$

In polluted airsheds, other direct sources also form OH through photodissociation, including nitrous acid:

$$HONO + h\nu(\lambda < 370 \text{ nm}) \rightarrow OH + NO, \quad (15)$$

and hydrogen peroxide (H_2O_2):

$$HOOH + h\nu(\lambda < 370 \text{ nm}) \rightarrow 2OH. \quad (16)$$

A very important "thermal" source of hydroxyl radicals, as well as NO_2, is the reaction of HO_2 with NO:

$$HO_2 + NO \rightarrow OH + NO_2. \quad (17)$$

This is a major chain propagation step in the overall reaction mechanism for ozone formation in photochemical air pollution. Because HO_2 is intimately tied to OH through reaction (17) and cycles such as that in Fig. 1.4, when NO is present the sources and sinks of HO_2 are, in effect, sources or sinks of the OH radical.

Sources of HO_2 include the reactions of O_2 with hydrogen atoms and formyl radicals, both of which are produced, for example, by the photodissociation of gaseous formaldehyde following absorption of solar actinic UV radiation.

$$HCHO + h\nu(\lambda < 370 \text{ nm}) \rightarrow H + HCO, \quad (18)$$

$$H + O_2 \xrightarrow{M} HO_2, \quad (19)$$

$$HCO + O_2 \rightarrow HO_2 + CO. \quad (20)$$

Another source of the hydroperoxyl radical is the abstraction of a hydrogen atom from alkoxy radicals by molecular oxygen:

$$RCH_2O + O_2 \rightarrow RCHO + HO_2. \quad (21)$$

The relative importance of these sources of OH and HO_2 radicals depends on the species present in the air mass, and hence on location and time of day. Figure 1.5, for example, shows the relative contributions as a function of time of day of three sources of OH/HO_2 in an urban air mass. In this case, nitrous acid is predicted to be the major OH source in the early morning hours, HCHO in mid-morning, and O_3 later in the day when its concentration has built up significantly (Winer, 1985; Winer and Biermann, 1994).

In summary, NO is now known to be converted to NO_2 during daylight hours in a reaction sequence initiated by OH attack on organics, and involving HO_2 and RO_2 free radicals. These peroxy radicals are the species that actually convert NO to NO_2 at ambient concentrations where the thermal oxidation of NO by O_2 is negligible.

c. Nighttime Chemistry of NO_2

The late 1970's saw the birth of a new aspect of atmospheric chemistry. Thus, in addition to ozone and photochemical oxidant formed in the daytime photooxidation of VOCs, there is an important *nighttime* chemistry, not only in polluted urban and suburban air environments, but also in relatively remote atmospheres.

In the late 1970's, a group of German researchers developed a novel instrument, the long-path length (e.g., 1–17 km) ultraviolet–visible differential optical absorption spectrometer, DOAS (Platt et al., 1979). Application of this instrument in field studies in remote

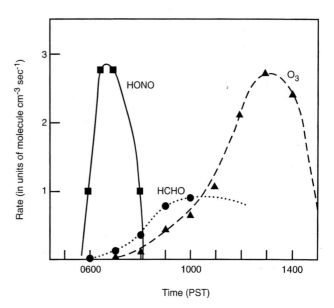

FIGURE 1.5 Predicted rates of generation of OH/HO_2 in a polluted urban atmosphere as a function of time of day for three free radical sources (adapted from Winer, 1985).

and polluted ambient air in Europe and the United States provided unequivocal evidence for the presence at night of two important "trace" nitrogenous species: gaseous nitrous acid (Perner and Platt, 1979) and the gaseous nitrate radical, NO_3 (Platt *et al.*, 1980). Both photodissociate rapidly and efficiently in daylight—hence the term "nighttime chemistry." The ambient levels, rates of formation, reactivities, and fates as well as the experimental details of current DOAS systems are discussed in detail in subsequent chapters.

(1) The NO_3 radical Briefly, the NO_3 radical is formed in the reaction

$$NO_2 + O_3 \rightarrow NO_3 + O_2 \quad (22)$$

and found at night at levels ranging from less than a few ppt (parts per trillion, 10^9) in remote regions to several hundred ppt in polluted atmospheres. It plays at least two major roles in the troposphere. Thus, it is nighttime sink for certain VOCs through addition as well as H-atom abstraction reactions

$$NO_3 + RH \rightarrow HNO_3 + R. \quad (23)$$

Furthermore, it reacts with NO_2 to form gaseous dinitrogen pentoxide, N_2O_5, in the equilibrium

$$NO_3 + NO_2 \leftrightarrow N_2O_5. \quad (24)$$

To date, there are no *direct* tropospheric measurements of N_2O_5 at the levels predicted to be in natural or polluted air masses. However, concentrations of N_2O_5 as high as 10–15 ppb have been calculated for the Los Angeles area using simultaneous measurements of ambient NO_3 and NO_2 and the equilibrium constant for reaction (24) (e.g., see Atkinson *et al.*, 1986).

In the troposphere N_2O_5 is an important nighttime source of nitric acid through its rapid hydrolysis on wet surfaces and aerosol particles:

$$N_2O_{5(g)} + H_2O \xrightarrow{\text{surfaces, aerosol particles}} 2HNO_3. \quad (25)$$

Furthermore, N_2O_5 plays an important role in key stratospheric heterogeneous processes (see Chapter 12).

(2) Gas-phase nitrous acid Gaseous HONO plays an important role in the chemistry of irradiated mixtures of VOC and NO_x in air, whether in smog chambers or in ambient atmospheres. Thus, it strongly absorbs actinic UV radiation and, at sunrise (Fig. 1.5), decomposes into OH radicals and NO with a high quantum efficiency:

$$HONO + h\nu(\lambda < 370 \text{ nm}) \rightarrow OH + NO. \quad (26)$$

Use of the long-path length DOAS technique has confirmed its presence in polluted ambient as well as relatively clean continental air masses at levels ranging from ~ 15 ppb at night in a highly polluted air mass down to sub-ppb levels in remote regions (see Chapter 11). It is found at much higher levels in indoor air environments having combustion sources such as gas or propane stoves (see Chapter 15).

Direct sources of ambient HONO, established unequivocally through use of the long-path length DOAS technique, include primary emissions, e.g., from light-duty motor vehicles having high levels of NO_x in exhaust gases (Pitts *et al.*, 1984). As discussed in Chapter 15, emissions from indoor combustion sources, e.g., gas-fired kitchen stoves and gas or propane heaters, can also produce high levels of HONO in poorly ventilated indoor air environments.

Interestingly, heterogeneous processes appear to be involved in HONO formation, certainly in smog chambers and indoor air environments and most likely on a variety of surfaces outdoors. It is produced from gaseous NO_2 and adsorbed water in a heterogeneous reaction on surfaces (see Chapter 7):

$$2NO_2 + H_2O \xrightarrow{\text{surface}} HONO_{(g)} + HNO_3(\text{ads?}). \quad (27)$$

4. Acid Deposition

a. Historical

The recognition of acid deposition, commonly called "acid rain," also has a long history. In England and Sweden, the presence of sulfur compounds and acids in polluted air and rain was recognized as early as the eighteenth century. Indeed, in 1692, Robert Boyle referred to "nitrous or salino-sulphurous spirits" in the air in his book *A General History of the Air* [see the excellent historical perspectives given by Brimblecombe (1978) and Cowling (1982)].

Remarkably, in 1872, a century before it became an international issue, a treatise on acid rain was published in England by Robert Angus Smith. Twenty years earlier, he had analyzed rain near Manchester and noted three types of areas as one moved from the city to the surrounding countryside:

> "that with carbonate of ammonia in the fields at a distance, that with sulfate of ammonia in the suburbs and that with sulphuric acid or acid sulphate, in the town."

In his 1872 book *Air and Rain: The Beginnings of a Chemical Climatology*, Smith coined the term *acid rain* and described many of the factors affecting it, such as coal combustion and the amount and frequency of precipitation. He also suggested experimental protocols to be followed in sample collection and analysis and described acid rain damage to plants and materials.

b. Overview of Acidic Rain and Fogs

Acid rain arises from the oxidation of SO_2 and NO_2 in the troposphere to form sulfuric and nitric acids, as well as other species, which are subsequently deposited at the earth's surface, either in precipitation (*wet deposition*) or in dry form (*dry deposition*). The contribution of organic acids has also been recognized recently (see Chapter 8). These oxidation and deposition processes can occur over relatively short distances from the primary pollutant sources or at distances of a 1000 km or more. Thus both short-range and long-range transport must be considered.

The gas-phase oxidation of both SO_2 and NO_2 is initiated by reaction with hydroxyl radicals:

$$OH + SO_2 \xrightarrow{M} HOSO_2 \xrightarrow{H_2O} H_2SO_4, \quad (28)$$

$$OH + NO_2 \xrightarrow{M} HNO_3. \quad (10)$$

In the case of SO_2, oxidation in the aqueous phase, present in the atmosphere in the form of aerosol particles, clouds, and fogs, is also important. Thus SO_2 from the gas phase dissolves in these water droplets and may be oxidized within the droplet by such species as H_2O_2, O_3, O_2, and free radicals. Oxidation of SO_2 on the surfaces of solids either present in the air or suspended in the water droplets is also possible. On the other hand, it is believed that HNO_3 is formed primarily by reaction (10) in the gas phase and subsequently dissolves in droplets.

These oxidation processes can lead to highly acidic fogs. For example, pH values as low as 1.69 have been measured in coastal regions of southern California (Jacob and Hoffmann, 1983). These high acidities, accompanied by high concentrations of other anions and cations, are likely due to evaporation of water from the fog droplets, leaving very high concentrations of ions in a strongly acidic liquid phase. Such acid fogs, whether in London or Los Angeles, are a major health concern because the droplets are sufficiently small to be efficiently inhaled (Hoffmann, 1984).

C. CHEMISTRY OF THE NATURAL TROPOSPHERE: REMOTE ATMOSPHERES

Although there is sometimes a tendency to think of the chemistry of the "natural" troposphere as somehow different from that of more polluted areas, it is essentially the same VOC–NO_x chemistry described earlier in this chapter. However, given that there are the significant biogenic sources of a wide variety of organics, the major difference lies in the NO_x levels, which are much lower in remote regions.

Because of these lower concentrations of NO, the self-reactions of HO_2 and RO_2 radicals can become competitive with NO reactions:

$$HO_2 + HO_2 \xrightarrow{M, H_2O} H_2O_2 + O_2, \quad (11)$$

$$HO_2 + RO_2 \rightarrow ROOH + O_2, \quad (12)$$

$$RO_2 + RO_2 \rightarrow products. \quad (29)$$

Because reactions (11), (12), and (29) do not oxidize NO to NO_2, they do not result in the formation of O_3 (via NO_2 photolysis). Indeed, an additional reaction of HO_2, that with O_3,

$$HO_2 + O_3 \rightarrow OH + 2O_2, \quad (30)$$

actually results in the destruction of ozone.

As developed in more detail in Chapter 6, the NO concentration at which other reactions such as (11), (12), and (29) occur at approximately the same rate as the HO_2/RO_2 + NO reaction is in the 10–50 ppt range. These concentrations are sufficiently low that they are encountered only in remote atmospheres, where the influence of anthropogenic emissions is minimized.

However, it is noteworthy that on a global level, O_3 levels have been increasing significantly over the past century, coinciding with the increase in fossil fuel usage and the associated increase in NO_x emissions (Bojkov, 1986; Volz and Kley, 1988; McKeen *et al.*, 1989; Anfossi *et al.*, 1991; Sandroni *et al.*, 1992; Marenco *et al.*, 1994; Anfossi and Sandroni, 1997). Figure 1.6 shows that O_3 levels appear to have been about 10–15 ppb a century ago, compared to 30–40 ppb measured around the world today. This is consistent with a shift in the

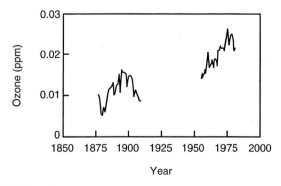

FIGURE 1.6 Typical tropospheric ozone concentrations in the 1800's and present values (adapted from Volz and Kley, 1988).

chemistry on a global scale from the self-reactions of HO_2 and RO_2 [reactions (11), (12), (29), and (30) with O_3] to their reaction with NO to generate NO_2 and subsequently O_3.

In short, the chemistry of remote regions differs primarily in the availability of NO_x.

D. CHEMISTRY OF THE STRATOSPHERE

The stratosphere is often referred to as the "ozone layer," because of the relatively high concentrations produced by photochemical reactions in this region of the atmosphere. Ozone, derived from the Greek word meaning "to smell," was first discovered by Schönbein in 1839. It has a rather pungent smell, which is sometimes noticeable around copy machines and laser printers that use discharge processes.

As described earlier, in the stratosphere, a steady-state concentration of O_3 is produced naturally by the Chapman cycle, reactions (1)–(4). Until about 1970, relatively little attention was paid to potential anthropogenic (i.e., man-made) perturbations of the stratosphere. At that time, Crutzen (1970) examined the potential role of NO and NO_2 formed in the stratosphere from reactions of N_2O that was originally generated at the earth's surface. Because N_2O is unreactive in the troposphere, it has a sufficiently long lifetime to end up in the stratosphere, where it can be converted into NO (see Chapter 12). Crutzen (1970) proposed that the NO and NO_2 formed from reactions of N_2O can then participate in a chain reaction that destroys O_3:

$$NO + O_3 \rightarrow NO_2 + O_2, \quad (8)$$

$$NO_2 + h\nu \rightarrow NO + O, \quad (31)$$

$$NO_2 + O \rightarrow NO + O_2. \quad (32)$$

Subsequently, Johnston (1971) suggested that NO_x emitted directly into the stratosphere from supersonic transport aircraft (SSTs) could decrease the steady-state concentration of O_3 via reactions (8) and (32), leading to increased UV radiation at the earth's surface. Although the number of SSTs that ultimately were produced was far less than originally anticipated, the same issue of NO_x destruction of O_3 has again been revisited with respect to the possible future production and use of high-speed civil transport (HSCT) aircraft (see Chapter 12).

In 1974, Cicerone and Stolarski suggested that if there were sources of atomic chlorine in the stratosphere, the following catalytic ozone destruction cycle could occur:

$$Cl + O_3 \rightarrow ClO + O_2 \quad (33)$$

$$ClO + O \rightarrow Cl + O_2 \quad (34)$$

$$\text{Net: } O_3 + O \rightarrow 2O_2$$

Shortly thereafter, Molina and Rowland (1974) published their seminal paper in which they showed that, because of the lack of removal mechanisms for these inert chemicals in the troposphere, chlorofluorocarbons (CFCs) are expected to reach the stratosphere. Once in the stratosphere, the CFCs are exposed to UV radiation in the region in which they absorb, leading to the production of chlorine atoms:

$$CF_2Cl_2 + h\nu \rightarrow Cl + CF_2Cl. \quad (35)$$

From the outset, their hypothesis caused worldwide concern because UV-B can damage DNA, cause skin cancer, including malignant melanoma in humans, and have long-term effects on the eye, in addition to its impact on a variety of ecosystems.

This foundation of the interaction between tropospheric processes and emissions and stratospheric chemistry established by Crutzen, Molina, and Rowland led to their sharing the 1995 Nobel Prize in Chemistry.

Until the mid-1980's, estimates of the reduction in the steady-state concentration of stratospheric O_3 were based largely on the application of large computer models that incorporated the known mechanisms and kinetics of individual chemical reactions. Because of natural variations in stratospheric ozone, unequivocal detection of the expected change is quite difficult. However, in 1985 a remarkable observation was made by a group of British researchers (Farman et al., 1985) who had been monitoring total column ozone in the Antarctic. Total column ozone is the integrated ozone measured in a vertical column throughout the atmosphere and is often expressed in Dobson units (DU), where 1 DU is equivalent to a column of O_3 of 10^{-5} m height at STP (1 atm pressure and 273.15 K). Since approximately 85–90% of the total column ozone is in the stratosphere, changes in column ozone are particularly sensitive to stratospheric ozone levels.

Figure 1.7 shows the total column ozone measured in October at one Antarctic location, Halley Bay, as a function of year. This includes the original Farman et al. (1985) data, as well as more recent data up to 1994 (Jones and Shanklin, 1995). It is clear that starting in the late 1970's, there was a dramatic drop in total column ozone at the end of the polar winter when sunrise occurs. Observation of such a rapid change is unprecedented and quite remarkable.

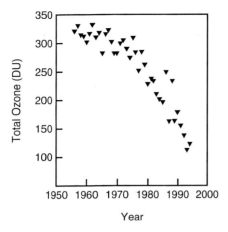

FIGURE 1.7 Average total column ozone measured in October at Halley Bay, Antarctica, from 1957 to 1994 (adapted from Jones and Shanklin, 1995).

After the first reports of this phenomenon, major field campaigns were launched, which clearly established a relationship between ozone destruction and chlorine chemistry. For example, Fig. 1.8 shows simultaneous aircraft measurements of ozone and the free radical ClO as the plane flew toward the South Pole. As it entered the polar vortex, a relatively well-contained air mass over Antarctica, O_3 dropped dramatically and ClO rose simultaneously. The key to the dramatic changes in O_3 occurring in this region now appears to be "heterogeneous" chemistry occurring on and in polar stratospheric clouds (PSC's), combined with the formation of a relatively well-contained air mass over the continent during the polar winter. This unique combination of chemistry and meteorology is discussed in more detail in Chapter 12.

In short, although the history of anthropogenic perturbations to the stratosphere is much shorter, it is clear that these are also important. Indeed, such perturbations are expected to affect the chemistry of the troposphere as well; for example, increased UV radiation will alter photochemistry at the earth's surface.

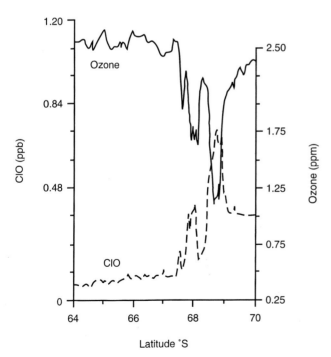

FIGURE 1.8 Measured concentrations of the chlorine monoxide free radical (ClO) as well as O_3 outside and inside the polar vortex on August 23, 1987 (adapted from Anderson, 1989).

E. GLOBAL CLIMATE CHANGE

Destruction of stratospheric ozone caused by relatively small atmospheric concentrations of chlorofluorocarbons has vividly illustrated the capacity of human activity to alter our atmosphere in a manner that has significant and far-ranging effects. There is similar concern for the effects of greenhouse gases on the earth's climate.

Solar radiation provides energy to the earth. Radiative balance is achieved through a combination of processes summarized in Fig. 1.9. A major process is the re-emission of radiation in the infrared by the earth's surface, so-called "terrestrial radiation." This infrared radiation is absorbed and re-emitted by a number of atmospheric gases, during which part of the energy is converted to thermal energy. The result of the balance in the incoming and outgoing radiation is a relatively "warm" troposphere, compared to what it would be without this natural "greenhouse effect."

It is known from measurements that the atmospheric concentrations of a number of greenhouse gases have increased significantly since the beginning of the industrial revolution. These include not only the well-recognized CO_2 but also a host of other trace gases, including O_3, as already discussed, and N_2O and CH_4 whose major sources are biogenic. Increases in such gases alter the radiation balance by trapping more of the terrestrial infrared radiation, which results in a larger amount of energy converted to thermal energy in the troposphere. More complex are the effects of aerosol particles. These impact the radiative properties of the earth both through direct effects such as scattering and, to a lesser extent, absorbing solar radiation and through indirect effects, by altering cloud properties.

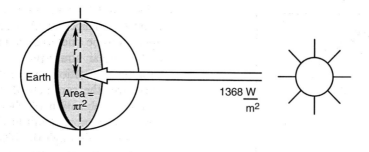

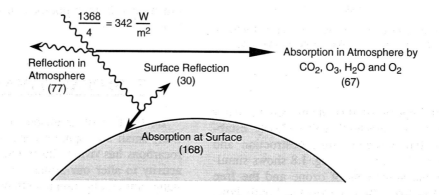

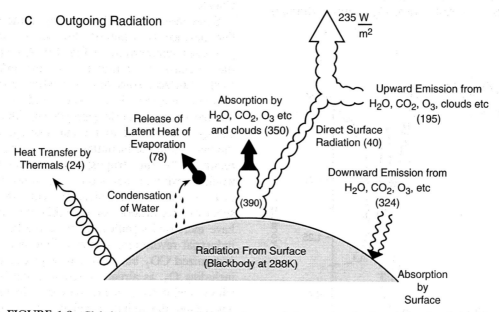

FIGURE 1.9 Global average mean radiation and energy balance per unit of earth's surface [adapted with permission from IPCC (1996) with numbers from Kiehl and Trenberth (1997)].

Significantly complicating predictions of the impact of human activities on global climate is incomplete understanding of feedbacks, which can be either positive or negative. As a result, there are currently large "error bars" on predictions of global climate change due to human activities, and this is a very active area of research. The chemical aspects of the atmosphere relevant to global climate issues are discussed in detail in Chapter 14.

F. INDOOR AIR POLLUTION

A major reason for understanding the chemistry of the atmosphere is the impact that changes can have on human health and well-being. With respect to effects due to direct inhalation of gases and particles, exposure occurs not only outdoors but also indoors as well. Indeed, the vast majority of time for most people is spent indoors. As a result, understanding the nature of the indoor atmosphere and human impacts on it is important as well.

A number of studies have documented that concentrations of some of the directly emitted species found in outdoor atmospheres can be quite high indoors if there are emission sources present such as combustion heaters, gas stoves, or tobacco smoke. In addition, there is evidence for chemistry analogous to that occurring outdoors taking place in indoor air environments, with modifications for different light intensities and wavelength distributions, shorter residence times, and different relative concentrations of reactants. In Chapter 15, we briefly summarize what is known about the chemical composition and chemistry of indoor atmospheres.

G. DISCUSSION TOPIC AND OZIPR MODEL

1. Discussion Topic: "Background Ozone"

It was accepted for a number of years in the atmospheric chemistry community that so-called "background" ozone was typically around 30–40 ppb. However, starting in the mid-1980's, a number of researchers examined the literature of a century earlier, shortly after ozone was discovered by Schönbein in 1839, and discovered several series of ozone measurements that had been made at different locations in the troposphere. Some of the papers describing this research include the papers by Bojkov (1986), Volz and Kley (1988), and Anfossi et al. (1991).

Some Points for Discussion

a. What techniques were used for measuring ozone in the nineteenth century? What were the likely interferences and/or parameters that could affect the results and how serious were they likely to be?

b. At what locations were measurements made for a period of years?

c. Given the potential interferences, what are the typical levels of tropospheric O_3 in the nineteenth century?

d. Given typical tropospheric O_3 levels today, discuss possible causes for the increase.

e. Name as many impacts as you can think of that might be associated with this increase in O_3 on a global basis. Be sure to include effects on the chemistry of the atmosphere as well as effects on health, global change, etc.

f. Some of the ozone levels reported in these papers are in mPa. What is 1 mPa in the more commonly used units of ppb? (See Chapter 2 for a discussion of units and conversions.)

2. OZIPR Model

The OZIPR model is described in Appendix III. It is a box model that has comprehensive chemical submodels that lump organics in two different ways. It allows for variation of meteorology through time-dependent mixing heights, temperatures, etc.

Familiarize yourself with this model and explore the effects of meteorology on the predicted peak ozone concentrations by varying the initial mixing height and its rate of change with time.

Examine the effects of photochemistry by varying the date from June 21 to December 21 and by varying the latitude.

References

Allègre, C. J., and S. H. Schneider, "The Evolution of the Earth," *Sci. Am.*, 66–75, October (1994).

Anderson, J. G., "Free Radicals in the Earth's Atmosphere: Measurement and Interpretation," in *Ozone Depletion, Greenhouse Gases and Climate Change*, National Academy Press, Washington, DC, pp. 56–65, 1989.

Anfossi, D., S. Sandroni, and S. Viarengo, "Tropospheric Ozone in the Nineteenth Century: The Moncalieri Series," *J. Geophys. Res.*, 96, 17349–17352 (1991).

Anfossi, D., and S. Sandroni, "Short Communication—Ozone Levels in Paris One Century Ago," *Atmos. Environ.*, 31, 3481–3482 (1997).

Atkinson, R., A. M. Winer, and J. N. Pitts, Jr., "Estimation of Nighttime N_2O_5 Concentrations from Ambient NO_2 and NO_3 Radical Concentrations and the Role of N_2O_5 in Nighttime Chemistry," *Atmos. Environ.*, 20, 331–339 (1986).

Blacet, F. E., "Photochemistry in the Lower Atmosphere," *Ind. Eng. Chem.*, 44, 1339–1342 (1952).

Bojkov, R. D., "Surface Ozone during the Second Half of the Nineteenth Century," *J. Am. Meteorol. Soc.*, 25, 343–352 (1986).

Brimblecombe, P., "Interest in Air Pollution among Early Fellows of the Royal Society," *Notes Rec. R. Soc., 32,* 123–129 (1978).

Cicerone, R. J., and R. S. Stolarski, "Stratospheric Chlorine: A Possible Sink for Ozone," *Can. J. Chem., 52,* 1610–1615 (1974).

Cochran, L. S., R. A. Pielke, and E. Kovács, "Selected International Receptor-Based Air Quality Standards," *J. Air Waste Manage. Assoc., 42,* 1567–1572 (1992).

Cowling, E. B., "Acid Precipitation in Historical Perspective," *Environ. Sci. Technol., 16,* 110A–123A (1982).

Crutzen, P. J., "The Influence of Nitrogen Oxides on the Atmospheric Ozone Content," *Q. J. R. Meteorol. Soc., 96,* 320–325 (1970).

Evelyn, J., *Fumifugium: Or the Inconvenience of the Aer and Smoke of London Dissipated, Together with Some Remedies Humbly Proposed,* Bedel and Collins, London, 1661.

Farman, J. C., B. G. Gardiner, and J. D. Shanklin, "Large Losses of Total Ozone in Antarctica Reveal Seasonal ClO_x/NO_x Interaction," *Nature, 315,* 207–210 (1985).

Finlayson-Pitts, B. J., and J. N. Pitts, Jr., "The Chemical Basis of Air Quality: Kinetics and Mechanisms of Photochemical Air Pollution and Application to Control Strategies," *Adv. Environ. Sci. Technol., 7,* 76–162 (1977).

Firket, J., "Fog along the Meuse Valley," *Trans. Faraday Soc., 32,* 1192–1197 (1936).

Goodhill, V., "Maimonides—Modern Medical Relevance," XXVI Wherry Memorial Lecture, Transactions of the American Academy of Ophthalmology and Otolaryngology, p. 463, May–June 1971.

Haagen-Smit, A. J., "Chemistry and Physiology of Los Angeles Smog," *Ind. Eng. Chem., 44,* 1342–1346 (1952).

Haagen-Smit, A. J., and M. M. Fox, "Photochemical Ozone Formation with Hydrocarbons and Automobile Exhaust," *J. Air Pollut. Control Assoc., 4,* 105–108, 136 (1954).

Haagen-Smit, A. J., and M. M. Fox, "Automobile Exhaust and Ozone Formation," *SAE Trans., 63,* 575–580 (1955).

Haagen-Smit, A. J., and M. M. Fox, "Ozone Formation in Photochemical Oxidation of Organic Substances," *Ind. Eng. Chem., 48,* 1484–1487 (1956).

Haagen-Smit, A. J., E. F. Darley, M. Zaitlin, H. Hull, and W. Noble, "Investigation on Injury to Plants from Air Pollution in the Los Angeles Area," *Plant Physiol., 27,* 18–34 (1952).

Haagen-Smit, A. J., C. E. Bradley, and M. M. Fox, "Ozone Formation in Photochemical Oxidation of Organic Substances," *Ind. Eng. Chem., 45,* 2086–2089 (1953).

Heicklen, J., K. Westberg, and N. Cohen, Center for Air Environmental Studies, Report No. 115-69, 1969.

Hoffmann, M. R., "Response to Comment on 'Acid Fog,'" *Environ. Sci. Technol., 18,* 61–64 (1984).

IPCC, Intergovernmental Panel on Climate Change, Contribution of Working Group I to the Second Assessment Report (J. T. Houghton, L. G. Meira Filho, B. A. Callander, N. Harris, A. Kattenberg, and K. Maskell, Eds.), *Climate Change 1995,* Cambridge Univ. Press, Cambridge, UK, 1996.

Jacob, D. J., and M. R. Hoffmann, "A Dynamic Model for the Production of H^+, NO_3^-, and SO_4^{2-} in Urban Fog," *J. Geophys. Res., 88,* 6611–6621 (1983).

Johnston, H. S., "Reduction of Stratospheric Ozone by Nitrogen Oxide Catalysts from Supersonic Transport Exhaust," *Science, 173,* 517–522 (1971).

Jones, A. E., and J. D. Shanklin, "Continued Decline of Total Ozone over Halley, Antarctica, Since 1985," *Nature, 376,* 409–411 (1995).

Kasting, J. F., "Earth's Early Atmosphere," *Science, 259,* 920–926 (1993).

Kiehl, J. T., and K. E. Trenberth, "Earth's Annual Global Mean Energy Budget," *Bull. Am. Meteorol. Soc., 78,* 197–208 (1997).

Kirchner, J. W., "The Gaia Hypothesis: Can It Be Tested?", *Rev. Geophys., 27,* 223–235 (1989).

Lenton, T. M., "Gaia and Natural Selection," *Nature, 394,* 439–447 (1998).

Levy, H., "Normal Atmosphere: Large Radical and Formaldehyde Concentrations Predicted," *Science, 173,* 141–143 (1971).

Lodge, J. P., Jr., Ed., *The Smoke of London. Two Prophecies*: Evelyn, J., 1661. *Fumifugium: Or the Inconvenience of the Aer and Smoke of London Dissipated*, and Barr, R., date unknown, *The Doom of London,* Maxwell Reprint Co., Elmsford, New York, 1969.

Lovelock, J. E., "Geophysiology, the Science of Gaia," *Rev. Geophys., 27,* 215–222 (1989).

Mamane, Y. "Air Pollution Control in Israel during the First and Second Century," *Atmos. Environ., 21,* 1861–1863 (1987).

Marenco, A., H. Gouget, P. Nedelec, J.-P. Pages, and F. Karcher, "Evidence of a Long-Term Increase in Tropospheric Ozone from Pic du Midi Data Series: Consequences: Positive Radiative Forcing," *J. Geophys. Res., 99,* 16617–16632 (1994).

McKeen, S., D. Kley, and A. Volz, "The Historical Trend of Tropospheric Ozone over Western Europe: A Model Perspective," in *Ozone in the Atmosphere* (R. D. Bojkov and P. Fabian, Eds.), A. Deepak Publishing, Hampton, VA, 1989.

Middleton, J. T., J. B. Kendrick, Jr., and H. W. Schwalm, "Injury to Herbaceous Plants by Smog or Air Pollution," *U.S.D.A. Plant Dis. Rep., 34,* 245–252 (1950).

Molina, M. J., and F. S. Rowland, "Stratospheric Sink for Chlorofluoromethanes: Chlorine Atom-Catalysed Destruction of Ozone," *Nature, 249,* 810–812 (1974).

Perner, D., and U. Platt, "Detection of Nitrous Acid in the Atmosphere by Differential Optical Absorption," *Geophys. Res. Lett., 6,* 917–920 (1979).

Platt, U., D. Perner, and H. W. Pätz, "Simultaneous Measurement of Atmospheric CH_2O, O_3, and NO_2 by Differential Optical Absorption," *J. Geophys. Res., 84,* 6329–6335 (1979).

Platt, U., D. Perner, A. M. Winer, G. W. Harris, and J. N. Pitts, Jr., "Detection of NO_3 in the Polluted Troposphere by Differential Optical Absorption," *Geophys. Res. Lett., 7,* 89–92 (1980).

Pitts, J. N., Jr., H. W. Biermann, A. M. Winer, and E. C. Tuazon, "Spectroscopic Identification and Measurement of Gaseous Nitrous Acid in Dilute Auto Exhaust," *Atm. Environ., 18,* 847–854 (1984).

Rouéché, B., *Eleven Blue Men,* pp. 173–191, New Berkley Medallian Edition, New York, 1965.

Sandroni, S., D. Anfossi, and S. Viarengo, "Surface Ozone Levels at the End of the Nineteenth Century in South America," *J. Geophys. Res., 97,* 2535–2539 (1992).

Schneider, S. H., "Debating Gaia," *Environment, 32,* 5–9 & 29–32 (1990).

Stedman, D. H., E. D. Morris, Jr., E. E. Daby, H. Niki, and B. Weinstock, "The Role of OH Radicals in Photochemical Smog Reactions," 160th National Meeting of the American Chemical Society, Chicago, IL, Sept. 14–18, 1970.

Volz, A., and D. Kley, "Evaluation of the Montsouris Series of Ozone Measurements Made in the Nineteenth Century," *Nature, 332,* 240–242 (1988).

Weinstock, B., "Carbon Monoxide: Residence Time in the Atmosphere," *Science, 166,* 224–225 (1969).

Wilkins, E. T., "Air Pollution and the London Fog of December 1952," *J. R. Sanitary Inst., 74,* 1–21 (1954).

Winer, A. M., "Air Pollution Chemistry," in *Handbook of Air Pollution Analysis* (R. M. Harrison and R. Perry, Eds.), 2nd ed., Chap. 3. Chapman and Hall, London, 1985.

Winer, A. M., and H. W. Biermann, "Long Pathlength Differential Optical Absorption Spectroscopy (DOAS) Measurements of Gaseous HONO, NO_2, and HCHO in the California South Coast Air Basin," *Res. Chem. Intermed., 20,* 423–445 (1994).

CHAPTER 2

The Atmospheric System

As discussed in Chapter 1, much of our understanding of the chemistry of our atmosphere is based on early studies of air pollution; these are often treated in the context of an overall "system." This approach starts with the various sources of anthropogenic and natural emissions and tracks the resulting pollutants through their atmospheric transport, transformations, and ambient concentrations—on local, regional, and global scales—to their ultimate chemical and physical fates, including their impacts on our health and environment.

Figure 2.1 is a simplified diagram illustrating the major elements. *Primary* pollutants are defined as those emitted directly into the air, e.g., SO_2, NO, CO, Pb, organics [including HAPS (hazardous air pollutants)], and combustion-generated particulate matter (PM). Sources may be anthropogenic, biogenic, geogenic, or some combination thereof. Once in the atmosphere, they are subjected to dispersion and transport, i.e., meteorology, and simultaneously to chemical and physical transformations into gaseous and particulate *secondary* pollutants; the latter are defined as those formed from reactions of the primary pollutants in air. Both primary and secondary pollutants are removed at the earth's surface via wet or dry deposition and, in the processes of transport, transformation, and deposition, can impact a variety of receptors, for example, humans, animals, aquatic ecosystems, forests and agricultural crops, and materials.

From a detailed knowledge of the emissions, topography, meteorology, chemistry, and deposition processes, one can develop mathematical models that predict the concentrations of primary and secondary pollutants as a function of time at various locations. Depending on the particular model, these may describe pollutant concentrations over a variety of scales:

- In a plume from a specific point source (plume models)
- In an air basin from a combination of diverse mobile and stationary sources (airshed models)
- Over a large geographical area downwind from a group of sources (long-range transport and regional models)
- Over the entire earth (global models)

To test these models, their predictions must be compared to the observed concentrations of various species; model inputs are adjusted to obtain acceptable agreement between the observed and predicted values. These models can then be used, in combination with the documented impacts on receptors, to develop health and/or environmental risk assessments and various control strategy options.

Finally, through legislative and administrative action, health-protective and cost-effective risk-management decisions can be made, and regulatory actions implemented, that directly affect the starting point of our atmospheric system, that is, the primary emissions and their sources.

To place the remainder of this book on atmospheric chemistry in perspective, the various components of our "atmospheric system" are treated briefly next.

A. EMISSIONS

In describing a given air mass and the chemical reactions occurring therein, one must consider both natural and anthropogenic sources of primary emissions and evaluate their relative importance. Thus the impact on air quality of natural emissions can be an important issue because cost-effective control strategies must take into account the relative strengths of emissions from all sources, not just those of anthropogenic origin. However, it is not only the relative amounts of total emissions that must be considered but also the chemical nature of the emissions, e.g., their reactivities and their temporal and spatial distributions.

Emissions inventories are typically obtained by combining the rate of emissions from various sources (the "emission factors") with the number of each type of source and the time over which the emissions occur.

Inventories are compiled in various formats. For example, they can be assembled for various individual anthropogenic processes such as refining, or natural processes such as volcanic eruptions, in which emissions of all of the relevant species associated with that event are estimated. Alternatively, and more commonly, emissions inventories are compiled by species, showing the various sources that contribute to the total emissions of each.

Emission factors for various sources in the United States have been published by the Environmental Protection Agency in the form of the document *AIRCHIEF*, short for the *Air Clearing House for Inventories and Emission Factors*. Such data are available on CD-ROM as well as on-line through the EPA Web site (see Appendix IV). In Europe, the Commission of the European Communities has published a handbook of emission factors as well (e.g., see CEC (1988, 1989, 1991), McInnes (1996), and Web site in Appendix IV). Emissions inventories and emission factors for Europe are also found in the volume edited by Fenger *et al.* (1998).

On a global scale, emissions inventories for a variety of species are currently under development under the

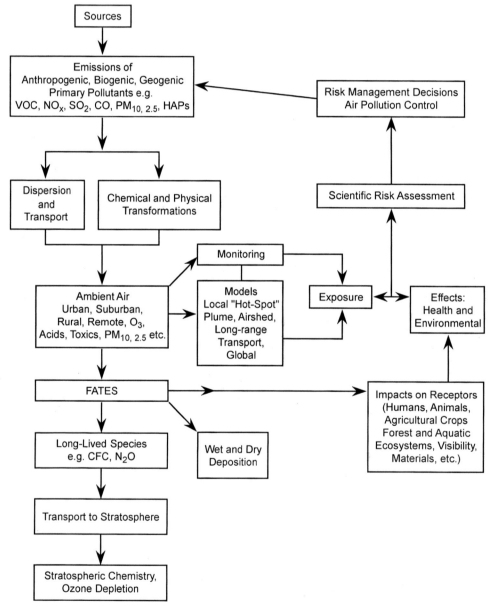

FIGURE 2.1 The atmospheric air pollution system.

auspices of the International Global Atmospheric Chemistry Project (IGAC), and various available inventories are described by Graedel *et al.* (1993). Data for some of the major pollutants follow.

1. Oxides of Nitrogen

We shall follow here the convention in current use that defines NO_x as the sum of ($NO + NO_2$) and NO_y as the sum of all reactive nitrogen-containing species, e.g., $NO_y = (NO + NO_2 + HNO_3 + PAN + HONO + NO_3 + N_2O_5 +$ organic nitrates etc.). By far the most significant species emitted by anthropogenic processes is nitric oxide, produced when N_2 and O_2 in air react during high-temperature combustion processes. In addition, some NO_x is formed from nitrogen in the fuel. Smaller amounts of NO_2 are produced by the further oxidation of NO; trace amounts of other nitrogenous species such as HNO_3 are also formed.

The fraction of the total that is emitted as NO clearly depends on the conditions associated with the specific combustion process. While most (typically > 90%) of the NO_x emitted is believed to be in the form of NO, the fraction of NO_2 can vary from less than 1% to more than 30% (e.g., Lenner, 1987).

Figure 2.2 shows the contribution of various sources to the total anthropogenic NO_x emissions, 23×10^6 short tons, or 21 Tg (expressed as NO_2), in the United States in 1996 (1 Tg = 1 teragram = 10^{12} g and one short ton = 0.907×10^6 g). This can be compared to total *global* anthropogenic emissions of approximately 72 Tg of NO_x (expressed as NO_2) (Müller, 1992).

Figure 2.3 shows the trend in NO_x emissions for North America, Europe, the USSR, and Asia from 1970 to 1986 (Hameed and Dignon, 1992). While those

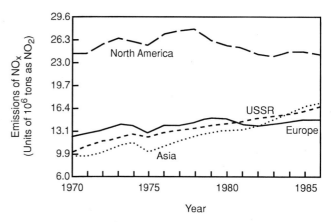

FIGURE 2.3 NO_x emissions in million tons of equivalent NO_2 for the period 1970 to 1986 for Asia, Europe, North America, and the USSR (from Hameed and Dignon, 1992).

of North America and Europe have decreased or leveled off, those from Asia and the USSR increased significantly, a trend that has continued. Figure 2.4a shows the geographical pattern of the emission flux of NO_x in Asia in 1987 (Akimoto and Narita, 1994). Clearly, Japan and China are major contributors to the flux of NO_x in this region, with the City of Tokyo having the highest emission flux rate.

There are also significant natural sources of oxides of nitrogen, in particular nitric oxide, which is produced by biomass burning as well as by soils where nitrification, denitrification, and the decomposition of nitrite (NO_2^-) contribute to NO production. Figure 2.4b, for example, shows the relative emission rates for biogenically produced NO in the United States in 1990 (EPA, 1995).

Another important natural source is NO_x produced by lightning, with recent estimates in the range of 10–33 Tg yr^{-1} as NO_2 (Flatøy and Hov, 1997; Price *et al.*, 1997a, 1997b; Wang *et al.*, 1998). By comparison to the estimated emissions from biomass burning and continental biogenic sources (Table 2.1), it is seen that lightning is quite important.

There is also some NO produced from the oxidation of NH_3 by photochemical processes in oceans and by some terrestrial plants (e.g., Wildt *et al.*, 1997).

Table 2.1 gives an estimate for global-scale natural and anthropogenic emissions of NO_x as well as of CO, CH_4, and VOC (Müller, 1992). It is seen that biomass burning and biogenic emissions of NO are comparable and together equal to about half of the anthropogenic emissions.

Nitrous oxide (N_2O, "laughing gas") is also produced by biological processes and, to a lesser extent, by anthropogenic processes (see Chapter 14.B.2c). While

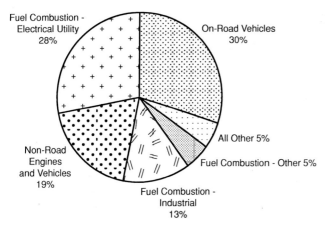

FIGURE 2.2 Contribution of various sources to total anthropogenic NO_x emissions in the United States in 1996 (from EPA, 1999).

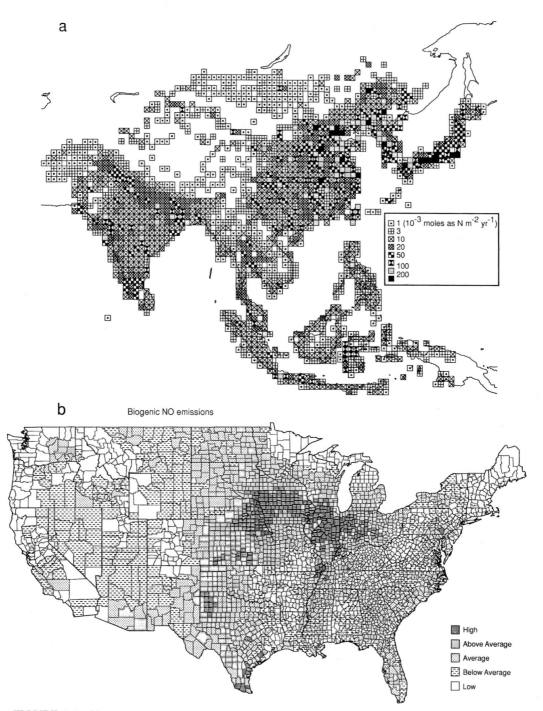

FIGURE 2.4 (a) Pattern of 1987 annual emission flux of NO_x in Asia (in units of millimoles as N per m^2 per year) (from Akimoto and Narita, 1994). (b) Estimated relative rates of biogenic emissions of NO in the United States in 1990 (from EPA, 1995).

biomass burning is the major anthropogenic source, there are a variety of smaller sources, including motor vehicles. Interestingly, the N_2O emissions from current catalyst-equipped cars appear to be higher than from noncatalyst-equipped vehicles (Berges *et al.*, 1993), but total NO_x emissions are much lower. N_2O is inert in the troposphere and is ultimately transported to the stratosphere, where it acts as a major source of NO_x (see Chapters 1 and 12).

2. Volatile Organic Compounds (VOC)

Historically, organics in the troposphere have been measured as non-methane hydrocarbons (NMHC). As

A. EMISSIONS

TABLE 2.1 Global Emission Estimates for CO, NO_x, CH_4, and VOC from Both Anthropogenic and Natural Sources (in Tg/yr)[a]

	Anthropogenic sources	Biomass burning	Continental biogenic sources[b]	Oceans	Total
CO	383	730	165	165	1440
NO_x[c]	72	18	22	0.01	122
CH_4	132	54	310	10	506
VOC	98	51	500	30–300	750

[a] Source: Müller (1992).
[b] Includes animal, microbial, and foliage emissions.
[c] Expressed as NO_2.

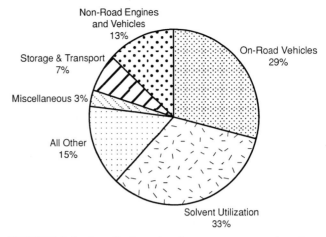

FIGURE 2.5 Contribution of various sources to total anthropogenic VOC emissions in the United States in 1996 (from EPA, 1997).

discussed elsewhere in this book, methane oxidizes relatively slowly in the troposphere and hence on a short-term basis (e.g., hours) does not contribute significantly to ozone formation compared to other organics (with some exceptions, e.g., CFCs). As a result, in terms of smog formation in urban and regional areas, methane has been excluded from consideration and controls have focused on the suite of larger organics.

However, with the recognition that a variety of organics (e.g., aldehydes) and not just hydrocarbons are important in the troposphere, alternative terminologies have been used to describe organics other than methane. These include the most commonly used term VOC (volatile organic compounds) as well as ROG (reactive organic gases) and NMOC (non-methane organic compounds). We use the term VOC, recognizing that certain volatile species such as CFCs are not included.

Figure 2.5 shows the distribution of anthropogenic sources of VOCs in the United States in 1996, a total of 19×10^6 short tons, or 17 Tg (EPA, 1997). Figure 2.6 shows VOC emissions in Europe (Friedrich and Obermeier, 1999).

Over the past decade, it has become clear that, on a global scale, biogenic processes also release substantial quantities of reactive hydrocarbons such as isoprene and α-pinene, in addition to methane and other organics, including oxygenated species such as methanol, 2-methyl-3-buten-2-ol, hexenol, acetone, and formic and acetic acids (see, for example, Fehsenfeld et al., 1992; Golden et al., 1993; Monson et al., 1995; Singh et al., 1995; Fall, 1999; Guenther, 1999); see Chapter 6.J.1.

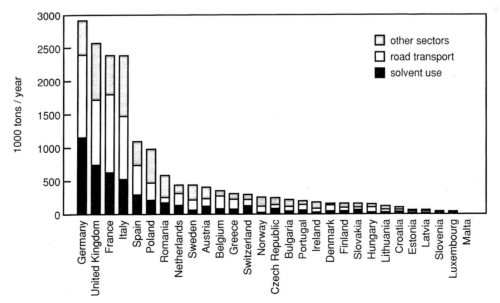

FIGURE 2.6 Annual 1990 VOC emissions in Europe [adapted from McInnes (1996) and Friedrich and Obermeier (1999)].

TABLE 2.2 Estimates of Global VOC Emissions Rates (Tg of C yr^{-1})[a]

Source	Isoprene	Monoterpene	Other more reactive organics[b]	Other less reactive organics[c]	Total VOC
Woods	372	95	177	177	821
Crops	24	6	45	45	120
Shrubs	103	25	33	33	194
Ocean	0	0	2.5	2.5	5
Other	4	1	2	2	9
All	503	127	260	260	1150

[a] From Guenther et al. (1995).
[b] Defined as having lifetimes < 1 day under typical tropospheric conditions.
[c] Defined as having lifetimes of > 1 day.

Table 2.2 shows one estimate of the global annual organic emissions from biogenic and oceanic sources and biomass burning. Such estimates are highly uncertain because the factors controlling biogenic emissions are complex, leading to large variations with time of day, season, geographical location, etc. (Lamb et al., 1993; Geron et al., 1994; Monson et al., 1995). Illustrative of this is the significant difference in the VOC emission estimates between Tables 2.2 and 2.1. Some sources are not included in these estimates, such as small amounts of isoprene produced by phytoplankton in oceans (e.g., Bonsang et al., 1992; Moore et al., 1994). A biogenics emission inventory for the United States is given by Pierce et al. (1998). The role of biogenic VOC in tropospheric chemistry is discussed in Chapter 6. For a review of VOC in natural and polluted atmospheres, see the book edited by Hewitt (1999).

3. Carbon Monoxide

Carbon monoxide is produced by the incomplete combustion of fossil fuels, and in major urban areas of developed nations a major source is the exhaust from light-duty motor vehicles (LDMV). Figure 2.7 shows the distribution of sources within the United States in 1996. A total of 89×10^6 short tons, or 81 Tg, of CO were emitted, about 60% of which comes from highway vehicles (EPA, 1997).

Natural sources of CO include CO from biomass burning and the oxidation of organics such as methane and isoprene, CO from biological processes in soils, CO from vegetation and termites, and CO from the ocean.

4. Sulfur Compounds

Ninety percent or more of the sulfur in fossil fuels is emitted in the form of sulfur dioxide (SO_2) during combustion, the remainder being primarily in the form of sulfates. Therefore, its emissions can be calculated for a given source with some accuracy from the rate of fuel consumption and the percentage of sulfur in the fuel. For example, Fig. 2.8 shows the SO_2 emissions from a number of countries as a function of total fuel consumption (Kato and Akimoto, 1992). Clearly, there is a relatively linear relationship between the two. In the United States in 1996, a total of 19×10^6 short tons, or 17 Tg, of SO_2 were emitted by anthropogenic sources, with the distribution of sources shown in Fig. 2.9 (EPA, 1997).

Figure 2.10 shows the trend in SO_2 emissions from North America, Europe, the USSR, and Asia from 1970 to 1986 (Hameed and Dignon, 1992). While SO_2 emissions from North America have decreased significantly, those from Europe have remained about the same and in the USSR have increased somewhat. However, SO_2 emissions from Asia have increased dramatically due to the increased combustion of fossil fuels, particularly from the use of coal in China and from biomass burning in Southeast Asia (Arndt et al., 1997). For example, Fig. 2.11 shows the 1987 annual emission fluxes of SO_2 in Asia, where China is seen to be a major source (Akimoto and Narita, 1994). Emissions from ships are also thought to contribute significantly (e.g., Streets et al., 1997; Corbett and Fishbeck, 1997).

Natural emissions of sulfur compounds to the atmosphere occur from a variety of sources, including volcanic eruptions, sea spray, and a host of biogenic processes (e.g., Aneja, 1990). Most of the volcanic sulfur is emitted as SO_2, with smaller and highly variable amounts of hydrogen sulfide and dimethyl sulfide (CH_3SCH_3). Sea spray contains sulfate, some of which is carried over land masses.

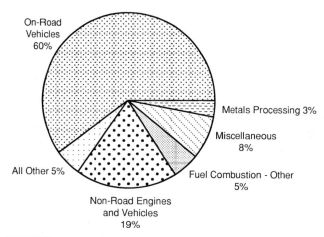

FIGURE 2.7 Contribution of various sources to the total anthropogenic CO emissions in the United States in 1996 (from EPA, 1997).

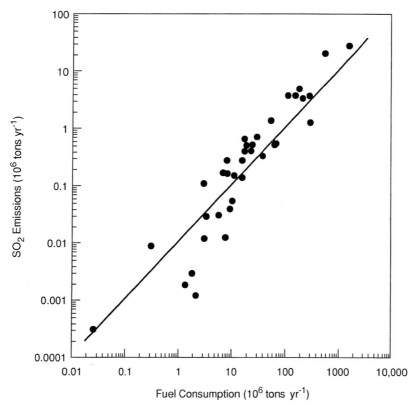

FIGURE 2.8 SO$_2$ emissions in many different regions as a function of rate of fuel consumption. Data for Europe and the United States are for 1980, and those for Asia are for 1987 (from Kato and Akimoto, 1992).

Biogenic processes, however, emit reduced forms of sulfur, including dimethyl sulfide and hydrogen sulfide, with lesser amounts of carbon disulfide (CS$_2$), dimethyl disulfide (CH$_3$SSCH$_3$), carbonyl sulfide (COS), and methyl mercaptan (CH$_3$SH). These reduced sulfur compounds are then oxidized in the atmosphere as described in detail in Chapter 8.E.

One estimate of the global emissions of sulfur compounds from both anthropogenic and natural sources is given in Table 2.3 (Spiro *et al.*, 1992).

5. Total Suspended Particles (TSP), PM10, and PM2.5

Air quality standards for particulate matter in the United States were expressed some years ago in terms of the mass of total suspended particulate matter (TSP).

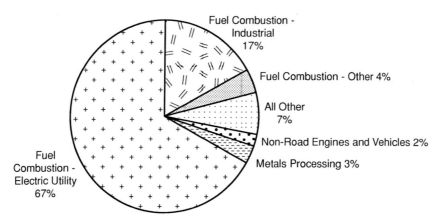

FIGURE 2.9 Contribution of various sources to total anthropogenic SO$_2$ emissions in the United States in 1996 (from EPA, 1997).

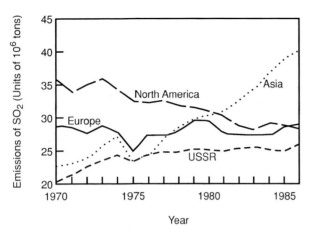

FIGURE 2.10 SO_2 emissions in million tons of equivalent SO_2 for the period 1970 to 1986 for Asia, Europe, North America, and the USSR (from Hameed and Dignon, 1992).

The standard was then changed to mass of suspended particulate matter less than 10 μm in size, commonly called PM10 or PM_{10}, and more recently was modified to include particulate matter less than 2.5 μm in diameter, PM2.5 or $PM_{2.5}$.

The rationale for basing air quality standards on smaller particles is evident from an examination of Fig. 2.12, a diagram of the human respiratory tract. Larger particles that are inhaled are removed in the head or upper respiratory tract. The respiratory system from the nose through the tracheobronchial region is covered with a layer of mucus that is continuously moved upward by the motion of small hairlike projections called cilia. Large particles deposit on the mucus, are moved up, and are ultimately swallowed.

On the other hand, particles from fossil fuel combustion and gas-to-particle conversion are generally much smaller ($<2.5\text{-}\mu$m diameter) and fall in the *respirable* size range. These particles can reach the alveolar region where gas exchange occurs. This region is not coated with a protective mucus layer, and here the clearance time for deposited particles is much greater than in the upper respiratory tract; hence the potential for health effects is much greater (Phalen, 1984).

Figure 2.13 shows the deposition of particles in various regions of the respiratory tract as a function of particle diameter (Phalen, 1984; Phalen *et al.*, 1991; Yeh *et al.*, 1996). The deposition fraction of PM_{10} in the pulmonary and tracheobronchial regions can be quite large, so it is not surprising that health effects could be associated with these particles. Deposition in the upper portions of the respiratory system is dominated primarily by the large particles, which are readily taken out in the nose and upper airways.

The deposition curves shown in Fig. 2.13 can be understood in terms of the major mechanisms of removal of particles in the respiratory tract: diffusion, sedimentation, and inertial impaction (see Chapter 9). The smallest particles undergo rapid Brownian diffusion, which carries them to the lung surface where they can be taken up; this is responsible for the large deposition in the pulmonary region seen in Fig. 2.13 for particle sizes below about 0.5 μm. Gravitational settling, i.e., sedimentation, is also an important mechanism of deposition both in the pulmonary region and in the tracheobronchial region. In both cases, the airways are relatively small so that the particle does not have large distances to travel before reaching a surface. The third mechanism, impaction, occurs when the airstream in which the particle is suspended changes direction due to a bifurcation in the lung, but the inertia of the particle carries it forward to impact the lung surface.

There has been great interest in airborne particulate matter recently due to the results of a number of epidemiological studies showing a correlation between increased mortality and levels of airborne particles. Figure 2.14 shows one such correlation reported by Dockery *et al.* (1993). A clear relationship between mortality rates and the concentration of fine particles $PM_{2.5}$, as well as with particle sulfate, is seen. Since sulfate is found primarily in fine particles, these observations are not independent. Schwartz *et al.* (1996) report a 1.5% increase in total daily mortality with an increase of 10 μg m^{-3} in $PM_{2.5}$. Deaths due to chronic obstructive pulmonary disease increased by 3.3% and those to ischemic heart disease by 2.1%.

What is somewhat puzzling, but certainly intriguing, is that the cities included in the studies in Fig. 2.14, as well as more recent ones where these findings have been corroborated, are quite disparate in terms of location and the types of air pollutants that would be expected to predominate in each region, yet a consistent relationship is found. Most such epidemiological studies to date are consistent with this finding. This suggests either that the health effects associated with particles are independent of their chemical composition or that there is some common chemical component. In addition, recent analysis of these studies also finds a correlation with other air pollutants as well and suggests that more than one pollutant may be involved (Lipfert and Wyzga, 1995). This issue is a fascinating one that clearly requires much more research on the formation, chemical composition, and effects of particles and associated air pollutants (e.g., see Phalen and McClellan, 1995; Dockery and Schwartz, 1995; Bascom *et al.*, 1996a,b; Wilson and Suh, 1997; and the review by Vedal, 1997).

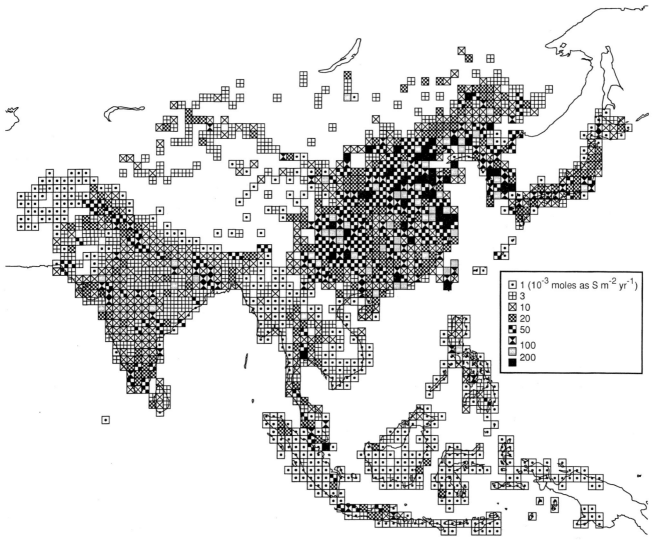

FIGURE 2.11 Patterns of 1987 annual emission flux of SO_2 in Asia (in units of millimoles as S per m^2 per year) (from Akimoto and Narita, 1994).

Interestingly, these fine particles not only are of great concern from the point of view of health effects but also are responsible for most of the light scattering, that is, visibility reduction. Thus an improvement in visibility in areas impacted by air pollution may be accompanied by a reduction in the total particle mass deposition in the alveolar region of the respiratory system as well.

In the United States in 1996, the total emissions of PM_{10} were 31×10^6 short tons per year, or 28 Tg per year (EPA, 1997). Fugitive dust sources such as unpaved roads make up $\sim 90\%$ of the total PM_{10} emissions. Figure 2.15 shows sources of PM_{10} in the United States in 1996 split into (a) nonfugitive dust sources ($\sim 10\%$ of the total) and (b) fugitive dust sources (EPA, 1997).

Globally, anthropogenic emissions of PM_{10} have been estimated to be 345 Tg yr^{-1} without including secondary nitrate and organics (Wolf and Hidy, 1997)

TABLE 2.3 Estimates of Global Emissions of Sulfur Compounds in 1980 (Tg yr^{-1} as S)[a]

Source	Sulfur emissions (Tg yr^{-1})
Fuel combustion/industrial activities	77.6
Biomass burning	2.3
Volcanic eruptions	9.6
Marine biosphere	11.9
Terrestrial biosphere	0.9
Total	102.2

[a] Source: Spiro *et al.* (1992).

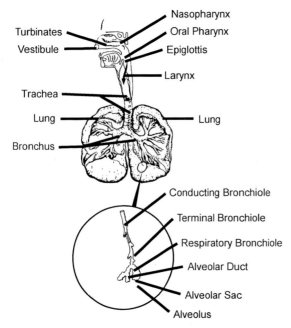

FIGURE 2.12 Schematic diagram of human respiratory tract. (From Hinds, W. C. *Aerosol Technology.* Copyright © 1982 John Wiley & Sons, Inc. Reprinted by permission of John Wiley & Sons, Inc.)

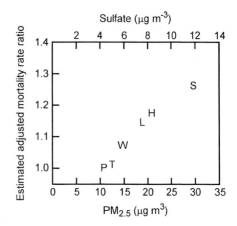

FIGURE 2.14 Estimated adjusted mortality rate ratios, taking the least polluted city, Portage, Wisconsin (P), as 1.0. T = Topeka, Kansas; W = Watertown, Massachusetts; L = St. Louis, Missouri; H = Harriman, Tennessee; S = Steubenville, Ohio. (Adapted from Dockery *et al.,* 1993.)

and are projected to increase by about a factor of two from 1990 to 2040.

The composition and sources of particles are discussed in detail in Chapter 9. Major natural sources of particles include terrestrial dust caused by winds, sea spray, biogenic emissions, volcanic eruptions, and wildfires. As with anthropogenic particulate emissions, particle size is important. Operationally speaking, terrestrial dust is generally in the size range $\sim > 10$ μm in diameter and is primarily composed of crustal elements, including silicon, aluminum, iron, sodium, potassium, calcium, and magnesium. Because particles from mechanical erosion processes tend to be quite large, they have been of less concern from the point of view of health effects since they tend to be removed in the upper respiratory system.

Particles are generated at the surface of the ocean by the bursting of bubbles and some of these are

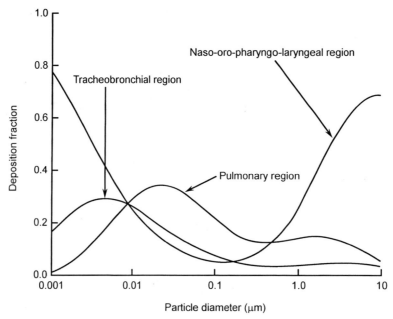

FIGURE 2.13 Calculated deposition of particles in various regions of the lung for polydisperse aerosol ($\sigma_g = 2.5$; see Chapter 9.A.2) (adapted from Yeh *et al.,* 1996).

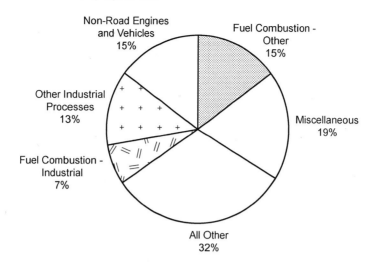

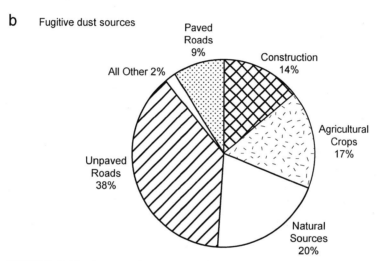

FIGURE 2.15 Sources of PM_{10} emissions in the United States in 1996 (a) excluding fugitive dust sources; transportation sources accounted for 21% of the total, and (b) fugitive dust sources (adapted from EPA, 1997).

carried inland. While more than 50% tend to be > 3 μm, in diameter, recent studies suggest that there are also a number of particles produced in the submicron size range as well (see Chapter 9). The chemical composition of sea-generated particles includes the elements found in seawater (primarily chlorine, sodium, sulfur, magnesium, potassium, and calcium) and organic materials, perhaps including viruses, bacteria, and so on.

Biological emissions of particles may occur from plants and trees; additionally, volatile organics such as isoprene and terpenes can react in the air to form small particles.

Volcanic eruptions are highly variable but can produce large amounts of particles. For example, for the St. Augustine (Alaska) eruption in 1976, particulate emissions over the period of 1 year were estimated to be $\sim 6 \times 10^6$ metric tons (Stith *et al.*, 1978). Wildfires and biomass burning also produce significant particulate matter, most of it in the respirable size range from ~ 0.1 to 1 μm. Elemental carbon and organics form the majority of these particles, with some minerals also being present.

6. Lead

Prior to the introduction of unleaded fuels, gasoline combustion in motor vehicles was by far the greatest

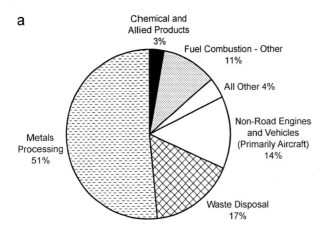

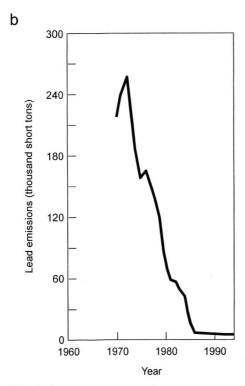

FIGURE 2.16 (a) Contribution of various sources to total anthropogenic Pb emissions in the United States in 1996. (b) Trend in lead emissions in the United States (from EPA, 1995, 1997).

source of atmospheric lead. However, as seen in Fig. 2.16a, this is no longer the case in the United States and most likely in other countries where unleaded gasoline is in widespread use. Metals processing now accounts for more than half of the total. As seen in Fig. 2.16b, emissions of lead have dropped dramatically over the past two decades, due largely to the phase out of its use in gasoline (EPA, 1995, 1997).

B. METEOROLOGY

Clearly, the concentrations of pollutants in ambient air, and hence their impacts, are determined not only by their rates of emissions but also by the nature and efficiencies of their chemical and physical "*sinks*," e.g., chemical transformations, as well as wet and dry deposition to the earth's surface. To a large extent, these competing processes are affected not only by direct dispersion and transport but also by such meteorological factors as temperature, sunlight intensity, and the presence of temperature inversions as well as clouds and fogs.

For example, during the severe air pollution episodes in London (Table 1.1), meteorological conditions were such that the pollutants were effectively contained in relatively small volumes, leading to high pollutant concentrations. Furthermore, fog water droplets provide an aqueous medium for the liquid phase of SO_2 to sulfate conversion. Additionally, the cycle of condensation of water vapor on aerosols at low temperatures, followed by evaporation during the day, is thought to be a major factor controlling the concentrations of pollutants within the droplets (Munger *et al.*, 1983).

Indeed, as noted in Chapter 1, fogs with pH values < 2 have been observed in heavily populated coastal cities in southern California (Jacob and Hoffmann, 1983; Hoffmann, 1984; Jacob *et al.*, 1985), and a variety of studies have also shown that fogs in the pH range of 2–3 are common throughout the world.

We discuss briefly in the following sections some meteorological parameters that are useful in the context of atmospheric chemistry.

1. Lapse Rate: Temperature and Altitude

As seen in Fig. 1.1, in the lowest 10 km of the earth's atmosphere, the air temperature generally decreases with altitude. The rate of this temperature change with altitude, the "lapse rate," is by definition the negative of the change in temperature with altitude, i.e., $-dT/dz$. Averaging over time and large geographic regions has shown that within the lowest 10 km of the atmosphere, the lapse rate is usually positive and is typically 6–7 degrees per km. A temperature inversion is said to exist when the lapse rate is negative.

The origin of the lapse rate can be understood on the basis of fundamental thermodynamics. That is, under the assumptions of a dry air parcel rising adiabatically in the atmosphere, the temperature is expected to fall about 10 degrees per kilometer increase in altitude. This drop in temperature is defined as a positive lapse rate.

BOX 2.1
DERIVATION OF LAPSE RATE IN TROPOSPHERE

To understand the origin of this lapse rate, it is convenient to consider first a volume (V) of dry air that is sufficiently large that exchange of molecules across its boundaries is negligible but sufficiently small that there is no significant heat exchange via entrainment/detrainment of large "blobs" of air from the surrounding environment.

Air can be treated as an ideal gas and hence follows the ideal gas law:

$$PV = nRT, \quad \text{(A)}$$

where P is the air pressure, n the number of moles of gas in the air parcel, and T the temperature (in K). The air pressure is due to the weight, i.e., gravitational force, of the column of air above it:

$$P = F/A = mg/A = \rho V g/A, \quad \text{(B)}$$

where m is the mass of air of density ρ in a column of air of area A and volume V and g is the acceleration due to gravity (9.8 m s^{-2}). Using the relationship $V = Az$, where z is the altitude, we may write the variation in pressure with z (expressed as a positive number), as

$$dP/dz = -d/dz\,[\rho V g/A]$$
$$= -d/dz\,[\rho(Az)g/A] = -\rho g. \quad \text{(C)}$$

However, using the ideal gas law and the definition of density, $\rho = (n/V)\text{MW}_a$, where MW_a is the molecular weight of air (29 × 10^{-3} kg mol^{-1}), we may rewrite Eq. (C) in the form

$$dP/dz = -(n/V)\text{MW}_a g = -(P/RT)\text{MW}_a g. \quad \text{(D)}$$

This relationship for the variation of pressure with altitude can be converted into that between temperature and altitude as shown in Eq. (N) below for a parcel of dry air that rises without heat exchange occurring between it and the surrounding air; this is known as an adiabatic process.

The first law of thermodynamics is

$$dU = dq + dw, \quad \text{(E)}$$

where U is the internal energy of the gas, q is the heat going in or out of the system, and w is the work performed on or by the gas. Heat and work are defined such that heat going *into* the system and work done *on* the system are positive quantities. By definition, an adiabatic process is one for which $dq = 0$, so that $dU = dw$.

The change in internal energy, dU, of an ideal gas is given by

$$dU = nc_v\,dT, \quad \text{(F)}$$

where c_v is the molar heat capacity at constant volume, i.e., the amount of heat required to raise the temperature of the gas 1°C if it is heated at constant volume. (Under the latter conditions, any heat going into the gas must go into its internal energy since no work of expansion is done at constant volume.) Hence

$$dw = nc_v\,dT. \quad \text{(G)}$$

The work done by an expanding gas is defined as $dw = -P\,dV$. From the ideal gas law,

$$d(PV) = d(nRT) \quad \text{(H)}$$

and

$$P\,dV + V\,dP = nR\,dT \quad \text{(I)}$$

so that

$$dw = -P\,dV = V\,dP - nR\,dT. \quad \text{(J)}$$

Combining Eqs. (G) and (J) and rearranging, one obtains:

$$n(c_v + R)\,dT = V\,dP. \quad \text{(K)}$$

For an ideal gas, however, $(c_p - c_v) = R$, so that Eq. (K), with application of the ideal gas law for V, becomes

$$nc_p\,dT = (nRT/P)\,dP \quad \text{(L)}$$

or

$$dT/dP = RT/c_p P. \quad \text{(M)}$$

Combining Eqs. (D) and (M), one obtains the temperature–altitude profile, or lapse rate, for a dry adiabatic gas:

$$dT/dz = [dT/dP]/[dz/dP] = -\text{MW}_a g/c_p. \quad \text{(N)}$$

Substituting the value of MW_a and c_p (29 J K^{-1} mol^{-1}) for air, the dry adiabatic lapse rate Γ_d is given by

$$\Gamma_d = -dT/dz = 9.8 \times 10^{-3}\ \text{K m}^{-1}. \quad \text{(O)}$$

In reality, measured lapse rates are ≈ 6–7°C per km. This is due to the fact that atmospheric air is not dry but contains significant amounts of water vapor that also cools as the air parcel rises. When it reaches saturation, it condenses and releases its heat of vaporization, which warms the air somewhat, resulting in a less steep drop in temperature with altitude than expected for a dry air parcel.

2. Potential Temperature

A concept that is very useful in relating meteorological conditions to the mixing and transport of air pollutants is that of *potential temperature*. Potential temperature (θ) is defined as the temperature an air parcel of temperature T and pressure P would have if it were expanded or compressed under adiabatic conditions to some reference pressure P_0.

Adiabatic expansion or compression of an air mass maintains a constant potential temperature. From the definition of entropy, S, as $dS = dq_{rev}/T$, these processes are also constant-entropy processes since no heat is exchanged between the system (i.e., the air parcel) and its surroundings. Hence the term "isentropic" is used to describe processes that occur without a change in potential temperature.

Potential temperature is a very useful parameter in several ways. First, air pollutants or trace gases within an air parcel having a constant value of θ can be assumed to be well mixed within that air parcel. Thus only limited numbers of measurements within the parcel are necessary to characterize its composition.

Second, air parcels tend to conserve their potential temperature; i.e., air parcels tend to move along lines of constant θ. Thus the potential temperature becomes a sort of tracer for the history of air parcels.

For example, a number of field campaigns have been carried out over the North Atlantic Ocean off the coast of Nova Scotia, Canada. Under some conditions, relatively high levels of O_3 were observed whose source was unknown. However, in some instances the air parcel containing the higher ozone concentrations could be tracked upwind over northern Canada (Berkowitz et al., 1995). Figure 2.17 shows the isentropic surface (i.e., surface of constant potential temperature) leading back from the sampling point to higher altitudes over the Arctic region. This surface suggests that the source of the higher ozone over the North Atlantic on that particular occasion was upper tropospheric/lower stratospheric air that had traveled from the Arctic regions.

3. Temperature Inversions

In a "normal" troposphere that has a positive lapse rate, i.e., where the temperature is falling with altitude, warm air close to the earth's surface, being less dense, rises and is replaced by cooler air from higher elevations. This results in mixing within the troposphere.

In some situations, however, the temperature of the air, at some height within the troposphere, may start to rise with increasing altitude before reversing itself again; that is, the lapse rate changes from positive to negative to positive (Fig. 2.18). This region, with a

---

BOX 2.2

DERIVATION OF EXPRESSION FOR POTENTIAL TEMPERATURE

Recall (see any standard physical chemistry text) that for adiabatic expansions or compressions of an ideal gas, there are several relationships between P, V, and T that hold; e.g., PV^γ = constant, where γ is the ratio of the heat capacities at constant pressure and volume, i.e., $\gamma = c_p/c_v$. Most useful in the context of potential temperature is $TP^{\gamma/\gamma-1}$ = constant. Applying this latter relationship,

$$TP^{(\gamma/\gamma-1)} = \theta P_0^{(\gamma/\gamma-1)}; \qquad (P)$$

i.e.,

$$\theta = T(P/P_0)^{\gamma/\gamma-1} = T(P/P_0)^{c_p/R}, \qquad (Q)$$

where $c_p/R = 3.5$ for dry air. Rearranging Eq. (Q) gives

$$T = \theta(P/P_0)^{R/C_p} = \theta(P/P_0)^{0.286}. \qquad (R)$$

A plot of $p^{0.286}$ versus temperature for a given θ is a straight line of constant potential temperature. Such plots are known as pseudoadiabatic charts when plotted with an inverted pressure scale so that pressure increases from top to bottom.

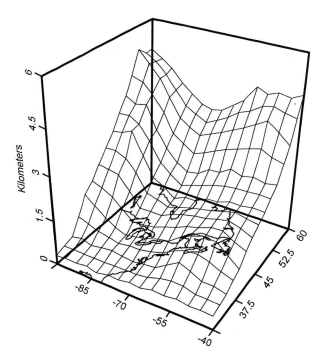

FIGURE 2.17 Isentropic surface at 300 K (i.e., surface of constant potential temperature) over eastern North America on September 8, 1992 (from Berkowitz et al., 1995).

negative lapse rate, is known as an inversion layer. In effect, it acts as a "lid" on an air mass because the cooler air underneath it, being more dense, will not rise through it. In effect, pollutants trapped below the inversion layer are not mixed rapidly throughout the entire troposphere but are confined to the much smaller volume beneath the inversion layer; this generally results in much higher ground-level concentrations of species that are emitted at the surface.

The formation of thermal inversions is one of the most important meteorological factors contributing to air pollution problems in urban areas. There are two major sources of thermal inversions. Radiation (or ground) inversions are caused by the rapid cooling of the earth's surface, along with the layer of air immediately above it, by the emission of infrared radiation immediately after sunset. On calm nights, this cooling may be sufficiently rapid that the layer of air adjacent to the surface becomes cooler than the air above; that is, an inversion forms. This can persist until sufficient heating of the surface and the air above it occurs to "break" the inversion at dawn. With this type of inversion, the inversion height—the distance from the earth's surface to the point at which the lapse rate reverses—is often quite small. For example, in the 1952 London smog episode, inversion heights as low as 150 ft were observed in some locations.

Overhead (or subsidence) inversions associated with photochemical air pollution are caused by the sinking motion of air masses as they pass over the continent. This leads to compression and heating of the air immediately below, resulting in a change in the lapse rate, that is, to the formation of an inversion layer. The inversion height is significantly higher than in the case of radiation inversions; for example, ~ 1500 ft would represent a relatively low subsidence inversion height.

Interestingly, the vertical distribution of photochemical oxidant may not be such that it falls off rapidly at the inversion layer. In fact, in a classic series of experiments, Edinger and co-workers (1972, 1973) showed that oxidant concentrations in the Los Angeles air basin could be higher within the inversion layer than at ground level. Thus, Fig. 2.19, for example, shows one temperature and oxidant profile for June 20, 1970, over Santa Monica, California, a city adjacent to the Pacific Ocean. Several "layers" of oxidant (mainly O_3) exist

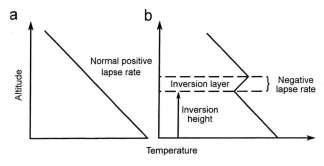

FIGURE 2.18 Variation of temperature with altitude within the troposphere: (a) normal lapse rate; (b) change in lapse rate from positive to negative, characteristic of a thermal inversion.

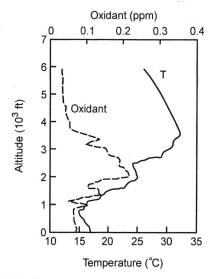

FIGURE 2.19 Temperature and oxidant profiles at 1:28 p.m. over Santa Monica, California, on June 20, 1970 (from Edinger, 1973).

within the inversion layer, reaching concentrations as high as 0.2 ppm, compared to a ground-level concentration of ~0.1 ppm.

The reason for this phenomenon is that the mountain slopes surrounding the Los Angeles air basin become heated by the sun; the layer of air in contact with the slopes is also heated and moves up the slopes. When the inversion layer is deep and strong, much of this rising polluted air does not get sufficiently warm to penetrate the inversion completely, so it moves out and away from the slopes and remains within the inversion layer; multiple pollution peaks within the inversion layer result.

Because of such meteorological phenomena, pollutants can be trapped aloft and transported over large distances, undergoing essentially no deposition at the surfaces. Clearly, surface measurements are not adequate to document such high-level transport.

The *boundary layer* is the lowest part of the atmosphere, closest to the earth's surface. Stull (1988) defines the boundary layer as "that part of the troposphere that is directly influenced by the presence of the earth's surface, and responds to surface forcings [such as heat transfer, pollutant emissions, evaporation etc.] with a timescale of about an hour or less." Typical boundary layer heights range from ~100 to 3000 m in altitude. The rest of the overlying troposphere is called the *free troposphere*.

Figure 2.20 summarizes the role of inversions and the boundary layer in terms of typical changes in mixing of the atmosphere close to the earth's surface at various times of the day (Stull, 1988). At midday, there is generally a reasonably well-mixed convective layer lying above the surface layer into which the direct emissions are injected. As the sun goes down, radiative cooling results in the formation of a stable nocturnal boundary layer, corresponding to a radiation inversion. Above this is a residual layer that contains the species that were well-mixed in the boundary layer during the day but that do not mix rapidly during the night with either the nocturnal boundary layer below or the free troposphere above. At sunrise, heating of the earth's surface results in mixing of the contents of the nocturnal boundary layer and the residual layer above it. Clearly, such meteorological changes can have significant impacts on the spatial distribution of pollutants emitted at the earth's surface, leading to chemistry that varies both spatially (in 3 dimensions) and diurnally. Coupling between the transport processes and chemistry on local to global scales is discussed by Kley (1997).

In summary, as we shall see throughout this book, meteorological parameters are extremely important, not only in determining the dispersion and transport of pollutants but also in determining their chemistry. The reader is encouraged to consult meteorology texts for a much more detailed treatment of this subject.

C. REMOVAL FROM THE ATMOSPHERE: WET AND DRY DEPOSITION

While the central thrust of this book involves the chemical transformations occurring in air, it is clear

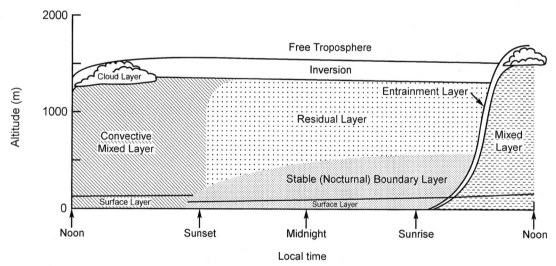

FIGURE 2.20 Schematic of mixing processes in atmosphere close to the earth's surface as a function of time of day. (Adapted, with kind permission from Kluwer Academic Publishers, from R. B. Stull, 1988, *An Introduction to Boundary Layer Meteorology*, Fig. 1.7. © 1988 by Kluwer Academic Publishers.)

that physical removal at the earth's surface is important for many primary and secondary pollutants.

Meteorology, however, also plays important roles other than mere transport of pollutants as they react. As we have seen, the formation of acids from their precursors involves reactions not only in the gas phase but also in the liquid phase (e.g., in clouds, fogs, and dews) and perhaps on the surfaces of solids as well. Both particles and gases can be deposited at the earth's surface in two ways, termed *dry deposition* or *wet deposition*, depending on the phase in which a species strikes the earth's surface and is taken up. Thus pollutants may be dissolved in clouds, fog, rain, or snow; when these water droplets impact the earth's surface (which includes not only soil but grass, trees, buildings, etc.), it is termed *wet deposition*. However, pollutants in the form of either gases or small particles can also be transported to ground level and absorbed and/or adsorbed by materials there without first being dissolved in atmospheric water droplets; this is called *dry deposition*. It should be noted that the surface itself may be wet or dry; the term dry deposition only refers to the mechanism of transport to the surface, not to the nature of the surface itself.

Because of the highly variable nature of precipitation events, quantitatively estimating wet deposition of pollutants is difficult. In addition to meteorological factors, parameters such as the solubility of the pollutant in ice, snow, and rain and how this varies with temperature and pH, the size of the water droplets, and the number present must also be considered; for example, snow may scavenge some species more efficiently than rain.

As an approximation, the rate of wet deposition of a pollutant is sometimes taken as λC, where C is the pollutant concentration and λ is known as a washout coefficient which is proportional to the precipitation intensity (Shaw, 1984).

Dry deposition can also be a very important mechanism for removing pollutants from the atmosphere in the absence of precipitation. Indeed, even in such places as eastern England, the ratio of dry to wet removal of SO_2 has been estimated to be $\sim$ 2:1 (Davies and Mitchell, 1983). If this is the case, then in arid and semiarid regions such as much of the western United States dry deposition is clearly important.

Dry deposition is usually characterized by a *deposition velocity*, V_g. The net flux (F) of a species to the surface is proportional to the concentration of that species in air, $[S]$; i.e., $F \propto [S]$. The deposition velocity is just the proportionality constant relating flux and concentration, i.e.,

$$F = -V_g[S]$$

or

$$V_g = -F/[S], \qquad (S)$$

where $[S]$ is the concentration at some reference height z. By convention, the deposition velocity is a positive number, but fluxes toward the surface are taken as negative; hence the negative sign in (S). The amount of the species deposited per unit area per second in a geographical location, that is, the flux, can be calculated if the deposition velocity and the pollutant concentration are known. The deposition velocity is also frequently related to a resistance r:

$$V_g = 1/r. \qquad (T)$$

By analogy to electrical systems, the resistance r can be thought of as consisting of several components. For convenience, three such components are often defined, a surface resistance (r_{surf}), which depends on the affinity of the surface for the species, a boundary layer resistance (r_{boun}), which depends on the molecular diffusivity of the gas in air, and a gas-phase resistance (r_{gas}), which depends on the micrometeorology that transports the gas to the surface:

$$r = r_{gas}(z) + r_{boun} + r_{surf}. \qquad (U)$$

The gas-phase resistance depends on the height (z) above the earth's surface, as does the concentration of the pollutant; as a result the deposition velocity is also a function of height.

Figure 2.21 schematically depicts the dry deposition of a pollutant to a typical surface in the form of resistances (Lovett, 1994; Wesely and Hicks, 1999). In this case, the surface resistance r_{surf} has been broken down even further into a combination of parallel and series resistances (r_s, r_m, r_{ct}, r_{soil}, r_{water}, etc.). Since leaves may absorb pollutants either through stomata or through the cuticles, the absorption into the leaf is represented by two parallel resistances, r_{ct} for the cuticular resistance and r_s for the stomatal resistance, which is in series with a mesophyllic resistance r_m. Also shown are resistances for uptake into the lower part of the plant canopy and into water, soil, or other surfaces.

The relative importance of gas-phase and surface resistances depends on the nature of the pollutant and the surface as well as the meteorology (Shaw, 1984; Unsworth *et al.*, 1984; Chameides, 1987; Wesely and Hicks, 1999). The gas-phase resistance (r_{gas}) is determined by the vertical eddy diffusivity, which depends on the evenness of the surface and the meteorology, for example, wind speed, solar surface heating, and so on. The surface resistance (r_{surf}) depends on the detailed characteristics of the surface (e.g., type, whether

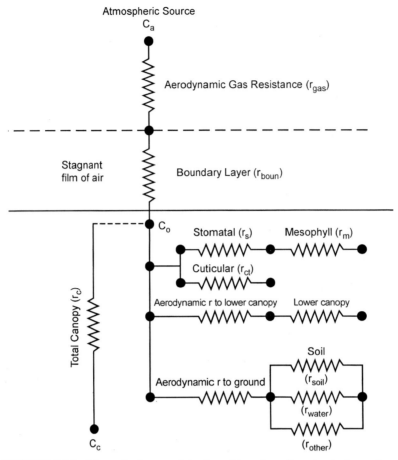

FIGURE 2.21 Schematic diagram of dry deposition of a pollutant onto the surface of a leaf, using the concept of resistances (r) for deposition. (Adapted from Weseley and Hicks, 1999, *Atmospheric Environment*, with permission from Elsevier Science.)

leaf, building soil, snow, wetness, etc.; see Dasch, 1985; Vandenberg and Knoerr, 1985) as well as on the nature of the pollutant being deposited. The resistance r_{boun} in the film of air immediately adjacent to the surface depends on the shape of the surface and the molecular diffusivity of the depositing species, whereas the resistances r_{ct} and r_s reflect the resistance to adsorption by the surface, which depends both on the nature of the surface itself and on the depositing species. For highly reactive gases, the surface resistance may be sufficiently small so that transport to the surface becomes rate limiting; for example, the surface resistance for deposition of HNO_3 on grass during the day has been shown to be approximately zero (Huebert and Robert, 1985). Similarly, Chameides (1987) has shown that the dynamical resistance determines essentially entirely the uptake into dew of highly soluble species such as HNO_3, whereas for less soluble compounds such as SO_2 and O_3, the surface resistance plays an important role.

Because of the effects of physical, chemical, biological, and meteorological parameters on the "resistances," the deposition velocity (V_g) defined by Eq. (T) also depends on these. As a result, deposition velocities reported for various pollutants show a wide range, from several hundredths to several cm s^{-1}, depending on the conditions during the measurement. For example, the mean value of V_g for SO_2 over grass in one study was 0.56 cm s^{-1} for dry grass but 0.93 cm s^{-1} for wet grass (Davies and Mitchell, 1983). A diurnal variation in deposition velocities has been reported, with daytime values being greater than those at night (Hicks et al., 1983); presumably both r_{gas} and r_{surf} increase under nighttime conditions. For example, peak values of O_3 deposition of up to 1 cm s^{-1} during midday but less than 0.1 cm s^{-1} in the evening have been reported (Droppo, 1985).

For particles, V_g depends on the particle size; thus a minimum in V_g is generally observed at particle diameters around 0.5 μm, ~ 0.01 cm s^{-1} being a typical

TABLE 2.4 Typical Values of Dry Deposition Velocities $(V_g)^a$

	Surface	Typical range for V_g^b
Larger particles (> 2 μm)	Exterior surfaces	0.5–2
Fine particles (≤ 2 μm)	Exterior surfaces	< 0.5
Gases		
SO_2	Exterior surfaces and interior of leaves	0.2–1 for dry foliage, stomata open, > 1 for a wet surface
HNO_3	Primarily exterior surfaces, but also interior of leaves	1–5, perhaps greater
NO_2	Primarily interior of leaves, also exterior surfaces	0.1–0.5 when stomata open
NH_3	Exterior surfaces and interior of leaves	0.5–5, with highest values in humid conditions
O_3	Primarily interior of leaves, also exterior surfaces	0.1–0.8

[a] From Lovett (1994).
[b] V_g given in cm s^{-1}.

minimum value for grasses and grasslike surfaces (McMahon and Denison, 1979; Hicks and Garland, 1983).

Table 2.4 gives some typical ranges of deposition velocities reported for various pollutants and surfaces (Lovett, 1994). The wide ranges given in Table 2.4 reflect a combination of experimental uncertainties as well as real differences due to meteorology, nature of the surface, diurnal variations, and so on. A discussion of these is given by Wesely and Hicks (1999). For example, Quinn and Ondov (1998) report that deposition velocities of certain elements in fine particles may increase by a factor of four or more at higher relative humidities (RH) compared to lower RH, due to the growth in size of the particles at higher RH. The overall uncertainty in the appropriate value of the deposition velocity to use under a given set of circumstances can thus be quite large; for example, Sievering (1984) estimated the uncertainty in the dry deposition of small particles (diameter < 1 μm) to water surfaces to be about two orders of magnitude!

For further details on dry deposition, its measurement, and typical values of V_g for various pollutants, the reader is referred to the articles and books by McMahon and Denison (1979), Georgii and Pankrath (1982), Pruppacher *et al.* (1983), Hicks (1984), Dabberdt *et al.* (1993), Wesely (1989), Erisman *et al.* (1994), Foken *et al.* (1995), Lovett (1994), Gao and Wesely (1995), and Wesely and Hicks (1999).

D. TYPICAL AMBIENT CONCENTRATIONS AND AIR QUALITY STANDARDS

1. Units of Concentrations and Conversions

A number of different units are used in expressing the concentrations of various species in the atmosphere; we shall review these here. For a review of IUPAC recommendations, see Schwartz and Warneck (1995).

a. Parts per Million, Parts per Hundred Million, Parts per Billion, and Parts per Trillion

For gas-phase species, the most commonly used units are parts per million (ppm), parts per hundred million (pphm), parts per billion (ppb), and parts per trillion (ppt). These units express the number of molecules of pollutant found in a million (10^6), a hundred million (10^8), a billion (an American billion (10^9)), or a trillion (10^{12}) molecules of air, respectively. It should be noted that, although these are commonly used, confusion may arise in that in some European countries a billion means 10^{12} and a trillion means 10^{18}. As a result, in some journals when these units are used, they are accompanied by a definition, e.g., ppb (parts in 10^9). Care must also be taken to ensure that ppt means parts per trillion and not parts per thousand. The latter unit is often used in isotope studies and is denoted parts per mille, or ‰.

Alternatively, because numbers of molecules (or moles) are proportional to their volumes according to the ideal gas law ($PV = nRT$), these units may be thought of as the number of volumes of the pollutant found in 10^6, 10^8, 10^9, or 10^{12} volumes of air, respectively.

This ratio of moles, molecules, or volumes of the species to the number of moles, molecules, or volumes of dry air is more commonly known as the *mixing* ratio. The use of mixing ratios is widespread for expressing the relative amounts of a species at various altitudes throughout the atmosphere. Since the total air pressure and hence total concentration of air decreases with altitude, a constant mixing ratio does not imply a constant *concentration* (i.e., as expressed in moles or molecules per unit volume). Although not expressly stated in most cases, mixing ratios are usually referenced to dry air. If water vapor is included, the mixing ratio would vary with humidity, which could induce a variation of a few percent. Although we shall use the term concentration frequently, if the units are ppm, ppb, or ppt, it should be understood that this is really a mixing ratio. Some journals emphasize this by writing these units as ppm (v:v), etc.

For example, a background concentration of O_3 may be 0.04 ppm; this is 4 pphm, 40 ppb, or 40,000 ppt. Thus in 10^8 molecules of air, only 4 are O_3; alternatively, in every 10^8 volumes of air, only 4 volumes are due to O_3. While it is most convenient to express the concentration of O_3 in ppm, pphm, or ppb, other important atmospheric species can be present in much smaller concentrations. For example, the hydroxyl free radical (OH), which, as we saw, drives the daytime chemistry of both the clean and polluted troposphere, is believed to have typical concentrations of only < 0.1 ppt. Hence either ppt or an alternate unit discussed in the next section (number per cm^3) is used.

Concentrations of pollutants in ambient air are normally sufficiently small that ppm is the largest unit in use. However, pollutant concentrations in stacks or exhaust trains prior to mixing and dilution with air are much higher, and percent (i.e., parts per hundred) is sometimes used in this case. For example, carbon monoxide concentrations in automobile exhaust are measured in percentages, reflecting the numbers of CO molecules (or volumes) per 100 molecules (or volumes) of exhaust.

Finally, it is important to remember that the number of molecules, or volumes, of a given gaseous species forms the basis of units in atmospheric chemistry; in water chemistry, mass rather than volume is used as the basis for expressing concentrations in ppm, and so on.

b. Number per Cubic Centimeter

A second type of concentration unit is generally used for species such as free radicals (e.g., OH) present at sub-ppt levels. It is the number of molecules, atoms, or free radicals present in a given volume of air, usually a cubic centimeter (cm^3). One can convert from units of ppm, pphm, ppb, or ppt to units of number per cm^3 using the ideal gas law. Thus the number of moles per L in air at 1 atm pressure and 25°C (298 K) is given by

$$n/V = P/RT$$
$$= 1 \text{ atm}/[0.08206 \text{ (L atm/K mol)} \times 298 \text{ K}]$$
$$= 0.0409 \text{ mol/L}$$

Converting to units of molecules per cm^3, one obtains

$$n/V = 0.0409 \text{ mol/L} \times 10^{-3} \text{ (L/cm}^3) \times 6.02 \times 10^{23}$$
$$\text{molecules/mol} = 2.46 \times 10^{19} \text{ molecules cm}^{-3}$$

From the definition of ppm as the number of pollutant molecules per 10^6 molecules of air, 1 ppm corresponds to $2.46 \times 10^{19} \times 10^{-6} = 2.46 \times 10^{13}$ molecules per cm^3 at 25°C and 1 atm total pressure. It follows that a concentration of the OH radical in polluted air

TABLE 2.5 Conversion between Units of Concentration in ppm, pphm, ppb, ppt, and Molecules cm^{-3}, Assuming 1 atm Pressure and 25°C[a]

Parts per	Unit	Molecules, atoms, or radicals per cm^3
10^6	1 ppm	2.46×10^{13}
10^8	1 pphm	2.46×10^{11}
10^9	1 ppb	2.46×10^{10}
10^{12}	1 ppt	2.46×10^7

[a] 1 ppm in units of mass per cubic meter = $40.9 \times$ (MW) $\mu g \text{ m}^{-3}$.

of 0.1 ppt is $2.46 \times 10^{19} \times 10^{-12} \times 0.1 = 2.46 \times 10^6$ molecules per cm^3. As we shall see, the peak OH concentration is much smaller so the use of fractional ppt units becomes inconvenient. Units of number per cm^3 are more commonly used in such cases.

Table 2.5 summarizes the relationship between these units at 1 atm pressure and 25°C. Of course, corrections must be made if the temperature or pressure differs significantly.

c. Micrograms per Cubic Meter

A third unit of measurement for gaseous species is mass per unit volume, usually 10^{-6} g per cubic meter ($\mu g \text{ m}^{-3}$). Since 1 atm at 25°C contains 4.09×10^{-2} mol L^{-1}, 1 ppm must contain $(4.09 \times 10^{-2}) \times 10^{-6}$ or 4.09×10^{-8} mol L^{-1} or 4.09×10^{-5} mol m^{-3}. If the molecular weight of the pollutant is MW grams per mole, then 1 ppm in units of mass per m^3 is $(4.09 \times 10^{-5}) \times $ (MW) g m^{-3} or $(40.9) \times$ (MW) $\mu g \text{ m}^{-3}$.

Returning to the example of a background O_3 of 0.04 ppm, this concentration in $\mu g \text{ m}^{-3}$ is $40.9 \times 48 \times 0.04$ (where 48 g mol^{-1} is the molecular weight of O_3), or 79 $\mu g \text{ m}^{-3}$. If the pollutant were 0.04 ppm SO_2 (MW = 64) rather than O_3, the concentration of SO_2 would be $(40.9 \times 64) \times 0.04$, or 105 $\mu g \text{ m}^{-3}$.

The conversion between $\mu g \text{ m}^{-3}$ and ppm, pphm, ppb, and ppt can be summarized as follows:

$$\mu g \text{ m}^{-3} = \text{ppm} \times 40.9(\text{MW})$$
$$= \text{pphm} \times 0.409(\text{MW})$$
$$= \text{ppb} \times 0.0409(\text{MW})$$
$$= \text{ppt} \times (4.09 \times 10^{-5})(\text{MW})$$

For atmospheric particulate matter, concentrations are expressed in mass per unit volume, commonly $\mu g \text{ m}^{-3}$, or in the number of particles per unit volume, for example, per cm^3.

TABLE 2.6 National Ambient Air Quality Standards (NAAQS) Set by the United States Federal Government[a]

Pollutant	Primary Concentration	Primary Time	Secondary Concentration	Secondary Time
CO	9.0 ppm	8 h		
	35.0 ppm	1 h		
SO_2	0.03 ppm	Annual average		
	0.14 ppm	24 h	0.5 ppm	3 h
O_3[b]	0.12 ppm	1 h	0.12 ppm	1 h
	0.08 ppm	8 h	0.08 ppm	8 h
NO_2	0.053 ppm	Annual arithmetic mean	0.053 ppm	Annual average
Particulate matter				
PM_{10}	50 $\mu g\,m^{-3}$	Annual arithmetic mean	50 $\mu g\,m^{-3}$	Annual arithmetic mean
	150 $\mu g\,m^{-3}$	24 h	150 $\mu g\,m^{-3}$	24 h
$PM_{2.5}$	15 $\mu g\,m^{-3}$	Annual arithmetic mean	15 $\mu g\,m^{-3}$	Annual arithmetic mean
	65 $\mu g\,m^{-3}$	24 h	65 $\mu g\,m^{-3}$	24 h
Lead	1.5 $\mu g\,m^{-3}$	Quarterly average	1.5 $\mu g\,m^{-3}$	Quarterly average

[a] NAAQS in effect in 1998; see Web Site listed in Appendix IV.

[b] The 1-h O_3 standard applies only to areas designated as nonattainment areas when the new 8-h standard was adopted in July 1997.

2. Criteria and Noncriteria Pollutants and Air Quality Standards

A number of pollutants have documented effects on people, plants, or materials at concentrations, or approaching those, found in polluted air (e.g., see Bascom *et al.*, 1996a,b). In the United States, seven of them are known as *criteria pollutants* and national ambient air quality standards (NAAQS) have been set for them "to protect public health and welfare."

The U.S. NAAQS and the recommended values or limits set by some other countries and the World Health Organization are given in Tables 2.6 and 2.7. (Note that the data in Table 2.7 were compiled in the early 1990s and there may therefore be some changes since then.) Ambient air quality standards have two

TABLE 2.7 Some Ambient Air Quality Standards (in ppm) for Some Gases[a]

Country	CO	SO_2	O_3	NO_2
WHO[b]	26	0.13	0.08	0.21
Australia[c]	30	0.17	0.12	0.16
New Zealand	35	0.05 (24 h)	0.06	0.11
Canada	13	0.17	0.08	0.21
Mexico	13 (8 h)	0.13 (24 h)	0.11	0.21
United States	35	0.14 (24 h)	0.08 (8 h) 0.12 (1 h)	0.05 (annual average)
Austria	35	0.08 (30 min)	0.06 (30 min)	0.11 (30 min)
Switzerland	7 (24 h)	0.04 (30 min)	0.06	0.05 (30 min)
Netherlands	35	0.32	0.06	0.09
Israel	52 (30 min)	0.19 (30 min)	0.12 (30 min)	0.50 (30 min)
Poland	4 (30 min)	0.23 (30 min)	0.05 (30 min)	0.27 (30 min)
Soviet Union	18	—	0.05	0.05
Hungary	9 (30 min)	0.19 (30 min)	0.03 (30 min)	0.05 (30 min)
Japan	20 (8 h)	0.10	0.06	0.04
South Africa	—	0.30	0.12	0.57

[a] As of 1992 reported by Cochran *et al.* (1992) except for United States, 1997, standards; 1-h short-term standards reported unless otherwise indicated.

[b] WHO = World Health Organization Standards.

[c] For Victoria.

components: a concentration and a time. For example, the U.S. NAAQS for CO is 35 ppm for 1 h. This means that to avoid deleterious effects, the CO concentration should not exceed 35 ppm for more than 1 h. The data in Table 2.7 are those for short exposure times, although standards also exist for longer times in many cases (see Cochran et al., 1992).

Table 2.8 illustrates the severity of air pollution problems in 20 so-called "megacities" around the world. Megacities were defined in this study as urban areas currently having, or anticipated to have by the year 2000, populations of $\geq$ 10 million. In most cases, the World Health Organization guidelines (Tables 2.7) are exceeded by more than a factor of two for at least one of the air pollutants shown (Mage et al., 1996).

The United States sets two types of NAAQS, primary and secondary. They are based on information contained in air quality criteria documents that contain a wealth of information on all aspects of the criteria pollutants, as do the documents by the World Health Organization. These should be consulted for detailed information and references regarding pollutant sources, ambient levels, chemical transformations, effects, and so on.

Note that the definition of a U.S. *primary* ambient air quality standard is one designed "to protect the public health"—with "an adequate margin of safety." These standards are set to protect even the most susceptible groups in the population, including those with cardiac and respiratory disease and newborn infants whose defense systems are not well developed.

Secondary NAAQS are set to protect "public welfare." This includes economic losses due to damage to agricultural crops, forests, and materials as well as aesthetic effects, including visibility degradation.

In addition to the criteria pollutants, a wide variety of trace gaseous and particulate species are present in the polluted troposphere (Finlayson-Pitts and Pitts, 1997). Table 2.9 shows some of these gaseous noncriteria pollutants identified in photochemical air pollution and gives typical concentrations under conditions ranging from those in remote areas to severely polluted urban air (see also Chapter 11).

Although their peak concentrations are usually only in the ppb, or even ppt, range, taken together they can form a substantial fraction of the concentration of their copollutant ozone, a criteria pollutant for which an air quality standard has been set and which, in southern California, is used as the basis for smog alerts. For example, Fig. 2.22 shows the maximum concentrations of some of these secondary pollutants determined using FT-IR spectroscopy over a kilometer path length dur-

TABLE 2.8 Air Pollution Problems in Some Megacities[a]

City	SO$_2$	CO	NO$_2$	O$_3$	Pb	PM
Bangkok	x		x	x	xx	xxx
Beijing	xxx	x	x	xx	x	xxx
Bombay	x	x	x		x	xxx
Buenos Aires					x	xx
Cairo		xx			xxx	xxx
Calcutta	x		x		x	xxx
Delhi	x	x	x		x	xxx
Jakarta	x	xx	x	xx	xx	xxx
Karachi	x				xxx	xxx
London	x	xx	x	x	x	x
Los Angeles	x	xx	xx	xxx	x	xx
Manila	x				xx	xxx
Mexico City	xxx	xxx	xx	xxx	xx	xxx
Moscow		xx	xx		x	xx
New York	x	xx	x	xx	x	x
Rio de Janeiro	xx	x			x	xx
São Paulo	x	xx	xx	xxx	x	xx
Seoul	xxx	x	x	x	x	xxx
Shanghai	xx					xxx
Tokyo	x	x	x	xxx		x

[a] From Mage et al. (1996) and personal communication; xxx means the WHO guidelines for that pollutant were exceeded by more than a factor of two, xx means they were exceeded by up to a factor of two, x means WHO guidelines are normally met, and blank means there are not sufficient data.

TABLE 2.9 Some Noncriteria Pollutants and Typical Concentration Ranges Observed from Remote Areas to Severe Air Pollution Episodes in Urban Areas

Noncriteria pollutant	Typical range of concentrations
Nitrogenous species	
Nitric oxide	0.02–2000 ppb
Peroxyacetyl nitrate	0.05–70 ppb
Nitric acid	< 0.1–50 ppb
Nitrous acid	< 0.03–10 ppb
Nitrate radical	< 5–430 ppt
Dinitrogen pentoxide[a]	≤ 15 ppb
Ammonia	< 0.02–100 ppb
Oxygenated organics	
Formaldehyde	< 0.5–75 ppb
Formic acid	≤ 20 ppb
Methanol	≤ 40 ppb

[a] Calculated from ambient NO_3 and NO_2 concentrations.

ing a severe photochemical air pollution episode in October 1978 in Claremont, California, a city approximately 30 miles east and generally downwind of central Los Angeles (Tuazon et al., 1981). Also shown is the observed peak ozone level, with the California Air Quality Standard (CAQS) for O_3 and the first-, second-, and third-stage alert levels. These alert levels, or *episode criteria*, are set by the State of California to define the concentrations at which short-term exposures may affect susceptible portions of the population; various control actions are required when each episode level is reached. It is seen that the sum of the peak concentrations of the toxic and/or acidic noncriteria pollutants, formaldehyde, nitric acid, PAN, and formic acid, totaled almost 40% of the peak ozone concentration.

It should be noted that most measurements of air pollutants made for regulatory purposes, i.e., for effects on health, are made close to the earth's surface. However, there can be significant changes in pollutant concentrations even over relatively small distances. For example, Micallef and Colls (1998) measured particulate matter at different heights above the ground in a street canyon in a city and found that the concentrations of PM_{10} were typically about 35% higher at 2.88-m elevation than at 0.8-m elevation. Thus, reported measurements of air pollutants may depend on the sampling height, as may actual exposures (e.g., of children compared to adults).

E. EFFECTS ON VISIBILITY AND MATERIALS

The loss of visibility that commonly accompanies high pollutant levels is perhaps the aspect of air pollution most obvious to the public; it is due to the scattering and, to a lesser extent, absorption of light by pollutants. The total light extinction—the sum of scattering and absorption by gases and particles—and hence visibility reduction depend on both the wavelength of the light and the scattering angle, that is, the position of the sun, so that "haze" due to air pollution may appear to have different colors and densities, depending on the conditions.

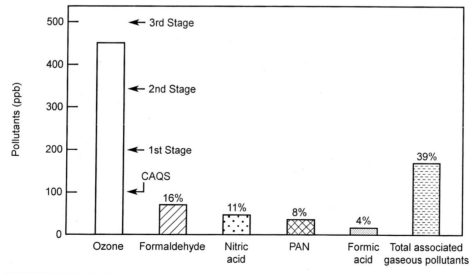

FIGURE 2.22 Maximum concentrations of some trace pollutants and the percentage of the peak ozone they form when summed. Also shown is the California Air Quality Standard for O_3 and the various alert levels (from Tuazon et al., 1981).

Of the gaseous air pollutants, only NO_2 absorbs visible light to a significant extent and thus contributes to visibility reduction. It is an orange-brown gas that absorbs radiation strongly at $\lambda < 430$ nm; hence it acts as a filter for blue light. The brownish color of many polluted urban areas and the accompanying spectacular sunsets are at least partly due to the presence of NO_2. For example, NO_2 has been found to contribute to the "brown cloud" over Albuquerque, New Mexico (Franzblau et al., 1993). However, the contribution of NO_2 to the total extinction is generally small, for example, < 10%.

Scattering of light by gaseous air pollutants is negligible compared to the Rayleigh scattering by O_2 and N_2 because of the very low concentrations of pollutants compared to those of the major components of air.

Thus, particulate matter suspended in air is generally responsible for the majority of light scattering and absorption and hence for visibility reduction associated with air pollution (see Chapter 9). Small particles with diameters of ~ 0.1–1 μm contribute the greatest amount to light scattering and hence visibility reduction is closely related to the fine-particle concentration. The size range that has the potential for maximum health effects also contributes the greatest amount to aesthetic effects. It is believed that scattering by particles is responsible for something in the range of 50–95% of the total extinction due to air pollutants, depending on the conditions.

Elemental carbon is believed to be the major significant light-absorbing species in particles. In remote regions, absorption may only account for ~ 5–10% of the total extinction, whereas in urban areas, its contribution is greater, up to ~ 50%.

In terms of chemical composition, visibility reduction due to particulate matter is generally related most closely to scattering by sulfate and nitrate and absorption and scattering by elemental carbon. Relative humidity (RH) is also an important factor, with a significant reduction in visibility occurring as the RH increases from 50 to 90%.

Effects on materials vary widely, depending on the composition of the material and on the particular air pollutant. For example, rubber products and polymers are especially susceptible to O_3, whereas metals such as iron are particularly sensitive to acids. Indeed, Haagen-Smit and co-workers (1959) used the cracking of rubber by O_3 to measure its concentration in air in the "early" days of photochemical air pollution. A detailed description of the effects of air pollutants on various types of materials is found in the article by Graedel and McGill (1986).

F. ECONOMICS

While it is clear that there are economic costs associated with air pollution as well as with measures taken to clean it up or prevent it, these costs are not always well defined. This is especially true in the case of the costs associated with its effects on human health, plants, materials, and esthetic effects. Some discussion of this aspect of air pollution is found in the articles by Adams et al. (1989), Hall et al. (1992), and Lin et al. (1994).

G. ATMOSPHERIC CHEMISTRY: RISK ASSESSMENTS AND PUBLIC POLICIES FOR AIR POLLUTION CONTROL

Over the past several decades there have been increasingly strong demands in Europe and the United States that atmospheric chemistry research be directly useful in developing scientific risk assessments and public policies. For example, one component of the EUROTRAC program (European Experiment on Transport and Transformation of Environmentally Relevant Trace Constituents in the Troposphere) "... is expected to assimilate the scientific results from EUROTRAC and present them in a condensed form, together with recommendations where appropriate, so that they are suitable for use by those responsible for environmental planning and management in Europe" (EUROTRAC, 1993).

An important aspect of this translation of science into public policy has been the separation of risk *assessment* and risk *management* (National Research Council, 1994). Thus, for major areas of public concern that involve atmospheric chemistry, scientific *risk assessments* have been carried out that review and summarize the current state of the science in a particular area, along with the uncertainties. These documents can then be used by those involved in *risk management* to make knowledgeable public policy decisions. Examples include the regular reviews of stratospheric ozone depletion published every few years by the World Meteorological Organization (e.g., WMO, 1995) and of global climate issues published by the International Panel on Climate Change (e.g., see IPCC, 1996). A similar effort known as NARSTO (North American Research Strategies for Tropospheric Ozone) is underway for tropospheric ozone and particles and represents a major coordinated scientific effort between Canada, Mexico, and the United States to further understand, and deal with, the transport of ozone across their borders. As an example, we conclude in Chapter

16 with a brief treatment of the central role of atmospheric chemistry in developing scientific risk assessments for airborne hazardous air pollutants (HAPs).

H. PROBLEMS

1. Measuring air pollution due to particles by mass can be quite misleading in terms of their ultimate impacts, for example on health. For example, the results of some laboratory studies show that ultrafine particles cause inflammatory responses while larger particles with the same chemical composition do not. Calculate how many particles with diameter 0.2 μm would need to be collected on a "hi-vol" sampler (which is essentially just a filter) to have the same mass as one particle of diameter 20 μm. Assume they have the same chemical composition and hence equal densities.

2. Derive the relationship between 1 ppt at 298 K and 1 atm pressure and the concentration of a species in μg m^{-3} from first principles.

3. The most stringent 1-h AQS for CO shown in a review of such standards by Cochran *et al.* [*J. Air Waste Manage.*, 42, 1567 (1992)] is 15,000 μg m^{-3} (Canada) and the least stringent is 40,000 μg m^{-3} (adopted by a number of countries, including the United States). What is this range of AQS in units of ppm?

4. Free radicals such as OH are present in such small concentrations that their concentrations are frequently given in units of molecules cm^{-3} rather than ppm, ppb, etc. A typical OH concentration in the lower troposphere is 5×10^5 radicals cm^{-3}. What is this concentration in terms of the mixing ratio unit ppt, assuming 298 K and 1 atm pressure?

5. You measure an air temperature of 260 K at a pressure of 600 mbar. What is the potential temperature, assuming a reference pressure of 1000 mbar?

6. An air parcel has a temperature of 7°F and a pressure of 450 Torr. What is its potential temperature (θ) if the reference pressure is 1 atm? Take c_p for air to be 29.1 J K^{-1} mol^{-1}.

7. Using the dry adiabatic lapse rate, by how much would you expect the temperature to change from the earth's surface to an altitude of 1000 feet, which, as seen in Fig. 2.19, sometimes corresponds to the bottom of the inversion layer in the Los Angeles area?

8. Estimate the mass of the atmosphere, given that the area of the surface of the earth is about 5.1×10^{14} m^2. Assume a uniform pressure of 1 atm at the earth's surface. Compare this to the mass of the earth itself, 6.0×10^{24} kg.

References

Adams, R. M., J. D. Glyer, S. L. Johnson, and B. A. McCarl, "A Reassessment of the Economic Effects of Ozone on U.S. Agriculture," *J. Air Pollut. Control Assoc.*, 39, 960–968 (1989).

Akimoto, H., and H. Narita, "Distribution of SO_2, NO_x, and CO_2 Emissions from Fuel Combustion and Industrial Activities in Asia with 1° × 1° Resolution," *Atmos. Environ.*, 28, 213–225 (1994).

Aneja, V. P., "Natural Sulfur Emissions into the Atmosphere," *J. Air Waste Manage. Assoc.*, 40, 469–476 (1990).

Arndt, R. L., G. R. Carmichael, D. G. Streets, and N. Bhatti, "Sulfur Dioxide Emissions and Sectorial Contributions to Sulfur Deposition in Asia," *Atmos. Environ.*, 31, 1553–1572 (1997).

Bascom, R., P. A. Bromberg, D. L. Costa, R. Devlin, D. W. Dockery, M. W. Frampton, W. Lambert, J. M. Samet, F. E. Speizer, and M. Utell, "Health Effects of Outdoor Air Pollution, Part 1," *Am. J. Respir. Crit. Care Med.*, 153, 3–50 (1996a).

Bascom, R., P. A. Bromberg, D. L. Costa, R. Devlin, D. W. Dockery, M. W. Frampton, W. Lambert, J. M. Samet, F. E. Speizer, and M. Utell, "Health Effects of Outdoor Air Pollution, Part 2," *Am. J. Respir. Crit. Care Med.*, 153, 477–498 (1996b).

Berges, M. G. M., R. M. Hofmann, D. Scharffe, and P. J. Crutzen, "Nitrous Oxide Emissions from Motor Vehicles in Tunnels and Their Global Extrapolation," *J. Geophys. Res.*, 98, 18527–18531 (1993).

Berkowitz, C. M., K. M. Busness, E. G. Chapman, J. M. Thorp, and R. D. Saylor, "Observations of Depleted Ozone within the Boundary Layer of the Western North Atlantic," *J. Geophys. Res.*, 100, 11483–11496 (1995).

Bonsang, B., C. Polle, and G. Lambert, "Evidence for Marine Production of Isoprene," *Geophys. Res. Lett.*, 19, 1129–1132 (1992).

CEC (Commission of the European Communities), "Emission Factors," DG XI CORINAIR, TNO Report No. 88-355, C. Veldt and A. Bakkum, Apeldoorn, The Netherlands, 1988.

CEC (Commission of the European Communities), "Environment and Quality of Life," CORINAIR Working Group on Emission Factors for Calculating 1985 Emissions for Road Traffic, Vol. 1: Methodology and Emission Factors, Final Report, EUR 12260 EN, 1989.

CEC (Commission of the European Communities), "Default Emission Factors Handbook," 2nd ed., CORINAIR Inventory Project, CITEPA, Paris, July 1991.

Chameides, W. L., "Acid Dew and the Role of Chemistry in the Dry Deposition of Reactive Gases to Wetted Surfaces," *J. Geophys. Res.*, 92, 11895–11908 (1987).

Cochran, L. S., R. A. Pielke, and E. Kovács, "Selected International Receptor-Based Air Quality," *J. Air Waste Manage. Assoc.*, 42, 1567–1572 (1992).

Corbett, J. J., and P. Fischbeck, "Emissions from Ships," *Science*, 278, 823–824 (1997).

Dabberdt, W. F., D. H. Lenschow, T. W. Horst, P. R. Zimmerman, S. P. Oncley, and A. C. Delany, "Atmosphere–Surface Exchange Measurements," *Science*, 260, 1472–1481 (1993).

Dasch, J. M., "Direct Measurement of Dry Deposition to a Polyethylene Bucket and Various Surrogate Surfaces," *Environ. Sci. Technol.*, 19, 721–725 (1985).

Davies, T. D., and J. R. Mitchell, "Dry Deposition of Sulphur Dioxide onto Grass in Rural Eastern England with Some Comparisons with Other Forms of Sulphur Deposition," in *Precipitation Scavenging, Dry Deposition, and Resuspension*, Vol. 2, *Dry Deposition and Resuspension* (H. R. Pruppacher, R. G. Semonin, and W. G. Slinn, Eds.), pp. 795–804, Elsevier, New York, 1983.

Dockery, D. W., and J. Schwartz, "Particulate Air Pollution and Mortality: More Than the Philadelphia Story," *Epidemiology, 6,* 629–632 (1995).

Dockery, D. W., C. A. Pope III, X. Xu, J. D. Spengler, J. H. Ware, M. E. Fay, B. G. Ferris, and F. E. Speizer, "An Association between Air Pollution and Mortality in Six U.S. Cities," *N. Engl. J. Med., 329,* 1753–1759 (1993).

Droppo, J. G., Jr., "Concurrent Measurements of Ozone Dry Deposition Using Eddy Correlation and Profile Flux Methods," *J. Geophys. Res., 90,* 2111–2118 (1985).

Edinger, J. G., "Vertical Distribution of Photochemical Smog in Los Angeles Basin," *Environ. Sci. Technol., 7,* 247–252 (1973).

Edinger, J. G., M. H. McCutchan, P. R. Miller, B. C. Ryan, M. J. Schroeder, and J. V. Behar, "Penetration and Duration of Oxidant Air Pollution in the South Coast Air Basin of California," *J. Air Pollut. Control Assoc., 22,* 882–886 (1972).

EPA, U.S. Environmental Protection Agency, "National Air Pollutant Emission Trends, 1900–1994," EPA-454/R-95-011, October 1995.

EPA, U.S. Environmental Protection Agency, "National Air Pollutant Emission Trends, 1900–1996," EPA-454/R-97-011, December 1997.

Erisman, J. W., A. Van Pul, and P. Wyers, "Parameterization of Surface Resistance for the Quantification of Atmospheric Deposition of Acidifying Pollutants and Ozone," *Atmos. Environ., 28,* 2595–2607 (1994).

"EUROTRAC Annual Report, 1993," P. Borrell, Coordinator, International Scientific Secretariat, Fraunhofer Institute, Kreuzeckbahnstrasse 19, D-82467, Garmisk-Partenkirchen, Germany.

Fall, R., "Biogenic Emissions of Volatile Organic Compounds from Higher Plants," in *Reactive Hydrocarbons in the Atmosphere* (C. N. Hewitt, Ed.), Chap. 2, Academic Press, San Diego, 1999.

Fehsenfeld, F., J. G. Calvert, R. Fall, P. Goldan, A. B. Guenther, C. N. Hewitt, B. Lamb, S. Liu, M. Trainer, H. Westberg, and P. Zimmerman, "Emissions of Volatile Organic Compounds from Vegetation and the Implications for Atmospheric Chemistry," *Global Biogeochem. Cycles, 6,* 389–430 (1992).

Fenger, J., O. Hertel, and F. Palmgren, *Urban Air Pollution—European Aspects,* Kluwer Academic Publishers, Dordrecht, The Netherlands, 1998.

Finlayson-Pitts, B. J., and J. N. Pitts, Jr., "Tropospheric Air Pollution: Ozone, Airborne Toxics, Polycyclic Aromatic Hydrocarbons, and Particles," *Science, 276,* 1045–1052 (1997).

Flatøy, F., and O. Hov, "NO_x from Lightning and the Calculated Chemical Composition of the Free Troposphere," *J. Geophys. Res., 102,* 21373–21381 (1997).

Foken, Th., R. Dlugi, and G. Kramm, "On the Determination of Dry Deposition and Emission of Gaseous Compounds at the Biosphere–Atmosphere Interface," *Meteorol. Z., 4,* 91–118 (1995).

Franzblau, E., C. J. Popp, E. W. Prestbo, N. A. Marley, and J. S. Gaffney, "Remote Measurement of NO_2 in the Brown Cloud over Albuquerque, New Mexico," *Environ. Monit. Assess., 24,* 231–242 (1993).

Friedrich, R., and A. Obermeier, "Anthropogenic Emissions of Volatile Organic Compounds," in *Reactive Hydrocarbons in the Atmosphere* (C. N. Hewitt, Ed.), Chap. 1, Academic Press, San Diego, 1999.

Gao, W., and M. L. Wesely, "Modeling Gaseous Dry Deposition over Regional Scales with Satellite Observations. I. Model Development," *Atmos. Environ., 29,* 727–737 (1995).

Georgii, H.-W., and J. Pankrath, Eds., *Deposition of Atmospheric Pollutants,* Proceedings of a Colloquium held at Oberursel/Taunus, West Germany, November 9–11, 1981, Reidel, Dordrecht, Holland, 1982.

Geron, C. D., A. B. Guenther, and T. E. Pierce, "An Improved Model for Estimating Emissions of Volatile Organic Compounds from Forests in the Eastern United States," *J. Geophys. Res., 99,* 12773–12791 (1994).

Golden, P. D., W. C. Kuster, F. C. Fehsenfeld, and S. A. Montzka, "The Observation of a C_5 Alcohol Emission in a North American Pine Forest," *Geophys. Res. Lett., 20,* 1039–1042 (1993).

Graedel, T. E., and R. McGill, "Degradation of Materials in the Atmosphere," *Environ. Sci. Technol., 20,* 1093–1100 (1986).

Graedel, T. E., T. S. Bates, A. F. Bouwman, D. Cunnold, J. Dignon, I. Fung, D. J. Jacob, B. K. Lamb, J. A. Logan, G. Marland, P. Middleton, J. M. Pacyna, M. Placet, and C. Veldt, "A Compilation of Inventories of Emissions to the Atmosphere," *Global Biogeochem. Cycles, 7,* 1–26 (1993).

Guenther, A., C. N. Hewitt, D. Erickson, R. Fall, C. Geron, T. Graedel, P. Harley, L. Klinger, M. Lerdau, W. A. McKay, T. Pierce, B. Scholes, R. Steinbrecher, R. Tallamraju, J. Taylor, and P. Zimmerman, "A Global Model of Natural Volatile Organic Compound Emissions," *J. Geophys. Res., 100,* 8873–8892 (1995).

Guenther, A., "Modeling Biogenic Volatile Organic Compound Emissions to the Atmosphere," in *Reactive Hydrocarbons in the Atmosphere* (C. N. Hewitt, Ed.), Chap. 3, Academic Press, San Diego, 1999.

Haagen-Smit, A. J., M. F. Brunelle, and J. W. Haagen-Smit, "Ozone Cracking in the Los Angeles Area," *Rubber Chem. Technol., 32,* 1134–1142 (1959).

Hall, J. V., A. M. Winer, M. T. Kleinman, F. W. Lurmann, V. Brajer, and S. Colome, "Valuing the Health Benefits of Clean Air," *Science, 255,* 812–817 (1992).

Hameed, S., and J. Dignon, "Global Emissions of Nitrogen and Sulfur Oxides in Fossil Fuel Combustion," *J. Air Waste Manage. Assoc., 42,* 159–163 (1992).

Hewitt, C. N., Ed., *Reactive Hydrocarbons in the Atmosphere,* Academic Press, San Diego, 1999.

Hicks, B. B., Ed., *Deposition both Wet and Dry,* Acid Precipitation Series, Vol. 4 (J. I. Teasley, Series Ed.), Butterworth, Stoneham, MA, 1984.

Hicks, B. B., and J. A. Garland, "Overview and Suggestions for Future Research on Dry Deposition," in *Precipitation Scavenging, Dry Deposition, and Resuspension," Vol. 2, Dry Deposition and Resuspension* (H. R. Pruppacher, R. G. Semonin, and W. G. Slinn, Eds.), pp. 1429–1433, Elsevier, New York, 1983.

Hicks, B. B., M. L. Wesely, R. L. Coulter, R. L. Hart, J. L. Durham, R. E. Speer, and D. H. Stedman, "An Experimental Study of Sulfur Deposition to Grassland," in *Precipitation Scavenging, Dry Deposition, and Resuspension," Vol. 2, Dry Deposition and Resuspension* (H. R. Pruppacher, R. G. Semonin, and W. G. Slinn, Eds.), pp. 933–942, Elsevier, New York, 1983.

Hoffmann, M. R., "Response to Comment on 'Acid Fog,'" *Environ. Sci. Technol., 18,* 664 (1984).

Hinds, W. C., *Aerosol Technology,* Wiley, New York, 1982.

Huebert, B. J., and C. H. Robert, "The Dry Deposition of Nitric Acid to Grass," *J. Geophys. Res., 90,* 2085–2090 (1985).

IPCC, Intergovernmental Panel on Climate Change, Contribution of Working Group I to the Second Assessment Report, (J. T. Houghton, L. G. Meira Filho, B. A. Callander, N. Harris, A. Kattenberg, and K. Maskell, Eds.), *Climate Change 1995,* Cambridge University Press, Cambridge, UK, 1996.

Jacob, D. J., and M. R. Hoffmann, "A Dynamic Model for the Production of H^+, NO_3^-, and SO_4^{2-} in Urban Fog," *J. Geophys. Res., 88,* 6611–6621 (1983).

Jacob, D. J., J. M. Waldman, J. W. Munger, and M. R. Hoffmann, "Chemical Composition of Fogwater Collected along the California Coast," *Environ. Sci. Technol., 19*, 730–736 (1985).

Kato, N., and H. Akimoto, "Anthropogenic Emissions of SO_2 and NO_x in Asia: Emission Inventories," *Atmos. Environ., 26A*, 2997–3017 (1992).

Kley, D., "Tropospheric Chemistry and Transport," *Science, 276*, 1043–1045 (1997).

Lamb, B., D. Gay, H. Westberg, and T. Pierce, "A Biogenic Hydrocarbon Emission Inventory for the U.S.A. Using a Simple Forest Canopy Model," *Atmos. Environ., 27*, 1673–1690 (1993).

Lenner, M., "Nitrogen Dioxide in Exhaust Emissions from Motor Vehicles," *Atmos. Environ., 21*, 37–43 (1987).

Lin, X., K. R. Polenske, and K. Robinson, "Economic Impact Analyses in U.S. State and Local Air Pollution Control Agencies: A Survey," *J. Air Waste Manage. Assoc., 44*, 134–140 (1994).

Lipfert, F. W., and R. E. Wyzga, "Air Pollution and Mortality: Issues and Uncertainties," *J. Air Waste Manage. Assoc., 45*, 949–966 (1995).

Lovett, G. M., "Atmospheric Deposition of Nutrients and Pollutants in North America: An Ecological Perspective," *Ecol. Appl., 4*, 629–650 (1994).

Mage, D., G. Ozolins, P. Peterson, A. Webster, R. Orthofer, V. Vandeweerd, and M. Gwynne, "Urban Air Pollution in Megacities of the World," *Atmos. Environ., 30*, 681–686 (1996).

McInnes, G., Ed., *Joint EMEF / CORINAIR Atmospheric Emission Inventory Guidebook*, 1st ed., European Environmental Agency, Copenhagen, Denmark, 1996.

McMahon, T. A., and P. J. Denison, "Empirical Atmospheric Deposition Parameters—A Survey," *Atmos. Environ., 13*, 571–585 (1979).

Micallef, A., and J. J. Colls, "Variation in Airborne Particulate Matter Concentration over the First Three Metres from Ground in a Street Canyon: Implications for Human Exposure," *Atmos. Environ., 32*, 3795–3799 (1998).

Monson, R. K., M. T. Lerdau, T. D. Sharkey, D. S. Schimel, and R. Fall, "Biological Aspects of Constructing Volatile Organic Emission Inventories," *Atmos. Environ., 29*, 2989–3002 (1995).

Moore, R. M., D. E. Oram, and S. A. Penkett, "Production of Isoprene by Marine Phytoplankton Cultures," *Geophys. Res. Lett., 21*, 2507–2510 (1994).

Müller, J.-F. "Geographical Distribution and Seasonal Variation of Surface Emissions and Deposition Velocities of Atmospheric Trace Gases," *J. Geophys. Res., 97*, 3787–3804 (1992).

Munger, J. W., D. J. Jacob, J. M. Waldman, and M. R. Hoffmann, "Fogwater Chemistry in an Urban Atmosphere," *J. Geophys. Res., 88*, 5109–5121 (1983).

National Research Council, *Science and Judgment in Risk Assessment*, Committee on Risk Assessment of Hazardous Air Pollutants, National Academy Press, Washington, DC, 1994.

Phalen, R. F., *Inhalation Studies: Foundations and Techniques*, CRC Press, Boca Raton, FL, 1984.

Phalen, R. F., and R. O. McClellan, "PM-10 Research Needs," *Inhalat. Toxicol., 7*, 773–779 (1995).

Phalen, R. F., R. G. Cuddihy, G. L. Fisher, O. R. Moss, R. B. Schlesinger, D. L. Swift, and H.-C. Yeh, "Main Features of the Proposed NCRP Respiratory Tract Model," *Radiat. Protect. Dosim., 38*, 179–184 (1991).

Pierce, T., C. Geron, L. Bender, R. Dennis, G. Tonnesen, and A. Guenther, "Influence of Increased Isoprene Emissions on Regional Ozone Modeling," *J. Geophys. Res., 103*, 25611–25629 (1998).

Price, C., J. Penner, and M. Prather, "NO_x from Lightning. 1. Global Distribution Based on Lightning Physics," *J. Geophys. Res., 102*, 5929–5941 (1997a).

Price, C., J. Penner, and M. Prather, "NO_x from Lightning. 2. Constraints from the Global Atmospheric Electric Circuit," *J. Geophys. Res., 102*, 5943–5951 (1997b).

Pruppacher, H. R., R. G. Semonin, and W. G. N. Slinn, *Precipitation Scavenging, Dry Deposition, and Resuspension*, Vol. 2, *Dry Deposition and Resuspension*, Proceedings of the Fourth International Conference, Santa Monica, California, November 29–December 3, 1982, Elsevier, New York, 1983.

Quinn, T. L., and J. M. Ondov, "Influence of Temporal Changes in Relative Humidity on Dry Deposition Velocities and Fluxes of Aerosol Particles Bearing Trace Elements," *Atmos. Environ., 32*, 3467–3479 (1998).

Schwartz, S. E., and P. Warneck, "Units for Use in Atmospheric Chemistry (IUPAC Recommendations 1995)," *Pure Appl. Chem., 67*, 1377–1406 (1995).

Schwartz, J., D. W. Dockery, and L. M. Neas, "Is Daily Mortality Associated Specifically with Fine Particles?" *J. Air Waste Manage. Assoc., 46*, 927–939 (1996).

Shaw, R. W., "The Atmosphere as Delivery Vehicle and Reaction Chamber for Acid Precipitation," in *Meteorological Aspects of Acid Rain* (C. M. Bhumralker, Ed.), Acid Precipitation Series (J. L. Teasley, Series Ed.), Vol. 1, pp. 33–55, Butterworth, Stoneham, MA, 1984.

Sievering, H., "Small-Particle Dry Deposition on Natural Waters: Modeling Uncertainty," *J. Geophys. Res., 89*, 9679–9681 (1984).

Singh, H. B., M. Kanakidou, P. J. Crutzen, and D. J. Jacob, "High Concentrations and Photochemical Fate of Oxygenated Hydrocarbons in the Global Troposphere," *Nature, 378*, 50–54 (1995).

Spiro, P. A., D. J. Jacob, and J. A. Logan, "Global Inventory of Sulfur Emissions with a $1° \times 1°$ Resolution," *J. Geophys. Res., 97*, 6023–6036 (1992).

Stith, J. L., P. V. Hobbs, and L. F. Radke, "Airborne Particle and Gas Measurements in the Emissions from Six Volcanoes," *J. Geophys. Res., 83*, 4009–4017 (1978).

Streets, D. G., G. R. Carmichael, and R. L. Arndt, "Sulfur Dioxide Emissions and Sulfur Deposition from International Shipping in Asian Waters," *Atmos. Environ., 31*, 1573–1582 (1997).

Stull, R. B. *An Introduction to Boundary Layer Meteorology*, Kluwer Academic, Dordrecht/Norwell, MA, 1988.

Tuazon, E. C., A. M. Winer, and J. N. Pitts, Jr., "Trace Pollutant Concentrations in a Multiday Smog Episode in the California South Coast Air Basin by Long Path Length Fourier Transform Infrared Spectroscopy," *Environ. Sci. Technol., 15*, 1232–1237 (1981).

Unsworth, M. H., A. S. Heagle, and W. W. Heck, "Gas Exchange in Open-Top Field Chambers. I. Measurement and Analysis of Atmospheric Resistances to Gas Exchange," *Atmos. Environ., 18*, 373–380 (1984).

Vandenberg, J. J., and K. R. Knoerr, "Comparison of Surrogate Surface Techniques for Estimation of Sulfate Dry Deposition," *Atmos. Environ., 19*, 627–635 (1985).

Vedal, S., "Ambient Particles and Health: Lines That Divide," *J. Air Waste Manage. Assoc., 47*, 551–581 (1997).

Wang, Y., D. J. Jacob, and J. A. Logan, "Global Simulation of Tropospheric O_3–NO_x–Hydrocarbon Chemistry. 1. Model Formulation," *J. Geophys. Res., 103*, 10713–10725 (1998).

Wesely, M. L., "Parameterization of Surface Resistances to Gaseous Dry Deposition in Regional-Scale Numerical Models," *Atmos. Environ., 23*, 1293–1304 (1989).

Wesely, M. L., and B. B. Hicks, "Recent Progress in Dry Deposition Studies," *Atmos. Environ.* (1999). (In press.)

Wildt, J., D. Kley, A. Rockel, P. Rockel, and H. J. Segschneider, "Emission of NO from Several Higher Plant Species," *J. Geophys. Res., 102*, 5919–5927 (1997).

Wilson, W. E., and H. H. Suh, "Fine Particles and Coarse Particles: Concentration Relationships Relevant to Epidemiologic Studies," *J. Air Waste Manage. Assoc., 47,* 1238–1249 (1997).

Wolf, M. E., and G. M. Hidy, "Aerosols and Climate: Anthropogenic Emissions and Trends for 50 Years," *J. Geophys. Res., 102,* 11113–11121 (1997).

World Meteorological Organization (WMO), "Scientific Assessment of Ozone Depletion: 1994," Global Ozone Research and Monitoring Project, Report No. 37, published February 1995; 1998 update, Report No. 44, published February, 1999.

Yeh, H.-S., R. G. Cuddihy, R. F. Phalen, and I.-Y. Chang, "Comparisons of Calculated Respiratory Tract Deposition of Particles Based on the Proposed NCRP Model and the New ICRP66 Model," *Aerosol Sci. Technol., 25,* 134–140 (1996).

CHAPTER 3

Spectroscopy and Photochemistry
Fundamentals

The chemistry of the atmosphere is driven to a very large measure by light, i.e., by photochemical processes. For example, the generation of the OH free radical in remote areas occurs primarily via the photolysis of O_3 in the presence of water vapor. Thus, the fundamental spectroscopy and photochemistry of atmospheric species is central to understanding the chemistry. In addition, as discussed in Chapter 11, the interaction of molecules with light is used extensively for detecting and measuring both trace species and urban air pollutants. As we shall see in Chapter 14, the interaction of gases and particles with infrared radiation plays a central role in the earth's climate.

In this chapter, we give a brief overview of the fundamentals of spectroscopy and photochemistry needed in atmospheric chemistry; for detailed treatments, see Calvert and Pitts (1966), Okabe (1978), Turro (1978), Wayne (1988), and Gilbert and Baggott (1991). Specifics for individual molecules are found in Chapter 4. Excellent treatments of atmospheric radiation are given by Liou (1980), Goody and Yung (1989), and Lenoble (1993).

A. BASIC PRINCIPLES

1. Molecular Energy Levels and Absorption and Emission Spectroscopy

We shall deal first with the simplest system, that of diatomic molecules, and then extrapolate to polyatomic systems.

Atoms in molecules undergo a variety of motions relative to each other. As illustrated in Fig. 3.1 for the water molecule, these can be separated into vibrational motions involving the various chemical bonds in the molecule, rotation of the molecule as a whole, and translational motion of the molecule, i.e., movement in the three coordinates x, y, and z. However, since the interaction of light with the molecule only directly changes the vibrational and rotational motions, we shall not consider translation further here. Of course, once the molecule has absorbed the light initially, the energy may also be converted into translational energy, at least in part, leading to an increase in temperature.

In addition to inducing changes in the positions of atoms within the molecule, the absorption of light can lead to changes in electron distribution. In contrast to vibrational and rotational changes, such electronic transitions typically require sufficient energy (in the ultraviolet and visible regions) that they can lead to breaking of chemical bonds; i.e., photochemistry can occur. It is this latter process that, in general, is of most interest in atmospheric chemistry.

a. *Diatomic Molecules*

(1) Vibrational energy and transitions   As seen in Fig. 3.2a, the bond between the two atoms in a diatomic molecule can be viewed as a vibrating spring in which, as the internuclear distance changes from the equilibrium value r_e, the atoms experience a force that tends to restore them to the equilibrium position. The *ideal*, or harmonic, oscillator is defined as one that obeys Hooke's law; that is, the restoring force F on the atoms in a diatomic molecule is proportional to their displacement from the equilibrium position.

Substitution of the potential energy for this harmonic oscillator into the Schrödinger wave equation gives the allowed vibrational energy levels, which are quantified and have energies E_v given by

$$E_v = h\nu_{vib}(v + 1/2), \qquad (A)$$

where ν_{vib} is a constant characteristic of the molecule and is related to the strength of the bond and the

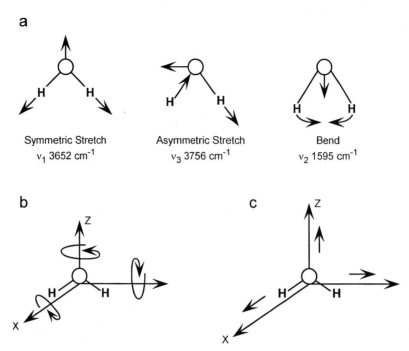

FIGURE 3.1 (a) Internal vibrations of the bonds in the water molecule, (b) rotational motion of water, and (c) translation of the water molecule.

reduced mass of the molecule. The *vibrational quantum number* v can have the integral values 0, 1, 2, ... Thus the vibrational energy levels of this ideal oscillator are equally spaced.

However, as seen in Fig. 3.2, this idealized harmonic oscillator (Fig. 3.2b) is satisfactory only for low vibrational energy levels. For real molecules, the potential energy rises sharply at small values of r, when the atoms approach each other closely and experience significant charge repulsion; furthermore, as the atoms move apart to large values of r, the bond stretches until it ultimately breaks and dissociation occurs (Fig. 3.2c).

To obtain the allowed energy levels, E_v, for a real diatomic molecule, known as an *anharmonic* oscillator, one substitutes the potential energy function describing the curve in Fig. 3.2c into the Schrödinger equation; the allowed energy levels are

$$E_v = h\nu_{vib}(v + 1/2) - h\nu_{vib}x_e(v + 1/2)^2 + h\nu_{vib}y_e(v + 1/2)^3 + \cdots . \quad (B)$$

Once again v is the vibrational quantum number with allowed values of 0, 1, 2,..., and x_e and y_e are anharmonicity constants characteristic of the molecule.

Equation (B) is often expressed in wavenumbers, ω_{vib}; the allowed energy states, $\bar{E}_v$, in units of wavenumbers (cm^{-1}) become

$$\bar{E}_v = \omega_e(v + 1/2) - \omega_e x_e(v + 1/2)^2 + \omega_e y_e(v + 1/2)^3 + \cdots . \quad (C)$$

Note that throughout this book we use a bar over a parameter (e.g., $\bar{E}$) if it is expressed in units of wavenumbers. Values for ω_e, x_e, and y_e for a number of diatomic molecules are found in Herzberg's classic *Molecular Spectra and Molecular Structure. I. Spectra of Diatomic Molecules* (1950) and in Huber and Herzberg (1979).

An important consequence of using the potential energy for a real molecule in the Schrödinger equation is that the vibrational energy levels become more closely spaced with increasing quantum number v (Fig. 3.2c versus 3.2b).

When exposed to electromagnetic radiation of the appropriate energy, typically in the infrared, a molecule can interact with the radiation and absorb it, exciting the molecule into the next higher vibrational energy level. For the ideal harmonic oscillator, the selection rules are $\Delta v = \pm 1$; that is, the vibrational energy can only change by one quantum at a time. However, for anharmonic oscillators, weaker overtone transitions due to $\Delta v = \pm 2, \pm 3$, etc. may also be observed because of their nonideal behavior. For polyatomic molecules with more than one fundamental vibration, e.g., as seen in Fig. 3.1a for the water molecule, both overtones and

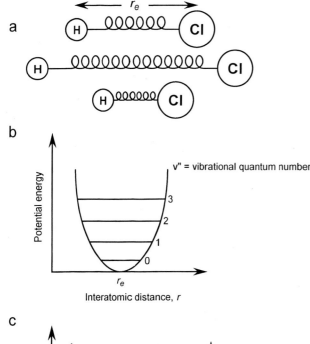

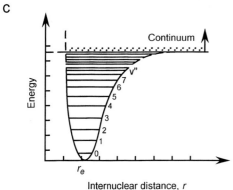

FIGURE 3.2 (a) Vibration of diatomic molecule, HCl, (b) potential energy of an ideal harmonic oscillator, and (c) an anharmonic oscillator described by the Morse function.

and H_2, whose dipole moments remain constant during vibration, do not.

Since at room temperature most molecules are in the $v'' = 0$ state, and $\Delta v = \pm 1$ is by far the strongest transition, most molecules go from $v'' = 0$ to $v' = 1$. (We follow the Herzberg convention that the quantum number of the upper state is designated by a prime, and the lower state by a double prime; e.g., in this example, $v'' = 0 \to v' = 1$.) As a result, a single vibrational absorption band (with associated rotational structure) is normally observed in the infrared, the region corresponding to the energy level differences given by Eq. (C).

(2) Rotational energy and transitions If a molecule has a permanent dipole moment, its rotation in space produces an oscillating electric field; this can also interact with electromagnetic radiation, resulting in light absorption.

In the idealized case for rotation of a diatomic molecule, one assumes the molecule is analogous to a dumbbell with the atoms held at a fixed distance r from each other; that is, it is a *rigid rotor*. The simultaneous vibration of the molecule is ignored, as is the increase of internuclear distance at high rotational energies arising from the centrifugal force on the two atoms.

For this idealized case, the rotational energy levels (in cm^{-1}) are given by

$$\overline{E}_r = \overline{B}(J)(J+1) \text{ cm}^{-1}, \qquad \text{(D)}$$

where $\overline{B}$, the rotational constant characteristic of the molecule, is given by

$$\overline{B} = h/8\pi^2 Ic. \qquad \text{(E)}$$

I is the moment of inertia of the molecule, given by $I = \mu r^2$, where μ is the reduced mass defined by $\mu^{-1} = [(M_A^{-1}) + (M_B^{-1})]$, M_A and M_B are the atomic masses, and r is the fixed, internuclear distance. J is the rotational quantum number; its allowed values are 0, 1, 2, ...

For a real rotating diatomic molecule, known as a nonrigid rotor, Eq. (D) becomes

$$\overline{E}_r = \overline{B}(J)(J+1) - \overline{D}(J)^2(J+1)^2. \qquad \text{(F)}$$

The constant $\overline{D}$ is characteristic of the diatomic molecule and is much smaller than $\overline{B}$; generally, $\overline{D} \approx 10^{-4}\overline{B}$. The second term in Eq. (F) generally becomes important at large values of J when centrifugal force increases the separation between atoms.

Because of the requirement of a permanent dipole moment, only heteronuclear molecules can absorb radiation and change their rotational energy. For the

combination bands (i.e., those that are the sum of two or more fundamental vibrations) occur.

Because the vibrational energy level spacing is relatively large (typically of the order of 10^3 cm^{-1}) compared to their thermal energy, most molecules at room temperature are in their lowest vibrational energy level and light absorption normally occurs from $v = 0$.

For a purely vibrational transition, the selection rule for absorption of light requires that there be a changing dipole moment during the vibration. This oscillating dipole moment produces an electric field that can interact with the oscillating electric and magnetic fields of the electromagnetic radiation. Thus heteronuclear diatomic molecules such as NO, HCl, and CO absorb infrared radiation and undergo vibrational transitions, whereas homonuclear diatomic molecules such as O_2

idealized case of a rigid rotor, the selection rule is $\Delta J = \pm 1$. For the energy levels given by Eq. (D), the energy level splitting between consecutive rotational energy levels is given by

$$\Delta \overline{E}_r = 2\overline{B}J', \quad (G)$$

where J' is the quantum number of the *upper* rotational state involved in the transition. Thus the spacing between rotational energy levels increases with increasing rotational quantum number. Splittings are small compared to those between vibrational energy levels, typically of the order of 10 cm^{-1} in the lower levels; this corresponds to absorption in the microwave region. Indeed, these spacings are sufficiently small that the population of the rotational energy levels above $J = 0$ is significant at room temperature because the thermal energy available is sufficient to populate the higher rotational levels.

The Boltzmann expression can be used to calculate the relative populations of molecules in any rotational state J compared to the lowest rotational state $J = 0$ at temperature T (K):

$$N_J/N_0 = (2J + 1)e^{-E_J/kT}. \quad (H)$$

In Eq. (H), k is the Boltzmann constant (1.381×10^{-23} J K^{-1}) and E_J is the energy of the Jth rotational level given by Eq. (D) for the ideal rigid rotor, or Eq. (F) for the nonideal case.

The exponential energy factor in Eq. (H) gives decreasing populations with increasing J, but the degeneracy factor $(2J + 1)$ works in the opposite direction. As a result, rotational populations increase initially with increasing J, reach a peak, and subsequently decrease.

The combination of increased spacing between energy levels as J increases and the significant population of molecules in higher rotational energy levels means that the absorption of microwave radiation occurs from a number of different initial states, resulting in a series of absorption lines, rather than a single line as seen in the pure vibrational infrared spectra of diatomic molecules. From the spacing of these lines, the rotational constant $\overline{B}$, and hence the moment of inertia and the internuclear spacing, can be obtained using Eq. (G).

(3) Vibration–rotation Molecules, of course, vibrate and rotate simultaneously. It is a good approximation that the total energy of the molecule (excluding translation) is the sum of the vibrational (V), rotational (R), and electronic (E) energy of a molecule; that is, $E_{\text{total}} = E_V + E_R + E_E$. For the case where there is no

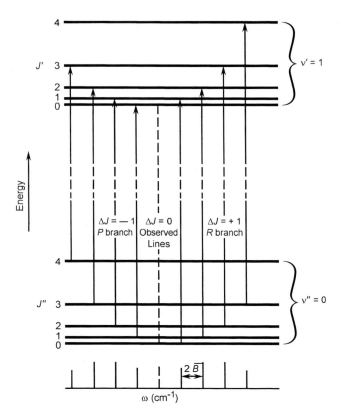

FIGURE 3.3 Schematic diagram of energy levels involved in HCl vibration–rotation transitions at room temperature (from Herzberg, 1950).

electronic energy change for an ideal harmonic oscillator–rigid rotor, the selection rules $\Delta v = \pm 1$ and $\Delta J = \pm 1$ apply. At room temperature, $v'' = 0 \to v' = 1$ is the only significant vibrational transition in absorption, but, as discussed earlier, a variety of rotational transitions can occur.

Figure 3.3 shows some of these possible transitions for HCl. Those with $\Delta J = +1$ are known as the R branch and occur at the high-energy side of the hypothetical transition $\Delta v = 1$, $\Delta J = 0$ (this is not allowed because of the selection rule, $\Delta J = \pm 1$). Those with $\Delta J = -1$ on the low-frequency side of the hypothetical transition form the P branch. Figure 3.4 shows the absorption spectrum of HCl at room temperature, with the rotational transitions responsible for each line. The relative intensities of the lines reflect the relative populations of the absorbing rotational levels; the peaks are doublets due to the separate absorptions of the two chlorine isotopes, that is, H^{35}Cl and H^{37}Cl, which have different reduced masses and hence values of the rotational constant $\overline{B}$.

(4) Electronic energy and transitions The electronic states of a diatomic molecule are described by several

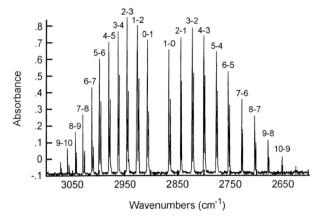

FIGURE 3.4 Vibration–rotation spectrum of 0.18 Torr HCl at room temperature using a path length of 19.2 m. Resolution is 0.25 cm^{-1}. The rotational transitions are shown as (initial J, final J) (from B. J. Finlayson-Pitts and S. N. Johnson, unpublished data).

molecular quantum numbers, Λ, S, and Ω. Λ is the component of the total electronic orbital angular momentum L along the internuclear axis and can be determined from the electronic spectrum of the molecule (see Herzberg, 1950). Allowed values of Λ of 0, 1, 2, and 3 correspond to electronic states designated as Σ, Π, Δ, and Φ, respectively.

The spin quantum number S represents the net spin of the electrons. It has an integral value or zero for even numbers of electrons and half-integral value for odd numbers. The *multiplicity* of a molecular state is defined as $(2S + 1)$ and is written as a superscript to the left of the symbol corresponding to Λ. Values of S of 0, 1/2, and 1, corresponding to multiplicities of 1, 2, and 3, are referred to as singlet, doublet, and triplet states, respectively. While most stable molecules have ground singlet states, one important exception is O_2, where the ground state is a triplet. As discussed in Chapter 4, this has important implications for its spectroscopy and photochemistry.

For many molecules the quantum number Ω is defined and is given by the vector sum of Λ and Σ,

$$\Omega = |\Lambda + \Sigma|,$$

where Σ is the vector component of S in the direction of the internuclear axis. Σ can have the values $+S$, $S - 1, \ldots, -S$ and can be positive, negative, or zero.

Two other symbols are used in designating electronic states according to their symmetry. For homonuclear diatomic molecules, states are designated "g" or "u" as a subscript to the right of the Λ symbol, depending on whether or not the wavefunction describing the molecular state changes sign when reflected through the center of symmetry of the molecule. If it does change sign, it is designated "u" (for ungerade = uneven); if it does not, it is designated "g" (for gerade = even).

Finally, the symbols + and −, written as superscripts to the Λ symbol, refer to two types of sigma states, Σ^+ and Σ^-. If the wavefunction is unaltered by reflection through a plane passing through the two nuclei, the state is positive (+); if it changes sign, it is negative (−).

The selection rules for electronic transitions are not as clear-cut as in the case of vibration and rotation. In the case of molecules consisting of relatively light nuclei, which is the case for many molecules of tropospheric interest, the selection rules

$$\Delta \Lambda = 0, \pm 1 \quad \text{and} \quad \Delta S = 0$$

apply. Thus transitions between states of unlike multiplicity (e.g., singlet → triplet) are "forbidden" but in some cases may occur with a relatively small probability, most notably with the oxygen molecule, which has a triplet ground state. In terms of the symmetry of the wavefunctions, u ↔ g transitions are allowed but u ↔ u and g ↔ g are forbidden. In addition, Σ^+ states cannot combine with Σ^- states; that is, $\Sigma^+ \leftrightarrow \Sigma^+$ and $\Sigma^- \leftrightarrow \Sigma^-$ transitions are allowed but $\Sigma^+ \leftrightarrow \Sigma^-$ are forbidden. Table 3.1 summarizes these selection rules for molecules with light nuclei.

Upon absorption of light of an appropriate wavelength, a diatomic molecule can undergo an *electronic* transition, along with simultaneous vibrational and rotational transitions. In this case, there is no restriction on Δv. That is, the selection rule $\Delta v = \pm 1$ valid for purely vibrational and vibrational–rotational transitions no longer applies; thus numerous vibrational transitions can occur. If the molecule is at room temperature, it will normally be in its lower state, $v'' = 0$; hence transitions corresponding to $v'' = 0$ to $v' = 0$,

TABLE 3.1 Allowed[a] Electronic Transitions of Diatomic Molecules Having Light Nuclei[b]

Homonuclear diatomic (equal nuclear charge)	Heteronuclear diatomic (unequal nuclear charge)
$\Sigma_g^+ \leftrightarrow \Sigma_u^+$	$\Sigma^+ \leftrightarrow \Sigma^+$
$\Sigma_g^- \leftrightarrow \Sigma_u^-$	$\Sigma^- \leftrightarrow \Sigma^-$
$\Pi_g \leftrightarrow \Sigma_u^+$, $\Pi_u \leftrightarrow \Sigma_g^+$	$\Pi \leftrightarrow \Sigma^+$
$\Pi_g \leftrightarrow \Sigma_u^-$, $\Pi_u \leftrightarrow \Sigma_g^-$	$\Pi \leftrightarrow \Sigma^-$
$\Pi_g \leftrightarrow \Pi_u$	$\Pi \leftrightarrow \Pi$
$\Pi_g \leftrightarrow \Delta_u$, $\Pi_u \leftrightarrow \Delta_g$	$\Pi \leftrightarrow \Delta$
$\Delta_g \leftrightarrow \Delta_u$	$\Delta \leftrightarrow \Delta$

[a] Presuming that the rule $\Delta S = 0$ is obeyed.
[b] Source: Herzberg (1950), p. 243.

1, 2, 3,... in the upper electronic state are usually observed.

The rotational selection rule is $\Delta J = 0, \pm 1$, except for the case of a transition involving $\Omega = 0$ for both the upper and lower states. Thus three sets of lines (known as the P, Q, and R branches) corresponding to $\Delta J = -1, 0$, and $+1$, respectively, are observed for each band arising from a particular vibrational transition. Figure 3.5 illustrates these transitions schematically. However, if $\Omega = 0$ for both upper and lower states ($^1\Sigma \to {}^1\Sigma$ transition), the rotational selection rule is $\Delta J = \pm 1$, and the Q branch does not appear. For further details, see Herzberg (1945, 1950, 1967).

As a general rule, the most probable, that is, most intense, vibrational transitions within a given electronic transition will be those in which the vibrational probabilities are maximum in both the initial and final states. An important restriction is that only *vertical* transitions are allowed. This is a consequence of the Franck–Condon principle, which states that the time for an electronic transition to occur (typically 10^{-15} s) is so short relative to the time it takes for one vibration ($\sim 10^{-13}$ s) that the internuclear distance remains essentially constant during the electronic transition.

Figure 3.6 shows the Morse potential energy curves for two hypothetical electronic states of a diatomic molecule, the vibrational energy levels for each, and the shape of the vibrational wavefunctions (ψ) within

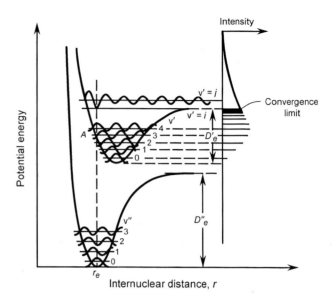

FIGURE 3.6 Potential energy curves for the ground state and an electronically excited state of a hypothetical diatomic molecule. Right-hand side shows relative intensities expected for absorption bands (from Calvert and Pitts, 1966).

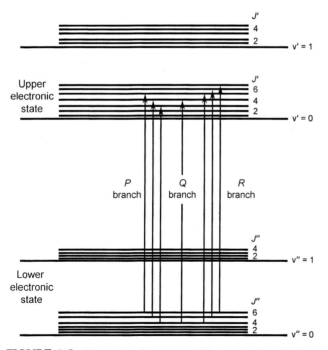

FIGURE 3.5 Schematic of some possible rotational and vibrational transitions involved during an electronic transition of a diatomic molecule from the ground electronic state.

each vibrational energy level. At room temperature, most molecules will originate in $v'' = 0$; the vertical line is at the midpoint of $v'' = 0$ since the probability of finding the molecule at $r = r_e$ is a maximum here.

The probability of a particular transition from $v'' = 0$ to an upper vibrational level v' is determined largely by the product of the wavefunctions for the two states, $\psi_{v'}\psi_{v''}$. A qualitative examination of the wavefunctions in the upper state, $\psi_{v'}$, in Fig. 3.6 shows that $\psi_{v'}$, and hence the product $\psi_{v'}\psi_{v''}$, is a maximum around $v' = 4$; thus the vibrational transition corresponding to $v'' = 0 \to v' = 4$ is expected to be the most intense. On the other hand, the wavefunction at $v' = 0$ is very small; hence the $v'' = 0 \to v' = 0$ transition should be weak. The right side of Fig. 3.6 shows the corresponding intensities expected for the various absorption lines in this electronic transition.

The potential energy curves of excited electronic states need not have potential energy minima, such as those shown in Fig. 3.6. Thus Fig. 3.7 shows two hypothetical cases of *repulsive states* where no minima are present. Dissociation occurs immediately following light absorption, giving rise to a spectrum with a structureless continuum. Transition *a* represents the case where dissociation of the molecule AB produces the atoms A and B in their ground states, and transition *b* the situation where dissociation produces one of the atoms in an electronically excited state, designated A*.

Some molecules may have a number of excited electronic states, some of which have potential minima, as

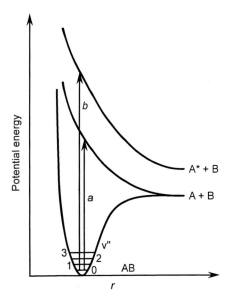

FIGURE 3.7 Potential energy curves for a hypothetical diatomic molecule showing electronic transitions to two repulsive excited states having no minima. A* is an electronically excited atom.

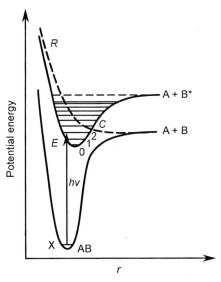

FIGURE 3.8 Potential energy curves for the ground state and two electronically excited states in a hypothetical diatomic molecule. Predissociation may occur when the molecule is excited into higher vibrational levels of the state E and crosses over to repulsive state R at the point C (from Okabe, 1978).

in Fig. 3.6, and some of which are wholly repulsive, as in Fig. 3.7. In this case, depending on the wavelength absorbed (i.e., the electronically excited state reached), the molecule may dissociate or undergo one of the photophysical processes described below.

A simplified, hypothetical example of this situation is shown in Fig. 3.8. If the molecule absorbs light corresponding to energies insufficient to produce vibrational energy levels above $v' = 2$ in the excited state E, a structured absorption (and emission) spectrum is observed. However, if the photon energy is greater than that required to produce $A + B$, that is, greater than the bond dissociation energy, then the molecule may be excited either into the repulsive state R, from which it immediately dissociates into ground-state atoms $A + B$, or alternatively into vibrational levels $v' > 2$ of excited state E. In the latter case, the excited molecule may undergo one of the photophysical processes discussed below (fluorescence, deactivation, etc.) or it may cross over from state E into the repulsive state R (point C in Fig. 3.8) and dissociate. This phenomenon is known as *predissociation*. In a case such as that in Fig. 3.8, the absorption spectrum would be expected to show well-defined rotational and vibrational structure up to a certain energy corresponding to the transition $v'' = 0 \rightarrow v' = 2$. For transitions to higher vibrational levels, the rotational structure becomes blurred, and a *predissociation spectrum* is observed. I_2 is an example of a molecule with both a low-lying repulsive electronically excited state and bound excited states; excitation from the ground $^1\Sigma$ state into the repulsive $^1\Pi$ state results in dissociation into the ground state iodine atoms. On the other hand, excitation into the $B^3\Pi$ state below the dissociation limit gives electronically excited I_2, which returns to the ground state via light emission or crosses into the repulsive $^1\Pi$ state and predissociates (Okabe, 1978).

Finally, if the incident photon is sufficiently energetic and the appropriate selection rules are obeyed, the excited molecule may be produced in vibrational levels of state E that are sufficiently high that the molecule immediately dissociates into $A + B^*$, where B^* is an electronically excited state.

b. Polyatomic Molecules

The principles discussed for diatomic molecules generally apply to polyatomic molecules, but their spectra are much more complex. For example, instead of considering rotation only about an axis perpendicular to the internuclear axis and passing through the center of mass, for nonlinear molecules, one must think of rotation about three mutually perpendicular axes as shown in Fig. 3.1b. Hence we have three rotational constants $\overline{A}$, $\overline{B}$, and $\overline{C}$ with respect to these three principal axes.

Furthermore, polyatomic molecules consisting of n atoms have $3n - 6$ vibrational degrees of freedom (or $3n - 5$ in the special case of a linear polyatomic molecule), instead of just one as in the case of a diatomic molecule. Some or all of these may absorb infrared radiation, leading to more than one infrared absorption band. In addition, overtone bands ($\Delta v > 1$)

and combination bands (absorptions corresponding to the sum of two or more of the fundamental vibrations) are much more common.

An atmospherically relevant example of just how complex infrared spectra of polyatomic molecules can be is water vapor. A multitude of observed transitions occurs in the regions $\sim 1300-2000$ cm^{-1} and $3000-4000$ cm^{-1}. This, combined with the relatively high, and often rapidly changing, concentrations of water vapor, makes long-path-length infrared spectroscopic determination of trace atmospheric species in ambient air very difficult in these wavelength regions. Similarly, absorption of CO_2 renders the region from ~ 2230 to ~ 2390 cm^{-1} unusable for ambient air studies.

2. Fates of Electronically Excited Molecules

Once a molecule is excited into an electronically excited state by absorption of a photon, it can undergo a number of different primary processes. *Photochemical* processes are those in which the excited species dissociates, isomerizes, rearranges, or reacts with another molecule. *Photophysical* processes include *radiative* transitions in which the excited molecule emits light in the form of fluorescence or phosphorescence and returns to the ground state and *nonradiative* transitions in which some or all of the energy of the absorbed photon is ultimately converted to heat.

a. Photophysical Processes

Photophysical processes are often displayed in the form of the Jablonski-type energy level diagram shown in Fig. 3.9. The common convention is that singlet states are labeled S_0, S_1, S_2, and so on and the triplets are labeled T_1, T_2, T_3, and so on, in order of increasing energy. Vibrational and rotational states are shown as being approximately equally spaced only for clarity of presentation. Radiative transitions, for example, fluorescence (F) and phosphorescence (P), are shown as solid lines, and nonradiative transitions as wavy lines. Vertical distances between the vibrational–rotational levels of the singlet ground state, S_0, and the two electronically excited states, the first excited singlet, S_1, and its triplet, T_1, correspond to their energy gaps.

Fluorescence is defined as the emission of light due to a transition between states of like multiplicity, for example, $S_1 \rightarrow S_0 + h\nu$. This is an allowed transition, and hence the lifetime of the upper state with respect to fluorescence is usually short, typically $10^{-6}-10^{-9}$ s. For example, the fluorescence lifetime of OH in the electronically excited $A^2\Sigma^+$ state is ~ 0.7 μs (McDer-

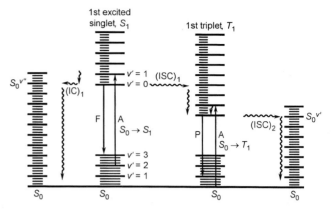

FIGURE 3.9 Jablonski diagram illustrating photophysical radiative and nonradiative transitions. S_0 = ground singlet state, S_1 = first excited singlet state, T_1 = first triplet state, A = absorption of light, F = fluorescence, P = phosphorescence, IC = internal conversion, ISC = intersystem crossing. Radiative transitions are shown by solid lines, and nonradiative transitions by wavy lines. Photochemical processes are not indicated.

mid and Laudenslager, 1982; Crosley, 1989). Chemiluminescence is similar to fluorescence except that the excited state is generated in a chemical reaction.

Phosphorescence is defined as the emission of light due to a transition between states of different spin multiplicities. Because this is theoretically not an allowed transition for an ideal unperturbed molecule, phosphorescence lifetimes tend to be relatively long, typically $10^{-3}-10^{-2}$ s.

Intersystem crossing (ISC) is the intramolecular crossing from one state to another of different multiplicity without the emission of radiation. In Fig. 3.9 (ISC)$_1$ shows the transfer from the first excited singlet state S_1 to the first excited triplet state T_1. Since the process is horizontal, the total energy remains the same and the molecule initially is produced in upper vibrational and rotational levels of T_1, from which it is deactivated as shown by the vertical wavy line. Similarly, (ISC)$_2$ shows the intersystem crossing from T_1 to upper vibrational and rotational states of the ground state S_0, from which vibrational deactivation to $v'' = 0$ then occurs.

Internal conversion (IC) is the intramolecular crossing of an excited molecule from one state to another of the same multiplicity without the emission of radiation. As seen in Fig. 3.9, the horizontal wavy line (IC)$_1$ represents internal conversion from the lowest excited singlet state S_1 to high vibrational levels of the ground state S_0; this is generally followed by vibrational deactivation to $v'' = 0$.

Intramolecular photophysical processes available to an excited molecule shown in Fig. 3.9 predominate at low pressures where collisions with other molecules are relatively infrequent. However, at 1 atm pressure, or in the liquid state, the excited molecule can undergo many collisions with ground-state molecules; this can lead to collisional deactivation of the excited species by several paths. For example, an electronically excited molecule, A*, in the $S_1^{v'>0}$ state could undergo a series of collisions, be vibrationally (and rotationally) deactivated, and fall into the $S_1^{v'=0}$ state. The energy lost by A* is carried off as translational energy of the ground-state collision partner, B. From here, A ($S_1^{v'=0}$) can undergo photophysical or photochemical processes. Alternatively, *energy transfer* from A* to the collision partner can occur in which the excitation energy appears as excess vibrational, rotational, and/or electronic energy of molecule B.

Collisional deactivation and energy transfer play important roles in tropospheric chemistry. For example, electronically excited SO_2 in the $^3B_1$ state can be deactivated by O_2 (as well as by N_2 and H_2O) to the ground ($^1A_1$) state, with part of this process occurring via triplet–triplet energy transfer to generate singlet electronically excited states of O_2:

$$SO_2(^3B_1) + O_2(^3\Sigma_g^-) \rightarrow SO_2(^1A_1) + O_2(^1\Sigma_g^+, ^1\Delta_g). \quad (1)$$

Note that in the transfer of electronic energy between an excited atom or molecule and a second atom or molecule, the Wigner spin conservation rule generally applies. This states that the overall spin angular momentum of the system should not change during the energy transfer (see Herzberg for details). Because O_2 has the unusual property of having a ground triplet state, energy transfer from triplet collision partners can produce the reactive singlet states of molecular oxygen. Indeed this is the mode of action in some photodynamic therapies in medicine.

Similarly, collisional deactivation is an important factor in trying to detect and measure various gaseous species in the troposphere using the technique of induced fluorescence. For example, as discussed in Chapter 11, induced fluorescence is one of the techniques applied to determine the concentration of OH free radicals in the troposphere. The OH is excited to the $A^2\Sigma^+$ state, from which it fluoresces as it returns to the ground state. However, collisional deactivation of excited OH by O_2 and N_2 is significant at 1 atm pressure; this reduces the emitted light relative to interfering signals. Expansion of the air sample to lower pressures reduces this quenching and increases the overall sensitivity (see Chapter 11).

b. Photochemical Processes

In contrast to the photo*physical* processes just described, photochemical processes produce new chemical species. Such processes can be characterized by the type of chemistry induced by light absorption: photodissociation, intramolecular rearrangements, photoisomerization, photodimerization, hydrogen atom abstraction, and photosensitized reactions.

Of these, photodissociation is by far the most pervasive and important in atmospheric chemistry. For example, the photodissociation of NO_2 into ground-state oxygen atoms,

$$NO_2(X^2A_1) + h\nu \ (290 < \lambda < 430 \text{ nm}) \rightarrow$$
$$NO(X^2\Pi) + O(^3P), \quad (2)$$

followed by the reaction of $O(^3P)$ with O_2, is the sole known source of anthropogenically produced O_3 in the troposphere.

The reader will encounter numerous other examples of photodissociation throughout this text, so it will not be treated further here. However, as will become obvious in examining the chemistry of both the troposphere and stratosphere in later chapters, it is photochemistry that indeed drives the chemistry of the atmosphere.

c. Quantum Yields

The relative efficiencies of the various photophysical and photochemical primary processes are described in terms of quantum yields, ϕ. The primary quantum yield, ϕ, for the ith process, either photophysical or photochemical, is given by Eq. (I):

$$\phi_i = \frac{\text{Number of excited molecules proceeding by process } i}{\text{Total number of photons absorbed}} \quad (I)$$

For example, the nitrate radical, which plays an important role in nighttime chemistry (see Chapter 6), absorbs light in the red region of the visible (600–700 nm). The electronically excited state formed on light absorption can dissociate into either $NO_2 + O$ or into $NO + O_2$, or it can fluoresce:

$$NO_3 + h\nu \rightarrow NO_3^*, \quad (3)$$
$$NO_3^* \rightarrow NO_2 + O, \quad (4a)$$
$$\rightarrow NO + O_2, \quad (4b)$$
$$\rightarrow NO_3 + h\nu. \quad (4c)$$

The primary quantum yields for each process are defined as follows:

$$\phi_{4a} = \frac{\text{Number of NO}_2 \text{ or O}(^3\text{P}) \text{ formed in the primary process}}{\text{Number of photons absorbed by NO}_3},$$

$$\phi_{4b} = \frac{\text{Number of NO or O}_2 \text{ formed in the primary process}}{\text{Number of photons absorbed by NO}_3},$$

$$\phi_{4c} = \frac{\text{Number of photons emitted by NO}_3}{\text{Number of photons absorbed by NO}_3}.$$

ϕ_{4c} is also known as the fluorescence quantum yield, ϕ_f.

By definition, the sum of the primary quantum yields for all photochemical and photophysical processes taken together must add up to unity, i.e.,

$$\Sigma(\phi_f + \phi_p + \phi_{deact} + \ldots \phi_a + \phi_b + \ldots) = 1.0,$$

where ϕ_f, ϕ_p, and ϕ_{deact} are the primary quantum yields for the photophysical processes of fluorescence, phosphorescence, and collisional deactivation, respectively, and ϕ_a, ϕ_b, and so on are the primary quantum yields for the various possible photochemically reactive decomposition paths of the excited molecule.

For example, in the case of NO_3, ϕ_{4a} is, within experimental error, 1.0 up to 585 nm and then decreases to zero at 635 nm. As path (4a) falls off above 585 nm, path (4b), ϕ_{4b} increases to a peak of 0.36 at approximately 595 nm and then also decreases at longer wavelengths (Orlando et al., 1993; Davis et al., 1993; Johnston et al., 1996). As the quantum yields for both (4a) and (4b) decline, fluorescence, (4c) is observed (Nelson et al., 1983; Ishiwata et al., 1983), increasing toward unity at ~ 640 nm. In short, at different wavelengths the contribution of the various processes, (4a), (4b), and (4c), varies, but always consistent with their sum being unity.

While the aim of photochemical studies is generally to measure primary quantum yields, this is not always experimentally feasible. For example, NO reacts rapidly with NO_3 to form NO_2. Thus determination of ϕ_{4a} or ϕ_{4b} by measuring the NO and NO_2 formed can be complicated by this secondary reaction of NO with NO_3, and the measured yields of NO and NO_2 may not reflect the efficiency of the primary photochemical processes.

In some cases, then, the *overall* quantum yield, rather than the *primary* quantum yield, is reported. The overall quantum yield for a particular product A, usually denoted by Φ_A, is defined as the number of molecules of the product A formed per photon absorbed. Because of the potential contribution of secondary chemistry to the formation of stable products, the overall quantum yield of a particular product may exceed unity. Indeed, in chain reactions, overall quantum yields for some products may be of the order of 10^6 or more.

B. ABSORPTION OF LIGHT

1. Basic Relationships

Light has both wave-like and particle-like properties. As a wave, it is a combination of oscillating electric and magnetic fields perpendicular to each other and to the direction of propagation (Fig. 3.10). The distance between consecutive peaks is the wavelength, λ, and the number of complete cycles passing a fixed point in 1 s is the frequency, ν. They are inversely proportional through the relationship

$$\lambda = c/\nu, \qquad (J)$$

where c is the speed of light in a vacuum, 2.9979×10^8 m s^{-1}.

Considered as a particle, the energy of a quantum of light E is

$$E = h\nu = hc/\lambda, \qquad (K)$$

where h is Planck's constant, 6.6262×10^{-34} J s per quantum, and the frequency ν is in s^{-1}. In the visible and ultraviolet regions of the spectrum, wavelength is commonly expressed in nanometer units, 1 nm = 10^{-9} m. In the older literature, units of angstroms, 1 Å = 10^{-10} m, are also found.

In the infrared region both microns [1 micron = 1 micrometer (μm) = 10^{-6} m] and wavenumbers ω (in cm^{-1}) are employed; ω is the reciprocal of the wave-

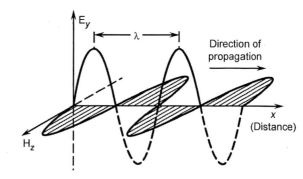

FIGURE 3.10 The instantaneous electric (E_y) and magnetic (H_z) field strength vectors of a plane-polarized light wave as a function of position along the axis of propagation (x) (from Calvert and Pitts, 1966).

length λ expressed in centimeters. It is directly related to energy through the Planck relationship,

$$E = hc\omega, \qquad (L)$$

and today is generally the unit of choice in infrared spectroscopy.

Since chemists often deal experimentally with moles rather than molecules, a convenient unit is a mole of quanta, defined as 1 einstein. The energy of 1 einstein of light of wavelength λ in nm is

$$\begin{aligned} E &= (6.02 \times 10^{23})h\nu = 6.02 \times 10^{23} hc/\lambda, \\ &= 1.196 \times 10^5/\lambda \text{ kJ einstein}^{-1}, \\ &= 2.859 \times 10^4/\lambda \text{ kcal einstein}^{-1}. \end{aligned} \qquad (M)$$

Another unit used in photochemistry to express the energy of a quantum of radiation is the electron volt; 1 eV = 96.49 kJ mol^{-1} = 23.06 kcal mol^{-1}. Thus for λ in nm

$$E = hc/\lambda = 1.240 \times 10^3/\lambda \text{ eV}. \qquad (N)$$

To put these energies and wavelengths in perspective, Table 3.2 gives some typical wavelengths, frequencies, wavenumbers, and energies of various regions of the electromagnetic spectrum. The region of most direct interest in tropospheric photochemistry ranges from the visible at $\sim$ 700 nm to the near-ultraviolet at $\sim$ 290 nm, the short-wavelength cutoff of the stratospheric ozone layer. The corresponding energies [Eq. (M)], 170.9 and 412.4 kJ einstein^{-1} (or 40.8 and 98.6 kcal einstein^{-1}), are sufficient to break chemical bonds ranging from, for example, the weak O_2—O bond in ozone, $\sim$ 100 kJ mol^{-1} ($\sim$ 25 kcal mol^{-1}), to the moderately strong C—H bond in formaldehyde, $\sim$ 368 kJ mol^{-1} ($\sim$ 88 kcal mol^{-1}).

Other spectral regions are also important because the detection and quantification of small concentrations of labile molecular, free radical, and atomic species of tropospheric interest both in laboratory studies and in ambient air are based on a variety of spectroscopic techniques that cover a wide range of the electromagnetic spectrum. For example, the relevant region for infrared spectroscopy of stable molecules is generally from $\sim$ 500 to 4000 cm^{-1} (20–2.5 μm), whereas the detection of atoms and free radicals by resonance fluorescence employs radiation down to 121.6 nm, the Lyman α line of the H atom.

Table 3.3 gives some relationships between commonly used energy units. Today the SI system of units is in general use, although much of the data in the literature is in the older units. Thus we use both types of units for energy, that is calories or kilocalories and joules or kilojoules, where 1 cal = 4.184 J.

2. The Beer–Lambert Law

The basis for the measurement of the strength of light absorption by a molecule at various wavelengths is shown in Fig. 3.11. A parallel monochromatic light beam of wavelength λ and power P_0 or intensity I_0, defined as the energy per second striking a unit area,

TABLE 3.2 Typical Wavelengths, Frequencies, Wavenumbers, and Energies of Various Regions of the Electromagnetic Spectrum

Name	Typical wavelength or range of wavelengths (nm)	Typical range of frequencies ν (s^{-1})	Typical range of wavenumbers ω (cm^{-1})	Typical range of energies (kJ einstein^{-1})[a]
Radiowave	$\sim 10^8$–10^{13}	$\sim 3 \times 10^4$–3×10^9	10^{-6}–0.1	$\sim 10^{-3}$–10^{-8}
Microwave	$\sim 10^7$–10^8	$\sim 3 \times 10^9$–3×10^{10}	0.1–1	$\sim 10^{-2}$–10^{-3}
Far-infrared	$\sim 10^5$–10^7	$\sim 3 \times 10^{10}$–3×10^{12}	1–100	$\sim 10^{-2}$–1
Near-infrared	$\sim 10^3$–10^5	$\sim 3 \times 10^{12}$–3×10^{14}	10^2–10^4	~ 1–10^2
Visible				
Red	700	4.3×10^{14}	1.4×10^4	1.7×10^2
Orange	620	4.8×10^{14}	1.6×10^4	1.9×10^2
Yellow	580	5.2×10^{14}	1.7×10^4	2.1×10^2
Green	530	5.7×10^{14}	1.9×10^4	2.3×10^2
Blue	470	6.4×10^{14}	2.1×10^4	2.5×10^2
Violet	420	7.1×10^{14}	2.4×10^4	2.8×10^2
Near-ultraviolet	400–200	$(7.5$–$15.0) \times 10^{14}$	$(2.5$–$5) \times 10^4$	$(3.0$–$6.0) \times 10^2$
Vacuum ultraviolet	$\sim$ 200–50	$(1.5$–$6.0) \times 10^{15}$	$(5$–$20) \times 10^4$	$\sim (6.0$–$24) \times 10^2$
X-Ray	$\sim$ 50–0.1	$\sim (0.6$–$300) \times 10^{16}$	$(0.2$–$100) \times 10^6$	$\sim 10^3$–10^6
γ-Ray	≤ 0.1	$\sim 3 \times 10^{18}$	$\geq 10^8$	$> 10^6$

[a] For kcal einstein^{-1}, divide by 4.184 (1 cal = 4.184 J).

TABLE 3.3 Some Relationships between Commonly Used Energy Units

(kJ mol^{-1})
- ×0.2390 = kcal mol^{-1}
- ×0.0104 = eV
- ×83.59 = cm^{-1}

(kcal mol^{-1})
- ×4.184 = kJ mol^{-1}
- ×0.04336 = eV
- ×349.8 = cm^{-1}

(cm^{-1})
- ×1.196 × 10^{-2} = kJ mol^{-1}
- ×2.859 × 10^{-3} = kcal mol^{-1}
- ×1.240 × 10^{-4} = eV

(eV)
- ×96.49 = kJ mol^{-1}
- ×23.06 = kcal mol^{-1}
- ×8.064 × 10^{3} = cm^{-1}

passes through a sample of length l (cm) and concentration C (mol L^{-1}). If this wavelength is absorbed by the sample, the power of the beam exiting the sample is reduced to P (or I). The transmittance (T) is defined as I/I_0 (or P/P_0).

As the light passes through a thickness of sample dl, it undergoes a fractional reduction in intensity proportional to the absorbing path length, i.e.,

$$(dI/I \propto) - (dl), \quad (O)$$

where the negative sign reflects the reduction in intensity with an increase in path length. Since the constant of proportionality must involve the concentration (C)

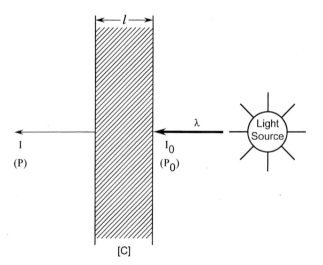

FIGURE 3.11 Schematic diagram of experimental approach to the Beer–Lambert law.

of the absorbing substance, this integrates to

$$\ln(I/I_0) = -kCl. \quad (P)$$

The most commonly used form of this Beer–Lambert law involves logarithms to the base 10:

$$\log(I_0/I) = \varepsilon Cl. \quad (Q)$$

C is in units of mol L^{-1}, l is in cm, and the constant of proportionality ε (L mol^{-1} cm^{-1}) is known as the molar absorptivity or molar extinction coefficient. The dimensionless quantity $\log(I_0/I)$ is known as the absorbance, A, which is related to the transmittance by $A = \log(I_0/I) = -\log T$.

Most commercial spectrometers report absorbance, as defined in Eq. (Q), versus wavelength. This is very important to recognize, since as we will see later, calculations of the rate of light absorption in the atmosphere require the use of absorption coefficients to the base e rather than to the base 10. While the recent atmospheric chemistry literature reports absorption cross sections to the base e, most measurements of absorption coefficients reported in the general chemical literature are to the base 10. If these are to be used in calculating photolysis rates in the atmosphere, the factor of 2.303 must be taken into account.

In gas-phase tropospheric chemistry, the most common units for concentration, N, are molecules cm^{-3} and for path length, l, units of cm. The form of the Beer–Lambert law is then

$$\ln(I_0/I) = \sigma Nl \quad (R)$$

or

$$I/I_0 = \exp(-\sigma Nl), \quad (S)$$

where it must again be emphasized that σ, known as the absorption cross section, must have been measured with the appropriate form of the Beer–Lambert law to the base e. The dimensionless exponent σNl is often referred to as the "optical depth."

In the past, some absorption coefficients for gases have been reported with concentrations in units of atmospheres, so that the absorption coefficient is in units of atm^{-1} cm^{-1}. Since the pressure depends on temperature, the latter (usually 273 or 298 K) were also reported.

For most tropospheric situations involving gaseous species, the Beer–Lambert law is an accurate method for treating light absorption; similar considerations apply to nonassociated molecules in dilute solution. However, under laboratory conditions with relatively high concentrations of the absorbing species, deviations may arise from a variety of factors, including concentration- and temperature-dependent association or dissociation reactions, deviations from the ideal gas law, and saturation of very narrow lines with increasing

TABLE 3.4 Conversion Factors for Changing Absorption Coefficients from One Set of Units to Another

Both units in either logarithmic base e or base 10

$(cm^2\ molecule^{-1})\begin{cases} \times\ 2.69 \times 10^{19} = (atm\ at\ 273\ K)^{-1}\ (cm^{-1}) \\ \times\ 2.46 \times 10^{19} = (atm\ at\ 298\ K)^{-1}\ (cm^{-1}) \\ \times\ 3.24 \times 10^{16} = (Torr\ at\ 298\ K)^{-1}\ (cm^{-1}) \\ \times\ 6.02 \times 10^{20} = L\ mol^{-1}\ cm^{-1} \end{cases}$

$(atm\ at\ 298\ K)^{-1}\ (cm^{-1})\begin{cases} \times\ 4.06 \times 10^{-20} = cm^2\ molecule^{-1} \\ \times\ 1.09 = (atm\ at\ 273\ K)^{-1}\ (cm^{-1}) \end{cases}$

$(L\ mol^{-1}\ cm^{-1})\begin{cases} \times\ 4.46 \times 10^{-2} = (atm\ at\ 273\ K)^{-1}\ (cm^{-1}) \\ \times\ 4.09 \times 10^{-2} = (atm\ at\ 298\ K)^{-1}\ (cm^{-1}) \\ \times\ 5.38 \times 10^{-5} = (Torr\ at\ 298\ K)^{-1}\ (cm^{-1}) \\ \times\ 1.66 \times 10^{-21} = cm^2\ molecule^{-1} \end{cases}$

Change of both logarithmic base and units

$(cm^2\ molecule^{-1}),\ base\ e\begin{cases} \times\ 1.17 \times 10^{19} = (atm\ at\ 273\ K)^{-1}\ (cm^{-1}),\ base\ 10 \\ \times\ 1.07 \times 10^{19} = (atm\ at\ 298\ K)^{-1}\ (cm^{-1}),\ base\ 10 \\ \times\ 1.41 \times 10^{16} = (Torr\ at\ 298\ K)^{-1}\ (cm^{-1}),\ base\ 10 \\ \times\ 2.62 \times 10^{20} = L\ mol^{-1}\ cm^{-1},\ base\ 10 \end{cases}$

$(L\ mol^{-1}\ cm^{-1}),\ base\ 10\begin{cases} \times\ 3.82 \times 10^{-21} = cm^2\ molecule^{-1},\ base\ e \\ \times\ 0.103 = (atm\ at\ 273\ K)^{-1}\ (cm^{-1}),\ base\ e \\ \times\ 9.42 \times 10^{-2} = (atm\ at\ 298\ K)^{-1}\ (cm^{-1}),\ base\ e \end{cases}$

$(atm\ at\ 273\ K)^{-1}\ (cm^{-1}),\ base\ 10\begin{cases} \times\ 8.57 \times 10^{-20} = cm^2\ molecule^{-1},\ base\ e \\ \times\ 51.6 = L\ mol^{-1}\ cm^{-1},\ base\ e \end{cases}$

$(Torr\ at\ 298\ K)^{-1}\ (cm^{-1}),\ base\ 10\begin{cases} \times\ 7.11 \times 10^{-17} = cm^2\ molecule^{-1},\ base\ e \\ \times\ 4.28 \times 10^4 = L\ mol^{-1}\ cm^{-1},\ base\ e \end{cases}$

$(atm\ at\ 298\ K)^{-1}\ (cm^{-1}),\ base\ 10\begin{cases} \times\ 9.35 \times 10^{-20} = cm^2\ molecule^{-1},\ base\ e \\ \times\ 2.51 = (atm\ at\ 273\ K)^{-1}\ (cm^{-1}),\ base\ e \end{cases}$

concentrations, i.e., increasing pressures. Particularly important is the situation in which a "monochromatic" analyzer beam actually has a bandwidth that is broad relative to very narrow lines of an absorbing species. In this case, which is often encountered in the infrared, for example, the Beer–Lambert law is nonlinear. Clearly, to be on the safe side, it is good practice to verify the linearity of $\ln(I_0/I)$ plots as a function of absorber concentration when experimentally determining absorption coefficients.

Table 3.4 gives conversion factors for converting absorption coefficients from one set of units to another and for changing between logarithms to the base 10 and base e.

C. ATMOSPHERIC PHOTOCHEMISTRY

1. Solar Radiation and Its Transmission through the Atmosphere

a. The Sun and Its Relationship to the Earth: Some Important Definitions for Atmospheric Chemistry

The sun can be considered a spherical light source of diameter 1.4×10^6 km located 1.5×10^8 km from the earth's surface. Incoming direct sunlight at the earth's surface is treated as a beam with an angle of collimation of $\sim 0.5°$ and thus is essentially parallel to $\pm 0.25°$.

The total intensity of sunlight outside the earth's atmosphere is characterized by the solar constant, defined as the total amount of light received per unit area normal to the direction of propagation of the light; the mean value is 1368 W m^{-2}, although variations from this mean are common (Lean, 1991).

Of more direct interest for atmospheric photochemistry is the solar flux per unit interval of wavelength. Values up to approximately 400 nm are provided by Atlas 3 (see Web site in Appendix IV) and from 400 nm on by Neckel and Labs (1984). Figure 3.12 shows the solar flux as a function of wavelength outside the atmosphere and at sea level for a solar zenith angle of 0° (Howard *et al.*, 1960).

Outside the atmosphere, the solar flux approximates blackbody emission at ~ 5770 K. However, light absorption or scattering by atmospheric constituents modifies the spectral distribution. The attenuation due to the presence of various naturally occurring atmospheric constituents is shown by the hatched areas in Fig. 3.12.

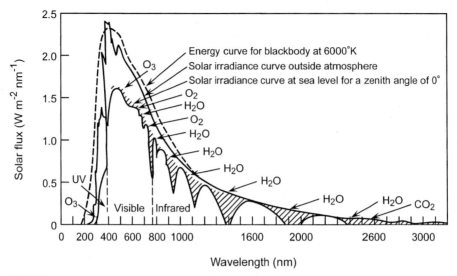

FIGURE 3.12 Solar flux outside the atmosphere and at sea level, respectively. The emission of a blackbody at 6000 K is also shown for comparison. The species responsible for light absorption in the various regions (O_3, H_2O, etc.) are also shown (from Howard et al., 1960).

Figure 3.13 shows the altitude corresponding to maximum light absorption by atomic and molecular oxygen and nitrogen and by O_3 as a function of wavelength up to $\lambda = 300$ nm with the sun directly overhead (Friedman, 1960).

Because of the presence of these absorbing species in the upper atmosphere, only light of $\lambda > 290$ nm is available for photochemical reactions in the troposphere. It is often expressed as the integrated radiation coming from all directions to a sphere and is referred to as *actinic radiation*, although in the strictest sense, "actinic" means "capable of causing photochemical reactions."

The ultraviolet region, $\lambda < 400$ nm, is often divided into what is known as the UV-A region from 315 to 400 nm, the UV-B region from 280 to 315 nm, and the UV-C region from 200 to 280 nm.

The effect of light scattering and absorption by atmospheric constituents on the intensity and wavelength distribution of sunlight at the earth's surface depends on both the nature and concentration of the gases and particles as well as the path length through

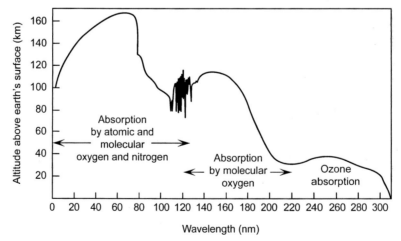

FIGURE 3.13 Approximate regions of maximum light absorption of solar radiation in the atmosphere by various atomic and molecular species as a function of altitude and wavelength with the sun overhead (from Friedman, 1960).

which the light passes, as expected from the Beer–Lambert law. The path length, that is, the distance from the outer reaches of the atmosphere to an observer on the earth's surface, is a function of the angle of the sun and hence time of day, latitude, and season. In addition, reflection of light from the earth's surface alters the light intensity at any given point in the atmosphere, as does the presence of clouds.

The angle of the sun relative to a fixed point on the surface of the earth is characterized by the solar zenith angle θ, defined, as shown in Fig. 3.14, as the angle between the direction of the sun and the vertical. Thus a zenith angle of zero corresponds to an overhead, noonday sun, and a zenith angle of $\sim 90°$ approximates sunrise and sunset. The greater the zenith angle, the longer is the path length through the atmosphere and hence the greater the reduction in solar intensity by absorption and scattering processes.

The path length L for direct solar radiation traveling through the earth's atmosphere to a fixed point on the earth's surface can be estimated geometrically using Fig. 3.14. This "flat earth" approximation is accurate for zenith angles $< 60°$. One can approximate L using

$$\cos \theta \cong h/L \tag{T}$$

or

$$L \cong h/\cos \theta \cong h \sec \theta. \tag{U}$$

A common term used to express the path length traversed by solar radiation to reach the earth's surface is the air mass, m, defined as

$$m = \frac{\text{Length of path of direct solar radiation through the atmosphere}}{\text{Length of vertical path through the atmosphere}}. \tag{V}$$

With reference to Fig. 3.14, for zenith angles less than 60°,

$$m \cong L/h \cong \sec \theta. \tag{W}$$

At larger angles, corrections for curvature of the atmosphere and refraction must be made to L and m.

Table 3.5 shows values of the air mass at various zenith angles θ, either estimated using $m = \sec \theta$ or corrected for curvature of the atmosphere and for refraction; it is seen that only for $\theta > 60°$ does this correction become significant.

b. Solar Spectral Distribution and Intensity in the Troposphere

When the radiation from the sun passes through the earth's atmosphere, it is modified both in intensity and

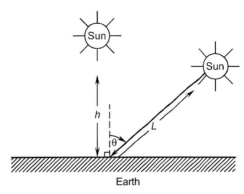

FIGURE 3.14 Definition of solar zenith angle θ at a point on the earth's surface.

in spectral distribution by absorption and scattering by gases as well as by particulate matter. As a result, the actual actinic flux to which a given volume of air is exposed is affected by the zenith angle (i.e., time of day, latitude, and season), by the extent of surface reflections, and by the presence of clouds. Madronich (1993) discusses these variables, with particular emphasis on the effects on the UV reaching the earth's surface.

To estimate the solar flux available for photochemistry in the troposphere then, one needs to know not only the flux outside the atmosphere but also the extent of light absorption and scattering within the atmosphere. We discuss here the actinic flux $F(\lambda)$ at the earth's surface; the effects of elevation and of height above the surface are discussed in Sections C.2.d and C.2.e.

The reduction in solar intensity due to scattering and absorption can be estimated using a form of the Beer–Lambert law:

$$I/I_0 = e^{-tm}. \tag{X}$$

In Eq. (X), I_0 is the light intensity at a given wavelength incident at the top of the atmosphere and I is the intensity of the light transmitted to the earth's surface; t is the total attenuation coefficient described below and m is the air mass as defined earlier. For the sun directly overhead (i.e., zenith angle $\theta = 0$) the air mass is unity ($m = 1.0$); the attenuation coefficient then reflects the minimum possible attenuation by the atmosphere. As θ increases until the sun is on the horizon (i.e., sunset or sunrise), m also increases (Table 3.5); thus the attenuation of the sunlight increases due to the increased path length in the atmosphere through which the light must travel to reach the earth's surface.

The attenuation coefficient, t, represents a combination of light scattering and absorption by gases and

> **BOX 3.1**
>
> # CALCULATION OF SOLAR ZENITH ANGLE
>
> The solar zenith angle can be calculated in the following manner for any particular location (i.e., latitude and longitude), day of the year (d_n), and time of day as described by Spencer (1971) and Madronich (1993). First, one needs to calculate what is known as the local hour angle (t_h), which is defined as the angle (in radians) between the meridian of the observer and that of the sun:
>
> $$t_h \text{ (in radians)} = \pi[(GMT/12) - 1 + (\text{longitude}/180)] + EQT,$$
>
> where GMT is Greenwich mean time converted from the local time, longitudes (in degrees) west of the Greenwich meridian are negative, and EQT is the "equation of time," given by
>
> $$\begin{aligned} EQT = & \, 7.5 \times 10^{-5} + 1.868 \times 10^{-3} \cos N \\ & - 3.2077 \times 10^{-2} \sin N \\ & - 1.4615 \times 10^{-2} \cos 2N \\ & - 4.0849 \times 10^{-2} \sin 2N. \end{aligned}$$
>
> where N is defined as
>
> $$N \text{ (in radians)} = 2\pi d_n/365.$$
>
> The day of the year, d_n, is defined as the day number (0–364), with 0 corresponding to January 1 and 364 to December 31.
>
> The second derived parameter that is needed for calculating the solar zenith angle at a particular time and place is the solar declination, δ, defined as the angle between the direction of the sun and the equatorial plane of the earth. The value of δ, which is 0° at the spring and fall equinoxes and falls between $+23.45°$ (June 21) and $-23.45°$ (December 21), can be calculated from the following:
>
> $$\begin{aligned} \delta \text{ (in radians)} = & \, 6.918 \times 10^{-3} - 0.399912 \cos N \\ & + 0.070257 \sin N - 6.758 \times 10^{-3} \\ & \times \cos 2N + 9.07 \times 10^{-4} \\ & \times \sin 2N - 2.697 \times 10^{-3} \cos 3N \\ & + 1.480 \times 10^{-3} \sin 3N. \end{aligned}$$
>
> The solar zenith angle (θ) for that particular time and place is then determined from:
>
> $$\cos \theta = \sin \delta \sin(\text{latitude}) + \cos \delta \cos(\text{latitude}) \cos t_h,$$
>
> where δ and t_h are calculated as already described and latitudes north of the equator (expressed in radians) are positive and south are negative. If all of the input parameters are in radians, θ is also obtained in radians and can be converted to degrees using 1 rad = 57.296°. For example, at Los Angeles, California (latitude = 34.03°N, longitude = 118.14°W) on September 21 at noon PST, GMT = 20.0, $N = 4.53$, EQT = 0.0301, $t_h = 0.0626$, $\delta = 0.0179$ rad, and $\cos \theta = 0.837$, giving a solar zenith angle of 0.579 rad, or 33°.

particles and is actually a sum of four terms,

$$t = t_{sg} + t_{ag} + t_{sp} + t_{ap}, \quad (Y)$$

where sg = light <u>s</u>cattering by <u>g</u>ases, ag = light <u>a</u>bsorption by <u>g</u>ases, sp = light <u>s</u>cattering by <u>p</u>articles, and ap = light <u>a</u>bsorption by <u>p</u>articles.

Gases scatter light by molecular, or Rayleigh, scattering. The intensity, $I(\lambda, \Theta)$ of light of wavelength λ scattered at an angle θ to the direction of incident light is determined by a number of factors. These include the incident light intensity, the angle Θ, the distance from the scattering molecule, and the index of refraction and size of the scattering molecule. In addition, and most importantly, Rayleigh scattering varies inversely with the fourth power of the wavelength.

Making the simplifying assumptions of a homogeneous atmosphere of fixed height of 7.996×10^5 cm and of uniform temperature and pressure throughout, Rayleigh scattering can be simplified for application to the atmosphere; as discussed in detail by Leighton (1961), the attenuation coefficient for scattering by gases, t_{sg}, becomes

$$t_{sg} = 1.044 \times 10^5 (n_{0\lambda} - 1)^2 / \lambda^4, \quad (Z)$$

where $n_{0\lambda}$ is the index of refraction of air at wavelength λ and the pressure and temperature of interest.

TABLE 3.5 Values of the Air Mass m at the Earth's Surface for Various Zenith Angles: (a) Calculated from $m = \sec \theta$ and (b) Corrected for Atmospheric Curvature and for Refraction

Zenith angle θ (deg)	$m = \sec \theta$	Air mass (m)
0	1.00	1.00
10	1.02	1.02
20	1.06	1.06
30	1.15	1.15
40	1.31	1.31
50	1.56	1.56
60	2.00	2.00
70	2.92	2.90
78	4.81	4.72
86	14.3	12.4

Source: Demerjian *et al.* (1980).

The dependence of Rayleigh scattering on λ^{-4} is evident in Fig. 3.15, which shows the attenuation coefficient for Rayleigh scattering as a function of wavelength from 290 to 700 nm; shorter wavelengths (i.e., in the blue ultraviolet region) are scattered much more strongly than the longer wavelengths.

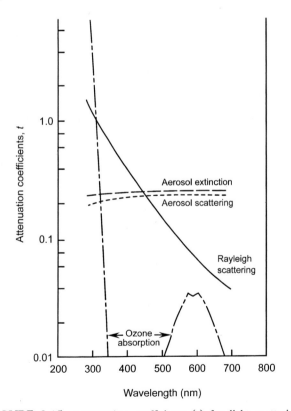

FIGURE 3.15 Attenuation coefficients (t) for light scattering (Rayleigh scattering) and absorption (ozone absorption) by gases and for scattering and scattering plus absorption (aerosol extinction) by particles [from Peterson (1976) and Demerjian *et al.* (1980)].

In the atmosphere, light absorption in the ultraviolet region is predominantly due to O_3 and this is predominantly in the stratosphere (Figs. 3.12 and 3.13). Since the absorption coefficients (σ) of O_3 are reasonably well established, a variant of the Beer–Lambert law can be applied to determine how much of the incident light is absorbed by O_3:

$$\frac{I}{I_0} = e^{-\sigma A m}. \qquad \text{(AA)}$$

A is the effective column O_3 (molecules cm^{-2}), σ its absorption cross section at that wavelength, and m the air mass. One needs to know, in addition to σ, the O_3 concentration as a function of altitude (z), that is,

$$A = \int_{z=0}^{\infty} O_3(z)\, dz.$$

Using the published absorption coefficients (σ) as a function of wavelength, one can then apply the Beer–Lambert law to calculate the intensity of light transmitted through such a vertical column to the earth's surface or to an altitude z. The resulting attenuation coefficients for O_3 are shown in Fig. 3.15 for an overhead sun. Clearly, O_3 is responsible for most of the attenuation of light directly from the sun of $\lambda < 310$ nm reaching the earth's surface.

This region of the spectrum around 300 nm is a crucial one for tropospheric photochemistry in both clean and polluted atmospheres. As we have indicated earlier, it is here that species such as ozone and aldehydes photolyze to produce atoms and free radicals critical to the chemistry of the troposphere.

Scattering and absorption of light by particulate matter are much more complex and will not be treated in detail here. Clearly, the size distribution and chemical composition, as well as the concentration of the particles, are very important in determining the extent of light scattering and absorption. Since these parameters will vary significantly geographically, seasonally, and diurnally, accurately estimating their impact on light intensities at a particular location at the earth's surface is difficult. Simplifications for the attenuation coefficient for scattering by particles such as

$$t_{sp} = b/\lambda^n \qquad \text{(BB)}$$

are often made, where b depends on the concentration of particles and n on their size; for example, n decreases from ~4 to 0 as the particle size increases (Leighton, 1961).

One estimate of the attenuation coefficients for light scattering by particles, t_{sp}, is also given in Fig. 3.15

(Demerjian et al., 1980). Also shown are these researchers' estimates of total scattering plus absorption due to particulate matter, known as the *aerosol extinction*:

$$t_{sp} + t_{ap} = \text{Aerosol extinction.}$$

In this case, the radii of the particles were assumed to fall between 0.01 and 2.0 μm; the peak in the number versus size distribution was at 0.07 μm.

Given estimated values for the attenuation coefficients for scattering and absorption of light by gases and particles (i.e., t_{sg}, t_{ag}, t_{sp}, and t_{ap}), one can calculate from Eq. (X) the fraction of the direct solar intensity incident on the top of the atmosphere that is transmitted to the earth's surface at any given wavelength. However, when one considers the *actual* light intensity that reaches a given volume of gas in the troposphere, one must take into account not only this direct solar radiation but also two other sources of indirect light: (1) light, either from the sun or reflected from the earth's surface, that is scattered to the volume by gases or particles, known as diffuse solar radiation or sky radiation, and (2) light that is reflected from the earth's surface. These are illustrated in Fig. 3.16.

Estimating the intensity of the scattered light at a given point in the atmosphere is difficult because of the substantial uncertainties and variability involved in the factors that contribute to light scattering; for example, the size distribution, concentration, and composition of particles, which to a large extent cause this scattering, are highly variable geographically and temporally and are not always well known for a particular point in space and time.

The amount of light reflected from the earth's surface to a volume of air clearly depends on the type of surface, as well as the wavelength of light; thus snow is highly reflecting, whereas black lava rock reflects very little of the incident radiation. The term used to describe the extent of this reflection is the *surface albedo*, which is the fraction of light incident on the surface that is reflected. Reflection can be specular, in which the angles of incidence and reflection are equal (e.g., a water surface at large zenith angles), or diffuse, in which light is reflected equally in all directions regardless of the angle of incidence (e.g., white rocks or buildings); the latter is known as "Lambertian" reflection. Table 3.6 gives some reported values of surface albedos for different types of surfaces. It should be noted that, as expected, albedos are wavelength dependent (e.g., see McLinden et al. (1997) for wavelength dependence of ocean albedos and Herman and Celarier (1997) for albedos in the UV from 340 to 380 nm).

One can thus estimate the total light intensity incident on a given volume of air in the troposphere due to direct solar radiation, scattering, and reflection. The light absorbed in that volume can then be calculated

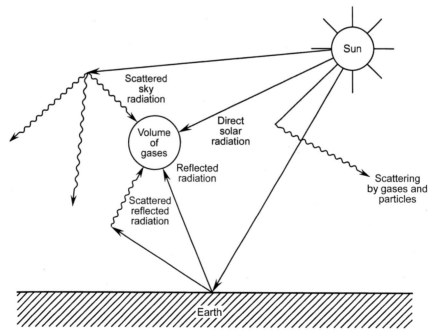

FIGURE 3.16 Different sources of radiation striking a volume of gas in the atmosphere. These sources are direction radiation from the sun, radiation scattered by gases and particles, and radiation reflected from the earth's surface.

TABLE 3.6 Some Typical Albedos for Various Types of Surfaces

Type of surface	Albedo	Reference
Snow	0.69	Angle *et al.*, 1992
	0.93[b]	Dickerson *et al.*, 1982
	0.9–1.0	Junkermann, 1994
Ocean	0.07[b]	Dickerson *et al.*, 1982
	0.06–0.08[a]	Eck *et al.*, 1987
Forests	0.06–0.18[b]	Dickerson *et al.*, 1982
	0.02[a]	Eck *et al.*, 1987
	0.17	Angle *et al.*, 1992
Fields and meadows	0.03–0.04[a]	Eck *et al.*, 1987
Desert	0.06–0.09[a]	Eck *et al.*, 1987
Salt flats	0.57–0.65[a]	Eck *et al.*, 1987

[a] Minimum reflectivities at 370 nm.
[b] Measured with respect to NO_2 photolysis.

FIGURE 3.17 Typical device (Eppley Laboratories Model 8-48) used to measure solar irradiance. The detector consists of a differential thermopile with the hot junction receivers blackened with flat black coating and the cold junction receivers whitened with $BaSO_4$ (photo supplied courtesy of G. L. Kirk, Eppley Laboratories).

using the Beer–Lambert law, if the concentrations and absorption coefficients of all absorbing species are known.

2. Calculating Photolysis Rates in the Atmosphere

a. Photolysis Rate Constant $[k_p]$, Radiance $[L(\lambda)]$, Actinic Flux $[F(\lambda)]$, and Irradiance $[E(\lambda)]$

The rate of photodissociation of a molecule, A, upon light absorption,

$$A + h\nu \rightarrow B + C, \quad (5)$$

can be described as a first-order process (see Chapter 5) with a rate constant, k_p, known as the photolysis rate constant:

$$d[A]/dt = -k_p[A]. \quad (CC)$$

In effect, k_p takes into account the intensity of available light that the molecule can absorb, the intrinsic strength of light absorption in that region by A, i.e., the absorption cross section σ, and the quantum yield for photodissociation, ϕ.

The light available to a molecule in air for absorption and photodissociation includes both direct and scattered and reflected radiation coming from all directions as described earlier and depicted in Fig. 3.16. The term *actinic flux* or *spherically integrated actinic flux*, denoted by $F(\lambda)$, is used to describe the total intensity of this light and is the quantity of interest in calculating k_p.

However, in practice, available light intensity is often measured using flat-plate devices such as the one shown in Fig. 3.17. Here light of a given wavelength that comes from all directions in a hemisphere and crosses the surface is measured. The net flux crossing the surface per unit area and time is known as the *irradiance*, $E(\lambda)$, and represents the flow of light across a flat plane rather than the total light coming from all directions that a molecule actually encounters in the atmosphere. Although actinic flux and irradiance are clearly related, they are not identical; for a detailed treatment of the actinic flux, irradiance, and radiance (defined next), see Madronich (1987).

An expression for the number of A molecules dissociating per unit volume per unit time is developed in Box 3.2. Comparing Eq. (NN) in Box 3.2 to Eq. (CC), the photolysis rate constant k_p must be given by

$$k_p = \int_\lambda \phi(\lambda)\sigma(\lambda) \int_\omega L(\lambda, \Theta, \phi) \, d\omega \, d\lambda$$
$$= \int_\lambda \phi(\lambda)\sigma(\lambda) F(\lambda) \, d\lambda, \quad (OO)$$

where $F(\lambda)$ is the spherically integrated actinic flux, Eq. (FF). We again stress that $\sigma(\lambda)$ is the absorption cross section to the base e, arising from the use of the differential form of the Beer–Lambert law to obtain Eq. (KK).

There are several approaches to measuring actinic fluxes and photolysis rate constants. One approach is to measure the rate of decay of a species such as NO_2 directly, so-called "chemical actinometry" (e.g. see Madronich *et al.*, 1983). Another approach is to measure the light intensity and convert this to an actinic flux.

BOX 3.2
RELATIONSHIPS BETWEEN RADIANCE, IRRADIANCE, ACTINIC FLUX, AND PHOTOLYSIS RATE CONSTANTS

To calculate k_p, let us take the case shown in Fig. 3.18a of light striking the top of a very thin layer of air of thickness dz. The light originates in a solid angle $d\omega$ and strikes the top surface of the thin layer at an angle θ to the vertical. The intensity of the incoming light at angle θ shown in Fig. 3.18a is known as the *radiance*, $L(\lambda, \theta, \phi)$. By definition, radiance is the number of photons (or energy) in the wavelength interval $d\lambda$ originating from a small solid angle $d\omega$ and striking a small surface area da in time dt at an angle θ to the vertical.

However, the net flux across the surface is, in effect, determined by the portion of the surface that is perpendicular to the incoming light beam. As seen in Fig. 3.18b, from simple geometric considerations the portion of the surface perpendicular to the incoming beam is $2x = (da)\cos\theta$. Thus the net photons (or energy), dP in a wavelength interval $d\lambda$ that originates in a solid angle $d\omega$ and at an angle θ to the normal and crosses a small surface area, da, in time dt is given by

$$dP = L(\lambda, \theta, \phi)\cos\theta\, da\, d\omega\, dt\, d\lambda. \quad \text{(DD)}$$

The *irradiance* $E(\lambda)$, which is directly measured by flat-plate devices, is by definition the total number of photons per unit surface area, time, and wavelength. Thus

$$E(\lambda) = \int (dP/da\, d\lambda\, dt) = \int L(\lambda, \theta, \phi)\cos\theta\, d\omega. \quad \text{(EE)}$$

The actinic flux $F(\lambda)$ is the total incident light intensity integrated over all solid angles, given by

$$F(\lambda) = \int_\omega L(\lambda, \theta, \phi)\, d\omega. \quad \text{(FF)}$$

Thus the irradiance, $E(\lambda)$, and the actinic flux, $F(\lambda)$, differ by the factor $\cos\theta$. Only for $\theta = 0°$, i.e., for a parallel beam of light perpendicular to the surface, are the irradiance and flux equal.

We now need to convert $d\omega$ into terms involving the spherical coordinates θ and ϕ. As shown in Fig. 3.19a, a given solid angle ω traces an area a on the surface of a sphere of radius r. When $a = r^2$, the solid angle ω is by definition 1 sr (sr = steradian). For the more general case of a surface area a subtended by the solid angle ω,

$$\omega \text{ (steradians)} = a/r^2$$

or

$$d\omega = da/r^2. \quad \text{(GG)}$$

As shown in Fig. 3.19b, for small changes in the angles θ and ϕ, there is a change in the surface area, da, on a sphere of radius r given by

$$da = (r\sin\theta\, d\phi)(r\, d\theta) = r^2 \sin\theta\, d\theta\, d\phi,$$

i.e.,

$$d\omega = da/r^2 = \sin\theta\, d\theta\, d\phi. \quad \text{(HH)}$$

Combining Eqs. (FF) and (HH), the actinic flux becomes

$$F(\lambda) = \int_\phi \int_\theta L(\lambda, \theta, \phi)\sin\theta\, d\theta\, d\phi. \quad \text{(II)}$$

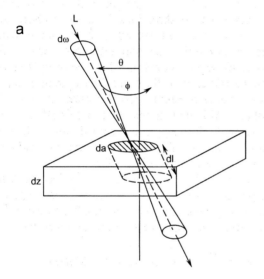
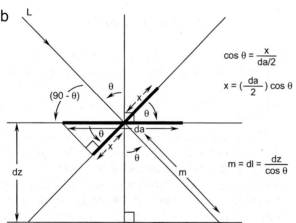

FIGURE 3.18 Typical light ray striking a thin layer of air in the atmosphere (adapted from Madronich, 1987).

The absolute change in intensity at a particular wavelength is thus given by $dI = \sigma(\lambda)[A]I\,dl$. This is just the number of photons absorbed as the light passes through the thin layer. I is the incident light, i.e., dP from Eq. (DD). If the quantum yield at this wavelength is $\phi(\lambda)$, then using $I = dP$ and substituting in for dP from Eq. (DD), the total number of molecules of A that photodissociate is given by

Number of A dissociating
$$= \phi(\lambda)\,dI = \phi(\lambda)\sigma(\lambda)[A]\,dP\,dl$$
$$= \phi(\lambda)\sigma(\lambda)[A]\,dl$$
$$\times \{L(\lambda,\theta,\phi)\cos\theta\,da\,d\omega\,dt\,d\lambda\}. \quad \text{(KK)}$$

Since the total path length for light absorption is given by $dl = dz/\cos\theta$ (Fig. 3.18b), this becomes

Number of A dissociating
$$= \phi(\lambda)\sigma(\lambda)[A]\,dz \times \{L(\lambda,\theta,\phi)\,da\,d\omega\,dt\,d\lambda\}. \quad \text{(LL)}$$

However, this represents the number of A molecules dissociating only due to light absorption in the wavelength interval $d\lambda$ and only for incident light over the solid angle $d\omega$ and surface area da. To obtain the total number of A dissociating, Eq. (LL) must be integrated over all wavelengths, solid angles, and surface areas

Total number of A dissociating
$$= \int_\lambda \int_\omega \int_a \phi(\lambda)\sigma(\lambda)[A]L(\lambda,\theta,\phi)\,dz\,da\,d\omega\,dt\,d\lambda$$
$$= [A]\,dt\left(\int_a dz\,da\right)$$
$$\times \left\{\int_\lambda \int_\omega \phi(\lambda)\sigma(\lambda)L(\lambda,\theta,\phi)\,d\omega\,d\lambda\right\}. \quad \text{(MM)}$$

The first integration over the surface area a is just the volume. Rearranging, Eq. (MM) becomes

$$\frac{\text{Total number of A dissociating/volume}}{dt}$$
$$= -\frac{d[A]}{dt}$$
$$= [A]\int_\lambda \int_\omega \phi(\lambda)\sigma(\lambda)L(\lambda,\theta,\phi)\,d\omega\,d\lambda$$
$$= [A]\int_\lambda \phi(\lambda)\sigma(\lambda)\int_\omega L(\lambda,\theta,\phi)\,d\omega\,d\lambda. \quad \text{(NN)}$$

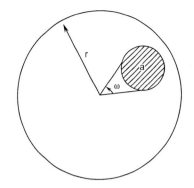

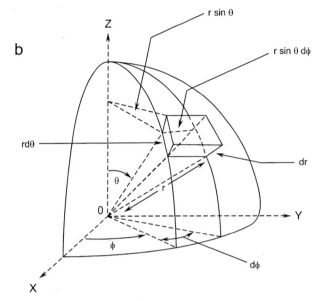

FIGURE 3.19 Conversion of solid angle ω to spherical coordinates.

Similarly, the irradiance $E(\lambda)$ is given by

$$E(\lambda) = \int_\omega L(\lambda,\theta,\phi)\cos\theta\,d\omega$$
$$= \int_\phi \int_\theta L(\lambda,\theta,\phi)\sin\theta\cos\theta\,d\theta\,d\phi. \quad \text{(JJ)}$$

Returning to Fig. 3.18a, as light passes through this thin layer of air, it can be absorbed by the molecules of A. From the differential form of the Beer–Lambert law, Eq. (O), the change in light intensity, dI, on passing through this small volume of air containing the absorber A is given by

$$dI/I \propto -(dl) = -k\,dl = -\sigma[A]\,dl,$$

where the negative sign indicates that the intensity decreases as the light passes through the sample.

It is important to note that the application of this form of the Beer–Lambert law, inherent in which are natural logarithms, means that the absorption cross section, σ, must be that to the base e.

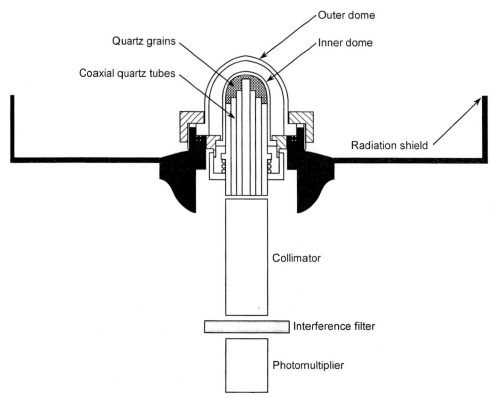

FIGURE 3.20 Schematic diagram of a 2π radiometer used to measure actinic fluxes (adapted from Junkermann *et al.*, 1989).

Light intensities can be measured using flat-plate radiometers such as that in Fig. 3.17. As discussed earlier, this measures the flux through a horizontal plane, and as a result there is a difference of $\cos \theta$ between this measured *irradiance* and the actinic flux (Box 3.2). Another approach is to use what is sometimes referred to as a 2π radiometer, which measures the light intensity striking half of a sphere. A typical example is shown in Fig. 3.20 (Junkermann *et al.*, 1989). The light collector consists of coaxial quartz tubes located inside a quartz dome, such that light is collected from all directions in a hemisphere equally well (hence the "2π" designation). Because light is collected from all directions within 2π sr, the $\cos \theta$ factor does not apply. The space between the quartz tubes and the dome is filled with quartz grains to scatter the light. The scattered light is transmitted by the quartz tubes through a set of interference filters to a detector. The filter–detector combination is chosen for the particular measurement of interest, e.g., O_3 or NO_2 photolysis. The shield is used to limit the field of view to exactly 2π sr. Two such detectors, one pointing up and one pointing down, can be used to cover the entire 4π sphere. A similar detector for aircraft use is described by Volz-Thomas *et al.* (1996).

b. Estimates of the Actinic Flux, $F(\lambda)$, at the Earth's Surface

There are a number of estimates of the actinic flux at various wavelengths and solar zenith angles in the literature (e.g., see references in Madronich, 1987, 1993). Clearly, these all involve certain assumptions about the amounts and distribution of O_3 and the concentration and nature (e.g., size distribution and composition) of particles which determine their light scattering and absorption properties. Historically, one of the most widely used data sets for actinic fluxes at the earth's surface is that of Peterson (1976), who recalculated these solar fluxes from 290 to 700 nm using a radiative transfer model developed by Dave (1972). Demerjian *et al.* (1980) then applied them to the photolysis of some important atmospheric species. In this model, molecular scattering, absorption due to O_3, H_2O, O_2, and CO_2, and scattering and absorption by particles are taken into account.

Madronich (1998) has calculated actinic fluxes using updated values of the extraterrestrial flux. In the 150- to 400-nm region, values from *Atlas* are used (see Web site in Appendix IV) whereas from 400 to 700 nm those of Neckel and Labs (1984) are used. In addition, the ozone absorption cross sections of Molina and Molina (1986) and the radiation scheme of Stamnes *et al.* (1988) were used. Other assumptions, e.g., the particle concentration and distribution, are the same as those of Demerjian *et al.* (1980).

In the "average" case for which the model calculations were carried out, absorption and scattering as the light traveled from the top of the atmosphere to the earth's surface were assumed to be due to O_3 (UV light absorption with column O_3 of 300 Dobson units), air molecules (scattering), and particles (scattering and absorption). The "best estimate" surface albedo varied from 0.05 in the 290- to 400-nm region to 0.15 in the 660- to 700-nm region. The surface was assumed to be what is known as an ideal "Lambert surface," meaning that it diffuses the incident light sufficiently well that it is reradiated equally in all upward directions, i.e., isotropically.

Table 3.7 gives the calculated actinic fluxes at the earth's surface as a function of zenith angle assuming the "best estimate" surface albedo. These data are plotted for six wavelength intervals as a function of zenith angle in Fig. 3.21. The initially small change in actinic flux with zenith angle as it increases from 0 to ~ 50° at a given wavelength followed by the rapid drop of intensity from 50 to 90° is due to the fact that the air mass m changes only gradually to ~ 50° but then increases much more rapidly to $\theta = 90°$ [see Table 3.5 and Eq. (X)]. At the shorter wavelengths at a fixed zenith angle, the rapid increase in actinic flux with wavelength is primarily due to the strongly decreasing O_3 absorption in this region.

The actinic fluxes calculated by Madronich (1998) for altitudes of 15, 25, and 40 km are collected in Tables 3.15 to 3.17.

c. *Effects of Latitude, Season, and Time of Day on $F(\lambda)$*

To estimate photolysis rates for a given geographical location, one must take into account the latitude and season, as well as the time of day.

The data in Table 3.7 are representative for the average earth–sun distance characteristic of early April and October. The orbit of the earth is slightly elliptical, so that there is a small change in the earth–sun distance, which causes a small change (< 3%) in the solar flux with season. Correction factors for this seasonal variation for some dates from Demerjian *et al.* (1980) are given in Table 3.8. As discussed by Madronich (1993), the correction factors for solar intensity can be calculated for any other date using the following

$$(R_0/R_n)^2 = 1.000110 + 0.034221 \cos N$$
$$+ 1.280 \times 10^{-3} \sin N$$
$$+ 7.19 \times 10^{-4} \cos 2N$$
$$+ 7.7 \times 10^{-5} \sin 2N,$$

where R_0 is the average earth–sun distance, R_n is the earth–sun distance on day d_n as defined earlier, $N = 2\pi d_n/365$ radians, and $(R_0/R_n)^2$ represents the correction factor.

Table 3.9 summarizes the solar zenith angles at latitudes of 20, 30, 40, and 50°N as a function of month and true solar time. True solar time, also known as apparent solar time or apparent local solar time, is defined as the time scale referenced to the sun crossing the meridian at noon. For example, at a latitude of 50°N at the beginning of January, two hours before the sun crosses the meridian corresponds to a true solar time of 10 a.m.; from Table 3.9, the solar zenith angle at this time is 77.7°.

To obtain the actinic flux at this time at any wavelength, one takes the fluxes in Table 3.7 listed under 78°; thus the flux in the 400- to 405-nm wavelength interval at 10 a.m. at 50°N latitude is 0.48×10^{15} photons cm^{-2} s^{-1}.

For other latitudes, dates, and times, the solar zenith angle can be calculated as described by Madronich (1993) and summarized earlier.

Afternoon values of θ are not given in Table 3.9 as the data are symmetrical about noon. Thus at a time of 2 p.m. at 50°N latitude, the flux would be the same as calculated for 10 a.m.

Figure 3.22 shows the solar angle θ as a function of true solar time for several latitudes and different times of the year. As expected, only for the lower latitudes at the summer solstice does the solar zenith angle approach 0° at noon. For a latitude of 50°N even at the summer solstice, θ is 27°.

Figure 3.23 shows the diurnal variation of the solar zenith angle as a function of season for Los Angeles, which is located at a latitude of 34.1°N. Clearly, the peak solar zenith angle varies dramatically with season.

These differences in light intensity, and in its diurnal variation at different latitudes and seasons, are critical because they alter the atmospheric chemistry at various geographical locations due to the fact that photochemistry is the major source of the free radicals such as OH that drive the chemistry.

d. *Effect of Surface Elevation on $F(\lambda)$*

The variation in the actinic flux with *surface* elevation is important because some of the world's major cities are located substantially above sea level. For example, Mexico City and Denver, Colorado, are at elevations of 2.2 and 1.6 km, respectively.

Table 3.10 shows the calculated percentage increase in the actinic flux at the earth's surface for an elevation of 1.5 km and atmospheric pressure of 0.84 atm (corresponding approximately to Denver) as a function of zenith angle for four wavelength intervals. In this calculation, it was assumed that the vertical O_3 and particle

TABLE 3.7 Actinic Flux Values $F(\lambda)$ at the Earth's Surface as a Function of Wavelength Interval and Solar Zenith Angle within Specific Wavelength Intervals for Best Estimate Surface Albedo Calculated by Madronich (1998)[a]

Wavelength interval (nm)	Exponent[b]	Solar zenith angle (deg)									
		0	10	20	30	40	50	60	70	78	86
		Actinic fluxes (photons cm^{-2} s^{-1})									
290–292	14	0.00	0.00	0.00	0.00	0.00	0.00	0.00	0.00	0.00	0.00
292–294	14	0.00	0.00	0.00	0.00	0.00	0.00	0.00	0.00	0.00	0.00
294–296	14	0.00	0.00	0.00	0.00	0.00	0.00	0.00	0.00	0.00	0.00
296–298	14	0.01	0.01	0.01	0.01	0.00	0.00	0.00	0.00	0.00	0.00
298–300	14	0.03	0.03	0.02	0.02	0.01	0.00	0.00	0.00	0.00	0.00
300–302	14	0.07	0.07	0.06	0.04	0.03	0.01	0.00	0.00	0.00	0.00
302–304	14	0.18	0.18	0.15	0.12	0.08	0.04	0.01	0.00	0.00	0.00
304–306	14	0.33	0.32	0.29	0.23	0.16	0.09	0.04	0.01	0.00	0.00
306–308	14	0.51	0.49	0.45	0.37	0.28	0.17	0.08	0.02	0.00	0.00
308–310	14	0.66	0.65	0.60	0.51	0.40	0.27	0.14	0.04	0.01	0.00
310–312	14	0.99	0.97	0.90	0.79	0.64	0.45	0.25	0.09	0.02	0.00
312–314	14	1.22	1.19	1.12	1.00	0.82	0.61	0.36	0.14	0.04	0.00
314–316	14	1.37	1.34	1.27	1.14	0.96	0.73	0.46	0.20	0.06	0.01
316–318	14	1.67	1.64	1.56	1.42	1.22	0.95	0.62	0.29	0.10	0.01
318–320	14	1.70	1.68	1.60	1.47	1.27	1.01	0.69	0.34	0.13	0.02
320–325	14	5.30	5.24	5.03	4.66	4.10	3.34	2.36	1.27	0.52	0.10
325–330	14	7.72	7.63	7.36	6.88	6.15	5.12	3.75	2.15	0.96	0.22
330–335	14	8.26	8.17	7.91	7.44	6.70	5.65	4.23	2.50	1.16	0.29
335–340	14	7.98	7.91	7.67	7.24	6.56	5.59	4.25	2.58	1.23	0.33
340–345	14	8.64	8.57	8.32	7.88	7.17	6.15	4.73	2.91	1.40	0.38
345–350	14	8.73	8.65	8.42	7.99	7.30	6.30	4.88	3.04	1.47	0.40
350–355	14	10.00	9.92	9.67	9.20	8.43	7.31	5.71	3.60	1.76	0.47
355–360	14	8.98	8.91	8.69	8.28	7.62	6.64	5.22	3.33	1.64	0.43
360–365	14	9.97	9.90	9.67	9.23	8.52	7.46	5.91	3.80	1.88	0.49
365–370	15	1.24	1.23	1.20	1.15	1.07	0.94	0.75	0.48	0.24	0.06
370–375	15	1.10	1.09	1.07	1.02	0.95	0.84	0.67	0.44	0.22	0.06
375–380	15	1.26	1.25	1.22	1.17	1.09	0.97	0.78	0.52	0.26	0.07
380–385	15	1.06	1.06	1.04	1.00	0.93	0.82	0.67	0.45	0.23	0.06
385–390	15	1.17	1.16	1.14	1.10	1.03	0.92	0.75	0.50	0.26	0.06
390–395	15	1.17	1.17	1.14	1.10	1.03	0.92	0.76	0.51	0.27	0.07
395–400	15	1.43	1.42	1.40	1.35	1.26	1.13	0.93	0.64	0.33	0.08
400–405	15	2.02	2.01	1.98	1.91	1.79	1.61	1.33	0.91	0.48	0.12
405–410	15	1.97	1.96	1.92	1.86	1.75	1.57	1.30	0.90	0.48	0.12
410–415	15	2.06	2.04	2.01	1.94	1.83	1.65	1.37	0.96	0.51	0.12
415–420	15	2.09	2.08	2.05	1.98	1.87	1.69	1.41	0.99	0.54	0.13
420–430	15	4.13	4.11	4.04	3.92	3.70	3.36	2.82	2.00	1.09	0.26
430–440	15	4.26	4.24	4.18	4.05	3.84	3.50	2.96	2.12	1.17	0.28
440–450	15	5.05	5.03	4.96	4.82	4.57	4.18	3.56	2.58	1.45	0.34
450–460	15	5.66	5.64	5.56	5.39	5.12	4.67	3.98	2.90	1.65	0.38
460–470	15	5.75	5.72	5.64	5.48	5.21	4.77	4.08	3.00	1.72	0.39
470–480	15	5.86	5.83	5.75	5.60	5.32	4.89	4.19	3.10	1.80	0.41
480–490	15	5.74	5.72	5.64	5.49	5.23	4.81	4.14	3.08	1.80	0.40
490–500	15	5.94	5.91	5.83	5.68	5.42	5.00	4.31	3.22	1.90	0.42
500–510	15	6.10	6.07	5.99	5.82	5.54	5.09	4.38	3.27	1.92	0.42
510–520	15	5.98	5.95	5.87	5.71	5.44	5.00	4.32	3.24	1.92	0.41
520–530	15	6.20	6.17	6.09	5.92	5.64	5.20	4.49	3.37	2.00	0.42
530–540	15	6.38	6.38	6.27	6.10	5.81	5.35	4.63	3.48	2.07	0.42
540–550	15	6.37	6.34	6.26	6.10	5.81	5.35	4.63	3.49	2.08	0.42
550–560	15	6.55	6.52	6.43	6.26	5.96	5.49	4.74	3.57	2.13	0.43
560–570	15	6.61	6.58	6.49	6.31	6.01	5.53	4.78	3.59	2.13	0.41
570–580	15	6.69	6.66	6.57	6.39	6.09	5.60	4.84	3.64	2.17	0.41
580–600	16	1.35	1.34	1.32	1.29	1.23	1.13	0.98	0.74	0.45	0.09
600–620	16	1.36	1.35	1.34	1.30	1.24	1.14	0.99	0.75	0.46	0.09
620–640	16	1.37	1.37	1.35	1.31	1.26	1.16	1.01	0.78	0.48	0.10
640–660	16	1.38	1.37	1.35	1.32	1.26	1.17	1.02	0.79	0.50	0.11
660–680	16	1.43	1.42	1.40	1.37	1.31	1.21	1.06	0.83	0.53	0.12
680–700	16	1.40	1.40	1.38	1.34	1.28	1.19	1.05	0.82	0.54	0.12

[a] The authors are grateful to Dr. Sasha Madronich for generously providing these calculations.
[b] This column lists the power of 10 by which all entries should be multiplied. For example, at $\theta = 0°$ the total actinic flux in the wavelength interval from 306 to 308 nm is 0.51×10^{14} photons cm^{-2} s^{-1}.

C. ATMOSPHERIC PHOTOCHEMISTRY

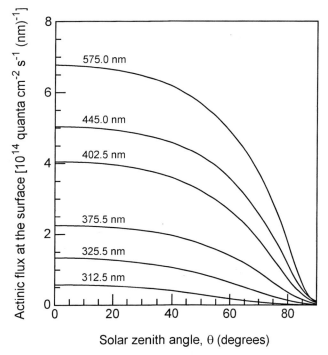

FIGURE 3.21 Calculated actinic flux centered on the indicated wavelengths at the earth's surface using best estimate albedos as a function of solar zenith angle (from Madronich, 1998).

those of Peterson (1976) and Demerjian *et al.* (1980), since they are relative values, they are not expected to differ significantly from those that would be derived with the Madronich (1998) actinic flux values.)

e. Effect of Height above Earth's Surface on $F(\lambda)$

Figure 3.24 shows the *relative* changes in the total actinic flux as a function of altitude from 0 to 15 km at solar zenith angles of 20, 50, and 78° and at wavelengths of 332.5 (part a), 412.5 (part b), and 575 nm (part c), respectively. Again, since these are relative changes, these results of Peterson (1976) and Demerjian *et al.* (1980) are not expected to be significantly different from those that would be obtained with the Madronich (1998) actinic flux estimates.

At the largest solar zenith angle shown, 78°, all of the curves show a decrease in the actinic flux from 15 km to lower altitudes. This occurs because at these large values of θ and hence long path lengths through the atmosphere, backscattering of the light increases as it passes through the atmosphere.

The calculated actinic flux typically increases significantly in the first few kilometers. This is partly due to scattering of light by particulate matter and to light absorption by tropospheric O_3 close to the surface. The effect of O_3 can be seen by comparing the total fluxes at 332.5 nm (Fig. 3.24a), where O_3 absorbs, to those at 575 nm (Fig. 3.24b), where it does not.

Peterson and co-workers have examined the percentage increase in total actinic flux, going from the surface to ~1 km; they estimate that at short wavelengths ($\lambda \leq 310$ nm), the increase is > 37.5% for all zenith angles. This increase in flux with altitude at short wavelengths could be particularly significant in photochemical smog formation. Thus pollutants trapped in an inversion layer aloft may be exposed to higher actinic fluxes than at ground level and photolyze more rapidly, hastening the formation of various secondary pollutants. The increased actinic flux with altitude close to the earth's surface is the basis for their suggestion that the presence of increased O_3 in, or close to, the inversion layer may be at least partially the result of the height dependence of $F(\lambda)$.

These predictions have been borne out experimentally in studies in which the rate of photolysis of NO_2 was measured from the surface to ~7.6-km altitude and found to increase with height by more than 50% (Kelley *et al.*, 1995; Volz-Thomas *et al.*, 1996).

f. Sensitivity of Calculated Actinic Fluxes to Input Values for Surface Albedo and Ozone and Particle Concentrations

As discussed earlier, the net actinic flux incident on a volume of air is sensitive to a number of parameters,

concentrations were the same but that the Rayleigh scattering was reduced due to the lowered pressure, i.e., lower gas concentrations. The increase in actinic flux in the UV is relatively small (< 5%) for zenith angles less than ~45°; at larger zenith angles, the change is less than 13%. For the longer wavelengths, it is small at all zenith angles. (Although these data are

TABLE 3.8 Correction Factors for Extraterrestrial Solar Flux Values Depending on Earth–Sun Distance at Various Times of the Year

Date	Correction factor	Date	Correction factor
Jan 1	1.033	Jul 1	0.966
Jan 15	1.032	Jul 15	0.967
Feb 1	1.029	Aug 1	0.970
Feb 15	1.024	Aug 15	0.974
Mar 1	1.018	Sep 1	0.982
Mar 15	1.011	Sep 15	0.989
Apr 1	1.001	Oct 1	0.998
Apr 15	0.993	Oct 15	1.006
May 1	0.984	Nov 1	1.015
May 15	0.978	Nov 15	1.022
Jun 1	0.971	Dec 1	1.027
Jun 15	0.968	Dec 15	1.031

Source: Demerjian *et al.* (1980).

TABLE 3.9 Tabulation of Solar Zenith Angles (deg) as a Function of True Solar Time and Month

	0400	0430	0500	0530	0600	0630	0700	0730	0800	0830	0900	0930	1000	1030	1100	1130	1200
Latitude 20°N																	
Jan 1							84.9	78.7	72.7	66.1	61.5	56.5	52.1	48.3	45.5	43.6	43.0
Feb 1						88.9	82.5	75.8	69.6	63.3	57.7	52.2	47.4	43.1	40.0	37.8	37.2
Mar 1						85.7	78.8	72.0	65.2	58.6	52.3	46.2	40.5	35.5	31.4	28.6	27.7
Apr 1					88.5	81.5	74.4	67.4	60.3	53.4	46.5	39.7	33.2	26.9	21.3	17.2	15.5
May 1					85.0	78.2	71.2	64.3	57.2	50.2	43.2	36.1	29.1	26.1	15.2	8.8	5.0
Jun 1				89.2	82.7	76.0	69.3	62.5	55.7	48.8	41.9	35.0	28.1	21.1	14.2	7.3	2.0
Jul 1				88.8	82.3	75.7	69.1	62.3	55.5	48.7	41.8	35.0	28.1	21.2	14.3	7.7	3.1
Aug 1					83.8	77.1	70.2	63.3	56.4	49.4	42.4	35.4	28.3	21.3	14.3	7.3	1.9
Sep 1					87.2	80.2	73.2	66.1	59.1	52.1	45.1	38.1	31.3	24.7	18.6	13.7	11.6
Oct 1						84.1	77.1	70.2	63.3	56.5	49.9	43.5	37.5	32.0	27.4	24.3	23.1
Nov 1						87.8	81.3	74.5	68.3	61.8	56.0	50.2	45.3	40.7	37.4	35.1	34.4
Dec 1							84.3	78.0	71.8	66.1	60.5	55.6	50.9	47.2	44.2	42.4	41.8
Latitude 30°N																	
Jan 1							89.4	83.7	78.3	73.2	68.4	64.1	60.4	57.3	55.0	53.5	53.0
Feb 1							86.2	80.3	74.6	69.1	64.1	59.4	55.3	51.9	49.3	47.7	47.2
Mar 1						87.5	81.1	74.9	68.8	62.9	57.4	52.2	47.5	43.5	40.3	38.4	37.7
Apr 1					87.2	81.4	74.9	68.5	62.1	55.8	49.6	43.8	38.2	33.3	29.3	26.5	25.5
May 1				88.9	82.7	76.3	69.9	63.4	56.9	50.4	44.0	37.6	31.4	25.6	20.4	16.5	15.0
Jun 1				85.3	79.2	73.1	66.8	60.4	54.0	47.5	41.0	34.5	28.1	21.7	15.7	10.5	8.0
Jul 1				84.7	78.7	72.5	66.3	60.0	53.5	47.1	40.6	34.1	27.7	21.2	15.1	9.6	6.9
Aug 1				87.2	81.0	74.7	68.3	61.9	55.4	48.9	42.4	36.0	29.7	23.6	18.1	13.7	11.9
Sep 1					85.9	79.4	72.9	66.4	60.0	53.6	47.3	41.2	35.5	30.2	25.8	22.8	21.6
Oct 1						85.1	78.7	72.4	66.2	60.1	54.4	48.8	43.8	39.5	36.1	33.9	33.1
Nov 1							84.6	78.4	72.8	67.1	62.0	57.0	53.0	49.3	46.7	44.9	44.4
Dec 1							88.7	82.8	77.3	72.2	67.3	63.0	59.2	56.0	53.7	52.2	51.8
Latitude 40°N																	
Jan 1								89.0	84.2	79.8	75.7	72.1	69.0	66.4	64.6	63.4	63.0
Feb 1								84.8	79.8	75.2	70.7	67.0	63.5	60.9	58.8	57.6	57.2
Mar 1							89.1	83.7	78.1	72.8	67.8	63.1	58.8	55.1	51.9	49.6	48.1	47.7
Apr 1						87.1	81.4	75.6	70.0	64.4	59.0	53.8	49.0	44.6	40.9	38.0	36.2	35.5
May 1				85.9	80.5	74.7	68.9	63.2	57.5	51.8	46.3	41.0	36.1	31.7	28.2	25.8	25.0
Jun 1			86.8	81.5	76.1	70.5	64.9	59.2	53.4	47.7	42.0	36.5	31.2	26.2	22.1	19.1	18.0
Jul 1			86.0	80.8	75.4	69.9	64.3	58.6	52.8	47.1	41.4	35.8	30.4	25.4	21.1	18.0	16.9
Aug 1			89.3	83.9	78.4	72.8	67.1	61.4	55.6	49.9	44.4	39.0	33.8	29.2	25.4	22.8	21.9
Sep 1					84.6	78.9	73.2	67.5	61.8	56.3	51.0	46.0	41.4	37.5	34.4	32.3	31.6
Oct 1						86.3	80.6	75.1	69.7	64.5	59.6	55.1	51.2	47.8	45.3	43.6	43.1
Nov 1							88.1	82.6	77.8	72.8	68.6	64.4	61.1	58.2	56.1	54.7	54.4
Dec 1								88.1	83.3	78.8	74.7	71.0	67.8	65.3	63.3	62.2	61.8
Latitude 50°N																	
Jan 1										86.5	83.2	80.2	77.7	75.7	74.2	73.3	73.0
Feb 1									89.5	85.3	81.5	78.0	74.9	72.2	70.0	68.5	67.5	67.2
Mar 1								86.3	81.8	77.4	73.3	69.6	66.2	63.2	60.9	59.1	58.0	57.7
Apr 1						86.6	81.8	76.9	72.2	67.6	63.2	59.1	55.4	52.0	49.3	47.2	45.9	45.5
May 1				87.8	83.2	78.5	73.7	68.9	64.1	59.4	54.7	50.3	46.2	42.5	39.4	37.0	35.5	35.0
Jun 1		86.7	82.4	78.0	73.3	68.6	63.8	59.0	54.2	49.5	44.9	40.6	36.6	33.1	30.4	28.6	28.0
Jul 1	89.7	85.7	81.5	77.1	72.5	67.8	63.0	58.2	53.4	48.7	44.1	39.7	35.6	32.1	29.3	27.5	26.9
Aug 1		89.7	85.4	80.8	76.2	71.4	66.6	61.8	57.0	52.4	47.9	43.7	39.9	36.6	34.1	32.5	31.9
Sep 1				88.3	83.6	78.8	74.0	69.2	64.6	60.1	55.9	52.0	48.5	45.7	43.5	42.1	41.6
Oct 1						87.6	82.8	78.2	73.8	69.6	65.7	62.1	59.1	56.5	54.7	53.5	53.1
Nov 1								87.1	83.0	79.0	75.5	72.3	69.5	67.3	65.7	64.7	64.4
Dec 1									89.2	85.5	82.1	79.1	76.5	74.5	73.0	72.1	71.8

Source: Peterson (1976) and Demerjian *et al.* (1980).

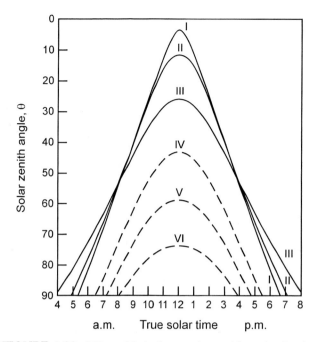

FIGURE 3.22 Effect of latitude on solar zenith angle. On the scale of true solar time, also called apparent solar time and apparent local solar time, the sun crosses the meridian at noon. The latitudes and seasons represented are as follows: I, 20°N latitude, summer solstice; II, 35°N latitude, summer solstice; III, 50°N latitude, summer solstice; IV, 20°N latitude, winter solstice; V, 35°N latitude, winter solstice; VI, 50°N latitude, winter solstice (from Leighton, 1961).

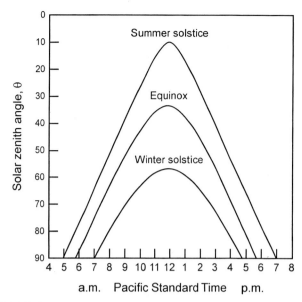

FIGURE 3.23 Relation between solar zenith angle and time of day at Los Angeles, California (from Leighton, 1961).

including surface reflection (i.e., albedo) and the concentrations of ozone and particulate matter, which scatter and/or absorb light.

(1) Surface albedo In Table 3.11 the calculated actinic fluxes are given for a surface albedo of 80%, which might correspond to the situation over snow, for example. As expected, the higher surface reflectivity leads to substantially higher actinic fluxes and hence enhanced photochemistry (see Problem 11).

(2) Total column ozone Since O_3 absorbs light primarily in the near-ultraviolet, a change in its concentration will have the greatest effect in this wavelength region. Table 3.12 shows the calculated percentage increase in actinic flux at the earth's surface for a 5% decrease in total column ozone (Madronich, 1998). Clearly, the UV flux is quite sensitive to changes in the O_3 concentration, with the greatest changes occurring at shorter wavelengths where the O_3 absorption cross sections are increasing sharply (see Chapter 4.B). This is particularly important since the total column abundance of O_3 can change by 10% or more within a season or latitude belt. In addition, decreases in stratospheric ozone due to chlorofluorocarbons (see Chapters 12 and 13) will impact the actinic flux and hence the photochemistry at the earth's surface.

TABLE 3.10 Percentage Increase in the Calculated Actinic Flux at a Surface Elevation of 1.5 km Using Best Estimate Albedos as a Function of Solar Zenith Angle and Selected Wavelengths[a] (Relative to Sea Level)[b]

Wavelength (nm)	Actinic flux increase (%)									
	0°[a]	10°	20°	30°	40°	50	60°	70°	78°	86°
340–345	2.1	2.3	2.6	3.2	4.2	5.7	8.1	11.4	12.4	7.5
400–405	0.9	0.9	1.1	1.5	2.1	3.0	4.6	7.6	10.9	6.7
540–550	0.2	0.2	0.2	0.4	0.5	0.9	1.4	2.4	4.3	4.7
680–700	0.02	0.02	0.05	0.1	0.2	0.3	0.5	1.0	1.7	2.8

[a] Zenith angles.
[b] Source: Peterson (1976) and Demerjian *et al.* (1980).

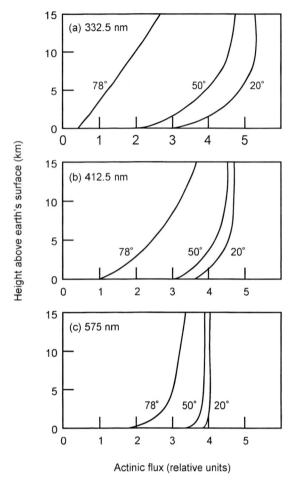

FIGURE 3.24 Calculated relative actinic flux using best estimate albedos as a function of height above the earth's surface for solar zenith angles θ of 20, 50, and 78°, respectively, at (a) 332.5, (b) 412.5, and (c) 575 nm [from Peterson (1976) and Demerjian et al. (1980)].

However, tropospheric ozone formed as an air pollutant by VOC–NO_x chemistry discussed throughout this book can also impact solar radiation reaching the earth's surface. For example, Frederick et al. (1993) reported that measurements of broadband UV in Chicago had a marginally significant negative correlation to surface O_3 concentrations under clear-sky conditions.

(3) Aerosol particles Table 3.13 shows the percentage change in the actinic flux calculated by Peterson (1976) and Demerjian et al. (1980) for two cases: (1) a particle concentration of zero, corresponding to a very clean atmosphere, and (2) a total particle concentration doubled compared to the base case. The actinic flux is predicted to increase if the total particle concentration is zero and decrease if it doubles (note, however, as discussed later, the sensitivity to the vertical distribution of particles and the relative importance of light scattering compared to absorption).

Figure 3.25 shows the results of one set of calculations of the effects of aerosol particles whose properties were judged to be characteristic of continental or urban situations, respectively, on the transmission of UV and visible radiation to the earth's surface (Erlick and Frederick, 1998). The ratio of the transmission with particles to that without is plotted in two wavelength regions, one in the UV and one in the visible. Two different relative humidity scenarios are shown. The "average summer relative humidity" was 70% RH in the boundary layer and 20% RH in the free troposphere. The high relative humidity case assumes 90% RH in the boundary layer and 30% in the free troposphere. (The RH in the stratosphere was taken to be 0% in both cases; see Chapter 12.)

The transmission of UV below $\sim$ 320 nm is particularly impacted by aerosol particles. This is primarily due to multiple scattering caused by the aerosol particles, which enhances the light absorption by O_3 in this region, since the effective absorption path length is increased (Erlick and Frederick, 1998). There is also a small contribution from assumed light absorption by aerosols (which, however, is highly uncertain; see Chapter 9). The increase in transmission with wavelength above 320 nm is due to decreased Mie light scattering by the particles, which depends on λ (see Chapter 9). It is evident that aerosol particles, particularly at high RH (which affects particle size by water uptake), can have significant impacts on the actinic flux at the earth's surface.

Model studies that incorporate both scattering and absorption of light by particles have shown that the vertical distribution and the relative importance of scattering versus absorption are critical in determining not only the magnitude but also the sign of the effect of particles on the actinic flux in the boundary layer and the associated photolysis rates for gases. For example, Dickerson et al. (1997) have shown that particles in the boundary layer which primarily scatter UV light lead to decreased actinic fluxes at the earth's surface but increased fluxes a few hundred meters above the surface. This leads to increased rates of photolysis of such species as NO_2 in the boundary layer. On the other hand, for aerosols that absorb strongly, the opposite effect occurs, reducing the actinic flux and photolysis rates of gases such as NO_2 (e.g., see Jacobson (1998) and Krotkov et al. (1998)).

It is these contrasting effects of aerosol particles, combined with uncertainties in the contribution of absorption due to O_3, that provide the largest uncertainties in calculations of actinic fluxes and photolysis rates in the boundary layer (e.g., Schwander et al., 1997). As a result, it is important to use the appropriate input

TABLE 3.11 Actinic Flux Values $F(\lambda)$ at the Earth's Surface as a Function of Wavelength Interval and Solar Zenith Angle within Specified Wavelength Intervals for a Surface Albedo of 80% Calculated by Madronich (1998)[a]

Wavelength interval (nm)	Exponent[b]	Solar zenith angle (degrees)									
		0	10	20	30	40	50	60	70	78	86
		Actinic fluxes (photons cm^{-2} s^{-1})									
290–292	14	0.00	0.00	0.00	0.00	0.00	0.00	0.00	0.00	0.00	0.00
292–294	14	0.00	0.00	0.00	0.00	0.00	0.00	0.00	0.00	0.00	0.00
294–296	14	0.01	0.01	0.00	0.00	0.00	0.00	0.00	0.00	0.00	0.00
296–298	14	0.03	0.03	0.02	0.01	0.01	0.00	0.00	0.00	0.00	0.00
298–300	14	0.08	0.08	0.06	0.04	0.02	0.01	0.00	0.00	0.00	0.00
300–302	14	0.19	0.18	0.15	0.11	0.06	0.03	0.01	0.00	0.00	0.00
302–304	14	0.49	0.47	0.40	0.30	0.19	0.09	0.03	0.00	0.00	0.00
304–306	14	0.91	0.88	0.77	0.60	0.41	0.23	0.09	0.02	0.00	0.00
306–308	14	1.42	1.37	1.22	0.99	0.71	0.43	0.19	0.05	0.01	0.00
308–310	14	1.90	1.84	1.66	1.38	1.04	0.67	0.33	0.10	0.02	0.00
310–312	14	2.87	2.79	2.55	2.16	1.67	1.13	0.61	0.21	0.05	0.01
312–314	14	3.57	3.47	3.20	2.76	2.19	1.54	0.88	0.35	0.10	0.01
314–316	14	4.04	3.94	3.65	3.19	2.58	1.87	1.13	0.49	0.16	0.02
316–318	14	4.97	4.85	4.52	3.99	3.27	2.43	1.53	0.71	0.25	0.04
318–320	14	5.10	4.99	4.67	4.15	3.44	2.60	1.70	0.84	0.32	0.05
320–325	14	15.96	15.65	14.72	13.19	11.11	8.61	5.82	3.09	1.32	0.26
325–330	14	23.28	22.85	21.58	19.50	16.64	13.16	9.19	5.15	2.39	0.58
330–335	14	24.81	24.38	23.08	20.95	18.00	14.39	10.22	5.89	2.84	0.75
335–340	14	23.85	23.45	22.24	20.25	17.50	14.10	10.14	5.96	2.94	0.83
340–345	14	25.65	25.23	23.96	21.86	18.95	15.34	11.12	6.60	3.28	0.94
345–350	14	25.75	25.33	24.08	22.01	19.13	15.55	11.33	6.77	3.37	0.96
350–355	14	29.31	28.85	27.45	25.13	21.90	17.87	13.09	7.88	3.93	1.13
355–360	14	26.16	25.75	24.52	22.48	19.64	16.08	11.84	7.17	3.59	1.03
360–365	14	28.87	28.43	27.09	24.88	21.78	17.89	13.23	8.06	4.05	1.15
365–370	15	3.57	3.52	3.35	3.08	2.70	2.23	1.65	1.01	0.51	0.14
370–375	15	3.15	3.10	2.96	2.72	2.39	1.98	1.48	0.91	0.46	0.13
375–380	15	3.57	3.52	3.36	3.10	2.73	2.26	1.69	1.05	0.53	0.15
380–385	15	3.01	2.97	2.84	2.62	2.31	1.92	1.44	0.90	0.46	0.13
385–390	15	3.30	3.25	3.11	2.87	2.54	2.11	1.59	1.00	0.51	0.14
390–395	15	3.29	3.24	3.10	2.87	2.54	2.12	1.60	1.01	0.52	0.14
395–400	15	4.00	3.94	3.77	3.49	3.09	2.59	1.97	1.25	0.64	0.17
400–405	15	5.52	5.44	5.21	4.82	4.28	3.59	2.74	1.74	0.90	0.24
405–410	15	5.35	5.28	5.06	4.69	4.17	3.50	2.68	1.71	0.89	0.24
410–415	15	5.56	5.49	5.26	4.88	4.34	3.65	2.80	1.80	0.94	0.25
415–420	15	5.64	5.56	5.33	4.95	4.41	3.72	2.86	1.85	0.97	0.26
420–430	15	11.07	10.93	10.48	9.74	8.70	7.35	5.68	3.70	1.95	0.51
430–440	15	11.35	11.20	10.76	10.01	8.95	7.59	5.89	3.87	2.05	0.53
440–450	15	13.38	13.21	12.69	11.83	10.60	9.01	7.03	4.65	2.49	0.64
450–460	15	14.40	14.21	13.67	12.75	11.44	9.75	7.64	5.09	2.75	0.70
460–470	15	14.53	14.35	13.81	12.89	11.58	9.89	7.78	5.22	2.84	0.71
470–480	15	14.73	14.55	14.00	13.08	11.77	10.07	7.95	5.35	2.93	0.72
480–490	15	14.39	14.21	13.68	12.79	11.52	9.88	7.81	5.28	2.90	0.70
490–500	15	14.82	14.63	14.09	13.18	11.88	10.20	8.09	5.49	3.03	0.72
500–510	15	14.63	14.45	13.92	13.03	11.75	10.09	8.01	5.44	3.00	0.70
510–520	15	14.29	14.12	13.60	12.73	11.49	9.88	7.86	5.36	2.97	0.68
520–530	15	14.77	14.59	14.06	13.17	11.89	10.23	8.14	5.56	3.09	0.69
530–540	15	15.17	15.04	14.44	13.52	12.21	10.51	8.36	5.71	3.16	0.69
540–550	15	15.10	14.92	14.38	13.47	12.17	10.48	8.34	5.71	3.16	0.68
550–560	15	15.22	15.04	14.49	13.58	12.27	10.57	8.43	5.77	3.20	0.68
560–570	15	15.33	15.14	14.60	13.67	12.36	10.64	8.48	5.79	3.20	0.65
570–580	15	15.48	15.30	14.74	13.81	12.48	10.75	8.57	5.86	3.23	0.65
580–600	16	3.11	3.07	2.96	2.78	2.51	2.17	1.73	1.19	0.66	0.13
600–620	16	3.08	3.04	2.93	2.75	2.49	2.15	1.72	1.19	0.67	0.13
620–640	16	3.09	3.06	2.95	2.77	2.52	2.18	1.76	1.23	0.70	0.15
640–660	16	3.02	2.99	2.88	2.71	2.47	2.15	1.74	1.23	0.72	0.16
660–680	16	3.06	3.02	2.92	2.75	2.50	2.18	1.78	1.27	0.75	0.17
680–700	16	2.99	2.96	2.86	2.70	2.46	2.15	1.75	1.26	0.76	0.17

[a] The authors are grateful to Dr. Sasha Madronich for generously providing these calculations.
[b] This column lists the power of 10 by which all entries should be multiplied. For example, at $\theta = 0°$ the total actinic flux in the wavelength interval from 306 to 308 nm is 1.42×10^{14} photons cm^{-2} s^{-1}.

TABLE 3.12 Percentage Increase in Actinic Fluxes at the Earth's Surface for a 5% Decrease in Total Column Ozone Calculated by Madronich (1998)[a]

Wavelength interval (nm)	Solar zenith angle (deg)									
	0	10	20	30	40	50	60	70	78	86
290–292	65.2	66.4	70.4	78.0	90.9	110.6	106.7	96.3	94.8	93.8
292–294	47.5	48.4	51.1	56.2	64.9	78.7	85.0	71.2	69.1	68.1
294–296	35.5	36.1	38.0	41.6	47.7	57.5	67.5	55.4	52.2	51.1
296–298	26.0	26.5	28.0	30.4	34.6	41.5	50.9	45.4	39.7	38.2
298–300	19.6	19.9	20.8	22.7	25.7	30.7	38.3	39.8	31.0	29.5
300–302	14.7	14.9	15.6	17.0	19.2	22.8	28.6	34.0	26.7	23.2
302–304	11.1	11.3	11.8	12.8	14.4	17.1	21.5	27.6	23.9	18.8
304–306	8.4	8.5	8.9	9.7	10.9	12.8	16.1	21.5	22.2	15.8
306–308	6.4	6.8	6.8	7.3	8.2	9.7	12.2	16.5	19.7	13.9
308–310	4.8	4.9	5.1	5.5	6.2	7.3	9.2	12.5	16.3	12.9
310–312	3.7	3.8	3.9	4.2	4.8	5.6	7.0	9.6	13.1	12.4
312–314	2.8	2.9	3.0	3.2	3.6	4.2	5.3	7.3	10.2	11.7
314–316	4.4	2.2	2.3	2.4	2.7	3.2	4.0	5.5	7.8	10.4
316–318	1.6	1.6	1.7	1.9	2.1	2.4	3.0	4.2	6.0	8.9
318–320	1.2	1.2	1.3	1.4	1.5	1.8	2.2	3.1	4.4	7.1
320–325	0.7	0.7	0.8	0.8	0.9	1.1	1.3	1.8	2.7	4.6
325–330	0.3	0.4	0.4	0.4	0.4	0.5	0.6	0.9	1.3	2.3

[a] We are grateful to Dr. Sasha Madronich for these calculations.

values for these parameters in carrying out photolysis calculations for particular locations. This is especially true for unusual situations such as high particle and/or surface-level ozone concentrations or unusual geography such as mountains, which shield the light at large solar zenith angles (e.g., Castro et al., 1997).

g. Effects of Clouds on $F(\lambda)$

All the calculated actinic fluxes discussed so far refer to a cloudless sky. The effects of clouds are complex, in that they reduce the direct radiation at the earth's surface from the sun but, at the same time, can increase the total actinic flux directly above the cloud due to scattering from the top surface of the cloud. Madronich (1987) has treated the case of large uniform clouds of various optical depths which, however, are sufficiently large that the cloud completely diffuses both the reflected and the transmitted light. The actinic flux above the cloud then becomes a combination of the incident light (a combination of direct sunlight and

TABLE 3.13 Percentage Change of Calculated Actinic Flux at the Earth's Surface Using Best Estimate Albedos as a Function of Solar Zenith Angle and Selected Wavelengths When Model Aerosol Concentrations Are Either Zero or Doubled[a]

Wavelength (nm)	Actinic flux change (%)									
	0°[b]	10°	20°	30°	40°	50°	60°	70°	78°	86°
340–345										
No aerosol	+8.2	+8.4	+8.8	+9.5	+10.7	+12.7	+16.1	+22.3	+26.5	+17.6
Double	−6.1	−6.3	−6.6	−7.3	−8.3	−10.1	−12.8	−16.1	−16.4	−12.5
400–405										
No aerosol	+5.8	+6.0	+6.4	+7.1	+8.3	+10.7	+15.3	+6.2	+46.8	+35.7
Double	−4.0	−4.1	−4.5	−5.3	−6.6	−8.8	−12.6	−19.4	−24.9	−15.9
540–550										
No aerosol	+0.9	+1.0	+1.2	+1.8	+2.9	+5.1	+10.4	+25.4	+67.1	+261
Double	−0.8	−0.9	−1.4	−2.2	−3.7	−6.4	−11.6	−21.4	−33.6	−27.4

[a] From Peterson (1976) and Demerjian et al. (1980); although these are based on actinic fluxes different from those in Table 3.7, the relative changes calculated here should be similar to those that would be derived using the model from which the data in Table 3.7 were derived; the changes in the particle concentration are relative to the base case shown in Table 3.7.
[b] Solar zenith angle.

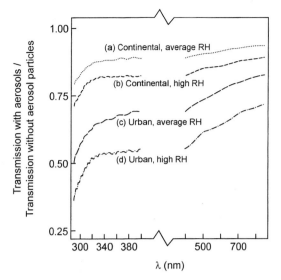

FIGURE 3.25 Calculated ratio of transmission of UV and visible light to the earth's surface in the presence of aerosol particles compared to that with no aerosol particles for typical continental aerosol particles at (a) average summer RH and (b) high summer RH and for urban aerosol particles with (c) average RH and (d) high RH. (Adapted from Erlick and Frederick, 1998.)

downward-directed diffuse light), the light that has undergone diffuse reflection at the top of the cloud, and the light that has passed through the cloud, undergone reflection at the surface, and then been transmitted back upward through the cloud.

Table 3.14 shows the results of some of Madronich's calculations of the actinic flux enhancements for two cases of a collimated direct beam of light striking the top of a cloud, first under typical summer conditions at

TABLE 3.14 Calculated Enhancements or Depressions of Actinic Fluxes above and below Perfectly Light-Diffusing Clouds of Different Optical Depths[a]

Conditions	Cloud optical depth	Above cloud	Below cloud
Summer[b]	0.0 (clear sky; no cloud)	1.09	1.09
	8	1.65	1.35
	128	2.70	0.19
Winter[c]	0.0 (clear sky; no cloud)	1.55	1.55
	8	1.59	0.81
	128	1.65	0.27

[a] From Madronich, 1987.
[b] Calculated for "typical" summer conditions of a solar zenith angle of 20° and a surface albedo of 5% with an incident light beam that is collimated.
[c] Calculated for "typical" winter conditions of a solar zenith angle of 70° and a surface reflectivity of 80% for a collimated incident light beam.

a solar zenith angle of 20° for the case of a 5% surface albedo and second, for typical winter conditions with a solar zenith angle of 70° and a surface albedo of 80%, which would be characteristic of snow, for example.

The cloudless case shown first for a small solar zenith angle and typical summertime conditions shows an enhancement due to reflections from the surface. The cloud with an optical depth of 8 corresponds to a total of 67% transmission of the light through the cloud, but essentially all of it is diffused by the cloud and is therefore not directly transmitted light. The cloud with an optical depth of 128 only transmits a total of 9% of the light, essentially all of which is again diffuse.

Under the typical summertime conditions, the thinner cloud shows an increase of 65% in the actinic flux above the cloud whereas the thicker cloud shows an increase of almost a factor of three, the maximum theoretically possible. This is due to scattering of diffuse light from the top of the cloud, as well as from the ground. As expected, below the thicker cloud, the total actinic flux is reduced, in this calculation, to 19% of the clear-sky value. However, for the thinner cloud of optical density 8, the actinic flux below the cloud is actually calculated to be *greater* than for the cloudless case. This occurs in the case of a small solar zenith angle and direct (rather than diffuse) incident light because the direct incident light is diffused as it traverses the cloud; as discussed earlier for the case of the actinic flux above a Lambertian surface, conversion of a direct to diffuse source leads to an enhancement in the actinic flux.

Similar trends are predicted for the winter case chosen, except that the unexpected below-cloud enhancement discussed is not seen.

Interestingly, in the air inside the cloud itself, particularly near the top of the cloud, there can be significant enhancements of the actinic flux due to this scattering phenomenon. The enhancements expected depend on a variety of factors, including the solar zenith angle, the amount of direct vs diffuse incident light, surface albedo, cloud optical depth, etc. Madronich (1987) suggests for a "typical" summer average that the enhancement factors vary linearly from 1.7 near the top of the cloud, to 1.0 (i.e., no enhancement) in the middle of the cloud, to 0.4 (i.e., a reduction in actinic flux) at the bottom of the cloud.

This behavior has been borne out experimentally. Figure 3.26, for example, shows some vertical measurements of the actinic flux below, in, and above a cloud (Vilà-Guerau de Arellano *et al.*, 1994). The dotted line shows the calculated actinic flux in the absence of clouds for these particular conditions. At the cloud

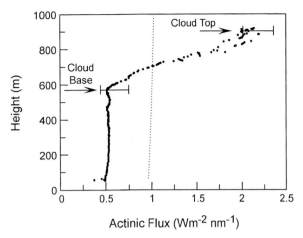

FIGURE 3.26 Vertical measurements of actinic fluxes below, in, and above a cloud. The dotted line shows calculated clear-sky values (from Vilà-Guerau de Arellano et al., 1994).

FIGURE 3.27 Dr. Wolfgang Junkermann prepares for a flight to measure actinic fluxes in Germany. The radiometer can be seen mounted by the wheel. Typical data are shown in Figure 3.28. (The authors are grateful to Dr. Junkermann for providing this photograph.)

base, the measured flux is 0.56 W m^{-2} nm^{-1}, compared to a calculated value for a cloudless sky of 0.93 W m^{-2} nm^{-1}. At the top of the cloud, the flux increased significantly to 2.1 W m^{-2} nm^{-1}. Inside the cloud itself, the flux increased linearly.

In short, while the flux under a cloud is generally less than the clear-sky value, inside and above the cloud it can be significantly larger, leading to enhanced rates of photolysis of photochemically active species. For example, the photolysis of O_3 to form electronically excited $O(^1D)$, followed by the reaction of the latter with water vapor, is a major source of OH in the troposphere. As a result, actinometric measurements are often made by measuring the rate of production of $O(^1D)$ directly, known as $J(O^1D)$, or, alternatively, using a light detector calibrated for this photolysis process. Junkermann (1994), for example, has used a hang glider (Fig. 3.27) equipped with a photoelectric detector to fly spiral flight paths from the top of a mountain to the valley floor in Germany while measuring vertical profiles of the light intensity coming into a 2π radiometer as described earlier (Fig. 3.20). Two such hemispherical detectors are used, one of which is downward facing and one of which is upward facing; the sum of the two gives the spherically integrated actinic flux, which is the parameter of interest for measuring total photolysis rates in the atmosphere. To relate this measurement of light intensity to $J(O^1D)$, a combination of optical filters and appropriate detectors is used (Junkermann et al., 1989).

Figure 3.28 gives a typical measurement of the upward component of $J(O^1D)$ as well as the total value during one flight made at a solar zenith angle of 62° (Junkerman et al., 1994). The values of $J(O^1D)$ are clearly reduced below the cloud, increase linearly inside the cloud, and are more than double the below-cloud values above the cloud top. As expected, the OH concentrations above the cloud are higher as well. For example, Mauldin et al. (1997) report OH concentrations of $(8-15) \times 10^6$ OH cm^{-3} above clouds compared to $(3-5) \times 10^6$ OH cm^{-3} in cloud-free regions. Similarly, Volz-Thomas et al. (1996) measured values of $J(NO_2)$ that were about twice as high above clouds compared to cloud-free days.

The intensity of all wavelengths is not affected equally by clouds. For example, Bordewijk et al. (1995) reported that the relationship between total solar radiation and UV in the 285- to 345-nm region measured at ground level is nonlinear, with relatively higher amounts of UV reaching the surface. Indeed, they suggest that even when the total solar radiation is decreased by 20% due to clouds, the UV intensity can be unchanged. Seckmeyer et al. (1996) also reported a wavelength dependence for radiation reaching the earth's surface through clouds. It has been shown that this dependence is not due to the properties of water in the clouds but rather to longer effective path lengths due to scattering (e.g., Kylling et al., 1997; Mayer et al., 1998). The increased path then gives a wavelength dependence through Rayleigh scattering and the enhanced light absorption by O_3 and particles. The backscattered light from clouds (which can be measured by satellites) not only has been reported to be wavelength dependent but also differs for high-level clouds compared to low- or mid-level clouds (e.g., see Wen and Frederick (1995) and Chapter 14.C).

One final interesting aspect of clouds and actinic fluxes is that inside the cloud droplets themselves, an

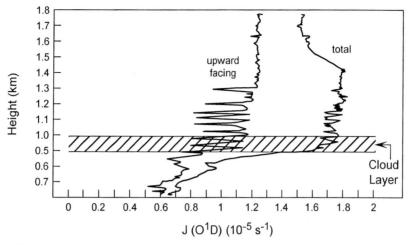

FIGURE 3.28 Vertical measurement of $J(O^1D)$ using an upward-facing detector and the total $J(O^1D)$ measured using both upward- and downward-facing detectors (from Junkermann, 1994).

increased actinic flux is expected compared to the surrounding air. As discussed in detail by Madronich (1987), there are several different effects that must be taken into account when a light beam strikes a water droplet in air. Initially, of course, some of the light will be reflected from the surface and not enter the drop itself. The portion of the incident beam that enters the droplet is subject to absorption, depending on the droplet composition and the wavelength of light, which also effectively reduces the actinic flux. However, counterbalancing these effects is the possibility of multiple internal reflections occurring at the inner droplet surface, which redirects the light beam back into the droplet. In addition, refraction of the light beam at the air–water interface as it enters the drop leads to an increased path length through the drop itself, in effect increasing the probability of light absorption (Beer–Lambert law).

The net enhancement factor for a droplet consisting of pure water can be as much as 1.6 (Madronich, 1987). Calculations by Ruggaber *et al.* (1997) suggest that the actinic flux inside cloud drops with a typical size distribution and dissolved particulate matter is more than a factor of two greater than in the cloud interstitial air. This effect of enhanced actinic flux inside droplets may be quite important for aqueous-phase photochemistry in fogs and clouds.

h. Comparison of Calculated Actinic Fluxes to Experimentally Measured Values

As described earlier, measurements of actinic fluxes are made using chemical actinometry, particularly the photolysis of NO_2 or O_3, or using flat-plate or 2π radiometers. Intercomparisons of such measurements have been made by a number of investigators, as well as comparison with calculated photolysis rates using published actinic fluxes such as those in Table 3.7. In general, there is good agreement between results obtained with different radiometers and calculated values, with the largest uncertainties generally being at shorter wavelengths (< 310 nm) and larger solar zenith angles (e.g., see Seckmeyer *et al.*, 1995; Kato *et al.*, 1997; and Halthore *et al.*, 1997).

Figure 3.29, for example, shows measurements of the photolysis rate of O_3, $J(O_3)$, made at the Mauna Loa Observatory on two different days, compared to model calculations of the photolysis rate constant (Shetter *et al.*, 1996). The two model calculations use different assumptions regarding the quantum yield for O_3 photolysis in the absorption "tail" beyond 310 nm (see Chapter 4.B). The measurements are in excellent agreement for the second day but somewhat smaller than the model calculations on the first.

Similarly, Fig. 3.30 shows measurements of $J(NO_2)$ at an altitude of 7–7.5 km as a function of solar zenith angle compared to a multidirectional model calculation (Volz-Thomas *et al.*, 1996). The agreement in this case is generally good. However, this is not always the case. For example, Fig. 3.31 shows some measurements of $J(NO_2)$ as a function of solar zenith angle made by different groups at different locations and using different techniques (Kraus and Hofzumahaus, 1998).

The reasons for discrepancies between various measurements and between the measured and model calculated values are not clear. Lantz *et al.* (1996) suggest that one factor that will affect instantaneous photolysis rates is cloud cover and that under some circumstances, the instantaneous photolysis rates may exceed

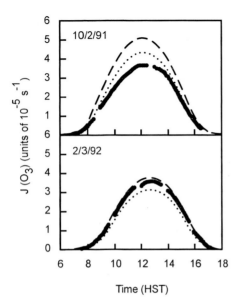

FIGURE 3.29 Measured rates of O_3 photolysis, $J(O_3)$, shown as heavy solid line, at Mauna Loa Observatory on two days (October 2, 1991, and February 3, 1992) compared to model calculations using two different assumptions (shown by the lighter dotted and dashed lines, respectively) for the quantum yield for O_3 photolysis at $\lambda > 310$ nm. (Adapted from Shetter et al., 1996.)

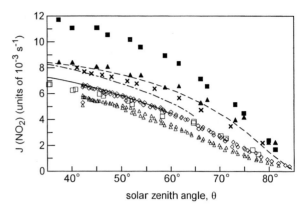

FIGURE 3.31 Some values of $J(NO_2)$ measured using different techniques as a function of solar zenith angle: ($\triangle$ $\diamond$) Kraus and Hofzumahaus (1998); ($\times$) Madronich et al. (1983); ($\square$) Brauers and Hofzumahaus (1992); ($\blacktriangle$) Shetter et al. (1992); ($\blacksquare$) Lantz et al. (1996); short dashed line is from Parrish et al. (1983); --- is from Dickerson et al. (1982) and solid line is from Müller and Schurath (1986). (Adapted from Kraus and Hofzumahaus, 1998.)

the clear-sky values. However, the average photolysis rate will not exceed the clear-sky value. This remains an area of active investigation.

i. Actinic Fluxes in the Stratosphere

Tables 3.15, 3.16, and 3.17 give Madronich's calculated actinic fluxes for altitudes of 15, 25, and 40 km, respectively. The reduction in actinic flux as the light travels through the atmosphere is very evident. Thus, at 40 km but not at 15 km, there is substantial light intensity at 200 nm. As discussed in Chapter 12, this is why chlorofluorocarbons, which absorb light in the 200-nm region, do not photodissociate until they reach the mid to upper stratosphere.

Figure 3.32 shows some calculated actinic fluxes in the stratosphere at 20-, 30-, 40-, and 50-km altitude at a solar zenith angle of 30° (DeMore et al., 1997) as well as at ground level. The surface albedo was assumed to be 0.3 and the aerosol concentrations typical of "moderate volcanic conditions."

The "window" in the stratosphere around 200 nm (where CFCs absorb) between the O_2 and O_3 absorptions (Fig. 3.13) is clearly evident in the actinic fluxes shown in Fig. 3.32. Figure 3.32 also clearly illustrates the tropospheric actinic cutoff of approximately 290 nm.

3. Procedure for Calculating Photolysis Rates

As discussed in Section 3.C.2a, the rate of loss of a molecule A from the troposphere by photolysis is given by Eq. (CC):

$$d[A]/dt = -k_p[A], \qquad (CC)$$

where k_p is the photolysis rate constant (s^{-1}) given by Eq. (OO):

$$k_p = \int_\lambda \phi(\lambda)\sigma(\lambda)F(\lambda)\,d\lambda. \qquad (OO)$$

The primary quantum yield $\phi(\lambda)$ represents the fraction of excited molecules that undergo photochemistry

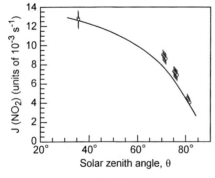

FIGURE 3.30 Values of $J(NO_2)$ at 7- to 7.5-km altitude as a function of solar zenith angle (θ) measured using 2π radiometers (circles) compared to a model calculated photolysis rate (solid line). (Adapted from Volz-Thomas et al., 1996.)

TABLE 3.15 Actinic Flux Values $F(\lambda)$ at an Altitude of 15 km above the Earth's Surface as a Function of Wavelength Interval and Solar Zenith Angle within Specified Wavelength Intervals for Best Estimate Surface Albedo Calculated by Madronich (1998)[a]

Wavelength interval (nm)	Exponent[b]	Solar zenith angle (deg)									
		0	10	20	30	40	50	60	70	78	86
		Actinic fluxes (photons cm^{-2} s^{-1})									
290–292	14	0.00	0.00	0.00	0.00	0.00	0.00	0.00	0.00	0.00	0.00
292–294	14	0.00	0.00	0.00	0.00	0.00	0.00	0.00	0.00	0.00	0.00
294–296	14	0.02	0.02	0.01	0.01	0.00	0.00	0.00	0.00	0.00	0.00
296–298	14	0.06	0.06	0.05	0.03	0.02	0.01	0.00	0.00	0.00	0.00
298–300	14	0.13	0.12	0.11	0.08	0.05	0.03	0.01	0.00	0.00	0.00
300–302	14	0.23	0.23	0.20	0.17	0.12	0.07	0.03	0.00	0.00	0.00
302–304	14	0.50	0.49	0.45	0.39	0.30	0.20	0.10	0.02	0.00	0.00
304–306	14	0.82	0.80	0.75	0.67	0.55	0.40	0.23	0.07	0.01	0.00
306–308	14	1.16	1.14	1.09	0.99	0.85	0.65	0.42	0.17	0.03	0.00
308–310	14	1.44	1.43	1.37	1.27	1.12	0.91	0.64	0.31	0.08	0.00
310–312	14	2.07	2.05	1.98	1.87	1.69	1.43	1.06	0.59	0.20	0.01
312–314	14	2.46	2.44	2.38	2.26	2.08	1.82	1.43	0.89	0.37	0.03
314–316	14	2.69	2.67	2.62	2.52	2.35	2.10	1.72	1.16	0.56	0.05
316–318	14	3.23	3.21	3.16	3.05	2.88	2.62	2.21	1.58	0.85	0.11
318–320	14	3.24	3.22	3.18	3.09	2.95	2.72	2.35	1.77	1.05	0.18
320–325	14	9.81	9.77	9.67	9.47	9.13	8.57	7.64	6.08	4.03	1.01
325–330	14	13.78	13.75	13.65	13.45	13.08	12.46	11.39	9.52	6.90	2.37
330–335	14	14.32	14.29	14.21	14.04	13.72	13.17	12.20	10.46	7.94	3.21
335–340	14	13.45	13.44	13.38	13.25	13.00	12.55	11.73	10.24	8.04	3.64
340–345	14	14.19	14.18	14.13	14.01	13.77	13.34	12.54	11.06	8.83	4.23
345–350	14	13.95	13.94	13.89	13.79	13.58	13.18	12.45	11.05	8.93	4.45
350–355	14	15.63	15.62	15.58	15.47	15.26	14.85	14.08	12.59	10.29	5.32
355–360	14	13.70	13.69	13.66	13.58	13.41	13.08	12.43	11.19	9.23	4.93
360–365	14	14.91	14.91	14.88	14.80	14.63	14.29	13.63	12.33	10.27	5.64
365–370	15	1.82	1.82	1.82	1.81	1.79	1.75	1.67	1.52	1.28	0.72
370–375	15	1.59	1.58	1.58	1.58	1.56	1.53	1.47	1.34	1.13	0.65
375–380	15	1.78	1.78	1.78	1.77	1.76	1.72	1.66	1.52	1.29	0.76
380–385	15	1.48	1.48	1.48	1.47	1.46	1.44	1.38	1.27	1.09	0.66
385–390	15	1.60	1.60	1.60	1.60	1.59	1.56	1.51	1.39	1.20	0.74
390–395	15	1.58	1.58	1.58	1.58	1.57	1.55	1.50	1.39	1.20	0.76
395–400	15	1.90	1.90	1.90	1.90	1.89	1.86	1.81	1.68	1.47	0.94
400–405	15	2.63	2.63	2.63	2.62	2.61	2.57	2.49	2.33	2.04	1.32
405–410	15	2.52	2.52	2.52	2.52	2.51	2.48	2.41	2.25	1.98	1.31
410–415	15	2.60	2.60	2.60	2.60	2.59	2.56	2.49	2.34	2.07	1.38
415–420	15	2.62	2.62	2.62	2.62	2.61	2.58	2.51	2.37	2.10	1.43
420–430	15	5.09	5.09	5.09	5.09	5.07	5.02	4.90	4.63	4.14	2.86
430–440	15	5.15	5.15	5.16	5.16	5.14	5.10	4.99	4.74	4.27	3.02
440–450	15	6.00	6.00	6.01	6.01	6.00	5.96	5.84	5.56	5.04	3.64
450–460	15	6.50	6.50	6.50	6.49	6.47	6.42	6.29	5.99	5.45	3.99
460–470	15	6.51	6.51	6.51	6.51	6.49	6.44	6.32	6.04	5.51	4.09
470–480	15	6.56	6.56	6.56	6.55	6.54	6.49	6.38	6.10	5.59	4.19
480–490	15	6.35	6.35	6.35	6.35	6.34	6.29	6.19	5.93	5.44	4.09
490–500	15	6.51	6.50	6.50	6.50	6.49	6.45	6.34	6.09	5.60	4.24
500–510	15	6.52	6.52	6.51	6.50	6.47	6.41	6.29	6.01	5.50	4.12
510–520	15	6.34	6.34	6.33	6.32	6.29	6.24	6.12	5.86	5.38	4.05
520–530	15	6.52	6.52	6.51	6.50	6.47	6.41	6.29	6.01	5.51	4.10
530–540	15	6.68	6.71	6.67	6.65	6.62	6.55	6.42	6.13	5.59	4.10
540–550	15	6.63	6.63	6.62	6.60	6.57	6.50	6.37	6.08	5.54	4.05
550–560	15	6.73	6.73	6.71	6.69	6.64	6.57	6.42	6.12	5.56	4.03
560–570	15	6.75	6.74	6.73	6.70	6.65	6.57	6.41	6.08	5.49	3.87
570–580	15	6.80	6.80	6.78	6.75	6.70	6.62	6.46	6.12	5.51	3.84
580–600	16	1.36	1.36	1.36	1.35	1.34	1.33	1.30	1.23	1.11	0.7
600–620	16	1.36	1.36	1.35	1.35	1.34	1.32	1.29	1.22	1.11	
620–640	16	1.36	1.36	1.36	1.35	1.34	1.33	1.30	1.24	1.14	
640–660	16	1.35	1.34	1.34	1.33	1.32	1.31	1.28	1.23	1.15	
660–680	16	1.38	1.38	1.37	1.37	1.35	1.34	1.31	1.26	1.1	
680–700	16	1.35	1.35	1.34	1.34	1.33	1.31	1.29	1.25		

[a] The authors are grateful to Dr. Sasha Madronich for generously providing these calculations.
[b] This column lists the power of 10 by which all entries should be multiplied. For example, at $\theta = 0°$ the wavelength interval from 306 to 308 nm is 1.16×10^{14} photons cm^{-2} s^{-1}.

TABLE 3.16 Actinic Flux Values $F(\lambda)$ at an Altitude of 25 km above the Earth's Surface as a Function of Wavelength Interval and Solar Zenith Angle within Specified Wavelength Intervals for Best Estimate Surface Albedo Calculated by Madronich (1998)[a]

Wavelength interval (nm)	Exponent[b]	Solar zenith angle (deg)									
		0	10	20	30	40	50	60	70	78	86
		Actinic fluxes (photons cm^{-2} s^{-1})									
290–292	14	0.06	0.05	0.05	0.03	0.02	0.01	0.00	0.00	0.00	0.00
292–294	14	0.11	0.11	0.09	0.07	0.05	0.03	0.01	0.00	0.00	0.00
294–296	14	0.21	0.20	0.18	0.15	0.11	0.07	0.03	0.00	0.00	0.00
296–298	14	0.34	0.33	0.31	0.27	0.21	0.14	0.07	0.02	0.00	0.00
298–300	14	0.46	0.45	0.43	0.38	0.32	0.24	0.14	0.05	0.01	0.00
300–302	14	0.58	0.57	0.55	0.50	0.44	0.35	0.23	0.10	0.02	0.00
302–304	14	0.94	0.93	0.90	0.84	0.75	0.63	0.46	0.24	0.07	0.00
304–306	14	1.22	1.21	1.18	1.12	1.02	0.89	0.69	0.42	0.16	0.01
306–308	14	1.48	1.46	1.43	1.37	1.27	1.13	0.93	0.62	0.29	0.02
308–310	14	1.63	1.62	1.59	1.53	1.44	1.31	1.11	0.81	0.45	0.04
310–312	14	2.17	2.16	2.12	2.06	1.96	1.81	1.57	1.21	0.74	0.12
312–314	14	2.47	2.46	2.42	2.36	2.26	2.11	1.87	1.50	1.01	0.23
314–316	14	2.62	2.61	2.58	2.53	2.43	2.29	2.07	1.71	1.23	0.38
316–318	14	3.10	3.09	3.06	3.00	2.91	2.76	2.52	2.13	1.61	0.61
318–320	14	3.09	3.08	3.06	3.01	2.92	2.79	2.58	2.23	1.74	0.78
320–325	14	9.35	9.33	9.27	9.15	8.95	8.61	8.06	7.13	5.84	3.20
325–330	14	13.21	13.19	13.13	13.01	12.78	12.40	11.75	10.62	9.04	5.82
330–335	14	13.79	13.78	13.73	13.63	13.43	13.09	12.49	11.42	9.93	6.90
335–340	14	13.03	13.02	12.98	12.90	12.75	12.47	11.96	11.04	9.73	7.13
340–345	14	13.79	13.78	13.75	13.68	13.54	13.26	12.76	11.84	10.52	7.89
345–350	14	13.59	13.58	13.56	13.50	13.37	13.12	12.66	11.79	10.52	8.01
350–355	14	15.27	15.26	15.24	15.18	15.05	14.80	14.31	13.37	12.00	9.25
355–360	14	13.41	13.41	13.39	13.35	13.25	13.04	12.63	11.85	10.68	8.32
360–365	14	14.63	14.62	14.61	14.57	14.47	14.27	13.85	13.03	11.78	9.26
365–370	15	1.79	1.79	1.79	1.78	1.77	1.75	1.70	1.60	1.46	1.15
370–375	15	1.56	1.56	1.56	1.56	1.55	1.53	1.49	1.41	1.28	1.02
375–380	15	1.75	1.75	1.75	1.75	1.74	1.72	1.68	1.59	1.46	1.17
380–385	15	1.46	1.46	1.46	1.46	1.45	1.44	1.40	1.33	1.22	0.99
385–390	15	1.59	1.59	1.59	1.58	1.58	1.56	1.53	1.46	1.34	1.09
390–395	15	1.57	1.57	1.57	1.57	1.56	1.55	1.52	1.45	1.34	1.09
395–400	15	1.88	1.88	1.89	1.89	1.88	1.87	1.83	1.75	1.62	1.33
400–405	15	2.61	2.61	2.61	2.61	2.60	2.58	2.53	2.42	2.24	1.86
405–410	15	2.50	2.51	2.51	2.51	2.50	2.48	2.44	2.34	2.17	1.80
410–415	15	2.59	2.59	2.59	2.59	2.58	2.57	2.52	2.42	2.26	1.88
415–420	15	2.61	2.61	2.61	2.61	2.60	2.59	2.55	2.45	2.29	1.92
420–430	15	5.06	5.06	5.07	5.07	5.06	5.04	4.97	4.79	4.48	3.78
430–440	15	5.13	5.13	5.14	5.14	5.14	5.12	5.06	4.89	4.59	3.90
440–450	15	5.98	5.99	5.99	6.00	6.00	5.98	5.91	5.73	5.40	4.62
450–460	15	6.48	6.48	6.48	6.48	6.47	6.44	6.36	6.16	5.81	4.98
460–470	15	6.50	6.50	6.50	6.50	6.49	6.47	6.39	6.20	5.85	5.04
470–480	15	6.54	6.54	6.55	6.55	6.54	6.52	6.45	6.26	5.93	5.11
480–490	15	6.34	6.34	6.35	6.35	6.35	6.32	6.26	6.08	5.76	4.97
490–500	15	6.50	6.50	6.50	6.50	6.50	6.48	6.42	6.25	5.92	5.11
500–510	15	6.51	6.51	6.51	6.50	6.48	6.44	6.36	6.17	5.83	4.99
510–520	15	6.34	6.33	6.33	6.32	6.31	6.27	6.19	6.01	5.69	4.87
520–530	15	6.52	6.52	6.51	6.50	6.48	6.45	6.37	6.18	5.84	4.97
530–540	15	6.68	6.70	6.67	6.66	6.64	6.60	6.51	6.32	5.96	5.03
540–550	15	6.63	6.63	6.62	6.61	6.59	6.55	6.46	6.27	5.91	4.97
550–560	15	6.73	6.73	6.72	6.70	6.67	6.62	6.52	6.31	5.94	4.97
560–570	15	6.75	6.75	6.73	6.72	6.68	6.63	6.52	6.31	5.91	4.87
570–580	15	6.81	6.80	6.79	6.77	6.74	6.68	6.58	6.35	5.95	4.87
580–600	16	1.36	1.36	1.36	1.36	1.35	1.34	1.32	1.27	1.20	0.99
600–620	16	1.36	1.36	1.35	1.35	1.34	1.33	1.31	1.26	1.19	0.98
620–640	16	1.36	1.36	1.36	1.35	1.35	1.33	1.32	1.28	1.21	1.02
640–660	16	1.35	1.34	1.34	1.34	1.33	1.32	1.30	1.26	1.20	1.03
660–680	16	1.38	1.38	1.37	1.37	1.36	1.34	1.32	1.28	1.22	1.07
680–700	16	1.35	1.35	1.34	1.34	1.33	1.31	1.30	1.26	1.21	1.07

[a] The authors are grateful to Dr. Sasha Madronich for generously providing these calculations.
[b] This column lists the power of 10 by which all entries should be multiplied. For example, at $\theta = 0°$ the total actinic flux in the wavelength interval from 306 to 308 nm is 1.48×10^{14} photons cm^{-2} s^{-1}.

TABLE 3.17 Actinic Flux Values $F(\lambda)$ at an Altitude of 40 km above the Earth's Surface as a Function of Wavelength Interval and Solar Zenith Angle within Specified Wavelength Intervals for Best Estimate Surface Albedo Calculated by Madronich (1998)[a]

Wavelength interval (nm)	Exponent[b]	Solar zenith angle (deg)									
		0	10	20	30	40	50	60	70	78	86
		Actinic fluxes (photons cm^{-2} s^{-1})									
202–205	14	0.03	0.02	0.02	0.02	0.02	0.02	0.02	0.02	0.01	0.00
205–210	14	0.06	0.06	0.06	0.06	0.06	0.06	0.05	0.04	0.03	0.01
210–215	14	0.14	0.14	0.14	0.14	0.13	0.12	0.11	0.09	0.06	0.01
215–220	14	0.15	0.15	0.15	0.14	0.13	0.12	0.10	0.07	0.04	0.00
220–225	14	0.17	0.16	0.16	0.15	0.14	0.12	0.09	0.05	0.02	0.00
225–230	14	0.11	0.11	0.11	0.10	0.09	0.07	0.05	0.02	0.00	0.00
230–235	14	0.08	0.08	0.08	0.07	0.06	0.04	0.02	0.01	0.00	0.00
235–240	14	0.06	0.05	0.05	0.04	0.03	0.02	0.01	0.00	0.00	0.00
240–245	14	0.05	0.04	0.04	0.03	0.02	0.01	0.01	0.00	0.00	0.00
245–250	14	0.03	0.03	0.02	0.02	0.01	0.01	0.00	0.00	0.00	0.00
250–255	14	0.03	0.02	0.02	0.02	0.01	0.01	0.00	0.00	0.00	0.00
255–260	14	0.05	0.05	0.05	0.04	0.02	0.01	0.00	0.00	0.00	0.00
260–265	14	0.09	0.09	0.08	0.06	0.04	0.02	0.01	0.00	0.00	0.00
265–270	14	0.23	0.22	0.20	0.17	0.13	0.08	0.03	0.01	0.00	0.00
270–275	14	0.31	0.30	0.28	0.24	0.19	0.13	0.07	0.02	0.00	0.00
275–280	14	0.44	0.43	0.41	0.37	0.31	0.24	0.14	0.05	0.01	0.00
280–285	14	0.86	0.85	0.82	0.77	0.69	0.58	0.42	0.22	0.06	0.00
285–290	14	1.57	1.56	1.52	1.47	1.37	1.24	1.02	0.69	0.32	0.01
290–292	14	1.36	1.36	1.34	1.30	1.25	1.16	1.03	0.79	0.47	0.05
292–294	14	1.29	1.29	1.27	1.25	1.21	1.14	1.04	0.85	0.57	0.10
294–296	14	1.39	1.38	1.37	1.35	1.32	1.26	1.17	1.00	0.73	0.19
296–298	14	1.44	1.43	1.42	1.41	1.38	1.33	1.26	1.12	0.88	0.31
298–300	14	1.38	1.37	1.37	1.35	1.33	1.30	1.24	1.13	0.94	0.43
300–302	14	1.32	1.31	1.31	1.30	1.28	1.26	1.21	1.13	0.98	0.53
302–304	14	1.72	1.72	1.72	1.71	1.69	1.66	1.61	1.52	1.36	0.85
304–306	14	1.89	1.89	1.88	1.87	1.85	1.82	1.78	1.70	1.56	1.08
306–308	14	2.00	2.00	1.99	1.98	1.95	1.92	1.88	1.80	1.68	1.25
308–310	14	1.99	1.99	1.98	1.96	1.94	1.90	1.86	1.78	1.67	1.33
310–312	14	2.46	2.45	2.44	2.42	2.38	2.34	2.27	2.17	2.05	1.70
312–314	14	2.64	2.63	2.62	2.59	2.55	2.49	2.41	2.30	2.16	1.84
314–316	14	2.70	2.69	2.68	2.65	2.61	2.54	2.45	2.32	2.17	1.87
316–318	14	3.12	3.11	3.09	3.06	3.01	2.93	2.82	2.66	2.46	2.14
318–320	14	3.06	3.06	3.04	3.01	2.96	2.88	2.77	2.59	2.39	2.07
320–325	14	9.18	9.16	9.12	9.04	8.90	8.68	8.33	7.77	7.10	6.09
325–330	14	12.97	12.96	12.91	12.82	12.65	12.36	11.88	11.09	10.09	8.59
330–335	14	13.59	13.58	13.54	13.45	13.30	13.03	12.55	11.74	10.69	9.09
335–340	14	12.87	12.86	12.83	12.77	12.64	12.41	11.99	11.25	10.26	8.73
340–345	14	13.65	13.64	13.62	13.56	13.44	13.21	12.79	12.03	11.01	9.38
345–350	14	13.47	13.46	13.44	13.39	13.28	13.08	12.68	11.96	10.97	9.38
350–355	14	15.14	15.14	15.12	15.08	14.97	14.75	14.33	13.55	12.46	10.68
355–360	14	13.31	13.31	13.30	13.27	13.18	13.01	12.66	12.00	11.06	9.50
360–365	14	14.53	14.53	14.52	14.49	14.41	14.23	13.88	13.18	12.18	10.49
365–370	15	1.78	1.78	1.78	1.77	1.77	1.75	1.70	1.62	1.50	1.30
370–375	15	1.55	1.55	1.55	1.55	1.54	1.53	1.49	1.42	1.32	1.14
375–380	15	1.75	1.75	1.75	1.74	1.74	1.72	1.68	1.61	1.50	1.30
380–385	15	1.45	1.45	1.45	1.45	1.45	1.44	1.41	1.35	1.26	1.09
385–390	15	1.58	1.58	1.58	1.58	1.58	1.56	1.53	1.47	1.37	1.20
390–395	15	1.56	1.56	1.56	1.56	1.56	1.55	1.52	1.46	1.37	1.20
395–400	15	1.88	1.88	1.88	1.88	1.88	1.87	1.84	1.77	1.66	1.45
400–405	15	2.60	2.60	2.60	2.60	2.59	2.58	2.54	2.44	2.29	2.01
405–410	15	2.50	2.50	2.50	2.50	2.50	2.48	2.44	2.36	2.22	1.95
410–415	15	2.58	2.58	2.58	2.58	2.58	2.57	2.53	2.44	2.30	2.03
415–420	15	2.60	2.60	2.60	2.60	2.60	2.59	2.55	2.47	2.33	2.06
420–430	15	5.05	5.05	5.06	5.06	5.06	5.04	4.98	4.83	4.56	4.04

(continues)

TABLE 3.17 (continued)

Wavelength interval (nm)	Exponent[b]	Solar zenith angle (deg)									
		0	10	20	30	40	50	60	70	78	86
430–440	15	5.13	5.13	5.13	5.14	5.14	5.12	5.07	4.92	4.67	4.15
440–450	15	5.98	5.98	5.99	6.00	6.00	5.99	5.93	5.77	5.49	4.89
450–460	15	6.47	6.47	6.48	6.48	6.47	6.45	6.38	6.20	5.90	5.27
460–470	15	6.49	6.49	6.49	6.50	6.49	6.47	6.41	6.24	5.95	5.33
470–480	15	6.54	6.54	6.54	6.55	6.54	6.53	6.47	6.31	6.02	5.41
480–490	15	6.34	6.34	6.34	6.35	6.35	6.33	6.28	6.13	5.86	5.28
490–500	15	6.50	6.49	6.50	6.50	6.50	6.49	6.44	6.30	6.03	5.44
500–510	15	6.51	6.51	6.50	6.50	6.48	6.46	6.39	6.23	5.96	5.38
510–520	15	6.33	6.33	6.33	6.32	6.31	6.28	6.22	6.08	5.82	5.26
520–530	15	6.52	6.52	6.51	6.51	6.49	6.47	6.40	6.26	6.00	5.43
530–540	15	6.68	6.71	6.67	6.66	6.65	6.62	6.56	6.41	6.15	5.57
540–550	15	6.63	6.63	6.63	6.62	6.60	6.58	6.51	6.37	6.11	5.55
550–560	15	6.73	6.73	6.72	6.71	6.69	6.65	6.58	6.43	6.16	5.60
560–570	15	6.76	6.75	6.74	6.73	6.71	6.67	6.59	6.44	6.18	5.62
570–580	15	6.82	6.81	6.80	6.79	6.76	6.73	6.65	6.50	6.24	5.68
580–600	16	1.36	1.36	1.36	1.36	1.35	1.35	1.33	1.30	1.25	1.14
600–620	16	1.36	1.36	1.36	1.35	1.35	1.34	1.32	1.29	1.24	1.13
620–640	16	1.36	1.36	1.36	1.35	1.35	1.34	1.33	1.30	1.25	1.14
640–660	16	1.35	1.34	1.34	1.34	1.33	1.32	1.30	1.27	1.23	1.12
660–680	16	1.38	1.38	1.37	1.37	1.36	1.35	1.33	1.29	1.25	1.14
680–700	16	1.35	1.35	1.34	1.34	1.33	1.32	1.30	1.27	1.22	1.12

[a] The authors are grateful to Dr. Sasha Madronich for generously providing these calculations.
[b] This column lists the power of 10 by which all entries should be multiplied. For example, at $\theta = 0°$ the total actinic flux in the wavelength interval from 306 to 308 nm is 2.0×10^{14} photons cm^{-2} s^{-1}.

(as opposed to a photophysical process such as fluorescence or energy transfer). For example, once NO_2 has absorbed light and is in an electronically excited state, it can either dissociate or energy transfer to other molecules in air, most commonly N_2 or O_2, and return to the ground state:

$$NO_2 + h\nu \rightarrow NO_2^* \quad (6)$$

$$NO_2^* \rightarrow NO + O(^3P) \quad (7)$$

$$NO_2^* + M\ (N_2, O_2) \rightarrow NO_2 + M. \quad (8)$$

Only reaction (7) leads to the removal of NO_2 via photochemistry and hence the quantum yield for reaction (7) is needed to calculate the photolysis rate. Data on both primary quantum yields and absorption cross section $\phi(\lambda)$, characteristic of each molecule, are found in Chapter 4.

It must again be stressed that the absorption cross sections, $\sigma(\lambda)$, used to calculate photolysis rates are to the base e, not base 10, even though the latter is what has often been measured and reported in the literature in the past.

The actinic flux $F(\lambda)$, describing the intensity of light available to the molecule for absorption, depends on many factors, including geographical location, time, season, presence or absence of clouds, and the total amount of O_3 and particles in the air which scatter light as it passes through the atmosphere. At the earth's surface, however, the actinic flux estimates and associated data of Madronich (1998) in Table 3.7 are commonly used to estimate rates and lifetimes of species with respect to photolysis under cloudless conditions.

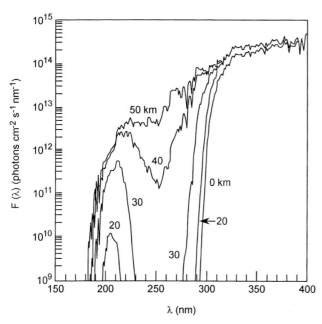

FIGURE 3.32 Calculated actinic fluxes as a function of altitude for a solar zenith angle of 30° and a surface albedo of 0.3. (From DeMore et al., 1997.)

Because the actinic flux data are reported as averages over certain wavelength intervals, rather than integrating over Eq. (OO) in a continuous manner, in practice one calculates the sum of the product $\phi(\lambda)\sigma(\lambda)F(\lambda)$ over discrete wavelength intervals $\Delta\lambda$. The intervals are chosen to match the available flux data; for example, in Table 3.7, actinic fluxes are reported as averages over 2-nm intervals from 290 to 320 nm, which is important for the O_3 absorption, 5-nm intervals from 320 to 420 nm, 10-nm intervals from 420 to 580 nm, and 20-nm intervals from 580 to 700 nm. Since the primary quantum yield, $\phi(\lambda)$, and the absorption cross section, $\sigma(\lambda)$, are not normally reported over identical intervals, representative averages of these parameters over the same intervals for which the actinic flux data are reported must be calculated from the literature data.

In the most commonly used form, then, Eq. (OO) becomes:

$$k_p(s^{-1}) = \sum_{\lambda=290\,nm}^{\lambda_i} \phi_{av}(\lambda)\sigma_{av}(\lambda)F_{av}(\lambda), \quad (PP)$$

where $\phi_{av}(\lambda)$ is the primary quantum yield for the photolysis of the molecules averaged over the wavelength interval $\Delta\lambda$, centered at λ, $\sigma_{av}(\lambda)$ is the absorption cross section, *base e*, averaged over the wavelength interval $\Delta\lambda$, centered at λ, and $F_{av}(\lambda)$ is the actinic flux in photons cm^{-2} s^{-1} summed over the wavelength interval $\Delta\lambda$, centered at λ, at a solar zenith angle θ (Table 3.7) corrected for season (Table 3.8). If desired, corrections for surface elevation, altitude, etc. can be included. Note that the values in the tables are the *total* actinic fluxes over the wavelength intervals given. They are *not* per nm.

The sum (or integral if Eq. (OO) is used) is carried out from the lower limit of wavelengths in the troposphere, 290 nm, to some wavelength λ_i at which either the primary quantum yield or the absorption cross section becomes negligible.

Experimentally, while the determination of absorption cross sections is fairly straightforward, measuring primary quantum yields is not, due to interference from rapid secondary reactions. As a result, in cases where quantum yield data are not available, calculations of *maximum* rates of photolysis are often carried out in which it is assumed that $\phi(\lambda) = 1.0$. It should be emphasized in such cases that this represents only a maximum rate constant for photolysis; the true rate constant may be much smaller, even zero, if photophysical fates of the excited molecule such as fluorescence or quenching predominate.

TABLE 3.18 Absorption Cross Sections, $\sigma(\lambda)$, and Primary Quantum Yields, $\phi(\lambda)$, for Reactions (9a) and (9b) for CH_3CHO at Room Temperature and 1 atm in Air[a]

Wavelength, λ (nm)	Absorption cross section, σ (10^{-20} cm^2 molecule^{-1} s^{-1})	Quantum yield	
		(9a)	(9b)
290	4.89	0.53	0.01
292	4.68		
294	4.33		
295	4.27	0.48	0.0
296	4.24		
298	4.41		
300	4.16	0.43	
302	3.86		
304	3.48		
305	3.42	0.37	
306	3.42		
308	3.33		
310	2.93		
312	2.53		
314	2.44		
315	2.20	0.17	
316	2.04		
318	1.98		
320	1.72	0.10	
325	1.14	0.04	
330	0.688	0.0	

[a] Atkinson *et al.* (1997).

4. Example: Photolysis of Acetaldehyde at the Earth's Surface

To illustrate the application of Eqs. (OO) and (PP), let us calculate the rate of photolysis of acetaldehyde. Aldehydes such as CH_3CHO play an important role in tropospheric chemistry because they photodissociate to produce free radicals. In the case of acetaldehyde, there are four possible sets of products:

$$CH_3CHO + h\nu \rightarrow CH_3 + HCO, \quad (9a)$$
$$\rightarrow CH_4 + CO, \quad (9b)$$
$$\rightarrow H + CH_3CO, \quad (9c)$$
$$\rightarrow H_2 + CH_2CO. \quad (9d)$$

For the purposes of illustration, the rate of photolysis will be calculated for conditions of a cloudless day at the earth's surface at 30°N latitude six hours after noon on July 1.

From Eq. (PP), we need the primary quantum yields for each of the processes (9a)–(9d), the absorption cross sections, *base e*, and the actinic flux values, $F(\lambda)$. Table 3.18 gives IUPAC recommendations (Atkinson *et al.*, 1997) for the absorption cross sections and primary quantum yields for CH_3CHO. Primary quantum

TABLE 3.19 Calculated Photolysis Rate Constants for CH_3CHO Photolysis at 30°N Latitude Six Hours after Noon on July 1

Wavelength interval, $\Delta\lambda$ (nm)	Actinic flux, $F_{av}(\lambda)$ (10^{14} photons $cm^{-2}\,s^{-1}$)	Absorption cross section, $\sigma_{av}(\lambda)$ ($10^{-20}\,cm^2$ molecule^{-1})	Quantum yield for reactions (9a) and (9b) $\phi_{av}^{9a}(\lambda)$	$\phi_{av}^{9b}(\lambda)$	$\phi_{av}^{9a}\sigma_{av}DF_{av}$ ($10^{-6}\,s^{-1}$)^a	$\phi_{av}^{9b}\sigma_{av}DF_{av}$ ($10^{-6}\,s^{-1}$)^a
290–292	0	4.78	0.52	0.01	0	0
292–294	0	4.51	0.50	0	0	0
294–296	0	4.27	0.48	0	0	0
296–298	0	4.33	0.46	0	0	0
298–300	0	4.29	0.44	0	0	0
300–302	0	4.01	0.42	0	0	0
302–304	0	3.67	0.40	0	0	0
304–306	0	3.42	0.37	0	0	0
306–308	0	3.38	0.33	0	0	0
308–310	0.01	3.13	0.27	0	0.008	0
310–312	0.02	2.73	0.25	0	0.013	0
312–314	0.04	2.49	0.21	0	0.020	0
314–316	0.06	2.20	0.17	0	0.022	0
316–318	0.10	2.01	0.14	0	0.027	0
318–320	0.13	1.85	0.11	0	0.026	0
320–325	0.52	1.43	0.07	0	0.050	0
325–330	0.96	0.914	0.02	0	0.017	0

Totals^a: $\Sigma\phi_{av}^{9a}\sigma_{av}DF_{av} = 0.183 \times 10^{-6}\,s^{-1} = k_p^{9a}$ $\Sigma\phi_{av}^{9b}\sigma_{av}DF_{av} = 0\,s^{-1} = k_p^{9b}$

^{a} D = earth–sun correction distance of 0.966.

yields for (9c) and (9d) are not given because they are sufficiently small that they can be ignored. Hence we only need to consider reactions (9a) and (9b).

As is typical for such data, values at *specific* wavelengths are given, rather than averages over the wavelength *intervals* for which the actinic flux data are provided in Table 3.7. Hence it is necessary to use the data in Table 3.18 to obtain averages over the appropriate wavelength intervals. One can plot the data in Table 3.18 as a function of wavelength and integrate over each interval to obtain the appropriate average. Indeed, the absorption cross sections are given every nm by Atkinson *et al.* (1997). In the case of slowly varying functions, a reasonable approximation can be obtained by using a simple numerical average of the values at each end of the wavelength interval. The averages over the appropriate intervals obtained in this manner are shown in Table 3.19.

To use the actinic flux data in Table 3.7, we need to know the solar zenith angle at 30°N latitude six hours after noon on July 1. We can obtain this from Table 3.9. Since the values are symmetrical about noon, we use the data for 0600 on July 1 at 30°N latitude, which from Table 3.9 corresponds to a solar zenith angle of 78.7°. To obtain the actinic fluxes corresponding to this solar zenith angle, one returns to Table 3.7 and interpolates between the 78° and 86° values given in the last two columns. For simplicity in presenting these calculations, we use the values given for 78°. A correction for the earth–sun distance must also be made using the data of Table 3.8. For July 1, this correction factor is 0.966; i.e., the actinic flux values in Table 3.7 must be multiplied by D = 0.966.

Table 3.19 summarizes these corrected values of $F(\lambda)$ for the wavelength region of interest for CH_3CHO.

Once the actinic fluxes, quantum yields, and absorption cross sections have been summarized as in Table 3.19, the individual products $\phi_{av}(\lambda)\sigma_{av}(\lambda)F(\lambda)$ for each wavelength interval can be calculated and summed to give k_p. Note that the individual reaction channels (9a) and (9b) are calculated separately and then added to get the total photolysis rate constant for the photolysis of acetaldehyde. However, the rate constants for the individual channels are also useful in that (9a) produces free radicals that will participate directly in the NO to NO_2 conversion and hence in the formation of O_3, etc., while (9b) produces relatively unreactive stable products.

In this case, the photolysis rate constant for reaction (9b) is zero because the quantum yield drops off rapidly to zero above 290 nm, and at this large solar zenith angle, the actinic flux is negligible at wavelengths below 300 nm. Hence only reaction (9a), leading to free radicals, contributes to the photolysis of CH_3CHO under these conditions. Adding up the individual contributions at each wavelength interval, one obtains $k_p^{tot} = k_p^{9a} = 1.83 \times 10^{-7}$ s^{-1}. Hence the rate of loss of acetaldehyde by photolysis is given by:

$$\frac{d[CH_3CHO]}{dt} = -k_p[CH_3CHO],$$

where $k_p = 1.83 \times 10^{-7}$ s^{-1}, and in this particular case, the loss proceeds entirely via one channel, (9a).

As described in Chapter 5, the *natural lifetime* for acetaldehyde with respect to photolysis under these conditions can be calculated from k_p for the overall reaction. The natural lifetime, τ, is defined as the time for the concentration of CH_3CHO to fall to $1/e$ of its initial value, where e is the base of natural logarithms. The natural lifetime of acetaldehyde under these conditions is therefore given by $\tau = 1/k_p = 5.5 \times 10^6$ s = 63 days. Of course, these conditions do not exist for 63 days, so the lifetime is hypothetical. However, it does provide a sort of "back-of the envelope" method of assessing the relative rapidity of loss of the compound by photolysis compared to other processes, such as reaction with OH.

D. PROBLEMS

For the following problems, where necessary use absorption cross sections and quantum yield data found in the relevant sections of Chapter 4.

1. Following the procedure outlined in Section 3.C.1a, calculate the solar zenith angle for your city or town at the following times: (a) noon on January 1; (b) 8:00 a.m. on March 15 ("Beware the ides of March..."); (c) noon on June 21; (d) 3:30 p.m. on September 1; (e) 9:00 a.m. on December 21. The latitudes and longitudes for various locations can be found, for example, in the *Rand McNally International Atlas*.

2. Following the procedure in Section 3.C.2c, calculate the earth-sun correction factors for the actinic flux on the following dates: (a) January 1; (b) February 25; (c) April 15; (d) June 8; (e) September 1; (f) November 15. You can check some of these answers against Table 3.8.

3. Calculate the photolysis rate constant, k_p, for the photolysis of Cl_2 at solar zenith angles of (a) 10, (b) 30, (c) 50, (d) 70, and (e) 86°, respectively, at the surface of the earth on January 1.

4. Calculate the photolysis rate constants, k_p, for each of the two photolysis paths as well as the overall photolysis rate constant for the photolysis of $ClONO_2$ at solar zenith angles of (a) 0, (b) 40, and (c) 78°, respectively, at the surface of the earth on February 25 (see Problem 2 for the earth-sun correction factor). Assume the quantum yield for production of $Cl + NO_3$ is 0.65 and that for $ClO + NO_2$ is 0.35 from 290 to 320 nm and that for $Cl + NO_3$ is 1.0 from 320 nm to longer wavelengths.

5. Calculate the photolysis rate constant, k_p, for the photolysis of methyl nitrate (CH_3ONO_2) at solar zenith angles of (a) 10, (b) 30, (c) 50, (d) 70, and (e) 86°, respectively, at the surface of the earth on June 8 (see Problem 2 for the earth-sun correction factor).

6. Choose one of the following molecules: (a) $CH_3CH_2ONO_2$, (b) 2-propyl nitrate, (c) PAN, (d) $ClONO_2$, (e) $BrONO_2$, (f) HOCl, (g) HOBr, (h) H_2O_2, (i) CH_3OOH, (j) N_2O_5, (k) HNO_3, (l) HO_2NO_2, (m) CH_3I. Calculate the photolysis rate constant at 11:00 a.m. true solar time at 30°N on March 1 at the earth's surface. Assume 298 K unless data are not given for that temperature, in which case use the closest temperature for which data are given. Use the data given for the "best estimate" surface albedo. If there are missing parameters, e.g., quantum yields, state clearly the assumptions you are making. If the solar zenith angle is within a degree of those for which data are given, you may use the flux values given without interpolation.

7. For the molecules given in Problem 6, calculate the photolysis rate constant at the earth's surface at 50°N on April 1 at a time 1.5 h before the sun crosses the meridian. Assume 298 K unless data are not given for that temperature, in which case use the closest temperature for which data are given. Use the data given for the "best estimate" surface albedo. If there are missing parameters, e.g., quantum yields, state clearly the assumptions you are making. If the solar zenith angle is within a degree of those for which data are given, you may use the flux values given without interpolation.

8. Use the data in Tables 3.7, 3.15, 3.16, and 3.17 to calculate the ratio of the actinic flux for "best estimate" albedo at 40, 25, and 15 km to that at 0 km for the wavelength intervals 300–302, 400–405, and 500–510 nm, respectively. Comment on the reason for the differences between the three wavelength regions.

9. Use the data in Tables 3.7 and 3.11 to calculate the ratio of the actinic flux at the earth's surface for an 80% surface albedo compared to the "best estimate" albedo at solar zenith angles of 0 and 78° for the following wavelength regions: 298–300, 318–320, and 400–405 nm. Comment on the expected effects on photochemistry in the boundary layer.

10. Comment on why the percentage increase in actinic flux at the surface for a 5% decrease in O_3 is greater at larger solar zenith angles (Table 3.12).

11. The surface albedo can have a large effect on the total light available for photolysis in the troposphere. Calculate the factor by which the photolysis of H_2O_2 would increase at a solar zenith angle of 60° on December 1 over snow with a surface albedo of 80% compared to a normal "best estimate" surface albedo.

References

Angle, R. P., M. Brennand, and H. S. Sandhu, "Surface Albedo Measurements at 53°N Latitude," *Atmos. Environ., 26A,* 1545–1547 (1992).

Atkinson, R., D. L. Baulch, R. A. Cox, R. F. Hampson, J. A. Kerr, M. J. Rossie, and J. Troe, "Evaluated Kinetic and Photochemical Data for Atmospheric Chemistry. Supplement V. IUPAC Subcommittee on Gas Kinetic Data Evaluation for Atmospheric Chemistry," *J. Phys. Chem. Ref. Data, 26,* 521–1011 (1997).

Bordewijk, J. A., H. Slaper, H. A. J. M. Reinen, and E. Schlamann, "Total Solar Radiation and the Influence of Clouds and Aerosols on the Biologically Effective UV," *Geophys. Res. Lett., 22,* 2151–2154 (1995).

Brauers, T., and A. Hofzumahaus, "Latitudinal Variation of Measured NO_2 Photolysis Frequencies over the Atlantic Ocean between 50°N and 30°S," *J. Atmos. Chem., 15,* 269–282 (1992).

Calvert, J. G., and J. N. Pitts, Jr., *Photochemistry,* Wiley, New York, 1966.

Castro, T., L. G. Ruiz-Suarez, J. C. Ruiz-Suarez, M. J. Molina, and M. Montero, "Sensitivity Analysis of a UV Radiation Transfer Model and Experimental Photolysis Rates of NO_2 in the Atmosphere of Mexico City," *Atmos. Environ., 31,* 609–620 (1997).

Crosley, D. R., "Rotational and Translational Effects in Collisions of Electronically Excited Diatomic Hydrides," *J. Phys. Chem., 93,* 6273–6282 (1989).

Dave, J. V., *Development of Programs for Computing Characteristics of Ultraviolet Radiation,* Final Report under Contract NAS 5-21680, NASA Report CR-139134, National Aeronautics and Space Administration, Goddard Space Flight Center, Greenbelt, MD, NTIS No. N75-10746/6SL, 1972.

Davis, H. F., B. Kim, H. S. Johnston, and Y. T. Lee, "Dissociation Energy and Photochemistry of NO_3," *J. Phys. Chem., 97,* 2172–2180 (1993).

Demerjian, K. L., K. L. Schere, and J. T. Peterson, "Theoretical Estimates of Actinic (Spherically Integrated) Flux and Photolytic Rate Constants of Atmospheric Species in the Lower Troposphere," *Adv. Environ. Sci. Technol., 10,* 369–459 (1980).

DeMore, W. B., S. P. Sander, D. M. Golden, R. F. Hampson, M. J. Kurylo, C. J. Howard, A. R. Ravishankara, C. E. Kolb, and M. J. Molina, "Chemical Kinetics and Photochemical Data for Use in Stratospheric Modeling," in JPL Publication 97-4, Jet Propulsion Laboratory, Pasadena, CA, January 15, 1997.

Dickerson, R. R., D. H. Stedman, and A. C. Delany, "Direct Measurements of Ozone and Nitrogen Dioxide Photolysis Rates in the Troposphere," *J. Geophys. Res., 87,* 4933–4946 (1982).

Dickerson, R. R., S. Kondragunta, G. Stenchikov, K. L. Civerolo, B. G. Doddridge, and B. N. Holben, "The Impact of Aerosols on Solar Ultraviolet Radiation and Photochemical Smog," *Science, 278,* 827–830 (1997).

Eck, T. F., P. K. Bhartia, P. H. Hwang, and L. L. Stowe, "Reflectivity of Earth's Surface and Clouds in Ultraviolet from Satellite Observations," *J. Geophys. Res., 92,* 4287–4296 (1987).

Erlick, C., and J. E. Frederick, "Effects of Aerosols on the Wavelength Dependence of Atmospheric Transmission in the Ultraviolet and Visible. 2. Continental and Urban Aerosols in Clear Skies," *J. Geophys. Res., 103,* 23275–23285 (1998).

Frederick, J. E., A. E. Koob, A. D. Alberts, and E. C. Weatherhead, "Empirical Studies of Tropospheric Transmission in the Ultraviolet: Broadband Measurements," *J. Appl. Meteorol., 32,* 1883–1892 (1993).

Friedman, H., in *Physics of the Upper Atmosphere* (J. A. Ratcliffe, Ed.), Academic Press, New York, 1960.

Gilbert, A., and J. Baggott, *Essentials of Molecular Photochemistry,* CRC Press, Boca Raton, FL, 1991.

Goody, R. M., and Y. L. Yung, *Atmospheric Radiation—Theoretical Basis,* Oxford Univ. Press, New York, 1989.

Halthore, R. N., S. E. Schwartz, J. J. Michalsky, G. P. Anderson, R. A. Ferrare, B. N. Holben, and H. M. Ten Brink, "Comparison of Model Estimated and Measured Direct-Normal Solar Irradiance," *J. Geophys. Res., 102,* 29991–30002 (1997).

Herman, J. R., and E. A. Celarier, "Earth Surface Reflectivity Climatology at 340–380 nm from TOMS Data," *J. Geophys. Res., 102,* 28003–28011 (1997).

Herzberg, G. *Molecular Spectra and Molecular Structure, I. Spectra of Diatomic Molecules,* 1950; *II. Infrared and Raman Spectra of Polyatomic Molecules,* 1945; *III. Electronic Spectra and Electronic Structure of Polyatomic Molecules,* Van Nostrand, Princeton, NJ, 1967.

Howard, J. N., J. I. F. King, and P. R. Gast, "Thermal Radiation," *Handbook of Geophysics,* Chap. 16. Macmillan, New York (1960).

Huber, K., and G. Herzberg, *Molecular Spectra and Molecular Structure. IV. Constants of Diatomic Molecules,* Van Nostrand, Princeton, NJ, 1979.

Ishiwata, T., I. Fujiwara, Y. Naruge, K. Obi, and I. Tanaka, "Study of NO_3 by Laser-Induced Fluorescence," *J. Phys. Chem., 87,* 1349–1352 (1983).

Jacobson, M. Z., "Studying the Effects of Aerosols on Vertical Photolysis Rate Coefficient and Temperature Profiles over an Urban Airshed," *J. Geophys. Res., 103,* 10593–10604 (1998).

Johnston, H. S., H. F. Davis, and Y. T. Lee, "NO_3 Photolysis Product Channels: Quantum Yields from Observed Energy Thresholds," *J. Phys. Chem., 100,* 4713–4723 (1996).

Junkermann, W. "Measurements of the $J(O^1D)$ Actinic Flux Within and Above Stratiform Clouds and Above Snow Surfaces", *Geophys. Res. Lett. 21,* 793–796 (1994).

Junkermann, W., U. Platt, and A. Volz-Thomas, "A Photoelectric Detector for the Measurement of Photolysis Frequencies of Ozone and Other Atmospheric Molecules," *J. Atmos. Chem., 8,* 203–227 (1989).

Kato, S., T. P. Ackerman, E. E. Clothiaux, J. H. Mather, G. G. Mace, M. L. Wesely, F. Murcray, and J. Michalsky, "Uncertainties in Modeled and Measured Clear-Sky Surface Shortwave Irradiances," *J. Geophys. Res., 102,* 25881–25898 (1997).

Kelley, P., R. R. Dickerson, W. T. Luke, and G. L. Kok, "Rate of NO_2 Photolysis from the Surface to 7.6 km Altitude in Clear-Sky and Clouds," *Geophys. Res. Lett., 22,* 2621–2624 (1995).

Kraus, A., and A. Hofzumahaus, "Field Measurements of Atmospheric Photolysis Frequencies for O_3, NO_2, HCHO, CH_3CHO, H_2O_2, and HONO by UV Spectroradiometry," *J. Atmos. Chem., 31,* 161–180 (1998).

Krotkov, N. A., P. K. Bhartia, J. R. Herman, V. Fioletov, and J. Kerr, "Satellite Estimation of Spectral Surface UV Irradiance in the Presence of Tropospheric Aerosols. 1. Cloud-Free Case," *J. Geophys. Res., 103,* 8779–8793 (1998).

Kylling, A., A. Albold, and G. Seckmeyer, "Transmittance of a Cloud Is Wavelength-Dependent in the UV-Range: Physical Interpretation," *Geophys. Res. Lett., 24*, 397–400 (1997).

Lantz, K. O., R. E. Shetter, C. A. Cantrell, S. J. Flocke, J. G. Calvert, and S. Madronich, "Theoretical, Actinometric, and Radiometric Determinations of the Photolysis Rate Coefficient of NO_2 during the Mauna Loa Observatory Photochemistry Experiment 2," *J. Geophys. Res., 101*, 14613–14629 (1996).

Lean, J., "Variations in the Sun's Radiative Output," *Rev. Geophys., 29*, 505–535 (1991).

Leighton, P. A., *Photochemistry of Air Pollution*, Academic Press, New York, 1961.

Lenoble, J., *Atmospheric Radiative Transfer*, A. Deepak Publishing, Hampton, VA, 1993.

Liou, K.-N., *An Introduction to Atmospheric Radiation*, Academic Press, New York, 1980.

Madronich, S. "Photodissociation in the Atmosphere. 1. Actinic Flux and the Effects of Ground Reflections and Clouds," *J. Geophys. Res., 92*, 9740–9752 (1987).

Madronich, S., "The Atmosphere and UV-B Radiation at Ground Level," in *Environmental UV Photobiology* (A. R. Young, Ed.), pp. 1–39, Plenum, New York, 1993.

Madronich, S., personal communication (1998). We are most grateful to Dr. Madronich for providing the results of his calculations for use by the atmospheric chemistry community.

Madronich, S., D. R. Hastie, B. A. Ridley, and H. I. Schiff, "Measurement of the Photodissociation Coefficient of NO_2 in the Atmosphere. I. Method and Surface Measurements," *J. Atmos. Chem., 1*, 3–25 (1983).

Mauldin, R. L., III, S. Madronich, S. J. Flocke, and F. L. Eisele, "New Insights on OH: Measurements around and in Clouds," *Geophys. Res. Lett., 24*, 3033–3036 (1997).

Mayer, B., A. Kylling, S. Madronich, and G. Seckmeyer, "Enhanced Absorption of UV Radiation Due to Multiple Scattering in Clouds: Experimental Evidence and Theoretical Explanation," *J. Geophys. Res., 103*, 31241–31254 (1998).

McDermid, I. S., and J. B. Laudenslager, "Radiative Lifetimes and Quenching Rate Coefficients for Directly Excited Rotational Levels of $OH(A^2\Sigma^+, v' = 0)$, *J. Chem. Phys., 76*, 1824–1831 (1982).

McLinden, C. A., J. C. McConnell, E. Griffoen, C. T. McElroy, and L. Pfister, "Estimating the Wavelength-Dependent Ocean Albedo under Clear-Sky Conditions Using NASA ER 2 Spectroradiometer Measurements," *J. Geophys. Res., 102*, 18801–18811 (1997).

Molina, L. T., and M. J. Molina, "Absolute Absorption Cross Sections of Ozone in the 185- to 350-nm Wavelength Range," *J. Geophys. Res., 91*, 14501–14508 (1986).

Müller, R., and U. Schurath, "Entwicklung eines Gerätes zur kontinuierlichen Messung der Photodissoziations-Geschwindigkeit von Aldehyden in der Atmosphäre durch Nachweis des erzeugten CO," Gesellschaft für Strahlen-und Umweltforschung mbH München, Abschlussbericht Vorhaben KBF 53 (1986).

Neckel, H., and D. Labs, "The Solar Radiation between 3300 and 12500 Å," *Solar Phys., 90*, 205–258 (1984).

Nelson, H. H., L. Pasternack, and J. R. McDonald, "Laser-Induced Excitation and Emission Spectra of NO_3," *J. Phys. Chem., 87*, 1286–1288 (1983).

Okabe, H. *Photochemistry of Small Molecules*, Wiley, New York, 1978.

Orlando, J. J., G. S. Tyndall, G. K. Moortgat, and J. G. Calvert, "Quantum Yields for NO_3 Photolysis between 570 and 635 nm," *J. Phys. Chem., 97*, 10996–11000 (1993).

Parrish, D. D., P. C. Murphy, D. L. Albritton, and F. C. Fehsenfeld, "The Measurement of the Photodissociation Rate of NO_2 in the Atmosphere," *Atmos. Environ., 17*, 1365–1379 (1983).

Peterson, J. T., "Calculated Actinic Fluxes (290–700 nm) for Air Pollution Photochemistry Applications," U.S. Environmental Protection Agency Report No. EPA-600/4-76-025, June 1976.

Ruggaber, A., R. Dlugi, A. Bott, R. Forkel, H. Herrmann, and H.-W. Jacobi, "Modelling of Radiation Quantities and Photolysis Frequencies in the Aqueous Phase in the Troposphere," *Atmos. Environ., 31*, 3137–3150 (1997).

Schwander, H., P. Koepke, and A. Ruggaber, "Uncertainties in Modeled UV Irradiances Due to Limited Accuracy and Availability of Input Data," *J. Geophys. Res., 102*, 9419–9429 (1997).

Seckmeyer, G., B. Mayer, G. Bernhard, R. L. McKenzie, P. V. Johnston, M. Kotkamp, C. R. Booth, T. Lucas, T. Mestechkina, C. R. Roy, H. P. Gies, and D. Tomlinson, "Geographical Differences in the UV Measured by Intercompared Spectroradiometers," *Geophys. Res. Lett., 22*, 1889–1892 (1995).

Seckmeyer, G., R. Erb, and A. Albold, "Transmittance of a Cloud is Wavelength-Dependent in the UV-Range," *Geophys. Res. Lett., 23*, 2753–2755 (1996).

Shetter, R. E., A. H. McDaniel, C. A. Cantrell, S. Madronich, and J. G. Calvert, "Actinometer and Eppley Radiometer Measurements of the NO_2 Photolysis Rate Coefficient during the Mauna Loa Observatory Photochemistry Experiment," *J. Geophys. Res., 97*, 10349–10359 (1992).

Shetter, R. E., C. A. Cantrell, K. O. Lantz, S. J. Flocke, J. J. Orlando, G. S. Tyndall, T. M. Gilpin, C. A. Fischer, S. Madronich, and J. G. Calvert, "Actinometric and Radiometric Measurement and Modeling of the Photolysis Rate Coefficient of Ozone to $O(^1D)$ during Mauna Loa Observatory Photochemistry Experiment. 2," *J. Geophys. Res., 101*, 14631–14641 (1996).

Spencer, J. W., "Fourier Series Representation of the Position of the Sun," *Search, 2*, 172 (1971).

Stamnes, K., S.-C. Tsay, W. Wiscombe, and K. Jayaweera, "Numerically Stable Algorithm for Discrete-Ordinate-Method Radiative Transfer in Multiple Scattering and Emitting Layered Media," *Appl. Opt., 27*, 2502–2509 (1988).

Turro, N. J., *Modern Molecular Photochemistry*, Benjamin/Cummings, Menlo Park, CA, 1978.

Vilà-Guerau de Arellano, J., P. G. Duynkerke, and M. van Weele, "Tethered-Balloon Measurements of Actinic Flux in a Cloud-Capped Marine Boundary Layer," *J. Geophys. Res., 99*, 3699–3705 (1994).

Volz-Thomas, A., A. Lerner, H.-W. Pätz, M. Schultz, D. S. McKenna, R. Schmitt, S. Madronich, and E. P. Röth, "Airborne Measurements of the Photolysis Frequency of NO_2," *J. Geophys. Res., 101*, 18613–18627 (1996).

Wayne, R. P., *Principles and Applications of Photochemistry*, Oxford University Press, Oxford, 1988.

Wen, G., and J. E. Frederick, "The Effects of Horizontally Extended Clouds on Backscattered Ultraviolet Sunlight," *J. Geophys. Res., 100*, 16387–16393 (1995).

CHAPTER

4

Photochemistry of Important Atmospheric Species

Absorption of sunlight induces photochemistry and generates a variety of free radicals that drive the chemistry of the troposphere as well as the stratosphere. This chapter focuses on the absorption spectra and photochemistry of important atmospheric species. These data can be used in conjunction with the actinic fluxes described in the preceding chapter to estimate rates of photolysis of various molecules as well as the rate of generation of photolysis products, including free radicals, from these photochemical processes.

There are several highly useful sources of data on the absorption spectra and photochemistry of atmospheric species. NASA publishes on a regular basis a summary of kinetics and photochemical data directed to stratospheric chemistry (DeMore et al., 1997). However, much of the data is also relevant to the troposphere. This document can be obtained from the Jet Propulsion Laboratory in Pasadena, California. Alternatively, the data are available through the Internet (see Appendix IV). IUPAC also publishes regularly in *The Journal of Physical Chemical Reference Data* a summary directed more toward tropospheric chemistry (Atkinson et al., 1997a, 1997b). Finally, Nölle et al. (1999) have made available a CD-ROM containing the UV–visible spectra of species of atmospheric interest.

We do not attempt a comprehensive treatment of the literature on each of the compounds discussed herein. With apologies to our colleagues whose work may not be explicitly cited, we shall rely on these exhaustive evaluations carried out by NASA (DeMore et al., 1997) and IUPAC (Atkinson et al., 1997a, 1997b) whenever possible. The reader should consult these evaluations, in addition to the original literature after 1998, for details and more recent studies.

A. MOLECULAR OXYGEN

The absorption of light by both molecular oxygen and ozone is a strong determinant of the intensity and wavelength distribution that reaches the troposphere and stratosphere and hence is available to cause photochemical reactions. O_2 absorbs light particularly strongly in the ultraviolet at wavelengths below ~ 200 nm, giving rise to the term "vacuum ultraviolet" for this region; experimental studies involving light in this wavelength range require an optical path from which air has been removed.

1. Absorption Spectra

The potential energy curves for the ground state and for the first four electronically excited states of O_2 are shown in Fig. 4.1. The ground state, $X^3\Sigma_g^-$, is unusual in that it is a triplet; as a result, only transitions to upper triplet states are spin-allowed. The transition from the ground state to the $A^3\Sigma_u^+$ state is also theoretically forbidden because it involves a $(+) \to (-)$ transition (see Chapter 3.A.1); however, this $X^3\Sigma_g^- \to A^3\Sigma_u^+$ transition does occur weakly, resulting in weak absorption bands ($\sigma \leq 10^{-23}$ cm^2 molecule^{-1}) known as the Herzberg continuum at wavelengths between ~ 190 and 300 nm. Further details on the spectroscopy can be found in Slanger and Cosby (1988).

The absorption cross sections between 205 and 240 nm recommended by the NASA evaluation (DeMore et al., 1997) are shown in Table 4.1.

The $X^3\Sigma_g^- \to B^3\Sigma_u^-$ transition is allowed and as seen in Figs. 4.2 and 4.3 results in an absorption in the 130- to 200-nm region known as the Schumann–Runge system. The banded structure from about 175 to 200 nm corresponds to transitions from $v'' = 0$ as well as $v'' = 1$ (i.e., hot bands) of the ground $X^3\Sigma_g^-$ state to different vibrational levels of the upper state.

The upper $^3\Sigma_u^-$ state is crossed by the repulsive $^3\Pi_u$ state (Fig. 4.1) at $\sim v' = 4$, providing a mechanism for the production of two ground-state O(^3P) atoms from the $B^3\Sigma_u^-$ state. The absorption spectrum becomes continuous at ~ 175 nm, with a strong absorption

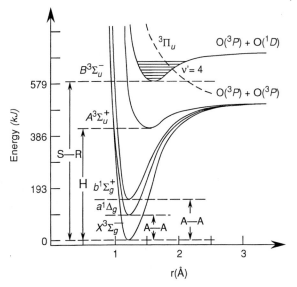

FIGURE 4.1 Potential energy curves for ground and first four excited states of O_2. S–R = Schumann–Runge system, H = Herzberg continuum, A–A = atmospheric bands (adapted from Gaydon, 1968).

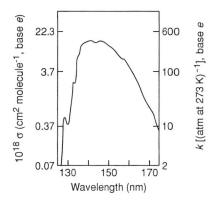

FIGURE 4.2 Absorption coefficients for O_2 in the Schumann–Runge continuum. Note log scale. (Adapted from Inn, 1955.)

down to ~ 130 nm. This continuum is believed to be due to dissociation of the $B^3\Sigma_u^-$ state to $O(^3P) + O(^1D)$. Below 130 nm, a banded absorption again appears, as seen in Fig. 4.4; at wavelengths below 133.2 nm, there is sufficient energy to produce $O(^1S)$ atoms. There is a minimum in the absorption at 121.6 nm,

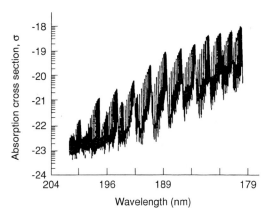

FIGURE 4.3 Semilogarithmic plot (base 10) of measured absorption coefficients in terms of the absorption cross section, σ (cm^2 $molecule^{-1}$), base e, for O_2 at 300 K in the 179.3- to 201.5-nm region. The structure seen for $\sigma > 10^{-22}$ cm^2 $molecule^{-1}$ is real; at smaller cross sections, some noise is present (adapted from Yoshino et al., 1992).

TABLE 4.1 Absorption Cross Sections (Base e) for O_2 between 205 and 240 nm[a]

Wavelength (nm)	$10^{24}\sigma$ (cm^2 $molecule^{-1}$)	Wavelength (nm)	$10^{24}\sigma$ (cm^2 $molecule^{-1}$)
205	7.35	223	3.89
206	7.13	224	3.67
207	7.05	225	3.45
208	6.86	226	3.21
209	6.68	227	2.98
210	6.51	228	2.77
211	6.24	229	2.63
212	6.05	230	2.43
213	5.89	231	2.25
214	5.72	232	2.10
215	5.59	233	1.94
216	5.35	234	1.78
217	5.13	235	1.63
218	4.88	236	1.48
219	4.64	237	1.34
220	4.46	238	1.22
221	4.26	239	1.10
222	4.09	240	1.01

[a] From DeMore et al. (1997) recommendations based on Yoshino et al., Planet. Space Sci., 36, 1469 (1988).

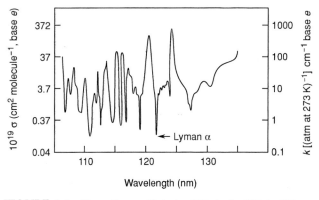

FIGURE 4.4 Absorption coefficients of O_2 in the 105- to 130-nm region. The O_2 absorption line corresponding to the Lyman α line of the H atom is shown by the arrow. Note the log scale: Absorption coefficient k on right is for units of [atm at 273 K]$^{-1}$ cm^{-1}. (Adapted from Inn, 1955.)

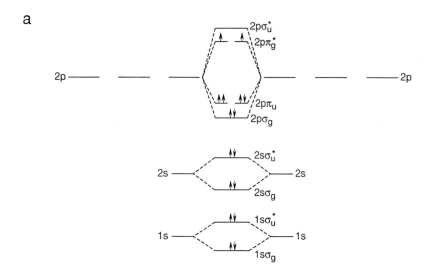
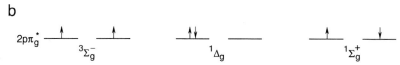

FIGURE 4.5 (a) Molecular orbital diagram of ground $X^3\Sigma_g^-$ state of O_2. (b) Comparison of highest occupied $2p\pi_g^*$ MO for the ground state, $X^3\Sigma_g^-$, and the electronically excited $a^1\Delta_g$ and $b^1\Sigma_g^+$ states.

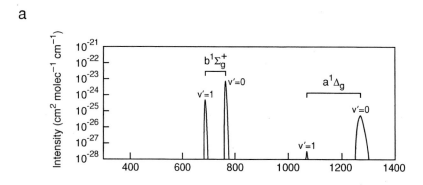
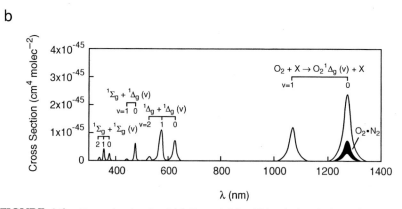

FIGURE 4.6 Absorption bands of (a) O_2 and (b) collision-induced absorptions of O_2 with O_2 and for the 1.26-μm band with N_2, respectively (adapted from Solomon et al., 1998).

coincident with the Lyman α line of the hydrogen atom. This fortunate coincidence allows O_2 to be used in laboratory studies of atomic hydrogen reactions as a filter for Lyman α radiation. For example, if one wishes to detect H atoms by resonance fluorescence or resonance absorption at 121.6 nm (Chapter 5.B), radiation at other wavelengths that might interfere (e.g., from oxygen atoms at 130.5 nm) can be selectively blocked by using a stream of O_2 in front of the detector.

In addition to the absorptions in the ultraviolet, O_2 also has very weak absorptions in the red (762 and 688 nm) and infrared (1.27 and 1.06 μm), known as the atmospheric oxygen bands. These produce O_2 in the singlet excited states $a^1\Delta_g$ and $b^1\Sigma_g^+$, respectively (collectively referred to as *singlet oxygen*). Figure 4.5 shows the difference between these two forms of singlet oxygen and the ground state in terms of a molecular orbital diagram. The two bands at 1.27 and 1.06 μm correspond to transitions to two different vibrational levels, $v' = 0$ (1.27 μm) and $v' = 1$ (1.06 μm) within the $a^1\Delta_g$ state. Similarly, those at 762 and 688 nm correspond to transitions to $v' = 0$ and $v' = 1$, respectively, in the $b^1\Sigma_g^+$ state. See Wayne (1994) for a review of singlet oxygen in the atmosphere.

In addition to these absorptions of O_2 due to transitions between various electronic states, absorptions attributed to van der Waals molecules as well as to collision-induced absorptions have also been reported. van der Waals molecules, or dimers, $(O_2)_2$, are weakly bound species that have been observed spectroscopically at low temperatures. For example, Long and Ewing (1973) investigated the infrared and visible spectra of oxygen at temperatures of ~ 90 K and reported structured absorptions attributable to the dimer, where the binding energy of the dimer was 0.53 kcal mol^{-1}. In addition to the narrower absorptions by such molecules, there are also broad absorptions assigned to collision-induced transitions in O_2; that is, in the brief span of a collision, a dipole moment is induced that leads to a transition that would not occur in the absence of the perturbation associated with the collision. Because of the short collision time, the uncertainty in the energy, i.e., width of the absorption, is correspondingly large, leading to broad, structureless absorptions (e.g., Blickensderfer and Ewing, 1969). A review of the van der Waals and collision-induced absorptions is given by Solomon *et al.* (1998). In any case, because of the relatively large O_2 concentration in air, such complexes have the potential to exist in significant amounts compared to other trace gases (e.g., Calo and Narcisi, 1980; Perner and Platt, 1980).

Figure 4.6 shows the absorption bands due to (a) O_2 and (b) collision complexes of O_2 (Solomon *et al.*, 1998; Greenblatt *et al.*, 1990). The absorption cross sections for the bands in the 455- to 830-nm region have been reported by Newnham and Ballard (1998). These oxygen complexes also absorb in the infrared (e.g., Orlando *et al.*, 1991). The transitions for the collision complexes are from the ground $X^3\Sigma_u^-$ state to the upper states shown on the figure. While most of the bands are due to $O_2 \cdot O_2$ collisions, Solomon *et al.* (1998) postulate that $O_2 \cdot N_2$ at 1.26 μm may also play a role. These collision-induced absorptions absorb a small amount of incoming solar radiation and hence play a role in the earth's radiation balance (e.g., see Murcray *et al.*, 1997; Pfeilsticker *et al.*, 1997; Solomon *et al.*, 1998; and Chapter 14). Perner and Platt (1980) speculated that, depending on their quenching rates in the atmosphere, they could contribute to atmospheric oxidations. Finally, because of the amount of O_2 in air, one must take into account these absorptions when using visible absorption spectroscopy to measure species in these spectral regions (e.g., Volkamer *et al.*, 1998; and Chapter 11.A.1d).

2. Photochemistry

Table 4.2 summarizes the threshold wavelengths for production of ground-state oxygen atoms, $O(^3P)$, as well as electronically excited $O(^1D)$ and $O(^1S)$ atoms. Dissociation of O_2 in the 175- to 242-nm region to produce atoms is particularly important in the stratosphere because it is the only significant source of O_3 via reactions (1) and (2) (see Chapter 12):

$$O_2 + h\nu \rightarrow 2O, \quad (1)$$

$$O + O_2 \xrightarrow{M} O_3. \quad (2)$$

Because 242.4 nm is the wavelength threshold for dissociation of O_2, absorption of longer wavelengths is not expected to result in O_3 formation. However, irradiation of O_2 at 248 nm has been observed to generate O_3 (Slanger *et al.*, 1988; Miller *et al.*, 1994). It appears to involve the initial generation of small amounts of O_3

TABLE 4.2 Threshold Wavelengths for the Production of Ground-State or Electronically Excited Oxygen Atoms from O_2 Photolysis[a]

Electronic state of oxygen atoms[b]	Threshold wavelength (nm)
$O(^3P) + O(^3P)$	242.4
$O(^3P) + O(^1D)$	175.0
$O(^3P) + O(^1S)$	133.2

[a] From Okabe (1978).
[b] $O(^3P)$ is the ground-state species.

through mechanisms as yet not well understood, perhaps via hot-band absorption by $v'' = 1$ of the ground state of O_2, followed by dissociation (Freeman et al., 1990). Once a small amount of O_3 is formed via reactions (1) and (2), it photolyzes in part (see later) to generate O_2 in its ground electronic state, some of which is vibrationally excited, $O_2(v'')$. Mechanisms that have been proposed include the direct absorption of light by this vibrationally excited O_2 to photodissociate (Slanger et al., 1988) or, alternatively, reaction with O_2 to give O_3 and $O(^3P)$, which then forms O_3 via reaction (2) as proposed by Wodtke and co-workers (Price et al., 1993; Rogaski et al., 1993; Miller et al., 1994). In the latter case, levels $v'' \geq 26$ are sufficiently energetic to allow reaction (3) to proceed:

$$O_2(v'' \geq 26) + O_2 \rightarrow O_3 + O(^3P). \quad (3)$$

Supporting this mechanism have been measurements of the production of $O_2(v'' \geq 26)$ in the photolysis of O_3 at shorter wavelengths (< 243 nm) (e.g., Miller et al., 1994; Syage, 1995, 1996a, 1996b; Stranges et al., 1995; Wilson et al., 1997). This is an important issue in the chemistry of the upper atmosphere, in that models have consistently underpredicted O_3 at altitudes of 40–80 km, suggesting that there is an as yet unrecognized source of O_3 at these altitudes.

Although light absorption to excite O_2 directly into the $b^1\Sigma_g^+$ and $a^1\Delta_g$ states is very weak, there are other sources of singlet oxygen, including energy transfer (e.g., from electronically excited NO_2^*), O_3 photolysis, and exothermic chemical reactions. When $O_2(a^1\Delta_g)$ and $O_2(b^1\Sigma_g^+)$ are formed in such processes, they do not readily undergo radiative transitions to the ground state because, as in absorption, the processes are forbidden. Thus the radiative lifetime of the $b^1\Sigma_g^+$ state, which lies 37.5 kcal above the ground state, is ~ 12 s (Wallace and Hunten, 1968), whereas that for the $a^1\Delta_g$ state, 22.5 kcal above the ground state, is ~ 67 min (Slanger and Cosby, 1988). As a result, when these electronic states are formed in the atmosphere, they are primarily collisionally deactivated to ground-state O_2.

Under low-pressure conditions in the laboratory or in the upper atmosphere where collisional deactivation is slow, weak radiative transitions from these two excited singlet states to the ground state are observed. The (0, 0) emission bands of the $b^1\Sigma_g^+ \rightarrow X^3\Sigma_g^-$ and the $a^1\Delta_g \rightarrow X^3\Sigma_g^-$ transitions occur at 761.9 and 1269 nm, respectively. The 761.9-nm band due to $b^1\Sigma_g^+$ is often observed in systems containing the $a^1\Delta_g$ state because of the energy pooling reaction:

$$O_2(a^1\Delta_g) + O_2(a^1\Delta_g) \rightarrow O_2(X^3\Sigma_g^-) + O_2(b^1\Sigma_g^+). \quad (4)$$

B. OZONE

O_3 plays a central role in tropospheric chemistry. Not only is it a highly reactive and toxic species, but it absorbs both infrared and ultraviolet light, contributing to the "greenhouse effect" (see Chapter 14) and providing protection from exposure to damaging UV. In the process of absorbing in the UV, it generates electronically excited oxygen atoms that react to form OH, a ubiquitous atmospheric oxidant.

1. Absorption Spectra

Absorption spectra of O_3 are shown in Figs. 4.7 and 4.8. The strongest absorption, in the 200- to 300-nm region, is known as the Hartley bands. It is this absorption that is responsible for the so-called "actinic cutoff"

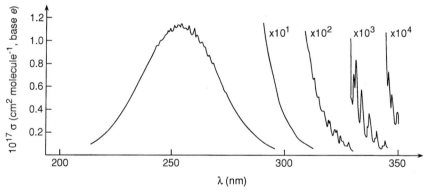

FIGURE 4.7 UV absorption of O_3 at room temperature in the Hartley and Huggins bands. At the longer wavelengths, each curve has been expanded by the factor shown. (Adapted from Daumont et al., 1989.)

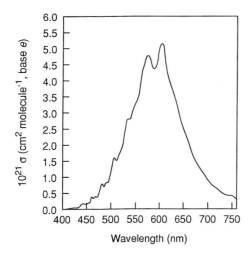

FIGURE 4.8 Absorption cross sections of O_3 in the Chappuis band at room temperature (adapted from Burkholder and Talukdar, 1994).

in the troposphere at 290 nm; thus O_3 in the stratosphere strongly absorbs $\lambda < 290$ nm, limiting the UV reaching the earth's surface.

The Huggins bands are in the 300- to 360-nm region, whereas the 440- to 850-nm region represents the Chappuis bands. As seen in Figs. 4.7 and 4.8, both absorptions are much weaker than the Hartley bands.

Table 4.3 shows the ozone absorption cross sections (base e) at 298 and at 226 K measured by Molina and Molina (1986), with which recent studies are in agreement, particularly below 240 nm (e.g., see Malicet et al., 1995; DeMore et al., 1997). The cross sections below 275 nm are not very sensitive to temperature over the 226–298 K temperature range, typically < 1%, but the temperature dependence becomes significant above 275 nm. Table 4.4 shows the temperature dependence of these absorption cross sections at longer wavelengths, averaged over the spectral regions shown.

2. Photochemistry

The photolysis of O_3 produces molecular oxygen and atomic oxygen, either or both of which may be in electronically excited states, depending on the excitation energy. Table 4.5 shows the wavelength threshold below which each combination of products may be formed. For example, the primary photochemical process

$$O_3 + h\nu \rightarrow O_2(^1\Delta_g) + O(^1D) \quad (5)$$

in principle requires light corresponding to $\lambda = 310$ nm. (However, as discussed below, production of $O(^1D)$ has been observed experimentally out to at least 336 nm.)

In addition to energetic considerations, however, there are other factors such as spin conservation that also determine the importance of various sets of products. As discussed in Chapter 3.A, since the ground state of O_3 is a singlet, dissociation into either two singlet states (e.g., reaction (5)) or into two triplet states is expected to predominate. However, as discussed shortly, both hot-band absorption by rovibrationally excited O_3 and by a spin-forbidden process are believed to contribute significantly to the atmospheric photochemistry of O_3.

The most important aspect of O_3 photochemistry for the troposphere is the yield and wavelength dependence of $O(^1D)$ production in reaction (5) since it is a source of hydroxyl free radicals via its reaction with water vapor:

$$O(^1D) + H_2O_{(g)} \rightarrow 2OH, \quad (6)$$

$$k_6 = 2.2 \times 10^{-10} \text{ cm}^3 \text{ molecule}^{-1} \text{ s}^{-1}$$

(DeMore et al., 1997). Only about 5% of the $O(^1D)$–H_2O interaction results in deactivation of the $O(^1D)$ to the ground state $O(^3P)$, and < 1% in the alternate set of products $H_2 + O_2$ (DeMore et al., 1997).

The very fast reaction of $O(^1D)$ with water vapor occurs in competition with its deactivation by air:

$$O(^1D) + M \rightarrow O(^3P) + M, \quad (7)$$
$$k_7 = 2.6 \times 10^{-11} \text{ cm}^3 \text{ molecule}^{-1} \text{ s}^{-1} \text{ for M} = N_2,$$
$$= 4.0 \times 10^{-11} \text{ cm}^3 \text{ molecule}^{-1} \text{ s}^{-1} \text{ for M} = O_2$$

(DeMore et al., 1997). In 1 atm of air at 50% relative humidity and 298 K, approximately 10% of the $O(^1D)$ produced via reaction (5) reacts with water vapor to form hydroxyl radicals.

The wavelength dependence of the quantum yield for $O(^1D)$ production has been somewhat controversial. The recommended value between 290 and 305 nm is $\phi[O(^1D)] = 0.95$, decreasing slightly at shorter wavelengths to values in the range 0.85–0.90 (see review in DeMore et al., 1997). Talukdar et al. (1998) reported yields of $O(^1D)$ of 0.89 ± 0.04 (2σ), independent of temperature from 203 to 320 K, consistent with the production of small amounts of $O(^3P)$ in this region (e.g., Brock and Watson, 1980). The falloff in the quantum yield at shorter wavelengths is unusual and not well understood in terms of the spectroscopy and electronic states of O_3 (e.g., Steinfeld et al., 1987; Wayne, 1987). Wayne (1987) suggests that the upper electronic states initially populated on light absorption have a curve crossing with a repulsive state that correlates with the ground-state products $O(^3P) + O_2(^3\Sigma_g^-)$ and that the relative positions of the states are such

TABLE 4.3 Ozone Absorption Cross Sections (Base e)[a]

Wavelength (nm)	$10^{20}\sigma$ (cm^2 molecule^{-1})		Wavelength (nm)	$10^{20}\sigma$ (cm^2 molecule^{-1})	
	$T = 226$ K	$T = 298$ K		$T = 226$ K	$T = 298$ K
185.0	64.37	65.37	262.0	1064	1057
186.0	62.59	61.87	263.0	1016	1022
187.0	59.33	59.41	264.0	1013	1006
188.0	56.55	56.59	265.0	961.2	965.7
189.0	54.63	54.24	266.0	946.7	948.5
190.0	51.63	51.14	267.0	877.9	884.1
191.0	48.42	48.80	268.0	872.2	875.4
192.0	45.95	46.06	269.0	803.8	810.4
193.0	43.12	43.36	270.0	796.1	798.0
194.0	40.88	40.66	271.0	736.7	741.5
195.0	38.27	38.64	272.0	710.7	714.7
196.0	36.42	36.73	273.0	666.0	669.8
197.0	34.63	35.00	274.0	601.3	614.0
198.0	33.33	33.49	275.0	587.1	591.3
199.0	32.13	32.09	276.0	537.9	545.0
200.0	31.45	31.54	277.0	504.0	509.6
201.0	31.26	31.15	278.0	461.7	466.8
202.0	31.56	31.79	279.0	424.5	432.6
203.0	32.55	32.51	280.0	398.3	400.1
204.0	34.00	33.65	281.0	364.2	367.3
205.0	36.23	35.85	282.0	322.0	325.0
206.0	38.87	38.55	283.0	299.0	302.5
207.0	42.39	42.00	284.0	264.4	271.2
208.0	46.84	46.40	285.0	239.8	246.5
209.0	51.88	51.18	286.0	219.5	223.8
210.0	58.06	57.16	287.0	197.8	203.4
211.0	65.28	64.02	288.0	168.7	175.0
212.0	73.12	71.94	289.0	153.5	158.5
213.0	82.58	81.04	290.0	136.5	141.8
214.0	92.55	90.96	291.0	122.5	128.5
215.0	104.1	102.3	292.0	104.3	111.1
216.0	116.9	114.6	293.0	93.75	100.2
217.0	131.4	128.7	294.0	81.48	87.11
218.0	146.4	143.9	295.0	72.70	77.53
219.0	163.8	160.1	296.0	62.02	67.27
220.0	179.9	178.5	297.0	54.69	59.55
221.0	200.0	198.2	298.0	47.14	51.24
222.0	221.7	220.0	299.0	42.31	45.51
223.0	244.3	242.9	300.0	36.16	39.64
224.0	268.8	268.4	301.0	31.28	34.63
225.0	296.3	294.3	302.0	27.96	30.73
226.0	323.9	322.6	303.0	23.25	26.50
227.0	354.2	351.3	304.0	21.56	24.01
228.0	385.7	382.9	305.0	17.71	20.15
229.0	416.4	414.1	306.0	16.03	18.08
230.0	450.6	447.6	307.0	13.67	15.65
231.0	485.9	481.4	308.0	12.02	13.64
232.0	523.0	518.1	309.0	10.64	12.43
233.0	558.9	554.9	310.0	8.637	10.20
234.0	592.3	589.1	311.0	7.925	9.260
235.0	634.9	631.8	312.0	6.697	7.947
236.0	677.2	672.2	313.0	5.691	6.883
237.0	714.2	709.4	314.0	5.334	6.294
238.0	753.2	748.6	315.0	4.186	5.199
239.0	784.3	781.9	316.0	3.896	4.792
240.0	836.7	831.4	317.0	3.351	4.146
241.0	864.0	860.3	318.0	3.063	3.757
242.0	901.6	897.1	319.0	1.999	2.765
243.0	939.6	933.3	320.0	2.859	3.243
244.0	975.2	971.7	321.0	1.368	2.041
245.0	1007	993.2	322.0	2.117	2.435
246.0	1042	1033	323.0	1.529	1.983
247.0	1058	1047	324.0	0.7852	1.250
248.0	1079	1071	325.0	1.486	1.727
249.0	1124	1112	326.0	0.7276	1.105
250.0	1134	1124	328.0	1.158	1.300
251.0	1123	1114	330.0	0.2854	0.4923
252.0	1165	1155	332.0	0.2415	0.4347
253.0	1149	1140	334.0	0.3788	0.5343
254.0	1169	1159	336.0	0.0871	0.1875
255.0	1174	1161	338.0	0.1311	0.2086
256.0	1158	1154	340.0	0.1549	0.2082
257.0	1147	1139	342.0	0.0285	0.0776
258.0	1130	1124	344.0	0.1271	0.1419
259.0	1151	1145	346.0	0.0195	0.0641
260.0	1086	1080	348.0	0.0181	0.0386
261.0	1095	1094	350.0	0.0076	0.0294

[a] From Molina and Molina (1986).

TABLE 4.4 Ozone Absorption Cross Sections[a,b] as a Function of Temperature Averaged over the Spectral Intervals Shown

Wavelength range (nm)	Parameters		
	a	b	c
277.778–281.690	4.0293×10^2	$+4.3819 \times 10^{-2}$	0
281.690–285.714	2.7776×10^2	$+6.3125 \times 10^{-2}$	0
285.714–289.855	1.8417×10^2	-9.6665×10^{-2}	2.1026×10^{-4}
289.855–294.118	1.1300×10^2	-1.0700×10^{-1}	3.2697×10^{-4}
294.118–298.507	6.5087×10	-8.0018×10^{-2}	2.2679×10^{-4}
298.507–303.030	3.6161×10	-6.7156×10^{-2}	3.3314×10^{-4}
303.030–307.692	1.9615×10	-4.4193×10^{-2}	2.0338×10^{-4}
307.692–312.5	1.0459×10	-2.8831×10^{-2}	1.3909×10^{-4}
312.5–317.5	5.4715	-2.0092×10^{-2}	9.8870×10^{-5}
317.5–322.5	2.7569	-1.0067×10^{-2}	2.9515×10^{-5}
322.5–327.5	1.3527	-5.7513×10^{-3}	1.1088×10^{-5}
327.5–332.5	6.9373×10^{-1}	-2.9792×10^{-3}	3.1038×10^{-6}
332.5–337.5	3.2091×10^{-1}	-1.9502×10^{-3}	5.6456×10^{-6}
337.5–342.5	1.4484×10^{-1}	-1.1025×10^{-3}	2.8818×10^{-6}
342.5–347.5	7.5780×10^{-2}	-5.7359×10^{-4}	1.6055×10^{-6}

[a] $\sigma(O_3, T) = a + b(T - 230) + C(T - 230)^2$; T is in K; $\sigma(O_3)$ is in units of 10^{-20} cm^2 molecule^{-1} (base e).
[b] From Molina and Molina (1986).

that higher energies are needed for this to occur. Theoretical studies by Banichevich et al. (1993) suggest that there are a number of excited singlet and triplet states that are only weakly bound or are repulsive and correlate with $O(^3P) + O_2(^3\Sigma_g^-)$. As discussed in Section 4.A.2, photolysis at shorter wavelengths, e.g., 226 nm, produces in part ground electronic state O_2 that is in highly vibrationally excited states.

As discussed earlier, the spin-allowed process that produces $O(^1D)$ and $O_2(^1\Delta_g)$ has an apparent energetic threshold corresponding to 310 nm. However, significant yields of $O(^1D)$ have been observed at wavelengths longer than this in a number of experiments (e.g., Arnold et al., 1977; Trolier and Wiesenfeld, 1988; Armerding et al., 1995; Takahashi et al., 1996a, 1996b, 1997; Ball et al., 1997; Silvente et al., 1997; Talukdar et al., 1997, 1998; Ravishankara et al., 1998). For example, Fig. 4.9 shows the quantum yields of $O(^1D)$ from 300 to 330 nm measured in three of these studies (Armerding et al., 1995; Ball et al., 1997; Talukdar et al., 1998). Silvente et al. (1997) observe substantial yields (~0.05) of $O(^1D)$ out to ~336 nm, well beyond the energetic threshold for reaction (5).

It appears that there are two processes that contribute to this $O(^1D)$ production. The first is so-called "hot-band absorption" by O_3 in which the additional energy comes from internal vibrational and rotational energy (Michelsen et al., 1994), a phenomenon that is well established in the case of NO_2 photodissociation (see later). This is believed to be responsible for $O(^1D)$

TABLE 4.5 Wavelength Threshold (nm) Below Which Indicated Reactions Are Energetically Possible in the Photolysis of O_3[a]

Electronic state of oxygen atom	Electronic state of molecular O_2				
	$^3\Sigma_g^-$	$^1\Delta_g$	$^1\Sigma_g^+$	$^3\Sigma_u^+$	$^3\Sigma_u^-$
^3P	1180	612	463	230	173
^1D	411	310	267	168	136
^1S	237	199	181	129	109

[a] From Wayne (1987) and Okabe (1978).

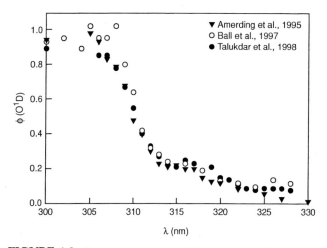

FIGURE 4.9 Some measurements of the quantum yields for production of $O(^1D)$ in the photolysis of O_3 at 298 K.

production in the region from approximately 306 to 324 nm. As expected, if this is the case, the quantum yield in this region decreases with temperature, since the Boltzmann population of the absorbing, vibrationally excited state of O_3 decreases; for example, at 320 nm and 298 K, $\phi[O(^1D)] = 0.15$, but decreases to 0.06 at 203 K (Talukdar et al., 1998).

The second process generating small amounts of $O(^1D)$ is thought to be the spin-forbidden reaction

$$O_3 + h\nu \rightarrow O(^1D) + O_2(^3\Sigma_g^-). \quad (8)$$

This spin-forbidden process appears to be the major source of $O(^1D)$ beyond 325 nm, particularly at low temperatures where the contribution of vibrationally excited O_3 is minimized. Thus, in contrast to the decreasing yields of $O(^1D)$ with temperature in the 306- to 324-nm region, yields in the 325- to 329-nm region are relatively constant with temperature; at lower temperatures, the quantum yield in this region approaches 0.06, which has therefore been assigned by Talukdar et al. (1998) as the upper limit for the production of $O(^1D)$ in this region by reaction (8).

Table 4.6 summarizes the wavelength and temperature dependence for $O(^1D)$ production recommended by Talukdar et al. (1998). Beyond 329 nm, the quantum yield from the spin-forbidden process appears to be $\sim 0.05-0.06$ (Silvente et al., 1997; Talukdar et al., 1998). The absorption cross sections of O_3 become sufficiently small beyond 360 nm that $O(^1D)$ production beyond this wavelength is not expected to be significant for atmospheric applications.

Ball et al. (1993) and Ball and Hancock (1995) measured relative yields of $O_2(^1\Delta_g)$, the other product expected in the spin-allowed reaction (5), and found that it too was produced at significant yields. For example, Fig. 4.10 shows the measured yields of $O_2(^1\Delta_g)$ normalized to $\phi[O(^1D)] = 0.95$ in the 290- to 300-nm region. The yields are similar in magnitude to those for $O(^1D)$. If reaction (8) is responsible for $O(^1D)$ production in the region beyond 325 nm, the yield of $O_2(^1\Delta_g)$ might also be expected to fall to zero. However, $O_2(^1\Delta_g)$ may also be generated by other paths, for example, that giving $O(^3P) + O_2(^1\Delta_g)$, which requires light of 612 nm or shorter wavelengths (Table 4.5).

This is consistent with studies of absorption in the Huggins bands, in which the products appear to be $O_2(^1\Delta_g)$ and ground-state oxygen atoms (Okabe, 1978). Thus, the spin-forbidden process (9) to give electronically excited O_2 and ground-state oxygen atoms must be occurring:

$$O_3 + h\nu \rightarrow O_2(^1\Delta_g \text{ and/or } ^3\Sigma_g^+) + O(^3P). \quad (9)$$

In the Chappuis region (440-850 nm), the products appear to be the ground-state species (Okabe, 1978):

$$O_3 + h\nu \rightarrow O_2(X^3\Sigma_g^-) + O(^3P). \quad (10)$$

In both these regions, the absorption coefficient is one to two orders of magnitude less than that at 300

TABLE 4.6 Parameterization of Quantum Yields for $O(^1D)$ Production from O_3 Photolysis in the 306- to 329-nm Region at Various Temperatures[a]

Wavelength (nm)	A	B
306	0.80	9.84
307	0.78	1.44
308	0.87	53.1
309	0.76	73.9
310	1.31	305.5
311	2.37	600
312	5.8	925.9
313	11.4	1191
314	20.1	1423
315	26.4	1514
316	26.8	1512
317	26.8	1542
318	28.33	1604
319	30.6	1604
320	44.4	1866
321	50.2	1931
322	27.8	1882
323	74.1	2329
324	868	3085
325	0.37	689
326	0.24	619
327	0.068	258
328	26.16	2131
329	0.15	470

[a] Using the quantum yield expression recommended by Talukdar et al., 1998: $\phi = 0.06 + Ae^{-B/T}$.

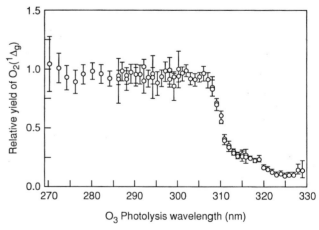

FIGURE 4.10 Relative yield of $O_2(^1\Delta_g)$ from the photolysis of O_3 (adapted from Ball et al., 1993).

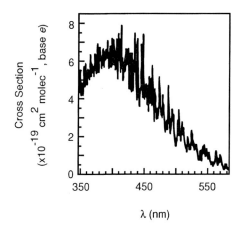

FIGURE 4.11 High-resolution (0.003 nm) absorption spectrum of NO_2 at room temperature (adapted from Harder et al., 1997).

nm. In addition, they produce $O(^3P)$ rather than $O(^1D)$. Under conditions typical of urban and suburban areas where sufficient NO_2 is present, the production of $O(^3P)$ from photolysis of O_3 will be much less than that from NO_2; in addition, since the $O(^3P)$ reacts with O_2 to regenerate O_3, this leads to no net loss of ozone. As a result, the photolysis processes generating $O(^3P)$ are not significant in tropospheric chemistry.

C. NITROGEN DIOXIDE

1. Absorption Spectra

Figure 4.11 shows a high-resolution (0.003 nm) absorption spectrum of NO_2 (Harder et al., 1997) and Tables 4.7 and 4.8 show the recommended absorption cross sections (DeMore et al., 1997). Those in the 202- to 274-nm region are temperature independent. At the longer wavelengths there appears to be a small temperature dependence parameterized as shown in Table 4.8.

As discussed in Chapter 11, NO_2 is frequently measured in the atmosphere using UV–visible absorption spectroscopy. Because of the large number of possible absorbing species in the atmosphere, NO_2 is usually measured using the peak-to-valley absorbance of selected peaks, rather than peak to zero absorbance. While reducing the temperature from 298 K only slightly affects the peak absorption cross sections, it does decrease the minima in the spectrum, thus increasing the peak-to-valley absorbance differences used for measuring NO_2 in the atmosphere by differential optical absorption spectrometry (e.g., Davidson et al., 1988; Harwood and Jones, 1994; Vandaele et al., 1996; Harder et al., 1997). This is expected since lowering the temperature will lower the population of the higher rotational energy levels in the ground state, leading to a smaller range of energies and wavelengths absorbed in a given transition. In addition to the temperature dependence, a pressure dependence of the absorption cross sections has been observed in high-resolution studies (Harder et al., 1997), which is important for application in the upper atmosphere.

One of the complications in measuring NO_2 absorption cross sections, particularly at lower temperatures and higher concentrations, is the presence of the dimer N_2O_4 in equilibrium with it:

$$2NO_2 \leftrightarrow N_2O_4. \qquad (11, -11)$$

TABLE 4.7 Averaged Absorption Cross Sections (Base e) from 202 to 274 nm for NO_2 at 298 K[a]

λ (nm)	$10^{20}\sigma$[b] (cm^2 $molecule^{-1}$)	λ (nm)	$10^{20}\sigma$[b] (cm^2 $molecule^{-1}$)
202.02–204.08	41.45	232.56–235.29	16.89
204.08–206.19	44.78	235.29–238.09	16.18
206.19–208.33	44.54	238.09–240.96	8.812
208.33–210.53	46.41	240.96–243.90	7.472
210.53–212.77	48.66	243.90–246.91	3.909
212.77–215.06	48.18	246.91–250.00	2.753
215.06–217.39	50.22	250.00–253.17	2.007
217.39–219.78	44.41	253.17–256.41	1.973
219.78–222.22	47.13	256.41–259.74	2.111
222.22–224.72	37.72	259.74–263.16	2.357
224.72–227.27	39.29	263.16–266.67	2.698
227.27–229.89	27.40	266.67–270.27	3.247
229.89–232.56	27.78	270.27–273.97	3.785

[a] From DeMore et al. (1997) recommendation based on Schneider et al., J. Photochem. Photobiol. A., 40, 195, (1987).
[b] Independent of temperature.

TABLE 4.8 Recommended Averaged Absorption Cross Sections (Base e) for NO_2 from 274 to 422 nm at 0°C and Their Temperature Dependence[a]

λ (nm)	$10^{20}\sigma$ (cm^2 molecule^{-1})	$10^{22}a$[b] (cm^2 molecule^{-1} degree^{-1})
273.97–277.78	5.03	0.075
277.78–281.69	5.88	0.082
281.69–285.71	7.00	−0.053
285.71–289.85	8.15	−0.043
289.85–294.12	9.72	−0.031
294.12–298.51	11.54	−0.162
298.51–303.03	13.44	−0.284
303.03–307.69	15.89	−0.357
307.69–312.50	18.67	−0.536
312.5–317.5	21.53	−0.686
317.5–322.5	24.77	−0.786
322.5–327.5	28.07	−1.105
327.5–332.5	31.33	−1.355
332.5–337.5	34.25	−1.277
337.5–342.5	37.98	−1.612
342.5–347.5	40.65	−1.890
347.5–352.5	43.13	−1.219
352.5–357.5	47.17	−1.921
357.5–362.5	48.33	−1.095
362.5–367.5	51.66	−1.322
367.5–372.5	53.15	−1.102
372.5–377.5	55.08	−0.806
377.5–382.5	56.44	−0.867
382.5–387.5	57.57	−0.945
387.5–392.5	59.27	−0.923
392.5–397.5	58.45	−0.738
397.5–402.5	60.21	−0.599
402.5–407.5	57.81	−0.545
407.5–412.5	59.99	−1.129
412.5–417.5	56.51	0.001
417.5–422.5	58.12	−1.208

[a] From DeMore et al. (1997) recommendation based on Davidson et al. (1988); see references.
[b] This is the temperature coefficient of $\sigma(t) = \sigma(0°) + at$, where t is in °C.

TABLE 4.9 Average Cross Sections (Base e) for N_2O_4[a,b]

Wavelength (nm)	$10^{20}\sigma$ (cm^2 molecule^{-1})	
	253 K	213 K
320	43.87	40.46
325	52.93	50.32
330	60.89	58.97
335	65.57	64.23
340	66.46	64.29
345	63.07	60.1
350	55.87	50.25
355	47.15	37.64
360	37.93	25.44
365	30.84	16.04
370	25.39	9.67
375	22.02	5.61
380	20.51	3.01
385	18.80	1.62
390	18.24	0.84
395	16.78	0.52
400	17.16	0.19
405	15.47	
410	15.01	
415	13.87	
420	13.66	
425	12.31	
430	10.91	

[a] Averaged over 5-nm intervals centered on the wavelength shown.
[b] From Harwood and Jones (1994).

As seen from the data in Table 4.9, N_2O_4 also absorbs strongly in this region, so care must be taken in both spectroscopic and kinetic studies to take this into account. For low concentrations of NO_2 at room temperature, significant amounts of the dimer are not present, but this is not the case at higher concentrations. For example, using the recommended value of $K = 2.5 \times 10^{-19}$ cm^3 molecule^{-1} at 298 K for reaction (11, −11) (DeMore et al., 1997), the pressure of N_2O_4 in equilibrium with 1.0 Torr of NO_2 (equivalent to 1318 ppm) is 8.1×10^{-3} Torr, i.e., 0.81%, but for 1 ppb NO_2 (7.6×10^{-7} Torr), it is only 6×10^{-9} ppb, i.e., 6×10^{-7}%!

This equilibrium means that great caution must be exercised in extrapolating both kinetic and mechanistic studies from Torr concentrations of NO_2 to ambient ppt–ppb levels, since N_2O_4 may contribute significantly under some conditions, especially high concentrations and lower temperatures. In other cases, it is not a problem. For example, the reaction of hydrogen atoms with NO_2 is used as a source of OH in fast flow discharge systems. The reservoir holding NO_2 is often at relatively high concentrations and contains significant amounts of N_2O_4. However, the equilibrium (11, −11) shifts sufficiently rapidly that the N_2O_4 dissociates to form NO_2 so that the initial presence of N_2O_4 is not a problem.

2. Photochemistry

Nitrogen dioxide photodissociates at $\lambda < 420$ nm to give nitric oxide and an oxygen atom:

$$NO_2 + h\nu \rightarrow NO + O. \quad (12)$$

The quantum yield for oxygen atom production in (12) has been studied extensively because of its role as the only significant anthropogenic source of O_3 in the troposphere via (12) followed by (2):

$$O + O_2 \xrightarrow{M} O_3. \quad (2)$$

TABLE 4.10 Calculated Wavelengths (nm) for NO$_2$ Photolysis Below Which the Fragments Shown Can Be Produced[a,b]

NO	Oxygen atoms		
	^3P	^1D	^1S
$X^2\Pi$	397.8	243.9	169.7
$A^2\Sigma^+$	144.2	117.4	97

[a] From Okabe (1978).
[b] Assuming no contribution from internal energy of the molecule.

Table 4.10 gives the calculated wavelengths below which it is energetically possible to produce the fragments in each of the electronic states shown if there is no contribution from internal energy of the molecule. The threshold wavelength for production of ground-state NO and O atoms is 397.8 nm.

Figure 4.12 shows the primary quantum yields for the production of NO in reaction (12). The quantum yield is within experimental error of 1 up to 395 nm, declining slightly to 0.82 at the theoretical threshold for dissociation at 397.8 nm. This has been attributed to the formation of a nondissociative excited state of NO$_2$. In ambient air, electronically excited NO$_2$ which does not dissociate to form O(^{3}P) is collisionally deactivated. When O$_2$ is the collision partner, energy transfer may occur a fraction of the time to form O$_2$($^1\Delta_g$) (Jones and Bayes, 1973a):

$$NO_2^* + O_2 \rightarrow NO_2 + O_2(^1\Delta_g). \quad (13)$$

Figure 4.12 shows that significant photodissociation continues at longer wavelengths (Roehl et al., 1994). The nonzero quantum yield for reaction (12) between 397.8 and 420 nm is believed to be due to a contribution from internal vibrational and rotational energy as well as to a small contribution from collisions. Quantum yields calculated assuming that the vibration–rotation internal energy of the molecule is available to overcome the photon energy deficiency (dotted line in Fig. 4.12) as well as the energy acquired in collisions of the excited NO$_2$ with other molecules (dashed line) are shown (Roehl et al., 1994). A good fit to the data is obtained, supporting the importance of internal as well as collisional energy for photodissociation in the 398- to 420-nm region.

A two-photon process at 435 nm to generate O(^{1}D) has been observed in laboratory studies (Crowley and Carl, 1997). However, the combination of relatively low light intensities and high pressures which quench excited NO$_2$ make this unimportant in the atmosphere.

TABLE 4.11 Quantum Yields for NO$_2$ Photolysis[a]

Wavelength (nm)	ϕ_{12}	Wavelength (nm)	ϕ_{12}
< 285	1.000	393	0.953
290	0.999	394	0.950
295	0.998	395	0.942
300	0.997	396	0.922
305	0.996	397	0.870
310	0.995	398	0.820
315	0.994	399	0.760
320	0.993	400	0.695
325	0.992	401	0.635
330	0.991	402	0.560
335	0.990	403	0.485
340	0.989	404	0.425
345	0.988	405	0.350
350	0.987	406	0.290
355	0.986	407	0.225
360	0.984	408	0.185
370	0.981	409	0.153
375	0.979	410	0.130
380	0.975	411	0.110
381	0.974	412	0.094
382	0.973	413	0.083
383	0.972	414	0.070
384	0.971	415	0.059
385	0.969	416	0.048
386	0.967	417	0.039
387	0.966	418	0.030
388	0.964	419	0.023
389	0.962	420	0.018
390	0.960	421	0.012
391	0.959	422	0.008
392	0.957	423	0.004
		424	0.00

[a] From DeMore et al. (1997) recommendations based on Gardner et al., J. Geophys. Res., 92, 6642 (1987); Harker et al., Chem. Phys. Lett., 50, 394 (1977); Jones and Bayes, 1973b and Davenport, Report # FAA-EQ-78-14, U.S. Federal Aviation Administration, Washington, D.C. (1978).

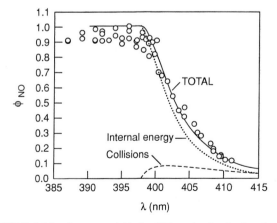

FIGURE 4.12 Quantum yields for NO production in the photolysis of NO$_2$ at 298 K. Calculated quantum yields due to internal energy (dotted line), the calculated dissociation due to collision (dashed line), and the sum of these two calculations (solid line) are also shown (adapted from Roehl et al., 1994).

Table 4.11 summarizes the recommended quantum yields in the 285- to 424-nm region (DeMore *et al.*, 1997).

D. NITRIC ACID

Figure 4.13 shows the absorption spectrum of gaseous HNO_3 at room temperature (Burkholder *et al.*, 1993) and Table 4.12 shows the recommended absorption cross sections at 298 K (DeMore *et al.*, 1997).

The quantum yield for photodissociation, reaction (14), is approximately 1 from 200 to 315 nm (DeMore *et al.*, 1997):

$$HNO_3 + h\nu \rightarrow OH + NO_2. \quad (14)$$

At shorter wavelengths in the vacuum ultraviolet, the path to give O + HONO appears to become important.

Donaldson *et al.* (1997) have proposed that absorption in the visible due to OH vibrational overtones could be important in the lower stratosphere at large solar zenith angles. Transfer of energy from the absorbing mode to the HO—NO_2 bond may then cause dissociation, as observed, for example, in HOCl (e.g.,

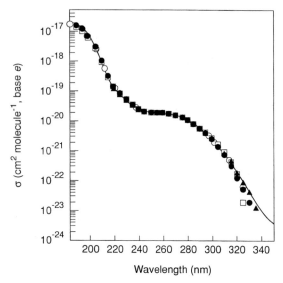

FIGURE 4.13 Absorption spectrum of HNO_3 (solid line) at 298 K. Symbols represent data reported in previous studies from other laboratories. (Adapted from Burkholder *et al.*, 1993.)

Barnes *et al.*, 1997). The energy needed for HO—NO_2 bond cleavage is slightly larger than $5\nu_{OH}$, but a contribution from rotational energy may make this feasible.

TABLE 4.12 Absorption Cross Sections (Base e) for HNO_3 Vapor at 298 K[a]

Wavelength (nm)	$10^{20}\sigma$ (cm^2 molecule^{-1})	Wavelength (nm)	$10^{20}\sigma$ (cm^2 molecule^{-1})	Wavelength (nm)	$10^{20}\sigma$ (cm^2 molecule^{-1})
190	1360	244	2.16	298	0.316
192	1225	246	2.06	300	0.263
194	1095	248	2.00	302	0.208
196	940	250	1.97	304	0.167
198	770	252	1.96	306	0.133
200	588	254	1.95	308	0.105
202	447	256	1.95	310	0.0814
204	328	258	1.93	312	0.0628
206	231	260	1.91	314	0.0468
208	156	262	1.87	316	0.0362
210	104	264	1.83	318	0.0271
212	67.5	266	1.77	320	0.0197
214	43.9	268	1.70	322	0.0154
216	29.2	270	1.62	324	0.0108
218	20.0	272	1.53	326	0.00820
220	14.9	274	1.44	328	0.00613
222	11.8	276	1.33	330	0.00431
224	9.61	278	1.23	332	0.00319
226	8.02	280	1.12	334	0.00243
228	6.82	282	1.01	336	0.00196
230	5.75	284	0.909	338	0.00142
232	4.87	286	0.807	340	0.00103
234	4.14	288	0.709	342	0.00086
236	3.36	290	0.615	344	0.00069
238	2.93	292	0.532	346	0.00050
240	2.58	294	0.453	348	0.00042
242	2.34	296	0.381	350	0.00042

[a] From DeMore *et al.* (1997) recommendation based on Burkholder *et al.* (1993); see these for the temperature dependence, which can be significant in the atmosphere.

At present, the quantitative data on absorption strengths etc. needed to fully evaluate this possibility are not available.

E. NITROUS ACID

As discussed in Chapter 1, nitrous acid is important in tropospheric chemistry because it photolyzes to form OH:

$$HONO + h\nu \rightarrow OH + NO. \qquad (15)$$

Measuring experimentally the quantum yields for reaction (15) as well as the absorption cross sections has been difficult because of contamination of the HONO by other, strongly absorbing species, particularly NO_2, necessitating corrections for their contributions to the absorption. However, it is believed, based on studies by Cox and Derwent (1976/1977) that the quantum yield for reaction (15) is 1.0 at $\lambda < 400$ nm.

Figure 4.14 shows the absorption spectrum of gas-phase nitrous acid and Table 4.13 gives absorption cross sections (Bongartz et al., 1991) corrected by the factor of 0.855 as recommended by Bongartz et al. (1994).

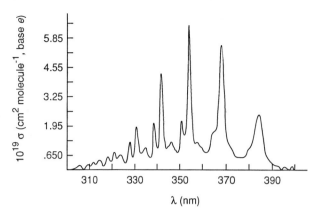

FIGURE 4.14 Absorption spectrum of HONO at 277 K (adapted from Bongartz et al., 1991). Note that the absolute values of the cross sections shown here should be multiplied by 0.855 as recommended by Bongartz et al. (1994).

TABLE 4.13 Recommended Absorption Cross Sections (Base e) for HONO[a]

Wavelength (nm)	$10^{20} \sigma$ (cm^2 molecule^{-1})	Wavelength (nm)	$10^{20} \sigma$ (cm^2 molecule^{-1})	Wavelength (nm)	$10^{20} \sigma$ (cm^2 molecule^{-1})
310	1.1	339	16.1	368	44.5
311	1.6	340	8.6	369	33.2
312	2.4	341	14.5	370	15.2
313	1.9	342	33.0	371	9.7
314	3.1	343	12.7	372	8.6
315	2.6	344	8.3	373	6.6
316	1.2	345	9.3	374	5.3
317	2.7	346	10.5	375	4.5
318	4.8	347	8.9	376	4.5
319	3.1	348	7.8	377	4.3
320	4.2	349	6.8	378	5.0
321	6.7	350	9.6	379	6.8
322	4.2	351	18.1	380	8.2
323	4.4	352	13.3	381	9.7
324	6.1	353	16.3	382	13.6
325	4.3	354	49.7	383	18.0
326	2.5	355	31.1	384	20.6
327	5.6	356	12.1	385	17.4
328	10.0	357	10.0	386	11.5
329	5.2	358	10.3	387	7.7
330	9.5	359	8.9	388	4.8
331	15.3	360	7.7	389	2.9
332	7.4	361	7.1	390	2.3
333	6.5	362	6.8	391	1.7
334	8.2	363	8.2	392	1.3
335	8.2	364	12.5	393	0.94
336	6.2	365	14.4	394	0.5
337	4.5	366	15.6	395	0.86
338	8.6	367	25.8	396	0.3

[a] From DeMore et al. (1997) recommendation based on Bongartz et al. (1991), but multiplied by 0.855 as recommended by Bongartz et al. (1994). The value at 354 nm is in excellent agreement with the measured value of $(5.02 \pm 0.76) \times 10^{-19}$ cm^2 molecule^{-1} of Pagsberg et al. (1997).

F. PEROXYNITRIC ACID

Peroxynitric acid (HO_2NO_2) is a relatively unstable species that is important as a reservoir for NO_2 at lower temperatures via the reversible reaction (16, −16):

$$HO_2NO_2 \leftrightarrow HO_2 + NO_2. \quad (16, -16)$$

Figure 4.15 and Table 4.14 give the absorption cross sections. They do not show a significant temperature dependence down to 253 K (Singer et al., 1989).

There are a variety of potential photodissociation pathways, including the following:

$$HO_2NO_2 + h\nu \rightarrow HO_2 + NO_2, \quad (17a)$$
$$\rightarrow HO_2 + NO + O, \quad (17b)$$
$$\rightarrow OH + NO_2 + O, \quad (17c)$$
$$\rightarrow OH + NO + O_2, \quad (17d)$$
$$\rightarrow OH + NO_3. \quad (17e)$$

TABLE 4.14 Recommended Absorption Cross Sections (Base e) for Gaseous HO_2NO_2[a]

Wavelength (nm)	$10^{20}\sigma$ (cm^2 molecule^{-1})	Wavelength (nm)	$10^{20}\sigma$ (cm^2 molecule^{-1})
190	1010	260	28.5
195	816	265	23.0
200	563	270	18.1
205	367	275	13.4
210	239	280	9.3
215	161	285	6.2
220	118	290	3.9
225	93.5	295	2.4
230	79.2	300	1.4
235	68.2	305	0.9
240	58.1	310	0.5
245	48.9	315	0.3
250	41.2	320	0.2
255	35.0	325	0.1

[a] From DeMore et al. (1997) recommendation based on Molina and Molina (1981) and Singer et al. (1989).

At 248 nm, the quantum yield for OH is 0.34 ± 0.16, and the formation of electronically excited NO_2^* with yields of < 30% was observed (MacLeod et al., 1988). Reaction (17a), where some of the NO_2^* is formed in an electronically excited state, and reaction (17e), which is the most direct path to OH, are thought to be the important paths. In the absence of additional studies, MacLeod et al. (1988) recommend for application to atmospheric situations $\phi_{17a} = 0.65$ and $\phi_{17e} = 0.35$.

G. NITRATE RADICAL

The nitrate radical, NO_3, is an important intermediate in nightime chemistry. Its spectroscopy, photochemistry, and chemistry are reviewed in detail by Wayne et al. (1991) and by Atkinson (1991).

As shown in Figure 4.16, NO_3 is unusual in that it absorbs strongly in the red region (620–670 nm) of the visible spectrum, unlike most atmospherically important species whose absorptions typically fall in the UV. Its absorption in this region is banded, which allows its detection and measurement using spectroscopic techniques (see Sections A.1d and A.4a in Chapter 11).

Table 4.15 gives the absorption cross sections and quantum yields at 298 K. A number of studies report increased values at 662 nm at lower temperatures (e.g., Ravishankara and Mauldin, 1986; Sander, 1986; Yokelson et al., 1994), while one (Cantrell et al., 1987) finds no change. This is important, since these cross sections are used to derive absolute concentrations of NO_3 in the atmosphere, where the temperature during measurement can vary considerably.

a
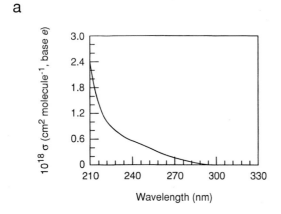

b
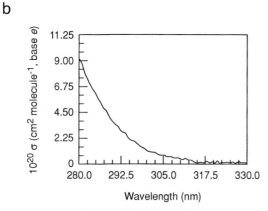

FIGURE 4.15 Absorption spectra of HO_2NO_2 at 298 K in the (a) 210- to 300-nm and (b) 280- to 330-nm regions (adapted from Singer et al., 1989).

There are two possible decomposition pathways for NO$_3$:

$$NO_3 + h\nu \rightarrow NO_2 + O(^3P), \quad (18a)$$
$$\rightarrow NO + O_2. \quad (18b)$$

Figure 4.17 shows the energetics of these pathways, including the possible formation of electronically excited singlet states of O$_2$. The threshold for (18a) is 585.5 nm (Johnston et al., 1996). While reaction (18b) is close to thermoneutral overall, there is a substantial energy barrier to the dissociation, 47.3 ± 0.8 kcal mol^{-1}; the threshold observed for this reaction is 594.5 nm (Johnston et al., 1996).

Figure 4.18 shows the results of a reevaluation of the quantum yields as a function of wavelength based on the observed energy thresholds for each channel and on consideration of the potential contributions of rotational and vibrational energy to the dissociation of NO$_3$ (Davis et al., 1993; Johnston et al., 1996). These are in excellent agreement with the experiments of Orlando et al. (1993), except for the region from 595 to 635 nm, where the experimental quantum yields for (18a) are larger than those predicted based on the model of Johnston et al. (1996). The reasons for the discrepancy in this region are not clear, but Johnston and co-workers suggest several possibilities, including contributions from an as yet unrecognized low-lying electronic state.

ϕ_{18a} is ~1 from 570 to 585 nm and then decreases gradually to zero at 635 nm. On the other hand, as (18a) decreases at longer wavelengths, ϕ_{18b} first increases to a peak value of ~0.36 at 595 nm and then drops off at longer wavelengths. Also shown is the increase in fluorescence quantum yield from zero at 595 nm. As expected (Chapter 3.A.2), as the quantum yields for photochemical channels decline, that for the photophysical process of fluorescence increases.

For a solar zenith angle of 0°, the photolysis rate constants are estimated to be in the range 0.17–0.19 s^{-1} for (18a) and 0.016–0.020 s^{-1} for (18b) at the earth's surface in the absence of clouds (Orlando et al., 1993; Johnston et al., 1996). The O(^{3}P) that is formed in the predominant path will add to O$_2$ to generate O$_3$, which can then react with NO$_2$ to regenerate NO$_3$.

H. DINITROGEN PENTOXIDE

Figure 4.19 and Table 4.16 show the absorption spectrum and cross sections for N$_2$O$_5$ at room temperature (Harwood et al., 1993, 1998). It should be noted that while these are in excellent agreement with the recommendations of DeMore et al. (1997) above 260 nm, the Harwood et al. (1998) values are smaller than the recommended ones below this.

The photolysis produces NO$_3$ with a quantum yield of approximately unity above 300 nm (DeMore et al., 1997). Harwood et al. (1998) report quantum yields for NO$_3$ of 0.96 ± 0.15 at 308 nm and 1.03 ± 0.15 at 352.5 nm. However, at 248 nm, the quantum yield for NO$_3$ drops to 0.64 ± 0.10; O(^{3}P) production also occurs at 248 nm, with a quantum yield of 0.72 ± 0.17 (Ravishankara et al., 1986). The O(^{3}P) generation could be due to the subsequent dissociation of an initially produced electronically excited NO$_2^*$ (Oh et al., 1986), to reaction (19b), or to the dissociation of vibrationally excited NO$_3$ (Harwood et al., 1998):

$$N_2O_5 + h\nu \rightarrow NO_3 + NO_2^*, \quad (19a)$$
$$\rightarrow NO_3 + NO + O. \quad (19b)$$

I. NITROUS OXIDE

Figure 4.20 shows the absorption spectrum of N$_2$O at room temperature and Table 4.17 the absorption cross sections. It does not absorb significantly in the "actinic" region above 290 nm, but at much shorter wavelengths found in the stratosphere, absorbs and dissociates with unit quantum yield to N$_2$ and electronically excited O(^{1}D):

$$N_2O + h\nu \rightarrow N_2 + O(^1D). \quad (20)$$

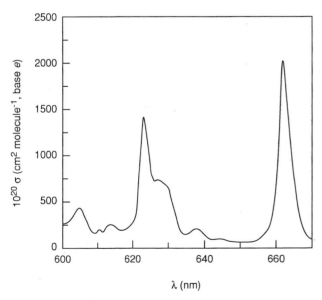

FIGURE 4.16 Absorption spectrum of NO$_3$ at 298 K [adapted from DeMore et al., 1997 based on data from Ravishankara and Mauldin (1986), Sander (1986), and Canosa-Mas et al. (1987)].

TABLE 4.15 Absorption Cross Sections (Base e) and Quantum Yields for NO_3 at 298 K[a]

Wavelength (nm)	$10^{20}\sigma$ (cm² molecule⁻¹)	Quantum yield $NO_2 + O$	Quantum yield $NO + O_2$	Wavelength (nm)	$10^{20}\sigma$ (cm² molecule⁻¹)	Quantum yield $NO_2 + O$	Quantum yield $NO + O_2$
585		0.983	0.0	628	702	0.0339	0.074
586		0.967	0.015	629	672	0.0294	0.070
587		0.943	0.039	630	638	0.0264	0.065
588		0.885	0.097	631	470	0.0236	0.058
589		0.854	0.128	632	344	0.0195	0.051
590		0.793	0.190	633	194	0.0177	0.047
591		0.763	0.220	634	142	0.0161	0.043
592		0.734	0.249	635	128	0.0146	0.037
593		0.680	0.303	636	159	0.0119	0.032
594		0.654	0.328	637	191	0.0107	0.029
595		0.608	0.359	638	193	0.0096	0.027
596		0.587	0.357	639	162	0.0086	0.024
597		0.567	0.318	640	121	0.0072	0.020
598		0.531	0.323	641	99		
599		0.509	0.314	642	91		
600	258	0.472	0.291	643	93		
601	263	0.438	0.296	644	92		
602	302	0.415	0.291	645	85		
603	351	0.371	0.283	646	72		
604	413	0.351	0.280	647	69		
605	415	0.323	0.264	648	60		
606	322	0.296	0.271	649	51		
607	225	0.280	0.268	650	49		
608	170	0.259	0.250	651	52		
609	153	0.238	0.248	652	55		
610	192	0.226	0.236	653	61		
611	171	0.210	0.205	654	76		
612	202	0.193	0.200	655	93		
613	241	0.181	0.190	656	131		
614	242	0.166	0.166	657	172		
615	210	0.147	0.166	658	222		
616	190	0.137	0.160	659	356		
617	189	0.124	0.141	660	658		
618	208	0.108	0.143	661	1308		
619	229	0.0993	0.139	662	2000		
620	292	0.0897	0.131	663	1742		
621	450	0.0769	0.127	664	1110		
622	941	0.0704	0.122	665	752		
623	1407	0.0643	0.117	666	463		
624	1139	0.0552	0.106	667	254		
625	796	0.0487	0.0985	668	163		
626	703	0.0442	0.092	669	113		
627	715	0.0393	0.085	670	85		

[a] Absorption cross sections from DeMore et al. (1997) recommendation based on Ravishankara and Mauldin (1986), Sander (1986), and Canosa-Mas et al. (1987); quantum yields from Johnston et al. (1996).

It is this process, followed by reaction (21a)

$$O(^1D) + N_2O \rightarrow 2NO, \quad (21a)$$
$$\rightarrow N_2 + O_2, \quad (21b)$$

that is primarily responsible for the production of reactive oxides of nitrogen in the stratosphere in a "natural" atmosphere (see Chapter 12).

J. ORGANIC NITRATES AND PEROXYACETYL NITRATE

1. Organic Nitrates

Figure 4.21 shows the absorption spectrum of some simple alkyl nitrates and Tables 4.18, 4.19, and 4.20 give some typical absorption cross sections (Roberts and Fajer, 1989; Turberg et al., 1990; Clemitshaw et al.,

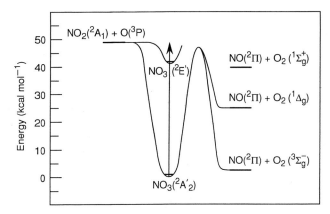

FIGURE 4.17 Energetics of NO$_3$ photodissociation (adapted from Davis et al., 1993).

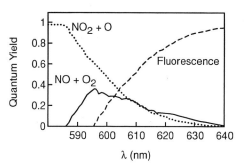

FIGURE 4.18 Quantum yields for NO$_3$ photolysis: dotted line, NO$_3 \to$ NO$_2$ + O; solid line, NO$_3 \to$ NO + O$_2$; dashed line, fluorescence quantum yields (adapted from Johnston et al., 1996).

1997). The absorption drops off strongly in the actinic region, with the result that they are not photodissociated rapidly in the troposphere (Roberts and Fajer, 1989; Turberg et al., 1990; Clemitshaw et al., 1997).

There are several different possible primary processes, depending on the photolysis wavelength. For example, ethyl nitrate in principle has three paths:

$$C_2H_5ONO_2 + h\nu \to C_2H_5O + NO_2, \quad (22a)$$
$$\to CH_3CHO + HONO, \quad (22b)$$
$$\to C_2H_5ONO + O. \quad (22c)$$

While the relative importance of the various paths is not well established, it is expected that dissociation to the alkoxy radical, RO, and NO$_2$ will predominate. Luke et al. (1989) experimentally measured rates of photolysis of simple alkyl nitrates and compared them to rates calculated using the procedures outlined in Chapter 3.C.2. Figure 4.22 compares the experimentally determined values of the photolysis rate constants (k_p) for ethyl and n-propyl nitrate with the values calculated assuming a quantum yield for photodissociation of unity. The good agreement suggests that the quantum yield for photodissociation of the alkyl nitrates indeed approaches 1.0.

2. Peroxyacetyl Nitrate

Peroxyacetyl nitrate, CH$_3$C(O)OONO$_2$, commonly referred to as PAN, is important as a means of transporting NO$_x$ over large distances. It is quite stable at low temperatures but decomposes at higher temperatures to release NO$_2$. Hence whether photolysis can compete with the thermal decomposition is of great interest. Figure 4.23 shows the absorption spectrum of PAN, CH$_3$CO$_3$NO$_2$, and Table 4.21 the absorption cross sections (Talukdar et al., 1995).

Photolysis by several pathways is possible:

$$CH_3C(O)OONO_2 \to CH_3C(O)O + NO_3, \quad (23a)$$
$$\to CH_3C(O)OO + NO_2, \quad (23b)$$
$$\to CH_3C(O) + O_2 + NO_2. \quad (23c)$$

The wavelength thresholds for these at 298 K are 1025, 990, and 445 nm, respectively. Mazely et al. (1995, 1997) have measured the quantum yields for production of NO$_2$ and NO$_3$ at 248 nm and find $\phi(NO_2) = 0.83 \pm 0.09$ and $\phi(NO_3) = 0.3 \pm 0.1$. The quantum yields at longer wavelengths have not been reported.

K. SULFUR DIOXIDE AND SULFUR TRIOXIDE

1. SO$_2$

As seen in Fig. 4.24, SO$_2$ absorbs light strongly up to ~ 300 nm, with a much weaker absorption from 340 to

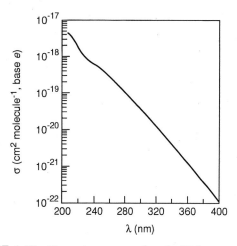

FIGURE 4.19 Absorption cross sections for N$_2$O$_5$ at room temperature (adapted from Harwood et al., 1998).

TABLE 4.16 Absorption Cross Sections (Base e) for N_2O_5[a]

Wavelength (nm)	$10^{20}\sigma$ (cm² molecule⁻¹)	Wavelength (nm)	$10^{20}\sigma$ (cm² molecule⁻¹)	Wavelength (nm)	$10^{20}\sigma$ (cm² molecule⁻¹)
208	418	272	14.9	336	0.462
210	380	274	13.7	338	0.412
212	335	276	12.4	340	0.368
214	285	278	11.4	342	0.328
216	236	280	10.5	344	0.293
218	196	282	9.59	346	0.262
220	165	284	8.74	348	0.234
222	140	286	7.94	350	0.210
224	119	288	7.20	352	0.188
226	105	290	6.52	354	0.167
228	92.6	292	5.88	356	0.149
230	83.8	294	5.29	358	0.133
232	76.9	296	4.75	360	0.120
234	70.8	298	4.26	362	0.107
236	65.8	300	3.81	364	0.0958
238	61.4	302	3.40	366	0.0852
240	57.1	304	3.03	368	0.0763
242	53.1	306	2.70	370	0.0685
244	49.3	308	2.40	372	0.0613
246	45.6	310	2.13	374	0.0545
248	41.9	312	1.90	376	0.0484
250	38.6	314	1.68	378	0.0431
252	35.5	316	1.49	380	0.0383
254	32.6	318	1.33	382	0.0341
256	29.9	320	1.18	384	0.0305
258	27.5	322	1.05	386	0.0273
260	25.2	324	0.930	388	0.0242
262	23.1	326	0.826	390	0.0215
264	21.1	328	0.735	392	0.0193
266	19.4	330	0.654	394	0.0172
268	17.8	332	0.582	396	0.0150
270	16.2	334	0.518	398	0.0134

[a] From Harwood et al. (1998).

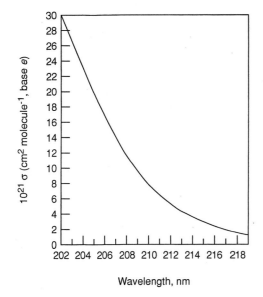

FIGURE 4.20 Absorption cross sections for N_2O at room temperature (adapted from Selwyn et al., 1977).

400 nm. The strong absorption peaking at ~290 nm forms two singlet excited states of SO_2,

$$SO_2(X^1A_1) + h\nu(240 < \lambda < 330 \text{ nm}) \rightarrow SO_2(^1A_2, {}^1B_1), \quad (24)$$

while the weak absorption involves a spin-forbidden transition to a triplet state:

$$SO_2(X^1A_1) + h\nu(340 < \lambda < 400 \text{ nm}) \rightarrow SO_2({}^3B_1). \quad (25)$$

SO_2 can dissociate to $SO + O$ only at wavelengths below 218 nm:

$$SO_2(X^1A_1) + h\nu(\lambda < 218 \text{ nm}) \rightarrow SO(^3\Sigma_u^-) + O(^3P). \quad (26)$$

Thus photodissociation does not occur in the tropo-

TABLE 4.17 Absorption Cross Sections (Base e) for N_2O at 296–302 K[a]

Wavelength (nm)	$10^{20}\sigma$ (cm² molecule⁻¹)	Wavelength (nm)	$10^{20}\sigma$ (cm² molecule⁻¹)	Wavelength (nm)	$10^{20}\sigma$ (cm² molecule⁻¹)
173	11.3	196	6.82	219	0.115
174	11.9	197	6.10	220	0.0922
175	12.6	198	5.35	221	0.0739
176	13.4	199	4.70	222	0.0588
177	14.0	200	4.09	223	0.0474
178	13.9	201	3.58	224	0.0375
179	14.4	202	3.09	225	0.0303
180	14.6	203	2.67	226	0.0239
181	14.6	204	2.30	227	0.0190
182	14.7	205	1.95	228	0.0151
183	14.6	206	1.65	229	0.0120
184	14.4	207	1.38	230	0.00955
185	14.3	208	1.16	231	0.00760
186	13.6	209	0.980	232	0.00605
187	13.1	210	0.755	233	0.00478
188	12.5	211	0.619	234	0.00360
189	11.7	212	0.518	235	0.00301
190	11.1	213	0.421	236	0.00240
191	10.4	214	0.342	237	0.00191
192	9.75	215	0.276	238	0.00152
193	8.95	216	0.223	239	0.00123
194	8.11	217	0.179	240	0.00101
195	7.57	218	0.142		

[a] From DeMore *et al.* (1997) recommendation based on Selwyn *et al.* (1977); cross sections from 173–209 nm at 302 K and from 209–240 nm at 296 K.

sphere, where only wavelengths of 290 nm and above are present.

For a more detailed description of quenching of excited SO_2 in the atmosphere after light absorption, see Calvert and Stockwell (1984).

2. SO_3

As discussed in Chapter 8, SO_3 reacts very rapidly with gaseous water, ultimately forming sulfuric acid. However, in the upper atmosphere, where water concentrations are small, photolysis could potentially compete with the reaction with water. Some of the absorption cross sections for SO_3 at room temperature measured by Burkholder and McKeen (1997) from 195 to 330 nm are given in Table 4.22. Based on these, they predict photolysis will be as fast or faster than reaction with H_2O only at altitudes of > 40 km.

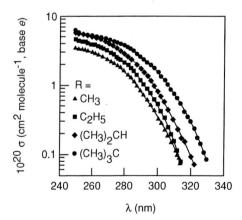

FIGURE 4.21 Absorption cross sections of methyl, ethyl, 2-propyl, and *tert*-butyl nitrates (adapted from Roberts and Fajer, 1989).

TABLE 4.18 Recommended Absorption Cross Sections (Base e) of Methyl Nitrate (CH_3ONO_2) at Room Temperature[a]

Wavelength (nm)	$10^{20}\sigma$ (cm² molecule⁻¹)	Wavelength (nm)	$10^{20}\sigma$ (cm² molecule⁻¹)
270	2.37	305	0.211
275	1.99	310	0.139
280	1.57	315	0.0626
285	1.17	320	0.0312
290	0.836	325	0.0144
295	0.559	330	0.0066
300	0.356		

[a] IUPAC recommendation (Atkinson *et al.*, 1997a) based on Roberts and Fajer (1989) and Rattigan *et al.* (1992).

TABLE 4.19 Recommended Absorption Cross Sections (Base e) of Ethyl Nitrate ($C_2H_5ONO_2$) at Room Temperature[a]

Wavelength (nm)	$10^{20}\sigma$ (cm² molecule⁻¹)	Wavelength (nm)	$10^{20}\sigma$ (cm² molecule⁻¹)
185	1710	260	4.1
188	1760	265	3.7
190	1710	270	3.2
195	1490	275	2.8
200	1140	280	2.3
205	738	285	1.8
210	400	290	1.3
215	195	295	0.85
220	91	300	0.54
225	45	305	0.32
230	24	310	0.18
235	13	315	0.091
240	8.0	320	0.045
245	5.6	325	0.023
250	4.7	330	0.011
255	4.3		

[a] IUPAC recommendation (Atkinson et al., 1997a), based on Roberts and Fajer (1989) and Turberg et al. (1990); the values of Clemitshaw et al. (1997) are in excellent agreement with these.

TABLE 4.20 Recommended Absorption Cross Sections (Base e) of 2-Propyl Nitrate [$(CH_3)_2CHONO_2$] at Room Temperature[a]

Wavelength (nm)	$10^{20}\sigma$ (cm² molecule⁻¹)	Wavelength (nm)	$10^{20}\sigma$ (cm² molecule⁻¹)
185	1790	260	4.17
188	1810	265	3.90
190	1790	270	3.41
195	1610	275	2.94
200	1260	280	2.34
205	867	285	1.88
210	498	290	1.42
215	247	295	1.02
220	132	300	0.685
225	64.9	305	0.436
230	37.4	310	0.259
235	17.8	315	0.149
240	10.0	320	0.0784
245	6.74	325	0.0422
250	4.90	330	0.0238
255	4.47	335	0.0126

[a] From 185 to 215 nm, values are IUPAC recommendation (Atkinson et al., 1997a) based on Roberts and Fajer (1989) and Turberg et al. (1990); from 220 to 335 nm, values are from Clemitshaw et al. (1997).

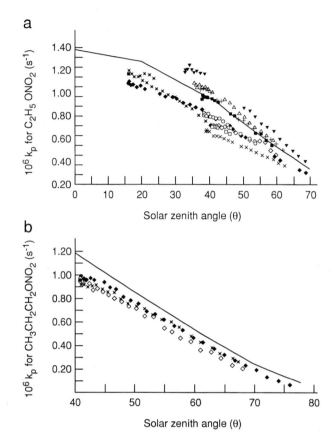

FIGURE 4.22 Experimental values of the photolysis rate constant, k_p, for (a) ethyl nitrate and (b) n-propyl nitrate as a function of zenith angle compared to calculated values shown by the solid lines. Different symbols represent different measurement days (adapted from Luke et al., 1989).

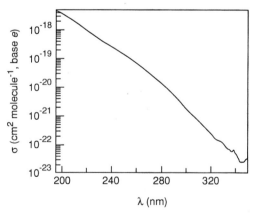

FIGURE 4.23 Absorption spectrum of PAN at room temperature (adapted from Talukdar et al., 1995).

TABLE 4.21 Absorption Cross Sections (Base e) of PAN at 298 K[a]

Wavelength (nm)	$10^{20} \sigma$ (cm^2 molecule^{-1})	Wavelength (nm)	$10^{20} \sigma$ (cm^2 molecule^{-1})
200	361	280	1.46
210	196	290	0.537
220	89.7	300	0.189
230	45.8	310	0.0666
240	24.4	320	0.0252
250	12.9	330	0.0106
260	6.85	340	0.00502
270	3.37	350	0.00165

[a] From Talukdar *et al.* (1995); see original reference for temperature dependence.

L. HYDROGEN PEROXIDE AND ORGANIC HYDROPEROXIDES

Figure 4.25 shows the absorption cross sections for H$_2$O$_2$ and methyl hydroperoxide at room temperature, and Table 4.23 summarizes these data (Vaghjiani and Ravishankara, 1989; DeMore *et al.*, 1997). Although the absorption is dropping off rapidly at wavelengths above the actinic cutoff of 290 nm, photolysis is still a significant loss process for these peroxides in the troposphere. In both cases, OH is the major product formed, with a product quantum yield of 2 for H$_2$O$_2$, corresponding to a photodissociation quantum yield of 1 at wavelengths > 222 nm (e.g., Vaghjiani and Ravishankara, 1990; Vaghjiani *et al.*, 1992),

$$H_2O_2 + h\nu \rightarrow 2OH, \quad (27)$$

and a quantum yield of 1 at 248 nm for CH$_3$OOH (Vaghjiani and Ravishankara, 1990; Thelen *et al.*, 1993)

$$CH_3OOH + h\nu \rightarrow CH_3O + OH. \quad (28)$$

M. ALDEHYDES AND KETONES

Figures 4.26, 4.27, and 4.28 show typical UV absorption spectra for some simple aldehydes and ketones (Rogers, 1990; Martinez *et al.*, 1992; see also Cronin and Zhu, 1998, for *n*-pentanal). Formaldehyde stands out from the higher aldehydes and ketones in that it has a highly structured spectrum and furthermore, the absorption extends out to longer wavelengths. The latter difference is particularly important because the solar intensity increases rapidly with wavelength here (Chapter 3.C.1) and hence the photolysis rate constant for HCHO and the rate of production of free radicals

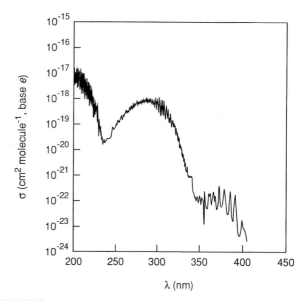

FIGURE 4.24 Absorption cross section of SO$_2$ (adapted from Manatt and Lane, 1993).

from this photolysis are much higher than for the larger molecules.

The absolute values of the absorption cross sections of HCHO have been somewhat controversial. This appears to be due to a lack of sufficient resolution in some studies; as discussed in Chapter 3.B.2, if the spectral resolution is too low relative to the bandwidth, nonlinear Beer–Lambert plots result. The strongly banded structure means that calculations of the photolysis rate constant require actinic flux data that have much finer resolution than the 2- to 5-nm intervals for which these flux data are given in Chapter 3 or, alternatively, that the measured absorption cross sections must be appropriately averaged. One significant advantage of the highly structured absorption of HCHO is that it can be used to measure low concentrations of this important aldehyde in the atmosphere by UV absorption (see Sections A.1d and A.4f in Chapter 11.).

TABLE 4.22 Absorption Cross Sections (Base e) for SO$_3$[a]

Wavelength (nm)	$10^{20} \sigma$ (cm^2 molecule^{-1})	Wavelength (nm)	$10^{20} \sigma$ (cm^2 molecule^{-1})
200	73.2	270	0.416
210	46.4	280	0.195
220	19.1	290	0.0907
230	7.49	300	0.0424
240	3.69	310	0.0192
250	1.80	320	0.00809
260	0.867	330	0.00363

[a] From Burkholder and McKeen (1997).

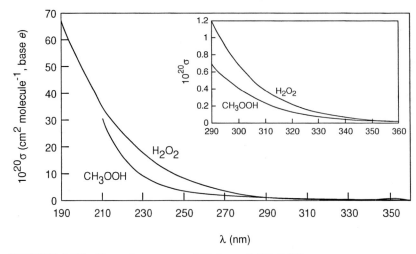

FIGURE 4.25 Absorption spectra of H_2O_2 and CH_3OOH at room temperature (data for H_2O_2 from DeMore *et al.*, 1997 recommendation, and for CH_3OOH from Vaghjiani and Ravishankara, 1989).

TABLE 4.23 Recommended Absorption Cross Sections (Base e) of H_2O_2 and CH_3OOH at Room Temperature[a]

Wavelength (nm)	$10^{20}\sigma$ (cm² molecule⁻¹) H_2O_2	$10^{20}\sigma$ (cm² molecule⁻¹) CH_3OOH	Wavelength (nm)	$10^{20}\sigma$ (cm² molecule⁻¹) H_2O_2	$10^{20}\sigma$ (cm² molecule⁻¹) CH_3OOH
190	67.2		280	2.0	1.09
195	56.4		285	1.5	0.863
200	47.5		290	1.2	0.691
205	40.8		295	0.90	0.551
210	35.7	31.2	300	0.68	0.413
215	30.7	20.9	305	0.51	0.313
220	25.8	15.4	310	0.39	0.239
225	21.7	12.2	315	0.29	0.182
230	18.2	9.62	320	0.22	0.137
235	15.0	7.61	325	0.16	0.105
240	12.4	6.05	330	0.13	0.079
245	10.2	4.88	335	0.10	0.061
250	8.3	3.98	340	0.07	0.047
255	6.7	3.23	345	0.05	0.035
260	5.3	2.56	350	0.04	0.027
265	4.2	2.11	355		0.021
270	3.3	1.70	360		0.016
275	2.6	1.39	365		0.012

[a] H_2O_2 data from DeMore *et al.* (1997) recommendation based on Molina and Molina (1981), Vaghijiani and Ravishankara (1989); Lin *et al.*, *Geophys. Res. Lett.*, 5, 113 (1978); and Nicovich and Wine, *J. Geophys. Res.*, 93, 2417 (1988); CH_3OOH data from Vaghjiani and Ravishankara (1989).

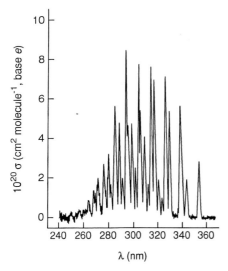

FIGURE 4.26 Absorption spectrum of HCHO at room temperature (adapted from Rogers, 1990).

Table 4.24 summarizes absorption cross sections for HCHO averaged over 2.5-nm intervals centered at the wavelengths shown.

HCHO photolyzes to form two sets of products, one free radical in nature and one set of stable products:

$$\text{HCHO} + h\nu \rightarrow \text{H} + \text{HCO}, \quad (29a)$$

$$\rightarrow \text{H}_2 + \text{CO}. \quad (29b)$$

Table 4.25 shows the recommended quantum yields for these two channels (Atkinson et al., 1997a).

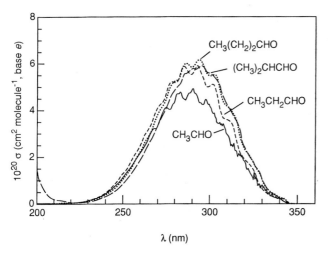

FIGURE 4.27 Absorption spectra for some simple aldehydes (adapted from Martinez et al., 1992).

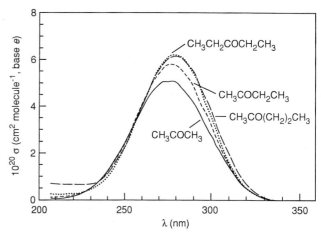

FIGURE 4.28 Absorption spectra for some simple ketones (adapted from Martinez et al., 1992).

Table 4.26 summarizes the absorption cross sections for the relatively unstructured absorptions by other small aldehydes and ketones (Martinez et al., 1992). For acetaldehyde, there are three possible decomposi-

TABLE 4.24 Recommended Absorption Cross Sections (Base e) for HCHO at 223 and 293 K[a,b]

Wavelength (nm)	$10^{20} \sigma$ (cm^2 molecule^{-1})	
	223 K	293 K
301.25	1.38	1.36
303.75	4.67	4.33
306.25	3.32	3.25
308.75	2.27	2.22
311.25	0.758	0.931
313.75	3.65	3.40
316.25	4.05	3.89
318.75	1.66	1.70
321.25	1.24	1.13
323.75	0.465	0.473
326.25	5.06	4.44
328.75	2.44	2.29
331.25	1.39	1.28
333.75	0.0926	0.123
336.25	0.127	0.113
338.75	3.98	3.36
341.25	0.805	0.936
343.75	1.44	1.26
346.25	0.00339	0.071
348.75	0.00905	0.0397
351.25	0.169	0.235
353.75	1.83	1.55
356.25	0.0354	0.125

[a] 2.5-nm interval centered at given λ.
[b] IUPAC recommendation (Atkinson et al., 1997a) and DeMore et al. (1997) recommendation based on Cantrell et al. (1990).

tion channels:

$$CH_3CHO + h\nu \rightarrow CH_3 + HCO, \quad (30a)$$
$$\rightarrow CH_4 + CO, \quad (30b)$$
$$\rightarrow CH_3CO + H. \quad (30c)$$

The quantum yield for (30c) is < 0.01 at all wavelengths above 290 nm and hence can be ignored. Table 4.27 summarizes the recommended quantum yields for (30a) and (30b) (Atkinson et al., 1997a).

There are fewer data available on the photochemistry of larger aldehydes, but it appears that when such compounds are present in air, they may also serve as significant free radical sources. For example, Cronin and Zhu (1998) have measured the yields of HCO from the photolysis of n-pentanal, presumably due to the channel giving n-C_4H_9 + HCO. The HCO yield varied from 0.058 at 280 nm to a peak of 0.20 at 315 nm, decreasing to 0.087 at 330 nm. These yields are sufficiently large that the rate of free radical production from the photolysis of n-pentanal is calculated to be about twice that of CH_3CHO. A variety of stable products such as CH_3CHO were also observed, suggesting the contribution of other photolysis channels (e.g., the Norrish Type II process (Calvert and Pitts, 1966), giving C_3H_6 + CH_3CHO) that will further contribute to free radical formation via subsequent photolysis of such products.

As discussed in Chapter 6.J, acetone photochemistry is of interest because this ketone is distributed globally, has both biogenic and anthropogenic sources, and has been proposed to be a significant source of free radicals in the upper troposphere. The absorption cross sections of acetone (as well as other aldehydes and ketones) are temperature dependent at the longer wavelenths, which is important for application to the colder upper troposphere. Figure 4.29, for example, shows the absorption cross sections of acetone at 298 and 261 K, respectively (Hynes et al., 1992; see also Gierczak et al., 1998).

Acetone has two potential paths for photochemical decomposition:

$$CH_3COCH_3 + h\nu \rightarrow CH_3CO + CH_3, \quad (31a)$$
$$\rightarrow 2CH_3 + CO. \quad (31b)$$

The wavelength thresholds are 338 nm for (31a) and 299 nm for (31b). Thus, (31a) is expected to predominate at the earth's surface. As might be expected, once excited, acetone can be collisionally quenched in competition with decomposition, and hence the quantum yields decrease with increasing total pressure. Figure 4.30, for example, shows the measured quantum yields for the decomposition of acetone at 760 Torr total pressure and the values extrapolated to zero pressure (Gierczak et al., 1998). In the tropospherically important wavelength region, the yields are small beyond about 330 nm at 1 atm pressure.

Gierczak et al. (1998) have also measured the temperature dependence for the absorption cross sections in addition to the quantum yields as a function of pressure and temperature. They have used these data, combined with the kinetics of the OH–acetone reaction, which is the other major removal process, to calculate the contributions of the OH reactions and of photolysis to the loss of acetone in the atmosphere as a function of altitude. Figure 4.31 shows that photolysis is a significant, but not the major, contributor at the

TABLE 4.25 Recommended Quantum Yields for Photolysis of HCHO[a]

Wavelength (nm)	H + HCO	H_2 + CO
240	0.27	0.49
250	0.29	0.49
260	0.30	0.49
270	0.38	0.43
280	0.57	0.32
290	0.73	0.24
300	0.78	0.21
301.25	0.749	0.251
303.75	0.753	0.247
306.25	0.753	0.247
308.75	0.748	0.252
311.25	0.739	0.261
313.75	0.724	0.276
316.25	0.684	0.316
318.75	0.623	0.368
321.25	0.559	0.423
323.75	0.492	0.480
326.25	0.420	0.550
328.75	0.343	0.634
331.25	0.259	0.697
333.75	0.168	0.739
336.25	0.093	0.728
338.75	0.033	0.667
341.25	0.003	0.602
343.75	0.001	0.535
346.25	0	0.469
348.75	0	0.405
351.25	0	0.337
353.75	0	0.265
356.25	0	0.197

[a] IUPAC recommendations from 240 to 300 nm (Atkinson et al., 1997a) and NASA recommendations from 301 to 356 nm (DeMore et al., 1997), where the latter are for 2.5 nm intervals centered on the indicated wavelength based on Horowitz and Calvert, Int. J. Chem. Kinet., 10, 805 (1978); Moortgat and Warneck, J. Chem. Phys., 70, 3639 (1979); and Moortgat et al., ibid., 78, 1185 (1983).

TABLE 4.26 Absorption Cross Sections^a (Base e) for Acetaldehyde, Propionaldehyde, n-Butyraldehyde, Acetone, and 2-Butanone at 300 K^a

Wavelength (nm)	$10^{20} \sigma$ (cm^2 molecule^{-1})				
	CH$_3$CHO	CH$_3$CH$_2$CHO	n-CH$_3$CH$_2$CH$_2$CHO	CH$_3$COCH$_3$	CH$_3$COCH$_2$CH$_3$
210	0.049	0.057	0.048	0.104	0.160
218	0.052	0.080	0.078	0.163	0.225
230	0.151	0.163	0.135	0.533	0.534
238	0.375	0.407	0.320	1.09	1.03
250	1.13	1.29	1.06	2.47	2.45
258	1.99	2.25	1.96	3.61	3.74
270	3.42	4.12	3.85	4.91	5.40
280	4.50	5.16	5.24	5.05	5.74
290	4.89	5.56	5.85	4.19	4.94
295	4.27	5.57	6.10	3.52	4.08
300	4.16	5.04	5.48	2.77	3.30
305	3.42	4.32	5.14	2.11	2.33
310	2.93	3.60	4.12	1.41	1.58
315	2.20	2.77	3.50	0.858	0.896
320	1.72	1.83	2.28	0.467	0.457
325	1.14	1.30	1.71	0.205	0.189
330	0.688	0.575	0.878	0.067	0.067
335	0.350	0.325	0.491	0.017	0.020
340	0.150	0.155	0.217	0.005	0.005
345	0.021	0.025	0.042	0.002	0.001
350	0.008	0.010	0.015	0.001	0.000
355	0.004	0.002	0.006	0.000	0.000
360	0.003	0.000	0.002		

^{a} Adapted from Martinez et al. (1992); for values every 1 nm, see original reference.

earth's surface. However, it becomes increasingly important at higher altitudes due to the decreased quenching at lower pressures and hence increased photodissociation quantum yields. The OH reaction becomes less important because of the decreasing OH concentrations and slower kinetics at the lower temperatures.

N. CHLORINE NITRATE (ClONO$_2$) AND BROMINE NITRATE (BrONO$_2$)

Chlorine nitrate and bromine nitrate are recognized as key species in the chemistry of the stratosphere. In

TABLE 4.27 Recommended Quantum Yields for CH$_3$CHO Photolysis at 298 K^a

Wavelength (nm)	CH$_3$ + HCO	CH$_4$ + CO
260	0.31	0.46
270	0.39	0.31
280	0.58	0.05
290	0.53	0.01
295	0.48	0.00
300	0.43	
305	0.37	
315	0.17	
320	0.10	
325	0.04	
330	0.00	

^{a} IUPAC recommendation (Atkinson et al., 1997a) based on Horowitz and Calvert, J. Phys. Chem., 86, 3105 (1982) and Meyrahn et al., in Atmospheric Trace Constituents, F. Herbert, Ed., Mainz, Germany (1981).

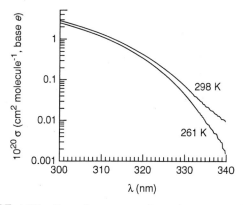

FIGURE 4.29 Absorption cross sections of acetone at 298 and 261 K (adapted from Hynes et al., 1992).

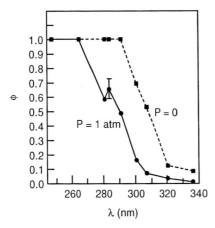

FIGURE 4.30 Measured quantum yields for acetone photodissociation as a function of wavelength at 1 atm total pressure and extrapolated to zero total pressure (adapted from Gierczak et al., 1998).

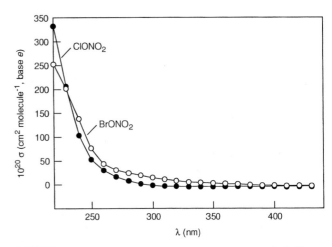

FIGURE 4.32 Absorption spectra of ClONO$_2$ and BrONO$_2$ at room temperature (based on data in DeMore et al., 1997, Burkholder et al., 1994, and Deters et al., 1998).

addition, as discussed in Chapter 6.A, there is an increasing recognition that since atomic chlorine and bromine may play key roles in the chemistry of the marine boundary layer, they may also be important in the troposphere.

Figure 4.32 shows the absorption spectra of these two nitrates at room temperature, and Table 4.28 summarizes the absorption cross sections (Burkholder et al., 1994, 1995; DeMore et al., 1997; Deters et al., 1998).

There are several feasible photolysis routes for these nitrates; e.g., for chlorine nitrate:

$$ClONO_2 + h\nu \rightarrow ClO + NO_2 \quad (\lambda \leq 1066 \text{ nm}), \tag{32a}$$

$$\rightarrow Cl + NO_3 \quad (\lambda \leq 6970 \text{ nm}), \tag{32b}$$

$$\rightarrow O + ClONO \quad (\lambda \leq 426 \text{ nm}), \tag{32c}$$

$$\rightarrow Cl + O + NO_2 \quad (\lambda \leq 314 \text{ nm}). \tag{32d}$$

A variety of experimental measurements indicated that the last two channels giving oxygen atoms are not important and that the first two predominate over the wavelengths of atmospheric interest (see review in DeMore et al., 1997). Table 4.29 summarizes some of the measurements of the species expected from reactions (32a)–(32d). These measurements are in agreement with studies of ClONO$_2$ photolysis using broadband light sources, in which the branching ratio for (32a) above 300 nm was reported to be 0.44 ± 0.08 and that for (32b) 0.56 ± 0.08; between 200 and 300 nm, these branching ratios were 0.61 ± 0.20 and 0.39 ± 0.20, respectively (Nickolaisen et al., 1996). Although pressure dependence was reported in the latter studies, other studies find no pressure dependence up to pressures of ~100 Torr or differences using various bath gases such as He and N$_2$ (Tyndall et al., 1997; Goldfarb et al., 1997). Interestingly, the ClO radicals generated in (32a) are vibrationally excited (Tyndall et al., 1997).

The photochemistry of bromine nitrate is expected to have similar reaction channels:

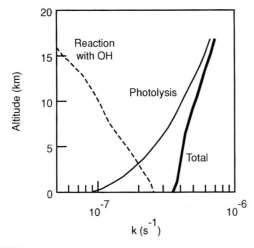

FIGURE 4.31 Calculated first-order rate constants for loss of CH$_3$COCH$_3$ due to reaction with OH (i.e., k[OH]) or photolysis (i.e., k_p) as a function of altitude (adapted from Gierczak et al., 1998).

TABLE 4.28 Recommended Absorption Cross Sections (Base e) of $ClONO_2$ and $BrONO_2$ at Room Temperature[a]

Wavelength (nm)	$10^{20} \sigma$ (cm² molecule⁻¹)	
	$ClONO_2$	$BrONO_2$
210	314	362
220	332	254
230	208	204
240	105	140
250	56.0	78.9
260	33.8	47.1
270	20.5	33.7
280	11.9	27.8
290	6.41	23.1
300	3.13	18.8
310	1.60	14.7
320	0.831	11.6
330	0.466	9.70
340	0.302	8.47
350	0.229	7.35
360	0.200	6.28
370	0.159	5.11
380	0.121	4.01
390	0.0909	3.05
400	0.0638	2.31
410	0.0444	1.81
420	0.0316	1.49
430	0.0189	1.28
440		1.10
450		0.925
460		0.740
470		0.553
480		0.394
490		0.255
500		0.165

[a] Recommended (DeMore *et al.*, 1997) $ClONO_2$ absorption cross sections are from Burkholder *et al.* (1994), and $BrONO_2$ absorption cross sections are from Deters *et al.* (1998), which are in excellent agreement with Burkholder *et al.* (1995).

$$BrONO_2 + h\nu \rightarrow BrO + NO_2 \quad (\lambda \leq 1076 \text{ nm}), \tag{33a}$$

$$\rightarrow Br + NO_3 \quad (\lambda \leq 879 \text{ nm}), \tag{33b}$$

$$\rightarrow O + BrONO \quad (\lambda \leq 391 \text{ nm}), \tag{33c}$$

$$\rightarrow Br + O + NO_2 \quad (\lambda \leq 344 \text{ nm}). \tag{33d}$$

NO_3 production has been observed in this system (Deters *et al.*, 1998; Harwood *et al.*, 1998). The quantum yields for NO_3 at 248, 308, and 352.5 nm have been reported to be 0.28 ± 0.09, 1.01 ± 0.35, and 0.92 ± 0.43, respectively, indicating that channel (33b) is a

TABLE 4.29 Some Product Quantum Yield Measurements in the Photolysis of $ClONO_2$[a]

λ (nm)	Cl[a]	NO_3[b]	ClO[a]	NO_2	O[a]
193.2	0.53	0.18	0.29	0.36[c]	0.37
222.0	0.46	—	0.64	—	0.17
248.25	0.41	0.60	0.39	0.46[c]	< 0.10
308.15	0.67[c,d]	0.67			
	0.64	0.67[d]	0.37	0.33[d]	< 0.05[a,e]
	0.80[e]		0.28[e]		
352.5	—	0.93	—	—	—

[a] Goldfarb *et al.* (1997) unless otherwise indicated.
[b] Yokelson *et al.* (1997).
[c] Minton *et al.* (1992); ratios converted to absolute values assuming a photodissociation quantum yield of 1 and only two channels.
[d] Moore *et al.* (1995); ratios converted to absolute values assuming a photodissociation quantum yield of 1 and only two channels.
[e] Tyndall *et al.* (1997).

major pathway in the atmosphere; at 248 nm, the quantum yield of oxygen atoms was ~ 0.2, that for Br ~ 0.5, and that for BrO ~ 0.5 (Harwood *et al.*, 1998).

O. HCl AND HBr

Figure 4.33 shows the absorption cross sections of HCl and HBr at room temperature (DeMore *et al.*, 1997; Huebert and Martin, 1968). Neither absorb above 290 nm, so their major tropospheric fates are deposition or reaction with OH. Even in the stratosphere, photolysis is sufficiently slow that these hydrogen halides act as temporary halogen reservoirs (see Chapter 12).

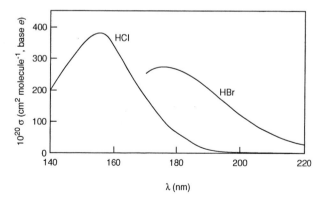

FIGURE 4.33 Absorption spectra of HCl and HBr at room temperature (based on data in DeMore *et al.*, 1997, and Huebert and Martin, 1968).

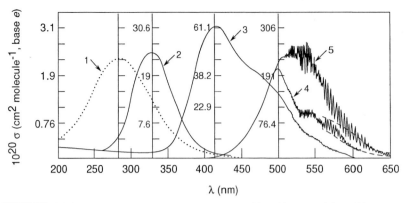

FIGURE 4.34 Absorption spectra of the halogens: (1) $F_2(g)$, 25°C; (2) $Cl_2(g)$, 18°C; (3) $Br_2(g)$, 25°C; (4) $I_2(g)$, 70–80°C; (5) $I_2(g)$ plus 1 atm air, 70–80°C (adapted from Calvert and Pitts, 1966).

P. THE HALOGENS

Figure 4.34 compares the absorption spectra of the diatomic halogens, F_2, Cl_2, Br_2, and I_2. Cl_2 is of particular recent interest in the troposphere in that levels up to ~ 150 ppt have been observed in marine areas (Keene *et al.*, 1993; Pszenny *et al.*, 1993; Spicer *et al.*, 1998). Table 4.30 summarizes the absorption cross sections of Cl_2, Br_2, and BrCl (DeMore *et al.*, 1997; Maric *et al.*, 1994; Hubinger and Nee, 1995). These diatomics all dissociated with a photodissociation quantum yield of 1 (Calvert and Pitts, 1966).

Q. ClO, BrO, AND IO

Figures 4.35, 4.36, and 4.37 show the absorption spectra of the free radicals ClO, BrO, and IO, respectively (Wahner *et al.*, 1988; DeMore *et al.*, 1997; Laszlo *et al.*, 1995). All have beautifully banded structures at longer wavelengths and large absorption cross sections, which allows one to measure these species in laboratory and atmospheric systems using differential optical absorption spectrometery (DOAS) (see Chapter 11.A.1d). However, as in the case of HCHO, adequate resolution is an important factor in obtaining accurate cross sections.

Dissociation to the constituent atoms is clearly the only possible photochemical decomposition path. The photochemistry of ClO has been studied as a function of wavelength from 237 to 270 nm by Schmidt *et al.* (1998a) using REMPI (resonance-enhanced multiphoton ionization) to follow ClO, chlorine, and oxygen atoms. ClO dissociates to chlorine and oxygen atoms with a quantum yield of 1. At wavelengths greater than 263.4 nm, where the absorption is banded, the oxygen atoms are generated in the ground $O(^3P)$ state, whereas at shorter wavelengths where absorption is continuous, $O(^1D)$ is formed.

IO absorbs strongly into the visible region, making its photochemistry in the troposphere potentially quite interesting. Table 4.31 gives the average absorption cross sections for both BrO and IO over 5-nm intervals (DeMore *et al.*, 1997; Laszlo *et al.*, 1995; Cox and Coker, 1983), which can be used for calculating its lifetime with respect to photolysis. Laszlo *et al.* (1995) calculate a lifetime for IO of only 3.7 s at a solar zenith angle of 40°, suggesting that photolysis will predominate in its loss processes *if* the quantum yield for photodissociation is 1.0 as assumed in these calculations.

R. ClOOCl

As discussed in Chapter 12, the ClO dimer is a central species in the chemistry of the Antarctic ozone hole. Table 4.32 gives the recommended absorption cross sections (DeMore *et al.*, 1997). The photodissociation can, in principle, proceed by two paths:

$$ClOOCl + h\nu \rightarrow Cl + ClOO, \qquad (34a)$$
$$\rightarrow 2ClO. \qquad (34b)$$

The reaction is believed to proceed predominantly via (34a) to generate chlorine atoms (Molina *et al.*, 1990). However, several groups have reported that the quantum yield for this channel is less than 1.0. For example, Schmidt *et al.* (1998b) used REMPI with time-of-flight mass spectrometry (TOF-MS) to follow the production of oxygen and chlorine atoms as well as ClO in vibrational levels up to $v'' = 5$ in the photolysis of the dimer. At a photolysis wavelength of 250 nm, the

TABLE 4.30 Absorption Cross Sections (Base e) of Cl_2, Br_2, and BrCl at 298 K[a]

Wavelength (nm)	$10^{20}\sigma$ (cm^2 molecule^{-1})		
	Cl_2	Br_2	BrCl
260	0.20	0.31	2.5
270	0.82	0.19	1.5
280	2.6	0.13	1.2
290	6.2	0.10	0.63
300	11.9	0.09	0.61
310	18.5	0.12	1.2
320	23.7	0.15	2.8
330	25.5	0.35	7.4
340	23.5	1.1	14.2
350	18.8	3.4	22.9
360	13.2	8.4	33.3
370	8.4	16.8	38.7
380	5.0	28.7	38.5
390	2.9	42.1	34.7
400	1.8	53.0	28.2
410	1.3	64.1	21.9
420	0.96	64.8	16.9
430	0.73	60.5	14.2
440	0.54	53.4	12.4
450	0.38	48.1	11.1
460	0.26	44.3	9.6
470	0.16	41.5	8.0
480		36.6	6.8
490		33.5	5.0
500		30.2	3.8
510		24.7	3.1
520		19.5	2.3
530		15.4	1.5
540		12.1	0.96
550		9.5	0.76
560		7.0	0.31
570		5.3	
580		3.6	
590		2.5	
600		1.1	

[a] From DeMore et al. (1997) recommendation based on Maric et al., J. Photochem. Photobiol. A., 70, 205 (1993) for Cl_2, and Maric et al. (1994) and Hubinger and Nee (1995) for Br_2 and BrCl.

quantum yield for chlorine atom production was measured to be 0.65 ± 0.15, but ClO was not observed. Assuming that all of the excited dimer dissociates, this suggests that the production of ClO in vibrational levels $v'' > 5$ accounted for about a third of the reaction. Moore et al. (1999) used TOF-MS to follow the products of photolysis (including Cl atoms, ClO, and O_2) of the dimer at 248 and 308 nm, respectively. At 248 nm, chlorine atom production from two primary processes was observed and attributed to reaction (34a), along with a concerted process producing $2Cl + O_2$ directly. They also observed the production of ClO at both 248 and 308 nm. At 248 nm, the relative Cl and ClO production was 0.88:0.12 and at 308, it was 0.90:0.10. Their studies suggest that under stratospheric conditions, the quantum yield for the channel producing chlorine atoms is 0.9 ± 0.1 and that for producing ClO is 0.1 ± 0.1 with an upper limit of 0.31, which is not inconsistent with the work of Schmidt et al. (1998b).

S. OClO

Figure 4.38 shows the absorption spectrum of OClO at 204 K (Wahner et al., 1987; DeMore et al., 1997). The photochemistry, which is complex, is reviewed by DeMore et al. (1997), with the recommendation that a quantum yield of 1 be adopted for reaction (35):

$$OClO + h\nu \rightarrow O + ClO. \quad (35)$$

The absorption spectrum for the analogous bromine compound, OBrO, is also highly structured, extending from ~400 to 650 nm (e.g., Miller et al., 1997).

T. HOCl, HOBr, AND HOI

Figure 4.39 shows the absorption spectrum of HOCl and Table 4.33 summarizes the absorption cross sections, for which most measurements are in relatively good agreement. The major products are OH + Cl, with a quantum yield of 1 above 290 nm (e.g., Vogt and Schindler, 1992; Schindler et al., 1997):

$$HOCl + h\nu \rightarrow OH + Cl. \quad (36)$$

Barnes et al. (1998) have measured the yield of OH from HOCl photolysis and find, in addition to the strong absorption shown in Fig. 4.39, a weak absorption feature at 380 nm due to excitation to the lowest triplet state. Although the absorption cross section of this weak absorption is only 4×10^{-21} cm^2 molecule^{-1}, its contribution lowers the calculated stratospheric lifetime of HOCl by ~10–20%.

Figure 4.40 shows measurements of the HOBr absorption cross sections (Benter et al., 1995; Orlando and Burkholder, 1995; Barnes et al., 1996; Rattigan et al., 1996; Deters et al., 1996; Ingham et al., 1998). As is evident from this figure, there is significant disagreement in the absorption cross sections, particularly for the weak absorption band at wavelengths beyond 400 nm. However, this appears to be a real feature of the HOBr absorption. Thus, Sinha and co-workers (Barnes et al., 1996) monitored the production of OH as a function of photolysis wavelength in a mixture containing HOBr. Production of OH in the 440- to 540-nm region was observed, indicative of a weak absorption band here, possibly due to a transition from the ground state to a dissociative triplet state. Although it is a

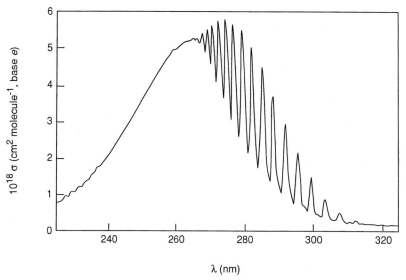

FIGURE 4.35 Absorption spectrum of ClO (adapted from DeMore *et al.*, 1997).

much weaker absorption than at the shorter wavelengths, in this region the solar flux is increasing significantly (see Chapter 3.C.1). Barnes *et al.* (1996) and Ingham *et al.* (1998) calculate that at large solar zenith angles, this weak absorption could decrease the lifetime with respect to photolysis by almost a factor of two.

Because of the uncertainties in the absolute cross sections for HOBr, both the DeMore *et al.* (1997) recommendation and the more recent values of Ingham *et al.* (1998) are given in Table 4.34.

The process again appears to be primarily dissociation to OH + Br, with a quantum yield for bromine atoms of >0.95 at 363 nm (Benter *et al.*, 1995).

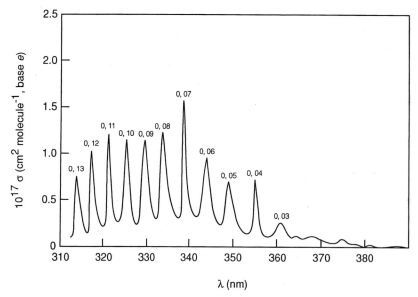

FIGURE 4.36 Absorption spectrum of BrO at room temperature (adapted from Wahner *et al.*, 1988).

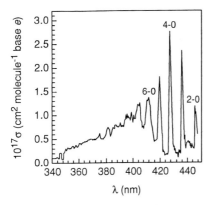

FIGURE 4.37 Absorption spectrum of IO at room temperature (adapted from Laszlo et al., 1995).

Figure 4.41 shows one set of absorption cross sections for HOI (Bauer et al., 1998). This dissociates to OH + I, with a quantum yield for OH production of ~1 at 355 nm (Bauer et al., 1998).

U. NITROSYL CHLORIDE (ClNO) AND NITRYL CHLORIDE (ClNO$_2$)

Figures 4.42 and 4.43 show the absorption spectra of nitrosyl chloride (ClNO) and nitryl chloride (ClNO$_2$), respectively (Roehl et al., 1992; Ganske et al., 1992), and Tables 4.35 and 4.36 list the recommended room temperature absorption cross sections (DeMore et al., 1997). In both cases, chlorine atoms are produced with unit quantum yield:

$$\text{ClNO} + h\nu \rightarrow \text{Cl} + \text{NO}, \quad (37)$$

$$\text{ClNO}_2 + h\nu \rightarrow \text{Cl} + \text{NO}_2. \quad (38)$$

V. HALOGENATED METHANES AND ETHANES

Figures 4.44 and 4.45 show absorption spectra of some simple chlorofluoro-methanes and ethanes, respectively (Hubrich and Stuhl, 1980). Tables 4.37 and 4.38 give the recommended absorption cross sections for some of these compounds (DeMore et al., 1997). None of these compounds absorb in the actinic region above 290 nm, but do around 180–200 nm, wavelengths only found in the stratosphere. As discussed in Chapter 12, it is photolysis at these short wavelengths to generate atomic chlorine that is responsible, along with bromine and perhaps in some cases, iodine atoms, for the chain destruction of stratospheric ozone.

TABLE 4.31 Absorption Cross Sections (Base e) Averaged over 5-nm Intervals in BrO and IO[a]

Wavelength region (nm)	$10^{18} \sigma_{av}$ (cm^2 molecule^{-1}) BrO	IO
300–305	2.00	
305–310	2.59	
310–315	4.54	
315–320	3.91	
320–325	6.00	
325–330	7.53	
330–335	6.28	
335–340	5.89	
340–345	5.15	
345–350	3.99	1.5
350–355	2.28	2.9
355–360	1.72	3.3
360–365	1.61	3.7
365–370	0.92	4.2
370–375	0.51	4.9
375–380		3.8
380–385		5.6
385–390		6.0
390–395		6.7
395–400		7.6
400–405		7.8
405–410		9.4
410–415		9.3
415–420		11.1
420–425		8.3
425–430		11.4
430–435		6.1
435–440		11.7
440–445		3.0
445–450		14.1
450–455		4.0
455–460		10.0
460–465		4.2
465–470		2.8

[a] From DeMore et al. (1997) recommendation based on Laszlo et al. (1995) and Cox and Coker (1983).

Tables 4.39 and 4.40 give the recommended absorption cross sections of some hydrochlorofluorocarbons at room temperature, and those of some brominated organics are found in Table 4.41 (DeMore et al., 1997).

Finally, alkyl iodides, some of which have natural sources, are of interest for both the troposphere and stratosphere. Figure 4.46 shows the absorption spectra of some simple alkyl iodides and Table 4.42 the absorption cross sections for CH$_3$I (Roehl et al., 1997).

W. PROBLEMS

1. Small amounts of O$_3$ are generated by photolysis of O$_2$ at 248 nm, even though this is beyond the

TABLE 4.32 Recommended Absorption Cross Sections (Base e) of ClOOCl at 200–250 K[a]

λ (nm)	$10^{20}\sigma$ (cm^2 molecule^{-1})	λ (nm)	$10^{20}\sigma$ (cm^2 molecule^{-1})	λ (nm)	$10^{20}\sigma$ (cm^2 molecule^{-1})	λ (nm)	$10^{20}\sigma$ (cm^2 molecule^{-1})
190	565.0	256	505.4	322	23.4	388	1.4
192	526.0	258	463.1	324	21.4	390	1.3
194	489.0	260	422.0	326	19.2	392	1.2
196	450.0	262	381.4	328	17.8	394	1.1
198	413.0	264	344.6	330	16.7	396	1.0
200	383.5	266	311.6	332	15.6	398	0.92
202	352.9	268	283.3	334	14.4	400	0.85
204	325.3	270	258.4	336	13.3	402	0.78
206	298.6	272	237.3	338	13.1	404	0.71
208	274.6	274	218.3	340	12.1	406	0.65
210	251.3	276	201.6	342	11.5	408	0.60
212	231.7	278	186.4	344	10.9	410	0.54
214	217.0	280	172.5	346	10.1	412	0.50
216	207.6	282	159.6	348	9.0	414	0.46
218	206.1	284	147.3	350	8.2	416	0.42
220	212.1	286	136.1	352	7.9	418	0.38
222	227.1	288	125.2	354	6.8	420	0.35
224	249.4	290	114.6	356	6.1	422	0.32
226	280.2	292	104.6	358	5.8	424	0.29
228	319.5	294	95.4	360	5.5	426	0.27
230	365.0	296	87.1	362	4.5	428	0.25
232	415.4	298	79.0	364	4.1	430	0.23
234	467.5	300	72.2	366	3.8	432	0.21
236	517.5	302	65.8	368	3.5	434	0.19
238	563.0	304	59.9	370	3.2	436	0.17
240	600.3	306	54.1	372	2.9	438	0.16
242	625.7	308	48.6	374	2.7	440	0.15
244	639.4	310	43.3	376	2.4	442	0.13
246	642.6	312	38.5	378	2.2	444	0.12
248	631.5	314	34.6	380	2.1	446	0.11
250	609.3	316	30.7	382	1.9	448	0.10
252	580.1	318	28.0	384	1.7	450	0.09
254	544.5	320	25.6	386	1.6		

[a] From DeMore et al. (1997) recommendation based on Cox and Hayman, *Nature*, *332*, 796 (1988), DeMore and Tschvikow-Roux, *J. Phys. Chem.*, *94*, 5856 (1990), Permien et al., Air Pollution Report #17, Environmental Research Program of the CEC, Brussels (1988) and Burkholder et al., *J. Phys. Chem.*, *94*, 687 (1990).

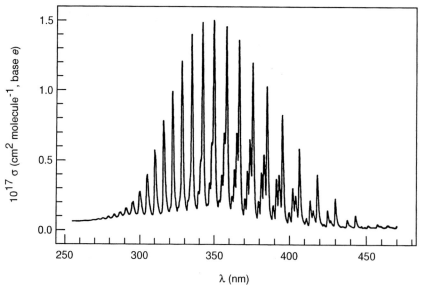

FIGURE 4.38 Absorption spectrum of OClO at 204 K (adapted from Wahner et al., 1987).

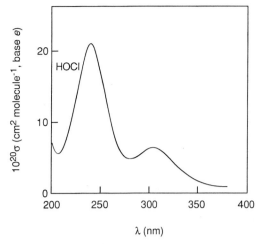

FIGURE 4.39 Absorption spectrum of HOCl (adapted from Burkholder, 1993).

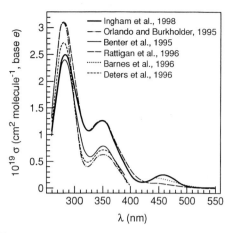

FIGURE 4.40 Absorption spectra of HOBr (adapted from Ingham et al., 1998).

TABLE 4.33 Absorption Cross Sections of HOCl at Room Temperature[a]

Wavelength (nm)	$10^{20}\sigma$ (cm^2 molecule^{-1})	Wavelength (nm)	$10^{20}\sigma$ (cm^2 molecule^{-1})	Wavelength (nm)	$10^{20}\sigma$ (cm^2 molecule^{-1})
200	7.1	262	9.3	322	4.6
202	6.1	264	8.3	324	4.3
204	5.6	266	7.4	326	4.2
206	5.4	268	6.6	328	3.8
208	5.5	270	6.0	330	3.5
210	5.7	272	5.5	332	3.3
212	6.1	274	5.2	334	3.1
214	6.6	276	4.9	336	2.7
216	7.5	278	4.8	338	2.5
218	8.4	280	4.7	340	2.4
220	9.7	282	4.8	342	2.1
222	10.9	284	4.8	344	1.8
224	12.2	286	4.9	346	1.8
226	13.5	288	5.1	348	1.7
228	15.0	290	5.3	350	1.5
230	16.4	292	5.4	352	1.3
232	17.7	294	5.6	354	1.3
234	18.7	296	5.8	356	1.2
236	19.7	298	5.9	358	1.0
238	20.3	300	6.0	360	0.8
240	20.7	302	6.0	362	1.0
242	21.0	304	6.1	364	1.0
244	20.5	306	6.0	366	0.9
246	19.6	308	6.0	368	0.8
248	18.6	310	5.9	370	0.8
250	17.3	312	5.7	372	1.0
252	15.9	314	5.6	374	0.8
254	14.6	316	5.4	376	0.8
256	13.2	318	5.1	378	0.6
258	11.8	320	4.9	380	0.8
260	10.5				

[a] From DeMore et al. (1997) recommendation based on Burkholder (1993).

TABLE 4.34 Absorption Cross Sections of HOBr

λ (nm)	$10^{20}\sigma$ (cm² molecule⁻¹)	λ (nm)	$10^{20}\sigma$ (cm² molecule⁻¹)	λ (nm)	$10^{20}\sigma$ (cm² molecule⁻¹)
240	6.7	330	11.0 (10.8)	420	1.3 (0.967)
245	5.2	335	11.5 (11.3)	425	1.1 (1.00)
250	6.7	340	12.0 (11.9)	430	0.92 (1.15)
255	9.9	345	12.3 (12.3)	435	0.84 (1.40)
260	14.1 (10.5)	350	12.5 (12.4)	440	0.74 (1.68)
265	18.9 (14.6)	355	12.2 (12.2)	445	0.71 (1.96)
270	23.9 (18.7)	360	11.6 (11.5)	450	0.67 (2.18)
275	28.0 (22.1)	365	10.7 (10.5)	455	0.65 (2.28)
280	30.4 (24.3)	370	9.6 (9.32)	460	0.61 (2.28)
285	30.8 (25.0)	375	8.4 (8.00)	465	0.53 (2.14)
290	28.7 (24.0)	380	7.4 (6.66)	470	0.49 (1.91)
295	25.2 (21.9)	385	6.2 (5.38)	475	0.40 (1.62)
300	20.9 (19.1)	390	5.1 (4.22)	480	0.34 (1.30)
305	16.8 (16.2)	395	4.1 (3.24)	485	0.28 (0.993)
310	13.8 (13.6)	400	3.3 (2.43)	490	0.21 (0.723)
315	11.8 (11.8)	405	2.6 (1.80)	495	0.14 (0.502)
320	10.8 (10.8)	410	2.0 (1.36)	500	0.09 (0.334)
325	10.6 (10.5)	415	1.6 (1.08)	505	0.05 (0.212)

[a] From DeMore *et al.* (1997) recommendation based on Rattigan *et al.* (1996); values in parentheses are from Ingham *et al.* (1998).

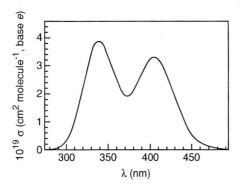

FIGURE 4.41 Absorption spectrum of HOI (adapted from Bauer *et al.*, 1998).

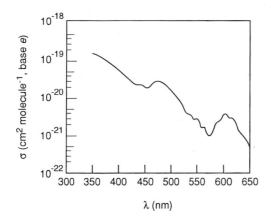

FIGURE 4.42 Absorption spectrum of ClNO (adapted from Roehl *et al.*, 1992).

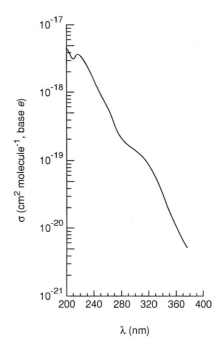

FIGURE 4.43 Absorption spectrum of ClNO$_2$ at 298 K (adapted from Ganske *et al.*, 1992).

TABLE 4.35 Recommended Absorption Cross Sections (Base e) of ClNO at Room Temperature[a]

Wavelength (nm)	$10^{20}\sigma$ (cm^2 molecule^{-1})	Wavelength (nm)	$10^{20}\sigma$ (cm^2 molecule^{-1})	Wavelength (nm)	$10^{20}\sigma$ (cm^2 molecule^{-1})	Wavelength (nm)	$10^{20}\sigma$ (cm^2 molecule^{-1})
190	4320	246	45.2	302	10.3	370	11.0
192	5340	248	37.7	304	10.5	375	9.95
194	6150	250	31.7	306	10.8	380	8.86
196	6480	252	27.4	308	11.1	385	7.82
198	6310	254	23.7	310	11.5	390	6.86
200	5860	256	21.3	312	11.9	395	5.97
202	5250	258	19.0	314	12.2	400	5.13
204	4540	260	17.5	316	12.5	405	4.40
206	3840	262	16.5	318	13.0	410	3.83
208	3210	264	15.3	320	13.4	415	3.38
210	2630	266	14.4	322	13.6	420	2.89
212	2180	268	13.6	324	14.0	425	2.45
214	1760	270	12.9	326	14.3	430	2.21
216	1400	272	12.3	328	14.6	435	2.20
218	1110	274	11.8	330	14.7	440	2.20
220	896	276	11.3	332	14.9	445	2.07
222	707	278	10.7	334	15.1	450	1.87
224	552	280	10.6	336	15.3	455	1.79
226	436	282	10.2	338	15.3	460	1.95
228	339	284	9.99	340	15.2	465	2.25
230	266	286	9.84	342	15.3	470	2.50
232	212	288	9.71	344	15.1	475	2.61
234	164	290	9.64	346	15.1	480	2.53
236	120	292	9.63	348	14.9	485	2.33
238	101	294	9.69	350	14.2	490	2.07
240	82.5	296	9.71	355	13.6	495	1.78
242	67.2	298	9.89	360	12.9	500	1.50
244	55.2	300	10.0	365	12.0		

[a] From DeMore et al. (1997) recommendation based on Roehl et al. (1992) from 350–500 nm and Tyndall et al., J. Photochem., 36, 133 (1987) from 190–350 nm.

TABLE 4.36 Recommended Absorption Cross Sections (Base e) of Gaseous ClNO$_2$ at 298 K[a]

Wavelength (nm)	$10^{20}\sigma$ (cm^2 molecule^{-1})	Wavelength (nm)	$10^{20}\sigma$ (cm^2 molecule^{-1})
200	468	280	22.0
210	320	290	17.3
216	348	300	14.9
220	339	310	12.1
230	226	320	8.87
240	133	330	5.84
250	90.6	340	3.54
260	61.3	350	2.04
270	35.3	360	1.15
		370	0.69

[a] From DeMore et al. (1997) recommendation based on Ganske et al. (1992).

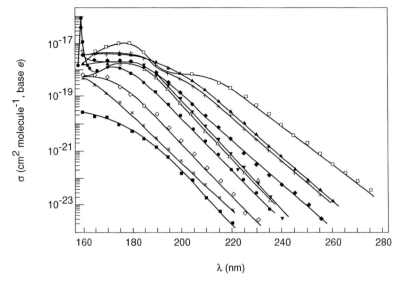

FIGURE 4.44 Absorption cross sections of the halogenated methanes at 298 K: +, $CHCl_3$; △, $CHCl_2F$; ×, $CHClF_2$; ◆, CH_2Cl_2; ◇, CH_2ClF; ●, CH_3Cl; □, CCl_4; ▲, CCl_3F (F-11); ▼, CCl_2F_2 (F-12); ■, $CClF_3$ (adapted from Hubrich and Stuhl, 1980).

TABLE 4.37 Recommended Absorption Cross Sections of Some Chlorofluorocarbons (Units of 10^{-20} cm^2 molecule^{-1}, Base e) at Room Temperature[a]

Wavelength (nm)	CCl_4	CCl_3F (CFC-11)	CCl_2F_2 (CFC-12)	CH_3Cl	$CHClF_2$ (HCFC-22)	CH_3CCl_3
170		316	124			
180	772	308	173		1.91	
190	144	178	62.8	12.7	0.245	192
200	64.8	64.7	8.84	1.76	0.032	81.0
210	46.6	15.4	0.80	0.206		24.0
220	17.0	2.42	0.068			4.15
230	4.07	0.35	0.0055			0.700
240	0.784	0.0464	0.00029			0.102
250	0.183	0.00661				
260	0.0253	0.00147				
270	0.0061					

[a] From DeMore et al. (1997) recommendations based on Hubrich and Stuhl (1980), Hubrich et al., *Ber. Bunsenges. Phys. Chem.*, **81**, 437 (1977), Vanlgethem-Meurée et al., *Bull. Cl. Sci., Acad. R. Belg.*, **64**, 31, 42 (1978); *Geophys. Res. Lett.*, **6**, 451 (1979), Green and Wayne, *J. Photochem.*, **6**, 375 (1976/77) and Simon et al., *J. Atmos. Chem.*, **7**, 107 (1988).

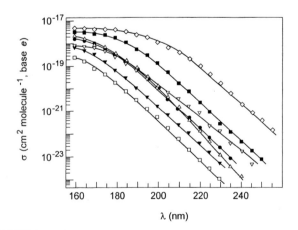

FIGURE 4.45 Absorption spectra of some chlorinated ethanes vs the wavelength at 298 K: ◇, CH_3CCl_3; ▽, CF_3CH_2Cl; △, CH_3CH_2Cl; □, CF_3CClF_2; ▼, CH_3CClF_2; ●, $CClF_2CClF_2$; ■, CCl_2FCClF_2 (adapted from Hubrich and Stuhl, 1980).

TABLE 4.38 Recommended Absorption Cross Sections of Some Chlorofluorocarbons (Units of 10^{-20} cm^2 molecule^{-1}, Base e) at Room Temperature[a]

Wavelength (nm)	$CF_2ClCFCl_2$ (CFC-113)	CF_2ClCF_2Cl (CFC-114)	CF_3CF_2Cl (CFC-115)
180		26	1.45
190	64.5	5.35	0.27
200	12.5	0.80	0.044
210	1.8	0.104	
220	0.220	0.012	
230	0.027		

[a] From DeMore et al. (1997) recommendations based on Simon et al., *Annales Geophysicae*, **6**, 239 (1988).

TABLE 4.39 Recommended Absorption Cross Sections (Base e) of Hydrochlorofluoroethanes at 298 K[a]

Wavelength (nm)	$10^{20}\sigma$ (cm² molecule⁻¹)			
	CH_3CFCl_2 (HCFC-141b)	CH_3CF_2Cl (HCFC-142b)	CF_3CHCl_2 (HCFC-123)	CF_3CHFCl (HCFC-124)
190	83.8	0.94	59.0	0.77
192	64.1	0.66	44.5	0.55
194	47.4	0.46	32.9	0.39
196	34.0	0.31	23.6	0.27
198	23.8	0.21	16.9	0.18
200	16.4	0.14	11.9	0.13
202	11.1	0.09	8.3	0.086
204	7.4	0.061	5.7	0.060
206	4.9	0.039	4.0	0.040
208	3.2	0.026	2.7	0.027
210	2.1	0.017	1.8	0.019
212	1.4	0.010	1.3	0.012
214	0.89	0.007	0.87	0.008
216	0.57	0.004	0.61	0.006
218	0.37	0.003	0.40	0.004
220	0.24	0.002	0.28	0.003

[a] From DeMore *et al.* (1997) recommendations based on Fahr *et al., J. Geophys. Res., 98,* 20467 (1993) for -141b, Gillotay and Simon, *J. Atmos. Chem., 12,* 269 (1991); *ibid, 13,* 289 (1991), and Orlando *et al., J. Geophys. Res., 96,* 5013 (1991) for -142b, -123, and -124.

TABLE 4.40 Recommended Absorption Cross Sections (Base e) of $CF_3CF_2CHCl_2$ (HCFC-225ca) and CF_2ClCF_2CHFCl (HCFC-225cb)[a]

Wavelength (nm)	$10^{20}\sigma$ (cm² molecule⁻¹)	
	$CF_3CF_2CHCl_2$ (HCFC-225ca)	CF_2ClCF_2CHFCl (HCFC-225cb)
160	269	188
165	197	145
170	183	91
175	191	47
180	177	21
185	129	9.1
190	74	3.5
195	37	1.4
200	16	0.63
205	6.9	0.33
210	2.9	0.25
215	1.2	
220	0.46	
225	0.17	
230	0.065	
235	0.025	
239	0.011	

[a] From DeMore *et al.* (1997) recommendations based on Braun *et al., J. Geophys. Res., 96,* 13009 (1991).

TABLE 4.41 Recommended Absorption Cross Sections (Base e) for Some Bromine Compounds at Room Temperature[a]

λ (nm)	$10^{20}\sigma$ (cm² molecule⁻¹)				λ (nm)	$10^{20}\sigma$ (cm² molecule⁻¹)			
	CH_3Br	$CHBr_3$	CF_2ClBr (Halon-1211)	CF_3Br (Halon-1301)		CH_3Br	$CHBr_3$	CF_2ClBr (Halon-1211)	CF_3Br (Halon-1301)
190	44	399	47	6.4	278		12	0.15	
192	53	360	58	7.5	280		9.9	0.1	
194	62	351	70	8.5	282		7.8	0.079	
196	69	366	83	9.5	284		6.1	0.058	
198	76	393	96	10.4	286		4.81	0.043	
200	79	416	112	11.2	288		3.75	0.031	
202	80	433	118	11.8	290		2.88		
204	79	440	121	12.2	292		2.22		
206	77	445	122	12.4	294		1.70		
208	73	451	121	12.4	296		1.28		
210	67	468	117	12.0	298		0.951		
212	61	493	112	11.4	300		0.719		
214	56	524	106	10.7	302		0.530		
216	49	553	98	9.8	304		0.394		
218	44	574	90	8.8	306		0.298		
220	38	582	81	7.7	308		0.226		
222	32	578	72	6.7	310		0.171		
224	28	558	64	5.7	312		0.127		
226	23	527	56	4.7	314		0.0952		
228	19	487	49	3.8	316		0.0712		
230	15	441	42	3.1	318		0.0529		
232	12	397	36	2.4	320		0.0390		
234	9.9	362	31	1.9	322		0.0289		
236	7.6	324	26	1.4	324		0.0215		
238	5.9	295	22	1.1	326		0.0162		
240	4.5	273	18	0.81	328		0.0121		
242	3.3	253	15	0.59	330		0.0092		
244	2.5	234	12	0.43	332		0.0069		
246	1.8	214	10	0.31	334		0.0052		
248	1.3	194	8.0	0.22	336		0.0040		
250	0.96	174	6.5	0.16	338		0.0031		
252	0.69	158	5.1	0.11	340		0.0024		
254	0.49	136	4.0	0.076	342		0.0018		
256	0.34	116	3.2	0.053	344		0.0013		
258	0.23	99	2.4	0.037	346		0.0010		
260	0.16	83	1.9	0.026	348		0.00080		
262		69	1.4	0.018	350		0.00064		
264		57	1.1	0.012	352		0.00054		
266		47	0.84	0.009	354		0.00046		
268		38	0.63	0.006	356		0.00032		
270		31	0.48		358		0.00024		
272		25	0.36		360		0.00017		
274		20	0.27		362		0.00013		
276		16	0.20						

[a] From DeMore et al. (1997) recommendations based on Gillotay and Simon, *Annales Geophysicae, 6,* 211 (1988) for CH_3Br, Gillotay and Simon, *J. Atmos. Chem., 8,* 41 (1989) and Burkholder et al., *J. Geophys. Res., 96,* 5025 (1991) for Halon-1211 and -1301.

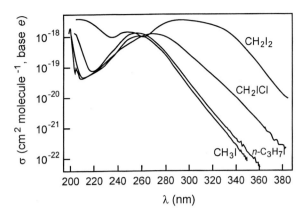

FIGURE 4.46 Absorption spectra of some simple alkyl iodides (adapted from Roehl *et al.*, 1997).

energetic threshold for O_2 dissociation. (a) Using the heats of formation in Appendix I, calculate the maximum wavelength at which O_2 can photodissociate. (b) Given that the force constant for the bond in O_2 is $k = 1140$ N m^{-1}, show that if ground-state O_2 has one vibrational quantum of energy, this will be sufficient to cause dissociation on absorption of 248-nm light. Assume O_2 behaves like a simple harmonic oscillator.

2. Using the force constant for O_2 given in Problem 1 and again assuming a simple harmonic oscillator behavior, calculate the highest possible vibrational level that ground electronic state O_2 could be produced upon photolysis of O_3 at 290 nm if the oxygen atom generated is (a) ground-state $O(^3P)$ or (b) electronically excited $O(^1D)$. Using your calculations in Problem 1b, calculate the vibrational energy level in which O_2 can be generated by O_3 photolysis at 243 nm. Compare to $v' = 26$ discussed in the text and comment on potential reasons for any differences.

3. Methyl isothiocyanate (CH_3NCS) is used as a soil fumigant. It is also the toxic substance generated when metam sodium, $CH_3NHCS_2^-Na^+$, reacts with water, as was the case near Dunsmuir, California, in the early 1990s, when a train carrying metam sodium derailed into a river. Alvarez and Moore (1994) have measured its absorption cross sections in the actinic region above 290 nm:

Wavelength range (nm)	(units of 10^{-20} cm^2 molecule^{-1}, base e)
295–300	1.70
300–305	1.23
305–310	0.852
310–315	0.563
315–320	0.348
320–325	0.196
325–330	0.096
330–335	0.040
335–340	0.017

Using these data and assuming a quantum yield of unity for the subsequent process forming methyl isocyanide

$$CH_3NCS + h\nu \rightarrow CH_3NC + S,$$

estimate the photolysis rate constant and lifetime for this compound at a latitude of 40°N on July 1 two hours before noon.

4. The absorption spectra of bromine compounds are generally shifted toward longer wavelengths compared to those of chlorine compounds.

(a) Use the absorption cross sections for Cl_2, BrCl, and Br_2 in Table 4.30 to calculate the factors by which the photolysis rate increases for BrCl and Br_2 compared to Cl_2 at a solar zenith angle of 20° at the earth's surface on December 15.

(b) Do the same calculation for a solar zenith angle of 86°. Comment on possible reasons for any differences.

5. The absorption spectra of iodine compounds are generally shifted toward longer wavelengths compared to those of both bromine and chlorine compounds.

(a) Use the absorption cross sections for IO and BrO in Table 4.31 to calculate the factor by which the photolysis rate increases for IO compared to BrO at a solar zenith angle of 20° at the earth's surface on July 1. Assume quantum yields of unity.

(b) Do the same calculation for a solar zenith angle of 86°. Comment on possible reasons for the differences.

6. Burkholder and McKeen (1997) report, based on their measurements of the absorption cross sections for

TABLE 4.42 Absorption Cross Sections (Base e) of CH_3I at 298 K^a

Wavelength (nm)	$10^{20}\sigma$ (cm^2 molecule^{-1})	Wavelength (nm)	$10^{20}\sigma$ (cm^2 molecule^{-1})
200	181	280	26.9
205	9.05	285	14.6
210	4.27	290	7.54
215	5.35	295	3.79
220	6.92	300	1.98
225	9.05	305	1.07
230	12.4	310	0.603
235	20.8	315	0.352
240	37.9	320	0.212
245	64.3	325	0.122
250	93.0	330	0.0724
255	112	335	0.0415
260	113	340	0.0225
265	96.6	345	0.0131
270	70.9	350	0.00738
275	45.9		

^{a} From Roehl *et al.* (1997).

SO_3 (Table 4.22), that photolysis of this compound will become important at higher altitudes, sufficiently so to compete with its reaction with water to form sulfuric acid. Compare the SO_3 photolysis rate for an overhead sun for an altitude of 0 km to its rate of hydrolysis assuming a collision-controlled value for the effective bimolecular rate constant of 10^{-10} cm^3 molecule^{-1} s^{-1} and 50% RH at 298 K at the earth's surface.

References

Alvarez, R. A., and C. B. Moore, "Quantum Yield for Production of CH_3NC in the Photolysis of CH_3NCS," *Science*, 263, 205–207 (1994).

Armerding, W., F. J. Comes, and B. Schülke, "$O(^1D)$ Quantum Yields of Ozone Photolysis in the UV from 300 nm to Its Threshold and at 355 nm," *J. Phys. Chem.*, 99, 3137–3143 (1995).

Arnold, I., F. J. Comes, and G. K. Moortgat, "Laser Flash Photolysis: Quantum Yield of $O(^1D)$ Formation from Ozone," *Chem. Phys.*, 24, 211–217 (1977).

Atkinson, R. "Kinetics and Mechanisms of the Gas-Phase Reactions of the NO_3 Radical with Organic Compounds," *J. Phys. Chem. Ref. Data*, 20, 459–507 (1991).

Atkinson, R., D. L. Baulch, R. A. Cox, R. F. Hampson, Jr., J. A. Kerr, M. J. Rossi, and J. Troe, "Evaluated Kinetic and Photochemical Data for Atmospheric Chemistry: Supplement VI. IUPAC Subcommittee on Gas Kinetic Data Evaluation for Atmospheric Chemistry," *J. Phys. Chem. Ref. Data*, 26, 1329–1499 (1997a).

Atkinson, R., D. L. Baulch, R. A. Cox, R. F. Hampson, Jr., J. A. Kerr, M. J. Rossi, and J. Troe, "Evaluated Kinetic, Photochemical, and Heterogeneous Data for Atmospheric Chemistry: Supplement V. IUPAC Subcommittee on Gas Kinetic Data Evaluation for Atmospheric Chemistry," *J. Phys. Chem. Ref. Data*, 26, 521–1011 (1997b).

Ball, S. M., G. Hancock, I. J. Murphy, and S. P. Tayner, "The Relative Quantum Yields of $O_2(^1\Delta_g)$ from the Photolysis of Ozone in the Wavelength Range 270 nm < λ < 329 nm," *Geophys. Res. Lett.*, 20, 2063–2066 (1993).

Ball, S. M., and G. Hancock, "The Relative Quantum Yields of $O_2(a^1\Delta_g)$ from the Photolysis of Ozone at 227 K," *Geophys. Res. Lett.*, 22, 1213–1216 (1995).

Ball, S. M., G. Hancock, S. E. Martin, and J. C. Pinot de Moira, "A Direct Measurement of the $O(^1D)$ Quantum Yields from the Photodissociation of Ozone between 300 and 328 nm," *Chem. Phys. Lett.*, 264, 531–538 (1997).

Banichevich, A., S. D. Peyerimhoff, and F. Grein, "Potential Energy Surfaces of Ozone in Its Ground State and in the Lowest-Lying Eight Excited States," *Chem. Phys.*, 178, 155–188 (1993).

Barnes, R. J., M. Lock, J. Coleman, and A. Sinha, "Observation of a New Absorption Band of HOBr and Its Atmospheric Implications," *J. Phys. Chem.*, 100, 453–457 (1996).

Barnes, R. J., G. Dutton, and A. Sinha, "Unimolecular Dissociation of HOCl near Threshold: Quantum State and Time-Resolved Studies," *J. Phys. Chem. A*, 101, 8374–8377 (1997).

Barnes, R. J., A. Sinha, and H. A. Michelsen, "Assessing the Contribution of the Lowest Triplet State to the Near-UV Absorption Spectrum of HOCl," *J. Phys. Chem. A*, 102, 8855–8859 (1998).

Bauer, D., T. Ingham, S. A. Carl, G. K. Moortgat, and J. N. Crowley, "Ultraviolet-Visible Absorption Cross Sections of Gaseous HOI and Its Photolysis at 355 nm," *J. Phys. Chem. A*, 102, 2857–2864 (1998).

Benter, T., C. Feldmann, U. Kirchner, M. Schmidt, S. Schmidt, and R. N. Schindler, "UV/VIS-Absorption Spectra of HOBr and CH_3OBr; $Br(^2P_{3/2})$ Atom Yields in the Photolysis of HOBr," *Ber. Bunsenges. Phys. Chem.*, 99, 1144–1147 (1995).

Blickensderfer, R. P., and G. E. Ewing, "Collision-Induced Absorption Spectrum of Gaseous Oxygen at Low Temperatures and Pressures. II. The Simultaneous Transitions $^1\Delta_g + {}^1\Delta_g \leftarrow {}^3\Sigma_g^- + {}^3\Sigma_g^-$ and $^1\Delta_g + {}^3\Sigma_g^+ \leftarrow {}^3\Sigma_g^- + {}^3\Sigma_g^-$," *J. Chem. Phys.*, 51, 5284–5289 (1969).

Bongartz, A., J. Kames, F. Welter, and U. Schurath, "Near-UV Absorption Cross Sections and Trans/Cis Equilibrium of Nitrous Acid," *J. Phys. Chem.*, 95, 1076–1082 (1991).

Bongartz, A., J. Kames, U. Schurath, Ch. George, Ph. Mirabel, and J. L. Ponche, "Experimental Determination of HONO Mass Accommodation Coefficients Using Two Different Techniques," *J. Atmos. Chem.*, 18, 149 (1994).

Brock, J. C., and R. T. Watson, "Ozone Photolysis: Determination of the $O(^3P)$ Quantum Yield at 266 nm," *Chem. Phys. Lett.*, 71, 371–375 (1980).

Burkholder, J. B., "Ultraviolet Absorption Spectrum of HOCl," *J. Geophys. Res.*, 98, 2963–2974 (1993).

Burkholder, J. B., R. K. Talukdar, A. R. Ravishankara, and S. Solomon, "Temperature Dependence of the HNO_3 UV Absorption Cross Sections," *J. Geophys. Res.*, 98, 22937–22948 (1993).

Burkholder, J. B., and R. K. Talukdar, "Temperature Dependence of the Ozone Absorption Spectrum over the Wavelength Range 410 to 760 nm," *Geophys. Res. Lett.*, 21, 581–584 (1994).

Burkholder, J. B., R. K. Talukdar, and A. R. Ravishankara, "Temperature Dependence of the $ClONO_2$ UV Absorption Spectrum," *Geophys. Res. Lett.*, 21, 585–588 (1994).

Burkholder, J. B., A. R. Ravishankara, and S. Solomon, "UV/Visible and IR Absorption Cross Sections of $BrONO_2$," *J. Geophys. Res.*, 100, 16793–16800 (1995).

Burkholder, J. B., and S. McKeen, "UV Absorption Cross Sections for SO_3," *Geophys. Res. Lett.*, 24, 3201–3204 (1997).

Calo, J. M., and R. S. Narcisi, "van der Waals Molecules—Possible Roles in the Atmosphere," *Geophys. Res. Lett.*, 7, 289–292 (1980).

Calvert, J. G., and J. N. Pitts, Jr., *Photochemistry*, Wiley, New York, 1966.

Calvert, J. G., and W. R. Stockwell, "Mechanism and Rates of the Gas Phase Oxidations of Sulfur Dioxide and Nitrogen Oxides in the Atmosphere," in SO_2, NO and NO_2 *Oxidation Mechanisms: Atmospheric Considerations* (J. G. Calvert, Ed.), Acid Precipitation Series, Vol. 3, pp. 1–62 (J. I. Teasley, Series Editor), Butterworth, Stoneham, MA, 1984.

Canosa-Mas, C. E., M. Fowles, P. J. Houghton, and R. P. Wayne, "Absolute Absorption Cross Section Measurements on NO_3," *J. Chem. Soc. Faraday Trans. II*, 83, 1465–1474 (1987).

Cantrell, C. A., J. A. Davidson, R. E. Shetter, B. A. Anderson, and J. G. Calvert, "The Temperature Invariance of the NO_3 Absorption Cross Section in the 662-nm Region," *J. Phys. Chem.*, 91, 5858–5863 (1987).

Cantrell, C. A., J. A. Davidson, A. H. McDaniel, R. E. Shetter, and J. G. Calvert, "Temperature-Dependent Formaldehyde Cross Sections in the Near-Ultraviolet Spectral Region," *J. Phys. Chem.*, 94, 3902–3908 (1990).

Clemitshaw, K. C., J. Williams, O. V. Rattigan, D. E. Shallcross, K. S. Law, and R. A. Cox, "Gas-Phase Ultraviolet Absorption Cross-Sections and Atmospheric Lifetimes of Several C_2–C_5 Alkyl Nitrates," *J. Photochem. Photobiol. A*, 102, 117–126 (1997).

Cox, R. A., and R. G. Derwent, "The Ultraviolet Absorption Spectrum of Gaseous Nitrous Acid," *J. Photochem.*, 6, 23–34 (1976/77).

Cox, R. A., and G. B. Coker, "Absorption Cross Sections and Kinetics of IO in the Photolysis of CH_3I in the Presence of Ozone," *J. Phys. Chem.*, 87, 4478–4484 (1983).

Cronin, T. J., and L. Zhu, "Dye Laser Photolysis of *n*-Pentanal from 280 to 330 nm," *J. Phys. Chem. A, 102,* 10274–10279 (1998).

Crowley, J. N., and S. A. Carl, "OH Formation in the Photoexcitation of NO_2 beyond the Dissociation Threshold in the Presence of Water Vapor," *J. Phys. Chem. A, 101,* 4178–4184 (1997).

Daumont, D., A. Barbe, J. Brion, and J. Malicet, "New Absolute Absorption Cross Section of O_3 in the 195–350 nm Spectral Range," in *Ozone in the Atmosphere* (R. D. Bojkov and P. Fabian, Eds.), Deepak Publishing, Hampton, VA, 1989.

Davidson, J. A., C. A. Cantrell, A. H. McDaniel, R. E. Shetter, S. Madronich, and J. G. Calvert, "Visible–Ultraviolet Absorption Cross Sections for NO_2 as a Function of Temperature," *J. Geophys. Res., 93,* 7105–7112 (1988).

Davis, H. F., B. Kim, H. S. Johnston, and Y. T. Lee, "Dissociation Energy and Photochemistry of NO_3," *J. Phys. Chem., 97,* 2172–2180 (1993).

DeMore, W. B., S. P. Sander, D. M. Golden, R. F. Hampson, M. J. Kurylo, C. J. Howard, A. R. Ravishankara, C. E. Kolb, and M. J. Molina, "Chemical Kinetics and Photochemical Data for Use in Stratospheric Modeling. Evaluation No. 12," Jet Propulsion Laboratory, Pasadena, CA, January 15, 1997.

Deters, B., J. P. Burrows, S. Himmelmann, and C. Blindauer, "Gas Phase Spectra of HOBr and Br_2O and Their Atmospheric Significance," *Ann. Geophys., 14,* 468–475 (1996).

Deters, B., J. P. Burrows, and J. Orphal, "UV–Visible Absorption Cross Sections of Bromine Nitrate Determined by Photolysis of $BrONO_2/Br_2$ Mixtures," *J. Geophys. Res., 103,* 3563–3570 (1998).

Donaldson, D. J., G. J. Frost, K. H. Rosenlof, A. F. Tuck, and V. Vaida, "Atmospheric Radical Production by Excitation of Vibrational Overtones via Absorption of Visible Light," *Geophys. Res. Lett., 24,* 2651–2654 (1997).

Freeman, D. E., K. Yoshino, and W. H. Parkinson, "Formation of Ozone by Irradiation of Oxygen at 248 Nanometers," *Science, 250,* 1432–1433 (1990).

Ganske, J. A., H. N. Berko, and B. J. Finlayson-Pitts, "Absorption Cross Sections for Gaseous $ClNO_2$ and Cl_2 at 298 K: Potential Organic Oxidant Source in the Marine Troposphere," *J. Geophys. Res., 97,* 7651–7656 (1992).

Gaydon, A. G., *Dissociation Energies and the Spectra of Diatomic Molecules,* 3rd ed., Chapman and Hall, London, 1968.

Gierczak, T., J. B. Burkholder, S. Bauerle, and A. R. Ravishankara, "Photochemistry of Acetone under Tropospheric Conditions," *Chem. Phys., 231,* 229–244 (1998).

Goldfarb, L., A.-M. Schmoltner, M. K. Gilles, J. B. Burkholder, and A. R. Ravishankara, "Photodissociation of $ClONO_2$: 1. Atomic Resonance Fluorescence Measurements of Product Quantum Yields," *J. Phys. Chem. A, 101,* 6658–6666 (1997).

Greenblatt, G. D., J. J. Orlando, J. B. Burkholder, and A. R. Ravishankara, "Absorption Measurements of Oxygen between 330 and 1140 nm," *J. Geophys. Res., 95,* 18577–18582 (1990).

Harder, J. W., J. W. Brault, P. V. Johnston, and G. H. Mount, "Temperature Dependent NO_2 Cross Sections at High Spectral Resolution," *J. Geophys. Res., 102,* 3861–3879 (1997).

Harwood, M. H., R. L. Jones, R. A. Cox, E. Lutman, and O. V. Rattigan, "Temperature-Dependent Absorption Cross Sections of N_2O_5," *J. Photochem. Photobiol. A: Chem., 73,* 167–175 (1993).

Harwood, M. H., and R. L. Jones, "Temperature Dependent Ultraviolet-Visible Absorption Cross Sections of NO_2 and N_2O_4: Low-Temperature Measurements of the Equilibrium Constant for $2NO_2 \leftrightarrow N_2O_4$," *J. Geophys. Res., 99,* 22955–22964 (1994).

Harwood, M. H., J. B. Burkholder, and A. R. Ravishankara, "Photodissociation of $BrONO_2$ and N_2O_5: Quantum Yields for NO_3 Production at 248, 308, and 352.5 nm," *J. Phys. Chem. A, 102,* 1309–1317 (1998).

Hubinger, S., and J. B. Nee, "Absorption Spectra of Cl_2, Br_2 and BrCl between 190 and 600 nm," *J. Photochem. Photobiol. A, 86,* 1–7 (1995).

Hubrich, C., and F. Stuhl, "The Ultraviolet Absorption of Some Halogenated Methanes and Ethanes of Atmospheric Interest," *J. Photochem., 12,* 93–107 (1980).

Huebert, B. J., and R. M. Martin, "Gas-Phase Far-Ultraviolet Absorption Spectrum of Hydrogen Bromide and Hydrogen Iodide," *J. Phys Chem., 72,* 3046–3048 (1968).

Hynes, A. J., E. A. Kenyon, A. J. Pounds, and P. H. Winer, "Temperature Dependent Absorption Cross-Sections for Acetone and *n*-Butanone—Implications for Atmospheric Lifetimes," *Spectrochim. Acta, 48A,* 1235–1242 (1992).

Ingham, T., D. Bauer, J. Landgraf, and J. N. Crowley, "Ultraviolet–Visible Absorption Cross Sections of Gaseous HOBr," *J. Phys. Chem. A, 102,* 3293–3298 (1998).

Inn, E. C. Y., "Vacuum Ultraviolet Spectroscopy," *Spectrochim. Acta, 7,* 65–87 (1955).

Johnston, H. S., H. F. Davis, and Y. T. Lee, "NO_3 Photolysis Product Channels: Quantum Yields from Observed Energy Thresholds," *J. Phys. Chem., 100,* 4713–4723 (1996).

Jones, I. T. N., and K. D. Bayes, "Formation of $O_2(a^1\Delta_g)$ by Electronic Energy Transfer in Mixtures of NO_2 and O_2," *J. Chem. Phys., 59,* 3119–3127 (1973a).

Jones, I. T. N., and K. D. Bayes, "Photolysis of Nitrogen Dioxide," *J. Chem. Phys., 59,* 4836–4844 (1973b).

Keene, W. C., J. R. Maben, A. A. P. Pszenny, and J. N. Galloway, "A Measurement Technique for HCl and Cl_2 in the Marine Boundary Layer," *Environ. Sci. Technol., 27,* 866–874 (1993).

Laszlo, B., M. J. Kurylo, and R. E. Huie, "Absorption Cross Sections, Kinetics of Formation, and Self-Reaction of the IO Radical Produced via the Laser Photolysis of $N_2O/I_2/N_2$ Mixtures," *J. Phys. Chem., 99,* 11701–11707 (1995).

Long, C. A., and G. E. Ewing, "Spectroscopic Investigation of van der Waals Molecules. I. The Infrared and Visible Spectra of $(O_2)_2^*$," *J. Chem. Phys., 58,* 4824–4834 (1973).

Luke, W. T., R. R. Dickerson, and L. J. Nunnermacker, "Direct Measurements of the Photolysis Rate Coefficients and Henry's Law Constants of Several Alkyl Nitrates," *J. Geophys. Res., 94,* 14905–14921 (1989).

Malicet, J., D. Daumont, J. Charbonnier, C. Parisse, A. Chakir, and J. Brion, "Ozone UV Spectroscopy. II. Absorption Cross-Sections and Temperature Dependence," *J. Atmos. Chem., 21,* 263–273 (1995).

Manatt, S. L., and A. L. Lane, "A Compilation of the Absorption Cross Sections of SO_2 from 106 to 403 nm," *J. Quant. Spectrosc. Radiat. Transfer, 50,* 267–276 (1993).

Maric, D., J. P. Burrows, and G. K. Moortgat, "A Study of the UV–Visible Absorption Spectra of Br_2 and BrCl," *J. Photochem. Photobiol., 83,* 179–192 (1994).

Martinez, R. D., A. A. Buitrago, N. W. Howell, C. H. Hearn, and J. A. Joens, "The Near U.V. Absorption Spectra of Several Aliphatic Aldehydes and Ketones at 300 K," *Atmos. Environ., 26A,* 785–792 (1992).

Mazely, T. L., R. R. Friedl, and S. P. Sander, "Production of NO_2 from Photolysis of Peroxyacetyl Nitrate," *J. Phys. Chem., 99,* 8162–8169 (1995).

Mazely, T. L., R. R. Friedl, and S. P. Sander, "Quantum Yield of NO_3 from Peroxyacetyl Nitrate Photolysis," *J. Phys. Chem. A, 101,* 7090–7097 (1997).

MacLeod, H., G. P. Smith, and D. M. Golden, "Photodissociation of Pernitric Acid (HO_2NO_2) at 248 nm," *J. Geophys. Res., 93,* 3813–3823 (1988).

Michelsen, H. A., R. J. Salawitch, P. O. Wennberg, and J. G. Anderson, "Production of $O(^1D)$ from Photolysis of O_3," *Geophys. Res. Lett., 21,* 2227–2230 (1994).

Miller, C. E., S. L. Nickolaisen, J. S. Francisco, and S. P. Sander, "The OBrO C($^2A_2$) ← x($^2B_1$) Absorption Spectrum," *J. Chem. Phys., 107,* 2300–2307 (1997).

Miller, R. L., A. G. Suits, P. L. Houston, R. Toumi, J. A. Mack, and A. M. Wodtke, "The 'Ozone Deficit' Problem: O_2(X, $v > 26$) + O(^{3}P) from 226-nm Ozone Photodissociation," *Science, 265,* 1831–1838 (1994).

Minton, T. K., C. M. Nelson, T. A. Moore, and M. Okumura, "Direct Observation of ClO from Chlorine Nitrate Photolysis," *Science, 258,* 1342–1345 (1992).

Molina, L. T., and M. J. Molina, "UV Absorption Cross Sections of HO_2NO_2 Vapor," *J. Photochem., 15,* 97–108 (1981).

Molina, L. T., and M. J. Molina, "Absolute Absorption Cross Sections of Ozone in the 185–350 nm Wavelength Range," *J. Geophys. Res., 91,* 14501–14508 (1986).

Molina, M. J., A. J. Colussi, L. T. Molina, R. N. Schindler, and T.-L. Tso, "Quantum Yield of Chlorine-Atom Formation in the Photodissociation of Chlorine Peroxide (ClOOCl) at 308 nm," *Chem. Phys. Lett., 173,* 310–315 (1990).

Moore, T. A., M. Okumura, M. Tagawa, and T. K. Minton, "Dissociation Dynamics of $ClONO_2$ and Relative Cl and O Product Yields following Photoexcitation at 308 nm," *Faraday Discuss., 100,* 295–307 (1995).

Moore, T. A., M. Okumura, J. W. Seale, and T. K. Minton, "UV Photolysis of ClOOCl," *J. Phys. Chem. A, 103,* 1691–1695 (1999).

Murcray, F. J., A. Goldman, J. C. Landry, and T. M. Stephen, "O_2 Continuum: A Possible Explanation for the Discrepancies between Measured and Modeled Shortwave Surface Irradiances," *Geophys. Res. Lett., 24,* 2315–2317 (1997).

Newnham, D. A., and J. Ballard, "Visible Absorption Cross Sections and Integrated Absorption Intensities of Molecular Oxygen (O_2 and O_4)," *J. Geophys. Res., 103,* 28801–28816 (1998).

Nickolaisen, S. L., S. P. Sander, and R. R. Friedl, "Pressure-Dependent Yields and Product Branching Ratios in the Broadband Photolysis of Chlorine Nitrate," *J. Phys. Chem., 100,* 10165–10178 (1996).

Nölle, A., F. Pätzold, S. Pätzold, R. Meller, G. K. Moortgat, E. P. Röth, R. Ruhnke, and H. Keller-Rudek, "UV/Vis Spectra of Atmospheric Constituents," Version 1, ISBN 3-89100-030-8 available as CD-ROM (see Appendix IV) (1999).

Oh, D., W. Sisk, A. Young, and H. Johnston, "Nitrogen Dioxide Fluorescence from N_2O_5 Photolysis," *J. Chem. Phys., 85,* 7146–7158 (1986).

Okabe, H. *Photochemistry of Small Molecules,* Wiley, New York, 1978.

Orlando, J. J., G. S. Tyndall, K. E. Nickerson, and J. G. Calvert, "The Temperature Dependence of Collision-Induced Absorption by Oxygen near 6 μm," *J. Geophys. Res., 96,* 20755–20760 (1991).

Orlando, J. J., G. S. Tyndall, G. K. Moortgat, and J. G. Calvert, "Quantum Yields for NO_3 Photolysis between 570 and 635 nm," *J. Phys. Chem., 97,* 10996–11000 (1993).

Orlando, J. J., and J. B. Burkholder, "Gas-Phase UV/Visible Absorption Spectra of HOBr and Br_2O," *J. Phys. Chem., 99,* 1143–1150 (1995).

Pagsberg, P., E. Bjergbakke, E. Ratajczak and A. Sillesen, "Kinetics of the Gas Phase Reaction OH + NO (+M) → HONO(+M) and the Determination of the UV Absorption Cross Sections of HONO," *Chem. Phys. Lett., 272* 383–390 (1997).

Perner, D., and U. Platt, "Absorption of Light in the Atmosphere by Collision Pairs of Oxygen $(O_2)_2$," *Geophys. Res. Lett., 7,* 1053–1056 (1980).

Pfeilsticker, K., F. Erle, and U. Platt, "Notes and Correspondence: 'Absorption of Solar Radiation by Atmospheric O_4,'" *J. Atmos. Sci., 54,* 933–939 (1997).

Price, J. M., J. A. Mack, C. A. Rogaski, and A. M. Wodtke, "Vibrational-State-Specific Self-Relaxation Rate Constant. Measurements of Highly Vibrationally Excited O_2($v = 19-18$)," *Chem. Phys., 175,* 83–98 (1993).

Pszenny, A. A. P., W. C. Keene, D. J. Jacob, S. Fan, J. R. Maben, M. P. Zetwo, M. Springer-Young, and J. N. Galloway, "Evidence of Inorganic Chlorine Gases Other Than Hydrogen Chloride in Marine Surface Air," *Geophys. Res. Lett., 20,* 699–702 (1993).

Rattigan, O., E. Lutman, R. L. Jones, R. A. Cox, K. Clemitshaw, and J. Williams, "Temperature-Dependent Absorption Cross Sections of Gaseous Nitric Acid and Methyl Nitrate," *J. Photochem. Photobiol. A, 66,* 313–326 (1992); *corrigendum, ibid, 69,* 125–126 (1992).

Rattigan, O. V., D. J. Lary, R. L. Jones, and R. A. Cox, "UV–Visible Absorption Cross Sections of Gaseous Br_2O and HOBr," *J. Geophys. Res., 101,* 23021–23033 (1996).

Ravishankara, A. R., and R. L. Mauldin, III, "Temperature Dependence of the NO_3 Cross Section in the 662-nm Region," *J. Geophys. Res., 91,* 8709–8712 (1986).

Ravishankara, A. R., G. Hancock, M. Kawasaki, and Y. Matsumi, "Photochemistry of Ozone: Surprises and Recent Lessons," *Science, 280,* 60–61 (1998).

Roberts, J. M., and R. W. Fajer, "UV Absorption Cross Sections of Organic Nitrates of Potential Atmospheric Importance and Estimation of Atmospheric Lifetimes," *Environ. Sci. Technol., 23,* 945–951 (1989).

Roehl, C. M., J. J. Orlando, and J. G. Calvert, "The Temperature Dependence of the UV–Visible Absorption Cross Sections of NOCl," *J. Photochem. Photobiol. A: Chem., 69,* 1–5 (1992).

Roehl, C. M., J. J. Orlando, G. S. Tyndall, R. E. Shetter, G. J. Vázquez, C. A. Cantrell, and J. G. Calvert, "Temperature Dependence of the Quantum Yields for the Photolysis of NO_2 near the Dissociation Limit," *J. Phys. Chem., 98,* 7837–7843 (1994).

Roehl, C. M., J. B. Burkholder, G. K. Moortgat, A. R. Ravishankara, and P. J. Crutzen, "Temperature Dependence of UV Absorption Cross Sections and Atmospheric Implications of Several Alkyl Iodides," *J. Geophys. Res., 102,* 12819–12829 (1997).

Rogaski, C. A., J. M. Price, J. A. Mack, and A. M. Wodtke, "Laboratory Evidence for a Possible Non-LTE Mechanism of Stratospheric Ozone Formation," *Geophys. Res. Lett., 20,* 2885–2888 (1993).

Rogers, J. D., "Ultraviolet Absorption Cross Sections and Atmospheric Photodissociation Rate Constants of Formaldehyde," *J. Phys. Chem., 94,* 4011–4015 (1990).

Sander, S. P., "Temperature Dependence of the NO_3 Absorption Spectrum," *J. Phys. Chem., 90,* 4135–4142 (1986).

Schindler, R. N., M. Liesner, S. Schmidt, U. Kirchner, and Th. Benter, "Identification of Nascent Products Formed in the Laser Photolysis of CH_3OCl and HOCl at 308 nm and around 235 nm. Total Cl-Atom Quantum Yields and the State and Velocity Distributions of Cl(^2P_j)," *J. Photochem. Photobiol. A: Chem., 107,* 9–19 (1997).

Schmidt, S., Th. Benter, and R. N. Schindler, "Photodissociation Dynamics of ClO Radicals in the Range (237 < λ < 270) nm and at 205 nm and the Velocity Distribution of O(^{1}D) Atoms," *Chem. Phys. Lett., 282,* 292–298 (1998a).

Schmidt, S., R. N. Schindler, and T. Benter, "Photodissociation Dynamics of ClO and ClOOCl: Branching Ratios, Kinetic Energy and Quantum Yield of Primary Photoproducts," Presented at the XXIII Informal Conference on Photochemistry, May 10–15, 1998b, Pasadena, CA.

Selwyn, G., J. Podolske, and H. S. Johnston, "Nitrous Oxide Ultraviolet Absorption Spectrum at Stratospheric Temperatures," *Geophys. Res. Lett., 4,* 427–430 (1977).

Silvente, E., R. C. Richter, M. Zheng, E. S. Saltzman, and A. J. Hynes, "Relative Quantum Yields for O^1D Production in the

Photolysis of Ozone between 301 and 336 nm: Evidence for the Participation of a Spin-Forbidden Channel," *Chem. Phys. Lett.*, 264, 309–315 (1997).

Singer, R. J., J. N. Crowley, J. P. Burrows, W. Schneider, and G. K. Moortgat, "Measurement of the Absorption Cross-Section of Peroxynitric Acid between 210 and 330 nm in the Range 253–298 K," *J. Photochem. Photobiol. A: Chem.*, 48, 17–32 (1989).

Slanger, T. G., and P. C. Cosby, "O_2 Spectroscopy below 5.1 eV," *J. Phys. Chem.*, 92, 267–282 (1988).

Slanger, T. G., L. E. Jusinski, G. Black, and G. E. Gadd, "A New Laboratory Source of Ozone and Its Potential Atmospheric Implications," *Science*, 241, 945–950 (1988).

Solomon, S., R. W. Portmann, R. W. Sanders, and J. S. Daniel, "Absorption of Solar Radiation by Water Vapor, Oxygen, and Related Collision Pairs in the Earth's Atmosphere," *J. Geophys. Res.*, 103, 3847–3858 (1998).

Spicer, C. W., E. G. Chapman, B. J. Finlayson-Pitts, R. A. Plastridge, J. M. Hubbe, J. D. Fast, and C. M. Berkowitz, "Unexpectedly High Concentrations of Molecular Chlorine in Coastal Air," *Nature*, 394, 353–356 (1998).

Steinfeld, J. I., S. M. Adler-Golden, and J. W. Gallagher, "Critical Survey of Data on the Spectroscopy and Kinetics of Ozone in the Mesosphere and Thermosphere," *J. Phys. Chem. Ref. Data*, 16, 911–951 (1987).

Stranges, D., X. Yang, J. D. Chesko, and A. G. Suits, "Photodissociation of Ozone at 193 nm by High-Resolution Photofragment Translational Spectroscopy," *J. Chem. Phys.*, 102, 6067–6077 (1995).

Syage, J. A., "Photofragment Imaging of Ozone Photodissociation: $O_3 \rightarrow O(^3P_j) + O_2(X, v)$ at 226 nm," *J. Phys. Chem.*, 99, 16530–16533 (1995).

Syage, J. A., "Photofragment Imaging of Ozone Photodissociation: $O_3 \rightarrow O(^3P_j) + O_2(X, v)$," *J. Phys. Chem.*, 100, 13885 (1996a).

Syage, J. A., "Photofragment Imaging by Sections for Measuring State-Resolved Angle-Velocity Differential Cross Sections," *J. Chem. Phys.*, 105, 1007–1022 (1996b).

Takahashi, K., M. Kishigami, Y. Matsumi, M. Kawasaki, and A. J. Orr-Ewing, "Observation of the Spin-Forbidden $O(^1D) + O_2(X^3\Sigma_g^-)$ Channel in the 317–327 nm Photolysis of Ozone," *J. Chem. Phys.*, 105, 5290–5293 (1996a).

Takahashi, K., Y. Matsumi, and M. Kawasaki, "Photodissociation Processes of Ozone in the Huggins Band at 308–326 nm: Direct Observation of $O(^1D_2)$ and $O(^3P_j)$ Products," *J. Phys. Chem.*, 100, 4084–4089 (1996b).

Takahashi, K., M. Kishigami, N. Taniguchi, Y. Matsumi, and M. Kawasaki, "Photofragment Excitation Spectrum for $O(^1D)$ from the Photodissociation of Jet-Cooled Ozone in the Wavelength Range 305–329 nm," *J. Chem. Phys.*, 106, 6390–6397 (1997).

Talukdar, R. K., J. B. Burkholder, A.-M. Schmoltner, J. M. Roberts, R. R. Wilson, and A. R. Ravishankara, "Investigation of the Loss Processes for Peroxyacetyl Nitrate in the Atmosphere: UV Photolysis and Reaction with OH," *J. Geophys. Res.*, 100, 14163–14173 (1995).

Talukdar, R. K., M. K. Gilles, F. Battin-Leclerc, A. R. Ravishankara, J.-M. Fracheboud, J. J. Orlando, and G. S. Tyndall, "Photolysis of Ozone at 308 and 248 nm: Quantum Yield of $O(^1D)$ as a Function of Temperature," *Geophys. Res. Lett.*, 24, 1091–1094 (1997).

Talukdar, R. K., C. A. Longfellow, M. K. Gilles, and A. R. Ravishankara, "Quantum Yields of $O(^1D)$ in the Photolysis of Ozone between 289 and 329 nm as a Function of Temperature," *Geophys. Res. Lett.*, 25, 143–146 (1998).

Thelen, M.-A., P. Felder, and J. R. Huber, "The Photofragmentation of Methyl Hydroperoxide CH_3OOH at 193 and 248 nm in a Cold Molecular Beam," *Chem. Phys. Lett.*, 213, 275–281 (1993).

Trolier, M., and J. R. Wiesenfeld, "Relative Quantum Yield of $O(^1D_2)$ Following Ozone Photolysis between 275 and 325 nm," *J. Geophys. Res.*, 93, 7119–7124 (1988).

Turberg, M. P., D. M. Giolando, C. Tilt, T. Soper, S. Mason, M. Davies, P. Klingensmith, and G. A. Takacs, "Atmospheric Photochemistry of Alkyl Nitrates," *J. Photochem. Photobiol. A: Chem.*, 51, 281–292 (1990).

Tyndall, G. S., C. S. Kegley-Oswen, J. J. Orlando, and J. G. Calvert, "Quantum Yields for $Cl(^2P_{3/2,1/2})$, ClO, and $O(^3P)$ in the Photolysis of Chlorine Nitrate at 308 nm," *J. Chem. Soc., Faraday Trans.*, 93, 2675–2682 (1997).

Vaghjiani, G. L., and A. R. Ravishankara, "Absorption Cross Sections of CH_3OOH, H_2O_2 and D_2O_2 Vapors between 210 and 365 nm at 297 K," *J. Geophys. Res.*, 94, 3487–3492 (1989).

Vaghjiani, G. L., and A. R. Ravishankara, "Photodissociation of H_2O_2 and CH_3OOH at 248 nm and 298 K: Quantum Yields for OH, $O(^3P)$ and $H(^2S)$," *J. Chem. Phys.*, 92, 996–1003 (1990).

Vaghjiani, G. L., A. A. Turnipseed, R. F. Warren, and A. R. Ravishankara, "Photodissociation of H_2O_2 at 193 and 222 nm: Products and Quantum Yields," *J. Chem. Phys.*, 96, 5878–5886 (1992).

Vandaele, A. C., C. Hermans, P. C. Simon, M. Van Roozendael, J. M. Guilmot, M. Carleer, and R. Colin, "Fourier Transform Measurement of NO_2 Absorption Cross-Section in the Visible Range at Room Temperature," *J. Atmos. Chem.*, 25, 289–305 (1996).

Vogt, R., and R. N. Schindler, "Product Channels in the Photolysis of HOCl," *J. Photochem. Photobiol. A: Chem.*, 66, 133–140 (1992).

Volkamer, R., T. Etzkorn, A. Geyer, and U. Platt, "Correction of the Oxygen Interference with UV Spectroscopic (DOAS) Measurements of Monocyclic Aromatic Hydrocarbons in the Atmosphere," *Atmos. Environ.*, 32, 3731–3747 (1998).

Wahner, A., G. S. Tyndall, and A. R. Ravishankara, "Absorption Cross Sections for OClO as a Function of Temperature in the Wavelength Range 240–480 nm," *J. Phys. Chem.*, 91, 2734–2738 (1987).

Wahner, A., A. R. Ravishankara, S. P. Sander, and R. R. Friedl, "Absorption Cross Section of BrO between 312 and 385 nm at 298 and 223 K," *Chem. Phys. Lett.*, 152, 507–512 (1988).

Wallace, L., and D. M. Hunten, "Dayglow of the Oxygen A Band," *J. Geophys. Res.*, 73, 4813–4834 (1968).

Wayne, R. P., "The Photochemistry of Ozone," *Atmos. Environ.*, 21, 1683–1694 (1987).

Wayne, R. P., "Singlet Oxygen in the Environmental Sciences," *Res. Chem. Intermed.*, 20, 395–422 (1994).

Wayne, R. P., I. Barnes, P. Biggs, J. P. Burrows, C. E. Canosa-Mas, J. Hjorth, G. Lebras, G. K. Moortgat, D. Perner, G. Poulet, G. Restelli, and H. Sidebottom, "The Nitrate Radical: Physics, Chemistry, and the Atmosphere," *Atmos. Environ.*, 25A, 1–203 (1991).

Wilson, R. J., J. A. Mueller, and P. L. Houston, "Speed-Dependent Anisotropy Parameters in the UV Photodissociation of Ozone," *J. Phys. Chem. A*, 101, 7593–7599 (1997).

Yokelson, R. J., J. B. Burkholder, R. W. Fox, R. K. Talukdar, and A. R. Ravishankara, "Temperature Dependence of the NO_3 Absorption Spectrum," *J. Phys. Chem.*, 98, 13144–13150 (1994).

Yokelson, R. J., J. B. Burkholder, R. W. Fox, and A. R. Ravishankara, "Photodissociation of $ClONO_2$: 2. Time-Resolved Absorption Studies of Product Quantum Yields," *J. Phys. Chem. A*, 101, 6667–6678 (1997).

Yoshino, K., J. R. Esmond, A. S.-C. Cheung, D. E. Freeman, and W. H. Parkinson, "High Resolution Absorption Cross Sections in the Transmission Window Region of the Schumann–Runge Bands and Herzberg Continuum of O_2," *Planet. Space Sci.*, 40, 185–192 (1992).

CHAPTER 5

Kinetics and Atmospheric Chemistry

Understanding the kinetics of reactions of various species in the atmosphere is critical for assessing their atmospheric fates. As such, reaction kinetics form a key component of risk assessments of airborne toxic chemicals, for example (see Chapter 16). In addition, it greatly simplifies the number of reactions that must be considered in assessing the atmospheric fates of a particular compound of interest. In the case of organics, for example, there are, in principle, many potential reactions that could occur in the atmosphere. Given the thousands of organics found in air, the total number of reactions that must be considered quickly becomes intractable unless there is some means to pare the list down. Knowing the kinetics of reaction of various classes of compounds with OH, O_3, etc. allows one to estimate lifetimes under typical atmospheric conditions and hence to rule out those reactions that are too slow to be significant, allowing one to concentrate on the most important reactions.

In this chapter we briefly review the fundamental kinetics needed for application to atmospheric systems and discuss some of the most common methods for determining rate constants in the laboratory. This includes so-called "heterogeneous" reactions, whose importance in the stratosphere is now well established and which are increasingly being recognized as important in tropospheric systems. For a review of these areas, see Molina *et al.* (1996).

A. FUNDAMENTAL PRINCIPLES OF GAS-PHASE KINETICS

1. Definitions

a. Elementary vs Overall Reactions

Elementary reactions are defined as those that cannot be broken down into two or more simpler reactions. Generally, they consist of one or two reactant species and are referred to as unimolecular and bimolecular processes, respectively. However, there are a number of important gas-phase processes in which three different species participate; these are termolecular reactions. In the troposphere they usually involve N_2 and/or O_2 as one of the three participants; the role of the third molecule is generally to act as an "inert gas" that stabilizes the energy-rich intermediate formed from the recombination of two species by siphoning off the excess energy, thus preventing dissociation back into the reactants. In such cases, rather than being specific as to the colliding third body, the symbol "M" is used.

Examples of these three classes of gas-phase reactions are

Unimolecular. The thermal decomposition of PAN:

$$CH_3\overset{O}{\overset{\|}{C}}OONO_2 \longrightarrow CH_3\overset{O}{\overset{\|}{C}}OO + NO_2. \quad (1)$$

Bimolecular. Formation of the gaseous nitrate radical:

$$O_3 + NO_2 \rightarrow NO_3 + O_2. \quad (2)$$

Termolecular. The formation of ozone by the reaction of a ground-state oxygen atom, $O(^3P)$, with O_2:

$$O(^3P) + O_2 + M \rightarrow O_3 + M. \quad (3)$$

Because M does not enter into the reaction chemically, such reactions are usually written with "M" above the arrow, although in some cases it is omitted entirely and simply understood to be present:

$$O(^3P) + O_2 \overset{M}{\rightarrow} O_3.$$

Each of these types of elementary processes will be treated in some detail in subsequent sections of this chapter.

While two-body collisions are common in the gas phase, three-body collisions are much less probable and four-body collisions can essentially be ignored because of their low probability. Thus the majority of the reactions we deal with in the atmosphere are bimolecular, with a lesser number being termolecular or unimolecular.

An *overall* reaction includes two or more elementary reactions; indeed there is no limit to the number of reactants or elementary reactions comprising an overall reaction. Thus if a single reaction step *as written* has four or more reactants, it cannot be an elementary process, and it must occur via two or more consecutive steps. If a reaction step contains two to three reactants, it may, or may not, be an elementary reaction.

b. Rate Laws, Reaction Order, and the Rate Constant

In studying any particular reaction, one does not really know *a priori* if it is an elementary reaction or not, unless it involves four or more species, in which case it cannot be elementary. Determination of the *rate law* for the reaction is the first step in assessing whether it could be elementary.

The rate of a reaction is defined as the change in the concentration of a reactant or product with time. For simple reactions occurring with unit stoichiometry, the rate expressed in terms of reactant disappearance is the same as the rate in terms of product formation. For example, for reaction (4), the reaction of ozone with nitric oxide,

$$NO + O_3 \rightarrow NO_2 + O_2, \quad (4)$$

the rate is defined as

$$\text{Rate} = \frac{-d[NO]}{dt} = \frac{-d[O_3]}{dt}$$
$$= \frac{+d[NO_2]}{dt} = \frac{+d[O_2]}{dt}. \quad (A)$$

For reactions of the more general form

$$aA + bB \rightarrow cC + dD, \quad (5)$$

where the stoichiometric coefficients a, b, c, and d are not all unity, the rate in terms of disappearance of A may not be equal to the rate in terms of disappearance of B or the appearance of C or D. To take such differences in stoichiometry into account, the rate of the generalized reaction (5) is defined by convention as

$$\text{Rate} = -\frac{1}{a}\frac{d[A]}{dt} = -\frac{1}{b}\frac{d[B]}{dt}$$
$$= +\frac{1}{c}\frac{d[C]}{dt} = +\frac{1}{d}\frac{d[D]}{dt}. \quad (B)$$

For example, in the thermal oxidation of NO by oxygen,

$$2NO + O_2 \rightarrow 2NO_2, \quad (6)$$

two molecules of NO disappear for each molecule of O_2 reacted, and the rate of loss of NO is twice that of O_2:

$$\text{Rate} = -\frac{1}{2}\frac{d[NO]}{dt} = \frac{-d[O_2]}{dt} = +\frac{1}{2}\frac{d[NO_2]}{dt}.$$

While this convention is now widely used, it was not in some early kinetic studies. Thus one must be careful to note exactly how the rate is defined so that the reported rate constants are interpreted and applied correctly. In systems of atmospheric interest, the rate law or rate expression for a reaction, either elementary or overall, is the equation expressing the dependence of the rate on the concentrations of reactants. In a few reactions (mainly those in solution), products may also appear in the rate law.

For the general *overall* reaction (5), the rate law has the form

$$\text{Rate} = k[A]^m[B]^n[C]^p[D]^q,$$

where, depending on the mechanism of the reaction, m, n, p, and q may be zero, integers, or fractions. As noted earlier, in most gas-phase atmospheric reactions, the exponents of the product concentration (i.e., p and q) are zero and the rate laws involve only the reactant species. It is important to stress here that in contrast to elementary reactions, in *overall* reactions the exponents in the rate laws (e.g., m, n, p, and q) *do not* necessarily bear a relationship to the stoichiometric coefficients of the reaction (e.g., a, b, c, and d).

The importance of distinguishing between elementary and overall reactions comes in formulating rate laws. For *elementary reactions only*, the rate law may be written directly from the stoichiometric equation. Thus for the general *elementary* gas-phase reaction

$$aA + bB \rightarrow cC + dD, \quad (7)$$
$$\text{Rate} = k[A]^a[B]^b,$$

where $(a + b) \leq 3$ by definition of an elementary reaction. For example, the rate expression for the elementary reaction (4) is given by

$$\text{Rate} = k_4[NO][O_3].$$

The *rate constant*, k, is simply the constant of proportionality in the expression relating the rate of a reaction to the concentrations of reactants and/or products, each expressed with the appropriate exponent. The *order* of a reaction is defined as the sum of the exponents in the rate law. Thus reaction (4) is $(1 + 1) =$ second order. The order with respect to each species appearing in the rate law is the exponent of the concentration of that species; thus reaction (4) is first order in both O_3 and NO.

The basis of predicting rate laws for elementary reactions from the stoichiometric equation lies in the fact that they must occur during a single collision (although the *probability* of reaction during any one collision is equal to or less than unity). Thus doubling the concentration of O_3 in reaction (4) will double the number of collisions per second of O_3 with NO. Assuming the probability of reaction per collision remains constant, then the number of O_3 and NO molecules reacting, and O_2 and NO_2 formed per unit time (i.e., the rate), must double.

The thermal oxidation of NO by molecular oxygen, reaction (6), is another example where the stoichiometry and the molecularity of the reaction are directly related, and the rate law is

$$\text{Rate} = k_6^{III}[NO]^2[O_2].$$

Thus the rate is proportional to the first power of the oxygen concentration and the square of the nitric oxide concentration and the reaction order is $1 + 2 = 3$. However, in the troposphere, the O_2 concentration is always so large relative to NO that it is effectively constant and thus can be incorporated into the rate constant k_6^{III}. The rate law is now written

$$\text{Rate} = k_6^{bi}[NO]^2,$$

and the reaction is referred to as *pseudo-second-order*. We adopt the convention of writing a third-order rate constant as k^{III}, and a pseudo-second-order rate constant as k^{bi}, as illustrated in the preceding equations.

The rate law and the reaction order can often be used to show that a reaction cannot be an elementary reaction since, in the latter case, the exponents must be integers and the overall reaction order must be ≤ 3. However, it should be noted that these kinetic parameters cannot be used to confirm that a particular reaction *is* elementary; they can only indicate that the kinetic data do not rule out the possibility that the reaction is elementary.

In gas-phase reactions, concentrations are usually expressed in molecules cm^{-3} and time in seconds, the convention we employ in this book. Thus the units of k are as follows: first order, s^{-1}; second order, cm^3 molecule^{-1} s^{-1}; third order, cm^6 molecule^{-2} s^{-1}.

Concentrations of gaseous pollutants are often expressed in terms of parts per million (ppm) by volume, and time is expressed in minutes. Use of these concentration units must be reflected in the units used for the rate constants as well; for example, second-order rate constants are in units of ppm^{-1} min^{-1}. Occasionally, gas concentrations are given in units of mol L^{-1} or in units of pressure such as Torr, atmospheres, or Pascals; these can be converted to the more conventional units

TABLE 5.1 Some Common Conversion Factors for Gas-Phase Reactions

Concentrations[a]
1 mol L^{-1} = 6.02×10^{20} molecules cm^{-3}
1 ppm = 2.46×10^{13} molecules cm^{-3}
1 ppb = 2.46×10^{10} molecules cm^{-3}
1 ppt = 2.46×10^{7} molecules cm^{-3}
1 atm = 760 Torr = 4.09×10^{-2} mol L^{-1}
    = 2.46×10^{19} molecules cm^{-3}

Second-order rate constants
cm^3 molecule^{-1} s^{-1} × 6.02×10^{20} = L mol^{-1} s^{-1}
ppm^{-1} min^{-1} × 4.08×10^{5} = L mol^{-1} s^{-1}
ppm^{-1} min^{-1} × 6.77×10^{-16} = cm^3 molecule^{-1} s^{-1}
atm^{-1} s^{-1} × 4.06×10^{-20} = cm^3 molecule^{-1} s^{-1}

Third-order rate constants
cm^6 molecule^{-2} s^{-1} × 3.63×10^{41} = L^2 mol^{-2} s^{-1}
ppm^{-2} min^{-1} × 9.97×10^{12} = L^2 mol^{-2} s^{-1}
ppm^{-2} min^{-1} × 2.75×10^{-29} = cm^6 molecule^{-2} s^{-1}

[a] The concentrations ppm, ppb, and ppt are relative to air at 1 atm and 25°C, where 1 atm = 760 Torr total pressure.

in tropospheric chemistry using the ideal gas law. Table 5.1 gives some common conversion factors for gas-phase concentrations and rate constants at 1 atm pressure (760 Torr total pressure) and 25°C.

For solution-phase reactions, we use concentration units of mol L^{-1}, with units for the corresponding rate constants of L mol^{-1} s^{-1} (second order) and L^2 mol^{-2} s^{-1} (third order).

c. Half-Lives and Lifetimes

A rate constant is a quantitative measure of how fast reactions proceed and therefore is an indicator of how long a given set of reactants will survive in the atmosphere under a particular set of reactant concentrations. However, the rate constant per se is not a parameter that by itself is readily related to the average length of time a species will survive in the atmosphere before reacting. More intuitively meaningful parameters are the *half-life* ($t_{1/2}$) or the *natural lifetime* (τ), the latter usually referred to simply as "lifetime," of a pollutant with respect to reaction with a labile species such as OH or NO_3 radicals.

The half-life ($t_{1/2}$) is defined as the time required for the concentration of a reactant to fall to one-half of its initial value, whereas the lifetime is defined as the time it takes for the reactant concentration to fall to $1/e$ of its initial value (e is the base of natural logarithms, 2.718). Both $t_{1/2}$ and τ are directly related to the rate constant and to the concentrations of any other reactants involved in the reactions. These relationships are given in general form in Table 5.2 for first-, second-, and third-order reactions and are derived in Box 5.1.

TABLE 5.2 Relationships between the Rate Constant, Half-Lives, and Lifetimes for First-, Second-, and Third-Order Reactions

Reaction order	Reaction	Half-life of A	Lifetime of A
First	(1) $A \xrightarrow{k_1}$ Products	$t_{1/2}^A = 0.693/k_1$	$\tau^A = 1/k_1$
Second	(2) $A + B \xrightarrow{k_2}$ Products	$t_{1/2}^A = 0.693/k_2[B]$	$\tau^A = 1/k_2[B]$
Third	(3) $A + B + C \xrightarrow{k_3}$ Products	$t_{1/2}^A = 0.693/k_3[B][C]$	$\tau^A = 1/k_3[B][C]$

A relevant example is the use of lifetimes to characterize the reactivity of organics. Compressed natural gas (CNG), for example, is a widely used fuel whose major component is methane, CH_4. The only known significant chemical loss process for CH_4 is reaction with OH:

$$CH_4 + OH \rightarrow CH_3 + H_2O,$$
$$k_8^{298\,K} = 6.3 \times 10^{-15} \text{ cm}^3 \text{ molecule}^{-1} \text{ s}^{-1}. \quad (8)$$

Taking a typical average, daytime OH concentration of 1×10^6 radicals cm^{-3}, the lifetime of CH_4 with respect to this removal process is

$$\tau_{OH}^{CH_4} = \frac{1}{k_8[OH]}$$
$$= \frac{1}{[6.3 \times 10^{-15} \text{ cm}^3 \text{ molecule}^{-1} \text{ s}^{-1} \times (1 \times 10^6 \text{ radicals cm}^{-3})]}$$
$$= 1.59 \times 10^8 \text{ s} = 5 \text{ years}.$$

Propane is another organic used widely as a fuel. It also reacts with OH:

$$C_3H_8 + OH \rightarrow C_3H_7 + H_2O,$$
$$k_9^{298\,K} = 1.1 \times 10^{-12} \text{ cm}^3 \text{ molecule}^{-1} \text{ s}^{-1}. \quad (9)$$

Assuming the same OH concentration, one calculates

$$\tau_{OH}^{C_3H_8} = 9.1 \times 10^5 \text{ s} = 10.5 \text{ days}.$$

These calculations illustrate why regulatory agencies have concentrated on controlling "non-methane hydrocarbons." Because methane reacts so slowly in the troposphere, it is generally not of concern from the point of view of ozone formation in urban areas, i.e., over the time scale of hours to a few days. It is also the reason that CH_4 is the only organic to survive long enough in the troposphere to cross the tropopause and enter the stratosphere in significant concentrations (see Chapter 12). On the other hand, propane reacts sufficiently quickly that it can contribute to local and regional photochemical smog formation, as suggested in Mexico City (Blake and Rowland, 1995).

Two points should be made about such calculations of tropospheric lifetimes. First, they are valid only for the specified reaction; if there are other competing loss processes such as photolysis, the actual overall lifetime will be shortened accordingly. On the other hand, for a species such as CH_4, which does not photolyze or react significantly with other atmospheric species such as O_3 or NO_3, $\tau_{OH}^{CH_4}$ is indeed close to the overall lifetime of CH_4.

Second, in bi- and termolecular reactions, $t_{1/2}$ and τ depend on the concentration of other reactants; this is particularly important when interpreting atmospheric lifetimes. For example, as discussed earlier, reaction with the OH radical is a major fate of most organics during daylight in both the clean and polluted troposphere. However, the actual concentrations of OH at various geographical locations and under a variety of conditions are highly variable; for example, its concentration varies diurnally since it is produced primarily by photochemical processes. Finally, the concentration of OH varies with altitude as well, so the lifetime will depend on where in the troposphere the reaction occurs.

Thus when a lifetime of an organic in the atmosphere is cited with respect to OH attack, one should examine carefully the concentration of OH that was *assumed* in arriving at that lifetime; the substantial uncertainties in these estimated lifetimes that arise from the uncertainties in the estimated atmospheric OH concentrations should be clearly recognized.

2. Termolecular Reactions and Pressure Dependence of Rate Constants

Termolecular elementary reactions, whose rates depend on the total pressure, are important in the atmosphere. Examples include the formation of O_3,

$$O(^3P) + O_2 + M \rightarrow O_3 + M, \quad (3)$$

> **BOX 5.1**
>
> **DERIVATION OF HALF-LIVES AND LIFETIMES FROM KINETICS**
>
> The expressions for half-lives and lifetimes in Table 5.2 can be readily derived from the rate laws. For a first-order reaction of a pollutant species A, the rate law for the reaction
>
> $$A \xrightarrow{k_1} \text{Products}$$
>
> is given by
>
> $$\frac{-d[A]}{dt} = k_1[A].$$
>
> Rearranging, this becomes
>
> $$\frac{-d[A]}{[A]} = k_1 \, dt.$$
>
> Integrating from time $t = 0$ when the initial concentration of A is $[A]_0$ to time t when the concentration is $[A]$, one obtains
>
> $$\ln \frac{[A]}{[A]_0} = -k_1 t.$$
>
> After one half-life (i.e., at $t = t_{1/2}$) by definition $[A] = 0.5[A]_0$. Substituting into the integrated rate expression, one obtains
>
> $$t_{1/2} = -\frac{\ln 0.5}{k_1} = \frac{0.693}{k_1}.$$
>
> For second- and third-order reactions, *if one assumes the concentrations of the reactants other than A are constant with time*, the derivation is the same except that k is replaced by $k[B]$ (second order) or $k[B][C]$ (third order).
>
> In most practical situations, however, the concentration of at least one of the other reactants is not constant but changes with time due to reactions, fresh injections of pollutants, and so on. As a result, using half-lives (or lifetimes) of a pollutant with respect to second- or third-order reactions is an approximation that involves *assumed* constant concentrations of the other reactants. These half-lives for bimolecular and termolecular reactions are thus directly affected by the concentrations of the other reactant.
>
> Derivation of the relationship between the rate constant k and the lifetime τ follows that for $t_{1/2}$, except that, from the definition of τ, at $t = \tau$, $[A] = [A]_0/e$.

and the oxidations of SO_2 and NO_2 via gas-phase OH reactions:

$$OH + SO_2 \xrightarrow{M} HOSO_2, \qquad (10)$$

$$OH + NO_2 \xrightarrow{M} HONO_2. \qquad (11)$$

The reason for the pressure dependence of termolecular reactions can be seen by taking reaction (3) as an example. The exothermic bond formation between $O(^3P)$ and O_2 releases energy that must be removed to form a stable O_3 molecule; if the energy remains as internal energy, the O_3 will quickly fly apart to re-form $O + O_2$. The third molecule, M, is any molecule that stabilizes the excited $(O_3)^*$ intermediate by colliding with it and removing some of its excess internal energy. Treating reaction (3) as an elementary reaction

$$\text{Rate} = k_3[O][O_2][M] = +\frac{d[O_3]}{dt},$$

one might expect the rate to increase with the concentration or pressure of the third body M. However, there clearly must be some limit since the rate cannot increase to infinity but only to some upper limit determined by how fast the two reactive species can combine chemically. As a result, one might intuitively expect the rates of reactions such as (3), (10), and (11) to increase initially as the pressure of M is increased from zero and then to plateau at some limiting value at high pressures.

Let us take the reaction (10) of OH with SO_2 as an example of a termolecular reaction of atmospheric interest and examine how its pressure dependence is established. It is common in kinetic studies to follow the decay of one reactant in an excess of the second reactant. In the case of reaction (10), the decay of OH is followed in the presence of excess SO_2 and the third body M, where M is an inert "bath" gas such as He,

Ar, or N_2. Since it is assumed to be an elementary reaction, the rate law for reaction (10) can be written for low pressures:

$$\frac{-d[OH]}{dt} = k_{10}^{III}[OH][SO_2][M].$$

If [M] is constant, k_{10}^{III} and [M] can be combined to form an effective bimolecular rate constant, $k_{10}^{bi} = k_{10}^{III}[M]$:

$$\frac{-d[OH]}{dt} = k_{10}^{III}[OH][SO_2][M] = k_{10}^{bi}[OH][SO_2].$$

Since SO_2 is in great excess, its concentration does not change significantly even when all the OH has reacted and hence it remains approximately constant throughout the reaction at its initial value, $[SO_2]_0$. Rearranging the rate law and integrating from time $t = 0$ when the initial concentration of OH is $[OH]_0$ to time t when the OH concentration is [OH], one obtains

$$\ln \frac{[OH]}{[OH]_0} = -k_{10}^{bi}[SO_2]_0 t.$$

Since the initial concentration of OH, $[OH]_0$, is a constant, a plot of ln[OH] against reaction time t should be a straight line with slope or decay rate given by

$$\text{Decay rate } (s^{-1}) = -k_{10}^{bi}[SO_2]_0.$$

A plot of these decay rates against $[SO_2]_0$ should thus be linear, with the slopes increasing with pressure since k_{10}^{bi} depends on [M].

Figure 5.1 shows such a plot of the absolute values of the observed OH decay rates against $[SO_2]_0$ at total pressures of Ar from 50 to 402 Torr (Atkinson et al., 1976). As expected, the decay rates are linear with $[SO_2]_0$ and increase with the pressure of M.

To obtain the termolecular rate constant k_{10}^{III}, the effective bimolecular rate constant $k_{10}^{bi} = k_{10}^{III}[M]$ is plotted in Fig. 5.2 as a function of total pressure (i.e., of [M]). As expected from the earlier discussion, k_{10}^{bi} increases with [M] at low pressures but approaches a plateau at higher pressures.

Termolecular reactions can be treated, as a first approximation, as if they consist of several elementary steps, for example, for reaction (10),

$$OH + SO_2 \underset{k_b}{\overset{k_a}{\rightleftharpoons}} HOSO_2^*, \quad (12, -12)$$

$$HOSO_2^* + M \overset{k_c}{\rightleftharpoons} HOSO_2 + M. \quad (13)$$

$HOSO_2^*$ is the exited OH–SO_2 adduct that contains the excess internal energy from bond formation in (12), and $HOSO_2$ is the stabilized adduct resulting when some of this internal energy is removed by a collision with M.

If the system is treated as if the concentration of the energized adduct ($HOSO_2^*$) remains constant with time, then its rates of formation and loss are equal. These rates can be written from Eqs. (12), (−12), and (13) since these are assumed to be elementary reactions. Thus

$$\frac{d[HOSO_2^*]}{dt} = 0 = k_a[OH][SO_2] - k_b[HOSO_2^*]$$
$$- k_c[HOSO_2^*][M].$$

This is an example of the *steady-state approximation*, widely employed in gas-phase kinetics and mechanistic studies.

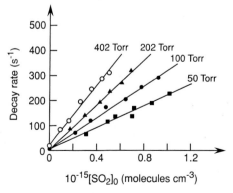

FIGURE 5.1 Plots of the OH decay rates against the initial SO_2 concentration at total pressures of Ar from 50 to 402 Torr (adapted from Atkinson et al., 1976).

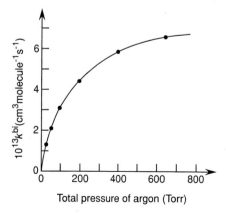

FIGURE 5.2 Plot of k_{10}^{bi} against total pressure for M = Ar for the reaction of OH with SO_2 (adapted from Atkinson et al., 1976).

Rearranging, an expression for [HOSO$_2^*$] is obtained in terms of the reactants OH and SO$_2$:

$$[\text{HOSO}_2^*] = \frac{k_a[\text{OH}][\text{SO}_2]}{k_b + k_c[\text{M}]}.$$

The rate of the reaction in terms of product formation is given by

$$\frac{d[\text{HOSO}_2]}{dt} = k_c[\text{M}][\text{HOSO}_2^*],$$

$$= k_c[\text{M}]\frac{k_a[\text{OH}][\text{SO}_2]}{k_b + k_c[\text{M}]},$$

$$= \left\{\frac{k_a k_c[\text{M}]}{k_b + k_c[\text{M}]}\right\}[\text{OH}][\text{SO}_2].$$

In this form, k_{10}^{bi} can be rationalized by the combination of rate constants and [M] given in brackets. Alternatively, $(k_{10}^{bi})^{-1}$ is given by

$$\frac{1}{k_{10}^{bi}} = \frac{k_b + k_c[\text{M}]}{k_a k_c[\text{M}]} = \frac{k_b}{k_a k_c[\text{M}]} + \frac{1}{k_a}.$$

At "infinite" pressure, where $1/[\text{M}] = 0$, the rate constant should have its high-pressure limiting value. It is seen that this high-pressure limiting value, k_{10}^∞, is equal to k_a. One would also qualitatively expect $k_{10}^\infty = k_a$ from the reaction scheme consisting of (12), (−12), and (13); thus in the limit of infinite pressure, all the energized adducts formed in (12) will be stabilized in (13) and none will have a chance to decompose back to reactants via (−12). In this case, the rate constant will just be that for formation of HOSO$_2^*$, that is, k_a.

This approximate treatment of termolecular reactions can also be used to examine how the third-order, low-pressure rate constant k^{III} relates to the rate constants k_a, k_b, and k_c for the elementary reactions assumed to be involved. As [M] approaches zero, k_{10}^{bi} approaches $k_a k_c[\text{M}]/k_b$, so that k_{10}^{III} is given by

$$k_{10}^{III} = \frac{k_a k_c}{k_b}.$$

Many addition reactions such as the OH–SO$_2$ reaction are in the falloff region between second and third order in the range of total pressures encountered from the troposphere through the stratosphere. Troe and co-workers have carried out extensive theoretical studies of addition reactions and their reverse unimolecular decompositions as a function of pressure (e.g., see Troe, 1979, 1983). In this work they have developed expressions for the rate constants in the falloff region; these are now most commonly used to derive the limiting low- and high-pressure rate constants from experimental data as well as to report the temperature and pressure dependence of termolecular reactions in compilations of kinetic data.

Equation (C) gives the most commonly used form of the rate constant expression of Troe and co-workers:

$$k = \frac{k_0[\text{M}]}{1 + k_0[\text{M}]/k_\infty} F_C^{\{1 + [\log_{10} k_0[\text{M}]/k_\infty]^2\}^{-1}}. \quad (C)$$

In equation (C), k_0 (or k^{III} as used earlier) is the low-pressure limiting rate constant and k_∞ is the high-pressure limiting rate constant. F_C is known as the broadening factor of the falloff curve; its actual value depends on the particular reaction and can be calculated theoretically. Troe (1979) suggests that for reactions under atmospheric conditions, the value of F_C will be ~ 0.7–0.9, independent of temperature. However, values as low as 0.4 are often observed. The NASA evaluations of stratospheric reactions (DeMore et al., 1997) take $F_C = 0.6$ for all reactions. The IUPAC evaluation (Atkinson et al., 1997a,b) does not restrict F_C to 0.6. However, it is important to note that the values of k_0 and k_∞ will depend on the value of F_C used to match the experimental data. For example, for reaction (11)

$$\text{OH} + \text{NO}_2 \xrightarrow{M} \text{HONO}_2, \quad (11)$$

the JPL evaluation recommends, at 300 K, $k_0 = 2.5 \times 10^{-30}$ cm^6 molecule^{-2} s^{-1} and $k_\infty = 1.6 \times 10^{-11}$ cm^2 molecule^{-1} s^{-1} with $F_C = 0.6$, whereas the IUPAC recommendation is $k_0 = 2.6 \times 10^{-30}$ and $k_\infty = 7.5 \times 10^{-11}$ with $F_C = 0.41$ (Atkinson et al., 1997b). (Note, however, that more recent studies discussed in Chapter 7.B.1 suggest both of these recommendations may give systematically high rate constants, especially below 240 K.)

The temperature dependence of k arises primarily in the temperature dependence of k_0 and k_∞, which are expressed in terms of their values at 300 K, k_0^{300} and k_∞^{300}. Thus

$$k_0^T = k_0^{300}\left(\frac{T}{300}\right)^{-n}$$

and

$$k_\infty^T = k_\infty^{300}\left(\frac{T}{300}\right)^{-m}.$$

Thus, there are a number of parameters (k_0^{300}, k_∞^{300}, n, m, and F_C) that must be known in order to calculate the rate constant throughout the falloff region. Experimentally, these are determined by obtaining a best fit to the experimentally measured values of k using a calculated or assumed value of $F_C = 0.6$, so that rate constants for the termolecular reactions can be calculated as a function of pressure and temperature. For example, for the reaction (10) of OH with SO_2 discussed earlier, $k_0^{300} = (3.0 \pm 1.0) \times 10^{-31}$ cm^6 $molecule^{-2}$ s^{-1}, $n = (3.3 \pm 1.5)$, $k_\infty^{300} = (1.5 \pm 0.5) \times 10^{-12}$ cm^{-3} $molecule^{-1}$ s^{-1}, and $m = 0$ (DeMore et al., 1997). At 300 K and 760 Torr pressure, $[M] = 2.46 \times 10^{19}$ molecules cm^{-3} and $k_0[M] = 7.4 \times 10^{-12}$ cm^3 $molecule^{-1}$ s^{-1}. Thus the value of the rate constant under these conditions is calculated as

$$k = \left[\frac{7.4 \times 10^{-12}}{1 + (7.4 \times 10^{-12})/(1.5 \times 10^{-12})}\right]$$
$$\times 0.6^{\{1 + (\log[(7.4 \times 10^{-12})/(1.5 \times 10^{-12})])^2\}^{-1}},$$
$$= [1.25 \times 10^{-12}]0.6^{0.68}$$
$$= 8.8 \times 10^{-13} \text{ cm}^3 \text{ molecule}^{-1} \text{ s}^{-1}.$$

However, under conditions typical of the lower stratosphere (~ 20 km), the temperature and pressure are much lower. Let us calculate k_{10} for conditions where the temperature is 219 K and the total pressure is 39 Torr. First we need to recognize that [M] changes not only with pressure but also with temperature according to the ideal gas law: $[M] = P/RT$. Under these conditions, $[M] = 1.7 \times 10^{18}$ molecules cm^{-3}. The low- and high-pressure limiting rate constants at this temperature are given by Eqs. (D) and (E):

$$k_0^{219} = 3.0 \times 10^{-31}\left(\frac{219}{300}\right)^{-3.3}$$
$$= 8.5 \times 10^{-31} \text{ cm}^6 \text{ molecule}^{-2} \text{ s}^{-1}, \quad (D)$$
$$k_\infty^{219} = 1.5 \times 10^{-12}\left(\frac{219}{300}\right)^{-0}$$
$$= 1.5 \times 10^{-12} \text{ cm}^3 \text{ molecule}^{-1} \text{ s}^{-1}. \quad (E)$$

From Eq. (C), the rate constant at 219 K and 39 Torr pressure is

$$k = \left[\frac{1.5 \times 10^{-12}}{(1 + 1.5 \times 10^{-12})/(1.5 \times 10^{-12})}\right]$$
$$\times 0.6^{[1 + (\log[(1.4 \times 10^{-12})/(1.5 \times 10^{-12})])^2]^{-1}},$$
$$= [7.5 \times 10^{-13}]0.6^{0.999}$$
$$= 4.5 \times 10^{-13} \text{ cm}^3 \text{ molecule}^{-1} \text{ s}^{-1}.$$

This is almost a factor of two smaller than at 300 K and 760 Torr pressure.

In summary, rate constants for addition reactions in the atmosphere can be estimated as a function of temperature and pressure if values are available for the low- and high-pressure limiting rate constants as a function of temperature, that is, if k_0^{300}, k_∞^{300}, n, m, and F_C are known.

At first glance, it might appear that the vast majority of the bimolecular reactions with which one deals in the troposphere are simple concerted reactions, that is, during the collision of the reactants there is a reorganization of the atoms, leading directly to the formation of the products. However, it has become increasingly apparent in recent years that some important reactions that *appeared* to be concerted exhibit characteristics such as pressure dependencies that are not consistent with a direct concerted process.

A classic case is the reaction of OH with CO:

$$OH + CO \rightarrow H + CO_2. \qquad (14)$$

This reaction appears to be an elementary bimolecular reaction involving a simple transfer of an oxygen atom from OH to CO. In accord with the definition of an elementary reaction, one can imagine that it occurs during one collision of an OH radical with a CO molecule.

A number of studies of the kinetics of this reaction were carried out in the 1960s and the early 1970s, and the room temperature rate constants, measured at total pressures up to ~ 200 Torr in inert gases such as He, Ar, and N_2, were generally in good agreement with $k_{14} \sim 1.5 \times 10^{-13}$ cm^3 $molecule^{-1}$ s^{-1} at room temperature. In fact, this reaction was often used to test whether a newly constructed kinetic apparatus was functioning properly.

However, a variety of studies since the mid-1970s has established that it is not, in fact, a simple bimolecular reaction as implied by reaction (14) but rather involves the formation of an excited HOCO* intermediate (e.g., see Fulle et al., 1996; Golden et al., 1998; and references therein):

$$OH + CO \underset{k_{-a}}{\overset{k_a}{\rightleftharpoons}} (HOCO)^* \overset{M}{\leftrightarrow} HOCO,$$
$$\overset{k_b}{\longrightarrow} OH + CO,$$
$$\overset{k_c}{\longrightarrow} H + CO_2. \qquad (15)$$

In (15), HOCO is the radical adduct of OH + CO, and HOCO* is the adduct containing excess internal energy resulting from the energy released by bond formation between OH and CO. As described earlier, M is any molecule or atom that collides with the HOCO*, removing some of its excess energy; in practice, it is usually an inert bath gas such as He or Ar that is present in great excess over the reactants.

Reactions such as (15), which proceed with the formation of a bound adduct between the reactants, are known as indirect or nonconcerted reactions. The

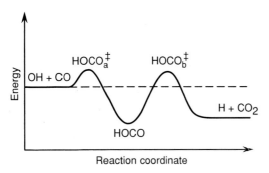

FIGURE 5.3 Typical potential energy diagram proposed for reaction of OH with CO (adapted from Mozurkewich et al., 1984). Note the well corresponding to formation of the (HOCO) intermediate.

adduct is "stable" in the sense that it corresponds to a well on the potential energy surface connecting the reactants and products (Fig. 5.3); as such it has a finite lifetime and should be capable of being detected using appropriate techniques. Because of the complex nature of the mechanism, such reactions can exhibit a relatively complex temperature dependence. In addition, if the rate of collisional stabilization of the excited adduct is comparable to its rate of decomposition, a pressure dependence may result, as in the OH + CO reaction. The distinction between bimolecular and termolecular reactions blurs in such cases. In any event, the OH + CO rate constant at 1 atm in air is now thought to be $\sim 2.4 \times 10^{-13}$ cm^3 molecule^{-1} s^{-1}, significantly greater than the low-pressure value, 1.5×10^{-13} cm^3 molecule^{-1} s^{-1}, that had been widely accepted at one time.

3. Temperature Dependence of Rate Constants

a. Arrhenius Expression

The temperature dependence of many rate constants can be fit over a relatively narrow temperature range by the exponential *Arrhenius equation*

$$k = Ae^{-E_a/RT}, \tag{F}$$

where **R** is the gas constant and the temperature T is in kelvin (K = °C + 273.15). A, the preexponential factor, and E_a, the activation energy, are parameters characteristic of the particular reaction.

To a first approximation over the relatively small temperature range encountered in the troposphere, A is found to be independent of temperature for many reactions, so that a plot of ln k versus T^{-1} gives a straight line of slope $-E_a/\mathbf{R}$ and intercept equal to ln A. However, the Arrhenius expression for the temperature dependence of the rate constant is empirically based. As the temperature range over which experiments could be carried out was extended, nonlinear Arrhenius plots of ln k against T^{-1} were observed for some reactions. This is not unexpected when the predictions of the two major kinetic theories in common use today, collision theory and transition state theory, are considered. A brief summary of the essential elements of these is found in the following sections, as we refer to them periodically throughout the text.

For many reactions, the temperature dependence of A is small (e.g., varies with $T^{1/2}$) compared to the exponential term so that Eq. (F) is a good approximation, at least over a limited temperature range. For some reactions encountered in tropospheric chemistry, however, this is not the case. For example, for reactions in which the activation energy is small or zero, the temperature dependence of A can become significant. As a result, the Arrhenius expression (F) is not appropriate to describe the temperature dependence, and the form

$$k = BT^n e^{-E_a/RT} \tag{G}$$

is frequently used, where B is a temperature-independent constant characteristic of the reaction and n is a number adjusted to provide a best fit to the data.

While most reactions with which we deal in atmospheric chemistry increase in rate as the temperature increases, there are several notable exceptions. The first is the case of termolecular reactions, which generally slow down as the temperature increases. This can be rationalized qualitatively on the basis that the lifetime of the excited bimolecular complex formed by two of the reactants with respect to decomposition back to reactants decreases as the temperature increases, so that the probability of the excited complex being stabilized by a collision with a third body falls with increasing temperature.

An alternate explanation can be seen by treating termolecular reactions as the sum of bimolecular reactions, as was illustrated in Section A.2 for the OH + SO$_2$ + M reaction. Recall that the third-order, low-pressure rate constant k^{III} can be expressed as the product of the three rate constants k_a, k_b, and k_c for the three individual reaction steps (12), (-12), and (13):

$$k_{10}^{III} = \frac{k_a k_c}{k_b}.$$

Expressing each of the component rate constants in the Arrhenius form, k_{10}^{III} becomes

$$k_{10}^{III} = A_{10} e^{-E_{10}/RT} = \left(\frac{A_a A_c}{A_b}\right) e^{-(E_a + E_c - E_b)/RT}. \tag{H}$$

Thus the activation energy for the reaction, E_{10}, is a combination of the activation energies for the individual steps, $(E_a + E_c - E_b)$. If $E_b > (E_a + E_c)$, that is, if the activation energy for decomposition of the energized adduct $(HOSO_2)^*$ back to reactants is greater than the sum of $E_a + E_c$, then the effective activation energy E_{10} for the termolecular reaction becomes negative and the rate constant decreases as the temperature increases.

We now briefly consider the predictions of collision theory and transition state theory for the expected form of the rate constant and its temperature dependence.

b. Predictions of Collision Theory and Transition State Theory

Collision theory is based on the concept that molecules behave like hard spheres; during a collision of two species, a reaction may occur. To estimate a rate constant for a bimolecular reaction between reactants A and B based on this theory, one needs first to calculate the number of collisions occurring in a unit volume per second (Z_{AB}) when the two species, A and B, having radii r_A and r_B, are present in concentrations N_A and N_B, respectively. From gas kinetic theory, this can be shown to be given by Eq. (I):

$$Z_{AB} = (r_A + r_B)^2 \left(\frac{8\pi \mathbf{k} T}{\mu}\right)^{1/2} N_A N_B. \tag{I}$$

μ is the reduced mass of A and B [$\mu = m_A m_B/(m_A + m_B)$], where m is the mass of the individual meucles, $\mathbf{k}$ is the Boltzmann constant (1.38×10^{-23} J K^{-1}), and T is the temperature in kelvin. In deriving Eq. (I), it is assumed that A and B are hard spheres and only collide when their centers come within a distance ($r_A + r_B$) of each other.

In considering reactions between colliding molecules, one must take into account two additional factors. First, different collisions will occur with different amounts of energy depending on the speed of the molecules as they collide. Most reactions are expected to have an energy barrier that must be surmounted for a reaction to occur. This energy barrier arises from the net effect of simultaneous bond breaking and formation; at the transition state in the reaction, the energy released from bond formation is generally less than that required for bond breaking. The difference, which is the energy barrier, must be supplied in other ways if reaction is to take place.

Second, even if the reactants collide with enough energy to surmount the energy barrier, they may not react if they are not in the proper orientation with respect to each other. The importance of this so-called *steric factor* can be illustrated using the reaction of ground-state oxygen atoms with the hydroxyl free radical. If the oxygen atom collides with the oxygen atom end of the OH, the orientation is correct for an overall reaction to $O_2 + H$:

$$O + O\text{—}H \to [O\text{····}O\text{····}H] \to O_2 + H.$$

However, if it collides with the hydrogen atom end, no net chemical change will result, although, in fact, an exchange reaction (i.e., exchange of the oxygen atoms) may occur:

$$O + H\text{—}O \to [O\text{····}H\text{····}O] \to$$
$$OH + O \quad \text{(no net chemical change)}.$$

To take into account the energy requirement, one can modify the result in Eq. (I) by calculating only the number of collisions between A and B that have a certain minimum energy, E_0. In the simplest approach, it is assumed that no reaction occurs if the energy of the colliding pair is less than E_0, and reaction occurs 100% of the time for energies $\geq E_0$. Alternatively, it can be assumed that for energies $> E_0$ the probability of reaction increases as the relative collision energy increases.

To take into account this dependence on energy, the concept of A and B being hard spheres with radii r_A and r_B can be modified. Let $\sigma_{AB} = \pi(r_A + r_B)^2$ be the *collisional* cross section for hard-sphere collisions between A and B and σ_R the *reaction* cross section for reaction between A and B. At energies $< E_0$, $\sigma_R = 0$; that is, no reaction occurs. For energies $> E_0$, σ_R could be taken as a constant (e.g., equal to σ_{AB}), which would correspond to assuming all collisions having energies above the threshold energy leading to reaction. The alternate approach of an increase in reaction probability with increasing energy above the threshold corresponds to assuming an expression for σ_R that is a function of total energy E, for example,

$$\sigma_R = \sigma_{AB}\left(1 - \frac{E_0}{E}\right) \quad \text{for } E \geq E_0, \tag{J}$$
$$= 0 \quad \text{for } E < E_0.$$

For the form of σ_R shown in Eq. (J), integrating over all total energies from 0 to infinity, the rate of reactive collisions becomes

$$Z_R = \sigma_{AB}\left(\frac{8\mathbf{k}T}{\pi\mu}\right)^{1/2} e^{-E_0/\mathbf{k}T} N_A N_B. \tag{K}$$

The rate constant k can thus be identified with Eq. (L):

$$k = \sigma_{AB}\left(\frac{8\mathbf{k}T}{\pi\mu}\right)^{1/2} e^{-E_0/\mathbf{k}T}. \tag{L}$$

Equation (L), however, does not take into account the need for proper orientation of the colliding molecules in order for a reaction to occur. This is commonly done by including an extra factor, P (the steric factor), which is the probability that the colliding molecules will have the correct orientation. This leads to Eq. (M):

$$k = P\sigma_{AB}\left(\frac{8\mathbf{k}T}{\pi\mu}\right)^{1/2} e^{-E_0/\mathbf{k}T}. \quad \text{(M)}$$

By comparison of (M) and (F), it can be seen that the preexponential factor A in the Arrhenius equation can be identified with $P\sigma_{AB}(8\mathbf{k}T/\pi\mu)^{1/2}$ and the activation energy, E_a, with the threshold energy E_0. It is important to note that collision theory predicts that the preexponential factor should indeed be dependent on temperature ($T^{1/2}$). The reason so many reactions appear to follow the Arrhenius equation with A being temperature independent is that the temperature dependence contained in the exponential term normally swamps the smaller $T^{1/2}$ dependence. However, for reactions where E_a approaches zero, the temperature dependence of the preexponential factor can be significant.

Collision theory is used mainly as a reference for the efficiency of reactions. Thus, at a temperature of 25°C (298 K), the rate constant for a reaction between two molecules each having a radius of 0.2 nm and a molecular weight of 50 g mol^{-1} would, according to Eq. (M), be 2.5×10^{-10} cm^3 molecule^{-1} s^{-1} for $P = 1$ and $E_0 = 0$. That is, when there are no steric or energy barriers to the reaction, the rate constant should be of the order of 10^{-10} cm^3 molecule^{-1} s^{-1}. A reaction with a rate constant of 10^{-15} cm^3 molecule^{-1} s^{-1} is therefore one that goes in approximately every 10^5 collisions.

Transition state theory is more commonly applied today than collision theory. It is especially useful in examining reactions in solution and avoids the problem of introducing arbitrary factors such as the steric factor to take into account steric requirements.

Transition state theory treats a reacting system thermodynamically. Let us again take a bimolecular reaction between A and B. Transition state theory assumes that as A and B collide and start to react, they form a species called the activated complex, which corresponds to the A–B adduct at the peak of the energy hill lying between reactants and products. This activated complex is thus in a "transition state" and can either fall back to reactants or go on to form products. The activated complex is normally indicated with a double dagger symbol, $AB^{\neq}$. The reaction can thus be given as

$$A + B \underset{k_r}{\overset{k_f}{\rightleftharpoons}} AB^{\neq} \xrightarrow{k_p} \text{Products}.$$

Assuming the activated complex is in equilibrium with reactants, one can define the equilibrium constant $K^{\neq}$:

$$K^{\neq} = \frac{k_f}{k_r} = \frac{[AB^{\neq}]}{[A][B]}. \quad \text{(N)}$$

$K^{\neq}$ is related to the standard free energy change in going from reactants to the transition state, $\Delta G^{0\neq}$ by the usual thermodynamic relationship:

$$K^{\neq} = e^{-\Delta G^{0\neq}/RT} = (e^{-\Delta H^{0\neq}/RT})(e^{\Delta S^{0\neq}/R}). \quad \text{(O)}$$

$\Delta H^{\circ\neq}$ and $\Delta S^{\circ\neq}$ are the standard enthalpy and entropy changes in going from reactants to the *transition state* (not to products).

The reaction rate is determined by the rate at which $AB^{\neq}$ forms products

$$\text{Rate of product formation} = \nu[AB^{\neq}].$$

where ν is the frequency with which $AB^{\neq}$ breaks up into products. Substituting for $[AB^{\neq}]$ from Eq. (N) and for $K^{\neq}$ from (O), one obtains

$$\text{Rate} = \nu K^{\neq}[A][B],$$
$$= \nu(e^{\Delta S^{\circ\neq}/R})(e^{-\Delta H^{\circ\neq}/RT})[A][B]. \quad \text{(P)}$$

The enthalpy change $\Delta H^{\circ\neq}$ is related to the energy change in going from reactants to the transition state, that is, to the activation energy. The frequency ν of breakup of the activated complex into products is often approximated by $\nu = \mathbf{k}T/h$, where $\mathbf{k}$ and h are the Boltzmann and Planck constants, respectively. Comparison of Eq. (P) to the rate equation shows that

$$k = \frac{\mathbf{k}T}{h}(e^{\Delta S^{\circ\neq}/R})(e^{-\Delta H^{\circ\neq}/RT}). \quad \text{(Q)}$$

Again the preexponential factor is seen to be temperature dependent, but for large activation energies, the exponential term dominates the temperature dependence of the rate constant.

The preexponential factor involves the entropy change in going from reactants to the transition state; the more highly ordered and tightly bound is the transition state, the more negative $\Delta S^{\circ\neq}$ will be and the lower the preexponential factor will be. Transition state theory thus automatically takes into account the effect of steric factors on rate constants, in contrast to collision theory.

An alternate form of the rate constant predicted by transition state theory using a statistical mechanical approach for the equilibrium constant $K^{\neq}$ is Eq. (R):

$$k = \left(\frac{\mathbf{k}T}{h}\right)\left(\frac{q_{AB^{\neq}}}{q_A q_B}\right)e^{-E^{\neq}/RT}. \quad \text{(R)}$$

Here the q's are partition functions for the reactants (q_A and q_B) and the transition state ($q_{AB}^{\neq}$), and $E^{\neq}$ is the energy difference between the reactants and the transition state.

The partition functions include contributions from translational, rotational, vibrational, and electronic degrees of freedom,

$$q_{trans} q_{rot} q_{vib} q_{elect}.$$

Those for the reactants can be evaluated using conventional techniques discussed in physical chemistry texts; estimating the partition functions for the transition state requires making assumptions concerning the nature of the transition state.

Transition state theory can be used to test reaction dynamics on a molecular scale. Thus one can hypothesize a spatial configuration of the atoms in the transition state and from this calculate $\Delta S^{\circ \neq}$; the predicted rate constant can then be compared to that observed. If the agreement is not acceptable, the molecular configuration of the transition state can be adjusted until such agreement is obtained. Assuming this molecular configuration approximates the actual form of the intermediate in the reaction, one can learn something about the chemical dynamics of the reaction.

An example of the application of transition state theory to atmospheric reactions is the reaction of OH with CO. As discussed earlier, this reaction is now believed to proceed by the formation of a radical adduct HOCO, which can decompose back to reactants or go on to form the products $H + CO_2$. For complex reactions such as this, transition state theory can be applied to the individual reaction steps, that is, to the steps shown in reaction (15). Figure 5.3 shows schematically the potential energy surface proposed for this reaction (Mozurkewich et al., 1984). The adduct HOCO, corresponding to a well on the potential energy surface, can either decompose back to reactants via the transition state shown as $HOCO_a^{\neq}$ or form products via transition state $HOCO_b^{\neq}$.

c. Example of Importance of Temperature Dependence: PAN Decomposition

As we have seen in Chapter 1, peroxyacetyl nitrate (PAN) is a powerful lachrymator and severe plant phytotoxicant formed in irradiated VOC–NO_x mixtures from the reaction of peroxyacetyl radicals with NO_2:

$$\underset{CH_3COO}{\overset{O}{\|}} + NO_2 \longrightarrow \underset{CH_3COONO_2}{\overset{O}{\|}}. \quad (16)$$

PAN is thermally unstable, however, and decomposes at higher temperatures to reform peroxyacetyl radicals and NO_2, that is, the reverse of reaction (16):

$$\underset{CH_3COONO_2}{\overset{O}{\|}} \overset{\Delta}{\longrightarrow} \underset{CH_3COO}{\overset{O}{\|}} + NO_2. \quad (-16)$$

The rate constant for PAN decomposition (k_{-16}) is strongly temperature dependent. As a result, this is important in the atmosphere because PAN acts as a reservoir for NO_2. At low temperatures, much of the total NO_x may be tied up in PAN; as the atmosphere warms, due to either diurnal temperature variations or the air mass being transported into warmer regions, the rate of PAN decomposition increases. This releases NO_2, which can then form other secondary pollutants such as O_3 and HNO_3.

The rate constant k_{-16} can be fit to the Arrhenius form (see, for example, Grosjean et al., 1994)

$$k_{-16} \, (s^{-1}) = 1.58 \times 10^{16} e^{-(112.5 \, kJ/mol)/RT}$$

over the temperature range normally encountered in the troposphere. At room temperature (25°C = 77°F) k_{-16} is 3.0×10^{-4} s^{-1}, giving a natural lifetime with respect to decomposition of 55 min. However, at 0°C (32°F), k_{-16} is 4.8×10^{-6} s^{-1}, corresponding to a lifetime of ~57 h. At 35°C (95°F), the rate constant k_{-16} (1.3×10^{-3} s^{-1}) is larger than that at 0°C by almost three orders of magnitude and the lifetime for PAN is correspondingly shorter, only ~12 min! Clearly, the temperature dependence of rate constants can be extremely important in determining the lifetimes and fates of certain species in the atmosphere, as well as their contribution to secondary pollutant formation.

B. LABORATORY TECHNIQUES FOR DETERMINING ABSOLUTE RATE CONSTANTS FOR GAS-PHASE REACTIONS

In this section we discuss the major experimental methods used to determine *absolute* rate constants for gas-phase reactions relevant to atmospheric chemistry. These include fast-flow systems (FFS), flash photolysis (FP), static reaction systems, and pulse radiolysis. The determination of *relative* rate constants is discussed in Section C.

In general, we use simple bimolecular reactions of the type

$$A + B \rightarrow Products \quad (17)$$

as illustrations. However, the techniques can be modified to study termolecular reactions, as discussed earlier, as well as unimolecular reactions.

To study the reaction kinetics of a relatively reactive species A with a reactant B, one normally follows the loss of small amounts of A in the presence of a great excess of B. This requires then that one be able first to generate A and second to monitor its concentration as a function of time. Ideally, to fully elucidate the reaction mechanism, one would also monitor the concentrations of intermediates and products. As we shall see, in practice, for many reactions this proves to be much more difficult than to simply determine the rate constant itself.

1. Kinetic Analysis

The rate law for a simple bimolecular reaction such as (17) is given by

$$\frac{-d[A]}{dt} = k_{17}[A][B]. \qquad (S)$$

If a small concentration of A is generated in a great excess of B, then even if (17) is allowed to go to completion, the concentration of B will remain essentially constant at its initial concentration $[B]_0$. Integrating (S) and treating $[B]_0$ as constant, one obtains

$$\ln \frac{[A]}{[A]_0} = -k_{17}[B]_0 t. \qquad (T)$$

That is, A decays exponentially with time determined by $(k_{17}[B]_0)$, as if it were a first-order reaction. Thus under these so-called *pseudo-first-order* conditions, a plot of $\ln[A]$ against time for a given value of $[B]_0$ should be linear with a slope equal to $(-k_{17}[B]_0)$. These plots are carried out for a series of concentrations of $[B]_0$ and the values of the corresponding decays determined. Finally, the absolute rate constant of interest, k_{17}, is the slope of a plot of the absolute values of these decay rates against the corresponding values of $[B]_0$. Some examples are discussed below.

As we have seen earlier, even third-order reactions can be reduced to pseudo-first-order reactions by keeping the concentrations of all species except A constant and in great excess compared to A. This technique of using pseudo-first-order conditions is by far the most common technique for determining rate constants. Not only does it require monitoring only one species, A, as a function of time, but even absolute concentrations of A need not be measured. Because the ratio $[A]/[A]_0$ appears in Eq. (T), the measurement of any parameter that is *proportional* to the concentration of A will suffice in determining k_{17}, since the proportionality constant between the parameter and [A] cancels out in Eq. (T). For example, if A absorbs light in a convenient spectral region and Beer's law is obeyed, then the absorbance (Abs) of a given concentration of A, N (number cm^{-3}), is given by

$$\ln \frac{I_0}{I} = \text{Abs} = \sigma N l, \qquad (U)$$

where I_0 and I are the intensities of the incident and transmitted light, respectively, l is the optical path length, and σ is the absorption cross section of A (to the base e)

Substituting into Eq. (T) for $[A] = N = \text{Abs}/\sigma l$, one obtains

$$\ln \frac{(\text{Abs}/\sigma l)}{(\text{Abs}/\sigma l)_0} = \ln \frac{(\text{Abs})}{(\text{Abs})_0} = -k_{17}[B]_0 t, \qquad (V)$$

where (Abs) and (Abs)$_0$ are the absorbance of the light by A at times t and $t = 0$, respectively. For example, O_3 has a strong absorption at 254 nm, which can be used to monitor its concentration.

This ability to monitor a parameter that is *proportional* to concentration, rather than the absolute concentration itself, affords a substantial experimental advantage in most kinetic studies, since determining absolute concentrations of atoms and free radicals is often difficult.

This pseudo-first-order kinetic analysis is generally applied regardless of the experimental system used.

2. Fast-Flow Systems

Fast-flow systems (FFS) consist of a flow tube typically 2- to 5-cm in diameter in which the reactants A and B are mixed in the presence of a large amount of an inert "bath gas" such as He or Ar. As the mixture travels down the flow tube at relatively high linear flow speeds (typically 1000 cm s^{-1}), A and B react. The decay of A along the length of the flow tube, that is, with time, is followed and Eq. (T) applied to obtain the rate constant of interest.

The term *fast flow* comes from the high flow speeds. In most of these systems, discharges are used to generate A or another species that is a precursor to A; hence the term fast-flow discharge system (FFDS) is also commonly applied. Since fast-flow discharge systems have been applied in many kinetic and mechanistic studies relevant to tropospheric chemistry (e.g., see Howard, 1979; Kaufman, 1984), we concentrate on them. However, all fast-flow systems rely on the same experimental and theoretical principles.

Figure 5.4 is a schematic diagram of a typical fast-flow discharge system. The reactive species is generated in a microwave discharge and enters at the upstream

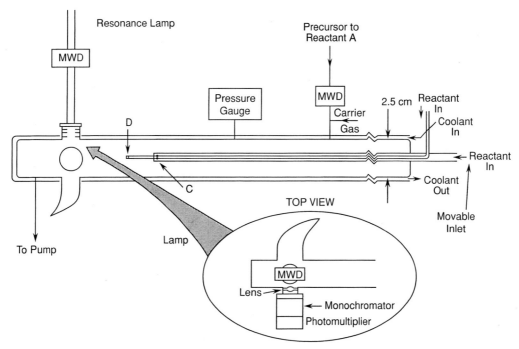

FIGURE 5.4 Schematic diagram of a fast-flow discharge system (adapted from Beichert et al., 1995). MWD = microwave discharge used to generate the reaction species from a precursor and to generate light of the appropriate wavelength to measure it using the resonance lamp.

end of the flow tube, and the second reactant is added through one port, C or D, of a movable inlet. The detector is fixed at the downstream end of the flow tube and the reaction time is varied by moving the mixing point for C and D (i.e., the movable inlet) relative to the fixed detector.

For example, OH can be generated by the reaction of H with NO_2:

$$H + NO_2 \rightarrow OH + NO. \quad (18)$$

Atomic hydrogen is formed by discharging dilute H_2/He mixtures. Then these H atoms are converted to OH by the addition of excess NO_2, e.g., through port C. By adjusting the concentration of NO_2 appropriately, essentially all of the H can be converted to OH before the second reactant is added through port D.

Table 5.3 shows some typical sources of reactive species of atmospheric interest used in FFDS, while Table 5.4 shows some of the methods used to detect them.

Under conditions where the *plug flow* assumption is valid, that is, concentration gradients are negligible so that the linear flow velocity of the carrier gas is the same as that of the reactants, the time (t) for A and B to travel a distance d along the flow tube is given by

$$t = \frac{d}{v}. \quad (W)$$

Here v is the linear flow speed, which can be calculated from the cross-sectional area of the flow tube (A_R), the total pressure (P) in the flow tube, the temperature (T), and the molar flow rates (dn/dt) of the reactants and the diluent gas:

$$v = \frac{RT}{A_R P}\left(\frac{dn}{dt}\right). \quad (X)$$

TABLE 5.3 Some Typical Sources of Reactive Species in Fast-Flow Discharge Systems

Reactive species	Source
OH	$H_2 \xrightarrow{MWD} 2H$
	$H + NO_2 \rightarrow OH + NO$
	$F_2 \xrightarrow{MWD} 2F$
	$F + H_2O \rightarrow OH + HF$
Cl	$F_2 \xrightarrow{MWD} 2F$
	$F + HCl \rightarrow Cl + HF$
NO_3	$F_2 \xrightarrow{MWD} 2F$
	$F + HNO_3 \rightarrow NO_3 + HF$
HO_2	$F_2 \xrightarrow{MWD} 2F$
	$F + H_2O_2 \rightarrow HO_2 + HF$
$O(^3P)$	$N_2 \xrightarrow{MWD} 2N$
	$N + NO \rightarrow O(^3P) + N_2$

TABLE 5.4 Some Typical Detection Systems Used for Reactive Species in Fast-Flow Systems

Technique	Typical application
Resonance fluorescence or light absorption	OH, 308 nm Cl, 135 nm O(^{3}P), 130.5 nm H(^{2}S), 121.6 nm CH$_3$O, 303.9 nm NO$_3$, 662.9 nm
Chemiluminescence	O(^{3}P): $O + NO \xrightarrow{M} NO_2^* + M$ H(^{2}S): $H + NO \xrightarrow{M} HNO^* + M$
Mass spectrometry	OH R HO$_2$ Stable products
Electron paramagnetic resonance (EPR)	OH
Laser magnetic resonance (LMR)	OH, HO$_2$

At typical linear flow speeds of 1000 cm s^{-1}, 1 cm along the tube corresponds to ~ 1-ms reaction time. Thus a flow tube of length 1 m can be used to study reactions at reaction times up to 100 ms.

Total pressures in most FFDS have typically been in the range 0.5–10 Torr where rapid diffusion across the flow tube ensures a relatively flat concentration profile of the reactants, so that Eq. (W) is valid. Maintaining the discharge used to generate atoms and free radicals is also difficult above a few Torr total pressure. The lower end of the pressure range is determined by the need to maintain viscous flow and to avoid significant axial concentration gradients. The latter may arise because of the lower concentrations of reactants at the downstream end of the flow tube; these can cause the true flow speed of the reactants to be greater than the calculated linear flow speed, due to their axial diffusion. Techniques for estimating errors due to such factors and for correcting measured rate constants for them are discussed in detail by Mulcahy (1973), Brown (1978), and Lambert et al. (1985).

However, flow tube systems for use at much higher pressures, up to several hundred Torr, have also been designed and applied to reactions of atmospheric interest (e.g., see Keyser, 1984; Abbatt et al., 1990, 1992; Seeley et al., 1993; and Donahue et al., 1996a). At these higher pressures, the velocity and radical axial and radial concentration profiles are experimentally determined and the full continuity equation describing the concentration profiles is solved.

A major factor in many FFDS studies is diffusion of the reactive species accompanied by their loss at the walls of the flow tube. Unfortunately, OH radicals are particularly sensitive to removal by wall reactions. While the mechanism and products of these wall reactions are unknown, it has been established that the rate of loss at the walls can be minimized by using various flow tube wall coatings or treatments. These include substances such as teflon or halocarbon waxes, which simply cover the entire surface so the incoming reactive species are only exposed to relatively unreactive carbon–halogen bonds, or treatment with boric or phosphoric acids. While such treatments have been shown to lower the rates of removal at the walls, *why* they do so is not clear.

Fortunately, the kinetics of the wall loss, measured from the decay of the reactive species in the absence of added reactant, are generally observed to be first order, so that corrections for these processes can be readily incorporated into the kinetic analyses. When these wall losses are significant, the integrated form of the rate expression (T) for reaction (17) of A + B becomes

$$\ln \frac{[A]}{[A]_0} = -(k_{17}[B]_0 + k_w)t, \qquad (Y)$$

where k_w is the observed first-order loss of A at the walls of the flow tube in the absence of B. The rate constant k_{17} can then be extracted from the slopes of plots of the pseudo-first-order rates of decay, $R = (k_{17}[B]_0 + k_w)$, against $[B]_0$.

An example is shown in Figs. 5.5 and 5.6 for the reaction of OH with nitrosyl chloride, ClNO. Figure 5.5 shows the decay of OH resonance fluorescence emission intensity (proportional to the OH concentration) as a function of reaction time in a fast-flow discharge system at ~ 1 Torr total pressure as the concentration of ClNO is increased from 0 to 14.1 × 10^{13} molecules cm^{-3}. As expected from Eq. (Y), the absolute value of the slope of the lines increases as [ClNO]$_0$ increases. Figure 5.6 shows the plot of the absolute values of these slopes against [ClNO]$_0$. The slope of this plot gives the rate constant for the reaction of OH with ClNO,

$$OH + ClNO \rightarrow Products, \qquad (19)$$

under these conditions, which in this case gives $k_{19} = 5.6 \times 10^{-13}$ cm^3 molecule^{-1} s^{-1}. The nonzero decay of OH when the ClNO concentration is zero is due to loss of OH at the walls of the flow tube.

These wall reactions can be a problem in FFDS studies. To avoid unrecognized interferences in the data associated with these heterogeneous reactions, as well as other secondary reactions, it is generally recommended that flow tube studies of a particular reaction be carried out using as many different wall coatings as possible. In addition, the use of different carrier gases

B. LABORATORY TECHNIQUES FOR DETERMINING ABSOLUTE RATE CONSTANTS FOR GAS-PHASE REACTIONS

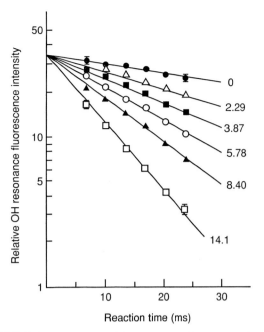

FIGURE 5.5 Typical plot of OH resonance fluorescence intensity as a function of reaction time in the presence of increasing concentrations of ClNO (in units of 10^{13} molecules cm^{-3}) at 373 K (adapted from Finlayson-Pitts *et al.*, 1986).

and flow tubes of different diameters is recommended.

Flow tubes have also been used in combination with such techniques as mass spectrometry and FTIR for product studies. For example, high-pressure flow tubes with a White cell and FTIR at the downstream end

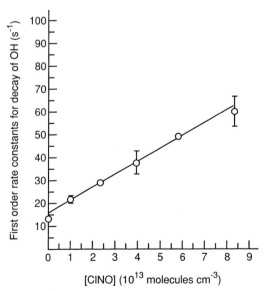

FIGURE 5.6 Typical plot of observed first-order rate constants for the decay of OH as a function of the initial ClNO concentration at 373 K (adapted from Finlayson-Pitts *et al.*, 1986).

have been used with modulation of the reactants to obtain mechanistic information. In this approach, the radical source is modulated, and changes in the spectra with the source on and off are used to identify and quantify products (Donahue *et al.*, 1996b).

3. Flash Photolysis Systems

As the name implies, this technique relies on flash photolysis to generate the reactive species A. In one of the most common configurations, resonance or induced fluorescence is used to monitor the decay of A—hence the name *flash photolysis–resonance fluorescence* (FP-RF). Since lasers are now frequently used as the photolysis source, the term *laser flash photolysis–resonance fluorescence* (LFP-RF) is also used.

Figure 5.7 is a schematic diagram of a typical FP-RF apparatus used to study chlorine atom reactions (Nicovich and Wine, 1996). For example, the fourth harmonic at 266 nm from an Nd:YAG laser can be used to generate chlorine atoms from the dissociation of phosgene, $COCl_2$. After a preset time following the photolytic flash, the time decay of the reactive species is monitored using the fluorescence excited by a resonance lamp. Since B is present in concentrations in great excess compared to A, care must be taken to avoid impurities that may react with A or photolyze to produce reactive species that do. A restriction on the nature of B is that it must not photolyze significantly itself; reactions of such species as NO_2 and O_3, which dissociate to produce highly reactive oxygen atoms, are often difficult to study with this technique. In addition, care must be taken to avoid the buildup of reaction products or of photolysis products in the photolysis cell, since some of these can photolyze and produce interfering secondary reactions. This is usually accomplished by using a slow flow of gas through the cell.

The limitations on the total pressure in the FP-RF cell are far less severe than those for FFDS. The lower end of the pressure range that can be used is determined by the need to minimize diffusion of the reactants out of the viewing zone. The upper end is determined primarily by the need to minimize both the absorption of the flash lamp radiation by the carrier gas and the quenching of the excited species being monitored by RF. In practice, pressures of ~5 Torr up to several atmospheres are used. The kinetic analysis is again typically pseudo-first-order with the "stable" reactant molecule B in great excess over the reactive species as outlined earlier. Table 5.5 gives some typical sources of reactive species used in FP-RF systems.

An example of the use of FP-RF to study the kinetics of an atmospherically relevant reaction is found in Fig. 5.8 (Stickel *et al.*, 1992). Chlorine atoms were

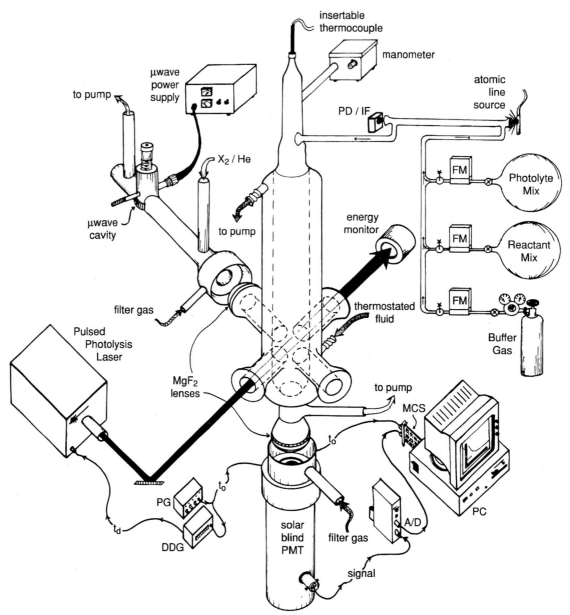

FIGURE 5.7 Laser photolysis resonance fluorescence apparatus for studying the kinetics of gas-phase reactions of H, O, Cl, and Br atoms with atmospheric trace gases. A/D, amplifier/discriminator; DDG, digital delay generator; FM, flow meter; IF, interference filter; MCS, multichannel scaler; PD, photodiode array detector; PG, pulse generator; PMT, photomultiplier. (Graciously provided by J. M. Nicovich and P. H. Wine, Georgia Institute of Technology.).

formed by laser photolysis of $COCl_2$ at 266 nm and detected using resonance fluorescence in the 135- to 140-nm region. As expected, the decay of Cl in the presence of a great excess of CH_3SCH_3 (DMS) is exponential (Fig. 5.8a), and slopes of such decays are linear with the concentration of DMS (Fig. 5.8b). From the slope of the line in Fig. 5.8b, the rate constant at this temperature and pressure was determined to be $(k = 2.71 \pm 0.09) \times 10^{-10}$ cm^3 $molecule^{-1}$ s^{-1}.

4. Pulse Radiolysis

Pulse radiolysis has been used in a number of kinetic studies, for example by the group at RISØ National Laboratory (Denmark) using the Febetron field emission accelerator facilities (e.g., see Nielsen and Sehested, 1993; Pagsberg et al., 1995; and Wallington et al., 1998). A short (30 ns) pulse of high-energy (~ 2 MeV) electrons impacts a reaction cell containing an

TABLE 5.5 Some Typical Sources of Reactive Species in FP – RF Systems

Reactive species	Source
OH	H_2O
	Reactions of $O(^1D)$, e.g., O_3, N_2O
	e.g., $O_3 + h\nu \rightarrow O(^1D) + O_2$
	$O(^1D) + H_2 \rightarrow OH + H$
	or $O(^1D) + H_2O \rightarrow 2OH$
	HNO_3
	H_2O_2
Cl	$COCl_2$
$O(^3P)$	O_2
H	Alkanes, e.g., C_3H_8
RO	RONO

atom or free radical source and reactant of interest. For example, if SF_6 is used as the bath gas, F atoms are generated, which can then undergo secondary reactions to form other reactant free radicals. For example, alkyl radicals can be generated from the F + RH reaction, and if carried out in the presence of O_2, RO_2 is formed (e.g., see Nielsen and Sehested, 1993; and Wallington et al., 1998). Hydrogen atoms can be generated by the electron bombardment of H_2, H, and OH using H_2O and O from CO_2 radiolysis.

The reaction cell has a White cell optical system (see Chapter 11.A.1c) with a pulsed xenon lamp light source. Once the radicals are formed, they are detected by their absorptions in the UV using the Xe lamp and a monochromator–photomultiplier or photodiode array detector. Thus the absorption spectra of the free radicals generated in the system can be measured and the absorption at a particular wavelength used to follow their reaction kinetics.

5. Cavity Ring Down Method

The cavity ring down method was first described by O'Keefe and Deacon (1988) and has been reviewed by Paul and Saykally (1997) and Scherer et al. (1997). This technique is based on the sequential loss of light intensity as a light pulse repeatedly traverses the length of a cell during multiple reflections between two mirrors. Loss of intensity occurs both during reflection at the mirrors and, if an absorbing gas is present, by its absorption as well. As will be seen shortly, the change in the time profile of the light transmitted through mirror B when an absorber is present can be used to follow the concentration of the absorber and hence to carry out kinetic studies.

Figure 5.9 is a schematic diagram of a typical cavity ring down apparatus. A laser pulse enters a reaction cell that has two highly reflecting mirrors. The distance between the mirrors, L, must be large compared to the pulse width to avoid multibeam interference in the cell. After traveling through the length of the cell the first time, the laser pulse strikes mirror B. If the reflectivity of the mirrors is R (defined as the fraction of light reflected), then $(1 - R)$ is the fraction of light lost by reflection at this surface. If the incident light intensity is I', then the intensity of light lost by the reflection is $(1 - R)I'$. Assuming that both mirrors A and B have the same reflectivity, then in one round trip in the cell, from mirror B to mirror A and back, the lost intensity is $dI \sim 2(1 - R)I'$. The time to make this one round trip is t_r (where $t_r = 2L/c$, and c is the speed of light). Under these conditions,

$$\frac{dI}{dt} \approx \frac{-2(1-R)I'}{t_r} = \frac{-(1-R)I'}{t_r/2},$$

or

$$\frac{I}{I_0} = \exp\left[-t\frac{(1-R)}{t_r/2}\right] = \exp\left[\frac{-t}{L/[c(1-R)]}\right]$$

$$= \exp\left[-\frac{t}{\tau_0}\right], \quad (Z)$$

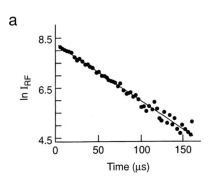

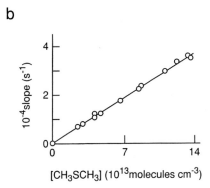

FIGURE 5.8 (a) Typical decay of resonance fluorescence from atomic chlorine in the presence of CH_3SCH_3 (8.6×10^{13} molecules cm^{-3}) at 297 K and in 50 Torr N_2 as the carrier gas (adapted from Stickel et al., 1992). (b) Typical pseudo-first-order plot of slopes of plots such as those in part (a) against the initial concentration of CH_3SCH_3 (adapted from Stickel et al., 1992).

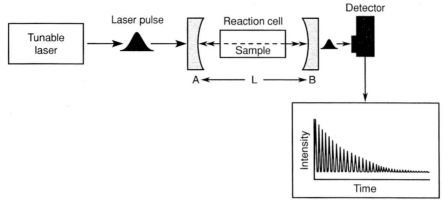

FIGURE 5.9 Schematic diagram of a cavity ring down apparatus (adapted from Paul and Saykally, 1997).

where τ_0 is a constant known as the cavity ring down time in the absence of an absorbing gas. Figure 5.9 also illustrates the resulting measured signal intensity as a function of time, from which the cavity ring down time is obtained.

In the presence of an absorbing gas with absorption cross section σ and concentration N, the fractional change in light intensity is given by $A = (1 - e^{-\sigma NL})$ (see Problem 10). This change in intensity occurs in addition to the $(1 - R)$ loss on the mirror and Eq. (Z) becomes

$$\frac{I}{I_0} = \exp\left\{-t\left[\frac{(1-R)}{t_r/2} + \frac{A}{t_r/2}\right]\right\} = \exp\left[\frac{-t}{\tau}\right], \quad \text{(AA)}$$

where τ is the cavity ring down time in the presence of the gas. It can be shown (see Problem 10) that the absorption (A) of the gas is therefore related to τ and τ_0, by

$$A = \frac{t_r}{2}\left[\frac{1}{\tau} - \frac{1}{\tau_0}\right]. \quad \text{(BB)}$$

Thus measurement of the cavity ring down times in the presence and absence of the absorbing gas allows the absorption of the gas to be obtained, and from $A = (1 - e^{-\sigma NL}) \sim \sigma NL$, the concentration of the gas determined. These concentrations as a function of time therefore allow the reaction kinetics of the absorbing species to be determined.

The advantages of the cavity ring down technique are its high sensitivity (fractional absorptions per pass of $\sim 10^{-6}$) and good time resolution ($\sim 10^{-5}$ s). In addition, it has the usual advantages associated with absorption spectrometry—e.g., most species can be detected through light absorption at specific wavelengths—and is experimentally relatively simple to carry out.

Typical examples of its application to processes of interest for atmospheric chemistry are the measurement of the kinetics of the reaction of the vinoxy radical with O_2 (Zhu and Johnston, 1995) and the kinetics of the $C_2H_5 + C_2H_5$ and $C_2H_5O_2 + C_2H_5O_2$ reactions (Atkinson and Hudgens, 1997). In addition, it has been shown to be useful for probing surface processes as well when combined with total internal reflection techniques (Pipino et al., 1997).

6. Static Techniques

Some reactive species of atmospheric interest such as O_3 are not so highly reactive that they must be generated and studied on subsecond time scales, as is true for atoms and free radicals. In these cases, where the reactions are generally much slower than those of atoms and free radicals such as OH, the experimental techniques used for determining absolute rate constants are often static in nature and do not require *in situ* generation of the reactive species. Taking O_3 as an example, mixtures of O_3 in O_2 can be generated using commercial ozonizers, which are either electrical discharges or UV lights. This mixture is injected into a cell or chamber with the reactant of interest, and its decay followed. Monitoring O_3 accurately and inexpensively *in situ* is reasonably straightforward because it absorbs strongly at 253.7 nm, a stable, inexpensive, and readily available mercury resonance line. Alternatively, samples of the reaction mixture can be withdrawn and the O_3 analyzed with a commercial O_3 monitoring instrument.

Ozone decomposes on surfaces, the rate depending on the nature of the particular surface and whether it has been previously "conditioned" by exposure to O_3. While this heterogeneous decomposition is much slower than the wall loss of OH, the homogeneous gas-phase

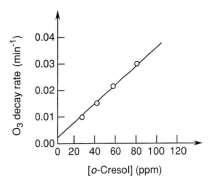

FIGURE 5.10 Plots of ozone pseudo-first-order decay rate constant as a function of the o-cresol concentration using U.S. EPA protocol for determining O_3 rate constants (adapted from Pitts et al., 1981).

reactions of O_3 of interest are also slower than those of OH. However, the kinetics are essentially the same, i.e., Eq. (Y) applies where A is O_3 and B the stable reactant of interest.

Figure 5.10 shows the results of a typical experiment. An FEP Teflon reaction chamber was subdivided into two chambers using metal rods, and O_3 was then introduced into one side of the chamber at concentrations that gave ~1 ppm in the entire chamber when mixed. The organic in excess concentration was injected into the other side, both the organic and O_3 being in ultrahigh-purity air. After removal of the metal barriers and mixing, O_3 was monitored as a function of time using a commercial chemiluminescence ozone analyzer. In this particular case, a rate constant for the reaction of O_3 with o-cresol of 2.55×10^{-19} cm^3 molecule^{-1} s^{-1} was obtained (Pitts et al., 1981).

Such experiments can also be carried out with O_3 in great excess; however, a technique must be available for following the concentration of the reactant X with time, and corrections may have to be made for changing ozone concentrations due to wall losses during the experiment. In addition, interferences from secondary reactions are more likely under these conditions.

C. LABORATORY TECHNIQUES FOR DETERMINING RELATIVE RATE CONSTANTS FOR GAS-PHASE REACTIONS

Many of the rate constants for gas-phase reactions of atmospheric interest reported in the literature were actually determined not as *absolute* values but rather as a *ratio of rate constants*. Thus if the absolute value for one of the rate constants has been determined independently, the second one can then be calculated from the experimentally determined ratio.

In the simplest case, determining relative rate constants for a reactive species A is based on a competition between two reactions:

$$A + X_1 \xrightarrow{k_1} P_1, \quad (20)$$

$$A + X_2 \xrightarrow{k_2} P_2. \quad (21)$$

X_1 and X_2 are the reactants that compete for A, and P_1 and P_2 are the respective products of these reactions. By monitoring the concentrations of the reactants X_1 and X_2, the concentrations of the products P_1 and P_2, or the change in the concentration of one of these as the second reactant is added with time, one can obtain the rate constant ratio k_1/k_2.

For example, if X_1 and X_2 are monitored, the relevant rate laws are as follows:

$$\frac{-d[X_1]}{dt} = k_1[A][X_1],$$

$$\frac{-d[X_2]}{dt} = k_2[A][X_2].$$

Rearranging one obtains

$$\frac{-d\ln[X_1]}{dt} = k_1[A], \quad (CC)$$

$$\frac{-d\ln[X_2]}{dt} = k_2[A]. \quad (DD)$$

Combining (CC) and (DD) to eliminate [A] yields Eq. (EE):

$$[A] = -\frac{1}{k_1}\frac{d\ln[X_1]}{dt} = -\frac{1}{k_2}\frac{d\ln[X_2]}{dt}. \quad (EE)$$

Integrating from time $t = 0$ when the initial concentrations are $[X_1]_0$ and $[X_2]_0$, respectively, to time t when the concentrations are $[X_1]_t$ and $[X_2]_t$ gives Eq. (FF)

$$\ln\frac{[X_1]_0}{[X_1]_t} = \frac{k_1}{k_2}\ln\frac{[X_2]_0}{[X_2]_t}. \quad (FF)$$

Thus the concentrations of X_1 and X_2 as a function of reaction time, plotted as given by Eq. (FF) (i.e., $\ln([X_1]_0/[X_1]_t)$ versus $\ln([X_2]_0/[X_2]_t)$), can be used to derive the rate constant ratio k_1/k_2. If an absolute value is known for one of the two rate constants from independent studies, then an absolute value for the second one can be obtained.

For example, Figure 5.11 shows typical results from a relative rate experiment on the reaction of chlorine atoms with some simple alkanes (Beichert et al., 1995). The chlorine atoms in this case were produced by the

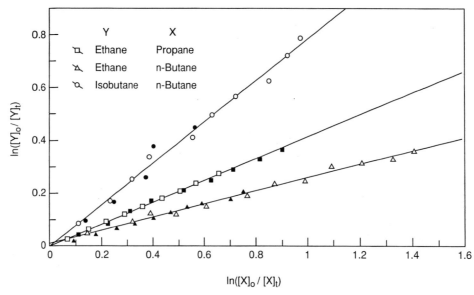

FIGURE 5.11 Plots of the relative decays of pairs of organics (Eq. (FF)) in the presence of chlorine atoms at room temperature (adapted from Beichert *et al.*, 1995).

blacklamp photolysis of Cl_2. The reaction vessel was a 50-L Teflon collapsible chamber into which Cl_2 and the alkanes were introduced as a dilute mixture in air. The mixture was sampled periodically into a gas-sampling valve interfaced to a gas chromatograph so that the concentrations of the alkanes were measured as a function of photolysis time.

The most common sources of OH used for relative rate studies include the photolysis of HONO (e.g., see Cox, 1975) or alternatively methyl nitrite (CH_3ONO) in air in the presence of NO (Atkinson *et al.*, 1981):

$$CH_3ONO + h\nu \rightarrow CH_3O + NO, \quad (22)$$

$$CH_3O + O_2 \rightarrow HCHO + HO_2, \quad (23)$$

$$HO_2 + NO \rightarrow OH + NO_2. \quad (24)$$

Determination of OH relative rate constants for compounds that photolyze significantly in actinic radiation requires a nonphotolytic source of OH. Three such OH sources are $H_2O_2-NO_2-CO$ mixtures (Campbell *et al.*, 1975, 1979; Audley *et al.*, 1982), the thermal decomposition of HO_2NO_2 in the presence of NO (Barnes *et al.*, 1982), and O_3–hydrazine reactions (Tuazon *et al.*, 1983) or O_3–alkane reactions in the dark (Finlayson-Pitts *et al.*, 1993). However, in these cases, the reactant must not react with O_3, HO_2, or H_2O_2, and care must be taken in interpreting the data since these systems have the potential of being rather complex. Indeed, the rate constants derived have not always agreed well with literature values. Until the general features of the mechanisms involved in the production of OH in these systems have been fully elucidated, the simultaneous production of other highly reactive species, and hence possible interfering secondary reactions, cannot be firmly ruled out.

Relative rate techniques have the advantage that such relative measurements can be made with greater precision than absolute rate constant measurements because only relative, not absolute, concentrations of X_1 and X_2 need be measured. Indeed, precisions of 5% or better are common using these techniques. Note, however, that increased *precision* does not *necessarily* imply increased accuracy.

Another advantage is that the species A, which is frequently a highly reactive free radical such as OH which is difficult to measure, need not be monitored in such experiments; only X_1 and X_2, which are usually stable and easily measured molecules, such as hydrocarbons, are followed. Finally, relative rate experiments can often be carried out under conditions directly relevant to the atmosphere, e.g., low concentrations of the reactants in high pressures of air.

The accuracy of the results, however, depends critically on knowing enough of the mechanistic details of the reaction system to be sure that the kinetic analysis, which is not always straightforward in complex systems, is valid. Furthermore, obtaining an accurate rate constant from the rate constant ratio (k_1/k_2) requires accurate knowledge of the second, reference rate constant.

D. REACTIONS IN SOLUTION

1. Interactions of Gaseous Air Pollutants with Atmospheric Aqueous Solutions

Because of the gaseous nature of many of the important primary and secondary pollutants, the emphasis in kinetic studies of atmospheric reactions historically has been on gas-phase systems. However, it is now clear that reactions that occur in the liquid phase and on the surfaces of solids and liquids play important roles in such problems as stratospheric ozone depletion (Chapters 12 and 13), acid rain, and fogs (Chapters 7 and 8) and in the growth and properties of aerosol particles (Chapter 9). We therefore briefly discuss reaction kinetics in solution in this section and "heterogeneous" kinetics in Section E.

The aqueous phase that serves as a reaction medium in the atmosphere is present in the form of clouds, fogs, rain, and particulate matter consisting of either an aqueous solution containing pollutants or a film of water surrounding an insoluble core (see Chapter 9). For example, at typical relative humidities, ~30–50% of the aerosol mass is due to water (Graedel and Weschler, 1981). However, many of the species that are believed to react in such atmospheric solutions, for example, SO_2, O_3, H_2O_2, and NO_x, are emitted or formed in the gas phase. Before reactions can occur in solution, then, several steps illustrated in Fig. 5.12 must first take place:

(1) Diffusion of the gas to the surface of the droplet
(2) Transport of the gas across the air–water interface
(3) Diffusion of the solvated species into the bulk phase of the droplet
(4) Reaction of the species in the aqueous phase or at the interface itself (see Section E.1).

Diffusion of gases is fast relative to diffusion in the aqueous phase; i.e., step 1 is fast relative to step 3. Thus diffusion coefficients for gases at 1 atm pressure are ~ 0.1–1 cm^2 s^{-1}, whereas in liquids they are ~ 10^{-5} cm^2 s^{-1} for small molecules. As discussed in detail by Schwartz and Freiberg (1981), gas-phase diffusion, in most (but not all) cases, will not be the slowest (i.e., rate-determining) step.

Gases dissolve in aqueous solution to various extents, depending on the nature of the gas. At sufficiently long times, an equilibrium can be established between the gas- and liquid-phase concentrations, which is described by Henry's law:

$$[X] = H_X P_X, \qquad (GG)$$

where $[X]$ is the equilibrium concentration of X in solution (in M = mol L^{-1}), P_X is the gas-phase equilibrium pressure (in atm), and H_X is the Henry's law constant (in M atm^{-1}). Table 5.6 shows some values of H for some species of interest dissolving in aqueous solutions at 25°C (Schwartz, 1984a). They range from ~ 10^{-3} M atm^{-1} for relatively insoluble gases such as O_2 to ~ 10^5 M atm^{-1} for highly soluble gases such as H_2O_2 and HNO_3.

Henry's law can be applied to predict solution concentrations only if certain conditions are met. Thus it

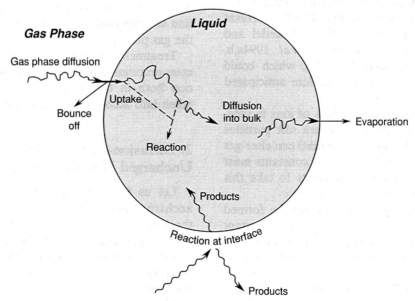

FIGURE 5.12 Schematic diagram of uptake and reaction of gases in liquids.

TABLE 5.6 Henry's Law Coefficients (H) of Some Atmospheric Gases Dissolving in Liquid Water at 25°C

Gas	H (mol L^{-1} atm^{-1})	Reference[e]
O_2	1.3×10^{-3}	Loomis, 1928
NO	1.9×10^{-3}	Loomis, 1928
C_2H_4	4.9×10^{-3}	Loomis, 1928
NO_2[a]	1×10^{-2}	Schwartz and White, 1983
O_3	$(0.82-1.3) \times 10^{-2}$	Briner and Perrottet, 1939
N_2O	2.5×10^{-2}	Loomis, 1928
CO_2[b]	3.4×10^{-2}	Loomis, 1928
SO_2[b]	1.22	Maahs, 1982
HONO[b]	49	Schwartz and White, 1981
NH_3[b]	62	Van Krevelen et al., 1949
H_2CO	6.3×10^3	Blair and Ledbury, 1925
H_2O_2	$(0.7-1.0) \times 10^5$	Martin and Damschen, 1981
	1.4×10^{5}[c]	Yoshizumi et al., 1984
	6.9×10^{4}[d]	Hwang and Dasgupta, 1985
HNO_3	2.1×10^5	Schwartz and White, 1981
HO_2	$(1-3) \times 10^3$	Schwartz, 1984b
OH	30	Golden et al., 1990; Hanson et al., 1992
PAN	5	Holdren et al., 1984
CH_3SCH_3	0.48-0.56	Dacey et al., 1984

Source: Adapted from Schwartz (1984a).
[a] Physical solubility; reacts with liquid water.
[b] Physical solubility exclusive of acid–base equilibria.
[c] At 20°C.
[d] Temperature dependence also reported as $H = \exp[7.92 \times 10^3/T \text{ (K)} - 15.44]$.
[e] See Table 8.1 for references and additional data.

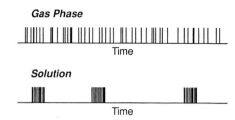

FIGURE 5.13 Patterns of A–B collisions expected in the gas phase and in solution (adapted from Adamson, 1973).

assumes that there are no irreversible chemical reactions that are so fast that the equilibrium cannot be established. It also assumes that the surface of the droplet is an unimpeded air–water interface; as discussed in more detail in Chapter 9, some atmospheric aerosols may have an organic surface film (e.g., Husar and Shu, 1975; Chang and Hill, 1980; Graedel and Weschler, 1981; Gill et al., 1983; Pankow et al., 1994a,b, 1997; Goss and Schwarzenbach, 1998), which could alter the establishment of the equilibrium anticipated by Henry's law.

In addition, the high concentrations of ions in solutions of high ionic strength such as sea salt particles (especially near their deliquescence point) can alter gas solubility. In this case, the Henry's law constants must be modified using Setchenow coefficients to take this effect into account (e.g., Kolb et al., 1997).

Ionized and/or hydrolyzed species may be formed from the dissolved gas in some cases. An important example is SO_2, which dissolves to set up equilibria involving HSO_3^- and SO_3^{2-} in a manner similar to CO_2 (see Chapter 8).

One major difference between reactions in the gas phase and in solution is the presence of solvent molecules in the latter case. In the liquid phase, molecules are in close contact, with the space between molecules being ~10% of the distance between their centers. Reactants thus have a number of nearest neighbors, in the range of ~4–12, with which they can collide. The reactants can then be thought of as existing in a solvent "cage," in which they undergo many collisions before breaking out of that particular environment. If two participants required for a chemical reaction diffuse into such a solvent cage, they will then be held together for a period of time and undergo a number of collisions with each other; such a series of collisons is known as an *encounter*. Because of this *cage effect*, highly reactive species such as atoms and free radicals that are formed in the cage, for example, by photolysis, have a much higher efficiency of recombination than if they were in the gas phase.

Compared to the gas phase, then, reactants take longer to diffuse together, but once they find themselves as nearest neighbors, they undergo a series of collisions rather than separating after one collision; this difference is illustrated in Fig. 5.13 (Adamson, 1973). As a result, for neutral nonpolar reactants (as opposed to ions; see later), the rate constants in solution are expected to be approximately equal to those in the gas phase.

Treatment of systems in which gas-phase diffusion, mass accommodation, liquid phase diffusion, and reaction both in the bulk and at the interface must be taken into account is discussed in Section E.1.

2. Diffusion-Controlled Reactions of Uncharged Nonpolar Species in Solution

Let us first consider a very fast reaction between uncharged nonpolar reactants in solution. In this case, the rate is controlled by the number of encounters. Once A and B diffuse into the same solvent cage, they will react; hence the rate of these diffusion-controlled reactions is determined by how fast A and B diffuse together in solution.

Fick's first law describes the rate of diffusion of a species A in solution across an area E in the direction of the x axis, for example. The rate of diffusion, $J = dn/dt$ (in molecules s^{-1}), is given, according to Fick's first law, by

$$J = \frac{dn}{dt} = -DE\frac{\delta[N_A]}{\delta x},$$

where dn is the amount of A crossing the area E (cm^2) in time dt, D is the diffusion coefficient (in units of cm^2 s^{-1}), and $\delta[N_A]/\delta x$ is the gradient (in units of molecules cm^{-4}) of the concentration of A in the x direction.

Starting with Fick's first law, one can calculate for a solution of two reactants A and B the frequency of A–B encounters, which is in effect the reaction rate constant for diffusion-controlled reactions. This is given by the following, in units of L mol^{-1} s^{-1}:

$$k = 4\pi r_{AB} D_{AB} (6.02 \times 10^{20}). \quad \text{(HH)}$$

r_{AB} is the distance between the centers of the molecules when they react and $D_{AB} = (D_A + D_B)$, where D_A and D_B are the diffusion coefficients of A and B, respectively. For typical values of $r_{AB} = 0.4$ nm and $D_A = D_B = 2 \times 10^{-5}$ cm^2 s^{-1}, a rate constant of $\sim 10^{10}$ L mol^{-1} s^{-1} is obtained. In solution then, diffusion-controlled bimolecular reactions between uncharged species occur with rate constants $\sim 10^{10}$ L mol^{-1} s^{-1}. For reactions with significant activation energies and/or steric requirements, the rate constants are correspondingly lower.

In this case of uncharged, nonpolar reactions, there is little interaction between the reactants and the solvent. As a result, the solvent does not play an important role in the kinetics per se, except through its role in determining the solubility of reactive species and cage effects. The rate constants for such reactions therefore tend to be similar to those for the same reactions occurring in the gas phase. Thus, as we saw earlier, diffusion-controlled reactions in the gas phase have rate constants of $\sim 10^{-10}$ cm^3 molecule^{-1} s^{-1}, which in units of L mol^{-1} s^{-1} corresponds to $\sim 6 \times 10^{10}$ L mol^{-1} s^{-1}, about equal to (usually slightly greater than) that for diffusion-controlled reactions in solution.

3. Reactions of Charged Species in Solution

If the reactants are ionic with opposite charges, the rate constant can be greater than 10^{10} L mol^{-1} s^{-1} due to the favorable attractive forces. For example, the rate constant for the reaction of H$^+$ with OH$^-$ in aqueous solutions at 25°C is $\sim 10^{11}$ L mol^{-1} s^{-1}. On the other hand, the electrostatic repulsion between ions of like sign can significantly slow their reaction. Similarly, if the reactants are polar molecules, electrostatic forces between them and the solvent may come into play.

For ions and polar molecules, the nature of the solvent is an important factor in solution-phase reactions. Following the derivation of Laidler and Meiser (1982), we first consider the reaction between two ions A and B with charges $Z_A e$ and $Z_B e$, respectively, where e is unit electronic charge and Z_A and Z_B are the number of unit charges on the ions, i.e., are whole positive or negative numbers. The electrostatic force (F) between these two ions separated by a distance r in a vacuum is given by Coulomb's law,

$$F = \frac{Z_A Z_B e^2}{4\pi\varepsilon_0 r^2}, \quad \text{(II)}$$

where $\varepsilon_0 = 8.85 \times 10^{-12}$ C^2 N^{-1} m^{-2}, the *permittivity* of a vacuum. However, if the ions are immersed in a solvent, having a dielectric constant ε, the electrostatic force between them is modified by the properties of the solvent. Equation (II) thus becomes

$$F = \frac{Z_A Z_B e^2}{4\pi\varepsilon_0 \varepsilon r^2}. \quad \text{(JJ)}$$

The higher the solvent dielectric constant ε, the more the electrostatic force between the ions is reduced. From this expression for the force between two ions, one can calculate the work done to bring the two ions from infinite distance to the distance necessary to react, $r = d_{AB}$; this is equal to the change in free energy due to the electrostatic forces as the ions approach each other, ΔG_{es}. The total free energy change in bringing the ions together is the sum of this electrostatic term and a nonelectrostatic one, ΔG_0:

$$\Delta G_{TOT} = \Delta G_0 + \Delta G_{es},$$

$$= \Delta G_0 + \frac{(6.02 \times 10^{23})Z_A Z_B e^2}{4\pi\varepsilon_0 \varepsilon d_{AB}}. \quad \text{(KK)}$$

(Avogadro's number is included in the electrostatic term to convert to units of per mole rather than per molecule.)

As seen in Section A.3.b, the transition state form of the rate constant is given by

$$k = \frac{\mathbf{k}T}{h} e^{-\Delta G^{\circ \neq}/\mathbf{R}T}.$$

The free energy of activation $\Delta G^{\circ \neq}$ for bringing two ions to the necessary distance d_{AB} in order to react is given by equation (KK). Thus the natural logarithm of the rate constant becomes

$$\ln k = \ln \frac{\mathbf{k}T}{h} - \frac{\Delta G_0}{\mathbf{R}T} - \frac{(6.02 \times 10^{23})Z_A Z_B e^2}{4\pi\varepsilon_0 \varepsilon d_{AB} \mathbf{R}T},$$

$$\ln k = \ln k_0 - \frac{Z_A Z_B e^2}{4\pi\varepsilon_0 \varepsilon d_{AB} \mathbf{k}T}, \quad \text{(LL)}$$

where

$$\ln k_0 = \ln \frac{kT}{h} - \frac{\Delta G_0}{RT}$$

and the Boltzmann constant **k** has been substituted in the electrostatic term for $R/(6.02 \times 10^{23}) = k$. The term k_0 is the rate constant for the ion reactions in a medium where $\varepsilon = \infty$, that is, when the electrostatic forces have become zero.

Equation (LL) predicts that the rate constant for the reaction between two ions in solution will depend on the dielectric constant ε, and hence the nature of the solvent. A plot of $\ln k$ against $1/\varepsilon$ should be a straight line with slope of $-(Z_A Z_B e^2 / 4\pi \varepsilon_0 d_{AB} kT)$. From the slope, d_{AB} can be obtained. Experimentally, it is found that Eq. (LL) is indeed followed in many cases.

In the atmosphere, most solutions available for reaction are aqueous. The dielectric constant for water at 25°C is $\varepsilon = 78.3$.

A second important factor for reactions in solution between ions or polar molecules is the ionic strength (I) of the solution. I is defined as

$$I = \frac{1}{2} \sum C_i Z_i^2,$$

where C_i is the molar concentration of the ith ion and Z_i is its charge. For a 1 M solution of $NaNO_3$, for example, $I = 1$ M, whereas for a 1 M solution of Na_2SO_4, $I = 3$ M. ($C_{Na^+} = 2$, $Z_{Na^+} = +1$, $C_{SO_4^{2-}} = 1$, $Z_{SO_4^{2-}} = -2$.)

In solution thermodynamics, the concentration (C) of ions is replaced by their activity, a, where $a = C\gamma$ and γ is the activity coefficient that takes into account nonideal behavior due to ion–solvent and ion–ion interactions. The Debye–Hückel limiting law predicts the relationship between the ionic strength of a solution and γ for an ion of charge Z in dilute solutions:

$$\log \gamma = -BZ^2 I^{1/2}. \quad \text{(MM)}$$

B is a constant that depends on the properties of the solution, for example, on its dielectric constant, and on the temperature. For water at 25°C, $B = 0.51$ $L^{1/2}$ $mol^{-1/2}$. The Debye–Hückel limiting law applies only for solutions of low ionic strength, for example, below ~ 0.01 M for 1:1 electrolytes, such as $NaNO_3$, and below ~ 0.001 M for electrolytes of higher charge.

The influence of ionic strength on solution rate constants can be anticipated by again resorting to transition state theory. For the reaction

$$A + B \rightleftharpoons X^{\neq} \xrightarrow{k_p} \text{Products},$$

the rate of the reaction is given by $k_p [X^{\neq}]$, where $[X^{\neq}]$ is the concentration of the activated complex in the transition state. The concentration of the activated complexes can be obtained from the equilibrium assumed between the reactants and the transition state:

$$K^{\neq} = \frac{a_{X^{\neq}}}{a_A a_B} = \frac{\gamma_X^{\neq} [X^{\neq}]}{\gamma_A [A] \gamma_B [B]}.$$

Thus

$$\text{Rate} = k_p K^{\neq} \frac{\gamma_A \gamma_B [A][B]}{\gamma_X^{\neq}},$$

$$= k[A][B],$$

where the reaction rate constant is given by

$$k = k_p K^{\neq} \frac{\gamma_A \gamma_B}{\gamma_X^{\neq}}.$$

Thus

$$\log k = \log(k_p K^{\neq}) + \log \frac{\gamma_A \gamma_B}{\gamma_X^{\neq}}.$$

Using the Debye–Hückel limiting law for the relationship between the activity coefficients γ and the ionic strength of the solution, one finds

$$\log k = \log k_0 + \log \gamma_A + \log \gamma_B - \log \gamma_{X^{\neq}},$$

$$= \log k_0 + 2B Z_A Z_B I^{1/2}.$$

With $B = 0.51$ $L^{1/2}$ $mol^{-1/2}$ for aqueous solutions at 25°C, this becomes

$$\log k = \log k_0 + 1.02 Z_A Z_B I^{1/2}. \quad \text{(NN)}$$

Thus a plot of $\log k$ against $I^{1/2}$ should give straight lines of slope $1.02 Z_A Z_B$ and intercepts of $\log k_0$. The constant k_0 is seen to be the rate constant in a solution of zero ionic strength, that is, at infinite dilution. Equation (NN) also predicts that reactions between ions of the same sign should speed up as the ionic strength increases, whereas reactions between oppositely charged ions should slow down with increasing ionic strength. For reactions between an ion and an uncharged molecule ($Z_B = 0$), ionic strength should not alter the rate constant. These relationships have been confirmed for solutions that are sufficiently dilute so that the Debye–Hückel law is applicable (Fig. 5.14). As might be expected, deviations are observed at higher ionic strengths (e.g., see the text by Benson, 1960, for a more detailed discussion).

This effect of ionic strength on solution rate constants is very important in studying reactions relevant to atmospheric chemistry. Thus care must be taken to study the effects of ionic strength over a range that

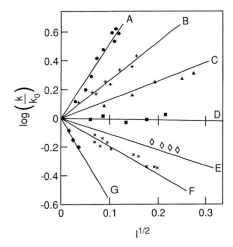

FIGURE 5.14 Variation of rate constant with ionic strength (I) of the solution for reactants having different charges. Reactions: (A) $[Co(NH_3)_5Br]^{2+} + Hg^{2+} + H_2O \rightarrow [Co(NH_3)_5(H_2O)]^{3+} + (HgBr)^+$; (B) $S_2O_8^{2-} + I^- \rightarrow$ (intermediates) $\rightarrow I_3^- + 2SO_4^{2-}$ (not balanced); (C) $[O_2NNCOOEt]^- + OH^- \rightarrow N_2O + CO_3^{2-} + EtOH$; (D) cane sugar $+ OH^- \rightarrow$ invert sugar (hydrolysis reaction); (E) $H_2O_2 + H^+ + Br^- \rightarrow H_2O + 1/2 Br_2$ (not balanced); (F) $[Co(NH_3)_5Br]^{2+} + OH^- \rightarrow [Co(NH_3)_5(OH)]^{2+} + Br^-$; (G) $Fe^{2+} + Co(C_2O_4)_3^{3-} \rightarrow Fe^{3+} + [Co(C_2O_4)_3]^{4-}$. (Adapted from Benson, 1960.)

approximates those found in the atmosphere. Aerosols in polluted urban areas can be highly concentrated solutions with ionic strengths in the range of 8–19 M (Stelson and Seinfeld, 1981); reactions in these solutions will not follow the "ideal" relationships discussed earlier. On the other hand, cloud water and rainwater in clean areas contain much lower solute concentrations; for example, from the ionic composition of precipitation samples in the maritime area of Cape Grim, Australia (Ayers, 1982), the ionic strength can be calculated to be ~10^{-3} M.

4. Experimental Techniques Used for Studying Solution Reactions

The approaches to studying reaction kinetics in the liquid phase are analogous to those in the gas phase, that is, the use of various spectroscopic techniques to follow the loss of one reactant in the presence of a large excess of the second reactant. UV–visible spectroscopy is a primary tool for following both stable species and radicals in solution.

As discussed in Chapters 7, 8, and 9, there are a number of free radical species whose reactions in the aqueous phase drive the chemistry of clouds and fogs. These include OH, HO_2, NO_3, halogen radicals such as Cl_2^-, sulfur oxide radicals, and RO_2. Generation of these radicals in the liquid phase for use in kinetic studies is typically carried out using either flash photolysis or pulse radiolysis.

Flash photolysis can be carried out using either broadband light sources in the 200- to 300-nm range, with pulse durations of the order of microseconds, or lasers with specific wavelengths and pulse durations of nanoseconds to femtoseconds. The advantages of lasers lie in the use of specific wavelengths, which minimizes the simultaneous photolysis of reactants or products at other wavelengths that can occur with broadband light sources, and in the availability of higher light intensities to generate larger radical concentrations. Excimer lasers have proven especially useful: ArF at 193 nm, KrF at 248 nm, XeCl at 308 nm, and XeF at 351 nm.

Pulse radiolysis relies on the interaction of high-energy ionizing radiation to generate free radicals during a short (μs to ns) radiation pulse (see Pagsberg et al., 1995). The hydroxyl radical can be easily generated through processes described shortly, and other species by secondary reactions of OH. With water as a solvent, bombardment with ionizing radiation generates electrons (e^-), H_2O^+, and excited water molecules, H_2O^*. The electrons either form hydrated electrons, e_{aq}^-, or ionize additional water molecules. The hydroxyl radical is generated from reactions of H_2O^+ and H_2O^*:

$$H_2O^+ + H_2O \rightarrow OH + H_3O^+, \qquad (25)$$

$$H_2O^* \rightarrow H + OH, \qquad (26a)$$

$$\rightarrow H_2 + O. \qquad (26b)$$

Higher OH yields can be obtained if the solution contains N_2O:

$$e_{aq}^- + N_2O \rightarrow N_2 + O^-, \qquad (27)$$

$$O^- + H_2O \rightarrow OH + OH^-. \qquad (28)$$

To generate other free radicals, OH can be reacted with other species (e.g., see Zellner and Herrmann, 1995). For example,

$$OH + HSO_3^- \rightarrow H_2O + SO_3^-, \qquad (29)$$

$$OH + HCO_3^- \rightarrow H_2O + CO_3^-, \qquad (30)$$

$$OH + Cl^- \leftrightarrow ClOH^-, \qquad (31, -31)$$

$$ClOH^- + H^+ \leftrightarrow Cl + H_2O, \qquad (32, -32)$$

$$Cl + Cl^- \rightarrow Cl_2^-. \qquad (33)$$

In the case of the chlorine reactions, the involvement of H^+ in the second step means that these reactions are efficient sources of chlorine atoms only at a pH less than about 4. Br^- can be converted to Br atoms in reactions analogous to those for chlorine, but in this case, generation of atomic bromine occurs up to a pH of about 11 (Zellner and Herrmann, 1995).

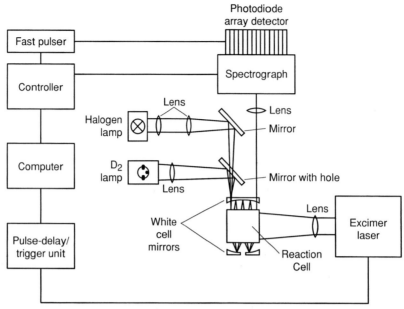

FIGURE 5.15 Schematic of typical apparatus used to study kinetics of reaction in the liquid phase (adapted from Zellner and Herrmann, 1995).

The reaction of OH with nitrate ion is endothermic so OH cannot be used to generate NO_3 in solution. It is often generated by reactions such as

$$S_2O_8^{2-} + h\nu \rightarrow 2SO_4^-, \quad (34)$$

$$SO_4^- + NO_3^- \leftrightarrow SO_4^{2-} + NO_3, \quad (35, -35)$$

or the photolysis of ceric ammonium nitrate solutions at 350 nm (e.g., Alfassi *et al.*, 1993):

$$Ce^{IV}(NO_3)_6^{2-} + h\nu \rightarrow NO_3 + Ce^{III}(NO_3)_5^{2-}. \quad (36)$$

Figure 5.15 is a schematic diagram of a typical apparatus used to study aqueous-phase kinetics (Zellner and Herrmann, 1995). An excimer laser is used to generate the free radical species of interest which is located in a cavity with White cell optics used to obtain long total path lengths (e.g., 60 cm) using a shorter base path (see Chapter 11.A.1c). The time dependence of the concentrations of the free radical reactant can be followed using its absorption of light from a halogen or D_2 lamp. In this particular apparatus, a photodiode array detector is used so that a range of wavlengths can be followed, rather than a single wavelength. This allows several different reactants and products to be monitored simultaneously, or if only one absorbing species is present, its absorption spectrum can be obtained.

E. LABORATORY TECHNIQUES FOR STUDYING HETEROGENEOUS REACTIONS

"Heterogeneous" reactions in the context of atmospheric chemistry are defined as those between gases and either solids or liquids. Heterogeneous reactions have been suggested for decades as being important in the atmosphere. However, historically, forays into this area by a variety of scientists usually concluded in a rapid retreat, due to highly variable results that were (and, unfortunately, continue to be) common. The recognition of the importance of condensed phases in the oxidation of SO_2 to sulfuric acid (see Chapter 8.C.3) and the discovery of the dramatic Antarctic ozone "hole" (see Chapter 12.C.4) reinforced the key role of heterogeneous reactions, and the atmospheric chemistry community has tackled this area anew. As a result, while we are still in the "dark ages" in terms of understanding kinetics and mechanisms of these processes on a molecular level, a great deal of progress has been made in the past decade.

There are many different types of surfaces available for reactions in the atmosphere. In the stratosphere, these include ice crystals, some containing nitric acid, liquid sulfuric acid–water mixtures, and ternary solutions of nitric and sulfuric acids and water. In the troposphere, liquid particles containing sulfate, nitrate, organics, trace metals, and carbon are common. Sea salt particles dominate in marine areas. In addition, there are large episodic sources of particles emitted directly into both the troposphere and stratosphere, such as rocket exhausts where particles containing carbon soot, alumina, and metal oxides can be emitted in large quantities.

Before we describe some of the common techniques used to study the kinetics and mechanisms of heterogeneous reactions, a few words regarding the difficulties in this area are appropriate. To put these in perspective, consider first the current state of understanding of gas-phase kinetics. There are a number of both absolute and relative rate techniques available for studying gas-phase reactions, and the methodologies for preparing reactants and measuring products are generally quite well developed. As a result, agreement to within ~15% on gas phase reaction rate constants measured using different techniques and by different laboratories is now common. In addition, understanding of the reactions at a molecular level is generally good.

In contrast, in most heterogeneous reactions, we really do not even understand what one of the reactants, the surface, looks like on a molecular level; i.e., the condensed-phase molecule and its environment that the incoming gaseous reactant encounters is not well characterized. An example of our incomplete understanding of the nature of surfaces is controversy about effective surface areas for ice and whether ice surfaces prepared in the laboratory for example, are porous or not (e.g., see Keyser *et al.*, 1993; and Hanson and Ravishankara, 1993a).

In addition, how the surface changes during reaction and how such changes affect both the reactivity and mechanism are unclear. One well-recognized phenomenon is that of saturation of the surface, i.e., complete reaction of the surface species so that the reaction comes to a halt. However, there are other potentially important aspects as well. For example, it is well known in surface science that adsorbates can lead to restructuring of the surface (Somorjai, 1994), yet this phenomenon has yet to be demonstrated on surfaces of atmospheric interest. This is in part due to the incompatibility between atmospheric conditions (air at pressures up to 1 atm and containing significant amounts of water vapor) and the ultrahigh-vacuum conditions typical of surface science studies. However, marriage of these two fields will be ultimately needed for a complete understanding of heterogeneous processes in the atmosphere, and progress is being made with some systems (e.g., Hemminger, 1999).

Different terms and symbols have been used in the literature to differentiate reversible, physical uptake from irreversible uptake via chemical reactions. Additional confusion arises from the fact that the observed uptake of a species from the gas phase is usually a *net* uptake affected by a number of factors such as changes in the surface during the uptake (e.g., due to saturation) and reevaporation into the gas phase due to limited solubility of the species. However, the most common terminology now in use is the following:

Surface reaction probability (γ_{rxn}) is the net fraction of gas-condensed phase collisions that leads to the irreversible uptake of the gas due to chemical reaction. The symbol γ_{rxn} (or sometimes ϕ) is most commonly used for reaction probabilities.

Mass accommodation coefficient (α) is the fraction of gas-condensed phase collisions that result in uptake of the gas by the condensed phase:

$$\alpha = \frac{\text{Number of gas molecules taken up by the surface}}{\text{Number of gas-surface collisions}}.$$

(OO)

This is the probability that a molecule that strikes the surface will cross the interface into the condensed phase. It does *not* represent the net uptake; i.e., it does not include the reverse effect of reevaporation from the condensed phase into the gas phase. α is also sometimes referred to as a sticking coefficient for uptake on solid surfaces.

Net collisional uptake probability (γ_{net}) is the net rate of uptake of the gas normalized to the rate of gas-surface collisions and since this is what is measured, is also often referred to as γ_{meas}.

The net loss of the gaseous species in the presence of a condensed phase having known volume and surface area is what is typically measured in laboratory studies. To obtain information on the fundamental processes contributing to this net measured uptake, all of the processes shown schematically in Fig. 5.12 must be taken into account: diffusion of the gas to the surface, uptake, diffusion in the liquid phase, and reaction either in the bulk or at the interface itself.

We therefore first briefly discuss the analysis of systems that involve diffusion in the gas and liquid phases, uptake, and reaction in the bulk liquid or at the interface. Following that, we give a brief description of some of the most common methods used to measure mass accommodation coefficients and reaction kinetics for heterogeneous atmospheric reactions. Included are some new approaches that appear to be especially promising. For a review of this area, see Kolb *et al.* (1995, 1997).

1. Analysis of Systems with Gas- and Liquid-Phase Diffusion, Mass Accommodation, and Reactions in the Liquid Phase or at the Interface

Both for laboratory measurements and for atmospheric processes, the uptake of a gas into a liquid followed by reaction involves a number of different physical (e.g., diffusion and uptake at the interface) as well as chemical processes. These were depicted in Fig. 5.12. We treat here in more detail the individual steps and how the net uptake of a gas into solution is determined by these steps.

1. *Transport of the gas to the surface and the initial interaction.* The first step in heterogeneous reactions involving the uptake and reaction of gases into the liquid phase is diffusion of the gas to the interface. At the interface, the gas molecule either bounces off or is taken up at the surface. These steps involve, then, gaseous diffusion, which is determined by the gas-phase diffusion coefficient (D_g) and the gas-surface collision frequency given by kinetic molecular theory.

2. *Uptake at the interface.* If the gas molecule is taken up at the surface, it enters the interface region and then the bulk. The efficiency of uptake involving crossing the interface is described by the *mass accommodation coefficient* (α) defined earlier. Molecular-level mechanisms by which gas molecules are taken up into liquids are discussed elsewhere (e.g., see Davidovits *et al.*, 1995; Taylor *et al.*, 1996, 1997; and Nathanson *et al.*, 1996).

3. *Diffusion into the bulk.* This is determined by the diffusion coefficient in the liquid (D_l). Diffusion within the bulk aqueous phase is much slower than gas-phase diffusion and can be rate-limiting under conditions of high reactant concentrations where the rate of the chemical reaction is high. This appears to have been a problem in some experimental studies of some aqueous-phase reactions relevant to the atmosphere where either bulk solutions or large droplets and reactant concentrations higher than atmospheric were used (Freiberg and Schwartz, 1981).

4. *Henry's law equilibrium.* If there is no reaction in the liquid phase (or it is slow relative to uptake and diffusion), the gas-liquid system eventually comes to equilibrium, which can usually be described by Henry's law discussed earlier. This does not reflect a lack of uptake of the gas at equilibrium but rather equal rates of uptake and evaporation; i.e., it is a *dynamic* equilibrium (see Problem 12). The equilibrium between the gas-and liquid-phase concentrations is characterized by the Henry's law constant, H (mol L^{-1} atm^{-1}), where $H = [X]/P_X$.

5. *Reaction in the bulk.* Reaction can occur in solution close to the surface or throughout the bulk of the liquid phase, depending on the speed of the reaction compared to diffusion. We shall see that whether the reaction occurs close to the surface or throughout the bulk has important implications for the kinetics, since in the former case, the reaction depends on the particle surface area, whereas in the latter it depends on the particle volume.

We shall treat here reactions occurring in the condensed phase as if they are first-order, irreversible reactions with a rate constant k. Of course, this also applies to second-order reactions with the second reactant, B, in great excess, since they are then pseudo-first-order with $k' = k[B]$.

6. *Reactions at the interface.* There has been increasing recognition that reactions may also occur at the interface itself. That is, species such as SO$_2$, NH$_3$, and organics do not simply cross the interface by physical transport but rather form unique chemical species at the interface (e.g., Donaldson *et al.*, 1995; Allen *et al.*, 1999; Donaldson, 1999; Donaldson and Anderson, 1999). These unique interface species can then react at the surface without actually being taken up into the bulk of the solution. Although relatively little is currently known at a molecular level about such processes, reactions in this "fourth phase" may prove to be very important in atmospheric processes, for example in the generation of HONO in the NO$_2$ reaction with water at surfaces (see Chapter 7.B.3b).

The measured, net uptake of a gas into a liquid phase can be related to these various processes, i.e., to D_g, α, D_l, H, and k. However, concentrations in both the gas and liquid phases as well as the volumes and surface areas available for reaction are quite different in laboratory studies compared to the atmospheric situation. We shall therefore also examine how net gas uptake measured in laboratory studies can be related to uptake under atmospheric conditions.

The physical and chemical processes occurring in a gas-liquid system are often treated in terms of a resistance model described in Box 5.2. As discussed there, the net uptake of gas (γ_{net}) can be treated under some conditions in terms of "conductances," Γ, normalized to the rate of gas-surface collisions. Individual conductances are associated with gas-phase diffusion to the surface (Γ_g), mass accommodation across the interface (α), solubility (Γ_{sol}), and finally, reaction in the bulk aqueous phase (Γ_{rxn}). This leads to Eq. (QQ):

$$\frac{1}{\gamma_{\text{net}}} = \frac{1}{\Gamma_g} + \frac{1}{\alpha} + \frac{1}{\Gamma_{\text{rxn}} + \Gamma_{\text{sol}}} \qquad \text{(QQ)}$$

BOX 5.2
RESISTANCE MODEL OF GAS–DROPLET INTERACTIONS

We shall treat the individual processes in terms of the rate of transfer of gas across a surface of unit area per second. However, this rate will be expressed relative to the number of gas–surface collisions per second, given according to kinetic molecular theory by

Number of collisions per unit area second of gas with surface

$$= \frac{N_g u_{av}}{4} = \frac{N_g}{4}\sqrt{\frac{8RT}{\pi M}} = N_g\sqrt{\frac{RT}{2\pi M}}. \quad \text{(PP)}$$

In Eq. (PP), N_g is the gas concentration (molecules cm^{-3}), u_{av} is the average molecular speed in the gas phase, R is the gas constant (J K^{-1} mol^{-1}), T the temperature (K), and M is the molecular weight (kg) of the gas. The normalized rates, i.e., divided by the rate of gas–surface collisions in Eq. (PP), will be referred to as "conductances," Γ, for reasons that will become apparent shortly. However, the reader should keep in mind that these "conductances" just reflect the speeds of the individual processes.

The uptake of a gas into a liquid followed by its reaction can be described by a series of coupled differential equations (Danckwerts, 1951, 1970), which can be solved exactly only for some specific cases.

However, under many conditions the individual processes can be treated as if they are not coupled. In this case, an approximation that has found widespread use (e.g., see Schwartz and Freiberg, 1981; Schwartz, 1986; and Kolb et al., 1995, 1997), and that helps to assess the relative importance of each of the terms, is to treat the individual processes in terms of an electrical circuit (Fig. 5.16). Dimensionless "conductances," Γ [where conductance = (resistance)$^{-1}$], associated with each process reflect rates normalized to the rate of gas–surface collisions, and the corresponding "resistances" are given by $1/\Gamma$. The net, overall measured resistance, $(\gamma_{net})^{-1}$, is then related to the individual resistances (see Problem 7) by

$$\frac{1}{\gamma_{net}} = \frac{1}{\Gamma_g} + \frac{1}{\alpha} + \frac{1}{\Gamma_{rxn} + \Gamma_{sol}}. \quad \text{(QQ)}$$

As already discussed, γ_{net} is a net probability normalized to the number of gas–surface collisions and is the parameter actually measured in experiments (and hence also often referred to as γ_{meas}). In Eq. (QQ), each conductance represents one of the processes involved; i.e., Γ_g involves the conductance for gas-phase diffusion, Γ_{rxn} that for reaction in the aqueous phase, and Γ_{sol} that for solubility and diffusion into the bulk. Each of the terms has been normalized and made unitless by dividing by the rate of gas–surface collisions, Eq. (PP), except for α, which by definition is already normalized to this parameter.

Let us now examine each of these terms individually.

Diffusion of the gas to the surface (Γ_g). As described by Fuchs and Sutugin (1970, 1971) in their comprehensive treatment of highly dispersed aerosols, the rate of transfer of mass to the surface of a spherical particle by diffusion of a gas is described by

$$x^2 \frac{dN}{dx} = \text{constant}, \quad \text{(RR)}$$

where N is the gas concentration a distance x from the center of the particle and the constant depends on the boundary conditions at the surface of the particle. Take the case where the Knudsen number, K_n, defined as the ratio of the mean free path of the gas to the radius of the particle (a), approaches 0 (i.e., the mean free path is small compared to the particle size); all molecules colliding with the surface are taken up at the surface and the boundary condition is $(N)_{x=a} = 0$. Fuchs and Sutugin show that Eq. (RR) can be integrated to obtain the gas concentration as a function of x, $N = N_\infty(1 - a/x)$, where N_∞ is the gas concentration at $x = \infty$, which can be taken as N_g for a constant gas-phase concentration. They also show that for small Knudsen numbers where $K_n \to 0$, the rate of diffusion of the gas to the surface of the particle of radius a is given by

Number of molecules per second diffusing to the surface $= 4\pi D_g a N_g.$ (SS)

The surface area of a spherical particle of radius a is $4\pi a^2$. Thus, the rate of diffusion to the surface

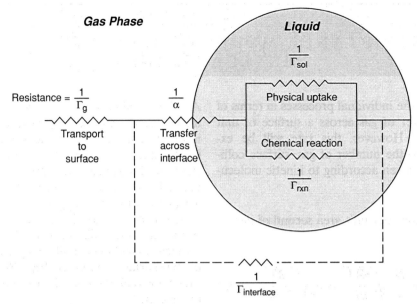

FIGURE 5.16 Schematic of resistance model for diffusion, uptake, and reaction of gases with liquids. Γ_g represents the transport of gases to the surface of the particle, α the mass accommodation coefficient for transfer across the interface, Γ_{sol} the solubilization and diffusion in the liquid phase, Γ_{rxn} the bulk liquid-phase reaction, and $\Gamma_{interface}$ the reaction of the gas at the interface.

per unit surface area is given by

Rate of diffusion to surface per cm² per second

$$= 4\pi D_g a N_g / 4\pi a^2 = 2 D_g N_g / d, \quad \text{(TT)}$$

where $d = 2a$ is the particle diameter. Normalizing to the rate of collisions, Γ_g becomes

$$\Gamma_g = 8 D_g / u_{av} d. \quad \text{(UU)}$$

As discussed by Fuchs and Sutugin (1970, 1971) and Motz and Wise (1960), in this continuous regime, distortion of the Boltzmann velocity distribution in the region close to the surface occurs if there is rapid uptake. In effect, the normal thermal velocity distribution is distorted so that the effective speed toward the surface is higher. In the case of a surface where the uptake occurs on every collision, the net speed toward the surface is effectively doubled. This adds an additional term to the rate of transfer of the gas to the surface, which when normalized using Eq. (PP), gives an additional "resistance" term of $-1/2$. The overall normalized conductance is therefore given by

$$\frac{1}{\Gamma_g} = \frac{u_{av} d}{8 D_g} - \frac{1}{2}. \quad \text{(VV)}$$

This applies to diffusion to a planar surface in the continuous regime, when the Knudsen number is small, and is the expression for gas diffusion most often encountered in the atmospheric chemistry literature.

For the other extreme of the free molecular regime where $K_n \to \infty$, the particle radius is small compared to the mean free path. In this case, the thermal velocity distribution of the gas is not distorted by uptake at the surface. In effect, the gas molecules do not "see" the small particles. For this case, Fuchs and Sutugin (1970, 1971) show that for diffusion to a spherical particle of radius a

Number of molecules diffusing to surface per second

$$= \pi a^2 \alpha_c u_{av} N_g, \quad \text{(WW)}$$

where α_c is the probability of uptake at the surface and u_{av} is the mean thermal velocity.

For intermediate regimes of K_n, which are common both in the atmosphere and in many laboratory studies, exact calculations are not readily carried out. Fuchs and Sutugin (1970, 1971) suggest the form

Number of molecules diffusing to surface per second

$$= 4\pi D_g a N_g / (1 + \lambda K_n). \quad \text{(XX)}$$

As $K_n \to 0$, Eq. (XX) approaches Eq. (SS), as expected. Values of λ are provided in the literature (Fuchs and Sutugin, 1970, 1971); as an approximation,

$$\lambda \approx \frac{1.33 + 0.71/K_n}{1 + 1/K_n}. \quad \text{(YY)}$$

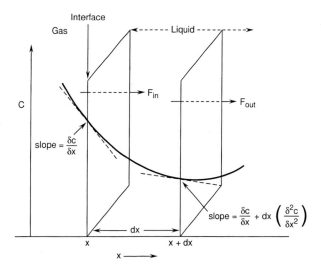

FIGURE 5.17 Schematic diagram for treatment of diffusion of a species in one dimension in a liquid.

Uptake across the interface into solution (α). By definition, this is described by the mass accommodation coefficient, α, and $1/\alpha$ is the "interfacial resistance."

Solubility and diffusion in the liquid phase (Γ_{sol}). Consider the diffusion of a dissolved species in one dimension into the bulk solution from the interface region. The concentration of the species in the liquid will depend on time as well as on the distance from the interface. It is assumed first that no reaction is taking place in the aqueous phase.

Figure 5.17 shows a general case of diffusion in the x direction across a plane of unit area. The concentration of the diffusing species is taken as c. The net flux in at x, F_{in} (molecules cm^2 s^{-1}), is given by

$$F_{in} = -D_l \frac{\delta c}{\delta x}, \quad \text{(ZZ)}$$

where D_l is the diffusion coefficient in the liquid phase (units of cm^2 s^{-1}) and ($\delta c/\delta x$) is the concentration gradient at x. The concentration gradient (i.e., slope of the concentration versus distance) is also changing with distance. At a distance dx from the position x, this concentration gradient is given by $[(\delta c/\delta x) + dx(\delta^2 c/\delta x^2)]$ and the net flux across the plane at dx, i.e., *out* of the volume bounded by x and $(x + dx)$, is given by

$$F_{out} = +D_l \left[\frac{\delta c}{\delta x} + dx \frac{\delta^2 c}{\delta x^2} \right]. \quad \text{(AAA)}$$

The net flux in the region bounded by x and $(x + dx)$ is the combination of the net flux in and out, i.e.,

$$F_{net} = dx \left(\frac{\delta c}{\delta t} \right)$$
$$= -D_l \frac{\delta c}{\delta x} + D_l \left[\frac{\delta c}{\delta x} + dx \frac{\delta^2 c}{\delta x^2} \right], \quad \text{(BBB)}$$

which reduces to

$$\frac{\delta c}{\delta t} = D_l \frac{\delta^2 c}{\delta x^2}. \quad \text{(CCC)}$$

The rate of transfer (Rt, in units of molecules or moles per cm^2 per second) of the species across a plane at $x = 0$, is given by

$$Rt = -D_l \left(\frac{\delta c}{\delta x} \right)_{x=0}. \quad \text{(DDD)}$$

The concentration gradient $(\delta c/\delta x)_{x=0}$ at the surface depends on time because as uptake occurs, reevaporation back to the gas phase becomes increasingly important and the magnitude of the concentration gradient decreases.

Equation (CCC) can be solved to obtain the rate, Rt, under certain boundary conditions. Take the case where the concentration in the bulk liquid is given by $c_{l,bulk}$ at time $t = 0$ as well as at $x = \infty$ for times $t > 0$. It is assumed that there is a thin layer at the surface that contains the dissolved species in equilibrium with the gas immediately adjacent to the surface. This interface concentration is denoted as $c_{l,interface}$. Under these conditions, Eq. (CCC) can be solved (see Danckwerts, 1970, pp. 31–33) to obtain the rate of transfer per unit surface area after exposure time t, as

$$\text{Rate} = (c_{l,interface} - c_{l,bulk})\sqrt{\frac{D_l}{\pi t}}. \quad \text{(EEE)}$$

As intuitively expected, the rate of transfer across the interface depends on the difference in the liquid-phase concentrations at the interface and in the bulk and on the diffusion coefficient in the liquid. In addition, it depends inversely on the time of exposure of the liquid to the gas because of the increasing importance of reevaporation back to the gas phase at longer times. When $c_{l,bulk} = 0$, Eq. (EEE) becomes

$$\text{Rate} = (c_{l,interface})\sqrt{\frac{D_l}{\pi t}}. \quad \text{(FFF)}$$

The dissolved species at the interface is considered to occur in a thin layer of thickness $(D_l t)^{1/2}$

and is also assumed to be in Henry's law equilibrium with the gas immediately adjacent to the interface. While Henry's law is commonly expressed as $H = [X]/P_X$, with the Henry's law constant H in units of mol L^{-1} atm^{-1}, it can also be expressed in unitless form if the gas-phase concentration is expressed in units of mol L^{-1} instead of in units of pressure. With this conversion (see Problem 5), the liquid-phase concentration in the interface region, $N_l = c_{l,\text{interface}}$, is related to the gas-phase concentration immediately adjacent to the interface, N_g, by $N_l = N_g H \mathbf{R} T$ (where $\mathbf{R}$ is the gas constant). Substituting into Eq. (FFF), the rate of transfer of the gas is given by

$$\text{Rate} = N_g H \mathbf{R} T \sqrt{\frac{D_l}{\pi t}}. \quad \text{(GGG)}$$

Normalizing the rate of gas–surface collisions using Eq. (PP), one obtains

$$\Gamma_{\text{sol}} = \frac{4H\mathbf{R}T}{u_{\text{av}}} \sqrt{\frac{D_l}{\pi t}}. \quad \text{(HHH)}$$

Note that in this case, Γ_{sol} decreases with increasing time of exposure of the liquid to the gas. This reflects reevaporation from the liquid becoming increasingly important as the concentration of the dissolved species increases.

If the liquid layer is very thin, as is the case for some particles in the atmosphere, the interface layer with thickness $(D_l t)^{1/2}$ may comprise the entire particle, in which case the liquid is in Henry's law equilibrium with the gas, there is no net uptake, and Eq. (HHH) is not applicable. Similarly, at very long reaction times, i.e., as $t \to \infty$, $\Gamma_{\text{sol}} \to 0$. That is, at very long exposure times, there is no net uptake because the system has come to equilibrium and the rates of uptake and reevaporation are equal.

Reaction in the liquid phase (Γ_{rxn}). Now consider the case where an irreversible, first-order reaction with rate constant k (s^{-1}) takes place, in addition to diffusion and solubilization. Equation (CCC) becomes

$$\frac{\delta c}{\delta t} = D_l \frac{\delta^2 c}{\delta x^2} - kc. \quad \text{(III)}$$

This can be solved assuming that the concentration in the bulk liquid is 0 inside the drop at time $t = 0$ as well as at $x = \infty$ at longer times, and that the concentration in the thin layer at the interface is in Henry's law equilibrium with the adjacent gas (Danckwerts, 1970, pp. 33–37). Under these conditions, the rate of transfer of molecules across a plane under conditions where $kt \gg 1$ becomes independent of time and is given by

$$\text{Rate} = c_{l,\text{interface}} \sqrt{D_l k}. \quad \text{(JJJ)}$$

The normalized rate for reaction is then given by

$$\Gamma_{\text{rxn}} = \frac{4H\mathbf{R}T}{u_{\text{av}}} \sqrt{D_l k} \quad \text{(KKK)}$$

(see Problem 6). This applies to irreversible reactions or to those where the solubility of the reaction product is very large.

Let us now return briefly to the question of the relationship between net reactive uptake coefficients measured in laboratory systems, where the liquid films are generally quite thick, and particles in the atmosphere, which can be quite small and hence effectively have thin liquid films. A measure of the distance from the interface in which the reaction occurs is the *diffuso-reactive length*, l, which is defined as

$$l = \sqrt{\frac{D_l}{k}}. \quad \text{(LLL)}$$

Associated with this is the *diffuso-reactive parameter*, q, where

$$q = a\sqrt{\frac{k}{D_l}} = \frac{a}{l} \quad \text{(MMM)}$$

and a is the particle radius. Using an approach similar to that discussed earlier, Hanson *et al.* (1994) showed that the effective reactive uptake coefficient for small drops, γ_e, is related to that measured in thick laboratory films, γ_{meas}, by

$$\frac{1}{\gamma_e} = \frac{1}{\alpha} + \frac{1}{\gamma_{\text{meas}}} \cdot \frac{1}{(\coth q - 1/q)}. \quad \text{(NNN)}$$

For large values of α, then,

$$\gamma_e \approx \gamma_{\text{meas}}(\coth q - 1/q) \quad \text{(OOO)}$$

and the factor $(\coth q - 1/q)$ is the correction factor that must be applied in extrapolating laboratory measurements to particles in the atmosphere. When q is large, i.e., the diffuso-reactive length is small compared to the size of the particle, the correction factor is ~ 1 and the values for the reactive uptake coefficient measured in the laboratory can be applied to atmospheric particles. However, when q is small, i.e., the diffuso-reactive length is about the same as, or greater than, the particle radius, reac-

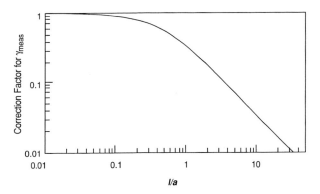

FIGURE 5.18 Correction factors for the measured uptake coefficient, γ_{meas}, as a function of the ratio of the diffuso-reactive length (l) to the droplet radius (a) (adapted from Hanson et al., 1994).

tion occurs throughout the particle volume. In this case, the correction factor becomes significant and γ_e is less than γ_{meas}. Figure 5.18 shows the values of this correction factor for various values of l/a (Hanson et al., 1994). Clearly, this correction factor can be quite large for small particle sizes/large diffuso-reactive lengths. Theoretical treatment suggests that the correction term may also vary with the electrical charge on the particle (Aikin and Pesnell, 1998), although it is not clear if this effect is significant for atmospheric droplets.

If the reaction is slow compared to diffusion so that the diffuso-reactive length is about the same or larger than the particle size, reaction takes place throughout the entire volume of the particle. On the other hand, if reaction is very fast compared to diffusion, i.e., the diffuso-reactive length is small compared to the size of the particle, reaction occurs close to the interface; in this case, it is the surface area, rather than the volume, that determines the magnitude of the reactive uptake (e.g., see Fried et al., 1994).

As seen in Box 5.2, Eq. (QQ) can be shown to be equivalent to Eq. (PPP):

$$\frac{1}{\gamma_{net}} = \frac{1}{\Gamma_g} + \frac{1}{\alpha} + \frac{1}{\dfrac{4HRT}{u_{av}}\left[\sqrt{\dfrac{D_1}{\pi t}} + \sqrt{D_1 k}\right]}, \quad \text{(PPP)}$$

where the symbols are as defined earlier, **R** is the gas constant and u_{av} is the average molecular speed in the gas phase.

There are several limiting cases of Eq. (PPP) of interest in atmospheric systems:

1. *Fast gas transport, high solubility, and/or fast reaction.* In this case, $1/\gamma_{net}$ approaches $1/\alpha$; i.e., the maximum value for the measured uptake approaches the mass accommodation coefficient.

2. *Fast gas transport, low solubility, and fast reaction.* In this case, $\Gamma_{sol} \ll \Gamma_{rxn}$ and Eq. (PPP) becomes

$$\frac{1}{\gamma_{net}} = \frac{1}{\gamma_{meas}} = \frac{1}{\alpha} + \frac{u_{av}}{4HRT\sqrt{D_1 k}}, \quad \text{(QQQ)}$$

where γ_{meas} is the measured uptake coefficient for the gas.

3. *High solubility (and/or short exposure times), no reaction.* In this case, the rate of mass transfer due to diffusion and solubility is large and Eq. (PPP) becomes

$$\frac{1}{\gamma_{net}} = \frac{1}{\gamma_{meas}} = \frac{1}{\Gamma_g} + \frac{1}{\alpha}. \quad \text{(RRR)}$$

That is, the net measured uptake measures the mass accommodation coefficient, corrected for the rate of transport of the gas to the surface.

4. *Gas transport and mass accommodation fast, solubility low, and slow reaction.* In this case, γ_{meas} is determined by the solubility and slow reaction and is given by

$$\gamma_{net} = \gamma_{meas} = \frac{4HRT\sqrt{D_1}}{u_{av}}\left[\sqrt{\frac{1}{\pi t}} + \sqrt{k}\right]. \quad \text{(SSS)}$$

However, because solubility and reaction are not decoupled processes in this case, this does not represent an exact solution and is valid only under conditions where $(kt)^{1/2} < 1$ (Kolb et al., 1995).

5. *Low solubility and no reaction.* In this case, equation (PPP) becomes

$$\frac{1}{\gamma_{net}} = \frac{1}{\gamma_{meas}} = \frac{1}{\Gamma_g} + \frac{1}{\alpha} + \frac{u_{av}\sqrt{\pi t}}{4HRT\sqrt{D_1}}. \quad \text{(TTT)}$$

Figure 5.19 illustrates the application of Eq. (PPP) to the reaction of O_3 with I^- in solution. In this case, reaction is faster than the diffusion solubility term so the latter can be neglected. In addition, gas-phase diffusion is fast and can be neglected. Hence Eq. (PPP) reduces to Eq. (QQ). The rate constant in this case is a pseudo-first-order rate constant, $k = k'[a_{I^-}]$, where k' is the second-order rate constant for the solution-phase

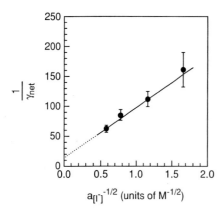

FIGURE 5.19 Plot of measured $(\gamma_{net})^{-1}$ for the uptake of O_3 into iodide-containing aqueous solutions against the (activity of iodide ion)$^{-1/2}$ at 277 K (adapted from Hu *et al.*, 1995).

TABLE 5.7 Characteristic Times Associated with the Uptake of Gases into Liquid Particles and Their Reaction in the Bulk Phase[a]

Process	Expression for characteristic time[b]
Gas-phase diffusion	$a^2/\pi^2 D_g$
Establishing equilibrium at the gas–liquid interface[c]	$D_l(4HRT/\alpha u_{av})^2$
Aqueous-phase diffusion	$a^2/\pi^2 D_l$
Aqueous-phase reaction	$1/k$
Aqueous-phase reaction relative to gas phase concentrations	$1/kHRT$
Relative importance of interfacial mass transport limitation to gas-phase mass transport limitation	$4D_g/a\alpha u_{av}$
Relative importance of aqueous-phase reaction to aqueous-phase diffusion	$ak^{1/2}/D_l^{1/2}$

[a] Adapted from Schwartz (1984a), and Shi and Seinfeld (1991).
[b] a = particle radius, D_g = gas-phase diffusion coefficient, D_l = liquid-phase diffusion coefficient, H = Henry's law constant, α = mass accommodation coefficient, u_{av} = mean thermal speed, and k = first-order aqueous-phase rate constant.
[c] For semi-infinite phase geometry.

reaction of O_3 with I^- and a_{I^-} is the activity of iodide ion in solution. A plot of $1/\gamma_{net}$, where γ_{net} is the measured uptake coefficient, against $k^{-1/2} = (k'a_{I^-})^{-1/2}$, i.e., as a function of $(a_{I^-})^{-1/2}$, should give a straight line from whose slope the second-order rate constant k' can be obtained if the Henry's Law constant (H) and liquid-phase diffusion coefficient (D_l) are known. The intercept corresponds to $1/\alpha$. As expected the plot of $1/\gamma_{net}$ against $(a_{I^-})^{-1/2}$ is linear and the value of k' derived from the slope, 4×10^9 L mol^{-1} s^{-1}, is in reasonable agreement with literature values (see Problem 9).

Another means of comparing the relative speeds of the various steps represented in Eq. (PPP) involves using the "characteristic time" for each process. Table 5.7 summarizes the parameters involved in each of these characteristic times. These allow an initial assessment of the relative speeds so that those steps which are rate-determining under a particular set of conditions can be readily identified.

6. *Reactions at the interface.* There is increasing evidence that reactions occur not only in the bulk but also at the interface itself. For example, Hanson and Ravishankara (1994) reported evidence that the stratospherically important reaction of $ClONO_2$ with HCl (see Chapter 12) occurs both at the surface and in the bulk. Similarly, Hu *et al.* (1995) investigated the uptake and reaction of Cl_2 into aqueous droplets containing bromide ion. Figure 5.20 shows a plot of $1/\gamma_{net}$, where γ_{net} is the net measured uptake coefficient, against $(a_{Br^-})^{-1/2}$ and, for comparison, the line expected if Cl_2 reacted with Br^- in the bulk phase with a rate constant of 7.7×10^9 L mol^{-1} s^{-1} as reported earlier in the literature, assuming a mass accommodation coefficient of $\alpha = 1$. Clearly, the measured rates of uptake exceed those expected, which is all the more surprising since

the rate constant represented by the dashed line is already close to the diffusion-controlled limit. In addition, the plot is not linear as expected from Eq. (PPP), particularly at higher concentrations of bromide.

Hu *et al.* (1995) interpret this as evidence for an additional reaction channel corresponding to a rapid reaction at the interface itself. Several different approaches have been taken to introduce this additional reaction dimension into the resistance model. For example, Hu *et al.* (1995) add another resistance term to Fig. 5.16, $1/\Gamma_{interface}$, shown by the dashed lines in that figure. Their data suggest that $1/\Gamma_{interface}$ is of the form $[1 + k''a_X]/(k''a_X p_s)$, where k'' is a measure of the

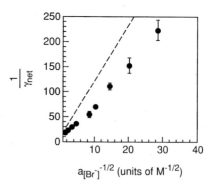

FIGURE 5.20 Plot of $(\gamma_{net})^{-1}$ for the uptake of Cl_2 into aqueous solutions containing various concentrations of bromide ion as a function of the (activity of bromide ion)$^{-1/2}$ at 293 K (adapted from Hu *et al.*, 1995).

surface reaction rate, a_X is the activity of the solution-phase reactant, and p_s is a probability parameter between 0 and 1. Another approach is described by Hanson (1997, 1998).

Similarly, there is some experimental evidence for surface reactions of organics. For example, Schweitzer *et al.* (1998) report evidence for a surface reaction (probably protonation) of glyoxal, $(CHO)_2$, on acid surfaces at temperatures below 273 K.

Although this area of reactions at interfaces is relatively new and not well understood, it may potentially be more significant than previously recognized. Because of the unique characteristics of such processes both kinetically and mechanistically compared to bulk aqueous-phase or gas-phase reactions, we suggest the term "fourth phase" be used to describe this chemistry at gas–liquid interfaces in the atmosphere.

In short, when treating the uptake of gases into particles, clouds, and fogs in the atmosphere and their reactions either at the interface or in the bulk, one must take into account all of the processes depicted in Fig. 5.12. While exact solutions for the series of coupled differential equations describing the individual steps are not always possible, approximate solutions have been derived for most situations of atmospheric interest in which the various steps can be treated as decoupled processes. In extrapolating values for the various steps derived from laboratory studies to particles in the atmosphere, one must take into account differences in conditions, including particle size. Summing up, if the fundamental parameters such as the Henry's law constants, diffusion coefficients, and rate constants are known, extrapolation to the atmosphere can be carried out reliably and reasonably accurately.

We now turn to a brief description of typical laboratory techniques used to determine kinetic parameters that characterize heterogeneous reactions in the atmosphere.

2. Knudsen Cells

Much of the data on heterogeneous reactions in the atmosphere, particularly the earliest work, were generated using Knudsen cells (Golden *et al.*, 1973; Caloz *et al.*, 1997; Fenter *et al.*, 1997). Figure 5.21 is a schematic diagram of a Knudsen cell. Gases flow into the cell, which has an orifice of known size connected to a low-pressure system. Gases exiting the Knudsen cell through this orifice are detected and measured, usually by mass spectrometry. When the gas is exposed to a surface that takes up the gas, the concentration of the gas in the cell and hence the amount exiting the orifice decrease. From the change, the net uptake of the gas by the surface can be determined in the following manner.

Let the flow of molecules into the Knudsen cell be F (molecules s^{-1}). In the absence of the reactive surface, these molecules are removed when they strike the escape aperture into the mass spectrometer. Let k_{esc} be the effective first-order rate constant (s^{-1}) for escape of the gas from the cell through this orifice, which can be measured experimentally. Alternatively, k_{esc} can be calculated from kinetic molecular theory since the number of collisions per second, J_s, of a gas on a

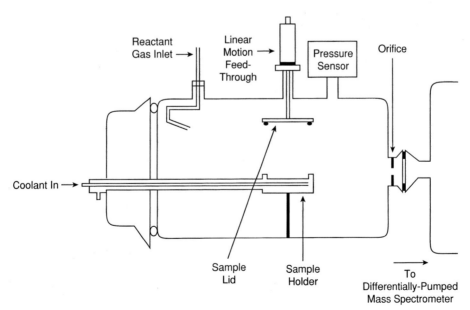

FIGURE 5.21 Schematic of a Knudsen cell.

surface of area A_s, or equivalently, J_h a hole of surface area A_h, is given by

$$J_h = A_h(N/V)(u_{av}/4), \quad \text{(UUU)}$$

i.e.,

$$k_{esc} = \frac{J_h}{N} = \frac{A_h}{V}(u_{av}/4). \quad \text{(VVV)}$$

Here u_{av} is the average speed of the molecules of molecular weight M at temperature T and is given by $u_{av} = (8kT/\pi M)^{1/2}$, where $k = 1.381 \times 10^{-23}$ J K^{-1}, N is the number of molecules in the cell, and V is the cell volume.

Under steady-state conditions, the flow into the cell is balanced by escape from the orifice, and the number of gas molecules in the cell in the absence of the reactive surface, N_0, remains constant, i.e.,

$$\frac{dN_0}{dt} = F - k_{esc}N_0 = 0.$$

When the gas is exposed to the reactive surface, some of the gas-phase molecules are removed with an effective first-order rate constant k_r, thus reducing the number of molecules in the cell to N_r. A new steady-state is established such that

$$\frac{dN_r}{dt} = F - k_{esc}N_r - k_r N_r = 0. \quad \text{(WWW)}$$

Combining these two equations and rearranging, one obtains

$$k_r = k_{esc}\left(\frac{N_0 - N_r}{N_r}\right). \quad \text{(XXX)}$$

The first-order rate constant k_r (s^{-1}) for the heterogeneous reaction is related to the rate of gas–surface collisions, J_s, and the fraction of those collisions that lead to uptake, γ_{net}, since $k_r N_r = \gamma_{net} J_s$. Since J_s is also equal to $A_s(N_r/V)(u_{av}/4)$, then

$$\gamma_{net} J_s = \gamma_{net} A_s(N_r/V)(u_{av}/4) = k_r N_r,$$

i.e.,

$$k_r = \frac{\gamma_{net} A_s}{V}(u_{av}/4). \quad \text{(YYY)}$$

Combining Eqs. (VVV), (XXX), and (YYY), the net uptake probability is given by

$$\gamma_{net} = \frac{A_h}{A_s}\frac{N_0 - N_r}{N_r}. \quad \text{(ZZZ)}$$

Since the ratio of the number of gas molecules in the cell in the presence and absence of the reactive surface

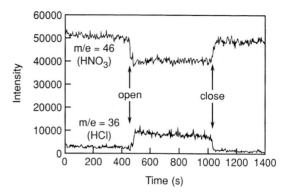

FIGURE 5.22 Typical uptake of gaseous HNO$_3$ by solid NaCl (monitored at $m/e = 46$) and increase in the gaseous HCl product (monitored at $m/e = 36$) (adapted from Beichert and Finlayson-Pitts, 1996).

is needed to calculate γ_{net}, the measurement need not be one of absolute concentrations but rather just relative values. Hence the relative signals measured using mass spectrometry in the presence and absence of the reactive surface are typically used to obtain γ_{net}.

Knudsen cells are operated at low pressures, typically < 10 mTorr, so that the mean free path (L) of the gas molecules is much larger than the diameter of the escape orifice. Since $L = 1/[2^{0.5}\pi d^2(N/V)]$, where d is the molecular diameter of the gas, at 10 mTorr N$_2$, $L = 0.5$ cm, for example.

Figure 5.22 shows the results of a typical Knudsen cell experiment for the uptake of HNO$_3$ by crystalline NaCl powder. The parent peak at $m/e = 63$ is weak, so the peak at $m/e = 46$ is followed instead. Both the loss of HNO$_3$ and the release of the product HCl into the gas phase can be followed and are seen to occur simultaneously as expected for the reaction

$$HNO_{3(g)} + NaCl_{(s)} \rightarrow HCl_{(g)} + NaNO_{3(s)}. \quad (37)$$

In Knudsen cell studies, as in other techniques, care must be taken to avoid surface "saturation," i.e., complete reaction of the surface so that one is no longer measuring uptake by the original surface. For example, at 10 mTorr there are approximately 10^{18} collisions per cm^2 per second of the gas with the surface. For typical areas of reactive surface that are convenient to use in these systems, the surface will have of the order of 10^{16} surface sites. Thus if every collision leads to uptake and there is no reevaporation from the surface, it will be completely reacted in only 10 ms!

Another important factor to recognize is that the *net* uptake coefficient determined using Knudsen cells may not represent the true uptake or trapping of the gas by the surface if reevaporation into the gas phase occurs, which must be taken into account in such cases. In principle, the mass accommodation coefficient is the uptake measured as $t \rightarrow 0$ or, equivalently, as the aqueous-phase concentration of the gaseous species of interest approaches zero.

For some typical examples of the applications of Knudsen cells to atmospheric reactions, see Quinlan *et al.* (1990), Fenter *et al.* (1994), Beichert and Finlayson-Pitts (1996), and De Haan and Finlayson-Pitts (1997).

3. Flow Tube Studies

As discussed earlier, flow tubes have been applied for many years to obtaining absolute rate constants for a variety of gas-phase reactions, especially with highly reactive free radical intermediates such as OH and Cl. More recently, the same approach has been applied to studying reactions of gases with both solid and liquid surfaces (e.g., McMurry and Stolzenburg, 1987).

The flow tube walls can be coated with the condensed phase of interest, and the gaseous reactant added through a movable injector. As the distance between the injector tip and the detector is increased, the gas is increasingly removed at the walls of the flow tube, and a pseudo-first-order rate constant for removal of the gas, k_s, is measured. From Eq. (UUU), the number of collisions per second per unit area is given by $(N/V)(u_{av}/4)$, where (N/V) is the concentration of the gas and u_{av} the mean molecular speed. If the net uptake probability is γ_{net}, the number of gas molecules removed at the surface per second in a flow tube of radius r and length l is $\gamma_{net}(N/V)(u_{av}/4)(2\pi rl)$ and the change in the *concentration* per second is $d(N/V)/dt = \gamma_{net}(N/V)(u_{av}/4)(2\pi rl)/(\pi r^2 l) = (\gamma_{net} u_{av}/2r)(N/V)$. Thus the first-order rate constant for loss of the gas at the surface is $k_s = \gamma_{net} u_{av}/2r$; i.e., the net uptake probability γ_{net} is given by

$$\gamma_{net} = \frac{2rk_s}{u_{av}},$$

where r is the flow tube radius and u_{av} is the mean thermal speed of the molecules.

It is often inconvenient and/or experimentally impossible to coat the walls of the flow tube with the condensed phase, e.g., for horizontally mounted flow tubes. In this case, the liquid can be held in a rectangular container on the bottom of the flow tube. While the principle of the experiment is the same, corrections for only a portion of the surface area being reactive must be made. The same approach has been applied to studying the reactions of gases with solids. If the solid sample is in the form of a powder, there are usually multilayers of the crystalline grains in the sample container, which makes determination of the effective surface area available for reaction much more complex. For some typical applications of flow tubes to studying heterogeneous reaction kinetics, see Hanson and Ravishankara (1993b), Zhang *et al.* (1994), and Leu *et al.* (1995).

4. Falling-Droplet Apparatus

A technique for obtaining the mass accommodation coefficients for the uptake of gases into liquid droplets is the falling-droplet apparatus, which has been applied to a number of atmospherically relevant species (e.g., see Gardner *et al.*, 1987; Worsnop *et al.*, 1989; Nathanson *et al.*, 1996; and Robinson *et al.*, 1998). Figure 5.23 shows a schematic of a typical falling-droplet apparatus. The droplets are generated using a vibrating orifice generator and are ejected into a flow tube at linear flow speeds of about 1500–4500 cm s^{-1}. As the stream of droplets flows down the tube, it interacts with the gas of interest, which can be added at various positions along the length of the flow tube. The gas concentration is measured using techniques such as mass spectrometry or tunable diode infared laser spectroscopy at the downstream end of the flow tube after it has interacted with the droplets of known surface area for a known time; alternatively, the droplets can be collected and their composition determined (e.g.,

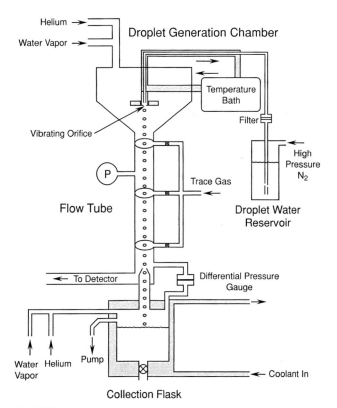

FIGURE 5.23 Schematic diagram of typical falling-droplet apparatus used for studying heterogeneous atmospheric reactions (adapted from Jayne *et al.*, 1992).

Ponche *et al.*, 1993). The droplet flow can be turned off and on to measure the change in the gas concentration caused by the droplets, or alternatively, the change in the gas concentration when the droplet surface area is changed can be measured. From the change in the gas concentration, the uptake of the gas by the liquid can then be extracted in the following manner.

If the flow of the carrier gas (e.g., He) is given by F_g (cm^3 s^{-1}) and Δn is the change in the trace gas concentration due to uptake by the droplets, then the number of gas molecules taken up per second is just $F_g \Delta n$. The number of gas-droplet collisions per second per unit area is given (Eq. PP) as $J' = N_g u_{av}/4$, where N_g is the number of gas molecules per unit volume and u_{av} is the mean molecular (thermal) speed. If A_d is the surface area of one droplet and there are N^* droplets to which the gas is exposed, then the total available surface area is $(N^* A_d)$, the total number of gas-droplet collisions is $J' = (N^* A_d) N_g u_{av}/4$, and the *measured* mass accommodation coefficient becomes

$$\gamma_{obs} = \frac{F_g \Delta n}{(N^* A_d N_g u_{av})/4}.$$

While the experiments are thus conceptually straightforward, this is not always the case with respect to the interpretation and extraction of the true mass accommodation coefficient because of the simultaneous occurrence of all of the processes depicted in Fig. 5.12. The approach to extracting α from the measurements of the net gas uptake was treated above in Section E.1.

5. Bubble Apparatus

For gas–liquid combinations with relatively small uptake coefficients ($\sim 10^{-4}$–10^{-7}), longer interaction times between the gas and liquid are needed than can be obtained with the falling-droplet apparatus. These are provided in a bubble apparatus, a typical example of which is shown in Fig. 5.24. The gas of interest as a mixture with an inert carrier gas is introduced as a stream of bubbles into the liquid of interest. The interaction time is varied by moving the gas injector relative to the surface. The composition of the gas exiting the top of the liquid is measured as a function of the interaction time (typically 0.1–1 s), e.g., by mass spectrometry. The interaction time is limited by the depth in the liquid at which the bubbles are injected and their buoyancy. Longer interaction times and better control over them have been achieved using a modified apparatus in which the bubbles are generated and transported horizontally (Swartz *et al.*, 1997).

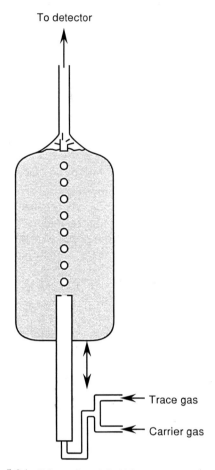

FIGURE 5.24 Schematic of bubble apparatus (adapted from Shorter *et al.*, 1995).

6. Aerosol Chambers

Several aerosol chambers have been applied recently to studying aerosol kinetics and mechanisms. Figure 5.25 shows one type of design that is a variant on flow tubes used extensively in studying gas-phase reactions. The aerosol travels along the length of a laminar flow reactor, with gases added through a movable injector, and changes in the gas and/or particles are followed as a function of reaction time, typically of the order of seconds. As a result, these have been applied to studying fairly fast heterogeneous reactions (e.g., see Fried *et al.*, 1994; and Lovejoy and Hanson, 1995). Alternate designs used to study slower reactions on time scales of the order of minutes have also been implemented (e.g., see Karlsson and Ljungström, 1995).

Static aerosol chambers have been developed for studying slower heterogeneous reactions (e.g., Zetzsch and Behnke, 1992; Anthony *et al.*, 1995; De Haan *et al.*, 1999). Figures 5.26 and 5.27 show one such system (De Haan *et al.*, 1999). It consists of a 561-L stainless steel

E. LABORATORY TECHNIQUES FOR STUDYING HETEROGENEOUS REACTIONS

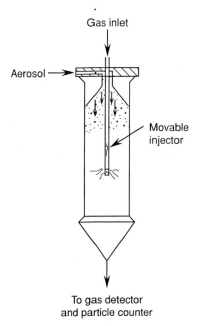

FIGURE 5.25 Schematic diagram of aerosol apparatus (adapted from Fried *et al.*, 1994).

and aluminum chamber whose walls are coated with halocarbon wax to provide a relatively unreactive surface. The top has a series of windows that can be either quartz or borosilicate glass, so that photolysis of the gas–particle mixtures can be studied. The chamber can be heated and pumped. It is equipped with two long-path optical systems, one in the UV–visible (for differential optical absorption spectrometry, DOAS) and one for FTIR (see Chapter 11.A.1c). An atmospheric pressure ionization mass spectrometer (API-MS) is also interfaced to the chamber to measure gases such as Cl_2 and Br_2, which cannot be measured using DOAS or FTIR. An aerosol generator and differential mobility analyzer are attached for generating particles and determining their size distribution (see Chapter 11.B.2b). This has been applied, for example, to studies of Cl_2 generation from the photolysis of O_3 in the presence of deliquesced sea salt particles (Oum *et al.*, 1998).

7. Liquid Jet Apparatus

Figure 5.28 shows a liquid jet apparatus used to study the uptake of HONO on water (Bongartz *et al.*, 1994). Water is forced through a capillary to form a jet with a diameter of the order of 100 μm. The jet flows

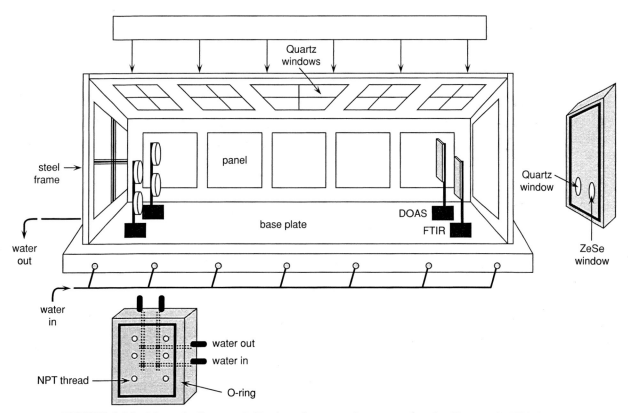

FIGURE 5.26 Schematic diagram of side view of an aerosol apparatus (see De Haan *et al.*, 1999, for a description of the chamber).

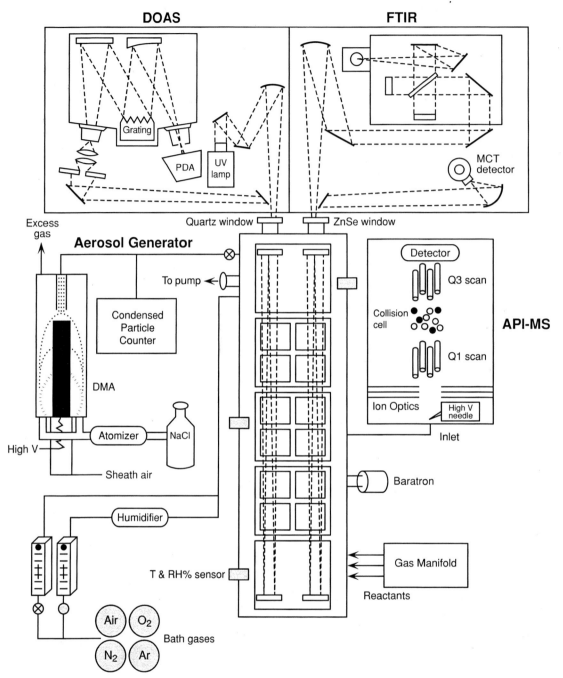

FIGURE 5.27 Top view of aerosol apparatus (De Haan *et al.*, 1999).

through an atmosphere of air containing the gas of interest, which is taken up by the liquid stream. The liquid is collected and analyzed to obtain the gas uptake. The reaction time, typically 0.1–1 ms, is varied by varying the length of the liquid jet exposed to the gas prior to collection. This type of experiment is related to the falling-droplet approach except that it utilizes the first millimeter of flow before the breakup into droplets. While the experimental approach is again relatively simple, the analysis is quite complex in that the time-dependent modeling must be carried out which incorporates the gas- and liquid-phase diffusion processes and chemical reactions in the liquid phase as well as the uptake of the gas at the interface. The value of the mass accommodation coefficient giving the best fit to the data is then obtained.

Another promising approach is the use of thin liquid jets a few micrometers in diameter combined with vacuum techniques that allow the application of surface science techniques such as photoelectron spec-

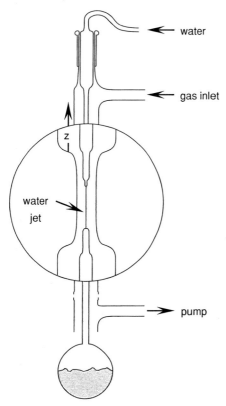

FIGURE 5.28 Schematic diagram of the liquid jet apparatus (adapted from Bongartz *et al.*, 1994).

troscopy to liquid surfaces (e.g., Faubel *et al.*, 1988, 1997). The use of high-vacuum electron spectroscopies on volatile liquids is limited by their vapor pressure, since the liquid evaporates during the measurement. In addition, the distance between the liquid surface and the spectrometer slit has to be sufficiently small that the electrons can be appropriately sampled. As a result, there are practical limits on the combination of (vapor pressure × distance) to < 1 Torr mm for X-ray photoelectron spectroscopy (XPS) and < 0.1 Torr mm for ultraviolet photoelectron spectroscopy (UPS) (Faubel *et al.*, 1997). A liquid jet apparatus using UPS that meets these limitations has been applied to liquid water, with the spectra suggesting orientation of surface water molecules with the hydrogen atoms pointing outward (Faubel *et al.*, 1997). Although such techniques have not yet been applied to systems of atmospheric interest, they could potentially be quite useful in exploring such surfaces.

8. DRIFTS

While the majority of techniques for studying heterogeneous reactions monitor changes in the gas-phase concentration, some focus on changes in the condensed phase. One such technique is diffuse reflectance infrared Fourier transform spectroscopy (DRIFTS). This technique has been used in the past as a method for analyzing solids. However, it has also proven very useful for studying the kinetics and mechanisms of the reactions of infrared-transparent solids such as NaCl, the major component of sea salt aerosols as well as synthetic sea salt (Vogt and Finlayson-Pitts, 1994, 1995; Langer *et al.*, 1997).

Figure 5.29 is a schematic diagram of a DRIFTS apparatus that has been applied to studying the reactions of the components of sea salt particles with various oxides of nitrogen. As the reactions occur, nitrate, which absorbs strongly in the infrared, is formed on the salt surface. Since the reactant solids do not absorb in the infrared, the increase in nitrate with time can be readily followed and used to obtain reaction probabilities.

9. Surface Science Techniques

Traditionally, surface science methodologies have not been applied to problems of atmospheric interest because of the incompatibility of atmospheric conditions and the ultrahigh-vacuum conditions commonly used in the surface sciences. However, with the increasing recognition of the need to understand heterogeneous atmospheric processes at the molecular level, more attention is being focused on the application of these techniques to problems of atmospheric interest. Thus, X-ray photoelectron spectroscopy (XPS), Auger electron spectroscopy, and ultraviolet photoelectron spectroscopy (UPS) have been applied to both solids and liquids of atmospheric interest. For example, Hemminger (1999) reviews the application of surface science techniques, including XPS and TEM, to studying the reactions of NaCl found in sea salt and to Al_2O_3 found in soils. Similarly, Fairbrother *et al.* (1996) described a UHV apparatus having several differential pumping systems and permitting the application of both Auger and XPS to the surface of liquid H_2SO_4–H_2O mixtures at room temperature. As described earlier, Faubel *et al.* (1997) have reported the UPS spectrum of the surface of liquid water using thin liquid jets.

One approach is to carry out reactions of interest in attached chambers under conditions approaching atmospheric and then do the actual surface analyses under ultrahigh-vacuum conditions. For example, XPS has been used to follow the formation of nitrate on the surface of NaCl exposed to HNO_3 (Laux *et al.*, 1994, 1996; Vogt *et al.*, 1996; Hemminger, 1999). Figure 5.30 shows the apparatus used to "dose" known quantities of HNO_3 onto the NaCl surface (Laux *et al.*, 1994). After each dose, the loss of Cl and uptake of N and O

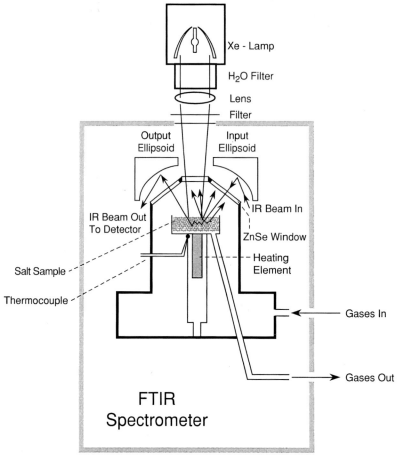

FIGURE 5.29 Schematic of DRIFTS apparatus (adapted from Vogt and Finlayson-Pitts, 1994).

are measured in an adjacent chamber using XPS. Typical time profiles are shown in Fig. 5.31. That the nitrogen and oxygen are in the oxidation state expected for NO_3^- can be confirmed from the binding energies observed for these elements. The reaction probabilities are calculated from the ratio of the number of surface nitrate ions formed, measured using XPS, to the total number of gas–solid collisions, calculated from the length and diameter of the capillary on the doser. Mechanistic information can also be obtained by exposing the crystal to other species such as water vapor in a separate chamber attached to the XPS (e.g., see Laux et al., 1996) and, as discussed in Chapter 11, by the use of TEM (Allen et al., 1996).

Nonlinear optical spectroscopies such as second harmonic generation (SHG) and sum frequency generation (SFG) are finding increasing use in probing species at interfaces (e.g., Eisenthal, 1996). For example, SHG was used by Donaldson et al. (1995) to detect a surface-bound SO_2 species, and SFG has been applied to elucidate the structure of dimethyl sulfoxide at liquid interfaces (Allen et al., 1999). These techiques are very promising for providing molecular-level understanding of liquid interfaces of interest in the atmosphere.

10. Other Methods

The foregoing methods are certainly not exclusive, and many other techniques such as cloud chambers (e.g., see Miller et al., 1987) and fluidized bed reactors have also been applied to following the kinetics of heterogeneous reactions relevant to the atmosphere. However, due to space limitations, these will not be treated in detail here.

F. COMPILATIONS OF KINETIC DATA FOR ATMOSPHERIC REACTIONS

Fortunately, there are several compendia of kinetic data applicable for atmospheric reactions that are

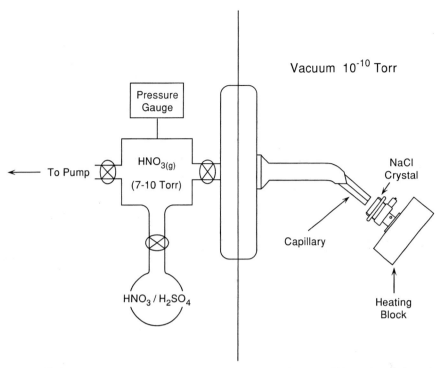

FIGURE 5.30 Schematic diagram of XPS apparatus used for gas–solid reactions (adapted from Laux *et al.*, 1994).

widely available. IUPAC carries out critical evaluations of kinetic data on a wide variety of reactions of atmospheric interest and makes recommendations for the most reliable kinetic parameters to be used. The most recent ones include data on both gas-phase and heterogeneous reactions and are published in the readily available *Journal of Physical Chemical Reference Data* (Atkinson *et al.*, 1997a,b).

NASA also carries out kinetic evaluations for reactions of interest in the stratosphere (although many of them are also important in the troposphere) (DeMore *et al.*, 1997). This document can be obtained from the Jet Propulsion Laboratory (California Institute of Technology, Document Distribution, MS 512-110, 4800 Oak Grove Drive, Pasadena, CA 91109) or through the Internet (see Appendix IV).

Finally, the National Institute of Standards and Technology (NIST) in the United States has several chemical kinetics databases that are available for purchase from the Office of Standard Reference Data at NIST. The NIST Standard Reference Data Base 17 gives gas-phase rate constants through 1993 and Data Base 40 gives solution-phase data through 1992. In addition, aqueous-phase data are available through the Radiation Chemistry Data Center of the Notre Dame Radiation Laboratory (http://www.rcdc.nd.edu/).

In addition to these highly useful data sets, periodically there are reviews directed to the reactions of one particular species (e.g., OH, NO_3, or O_3) or group of compounds (e.g., RO_2 radicals). These are referenced in the appropriate sections of Chapter 6. For example, a review of the gas-phase tropospheric chemistry of

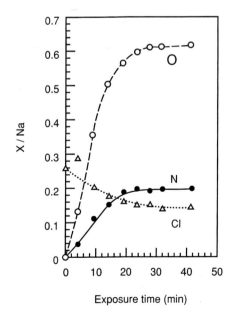

FIGURE 5.31 Change in surface concentrations of N(1s), O(1s), and Cl(2p) as a function of exposure time of an NaCl(100) single crystal to gaseous HNO_3 (adapted from Laux *et al.*, 1996).

organic compounds was published by Atkinson in 1994 and of alkanes and alkenes in 1997 (Atkinson, 1994, 1997); these include recommendations on rate constants as well as reaction mechanisms and products.

G. PROBLEMS

1. The rate constants for the reactions of propene with OH, O_3, and NO_3 at 1 atm pressure and 298 K are 2.6×10^{-11}, 1.0×10^{-17}, and 9.5×10^{-15} cm^3 molecule^{-1} s^{-1}, respectively. Typical peak concentrations at the surface in a moderately polluted atmosphere are 5×10^6 OH cm^{-3}, 100 ppb O_3, and 50 ppt NO_3. Assume 298 K and 1 atm pressure. (a) What are the half-lives of propene with respect to removal from the atmosphere by each of these species? (b) What is the half-life of propene with respect to all three acting simultaneously (which does not actually happen since they peak at different times of the day; see Chapter 6). (c) Repeat parts (a) and (b) but calculate the lifetimes instead of the half-lives.

2. Much of the total oxides of nitrogen in the Arctic at surface level in the winter is tied up in the form of PAN. (a) Use the kinetic parameters in the text to calculate the lifetime of PAN with respect to thermal decomposition under typical temperatures of $-35°C$. (b) What would the temperature have to be to give it a lifetime of half a day?

3. Table 5.8 gives some of the kinetic parameters for some termolecular reactions of atmospheric interest. Choose one of these reactions and do the following:

(a) Calculate the overall effective bimolecular rate constant at a typical surface pressure of 750 Torr and temperature of 300 K.

(b) Calculate the effective overall rate constant at the following total pressures: 0, 150, 300, 450, 600, and 750 Torr. Assume the temperature is 300 K. Plot the rate constant as a function of pressure and explain the shape of the curve on a molecular level.

(c) Some cities such as Mexico City are at higher elevations (~ 2200 m) where the air pressure is lower. Calculate the rate constant at a pressure of 630 Torr, assuming a temperature of 300 K. Comment on the direction of change in the rate constant on a molecular level.

(d) Temperatures as low as $-50°F$ can be found at some polar locations at the earth's surface, while temperatures as high as 120°F are found in desert regions. Calculate the following, assuming a pressure of 750 Torr:

   i. The values of k_0 and k_∞ at these two temperatures. Do the directions of the change in these two limiting rate constants as a function of temperature make sense on a molecular level? Explain why or why not.

   ii. Calculate the overall rate constant at these two temperature extremes and comment on the reason for the direction of the change, on a molecular level, in going from low to high temperatures.

TABLE 5.8 Some Kinetic Parameters for Some Termolecular Reactions[a]

Reaction	k_0^{300} [b] (cm^6 molecule^{-2} s^{-1})	n	k_∞^{300} [b] (cm^3 molecule^{-1} s^{-1})	m
$O + O_2 \xrightarrow{M} O_3$	6.0 (-34)	2.3	—	—
$H + O_2 \xrightarrow{M} HO_2$	5.7 (-32)	1.6	7.5 (-11)	0
$OH + NO_2 \xrightarrow{M} HNO_3$[c]	2.5 (-30)	4.4	1.6 (-11)	1.7
$HO_2 + NO_2 \xrightarrow{M} HO_2NO_2$	1.8 (-31)	3.2	4.7 (-12)	1.4
$NO_2 + NO_3 \xrightarrow{M} N_2O_5$	2.2 (-30)	3.9	1.5 (-12)	0.7
$CH_3 + O_2 \xrightarrow{M} CH_3O_2$	4.5 (-31)	3.0	1.8 (-12)	1.7
$C_2H_5 + O_2 \xrightarrow{M} C_2H_5O_2$	1.5 (-28)	3.0	8.0 (-12)	0
$CH_3C(O)O_2 + NO_2 \xrightarrow{M}$ PAN	9.7 (-29)	5.6	9.3 (-12)	1.5
$ClO + NO_2 \xrightarrow{M} ClONO_2$	1.8 (-31)	3.4	1.5 (-11)	1.9
$BrO + NO_2 \xrightarrow{M} BrONO_2$	5.2 (-31)	3.2	6.9 (-12)	2.9
$OH + SO_2 \xrightarrow{M} HOSO_2$	3.0 (-31)	3.3	1.5 (-12)	0

[a] Data from DeMore et al. (1997).
[b] 6.0 (-34) ≡ 6.0×10^{-34}.
[c] See discussion of more recent kinetics studies of this reaction in Chapter 7.B.1.

4. Use the data in Fig. 5.11 to calculate the rate constants for the reactions of chlorine atoms with (a) ethane and (b) isobutane. The rate constants for the reactions of Cl with propane and n-butane are 1.2×10^{-10} and 2.1×10^{-10} cm^3 molecule^{-1} s^{-1}, respectively.

5. Show that the unitless form of Henry's law leads to $N_l = N_g H\mathbf{R}T$.

6. Show that the expression for Γ_{rxn} given in Eq. (KKK) can be obtained from the expression in Eq. (JJJ) for the rate of transfer of molecules across a plane.

7. Show that Eq. (QQ) applies for the electrical circuit analogy in Fig. 5.16 with the individual conductances and resistances defined as shown (but excluding the interface reaction).

8. Show that Eq. (XX) reduces to Eq. (SS) for small values of K_n.

9. Use the data of Hu et al. (1995) in Fig. 5.19 to derive the second-order rate constant for the $O_3 + I^-$ reaction in the liquid phase assuming that solubility and gas-phase diffusion are not limiting factors. Also derive a value for the mass accommodation coefficient for O_3 based on these data. The Henry's law constant for O_3 can be taken to be 0.02 M atm^{-1}, the temperature is 277 K, and the diffusion coefficient in the liquid phase 1.3×10^{-5} cm^2 s^{-1}.

10. (a) Show that the fractional change in light intensity due to a concentration N molecules cm^{-3} of an absorbing gas with absorption cross section σ is $(1 - e^{-\sigma NL})$. (b) Show that Eq. (BB) describes the absorption of light by a gas molecule in the cavity ring down method. (c) Comment on why the cavity ring down method is able to measure such small absorptions compared to standard absorption spectrometry.

11. You are using a Knudsen cell with a circular reactive surface of diameter 4.9 cm. If you anticipate that the uptake coefficient for a reaction of interest is ~ 0.01, what diameter should the aperture to the mass spectrometer be to observe a drop in the reactant signal of 50% on exposure to the reactive surface?

12. When a liquid becomes saturated with a gas, i.e., has reached Henry's law equilibrium, the rates of uptake and reevaporation are equal. Use Eq. (PP), the mass accommodation coefficient α, and the relationship between the gas-phase concentration N_g and the liquid-phase concentration N_l developed in Problem 5 to show that

$$\frac{N_g \alpha u_{av}}{4} = \frac{N_l \alpha u_{av}}{4H\mathbf{R}T}.$$

Note that this suggests that the term $(\alpha/H\mathbf{R}T)$, the "evaporation coefficient" for dissolved gases moving from the liquid to the gas phase, is analogous to α, which reflects the efficiency of uptake across the interface from the gas to the condensed phase (e.g., see Kolb et al., 1997).

13. The low and high pressure limiting rate constants for the reaction of ClO with NO$_2$ to form chlorine nitrate at 300 K, ClO + NO$_2$ + M $\rightleftarrows$ ClONO$_2$ + M (1, −1) are given by $k_0 = 1.8 \times 10^{-31}$ cm^6 molecule^{-2} s^{-1} and $k_\infty = 1.5 \times 10^{-11}$ cm^3 molecule^{-1} s^{-1} (DeMore et al., 1997). The kinetics of its thermal decomposition have also been measured by Anderson and Fahey [*J. Phys. Chem.*, **94** 644 (1990)] who report $k_{-1} = 10^{-6.16} \exp(-90.7 \text{ kJ mol}^{-1}/\mathbf{R}T)$ in units of cm^3 molecule^{-1} s^{-1}. Calculate (a) the effective bimolecular rate constant k_1 at 298 K, (b) the rate constant k_{-1} for the thermal decomposition at 298 K and 1 atm pressure, and (c) the lifetime with respect to the thermal decomposition. (d) Compare its rate of thermal decomposition to its photolysis rates at the earth's surface (see Problem 3-4).

References

Abbatt, J. P. D., K. L. Demerjian, and J. G. Anderson, "A New Approach to Free-Radical Kinetics: Radially and Axially Resolved High-Pressure Discharge Flow with Results for OH + (C$_2$H$_6$, C$_3$H$_8$, n-C$_4$H$_{10}$, n-C$_5$H$_{12}$ → Products at 297 K," *J. Phys. Chem.*, **94**, 4566–4575 (1990).

Abbatt, J. P. D., F. F. Fenter, and J. G. Anderson, "High-Pressure Discharge Flow Kinetics Study of OH + CH$_3$SCH$_3$, CH3SSCH$_3$ → Products from 297 to 368 K," *J. Phys. Chem.*, **96**, 1780–1785 (1992).

Adamson, A. W., *A Textbook of Physical Chemistry*, Academic Press, New York, 1973.

Aikin, A. C., and W. D. Pesnell, "Uptake Coefficient of Charged Aerosols—Implications for Atmospheric Chemistry," *Geophys. Res. Lett.*, **25**, 1309–1312 (1998).

Alfassi, Z. B., S. Padmaja, P. Neta, and R. E. Huie, "Rate Constants for Reactions of NO$_3$ Radicals with Organic Compounds in Water and Acetonitrile," *J. Phys. Chem.*, **97**, 3780–3782 (1993).

Allen, H. C., J. M. Laux, R. Vogt, B. J. Finlayson-Pitts, and J. C. Hemminger, "Water-Induced Reorganization of Ultrathin Nitrate Films on NaCl—Implications for the Tropospheric Chemistry of Sea Salt Particles," *J. Phys. Chem.*, **100**, 6371–6375 (1996).

Allen, H. C., D. E. Gragson, and G. L. Richmond, "Molecular Structure and Adsorption of Dimethyl Sulfoxide at the Surface of Aqueous Solutions," *J. Phys. Chem. B*, **103**, 660–666 (1999).

Anthony, S. E., R. T. Tisdale, R. S. Disselkamp, and M. A. Tolbert, "FTIR Studies of Low Temperature Sulfuric Acid Aerosols," *Geophys. Res. Lett.*, **22**, 1105–1108 (1995).

Atkinson, D. B., and J. W. Hudgens, "Chemical Kinetic Studies Using Ultraviolet Cavity Ring-Down Spectroscopic Detection: Self-Reaction of Ethyl and Ethylperoxy Radicals and the Reaction O$_2$ + C$_2$H$_5$ → C$_2$H$_5$O$_2$," *J. Phys. Chem. A*, **101**, 3901–3909 (1997).

Atkinson, R., R. Perry, and J. N. Pitts, Jr., "Rate Constants for the Reactions of the OH Radical with NO$_2$ (M = Ar and N$_2$) and SO$_2$ (M = Ar)," *J. Chem. Phys.*, **65**, 306–310 (1976).

Atkinson, R., W. P. L. Carter, A. M. Winer, and J. N. Pitts, Jr., "An Experimental Protocol for the Determination of OH Radical Rate Constants with Organics Using Methyl Nitrite Photolysis as an OH Radical Source," *J. Air Pollut. Control Assoc.*, **31**, 1090–1092 (1981).

Atkinson, R., "Gas-Phase Tropospheric Chemistry of Organic Compounds," *J. Phys. Chem. Ref. Data,* Monograph No. 2, American Chemical Society and the American Institute of Physics, Woodbury, NY, 1994.

Atkinson, R., "Gas-Phase Tropospheric Chemistry of Volatile Organic Compounds: 1. Alkanes and Alkenes," *J. Phys. Chem. Ref. Data, 26,* 215–290 (1997).

Atkinson, R., D. L. Baulch, R. A. Cox, R. F. Hampson, Jr., J. A. Kerr, M. J. Rossi, and J. Troe, "Evaluated Kinetic, Photochemical, and Heterogeneous Data for Atmospheric Chemistry: Supplement V, IUPAC Subcommittee on Gas Kinetic Data Evaluation for Atmospheric Chemistry," *J. Phys. Chem. Ref. Data, 26,* 521–1011 (1997a).

Atkinson, R., D. L. Baulch, R. A. Cox, R. F. Hampson, Jr., J. A. Kerr, M. J. Rossi, and J. Troe, "Evaluated Kinetics and Photochemical Data for Atmospheric Chemistry: Supplement VI," *J. Phys. Chem. Ref. Data, 26,* 1329–1499 (1997b).

Audley, G. J., D. L. Baulch, L. M. Campbell, D. J. Waters, and G. Watling, "Gas-Phase Reactions of Hydroxyl Radicals with Alkyl Nitrite Vapours in $H_2O_2 + NO_2 + CO$ Mixtures," *J. Chem. Soc., Faraday Trans. 1, 78,* 611–617 (1982).

Ayers, G. P., "The Chemical Composition of Precipitation: A Southern Hemisphere Perspective," in *Atmospheric Chemistry* (E. D. Goldberg, Ed.), pp. 41–56, Springer-Verlag, New York, 1982.

Barnes, L., V. Bastian, K. H. Becker, E. H. Fink, and F. Zabel, "Reactivity Studies of Organic Substances towards Hydroxyl Radicals under Atmospheric Conditions," *Atmos. Environ., 16,* 545–550 (1982).

Beichert, P., L. Wingen, J. Lee, R. Vogt, M. J. Ezell, M. Ragains, R. Neavyn, and B. J. Finlayson-Pitts, "Rate Constants for the Reactions of Chlorine Atoms with Some Simple Alkanes at 298 K: Measurement of a Self-Consistent Set Using both Absolute and Relative Rate Methods," *J. Phys. Chem., 99,* 13156–13162 (1995).

Beichert, P., and B. J. Finlayson-Pitts, "Knudsen Cell Studies of the Uptake of Gaseous HNO_3 and Other Oxides of Nitrogen on Solid NaCl: The Role of Surface-Adsorbed Water," *J. Phys. Chem., 100,* 15218–15228 (1996).

Benson, S. W., The *Foundations of Chemical Kinetics,* McGraw-Hill, New York, 1960.

Blake, D. R., and F. S. Rowland, "Urban Leakage of Liquified Petroleum Gas and Its Impact on Mexico City Air Quality," *Science, 269,* 953–956 (1995).

Bongartz, A., J. Kames, U. Schurath, Ch. George, Ph. Mirabel, and J. L. Ponche, "Experimental Determination of HONO Mass Accommodation Coefficients Using Two Different Techniques," *J. Atmos. Chem., 18,* 149–169 (1994).

Brown, R. L., "Tubular Flow Reactors with First-Order Kinetics," *J. Res. Natl. Bur. Stand., 83,* 1–8 (1978).

Caloz, F., F. F. Fenter, K. D. Tabor, and M. J. Rossi, "I: Design and Construction of a Knudsen-Cell Reactor for the Study of Heterogeneous Reactions over the Temperature Range 130–750 K, Performances and Limitations," *Rev. Sci. Instrum., 68,* 3172–3179 (1997).

Campbell, I. M., B. J. Handy, and R. M. Kirby, "Gas Phase Chain Reaction of $H_2O_2 + NO_2 + CO$," *J. Chem. Soc. Faraday Trans. 1, 71,* 867–874 (1975).

Campbell, I. M., and P. E. Parkinson, "Mechanism and Kinetics of the Chain Reaction in $H_2O_2 + NO_2 + CO$ Systems," *J. Chem. Soc., Faraday Trans. 1, 75,* 2048–2059 (1979).

Chang, D. P. Y., and R. C. Hill, "Retardation of Aqueous Droplet Evaporation by Air Pollutants," *Atmos. Environ., 14,* 803–807 (1980).

Cox, R. A., "The Photolysis of Gaseous Nitrous Acid—A Technique for Obtaining Kinetic Data on Atmospheric Photooxidation Reactions," *Int. J. Chem. Kinet. Symp., 1,* 379–398 (1975).

Danckwerts, P. V., "Absorption by Simultaneous Diffusion and Chemical Reaction into Particles of Various Shapes and into Falling Drops," *Trans. Faraday Soc., 47,* 1014–1023 (1951).

Danckwerts, P. V., *Gas–Liquid Reactions* (J. J. Carberry, M. S. Peters, W. R. Schowalter, and J. Wei, Eds.), Chemical Engineering Series, McGraw-Hill, New York, 1970.

Davidovits, P., J. H. Hu, D. R. Worsnop, M. S. Zahniser, and C. E. Kolb, "Entry of Gas Molecules into Liquids," *Faraday Discuss., 100,* 65–82 (1995).

De Haan, D. O., and B. J. Finlayson-Pitts, "Knudsen Cell Studies of the Reaction of Gaseous Nitric Acid with Synthetic Sea Salt at 298 K," *J. Phys. Chem. A, 101,* 9993–9999 (1997).

De Haan, D. O., T. Brauers, K. Oum, J. Stutz, T. Nordmeyer, and B. J. Finlayson-Pitts, "Heterogeneous Chemistry in the Troposphere: Experimental Approaches and Applications to the Chemistry of Sea Salt Particles," *Int. Rev. Phys. Chem.,* in press (1999).

DeMore, W. B., S. P. Sander, D. M. Golden, R. F. Hampson, M. J. Kurylo, C. J. Howard, A. R. Ravishankara, C. E. Kolb, and M. J. Molina, "Chemical Kinetics and Photochemical Data for Use in Stratospheric Modeling," in JPL Publication 97-4, Jet Propulsion Laboratory, Pasadena, California, January 15, 1997.

Donahue, N. M., J. S. Clark, K. L. Demerjian, and J. G. Anderson, "Free-Radical Kinetics at High Pressure: A Mathematical Analysis of the Flow Reactor," *J. Phys. Chem., 100,* 5821–5838 (1996a).

Donahue, N. M., K. L. Demerjian, and J. G. Anderson, "Reaction Modulation Spectroscopy: A New Approach to Quantifying Reaction Mechanisms," *J. Phys. Chem., 100,* 17855–17861 (1996b).

Donaldson, D. J., "Adsorption of Gases at the Air-Water Interface. I. NH_3," *J. Phys. Chem. A, 103,* 62–70 (1999).

Donaldson, D. J., J. A. Guest, and M. C. Goh, "Evidence for Adsorbed SO_2 at the Aqueous–Air Interface," *J. Phys. Chem., 99,* 9313–9315 (1995).

Donaldson, D. J., and D. Anderson, "Adsorption of Atmospheric Gases at the Air–Water Interface. 2. C_1–C_4 Alcohols, Acids and Acetone," *J. Phys. Chem. A, 103,* 871–876 (1999).

Eisenthal, K. B., "Photochemistry and Photophysics of Liquid Interfaces by Second Harmonic Spectroscopy," *J. Phys. Chem., 100,* 12997–13006 (1996).

Fairbrother, D. H., H. Johnston, and G. Somorjai, "Electron Spectroscopy Studies of the Surface Composition in the H_2SO_4–H_2O Binary System," *J. Phys. Chem., 100,* 13696–13700 (1996).

Faubel, M., S. Schlemmer, and J. P. Toennies, "A Molecular Beam Study of the Evaporation of Water from a Liquid Jet," *Z. Phys. D: At., Mol. Clusters, 10,* 269–277 (1988).

Faubel, M., B. Steiner, and J. P. Toennies, "Photoelectron Spectroscopy of Liquid Water, Some Alcohols, and Pure Nonane in Free Micro Jets," *J. Chem. Phys., 106,* 9013–9031 (1997).

Fenter, F. F., F. Caloz, and M. J. Rossi, "Kinetics of Nitric Acid Uptake by Salt," *J. Phys. Chem., 98,* 9801–9810 (1994).

Fenter, F. F., F. Caloz, and M. J. Rossi, "II: Simulation of Flow Conditions in Low-Pressure Reactors (Knudsen Cells) Using a Monte-Carlo Technique," *Rev. Sci. Instrum., 68,* 3180–3186 (1997).

Finlayson-Pitts, B. J., M. J. Ezell, and C. E. Grant, "Temperature Dependence of the OH + ClNO Reaction: Evidence for Two Competing Reaction Channels," *J. Phys. Chem., 90,* 17–19 (1986).

Finlayson-Pitts, B. J., S. K. Hernandez, and H. N. Berko, "A New Dark Source of the Gaseous Hydroxyl Radical for Relative Rate Measurements," *J. Phys. Chem., 97,* 1172–1177 (1993).

Freiberg, J. E., and S. E. Schwartz, "Oxidation of SO_2 in Aqueous Droplets: Mass-Transport Limitation in Laboratory Studies and the Ambient Atmosphere," *Atmos. Environ., 15,* 1145–1154 (1981).

Fried, A., B. E. Henry, J. G. Calvert, and M. Mozurkewich, "The Reaction Probability of N_2O_5 with Sulfuric Acid Aerosols at Stratospheric Temperatures and Compositions," *J. Geophys. Res., 99,* 3517–3532 (1994).

Fuchs, N. A., and A. G. Sutugin, *Highly Dispersed Aerosols,* Ann Arbor Science Publishers, Inc., Ann Arbor, MI, 1970.

Fuchs, N. A., and A. G. Sutugin, "High-Dispersed Aerosols," in *Topics in Current Aerosol Research* (G. M. Hidy and J. R. Brock, Eds.), pp. 1–60, Pergamon, New York, 1971.

Fulle, D., H. F. Hamann, H. Hippler, and J. Troe, "High Pressure Range of Addition Reactions of HO. II. Temperature and Pressure Dependence of the Reaction HO + CO ↔ HOCO → H + CO_2," *J. Chem. Phys.,* 105, 983–1000 (1996).

Gardner, J. A., L. R. Watson, Y. G. Adewuyi, P. Davidovits, M. S. Zahniser, D. R. Worsnop, and C. E. Kolb, "Measurement of the Mass Accommodation Coefficient of $SO_2(g)$ on Water Droplets," *J. Geophys. Res.,* 92, 10887–10895 (1987).

Gill, P. S., T. E. Graedel, and C. J. Weschler, "Organic Films on Atmospheric Aerosol Particles, Fog Droplets, Cloud Droplets, Raindrops, and Snowflakes," *Rev. Geophys. Space Phys.,* 21, 903–920 (1983).

Golden, D. M., G. N. Spokes, and S. W. Benson, "Very Low-Presssure Pyrolysis (VLPP): A Versatile Kinetic Tool," *Angew. Chem., Int. Ed. Engl.,* 12, 534–546 (1973).

Golden, D. M., G. P. Smith, A. B. McEwen, C.-L. Yu, B. Eiteneer, M. Frenklach, G. L. Vaghjiani, A. R. Ravishankara, and F. P. Tully, "OH(OD) + CO: Measurements and an Optimized RRKM Fit," *J. Phys. Chem. A,* 102, 8598–8606 (1998).

Goss, K.-U., and R. P. Schwarzenbach, "Gas/Solid and Gas/Liquid Partitioning of Organic Compounds: Critical Evaluation of the Interpretation of Equilibrium Constants," *Environ. Sci. Technol.,* 32, 2025–2032 (1998).

Graedel, T. E., and C. J. Weschler, "Chemistry within Aqueous Atmospheric Aerosols and Raindrops," *Rev. Geophys. Space Phys.,* 19, 505–539 (1981).

Grosjean, D., E. Grosjean, and E. L. Williams, II, "Thermal Decomposition of PAN, PPN, and Vinyl-PAN," *J. Air Waste Manage. Assoc.,* 44, 391–396 (1994).

Hanson, D. R., and A. R. Ravishankara, "Comment on Porosities of Ice Films Used to Simulate Stratospheric Cloud Surfaces-Response," *J. Phys. Chem.,* 97, 2802–2803 (1993a).

Hanson, D. R., and A. R. Ravishankara, "Uptake of HCl and HOCl onto Sulfuric Acid: Solubilities, Diffusivities, and Reaction," *J. Phys. Chem.,* 97, 12309–12319 (1993b).

Hanson, D. R., and A. R. Ravishankara, "Reactive Uptake of $ClONO_2$ onto Sulfuric Acid Due to Reaction with HCl and H_2O," *J. Phys. Chem.,* 98, 5728–5735 (1994).

Hanson, D. R., A. R. Ravishankara, and S. Solomon, "Heterogeneous Reactions in Sulfuric Acid Aerosols: A Framework for Model Calculations," *J. Geophys. Res.,* 99, 3615–3629 (1994).

Hanson, D. R., "Surface-Specific Reactions on Liquids," *J. Phys. Chem. B,* 101, 4998–5001 (1997).

Hanson, D. R., "Reaction of $ClONO_2$ with H_2O and HCl in Sulfuric Acid and $HNO_3/H_2SO_4/H_2O$ Mixtures," *J. Phys. Chem. A,* 102, 4794–4807 (1998).

Hemminger, J. C., "Heterogeneous Chemistry in the Troposphere: A Modern Surface Chemistry Approach to the Study of Fundamental Processes," *Int. Rev. Phys. Chem.,* in press (1999).

Howard, C. J., "Kinetic Measurements Using Flow Tubes," *J. Phys. Chem.,* 83, 3–8 (1979).

Hu, J. H., Q. Shi, P. Davidovits, D. R. Worsnop, M. S. Zahniser, and C. E. Kolb, "Reactive Uptake of $Cl_2(g)$ and $Br_2(g)$ by Aqueous Surfaces as a Function of Br^- and I^- Ion Concentration: The Effect of Chemical Reaction at the Interface," *J. Phys. Chem.,* 99, 8768–8776 (1995).

Husar, R. B., and W. R. Shu, "Thermal Analysis of the Los Angeles Smog Aerosol," *J. Appl. Meteorol.,* 14, 1558–1565 (1975).

Jayne, J. T., S. X. Duan, P. Davidovits, D. R. Worsnop, M. S. Zahniser, and C. E. Kolb, "Uptake of Gas-Phase Aldehydes by Water Surfaces," *J. Phys. Chem.,* 96, 5452–5460 (1992).

Jet Propulsion Laboratory, "Chemical Kinetics and Photochemical Data for Use in Stratospheric Modeling," Evaluation #12, JPL Publication 97-4, 1997.

Karlsson, R., and E. Ljungström, "Nitrogen Dioxide and Sea Salt Particles—A Laboratory Study," *J. Aerosol Sci.,* 26, 39–50 (1995).

Kaufman, F., "Kinetics of Elementary Radical Reactions in the Gas Phase," *J. Phys. Chem.,* 88, 4909–4917 (1984).

Keyser, L. T., "High-Pressure Flow Kinetics. A Study of the OH + HCl Reaction from 2 to 100 Torr," *J. Phys. Chem.,* 88, 4750–4758 (1984).

Keyser, L. F., M.-T. Leu, and S. B. Moore, "Comment on Porosities of Ice Films Used to Simulate Stratospheric Cloud Surfaces," *J. Phys. Chem.,* 97, 2800–2801 (1993).

Kolb, C. E., D. R. Worsnop, M. S. Zahniser, P. Davidovits, L. F. Keyser, M.-T. Leu, M. J. Molina, D. R. Hanson, A. R. Ravishankara, L. R. Williams, and M. A. Tolbert, "Laboratory Studies of Atmospheric Heterogeneous Chemistry," in *Progress and Problems in Atmospheric Chemistry,* (J. R. Barker, Ed.), Chap. 18, Advanced Series in Physical Chemistry (C.-Y. Ng, Ed.), Vol. 3, World Scientific, Singapore, 1995.

Kolb, C. E., J. T. Jayne, D. R. Worsnop, and P. Davidovits, "Solubility Data Requirements and New Experimental Methods in Atmospheric Aerosol Research," *Pure Appl. Chem.,* 69, 959–968 (1997).

Laidler, K. J., and J. H. Meiser, *Physical Chemistry,* Benjamin/Cummings, Menlo Park, CA, 1982.

Lambert, M., C. M. Sadowski, and T. Carrington, "Uses of the Transit Time Distribution in Kinetic Flow Systems," *Int. J. Chem. Kinet.,* 7, 685–708 (1985).

Langer, S., R. S. Pemberton, and B. J. Finlayson-Pitts, "Diffuse Reflectance Infrared Studies of the Reaction of Synthetic Sea Salt Mixtures with NO_2: A Key Role for Hydrates in the Kinetics and Mechanism," *J. Phys. Chem. A,* 101, 1277–1286 (1997).

Laux, J. M., J. C. Hemminger, and B. J. Finlayson-Pitts, "X-ray Photoelectron Spectroscopic Studies of the Heterogeneous Reaction of Gaseous Nitric Acid with Sodium Chloride: Kinetics and Contribution to the Chemistry of the Marine Troposphere," *Geophys. Res. Let.,* 21, 1623–1626 (1994).

Laux, J. M., T. F. Fister, B. J. Finlayson-Pitts, and J. C. Hemminger, "X-ray Photoelectron Spectroscopy Studies of the Effects of Water Vapor on Ultrathin Nitrate Layers on NaCl," *J. Phys. Chem.,* 100, 19891–19897 (1996).

Leu, M.-T., R. S. Timonen, L. F. Keyser, and Y. L. Yung, "Heterogeneous Reactions of $HNO_{3(g)} + NaCl_{(s)} \rightarrow HCl_{(g)} + NaNO_{3(s)}$ and $N_2O_{5(g)} + NaCl_{(s)} \rightarrow ClNO_{2(g)} + NaNO_{3(s)}$," *J. Phys. Chem.,* 99, 13203–13212 (1995).

Lovejoy, E. R., and D. R. Hanson, "Measurement of the Kinetics of Reactive Uptake by Submicron Sulfuric Acid Particles," *J. Phys. Chem.,* 99, 2080–2087 (1995).

McMurry, P. H., and M. R. Stolzenburg, "Mass Accommodation Coefficients from Penetration Measurements in Laminar Tube Flow," *Atmos. Environ.,* 21, 1231–1234 (1987).

Miller, D. F., D. Lamb, and A. W. Gertler, "SO_2 Oxidation in Cloud Drops Containing NaCl or Sea Salt as Condensation Nuclei," *Atmos. Environ.,* 21, 991–993 (1987).

Molina, M. J., L. T. Molina, and D. M. Golden, "Environmental Chemistry (Gas and Gas–Solid Interactions): The Role of Physical Chemistry," *J. Phys. Chem.,* 100, 12888–12896 (1996).

Motz, H., and H. Wise, "Diffusion and Heterogeneous Reaction. III. Atom Recombination at a Catalytic Boundary," *J. Chem. Phys.,* 32, 1893–1894 (1960).

Mozurkewich, M., J. J. Lamb, and S. W. Benson, "Negative Activation Energies and Curved Arrhenius Plots. 2. OH + CO," *J. Phys. Chem.,* 88, 6435–6441 (1984).

Mulcahy, M. F. R., *Gas Kinetics,* Halsted, New York, 1973.

Nathanson, G. M., P. Davidovits, D. R. Worsnop, and C. E. Kolb, "Dynamics and Kinetics at the Gas–Liquid Interface," *J. Phys. Chem.,* 100, 13007–13020 (1996).

Nicovich, J. M., and P. H. Wine, personal communication (1996).

Nielsen, O. J., and J. Sehested, "Upper Limits for the Rate Constants of the Reactions of CF_3O_2 and CF_3O Radicals with Ozone at 295 K," *Chem. Phys. Lett., 213*, 433–441 (1993).

O'Keefe, A., and D. A. G. Deacon, "Cavity Ring-Down Optical Spectrometer for Absorption Measurements Using Pulsed Laser Sources," *Rev. Sci. Instrum, 59*, 2544–2551 (1988).

Oum, K. W., M. J. Lakin, D. O. De Haan, T. Brauers, and B. J. Finlayson-Pitts, "Formation of Molecular Chlorine from the Photolysis of Ozone and Aqueous Sea-Salt Particles," *Science, 279*, 74–77 (1998).

Pagsberg, P., O. J. Nielsen, and C. Anastasi, "Gas Phase Studies in Atmospheric Chemistry Using Pulse Radiolysis and Transient Absorption Spectroscopy," in *Spectroscopy in Environmental Science* (R. J. H. Clark and R. E. Hester, Eds.), Chap. 6, Wiley, 1995.

Pankow, J. F., "An Absorption Model of Gas/Particle Partitioning of Organic Compounds in the Atmosphere," *Atmos. Environ., 28*, 185–188 (1994a).

Pankow, J. F., "An Absorption Model of the Gas/Aerosol Partitioning Involved in the Formation of Secondary Organic Aerosol," *Atmos. Environ., 28*, 189–193 (1994b).

Pankow, J. F., "Partitioning of Semi-Volatile Organic Compounds to the Air/Water Interface," *Atmos. Environ., 31*, 927–929 (1997).

Paul, J. B., and R. J. Saykally, "Cavity Ringdown Laser Absorption Spectroscopy," *Anal. Chem., 69*, A287–A292 (1997).

Pipino, A. C. R., J. W. Hudgens, and R. E. Huie, "Evanescent Wave Cavity Ring-Down Spectroscopy for Probing Surface Processes," *Chem. Phys. Lett., 280*, 104–112 (1997).

Pitts, J. N., Jr., A. M. Winer, D. R. Fitz, S. M. Aschmann, and R. Atkinson, "Experimental Protocol for Determining Ozone Reaction Rate Constants," U.S. Environmental Protection Agency, Report No. EPA-600/S3-81-024, May 1981.

Ponche, J. L., Ch. George, and Ph. Mirabel, "Mass Transfer at the Air/Water Interface: Mass Accommodation Coefficients of SO_2, HNO_3, NO_2, and NH_3," *J. Atmos. Chem., 16*, 1–21 (1993).

Quinlan, M. A., C. M. Reihs, D. M. Golden, and M. A. Tolbert, "Heterogeneous Reactions on Model Polar Stratospheric Cloud Surfaces: Reaction of N_2O_5 on Ice and Nitric Acid Trihydrate," *J. Phys. Chem., 94*, 3255–3260 (1990).

Robinson, G. N., D. R. Worsnop, J. T. Jayne, C. E. Kolb, E. Swartz, and P. Davidovits, "Heterogeneous Uptake of HCl by Sulfuric Acid Solutions," *J. Geophys. Res., 103*, 25371–25381 (1998).

Scherer, J. J., J. B. Paul, A. O'Keefe, and R. J. Saykally, "Cavity Ringdown Laser Absorption Spectroscopy: History, Development, and Application to Pulsed Molecular Beams," *Chem. Rev., 97*, 25–51 (1997).

Schwartz, S. E., and J. E. Freiberg, "Mass-Transport Limitation to the Rate of Reaction of Gases in Liquid Droplets: Application to Oxidation of SO_2 in Aqueous Solutions," *Atmos. Environ., 15*, 1129–1144 (1981).

Schwartz, S. E. "Gas–Aqueous Reactions of Sulfur and Nitrogen Oxides in Liquid Water Clouds," in *SO_2, NO, and NO_2 Oxidation Mechanisms: Atmospheric Considerations* (J. G. Calvert, Ed.), pp. 173–208 and references therein, Acid Precipitation Series, Vol. 3 (J. I. Teasley, Series Ed.), Butterworth, Stoneham, MA, 1984a.

Schwartz, S. E., "Mass-Transport Considerations Pertinent to Aqueous Phase Reactions of Gases in Liquid-Water Clouds," *NATO ASI Series, G6*, 416–471 (1986), and in *Chemistry of Multiphase Atmospheric Systems* (W. Jaeschke, Ed.), pp. 415–471, Springer-Verlag, New York, 1986.

Schweitzer, F., L. Magi, P. Mirabel, and C. George, "Uptake Rate Measurements of Methanesulfonic Acid and Glyoxal by Aqueous Droplets," *J. Phys. Chem. A, 102*, 593–600 (1998).

Seeley, J. V., J. T. Jayne, and M. J. Molina, "High Pressure Fast-Flow Technique for Gas Phase Kinetics Studies," *Int. J. Chem. Kinet., 25*, 571–594 (1993).

Shi, B., and J. H. Seinfeld, "On Mass Transport Limitation to the Rate of Reaction of Gases in Liquid Droplets," *Atmos. Environ., 25A*, 2371–2383 (1991).

Shorter, J. A., W. J. De Bruyn, J. Hu, E. Swartz, P. Davidovits, D. R. Worsnop, M. S. Zahniser, and C. E. Kolb, "Bubble Column Apparatus for Gas–Liquid Heterogeneous Chemistry Studies," *Environ. Sci. Technol., 29*, 1171–1178 (1995).

Somorjai, G. A., *Introduction to Surface Chemistry and Catalysis*, Wiley, New York, 1994.

Stelson, A. W., and J. H. Seinfeld, "Chemical Mass Accounting of Uban Aerosol," *Environ. Sci. Technol., 15*, 671–679 (1981).

Stickel, R. E., J. M. Nicovich, S. Wang, Z. Zhao, and P. H. Wine, "Kinetic and Mechanistic Study of the Reaction of Atomic Chlorine with Dimethyl Sulfide," *J. Phys. Chem., 96*, 9875–9883 (1992).

Swartz, E., J. Boniface, I. Tchertkov, O. V. Rattigan, D. V. Robinson, P. Davidovits, D. R. Worsnop, J. T. Jayne, and C. E. Kolb, "Horizontal Bubble Train Apparatus for Heterogeneous Chemistry Studies: Uptake of Gas-Phase Formaldehyde," *Environ. Sci. Technol., 31*, 2634–2641 (1997).

Taylor, R. S., L. X. Dang, and B. C. Garrett, "Molecular Dynamics Simulations of the Liquid/Vapor Interface of SPC/E Water," *J. Phys. Chem., 100*, 11720–11725 (1996).

Taylor, R. S., D. Ray, and B. C. Garrett, "Understanding the Mechanism for the Mass Accommodation of Ethanol by a Water Droplet," *J. Phys. Chem. B, 101*, 5473–5476 (1997).

Troe, J., "Predictive Possibilities of Unimolecular Rate Theory," *J. Phys. Chem., 83*, 114–126 (1979).

Troe, J., "Theory of Thermal Unimolecular Reactions in the Fall-Off Range. Strong Collision Rate Constants," *Ber. Bunsenges. Phys. Chem., 87*, 161–169 (1983).

Tuazon, E. C., W. P. L. Carter, R. Atkinson, and J. N. Pitts, Jr., "The Gas Phase Reaction of Hydrazine and Ozone: A Nonphotolytic Source of OH Radicals for Measurement of Relative OH Radical Rate Constants," *Int. J. Chem. Kinet., 15*, 619–629 (1983).

Vogt, R., and B. J. Finlayson-Pitts, "A Diffuse Reflectance Infrared Fourier Transform Spectroscopic (DRIFTS) Study of the Surface Reaction of NaCl with Gaseous NO_2 and HNO_3," *J. Phys. Chem., 98*, 3747–3755 (1994); *J. Phys. Chem., 99*, 13052 (1995).

Vogt, R., C. Elliott, H. C. Allen, J. M. Laux, J. C. Hemminger, and B. J. Finlayson-Pitts, "Some New Laboratory Approaches to Studying Tropospheric Heterogeneous Reactions," *Atmos. Environ., 30*, 1729–1737 (1996).

Wallington, T. J., A. Guschin, T. N. N. Stein, J. Platz, J. Sehested, L. K. Christensen, and O. J. Nielsen, "Atmospheric Chemistry of $CF_3CH_2OCH_2CF_3$: UV Spectra and Kinetic Data for $CF_3CH(\cdot)OCH_2CF_3$ and $CF_3CH(OO\cdot)OCH_2CF_3$ Radicals and Atmospheric Fate of $CF_3CH(O\cdot)OCH_2CF_3$ Radicals," *J. Phys. Chem. A, 102*, 1152–1161 (1998).

Worsnop, D. R., M. S. Zahniser, C. E. Kolb, J. A. Gardner, L. R. Watson, J. M. Van Doren, J. T. Jayne, and P. Davidovits, "Temperature Dependence of Mass Accommodation of SO_2 and H_2O_2 on Aqueous Surfaces," *J. Phys. Chem., 93*, 1159–1172 (1989).

Zellner, R., and H. Herrmann, "Free Radical Chemistry of the Aqueous Atmospheric Phase," in *Spectroscopy in Environmental Science* (R. J. H. Clark and R. E. Hester, Eds.), Chapter 9, pp. 381–451, Wiley, New York, 1995.

Zetzsch, C., and W. Behnke, "Heterogeneous Photochemical Sources of Atomic Chlorine in the Troposphere," *Ber. Bunsenges. Phys. Chem., 96*, 488–493 (1992).

Zhang, R., J. T. Jayne, and M. J. Molina, "Heterogeneous Interactions of $ClONO_2$ and HCl with Sulfuric Acid Tetrahydrate: Implications for the Stratosphere," *J. Phys. Chem., 98*, 867–874 (1994).

Zhu, L., and G. Johnston, "Kinetics and Products of the Reaction of the Vinoxy Radical with O_2," *J. Phys. Chem., 99*, 15114–15119 (1995).

CHAPTER

6

Rates and Mechanisms of Gas-Phase Reactions in Irradiated Organic – NO_x – Air Mixtures

The myriad organics found in the troposphere, in combination with a variety of potential oxidizing species, can easily conjure up a nightmare in terms of the chemistry and kinetics needed to fully understand the gas-phase chemistry of organic–NO_x mixtures in the lower atmosphere. Fortunately, through studies over the past three decades, a great deal has been learned regarding which oxidants are important for each class of organics. One can use this information to examine the most important types of reactions, which once understood, can be readily extrapolated to other organic compounds.

Let us first review the oxidants that have been recognized for some time as important in the troposphere, as well as atomic chlorine, for which there is increasing evidence of a contribution in marine regions.

A. SOURCES OF OXIDANTS IN THE TROPOSPHERE: OH, O_3, NO_3, HO_2, AND Cl

The major recognized oxidants for organics in the troposphere are OH and O_3, with a contribution from NO_3 at night. The hydroperoxyl free radical, HO_2, as we shall see, also reacts readily with aldehydes but does not significantly impact tropospheric chemistry due to a rapid reverse reaction. It is, of course, responsible for converting NO to NO_2, ultimately leading to the production of O_3, as well as a whole host of compounds included under the umbrella of NO_y. The HO_2 + NO reaction also generates OH, so understanding the sources of HO_2 is important in understanding sources of OH. Because there is increasing evidence for the production of atomic chlorine in marine areas (although its source is unknown), we shall briefly discuss this as well.

1. OH

The major source of OH in remote areas is the photolysis of O_3 to electronically excited $O(^1D)$, followed by its reaction with water vapor:

$$O_3 + h\nu(\lambda \lesssim 336 \text{ nm}) \rightarrow O(^1D) + O_2, \quad (1)$$

$$O(^1D) + H_2O \rightarrow 2OH, \quad (2a)$$

$$O(^1D) \xrightarrow{M} O(^3P). \quad (2b)$$

It should be noted that only a portion of the $O(^1D)$ formed generates OH via reaction (2a); the remainder is deactivated to ground-state $O(^3P)$, reaction (2b), which then re-forms O_3. For example, at 50% RH and 300 K at the earth's surface, about 10% of the $O(^1D)$ formed generates OH. As a result, as discussed later in this chapter, the relative importance of (2a) decreases at higher altitudes due to the decrease in water vapor. This is also an important source in polluted areas, where, however, there are additional sources as well. These include the photolysis of gaseous nitrous acid (HONO) and hydrogen peroxide (H_2O_2):

$$HONO + h\nu(\lambda < 400 \text{ nm}) \rightarrow OH + NO, \quad (3)$$
$$H_2O_2 + h\nu(\lambda < 370 \text{ nm}) \rightarrow 2OH. \quad (4)$$

In addition, in the presence of NO concentrations larger than ~10 ppt, sources of HO_2 are, in effect, sources of OH:

$$HO_2 + NO \rightarrow OH + NO_2. \quad (5)$$

Because most of the OH sources are photolytic in nature (exceptions being production via the reactions of O_3 with alkenes, free radicals produced by the thermal decomposition of compounds such as PAN or peroxynitric acid, HO_2NO_2, or the nighttime reactions of NO_3), OH is a major oxidant primarily during daylight hours.

2. O_3

Tropospheric O_3 is known to be formed upon photolysis of NO_2 in air, first suggested by Blacet (1952):

$$NO_2 + h\nu(\lambda \leq 420 \text{ nm}) \rightarrow NO + O(^3P), \quad (6)$$

$$O(^3P) + O_2 \xrightarrow{M} O_3. \quad (7)$$

While elevated NO_x levels are clearly associated with anthropogenic emissions (see Chapter 2), there are also small concentrations due to natural processes. Hence small concentrations of O_3 are also formed via the reactions of natural VOC and NO_x discussed later in this chapter. Finally, the periodic intrusion of stratospheric air with its relatively high concentrations of O_3 provides an additional source of tropospheric ozone.

Although the photolysis of NO_2 is the major source of O_3, ozone is sufficiently long-lived that it can be transported downwind and survive into the nighttime hours. Hence it is a player in tropospheric chemistry throughout the day and night.

3. NO_3

The nitrate radical is formed by the reaction of NO_2 with O_3:

$$NO_2 + O_3 \rightarrow NO_3 + O_2. \quad (8)$$

As discussed in Chapter 4, NO_3 only exists in sufficient concentrations to play a role in nighttime chemistry, due to its strong absorption of light in the visible and subsequent photodissociation.

4. HO_2

Any reactions that produce H or HCO in the troposphere act as HO_2 sources:

$$H + O_2 \xrightarrow{M} HO_2, \quad (9)$$

$$HCO + O_2 \rightarrow HO_2 + CO. \quad (10)$$

Thus, formaldehyde photolysis is a major source of HO_2 during the day:

$$HCHO + h\nu(\lambda \leq 370 \text{ nm}) \rightarrow H + HCO. \quad (11)$$

The photolysis of higher aldehydes, RCHO, also forms HCO (see Chapter 4) and hence HO_2:

$$RCHO + h\nu \rightarrow R + HCO. \quad (12)$$

However, as discussed in Chapter 4, the absorption spectrum of higher aldehydes cuts off at shorter wavelengths than formaldehyde. This, combined with higher quantum yields for radical production in the 290- to 340-nm range and the fact that HCHO produces $2HO_2$ essentially immediately upon dissociation, makes the photolysis of aldehydes larger than formaldehyde less important at equal concentrations of the aldehydes.

The reactions of some alkoxy radicals generated in the VOC oxidation sequence, for example those of CH_3O and C_2H_5O radicals discussed in more detail later, with O_2 also generate HO_2:

$$RCH_2O + O_2 \rightarrow RCHO + HO_2. \quad (13)$$

The alkoxy radical originates in the oxidation of a VOC to an alkyl radical by any of the oxidants described here or, alternatively, in the thermal decomposition of species such as PAN which gives the CH_3 radical. In short, HO_2 is a natural consequence of the oxidation of organics.

Finally, the decomposition of peroxynitric acid, HO_2NO_2, which is strongly temperature dependent, generates HO_2 directly:

$$HO_2NO_2 \leftrightarrow HO_2 + NO_2. \quad (14, -14)$$

5. Cl

In marine areas, wave action generates airborne droplets of seawater from which the water can evaporate, leaving a suspended particle of the dissolved solids. Because this is mainly NaCl, the possibility exists for the generation of atomic chlorine via the reactions of NaCl with gaseous species such as N_2O_5 or $ClONO_2$ (see reviews by Finlayson-Pitts, 1993; Graedel and Keene, 1995; Andreae and Crutzen, 1997; Finlayson-Pitts and Pitts, 1997; De Haan et al., 1999; and Hemminger, 1999), e.g.,

$$N_2O_{5(g)} + NaCl_{(s)} \rightarrow ClNO_{2(g)} + NaNO_{3(s)}, \quad (15)$$

$$ClONO_{2(g)} + NaCl_{(s)} \rightarrow Cl_{2(g)} + NaNO_{3(s)}. \quad (16)$$

(Chlorine nitrate is formed from the reaction of ClO with NO_2.) These reactions also occur when NaCl is in the aqueous phase, in competition with the hydrolysis of N_2O_5 and $ClONO_2$, i.e., above the deliquescence point of NaCl in sea salt (e.g., Behnke et al., 1997). Photolysis then generates chlorine atoms, e.g.,

$$ClNO_2 + h\nu \rightarrow Cl + NO_2, \quad (17)$$

$$Cl_2 + h\nu \rightarrow 2Cl. \quad (18)$$

It is likely that there are as yet ill-defined aqueous-phase reactions in the airborne seawater droplets that release photochemically labile chlorine gases. For example, Oum et al. (1998a) have shown that Cl_2 is formed when sea salt aerosols above their deliquescence point are irradiated at 254 nm in the presence of O_3, generating OH which initiates Cl^- oxidation.

Such reactions may also be important in other situations in the troposphere. For example, Shaw (1991) has observed salt particles as far as 900 km inland in Alaska, and chloride salts are used on many roads in cold climates in the wintertime. In addition, in the plumes from oil well burning in Kuwait, salt particles were observed, due to the brine that was mixed with the oil in the wells (e.g., see Cahill et al., 1992).

Direct evidence for the potential importance of Cl as an organic oxidant comes from recent measurements of inorganic chlorine-containing species other than HCl in the marine troposphere in midlatitudes (Keene et al., 1993; Pszenny et al., 1993). In particular, Cl_2 has been identified using atmospheric pressure ionization mass spectrometry (API-MS) in a coastal region (Spicer et al., 1998). Interestingly, the concentrations, up to 150 ppt, are much higher than can be explained by any known chemistry, again highlighting the contribution of some as yet unidentified chemistry in the marine boundary layer. During the day, any Cl_2 formed will absorb strongly in the 300- to 400-nm region (Chapter 4), and dissociate, generating atomic chlorine.

Indirect evidence for the involvement of atomic chlorine in the chemistry of marine atmospheres comes from the measurement of simple organics, where their relative rates of decay frequently cannot be matched assuming attack only by OH (Wingenter et al., 1996). Estimates of the peak concentrations of atomic chlorine range from $\sim 10^3$ to 10^6 radicals cm^{-3} in the marine boundary layer (e.g., Pszenny et al., 1993; Singh et al., 1996a). However, on a global scale, the concentrations are likely much smaller. For example, Rudolph et al. (1996) and Singh et al. (1996b) have examined the budget for tetrachloroethene (TCE), which reacts relatively rapidly with Cl compared to OH. The measured atmospheric concentrations of TCE are consistent with the known emissions and removal solely by reaction with OH, from which an upper limit for the *global annual average* Cl atom concentrations was estimated to be $\leq 10^3$ atoms cm^{-3}. However, most of it is in the marine boundary layer (MBL) so that these averaged values may not be inconsistent with peak MBL concentrations of 10^4–10^6 atoms cm^{-3}.

In short, while there is evidence that atomic chlorine is generated from sea salt reactions and contributes to organic oxidations in the marine boundary layer, the nature and strength of the sources remain to be elucidated.

B. LIFETIMES OF TYPICAL ORGANICS IN THE TROPOSPHERE

To pare the list of VOC oxidations down to the most important processes, we can calculate the effective lifetimes of organics with respect to reactions with each of the oxidants listed in the previous section. Since these natural lifetimes are defined as $\tau = 1/k_p[X]$, we also need to assume an average concentration for the oxidant, [X]. We can therefore take a typical organic from each of the major classes (alkane, alkene, aromatic, etc.) and compare the individual lifetimes for reaction with OH, O_3, NO_3, etc. Those reactions having very long lifetimes are insignificant with respect to their contribution to tropospheric chemistry and hence can be ignored for the purposes of this discussion.

Table 6.1 shows such a set of calculated lifetimes for the oxidants discussed in Section A. The most significant reactions are as follows:

Alkanes: OH, to a lesser extent NO_3, and in the marine boundary layer (MBL) Cl

Alkenes: OH, O_3, NO_3, and to a lesser extent Cl (MBL)

Alkynes: OH, Cl (MBL)

Aromatics: OH, Cl (MBL)

Aldehydes: OH, NO_3, (HO_2), Cl (MBL)

TABLE 6.1 Estimated Lifetimes of Representative Organics in the Troposphere[a]

Organic	OH (1×10^6 cm^{-3})	O_3 (100 ppb)	NO_3 (50 ppt)	HO_2 (2×10^8 cm^{-3}, 8 ppt)	Cl (1×10^4 cm^{-3})
n-Butane	5 days	≥ 1300 yr	205 days		5 days
trans-2-Butene	4.3 h	36 min	35 min		~4 days
Acetylene	14 days	≥ 400 days	≥ 188 days		~22 days[c]
Toluene	2 days	≥ 400 days	138 days[d]		20 days
HCHO	1.2 days	≥ 463 days	16 days	18/h[b]	16 days

[a] $\tau = 1/k_p$[oxidant] = time for the organic to fall to $1/e$ of its initial value; except as shown here, rate constants are found in text.

[b] Note: This is only for the forward reaction. Since the adduct decomposes back to reactants under most atmospheric conditions. The effective atmospheric lifetime is much longer.

[c] Based on k (Cl + C_2H_2) = 5.3×10^{-11} cm^3 molecule^{-1} s^{-1} from $k_0 = 5.7 \times 10^{-30}$ cm^6 molecule^{-2} s^{-1}, $k_\infty = 2.3 \times 10^{-10}$ cm^3 molecule^{-1} s^{-1} and $F_c = 0.6$ (Atkinson et al., 1997a).

[d] Using $k = 6.8 \times 10^{-17}$ cm^3 molecule^{-1} s^{-1} (Atkinson, 1994).

The HO_2–aldehyde reaction is in parentheses because, as we shall see later, it is a reversible reaction that is sufficiently fast in the reverse direction under typical tropospheric conditions that no overall reaction, in effect, occurs.

Each of these reactions, both the kinetics and mechanisms, are discussed in the remainder of this chapter.

C. REACTIONS OF ALKANES

1. Hydroxyl Radical (OH)

The hydroxyl radical has a strong tendency to abstract a hydrogen atom whenever possible, forming the thermodynamically stable water molecule. In the case of alkanes, the reaction is therefore

$$OH + RH \rightarrow R + H_2O. \quad (19)$$

Table 6.2 summarizes rate constants for some OH–alkane reactions; for recent recommendations for other alkanes, see Atkinson (1994, 1997a) and Atkinson *et al.* (1997a).

The temperature dependence is given in the form $k = BT^n e^{-C/T}$, where n is usually taken as 2, except for CH_4. This fit procedure provides curvature in the Arrhenius plot similar to that which is observed. Donahue *et al.* (1998a) propose an alternate form of the

TABLE 6.2 Rate Constants and Temperature Dependence[a,b] for Reaction of OH Radicals with Some Alkanes

Alkane	k (10^{-12} cm^3 molecule^{-1} s^{-1}) at 298 K	B (10^{-18} cm^3 molecule^{-1} s^{-1})	C (K)	n	D (10^{-9} K cm^3 s^{-1})	F (K)
Methane	0.00618	0.0965	1082	2.58		
Ethane	0.254	15.2	498	2	1.24	1042
Propane	1.12	15.5	61	2	1.32	616
n-Butane	2.44	16.9	−145	2	1.68	456
2-Methylpropane	2.19	11.6	−225	2	0.75	257
n-Pentane	4.0	24.4	−183	2	2.46	414
2-Methylbutane	3.7					
2,2-Dimethylpropane	0.85	18.0	189	2		
n-Hexane	5.45	15.3	−414	2	2.10	284
2-Methylpentane	5.3					
3-Methylpentane	5.4					
2,3-Dimethylbutane	5.8	12.4	−494	2		
n-Heptane	7.0	15.9	−478	2		
2,2-Dimethylpentane	3.4					
2,2,3-Trimethylbutane	4.2	8.5	−516	2		
n-Octane	8.7	27.6	−378	2		
2,2,4-Trimethylpentane	3.6	20.8	−196	2		
2,2,3,3-Tetramethylbutane	1.05	19.1	144	2		
n-Nonane	10.0	25.1	−447	2		
n-Decane	11.2	31.3	−416	2		
n-Undecane	12.9					
n-Dodecane	13.9					
n-Tridecane	16					
n-Tetradecane	18					
n-Pentadecane	21					
n-Hexadecane	23					
Cyclopropane	0.084					
Cyclobutane	1.5					
Cyclopentane	5.02 (4.8)[c]	25.7	−235	2	1.97	253
Cyclohexane	7.21 (7.2)[c]	28.8	−309	2	2.36	227
Cycloheptane	13				4.25	256
Methylcyclohexane	10 (9.4)[c]					

[a] The parameters B and C give the temperature dependence in the form $k = BT^n e^{-C/T}$, where $C = E_a/R$. From Atkinson (1997a).

[b] The parameters D and $F = E_a/R$ are for the temperature dependence in the form recommended by Donahue *et al.* (1998a): $k(T) = De^{-F/T}/[T(1 - e^{-1.44v_1/T})^2(1 - e^{-1.44v_2/T})]$, where two bends at $v_1 = 300$ cm^{-1} and one bend at $v_2 = 500$ cm^{-1} are treated explicitly.

[c] Kramp and Paulson (1998).

temperature dependence consistent with a simplified form of transition state theory:

$$k(T) = \frac{De^{-F/T}}{T(1 - e^{-1.44\nu_1/T})^2(1 - e^{-1.44\nu_2/T})},$$

where ν_1 is the degenerate C–H–O bend frequency (cm^{-1}) and ν_2 is the H–O–H bend frequency (cm^{-1}). Table 6.2 also shows the parameters D and $F = E_a/R$ for this form of the temperature dependence using $\nu_1 = 300$ cm^{-1} and $\nu_2 = 500$ cm^{-1}.

The first thing that stands out in Table 6.2 is that the OH–CH$_4$ rate constant, 6.2×10^{-15} cm^3 molecule^{-1} s^{-1}, is much smaller than those for the higher alkanes, a factor of 40 below that for ethane. This relatively slow reaction between OH and CH$_4$ is the reason that the focus is on "non-methane hydrocarbons" (NMHC) in terms of ozone control in urban areas. Thus, even at a typical peak OH concentration of 5×10^6 molecules cm^{-3}, the calculated lifetime of CH$_4$ at 298 K is 373 days, far too long to play a significant role on urban and even regional scales. Clearly, however, this reaction is important in the global troposphere (see Chapter 14.B.2b).

Second, the room temperature rate constants increase with increasing size and complexity of the alkane and are of the order of 10^{-11} cm^3 molecule^{-1} s^{-1} for the largest alkanes. To put this in perspective, a diffusion-controlled reaction, i.e., one that occurs on every collision of the reactants, is of the order of $\sim (3-5) \times 10^{-10}$ cm^3 molecule^{-1} s^{-1}. Thus for the larger alkanes, reaction occurs in approximately one in 10 collisions, which is quite a fast process.

As discussed in Chapter 5, kinetic theories predict that the preexponential factor should have a temperature dependence that manifests itself in curved Arrhenius plots if the reactions are studied over a sufficiently broad temperature range. This is the case for OH–alkane reactions, where there has been great interest in the high-temperature kinetics for combustion systems. Table 6.2 also shows the temperature dependence for the OH reactions in the form $k = BT^n e^{-C/T}$, where $C = E_a/R$ and in the form recommended by Donahue et al. (1998a).

The C–H bond strength is largest for primary C–H bonds at ~ 101 kcal mol^{-1}, decreasing to ~ 98 kcal mol^{-1} for secondary and ~ 96 kcal mol^{-1} for tertiary C–H bonds (Lide, 1998–1999). Hence one expects that, all else being equal, a tertiary C–H will react faster than a secondary C–H, which in turn will react faster than a primary C–H. Greiner (1970), whose measurements of the absolute rate constants for OH reactions in the mid-1960s provided the first clue of the potential importance of OH in the troposphere, suggested that the rate constant for the overall reaction, k^{tot}, could be treated as the sum of contributions from each type of abstractable hydrogen in the following manner:

$$k^{\text{tot}} = N_p k_p + N_s k_s + N_t k_t. \quad (A)$$

k_p, k_s, and k_t represent rate constants for the abstraction of primary, secondary, and tertiary hydrogens, respectively, and N_p, N_s, and N_t are the corresponding numbers of each kind of hydrogen. If the rate constants (k^{tot}) for a number of simple alkanes are known, the experimental data can be fit to obtain best values for k_p, k_s, and k_t. These can then be used to predict the rate constant for the reaction of OH with an alkane where experimental measurements have not been made.

This type of structure–reactivity relationship (SRR) works reasonably well for the simple alkanes. However, clearly one would expect that the nature of adjacent groups would also have an effect on the rate constant, albeit a smaller one than the type of C–H bond. A variant of the approach in Eq. (A) is to use rate constants per primary, secondary, or tertiary *group* modified by factors reflecting the adjacent groups. In this case,

$$k'_p(\text{CH}_3\text{X}) = k_p^* F(\text{X}),$$
$$k'_s(\text{CH}_2\text{XY}) = k_s^* F(\text{X})F(\text{Y}),$$
$$k'_t(\text{CHXYZ}) = k_t^* F(\text{X})F(\text{Y})F(\text{Z}),$$

where k_p^*, k_s^*, and k_t^* are the rate constants for the CH$_3$– *group*, the –CH$_2$– *group*, and the >CH– *group*, respectively, and the F factors reflect how the rate constants for the individual groups are modified by adjacent groups, X, Y, and Z. Recommended values (Kwok and Atkinson, 1995; Atkinson, 1997a) of these group rate constants are $k_p^* = 1.36 \times 10^{-13}$, $k_s^* = 9.34 \times 10^{-13}$, and $k_t^* = 1.94 \times 10^{-12}$ cm^3 molecule^{-1} s^{-1} at 298 K. The modifying F parameters at 298 K are taken as 1.00 for X = CH$_3$ and 1.23 for all of the other simple alkyl groups, –CH$_2$–, >CH–, and >C<. Other correction factors must be included for cyclic compounds (Atkinson, 1997a).

For example, the rate constant for the reaction of OH with 2-methylbutane,

$$\text{OH} + \text{CH}_3-\underset{\underset{\text{CH}_3}{|}}{\overset{\overset{\text{H}}{|}}{\text{C}}}-\text{CH}_2\text{CH}_3 \rightarrow \text{H}_2\text{O} + \text{R},$$

can be estimated as follows:

$$k^{\text{tot}} = 2[k_p^* F(>\text{CH}-)]$$
$$+ k_t^*[F(\text{CH}_3-)F(\text{CH}_3-)F(-\text{CH}_2-)]$$

$$+ k_s^*[F(>CH-)F(CH_3-)] + k_p^*F(-CH_2-)$$
$$= 2(1.36 \times 10^{-13})[1.23]$$
$$+ (1.94 \times 10^{-12})[1.0 \times 1.0 \times 1.23]$$
$$+ (9.34 \times 10^{-13})[1.23 \times 1.00]$$
$$+ (1.36 \times 10^{-13})(1.23)$$
$$= (0.335 + 2.39 + 1.15 + 0.167) \times 10^{-12}$$
$$= 4.0 \times 10^{-12} \text{ cm}^3 \text{ molecule}^{-1} \text{ s}^{-1}.$$

This is within 10% of the recommended value (Table 6.2) of 3.7×10^{-12} cm³ molecule⁻¹ s⁻¹.

A similar approach can be used for estimating rate constants with a variety of other organics such as alkenes, alcohols, and nitrates, although the agreement is not as good in many cases (Kwok and Atkinson, 1995).

In addition to this type of empirical approach, there are several other approaches that are related more directly to specific properties of the organic, such as the C–H bond dissociation enthalpies (Heicklen, 1981; Jolly et al., 1985), ionization energy (Gaffney and Levine, 1979), or NMR shifts (Hodson, 1988). In addition, molecular orbital calculations (Klamt, 1993) and transition state theory (Cohen and Benson, 1987) have been applied.

2. Nitrate Radical (NO_3)

Like OH, the nitrate radical also abstracts a hydrogen atom, forming nitric acid and an alkyl radical:

$$NO_3 + RH \rightarrow HNO_3 + R. \quad (20)$$

The kinetics and mechanisms of nitrate radical reactions with alkanes and a variety of other organics relevant to the atmosphere are discussed in detail in two excellent reviews by Wayne et al. (1991) and Atkinson (1991). The kinetics of the NO_3–alkane reactions are summarized in Table 6.3, where it can be seen that, with the exception of methane, they are in the range 10^{-18}–10^{-16} cm³ molecule⁻¹ s⁻¹.

While these reactions are much slower than the corresponding OH reactions, the nighttime peak concentrations of NO_3 under some conditions are much larger than those of OH during the day, ~400 ppt vs 0.4 ppt. Even given the differences in concentration, however, as seen from the lifetimes in Table 6.1, the nitrate radical reaction is still relatively slow. While the removal of the alkanes by NO_3 is thus not expected to be very significant under most tropospheric conditions, reaction (20) can contribute to HNO_3 formation and the removal of NO_x from the atmosphere.

The rate constants for the simple alkanes can be empirically fit by assigning rate constants to the primary and secondary C–H groups, along with sub-

TABLE 6.3 Rate Constants at Room Temperature for Reaction of NO_3 Radicals with Alkanes[a]

Alkane	k (10^{-17} cm³ molecule⁻¹ s⁻¹)
Methane	< 0.1
Ethane	0.14[b]
Propane	1.7[b]
n-Butane	4.59
2-Methylpropane	10.6
n-Pentane	8.7
2-Methylbutane	16
n-Hexane	11
2,3-Dimethylbutane	44
Cyclohexane	14
n-Heptane	15
n-Octane	19
n-Nonane	23

[a] From Atkinson (1997a).
[b] Estimated from group contributions as described in text.

stituent factors F used to modify these rate constants. The best fit to the data at 298 K are found for the following values (Atkinson, 1997a):

$$k_p^* = 7.0 \times 10^{-19}, \quad F(CH_3-) = 1.00,$$
$$k_s^* = 1.22 \times 10^{-17}, \quad F(-CH_2-) = 1.67.$$

However, this approach is not accurate for branched chain alkanes (Atkinson, 1997).

3. Chlorine Atoms (Cl)

Table 6.4 summarizes the rate constants for the reactions of chlorine atoms with alkanes. Structure-reactivity relationships have again been developed for

TABLE 6.4 Rate Constants for the Reactions of Cl Atoms with Alkanes[a]

Alkane	$k^{298 K}$ (10^{-11} cm³ molecule⁻¹ s⁻¹)	A (10^{-12} cm³ molecule⁻¹ s⁻¹)	E_a/R (K)
Methane	0.010	9.6	1350
Ethane	5.9	81	95
Propane	13.7	120	−40
n-Butane	21.8	218	0
Isobutane	14.3		
n-Pentane	28		
n-Hexane	34		
n-Heptane	39		
n-Octane	46		
n-Nonane	48		
n-Decane	55		

[a] From Atkinson (1997a); temperature dependence is in form $k = Ae^{-E_a/RT}$.

298 K (Atkinson, 1997a):

$k_p^* = 3.5 \times 10^{-11}$, $F(CH_3-) = 1.00$,

$k_s^* = 9.3 \times 10^{-11}$, $F(-CH_2-) = F(>CH-)$
$= F(>C<) = 0.79$,

$k_t^* = 6.8 \times 10^{-11}$.

The reaction involves abstraction of a hydrogen atom to form HCl:

$$Cl + RH \rightarrow HCl + R. \quad (21)$$

D. REACTIONS OF ALKYL (R), ALKYLPEROXY (RO_2), AND ALKOXY (RO) RADICALS IN AIR

As already seen, the reactions of OH, NO_3, and Cl with alkanes generate alkyl radicals. We shall see in the subsequent sections of this chapter that the production of alkyl radicals of various types is a general characteristic of organic oxidations. Here we will trace the atmospheric fates of typical alkyl radicals formed in OH, NO_3, and Cl atom oxidations of alkanes. However, the principles are general and can be applied to those formed by the reactions of other organics such as alkenes as well.

1. Alkyl Radicals (R)

The only fate of alkyl radicals in air is reaction with O_2:

$$R + O_2 \xrightarrow{M} RO_2. \quad (22)$$

For the case where R is a methyl group, the reaction is in the falloff region at 1 atm, and for C_2H_5, it is close to the high-pressure limit. For C_3H_7 and above, a value for the effective second-order rate constant, k_∞, of $\sim(0.8-2) \times 10^{-11}$ cm^3 molecule^{-1} s^{-1} means that the lifetime of an alkyl radical at $\sim$1 atm in air is $\sim$10 ns.

A second possible channel, in which the alkene and HO_2 are formed from decomposition of the excited RO_2^* intermediate before it is stabilized, is very small; thus, this channel is < 1% for $C_2H_5 + O_2 \rightarrow (C_2H_5O_2)^* \rightarrow C_2H_4 + HO_2$ and < 0.13% for $C_3H_7 + O_2 \rightarrow (C_3H_7O_2)^* \rightarrow C_3H_6 + HO_2$ at 1 atm pressure and room temperature but becomes important at higher temperatures, e.g., in combustion systems (Kaiser, 1995; Benson, 1996; Kaiser and Wallington, 1996a).

2. Alkylperoxy Radicals (RO_2)

Alkylperoxy radicals (RO_2) in the atmosphere react primarily with NO, HO_2, and other RO_2. Reaction with the nitrate radical, NO_3, at night, has been recently recognized as being important as well. A less important reaction is that with NO_2. There are two excellent reviews of RO_2 chemistry by Lightfoot et al. (1992) and Wallington et al. (1992), including the thermochemistry, spectroscopy, kinetics, and mechanisms. The reader should consult these for references to the original work in this area.

a. Reactions with NO

The reactions of RO_2 with NO are quite fast and do not vary significantly with the nature of the alkyl group (Wallington et al., 1992; Lightfoot et al., 1992; Masaki et al., 1994; Maricq and Szente, 1996; Eberhard and Howard, 1996, 1997). The recommended rate constants at 298 K are 7.5×10^{-12} cm^3 molecule^{-1} s^{-1} for R = CH_3 and $(8-9) \times 10^{-12}$ cm^3 molecule^{-1} s^{-1} for all other R groups (Atkinson, 1997a; Eberhard and Howard, 1997).

The major path produces alkoxy radicals (RO) and NO_2:

$$RO_2 + NO \rightarrow RO + NO_2. \quad (23a)$$

There is a second path for the larger RO_2 radicals corresponding to addition of the NO and followed by isomerization to form an alkyl nitrate:

$$RO_2 + NO \rightarrow RONO_2. \quad (23b)$$

The mechanism is postulated to be (Darnall et al., 1976)

$$RO_2 + NO \rightarrow (ROONO)^* \rightarrow RO + NO_2, \quad (23a)$$
$$\downarrow$$
$$(RONO_2)^* \xrightarrow{M} RONO_2. \quad (23b)$$

Hence one would expect the second channel to depend on the size of the radical and to show some pressure dependence as well as a negative temperature dependence, all of which are found to be the case.

Table 6.5 shows the fraction of the total reaction forming the stable alkyl nitrate, i.e., the branching ratio $k_{23b}/(k_{23a} + k_{23b})$ for some simple alkylperoxy radicals at 1 atm and room temperature. The ratio is smaller for primary and tertiary RO_2; in general, the ratio k_{23b}/k_{23a} for primary and tertiary alkylperoxy radicals is about 40 and 30%, respectively, of that for secondary RO_2 (Atkinson, 1997a).

It can be seen from the data in Table 6.5 that the fraction of the total RO_2 + NO reaction that forms the stable alkyl nitrate at 1 atm can be substantial, as much as 35% for the larger radicals. (However, more recent measurements by Aschmann *et al.* (1999) give lower nitrate yields for the hexyl, heptyl, and octyl alkylperoxy radicals, ranging from 15% for the sum of (2-hexyl + 3-hexyl) to 24% for the sum of (2-octyl + 3-octyl + 4-octyl RO_2.) It is thought that such reactions may be significant sources of such alkyl nitrates and perhaps form part of the "missing NO_y" (see Section E.3).

TABLE 6.5 Yields of $RONO_2$ in RO_2 + NO Reactions at Room Temperature and 1 atm[a]

R	Branching ratio = $k_{23b} / (k_{23a} + k_{23b})$
Ethane	
Ethyl	≤0.014
Propane	
1-Propyl	0.020
2-Propyl	0.05
n-Butane	
1-Butyl	≤0.04
2-Butyl	0.083
Isobutane	
2-Methyl-1-propyl	0.075
tert-Butyl	0.18
n-Pentane	
1-Pentyl	0.06
2-Pentyl	0.13
3-Pentyl	0.12
Isopentane	
2-Methyl-1-butyl	0.040
2-Methyl-2-butyl	0.044–0.056
2-Methyl-3-butyl	0.074–0.15
3-Methyl-1-butyl	0.043
n-Pentane	
n-Pentyl	0.51
n-Hexane	
1-Hexyl	0.12
2-Hexyl	0.22[b]
3-Hexyl	0.22[b]
2-Methylpentane	
2-Methyl-2-pentyl	0.035
3-Methylpentane	
3-Methyl-2-pentyl	0.14–0.16
n-Heptane	
1-Heptyl	0.20
2-Heptyl	0.32[b]
3-Heptyl	0.31[b]
4-Heptyl	0.29[b]
n-Octane	
1-Octyl	0.36
2-Octyl	0.35[b]
3-Octyl	0.34[b]
4-Octyl	0.32[b]

[a] Adapted from Lightfoot *et al.* (1992).
[b] Aschmann *et al.* (1999) have measured 0.15 for the sum of (2-hexyl + 3-hexyl), 0.19 for (2-heptyl + 3-heptyl + 4-heptyl), and 0.24 for (2-octyl + 3-octyl + 4-octyl).

There is evidence that in some cases, the alkoxy radical formed in the RO_2 + NO reaction contains sufficient excess energy that it can decompose under atmospheric conditions. This is the case, for example, for some of the alkoxy radicals formed in the oxidation of alternate CFCs (see Chapter 13.D.2a). It has also been postulated for the alkoxy radical formed from the NO reaction with $HOCH_2CH_2O_2$, formed in the OH + C_2H_4 reaction (Orlando *et al.*, 1998). In the latter case, about 25% of the excited $(HOCH_2CH_2O)^*$ decomposes to HCHO + CH_2OH, with the remainder being stabilized. The stabilized radicals then decompose to HCHO + CH_2OH or react with O_2.

b. Reactions with HO_2 and RO_2

Alternate fates of RO_2 are reactions with HO_2 or with other RO_2 radicals:

$$RO_2 + HO_2 \rightarrow ROOH + O_2, \quad (24a)$$
$$\rightarrow \text{Carbonyl compound} + H_2O + O_2, \quad (24b)$$
$$\rightarrow ROH + O_3, \quad (24c)$$
$$RO_2 + R'O_2 \rightarrow \text{Products}. \quad (25)$$

The reaction of simple alkylperoxy radicals with HO_2 is believed to occur primarily by path (24a) to form the hydroperoxide, although for more complex RO_2 radicals there is some evidence for some contribution of the other paths. The rate constant (k_{24}) at room temperature is 5.2×10^{-12} for R = CH_3, increasing to 7.7×10^{-12} for C_2H_5 and $\sim 1.5 \times 10^{-11}$ cm^3 $molecule^{-1}$ s^{-1} for the larger radicals (Lightfoot *et al.*, 1992; Fenter *et al.*, 1993; Atkinson, 1997a).

The self-reaction of RO_2 radicals, i.e., RO_2 + RO_2, or their reaction with alkylperoxy radicals of different structure, i.e., RO_2 + $R'O_2$ as shown in reaction (25), is complex. In principle, one expects an excited intermediate $(ROOOOR)^*$ to be formed, which could decompose by a number of different paths; i.e., the overall reaction could be one of the following paths:

$$RO_2 + RO_2 \rightarrow 2RO + O_2, \quad (25a)$$
$$\rightarrow ROH + RCHO + O_2, \quad (25b)$$
$$\rightarrow ROOR + O_2. \quad (25c)$$

The term "branching ratio" is used to describe the relative importance of each path. Thus the branching ratio for (25a) is the fraction of the total reaction proceeding by this path and is given by $k_{25a}/(k_{25a} + k_{25b} + k_{25c})$. Clearly, the sum of the branching ratios for a reaction must add up to 1.0.

Table 6.6 gives some of the values of the recommended rate constants for the self-reaction, k_{25}, and

TABLE 6.6 Recommended Rate Constants and Branching Ratios at Room Temperature for the Self-Reactions of Some RO_2 Radicals[a]

RO_2	k_{25}^{298K} (cm^3 molecule^{-1} s^{-1})	Branching ratios		
		(25a) (2RO + O$_2$)	(25b) (ROH + RCHO + O$_2$)	(25c) (ROOR + O$_2$)
CH_3O_2	3.7×10^{-13}	0.33 ± 0.05[b]	~ 0.67[f]	Minor
		0.30 ± 0.08[c]	~ 0.70[f]	
		0.41 ± 0.04[d]		< 0.006[d]
$HOCH_2CH_2O_2$	2.3×10^{-12}	0.50[e]	0.50	
$C_2H_5O_2$	6.4×10^{-14}	0.63 ± 0.06[b,c]	0.32[g]	0.05[c]
$n\text{-}C_3H_7O_2$	3×10^{-13}			
$i\text{-}C_3H_7O_2$	1×10^{-15}	0.56	0.44	0
$HOCH(CH_3)CH(CH_3)O_2$	6.7×10^{-13e}	~ 0.2[e]		
$t\text{-}C_4H_9O_2$	2×10^{-17}			
$HOC(CH_3)_2C(CH_3)_2O_2$	4×10^{-15e}	1.0[e]		

[a] Rate constants from IUPAC recommendations (Atkinson et al., 1997a).
[b] Lightfoot et al. (1992).
[c] Wallington et al. (1992).
[d] Tyndall et al. (1998).
[e] Boyd and Lesclaux (1997).
[f] Calculated assuming path c is negligible.
[g] Calculated from the branching yields of the other two paths.

where there are available data, the branching ratios. For the simple alkylperoxy radicals, the contribution of path (c) to form a peroxide and O_2 is generally small, whereas both paths (a) and (b) contribute significantly at room temperature. The self-reaction of primary RO_2 occurs with the highest rate constant, $\sim 10^{-13}$, that of secondary RO_2 is slower at $\sim 10^{-15}$, and that of tertiary RO_2 is even slower at 10^{-17} cm^3 molecule^{-1} s^{-1}. Kirchner and Stockwell (1996) have parameterized the rate constants for the self-reactions in the form

$$k \text{ (cm}^3 \text{ molecule}^{-1} \text{ s}^{-1}) = 2 \times 10^{-14} \exp\left[3.8A - 5\alpha + \frac{3N}{1 + 0.02N^2}\right],$$

where $A = 0$ for simple alkyl radicals and 1 if there are additional oxygen atoms in the alkyl group, α is the number of R or RO groups attached to the $-COO$ group, and N is the number of carbon atoms in the RO_2. This gives values within about a factor of three of measured rate constants and can be used for $N \leq 7$. For larger RO_2, the value calculated with $N = 7$ is recommended since the rate constants do not increase significantly beyond that. This formulation is not appropriate for use for peroxy radicals derived from the reactions of OH with terminal alkenes or for those with functional groups containing oxygen attached to the terminal carbon, e.g., $(HO)(R_1)(R_2)COO$ (Kirchner and Stockwell, 1996).

Atkinson (1997a) recommends values of 2.5×10^{-13}, 5×10^{-15}, and 2×10^{-17} cm^3 molecule^{-1} s^{-1} for the self-reaction of primary, secondary, and tertiary RO_2 radicals, respectively, with uncertainties of at least a factor of five. For the cross-reaction of two different radicals R_aO_2 and R_bO_2, the recommendation is $k \sim 2(k_a k_b)^{0.5}$, based on Madronich and Calvert (1990), where k_a and k_b are the self-reaction rate constants for each of the two radicals. This generally gives rate constants within a factor of two of the experimental value for the cases where such rate constants are available. However, in some cases when the RO_2 self-reaction is fast, e.g., for $CH_3C(O)O_2$, the cross combination reactions with CH_3O_2 or $C_2H_5O_2$ also tend to be fast (e.g., Villenave and Lesclaux, 1996).

c. Reactions with NO_3

Simple alkyl peroxy radicals have been shown to react with NO_3, e.g., for CH_3O_2:

$$CH_3O_2 + NO_3 \rightarrow CH_3O + NO_2 + O_2. \quad (26)$$

The rate constant is in the range of $(1-3) \times 10^{-12}$ cm^3 molecule^{-1} s^{-1} at 298 K (e.g., Biggs et al., 1994a, 1994b; Kukui et al., 1995; Daële et al., 1995), similar to that for the $CH_3C(O)O_2$ reaction with NO_3 (4×10^{-12} cm^3 molecule^{-1} s^{-1}; Canosa-Mas et al., 1996). Although there are few data on the range of possible NO_3^- peroxy radical reactions, as discussed in more detail later, it appears that such reactions may be very important at night (e.g., Platt et al., 1990; Canosa-Mas et al., 1996; Kirchner and Stockwell, 1996).

d. Reactions with NO_2

Alkylperoxy radicals can also react with NO_2 in a three-body reaction to form a peroxynitrate, RO_2NO_2:

$$RO_2 + NO_2 \xrightarrow{M} ROONO_2.$$

This reaction is in the falloff region at 1 atm for $R = CH_3$ and C_2H_5 but for larger radicals is effectively in the high-pressure limit with a rate constant $\sim 9 \times 10^{-12}$ cm^3 molecule^{-1} s^{-1}. The reaction is reversible in that the peroxynitrates thermally decompose in the reverse of this reaction. As we shall see, this reaction of RO_2 with NO_2 is not significant compared to its reactions with NO, HO_2, or other RO_2 under most tropospheric conditions.

e. Fate of RO_2 under Typical Tropospheric Conditions

To compare the relative importance of these potential atmospheric fates of RO_2 under typical polluted conditions, and particularly the relative importance of the NO reaction, let us take the $C_2H_5O_2$ radical as an example. The lifetime of $C_2H_5O_2$ with respect to reaction with NO, HO_2, or $C_2H_5O_2$ at peak concentrations of 20 ppb, 40 ppt, and 40 ppt, respectively, can then be calculated from $\tau = 1/k[X]$ as 0.2, 1.3×10^2 s and 1.6×10^4 s, respectively. (Note that the NO and HO_2/RO_2 peaks will not occur simultaneously.) At night, with an NO_3 concentration of 100 ppt, the lifetime would be ~ 135 s, assuming a rate constant of 3×10^{-12} cm^3 molecule^{-1} s^{-1}. In short, in areas impacted by anthropogenic emissions, the reaction of RO_2 with NO will predominate.

One can then ask under what conditions the reaction of RO_2 with NO will be equivalent in rate to that with HO_2. To make this calculation, one can estimate the concentration of NO at which $k_{23}[NO] = k_{24}[HO_2]$. Since $k_{23} \sim k_{24} \sim 1 \times 10^{-11}$ cm^3 molecule^{-1} s^{-1}, this suggests that the removal of RO_2 by HO_2 will compete with that by NO when the concentrations of these two, HO_2 and NO, are equal. Typical peak HO_2 concentrations in polluted areas are believed to be of the order of 10^9 cm^{-3}, corresponding to ~ 40 ppt. "Average" HO_2 concentrations may more typically be $(1-2) \times 10^8$ cm^{-3}, corresponding to concentrations of 4–8 ppt. Such equivalent small concentrations of NO are found in the remote troposphere and, under such conditions, reactions of RO_2 with HO_2 or other RO_2 can become quite important (see Section J.2b).

In addition, the reaction of RO_2 with NO_3 at night can become significant. For example, at HO_2 of 4 ppt and NO_3 of 10 ppt, which have been reported in a relatively clean marine region (Carslaw et al., 1997), the lifetimes of RO_2 relative to reaction with these species are comparable at $\sim 10^3$ s. Kirchner and Stockwell (1996) predict that in more polluted areas, the $RO_2 + NO_3$ reactions may be important. For example, under conditions chosen to be representative of aged polluted air masses mixing with rural air masses, they calculate that 77% of the total RO_2 at night reacts with NO_3; overall (both day and night), however, only 0.66% is removed by reaction with NO_3.

The relative magnitude of the $RO_2 + NO$ reaction compared to the $RO_2 + HO_2$ or RO_2 reactions is a critical factor in ozone formation in the troposphere. As discussed in more detail in Section J, if $RO_2 + NO$ predominates, NO_2 is formed and through its photolysis to $O(^3P)$, O_3 is ultimately generated. On the other hand, the $RO_2 + HO_2/RO_2$ reactions can lead to the formation of stable products such as ROOH, without conversion of an NO to NO_2. In this case, no O_3 is formed and indeed, through its photolysis to form OH radicals and subsequent reactions with HO_2 and OH, destruction of O_3 occurs.

3. Alkoxy Radicals (RO)

The reactions of RO_2 with NO and with RO_2 generate alkoxy radicals (RO). Alkoxy radicals have several possible atmospheric fates, depending on their particular structure. These include reaction with O_2, decomposition, and isomerization; as we shall see, reactions with NO and NO_2 are unlikely to be important under most tropospheric conditions. Atkinson et al. (1995b) and Atkinson (1997b) have reviewed reactions of alkoxy radicals and β-hydroxyalkyl radicals:

- Reaction with O_2: If the carbon to which the alkoxy oxygen is attached also has a hydrogen, this H can be abstracted to give HO_2 and a carbonyl compound:

$$\cdot O - \overset{H}{\underset{|}{C}} - \xrightarrow[HO_2]{O_2} O = C \diagup \qquad (27)$$

This reaction may proceed via the initial addition of O_2 to give a trioxy intermediate ($-C-OOO\cdot$) followed by rearrangement and elimination of HO_2 (e.g., Jungkamp et al., 1997). Atkinson (1997a, 1997b) recommends rate constants of 9.5×10^{-15} and 8×10^{-15} cm^3 molecule^{-1} s^{-1} at 298 K for primary RCH_2O and secondary R_1R_2CHO radicals, respectively.

- Scission of a C–C bond:

$$CH_3 \overset{a}{\underset{}{\}}} \underset{H}{\overset{O}{\underset{|}{C}}} \overset{b}{\underset{}{\}}} CH_2CH_3 \xrightarrow{a} CH_3 + \underset{H}{\overset{O}{\diagup}} CCH_2CH_3$$

$$\downarrow b$$

$$CH_3C \underset{H}{\overset{O}{\diagup}} + CH_2CH_3 \qquad (28)$$

Scission to produce the larger alkyl radical dominates. For example, in the case of the 2-butoxy radical in Eq. (28), path (b) dominates over path (a) by more than two orders of magnitude (see Atkinson, 1997b, and references therein).

- If the alkoxy radical has four or more carbon atoms and can form a 6-membered transition state via the alkoxy oxygen abstracting H from a C–H bond, an intramolecular isomerization may occur:

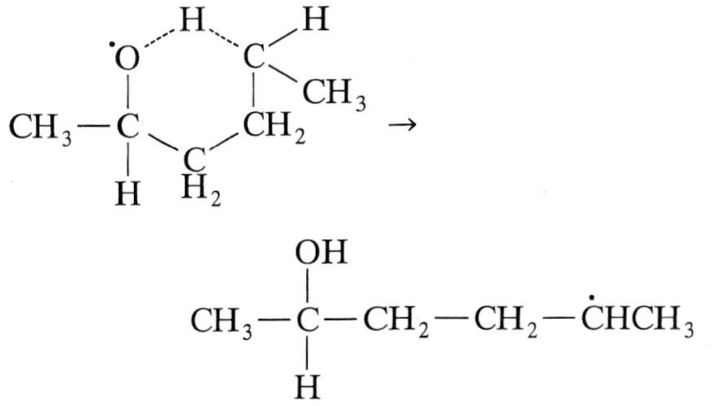

$$CH_3-\underset{H}{\overset{\cdot O \cdots H}{C}}\underset{H_2}{\overset{C}{-}}\underset{C}{\overset{CH_3}{-}}CH_2 \rightarrow$$

$$CH_3-\underset{H}{\overset{OH}{\underset{|}{C}}}-CH_2-CH_2-\dot{C}HCH_3 \quad (29)$$

Structure–reactivity relationships have also been developed for these isomerizations. For H abstraction from a CH_3–X, CH_2–XY, or CH–XYZ, respectively, through a 6-membered transition state, rate constants of 1.6×10^5, 1.6×10^6, and 4×10^6 s^{-1}, respectively, are recommended (Atkinson, 1997a, 1997b). These are corrected for neighboring group effects using $F(-CH_3) = 1.0$, $F(-CH_2-) = F(>CH-) = F(>C<) = 1.27$, and $F(-OH) = 4.3$.

- Reaction with NO or NO_2:

$$RO + NO \xrightarrow{M} RONO, \quad (30)$$

$$RO + NO_2 \xrightarrow{M} RONO_2. \quad (31)$$

The latter reactions with NO and NO_2 are less likely than the first three possibilities. With $k_{30} \sim k_{31} \sim 3 \times 10^{-11}$ cm^3 molecule^{-1} s^{-1} at 298 K and 1 atm pressure, the pseudo-first-order rate of reaction, k_{30}[NO] or k_{31}[NO$_2$], is ~ 75 s^{-1} at 100 ppb of NO or NO_2. For a species such as $(CH_3)_3CO^\cdot$ that can only decompose (and then only relatively slowly), these reactions could be responsible for $\sim 10\%$ of the alkoxy radical reaction at the relatively high NO and NO_2 concentrations of 100 ppb.

Table 6.7 compares the estimated rates of reaction with O_2, decomposition, and isomerization for some alkoxy radicals with different structures. It is important to recognize that there is a great deal of uncertainty in many of these estimates, and this is an area that clearly requires more research. However, given these caveats, it is clear that where isomerization is possible, it usually predominates at room temperature. (Note, however, that this will be slower at the lower temperatures found at higher altitudes.) When isomerization is not feasible, e.g., for the smaller alkoxy radicals or for branched species, reaction with O_2 is always significant and usually predominates.

As the size of the alkane increases, so does the relative importance of isomerization. For example, consider the competition between isomerization and reaction with O_2 for the n-pentoxy radical:

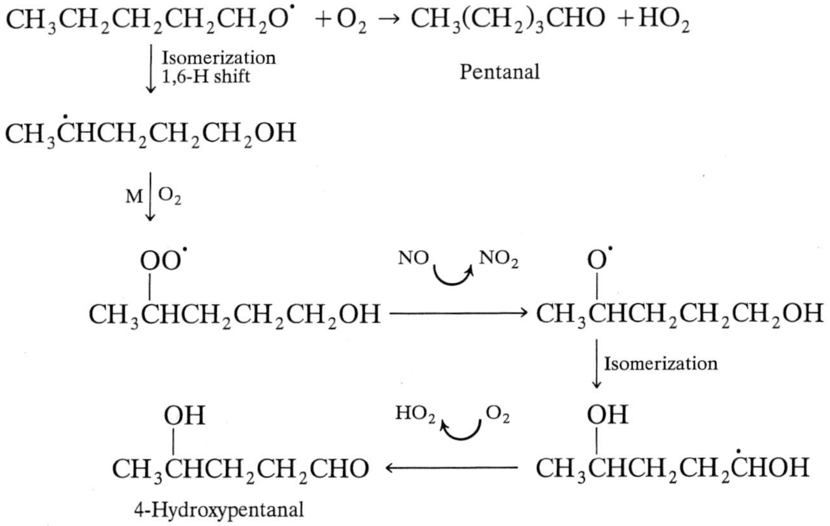

$$CH_3CH_2CH_2CH_2CH_2O^\cdot + O_2 \rightarrow CH_3(CH_2)_3CHO + HO_2$$

Isomerization 1,6-H shift ↓   Pentanal

$$CH_3\dot{C}HCH_2CH_2CH_2OH$$

↓ M, O_2

$$\underset{CH_3CHCH_2CH_2CH_2OH}{\overset{OO^\cdot}{|}} \xrightarrow{NO \curvearrowright NO_2} \underset{CH_3CHCH_2CH_2CH_2OH}{\overset{O^\cdot}{|}}$$

↓ Isomerization

$$\underset{CH_3CHCH_2CH_2CHO}{\overset{OH}{|}} \xleftarrow{HO_2 \curvearrowright O_2} \underset{CH_3CHCH_2CH_2\dot{C}HOH}{\overset{OH}{|}}$$

4-Hydroxypentanal

TABLE 6.7 Rates (s^{-1}) of Alkoxy Radical Reactions at 298 K and 1 atm Air[a]

RO	Decomposition	Reaction with O_2[b]	Isomerization
$CH_3O \cdot$	5.3×10^{-2}	1×10^4	
$C_2H_5O \cdot$	0.3	5×10^4	
$n\text{-}C_4H_9O \cdot$	5.8×10^2	5×10^4	2.0×10^5
$CH_3\text{—}\underset{\underset{O\cdot}{\vert}}{\overset{\overset{H}{\vert}}{C}}\text{—}CH_2CH_3$	2.3×10^4	4×10^4	
$(CH_3)_3CO \cdot$	1×10^3		
$CH_3\text{—}\underset{\underset{O\cdot}{\vert}}{\overset{\overset{H}{\vert}}{C}}\text{—}CH_2CH_2CH_3$	1.7×10^4	4×10^4	2×10^5
$CH_3CH_2\text{—}\underset{\underset{O\cdot}{\vert}}{\overset{\overset{H}{\vert}}{C}}\text{—}CH_2CH_3$	1.6×10^4	4×10^4	
$CH_3\text{—}\underset{\underset{O\cdot}{\vert}}{\overset{\overset{H}{\vert}}{C}}\text{—}(CH_2)_3CH_3$	2.8×10^4	4×10^4	2×10^6
$CH_3CH_2\text{—}\underset{\underset{O\cdot}{\vert}}{\overset{\overset{H}{\vert}}{C}}\text{—}(CH_2)_2CH_3$	3.4×10^4	4×10^4	2×10^5
$(CH_3)_3COCH_2O \cdot$	1.1×10^{-3}	3.8×10^6	2.0×10^5
$CH_3\text{—}\underset{\underset{CH_3}{\vert}}{\overset{\overset{CH_3}{\vert}}{C}}\text{—}CH_2O \cdot$	9.8×10^3	2.4×10^4	$\geq 7 \times 10^4$

[a] Adapted from Atkinson (1994, 1997a, 1997b) and Atkinson et al. (1995b).
[b] Shown as $k[O_2]$ (s^{-1}), based on recommended rate constants for RO + O_2 (Atkinson, 1997a).

The last step, forming 4-hydroxypentanal, which overall is net hydrogen abstraction from an α-hydroxy radical to form a carbonyl compound and HO_2, is characteristic of this type of radical; it is discussed later in more detail with respect to the reaction of the smallest α-hydroxy radical, CH_2OH, with O_2 to form HCHO + HO_2.

The relative importance of the two pathways is reflected in the ratio of the yields of the hydroxycarbonyls compared to the carbonyl products with the same number of carbons as the parent compound as well as in the absolute yields of the latter products. Thus the carbonyl product yields decrease from 70% for the n-butane reaction to 0.7% for the n-octane reaction; at the same time, the ratio of hydroxycarbonyl products to carbonyls increases from 0.14 to 50 (Kwok et al., 1996a). In short, isomerization becomes increasingly more important for the larger alkoxy radicals.

A study by Eberhard et al. (1995) illustrates the differences in the importance of the various paths that are associated with what may appear at first glance to be small changes in structure. They assessed the relative importance of the various reaction paths for the 2-hexoxy and 3-hexoxy radicals expected from the abstraction of a hydrogen in either the 2- or 3-position, respectively (e.g., by OH or NO_3), followed by addition of O_2 and reaction with NO. Based on the final product yields, they estimated the branching ratios for each possible reaction path. Figure 6.1 shows the possible reactions of each of these alkoxy radicals and the branching ratios they calculate, which are in reasonable agreement with the estimates shown in Table 6.7.

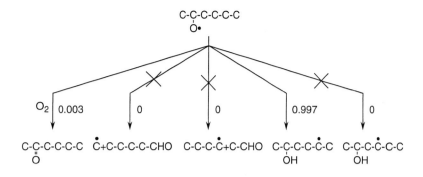

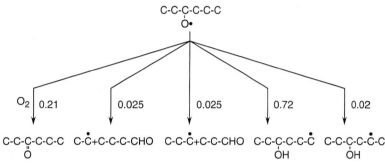

FIGURE 6.1 Possible fates of the 2-hexoxy and 3-hexoxy free radicals (adapted from Eberhard et al., 1995).

It should be noted that some alkoxy radicals derived from oxygenated compounds may have some unique chemistry in addition to these reactions. For example, Tuazon et al. (1998b) have shown that alkoxy radicals derived from OH + acetate reactions, i.e., $R_1C(O)OCH(O^{\cdot})R_2$, rearrange and decompose to RCO and the acid $RC(O)OH$.

4. Summary of R, RO_2, and RO Radical Reactions in the Troposphere

In summary, alkyl radicals, R, are all converted to alkylperoxy radicals, RO_2, in the troposphere. RO_2 reacts with NO, except in very remote regions where the concentration of NO is of the order of ~40 ppt or less. The reaction converts NO to NO_2 and hence acts as a source of O_3 via the NO_2 photolysis. For the larger alkylperoxy radicals, a substantial portion of the RO_2 + NO reaction also gives the stable alkyl nitrate, $RONO_2$.

As discussed in more detail in Section J.2.b, in remote regions where the NO concentrations are very low, reactions of RO_2 with other peroxy radicals such as HO_2 and RO_2 occur. Finally, under very high concentrations of NO_2, RO_2 may react with NO_2.

The alkoxy radical, RO, generated by RO_2 reactions can react with O_2, decompose, or undergo an isomerization, depending on the structure of the radical. At very high NO_x concentrations for a few alkoxy radicals, reactions with NO and NO_2 may also occur to a small extent. Figure 6.2 summarizes these possibilities.

E. REACTIONS OF ALKENES (INCLUDING BIOGENICS)

1. Hydroxyl Radical (OH)

As for other organics in the atmosphere, the OH radical is a major oxidant for alkenes. Table 6.8 gives the rate constants for some OH–alkene reactions as well as their temperature dependence in Arrhenius form. Several points are noteworthy: (1) the reactions are very fast, approaching 10^{-10} cm^3 molecule^{-1} s^{-1} for the larger alkenes; (2) the rate constants have a pressure dependence; (3) the apparent Arrhenius activation energies are "negative."

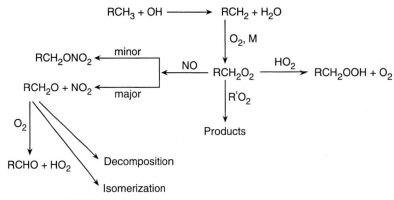

FIGURE 6.2 Summary of alkane oxidation by OH in air.

TABLE 6.8 Rate Constants and Temperature Dependence[a] for the Reactions of OH Radicals with Alkenes[e] at 1 atm Total Pressure of Air[b]

Alkene	k^c (10^{-12} cm^3 molecule^{-1} s^{-1})	A (10^{-12} cm^3 molecule^{-1} s^{-1})	E_a/R (K)
Ethene	8.52	1.96	−438
Propene	26.3	4.85	−504
1-Butene	31.4	6.55	−467
cis-2-Butene	56.4	11.0	−487
trans-2-Butene	64.0	10.1	−550
2-Methylpropene	51.4	9.47	−504
1-Pentene	31.4		
cis-2-Pentene	65		
trans-2-Pentene	67		
Cyclopentene	67		
3-Methyl-1-butene	31.8	5.32	−533
2-Methyl-1-butene	61		
2-Methyl-2-butene	86.9	19.2	−450
1-Hexene	37		
Cyclohexene	67.7		
1-Heptene	40		
trans-2-Heptene	68		
Cycloheptene	74		
1,3-Butadiene	66.6	14.8	−448
2-Methyl-1,3-butadiene (isoprene)	101	25.4	−410
Camphene	53		
2-Carene	80		
Limonene	171		
α-Phellandrene	313		
β-Phellandrene	168		
α-Pinene	53.7	12.1	−444
β-Pinene	78.9	23.8	−357
α-Terpinene	363		
γ-Terpinene	177		
Terpinolene	225		
Methyl vinyl ketone	18.8[d]		
Methacrolein	33.5[d]		

[a] $k = Ae^{-E_a/RT}$; valid only for the 250–425 K range.
[b] From Atkinson (1997a).
[c] High-pressure limiting rate constants (k_∞) except for C_2H_4 and C_3H_6.
[d] From Atkinson (1994).
[e] See Fig. 6.22 for structures of biogenics.

These observations are consistent with the primary reaction path being addition of OH to the double bond to form an adduct or intermediate that can decompose back to the reactants, or be stabilized:

$$\text{OH} + \underset{}{>}\text{C}=\text{C}\underset{}{<} \rightleftharpoons \left[\underset{}{>}\text{C}\overset{\text{OH}}{\underset{\vdots}{=}}\text{C}\underset{}{<} \right] \quad (32, -32)$$

$$\left[\underset{}{>}\text{C}\overset{\text{OH}}{\underset{\vdots}{=}}\text{C}\underset{}{<} \right] \xrightarrow{M} -\overset{\text{OH}}{\underset{|}{\text{C}}}-\dot{\text{C}}- \quad (33)$$

This mechanism has been confirmed by the mass spectrometric observation of the OH–alkene adducts themselves (e.g., Morris *et al.*, 1971; Hoyermann and Sievert, 1983). Only at low pressures for the smaller alkenes is decomposition back to reactants significant.

However, this does illustrate the importance of understanding the fundamental mechanisms in order to extrapolate to atmospheric conditions reliably. A number of experimental techniques used for studying gas-phase kinetics and mechanisms require low pressures and, under these conditions, decomposition of the OH–alkene adduct can predominate. As long as the fundamental mechanisms are understood and the kinetics determined as a function of pressure, extrapolation to atmospheric conditions is possible. Clearly, confirmation using studies at atmospheric pressure is also important.

Hydrogen atom abstraction can occur to a small extent, particularly with larger and more highly branched compounds. However, the contribution of this path is, overall, relatively small. For example, for the reaction with 3-methyl-1-butene, where there is a weaker allylic C–H bond, ~5–10% of the reaction proceeds by abstraction at 1 atm in air (Atkinson *et al.*, 1998).

For all but the two smallest alkenes, ethene and propene, the rate constants are at their high-pressure limits at 1 atm, and even for these two compounds, the effective rate constant is within ~10% of k_∞.

In the case of unsymmetrical alkenes, the OH radical can add to either end of the double bond. There is evidence that, as expected, it preferentially adds to form the secondary radical. For example, for the propene reaction (Cvetanovic, 1976), ~65% of the adducts formed correspond to (34a) and 35% to (34b):

$$\text{OH} + \text{CH}_3\text{CH}=\text{CH}_2 \rightarrow \text{CH}_3\dot{\text{C}}\text{H}-\text{CH}_2\text{OH}, \quad (34a)$$

$$\rightarrow \text{CH}_3\text{CH}(\text{OH})-\dot{\text{C}}\text{H}_2. \quad (34b)$$

Because the β-hydroxyalkyl radicals formed are substituted alkyl radicals, they react with O_2 to form alkylperoxy radicals, e.g.,

$$\text{CH}_3\dot{\text{C}}\text{H}-\text{CH}_2\text{OH} + \text{O}_2 \rightarrow \text{CH}_3\overset{\overset{\text{OO}^\bullet}{|}}{\text{CH}}-\text{CH}_2\text{OH}. \quad (35)$$

These β-hydroxyalkyperoxy radicals undergo the same reaction discussed earlier for RO_2 radicals, i.e., reaction primarily with NO,

$$\text{CH}_3\overset{\overset{\text{OO}^\bullet}{|}}{\text{CH}}-\text{CH}_2\text{OH} + \text{NO}$$

$$\rightarrow \text{CH}_3\overset{\overset{\text{O}^\bullet}{|}}{\text{CH}}-\text{CH}_2\text{OH} + \text{NO}_2 \quad (36a)$$

$$\rightarrow \text{CH}_3\overset{\overset{\text{ONO}_2}{|}}{\text{CH}}-\text{CH}_2\text{OH}, \quad (36b)$$

with rate constants that are about the same as those for simple alkylperoxy radicals. The hydroxy nitrate yields for some simple C_4–C_6 alkenes are ~2–6%, about half those for simple alkyl radicals (Table 6.5) (e.g., O'Brien *et al.*, 1998; Chen *et al.*, 1998). At low concentrations of NO, these RO_2 radicals also react with HO_2 and other RO_2 radicals (e.g., see Hatakeyama *et al.*, 1995; and Tuazon *et al.*, 1998a).

There is increasing evidence that for some reactions, the RO_2 + NO reaction produces a fraction of the alkoxy radicals with sufficient energy that they decompose immediately. For example, Orlando *et al.* (1998) observed that in the reaction of OH with C_2H_4, approximately 25% of the $HOCH_2CH_2O$ radicals generated in the reaction of $HOCH_2CH_2OO$ with NO decomposed before they could be collisionally stabilized. Similar observations have been made for RO_2 from the reactions of alternate CFCs (see Chapter 13).

The β-hydroxyalkoxy radicals such as that formed in (36a) can, in principle, react as discussed earlier, i.e., with O_2, decompose, or, for the larger radicals, undergo intramolecular isomerization. For the small alkenes ($\leq C_4$), decomposition appears to be the most important process (e.g., Tuazon *et al.*, 1998a). For example, for the β-hydroxyalkoxy radical formed in (36a), the primary fate is

$$\text{CH}_3\overset{\overset{\text{O}^\bullet}{|}}{\text{CH}}-\text{CH}_2\text{OH} \rightarrow \text{CH}_3\text{CHO} + \dot{\text{C}}\text{H}_2\text{OH}. \quad (37)$$

In the case of the stabilized alkoxy radical formed in the ethene reaction, decomposition and reaction with

O_2 are both important at 1 atm in air (Niki *et al.*, 1981; Orlando *et al.*, 1998):

$$CH_2(OH)CH_2\dot{O} + O_2 \rightarrow CH_2(OH)CHO + HO_2, \quad (38)$$

$$CH_2(OH)CH_2\dot{O} \rightarrow HCHO + \dot{C}H_2OH. \quad (39)$$

For the β-hydroxyalkoxy radicals formed from alkenes C_5 and larger, there is experimental evidence that isomerization starts to dominate (e.g., see Atkinson *et al.*, 1995d; Kwok *et al.*, 1996b). Thus, isomerization followed by reaction with O_2, NO, etc., ultimately leads to the formation of dihydroxycarbonyl compounds. For the reaction of OH with 1-butene, for example, isomerization of one of the alkoxy radicals ultimately leads to 3,4-dihydroxybutanal in competition with its decomposition and reaction with O_2:

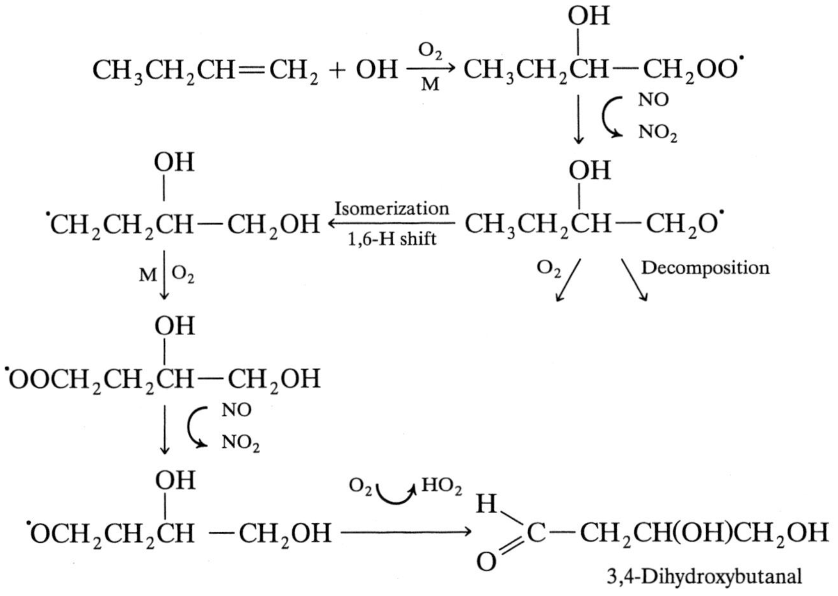

characteristic manner with O_2:

$$\cdot CH_2OH + O_2 \rightarrow HCHO + HO_2,$$
$$k = 9.4 \times 10^{-12} \text{ cm}^3 \text{ molecule}^{-1} \text{ s}^{-1}$$
$$\text{(Atkinson } et\ al.\text{, 1997a)}. \quad (40)$$

The larger α-hydroxy radicals react in a similar manner:

$$\dot{C}R_1R_2OH + O_2 \rightarrow R_1R_2C{=}O + HO_2. \quad (41)$$

Interestingly, the mechanism appears to involve the initial addition of O_2 to the carbon radical, followed by isomerization and decomposition:

$$\cdot CH_2OH + O_2 \rightarrow \cdot OOCH_2OH$$
$$\leftrightarrow HOOCH_2O \rightarrow HO_2 + HCHO. \quad (40)$$

$$CH_3CH_2CH{=}CH_2 + OH \xrightarrow[M]{O_2} CH_3CH_2\underset{\underset{OH}{|}}{CH}{-}CH_2OO\cdot$$

$$\downarrow \text{NO} / \text{NO}_2$$

$$\cdot CH_2CH_2\underset{\underset{OH}{|}}{CH}{-}CH_2OH \xleftarrow[\text{1,6-H shift}]{\text{Isomerization}} CH_3CH_2\underset{\underset{OH}{|}}{CH}{-}CH_2O\cdot$$

$$M \downarrow O_2 \qquad \qquad O_2 \nearrow \qquad \searrow \text{Decomposition}$$

$$\cdot OOCH_2CH_2\underset{\underset{OH}{|}}{CH}{-}CH_2OH$$

$$\downarrow \text{NO}/\text{NO}_2$$

$$\cdot OCH_2CH_2\underset{\underset{OH}{|}}{CH}{-}CH_2OH \xrightarrow{O_2 \curvearrowright HO_2} \underset{O}{\overset{H}{\diagup}}C{-}CH_2CH(OH)CH_2OH$$

3,4-Dihydroxybutanal

The yields of such dihydroxycarbonyl products have been measured to increase continuously from 0.04 for 1-butene to 0.6 for the 1-octene reaction (Kwok *et al.*, 1996b). The low yield for 1-butene reflects the fact that only one of the two possible alkoxy radicals formed can undergo isomerization via a 6-membered transition state.

The carbon-centered hydroxy-containing radical $\dot{C}H_2OH$ formed in reactions (37) and (39) is encountered frequently as an intermediate in tropospheric organic oxidations. It is commonly referred to as an α-hydroxy radical. These types of radicals react in a

For example, no deuterium isotope effect is observed on the rate constant when CH_2OD is substituted for CH_2OH (Grotheer *et al.*, 1988; Pagsberg *et al.*, 1989), as would be expected if a direct hydrogen abstraction were occurring.

Similar principles apply to more complex alkenes. One such case is that of isoprene, an important biogenically produced hydrocarbon:

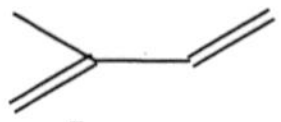

Isoprene

Figure 6.3 shows some of the major pathways in the

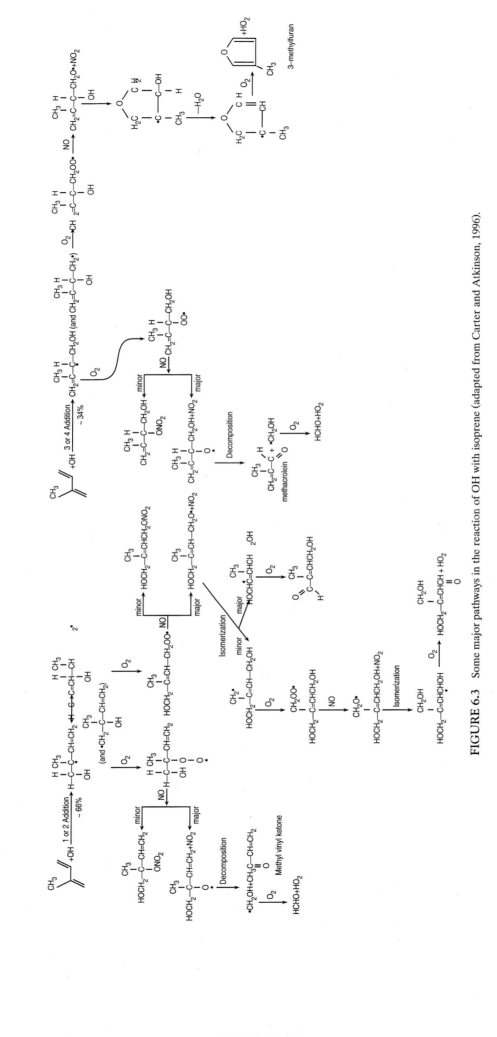

FIGURE 6.3 Some major pathways in the reaction of OH with isoprene (adapted from Carter and Atkinson, 1996).

OH–isoprene reaction. Detailed mechanisms have been developed by several groups (e.g., Paulson and Seinfeld, 1992b; Carter and Atkinson, 1996; Zimmermann and Poppe, 1996). Addition to the 1- (or 2-) position gives methyl vinyl ketone, whereas OH addition to the 3- (or 4-) position yields methacrolein. (Note that the O_3 reaction also generates these products, e.g., Aschmann and Atkinson, 1994.) Small yields of 3-methylfuran have also been reported (e.g., Atkinson et al., 1989; Tuazon and Atkinson, 1990; Paulson et al., 1992a). However, together these account for only about 60% of the total reaction (Tuazon and Atkinson, 1990; Paulson et al., 1992a; Carter and Atkinson, 1996). There is experimental evidence for another path in which isomerization of the double bond to the 2,3-position occurs after the initial addition of the OH; subsequent addition of O_2 and the usual reactions of RO_2 and RO give the hydroxycarbonyls $CHOC(CH_3)=CHCH_2OH$ and $HOCH_2C(CH_3)=CHCHO$ (Kwok et al., 1995).

As shown in Fig. 6.3 and as expected by analogy to the chemistry of simple alkenes, organic nitrates are formed in small yields from the reaction of the peroxy radicals with NO_2. For example, Chen et al. (1998) report the formation of seven different organic nitrates in this system, with a total yield of 4.4%. This formation of organic nitrates may be important as a minor NO_x sink in forested regions, however. For example, Starn et al. (1998a,b) estimated that the isoprene nitrates comprise 0.5–2% of the NO_y during the day but up to 9% at night in a rural area in the state of Tennessee.

2. Ozone (O_3)

The room temperature rate constants for the reactions of O_3 with some alkenes are given in Table 6.9. While the values are many orders of magnitude smaller than those for the corresponding OH reactions, the fact that tropospheric ozone concentrations are so much larger makes these reactions a significant removal process for the alkenes.

Despite many decades of research, the mechanisms of ozone–alkene reactions in the gas phase are still not well understood, certainly not as well understood as the corresponding reactions in the condensed phases. Figure 6.4 compares the overall characteristics of these reactions in the gas and condensed phases.

It is well known that the initial step in the reaction is the addition of O_3 across the double bond to form what is known as a primary ozonide, or molozonide:

$$O_3 + \underset{R_2}{\overset{R_1}{}}C=C\underset{R_4}{\overset{R_3}{}} \rightarrow \underset{R_2}{\overset{R_1}{}}C\overset{O-O-O}{\underset{\text{Primary ozonide}}{||}}C\underset{R_4}{\overset{R_3}{}}$$

(42)

This primary ozonide is not stable. One of the two peroxy O–O bonds (marked a or b in Eq. 43) and the C–C bond (marked c) cleave simultaneously to give an aldehyde or ketone and an intermediate called the Criegee intermediate, named after the German chemist who originally proposed this mechanism (e.g., see Criegee, 1975):

$$\underset{R_2}{\overset{R_1}{}}C\overset{a\overset{O-O}{||}b}{\underset{c}{||}}C\underset{R_4}{\overset{R_3}{}} \begin{array}{c} \overset{a}{\nearrow} R_1R_2C=O + R_3R_4\dot{C}OO\cdot \overset{(+)}{(-)} \\ \text{Criegee intermediate} \\ \overset{b}{\searrow} R_3R_4C=O + R_1R_2\dot{C}OO\cdot \overset{(+)}{(-)} \\ \text{Criegee intermediate} \end{array}$$

(43)

TABLE 6.9 Rate Constants and Temperature Dependence[a] for the Gas-Phase Reactions of O_3 with Some Alkenes[b]

Alkene	k (10^{-18} cm^3 molecule^{-1} s^{-1})	A (10^{-15} cm^3 molecule^{-1} s^{-1})	E_a/R (K)
Ethene	1.6	9.14	2580
Propene	10.1	5.51	1878
1-Butene	9.64	3.36	1744
2-Methylpropene	11.3	2.70	1632
cis-2-Butene	125	3.22	968
trans-2-Butene	190	6.64	1059
1-Pentene	10.0		
Cyclopentene	570	1.8	350
2-Methyl-2-butene	403	6.51	829
1-Hexene	11.0		
Cyclohexene	81.4	2.88	1063
cis-3-Methyl-2-pentene	450		
trans-3-Methyl-2-pentene	560		
2,3-Dimethyl-2-butene	1130	3.03	294
1,3-Butadiene	6.3	13.4	2283
2-Methyl-1,3-butadiene	12.8	7.86	1913
Myrcene	470		
2-Carene	230		
3-Carene	37		
Limonene	200		
α-Phellandrene	2980		
β-Phellandrene	47		
α-Pinene	86.6	1.01	732
β-Pinene	15		
α-Terpinene	2.1×10^4		
γ-Terpinene	140		
Terpinolene	1880		
Methyl vinyl ketone	5.6[c]		
Methacrolein	1.2[c]		

[a] $k = Ae^{-E_a/RT}$.
[b] From Atkinson (1997a) and Atkinson et al. (1997a); for structures of biogenics, see Fig. 6.22.
[c] Average of Grosjean and Grosjean (1998a) and Neeb et al. (1998b).

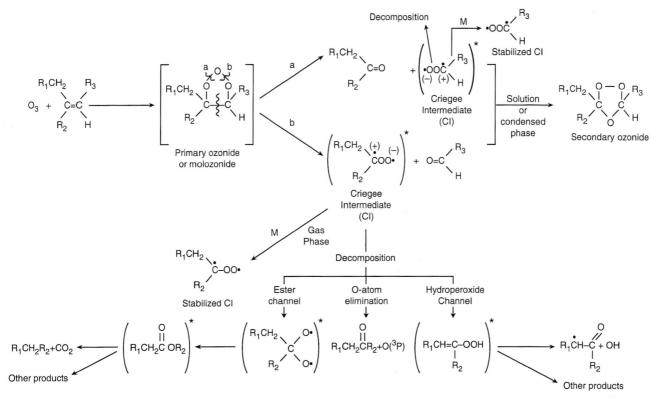

FIGURE 6.4 Overall mechanisms of O_3–alkene reactions in the gas and condensed phase, respectively.

In the liquid phase, the Criegee intermediates have been assumed to be zwitterions and hence the term "Criegee zwitterion" is commonly used. In the gas phase, the structure is usually written as a biradical (although it may really be more zwitterionic in character; e.g., see Cremer *et al.*, 1993). Hence "Criegee biradical" is frequently used for this gaseous intermediate. Sander (1990), Bunnelle (1991), and Cremer *et al.* (1993) give a more detailed discussion of the structure and properties of the Criegee intermediate.

In any event, in the solvent cage in which they are formed in the liquid phase, or for higher molecular weight alkenes condensed on surfaces, the two fragments formed by decomposition of the primary ozonide are held in close proximity and recombine to form a secondary ozonide:

$$R_1R_2C=O + R_3R_4\dot{C}OO\cdot$$
$$R_3R_4C=O + R_1R_2\dot{C}OO\cdot$$
$$\rightarrow \underset{R_2}{\overset{R_1}{C}}\underset{O}{\overset{O-O}{\diagup}}\underset{R_4}{\overset{R_3}{C}} \quad (44)$$

In addition to the effects of close proximity, the condensed phases act to remove excess energy in the fragments.

In the gas phase, however, two effects come into play. First, the two fragments formed on decomposition of the primary ozonide fly apart rapidly. As a result, the formation of secondary ozonides by recombination, reaction (44), does not occur to a significant extent in the gas phase [unless the mixture is doped with large quantities of an aldehyde or ketone to trap the Criegee intermediate (e.g., Neeb *et al.*, 1998a) or very high concentrations are used (e.g., Fajgar *et al.* (1996); Griesbaum *et al.* (1998)]. Second, there is no efficient mechanism for removal of excess energy from the carbonyl compound and the Criegee intermediate. The Criegee intermediate contains excess energy and either can be stabilized or decompose in a variety of ways. For example, for the two possible Criegee intermediates produced in the O_3–propene reaction, the following paths are possible (where the branching ratios are the IUPAC evaluation recommendations (Atkinson *et al.*, 1997a):

$$(H\dot{C}HOO\cdot)^* + M \xrightarrow{0.37} H\dot{C}HOO\cdot + M, \quad (45a)$$
$$\xrightarrow{0.12} HCO + OH, \quad (45b)$$
$$\xrightarrow{0.38} CO + H_2O, \quad (45c)$$
$$\xrightarrow{0.13} CO_2 + H_2, \quad (45d)$$
$$\xrightarrow{0} CO_2 + 2H, \quad (45e)$$
$$\xrightarrow{0} HCOOH. \quad (45f)$$

Neeb *et al.* (1998a) report branching ratios for (45a) of 0.50, (45b) + (45c) of 0.23, (45d) + (45e) of 0.23, and (45f) of 0.04.

$$(CH_3\dot{C}HOO\cdot)^* + M \xrightarrow{0.15} CH_3\dot{C}HOO\cdot, \quad (46a)$$
$$\xrightarrow{0.54} CH_3 + CO + OH, \quad (46b)$$
$$\longrightarrow CH_3 + CO_2 + H, \quad (46c)$$
$$\xrightarrow{(0.17 = c + d)} HCO + CH_3O, \quad (46d)$$
$$\xrightarrow{0.14} CH_4 + CO_2, \quad (46e)$$
$$\xrightarrow{0} CH_3OH + CO. \quad (46f)$$

Table 6.10 gives the ranges of observed yields of the stabilized Criegee intermediates at 1 atm pressure in air and at room temperature. Clearly, significant decomposition of the intermediates occurs under typical tropospheric conditions.

The fraction of the primary ozonide decomposition that goes by path a in reaction (43) vs path b for unsymmetrical alkenes has been determined from the product yields in a number of studies (e.g., see series of papers by Grosjean and co-workers and Atkinson and co-workers). One might expect, based on thermochemical arguments, that the decomposition giving the more stable, i.e., more highly substituted, biradical would be preferred, and this is indeed what has been observed (e.g., see Horie and Moortgat, 1991; Grosjean *et al.*, 1994c; Atkinson *et al.*, 1995c; and Grosjean and Grosjean, 1997, 1998a). Atkinson (1997a) recommends, based on the literature, the following branching ratios:

$$O_3 + R_1CH{=}CH_2 \xrightarrow{0.5} (R_1\dot{C}HOO\cdot)^* + HCHO,$$
$$\xrightarrow{0.5} R_1CHO + (H\dot{C}HOO\cdot)^*,$$

TABLE 6.10 Yields of Stabilized Criegee Intermediates at Room Temperature and 1 atm Air[a]

Alkene	Yield
Ethene	0.35–0.47
Propene	0.25–0.44
2-Methylpropene	0.17
cis-2-Butene	0.18
trans-2-Butene	0.19–0.42
2,3-Dimethyl-2-butene	0.30
1-Octene	0.22
Cyclopentene	0.05
Cyclohexene	0.03
Cycloheptene	0.03
1-Methylcyclohexene	0.10
Methylenecyclohexane	0.22
α-Pinene	0.13
β-Pinene	0.25

[a] From Hatakeyama and Akimoto (1994) and Atkinson (1994, 1997a); see these for original references.

$$O_3 + R_1R_2C{=}CH_2 \xrightarrow{0.65} (R_1R_2\dot{C}OO\cdot)^* + HCHO,$$
$$\xrightarrow{0.35} R_1C(O)R_2 + (H\dot{C}HOO\cdot)^*,$$
$$O_3 + R_1R_2C{=}CHR_3 \xrightarrow{0.65} (R_1R_2\dot{C}OO\cdot)^* + R_3CHO,$$
$$\xrightarrow{0.35} R_1C(O)R_2 + (R_3\dot{C}HOO\cdot)^*.$$

While free radicals were reported in ozone–alkene reactions as long ago as 1958 by Saltzman, the importance and magnitude of these processes have only relatively recently been appreciated by the atmospheric chemistry community. Of greatest significance is the production of OH in these reactions, which was first observed in these systems in a vibrationally excited state, $OH^{\neq}$, probably from the initial production of H atoms, followed by the well-known reaction $H + O_3 \rightarrow OH^{\neq} + O_2$ (Finlayson *et al.*, 1972). Rapid secondary reactions of OH with the parent alkene complicates interpretation of the original O_3–alkene mechanism and products. In addition, in kinetic studies where the loss of the alkene is followed, the derived rate constants may be too large unless the contribution of OH secondary reactions is taken into account or it is scavenged from the system. However, more important than the implications for laboratory studies are those for the role of ozone–alkene reactions in tropospheric chemistry. Thus, in addition to producing carbonyl compounds that can photolyze to generate free radicals, ozone–alkene reactions contribute directly to the generation of free radicals via reactions such as (45) and (46). This will be especially important at night when there are no photolytic sources of OH.

Table 6.11 gives some measured yields of OH in some ozone–alkene reactions. These have been determined by trapping the OH, e.g., using a large excess of cyclohexane (e.g., Grosjean *et al.*, 1994c; Grosjean and Grosjean, 1997) or 2-butanol (e.g., Chew and Atkinson, 1996), and measuring the major "OH-trap" products such as cyclohexanone and cyclohexanol (Atkinson *et al.*, 1992; Aschmann *et al.*, 1997a) and butanone, respectively (e.g., Chew and Atkinson, 1996), under conditions where the formation of OH via the $HO_2 + O_3$ reaction is minimized. Other approaches include trapping by CO to form CO_2 (e.g., Gutbrod *et al.*, 1997b) or following the loss of pairs of tracer organics that react with OH (e.g., Schäfer *et al.*, 1997; Paulson *et al.*, 1997; Marston *et al.*, 1998). In addition to the indirect measurements of OH, a direct measurement of its production using laser-induced fluorescence has been carried out for the reactions of O_3 with ethene, *trans*-2-butene, 2,3-dimethyl-2-butene, and isoprene at 4- to 6-Torr pressures in N_2. Yields of OH from 0.4 to 0.8 were observed by Donahue *et al.* (1998b).

In short, O_3–alkene reactions generate OH, with yields approaching unity in the case of highly branched alkenes.

The mechanism of decomposition of the Criegee intermediates is believed to occur via several reaction channels shown for the $[(R_1CH_2)(R_2)\dot{C}HO\dot{O}]^*$ Criegee intermediate in Fig. 6.4. The oxygen-atom elimination channel for simple alkenes is not believed to be important. However, the ester and hydroperoxide channels are important and explain the production of free radicals such as OH. Theoretical calculations have shed some light on this (e.g., Gutbrod *et al.*, 1996, 1997a;

Anglada *et al.*, 1996). Figure 6.5, for example, shows the calculations of Gutbrod *et al.* (1996) for the reaction of the Criegee intermediate $\dot{C}(CH_3)_2O\dot{O}$, formed in the reaction of O_3 with 2,3-dimethyl-2-butene. The path that forms the dioxinane intermediate has a higher activation barrier than that forming the unsaturated hydroperoxide, and the latter channel that leads to OH formation is predicted to predominate. The key to OH production is predicted to be an alkyl group in a *syn* position so that it can interact with the terminal oxygen (Gutbrod *et al.*, 1996, 1997a):

$$\begin{array}{c} H_2C \overset{H}{\diagdown} O \\ \diagdown \\ H_3C C\!\!-\!\!O \end{array}$$

Grosjean *et al.* (1994c) propose that the hydroperoxide channel may also be responsible for the formation of hydroxycarbonyl and/or dicarbonyl products:

$$CH_3\dot{C}HOO\cdot \rightarrow CH_2\!=\!CHOOH$$
$$\rightarrow [CH(O)CH_2OH]^*$$
$$\overset{M}{\rightarrow} CH(O)CH_2OH,$$
$$\rightarrow H_2 + (CHO)_2. \quad (47)$$

TABLE 6.11 Yields of OH from Gas-Phase O_3 – Alkene Reactions at 1 atm Pressure[a]

Alkene	OH yield
Ethene	0.12,[b] 0.08[h]
Propene	0.33,[a] 0.18[h]
1-Butene	0.41[a]
1-Pentene	0.37[f]
1-Hexene	0.32[f]
1-Heptene	0.27[f]
1-Octene	0.18[f]–0.45[c]
cis-2-Butene	0.41,[a] 0.17[h]
trans-2-Butene	0.64,[a] 0.24[h]
Cyclopentene	0.61[f]
Cyclohexene	0.68[a]
1-Methylcyclohexene	0.90[f]
2-Methylpropene	0.84[a]
2-Methyl-1-butene	0.83[a]
2-Methyl-2-butene	0.89[a]
2,3-dimethyl-2-butene	0.5[h]–1.0[a,d]
Limonene	0.86[b]
Myrcene	1.15[b]
α-Pinene	0.70–0.85[b,g,i]
β-Pinene	0.35[b]
Terpinolene	1.03[b]
Camphene	≤0.18[b]
1,3-Butadiene	0.08[a]
Isoprene	0.19–0.27[b,g,e,h]

[a] From Atkinson and Aschmann (1993) and Atkinson (1997a); Donahue *et al.* (1998) have reported OH yields of 0.4–0.8 at low pressures (4–6 Torr) in N_2 from the ethene, *trans*-2-butene, 2,3-dimethyl-2-butene, and isoprene reactions; see Fig. 6.22 for biogenics structures.
[b] From Atkinson *et al.* (1992).
[c] From Paulson and Seinfeld (1992a).
[d] From Niki *et al.* (1987).
[e] From Paulson *et al.* (1992b).
[f] From Atkinson *et al.* (1995a).
[g] Paulson *et al.* (1998); OH yield of 0.25 from O_3 + isoprene supersedes Paulson *et al.* (1992b) yield of 0.68.
[h] From Gutbrod *et al.* (1997b); the OH yields reported using this CO trapping method tend to be lower than those measured using other techniques.
[i] Chew and Atkinson (1996).

The generation of OH in O_3–alkene reactions has important implications for tropospheric chemistry. Thus the O_3–alkene reactions could be important free radical sources at dusk and during the night when photolytic sources of OH are minimal (e.g., Paulson and Orlando, 1996; Bey *et al.*, 1997; Paulson *et al.*, 1998). For example, Paulson and Orlando (1996) predicted that ~10–15% of the total radical production may be from O_3–alkene reactions in a typical rural area in the southeastern United States. As seen in Fig. 6.6, this reaction is expected to be most important at night.

There are a number of potential atmospheric fates of the stabilized Criegee intermediate. These include reactions with water vapor, SO_2, NO, NO_2, CO, and aldehydes and ketones. In the latter cases, the secondary ozonides are formed as discussed earlier. Neeb *et al.* (1998a) propose that the secondary ozonides are formed with sufficient excess energy that they may partly decompose rather than being stabilized; for the $CH_3CHO + \dot{C}H_2O\dot{O}$ reaction, for example, formation of both the stabilized secondary ozonide and HCHO and CO_2 was observed by Neeb *et al.* (1998a). Similarly, the reaction of the ethene Criegee intermediate, $\dot{C}H_2O\dot{O}$, with HCHO, observed to generate HCHO, HCOOH, and CO, was proposed to occur via an excited secondary ozonide that completely decomposes to

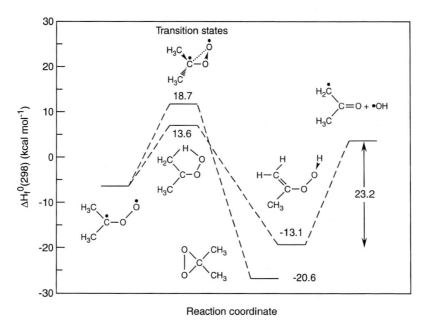

FIGURE 6.5 Calculated energetics of two possible reaction pathways for the Criegee intermediate $(CH_3)_2\dot{C}OO\cdot$ (adapted from Gutbrod *et al.*, 1996).

(HCHO + HCOOH), (HCHO + CO + H_2O), or (HCHO + HCO + OH).

The reaction with water vapor has a number of potential paths:

$$R_1H\dot{C}OO\cdot + H_2O \rightarrow R_1C(O)OH + H_2O, \quad (48a)$$

$$\xrightarrow{M} R_1\underset{\underset{OH}{|}}{\overset{\overset{H}{|}}{C}}OOH \rightarrow RC(O)OH + H_2O, \quad (48b)$$

$$\rightarrow R_1CHO + H_2O_2. \quad (48c)$$

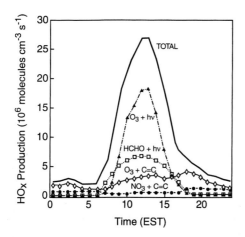

FIGURE 6.6 Calculated rates of HO_x radical generation from various sources for a rural forested site in the southeastern United States (adapted from Paulson and Orlando, 1996).

The formation of carboxylic acids in these reactions may be a significant source of organic acids in air. Reaction (48a) appears to occur by at least two mechanisms. Thus, the addition of isotopically labeled $H_2^{18}O$ leads to the formation of carboxylic acid, some of which contains the ^{18}O label and some of which does not (Hatakeyama *et al.*, 1981), suggesting mechanisms such as the following:

$$R_1\dot{C}HOO\cdot + H_2^{18}O$$
$$\rightarrow [R_1\dot{C}HOO \cdot H_2^{18}O]^*$$
$$\rightarrow R_1CO^{18}OH, R_1C^{18}OOH + H_2O$$
$$\rightarrow R_1COOH + H_2^{18}O.$$

Analogous trapping reactions for the Criegee intermediate using such reactants as H_2O_2, CH_3OH, C_2H_5OH, and HCOOH have been observed for $\dot{C}H_2O\dot{O}$ from the ethene–O_3 reaction (e.g., Neeb *et al.*, 1996; Wolff *et al.*, 1997).

While it is often assumed that only the stabilized form of the Criegee intermediate undergoes such reactions, Moortgat and co-workers (Horie *et al.*, 1994) have suggested that in the ethene reaction, it is the excited form that reacts with water vapor to form the acid, while the reaction of the stabilized Criegee intermediate leads to a hydroxyhydroperoxide, corresponding to reaction (48b).

The formation of hydroxyhydroperoxides has only relatively recently been recognized, probably due to the difficulty in analyzing for these reactive species. An additional complication is that these hydroxyhydroperoxides are known to decompose in solution to H_2O_2 and the carbonyl compounds. This decomposition, reaction (48c), which can occur during analysis in solution or on the walls of the reaction chamber, may be responsible for the highly varying yields of H_2O_2 and hydroxyhydroperoxides reported in various studies (e.g., see Gäb et al., 1985, 1995; Hatakeyama and Akimoto, 1994; Horie et al., 1994; Hatakeyama et al., 1993; Hewitt and Kok, 1991; Simonaitis et al., 1991; Wolff et al., 1997; Neeb et al., 1997; and Sauer et al., 1999). At any rate, their formation is now accepted, although the yields as a function of water vapor concentration under typical atmospheric conditions are not well established for a variety of alkenes.

For example, in the isoprene reaction with O_3, H_2O_2 has been observed as a product, with its yield increasing as the relative humidity increased from 1 to 9% (Sauer et al., 1999). The yields of methacrolein and methyl vinyl ketone also increased simultaneously, as expected if they were also formed in part from the decomposition of an intermediate hydroxyhydroperoxide.

The reaction with SO_2 has been thought in the past to proceed via the direct formation of SO_3:

$$R_1\dot{C}HOO\cdot + SO_2 \rightarrow R_1CHO + SO_3. \quad (49)$$

However, as discussed in detail by Hatakeyama and Akimoto (1994), there is increasing evidence for the formation of an addition complex such as that proposed by Martinez and Herron (1981) and observed earlier by mass spectrometry by Schulten and Schurath (1975):

$$R_1\dot{C}HOO\cdot + SO_2 \rightarrow \begin{array}{c} R_1 \\ \diagdown \\ H \end{array} C \begin{array}{c} O-O \\ \diagup \quad \diagdown \\ \diagdown \quad \diagup \\ O \end{array} S=O \quad (50)$$

In the presence of water vapor, the adduct then reacts with H_2O to generate sulfuric acid.

Other reactions proposed for the Criegee intermediate include

$$R_1\dot{C}HOO\cdot + NO \rightarrow R_1CHO + NO_2, \quad (51)$$

$$R_1\dot{C}HOO\cdot + NO_2 \rightarrow R_1CHO + NO_3, \quad (52)$$

$$R_1\dot{C}HOO\cdot + CO \rightarrow R_1CHO + CO_2. \quad (53)$$

However, the relative importance of these reactions is highly uncertain. Table 6.12 gives one estimate of the range of possible rate constants for these reactions of the Criegee intermediate and the calculated lifetimes of the intermediate under some typical atmospheric conditions. Reaction with water vapor is expected to be the major fate of the Criegee intermediate, with per-

TABLE 6.12 Range of Reported Rate Constants for the Reactions of the Criegee Intermediate with Some Gases[a] and Associated Lifetimes of the Criegee Intermediate under Polluted Tropospheric Conditions

Reactant	k (cm^3 molecule^{-1} s^{-1})	τ_{CI}[b] (s)
H_2O	$1 \times 10^{-15} - 2 \times 10^{-19}$	0.003–400
CO	1×10^{-14}	8
HCHO	$2 \times 10^{-12} - 2 \times 10^{-16}$	$20-(2 \times 10^5)$
NO	7×10^{-12}	0.1
NO_2	$7 \times 10^{-13} - 1 \times 10^{-17}$	$1-10^5$
SO_2	$1.7 \times 10^{-11} - 3 \times 10^{-15}$	0.05–270

[a] Adapted from Hatakeyama and Akimoto (1994).
[b] Lifetime of Criegee intermediate assuming the following concentrations: H_2O, 4×10^{17} molecules cm^{-3}, corresponding to a relative humidity of ~50% at 298 K; CO, 0.5 ppm; HCHO, 1 ppb; NO, NO_2, SO_2, 50 ppb.

haps some contribution by reaction with NO and/or SO_2 in more polluted atmospheres.

It should be noted that while the mechanism outlined in this section describes the overall features of O_3–alkene chemistry, there are also other minor paths as well. For example, small yields of epoxides that appear to be formed in the primary reaction have been observed as products of the reactions of some dienes and cycloalkenes (e.g., see Paulson et al., 1992b; and Atkinson et al., 1994a, 1994b). The reader should consult the rather extensive ozone literature for further details on both the condensed- and gas-phase reactions.

As discussed in Chapter 9, there are a variety of low-volatility organics, some of which are unsaturated, which are found in or on particles in the atmosphere, and these would be expected to undergo heterogeneous reactions with O_3. Consistent with this, de Gouw and Lovejoy (1998) report reaction probabilities of $\sim 10^{-2}-10^{-3}$ for O_3 on organic liquids containing unsaturated sites.

3. Nitrate Radical (NO_3)

Table 6.13 gives the rate constants and temperature dependence for the reactions of the nitrate radical, NO_3, with some alkenes. The room temperature rate constants span about six orders of magnitude, approaching diffusion controlled for some of the biogenics. Those activation energies that have been measured are relatively small (or negative by a small amount), and the rate constants increase with increasing alkyl substitution at the double bond. This, plus the much larger values compared to the rate constants for reactions with alkanes where H-abstraction occurs, suggests

TABLE 6.13 Room Temperature Rate Constants and Temperature Dependence[a] for the Gas-Phase Reactions of the NO_3 Radical with Some Alkenes[b]

Alkene	k (cm^3 molecule^{-1} s^{-1}) at 298 K	A (cm^3 molecule^{-1} s^{-1})	E_a/R (K)
Ethene	2.1×10^{-16}	c	c
Propene	9.5×10^{-15}	4.6×10^{-13}	1156
1-Butene	1.4×10^{-14}	3.14×10^{-13}	938
2-Methylpropene	3.3×10^{-13}		
cis-2-Butene	3.5×10^{-13}		
trans-2-Butene	3.9×10^{-13}	d	d
2-Methyl-2-butene	9.4×10^{-12}		
2,3-Dimethyl-2-butene	5.7×10^{-11}		
1,3-Butadiene	1.0×10^{-13}		
2-Methyl-1,3-butadiene (isoprene)	6.8×10^{-13}	3.03×10^{-12}	446
Cyclopentene	5.3×10^{-13}		
Cyclohexene	5.9×10^{-13}	1.05×10^{-12}	174
Cycloheptene	4.8×10^{-13}		
Camphene	6.2×10^{-13}[f]	3.1×10^{-12}	481
2-Carene	1.9×10^{-11}		
3-Carene	9.1×10^{-12}		
Limonene	1.2×10^{-11}		
α-Pinene	5.9×10^{-12}[f]	3.5×10^{-13}	−841
β-Pinene	2.1×10^{-12}[f]	1.6×10^{-10}	1248
α-Phellandrene	7.3×10^{-11}		
β-Phellandrene	8.0×10^{-12}		
α-Terpinene	1.4×10^{-10}		
γ-Terpinene	2.9×10^{-11}		
Terpinolene	9.7×10^{-11}		
Methyl vinyl ketone	$< 6 \times 10^{-16}$[d]		
Methacrolein	3.3×10^{-15}[e]		

[a] $k = Ae^{-E_a/RT}$.
[b] From Atkinson (1994) and Atkinson et al. (1997a).
[c] Recommended temperature dependence is $k = 4.88 \times 10^{-18} T^2 e^{-2282/T}$ cm^3 molecule^{-1} s^{-1} from 290 to 523 K.
[d] Dlugokencky and Howard (1989) report a curved Arrhenius plot that can be matched from 204 to 378 K by $k = (1.78 \pm 0.36) \times 10^{-12} \exp[(-530 \pm 100)/T] + (1.28 \pm 0.26) \times 10^{-14} \exp[(570 \pm 110)/T]$; Atkinson (1997a) recommends $k = 1.22 \times 10^{-18} e^{382/T}$ cm^3 molecule^{-1} s^{-1} from 204 to 378 K.
[e] From Chew et al. (1998).
[f] From Martinez et al. (1998).

that NO_3 undergoes electrophilic addition to the double bond in much the same fashion as OH:

$$NO_3 + \quad \! \text{>C=C<} \rightarrow \text{—C(ONO}_2\text{)—Ċ—} \quad (54)$$

However, there is no pressure dependence of the rate constants over the range from about 1 Torr up to 1 atm, suggesting that the adduct does not decompose significantly back to reactants under atmospheric conditions.

This mechanism is consistent with the observation of significant yields of epoxide products and NO_2 for some alkenes (Olzmann et al., 1994). For example, Fig. 6.7 shows the infrared spectrum of the minor products from the reaction of NO_3 with 2,3-dimethyl-2-butene at 740 Torr in air, after the spectra of the reactants and the major product, acetone, as well as of NO_2 and HNO_3 have been subtracted out. Bands due to the epoxide (in ~17% yield) are clearly seen, as well as bands due to 2,3-dinitroxy-2,3-dimethyl-2-butane. Together with acetone, these compounds account for 75% of the reacted alkene (Skov et al., 1994).

Figure 6.8 summarizes the mechanism of the reaction of NO_3 with 2,3-dimethyl-2-butene (Skov et al., 1994; Olzmann et al., 1994). The addition of the NO_3 to a double bond is about 20 kcal mol^{-1} exothermic, so that the initially formed adduct is excited and either can be stabilized or decompose by breaking the weaker CO–NO_2 bond to form the epoxide and NO_2. The stabilized adduct can also decompose in a similar manner, in competition with its reaction with O_2. Under

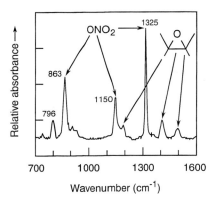

FIGURE 6.7 Infrared spectrum of products of the reaction of 2,3-dimethyl-2-butene with NO_3 (spectra of acetone, NO_2, and HNO_3 have been subtracted out) (adapted from Skov *et al.*, 1994).

surface-level tropospheric conditions, 1 atm total pressure in air, the epoxide yields are small but not insignificant. For example, epoxide yields of 7, 12, 18, and 28% from the reactions of isobutene, *trans*-2-butene, 1-butene, and propene, respectively, have been reported (Berndt and Böge, 1995).

The lifetime of the excited NO_3-alkene adducts is sufficiently long that rotation about the C–C bond leads to the same yields of *trans*- and *cis*-epoxides regardless of the configuration of the reactant alkene; for example, the reactions of both *cis*- and *trans*-2-butene give about 80% of the *trans* form of the product epoxide and 20% of the *cis* form (Benter *et al.*, 1994).

The peroxy radical formed in Fig. 6.8 is expected to react in air as discussed earlier, i.e., with NO, HO_2, RO_2, or NO_2. Since NO_3 itself reacts rapidly with NO, large concentrations of NO will not be present at the

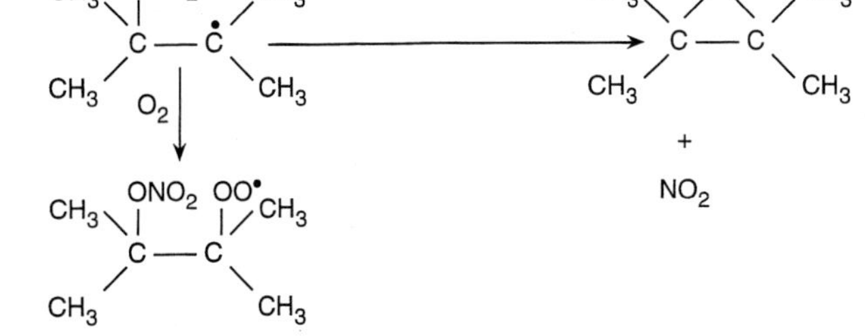

FIGURE 6.8 Mechanism of the NO_3 reaction with 2,3-dimethyl-2-butene (adapted from Skov *et al.*, 1994).

same time as NO_3. However, there may still be sufficient amounts for this to be a significant fate of the peroxy radical. Figure 6.9 traces the expected fates of this radical through to stable organic products. The dinitroxybutane is formed from the alkoxy radical–NO_2 reaction and acetone from the alkoxy radical decomposition. The observation of smaller amounts of other aldehydes and formaldehyde also suggests a small contribution from a 1,4-H shift and subsequent reactions of the alkyl radical thus formed (Hjorth *et al.*, 1990).

The results of laboratory studies of the mechanism of NO_3 radical reactions with alkenes tend to be slanted toward observing polyfunctional organics containing more than one nitro group. The reason for this is that the thermal decomposition of N_2O_5,

$$N_2O_5 \leftrightarrow NO_3 + NO_2, \qquad (55, -55)$$

is often used as the source of NO_3. As a result, relatively high concentrations of NO_2 may also be present, and depending on the concentration regimes employed, these can trap the alkylperoxy and alkoxy radicals more efficiently than would normally be the case in the atmosphere. On the other hand, sources such as the reaction of fluorine atoms with HNO_3,

$$F + HNO_3 \rightarrow HF + NO_3, \qquad (56)$$

are normally used at low pressures (~ Torr) and in the absence of air because such conditions are needed to generate the halogen atoms. Under these conditions, the products may be quite different from those at 1 atm in air. For example, the yields of epoxides are much greater, typically approaching 100%, under these low-pressure, low-O_2 conditions (Skov *et al.*, 1994; Berndt and Böge, 1995).

However, it is interesting that vicinal dinitrates have been measured in air with higher concentrations at night, as would be expected if nitrate radical reactions were the source. For example, Schneider *et al.* (1998) measured the diurnal variation of a series of alkyl nitrates in rural air in Germany and found higher nighttime levels of vicinal dinitrates. They attributed these to the addition of NO_3 to double bonds, formation of the alkylperoxy radical, and then reaction with NO, with the minor channel in the RO_2 + NO reaction leading to a second nitrate group on the adjacent carbon.

As seen from the rate constants in Table 6.13, the reactions of biogenic hydrocarbons with NO_3 are quite fast, so this reaction is expected to be a major fate of these biogenics at night and to also contribute to the removal of NO_x. The lifetime of isoprene with respect to reaction with NO_3 at 50 ppt, for example, is only about 20 min.. Addition of NO_3 occurs primarily at the 1-position and leads in the absence of NO to the

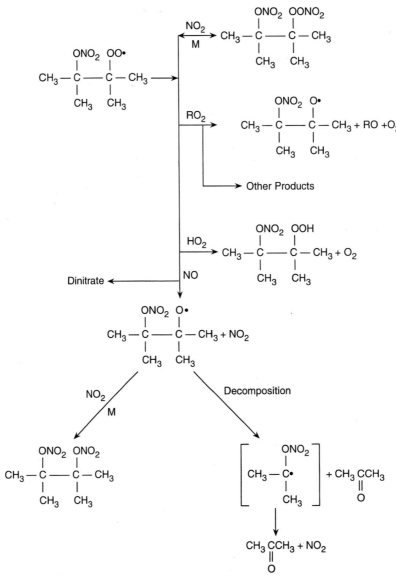

FIGURE 6.9 Expected atmospheric fates of peroxy radical formed in NO$_3$ addition to 2,3-dimethyl-2-butene.

formation of 4-nitroxy-3-methyl-2-butenal as a major product (e.g., Jay and Stieglitz, 1989; Barnes *et al.*, 1990; Skov *et al.*, 1992; Kwok *et al.*, 1996c; Berndt and Böge, 1997):

4-Nitroxy-3-methyl-2-butenal

Also formed are a variety of hydroxynitrates, nitrooxyhydroperoxides, and hydroxycarbonyls anticipated from the reactions of alkylperoxy and alkoxy radicals formed after the initial addition of NO$_3$ to the double bond (e.g., Kwok *et al.*, 1996c). The yield of 4-nitroxy-3-methyl-2-butenal decreases with increasing NO concentrations, while the yields of methyl vinyl ketone and methacrolein increase (Berndt and Böge, 1997). This has been attributed to the decomposition of the alkoxy radicals formed after 1,2-addition, e.g.,

Methyl vinyl ketone

and after 3,4-addition, respectively:

[structure] ONO$_2$ → [structure] + HCHO + NO$_2$.
Methacrolein

In addition, in the nighttime atmosphere, reaction of the RO$_2$ radicals with NO$_3$ may occur as discussed earlier. As a result, the products of the isoprene–NO$_3$ reaction in the atmosphere will depend on the concentrations of NO, NO$_3$, HO$_2$, and RO$_2$.

Field studies suggest that the nitrate radical reaction can also be a major contributor to isoprene decay at night, as well as contributing to the formation of organic nitrates in air. For example, Starn et al. (1998b) found that when the product of NO$_2$ and O$_3$ (which form NO$_3$) was high in a forested region in the southeastern United States, isoprene often decayed rapidly at dusk. This reaction of NO$_3$ with isoprene was estimated to be the major sink for NO$_3$ under some conditions in this area.

The reader is referred to two excellent reviews of NO$_3$ chemistry by Wayne et al. (1991) and Atkinson (1991) for further details.

The plethora of multifunctional products anticipated from nitrate radical reactions with alkenes (Fig. 6.8) may well be in part responsible for what is commonly referred to as "missing NO$_y$." In brief, one can measure total NO$_y$ using techniques such as the "master blaster" described in Chapter 11, in which one converts all oxygenated nitrogen-containing species into NO using a Au/CO converter and then measures the total NO. At the same time, one can measure various individual components of NO$_y$ (NO, NO$_2$, HNO$_3$, PAN, etc.) separately and determine whether their sum accounts for the observed NO$_y$. As discussed in detail in Chapter 11.A.4a, the sum of the individual compounds is often less than the measured NO$_y$, and this difference, which is not accounted for, is termed "missing NO$_y$." Given the mixture of potential nitrogen-containing organics formed in the NO$_3$ radical reactions (e.g., Figs. 6.8 and 6.9), as well as in NO$_2$ reactions with organic free radicals (see earlier), it is perhaps not surprising that in many air masses species other than NO$_x$, HNO$_3$, PAN, and particulate nitrate are present at significant concentrations. It remains a challenge to identify and measure such species in a specific and sensitive manner.

4. Chlorine Atoms (Cl)

Table 6.14 gives the rate constants for the reactions of chlorine atoms with some simple alkenes and some biogenic hydrocarbons. As expected, the reactions are

TABLE 6.14 Some Reported Values of the Rate Constants for the Reactions of Chlorine Atoms with Some Alkenes at 1 atm Total Pressure and 298 K

Alkene	k (10^{-10} cm^3 molecule^{-1} s^{-1})
Ethene	0.99[a]
	1.2[b]
	1.2[c]
Propene	2.3[a]
	2.7[b]
	2.5[d]
1-Butene	2.2[a]
1,3-Butadiene	4.2[a]
	5.6[e]
Isoprene	4.6[h]
	5.5[e]
	2.8[f]
(1R)-(+)-α-Pinene	4.8[g]
(1S)-(−)-α-Pinene	4.6[g]
3-Carene	5.6[g]
Myrcene	6.6[g]
p-Cymene	2.1[g]
Methyl vinyl ketone	2.0[g]

[a] Stutz et al. (1998).
[b] Atkinson and Aschmann (1985), corrected to k(Cl + n-butane) = 2.18×10^{-10} cm^3 molecule^{-1} s^{-1}.
[c] Kaiser and Wallington (1998), based on reported values of k_0 and k_∞.
[d] Kaiser and Wallington (1996).
[e] Bierbach et al. (1996).
[f] Bedjanian et al. (1998); this is the value of k_∞.
[g] Finlayson-Pitts et al. (1999).
[h] Ragains and Finlayson-Pitts (1997).

all very fast, approaching the collision-controlled regime.

The reaction proceeds primarily by addition to a double bond to form a chlorine-containing alkyl radical, which then adds O$_2$ to ultimately generate characteristic chlorine-containing oxygenated products. For example, 1-chloro-3-methyl-3-buten-2-one is formed in the absence of NO from the addition of chlorine atoms to the 4-position of isoprene, followed by secondary reactions of the alkyl radical with O$_2$ etc. This product might therefore be expected to be a reaction product in the low-NO$_x$ environment of the marine boundary layer (Nordmeyer et al., 1997).

A small portion of the reaction proceeds by what appears to be abstraction of the allylic hydrogen atom from the weaker C–H bond (e.g., Lee and Rowland, 1977). In the case of the isoprene reaction, for example, about 15% of the overall reaction at 1 atm pressure and room temperature proceeds by abstraction (Ragains and Finlayson-Pitts, 1997). It should be noted, however, that this net hydrogen abstraction may not be a simple abstraction per se but rather proceed via an addition–elimination reaction (e.g., Kaiser and Wallington, 1996b; Ragains and Finlayson-Pitts, 1997).

5. Nitrogen Dioxide (NO_2)

NO_2 can add to the double bond of conjugated dienes, but the reaction is generally slow, ranging from 1×10^{-20} cm^3 molecule^{-1} s^{-1} for 2,3-dimethyl-2-butene to 1.3×10^{-17} cm^3 molecule^{-1} s^{-1} for α-phellandrene (see Atkinson, 1997a, for a review). Addition of NO_2 to one of the double bonds generates an alkyl radical that reacts as expected, i.e., adds O_2 to form RO_2 etc. This reaction is not likely to be significant under most atmospheric conditions; for example, at 0.1 ppm NO_2, the lifetime of α-phellandrene is about 9 h, much longer than its lifetimes with respect to reaction with OH and O_3 (see Problem 9). However, it may be important in some laboratory systems where high NO_x concentrations are used. For example, it may be responsible for the observed generation of OH in the dark reaction of isoprene and other conjugated dienes (e.g., Atkinson et al., 1984) with NO_2 in the presence of NO in the dark (Tuazon and Atkinson, 1990) (see Problem 10). Harrison et al. (1998) have also proposed that it could be responsible for some of the observed NO to NO_2 conversion during some air pollution episodes at night in London.

F. REACTIONS OF ALKYNES

1. Hydroxyl Radical (OH)

The only significant loss of alkynes is reaction with OH, for which a pressure dependence is observed. Table 6.15 gives the high-pressure limiting rate constants for the OH reactions with acetylene, propyne, 1-butyne, and 2-butyne. The reaction of acetylene approaches the high-pressure limit at several thousand Torr (see Problem 5). However, for the larger alkynes, the reactions are essentially at the high-pressure limit at 1 atm (and room temperature).

TABLE 6.15 High-Pressure Limiting Rate Constants (k_∞) for the Reaction of OH Radicals with Alkynes at 298 K[a]

Alkyne	k_∞(OH) (10^{-12} cm^3 molecule^{-1} s^{-1})
Acetylene	0.90
Propyne	5.9
1-Butyne	8.0
2-Butyne	27.4

[a] From Atkinson (1994) and Atkinson et al. (1997a).

The magnitude of the rate constants, their observed pressure dependence, and the products of the reactions are consistent with the mechanism involving the initial addition of OH to the triple bond. For example, the OH–1-butyne reaction at 298 K is about a factor of three faster than the reaction with n-butane (see Table 6.2), despite the fact that it has fewer abstractable hydrogens and the $\equiv$C–H bond is much stronger than a primary –C–H bond ($\sim$ 125 vs $\sim$ 100 kcal mol^{-1}). In addition, a pressure dependence is not consistent with a simple hydrogen atom abstraction (see Chapter 5.A.2).

The reactions give as major products the corresponding dicarbonyls; i.e., acetylene gives glyoxal [$(CHO)_2$], propyne gives methylglyoxal [CH_3COCHO], and 2-butyne gives biacetyl [$(CH_3CO)_2$] (Schmidt et al., 1985; Hatakeyama et al., 1986). This is consistent with the following reaction sequence:

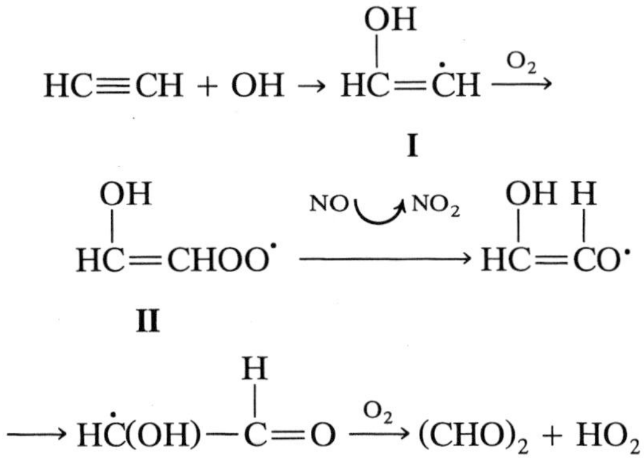

$$HC\equiv CH + OH \rightarrow HC\overset{OH}{=}\overset{\cdot}{CH} \xrightarrow{O_2}$$
$$\mathbf{I}$$

$$HC\overset{OH}{=}CHOO\cdot \xrightarrow{NO \; NO_2} HC\overset{OH}{=}\overset{H}{CO}\cdot$$
$$\mathbf{II}$$

$$\rightarrow H\overset{\cdot}{C}(OH)-\overset{H}{C}=O \xrightarrow{O_2} (CHO)_2 + HO_2 \quad (57)$$

Formic acid is another product, possibly from the rearrangement and decomposition of the intermediate **II** (Hatakeyama et al., 1986).

Siese and Zetzsch (1995) and Bohn and Zetzsch (1998) have studied the OH–C_2H_2 reaction using FP-RF (see Chapter 5.B.3) and observed biexponential decays of OH. They propose that the adduct **I** has two channels in its reaction with O_2, rather than one as shown above, and that one of the two generates OH and glyoxal, a small portion of which is excited and decomposes to HCO.

The vinoxy radical has also been observed as an intermediate (Schmidt et al., 1985), suggesting that a portion of the initial adduct isomerizes:

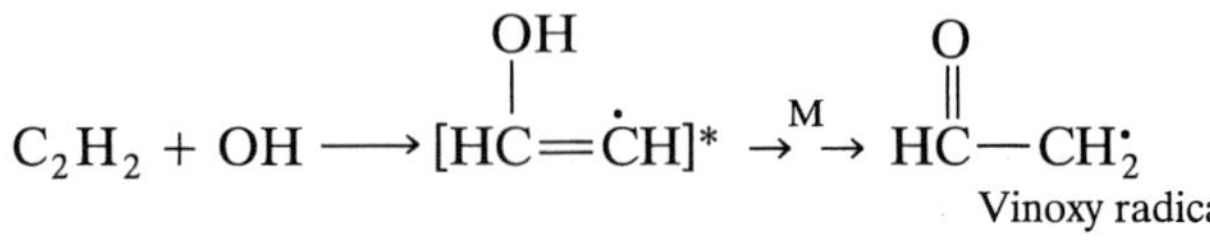

$$C_2H_2 + OH \longrightarrow [HC\overset{OH}{=}\overset{\cdot}{CH}]^* \xrightarrow{M} HC\overset{O}{-}\overset{\cdot}{CH_2}$$
$$\text{Vinoxy radical}$$
$$(58)$$

The vinoxy radical reacts rapidly with O_2 with a high-pressure limiting rate constant $k_\infty = (1.9 \pm 0.2) \times 10^{-13}$ cm^3 molecule^{-1} s^{-1} at $P \geq 400$ Torr (Zhu and Johnston, 1995). While the reaction leads to glyoxal formation, the production of glyoxal is delayed compared to the disappearance of the vinoxy radical, suggesting that a long-lived adduct with O_2 is first formed and then subsequently decomposes, in part to glyoxal,

$$CH(O)\dot{C}H_2 + O_2 \rightarrow [CH(O)CH_2O\dot{O}] \rightarrow \rightarrow$$
$$(CHO)_2 + OH, \quad (59)$$

perhaps along with the regeneration of OH, which Schmidt *et al.* (1985) also observed in the presence of O_2. However, Zhu and Johnston (1995) observed the yield of glyoxal from reaction (59) to be only 15%, whereas Hatakeyama *et al.* (1986) measured glyoxal yields in the OH–C_2H_2 reaction of ~70%. Thus reaction (59) cannot be the major path for glyoxal formation in the OH + C_2H_2 reaction. Another reaction proposed for the vinoxy radical (Gutman and Nelson, 1983) is reaction (60):

$$CH(O)CH_2 \cdot + O_2 \rightarrow OH + HCHO + CO. \quad (60)$$

In short, while the overall features of OH–alkyne reactions are understood, more research needs to be done, especially on the alkynes larger than acetylene.

G. REACTIONS OF SIMPLE AROMATIC HYDROCARBONS

1. Hydroxyl Radical (OH)

The reactions of simple aromatic hydrocarbons with OH provide a classic example of how kinetics can be used to elucidate reaction mechanisms. Figure 6.10 shows a semilogarithmic plot of the decay of OH in the presence of a great excess of toluene from 298 to 424 K at ~100 Torr total pressure in argon. While one would expect such plots to be linear (Chapter 5.B.1), this is only observed to be the case at temperatures below 325 K and above about 380 K; at the intermediate temperatures, the plots are clearly curved.

Even more unusual behavior is observed for the temperature dependence of the rate constant. Figure 6.11 shows these data in Arrhenius form for the reactions of toluene and 1,2,3-trimethylbenzene. At the higher temperatures, the Arrhenius plot is linear with a normal activation energy (i.e., the rate constant increases with increasing temperature). However, as the temperature is lowered, there is a sharp discontinuity in the plot and at lower temperatures the temperature dependence is reversed; i.e., the rate constants decrease with increasing temperature.

These results are now known to reflect the occurrence of two, quite different mechanisms over this temperature range. At the higher temperatures, ab-

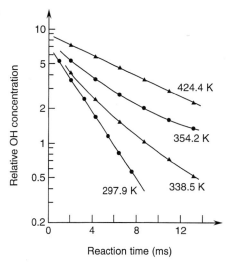

FIGURE 6.10 Semilogarithmic plots of OH decays as a function of reaction time in the presence of a great excess of toluene (~5 × 10^{13} cm^{-1}) at temperatures from 297.9 to 424.4 K and in ~100 Torr argon (adapted from Perry *et al.*, 1977).

straction of a hydrogen atom occurs, in the case of the substituted benzenes from the alkyl side chains, where the C–H bonds are weaker (~85 kcal mol^{-1} vs ~110 kcal mol^{-1} for C–H bonds in benzene):

$$OH + C_6H_5CH_3 \longrightarrow C_6H_5CH_2 + H_2O \quad (61)$$

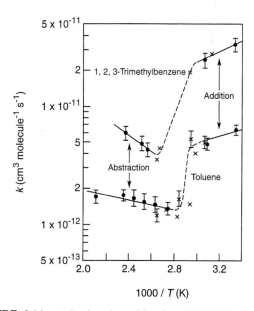

FIGURE 6.11 Arrhenius plots of log k vs 1000/T for the reaction of OH with toluene and 1,2,3-trimethylbenzene: (●) exponential OH decays observed; (×) nonexponential OH decays observed (adapted from Perry *et al.*, 1977).

The temperature dependence is that expected for a hydrogen abstraction; i.e., the rate constant increases with temperature (left side of Fig. 6.11).

As the temperature is lowered, the abstraction slows down but another reaction, addition of OH to the aromatic ring, takes place:

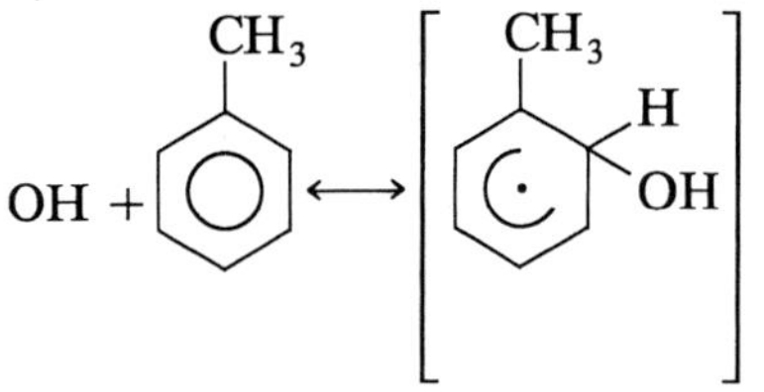

$$OH + C_6H_5CH_3 \leftrightarrow [C_6H_5(CH_3)(H)(OH)\cdot] \quad (62, -62)$$

Theoretical calculations support the expectation that the preferred site of initial OH attack is *ortho* to the methyl group (Andino *et al.*, 1996), but addition to the other positions also occurs. If the OH–aromatic adduct, which contains ~ 18 kcal mol^{-1} excess energy, is not stabilized, it decomposes back to reactants, reaction (-62). The existence of the adduct in the case of the OH–benzene reaction has been observed spectroscopically (Fritz *et al.*, 1985; Knispel *et al.*, 1990; Markert and Pagsberg, 1993; Bjergbakke *et al.*, 1996). As expected for such a mechanism, the rate constants at temperatures below ~ 300 K exhibit a pressure dependence at lower pressures. At higher temperatures, the rate of decomposition of the excited adduct back to reactants is higher, so the net contribution of adduct formation to the overall reaction is small compared to H-abstraction.

Conversely, at the lower temperatures, the rate constant for H-abstraction is small while, at the same time, the rate of adduct decomposition is lowered. As a result, at the lower temperatures (right side of Fig. 6.11), adduct formation predominates and a "negative" temperature dependence, as well as a dependence on pressure is observed for the overall rate constant. In the intermediate region, both addition and abstraction are occurring at significant rates, leading to the curved OH decay plots in Fig. 6.10 and the discontinuities in the Arrhenius plots of Fig. 6.11.

Table 6.16 shows the room temperature rate constants for the reactions of OH with some simple aromatics as well as the branching ratio for abstraction, i.e., the ratio $k_{61}/(k_{61} + k_{62})$. Abstraction accounts for less than about 10% of the reaction at room temperature for those alkylbenzenes studied to date. It is noteworthy that the reactions are all quite fast, even that for benzene being within approximately two orders of magnitude of diffusion controlled.

The products of the abstraction path are easily predictable, based on our understanding of the fates of alkyl radicals produced in alkane reactions (see Sections C and D). For example, in the case of toluene, the alkyl radical formed in reaction (61) adds O_2 and ultimately forms benzaldehyde and smaller amounts of benzyl nitrate, as shown in Fig. 6.12.

The greatest uncertainty in the mechanism is the fate of the stabilized OH–aromatic adduct. It had been assumed that it would react with O_2 and ultimately lead to the formation of oxygenated compounds. However, there is also some evidence that the kinetics of the reaction of the adduct with O_2 may be sufficiently slow at room temperature that the reaction with NO_2 may be competitive at the relatively high concentrations of NO_2 used in laboratory studies of these reactions. The rate constants at room temperature for the reactions with O_2 and NO_2 have been reported to be $\sim 5 \times 10^{-16}$ and $\sim 3 \times 10^{-11}$ cm^3 molecule^{-1} s^{-1}, respectively (Knispel *et al.*, 1990; Goumri *et al.*, 1991), in which case the removal of the adduct by these two pathways at 1 atm in air would be equal only at the very high NO_2 concentration of ~ 3.5 ppm. While these kinetic data suggest that the adduct will react with O_2 under atmospheric conditions, in many labora-

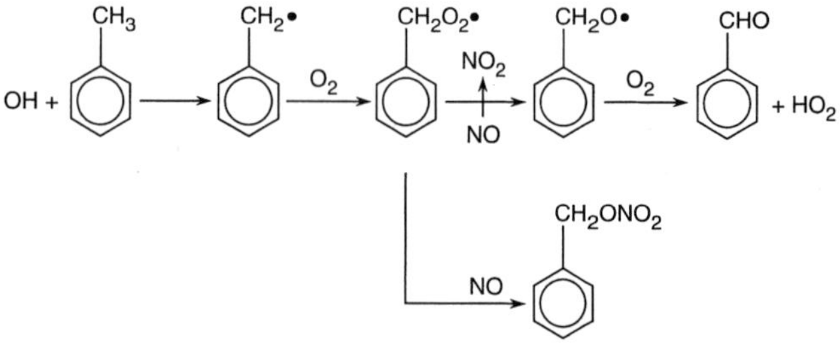

TABLE 6.16 Rate Constants at Room Temperature for OH – Aromatic Hydrocarbon Reactions and Branching Ratios for the Abstraction Reaction[a]

Aromatic	k (10^{-12} cm^3 molecule^{-1} s^{-1})	Branching ratio for abstraction
Benzene	1.2	0.05
Toluene	6.0 (5.5)[b]	0.07–0.12
Ethylbenzene	7.1	
o-Xylene	13.7	0.05–0.10
m-Xylene	23.6 (22.0)[b]	0.04
p-Xylene	14.3	0.08
n-Propylbenzene	6.0	
Isopropylbenzene	6.5	
o-Ethyltoluene	12.3	
m-Ethyltoluene	19.2	
p-Ethyltoluene	12.1	
1,2,3-Trimethylbenzene	32.7	0.06
1,3,5-Trimethylbenzene	57.5 (57.3)[b]	0.03
tert-Butylbenzene	4.6	

[a] From Atkinson (1994).
[b] Numbers in parentheses from Kramp and Paulson (1998).

FIGURE 6.12 Mechanism of formation of benzyl nitrate and benzaldehyde from reaction from the methyl side group of toluene.

tory studies where higher NO_x levels are used, a significant portion of the OH–aromatic adducts may be trapped by NO_2, rather than reacting with O_2. As a result, the products from some laboratory studies may not be directly applicable to ambient air.

More recent measurements (Bjergbakke et al., 1996) report the observation of the $HO-C_6H_6O_2$ peroxy radical by UV absorption (which has been controversial; see Koch, 1997, and Pagsberg, 1997). This study reports rate constants for the $OH-C_6H_6$ adduct of $\sim 5 \times 10^{-13}$ cm^3 $molecule^{-1}$ s^{-1} with O_2 and 1×10^{-11} cm^3 $molecule^{-1}$ s^{-1} with NO_2, in which case reaction of the adduct with NO_2 is not expected to be important. This is clearly an area that warrants further study in order to be able to extrapolate reliably the results of laboratory studies to atmospheric conditions.

Species such as cresol are known to be products of the toluene oxidation and have commonly been assumed to be formed via reactions such as (63)

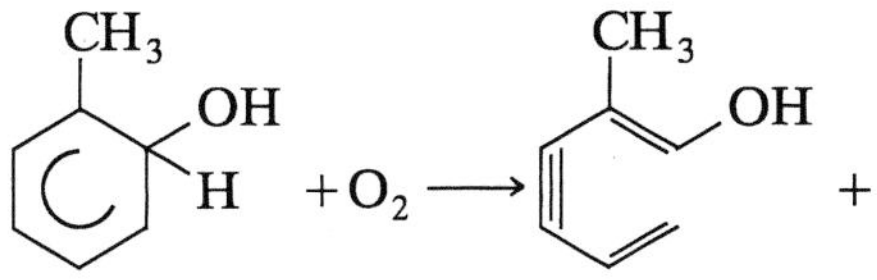

The actual mechanism may not be as simple as implied by this equation, however. For example, Narita and Tezuka (1982) have shown that, in the solution phase oxidation at least, some of the cresol product contains an oxygen atom from the O_2. In addition, in the case of reaction of benzene, phenol may be formed by direct displacement of a hydrogen atom by OH (Bjergbakke et al., 1996; Koch, 1997; Pagsberg, 1997).

There are also data suggesting that O_2 reacts with the OH–aromatic adduct by abstraction of the hydroxyl hydrogen to give an epoxide that photolyzes to the alcohol. Such epoxides are known to be in equilibrium with the isomeric oxepins. For the benzene reaction, for example, the formation of phenol may occur via the following mechanism:

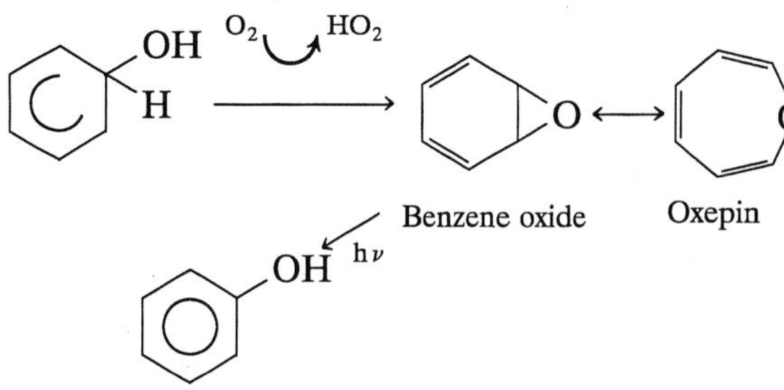

For example, Klotz et al. (1997, 1998) have shown that benzene oxide/oxepin photolyzes in sunlight to give phenol with a yield of 43.2 ± 4.5%. This reaction mechanism is therefore feasible for the formation of phenol in the benzene–OH reaction. However, photolysis of toluene 1,2-oxide/2-methyloxepin gave o-cresol only in small yields, 2.7 ± 2.2% (Klotz et al., 1998); this suggests that cresols formed in the OH–toluene reaction come primarily from the direct reaction (63) of the OH adduct with O_2, in contrast to the conclusions of Moschonas et al. (1999).

A variety of smaller multifunctional oxygenated compounds are also found as products of the gas-phase OH–aromatic reactions. Table 6.17 shows the yields of the smallest dicarbonyl compounds from these reactions, which, while small, are not insignificant. In addition to these products, a variety of other multifunctional compounds are typically found, the numbers, types, and concentrations of these products depending on the analytical methodologies used, the reaction conditions, and the skill and imagination of the experimentalist! Table 6.18, for example, shows some products observed in the photooxidation of toluene in air where the loss is due to attack by OH (Dumdei et al., 1988). In this particular study, ~44% of the reacted toluene could be accounted for by the products shown in Table 6.18.

Similarly, Yu et al. (1997) and Yu and Jeffries (1997) report a total of 50 products from the oxidation of

TABLE 6.17 Fractional Yields of Dicarbonyl Compounds from the OH Reaction with Some Aromatic Hydrocarbons at 1 atm Total Pressure and 298 K[a]

Aromatic	Glyoxal $(CHO)_2$	Methylglyoxal CH_3COCHO	Biacetyl $(CH_3CO)_2$
Benzene	0.21		
Toluene	0.08–0.15	0.08–0.15	
o-Xylene	0.03–0.09	0.12–0.25	0.09–0.26
m-Xylene	0.09–0.13	0.27–0.42	
p-Xylene	0.12–0.24	0.11	
1,2,3-Trimethylbenzene	0.06–0.07	0.15–0.18	0.32–0.45
1,2,4-Trimethylbenzene	0.05–0.08	0.36	0.05–0.11
1,3,5-Trimethylbenzene		0.60–0.64	

[a] From Atkinson (1994).

TABLE 6.18 Some Products Observed in the Photooxidation of Toluene[a]

Species	Percentage yield
$CH_3C(O)CHO$	7.7
$(CHO)_2$	5.8
Methylbutenedial	5.8
C_6H_5CHO	5.0
(Hydroxymethyl)butenedial	3.4
Peroxyacetyl nitrate	3.3
Oxoheptadienal	2.7
CH_3COOH	1.8
HCHO	1.0
Hexadienal	1.0
Hydroxyoxoheptadienal	1.0
Hydroxydioxohexenal	0.8
Dioxohexenal	0.7
Hydroxyoxohexenal	0.6
Hydroxyhexadienal	0.6
Methyl vinyl ketone	0.5
Methylfuran	0.5
Hydroxyoxobutanal	0.5
Hexadienedial	0.3
Hydroxymethyl vinyl ketone	0.3
Butenedial	0.3
Hydroxybutenedial	0.2
CH_3CHO	0.2
Pentadienal	0.1
Acrolein	0.06
Pyruvic acid	0.06
Total	44.2

[a] From Dumdei *et al.* (1988); see also Yu *et al.* (1997).

study (Bartolotti and Edney, 1995). Kwok *et al.* (1997) have studied the reactions of OH with *o*-, *m*-, and *p*-xylene and identified a variety of open-chain multifunctional products, including

$$HC(O)CH=CHCHO,$$
$$CH_3C(O)CH=CHCHO,$$
$$CH_3C(O)CH=CHC(O)CH_3,$$
$$CH_3C(O)C(CH_3)=CHCH=CHCHO,$$

and their isomers. It is typical of such aromatic oxidations that only ~40–70% of the reacted parent organic can be accounted for in measured products.

Figure 6.13 shows one postulated reaction sequence taking the adduct to methylglyoxal and butenedial. This represents just one of many possible reaction paths. OH can also add to the ring at the *meta* or *para* positions, in addition to the *ortho* position shown in Eq. (62). The addition of O_2, shown in Fig. 6.13 to occur at the 1-position relative to the OH group, could also occur at the 3- or 5-positions. Ring-cleavage products are also observed in the absence of NO_x (Atkinson and Aschmann, 1994; Seuwen and Warneck, 1996). Subsequent cyclization of the peroxy radical to form allylically stabilized five-membered bicyclic radicals is expected to be the most energetically favored, with the formation of nonallylically stabilized biradicals being endothermic (Andino *et al.*, 1996).

Theoretical studies also suggest that some of the peroxy radicals formed by addition of O_2 to the OH–aromatic adducts may react with NO in competition with cyclization (Andino *et al.*, 1996), generating NO_2 and an aromatic peroxy radical that may subsequently form phenolic and unsaturated derivatives through reactions with O_2:

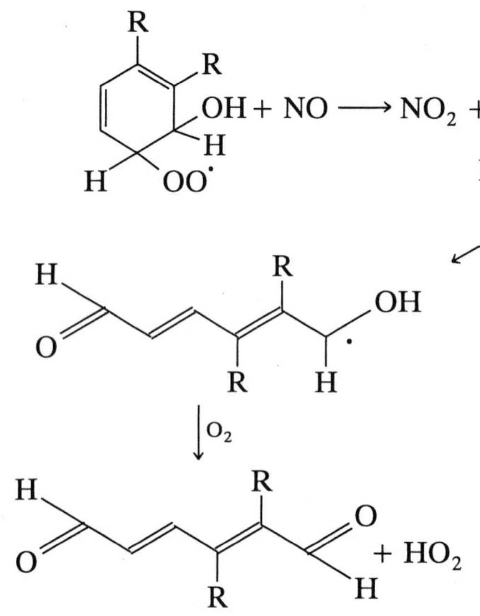

toluene in air in the presence of NO_x. These include additional compounds from those in Table 6.18, such as benzoquinones and epoxides, the latter hypothesized to be formed from an epoxide type of structure predicted theoretically for the OH–aromatic–O_2 adduct in one

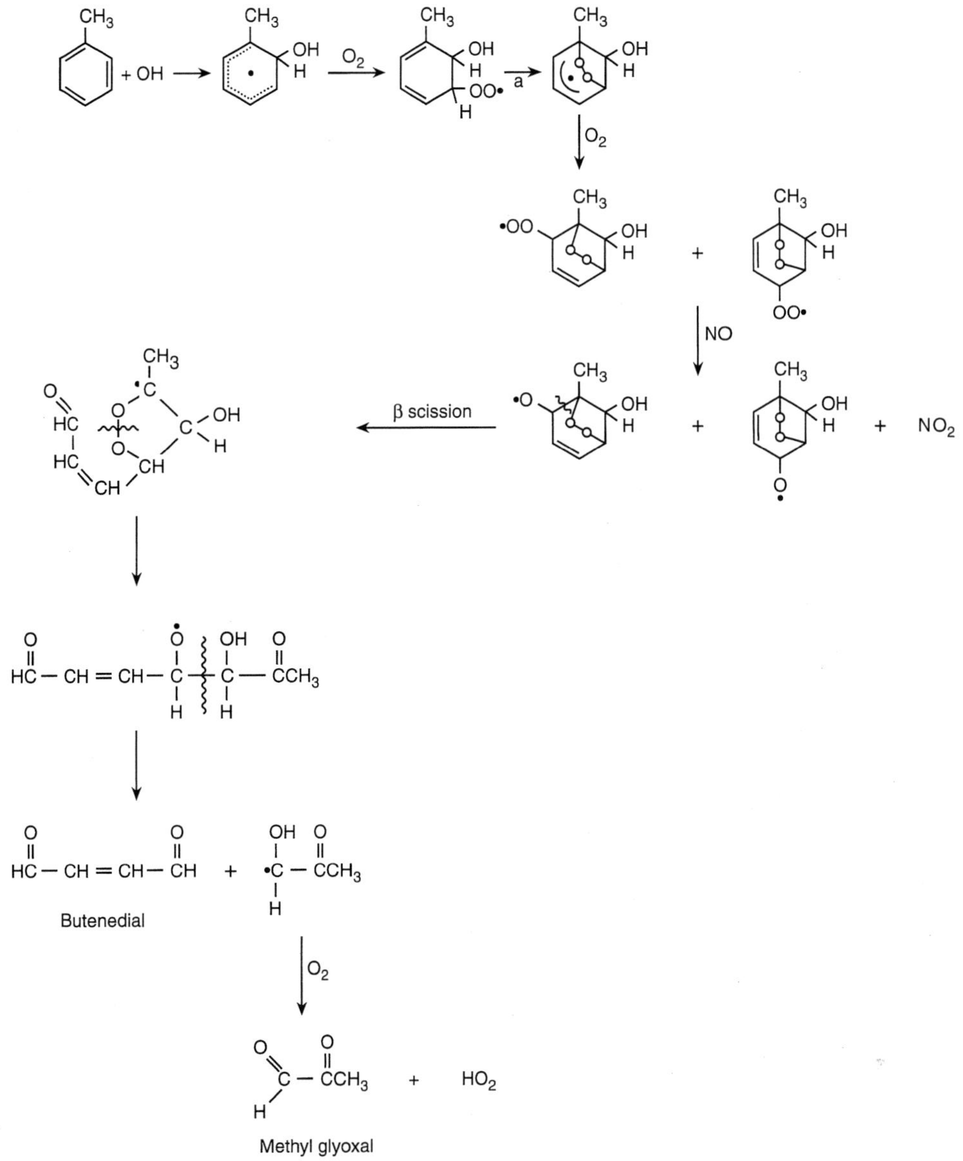

FIGURE 6.13 One postulated fate of the OH–toluene adduct in air.

The simplest phenoxy radical, C_6H_5O, does not react with O_2 ($k < 5 \times 10^{-21}$ cm^3 molecule^{-1} s^{-1}) but does react with NO ($k = 1.9 \times 10^{-12}$ cm^3 molecule^{-1} s^{-1}) and with NO_2 ($k = 2.1 \times 10^{-12}$ cm^3 molecule^{-1} s^{-1}), suggesting that reactions with NO_x will be its primary fate in the troposphere (Platz et al., 1998b). However, this may not be the case for the larger, hydroxylated phenoxy radicals from the OH–aromatic–O_2–NO reaction sequence.

In short, there are a multitude of potential reaction pathways that must be considered in OH–aromatic reactions, and the details of the mechanism remain to be elucidated.

The formation of multifunctional, highly reactive "products" such as butenedial may be responsible for much of the missing carbon in these reactions. Such products will also react rapidly with OH, with O_3, and, when present, with NO_3, as well as photolyze. For example, Fig. 6.14 shows the infrared spectrum obtained after a mixture of the cis and trans forms of butenedial were photolyzed using fluorescent lamps with wavelengths $320 \le \lambda \le 480$ nm and the absorptions due to the products CO, CO_2, HCHO, and HCOOH were subtracted out (Bierbach et al., 1994). Most of the remaining bands are due to the product 3H-furan-2-one formed from an intramolecular rearrangement of butenedial.

Reaction of butenedial with OH was also shown to give maleic anhydride as a major product (along with glyoxal). Figure 6.15 shows possible mechanisms for

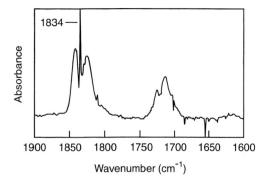

FIGURE 6.14 Infrared spectrum due primarily to 3-H-furan-2-one formed in the photolysis of butenedial after subtraction of bands due to CO, CO_2, HCHO, and HCOOH (adapted from Bierbach et al., 1994).

formation of maleic anhydride from butenedial, which is itself formed in the toluene–OH oxidation.

Further complicating mechanistic understanding is the possibility of different isomers of some of the multifunctional products. For example, one of the products of the *p*-xylene–OH reaction is 3-hexene-2,5-dione:

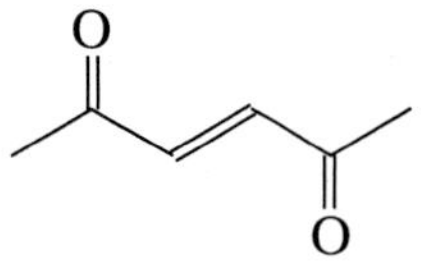

As shown in Fig. 6.16, it is expected to react with OH to form 4-hydroxyhexane-2,3,5-trione, which, as Wiesen et al. (1995) have shown, exists in equilibrium with the enediol form. Further reaction of the enediol with OH forms the trione shown in Fig. 6.16. Wiesen et al. (1995) suggest that such polyketones could account for 25–30% of the carbon, which would improve the carbon balance in OH–aromatic reactions considerably.

In short, the mechanism of OH–aromatic reactions remains today one of the least understood areas in tropospheric chemistry.

2. Nitrate Radical (NO_3)

As seen in Table 6.1, the reactions of the nitrate radical with the simple aromatic hydrocarbons are generally too slow to be important in the tropospheric decay of the organic. However, one of the products of the aromatic reactions, the cresols, reacts quite rapidly with NO_3. *o*-Cresol, for example, reacts with NO_3 with a room temperature rate constant of 1.4×10^{-11} cm^3 molecule^{-1} s^{-1}, giving a lifetime for the cresol of only ~1 min at 50 ppt NO_3. This rapid reaction is effectively an overall hydrogen abstraction from the phenolic OH (although the mechanism is likely not a direct abstraction process; e.g., Atkinson, 1994):

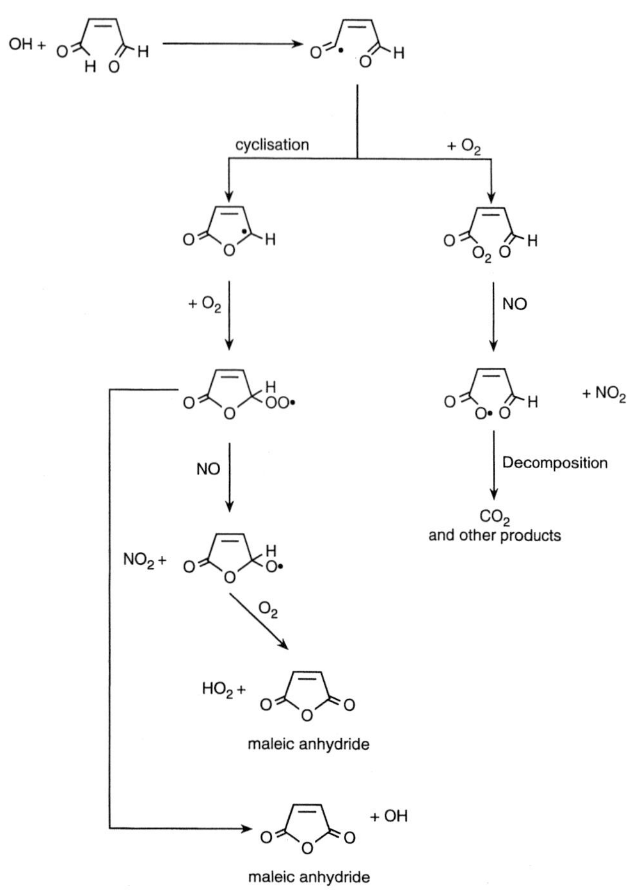

FIGURE 6.15 Postulated reaction scheme from the reaction of *cis*-butenedial with OH (adapted from Bierbach et al., 1994).

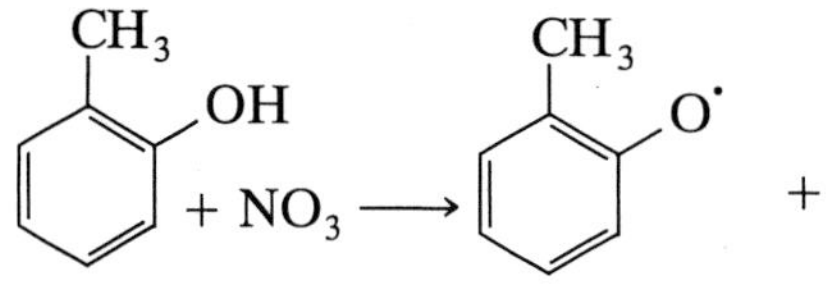

(64)

3. Chlorine Atoms (Cl)

Chlorine atoms react with aromatic hydrocarbons, but only at a significant rate with those having saturated side chains from which the chlorine atom can abstract a hydrogen or unsaturated side chains to which it can add. For example, the rate constant for the Cl atom reaction with benzene is 1.3×10^{-15} cm^3 molecule^{-1} s^{-1} (Shi and Bernhard, 1997). On the other hand, the rate constants for the reactions with toluene and *p*-xylene are 0.59×10^{-10} and 1.5×10^{-10} cm^3 molecule^{-1} s^{-1}, respectively (Shi and Bernhard, 1997), and that for reaction with *p*-cymene is 2.1×10^{-10} cm^3 molecule^{-1} s^{-1} (Finlayson-Pitts et al., 1999). Hence

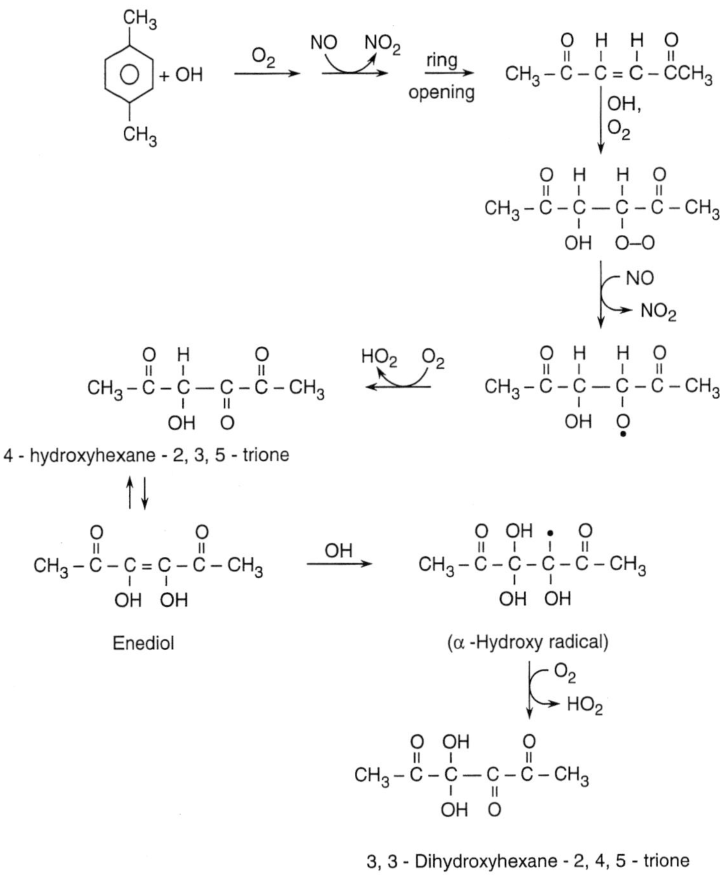

FIGURE 6.16 Formation of polyketones in the OH reaction with *p*-xylene (adapted from Wiesen *et al.*, 1995).

reactions of aromatic hydrocarbons with chlorine atoms will be significant primarily for those species having reactive groups attached to the ring.

H. REACTIONS OF OXYGEN-CONTAINING ORGANICS

1. Reactions of OH, NO_3, and Cl

Table 6.19 gives the room temperature rate constants for the reactions of some oxygen-containing organics with OH as well as with NO_3 and with chlorine atoms. As expected, the OH reactions are reasonably fast, ranging from $\sim 10^{-13}$ cm^3 molecule^{-1} s^{-1} with acetone to more than 10^{-11} cm^3 molecule^{-1} s^{-1} with the aldehydes, furan, and, in general, compounds having an alkyl group larger than $-CH_3$. Chlorine atoms also react very quickly. The NO_3 reactions in most cases are sufficiently slow that OH is the only significant tropospheric oxidant on a global basis for these oxygenated compounds. However, NO_3 can play a role in aldehyde oxidations, especially at night when OH levels are very low.

Aldehydes. Reactions with aldehydes occur by abstraction of the relatively weak (~ 87 kcal mol^{-1}) aldehydic hydrogen:

$$OH(NO_3, Cl) + RCHO$$
$$\rightarrow RCO + H_2O\ (HNO_3, HCl). \quad (65)$$

The RCO radical produced then adds O_2 as expected. For example, for acetaldehyde, reaction (65) is followed by

$$CH_3\overset{O}{\overset{\|}{C}} + O_2 \rightarrow CH_3\overset{O}{\overset{\|}{C}}-OO \xrightarrow{NO} CH_3\overset{O}{\overset{\|}{C}}-O + NO_2, \quad (66)$$

$$\xrightarrow{NO_2} CH_3\overset{O}{\overset{\|}{C}}-OONO_2,$$
$$\text{PAN} \quad (67)$$

$$CH_3\overset{O}{\overset{\|}{C}}O \longrightarrow CH_3 + CO_2. \quad (68)$$

TABLE 6.19 Room Temperature Rate Constants (cm^3 molecule^{-1} s^{-1}) for the Reactions of Some Oxygen-Containing Organics[a]

Organic	OH	NO$_3$	Cl
Aldehydes			
HCHO	9.2×10^{-12}	5.8×10^{-16}	7.3×10^{-11} [w]
CH$_3$CHO	1.6×10^{-11}	2.7×10^{-15}	7.2×10^{-11}
CH$_3$CH$_2$CHO	2.0×10^{-11}		1.2×10^{-10}
(CHO)$_2$	1.1×10^{-11}		
CH$_2$=C(CH$_3$)—CHO (Methacrolein)	3.4×10^{-11}	$<8 \times 10^{-15}$ [e]	
Ketones			
CH$_3$COCH$_3$	2.2×10^{-13}	$<3 \times 10^{-17}$	3.5×10^{-12}
CH$_3$COCH$_2$—CH(CH$_3$)—CH$_3$	1.4×10^{-11}		
CH$_2$=CH—C(O)—CH$_3$ (Methyl vinyl ketone)	1.9×10^{-11}	$<1.2 \times 10^{-16}$ [e]	2.0×10^{-10} [x]
Alcohols			
CH$_3$OH	9.3×10^{-13}	2.4×10^{-16}	5.5×10^{-11} [c]
C$_2$H$_5$OH	3.2×10^{-12}	$<2 \times 10^{-15}$	9.4×10^{-11}
n-C$_3$H$_7$OH	5.5×10^{-12}		1.5×10^{-10}
(CH$_2$OH)$_2$	1.5×10^{-11} [k]		
CH$_3$CH(OH)CH$_2$OH	2.2×10^{-11} [k]		
CH$_3$—C(CH$_3$)(OH)—CH=CH$_2$ (2-Methyl-3-buten-2-ol)	6.4×10^{-11} [f]	1.2×10^{-14} [e]	
Acids			
HCOOH	4.5×10^{-13}		2.0×10^{-13}
CH$_3$COOH	8×10^{-13}		2.8×10^{-14}
CH$_3$CH$_2$COOH	1.2×10^{-12}		
Epoxides			
CH$_3$CH$_2$CH(O)CH$_2$	1.9×10^{-12} [b]		
Furans			
Furan	4.2×10^{-11} [c]	1.0×10^{-12} [t]	
2-Methylfuran	6.2×10^{-11} [c]	2.6×10^{-11} [t]	
3-Methylfuran	9.3×10^{-11}	$(1.3–2.9) \times 10^{-11}$ [u,t]	
2-Ethylfuran	1.1×10^{-10} [c]		
2,5-Dimethylfuran	1.3×10^{-10} [c]	5.8×10^{-11} [t]	
Ethers			
CH$_3$OCH$_3$	3.0×10^{-12}		
C$_2$H$_5$OC$_2$H$_5$	1.3×10^{-11}		
(CH$_3$CH$_2$CH$_2$CH$_2$)$_2$O	2.9×10^{-11} [j]		
CH$_3$OC(CH$_3$)$_3$ (MTBE)	3.1×10^{-12} [g]		
CH$_3$CH(OH)CH$_2$OCH$_3$	2.1×10^{-11} [k]		
1,3,5-Trioxane (C$_3$H$_6$O$_3$)	6.0×10^{-12} [l]		1.0×10^{-10} [l]
CH$_3$OCH$_2$OCH$_3$	4.8×10^{-12} [m]		
CH$_3$OCH$_2$CH$_2$OCH$_3$	$(2.7–4.1) \times 10^{-11}$ [n]		
CH$_3$CH$_2$OC(CH$_3$)$_3$ (ETBE)	8.8×10^{-12} [g]	8.71×10^{-15} [d]	

(continues)

H. REACTIONS OF OXYGEN-CONTAINING ORGANICS

TABLE 6.19 (*Continued*)

Organic	OH	NO$_3$	Cl
Ethers (*Continued*)			
(CH$_3$)$_3$C—O—C(CH$_3$)$_3$ Di-*tert*butyl ether	$3.7 \times 10^{-12\,h}$	$2.8 \times 10^{-16\,h}$	$1.4 \times 10^{-10\,h}$
Formates, acetates, and other esters			
HC(O)OCH$_3$	$1.7 \times 10^{-13\,n}$		
HC(O)OC$_2$H$_5$	$8.5 \times 10^{-13\,n}$		
HC(O)OC$_4$H$_9$	$3.5 \times 10^{-12\,n}$		
HC(O)C(CH$_3$)$_3$	$7.5 \times 10^{-13\,n}$		
CH$_3$C(O)OCH$_3$	$3.2 \times 10^{-13\,o}$		
C$_2$H$_5$C(O)OCH$_3$	$8.3 \times 10^{-13\,s}$		
n-C$_4$H$_9$C(O)OCH$_3$	$4.8 \times 10^{-12\,s}$		
CH$_3$C(O)OC$_2$H$_5$	$1.7 \times 10^{-12\,o,v}$		
CH$_3$C(O)OC$_4$H$_9$	$5.5 \times 10^{-12\,o,v}$		
CH$_3$C(O)OCH(CH$_3$)$_2$	$3.8 \times 10^{-12\,p,v}$		
CH$_3$C(O)OC(CH$_3$)$_3$	$5.6 \times 10^{-13\,p,v}$		
C$_2$H$_5$OCH$_2$CH$_2$C(O)OC$_2$H$_5$	$2.3 \times 10^{-11\,q}$		
CH$_3$OC(O)(CH$_2$)$_x$C(O)OCH$_3$			
$x = 2$	$1.4 \times 10^{-12\,r}$		
$x = 3$	$3.3 \times 10^{-12\,r}$		
$x = 4$	$8.4 \times 10^{-12\,r}$		
Glycol ethers (ROCH$_2$CH$_2$OH)			
R = CH$_3$	$(1.1-1.4) \times 10^{-11\,i,m}$		
R = C$_2$H$_5$	$(1.5-1.9) \times 10^{-11\,i,m}$		
R = C$_3$H$_7$	$1.6 \times 10^{-11\,i}$		
R = C$_4$H$_9$	$1.9 \times 10^{-11\,i}$		

[a] Unless otherwise stated, from Atkinson *et al.* (1997a) or Atkinson (1994). The additional references cited are not intended to be comprehensive, but merely illustrative of studies of these reactions. [b] From Wallington *et al.* (1988). [c] From Bierbach *et al.* (1992). [d] From Langer and Ljungström (1994). [e] From Rudich *et al.* (1996). [f] From Rudich *et al.* (1995). [g] From Teton *et al.* (1996). [h] From Langer *et al.* (1996). [i] From Stemmler *et al.* (1996). [j] From Kramp and Paulson (1998). [k] From Aschmann and Atkinson (1998a). [l] From Platz *et al.* (1998a). [m] From Porter *et al.* (1997). [n] From Le Calvé *et al.* (1997a). [o] From El Boudali *et al.* (1996). [p] From Le Calvé *et al.* (1997b). [q] From Baxley *et al.* (1997). [r] From Aschmann and Atkinson (1998b). [s] From Le Calvé *et al.* (1997c). [t] From Kind *et al.* (1996). [u] From Alvarado *et al.* (1996). [v] From Picquet *et al.* (1998). [w] From DeMore *et al.* (1997) [x] From Finlayson-Pitts *et al.* (1999).

Ketones. Reactions of ketones are similar to those of alkanes, with abstraction by OH, NO$_3$, and Cl occurring from the alkyl chains. In the case of acetone, for example, these reactions generate the radical CH$_3$COCH$_2$O$_2$ in air. As for typical RO$_2$, this reacts rapidly with NO ($k^{298} = 8 \times 10^{-12}$ cm^3 molecule^{-1} s^{-1}) and with NO$_2$ ($k^{298} = 6.4 \times 10^{-12}$ cm^3 molecule^{-1} s^{-1}) (Sehested *et al.*, 1998b), giving calculated lifetimes with respect to reaction with 100 ppt each of NO and NO$_2$ of about 1 min. In remote areas where the NO$_x$ concentrations can be much smaller (see later and Chapter 11), reactions of CH$_3$COCH$_2$O$_2$ with HO$_2$ and other RO$_2$ may also become important.

Photolysis of acetone also contributes significantly to its loss and is the dominant loss process in the upper troposphere (Gierczak *et al.*, 1998; see Section J.3).

The peroxynitrate CH$_3$COCH$_2$OONO$_2$ formed from the NO$_2$ reaction thermally decomposes, with a rate constant of ~ 3 s^{-1} at 700 Torr and 295 K. Sehested *et al.* (1998b) suggest that its lifetime with respect to thermal decomposition is sufficiently small even at the lower temperatures of the upper troposphere that this species cannot participate in long-range transport of NO$_x$, as is the case for PAN, and that in the upper troposphere, reaction of CH$_3$COCH$_2$OONO$_2$ with OH and photolysis will be major fates.

Alcohols. Reactions of the simple alcohols in air are of particular interest because of their use as alternate fuels, including as blends with gasoline (see Chapter 16.D). There are two possible abstraction sites for the reaction with methanol, the alcohol O—H or the alkyl

C–H. Since the alcohol O–H bond is stronger (~104 kcal mol^{-1}) than the alkyl C–H bond (~96 kcal mol^{-1}) (Lide, 1998–1999), one would expect that abstraction from the alkyl group would predominate, e.g., reaction (69a) would be faster than reaction (69b):

$$CH_3OH + OH \rightarrow CH_2OH + H_2O, \quad (69a)$$
$$\rightarrow CH_3O + H_2O. \quad (69b)$$

In agreement with these expectations, reaction (69a) does indeed appear to account for ~85% of the overall reaction (McCaulley et al., 1989).

In the case of ethanol, three different hydrogen abstractions are possible:

$$CH_3CH_2OH + OH \rightarrow CH_3\dot{C}HOH + H_2O, \quad (70a)$$
$$\rightarrow \cdot CH_2CH_2OH + H_2O, \quad (70b)$$
$$\rightarrow CH_3CH_2O\cdot + H_2O. \quad (70c)$$

Abstraction of the secondary hydrogen, reaction (70a), accounts for about 90% of the overall reaction, with the remainder split about equally between (70b) and (70c) (Atkinson et al., 1997a). Similarly, the reaction of OH with 2-butanol and 2-pentanol proceeds predominantly by abstraction of the alkyl hydrogen of the –CH(OH) group (Chew and Atkinson, 1996; Baxley and Wells, 1998).

Carboxylic acids. The mechanism of the OH reaction with carboxylic acids is interesting in that it appears to undergo both addition to form a hydrogen-bonded complex, and abstraction. For the reactions with formic and acetic acids, for example, Singleton, Paraskevopoulos, and co-workers (Jolly et al., 1986; Singleton et al., 1988, 1989) proposed the following mechanism for the formation and reaction of the hydrogen-bonded complex:

$$OH + CH_3C\begin{smallmatrix}OH\\\\O\end{smallmatrix} \longleftrightarrow \left[CH_3C\begin{smallmatrix}OH\\\\O\cdots H-O\end{smallmatrix}\right]$$

Hydrogen-bonded complex
$$\downarrow$$

$$CH_3C\begin{smallmatrix}O\cdot\\\\O\end{smallmatrix} + H_2O \longleftarrow \left[CH_3C\begin{smallmatrix}O\cdots H\\\\O\cdots H\end{smallmatrix}O\right]$$

Thus, it is the carboxylic hydrogen that is ultimately abstracted by this channel. This is consistent with the decrease in the room temperature rate constant upon deuterium substitution from 5.2×10^{-11} to 4.9×10^{-11} to 1.4×10^{-11} cm^3 molecule^{-1} s^{-1} for OH + CH$_3$COOH vs OH + CD$_3$COOH vs OH + CD$_3$COOD. Similarly, the rate constants for OH + HCOOH and DCOOH are identical (Wine et al., 1985). That is, only deuteration of the acidic hydrogen has a large effect on the rate constant. The acid dimer also reacts about two orders of magnitude slower than the monomer, because the carboxylic hydrogen is tied up in hydrogen bonding in the dimers:

$$CH_3-C\begin{smallmatrix}O\cdots H\\\\O\cdots H\end{smallmatrix}\begin{smallmatrix}O\\\\O\end{smallmatrix}C-CH_3$$

In competition with the formation and subsequent reaction of the OH–acid complex, there is also a direct hydrogen abstraction channel:

$$OH + CH_3COOH \rightarrow H_2O + \dot{C}H_2COOH.$$

In the case of acetic acid, this channel is small at room temperature. The relative rates of the two channels for formic, acetic, and propionic acids are discussed in detail by Singleton et al. (1989).

Overall, given the slowness of the reactions (lifetime ~26 days for HCOOH at [OH] = 1×10^6 radicals cm^{-3}) and the high solubilities and "stickiness" of these acids, they are likely removed primarily by wet and dry deposition rather than by reaction with OH.

OH radical reactions with other oxygen-containing organics such as ethers and glycol ethers, which are used as gasoline additives, solvents, and intermediates in chemical manufacturing, occur by hydrogen abstraction, as expected (e.g., see Wells et al. (1996) for the reaction of 2-ethoxy acetate, Stemmler et al. (1997) for the reaction of 2-butoxyethanol, and Tuazon et al. (1998b) for reactions of acetates). For example, Kerr and co-workers (Eberhard et al., 1993) report that the OH-radical reaction with diethyl ether gives ethyl formate as the major product, with smaller amounts of ethyl acetate and acetaldehyde, as well as some other products that can be rationalized by the mechanisms discussed earlier (see Problem 6). The mechanism of the NO$_3$ radical reactions with furan and tetramethylfuran is treated by Berndt et al. (1997).

2. Hydroperoxyl Radical (HO$_2$)

Interestingly, while HO$_2$ is fairly unreactive toward most organics, it does react quite readily with HCHO and certain other aldehydes such as CH$_3$CHO, CH$_3$COCHO, and (CHO)$_2$:

$$HO_2 + H_2C=O \leftrightarrow OOCH_2OH \quad (71, -71)$$

followed by

$$OOCH_2OH \leftrightarrow HOOCH_2O \rightarrow HCHO + HO_2.$$

Theoretical studies suggest reaction (71) occurs by a concerted process in which the H from HO_2 is transferred simultaneously to the oxygen of $C=O$, while the terminal oxygen of HO_2 adds to the carbon (Evleth *et al.*, 1993). The forward rate constant, k_{71}, is 7.9×10^{-14} cm^3 molecule^{-1} s^{-1}, leading to a lifetime of HCHO with respect to HO_2 of only 7 h at $[HO_2] = 5 \times 10^8$ radicals cm^{-3}. However, the reverse decomposition is also fast, $k_{-71} = 150$ s^{-1} at 298 K. If the rate constant for the reaction of the peroxy radical formed in (71) with NO is $\sim 8 \times 10^{-12}$ cm^3 molecule^{-1} s^{-1},

$$OOCH_2OH + NO \rightarrow OCH_2OH + NO_2, \quad (72)$$

then even at 100 ppb NO, found only in highly polluted areas, the removal of $OOCH_2OH$ in reaction (72) would only be 13% of that via its decomposition, reaction (−71).

In short, while the addition of HO_2 to aldehydes is fast, the decomposition of the adduct back to reactants is sufficiently fast compared to alternate reactions that little net loss of the aldehyde occurs under most conditions.

I. REACTIONS OF NITROGENOUS ORGANICS

1. Peroxyacetyl Nitrate and Its Homologs

Shortly after the discovery of photochemical air pollution, Stephens and his co-workers (Stephens *et al.*, 1956; Stephens, 1987) applied long-path infrared spectroscopy to identifying and measuring products in the photooxidation of organic–NO_x mixtures. In the photooxidations of 3-methylheptane and, to a larger extent, 2,3-butanedione, a set of infrared bands that could not be assigned to known products was observed. These were assigned to a previously unobserved species, which was initially called "compound X." It was ultimately shown to be peroxyacetyl nitrate (PAN):

$$\overset{O}{\underset{}{\overset{\|}{CH_3C}}}OONO_2$$
PAN

For a detailed discussion of the discovery of PAN in the context of early studies of photochemical air pollution, the reader is referred to an article by Stephens (1987).

PAN is the simplest member of a series of compounds known as peroxyacyl nitrates, having the structure

$$\overset{O}{\underset{}{\overset{\|}{RC}}}OONO_2$$

The nomenclature for this homologous series is somewhat confused. The term "PANs" has been used historically to denote peroxyacyl nitrates, and this terminology continues to be used extensively in the literature, despite the lack of adherence to traditional IUPAC rules of nomenclature. Because the PANs can be considered to be mixed anhydrides of carboxylic acids and nitric acid, another suggestion (Roberts, 1990) has been "peroxyacetic nitric anhydride" for $CH_3C(O)OONO_2$ and "peroxy carboxylic nitric anhydrides" for the whole class of compounds. Although it does not follow the IUPAC rules, it would be consistent with the widespread use of the name "PAN" but also reflect the structure more accurately. Table 6.20 shows the structures and commonly used names of some PANs that have been observed in the atmosphere and/or in laboratory studies.

PAN is known to play an important role in tropospheric chemistry. As discussed in this section, its thermal decomposition releases both NO_2 and an organic free radical, so that it can act as an NO_x reservoir and ultimately as a source of OH in the dark. In addition, PAN is a strong lachrymator (eye irritant), is mutagenic in certain bacterial assays, and is phytotoxic to plants. Because of these broad effects on a variety of systems, its formation and reactions have been studied in some detail.

As seen earlier in this chapter, the oxidation of organics produces a variety of free radicals, some of which can form peroxyacyl nitrates. For example, acetaldehyde is a classic precursor to the PAN through the following reaction sequence:

$$CH_3CHO + OH \rightarrow CH_3CO + H_2O, \quad (73)$$
$$CH_3CO + O_2 + M \rightarrow CH_3C(O)OO + M, \quad (74)$$
$$CH_3C(O)OO + NO_2 + M \leftrightarrow CH_3C(O)OONO_2 + M. \quad (75, -75)$$

As expected for a termolecular reaction, reaction (74) proceeds through the formation of an excited intermediate that can decompose back to reactants at lower pressures. At 298 K and 1 atm pressure, $k_{74} = 3.2 \times 10^{-12}$ cm^3 molecule^{-1} s^{-1} (e.g., Tyndall *et al.*, 1997), about half that for other alkyl radical reactions with O_2 discussed earlier.

A key feature of reaction (75) is that it is reversible; i.e., PAN thermally decomposes back to peroxyacetyl

TABLE 6.20 Structures and Commonly Used Names for Some Peroxyacyl Nitrates Found in the Atmosphere and/or in Laboratory Studies

Name	Acronym	Structure
Peroxyacetyl nitrate (peroxyacetic nitric anhydride)	PAN	$CH_3\overset{\overset{O}{\|\|}}{C}OONO_2$
Peroxypropionyl nitrate	PPN	$CH_3CH_2\overset{\overset{O}{\|\|}}{C}OONO_2$
Peroxy-n-butyryl nitrate	PnBN	$CH_3CH_2CH_2\overset{\overset{O}{\|\|}}{C}OONO_2$
Peroxybenzoyl nitrate	PBzN	$C_6H_5\overset{\overset{O}{\|\|}}{C}OONO_2$
Peroxymethacryloyl nitrate	MPAN	$CH_2=\underset{\underset{}{}}{\overset{\overset{H_3C}{\|}}{C}}-\overset{\overset{O}{\|\|}}{C}OONO_2$

radicals and NO$_2$. Indeed, it is this thermal decomposition that accounts for most of the loss of PAN under typical conditions in the troposphere.

The simplest compound in the series, PAN, is also the one found in the highest concentrations in the atmosphere. While PAN, PPN, PnBN, PBzN, and MPAN have all been identified in ambient air, PAN predominates by a large margin, and it is only for PAN and, to a lesser extent, PPN that a large number of measurements have been reported (e.g., see Gaffney et al., 1989; Roberts, 1990; Altshuller, 1993; Williams et al., 1993; Grosjean et al., 1993a, 1993b; Kleindienst, 1994; and Chapter 11.A.4g). Measurements in a number of locations around the world typically show PPN/PAN ratios of ~0.1–0.3. In highly polluted air, PAN has been observed to reach concentrations as high as 70 ppb.

The oxidation of isoprene, one of the most common biogenic hydrocarbons, is expected to lead to the formation of some MPAN. Thus, as treated earlier, one of the major products of isoprene oxidation is methacrolein. Its subsequent oxidation by OH is expected to form MPAN:

$$CH_2=\underset{Methacrolein}{\overset{CH_3}{\underset{|}{C}}}-\overset{O}{\underset{\|}{C}}-H \xrightarrow{OH \quad H_2O} CH_2=\overset{CH_3}{\underset{|}{C}}-\overset{O}{\underset{\|}{C}}\cdot \downarrow O_2$$

$$CH_2=\underset{MPAN}{\overset{CH_3}{\underset{|}{C}}}-\overset{O}{\underset{\|}{C}}OONO_2 \xleftarrow{NO_2} CH_2=\overset{CH_3}{\underset{|}{C}}-\overset{O}{\underset{\|}{C}}OO\cdot$$

(76)

Like the other higher PANs, MPAN is expected to be formed in relatively small yields and indeed, this compound has been identified and measured in rural areas where isoprene is present (e.g., see Bertman and Roberts, 1991; Williams et al., 1997; Nouaime et al., 1998; and Roberts et al., 1998). In one study in the rural southeastern United States, the MPAN concentration was found to track that of methacrolein but with a delay of several hours (Fig. 6.17); Nouaime et al. (1998) attributed this positive correlation to a more rapid formation of methacrolein compared to its rate of reaction to form MPAN.

Of all the possible fates for PANs in the atmosphere, thermal decomposition is usually the most important; e.g., for PAN

$$CH_3C(O)OONO_2 \xrightarrow{\Delta} CH_3C(O)OO + NO_2. \quad (77)$$

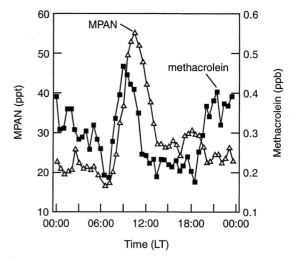

FIGURE 6.17 Diurnal variation of mean mixing ratios of MPAN and its precursor methacrolein measured near Nashville, Tennessee (adapted from Nouaime et al., 1998).

Table 6.21 gives the Arrhenius parameters for the decomposition of various PANs and the rate constants and corresponding lifetimes for PAN. The decomposition is strongly temperature dependent, with long lifetimes, of the order of a year or more at low temperatures of ~215 K, and very short lifetimes, < 1 h, at the higher temperatures around 298 K.

The effective rate of decomposition depends on the ratio of NO to NO_2. Thus the acetylperoxy radical formed in the decomposition must be removed by further reaction, e.g., with NO,

$$CH_3C(O)OO + NO \rightarrow CH_3C(O)O + NO_2, \quad (78)$$

before it reacts back with NO_2 to regenerate PAN, reaction (75), in which case there has been no net reaction (see Problem 8). As a result, the ratio of rate constants k_{75}/k_{78} as well as the ratio of NO_2 to NO concentrations is important. This rate constant ratio has been measured to be $k_{75}/k_{78} \sim 0.4$ for both PAN and peroxypropionyl nitrate (e.g., see Seefeld et al., 1997; Seefeld and Kerr, 1997; and Sehested et al., 1998b). The $CH_3C(O)O$ radical rapidly decomposes to $CH_3 + CO_2$. In regions where NO levels are low, $CH_3C(O)OO$ can also be removed by reactions with other $RC(O)OO$, RO_2, HO_2, and NO_3, which then become important in determining the lifetime of PAN (Madronich and Calvert, 1990; Stockwell et al., 1995).

This strong temperature dependence has important implications for the role of PAN, and its higher homologs, in the troposphere. Thus, when it is formed at lower temperatures or is transported into colder regions, it is stabilized and acts as an NO_x reservoir. When an air mass containing PAN is transported into warmer regions, however, the PAN decomposes, releasing both the NO_2 and the acetylperoxy free radical. If sufficient NO is present, the organic free radical reacts in the following manner:

$$CH_3C(O)OO + NO \rightarrow CH_3C(O)O + NO_2, \quad (78)$$
$$CH_3C(O)O \rightarrow CH_3 + CO_2, \quad (79)$$

TABLE 6.21 Rate Constants and Temperature Dependence[a] for the Thermal Decomposition of Some PANs

Compound	Temperature dependence	$k^{298\ K}$ (10^{-4} s^{-1})	τ^b (min)
$CH_3\overset{O}{\overset{\|}{C}}OONO_2$ PAN	$k_0 = 4.9 \times 10^{-3} \exp(-12{,}100/T)[N_2]$ $k_\infty = 5.4 \times 10^{16} \exp(-13{,}830/T)$ $F_c = 0.3$	3.3	50
$CH_3CH_2\overset{O}{\overset{\|}{C}}OONO_2$ PPN	$2 \times 10^{15} \exp(-12{,}800/T)$	4.4	38
$CH_3CH_2CH_2\overset{O}{\overset{\|}{C}}OONO_2$ PnBN	$3.2 \times 10^{18} \exp(-15{,}150/T)$	2.7	62
$i\text{-}C_3H_7\overset{O}{\overset{\|}{C}}OONO_2$ PiBN	$1.6 \times 10^{18} \exp(-15{,}000/T)$	2.2	76
$n\text{-}C_4H_9\overset{O}{\overset{\|}{C}}OONO_2$	$1.3 \times 10^{12} \exp(-10{,}870/T)$	1.8	93
$i\text{-}C_4H_9\overset{O}{\overset{\|}{C}}OONO_2$	$2.5 \times 10^{9} \exp(-15{,}800/T)$	2.4	69
$CH_2{=}\overset{H_3C}{\overset{\|}{C}}{-}\overset{O}{\overset{\|}{C}}OONO_2$ MPAN	$1.6 \times 10^{16} \exp(-13{,}488/T)$	3.5	48

[a] For PAN and PPN, from Atkinson et al. (1997a); for MPAN, from Roberts and Bertman (1992); for PnBN, PiBN, and n- and i-$C_4H_9C(O)OONO_2$, from Grosjean et al. (1994a, 1994b) over a limited temperature range (291–299 K). Note that PAN is not in the high-pressure limit at 298 K and 1 atm, whereas the higher homologs are.

[b] Assuming no back reaction with NO_2 (see text).

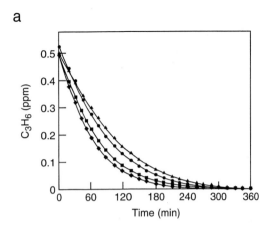

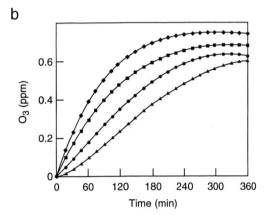

FIGURE 6.18 Concentration–time profiles in an environmental chamber for (a) propene and (b) O_3 as a function of increasing initial concentrations of PAN. Temperature ~30°C, relative humidity ~60%, [NO] = [NO_2] = 0.26 ppm, [C_3H_6] = 0.5 ppm. (▲) No added PAN; (●) 0.06 ppm PAN added; (■) 0.13 ppm PAN added; (♦) 0.26 ppm added PAN (adapted from Carter et al., 1981a).

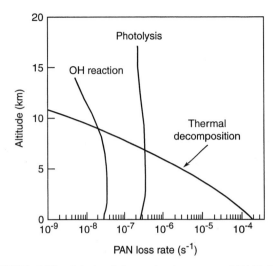

FIGURE 6.19 Calculated first-order loss rates of PAN due to thermal decomposition, OH reaction, and photolysis as a function of altitude (assuming diurnally averaged actinic fluxes for 30°N, July 4) (adapted from Talukdar et al., 1995).

followed by the reaction of CH_3 with O_2, etc., to form HCHO and HO_2, or HCHO, CH_3OH, and CH_3OOH under low-NO_x conditions. As a result of this chemistry, it was suggested by Hendry and Kenley (1979) that PAN should act as an accelerator for photochemical smog formation, and indeed this is the case. Figure 6.18, for example, shows that in smog chamber studies, the loss of propene and the formation of O_3 are accelerated by the addition of PAN (Carter et al., 1981a).

At lower temperatures such as those found in the Arctic, the majority of NO_y is often present in the form of PAN because of its thermal stability. For example, 50–90% of NO_y may exist in the form of PAN in the Arctic spring (Bottenheim et al., 1986; Barrie and Bottenheim, 1991; Jaffe, 1993). It is also interesting that, in relatively "clean," low-NO_x regions at higher temperatures, PANs may also tie up significant amounts of NO_x (Singh et al., 1985; Madronich and Calvert, 1990).

The reaction of OH with PAN and its homologs appears to be sufficiently slow, $k < 3 \times 10^{-14}$ cm³ molecule^{-1} s^{-1} at 298 K (Talukdar et al., 1995), that it cannot compete with thermal decomposition. For example, Figure 6.19 shows the calculated loss rates for PAN in the atmosphere as a function of altitude (Talukdar et al., 1995). Photolysis (see Chapter 4.J) only becomes important above 5 km, and the reaction with OH does not compete at any altitude.

For a detailed treatment of peroxyacyl nitrates, see the recent reviews by Gaffney et al., 1989; Roberts, 1990; Altshuller, 1993; and Kleindienst, 1994.

2. Alkyl Nitrates and Nitrites

a. Alkyl Nitrates

As discussed earlier in this chapter, alkyl nitrates ($RONO_2$) are expected to be formed as minor products of the RO_2 + NO reaction, especially in the case of larger alkylperoxy radicals where as much as ~30–35% of the reaction may give stabilized alkyl nitrates rather than RO + NO_2. It is noteworthy that while NO_x levels tend to be low in "clean" areas, modeling studies by Madronich and Calvert (1990) suggest that a significant fraction (~30–70%) of NO_y may be tied up as various alkyl nitrates (including multifunctional nitrates) in these regions.

The absorption cross sections and photochemistry of alkyl nitrates are discussed in Chapter 4.J, where the

TABLE 6.22 Room Temperature Rate Constants for the Reactions of OH with Some Simple Alkyl Nitrates at 298 K[a]

Alkyl nitrate	k^{298K} (10^{-13} cm^3 molecule^{-1} s^{-1})
CH$_3$ONO$_2$	0.35, 0.30,[b] 0.24[c]
C$_2$H$_5$ONO$_2$	4.9, 2.0,[b] 1.8[c]
n-C$_3$H$_7$ONO$_2$	7.3
i-C$_3$H$_7$ONO$_2$	4.9, 2.9[c]
CH$_3$CH(ONO$_2$)CH$_2$CH$_3$	9.2
CH$_3$CH(ONO$_2$)CH$_2$CH$_2$CH$_3$	18.5
CH$_3$CH$_2$CH(ONO$_2$)CH$_2$CH$_3$	11.2
CH$_3$CH$_2$CH(ONO$_2$)(CH$_2$)$_4$CH$_3$	38.8

[a] From Atkinson et al. (1997a) and Atkinson (1994).
[b] Kakesu et al. (1997), from 304 to 310 K for CH$_3$ONO$_2$ and from 298 to 310 K for C$_2$H$_5$ONO$_2$.
[c] Talukdar et al. (1997).

photolysis appears to occur via breaking the O–NO$_2$ bond. However, the absorption cross sections are not large, and hence the lifetimes with respect to photolysis are quite long, of the order of a week or more at the earth's surface and several days at higher altitudes where the solar flux is larger (e.g., see Clemitshaw et al., 1997).

Reaction with OH is, however, reasonably rapid as might be expected and is of the same order of magnitude as the OH–alkane reactions. Table 6.22, for example, shows the room temperature rate constants for the reactions of OH at 298 K with some alkyl nitrates. With 2-butyl nitrate as an example, the lifetime with respect to the OH reaction with OH at 1×10^6 radicals cm^{-3} is about 13 days, comparable to the photolysis rate. As with the alkanes, abstraction of a hydrogen atom occurs to form an alkyl radical, whose fate is the same as discussed in Section 6.D earlier.

b. Alkyl Nitrites

Alkyl nitrites (RONO) absorb light strongly in the actinic region, dissociating to form RO + NO. Because of this rapid photolysis, other reactions such as that with OH cannot compete, and alkyl nitrites have not been generally observed in the troposphere at significant concentrations.

3. Amines, Nitrosamines, and Hydrazines

a. Amines

Aliphatic amines are emitted from a variety of sources, including feedlots, sewage treatment, waste incineration, and industrial activities (e.g., Schade and Crutzen, 1995). They have also been measured in nonurban areas (e.g., see Van Neste et al., 1987; Gorzelska and Galloway, 1990; and Eisele and Tanner, 1990). Because they do not absorb light in the actinic region (Calvert and Pitts, 1966), they are not removed by photolysis. Hence reactions with atmospheric oxidants such as OH and O$_3$ are the major removal processes for these organic nitrogen compounds. Reaction with HNO$_3$ may also occur in polluted urban areas.

Table 6.23 gives the rate constants for the reactions of OH and O$_3$ with some simple alkyl amines at room

TABLE 6.23 Room Temperature Rate Constants and Estimated Atmospheric Lifetimes for the Gas-Phase Reactions of Some Alkyl Amines and Amides with OH[a] and O$_3$[b]

Amine or amide	OH		O$_3$	
	k (10^{-11} cm^3 molecule^{-1} s^{-1})	τ^c (h)	k (10^{-18} cm^3 molecule^{-1} s^{-1})	τ^d
CH$_3$NH$_2$	2.2,[a] 1.7[e]	13	0.0074	~2 yr
C$_2$H$_5$NH$_2$	2.8, 2.4[e]	10		
(CH$_3$)$_2$NH	6.5[a,e]	4	1.67	67 h
(CH$_3$)$_3$N	6.1,[a] 3.6[e]	5	7.84	14 h
(CH$_3$)$_3$CNH$_2$	1.2[f]	23		
CH$_3$NHC(O)CH$_3$	0.52[g]	53		
CH$_3$NHC(O)C$_2$H$_5$	0.76[g]	37		
(CH$_3$)$_2$NC(O)CH$_3$	1.4[g]	20		

[a] From Atkinson (1989).
[b] From Tuazon et al. (1994).
[c] Assuming [OH] = 1×10^6 radicals cm^{-3}.
[d] Assuming [O$_3$] = 100 ppb = 2.5×10^{12} molecules cm^{-3}.
[e] Carl and Crowley (1998).
[f] Koch et al. (1996).
[g] Koch et al. (1997).

temperatures and the associated atmospheric lifetimes under typical, moderately polluted tropospheric conditions. As might be expected, OH reacts quite rapidly with the simple amines. Perhaps somewhat surprising, however, are the relatively rapid reactions of di- and trimethylamine with O_3.

The N–H bond strength in CH_3NH_2 is ~100 kcal mol^{-1}, stronger than the C–H bond at 93 kcal mol^{-1}. Similarly, in $(CH_3)_2NH$, the N–H bond strength is ~92 kcal mol^{-1}, compared to 87 kcal mol^{-1} for the C–H bond. One might expect abstraction of a C–H hydrogen to predominate in these reactions, and this appears to be the case. For example, for $(CH_3)_2NH$, Lindley *et al.* (1979) report that 63% of the reaction occurs by (80a),

$$OH + (CH_3)_2NH \rightarrow H_2O + CH_2NHCH_3, \quad (80a)$$

and the remainder by (80b):

$$OH + (CH_3)_2NH \rightarrow H_2O + (CH_3)_2N \quad (80b)$$

The subsequent reactions of the alkyl radical formed in reaction (80a) and the dimethylamino radical in (80b) are expected to be as follows:

$$(CH_3)_2N + O_2 \rightarrow \text{Products}$$
$$[\text{e.g., } CH_2=NCH_3 + HO_2 \text{ or } (CH_3)_2NOO?], \quad (81)$$

$$(CH_3)_2N + NO \xrightarrow{M} (CH_3)_2NNO, \quad (82)$$

$$(CH_3)_2N + NO_2 \rightarrow (CH_3)_2NNO_2, \quad (83a)$$

$$\rightarrow HONO + CH_2=NCH_3, \quad (83b)$$

$$(CH_3)_2N + (CH_3)_2NNO_2 \rightarrow$$
$$(CH_3)_2NN(CH_3)_2 + NO_2, \quad (84)$$

$$2(CH_3)_2N \rightarrow (CH_3)_2NN(CH_3)_2, \quad (85)$$

$$CH_2NHCH_3 + O_2 \longrightarrow$$

$$OOCH_2NHCH_3 \xrightarrow{NO \quad NO_2} OCH_2NHCH_3$$

$$\downarrow O_2, HO_2$$

$$\underset{H}{\overset{O}{\parallel}}C-N\underset{CH_3}{\overset{H}{\diagdown}}$$

$$(86)$$

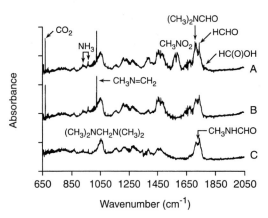

FIGURE 6.20 Infrared spectrum of the products of reaction of 19 ppm $(CH_3)_2NH$ with 5.5 ppm O_3 at room temperature and in 740 Torr air: (A) product spectrum; (B) after subtraction of $(CH_3)_2NCHO$, CH_3NO_2, HCHO, and HCOOH absorptions; (C) after subtraction of $(CH_3)N=CH_2$ and CO_2 absorptions (adapted from Tuazon *et al.*, 1994).

Most of the products predicted by this reaction scheme have been observed in laboratory studies, including dimethylnitramine [$(CH_3)_2NNO_2$], HONO, tetramethylhydrazine [$(CH_3)_2NN(CH_3)_2$], and small amounts of methylformamide as products of the photooxidation of dimethylamine (e.g., see Pitts *et al.*, 1978; and Tuazon *et al.*, 1978). In addition, HCHO, CO, and small amounts of dimethylformamide were identified in these studies.

It is noteworthy that the dimethylamino radical reaction with O_2 is about a factor of ~10^6–10^7 slower than its reactions with NO and NO_2. For example, Lindley *et al.* (1979) measured the ratio of rate constants $k_{81}/k_{82} = 1.5 \times 10^{-6}$ and $k_{81}/k_{83a} = 3.9 \times 10^{-7}$. Thus, at ~10 ppb NO_x, reactions of the nitrogen-centered radical with NO and NO_2, in addition to O_2, become important. This is perhaps not surprising, given that NH_2 radicals also react extremely slowly with O_2, with an upper limit of 6×10^{-21} cm^3 molecule^{-1} s^{-1} (Tyndall *et al.*, 1991).

There are few kinetic or product studies of the reactions of O_3 with simple amines. Figure 6.20 shows the infrared spectrum of the products of the reaction of dimethylamine with O_3 (Tuazon *et al.*, 1994). The major products observed are $CH_3N=CH_2$, CH_3NO_2, HCHO, CH_3NHCHO, $(CH_3)_2NCHO$, $(CH_3)_2NCH_2N(CH_3)_2$, HCOOH, and CO_2. The major nitrogen-containing products were explained by Tuazon and co-workers by the following mechanism: The

$$(CH_3)_2NH + O_3 \longrightarrow \left[\begin{array}{c} O \\ \uparrow \\ (CH_3)_2NH \end{array} \right]^* + O_2$$

$$\downarrow$$

$$[(CH_3)_2NOH]^* \longrightarrow (CH_3)_2N + OH$$

$$\swarrow_{O_3} \quad \downarrow_{-H_2O} \quad \searrow_{O_2}^{HO_2}$$

$$(CH_3)_2NO + O_2 + OH \quad CH_3N=CH_2$$

$$\downarrow_{O_3}$$

$$CH_3NO_2 + O_2 + CH_3 \longrightarrow \quad \longrightarrow HCHO$$

(87)

CH_3NHCHO arises from the reaction of the OH generated in the system with the parent amine to form the CH_2NHCH_3 alkyl radical, followed by reaction (86). N,N,N',N'-Tetramethyldiaminomethane, $(CH_3)_2NCH_2N(CH_3)_2$, seen in Fig. 6.20C, was shown to be formed in the reaction of dimethylamine with the product HCHO:

$$2(CH_3)_2NH + HCHO \rightarrow$$
$$(CH_3)_2NCH_2N(CH_3)_2 + H_2O. \quad (88)$$

Further reaction of this product with O_3 was shown to give $(CH_3)_2NCHO$, observed as a product in Fig. 6.20A.

It should be noted, however, that in studies of amine photooxidations, it is generally true that a significant fraction of the reacted parent amine remains unaccounted for in the identified products. Clearly, the mechanisms and products are complex and warrant further investigation.

b. Nitrosamines

One difficulty in studying the photooxidation of amines is the rapid reaction in the dark with nitrous acid to form nitrosamines (Hanst *et al.*, 1977; Pitts *et al.*, 1978):

$$R_2NH + HONO \leftrightarrow R_2NNO + H_2O. \quad (89)$$

As discussed in Chapter 7.B.3, NO_2 undergoes a surface reaction with water, which is perhaps enhanced at the air–water interface, forming HONO:

$$2NO_2 + H_2O \xleftrightarrow{\text{surface}} HONO + HNO_3. \quad (90)$$

Thus, in the course of preparing reactant mixtures for photooxidation studies under typical atmospheric conditions where both NO_2 and water vapor are present, it is essentially impossible to avoid the production of some HONO, and in the case of studies of amine reactions, some nitrosamines. However, this too is quite relevant, since nitrosamines are carcinogenic in experimental animals. In addition, there are a number of sources that emit nitrosamines directly into the air, including leather tanneries, rocket fuels, tire and amine factories, and tobacco smoke (e.g., see Fine, 1980).

The major atmospheric fate of the *N*-nitrosamines such as *N*-nitrosodimethylamine is photolysis (Tuazon *et al.*, 1984):

$$(CH_3)_2NNO + h\nu \rightarrow (CH_3)_2N + NO. \quad (91)$$

The dimethylamino radical then reacts as described earlier.

c. Hydrazines

Hydrazines see widespread use as fuels, for example, in the space shuttle and as a source of emergency power in the F-16 fighter plane. As a result of the industrial and fuel uses of hydrazines, with their accompanying transport and storage, some emissions to the atmosphere occur and hence there is interest in their atmospheric reactions.

Hydrazines do not photolyze in the actinic UV, but reactions with OH and O_3 must be considered. The rate constants for reaction of OH with N_2H_4 and CH_3NHNH_2 are $(6.1 \pm 1.0) \times 10^{-11}$ and $(6.5 \pm 1.3) \times 10^{-11}$ cm^3 molecule^{-1} s^{-1}, respectively, essentially independent of temperature over the range 298–424 K (Harris *et al.*, 1979). At an OH concentration of 1×10^6 cm^{-3}, the lifetimes of both N_2H_4 and CH_3NHNH_2 will be ~4–5 h. Harris and co-workers (1979) estimate that the rate constant for the reaction of OH with 1,1-dimethylhydrazine is ~$(5 \pm 2) \times 10^{-11}$ cm^3 molecule^{-1} s^{-1}, so that its lifetime with respect to OH will be similar, ~6 h.

Reaction with O_3 is also relatively fast. Tuazon *et al.* (1981) estimate that the rate constant for the N_2H_4–O_3 reaction at 294–297 K is ~1×10^{-16} cm^3 molecule^{-1} s^{-1}. This corresponds to a lifetime of about 1 h at an O_3 concentration of 0.1 ppm. The rate constants for the O_3–CH_3NHNH_2 and O_3–$(CH_3)_2NNH_2$ reactions were too fast to measure under their experimental conditions; the reactions of ~1–3 ppm O_3 with 0.4–4 ppm CH_3NHNH_2 and ~2 ppm O_3 with ~0.2–2 ppm $(CH_3)_2NNH_2$ were complete in less than 2–3 min.

From these data, the rate constants must be $>10^{-15}$ cm^3 molecule^{-1} s^{-1}, and the lifetimes of these two hydrazines must be less than 7 min at 0.1 ppm O_3.

The mechanism of the reaction of hydrazines with O_3 has been investigated using FTIR (Tuazon *et al.*, 1981; Carter *et al.*, 1981b). In the case of N_2H_4, the major product was H_2O_2, and N_2O appeared as a minor product; these are consistent with the following mechanism:

Initiation

$$H_2NNH_2 + O_3 \rightarrow H_2N-\dot{N}H + OH + O_2 \quad (92)$$

Propagation

$$H_2N-\dot{N}H + O_2 \rightarrow HN=NH + HO_2 \quad (93)$$

$$HN=NH + O_3 \rightarrow HN=\dot{N} + OH + O_2 \quad (94)$$

$$H_2NNH_2 + OH \rightarrow H_2N-\dot{N}H + H_2O \quad (95)$$

Product Formation

$$HN=\dot{N} \rightarrow H + N_2 \quad (96)$$

$$H + O_2 \xrightarrow{M} HO_2 \quad (97)$$

$$HO_2 + HO_2 \xrightarrow{M, H_2O} H_2O_2 + O_2 \quad (98)$$

According to this mechanism, most of the nitrogen in the hydrazine would form N_2, which would not have been detected in this system. The diazene HN=NH would be expected to react with OH radicals:

$$HN=NH + OH \rightarrow HN=\dot{N} + H_2O, \quad (99)$$

and ultimately form N_2 and H_2O_2 via reactions (96)–(98).

Interestingly, this reaction has been used as a non-photolytic OH source for kinetic studies because of the production of OH in the initial reaction (92) between O_3 and N_2H_4 (Tuazon *et al.*, 1983a).

Figure 6.21 shows an FTIR spectrum taken during the studies by Tuazon *et al.* (1981) of the reaction of CH_3NHNH_2 with O_3 (the absorption bands from NH_3 which form from the slow decay of the hydrazine in the dark have been subtracted from the spectra). After initial injection of O_3, with the hydrazine present in excess, the observed products were methyl hydroperoxide (CH_3OOH), diazomethane (CH_2N_2), H_2O_2, methyldiazene ($CH_3N=NH$), HCHO, CH_3OH, and traces of CH_3ONO_2. After a second injection of O_3 into the system so that O_3 was in excess, $CH_3N=NH$ and CH_2N_2 disappear, and higher yields of CH_3OOH, CH_3OH, and HCHO result. Ninety-two percent of the initial carbon atoms could be accounted for in the

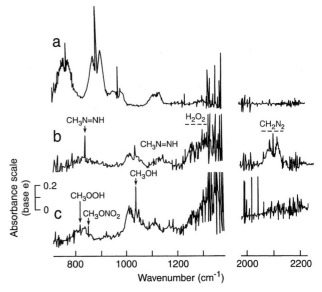

FIGURE 6.21 FTIR spectra taken during reaction of CH_3NHNH_2 with O_3: (a) 3.68 ppm CH_3NHNH_2 before reaction; (b) 2 min after injection of 2.8 ppm O_3; (c) 2.8 ppm O_3 injected 38 min after first injection (spectrum taken 2 min after second injection). NH_3 absorptions have been subtracted from (a) and (b), and both NH_3 and O_3 absorptions from (c) (adapted from Tuazon *et al.*, 1981).

observed products, but at least 95% of the initial nitrogen in CH_3NHNH_2 could not be found, indicating it likely formed N_2.

The reaction mechanism proposed by Tuazon *et al.* (1981, 1982) to explain these results is analogous to that for N_2H_4. In this case, the intermediate $CH_3N=NH$ formed in the reaction analogous to (93) was identified by FTIR. Reactions (100a) and (100b) explain the formation of CH_2N_2:

$$CH_3N=NH + O_3 \xrightarrow{O_2} CH_2N_2 + OH + O_2 + HO_2, \quad (100a)$$

$$CH_3N=NH + O_3 \rightarrow CH_2N_2 + H_2O + O_2. \quad (100b)$$

In excess O_3, CH_2N_2 can react via (100c):

$$CH_2N_2 + O_3 \rightarrow HCHO + O_2 + N_2. \quad (100c)$$

The formation of CH_3OOH, HCHO, CH_3OH, and CH_3ONO_2 is expected from secondary reactions of the methyl radical as discussed earlier in this chapter.

The reaction of 1,1-dimethylhydrazine with O_3 gave the carcinogen *N,N*-dimethylnitrosamine as the major product in ~60% yields within 2- to 3-min reaction time (Tuazon *et al.*, 1981). Minor products were HCHO, H_2O_2, HONO, and perhaps NO_x. For a discussion of the complex mechanisms, the reader should consult the original reference.

NO_2 has also been shown to react with 1,1-dimethylhydrazine in air, forming HONO and tetramethyltetrazine-2, $(CH_3)_2NN=NN(CH_3)_2$ (Tuazon et al., 1983b). The reaction is also proposed to involve abstraction of a hydrogen from the weak N–H bond by NO_2, forming HONO. The tetramethyltetrazine-2 is hypothesized to be formed by the addition of NO_2 to the $(CH_3)_2NNH$ radical, followed by decomposition to $(CH_3)_2N_2$ + HONO and the self-recombination of the $(CH_3)_2N_2$ radicals (Tuazon et al., 1982). The apparent overall rate constant for the reaction was 2.3×10^{-17} cm^3 molecule^{-1} s^{-1} so that the lifetime of 1,1-dimethylhydrazine at an NO_2 concentration of 0.1 ppm would be ~5 h. Since the lifetimes with respect to 0.1 ppm O_3 or 1×10^6 OH radicals cm^{-3} are ~7 min and 6 h, respectively, the reaction of NO_2 can contribute to the atmospheric reactions of the hydrazine only at low O_3 levels.

Hydrazine, monomethylhydrazine, and asymmetrical dimethylhydrazine have also been shown to react rapidly ($k > 10^{-15}$ cm^3 molecule^{-1} s^{-1}) with HNO_3 in the gas phase to form the corresponding hydrazinium nitrate aerosols (Tuazon et al., 1982).

J. CHEMISTRY OF REMOTE REGIONS

Gas-phase chemistry in remote areas is, in most cases, analogous to that in more polluted regions. The major difference is in lower NO_x emissions and hence concentrations. In addition, in continental regions, there are substantial emissions of biogenic organics, many of which are highly reactive toward OH, O_3, NO_3, and Cl atoms and in oceanic regions, dimethyl sulfide (DMS), which reacts with OH, NO_3, and Cl atoms.

As discussed briefly in Chapter 1 and in more detail in Chapter 14, it is unlikely that there are any regions at the earth's surface that have not been significantly impacted by anthropogenic emissions. Even over the central Atlantic and Pacific Oceans, for example, significant contributions to the chemistry from anthropogenic emissions are often observed (e.g., Parrish et al., 1993b; Dickerson et al., 1995). One means of testing for anthropogenic emissions is the use of a correlation between CO and O_3. CO is generated primarily from combustion processes in vehicles and industrial processes (e.g., Fig. 2.7 and Table 2.1), and its chemistry is the relatively slow reaction with OH (lifetime of ~90 days at an OH concentration of 5×10^5 cm^{-3}). As a result, as an air mass containing VOC and NO_x ages and undergoes the photochemical reactions discussed earlier, O_3 is formed. Such air masses therefore often have both increased O_3 and CO, with $\Delta[O_3]/\Delta[CO]$ ~ 0.3–0.4 being typical (e.g., Dickerson et al., 1989; Chin et al., 1994; Wang et al., 1996; Harris et al., 1998; Parrish et al., 1993b, 1998; Kajii et al., 1998). This relationship is impacted, of course, by other factors as well such as the production of CO in the oxidation of VOC and by the deposition of O_3 during transport, for which corrections can be estimated (e.g., Chin et al., 1994).

In this section, we discuss briefly the distinguishing chemistry associated with remote regions of the troposphere, including continental and marine areas, focusing primarily on regions of minimal anthropogenic influence. There is increasing evidence of interesting chemistry in the upper troposphere, which is also discussed. Finally, we briefly treat some unusual chemistry occurring in the Arctic, which is characterized by long periods of darkness and low temperatures during winter followed by long periods of sunlight.

1. Emissions of Biogenic Organics

As discussed briefly in Chapter 2, there are substantial biogenic emissions of VOC. (We exclude a discussion of methane, since it is of low reactivity and its chemistry is quite straightforward; see Chapter 14 for a discussion of its implications for global climate change.) On a global basis, the total may exceed anthropogenic emissions by as much as an order of magnitude (see Tables 2.1 and 2.2 and the volume edited by Hewitt, 1999). In urban areas, of course, their contributions are relatively less important. For example, Benjamin et al. (1997) report that biogenic hydrocarbons in the South Coast Air Basin of southern California are ~10% of the total VOC emission inventory on a typical summer day. The contribution of biogenic organics to urban O_3 formation is of course variable, depending on the particular locations (e.g., Chameides et al., 1988; Roselle et al., 1991; Pierce et al., 1998; see also Chapter 16).

Several thousand different biogenic VOCs have been identified (e.g., Isidorov et al., 1985; Graedel et al., 1986; Puxbaum, 1997; Helas et al., 1997; Fall, 1999). The most well known are ethene, isoprene, and the monoterpenes emitted by terrestrial plants. As discussed in this section, it has been increasingly recognized that there are biogenic emissions of oxygen-containing organics, from small alcohols such as methanol to larger aldehydes, ketones, and alcohols such as 2-methyl-3-buten-2-ol. Figures 6.22 and 6.23 summarize some of the VOC that have been observed and possible mechanisms of production (Fall, 1999). Different compounds are produced in different parts of the plant and by different physiological processes; the

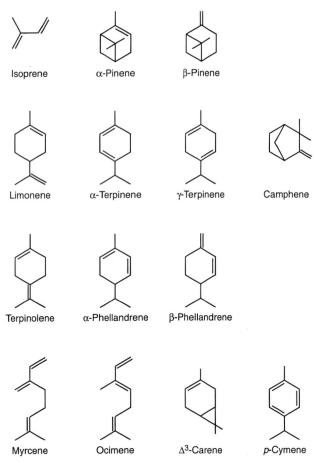

FIGURE 6.22 Chemical structures of some biogenically emitted hydrocarbons.

present understanding of these processes is reviewed by Fall (1999).

Ethene is a plant hormone emitted at a rate of several teragrams per year from plants; smaller amounts are emitted by soils and oceans (Rudolph, 1997). Interestingly, the emissions of ethene as well as some other VOCs (see later) increase significantly (by as much as factors of one to two orders of magnitude) when the plant is stressed, for example by mechanical means, high temperatures, or lack of water.

Isoprene is the major single, non-methane VOC emitted by plants:

Isoprene

Significant amounts of larger hydrocarbons are also generated by plants and emitted to the atmosphere. The larger hydrocarbon compounds generally fall under the classification of isoprenoids, or terpenoids, consisting of groups of 5-carbon isoprene type units (although they are not formed from isoprene). The monoterpenes are the C_{10} compounds, sesquiterpenes the C_{15} compounds, diterpenes the C_{20} compounds, triterpenes the C_{30} compounds, and tetraterpenes the C_{40} compounds.

Table 6.24, for example, shows one estimate of the annual global emissions of isoprene, other monoterpenes, and VOCs as well as methane (Guenther, 1999). Emissions of isoprene are believed to be about four times those of the other monoterpenes and about equal to all other VOCs.

Hardwood species such as oaks, poplars, aspen, and ironwood are generally isoprene emitters. However, even within plant families, not all species are isoprene emitters. For example, while North American oaks emit isoprene, many European oak species do not. For example, Steinbrecher et al. (1997) measured the emissions of isoprene and monoterpenes from five species of Mediterranean oak; two of them were strong isoprene emitters, whereas the other three did not emit significant amounts. Similarly, Kesselmeier et al. (1998) measured emissions of isoprene and monoterpenes from a Holm oak and a white oak growing side by side; the white oak was a strong isoprene emitter, whereas the Holm oak was a strong monoterpene emitter.

In addition to deciduous trees such as some oaks, isoprene is also emitted from other plants, including shrubs, gorse, vines, ferns, and plants characteristic of tropical savannas and peatlands characteristic of boreal regions (e.g., Guenther et al., 1996a,b; Cao et al., 1997; Owen et al., 1998; Janson and De Serves, 1998; Guenther, 1999; Fall, 1999). Isoprene emissions have been reported to be correlated with successional patterns; for example, Klinger et al. (1998) report higher isoprene emissions in the early to mid-successional savanna ecosystems, compared to the later, rainforest development. Other sources (all believed to be relatively small) include marine phytoplankton (e.g., Bonsang et al., 1992; Moore et al., 1994; Milne et al., 1995; Broadgate et al., 1997), bacteria and fungi (e.g., Kuzma et al., 1995), the breath of humans and other mammals (e.g., Gelmont et al., 1981; Mendis et al., 1994; Phillips et al., 1994; Jones et al., 1995; Foster et al., 1996; Sharkey, 1996; Fenske and Paulson, 1999), and some industrial processes (e.g., Guenther, 1999).

Isoprene production by plants is very sensitive to light as well as temperature (e.g., Sanadze and Kalandadze, 1966; Tingey et al., 1979). Figure 6.24a shows the production of isoprene from an aspen leaf when light in the photosynthetically active range is turned on and off (Monson et al., 1991; Fall, 1999). When the light is switched on, isoprene emissions rise and when it is turned off, fall even more rapidly. Figure 6.24b shows

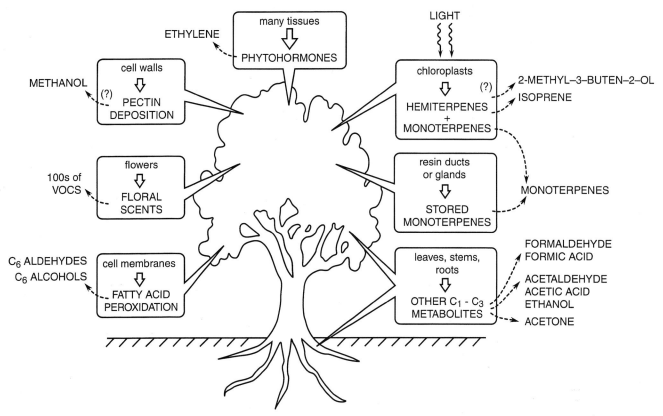

FIGURE 6.23 Schematic diagram of mechanisms of production of some biogenically emitted VOCs (adapted with permission from Fall, 1999).

the emissions as a function of this light intensity; the emissions rise rapidly and approach (but do not reach) a plateau (Monson et al., 1992; Fall, 1999). Finally, Fig. 6.25 shows the effect of temperature on the isoprene emission rate; emissions rise to a temperature of ~40–45 °C and then rapidly fall (e.g., Monson et al., 1992; Sharkey and Singsaas, 1995; Fall and Wildermuth, 1998; Fall, 1999). These influences of light and temperature are believed to be major factors in the variable emissions of isoprene measured in field studies (e.g., Lamb et al., 1996; Guenther et al., 1996b; Geron et al., 1997; Owen et al., 1997; Guenther and Hills, 1998; Drewitt et al., 1998).

This dependence on light levels and temperature is believed to be due to the mechanism of production of isoprene in the plant, which involves the enzyme isoprene synthetase and dimethylallyl diphosphate (DMAPP) as a precursor to isoprene (e.g., see Silver and Fall, 1995; and Monson et al., 1995). Either the enzyme, the formation of DMAPP, or both may be light sensitive (Wildermuth and Fall, 1996). The temperature effect has been attributed to effects on the enzyme, increasing its activity initially and then leading to irreversible denaturation (and/or possibly membrane damage) (Fall and Wildermuth, 1998).

Conifers tend to be sources of a variety of larger terpenoids, a major component of which are the C_{10} monoterpenes (Table 6.24). The structures of some of these are shown in Fig. 6.22. The mechanisms of the formation of these hydrocarbons in plants are closely linked, so that emissions of more than one monoterpene often occur together. For example, geranyl

TABLE 6.24 Estimated Global Annual Biogenic VOC Emissions (Tg yr^{-1})[a]

Source	Isoprene	Monoterpenes	Other VOCs[b]
Canopy foliage	460	115	500
Terrestrial ground cover and soils	40	13	50
Flowers	0	2	2
Ocean and freshwater	1	<0.001	10
Animals, humans, and insects	0.003	<0.001	0.003
Anthropogenic (including biomass burning)	0.01	1	93
Total	~500	~130	~650

[a] From Guenther (1999) and references therein.
[b] Other VOCs include all volatile organic compounds other than methane, isoprene, and monoterpenes.

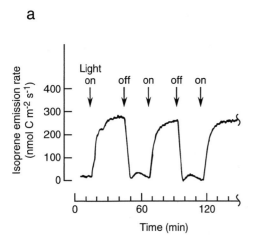

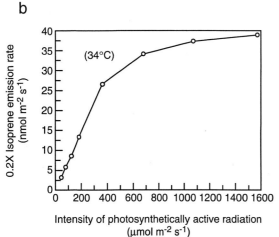

FIGURE 6.24 Effect of light on isoprene emission rate from (a) aspen leaf and (b) velvet bean leaf (adapted from Monson et al., 1991, 1992; and Fall, 1999).

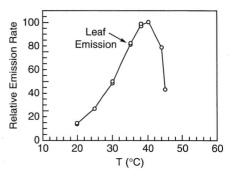

FIGURE 6.25 Effect of willow leaf temperature on isoprene emission rate (adapted from Fall and Wildermuth, 1998; and Fall, 1999).

diphosphate is a precursor for α- and β-pinene, limonene, and myrcene, whose generation involves the enzyme limonene synthase (e.g., see Fall, 1999). In contrast to isoprene, which does not have a reservoir in leaves in many cases, monoterpenes are generated and stored in the plant prior to emission so that in general, their emission to the atmosphere is not as closely tied to short-term controls over their biosynthesis (e.g., Monson et al., 1995; Fall, 1999).

Emissions of monoterpenes have been observed from a variety of plants, including pines (e.g., Juuti et al., 1990; Guenther et al., 1994; Street et al., 1997; Staudt et al., 1997), resin in pine forests (e.g., Pio and Valente, 1998), spruce (Street et al., 1996), some deciduous trees such as oaks (e.g., Benjamin et al., 1996; Street et al., 1997; Kesselmeier et al., 1998), and gorse (e.g., Cao et al., 1997). Interestingly, as for ethene, increased emissions have been observed when plants are stressed (Fall, 1999). For example, Juuti et al. (1990) report that the monoterpene emission rates from a Monterey pine increased by factors of 10–50 during rough handling.

As for isoprene, emission rates of the monoterpenes increase with temperature, although different plant species exhibit different temperature sensitivities and different compounds can also show different dependencies on temperature (e.g., Tingey et al., 1980; Loreto et al., 1996; Owen et al., 1997; Schween et al., 1997; Drewitt et al., 1998). The temperature dependence of monoterpene emissions is often taken into account by multiplying the base emission rate at a reference temperature T_s by the factor $e^{[\beta(T-T_s)]}$, where T is the leaf temperature and β is a coefficient that reflects the temperature sensitivity of emissions (e.g., Guenther et al., 1993). Light also affects monoterpene emissions but does not appear to be as significant as for isoprene (e.g., Tingey et al., 1980; Loreto et al., 1996; Guenther et al., 1996b).

Larger hydrocarbons such as the C_{15} sesquiterpenes also have biogenic sources such as sage (Arey et al., 1995).

As shown in Table 6.24, oceans and freshwater are not believed to be major sources of VOC to the atmosphere. As indicated earlier, isoprene is thought to be generated in small amounts in the oceans by marine phytoplankton. In addition, a variety of small hydrocarbons have been identified both in seawater and in the air above it, including alkanes (ethane, propane, n-butane, isobutane, n-pentane, isopentane, and n-hexane), alkenes (ethene, propene, 1- and 2-butene, isobutene, 1-pentene, and 1-hexene), and acetylene (e.g., Arlander et al., 1990; Rudolph and Johnen, 1990; Bonsang et al., 1991; Plass-Dülmer et al., 1993). While some of these may be due to long-range transport from the continents (Rudolph and Johnen, 1990), it appears that the ocean is indeed a source of most, if not all, of these light hydrocarbons. A major organic found in ocean areas is dimethyl sulfide (DMS), whose oxidation products are believed to play a significant role in particle formation and hence radiative properties in the marine boundary layer (see Chapters 8.E.1 and 14.C). DMS also plays a major role in determining the lifetime and fate of NO$_3$ (Carslaw *et al.*, 1997).

In addition to hydrocarbons, biogenic processes also produce a number of oxygen-containing organics. One of the most important appears to be 2-methyl-3-buten-2-ol (MBO), first identified in a forested area by Goldan *et al.* (1993):

Direct emissions of this compound from loblolly pine as well as from lodgepole and ponderosa pines were subsequently demonstrated (Guenther *et al.*, 1996b; Harley *et al.*, 1998). The concentrations of MBO measured by Goldan *et al.* (1993) were factors of 4–7 times those of isoprene, indicating the potential importance of this compound in the chemistry of remote regions. Its emissions appear to be regulated by temperature and light, similar to that of isoprene. This may be due to formation from a common precursor such as DMAPP (Fall, 1999).

Figure 6.26 shows the structures of some oxygen-containing compounds for which there is evidence of direct biogenic emissions. (3Z)-Hexenol and (3Z)-

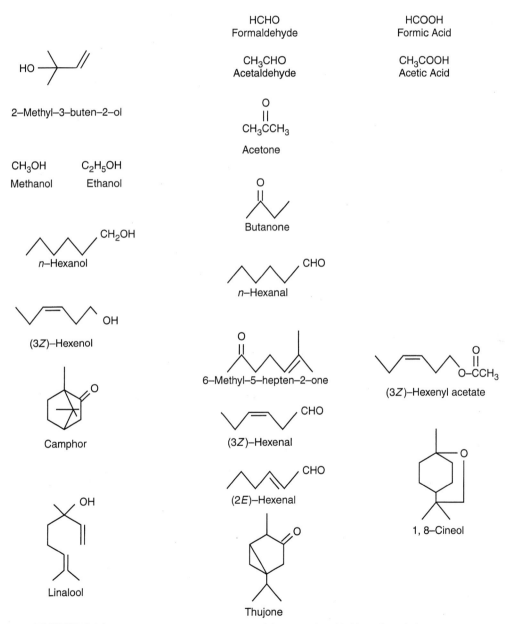

FIGURE 6.26 Structures of some oxygen-containing organics with biogenic emission sources.

hexenyl acetate, for example, are emitted by a number of plant species. For example, Arey *et al.* (1991a) identified emissions of these compounds from more than a dozen different agricultural plants in California as well as from Valley Oak and Whitethorn, and Kirstine *et al.* (1998) reported emissions from grass and clover. Such emissions are enhanced by mechanical damage. For example, Kirstine *et al.* (1998) reported that the emissions of (3Z)-hexenol and (3Z)-hexenyl acetate increased by three orders of magnitude when the grass and clover were mowed! A second, even larger, emission of these and other organics during the subsequent drying of the plants has been observed (de Gouw *et al.*, 1999).These compounds as well as other volatile C_6 aldehydes and alcohols are generated by the biochemical degradation of α-linolenic acid in the plants (Fall, 1999). Similarly, the breakdown of linoleic acid gives *n*-hexanal and *n*-hexanol.

Other examples of emissions of larger oxygen-containing organics include camphor, cineole, and thujone which are emitted by California sagebrush (Arey *et al.*, 1995), cineole from pines and eucalyptus (Staudt *et al.*, 1997), the unsaturated alcohol linalool from the blossoms of Valencia orange trees (Arey *et al.*, 1991b) and from certain pines (Kesselmeier *et al.*, 1997; Staudt *et al.*, 1997), and eucalyptol from grass and clover (e.g., Kirstine *et al.*, 1998). Ciccioli *et al.* (1997) measured emissions in a Mediterranean region known as the Mediterranean Pseudosteppe and found that not only isoprene but also a range of C_6–C_{10} aldehydes, linalool, and acetic acid were emitted by the vegetation.

In addition to these larger VOCs, there are biogenic sources of a wide variety of small alcohols, aldehydes, ketones, and acids. For example, emissions of methanol and acetone have been reported from plant leaves, grass, and clover (e.g., MacDonald and Fall, 1993; Nemecek-Marshall *et al.*, 1995; Fall and Benson, 1996; Kirstine *et al.*, 1998). Table 6.25 shows some of the compounds measured in grass and clover emissions (Kirstine *et al.*, 1998). Clearly, a wide variety of oxygen-containing species are emitted from this one source alone. Direct emissions of formaldehyde, acetaldehyde, and formic and acetic acids have been observed from oaks and pines (Kesselmeier *et al.*, 1997).

Consistent with the direct observation of the emissions are field measurements in remote areas. For example, Fehsenfeld *et al.* (1992) measured the composition of VOC at two rural locations in the United States, summarized in Fig. 6.27. Alcohols and carbonyl compounds comprise 40–70% of the total. Of these, a significant portion appear to be direct emissions, with methanol being a major contributor to the oxygen-containing portion. Similarly, Singh *et al.* (1995) reported relatively high concentrations of methanol and acetone in the free troposphere, at least a portion of which may be due to biogenic emissions. For example, based on a comparison of measurements and model predictions, Wang *et al.* (1998) propose that biogenic emissions account for about 40% of the acetone on a global basis.

TABLE 6.25 Some Oxygen-Containing Organics Observed in Emissions from Grass and Clover[a]

Compound	Percentage of total carbon	
	Grass	Clover
Methanol	11–15	15.1
Ethanol	16–21	0.4
Acetaldehyde	13–16	0.3
Acetone	11–16	22.9
Propanol	2	< 0.1
2-Methyl-2-propenal	1	< 0.1
2,3-Butanedione	1	< 0.1
Butanone	0.4–2	50.0
3-Methylbutanal	1	< 0.1
Pentanal	1	< 0.1
2-Pentanone	0.2–0.8	< 0.1
Hexanal	4	< 0.1
4-Methyl-2-pentanone	0.5–1.1	< 0.1
Benzaldehyde	0.9–1.9	< 0.1
Nonanal	0.3–1.1	< 0.1
Eucalyptol	1	0.1

[a] Adapted from Kirstine *et al.* (1998).

As is expected, there are also a variety of oxygen-containing organics found in rural and remote areas, which are oxidation products of the directly emitted biogenics. For example, in areas with significant isoprene emissions, the oxidation products methyl vinyl ketone (MVK), methacrolein (MACR), and 3-methylfuran are also typically present (e.g., Yokouchi *et al.*, 1993; Montzka *et al.*, 1993, 1995; Biesenthal *et al.*, 1998; Helmig *et al.*, 1998b). Biesenthal and Shepson (1997) suggest that MVK and MACR may also be generated by automobile exhaust, based on the correlation of these compounds with CO in an urban area.

Indeed, separating out direct emissions and the formation by oxidation in air of other biogenics is not straightforward. For example, 6-methyl-5-hepten-2-one (Fig. 6.26) has been reported in air in different locations by a number of groups (e.g., Ciccioli *et al.*, 1993a, 1993b; König *et al.*, 1995; Helmig *et al.*, 1996). However, the reaction of O_3 with organics containing the structural group $(CH_3)_2C=CHCH_2CH_2C(CH_3)=C-$ also gives this compound (e.g., Fruekilde *et al.*, 1998), as expected from the earlier discussion of mechanisms of ozonolysis and studies of struc-

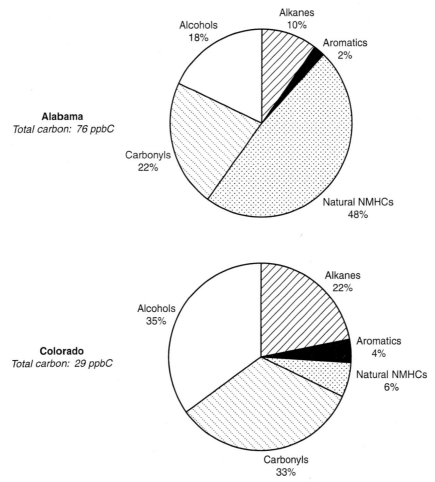

FIGURE 6.27 Distribution of organics observed in rural areas in Alabama and Colorado, respectively (adapted from Fehsenfeld et al., 1992).

turally similar compounds such as linalool, $(CH_3)_2C=CHCH_2CH_2C(CH_3)(OH)CH=CH_2$ (e.g., Shu et al., 1997). For example, ozonolysis of squalene (Fig. 6.28) was demonstrated to form gaseous 6-methyl-5-hepten-2-one, acetone, and geranyl acetone, respectively. 4-Oxopentanal was also formed from the further oxidation of 6-methyl-5-hepten-2-one (Grosjean et al., 1996; Smith et al., 1996; Fruekilde et al., 1998). These compounds were also observed when leaves of common vegetation found in the Mediterranean were exposed to O_3. Furthermore, these products could be formed from the reaction of glass wool that had been in contact with human skin, which also contains squalene as a lipid; such observations suggest the importance of avoiding contamination of samples during measurements of biogenic organics.

In short, while a variety of oxygen-containing biogenic organics have been observed to be generated from plants and most are likely direct emissions, care must be taken to distinguish such direct emissions from possible formation from oxidation of larger biogenic hydrocarbons and/or, in some cases, contamination during sample handling.

2. Chemistry

a. Biogenic Hydrocarbons

Although some of the biogenic VOCs are relatively simple compounds such as ethene, most are quite complex in structure (e.g., Figs. 6.22 and 6.26). Furthermore, they tend to be unsaturated, often with multiple double bonds. As a result, they are very reactive (see Chapter 16.B) with OH, O_3, NO_3, and Cl atoms (e.g., Atkinson et al., 1995a). In addition, because they are quite large and of relatively low volatility, their polar oxidation products are even less volatile. This makes elucidating reaction mechanisms and quantifying product yields quite difficult. For a review of this area, see Atkinson and Arey (1998).

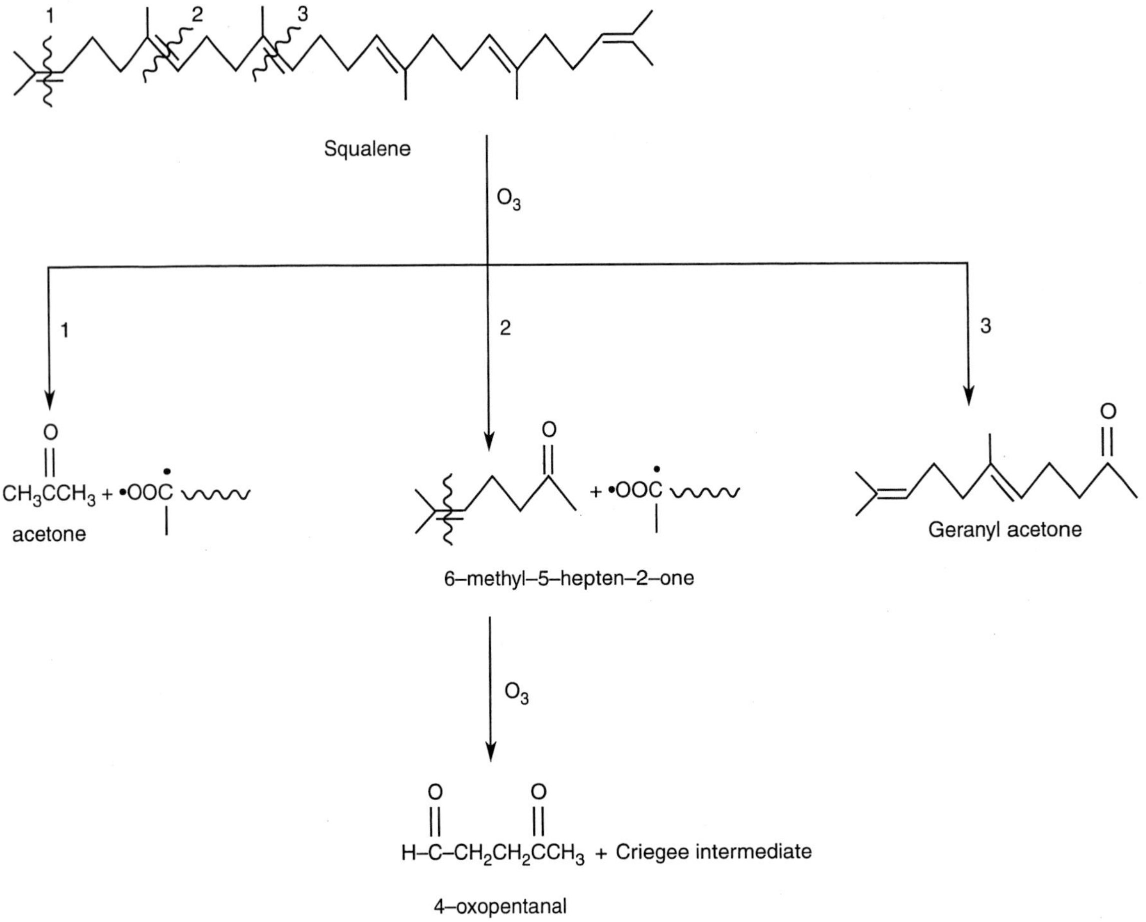

FIGURE 6.28 Oxidation of squalene to 6-methyl-5-hepten-2-one, acetone, geranyl acetone, and 4-oxopentanal (adapted from Fruekilde et al., 1998).

Some of the reactions, e.g., that of isoprene with OH and NO_3, were discussed earlier in this chapter. Table 6.26 summarizes some of the major products observed in the gas-phase reactions of several other biogenic hydrocarbons with OH and O_3 (Atkinson, 1997a). These products are anticipated, based on the mechanisms described earlier in this chapter. As also expected, the yields of these major products generally do not account for 100% of the reactant lost, and there are a number of other products, including multifunctional species, that are also formed. As an example, the formation of more than 30 individual products has been observed from the reaction of α-pinene with O_3 in air, some of which are unidentified, and the same is true for the Δ^3-carene reaction (Yu et al., 1998). Products included hydroxy oxoacids, hydroxy dicarbonyls, and dicarbonyls. The formation of low-volatility products that form particles (e.g., Hoffmann et al., 1998; Jang and Kamens, 1999) is likely responsible for a significant fraction of the products missing from the gas phase. For example, Jang and Kamens (1999) have identified a variety of multifunctional oygenated products in aerosol particles from the α-pinene reaction with O_3, including diacids such as 2,2-dimethylcyclobutane-1,3-dicarboxylic acid and pinic acid. These diacids have sufficiently low vapor pressures that they are believed to contribute significantly to the formation of particles in this system.

This is supported by studies of the aerosol composition in forested areas. For example, Kavouras et al. (1998) identified cis- and trans-pinonic acids as well as pinonaldehyde and nopinone in particles in a forest in Portugal. The diurnal variations of the pinonic acids and formic acid were similar, peaking in the afternoon as expected if they were formed by the reaction of O_3 with α-pinene. On the other hand, the concentrations of pinonaldehyde, expected from the oxidation of α-pinene by OH, O_3, and NO_3, and nopinone, from the oxidation of β-pinene, were the smallest in the after-

TABLE 6.26 Some Products and Their Yields Observed in the Gas-Phase Reactions of Some Biogenic Hydrocarbons with OH and O_3[a]

Biogenic	Structure	Product	OH reaction yield	O_3 reaction yield
Limonene		(structure with CHO)	0.29 ± 0.06[b]	
		(cyclohexene ketone)	0.20 ± 0.03[b]	≤ 0.04[b]
		HCHO		0.10[e]
β-Phellandrene		(cyclohexenone)	0.29 ± 0.07[c]	0.29 ± 0.06[c]
α-Pinene		(structure with CHO)	0.28 ± 0.05[b] 0.56 ± 0.04[d]	0.143 ± 0.024[g]
		CH_3COCH_3	0.11 ± 0.03[f]	
β-Pinene		(ketone)	0.27 ± 0.04[b] 0.79 ± 0.08[d]	0.23 ± 0.05[b] 0.22[e]
		HCHO	0.54 ± 0.05[d]	0.42[e]
		CH_3COCH_3	0.085 ± 0.018[f]	
Sabinene		(ketone)	0.17 ± 0.03[b]	0.50 ± 0.09[b]
Terpinolene		(cyclohexenone)	0.26 ± 0.06[b]	0.40 ± 0.06[b]
		(structure with CHO)	0.08 ± 0.02[b]	

[a] Adapted from Atkinson (1997a).
[b] Hakola et al. (1994).
[c] Hakola et al. (1993).
[d] Hatakeyama et al. (1991); yields may be too high due to contribution from other products to IR absorption.
[e] Grosjean et al. (1993a).
[f] Aschmann et al. (1998).
[g] Alvarado et al. (1998b).

noon, likely reflecting their simultaneous removal by reaction with OH and/or photolysis.

The atmospheric oxidation of biogenic hydrocarbons appears to be a significant source of CO on a global level, accounting for ~10–20% of the total (e.g., Hatakeyama et al., 1991; Miyoshi et al., 1994; Röckmann et al., 1998).

Nighttime reactions of biogenics with NO_3 are also important. As seen in Table 6.13, many of these reactions are very fast, leading to relatively short lifetimes of the biogenics at night. For α-pinene, for example, the lifetime with respect to reaction with 50 ppt NO_3 is only ~2 min. Again the reactions produce a number of multifunctional products, in this case many of them containing the nitrate group. For example, Wängberg et al. (1997) found total organic nitrate yields of about 14% in the NO_3–α-pinene reaction, which included 3-oxopinane 2-nitrate (3% yield) and 2-hydroxypinane 3-nitrate (5% yield), as well as the major product pinonaldehyde (62% yield):

Pinonaldehyde   3-Oxopinane 2-nitrate

2-Hydroxypinane 3-nitrate

It is interesting that the reactions of α-terpinene and α-phellandrene with NO_3 also gave the corresponding aromatic, p-cymene, in significant yields (6 and 22%, respectively), suggesting an additional path producing $NO_2 + H_2O$ + aromatic (Berndt et al., 1996).

Since 2-methyl-3-buten-2-ol is now recognized as being emitted in significant quantities, its kinetics and mechanisms of oxidation are also of interest. Reaction with OH is fast, with the rate constant reported to be in the range of $(4-7) \times 10^{-11}$ cm^3 molecule^{-1} s^{-1} (Rudich et al., 1995; Fantechi et al., 1998a; Ferranto et al., 1998). As expected, the reaction proceeds primarily by addition to the double bond. The major products of the reaction in air are $CHOCH_2OH$ (glycoaldehyde), acetone, and HCHO (Grosjean and Grosjean, 1995; Fantechi et al., 1998b; Ferranto et al., 1998) (see Problem 12).

2-Methyl-3-buten-2-ol also reacts with O_3 and NO_3, with the rate constants for these reactions being 1×10^{-17} and $\sim(1-2) \times 10^{-14}$ cm^3 molecule^{-1} s^{-1}, respectively (Grosjean and Grosjean, 1994; Rudich et al., 1996; Hallquist et al., 1996; Fantechi et al., 1998a). Major products of the O_3 reaction have been reported to be acetone, HCHO, HCOOH, and possibly 2-hydroxy-2-methylpropanal. Major products of the NO_3 reaction are acetone and organic nitrates, peroxy nitrates, and carbonyl nitrates (e.g., Fantechi et al., 1998b). These are consistent with the mechanisms of reaction of O_3 and NO_3 discussed earlier in this chapter.

As anticipated from their structures, the products formed from the initial reactions of the biogenics (Table 6.26) are also highly reactive with OH, O_3, and NO_3. For example, the rate constants (Alvarado et al., 1998a) for the reaction of pinonaldehyde with OH, O_3, and NO_3 are 4.8×10^{-11}, $< 2 \times 10^{-20}$, and 2×10^{-14} cm^3 molecule^{-1} s^{-1}, respectively (see also Glasius et al., 1997; and Hallquist et al., 1997), leading to a lifetime of ~6 h with respect to the OH reaction (at an [OH] of 1×10^6 radicals cm^{-3}) and ~11 h with respect to the reaction with 50 ppt NO_3. Photolysis is also important, with a photolysis lifetime of ~3 h for 50°N on July 1, assuming a quantum yield of one (Hallquist et al., 1997).

Again, the reaction products are complex. For example, the NO_3 reaction has been reported to give 3-acetyl-2,2-dimethylcyclobutane acetylperoxynitrate (Wängberg et al., 1997; Nozière and Barnes, 1998):

3-Acetyl-2,2-dimethylcyclobutane acetylperoxynitrate

A review of the gas-phase oxidation products is given by Calogirou et al. (1999).

The chemistry of dimethyl sulfide, a major biogenic in marine areas, is discussed in Chapter 8.E.1.

b. General Remote Boundary Layer Chemistry

The most universal characteristic of remote regions compared to those clearly subject to anthropogenic influences is the low NO_x (see Crutzen, 1995, for a review). Under these conditions, OH is generated by the photolysis of O_3 to $O(^1D)$, followed by its reaction with water vapor, which occurs in competition with deactivation to $O(^3P)$:

$$O_3 + h\nu \rightarrow O(^1D) + O_2, \tag{101}$$

$$O(^1D) + H_2O \rightarrow 2OH, \tag{102a}$$

$$+ M \rightarrow O(^3P) + M. \tag{102b}$$

In remote marine regions where there are not significant sources of large, biogenic VOCs, OH is removed by reaction with CO and CH_4:

$$OH + CO \xrightarrow{O_2} HO_2 + CO_2, \quad (103)$$

$$OH + CH_4 \xrightarrow{O_2} CH_3O_2 + H_2O. \quad (104)$$

Both HO_2 and OH can react with O_3:

$$HO_2 + O_3 \rightarrow OH + 2O_2, \quad (105)$$

$$OH + O_3 \rightarrow HO_2 + O_2. \quad (106)$$

In the absence of NO, radical–radical reactions of HO_2 and of RO_2 occur. The self-reaction of HO_2 is both pressure and water concentration dependent:

$$HO_2 + HO_2 \xrightarrow{M, H_2O} H_2O_2 + O_2. \quad (107)$$

As indicated by the involvement of water vapor and an inert third body, this reaction has several channels (see DeMore et al., 1997, for a review). There is both a bimolecular channel, which is pressure independent, and a termolecular channel, which is pressure dependent. In addition, the rate constant increases in the presence of gaseous water, suggesting that the reaction proceeds through a mechanism such as

$$HO_2 + H_2O \leftrightarrow (HO_2 \cdot H_2O),$$
$$HO_2 + (HO_2 \cdot H_2O) \rightarrow \rightarrow H_2O_2 + O_2 + H_2O,$$
$$2(HO_2 \cdot H_2O) \rightarrow H_2O_2 + O_2 + 2H_2O.$$

The binding energy of the $HO_2 \cdot H_2O$ complex has been calculated to be 6.9 kcal mol^{-1} (Aloisio and Francisco, 1998). The recommended overall rate constant (in units of cm^3 molecule^{-1} s^{-1}) for reaction (107) is given by (DeMore et al., 1997)

$$k_{107} = [2.3 \times 10^{-13} e^{600/T} + 1.7 \times 10^{-33}[M]e^{1000/T}]$$
$$\times \{1 + 1.4 \times 10^{-21}[H_2O]e^{2200/T}\}.$$

At 1 atm pressure, 298 K, and 50% relative humidity, $k_{107} = 5.5 \times 10^{-12}$ cm^3 molecule^{-1} s^{-1} (see Problem 13). The pressure and water vapor dependences are quite significant. For example, Stockwell (1995) points out that the relative error can be as much as 75% near the earth's surface and 30% at 10 km, leading to underestimates of the rate of formation of H_2O_2 and overestimates of the rates of formation of organic peroxides (formed from $HO_2 + RO_2$; see the following) and of O_3.

In addition to the HO_2 self-reaction, there are also HO_2–RO_2 and RO_2–RO_2 reactions:

$$HO_2 + CH_3O_2 \rightarrow CH_3OOH + O_2, \quad (108)$$

$$CH_3O_2 + CH_3O_2 \rightarrow CH_3OH + HCHO + O_2, \quad (109a)$$
$$\rightarrow 2CH_3O + O_2. \quad (109b)$$

Although reaction (108) is generally accepted to represent the major, if not sole, reaction path, it has been suggested by Ayers et al. (1997), based on measurements of HCHO in clean marine air, that a portion may proceed by an alternate path to produce HCHO + H_2O + O_2. As discussed earlier, path (109a) is the major path in the CH_3O_2 self-reaction at room temperature, with a branching ratio of ~ 0.7 and the remainder occurring via (109b); however, k_{109a}/k_{109b} is temperature dependent, with the relative importance of reaction (109b) decreasing at lower temperatures. For example, the recommended temperature dependence for these channels gives a branching ratio for (109b) of ~ 0.13 at 245 K compared to 0.3 at 298 K (Atkinson et al., 1997a).

The result of this chemistry is the photochemical destruction of O_3 and the formation of peroxides.

On the other hand, when sufficient concentrations of NO are present, HO_2 and RO_2 both react with NO:

$$HO_2 + NO \rightarrow OH + NO_2, \quad (110)$$

$$CH_3O_2 + NO \rightarrow CH_3O + NO_2, \quad (111)$$

As discussed earlier, the NO_2 then photolyzes to $O(^3P)$, which adds to O_2 to form O_3. Under these conditions, O_3 will be formed. The concentration of NO at which this crossover from ozone destruction to ozone formation occurs is central to the chemistry of both remote and polluted regions.

At night, there can be significant concentrations of NO_3 radicals present, along with HO_2 and RO_2. HO_2 reacts with NO_3,

$$HO_2 + NO_3 \rightarrow OH + NO_2 + O_2, \quad (112)$$

with a recommended rate constant of 3.5×10^{-12} cm^3 molecule^{-1} s^{-1} at 298 K (DeMore et al., 1997). Given that this rate constant is similar to that for the HO_2 self-reaction under typical tropospheric conditions near the surface, this reaction can be a significant contributor to the removal of HO_2 at night. It accomplishes the same thing as NO, i.e., converts HO_2 to OH and generates NO_2.

In addition to gas-phase chemistry, aqueous-phase chemistry discussed in Chapter 8.C.3 taking place in clouds can also be important in remote regions. For example, modeling studies by Lelieveld and Crutzen (1990) suggest that clouds may decrease the net production of O_3 by uptake of HO_2, dissociation to H^+ + O_2^-, and reaction of O_3 with O_2^- in cloud droplets.

A test of our understanding of the chemistry of remote regions thus requires measurements of not only

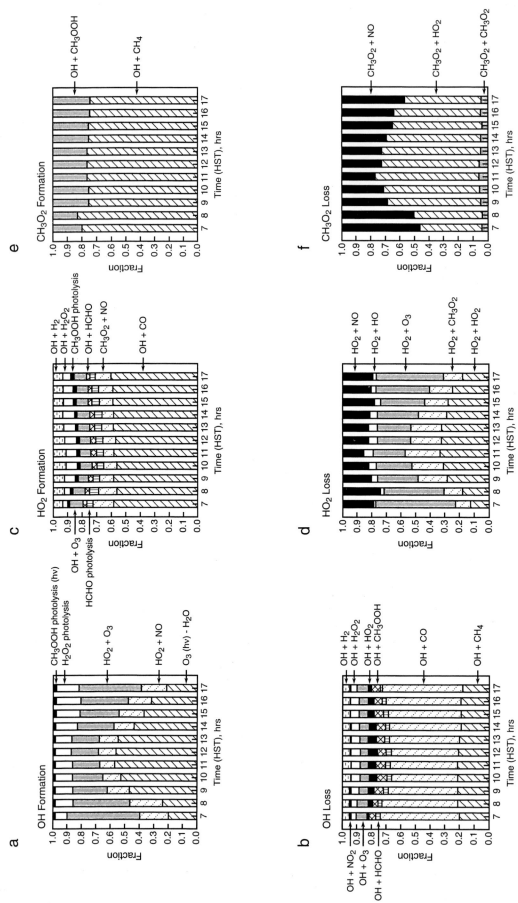

FIGURE 6.29 Estimated rates of formation and loss of OH, HO_2, and CH_3O_2 in what is primarily free tropospheric air (adapted from Cantrell et al., 1996).

stable species such as O_3, CO, CH_4, and NO but also free radicals such as HO_2, RO_2, and OH. Such studies have been carried out in a number of locations (see Chapter 11.A.4), including remote regions. For example, Cantrell *et al.* (1996) measured total peroxy radical concentrations ($HO_2 + RO_2$) at the Mauna Loa Observatory and compared them to model calculations. On one day in which mostly free tropospheric air reached the sampling site, the measurements and model calculations were in relatively good agreement. However, on other days, the measured values were significantly smaller than predicted. This was hypothesized as being due to the possible loss of peroxy radicals on aerosol particles and/or reaction of OH with unidentified organics.

Figure 6.29 shows the estimated rates of removal and formation of OH, HO_2, and CH_3O_2 for the day having primarily free tropospheric air at the sampling site. The photolysis of O_3 to $O(^1D)$ is estimated to be a major source of OH, but not the only source, with its contribution varying from ~20 to 55% of the total rate of OH production. The major source of HO_2 is the reaction of OH with CO, which is also the major sink for OH. Interestingly, the production of CH_3O_2 is predicted to occur to a small, but significant, extent from the reaction of OH with CH_3OOH formed in the $HO_2 + CH_3O_2$ reaction.

Figure 6.30 shows the predicted rates of O_3 formation and loss for the day corresponding to Fig. 6.29. The net is a loss of O_3; i.e., the air mass is in the low-NO_x regime.

Penkett, Ayers, Galbally, and co-workers have measured [$HO_2 + RO_2$] radicals using the radical amplifier technique described in Chapter 11.A.4, peroxides, O_3, and the photolysis rate of O_3, $J(O^1D)$, in remote marine air. Assuming that reaction (109) is sufficiently slow compared to reaction (108) that it can be ignored, steady-state analysis can be applied to OH, HO_2, and RO_2 to give the following expressions (Penkett *et al.*, 1997):

$$d[OH]/dt = (2f)J(O^1D)[O_3] \\ + k_{105}[HO_2][O_3] - k_{103}[OH][CO] \\ - k_{104}[OH][CH_4] \approx 0. \quad (B)$$

In Eq. (B), f is the fraction of $O(^1D)$ that reacts with water vapor to form OH,

$$f = k_{102b}[H_2O]/\{k_{102a}[H_2O] + k_{102b}[M]\},$$

and is typically about 0.1.

$$d[HO_2]/dt = k_{103}[CO][OH] \\ - k_{105}[HO_2][O_3] \\ - 2k_{107}[HO_2]^2 \\ - k_{108}[HO_2][CH_3O_2] \approx 0, \quad (C)$$

$$d[CH_3O_2]/dt = k_{104}[OH][CH_4] \\ - k_{108}[CH_3O_2][HO_2] \approx 0. \quad (D)$$

The net change in the free radical concentration (i.e., OH + HO_2 + CH_3O_2) is given by the sum of Eqs. (B)–(D):

$$d[OH + HO_2 + CH_3O_2]/dt \\ = (2f)J(O^1D)[O_3] - 2k_{107}[HO_2]^2 \\ - 2k_{108}[HO_2][CH_3O_2]. \quad (E)$$

Under steady-state conditions, the net change is 0.

If the [HO_2]/{[HO_2] + [RO_2]} ratio is a constant, α, then a rate constant representing both the $HO_2 + HO_2$ and $HO_2 + CH_3O_2$ reactions, k_{sum}, can be formulated:

$$k_{sum} = \alpha^2 k_{107} + \alpha(1 - \alpha)k_{108}. \quad (F)$$

Setting the right-hand side of Eq. (E) to 0 and using the composite rate constant defined by Eq. (F), one can show (see Problem 11) that

$$\{[HO_2] + [RO_2]\} = \{fJ(O^1D)[O_3]/k_{sum}\}^{0.5}. \quad (G)$$

That is, the peroxy radical concentrations in the absence of NO should vary with the square root of $J(O^1D)$ under conditions of relatively constant O_3 concentrations.

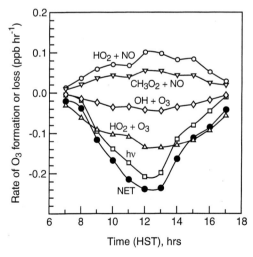

FIGURE 6.30 Calculated rates of processes leading to the formation and loss of O_3 for conditions of primarily free tropospheric air (see also Fig. 6.29) (adapted from Cantrell *et al.*, 1996).

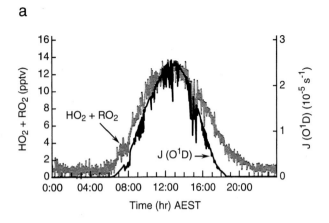

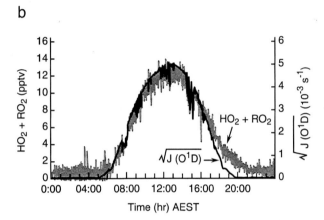

FIGURE 6.31 Measured peroxy radical concentrations ([HO$_2$] + [RO$_2$]) and ozone photolysis rate $J(O^1D)$ (a) or $\{J(O^1D)\}^{0.5}$ (b) in clean marine air at Cape Grim, Tasmania (adapted from Penkett *et al.*, 1997).

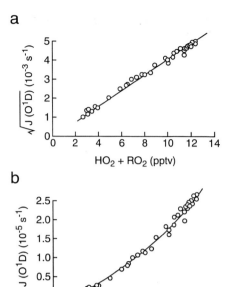

FIGURE 6.32 Measured concentrations of HO$_2$ + RO$_2$ as a function of (a) square root of ozone photolysis rate $\{J(O^1D)\}^{0.5}$ and (b) $J(O^1D)$ for a clean marine air mass at Cape Grim, Tasmania. Lines are guides for the eye only (adapted from Penkett *et al.*, 1997).

Figure 6.31 shows plots of measured peroxy radical concentrations for low-NO$_x$ conditions at Cape Grim, Tasmania (Penkett *et al.*, 1997). Overlaid are plots of $J(O^1D)$ and $\{J(O^1D)\}^{0.5}$. The plot of $\{J(O^1D)\}^{0.5}$ provides a better match. The slower decay in the peroxy radical concentration at dusk is due to the slow decay due to self-reactions, with some contribution from the CH$_3$O$_2$ + O$_3$ reaction ($k \sim 1 \times 10^{-17}$ cm^3 molecule^{-1} s^{-1}; Tyndall *et al.*, 1998) and perhaps a small contribution from deposition (Monks *et al.*, 1996).

The relationship between $\{J(O^1D)\}^{0.5}$ and $\{[HO_2] + [RO_2]\}$ is better illustrated in Fig. 6.32, which shows plots of both the square root and first power of $J(O^1D)$ against the measured peroxy radical concentrations; the square root plot is linear, while the first-order plot curves significantly.

This chemistry suggests that under low-NO$_x$ conditions, there should be net photochemical destruction of O$_3$ accompanied by the formation of peroxides. That is,
there should be a negative correlation between O$_3$ and peroxides, which has indeed been observed to be the case in remote regions (e.g., see Penkett *et al.*, 1995).

The relationship between the peroxy radical concentration and the ozone photolysis rate constant for these higher NO conditions can be again approximated using steady-state analysis (Penkett *et al.*, 1997; Carpenter *et al.*, 1997). While OH is recycled in its reactions with CO and CH$_4$ via HO$_2$, it is permanently removed at higher NO$_x$ concentrations by the reaction of OH with NO$_2$, forming nitric acid:

$$\text{OH} + \text{NO}_2 \xrightarrow{M} \text{HONO}_2. \quad (113)$$

Thus the steady-state concentration of OH is given by

$$[\text{OH}] = \frac{(2f)J(O^1D)[O_3]}{k_{113}[NO_2]}, \quad (H)$$

where k_{113} is the effective second-order rate constant. However, the OH concentration is also determined by its interconversion with HO$_2$ and CH$_3$O$_2$. Treating the OH + CO and HO$_2$ + NO as one cycle and OH + CH$_4$ and CH$_3$O$_2$ + NO as another,

$$k_{103}[\text{OH}][\text{CO}] = k_{110}[\text{HO}_2][\text{NO}], \quad (I)$$

$$k_{104}[\text{OH}][\text{CH}_4] = k_{111}[\text{CH}_3\text{O}_2][\text{NO}]$$

$$\cong k_{110}[\text{CH}_3\text{O}_2][\text{NO}], \quad (J)$$

since the rate constants for reactions (111) and (110) are similar. Adding (I) and (J), one obtains

$$\{[HO_2] + [CH_3O_2]\} = \frac{[OH]}{k_{110}[NO]}\{k_{103}[CO] + k_{104}[CH_4]\}, \quad (K)$$

and using Eq. (H) for [OH], this becomes:

$$\{[HO_2] + [CH_3O_2]\} = \frac{(2f)J(O^1D)[O_3]}{k_{113}[NO_2]}\left\{\frac{k_{103}[CO] + k_{104}[CH_4]}{k_{110}[NO]}\right\}. \quad (L)$$

That is, in the presence of sufficient quantities of NO that HO_2 and CH_3O_2 react primarily with NO rather than with each other, the total concentration of peroxy radicals should vary directly with the photolysis rate constant for O_3 rather than with its square root as was the case at low NO.

Figure 6.33 shows a plot of the total peroxy radical concentrations measured at Mace Head, Ireland, as a function of either the square root or the first power of $J(O^1D)$ (Carpenter et al., 1997). Consistent with Eq. (L), the concentrations vary with the first power under these conditions.

From such studies, Carpenter et al. (1997) conclude that the crossover point between O_3 destruction and formation occurs at NO concentrations of ~55 ± 30

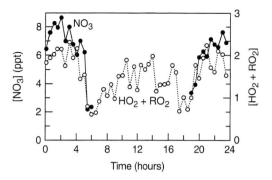

FIGURE 6.34 Measured diurnal variation of NO_3 and HO_2 + RO_2, respectively, at Weybourne, U.K. (adapted from Carslaw et al., 1997).

ppt at Mace Head, Ireland, in the late spring and 23 ± 20 ppt at Cape Grim, Tasmania, during the summer.

As discussed earlier in this chapter, NO_3 drives nighttime chemistry. Through its reactions with organics, it would be expected to generate HO_2 and RO_2 at night and hence NO_3 and peroxy radical concentrations should be related. Figure 6.34 shows one set of measurements of these species in a coastal marine boundary layer at Weybourne, U.K. (Carslaw et al., 1997). The temporal profile of peroxy radicals at night follows that of NO_3; during the day, there are additional sources, of course, through OH and O_3 reactions. Carslaw et al. (1997) suggest that the reactions of HO_2, and perhaps $CH_3SCH_2O_2$ (from DMS oxidation), with NO_3 at night may also be important.

3. Upper Troposphere

While a great deal is known about the chemistry of the lower troposphere, particularly the boundary layer, as well as the stratosphere (see Chapters 12 and 13), much less is known about the upper troposphere, the region between the two. This region has attracted increasing attention for a number of reasons, including the potential impact of commercial aircraft. Of particular concern is understanding the formation and fate of O_3, whose concentration in this region is important for its role as a greenhouse gas (see Chapter 14.B), in addition to its role in the photochemical reactions.

Measurements made in this region have raised questions regarding our understanding of the chemistry involved, as well as the transport processes that can affect ozone in this region (e.g., Suhre et al., 1997). Figure 6.35, for example, shows one set of measurements of OH as a function of solar zenith angle at an altitude of 11.8 km near Hawaii (Wennberg et al., 1998). Also shown are model predictions based on the

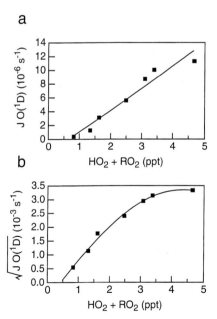

FIGURE 6.33 Measured concentrations of HO_2 + RO_2 as a function of (a) ozone photolysis rate, $J(O^1D)$, or (b) $\{J(O^1D)\}^{0.5}$ under polluted conditions at Mace Head, Ireland. Lines are guides for the eye (adapted from Carpenter et al., 1997).

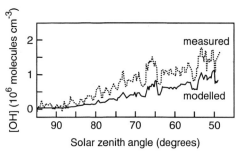

FIGURE 6.35 Measured OH concentrations at an altitude of 11.8 km near Hawaii and concentrations predicted using simple chemistry (adapted from Wennberg *et al.*, 1998).

simple O_3–H_2O–OH–CO–CH_4 chemistry outlined earlier. The measured concentrations of OH are significantly larger than predicted by the model, indicating that there are additional sources of OH and other free radicals.

In general, for a compound to reach the upper troposphere in sufficient concentrations to impact the chemistry, it must not react rapidly in the lower troposphere. However, more reactive compounds can be rapidly transported (on the time scale of a few minutes) from the surface to the upper troposphere through convective events (e.g., thunderstorms) (e.g., Gidel, 1983; Chatfield and Crutzen, 1984; Chatfield and Alkezweeny, 1990; Pickering *et al.*, 1992; Wang *et al.*, 1995; Kley *et al.*, 1996; Mahlman, 1997; Kley, 1997; Jaeglé *et al.*, 1998a; Talbot *et al.*, 1998). As a result, some compounds with relatively short tropospheric lifetimes can be carried into the upper troposphere and act as free radical sources.

The additional source of HO_x free radicals has been proposed to be the photolysis of compounds such as CH_3COCH_3, HCHO, CH_3OOH, and H_2O_2 carried into the upper troposphere by such convective transport (e.g., Chatfield and Crutzen, 1984; Singh *et al.*, 1995; Arnold *et al.*, 1997a, 1997b; McKeen *et al.*, 1997; Prather and Jacob, 1997; Jaeglé *et al.*, 1997, 1998a; Folkins *et al.*, 1998; Lee *et al.*, 1998). Thus, inclusion of acetone photolysis improves the agreement between the measured and modeled values of OH. Figure 6.36, for example, shows the measured and calculated values of OH with and without acetone photolysis in the model as well as calculated rates of production of HO_x from the photolysis of O_3 and acetone, respectively (Wennberg *et al.*, 1998). At higher altitudes where acetone photolysis predominates, the agreement between the measurements and model is good. At lower altitudes, however, measured OH is still larger than the predicted values even with acetone photolysis, indicat-

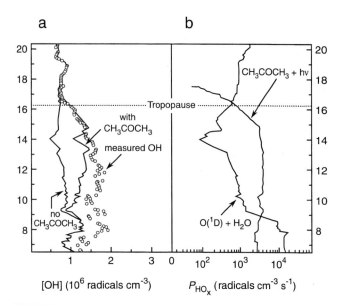

FIGURE 6.36 (a) Measured OH concentrations as a function of altitude and model-predicted concentrations without acetone photolysis and with acetone photolysis, respectively. (b) Calculated rates of HO_x production from O_3 and acetone photolysis, respectively, as a function of altitude. (Adapted from Wennberg *et al.*, 1998.)

ing a significant contribution of other species, perhaps compounds such as CH_3OOH, as well.

Based on this chemistry, the production rate of O_3 is expected to be very sensitive to the NO concentration, increasing with NO (see also Chapter 16 for a discussion of the dependence of O_3 generation on NO_x). In this context, Folkins *et al.* (1998) suggest that acetone is likely the major contributor to enhanced ozone production in the upper troposphere, since increased CH_3OOH and H_2O_2 concentrations at 9- to 12-km altitude were observed only at very small NO concentrations, indicative of clean marine boundary layer air; under such low NO_x conditions, destruction rather than production of O_3 is expected.

Another significant uncertainty in our understanding of the chemistry of the upper troposphere involves oxides of nitrogen. The measured NO_x/NO_y ratios have been observed to be higher than expected based on model predictions (e.g., Jaeglé *et al.*, 1998b). The HNO_3/NO_x ratio in the free troposphere also appears to be smaller than expected (e.g., Liu *et al.*, 1992; Chatfield, 1994). One possible explanation is that there is some unrecognized chemistry reconverting NO_y back to NO_x (e.g., Chatfield, 1994) that would bring the measurements and model predictions into better agreement (e.g., Hauglustaine *et al.*, 1996). However, it has also been suggested that these observations may reflect additional injection of NO_x by convective transport, by lightning (e.g., McKeen *et al.*, 1997; Prather and Jacob,

1997; Jaeglé et al., 1998b; Brunner et al., 1998; Dias-Lalcaca et al., 1998), or by uptake of HNO_3 in clouds. Another possibility is errors in the kinetics for NO_x and NO_y reactions in the models. For example, those for the $OH + NO_2$ and $OH + HNO_3$ reactions have been recently revised and bring the models and measurements into better agreement (see Chapters 7.B.1 and 7.E.2 and Problem 7.9).

Injection of air from the stratosphere into the upper troposphere is also very important under some conditions for determining the concentrations of various species in this region. For example, Suhre et al. (1997) measured high O_3 concentrations in the upper equatorial troposphere due to input from the stratosphere. Similarly, Dias-Lalcaca et al. (1998) carried out measurements of NO, NO_2, and O_3 in the tropopause region on flights of commercial passenger aircraft. Regions of very high ozone, up to 450 ppb, were encounted simultaneously with higher NO_x concentrations. These were attributed to air of stratospheric origin.

There has been a great deal of research activity on the effects of subsonic aircraft in the upper troposphere, with respect to impacts both on the chemistry and on the radiation balance through effects on clouds and O_3 (e.g., see April 15, May 1, and May 15, 1998, issues of *Geophysical Research Letters* and the July 27, 1998, issue of *Atmospheric Environment*). Aircraft emit a variety of pollutants, including NO_x, SO_2, and particles whose concentrations have provided "exhaust signatures" in some studies (e.g., Schlager et al., 1997; Hofmann et al., 1998).

Of particular concern is the impact of oxides of nitrogen emissions on O_3 (e.g., see Ehhalt et al., 1992; and Ehhalt and Rohrer, 1995). As is typical of combustion systems, NO_x emissions are primarily in the form of NO (e.g., Schulte et al., 1997). As discussed earlier in this chapter and elsewhere in this book (e.g., see Chapter 16), the impact of added NO_x on the generation of O_3 depends on existing levels. At low NO levels, added NO leads to increased O_3 formation. However, at sufficiently high NO_x, OH reacts with NO_2 to form HNO_3, effectively removing NO_x from the system and terminating ozone production. The level at which this occurs in the upper troposphere is ≈ 300 ppt NO (e.g., see Wennberg et al., 1998; and Grooß et al., 1998).

There is evidence from laboratory studies that heterogeneous reactions on sulfate particles may be important in the upper troposphere as well. For example, HCHO uptake into sulfuric acid solutions or ternary mixtures of sulfuric and nitric acids and water has been observed in laboratory studies (e.g., Tolbert et al., 1993; Jayne et al., 1996; Iraci and Tolbert, 1997). In sulfuric acid, the effective Henry's law constant at the low temperatures found in the upper troposphere and lower stratosphere is large, $\sim 10^6 - 10^7$ M atm^{-1}, and polymerization of the HCHO occurs as its concentration increases (Iraci and Tolbert, 1997). When HNO_3 is present, a reaction occurs that generates HONO and HCOOH:

$$HCHO + HNO_3 \rightarrow HONO + HCOOH. \quad (114)$$

NO_2 is also formed, perhaps by the subsequent reaction of HONO with HNO_3 (which is the reverse of the surface hydrolysis of NO_2 discussed in Chapter 7.B.3 thought to be a significant source of HONO in the troposphere):

$$HONO + HNO_3 \rightarrow 2NO_2 + H_2O. \quad (115)$$

These products were observed at room temperature, although their formation at the lower temperatures found in the upper troposphere could not be confirmed (Iraci and Tolbert, 1997). Such reactions may contribute to a conversion of HNO_3 to NO_x proposed by Chatfield (1994).

Similarly, the uptake of acetone into sulfuric acid–water solutions has been reported (Duncan et al., 1998), with the formation of 4-methyl-3-penten-2-one and trimethylbenzene at temperatures above 200 K and 75 wt% H_2SO_4.

In short, although relatively little is known about the possibility of heterogeneous chemistry of organics in the upper troposphere, the results of initial laboratory studies suggest that this may be important.

4. Arctic

Atmospheric chemistry in the Arctic has been the subject of studies for many years, in part because of the observation of "Arctic haze" decades ago. This haze is composed of particles with significant amounts of sulfate, about half of which is due to long-range transport from other regions, particularly Eurasia during the winter (e.g., Barrie and Bottenheim, 1991; Polissar et al., 1998a, 1998b).

As might be expected, levels of most pollutants are quite low in the Arctic when air is not being transported from populated regions. For example, during the spring in Alaska during periods of southerly wind flow, surface-level concentrations of small (C_2-C_5) hydrocarbons are 8 ppb C, O_3 is $\sim 20-40$ ppb, NO is <10 ppt, NO_x is ~ 30 ppt, and NO_y is ~ 400 ppt (e.g., Doskey and Gaffney, 1992; Honrath and Jaffe, 1992; Sandholm et al., 1992; Beine et al., 1996). PAN tends to be a larger portion of NO_y than normally expected for "clean" regions, often in the range of 50–90% (e.g., Bottenheim et al., 1986; Barrie and Bottenheim, 1991; Jaffe, 1993; Jaffe et al., 1997). It forms an increasingly larger fraction of NO_y at higher altitudes due to stabilization with respect to thermal decomposition at lower temperatures (see Chapter 5.A.3c). For example, PAN was reported to be about 10% of NO_y near the surface, increasing to 45% at altitudes of 4.5–6.1 km over the Arctic (Sandholm *et al.*, 1992; Singh *et al.*, 1992).

An unusual phenomenon was reported in the Arctic in the mid-1980s. Ozone measured at ground level was observed to decrease rapidly to small concentrations, at times near zero (Bottenheim *et al.*, 1986; Oltmans and Komhyr, 1986). As seen in Fig. 6.37, an increase in bromide ion collected on filters (f-Br) was inversely correlated with the O_3 decrease (Barrie *et al.*, 1988; Oltmans *et al.*, 1989; Sturges *et al.*, 1993; Lehrer *et al.*, 1997); this could reflect either particle bromide or a "sticky" gas such as HBr that could be collected on the filter simultaneously. This correlation suggested that the loss of ozone was due to gas-phase chain reactions involving bromine such as

$$Br + O_3 \rightarrow BrO + O_2, \qquad (116)$$
$$BrO + BrO \rightarrow 2Br + O_2, \qquad (117a)$$
$$\rightarrow Br_2 + O_2, \qquad (117b)$$
$$Br_2 + h\nu \rightarrow 2Br, \qquad (118)$$
$$BrO + h\nu \rightarrow Br + O. \qquad (119)$$

Subsequently, additional reactions for the chain destruction of ozone were suggested, including one involving ClO (Le Bras and Platt, 1995):

$$ClO + BrO \rightarrow Br + Cl + O_2, \qquad (120a)$$
$$\rightarrow BrCl + O_2, \qquad (120b)$$
$$\rightarrow OClO + Br, \qquad (120c)$$
$$BrCl + h\nu \rightarrow Br + Cl, \qquad (121)$$
$$OClO + h\nu \rightarrow O + ClO. \qquad (122)$$

Such chain reactions imply low NO_x concentrations, which would otherwise terminate the chains by reacting with BrO and ClO to form $BrONO_2$ and $ClONO_2$, respectively. Unless the nitrates could be recycled rapidly back to active forms of the halogens, they would terminate the chain. NO_x concentrations do indeed appear to be small during O_3 depletion episodes. Beine *et al.* (1997), for example, measured $NO_x < 4.5$ ppt during O_3 depletion episodes in Norway.

In addition, mechanisms for regeneration of photochemically active bromine that involve aerosol particles or reactions on the snowpack have been proposed. For example, McConnell *et al.* (1992) and Tang and McConnell (1996) proposed that HBr and organobromine compounds could be converted to Br_2 through adsorption and reaction on ice and aerosol particles. Fan and Jacob (1992) suggested that HOBr, formed by the reaction of BrO with HO_2,

$$BrO + HO_2 \rightarrow HOBr + O_2, \qquad (123)$$

or by hydrolysis of bromine nitrate,

$$BrONO_2 + H_2O \rightarrow HOBr + HNO_3, \qquad (124)$$

would form Br_2 by reaction with Br^- in aerosol particles:

$$HOBr + Br^- + H^+ \rightarrow Br_2 + H_2O. \qquad (125)$$

Aranda *et al.* (1997) have shown in laboratory studies that CH_3O_2 also reacts with BrO in a manner analogous to the HO_2 reaction (123), with a rate constant at 298 K of 5.7×10^{-12} cm^3 molecule^{-1} s^{-1}. About 80% of the reaction generates HOBr + CH_2O_2, with the remainder forming Br + CH_3O + O_2. In either case, photochemically active bromine species are regenerated.

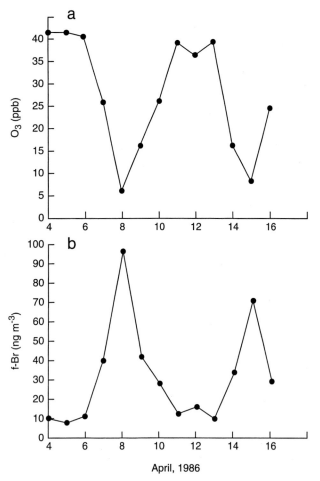

FIGURE 6.37 (a) Surface-level O_3 at Alert, Canada, and (b) filter-collected bromide (f-Br) during an ozone depletion episode (adapted from Barrie *et al.*, 1988).

Crutzen and co-workers (Sander and Crutzen, 1996; Vogt *et al.*, 1996) have developed a model for chemistry in the marine boundary layer at midlatitudes, in which autocatalytic cycles involving sea salt particles generate photochemically active gases such as BrCl, Br_2, and Cl_2. It is likely that such chemistry also occurs in the Arctic as well. In these cycles, reactions (125) and (126) in the condensed phase,

$$HOBr + Cl^- + H^+ \rightarrow BrCl + H_2O, \quad (126)$$

followed by reactions such as (127)–(131), also occurring in the condensed phase,

$$BrCl + Br^- \leftrightarrow Br_2Cl^-, \quad (127)$$

$$Br_2Cl^- \leftrightarrow Br_2 + Cl^-, \quad (128)$$

$$BrCl + H_2O \leftrightarrow HOBr + Cl^- + H^+, \quad (129)$$

$$BrCl + Cl^- \rightarrow BrCl_2^-, \quad (130)$$

$$BrCl_2^- + H_2O \rightarrow HOBr + H^+ + 2Cl^-, \quad (131)$$

lead to the generation of photochemically active halogens in the gas phase.

Much of this chemistry has been confirmed experimentally. For example, Kirchner *et al.* (1997) showed that HOBr reacts at 240 K with the surface of ice that has been doped with sea salt, generating gaseous BrCl and Br_2. The chemistry is similar to that at midlatitudes, only in this case occurs in a quasi-liquid layer on the ice surface. Figure 6.38 is a schematic diagram of this chemistry, for which many of the reaction kinetics in aqueous solution have been reported (e.g., Wang *et al.*, 1994). Similar chemistry occurs in the interactions of HOBr with HCl on ice (Abbatt, 1994) and with aqueous aerosols of NaCl (Abbatt and Waschewsky, 1998) as well as in the reaction of O_3 with frozen seawater ice in the dark (Oum *et al.*, 1998b); in the latter case, O_3 oxidizes Br^- to $BrO^-/HOBr$ and the chemistry then leads to the generation of gaseous Br_2 through the chemistry in Fig. 6.38.

The central point is that photochemically active bromine, and perhaps chlorine (see following), compounds are generated that lead to the chain destruction of gaseous O_3 at ground level at polar sunrise.

However, what remains unknown is the source of the original bromine that initiates the chemistry. There have been a number of hypotheses, including the photolysis of bromoform which is generated by biological processes in the ocean (Barrie *et al.*, 1988) or reactions of sea salt, either suspended in the air or deposited on, or associated with, the snowpack. These include photolysis of $BrNO_2$ formed from the reaction of sea salt particles with N_2O_5 (Finlayson-Pitts *et al.*, 1990), the

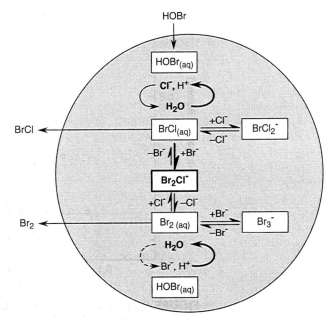

FIGURE 6.38 Schematic diagram of HOBr chemistry with sea salt particles/ice (graciously provided by T. Benter).

formation of HOBr from the oxidation of bromide by HSO_5^- (Mozurkewich, 1995), or the reaction of O_3 with bromide in frozen seawater ice (Oum *et al.*, 1998b; De Haan *et al.*, 1999).

Supporting a seawater source for the halogens is the observation by Shepson and co-workers of significant amounts of as yet unidentified photolyzable chlorine as well as bromine compounds in the spring in the Arctic (Impey *et al.*, 1997a, 1997b). In addition, Platt and co-workers have detected both BrO and ClO at the surface during ozone depletion events (Platt and Hausmann, 1994; Hausmann and Platt, 1994; Tuckermann *et al.*, 1997).

It is also clear that during periods of low surface ozone, chlorine atoms are a major reactant for hydrocarbons (e.g., Jobson *et al.*, 1994; Solberg *et al.*, 1996; Ariya *et al.*, 1998). Figure 6.39, for example, shows the measured ratios of isobutane, *n*-butane, and propane during an ozone depletion event (Jobson *et al.*, 1994). These particular pairs of hydrocarbons were chosen to differentiate chlorine atom chemistry from OH reactions. Thus isobutane and propane have similar rate constants for reaction with Cl but different rate constants for reaction with OH. If chlorine atoms are responsible for the loss of these organics, their ratio should remain relatively constant in the air mass, as indicated by the line marked "Cl." Similarly, isobutane and *n*-butane have similar rate constants for removal by OH but different rate constants for reactions with

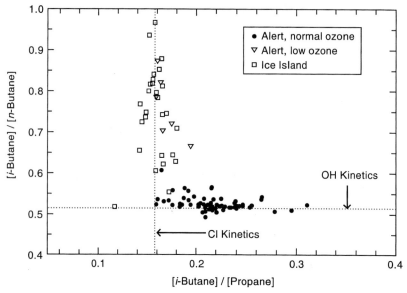

FIGURE 6.39 Decay of some simple alkanes during an episode of surface-level O_3 depletion at Alert, Canada, and at an ice island upwind. Also shown are the relative alkane concentrations for a "normal" time without O_3 depletion. The dotted lines show the ratios expected if the alkanes are being removed by Cl or OH, respectively (adapted from Jobson *et al.*, 1994).

chlorine atoms, and their ratio should remain constant if reaction with OH is the major removal process. As seen in Fig. 6.39, during ozone depletion events at Alert, Canada, and at an ice island upwind, the removal rates are consistent with chlorine atom reactions. On the other hand, in "normal ozone" conditions, the hydrocarbon loss is indicative of OH kinetics.

Aldehydes and ketones have also been measured during these ozone depletion events. Although these are expected to be generated from the oxidation of organics by chlorine atoms, the levels are not completely consistent with this as the only source. For example, Shepson *et al.* (1996) report that much higher concentrations of HCHO are observed than expected from such oxidation processes. High concentrations have been measured in the interstitial air in the snowpack in the presence of sunlight, suggesting that photochemical processes in the snowpack generate HCHO (Sumner and Shepson, 1999).

It is interesting that recent satellite observations (e.g., Richter *et al.*, 1998; Chance, 1998; Wagner and Platt, 1998; Hegels *et al.*, 1998) indicate that BrO is formed over large parts of the Arctic during spring, coincident with retreating sea ice and increasing radiation. Hopper *et al.* (1998) report that ozone depletion occurs in a layer above the sea ice, both observations again indicative of a seawater source for the bromine.

In short, a bromine-catalyzed destruction of surface-level ozone occurs over a wide area in the Arctic at polar sunrise. The mechanism likely involves regeneration of photochemically active bromine via heterogeneous reactions on aerosol particles, the snowpack, and/or frozen seawater. The source of the bromine is likely sea salt, but the nature of the reactions initiating this ozone loss remains to be identified. For a review, see the volume edited by Niki and Becker (1993) and an issue of *Tellus* (Barrie and Platt, 1997).

K. ATMOSPHERIC CHEMISTRY AND BIOMASS BURNING

Biomass burning is used for a variety of purposes, including clearing of forests for agricultural and grazing purposes, control of weeds and other unwanted plants, and removal of agricultural waste and stubble after harvest. Biomass fuels are also used for heating, cooking, and producing charcoal (Crutzen and Andreae, 1990; Levine *et al.*, 1995). While the burning of savannas has been the focus of many studies and is believed to contribute about 40% of the total carbon released by biomass burning each year, the other contributions, including burning of boreal forests, are also substantial during some time periods (Levine *et al.*, 1995). Indeed, there is evidence from changes in composition of the Greenland ice cores (see Chapter 14.E) for burning of boreal vegetation at high latitudes over the past approximately 800 years (e.g., Savarino and Legrand, 1998).

Emissions from biomass burning include a wide range of gases and particles, in quantities that in some cases can be significant not only locally and regionally but also on a global scale. Table 6.27, for example, shows one estimate of contributions of biomass burning to the global emissions of some of the emissions (Andreae, 1991). Clearly, biomass burning is important on a global scale, contributing to ozone formation over wide regions (see later) and through both ozone formation and particle emissions, impacting radiative properties regionally and hence potentially climate (e.g., Penner et al., 1992; Kaufman and Fraser, 1997; Christopher et al., 1996; Hobbs et al., 1997; Portmann et al., 1997).

The production of a wide range of compounds is associated with biomass burning. These include not only CO, CO_2, and NO_x but a variety of light hydrocarbons, CH_3Cl and CH_3Br, polycyclic aromatic hydrocarbons (see Chapter 10), HCHO, formic and acetic acids, CH_3OH, phenol, 2-hydroxyethanal, H_2O_2, methyl hydroperoxide, hydroxymethyl hydroperoxide, 1-hydroxyethyl hydroperoxide, NH_3, HCN, CH_3CN, HNO_3, PAN, H_2, some sulfur compounds such as SO_2 and COS, and particles (e.g., LeBel et al., 1988; Hegg et al., 1990; Cofer et al., 1990; Levine, 1991; Rogers et al., 1991; Andreae, 1991; Rudolph et al., 1992; Laursen et al., 1992; Reinhardt and Ward, 1995; Kuhlbusch and Crutzen, 1995; Yokelson et al., 1996, 1997; Lee et al., 1997, 1998; Nichol, 1998; Mauzerall et al., 1998; Crutzen et al., 1998).

Fires go through several different phases of burning, with emissions of different compounds tending to be associated with each phase. Most commonly, two phases, a flaming stage followed by a smoldering stage, have been considered. However, Yokelson et al. (1996) have suggested, based on their laboratory measurements of emissions from the burning of biomass materials, that a third pyrolysis stage, associated with unburned pyrolysis products from heating of the fuel, should also be considered.

The initial, hot, flaming stage of the fire is associated with emissions of CO_2, NO_x, N_2O, SO_2, and black/elemental carbon; during the second phase, CO, CH_4, NH_3, HCN, CH_3CN, and some light hydrocarbons such as ethane and acetylene are emitted (Fig. 6.40). Emissions of individual compounds are usually expressed as enhancement or emission factors relative to CO_2, defined as Δcompound/ΔCO_2, or relative to CO, Δcompound/ΔCO, where Δcompound is the difference between the concentration in the smoke plume and its background concentration outside the plume. (Corrections for the loss of CO by chemical reactions and for changes in the ratio due to dilution with air outside the plume, which contains different mixing ratios, must be taken into account; e.g., see Mauzerall et al., 1998.)

TABLE 6.27 Estimated Emissions from Biomass Burning Compared to Global Emissions[a]

Species	Biomass burning (Tg of element / yr)	All sources (Tg of element / yr)	Biomass burning (%)
Carbon dioxide (gross from combustion)	3500	8700	40
Carbon dioxide (net from deforestation)	1800	7000	26
Carbon monoxide	350	1100	32
Methane	38	380	10
Nonmethane hydrocarbons[b]	24	100	24
Nitrous oxide	0.8	13	6
NO_x	8.5	40	21
Ammonia	5.3	44	12
Nitrous oxide	0.8	13	6
Sulfur gases	2.8	150	2
Carbonyl sulfide	0.09	1.4	6
Methyl chloride	0.51	2.3	22
Hydrogen	19	75	25
Tropospheric ozone[c]	420	1100	38
Total particulate matter	104	1530	7
Particulate organic carbon	69	180	39
Elemental carbon (black soot)	19	<22	>86

[a] From Andreae (1991); see for original references.
[b] Excluding isoprene and terpenes.
[c] Formed from reactions in air due to biomass burning.

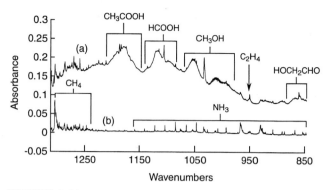

FIGURE 6.41 Infrared spectra of gases from burning a hardwood (a) when glowing combustion and white smoke are observed and (b) ~1 min later when there is no longer white smoke (adapted from Yokelson *et al.*, 1997).

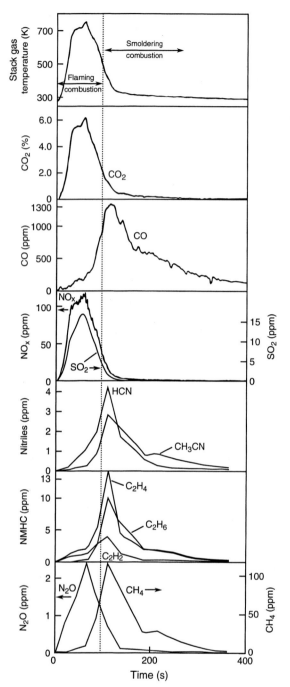

FIGURE 6.40 Stack gas temperature and emissions of various species from an experimental burning of grass in Venezuela (adapted from Crutzen and Andreae, 1990).

The use of *in situ* techniques such as FTIR in both laboratory and field studies has also permitted measurement of a variety of compounds associated with biomass burning that are difficult to analyze by traditional methods. For example, as seen in Fig. 6.41, Yokelson *et al.* (1996, 1997) have identified formic and acetic acids as well as methanol and 2-hydroxyethanal as products of burning during the smoldering phase in a laboratory study using FTIR. Remote infrared spectroscopy has also been used in field studies to detect such compounds as formic acid, methanol, ethene, and ammonia (Worden *et al.*, 1997), and remote sensing from satellites has been used to detect NO_2 and HCHO (Thomas *et al.*, 1998).

In addition to the compounds that are directly emitted, there are a number of compounds formed in the plume from the VOC–NO_x chemistry described earlier. In particular, the emissions from biomass burning have been observed to form elevated concentrations of O_3 over large regions (e.g., Fishman *et al.*, 1986, 1991; Logan and Kirchhoff, 1986; Cros *et al.*, 1992; Andreae *et al.*, 1994; Kim and Newchurch, 1996, 1998; Levy *et al.*, 1997; Mauzerall *et al.*, 1998; Jonquières *et al.*, 1998; Poppe *et al.*, 1998).

Mauzerall *et al.* (1998) measured a variety of species in plumes from biomass burning over the tropical South Atlantic Ocean as a function of the age of the plume and determined their enhancement ratios relative to both CO and CO_2. Species that are directly emitted by the biomass burning or formed rapidly are expected to be strongly correlated with CO at short times, but less so at longer times. CH_4, for example, was strongly correlated with CO, with a relatively constant enhancement ratio using CO_2 as the reference compound and a slight increase with age using CO, due to the small loss of CO. Similarly, HCHO and possibly H_2O_2 were correlated with CO in a manner suggesting direct emissions of these compounds, in agreement with measurements made by Lee *et al.* over the South Atlantic, Brazil, and southern Africa (1997, 1998).

On the other hand, species whose correlation with CO increases as the plume ages and with enhancement ratios that also increase with plume age are likely generated by chemical reactions in the plume. This was the case for O_3, as expected from known VOC–NO_x chemistry, with the enhancement ratio using CO as the reference increasing from 0.15 to 0.74 as the plume aged. Acetone also appeared to be formed as a secondary oxidation product, as did CH_3OOH.

Interconversion of various forms of oxides of nitrogen appeared to be occurring in the plume. Thus the enhancement ratio $\Delta NO_x/\Delta CO$ did not change substantially with the plume age, despite the fact that known chemistry of the primary emissions must be occurring. Mauzerall *et al.* (1998) postulate that continuing PAN decomposition to NO_2 helped to maintain this ratio at a relatively constant value. This was supported by measurements showing that in one fresh plume, PAN mixing ratios were larger than those of NO_x (by ~25%) and about a factor of 4 times that of HNO_3. However, after 5 days, PAN levels are about a third of those of HNO_3. It appears that PAN was the major form of NO_y initially but as the plume aged, it decomposed to NO_2, which was then ultimately oxidized to HNO_3.

Figure 6.42 shows the fraction of HO_x production due to O_3 photolysis as well as the contributions from photolysis of H_2O_2, CH_3OOH, HCHO, and acetone at different altitudes in a plume from biomass burning, based on measurements of these compounds (Lee *et al.*, 1998). O_3 photolysis becomes less important at higher altitudes due to lower water vapor concentrations which are needed to react with the $O(^1D)$ produced initially in order to generate OH. Formaldehyde becomes increasingly important at higher altitudes, with a significant contribution from H_2O_2 and in some cases, at high altitudes particularly, from CH_3OOH. Acetone is a minor contributor to the HO_x production.

Folkins *et al.* (1997) directly measured OH and HO_2 using an aircraft platform while descending through a plume from biomass burning. The measured concentrations of HO_x and the associated calculated rate of production of O_3 were much larger than expected based on the known chemistry. They suggested that, as for the clean upper troposphere, there must be an additional source of HO_x in biomass plumes.

In short, there is a rich and varied chemistry associated with the emissions from biomass burning, which have effects on local, regional, and global scales. A review of various aspects of biomass burning is found in the volume edited by Levine (1991).

L. PROBLEMS

1. The individual lifetimes of *trans*-2-butene with respect to typical peak concentrations of OH, O_3, and NO_3 are given in Table 6.1. (a) Develop an expression

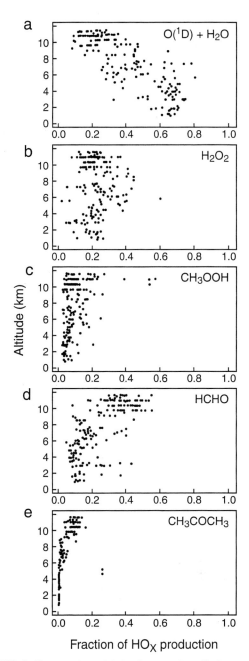

FIGURE 6.42 Fraction of HO_x from each radical source as a function of altitude over the tropical Atlantic Ocean (adapted from Lee *et al.*, 1998).

for the overall "instantaneous" lifetime of an organic with respect to these three oxidants acting together at the same time. (b) Calculate the overall lifetime of this compound at concentrations of OH of 4×10^4 molecules cm^{-3}, of O_3 of 50 ppb, and of NO_3 of 100 ppt, which might be typical of a nighttime situation, using the rate constants given in this chapter.

2. Use the structure–reactivity relationship approach to calculate the rate constants for the reactions of OH with the following compounds and calculate the percentage difference from the recommended values in Table 6.2: (a) ethane, (b) n-butane, (c) 2-methylpentane, (d) 2,2-dimethylpentane, (e) 2,2,3-trimethylbutane, (f) n-nonane, (g) n-decane.

3. Use the structure-reactivity relationship approach to calculate the rate constants for the reactions of NO_3 with the following compounds and calculate the percentage difference from the recommended values in Table 6.3: (a) n-butane, (b) 2-methylbutane, (c) n-hexane, (d) 2,3-dimethylbutane, (e) n-octane, (f) n-nonane.

4. Two of the products of the OH reaction with limonene that have been observed are the following:

Show mechanisms that would form these products.

5. The OH–C_2H_2 reaction is not quite at the high-pressure limit at 1 atm. The OH–acetylene reaction occurs by addition of the OH to the triple bond. For the reaction at 300 K, $k_0 = 5.0 \times 10^{-30}$ cm^6 molecule^{-2} s^{-1}, $k_\infty = 9.0 \times 10^{-13}$ cm^3 molecule^{-1} s^{-1}, and $F_C = 0.62$. (a) Calculate and plot the rate constant as a function of pressure up to 3000 Torr at 300 K. (b) What is the rate constant at 760 Torr and 300 K? How much different is this from the high-pressure limiting value?

6. Develop a mechanism for the photooxidation of diethyl ether in the presence of NO_x that explains why Eberhard et al. (1993) observed ethyl formate, ethyl acetate, and acetaldehyde as products.

7. Calculate the lifetimes of PPN with respect to thermal decomposition at temperatures of $-30, 32, 70,$ and 120°F, respectively, assuming there is sufficient NO present to prevent the back reaction with NO_2 to regenerate PPN.

8. (a) It is pointed out in the text that the effective rate of decomposition of PAN depends on how much NO is present, relative to NO_2, to prevent the back reaction to regenerate PAN. Show that the lifetime of PAN with respect to its thermal decomposition as a function of the NO and NO_2 concentration and the rate constants k_{75}, k_{-75}, and k_{78} is given by $\tau = (1/k_{-75})(1 + k_{75}[NO_2]/k_{78}[NO])$. Assume that the acetylperoxy radical is in a steady state and that reactions 75, -75, and 78 represent the entire reaction system. (b) Calculate the lifetimes with respect to the thermal decomposition of PAN at 25°C at ratios of $[NO]/[NO_2]$ of 10, 1, and 0.1, respectively.

9. Use the rate constants provided in this chapter to show whether or not the reaction of NO_2 with α-phellandrene could compete with the OH, O_3, and NO_3 reactions for the removal of this compound from the atmosphere.

10. The OH radical has been observed to be formed in a dark reaction when isoprene, high concentrations of NO_2, and NO are present. Postulate a mechanism for this OH generation.

11. Show that Eq. (G), which indicates that the peroxy radical concentration should vary with the square root of the O_3 photolysis rate, can be derived from Eqs. (B)–(F).

12. Ferronato et al. (1998) have reported HCHO, $CHOCH_2OH$, and CH_3COCH_3 as major products of the reaction of the biogenically emitted 2-methyl-3-buten-2-ol with OH in air. Develop a likely mechanism for the formation of these products.

13. (a) Calculate and plot the rate constant for the HO_2 self-reaction as a function of relative humidity at 298 K in increments of 10% from 0 to 100% at 1 atm pressure. (b) By what factor will the rate constant for HO_2 self-reaction change from the surface at a pressure of 1 atm, temperature of 298 K, and 50% RH to the lower stratosphere, assuming a pressure of 100 Torr, a temperature of 220 K, and a water vapor mixing ratio of 5 ppm?

14. Assess the atmospheric fates of 1-pentene by answering the following questions:

(a) What are the potentially important tropospheric oxidants you have to consider, and what are typical concentrations of each oxidant under moderately polluted conditions?

(b) Calculate the lifetime of the organic with respect to each of the oxidants in (a). If the rate constant(s) you need are not available, estimate them based on those for analogous compounds.

(c) Show the mechanism of reaction of the organic with each of the important oxidants, following through to stable products in each case. Be sure to consider all possibilities. Be sure to consider all possible fates of reactive intermediates, and when some can be ruled out, state so explicitly.

(d) Suppose the oxidation occurs in a remote area where [NO] is close to 0. Trace the mechanism of the OH reaction through to one stable product under these conditions.

References

Abbatt, J., "Heterogeneous Reaction of HOBr with HBr and HCl on Ice Surfaces at 228 K," *Geophys. Res. Lett.*, **21**, 665–668 (1994).

Abbatt, J. P. D., and G. C. G. Waschewsky, "Heterogeneous Interactions of HOBr, HNO_3, O_3, and NO_2 with Deliquescent NaCl

Aerosols at Room Temperature," *J. Phys. Chem. A, 102,* 3719–3725 (1998).

Aloisio, S., and J. S. Francisco, "Existence of a Hydroperoxy and Water ($HO_2 \cdot H_2O$) Radical Complex," *J. Phys. Chem. A., 102,* 1899–1902 (1998).

Altshuller, A. P., "PAN's in the Atmosphere," *J. Air Waste Manage. Assoc., 43,* 1221–1230 (1993).

Alvarado, A., R. Atkinson, and J. Arey, "Kinetics of the Gas-Phase Reactions of NO_3 Radicals and O_3 with 3-Methylfuran and the OH Radical Yield from the O_3 Reaction," *Int. J. Chem. Kinet., 28,* 905–909 (1996).

Alvarado, A., J. Arey, and R. Atkinson, "Kinetics of the Gas-Phase Reactions of OH and NO_3 Radicals and O_3 with the Monoterpene Reaction Products Pinonaldehyde, Caronaldehyde, and Sabinaketone," *J. Atmos. Chem., 31,* 281–297 (1998a).

Alvarado, A., E. C. Tuazon, S. M. Aschmann, R. Atkinson, and J. Arey, "Products of the Gas-Phase Reactions of $O(^3P)$ Atoms and O_3 with α-Pinene and 1,2-Dimethyl-1-cyclohexene," *J. Geophys. Res., 103,* 25541–25551 (1998b).

Andino, J. M., J. N. Smith, R. C. Flagan, W. A. Goddard, III, and J. H. Seinfeld, "Mechanism of Atmospheric Photooxidation of Aromatics: A Theoretical Study," *J. Phys. Chem., 100,* 10967–10980 (1996).

Andreae, M. O., "Biomass Burning: Its History, Use, and Distribution and Its Impact on Environmental Quality and Global Climate," in *Global Biomass Burning: Atmospheric, Climatic, and Biospheric Implications* (J. S. Levine, Ed.), pp. 1–21, MIT Press, Cambridge, MA, 1991.

Andreae, M. O., B. E. Anderson, D. R. Blake, J. D. Bradshaw, J. E. Collins, G. L. Gregory, G. W. Sachse, and M. C. Shipham, "Influence of Plumes from Biomass Burning on Atmospheric Chemistry over the Equatorial and Tropical South Atlantic during CITE 3," *J. Geophys. Res., 99,* 12793–12808 (1994).

Andreae, M. O., and P. J. Crutzen, "Atmospheric Aerosols: Biogeochemical Sources and Role in Atmospheric Chemistry," *Science, 276,* 1052–1058 (1997).

Anglada, J. M., J. M. Bofill, S. Olivella, and A. Solé, "Unimolecular Isomerizations and Oxygen Atom Loss in Formaldehyde and Acetaldehyde Carbonyl Oxides. A Theoretical Investigation," *J. Am. Chem. Soc., 118,* 4636–4647 (1996).

Aranda, A., G. Le Bras, G. La Verdet, and G. Poulet, "The BrO + CH_3O_2 Reaction: Kinetics and Role in the Atmospheric Ozone Budget," *Geophys. Res. Lett., 24,* 2745–2748 (1997).

Arey, J., A. M. Winer, R. Atkinson, S. M. Aschmann, W. D. Long, and C. L. Morrison, "The Emission of (Z)-3-Hexen-1-ol, (Z)-3-Hexenylacetate and Other Oxygenated Hydrocarbons from Agricultural Plant Species," *Atmos. Environ., 25A,* 1063–1075 (1991a).

Arey, J., S. B. Corchnoy, and R. Atkinson, "Emission of Linalool from Valencia Orange Blossoms and Its Observation in Ambient Air," *Atmos. Environ., 25A,* 1377–1381 (1991b).

Arey, J., D. E. Crowley, M. Crowley, M. Resketo, and J. Lester, "Hydrocarbon Emissions from Natural Vegetation in California's South Coast Air Basin," *Atmos. Environ., 29,* 2977–2988 (1995).

Ariya, P. A., B. T. Jobson, R. Sander, H. Niki, G. W. Harris, J. G. Hopper, and K. G. Anlauf, "Measurements of $C_2–C_7$ Hydrocarbons during the Polar Sunrise Experiment 1994: Further Evidence for Halogen Chemistry in the Troposphere," *J. Geophys. Res., 103,* 13169–13180 (1998).

Arlander, D. W., D. R. Cronn, J. C. Farmer, F. A. Menzia, and H. H. Westberg, "Gaseous Oxygenated Hydrocarbons in the Remote Marine Troposphere," *J. Geophys. Res., 95,* 16391–16403 (1990).

Arnold, F., J. Schneider, K. Gollinger, H. Schlager, P. Schulte, D. E. Hagen, P. D. Whitefield, and P. van Velthoven, "Observation of Upper Tropospheric Sulfur Dioxide- and Acetone-Pollution: Potential Implications for Hydroxyl Radical and Aerosol Formation," *Geophys. Res. Lett., 24,* 57–60 (1997a).

Arnold, F., V. Bürger, B. Droste-Fanke, F. Grimm, A. Krieger, J. Schneider, and T. Stilp, "Acetone in the Upper Troposphere and Lower Stratosphere: Impact on Trace Gases and Aerosols," *Geophys. Res. Lett., 24,* 3017–3020 (1997b).

Aschmann, S. M., and R. Atkinson, "Formation Yields of Methyl Vinyl Ketone and Methacrolein from the Gas-Phase Reaction of O_3 with Isoprene," *Environ. Sci. Technol., 28,* 1539–1542 (1994).

Aschmann, S. M., A. A. Chew, J. Arey, and R. Atkinson, "Products of the Gas-Phase Reaction of OH Radicals with Cyclohexane: Reactions of the Cyclohexoxy Radical," *J. Phys. Chem. A., 101,* 8042–8048 (1997a).

Aschmann, S. M., Y. Shu, J. Arey, and R. Atkinson, "Products of the Gas-Phase Reactions of cis-3-Hexen-1-ol with OH Radicals and O_3," *Atmos. Environ., 31,* 3551–3560 (1997b).

Aschmann, S. M., and R. Atkinson, "Kinetics of the Gas-Phase Reactions of the OH Radical with Selected Glycol Ethers, Glycols, and Alcohols," *Int. J. Chem. Kinet., 30,* 533–540 (1998a).

Aschmann, S. M., and R. Atkinson, "Rate Constants for the Gas-Phase Reactions of Selected Dibasic Esters with the OH Radical," *Int. J. Chem. Kinet., 30,* 471–474 (1998b).

Aschmann, S. M., A. Reissell, R. Atkinson, and J. Arey, "Products of the Gas Phase Reactions of the OH Radical with α- and β-Pinene in the Presence of NO," *J. Geophys. Res., 103,* 25553–25561 (1998).

Aschmann, S. M., E. S. C. Kwok, R. Atkinson, and J. Arey, unpublished data (1999).

Atkinson, R., S. M. Aschmann, A. M. Winer, and J. N. Pitts, Jr., "Gas Phase Reaction of NO_2 with Alkenes and Dialkenes," *Int. J. Chem. Kinet., 16,* 697–706 (1984).

Atkinson, R., and S. M. Aschmann, "Kinetics of the Gas Phase Reaction of Cl Atoms with a Series of Organics at 296 ± 2 K and Atmospheric Pressure," *Int. J. Chem. Kinet., 17,* 33–41 (1985).

Atkinson, R. "Kinetics and Mechanisms of the Gas-Phase Reactions of the Hydroxyl Radical with Organic Compounds," *J. Phys. Chem. Data Monograph No. 1,* 1–246 (1989).

Atkinson, R., S. M. Aschmann, E. C. Tuazon, J. Arey, and B. Zielinska, "Formation of 3-Methylfuran from the Gas-Phase Reaction of OH Radicals with Isoprene and the Rate Constant for Its Reaction with the OH Radical," *Int. J. Chem. Kinet., 21,* 593–604 (1989).

Atkinson, R., "Kinetics and Mechanisms of the Gas-Phase Reactions of the NO_3 Radical with Organic Compounds," *J. Phys. Chem. Ref. Data, 20,* 459–507 (1991).

Atkinson, R., S. M. Aschmann, J. Arey, and B. Shorees, "Formation of OH Radicals in the Gas Phase Reactions of O_3 with a Series of Terpenes," *J. Geophys. Res., 97,* 6065–6073 (1992).

Atkinson, R., and S. M. Aschmann, "OH Radical Production from the Gas-Phase Reactions of O_3 with a Series of Alkenes under Atmospheric Conditions," *Environ. Sci. Technol., 27,* 1357–1363 (1993).

Atkinson, R., "Gas Phase Tropospheric Chemistry of Organic Compounds," *J. Phys. Chem. Ref. Data, Monogr. No. 2,* 11–216 (1994).

Atkinson, R., J. Arey, S. M. Aschmann, and E. C. Tuazon, "Formation of $O(^3P)$ Atoms and Epoxides from the Gas-Phase Reaction of O_3 with Isoprene," *Res. Chem. Intermed., 20,* 385–394 (1994a).

Atkinson, R., and S. M. Aschmann, "Products of the Gas-Phase Reactions of Aromatic Hydrocarbons: Effect of NO_2 Concentration," *Int. J. Chem. Kinet., 26,* 929–944 (1994).

Atkinson, R., S. M. Aschmann, J. Arey, and E. C. Tuazon, "Formation Yields of Epoxides and $O(^3P)$ Atoms from the Gas-Phase Reactions of O_3 with a Series of Alkenes," *Int. J. Chem. Kinet., 26,* 945–950 (1994b).

Atkinson, R., J. Arey, S. M. Aschmann, S. B. Corchnoy, and Y. Shu, "Rate Constants for the Gas-Phase Reactions of cis-3-Hexen-1-ol, cis-3-Hexenylacetate, trans-2-Hexenal, and Linalool with OH and NO_3 Radicals and O_3 at 296 ± 2 K, and OH Radical Formation Yields from the O_3 Reactions," Int. J. Chem. Kinet., 27, 941–955 (1995a).

Atkinson, R., E. S. C. Kwok, J. Arey, and S. M. Aschmann, "Reactions of Alkoxy Radicals in the Atmosphere," Faraday Discuss., 100, 23–37 (1995b).

Atkinson, R., E. C. Tuazon, and S. M. Aschmann, "Products of the Gas-Phase Reactions of O_3 with Alkenes," Environ. Sci. Technol., 29, 1860–1866 (1995c).

Atkinson, R., E. C. Tuazon, and S. M. Aschmann, "Products of the Gas-Phase Reactions of a Series of 1-Alkenes and 1-Methylcylohexene with the OH Radical in the Presence of NO," Environ. Sci. Technol., 29, 1674–1680 (1995d).

Atkinson, R., "Gas-Phase Tropospheric Chemistry of Volatile Organic Compounds: 1. Alkanes and Alkenes," J. Phys. Chem. Ref. Data, 26, 215–290 (1997a).

Atkinson, R., "Atmospheric Reactions of Alkoxy and β-Hydroxyalkoxy Radicals," Int. J. Chem. Kinet., 29, 99–111 (1997b).

Atkinson, R., D. L. Baulch, R. A. Cox, R. F. Hampson, Jr., J. A. Kerr, M. J. Rossi, and J. Troe, "Evaluated Kinetic, Photochemical, and Heterogeneous Data for Atmospheric Chemistry. 5. IUPAC Subcommittee on Gas Kinetic Data Evaluation for Atmospheric Chemistry," J. Phys. Chem. Ref. Data, 26, 521–1011 (1997a).

Atkinson, R., D. L. Baulch, R. A. Cox, R. F. Hampson, Jr., J. A. Kerr, M. J. Rossi, and J. Troe, "Evaluated Kinetic and Photochemical Data for Atmospheric Chemistry: Supplement VI—IUPAC Subcommittee on Gas Kinetic Data Evaluation for Atmospheric Chemistry," J. Phys. Chem. Ref. Data, 26, 1329–1499 (1997b).

Atkinson, R., and J. Arey, "Atmospheric Chemistry of Biogenic Organic Compounds," Accounts Chem. Res., 31, 574–583 (1998).

Atkinson, R., E. C. Tuazon, and S. M. Aschmann, "Products of the Gas-Phase Reaction of the OH Radical with 3-Methyl-1-butene in the Presence of NO," Int. J. Chem. Kinet., 30, 577–587 (1998).

Ayers, G. P., R. W. Gillett, H. Granek, C. de Serves, and R. A. Cox, "Formaldehyde Production in Clean Marine Air," Geophys. Res. Lett., 24, 401–404 (1997).

Barnes, I., V. Bastian, K. H. Becker, and Z. Tong, "Kinetics and Products of the Reactions of NO_3 with Monoalkenes, Dialkenes and Monoterpenes," J. Phys. Chem., 94, 2413–2419 (1990).

Barrie, L. A., J. W. Bottenheim, R. C. Schnell, P. J. Crutzen, and R. A. Rasmussen, "Ozone Destruction and Photochemical Reactions at Polar Sunrise in the Lower Arctic Atmosphere," Nature, 334, 138–141 (1988).

Barrie, L. A., and J. W. Bottenheim, "Sulphur and Nitrogen Pollution in the Arctic Atmosphere," in Pollution of the Arctic Atmosphere (W. T. Sturges, Ed.), pp. 155–182, Elsevier, Amsterdam/New York, 1991, and references therein.

Barrie, L., and U. Platt, "Arctic Tropospheric Chemistry: An Overview," Tellus, 49B, 450–454 (1997).

Bartolotti, L., and E. O. Edney, "Density Functional Theory Derived Intermediates from the OH Initiated Atmospheric Oxidation of Toluene," Chem. Phys. Lett., 245, 119–122 (1995).

Baxley, J. S., M. V. Henley, and J. R. Wells, "The Hydroxyl Radical Reaction Rate Constant and Products of Ethyl 3-Ethoxypropionate," Int. J. Chem. Kinet., 29, 637–644 (1997).

Baxley, J. S., and J. R. Wells, "The Hydroxyl Radical Reaction Rate Constant and Atmospheric Transformation Products of 2-Butanol and 2-Pentanol," Int. J. Chem. Kinet., 30, 745–752 (1998).

Bedjanian, Y., G. Laverdet, and G. Le Bras, "Low-Pressure Study of the Reaction of Cl Atoms with Isoprene," J. Phys. Chem., 102, 953–959 (1998).

Behnke, W., C. George, V. Scheer, and C. Zetzsch, "Production and Decay of $ClNO_2$ from the Reaction of Gaseous N_2O_5 with NaCl Solution: Bulk and Aerosol Experiments," J. Geophys. Res., 102, 3795–3804 (1997).

Beine, H. J., D. A. Jaffe, D. R. Blake, E. Atlas, and J. Harris, "Measurements of PAN, Alkyl Nitrates, Ozone, and Hydrocarbons during Spring in Interior Alaska," J. Geophys. Res., 101, 12613–12619 (1996).

Beine, H. J., D. A. Jaffe, F. Stordal, M. Engardt, S. Solbert, N. Schmidbauer, and K. Holmèn, "NO_x during Ozone Depletion Events in the Arctic Troposphere at Ny-Ålesund, Svalbard," Tellus, 49B, 556–565 (1997).

Benjamin, M. T., M. Sudol, L. Bloch, and A. M. Winer, "Low-Emitting Urban Forests: A Taxonomic Methodology for Assigning Isoprene and Monoterpene Emission Rates," Atmos. Environ., 30, 1437–1452 (1996).

Benjamin, M. T., M. Sudol, D. Vorsatz, and A. M. Winer, "A Spatially and Temporally Resolved Biogenic Hydrocarbon Emissions Inventory for the California South Coast Air Basin," Atmos. Environ., 31, 3087–3100 (1997).

Benson, S. W., "Some Observations on the Thermochemistry and Kinetics of Peroxy Radicals," J. Phys. Chem., 100, 13544–13547 (1996).

Benter, Th., M. Liesner, R. N. Schindler, H. Skov, J. Hjorth, and G. Restelli, "REMPI-MS and FTIR Study of NO_2 and Oxirane Formation in the Reactions of Unsaturated Hydrocarbons with NO_3 Radicals," J. Phys. Chem., 98, 10492–10496 (1994).

Berndt, T., and O. Böge, "Products and Mechanism of the Reaction of NO_3 with Selected Acyclic Monoalkenes," J. Atmos. Chem., 21, 275–291 (1995).

Berndt, T., O. Böge, I. Kind, and W. Rolle, "Reaction of NO_3 Radicals with 1,3-Cyclohexadiene, α-Terpinene, and α-Phellandrene: Kinetics and Products," Ber. Bunsenges. Phys. Chem., 100, 462–469 (1996).

Berndt, T., and O. Böge, "Gas-Phase Reaction of NO_3 Radicals with Isoprene: A Kinetic and Mechanistic Study," Int. J. Chem. Kinet., 29, 755–765 (1997).

Berndt, T., O. Böge, and W. Rolle, "Products of the Gas-Phase Reactions of NO_3 Radicals with Furan and Tetramethylfuran," Environ. Sci. Technol., 31, 1157–1162 (1997).

Bertman, S. B., and J. M. Roberts, "A PAN Analog from Isoprene Photooxidation," Geophys. Res. Lett., 18, 1461–1464 (1991).

Bey, I., B. Aumont, and G. Toupance, "The Nighttime Production of OH Radicals in the Continental Troposphere," Geophys. Res. Lett., 24, 1067–1070 (1997).

Bierbach, A., I. Barnes, and K. H. Becker, "Rate Coefficients for the Gas-Phase Reactions of Hydroxyl Radicals with Furan, 2-Methylfuran, 2-Ethylfuran, and 2,5-Dimethylfuran at 300 ± 2 K," Atmos. Environ., 26A, 813–817 (1992).

Bierbach, A., I. Barnes, K. H. Becker, and E. Wiesen, "Atmospheric Chemistry of Unsaturated Carbonyls: Butenedial, 4-Oxo-2-pentenal, 3-Hexene-2,5-dione, Maleic Anhydride, 3H-Furan-2-one, and 5-Methyl-3H-furan-2-one," Environ. Sci. Technol., 28, 715–729 (1994).

Bierbach, A., I. Barnes, and K. H. Becker, "Rate Coefficients for the Gas-Phase Reactions of Bromine Radicals with a Series of Alkenes, Dienes, and Aromatic Hydrocarbons at 298 ± 2 K," Int. J. Chem. Kinet., 28, 565–577 (1996).

Biesenthal, T. A., and P. B. Shepson, "Observations of Anthropogenic Inputs of the Isoprene Oxidation Products Methyl Vinyl Ketone and Methacrolein to the Atmosphere," Geophys. Res. Lett., 24, 1375–1378 (1997).

Biesenthal, T. A., J. W. Bottenheim, P. B. Shepson, S.-M. Li, and P. C. Brickell, "The Chemistry of Biogenic Hydrocarbons at a Rural Site in Eastern Canada," *J. Geophys. Res., 103,* 25487–25498 (1998).

Biggs, P., C. E. Canosa-Mas, J.-M. Fracheboud, D. E. Shallcross, and R. P. Wayne, "Investigation into the Kinetics and Mechanism of the Reaction of NO_3 with HCO_2 at 298 K and 2.5 Torr: A Potential Source of OH in the Nighttime Troposphere?" *J. Chem. Soc., Faraday Trans., 90,* 1205–1210 (1994a).

Biggs, P., C. E. Canosa-Mas, J.-M. Fracheboud, D. E. Shallcross, and R. P. Wayne, "Investigation into the Kinetics and Mechanism of the Reaction of NO_3 with CH_3 and CH_3O at 298 K between 0.6 and 8.5 Torr: Is There a Chain Decomposition Mechanism in Operation?" *J. Chem. Soc., Faraday Trans., 90,* 1197–1204 (1994b).

Bjergbakke, E., A. Sillesen, and P. Pagsberg, "UV Spectrum and Kinetics of Hydroxycyclohexadienyl Radicals," *J. Phys. Chem., 100,* 5729–5736 (1996).

Blacet, F. E., "Photochemistry in the Lower Atmosphere," *Ind. Eng. Chem., 44,* 1339–1342 (1952).

Bohn, B., and C. Zetzsch, "Formation of HO_2 from OH and C_2H_2 in the Presence of O_2," *J. Chem. Soc., Faraday Trans., 94,* 1203–1210 (1998).

Bonsang, B., D. Martin, G. Lambert, M. Kanakidou, L. C. Le Roulley, and G. Sennequier, "Vertical Distribution of Nonmethane Hydrocarbons in the Remote Marine Boundary Layer," *J. Geophys. Res., 96,* 7313–7324 (1991).

Bonsang, B., C. Polle, and G. Lambert, "Evidence for Marine Production of Isoprene," *Geophys. Res. Lett., 19,* 1129–1132 (1992).

Bottenheim, J. W., A. G. Gallant, and K. A. Brice, "Measurements of NO_y Species and O_3 at 82°N Latitude," *Geophys. Res. Lett., 13,* 113–116 (1986).

Boyd, A. A., and R. Lesclaux, "The Temperature Dependence of the Rate Coefficients for β-Hydroxyperoxy Radical Self Reactions," *Int. J. Chem. Kinet., 29,* 323–331 (1997).

Broadgate, W. J., P. S. Liss, and S. A. Penkett, "Seasonal Emissions of Isoprene and Other Reactive Hydrocarbon Gases from the Ocean," *Geophys. Res. Lett., 24,* 2675–2678 (1997).

Brunner, D., J. Staehelin, and D. Jeker, "Large-Scale Nitrogen Oxide Plumes in the Tropopause Region and Implications for Ozone," *Science, 282,* 1305–1309 (1998).

Bunnelle, W. H., "Preparation, Properties, and Reactions of Carbonyl Oxides," *Chem. Rev., 91,* 335–362 (1991).

Cahill, T. A., K. Wilkinson, and R. Schnell, "Composition Analyses of Size-Resolved Aerosol Samples Taken from Aircraft Downwind of Kuwait, Spring, 1991," *J. Geophys. Res., 97,* 14513–14520 (1992).

Calogirou, A., B. R. Larsen, and D. Kotzias, "Gas-Phase Terpene Oxidation Products: A Review," *Atmos. Environ., 33,* 1428–1439 (1999).

Calvert, J. G., and J. N. Pitts, Jr., *Photochemistry,* Wiley, New York, 1966.

Canosa-Mas, C. E., M. D. King, R. Lopez, C. J. Percival, R. P. Wayne, D. E. Shallcross, J. A. Pyle, and V. Daële, "Is the Reaction between $CH_3C(O)O_2$ and NO_3 Important in the Nighttime Troposphere?" *J. Chem. Soc., Faraday Trans., 92,* 2211–2222 (1996).

Cantrell, C. A., R. E. Shetter, T. M. Gilpin, J. G. Calvert, F. L. Eisele, and D. J. Tanner, "Peroxy Radical Concentrations Measured and Calculated from Trace Gas Measurements in the Mauna Loa Observatory Photochemistry Experiment 2," *J. Geophys. Res., 101,* 14653–14664 (1996).

Cao, X.-L., C. Boissard, A. J. Juan, C. N. Hewitt, and M. Gallagher, "Biogenic Emissions of Volatile Organic Compounds from Gorse (*Ulex europaeus*): Diurnal Emission Fluxes at Kelling Heath, England," *J. Geophys. Res., 102,* 18903–18915 (1997).

Carl, S. A., and J. N. Crowley, "Sequential Two (Blue) Photon Absorption by NO_2 in the Presence of H_2 as a Source of OH in Pulsed Photolysis Kinetic Studies: Rate Constants for Reaction of OH with CH_3NH_2, $(CH_3)_2NH$, $(CH_3)_3N$, and $C_2H_5NH_2$ at 298 K," *J. Phys. Chem. A., 102,* 8131–8141 (1998).

Carpenter, L. J., P. S. Monks, B. J. Bandy, S. A. Penkett, I. E. Galbally, and C. P. Meyer, "A Study of Peroxy Radicals and Ozone Photochemistry at Coastal Sites in the Northern and Southern Hemispheres," *J. Geophys. Res., 102,* 25417–25427 (1997).

Carslaw, N., L. J. Carpenter, J. M. C. Plane, B. J. Allan, R. A. Burgess, K. C. Clemitshaw, H. Coe, and S. A. Penkett, "Simultaneous Observations of Nitrate and Peroxy Radicals in the Marine Boundary Layer," *J. Geophys. Res., 102,* 18917–18933 (1997).

Carter, W. P. L., A. M. Winer, and J. N. Pitts, Jr., "Effect of Peroxyacetyl Nitrate on the Initiation of Photochemical Smog," *Environ. Sci. Technol., 15,* 831–834 (1981a).

Carter, W. P. L., E. C. Tuazon, A. M. Winer, and J. N. Pitts, Jr., "Gas Phase Reactions of N,N-Dimethylhydrazine with Ozone and NO_x in Simulated Atmospheres," in *N-Nitroso Compounds"* (R. A. Scanlan and S. R. Tannenbaum, Eds.), pp. 117–131, ACS Symposium Series No. 174, Am. Chem. Soc., Washington, DC, 1981b.

Carter, W. P. L., and R. Atkinson, "Development and Evaluation of a Detailed Mechanism for the Atmospheric Reactions of Isoprene and NO_x," *Int. J. Chem. Kinet., 28,* 497–530 (1996).

Chameides, W. L., R. W. Lindsay, J. Richardson, and C. S. Kiang, "The Role of Biogenic Hydrocarbons in Urban Photochemical Smog: Atlanta as a Case Study," *Science, 241,* 1473–1475 (1988).

Chance, K., "Analysis of BrO Measurements from the Global Ozone Monitoring Experiment," *Geophys. Res. Lett., 25,* 3335–3338 (1998).

Chatfield, R., and P. Crutzen, "Sulfur Dioxide in Remote Oceanic Air: Cloud Transport of Reactive Precursors," *J. Geophys. Res., 89,* 7111–7132 (1984).

Chatfield, J. K. S., and A. J. Alkezweeny, "Tracer Study of Vertical Exchange by Cumulus Clouds," *J. Geophys. Res., 95,* 18473–18488 (1990).

Chatfield, R. B., "Anomalous HNO_3/NO_x Ratio of Remote Tropospheric Air: Conversion of Nitric Acid to Formic Acid and NO_x?" *Geophys. Res. Lett., 21,* 2705–2708 (1994).

Chen, X., D. Hulbert, and P. B. Shepson, "Measurement of the Organic Nitrate Yield from OH Reaction with Isoprene," *J. Geophys. Res., 103,* 25563–25568 (1998).

Chew, A. A., and R. Atkinson, "OH Radical Formation Yields from the Gas-Phase Reactions of O_3 with Alkenes and Monoterpenes," *J. Geophys. Res., 101,* 28649–28653 (1996).

Chew, A. A., R. Atkinson, and S. M. Aschmann, "Kinetics of the Gas-Phase Reactions of NO_3 Radicals with a Series of Alcohols, Glycol Ethers, Ethers and Chloroalkenes," *J. Chem. Soc., Faraday Trans., 94,* 1083–1089 (1998).

Chin, M., D. J. Jacob, J. W. Munger, D. D. Parrish, and B. G. Doddridge, "Relationship of Ozone and Carbon Monoxide over North America," *J. Geophys. Res., 99,* 14565–14573 (1994).

Christopher, S. A., D. V. Kliche, J. Chou, and R. M. Welch, "First Estimates of the Radiative Forcing of Aerosols Generated from Biomass Burning Using Satellite Data," *J. Geophys. Res., 101,* 21265–21273 (1996).

Ciccioli, P., E. Brancaleoni, M. Frattoni, A. Cecinato, and A. Brachetti, "Ubiquitous Occurrence of Semi-Volatile Carbonyl Compounds in Tropospheric Samples and Their Possible Sources," *Atmos. Environ., 27A,* 1891–1901 (1993a).

Ciccioli, P., E. Brancaleoni, A. Cecinato, R. Sparapani, and M. Frattoni, "Identification and Determination of Biogenic and

Anthropogenic Volatile Organic Compounds in Forest Areas of Northern and Southern Europe and a Remote Site of the Himalaya Region by High-Resolution Gas Chromatography–Mass Spectrometry," *J. Chromatogr., 643,* 55–69 (1993b).

Ciccioli, P., C. Fabozzi, E. Brancaleoni, A. Cecinato, M. Frattoni, S. Cieslik, D. Kotzias, G. Seufert, P. Foster, and R. Steinbrecher, "Biogenic Emission from the Mediterranean Pseudosteppe Ecosystem Present in Castelporziano," *Atmos. Environ., 31,* 167–175 (1997).

Clemitshaw, K. C., J. Williams, O. V. Rattigan, D. E. Shallcross, K. S. Law, and R. A. Cox, "Gas-Phase Ultraviolet Absorption Cross-Sections and Atmospheric Lifetimes of Several C_2–C_5 Alkyl Nitrates," *J. Photochem. Photobiol A: Chem., 102,* 117–126 (1997).

Cofer, W. R., III, J. S. Levine, E. L. Winstead, P. J. LeBel, A. M. Koller, Jr., and C. R. Hinkle, "Trace Gas Emissions from Burning Florida Wetlands," *J. Geophys. Res., 95,* 1865–1870 (1990).

Cohen, N., and S. W. Benson, "Empirical Correlations for Rate Coefficients for Reactions of OH with Haloalkanes," *J. Phys. Chem., 91,* 162–170 (1987).

Cremer, D., J. Gauss, E. Kraka, J. F. Stanton, and R. J. Bartlett, "A CCSD(T) Investigation of Carbonyl Oxide and Dioxirane. Equilibrium Geometries, Dipole Moments, Infrared Spectra, Heats of Formation, and Isomerization Energies," *Chem. Phys. Lett., 209,* 547–556 (1993).

Criegee, R., "Mechanisms of Ozonolysis," *Angew. Chem., Int. Ed. Engl., 14,* 745–752 (1975).

Cros, B., D. Nganga, A. Minga, J. Fishman, and V. Brackett, "Distribution of Tropospheric Ozone at Brazzaville, Congo, Determined from Ozonesonde Measurements," *J. Geophys. Res., 97,* 12869–12875 (1992).

Crutzen, P. J., and M. O. Andreae, "Biomass Burning in the Tropics: Impact on Atmospheric Chemistry and Biogeochemical Cycles," *Science, 250,* 1669–1678 (1990).

Crutzen, P. J., "Overview of Tropospheric Chemistry: Developments during the Past Quarter Century and a Look Ahead," *Faraday Discuss., 100,* 1–21 (1995).

Crutzen, P. J., N. F. Elansky, M. Hahn, G. S. Golitsyn, C. A. M. Brenninkmeijer, D. H. Scharffe, I. B. Belikov, M. Maiss, P. Bergamaschi, T. Röckmann, A. M. Grisenko, and V. M. Sevostyanov, "Trace Gas Measurements between Moscow and Vladivostok Using the Trans-Siberian Railroad," *J. Atmos. Chem., 29,* 179–194 (1998).

Cvetanovic, R. J., "Chemical Kinetic Studies of Atmospheric Interest," 12th International Symposium on Free Radicals, Laguna Beach, CA, January 4–9, 1976.

Daële, V., G. Laverdet, G. Le Bras, and G. Poulet, "Kinetics of the Reactions $CH_3O + NO$, $CH_3O + NO_3$, and $CH_3O_2 + NO_3$," *J. Phys. Chem., 99,* 1470–1477 (1995).

Darnall, K. R., W. P. L. Carter, A. M. Winer, A. C. Lloyd, and J. N. Pitts, Jr., "Importance of $RO_2 + NO$ in Alkyl Nitrate Formation from C_4–C_6 Alkane Photooxidations under Simulated Atmospheric Conditions," *J. Phys. Chem., 80,* 1948–1950 (1976).

de Gouw, J. A., and E. R. Lovejoy, "Reactive Uptake of Ozone by Liquid Organic Compounds," *Geophys. Res. Lett., 25,* 931–934 (1998).

de Gouw, J. A., C. J. Howard, T. G. Custer, and R. Fall, "Emissions of Volatile Organic Compounds from Cut Grass and Clover Are Enhanced during the Drying Process," *Geophys. Res. Lett., 26,* 811–814 (1999).

De Haan, D., T. Brauers, K. Oum, J. Stutz, T. Nordmeyer, and B. J. Finlayson-Pitts, "Heterogeneous Chemistry in the Troposphere: Experimental Approaches and Applications to the Chemistry of Sea Salt Particles," *Int. Rev. Phys. Chem.,* in press (1999).

DeMore, W. B., S. P. Sander, D. M. Golden, R. F. Hampson, M. J. Kurylo, C. J. Howard, A. R. Ravishankara, C. E. Kolb, and M. J. Molina, "Chemical Kinetics and Photochemical Data for Use in Stratospheric Modeling," in JPL Publication 97-4, Jet Propulsion Laboratory, Pasadena, CA, January 15, 1997.

Dias-Lalcaca, P., D. Brunner, W. Imfeld, W. Moser, and J. Staehelin, "An Automated System for the Measurement of Nitrogen Oxides and Ozone Concentrations from a Passenger Aircraft: Instrumentation and First Results of the NOXAR Project," *Environ. Sci. Technol., 32,* 3228–3236 (1998).

Dickerson, R. R., B. S. Gockel, W. T. Luke, D. P. McNamara, L. J. Nunnermacker, K. E. Pickering, J. P. Greenberg, and P. R. Zimmerman, "Profiles of Photochemically Active Trace Gases in the Troposphere," in *Ozone in the Atmosphere, Proceedings of the Quadrennial Ozone Symposium 1988 and Tropospheric Ozone Workshop, Göttingen, FRG, August 4–13* (R. D. Bojkov and P. Fabian, Eds.), pp. 463–466, A. Deepak Publishing, Hampton, VA, 1989.

Dickerson, R. R., B. G. Doddridge, P. Kelley, and K. P. Rhoads, "Large-Scale Pollution of the Atmosphere over the Remote Atlantic Ocean: Evidence from Bermuda," *J. Geophys. Res., 100,* 8945–8952 (1995).

Dlugokencky, E. J., and C. J. Howard, "Studies of NO_3 Radical Reactions with Some Atmospheric Organic Compounds at Low Pressures," *J. Phys. Chem., 93,* 1091–1096 (1989).

Donahue, N. M., J. G. Anderson, and K. L. Demerjian, "New Rate Constants for Ten OH–Alkane Reactions from 300 to 400 K: An Assessment of Accuracy," *J. Phys. Chem. A., 102,* 3121–3126 (1998a).

Donahue, N. M., J. H. Kroll, J. G. Anderson, and K. L. Demerjian, "Direct Observation of OH Production from the Ozonolysis of Olefins," *Geophys. Res. Lett., 25,* 59–62 (1998b).

Doskey, P. V., and J. S. Gaffney, "Non-Methane Hydrocarbons in the Arctic Atmosphere at Barrow, Alaska," *Geophys. Res. Lett., 19,* 381–384 (1992).

Drewitt, G. B., K. Curren, D. G. Steyn, T. J. Gillespie, and H. Niki, "Measurement of Biogenic Hydrocarbon Emissions from Vegetation in the Lower Fraser Valley, British Columbia," *Atmos. Environ., 32,* 3457–3466 (1998).

Dumdei, B. E., D. V. Kenny, P. B. Shepson, T. E. Kleindienst, C. M. Nero, L. T. Culpitt, and L. D. Claxton, "MS/MS Analysis of the Products of Toluene Photooxidation and Measurement of Their Mutagenic Activity," *Environ. Sci. Technol., 22,* 1493–1498 (1988).

Duncan, J. L., L. R. Schindler, and J. T. Roberts, "A New Sulfate-Mediated Reaction: Conversion of Acetone to Trimethylbenzene in the Presence of Liquid Sulfuric Acid," *Geophys. Res. Lett., 25,* 631–634 (1998).

Eberhard, J., C. Müller, D. W. Stocker, and J. A. Kerr, "The Photo-oxidation of Diethyl Ether in Smog Chamber Experiments Simulating Tropospheric Conditions: Product Studies and Proposed Mechanism," *Int. J. Chem. Kinet., 25,* 639–649 (1993).

Eberhard, J., C. Müller, D. W. Stocker, and J. A. Kerr, "Isomerization of Alkoxy Radicals under Atmospheric Conditions," *Environ. Sci. Technol., 29,* 232–241 (1995).

Eberhard, J., and C. J. Howard, "Temperature-Dependent Kinetics Studies of the Reactions of $C_2H_5O_2$ and n-$C_3H_7O_2$ Radicals with NO," *Int. J. Chem. Kinet., 28,* 731–740 (1996).

Eberhard, J., and C. J. Howard, "Rate Coefficients for the Reactions of Some C_3 to C_5 Hydrocarbon Peroxy Radicals with NO," *J. Phys. Chem. A, 101,* 3360–3366 (1997).

Ehhalt, D. H., F. Rohrer, and A. Wahner, "Sources and Distribution of NO_x in the Upper Troposphere at Northern Mid-Latitudes," *J. Geophys. Res., 97,* 3725–3738 (1992).

Ehhalt, D., and F. Rohrer, "The Impact of Commercial Aircraft on Tropospheric Ozone," in *The Chemistry of the Atmosphere, Proceedings of the 7th BOC Priestley Conference.,* Lewisburg, Pennsylvania, U.S.A. (A. R. Bandy, Ed.), The Royal Society of Chemistry, Cambridge, 1995.

Eisele, F. L., and D. J. Tanner, "Identification of Ions in Continental Air," *J. Geophys. Res., 95,* 20539–20550 (1990).

El Boudali, A., S. Le Calvé, G. Le Bras, and A. Mellouki, "Kinetic Studies of OH Reactions with a Series of Acetates," *J. Phys. Chem., 100,* 12364–12368 (1996).

Evleth, E. M., C. F. Melius, M. T. Rayez, J. C. Rayez, and W. Forst, "Theoretical Characterization of the Reaction of HO_2 with Formaldehyde," *J. Phys. Chem., 97,* 5040–5045 (1993).

Fajgar, R., J. Vítek, Y. Haas, and J. Pola, "Observations of Secondary 2-Butene Ozonide in the Ozonation of Trans-2-butene in the Gas Phase," *Tetrahedron Lett., 37,* 3391–3394 (1996).

Fall, R., and A. A. Benson, "Leaf Methanol—The Simplest Natural Product from Plants," *Trends Plant Sci., 1,* 296–301 (1996).

Fall, R., and M. C. Wildermuth, "Isoprene Synthase: From Biochemical Mechanism to Emission Algorithm," *J. Geophys. Res., 103,* 25599–25609 (1998).

Fall, R., "Biogenic Emissions of Volatile Organic Compounds from Higher Plants," Ch. 2, pp. 41–96. In *Reactive Hydrocarbons in the Atmosphere,* (C. N. Hewitt, Ed.). Academic Press, San Diego, 1999.

Fan, S.-M., and D. J. Jacob, "Surface Ozone Depletion in Arctic Spring Sustained by Bromine Reactions on Aerosols," *Nature, 359,* 522–524 (1992).

Fantechi, G., N. R. Jensen, J. Hjorth, and J. Peeters, "Determination of the Rate Constants for the Gas-Phase Reactions of Methyl Butenol with OH Radicals, Ozone, NO_3 Radicals, and Cl Atoms," *Int. J. Chem. Kinet., 30,* 589–594 (1998a).

Fantechi, G., N. R. Jensen, J. Hjorth, and J. Peeters, "Mechanistic Studies of the Atmospheric Oxidation of Methyl Butenol by OH Radicals, Ozone, and NO_3 Radicals," *Atmos. Environ., 32,* 3547–3556 (1998b).

Ferronato, C., J. J. Orlando, and G. S. Tyndall, "Rate and Mechanism of the Reactions of OH and Cl with 2-Methyl-3-buten-2-ol," *J. Geophys. Res., 103,* 25579–25586 (1998).

Fehsenfeld, F., J. Calvert, R. Fall, P. Goldan, A. B. Guenther, C. N. Hewitt, B. Lamb, S. Liu, M. Trainer, H. Westberg, and P. Zimmerman, "Emissions of Volatile Organic Compounds from Vegetation and the Implications for Atmospheric Chemistry," *Global Biogeochem. Cycl., 6,* 389–430 (1992).

Fenske, J. D., and S. E. Paulson, "Human Breath Emissions of VOCs," *J. Air Waste Manage. Assoc., 49,* 594–598 (1999).

Fenter, F. F., V. Catoire, R. Lesclaux, and P. D. Lightfoot, "The Ethylperoxy Radical: Its Ultraviolet Spectrum, Self-Reaction, and Reaction with HO_2, Each Studied as a Function of Temperature," *J. Phys. Chem., 97,* 3530–3538 (1993).

Fine, D. H., "*N*-Nitroso Compounds in the Environment," *Adv. Environ. Sci. Technol., 10,* 39–123 (1980).

Finlayson, B. J., J. N. Pitts, Jr., and H. Akimoto, "Production of Vibrationally Excited OH in Chemiluminescent Ozone–Olefin Reactions," *Chem. Phys. Lett., 12,* 495–498 (1972).

Finlayson-Pitts, B. J., F. E. Livingston, and H. N. Berko, "Ozone Destruction and Bromine Photochemistry at Ground Level in the Arctic Spring," *Nature, 343,* 622–625 (1990).

Finlayson-Pitts, B. J., "Chlorine Atoms as a Potential Tropospheric Oxidant in the Marine Boundary Layer," *Res. Chem. Intermed., 19,* 235–249 (1993).

Finlayson-Pitts, B. J., and J. N. Pitts, Jr., "Tropospheric Air Pollution: Ozone, Airborne Toxics, Polycyclic Aromatic Hydrocarbons, and Particles," *Science, 276,* 1045–1052 (1997).

Finlayson-Pitts, B. J., C. J. Keoshian, B. Buehler, and A. A. Ezell, "Kinetics of Reaction of Chlorine Atoms with Some Biogenic Organics," *Int. J. Chem. Kinet., 31,* 491–499 (1999).

Fishman, J., P. Minnis, and H. G. Reichle, Jr., "Use of Satellite Data to Study Tropospheric Ozone in the Tropics," *J. Geophys. Res., 91,* 14451–14465 (1986).

Fishman, J., K. Fakhruzzaman, B. Cros, and D. Nganga, "Identification of Widespread Pollution in the Southern Hemisphere Deduced from Satellite Analyses," *Science, 252,* 1693–1696 (1991).

Folkins, I., P. O. Wennberg, T. F. Hanisco, J. G. Anderson, and R. J. Salawitch, "OH, HO_2, and NO in Two Biomass Burning Plumes: Sources of HO_x and Implications for Ozone Production," *Geophys. Res. Lett., 24,* 3185–3188 (1997).

Folkins, I., R. Chatfield, H. Singh, Y. Chen, and B. Heikes, "Ozone Production Efficiencies of Acetone and Peroxides in the Upper Troposphere," *Geophys. Res. Lett., 25,* 1305–1308 (1998).

Foster, W. M., L. Jiang, P. T. Stetkiewicz, and T. H. Risby, "Breath Isoprene: Temporal Changes in Respiratory Output after Exposure to Ozone," *J. Appl Physiol., 80,* 706–710 (1996).

Fritz, B., V. Handwerk, M. Preidel, and R. Zellner, "Direct Detection of Hydroxy-Cyclohexadienyl in the Gas Phase by cw UV Laser Absorption," *Ber. Bunsenges. Phys. Chem., 89,* 343–344 (1985).

Fruekilde, P., J. Hjorth, N. R. Jensen, D. Kotzias, and B Larsen, "Ozonolysis at Vegetation Surfaces: A Source of Acetone, 4-Oxopentanal, 6-Methyl-5-hepten-2-one, and Geranyl Acetone in the Troposphere," *Atmos. Environ., 32,* 1893–1902 (1998).

Gäb, S., E. Hellpointner, W. V. Turner, and F. Korte, "Hydroxymethyl hydroperoxide and Bis(hydroxymethyl)-peroxide from Gas Phase Ozonolysis of Naturally Occurring Alkenes," *Nature, 316,* 535–536 (1985).

Gäb, S., W. V. Turner, S. Wolff, K. H. Becker, L. Ruppert, and K. J. Brockmann, "Formation of Alkyl and Hydroxyalkyl Hydroperoxides on Ozonolysis in Water and in Air," *Atmos. Environ., 29,* 2401–2407 (1995).

Gaffney, J. S., and S. Z. Levine, "Predicting Gas-Phase Organic Molecule Reaction Rates Using Linear Free Energy Correlations. I. $O(^3P)$ and OH Addition and Abstraction Reactions," *Int. J. Chem. Kinet., 11,* 1197–1209 (1979).

Gaffney, J. S., N. A. Marley, and E. W. Prestbo, "Peroxyacyl Nitrates (PAN's): Their Physical and Chemical Properties," in *The Handbook of Environmental Chemistry,* (O. Hutzinger, Ed.), Vol. 4, Part B, pp. 1–38, Springer-Verlag, Berlin, 1989.

Gelmont, D., R. A. Stein, and J. F. Mead, "Isoprene—The Main Hydrocarbon in Human Breath," *Biochem. Biophys. Res. Commun., 99,* 1456–1460 (1981).

Geron, C. D., D. Nie, R. R. Arnts, T. D. Sharkey, E. L. Singsaas, P. J. Vanderveer, A. Guenther, J. E. Sickles, II, and T. E. Kleindienst, "Biogenic Isoprene Emission: Model Evaluation in a Southeastern United States Bottomland Deciduous Forest," *J. Geophys. Res., 102,* 18889–18901 (1997).

Gidel, L. T., "Cumulus Cloud Transport of Transient Tracers," *J. Geophys. Res., 88,* 6587–6599 (1983).

Gierczak, T., J. B. Burkholder, S. Bauerle, and A. R. Ravishankara, "Photochemistry of Acetone under Tropospheric Conditions," *Chem. Phys., 231,* 229–244 (1998).

Glasius, M., A. Calogirou, N. R. Jensen, J. Hjorth, and C. J. Nielsen, "Kinetic Study of Gas-Phase Reactions of Pinonaldehyde and Structurally Related Compounds," *Int. J. Chem. Kinet., 29,* 527–533 (1997).

Goldan, P. D., W. C. Kuster, F. C. Fehsenfeld, and S. A. Montzka, "The Observation of a C_5 Alcohol Emission in a North American Pine Forest," *Geophys. Res. Lett., 20,* 1039–1042 (1993).

Gorzelska, K., and J. N. Galloway, "Amine Nitrogen in the Atmospheric Environment over the North Atlantic Ocean," *Global Biogeochem. Cycl., 4,* 309–333 (1990).

Goumri, A., J. F. Pauwels, and P. Devolder, "Rate of the OH + C_6H_6 + He Reaction in the Fall-Off Range by Discharge Flow and OH Resonance Fluorescence," *Can. J. Chem., 69,* 1057–1064 (1991).

Graedel, T. E., D. T. Hawkins, and L. D. Claxton, *Atmospheric Chemical Compounds—Sources, Occurrence, and Bioassay*, Academic Press, San Diego, 1986.

Graedel, T. E., and W. C. Keene, "Tropospheric Budget of Reactive Chlorine," *Global Biogeochem. Cycl.*, 9, 47–77 (1995).

Greiner, N. R., "Hydroxyl Radical Kinetics by Kinetic Spectroscopy. VI. Reactions with Alkanes in the Range 300–500 K," *J. Chem. Phys.*, 53, 1070–1076 (1970).

Griesbaum, K., V. Miclaus, and I. C. Jung, "Isolation of Ozonides from Gas-Phase Ozonolyses of Terpenes," *Environ. Sci. Technol.*, 32, 647–649 (1998).

Grooß, J.-U., C. Brühl, and T. Peter, "Impact of Aircraft Emissions on Tropospheric and Stratospheric Ozone. Part I: Chemistry and 2-D Model Results," *Atmos. Environ.*, 32, 3173–3184 (1998).

Grosjean, D., E. L. Williams, II, E. Grosjean, J. M. Andino, and J. H. Seinfeld, "Atmospheric Oxidation of Biogenic Hydrocarbons: Reaction of Ozone with β-Pinene, d-Limonene, and *trans*-Caryophyllene," *Environ. Sci. Technol.*, 27, 2754–2758 (1993a).

Grosjean, D., E. L. Williams, II, and E. Grosjean, "Peroxyacyl Nitrates at Southern California Mountain Forest Locations," *Environ. Sci. Technol.*, 27, 110–121 (1993b).

Grosjean, D., E. L. Williams, II, and E. Grosjean, "Ambient Levels of Peroxy-*n*-butyryl Nitrate at a Southern California Mountain Forest Smog Receptor Location," *Environ. Sci. Technol.*, 27, 326–331 (1993c).

Grosjean, E., and D. Grosjean, "Rate Constants for the Gas-Phase Reactions of Ozone with Unsaturated Aliphatic Alcohols," *Int. J. Chem. Kinet.*, 26, 1185–1191 (1994).

Grosjean, D., E. Grosjean, and E. L. Williams, "Formation and Thermal Decomposition of Butyl-Substituted Peroxyacyl Nitrates: n-$C_4H_9C(O)OONO_2$ and i-$C_4H_9C(O)OONO_2$," *Environ. Sci. Technol.*, 28, 1099–1105 (1994a).

Grosjean, D., E. Grosjean, and E. L. Williams, "Thermal Decomposition of C_3-Substituted Peroxyacyl Nitrates," *Res. Chem. Intermed.*, 20, 447–461 (1994b).

Grosjean, D., E. Grosjean, and E. L. Williams, "Atmospheric Chemistry of Olefins: A Product Study of the Ozone–Alkene Reaction with Cyclohexane Added to Scavenge OH," *Environ. Sci. Technol.*, 28, 186–196 (1994c).

Grosjean, D., and E. Grosjean, "Carbonyl Products of the Ozone-Unsaturated Alcohol Reaction," *J. Geophys. Res.*, 100, 22815–22820 (1995).

Grosjean, E., D. Grosjean, and J. H. Seinfeld, "Gas-Phase Reaction of Ozone with *trans*-2-Hexenal, *trans*-2-Hexenyl Acetate, Ethylvinyl Ketone, and 6-Methyl-5-hepten-2-one," *Int. J. Chem. Kinet.*, 28, 373–382 (1996).

Grosjean, E., and D. Grosjean, "Gas Phase Reaction of Alkenes with Ozone: Formation Yields of Primary Carbonyls and Biradicals," *Environ. Sci. Technol.*, 31, 2421–2427 (1997).

Grosjean, E., and D. Grosjean, "The Gas-Phase Reaction of Alkenes with Ozone: Formation Yields of Carbonyls from Biradicals in Ozone–Alkene–Cyclohexane Experiments," *Atmos. Environ.*, 32, 3393–3402 (1998a).

Grosjean, E., and D. Grosjean, "Rate Constants for the Gas-Phase Reaction of Ozone with Unsaturated Oxygenates," *Int. J. Chem. Kinet.*, 30, 21–29 (1998b).

Grotheer, H.-H., G. Riekert, D. Walter, and Th. Just, "Non-Arrhenius Behavior of the Reaction of Hydroxymethyl Radicals with Molecular Oxygen," *J. Phys. Chem.*, 92, 4028–4030 (1988).

Guenther, A. B., P. R. Zimmerman, P. C. Harley, R. K. Monson, and R. Fall, "Isoprene and Monoterpene Emission Rate Variability: Model Evaluations and Sensitivity Analyses," *J. Geophys. Res.*, 98, 12609–12617 (1993).

Guenther, A., P. Zimmerman, and M. Wildermuth, "Natural Volatile Organic Compound Emission Rate Estimates for U.S. Woodland Landscapes," *Atmos. Environ.*, 28, 1197–1210 (1994).

Guenther, A., L. Otter, P. Zimmerman, J. Greenberg, R. Scholes, and M. Scholes, "Biogenic Hydrocarbon Emissions from Southern Africa Savannas," *J. Geophys. Res.*, 101, 25859–25865 (1996a).

Guenther, A., P. Zimmerman, L. Klinger, J. Greenberg, C. Ennis, K. Davis, W. Pollock, H. Westberg, G. Allwine, and C. Geron, "Estimates of Regional Natural Volatile Organic Compound Fluxes from Enclosure and Ambient Measurements," *J. Geophys. Res.*, 101, 1345–1359 (1996b).

Guenther, A. B., and A. J. Hills, "Eddy Covariance Measurement of Isoprene Fluxes," *J. Geophys. Res.*, 103, 13145–13152 (1998).

Guenther, A., "Modeling Biogenic Volatile Organic Compound Emissions to the Atmosphere," Ch. 3, pp. 97–118. In *Reactive Hydrocarbons in the Atmosphere*, Academic Press, San Diego, 1999.

Gutbrod, R., R. N. Schindler, E. Kraka, and D. Cremer, "Formation of OH Radicals in the Gas Phase Ozonolysis of Alkenes: The Unexpected Role of Carbonyl Oxides," *Chem. Phys. Lett.*, 252, 221–229 (1996).

Gutbrod, R., E. Kraka, R. N. Schindler, and D. Cremer, "Kinetic and Theoretical Investigation of the Gas-Phase Ozonolysis of Isoprene: Carbonyl Oxides as an Important Source for OH Radicals in the Atmosphere," *J. Am. Chem. Soc.*, 119, 7330–7342 (1997a).

Gutbrod, R., S. Meyer, M. M. Rahman, and R. N. Schindler, "On the Use of CO as Scavenger for OH Radicals in the Ozonolysis of Simple Alkenes and Isoprene," European Community (LABVOC), Project No. EV5V-CT91-0038, Wiley, New York, 1997b.

Gutman, D., and H. H. Nelson, "Gas-Phase Reactions of the Vinoxy Radical with O_2 and NO," *J. Phys. Chem.*, 87, 3902–3905 (1983).

Hakola, H., B. Shorees, J. Arey, and R. Atkinson, "Product Formation from the Gas-Phase Reactions of OH Radicals and O_3 with β-Phellandrene," *Environ. Sci. Technol.*, 27, 278–283 (1993).

Hakola, H., J. Arey, S. M. Aschmann, and R. Atkinson, "Product Formation from the Gas-Phase Reactions of OH Radicals and O_3 with a Series of Monoterpenes," *J. Atmos. Chem.*, 18, 75–102 (1994).

Hallquist, M., S. Langer, E. Ljungström, and I. Wängberg, "Rates of Reaction between the Nitrate Radical and Some Unsaturated Alcohols," *Int. J. Chem. Kinet.*, 28, 467–474 (1996).

Hallquist, M., I. Wängberg, and E. Ljungström, "Atmospheric Fate of Carbonyl Oxidation Products Originating from α-Pinene and Δ^3-Carene: Determination of Rate of Reaction with OH and NO_3 Radicals, UV Absorption Cross Sections, and Vapor Pressures," *Environ. Sci. Technol.*, 31, 3166–3172 (1997).

Hanst, P. L., J. W. Spence, and M. Miller, "Atmospheric Chemistry of *N*-Nitroso Dimethylamine," *Environ. Sci. Technol.*, 11, 403–405 (1977).

Harley, P., V. Fridd-Stroud, J. Greenberg, A. Guenther, and P. Vasconellos, "Emission of 2-Methyl-3-buten-2-ol by Pines: A Potentially Large Natural Source of Reactive Carbon to the Atmosphere," *J. Geophys. Res.*, 103, 25479–25486 (1998).

Harris, G. W., R. Atkinson, and J. N. Pitts, Jr., "Kinetics of the Reactions of the OH Radical with Hydrazine and Methylhydrazine," *J. Phys. Chem.*, 83, 2557–2559 (1979).

Harris, J. M., S. J. Oltmans, E. J. Dlugokencky, P. C. Novelli, B. J. Johnson, and T. Mefford, "An Investigation into the Source of the Springtime Tropospheric Ozone Maximum at Mauna Loa Observatory," *Geophys. Res. Lett.*, 25, 1895–1898 (1998).

Harrison, R. M., J. P. Shi, and J. L. Grenfell, "Novel Nighttime Free Radical Chemistry in Severe Nitrogen Dioxide Pollution Episodes," *Atmos. Environ.*, 32, 2769–2774 (1998).

Hatakeyama, S., H. Bandow, M. Okuda, and H. Akimoto, "Reactions of CH_2OO and $CH_2(^1A_1)$ with H_2O in the Gas Phase," *J. Phys. Chem.*, 85, 2249–2254 (1981).

Hatakeyama, S., N. Washida, and H. Akimoto, "Rate Constants and Mechanisms for the Reaction of OH(OD) Radicals with Acetylene, Propyne, and 2-Butyne in Air at 297 ± 2 K," *J. Phys. Chem.*, *90*, 173–178 (1986).

Hatakeyama, S., K. Izumi, T. Fukuyama, H. Akimoto, and N. Washida, "Reactions of OH with α-Pinene and β-Pinene in Air: Estimate of Global CO Production from the Atmospheric Oxidation of Terpenes," *J. Geophys. Res.*, *96*, 947–958 (1991).

Hatakeyama, S., H. Lai, S. Gao, and K. Murano, "Production of Hydrogen Peroxide and Organic Hydroperoxides in the Reactions of Ozone with Natural Hydrocarbons in Air," *Chem Lett.*, 1287–1290 (1993).

Hatakeyama, S., and H. Akimoto, "Reactions of Criegee Intermediates in the Gas Phase," *Res. Chem. Intermed.*, *20*, 503–524 (1994).

Hatakeyama, S., H. Lai, and K. Murano, "Formation of 2-Hydroxyethyl Hydroperoxide in an OH-Initiated Reaction of Ethylene in Air in the Absence of NO," *Environ. Sci. Technol.*, *29*, 833–835 (1995).

Hauglustaine, D. A., B. A. Ridley, S. Solomon, P. G. Hess, and S. Madronich, "HNO_3/NO_x Ratio in the Remote Troposphere during MLOPEX 2: Evidence for Nitric Acid Reduction on Carbonaceous Aerosols?" *Geophys. Res. Lett.*, *23*, 2609–2612 (1996).

Hausmann, M., and U. Platt, "Spectroscopic Measurement of Bromine Oxide and Ozone in the High Arctic during Polar Sunrise Experiment 1992," *J. Geophys. Res.*, *99*, 25399–25413 (1994).

Hegels, E., P. J. Crutzen, T. Klüpfel, D. Perner, and J. P. Burrows, "Global Distribution of Atmospheric Bromine-Monoxide from GOME on Earth Observing Satellite ERS-2," *Geophys. Res. Lett.*, *25*, 3127–3130 (1998).

Hegg, D. A., L. F. Radke, P. V. Hobbs, R. A. Rasmussen, and P. J. Riggan, "Emissions of Some Trace Gases from Biomass Fires," *J. Geophys. Res.*, *95*, 5669–5675 (1990).

Heicklen, J., "The Correlation of Rate Coefficients for H-Atom Abstraction by HO Radicals with C–H Bond Dissociation Enthalpies," *Int. J. Chem. Kinet.*, *13*, 651–665 (1981).

Helas, G., S. Slanina, and R. Steinbrecher, Eds., *Biogenic Volatile Organic Compounds in the Atmosphere*, SPB Academic Publishing, Amsterdam, 1997.

Helmig, D., W. Pilock, J. Greenberg, and P. Zimmerman, "Gas Chromatography Mass Spectrometry Analysis of Volatile Organic Trace Gases at Mauna Loa Observatory, Hawaii," *J. Geophys. Res.*, *101*, 14697–14710 (1996).

Helmig, D., B. Balsley, K. Davis, L. R. Kuck, M. Jensen, J. Bognar, T. Smith, Jr., R. V. Arrieta, R. Rodriquez, and J. W. Birks, "Vertical Profiling and Determination of Landscape Fluxes of Biogenic Nonmethane Hydrocarbons within the Planetary Boundary Layer in the Peruvian Amazon," *J. Geophys. Res.*, *103*, 25519–25532 (1998a).

Helmig, D., J. Greenberg, A. Guenther, P. Zimmerman, and C. Geron, "Volatile Organic Compounds and Isoprene Oxidation Products at a Temperate Deciduous Forest Site," *J. Geophys. Res.*, *103*, 22397–22414 (1998b).

Hemminger, J. C., "Heterogeneous Chemistry in the Troposphere: A Modern Surface Chemistry Approach to the Study of Fundamental Processes," *Int. Rev. Phys. Chem.*, in press (1999).

Hendry, D. G., and R. A. Kenley, "Atmospheric Chemistry of Peroxynitrates," in *Nitrogenous Air Pollutants* (D. Grosjean, Ed.), pp. 137–148, Ann Arbor Science, Ann Arbor, MI, 1979.

Hewitt, C. N., and G. L. Kok, "Formation and Occurrence of Organic Hydroperoxides in the Troposphere: Laboratory and Field Observations," *J. Atmos. Chem.*, *12*, 181–194 (1991).

Hewitt, C. N., Ed., *Reactive Hydrocarbons in the Atmosphere*, Academic Press, San Diego, 1999.

Hjorth, J., C. Lohse, C. J. Nielsen, H. Skov, and G. Restelli, "Products and Mechanisms of the Gas-Phase Reactions between NO_3 and a Series of Alkenes," *J. Phys. Chem.*, *94*, 7494–7500 (1990).

Hobbs, P. V., J. S. Reid, R. A. Kotchenruther, R. J. Ferek, and R. Weiss, "Direct Radiative Forcing by Smoke from Biomass Burning," *Science*, *275*, 1776–1778 (1997).

Hodson, J., "The Estimation of the Photodegradation of Organic Compounds by Hydroxyl Radical Reaction Rate Constants Obtained from Nuclear Magnetic Resonance Spectroscopy Chemical Shift Data," *Chemosphere*, *17*, 2339–2348 (1988).

Hoffmann, T., R. Bandur, U. Marggraf, and M. Linscheid, "Molecular Composition of Organic Aerosols Formed in α-Pinene/O_3 Reaction: Implications for New Particle Formation Processes," *J. Geophys. Res.*, *103*, 25569–25578 (1998).

Hofmann, D. J., R. S. Stone, M. E. Wood, T. Deshler, and J. M. Harris, "An Analysis of 25 Years of Balloonborne Aerosol Data in Search of a Signature of the Subsonic Commercial Aircraft Fleet," *Geophys. Res. Lett.*, *25*, 2433–2436 (1998).

Honrath, R. E., and D. A. Jaffe, "The Seasonal Cycle of Nitrogen Oxides in the Arctic Troposphere at Barrow, Alaska," *J. Geophys. Res.*, *97*, 20615–20630 (1992).

Hopper, J. F., L. A. Barrie, A. Silis, W. Hart, A. J. Gallant, and H. Dryfhout, "Ozone and Meteorology during the 1994 Polar Sunrise Experiment," *J. Geophys. Res.*, *103*, 1481–1492 (1998).

Horie, O., and G. K. Moortgat, "Decomposition Pathways of the Excited Criegee Intermediates in the Ozonolysis of Simple Alkenes," *Atmos. Environ.*, *25A*, 1881–1896 (1991).

Horie, O., P. Neeb, S. Limbach, and G. K. Moortgat, "Formation of Formic Acid and Organic Peroxides in the Ozonolysis of Ethene with Added Water Vapour," *Geophys. Res. Lett.*, *21*, 1523–1526 (1994).

Hoyermann, K., and R. Sievert, "Elementarreaktionen in der Oxidation von Alkenen," *Ber. Bunsenges. Phys. Chem.*, *87*, 1027–1031 (1983).

Impey, G. A., P. B. Shepson, D. R. Hastie, and L. A. Barrie, "Measurement Technique for the Determination of Photolyzable Chlorine and Bromine in the Atmosphere," *J. Geophys. Res.*, *102*, 15999–16004 (1997a).

Impey, G. A., P. B. Shepson, D. R. Hastie, L. A. Barrie, and K. G. Anlauf, "Measurements of Photolyzable Chlorine and Bromine during the Polar Sunrise Experiment 1995," *J. Geophys. Res.*, *102*, 16005–16010 (1997b).

Iraci, L. T., and M. A. Tolbert, "Heterogeneous Interaction of Formaldehyde with Cold Sulfuric Acid: Implications for the Upper Troposphere and Lower Stratosphere," *J. Geophys. Res.*, *102*, 16099–16107 (1997).

Isidorov, V. A., I. G. Zenkevich, and B. V. Ioffe, "Volatile Organic Compounds in the Atmosphere of Forests," *Atmos. Environ.*, *19*, 1–8 (1985).

Jaeglé, L., D. J. Jacob, P. O. Wennberg, C. M. Spivakovsky, T. F. Hanisco, E. J. Lanzendorf, E. J. Hintsa, D. W. Fahey, E. R. Keim, M. H. Proffitt, E. L. Atlas, F. Flocke, S. Schauffler, C. T. McElroy, C. Midwinter, L. Pfister, and J. C. Wilson, "Observed OH and HO_2 in the Upper Troposphere Suggest a Major Source from Convective Injection of Peroxides," *Geophys. Res. Lett.*, *24*, 3181–3184 (1997).

Jaeglé, L., D. J. Jacob, W. H. Brune, D. Tan, I. C. Faloona, A. J. Weinheimer, B. A. Ridley, T. L. Campos, and G. W. Sachse, "Sources of HO_x and Production of Ozone in the Upper Troposphere over the United States," *Geophys. Res. Lett.*, *25*, 1709–1712 (1998a).

Jaeglé, L., D. J. Jacob, Y. Wang, A. J. Weinheimer, B. A. Ridley, T. L. Campos, G. W. Sachse, and D. E. Hagen, "Sources and Chemistry of NO_x in the Upper Troposphere over the United States," *Geophys. Res. Lett.*, *25*, 1705–1708 (1998b).

Jaffe, D., "The Relationship between Anthropogenic Nitrogen Oxides and Ozone Trends in the Arctic Troposphere," in *The Tropospheric Chemistry of Ozone in the Polar Regions* (H. Niki and

K. H. Becker, Eds.), NATO ASI Series, Series I: Global Environmental Change, Vol. 7, pp. 105–115, 1993.

Jaffe, D. A., T. K. Berntsen, and I. S. A. Isaksen, "A Global Three-Dimensional Chemical Transport Model. 2. Nitrogen Oxides and Nonmethane Hydrocarbon Results," *J. Geophys. Res.*, 102, 21281–21296 (1997).

Jang, M., and R. M. Kamens, "Newly Characterized Products and Composition of Secondary Aerosols from the Reaction of α-Pinene with Ozone," *Atmos. Environ.*, 33, 459–474 (1999).

Janson, R., and C. De Serves, "Isoprene Emissions from Boreal Wetlands in Scandinavia," *J. Geophys. Res.*, 103, 25513–25517 (1998).

Jay, K., and L. Stieglitz, "The Gas Phase Addition of NO_x to Olefins," *Chemosphere*, 19, 1939–1950 (1989).

Jayne, J. T., D. R. Worsnop, C. E. Kolb, E. Swartz, and P. Davidovits, "Uptake of Gas-Phase Formaldehyde by Aqueous Acid Surfaces," *J. Phys. Chem.*, 100, 8015–8022 (1996).

Jobson, B. T., H. Niki, Y. Yokouchi, J. Bottenheim, F. Hopper, and R. Leaitch, "Measurements of C_2–C_6 Hydrocarbons during the Polar Sunrise 1992 Experiment: Evidence for Cl Atoms and Br Atom Chemistry," *J. Geophys. Res.*, 99, 25355–25368 (1994).

Jolly, G. S., G. Paraskevopoulos, and D. L. Singleton, "Rates of OH Radical Reactions. XII. The Reactions of OH with c-C_3H_6, c-C_5H_{10}, and c-C_7H_{14}. Correlation of Hydroxyl Rate Constants with Bond Dissociation Energies," *Int. J. Chem. Kinet.*, 17, 1–10 (1985).

Jolly, G. S., P. J. McKenney, D. L. Singleton, G. Paraskevopoulos, and A. R. Brossard, "Rate Constant and Mechanism for the Reaction of Hydroxyl Radical with Formic Acid," 90, 6557–6562 (1986).

Jones, A. W., V. Lagesson, and C. Tagesson, "Origins of Breath Isoprene," *J. Clin. Pathol.*, 48, 979–980 (1995).

Jonquières, I., A. Marenco, A. Maalej, and F. Rohrer, "Study of Ozone Formation and Transatlantic Transport from Biomass Burning Emissions over West Africa during the Airborne Tropospheric Ozone Campaigns TROPOZ I and TROPOZ II," *J. Geophys. Res.*, 103, 19059–19073 (1998).

Jungkamp, T. P. W., J. N. Smith, and J. H. Seinfeld, "Atmospheric Oxidation Mechanism of n-Butane: The Fate of Alkoxy Radicals," *J. Phys. Chem. A*, 101, 4392–4401 (1997).

Juuti, S., J. Arey, and R. Atkinson, "Monoterpene Emission Rate Measurements from a Monterey Pine," *J. Geophys. Res.*, 95, 7515–7519 (1990).

Kaiser, E. W., "Temperature and Pressure Dependence of the C_2H_4 Yield from the Reaction $C_2H_5 + O_2$," *J. Phys. Chem.*, 99, 707–711 (1995).

Kaiser, E. W., and T. J. Wallington, "Formation of C_3H_6 from the Reaction $C_3H_7 + O_2$ and C_2H_3Cl from $C_2H_4Cl + O_2$ at 297 K," *J. Phys. Chem.*, 100, 18770–18774 (1996a).

Kaiser, E. W., and T. J. Wallington, "Pressure Dependence of the Reaction $Cl + C_3H_6$," *J. Phys. Chem.*, 100, 9788–9793 (1996b).

Kaiser, E. W., and T. J. Wallington, "Comment on 'Inverse Kinetic Isotope Effect in the Reaction of Atomic Chlorine with C_2H_4 and C_2D_4,'" *J. Phys. Chem. A*, 102, 6054–6055 (1998).

Kajii, Y., K. Someno, H. Tanimoto, J. Hirokawa, and H. Akimoto, "Evidence for the Seasonal Variation of Photochemical Activity of Tropospheric Ozone: Continuous Observation of Ozone and CO at Happo, Japan," *Geophys. Res. Lett.*, 25, 3505–3508 (1998).

Kakesu, M., H. Bandow, N. Takenaka, Y. Maeda, and N. Washida, "Kinetic Measurements of Methyl and Ethyl Nitrate Reactions with OH Radicals," *Int. J. Chem. Kinet.*, 29, 933–941 (1997).

Kaufman, Y. J., and R. S. Fraser, "The Effect of Smoke Particles on Clouds and Climate Forcing," *Science*, 277, 1636–1639 (1997).

Kavouras, I. G., N. Mihalopoulos, and E. G. Stephanou, "Formation of Atmospheric Particles from Organic Acids Produced by Forests," *Nature*, 395, 683–686 (1998).

Keene, W. C., J. R. Maben, A. A. P. Pszenny, and J. N. Galloway, "Measurement Technique for Inorganic Chlorine Gases in the Marine Boundary Layer," *Environ. Sci. Technol.*, 27, 866–874 (1993).

Kesselmeier, J., K. Bode, U. Hofmann, H. Müller, L. Schäfer, A. Wolf, P. Ciccioli, E. Brancaleoni, A. Cecinato, M. Frattoni, P. Foster, C. Ferrari, V. Jacob, J. L. Fugit, L. Dutaur, V. Simon, and L. Torres, "Emission of Short Chained Organic Acids, Aldehydes, and Monoterpenes from *Quercus ilex* L. and *Pinus pinea* L. in Relation to Physiological Activities, Carbon Budget, and Emission Algorithms," *Atmos. Environ.*, 31, 119–133 (1997).

Kesselmeier, J., K. Bode, L. Schäfer, G. Schebeske, A. Wolf, E. Brancaleoni, A. Cecinato, P. Ciccioli, M. Frattoni, L. Dutaur, J. L. Fugit, V. Simon, and L. Torres, "Simultaneous Field Measurements of Terpene and Isoprene Emissions from Two Dominant Mediterranean Oak Species in Relation to a North American Species," *Atmos. Environ.*, 32, 1947–1953 (1998).

Kim, J. H., and M. J. Newchurch, "Climatology and Trends of Tropospheric Ozone over the Eastern Pacific Ocean: The Influences of Biomass Burning and Tropospheric Dynamics," *Geophys. Res. Lett.*, 23, 3723–3726 (1996).

Kim, J. H., and M. J. Newchurch, "Biomass-Burning Influence on Tropospheric Ozone over New Guineas and South America," *J. Geophys. Res.*, 103, 1455–1461 (1998).

Kind, I., T. Berndt, O. Böge, and W. Rolle, "Gas-Phase Rate Constants for the Reaction of NO_3 Radicals with Furan and Methyl-Substituted Furans," *Chem. Phys. Lett.*, 256, 679–683 (1996).

Kirchner, F., and W. R. Stockwell, "Effect of Peroxy Radical Reactions on the Predicted Concentrations of Ozone, Nitrogenous Compounds, and Radicals," *J. Geophys. Res.*, 101, 21007–21022 (1996); correction, *ibid.*, 102, 10871 (1997).

Kirchner, U., Th. Benter, and R. N. Schindler, "Experimental Verification of Gas Phase Bromine Enrichment in Reactions of HOBr with Sea Salt Doped Ice Surfaces," *Ber. Bunsenges. Phys. Chem.*, 101, 975–977 (1997).

Kirstine, W., I. Galbally, Y. Ye, and M. Hooper, "Emissions of Volatile Organic Compounds (Primarily Oxygenated Species) from Pasture," *J. Geophys. Res.*, 103, 10605–10619 (1998).

Klamt, A. "Estimation of Gas-Phase Hydroxyl Radical Rate Constants of Organic Compounds from Molecular Orbital Calculations," *Chemosphere*, 26, 1273–1289 (1993).

Kleindienst, T. A., "Recent Developments in the Chemistry and Biology of Peroxyacetyl Nitrate," *Res. Chem. Intermed.*, 20, 335–384 (1994).

Kley, D., P. J. Crutzen, H. G. J. Smit, H. Vömel, S. J. Oltmans, H. Grassl, and V. Ramanathan, "Observations of Near-Zero Ozone Concentrations over the Convective Pacific: Effects on Air Chemistry," *Science*, 274, 230–233 (1996).

Kley, D., "Tropospheric Chemistry and Transport," *Science*, 276, 1043–1045 (1997).

Klinger, L. F., J. Greenberg, A. Guenther, G. Tyndall, P. Zimmerman, M. M'Bangui, J. M. Moutsamboté, and D. Kenfack, "Patterns in Volatile Organic Compound Emissions along a Savanna-Rainforest Gradient in Central Africa," *J. Geophys. Res.*, 103, 1443–1454 (1998).

Klotz, B., I. Barnes, K. H. Becker, and B. T. Golding, "Atmospheric Chemistry of Benzene Oxide/Oxepin," *J. Chem. Soc., Faraday Trans.*, 93, 1507–1516 (1997).

Klotz, B., I. Barnes, and K. H. Becker, "New Results on the Atmospheric Photooxidation of Simple Alkylbenzenes," *Chem. Phys.*, 231, 289–301 (1998).

Knispel, R., R. Koch, M. Siese, and C. Zetzsch, "Adduct Formation of OH Radicals with Benzene, Toluene, and Phenol and Consecutive Reactions of the Adducts with NO_x and O_2," *Ber. Bunsenges. Phys. Chem.*, 94, 1375–1379 (1990).

Koch, R., H.-U. Krüger, M. Elend, W.-U. Palm, and C. Zetzsch, "Rate Constants for the Gas-Phase Reaction of OH with Amines: *tert*-Butyl Amine, 2,2,2-Trifluoroethyl Amine, and 1,4-Diazabicyclo[2.2.2]octane," *Int. J. Chem. Kinet.*, 28, 807–815 (1996).

Koch, R., "Comment on 'UV Spectrum and Kinetics of Hydroxycyclohexadienyl Radicals,'" *J. Phys. Chem. B*, 101, 293 (1997).

Koch, R., W.-U. Palm, and C. Zetzsch, "First Rate Constants for Reactions of OH Radicals with Amides," *Int. J. Chem. Kinet.*, 29, 81–87 (1997).

König, G., M. Brunda, H. Puxbaum, C. Hewitt, C. Duckham, and J. Rudolph, "Relative Contribution of Oxygenated Hydrocarbons to the Total Biogenic VOC Emission of Selected Mid-European Agricultural and Natural Plant Species," *Atmos. Environ.*, 29A, 861–874 (1995).

Kramp, F., and S. E. Paulson, "On the Uncertainties in the Rate Coefficients for OH Reactions with Hydrocarbons, and the Rate Coefficients of the 1,3,5-Trimethylbenzene and *m*-Xylene Reactions with OH Radicals in the Gas Phase," *J. Phys. Chem. A*, 102, 2685–2690 (1998).

Kuhlbusch, T. A. J., and P. J. Crutzen, "Toward a Global Estimate of Black Carbon in Residues of Vegetation Fires Representing a Sink of Atmospheric CO_2 and a Source of O_2," *Global Biogeochem. Cycl.*, 9, 491–501 (1995).

Kukui, A. S., T. P. W. Jungkamp, and R. N. Schindler, "Aldehyde Formation in the Reaction of Methoxy Radicals with NO_3," *Ber. Bunsenges. Phys. Chem.*, 99, 1565–1567 (1995).

Kuzma, J., M. Nemecek-Marshall, W. H. Pollock, and R. Fall, "Bacteria Produce the Volatile Hydrocarbon Isoprene," *Curr. Microbiol.*, 30, 97–103 (1995).

Kwok, E. S. C., and R. Atkinson, "Estimation of Hydroxyl Radical Reaction Rate Constants for Gas-Phase Organic Compounds Using a Structure–Reactivity Relationship: An Update," *Atmos. Environ.*, 29, 1685–1695 (1995).

Kwok, E. S. C., R. Atkinson, and J. Arey, "Observation of Hydroxycarbonyls from the OH Radical Initiated Reaction of Isoprene," *Environ. Sci. Technol.*, 29, 2467–2469 (1995).

Kwok, E. S. C., J. Arey, and R. Atkinson, "Alkoxy Radical Isomerization in the OH Radical-Initiated Reactions of C_4–C_8 *n*-Alkanes," *J. Phys. Chem.*, 100, 214–219 (1996a).

Kwok, E. S. C., R. Atkinson, and J. Arey, "Isomerization of β-Hydroxyalkoxy Radicals Formed from the OH Radical-Initiated Reactions of C_4–C_8 1-Alkenes," *Environ. Sci. Technol.*, 30, 1048–1052 (1996b).

Kwok, E. S. C., S. M. Aschmann, J. Arey, and R. Atkinson, "Product Formation from the Reaction of the NO_3 Radical with Isoprene and Rate Constants for the Reactions of Methacrolein and Methyl Vinyl Ketone with the NO_3 Radical," *Int. J. Chem. Kinet.*, 28, 925–934 (1996c).

Kwok, E. S. C., S. M. Aschmann, R. Atkinson, and J. Arey, "Products of the Gas-Phase Reactions of *o*-, *m*-, and *p*-Xylene with the OH Radical in the Presence and Absence of NO_x," *J. Chem. Soc., Faraday Trans.*, 93, 2847–2854 (1997).

Lamb, B., T. Pierce, D. Baldocchi, E. Allwine, S. Dilts, H. Westberg, C. Geron, A. Guenther, L. Klinger, P. Harley, and P. Zimmerman, "Evaluation of Forest Canopy Models for Estimating Isoprene Emissions," *J. Geophys. Res.*, 101, 22787–22797 (1996).

Langer, S., and E. Ljungström, "Reaction of the Nitrate Radical with Some Potential Automotive Fuel Additives. A Kinetic and Mechanistic Study," *J. Phys. Chem.*, 98, 5906–5912 (1994).

Langer, S., E. Ljungström, I. Wängberg, T. J. Wallington, and O. J. Nielsen, "Atmospheric Chemistry of Di-*tert*-butyl Ether—Rates and Products of the Reactions with Chlorine Atoms, Hydroxyl Radicals, and Nitrate Radicals," *Int. J. Chem. Kinet.*, 28, 299–306 (1996).

Laursen, K. K., P. V. Hobbs, L. F. Radke, and R. A. Rasmussen, "Some Trace Gas Emissions from North American Biomass Fires with an Assessment of Regional and Global Fluxes from Biomass Burning," *J. Geophys. Res.*, 97, 20687–20701 (1992).

LeBel, P. J., W. R. Cofer, III, J. S. Levine, S. A. Vay, and P. D. Roberts, "Nitric Acid and Ammonia Emissions from a Mid-Latitude Prescribed Wetlands Fire," *Geophys. Res. Lett.*, 15, 792–795 (1988).

Le Bras, G., and U. Platt, "A Possible Mechanism for Combined Chlorine and Bromine Catalyzed Destruction of Tropospheric Ozone in the Arctic," *Geophys. Res. Lett.*, 22, 599–602 (1995).

Le Calvé, S., G. Le Bras, and A. Mellouki, "Temperature Dependence for the Rate Coefficients of the Reactions of the OH Radical with a Series of Formates," *J. Phys. Chem. A.*, 101, 5489–5493 (1997a).

Le Calvé, S., G. Le Bras, and A. Mellouki, "Kinetic Studies of OH Reactions with *Iso*-Propyl, *Iso*-Butyl, *Sec*-Butyl, and *Tert*-Butyl Acetate," *Int. J. Chem. Kinet.*, 29, 683–688 (1997b).

Le Calvé, S., G. Le Bras, and A. Mellouki, "Kinetic Studies of OH Reactions with a Series of Methyl Esters," *J. Phys. Chem. A*, 101, 9137–9141 (1997c).

Lee, F. S. C., and F. S. Rowland, "Thermal Chlorine-38 Reactions with Propene," *J. Phys. Chem.*, 81, 1222–1229 (1977).

Lee, M., B. G. Heikes, D. J. Jacob, G. Sachse, and B. Anderson, "Hydrogen Peroxide, Organic Hydroperoxide, and Formaldehyde as Primary Pollutants from Biomass Burning," *J. Geophys. Res.*, 102, 1301–1309 (1997).

Lee, M., B. G. Heikes, and D. J. Jacob, "Enhancements of Hydroperoxides and Formaldehyde in Biomass Burning Impacted Air and Their Effect on Atmospheric Oxidant Cycles," *J. Geophys. Res.*, 103, 13201–13212 (1998).

Lehrer, E., D. Wagenbach, and U. Platt, "Aerosol Chemical Composition during Tropospheric Ozone Depletion at Ny Ålesund/Svalbard," *Tellus*, 49B, 486–495 (1997).

Lelieveld, J., and P. J. Crutzen, "Influences of Cloud Photochemical Processes on Tropospheric Ozone," *Nature*, 343, 227–233 (1990).

Levine, J. S., Ed., *Global Biomass Burning: Atmospheric, Climatic, and Biospheric Implications*, The MIT Press, Cambridge, Massachusetts, 1991.

Levine, J. S., W. R. Cofer, III, D. R. Cahoon, Jr., and E. L. Winstead, "Biomass Burning: A Driver for Global Change," *Environ. Sci. Technol.*, 29, 120A–125A (1995).

Levy, H., II, P. S. Kasibhatla, W. J. Moxim, A. A. Klonecki, A. I. Hirsch, S. J. Oltmans, and W. L. Chameides, "The Global Impact of Human Activity on Tropospheric Ozone," *Geophys. Res. Lett.*, 24, 791–794 (1997).

Lide, D. R., Ed., *CRC Handbook of Chemistry and Physics*, 79th ed., CRC Press, Boca Raton, FL, 1998–1999.

Lightfoot, P. D., R. A. Cox, J. N. Crowley, M. Destriau, G. D. Hayman, M. E. Jenkin, G. K. Moortgat, and F. Zabel, "Organic Peroxy Radicals: Kinetics, Spectroscopy and Tropospheric Chemistry," *Atmos. Environ.*, 26A, 1805–1961 (1992).

Lindley, C. R. C., J. G. Calvert, and J. H. Shaw, "Rate Studies of the Reactions of the $(CH_3)_2N$ Radical with O_2, NO, and NO_2," *Chem. Phys. Lett.*, 67, 57–62 (1979).

Liu, S. C., M. Trainer, M. A. Carroll, G. Hubler, D. D. Montzka, R. B. Norton, B. A. Ridley, J. G. Walega, E. L. Atlas, B. G. Heikes, B. J. Huebert, and W. Warren, "A Study of the Photochemistry and Ozone Budget during the Mauna Loa Observatory Photochemistry Experiment," *J. Geophys. Res.*, 97, 10463–10471 (1992).

Logan, J. A., and V. W. J. H. Kirchhoff, "Seasonal Variations of Tropospheric Ozone at Natal, Brazil," *J. Geophys. Res.*, 91, 7875–7881 (1986).

Loreto, F., P. Ciccioli, A. Cecinato, E. Brancaleoni, M. Frattoni, and D. Tricoli, "Influence of Environmental Factors and Air Composition on the Emission of α-Pinene from *Quercus ilex* Leaves," *Plant Physiol.*, 110, 267–275 (1996).

MacDonald, R. C., and R. Fall, "Detection of Substantial Emissions of Methanol from Plants to the Atmosphere," *Atmos. Environ.*, 27A, 1709–1713 (1993).

Madronich, S., and J. G. Calvert, "Permutation Reactions of Organic Peroxy Radicals in the Troposphere," *J. Geophys. Res.*, 95, 5697–5715 (1990).

Mahlman, J. D., "Dynamics of Transport Processes in the Upper Troposphere," *Science*, 276, 1079–1083 (1997).

Maricq, M. M., and J. J. Szente, "Kinetics of the Reaction between Ethylperoxy Radicals and Nitric Oxide," *J. Phys. Chem.*, 100, 12374–12379 (1996).

Markert, F., and P. Pagsberg, "UV Spectra and Kinetics of Radicals Produced in the Gas Phase Reactions of Cl, F, and OH with Toluene," *Chem. Phys. Lett.*, 209, 445–454 (1993).

Marston, G., C. D. McGill, and A. R. Rickard, "Hydroxyl-Radical Formation in the Gas-Phase Ozonolysis of 2-Methylbut-2-ene," *Geophys. Res. Lett.*, 25, 2177–2180 (1998).

Martinez, E., B. Cabanas, A. Aranda, and P. Martin, "Kinetics of the Reactions of NO_3 Radical with Selected Monoterpenes: A Temperature Dependence Study," *Environ. Sci. Technol.*, 32, 3730–3734 (1998).

Martinez, R. R. I., and J. T. Herron, "Gas-Phase Reaction of SO_2 with a Criegee Intermediate in the Presence of Water Vapor," *J. Environ. Health Sci.*, A16, 623–636 (1981).

Masaki, A., S. Tsunashima, and N. Washida, "Rate Constant for the Reaction of CH_3O_2 with NO," *Chem. Phys. Lett.*, 218, 523–528 (1994).

Mauzerall, D. L., J. A. Logan, D. J. Jacob, B. E. Anderson, D. R. Blake, J. D. Bradshaw, B. Heikes, G. W. Sachse, H. Singh, and B. Talbot, "Photochemistry in Biomass Burning Plumes and Implications for Tropospheric Ozone over the Tropical South Atlantic," *J. Geophys. Res.*, 103, 8401–8423 (1998); correction, 103, 19281–19282 (1998).

McCaulley, J. A., N. Kelly, M. F. Golde, and F. Kaufman, "Kinetic Studies of the Reactions of F and OH with CH_3OH," *J. Phys. Chem.*, 93, 1014–1018 (1989).

McConnell, J. C., G. S. Henderson, L. Barrie, J. Bottenheim, H. Niki, C. H. Langford, and E. M. J. Templeton, "Photochemical Bromine Production Implicated in Arctic Boundary-Layer Ozone Depletion," *Nature*, 355, 150–152 (1992).

McKeen, S. A., T. Gierczak, J. B. Burkholder, P. O. Wennberg, T. F. Hanisco, E. R. Keim, R.-S. Gao, S. C. Liu, A. R. Ravishankara, and D. W. Fahey, "The Photochemistry of Acetone in the Upper Troposphere: A Source of Odd-Hydrogen Radicals," *Geophys. Res. Lett.*, 24, 3177–3180 (1997).

Mendis, S., P. A. Sobotka, and D. E. Euler, "Pentane and Isoprene in Expired Air from Humans: Gas-Chromatographic Analysis of Single Breath," *Clin. Chem.*, 40, 1485–1488 (1994).

Milne, P. J., D. D. Riemer, R. G. Zika, and L. E. Brand, "Measurement of Vertical Distribution of Isoprene in Surface Seawater, Its Chemical Fate, and Its Emission from Several Phytoplankton Monocultures," *Mar. Chem.*, 48, 237–244 (1995).

Miyoshi, A., S. Hatakeyama, and N. Washida, "OH Radical-Initiated Photooxidation of Isoprene: An Estimate of Global CO Production," *J. Geophys. Res.*, 99, 18779–18787 (1994).

Monks, P. S., L. J. Carpenter, S. A. Penkett, and G. P. Ayers, "Night-Time Peroxy Radical Chemistry in the Remote Marine Boundary Layer over the Southern Ocean," *Geophys. Res. Lett.*, 23, 535–538 (1996).

Monson, R. K., A. J. Hills, P. R. Zimmerman, and R. Fall, "Studies of the Relationship between Isoprene Emission Rate and CO_2 or Photon-Flux Density Using a Real-Time Isoprene Analyzer," *Plant Cell Environ.*, 14, 517–523 (1991).

Monson, R. K., C. H. Jaeger, W. W. Adams, III, E. M. Driggers, G. M. Silver, and R. Fall, "Relationships among Isoprene Emission Rate, Photosynthesis, and Isoprene Emission Rate as Influenced by Temperature," *Plant Physiol.*, 98, 1175–1180 (1992).

Monson, R. K., M. T. Lerdau, T. D. Sharkey, D. S. Schimel, and R. Fall, "Biological Aspects of Constructing Volatile Organic Compound Emission Inventories," *Atmos. Environ.*, 29, 2989–3002 (1995).

Montzka, S. A., M. Trainer, P. D. Goldan, W. C. Kuster, and F. C. Fehsenfeld, "Isoprene and Its Oxidation Products, Methyl Vinyl Ketone and Methacrolein, in the Rural Troposphere," *J. Geophys. Res.*, 98, 1101–1111 (1993).

Montzka, S. A., M. Trainer, W. M. Angevine, and F. C. Fehsenfeld, "Measurements of 3-Methyl Furan, Methyl Vinyl Ketone, and Methacrolein at a Rural Forested Site in the Southeastern United States," *J. Geophys. Res.*, 100, 11393–11401 (1995).

Moore, R. M., D. E. Oram, and S. A. Penkett, "Production of Isoprene by Marine Phytoplankton Cultures," *Geophys. Res. Lett.*, 21, 2507–2510 (1994).

Morris, E. D., Jr., D. H. Stedman, and H. Niki, "Mass Spectrometric Study of the Reactions of the Hydroxyl Radical with Ethylene, Propylene, and Acetaldehyde in a Discharge Flow System," *J. Am. Chem. Soc.*, 93, 3570–3572 (1971).

Moschonas, N., D. Danalatos, and S. Glavas, "The Effect of O_2 and NO_2 on the Ring Retaining Products of the Reaction of Toluene with Hydroxyl Radicals," *Atmos. Environ.*, 33, 111–116 (1999).

Mozurkewich, M., "Mechanisms for the Release of Halogens from Sea-Salt Particles by Free Radical Reactions," *J. Geophys. Res.*, 100, 14199–14207 (1995).

Narita, N., and T. Tezuka, "On the Mechanism of Oxidation of Hydroxycyclohexadienyl Radicals with Molecular Oxyen," *J. Am. Chem. Soc.*, 104, 7316–7318 (1982).

Neeb, P., O. Horie, and G. K. Moortgat, "Gas-Phase Ozonolysis of Ethene in the Presence of Hydroxylic Compounds," *Int. J. Chem. Kinet.*, 28, 721–730 (1996).

Neeb, P., F. Sauer, O. Horie, and G. K. Moortgat, "Formation of Hydroxymethyl Hydroperoxide and Formic Acid in Alkene Ozonolysis in the Presence of Water Vapour," *Atmos. Environ.*, 31, 1417–1423 (1997).

Neeb, P., O. Horie, and G. K. Moortgat, "The Ethene–Ozone Reaction in the Gas Phase," *J. Phys. Chem. A*, 102, 6778–6785 (1998a).

Neeb, P., A. Kolloff, S. Koch, and G. K. Moortgat, "Rate Constants for the Reactions of Methylvinyl Ketone, Methacrolein, Methacrylic Acid, and Acrylic Acid with Ozone," *Int. J. Chem. Kinet.*, 30, 769–776 (1998b).

Nemecek-Marshall, M., R. C. MacDonald, J. J. Franzen, C. L. Wojciechowski, and R. Fall, "Methanol Emission from Leaves," *Plant Physiol.*, 108, 1359–1368 (1995).

Nichol, J., "Smoke Haze in Southeast Asia: A Predictable Recurrence," *Atmos. Environ.*, 32, 2715–2716 (1998).

Niki, H., P. D. Maker, C. M. Savage, and L. P. Beitenbach, "An FTIR Study of Mechanisms for the HO Radical Initiated Oxidation of C_2H_4 in the Presence of NO: Detection of Glycolaldehyde," *Chem. Phys. Lett.*, 80, 499–503 (1981).

Niki, H., P. D. Maker, C. M. Savage, and L. P. Breitenbach, "FTIR Spectroscopic Study of the Mechanism for the Gas-Phase Reaction between Ozone and Tetramethylethylene," *J. Phys. Chem.*, 91, 941–946 (1987).

Niki, H., and K. H. Becker, Eds., *The Tropospheric Chemistry of Ozone in the Polar Regions*, NATO ASI Series I: Global Environmental Change, Vol. 7, Springer-Verlag, Berlin, 1993.

Nordmeyer, T., W. Wang, M. L. Ragains, B. J. Finlayson-Pitts, C. W. Spicer, and R. A. Plastridge, "Unique Products of the Reaction of Isoprene with Atomic Chlorine: Potential Markers of Chlorine Atom Chemistry," *Geophys. Res. Lett.*, 24, 1615–1618 (1997).

Nouaime, G., S. B. Bertman, C. Seaver, D. Elyea, H. Huang, P. B. Shepson, T. K. Starn, D. D. Riemer, R. G. Zika, and K. Olszyna,

"Sequential Oxidation Products from Tropospheric Isoprene Chemistry: MACR and MPAN at a NO_x-Rich Forest Environment in the Southeastern United States," *J. Geophys. Res., 103,* 22463–22471 (1998).

Nozière, B., and I. Barnes, "Evidence for Formation of a PAN Analogue of Pinonic Structure and Investigation of Its Thermal Stability," *J. Geophys. Res., 103,* 25587–25597 (1998).

O'Brien, J. M., E. Czuba, D. R. Hastie, J. S. Francisco, and P. B. Shepson, "Determination of the Hydroxy Nitrate Yields from the Reaction of C_2–C_6 Alkenes with OH in the Presence of NO," *J. Phys. Chem. A, 102,* 8903–8908 (1998).

Oltmans, S. J., and W. D. Komhyr, "Surface Ozone Distributions and Variations from 1973–1984 Measurements at the NOAA Geophysical Monitoring for Climatic Change Baseline Observatories," *J. Geophys. Res., 91,* 5229–5236 (1986).

Oltmans, S. J., R. C. Schnell, P. J. Sheridan, R. E. Peterson, S.-M. Li, J. W. Winchester, P. P. Tans, W. T. Sturges, J. D. Kahl, and L. A. Barrie, "Seasonal Surface Ozone and Filterable Bromine Relationship in the High Arctic," *Atmos. Environ., 23,* 2431–2441 (1989).

Olzmann, M., Th. Benter, M. Liesner, and R. N. Schindler, "On the Pressure Dependence of the NO_2 Product Yield in the Reaction of NO_3 Radicals with Selected Alkenes," *Atmos. Environ., 28,* 2677–2683 (1994).

Orlando, J. J., G. S. Tyndall, M. Bilde, C. Ferronato, T. J. Wallington, L. Vereecken, and J. Peeters, "Laboratory and Theoretical Study of the Oxy Radicals in the OH- and Cl-Initiated Oxidation of Ethene," *Phys. Chem. A., 102,* 8116–8123 (1998).

Oum, K. W., M. J. Lakin, D. O. DeHaan, T. Brauers, and B. J. Finlayson-Pitts, "Formation of Molecular Chlorine from the Photolysis of Ozone and Aqueous Sea-Salt Particles," *Science, 279,* 74–77 (1998a).

Oum, K. W., M. J. Lakin, and B. J. Finlayson-Pitts, "Bromine Activation in the Troposphere by the Dark Reaction of O_3 with Seawater Ice," *Geophys. Res. Lett., 25,* 3923–3926 (1998b).

Owen, S., C. Boissard, R. A. Street, S. C. Duckham, O. Csiky, and C. N. Hewitt, "Screening of 18 Mediterranean Plant Species for Volatile Organic Compound Emissions," *Atmos. Environ., 31,* 101–117 (1997).

Owen, S. M., C. Boissard, B. Hagenlocher, and C. N. Hewitt, "Field Studies of Isoprene Emissions from Vegetation in the Northwest Mediterranean Region," *J. Geophys. Res., 103,* 25499–25511 (1998).

Pagsberg, P., J. Munk, C. Anastasi, and V. Simpson, "UV Spectrum of CD_2OD and Its Reactions with O_2, NO, and NO_2," *Chem. Phys. Lett., 157,* 271–276 (1989).

Pagsberg, P., "Reply to Comment on 'UV Spectrum and Kinetics of Hydroxycyclohexadienyl Radicals,'" *J. Phys. Chem. B, 101,* 294 (1997).

Parrish, D. D., M. P. Buhr, M. Trainer, R. B. Norton, J. P. Shimshock, F. C. Fehsenfeld, K. G. Anlauf, J. W. Bottenheim, Y. Z. Tang, H. A. Wiebe, J. M. Roberts, R. L. Tanner, L. Newman, V. C. Bowersox, K. J. Olszyna, E. M. Bailey, M. O. Rodgers, T. Wang, H. Berresheim, U. K. Roychowdhury, and K. L. Demerjian, "The Total Reactive Oxidized Nitrogen Levels and the Partitioning between the Individual Species at Six Rural Sites in Eastern North America," *J. Geophys. Res., 98,* 2927–2939 (1993a).

Parrish, D. D., J. S. Holloway, M. Trainer, P. C. Murphy, G. L. Forbes, and F. C. Fehsenfeld, "Export of North American Ozone Pollution to the North Atlantic Ocean," *Science, 259,* 1436–1439 (1993b).

Parrish, D. D., M. Trainer, J. S. Holloway, J. E. Yee, M. S. Warshawsky, F. C. Fehsenfeld, G. L. Forbes, and J. L. Moody, "Relationships between Ozone and Carbon Monoxide at Surface Sites in the North Atlantic Region," *J. Geophys. Res., 103,* 13357–13376 (1998).

Paulson, S. E., and J. H. Seinfeld, "Atmospheric Photochemical Oxidation of 1-Octene: OH, O_3, and $O(^3P)$ Reactions," *Environ. Sci. Technol., 26,* 1165–1173 (1992a).

Paulson, S. E., and J. H. Seinfeld, "Development and Evaluation of a Photooxidation Mechanism for Isoprene," *J. Geophys. Res., 97,* 20703–20715 (1992b).

Paulson, S. E., R. C. Flagan, and J. H. Seinfeld, "Atmospheric Photooxidation of Isoprene. 1. The Hydroxyl Radical and Ground State Atomic Oxygen Reactions," *Int. J. Chem. Kinet., 24,* 79–101 (1992a).

Paulson, S. E., R. C. Flagan, and J. H. Seinfeld, "Atmospheric Photooxidation of Isoprene. 2. The Ozone–Isoprene Reaction," *Int. J. Chem. Kinet., 24,* 103–125 (1992b).

Paulson, S. E., and J. J. Orlando, "The Reactions of Ozone with Alkenes: An Important Source of HO_x in the Boundary Layer," *Geophys. Res. Lett., 23,* 3727–3730 (1996).

Paulson, S. E., A. D. Sen, P. Liu, J. D. Fenske, and M. J. Fox, "Evidence for Formation of OH Radicals from the Reaction of O_3 with Alkenes in the Gas Phase," *Geophys. Res. Lett., 24,* 3193–3196 (1997).

Paulson, S. E., M. Chung, A. D. Sen, and G. Orzechowska, "Measurement of OH Radical Formation from the Reaction of Ozone with Several Biogenic Alkenes," *J. Geophys. Res., 103,* 25533–25539 (1998).

Penkett, S. A., B. J. Bandy, C. E. Reeves, D. McKenna, and P. Hignett, "Measurements of Peroxides in the Atmosphere and Their Relevance to the Understanding of Global Tropospheric Chemistry," *Faraday Discuss., 100,* 155–174 (1995).

Penkett, S. A., P. S. Monks, L. J. Carpenter, K. C. Clemitshaw, G. P. Ayers, R. W. Gillett, I. E. Galbally, and C. P. Meyer, "Relationships between Ozone Photolysis Rates and Peroxy Radical Concentrations in Clean Marine Air over the Southern Ocean," *J. Geophys. Res., 102,* 12805–12817 (1997).

Penner, J. E., R. E. Dickinson, and C. A. O'Neill, "Effects of Biomass Burning Aerosol on Global Radiation Budget," *Science, 256,* 1432–1434 (1992).

Perry, R. A., R. Atkinson, and J. N. Pitts, Jr., "Kinetics and Mechanism of the Gas Phase Reaction of OH Radicals with Aromatic Hydrocarbons over the Temperature Range 296–473 K," *J. Phys. Chem., 81,* 296–303 (1977).

Phillips, M., J. Greenberg, and J. Awad, "Metabolic and Environmental Origins of Volatile Organic Compounds in Breath," *J. Clin. Pathol., 47,* 1052–1053 (1994).

Pickering, K. E., A. M. Thompson, J. R. Scala, W.-K. Tao, R. R. Dickerson, and J. Simpson, "Free Tropospheric Ozone Production Following Entrainment of Urban Plumes into Deep Convection," *J. Geophys. Res., 97,* 17985–18000 (1992).

Picquet, B., S. Heroux, A. Chebbi, J.-F. Doussin, R. Durand-Jolibois, A. Monod, H. Loirat, and P. Carlier, "Kinetics of the Reactions of OH Radicals with Some Oxygenated Volatile Organic Compounds under Simulated Atmospheric Conditions," *Int. J. Chem. Kinet., 30,* 839–847 (1998).

Pierce, T., C. Geron, L. Bender, R. Dennis, G. Tonnesen, and A. Guenther, "Influence of Increased Isoprene Emissions on Regional Ozone Modeling," *J. Geophys. Res., 103,* 25611–25629 (1998).

Pio, C. A., and A. A. Valente, "Atmospheric Fluxes and Concentrations of Monoterpenes in Resin-Tapped Pine Forests," *Atmos. Environ., 32,* 683–691 (1998).

Pitts, J. N., Jr., D. Grosjean, K. Van Cauwenberghe, J. P. Schmid, and D. R. Fitz, "Photooxidation of Aliphatic Amines under Simulated Atmospheric Conditions: Formation of Nitrosamines, Nitramines, Amides, and Photochemical Oxidant," *Environ. Sci. Technol., 12,* 946–953 (1978).

Plass-Dülmer, C., A. Khedim, R. Koppmann, F. J. Johnen, J. Rudolph, and H. Kuosa, "Emissions of Light Nonmethane Hydrocarbons from the Atlantic into the Atmosphere," *Global Biogeochem. Cycl., 7,* 211–228 (1993).

Platt, U., G. Le Bras, G. Poulet, J. P. Burrows, and G. Moortgat, "Peroxy Radicals from Nighttime Reaction of NO_3 with Organic Compounds," *Nature, 348,* 147–149 (1990).

Platt, U., and M. Hausmann, "Spectroscopic Measurement of the Free Radicals NO_3, BrO, IO, and OH in the Troposphere," *Res. Chem. Intermed., 20,* 557–578 (1994).

Platz, J., L. K. Christensen, J. Sehested, O. J. Nielsen, T. J. Wallington, C. Sauer, I. Barnes, K. H. Becker, and R. Vogt, "Atmospheric Chemistry of 1,3,5-Trioxane: UV Spectra of c-$C_3H_5O_3$(·) and (c-$C_3H_5O_3)O_2$(·) Radicals, Kinetics of the Reactions of (c-$C_3H_5O_3)O_2$(·) Radicals with NO and NO_2, and Atmospheric Fate of the Alkoxy Radical (c-$C_3H_5O_3$)O(·)," *J. Phys. Chem. A, 102,* 4829–4838 (1998a).

Platz, J., O. J. Nielsen, T. J. Wallington, J. C. Ball, M. D. Hurley, A. M. Straccia, W. F. Schneider, and J. Sehested, "Atmospheric Chemistry of the Phenoxy Radical, C_6H_5O(·): UV Spectrum and Kinetics of Its Reaction with NO, NO_2, and O_2," *J. Phys. Chem. A, 102,* 7964–7974 (1998b).

Polissar, A. V., P. K. Hopke, W. C. Malm, and J. F. Sisler, "Atmospheric Aerosol over Alaska. 1. Spatial and Seasonal Variability," *J. Geophys. Res., 103,* 19035–19044 (1998a).

Polissar, A. V., P. K. Hopke, P. Paatero, W. C. Malm, and J. F. Sisler, "Atmospheric Aerosol over Alaska. 2. Elemental Composition and Sources," *J. Geophys. Res., 103,* 19045–19057 (1998b).

Poppe, D., R. Koppmann, and J. Rudolph, "Ozone Formation in Biomass Burning Plumes: Influence of Atmospheric Dilution," *Geophys. Res. Lett., 25,* 3823–3826 (1998).

Porter, E., J. Wenger, J. Treacy, H. Sidebottom, A. Mellouki, S. Téton, and G. Le Bras, "Kinetic Studies on the Reactions of Hydroxyl Radicals with Diethers and Hydroxyethers," *J. Phys. Chem. A, 101,* 5770–5775 (1997).

Portmann, R. W., S. Solomon, J. Fishman, J. R. Olson, J. T. Kiehl, and B. Briegleb, "Radiative Forcing of the Earth's Climate System Due to Tropical Tropospheric Ozone Production," *J. Geophys. Res., 102,* 9409–9417 (1997).

Prather, M. J., and D. J. Jacob, "A Persistent Imbalance in HO_x and NO_x Photochemistry of the Upper Troposphere Driven by Deep Tropical Convection," *Geophys. Res. Lett., 24,* 3189–3192 (1997).

Pszenny, A. A. P., W. C. Keene, D. J. Jacob, S. Fan, J. R. Maben, M. P. Zetwo, M. Springer-Young, and J. N. Galloway, "Evidence of Inorganic Chlorine Gases Other Than Hydrogen Chloride in Marine Surface Air," *Geophys. Res. Lett., 20,* 699–702 (1993).

Puxbaum, H., "Biogenic Emissions of Alcohols, Ester, Ether, and Higher Aldehydes," in *Biogenic Volatile Organic Compounds in the Atmosphere* (G. Helas, J. Slanina, and R. Steinbrecher, Eds.), pp. 79–99, SPB Academic Publishing, Amsterdam, 1997.

Ragains, M. L., and B. J. Finlayson-Pitts, "Kinetics and Mechanism of the Reaction of Cl Atoms with 2-Methyl-1,3-butadiene (Isoprene) at 298 K," *J. Phys. Chem. A., 101,* 1509–1517 (1997).

Reinhardt, T., and D. E. Ward, "Factors Affecting Methyl Chloride Emissions from Forest Biomass Combustion," *Environ. Sci. Technol., 29,* 825–832 (1995).

Richter, A., F. Wittrock, M. Eisinger, and J. P. Burrows, "GOME Observations of Tropospheric BrO in Northern Hemispheric Spring and Summer 1997," *Geophys. Res. Lett., 25,* 2683–2686 (1998).

Roberts, J. M., "The Atmospheric Chemistry of Organic Nitrates," *Atmos. Environ., 24A,* 243–287 (1990).

Roberts, J. M., and S. B. Bertman, "The Thermal Decomposition of Peroxyacetic Nitric Anhydride (PAN) and Peroxymethacrylic Nitric Anhydride (MPN)," *Int. J. Chem. Kinet., 24,* 297–307 (1992).

Roberts, J. M., J. Williams, K. Baumann, M. P. Buhr, P. D. Goldan, J. Holloway, G. Hübler, W. C. Kuster, S. A. McKeen, T. B. Ryerson, M. Trainer, E. J. Williams, F. C. Fehsenfeld, S. B. Bertman, G. Nouaime, C. Seaver, G. Grodzinsky, M. Rodgers, and V. L. Young, "Measurements of PAN, PPN, and MPAN Made during the 1994 and 1995 Nashville Intensives of the Southern Oxidant Study: Implications for Regional Ozone Production from Biogenic Hydrocarbons," *J. Geophys. Res., 103,* 22473–22490 (1998).

Röckmann, T., C. A. M. Brenninkmeijer, P. Neeb, and P. J. Crutzen, "Ozonolysis of Nonmethane Hydrocarbons as a Source of the Observed Mass Independent Oxygen Isotope Enrichment in Tropospheric CO," *J. Geophys. Res., 103,* 1463–1470 (1998).

Rogers, C. F., J. G. Hudson, B. Zielinska, R. L. Tanner, J. Hallett, and J. G. Watson, "Cloud Condensation Nuclei from Biomass Burning," in *Global Biomass Burning: Atmospheric, Climatic, and Biospheric Implications* (J. S. Levine, Ed.), Chap. 53, pp. 431–438, MIT Press, Cambridge, MA, 1991.

Roselle, S. J., T. E. Pierce, and K. L. Schere, "The Sensitivity of Regional Ozone Modeling to Biogenic Hydrocarbons," *J. Geophys. Res., 96,* 7371–7394 (1991).

Rudich, Y., R. K. Talukdar, J. B. Burkholder, and A. R. Ravishankara, "Reaction of Methylbutene with Hydroxyl Radical: Mechanism and Atmospheric Implications," *J. Phys. Chem., 99,* 12188–12194 (1995).

Rudich, Y., R. K. Talukdar, R. W. Fox, and A. R. Ravishankara, "Rate Coefficients for Reactions of NO_3 with a Few Olefins and Oxygenated Olefins," *J. Phys. Chem., 100,* 5374–5381 (1996).

Rudolph, J., and F. J. Johnen, "Measurements of Light Atmospheric Hydrocarbons over the Atlantic in Regions of Low Biological Activity," *J. Geophys. Res., 95,* 20583–20591 (1990).

Rudolph, J., A. Khedim, and B. Bonsang, "Light Hydrocarbons in the Tropospheric Boundary Layer over Tropical Africa," *J. Geophys. Res., 97,* 6181–6186 (1992).

Rudolph, J., R. Koppmann, and Ch. Plass-Dülmer, "The Budgets of Ethane and Tetrachloroethene—Is There Evidence for an Impact of Reactions with Chlorine Atoms in the Troposphere?" *Atmos. Environ., 30,* 1887–1894 (1996).

Rudolph, J., "Biogenic Sources of Atmospheric Alkenes and Acetylene," in *Biogenic Volatile Organic Carbon Compounds in the Atmosphere* (G. Helas, J. Slanina, and R. Steinbrecher, Eds.), pp. 53–65, SPB Academic Publishing, Amsterdam, 1997.

Saltzman, B. E., "Kinetic Studies of Formation of Atmospheric Oxidants," *Ind. Eng. Chem., 50,* 677–682 (1958).

Sanadze, G. A., and A. N. Kalandadze, "Light and Temperature Curves of the Evolution of C_5H_8," *Fiz. Rast, Moscow, 13,* 411–413 (1966).

Sander, W., "Carbonyl Oxides: Zwitterions or Diradicals?" *Angew. Chem., Int. Ed. Engl., 29,* 344–354 (1990).

Sander, R., and P. Crutzen, "Model Study Indicating Halogen Activation and Ozone Destruction in Polluted Air Masses Transported to the Sea," *J. Geophys. Res., 101,* 9121–9138 (1996).

Sandholm, S. T., J. D. Bradshaw, G. Chen, H. B. Singh, R. W. Talbot, G. L. Gregory, D. R. Blake, G. W. Sachse, E. V. Browell, J. D. W. Barrick, M. A. Shipham, A. S. Bachmeier, and D. Owen, "Summertime Tropospheric Observations Related to N_xO_y Distributions and Partitioning over Alaska: Arctic Boundary Layer Expedition 3A," *J. Geophys. Res., 97,* 16481–16509 (1992).

Sauer, F., C. Schäfer, P. Neeb, O. Horie, and G. K. Moortgat, "Formation of Hydrogen Peroxide in the Ozonolysis of Isoprene and Simple Alkenes under Humid Conditions," *Atmos. Environ., 33,* 229–241 (1999).

Savarino, J., and M. Legrand, "High Northern Latitude Forest Fires and Vegetation Emissions over the Last Millenium Inferred from the Chemistry of a Central Greenland Ice Core," *J. Geophys. Res.*, 103, 8267–8279 (1998).

Schade, G. W., and P. J. Crutzen, "Emission of Aliphatic Amines from Animal Husbandry and Their Reactions: Potential Source of N_2O and HCN," *J. Atmos. Chem.*, 22, 319–346 (1995).

Schäfer, C., O. Horie, J. N. Crowley, and G. K. Moortgat, "Is the Hydroxyl Radical Formed in the Gas-Phase Ozonolysis of Alkenes?" *Geophys. Res. Lett.*, 24, 1611–1614 (1997).

Schlager, H., P. Konopka, P. Schulte, U. Schumann, H. Ziereis, F. Arnold, M. Klemm, D. E. Hagen, P. D. Whitefield, and J. Ovarlez, "*In Situ* Observations of Air Traffic Emission Signatures in the North Atlantic Flight Corridor," *J. Geophys. Res.*, 102, 10739–10750 (1997).

Schmidt, V., G. Y. Zhu, K. H. Becker, and E. H. Fink, "Study of OH Reactions at High Pressures by Excimer Laser Photolysis–Dye Laser Fluoresence," *Ber. Bunsenges. Phys. Chem.*, 89, 321–322 (1985).

Schneider, M., O. Luxenhofer, A. Deissler, and K. Ballschmiter, "C_1–C_{15} Alkyl Nitrates, Benzyl Nitrate, and Bifunctional Nitrates: Measurements in California and South Atlantic Air and Global Comparison Using C_2Cl_4 and $CHBr_3$ as Marker Molecules," *Environ. Sci. Technol.*, 32, 3055–3062 (1998).

Schulte, P., H. Schlager, H. Ziereis, U. Schumann, S. L. Baughcum, and F. Deidewig, "NO_x Emission Indices of Subsonic Long-Range Jet Aircraft at Cruise Altitude: *In Situ* Measurements and Predictions," *J. Geophys. Res.*, 102, 21431–21442 (1997).

Schulten, H.-R., and U. Schurath, "Analysis of Aerosols from the Ozonolysis of 1-Butene by High-Resolution Field Desorption Mass Spectrometry," *J. Phys. Chem.*, 79, 51–57 (1975).

Schween, J. H., R. Dlugi, C. N. Hewitt, and P. Foster, "Determination and Accuracy of VOC-Fluxes above the Pine/Oak Forest at Castelporziano," *Atmos. Environ.*, 31, 199–215 (1997).

Seefeld, S., and J. A. Kerr, "Kinetics of the Reactions of Propionylperoxy Radicals with NO and NO_2: Peroxypropionyl Nitrate Formation under Laboratory Conditions Related to the Troposphere," *Environ. Sci. Technol.*, 31, 2949–2953 (1997).

Seefeld, S., D. J. Kinnison, and J. Alistair Kerr, "Relative Rate Study of the Reactions of Acetylperoxy Radicals with NO and NO_2: Peroxyacetyl Nitrate Formation under Laboratory Conditions Related to the Troposphere," *J. Phys. Chem. A*, 101, 55–59 (1997).

Sehested, J., L. K. Christensen, T. Møgelberg, O. J. Nielsen, T. J. Wallington, A. Guschin, J. J. Orlando, and G. S. Tyndall, "Absolute and Relative Rate Constants for the Reactions $CH_3C(O)O_2$ + NO and $CH_3C(O)O_2$ + NO_2 and Thermal Stability of $CH_3C(O)O_2NO_2$," *J. Phys. Chem. A*, 102, 1779–1789 (1998a).

Sehested, J., L. K. Christensen, O. J. Nielsen, M. Bilde, T. J. Wallington, W. F. Schneider, J. J. Orlando, and G. S. Tyndall, "Atmospheric Chemistry of Acetone: Kinetic Study of the $CH_3C(O)CH_2O_2$ + NO/NO_2 Reactions and Decomposition of $CH_3C(O)CH_2O_2NO_2$," *Int. J. Chem. Kinet.*, 30, 475–489 (1998b).

Seuwen, R., and P. Warneck, "Oxidation of Toluene in NO_x Free Air: Product Distribution and Mechanism," *Int. J. Chem. Kinet.*, 28, 315–332 (1996).

Sharkey, T. D., and E. L. Singsaas, "Why Plants Emit Isoprene," *Nature*, 374, 769 (1995).

Sharkey, T. D., "Isoprene Synthesis by Plants and Animals," *Endeavor*, 20, 74–78 (1996).

Shaw, G., "Aerosol Chemical Components in Alaska Air Masses. 2. Sea Salt and Marine Product," *J. Geophys. Res.*, 96, 22369–22372 (1991).

Shepson, P. B., A.-P. Sirju, J. F. Hopper, L. A. Barrie, V. Young, H. Niki, and H. Dryfhout, "Sources and Sinks of Carbonyl Compounds in the Arctic Ocean Boundary Layer: Polar Ice Floe Experiment," *J. Geophys. Res.*, 101, 21081–21089 (1996).

Shi, J. C., and M. J. Bernhard, "Kinetic Studies of Cl-Atom Reactions with Selected Aromatic Compounds Using the Photochemical Reaction–FTIR Spectroscopy Technique," *Int. J. Chem. Kinet.*, 29, 349–358 (1997).

Shu, Y., E. S. C. Kwok, E. C. Tuazon, R. Atkinson, and J. Arey, "Products of the Gas-Phase Reactions of Linalool with OH Radicals, NO_3 Radicals, and O_3," *Environ. Sci. Technol.*, 31, 896–904 (1997).

Siese, M., and C. Zetzsch, "Addition of OH to Acetylene and Consecutive Reactions of the Adduct with O_2," *Z. Phys. Chem.*, 188, 75–89 (1995).

Silver, G. M., and R. Fall, "Characterization of Aspen Isoprene Synthase, an Enzyme Responsible for Leaf Isoprene Emission to the Atmosphere," *J. Biol. Chem.*, 270, 13010–13016 (1995).

Simonaitis, R., K. J. Olszyna, and J. F. Meagher, "Production of Hydrogen Peroxide and Organic Peroxides in the Gas Phase Reactions of Ozone with Natural Alkenes," *Geophys. Res. Lett.*, 18, 9–12 (1991).

Singh, H. B., L. J. Salas, B. A. Ridley, J. D. Shetter, N. M. Donahue, F. C. Fehsenfeld, D. W. Fahey, D. D. Parish, E. J. Williams, S. C. Liu, G. Hubler, and P. C. Murphy, "Relationship between Peroxyacetyl Nitrate and Nitrogen Oxides in the Clean Troposphere," *Nature*, 318, 347–349 (1985).

Singh, H. B., D. Herlth, D. O'Hara, K. Zahnle, J. D. Bradshaw, S. T. Sandholm, R. Talbot, P. J. Crutzen, and M. Kanakidou, "Relationship of Peroxyacetyl Nitrate to Active and Total Odd Nitrogen at Northern High Latitudes: Influence of Reservoir Species on NO_x and O_3," *J. Geophys. Res.*, 97, 16523–16530 (1992).

Singh, H. B., M. Kanakidou, P. J. Crutzen, and D. J. Jacob, "High Concentrations and Photochemical Fate of Oxygenated Hydrocarbons in the Global Troposphere," *Nature*, 378, 50–54 (1995).

Singh, H. B., G. L. Gregory, A. Anderson, E. Browell, G. W. Sachse, D. D. Davis, J. Crawford, J. D. Bradshaw, R. Talbot, D. R. Blake, D. Thornton, R. Newell, and J. Merrill, "Low Ozone in the Marine Boundary Layer of the Tropical Pacific Ocean: Photochemical Loss, Chlorine Atoms, and Entrainment," *J. Geophys. Res.*, 101, 1907–1917 (1996a).

Singh, H. B., A. N. Thakur, Y. E. Chen, and M. Kanakidou, "Tetrachloroethylene as an Indication of Low Cl Atom Concentrations in the Troposphere," *Geophys. Res. Lett.*, 23, 1529–1532 (1996b).

Singleton, D. L., G. Paraskevopoulos, R. S. Irwin, G. S. Jolly, and D. J. McKenney, "Rate and Mechanism of the Reaction of Hydroxyl Radicals with Formic and Deuterated Formic Acids," *J. Am. Chem. Soc.*, 110, 7786–7790 (1988).

Singleton, D. L., G. Paraskevopoulos, and R. S. Irwin, "Rates and Mechanism of the Reactions of Hydroxyl Radicals with Acetic, Deuterated Acetic, and Propionic Acids in the Gas Phase," *J. Am. Chem. Soc.*, 111, 5248–5251 (1989).

Skov, H., J. Hjorth, C. Lohse, N. R. Jensen, and G. Restelli, "Products and Mechanisms of the Reactions of the Nitrate Radical (NO_3) with Isoprene, 1,3-Butadiene and 2,3-Dimethyl-1,3-butadiene in Air," *Atmos. Environ.*, 26A, 2771–2783 (1992).

Skov, H., Th. Benter, R. N. Schindler, J. Hjorth, and G. Restelli, "Epoxide Formation in the Reactions of the Nitrate Radical with 2,3-Dimethyl-2-butene, *cis*- and *trans*-2-Butene, and Isoprene," *Atmos. Environ.*, 28, 1583–1592 (1994).

Smith, A. M., E. Rigler, E. S. C. Kwok, and R. Atkinson, "Kinetics and Products of the Gas-Phase Reactions of 6-Methyl-5-hepten-2-one and *trans*-Cinnamaldehyde with OH and NO_3 Radicals and O_3 at 296 ± 2K," *Environ. Sci. Technol.*, 30, 1781–1785 (1996).

Solberg, S., N. Schmidtbauer, A. Semb, F. Stordal, and O. Hov, "Boundary-Layer Ozone Depletion As Seen in the Norwegian Arctic in Spring," *J. Atmos. Chem., 23*, 301–332 (1996).

Spicer, C. E., E. G. Chapman, B. J. Finlayson-Pitts, R. A. Plastridge, J. M. Hubbe, J. D. Fast, and C. M. Berkowitz, "Unexpectedly High Concentrations of Molecular Chlorine in Coastal Air," *Nature, 394*, 353–356 (1998).

Starn, T. K., P. B. Shepson, S. B. Bertman, J. S. White, B. G. Splawn, D. D. Riemer, R. G. Zika, and K. Olszyna, "Observations of Isoprene Chemistry and Its Role in Ozone Production at a Semirural Site during the 1995 Southern Oxidants Study," *J. Geophys. Res., 103*, 22425–22435 (1998a).

Starn, T. K., P. B. Shepson, S. B. Bertman, D. D. Riemer, R. G. Zika, and K. Olszyna, "Nighttime Isoprene Chemistry at an Urban-Impacted Forest Site," *J. Geophys. Res., 103*, 22437–22447 (1998b).

Staudt, M., N. Bertin, U. Hansen, G. Seufert, P. Ciccioli, P. Foster, B. Frenzel, and J.-L. Fugit, "Seasonal and Diurnal Patterns of Monoterpene Emissions from *Pinus pinea (L.)*, under Field Conditions," *Atmos. Environ., 31*, 145–156 (1997).

Steinbrecher, R., K. Hauff, R. Rabong, and J. Steinbrecher, "Isoprenoid Emission of Oak Species Typical for the Mediterranean Area: Source Strength and Controlling Variables," *Atmos. Environ., 31*, 79–88 (1997).

Stemmler, K., D. J. Kinnison, and J. Alistair Kerr, "Room Temperature Rate Coefficients for the Reactions of OH Radicals with Some Monomethylene Glycol Monoalkyl Ethers," *J. Phys. Chem., 100*, 2114–2116 (1996).

Stemmler, K., W. Mengon, D. J. Kinnison, and J. A. Kerr, "OH Radical-Initiated Oxidation of 2-Butoxyethanol under Laboratory Conditions Related to the Troposphere: Product Studies and Proposed Mechanism," *Environ. Sci. Technol., 31*, 1496–1504 (1997).

Stephens, E. R., W. E. Scott, P. L. Hanst, and R. C. Doerr, "Recent Developments in the Study of the Organic Chemistry of the Atmosphere," *J. Air Pollut. Contr. Assoc., 6*, 159–165 (1956).

Stephens, E. R., "Smog Studies of the 1950's," *EOS, 68*, 89, 91–93 (1987).

Stockwell, W. R., "On the $HO_2 + HO_2$ Reaction: Its Misapplication in Atmospheric Chemistry Models," *J. Geophys. Res., 100*, 11695–11698 (1995).

Stockwell, W. R., J. B. Milford, D. F. Gao, and Y. J. Yang, "The Effect of Acetyl Peroxy-Peroxy Radical Reactions on Peroxyacetyl Nitrate and Ozone Concentrations," *Atmos. Environ., 29*, 1591–1599 (1995).

Street, R. A., S. C. Duckham, and C. N. Hewitt, "Laboratory and Field Studies of Biogenic Volatile Organic Compound Emissions from Sitka Spruce (*Picea sitchensis* Bong.) in the United Kingdom," *J. Geophys. Res., 101*, 22799–22806 (1996).

Street, R. A., S. Owen, S. C. Duckham, C. Boissard, and C. N. Hewitt, "Effect of Habitat and Age on Variations in Volatile Organic Compound (VOC) Emissions from *Quercus ilex* and *Pinus pinea*," *Atmos. Environ., 31*, 89–100 (1997).

Sturges, W. T., R. C. Schnell, and G. S. Dutton, "Spring Measurements of Tropospheric Bromine at Barrow, Alaska," *Geophys. Res. Lett., 20*, 201–204 (1993).

Stutz, J., M. J. Ezell, A. A. Ezell, and B. J. Finlayson-Pitts, "Rate Constants and Kinetic Isotope Effects in the Reactions of Atomic Chlorine with *n*-Butane and Simple Alkenes at Room Temperature," *J. Phys. Chem., 102*, 8510–8519 (1998).

Suhre, K., J.-P. Cammas, P. Nédelec, R. Rosset, A. Marenco, and H. G. J. Smit, "Ozone-Rich Transients in the Upper Equatorial Atlantic Troposphere," *Nature, 388*, 661–663 (1997).

Sumner, A. L., and P. B. Shepson, "Snowpack Production of Formaldehyde and Its Effect on the Arctic Troposphere," *Nature, 398*, 230–233 (1999).

Talbot, R. W., J. E. Dibb, and M. B. Loomis, "Influence of Vertical Transport on Free Tropospheric Aerosols over the Central USA in Springtime," *Geophys. Res. Lett., 25*, 1367–1370 (1998).

Talukdar, R. K., J. B. Burkholder, A.-M. Schmoltner, J. M. Roberts, R. R. Wilson, and A. R. Ravishankara, "Investigation of the Loss Processes for Peroxyacetyl Nitrate in the Atmosphere: UV Photolysis and Reaction with OH," *J. Geophys. Res., 100*, 14163–14173 (1995).

Talukdar, R. K., S. C. Herndon, J. B. Burkholder, J. M. Roberts, and A. R. Ravishankara, "Atmospheric Fate of Several Alkyl Nitrates. Part 1: Rate Coefficients of the Reactions of Alkyl Nitrates with Isotopically Labelled Hydroxyl Radicals," *J. Chem. Soc., Faraday Trans., 93*, 2787–2796 (1997).

Tang, T., and J. C. McConnell, "Autocatalytic Release of Bromine from Arctic Snow Pack during Polar Sunrise," *Geophys. Res. Lett., 23*, 2633–2636 (1996).

Teton, S., A. Mellouki, G. Le Bras, and H. Sidebottom, "Rate Constants for Reactions of OH Radicals with a Series of Asymmetrical Ethers and *tert*-Butyl Alcohol," *Int. J. Chem. Kinet., 28*, 291–297 (1996).

Thomas, W., E. Hegels, S. Slijkhuis, R. Spurr, and K. Chance, "Detection of Biomass Burning Combustion Products in Southeast Asia from Backscatter Data Taken by the GOME Spectrometer," *Geophys. Res. Lett., 25*, 1317–1329 (1998).

Tingey, D. T., M. Manning, L. C. Grothaus, and W. F. Burns, "The Influence of Light and Temperature on Isoprene Emission Rates from Live Oak," *Physiol. Plant., 47*, 112–118 (1979).

Tingey, D. T., M. Manning, L. C. Grothaus, and W. F. Burns, "Influence of Light and Temperature on Monoterpene Emission Rates from Slash Pine," *Plant Physiol., 65*, 797–801 (1980).

Tolbert, M. A., J. Pfaff, I. Jayaweera, and M. J. Prather, "Uptake of Formaldehyde by Sulfuric Acid Solutions: Impact on Stratospheric Ozone," *J. Geophys. Res., 98*, 2957–2962 (1993).

Tuazon, E. C., A. M. Winer, R. A. Graham, J. P. Schmid, and J. N. Pitts, Jr., "Fourier Transform Infrared Detection of Nitramines in Irradiated Amine–NO_x Systems," *Environ. Sci. Technol., 12*, 954–958 (1978).

Tuazon, E. C., W. P. L. Carter, A. M. Winer, and J. N. Pitts, Jr., "Reactions of Hydrazines with Ozone under Simulated Atmospheric Conditions," *Environ. Sci. Technol., 15*, 823–828 (1981).

Tuazon, E. C., W. P. L. Carter, R. V. Brown, R. Atkinson, A. M. Winer, and J. N. Pitts, Jr., "Atmospheric Reaction Mechanisms of Amine Fuels," Report No. ESL-TR-82-17, U.S. Air Force Engineering and Services Center, Tyndall Air Force Base, FL, March 1982.

Tuazon, E. C., W. P. L. Carter, R. Atkinson, and J. N. Pitts, Jr., "The Gas-Phase Reaction of Hydrazine and Ozone: A Nonphotolytic Source of OH Radicals for Measurement of Relative OH Radical Rate Constants," *Int. J. Chem. Kinet., 15*, 619–629 (1983a).

Tuazon, E. C., W. P. L. Carter, R. V. Brown, A. M. Winer, and J. N. Pitts, Jr., "Gas-Phase Reaction of 1,1-Dimethylhydrazine with Nitrogen Dioxide," *J. Phys. Chem., 87*, 1600–1605 (1983b).

Tuazon, E. C., W. P. L. Carter, R. Atkinson, A. M. Winer, and J. N. Pitts, Jr., "Atmospheric Reactions of *N*-Nitrosodimethylamine and Dimethylnitramine," *Environ. Sci. Technol., 18*, 49–54 (1984).

Tuazon, E. C., and R. Atkinson, "A Product Study of the Gas-Phase Reaction of Isoprene with the OH Radical in the Presence of NO_x," *Int. J. Chem. Kinet., 22*, 1221–1236 (1990).

Tuazon, E. C., R. Atkinson, S. M. Aschmann, and J. Arey, "Kinetics and Products of the Gas-Phase Reactions of O_3 with Amines and Related Compounds," *Res. Chem. Intermed., 20*, 303–320 (1994).

Tuazon, E. C., S. M. Aschmann, J. Arey, and R. Atkinson, "Products of the Gas-Phase Reactions of a Series of Methyl-Substituted Ethenes with the OH Radical," *Environ. Sci. Technol., 32*, 2106–2112 (1998a).

Tuazon, E. C., S. M. Aschmann, R. Atkinson, and W. P. L. Carter, "The Reactions of Selected Acetates with the OH Radical in the

Presence of NO: Novel Rearrangement of Alkoxy Radicals of Structure RC(O)OCH(O)R'," *J. Phys. Chem. A, 102,* 2316–2321 (1998b).

Tuckermann, M., R. Ackermann, C. Gölz, H. Lorenzen-Schmidt, T. Senne, J. Stutz, B. Trost, W. Unold, and U. Platt, "DOAS-Observation of Halogen Radical-Catalysed Arctic Boundary Layer Ozone Destruction during the ARCTOC-Campaigns 1995 and 1996 in Ny-Ålesund, Spitsbergen," *Tellus, 49B,* 533–555 (1997).

Tyndall, G. S., J. J. Orlando, K. E. Nickerson, C. A. Cantrell, and J. G. Calvert, "An Upper Limit for the Rate Coefficient of the Reaction of NH_2 Radicals with O_2 Using FTIR Product Analysis," *J. Geophys. Res., 96,* 20761–20768 (1991).

Tyndall, G. S., J. J. Orlando, T. J. Wallington, and M. D. Hurley, "Pressure Dependence of the Rate Coefficients and Product Yields for the Reaction of CH_3CO Radicals with O_2," *Int. J. Chem. Kinet., 29,* 655–663 (1997).

Tyndall, G. S., T. J. Wallington, and J. C. Ball, "FTIR Product Study of the Reactions $CH_3O_2 + CH_3O_2$ and $CH_3O_2 + O_3$," *J. Phys. Chem. A, 102,* 2547–2554 (1998).

Van Neste, A., and R. A. Duce, "Methylamines in the Marine Atmosphere," *Geophys. Res. Lett., 14,* 711–714 (1987).

Villenave, E., and R. Lesclaux, "Kinetics of the Cross Reactions of CH_3O_2 and $C_2H_5O_2$ Radicals with Selected Peroxy Radicals," *J. Phys. Chem., 100,* 14372–14382 (1996).

Vogt, R., P. Crutzen, and R. Sander, "A Mechanism for Halogen Release from Sea-Salt Aerosol in the Remote Marine Boundary Layer," *Nature, 383,* 327–330 (1996).

Wagner, T., and U. Platt, "Satellite Mapping of Enhanced BrO Concentrations in the Troposphere," *Nature, 395,* 486–490 (1998).

Wallington, T. J., P. Dagaut, and M. J. Kurylo, "Correlation between Gas-Phase and Solution-Phase Reactivities of Hydroxyl Radicals Towards Saturated Organic Compounds," *J. Phys. Chem., 92,* 5024–5028 (1988).

Wallington, T. J., P. Dagaut, and M. J. Kurylo, "Ultraviolet Absorption Cross Sections and Reaction Kinetics and Mechanisms for Peroxy Radicals in the Gas Phase," *Chem. Rev., 92,* 667–710 (1992).

Wang, C., P. J. Crutzen, V. Ramanathan, and S. F. Williams, "The Role of a Deep Convective Storm over the Tropical Pacific Ocean in the Redistribution of Atmospheric Chemical Species," *J. Geophys. Res., 100,* 11509–11516 (1995).

Wang, T. X., M. D. Kelley, J. N. Cooper, R. C. Beckwith, and D. W. Margerum, "Equilibrium, Kinetic, and UV-Spectral Characteristics of Aqueous Bromine Chloride, Bromine and Chlorine Species," *Inorg. Chem., 33,* 5872–5878 (1994).

Wang, T., M. A. Carroll, G. M. Albercook, K. R. Owens, K. A. Duderstadt, A. N. Markevitch, D. D. Parrish, J. S. Holloway, F. C. Fehsenfeld, G. Forbes, and J. Ogren, "Ground-Based Measurements of NO_x and Total Reactive Oxidized Nitrogen (NO_y) at Sable Island, Nova Scotia, during the NARE 1993 Summer Intensive," *J. Geophys. Res., 101,* 28991–29004 (1996).

Wang, Y., J. A. Logan, and D. J. Jacob, "Global Simulation of Tropospheric O_3-NO_x-Hydrocarbon Chemistry. 2. Model Evaluation and Global Ozone Budget," *J. Geophys. Res., 103,* 10727–10755 (1998).

Wängberg, I., I. Barnes, and K. H. Becker, "Product and Mechanistic Study of the Reaction of NO_3 Radicals with α-Pinene," *Environ. Sci. Technol., 31,* 2130–2135 (1997).

Wayne, R. P., I. Barnes, P. Biggs, J. P. Burrows, C. E. Canosa-Mas, J. Hjorth, G. Le Bras, G. K. Moortgat, D. Perner, G. Poulet, G. Restelli, and H. Sidebottom, "The Nitrate Radical: Physics, Chemistry, and the Atmosphere," *Atmos. Environ., 25A,* 1–203 (1991).

Wells, J. R., F. L. Wiseman, D. C. Williams, J. S. Baxley, and D. F. Smith, "The Products of the Reaction of the Hydroxyl Radical with 2-Ethoxyethyl Acetate," *Int. J. Chem. Kinet., 28,* 475–480 (1996).

Wennberg, P. O., T. F. Hanisco, L. Jaeglé, D. J. Jacob, E. J. Hintsa, E. J. Lanzendorf, J. G. Anderson, R.-S. Gao, E. R. Keim, S. G. Donnelly, L. A. Del Negro, D. W. Fahey, S. A. McKeen, R. J. Salawitch, C. R. Webster, R. D. May, R. L. Herman, M. H. Proffitt, J. J. Margitan, E. L. Atlas, S. M. Schauffler, F. Flocke, C. T. McElroy, and T. P. Bui, "Hydrogen Radicals, Nitrogen Radicals, and the Production of O_3 in the Upper Troposphere," *Science, 279,* 49–53 (1998).

Wiesen, E., I. Barnes, and K. H. Becker, "Study of the OH-Initiated Degradation of the Aromatic Photooxidation Product 3,4-Dihydroxy-3-hexene-2,5-dione," *Environ. Sci. Technol., 29,* 1380–1386 (1995).

Wildermuth, M. C., and R. Fall, "Light-Dependent Isoprene Emission. Characterization of a Thylakoid-Bound Isoprene Synthase in *Salix discolor* Chloroplasts," *Plant Physiol., 112,* 171–182 (1996).

Williams, E. L., E. Grosjean, and D. Grosjean, "Ambient Levels of the Peroxyacyl Nitrates PAN, PPN, and MPAN in Atlanta, Georgia," *J. Air Waste Manage. Assoc., 43,* 873–879 (1993).

Williams, J., J. M. Roberts, F. C. Fehsenfeld, S. B. Bertman, M. P. Buhr, P. D. Goldan, G. Hübler, W. C. Kuster, T. B. Ryerson, M. Trainer, and V. Young, "Regional Ozone from Biogenic Hydrocarbons Deduced from Airborne Measurements of PAN, PPN, and MPAN," *Geophys. Res. Lett., 24,* 1099–1102 (1997).

Wine, P. H., R. J. Astalos, and R. L. Mauldin III, "Kinetic and Mechanistic Study of the OH + HCOOH Reaction," *J. Phys. Chem., 89,* 2620–2624 (1985).

Wingenter, O. W., M. K. Kubo, N. J. Blake, T. W. Smith, D. R. Blake, and F. S. Rowland, "Hydrocarbon and Halocarbon Measurements as Photochemical and Dynamical Indicators of Atmospheric Hydroxyl, Atomic Chlorine, and Vertical Mixing Obtained during Lagrangian Flights," *J. Geophys. Res., 101,* 4331–4340 (1996).

Wolff, S., A. Boddenberg, J. Thamm, W. V. Turner, and S. Gäb, "Gas-Phase Ozonolysis of Ethene in the Presence of Carbonyl-Oxide Scavengers," *Atmos. Environ., 31,* 2965–2969 (1997).

Worden, H., R. Beer, and C. P. Rinsland, "Airborne Infrared Spectroscopy of 1994 Western Wildfires," *J. Geophys. Res., 102,* 1287–1299 (1997).

Yokelson, R. J., D. W. T. Griffith, and D. E. Ward, "Open-Path Fourier Transform Infrared Studies of Large-Scale Laboratory Biomass Fires," *J. Geophys. Res., 101,* 21067–21080 (1996).

Yokelson, R. J., R. Susott, D. E. Ward, J. Reardon, and D. W. T. Griffith, "Emissions from Smoldering Combustion of Biomass Measured by Open-Path Fourier Transform Infrared Spectroscopy," *J. Geophys. Res., 102,* 18865–18877 (1997).

Yokouchi, Y., H. Bandow, and H. Akimoto, "Development of Automated Gas Chromatographic–Mass Spectrometric Analysis for Natural Volatile Organic Compounds in the Atmosphere," *J. Chromatogr., 642,* 401–407 (1993).

Yu, J., H. E. Jeffries, and K. G. Sexton, "Atmospheric Photooxidation of Alkylbenzenes. I. Carbonyl Product Analyses," *Atmos. Environ., 31,* 2261–2280 (1997).

Yu, J., and H. E. Jeffries, "Atmospheric Photooxidation of Alkylbenzenes. II. Evidence of Formation of Epoxide Intermediates," *Atmos. Environ., 31,* 2281–2287 (1997).

Yu, J., R. C. Flagan, and J. H. Seinfeld, "Identification of Products Containing –COOH, –OH, and –C=O in Atmospheric Oxidation of Hydrocarbons," *Environ. Sci. Technol., 32,* 2357–2370 (1998).

Zhu, L., and G. Johnston, "Kinetics and Products of the Reaction of the Vinoxy Radical with O_2," *J. Phys. Chem., 99,* 15114–15119 (1995).

Zimmermann, J., and D. Poppe, "A Supplement for the RADM2 Chemical Mechanism: The Photooxidation of Isoprene," *Atmos. Environ., 30,* 1255–1269 (1996).

CHAPTER 7

Chemistry of Inorganic Nitrogen Compounds

Oxides of nitrogen play a central role in essentially all facets of atmospheric chemistry. As we have seen, NO_2 is key to the formation of tropospheric ozone, contributing to acid deposition (some are toxic to humans and plants), and forming other atmospheric oxidants such as the nitrate radical. In addition, in the stratosphere their chemistry and that of halogens interact closely to control the chain length of ozone-destroying reactions.

As discussed in Chapter 2.A.1, most of the primary emissions of NO_x ($= NO + NO_2$) are in the form of nitric oxide, NO. The overall oxidation sequence is conversion of NO to NO_2, which is ultimately converted to HNO_3 and other oxidized forms such as PAN. Even in the case of PAN, the end product is ultimately HNO_3 since PAN can decompose back to NO_2 (see Chapter 6.I). While there has been speculation that there are processes that can convert HNO_3 back into reactive forms, which could be important in both the troposphere and stratosphere (e.g., see Chatfield, 1994; Hauglustaine *et al.*, 1996; and Lary *et al.*, 1997), none have been confirmed to date to occur in the atmosphere.

Figure 7.1, for example, shows the concentrations of the major nitrogen-containing products as a function of reaction time during a typical smog chamber experiment. Curve I is the concentration of $NO + NO_2$ expected if no reaction occurred and the concentrations decreased only due to dilution during the experiment. Curve III is the sum of the measured gas-phase concentrations of ($NO + NO_2 + PAN + HNO_3$). The difference between the measured concentrations and those expected from the concentrations of the initial reactants increases significantly during the run. However, in separate studies, the rate of loss of HNO_3 to the chamber walls was determined. Using this rate, Spicer (1983) calculated the amount of HNO_3 expected to be adsorbed on the chamber walls; this is shown by the shaded area between curves II and III. When this adsorbed HNO_3 is taken into account, about 90% of the nitrogen can be accounted for at the end of the run.

The ratio of PAN to HNO_3 in such experiments depends on a number of factors, especially the initial VOC/NO_x ratio. Figure 7.2 shows the ratio of the final concentrations of PAN to HNO_3 (sum of gaseous and adsorbed) as a function of the initial VOC/NO_x ratio. The increase in PAN relative to HNO_3 is due to increasing concentrations of the $CH_3C(O)OO$ radicals that form PAN as the VOC concentrations increase. This makes the $CH_3C(O)OO + NO_2$ reaction more competitive with the $OH + NO_2$ reaction, the major HNO_3 source in this system. At VOC/NO_x ratios of ~5-10 typical of urban areas, smog chamber experiments suggest that HNO_3 should exceed PAN by factors of ~2-5. However, the actual ratios depend on the particular conditions, e.g., chemical composition, temperature, and light. As discussed in Chapter 6, PAN can under some conditions (e.g., the Arctic) constitute the major portion of NO_y. (Recall NO_y is defined as $NO_x + HNO_3 + 2N_2O_5 + NO_3$ + organic nitrates + particulate nitrate + ··· .)

Nitric acid undergoes both wet and dry deposition rapidly and can be neutralized by ammonia, the major gaseous base found in the atmosphere. As discussed in Section E.2, the neutralization reaction is an equilibrium reaction so that by itself, this does not result in permanent removal from the atmosphere. However, as seen in this chapter and in Chapter 9, this acid-base reaction has some important implications for visibility in the atmosphere and for the nitrate concentrations found in respirable particles.

While nitric acid is one of the major contributors to acid deposition (more colloquially "acid rain"), we treat its chemistry separately from that of sulfuric and organic acids discussed in the following chapter. The reason for treating it first is that the chemistry of

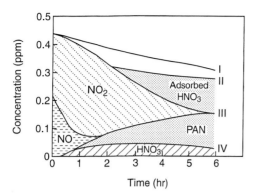

FIGURE 7.1 Cumulative plot of oxidized nitrogen compounds for a typical smog chamber run. Initial concentrations were 4.55 ppmC VOC, 0.21 ppm NO, and 0.23 ppm NO_2 (from Spicer, 1983).

oxides of nitrogen is rich and varied, with its contribution to acid deposition being only one facet. Thus, as NO is converted ultimately to HNO_3, important nitrogenous intermediates are formed along the way. There is some fascinating chemistry (and toxicology) associated with oxides of nitrogen and, as we shall see, a number of important questions remain unresolved.

A. OXIDATION OF NO TO NO_2 AND THE LEIGHTON RELATIONSHIP

In the early 1950s, the major "ingredients" in photochemical air pollution had been identified by Haagen-Smit and co-workers as VOC and NO_x, and the photolysis of NO_2 had been identified by Blacet as the source of the high ozone levels (see Chapter 1.B.3). Initially, the atmospheric conversion of emitted NO to NO_2 was thought to be due to its reaction with O_2:

$$2NO + O_2 \rightarrow 2NO_2. \quad (1)$$

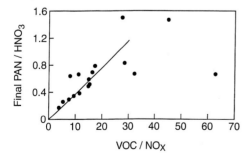

FIGURE 7.2 Ratio of final concentrations of PAN to HNO_3 (PAN/HNO_3) versus initial VOC/NO_x in a series of smog chamber experiments. The HNO_3 includes both that in the gas phase and that estimated to be adsorbed on chamber walls (from Spicer, 1983).

Indeed, this reaction can be easily demonstrated by mixing relatively high (i.e., ~Torr) concentrations of NO with air. The colorless NO is rapidly converted to the brown-orange NO_2, and one can feel the reaction vessel warm as the reaction exothermicity is released.

Despite the fact that reaction (1) is often cited, erroneously, as responsible for the NO to NO_2 conversion in the atmosphere, elementary reaction kinetics can be used to demonstrate that this cannot be the case. Even in a highly polluted atmosphere, the conversion of NO to NO_2 occurs over a period of several hours. Reaction (1) is kinetically second order in NO in both the gas and liquid phases (e.g., DeMore et al., 1997; Lewis and Deen, 1994). Following the conventions discussed in Chapter 5.A.1, the rate law for reaction (1) can be written as follows:

$$\frac{1}{2}\frac{d[NO_2]}{dt} = k_1[NO]^2[O_2].$$

As a result of the reaction being second order in NO, the rate of oxidation of NO to NO_2 decreases by a factor of 100 as the NO concentration falls by a factor of 10.

The recommended value of the third-order rate constant at room temperature is $k_1 = 2.0 \times 10^{-38}$ cm^6 $molecule^{-2}$ s^{-1} (Atkinson et al., 1997a). At 1 Torr NO, for example, the initial rate of oxidation in 1 atm air is about 40% per minute, whereas at 1 mTorr (1.3 ppm at 1 atm, 298 K), it is only 0.04% per minute (see also Problem 1). Thus, at a concentration of NO of even 0.1 ppm, found as a peak concentration in some polluted areas, the rate is too slow to be consistent with observed conversion to NO_2 on a time scale of hours.

However, the second-order nature of reaction (1) does provide a qualitative diagnostic for emissions of NO from certain power plants, smelters, etc. Occasionally, an orange-brown plume characteristic of NO_2 can be seen starting a short distance above the stack exit. In such a case, the concentration of the NO exiting the stack is sufficiently high that it is being rapidly oxidized to NO_2 by O_2 via reaction (1) (see Problem 2). At lower concentrations, oxidation of the NO by O_3 at the edges of the plume can also be important, giving oxidation rates as high as 20% per minute (e.g., see Cheng et al., 1986).

In most cases, however, such a plume is not visible because the NO concentrations are sufficiently small that reaction (1) is very slow. Once this was recognized in the 1950s, the puzzle was to identify the reactions responsible for the NO to NO_2 conversion. As discussed in Chapters 1.B and 6, it is now known that hydroperoxy and alkylperoxy free radicals are the oxidizing agents:

$$HO_2 + NO \rightarrow OH + NO_2, \quad (2)$$

$$RO_2 + NO \rightarrow RO + NO_2. \quad (3)$$

Sources of HO_2 and RO_2 are discussed in Chapter 6.

In a hypothetical atmosphere containing only NO, NO_2, and air, that is, no organics, the reactions controlling the concentrations of NO and NO_2 are (4), (5), and (6):

$$NO_2 + h\nu(\lambda \leq 420 \text{ nm}) \rightarrow NO + O(^3P), \quad (4)$$

$$O(^3P) + O_2 \xrightarrow{M} O_3, \quad (5)$$

$$O_3 + NO \rightarrow NO_2 + O_2. \quad (6)$$

The Leighton relationship, named after Philip Leighton, who wrote the first definitive monograph on air pollution in 1961, is given by

$$\frac{[O_3][NO]}{[NO_2]} = \frac{k_4}{k_6} \quad \text{or} \quad \frac{[NO_2]}{[NO]} = [O_3]\frac{k_6}{k_4}. \quad (A)$$

According to this relationship, also referred to frequently as the "photostationary state," the ratio of concentrations of O_3, NO, and NO_2 should be a constant given by the ratio of rate constants for photolysis of NO_2 and for the reaction of NO with O_3. Since k_4 [usually referred to as $k_p(NO_2)$] changes with the solar zenith angle, this ratio of concentrations is also expected to change during the day.

In general, the Leighton relationship is expected to hold when reactions (4) and (6) are the major loss processes for NO_2 and O_3 [reaction (5) is essentially always the loss process for $O(^3P)$]. Under such circumstances, the Leighton relationship can be used in computer models of tropospheric chemistry to minimize computation time. Thus, instead of carrying out numerical integration procedures separately to obtain $[O_3]$, [NO], and $[NO_2]$, if two of the three concentrations are known, one can obtain the third by using Eq. (A).

Figure 7.3 illustrates a test of Eq. (A) carried out by measuring NO, NO_2, and O_3 simultaneously (Ritter et al., 1979). Rearranging Eq. (A), one obtains

$$k_6[O_3][NO]/k_4[NO_2] = 1.0. \quad (B)$$

Thus, if O_3, NO, and NO_2 are measured simultaneously, with the values of k_6 and k_4 being well known, the left-hand side of Eq. (B) can be calculated and compared to the value of unity, which is expected if the Leighton relationship holds. Alternatively, the logarithm of the left-hand side should be zero. Figure 7.3 shows the logarithm of the left side of Eq. (B) as a function of time over a 2-h period midday in a rural

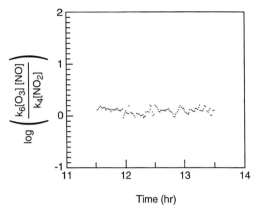

FIGURE 7.3 Test of photostationary state in rural Michigan on June 4, 1977, 11:30 to 13:30 hours (adapted from Ritter et al., 1979).

area in Michigan (U.S.) (Ritter et al., 1979). The values do indeed lie close to zero, although deviations from this are clearly seen. This is not unexpected, given the assumptions made in deriving Eqs. (A) and (B).

Deviations occur when the conversion of NO to NO_2 occurs by reactions other than that with O_3, i.e., by reactions (2) and (3) with HO_2 and RO_2 (e.g., Parrish et al., 1986; Carroll and Thompson, 1995); Eq. (A) then becomes

$$\frac{[NO_2]}{[NO]} = \frac{1}{k_4}\{k_6[O_3] + k_2[HO_2] + \sum k_3[RO_2]\}.$$

Deviations can also occur when loss processes for O_3 other than reaction (6) become significant so that O_3 is no longer in a steady state. These additional loss processes may include photolysis of O_3 as well as reactions with NO_2, alkenes, and the radicals HO_2 and OH. At sunset and sunrise, deviations are expected because the rate of photolysis of NO_2 is sufficiently small that steady-state assumptions are not valid (Calvert and Stockwell, 1983; Parrish et al., 1986). Finally, this ratio can be perturbed by fluxes of NO from the surface and micrometeorological effects (Carroll and Thompson, 1995).

B. OXIDATION OF NO_2

1. Daytime Gas-Phase Reaction with OH

NO_2 reacts readily with OH, forming nitric acid (shown as $HONO_2$ to emphasize the source, but as HNO_3 in most cases throughout this book):

$$NO_2 + OH \xrightarrow{M} HONO_2. \quad (7)$$

The reaction is termolecular, with recommended values (DeMore et al., 1997) of $k_0 = 2.5 \times 10^{-30}$ cm^6 molecule^{-2} s^{-1}, $n = 4.4$ for its temperature dependence, $k_\infty = 1.6 \times 10^{-11}$ cm^3 molecule^{-1} s^{-1} at 300 K, and $m = 1.7$ for its temperature dependence (see Chapter 5.A.2). These values of the low- and high-pressure limiting rate constants can be used as described in Chapter 5.A.2 to calculate an effective second-order rate constant at 300 K and 1 atm of 8.7×10^{-12} cm^3 molecule^{-1} s^{-1} (see Problem 4). At an OH concentration of 2×10^6 radicals cm^{-3}, this translates to a pseudo-first-order rate constant of 1.7×10^{-5} s^{-1}, and a lifetime of NO$_2$ with respect to this reaction of about 16 h.

The 1997 recommendations for the OH + NO$_2$ rate constants (DeMore et al., 1997; Atkinson et al., 1997a, 1997b) may be systematically high (e.g., Donahue et al., 1997) at temperatures below ~240 K. Thus, recent measurements at temperatures characteristic of the upper troposphere give rate constants that are smaller than the recommendations by ~10–30% (Brown et al., 1999a; Dransfield et al., 1999). In addition, O$_2$ appears to be only about 70% as efficient a third body as N$_2$ in the termolecular reaction. Using a modified form of the semiempirical equation for the rate constant in the falloff region (Chapter 5, Eq. (C)), which takes into account the variable collision efficiency β,

$$k = \left[\frac{k_0 \beta [M]}{1 + \frac{k_0 \beta [M]}{k_\infty}} \right] 0.6^{\{1 + [\log(k_0 \beta [M]/k_\infty)]^2\}^{-1}}.$$

Brown et al. (1999a) recommend $k_0^{300} = 2.47 \times 10^{-30}$ cm^6 molecule^{-2} s^{-1} with $n = 2.97$, $k_\infty^{300} = 1.45 \times 10^{-11}$ cm^3 molecule^{-1} s^{-1} with $m = 2.77$, and β(N$_2$) = 1.0, β(O$_2$) = 0.7, and β(air) = 0.94. The smaller values calculated for lower temperatures using this expression are important since the OH + NO$_2$ reaction converts NO$_x$ to NO$_y$; thus smaller values of this rate constant lead to increased model values of NO$_x$/NO$_y$ in the upper troposphere and lower stratosphere.

This reaction is primarily a daytime reaction because most OH sources are photolytic in nature. As a result, the NO$_2$ reaction with OH competes with NO$_2$ photolysis, reaction (4). As discussed in Chapter 3, a typical value of the photolysis rate constant for NO$_2$ would be $k_p = k_4 \simeq 7 \times 10^{-3}$ s^{-1} at a solar zenith angle of ~50° (e.g., see Fig. 3.31). Thus, the reaction with OH is not usually a dominant loss process for NO$_2$, but it is still sufficiently fast to form significant amounts of HNO$_3$ during the day, particularly in polluted regions with relatively large NO$_2$ concentrations.

2. Nighttime Reactions to Form NO$_3$ and N$_2$O$_5$

NO$_2$ also reacts with O$_3$, forming the nitrate radical, NO$_3$:

$$NO_2 + O_3 \rightarrow NO_3 + O_2. \quad (8)$$

The rate constant for this reaction at 298 K is relatively small, $k_8 = 3.2 \times 10^{-17}$ cm^3 molecule^{-1} s^{-1} (DeMore et al., 1997). However, at an O$_3$ concentration of 100 ppb, still near or below the air quality standards or guidelines of most countries (see Table 2.7), the lifetime of NO$_2$ with respect to this reaction is only 3.5 h.

It was not until the late 1970s that the importance of the nitrate radical was recognized when it was first reported by Noxon and co-workers (1978) in terms of its total column abundance, i.e., the concentration integrated through a column extending through the atmosphere from the earth's surface (see Chapter 11.A.4a). NO$_3$ was subsequently confirmed to be in the troposphere by Noxon et al. (1980) and by Platt and co-workers (1980, 1984) in polluted atmospheres and rural continental air.

As discussed in detail in Chapter 6, NO$_3$ is now recognized to be a major contributor to the chemistry of organics in the troposphere at night. Because it absorbs in the red region of the spectrum (Chapter 4.G), it photolyzes rapidly during the day so that its chemistry other than photolysis is essentially restricted to the dark hours.

In addition to reacting with organics, NO$_3$ also reacts with NO$_2$, forming dinitrogen pentoxide in a reversible, equilibrium reaction:

$$NO_3 + NO_2 \overset{M}{\leftrightarrow} N_2O_5. \quad (9, -9)$$

The forward reaction is a three-body reaction with values of $k_0 = 2.2 \times 10^{-30}$ cm^6 molecule^{-2} s^{-1} and $k_\infty = 1.5 \times 10^{-12}$ cm^3 molecule^{-1} s^{-1} at 300 K with $F_C = 0.6$ (see Chapter 5.A.2 for discussion of this factor) recommended by DeMore et al. (1997). Atkinson et al. (1997b) recommend $k_0 = 2.8 \times 10^{-30}$ cm^6 molecule^{-2} s^{-1} and $k_\infty = 2.0 \times 10^{-12}$ cm^3 molecule^{-1} s^{-1} with $F_C = 0.45$. The recommended value of the equilibrium constant $K_{9,-9}$ is 2.9×10^{-11} cm^3 molecule^{-1} at 298 K, with an uncertainty of $\pm 30\%$ (DeMore et al., 1997). This equilibrium constant has been the subject of numerous experimental studies, which have yielded results that ranged over a factor of two at room temperature. For example, a study by Wängberg et al. (1997) subsequent to the NASA and IUPAC recommendations reports a value of 2.34×10^{-11} cm^3 molecule^{-1}, about 20% smaller but within the relatively large uncertainty of the recommended value.

In any case, because of this equilibrium, sinks for N_2O_5 such as hydrolysis (see later) are, in essence, also sinks for NO_3 as well.

The NO_2–NO_3 reaction may also have a small contribution from a two-body channel:

$$NO_3 + NO_2 \rightarrow NO + NO_2 + O_2. \quad (10)$$

However, if it occurs, it appears to be minor. Thus, based on a review of the relevant studies reported in the literature, DeMore et al. (1997) suggest that $k_{10} = 4.5 \times 10^{-14} e^{-1260/T} = 6.6 \times 10^{-16}$ cm^3 molecule^{-1} s^{-1} at 298 K. This can be compared to an effective second-order rate constant for reaction (9) at 1 atm of 1.3×10^{-12} cm^3 molecule^{-1} s^{-1}. In short, the two-body reaction is more than three orders of magnitude slower than the termolecular process at 1 atm pressure.

3. Reactions of NO and NO_2 with Water and Alcohols

a. Uptake Into and Reaction with Liquid Water

It is known from studies carried out over many decades that oxides of nitrogen at high concentrations dissolve in aqueous solution and react to form species such as nitrate and nitrite. With the focus on acid deposition and the chemistry leading to the formation of nitric and sulfuric acids during the 1970s and 1980s, a great deal of research was carried out on these reactions at much lower concentrations relevant to atmospheric conditions (for reviews, see Schwartz and White, 1981, 1983; and Schwartz, 1984).

Through these studies, it was concluded that absorption of NO and NO_2 into the aqueous phase in the form of clouds and fogs in the atmosphere and their subsequent oxidation are not significant under typical atmospheric conditions. The major reasons for this are that NO and NO_2 are not highly soluble and, in addition, the reactions are kinetically rather slow due to the dependence of the rates on the square of the reactant concentration. As a result, like the oxidation of NO by O_2, the reactions slow down dramatically when the reactant concentrations are lowered to atmospheric levels. Let us take a brief look at these issues.

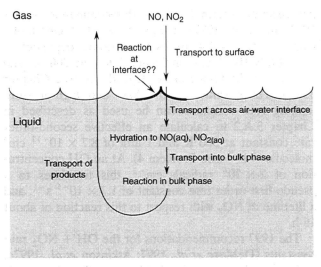

FIGURE 7.4 Processes involved in uptake of NO and NO_2 into the aqueous phase in fogs and clouds and subsequent oxidation.

Figure 7.4 illustrates the processes that must be taken into account when gaseous oxides of nitrogen interact with liquid water. The gas must first be transported to the liquid surface. It must then be taken up into the liquid and then diffuse away from the interface. Oxidation may occur in the bulk solution or, as the evidence increasingly suggests, at the interface itself.

Table 7.1 shows the major reactions of interest in this system. As discussed in Chapter 5, the Henry's law constant for a species X, H_X, is in effect the equilibrium constant for the gas–solution equilibrium:

$$X_{(g)} \leftrightarrow X_{(aq)}. \quad (11, -11)$$

That is,

$$H_X \text{ (mol L}^{-1} \text{ atm}^{-1}) = [X_{(aq)}]/P_X. \quad (C)$$

To put the values in Table 7.1 in perspective, highly soluble gases have Henry's law constants of the order of 10^5 and relatively insoluble gases have values of

TABLE 7.1 Some Rate and Equilibrium Constants of Aqueous-Phase Reactions of NO and NO_2[a]

Reaction	Rate or equilibrium expression	Value of rate or equilibrium constant
$NO_{(g)} \leftrightarrow NO_{(aq)}$	$H_{NO} = [NO]/p_{NO}$	1.93×10^{-3} M atm^{-1}
$NO_{2(g)} \leftrightarrow NO_{2(aq)}$	$H_{NO_2(aq)} = [NO_2]/p_{NO_2}$	1.0×10^{-2} M atm^{-1}
$2NO_{2(aq)} \leftrightarrow N_2O_{4(aq)}$	$K = [N_2O_4]/[NO_2]^2$	7×10^4 M^{-1}
$NO_{(aq)} + NO_{2(aq)} \leftrightarrow N_2O_{3(aq)}$	$K = [N_2O_3]/[NO][NO_2]$	3×10^4 M^{-1}
$2NO_{2(aq)} + H_2O_{(l)} \rightarrow 2H^+ + NO_2^- + NO_3^-$	$R_{12} = k_{12}[NO_2]^2$	7×10^7 M^{-1} s^{-1}
$NO_{(aq)} + NO_{2(aq)} + H_2O_{(l)} \rightarrow 2H^+ + 2NO_2^-$	$R_{13} = k_{13}[NO][NO_2]$	3×10^7 M^{-1} s^{-1}

[a] From Schwartz (1984).

$\sim 10^{-3}$ mol L^{-1} atm^{-1}. Thus NO and NO$_2$ are at the low end of solubilities of gases of atmospheric interest.

The two reactions that have been primarily considered are the following:

$$2NO_{2(g)} + H_2O_{(l)} \leftrightarrow 2H^+_{(aq)} + NO^-_{3(aq)} + NO^-_{2(aq)}, \quad (12, -12)$$

$$NO_{(g)} + NO_{2(g)} + H_2O_{(l)} \leftrightarrow 2H^+_{(aq)} + 2NO^-_{2(aq)}. \quad (13, -13)$$

Figure 7.5 shows a range for the rates of these reactions for NO, NO$_2$, and liquid water concentrations typical of the atmosphere. The rates drop off steeply with the concentrations of the gases involved, and even for high levels found in quite polluted areas, they are not sufficiently fast to contribute significantly to the aqueous-phase acidity.

In short, the uptake of NO and NO$_2$ into clouds and fogs, followed by their oxidation in the liquid phase, is not believed to contribute significantly to the formation of nitrate and acidity in fogs and clouds under most conditions. The major source of HNO$_3$ in fogs, clouds, and particles is thought to be oxidation of NO$_x$ to gaseous HNO$_3$, followed by its facile uptake into the condensed phase.

There are some intriguing observations, however, with regard to the possibility of reactions at the air–water interface that have unique kinetics and perhaps mechanisms (see also discussion in the following section). For example, Mertes and Wahner (1995) used a liquid jet apparatus (see Chapter 5.E.7) to study the uptake of NO$_2$ into a jet of water. The loss of NO$_2$ from the gas phase after passing over the jet was followed using differential optical absorption spectrometry (DOAS; see Chapter 11.A.1d); the formation of NO$_3^-$ and NO$_2^-$ in the aqueous phase was determined using a wet chemical technique. The uptake of NO$_2$ and formation of products were much faster than expected, and the apparent Henry's law constant for NO$_2$ was about an order of magnitude larger than that shown in Table 7.1. The reaction order in NO$_2$ was $\sim(1.4 \pm 0.3)$, based on a limited number of experiments and over less than an order of magnitude in concentration.

These observations could be reconciled with the extensive work on the interaction of NO$_2$ with bulk solutions if reaction (12) is much faster at the interface than in the bulk. The existence of such an enhanced reaction is also suggested by work using a falling-droplet apparatus (Ponche et al., 1993) and cloud and smog chambers discussed in the following section, where the reaction order in NO$_2$ was measured to be one, rather than two as in the bulk-phase reaction.

As discussed in Chapter 8, enhanced reactions of SO$_2$ at the interface have also been observed (Jayne et al., 1990). Surface second harmonic generation (SHG) experiments (Donaldson et al., 1995) subsequently identified a unique adsorbed SO$_2$ species at the air–water interface that may be involved in this enhanced reaction. Such SHG work on the uptake and reaction of NO$_2$ on water would clearly also be of value in understanding the kinetic anomalies. In addition, the use of sum frequency generation (SFG) spectroscopy, which in effect allows one to obtain the infrared spectrum of species present at interfaces, may shed some light on such reactions.

b. "Heterogeneous" Reaction of NO$_2$ with Water Vapor

Related to the uptake and reaction of NO$_2$ into liquid water and at the interface is a so-called heterogeneous "dark" reaction of gaseous NO$_2$ with water vapor to form nitrous acid, HONO. Potential formation processes and reactions of HONO in the atmosphere have been reviewed by Lammel and Cape (1996). This is a fascinating reaction in that, despite decades of research, the mechanism is still not understood. It occurs on a variety of surfaces, including water and acid surfaces (e.g., Kleffmann et al., 1998) and, as discussed in this chapter, on soot as well.

The reaction is usually assumed to be stoichiometrically the same as reaction (12) above:

$$2NO_2 + H_2O \xrightarrow{\text{surface}} HONO + HNO_3. \quad (14)$$

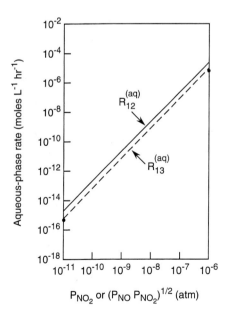

FIGURE 7.5 Rates of $2NO_{2(g)} + H_2O_{(l)} \rightarrow 2H^+ + NO_3^- + NO_2^-$ (R_{12}) and of $NO_{(g)} + NO_{2(g)} + H_2O_{(l)} \rightarrow 2H^+ + 2NO_2^-$ (R_{13}) as a function of partial pressure of NO$_2$ at 25°C, where the rates are in terms of the formation of aqueous-phase products in mol L^{-1} h^{-1} (adapted from Schwartz, 1984).

However, unlike the bulk liquid-phase reaction, there are many inconsistencies and uncertainties in the kinetics and mechanism. For example, the observed kinetics depend on the surface-to-volume ratio (S/V) of the reaction vessel, clearly indicating a heterogeneous reaction on the surface. While HONO is always produced, the formation of equivalent amounts of gaseous HNO_3 has not been observed. It is frequently argued that the nitric acid remains adsorbed to the walls of the reactor, a reasonable explanation given the very "sticky" nature of HNO_3 and its high mass accommodation coefficient and solubility in water. For example, Svensson et al. (1987) rinsed a piece of Teflon from the chamber after the reaction and analyzed for nitrate. They report that if an equivalent amount of nitrate was adsorbed on all of the reactor surfaces, reaction (14) would indeed represent the reaction and mass balance on the nitrogen would be achieved.

While the bulk reaction (12) is second order in NO_2, the heterogeneous reaction (14) is first order in NO_2 and first order in water vapor (Sakamaki et al., 1983; Pitts et al., 1984a; Svensson et al., 1987; Jenkin et al., 1988; Bambauer et al., 1994; Mertes and Wahner, 1995; Schwartz and Lee, 1995; Novakov, 1995; Kleffmann et al., 1998). The heterogeneous reaction is enhanced by light in the actinic region (Akimoto et al., 1987). Figure 7.6, for example, shows the enhancement in the rate constant as a function of the photolysis rate for NO_2 (k_{p,NO_2}), used as a means to measure the light intensity, at 4 and 20 ppm NO_2 in a smog chamber. Clearly, quite large enhancements, up to about an order of magnitude, are observed.

In investigating this reaction, Svensson et al. (1987) measured the uptake of water by a film of FEP Teflon using an electrobalance and concluded that there are already several monolayers of water on the surface at water vapor concentrations corresponding to 5% relative humidity. Thus even at what are very low water vapor concentrations in the atmosphere, there is sufficient surface-adsorbed water to provide both an aqueous phase and an interface for reaction.

Isotope studies also provide some intriguing hints at the mechanism. As seen in Fig. 7.7, when isotopically labeled $H_2^{18}O$ is used at concentrations at or above $\sim 5 \times 10^3$ ppm, $H^{18}ONO$ is formed initially (Sakamaki et al., 1983). Below this concentration, both $H^{18}ONO$ and $HON^{18}O$ are formed, suggesting that there may be more than one reaction path (Svensson et al., 1987).

It is possible that N_2O_4 may play a role in this formation of HONO from NO_2, at least at low temperatures. For example, Wang and Koel (1998) adsorbed NO_2 on ice at 86 K and carried out temperature-programmed desorption studies (TPD) in which species desorbing at various temperatures were identified using mass spectrometry. The desorption of HONO and HNO_3 was observed as the temperature increased to ~ 140 K. While N_2O_4 might be expected to be relatively more important at such low temperatures due to the shift in the $2NO_2 \leftrightarrow N_2O_4$ equilibrium, an enhancement of N_2O_4 relative to NO_2 has also been observed on a porous glass surface at room temperature (Barney and Finlayson-Pitts, unpublished data, 1999). Hence the possible role of N_2O_4 as an intermediate in HONO formation needs to be considered.

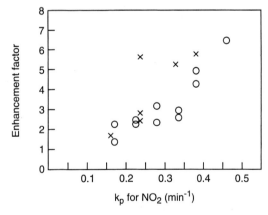

FIGURE 7.6 Enhancement in the rate of HONO formation from NO_2 and H_2O (○) and in the rate of CH_3ONO formation from CH_3OH and NO_2 (×) as a function of light intensity expressed as k_p for NO_2 (min^{-1}). No enhancement corresponds to a factor of 1.0 (adapted from Akimoto and Takagi, 1986, and Akimoto et al., 1987).

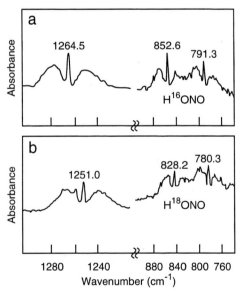

FIGURE 7.7 Infrared spectra of HONO formed in the reaction of 50 ppm NO_2 and 4700 ppm H_2O: (a) $H^{16}ONO$ from the reaction with $H_2^{16}O$; (b) $H^{18}ONO$ from the reaction with $H_2^{18}O$ (adapted from Sakamaki et al., 1983).

The production of NO has also been observed in this heterogeneous NO_2–H_2O reaction (Sakamaki et al., 1983; Pitts et al., 1984a; Svensson et al., 1987). In addition, recent studies show the formation of N_2O at longer times, both in the absence of SO_2 (e.g., Wiesen et al., 1995) and in its presence (e.g., Eriksson and Johansson, 1991; Pires et al., 1996; Pires and Rossi, 1995, 1997). While the mechanism of formation of N_2O is not clear, it is thought to involve secondary reactions of HONO (e.g., Kleffmann et al., 1994; see later). Indeed, this heterogeneous hydrolysis of NO_2 to HONO occurs in exhaust from combustion systems and is responsible for the artifact formation of N_2O reported in such samples (e.g., Muzio and Kramlich, 1988; Muzio et al., 1989).

A heterogeneous reaction of NO_2 has also been observed to occur with soot (e.g., Tabor et al., 1994; Chughtai et al., 1990, 1994; Kalberer et al., 1996; Rogaski et al., 1997; Gerecke et al., 1998; Ammann et al., 1998). NO has been observed as a major product in most studies, but large yields of HONO have also been reported by Gerecke et al. (1998) and Ammann et al. (1998). The reaction to form HONO appears to be quite fast, with initial uptake coefficients for NO_2 in the range of $\sim 10^{-1}$–10^{-4}, depending on the conditions such as reaction time, NO_2 concentration, and type of soot. The initial rate of HONO production was fast, decreasing at longer reaction times (Ammann et al., 1998; Gerecke et al., 1998).

The yields of HONO from the soot reaction have been observed to be greater than the 50% expected from the stoichiometry of reaction (14). For example, Gerecke et al. (1998) report HONO yields from 68 to 93%, depending on where in the flame the soot was collected and the fuel used to form the soot; this determines the nature of the soot surface, with the yield of HONO decreasing with distance from the flame base. As the HONO yield decreased, that of NO increased. In addition, Ammann et al. (1998) did not detect either gas-phase HNO_3 nor the amounts of particle-bound nitrogen that would be expected if the HNO_3 remained on the surface of the soot. All of these observations led these groups to propose an additional route to HONO formation involving the reduction of NO_2 by a surface group:

NO_2 + surface reduced site + H_2O →

HONO + surface oxidized site + OH^-

Gerecke et al. (1998) suggest that hydrogen at the surface may be involved in the reduction.

In short, the reduction of NO_2 on soot is quite fast and forms HONO in a process that may occur in parallel with reaction (14). This reaction may be particularly important in polluted urban areas as well as in the upper troposphere where soot from commercial aircraft is injected into the atmosphere.

There is also evidence from field studies for the generation of HONO, first detected unequivocally in the atmosphere by Perner and Platt (1979), at surfaces. Interpretation of such studies is complex due to the contributions of meteorology and the uncertainty in the nature of the available surfaces for reaction (14) (e.g., Lammel, 1996; Hjorth et al., 1996). However, in semirural areas, the surface appears to be a net source of HONO at concentrations of NO_2 above 10 ppb (Harrison and Kitto, 1994; Harrison et al., 1996).

HONO also undergoes deposition at surfaces in competition with its formation by the NO_2 heterogeneous reaction with water. For example, the mass accommodation coefficient for HONO on water has been reported to be in the range of 4×10^{-3} to ~ 0.15 over temperatures from 278 to 297 K (e.g., Kirchner et al., 1990; Bongartz et al., 1994; Mertes and Wahner, 1995). Thus aqueous particles and surfaces having adsorbed water can also act as a sink for gaseous HONO. This is consistent with the observations of Harrison et al. (1996) on the direction of HONO fluxes from the surface at various concentrations of NO_2; at NO_2 concentrations below 10 ppb in rural areas, surfaces were observed to be a net sink of HONO (e.g., see Harrison et al., 1996; and Harrison and Peak, 1997).

Notholt et al. (1992) and Andrés-Hernández et al. (1996) measured HONO, NO, NO_2, and aerosol surface areas at both urban and nonurban locations. They observed that at Ispra, Italy, HONO concentrations tended to correlate with NO_2, NO, and aerosol surface areas. Such studies support the formation of HONO from heterogeneous reactions of NO_2 at the surfaces of aerosol particles, fogs, buildings, and the ground.

Lammel and Perner (1988) measured gas-phase HONO using DOAS near Mainz (Germany) and, during the same period, collected aerosols using a low-pressure cascade impactor. Analysis of the aerosols for nitrite was carried out and the concentrations were found to be in great excess of that expected based on a Henry's law equilibrium. They suggested that the aerosols must be acting as a source of HONO. It is interesting that Clemens et al. (1997) also found that particles incubated in a reactor were a source of N_2O, perhaps from the reactions of HONO in or on the particles, as discussed earlier.

Another reaction that has been suggested as a source of HONO in the atmosphere is that of NO, NO_2, and water:

$$NO + NO_2 + H_2O \xrightleftharpoons{\text{surface}} 2HONO. \quad (15)$$

As for the NO_2 reaction (14), this reaction is slow if all reactants are in the gas phase, but a surface-catalyzed reaction is possible. Calvert and co-workers (1994) have reviewed possible mechanisms for HONO formation in the troposphere and, based on an analysis of ambient air data for NO, NO$_2$, H$_2$O, and particle concentrations, suggest that reaction (15) may be a good representation of a major HONO source. They hypothesized that the mechanism of the reaction may involve the initial formation of N$_2$O$_3$ from the reaction of NO with NO$_2$ and that N$_2$O$_3$ then reacts with water on surfaces to form HONO. However, a variety of laboratory studies on the formation of HONO from the reaction of NO$_2$ at surfaces find no effect of adding NO, indicating that reaction (15) is unlikely to be important (e.g., Pitts *et al.*, 1984a; Wiesen *et al.*, 1995; Gerecke *et al.*, 1998; Kleffmann *et al.*, 1998).

Understanding the kinetics and mechanisms of the production of HONO from NO$_2$ reaction is very important in that it appears to be a major source of HONO in smog chambers, inside homes, from automobiles (see Chapter 15), and perhaps in the troposphere at large. Clearly, more studies are warranted, especially those that can elucidate the nature and concentrations of species at the interface itself.

c. Reaction with Alcohols

A reaction of NO$_2$ similar to reaction (14) occurs with alcohols. The reaction with methanol, for example, produces methyl nitrite:

$$2NO_2 + CH_3OH \rightarrow CH_3ONO + HNO_3. \quad (16)$$

The *homogeneous* gas-phase reaction is slow, depending on the square of the NO$_2$ concentration and with a third-order rate constant at room temperature of $k_{16} = (5.7 \pm 0.6) \times 10^{-37}$ cm^6 molecule^{-2} s^{-1} (Niki *et al.*, 1982; Koda *et al.*, 1985).

However, on surfaces the reaction is much faster. For example, Takagi *et al.* (1986) studied this reaction in the presence of different surfaces, including quartz, Pyrex glass, stainless steel, and PFA (tetrafluoroethylene–perfluoroalkyl vinyl ether copolymer) coatings, and found the reaction rate decreased in going from stainless steel to Pyrex glass to a PFA coating to quartz. The CH$_3$OH–NO$_2$ reaction (16) was found to have a reaction order with respect to NO$_2$ of 1.0–1.5, depending on the nature of the surface. As shown in Fig. 7.6, like the reaction with H$_2$O, the reaction can also be photoenhanced (Akimoto and Takagi, 1986).

This particular reaction may be important when sampling exhaust from methanol-fueled combustion sources. For example, significant concentrations (46–69 ppm) of methyl nitrite have been reported in the exhaust of a methanol-fueled bus (Hanst and Stephens, 1989). Such large concentrations would be of concern if the use of methanol became widespread, since methyl nitrite acts as a free radical source via its photolysis and hence contributes to the NO to NO$_2$ conversion and ultimately to O$_3$ formation:

$$CH_3ONO + h\nu \rightarrow CH_3O + NO, \quad (17)$$

$$CH_3O + O_2 \rightarrow HCHO + HO_2, \quad (18)$$

$$HO_2 + NO \rightarrow OH + NO_2, \quad (19)$$

$$HCHO + h\nu \rightarrow H + CHO, \quad (20)$$

$$H + O_2 \xrightarrow{M} HO_2 \quad (21)$$

$$CHO + O_2 \rightarrow HO_2 + CO, \quad (22)$$

$$NO_2 + h\nu \rightarrow NO + O(^3P), \quad (4)$$

$$O(^3P) + O_2 \xrightarrow{M} O_3. \quad (5)$$

However, if one applies the heterogeneous kinetics reported by Takagi *et al.* (1986), one can calculate that methyl nitrite concentrations of the order measured in the exhaust sample could arise by reactions of unburned methanol and NO$_2$ on the walls of the sampling bag prior to analysis (Finlayson-Pitts *et al.*, 1992).

In short, while such heterogeneous reactions are not at all well understood, they may play very important roles in laboratory apparatus, in sampling systems, and perhaps in ambient air, where a variety of surfaces are available.

4. Other Reactions of NO$_2$

a. Organics

In principle, NO$_2$ can abstract a hydrogen atom from organics to form nitrous acid, HONO. For example, Pryor and Lightsey (1981) suggest that NO$_2$ at low concentrations in solution abstracts from the weak allylic C–H bond:

$$NO_2 + {>}C{=}\overset{|}{C}{-}\overset{|}{C}H_2 \longrightarrow {>}C{=}C{=}C{<}_H + HONO \quad (23)$$

There is also some evidence for a similar reaction of NO$_2$ with phospholipids adsorbed to a glass surface (Lai and Finlayson-Pitts, 1991). However, such reactions do not appear to be important in ambient air.

b. Sea Salt Particles and Mineral Oxides

Airborne sea salt particles are generated by wave action, which produces small droplets of seawater (Blanchard, 1985). As the droplets move inland, water can evaporate from the droplets, leaving a solid suspended particle containing the solids that were originally in the ocean water.

Hence, other reactions of NO_2 that could potentially occur in the atmosphere include the reaction with components of sea salt particles such as NaCl and NaBr (e.g., Robbins *et al.*, 1959; Cadle and Robbins, 1960; Schroeder and Urone, 1974; Chung *et al.*, 1978; Sverdrup and Kuhlman, 1980; Finlayson-Pitts, 1983; Zetzsch, 1987; Mamane and Gottlieb, 1990; Winkler *et al.*, 1991; Junkermann and Ibusuki, 1992; Vogt and Finlayson-Pitts, 1994; Karlsson and Ljungström, 1995; Vogt *et al.*, 1996; Peters and Ewing, 1996; De Haan *et al.*, 1999):

$$2NO_{2(g)} + NaCl_{(s)} \rightarrow ClNO_{(g)} + NaNO_{3(s)}, \quad (24)$$

$$2NO_{2(g)} + NaBr_{(s)} \rightarrow BrNO_{(g)} + NaNO_{3(s)}. \quad (25)$$

Both gaseous products, nitrosyl chloride and nitrosyl bromide, absorb light strongly in the visible (Chapter 4) and hence photolyze readily at dawn, producing atomic chlorine and bromine, respectively.

However, these reactions are also second order in NO_2 and appear to be too slow at atmospheric NO_2 levels to be important (Vogt and Finlayson-Pitts, 1994; Peters and Ewing, 1996). There is one aspect of the mechanism that is quite interesting, however, in that the mechanism appears to be at least in part a stepwise process involving the formation of a radical anion intermediate, $\{Cl \cdots NO_2\}^-$ in the solid, which has been identified by electron paramagnetic resonance (EPR) in both the NaCl and NaBr reactions at room temperature (Wan *et al.*, 1996). This intermediate appears to be remarkably stable and may be responsible for synergistic health effects observed when rats were exposed to a combination of NO_2 and NaCl aerosols (Last and Warren, 1987).

NO_2 has also been observed to be taken up on mineral oxides of types commonly found in particles in the atmosphere. For example, Miller and Grassian (1998) exposed powders of Al_2O_3 and TiO_2 to NO_2 in both the presence and the absence of water on the surface. At low NO_2 concentrations (e.g., 5 mTorr), only NO_2 which was chelated to the metal ion was observed using FTIR for both dry and hydrated oxides. While nitrate was observed at higher NO_2 concentrations, these are much larger than would be encountered in the atmosphere. Whether the chelated NO_2 on the surface can react at a significant rate with various atmospheric gases is not clear.

In short, while kinetics are very important in determining the importance of reactions in the atmosphere, other aspects such as the formation of reactive intermediates, and the physical and chemical nature of reaction surfaces, should also be taken into account.

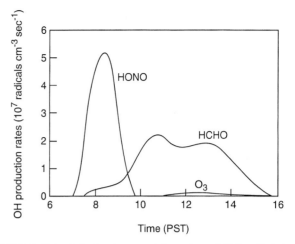

FIGURE 7.8 Calculated rates of formation of OH radical from photolysis of HONO, O_3, and HCHO at Long Beach, California, on December 10, 1987 (adapted from Winer and Biermann, 1994).

C. ATMOSPHERIC CHEMISTRY OF HONO

Nitrous acid, HONO, is of great interest because of its absorption of light in the actinic region and rapid photolysis to generate OH radicals (see Chapter 4.E). For example, on the basis of measurements of HONO, HCHO, and O_3 made in Long Beach, California, Winer and Biermann (1994) calculated OH radical formation rates due to photolysis of each of these species. Photolysis of HCHO gives H and CHO; these radicals both react with O_2 to give HO_2, which reacts further with NO to generate OH. Ozone photolysis gives OH through the generation of electronically excited $O(^1D)$ and its reaction with water. Figure 7.8 shows the calculated rates of OH formation from each precursor during the day. Clearly under these particular conditions, where the nighttime HONO concentration reached almost 15 ppb, HONO photolysis is by far the major source of OH in the early morning hours. In fact, Winer and Biermann calculate that in terms of the integrated OH production, HONO photolysis is comparable to that from HCHO and greatly exceeds that due to O_3 photolysis under these particular conditions.

In addition to its importance as an OH source and hence a driving force behind the chemistry at sunrise, there is concern about the formation of relatively high levels indoors, including inside automobiles, and the accompanying potential for health effects (see Chapter 15.B.2).

1. Formation of HONO

As already discussed, a major source of HONO is believed to be heterogeneous reactions of NO_2, including that with water adsorbed on various surfaces and with reactive sites on the surface of soot. During the day, it can also be formed by the reaction of OH with NO:

$$\text{OH} + \text{NO} \xrightarrow{M} \text{HONO}. \quad (26)$$

This termolecular reaction is in the falloff region between second and third order at 1 atm pressure and 298 K. It has a low-pressure limiting rate constant of $k_0 = 7.0 \times 10^{-31}$ cm^6 molecule^{-2} s^{-1} and a high-pressure limiting rate constant of $k_\infty = 3.6 \times 10^{-11}$ cm^3 molecule^{-1} s^{-1} at 300 K (DeMore et al., 1997). Since most OH sources require photolysis of some precursor, this reaction would be expected to be most important during the daylight hours. However, because HONO photolyzes so rapidly during the day, significant concentrations are not generated (see Problem 6).

Zhu and co-workers (1993) observed the formation of HONO in an environmental chamber during the decay of peroxynitric acid, HO$_2$NO$_2$, and suggested that HO$_2$NO$_2$, formed in the HO$_2$ + NO$_2$ reaction,

$$\text{HO}_2 + \text{NO}_2 \xrightarrow{M} \text{HO}_2\text{NO}_2, \quad (27)$$

reacted heterogeneously on the walls of the reactor to form HONO:

$$\text{HO}_2\text{NO}_2 \xrightarrow{\text{surface}} \text{HONO} + \text{O}_2. \quad (28)$$

Figure 7.9, for example, shows the decay of HO$_2$NO$_2$ and the formation of HONO and HNO$_3$ in their chamber. The peroxynitric acid was generated by reaction (27), where the HO$_2$ was formed by the bromine atom initiated oxidation of formaldehyde in air. Zhu et al.

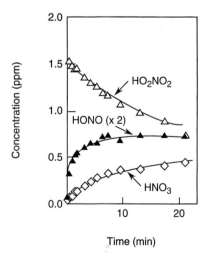

FIGURE 7.9 Concentration-time profiles of HO$_2$NO$_2$, HONO, and HNO$_3$ after 30-s irradiation of a mixture containing Br$_2$ (20 ppm), HCHO (3.9 ppm), and NO$_2$ (6.6 ppm) in 700 Torr of air (adapted from Zhu et al., 1993).

also considered as a potential HONO source a possible contribution from a minor channel in the homogeneous reaction of HO$_2$ with NO$_2$:

$$\text{HO}_2 + \text{NO}_2 \rightarrow \text{HONO} + \text{O}_2. \quad (29)$$

However, this reaction has not been observed to generate significant amounts of HONO, and only an upper limit for the rate constant has been derived. In a recent study, for example, Tyndall et al. (1995) found $k_{29} < 5 \times 10^{-16}$ cm^3 molecule^{-1} s^{-1} at room temperature, in agreement with earlier work of Graham et al. (1978). The observed generation of HONO in the studies of Zhu et al. is thus inconsistent with reaction (29). Further evidence for a heterogeneous source of the HONO in their system was the wide variation of the HONO yield from experiment to experiment over a period of time, consistent with a heterogeneous reaction in which the nature of the surfaces varied due to the introduction of a range of reactants and products.

It is noteworthy that measurements of OH, HO$_2$, NO, NO$_2$, and ClO in the lower stratosphere also suggest there is some as yet unrecognized source of HONO in that region. The measurements can be adequately modeled assuming that reaction (28) is the HONO source, with particles providing the surface (Salawitch et al., 1994).

In addition to what appears to be a heterogeneous chemical source for HONO, it has also been shown to be emitted directly from combustion systems. For example, it has been measured in the exhaust of non-catalyst-equipped automobiles (Pitts et al., 1984b), from natural gas combustion in a kitchen stove, and in the emissions from kerosene and propane space heaters (e.g., Pitts et al., 1985, 1989; Brauer et al., 1990; Febo and Perrino, 1991, 1995; Spicer et al., 1993; Vecera and Dasgupta, 1994).

In short, while HONO is believed to play an important role in atmospheric chemistry and perhaps be of concern from the point of view of health effects, much remains to be learned about its sources both indoors and outdoors.

2. Atmospheric Fates of HONO

As discussed earlier in this chapter and in Chapter 4, HONO photolyzes rapidly in the actinic region, generating OH and NO with unit quantum yield (DeMore et al., 1997):

$$\text{HONO} + h\nu \rightarrow \text{OH} + \text{NO}. \quad (30)$$

It can also react with OH,

$$\text{HONO} + \text{OH} \rightarrow \text{H}_2\text{O} + \text{NO}_2, \quad (31)$$

with a room temperature rate constant of 4.5×10^{-12} cm^3 molecule^{-1} s^{-1} (DeMore et al., 1997). However, given that the first-order photolysis rate constant is $\sim 1.3 \times 10^{-3}$ s^{-1} at a solar zenith angle of 40° (Calvert et al., 1994), even at a peak OH concentration of 1×10^7 radicals cm^{-3}, this reaction will be too slow to compete with the loss by photolysis.

There are some intriguing observations that suggest that HONO can undergo some as yet unrecognized reactions, at least in laboratory systems, and it seems likely in air as well. The self-reaction of gaseous HONO to form NO + NO$_2$ + H$_2$O (i.e., the reverse of reaction (15)) has been observed in laboratory systems (e.g., Ten Brink and Spoelstra, 1998) and treated theoretically (e.g., Mebel et al., 1998). This gaseous reaction is too slow to be important in the atmosphere. However, Kleffmann et al. (1994, 1998) have observed the formation of nitrous oxide, N$_2$O, during the decay of HONO in a laboratory system (Fig. 7.10), in a reaction that appears to occur on the reactor surface. While they suggest the overall reaction can be represented by (32),

$$4\text{HONO} \rightarrow \text{N}_2\text{O} + 2\text{HNO}_3 + \text{H}_2\text{O}, \quad (32)$$

this clearly cannot be an elementary reaction, and the mechanism remains unclear.

Nitrous acid/nitrite can also be oxidized in the aqueous solutions found in the atmosphere in the form of fogs, clouds, and particles. Nitrite is well known to be slowly oxidized in the dark to nitrate by dissolved oxygen in the liquid phase. However, it has been reported that the rate of this oxidation increases remarkably during freezing of the solution containing the nitrite (Takenaka et al., 1992, 1996). Figure 7.11, for example, shows the rate of nitrate formation in a nitrite solution at 25°C and in one with the cooling bath at -21°C (Takenaka et al., 1992). This unusual phenomenon has also been observed with respect to the oxidation of sulfur compounds in a cloud chamber (Finnegan et al., 1991; Finnegan and Pitter, 1991; and Chapter 8). Takenaka et al. (1996) propose that this acceleration is primarily due to concentration of nitrite and H$^+$ in solution during the freezing process, by a factor of 2.4×10^3 at -3°C, followed by the known oxidation by O$_2$. This is an area that clearly needs further investigation, as it has significant implications for the chemistry of freezing cloud and fog droplets in the atmosphere.

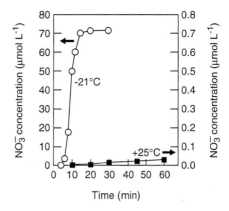

FIGURE 7.11 Formation of NO$_3^-$ in solutions of nitrite at pH 4.0 and either 25°C (right axis) or -21°C (left axis). Note the different scales for the rate of formation of nitrate. (Adapted from Takenaka et al., 1992.)

Free radical oxidation of nitrite by OH, for example, can also occur. This reaction (which will occur during daylight hours when photolytic sources of OH are present) and the chemistry of associated nitrogen oxides in solution have been studied by Løgager and Sehested (1993):

$$\text{OH} + \text{NO}_2^- \rightarrow \text{OH}^- + \text{NO}_2 \quad (33)$$

$$k_{33} = (6 \pm 1) \times 10^9 \text{ L mol}^{-1} \text{ s}^{-1}.$$

Uptake of HONO onto ice and H$_2$SO$_4$–H$_2$O solutions has been observed (e.g., Zhang et al., 1996; Fenter and Rossi, 1996). The uptake onto ice at temperatures from 180 to 200 K occurs with a mass accommodation coefficient (α) of $\sim 1 \times 10^{-3}$ and is reversible. Uptake onto sulfuric acid solutions depends on the concentration of the solution, with α varying from $\sim 10^{-4}$ at 55 wt% H$_2$SO$_4$ to $\sim 10^{-1}$ at 95 wt% H$_2$SO$_4$. Interestingly, in the presence of HCl, HONO undergoes a heterogeneous reaction on these surfaces to form gaseous ClNO. The reaction probability for HONO with frozen HCl solutions was measured to be ~ 0.1 (Fenter and Rossi, 1996). The reaction was slower on sulfuric acid solutions; Fenter and Rossi (1996) report that the reactive uptake is less than approxi-

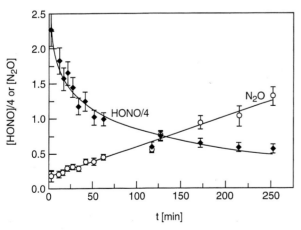

FIGURE 7.10 Observed formation of N$_2$O during the decay of HONO in a laboratory system (adapted from Kleffmann et al., 1994).

mately 10^{-3} even under optimized conditions of 60 wt% H_2SO_4, whereas Zhang et al. (1996) report values about an order of magnitude larger, in the range of 0.01–0.02, for 60–71 wt% H_2SO_4. Whether or not these heterogeneous reactions of HONO are important in the upper troposphere or stratosphere is not clear.

It is noteworthy that a similar reaction of HONO with HCl occurs, apparently, mainly in the gas phase (Wingen et al., 1999):

$$HONO + HCl \rightarrow ClNO + H_2O. \quad (34)$$

Although it is relatively slow ($k \sim 10^{-19}$ cm^3 molecule^{-1} s^{-1} at room temperature measured in a 560-L chamber), it has proven useful as a titration for HONO in order to measure absolute concentrations of this species.

As discussed in 7.B.3b above, deposition can also be a significant removal process for HONO.

D. REACTIONS OF NO_3 AND N_2O_5

1. Reactions of NO_3

Because of the rapid photolysis of NO_3 during the day (see Chapter 4.G), competing reactions of NO_3 are important primarily at night. Some of the most important dark reactions are discussed in the following sections. For detailed treatments of nitrate radical chemistry, the reader should consult the extensive reviews by Wayne et al. (1991) and Atkinson (1991).

a. Reactions with NO_2 and NO

One of the most important reactions of NO_3 is that with NO_2 to form N_2O_5, discussed earlier. This is the only known source of N_2O_5 in the atmosphere, and because NO_3 is only present at significant concentrations at night due to its rapid photolysis, the formation of N_2O_5 is restricted to the dark.

NO_3 also reacts quite rapidly with NO, however, and this appears to limit its lifetime in air under many circumstances. The reaction is fast, $k_{35} = 2.6 \times 10^{-11}$ cm^3 molecule^{-1} s^{-1} at 298 K (DeMore et al., 1997):

$$NO_3 + NO \rightarrow 2NO_2. \quad (35)$$

For example, with NO at 1 ppb, the lifetime of NO_3 with respect to this reaction can be calculated from $\tau^{NO_3} = 1/k_{35}[NO]$ to be only 2 s. As a result, NO and NO_3 do not coexist with both at high concentrations, as is the case for NO and O_3, and for the same reason.

b. Reactions with Organics

In Chapter 6, we saw that the reactions of organics with NO_3 are important in the troposphere. In general, NO_3 reactions are similar to those of the OH radical. That is, it abstracts from saturated hydrocarbons and aldehydes and adds to unsaturated hydrocarbons. As discussed in Chapter 6, the reactions with aromatic hydrocarbons are generally too slow to be important in the troposphere; the exceptions are particular compounds such as the cresols where the reaction is rapid.

Many of these reactions with organics are sufficiently fast to be an important sink of the organics at night, often rivaling in magnitude the loss by reaction with OH during the day (e.g., Smith et al., 1995). In addition, they provide a means of generating organic free radicals and subsequently OH at night when photolytic free radical sources are not available (e.g., Platt et al., 1990; Aliwell and Jones, 1998).

c. Thermal Decomposition

Several studies have suggested that NO_3 thermally decomposes at a sufficient rate to provide a significant sink for NO_3 in the troposphere (Cantrell et al., 1985; Johnston et al., 1986; Davidson et al., 1990):

$$NO_3 \overset{M}{\rightarrow} NO + O_2. \quad (36)$$

Johnston et al. (1986), for example, examined kinetic data from laboratory studies reported in the literature and estimated a first-order rate constant at 1 atm pressure at room temperature of $(3 \pm 2) \times 10^{-3}$ s^{-1}, corresponding to a lifetime for NO_3 with respect to decomposition of about 6 min. Subsequent attempts by Davidson et al. (1990) to measure this decomposition directly were complicated by a contribution from a wall-catalyzed decomposition of NO_3. However, they suggest that their data "support this conclusion weakly," with their data being "not inconsistent" with the rate constant suggested earlier by Johnston et al. (1986).

This suggested rapid thermal decomposition of NO_3 is somewhat controversial, however. Thus, a modeling study by Russell et al. (1986) carried out to examine nighttime chemistry found that such a rapid decomposition was inconsistent with atmospheric observations. Figure 7.12, for example, shows NO_3 radical concentrations measured in the California desert northeast of Los Angeles (Platt et al., 1984) and the predicted concentrations of Russell et al. (1986) first without taking into account the proposed thermal decomposition of NO_3 and then including it. Clearly, the case where the thermal decomposition is included in the calculations underpredicts the NO_3 by a large amount.

The activation barrier to such a decomposition based on recent studies also appears to be too high, 47.3 ± 0.8 kcal mol^{-1} (Davis et al., 1993), for this reaction to be fast.

Finally, chamber studies of the reactions of NO_2 and O_3 at various relative humidities from 8 to 70% were carried out by Mentel et al. (1996) and the con-

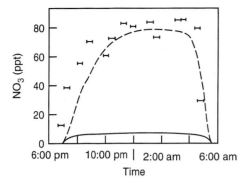

FIGURE 7.12 Comparison of measured (Platt *et al.*, 1984) and predicted nitrate radical concentrations at Edwards Air Force Base, California, May 23–24, 1982. The dashed line is the model prediction without the unimolecular decomposition of NO_3 and the solid line is that with the decomposition (adapted from Russell *et al.*, 1986).

centrations of NO_2, O_3, N_2O_5, HNO_3, and NO followed with time. Comparison of the experimental data to model predictions suggested that an upper bound for the rate constant for the NO_3 unimolecular decomposition is $k_{36} < 1.4 \times 10^{-4}$ s^{-1}.

d. Reaction with Water

It is difficult to separate out the uptake and/or reactions of NO_3 with water and those of N_2O_5. As discussed in the next section, there is an abundance of evidence that N_2O_5 is taken up by aqueous droplets and surfaces in both the troposphere and stratosphere and hydrolyzes to form HNO_3. However, it appears that NO_3 may also be taken up, and in this case, may act as a strong oxidant in solution (see, for example, Chameides, 1986a, 1986b; and Pedersen, 1995).

Figure 7.13, for example, shows the calculated lifetimes of NO_3 based on measurements made in a series of studies carried out at various locations in the Cali-

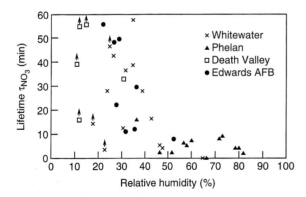

FIGURE 7.13 Calculated lifetime of NO_3 as a function of relative humidity at several locations in California. The points marked with an upward arrow indicate that these represent lower limits to the lifetimes (adapted from Platt *et al.*, 1984).

fornia desert (Platt *et al.*, 1984). The lifetimes were calculated based on the following reaction scheme:

$$NO_2 + O_3 \rightarrow NO_3 + O_2, \qquad (8)$$

$$NO_3 \xrightarrow{k_{loss}} \text{Loss of } NO_3.$$

Under conditions where the NO_3 concentration remained constant for a period of time, a steady state was assumed, i.e.,

$$d[NO_3]/dt = 0 = k_8[NO_2][O_3] - k_{loss}[NO_3],$$

from which the effective first-order rate constant for loss of NO_3, k_{loss}, could be calculated. The lifetime of NO_3 is then given by $\tau^{NO_3} = 1/k_{loss} = [NO_3]/k_8[NO_2][O_3]$. Since k_8 is known (DeMore *et al.*, 1997; Atkinson *et al.*, 1997a, 1997b), simultaneous measurements of NO_2, O_3, and NO_3 allow one to calculate τ^{NO_3}.

It is striking that the lifetimes fall to very small values as the relative humidity (RH) approaches 50%. At larger values of RH, not only is the gas-phase concentration of water larger, but condensation to form an aqueous liquid film on surfaces becomes more important. One cannot distinguish from data whether NO_3 is itself being taken up into a liquid film, whether N_2O_5 is being removed, or whether both processes are occurring.

Evidence for the uptake of NO_3 by aqueous solutions has been sought in both laboratory and field studies. A lower limit for the mass accommodation coefficient for NO_3 on liquid water of $> 2.5 \times 10^{-3}$ was reported by Thomas *et al.* (1989) and Mihelcic *et al.* (1993). Li *et al.* (1993) followed the formation of particulate nitrate in a rural area and, by comparing their measurements to model predictions, suggested that the mass accommodation coefficient for NO_3 on aqueous $(NH_4)_2SO_4$–NH_4HSO_4–H_2SO_4 aerosols is approximately unity, i.e.; NO_3 is taken up into the particle on every collision.

However, Rudich *et al.* (1996a, 1996b), Imamura *et al.* (1997), and Thomas *et al.* (1998) have measured the uptake of NO_3 on pure water as well as on aqueous solutions containing Cl^-, Br^-, I^-, NO_2^-, HSO_3^-, SO_3^{2-}, $HCOO^-$, CH_3COO^-, and OH^-. The uptake coefficient for pure water at 273 K was reported to be 2.0×10^{-4} (Rudich *et al.*, 1996a). Larger values were obtained with the salt solutions due to reactions of NO_3 in solution; for example, a lower limit of 2×10^{-3} was measured for the uptake on 0.1 M NaCl at 293 K (Thomas *et al.*, 1998), in good agreement with the value of 2.7×10^{-3} at 273 K reported by Rudich *et al.* (1996a).

In addition to the uptake of NO_3 from the gas phase, NO_3 can also be directly generated in the aqueous phase by reaction of nitrate ion with SO_4^-, the sulfate(VI) radical anion (Zellner and Herrmann, 1995):

$$NO_3^- + SO_4^- \rightarrow NO_3 + SO_4^{2-}. \quad (37)$$

The chemistry of NO_x–SO_x systems is discussed in more detail in the following chapter.

The hydrolysis of NO_3 has been proposed to generate OH:

$$NO_{3(aq)} + H_2O_{(l)} \rightarrow HNO_{3(aq)} + OH_{(aq)}. \quad (38)$$

As a result, uptake of NO_3 at night followed by its hydrolysis could provide a source of OH in these droplets at night, supplementing the photolytic sources during the day. The potential contribution of this reaction depends on its rate constant. For example, Rudich et al. (1998) suggest that if the rate constant at 298 K for reaction (38) is 6 L mol^{-1} s^{-1} as suggested by their earlier experiments (Rudich et al., 1996a), the generation of OH in such droplets in the remote troposphere at night by the NO_3 hydrolysis could be similar to that from OH uptake from the gas phase during the day. However, Thomas et al. (1998) report an upper limit for the rate constant for reaction (38) that is two orders of magnitude smaller. In this case, this reaction will not be important and removal via the hydrolysis of N_2O_5 in equilibrium with NO_3 becomes the major heterogeneous hydrolysis process.

At any rate, in addition to the hydrolysis reaction, there are a variety of additional oxidations in atmospheric aqueous solutions that can be carried out by NO_3. For example, NO_3 can oxidize S(IV) (see Chapter 8; Chameides, 1986a, 1986b; Huie et al., 1995; Rudich et al., 1998) as well as halogens in sea salt. In the case of S(IV), the initiation reactions are

$$NO_{3(aq)} + HSO_{3(aq)}^- \rightarrow NO_{3(aq)}^- + H_{(aq)}^+ + SO_{3(aq)}^-, \quad (39)$$

$k_{39} = 1.7 \times 10^9$ L mol^{-1} s^{-1} (Exner et al., 1992),

and

$$NO_{3(aq)} + SO_{3(aq)}^{2-} \rightarrow NO_{3(aq)}^- + SO_{3(aq)}^-, \quad (40)$$

$k_{40}(278\,K) = 3 \times 10^8$ L mol^{-1} s^{-1} (Exner et al., 1992).

The SO_3^- radical anion is then oxidized to sulfate, i.e., S(VI), in a series of subsequent steps. These are analogous to the OH-initiated oxidation of S(IV) in solution discussed in more detail in the following chapter.

In the case of chloride and bromide ions, the initiation reactions are

$$NO_{3(aq)} + Cl_{(aq)}^- \rightarrow NO_{3(aq)}^- + Cl_{(aq)}, \quad (41)$$

$k_{41} = 9.3 \times 10^6$ L mol^{-1} s^{-1} (Exner et al., 1992),

and

$$NO_{3(aq)} + Br_{(aq)}^- \rightarrow NO_{3(aq)}^- + Br_{(aq)}, \quad (42)$$

$k_{42} = 3.7 \times 10^8$ L mol^1 s^{-1} (Rudich et al., 1996a).

Again, the chemistry is analogous to the OH-initiated oxidation of chloride and bromide in solution. The subsequent aqueous-phase chemistry for the halogens is summarized in the following chapter.

Of course, in the atmosphere S(IV) and halogens will be present simultaneously and their chemistries become intertwined. In addition, other oxidants such as O_3, H_2O_2, and OH are present in the gas phase and can be taken up into solution. Subsequent photochemical reactions of O_3 and H_2O_2 generate a variety of free radicals, including OH and HO_2. This chemistry is summarized in the following chapter.

Figure 7.14 shows the calculated ratio of S(IV) oxidation with the uptake and reaction of NO_3 to that without the NO_3 contribution as a function of the chloride concentration in particles (Rudich et al., 1998). For reference, the saturation concentration of Cl^- in sea salt particles (i.e., at the deliquescence point) is ~6 M at room temperature. Under the assumptions of these particular calculations, the rate of aqueous-phase oxidation of S(IV) is estimated to increase by as much as 25% when NO_3 chemistry is taken into account. This uptake and reaction of NO_3 also decrease its gas-phase concentrations.

In addition to electron-transfer reactions such as (39)–(42), NO_3 can also react in solution with a variety of organics. For example, addition to the double bond in alkenes is quite fast, with rate constants of the order of 10^9 L mol^{-1} s^{-1} at room temperature; the reactions

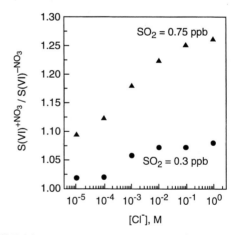

FIGURE 7.14 Model-estimated increase in S(IV) oxidation in the aqueous phase of sea salt particles due to the uptake and reactions of NO_3. O_3 taken as 40 ppb, NO_2 as 0.1 ppb, and H_2O_2 as 0.05 ppb. The Y-axis is the calculated ratio of oxidized sulfur, S(VI), formed in droplets when NO_3 chemistry is included to that when it is not (adapted from Rudich et al., 1998).

with alkylbenzenes are also fast ($k \sim 10^7$–10^8 L mol^{-1} s^{-1}) and likely proceed via electron transfer from the aromatic ring (Alfassi *et al.*, 1993; Huie, 1995).

NO$_3$ has also been observed to react with solid NaCl and KBr (Seisel *et al.*, 1997). The reactions are believed to first generate atomic chlorine and bromine, respectively, i.e.,

$$NO_{3(g)} + NaCl_{(s)} \rightarrow Cl + NaNO_{3(s)}, \quad (43)$$

$$NO_{3(g)} + NaBr_{(s)} \rightarrow Br + NaNO_{3(s)}. \quad (44)$$

The products observed were HCl and HBr + Br$_2$, respectively (Seisel *et al.*, 1997). These are likely due to the presence of small amounts of water on the salt surfaces (e.g., Beichert and Finlayson-Pitts, 1996). The reaction probabilities at room temperature were reported to be $\gamma_{43} = 4.9 \times 10^{-2}$ and $\gamma_{44} = 0.16$, respectively (Seisel *et al.*, 1997).

2. Reactions of N$_2$O$_5$

It is interesting that while N$_2$O$_5$ is believed to play a significant role in tropospheric chemistry, it has never been directly measured in the troposphere. However, using the measured concentrations of NO$_2$ and NO$_3$ and the equilibrium constant for the reaction in which it is formed,

$$NO_3 + NO_2 \underset{}{\overset{M}{\leftrightarrow}} N_2O_5. \quad (9, -9)$$

peak concentrations of N$_2$O$_5$ up to about 10–15 ppb in a polluted atmosphere have been calculated (Atkinson *et al.*, 1986). It is interesting that even at a rural site in the Baltic Sea region, however, the calculated mean N$_2$O$_5$ concentration has been estimated to be ~1 ppb, based on measurements of NO$_3$ and NO$_2$ (Heintz *et al.*, 1996). The major loss process in the troposphere appears to be hydrolysis, although other reactions may contribute to a small extent.

a. Hydrolysis

Gaseous N$_2$O$_5$ reacts with water to form HNO$_3$ both in the gas phase and on surfaces. The reaction of N$_2$O$_5$ in the gas phase is sufficiently slow in most studies that only an upper limit to the rate constant of $\sim 10^{-21}$ cm^3 molecule^{-1} s^{-1} has been determined (e.g., see Atkinson *et al.*, 1986; Hjorth *et al.*, 1987; and Sverdrup *et al.*, 1987). However, Mentel *et al.* (1996) and Wahner *et al.* (1998a) have used a large (250 m^3) chamber to study both the gas-phase and surface hydrolysis of N$_2$O$_5$ and report that the gas-phase reaction has both bimolecular and termolecular components:

$$N_2O_{5(g)} + H_2O_{(g)} \rightarrow 2HNO_{3(g)}, \quad (45)$$

$$N_2O_{5(g)} + 2H_2O_{(g)} \rightarrow 2HNO_{3(g)} + H_2O_{(g)}, \quad (46)$$

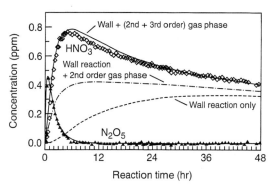

FIGURE 7.15 Measured loss of gaseous N$_2$O$_5$ in the presence of 8100 ppm H$_2$O and formation of HNO$_3$ as a function of reaction time in a large chamber (250 m^3) and model-predicted HNO$_3$ for the heterogeneous wall hydrolysis of N$_2$O$_5$ for the combinations of wall loss plus bimolecular gas phase (N$_2$O$_5$ + H$_2$O) reaction and wall loss plus bimolecular and termolecular (N$_2$O$_5$ + 2H$_2$O) reaction (adapted from Wahner *et al.*, 1998a).

where $k_{45} = 2.5 \times 10^{-22}$ cm^3 molecule^{-1} s^{-1} and $k_{46} = 1.8 \times 10^{-39}$ cm^6 molecule^{-2} s^{-1}. Figure 7.15, for example, shows the loss of gaseous N$_2$O$_5$ and formation of HNO$_3$ as a function of time compared to model predictions using three different assumptions. The lowest curve assumed that HNO$_3$ was only formed from the wall reaction of N$_2$O$_5$. The next curve assumed that in addition, the bimolecular, gas-phase reaction (45) occurred. The third curve, which matches the HNO$_3$ data quite well, assumes that in addition to the wall and bimolecular reactions, there is also a termolecular reaction (46). The implications of a possible termolecular contribution whose rate is proportional to [H$_2$O]^2 are significant; for example, Wahner *et al.* (1998a) suggest that at 50% RH and 291 K, the lifetime of N$_2$O$_5$ with respect to hydrolysis decreases to about 1.5 h from 4.5 h when the termolecular reaction is included.

N$_2$O$_5$ hydrolysis is well known to be catalyzed by surfaces; i.e., the reaction occurs rapidly on the aqueous films found on many surfaces:

$$N_2O_{5(g)} + H_2O_{(l)} \rightarrow 2HNO_{3(aq)}. \quad (47)$$

This is believed to be a major source of atmospheric HNO$_3$. For example, Ljungström and Hallquist (1996) report that the calculated rates of NO$_3$ formation and the measured rates of wet deposition of nitrate at a site in Sweden were very similar; they intepreted this as being due to the formation of N$_2$O$_5$, followed by its hydrolysis (although the direct uptake and reaction of NO$_3$ presumably could have contributed as well).

There have been many studies of reaction (47) using sulfuric acid or sulfate aerosols (e.g., ammonium sulfate) of various compositions and over a range of temperatures. The reaction probability (γ_{47}) for (47) falls in the range of 0.05–0.16 over H_2SO_4 concentrations from 39 to 96% (w:w) and from 213 to 298 K (e.g., see Lovejoy and Hanson, 1995; and DeMore et al., 1997). Hu and Abbatt (1997) have shown that γ_{47} for hydrolysis on both sulfuric acid and ammonium sulfate particles decreases from ~0.05 to 0.02 as the relative humidity falls. However, Wahner et al. (1998b) have reported much smaller values on $NaNO_3$ aerosols at low RH, $\gamma_{47} = 0.0018$ at 48% RH and 0.0032 at 62% RH, which they attributed to increased importance of the reverse reaction of NO_2^+ with NO_3^- to regenerate N_2O_5.

The uptake and hydrolysis of N_2O_5 on pure liquid water appears to be about the same as for sulfuric acid–water mixtures, ranging from about 0.01 to 0.06 over the temperature range from 262 to 293 K (e.g., Van Doren et al., 1990; Kirchner et al., 1990; George et al., 1994). It is also noteworthy that similar reactions of N_2O_5 occur with anionic and cationic water clusters (Wincel et al., 1994, 1995).

Dentener and Crutzen (1993) have modeled the impact on a global scale of the uptake and hydrolysis of N_2O_5 and NO_3 into aerosols in the troposphere. Both species were assumed to be taken up and hydrolyzed with equal reaction probabilities of 0.1. Figure 7.16 shows the fraction of HNO_3 (f_{HNO_3}) calculated to be formed in the month of January by the combination of reactions (47) and (48),

$$N_2O_{5(g)} + H_2O_{(l)} \to 2HNO_{3(aq)}, \quad (47)$$

$$NO_{3(g)} + H_2O_{(l)} \to \to HNO_{3(aq)}, \quad (48)$$

compared to the sum of its formation by the sum of reactions [(7) + (47) + (48)]. In much of the Northern Hemisphere, more than half, and as much as 90%, of the nitric acid formed is predicted to occur via these "heterogeneous" pathways.

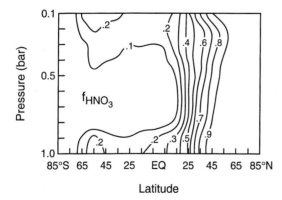

FIGURE 7.16 Model-predicted fraction of HNO_3 (f_{HNO_3}) produced by heterogeneous N_2O_5 and NO_3 hydrolysis in January (adapted from Dentener and Crutzen, 1993).

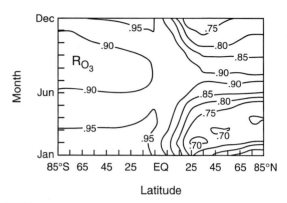

FIGURE 7.17 Model-predicted ratio, R_{O_3}, of O_3 concentrations with aerosol reactions included those without for all months (adapted from Dentener and Crutzen, 1993).

Furthermore, because these reactions result in the effective removal of NO_x from ozone production, by removing NO_2, the model also predicts that O_3 concentrations will decrease. Figure 7.17, for example, shows the model results for the ratio of O_3 (R_{O_3}) with the heterogeneous removal of NO_3 and N_2O_5 included to that without these aerosol reactions. In some locations, the O_3 concentrations are predicted to be as much as 30% lower than they would have been in the absence of the heterogeneous reactions. Because O_3 is also the major OH source on a global scale, via its photolysis to electronically excited oxygen atoms, $O(^1D)$, which react in part with gas-phase water, this also decreases the predicted OH levels.

In urban airsheds, the hydrolysis of N_2O_5 and NO_3 can be even more important as a source of HNO_3 due to their higher concentrations in more polluted regions.

The implications of reaction (47) in the stratosphere are discussed in detail in Chapter 12. However, suffice it to say that it also plays a critical role in that region of the atmosphere as well.

b. Other Reactions

While the hydrolysis of N_2O_5 is believed to represent its major loss process, there are other possibilities that have potentially interesting implications under certain conditions. For example, N_2O_5 reacts with the components of sea salt particles such as NaCl, NaBr, and NaI to form nitryl chloride, nitryl bromide, and nitryl iodide, respectively (e.g., Finlayson-Pitts et al., 1989a, 1989b; Behnke and Zetzsch, 1990; Zetzsch and Behnke, 1992; Junkermann and Ibusuki, 1992; George et al., 1994; Behnke et al., 1994, 1997; Leu et al., 1995; Fenter et al., 1996; Barnes et al., 1991; Schweitzer et al., 1998):

$$NaCl_{(s,aq)} + N_2O_{5(g)} \to ClNO_{2(g)} + NaNO_{3(s)}, \quad (49)$$

$$NaBr_{(s,aq)} + N_2O_{5(g)} \to BrNO_{2(g)} + NaNO_{3(s)}, \quad (50)$$

$$NaI_{(s,aq)} + N_2O_{5(g)} \to INO_{2(g)} + NaNO_{3(s)}. \quad (51)$$

Indeed, reactions (50) and (51) represented the first syntheses of gaseous BrNO$_2$ (Finlayson-Pitts *et al.*, 1989b) and INO$_2$, respectively (Barnes *et al.*, 1991). The importance of reactions (49)–(51) lies in the fact that the gaseous nitryl halides produced all absorb light in the actinic region, giving reactive halogen atoms (see Chapter 4.U). While the relative importance of these reactions as a source of halogen atoms is uncertain, they clearly could be important in the marine boundary layer as well as some distance inland from coastal marine areas. In addition, Michelangeli *et al.* (1991) have suggested they may occur in the stratosphere after the eruption of alkaline volcanoes such as El Chichon. Finally, some unique (hopefully!) situations such as the oil well burning in Kuwait produced plumes containing large concentrations of salt in which such reactions can potentially occur (e.g., Cahill *et al.*, 1992). The reader is referred to reviews by Finlayson-Pitts (1993), Graedel and Keene (1995), and De Haan *et al.* (1999) for further discussion of this issue.

It should be noted that these reactions are analogous to the reactions of N$_2$O$_5$ with HCl on ice discussed in Chapter 12.

A potential reaction of N$_2$O$_5$ that, however, has not been reported to date in atmospherically relevant systems is the addition of N$_2$O$_5$ across the double bond of alkenes:

$$N_2O_5 + >\!C\!=\!C\!< \longrightarrow -\underset{NO_2}{\overset{|}{C}}-\underset{ONO_2}{\overset{|}{C}}-$$

This reaction has been observed at much higher concentrations in laboratory studies (Stevens and Emmons, 1957; Lai and Finlayson-Pitts, 1991), and there is no reason *a priori* that it could not occur at night in the troposphere, particularly in polluted areas where relatively high concentrations of N$_2$O$_5$ can be formed (Atkinson *et al.*, 1986). Clearly, this is an area for further study.

E. ATMOSPHERIC CHEMISTRY OF HNO$_3$

1. Formation

The formation of HNO$_3$ has been discussed earlier and includes the following reactions:

$$NO_2 + OH \overset{M}{\rightarrow} HONO_2 \qquad (7)$$

$$N_2O_{5(g)} + H_2O_{(g, l)} \rightarrow 2HNO_{3(g, aq)} \qquad (47)$$

$$NO_{3(aq)} + H_2O_{(l)} \rightarrow HNO_{3(aq)} + OH_{(aq)} \qquad (38)$$

$$NO_3 + RH \rightarrow HNO_{3(g)} + R. \qquad (52)$$

In reaction (52), the organic can be any species with abstractable hydrogen atoms, including alkanes, aldehydes, etc.

2. Tropospheric Fates

Nitric acid is a "sticky" molecule and readily adsorbs to surfaces, particularly if there is water on the surface, which is essentially always the case in the troposphere. Because of this, it undergoes dry and wet deposition rapidly, with deposition velocities in the range 1–5 cm s^{-1}, at the highest end of the range of those measured for various gases in the troposphere (see Chapter 2.C for a discussion of deposition). In addition, theoretical predictions indicate that it should form a complex in the gas phase with water, with about 1% of the nitric acid being complexed at the surface (Tao *et al.*, 1996); such complexes have been observed in the laboratory using microwave spectroscopy (Canagaratna *et al.*, 1998).

Because of this high deposition velocity, dry deposition of HNO$_3$ can be responsible for much of the removal of inorganic nitrogen from the troposphere. For example, Russell and co-workers (1993) applied an airshed model which included removal by dry deposition to the Los Angeles air basin and calculated that about 40% of the total flux of nitrogen to the surface was in the form of HNO$_3$ for the particular day that was modeled. NH$_3$ deposition was the next highest contributor, responsible for about 25% of the nitrogen deposition. On the other hand, Nielsen and co-workers (1996), using the measured concentrations and deposition velocities shown in Table 7.2, estimate that in an open land area near Copenhagen (Denmark), HNO$_3$ was only responsible for about 16% of the flux, NO$_2$ for 34%, and unidentified NO$_y$ compounds (see later) for about 26%.

As discussed in Chapter 4.D, the absorption cross sections of HNO$_3$ decrease rapidly in the actinic region and hence its photolysis is slow relative to other losses (see Problem 5). In addition, the reaction with OH is relatively slow at typical tropospheric temperatures:

$$OH + HNO_3 \rightarrow H_2O + NO_3. \qquad (53)$$

While the abstraction of a hydrogen atom is accepted as the major reaction path, a minor contribution from a channel producing H$_2$O$_2$ + NO$_2$ cannot be ruled out with certainty. The kinetics of reaction (53) are most interesting in that, although it appears reasonable that it is a direct hydrogen atom abstraction, numerous kinetic studies have reported a negative temperature dependence and a small pressure dependence (DeMore *et al.*, 1997). Such kinetic behavior is indicative of the formation of a complex between OH and HNO$_3$ (see

TABLE 7.2 Distribution of NO_y Compounds in an Open Land Area near Copenhagen and Their Estimated Contribution to Dry Deposition of Nitrogen[a]

Compound	Atmospheric concentration (μg N m^{-3}) (mean $\pm 2\sigma$)	Typical dry deposition velocities (v_d) (cm s^{-1})	Percentage of dry deposition due to each compound
NO	0.70 ± 0.17	0.15[b]	8
NO_2	2.3 ± 0.4	0.2[a]	35
HNO_3	0.17 ± 0.03	1.3[b]	16
Inorganic nitrate	0.44 ± 0.12	0.1[b]	3
PAN and PPN	0.32 ± 0.04	0.44[b]	12
Residual gas NO_y[d]	0.3 ± 0.2	1.0[c]	26
Total NO_y	4.3 ± 0.9		100

[a] From Nielsen *et al.* (1996).
[b] From Hertel *et al.* (1994).
[c] Deposition velocities assumed to be the following (units of cm s^{-1}): N_2O_5, 1.3; alkyl nitrates, 0.44; bifunctional organic nitrates, 0.44–1.30 cm s^{-1}.
[d] Unidentified NO_y compounds, assumed to consist of N_2O_5, alkyl nitrates, and bifunctional organic nitrates.

Chapter 5.A.2), suggesting that the complex then decomposes to $H_2O + NO_3$.

The kinetics of the reaction have been expressed as a combination of a pressure-independent term and a pressure-dependent term (DeMore *et al.*, 1997):

$$k(M,T) = k_0 + \left[k_3[M]/\{1 + k_3[M]/k_2\}\right]. \quad (D)$$

where $k_0 = 7.2 \times 10^{-15}(e^{785/T})$ cm^3 molecule^{-1} s^{-1}, $k_2 = 4.1 \times 10^{-16}(e^{1440/T})$ cm^3 molecule^{-1} s^{-1}, and $k_3 = 1.9 \times 10^{-33}(e^{725/T})$ cm^6 molecule^{-2} s^{-1}. Subsequent studies suggested a stronger pressure dependence than predicted by Eq. (D) at temperatures below ~250 K. Thus, Brown *et al.* (1999b) have fit data from several different studies, including their own, to the form

$$k = k_0 + \left(\frac{k_\Delta}{1 + \left(\frac{k_\Delta}{\beta k_c[M]}\right)}\right), \quad (E)$$

where k_0 is the low-pressure limiting rate constant for the bimolecular direct abstraction reaction, k_Δ is the difference between the low- and high-pressure limiting rate constants ($k_\Delta = k_\infty - k_0$), k_c is the termolecular rate constant for stabilization of the complex, and β is the relative efficiency for stabilization of the complex by various gases. The best-fit values of the kinetic parameters were $k_0 = 2.41 \times 10^{-14}(e^{460/T})$ cm^3 molecule^{-1} s^{-1}, $k_\Delta = 2.69 \times 10^{-17}(e^{2199/T})$ cm^3 molecule^{-1} s^{-1}, $k_c = 6.51 \times 10^{-34}(e^{1335/T})$ cm^6 molecule^{-2} s^{-1}, and $\beta = 1.0$ for N_2, O_2, and air (Brown *et al.*, 1999b).

In short, the OH + HNO_3 reaction likely proceeds in part via a direct abstraction and in part via complex formation, with decomposition of the complex to $H_2O + NO_3$. There is no experimental evidence for the existence of a minor channel producing $H_2O_2 + NO_2$, although a small contribution at lower temperatures cannot be ruled out.

The only gaseous base present in the atmosphere at significant concentrations is ammonia, NH_3, which reacts rapidly with HNO_3 to form ammonium nitrate:

$$NH_{3(g)} + HNO_{3(g)} \leftrightarrow NH_4NO_{3(s,aq)}. \quad (54, -54)$$

It is interesting that theoretical predictions (Nguyen *et al.*, 1997) suggest that in the gas phase, a nitric acid and ammonia molecule will interact by forming a strong hydrogen bond, rather than by proton transfer. Water solvent is needed for the latter, suggesting that the formation of solid NH_4NO_3 from the gases likely involves some water as well. Musin and Lin (1998) have also predicted the existence of two minor, nonionic reaction channels for the NH_3–HNO_3 reaction, one forming $H_2NNO_2 + H_2O$ and the other $H_2NONO + H_2O$.

The ammonium nitrate formed in reaction (54) can exist either as a solid particle or in solution, and since this reaction is an equilibrium, it can redissociate to form the reactants. The deliquescence point for NH_4NO_3 at 25°C is 62% RH; i.e., at a water vapor concentration corresponding to 62% RH, the solid particle dissolves to form a concentrated liquid solution.

Mozurkewich (1993) has treated this system in detail and recommends that the equilibrium constant for reactions (54, −54) when the ammonium nitrate is a solid can be calculated as a function of temperature from

$$\ln(K_{54,-54}) = 118.87 - 24084/T - 6.025 \ln(T), \quad (F)$$

where T is in Kelvin and $K_{54,-54}$ is in (nanobar)^2. Since 1 bar = 0.987 atm, this is within a few percent of the value expressed in units of (ppb)^2.

At water vapor concentrations above the deliquescence point, the equilibrium is that between the reactant gases and aqueous ammonium nitrate. As treated in detail by Mozurkewich, the equilibrium constant, $K^*_{54,-54}$, then depends on the solution concentrations or activities:

$$K^*_{54,-54} = P_{NH_3} P_{HNO_3}$$
$$= \{a_{NH_3}/H_{NH_3}\} \cdot \{a_{HNO_3}/H_{HNO_3}\}, \quad (G)$$

where the a parameters are the solution-phase activities for NH_3 and HNO_3, respectively, which clearly depend on the solution concentrations, and the H factors are the Henry's law constants for these gases. The activity coefficients needed to obtain the activities in solutions of varying concentrations can be obtained by using standard thermodynamic approaches outlined by Mozurkewich (1993).

Figure 7.18 gives the ratio $(K^*/K)_{54,-54}$ of the calculated equilibrium constants for solution-phase ammonium nitrate compared to the solid salt product at various temperatures and water activities. As the water activity, i.e., water vapor pressure above the solution, increases, the equilibrium constant falls. That is, at higher relative humidities, relatively less HNO_3 and NH_3 are found in the vapor phase at equilibrium. This may be why relatively more ammonium nitrate in particles collected on filters evaporates at lower RHs compared to higher ones.

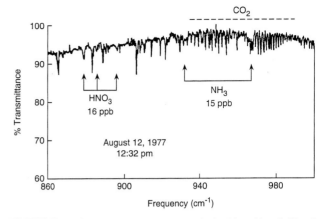

FIGURE 7.19 Infrared spectrum of air in Riverside, California, on August 12, 1977, showing the coexistence of the strong acid HNO_3 and the strong base NH_3 (graciously provided by E. C. Tuazon).

It is interesting that because of this equilibrium, the strong acid HNO_3 and the strong base NH_3 have been observed to coexist in air. Figure 7.19, for example, shows the infrared spectrum of air when both of these species could clearly be identified.

From such measurements, the product $(P_{HNO_3} P_{NH_3})$ measured in air has been calculated and compared to the calculated value based on the known equilibrium constants. Figure 7.20, for example, shows one comparison between the calculated and measured values in Rubidoux, California, about 85 km east of Los Angeles. The agreement between the two is quite good on the first day in particular (Hildemann et al., 1984). That the agreement is not perfect is not surprising, given that the particles undoubtedly contain a variety of other

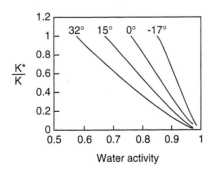

FIGURE 7.18 Ratio $(K^*/K)_{54,-54}$, where K^* and K are the equilibrium constants over the solution and over the solid at various temperatures (°C) (adapted from Mozurkewich, 1993).

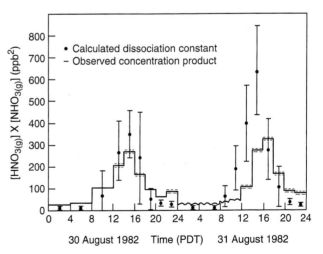

FIGURE 7.20 Experimentally measured concentration product $[HNO_3][NH_3]$ (in units of ppb^2) and calculated value of $K = [HNO_3][NH_3]$ for two days at Rubidoux, California (adapted from Hildemann et al., 1984).

species as well (see Chapter 9), and hence the thermodynamics should take these into account rather than treating it as a pure HNO_3–NH_3–NH_4NO_3 system.

The formation of ammonium nitrate has some interesting implications for visibility reduction. In the Los Angeles air basin, for example, the major NO_x sources are at the western, upwind end of the air basin. Approximately 40 mi east in the vicinity of the city of Chino, there is a large agricultural area that has significant emissions of ammonia. The geographical distribution of the estimated emissions of NO_x and NH_3 in 1982 are shown in Fig. 7.21. (It should be noted that vehicles with three-way catalysts appear to be significant sources of NH_3 as well, so that this distribution has likely changed with the introduction of such controls on vehicles; Fraser and Cass, 1998.) As a result, under typical meteorological conditions, air is carried inland during the day, with NO_x being oxidized to HNO_3 as the air mass moves downwind. When it reaches the agricultural area, the HNO_3 reacts with gaseous NH_3 to form ammonium nitrate. As discussed in Chapter 9, the particles formed by such gas-to-particle conversion processes are in a size range where they scatter light efficiently, giving the appearance of a very hazy or smoggy atmosphere even though other manifestations of smog such as ozone levels may not be highly elevated. A result of this reaction is that particle nitrate levels measured further downwind, e.g., at Rubidoux, tend to be elevated. Table 7.3, for example, shows the distribution of nitrogen for Rubidoux and at West Covina, a location upwind. The large increase in particulate nitrate after the air mass passes over the region of high ammonia emissions is clear (Spicer, 1982).

Similar "titrations" of gaseous HNO_3 with NH_3 to form new NH_4NO_4 particles and/or to enhance the growth of existing particles as the air mass passes over rural and agricultural areas have been observed in other locations as well, such as the Fraser River Valley east of Vancouver, in rural Ontario, Canada (e.g., Barthelmie and Pryor, 1998; Makar et al., 1998), and in Phoenix, Arizona (Watson et al., 1994).

Finally, in some areas with unique chemical compositions, HNO_3 may have the opportunity to react with other species as well. For example, it reacts relatively rapidly with NaCl, the major component of sea salt particles:

$$HNO_{3(g)} + NaCl_{(s)} \rightarrow HCl_{(g)} + NaNO_{3(s)}. \quad (55)$$

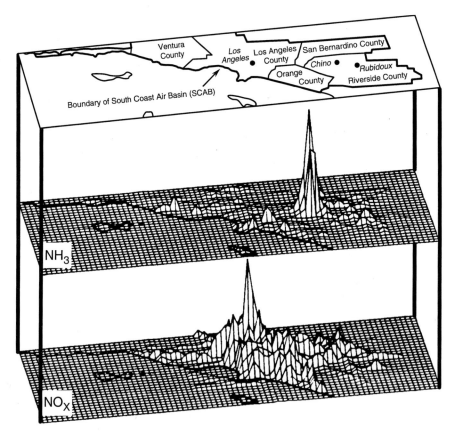

FIGURE 7.21 Estimated daily emissions of NH_3 and NO_x in the Los Angeles area in 1982 (adapted from Russell and Cass, 1986).

TABLE 7.3 Oxidized Nitrogen Distribution at Selected Locations[a] (Percentage of Total)

Location	West Covina, CA	Claremont, CA[d]	Rubidoux, CA	Phoenix, AZ	St. Louis, MO	Quail Farm, NJ
Year:	1973	1979	1976			1979
Type:	Urban	Urban	Suburban	Urban	Urban	Rural
Distance from source:	40 km	50 km	85 km	0 km	0 km	80 km
Source:	Los Angeles	Los Angeles	Los Angeles			Philadelphia
NO	23	5	< 3	45	33	11
NO_2	57	69	78	51	45	59
HNO_3	6	12	b	b	12	17
PAN	13	6	10	2	9	11
Particulate NO_3^-	0.5[c]	7	12	2	1[c]	—
PAN/HNO_3	2.2	0.2–0.5[c,e,f]			0.8	0.6

Source: Adapted from Spicer (1982).
[a] Based on 8-h daylight samples unless otherwise specified.
[b] Generally below instrument detection limit.
[c] 24-h averages.
[d] Only 8 days of data available.
[e] Tuazon *et al.* (1981).
[f] Ratio of peak concentrations was 0.3–0.8.

Measurements of the reaction probability for reaction (55) are somewhat controversial. The reaction is quite slow on single crystals, with a reaction probability $\leq 4 \times 10^{-4}$ (Laux *et al.*, 1994). Small crystals or ground powders react much faster, with a reaction probability of about $(1-2) \times 10^{-2}$ (Leu *et al.*, 1995; Fenter *et al.*, 1996; Beichert and Finlayson-Pitts, 1996). Water adsorbed to defects, steps, or edges of the crystal plays a major role in determining the reactivity, and the reaction has been hypothesized to occur in water bound to certain sites on the surface (Beichert and Finlayson-Pitts, 1996). Thus, HNO_3 is taken up into a "solution" on the surface and acidifies it to the point that HCl degasses. As HNO_3 continues to be taken up into this aqueous portion of the surface, nitrate is removed by precipitation of $NaNO_3$ and H^+ in the form of gaseous HCl. This is consistent with the lack of an apparent surface saturation for the reactions of NaCl powders, in contrast to single crystals under high vacuum, which are known not to hold adsorbed water.

On the other hand, Davies and Cox (1998) have reported a dependence of the reaction probability on the gas-phase concentration of HNO_3 as well as water vapor. The reaction probabilities they report are also one to two orders of magnitude smaller than those of the other powder studies. They interpret their data in terms of a modified version of the Beichert and Finlayson-Pitts (1996) mechanism, in which HNO_3 first physisorbs to the surface and then diffuses to a reactive site that holds water. A modified Langmuir reaction scheme has been proposed by Ghosal and Hemminger (1999), which explains the observed dependence on HNO_3.

If the relative humidity is above the deliquescence point of NaCl (75% at 25°C), the sea salt particles are concentrated salt solutions. (It should be noted that there is a hysteresis associated with decreasing relative humidities; i.e., once the particle is a solution, it does not "dry out" until RHs well below the deliquescence point are reached.) The thermodynamics of this system has been treated in detail (e.g., see Clegg and Brimblecombe, 1988; Brimblecombe and Clegg, 1988; and Tang *et al.*, 1988). In essence, HNO_3 is absorbed into solution and reacts to form HCl. Once the solution becomes sufficiently acidified, e.g., pH ~1–2, the HCl degasses, leaving nitrate ions in place of the chloride.

Such reactions, of course, will be important only in regions where the particles contain significant salt concentrations as discussed earlier and in Chapter 6.

Another potential reaction of HNO_3 is with soot particles. This is of particular interest for the upper troposphere and lower stratosphere, since soot emissions from commercial aircraft occur in these regions. Conversion of HNO_3 on soot particles back to NO and NO_2 would decrease the HNO_3/NO_x ratios and bring modeled values into closer agreement with the measurements (e.g., Chatfield, 1994; Hauglustaine *et al.*, 1996; Lary *et al.*, 1997). The uptake and reaction of HNO_3 on soot have been studied by several groups. Rogaski *et al.* (1997) report the results of Knudsen cell studies where a rapid uptake of HNO_3 (0.5–10 mTorr) at room temperature was observed. The reaction probability was measured to be 0.038 ± 0.008, with about two-thirds of the HNO_3 that was taken up being converted to NO and NO_2. However, Choi and Leu (1998) report that at lower pressures of HNO_3 (~10^{-7} Torr) and temperatures (220 K), characteristic of the upper troposphere and lower stratosphere, only physical adsorption of HNO_3 occurs on soot. They propose a bimolecular reaction of adjacent HNO_3 molecules at higher HNO_3 concentrations, forming NO_2, NO, and H_2O, along with a surface-bound oxygen that ultimately reacts with the carbon surface to form CO_2. Such a process would not be expected to occur, however, at the lower concentrations found in the atmosphere.

In short, although there are relatively few laboratory studies of the HNO_3–soot reaction, it appears that reduction of HNO_3 at the low concentrations found in the atmosphere is not likely to be an important process.

Given the sticky nature of HNO_3, it is likely that it is also taken up on other types of particles in the atmosphere, and depending on the nature of the surface, may react further. For example, Tabazadeh et al. (1998) have proposed that uptake on mineral particles and particles from biomass burning could be important in the upper troposphere, leading to reactions such as

$$CaCO_3 + 2HNO_3 \rightarrow Ca(NO_3)_2 + CO_2 + H_2O \quad (56)$$

and

$$(NH_4)^+ (HCOO^-)_{(aq)} + HNO_{3(g)} \rightarrow$$
$$NH_4NO_{3(aq)} + HCOOH_{(g)}, \quad (57)$$

where the latter is essentially an acid displacement reaction.

F. "MISSING" NO_y

The issue of "missing NO_y" is discussed in detail in Chapter 11.A.4. Briefly, composite measurements of the total nitrogen compounds in air have been made by passing the air sample over a reducing catalyst to form NO, which is then measured using conventional methods such as chemiluminescence. The term "NO_y" is used to denote this sum, i.e.,

$$NO_y = NO_x + PAN + HNO_3$$
$$+ NO_3^- \text{ (Particles)} + 2N_2O_5$$
$$+ \text{Other reactive N compounds}\ldots,$$

where

$$NO_x = NO + NO_2.$$

A number of field studies have been carried out in which individual nitrogen compounds have been measured simultaneously with NO_y to determine whether the individual species add up to the measured NO_y. In some cases there is indeed an approximate mass balance, whereas in others the sum of the individual species is less than the total measured NO_y. This discrepancy is often referred to as "missing NO_y." Compounds that may contribute to the shortfall, when it exists, are discussed in Chapters 6.E.3 and 11.A.4.

G. AMMONIA (NH_3)

Ammonia is the only significant gaseous base in the atmosphere and hence, as discussed earlier, plays an important role in neutralizing acids in the atmosphere. It is produced primarily by livestock wastes and fertilizer; smaller amounts are believed to be generated by industrial activities and natural processes in the soil (Schlesinger and Hartley, 1992; Langford et al., 1992). In addition, there may be some previously unrecognized sources such as burning of agricultural wastes, e.g., straw (Lee and Atkins, 1994). Vehicles equipped with three-way catalysts also emit ammonia; indeed, Fraser and Cass (1998) estimate that in the Los Angeles area ammonia emissions from vehicles in 1993 were comparable to those from the decomposition of livestock waste in dairy regions. [For a summary of emissions, chemistry, deposition, and impacts of ammonia, see *Atmos. Environ.*, 32, 269–594 (1998).]

In air, NH_3 neutralizes acids, forming ammonium nitrate in the case of HNO_3 as described earlier and ammonium sulfate and bisulfate in the case of H_2SO_4. As was the case with HNO_3, ammonia undergoes dry deposition readily so that much of the removal from the atmosphere is nonchemical in nature.

NH_3 also reacts with OH, although this reaction is relatively slow:

$$NH_3 + OH \rightarrow NH_2 + H_2O. \quad (58)$$

The room temperature rate constant is 1.6×10^{-13} cm^3 molecule^{-1} s^{-1} (DeMore et al., 1997; Atkinson et al., 1997a, 1997b), giving an estimated lifetime for NH_3 of 72 days with respect to reaction with OH at a typical average daytime concentration of 1×10^6 radicals cm^{-3}.

The NH_2 radical can, in principle, react with O_2, NO, NO_2, or O_3:

$$NH_2 + O_2 \rightarrow$$
$$\text{Products}(NH_2O_2, NO + H_2O, OH + HNO), \quad (59)$$
$$k_{59} \leq 6 \times 10^{-21} \text{ cm}^3 \text{ molecule}^{-1} \text{ s}^{-1},$$
$$NH_2 + NO \rightarrow N_2H + OH, \quad (60a)$$
$$\rightarrow N_2 + H_2O, \quad (60b)$$
$$k_{60} = 1.6 \times 10^{-11} \text{ cm}^3 \text{ molecule}^{-1} \text{ s}^{-1},$$
$$NH_2 + NO_2 \rightarrow N_2O + H_2O, \quad (61a)$$
$$\rightarrow NH_2O + NO, \quad (61b)$$
$$k_{61} = 2.0 \times 10^{-11} \text{ cm}^3 \text{ molecule}^{-1} \text{ s}^{-1},$$
$$NH_2 + O_3 \rightarrow \text{Products}, \quad (62)$$
$$k_{62} = 1.7 \times 10^{-13} \text{ cm}^3 \text{ molecule}^{-1} \text{ s}^{-1}.$$

The rate constants shown are those at 298 K recommended by Atkinson *et al.* (1997a, 1997b). At concentrations of NO and NO_2 of 1 ppb, respectively, and of O_3 of 100 ppb, which might be found in moderately polluted atmospheres, the lifetime of NH_2 with respect to reaction with O_2 is $\geq$ 30 s and those with respect to reaction with NO, NO_2, and O_3 are about 2–3 s. It appears likely, therefore, that the amidogen radical, NH_2, once formed, will not react with O_2 as one might expect at first glance, but rather with other atmospheric constituents.

The branching ratios for the reactions with NO and NO_2 have been studied. Park and Lin (1997a) and Wolf *et al.* (1997) report that the branching ratio for (60a) is 0.1 at room temperature. Park and Lin (1996, 1997b) measured the branching ratio for the production of N_2O in the NH_2 + NO_2 reaction (61a) to be 0.19 over the temperature range from 300 to 990 K.

H. PROBLEMS

1. For the oxidation of NO by O_2,

$$2NO + O_2 \rightarrow 2NO_2$$

the third-order rate constant at 298 K is $k_1 = 2.0 \times 10^{-38}$ cm^6 $molecule^{-2}$ s^{-1}. (a) What is the effective second-order rate constant at 1 atm in air? (b) Calculate the half-life for conversion of NO to NO_2 at 298 K and NO concentrations of 0.1, 10, and 10,000 ppm, respectively, at the earth's surface.

2. You see a brown plume starting about 5 m above the exit of a power plant stack and want to estimate the concentration of NO that must be in the plume initially. Assume your eye can detect an effective absorbance (base 10) due to NO_2 of 0.05 and that all of the NO_2 is formed by the thermal oxidation of NO by O_2 (see Problem 1). Assume the stack is 3 m in diameter and that this determines the effective path length through the plume. Also assume that the vertical plume speed is 1 m s^{-1} and the temperature is 298 K. Take the absorption cross section of NO_2 to be its peak value in the visible as given in Chapter 4. Estimate the approximate concentration of NO leaving the stack that gives this intensity of absorbance due to NO_2. Use the kinetic data given in Problem 1.

3. Derive the Leighton relationship, Eq. (A), in terms of the ratio $[NO_2]/[NO]$ and its modification for the oxidation of NO by HO_2 and RO_2.

4. (a) Calculate the rate constant for the OH + NO_2 reaction at 1 atm and 300 K, using the DeMore *et al.* (1997) recommended values of k_0 and k_∞ given in the text.

(b) By how much does the rate constant change when the conditions are 250 K and 2 Torr total pressure, as might be found at 40 km in the stratosphere, for example?

5. (a) Using the data in Fig. 4.13 and Table 4.12, calculate the lifetime of HNO_3 with respect to photolysis at a solar zenith angle of 0° at the earth's surface and at an altitude of 40 km on July 1. Comment on the significance of the difference in photolysis rates and lifetimes.

(b) A typical peak OH concentration in the troposphere could be about 5×10^6 radicals cm^{-3} at noon and about 1×10^5 cm^{-3} at sunrise or sunset. Using the kinetics for the OH + HNO_3 reaction in Eq. (D), calculate the lifetime of HNO_3 with respect to this reaction at noon and at sunrise or sunset, respectively, at 298 K. How does this compare to the rate of photolysis at a solar zenith angle of 0°?

6. Using the kinetics for the OH + NO reaction discussed in this chapter, estimate the steady-state concentration of HONO that would exist at noon at the earth's surface if the OH radical concentration is 5×10^6 radicals cm^{-3}, the NO concentration is 1 ppb, and the photolysis rate constant for HONO is 1.4×10^{-3} s^{-1}.

7. Use the kinetics parameters for the OH + NO_2 reaction reported by Brown *et al.* (1999a) to compare the overall rate constants for this reaction at temperatures and pressures of (a) 300 K and 1 atm and (b) 220 K and 20 Torr to those calculated using the DeMore *et al.* (1997) recommendation (see Problem 4).

8. Compare the values for the rate constant of the OH + HNO_3 reaction at 220 K and 100 Torr total pressure using Eq. (D) based on the DeMore *et al.* (1997) recommendations to that of Eq. (E) based on subsequent studies (Brown *et al.*, 1999b).

9. As discussed in Chapter 6.J.3, the ratio NO_x/NO_y measured in the upper troposphere is larger than predicted by many models. One potential factor is uncertainties in the reaction kinetics involving these species. Consider only the following two reactions:

$$OH + NO_2 \xrightarrow{M} HNO_3 \quad (7)$$

$$OH + HNO_3 \rightarrow H_2O + NO_3 \quad (53)$$

Assume HNO_3 is in a steady state. Calculate the ratio $[NO_2]/[HNO_3]$ at a temperature of 300 K and 1 atm pressure using the DeMore *et al.* (1997) recommendations and those of Brown *et al.* (1999a, 1999b). For (a) $T = 300$ K and $P = 1$ atm and (b) $T = 220$ K and $P = 150$ Torr, characteristic of the lower stratosphere/upper troposphere, would the revised kinetics be expected to bring the measurements and models into better agreement?

10. The rate of photolysis of NO_2 at Raleigh, North Carolina (35.8°N, 78.6°W) between 8 and 9 a.m. on a clear day, October 22, 1975, has been measured to be $J(NO_2) = 3.5 \times 10^{-3}$ s^{-1} (Demerjian et al., 1980). Use the OZIPR model (see Appendix III) to look at the rate of change of the NO_2 concentration under these conditions. For the purposes of this calculation, set the initial NO_2 concentration to 100 ppb and make all of the initial NO_x in the form of NO_2. Modify the inputs so that no dilution is occurring and there is no deposition of NO_2. Be sure to change the output concentrations printed so that NO, NO_2, and O_3 are shown explicitly.

a. Based on the value for $J(NO_2)$ given above and applying simple kinetics, approximately what concentration of NO_2 do you expect after this 1 h of photolysis?

b. Compare your calculation in part (a) to the predicted change in NO_2. How do they compare?

c. You should find that the model predicts relatively little NO_2 loss compared to your calculated loss in part (a). Why? What could you do to test this hypothesis? Try it! *Note:* This will require that you edit the meccm.rad file.

d. The Leighton relationship described in your text gives the relationship expected between O_3, NO, and NO_2 in a relatively simple system where a steady state is reached during photolysis. How well do your model calculations agree with the predictions of the Leighton relationship? Why might there be deviations?

References

Akimoto, H., and H. Takagi, "Formation of Methyl Nitrite in the Surface Reaction of Nitrogen Dioxide and Methanol. 2. Photoenhancement," *Environ. Sci. Technol.*, 20, 393–397 (1986).

Akimoto, H., H. Takagi, and F. Sakamaki, "Photoenhancement of the Nitrous Acid Formation in the Surface Reaction of Nitrogen Dioxide and Water Vapor: Extra Radical Source in Smog Chamber Experiments," *Int. J. Chem. Kinet.*, 19, 539–551 (1987).

Alfassi, Z. B., S. Padmaja, P. Neta, and R. E. Huie, "Rate Constants for Reactions of NO_3 Radicals with Organic Compounds in Water and Acetonitrile," *J. Phys. Chem.*, 97, 3780–3782 (1993).

Aliwell, S. R., and R. L. Jones, "Measurements of Tropospheric NO_3 at Midlatitude," *J. Geophys. Res.*, 103, 5719–5727 (1998).

Ammann, M., M. Kalberer, D. T. Jost, L. Tobler, E. Rössler, D. Piguet, H. W. Gäggeler, and U. Baltensperger, "Heterogeneous Production of Nitrous Acid on Soot in Polluted Air Masses," *Nature*, 395, 157–160 (1998).

Andrés-Hernández, M. D., J. Notholt, J. Hjorth, and O. Schrems, "A DOAS Study on the Origin of Nitrous Acid at Urban and Non-Urban Sites," *Atmos. Environ.*, 30, 175–180 (1996).

Atkinson, R., "Kinetics and Mechanisms of the Gas-Phase Reactions of the NO_3 Radical with Organic Compounds," *J. Phys. Chem. Ref. Data*, 20, 459–507 (1991).

Atkinson, R., A. M. Winer, and J. N. Pitts, Jr., "Estimation of Night-Time N_2O_5 Concentrations from Ambient NO_2 and NO_3 Radical Concentrations and the Role of N_2O_5 in Night-Time Chemistry," *Atmos. Environ.*, 20, 331–339 (1986).

Atkinson, R., D. L. Baulch, R. A. Cox, R. F. Hampson, Jr., J. A. Kerr, M. J. Rossi, and J. Troe, "Evaluated Kinetic, Photochemical, and Heterogeneous Data for Atmospheric Chemistry: Supplement V, IUPAC Subcommittee on Gas Kinetic Data Evaluation for Atmospheric Chemistry," *J. Phys. Chem. Ref. Data*, 26, 521–1011 (1997a).

Atkinson, R., D. L. Baulch, R. A. Cox, R. F. Hampson, Jr., J. A. Kerr, M. J. Rossi, and J. Troe, "Evaluated Kinetic and Photochemical Data for Atmospheric Chemistry: Supplement VI, IUPAC Subcommittee on Gas Kinetic Data Evaluation for Atmospheric Chemistry," *J. Phys. Chem. Ref. Data*, 26, 1329–1499 (1997b).

Bambauer, A., B. Brantner, M. Paige, and T. Novakov, "Laboratory Study of NO_2^- Reaction with Dispersed and Bulk Liquid Water," *Atmos. Environ.*, 28, 3225–3232 (1994).

Barnes, I., K. H. Becker, and J. Starcke, "Fourier Transform Infrared Spectroscopic Observations of Gaseous Nitrosyl Iodine, Nitryl Iodine, and Iodine Nitrate," *J. Phys. Chem.*, 95, 9736–9740 (1991).

Barthelmie, R. J., and S. C. Pryor, "Implications of Ammonia Emissions for Fine Aerosol Formation and Visibility Impairment—A Case Study from the Lower Fraser Valley, British Columbia," *Atmos. Environ.*, 32, 345–352 (1998).

Behnke, W., and C. Zetzsch, "Heterogeneous Photochemical Formation of Cl Atoms from NaCl Aerosol, NO_x and Ozone," *J. Aerosol Sci.*, 21, S229–S232 (1990).

Behnke, W., V. Scheer, and C. Zetzsch, "Production of $BrNO_2$, Br_2 and $ClNO_2$ from the Reaction between Sea Spray Aerosol and N_2O_5," *J. Aerosol Sci.*, 25, S277–S278 (1994).

Behnke, W., C. George, V. Scheer, and C. Zetzsch, "Production and Decay of $ClNO_2$ from the Reaction of Gaseous N_2O_5 with NaCl Solution: Bulk and Aerosol Experiments," *J. Geophys. Res.*, 102, 3795–3804 (1997).

Beichert, P., and B. J. Finlayson-Pitts, "Knudsen Cell Studies of the Uptake of Gaseous HNO_3 and Other Oxides of Nitrogen on Solid NaCl: The Role of Surface-Adsorbed Water," *J. Phys. Chem.*, 100, 15218–15228 (1996).

Blanchard, D. C., "The Oceanic Production of Atmospheric Sea Salt," *J. Geophys. Res.*, 90, 961–963 (1985).

Bongartz, A., J. Kames, U. Schurath, Ch. George, Ph. Mirabel, and J. L. Ponche, "Experimental Determination of HONO Mass Accommodation Coefficients Using Two Different Techniques," *J. Atmos. Chem.*, 18, 149–169 (1994).

Brauer, M., P. B. Ryan, H. H. Suh, P. Koutrakis, J. D. Spengler, N. P. Leslie, and I. H. Billick, "Measurements of Nitrous Acid inside Two Research Houses," *Environ. Sci. Technol.*, 24, 1521–1527 (1990).

Brimblecombe, P., and S. L. Clegg, "The Solubility and Behaviour of Acid Gases in the Marine Aerosol," *J. Atmos. Chem.*, 7, 1–18 (1988).

Brown, S. S., R. K. Talukdar, and A. R. Ravishankara, "Rate Constants for the Reaction $OH + NO_2 + M \rightarrow HNO_3 + M$ under Atmospheric Conditions," *Chem. Phys. Lett.*, 299, 277–284 (1999a).

Brown, S. S., R. K. Talukdar, and A. R. Ravishankara, "Reconsideration of the Rate Constant for the Reaction of Hydroxyl Radicals with Nitric Acid," *J. Phys. Chem. A* 103, 3031–3037 (1999b).

Cadle, R. D., and R. C. Robbins, "Kinetics of Atmospheric Chemical Reactions Involving Aerosols," *Disc. Faraday Soc.*, 30, 155–161 (1960).

Cahill, T. A., K. Wilkinson, and R. Schnell, "Composition Analyses of Size-Resolved Aerosol Samples Taken from Aircraft Downwind of Kuwait, Spring, 1991," *J. Geophys. Res.*, 97, 14513–14520 (1992).

Calvert, J. G., and W. R. Stockwell, "Deviations from the O_3–NO–NO_2 Photostationary State in Tropospheric Chemistry," *Can. J. Chem., 61,* 983–992 (1983).

Calvert, J. G., G. Yarwood, and A. M. Dunker, "An Evaluation of the Mechanism of Nitrous Acid Formation in the Urban Atmosphere," *Res. Chem. Intermed., 20,* 463–502 (1994).

Canagaratna, M., J. A. Phillips, M. E. Ott, and K. R. Leopold, "The Nitric Acid–Water Complex: Microwave Spectrum, Structure, and Tunneling," *J. Phys. Chem. A, 102,* 1489–1497 (1998).

Cantrell, C. A., W. R. Stockwell, L. G. Anderson, K. L. Busarow, D. Perner, A. Schmeltekopf, J. G. Calvert, and H. S. Johnston, "Kinetic Study of the NO_3–CH_2O Reaction and Its Possible Role in Nighttime Tropospheric Chemistry," *J. Phys. Chem., 89,* 139–146 and 4160 (1985).

Carroll, M. A., and A. M. Thompson, "NO_x in the Non-Urban Troposphere," in *Progress and Problems in Atmospheric Chemistry* (J. R. Barker, Ed.), Chapter 7, pp. 198–225, World Scientific, Singapore, 1995.

Chameides, W. L., "Possible Role of NO_3 in the Nighttime Chemistry of a Cloud," *J. Geophys. Res., 91,* 5331–5337 (1986a).

Chameides, W. L., "Reply," *J. Geophys. Res., 91,* 14571–14572 (1986b).

Chatfield, R. B. "Anomalous HNO_3/NO_x Ratio of Remote Tropospheric Air: Conversion of Nitric Acid to Formic Acid and NO_x?," *Geophys. Res. Lett., 21,* 2705–2708 (1994).

Cheng, L., E. Peake, D. Rogers, and A. Davis, "Oxidation of Nitric Oxide Controlled by Turbulent Mixing in Plumes from Oil Sands Extraction Plants," *Atmos. Environ., 20,* 1697–1703 (1986).

Choi, W., and M.-T. Leu, "Nitric Acid Uptake and Decomposition on Black Carbon (Soot) Surfaces: Its Implications for the Upper Troposphere and Lower Stratosphere," *J. Phys. Chem. A, 102,* 7618–7630 (1998).

Chughtai, A. R., W. F. Welch, Jr., M. S. Akhter, and D. M. Smith, "A Spectroscopic Study of Gaseous Products of Soot–Oxides of Nitrogen/Water Reactions," *Appl. Spectrosc., 44,* 294–298 (1990).

Chughtai, A. R., S. A. Gordon, and D. M. Smith, "Kinetics of the Hexane Soot Reaction with NO_2/N_2O_4 at Low Concentration," *Carbon, 32,* 405–416 (1994).

Chung, T. T., J. Dash, and R. J. O'Brien, "*In Situ* Studies of NaCl–Gas Reactions," *9th Int. Congr. Electron Microscop.,* 440–441 (1978).

Clegg, S. L., and P. Brimblecombe, "The Solubility and Behaviour of Acid Gases in the Marine Aerosol," *J. Atmos. Chem., 7,* 1–18 (1988).

Clemens, J., J. Burkhardt, and H. Goldbach, "Abiogenic Nitrous Oxide Formation on Aerosols," *Atmos. Environ., 31,* 2961–2964 (1997).

Davidson, J. A., C. A. Cantrell, R. E. Shetter, A. H. McDaniel, and J. G. Calvert, "The NO_3 Radical Decomposition and NO_3 Scavenging in the Troposphere," *J. Geophys. Res., 95,* 13963–13969 (1990).

Davies, J. A., and R. A. Cox, "Kinetics of the Heterogeneous Reaction of HNO_3 with NaCl: Effect of Water Vapor," *J. Phys. Chem. A, 102,* 7631–7642 (1998).

Davis, H. F., B. Kim, H. S. Johnston, and Y. T. Lee, "Dissociation Energy and Photochemistry of NO_3," *J. Phys. Chem., 97,* 2172–2180 (1993).

De Haan, D., T. Brauers, K. Oum, J. Stutz, T. Nordmeyer, and B. J. Finlayson-Pitts, " Heterogeneous Chemistry in the Troposphere: Experimental Approaches and Applications to the Chemistry of Sea Salt Particles," *Int. Rev. Phys. Chem.,* in press (1999).

DeMore, W. B., S. P. Sander, D. M. Golden, R. F. Hampson, M. J. Kurylo, C. J. Howard, A. R. Ravishankara, C. E. Kolb, and M. J. Molina, "Chemical Kinetics and Photochemical Data for Use in Stratospheric Modeling, Evaluation No. 12," JPL Publ. No. 97-4, January 15, 1997.

Demerjian, K. L., K. L. Schere, and J. T. Peterson, "Theoretical Estimates of Actinic (Spherically Integrated) Flux and Photolytic Rate Constants of Atmospheric Species in the Lower Troposphere," *Adv. Environ. Sci. Technol., 10,* 369–459 (1980).

Dentener, F. J., and P. J. Crutzen, "Reaction of N_2O_5 on Tropospheric Aerosols: Impact on the Global Distributions of NO_x, O_3, and OH," *J. Geophys. Res., 98,* 7149–7163 (1993).

Donahue, N. M., M. K. Dubey, R. Mohrschladt, K. L. Demerjian, and J. G. Anderson, "High-Pressure Flow Study of the Reactions OH + NO_x → $HONO_x$: Errors in the Falloff Region," *J. Geophys. Res., 102,* 6159–6168 (1997).

Donaldson, D. J., J. A. Guest, and M. C. Goh, "Evidence for Adsorbed SO_2 at the Aqueous Air Interface," *J. Phys. Chem., 99,* 9313–9315 (1995).

Dransfield, T. J., K. K. Perkins, N. M. Donahue, J. G. Anderson, M. M. Sprengnether, and K. L. Demerjian, "Temperature and Pressure Dependent Kinetics of the Gas-Phase Reaction of the Hydroxyl Radical with Nitrogen Dioxide," *Geophys. Res. Lett., 26,* 687–690 (1999).

Eriksson, P., and L.-G. Johansson, "The Formation of Sulfuric Acid, Nitrogen Monoxide, and Nitrous Acid on Gold in Air Containing Sub-ppm Concentrations of SO_2 and NO_2," *J. Electrochem. Soc., 138,* 1227–1233 (1991).

Exner, M., H. Herrmann, and R. Zellner, "Laser-Based Studies of Reactions of the Nitrate Radical in Aqueous Solution," *Ber. Bunsenges. Phys. Chem., 96,* 470–477 (1992).

Febo, A., and C. Perrino, "Prediction and Experimental Evidence for High Air Concentration of Nitrous Acid in Indoor Environments," *Atmos. Environ., 25A,* 1055–1061 (1991).

Febo, A., and C. Perrino, "Measurement of High Concentrations of Nitrous Acid Inside Automobiles," *Atmos. Environ., 29,* 345–351 (1995).

Fenter, F. F., F. Caloz, and M. J. Rossi, "Heterogeneous Kinetics of N_2O_5 Uptake on Salt, with a Systematic Study of the Role of Surface Presentation (for N_2O_5 and HNO_3)," *J. Phys. Chem., 100,* 1008–1019 (1996).

Fenter, F. F., and M. J. Rossi, "Heterogeneous Kinetics of HONO on H_2SO_4 Solutions and on Ice: Activation of HCl," *J. Phys. Chem., 100,* 13765–13775 (1996).

Finlayson-Pitts, B. J., "Reaction of NO_2 with NaCl and Atmospheric Implications of NOCl Formation," *Nature, 306,* 676–677 (1983).

Finlayson-Pitts, B. J., M. J. Ezell, and J. N. Pitts, Jr., "Formation of Chemically Active Compounds by Reactions of Atmospheric NaCl Particles with Gaseous N_2O_5 and $ClONO_2$," *Nature, 337,* 241–244 (1989a).

Finlayson-Pitts, B. J., F. E. Livingston, and H. N. Berko, "Synthesis and Identification by Infrared Spectroscopy of Gaseous Nitryl Bromide," *J. Phys. Chem., 93,* 4397–4400 (1989b).

Finlayson-Pitts, B. J., J. N. Pitts, Jr., and A. C. Lloyd, "Comment on 'A Study of the Stability of Methanol-Fueled Vehicle Emissions in Tedlar Bags,'" *Environ. Sci. Technol., 26,* 1668–1670 (1992).

Finlayson-Pitts, B. J., "Chlorine Atoms as a Potential Tropospheric Oxidant in the Marine Boundary Layer," *Res. Chem. Intermed., 19,* 235–249 (1993).

Finnegan, W. G., and R. L. Pitter, "Preliminary Study of Coupled Oxidation–Reduction Reactions of Included Ions in Growing Ice Crystals—Reply," *Atmos. Environ., 25,* 2912–2913 (1991).

Finnegan, W. G., R. L. Pitter, and L. G. Young, "Preliminary Study of Coupled Oxidation–Reduction Reactions of Included Ions in Growing Ice Crystals," *Atmos. Environ., 25,* 2531–2534 (1991).

Fraser, M. P., and G. R. Cass, "Detection of Excess Ammonia Emissions from In-Use Vehicles and the Implications for Fine Particle Control," *Environ. Sci. Technol., 32,* 1053–1057 (1998).

George, Ch., J. L. Ponche, Ph. Mirabel, W. Behnke, V. Scheer, and C. Zetzsch, "Study of the Uptake of N_2O_5 by Water and NaCl Solutions," *J. Phys. Chem., 98,* 8780–8784 (1994).

Gerecke, A., A. Thielmann, L. Gutzwiller, and M. J. Rossi, "The Chemical Kinetics of HONO Formation Resulting from Heterogeneous Interaction of NO_2 with Flame Soot," *Geophys. Res. Lett., 25,* 2453–2456 (1998).

Ghosal, S., and J. C. Hemminger, "Effect of Water on the HNO_3 Pressure Dependence of the Reaction between Gas-Phase HNO_3 and NaCl Surfaces," *J. Phys. Chem. A, 103,* 4777–4781 (1999).

Graedel, T. E., and W. C. Keene, "Tropospheric Budget of Reactive Chlorine," *Global Biogeochem. Cycl., 9,* 47–77 (1995).

Graham, R. A., A. M. Winer, and J. N. Pitts, Jr., "Pressure and Temperature Dependence of the Unimolecular Decomposition of HO_2NO_2," *J. Chem. Phys., 68,* 4505–4510 (1978).

Hanst, P. L., and E. R. Stephens, "Infrared Analysis of Engine Exhausts: Methyl Nitrite Formation from Methanol Fuel," *Spectroscopy, 4,* 33–38 (1989).

Harrison, R. M., and A.-M. N. Kitto, "Evidence for a Surface Source of Atmospheric Nitrous Acid," *Atmos. Environ., 28,* 1089–1094 (1994).

Harrison, R. M., J. D. Peak, and G. M. Collins, "Tropospheric Cycle of Nitrous Acid," *J. Geophys. Res., 101,* 14429–14439 (1996).

Harrison, R. M., and J. D. Peak, "Measurements of Concentration Gradients of HNO_2 and HNO_3 over a Semi-Natural Ecosystem: Discussion," *Atmos. Environ., 31,* 2891–2892 (1997).

Hauglustaine, D. A., B. A. Ridley, S. Solomon, P. G. Hess, and S. Madronich, "HNO_3/NO_x Ratio in the Remote Troposphere during MLOPEX 2: Evidence for Nitric Acid Reduction on Carbonaceous Aerosols?," *Geophys. Res. Lett., 23,* 2609–2612 (1996).

Heintz, F., U. Platt, H. Flentje, and R. Dubois, "Long-Term Observation of Nitrate Radicals at the Tor Station, Kap Arkona (Rügen)," *J. Geophys. Res., 101,* 22891–22910 (1996).

Hertel, O., J. Christensen, and Ø. Hov, "Modeling of the End Products of the Chemical Decomposition of DMS in the Marine Boundary Layer," *Atmos. Environ., 28,* 2431–2449 (1994).

Hildemann, L. M., A. G. Russell, and G. R. Cass, "Ammonia and Nitric Acid Concentrations in Equilibrium with Atmospheric Aerosols: Experiment vs. Theory," *Atmos. Environ., 18,* 1737–1750 (1984).

Hjorth, J., G. Ottobrini, F. Cappellani, and G. Restelli, "A Fourier Transform Infrared Study of the Rate Constant of the Homogeneous Gas-Phase Reaction $N_2O_5 + H_2O$ and Determination of Absolute Infrared Band Intensities of N_2O_5 and HNO_3," *J. Phys. Chem., 91,* 1565–1568 (1987).

Hjorth, J., J. Notholt, and M. D. Andrés-Hernández, "Reply to Lammel's Comment on 'A DOAS Study on the Origin of Nitrous Acid at Urban and Non-urban Sites,' by G. Lammel," *Atmos. Environ., 30,* 4103 (1996).

Hu, J. H., and J. P. D. Abbatt, "Reaction Probabilities for N_2O_5 Hydrolysis on Sulfuric Acid and Ammonium Sulfate Aerosols at Room Temperature," *J. Phys. Chem. A, 101,* 871–878 (1997).

Huie, R. E., "Free Radical Chemistry of the Atmospheric Aqueous Phase," in *Progress and Problems in Atmospheric Chemistry* (J. R. Barker, Ed.), pp. 374–419, World Scientific, Singapore, 1995.

Huie, R. E., and P. Neta, "Rate Constants for Some Oxidation of S(IV) by Radicals in Aqueous Solutions," *Atmos. Environ., 21,* 1743–1747 (1987).

Imamura, T., Y. Rudich, R. K. Talukdar, R. W. Fox, and A. R. Ravishankara, "Uptake of NO_3 on Water Solutions: Rate Coefficients for Reaction with Cloud Water Constituents," *J. Phys. Chem., 101,* 2316–2322 (1997).

Jayne, J. T., P. Davidovits, D. R. Worsnop, M. S. Zahniser, and C. E. Kolb, "Uptake of $SO_2(g)$ by Aqueous Surfaces as a Function of pH: The Effect of Chemical Reaction at the Interface," *J. Phys. Chem., 94,* 6041–6048 (1990).

Jenkin, M. E., R. A. Cox, and D. J. Williams, "Laboratory Studies of the Kinetics of Formation of Nitrous Acid from the Thermal Reaction of Nitrogen Dioxide and Water Vapor," *Atmos. Environ., 22,* 487–498 (1988).

Johnston, H. S., C. A. Cantrell, and J. G. Calvert, "Unimolecular Decomposition of NO_3 to Form NO and O_2 and a Review of N_2O_5/NO_3 Kinetics," *J. Geophys. Res., 91,* 5159–5172 (1986).

Junkermann, W., and T. Ibusuki, "FTIR Spectroscopic Measurements of Surface Bound Products of Nitrogen Oxides on Aerosol Surfaces—Implications for Heterogeneous HNO_2 Production," *Atmos. Environ., 26A,* 3099–3103 (1992).

Kalberer, M., K. Tabor, M. Ammann, Y. Parrat, E. Weingartner, D. Piguet, E. Rössler, D. T. Jost, A. Türler, H. W. Gäggeler, and U. Baltensperger, "Heterogeneous Chemical Processing of $^{13}NO_2$ by Monodisperse Carbon Aerosols at Very Low Concentrations," *J. Phys. Chem., 100,* 15487–15493 (1996).

Karlsson, R., and E. Ljunström, "Nitrogen Dioxide and Sea Salt Particles—A Laboratory Study," *J. Aerosol Sci., 26,* 39–50 (1995).

Kirchner, W., F. Welter, A. Bongartz, J. Kames, S. Schweighoefer, and U. Schurath, "Trace Gas Exchange at the Air/Water Interface: Measurements of Mass Accommodation Coefficients," *J. Atmos. Chem., 10,* 427–449 (1990).

Kleffmann, J., R. Kurtenback, and P. Wiesen, "Surface Catalyzed Conversion of NO_2 into HONO and N_2O: A New Source of Atmospheric N_2O?," in *Impact of Emissions from Aircraft and Spacecraft upon the Atmosphere*, Proceedings of an International Scientific Colloquium, Cologne, Germany, April 18–20, 1994, Porz-Wahnheide, pp. 146–157.

Kleffmann, J., K. H. Becker, and P. Wiesen, "Heterogeneous NO_2 Conversion Processes on Acid Surfaces: Possible Atmospheric Implications," *Atmos. Environ., 32,* 2721–2729 (1998).

Koda, S., K. Yoshikawa, J. Okada, and K. Akita, "Reaction Kinetics of Nitrogen Dioxide with Methanol in the Gas Phase," *Environ. Sci. Technol., 19,* 262–264 (1985).

Lai, C. C., and B. J. Finlayson-Pitts, "Reactions of Dinitrogen Pentoxide and Nitrogen Dioxide with 1-Palmitoyl-2-Oleoyl-sn-Glycero-3-Phosphocholine," *Lipids, 26,* 306–314 (1991).

Lammel, G., and D. Perner, "The Atmospheric Aerosol as a Source of Nitrous Acid in the Polluted Atmosphere," *J. Aerosol Sci., 19,* 1199–1202 (1988).

Lammel, G., "Comment on A DOAS Study on the Origin of Nitrous Acid at Urban and Non-urban Sites, by Andrés-Hernández *et al.* (1996)," *Atmos. Environ., 30,* 4101–4103 (1996).

Lammel, G., and J. N. Cape, "Nitrous Acid and Nitrite in the Atmosphere," *Chem. Soc. Rev., 25,* 361–369 (1996).

Langford, A. O., F. C. Fehsenfeld, J. Zachariassen, and D. S. Schimel, "Gaseous Ammonia Fluxes and Background Concentrations in Terrestrial Ecosystems of the United States," *Global Biogeochem. Cycl., 6,* 459–483 (1992).

Lary, D. J., A. M. Lee, R. Toumi, M. J. Newchurch, M. Pirre, and J. B. Renard, "Carbon Aerosols and Atmospheric Photochemistry," *J. Geophys. Res., 102,* 3671–3682 (1997).

Last, J. A., and D. L. Warren, "Synergistic Interaction between Nitrogen Dioxide and Respirable Aerosols of Sulfuric Acid or Sodium Chloride on Rat Lungs," *Toxicol. Appl. Pharmacol., 90,* 34–42 (1987).

Laux, J. M., J. C. Hemminger, and B. J. Finlayson-Pitts, "X-Ray Photoelectron Spectroscopic Studies of the Heterogeneous Reaction of Gaseous Nitric Acid with Sodium Chloride: Kinetics and Contribution to the Chemistry of the Marine Troposphere," *Geophys. Res. Lett., 21,* 1623–1626 (1994).

Lee, D. S., and D. H. F. Atkins, "Atmospheric Ammonia Emissions from Agricultural Waste Combustion," *Geophys. Res. Lett., 21,* 281–284 (1994).

Leighton, P. A., *Photochemistry of Air Pollution*, Academic Press, New York, 1961.

Leu, M. T., R. S. Timonen, L. F. Keyser, and Y. L. Yung, "Heterogeneous Reactions of $HNO_3(g) + NaCl(s) \rightarrow HCl(g) + NaNO_3(s)$ and $N_2O_5(g) + NaCl(s) \rightarrow ClNO_2(g) + NaNO_3(s)$," *J. Phys. Chem., 99,* 13203–13212 (1995).

Lewis, R. S., and W. M. Deen, "Kinetics of the Reaction of Nitric Oxide with Oxygen in Aqueous Solutions," *Chem. Res. Toxicol., 7,* 568–574 (1994).

Li, S.-M., K. G. Anlauf, and H. A. Wiebe, "Heterogeneous Nighttime Production and Deposition of Particle Nitrate at a Rural Site in North America during Summer 1988," *J. Geophys. Res., 98,* 5139–5157 (1993).

Ljungström, E., and M. Hallquist, "Nitrate Radical Formation Rates in Scandinavia," *Atmos. Environ., 30,* 2925–2932 (1996).

Løgager, T., and K. Sehested, "Formation and Decay of Peroxynitrous Acid: A Pulse Radiolysis Study," *J. Phys. Chem., 97,* 6664–6669 (1993).

Lovejoy, E. R., and D. R. Hanson, "Measurement of the Kinetics of Reactive Uptake by Submicron Sulfuric Acid Particles," *J. Phys. Chem., 99,* 2080–2087 (1995).

Makar, P. A., H. A. Wiebe, R. M. Staebler, S. M. Li, and K. Anlauf, "Measurement and Modeling of Particle Nitrate Formation," *J. Geophys. Res., 103,* 13095–13110 (1998).

Mamane, Y., and J. Gottlieb, "Heterogeneous Reactions of Nitrogen Oxides on Sea Salt and Mineral Particles—A Single Particle Approach," *J. Aerosol Sci., 21,* S225–S228 (1990).

Mebel, A. M., M. C. Lin, and C. F. Melius, "Rate Constant of the $HONO + HONO \rightarrow H_2O + NO + NO_2$ Reaction from ab Initio MO and TST Calculations," *J. Phys. Chem. A, 102,* 1803–1807 (1998).

Mentel, Th. F., D. Bleilebens, and A. Wahner, "A Study of Nighttime Nitrogen Oxide Oxidation in a Large Reaction Chamber—The Fate of NO_2, N_2O_5, HNO_3, and O_3 at Different Humidities," *Atmos. Environ., 30,* 4007–4020 (1996).

Mertes, S., and A. Wahner, "Uptake of Nitrogen Dioxide and Nitrous Acid on Aqueous Surfaces," *J. Phys. Chem., 99,* 14000–14006 (1995).

Michelangeli, D. V., M. Allen, and Y. L. Yung, "Heterogeneous Reactions with NaCl in the El Chichon Volcanic Aerosols," *Geophys. Res. Lett., 18,* 673–676 (1991).

Mihelcic, D., D. Klemp, P. Musgen, H. W. Patz, and A. Volz-Thomas, "Simultaneous Measurements of Peroxy and Nitrate Radicals at Schauinsland," *J. Atmos. Chem., 16,* 313–335 (1993).

Miller, T. M., and V. H. Grassian, "Heterogeneous Chemistry of NO_2 on Mineral Oxide Particles: Spectroscopic Evidence for Oxide-Coordinated and Water-Solvated Surface Nitrate," *Geophys. Res. Lett., 25,* 3835–3838 (1998).

Mozurkewich, M., "The Dissociation Constant of Ammonium Nitrate and Its Dependence on Temperature, Relative Humidity, and Particle Size," *Atmos. Environ., 27A,* 261–270 (1993).

Musin, R. N., and M. C. Lin, "Novel Bimolecular Reactions between NH_3 and HNO_3 in the Gas Phase," *J. Phys. Chem. A, 102,* 1808–1814 (1998).

Muzio, L. J., and J. C. Kramlich, "An Artifact in the Measurements of N_2O from Combustion Sources," *Geophys. Res. Lett., 15,* 1369–1372 (1988).

Muzio, L. J., M. E. Teague, J. C. Kramlich, J. A. Cole, J. M. McCarthy, and R. K. Lyon, "Errors in Grab Sample Measurements of N_2O from Combustion Sources," *JAPCA, 39,* 287–293 (1989).

Nguyen, M.-T., A. Jamka, R. Cazar, and F.-M. Tao, "Structure and Stability of the Nitric Acid–Ammonia Complex in the Gas Phase and in Water," *J. Chem. Phys., 106,* 8710–8717 (1997).

Nielsen, T., K. Pilegaard, A. H. Egelov, K. Granby, P. Hummelshøj, N. O. Jensen, and H. Skov, "Atmospheric Nitrogen Compounds: Occurrence, Composition, and Deposition," *Sci. Total Environ., 189/190,* 459–465 (1996).

Niki, H., P. D. Maker, C. M. Savage, and L. P. Breitenbach, "An FTIR Study of the Reaction between Nitrogen Dioxide and Alcohols," *Int. J. Chem. Kinet., 14,* 1199–1209 (1982).

Notholt, J., J. Hjorth, and F. Raes, "Formation of HNO_2 on Aerosol Surfaces during Foggy Periods in the Presence of NO and NO_2," *Atmos. Environ., 26A,* 211–217 (1992).

Novakov, T., Author's Reply to Schwartz and Lee's (1995) Comment on "Laboratory Study of NO_2 Reaction with Dispersed and Bulk Liquid Water," *Atmos. Environ., 29,* 2559–2560 (1995).

Noxon, J. F., R. B. Norton, and W. R. Henderson, "Observation of Atmospheric NO_3," *Geophys. Res. Lett., 5,* 675–678 (1978).

Noxon, J. F., R. B. Norton, and E. Marovich, "NO_3 in the Troposphere," *Geophys. Res. Lett., 7,* 125–128 (1980).

Park, J., and M. C. Lin, "Mass-Spectrometric Determination of Product Branching Probabilities for the $NH_2 + NO_2$ Reaction at Temperatures between 300 and 990 K," *Int. J. Chem. Kinet., 28,* 879–883 (1996).

Park, J., and M. C. Lin, "Laser-Initiated NO Reduction by NH_3: Total Rate Constant and Product Branching Ratio Measurements for the NH_2 + NO Reaction," *J. Phys. Chem. A, 101,* 5–13 (1997a).

Park, J., and M. C. Lin, "A Mass Spectrometric Study of the $NH_2 + NO_2$ Reaction," *J. Phys. Chem. A, 101,* 2643–2647 (1997b).

Parrish, D. D., M. Trainer, E. J. Williams, D. W. Fahey, G. Hübler, C. S. Eubank, S. C. Liu, P. C. Murphy, D. L. Albritton, and F. C. Fehsenfeld, "Measurements of the NO_x–O_3 Photostationary State at Niwot Ridge, Colorado," *J. Geophys. Res., 91,* 5361–5370 (1986).

Pedersen, T., "Nighttime Hydrogen Peroxide Production on Sulfuric-Acid-Aerosols Involving Nitrate and Sulfate Radicals," *Geophys. Res. Lett., 22,* 1497–1499 (1995).

Perner, D., and U. Platt, "Detection of Nitrous Acid in the Atmosphere by Differential Optical Absorption," *Geophys. Res. Lett., 6,* 917–920 (1979).

Peters, S. J., and G. E. Ewing, "The Reaction of NO_2 with NaCl(100)," *J. Phys. Chem., 100,* 14093–14102 (1996).

Pires, M., and M. J. Rossi, "The Heterogeneous Generation of N_2O from Exhaust Gases of Combustion: A Laboratory Study," *Geophys. Res. Lett., 22,* 3509–3512 (1995).

Pires, M., H. van den Bergh, and M. J. Rossi, "The Heterogeneous Formation of N_2O over Bulk Condensed Phases in the Presence of SO_2 at High Humidities," *J. Atmos. Chem., 25,* 229–250 (1996).

Pires, M., and M. J. Rossi, "The Heterogeneous Formation of N_2O in the Presence of Acidic Solutions: Experiments and Modeling," *Int. J. Chem. Kinet., 29,* 869–891 (1997).

Pitts, J. N., Jr., E. Sanhueza, R. Atkinson, W. P. L. Carter, A. M. Winer, G. W. Harris, and C. N. Plum, "An Investigation of the Dark Formation of Nitrous Acid in Environmental Chambers," *Int. J. Chem. Kinet., 16,* 919–939 (1984a).

Pitts, J. N., Jr., H. W. Biermann, A. M. Winer, and E. C. Tuazon, "Spectroscopic Identification and Measurement of Gaseous Nitrous Acid in Direct Auto Exhaust," *Atmos. Environ., 18,* 847–854 (1984b).

Pitts, J. N., Jr., T. J. Wallington, H. W. Biermann, and A. M. Winer, "Identification and Measurement of Nitrous Acid in an Indoor Environment," *Atmos. Environ., 19,* 763–767 (1985).

Pitts, J. N., Jr., H. W. Biermann, E. C. Tuazon, M. Green, W. D. Long, and A. M. Winer, "Time-Resolved Identification and Measurement of Indoor Air Pollutants by Spectroscopic Techniques: Nitrous Acid, Methanol, Formaldehyde, and Formic Acid," *J. Air Pollut. Control Assoc., 39,* 1344–1347 (1989).

Platt, U., D. Perner, A. M. Winer, G. W. Harris, and J. N. Pitts, Jr., "Detection of NO_3 in the Polluted Troposphere by Differential Optical Absorption," *Geophys. Res. Lett., 7,* 89–92 (1980).

Platt, U. F., A. M. Winer, H. W. Biermann, R. Atkinson, and J. N. Pitts, Jr., "Measurement of Nitrate Radical Concentrations in Continental Air," *Environ. Sci. Technol., 18,* 365–369 (1984).

Platt, U., G. Le Bras, G. Poulet, J. P. Burrows, and G. Moortgat, "Peroxy Radicals from Nighttime Reaction of NO_3 with Organic Compounds," *Nature, 348,* 147–149 (1990).

Ponche, J. L., C. George, and P. Mirabel, "Mass Transfer at the Air/Water Interface: Mass Accommodation Coefficients of SO_2, HNO_3, and NH_3," *J. Atmos. Chem., 16,* 1–21 (1993).

Pryor, W. A., and J. W. Lightsey, "Mechanisms of Nitrogen Dioxide Reactions: Initiation of Lipid Peroxidation and the Production of Nitrous Acid," *Science, 214,* 435–437 (1981).

Ritter, J. A., D. H. Stedman, and T. J. Kelly, "Ground-Level Measurements of Nitric Oxide, Nitrogen Dioxide and Ozone in Rural Air," in *Nitrogenous Air Pollutants: Chemical and Biological Implications,* pp. 325–343, Ann Arbor Science Publishers, Ann Arbor, MI., (1979).

Robbins, R. C., R. D. Cadle, and D. L. Eckhardt, "The Conversion of Sodium Chloride to Hydrogen Chloride in the Atmosphere," *J. Meteorol., 16,* 53–56 (1959).

Rogaski, C. A., D. M. Golden, and L. R. Williams, "Reactive Uptake and Hydration Experiments on Amorphous Carbon Treated with NO_2, SO_2, O_3, HNO_3, and H_2SO_4," *Geophys. Res. Lett., 24,* 381–384 (1997).

Rudich, Y., R. K. Talukdar, R. W. Fox, and A. R. Ravishankara, "Reactive Uptake on NO_3 on Pure Water and Ionic Solutions," *J. Geophys. Res., 101,* 21023–21031 (1996a).

Rudich, Y., R. K. Talukdar, T. Imamura, R. W. Fox, and A. R. Ravishankara, "Determination of Gas-Phase Diffusion Coefficients for NO_3 by Measuring Its Uptake on KI Solutions," *Chem. Phys. Lett., 261,* 467–473 (1996b).

Rudich, Y., R. K. Talukdar, and A. R. Ravishankara, "Multiphase Chemistry of NO_3 in the Remote Troposphere," *J. Geophys. Res., 103,* 16133–16143 (1998b).

Russell, A. G., and G. R. Cass, "Verification of a Mathematical Model for Aerosol Nitrate and Nitric Acid Formation and Its Use for Control Measure Evaluation," *Atmos. Environ., 20,* 2011–2025 (1986).

Russell, A. G., G. R. Cass, and J. H. Seinfeld, "On Some Aspects of Nighttime Atmospheric Chemistry," *Environ. Sci. Technol., 20,* 1167–1172 (1986).

Russell, A. G., D. A. Winner, R. A. Harley, K. F. McCue, and G. R. Cass, "Mathematical Modeling and Control of the Dry Deposition Flux of Nitrogen-Containing Air Pollutants," *Environ. Sci. Technol., 27,* 2772–2782 (1993).

Sakamaki, F., S. Hatakeyama, and H. Akimoto, "Formation of Nitrous Acid and Nitric Oxide in the Heterogeneous Dark Reaction of Nitrogen Dioxide and Water Vapor in a Smog Chamber," *Int. J. Chem. Kinet., 15,* 1013–1029 (1983).

Salawitch, R. J., S. C. Wofsy, P. O. Wennberg, R. C. Cohen, J. G. Anderson, D. W. Fahey, R. S. Gao, E. R. Keim, E. L. Woodbridge, R. M. Stimpfle, J. P. Koplow, D. W. Kohn, C. R. Webster, R. D. May, L. Pfister, E. W. Gottlieb, H. A. Michelsen, G. K. Yue, M. J. Prather, J. C. Wilson, C. A. Brock, H. H. Jonsson, J. E. Dye, D. Baumgardner, M. H. Proffitt, M. Loewenstein, J. R. Podolske, J. W. Elkins, G. S. Dutton, E. J. Hintsa, A. E. Dessler, E. M. Weinstock, K. K. Kelly, K. A. Boering, B. C. Daube, K. R. Chan, and S. W. Bowen, "The Diurnal Variation of Hydrogen, Nitrogen, and Chlorine Radicals: Implications for the Heterogeneous Production of HNO_2," *Geophys. Res. Lett., 21,* 2551–2554 (1994).

Schlesinger, W. H., and A. E. Hartley, "A Global Budget for Atmospheric NH_3," *Biogeochemistry, 15,* 191–211 (1992).

Schroeder, W. H., and P. Urone, "Formation of Nitrosyl Chloride from Salt Particles in Air," *Environ. Sci. Technol., 8,* 756–758 (1974).

Schwartz, S. E., and W. H. White, "Solubility Equilibria of the Nitrogen Oxides and Oxyacids in Dilute Aqueous Solution," *Adv. Environ. Sci. Eng., 4,* 1–45 (1981).

Schwartz, S. E., and W. H. White, "Kinetics of Reactive Dissolution of Nitrogen Oxides into Aqueous Solution," *Adv. Environ. Sci. Technol., 12,* 1–116 (1983).

Schwartz, S., "Gas–Aqueous Reactions of Sulfur and Nitrogen Oxides in Liquid-Water Clouds," in *SO_2, NO and NO_2 Oxidation Mechanisms: Atmospheric Considerations* (J. G. Calvert, Ed.), Acid Precipitation Series, Vol. 3, pp. 173–208 (J. I. Teasley, Series Ed.), Butterworth, Stoneham, MA, 1984.

Schwartz, S. E., and Y.-N. Lee, "Comment on 'Laboratory Study of NO_2 Reaction with Dispersed and Bulk Liquid Water,' by Bambauer *et al.* (1994)," *Atmos. Environ., 29,* 2557–2559 (1995).

Schweitzer, F., P. Mirabel, and C. George, "Multiphase Chemistry of N_2O_5, $ClNO_2$ and $BrNO_2$," *J. Phys. Chem. A., 102,* 3942–3952 (1998).

Seisel, S., F. Caloz, F. F. Fenter, H. van den Bergh, and M. J. Rossi, "The Heterogeneous Reaction of NO_3 with NaCl and KBr: A Nonphotolytic Source of Halogen Atoms," *Geophys. Res. Lett., 24,* 2757–2760 (1997).

Smith, N., J. M. C. Plane, C.-F. Nien, and P. A. Solomon, "Nighttime Radical Chemistry in the San Joaquin Valley," *Atmos. Environ., 29,* 2887–2897 (1995).

Spicer, C. W., "The Distribution of Oxidized Nitrogen in Urban Air," *Sci. Total Environ., 24,* 183–192 (1982).

Spicer, C. W., "Smog Chamber Studies of NO_x Transformation Rate and Nitrate/Precursor Relationships," *Environ. Sci. Technol., 17,* 112–120 (1983).

Spicer, C. W., D. V. Kenny, G. F. Ward, and I. H. Billick, "Transformations, Lifetimes, and Sources of NO_2, HONO, and HNO_3 in Indoor Environments," *J. Air Waste Manage. Assoc., 43,* 1479–1485 (1993).

Stevens, T. E., and W. D. Emmons, "The Dinitrogen Pentoxide-Olefin Reaction," *J. Amer. Chem. Soc., 79,* 6008–6014 (1957).

Svensson, R., E. Ljungström, and O. Lindqvist, "Kinetics of the Reaction between Nitrogen Dioxide and Water Vapor," *Atmos. Environ., 21,* 1529–1539 (1987).

Sverdrup, G. M., and M. R. Kuhlman, "Heterogeneous Nitrogen Oxide-Particle Reactions," in *Atmospheric Pollution,* (M. M. Benarie, Ed.), *Studies in Environmental Science,* Elsevier, Amsterdam, Vol. 8, pp. 245–248 (1980).

Sverdrup, G. M., C. W. Spicer, and G. F. Ward, "Investigation of the Gas Phase Reaction of Dinitrogen Pentoxide with Water Vapor," *Int. J. Chem. Kinet., 19,* 191–205 (1987).

Tabazadeh, A., M. Z. Jacobson, H. B. Singh, O. B. Toon, J. S. Lin, R. B. Chatfield, A. N. Thakur, R. W. Talbot, and J. E. Dibb, "Nitric Acid Scavenging by Mineral and Biomass Burning Aerosols," *Geophys. Res. Lett., 25,* 4185–4188 (1998).

Tabor, K. D., L. Gutzwiller, and M. J. Rossi, "Heterogeneous Chemical Kinetics of NO_2 on Amorphous Carbon at Ambient Temperature," *J. Phys. Chem., 98,* 6172–6186 (1994).

Takagi, H., S. Hatakeyama, H. Akimoto, and S. Koda, "Formation of Methyl Nitrite in the Surface Reaction of Nitrogen Dioxide and Methanol. 1. Dark Reaction," *Environ. Sci. Technol., 20,* 387–393 (1986).

Takenaka, N., A. Ueda, and Y. Maeda, "Acceleration of the Rate of Nitrite Oxidation by Freezing in Aqueous Solution," *Nature, 358,* 736–738 (1992).

Takenaka, N., A. Ueda, T. Daimon, H. Bandow, T. Dohmaru, and Y. Maeda, "Acceleration Mechanism of Chemical Reaction by Freezing: The Reaction of Nitrous Acid with Dissolved Oxygen," *J. Phys. Chem., 100,* 13874–13884 (1996).

Tang, I. N., H. R. Munkelwitz, and J. H. Lee, "Vapor–Liquid Equilibrium Measurements for Dilute Nitric Acid Solutions," *Atmos. Environ., 22,* 2579–2585 (1988).

Tao, F.-M., K. Higgins, W. Klemperer, and D. D. Nelson, "Structure, Binding Energy, and Equilibrium Constant of the Nitric Acid–Water Complex," *Geophys. Res. Lett., 23,* 1797–1800 (1996).

Ten Brink, H. M., and H. Spoelstra, "The Dark Decay of HONO in Environmental (Smog) Chambers," *Atmos. Environ., 32,* 247–251 (1998).

Thomas, K., D. Kley, D. Mihelcic, and A. Volz-Thomas, "Mass Accommodation Coefficient for NO_3 Radicals on Water: Implication for Atmospheric Oxidation Processes," International Conference on the Generation of Oxidants on Regional and Global Scales, Norwich, July 3–7, 1989.

Thomas, K., A. Volz-Thomas, D. Mihelcic, H. G. J. Smit, and D. Kley, "On the Exchange of NO_3 Radicals with Aqueous Solutions: Solubility and Sticking Coefficient," *J. Atmos. Chem., 29,* 17–43 (1998).

Tuazon, E. C., A. M. Winer, and J. N. Pitts, Jr., "Trace Pollutant Concentrations in a Multiday Smog Episode in the California South Coast Air Basin by Long Pathlength Fourier Transform Infrared Spectroscopy," *Environ. Sci. Technol., 15,* 1232–1237 (1981).

Tyndall, G. S., J. J. Orlando, and J. G. Calvert, "Upper Limit for the Rate Coefficient for the Reaction $HO_2 + NO_2 \rightarrow HONO + O_2$," *Environ. Sci. Technol., 29,* 202–206 (1995).

Van Doren, J. M., L. R. Watson, P. Davidovits, D. R. Worsnop, M. S. Zahniser, and C. E. Kolb, "Temperature Dependence of the Uptake Coefficients of HNO_3, HCl, and N_2O_5 by Water Droplets," *J. Phys. Chem., 94,* 3265–3269 (1990).

Vecera, Z., and P. K. Dasgupta, "Indoor Nitrous Acid Levels. Production of Nitrous Acid from Open-Flame Sources," *Int. J. Environ. Anal. Chem., 56,* 311–316 (1994).

Vogt, R., and B. J. Finlayson-Pitts, "A Diffuse Reflectance Infrared Fourier Transform Spectroscopic (DRIFTS) Study of the Surface Reaction of NaCl with Gaseous NO_2 and HNO_3," *J. Phys. Chem., 98,* 3747–3755 (1994); *J. Phys. Chem., 99,* 13052 (1995).

Vogt, R., C. Elliott, H. C. Allen, J. M. Laux, J. C. Hemminger, and B. J. Finlayson-Pitts, "Some New Laboratory Approaches to Studying Tropospheric Heterogeneous Reactions," *Atmos. Environ., 30,* 1729–1737 (1996).

Wahner, A., T. F. Mentel, and M. Sohn, "Gas-Phase Reaction of N_2O_5 with Water Vapor: Importance of Heterogeneous Hydrolysis of N_2O_5 and Surface Desorption of HNO_3 in a Large Teflon Chamber," *Geophys. Res. Lett., 25,* 2169–2172 (1998a).

Wahner, A., T. F. Mentel, M. Sohn, and J. Stier, "Heterogeneous Reaction of N_2O_5 on Sodium Nitrate Aerosol," *J. Geophys. Res., 103,* 31103–31112 (1998b).

Wan, J. K. S., J. N. Pitts, Jr., P. Beichert, and B. J. Finlayson-Pitts, "The Formation of Free Radical Intermediates in the Reactions of Gaseous NO_2 with Solid NaCl and NaBr: Atmospheric and Toxicological Implications," *Atmos. Environ., 30,* 3109–3113 (1996).

Wang, J., and B. E. Koel, "IRAS Studies of NO_2, N_2O_3 and N_2O_4 Adsorbed on Au(III) Surfaces and Reactions with Coadsorbed H_2O," *J. Phys. Chem. A., 102,* 8573–8579 (1998).

Wängberg, I., T. Etzkorn, I. Barnes, U. Platt, and K. H. Becker, "Absolute Determination of the Temperature Behavior of the $NO_2 + NO_3 + (M) \Leftrightarrow N_2O_5 + (M)$ Equilibrium," *J. Phys. Chem. A, 101,* 9694–9698 (1997).

Watson, J. G., J. C. Chow, F. W. Lurmann, and S. P. Musarra, "Ammonium Nitrate, Nitric Acid, and Ammonia Equilibrium in Wintertime Phoenix, Arizona," *J. Air Water Manage. Assoc., 44,* 405–412 (1994).

Wayne, R. P., I. Barnes, P. Biggs, J. P. Burrows, C. E. Canosa-Mas, J. Hjorth, G. Le Bras, G. K. Moortgat, D. Perner, G. Poulet, G. Restelli, and H. Sidebottom, "The Nitrate Radical: Physics, Chemistry and the Atmosphere," *Atmos. Environ., 25A,* 1–203 (1991).

Wiesen, P., J. Kleffmann, R. Kurtenbach, and K. H. Becker, "Mechanistic Study of the Heterogeneous Conversion of NO_2 into HONO and N_2O on Acid Sufaces," *Faraday Discuss., 100,* 121–127 (1995).

Wincel, H., E. Mereand, and A. W. Castleman, Jr., "Reactions of N_2O_5 with Protonated Water Clusters $H^+(H_2O)_n$, $n = 3$–30," *J. Phys. Chem., 98,* 8606–8610 (1994).

Wincel, H., E. Mereand, and A. W. Castleman, Jr., "Gas Phase Reactions of N_2O_5 with $X^-(H_2O)_n$, X = O, OH, O_2, HO_2, and O_3," *J. Phys. Chem., 99,* 1792–1798 (1995).

Winer, A. M., and H. W. Biermann, "Long Pathlength Differential Optical Absorption Spectroscopy (DOAS) Measurements of Gaseous HONO, NO_2, and HCHO in the California South Coast Air Basin," *Res. Chem. Intermed., 20,* 423–445 (1994).

Wingen, L. M., W. S. Barney, M. J. Lakin, T. Brauers, and B. J. Finlayson-Pitts, "A Unique Method for Laboratory Quantification of Gaseous Nitrous Acid (HONO) Using the Reaction $HONO + HCl \rightarrow ClNO + H_2O$," *J. Phys. Chem.,* submitted for publication (1999).

Winkler, T., J. Goschnick, and H. J. Ache, "Reactions of Nitrogen Oxides with NaCl as Model of Sea Salt Aerosol," *J. Aerosol Sci., 22,* S605–S608 (1991).

Wolf, M., D. L. Yang, and J. L. Durant, "A Comprehensive Study of the Reaction $NH_2 + NO \rightarrow$ Products: Reaction Rate Coefficients, Product Branching Fractions, and *ab Initio* Calculations," *J. Phys. Chem. A, 101,* 6243–6251 (1997).

Zellner, R., and H. Herrmann, "Free Radical Chemistry of the Aqueous Atmospheric Phase," in *Spectroscopy in Environmental Science* (R. J. H. Clark and R. E. Hester, Eds.), pp. 381–451, Wiley, New York, 1995.

Zetzsch, C., "Simulation of Atmospheric Photochemistry in the Presence of Solid Airborne Aerosols," in *Formation, Distribution and Chemical Transformation of Air Pollutants,* (R. Zellner, Ed.), p. 187, Monograph Vol. 104, VCH (1987).

Zetzsch, C., and W. Behnke, "Heterogeneous Photochemical Sources of Atomic Cl in the Troposphere," *Ber. Bunsenges. Phys. Chem., 96,* 488–493 (1992).

Zhang, R., M.-T. Leu, and L. F. Keyser, "Heterogeneous Chemistry of HONO on Liquid Sulfuric Acid: A New Mechanism of Chlorine Activation on Stratospheric Sulfate Aerosols," *J. Phys. Chem., 100,* 339–345 (1996).

Zhu, T., G. Yarwood, J. Chen, and H. Niki, "Evidence for the Heterogeneous Formation of Nitrous Acid from Peroxynitric Acid in Environmental Chambers," *Environ. Sci. Technol., 27,* 982–983 (1993).

CHAPTER 8

Acid Deposition
Formation and Fates of Inorganic and Organic Acids in the Troposphere

A. CONTRIBUTION OF H_2SO_4, HNO_3, HONO, AND ORGANIC ACIDS

Although the term *acid rain* has been used extensively in the popular literature to describe the formation and deposition of acids at the earth's surface, the terminology *acid deposition* is more commonly encountered in the scientific literature. The reason for this is that deposition of acids can occur either as *dry deposition* or as *wet deposition*. The former refers to the direct transport of acidic gases or small particles to the surface, followed by adsorption, without first being dissolved in an aqueous phase such as rain, clouds, or fog. Wet deposition, on the other hand, refers to the transport of acids to, and deposition on, surfaces (including soil, trees, grass, buildings, etc.) after the acids have been dissolved in an aqueous medium. It should be noted that the surface itself can be either wet or dry; the terms wet and dry deposition refer to the mechanism of transport to the surface, not to the nature of the surface itself.

Acid deposition has been recognized for centuries. For example, as described in the fascinating historical perspectives provided by Brimblecombe (1978) and Cowling (1982), Robert Boyle referred in 1692 to "nitrous or salino-sulphureous spirits" in air in his book *A General History of the Air*. In 1872, Robert Smith published in the United Kingdom a book entitled *Air and Rain: The Beginnings of a Chemical Climatology*. In it, he described many of the factors affecting what he termed "acid rain," including coal combustion and the amount and frequency of precipitation as well as the effects on plants and materials. It is fascinating that he actually analyzed rain near Manchester, U.K., and noted a change in the chemistry from the town center to remoter surrounding areas:

> "...carbonate of ammonia in the fields at a distance, ... sulphate of ammonia in the suburbs, and ... sulfuric acid or acid sulphate in the town."

Historically, the major acids believed to contribute to acid deposition in the troposphere have been sulfuric and nitric acids, formed by the oxidation in air of SO_2 and oxides of nitrogen, respectively. However, there is an increasing recognition that organic acids may contribute significantly to the total acid burden and indeed may represent the major acidic species even in polluted urban environments. In addition, since nitrous acid (HONO) is formed whenever NO_2 and water are present (see Chapter 7.B.3), its contribution to the total acidity, particularly to indoor air environments, has become of interest and concern.

The relative contributions of these acids to the total acid deposition depends, as one might expect, on the particular emission sources. For example, the ratio of H_2SO_4 to HNO_3 is typically 2:1 on the east coast of the United States, where there are significant sources of SO_2, but about 1:2 on the west coast, where NO_x emissions predominate. The contribution of HONO has not been well established, in part because it is difficult to measure in an accurate and specific manner (see Chapter 11.A.4a) so that the data base on its concentrations is somewhat limited. In addition, since it photolyzes rapidly, its concentrations and contributions to the total acidity during the day are not significant compared to sulfuric and nitric acids. It is only recently that the potential contribution of organic acids has also been recognized.

To treat the chemistry of oxides of nitrogen, which play such a central role in the chemistry of both the polluted and remote troposphere, in a consistent manner, we have discussed the formation and fates of

HNO₃ and HONO separately in Chapter 7. As a result, we shall deal in this chapter primarily with H_2SO_4 and organic acids. In addition, since much of the chemistry occurs in the aqueous phase, with oxidation of dissolved SO_2 initiated by a variety of free radical species, we shall also discuss the associated aqueous-phase chemistry of HO_x, carbonate/bicarbonate, and the halogens.

B. SOLUBILITY OF GASES IN RAIN, FOGS, AND CLOUDS: HENRY'S LAW AND AQUEOUS-PHASE EQUILIBRIA

The distribution of a species between the gas and the aquated forms due simply to physical solubility of the gas is described by the Henry's law constant, H_X (Fig. 8.1). The Henry's law constant (H_X) for a particular species X is, in effect, the equilibrium constant for the reaction

$$X_{(g)} \leftrightarrow X_{(aq)} \qquad H_x = [X]_{aq}/P_X.$$

Table 8.1 shows the values of these constants for some species of tropospheric interest. The most soluble gases have Henry's law constants of approximately 10^5 M atm^{-1}, whereas the least soluble have values about eight orders of magnitude smaller.

However, for species such as CO_2 and SO_2 whose aquated forms can react further, much larger amounts of the gases can be taken up into solution. For CO_2, for example, the bicarbonate ion (HCO_3^-) and the carbonate ion (CO_3^{2-}) are formed as the aquated form of CO_2 reacts:

$$CO_{2(g)} + H_2O_{(l)} \leftrightarrow CO_2 \cdot H_2O_{(aq)}$$
$$H_1 = 3.4 \times 10^{-2} \text{ M atm}^{-1}, \quad (1, -1)$$

$$CO_2 \cdot H_2O_{(aq)} \leftrightarrow HCO_3^- + H^+$$
$$K_2 = 4.3 \times 10^{-7} \text{ M}, \quad (2, -2)$$

$$HCO_3^- \leftrightarrow CO_3^{2-} + H^+$$
$$K_3 = 4.7 \times 10^{-11} \text{ M}. \quad (3, -3)$$

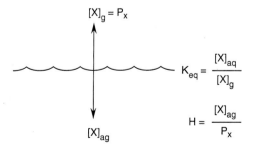

FIGURE 8.1 Henry's law applied to atmospheric systems.

(The Henry's law and equilibrium constants are from Sillén and Martell, 1964.) In such cases, the Henry's law constant reflects only the physical solubility (i.e., reaction (1, −1). A "pseudo-Henry's law constant," H^*, is often defined to take into account the increased uptake compared to that expected based on simple dissolution of the gas without further reaction. With CO_2 as an example, this pseudo-Henry's law constant is defined by Eq. (A):

$$H^*_{CO_2} = [CO_2 \cdot H_2O + HCO_3^- + CO_3^{2-}]/P_{CO_2}. \quad (A)$$

That is, H^* is the sum of all aqueous-phase forms of the species divided by its equilibrium gas-phase pressure. The usual expressions for the equilibrium constants $K_2 = [HCO_3^-][H^+]/[CO_2 \cdot H_2O]$ and $K_3 = [CO_3^{2-}][H^+]/[HCO_3^-]$ can be used to replace $[HCO_3^-]$ and $[CO_3^{2-}]$ in Eq. (A), resulting in the following expression (see Problem 1) for the pseudo-Henry's law constant as a function of pH:

$$H^*_{CO_2} = H_1\{1 + K_2/[H^+] + (K_2K_3)/[H^+]^2\}. \quad (B)$$

Equations (A) and (B) allow one to calculate at a given pH the value of $H^*_{CO_2}$ and hence the total concentration of the dissolved forms of CO_2 for a given gas-phase concentration. Alternatively, one can estimate the pH of the water for given gas- and aqueous-phase concentrations.

It should be noted that Henry's law describes the equilibrium situation and can only be applied when this is the case. The presence of surfactants on the surface, for example, has been shown in laboratory studies to reduce the apparent Henry's law constant in some cases (e.g., Anderson, 1992). In addition, there must be sufficient contact time between the gas and aqueous phase for equilibrium to be established.

Following this approach, one can calculate that the pH of atmospheric aqueous solutions in equilibrium with the current concentrations of CO_2 should be approximately 5.6, assuming other species such as ammonia are not available to neutralize the acids formed in reactions (2) and (3). However, over the past three decades or so, starting in Europe and particularly the Scandinavian countries, the formation of rain, clouds, and fogs with much higher acidities (lower pH) has been widely recognized. Rain with pHs of 4.5 is common worldwide, and even more acidic precipitation has been measured in many areas. Clouds with a pH in the mid-2 range and fogs with a pH as low as 1.69 (e.g., Aneja and Kim, 1993; Jacob and Hoffmann, 1983) have been observed. As discussed in this chapter, this apparent increase in acidity in going from rain to clouds to

TABLE 8.1 Henry's Law Coefficients (H) for Some Atmospheric Gases Dissolving in Liquid Water at 20–25°C[a]

Gas	H (mol L^{-1} atm^{-1})	Reference
O_2	1.3×10^{-3}	Loomis, 1928
NO	1.9×10^{-3}	Loomis, 1928
C_2H_4	4.9×10^{-3}	Loomis, 1928
NO_2[b]	1×10^{-2}	Schwartz and White, 1983
N_2O_4	1.4	Schwartz and White, 1981
N_2O_3	0.6	Schwartz and White, 1981
O_3	$(0.82–1.3) \times 10^{-2}$	Shorter et al., 1995; Briner and Perrottet, 1939
N_2O	2.5×10^{-2}	Loomis, 1928
CO_2[c]	3.4×10^{-2}	Loomis, 1928
SO_2[c]	1.2	Sillén and Martell, 1964; Maahs, 1982; Shorter et al., 1995
HONO[c]	49	Schwartz and White, 1981
NH_3[c]	62	Van Krevelen et al., 1949
HNO_3[d]	2.1×10^5	Schwartz and White, 1981
HO_2	$(1–3) \times 10^3$	Schwartz, 1984b
OH	30	Golden et al., 1990; Hanson et al., 1992
PAN	5	Holdren et al., 1984
CH_3SCH_3	0.48–0.56	Shorter et al., 1995; De Bruyn et al., 1995; Dacey et al., 1984
NO_3	0.6–1.8	Rudich et al., 1996; Thomas et al., 1998
CH_3SH	0.20	De Bruyn et al., 1995
H_2O_2	8.3×10^4	O'Sullivan et al., 1996
	1.1×10^5	Zhou and Lee, 1992; Staffelbach and Kok, 1993
CH_3OOH	3.1×10^2	O'Sullivan et al., 1996
$HOCH_2OOH$	1.7×10^6	O'Sullivan et al., 1996
	1.7×10^6	Staffelbach and Kok, 1993
	5×10^5	Zhou and Lee, 1992
$HOCH_2OOCH_2OH$	6×10^5	Zhou and Lee, 1992
$CH_3C(O)OOH$	8.4×10^2	O'Sullivan et al., 1996
C_2H_5OOH	3.4×10^2	O'Sullivan et al., 1996
H_2S	0.087	Shorter et al., 1995; De Bruyn et al., 1995
COS	0.022	Shorter et al., 1995; De Bruyn et al., 1995
CS_2	0.055	De Bruyn et al., 1995

[a] Adapted from Schwartz, 1984a; see also Web site in Appendix IV.
[b] Physical solubility; reacts with liquid water.
[c] Physical solubility; exclusive of acid–base equilibria.
[d] See Clegg and Brimblecombe (1988) and Brimblecombe and Clegg (1988) for temperature dependence.

fog likely reflects the different liquid water contents, i.e., in essence, the effects of concentration dilution.

C. OXIDATION OF SO_2

1. Field Studies

Combustion of fossil fuels containing sulfur produces SO_2. Because virtually all of the sulfur is emitted in the form of SO_2, with only small amounts in other forms such as H_2SO_4, sulfates, and SO_3, one can readily calculate SO_2 emissions from the sulfur content of the fuel. Once emitted, the gaseous SO_2 is oxidized in the plume itself or, after dilution with the surrounding air, to form H_2SO_4 and sulfates. It is these oxidation reactions that are the major focus of this chapter.

Numerous field studies of the rate of SO_2 oxidation in the troposphere have shown that the oxidation rate depends on a number of parameters. These include the presence of aqueous phase in the form of clouds and fogs, the concentration of oxidants such as H_2O_2 and

O_3, and sunlight intensity. For example, sulfur in cloudwater in the Los Angeles area over a two-year period from 1983 to 1985 was essentially all in the form of sulfate; at the same time, excess H_2O_2 was present in the cloudwater (Richards, 1995). Similarly, fogs and clouds have been shown to be associated with significant sulfate production (e.g., see Eatough et al., 1984; Jacob et al., 1987; Pandis et al., 1992; and Laj et al., 1997). SO_2 oxidation rates in a plume from a coal-fired power plant have been measured to be typically less than 0.5% h^{-1} in February but 1–3% at midday in June when the sunlight intensity and availability of oxidants are much greater (Lusis et al., 1978). The same effects of clouds and fogs, oxidants, and sunlight intensity on the rate of oxidation of SO_2 have been observed after the contents of the plume have been dispersed in air (e.g., see McMurry and Wilson, 1983; and Husain and Dutkiewicz, 1990). Indeed rates of oxidation as high as ~30% h^{-1} have been measured in ambient air in such locations as Budapest, Hungary (Mésaros et al., 1977), and St. Louis, Missouri (Breeding et al., 1976; Alkezweeny and Powell, 1977).

There are some sampling sites located on mountains that have been used to assess the relative amounts of SO_2 oxidation that occurs in clouds versus in the gas phase. In these cases, fixed sampling sites can be used and the concentrations of SO_2, sulfate, and associated species measured in the gas and particle phases as well as in the cloudwater as the cloud passes over the sampling site (e.g., see Saxena and Lin, 1990; Aneja and Kim, 1993; and Burkhard et al., 1995).

Husain and co-workers, for example (Husain, 1989; Husain et al., 1991; Husain and Dutkiewicz, 1992; Burkhard et al., 1995; Dutkiewicz et al., 1995), have developed techniques using Se, As, and Sb as tracers to follow the oxidation of SO_2 with time in clouds at Whiteface Mountain in New York State. The principle is based on the fact that the major sources of these metals are high-temperature combustion, e.g., of coal and oil. Thus these metals are found in particles (see Chapter 9) that act as condensation nuclei for cloud formation. These particles also contain sulfate formed from the gas-phase oxidation of SO_2. Because of the low vapor pressure of H_2SO_4, it becomes associated with particles, either by homogeneous condensation or by condensing out on preexisting particles (see Chapter 9.C). As a result, the metals can be used as tracers for sulfate formed in the gas phase, as opposed to sulfate formed by the uptake of SO_2 into cloudwater followed by oxidation.

Figure 8.2 depicts the principle of such an experiment. In air mass A below the cloud, there are sulfate and trace metals such as selenium in suspended particles, and SO_2 and oxidants in the gas phase. As the air

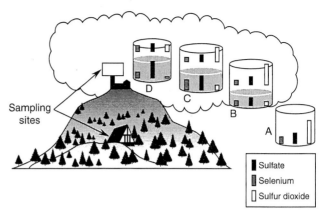

FIGURE 8.2 Schematic of cloud SO_2 oxidation studies (from Burkhard et al., 1995).

rises, the temperature falls and condensation to form cloudwater depicted in B results. The water content of the cloud increases with height through C and D. From B to D, not only is there sulfate in the cloudwater from the particles in A that served as condensation nuclei for formation of the cloud, but uptake of SO_2 and oxidants occurs, leading to in-cloud sulfate formation. As a result, gaseous SO_2 decreases and sulfate in the aqueous phase increases. Using this technique, Burkhard et al. showed the percentage of in-cloud oxidation to be quite large, up to ~50% of the total measured sulfate (Burkhard et al., 1994). Similar results have been obtained elsewhere using other techniques such as stable sulfur isotopes (e.g., see Tanaka et al., 1994).

As we shall see in the following sections, these observations are readily understood in terms of the kinetics and mechanisms of oxidation of SO_2. The oxidation of SO_2 occurs in solution and on the surfaces of solids as well as in the gas phase. Indeed, under many conditions typical of the troposphere, oxidation in the aqueous phase provided by clouds and fogs predominates, consistent with the observed dependence on these factors. The presence of oxidizers to react with the SO_2 is, of course, also a requirement; hence the dependence on O_3 (which is a useful surrogate for other oxidants as well) and sunlight, which is needed to generate significant oxidant concentrations.

Figure 8.3 summarizes the pathways that must be considered for SO_2 oxidation and for the deposition of sulfur compounds (Lamb et al., 1987). The focus of this chapter is on the chemistry converting SO_2 to sulfate in both the gas and condensed phases.

As we have seen in Chapter 7, the oxidation of NO_x to HNO_3 occurs to a large extent in the gas phase as well as by the hydrolysis of N_2O_5 on surfaces, and the acid is then taken up by dissolving in clouds and fogs;

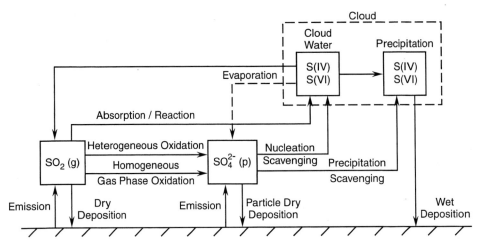

FIGURE 8.3 Summary of emission, oxidation, and deposition of S(IV) and S(VI). (Adapted from *Atmos. Environ. 21*, Lamb, D., Miller, D. F., Robinson, N. F., and Gertler, A. W. "The Importance of Liquid Water Concentration in the Atmospheric Oxidation of SO₂," pp. 2333–2344. Copyright 1987, with permission from Elsevier Science.)

in contrast, a large portion of the total SO$_2$ oxidation can occur in the condensed phase (e.g., see Pandis and Seinfeld, 1989a; Wurzler *et al.*, 1995; and Bergin *et al.*, 1996). Another important difference between nitric and sulfuric acids is in the stable form of these two acids in the atmosphere. The vapor pressure of HNO$_3$ is sufficiently large at tropospheric temperatures (e.g., 48 Torr at 20°C) that it exists in the gas phase at the ppt to ppb levels found in the lower atmosphere. Sulfuric acid, on the other hand, has a very low vapor pressure (9.9 × 10^{-6} Torr at 296 K) so that it exists in the condensed phase diluted with water to varying degrees (e.g., Roedel, 1979; Ayers *et al.*, 1980).

The vapor pressure of H$_2$SO$_4$ above solutions with water depends on the solution composition and the temperature. For example, the vapor pressure at 25°C varies from 2.6 × 10^{-9} Pa for a 54.1 wt% H$_2$SO$_4$–H$_2$O solution to 5.9 × 10^{-6} Pa for a 76.0 wt% solution (Marti *et al.*, 1997). The vapor pressures above solutions partially neutralized with ammonia are also reported by Marti *et al.* (1997); as discussed in Chapter 9.B.1, the vapor pressures of the partially neutralized solutions are orders of magnitude smaller than those of the acid. As a result, ammonia may play an important role in nucleation of gaseous sulfuric acid in the atmosphere to form new particles.

2. Oxidation in the Gas Phase

a. Hydroxyl Radical

The only significant oxidant for SO$_2$ in the gas phase is the OH radical:

$$\text{OH} + \text{SO}_2 \xrightarrow{\text{M}} \text{HOSO}_2. \quad (4)$$

This reaction is termolecular and is in the falloff region between second and third order at 1 atm pressure. The recommended high- and low-pressure limiting rate constants for reaction (4) at room temperature are $k_\infty = 2 \times 10^{-12}$ cm^3 molecule^{-1} s^{-1} and $k_0 = 4.0 \times 10^{-31}$ [N$_2$] cm^6 molecule^{-2} s^{-1} with $F_C = 0.45$ (see Chapter 5.A.2) (Atkinson *et al.*, 1997b) or, alternatively, $k_\infty = 1.5 \times 10^{-12}$ cm^3 molecule^{-1} s^{-1} and $k_0 = 3.0 \times 10^{-31}$ [M] cm^6 molecule^{-2} s^{-1} with $F_C = 0.6$ (DeMore *et al.*, 1997). The effective bimolecular rate constant at room temperature is thus $k_4^{bi} = (9.7$ or $8.8) \times 10^{-13}$ cm^3 molecule^{-1} s^{-1}, depending on which set of recommendations is used, and the corresponding lifetime of SO$_2$ with respect to OH at 1×10^6 radicals cm^{-3} is ~13 days.

The adduct free radical formed in reaction (4) [which has been detected directly using neutralization/reionization mass spectrometry (Egsgaard *et al.*, 1988)] subsequently reacts with O$_2$:

$$\text{HOSO}_2 + \text{O}_2 \xrightarrow{\text{M}} \text{HO}_2 + \text{SO}_3. \quad (5)$$

The exothermicity of reaction (5) has been estimated to be ~3.7 kcal mol^{-1} (e.g., Li and McKee, 1997).

Although the fate of the HOSO$_2$ adduct, and in particular the products of the HOSO$_2$–O$_2$ reaction, was controversial for a number of years, Stockwell and Calvert (1983) established that it must produce HO$_2$ since OH was regenerated in the presence of NO. This was subsequently confirmed by a number of researchers [e.g., see Meagher *et al.*, 1984; Margitan, 1984; Martin *et al.*, 1986; Gleason *et al.*, 1987; Gleason and Howard, 1988; and Anderson *et al.*, 1989] who measured the rate constant for reaction (5). This reaction is fast, with the current recommended rate constant being $k_5 = 4.3 \times 10^{-13}$ cm^3 molecule^{-1} s^{-1} at room temperature (Atkinson *et al.*, 1997b). The lifetime of the HOSO$_2$ adduct is thus expected to be only 0.5 μs at 1 atm in air.

While the SO$_3$ formed reacts rapidly with water, forming sulfuric acid,

$$SO_3 + H_2O \rightarrow \rightarrow H_2SO_4, \qquad (6)$$

the actual mechanism of reaction (6) was not well understood until recently. A combination of theoretical (Morokuma and Muguruma, 1994; Hoffmann and Schleyer, 1994) and experimental work (e.g., Reiner and Arnold, 1993; Kolb *et al.*, 1994; Phillips *et al.*, 1995; Lovejoy *et al.*, 1996; Jayne *et al.*, 1997) has established that SO$_3$ forms a complex with H$_2$O, but this does not appear to form H$_2$SO$_4$ directly. It appears likely that the H$_2$O–SO$_3$ complex reacts with a second water molecule, leading to H$_2$SO$_4$ formation via a cyclic six-membered transition state such as that shown in Fig. 8.4. An alternate mechanism that cannot be ruled out is the reaction of SO$_3$ with a water dimer. This leads to the reaction being second order in H$_2$O and first order in SO$_3$ (Lovejoy *et al.*, 1996; Jayne *et al.*, 1997).

While reaction of SO$_3$ with water vapor to form sulfuric acid is expected to be by far its major fate, there is also the possibility that some minor reaction paths could play a role under some circumstances. For example, SO$_3$ forms a complex with NH$_3$ and ultimately sulfamic acid (H$_2$NSO$_3$H), which rapidly forms dimers (Shen *et al.*, 1990; Lovejoy and Hanson, 1996). Jayne *et al.* (1997) measured a reaction probability of SO$_3$ of ~1 on surfaces that had adsorbed water, suggesting that such heterogeneous losses may become important at higher altitudes. In addition, as discussed in Chapter 4.K.2, the photolysis of SO$_3$ may compete with the reaction with water vapor in the upper stratosphere.

The production of HO$_2$ in reaction (5) is important to the overall chemistry of SO$_2$ oxidation in laboratory systems as well as in air. Thus in the presence of NO, the HO$_2$ reacts to regenerate OH:

$$HO_2 + NO \rightarrow OH + NO_2. \qquad (7)$$

As a result, the initial SO$_2$–OH reaction does not lead to the net loss of OH and a chain oxidation of SO$_2$ can result. Perhaps more important, the generation of HO$_2$ leads to increased H$_2$O$_2$ production. As discussed in Section C.3.e, this highly soluble gas is a major oxidant for S(IV) in the aqueous phase so that reaction (5) can affect not only gas-phase processes but also the oxidation in clouds and fogs (e.g., Stockwell, 1994).

b. Criegee Biradical

While OH is the major gas-phase oxidant for SO$_2$, Criegee biradicals may also contribute. This is so particularly at night when OH concentrations are small but significant concentrations of O$_3$ and alkenes may exist, generating the Criegee intermediate (see Chapter 6.E.2).

The first indication of a reaction between the Criegee intermediate and SO$_2$ came from studies by Cox and Penkett (1971, 1972), who showed that although the oxidation of SO$_2$ by O$_3$ alone was negligible, it was relatively fast in the presence of both ozone and alkenes. In addition, water vapor inhibited the SO$_2$ oxidation in this system. These observations can be understood in terms of the competition between the reactions of the Criegee intermediate with SO$_2$ and H$_2$O.

Since then, there have been a number of studies of this reaction, which have been summarized by Hatakeyama and Akimoto (1994). The mechanism appears to involve formation of an adduct that can either decompose to SO$_2$ and an isomerization product of the Criegee intermediate or, alternatively, react with a second SO$_2$ molecule to generate other products. For the Criegee intermediate formed in the ethene–ozone reaction, for example, the proposed reaction sequence is the following:

$$H\dot{C}HOO\cdot + SO_2 \rightarrow Adduct, \qquad (8)$$

$$Adduct \rightarrow HC(O)OH + SO_2, \qquad (9)$$

$$Adduct + SO_2 \rightarrow HCHO + SO_3 + SO_2. \qquad (10)$$

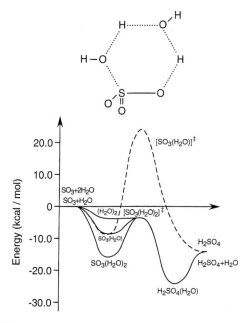

FIGURE 8.4 Predicted energetics for various mechanisms for the SO$_3$–H$_2$O reaction transition states are marked with a ‡ symbol (adapted from Morokuma and Muguruma, 1994).

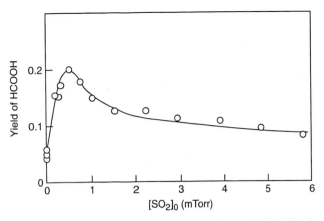

FIGURE 8.5 Yield of HCOOH as a function of added SO_2 in C_2H_4–O_3–SO_2 reactions (adapted from Hatakeyama and Akimoto, 1994).

Figure 8.5, for example, shows that the yield of formic acid first increases and then decreases as the SO_2 concentration increases, as expected from the competition of reactions (9) and (10) for the adduct. At the same time, as seen in Fig. 8.6, the increased yield of HCHO upon addition of SO_2 to the ethene–ozone reaction was equivalent to the consumption of SO_2 (Hatakeyama *et al.*, 1986).

Martinez and Herron (1981, 1983) proposed that the adduct has a cyclic structure, similar to organic secondary ozonides:

$$\begin{array}{c} H \\ \diagdown \\ C \\ H \diagup \diagdown \end{array} \begin{array}{c} O-O \\ \diagdown \\ S=O \\ O \diagup \end{array}$$

The reaction (10) of the adduct with SO_2 generates SO_3, which then reacts with water vapor to form H_2SO_4. However, the adduct itself also reacts directly with water vapor to form H_2SO_4 and HCHO (Martinez and Herron, 1981, 1983). Akimoto and co-workers have evaluated the relative rate constants for reaction of the HCHOO· –SO_2 adduct with H_2O compared to SO_2 to be $(0.6–2) \times 10^{-4}$ (Hatakeyama and Akimoto, 1994). However, at 298 K and 50% relative humidity, the water vapor concentration corresponds to about 4×10^{17} molecules cm^{-3} compared to $\sim 3 \times 10^{12}$ molecules cm^{-3} for SO_2 in a polluted atmosphere. As a result, reaction with water vapor is favored by more than an order of magnitude.

One can estimate the relative contribution of the Criegee intermediate (CI) to SO_2 oxidation in the gas phase in the troposphere. The absolute value of the rate constant for the reaction of the CI with SO_2 is not known, with estimates ranging from 1.7×10^{-11} to 3×10^{-15} cm^3 $molecule^{-1}$ s^{-1} (Hatakeyama and Akimoto, 1994). Using the highest value and a concentration of the CI of 1×10^5 molecules cm^{-3}, one obtains $\sim 10^{-6}$ s^{-1} for the first-order rate of removal of SO_2 by this reaction. This can be compared to the rate of removal of SO_2 by reaction with 1×10^6 OH radicals cm^3, which is also $\sim 10^{-6}$ s^{-1} using the effective bimolecular rate constant cited earlier. Using the lower estimates for the CI–SO_2 rate constant, which is more reasonable, would lower its contribution proportionately.

In short, the Criegee intermediate from alkene–ozone reactions can contribute, in principle, to the gas-phase oxidation of SO_2. In practice, it is likely less important than reaction with OH. In addition, as we shall see, even the OH–SO_2 gas-phase reaction is, under many conditions, swamped out by reactions occurring in the liquid phase found in clouds and fogs. As a result, the CI–SO_2 reaction may contribute in some circumstances but is unlikely to be a major contributor to SO_2 oxidation as a whole.

c. Other Gas-Phase Reactants

While one might expect that other tropospheric free radicals such as $O(^3P)$, HO_2, and RO_2 could react with SO_2 as well, such reactions are not significant.

Because the oxidants for SO_2 are generated in the VOC–NO_x system discussed in Chapter 6, the overall gas-phase mechanism for the oxidation of SO_2 to H_2SO_4 is quite complex. The reader is referred to the mechanism in the "RADM" model (*R*egional *A*cid *D*eposition *M*odel) for a treatment of the VOC–NO_x–SO_2 chemistry (Stockwell *et al.*, 1990; Gao *et al.*, 1996). It should be noted that an earlier version of this mechanism is given in some of the examples included with the OZIPR model discussed in Chapter 16 and whose applications are included with this book.

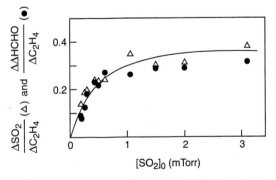

FIGURE 8.6 Loss of SO_2 and increase in HCHO as a function of SO_2 in the C_2H_4–O_3–SO_2 reactions. Triangle: $\Delta SO_2/\Delta C_2H_4$; solid circle: $\Delta\Delta HCHO/\Delta C_2H_4$ (adapted from Hatakeyama and Akimoto, 1994).

BOX 8.1
CALCULATING GAS-PHASE OXIDATION RATES IN PERCENT PER HOUR

For example, let us assume that the concentration of OH is constant. Then the rate of oxidation of SO_2 by OH,

$$SO_2 + OH \xrightarrow{M} HOSO_2, \quad (4)$$

is given by

$$-\frac{d[SO_2]}{dt} = k_4^{bi}[OH][SO_2], \quad (C)$$

where $k_4^{bi} = k_4[M]$. Integrating, one obtains

$$\frac{[SO_2]}{[SO_2]_0} = e^{-k_4^{bi}[OH]t}, \quad (D)$$

where $[SO_2]$ and $[SO_2]_0$ are the concentrations at time t and $t = 0$, respectively. The percentage change per hour in the ambient SO_2 concentration is thus given by

$$\% \, h^{-1} = -100\left(\frac{[SO_2]_{t=1h} - [SO_2]_0}{[SO_2]_0}\right). \quad (E)$$

Substituting into Eq. (E) from (D), one obtains

$$\% \, h^{-1} = 100(1 - e^{-k_4^{bi}[OH]t}), \quad (F)$$

where the time t is 1 h, expressed in the same time units used for the rate constant k_4. Given similar assumptions, the same type of relationship, of course, holds true for other oxidizing species as well.

---

d. Computing Oxidation Rates in the Atmosphere

The SO_2 oxidation rates in the units of % h^{-1} expected from the gas-phase reactions of SO_2 with OH and the Criegee intermediate can be computed as shown in Box 8.1 for comparison to rates of oxidation observed in field studies (see Section C.1), if the rate constants and the concentrations of OH and Criegee intermediates are known.

3. Oxidation in the Aqueous Phase

a. S(IV) Aqueous Equilibria

Sulfur dioxide gas dissolves in water to set up equilibria similar to those of CO_2:

$$SO_{2(g)} + H_2O \leftrightarrow SO_2 \cdot H_2O_{(aq)} \quad H_{11} = 1.242 \, M \, atm^{-1}, \quad (11, -11)$$

$$SO_2 \cdot H_2O_{(aq)} \leftrightarrow HSO_3^- + H^+ \quad K_{12} = 1.32 \times 10^{-2} \, M, \quad (12, -12)$$

$$HSO_3^- \leftrightarrow SO_3^{2-} + H^+ \quad K_{13} = 6.42 \times 10^{-8} \, M. \quad (13, -13)$$

As a result, *dissolved SO_2* really includes three chemical species: hydrated SO_2 ($SO_2 \cdot H_2O$), the bisulfite ion (HSO_3^-), and the sulfite ion (SO_3^{2-}). While HSO_3^- is commonly referred to as bisulfite, it should be noted that it can exist in two possible structures: $HOSO_2^-$ or HSO_3^-. Recent calculations indicate that the sulfonate form, HSO_3^-, is the most stable form (Brown and Barber, 1995). However, Hoffmann (1986) argues that the chemical reactivity of bisulfite in aqueous solutions is consistent with $HO-SO_2^-$, being the major reactive automer, which is in rapid equilibrium with HSO_3^-. While the formation of an $SO_2 \cdot H_2O$ complex similar to that shown in (11, −11) can, in principle, occur in the gas phase as well, the binding energy is low ($\sim 1.8-3.5$ kcal mol^{-1}; e.g., see Li and McKee, 1997; and Bishenden and Donaldson, 1998) and hence this complex is not important in the atmosphere.

The predominant form dissolved in solution depends on the acidity of the solution in which SO_2 dissolves. Figure 8.7 shows the concentrations of the three species as a function of pH; over the pH range typical of atmospheric droplets, 2–6, most of the dissolved SO_2 is in the form of bisulfite ion (HSO_3^-).

Because of the different forms in which dissolved SO_2 exists in solution, the oxidation state (i.e., +4) is often used to denote all these forms of SO_2 taken together, that is,

$$S(IV) = SO_2 \cdot H_2O + HSO_3^- + SO_3^{2-}.$$

The oxidized form of sulfur (i.e., sulfuric acid and sulfate) is in the +6 oxidation state and hence is commonly referred to as S(VI).

The individual reactions in the equilibria represented by (11)–(13) are relatively fast (Martin, 1984).

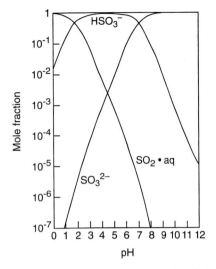

FIGURE 8.7 Mole fraction of sulfur species in solution at different acidities (adapted from Martin and Damschen, 1981).

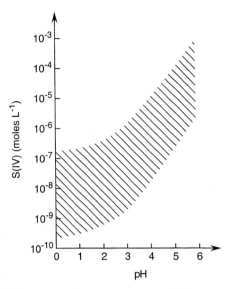

FIGURE 8.8 Range of expected aqueous S(IV) concentrations as a function of acidity for gas-phase SO_2 concentrations of 0.2–200 ppb (adapted from Martin, 1984).

For example, the rate constant for dissociation of hydrated SO_2, k_{12}, is 3.4×10^6 s^{-1} so that the half-life for dissociation of the hydrated SO_2 is only 0.2 μs. Similarly, the second ionization, reaction (13), occurs on time scales of less than a millisecond (Schwartz and Freiberg, 1981). Thus, regardless of which of the three species, $SO_2 \cdot H_2O$, HSO_3^-, or SO_3^{2-}, is the actual reactant in any particular oxidation, the equilibria will be reestablished relatively rapidly under laboratory conditions, and likely under atmospheric conditions as well. The latter is complicated by such factors as the size of the droplet, the efficiency with which gaseous SO_2 striking a droplet surface is absorbed, the chemical nature of the aerosol surface, and so on; for example, the presence of an organic surface film on the droplet could hinder the absorption of SO_2 from the gas phase.

As expected from the equilibria (11)–(13) and Le Chatelier's principle, the more acidic the droplet, the more equilibria will shift to the left, that is, the less the dissolved SO_2. Figure 8.8 shows the range of dissolved S(IV) concentrations expected in aqueous solutions that are in equilibrium with SO_2 in the gas phase at concentrations of 0.2–200 ppb and over a pH range of 0–6. It is seen that a wide range of concentrations, from $\sim 10^{-9}$ to 10^{-3} mol L^{-1}, of S(IV) is anticipated, depending on the pH and on the concentration of SO_2 in the gas phase. As expected, the aqueous-phase S(IV) concentration falls as the pH falls.

This dependence of the S(IV) concentrations on the pH of the droplet plays a critical role in determining which oxidant dominates the S(IV) oxidation. As discussed in more detail later, the rates of the various aqueous-phase reactions show different dependencies on pH. Some have rate coefficients that increase with increasing pH (e.g., O_3) whereas others (e.g., H_2O_2) show the opposite trend.

In the first case shown schematically in Fig. 8.9a, both the rate constant and solubility of S(IV) vary with pH in the same manner. As a result, the overall rate of production of S(IV), i.e., $k[S(IV)]$, by such reactions

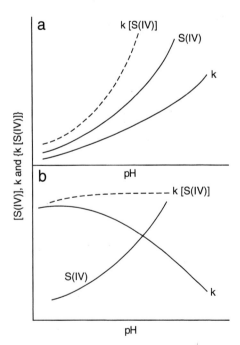

FIGURE 8.9 Schematic diagram of effect of pH on the rate constant k and on the concentration of dissolved S(IV) and its total rate of oxidation represented by $k[S(IV)]$ for two cases: (a) rate constant k decreases with pH; (b) k increases with pH.

typically shows a strong pH dependence, so that they represent important oxidation pathways for S(IV) only at higher pH values. This is the case for oxidation by O_3. Such oxidative pathways may initially contribute significantly to acid formation in a relatively neutral droplet; however, as more and more acid is formed, the rate of formation of S(VI) by these reactions will decrease. This *self-quenching* effect may lead either to a lowered net oxidation of SO_2 in the atmosphere or to other pathways that do not show this pH dependence becoming relatively more important.

On the other hand, if the rate constant decreases with increasing pH, as schematically shown in Fig. 8.9b, then the rate constant and solubility of S(IV) work in opposite directions. The net result in terms of S(VI) production may be, as in the case of H_2O_2, a relatively small dependence on pH. Such oxidations may thus contribute in a relatively constant fashion over the entire range of pH values of atmospheric interest.

Table 8.2 shows some of the important acid–base equilibria in aqueous phases in the atmosphere. In addition to the S(IV) and CO_2 equilibria, the dissociation of H_2O_2 and the HO_2 free radical (generated in reactions described later) generate H^+ and the reactive radical anions HO_2^- and O_2^-, respectively. Ammonia is the only gas-phase base found in the atmosphere. Uptake of strong acids such as HNO_3 and HCl also acidifies droplets, as do weak acids such as HONO and the organic acids.

The equilibria discussed earlier apply to SO_2 dissolved in pure water, and these have commonly been used for calculations of the concentrations of S(IV) in atmospheric droplets. However, a variety of measurements of the concentration of S(IV) in fog and cloudwater show that these concentrations are far in excess of what is expected based only on equilibria. Water droplets in the atmosphere, especially in or near urban areas, do not consist of pure water; they contain species such as aldehydes and Fe^{3+} that are known to form complexes in solution with the bisulfite or sulfite ions.

[For a detailed discussion of S(IV) complexes with transition metal ions and organics, see Huie and Peterson (1983) and Eatough and Hansen (1983).]

For example, aldehydes such as formaldehyde react with bisulfite and sulfite ions in solution (Boyce and Hoffmann, 1984):

$$H_2C=O + HSO_3^- \longleftrightarrow H-\overset{\overset{OH}{|}}{\underset{\underset{H}{|}}{C}}-\overset{\overset{O}{\|}}{\underset{\underset{O}{\|}}{S}}-O^- \quad (14)$$

$$H_2C=O + SO_3^{2-} \longleftrightarrow H-\overset{\overset{O^-}{|}}{\underset{\underset{H}{|}}{C}}-\overset{\overset{O}{\|}}{\underset{\underset{O}{\|}}{S}}-O^- \quad (15)$$

Table 8.3 shows the solubilities of some potentially important aldehydes in the form of the Henry's law constant (H) and the effective Henry's law constant (H^*) (Betterton and Hoffmann, 1988; Olson and Hoffmann, 1989). These aldehydes not only dissolve in aqueous solutions but also hydrate to form *gem*-diols (Buschmann *et al.*, 1980, 1982):

$$R_1R_2C=O + H_2O \leftrightarrow R_1R_2C(OH)_2.$$

This reaction is very rapid and, being acid catalyzed, occurs even more rapidly in acidic fogs and clouds than in neutral solutions. For example, at a pH of 3, the hydration of the small aldehydes occurs in about 1 s (Buschmann *et al.*, 1982). As seen from the data in Table 8.3, the equilibrium constant for this reaction, also known as the hydration constant $K_{hyd} = [R_1R_2C(OH)_2]_{aq}/[R_1R_2CHO]_{aq}$, varies from $\sim 10^3$ for HCHO to ~ 1 for CH_3CHO (and for higher straight-chain aldehydes; Buschmann *et al.*, 1980, 1982; Xu *et al.*, 1993; Sham and Joens, 1995). Thus, the aldehyde and diol exist in comparable concentrations in aqueous solutions for acetaldehyde and larger aldehydes, whereas HCHO will exist predominantly as $CH_2(OH)_2$.

TABLE 8.2 Some Important Acid–Base Equilibria in the Aqueous Phase in the Atmosphere

Aqueous-phase reaction	K^{298K}	Reference
$SO_2 \cdot H_2O_{(aq)} \leftrightarrow H^+ + HSO_3^-$	1.32×10^{-2}	Sillén and Martell, 1964
$HSO_3^- \leftrightarrow H^+ + SO_3^{2-}$	6.42×10^{-8}	Sillén and Martell, 1964
$CO_2 \cdot H_2O_{(aq)} \leftrightarrow H^+ + HCO_3^-$	4.3×10^{-7}	Sillén and Martell, 1964
$HCO_3^- \leftrightarrow H^+ + CO_3^{2-}$	4.7×10^{-11}	Sillén and Martell, 1964
$H_2O_2 \leftrightarrow H^+ + HO_2^-$	2.5×10^{-12}	Sauer, 1984
$HO_2 \leftrightarrow H^+ + O_2^-$	2.05×10^{-5}	Bielski, 1978; Bielski *et al.*, 1985
$NH_3 + H_2O \leftrightarrow NH_4^+ + OH^-$	1.8×10^{-5}	Zellner, 1995
$HONO \leftrightarrow H^+ + NO_2^-$	5.1×10^{-4}	Schwartz and White, 1981
$HCOOH \leftrightarrow H^+ + HCOO^-$	1.7×10^{-4}	Eigen *et al.*, 1964
$CH_3COOH \leftrightarrow H^+ + CH_3OO^-$	1.6×10^{-5}	Eigen *et al.*, 1964

TABLE 8.3 Intrinsic (H) and Effective (H^*) Henry's Law Constants and Hydration Constants (K_{hydr}) at 298 K for Some Aldehydes of Atmospheric Interest[a]

Aldehyde	H[b] (M atm^{-1})	H^*[c] (M atm^{-1})	K_{hydr}
HCHO	1.3	3.0×10^3	2.3×10^3
CH_3CHO	4.8	1.14×10	1.4
			1.13[e]
CH_3CH_2CHO			1.02[e]
$\overset{O}{\underset{\|\|}{HC}}-CHO$	≥1.4	≥3.0×10^5	2.2×10^5
$\overset{O}{\underset{\|\|}{CH_3C}}CHO$	1.4	3.7×10^3	2.7×10^3
$HOCH_2CHO$	4.1×10^3	4.1×10^4	10
C_6H_5CHO	—	3.7×10	—
CCl_3CHO	14	3.4×10^5	2.5×10^4
n-$CH_3CH_2CH_2CHO$			0.50[e]
n-$CH_3(CH_2)_4CHO$			0.49[f]

[a] Except where noted, from Betterton and Hoffmann (1988) and Olson and Hoffmann (1989).
[b] $H = [RCHO]_{aq}/[RCHO]_g$.
[c] $H^* = \{[RCHO]_{aq} + [RCH(OH)_2]_{aq}\}/[RCHO]_g = H(1 + K_{hydr})$.
[d] $K_{hydr} = [RCH(OH)_2]_{aq}/[RCHO]_{aq}$.
[e] Xu et al. (1993).
[f] Sham and Joens (1995).

Formation of the diols causes a significant change in the absorption spectra compared to the gas phase, with the shift to shorter wavelengths decreasing their photolysis in solution (Xu et al., 1993; Sham and Joens, 1995).

Because of this hydration, the total solubility, i.e., effective Henry's law constant, is larger than expected based on physical solubility alone. The data in Table 8.3 show that most aldehydes have quite large effective Henry's law constants (H^*), the exceptions being acetaldehyde and benzaldehyde. As a result of these high solubilities, significant concentrations can occur in fogs and clouds and hence be available to complex with S(IV).

For complex formation between aldehydes and S(IV) to be important in the troposphere, the aldehydes not only must have high solubility but also be present in air at significant concentrations and form stable adducts with S(IV) at a sufficiently fast rate that it can occur during the lifetime of a typical cloud or fog event. Table 8.4 gives the rate constants k_{14} and k_{15} for formation of the S(IV) complexes as well as the stability constants K_{14} and apparent stability constant K_{14}^{app}, defined as

$$K_{14} = \frac{[RCH(OH)SO_3^-]_{aq}}{[RCHO]_{aq}[HSO_3^-]_{aq}}$$

TABLE 8.4 Stability and Rate Constants for the Formation of S(IV) – Aldehyde Complexes in Aqueous Solution at 298 K

RCHO	K_{14}	K_{14}^{app}	k_{14} (L mol^{-1} s^{-1})	k_{15} (L mol^{-1} s^{-1})
HCHO	6.6×10^9	3.6×10^6	7.9×10^2	2.5×10^7
$CH_3\overset{O}{\underset{\|\|}{C}}CHO$	8.1×10^8	3.1×10^5	3.4×10^3	3.7×10^7
$HOCH_2CHO$	2.0×10^6	2.0×10^5	1.7	5.0×10^4
C_6H_5CHO	4.8×10^3	4.8×10^3	0.71	2.2×10^4

[a] From Olson and Hoffmann (1989).

and

$$K_{14}^{app} = \frac{[RCH(OH)SO_3^-]_{aq}}{\{[RCHO]_{aq} + [RCHO(OH)_2]_{aq}\}[HSO_3^-]_{aq}}$$

for various large aldehydes as well as formaldehyde (Olson and Hoffmann, 1989). As discussed in detail by Hoffmann and co-workers (e.g., see Olson and Hoffmann, 1988a, 1988b, 1989; Olson et al., 1988; and Betterton and Hoffmann, 1988), formaldehyde, glyoxal, hydroxyacetaldehyde, and, to a smaller extent, methylglyoxal all meet these criteria and hence can be important reservoirs for S(IV) in droplets, whereas acetaldehyde and benzaldehyde do not. Other multifunctional compounds such as glyoxylic acid [CHOC(O)OH] may also contribute in weakly acidic solutions (Olson and Hoffmann, 1988a, 1988b, 1989).

Although the S(IV)–aldehyde adducts are stable toward oxidation, one or more of the oxidation processes for HSO_3^- or SO_3^{2-} described below are likely to be much faster than adduct formation under typical fog and cloud conditions. For example, Fig. 8.10 shows the calculated times for complexing S(IV) with HCHO compared to the time for oxidation by H_2O_2 at different concentrations typical of various clouds and fogs as a function of pH (Rao and Collett, 1995). Even at the lowest H_2O_2 concentrations and highest HCHO concentrations, complexation only competes with oxidation at pH values above about 4.5. Thus the two processes, complexation and oxidation, are expected to occur in parallel, with oxidation being more important as the droplets become acidified.

The formation of such adducts, however, can play a significant role in determining the composition of clouds and fogs. Thus much larger aqueous-phase concentrations of aldehydes and S(IV) than predicted from simple Henry's law equilibria are possible if the adducts are formed. For example, Munger et al. (1986) measured the concentration of hydroxymethanesulfonate (HMSA), the HCHO–S(IV) adduct, in fog water as well as the aqueous-phase S(IV) and HCHO and found that in some cases, the fogs were supersaturated in S(IV) by as much as a factor of 10 compared to that expected on the basis of Henry's law. Similarly, Rao and Collett (1995) showed that most of the S(IV) in cloudwater samples collected in five different locations across the United States was in the form of HMSA, while there was excess HCHO, and interestingly, H_2O_2. Over the range of values of pH typically found in fogs and clouds, ~2–5, the HCHO–S(IV) adduct is primarily in the monovalent form shown in Eq. (14) because the $CH_2(OH)SO_3^-$ is a relatively weak acid, $pK_a = 11.7$ (Sørensen and Andersen, 1970; Jacob and Hoffmann, 1983; Deister et al., 1986). HMSA has also been measured in aerosol particles, presumably due to formation in cloudwater droplets followed by evaporation of water (e.g., Dixon and Aasen, 1999).

In short, complex formation involving S(IV) and aldehydes is now known to be important in a number of cases and must be considered in the chemistry of fogs and clouds.

With these caveats in mind concerning possible complex formation, we examine potential oxidants for S(IV) in solution. These include O_2, O_3, H_2O_2, free radicals such as OH and HO_2, and oxides of nitrogen (e.g., NO, NO_2, HONO, and HNO_3). Metal catalysis may play a role in some of these reactions.

There are two major factors to be considered in assessing the contribution of potential oxidants for S(IV) to the net aqueous-phase oxidation. The first is the aqueous-phase concentration of the species, and the second is the reaction kinetics, that is, the rate constant and its pH and temperature dependencies. As a first approximation to the aqueous-phase concentrations, Henry's law constants (Table 8.1) can be applied. It must be noted, however, that as discussed earlier for S(IV) this approach may lead to low estimates if complex formation occurs in solution. On the other hand, high estimates may result if equilibrium between the gas and liquid phases is not established, for example, if an organic film inhibits the gas-to-liquid transfer (see Section 9.C.2).

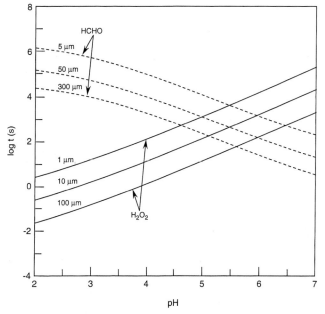

FIGURE 8.10 Calculated characteristic times for formation of the HCHO–S(IV) complex or oxidation by H_2O_2 under various conditions (adapted from Rao and Collett, 1995).

Note that the Henry's law constants given in Table 8.1 are generally for 25°C. At lower temperatures, the values will, of course, be larger due to the increased solubility. While the rate constants for most reactions, especially those in the liquid phase, decrease with decreasing temperature, the increased reactant concentrations tend to counterbalance this effect. Thus rates of reactant loss and product formation are not as sensitive to temperature as the rate constant alone.

b. Physical and Chemical Steps in Aqueous-Phase Oxidation

In the discussion that follows, we focus on the kinetic studies of S(IV) oxidations in aqueous solutions. However, it must be recognized that the oxidation itself is only one portion of a sequence of processes that leads from gas-phase SO_2 to aqueous-phase sulfate. The sequence of steps, depicted in Fig. 8.11, is as follows:

1. Transport of the gas to the surface of the droplet
2. Transfer of the gas across the air–liquid interface (note that the formation of unique surface species may occur; see below)
3. Formation of aqueous-phase equilibria of the dissolved species, for example, (11)–(13) in the case of SO_2
4. Transport of the dissolved species from the surface to the bulk aqueous phase of the droplet
5. Reaction in the droplet

Schwartz and Freiberg (1981) have calculated the rates of these processes for SO_2 and expressed them in terms of *characteristic times* τ, which for Step 5, chemical reaction, is the natural lifetime discussed in Section 5.A.1.c. For Steps 1–4, the characteristic time is the time to establish the appropriate steady state or equilibrium for the process involved; for example, for Step 1, it is the time to establish a steady-state concentration of the gas in the air surrounding the droplet. Seinfeld (1986) discusses in detail calculation procedures for these characteristic times. A brief summary of the results of Schwartz and Freiberg (1981) for Steps 1–4 is as follows:

(1) *Transport of gas to the surface*. Assuming mixing occurs by molecular diffusion rather than by mechanical or convective processes, the characteristic times for gas-phase diffusion to the surface are in the range 10^{-10}–10^{-4} s for droplets with radii from 10^{-5} to 10^{-2} cm, respectively.

(2) *Transfer of the gas across the air–liquid interface*. The time for the phase equilibrium to be established across the interface if no reaction is occurring depends on the Henry's law constant for the gas dissolving in the solution; the larger the value of this constant, the larger is the characteristic time for establishing equilibrium across the air–liquid interface because more of the gas must cross the interface in order for the equilibrium to be established. This characteristic time also depends on the mass accommodation coefficient (α), that is, the fraction of collisions with the surface that result in absorption of the molecule. Assuming this is unity (i.e., no "bounce-off"), the time to establish such equilibria for gases of atmospheric interest is of the order of $< 10^{-8}$–10^{-1} s over a droplet pH range of 2–6. Of course, if the molecule is absorbed into the droplet on only a small fraction of the collisions, this time will be much longer.

(3) *Formation of the S(IV)–H_2O equilibria*. As already discussed, this occurs on a time scale of milliseconds or less.

(4) *Transport of the dissolved species within the aqueous phase*. Because diffusion in liquids is much slower than in gases, the characteristic times for diffusion within the droplet itself are much greater (by about four orders of magnitude) than for diffusion of the gas to the droplet surface (again assuming mixing only by molecular diffusion). Thus the times are ~10^{-6}–1 s

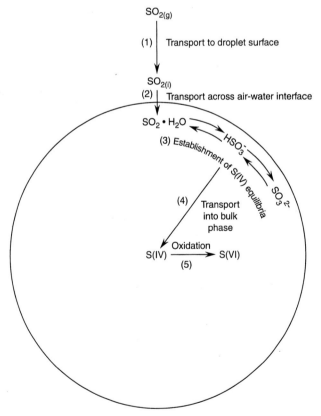

FIGURE 8.11 Schematic of steps involved in the transfer of SO_2 from the gas phase to the aqueous phase of an atmospheric water droplet and its oxidation in the liquid phase. $SO_{2(i)} = SO_2$ at the water–gas interface.

for droplets with radii from 10^{-5} to 10^{-2} cm, respectively.

Models for the uptake of gases into solution and their subsequent reaction are discussed in more detail in Chapter 5.E.1.

Table 8.5 shows the mass accommodation coefficients for SO_2, as well as for some other gases of tropospheric interest, on liquid water. It is seen that the uptake of most gases into liquid water is quite efficient. Interactions of gas molecules at the air–liquid interface may have additional implications other than the rate at which it is transferred into the aqueous phase. As discussed briefly in Chapter 5, there is increasing evidence that there are some species, and perhaps some chemistry, that are unique to the interface. For example, in the case of SO_2, Jayne *et al.* (1990a) observed that the uptake of SO_2 into water droplets was faster than expected based on the known kinetics in bulk solution. They suggested that a surface complex was formed between SO_2 and H_2O at the interface. Donaldson *et al.* (1995) subsequently observed such a species using second harmonic generation spectroscopy.

In short, while data are becoming increasingly available on the kinetics of uptake of gases into the aqueous

TABLE 8.5 Some Mass Accommodation Coefficients (α) for Gases of Tropospheric Interest on a Liquid Water Surface[a]

Gas	T (K)	α	Reference
SO_2	260–292	0.11	Worsnop *et al.*, 1989
		0.13	George *et al.*, 1992
O_3	292	$>2 \times 10^{-3}$	Utter *et al.*, 1992
OH	275	$>4 \times 10^{-3}$	Hanson *et al.*, 1992
HO_2	275	>0.02	Hanson *et al.*, 1992
		>0.2	Mozurkewich *et al.*, 1987
H_2O_2	273	0.18	Worsnop *et al.*, 1989
NO_2	298	$>1 \times 10^{-3}$	Ponche *et al.*, 1993
		1.5×10^{-3}	George *et al.*, 1992
NO_3	~290	$>2.5 \times 10^{-3}$	Mihelcic *et al.*, 1993
	273	2×10^{-4}	Rudich *et al.*, 1996
N_2O_5	271–282	0.04–0.06	Van Doren *et al.*, 1990
HONO	≈ 245, 297	0.05	Bongartz *et al.*, 1994
		0.06	George *et al.*, 1992
HNO_3	268–293	0.07–0.19	Van Doren *et al.*, 1990
		0.11	George *et al.*, 1992
NH_3	295	0.097	George *et al.*, 1992
		0.06	Ponche *et al.*, 1993; Bongartz *et al.*, 1995
CH_3OH	260–291	0.12–0.02	Jayne *et al.*, 1991
HCHO	260–270	0.04	Jayne *et al.*, 1992
CH_3COCH_3	260–285	0.066–0.013	Duan *et al.*, 1993
HCOOH	260–291	0.1–0.02	Jayne *et al.*, 1991
CH_3COOH	260–291	0.15–0.03	Jayne *et al.*, 1991
HCl	274–294	0.06–0.18	Van Doren *et al.*, 1990; George *et al.*, 1992
$CH_3S(O)CH_3$	273	0.10	De Bruyn *et al.*, 1994
$CH_3S(O)_2CH_3$	273	0.14	De Bruyn *et al.*, 1994
$CH_3S(O)_2OH$	273	0.13	De Bruyn *et al.*, 1994

[a] Source: DeMore *et al.* (1997).

phase found in clouds and fogs, there are some intriguing hints that there may also be some chemistry that is unique to the interface which might be thought of as a "fourth phase," in addition to gases, liquids, and solids. This is an area that remains to be explored.

As discussed in detail in Sections C.3.d and C.3.e, the fastest atmospheric reactions of SO_2 are believed to be with H_2O_2 and perhaps with O_3 at higher pH values. Under extreme conditions of large fog droplets (~ 10 μm) and very high oxidant concentrations, the chemical reaction times may approach those of diffusion, particularly in the aqueous phase. In this case, mass transport may become limiting. However, it is believed that under most conditions typical of the troposphere, this will not be the case and the chemical reaction rate will be rate determining in the S(IV) aqueous-phase oxidation.

Many experimental studies of the rates of oxidation of S(IV) in solution have used either bulk solutions or droplets that are very large compared to those found in the atmosphere. In addition, reactant concentrations in excess of atmospheric levels have often been used for analytical convenience. The use of large droplets increases the diffusion times, whereas higher reactant concentrations speed up the aqueous-phase chemical reaction rates. The combination of these two factors can lead to a situation where the rates of the diffusion processes, either of the gas to the droplet surface or more likely within the aqueous phase itself, become comparable to, or slower than, the chemical reaction rate. If this is not recognized, the observed rates may be attributed in error to the intrinsic chemical reaction rate.

In the atmosphere, suspended aqueous solutions are present in the form of aerosols, clouds, fogs, and rain. However, these have different liquid water contents (i.e., grams of $H_2O_{(l)}$ per cubic meter of air). As discussed in detail in Chapter 9, fine particles ($\lesssim 2$-μm diameter) emitted directly into the air or formed by chemical reactions can remain suspended for long periods of time. Many of these particles contain water, either in the form of dilute aqueous solutions or as thin films covering an insoluble core; as much as 50% of the mass may be liquid water. Since the total particulate mass in this size range per cubic meter of air can be as high as 10^{-4} g m^{-3} or more, the liquid water content due to these small particles is also of this order of magnitude.

Clouds, fogs, and rain, however, have much greater liquid water contents and thus have the potential for contributing more to atmospheric aqueous-phase oxidations. Clouds typically have liquid water contents of the order of ~ 1 g m^{-3}, with droplet diameters of the order of 5-50 μm; the number concentration and size distribution depend on the type of cloud. Fogs, on the other hand, have smaller liquid water contents (~ 0.1 g m^{-3}) and smaller droplet diameters, generally ~ 0.5-10 μm (Pruppacher and Klett, 1978). Raindrops are, of course, much larger than cloud or fog droplets, with diameters of ~ 0.2-3 mm and correspondingly large liquid water contents. However, because of their size, they remain suspended in the atmosphere for only minutes en route to the earth's surface, and hence the potential for oxidation processes to occur in raindrops is minimized.

While the volume of liquid water present is much larger in clouds and fogs than that in fine particles, the solute concentrations in the latter may be much higher, and this may serve to increase the rate of aqueous-phase oxidations. More importantly, these fine particles are believed to serve as sites for the condensation of water vapor, leading to the formation of fogs and clouds (Chapter 14.C.2).

The liquid water content of an air mass plays a role in determining the oxidation rate of SO_2 in aqueous atmospheric droplets. This can be seen from the expression developed in Box 8.2 for the rate of oxidation of SO_2 (in % h^{-1}) in the liquid phase.

Figure 8.12 shows the projected conversion of SO_2 to sulfate as a function of the volume of water per cubic meter of air available for conversion in the aqueous phase, covering a range typical of "haze particles," fogs, and clouds for atmospheric lifetimes which are typical for each (Lamb et al., 1987). As expected from Eq. (M), the conversion increases with the water available in the atmosphere. As we shall see, the aqueous-phase oxidation does indeed predominate in the atmosphere under many circumstances. Equations (G) and (M) apply as long as the partial pressure of SO_2 in the gas phase, P_{SO_2}, is measured simultaneously with the solution concentration of S(IV).

With these comments regarding the characteristics of atmospheric aqueous-phase oxidations in mind, we summarize the kinetics of S(IV) oxidation in solution by a series of individual potential atmospheric oxidants.

c. Oxidation by O_2 (Catalyzed and Uncatalyzed)

While oxidation of S(IV) in solution in the presence of O_2 has been known for many years, there has been considerable controversy concerning the rates, mechanisms, and effects of catalysts such as Fe^{3+} and Mn^{2+}, particularly under atmospheric conditions. However, studies over the past decade carried out in a number of laboratories, particularly those of Hoffmann and co-workers (e.g., Hoffmann and Boyce, 1983 and references therein) Martin and co-workers (1994 and references therein), have identified the various parameters that determine the overall rate of oxidation. As we shall see, the mechanism and kinetics are so complex that past confusion is understandable.

BOX 8.2
CALCULATING AQUEOUS-PHASE OXIDATION RATES IN PERCENT PER HOUR

The expression for the rate of oxidation in % h^{-1} in the liquid phase can be developed from a knowledge of the gas- and aqueous-phase reactant concentrations, the solution rate constant, the Henry's law constants (Table 8.1), and the liquid water content of air. In 1 m^3 of air, the rate of formation of S(VI) in the aqueous phase is given by

$$\frac{d[S(VI)]}{dt} \text{ (moles per m}^3 \text{ air)} = k[X][S(IV)]V, \quad (G)$$

where k is the solution-phase rate constant (L mol^{-1} s^{-1}), [X] and [S(IV)] are the aqueous-phase concentrations of the oxidant and S(IV), respectively, in units of moles per liter of solution, and V is the volume (L) of liquid water, that is, of aqueous solution available, per cubic meter of air. The rate of S(VI) formation is then expressed in moles per cubic meter of air per second.

To express this rate of oxidation in % h^{-1}, consistent with the units in which the results of field studies are often reported, one needs to divide this rate by the total number of moles of S(IV) per cubic meter of air, convert the unit time from s^{-1} to h^{-1}, and multiply by 100 to convert the fraction to percent. The gas-phase moles of SO$_2$ in a cubic meter of air is given, according to the ideal gas law, by

$$\left(\frac{n}{V}\right)_{SO_2} = \frac{1000 P_{SO_2}}{RT}, \quad (H)$$

where the factor of 1000 converts from L to m^3. The concentration of S(IV) in solution can be calculated using Henry's law in combination with a knowledge of the concentration of total dissolved S(IV) relative to dissolved SO$_2$ (i.e., SO$_2 \cdot$ H$_2$O). Thus the dissolved SO$_2$ concentration is given by

$$[SO_2]_{aq} = H_{SO_2} P_{SO_2}, \quad (I)$$

where H is the Henry's law constant based on physical solubility (Table 8.1) and P is the gas-phase pressure of SO$_2$. The total concentration of S(IV) in solution, taking into account the acid–base equilibria reactions (11)–(13), is then given by

$$[S(IV)]_{aq} = \eta H_{SO_2} P_{SO_2}, \quad (J)$$

where η is the ratio of the total dissolved S(IV) to that of dissolved SO$_2$. If there are V liters of liquid water per cubic meter of air, then the total number of moles of S(IV) contained in the atmospheric water droplets found in 1 m^3 of air becomes

$$\text{Moles of aqueous S(IV)} = \eta H_{SO_2} P_{SO_2} V. \quad (K)$$

The total number of moles of S(IV) in a cubic meter of air, including both gas and aqueous phases, is thus given by

Total S(IV) per m^3 of air
$$= \frac{1000 P_{SO_2}}{RT} + \eta H_{SO_2} P_{SO_2} V. \quad (L)$$

Thus, combining (G) and (L), the rate of oxidation of SO$_2$ in % h^{-1}, which occurs in aqueous solution in the atmosphere, is given by

$$\% \text{h}^{-1} = \left[\frac{100 k[X][S(IV)]V}{1000 P_{SO_2}/RT + \eta H_{SO_2} P_{SO_2} V}\right] \times 3600. \quad (M)$$

Studying the oxidation of S(IV) by O$_2$ in "pure" water without traces of catalysts or inhibitors has proven extremely difficult. Based on a compilation of many studies, Radojevic (1984) has recommended that the uncatalyzed rate of oxidation (in terms of the rate of sulfate formation) is given by

$$d[SO_4^{2-}]/dt = 0.32[SO_3^{2-}][H^+]^{1/2} \text{ (in mole L}^{-1} \text{ s}^{-1}). \quad (N)$$

Since it varies with the square root of the hydrogen ion concentration, it has a weak pH dependence. However, as we shall see, this uncatalyzed reaction is too slow to be of importance under typical tropospheric conditions.

Much more relevant to the aqueous phase in clouds and fogs in the atmosphere is the catalyzed oxidation of S(IV) by O$_2$. Both Fe^{3+} and Mn^{2+} catalyze the oxidation and as described in Chapter 9, both are common constituents of tropospheric aerosols even in remote

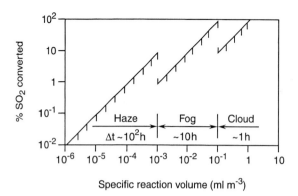

FIGURE 8.12 Percentage of SO_2 converted to sulfate after a time interval Δt in small "haze" particles, fogs, and clouds as a function of the aqueous reaction volume; note that the time intervals for each one are different, reflecting how long they typically last in the atmosphere (adapted from Lamb et al., 1987).

areas due to their generation from erosion of the earth's crust. Iron has a particularly rich chemistry because it forms a variety of complexes with OH^- and with various S(IV) aqueous forms (see, for example, Betterton, 1993; Brandt et al., 1994; and Millero et al., 1995). Figure 8.13 shows the calculated concentrations of various iron complexes as a function of pH in a solution containing 1×10^{-6} M Fe^{3+} and 1×10^{-5} M S(IV) at an ionic strength of 0.01. Hence elucidating the role of iron in the S(IV) oxidation has involved first understanding the nature of such complexes in solution. Further complicating the iron-catalyzed oxidation is that the mechanism changes from an ionic mechanism in the low-pH regime (0–3.6) to a free radical mechanism at higher pHs (4–7).

Table 8.6 shows two proposed mechanisms for the iron-catalyzed reaction at high acidities, in the pH range from 0 to 3.6. The recommended rate expression is given by (Martin, 1994):

$$-d[S(IV)]/dt = k[Fe(III)][S(IV)]/[H^+]$$

$$\text{(in mole L}^{-1}\text{ s}^{-1}\text{)}. \quad \text{(O)}$$

Fe(III) refers to the sum of all three-valent iron in solution, i.e., $Fe(III) = Fe^{3+} + FeOH^{2+} + Fe(OH)_2^+ + FeOHSO_3 + FeSO_3^+ +$ etc. Thus the rate of the iron-catalyzed reaction in the low-pH region decreases with increasing $[H^+]$. This means that it shows the behavior depicted in Fig. 8.9a; i.e., it is self-quenching. That is, as S(IV) is oxidized to the acid, the pH falls and the rate also decreases.

Despite the simplicity of the rate law implied by Eq. (O), the behavior of the kinetics is very sensitive to a variety of factors. Thus the reaction is inhibited not only by $[H^+]$ but also by the ionic strength (I) of the solution, by both S(IV) and S(VI), and, at high pH, by organics. Martin (1994) gives expressions for the dependence of the rate constant k in Eq. (O) on I, S(IV), and S(VI). The effect of ionic strength may be due to effects on the stability of complexes, whereas the sulfate is thought to complex one or more of the catalytic species in the reaction. Because of these complexities, the rate expression in Eq. (O) only applies for $[S(IV)] < 1 \times 10^{-5}$, $[Fe(III)] > 1 \times 10^{-7}$, $I < 0.01$, and $[S(VI)] < 1 \times 10^{-4}$ mol L^{-1}, where $k = 6.0$ s^{-1} (Martin, 1994).

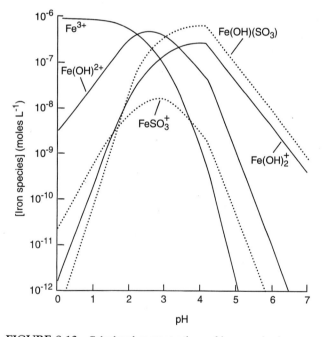

FIGURE 8.13 Calculated concentrations of iron species in aqueous solution for $[Fe(III)] = 1 \times 10^{-6}$ M, $[S(IV)] = 1 \times 10^{-5}$ M, and $I = 0.01$. The sulfur complexes are shown by the dotted lines (adapted from Martin, 1994).

TABLE 8.6 Some Proposed Mechanisms for the Catalyzed Oxidation of S(IV) in Aqueous Solutions

Hoffmann and Jacob (1984)
$Fe^{3+} + HSO_3^- \leftrightarrow FeSO_3^+ + H^+$
$FeSO_3^+ \rightarrow$ Internal redox
$\xrightarrow{fast}$ Addition of HSO_3^-
$\xrightarrow{fast}$ Addition of O_2
$\xrightarrow{fast}$ Formation of products

Conklin and Hoffmann (1988)
$HSO_3^- \leftrightarrow SO_3^{2-} + H^+$
$Fe^{3+} + H_2O \leftrightarrow FeOH^{2+} + H^+$
$FeOH^{2+} + SO_3^{2-} \leftrightarrow HOFeOSO_2$ complex
$HOFeOSO_2$ complex $+ O_2 \leftrightarrow O_2$ adduct
O_2 adduct $\xrightarrow[rearrangement]{acid-catalyzed} HOFe^{2+} + SO_3 + H_2O_2$
$SO_3 + H_2O \rightarrow 2H^+ + SO_4^{2-}$

In the high-pH (4–7) region, the reaction is first order in S(IV), but the dependence on the catalyst concentration is complex. Martin (1994) recommends the following rate expressions for pH around 4.0 and 5–6, respectively, where it is assumed that there is at least a trace of iron available for catalysis in the 5–6 pH range:

pH 4.0

$$-d[S(IV)]/dt = 1 \times 10^9 [Fe^{3+}]^2 [S(IV)]$$
$$(\text{in mol L}^{-1} \text{ s}^{-1}) \quad (P)$$

pH 5.0–6.0

$$-d[S(IV)]/dt = 1 \times 10^{-3} [S(IV)] \ (\text{in mol L}^{-1} \text{ s}^{-1})$$
$$(Q)$$

At pH > 4, the oxidation is inhibited by organics, suggesting a free radical mechanism. One proposed mechanism, which originates in the work of Bäckström (1934), is shown in Table 8.7. The inhibition occurs when the organics react with the sulfate radical ion, SO_4^-. This inhibition has also been seen in laboratory experiments using fogwater collected in Dübendorf, Switzerland, where oxidation rates for S(IV) were less than expected based on the kinetics of the iron-catalyzed oxidation (Kotronarou and Sigg, 1993).

The Mn^{2+}-catalyzed oxidation of SO_2 is also complex in both kinetics and mechanism. At pH 2, for example, the reaction is second order in Mn^{2+} and zero order in S(IV) (i.e., is independent of the dissolved sulfur concentration) when $[S(IV)] > 10^{-4}$ mol L^{-1} but first order in both Mn^{2+} and S(IV) at low concentrations, $[S(IV)] < 10^{-6}$ mol L^{-1}. Furthermore, the rate decreases with ionic strength at all S(IV) concentrations. Finally, there is a synergistic effect in the presence of both Mn^{2+} and Fe^{3+} (Martin, 1994).

In summary, the uncatalyzed oxidation of S(IV) occurs in aqueous solution but is very slow. However, given the ubiquitous occurrence of Fe^{3+} and Mn^{2+} (see Chapter 9), the uncatalyzed oxidation is likely irrelevant to atmospheric solutions. The catalyzed oxidations are complex in both kinetics and mechanism. We shall defer a comparison of their importance until other oxidation mechanisms are discussed. However, we shall see that the catalyzed oxidations are likely to contribute significantly to S(IV) oxidation in solution only at pH values near neutral, i.e., in the range of ~6–7. As the oxidation occurs and acid forms, the pH falls. The rapid falloff in the rate of the catalyzed oxidation with increasing $[H^+]$ then results in a rapid quenching of this path, as expected from Fig. 8.9a.

As discussed in Chapter 7, there is some evidence that freezing of aqueous solutions containing nitrite accelerates its oxidation to nitrate (Takenaka et al., 1992, 1996). A similar phenomenon has been reported in cloud chamber studies, where sulfide was observed to be oxidized to sulfate during ice crystal formation from expansion of droplets containing ionic salts (Finnegan et al., 1991; Finnegan and Pitter, 1991; Gross, 1991). Possible mechanisms are discussed by Finnegan and Pitter (1997).

d. Oxidation by O_3

While the Henry's law constant for ozone is fairly small (Table 8.1), there is sufficient ozone present in the troposphere globally to dissolve in clouds and fogs, hence presenting the potential for it to act as a S(IV) oxidant. Kinetic and mechanistic studies for the O_3–S(IV) reaction in aqueous solutions have been reviewed and evaluated by Hoffmann (1986), who shows that it can be treated in terms of individual reactions of the various forms of S(IV) in solution. That is, $SO_2 \cdot H_2O$, HSO_3^-, and SO_3^{2-} each react with O_3 by unique mechanisms and with unique rate constants, although in all cases the reactions can be considered to be a nucleophilic attack by the sulfur species on O_3.

Figure 8.14 shows the proposed mechanisms of reaction for each species. The overall rate of the S(IV) oxidation can then be represented by

$$-d[S(IV)]/dt = \{k_0 \alpha_0 + k_1 \alpha_1 + k_2 \alpha_2\}[S(IV)][O_3],$$
$$(R)$$

where α_0, α_1, and α_2 are the fractions of the total S(IV) in the form of $SO_2 \cdot H_2O$, HSO_3^-, and SO_3^{2-}, respectively. That is, $\alpha_0 = [SO_2 \cdot H_2O]/[S(IV)]$ etc., where S(IV) is the sum of $\{SO_2 \cdot H_2O + HSO_3^- +$

TABLE 8.7 Proposed Mechanism for the Catalyzed Oxidation of S(IV) in Aqueous Solutions of pH Range 4–7[a]

Reaction	Rate constant
$2Fe(OH)_2^+ \leftrightarrow Fe_2(OH)_4^{2+}$	K
$Fe_2(OH)_4^{2+} + OH^- \rightarrow OH + Fe(OH)_2 + Fe(OH)_2^+$	k_a
$Fe(OH)_2 \rightarrow Fe^{2+} + 2OH^-$	(fast)
$OH + HSO_3^- \rightarrow SO_3^- + H_2O$	k_b
$SO_3^- + O_2 \rightarrow SO_5^-$	k_c
$SO_5^- + SO_3^{2-} \rightarrow SO_4^{2-} + SO_4^-$	k_d
$SO_4^- + SO_3^{2-} \rightarrow SO_4^{2-} + SO_3^-$	k_e
$SO_4^- + Fe^{2+} \rightarrow Fe^{3+} + SO_4^{2-}$	k_f
$SO_4^- + \text{organic} \rightarrow SO_4^{2-} + \text{products}$	k_g
$SO_4^- \rightarrow \text{products}$	k_h

$k_b = 4.5 \times 10^9$ M^{-1} s^{-1}
$k_c = 1.5 \times 10^9$ M^{-1} s^{-1}
$k_d = 1.3 \times 10^7$ M^{-1} s^{-1}
$k_e = 2 \times 10^9$ M^{-1} s^{-1}
$k_f = 9.9 \times 10^8$ M^{-1} s^{-1}
k_g = see Neta et al. (1988) and Wine et al. (1989)
$k_h = 410$ s^{-1}

[a] Bäckström (1934), Huie and Neta (1987), and Martin (1994). Rate constants from Huie and Neta (1987), Neta et al. (1988), Tang et al., (1988), and Wine et al. (1989).

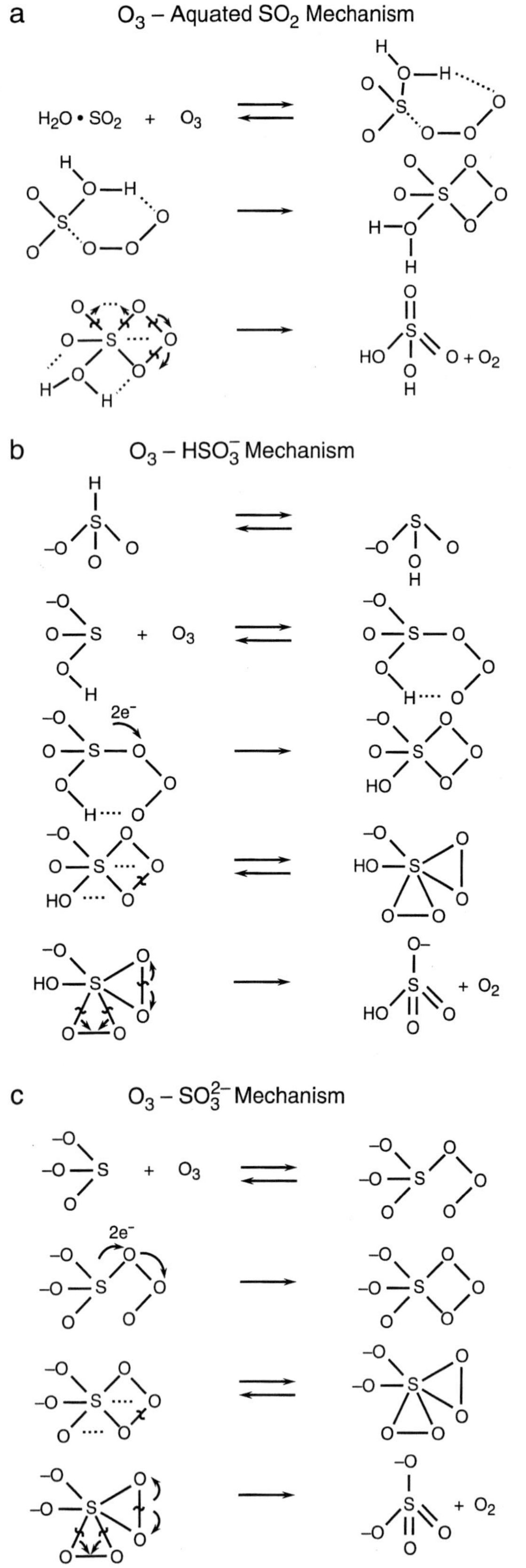

SO_3^{2-}}. Values for k_0, k_1, and k_2 are given in Table 8.8.

Figure 8.15 shows the pH dependence of the total rate of oxidation of S(IV) as well as the individual contributions from $SO_2 \cdot H_2O$, HSO_3^-, and SO_3^{2-}, respectively, under conditions where $[SO_2] = 20$ ppb and $[O_3] = 50$ ppb. The pH dependence arises from the dependence of the values of α in Eq. (R) on $[H^+]$. As discussed earlier and shown in Fig. 8.7, the fraction of total S(IV) that exists in each of the reactive forms depends on the pH since the equilibria (11)–(13) shift with pH. As a result, the three terms in Eq. (R) predominate in different pH regimes. As expected, Fig. 8.15 shows that the reaction with SO_3^{2-} predominates at high pH values, whereas that with $SO_2 \cdot H_2O$ and HSO_3^- predominates at lower pH values.

Overall, the rate of oxidation falls sharply as the pH drops. As a result, it is most important at relatively high pH values, e.g., before much oxidation has occurred. However, as S(IV) is oxidized to sulfuric acid and the pH falls, the rate decreases and the reaction in effect becomes self-quenching.

Lagrange and co-workers (1994) have examined the effect of added electrolytes on the kinetics and find that the rate constant in the pH range from 2.5 to 3.5 increases with ionic strength in a manner consistent with the following:

$$k = k_0 + F_j I_j,$$

where k_0 is the rate constant at zero ionic strength, I_j is the ionic strength, and F_j is a parameter that reflects the nature of the electrolyte. Interestingly, the nature of the electrolyte was shown to be important, with F_j for Na_2SO_4 being about 4 times larger than that for $NaClO_4$ and that for $NaCl$ being about twice as large. Lagrange *et al.* suggest that a free radical mechanism suggested by Penkett *et al.* (1979) is consistent with their data, rather than the mechanisms proposed in Fig. 8.14, i.e.,

$$2O_3 + OH^- \rightarrow OH + O_2^- + 2O_2,$$
$$OH + HSO_3^- \rightarrow HSO_3 + OH^-,$$

followed by reactions of HSO_3 discussed later.

FIGURE 8.14 Mechanism of oxidation of $SO_2 \cdot H_2O$, HSO_3^-, and SO_3^{2-} by O_3 in aqueous solution (adapted from Hoffmann, 1986).

TABLE 8.8 Rate Constants for Reactions of $SO_2 \cdot H_2O_{(aq)}$, HSO_3^-, and SO_3^{2-} with O_3 in Aqueous Solution at Room Temperature[a]

Reacting species	k (L mol^{-1} s^{-1})
$SO_2 \cdot H_2O_{(aq)}$	$k_0 = (2.4 \pm 1.1) \times 10^4$
HSO_3^-	$k_1 = (3.7 \pm 0.7) \times 10^5$
SO_3^{2-}	$k_2 = (1.5 \pm 0.6) \times 10^9$

[a] From Hoffmann (1986).

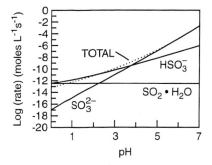

FIGURE 8.15 Rate of S(IV) oxidation by O_3 in the aqueous phase under conditions typical of a Los Angeles atmosphere, with SO_2 = 20 ppb and O_3 = 50 ppb (adapted from Hoffmann, 1986).

Since the ozone oxidation is important at higher pH values, this reaction is particularly important in sea salt particles generated from seawater, which has a pH of ~8. Thus SO_2 can be taken up into sea salt particles and oxidized by ozone to form sulfate. Because of the presence of buffering agents such as carbonates in the particles, this oxidation continues until the buffering agents are exhausted, at which point the pH falls and the rate of the ozone oxidation decreases; as discussed shortly, reactive chlorine and bromine species may also participate in S(IV) oxidation in sea salt particles (e.g., see Clarke and Radojevic, 1983; Clarke and Williams, 1983; Miller et al., 1987; Chameides and Stelson, 1992a, 1992b; Sievering et al., 1991, 1994, 1995; Clegg and Toumi, 1997; Keene et al., 1998).

e. Oxidation by H_2O_2 and Organic Peroxides

Hydrogen peroxide has been shown to oxidize S(IV) relatively rapidly in solution. Furthermore, because it is a highly soluble compound, even gas-phase concentrations in the low-ppb range, which are common, lead to significant concentrations in the liquid phase (e.g., Gunz and Hoffmann, 1990). For example, using the Henry's law constant for H_2O_2 of 1×10^5 M atm^{-1}, a 1-ppb gas-phase concentration would produce at equilibrium at 25°C an aqueous-phase concentration of 1×10^{-4} mol L^{-1}. This is approximately six orders of magnitude greater than the solution-phase concentrations of O_3 expected under ambient conditions! In addition to uptake from the gas phase, H_2O_2 can be formed in solution through the photooxidation of organics (e.g., Faust et al., 1997; Anastasio et al., 1997). This relatively large concentration anticipated in the solution phase is a significant factor in making H_2O_2 responsible for a major part of the oxidation of S(IV) in the aqueous phase (see later).

A second factor in favor of the importance of H_2O_2 in the S(IV) oxidation is the pH dependence of the rate coefficient. Figure 8.16 shows the results of some rate studies summarized by Martin and Damschen (1981), where k_0 is defined as

$$k_0 = \frac{1}{[H_2O_2][S(IV)]} \frac{d[S(IV)]}{dt} \quad (S)$$

and k_0 depends on [H$^+$] in a complex fashion (Overton, 1985).

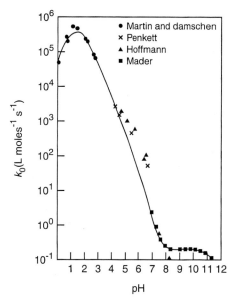

FIGURE 8.16 k_0 in expression $d[S(VI)]/dt = k_0[H_2O_2][S(IV)]$; effect of buffer removed and results converted to 25°C (adapted from Martin and Damschen, 1981).

The mechanism for this reaction was proposed by Hoffmann and Edwards (1975):

$$HSO_3^- + H_2O_2 \longleftrightarrow {}^-O{\!\!\!>}S-OOH + H_2O \quad (16a)$$
$${}^{}_{O}$$

$${}^-O{\!\!\!>}S-OOH + HA \longrightarrow H_2SO_4 + A^-, \quad (16b)$$
$${}^{}_{O}$$

where HA is an acid, or, alternatively,

$$HSO_3^- + HA \leftrightarrow H_2O \cdot SO_2 + A^-, \quad (17a)$$

which is followed by

$$A^- + H_2O_2 + H_2O \cdot SO_2 \longrightarrow {}^-O{\!\!\!>}S-OOH + HA \quad (17a)$$
$${}^{}_{O}$$

$${}^-O{\!\!\!>}S-OOH \longrightarrow HSO_4^-. \quad (18)$$
$${}^{}_{O}$$

At high pH values (which are not typical of most aqueous solutions in the atmosphere, however), H_2O_2 can also react with SO_3^{2-} (Lagrange et al., 1996). Other studies (McArdle and Hoffmann, 1983; Kunen et al.,

1983; Overton, 1985) are also in agreement with the data of Fig. 8.16. It is seen that in contrast to most of the other oxidations discussed earlier, k_0 decreases as the pH increases at pH $\geq$ 1.5. This is in the opposite direction to the S(IV) solubility (i.e., is qualitatively described by the situation depicted in Fig. 8.9b). The result is that, unlike the other reactions discussed so far, the overall rate of production of S(VI) from this reaction is relatively independent of pH over a wide pH range of interest in the atmosphere (see below). Interestingly, Lagrange et al. (1993) report that the reaction is catalyzed by chloride and ammonium ions.

The kinetics of the reaction can be expressed as

$$\frac{-d[H_2O_2]}{dt} = k[H^+][H_2O_2][S(IV)],$$

where k has been measured to be in the range (7.2 $\pm$ 2.0) $\times$ 10^7 to (9.6 $\pm$ 0.5) $\times$ 10^7 L^2 mol^{-2} s^{-1} (Lee et al., 1986; Lind et al., 1987). Typical H_2O_2 levels are up to a maximum of ~10 ppb in the gas phase and more typically ~1 ppb; in cloudwater, concentrations up to ~200 μM have been reported (e.g., see Sakugawa et al., 1990; Claiborn and Aneja, 1991; and Lee et al., 1993). The ubiquitous occurrence of H_2O_2, its large Henry's law constant, its high reactivity, and the pH dependence of the rate constant combine to make H_2O_2 the most important oxidant for S(IV) in the troposphere.

This is consistent with the results of a number of field studies. For example, the release of SO_2 into air entering a cloud base at Great Dunn Fell, England, resulted in the simultaneous increase in sulfate and decrease in H_2O_2 in the cloudwater (Clark et al., 1990), and H_2O_2 concentrations are often inversely related to SO_2 (e.g., Sakugawa and Kaplan, 1989).

It is not clear whether the reaction kinetics will be significantly altered by other species present in the aqueous phase. For example, Lee et al. (1986) showed that the kinetics of the H_2O_2-S(IV) reaction in freshly collected precipitation were only 15% below those measured in laboratory pure water. On the other hand, laboratory studies by Lagrange et al. (1993) suggest that the rate constant depends both on the ionic strength and on the nature of the electrolyte and that Fe^{2+} catalyzes the reaction.

Organic hydroperoxides have also been proposed as potential oxidants of S(IV) in solution (e.g., Graedel and Goldberg, 1983):

$$ROOH + HSO_3^- \rightarrow HSO_4^- + ROH, \quad (19)$$

$$k_{19} \simeq 7.2 \times 10^4 \text{ L mol}^{-1} \text{ s}^{-1}.$$

Both methyl hydroperoxide (CH_3OOH) and peracetic acid ($CH_3C(O)OOH$) have been shown to oxidize S(IV) in the aqueous phase (e.g., Lind and Lazrus, 1983; Lind et al., 1987). Lind and co-workers (1987) express the rate law for oxidation of S(IV) by H_2O_2, CH_3OOH, and $CH_3C(O)OOH$ in the form

$$\text{Rate (mol L}^{-1} \text{ s}^{-1}) = k[H^+][\text{peroxide}][S(IV)],$$

with k = 7.2 $\times$ 10^7 L^2 mol^{-2} s^{-1} for H_2O_2, 1.7 $\times$ 10^7 L^2 mol^{-2} s^{-1} for CH_3OOH, and 3.5 $\times$ 10^7 L^2 mol^{-2} s^{-1} for $CH_3C(O)OOH$. (The rate for $CH_3C(O)OOH$ was shown to contain an additional term, k [peroxide][S(IV)], where k = 610 L mol^{-1} s^{-1}, which becomes important above a pH of ~5.) As seen from the data in Table 8.1, the Henry's law constant for CH_3OOH is about three orders of magnitude less than that for H_2O_2 and that for $CH_3C(O)OOH$ is about two orders of magnitude less than for H_2O_2. Since their aqueous-phase concentrations are expected to be smaller than that of H_2O_2 they are not expected to contribute significantly compared to oxidation by H_2O_2. (See, for example, Kelly et al., 1985.)

Zhou and Lee (1992) measured the rate constant for the reaction of hydroxymethyl hydroperoxide, $HOCH_2OOH$, with S(IV) compared to H_2O_2 and found $k_{S(IV)+HOCH_2OOH}$ = {2.2 $\times$ 10^7[H$^+$]} L^2 mol^{-2} s^{-1} (based on $k_{S(IV)+H_2O_2}$ = {9.6 $\times$ 10^7[H$^+$]} L^2 mol^{-2} s^{-1} determined by Lee et al., 1986). As seen in Table 8.1, this hydroperoxide, which can be formed from the H_2O_2–HCHO reaction or the reaction of H_2O with the Criegee intermediate ($\dot{\text{H}}$CHOO·) (see Chapter 6), has a Henry's law constant that is even larger than that for H_2O_2. Because of uncertainties in its concentrations in fogs and clouds, it is not possible to make a firm estimate of its importance in the oxidation of S(IV). However, in one set of measurements made in Georgia, gas-phase $HOCH_2OOH$ constituted a large percentage of the total peroxides, with concentrations as high as 5 ppb (Lee et al., 1993). On the other hand, near Grand Canyon, Arizona, most of the total peroxide was measured to be H_2O_2 (Tanner and Schorran, 1995). Hence, $HOCH_2OOH$ may be important in S(IV) oxidation under some circumstances, depending on the relative amount compared to H_2O_2.

f. Oxidation by Oxides of Nitrogen

The oxides of nitrogen—NO, NO_2, NO_3, HONO, and HNO_3—have all been suggested as possible oxidizing agents for dissolved S(IV); however, the reactions of HNO_3 and NO at atmospheric concentrations are too slow to be significant (Lee and Schwartz, 1983; Martin, 1984; Schwartz, 1984a).

In aqueous solutions, nitrous acid reacts with S(IV) at a reasonable rate with a rate expresssion given by

(Martin, 1984):

$$-\frac{dS(IV)}{dt} = 142[H^+]^{1/2}[N(III)][S(IV)]. \quad (T)$$

However, the levels of gaseous HONO observed in polluted ambient air (~1–8 ppb) (see Chapter 11) taken with the Henry's law constant for HONO (Table 8.1) yield aqueous concentrations too low to contribute substantially to the aqueous-phase S(IV) oxidation.

For example, with a Henry's law constant for HONO of 49 M atm^{-1}, a gas-phase concentration of 1 ppb would result in a solution-phase concentration of only 4.9×10^{-8} mol L^{-1}, compared to an anticipated H$_2$O$_2$ solution-phase concentration of 10^{-4} mol L^{-1} at the same gas-phase concentration. The rate constants also favor the H$_2$O$_2$ reaction; at a pH of 3.0, that for oxidation of H$_2$O$_2$ is approximately a factor of 10^4 larger than that for reaction with HONO. Thus, the combination of concentrations and rate constants makes HONO unlikely to be a significant S(IV) oxidant in solution unless other oxidants such as O$_3$ or H$_2$O$_2$ are absent.

It is interesting, however, that the HONO–HSO$_3^-$ reaction has been shown to form a nitrene (HON:), which Mendiara and co-workers (1992) suggest could contribute to free radical formation in clouds and fogs.

Whether dissolved NO$_2$ contributes significantly to the S(IV) oxidation in solution is uncertain. As seen from the Henry's law constant in Table 8.1, NO$_2$ is relatively insoluble; thus at a gas-phase concentration of 10 ppb, the equilibrium aqueous-phase concentration is only 10^{-10} mol L^{-1}. However, Schwartz and co-workers (Schwartz, 1984a) have inferred from literature data that the rate constants for the NO$_2$ reaction with HSO$_3^-$ and SO$_3^{2-}$ may be sufficiently large, 3×10^5 and 1×10^7 L mol^{-1} s^{-1}, respectively, that the NO$_2$–S(IV) reaction could be significant. Littlejohn et al. (1993) suggest that the reaction is that of NO$_2$ with sulfite, leading to the formation of SO$_3^-$ and SO$_5^-$ radical anions. The NO$_2$–S(IV) reaction may also be catalyzed in solution by the presence of carbon particles (Schryer et al., 1983). Further support for an aqueous-phase oxidation of S(IV) by NO$_2$ comes from cloud chamber studies where significant sulfate production was observed when NO$_2$ was present even at concentrations as low as 5 ppb NO$_2$ (Gertler et al., 1984).

The reaction of NO$_3$ scavenged from the gas phase with S(IV) is discussed in Chapter 7 and, as seen there, may also be important (e.g., Chameides and Davis, 1983; Rudich et al., 1996, 1998).

g. Free Radical Reactions in Clouds and Fogs

As might be expected based on our knowledge of the gas-phase chemistry, there are a variety of free radicals in atmospheric aqueous systems as well and these too can participate in S(IV) oxidation in clouds and fogs. The free radicals arise either from absorption from the gas phase or, alternatively, from *in situ* production, largely from photochemical processes. As seen from the data in Table 8.5, uptake of OH and HO$_2$ from the gas phase is fast, with mass accommodation coefficients $> 4 \times 10^{-3}$ for OH and > 0.2 for HO$_2$ on liquid water at 275 K (Hanson et al., 1992; Mozurkewich et al., 1987). Uptake of species such as H$_2$O$_2$, which can photolyze to form OH, is also fast.

In addition, there are many photochemical processes in clouds and fogs that can produce reactive species such as peroxyl radicals, singlet oxygen, O$_2$($^1\Delta_g$), OH, HO$_2$, and H$_2$O$_2$. Thus, a variety of studies have detected the formation of such species upon irradiation of rainwater, cloudwater, and fogwater (e.g., Faust and Allen, 1992, 1993; Zuo and Hoigné, 1993; Faust et al., 1993; Anastasio et al., 1994; Faust, 1994; Arakaki et al., 1995; Arakaki and Faust, 1998). The actual reactions leading to the formation of these oxidants are not well established. Suggested mechanisms include the reaction of organics to form superoxide ion, O$_2^-$.

Zafiriou (1983), for example, suggested that absorption of light by organics, followed by intersystem crossing (ISC) to the triplet state (T) as described in Chapter 3, could occur. The subsequent reaction of the organic in a triplet state with O$_2$ could then give O$_2^-$:

$$\text{Organic chromophore (S}_0\text{)} + h\nu \rightarrow \text{Chromophore (S}_1\text{)}$$
$$\rightarrow \text{Chromophore (T}_1\text{)}, \quad (20)$$

$$\text{Chromophore (T}_1\text{)} + \text{O}_2 \rightarrow \text{O}_2^- + \text{Chromophore}^+. \quad (21)$$

While energy transfer between triplets and O$_2$ is well known, whether charge transfer can occur as shown in (21) is not clear.

Superoxide ion and HO$_2$ are closely coupled via reaction (22):

$$\text{O}_2^- + \text{H}^+ \leftrightarrow \text{HO}_2, \quad (22)$$
$$K_{eq} = 4.9 \times 10^4 \text{ L mol}^{-1}.$$

Production of H$_2$O$_2$ can then occur via

$$\text{O}_2^- + \text{HO}_2 \rightarrow \text{O}_2 + \text{HO}_2^- \xrightarrow[\text{fast}]{\text{H}^+} \text{H}_2\text{O}_2, \quad (23)$$
$$k_{23} = 9.7 \times 10^7 \text{ L mol}^{-1} \text{ s}^{-1},$$

or alternatively, as in the gas phase, by the self-reaction of HO$_2$:

$$2\text{HO}_2 \rightarrow \text{H}_2\text{O}_2 + \text{O}_2, \quad (24)$$
$$k_{24} = 8.3 \times 10^5 \text{ L mol}^{-1} \text{ s}^{-1}.$$

(The rate and equilibrium constants are from Bielski *et al.* (1978, 1985), which provides an excellent review of reactions of HO_2 and O_2^- in aqueous solutions.)

Production of superoxide can also occur via photolysis of iron–ligand (L) complexes, where L is oxalate, glyoxalate, or pyruvate, for example (Zuo and Hoigné, 1992, 1993; Zhu *et al.*, 1993; Erel *et al.*, 1993):

$$\text{Fe(III)}-\text{L} + h\nu \rightarrow \text{Fe(II)} + \text{L}^\bullet, \quad (25)$$

$$\text{L}^\bullet + \text{O}_2 \rightarrow \text{O}_2^{\bullet-} + \text{Oxidized ligand}. \quad (26)$$

Based on measurements of iron in both the (II) and (III) oxidation states and the anions in cloudwater and fogwater, Siefert *et al.* (1998) calculate that most of the Fe(III) is in the form of hydroxy species such as Fe(OH)_2^+, with much smaller amounts ($<10\%$) in the form of oxalate complexes such as $\text{Fe(oxalate)}_3^{3-}$.

Indeed, iron (and probably other transition metals as well) appears to play a major role in the production of free radicals in clouds and fogs [see, for example, reviews by Faust (1994) and Hoigné *et al.* (1994)]. Graedel, Weschler, and co-workers (Graedel and Weschler, 1981; Graedel *et al.*, 1985, 1986; Weschler *et al.*, 1986) proposed that photolysis of Fe(III) complexes may be a major source of OH in atmospheric droplets during the day. Thus, the monohydroxy complex Fe(OH)^{2+}, which is a major form of iron in solution at pHs commonly encountered in the troposphere (see Fig. 8.13), absorbs light in the 290- to 400-nm region (Fig. 8.17). It photolyzes to generate OH, with quantum yields of 0.14 at 313 nm and 0.017 at 360 nm (Faust and Hoigné, 1990) and 0.31 and 0.07 at 280 and 370 nm, respectively (Benkelberg and Warneck, 1995):

$$\text{Fe(OH)}^{2+} + h\nu \rightarrow \text{Fe}^{2+} + \text{OH}. \quad (27)$$

The combination of these significant quantum yields in the actinic region with substantial absorption coefficients means that the lifetime of Fe(OH)^{2+} is quite short, of the order of minutes, during the day and that reaction (27) is expected to be a major source of OH in aqueous atmospheric droplets (Benkelberg and Warneck, 1995; Siefert *et al.*, 1996).

Once formed, Fe^{2+} can be oxidized back to Fe^{3+}, for example, by the Fenton reaction involving H_2O_2, O_2^-, or HO_2 (Siefert *et al.*, 1996; Faust, 1994a,b; Arakaki and Faust, 1998), which again generate OH:

$$\text{Fe}^{2+} + \text{H}_2\text{O}_2 \rightarrow \text{Fe}^{3+} + \text{OH} + \text{OH}^-, \quad (28a)$$

$$\text{Fe(OH)}^+ + \text{H}_2\text{O}_2 \rightarrow \text{Fe(OH)}^{2+} + \text{OH} + \text{OH}^-, \quad (28b)$$

$$\text{Fe}^{2+} + \text{O}_2^- \xrightarrow{2\text{H}_2\text{O}} \text{Fe}^{3+} + \text{H}_2\text{O}_2 + 2\text{OH}^-, \quad (29)$$

$$\text{Fe}^{2+} + \text{HO}_2 + \text{H}^+ \rightarrow \text{Fe}^{3+} + \text{H}_2\text{O}_2. \quad (30)$$

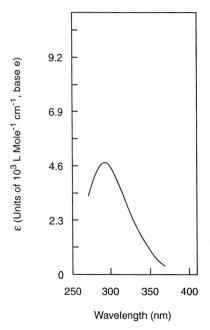

FIGURE 8.17 Absorption spectrum for Fe(OH)^{2+} (adapted from Benkelberg and Warneck, 1995).

Complexes of Fe^{2+} with oxalate, citrate, and phosphate also undergo similar reactions (Zepp *et al.*, 1992). In cloudwater, cycling between Fe^{2+} and Fe^{3+} occurs on a time scale of minutes (Faust, 1994a,b).

In agreement with such cycles, both Fe^{3+} and Fe^{2+} have been measured in airborne particles, fogs, clouds, and mineral dust particles; the percentage of the total iron that is in the reduced form, Fe^{2+}, varies widely but has been observed to be as high as $\sim 50\%$ (Dedik *et al.*, 1992; Zhuang *et al.*, 1992; Erel *et al.*, 1993) and can be even higher in cloudwater during the day (Behra and Sigg, 1990; Faust, 1996). For example, Siefert *et al.* (1998) measured the total concentration of Fe (as well as Mn, Cu, and Cr) in fogwater and cloudwater in three locations, including the oxidation states of the metals. In one sample collected during the day at Whiteface Mountain, New York, 63% of the total iron was present as Fe^{2+}. However, this was unusual, with most of the samples being below 50%. Fe(II) has also been shown to increase on irradiation of the particles (e.g., see Siefert *et al.*, 1996; Faust, 1996). Consistent with this, the concentration of Fe(II) during the day in mineral dust particles from North Africa and collected at Barbados was about twice that at night (Zhu *et al.*, 1997); Fe(II) was found to represent about 25% of the soluble iron but only a small portion (1.6%) of the total iron in the particles.

Table 8.9 shows some other potential photochemical sources of OH in aqueous solutions in the atmosphere.

TABLE 8.9 Summary of Photochemistry of Some Species in Aqueous Solutions

Photochemical dissociation	Quantum yield[a]	Reference
$O_3 + h\nu(\lambda \leq 336 \text{ nm}) \xrightarrow{H_2O} H_2O_2$	$\Phi = 0.48$ at 254 nm 0.23 at 310 nm	Gurol and Akata, 1996 Taube, 1957
$H_2O_2 + h\nu(\lambda \leq 380 \text{ nm}) \to 2OH$	$\Phi^{OH}(308 \text{ nm}) = 0.98$ $\Phi^{OH}(351 \text{ nm}) = 0.96$ $\Phi^{OH}(250 \text{ nm}) = 1.8$	Zellner et al., 1990; Zellner and Herrmann, 1995
$HONO + h\nu(\lambda \leq 390 \text{ nm}) \to OH + NO$	$\Phi^{OH}(280-390 \text{ nm}) = 0.35$	Fischer and Warneck, 1996
$HNO_3 + h\nu(\lambda \leq 320 \text{ nm}) \to OH + NO_2$	$\Phi \approx 0.1$[b]	Graedel and Weschler, 1981
$NO_2^- + h\nu(\lambda \leq 410 \text{ nm}) \xrightarrow{H_2O} NO + OH + OH^-$	$\Phi^{OH}(308 \text{ nm}) = 0.07$ $\Phi^{OH}(351 \text{ nm}) = 0.046$ $\Phi^{OH}(280 \text{ nm}) = 0.069$ $\Phi^{OH}(390 \text{ nm}) = 0.022$	Zellner et al., 1990 Zellner et al., 1990 Fischer and Warneck, 1996 Fischer and Warneck, 1996
$NO_3^- + h\nu(\lambda \leq 350 \text{ nm}) \to NO_2^- + O$ $\to NO_2 + O^- \xrightarrow{H_2O} OH^- + OH$	$\Phi^O(305 \text{ nm}) = 1.1 \times 10^{-3}$ $\Phi(305-313 \text{ nm}) = 0.013$	Warneck and Wurzinger, 1988; Zepp et al., 1987; Zellner et al., 1990; Zellner and Herrmann, 1995
$HO_2^- + h\nu(\lambda \leq 390 \text{ nm}) \xrightarrow{H_2O} OH + OH^- + O_2$		Treinin, 1970

[a] These are *effective quantum yields*, that is, those for photolysis and escape of the species from the solvent cage.
[b] Estimated.

As expected based on our knowledge of gas-phase chemistry, in addition to the Fenton type chemistry involving iron, photolysis of O_3, H_2O_2, HONO, and HNO_3 are all potential OH sources in clouds and fogs. In addition, the photolysis of nitrite, nitrate, and HO_2^- in aqueous solutions can also form OH. In short, there are many potential sources of OH in clouds and fogs.

Given that there are large sources of OH in atmospheric suspended droplets, oxidation of S(IV) by OH is expected. The proposed mechanism (see, for example, McElroy, 1986; Huie and Neta, 1987; Huie, 1995; and Buxton et al., 1996) is as follows (rate constants are in units of L mol s):

Initiation

$$OH + HSO_3^{2-} \to H_2O + SO_3^-, \quad (31)$$
$$k_{31} = (2.7-4.5) \times 10^9$$

(Buxton et al., 1996; Huie and Neta, 1987)

$$OH + SO_3^{2-} \to OH^- + SO_3^-, \quad (32)$$
$$k_{32} = 4.6 \times 10^9 \text{ (Buxton et al., 1996).}$$

Propagation

$$SO_3^- + O_2 \to SO_5^- \text{ (peroxymonosulfate radical)}, \quad (33)$$
$$k_{33} = (1.5-2.5) \times 10^9$$

(Huie and Neta, 1984; Buxton et al., 1996)

$$SO_5^- + HSO_3^- \to HSO_5^- + SO_3^-, \quad (34a)$$
$$k_{34a} = (0.83-2.5) \times 10^4$$

(Buxton et al., 1996; Huie and Neta, 1987)

$$\to HSO_4^- + SO_4^- \text{ (sulfate radical)}, \quad (34b)$$
$$k_{34b} = (0.034-7.5) \times 10^4$$

(Buxton et al., 1996; Huie and Neta, 1987)

$$SO_4^- + HSO_3^- \to HSO_4^- + SO_3^-, \quad (35)$$
$$k_{35} = (0.68-2) \times 10^9$$

(Buxton et al., 1996; Wine et al., 1989; Huie and Neta, 1987)

$$SO_4^- + SO_3^{2-} \to SO_3^- + SO_4^{2-}, \quad (36)$$
$$k_{36} = 5.5 \times 10^8$$

(Deister and Warneck, 1990)

$$SO_5^- + SO_5^- \to 2SO_4^- + O_2, \quad (37a)$$
$$k_{37} = (\sim 2.2-6) \times 10^8$$

(Buxton et al., 1996; Huie and Neta, 1987)

$$SO_4^- + H_2O \to OH + HSO_4^-, \quad (38)$$
$$k_{38}[H_2O] = 440 \text{ s}^{-1}$$

(Bao and Barker, 1996)

$$SO_5^- + SO_3^{2-} \xrightarrow{H^+} HSO_5^- + SO_3^-, \quad (39a)$$
$$k_{39a} = 3.6 \times 10^5$$

(Buxton et al., 1996)

$$\to SO_4^- + SO_4^{2-}, \quad (39b)$$
$$k_{39b} = 1.4 \times 10^5$$

(Buxton et al., 1996).

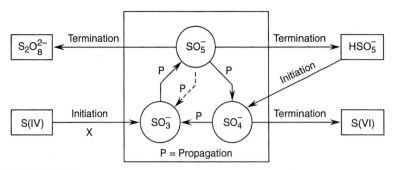

FIGURE 8.18 Summary of initiation, propagation, and termination steps in the free radical oxidation of S(IV) in solution. (Adapted from *J. Atmos. Chem. 20*, Sander R., Lelieveld, J., and Crutzen, P. J. "Modelling of the Nighttime Nitrogen and Sulfur Chemistry in Size Resolved Droplets of Orographic Cloud," Fig. 7, pp. 89–116. Copyright 1995, with kind permission from Kluwer Academic Publishers.)

Termination

$$SO_5^- + SO_5^- \rightarrow S_2O_8^{2-} + O_2, \quad (37b)$$

$$k_{37b} = (0.48 - 1.4) \times 10^8$$

(Buxton *et al.*, 1996; Huie and Neta, 1987)

$$SO_4^- + SO_4^- \rightarrow S_2O_8^{2-}, \quad (40)$$

$$k_{40} = 4.5 \times 10^8$$

(McElroy and Waygood, 1990)

$$SO_3^- + SO_3^- \rightarrow S_2O_6^{2-}, \quad (41)$$

$$k_{41} = 1.8 \times 10^8$$

(Waygood and McElroy, 1992)

$$SO_5^- + HO_2 \rightarrow HSO_5^- + O_2, \quad (42)$$

$$k_{42} = 5 \times 10^7$$

(Yermakov *et al.*, 1995).

It can be seen that there are some significant discrepancies in the rate constants measured for the individual reactions in this complex system, which affects how important this oxidation cycle is in atmospheric droplets. For example, the rate constant for the key propagation reactions (34a) and (34b), which occur in the pH range most commonly encountered in atmospheric droplets, was reported recently to be even smaller than shown above, at $k_{34} = (k_{34a} + k_{34b}) = 3.6 \times 10^3$ L mol^{-1} s^{-1} (Yermakov *et al.*, 1995). Clearly, additional work is needed in this area.

In addition, some of the key, free radical intermediates have other potential reactions that can impact the chemistry. The sulfate radical, SO_4^-, for example, reacts rapidly with aromatics in solution (e.g., Herrmann *et al.*, 1995; Huie, 1995), removing SO_4^- from the sequence shown.

The aqueous-phase free radical oxidation of S(IV) to S(VI) is summarized in Fig. 8.18 (Sander *et al.*, 1995) and Table 8.10. Oxidation is initiated by attack on the bisulfite ion, HSO_3^-, or the sulfite ion, SO_3^{2-}, by various species. The sulfite radical (SO_3^-), sulfate radical (SO_4^-), and peroxymonosulfate radical (SO_5^-) are the key intermediates involved in the chain propagation. Formation of sulfate, HSO_5^-, or $S_2O_8^{2-}$ leads to chain termination.

Initiation with X = OH has been discussed earlier. Table 8.11 summarizes some of the aqueous-phase HO_x chemistry in which OH is generated and reacts in the atmosphere. (Note that the rate constants for some of the aqueous phase reactions shown in Tables 8.10–8.16 depend on such factors as ionic strength; see Chapter 5.D.) Involved with this chemistry is that of bicarbonate/carbonate, since OH reacts with these species as well (Table 8.12). It is interesting that, in contrast to the high reactivity of OH toward S(IV) in aqueous solutions, direct reactions of HO_2/O_2^- with S(IV) do not appear to be important (Sedlak and Hoigné, 1994; Yermakov *et al.*, 1995).

The aqueous-phase and gas-phase chemistries of HO_x are sufficiently closely coupled that the chemistry shown in Tables 8.11 and 8.12 can affect gas-phase concentrations as well. For example, including the aqueous-phase chemistry in models of tropospheric ozone formation alters predicted O_3 concentrations, although whether the perturbation is significant is subject to some controversy (e.g., see Lelieveld and Crutzen, 1990; Jonson and Isaksen, 1993; Walcek *et al.*, 1997; Liang and Jacob, 1997).

Other species that can initiate this sulfur oxidation chemistry are NO_3 (discussed in Chapter 7.D.1) and Cl_2^-. The latter radical anion is formed in sea salt particles when atomic chlorine is generated and reacts with chloride ion. In addition, Vogt *et al.* (1996) have proposed that oxidation of SO_3^{2-} by HOCl and HOBr in sea salt particles may be quite important. Table 8.13 summarizes the aqueous-phase chlorine chemistry that occurs in sea salt particles and Table 8.14 the oxidation of S(IV) by reactive chlorine and bromine species in solution.

TABLE 8.10 Summary of Aqueous-Phase Chemistry of S(IV)

Reaction	Rate or equilibrium constant[a]	Reference
$SO_{2(g)} + H_2O \leftrightarrow SO_2 \cdot H_2O_{(aq)}$	$H = 1.24$ M atm^{-1}	b
$SO_2 \cdot H_2O_{(aq)} \leftrightarrow HSO_3^- + H^+$	$K = 1.32 \times 10^{-2}$ M	b
$HSO_3^- \leftrightarrow SO_3^{2-} + H^+$	$K = 6.42 \times 10^{-8}$ M	b
Initiation		
$X + HSO_3^- \to X^- + H^+ + SO_3^-$	(X = OH, NO$_3$, Cl$_2^-$, etc.; see text)	
$X + SO_3^{2-} \to X^- + SO_3^-$		
Propagation		
$SO_3^- + O_2 \to SO_5^-$	$(1.5–2.5) \times 10^9$	b
$SO_5^- + SO_3^{2-} \to SO_4^- + SO_4^{2-}$	5.5×10^5	b
$\phantom{SO_5^- + SO_3^{2-}} \to SO_2^{2-} + SO_4^-$	2.1×10^5	b
$SO_5^- + SO_5^- \to 2SO_4^- + O_2$	$(2.2–6) \times 10^8$	b
$SO_5^- + HSO_3^- \to HSO_5^- + SO_3^-$	$(0.83–2.5) \times 10^4$	b
$ \to HSO_4^- + SO_4^-$	$(0.034–7.5) \times 10^4$	b
$SO_4^- + SO_3^{2-} \to SO_3^- + SO_4^{2-}$	5.5×10^8	b
$SO_4^- + H_2O \to HSO_4^- + OH$	440	b
$SO_4^- + HSO_3^- \to HSO_4^- + SO_3^-$	$(0.68–2) \times 10^9$	b
$SO_4^- + H_2O_2 \to HSO_4^- + HO_2$	1.2×10^7	Wine et al., 1989
$SO_4^- + NO_3^- \to SO_4^{2-} + NO_3$	1.1×10^5	Wine et al., 1988
$HSO_4^- + OH \leftrightarrow H_2O + SO_4^-$	$k = 4.5 \times 10^5$ $K_{eq} \approx 10^3$	Huie, 1995
$SO_5^- + HO_2 \to HSO_5^- + O_2$	5×10^7	b
$HSO_3^- + O_2^- \to SO_3^- + HO_2^-$	3×10^4	Weinstein-Lloyd and Schwartz, 1991
Termination		
$SO_3^- + SO_3^- \to S_2O_6^{2-}$	1.8×10^8	Waygood and McElroy, 1992
$SO_4^- + SO_4^- \to S_2O_8^{2-}$	4.5×10^8	b
$SO_5^- + SO_5^- \to S_2O_8^{2-} + O_2$	$(0.48–1.4) \times 10^8$	b

[a] All in L–mol–s units.
[b] See references and discussion in Sections B and C.3g.

Bromine undergoes reactions similar to those of chlorine in the aqueous phase (Table 8.15). However, there are some important differences. For example, the oxidation of bromide ion by O_3, while not extremely fast, is still five orders of magnitude larger than that for the oxidation of Cl^-. As a result, O_3 reacts both at room temperature and just below freezing with solutions containing bromide to generate Br_2 (e.g., Oum et al., 1998b; Hirokawa et al., 1998). However, no reaction occurs in the dark with chloride-containing aqueous particles (e.g., Oum et al., 1998a; Hirokawa et al., 1998). A second difference is in the reactions

$$HOX^- + H^+ \leftrightarrow H_2O + Cl. \quad (43, -43)$$

Although the forward reactions are very fast for both HOCl$^-$ and HOBr$^-$, the reverse reaction for Br is very slow (1.4 s^{-1}) compared to that of Cl (2.5 × 10^5 s^{-1}). As a result, the equilibrium is far to the right for HOBr$^-$ even at high pH, whereas for HOCl$^-$, it is to the left except at low pH. Hence generation of reactive chlorine in solution occurs only in acidic solutions, whereas reactive bromine can be generated in neutral to somewhat basic solutions. This is important, because seawater and hence the sea salt droplet generated from it initially have a pH of approximately 8.

A third factor comes into play in bromine chemistry, which is that atmospheric solutions containing bromide and chloride are most typically formed from seawater. Wave action generates small airborne droplets of seawater, which thus initially contain the elements in the ratios found in seawater. The molar ratio of Br$^-$ to Cl$^-$ is ~1:650. However, despite the relatively small amounts of bromide relative to chloride, it plays a disproportionate role because of its reactivity and because its chemistry is closely intertwined with chloride ion chemistry. Table 8.16, for example, shows some of the interhalogen reactions of bromide and chloride. It can be seen that the chemistry preferentially generates Br$_2$ rather than Cl$_2$.

This is also true of the reactions of gaseous bromine compounds with solid chloride salts. For example, Mochida et al. (1998a) observed Br$_2$ and BrCl as products of the reaction of gaseous HOBr with solid NaCl

TABLE 8.11 Some Aqueous-Phase HO_x Chemistry

Reaction	k^{298K} (L mol^{-1} s^{-1})	Reference
$OH + O_3 \rightarrow HO_2 + O_2$	1.1×10^8	Neta et al., 1988
$OH + HO_2 \rightarrow H_2O + O_2$	1×10^{10}	Elliott and Buxton, 1992
$OH + O_2^- \rightarrow OH^- + O_2$	1×10^{10}	Elliott and Buxton, 1992
$OH + H_2O_2 \rightarrow H_2O + HO_2$	2.7×10^7	Buxton et al., 1988
$OH + OH \rightarrow H_2O_2$	5.5×10^9	Buxton et al., 1988
$OH + OH^- \rightarrow H_2O + O^-$	1.3×10^{10}	Buxton et al., 1988
$OH + HO_2^- \rightarrow HO_2 + OH^-$	7.5×10^9	Buxton et al., 1988
$OH + H_2 \rightarrow H_2O + H$	4.2×10^7	Buxton et al., 1988
$HO_2 + O_3 \rightarrow OH + 2O_2$	$< 1 \times 10^4$	
$HO_2 + HO_2 \rightarrow H_2O_2 + O_2$	8.3×10^5	Bielski, 1978
$HO_2 + O_2^- \rightarrow HO_2^- + O_2$	9.7×10^7	Bielski, 1978
$HO_2 + H_2O_2 \rightarrow OH + O_2 + H_2O$	0.5	Weinstein and Bielski, 1979
$O_2^- + O_3 \rightarrow O_2 + O_3^-$	1.5×10^9	Sehested et al., 1983; Zellner and Herrmann, 1995
$O_3^- + H_2O \xrightarrow{fast} OH^- + OH + O_2$		
$O_3^- + OH \rightarrow O_2^- + HO_2$	5.9×10^9	Buxton et al., 1988
$\rightarrow OH^- + O_3$	2.6×10^9	Buxton et al., 1988
$O_3^- + O^- \rightarrow 2O_2^-$	7.0×10^8	Buxton et al., 1988
$O_3^- \rightarrow O^- + O_2$	5.0×10^3	Neta et al., 1988
$O_3^- + H^+ \rightarrow OH + O_2$	9.0×10^{10}	Neta et al., 1988
$\rightarrow HO_3$	5.2×10^{10}	Bühler et al., 1984
$HO_3 \rightarrow OH + O_2$	1.1×10^5	Bühler et al., 1984
$\rightarrow H^+ + O_3^-$	3.7×10^4	Bühler et al., 1984
$O^- + H_2O \rightarrow OH^- + OH$	1.8×10^6	Buxton et al., 1988
$O^- + H_2O_2 \rightarrow O_2^- + H_2O$	$<5 \times 10^8$	Buxton et al., 1988
$O^- + HO_2^- \rightarrow O_2^- + OH^-$	4×10^8	Buxton et al., 1988
$O^- + O_2 \xrightarrow{H_2O} 2OH^- + O_2$	6×10^8	Buxton et al., 1988
$O^- + H_2 \rightarrow OH^- + H$	8×10^7	Buxton et al., 1988
$OH^- + O_3 \rightarrow HO_2^- + O_2$	48	Neta et al., 1988
$\rightarrow HO_2 + O_2^-$	70	Neta et al., 1988
$OH^- + O(^3P) \rightarrow HO_2^-$	4.2×10^8	Sauer et al., 1984
$HO_2^- + O_3 \rightarrow OH + O_2^- + O_2$	5.5×10^6	Neta et al., 1988
$HO_2^- + O(^3P) \rightarrow OH + O_2^-$	5.3×10^9	Sauer et al., 1984
$H + HO_2 \rightarrow H_2O_2$	1.0×10^{10}	Buxton et al., 1988
$H + O_2 \rightarrow HO_2$	2.1×10^{10}	Buxton et al., 1988
$H + O_3 \rightarrow OH + O_2$	3.7×10^{10}	Neta et al., 1988
$H + H_2O \rightarrow H_2 + OH$	10	Hartig and Getoff, 1982
$O(^3P) + O_2 \rightarrow O_3$	4×10^9	Kläning et al., 1984
$O(^3P) + H_2O_2 \rightarrow OH + HO_2$	1.6×10^9	Sauer et al., 1984

TABLE 8.12 Some Aqueous-Phase Carbonate/Bicarbonate Chemistry

Reaction	k (L mol^{-1} s^{-1})	Reference
$HCO_3^- + OH \rightarrow H_2O + CO_3^-$	8.5×10^6	Buxton et al., 1988
$OH + CO_3^{2-} \rightarrow OH^- + CO_3^-$	3.9×10^8	Buxton et al., 1988
$CO_3^- + O_2^- \rightarrow CO_3^{2-} + O_2$	6.5×10^8	Eriksen et al., 1985
$CO_3^- + H_2O_2 \rightarrow HCO_3^- + HO_2$	8.0×10^5	Behar et al., 1970
$CO_3^- + HO_2^- \rightarrow HCO_3^- + O_2^-$	5.6×10^7	Behar et al., 1970
$CO_3^- + OH \rightarrow$ product	3×10^9	Holcman et al., 1987
$CO_3^- + O_3^- \rightarrow CO_3^{2-} + O_3$	6×10^7	Holcman et al., 1987

TABLE 8.13 Some Aqueous-Phase Chlorine Chemistry

Reaction	Rate constant k^{298K} or equilibrium constant (L mol^{-1} s^{-1})	Reference
$Cl^- + O_3 \rightarrow ClO^- + O_2$	2.0×10^{-3}	Hoigné et al., 1985; Neta et al., 1988
$Cl^- + NO_3 \rightarrow NO_3^- + Cl$	1×10^7	Exner et al., 1992
$Cl^- + OH \rightarrow HOCl^-$	4.3×10^9	Jayson et al., 1973
$HOCl^- \rightarrow Cl^- + OH$	6.1×10^9	Jayson et al., 1973
$HOCl^- + H^+ \rightarrow H_2O + Cl$	1.45×10^{10}	Neta et al., 1988; de Violet, 1981
$Cl + H_2O \rightarrow HOCl^- + H^+$	2.5×10^5	Buxton et al., 1998
$Cl + Cl \rightarrow Cl_2$	8.8×10^7	Neta et al., 1988
$Cl + Cl^- \rightarrow Cl_2^-$	8.5×10^9	Buxton et al., 1998
$Cl_2^- \rightarrow Cl + Cl^-$	6.0×10^4	Buxton et al., 1998
$Cl_2^- + Cl_2^- \rightarrow Cl_2 + 2Cl^-$	2.0×10^9	Neta et al., 1988
$Cl_2^- + HO_2 \rightarrow 2Cl^- + O_2 + H^+$	$(1-4.5) \times 10^9$	Neta et al., 1988
$Cl_2^- + H_2O_2 \rightarrow 2Cl^- + HO_2 + H^+$	1.4×10^5	Neta et al., 1988
$Cl_2^- + H_2O \rightarrow 2Cl^- + H^+ + OH$	1.3×10^3	Buxton et al., 1998
$Cl_2^- + OH \rightarrow HOCl + Cl^-$	1.0×10^9	Wagner et al., 1986
$Cl_2^- + H \rightarrow H^+ + 2Cl^-$	7.0×10^9	Buxton et al., 1988
$Cl_2^- + OH^- \rightarrow HOCl^- + Cl^-$	4.5×10^7	Grigor'ev et al., 1987
$Cl_2 + H_2O \rightarrow HOCl + Cl^- + H^+$	22	Wang and Margerum, 1994
$Cl_2 + HO_2 \rightarrow H^+ + Cl_2^- + O_2$	1.0×10^9	Bjergbakke et al., 1981
$Cl_2 + Cl \rightarrow Cl_3$	5.3×10^8	Bunce et al., 1985
$HOCl + Cl^- + H^+ \rightarrow Cl_2 + H_2O$	1.8×10^4	Eigen and Kustin, 1962
$HOCl + O_2^- \rightarrow Cl^- + OH + O_2$	7.5×10^6	Long and Bielski, 1980

and Hirokawa et al. (1998) observed only Br_2 from the reaction of O_3 with a mixture of NaBr and NaCl at high relative humidity. Similarly, the reaction of gaseous Cl_2 with natural or synthetic sea salt gave only Br_2 as a product (Mochida et al., 1998b).

Excellent summaries of aqueous-phase chemistry are found in Huie (1995) and Zellner and Herrmann (1995). Extensive data bases for kinetics of aqueous-phase reactions are provided by the U.S. National Institute of Standards and Technology (see NIST, Ross et al., 1994) and by the University of Notre Dame Radiation Laboratory (see Appendix IV for Web site).

A variety of models incorporating the reactions of a variety of inorganics and organics, in addition to sulfur

TABLE 8.14 Aqueous-Phase Chemistry Involving S(IV) and Reactive Chlorine and Bromine Ions

Reaction	Rate constant k^{298K} (L mol^{-1} s^{-1})	Reference
Chlorine chemistry		
$Cl + Cl^- \rightarrow Cl_2^-$	8.5×10^9	Buxton et al., 1998
$Cl_2^- + HSO_3^- \rightarrow 2Cl^- + SO_3^- + H^+$	3.4×10^8	Huie and Neta, 1987
$Cl^- + SO_4^- \rightarrow SO_4^{2-} + Cl$	2.5×10^8	Wine et al., 1989; Huie and Clifton, 1990; Huie, 1995
$Cl^- + HSO_5^- \rightarrow HOCl + SO_4^{2-}$	1.4×10^{-3}	Fortnum et al., 1960; Clarke and Radojevic, 1983
$HOCl + SO_3^{2-} \rightarrow HSO_4^- + Cl^-$	$\geq 7.6 \times 10^{8a}$	Fogelman et al., 1989
Bromine chemistry[b]		
$Br + Br^- \rightarrow Br_2^-$	1.6×10^{10}	Scaiano et al., 1992
$Br_2^- + HSO_3^- \rightarrow 2Br^- + SO_3^- + H^+$	6.3×10^7	Shoute et al., 1991
$Br_2^- + SO_3^{2-} \rightarrow 2Br^- + SO_3^-$	2.2×10^8	Shoute et al., 1991
$Br^- + HSO_5^- \rightarrow HOBr + SO_4^{2-}$	1	Fortnum et al., 1960
$HOBr + SO_3^{2-} + H_2O \rightarrow SO_4^{2-} + Br^- + 2H^+$	$\geq 5 \times 10^{9\,a}$	Troy and Margerum, 1991

[a] Measured at pH of ~9–10; increases as pH decreases. See Vogt et al. (1996) for application to sea salt particles.
[b] See also Mozurkewich (1995).

TABLE 8.15 Some Aqueous-Phase Bromine Chemistry

Reaction	Rate constant k (L mol^{-1} s^{-1}) or equilibrium constant K_{eq}	Reference
$Br^- + O_3 \rightarrow BrO^- + O_2$	1.6×10^2	Haag and Hoigné, 1983
$BrO^- + H^+ \leftrightarrow HOBr$	$K_{eq} = 10^{8.8}$	Haag and Hoigné, 1983
$HOBr + Br^- + H^+ \rightarrow Br_2 + H_2O$	1.6×10^{10} L^2 mol^{-2} s^{-1}	Eigen and Kustin, 1962
$Br^- + OH \rightarrow HOBr^-$	1.1×10^{10}	Kläning and Wolff, 1985
$HOBr^- \rightarrow Br^- + OH$	3.3×10^7	Kläning and Wolff, 1985
$\rightarrow Br + OH^-$	4.2×10^6	Kläning and Wolff, 1985
$HOBr^- + H^+ \rightarrow H_2O + Br$	1.1×10^{10}	Kläning and Wolff, 1985
$Br + H_2O \rightarrow HOBr^- + H^+$	1.4 s^{-1}	Kläning and Wolff, 1985
$HOBr + Br^- \rightarrow Br_2 + OH^-$	$K_{eq} = 70$	Mamou et al., 1977
$Br + Br^- \rightarrow Br_2^-$	9×10^9	Nagarajan and Fessenden, 1985
	$K_{eq} = 1.1 \times 10^5$	Mamou et al., 1977
$Br_2^- + Br_2^- \rightarrow Br_2 + 2Br^-$	2×10^9	D'Angelantonio et al., 1988
$Br_2 + H_2O \rightarrow HOBr + Br^- + H^+$	1.1×10^2	Eigen and Kusten, 1962

compounds, have been developed for reactions in clouds and fogs. The reader is referred to papers by Seigneur and Saxena (1988), Pandis and Seinfeld (1989a, 1989b), Jacob et al. (1989), Lelieveld and Crutzen (1990), Walcek et al. (1990, 1997), Dennis et al., (1993), McHenry and Dennis (1994), Sander and Crutzen (1996), and Liang and Jacob (1997) for some examples of models used to examine the relative importance of various oxidation mechanisms.

h. Effect of Droplet Size on S(IV) Oxidation

The chemical composition of fogs, clouds, and particles (see Chapter 9) varies as a function of particle size. For example, Figure 8.19 shows the concentrations of the major cations and anions measured in small and large cloud droplets at La Jolla peak in southern California (Collett et al., 1994, 1999). The large drops are enriched in soil and sea salt derived species such as Mg^{2+}, Ca^{2+}, Na^+, and Cl^- whereas the smaller particles contain higher concentrations of sulfate and H^+, i.e., have a smaller pH. In a given sample, the difference in pH between large and small drops was usually less than one pH unit, although in some samples, the difference in pH was as much as 2. Similar results have been obtained at a variety of sampling sites and with different types of samplers (e.g., Bator and Collett, 1997), indicating this is a general phenomenon.

These data suggest that the larger fog and cloud droplets are formed on the larger, coarse particles, which contain largely elements associated with mechanical processes such as erosion and sea spray (see Chapter 9.A). Similarly, the smaller fog and cloud droplets arise from the smaller, accumulation mode particles, which contain species such as sulfate and ammonium ions. It is somewhat surprising that the drops retain in part the chemical signatures of the nuclei on which they were formed, given the number of chemical and physical interactions associated with fogs and clouds as they form and dissipate.

For oxidation of S(IV) by H_2O_2, which does not vary greatly over the pH range of atmospheric interest, such differences in droplet composition are not important. However, for the other oxidation mechanisms that do depend significantly on pH and/or other solute con-

TABLE 8.16 Some Aqueous-Phase Bromine/Chlorine Chemistry

Reaction	Rate constant k (L mol^{-1} s^{-1}) or equilibrium constant K_{eq}	Reference
$HOBr + Cl^- \rightarrow BrCl + OH^-$	$\gamma_{ice}^a = 0.001$–0.25	Abbatt, 1994; Kirchner et al., 1997
$BrCl + H_2O \leftrightarrow HOBr + Cl^- + H^+$	$k > 10^5$ s^{-1}	Wang et al., 1994
	$K_{eq} = 1.8 \times 10^{-5}$ M^2	Wang et al., 1994
$BrCl + Br^- \rightarrow Br_2Cl^-$	$k > 10^8$	Wang et al., 1994
	$K_{eq} = 1.8 \times 10^4$ M^{-1}	Wang et al., 1994
$Br_2Cl^- \leftrightarrow Br_2 + Cl^-$	$K_{eq} = 1.3$ M	Wang et al., 1994
$BrCl + Cl^- \leftrightarrow BrCl_2^-$	$K_{eq} = 6.0$ M^{-1}	Wang et al., 1994
$BrCl_2^- + H_2O \leftrightarrow HOBr + H^+ + 2Cl^-$	$K_{eq} = 3.0 \times 10^{-6}$ M^3	Wang et al., 1994

[a] Reaction probabilities for HOBr with chloride ion in ice; the larger values are for reaction with HCl due to acid catalysis.

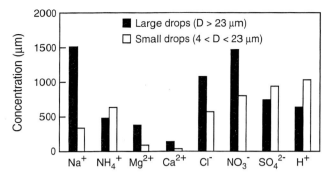

FIGURE 8.19 Concentrations of major anions and cations in coastal stratus clouds at La Jolla Peak, California, on July 6, 1993 (adapted from Collett et al., 1994).

centrations, e.g., by O_2 and O_3, these variations in composition with particle size can be important. Thus oxidation of S(IV) by O_2 and O_3 would be expected to be inhibited in the more acidic droplets.

In the vast majority of studies, the chemical composition of fogs and clouds is measured on bulk samples, i.e., collected without discrimination by size, and the relative importance of various oxidation mechanisms assessed based on this bulk composition. As discussed by Collett and co-workers (Collett et al., 1994; Bator and Collett, 1997), the use of the average cloudwater composition without taking into account the differences in chemistry in different drop sizes leads to an underestimation of the rate of S(IV) oxidation if the pH-dependent oxidation mechanisms contribute significantly. For example, model predictions by Gurciullo and Pandis (1997) suggest that the sulfate production could be underestimated by a factor of ~30 in the early stages of cloud formation. The size of the effect depends not only on the time but also especially on the concentrations of the oxidants H_2O_2 and O_3. The reason is that at higher pH values, O_3 can contribute significantly to the oxidation of S(IV) and this will therefore occur preferentially in the larger drops having lower concentrations of H^+. On the other hand, if the H_2O_2 concentration is large relative to O_3, the oxidation by the peroxide overwhelms the ozone reaction; in this case, there is little effect of particle size since the H_2O_2 oxidation is not highly dependent on pH.

Using bulk average composition to assess the chemistry of fogs and clouds can be misleading in other respects as well. For example, Pandis and Seinfeld (1991, 1992) show that the bulk mixture formed from drops with different pH values that are each in equilibrium with the gas phase does not itself conform to Henry's law, with the bulk mixture being supersaturated with respect to species such as weak acids and NH_3. Winiwarter et al. (1992) show that deviations in the opposite direction can result for highly soluble species in a closed system.

In short, the use of average bulk aqueous-phase composition to assess the importance of various oxidation pathways can lead to erroneous results if the composition of drops varies significantly with particle size.

i. Fog–Smog–Fog Cycles

Studies of the chemical composition of fogs have shown that they can be quite acidic and contain high concentrations of other cations and anions as well. For example, pH values in the range of 2–4 are not uncommon, and values as low as 1.69 have been observed (e.g., see Munger et al., 1983; Hoffmann and Jacob, 1984; Jacob et al., 1985; Fuzzi et al., 1984; Muir et al., 1986; Hering et al., 1987; and Erel et al., 1993).

The reason that the aqueous-phase concentrations in fogs can be so high is related in part to the liquid water content (LWC), which is a major difference between clouds and fogs. The liquid water content for fogs is typically of the order of 0.1 g m^{-3} of air, whereas that for clouds is about an order of magnitude higher. This small LWC in fogs corresponds to increased solute concentrations.

A second reason for the higher concentrations in fogwater is the relatively rapid condensation and evaporation of water vapor onto preexisting particles as the fog forms and then dissipates. Figure 8.20 is a schematic of this process. Aerosol particles, which typically contain sulfate and nitrate as well as a variety of other species, act as condensation nuclei for the condensation of water as the relative humidity increases during fog formation. As more water condenses, the particle components become increasingly diluted. Simultaneously, the solution can take up gases such as SO_2, HNO_3, and NH_3, and chemical reactions such as the oxidation of S(IV) to sulfate can occur in the droplet. Both the initial sulfate in the particle on which water condensed and sulfate formed by oxidation of S(IV) in the fog aqueous phase can contribute significantly to the total sulfate deposition. For example, Bergin et al. (1996) estimate that in a typical Arctic radiation fog, preexisting sulfate aerosol accounts for about 60% of the sulfate deposition and oxidation of S(IV) in the aqueous phase the remaining 40%.

Because these fog drops are much larger than the initial particles, gravitational settling is more important and acts to remove some of the particles (Pandis and Seinfeld, 1990). When the temperature rises, water begins to evaporate out of the fog droplet, concentrating the solutes. It is this process that likely led to the

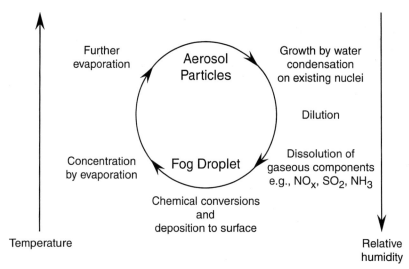

FIGURE 8.20 Schematic of role of atmospheric aerosols in fog formation and evaporation (adapted from Munger et al., 1983).

extremely low observed pH value of 1.69. This mechanism is consistent with fog–smog–fog cycles that are commonly observed in urban areas (Munger et al., 1983; Pandis et al., 1990). [However, it is interesting to note that in some cities with relatively good air quality, reduced fog formation has been observed and attributed to increased temperatures via the heat island effect (Sachweh and Koepke, 1995).]

Because such fogs can be so concentrated, there has been a concern regarding their effects on both human health and ecosystems. For example, Hoffmann (1984) has estimated the sulfate ion concentrations in the London fog during the 1952 smog episode. While the pollutants that were present and their concentrations are not well known, 4000 excess deaths occurred in a 5-day period (Chapter 1.B). Hoffmann and co-workers estimate that the sulfate concentration was approximately 11–46 milliequivalents per liter (mequiv L^{-1}) during this air pollution episode. This is of the same order of magnitude as those measured in acid fogs in other areas. Hence the health effects of acids and associated species in fogs continue to be of interest.

4. Oxidation on Surfaces

Adsorption of SO_2 on the surfaces of solids, followed by its oxidation on the surface, may provide a third route for the formation of sulfuric acid. The solid surfaces may be suspended in the gas phase or in atmospheric droplets, as discussed in detail by Chang and Novakov (1983).

It is known that SO_2 in air is oxidized on both graphite and soot particles, with water vapor enhancing the sulfate formation on the surfaces (Hulett et al., 1972; Novakov et al., 1974; Halstead et al., 1990). Whether this surface oxidation is enhanced by the presence of oxidants other than air (e.g., O_3 or NO_2) is not clear. For example, under certain conditions some researchers report an enhancement of the rate of uptake of SO_2 due to the presence of the copollutants O_3 and NO_2 (Cofer et al., 1981; Britton and Clarke, 1980); however, others have not observed such an effect (Baldwin, 1982; Halstead et al., 1990), particularly at low gas concentrations more closely approximating those in the atmosphere (Cofer et al., 1984).

Carbon surfaces of various types have been the subject of most studies. However, other types of surfaces, including alumina, fly ash, dust, MgO, V_2O_5, Fe_2O_3, and MnO_2, have also been shown to oxidize SO_2 and/or remove it from the gas phase (Hulett et al., 1972; Judeikis et al., 1978; Liberti et al., 1978; Barbaray et al., 1977, 1978; Halstead et al., 1990). As expected, the rate of removal depends on the nature of the particular surface, the presence of copollutants such as NO_2, and, as in the case of carbonaceous surfaces, the relative humidity. The increase with increasing water vapor suggests that oxidation of the SO_2 may occur in a thin film of water on the surface of the solid.

Carbonaceous particles suspended in aqueous solutions can also act as sites for efficient SO_2 oxidation. For example, Novakov and co-workers have studied the kinetics of such processes with respect to the concentrations of O_2, S(IV), and carbon particles as well as the pH and temperature dependencies (Brodzinsky et al., 1980; Chang et al., 1981; Benner et al., 1982).

Extrapolation of their results to atmospheric conditions led them to suggest that such heterogeneous reactions could be important in the aqueous-phase oxidation of SO_2.

It is difficult to evaluate quantitatively the importance of such heterogeneous reactions in the overall oxidation of S(IV). Their rates depend on the physical and chemical natures of the surfaces involved, including specific surface areas, the presence of defects and surface adsorbed water, etc., yet these are not well understood, especially for highly complex environmental gas–liquid–solid systems. For example, the rates of oxidation of SO_2 at 80% relative humidity on two different samples of fly ash obtained from two coal-fired power plants differed by more than an order of magnitude (Dlugi and Gusten, 1983). Even in laboratory systems the nature of relatively simple surfaces such as carbon depends on the history of the material.

However, the available evidence suggests that such heterogeneous reactions should be considered as potential contributors to the overall oxidation of S(IV), especially close to sources where particle concentrations and hence available surface areas are relatively high (see Chapter 9). Baldwin (1982), for example, estimates that the loss of gaseous SO_2 due to interactions with particle surfaces at a particle density of 100 $\mu g\ m^{-3}$ could be as high as 1% h^{-1}. Such rates may not be sustained for long periods of time due to saturation of the surface, if the surface is dry. However, Chang and Novakov (1983) point out that when the surface is wet, the active sites are constantly regenerated as the sulfate formed on the surface dissolves in the surrounding aqueous solution; given the availability of water in the atmosphere, it seems unlikely that any surfaces are really "dry."

Herring et al. (1996) have reported evidence for the heterogeneous oxidation of both SO_2 and NO_x on the surfaces of soil dust particles in the smoke plume from the 1991 Kuwait oil fires. The rate of SO_2 oxidation was estimated to be 6.5% h^{-1}.

In addition to the dark oxidation of S(IV) on surfaces, there may be photochemically induced processes as well. For example, irradiation of aqueous suspensions of solid α-Fe_2O_3 (hematite) containing S(IV) with light of $\lambda > 295$ nm resulted in the production of Fe(II) in solution (Faust and Hoffmann, 1986; Faust et al., 1989; Hoffmann et al., 1995). This reductive dissolution of the hematite has been attributed to the absorption of light by surface Fe(III)–S(IV) complexes, which leads to the generation of electron–hole pairs, followed by an electron transfer in which the adsorbed S(IV) is oxidized to the $SO_3^-\cdot$ radical anion. This initiates the free radical chemistry described earlier.

The photochemical reactions of such semiconductors in aqueous solutions may also influence S(IV) oxidation via the production of H_2O_2. The possibility of such photoassisted surface reactions in the atmosphere was first examined by Calvert in 1956. For example, Fe_2O_3 particles suspended in a bisulfite solution in the presence of O_2 rapidly oxidize the S(IV) to S(VI) in the presence of light, whereas no oxidation occurs in the dark (Frank and Bard, 1977). It has been suggested that this is due to the absorption of light by Fe_2O_3 and the migration of electrons in the conduction band to the particle surface where they react with O_2 and H^+ to form H_2O_2:

$$O_2 + 2e^-_{surface} + 2H^+ \xrightarrow{surface} H_2O_2. \quad (44)$$

Similar reactions have been observed more recently on a variety of solids, including ZnO, TiO_2, and desert sand; in the presence of organics, organic peroxides have also been observed (e.g., see Kormann et al., 1988). For a review of this area, the reader is referred to the review by Hoffmann et al. (1995).

5. Relative Importance of Various Oxidation Pathways for SO_2

Figure 8.21 shows one estimate of the relative importance of the oxidation of S(IV) by O_3 and H_2O_2, the Fe- and Mn-catalyzed O_2 oxidation, and the oxidation on a carbon surface at concentrations typically found in the atmosphere (Martin, 1984; Martin et al.,

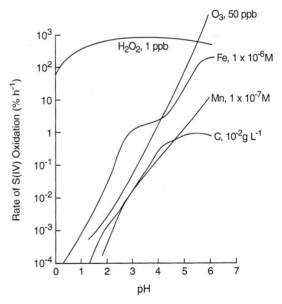

FIGURE 8.21 Estimated rates of oxidation of S(IV) in a hypothetical cloud with liquid water content of 1 mL m^{-3} (i.e., LWC of 1 g m^{-3}) based on 5 ppb gaseous SO_2 as a function of pH (adapted from Martin, 1984; Martin et al., 1991).

1991). It was assumed that there were no limitations on the rates of oxidation due to mass transport; as discussed in detail by Schwartz and Freiberg (1981), this assumption is justified except for very large droplets (>10 μm) and high pollutant concentrations (e.g., O_3 at 0.5 ppm) where the aqueous-phase reactions are very fast. It was also assumed that the aqueous phase present in the atmosphere was a cloud with a liquid water content (V) of 1 g m^{-3} of air. As seen earlier, the latter factor is important in the aqueous-phase rates of conversion of S(IV); thus the actual concentrations of iron, manganese, and so on in the liquid phase and hence the kinetics of the reactions depend on the liquid water content.

Only the oxidation by H_2O_2 is relatively independent of pH. This arises because the rate coefficient for the reaction and the solubility of S(IV) show opposite trends with pH (Fig. 8.9b). For the other species, the effects of the S(IV) solubility and the pH dependence of the kinetics work in the same direction (Fig. 8.9a), leading to a strong overall dependence on pH. The uncatalyzed oxidation of O_2 is not shown, because it is generally believed to be unimportant compared to the other mechanisms in real atmospheric droplets containing "impurities" such as metals that will act as catalysts.

The estimates in Fig. 8.21 show that H_2O_2 is expected to be the most important oxidant for S(IV) in clouds and fogs at pH < 4.5. At higher pH values, both O_3 and the iron-catalyzed O_2 oxidation can compete.

Figure 8.22 shows an estimate of the contributions to the oxidation of S(IV) by H_2O_2, by the iron-catalyzed O_2 oxidation, and by OH in both the gas and aqueous phases of a cloud (Jacob *et al.*, 1989). It is seen that H_2O_2 and the iron-catalyzed process predominate at night, but the gas-phase oxidation by OH becomes significant during the day when it is formed by photochemical processes. On the other hand, the contribution from oxidation by OH in the aqueous phase is relatively small.

D. ORGANIC ACIDS

While the focus in terms of acid deposition has been on sulfuric and nitric acids, it has been increasingly recognized that organic acids can also contribute significantly to the acidity of both the gas and aqueous phases in both urban and remote regions. A review of carboxylic acids in the atmosphere is given by Chebbi and Carlier (1996).

The major organic acids found in the gas phase are formic acid (HCOOH) and acetic acid (CH_3COOH), with smaller contributions from larger aliphatic acids and multifunctional acids such as pyruvic acid ($CH_3COCOOH$) and glyoxalic acid (COOHCHO) (e.g., see Kawamura *et al.*, 1985; and Khwaja, 1995). Formic and acetic acids have sufficiently high vapor pressures that they are found almost totally in the gas phase rather than in particles (e.g., see Kawamura *et al.*, 1985; Talbot *et al.*, 1988; Grosjean, 1989; and Khwaja, 1995). Concentrations of HCOOH in urban areas are typically a few ppb (e.g., see Dawson *et al.*, 1980; and summaries by Lawrence and Koutrakis, 1994; and Khwaja, 1995) although levels up to 45 ppb have been observed (Lawrence and Koutrakis, 1994). CH_3COOH is also present in urban areas in the low-ppb range, with reported concentrations as high as 15 ppb in the Los Angeles area (Grosjean, 1989).

In semirural and rural areas, the concentrations of HCOOH and CH_3COOH are somewhat smaller but still tend to be in the range around a ppb (e.g., Sanhueza *et al.*, 1996; Kumar *et al.*, 1996; Granby *et al.*, 1997a, 1997b). For example, at a rural site in Virginia, concentrations of formic and acetic acids are typically ~1 and 0.5 ppb, respectively, although peak formic acid concentrations as high as 10 ppb have been measured (Talbot *et al.*, 1995; Keene *et al.*, 1995). In the northern Congo, ground-level concentrations of ~0.5 ppb each have been measured for these two acids, but the levels were as high as ~3-4 ppb in the boundary layer above the surface (Helas *et al.*, 1992).

In marine remote regions, concentrations of formic and acetic acids are typically about 0.1-0.3 ppb, although much higher levels have been observed in stable layers of air at higher altitudes above the ocean (e.g., Chapman *et al.*, 1995).

Formic and acetic acids may constitute a large fraction of the *gas-phase* acidity. For example, Grosjean (1990) measured the concentrations of these two organic acids as well as the inorganic acids HNO_3 and HCl in southern California during a smog episode in

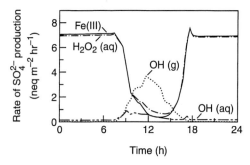

FIGURE 8.22 Calculated contributions to S(IV) oxidation in a cloud of the iron-catalyzed oxidation by O_2/Fe(III), by H_2O_2 and OH in solution, and by OH in the gas phase, OH(g), expressed in terms of rate of production of column S(IV) (adapted from Jacob *et al.*, 1989).

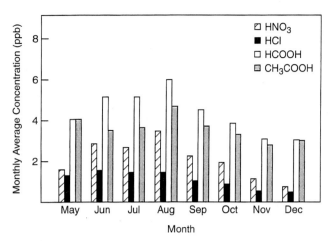

FIGURE 8.23 Average concentrations of gas-phase acids at eight sites in southern California in 1986 (adapted from Nolte et al., 1997).

August 1986. The two organic acids taken together represented 44–93% of the total gas-phase acids.

Similarly, Fig. 8.23 shows the monthly average concentrations of $HCOOH$, CH_3COOH, HCl, and HNO_3 measured at eight sites in southern California in 1986 (Nolte et al., 1997). Formic and acetic acids are seen to be the major acids, consistently exceeding HNO_3.

There are a number of potential sources, both primary and secondary, of formic and acetic acids in urban areas (e.g., see review by Chebbi and Carlier, 1996). Clearly the oxidation of organics can lead to the formation of acids. For example, as discussed in Chapter 6.E.2, the reaction of O_3 with alkenes generates a Criegee intermediate that can react with water vapor to generate a carboxylic acid. Aqueous-phase oxidations, e.g., of $HCHO$ to $HCOOH$, followed by evaporation from the condensed phase may also potentially contribute. Direct emissions (i.e., primary sources) include automobile exhaust, biomass combustion, stationary source emissions, for example, from vinegar manufacturing and cooking of food, and possibly natural emissions from vegetation (e.g., see Kawamura et al., 1985; Talbot et al., 1988, 1995; Dawson and Farmer, 1988; and Keene and Galloway, 1988).

The relative contributions of these sources depends, not surprisingly, on the particular location. For example, Granby et al. (1997a, 1997b) found that the concentrations of $HCOOH$ and CH_3COOH were similar in central Copenhagen and at a semirural site 30 km away, where the NO_y concentrations were an order of magnitude smaller. In addition, they correlated with the photochemical species such as O_3, suggesting that the acids were formed by chemical reactions (e.g., of O_3 with alkenes) during long-range transport. In southern California, these two acids were correlated with elemental carbon, used as an indicator of primary emissions, at upwind site close to the coast but not at downwind sites, suggesting that local primary emissions dominated upwind (Nolte et al., 1997). Conversely, the correlation between formic and acetic acids with both O_3 and particle sulfate increased downwind, suggesting a significant contribution to their formation from chemical reactions both in the gas phase and in solution (where most of the sulfate is formed). Similarly, in Venezuela, atmospheric oxidation of organics appears to be the major source of these acids, perhaps with some contribution from emissions from soils during the dry season (Sanhueza et al., 1996).

Grosjean (1989, 1992) estimated that about 40% of the $HCOOH$ and $\sim 75\%$ of the CH_3COOH measured at Claremont, just east of Los Angeles, are due to direct emissions. This is qualitatively consistent with carbon isotope measurements of $^{13}C/^{12}C$ of formic and acetic acids in rainwater collected in this area, which show $HCOOH$ is from a combination of direct emissions and secondary oxidation chemistry and CH_3COOH is primarily from direct emissions (Sakugawa and Kaplan, 1995).

A variety of dicarboxylic acids have been measured in air, including, for example, oxalic acid $[(COOH)_2]$, succinic acid $[HOOCCH_2CH_2COOH]$, and malonic acid $[HOOCCH_2COOH]$, as well as larger straight- and branched-chain carboxylic acids; unsaturated and aromatic acids such as phthalic acid are also observed in smaller concentrations (e.g., see Kawamura et al., 1996a, 1996b). Because of their lower vapor pressures, they are found predominantly in particles (see Chapter 9).

As discussed in Chapters 6 and 9, these dicarboxylic acids are believed to result in part from the oxidation of organics in air. However, Kawamura and Kaplan (1987) have also shown that automobile exhaust can be a significant source as well.

Organic acids can be removed by reaction with OH (see Chapter 6.H) as well as by wet or dry deposition. As a result, these acids are a common component of rain, clouds, fogs, and dews as would be expected from their large Henry's law constants, $\sim 10^3 - 10^4$ mol L^{-1} atm^{-1} (see Keene et al., 1995), and are found in the condensed phase from remote to highly polluted urban areas (e.g., see Norton, 1985; Keene and Galloway, 1986, 1988; Likens et al., 1987; Weathers et al., 1988; Muir, 1991; Sakugawa et al., 1993; Keene et al., 1995; and Khwaja et al., 1995).

In short, while the focus has been primarily on sulfuric and nitric acids as a source of acid deposition, it is clear that organic acids can also contribute significantly. The gas-phase concentrations of the simplest carboxylic acids, formic acid and acetic acid, are relatively high even in remote regions, of the order of a ppb. Both natural and anthropogenic sources have been proposed, but the nature of the individual sources and their relative contributions are not well established.

E. OXIDATION OF SULFUR COMPOUNDS OTHER THAN SO_2

As discussed in Chapter 2 and in more detail in Chapter 11, a variety of organic sulfur compounds in addition to inorganics such as H_2S and COS are emitted by biological sources. In the troposphere, they may ultimately be oxidized to SO_2 and H_2SO_4. However, the chemistry of these compounds tends to be complex, and a variety of partially oxidized sulfur compounds is formed first.

Table 8.17 summarizes the rate constants and estimated tropospheric lifetimes of some of these sulfur compounds with respect to reaction with OH or NO_3. The assumed concentrations of these oxidants chosen for the calculations are those characteristic of more remote regions, which are major sources of reduced sulfur compounds such as dimethyl sulfide (DMS). It is seen that OH is expected to be the most important sink for these compounds and that NO_3 may also be important, for example, for DMS oxidation (see also Chapter 6.J).

As discussed in Chapters 6 and 7, NaCl and other chloride salts in airborne sea salt particles are now believed to react to generate photochemically active compounds such as Cl_2 and $ClNO_2$ that photolyze to form chlorine atoms. Peak concentrations of atomic chlorine as high as 10^4–10^5 cm^{-3} in the early morning hours have been predicted (Keene *et al.*, 1993, 1996; Pszenny *et al.*, 1993; Wingenter *et al.*, 1996; Singh *et al.*, 1996; Tuckermann *et al.*, 1997; Spicer *et al.*, 1998). Chlorine atoms also react rapidly with reduced sulfur compounds and hence are potential oxidants for these organics in the early morning hours in coastal regions.

Finally, there has been a great deal of interest in the halogen oxides IO, BrO, and ClO as potential oxidants for organic sulfur compounds such as dimethyl sulfide. We therefore also discuss the current status of these reactions.

Dimethyl sulfide (DMS) is of particular interest since it is the major reduced sulfur compound produced in oceanic areas. In addition, as discussed in Chapter 14, it is believed to play a significant role in global climate issues; thus, its oxidation to SO_2 is expected to lead to the subsequent formation of sulfuric acid and sulfate particles by the chemistry that is the focus of this chapter. In addition, as we shall see, another oxidation product of DMS is methanesulfonic acid (CH_3SO_3H), which is highly soluble and has a sufficiently low vapor pressure that, like sulfuric acid, it exists primarily in the condensed phase (e.g., Ayers *et al.*, 1980; Clegg and Brimblecombe, 1985; Kreidenweis and Seinfeld, 1988a, 1988b). As a result, oxidation of DMS can generate condensation nuclei on which water condenses to form clouds. These clouds can scatter radiation back to space, lowering the surface temperature and hence the production of DMS, forming a feedback loop (Charlson *et al.*, 1987). Because of this potential feedback, there have been many recent studies examining the relationship between DMS, its oxidation products, and condensation nuclei (e.g., see Kreidenweis *et al.*, 1991; Berresheim *et al.*, 1993; Pandis *et al.*, 1994; Ferek *et al.*, 1995; Clarke *et al.*, 1996; and Bandy *et al.*, 1992, 1996). However, it is clear that understanding such feedbacks requires understanding the chemistry of DMS and, in particular, its role as a source of low-volatility products that can act as condensation nuclei.

As a result, we focus here on what is known about the tropospheric chemistry of DMS. As we shall see, the chemistry of even this relatively simple compound is complex, and much remains to be learned about its reaction mechanisms. For larger reduced sulfur com-

TABLE 8.17 Rate Constants and Lifetimes at Room Temperature for the Reactions of OH and NO_3 with Some Reduced Sulfur Compounds Emitted Biogenically

Compound	OH (day)		NO_3 (night)	
	k (cm^3 $molecule^{-1}$ s^{-1})[a]	τ_{OH}[b]	k (cm^3 $molecule^{-1}$ s^{-1})[a]	τ_{NO_3}[c]
CH_3SCH_3	6.5×10^{-12}	2 days	1.1×10^{-12}	10 h
CH_3SSCH_3	2.3×10^{-10}	1.2 h	0.7×10^{-12}	16 h
CH_3SH	3.3×10^{-11}	8 h	0.92×10^{-12}	12 h
H_2S	4.8×10^{-12}	2.4 days	$<1 \times 10^{-15}$	>1 yr
CS_2	4.7×10^{-12}	2.5 days	$<1 \times 10^{-15}$	>1 yr
COS	2.0×10^{-15}	16 yr	$<1 \times 10^{-16}$	>13 yr

[a] Rate constants at 1 atm in air at 298 K; from Atkinson *et al.* (1997a).
[b] $\tau_{OH} = 1/k[OH]$, where [OH] is taken to be 1×10^6 radicals cm^{-3}.
[c] $\tau_{NO_3} = 1/k[NO_3]$, where [NO_3] is taken to be 1 ppt = 2.5×10^7 radicals cm^{-3}.

pounds, including disulfides, even less is known. However, there are a number of organic sulfur intermediates formed in the DMS reactions that are common to the oxidations of other sulfur compounds as well. Hence some "educated guesses" as to the chemistry of other sulfur compounds can be made based on what is known about DMS chemistry as discussed in the following sections. For a review of dimethyl sulfide and other sulfur compounds in the atmosphere, see Berresheim et al. (1995).

1. Reactions of Dimethyl Sulfide (CH_3SCH_3)

a. Reaction with OH

The reaction of OH with DMS is believed to proceed by two channels, abstraction of a hydrogen from a methyl group and addition of OH to the sulfur atom:

$$OH + CH_3SCH_3 \rightarrow CH_3SCH_2 + H_2O, \quad (45a)$$

$$k_{45a} = 4.8 \times 10^{-12} \text{ cm}^3 \text{ molecule}^{-1} \text{ s}^{-1} \text{ at 298 K}$$

(Atkinson et al., 1997a)

$$OH + CH_3SCH_3 + M \leftrightarrow CH_3S(OH)CH_3 + M,$$
$$(45b, -45b)$$

$$k_{45b} = 1.7 \times 10^{-12} \text{ cm}^3 \text{ molecule}^{-1} \text{ s}^{-1}$$
$$\text{at 298 K, 1 atm}$$

(Atkinson et al., 1997a).

Thus, at 1 atm in air and 298 K, abstraction predominates. The addition channel (45b) would be expected to have a pressure dependence and a negative temperature dependence (see Chapter 5.A.2). Thus is consistent with the observation that the effective overall bimolecular rate constant in 1 atm of air decreases as the temperature increases from 250 to 310 K and that the fraction of the reaction that proceeds via (45a) increases from 0.24 to 0.87 over the same temperature range (e.g., Hynes et al., 1986).

Experimental studies of the reaction of OH with fully deuterated DMS (Hynes et al., 1995; Barone et al., 1996) give a bond strength for the adduct at 258 K of ~10–13 kcal mol^{-1}. In air, reaction (46) of the adduct with O_2 can occur in competition with its decomposition back to reactants [reaction (−45b)]:

$$CH_3S(OH)CH_3 + O_2 \rightarrow \text{Products.} \quad (46)$$

The rate constant (k_{46}) is independent of pressure from 100 to 700 Torr and temperature from 217 to 300 K and is reported to be $(8-10) \times 10^{-13}$ cm^3 molecule^{-1} s^{-1} (Hynes et al., 1995; Barone et al., 1996). At 1 atm in air, the first-order removal rate of the adduct by reaction with O_2 is thus 4×10^6 s^{-1}, similar to the unimolecular rate of decomposition of the adduct of 3.5×10^6 s^{-1} at 261 K (Hynes et al., 1986).

At least part of reaction (46) forms dimethyl sulfoxide (DMSO):

$$CH_3S(OH)CH_3 + O_2 \rightarrow CH_3SOCH_3 + HO_2. \quad (47)$$

Figure 8.24a, for example, shows the FTIR spectrum before the photolysis of mixtures of DMS in air with H_2O_2 as the OH source and the residual spectrum after 5 min of photolysis (Barnes et al., 1996). The reactants, as well as the product SO_2 have been subtracted out in Fig. 8.24b. Dimethyl sulfoxide (DMSO) as well as dimethyl sulfone, $CH_3SO_2CH_3$ (DMSO$_2$), and small amounts of COS are observed as products. DMSO is so reactive that it is rapidly converted into DMSO$_2$ in this system and hence both are observed in Fig. 8.24b. However, Barnes and co-workers calculate that the DMSO yield corrected for secondary oxidation is about the same as the fraction of the OH–DMS reaction that proceeds by addition under these conditions, i.e., that the major fate of the adduct is reaction (47). Turnipseed et al. (1996) measured the yield of HO_2 from reaction (47) to be 0.50 ± 0.15 at both 234 and 258 K, suggesting that there are other reaction paths than (47) as well. The mechanism of formation of COS is not clear but may involve the oxidation of thioformaldehyde ($H_2C=S$). The implications for the global budget of COS are discussed by Barnes et al. (1994b, 1996).

The radical formed by abstraction, (45a), is, in essence, an alkyl radical and is therefore expected to add O_2:

$$CH_3SCH_2 + O_2 + M \rightarrow CH_3SCH_2OO + M. \quad (48)$$

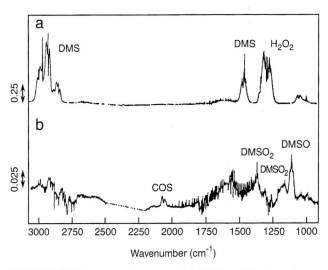

FIGURE 8.24 Infrared spectra of (a) DMS and H_2O_2 at 1 atm air before photolysis and (b) the product spectrum after 5-min irradiation and subtraction of peaks due to unreacted DMS and H_2O_2 and product SO_2 (adapted from Barnes et al., 1996).

The rate constant for reaction (48) has been reported as $(5.7 \pm 0.4) \times 10^{-12}$ cm^3 molecule^{-1} s^{-1} at 1 atm (Wallington et al., 1993) and somewhat smaller at lower pressures (Butkovskaya and Le Bras, 1994), consistent with similar reactions of alkyl radicals (see Chapter 6.D.1).

In the presence of NO, the alkylperoxy radical oxidizes NO to NO$_2$, forming an alkoxy radical, which can decompose to formaldehyde and the CH$_3$S radical (Wallington et al., 1993; Butkovskaya and Le Bras, 1994):

$$CH_3SCH_2OO + NO \rightarrow CH_3SCH_2O + NO_2, \quad (49)$$

$$CH_3SCH_2O \rightarrow CH_3S + HCHO. \quad (50)$$

The rate constant for reaction (49) has been estimated to be $\sim 1 \times 10^{-11}$ cm^3 molecule^{-1} s^{-1} at 298 K, similar to that of other RO$_2$ + NO reactions (Turnipseed et al., 1996).

An alternate potential fate for the CH$_3$SCH$_2$O radical is the well-known abstraction reaction of alkoxy radicals with O$_2$, forming methyl thioformate (Crutzen, 1983; Butkovskaya and Le Bras, 1994):

$$CH_3SCH_2O + O_2 \rightarrow CH_3SCHO + HO_2. \quad (51)$$

Infrared absorption bands attributable to methyl thioformate have been observed in the oxidation of DMS by OH in the absence of NO$_x$ but not when NO$_x$ was present (Barnes et al., 1996; Patroescu et al., 1999). Reaction (51) appears to be quite slow ($k < 1 \times 10^{-15}$ cm^3 molecule^{-1} s^{-1}), so that the dominant fate of CH$_3$SCH$_2$O is decomposition to HCHO + CH$_3$S; Turnipseed et al. (1996) measured the production of CH$_3$S in the reaction of OH with DMS and also suggest, based on its relatively high yield, that the thermal decomposition (50) predominates over reaction (51) with O$_2$ to form methyl thioformate. The methyl thioformate observed in laboratory systems in the absence of NO is thus likely due to cross reactions of CH$_3$SCH$_2$OO with itself or other RO$_2$ (Barnes et al., 1994a), and the abstraction channel in the OH + DMS reaction leads primarily to the formation of CH$_3$S. Methyl thioformate itself reacts rapidly with OH ($k = 1.1 \times 10^{-11}$ cm^3 molecule^{-1} s^{-1} at room temperature), with evidence from FTIR studies for both abstraction of the aldehydic hydrogen and addition to the sulfur; photolysis of methyl thioformate also occurs but is relatively slow, with a lifetime of ≥ 5.4 days (Patroescu et al., 1996).

Thus, in low-NO$_x$ environments, the CH$_3$SCH$_2$OO radical will react with HO$_2$ or other RO$_2$ radicals:

$$CH_3SCH_2OO + HO_2 \rightarrow CH_3SCH_2OOH + O_2, \quad (52a)$$

$$\rightarrow CH_3SCHO + H_2O + O_2. \quad (52b)$$

The fate of the CH$_3$S radical in the atmosphere is not clear but may include reaction with O$_2$, O$_3$, or NO$_2$. Table 8.18 summarizes the rate constants for these reactions at 298 K and the corresponding lifetimes under typical tropospheric conditions. Although only an upper limit can be placed on the rate constant for the O$_2$ reaction, it may still be the predominant reaction of CH$_3$S because of the large oxygen concentration in the atmosphere. However, this process forms a weakly bound adduct with a CH$_3$S–OO bond energy of about 11 kcal mol^{-1}, and decomposition back to reactants also occurs:

$$CH_3S + O_2 + M \leftrightarrow CH_3SOO + M. \quad (53, -53)$$

Evidence for formation of this adduct has been obtained in laboratory studies between 216 and 258 K, where CH$_3$S is observed to come to equilibrium in the presence of O$_2$ (Turnipseed et al., 1992). A contribution from the back reaction is difficult to avoid in experimental systems, making measurements of the true forward rate constant somewhat uncertain. Extrapolation of the measured kinetics to 298 K suggests that approximately 30–75% of the CH$_3$S would be in the form of the adduct at typical tropospheric temperatures of 298–275 K and 1 atm pressure (Turnipseed et al., 1992).

TABLE 8.18 Removal Rates of CH$_3$S in the Troposphere at 298 K[a]

Reaction of CH$_3$S	k(298 K) (cm^3 molecule^{-1} s^{-1})	[Reactant] (molecules cm^{-3})	τ(s)
CH$_3$S + O$_2$ → products	$< 6.0 \times 10^{-18}$	5.2×10^{18}	>0.03
CH$_3$S + O$_3$ → products	5.4×10^{-12}	9.8×10^{11} (40 ppb)	0.19
CH$_3$S + NO$_2$ → CH$_3$SO + NO	5.8×10^{-11}	2.5×10^{10} (1 ppb)	0.69
		2.5×10^8 (10 ppt)	69

[a] Rate constants from Tyndall and Ravishankara (1991), Dominé et al. (1992), and Turnipseed et al. (1993).

The fate of the CH_3SOO adduct is not known but by analogy to other peroxy radical reactions is expected to include reactions with NO and NO_2 (Turnipseed et al., 1993):

$$CH_3SOO + NO \rightarrow CH_3SO + NO_2, \quad (54)$$

$k_{54}(227-256\ K) = 1.1 \times 10^{-11}\ cm^3\ molecule^{-1}\ s^{-1}$

$$CH_3SOO + NO_2$$
$$\rightarrow CH_3SOONO_2\ (or\ CH_3SO + NO_3), \quad (55)$$

$k_{55}(227-246\ K) = 2.2 \times 10^{-11}\ cm^3\ molecule^{-1}\ s^{-1}.$

Another alternative is isomerization to CH_3SO_2 (Turnipseed and Ravishankara, 1993).

The data in Table 8.18 show that reactions of CH_3S with NO_2 and O_3 may also be important. This also appears to be the case with larger thio radicals such as C_2H_5S (Black et al., 1988). The reaction with NO_2 produces primarily $CH_3SO + NO$ (Barnes et al., 1987; Hatakeyama, 1989; Tyndall and Ravishankara, 1989; Dominé et al., 1990); a minor addition channel produces CH_3SNO_2, which has been observed in laboratory systems using FTIR (Barnes et al., 1987).

CH_3SO has been observed as a product of the reaction of CH_3S with O_3 (Dominé et al., 1992), suggesting that one channel is an oxygen atom transfer:

$$CH_3S + O_3 \rightarrow CH_3SO + O_2. \quad (56a)$$

The yield of CH_3SO at low (Torr) pressures is only 15%. However, the yield has not been determined at 1 atm. Since (56a) is highly exothermic ($\sim 59\ kcal\ mol^{-1}$), the CH_3SO may contain sufficient energy to decompose at low pressures to $CH_3 + SO$; quenching of an excited CH_3SO at 1 atm may lead to much greater yields of this radical. Other potential channels that are exothermic are the following:

$$CH_3S + O_3 \rightarrow CH_3 + SO + O_2, \quad (56b)$$
$$\rightarrow CH_2SO + H + O_2, \quad (56c)$$
$$\rightarrow CH_2SO + HO_2, \quad (56d)$$
$$\rightarrow CH_2S + OH + O_2, \quad (56e)$$
$$\rightarrow CH_3O + SO_2. \quad (56f)$$

Although Dominé et al. (1992) place an upper limit of 4% on the contribution of (56e), Barnes et al. (1996) point out that even a small production of CH_2S followed by its oxidation could be responsible for the small yields of COS they observe by FTIR.

The CH_3SO radical will be further oxidized under tropospheric conditions:

$$CH_3SO + O_3 \xrightarrow{13\%} CH_3S + 2O_2, \quad (57a)$$
$$\rightarrow CH_3SO_2\ (or\ CH_3 + SO_2) + O_2, \quad (57b)$$
$$\rightarrow CH_2SO_2 + H + O_2, \quad (57c)$$
$$\rightarrow Other\ products? \quad (57d)$$

$k_{57}(298\ K) = 6.0 \times 10^{-13}\ cm^3\ molecule^{-1}\ s^{-1}$

(Dominé et al., 1992)

$$CH_3SO + NO_2 \rightarrow CH_3SO_2\ (or\ CH_3 + SO_2) + NO, \quad (58)$$

$k_{58}(298\ K) = 1.2 \times 10^{-11}\ cm^3\ molecule^{-1}\ s^{-1}$

(Mellouki et al., 1988; Hatakeyama, 1989; Dominé et al., 1990).

There is also evidence for an addition reaction of CH_3SO with O_2 (e.g., Hatakeyama et al., 1989; Barone et al., 1995):

$$CH_3SO + O_2 + M \rightarrow CH_3S(O)OO + M. \quad (59)$$

For example, methylsulfinyl peroxynitrate, $CH_3S(O)OONO_2$, has been observed by FTIR (e.g., see Barnes et al., 1987; Hatakeyama, 1989; and Jensen et al., 1992), presumably from reaction (59) followed by reaction of the peroxy radical formed with NO_2.

By analogy to other reactions, there are several other fates of the $CH_3S(O)OO$ radical formed in (59) that might be expected to be important. These include decomposition or reaction with NO or other peroxy radicals (Barnes et al., 1987):

$$CH_3S(O)OO \rightarrow CH_3 + SO_3, \quad (60)$$
$$CH_3S(O)OO + NO \rightarrow CH_3S(O)O + NO_2. \quad (61)$$

The SO_3 formed in (60) will react with H_2O to form H_2SO_4 as described in detail earlier in this chapter, whereas $CH_3S(O)O$ formed in (61) may decompose or react further.

The CH_3SO_2 radical has a number of potential fates, including decomposition to $CH_3 + SO_2$ and reactions with NO_2, O_2, and O_3:

$$CH_3SO_2 + M \rightarrow CH_3 + SO_2 + M, \quad (62)$$

$k_{62}(298\ K) = 510\ s^{-1}$ at 1 Torr pressure

(Ray et al., 1996)

$$CH_3SO_2 + NO_2 \rightarrow CH_3SO_3 + NO, \quad (63)$$

$k_{63}(298\ K) = 2.2 \times 10^{-12}\ cm^3\ molecule^{-1}\ s^{-1}$

(Ray et al., 1996).

Only upper limits for rate constants for the reaction of CH_3SO_2 with O_2 and O_3 of $<6 \times 10^{-18}$ and $<8 \times 10^{-13}\ cm^3\ molecule^{-1}\ s^{-1}$ have been measured (Turnipseed and Ravishankara, 1993). *If the rate constant for decomposition measured at 1 Torr is the same at 1 atm, the decomposition is expected to predominate at 298 K.* As discussed by Ray et al. (1996), as the temperature decreases, the rate of the thermal decomposition will also decrease and some of the other reactions may become competitive.

However, Patroescu et al. (1999) observed in FTIR experiments in 1 atm air that the yield of methanesulfonic acid increased as the NO_x concentration increased. They attributed this observation to the addition of O_2 to CH_3SO_2, with secondary reactions involving NO leading to the formation of methanesulfonic acid, CH_3SO_2OH. Thus, at 1 atm in air, the reaction of CH_3SO_2 appeared to be much faster than its thermal decomposition. Reaction of the $CH_3SO_2(OO)$ adduct with NO_2 was postulated to give $CH_3SO_2OONO_2$, methanesulfonyl peroxynitrate, which was observed by FTIR (Patroescu et al., 1999).

The reaction of OH with dimethyl sulfoxide also appears to occur primarily by addition. Urbanski et al. (1998) have reported yields of CH_3 in this reaction of unity and suggest that the reaction is

$$OH + CH_3S(O)CH_3 \leftrightarrow [CH_3S(O)(OH)CH_3]^* \rightarrow$$
methanesulfinic acid

$$CH_3 + CH_3S(O)OH. \quad (64)$$

The rate constant at 298 K, based on the production of CH_3, was measured to be $k_{64} = (8.7 \pm 1.6) \times 10^{-11}$ cm^3 molecule^{-1} s^{-1} (Urbanski et al., 1998). The formation of methanesulfinic acid in the OH–dimethyl sulfide system has been reported by Sørensen et al. (1996), who proposed that it resulted from secondary reactions of the OH–DMS adduct rather than reaction (64). Further addition of OH to methanesulfinic acid followed by reaction with O_2 may then form methanesulfonic acid, $CH_3S(O)(O)OH$, which in the atmosphere is rapidly scavenged into particles (e.g., Clegg and Brimblecombe, 1985). The adduct may also react with O_2 and, by analogy to the OH–CH_3SCH_3 adduct, is expected to give $HO_2 + CH_3S(O)(O)CH_3$, dimethyl sulfone.

The oxidation of dimethyl sulfide (DMS) to dimethyl sulfoxide (DMSO) and the subsequent oxidation of the latter to methanesulfonic acid (MSA) have been observed in field studies. For example, one study in Antarctica which focused on the chemistry of dimethyl sulfide (Berresheim and Eisele, 1998) measured not only DMS but also a variety of its oxidation products, including DMSO, MSA, and dimethyl sulfone (Berresheim et al., 1998). The measured concentrations of DMSO were in agreement with model results if 80–100% of the OH + DMS reaction gave DMSO; as discussed earlier, the addition channel that leads to DMSO becomes relatively more important at the lower temperatures found in Antarctica. Furthermore, the MSA concentrations were generally consistent with its formation from the OH–DMSO reaction if 80% of the reaction generated methanesulfinic acid and 50% of this compound was oxidized to MSA (Davis et al., 1998).

Figure 8.25 summarizes the current understanding of the mechanism of DMS oxidation in the troposphere. For reviews, see Plane (1989), Turnipseed and Ravishankara (1993), Barnes et al. (1993, 1996), Barone et al. (1995), and Berresheim et al. (1995).

b. Reaction with the Nitrate Radical (NO_3)

The nitrate radical is also known to react rapidly with DMS:

$$CH_3SCH_3 + NO_3 \rightarrow CH_3SCH_2 + HNO_3, \quad (65)$$

$$k_{65} = 1.1 \times 10^{-12} \text{ cm}^3 \text{ molecule}^{-1} \text{ s}^{-1}$$

(Atkinson et al., 1997a).

In contrast to the OH reaction, which occurs during the day due to the photolytic sources of OH, reaction (65) is a nighttime reaction due to the rapid photolysis of NO_3 at dawn. It is interesting that while the *overall* reaction appears to correspond to a hydrogen atom abstraction rather than addition to the sulfur atom, the mechanism is believed to involve the initial formation of an adduct, followed by its decomposition to HNO_3 and the alkyl radical shown in reaction (65) (Jensen et al., 1992; Daykin and Wine, 1990; Butkovskaya and Le Bras, 1994). NO_2 is not formed in the NO_3–DMS reaction (Dlugokencky and Howard, 1988), ruling out an oxygen atom transfer either directly or via decomposition of the adduct.

c. Oxidation by Chlorine Atoms

Atomic chlorine reacts rapidly with DMS, with an overall rate constant of $(3.3 \pm 0.5) \times 10^{-10}$ cm^3 molecule^{-1} s^{-1} at 298 K and 700 Torr total pressure (Stickel et al., 1992). As is the case for the OH reaction, the chlorine atom reaction proceeds by two reaction channels, one an abstraction and the other addition to the sulfur atom:

$$Cl + CH_3SCH_3 \rightarrow HCl + CH_3SCH_2, \quad (66a)$$

$$\xrightarrow{M} CH_3S(Cl)CH_3. \quad (66b)$$

A potential reaction path producing $CH_3S + CH_3Cl$ can be ruled out based on the observation of very small yields of CH_3Cl, only 0.13% at 1 atm pressure (Langer et al., 1996), and Zhao et al. (1996) show a path producing CH_3 is also not important. Stickel et al. (1992) suggest that at 298 K and 1 atm pressure, the

E. OXIDATION OF SULFUR COMPOUNDS OTHER THAN SO_2

two paths, (66a) and (66b), are about equally important. The fate of the Cl–DMS adduct is not known.

d. Oxidation by Halogen Oxides: IO, BrO, and ClO

Halogen oxides are also potential reactants with DMS in the marine boundary layer (e.g., see Barnes et al., 1989). As discussed in Chapter 12, CH_3I as well as CH_3Cl and CH_3Br have natural oceanic sources. While CH_3Cl and CH_3Br do not absorb light significantly in the actinic region, CH_3I and other alkyl halides do (see Chapter 4.V), photolyzing to form iodine atoms:

$$CH_3I + h\nu \rightarrow CH_3 + I. \qquad (67)$$

The iodine atoms react with O_3, forming the IO radical, which can potentially oxidize DMS to DMSO, regenerating iodine atoms in a chain process:

$$IO + CH_3SCH_3 \rightarrow I + CH_3S(O)CH_3. \qquad (68)$$

The current recommendation for k_{68} at 298 K is 1.2×10^{-14} cm^3 molecule^{-1} s^{-1} (Atkinson et al., 1997a). The concentration of IO radicals is not well known, as it has only just been directly detected and measured in the troposphere at concentrations up to 6 ppt at a coastal site (Alicke et al., 1999). Using 10^8 cm^{-3}, one can calculate a lifetime for DMS with respect to reaction (68) of about 10 days, too slow to compete with OH, NO_3, and Cl atoms.

However, with the recent recognition of the potential importance of atomic chlorine and bromine under certain conditions in the Arctic at polar sunrise (e.g., see Barrie et al., 1988; and Niki and Becker, 1993), the potential for BrO and ClO chemistry has been reconsidered. As described in Chapter 6.J.4, at polar sunrise there is a rapid loss of ground-level O_3 that appears to be associated with reaction with atomic bromine and at the same time, there is evidence that chlorine atoms are playing a major role in the organic removal (Jobson et al., 1994). This is consistent with reactions of sea salt particles generating atomic bromine and chlorine, although the exact nature of the reactions and halogen atom precursors remains unknown.

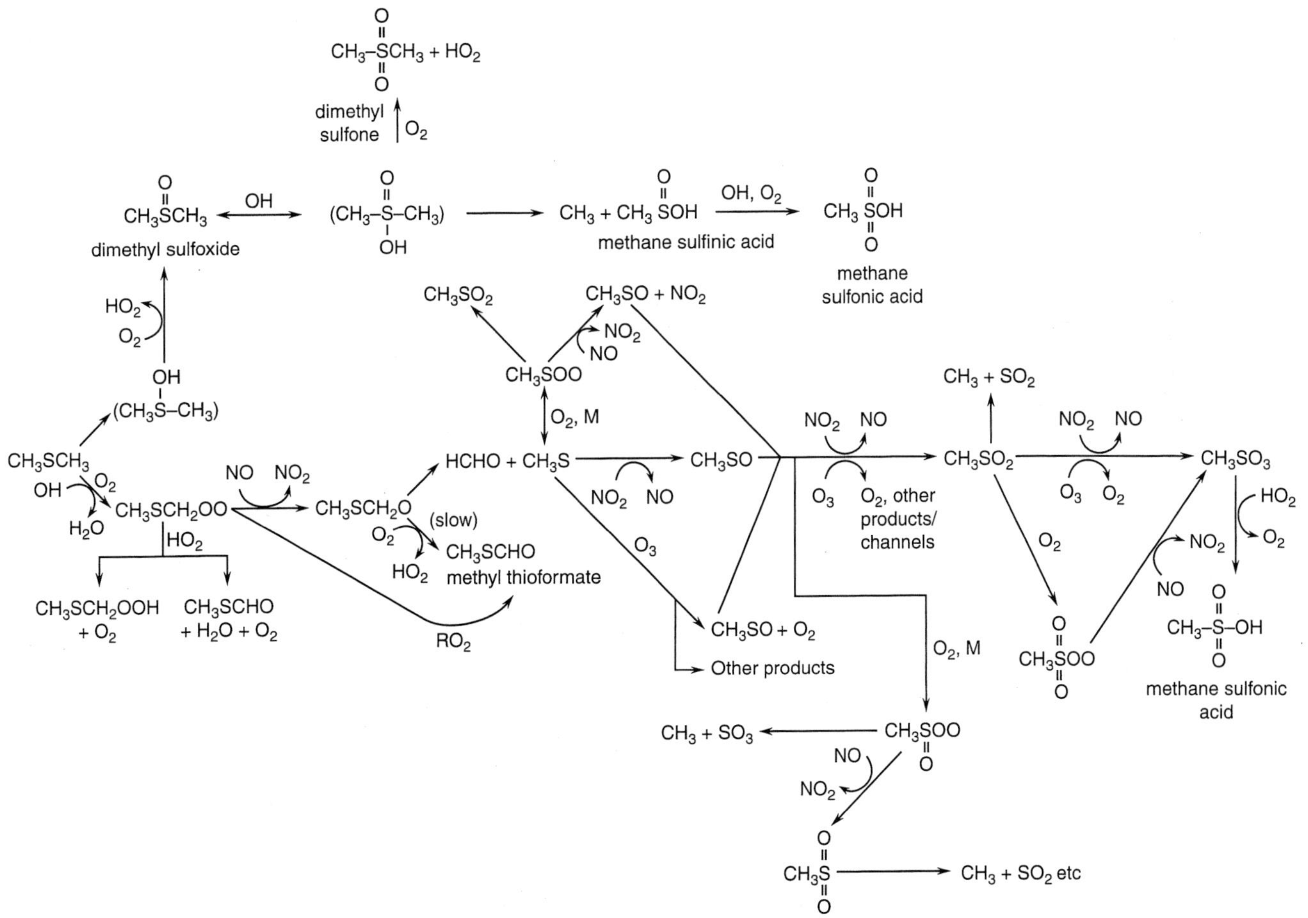

FIGURE 8.25 Overview of oxidation of DMS by OH in the troposphere (note that many of the reactions after the first step are the same in DMS reactions with SO_3, Cl, etc.).

The ClO + DMS reaction is quite slow, with a rate constant at 298 K of only 9.5×10^{-15} cm^3 molecule^{-1} s^{-1} (Barnes et al., 1991). Combined with the fact that Cl reacts with organics at essentially a diffusion-controlled rate, while the rate constant for reaction with O$_3$ is about an order of magnitude slower, the ClO + DMS reaction is not expected to be important in the troposphere.

The reaction with BrO, however, is potentially important under some conditions. As discussed in Chapter 11, UV-visible differential optical absorption spectrometry (DOAS) has been used to measure BrO under these conditions (e.g., see Hausmann et al., 1993; Platt and Hausmann, 1994; and Tuckermann et al., 1997). Concentrations up to about 30 ppt have been measured. The rate constant for the BrO–DMS reaction is 2.6×10^{-13} cm^3 molecule^{-1} s^{-1} at 298 K (Bedjanian et al., 1996) and gives with essentially unit yield DMSO and a bromine atom (Barnes et al., 1993; Bedjanian et al., 1996):

$$BrO + CH_3SCH_3 \rightarrow CH_3S(O)CH_3 + Br. \quad (69)$$

At 30 ppt BrO (7×10^8 cm^{-3}), the lifetime for the reaction of DMS with BrO is about 1.5 h and hence under these conditions can be significant. Obviously the situation in the Arctic at polar sunrise is unusual so that this represents the top end of estimates of the importance of BrO. However, model calculations suggest that it should be considered in other regions as well (Toumi, 1994).

e. Oxidation by Ozone in the Aqueous Phase

While the emphasis has been on oxidation of DMS and other reduced sulfur compounds in the gas phase, there is some indication that oxidation in the aqueous phase in clouds and fogs should also be considered. For example, Lee and Zhou (1994) have shown that DMS reacts with O$_3$ in aqueous solutions quite rapidly, with a rate constant at 288 K of 4×10^8 L mol^{-1} s^{-1}. They estimate that at 30 ppb O$_3$, a level found globally, the lifetime for in-cloud oxidation of DMS is about 3 days, of the same order of magnitude as that for the gas-phase oxidation by OH (see Table 8.17). Given the moderately high solubility of not only DMS but other sulfur compounds as well (see Henry's law constants in Table 8.1), this is clearly an area that warrants further research.

2. Dimethyl Disulfide (CH$_3$SSCH$_3$)

As seen from the data in Table 8.17, the OH reaction with CH$_3$SSCH$_3$ (DMDS) is almost two orders of magnitude faster than that with DMS. Because of its high reactivity, even small concentrations of DMDS may be significant. For example, Jefferson et al. (1998) measured H$_2$SO$_4$ and DMS in Antarctica and found that the DMS was insufficient to produce the levels of H$_2$SO$_4$ that were observed; they postulated that the oxidation of sulfur compounds such as DMDS may be responsible for the additional sulfuric acid formation.

The reaction of OH with dimethyl disulfide proceeds primarily by addition to a sulfur atom. As discussed by Abbatt et al. (1992), although one would expect abstraction of an H atom from the CH$_3$ group to occur as well at about the same rate as for DMS, such a channel is overwhelmed by the fast addition.

The initially formed adduct is thought to decompose:

$$OH + CH_3SSCH_3 \rightarrow [CH_3S(OH)SCH_3]$$
$$\rightarrow CH_3SOH + CH_3S. \quad (70)$$

The CH$_3$S and CH$_3$SOH then react in a manner analogous to the reactions discussed for the DMS oxidation, ultimately forming SO$_2$ and methanesulfonic acid (e.g., Hatakeyama and Akimoto, 1983; Barnes et al., 1994a).

The mechanism of the NO$_3$ radical reaction with DMDS is complex. The rate constants for the NO$_3$ reaction with CH$_3$SCH$_3$, CH$_3$SH, and CH$_3$SSCH$_3$ (Table 8.17) are similar and much larger than that for the reaction with H$_2$S. This suggests that by analogy with the dimethyl sulfide reaction, the initial step is addition to a sulfur atom to form an adduct that then reacts further:

$$CH_3SSCH_3 + NO_3 \rightarrow [CH_3S(NO_3)SCH_3] \rightarrow Products. \quad (71)$$

FTIR studies of the NO$_3$–DMDS reaction (Jensen et al., 1992; MacLeod et al., 1986) identified HCHO, SO$_2$, CH$_3$ONO$_2$, CH$_3$SO$_3$H, HNO$_3$, and CH$_3$SNO$_2$ as products of the reaction with NO$_3$. In addition, there were a number of FTIR bands that were unidentified. Presumably the decomposition of the adduct gives at least in part intermediates such as CH$_3$S and CH$_3$SO, which then go on to react as described earlier (see Fig. 8.25).

3. Methyl Mercaptan (CH$_3$SH)

The OH radical also reacts with CH$_3$SH by addition. The adduct appears to decompose at least in part to CH$_3$S (Hatakeyama and Akimoto, 1983; Grosjean,

1984), which then reacts as described earlier:

$$OH + CH_3SH \rightarrow [CH_3S(OH)H] \rightarrow CH_3S + H_2O. \quad (72)$$

The NO_3 radical reaction with CH_3SH gives a variety of products, including CH_3SO_3H, SO_2, CH_3ONO_2, HCHO, CH_3SNO_2, HNO_3, and CH_3SSCH_3 (Jensen et al., 1992; MacLeod et al., 1986). The mechanism proposed is addition to the sulfur atom, followed by decomposition:

$$NO_3 + CH_3SH \rightarrow [CH_3S(NO_3)H] \rightarrow CH_3S + HNO_3, \quad (73a)$$

$$\rightarrow SO_2 + \text{Other products}. \quad (73b)$$

The direct production of SO_2 in a reaction such as (73b) was suggested by the independence of the measured SO_2 yields on the oxygen concentration. NO_2 is not produced in significant yields (<5%) in the NO_3–CH_3SH reaction (Dlugokencky and Howard, 1988), ruling out a simple oxygen atom transfer or decomposition of the adduct to NO_2 and other products.

4. Hydrogen Sulfide (H_2S)

The reactions of both OH and NO_3 with H_2S appear to proceed by hydrogen atom abstraction:

$$OH + H_2S \rightarrow H_2O + SH, \quad (74)$$
$$NO_3 + H_2S \rightarrow HNO_3 + SH. \quad (75)$$

The SH radical then reacts further:

$$SH + O_2 \rightarrow \text{Products}, \quad (76)$$
$$k_{76} \leq 4 \times 10^{-19} \text{ cm}^3 \text{ molecule}^{-1} \text{ s}^{-1}$$
(Stachnik and Molina, 1987)
$$SH + O_3 \rightarrow HSO + O_2, \quad (77)$$
$$k_{77} = 3.7 \times 10^{-12} \text{ cm}^3 \text{ molecule}^{-1} \text{ s}^{-1}$$
(Atkinson et al., 1997a)
$$SH + NO_2 \rightarrow HSO + NO, \quad (78)$$
$$k_{78} = 5.8 \times 10^{-11} \text{ cm}^3 \text{ molecule}^{-1} \text{ s}^{-1}$$
(Atkinson et al., 1997a).

At concentrations of O_3 of 30 ppb and NO_2 of 0.1 ppb, the lifetimes of the SH radical with respect to reaction with O_2, O_3, and NO_2 are > 0.5, 0.4, and 7 s, respectively. Thus all three species may be involved in the atmospheric reactions of SH, with the latter two giving the SHO radical.

Possible atmospheric fates of the SHO radical reactions include

$$HSO + O_2 \rightarrow \text{Products}, \quad (79)$$
$$k_{79} \leq 2 \times 10^{-17} \text{ cm}^3 \text{ molecule}^{-1} \text{ s}^{-1}$$
(Lovejoy et al., 1987)
$$HSO + O_3 \rightarrow \text{Products}, \quad (80)$$
$$k_{80} = 1.1 \times 10^{-13} \text{ cm}^3 \text{ molecule}^{-1} \text{ s}^{-1}$$
(Atkinson et al., 1997a)
$$HSO + NO_2 \rightarrow HSO_2 + NO, \quad (81)$$
$$k_{81} = 9.6 \times 10^{-12} \text{ cm}^3 \text{ molecule}^{-1} \text{ s}^{-1}$$
(Lovejoy et al., 1987)

followed by

$$HSO_2 + O_2 \rightarrow \text{Products}, \quad (82)$$
$$k_{82} = 3.0 \times 10^{-13} \text{ cm}^3 \text{ molecule}^{-1} \text{ s}^{-1}$$
(Atkinson et al., 1997a).

The reaction (80) of HSO with O_3 has one channel generating $HS + 2O_2$ as well as others giving further oxidation of the HSO (Wang and Howard, 1990). The ultimate oxidation product in the atmosphere is SO_2.

5. Carbon Disulfide (CS_2)

The atmospheric sources and sinks of CS_2 and COS are reviewed by Chin and Davis (1993). Oxidation of CS_2 generates COS (discussed in the following section). COS is also generated by natural emissions, including photochemical production from organosulfur compounds in seawater (e.g., see Zepp and Andreae, 1994).

The reaction of OH with CS_2 is similar to the OH–CO reaction (see Chapter 5.A) in that the reaction proceeds by addition (e.g., Hynes et al., 1988; Diau and Lee, 1991a,b). An intermediate adduct is formed that can react with O_2:

$$OH + CS_2 + M \leftrightarrow HOCS_2 + M, \quad (83)$$
$$HOCS_2 + O_2 \rightarrow \text{Products}, \quad (84)$$
$$k_{84}(298 \text{ K}) = 3 \times 10^{-14} \text{ cm}^3 \text{ molecule}^{-1} \text{ s}^{-1}$$
(Atkinson et al., 1997a)

The recommended (Atkinson et al., 1997a, 1997b) low- and high-pressure limiting rate constants for reaction (83) are $k_0 = 8 \times 10^{-31}[N_2]$ cm^6 molecule^{-2} s^{-1} and $k_\infty = 8 \times 10^{-12}$ cm^3 molecule^{-1} s^{-1}, with $F_c = 0.8$ (see Chapter 5.A.2), giving an effective bimolecular rate constant at 298 K and 1 atm of 4.7×10^{-12} cm^3 molecule^{-1} s^{-1}. There is no experimental evidence for a direct channel forming HS + COS, and if it occurs, the rate constant must be less than 2×10^{-15} cm^3 molecule^{-1} s^{-1} (Murrells et al., 1990). The equilibrium constant for reaction (83) at 299 K is $K_{83} = 1.7 \times 10^{-17}$ cm^3 molecule^{-1} (Murrells et al., 1990).

As discussed in detail by Stickel et al. (1993), there are 26 potential sets of products of reaction (84) of the adduct with O_2 that are exothermic! They have measured the yield of COS to be 0.83 ± 0.08, of CO to be 0.16 ± 0.03, and of SO_2 to be 1.15 ± 0.10. Both the COS and about 75% of the SO_2 are produced rapidly, indicating that they either are formed directly from the reaction of the adduct with O_2, e.g., via

$$HOCS_2 + O_2 \rightarrow COS + HOSO, \quad (85a)$$
$$\rightarrow COS + H + SO_2, \quad (85b)$$
$$\rightarrow COS + HSO_2, \quad (85c)$$
$$\rightarrow COS + HO_2 + S, \quad (85d)$$
$$\rightarrow HO_2 + CO + S_2, \quad (85e)$$

or, alternatively, are formed from very fast secondary reactions of other primary products. CO is also produced rapidly, perhaps from such pathways as

$$HOCS_2 + O_2 \rightarrow CO + HOSO + S, \quad (85f)$$
$$\rightarrow CO + H + SO_2 + S. \quad (85g)$$

The formation of HO_2 in high yield, $95 \pm 15\%$ of the OH reacted, has also been observed using laser magnetic resonance (Lovejoy et al., 1990), indicating the importance of channels such as (85d) and (85e).

6. Carbonyl Sulfide (COS)

The reaction of OH with COS is significantly slower than the CS_2 reaction, with a rate constant at 298 K of 2.0×10^{-15} cm^3 molecule^{-1} s^{-1} (Atkinson et al., 1997a, 1997b). At an OH concentration of 1×10^6 radicals cm^{-3}, this gives a lifetime with respect to this reaction of about 16 years, far too long to be a significant removal path in the troposphere. Unlike the CS_2 reaction, the reaction of OH with COS is independent of pressure and of the presence of O_2 (e.g., see Cheng and Lee, 1986; and Wahner and Ravishankara, 1987). The reaction products are not known.

F. PROBLEMS

1. Derive Eq. (B), relating the pseudo-Henry's law constant to the true Henry's law constant for the physical solubility of CO_2 exclusive of other reactions and to the pH and equilibrium constants for reactions (2) and (3).

2. Using the kinetic and thermodynamic data given in the text, calculate the rates of sulfate formation in mol L^{-1} s^{-1} from the reaction of O_3 with each of the S(IV) species, i.e., $SO_2 \cdot H_2O$, HSO_3^-, and SO_3^{2-}, and the total rate of formation of S(VI) from these reactions at a pH of 3.0 and with a constant gas-phase concentration of SO_2 of 20 ppb and O_3 of 50 ppb (you can compare to the data in Fig. 8.15).

3. (a) What is the effective Henry's law constant for SO_2 for an aqueous solution of pH 5.0? (b) What is the total concentration of S(IV) in solution under these conditions if the gas-phase SO_2 concentration is 10 ppb?

4. Repeat Problem 3 for pH 2.0. Compare your answers to those for Problem 3 and explain based on the chemistry of S(IV) in solution.

5. Calculate the mole fractions of $SO_{2(aq)}$, HSO_3^-, and SO_3^{2-} in solution at a pH of 3.0, characteristic of many fogs, as well as the total concentration of S(IV) in solution for 20 ppb SO_2 in the gas phase.

6. Calculate the mole fractions of $SO_{2(aq)}$, HSO_3^-, and SO_3^{2-} in solution at a pH of 8.0, equal to that of seawater and therefore expected for airborne particles formed from seawater in clean regions as well as the total concentration of S(IV) in solution for 20 ppb SO_2 in the gas phase. Comment on the implications for the potential role of sea salt particles in the oxidation of SO_2 in the marine boundary layer.

7. How fast is the oxidation of SO_2, in percent per hour, in the gas phase for an OH concentration of 1×10^7 radicals cm^{-3}, a typical peak concentration in the lower troposphere? Comment on the significance of your calculated rate in light of observed rates of SO_2 oxidation of up to 30% h^{-1} in the troposphere.

8. Calculate the rate of oxidation of SO_2 in percent per hour by H_2O_2 in the aqueous phase for an SO_2 concentration of 10 ppb, a pH of 3.0, a temperature of 298 K, a constant gas-phase H_2O_2 concentration of 1 ppb, and a liquid water content (LWC) of (a) 0.1 g of liquid water per m^{-3} of air, typical of fogs, and (b) 1 g m^{-3}, typical of clouds. Why does the LWC affect the rate of oxidation of SO_2?

9. An air mass has an O_3 concentration of 60 ppb and cloud drops with pH 4.0. What must the concentration of gaseous H_2O_2 be so that the rate of oxidation of S(IV) in the cloud by these two oxidants is equal?

10. The acid dissociation constants for HCOOH, CH_3COOH, and HONO are 1.77×10^{-4}, 1.76×10^{-5}, and 4.6×10^{-4}, respectively. Assume that the liquid-phase acidity is provided by dissolution of gas-phase HNO_3. What would the gas-phase concentrations of HNO_3 need to be to give equal concentrations of the

dissociated and undissociated forms of each of these weak acids, i.e., to give [anion] = [undissociated acid].

11. Dimethyl sulfide is believed to be a significant source of sulfate in the marine boundary layer via its oxidation by OH, NO_3, and Cl. Compare the lifetimes for rates of oxidation of DMS by these three species, assuming peak concentrations of OH of 1×10^7 radicals cm^{-3}, NO_3 = 10 ppt, and Cl = 1×10^5 atoms cm^{-3}. (However, note that these oxidants peak at different times of the day; see Problem 12.)

12. NO_3 is an important oxidant only at night because of its rapid photolysis. At dawn, a pulse of chlorine atoms is expected as the photolyzable halogen precursors that built up overnight photolyze. OH is expected to peak around noon. Compare the lifetimes for oxidation of DMS by Cl and OH at dawn, where the concentrations of both species may be $\sim 1 \times 10^5$ cm^3.

References

Abbatt, J. P. D., F. F. Fenter, and J. G. Anderson, "High-Pressure Discharge Flow Kinetics Study of OH + CH_3SCH_3, CH_3SSCH_3 → Products from 297 to 368 K," *J. Phys. Chem., 96,* 1780–1785 (1992).

Abbatt, J., "Heterogeneous Reaction of HOBr with HBr and HCl on Ice Surfaces at 228 K," *Geophys. Res. Lett., 21,* 665–668 (1994).

Alicke, B., K. Hebestreit, J. Stutz, and U. Platt, "Iodine Oxide in the Marine Boundary Layer," *Nature, 397,* 572–573 (1999).

Alkezweeny, A. J., and D. C. Powell, "Estimation of Transformation Rate of SO_2 to SO_4 from Atmospheric Concentration Data," *Atmos. Environ., 11,* 179–182 (1977).

Anastasio, C., B. C. Faust, and J. M. Allen, "Aqueous Phase Photochemical Formation of Hydrogen Peroxide in Authentic Cloud Waters," *J. Geophys. Res., 99,* 8231–8248 (1994).

Anastasio, C., B. C. Faust, and C. J. Rao, "Aromatic Carbonyl Compounds as Aqueous-Phase Photochemical Sources of Hydrogen Peroxide in Acidic Sulfate Aerosols, Fogs, and Clouds. 1. Non-Phenolic Methoxybenzaldehydes and Methoxyacetophenones with Reductants (Phenols)," *Environ. Sci. Technol., 31,* 218–232 (1997).

Anderson, L. G., P. M. Gates, and C. R. Nold, "Mechanism of Atmospheric Oxidation of Sulfur Dioxide by Hydroxyl Radicals," in *Biogenic Sulfur in the Environment,* ACS Symposium Series No. 393 (E. S. Saltzman and W. J. Cooper, Eds.), pp. 437–449, 1989.

Anderson, M. A., "Influence of Surfactants on Vapor–Liquid Partitioning," *Environ. Sci. Technol., 26,* 2186–2191 (1992).

Aneja, V. P., and Kim, D.-S., "Chemical Dynamics of Clouds at Mt. Mitchell, North Carolina," *Air Waste, 43,* 1074–1083 (1993).

Arakaki, T., C. Anastasio, P. G. Shu, and B. C. Faust, "Aqueous-Phase Photoproduction of Hydrogen Peroxide in Authentic Cloud Waters: Wavelength Dependence, and the Effects of Filtration and Freeze–Thaw Cycles," *Atmos. Environ., 29,* 1697–1703 (1995).

Arakaki, T., and B. C. Faust, "Sources, Sinks, and Mechanisms of Hydroxyl Radical ($^{\cdot}$OH) Photoproduction and Consumption in Authentic Acidic Continental Cloud Waters from Whiteface Mountain, New York: The Role of the Fe(r) (r = II, III) Photochemical Cycle," *J. Geophys. Res., 103,* 3487–3504 (1998).

Atkinson, R., D. L. Baulch, R. A. Cox, R. F. Hampson, Jr., J. A. Kerr, M. J. Rossi, and J. Troe, "Evaluated Kinetic and Photochemical Data for Atmospheric Chemistry: Supplement VI—IUPAC Subcommittee on Gas Kinetic Data Evaluation for Atmospheric Chemistry," *J. Phys. Chem. Ref. Data, 26,* 1329–1499 (1997a).

Atkinson, R., D. L. Baulch, R. A. Cox, R. F. Hampson, Jr., J. A. Kerr, M. J. Rossi, and J. Troe, "Evaluated Kinetic, Photochemical, and Heterogeneous Data for Atmospheric Chemistry. 5. IUPAC Subcommittee on Gas Kinetic Data Evaluation for Atmospheric Chemistry," *J. Phys. Chem. Ref. Data, 26,* 521–1011 (1997b).

Ayers, G. P., R. W. Gillett, and J. L. Gras, "On the Vapor Pressure of Sulfuric Acid," *Geophys. Res. Lett., 7,* 433–436 (1980).

Bäckström, H. L. J., "Der Kettenmechanismus bie der Autoxydation von Natrium-Sulfitlosungen," *Z. Phys. Chem., 25B,* 122–138 (1934).

Baldwin, A. C., "Heterogeneous Reactions of Sulfur Dioxide with Carbonaceous Particles," *Int. J. Chem. Kinet., 14,* 269–277 (1982).

Bandy, A. R., D. L. Scott, B. W. Blomquist, S. M. Chen, and D. C. Thornton, "Low Yields of SO_2 from Dimethyl Sulfide Oxidation in the Marine Boundary Layer," *Geophys. Res. Lett., 19,* 1125–1127 (1992).

Bandy, A. R., D. C. Thornton, B. W. Blomquist, S. Chen, T. P. Wade, J. C. Ianni, G. M. Mitchell, and W. Nadler, "Chemistry of Dimethyl Sulfide in the Equatorial Pacific Atmosphere," *Geophys. Res. Lett., 23,* 741–744 (1996).

Bao, Z.-C., and J. R. Barker, "Temperature and Ionic Strength Effects on Some Reactions Involving Sulfate Radical [SO_4^-(aq)]," *J. Phys. Chem., 100,* 9780–9787 (1996).

Barbaray, B., J.-P. Contour, and G. Mouvier, "Sulfur Dioxide Oxidation over Atmospheric Aerosol-X-Ray Photoelectron Spectra of Sulfur Dioxide Adsorbed on V_2O_5 and Carbon," *Atmos. Environ., 11,* 351–356 (1977).

Barbaray, B., J.-P. Contour, and G. Mouvier, "Effects of Nitrogen Dioxide and Water Vapor on Oxidation of Sulfur Dioxide over V_2O_5 Particles," *Environ. Sci. Technol., 12,* 1294–1297 (1978).

Barnes, I., V. Bastian, and K. H. Becker, "FTIR Spectroscopic Studies of the $CH_3S + NO_2$ Reaction under Atmospheric Conditions," *Chem. Phys. Lett., 140,* 451–457 (1987).

Barnes, I., K. H. Becker, D. Martin, P. Carlier, G. Mouvier, J. L. Jourdain, G. Laverdet, and G. Le Bras, "Impact of Halogen Oxides on Dimethyl Sulfide Oxidation in the Marine Atmosphere," in *Biogenic Sulfur in the Environment,* Chapter 29, pp. 464–475, 1989.

Barnes, I., V. Bastian, K. H. Becker, and R. D. Overath, "Kinetic Studies of the Reactions of IO, BrO, and ClO with Dimethylsulfide," *Int. J. Chem. Kinet., 23,* 579–591 (1991).

Barnes, I., K. H. Becker, and R. D. Overath, "Oxidation of Organic Sulfur Compounds," in *The Tropospheric Chemistry of Ozone in the Polar Regions* (H. Niki and K. H. Becker, Eds.), NATO ASI Series, Vol. 17, pp. 371–383, Springer-Verlag, Berlin, 1993.

Barnes, I., K. H. Becker, and N. Mihalopoulos, "An FTIR Product Study of the Photooxidation of Dimethyl Disulfide," *J. Atmos. Chem., 18,* 267–289 (1994a).

Barnes, I., K. H. Becker, and I. Patroescu, "The Tropospheric Oxidation of Dimethyl Sulfide: A New Source of Carbonyl Sulfide," *Geophys. Res. Lett., 21,* 2389–2392 (1994b).

Barnes, I., K. H. Becker, and I. Patroescu, "FTIR Product Study of the OH Initiated Oxidation of Dimethyl Sulphide: Observation of Carbonyl Sulphide and Dimethyl Sulphoxide," *Atmos. Environ., 30,* 1805–1814 (1996).

Barone, S. B., A. A. Turnipseed, and A. R. Ravishankara, "Role of Adducts in the Atmospheric Oxidation of Dimethyl Sulfide," *Faraday Discuss., 100,* 39–54 (1995).

Barone, S. B., A. A. Turnipseed, and A. R. Ravishankara, "Reaction of OH with Dimethyl Sulfide (DMS). 1. Equilibrium Constant for OH + DMS Reaction and the Kinetics of the OH $\cdot$ DMS + O_2 Reaction," *J. Phys. Chem., 100,* 14694–14702 (1996).

Barrie, L. A., J. W. Bottenheim, R. C. Schnell, P. J. Crutzen, and R. A. Rasmussen, "Ozone Destruction and Photochemical Reactions at Polar Sunrise in the Lower Arctic Atmosphere," *Nature, 334,* 138–141 (1988).

Bator, A., and J. L. Collett, Jr., "Cloud Chemistry Varies with Drop Size," *J. Geophys. Res., 102,* 28071–28078 (1997).

Bedjanian, Yu., G. Poulet, and G. Le Bras, "Kinetic Study of the Reaction of BrO Radicals with Dimethylsulfide," *Int. J. Chem. Kinet., 28,* 383–389 (1996).

Behar, D., G. Czapski, and I. Duchovny, "Carbonate Radical in Flash Photolysis and Pulse Radiolysis of Aqueous Carbonate Solutions," *J. Phys. Chem., 74,* 2206–2210 (1970).

Behra, P., and L. Sigg, "Evidence for Redox Cycling of Iron in Atmospheric Water Droplets," *Nature, 344,* 419–421 (1990).

Benkelberg, H.-J., and P. Warneck, "Photodecomposition of Iron(III) Hydroxo and Sulfato Complexes in Aqueous Solution: Wavelength Dependence of OH and SO_4^- Quantum Yields," *J. Phys. Chem., 99,* 5215–5221 (1995).

Benner, W. H., R. Brodzinsky, and T. Novakov, "Oxidation of SO_2 in Droplets Which Contain Soot Particles," *Atmos. Environ., 16,* 1333–1339 (1982).

Bergin, M. H., S. N. Pandis, C. I. Davidson, J.-L. Jaffrezo, J. E. Dibb, A. G. Russell, and H. D. Kuhns, "Modeling of the Processing and Removal of Trace Gas and Aerosol Species by Arctic Radiation Fogs and Comparison with Measurements," *J. Geophys. Res., 101,* 14465–14478 (1996).

Berresheim, H., F. L. Eisele, D. J. Tanner, L. M. McInnes, D. C. R. Bell, and D. S. Covert, "Atmospheric Sulfur Chemistry and Cloud Condensation Nuclei (CCN) Concentrations over the Northeastern Pacific Coast," *J. Geophys. Res., 98,* 12701–12711 (1993).

Berresheim, H., P. H. Wine, and D. D. Davis, "Sulfur in the Atmosphere," in *Composition, Chemistry, and the Climate of the Atmosphere* (H. B. Singh, Ed.), pp. 251–307, Van Nostrand Reinhold, New York, 1995.

Berresheim, H., and F. L. Eisele, "Sulfur Chemistry in the Antarctic Troposphere Experiment: An Overview of Project SCATE," *J. Geophys. Res., 103,* 1619–1627 (1998).

Berresheim, H., J. W. Huey, R. P. Thorn, F. L. Eisele, D. J. Tanner, and A. Jefferson, "Measurements of Dimethyl Sulfide, Dimethyl Sulfoxide, Dimethyl Sulfone, and Aerosol Ions at Palmer Station, Antarctica," *J. Geophys. Res., 103,* 1629–1637 (1998).

Betterton, E. A., and M. R. Hoffmann, "Henry's Law Constants of Some Environmentally Important Aldehydes," *Environ. Sci. Technol., 22,* 1415–1418 (1988).

Betterton, E. A., "On the pH-Dependent Formation Constants of Iron(III) Sulfur(IV) Transient Complexes," *J. Atmos. Chem., 17,* 307–324 (1993).

Bielski, B. H. J., "Reevaluation of the Spectral and Kinetic Properties of HO_2 and O_2^- Free Radicals," *Photochem. Photobiol., 28,* 645–649 (1978).

Bielski, B. H. J., D. E. Cabelli, R. L. Arudi, and A. B. Ross, "Reactivity of HO_2/O_2^- Radicals in Aqueous Solution," *J. Phys. Chem. Ref. Data, 14,* 1041–1100 (1985).

Bishenden, E., and D. J. Donaldson, "*Ab-Initio* Study of $SO_2 + H_2O$," *J. Phys. Chem., 102,* 4638–4642 (1998).

Bjergbakke, E., S. Navaratnam, B. J. Parsons, and A. J. Swallow, "Reaction between HO_2 Chlorine in Aqueous Solution," *J. Am. Chem. Soc., 103,* 5926–5928 (1981).

Black, G., L. E. Jusinski, and R. Patrick, "Kinetics of the Reactions of C_2H_5S with NO_2, NO, and O_2 at 296 K," *J. Phys. Chem., 92,* 5972–5977 (1988).

Bongartz, A., J. Kames, U. Schurath, Ch. George, Ph. Mirabel, and J. L. Ponche, "Experimental Determination of HONO Mass Accommodation Coefficients Using Two Different Techniques," *J. Atmos. Chem., 18,* 149–169 (1994).

Bongartz, A., S. Schweighoefer, C. Roose, and U. Schurath, "The Mass Accommodation Coefficient of Ammonia on Water," *J. Atmos. Chem., 20,* 35–58 (1995).

Boyce, S. D., and M. R. Hoffmann, "Kinetics and Mechanism of the Formation of Hydroxymethanesulfonic Acid at Low pH," *J. Phys. Chem., 88,* 4740–4746 (1984).

Brandt, C., I. Fabian, and R. van Eldik, "Kinetics and Mechanism of the Iron(III)-Catalyzed Autoxidation of Sulfur(IV) Oxides in Aqueous Solution—Evidence for the Redox Cycling of Iron in the Presence of Oxygen and Modeling of the Overall Reaction Mechanism," *Inorg. Chem., 33,* 687–701 (1994).

Breeding, R. J., H. B. Klonis, J. P. Lodge, Jr., J. B. Pate, D. C. Sheesley, T. R. Englert, and D. R. Sears, "Measurements of Atmospheric Pollutants in the St. Louis Area," *Atmos. Environ., 10,* 181–194 (1976).

Brimblecombe, P., "Interest in Air Pollution among Early Fellows of the Royal Society," *Notes Rec. R. Soc., 32,* 123–129 (1978).

Brimblecombe, P., and S. L. Clegg, "The Solubility and Behaviour of Acid Gases in the Marine Aerosol," *J. Atmos. Chem., 7,* 1–18 (1988).

Briner, E., and E. Perrottet, "Détermination des Solubilités de l'Ozone dans l'Eau et dans Une Solution Aqueuse de Chlorure de Sodium: Calcul des Solubilités de l'Ozone Atmosphérique dans les Eaux," *Helv. Chim. Acta, 22,* 397–404 (1939).

Britton, L. G., and A. G. Clarke, "Heterogeneous Reactions of Sulfur Dioxide and SO_2/NO_2 Mixtures with a Carbon Soot Aerosol," *Atmos. Environ., 14,* 829–839 (1980).

Brodzinsky, R., S. G. Chang, S. S. Markowitz, and T. Novakov, "Kinetics and Mechanism for the Catalytic Oxidation of Sulfur Dioxide on Carbon in Aqueous Suspensions," *J. Phys. Chem., 84,* 3354–3358 (1980).

Brown, R. E., and F. Barber, "Ab Initio Studies of the Thermochemistry of the Bisulfite and the Sulfonate Ions and Related Compounds," *J. Phys. Chem., 99,* 8071–8075 (1995).

Bühler, R. E., J. Staehelin, and J. Hoigné, "Ozone Decomposition in Water Studied by Pulse Radiolysis. 1. HO_2/O_2^- and HO_3/O_3^- as Intermediates," *J. Phys. Chem., 88,* 2560–2564 (1984).

Bunce, N. J., K. U. Ingold, J. P. Landers, J. Lusztyk, and J. C. Scaiano, "Kinetic Study of the Photochlorination of 2,3-Dimethylbutane and Other Alkanes in Solution in the Presence of Benzene. First Measurements of the Absolute Rate Constants for Hydrogen Abstraction by the "Free" Chlorine Atom and the Chlorine Atom–Benzene π-Complex. Identification of These Two Species as the Only Hydrogen Abstractors in These Systems," *J. Am. Chem. Soc., 107,* 5464–5472 (1985).

Burkhard, E. G., V. A. Dutkiewicz, and L. Husain, "A Study of SO_2, SO_4^{2-}, and Trace Elements in Clear Air and Clouds above the Midwestern United States," *Atmos. Environ., 28,* 1521–1533 (1994).

Burkhard, E. G., B. M. Ghauri, V. A. Dutkiewicz, and L. Husain, "A Multielement Tracer Technique for the Determination of SO_2 Oxidation in Clouds," *J. Geophys. Res., 100,* 26051–26059 (1995).

Buschmann, H.-J., H.-H. Fuldner, and W. Knoche, "The Reversible Hydration of Carbonyl Compounds in Aqueous Solution. Part I. The Keto/Gem-Diol Equilibrium," *Ber. Bunsenges. Phys. Chem., 85,* 41–44 (1980).

Buschmann, H.-J., E. Dutkiewicz, and W. Knoche, "The Reversible Hydration of Carbonyl Compounds in Aqueous Solution. Part II. The Kinetics of the Keto/Gem-Diol Transition," *Ber. Bunsenges. Phys. Chem., 86,* 129–134 (1982).

Butkovskaya, N. I., and G. Le Bras, "Mechanism of the NO_3 + DMS Reaction by Discharge Flow Mass Spectrometry," *J. Phys. Chem., 98,* 2582–2591 (1994).

Buxton, G. V., M. Bydder, and G. A. Salmon, "The Equilibrium $Cl + Cl^- \leftrightarrow Cl_2^-$," *J. Chem. Soc. Faraday Trans., 94,* 653–657 (1998).

Buxton, G. V., C. L. Greenstock, W. P. Helman, and A. B. Ross, "Critical Review of Rate Constants for Reactions of Hydrated Electrons, Hydrogen Atoms, and Hydroxyl Radicals ($\cdot OH/ \cdot LO^-$) in Aqueous Solution," *J. Phys. Chem. Ref. Data, 17,* 513–886 (1988).

Buxton, G. V., S. McGowan, G. A. Salmon, J. E. Williams, and N. D. Wood, "A Study of the Spectra and Reactivity of Oxysulphur-Radical Anions Involved in the Chain Oxidation of S(IV): A Pulse and γ-Radiolysis Study," *Atmos. Environ., 30,* 2483–2493 (1996).

Calvert, J. G., "Photoactivated Surface Reactions," in *Air Pollution Foundation Report No. 15*, November 1956, pp. 91–112.

Chameides, W. L., and D. D. Davis, "The Coupled Gas-Phase/Aqueous-Phase Free Radical Chemistry of a Cloud," in *Precipitation Scavenging, Dry Deposition, and Resuspension* (H. R. Pruppacher, R. G. Semonin, and W. G. N. Slinn, Eds.), pp. 431–443, Elsevier, New York, 1983.

Chameides, W. L., and A. W. Stelson, "Aqueous-Phase Chemical Processes in Deliquescent Seasalt Aerosols," *Ber. Bunsenges. Chem., 96,* 461–470 (1992a).

Chameides, W. L., and A. W. Stelson, "Aqueous Phase Chemical Processes in Deliquescent Sea-Salt Aerosols: A Mechanism That Couples the Atmospheric Cycles of S and Sea Salt," *J. Geophys. Res., 97,* 20565–20580 (1992b).

Chang, S.-G., and T. Novakov, "Role of Carbon Particles in Atmospheric Chemistry," *Adv. Environ. Sci. Technol., 12,* 191–219 (1983).

Chang, S. G., R. Toossi, and T. Novakov, "The Importance of Soot Particles and Nitrous Acid in Oxidizing SO_2 in Atmospheric Aqueous Droplets," *Atmos. Environ., 15,* 1287–1292 (1981).

Chapman, E. G., D. V. Kenny, K. M. Busness, J. M. Thorp, and C. W. Spicer, "Continuous Airborne Measurements of Gaseous Formic and Acetic Acids over the Western North Atlantic," *Geophys. Res. Lett., 22,* 405–408 (1995).

Charlson, R. J., J. E. Lovelock, M. O. Andreae, and S. G. Warren, "Oceanic Phytoplankton, Atmospheric Sulfur, Cloud Albedo, and Climate," *Nature, 326,* 655–661 (1987).

Chebbi, A., and P. Carlier, "Carboxylic Acids in the Troposphere, Occurrence, Sources, and Sinks: A Review," *Atmos. Environ., 30,* 4233–4249 (1996).

Cheng, B.-M., and Y.-P. Lee, "Rate Constant of OH + OCS Reaction over the Temperature Range 255–483 K," *Int. J. Chem. Kinet., 18,* 1303–1314 (1986).

Chin, M., and D. D. Davis, "Global Sources and Sinks of OCS and CS_2 and Their Distributions," *Global Biogeochem. Cycl. 7,* 321–337 (1993).

Claiborn, C. S., and V. P. Aneja, "Measurements of Atmospheric Hydrogen Peroxide in the Gas Phase and in Cloud Water at Mt. Mitchell, North Carolina," *J. Geophys. Res., 96,* 18771–18787 (1991).

Clark, P. A., G. P. Gervat, T. A. Hill, A. R. W. Marsh, A. S. Chandler, T. W. Choularton, and M. J. Gay, "A Field Study of the Oxidation of SO_2 in Cloud," *J. Geophys. Res., 95,* 13985–13995 (1990).

Clarke, A. D., Z. Li, and M. Kitchy, "Aerosol Dynamics in the Equatorial Pacific Marine Boundary Layer: Microphysics, Diurnal Cycles, and Entrainment," *Geophys. Res. Lett., 23,* 733–736 (1996).

Clarke, A. G., and M. Radojevic, "Chloride Ion Effects on the Aqueous Oxidation of SO_2," *Atmos. Environ., 17,* 617–624 (1983).

Clarke, A. G., and P. T. Williams, "The Oxidation of Sulfur Dioxide in Electrolyte Droplets," *Atmos. Environ., 17,* 607–615 (1983).

Clegg, N. A., and R. Toumi, "Sensitivity of Sulphur Dioxide Oxidation in Sea Salt to Nitric Acid and Ammonia Gas Phase Concentrations," *J. Geophys. Res., 102,* 23241–23249 (1997).

Clegg, S. L., and P. Brimblecombe, "The Solubility of Methanesulphonic Acid and Its Implications for Atmospheric Chemistry," *Environ. Technol. Lett., 6,* 269–278 (1985).

Clegg, S. L., and P. Brimblecombe, "The Solubility and Behaviour of Acid Gases in the Marine Aerosol," *J. Atmos. Chem., 7,* 1–18 (1988).

Cofer, W. R., III, D. R. Schryer, and R. S. Rogowski, "The Oxidation of SO_2 on Carbon Particles in the Presence of O_3, NO_2, and N_2O," *Atmos. Environ., 15,* 1281–1286 (1981).

Cofer, W. R., III, D. R. Schryer, and R. S. Rogowski, "Oxidation of SO_2 by NO_2 and O_3 on Carbon: Implications to Tropospheric Chemistry," *Atmos. Environ., 18,* 243–245 (1984).

Collett, J. L., Jr., A. Bator, X. Rao, and B. B. Demoz, "Acidity Variations across the Cloud Drop Size Spectrum and Their Influence on Rates of Atmospheric Sulfate Production," *Geophys. Res. Lett., 21,* 2393–2396 (1994).

Collett, J. L., Jr., K. J. Hoag, D. E. Sherman, A. Bator, and L. W. Richards, "Spatial and Temporal Variations in San Joaquin Valley Fog Chemistry," *Atmos. Environ., 33,* 129–140 (1999).

Conklin, M. H., and M. R. Hoffmann, "Metal Ion–Sulfur(IV) Chemistry. 3. Thermodynamics and Kinetics of Transient Iron(III)–Sulfur(IV) Complexes," *Environ. Sci. Technol., 22,* 899–907 (1988).

Cowling, E. B., "Acid Precipitation in Historical Perspective," *Environ. Sci. Technol., 16,* 110A–123A (1982).

Cox, R. A., and S. A. Penkett, "Oxidation of SO_2 by Oxidants Formed in the Ozone–Olefin Reaction," *Nature, 230,* 321–322 (1971).

Cox, R. A., and S. A. Penkett, "Aerosol Formation from Sulfur Dioxide in the Presence of Ozone and Olefinic Hydrocarbons," *J. Chem. Soc., Faraday Trans. 1, 68,* 1735–1753 (1972).

Crutzen, P. J., "Atmospheric Interactions—Homogeneous Gas Reactions of C, N, and S Containing Compounds," in *The Major Biogeochemical Cycles and Their Interactions* (B. Bolin and R. B. Cook, Eds.), Chap. 3, pp. 67–114, Wiley, New York, 1983.

Dacey, J. W. H., S. G. Wakeham, and B. L. Howes, "Henry's Law Constants for Dimethylsulfide in Freshwater and Seawater," *Geophys. Res. Lett., 11,* 991–994 (1984).

D'Angelantonio, M., M. Venturi, and Q. G. Mulazzani, "A Re-examination of the Decay Kinetics of Pulse Radiolytically Generated Br_2^- Radicals in Aqueous Solution," *Radiat. Phys. Chem., 32,* 319–324 (1988).

Davis, D., G. Chen, P. Kasibhatla, A. Jefferson, D. Tanner, F. Eisele, D. Lenschow, W. Neff, and H. Berresheim, "DMS Oxidation in the Antarctic Marine Boundary Layer: Comparison of Model Simulations and Field Observations of DMS, DMSO, $DMSO_2$, H_2SO_4(g), MSA(g), and MSA(p)," *J. Geophys. Res., 103,* 1657–1678 (1998).

Dawson, G. A., J. C. Farmer, and J. L. Moyers, "Formic and Acetic Acids in the Atmosphere of the Southwest U.S.A.," *Geophys. Res. Lett., 7,* 725–728 (1980).

Dawson, G. A., and J. C. Farmer, "Soluble Atmospheric Trace Gases in the Southwestern United States. 2. Organic Species HCHO, HCOOH, and CH_3COOH," *J. Geophys. Res., 93,* 5200–5206 (1988).

Daykin, E. P., and P. H. Wine, "A Study of the Reactions of NO_3 Radicals with Organic Sulfides: Reactivity Trend at 298 K," *Int. J. Chem. Kinet., 22,* 1083–1094 (1990).

De Bruyn, W. J., J. A. Shorter, P. Davidovits, D. R. Worsnop, M. S. Zahniser, and C. E. Kolb, "Uptake of Gas Phase Sulfur Species Methanesulfonic Acid, Dimethylsulfoxide, and Dimethyl Sulfone by Aqueous Surfaces," *J. Geophys. Res., 99,* 16927–16932 (1994).

De Bruyn, W. J., E. Swartz, J. H. Hu, J. A. Shorter, P. Davidovits, D. R. Worsnop, M. S. Zahniser, and C. E. Kolb, "Henry's Law Solubilities and Setchenow Coefficients for Biogenic Reduced Sulfur Species Obtained from Gas–Liquid Uptake Measurements," *J. Geophys. Res., 100,* 7245–7251 (1995).

Dedik, A. N., P. Hoffmann, and J. Ensling, "Chemical Characterization of Iron in Atmospheric Aerosols," *Atmos. Environ., 26A,* 2545–2548 (1992).

Deister, U., R. Neeb, G. Helas, and P. Warneck, "Temperature Dependence of the Equilibrium $CH_2(OH)_2 + HSO_3^- = CH_2(OH)SO_3^- + H_2O$ in Aqueous Solution," *J. Phys. Chem., 90,* 3213–3217 (1986).

Deister, U., and P. Warneck, "Photooxidation of SO_3^{2-} in Aqueous Solution," *J. Phys. Chem., 94,* 2191–2198 (1990).

DeMore, W. B., S. P. Sander, D. M. Golden, R. F. Hampson, M. J. Kurylo, C. J. Howard, A. R. Ravishankara, C. E. Kolb, and M. J. Molina, "Chemical Kinetics and Photochemical Data for Use in Stratospheric Modeling. Evaluation No. 12," Jet Propulsion Laboratory Publication No. JPL 97-4, Pasadena, CA, January 15, 1997.

Dennis, R. L., J. N. McHenry, W. R. Barchet, F. S. Binkowski, and D. W. Byun, "Correcting RADM's Sulfate Underprediction: Discovery and Correction of Model Errors and Testing the Corrections through Comparisons against Field Data," *Atmos. Environ., 27A,* 975–997 (1993).

de Violet, Ph. F., "Polyhalide Radical Anions as Intermediates in Chemistry," *Rev. Chem. Intermed., 4,* 121–169 (1981).

Diau, E. W.-G., and Y.-P. Lee, "Kinetics of the Reactions of CS_2OH with O_2, NO, and NO_2," *J. Phys. Chem., 95,* 7726–7732 (1991a).

Diau, E. W.-G., and Y.-P. Lee, "Termolecular Rate Coefficients and the Standard Enthalpy of the Reaction $OH + CS_2 + M \rightarrow HOCS_2 + M$," *J. Phys. Chem., 95,* 379–386 (1991b).

Dixon, R. W., and H. Aasen, "Measurement of Hydroxymethanesulfonate in Atmospheric Aerosols," *Atmos. Environ., 33,* 2023–2029 (1999).

Dlugi, R., and H. Gusten, "The Catalytic and Photocatalytic Activity of Coal Fly Ashes," *Atmos. Environ., 17,* 1765–1771 (1983).

Dlugokencky, E. J., and C. J. Howard, "Laboratory Studies of NO_3 Radical Reactions with Some Atmospheric Sulfur Compounds," *J. Phys. Chem., 92,* 1188–1193 (1988).

Dominé, F., T. P. Murrells, and C. J. Howard, "Kinetics of the Reactions of NO_2 with CH_3S, CH_3SO, CH_3SS, and CH_3SSO at 297 K and 1 Torr," *J. Phys. Chem., 94,* 5839–5847 (1990).

Dominé, F., A. R. Ravishankara, and C. J. Howard, "Kinetics and Mechanisms of the Reactions of CH_3S, CH_3SO, and CH_3SS with O_3 at 300 K and Low Pressures," *J. Phys. Chem., 96,* 2171–2178 (1992).

Donaldson, D. J., J. A. Guest, and M. C. Goh, "Evidence for Adsorbed SO_2 at the Aqueous–Air Interface," *J. Phys. Chem., 99,* 9313–9315 (1995).

Duan, S. X., J. T. Jayne, P. Davidovits, D. R. Worsnop, M. S. Zahniser, and C. E. Kolb, "Uptake of Gas-Phase Acetone by Water Surfaces," *J. Phys. Chem., 97,* 2284–2288 (1993).

Dutkiewicz, V. A., E. G. Burkhard, and L. Husain, "Availability of H_2O_2 for Oxidation of SO_2 in Clouds in the Northeastern United States," *Atmos. Environ., 29,* 3281–3292 (1995).

Eatough, D. J., and L. D. Hansen, "Organic and Inorganic S(IV) Compounds in Airborne Particulate Matter," *Adv. Environ. Sci. Technol., 12,* 221–268 (1983).

Eatough, D. J., R. J. Arthur, N. L. Eatough, M. W. Hill, N. F. Mangelson, B. E. Richter, L. D. Hansen, and J. A. Cooper, "Rapid Conversion of $SO_{2(g)}$ to Sulfate in a Fog Bank," *Environ. Sci. Technol., 18,* 855–859 (1984).

Egsgaard, H., L. Carlsen, H. Florencio, T. Drewello, and H. Schwarz, "Experimental Evidence for the Gaseous HSO_3 Radical. The Key Intermediate in the Oxidation of SO_2 in the Atmosphere," *Chem. Phys. Lett., 148,,* 537–540 (1988).

Eigen, M., and K. Kustin, "The Kinetics of Halogen Hydrolysis," *J. Am. Chem. Soc., 84,,* 1355–1361 (1962).

Eigen, M., W. Kruse, G. Maass, and L. De Maeyer, "Rate Constants of Protolytic Reactions in Aqueous Solution," in *Progress in Reaction Kinetics* (G. Porter, Ed.), Vol. 2, Chap 6, Macmillan, New York, 1964.

Elliott, A. J., and G. V. Buxton, "Temperature Dependence of the Reactions $OH + O_2^-$ and $OH + HO_2$ in Water up to 200°C," *J. Chem. Soc. Faraday Trans., 88,* 2465–2470 (1992).

Erel, Y., S. O. Pehkonen, and M. R. Hoffmann, "Redox Chemistry of Iron in Fog and Stratus Clouds," *J. Geophys. Res., 98,* 18423–18434 (1993).

Eriksen, T. E., J. Lind, and G. Mereny, "On the Acid-Base Equilibrium of the Carbonate Radical," *Radiat. Phys. Chem., 26,* 197–199 (1985).

Exner, M., H. Herrmann, and R. Zellner, "Laser-Based Studies of Reactions of the Nitrate Radical in Aqueous Solution," *Ber. Bunsenges. Phys. Chem., 96,* 470–477 (1992).

Faust, B. C., and M. R. Hoffmann, "Photo-Induced Reductive Dissolution of α-Fe_2O_3 by Bisulfite," *Environ. Sci. Technol., 20,* 943–948 (1986).

Faust, B. C., M. R. Hoffmann, and D. W. Bahnemann, "Photocatalytic Oxidation of Sulfur Dioxide in Aqueous Suspensions of α-Fe_2O_3," *J. Phys. Chem., 93,* 6371–6381 (1989).

Faust, B. C., and J. Hoigné, "Photolysis of Fe (III)–Hydroxy Complexes as Sources of OH Radicals in Clouds, Fog, and Rain," *Atmos. Environ., 24A,* 79–89 (1990).

Faust, B. C., and J. M. Allen, "Aqueous-Phase Photochemical Sources of Peroxy Radicals and Singlet Molecular Oxygen in Clouds and Fog," *J. Geophys. Res., 97,* 12913–12926 (1992).

Faust, B. C., and J. M. Allen, "Aqueous-Phase Photochemical Formation of Hydroxyl Radical in Authentic Cloudwaters and Fogwaters," *Environ. Sci. Technol., 27,* 1221–1224 (1993).

Faust, B. C., C. Anastasio, J. M. Allen, and T. Arakaki, "Aqueous-Phase Photochemical Formation of Peroxides in Authentic Cloud and Fog Waters," *Science, 260,* 73–75 (1993).

Faust, B. C., "Photochemistry of Clouds, Fogs, and Aerosols," *Environ. Sci. Technol., 28,* 217A–222A (1994a).

Faust, B. C., "A Review of the Photochemical Redox Reactions of Iron(III) Species in Atmospheric, Oceanic, and Surface Waters: Influences on Geochemical Cycles and Oxidant Formation," in *Aquatic and Surface Photochemistry* (G. Helz, R. Zepp, and D. Crosby, Eds.), Chap. 1, pp. 3–37, Lewis, Boca Raton, FL, 1994b.

Faust, B. C., K. Powell, C. J. Rao, and C. Anastasio, "Aqueous-Phase Photolysis of Biacetyl (an α-Dicarbonyl Compound): A Sink for Biacetyl and a Source of Acetic Acid, Peroxyacetic Acid, Hydrogen Peroxide, and the Highly Oxidizing Acetylperoxyl Radical in Aqueous Aerosols, Fogs, and Clouds," *Atmos. Environ., 31,* 497–510 (1997).

Ferek, R. J., P. V. Hobbs, L. F. Radke, J. A. Herring, W. T. Sturges, and G. F. Cota, "Dimethyl Sulfide in the Arctic Atmosphere," *J. Geophys. Res., 100,* 26093–26104 (1995).

Finnegan, W. G., and R. L. Pitter, "Preliminary Study of Coupled Oxidation–Reduction Reactions of Included Ions in Growing Ice Crystals—Reply," *Atmos. Environ., 25,* 2912–2913 (1991).

Finnegan, W. G., R. L. Pitter, and L. G. Young, "Preliminary Study of Coupled Oxidation–Reduction Reactions of Included Ions in Growing Ice Crystals," *Atmos. Environ., 25A,* 2531–2534 (1991).

Finnegan, W. G., and R. L. Pitter, "Ion-Induced Charge Separations in Growing Single Ice Crystals: Effects on Growth and Interaction Processes," *J. Colloid Interface Sci., 189,* 322–327 (1997).

Fischer, M., and P. Warneck, "Photodecomposition of Nitrite and Undissociated Nitrous Acid in Aqueous Solution," *J. Phys. Chem., 100,* 18749–18756 (1996).

Fogelman, K. D., D. M. Walker, and D. W. Margerum, " Non-Metal Redox Kinetics: Hypochlorite and Hypochlorous Acid Reactions with Sulfite," *Inorg. Chem., 28,* 986–993 (1989).

Fortnum, D. H., C. J. Battaglia, S. R. Cohen, and J. O. Edwards, "The Kinetics of the Oxidation of Halide Ions by Monosubstituted Peroxides," *J. Am. Chem. Soc. 82,* 778–782 (1960).

Frank, S. N., and A. J. Bard, "Heterogeneous Photocatalytic Oxidation of Cyanide and Sulfite in Aqueous Solutions at Semiconductor Powers," *J. Phys. Chem., 81,* 1484–1488 (1977).

Fuzzi, S., R. A. Castillo, J. E. Kiusto, and G. G. Lala, "Chemical Composition of Radiation Fog Water at Albany, New York, and Its Relationship to Fog Microphysics," *J. Geophys. Res., 89,* 7159–7164 (1984).

Gao, D., W. R. Stockwell, and J. B. Milford, "Global Uncertainty Analysis of a Regional-Scale Gas-Phase Chemical Mechanism," *J. Geophys. Res., 101,* 9107–9119 (1996).

George, Ch., J. L. Ponche, and Ph. Mirabel, "Experimental Determination of Mass Accommodation Coefficient," in *Nucleation and Atmospheric Aerosols* (N. Fukuta and P. E. Wagner, Eds.), A. Deepak Publishing, Hampton, VA, 1992.

Gertler, A. W., D. F. Miller, D. Lamb, and U. Katz, "Studies of Sulfur Dioxide and Nitrogen Dioxide Reactions in Haze and Cloud," in *Chemistry of Particles, Fogs, and Rain* (J. L. Durham, Ed.), Acid Precipitation Series, Vol. 2, pp. 131–160 (J. I. Teasley, Series Ed.), Butterworth, Stoneham, MA, 1984.

Gleason, J. F., A. Sinha, and C. J. Howard, "Kinetics of the Gas-Phase Reaction $HOSO_2 + O_2 \rightarrow HO_2 + SO_3$," *J. Phys. Chem., 91,* 719–724 (1987).

Gleason, J. F., and C. J. Howard, "Temperature Dependence of the Gas-Phase Reaction $HOSO_2 + O_2 \rightarrow HO_2 + SO_3$," *J. Phys. Chem., 92,* 3414–3417 (1988).

Golden, D. M., V. M. Bierbaum, and C. J. Howard, "Comments on 'Reevaluation of the Bond-Dissociation Energies (ΔH_{DBE}) for H-OH, H-OOH, H-OO$^-$, H-O, H-OO$^-$ and H-OO'," *J. Phys. Chem., 94,* 5413–5415 (1990).

Graedel, T. E., and C. J. Weschler, "Chemistry within Aqueous Atmospheric Aerosols and Raindrops," *Rev. Geophys. Space Phys., 19,* 505–539 (1981).

Graedel, T. E., and K. I. Goldberg, "Kinetic Studies of Raindrop Chemistry. I. Inorganic and Organic Processes," *J. Geophys. Res., 88,* 10865–10882 (1983).

Graedel, T. E., C. J. Weschler, and M. L. Mandlich, "The Influence of Transition Metal Complexes on Atmospheric Droplet Acidity," *Nature, 317,* 240–242 (1985).

Graedel, T. E., M. L. Mandlich, and C. J. Weschler, "Kinetic Model Studies of Atmospheric Droplet Chemistry. 2. Homogeneous Transition Metal Chemistry in Raindrops," *J. Geophys. Res., 91,* 5205–5221 (1986).

Granby, K., C. S. Christensen, and C. Lohse, "Urban and Semi-Rural Observations of Carboxylic Acids and Carbonyls," *Atmos. Environ., 31,* 1403–1415 (1997a).

Granby, K., A. H. Egeløv, T. Nielsen, and C. Lohse, "Carboxylic Acids: Seasonal Variation and Relation to Chemical and Meteorological Parameters," *J. Atmos. Chem., 28,* 195–207 (1997b).

Grigor'ev, A. E., I. E. Makarov, and A. K. Pikaev, "Formation of Cl_2^- in the Bulk Solution during the Radiolysis of Concentrated Aqueous Solutions of Chlorides," *High Energy Chem., 21,* 99–102 (1987).

Grosjean, D., "Photooxidation of Methyl Sulfide, Ethyl Sulfide, and Methanethiol," *Environ. Sci. Technol., 18,* 460–468 (1984).

Grosjean, D., "Organic Acids in Southern California Air: Ambient Concentrations, Mobile Source Emissions, *in Situ* Formation, and Removal Processes," *Environ. Sci. Technol., 23,* 1506–1514 (1989).

Grosjean, D., "Liquid Chromatography Analysis of Chloride and Nitrate with 'Negative' Ultraviolet Detection: Ambient Levels and Relative Abundance of Gas-Phase Inorganic and Organic Acids in Southern California," *Environ. Sci. Technol., 24,* 77–81 (1990).

Grosjean, D., "Formic Acid and Acetic Acid: Emissions, Atmospheric Formation, and Dry Deposition at Two Southern California Locations," *Atmos. Environ., 26A,* 3279–3286 (1992).

Gross, G. W., "Preliminary Study of Coupled Oxidation–Reduction Reactions of Included Ions in Growing Ice Crystals—Discussion," *Atmos. Environ., 25A,* 2911–2913 (1991).

Gunz, D. W., and M. R. Hoffmann, "Atmospheric Chemistry of Peroxides: A Review," *Atmos. Environ., 24A,* 1601–1633 (1990).

Gurciullo, C. S., and S. N. Pandis, "Effect of Composition Variations in Cloud Droplet Populations on Aqueous-Phase Chemistry," *J. Geophys. Res., 102,* 9375–9385 (1997).

Gurol, M. D., and A. Akata, "Kinetics of Ozone Photolysis in Aqueous Solution," *AIChE J., 42,* 3283–3292 (1996).

Haag, W. R., and J. Hoigné, "Ozonation of Bromide-Containing Waters: Kinetics of Formation of Hypobromous Acid and Bromate," *Environ. Sci. Technol., 17,* 261–267 (1983).

Halstead, J. A., R. Armstrong, B. Pohlman, S. Sibley, and R. Maier, "Nonaqueous Heterogeneous Oxidation of Sulfur Dioxide," *J. Phys. Chem., 94,* 3261–3265 (1990).

Hanson, D. R., J. B. Burkholder, C. J. Howard, and A. R. Ravishankara, "Measurement of OH and HO_2 Radical Uptake Coefficients on Water and Sulfuric Acid Surfaces," *J. Phys. Chem., 96,* 4979–4985 (1992).

Hartig, K. J., and N. Getoff, "Reactivity of Hydrogen Atoms with Liquid Water," *J. Photochem., 18,* 29–38 (1982).

Hatakeyama, S., and H. Akimoto, "Reactions of OH Radicals with Methanethiol, Dimethyl Sulfide, and Dimethyl Disulfide in Air," *J. Phys. Chem., 87,* 2387–2395 (1983).

Hatakeyama, S., H. Kobayashi, Z.-Y. Lin, K. Takagi, and H. Akimoto, "Mechanism for the Reaction of Peroxymethylene with Sulfur Dioxide," *J. Phys. Chem., 90,* 4131–4135 (1986).

Hatakeyama, S., "Mechanisms for the Reaction of CH_3S with NO_2," in *Biogenic Sulfur in the Environment,* ACS Symposium Series No. 393 (E. S. Saltzman and W. J. Cooper, Eds.), Chap. 28, pp. 459–463, Am. Chem. Soc., Washington, DC, 1989.

Hatakeyama, S., and H. Akimoto, "Reactions of Criegee Intermediates in the Gas Phase," *Res. Chem. Intermed., 20,* 503–524 (1994).

Hausmann, M., J. Rudolf, and U. Platt, "Spectroscopic Measurement of Bromine Oxide, Ozone and Nitrous Acid in Alert," in *The Tropospheric Chemistry of Ozone in the Polar Regions: NATO ASI Series I: Global Environmental Change* (H. Niki and K. H. Becker, Eds.), Vol. 17, pp. 189–203, 1993.

Helas, G., H. Bingemer, and M. O. Andreae, "Organic Acids over Equatorial Africa: Results from DECAFE 88," *J. Geophys. Res., 97,* 6187–6193 (1992).

Hering, S. V., D. L. Blumenthal, R. L. Brewer, A. Gertler, M. Hoffmann, J. A. Kadlecek, and K. Pettus, "Field Intercomparison of Five Types of Fogwater Collectors," *Environ. Sci. Technol., 21,* 654–663 (1987).

Herring, J. A., R. J. Ferek, and P. V. Hobbs, "Heterogeneous Chemistry in the Smoke Plume from the 1991 Kuwait Oil Fires," *J. Geophys. Res., 101,* 14451–14463 (1996).

Herrmann, H., M. Exner, H.-W. Jacobi, G. Raabe, A. Reese, and R. Zellner, "Laboratory Studies of Atmospheric Aqueous-Phase

Free-Radical Chemistry: Kinetic and Spectroscopic Studies of Reactions of NO$_3$ and SO$_4^-$ Radicals with Aromatic Compounds," *Faraday Discuss., 100*, 129–153 (1995).

Hirokawa, J., K. Onaka, Y. Kajii, and H. Akimoto, "Heterogeneous Processes Involving Sodium Halide Particles and Ozone: Molecular Bromine Release in the Marine Boundary Layer in the Absence of Nitrogen Oxides," *Geophys. Res. Lett., 25*, 2449–2452 (1998).

Hoffmann, M. R., and J. O. Edwards, "Kinetics of the Oxidation of Sulfite by Hydrogen Peroxide in Acidic Solution," *J. Phys. Chem., 79*, 2096–2098 (1975).

Hoffmann, M. R., and S. D. Boyce, "Catalytic Autooxidation of Aqueous Sulfur Dioxide in Relationship to Atmospheric Systems," *Adv. Environ. Sci. Technol., 12*, 147–189 (1983).

Hoffmann, M. R., "Comment on Acid Fog," *Environ. Sci. Technol., 18*, 61–64 (1984).

Hoffmann, M. R., and D. J. Jacob, "Kinetics and Mechanisms of Catalytic Oxidation of Dissolved Sulfur Dioxide in Aqueous Solution: An Application to Nighttime Fog Water Chemistry," in *SO$_2$, NO, and NO$_2$ Oxidation Mechanisms: Atmospheric Considerations*, Acid Precipitation Series, Vol. 3, pp. 101–172 (J. I. Teasley, Series Ed.), Butterworth, Stoneham, MA, 1984.

Hoffmann, M. R., "On the Kinetics and Mechanism of Oxidation of Aquated Sulfur Dioxide by Ozone," *Atmos. Environ., 20*, 1145–1154 (1986).

Hoffmann, M., and P. Schleyer, "Acid Rain—Ab Initio Investigation of the H$_2$O · SO$_2$ Complex and Its Conversion into H$_2$SO$_4$," *J. Am. Chem. Soc., 116*, 4947–4952 (1994).

Hoffmann, M. R., S. T. Martin, W. Choi, and D. W. Bahnemann, "Environmental Applications of Semiconductor Photocatalysis," *Chem. Rev., 95*, 69–96 (1995).

Hoigné, J., H. Bader, W. R. Haag, and J. Staehelin, "Rate Constants of Reactions of Ozone with Organic and Inorganic Compounds in Water—III," *Water Res., 19*, 993–1004 (1985).

Hoigné, J., Y. Zuo, and L. Nowell, "Photochemical Reactions in Atmospheric Waters: Role of Dissolved Iron Species," in *Aquatic and Surface Photochemistry* (G. Helz, R. Zepp, and D. Crosby, Eds.), Chap. 4, pp. 75–84, Lewis, Boca Raton, FL, 1994.

Holcman, J., E. Bjergbakke, and K. Sehested, "The Importance of Radical–Radical Reactions in Pulse Radiolysis of Aqueous Carbonate/Bicarbonate," *Proc. 6th Tihany Symp. Radiat. Chem., 6*, 149–153 (1987).

Holcman, J., K. Sehested, E. Bjergbakke, and E. J. Hart, "Formation of Ozone in the Reaction between the Ozonide Radical Ion, O$_3^-$, and the Carbonate Radial Ion, CO$_3^-$, in Aqueous Alkaline Solutions," *J. Phys. Chem., 86*, 2069–2072 (1982).

Holdren, M. W., C. W. Spicer, and J. M. Hales, "Peroxyacetyl Nitrate Solubility and Decomposition Rate in Acidic Water," *Atmos. Environ., 18*, 1171–1173 (1984).

Huie, R. E., and N. C. Peterson, "Reaction of Sulfur(IV) with Transition-Metal Ions in Aqueous Solutions," *Adv. Environ. Sci. Technol., 12*, 117–146 (1983).

Huie, R. E., and P. Neta, "Chemical Behaviour of SO$_3^-$ and SO$_5^-$ Radicals in Aqueous Solutions," *J. Phys. Chem., 88*, 5665–5669 (1984).

Huie, R. E., and P. Neta, "Rate Constants for Some Oxidation of S(IV) by Radicals in Aqueous Solutions," *Atmos. Environ., 21*, 1743–1747 (1987).

Huie, R. E., and C. L. Clifton, "Temperature Dependence of the Rate Constants for Reactions of the Sulfate Radical, SO$_4^-$, with Anions," *J. Phys. Chem., 94*, 8561–8567 (1990).

Huie, R. E., "Free Radical Chemistry of the Atmospheric Aqueous Phase," in *Progress and Problems in Atmospheric Chemistry* (J. R. Barker, Ed.), Vol. 3, Chap. 10, pp. 374–419, World Scientific, Singapore, 1995.

Huie, R. E., and C. L. Clifton, "Kinetics of the Reaction of the Sulfate Radical with the Oxalate Anion," *Int. J. Chem. Kinet., 28*, 195–199 (1996).

Hulett, L. D., T. A. Carlson, B. R. Fish, and J. L. Durham, "Studies of Sulfur Compounds Adsorbed on Smoke Particles and Other Solids by Photoelectron Spectroscopy," in *Determination of Air Quality* (G. Mamantov and W. D. Shults, Eds.), pp. 179–187, Plenum, New York, 1972.

Husain, L., "A Technique for Determining In-Cloud Formation of SO$_4$," *Geophys. Res. Lett., 16*, 57–60 (1989).

Husain, L., and V. A. Dutkiewicz, "A Long-Term (1975–1988) Study of Atmospheric SO$_4^{2-}$: Regional Contributions and Concentration Trends," *Atmos. Environ., 24A*, 1175–1187 (1990).

Husain, L., V. A. Dutkiewicz, M. M. Husain, H. A. Khwaja, E. G. Burkhard, G. Mehmood, P. P. Parekh, and E. Canelli, "A Study of Heterogeneous Oxidation of SO$_2$ in Summer Clouds," *J. Geophys. Res., 96*, 18789–18805 (1991).

Husain, L., and V. A. Dutkiewicz, "Elemental Tracers for the Study of Homogeneous Gas Phase Oxidation of SO$_2$ in the Atmosphere," *J. Geophys. Res., 97*, 14635–14643 (1992).

Hynes, A. J., P. H. Wine, and D. H. Semmes, "Kinetics and Mechanism of OH Reactions with Organic Sulfides," *J. Phys. Chem., 90*, 4148–4156 (1986).

Hynes, A. J., P. H. Wine, and J. M. Nicovich, "Kinetics and Mechanism of the Reaction of OH with CS$_2$ under Atmospheric Conditions," *J. Phys. Chem., 92*, 3846–3852 (1988).

Hynes, A. J., R. B. Stoker, A. J. Pounds, T. McKay, J. D. Bradshaw, J. M. Nicovich, and P. H. Wine, "A Mechanistic Study of the Reaction of OH with Dimethyl-d_6 Sulfide. Direct Observation of Adduct Formation and the Kinetics of the Adduct Reaction with O$_2$," *J. Phys. Chem., 99*, 16967–16975 (1995).

Jacob, D. J., and M. R. Hoffmann, "A Dynamic Model for the Production of H$^+$, NO$_3^-$, and SO$_4^{2-}$ in Urban Fog," *J. Geophys. Res., 88C*, 6611–6621 (1983).

Jacob, D. J., J. M. Waldman, J. W. Munger, and M. R. Hoffmann, "Chemical Composition of Fogwater Collected along the California Coast," *Environ. Sci. Technol., 19*, 730–736 (1985).

Jacob, D. J., F. H. Shair, J. M. Waldman, J. W. Munger, and M. R. Hoffmann, "Transport and Oxidation of SO$_2$ in a Stagnant Foggy Valley," *Atmos. Environ., 21*, 1305–1314 (1987).

Jacob, D. J., E. W. Gottlieb, and M. J. Prather, "Chemistry of a Polluted Cloudy Boundary Layer," *J. Geophys. Res., 94*, 12975–13002 (1989).

Jayne, J. T., P. Davidovits, D. R. Worsnop, M. S. Zahniser, and C. E. Kolb, "Uptake of SO$_2$(g) by Aqueous Surfaces as a Function of pH: The Effect of Chemical Reaction at the Interface," *J. Phys. Chem., 94*, 6041–6048 (1990a).

Jayne, J. T., J. A. Gardner, P. Davidovits, D. R. Worsnop, M. S. Zahniser, and C. E. Kolb, "The Effect of H$_2$O$_2$ Content on the Uptake of SO$_2$(g) by Aqueous Droplets," *J. Geophys. Res., 95*, 20559–20563 (1990b).

Jayne, J. T., S. X. Duan, P. Davidovits, D. R. Worsnop, M. S. Zahniser, and C. E. Kolb, "Uptake of Gas-Phase Alcohol and Organic Acid Molecules by Water Surfaces," *J. Phys. Chem., 95*, 6329–6336 (1991).

Jayne, J. T., S. X. Duan, P. Davidovits, D. R. Worsnop, M. S. Zahniser, and C. E. Kolb, "Uptake of Gas-Phase Aldehydes by Water Surfaces," *J. Phys. Chem., 96*, 5452–5460 (1992).

Jayne, J. T., U. Pöschl, Y.-M. Chen, D. Dai, L. T. Molina, D. R. Worsnop, C. E. Kolb, and M. J. Molina, "Pressure and Temperature Dependence of the Gas-Phase Reaction of SO$_3$ and H$_2$O and the Heterogeneous Reaction of SO$_3$ and H$_2$O/H$_2$SO$_4$ Surfaces," *J. Phys. Chem. A, 101*, 10000–10011 (1997).

Jayson, G. G., B. J. Parsons, and A. J. Swallow, "Some Simple, Highly Reactive, Inorganic Chlorine Derivatives in Aqueous Solution," *J. Chem. Soc., Faraday Trans., 69*, 1597–1607 (1973).

Jefferson, A., D. J. Tanner, F. L. Eisele, and H. Berresheim, "Sources and Sinks of H_2SO_4 in the Remote Antarctic Marine Boundary Layer," *J. Geophys. Res.*, 103, 1639–1645 (1998).

Jensen, N. R., J. Hjorth, C. Lohse, H. Skov, and G. Restelli, "Products and Mechanisms of the Gas Phase Reactions of NO_3 with CH_3SCH_3, CD_3SCD_3, CH_3SH, and CH_3SSCH_3," *J. Atmos. Chem.*, 14, 95–108 (1992).

Jobson, B. T., H. Niki, Y. Yokouchi, J. Bottenheim, F. Hopper, and R. Leaitch, "Measurements of C_2–C_6 Hydrocarbons During the Polar Sunrise 1992 Experiment: Evidence for Cl Atoms and Br Atom Chemistry," *J. Geophys. Res.*, 99, 25,355–25,368 (1994).

Jonson, J. E., and I. S. A. Isaksen, "Tropospheric Ozone Chemistry: The Impact of Cloud Chemistry," *J. Atmos. Chem.*, 16, 99–122 (1993).

Judeikis, H. S., T. B. Stewart, and A. C. Wren, "Laboratory Studies of Heterogeneous Reactions of SO_2," *Atmos. Environ.*, 12, 1633–1641 (1978).

Kawamura, K., Ng, L.-L., and I. R. Kaplan, "Determination of Organic Acids (C_1–C_{10}) in the Atmosphere, Motor Exhaust, and Engine Oils," *Environ. Sci. Technol.*, 19, 1082–1086 (1985).

Kawamura, K., and I. R. Kaplan, "Motor Exhaust Emissions as a Primary Source for Dicarboxylic Acids in Los Angeles Ambient Air," *Environ. Sci. Technol.*, 21, 105–110 (1987).

Kawamura, K., and H. Kasukabe, "Source and Reaction Pathways of Dicarboxylic Acids, Ketoacids, and Dicarbonyls in Arctic Aerosols: One Year of Observations," *Atmos. Environ.*, 30, 1709–1722 (1996a).

Kawamura, K., R. Seméré, Y. Imai, Y. Fujii, and M. Hayashi, "Water Soluble Dicarboxylic Acids and Related Compounds in Antarctic Aerosols," *J. Geophys. Res.*, 101, 18721–18728 (1996b).

Keene, W. C., and J. N. Galloway, "Considerations Regarding Sources for Formic and Acetic Acids in the Troposphere," *J. Geophys. Res.*, 91, 14466–14474 (1986).

Keene, W. C., and J. N. Galloway, "The Biogeochemical Cycling of Formic and Acetic Acids through the Troposphere: An Overview of Current Understanding," *Tellus*, 40B, 322–334 (1988).

Keene, W. C., J. R. Maben, A. A. P. Pszenny, and J. N. Galloway, "Measurement Technique for Inorganic Chlorine Gases in the Marine Boundary Layer," *Environ. Sci. Technol.*, 27, 866–874 (1993).

Keene, W. C., B. W. Mosher, D. J. Jacob, J. W. Munger, R. W. Talbot, R. S. Artz, J. R. Maben, B. C. Daube, and J. N. Galloway, "Carboxylic Acids in Clouds at a High-Elevation Forested Site in Central Virginia," *J. Geophys. Res.*, 100, 9345–9357 (1995).

Keene, W. C., D. J. Jacob, and S.-M. Fan, "Reactive Chlorine: A Potential Sink for Dimethylsulfide and Hydrocarbons in the Marine Boundary Layer," *Atmos. Environ.*, 30, ii–iii (1996).

Keene, W. C., R. Sander, A. A. P. Pszenny, R. Vogt, P. J. Crutzen, and J. N. Galloway, "Aerosol pH in the Marine Boundary Layer: A Review and Model Evaluation," *J. Aerosol Sci.*, 29, 339–356 (1998).

Kelly, T. J., P. H. Daum, and S. E. Schwartz, "Measurements of Peroxides in Cloudwater and Rain," *J. Geophys. Res.*, 90, 7861–7871 (1985).

Khwaja, H. A., "Atmospheric Concentrations of Carboxylic Acids and Related Compounds at a Semiurban Site," *Atmos. Environ.*, 29, 127–139 (1995).

Khwaja, H. A., S. Brudnoy, and L. Husain, "Chemical Characterization of Three Summer Cloud Episodes at Whiteface Mountain," *Chemosphere*, 31, 3357–3381 (1995).

Kirchner, U., Th. Benter, and R. N. Schindler, "Experimental Verification of Gas Phase Bromine Enrichment in Reactions of HOBr with Sea Salt Doped Ice Surfaces," *Ber. Bunsenges. Phys. Chem.*, 101, 975–977 (1997).

Kläning, U. K., K. Sehested, and T. Wolff, "Ozone Formation in Laser Flash Photolysis of Oxoacids and Oxoanions of Chlorine and Bromine," *J. Chem. Soc., Faraday Trans. 1*, 80, 2969–2979 (1984).

Kläning, U. K., and T. Wolff, "Laser Flash Photolysis of HClO, ClO$^-$, HBrO, and BrO$^-$ in Aqueous Solution: Reactions of Cl- and Br-Atoms," *Ber. Bunsenges. Phys. Chem.*, 89, 243–245 (1985).

Kolb, C. E., J. T. Jayne, D. R. Worsnop, M. J. Molina, R. F. Meads, and A. A. Viggiano, "Gas Phase Reaction of Sulfur Trioxide with Water Vapor," *J. Am. Chem. Soc.*, 116, 10314–10315 (1994).

Kormann, C., D. W. Bahnemann, and M. R. Hoffmann, "Photocatalytic Production of H_2O_2 and Organic Peroxides in Aqueous Suspensions of TiO_2, ZnO, and Desert Sand," *Environ. Sci. Technol.*, 22, 798–806 (1988).

Kotronarou, A., and L. Sigg, "SO_2 Oxidation in Atmospheric Water: Role of Fe(II) and Effect of Ligands," *Environ. Sci. Technol.*, 27, 2725–2735 (1993).

Kreidenweis, S. M., and J. H. Seinfeld, "Nucleation of Sulfuric Acid–Water and Methanesulfonic Acid–Water Solution Particles: Implications for the Atmospheric Chemistry of Organosulfur Species," *Atmos. Environ.*, 22, 283–296 (1988a).

Kreidenweis, S. M., and J. H. Seinfeld, "Effect of Surface Tension of Aqueous Methanesulfonic Acid Solutions uon Nucleation and Growth of Aerosol," *Atmos. Environ.*, 22, 1499–1500 (1988b).

Kreidenweis, S. M., J. E. Penner, F. Yin, and J. H. Seinfeld, "The Effects of Dimethylsulfide upon Marine Aerosol Concentrations," *Atmos. Environ.*, 25A, 2501–2511 (1991).

Kumar, N., U. C. Kulshrestha, P. Khare, A. Saxena, K. M. Kumari, and S. S. Srivastava, "Measurements of Formic and Acetic Acid Levels in the Vapour Phase at Dayalbagh, Agra, India," *Atmos. Environ.*, 30, 3545–3550 (1996).

Kunen, S. M., A. L. Lazrus, G. L. Kok, and B. G. Heikes, "Aqueous Oxidation of SO_2 by Hydrogen Peroxide," *J. Geophys. Res.*, 88, 3671–3674 (1983).

Lagrange, J., C. Pallares, G. Wenger, and P. Lagrange, "Electrolyte Effects on Aqueous Atmospheric Oxidation of Sulphur Dioxide by Hydrogen Peroxide," *Atmos. Environ.*, 27A, 129–137 (1993).

Lagrange, J., C. Pallares, and P. Lagrange, "Electrolyte Effects on Aqueous Atmospheric Oxidation of Sulphur Dioxide by Ozone," *J. Geophys. Res.*, 99, 14595–14600 (1994).

Lagrange, J., C. Pallares, G. Wenger, and P. Lagrange, "Kinetics of Sulphur(IV) Oxidation by Hydrogen Peroxide in Basic Aqueous Solution," *Atmos. Environ.*, 30, 1013–1018 (1996).

Laj, P., S. Fuzzi, M. C. Facchini, G. Orsi, A. Berner, C. Kruisz, W. Wobrock, A. Hallberg, K. N. Bower, M. W. Gallagher, K. M. Beswick, R. N. Colvile, T. W. Choularton, P. Nason, and B. Jones, "Experimental Evidence for In-Cloud Production of Aerosol Sulphate," *Atmos. Environ.*, 31, 2503–2514 (1997).

Lamb, D., D. F. Miller, N. F. Robinson, and A. W. Gertler, "The Importance of Liquid Water Concentration in the Atmospheric Oxidation of SO_2," *Atmos. Environ.*, 21, 2333–2344 (1987).

Langer, S., B. T. McGovney, and B. J. Finlayson-Pitts, "The Dimethyl Sulfide Reaction with Atomic Chlorine and Its Implications for the Budget of Methyl Chloride," *Geophys. Res. Lett.*, 23, 1661–1664 (1996).

Lawrence, J. E., and P. Koutrakis, "Measurement of Atmospheric Formic and Acetic Acids: Methods Evaluation and Results from Field Studies," *Environ. Sci. Technol.*, 28, 957–964 (1994).

Lee, J. H., D. F. Leahy, I. N. Tang, and L. Newman, "Measurement and Speciation of Gas Phase Peroxides in the Atmosphere," *J. Geophys. Res.*, 98, 2911–2915 (1993).

Lee, Y.-N., and S. E. Schwartz, "Kinetics of Oxidation of Aqueous Sulfur(IV) by Nitrogen Dioxide," in *Precipitation Scavenging, Dry Deposition, and Resuspension*, (H. R. Pruppacher, R. G. Semonin, and W. G. N. Slinn, Eds.), Vol. 1, pp. 453–470, Elsevier, New York, 1983.

Lee, Y.-N., J. Shen, P. J. Klotz, S. E. Schwartz, and L. Newman, "Kinetics of Hydrogen Peroxide–Sulfur(IV) Reaction in Rainwater Collected at a Northeastern U.S. Site," *J. Geophys. Res., 91,* 13264–13274 (1986).

Lee, Y.-N., and X. Zhou, "Aqueous Reaction Kinetics of Ozone and Dimethylsulfide and Its Atmospheric Implications," *J. Geophys. Res., 99,* 3597–3605 (1994).

Lelieveld, J., and P. J. Crutzen, "Influences of Cloud Photochemical Processes on Tropospheric Ozone," *Nature, 343,* 227–233 (1990).

Li, W.-K., and M. L. McKee, "Theoretical Study of OH and H_2O Addition to SO_2," *J. Phys. Chem. A, 101,* 9778–9782 (1997).

Liang, J., and D. J. Jacob, "Effect of Aqueous Phase Cloud Chemistry on Tropospheric Ozone," *J. Geophys. Res., 102,* 5993–6001 (1997).

Liberti, A., D. Brocco, and M. Possanzini, "Adsorption and Oxidation of Sulfur Dioxide on Particles," *Atmos. Environ., 12,* 255–261 (1978).

Likens, G. E., W. C. Keene, J. M. Miller, and J. N. Galloway, "Chemistry of Precipitation from a Remote, Terrestrial Site in Australia," *J. Geophys. Res., 92,* 13299–13314 (1987).

Lind, J. A., and A. L. Lazrus, "Aqueous Phase Oxidation of Sulfur IV by Some Organic Peroxides," *EOS Trans., 64,* 670 (1983).

Lind, J. A., A. L. Lazrus, and G. L. Kok, "Aqueous Phase Oxidation of Sulfur(IV) by Hydrogen Peroxide, Methylhydroperoxide, and Peroxyacetic Acid," *J. Geophys. Res., 92,* 4171–4177 (1987).

Littlejohn, D., Y. Wang, and S.-G. Chang, "Oxidation of Aqueous Sulfite Ion by Nitrogen Dioxide," *Environ. Sci. Technol., 27,* 2162–2167 (1993).

Long, C. A., and B. H. Bielski, "Rate of Reaction of Superoxide Radical with Chloride-Containing Species," *J. Phys. Chem., 84,* 555–557 (1980).

Loomis, A. G., "Solubilities of Gases in Water," in *International Critical Tables*, Vol. III, pp. 255–261, McGraw-Hill, New York, 1928.

Lovejoy, E. R., N. S. Wang, and C. J. Howard, "Kinetic Studies of the Reactions of HSO with NO_2, NO, and O_2," *J. Phys. Chem., 91,* 5749–5755 (1987).

Lovejoy, E. R., T. P. Murrels, A. R. Ravishankara, and C. J. Howard, "Oxidation of CS_2 by Reaction with OH. 2. Yields of HO_2 and SO_2 in Oxygen," *J. Phys. Chem., 94,* 2386–2393 (1990).

Lovejoy, E. R., and D. R. Hanson, "Kinetics and Products of the Reaction $SO_3 + NH_3 + N_2$," *J. Phys. Chem., 100,* 4459–4465 (1996).

Lovejoy, E. R., D. R. Hanson, and L. G. Huey, "Kinetics and Products of the Gas Phase Reaction of SO_3 with Water," *J. Phys. Chem., 100,* 19911–19916 (1996).

Lusis, M. A., K. G. Anlauf, L. A. Barrie, and H. A. Wiebe, "Plume Chemistry Studies at a Northern Alberta Power Plant," *Atmos. Environ., 12,* 2429–2437 (1978).

Maahs, H. G., "Sulfur-Dioxide/Water Equilibria between 0° and 50°C: An Examination of Data at Low Concentrations," in *Heterogeneous Atmospheric Chemistry* (D. R. Schryer, Ed.), pp. 187–195, Geophysical Monograph 26, Am. Geophys. Union, Washington, DC, 1982.

MacLeod, H., S. M. Aschmann, R. Atkinson, E. C. Tuazon, J. A. Sweetman, A. M. Winer, and J. N. Pitts, Jr., "Kinetics and Mechanisms of the Gas Phase Reactions of the NO_3 Radical with a Series of Reduced Sulfur Compounds," *J. Geophys. Res., 91,* 5338–5346 (1986).

Mamou, A., J. Rabani, and D. Behar, "On the Oxidation of Aqueous Br^- by OH Radicals, Studied by Pulse Radiolysis," *J. Phys. Chem., 81,* 1447–1448 (1977).

Margitan, J. J., "Mechanism of the Atmospheric Oxidation of Sulfur Dioxide. Catalysis by Hydroxyl Radicals," *J. Phys. Chem., 88,* 3314–3318 (1984).

Marti, J. J., A. Jefferson, X. P. Cai, C. Richert, P. H. McMurry, and F. Eisele, "H_2SO_4 Vapor Pressure of Sulfuric Acid and Ammonium Sulfate Solutions," *J. Geophys. Res., 102,* 3725–3735 (1997).

Martin, D., J. L. Jourdain, and G. Le Bras, "Discharge Flow Measurements of the Rate Constants for the Reactions $OH + SO_2 + He$ and $HOSO_2 + O_2$ in Relation with the Atmospheric Oxidation of SO_2," *J. Phys. Chem., 90,* 4143–4147 (1986).

Martin, L. R., and D. E. Damschen, "Aqueous Oxidation of Sulfur Dioxide by Hydrogen Peroxide at Low pH," *Atmos. Environ., 15,* 1615–1621 (1981).

Martin, L. R., "Kinetic Studies of Sulfite Oxidation in Aqueous Solutions," in *SO_2, NO, and NO_2 Oxidation Mechanisms: Atmospheric Consideration*, Acid Precipitation Series, pp. 63–100 and references therein (J. I. Teasley, Series Ed.), Butterworth, Stoneham, MA, 1984.

Martin, L. R., M. W. Hill, A. F. Tai, and T. W. Good, "The Iron Catalyzed Oxidation of Sulfur(IV) in Aqueous Solution: Differing Effects of Organics at High and Low pH," *J. Geophys. Res., 96,* 3085–3097 (1991).

Martin, L. R., "Aqueous Sulfur(IV) Oxidation Revisited," in *Environmental Oxidants* (J. O. Nriagu and M. S. Simmons, Eds.), pp. 221–268, Wiley, New York, 1994.

Martinez, R. I., and J. T. Herron, "Gas-Phase Reaction of SO_2 with a Criegee Intermediate in the Presence of Water Vapor," *J. Environ. Health Sci., A16,* 623–636 (1981).

Martinez, R. I., and J. T. Herron, "Acid Precipitation: The Role of O_3–Alkene–SO_2 Systems in the Atmospheric Conversion of SO_2 to H_2SO_4 Aerosol," *J. Environ. Health Sci., A18,* 739–745 (1983).

McArdle, J. V., and M. R. Hoffmann, "Kinetics and Mechanism of the Oxidation of Aquated Sulfur Dioxide by Hydrogen Peroxide at Low pH," *J. Phys. Chem., 87,* 5425–5429 (1983).

McElroy, W. J., "The Aqueous Oxidation of SO_2 by OH Radicals," *Atmos. Environ., 20,* 323–330 (1986).

McElroy, W. J., and S. J. Waygood, "Kinetics of the Reaction of the SO_4^- Radical with SO_4^{2-}, $S_2O_8^{2-}$, H_2O, and Fe^{2+}," *J. Chem. Soc., Faraday Trans., 86,* 2557–2564 (1990).

McHenry, J. N., and R. L. Dennis, "The Relative Importance of Oxidation Pathways and Clouds to Atmospheric Ambient Sulfate Production As Predicted by the Regional Acid Deposition Model," *J. Appl. Meteorol., 33,* 890–905 (1994).

McMurry, P. H., and J. C. Wilson, "Droplet Phase (Heterogeneous) and Gas Phase (Homogeneous) Contributions to Secondary Ambient Aerosol Formation as Functions of Relative Humidity," *J. Geophys. Res., 88,* 5101–5108 (1983).

Meagher, J. F., K. J. Olszyna, and M. Luria, "The Effect of SO_2 Gas Phase Oxidation on Hydroxyl Smog Chemistry," *Atmos. Environ., 18,* 2095–2104 (1984).

Mellouki, A., J. L. Jourdain, and G. Le Bras, "Discharge Flow Study of the $CH_3S + NO_2$ Reaction Mechanism Using $Cl + CH_3SH$ as the CH_3S Source," *Chem. Phys. Lett., 148,* 231–236 (1988).

Mendiara, S. N., E. Ghibaudi, L. J. Perissinotti, and A. J. Colussi, "Free Radicals and Diradicals in the Reaction between Nitrous Acid and Bisulfite in Acid Aqueous Media," *J. Phys. Chem., 96,* 8089–8091 (1992).

Mésáros, E., D. J. Moore, and J. P. Lodge, Jr., "Sulfur Dioxide–Sulfate Relationships in Budapest," *Atmos. Environ., 11,* 345–349 (1977).

Mihelcic, D., D. Klemp, P. Msgen, W. W. Ptz, and A. Volz-Thomas, "Simultaneous Measurements of Peroxy and Nitrate Radicals at Schauinsland," *J. Atmos. Chem., 16,* 313–335 (1993).

Miller, D. F., D. Lamb, and A. W. Gertler, "SO_2 Oxidation in Cloud Drops Containing NaCl or Sea Salt as Condensation Nuclei," *Atmos. Environ., 21,* 991–993 (1987).

Millero, F. J., M. Gonzalez-Davila, and J. M. Santana-Casiano, "Reduction of Fe(III) with Sulfite in Natural Waters," *J. Geophys. Res., 100,* 7235–7244 (1995).

Mochida, M., H. Akimoto, H. van den Bergh, and M. J. Rossi, "Heterogeneous Kinetics of the Uptake of HOBr on Solid Alkali

Metal Halides at Ambient Temperature," *J. Phys. Chem. A, 102,* 4819–4828 (1998a).

Mochida, M., J. Hirokawa, Y. Kajii, and H. Akimoto, "Heterogeneous Reactions of Cl_2 with Sea Salts at Ambient Temperature: Implications for Halogen Exchange in the Atmosphere," *Geophys. Res. Lett., 25,* 3927–3930 (1998b).

Morokuma, K., and C. Muguruma, "Ab Initio Molecular Orbital Study of the Mechanism of the Gas Phase Reactions $SO_3 + H_2O$: Importance of the Second Water Molecule," *J. Am. Chem. Soc., 116,* 10316–10317 (1994).

Mozurkewich, M., P. H. McMurry, A. Gupta, and J. G. Calvert, "Mass Accommodation Coefficient for HO_2 Radicals on Aqueous Particles," *J. Geophys. Res., 92,* 4163–4170 (1987).

Mozurkewich, M., "Mechanisms for the Release of Halogens from Sea-Salt Particles by Free Radical Reactions," *J. Geophys. Res., 100,* 14199–14207 (1995).

Muir, P. S., K. A. Wade, B. H. Carter, T. V. Armentano, and R. A. Pribush, "Fog Chemistry at an Urban Midwestern Site," *J. Air Pollut. Control Assoc., 36,* 1359–1361 (1986).

Muir, P. S., "Fogwater Chemistry in a Wood-Burning Community, Western Oregon," *J. Air Waste Manage., 41,* 32–38 (1991).

Munger, J. W., D. J. Jacob, J. M. Waldman, M. R. Hoffmann, "Fogwater Chemistry in an Urban Atmosphere," *J. Geophys. Res., 88,* 5109–5121 (1983).

Munger, J. W., D. J. Jacob, and M. Hoffmann, "The Occurrence of Bisulfite–Aldehyde Addition Products in Fog- and Cloudwater," *J. Atmos. Chem., 1,* 335–350 (1984).

Munger, J. W., C. Tiller, and M. R. Hoffmann, "Identification of Hydroxymethanesulfonate in Fog Water," *Science, 231,* 247–249 (1986).

Murrells, T. P., E. R. Lovejoy, and A. R. Ravishankara, "Oxidation of CS_2 by Reaction with OH. 1. Equilibrium Constant for the Reaction $OH + CS_2 \leftrightarrow CS_2OH$ and the Kinetics of the $CS_2OH + O_2$ Reaction," *J. Phys. Chem., 94,* 2381–2386 (1990).

Nagarajan, V., and R. W. Fessenden, "Flash Photolysis of Transient Radicals. 1. X_2^- with X = Cl, Br, I, and SCN," *J. Phys. Chem., 89,* 2330–2335 (1985).

Neta, P., R. E. Huie, and A. B. Ross, "Rate Constants for Reactions of Inorganic Radicals in Aqueous Solution," *J. Phys. Chem. Ref. Data, 17,* 1027–1228 (1988).

Niki, H., and K. H. Becker, Eds., *The Tropospheric Chemistry of Ozone in the Polar Regions,* NATO ASI Series I: Global Environmental Change, Vol. 7, Springer-Verlag, Berlin, 1993.

NIST Standard Reference Data Base 40 NDRL/NIST Solution Kinetic Database, Version 2.0. Data compiled and evaluated by A. B. Ross, B. H. J. Bielski, G. V. Buxton, D. E. Cabeli, C. L. Greenstock, W. P. Helman, R. E. Huie, J. Grodkawski, and P. Neta. Database developed by W. G. Mallard, A. B. Moss, and W. P. Helman (1994).

Nolte, C. G., P. A. Solomon, T. Fall, L. G. Salmon, and G. R. Cass, "Seasonal and Spatial Characteristics of Formic and Acetic Acids Concentrations in the Southern California Atmosphere," *Environ. Sci. Technol., 31,* 2547–2553 (1997).

Norton, R. B., "Measurements of Formate and Acetate in Precipitation at Niwot Ridge and Boulder, Colorado," *Geophys. Res. Lett., 12,* 769–772 (1985).

Novakov, T., S. G. Chang, and A. B. Harker, "Sulfates as Pollution Particles: Catalytic Formation on Carbon (Soot) Particles," *Science, 186,* 259–261 (1974).

Olson, T. M., and M. R. Hoffmann, "Formation Kinetics, Mechanism, and Thermodynamics of Glyoxylic Acid–S(IV) Adducts," *J. Phys. Chem., 92,* 4246–4253 (1988a).

Olson, T. M., and M. R. Hoffmann, "Kinetics, Mechanism, and Thermodynamics of Glyoxal–S(IV) Adduct Formation," *J. Phys. Chem., 92,* 533–540 (1988b).

Olson, T. M., L. A. Torry, and M. R. Hoffmann, "Kinetics of the Formation of Hydroxyacetaldehyde–Sulfur(IV) Adducts at Low pH," *Environ. Sci. Technol., 22,* 1284–1289 (1988).

Olson, T. M., and M. R. Hoffmann, "Hydroxyalkylsulfonate Formation: Its Role as a S(IV) Reservoir in Atmospheric Water Droplets," *Atmos. Environ., 23,* 985–997 (1989).

O'Sullivan, D. W., M. Lee, B. C. Noone, and B. G. Heikes, "Henry's Law Constant Determinations for Hydrogen Peroxide, Methyl Hydroperoxide, Hydroxymethyl Hydroperoxide, Ethyl Hydroperoxide, and Peroxyacetic Acid," *J. Phys. Chem., 100,* 3241–3247 (1996).

Oum, K. W., M. J. Lakin, D. O. DeHaan, T. Brauers, and B. J. Finlayson-Pitts, "Formation of Molecular Chlorine from the Photolysis of Ozone and Aqueous Sea-Salt Particles," *Science, 279,* 74–75 (1998a).

Oum, K. W., M. J. Lakin, and B. J. Finlayson-Pitts, "Bromine Activation in the Troposphere by the Dark Reaction of O_3 with Seawater Ice," *Geophys. Res. Lett., 25,* 3923–3926 (1998b).

Overton, J. H., "Validation of the Hoffmann and Edwards' S(IV)–H_2O_2 Mechanism," *Atmos. Environ., 19,* 687–690 (1985).

Pandis, S. N., and J. H. Seinfeld, "Mathematical Modeling of Acid Deposition Due to Radiation Fog," *J. Geophys. Res., 94,* 12911–12923 (1989a).

Pandis, S. N., and J. H. Seinfeld, "Sensitivity Analysis of a Chemical Mechanism for Aqueous-Phase Atmospheric Chemistry," *94,* 1105–1126 (1989b).

Pandis, S. N., and J. H. Seinfeld, "The Smog–Fog–Smog Cycle and Acid Deposition," *J. Geophys. Res., 95,* 18489–18500 (1990).

Pandis, S. N., and J. H. Seinfeld, "Should Bulk Cloudwater or Fogwater Samples Obey Henry's Law?" *J. Geophys. Res., 96,* 10791–10798 (1991).

Pandis, S. N., and J. H. Seinfeld, Reply, *J. Geophys. Res., 97,* 6079–6081 (1992).

Pandis, S. N., J. H. Seinfeld, and C. Pilinis, "Heterogeneous Sulfate Production in an Urban Fog," *Atmos. Environ., 26A,* 2509–2522 (1992).

Pandis, S. N., L. M. Russell, and J. H. Seinfeld, "The Relationship between DMS Flux and CCN Concentration in Remote Marine Regions," *J. Geophys. Res., 99,* 16945–16957 (1994).

Patroescu, I. V., I. Barnes, and K. H. Becker, "FTIR Kinetic and Mechanistic Study of the Atmospheric Chemistry of Methyl Thiolformate," *J. Phys. Chem., 100,* 17207–17217 (1996).

Patroescu, I. V., I. Barnes, K. H. Becker, and N. Mihalopoulos, "FT-IR Product Study of the OH-Initiated Oxidation of DMS in the Presence of NO," *Atmos. Environ., 33,* 25–35 (1999).

Penkett, S. A., B. M. R. Jones, K. A. Brice, and A. E. J. Eggleton, "The Importance of Atmospheric Ozone and Hydrogen Peroxide in Oxidizing Sulphur Dioxide in Cloud and Rain Water," *Atmos. Environ., 13,* 123–137 (1979).

Phillips, J. A., M. Canagaratna, H. Goodfriend, and K. R. Leopold, "Microwave Detection of a Key Intermediate in the Formation of Atmospheric Sulfuric Acid: The Structure of H_2O–SO_3," *J. Phys. Chem., 99,* 501–504 (1995).

Plane, J. M. C., "Gas-Phase Atmospheric Oxidation of Biogenic Sulfur Compounds," in *Biogenic Sulfur in the Environment,* ACS Symposium Series No. 393 (E. S. Saltzman and W. J. Cooper, Eds.), Chap. 24, pp. 404–422, Am. Chem. Soc., Washington, DC, 1989.

Platt, U., and M. Hausmann, "Spectroscopic Measurement of the Free Radicals NO_3, BrO, IO, and OH in the Troposphere," *Res. Chem. Intermed., 20,* 557–578 (1994).

Ponche, J. L., Ch. George, and Ph. Mirabel, "Mass Transfer at the Air/Water Interface: Mass Accommodation Coefficients of SO_2, HNO_3, NO_2, and NH_3," *J. Atmos. Chem., 16,* 1–21 (1993).

Pruppacher, H. R., and J. D. Klett, *Microphysics of Clouds and Precipitation*, Reidel, Dordrecht, 1978.

Pszenny, A. A. P., W. C. Keene, D. J. Jacob, S. Fan, J. R. Maben, M. P. Zetwo, M. Springer-Young, and J. N. Galloway, "Evidence of Inorganic Chlorine Gases Other Than Hydrogen Chloride in Marine Surface Air," *Geophys. Res. Lett., 20*, 699–702 (1993).

Radojevic, M., "On the Discrepancy between Reported Studies of the Uncatalysed Aqueous Oxidation of SO_2 by O_2," *Environ. Technol. Lett., 5*, 549–566 (1984).

Rao, X., and J. L. Collett, Jr., "Behavior of S(IV) and Formaldehyde in a Chemically Heterogeneous Cloud," *Environ. Sci. Technol., 29*, 1023–1031 (1995).

Ray, A., I. Vassalli, G. Laverdet, and G. Le Bras, "Kinetics of the Thermal Decomposition of the CH_3SO_2 Radical and Its Reaction with NO_2 at 1 Torr and 298 K," *J. Phys. Chem., 100*, 8895–8900 (1996).

Reiner, Th., and F. Arnold, "Laboratory Flow Reactor Measurements of the Reaction $SO_3 + H_2O + M \rightarrow H_2SO_4 + M$: Implications for Gaseous H_2SO_4 and Aerosol Formation in the Plumes of Jet Aircraft," *Geophys. Res. Lett., 20*, 2659–2662 (1993).

Richards, L. W., "Airborne Chemical Measurements in Nighttime Stratus Clouds in the Los Angeles Basin," *Atmos. Environ., 29*, 27–46 (1995).

Roedel, W., "Measurement of Sulfuric Acid Saturation Vapor Pressure; Implications for Aerosol Formation by Heteromolecular Nucleation," *J. Aerosol Sci., 10*, 375–386 (1979).

Ross, A. B., and P. Neta, "Rate Constants for Reactions of Inorganic Radicals in Aqueous Solutions," NSRDS-NBS 65, U.S. Department of Commerce, Washington, DC, 1979.

Rudich, Y., R. K. Talukdar, and A. R. Ravishankara, "Reactive Uptake of NO_3 on Pure Water and Ionic Solutions," *J. Geophys. Res., 101*, 21023–21031 (1996).

Rudich, Y., R. K. Talukdar, and A. R. Ravishankara, "Multiphase Chemistry of NO_3 in the Remote Troposphere," *J. Geophys. Res., 103*, 16133–16143 (1998).

Sachweh, M., and P. Koepke, "Radiation Fog and Urban Climate," *Geophys. Res. Lett., 22*, 1073–1076 (1995).

Sakugawa, H., and I. R. Kaplan, "H_2O_2 and O_3 in the Atmosphere of Los Angeles and Its Vicinity: Factors Controlling Their Formation and Their Role as Oxidants of SO_2," *J. Geophys. Res., 94*, 12957–12973 (1989).

Sakugawa, H., I. R. Kaplan, W. Tsai, and Y. Cohen, "Atmospheric Hydrogen Peroxide," *Environ. Sci. Technol., 24*, 1452–1461 (1990).

Sakugawa, H., I. R. Kaplan, and L. S. Shepard, "Measurements of H_2O_2, Aldehydes, and Organic Acids in Los Angeles Rainwater: Their Sources and Deposition Rates," *Atmos. Environ., 27B*, 203–219 (1993).

Sakugawa, H., and I. R. Kaplan, "Stable Carbon Isotope Measurements of Atmospheric Organic Acids in Los Angeles, California," *Geophys. Res. Lett., 22*, 1509–1512 (1995).

Sander, R., J. Lelieveld, and P. J. Crutzen, "Modelling of the Nighttime Nitrogen and Sulfur Chemistry in Size Resolved Droplets of an Orographic Cloud," *J. Atmos. Chem., 20*, 89–116 (1995).

Sander, R., and P. J. Crutzen, "Model Study Indicating Halogen Activation and Ozone Destruction in Polluted Air Masses Transported to the Sea," *J. Geophys. Res., 101*, 9121–9138 (1996).

Sanhueza, E., L. Figueroa, and M. Santana, "Atmospheric Formic and Acetic Acids in Venezuela," *Atmos. Environ., 30*, 1861–1873 (1996).

Sauer, M. C., Jr., W. G. Brown, and E. J. Hart, "$O(^3P)$ Atom Formation by the Photolysis of Hydrogen Peroxide in Alkaline Aqueous Solutions," *J. Phys. Chem., 88*, 1398–1400 (1984).

Saxena, V. K., and N.-H. Lin, "Cloud Chemistry Measurements and Estimates of Acidic Deposition on an Above Cloudbase Coniferous Forest," *Atmos. Environ., 24A*, 329–352 (1990).

Scaiano, J. C., M. Barra, G. Calabrese, and R. Sinta, "Photochemistry of 1,2-Dibromoethane in Solution. A Model for the Generation of Hydrogen Bromide," *J. Chem. Soc., Chem. Commun.*, 1418–1419 (1992).

Schryer, D. R., R. S. Rogowski, and W. R. Cofer III, "The Reaction of Nitrogen Oxides with SO_2 in Aqueous Aerosols," *Atmos. Environ., 17*, 666 (1983).

Schwartz, S. E., "Gas–Aqueous Reactions of Sulfur and Nitrogen Oxides in Liquid-Water Clouds," in *SO_2, NO, and NO_2 Oxidation Mechanism: Atmospheric Considerations* (J. G. Calvert, Ed.), Acid Precipitation Series, Vol. 3, pp. 173–208 and references therein (J. I. Teasley, Series Ed.), Butterworth, Stoneham, MA, 1984a.

Schwartz, S. E., "Gas- and Aqueous-Phase Chemistry of HO_2 in Liquid Water Clouds," *J. Geophys. Res., 89*, 11589–11598 (1984b).

Schwartz, S. E., and J. E. Freiberg, "Mass-Transport Limitation to the Rate of Reaction of Gases in Liquid Droplets: Application to Oxidation of SO_2 in Aqueous Solutions," *Atmos. Environ., 15*, 1129–1144 (1981).

Schwartz, S. E., and W. H. White, "Solubility Equilibria of the Nitrogen Oxides and Oxyacids in Dilute Aqueous Solution," *Adv. Environ. Sci. Eng., 4*, 1–45 (1981).

Schwartz, S. E., and W. H. White, "Kinetics of Reactive Dissolution of Nitrogen Oxides into Aqueous Solution," *Adv. Environ. Sci. Technol., 12*, 1–116 (1983).

Sedlak, D. L., and J. Hoigné, "Oxidation of S(IV) in Atmospheric Water by Photooxidants and Iron in the Presence of Copper," *Environ. Sci. Technol., 28*, 1898–1906 (1994).

Sehested, K., J. Holcman, and E. J. Hart, "Rate Constants and Products of the Reactions of e_{aq}^-, O_2^-, and H with Ozone in Aqueous Solutions," *J. Phys. Chem., 87*, 1951–1954 (1983).

Seigneur, C., and P. Saxena, "A Theoretical Investigation of Sulfate Formation in Clouds," *Atmos. Environ., 22*, 101–115 (1988).

Seinfeld, J. H., *Atmospheric Chemistry and Physics of Air Pollution*, Wiley, New York, 1986.

Sham, Y. Y., and J. A. Joens, "Temperature Dependent Near-UV Molar Absorptivities of Several Small Aldehydes in Aqueous Solution," *Spectrochim. Acta, 51A*, 247–251 (1995).

Shen, G., M. Suto, and L. C. Lee, "Reaction Rate Constant of $SO_3 + NH_3$ in the Gas Phase," *J. Geophys. Res., 95*, 13981–13984 (1990).

Shorter, J. A., W. J. De Bruyn, J. Hu, E. Swartz, P. Davidovits, D. R. Worsnop, M. S. Zahniser, and C. E. Kolb, "Bubble Column Apparatus for Gas–Liquid Heterogeneous Chemistry Studies," *Environ. Sci. Technol., 29*, 1171–1178 (1995).

Shoute, L. C. T., Z. B. Alfassi, P. Neta, and R. E. Huie, "Temperature Dependence of the Rate Constants for Reaction of Dihalide and Azide Radicals with Inorganic Reductants," *J. Phys. Chem., 95*, 3238–3242 (1991).

Siefert, R. L., S. M. Webb, and M. R. Hoffmann, "Determination of Photochemically Available Iron in Ambient Aerosols," *J. Geophys. Res., 101*, 14441–14449 (1996).

Siefert, R. L., A. M. Johansen, M. R. Hoffmann, and S. O. Pehkonen, "Measurements of Trace Metal (Fe, Cu, Mn, Cr) Oxidation States in Fog and Stratus Clouds," *J. Air Waste Manage. Assoc., 48*, 128–143 (1998).

Sievering, H., J. Boatman, J. Galloway, W. Keene, Y. Kim, M. Luria, and J. Ray, "Heterogeneous Sulfur Conversion in Sea-Salt Aerosol Particles: The Role of Aerosol Water Content and Size Distribution," *Atmos. Environ., 25A*, 1479–1487 (1991).

Sievering, H., E. Gorman, Y. Kim, T. Ley, W. Seidl, and J. Boatman, "Heterogeneous Conversion Contribution to the Sulfate Observed over Lake Michigan," *Atmos. Environ., 28*, 367–370 (1994).

Sievering, H., E. Gorman, T. Ley, A. Pszenny, M. Springer-Young, J. Boatman, Y. Kim, C. Nagamoto, and D. Wellman, "Ozone Oxidation of Sulfur in Sea-Salt Aerosol Particles during the Azores Marine Aerosol and Gas Exchange Experiment," *J. Geophys. Res., 100*, 23075–23081 (1995).

Sillén, G. H., and A. E. Martell, *Stability Constants of Metal-Ion Complexes*, Special Publication 17, Chemical Society, London, 1964.

Singh, H. B., G. L. Gregory, A. Anderson, E. Browell, G. W. Sachse, D. D. Davis, J. Crawford, J. D. Bradshaw, R. Talbot, D. R. Blake, D. Thornton, R. Newell, and J. Merrill, "Low Ozone in the Marine Boundary Layer of the Tropical Pacific Ocean: Photochemical Loss, Chlorine Atoms, and Entrainment," *J. Geophys. Res., 101*, 1907–1917 (1996).

Sørensen, P. E., and V. S. Andersen, "The Formaldehyde–Hydrogen Sulphite System in Alkaline Aqueous Solution: Kinetics, Mechanism, and Equilibria," *Acta Chem. Scand., 24*, 1301–1306 (1970).

Sørensen, S., H. Falbe-Hansen, M. Mangoni, J. Hjorth, and N. R. Jensen, "Observation of DMSO and $CH_3S(O)OH$ from the Gas Phase Reaction between DMS and OH," *J. Atmos. Chem., 24*, 299–315 (1996).

Spicer, C. W., E. G. Chapman, B. J. Finlayson-Pitts, R. A. Plastridge, J. M. Hubbe, J. D. Fast, and C. M. Berkowitz, "Unexpectedly High Concentrations of Molecular Chlorine in Coastal Air," *Nature, 394*, 353–356 (1998).

Stachnik, R. A., and M. J. Molina, "Kinetics of the Reactions of SH Radicals with NO_2 and O_2," *J. Phys. Chem., 91*, 4603–4606 (1987).

Staffelbach, T. A., and G. L. Kok, "Henry's Law Constants for Aqueous Solutions of Hydrogen Peroxide and Hydroxymethyl Hydroperoxide," *J. Geophys. Res., 98*, 12713–12717 (1993).

Stickel, R. E., J. M. Nicovich, S. Wang, Z. Zhao, and P. H. Wine, "Kinetic and Mechanistic Study of the Reaction of Atomic Chlorine with Dimethyl Sulfide," *J. Phys. Chem., 96*, 9875–9883 (1992).

Stickel, R. E., M. Chin, E. P. Daykin, A. J. Hynes, P. H. Wine, and T. J. Wallington, "Mechanistic Studies of the OH-Initiated Oxidation of CS_2 in the Presence of O_2," *J. Phys. Chem., 97*, 13653–13661 (1993).

Stockwell, W. R., and J. G. Calvert, "The Mechanism of the OH–SO_2 Reaction," *Atmos. Environ., 17*, 2231–2235 (1983).

Stockwell, W. R., P. Middleton, J. S. Chang, and X. Tang, "The Second Generation Regional Acid Deposition Model Chemical Mechanism for Regional Air Quality Modeling," *J. Geophys. Res., 95*, 16343–16367 (1990).

Stockwell, W. R., "The Effect of Gas-Phase Chemistry on Aqueous-Phase Sulfur Dioxide Oxidation Rates," *J. Atmos. Chem., 19*, 317–329 (1994).

Takenaka, N., A. Ueda, and Y. Maeda, "Acceleration of the Rate of Nitrite Oxidation by Freezing in Aqueous Solution," *Nature, 358*, 736–738 (1992).

Takenaka, N., A. Ueda, T. Daimon, H. Bandow, T. Dohmaru, and Y. Maeda, "Acceleration Mechanism of Chemical Reaction by Freezing: The Reaction of Nitrous Acid with Dissolved Oxygen," *J. Phys. Chem., 100*, 13874–13884 (1996).

Talbot, R. W., K. M. Beecher, R. C. Harriss, and W. R. Cofer III, "Atmospheric Geochemistry of Formic and Acetic Acids at a Mid-Latitude Temperate Site," *J. Geophys. Res., 93*, 1638–1652 (1988).

Talbot, R. W., B. W. Mosher, B. G. Heikes, D. J. Jacob, J. W. Munger, B. C. Daube, W. C. Keene, J. R. Maben, and R. S. Artz, "Carboxylic Acids in the Rural Continental Atmosphere over the Eastern United States during the Shenandoah Cloud and Photochemistry Experiment," *J. Geophys. Res., 100*, 9335–9343 (1995).

Tanaka, N., D. M. Rye, Y. Xiao, and A. C. Lasaga, "Use of Stable Sulfur Isotope Systematics for Evaluating Oxidation Reaction Pathways and In-Cloud-Scavenging of Sulfur Dioxide in the Atmosphere," *Geophys. Res. Lett., 21*, 1519–1522 (1994).

Tang, Y., R. P. Thorn, R. L. Mauldin, III, and P. H. Wine, "Kinetics and Spectroscopy of the SO_4^- Radical in Aqueous Solution," *J. Photochem. Photobiol., 44*, 243–258 (1988).

Tanner, R. L., and D. E. Schorran, "Measurements of Gaseous Peroxides near the Grand Canyon—Implication for Summertime Visibility Impairment from Aqueous-Phase Secondary Sulfate Formation," *Atmos. Environ., 29*, 1113–1122 (1995).

Taube, H., "Photochemical Reactions of Ozone in Solution," *Trans. Faraday Soc., 53*, 656–665 (1957).

Thomas, K., A. Volz-Thomas, D. Mihelcic, H. G. J. Smit, and D. Kley, "On the Exchange of NO_3 Radicals with Aqueous Solutions: Solubility and Sticking Coefficient," *J. Atmos. Chem., 29*, 17–43 (1998).

Toumi, R., "BrO as a Sink for Dimethylsulphide in the Marine Atmosphere," *Geophys. Res. Lett., 21*, 117–120 (1994).

Treinin, A., "The Photochemistry of Oxyanions," *Isr. J. Chem., 8*, 103–113 (1970).

Troy, R. C., and D. W. Margerum, "Non-Metal Redox Kinetics—Hypochlorite and Hypochlorous Acid Reactions with Iodide and with Sulfite and the Hydrolysis of Bromosulfate," *Inorg. Chem., 30*, 3538–3543 (1991).

Tuckermann, M., R. Ackermann, C. Gölz, H. Lorenzen-Schmidt, T. Senne, J. Stutz, B. Trost, W. Unold, and U. Platt, "DOAS-Observation of Halogen Radical-Catalysed Arctic Boundary Layer Ozone Destruction during the ARCTOC-Campaigns 1995 and 1996 in Ny-Ålesund, Spitsbergen," *Tellus, 49B*, 533–555 (1997).

Turnipseed, A. A., S. B. Barone, and A. R. Ravishankara, "Observation of CH_3S Addition to O_2 in the Gas Phase," *J. Phys. Chem., 96*, 7502–7505 (1992).

Turnipseed, A. A., and A. R. Ravishankara, "The Atmospheric Oxidation of Dimethyl Sulfide: Elementary Steps in a Complex Mechanism," in *Dimethylsulphide: Oceans, Atmosphere, and Climate* (G. Restelli and G. Angeletti, Eds.), pp. 185–195, Kluwer Academic, Dordrecht/Norwell, MA, 1993.

Turnipseed, A. A., S. B. Barone, and A. R. Ravishankara, "Reactions of CH_3S and CH_3SOO with O_3, NO_2, and NO," *J. Phys. Chem., 97*, 5926–5934 (1993).

Turnipseed, A. A., S. B. Barone, and A. R. Ravishankara, "Reaction of OH with Dimethyl Sulfide. 2. Products and Mechanisms," *J. Phys. Chem., 100*, 14703–14713 (1996).

Tyndall, G. S., and A. R. Ravishankara, "Kinetics and Mechanism of the Reactions of CH_3S with O_2 and NO_2 at 298 K," *J. Phys. Chem., 93*, 2426–2435 (1989).

Tyndall, G. S., and A. R. Ravishankara, "Atmospheric Oxidation of Reduced Sulfur Species," *Int. J. Chem. Kinet., 23*, 483–527 (1991).

Urbanski, S. P., R. E. Stickel, and P. H. Wine, "Mechanistic and Kinetic Study of the Gas-Phase Reaction of the Hydroxyl Radical with Dimethyl Sulfoxide," *J. Phys. Chem. A, 102*, 10522–10529 (1998).

Utter, R. G., J. B. Burkholder, C. J. Howard, and A. R. Ravishankara, "Measurement of the Mass Accommodation Coefficient of Ozone on Aqueous Surfaces," *J. Phys. Chem., 96*, 4973–4979 (1992).

Van Doren, J. M., L. R. Watson, P. Davidovits, D. R. Worsnop, M. S. Zahniser, and C. E. Kolb, "Temperature Dependence of the Uptake Coefficients of HNO_3, HCl, and N_2O_5 by Water Droplets," *J. Phys. Chem., 94*, 3265–3269 (1990).

Van Krevelen, D. W., P. J. Hoftijzer, and F. J. Huntjens, "Composition and Vapor Pressures of Aqueous Solutions of Ammonia, Carbon Dioxide, and Hydrogen Sulfide," *Recl. Trav. Chim. Pays-Bas, 68*, 191–216 (1949).

Vogt, R., P. J. Crutzen, and R. Sander, "A Mechanism for Halogen Release from Sea-Salt Aerosol in the Remote Marine Boundary Layer," *Nature, 383*, 327–330 (1996).

Wagner, I., J. Karthauser, and H. Strehlow, "On the Decay of the Dichloride Anion Cl_2^- in Aqueous Solution," *Ber. Bunsenges. Phys. Chem.*, 90, 861–867 (1986).

Wahner, A., and A. R. Ravishankara, "The Kinetics of the Reaction of OH with COS," *J. Geophys. Res.*, 92, 2189–2194 (1987).

Walcek, C. J., W. R. Stockwell, and J. S. Chang, "Theoretical Estimates of the Dynamic, Radiative, and Chemical Effects of Clouds on Tropospheric Trace Gases," *Atmos. Res.*, 25, 53–69 (1990).

Walcek, C. J., H.-H. Yuan, and W. R. Stockwell, "The Influence of Aqueous-Phase Chemical Reactions on Ozone Formation in Polluted and Nonpolluted Clouds," *Atmos. Environ.*, 31, 1221–1237 (1997).

Wallington, T. J., T. Ellermann, and O. J. Nielsen, "Atmospheric Chemistry of Dimethyl Sulfide: UV Spectra and Self-Reaction Kinetics of CH_3SCH_2 and $CH_3SCH_2O_2$ Radicals and Kinetics of the Reactions $CH_3SCH_2 + O_2 \rightarrow CH_3SCH_2O_2$ and $CH_3SCH_2O_2 + NO \rightarrow CH_3SCH_2O + NO_2$," *J. Phys. Chem.*, 97, 8442–8449 (1993).

Wang, N. S., and C. J. Howard, "Kinetics of the Reactions of HS and HSO with O_3," *J. Phys. Chem.*, 94, 8787–8794 (1990).

Wang, T., and D. W. Margerum, "Kinetics of Reversible Chlorine Hydrolysis: Temperature Dependence and General Acid/Base-Assisted Mechanisms," *Inorg. Chem.*, 33, 1050–1055 (1994).

Wang, T. X., M. D. Kelley, J. N. Cooper, R. C. Beckwith, and D. W. Margerum, "Equilibrium, Kinetic, and UV-Spectral Characteristics of Aqueous Bromine Chloride, Bromine and Chlorine Species," *Inorg. Chem.*, 33, 5872–5878 (1994).

Warneck, P., and C. Wurzinger, "Product Quantum Yields for the 305-nm Photodecomposition of NO_3^- in Aqueous Solution," *J. Phys. Chem.*, 92, 6278–6283 (1988).

Waygood, S. J., and W. J. McElroy, "Spectroscopy and Decay Kinetics of the Sulfite Radical Anion in Aqueous Solution," *J. Chem. Soc. Faraday Trans.*, 88, 1525–1530 (1992).

Weathers, K. C., G. E. Likens, F. H. Bormann, S. H. Bicknell, B. T. Bormann, B. C. Daube, Jr., J. S. Eaton, J. N. Galloway, W. C. Keene, K. D. Kimball, W. H. McDowell, T. G. Siccama, D. Smiley, and R. A. Tarrant, "Cloudwater Chemistry from Ten Sites in North America," *Environ. Sci. Technol.*, 22, 1018–1026 (1988).

Weinstein, J., and B. H. J. Bielski, "Kinetics of the Interaction of HO_2 and O_2^- Radical with Hydrogen Peroxide. The Huber-Weiss Reaction," *J. Am. Chem. Soc.*, 101, 58–62 (1979).

Weinstein-Lloyd, J., and S. E. Schwartz, "Low-Intensity Radiolysis Study of Free-Radical Reactions in Cloudwater: H_2O_2 Production and Destruction," *Environ. Sci. Technol.*, 25, 791–800 (1991).

Weschler, C. J., M. L. Mandlich, and T. E. Graedel, "Speciation, Photosensitivity, and Reactions of Transition Metal Ions in Atmospheric Droplets," *J. Geophys. Res.*, 91, 5189–5204 (1986).

Wine, P. H., R. L. Mauldin, III, and R. P. Thorn, "Kinetics and Spectroscopy of the NO_3 Radical in Aqueous Ceric Nitrate–Nitric Acid Solutions," *J. Phys. Chem.*, 92, 1156–1162 (1988).

Wine, P. H., Y. Tang, R. P. Thorn, J. R. Wells, and D. D. Davis, "Kinetics of Aqueous Phase Reactions of the SO_4^- Radical with Potential Importance in Cloud Chemistry," *J. Geophys. Res.*, 94, 1085–1094 (1989).

Wingenter, O. W., M. K. Kubo, N. J. Blake, T. W. Smith, D. R. Blake, and F. S. Rowland, "Hydrocarbon and Halocarbon Measurements as Photochemical and Dynamical Indicators of Atmospheric Hydroxyl, Atomic Chlorine, and Vertical Mixing Obtained during Lagrangian Flights," *J. Geophys. Res.*, 101, 4331–4340 (1996).

Winiwarter, W., B. Brantner, and H. Puxbaum, "Comment on 'Should Bulk Cloudwater or Fogwater Samples Obey Henry's Law?' by S. N. Pandis and J. H. Seinfeld," *J. Geophys. Res.*, 97, 6075–6078 (1992).

Worsnop, D. R., M. S. Zahniser, C. E. Kolb, J. A. Gardner, L. R. Watson, J. M. Van Doren, J. T. Jayne, and P. Davidovits, "Temperature Dependence of Mass Accommodation of SO_2 and H_2O_2 on Aqueous Surfaces," *J. Phys. Chem.*, 93, 1159–1172 (1989).

Wurzler, S., A. I. Flossmann, H. R. Pruppacher, and S. E. Schwartz, "The Scavenging of Nitrate by Clouds and Precipitation," *J. Atmos. Chem.*, 20, 259–280 (1995).

Xu, H., P. J. Wentworth, N. W. Howell, and J. A. Joens, "Temperature Dependent Near-UV Molar Absorptivities of Aliphatic Aldehydes and Ketones in Aqueous Solution," *Spectrochim. Acta, Part A*, 49A, 1171–1178 (1993).

Yermakov, A. N., B. M. Zhitomirsky, G. A. Poskrebyshev, and S. I. Stoliarov, "Kinetic Study of SO_5^- and HO_2 Radical Reactivity in Aqueous Phase Bisulfite Oxidation," *J. Phys. Chem.*, 99, 3120–3127 (1995).

Zafiriou, O. C., "Natural Water Photochemistry," in *Chemical Oceanography* (J. P. Riley and R. Chester, Eds.), Vol. 8, pp. 339–379, Academic Press, London, 1983.

Zellner, R., M. Exner, and H. Herrmann, "Absolute OH Quantum Yields in the Laser Photolysis of Nitrate, Nitrite, and Dissolved H_2O_2 at 308 and 351 nm in the Temperature Range 278–353 K," *J. Atmos. Chem.*, 10, 411–425 (1990).

Zellner, R., and H. Herrmann, "Free Radical Chemistry of the Aqueous Atmospheric Phase," in *Spectroscopy in Environmental Science* (R. J. H. Clark and R. E. Hester, Eds.), Chap. 9, pp. 381–451, Wiley, New York, 1995.

Zepp, R. G., J. Hoigné, and H. Bader, "Nitrate-Induced Photooxidation of Trace Organic Chemicals in Water," *Environ. Sci. Technol.*, 21, 443–450 (1987).

Zepp, R. G., B. C. Faust, and J. Hoigné, "Hydroxyl Radical Formation in Aqueous Reactions (pH 3–8) of Iron(II) with Hydrogen Peroxide: The Photo-Fenton Reaction," *Environ. Sci. Technol.*, 26, 313–319 (1992).

Zepp, R. G., and M. O. Andreae, "Factors Affecting the Photochemical Production of Carbonyl Sulfide in Seawater," *Geophys. Res. Lett.*, 21, 2813–2816 (1994).

Zhao, Z., R. E. Stickel, and P. H. Wine, "Branching Ratios for Methyl Elimination in the Reactions of OD Radicals and Cl Atoms with CH_3SCH_3," *Chem. Phys. Lett.*, 251, 59–66 (1996).

Zhou, X., and Y.-N. Lee, "Aqueous Solubility and Reaction Kinetics of Hydroxymethyl Hydroperoxide," *J. Phys. Chem.*, 96, 265–272 (1992).

Zhu, X., J. M. Prospero, D. L. Savoie, F. J. Millero, R. G. Zika, and E. S. Saltzman, "Photoreduction of Iron(III) in Marine Mineral Aerosol Solutions," *J. Geophys. Res.*, 98, 9039–9046 (1993).

Zhu, X. R., J. M. Prospero, and F. J. Millero, "Diel Variability of Soluble Fe(II) and Soluble Total Fe in North African Dust in the Trade Winds at Barados," *J. Geophys. Res.*, 102, 21297–21305 (1997).

Zhuang, G., Z. Yi, R. A. Duce, and P. R. Brown, "Link between Iron and Sulphur Cycles Suggested by Detection of Fe(II) in Remote Marine Aerosols," *Nature*, 355, 537–539 (1992).

Zuo, Y., and J. Hoigné, "Formation of Hydrogen Peroxide and Depletion of Oxalic Acid in Atmospheric Water by Photolysis of Iron(III)–Oxalato Complexes," *Environ. Sci. Technol.*, 26, 1014–1022 (1992).

Zuo, Y., and J. Hoigné, "Evidence for Photochemical Formation of H_2O_2 and Oxidation of SO_2 in Authentic Fog Water," *Science*, 260, 71–73 (1993).

CHAPTER

9

Particles in the Troposphere

The most obvious characteristic of air pollution is the loss of visibility. This is primarily due to suspended airborne particles, which scatter light efficiently, giving the atmosphere a "hazy" appearance. As discussed in Chapter 14, scattering of light has direct effects on climate, altering the amount of solar radiation reaching the earth's surface; these suspended particles also contribute indirectly via their effects on clouds by acting as cloud condensation nuclei. Particles may also lead to heating of the lower atmosphere if they contain light absorbers such as elemental carbon, which strongly absorbs visible light, or mineral dusts, which absorb long-wavelength infrared (see Chapter 14.C.1). (For a series of papers on particle characteristics and their relationship to radiative forcing and climate, see the July 20, 1998, issue of *Journal of Geophysical Research*.) In addition to visibility and climate implications, these particles have significant health impacts which have been increasingly recognized (see Chapter 2.A.5). Finally, particles may act as sinks of reactive species such as HO_2, particularly in remote regions, hence affecting the chemistry of the gas phase as well (e.g., Saylor, 1997).

As a result of these widespread implications, understanding their direct sources, their formation from chemical reactions in air, their fates, and how their physical and chemical properties determine health and visibility impacts is critical. While many of the overall chemical and physical characteristics of particles have been elucidated, as we shall see in this chapter, there remain large gaps in our knowledge in areas central to policy and regulatory issues (National Research Council, 1998). As a result, this is a particularly active and rapidly evolving area of research in atmospheric chemistry.

A. PHYSICAL PROPERTIES

1. Some Definitions

Particles, or particulate matter, may be solid or liquid, with diameters between ~ 0.002 and ~ 100 μm. The lower end of the size range is not sharply defined because there is no accepted criterion at which a cluster of molecules becomes a particle. The upper end corresponds to the size of fine drizzle or very fine sand; these particles are so large that they quickly fall out of the atmosphere and hence do not remain suspended for significant periods of time. There are, of course, larger particles produced in the atmosphere (e.g., raindrops, ~ 1 mm, and hail, $\sim 1-20$ mm), but their rapid fallout precludes, for all practical purposes, their inclusion in the definition of atmospheric particles. As we shall see, the most important particles with respect to atmospheric chemistry and physics are in the 0.002- to 10-μm range.

Aerosols are defined as relatively stable suspensions of solid or liquid particles in a gas. Thus aerosols differ from particles in that an aerosol includes both the particles and the gas in which they are suspended. However, while this is the rigorous definition of aerosols, one should note that the term is often used in the atmospheric chemistry literature to denote just the particles.

Particles may be either directly emitted into the atmosphere or formed there by chemical reactions; we refer to these as primary and secondary particles, respectively. The relative importance of primary and secondary particles will clearly depend on the phenomena examined, the geographical location with its particular mix of emissions, and the atmospheric chemistry.

There are a number of properties of particles that are important for their role in atmospheric processes. These include, in addition to their number concentration, their mass, size, chemical composition, and aerodynamic and optical properties. Of these, size is the most important; it is related not only to the source of the particles (see later) but also to their effects on health, visibility, and climate.

Atmospheric particles are usually referred to as having a radius or a diameter, implying they are spherical. However, many particles in the atmosphere have quite

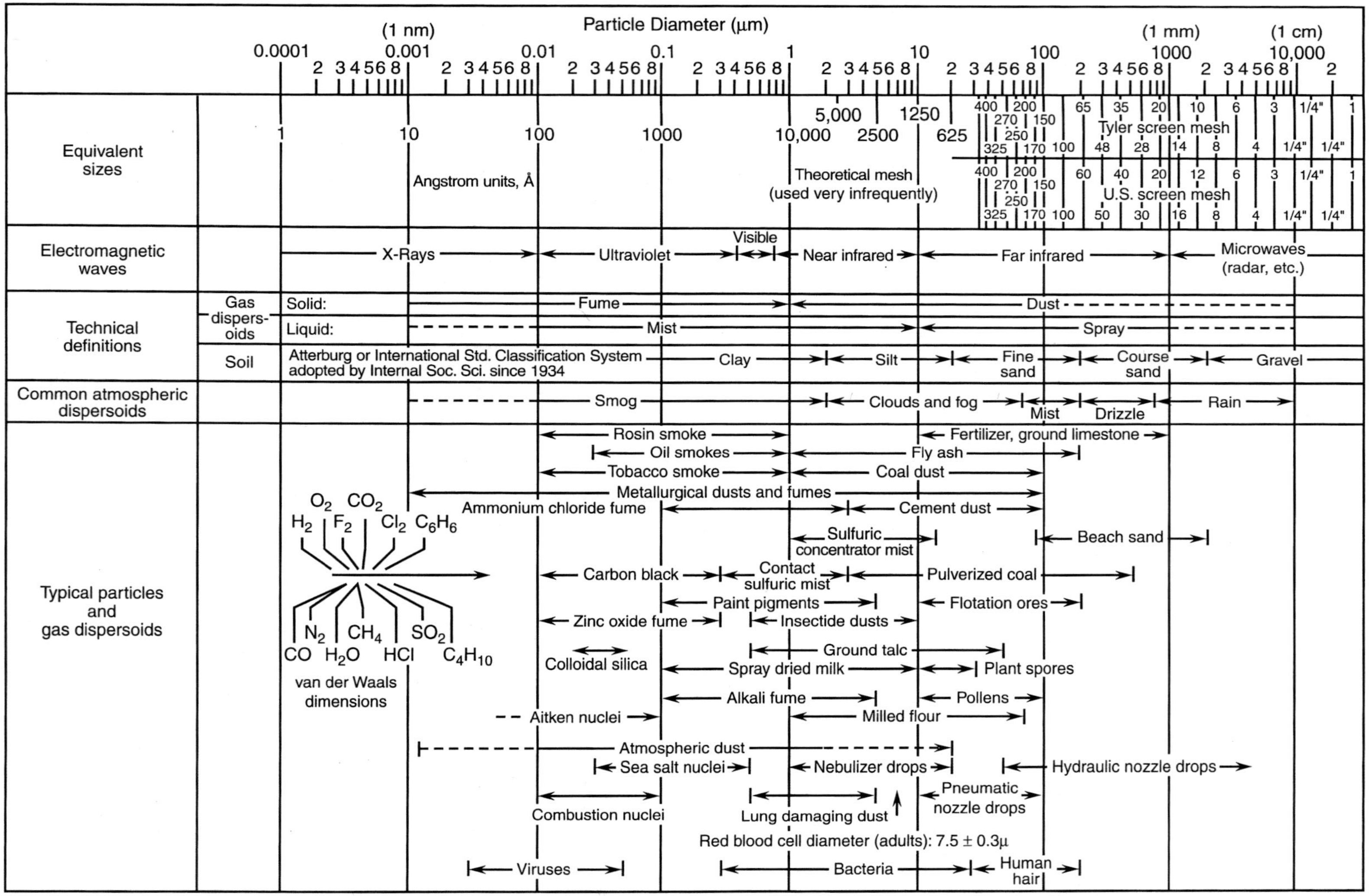

FIGURE 9.1 Some characteristics of particles and aerosols in ambient atmospheres and industrial settings (adapted from Lapple, 1961).

irregular shapes for which geometrical radii and diameters are not meaningful. Some means of expressing the size of such particles is essential since many important properties of the particle such as volume, mass, and settling velocity depend on the size. In practice, the size of such irregularly shaped particles is expressed in terms of some sort of *equivalent* or *effective* diameter that depends on a physical, rather than a geometrical, property.

There are several different types of effective diameters. One of the most commonly used is the *aerodynamic diameter*, D_a, which is defined as the diameter of a sphere of unit density (1 g cm^{-3}) that has the same terminal falling speed in air as the particle under consideration. This effective diameter is particularly useful because it determines the residence time in the air and it reflects the various regions of the respiratory system in which particles of different sizes become deposited. D_a is given by Eq. (A):

$$D_a = D_g k \sqrt{\frac{\rho_p}{\rho_0}}. \qquad (A)$$

D_g is the geometric diameter, ρ_p is the density of the particle, neglecting the buoyancy effects of air, ρ_0 is the reference density (1 g cm^{-3}), and k is a shape factor, which is 1.0 in the case of a sphere. Because of the effect of particle density on the aerodynamic diameter, a spherical particle of high density will have a larger aerodynamic diameter than its geometric diameter. However, for most substances, $\rho_p < 10$ so that the difference is less than a factor of ~ 3 (Lawrence Berkeley Laboratory, 1979). Particle densities are often lower than bulk densities of pure substances due to voids, pores, and cracks in the particles.

Throughout this chapter, we use the term *diameter* with the understanding that it is the aerodynamic diameter of the particle unless stated otherwise.

Another type of diameter commonly used is the *Stokes diameter*, D_s. This is defined as the diameter of a sphere that has the same density and settling velocity as the particle. Thus Stokes diameters are all based on settling velocities, whereas the aerodynamic diameter (D_a) also includes a standardized density of unity.

Figure 9.1 summarizes some of the characteristics of particles and aerosols encountered in both environmental and industrial atmospheres (Lapple, 1961).

This chapter treats those aerosol phenomena that are known or believed to be important in atmospheric chemistry. For treatment of related, but specialized, topics, a number of excellent references are available. The classic works on aerosol physics are *The Mechanics of Aerosols* by the late N. A. Fuchs (1964) and *Highly Dispersed Aerosols* (Fuchs and Sutugin, 1970). The chemical engineering aspects of aerosols are emphasized in S. K. Friedlander's book *Smoke, Dust, and Haze: Fundamentals of Aerosol Behavior* (1977); several topics, including photophoresis, thermophoresis, and coagulation, are included in *Aerosol Science*, edited by C. N. Davies (1966); and health considerations are treated in *Inhalation Studies* by R. F. Phalen (1984) and in *Pulmonary Toxicology of Respirable Particles* (Sanders et al., 1980). Finally, several works deal with laboratory generation and characterization of aerosols, including those by Hinds (1982), Mercer (1973), Willeke (1980), and Liu (1976).

2. Size Distributions

a. Number, Mass, Surface, and Volume Distributions

The atmosphere, whether in remote or urban areas, always contains significant concentrations of particles, up to 10^8 cm^{-3}. These may have diameters anywhere within the entire range from molecular clusters to ~ 100 μm. Because the size of atmospheric particles plays such an important role in both their chemistry and physics in the atmosphere as well as in their effects, it is important to know the distribution of sizes. We thus consider first how these size distributions are characterized.

An obvious way to express the distribution of particle sizes found in the atmosphere would be to plot, in the form of a histogram, the number (ΔN) of particles found in certain arbitrarily chosen intervals of diameter (ΔD), for example, from 0.002 to 0.01 μm, 1–10 μm, and so on. However, since in the atmosphere there tend to be a much greater number of small particles relative to large particles, a linear plot of ΔN against ΔD would give what would appear to be a narrow spike at the origin whose details could not be distinguished. This can be seen in Fig. 9.2, where typical impactor data from Wesolowski et al. (1980) for a sample consisting of Arizona road dust have been plotted; the original data in the form of a mass distribution have been converted to a number distribution using simplifying assumptions.

Figure 9.3 shows the same data but with the horizontal axis plotted as log D. This allows one to show a much larger particle size range than that which can be shown using a linear scale for the diameter.

A second problem in expressing the size distribution of aerosols is that the intervals of diameter over which it is experimentally convenient to measure the number of particles are not constant in terms of either D or log D. For example, as discussed in Chapter 11, a multistage impactor is often used to measure the number of particles in certain size ranges. The size intervals are

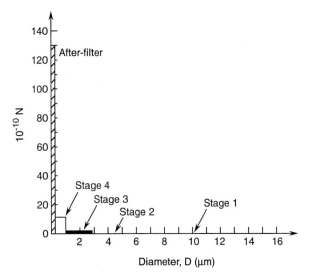

FIGURE 9.2 Plot of number of particles (N) against D (D = aerodynamic diameter) determined using a four-stage impactor with the cutoff points given in Table 9.1. It has been assumed that the particles are spherical with density 2.6 g cm^{-3}. Data from Wesolowski et al. (1980).

stage for a typical impactor are given in Table 9.1; also given are the diameter intervals ΔD as well as $\Delta \log D$ for each stage. It is seen that neither the intervals in terms of diameter nor those in terms of the logarithm of the diameter are equal. As a result, a plot of ΔN against D or $\log D$ will look something like those shown in Figs. 9.2 and 9.3.

These give a somewhat distorted picture of the size distribution, however, because the height of any bar, that is, the number of particles, depends on the width of the interval taken, that is, ΔD or $\Delta \log D$. To give a more physically descriptive picture of the size distribution, a modified plot of the number of particles normalized for the width of the diameter interval is used; that is, the number of particles per unit size interval is plotted on the vertical axis. With $\log D$ as the horizontal axis then, a normalized plot is one of $\Delta N / \Delta \log D$ against $\log D$, where ΔN is the number of particles in that interval of $\Delta \log D$; Fig. 9.4 is an example of this type of plot using the data in Figs. 9.2 and 9.3. The area under each rectangle then gives the number of particles in that size range.

Rather than showing histograms, one usually draws a smooth curve through the data. Figure 9.5, for example, shows one such curve of $\Delta N / \Delta \log D$ against $\log D$ for a typical urban model aerosol; to emphasize the wide range of numbers involved, a logarithmic scale has also been used for the vertical axis (Whitby and Sverdrup, 1980).

However, it is not only the number of particles in each size interval that is of interest but also how other properties such as mass, volume, and surface area are distributed among the various size ranges. For example, the U.S. Environmental Protection Agency's air quality

characterized by the 50% cutoff point for each stage, where the 50% cutoff point is defined as the diameter of spheres of unit density, 50% of which are collected by that stage of the impactor. For example, a typical set of 50% cutoff points is 8.0 (stage 1), 4.0 (stage 2), 1.5 (stage 3), and 0.5 μm (stage 4), respectively, with particles smaller than 0.5 μm being collected on an afterfilter. One might then use the approximation that each stage captures particles ranging from a diameter corresponding to midway between its cut point and that of the next higher and lower stages. With these assumptions, the ranges of particle diameters captured by each

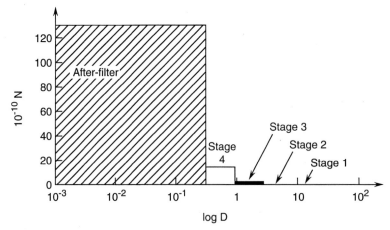

FIGURE 9.3 Plot of number of particles (N) against $\log D$ (D = aerodynamic diameter). Data same as those in Fig. 9.2.

TABLE 9.1 Typical 50% Cutoff Points and Approximate Range of Particle Diameters Captured by Each Stage in a Four-Stage Impactor[a]

Stage	50% cutoff point (μm)	Approximate range of particle diameters captured by that stage (μm)	ΔD (μm)	$\Delta \log D$ (μm)
1	8.0	14.0[b]–6.0	8.0	0.37
2	4.0	6.0–2.8	3.2	0.33
3	1.5	2.8–1.0	1.8	0.45
4	0.5	1.0–0.3[c]	0.7	0.52
Afterfilter	<0.5	0.3–0.001[c]	0.3	2.48

[a] From Wesolowski et al. (1980).
[b] Assuming a cutoff of 20 μm for the sampling probe.
[c] Assuming the afterfilter collects all particles >0.001-μm diameter.

standards for particles are expressed in terms of mass of particulate matter, with diameters <10 μm per unit volume of air, PM_{10}, and those with diameters <2.5 μm, $PM_{2.5}$ (see Chapter 2.D). It is thus important to know the mass distribution of atmospheric particulate matter. Similarly, surface and volume distributions are important when considering reactions of gases at the surface of particles or reactions occurring within the particles themselves, for example, the oxidation of SO_2 to sulfate.

Because of this need to know how the mass, surface, and volume are distributed among the various particle sizes, distribution functions for these parameters (i.e., mass, surface, and volume) are also commonly used for atmospheric aerosols in a manner analogous to the number distribution. That is, $\Delta m / \Delta \log D$, $\Delta S / \Delta \log D$, or $\Delta V / \Delta \log D$ is plotted against D on a logarithmic scale, where Δm, ΔS, and ΔV are the mass, surface area, and volume, respectively, found in a given size interval; again the area under these curves gives the total mass, surface, or volume in the interval considered. Figure 9.6 shows the surface and volume distributions for the number distribution shown in Fig. 9.5; also shown is the same number distribution for comparison, where the vertical axis is now linear (rather than logarithmic as in Fig. 9.5).

When the particle data are plotted as mass, surface, or volume distributions, an important characteristic of typical urban aerosols emerges clearly. As seen in the number distribution of Fig. 9.6a, there is a large peak at ~0.02 μm and a slight "knee" in the curve around 0.1 μm.

The volume distribution for the same aerosol (Fig. 9.6c) shows two strong peaks, one in the 0.1- to 1.0-μm range and the second in the 1- to 10-μm region, with a minimum in the 1- to 2-μm region. The surface distribution (Fig. 9.6b) shows a major peak in the vicinity of 0.1 μm, with smaller peaks in the region between 0.01 and 0.1 μm and between 1 and 10 μm.

The "knee" observed in the number distribution

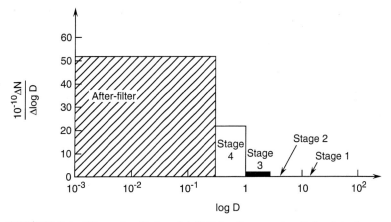

FIGURE 9.4 "Normalized" plot of $\Delta N / \Delta \log D$ versus $\log D$ for data shown in Figs. 9.2 and 9.3.

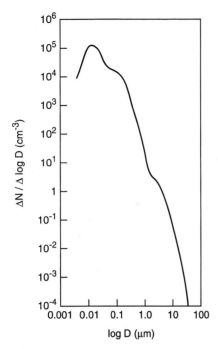

FIGURE 9.5 Plot of number distribution $\Delta N/\Delta \log D$ versus $\log D$ for a typical urban model aerosol (adapted from Whitby and Sverdrup, 1980).

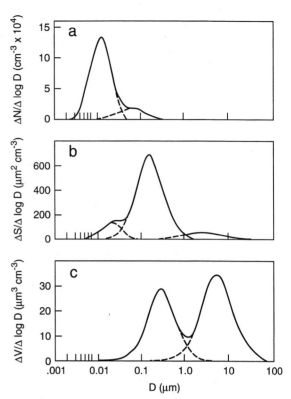

FIGURE 9.6 Number, surface, and volume distributions for a typical urban model aerosol (adapted from Whitby and Sverdrup, 1980).

suggests that the observed curve may be a combination of two different distributions, shown by the dashed lines in Fig. 9.6a. While the multimodal nature of atmospheric aerosols had been suggested earlier (e.g., Junge, 1963), Whitby and co-workers, in a classic set of papers (Whitby et al., 1972a, 1972b; Husar et al., 1972), were the first to establish and explore in detail the significance of such a bimodal distribution in terms of the origins, chemical characteristics, and removal processes of the two groups of particles of different sizes. Although instrument artifacts can be a confounding factor, it is now widely accepted that atmospheric aerosols usually occur in specific size groupings that are different in their origins and properties.

Indeed, based on the number, surface, and volume distributions shown in Fig. 9.6, Whitby and co-workers suggested that there were three distinct groups of particles contributing to this atmospheric aerosol. Particles with diameters >2.5 μm are identified as *coarse particles* and those with diameters <2.5 μm are called *fine particles*. The fine particle mode typically includes most of the total number of particles and a large fraction of the mass, for example, about one-third of the mass in nonurban areas and about one-half in urban areas. The fine particle mode can be further broken down into particles with diameters between ~ 0.08 and 1–2 μm, known as the accumulation range, and those with diameters between 0.01 and 0.08 μm, known as the transient or Aitken nuclei range.

As the technology for measuring small particles has improved (see Chapter 11.B), *ultrafine particles* have also been increasingly studied. While there is no fixed definition of these particles, they are usually taken to mean those with diameters less than 0.01 μm, i.e., $\lesssim 10$ nm.

In short, particles in the atmosphere are now frequently treated in terms of the four modes summarized in Fig. 9.7, which also shows the major sources and removal processes for each one. Although the vertical axis is not shown, it could in theory be any of the distributions discussed, that is, number, mass, surface, or volume.

Particles in the coarse particle range are usually produced by mechanical processes such as grinding, wind, or erosion. As a result, they are relatively large and hence settle out of the atmosphere by sedimentation, except on windy days, where fallout is balanced by reentrainment. Schmidt et al. (1998) also point out that particles generated in blizzards and sandstorms develop an electrostatic charge and that the electrostatic forces between the particles and the surface may be similar in magnitude to gravitational forces. These large particles

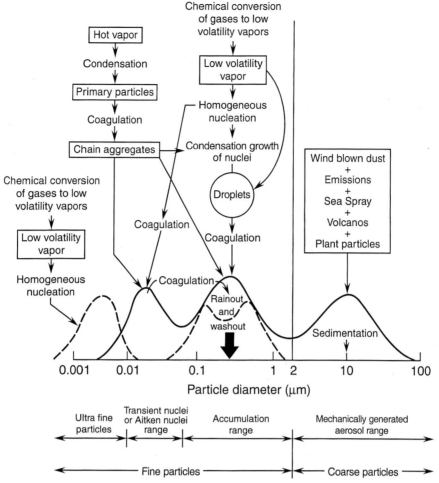

FIGURE 9.7 Schematic of an atmospheric aerosol size distribution showing four modes. The original hypothesis of Whitby and co-workers is shown by the solid, trimodal curves, and the fourth, ultrafine particle mode, as well as the two peaks sometimes observed in the accumulation mode are shown by the dashed lines (adapted from Whitby and Sverdrup, 1980).

can also be removed by washout. In the atmosphere, transport of coarse particles over long distances can occur, however, by convective processes. Chemically, their composition reflects their source, and hence one finds predominantly inorganics such as sand, sea salt, and so on in this range, although significant amounts of organics have also been reported associated with dust particles (e.g., Boon et al., 1998). Because the sources and sinks are different from those of the smaller modes, the occurrence of particles in this mode tends to be only weakly associated with the fine particle mode. The majority of biological particles, spores, pollens, and so on, tend to be in the coarse particle range. It should be noted, however, that this by no means indicates that compounds and elements associated with such mechanical processes are always found exclusively in the coarse particle range. For example, O'Dowd and Smith (O'Dowd and Smith, 1993b; Smith and O'Dowd, 1996) report that sea salt was the major component of all particles with radii above 0.05 μm over the northeast Atlantic Ocean.

While particles in the coarse particle mode are generally sufficiently large that they are removed relatively rapidly by gravitational settling, there are large-scale mechanisms of transport that can carry them long distances during some episodes. For example, there are many studies showing the transport of dust in larger particles from the Sahara Desert to the northwestern Mediterranean, Atlantic Ocean, Bermuda, the Barbados, Ireland, the Amazon Basin, and the United States (see, for example, papers by Prospero and Nees, 1977, 1986; Prospero et al., 1981; Talbot et al., 1986; Wolff et al., 1986; Savoie et al., 1989; Mateu et al., 1994; Artaxo and Hansson, 1995; Dentener et al., 1996; Jennings et al., 1996; Gatz and Prospero, 1996; Chiapello et al., 1997; Perry et al., 1997; Moulin et al., 1997; and Li-Jones and Prospero, 1998). Similarly, dust transported from Asia has been recorded on a regular basis over the Pacific (e.g., see Duce *et al.*, 1980; Uematsu *et al.*, 1985; and Zhang *et al.*, 1997) and in the Canadian Arctic (e.g., Kawamura *et al.*, 1996b). Asian dust has been observed, usually during the spring, at the Mauna Loa Observatory in Hawaii (e.g., see Shaw, 1980; Braaten and Cahill, 1986; Zieman *et al.*, 1995; and Holmes *et al.*, 1997); the elemental signature in particles in the size range 0.5- to 3.0-μm, particularly in terms of the Si/Fe and Ti/Fe ratios, is very similar to those measured during dust storms in Beijing, consistent with long-range transport of these particles (Braaten and Cahill, 1986). The mineralogy of the dust has proven useful in source determination of particles after long-range transport (e.g., see Leinen *et al.*, 1994; Merrill *et al.*, 1994; Avila *et al.*, 1997; and Caquineau *et al.*, 1998). While these "dust storms" are episodic in nature, such long-range transport of dust has been proposed to play a significant role in SO_2 and NO_x heterogeneous chemistry, in the photochemical cycles leading to O_3 formation (Dentener *et al.*, 1996), and, as discussed in Chapter 14, possibly in the radiative balance of the atmosphere as well.

Particles in the accumulation range with diameters from $\sim$0.08 to $\sim$1–2 μm typically arise from condensation of low-volatility vapors (e.g., following combustion) and from coagulation of smaller particles in the nuclei range either with themselves or, more likely, with the larger particles in the accumulation range. The coagulation rates for particles in the nuclei range with the larger particles in the accumulation range are usually much larger than for self-coagulation of the small particles; this occurs because of the high mobility of the smaller particles combined with the larger target area of the bigger particles.

Because of the nature of their sources, particles in the accumulation range generally contain far more organics than the coarse particles (other than biologically derived particles) as well as soluble inorganics such as NH_4^+, NO_3^-, and SO_4^{2-}.

While many particle distributions show one peak in the accumulation range, many instances have been observed in which there are two peaks. For example, as seen in Fig. 9.8, John and co-workers (1990) observed two peaks within the traditional accumulation mode, one at 0.2 ± 0.1 and one at 0.7 ± 0.2 μm, in studies of particles carried out in a relatively polluted urban area. This bimodal character of particles appears to occur quite frequently; for example, Hering *et al.* (1997) showed that during the summer in the Los Angeles area, two modes often occur, having diameters of about 0.26 μm (range from 0.10 to 0.39 μm) and 0.65 μm (range from 0.46 to 0.90 μm), respectively.

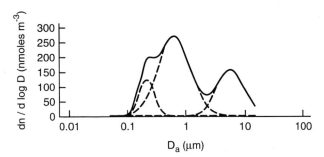

FIGURE 9.8 Typical size distribution of nitrate in southern California in 1987 fitted by the sum of three log-normal distributions with peaks at 0.2, 0.7, and 4.4 μm (adapted from John *et al.*, 1990).

Similarly, Fig. 9.9 shows a particle number size distribution for measurements made at a relatively clean site, Cheeka Peak, which is located approximately 2 km from the coast in Washington State at an altitude of 480 m (Quinn *et al.*, 1993). The data in Fig. 9.9 represent air masses arriving at the sampling site from aloft over Canada. Again two peaks are seen in the number distribution in the 0.1- to 1-μm range, one peaking at 0.15 and the other at 0.45 μm. A bimodal distribution of sulfate with mass mean diameters of 0.25 and 0.55 μm, respectively, has also been observed during the summer in Greenland (Kerminen *et al.*, 1998).

Hering and Friedlander (1982) made similar observations for particle sulfate and attributed the smallest mode (referred to as the "condensation" mode) to formation from gas-phase SO_2 oxidation and the larger modes (the "droplet" mode) to oxidation in the condensed phase. Meng and Seinfeld (1994) have shown that the droplet mode particles cannot arise from growth of the smaller, condensation mode particles and propose that the condensation mode particles are activated to form fog or cloud droplets, followed by chemical reactions and subsequent evaporation to form the "droplet" particles.

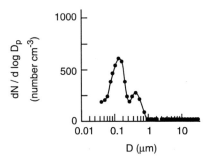

FIGURE 9.9 Particle number size distributions measured at Cheeka Peak, Washington, in 1991 (adapted from Quinn *et al.*, 1993).

Based on such observations and proposals, Ondov and Wexler (1998) have offered a modification of the Whitby accumulation mode, summarized in Fig. 9.10. High-temperature combustion sources produce particles in the Aitken nuclei and the accumulation mode. For example, the mass median diameters of particles emitted from a variety of combustion sources, including incinerators, coal- and oil-fired boilers, and automobiles and trucks, typically fall in the range of 0.05–0.35 μm (Hildemann et al., 1991a, 1991b; Ondov and Wexler, 1998). The relative numbers of particles produced in the Aitken nuclei range compared to the accumulation range depend on the nature of the combustion process (e.g., fuel and operating conditions) as well as the conditions of dilution. Figure 9.11, for example, shows the surface distribution of particles produced by the combustion of several organic compounds as well as by automobiles and by a burning candle. The "dirtier" flames (e.g., the candle and the acetone flame) produced significant numbers of particles in the accumulation mode, whereas the cleaner flames produce more Aitken nuclei (National Research Council, 1979).

Because the particles generated in high-temperature combustion processes contain hygroscopic compounds, water vapor can be taken up or evaporate, depending on the atmospheric conditions. In addition, the particles can be taken up into clouds and fogs. The condensed-phase water provides a medium for atmospheric reactions that generate low-volatility species; the best known example is the oxidation of SO_2 to sulfate (see Chapter 8.C.3). When the water evaporates, the remaining particle contains this additional material and hence has grown to a larger size (Fig. 9.10; see also Section B-3).

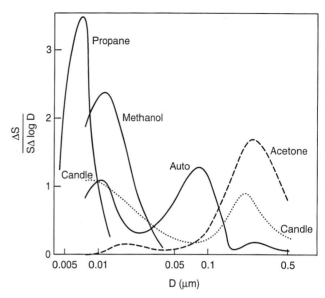

FIGURE 9.11 Surface distribution of particles from the combustion of several organics and from automobiles and a candle (adapted from National Research Council, 1979).

Two peaks in the accumulation mode may also result if the particles are externally mixed rather than internally mixed (the latter meaning that all particles have the same composition corresponding to a mixture of the various components). That is, the composition of individual particles may not be the same as the overall bulk particle composition, with some having more hygroscopic components than others. These are referred to as "externally mixed" particles; indeed single-particle analyses suggest that externally mixed particles are common (see Chapter 11.B.4). In this case, a

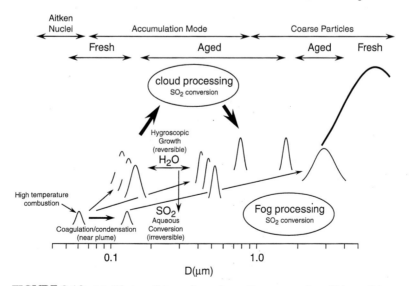

FIGURE 9.10 Modified particle modes and growth processes for sulfate particles involving aqueous-phase reactions in low altitude fogs and in higher altitude clouds upon advection of boundary-layer air upwards. (Adapted with permission from Ondov and Wexler, 1998. Copyright 1998 American Chemical Society.)

monodisperse (i.e., single mode) particle size distribution can develop into a bimodal distribution when the more hygroscopic particles take up increased amounts of water (e.g., see modeling studies of Kleeman *et al.*, 1997).

In short, growth of sulfate particles at least in the accumulation mode and the presence of two peaks are both believed to be largely controlled by interactions with water in the atmosphere, including the aqueous phase oxidation of SO_2 to sulfate.

Particles in the accumulation range tend to represent only a small portion of the total particle number (e.g., 5%) but a significant portion (e.g., 50%) of the aerosol mass. Because they are too small to settle out rapidly (see later), they are removed by incorporation into cloud droplets followed by rainout, or by washout during precipitation. Alternatively, they may be carried to surfaces by eddy diffusion and advection and undergo dry deposition. As a result, they have much longer lifetimes than coarse particles. This long lifetime, combined with their effects on visibility, cloud formation, and health, makes them of great importance in atmospheric chemistry.

The Aitken nuclei, with $0.01 < D < 0.08$ μm, arise from ambient-temperature gas-to-particle conversion as well as combustion processes in which hot, supersaturated vapors are formed and subsequently undergo condensation. These particles act as nuclei for the condensation of low-vapor-pressure gaseous species, causing them to grow toward the accumulation range; alternatively, these nuclei may grow larger by coagulation. This range contains most of the total number of particles but relatively little of the total mass because of their small size. The lifetime of these particles is short, sometimes on the order of minutes, due to their rapid coagulation.

Development of techniques to measure ultrafine particles with diameters less than 0.01 μm (10 nm) has established the presence of what might be considered a fourth mode, sometimes referred to as the *nucleation* mode. As discussed later, these ultrafine particles are generated by gas-to-particle conversion processes that are not yet well understood on a molecular level. Although they are sufficiently small that they do not contribute a large fraction of the total mass, they can be present at significant number concentrations. For example, in the Los Angeles area, the 24-h average concentration of ultrafine particles ($0.017 < D < 0.1$ μm) during the winter has been reported to be $\sim 10^4$ particles cm^{-3} (Hughes *et al.*, 1998). As discussed by Hughes and co-workers, although these particles do not contribute a large amount (only ~ 1 μg m^{-3}) to the total particulate mass, about 10^{11} of these ultrafine particles will be deposited in one day in the respiratory tract of a typical person breathing this air. Hence if the toxicological effects are determined primarily by the *number* of particles, rather than their mass, these ultrafine particles could ultimately prove to be quite important.

A good example of the complexity of size distributions is found in the work of Dodd *et al.* (1991). They measured the composition of various elements in particles collected in a rural area in Maryland as a function of size and observed in individual samples as many as four distinct modes between 0.09 and 1 μm. For example, in one particular sample, there were multiple peaks at ~ 1 μm containing Na, Fe, Ti, Mn, and Al, at 0.5–0.6 μm containing As, Sb, Se, V, La, and Ce, and at 0.28 μm containing S and a single peak at ~ 0.1 μm containing Na, Ga, Ti, Al, La, Ce, and Fe. Dodd and co-workers suggest the smallest particles represent primary emissions from high-temperature combustion sources, whereas the larger ones result from both secondary reactions (e.g., forming sulfate) and direct contributions from multiple sources.

Based on numerous size distributions measured in air, various categories of tropospheric aerosols have been proposed. Table 9.2, for example, shows a typical set of categories and some of their associated characteristics. However, these should be taken merely as examples rather than as fixed categories since many aerosols will display characteristics of more than one category.

b. Atmospheric Aerosols and Log-Normal Distributions

Ideally, one would like to describe various size distributions by some relatively simple mathematical function. Because there is no single theoretical basis for a particular function to describe atmospheric aerosols, various empirical matches have been carried out to the experimentally observed size distributions; some of these are discussed in detail elsewhere (e.g., see Hinds, 1982). Out of the various mathematical distribution functions for fitting aerosol data, the log-normal distribution (Aitchison and Brown, 1957; Patel *et al.*, 1976) has emerged as the mathematical function that most frequently provides a sufficiently good fit, and hence we briefly discuss its application to the size distribution of atmospheric aerosols.

Most readers will be familiar with the bell-shaped normal distribution plotted in Fig. 9.12. When applied to the size distribution of particles, for example, such a distribution is fully characterized by the arithmetic mean $\overline{D}$ and the standard deviation σ, where σ is defined such that 68% of the particles have sizes in the range $\overline{D} \pm \sigma$ In the log-normal distribution, the *logarithm* of the diameter D is assumed to have a normal distribution. (Either logarithms to the base 10 or loga-

TABLE 9.2 Some Typical Tropospheric Aerosols and Their Associated Properties[a]

Type	Typical number concentration (cm^{-3})	Typical characteristics
Marine	100–400	Three modes: $D_p < 0.1$, $0.1–0.6$, and >0.6 μm; most ($>95\%$) of particle mass but only 5–10% of total number in largest mode; large particles mainly sea salt aerosol, smallest are products (e.g., SO_4^{2-}) of gas (e.g., DMS) to particle conversion
Remote continental	$\sim 10^4$	Three modes centered at $D_p \sim 0.02$, 0.12, and 1.8 μm; includes products of gas-to-particle conversion and biological sources, e.g., pollens
Urban aerosol	$\geq 10^5$	Three modes: nuclei, accumulation, and coarse; larger particles contain crustal elements (Fe, Si, etc.), smaller contain nitrate, sulfate, ammonium, and elemental and organic carbon and are formed by combustion processes and gas-to-particle conversion
Nonurban continental	$\sim 10^3$	Three modes similar to urban aerosol
Desert	Variable depending on location	Most are large, $D_p > 1$ μm; contain desert soil elements such as Fe, Si, Al, and Ca
Polar	$\sim 15–150$	Typical $D_p \sim 0.15$ μm; contain gas-to-particle conversion products such as sulfate and ammonium
Free troposphere	~ 30	Monodisperse with $D_p \sim 0.2–0.5$ μm; nucleation major source; sulfate is major component
Biomass burning	$\sim 10^4$ close to source; $\leq 10^3$ downwind	Two modes often seen: $D_p \sim 0.1–0.4$ and $D_p > 2$ μm; smaller mode contains gas-to-particle conversion products (sulfate, nitrate, ammonium, organics) while larger mode has soil and ash particles

[a] From Pandis *et al.* (1995), Whitby and Sverdrup (1979), and Fitzgerald (1991).

rithms to the base e can be used, but since the latter is more common, we follow through the discussion using natural logarithms, ln.) This distribution is expressed as

$$\frac{dN}{d \ln D} = \frac{N_T}{\sqrt{2\pi} \ln \sigma_g} \exp\left[-\frac{(\ln D - \ln \overline{D}_{gN})^2}{2(\ln \sigma_g)^2}\right], \quad (B)$$

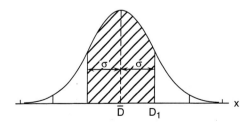

FIGURE 9.12 Meaning of standard deviation for a normal distribution. The hatched area represents 68% of total area under curve.

where N is the number of particles having diameters whose logarithms are between $\ln D$ and $\ln D + d \ln D$, N_T is the total number of particles, σ_g is known as the geometric standard deviation, and $\overline{D}_{gN}$ is the geometric number mean diameter, which for this distribution is equal to the number median diameter, defined as the diameter for which half the number of particles are smaller and half are larger. A typical plot of Eq. (B) is shown in Fig. 9.13, where $(dN/N_T)/d \ln D$ is plotted against the diameter on a logarithmic scale.

The geometric number mean diameter, $\overline{D}_{gN}$, is related to the arithmetic mean of ln(diameter):

$$\ln \overline{D}_{gN} = \frac{\Sigma n_j \ln d_j}{N_T}. \quad (C)$$

Here n_j is the number of particles in a group whose diameters are centered around d_j. Thus $\ln \overline{D}_{gN}$ is really a weighted value of $\ln d$, where the weighting is by the number of particles in that size interval.

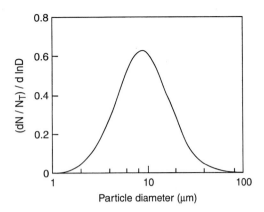

FIGURE 9.13 Frequency distribution curve (logarithmic size scale) (adapted from Hinds, 1982).

The meaning of the geometric standard deviation σ_g can be seen by referring to the meaning of standard deviation for a typical *normal* distribution. In this case, as shown in Fig. 9.12, the standard deviation is a measure of the spread about the mean ($\overline{D}$) and is defined such that there is a 68% probability of a particle having a diameter in the range $\overline{D} \pm \sigma$, and a 95% probability of it having a diameter in the range $\overline{D} \pm 2\sigma$. σ can be calculated for the normal distribution using Eq. (D),

$$\sigma = D_1 - \overline{D}, \quad (D)$$

where $\overline{D}$ is the diameter for which 50% of the particles have smaller diameters and 50% have larger diameters and $D_1 (= \overline{D} + \sigma)$ is the diameter for which there is an 84% (= 50 + 68/2) probability that a particle will have a diameter equal to or less than this value.

For a *log-normal* distribution, σ_g is still a measure of spread of the distribution, but it has a slightly different definition because $\ln D$, rather than D, is assumed to have a normal distribution. For a log-normal distribution σ_g is defined by Eq. (E):

$$\ln \sigma_g = \ln D_1 - \ln \overline{D} = \ln \frac{D_1}{\overline{D}}. \quad (E)$$

Thus

$$\sigma_g = \frac{D_1}{\overline{D}} \quad (F)$$

and 68% of the distribution is between D_g / σ_g and $D_g \sigma_g$. σ_g is a dimensionless parameter that must be >1.0 since $D_1 > \overline{D}$.

Although the log-normal distribution in Eq. (B) was given in terms of the distribution of the numbers of particles as a function of size, it can also be applied to the size distribution of the other properties of interest, that is, mass, surface, or volume. In these cases, the *average* size used to characterize the distribution is known as the geometric *mass* mean diameter, the geometric *surface* mean diameter, and geometric *volume* mean diameter, respectively. Like the geometric number mean size, the geometric mass mean diameter is given by

$$\ln \overline{D}_{gM} = \frac{\Sigma m_j \ln d_j}{M_T}, \quad (G)$$

where m_j is the mass of particles in group j having representative diameter d_j and M_T is the total mass in the sample, $M_T = \Sigma m_j$. Assuming that the particles are all perfectly smooth and spherical with density ρ, then the volume of each particle in group j is $v_j = \pi d_j^3 / 6$ and the total volume of the group j is $n_j(\pi d_j^3/6)$. The mass of particles in group j is then given by $m_j = n_j \rho (\pi d_j^3/6)$. Thus, the total mass M_T of all the particles in all groups is $M_T = \Sigma n_j \rho (\pi d_j^3/6)$. Equation (G) becomes

$$\ln \overline{D}_{gM} = \frac{(\pi \rho / 6) \Sigma (n_j d_j^3) \ln d_j}{(\pi \rho / 6) \Sigma n_j d_j^3} = \frac{\Sigma n_j d_j^3 \ln d_j}{\Sigma n_j d_j^3}. \quad (H)$$

Because mass and volume are directly related, with ρ as the constant of proportionality, the expression for the geometric volume mean diameter, $\overline{D}_{gV}$, is the same as that for $\overline{D}_{gM}$ given in Eq. (H).

Similarly, the geometric surface mean diameter, $\overline{D}_{gS}$, is defined as the average diameter weighted according to the surface areas s_j in the various groups:

$$\ln \overline{D}_{gS} = \frac{\Sigma s_j \ln d_j}{S_T} = \frac{\Sigma n_j d_j^2 \ln d_j}{\Sigma n_j d_j^2}, \quad (I)$$

using area = πd^2 and where S_T is the total surface area of all particles in all groups.

The surface mean diameter (as opposed to the *geometric surface* mean diameter) is also a very useful parameter, since

$$\overline{D}_S = \frac{\Sigma s_j d_j}{S_T} = \frac{\Sigma n_j d_j^3}{\Sigma n_j d_j^2} = \frac{6}{\rho} \frac{M_T}{S_T}. \quad (J)$$

Thus from the total mass (M_T) of the particle sample and its total surface area (S_T) (which can be determined experimentally), the surface mean diameter of the particles can be calculated if their density (ρ) is known.

An advantage of applying the log-normal distribution to atmospheric aerosols is that the value of the geometric standard deviation, σ_g, is the same for a given sample for all types of distributions—count, mass, surface, and volume. It is only the value of the geometric mean diameter that changes, depending on the

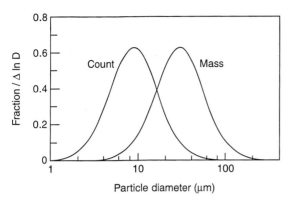

FIGURE 9.14 Count and mass distributions for a hypothetical log-normal sample. The spread, σ_g, of the two curves is seen to be the same, but the mean diameters associated with each are different (adapted from Hinds, 1982).

TABLE 9.3 Values of the Constant b in the Hatch–Choate Equations for Converting the Count Geometric Mean Diameter to Mass, Surface, or Volume Diameters

Type of mean diameter to be calculated	Geometric mean b^a	mean b^a
Mass	3	3.5
Surface	2	2.5
Volume	3	3.5

^{a} See Eq. (K).

property used for the distribution. This can be seen in Fig. 9.14, which shows a count and mass distribution for a hypothetical log-normal sample; the spread, σ_g, is seen to be the same for each distribution, but the geometric mean diameters, $\overline{D}_{gN}$ and $\overline{D}_{gM}$, are quite different.

In practice, when one measures the size distributions of aerosols using techniques discussed in Chapter 11, one normally measures one parameter, for example, number or mass, as a function of size. For example, impactor data usually give the mass of particles by size interval. From such data, one can obtain the geometric *mass* mean diameter (which applies only to the mass distribution), and σ_g, which, as discussed, is the same for all types of log-normal distributions for this one sample. Given the geometric mass mean diameter ($\overline{D}_{gM}$) in this case and σ_g, an important question is whether the other types of mean diameters (i.e., number, surface, and volume) can be determined from these data or if separate experimental measurements are required. The answer is that these other types of mean diameters can indeed be calculated for smooth spheres whose density is independent of diameter. The conversions are carried out using equations developed for fine-particle technology in 1929 by Hatch and Choate.

These Hatch–Choate equations are of the form

$$d/(\text{Number median diameter}) = \exp\left[b\left(\ln^2 \sigma_g\right)\right], \quad (K)$$

where d is the average diameter to be determined using a known value of the number median diameter, σ_g is the geometric standard deviation, and b is a constant whose value is determined by the type of average diameter, d, which is to be calculated. Recall that for a log-normal distribution, the number median diameter is equal to the geometric number mean diameter, $\overline{D}_{gN}$. Thus Eq. (K) becomes

$$d/\overline{D}_{gN} = \exp\left[b\left(\ln^2 \sigma_g\right)\right]. \quad (L)$$

Table 9.3 gives the values of b to be used for converting the count geometric mean diameter to the other types of geometric or mean diameters, respectively.

While we have concentrated here on the geometric means weighted by number, mass, surface, or volume, respectively, other types of "average" diameters are also sometimes cited in the literature. One commonly used is the diameter of average mass, which is defined as the diameter such that the mass of this particle multiplied by the total number of particles gives the total mass. The geometric mean diameters discussed here can also be converted to these types of diameters using a form of the Hatch–Choate equations. The reader is referred to the discussion by Hinds (1982) for definition of these other types of diameters and their conversion to the geometric mean diameters discussed here.

Whitby and co-workers have indicated that the many atmospheric aerosol size distributions that they have measured under a variety of conditions and at many locations can be fit reasonably well assuming three additive log-normal distributions corresponding to the Aitken nuclei range, the accumulation range, and the coarse particle range, respectively, as described earlier. Each of these log-normal distributions has its own characteristic value of σ_g as well as, of course, average diameters. For example, Fig. 9.6 contains the number, surface, and volume distributions for a typical urban aerosol; these were calculated to be consistent with the sum of two, or in the case of the surface distribution, three, additive components. These components are shown by the dashed lines in Fig. 9.6. From these separate distributions, geometric mean diameters and σ_g could be found assuming smooth spheres. Table 9.4 summarizes the parameters derived by Whitby and

TABLE 9.4 Summary of Parameters $\overline{D}_g$ and σ_g for the Three Additive Log-Normal Distributions Characterizing Data in Fig. 9.6[a]

Type of distribution used	Mode	Log-normal distribution	
		$\overline{D}^b$ (μm)	σ_g
Number	Aitken nuclei	$\overline{D}_{gN} = 0.013$	1.7
	Accumulation	$\overline{D}_{gN} = 0.069$	2.03
Surface	Aitken nuclei	$\overline{D}_{gS} = 0.023$	c
	Accumulation	$\overline{D}_{gS} = 0.19$	c
	Coarse particle	$\overline{D}_{gS} = 3.1$	2.15
Volume	Aitken nuclei	$\overline{D}_{gV} = 0.031$	c
	Accumulation	$\overline{D}_{gV} = 0.31$	c
	Coarse particle	$\overline{D}_{gV} = 5.7$	d

[a] From Whitby and Sverdrup (1980).
[b] Note that the diameter—number, surface, or volume—changes depending on the type of distribution used, whereas σ_g remains constant for each of the three modes. See text.
[c] Same as derived from the number distribution.
[d] Same as derived from the surface distribution.

Sverdrup (1980) based on the normal distributions in the three size ranges in Fig. 9.6.

The examples used so far are generally based on surface measurements. The particle concentration and size distribution also depend on altitude. Typical vertical distributions are discussed, for example, by Jaenicke (1992).

3. Particle Motion

One of the important properties of particles that contributes to both the observed size distribution and the number concentration of aerosols in the atmosphere is the motion they undergo when suspended in air. This includes gravitational settling and Brownian diffusion.

a. Gravitational Settling

In the free troposphere, particles are subjected to gravitational forces. They also can be subjected to electrical forces in nature as well as in the course of detection and measurement. When such forces are applied, the particle moves relative to the gas and hence is subjected to a resistance force. Stokes' law gives the force (F_R) acting on smooth spherical particles due to the laminar flow of air over them:

$$F_R = 3\pi \eta \nu D. \quad (M)$$

η is the gas viscosity, ν is the particle velocity relative to the gas, and D is the particle diameter. When a force such as gravity is applied to the particle, it speeds up until the frictional force equals the applied force; it then moves with a constant velocity known as the *terminal velocity*.

One can apply Stokes' law to atmospheric particles to calculate how fast they will settle out of the air when subjected to gravity alone. Thus the terminal settling velocity occurs when the frictional and gravitational forces are balanced, that is,

$$F_R = F_{gravity} = mg, \quad (N)$$

where m is the mass of the particle and g is the acceleration due to gravity (9.8 m s^{-2} at sea level). One can apply the relationship between mass, volume ($\pi D^3/6$), and density (ρ) of the particle. Equations (M) and (N) combine to give Eq. (O):

$$\frac{\pi D^3 \rho g}{6} = 3\pi \eta \nu D$$

or

$$\nu = \frac{D^2 \rho g}{18\eta} \text{ (for } D > 1.5 \text{ μm)}. \quad (O)$$

The settling velocity thus increases with the density of the particle and with the square of its diameter. In developing Eq. (O), the buoyancy effect of air, which tends to lower the effective particle density, has been ignored since it is much smaller than the particle density; it can be included if desired by replacing ρ by ($\rho_p - \rho_{air}$), where ρ_p is the particle density and ρ_{air} is the air density (1.2 × 10^{-3} g cm^{-3} at 20°C and 1 atm pressure).

The expression given in Eq. (O) for the terminal settling velocity only applies to particles with diameters >1.5 μm because its derivation is based on the assumption that the relative speed of the air at the surface of the particle is zero. However, as the particle becomes smaller, the air molecules appear less as a continuous fluid and more as discrete molecules separated by space through which the particles can "slip." The net effect is that the particles can move faster than predicted by Eq. (O) due to this slipping between gas molecules. To correct for this effect, a correction factor must be applied to the resistance force predicted by Stokes' law, Eq. (M). The correction factor, C, is a number greater than 1. Thus Eq. (M) is modified to

$$F_R = \frac{3\pi\eta v D}{C} \qquad (P)$$

and the settling velocity becomes

$$v = \frac{D^2 \rho g C}{18\eta}. \qquad (Q)$$

The correction factor, first derived by Millikan (1923), is given by Eq. (R):

$$C = 1 + \frac{l}{D}\left[2.514 + 0.800 \exp\left(-0.55\frac{D}{l}\right)\right]. \qquad (R)$$

This is often called the Cunningham correction factor, although Cunningham's original correction was of the form $1 + A(l/D)$ and did not include the exponential term. l is the mean free path between air molecule collisions, defined as the average distance traveled between collisions with another molecule; it can be easily calculated using simple kinetic molecular theory, and for air at 1 atm pressure and 20°C, l is 0.066 μm. For large particles, $D > 1.5$ μm; the term in brackets in Eq. (R) is <0.10 and hence Eq. (O) is a good approximation. For particles between ~0.1 and 1.5 μm, the third (exponential) term in Eq. (R) is small relative to 2.5, and hence the Cunningham correction factor can be approximated by

$$C = 1 + \frac{2.154 l}{D} \text{ (for 0.1 μm} < D < 1.5 \text{ μm).} \qquad (S)$$

Figure 9.15 shows the Cunningham slip correction factors for air at 1 atm pressure and 20°C; for the smallest particles, there is a significant correction to the speeds calculated using Eq. (O).

The assumptions inherent in the use of Stokes' law (e.g., relatively low speeds), which normally apply to atmospheric particles, are discussed in more detail by Hinds (1982) and Fuchs (1964). For our purposes, we

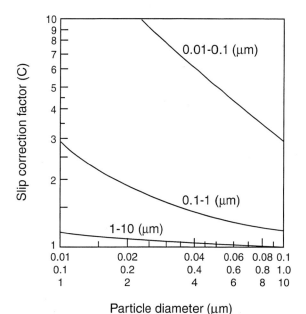

FIGURE 9.15 Cunningham correction factors to be applied in calculating terminal settling velocities of particles (adapted from Hinds, 1982).

need only recognize that the settling velocities for atmospheric particles calculated using Eq. (O) or (Q) are large for particles >10 μm in diameter. Figure 9.16 shows the settling velocities of spherical particles with $\rho = 1$ g cm^{-3} in still air at 0°C and 1 atm pressure as a function of particle diameter. It is seen that particles with diameters of the order of 10 μm or greater have settling velocities >0.1 cm s^{-1} and hence will settle out of the atmosphere relatively rapidly. However, those particles with diameters <1 μm will remain suspended for long periods of time and hence can participate in atmospheric transformations.

b. Brownian Diffusion

Small particles do not settle via gravity at a significant rate, but they do undergo Brownian diffusion. The classic example of Brownian diffusion is the random zigzag motion of smoke particles in air that can be observed because of light scattering by the particles. In fact, in the absence of convection, particles <0.1 μm in diameter are transported largely by Brownian diffusion, which is primarily responsible for the rapid coagulation of particles in the Aitken nuclei range.

The rates of Brownian diffusion can be quantified by considering a box of cross-sectional area 1 cm^2 having N particles cm^{-3} in one section (A) on the left, while the right section (B) is initially empty. Particles (or gas molecules) will always tend to diffuse from a region of higher concentrations to one of lower concentrations.

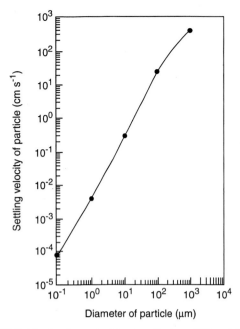

FIGURE 9.16 Settling velocities in still air at 0°C and 760 Torr pressure for particles having a density of 1 g cm^{-3} as a function of particle diameter. For spherical particles of unit density suspended in air near sea level, the Stokes law applies over a considerable range of particle sizes, where the line is straight, but a correction is required at the particle size extremes (adapted from LBL, 1979).

TABLE 9.5 Cumulative Deposition of Unit Density Particles onto a Horizontal Surface from Unit Aerosol Concentrations[a] during 100 s by Diffusion and Gravitational Settling[b]

Diameter (μm)	Cumulative deposition		Ratio diffusion/settling
	Diffusion (number cm^{-2})	Settling (number cm^{-2})	
0.001	2.5	6.5×10^{-5}	3.8×10^4
0.01	0.26	6.7×10^{-4}	390
0.1	2.9×10^{-2}	8.5×10^{-3}	3.4
1.0	5.9×10^{-3}	0.35	1.7×10^{-2}
10	1.7×10^{-3}	31	5.5×10^{-5}
100	5.5×10^{-4}	2500	2.2×10^{-7}

[a] This assumes an aerosol concentration of 1 particle cm^{-3} outside the gradient region.
[b] From Hinds (1982).

The rate at which they diffuse depends on the concentration gradient, dN/dx; the larger the gradient, the faster the rate of diffusion. This is the basis of the well-known Fick's first law of diffusion:

$$J = -\mathbf{D}\frac{dN}{dx}. \quad (T)$$

J is the flux of particles crossing a 1-cm^2 plane in 1 s (i.e., number cm^{-2} s^{-1}). The constant $\mathbf{D}$ is known as the diffusion coefficient and is simply the proportionality constant relating the flux to the concentration gradient. (Fick's first law applies, of course, not only to particles but also to gas and liquid molecules.)

Intuitively, one might expect that the rate of diffusion would increase with temperature and decrease with increasing gas viscosity and particle size. Indeed, this is observed to be the case because all these parameters contribute to the diffusion coefficient $\mathbf{D}$, given by

$$\mathbf{D} = \frac{kTC}{3\pi\eta D}. \quad (U)$$

k is the Boltzmann constant (1.38×10^{-23} J deg^{-1}), T the temperature (in K), C the Cunningham correction factor given by Eq. (R), η the viscosity of the gas in which the particles are suspended, and D the particle diameter. [The reader is referred to Hinds (1982) for a derivation of Eq. (U).] For larger particles where $C \sim 1$ (i.e., the slip correction is negligible), the rate of diffusion varies inversely with the particle diameter. For very small particles, the rate varies with $1/D^2$, which contributes to making diffusion a major transport mechanism for particles <0.1 μm. It is this Brownian diffusion that helps to carry small particles through the boundary layer to surfaces where they may stick on impact.

The relative importance of Brownian diffusion and gravitational settling in the deposition of particles may be seen by calculating the total deposition of particles onto a horizontal surface by these two processes in a given period of time under certain conditions. Table 9.5 shows the results of such a calculation for particle diameters from 0.001 to 100 μm, assuming spherical particles of unit density with a constant concentration of 1 particle cm^{-3} outside the gradient region; also shown is the ratio of the number of particles deposited by diffusion compared to the number deposited by gravitational settling. At a diameter of ~ 0.2 μm, the two mechanisms become equal, with diffusion greatly exceeding gravitational settling for particles in the Aitken nuclei range. (Note that if the *mass* deposited were calculated, the results would be quite different.)

Other factors also come into play in laboratory systems. For example, McMurry and Rader (1985) have shown that particle deposition at the walls of Teflon smog chambers is controlled by Brownian and turbulent diffusion for particles with $D_p < 0.05$ μm and by gravitational settling for particles with $D_p > 1.0$ μm. However, in the 0.05- to 1.0-μm range, the deposition is controlled by electrostatic effects; Teflon tends to hold an electrostatic charge, which leads to rapid deposition of charged particles and deposition rates almost an order of magnitude faster than expected if these effects of charge are not taken into account.

It should be kept in mind that these calculated rates of diffusion and gravitational settling are only applicable to still air. In fact, in the atmosphere the air is rarely still and is usually undergoing some degree of turbulent motion. In this case, the transport of particles becomes more complex and faster due to the velocity gradients and contorted patterns of air flow; however, a discussion of this is outside the scope of this book.

4. Light Scattering and Absorption and Their Relationship to Visibility Reduction

a. Light Scattering and Absorption

As discussed in Chapter 3, solar radiation passing through the atmosphere to the earth's surface is both scattered and absorbed by gases and particles. The intensity of radiation striking the surface can be expressed in the form of a Beer–Lambert law:

$$\frac{I}{I_0} = e^{-b_{ext}L}, \qquad \text{(V)}$$

where I_0 and I are the incident and transmitted light intensities, respectively, L is the path length of the light beam, and b_{ext} is known as the extinction coefficient and has units of (length)$^{-1}$. This extinction coefficient, representing the total reduction in light intensity due to scattering and absorption of light by gases and particles, is the sum of two terms,

$$b_{ext} = b_g + b_p, \qquad \text{(W)}$$

where b_g is the extinction due to gases and b_p that due to particles. Each of these terms can be broken down into contributions from light scattering and absorption so that Eq. (W) becomes

$$b_{ext} = b_{ag} + b_{sg} + b_{ap} + b_{sp}, \qquad \text{(X)}$$

where b_{ag} and b_{sg} are the light extinctions due to *a*bsorption and *s*cattering by *g*ases, and b_{ap} and b_{sp} are those due to *a*bsorption and *s*cattering by *p*articles.

Scattering and absorption of light by gases have already been discussed in Chapters 3 and 4. In terms of the absorption of visible light in the troposphere by gases, only NO_2 is believed to contribute significantly. However, light absorption by NO_2 is usually much less than the total light scattering and absorption by particles. Thus, when NO_2 is present in sufficient concentrations to absorb light, the atmosphere usually also contains relatively high concentrations of other pollu-

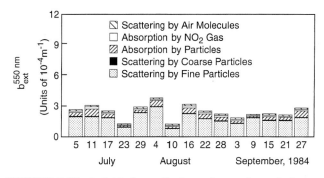

FIGURE 9.17 Individual contributions of scattering and absorption of light at 550 nm by gases and particles in Pasadena, California, during the summer of 1984 (adapted from Larson and Cass, 1989).

tants, including particles. As a result, light scattering and absorption by particles usually exceed light absorption by NO_2 in most situations. For example, in studies of Denver's "brown cloud," it was shown that of the total extinction excluding Rayleigh scattering by gases, 7% was due to light absorption by NO_2 and 93% was due to light scattering and absorption by particulate matter (Groblicki *et al.*, 1981; Waggoner *et al.*, 1983b).

Similarly, Fig. 9.17 shows the calculated individual contributions of various gases and particles to the scattering and absorption of light at 550 nm during a typical summer in Pasadena, California (Larson and Cass, 1989). Even in this area of relatively large NO_x sources, the contribution of NO_2 to the total light attention (expressed as b_{ext}) is small. For reasons to be discussed in more detail, scattering by fine particles predominates, with a smaller contribution due to light absorption by the particles.

Historically, the symbol b has been used both for the total extinction coefficient and for the individual contributions due to light absorption and scattering by particles and gases. However, the Radiation Commission of the International Association of Meteorology and Atmospheric Physics (IAMAP, 1978) has recommended use of the symbol σ instead. Thus the reader may find in the literature the terms σ_{ag}, σ_{sg}, σ_{ap}, and σ_{sp}, respectively, rather than b_{ag}, b_{sg}, b_{ap}, and b_{sp}. However, in keeping with the common practice in the literature in this area and to avoid confusion with light absorption cross sections by gases defined in Chapter 3, we use the symbol b.

The scattering of light by particles falls into three regions depending on the size of the particles relative to the wavelength (λ) of the light: (1) particle diameter $D \ll \lambda$, known as Rayleigh scattering, (2) $D \sim \lambda$, known as Mie scattering, and (3) $D \gg \lambda$. Since we are concerned at the earth's surface primarily with visible and near-ultraviolet light in the actinic region, that is, 290

$< \lambda < 750$ nm, the first case corresponds to particles with $D \lesssim 0.03$ μm and the third to particles with $D \gtrsim 10$ μm. Particles with sizes between these two extremes fall in the second category where $D \sim \lambda$; as we have seen, this is the most important size regime for atmospheric particles.

A common convention used when discussing light scattering as a function of particle size is to define a dimensionless size parameter α, which is the ratio of the circumference of the particle to the wavelength of the incident light:

$$\alpha = \frac{\pi D}{\lambda}. \quad (Y)$$

Very small particles ($\alpha \ll 3$) behave like gaseous molecules in scattering light and hence produce Rayleigh scattering, described in Chapter 3. Because the particles or molecules undergoing this type of scattering are small relative to the incident wavelength, the entire species is subjected at any instant of time to what appears to be a uniform electromagnetic field; this creates a dipole that oscillates with the changing electromagnetic field of the light wave and reradiates the energy in all directions. Thus Rayleigh scattering is symmetric in the forward and backward directions relative to the incident light beam and, as we saw in Chapter 3, varies as λ^{-4}.

Very large particles, on the other hand, that is, those with $D \gg \lambda$ ($\alpha \gg 3$), undergo geometric scattering, where the light beam refracted through the particle can be treated using classical optics. Between these two regimes where $D \sim \lambda$ ($\alpha \sim 3$), much more complex light scattering occurs, known as Mie scattering.

Because particles undergoing Mie scattering have dimensions of the same order as the wavelength of the incident light, the electromagnetic field of the light wave is not uniform over the entire particle at one instant of time, and a three-dimensional charge distribution is set up in the scattering particle. In 1908 Mie developed the solutions for scattering of light of wavelength λ by a homogeneous sphere of diameter D. As shown in Fig. 9.18, light is considered to be incident on the sphere and to be scattered at various angles θ to the direction of the unscattered beam. The incident and scattered beams are shown as the combination of two independent polarized beams: one (I_1) has its electric vector perpendicular to the scattering plane defined by the incident and the scattered beams, and the other (I_{11}) is parallel to it. The intensity of light at a distance R and a scattering angle θ from the particle is given by

$$I(\theta, R) = \frac{I_0 \lambda^2 (i_1 + i_{11})}{8\pi^2 R^2}, \quad (Z)$$

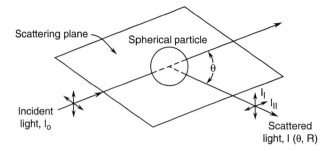

FIGURE 9.18 Diagram showing scattering angle, scattering plane, and the polarized components of scattered light. (From Hinds, W. C. *Aerosol Technology*. Copyright © 1982 John Wiley & Sons, Inc. Reprinted by permission of John Wiley & Sons, Inc.)

where I_0 is the intensity of the incident light beam (taken as unpolarized) and i_1 and i_{11} are known as the Mie intensity parameters for the perpendicular-polarized and parallel-polarized components of the scattered light, respectively. The Mie intensity parameters are a complex function of the refractive index (m) of the scatterer, the size parameter (α), and the scattering angle (θ). For further mathematical details and descriptions of Mie scattering, the reader is referred to the books by Van de Hulst (1957), Kerker (1969), and Bohren and Huffman (1983).

The refractive index of a material, m, is defined as the ratio of the speed of light (c) in a vacuum to that (v) in the material, that is, $m = c/v$. Because light travels more slowly in materials than in air, $m > 1$. The index of refraction, m_a, of materials that absorb light as well as scatter it is expressed in the form of a complex number (see Hinds, 1982)

$$m_a = m_r(1 - ai), \quad (AA)$$

where $i = \sqrt{-1}$, m_r is the real refractive index, and a is a constant that depends on the absorption coefficient of the material as well as to the wavelength. Values of the index of refraction of $\lambda = 589$ nm of some materials either found in the atmosphere or used to calibrate instruments that measure particle size using light scattering (see later) are given in Table 9.6. A typical refractive index for a dry aerosol that absorbs light in the atmosphere is $1.5 - 0.02i$ (Covert *et al.*, 1980).

Unlike Rayleigh scattering, which occurs equally in the forward and backward directions, Mie scattering is predominantly in the forward direction, except for the smallest particles. This can be seen in Fig. 9.19, which shows the Mie intensity parameters i_1 and i_{11} as a function of the scattering angle θ for three different values of the size parameter α defined by Eq. (Y), assuming the droplets are composed entirely of liquid water (i.e., $m = 1.333$). For $\alpha = 2.0$ and 10 (i.e., D/λ

A. PHYSICAL PROPERTIES

TABLE 9.6 Index of Refraction at 589 nm for Some Species Found in the Atmosphere or Used for Instrument Calibration[a]

Species	Index of refraction
Vacuum	1.0
Water vapor	1.00025
Air	1.00029
Water (liquid)	1.333
Ice	1.309
Rock salt	1.544
Sodium chloride in aqueous solutions	1.342–1.378[b]
Sulfuric acid in aqueous solutions	1.339–1.437[c]
Benzene	1.501
α-Pinene	1.465
d-Limonene	1.471
Nitrobenzene	1.550
Dioctyl phthalate	1.49
Oleic acid	1.46
Polystyrene latex	1.59
Carbon	$1.59 - 0.66i$[d,f]
Iron	$1.51 - 1.63i$[d]
Magnetite (Fe_3O_4)	$2.58 - 0.58i$[e]
Copper	$0.62 - 2.63i$[d]

[a] Data from the *Handbook of Chemistry and Physics*, unless otherwise noted (~20–25°C).
[b] For solution densities from 1.035 to 1.189.
[c] For solution densities from 1.028 to 1.811.
[d] From Hinds (1982).
[e] From Huffman and Stapp (1973).
[f] At $\lambda = 491$ nm.

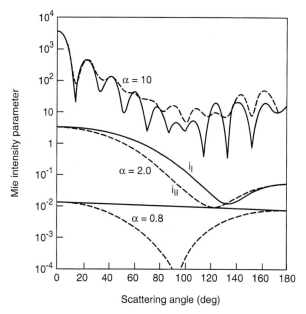

FIGURE 9.19 Mie intensity parameters versus scattering angle for water droplets ($m = 1.333$) having $\alpha = 0.8$, 2.0, and 10.0. Solid lines are i_1, and dashed lines are i_{11} (adapted from Hinds, 1982).

= 0.64 or 3.2), both i_1 and i_{11} fall from a maximum value at $\theta = 0°$ as θ increases. For smaller particles with $\alpha = 0.8$ (i.e., $D/\lambda = 0.26$), i_{11} initially falls as θ increases but then rises as θ approaches 180°, corresponding to backscattering; for these particles i_1 decreases only slightly from 0° to 180°. However, in all cases, i_1 and i_{11}, and hence the scattered light intensity [Eq. (Z)], show their maximum values at $\theta = 0°$, corresponding to forward light scattering.

The variation of scattered light intensity with θ as typified by Fig. 9.19 clearly becomes more complex as the particle size increases, with sharp oscillations seen at $\alpha = 10$. However, recall that this is for a spherical homogeneous particle of a fixed size and for monochromatic light (e.g., a laser); when the particle is irregular in shape, these oscillations are far less prominent. This is also true for a group of particles of various sizes, that is, a polydisperse aerosol, where the overall scattering observed is the sum of many different contributions from particles of various sizes. Finally, nonmonochromatic light and fluctuations in polarization also help to smooth out the oscillations.

The dependence of Mie scattering on particle size can be seen in Fig. 9.20, which shows the sum of the Mie intensity parameters ($i_1 + i_{11}$) as a function of the size parameter α for two scattering angles, $\theta = 30°$ and $\theta = 90°$, respectively. It is seen that Mie scattering generally increases with size over this range of values of α and, as seen in Fig. 9.19, scattering is more pronounced in the forward direction (i.e., at smaller values of θ).

It is noteworthy that in Fig. 9.20 the function becomes smooth and approaches a variation with D^6 as the size parameter decreases toward small values. This is expected, since in the limit of very small particles or molecules, Mie theory reduces to Rayleigh scattering, which, as seen in Chapter 3, varies with D^6.

It should be noted that scattering of light by particles can be measured using remote sensing techniques on satellites, from which such parameters as total aerosol optical thickness [i.e., the exponent ($b_{ext}L$) in $I = I_0 \exp(-b_{ext}L)$, Eq. (V)], albedo, etc. can be determined. However, as discussed in detail by Mishchenko *et al.* (1995), application of conventional Mie theory can lead to significant errors in the aerosol optical thickness if the particles are not spherical, as is assumed in development of Mie scattering theory.

Particles can also absorb light in the atmosphere; the radiant energy absorbed is then converted to heat. As discussed later, graphitic carbon is believed to be the species responsible for most of the light absorption occurring in typical urban atmospheres, although there

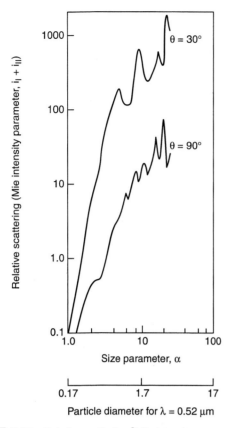

FIGURE 9.20 Relative scattering [Mie intensity parameter ($i_1 + i_{11}$)] versus size parameter for water droplets ($m = 1.33$) at scattering angles of 30° and 90° (adapted from Hinds, 1982).

an integrating nephelometer (Fig. 9.21). Light scattering through all angles (commonly known as b_{scat}) due to Rayleigh scattering by gases as well as scattering by particles is measured by this device (Ruby, 1985), i.e., $b_{scat} = b_{sg} + b_{sp}$. However, in the atmosphere under many conditions, $b_{sp} \gg b_{sg}$, so that b_{scat} measured using a nephelometer effectively measures the light scattering due to particles. In this instrument, air flows slowly through the measuring chamber where a flash lamp produces a pulse of white light. A photomultiplier measures the intensity of the scattered light. Rayleigh scattering by gases can be used to calibrate the instrument. For example, the scattering coefficient of He is calculated to be 3.0×10^{-7} m^{-1}, whereas that of clean air is 2.8×10^{-5} m^{-1} at 460 nm (Charlson et al., 1967).

Values of b_{sp}, the extinction coefficient for light scattering by particles, measured in ambient air using an integrating nephelometer (Fig. 9.21) range from $\sim 10^{-3}$ m^{-1} in highly polluted urban areas to $\sim 10^{-7}$ m^{-1} in remote locations (e.g., Nyeki et al., 1998). The extinction coefficient for particulate absorption, b_{ap}, is generally from $\sim 10^{-4}$ to $< 10^{-8}$ m^{-1} (Bodhaine, 1995; Nyeki et al., 1998). However, the relative contributions of light scattering and absorption at any particular location will clearly depend on the nature of the sources; this is discussed in more detail later in this chapter.

is evidence that organics may also contribute. Iron oxides such as hematite and magnetite also absorb light strongly (Table 9.6), their concentrations are not believed to be sufficiently high to contribute significantly to light absorption in urban areas compared to carbon. However, as discussed in Chapter 14, on a global scale such particles can make significant contributions to light absorption.

Light scattering by particles is usually measured with

b. Relationship of Light Scattering and Absorption to Visibility Reduction

One of the most evident manifestations of anthropogenic air pollution is the production of a *haze* which causes a reduction in visibility, that is, in *visual range*. Visual range is defined as the distance at which a black object can just be distinguished against the horizon. Two factors enter into visual range: visual acuity and contrast. In the daytime atmosphere, particles reduce the contrast perceived by an observer by scattering light from the object out of the line of sight to the

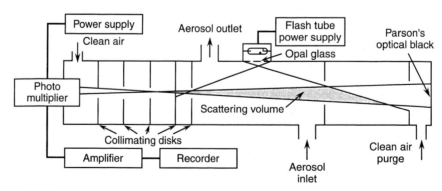

FIGURE 9.21 Schematic diagram of an integrating nephelometer (from Charlson et al., 1967).

observer's eyes; simultaneously, sunlight is scattered into the line of sight, making dark objects appear lighter. The result is a decrease in the contrast between the object and the horizon. At night, scattering of light out of the visual path decreases the contrast and hence the source intensity becomes a factor in visual range as well.

The Koschmieder equation has been shown to approximate the change in contrast of an object with distance away from an observer (Middleton, 1952); note that it has a form similar to that of the Beer–Lambert law:

$$\frac{C}{C_O} = e^{-b_{ext}L}. \quad (BB)$$

In Eq. (BB), C_O is the contrast relative to the horizon (or background) of an object seen at the observation point itself, that is, at a distance $L = 0$, and C is the contrast at the distance L. The contrast is defined as the ratio of the brightness of the object (B_O) to that of the horizon or background (B_H) minus one:

$$C = \frac{B_O}{B_H} - 1. \quad (CC)$$

For example, a black object at zero distance has a brightness of zero (e.g., it absorbs all the visible light) and hence has a C_O of -1.0. b_{ext} in Eq. (BB) is the total extinction as defined in Eq. (V). Observers typically can differentiate objects on the horizon if $C/C_O \sim 0.02-0.05$. A contrast of 0.02, corresponds, using Eq. (BB), to a visual range V_R of

$$V_R = \frac{\ln(C_O/C)}{b_{ext}} = \frac{3.9}{b_{ext}}. \quad (DD)$$

For a contrast of 0.05, $V_R = 3.0/b_{ext}$.

In a clean, particle-free atmosphere, some light scattering occurs due to the Rayleigh scattering by gases. For this scattering, $b_{sg} = 1.5 \times 10^{-5}$ m^{-1} integrated over the solar spectrum at sea level and 25°C (Ouimette et al., 1981). If light absorption by gases is negligible, as it usually is unless significant concentrations of NO_2 are present, then $b_{ext} = b_{sg} = 1.5 \times 10^{-5}$ m^{-1}, and the visual range is $\sim 200-260$ km for contrasts of 0.05–0.02. While this is an approximation that depends on the nature of the object and on the observer, it does give some idea of the visual range that can be expected in clean air.

Visual ranges can vary from hundreds of kilometers in remote areas to only a few kilometers in heavily polluted urban areas. In the latter case, most of the loss in visibility is due to light scattering, with some contribution from light absorption by suspended particulate matter.

It should be noted that the definition of visual range in Eq. (DD) is not always in accord with visual ranges reported from qualitative sightings of surrounding landmarks, as is done, for example, at airport observation towers. There are a number of factors that might influence this, such as the targets not being black or there being differences between various observers. In general, the airport visual ranges are less than those predicted from Eq. (DD) (Stevens et al., 1983; Waggoner, 1983; Lodge, 1983). Indeed, on the basis of airport visual range observations, Ozkaynak et al. (1985) suggest that the use of the coefficient 3.9 in Eq. (DD) is optimistic and that a value less than half that, 1.8, may be more appropriate in urban areas. Other methods of assessing visibility in the atmosphere are discussed by Richards et al. (1988, 1989).

Most of the light scattering by particles in the atmosphere is due to particles in the size range 0.1–1 μm as shown by calculations carried out during World War II for screening smoke particle sizes (Sinclair, 1950). This can be seen in Fig. 9.22, which shows the scattering coefficient of a single particle per unit volume as a function of the particle diameter for spheres with a refractive index of 1.50 and light of wavelength 550 nm. The portion of the total extinction coefficient due to particle scattering, b_{sp}, can be obtained by combining the curve in Fig. 9.22 with the particle volume size distribution, that is, by the curve of $\Delta V/\Delta \log D$ vs $\log D$.

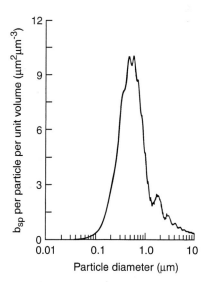

FIGURE 9.22 Scattering coefficient per particle divided by particle volume plotted as a function of diameter. The particles are assumed to be spheres of refractive index 1.50 and the light has $\lambda = 550$ nm (adapted from Waggoner and Charlson, 1976).

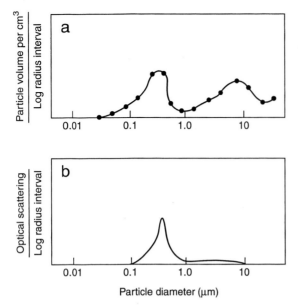

FIGURE 9.23 (a) Aerosol particle size distribution measured at Pomona during the 1972 State of California Air Resources Board ACHEX program. (b) Calculated optical scattering by particles, b_{sp}, for measured size distribution (adapted from Waggoner and Charlson, 1976).

For example, Fig. 9.23a shows the measured volume distribution of one ambient aerosol sample. When this volume distribution is multiplied by the size distribution of the scattering coefficient per unit volume in Fig. 9.22, one obtains the calculated curve for light scattering in Fig. 9.23b. It is seen that the particles in the 0.1- to 1-μm-diameter range, that is, in the accumulation mode, are clearly expected to predominate the light scattering.

This is supported by the correlation between the total aerosol volume of particles with diameters in the 0.1- to 1-μm range and the experimentally determined values of b_{sp} obtained using a nephelometer in many studies (e.g., Fig. 9.24). The slopes of lines such as that in Fig. 9.24, however, depend critically on the nature and history of the air mass and can vary by more than a factor of 10 from clean, nonurban air to highly polluted air in the vicinity of sources. For example, Sverdrup and Whitby (1980a) have shown that the ratio of submicron aerosol volume to b_{sp}, which corresponds to the slope of the line in Fig. 9.24, varies from 5 to 80, depending on the nature of the air mass. In addition, the correlation between b_{sp} and fine particles is usually not as clear-cut as seen in Fig. 9.24.

Consistent with the relationship between the aerosol fine particle volume and the particle scattering coefficient, a number of studies have shown that the fine particle mass and b_{sp} are also related. Figure 9.25 shows the scattering coefficient observed in studies in Denver, Colorado, by Groblicki and co-workers (1981) as a function of the observed mass in the fine and coarse particle ranges, respectively. It is seen that a good linear relationship exists between b_{sp} and the fine particle mass (FPM) but not between b_{sp} and the coarse particle mass. This has been observed in a number of areas ranging from pristine to industrial, with the ratio of the scattering coefficient to the fine particle mass concentration (b_{sp}/FPM) being approximately 3 in many areas (Waggoner et al., 1981; Conner et al., 1991).

Because of the dependence of Mie scattering on the refractive index and hence chemical composition of the particles, one would expect the light scattering coeffi-

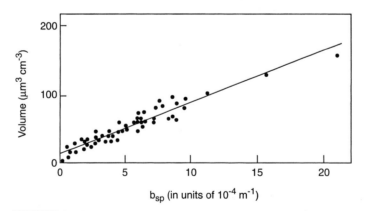

FIGURE 9.24 Plot of measured aerosol fine particle volume (including only those particles of 0.1- to 1.0-μm diameter) versus measured b_{sp}. Measurements were part of the State of California Air Resources Board ACHEX program (adapted from Waggoner and Charlson, 1976; data supplied by Dr. Clark of North American Rockwell).

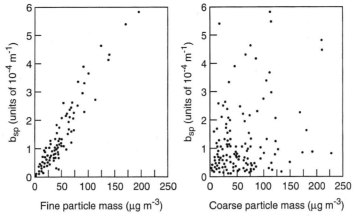

FIGURE 9.25 Correlation of b_{sp} with fine and coarse particulate mass (adapted from Groblicki *et al.*, 1981).

cient b_{sp} to depend on the particular particle composition. Many studies have probed this relationship, coming at it from several different directions. One common approach has been to measure the chemical composition of particles and then test for empirical relationships between light scattering and chemical composition.

In this case, a best fit of the measured b_{sp} to the chemical composition of the particles is often found using an equation of the form

$$b_{sp} = a_0 + \Sigma a_i M_i. \quad \text{(EE)}$$

where M_i is the mass concentration of the ith chemical species and a_i is the light scattering coefficient for that species per unit concentration. Such a relationship has been applied in many studies and has proven useful in identifying certain species (e.g., sulfate, nitrate, and organics) as being important in terms of light scattering (e.g., see Groblicki *et al.*, 1981; Appel *et al.*, 1985; Vossier and Macias, 1986; Solomon and Moyers, 1986; McMurry *et al.*, 1996; Eatough *et al.*, 1996; Laulainen and Trexler, 1997; and McInnes *et al.*, 1998). Values of a_i are usually in the range between 1 and 20, with a typical value of ~ 5 m^2 g^{-1}. Table 9.7 gives some values of a_i reported for various common aerosol constituents.

However, as discussed in detail by Sloane (1986) and White (1986), Eq. (EE) only holds if all of the species act independently. There are several reasons to believe that in many cases they do not. For example, if one species is removed from a particle, not only will its contribution $(a_i M_i)$ disappear but the particle size may

TABLE 9.7 Some Reported Values of the Light Scattering Coefficients (a_i) for Common Constituents of Atmospheric Particles[a]

Compound	a_i (m^2 g^{-1})	Reference
Elemental (black) carbon	0.45–1.4	Petzold *et al.*, 1997
Sulfate	5.2–13[b]	Howell and Huebert, 1998
	3–7	McMurry *et al.*, 1996
	2.1	Eatough *et al.*, 1996
	2.2–3.2	Hegg *et al.*, 1995
	3.2–13.5	White, 1986
Nitrate	1.8	Eatough *et al.*, 1996
	0.5–6.0	White, 1986
Organics	3–7	McMurry *et al.*, 1996
	1.1	Eatough *et al.*, 1996
	0–4.6	White, 1986

[a] For dry particles unless otherwise specified.
[b] At 80% RH.

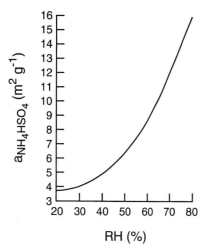

FIGURE 9.26 Calculated light scattering coefficient for NH_4HSO_4 droplets as a function of relative humidity (adapted from Sloane, 1986).

pending on their size and chemical composition, such particles can act as cloud condensation nuclei (see Chapter 14.C.2), taking up enough water to form cloud droplets.

The water content of aerosols varies widely. In the Los Angeles area, for example, a range of water contents in particles from a few percent up to 90 wt% has been observed (Ho et al., 1974; Sverdrup and Whitby, 1980a; Witz et al., 1988; Li et al., 1997), whereas in Sapporo, Japan, 0.4–3.2% of the fine particles ($D < 2$ μm) was reported to be water (Ohta et al., 1998). In accumulation mode particles over the North Atlantic Ocean, an average of 31% of the particle mass was water (Kim et al., 1995). The large amounts of water in particles in the high-humidity marine boundary layer have also been documented using a combination of atomic force microscopy (AFM) and transmission electron microscopy (TEM); thus, Pósfai et al. (1998) showed that the volume of particles measured under ambient conditions using AFM was up to four times that of the same particles under the vacuum conditions of the TEM measurements.

The uptake of water with increasing RH causes an increase in both mass and radius and a decrease in refractive index; the net effect of all these factors is an increase in light scattering. This can be seen in Fig. 9.27, where the light scattering coefficient, $b_{scat} = b_{sp}$, for some ambient aerosol particles measured using an integrating nephelometer was found to increase with the liquid water content of the aerosols.

The effect of water on aerosol light scattering is sometimes expressed as the *hygroscopic growth factor*,

also change. As seen from the earlier discussion of Mie scattering and from data such as those in Fig. 9.22, any such change in particle size will also change its contribution to light scattering.

Another potential effect is if one species affects the concentration of others due to chemical interactions. If the concentration of a species that is hygroscopic changes, the water content and average index of refraction for the particle may change simultaneously. For example, a reduction of about 7% in the index of refraction due to increased hydration can lead to a change in the scattering efficiency of as much as 34% (Hegg et al., 1993).

Figure 9.26 shows the calculated light scattering coefficient for NH_4HSO_4 ($a_{NH_4HSO_4}$), a common constitutent of atmospheric particles, as a function of relative humidity (Sloane, 1986). The contribution of this one compound to light scattering increases from 3.7 to 15.8 $m^2\ g^{-1}$ as the RH increases from 20 to 80%.

This key role of water in visibility reduction has been increasingly recognized as being very important. It is not surprising, given the changes in both the size and refractive index that water uptake causes.

Atmospheric aerosols are hygroscopic, taking up and releasing water as the RH changes (see also Section C.1) because some of the chemical components are themselves deliquescent in pure form. For example, sodium chloride, the major component of sea salt, deliquesces at 298 K at an RH of 75%, whereas ammonium sulfate, $(NH_4)_2SO_4$, and ammonium nitrate, NH_4NO_3, deliquesce at 80 and 62% RH, respectively. (See Table 9.16 for the deliquescence points of some common constituents of atmospheric particles.) De-

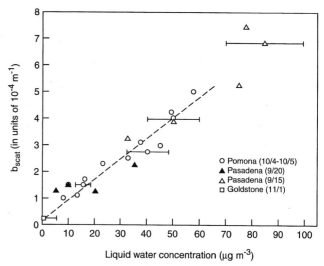

FIGURE 9.27 Comparison of light-scattering coefficient with the liquid water concentration in atmospheric aerosols at some locations in California in 1972 (adapted from Ho et al., 1974).

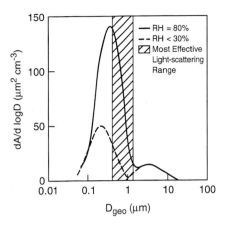

FIGURE 9.28 Size distribution of particle geometric cross section (*A*) as a function of geometric diameter for a typical rural aerosol (adapted from Hegg *et al.*, 1993).

defined as the ratio of b_{sp} at 80% RH to that at a low reference RH (e.g., Waggoner *et al.*, 1983a; Hegg *et al.*, 1993). Changes in light scattering due to changes in RH again reflect the effects on particle size and the index of refraction. Hegg and co-workers (1993) treat the effects on particle size in terms of two components, the first a change on the geometric cross section of the individual particles, and the second a change in the particle size distribution. Figure 9.28 shows the size distribution of the geometric cross section of particles under low RH conditions and at an RH of 80%. The initial distribution represents particles typical of a rural continental site having a number mean diameter of $D_N = 0.076$ μm, a geometric standard deviation of 2, and a number concentration of 2300 particles cm^{-3}. Increasing the relative humidity to 80% results in a larger fraction (41 vs 16%) of the particles being in the range that is most efficient for light scattering, giving a hygroscopic growth factor in this case of 4.2. Values in the range of 1–3 are typical (e.g., see Hegg *et al.*, 1996; Svenningsson *et al.*, 1992; Ten Brink *et al.*, 1996; and McInnes *et al.*, 1998). The increase in RH is thus seen in this case to have major effects primarily through this shift in the particle size distribution into a more effective range for light scattering. Hegg and co-workers (1993) point out that this effect is expected to be the largest for particles in rural areas, where the size distributions often peak below the efficient light scattering range and which can therefore take up water and grow into this range. On the other hand, aged urban particles often already exist with sizes in efficient light scattering ranges so that the uptake of water has relatively less effect. For example, McInnes *et al.* (1998) measured hygroscopic growth factors for particles (using a reference RH of 40%) at Sable Island, Nova Scotia, Canada. Particles in air masses from the northeastern United States that had been impacted by anthropogenic emissions had a hygroscopic growth factor of 1.7 ± 0.1, compared to 2.7 ± 0.4 for particles in marine air from the open ocean.

In short, the contribution of various components of particles to the scattering of light in the troposphere is complex, depending not only on the particular species of interest but also on its interactions with other constituents, on the initial particle size distribution, and on the relative humidity.

Significant advances have also been made in illustrating the effects of visibility reduction by generating simulated photographs calculated to mimic the scattering and absorption of light by particles typically found in the particular area of interest. Both ground-based photography (see, for example, Williams *et al.*, 1980; Malm *et al.*, 1983; Larson and Cass, 1989; Eldering *et al.*, 1993; and Molenar *et al.*, 1994) and satellite images (Eldering *et al.*, 1996) have been mimicked using this approach, both very effective ways to demonstrate the effects of particles on visibility reduction. Significant progress has also been made on developing grid-based models that incorporate both physical and chemical aerosol processes to predict effects on visibility (e.g., see Eldering and Cass, 1996).

Not only light scattering but also light absorption by particles can occur. It is generally agreed that the major contributor to light absorption in the visible region is *black* or *graphitic* carbon, often referred to as elemental carbon. On a practical level, this is the aerosol component that is insoluble in organic solvents and is not oxidized at temperatures below 400°C (Penner and Novakov, 1996). However, as discussed by Chang *et al.* (1982), carbon particles in the air are made up of a number of crystallites 2–3 nm in diameter, with each crystallite consisting of several carbon layers having the hexagonal structure of graphite, Fig. 9.29. Because of the presence of defects, dislocations, and discontinuities, there are unpaired electrons that constitute active sites in the carbon; during the formation of the carbon particle in combustion processes, these active sites can react with gases to incorporate other elements such as oxygen, nitrogen, and hydrogen into the structure. Thus *elemental*, *black*, or *graphitic* carbon found in atmospheric particles is not chunks of highly structured pure graphite, but rather is a related, but more complex, three-dimensional array of carbon with small amounts of other elements. As discussed later, the presence of polar groups on the surface is believed to play an important role in determining their properties, such as uptake of water.

Being black, this atmospheric carbon is a strong absorber of visible radiation. The specific mass absorption coefficient, b_{ap}, has been measured to be in the

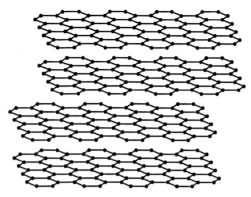

FIGURE 9.29 Structure of elemental carbon.

range from 2 to 20 $m^2\ g^{-1}$ (e.g., see Horvath, 1993; Liousse et al., 1993; Hitzenberger et al., 1996; Petzold et al., 1997; Kuhlbusch et al., 1998; and Moosmüller et al., 1998). Not surprisingly, the absorption depends on the wavelength of the light. Moosmüller et al. (1998) report that b_{ap} varies with $\lambda^{-2.7}$, whereas Horvath et al. (1997) report that the absorption coefficient for aerosol in Santiago, Chile, assumed to be due to elemental carbon, varied with $\lambda^{-0.92}$. Because of the nature of its sources, its contribution to light extinction varies geographically and temporally. For example, wood-burning fireplaces and diesels are major sources of elemental carbon, and areas with large numbers of these two sources generally have more graphitic carbon in the atmospheric aerosol, and hence more light absorption. Where wood burning is significant, more particulate graphite carbon would be expected in winter than in summer.

Absorption of light by carbon is expected to lead to heating of the atmosphere since the light energy is converted into thermal energy (see Chapter 14.C.1b). This is the opposite effect from scattering of light by particles back into the upper atmosphere. This heating effect would be expected to be most important in polluted urban areas (e.g., see Liu and Smith 1995; and Horvath, 1995).

Table 9.8 summarizes some measurements of the contribution of light absorption by particles, b_{ap}, for different types of locations from urban residential to remote. In urban areas, b_{ap} varies from ~13 to 42% of b_{ext}, whereas in very remote areas, the contribution of absorption is much less. This is not surprising since the combustion sources producing graphitic carbon tend to be in urban-industrial areas. However, it is noteworthy that soot has been found associated with many sulfate particles even over the remote oceans (Buseck and Pósfai, 1999).

There are other visible light absorbers than elemental carbon present in particles. These include carbonyl compounds from the oxidation of larger organics in the atmosphere (see Section C.2) and some soil components such as hematite (Fe_2O_3) (see Chapter 14.C.1b). Their contribution to total light extinction in various regions from urban to remote is not well understood, but it appears that it can be significant. For example, Malm et al. (1996) assessed the contributions of various particle components to scattering and absorption of light at the Lake Mead, Arizona, recreation area and at 18 sites in national parks in the western United States. The contributions of organic carbon, elemental carbon, and soil components to the total absorption of light by particles at Lake Mead were estimated to be approximately equal, at 18, 21, and 19% of the total absorption, respectively (the remainder was due to an unknown absorber or set of absorbers). At the national park sites, the contributions of organic and elemental carbon were again about equal, at 45 and 41% of the total absorption.

TABLE 9.8 Contribution of Light Absorption to Total Light Extinction Due to Particles in the Troposphere[a]

Type of atmosphere	Location	b_{ap} (m^{-1})	$b_{ap}/(b_{ap}+b_{sp})$	Reference
Remote	Mauna Loa, Hawaii	6×10^{-8}	0.069	Clarke and Charlson, 1985
Rural	Allegheny Mts./Laurel Hill	1.9×10^{-5}	0.13	Japar et al., 1986
	Shenandoah Valley/Blue Ridge Mts.	1.0×10^{-5}	0.051	Ferman et al., 1981
	Off East Coast of U.S. (Virginia)	$(0.01–1.0) \times 10^{-5}$	0.029–0.24	Novakov et al., 1997[b]
Urban/residential	Denver, Nov/Dec. 1978	6.6×10^{-5}	0.42	Groblicki et al., 1981
	Detroit, July 1981	2.4×10^{-5}	0.13	Wolff et al., 1982a,b
	Houston, Sept. 1980	3.0×10^{-5}	0.18	Dzubay et al., 1982
	Los Angeles Basin, Oct. 1980	6.4×10^{-5}	0.25	Pratsinis et al., 1984

[a] From Japar et al. (1986).
[b] At altitudes from 0.2 to 3 km.

Graphitic carbon also scatters light; in one study, its mass scattering coefficient was found to be approximately the same as that for sulfate (Appel et al., 1985). Because of its contribution to both scattering and absorption of light, graphitic carbon tends to play a proportionally much greater role in light extinction than its contribution to the particulate mass would suggest. For example, Horvath et al. (1997) report that 13% of the total light extinction in Santiago, Chile, was due to absorption by elemental carbon, but an additional 14% was due to scattering; as a result, more than a quarter of the total light extinction by aerosol particles could be attributed to elemental carbon.

This disproportionate effect on light extinction of elemental carbon compared to its mass fraction has been reported in a number of areas. For example, in Denver, graphitic carbon was found to represent 15% of the fine particle mass but contribute ~35% to the total light extinction (Groblicki et al., 1981); at Zilna Mesa, Arizona, during one sampling period, graphitic carbon comprised 1.4% of the total aerosol mass and 3.5% of the fine particle mass but was responsible for 15% of the light extinction (Ouimette and Flagan, 1982). Similarly, in one Los Angeles area study, graphitic carbon was found to represent ~8.5–10% of the fine particle mass but to account for ~14–21% of the total light extinction (Pratsinis et al., 1984).

The absorption of light by carbon is sufficiently strong that it has been used as a marker of long-range transport in unusual circumstances. Thus, the plumes from the burning of oil wells in Kuwait were detected at the Mauna Loa Observatory in Hawaii in the form of increased b_{ap} values. Before the oil well burning, b_{ap} was in the $(0.1–1) \times 10^{-7}$ m^{-1} range, whereas during the burning, values as high as 8.6×10^{-7} m^{-1} were measured (Bodhaine et al., 1992).

Table 9.9 summarizes the results of some recent studies of the contribution of various particle components to light scattering. The contribution of absorption, when measured, is also shown. Such data suggest that sulfate and organics are major contributors to light scattering, with the contribution of nitrate being more variable. Light absorption, even in these non-urban areas, appears to be significant as well. (It should be noted, however, that apportionment of light scattering to various components of the particles is sensitive to whether it is assumed that the particles are internally or externally mixed, which is not well established in most cases; e.g., see Malm and Kreidenweis, 1997.)

B. REACTIONS INVOLVED IN PARTICLE FORMATION AND GROWTH

The formation of secondary particulate matter in various size ranges by chemical reactions in the atmosphere may occur by a number of mechanisms. These include (1) reactions of gases to form low-vapor-pressure products (e.g., the oxidation of SO_2 to H_2SO_4 or the reaction of aromatics with OH to form multifunctional oxygenated products) followed by nucleation to form new particles or condensation on preexisting particles, along with some coagulation between particles, (2) reaction of gases on the surfaces of existing particles to form condensed-phase products (e.g., the reaction of gaseous HNO_3 with sea salt particles to form $NaNO_3$), and (3) chemical reactions within the aqueous phase in fogs, cloud, or aerosol particles (e.g., SO_2 oxidation to sulfate; see Chapter 8).

1. Nucleation, Condensation, and Coagulation

The condensation of a low-vapor-pressure species to form a new particle is known as *homogeneous nucleation*. Recall that the vapor pressure of a substance over the curved surface of a droplet is greater than over a flat surface of the same substance (e.g., see

TABLE 9.9 Some Reported Contributions (%) of Various Particle Compounds to Light Scattering and Absorption by Particles in the Troposphere

	Scattering					
Location	SO_4^{2-}	NO_3^-	Organics	Soil / coarse particles	Absorption	Reference
Meadview, Arizona[a]	25	3	32	12	28	Sisler and Malm, 1997
Meadview, Arizona[a]	40	—	≥48	—	—	McMurry et al., 1996
Bryce Canyon, Utah	31	16	20	10	22	Sisler and Malm, 1997
San Gorgonio Wilderness, California	21	34	25	9	21	Sisler and Malm, 1997
Fraser Valley, British Columbia, Canada[a]	29–39	16–35	22–33	1	—	Pryor et al., 1997
Canyonlands National Park, Utah	28	5	19	12	35	Eatough et al., 1996
Hopi Point, Grand Canyon[a]	39	—	≥50	—	—	McMurry et al., 1996

[a] Percentage contributions to b_{sp} only.

Adamson, 1973, and Chapter 14.C.2). The smaller the radius of the droplet, the higher is the vapor pressure over the droplet surface. For example, for pure water at 25°C, the vapor pressure is only 0.1% greater over a droplet of 1-μm radius compared to that for a flat surface but is 11% greater if the radius is 0.01 μm. (see Problem 1). This raises the question as to how homogeneous nucleation of even a single species can occur at all, since the first very small droplets formed would tend to evaporate rapidly. The explanation lies in the formation of molecular clusters of molecules that occur as molecules collide in the gas phase. When the system becomes supersaturated, the concentration of the condensable species increases, as does that of the clusters. The clusters grow by the sequential attachment of molecules until they reach a critical diameter (D^*) above which the droplets are stable and grow and below which they evaporate (Friedlander, 1977, 1983). The critical diameter is given by

$$D^* = \frac{4\gamma\bar{v}}{kT \ln s}, \quad \text{(FF)}$$

where γ is the surface tension of the chemical forming the particle, $\bar{v}$ is the molecular volume, k is the Boltzmann constant, T is the temperature (K), and s is the saturation ratio, defined as the ratio of the actual vapor pressure to the equilibrium vapor pressure at that temperature.

The term *binary homogeneous nucleation* is used to describe the formation of particles from two different gas-phase compounds such as sulfuric acid and water; such nucleation can occur when their individual concentrations are significantly smaller than the saturation concentrations needed for nucleation of the pure compounds. It is believed that in the atmosphere, formation of particles from low-volatility gases occurs not by condensation of a single species but rather by the formation and growth of molecular clusters involving at least two, and as described shortly, probably three or more different species.

A great deal of work on nucleation in the atmosphere has focused on sulfuric acid, since the oxidation of SO_2 is such a common atmospheric process and the acid product has low volatility. Similar considerations apply to methanesulfonic acid, which is formed by the oxidation of some organic sulfur compounds (see Chapter 8.E) and which is also found in particles because of its low vapor pressure (e.g., see Kreidenweis and Seinfeld, 1988a, 1988b). Even for the ostensibly simple case of sulfuric acid and water, the predicted and observed binary homogeneous nucleation rates are not in good agreement. For example, Fig. 9.30 shows both the theoretically predicted (Jaecker-Voirol and Mirabel, 1989) and the experimentally determined (Wyslouzil et al., 1991) concentrations of sulfuric acid needed to have

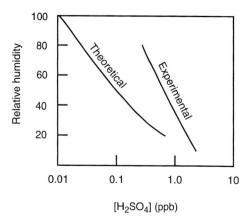

FIGURE 9.30 Theoretically predicted and experimentally measured concentrations of H_2SO_4 required for homogeneous nucleation of sulfuric acid at a rate of 1 particle cm^{-3} s^{-1} (adapted from Hoppel et al., 1994; based on theoretical calculations of Jaecker-Voirol and Mirabel (1989) and experimental data of Wyslouzil et al. (1991).

homogeneous nucleation into particles at various relative humidities (Hoppel et al., 1994). Clearly, there is a significant discrepancy between the predicted and measured concentrations.

As discussed in detail by Pandis et al. (1995), nucleation theory shows there is a critical sulfuric acid concentration above which binary nucleation of sulfuric acid and water should occur. Based on the work of Jaecker-Voirol and Mirabel (1989), they parameterize the theoretical dependence of this critical concentration needed to generate nuclei at a rate of 1 cm^{-3} s^{-1} on the relative humidity (RH, expressed in this case as a fraction, i.e., between 0 and 1) and temperature, T (in K):

$$[H_2SO_4]_{crit} \; (\mu g \; m^{-3})$$
$$= 0.16 \exp\{0.10T - 3.5RH - 27.7\}. \quad \text{(GG)}$$

Note that this predicts critical concentrations that are more than an order of magnitude below the experimentally observed concentrations of Wyslouzil et al. (1991) shown in Fig. 9.30 (see Problem 3). For a recent treatment of the binary homogeneous nucleation of H_2SO_4 and H_2O, see Kulmala et al. (1998).

The atmospheric situation is complicated by varying conditions of temperature, relative humidity, and concentrations of other gases such as NH_3 which can enhance nucleation rates over those expected for a well-mixed air mass at a fixed temperature and RH (e.g., see Nilsson and Kulmala, 1998). However, there is a general consensus that the observed rates of nucleation of H_2SO_4 often, indeed usually, exceed those expected from classical binary homogeneous nucleation theory. (Note that this is not always the case. For example, Pirjola et al. (1998) reported that the measured formation of nuclei in the Arctic boundary layer was consistent with homogeneous nucleation when the air was contaminated with SO_2 from upwind smelters; however, it was much greater than predicted theoretically for clean marine air.)

As discussed by Hoppel et al. (1994), it is likely that under atmospheric conditions, nucleation involves condensation onto preexisting clusters of molecules, or "prenucleation embryos," which are so small in size and number that they have not been measured. Ionic clusters, or neutral clusters formed from the recombination of positively and negatively charged clusters, could serve as embryos for nucleation of sulfuric acid and other species. For example, Turco et al. (1998) propose that recombination of positive ion clusters such as $H_3O^+(H_2O)_n$ with negative ion clusters such as $HSO_4^-(H_2SO_4)_m(H_2O)_q$ (which are known from mass spectrometric measurements to be ubiquitous in the atmosphere; see Chapter 11.A.2) may lead to stable large neutral cluster embryos. Hoppel and co-workers predict, for example, that a cluster of radius 1 nm could act as a nucleus for the condensation of sulfuric acid and water at 60% RH and an H_2SO_4 concentration of only 1 ppt, far below the concentrations shown in Fig. 9.30 for homogeneous nucleation.

The development of techniques to measure particles down to 3 nm in diameter (see Chapter 11.B) has provided some new insights into nucleation in the atmosphere. For example, McMurry, Eisele, and co-workers have measured simultaneously ultrafine particles and gaseous H_2SO_4 in a number of locations (e.g., Weber et al., 1995, 1996, 1997, 1998; Eisele and McMurry, 1997). Formation of ultrafine particles occurs at much smaller concentrations of gaseous H_2SO_4 than expected based on classical binary nucleation theory for sulfuric acid and water. In addition, the dependence on the concentration of gaseous H_2SO_4 is much smaller than expected theoretically. They propose that ammonia may assist in the nucleation process. Figure 9.31 shows the H_2SO_4 vapor pressure above the pure liquid and above a 1:1 mixture of NH_3 and H_2SO_4 (Marti et al., 1997a; Eisele and McMurry, 1997). The vapor pressure of H_2SO_4 is reduced by two to three orders of magnitude when NH_3 is present at equimolar levels. This large reduction in vapor pressure suggests that the reaction of NH_3, which is ubiquitous in the troposphere (see Chapter 11.A.4a), with H_2SO_4 on a molecular level may play a key role in the high observed rates of nucleation. This is consistent with the observation of high concentrations of the smallest measurable ultrafine particles downwind of a large penguin colony, which would be expected to be a significant source of NH_3 (Weber et al., 1998). Interestingly, the growth rate of ultrafine particles is often an order of magnitude

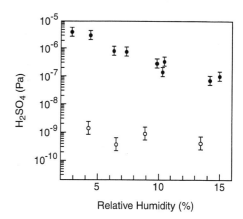

FIGURE 9.31 Vapor pressure of H_2SO_4 as a function of relative humidity for pure H_2SO_4 (●) and a 1:1 mixture of NH_3 and H_2SO_4 (○) (adapted from Marti et al., 1997a; and Eisele and McMurry, 1997).

larger than can be explained based solely on the uptake of gaseous H_2SO_4 by the particles, suggesting that other species such as organics contribute to the growth of ultrafine particles once nucleation to form new particles has occurred (e.g., see Marti et al., 1997b; and Weber et al., 1997). This is consistent with observations that organics can contribute up to 80% of the concentrations of condensation nuclei (CN) under some conditions (e.g., Rivera-Carpio et al., 1996). The uptake of inorganics such as HNO_3 and HCl combined with NH_3 has also been proposed to contribute to particle growth in some cases (e.g., Kerminen et al., 1997).

Similar observations have been made in the marine boundary layer where biogenically emitted sulfur compounds such as dimethyl sulfide (DMS) may serve as the source of gaseous H_2SO_4 and other low-volatility species, e.g., methanesulfonic acid, MSA (see Chapter 8.E). For example, Clarke et al. (1998) report measurements of ultrafine particle formation, DMS, SO_2, and gaseous H_2SO_4 in the tropical marine boundary layer that are consistent with the oxidation of DMS to SO_2 and then H_2SO_4, followed by nucleation. As in many other studies, the nucleation rate of H_2SO_4 was much larger than expected from classical nucleation theory. Interestingly, Weber et al. (1995) conclude from measurements of H_2SO_4, MSA, and ultrafine particles at the Mauna Loa Observatory that H_2SO_4 was the major precursor to ultrafine particles and that the contribution of MSA was small.

For a review of nucleation in the atmosphere, the reader is referred to *Nucleation and Atmospheric Aerosols* (Fukuta and Wagner, 1992; Kulmala and Wagner, 1996) and *Microphysics of Clouds and Precipitation* (Pruppacher and Klett, 1997).

Heterogeneous condensation is secondary aerosol formation by the scavenging of the low-vapor-pressure products onto preexisting particles. If the concentration of particles is sufficiently high, this dominates over the formation of new nuclei via homogeneous nucleation (e.g., Friedlander, 1978, 1980).

The condensation of low-volatility vapors on preexisting particles depends on a number of factors, including the rate of collisions of the gas with the surface, the probability of uptake per collision with the surface, i.e., the mass accommodation coefficient (see Chapter 5.E.1), the size of existing particles, and the difference in partial pressure of the condensing species between the air mass and the particle surface. While some of these parameters are reasonably well known, others are not. For example, mass accommodation coefficients for the complex surfaces found in the atmosphere are not well known. Indeed, the exact nature of the surfaces themselves, which determines the uptake and the partial pressures of gases at the surface, remains a research challenge.

However, despite the uncertainties in quantifying homogeneous new particle formation compared to condensation on preexisting particles, there are a variety of field data supporting the occurrence of both in the troposphere. Figure 9.32, for example, shows some particle measurements made from a ship off the coast of Washington State by Covert and co-workers (1992). Figure 9.32a shows the total particle surface area for particles with diameters >0.02 μm; this surface is needed for condensation of low-volatility vapors onto preexisting particles. It is seen that at about 1430 hours, the surface area decreased significantly, from about 22 to less than 5 μm^2 cm^{-3} of air. As the surface area fell to low values, the relative numbers of ultrafine particles, defined in this case as those with diameters <0.015 μm, increased (Fig. 9.32b). This is consistent with the formation of new particles by homogeneous condensation, since it occurred in the smallest size range and when the available surface area for condensation on existing particles was small. The subsequent growth in the number of slightly larger particles, with diameters in the 0.02- to 0.024-μm range (Fig. 9.32c) may then be due to heterogeneous condensation on the ultrafine particles. This causes an increase in these larger particles accompanied by a decrease in the ultrafine range, as seen in Fig. 9.32b.

Smog chamber studies have documented similar aerosol growth mechanisms. For example, in the photochemical oxidation of dimethyl sulfide, the formation and growth of particles in an initially particle-free system was observed. However, if seed particles with 34-nm mean size were present, an oscillation in the

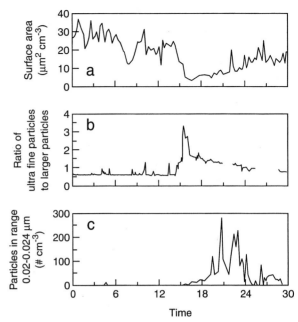

FIGURE 9.32 Some evidence from field studies along the coast of Washington State on April 22, 1991, both for new particle formation and for condensation on preexisting nuclei: (a) particle surface area for particles with $D > 0.02$ μm; (b) ratio of number of particles with $D < 0.015$ μm to those with $D > 0.015$ μm; (c) number concentration with diameters 0.02–0.024 μm (adapted from Covert *et al.*, 1992).

number of fine particles was observed, with bursts of nucleation occurring periodically (Flagan *et al.*, 1991). This was attributed to the heterogeneous condensation on the seed particles initially, but at too slow a rate to remove the low-volatility products. As the latter accumulate, homogeneous condensation occurred, forming a burst of new particles. Condensation on these new particle surfaces then occurs until they coagulate with the seed particles and the process begins anew.

For detailed discussions of the quantitative treatment of such condensation processes in the atmosphere, the reader is referred to articles by Pandis *et al.* (1995) and Kerminen and Wexler (1995).

Coagulation refers to the formation of a single particle via collision and sticking of two smaller particles. Small particles undergo relatively rapid Brownian motion, which leads to sufficient particle–particle collisions to cause such coagulation. The coagulation of smaller particles with much larger ones is similar to condensation of a gas on the larger particle and acts primarily to reduce the number of small particles, adding relatively little to the mass or size of the larger particles. Hence the larger mode will not show significant growth by such a mechanism. As expected, the rate of such processes depends on the diameter of the large particle, how rapidly the smaller particle is carried to it (i.e., the diffusion of the smaller particle), and the concentrations of the particles.

So-called self-coagulation, where the particles are approximately the same size, can, however, lead to changes in the size distribution of the aerosol particles. As one might expect, the rate of this process is a strong function of the particle concentration as well as the particle size.

Table 9.10 shows an estimate (Pandis et al., 1995) of the time scales for coagulation of smaller particles onto larger ones characteristic of various types of air masses. For comparison, typical time scales for condensation, dry deposition of the particles, and transport are also shown. (For discussions of dry deposition of particles, see, for example, Slinn (1982, 1993), Arimoto et al. (1987), and Main and Friedlander (1990)). As expected, condensation is fast, but coagulation is also significant on these time scales in some situations.

2. Reactions of Gases at Particle Surfaces

There are some well-known examples of reactions of gases with solids at the interface that are potentially important in the atmosphere. For example, the reactions of O_3 with adsorbed solid polycyclic aromatic hydrocarbons (PAH) are discussed in Chapter 10. Another example is the reaction of NaCl and NaBr in sea salt particles with gaseous oxides of nitrogen such as HNO_3 (see Chapter 6.J):

$$NaCl_{(s)} + HNO_{3(g)} \rightarrow HCl_{(g)} + NaNO_{3(s)}. \quad (1)$$

The replacement of chloride in sea salt particles by nitrate as well as by sulfate has been observed in many measurements in coastal areas using bulk filter samples (e.g., see Section C.1 and Chapter 6.J.2b) and, more recently, in single particles (e.g., Murphy et al., 1997; Gard et al., 1998). Even in these relatively simple cases, however, the nature of the reactions at the interface and how they contribute to particle growth and/or transformations is not clear. For example, in the reaction of NaCl with HNO_3 and other oxides of nitrogen such as NO_2, the initial reaction forms one or more unique surface nitrate species which do not alter the particle morphology significantly (Vogt and Finlayson-Pitts, 1994; Vogt et al., 1996; Allen et al., 1996). However, if exposed to water vapor even well below the deliquescence point and then dried, this surface nitrate reorganizes into small microcrystallites of $NaNO_3$ attached to the original salt surface (see Fig. 11.63 in Chapter 11). This generates very small particles of $NaNO_3$ and may be responsible for the observation of small particles in the marine boundary layer which are almost completely devoid of chloride (e.g., Mouri and Okada, 1993).

A further complication is the recent indication that even small amounts of strongly bound surface-adsorbed water may play a critical, indeed determining, role in the interaction of gases with surfaces traditionally thought to be solids. For example, in the $NaCl-HNO_3$ reaction, there is evidence that the reaction even in laboratory vacuum systems occurs on sites holding adsorbed water. As a result, the surface does not become saturated as one would expect for a solid surface, since the underlying reactant salt continues to dissolve in the surface water (Beichert and Finlayson-Pitts, 1996).

Another example of reactions at interfaces that is only now being recognized, due to the lack of suitable experimental techniques in the past, is that of species such as SO_2 and NO_2 at liquid interfaces. As discussed in Chapters 7 and 8, there is increasing evidence that the reactions of such species at the air–water interface can be fast relative to that in the bulk and may have unique reaction mechanisms compared to those in the bulk or gas phases. Given the paucity of data on such processes at the present time, they are generally not included in present models of aerosol growth. How-

TABLE 9.10 Typical Time Scales for Various Aerosol Fates[a]

Fate	Type of air mass			
	Urban	Remote marine	Free troposphere	Nonurban continental
Condensation	0.01–1 h	1–10 h	2–20 h	0.5–20 h
Coagulation of 0.03-μm-particles with larger particles	0.1–2 days	10–30 days	~50 days	1–5 days
Deposition				
0.03-μm particles	0.5–10 days	0.5–10 days	—	~1 month
0.3-μm particles	~1 month	~1 month	—	~1 month
Transport	2–5 days	1–2 weeks	3 days to 2 weeks	1–2 weeks

[a] From Pandis et al. (1995).

ever, as our understanding of such reactions at interfaces expands, their implications for the growth of aerosols in the troposphere will need to be critically assessed.

3. Reactions in the Aqueous Phase

We have seen in Chapter 8 that reactions in the aqueous phase present in the atmosphere in the form of clouds and fogs play a central role in the formation of sulfuric acid. Thus, an additional mechanism of particle formation and growth involves the oxidation of SO_2 (and other species as well) in such airborne aqueous media, followed by evaporation of the water to leave a suspended particle.

As discussed earlier, this process is believed to be responsible for the presence of more than one peak in the accumulation mode. For example, Hoppel and co-workers (1985, 1986, 1994) measured particle size distributions in and around clouds. Figure 9.33 shows a typical size distribution for in-cloud particles (the so-called *interstitial* particles) and for particles in the air mass just below the cloud (Hoppel *et al.*, 1994). The larger mode peaking at ~0.15 μm below the cloud is attributed to processing of the smaller particle mode by clouds in two ways (Hoppel *et al.*, 1986). First, the smaller particles are taken up by the cloud droplets. If the cloud evaporates before precipitating (which is most common), the agglomeration of smaller particles taken up into the droplet becomes a particle of larger size. Second, SO_2 can be absorbed into the droplet and oxidized to sulfate (see Chapter 8). When the cloud evaporates, it leaves a particle containing the newly formed sulfate. This has been proposed to explain the bimodal distribution of particles such as that in Fig. 9.33. In-cloud oxidation of SO_2 to sulfate has also been invoked to explain particles at ~0.1 μm in marine aerosols (e.g., Fitzgerald, 1991).

Similarly, Kerminen and Wexler (1995) have examined potential sources of the bimodal particle distribution often observed in the accumulation mode and suggested that it results from aqueous-phase oxidation of SO_2 to sulfate. They assume that the larger particles in such bimodal distributions are formed by the growth of particles in the smaller mode. In this case, to be consistent with atmospheric observations, the mechanism responsible for the two modes must occur on a time scale of hours (a day or so maximum) and must involve growth of the smaller particles preferentially over the larger ones. In addition, because the smaller mode does not get completely converted into the larger mode, only some of the smaller particles must grow or there must be a continuous source of these particles to replace those that grow into the larger mode. They show that only aqueous-phase oxidation of SO_2 in clouds or fogs is consistent with these requirements; growth by condensation of gases on existing particles or by coagulation of particles is too slow to be compatible with the atmospheric observations.

4. Relative Importance of Various Aerosol Growth Mechanisms

Different mechanisms of aerosol growth give rise to different so-called *growth laws*, which are expressions relating the change in particle size (e.g., volume or diameter) with time to the particle diameter. Because different mechanisms of particle formation give rise to different growth laws, one can test experimental data to see which mechanism or combination of mechanisms is consistent with the observations. For a more detailed discussion of this approach, see Friedlander (1977), Heisler and Friedlander (1977), McMurry and Wilson (1982), Pandis *et al.* (1995), and Kerminen and Wexler (1995).

C. CHEMICAL COMPOSITION OF TROPOSPHERIC AEROSOLS

As we have seen in our earlier discussion of the size distribution of tropospheric particles, the chemical components are not generally distributed equally among all sizes but, rather, tend to be found in specific size ranges characteristic of their source. Generally, the smallest ultrafine particles are produced by homogeneous nucleation and hence tend to contain secondary species such as sulfate and likely organics (see Section A.2). Particles in the Aitken nuclei range are produced

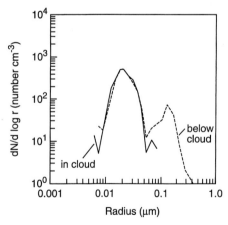

FIGURE 9.33 Size distribution of particles in clouds (solid line) and below the clouds (dashed line), showing two modes (adapted from Hoppel *et al.*, 1994).

by combustion processes, by coagulation of smaller particles, and by condensation of low-vapor-pressure products of gas-phase reactions. Hence these particles and accumulation mode particles tend to contain elements such as carbon and some trace metals such as V, which are characteristic of combustion, as well sulfates, nitrates, and polar organics. Finally, because mechanical processes are primarily responsible for coarse particles larger than about 2.5 μm, these larger particles typically contain elements in soil, sea salt, etc.

In the following, we illustrate these principles using data from studies in various locations and characteristic of various types of environments from polluted urban to the free troposphere.

1. Inorganic Species

a. Size Distribution

Table 9.11 shows the aerodynamic mass median diameter (MMD) for some typical inorganics that are common components of tropospheric particles. Also shown are the calculated crystal enrichment factors, EF_{crust}. These are a measure of the enrichment of the element in the airborne particles compared to that expected for the earth's crust, using aluminum as the reference element. Thus EF_{crust} for a particular element X is defined as

$$EF_{crust} = \{X_{air}/Al_{air}\}/\{X_{crust}/Al_{crust}\}, \quad (HH)$$

where "air" indicates the concentration in airborne particles and "crust" that in the earth's crust. A value of EF_{crust} of approximately unity indicates that the composition of the particles is consistent with that expected if they were formed by erosion of the earth's surface. In this case, one would also expect that they will fall in the coarse particle range, having typical diameters above about 2.5 μm.

The data in Table 9.11 are based on an extensive review of the literature through 1985 by Milford and Davidson (1985). Because they represent composites of many different studies carried out in many different locations by many different investigators, they will not match any particular sample of airborne particulate matter; indeed, for many of these elements, the size distribution is multimodal, which is not reflected in the median values shown in Table 9.11. On the other hand, such data demonstrate very clearly some characteristics of tropospheric particles that are common to many conditions.

For example, the most common elements in the earth's crust (Table 9.12 and Fig. 9.34) are O, Si, Al, Fe, Mg, Ca, Na, K, and Ti. These elements have MMDs > 3 μm and enrichment factors that are generally less than three (Table 9.11). That the enrichment

TABLE 9.11 Aerodynamic Mass Median Diameters of Tropospheric Particles Containing Various Elements Observed in a Number of Studies and Enrichment Factors[a]

Element	MMD (μm)	EF_{crust}[b]
W	0.43	19
Pb	0.55	1500
Hg	0.61	560
Se	0.68	3100
Cd	0.84	1900
Sb	0.86	1400
Br	0.89	1900
Ni	0.98	32
I	1.03	510
As	1.11	310
Cr	1.11	8.1
Zn	1.13	260
Cu	1.29	100
V	1.44	14
U	1.56	2.9
In	1.75	90
Ta	1.77	1.1
Cs	1.89	12
Mn	2.06	3.9
Eu	2.59	2.7
Co	2.63	3.5
Th	2.73	1.8
Sm	2.76	2.1
Cl	3.04	740
Ba	3.31	5.5
Fe	3.42	2.1
K	3.76	2.0
Na	3.78	4.4
Si	3.90	0.79
Sc	4.39	1.2
Al	4.54	1.0
Ca	4.64	2.8
Ce	5.10	2.6
Ga	6.00	2.5
Mg	6.34	2.4
Ti	6.52	1.4
Hf	7.65	2.0
Sr	11.9	1.5

[a] From Milford and Davidson (1985).
[b] Calculated crustal enrichment factors.

factors are not all unity suggests that there is some contribution of other sources to the airborne particles. For example, road dust is thought to be a major source of Ca since crushed limestone is often used in road paving materials.

On the other hand, in the literature reviewed for Table 9.11, particulate lead has an MMD of only 0.55 μm. This is consistent with the major source of lead to the atmosphere in the 1970s being combustion of leaded gasoline, the remainder being due mainly to smelting processes. As a result, Pb is primarily found in the accumulation mode and has a large enrichment factor

TABLE 9.12 Abundance in Earth's Crust of Some of the Most Common Crustal Elements and Those Commonly Found in Atmospheric Aerosols or Used as a Tracer[a]

	Weight percent	Atom percent
O	46.60	62.55
Si	27.72	21.22
Al	8.13	6.47
Fe	5.00	1.92
Mg	2.09	1.84
Ca	3.63	1.94
Na	2.83	2.64
K	2.59	1.42
Ti	0.44	
P	0.105	
Mn	0.0950	
S	0.0260	
C	0.0200	
V	0.0135	
La	0.0030	
N	0.0020	
Pb	0.0013	
Sm	0.0006	

[a] From Mason and Moore (1982).

of 1500. (Similar considerations are expected to apply to manganese in some polluted urban areas where (methylcyclopentadienyl)manganese tricarbonyl, MMT, is added to gasoline to improve octane and act as an antiknock agent (e.g., see Loranger and Zayed, 1997; and Wallace and Slonecker, 1997). Intermediate between these two extremes is vanadium, with an MMD of 1.44 μm and $EF_{crust} = 14$. Vanadium is produced by the combustion of fuel oil, which is expected to generate small particles, but is also found in the earth's crust (Table 9.12), which should lead to its being found in large particles as well. Clearly, the relative contribution of these two processes will depend on the particular location and conditions.

A word of caution is also in order with respect to assigning a particular particle to the fine or coarse particle modes. Since the size distributions can generally be described as log-normal, they do not have sharp cutoffs. A few particles at the top end of the fine mode distribution will have diameters larger than 2.5 μm and a few at the bottom end of the coarse mode will have diameters smaller than this. For example, as Lodge (1985) points out, for a coarse particle distribution with a geometric mean diameter of 15 μm and a geometric standard deviation of 3, about 5% of the particles will have diameters below the 2.5-μm fine particle cutoff. This may be responsible for observations that while Si and Ca dominate the coarse particle mode, they are also often found at significant levels in fine particles (e.g., see Katrinak et al., 1995).

In addition to crustal elements being found in airborne particles due to weathering processes, in marine areas one also finds particles characteristic of sea salt. Wave action entrains air and forms bubbles that rise to the surface. As they rise, dissolved organics may become adsorbed on them. The bubbles burst on reaching the surface, producing small droplets that are ejected into the air. Two types of drops have been distinguished—jet drops and film drops. Jet drops are produced from the jet of water that rises from the bottom of the collapsing bubble; film drops are produced from the bursting of the bubble water film. Particles with a wide range of sizes, from less than 0.1 μm to greater than 100 μm, are formed (Blanchard and Woodcock, 1980; Blanchard, 1985).

It might be assumed that initially the composition of the liquid drops would approximate that of seawater, given in Table 9.13. As evaporation of water in the droplet into the surrounding air mass occurs, the salt

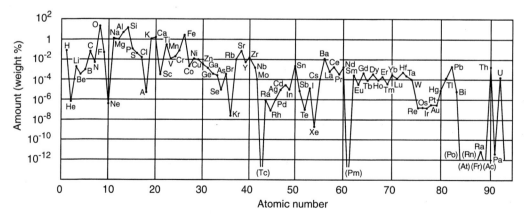

FIGURE 9.34 Amounts of elements (wt%) found in the earth's crust for atomic numbers up to 92 (adapted from Mason and Moore, 1982).

TABLE 9.13 Typical Sea Water Composition^a

Species	Concentration (mg L^{-1})
Al	1×10^{-2}
B	4.6
Ba	3×10^{-2}
Br	65
C	28
Ca	4×10^{2}
Cl	1.9×10^{4}
F	1.3
Fe	1×10^{-2}
I	6×10^{-2}
K	3.8×10^{2}
Mg	1.35×10^{3}
Mn	2×10^{-3}
Na	1.05×10^{4}
Pb	3×10^{-5}
Si	3
S	8.85×10^{2}
Sr	8.1
V	2×10^{-3}
Zn	1×10^{-2}

^{a} From *Handbook of Chemistry and Physics*. (Not all elements are shown.)

becomes more concentrated. If the relative humidity drops below the effluorescence point of NaCl (~43% at 25°C) (Tang, 1980), complete evaporation may occur, leaving a solid salt particle. Along with the salt is any organic material originally associated with the bubble. Thus, marine-derived sterols, fatty alcohols, and fatty acid salts have been shown to be enriched in the surface microlayer of the sea (the first 150 μm) compared to their concentrations in bulk seawater, particularly in bubble interfacial microlayer samples, which occur at the top 1 μm of the surface (Schneider and Gagosian, 1985). Since this top microlayer is believed to be involved in the aerosol formation process from the sea, adsorbed organics are expected on the surfaces of such sea salt particles. These organic layers may carry with them enriched concentrations of cations such as Fe^{3+} as well (Thomsen, 1983).

These sea salt particles may play a major role in the global distribution and fluxes of a number of elements. For example, Fogg and Duce (1985) suggest that they are a major source of atmospheric boron.

Salt particles whose chemical composition reflects that of seawater have been observed in many "clean air" locations. However, in a number of other cases, the ratio of the elemental mass of some elements to that of sodium differs from that in seawater. It has been suggested that ion fractionation may occur during particle formation at the surface; as discussed shortly, chemical reactions may also occur on the particle surface, releasing elements such as chlorine, bromine, and iodine to the gas phase.

As might be expected, the salt concentration and to some extent the size distribution depend on the meteorology, especially wind speed, which drives the wave action. For a detailed discussion of sea salt aerosol, the reader is referred to Blanchard and Woodcock (1980), Blanchard (1985), Erickson et al. (1986), Fitzgerald (1991), Gong and Barrie (1997), Gong et al. (1997), and O'Dowd et al. (1997).

Most of the elements found in coarse particles over land or sea are involatile and relatively chemically inert; however, for a few elements, such as the halogens, this is not true. As early as 1956, Junge noted in marine aerosols on the Florida coast that the particles contained nitrate and that the NO$_3^-$/Cl$^-$ ratio was highest when the wind direction was from the land and lowest when it was from the ocean; presumably the breeze from the land contained anthropogenic pollutants that could react with the particles (Junge, 1956). Since then, numerous investigators have observed such a chloride ion deficiency relative to the sodium concentration, and it has generally been ascribed to reactions of acids such as sulfuric and nitric acids with NaCl to produce gaseous HCl, for example,

$$HNO_{3(g)} + NaCl_{(s)} \rightarrow NaNO_{3(s)} + HCl_{(g)}, \quad (1)$$

$$H_2SO_4 + 2NaCl \rightarrow Na_2SO_4 + 2HCl. \quad (2)$$

The conditions under which HCl formed in acidified sodium chloride droplets would be expected to enter the gas phase have been treated by Clegg and Brimblecombe (1990). Cadle and co-workers (Robbins et al., 1959; Cadle and Robbins, 1960) observed that NaCl aerosols in the presence of 0.1–100 ppm NO$_2$ at relative humidities of 50–100% lost chloride ion from the particles. They ascribed this to the formation of nitric acid from NO$_2$, followed by reaction (1). Schroeder and Urone (1974) subsequently suggested that NO$_2$ could react directly with NaCl to produce gaseous nitrosyl chloride, ClNO, which they observed using infrared spectroscopy; stoichiometrically, this is represented as

$$2NO_{2(g)} + NaCl_{(s)} \rightarrow NaNO_{3(s)} + ClNO_{(g)}. \quad (3)$$

Since then, it has been established that a number of different gaseous oxides of nitrogen, including N$_2$O$_5$ and ClONO$_2$, also react rapidly with NaCl to generate NaNO$_3$ and gaseous products such as ClNO$_2$ and Cl$_2$. These reactions are discussed in Chapter 7 and reviewed by De Haan et al. (1999).

Similar reactions involving bromine and iodine in sea salt particles have also been suggested based on measurements of their concentrations in sea salt aerosols (Moyers and Duce, 1972a, 1972b; Duce et al.,

1973; Moyers and Colovos, 1974; Cicerone, 1981; Sturges and Harrison, 1986). Laboratory studies have confirmed such reactions. For example, gaseous NO_2, N_2O_5, and $ClONO_2$ react with NaBr to form BrNO, $BrNO_2$, and BrCl, respectively (e.g., Finlayson-Pitts and Johnson, 1988; Finlayson-Pitts et al., 1990; Berko et al., 1991; De Haan et al., 1999).

Figure 9.35 shows a typical set of mass size distributions for total suspended particles (TSP), Na, Cl, Al, V, NO_3^-, SO_4^{2-}, and NH_4^+ at Chichi in the Ogasawara (Bonin) Islands, about 1000 km southeast of the main island of Japan (Yoshizumi and Asakuno, 1986). As expected for a marine site such as this, Na and Cl from sea salt predominate, and both the TSP and Na and Cl components peak in the coarse particle range. Al is also found primarily in the larger particles and is attributed to a contribution from soil dust. On the other hand, vanadium, non-sea salt sulfate (nss-SO_4^{2-}), and ammonium are primarily in the fine particles. The vanadium levels are extremely low and likely reflect long-range transport of an air mass containing the products of combustion of fuel oil, which contains V; because it is likely associated with a combustion source, it would be expected in the fine particle mode, consistent with Fig. 9.35.

Sulfate shows two peaks, one in the coarse particle mode associated with sea salt, and one in the fine particle mode. The smaller peak is expected since the nss-SO_4^{2-} is formed from the oxidation of SO_2 and other sulfur-containing compounds such as dimethyl sulfide. The small size of particles containing ammonium reflects the uptake of the gaseous base NH_3 into the smaller acidic particles.

Similar data for sulfate have been reported in many studies. Figure 9.36, for example, shows overall average sulfate distributions measured in marine areas as well as at continental sites (Milford and Davidson, 1987). The marine data show two modes, a coarse mode associated with sea salt and a fine mode associated with gas-to-particle conversion. Sulfate in seawater, formed, for example, by the oxidation of sulfur-containing organics such as dimethyl sulfide, can be carried into the atmosphere during the formation of sea salt particles by processes described earlier and hence are found in larger particles. The continental data show only the fine particle mode, as expected for formation from the atmospheric oxidation of the SO_2 precursors.

Sulfate is a ubiquitous component of particles in the troposphere in both polluted urban areas and remote regions such as the Mauna Loa Observatory (e.g., see Johnson and Kumar, 1991). Indeed, sulfates can sometimes be the major component of tropospheric particles. For example, more than 90% of the total particles sampled in the upper troposphere were observed to contain sulfate (Sheridan et al., 1994), and TEM analysis of particles in the 0.3- to 2-μm range collected in rural Maryland showed that as much as 95% of the particles were sulfate (Mamane and Dzubay, 1986). Sulfur in the form of hydroxymethanesulfonate has also been observed in particles thought to be formed by in-cloud reactions followed by evaporation of water from the particles (Dixon and Aasen, 1999).

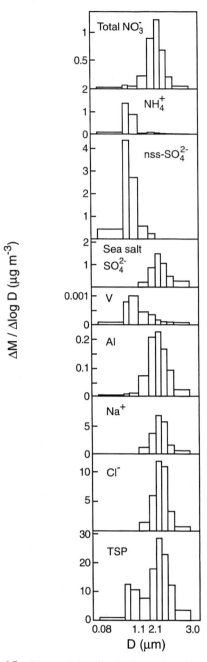

FIGURE 9.35 Observed size distributions of total suspended particles and some of the inorganic constituents of airborne particles at Chichi, Ogasawara Islands, Japan, in 1983 (adapted from Yoshizumi and Asakuno, 1986).

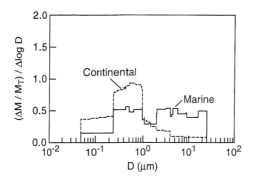

FIGURE 9.36 Average size distributions for sulfate in continental and marine aerosol. $\Delta M/M_T$ is the fraction of the total mass of sulfate (M_T) found in each particle size range (adapted from Milford and Davidson, 1987).

TABLE 9.14 Average Elemental and Nitrate Concentrations of Particles Measured at Ducke Reserve in the Amazon Basin[a]

	Concentration (ng m^{-3} ± sd)	
	Fine (<2 μm)	Coarse (>2 μm)
Mg	11.9 ± 3.8	20.9 ± 12.5
Al	8.37 ± 3.1	32.7 ± 13.8
Si	—	55.9 ± 29.8
P	3.87 ± 1.3	32.2 ± 12.9
S	259 ± 127	65.2 ± 23.4
Cl	6.06 ± 2.68	50.5 ± 43.4
K	161 ± 105	103 ± 36
Ca	3.94 ± 2.86	18.1 ± 11.3
Ti	0.89 ± 0.33	2.90 ± 1.37
V	1.14 ± 0.59	0.61 ± 0.35
Cr	0.76 ± 0.30	0.67 ± 0.28
Mn	0.30 ± 0.16	0.64 ± 0.31
Fe	6.53 ± 3.16	22.8 ± 12.5
Ni	0.54 ± 0.25	0.23 ± 0.11
Cu	0.36 ± 0.04	0.43 ± 0.21
Zn	1.61 ± 0.78	1.42 ± 0.86
Br	0.95 ± 0.51	1.28 ± 0.34
Sr	—	0.43 ± 0.22
Rb	0.52 ± 0.30	0.81 ± 0.52
Zr	—	0.40 ± 0.09
Pb	0.73 ± 0.29	0.46 ± 0.25
Mass (μg m^{-3})	6.75 ± 3.87	7.55 ± 1.60
Elemental carbon (ng m^{-3})	752 ± 258	—
NO$_3^-$ (ng m^{-3})	13.5 ± 5.1	111 ± 135

[a] From Artaxo et al. (1988).

Accurate and artifact-free determination of nitrate in particles is much more difficult than accurate determination of sulfate in particles because both positive and negative artifacts can occur. For example, gas-phase HNO$_3$ can be taken up by particles as the air mass being sampled is drawn over the collected particles, leading to a positive interference. On the other hand, since ammonium nitrate (which can dissociate to HNO$_3$ and NH$_3$; see Chapter 7.E.2) is believed to be one of the major forms of particle nitrate, volatilization of nitrate can occur as the NH$_4$NO$_3$ particles are being collected, leading to a negative interference (see, for example, Dunwoody, 1986; Milford and Davidson, 1987; and Sturges and Harrison, 1988).

Because of this possibility of artifacts in determining either the concentration of particulate nitrate or its size distribution, caution must be exercised in interpreting field data. However, it appears to be generally true that nitrate is found in both the fine and the coarse particle modes, with the relative amounts in each being highly variable. Milford and Davidson (1987) show that aerodynamic mass median diameters (MMD) from 0.3 to 4.2 μm have been reported for nitrate in particles in the literature, the larger ones generally being associated with marine regions. For example, the total nitrate size distribution measured in a relatively clean marine area (Fig. 9.35) shows that in this case, it is found primarily in the large-particle mode. This has been attributed to the reactions of NaCl and NaBr in sea salt particles with gaseous oxides of nitrogen such as HNO$_3$, NO$_2$, N$_2$O$_5$, and ClONO$_2$, all of which form NaNO$_3$ (see Chapter 7).

However, it is not only in marine areas that nitrate is found associated with larger particles. Table 9.14 for example, shows the composition of particles from the Amazon Basin where there was no significant sea salt contribution (Artaxo et al., 1988). Concentrations of nitrate are again seen to be much larger in the coarse particles. These data also illustrate again that soil components such as Si, Al, Mg, Fe, and Ca tend to be found mainly in the coarse particle mode, whereas sulfur (i.e., sulfate) is primarily in fine particles. Artaxo and co-workers have deconvoluted these elemental distributions using principal factor analysis into contributions from different sources, including soil, dust, and two other sources that appeared to be plant related.

On the other hand, nitrate is often a major component of the fine particle mode as well, especially in more polluted urban areas. Figure 9.37, for example, shows the frequency of observations of sulfate and nitrate in various particle sizes during a study in southern California in the summer of 1987 (John et al., 1990). Three distinct peaks are seen for both sulfate and nitrate, at 0.2 ± 0.1, 0.7 ± 0.2, and 4.4 ± 1.2 μm, respectively. Table 9.15 shows the concentrations of nitrate, sulfate, and ammonium in each size range, from which the predominance of nitrate, as well as sulfate and ammonium, in the accumulation mode is evident.

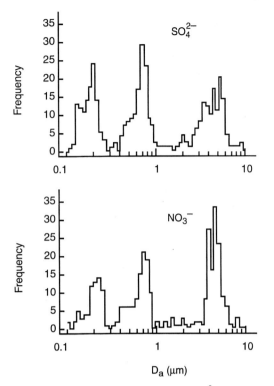

FIGURE 9.37 Frequency of observation of SO_4^{2-} and NO_3^- as a function of particle size during summer, 1987, in southern California (adapted from John *et al.*, 1990).

b. Trace Elements as Tracers

In a number of studies, the elemental composition of aerosol particles has been related to their source. Schroeder and co-workers (1987) review the sources, sinks, analysis, deposition, chemical forms, and global cycles of various trace elements. Because the relative amounts of trace elements vary for coal compared to oil-fueled power plants, for example, it has been suggested that certain elements or ratios of elements may serve as tracers for various sources (e.g., see Gordon, 1988; and Rahn and Lowenthal, 1984). Thus V and Ni are indicative of oil combustion, and elevated concentrations of elements such as As and Se are usually associated with coal burning and smelter operations (e.g., see Mosher and Duce, 1987; and Rabano *et al.*, 1989). The ratio of Mn to V and of Fe to Mg have been suggested as indicators of coal burning (Rahn, 1981; Lowenthal and Rahn, 1987; Parekh and Husain, 1987).

The relative concentrations of rare earth elements in fine particles have also been suggested as being indicative of emissions from oil-fired power plants and from refineries. For example, the ratio of lanthanum to samarium is much higher in emissions from these sources than found in particles formed from the earth's crust (Olmez and Gordon, 1985). Thus the La/Sm ratio was measured to be about 20 for refinery emissions and 28 for emissions from an oil-fired power plant compared to 5 from a coal-fired power plant and 9 from the earth's crust (Olmez and Gordon, 1985; Kitto *et al.*, 1992).

Rare earth isotopes have also been suggested as being useful for tracer studies. That is, they can be intentionally added to a source, such as flue gas or the fuel used in combustion, and their unique isotopic signature used to follow the plume as it disperses downwind. This approach, using the isotopes of Nd, for example, is discussed in more detail by Ondov and co-workers (Ondov and Kelly, 1991; Wight *et al.*, 1991; Ondov *et al.*, 1992; Lin *et al.*, 1992).

While there has been a great deal of interest in using the elemental composition of particles as tracers of sulfate and nitrate from sources located large distances upwind, caution must be exercised in using elements or ratios of their concentrations as indicators of such long-range transport. For example, element concentrations and ratios may change from that at the source as the air mass travels over downwind sources. For a discussion of this and some of the other potential problems, see Thurston and Laird (1985), Rahn and Lowenthal (1985), Lowenthal *et al.* (1988), Lowenthal and Rahn (1988), and Gordon (1988).

c. Source Apportionment Models

Because certain sources emit particles with characteristic elemental signatures, in principle one ought to be able to measure the composition of particles in the atmosphere and then "work backward" to calculate how much each source contributed to obtain the final, observed particle composition. This approach involves the use of receptor models, defined as "models that assess contributions from various sources based on observations at sampling sites (the receptors)" (Gordon, 1988).

There are a number of different types of source apportionment models, including the chemical mass balance method, factor analysis, multiple linear regression analysis, and Lagrangian modeling. The *chemical*

TABLE 9.15 Mean Concentrations of Ammonium, Nitrate, and Sulfate by Size in Particles in Southern California in 1987[a]

Peak size (μm)	NH_4^+	NO_3^-	SO_4^{2-}
0.2	75 ± 84	50 ± 90	24 ± 18
0.7	352 ± 283	219 ± 238	129 ± 66
4.4	64 ± 50	128 ± 75	29 ± 14

[a] From John *et al.* (1990).

mass balance method is based on the idea that the observed concentration of the ith element in the particles (OC_i) is the sum of contributions from each source:

$$OC_i = \sum_j c_{ij} s_j. \quad \text{(II)}$$

In Eq. (II), c_{ij} is the concentration of species i in the jth source and s_j is the source strength for source j (e.g., as mass m^{-3}) contributed by particles from that particular source. This assumes there are no loss processes for the element between source and receptor, e.g., reactions and/or volatilization. If the concentration profiles for the various contributing sources (i.e., the c_{ij}) are measured, as well as the total concentration OC_i at the receptor, the set of simultaneous equations represented by (II) can be solved to obtain s_j, the contribution of each source for that particular element.

Figure 9.38, for example, shows the application of the chemical mass balance approach to the fine particle fraction of particles collected at a location in Philadelphia (Dzubay *et al.*, 1988; Olmez *et al.*, 1988; Gordon, 1988). If the set of equations (II) fitted the data perfectly, the sum of the contributions of the various sources would be 100% for each element. Clearly, from the top frame, this is not the case for a number of elements, and both positive and negative deviations from 100% can be seen. However, the contributions of several sources are clear: Si and Fe from soil, Ni, V, and Ca from oil-fired power plants, Ti from a paint pigment plant, La, Ce, and Sm from a catalytic cracker, K, Zn, and Sn from an incinerator, Sb from an antimony roaster, and Pb and Br from motor vehicles.

This approach has been used to show, for example, that non-catalyst-equipped vehicles are the major source of fine particles in the area around São Paulo, Brazil (Alonso *et al.*, 1997), and in Mexico City (Vega *et al.*, 1997).

A second approach that is used extensively is *factor analysis*. This is based on examining the variations of the concentrations of various elements from their mean value. The data are then analyzed to find groups of elements whose changes as a group account for these variations. These groups of elements, which change in such a way as to indicate they are associated with each other, are known as "factors." In the factor analysis approach, the minimum number of factors or elemental groupings needed to explain the observed variations in the composition is sought. Figure 9.39, for example, shows the results of factor analysis applied to particles collected in the Amazon Basin (Artaxo and Hansson, 1995). Only two factors are needed to explain this data set, one a mineral dust component containing such elements as Si, Al, Ti, and Fe that peaks at a size of 2.0 μm, and the second a biological component containing P, S, K, Ca, Zn, and Sr. The biogenic-associated K, P, S, and Zn are distributed bimodally, as seen in Fig. 9.39. The biogenic component that has enhanced S, Sr, and Zn peaks at 0.25 μm while that with enhanced P and K is found in the 2- to 4-μm range.

This approach has also been applied to the organic component of particles. For example, Veltkamp *et al.* (1996) measured 18 organics in particles at Niwot Ridge, Colorado, as well as a variety of organics and inorganics in the gas phase. They identified seven factors, which include gas-phase internal combustion products (e.g., benzene), particle-phase oxygenated biogenic hydrocarbons (e.g., camphor), particle-phase anthropogenic products (e.g., *n*-heptadecanal), and particle-phase biogenic aldehydes (e.g., mid-molecular-weight *n*-aldehydes).

A third common approach is *multiple linear regression analysis*. In this approach, multiple linear regressions of the form

$$Y_i = a \text{ (tracer of source 1)}$$
$$+ b \text{ (tracer of source 2)} + \cdots + z \quad \text{(JJ)}$$

are carried out, where Y_i is the measured species concentration and the tracers for various sources are measured simultaneously. The regression coefficients $a, b, \ldots$ are then determined from the data. For example, Zweidinger *et al.* (1990) used multiple regression analysis to examine the use of volatile organic hydrocarbons (VOC) as mobile source tracers. They showed that there was a high correlation between certain VOCs and Pb, used as a tracer of mobile source emissions, and proposed that as Pb levels drop due to the increased use of nonleaded gasoline, these could be used to calculate the contribution of mobile sources to organics in fine particles. Swietlicki *et al.* (1996) also used this approach to elucidate the source of air pollution in Lund, Sweden.

Application of a combination of all of these approaches can be particularly powerful when the requisite input data are available. For example, Malm and Gebhart (1997) used a combination of these source apportionment techniques to show that in the Grand Canyon area, about 50% of the sulfur in particles was associated with sources that also emitted Se and hence was attributed to coal-fired power plants in the region. A strong association with bromine was also found, suggesting that wood smoke was also a significant source of sulfur in this region.

Finally, *Lagrangian* approaches to source apportionment have been used in some airsheds. For example,

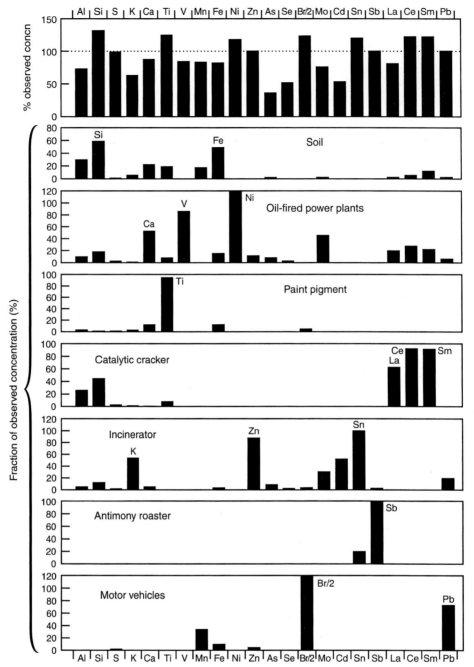

FIGURE 9.38 Example of chemical mass balance approach applied to fine particle fraction in Philadelphia in 1982 (adapted from Gordon, 1988).

Cass and co-workers (e.g., Rogge *et al.*, 1996; Gray and Cass, 1998; Kleeman and Cass, 1998) have used this method to assess the contribution of various sources of particles in the Los Angeles area. If the particle emissions from each source are known and assumed to remain in air as external mixtures, they can be tracked separately in the model as they are transported and react. Hence their contribution to the concentrations at a particular location can be calculated.

For further details of the application of receptor models, see Hopke (1985) and Gordon (1988).

d. Forms of Inorganics in Tropospheric Particles

The discussion thus far has focused on the elemental composition of particles, which can be determined using such techniques as neutron activation or X-ray fluorescence analysis (see Chapter 11). One of the questions of increasing interest is the chemical form in

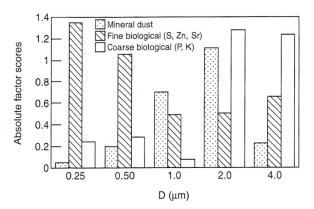

FIGURE 9.39 Analysis of particles as a fraction of particle size in the Amazon Basin using factor analysis (adapted from Artaxo and Hansson, 1995).

which these elements exist in particles and whether it is different in different particles. The terms *internal mixture* and *external mixture* are often used to describe whether two or more species are found together in one particle, i.e., as one "internal mixture," or, rather, are found in separate particles in the same sample, i.e., as an "external mixture."

Getting at this issue for most species has been difficult due primarily to uncertainties introduced during sampling. Thus, until recently, a large number of particles were generally collected, for example, on filters or impactor stages (see Chapter 11.B.1). Chemical analysis was then carried out on the filters or impactor stages. This has proven satisfactory for elemental analysis for most species such as metals, where either positive (e.g., due to adsorption or reactions of gases on existing particles) or negative artifacts (e.g., due to volatilization from the filter) are negligible.

This is not the case for certain critical species such as sulfate, nitrate, and the halogens, however, where artifacts can be severe. Even though analysis of particles after collection may show distinct separations of species into different particles, i.e., external mixtures, it is often difficult to show conclusively that this is the way they existed in the atmosphere. For example, as already discussed (also see Fig. 11.63 in Chapter 11), nitrate can be formed on the surface of salt particles by reaction of NaCl with various oxides of nitrogen, which might be thought of as an "internal" mixture. However, the subsequent exposure of these particles to water vapor followed by drying leads to the formation of individual microcrystallites of $NaNO_3$ and a fresh surface of NaCl, i.e., to an external mixture of $NaNO_3$ and NaCl. The same process may occur during sampling. For example, if a wet sea salt particle containing nitrate is collected on a filter and dried either during collection or subsequently during analysis, the internal mixture may separate during drying into separate crystals, i.e., into an external mixture of particles.

Because of these difficulties in sampling and analysis, relatively little is yet known about the actual chemical forms of inorganics in the atmosphere, except in some unique cases. For example, does sulfate exist as $(NH_4)_2SO_4$ and/or $(NH_4)HSO_4$ as suggested by the molar concentrations of NH_4^+, H^+, and SO_4^{2-} in some locations? One approach to this problem has involved the use of an instrument known as a humidograph (Charlson et al., 1974a, 1974b).

This method is based on the different deliquescence properties of various salts found in ambient air (Table 9.16). At the deliquescence point, the transition from a solid particle to a liquid droplet occurs suddenly, marking the onset of growth of the particle size. Figure 9.40, for example, shows the calculated change in particle size for four different sulfate aerosols, consisting of H_2SO_4, $(NH_4)HSO_4$, $(NH_4)_2SO_4$, and $(NH_4)_3H(SO_4)_2$ (letovicite), respectively, as a function of relative humidity (RH). At the deliquescence point, in each case the particle suddenly increases in size and then grows in a continuous fashion as the RH increases.

The deliquescence points of a variety of single-salt solutions as well as salt mixtures have been studied by a number of researchers (e.g., see Tang, 1976, 1996; Hänel and Zankl, 1979; Tang and Munkelwitz, 1993, 1994a, 1994b; Tang et al., 1978, 1995; and Wexler and Seinfeld, 1991). Tang and co-workers have established that for particles of a given size, the light scattered per unit weight of the dry salt does not depend strongly on the chemical constituents used to generate sulfate and nitrate aerosols but does depend strongly on particle

TABLE 9.16 Deliquescence Points of Some Salts Commonly Found in Ambient Air at 25°C[a]

Composition	Deliquescence Relative humidity (%)
$(NH_4)_2SO_4$	79.9
$(NH_4)HSO_4$	39.0
$(NH_4)_3H(SO_4)_2$	69.0
Na_2SO_4	84.2
$Na_2SO_4 \cdot 10H_2O$[b]	93.6
NH_4NO_3	61.8
NaCl	75.3
$NaNO_3$	74.3
NaCl–$NaNO_3$	68.0
KCl	84.2
NaCl–KCl	72.7

[a] From Tang and Munkelwitz (1993, 1994a, 1994b); note there is a hysteresis so that the RH at which the salts recrystallize as the RH drops (the effluorescence RH) is lower than these values.
[b] From Goldberg (1981).

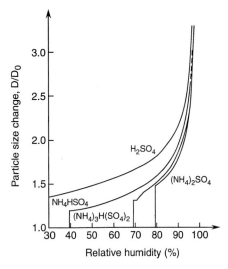

FIGURE 9.40 Calculated changes in particle size as a function of relative humidity at 25°C from particles with four different chemical compositions (adapted from Tang, 1980).

size as expected. They suggest that treating light scattering by such aerosols in terms of an external mixture rather than an internal mixture is satisfactory.

It should be noted that hysteresis occurs as the particles are dried out; i.e., the liquid solution does not form a solid particle as the water evaporates at the same RH as it went through the solid → liquid transition. Typically one must reach RHs 20–30% or more below the deliquescence point in order to "dry" the particle. Figure 9.41, for example, shows the uptake of water by solid $(NH_4)_2SO_4$ and its subsequent dehydration (Tang et al., 1995). At 80% RH the solid deliquesces but does not solidify (effluoresce) on drying until an RH of 37% is reached. (The presence of other species has been shown to increase this effluorescence

RH (e.g., Oatis et al., 1998). In addition, there is evidence that new metastable solid states can form in these droplets (Tang et al., 1995). Similar studies of ammonium bisulfate, NH_4HSO_4, have been carried out over a range of temperatures and relative humidities relevant to the atmosphere, and its phase diagram, including metastable states and a crystalline hydrate phase at low temperatures has been reported (Imre et al., 1997).

The uptake of water by the particles as the RH is increased can be followed by measuring the light scattering of the aerosol with an integrating nephelometer. Figure 9.42 shows the light scattering coefficient b_{sp} (as a ratio to that at 30%) as a function of RH for a pure H_2SO_4 aerosol and for one to which NH_3 has been added. b_{sp} increases monotonically for the hygroscopic H_2SO_4 particles. However, when NH_3 is added, converting the aerosol to $(NH_4)_2SO_4$, a sharp inflection point is seen at ~80% RH, as expected for ammonium sulfate. One can thus distinguish between H_2SO_4 and $(NH_4)_2SO_4$ by the dependence of the light scattering properties on RH, with and without added NH_3 (Charlson et al., 1974a). While this has the advantage of providing an in situ measurement of the chemical nature of the sulfates, the method responds mainly to those particle sizes that scatter light efficiently (i.e., ~0.1–1 μm); fortunately, this also appears to be in the range of greatest importance for health and visibility effects. When the particles are not pure sulfates, but mixtures with other species, the $(NH_4)_2SO_4$ must be at least 30–50 mol% to show a clear inflection point in the humidogram.

Using this humidograph technique, the presence of $H_2SO_4/(NH_4)HSO_4$ or $(NH_4)_2SO_4$ as a major component of ambient light scattering aerosols has been established in a number of locations. $(NH_4)_2SO_4$ was frequently observed, suggesting that there is often sufficient NH_3 present in ambient air to completely neu-

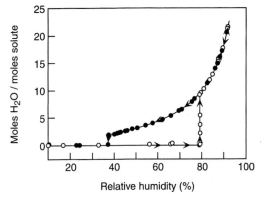

FIGURE 9.41 Uptake of water by $(NH_4)SO_4$ and its drying as a function of relative humidity at 25°3C (adapted from Tang et al., 1995).

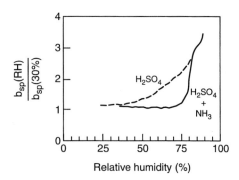

FIGURE 9.42 Humidogram for laboratory H_2SO_4 aerosol (dashed) and for the reaction product of H_2SO_4 and NH_3 (solid) (adapted from Charlson et al., 1974a).

tralize the sulfuric acid (Weiss et al., 1977; Waggoner et al., 1983a). A modified version of the humidograph, in which a heating and cooling cycle is included to differentiate various salts based on their thermal properties (also referred to as thermography), has also been applied to estimate the molar ratio of NH_4^+ to SO_4^{2-} (Cobourn et al., 1978; Weiss et al., 1982; Larson et al., 1982; Rood et al., 1985, 1987; Covert, 1988). This approach has also been used, in combination with laboratory measurements of the deliquescence point and thermodynamic considerations, to show that the aerosol particles in Riverside, California, were likely an internal mixture of $(NH_4)_2SO_4$, Na_2SO_4, and H_2O (Koloutsou-Vakakis and Rood, 1994).

The form of sulfate in particles has also been indirectly inferred from analysis of the cations in the particles, particularly the ammonium ion. For example, Turpin et al. (1997) measured NH_4^+ and SO_4^{2-} as well as H^+ and NO_3^- in particles collected at Meadview, Arizona. The ratio of NH_4^+ to SO_4^{2-} varied from 1.0 to 2.1, with an average of 1.46. Consistent with this degree of neutralization of the sulfuric acid was an average ratio of H^+ to SO_4^{2-} of 0.54.

Similarly, the relationship between H^+ and SO_4^{2-} has been shown to be an indicator of the acidity in particles at a variety of sites in Canada. For example, Brook et al. (1997) reported that the amounts of H^+ and SO_4^{2-} in particles were very strongly correlated in St. John, New Brunswick ($r = 0.92$), whereas at sites in Ontario and Quebec, the correlations were much weaker. The strong correlation between H^+ and SO_4^{2-} in St. John was attributed to local emissions of SO_2, followed by uptake into the fogs common to that area, and efficient oxidation in the aqueous phase to H_2SO_4 (see Chapter 8.C.3). At the other sites, there were larger concentrations of NH_3 to at least partially neutralize the sulfuric acid.

In short, some particles in air can become quite acidic. For example, the pH of particles in southern California was estimated to be 0.2 and below (Li et al., 1997).

The ratio of hydrogen to sulfur in particles has been used as an indicator of the degree of neutralization of sulfuric acid. For example, Cahill et al. (1996) measured hydrogen by proton elastic scattering analysis (PESA) and sulfur by particle-induced X-ray emission (PIXE) in particles collected at Shenandoah National Park, U.S. For particles with diameters from 0.069 to 0.34 μm, the ratio of H to S was consistent with $(NH_4)_2SO_4$. However, for those with diameters in the range from 0.34 to 1.15 μm, the ratio was highly variable, ranging from those corresponding to $(NH_4)_2SO_4$ to those expected for H_2SO_4. Similarly, using this approach, Day et al. (1997) found that the ratio of hydrogen to sulfur in particles at the Great Smoky Mountains National Park, U.S., was (5.4 ± 0.6) to 1.0, suggesting that the particles were slightly more acidic than NH_4HSO_4.

Sulfuric acid may exist not only inside particles but, in some cases, perhaps as a coating on the outside. For example, Cantrell et al. (1997) reported that heating particles collected in the Arctic during the spring to 130°C led to a decrease in their size but no change in their number concentration. They proposed that this was due to a coating of sulfuric acid on the particles, which volatilized at 130°C. (They argue that although an organic coating cannot be ruled out, it is less likely than a sulfuric acid coating.)

More recently, the advent of real-time, single-particle monitoring techniques has started to give some detailed information on the composition of single particles. Because these methodologies are undergoing development (see Chapter 11.B.4), there are many uncertainties still associated with them. However, initial results are intriguing. For example, negative ion laser ionization mass spectrometry measurements of particles in Colorado found that sulfate and nitrate were most commonly found in separate particles (Murphy and Thomson, 1997). Using the same technique, Noble and Prather (1996) show that particles in Riverside, California, could generally be broken down into four classes: organic, marine, soot, and inorganic.

Clearly, such real-time, single-particle techniques have great promise in terms of elucidating the chemical nature of different species in tropospheric particles and determining whether they exist as internal or external mixtures.

It should be noted that there are likely to be some aerosol particle components that are not readily detectable by the techniques in use now. For example, Kao and Friedlander (1995) have suggested that compounds such as H_2O_2 and free radicals that may be important toxicologically would have reacted prior to particle analysis and that species formed from such reactions, e.g., sulfate, may be used as markers of their presence.

e. Unique Sources of Particles

The foregoing discussion is intended to provide an overview of the most common characteristics of inorganic tropospheric particles. Obviously, there can be wide variation from such "typical" cases, especially under unusual circumstances.

For example, in some areas dust storms are prevalent and extremely high particle concentrations along with unique elemental signatures result. One such case is in central California, where particle concentrations from Owens (dry) Lake are highest in the United

TABLE 9.17 Individual Compounds Identified in the Ether-Extractable Organic Matter from Aerosol Particles Collected over the Southern Region of the North Atlantic[a]

Compound	Concentration (ng m^{-3} air at STP)	Compound	Concentration (ng m^{-3} air at STP)
I. Aliphatic hydrocarbons		IV. Organic acids	
n-Decane	0.55	Pelargonic acid	2.5
n-Undecane	1.9	Capric acid	3.5
n-Dodecane	0.92	Hendecanoic acid	6.1
n-Tridecane	0.85	Lauric acid	2.0
n-Tetradecane	1.2	Tridecanoic acid	1.6
n-Pentadecane	0.70	Myristic acid	1.3
n-Hexadecane	0.58	Pentadecanoic acid	3.8
n-Heptadecane	1.0	Palmitic acid	5.3
n-Octadecane	0.69	Margaric acid	2.1
n-Nonadecane	2.1	Stearic acid	2.4
n-Eicosane	0.01	Oleic acid	1.7
n-Heneicosane	0.61	Linoleic acid	3.1
n-Docosane	1.3	Linolenic acid	1.1
n-Tricosane	0.81	Nonadecanoic acid	1.1
n-Tetracosane	0.24	Arachidic acid	1.1
n-Pentacosane	0.10	Heneicosanoic acid	0.01
n-Hexacosane	0.39	Behenic acid	0.82
n-Heptacosane	0.01	Total	39.5
n-Octacosane	<0.01		
n-Nonacosane	<0.01	V. Organic bases	
n-Triacontane	<0.01	Quinoline	<0.01
Total	14.0	Isoquinoline	<0.01
		Aniline	<0.01
II. Polycyclic aromatic hydrocarbons[b]		Indole	0.20
Naphthalene	2.7	2-Methylindole	0.20
2-Methylnaphthalene	1.8	7-Methylindole	0.17
Acenaphthalene	0.38	2,3-Dimethylindole	0.14
Acenaphthene	0.25	2,5-Dimethylindole	0.15
Fluorene	0.31	2,4-Dimethylquinoline	0.41
Phenanthrene	0.05	2,6-Dimethylquinoline	0.02
Anthracene	0.49	2,8-Dimethylquinoline	0.78
2-Methylanthracene	2.2	α-Naphthylamine	<0.01
Fluoranthene	1.4	Benzo[h]quinoline	<0.01
Pyrene	0.24	Acridine	<0.01
Benzo[a]fluorene	0.78	Phenanthridine	<0.01
Chrysene	0.37	Total	2.1
Benzo[a]anthracene	0.28		
Benzo[a]pyrene	<0.01		
Benzo[e]pyrene	<0.01		
Perylene	<0.01		
Total	11.3		
III. Polar compounds			
Coumarine	0.04		
$peri$-Naphthenone	0.11		
Xanthone	0.22		
Anthrone	0.04		
Flavone	0.07		
Benzoanthrone	0.24		
Carbazole	0.04		
Total	0.76		

Total identified 67.7

[a] From Hahn (1980) and Ketseridis et al. (1976).
[b] See Chapter 10.

States during dust storms, as much as 1000 $\mu g\ m^{-3}$, with arsenic levels as high as 0.4 $\mu g\ m^{-3}$ (Reid et al., 1994).

Similarly, the plume from the burning of oil wells in Kuwait had an unusual composition, containing both soot and significant salt concentrations due to brine mixed in with the oil [see, for example, the special section in the *Journal of Geophysical Research*, Vol. 97 (D13), September 20, 1992].

Plumes from biomass burning can also have unique signatures. For example, organics, ammonium, potassium, sodium, nitrate, nitrite, sulfate, chloride, phosphate, elemental carbon, and the anions of organic acids (formate, acetate, oxalate, etc.) have all been measured in particles in the plumes from burning vegetation (e.g., see Cofer et al., 1988; Andreae et al., 1988; and Artaxo et al., 1994).

In short, while there are many common elements and size distributions in tropospheric particles found in many areas around the world, it is important to recognize that in some circumstances, the particle composition and size distributions may be unique.

2. Organics

While inorganics, particularly sulfates and nitrates, tend to be ubiquitous components of tropospheric particles, in many cases a large portion of the mass lies in organic compounds. The composition of the organics is complex, even in remote areas with negligible contributions from combustion processes. We first discuss typical organics found in remote areas, and then those characteristic of polluted urban areas. For a discussion of the history of this field and a collection of papers on this topic, see the Proceedings of the Fifth International Conference on Carbonaceous Particles in the Atmosphere [*J. Geophys. Res.*, 101, 19373–19627 (1996)].

a. Biogenically Derived Organics in Aerosol Particles

Table 9.17 shows some of the organics identified in nonurban aerosols. A wide variety of organics are found, including alkanes, alkenes, aromatics, fatty acids, alcohols, and organic bases.

n-Alkanes in the $C_{15}-C_{35}$ range are common components of nonurban aerosols. As seen in Fig. 9.43, there is a preference for compounds with *odd* numbers of carbon atoms. This is expressed in terms of the *carbon preference index* (CPI) (Cooper and Bray, 1963), which for alkanes is the sum of the odd carbon number alkanes over the sum of the even carbon number alkanes. It is often applied to restricted carbon number ranges to separate out different sources.

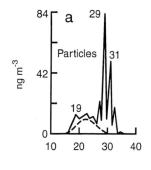
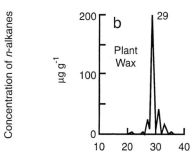
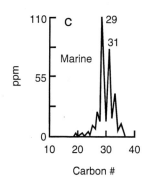

FIGURE 9.43 Concentrations of *n*-alkanes as a function of carbon number from (a) a particle sample collected in Jos, Nigeria (the dashed line represents the estimated concentrations from fossil fuel sources), (b) vegetation wax in the Jos area, and (c) particles collected over the Atlantic Ocean (adapted from Simoneit et al., 1988).

This can be illustrated, for example, by a study by Simoneit et al. (1988) of the organics in particles in Nigeria. As seen by comparing Fig. 9.43a to Fig. 9.43b, the $C_{26}-C_{35}$ range can be attributed largely to epicuticular waxes of vascular plants. The similarity in the distribution for particles collected over the Atlantic Ocean (Fig. 9.43c) was cited by Simoneit et al. (1988) as evidence for transport from continental areas. A CPI > 2 for $C_{23}-C_{35}$ *n*-alkanes is indicative of biogenic sources. For the sample shown in Fig. 9.43a, for example, the CPI for the $C_{25}-C_{35}$ region was 6.2. The same definition of CPI is applied to *n*-alkan-2-ones, which are also found in particles of biogenic origin.

Two other characteristics of organics in particles are also used to differentiate biogenic from anthropogenic sources (Simoneit, 1989). These are the unresolved to resolved component areas on gas chromatograms for the *n*-alkanes, referred to as U:R, and the carbon number of the compound present in the largest quantity in the homologous series, C_{max}. In Fig. 9.43a, for example, the dashed line is the estimated contribution of alkanes from fossil fuel sources in the Nigerian samples, where no odd carbon number preference is observed (i.e., CPI ~1). For this sample, the U:R ratio was 2.2, indicative of some contribution from fossil fuels (Simoneit *et al.*, 1988).

Also produced by biogenic processes are *n*-alkanoic acids and *n*-alkanols having a predominance of *even* carbon numbers. For these compounds, the CPI is defined as the ratio of *even* numbered carbon compounds to *odd* numbered compounds. The acids with fewer than 20 carbons (i.e., C_{10}–C_{18}) are believed to be from microbial sources, whereas the higher acids, C_{22}–C_{34}, are from vascular plant wax. Figure 9.44, for example, shows the *n*-alkanoic acids and *n*-alkanols from particles in a remote region in Nigeria compared to those found in the wax from local vegetation as well as to those in particles over the Atlantic Ocean. The particles from the remote region have a peak for the acids and alcohols at $C_{max} = C_{30}$, similar to that found in vegetation wax from that region (Figs. 9.44b and 9.44e). The CPI for the *n*-alkanoic acids in this sample was 4.3, and for the *n*-alkanols, 4.8, again reflecting the biogenic source of these acids. On the other hand, the alkanoic acids in particles from over the Atlantic peaked at C_{16}, reflecting a microbial source. The alkanoic acids may be, in part, oxidation products of the corresponding aldehydes released by the plants, since the aldehydes were, in this case, found in the plant wax but not in the aerosol particles (Simoneit *et al.*, 1988). [In other cases, e.g., Oregon, U.S., aldehydes and ketones were shown to be present both in the leaf wax and in airborne particles (e.g., Chen and Simoneit, 1994).

As discussed in Chapter 14.C, there is increasing evidence that such biogenically derived organics may play a role in climate by acting as cloud condensation nuclei. For example, Novakov *et al.* (1997) collected particles at a coastal site in Puerto Rico and at a site upwind over the Atlantic Ocean and showed that the organic component exceeded that due to sulfate. The organic material was water soluble and preliminary mass spectral analysis indicated the presence of fatty acids and carboxylic acids, consistent with the preceding discussion and Fig. 9.44. Similarly, Hegg *et al.* (1997) sampled particles off the mid-Atlantic coast of the United States and found that the carbonaceous component comprised ~50% of the dry weight and contributed between 4 and 18% of the total aerosol optical depth.

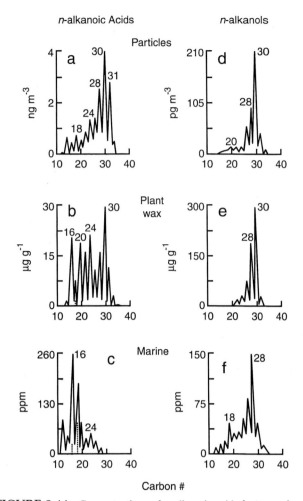

FIGURE 9.44 Concentrations of *n*-alkanoic acids (parts a–c) and *n*-alkanols (parts d–f) as a function of carbon number from (a, d) a particle sample collected in Jos, Nigeria, (b, e) vegetation wax in the Jos area, and (c, f) particles collected over the Atlantic Ocean (adapted from Simoneit *et al.*, 1988).

Similar biogenic organics and their distributions have been noted in a variety of locations, including the Amazon, China, and over the south Atlantic area (e.g., see Simoneit *et al.*, 1990, 1991a, 1991b). They are also found in urban areas but generally comprise a small fraction of the total aerosol mass; for example, they are about 1–3% of the total fine particulate matter in the Los Angeles area (Hildemann *et al.*, 1996).

Some monounsaturated fatty acids such as $C_{16:1}$ and $C_{18:1}$ (the first number is the number of carbon atoms and the second number is the number of double bonds) are also produced biogenically, in addition to some diunsaturated species (e.g., see Sicre *et al.*, 1990a). However, because they react fairly rapidly with tropospheric gases such as O_3, OH, and NO_3, their concentrations are highly variable and also dependent on storage conditions after sampling. A variety of dicarboxylic acids have been observed in aerosols in the Antarctic and attributed to the oxidation of unsaturated fatty acids emitted from marine microalgae (Kawamura *et al.*, 1996a). Also found in small amounts are *iso-* and *anteiso*-alkanoic acids (the latter have a CH_3 group on the third carbon from the end, rather than the second, as is the case for the *iso* compounds; i.e., the *anteiso* compounds are 3-methyl substituted and the *iso* compounds are 2-methyl substituted).

Figure 9.45 shows the structures of a few of the other organics found in tropospheric aerosols and believed to be of biogenic origin. The reader is referred to summaries by Simoneit (1989) and Graedel *et al.* (1986) for more detailed lists of compounds and descriptions of sources.

In short, biogenic processes produce complicated mixtures of organics that are structurally large and have sufficiently small vapor pressures that they are found primarily or exclusively in airborne particles. While the classes of compounds discussed are typical, there are a variety of other compounds found as well, depending on the particular location, time, etc. For

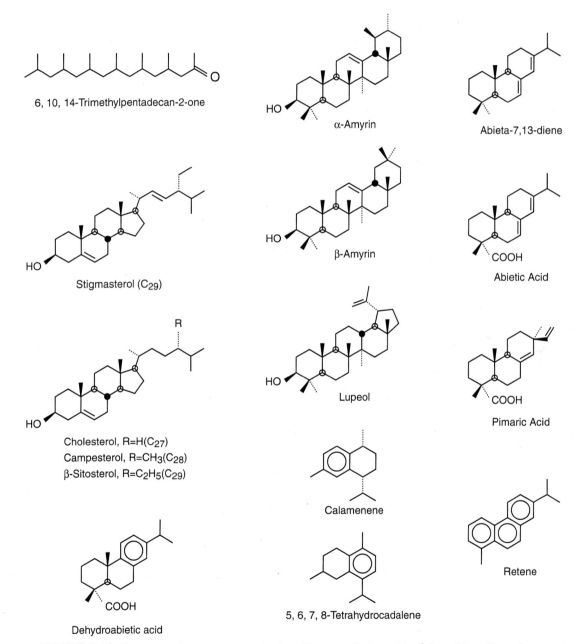

FIGURE 9.45 Some naturally occurring organics found in tropospheric particles (adapted from Simoneit *et al.*, 1988, 1989).

example, long-chain alkenones have been observed in New Zealand and attributed to an algal source but not on American Samoa (Sicre *et al.*, 1990a).

In addition to these *nonviable* organics, there are, of course, a whole host of viable species, such as fungi, bacteria, pollen, yeasts, and viruses, also present in the atmosphere. For more details of the variety of organics found in aerosols in various locations and their concentrations, the reader is referred to papers by Duce *et al.* (1983), Chen and Simoneit (1994), and Rogge *et al.* (1993d).

b. Organics in Anthropogenically Influenced and Aged Aerosol Particles

Particles collected in regions influenced by anthropogenic emissions and those in aged air masses where there has been ample opportunity for oxidation are even more complex than those in remote regions. Thus, they may contain not only the organics from biogenic emissions but also complex organics that either have been directly emitted from sources associated with human activities or have been formed in air from reactions of primary emissions.

(1) Direct emissions To fully appreciate the complexity of even those organics that are directly emitted from various sources in urban areas, the reader should consult the papers by Simoneit, Cass, and co-workers (e.g., see Standley and Simoneit, 1987; Simoneit *et al.*, 1988, 1993; Mazurek *et al.*, 1991; and Rogge *et al.*, 1993a, 1993b, 1993c, 1993e, 1994, 1996, 1997a, 1997b, 1998) in which they characterize organics from various sources. For example, automobiles and heavy-duty diesel trucks are shown to be sources of *n*-alkanes, *n*-alkanoic acids, aromatic aldehydes and acids, polycyclic aromatic hydrocarbons (PAH; see Chapter 10), oxidized PAH derivatives, steranes, pentacyclic triterpanes, and azanapthalenes (Rogge *et al.*, 1993a); abrasion of leaf surfaces a source of *n*-alkanes, *n*-alkanoic and *n*-alkenoic acids, *n*-alkanols, *n*-alkanals, and terpenoids (Rogge *et al.*, 1993b); combustion of natural gas home appliances a source of *n*-alkanes and *n*-alkanoic acids, PAH, oxidized PAH derivatives, and azaarenes (although total particle emissions from this source are fairly low) (Rogge *et al.*, 1993c); and wood smoke a source of compounds such as diterpenoid acids, retene, dehydroabietic acid (Fig. 9.45), and 13-isopropyl-5α-podocarpa-6,8,11,13-tetraen-16-oic acid (Standley and Simoneit, 1990; Mazurek *et al.*, 1991; Rogge *et al.*, 1993d). Other characteristic compounds from wood smoke include derivatives of 2-methoxyphenol (guaiacol) and 2,6-dimethoxyphenol (syringol) (Hawthorne *et al.*, 1988). Figure 9.46 shows

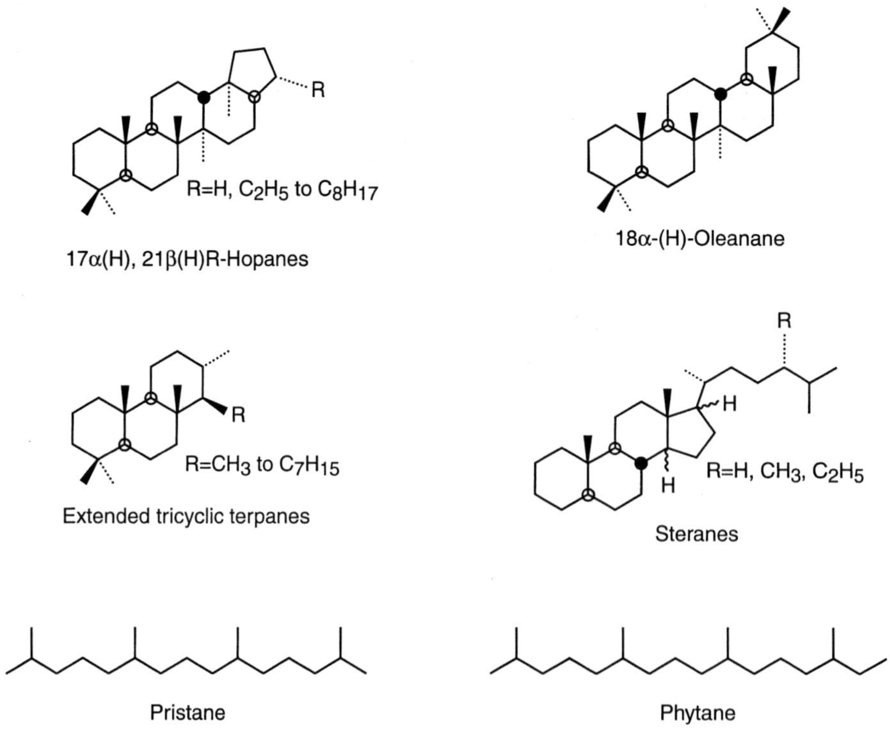

FIGURE 9.46 Some organics from petroleum that are found in tropospheric particles (adapted from Simoneit *et al.*, 1988).

some of the larger organics associated with petroleum that have been observed in tropospheric particles (Simoneit et al., 1988).

A particularly interesting example of direct emissions of particles into urban areas is that from cigarette smoke. Figure 9.47 shows the concentrations of some long-chain alkanes found in cigarette smoke (Fig. 9.47a) as well as from abrasion of leaves characteristic of the Los Angeles area (Fig. 9.47b) (Rogge et al., 1994). The "*i*" refers to the *iso* (i.e., 2-methyl) isomers while the "*a*" refers to the *anteiso* (i.e., 3-methyl) isomers. The distribution is quite different in cigarette smoke, with the i-C_{31} and a-C_{32} isomers being about equal in cigarette smoke but not in the leaf abrasion products.

In addition, the i-C_{33} component is relatively more important in the leaf abrasion particles. Similarly, Kavouras et al. (1998) have shown that in addition to the concentrations and ratios of the *iso*- and *anteiso*-alkanes, there are other markers associated with the alkanes in cigarette smoke. For example, the CPI for cigarette smoke was measured to be 3.44 compared to 1.41–1.66 for particles in urban and suburban areas. (The CPI for rural areas was 4.53, reflecting the contribution from plant waxes.)

Figure 9.47c shows the annual average concentrations for these compounds in particles from three locations in southern California, from which the contribution of cigarette smoke to the ambient particulate matter is evident. The signature of cigarette smoke has also been observed in outdoor air particles in Crete, Greece (Kavouras et al., 1998). Indeed, based on such data, Rogge et al. (1994) estimate that cigarette smoke was responsible for about 1% of the *outdoor* fine particle concentrations in the Los Angeles area!

(2) Secondary organic aerosol formation Recall from the discussion of the kinetics, mechanisms, and products of the gas-phase oxidation of organics (Chapter 6) that a wide variety of multifunctional organics can be formed by reactions of even relatively simple organics commonly found in urban air. Some of these have sufficiently low vapor pressures that they will exist primarily in the condensed phase, i.e., in the form of particles. In particular, the oxidation of the larger and cyclic alkenes, aromatic hydrocarbons, and the larger alkanes and cycloalkanes is expected to lead to condensed-phase multifunctional organic products (e.g., see Grosjean and Seinfeld, 1989; Grosjean, 1992) and this is indeed what is observed. For example, difunctionally substituted alkane derivatives of the type X–$(CH_2)_n$–X with $n = 1$–5 have been observed in a number of studies in the Los Angeles area. The substituent X can be –COOH, –CHO, –CH_2OH, –CH_2ONO, –COONO, or –$COONO_2$. Table 9.18 shows some of the difunctional species identified in submicron aerosols using high-resolution mass spectrometry (MS) of ambient particle samples introduced into the mass spectrometer by slow heating from 20 to 400°C; these compounds were present primarily in the submicron fraction of the particles, suggesting they were secondary in nature, that is, were formed from chemical reactions in the atmosphere (Schuetzle et al., 1975; Cronn et al., 1977). Similarly, the C_3–C_9 dicarboxylic acids measured in southern California appear to be predominantly secondary (Rogge et al., 1993d, 1996).

An idea of the complexity of the compounds found in air can be seen from the work of Yokouchi and

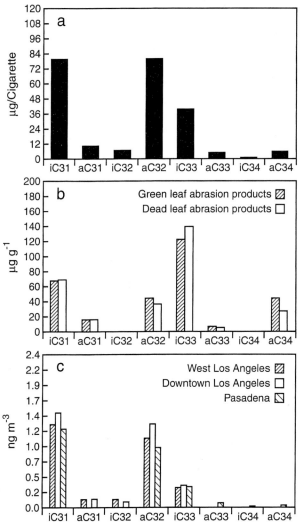

FIGURE 9.47 Distribution of 2-methyl (i = iso) and 3-methyl (a = anteiso) alkanes measured in (a) cigarette smoke, (b) particles from abrasion of leaves typical of the Los Angeles area, and (c) annual average concentrations measured in three locations in southern California in 1982 (adapted from Rogge et al., 1994).

TABLE 9.18 Some Difunctionally Substituted Alkane Derivatives Found in Submicron Ambient Particles in Urban Air[a]

Compound	n
$HOOC(CH_2)_nCOOH$	1–5
$HOOC(CH_2)_nCHO$	3–5
$HOOC(CH_2)_nCH_2OH$	3–5
$HOOC(CH_2)_nCH_2ONO$ or $CHO(CH_2)_nCH_2ONO_2$	3–5
$CHO(CH_2)_nCH_2OH$	3–5
$CHO(CH_2)_nCHO$	3–5
$HOOC(CH_2)_nCOONO$ or $CHO(CH_2)_nCOONO_2$	3–5
$CHO(CH_2)_nCOONO$	3, 4
$HOOC(CH_2)_nCOONO_2$	4, 5
$HOOC(CH_2)_nCH_2ONO_2$	3, 4

[a] From Schuetzle et al. (1975) and Cronn et al. (1977).

Ambe (1986), who analyzed for organic acids in particles collected in Tsukuba, Japan, about 60 km northeast of Tokyo by forming and analyzing their methyl esters. Figure 9.48 shows a typical chromatogram of the methyl esters of the organics in one sample, and Table 9.19 identifies the compounds associated with the numbered peaks in Fig. 9.48. It should be noted that this is not *all* of the organics in this sample, but rather only those that are highly polar and can be extracted into methanol. In one study in central Japan, dicarboxylic acids were shown to comprise as much as 30–50% of the total organic particulate matter and most of them were estimated to be formed by oxidation of precursors in air (Satsumabayashi et al., 1989, 1990). Once oxidized, however, these acids may be transported significant distances (e.g., see Sempere and Kawamura, 1996).

However, while a great many of these compounds are undoubtedly formed from reactions in air, some are also direct emissions. For example, Kawamura and Kaplan (1987) have identified a number of dicarboxylic acids in automobile exhaust, and Rogge et al. (1991) have identified C_4–C_8 dicarboxylic acids from meat cooking. In a study in Philadelphia, Lawrence and Koutrakis (1996a, 1996b) attribute dicarboxylic acids to a combination of combustion sources, photochemical formation, and biogenic sources.

Infrared spectroscopy applied to urban particles shows a complex mixture of organics and inorganics as expected, with evidence for multifunctional organics containing both carboxylic acid and organic nitrate groups, for example (e.g., O'Brien et al., 1975b; Mylonas et al., 1991; Blando et al., 1998). The presence of organic nitrates in tropospheric aerosols is not unique, however, to highly polluted urban areas. For example, they have also been observed in many rural and suburban areas such as in Denmark (e.g., Nielsen et al., 1995, 1998; see also Chapter 11). Also identified by IR are organics such as hydrocarbons, ketones, and esters (Gordon et al., 1988) as well as inorganics such as NH_4^+, NO_3^-, and SO_4^{2-} (e.g., see Cunningham et al., 1984; Pickle et al., 1990; Mylonas et al., 1991; and Blando et al., 1998). A typical example is found in Fig. 9.49.

Although one might expect the complex mixture shown in Fig. 9.49 in an urban atmosphere, the same is true of rural and remote areas. For example, Fig. 9.50 shows infrared spectra of particles with diameters from 0.5 to 1.0 μm collected in the Smoky Mountains,

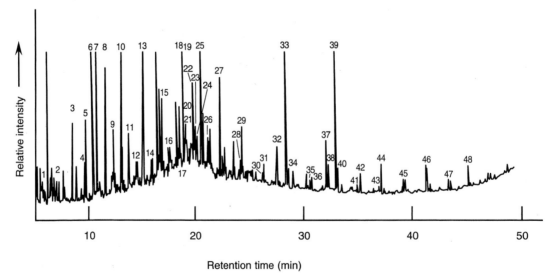

FIGURE 9.48 GC trace of methyl ester derivatives formed from polar organics in particles collected in Tsukuba, Japan (see Table 9.19) (adapted from Yokouchi and Ambe, 1986).

TABLE 9.19 Identification of Some of the Peaks of Polar Organic Acid Methyl Ester Derivatives Shown in Fig. 9.48[a]

Compound	Peak number
Oxalic acid	1
3-Hydroxyisovaleric acid	2
Malonic acid	3
2-Furoic acid	4
4-Oxopentanoic acid	5
Succinic acid	7
Methylsuccinic acid	8
Benzoic acid	9
Glutaric acid	10
2-Methylglutaric acid	11
Salicylic acid	12
Adipic acid	13
3-Methyladipic acid	14
Pimelic acid	15
Anisic acid	16
3-Hydroxybenzoic acid	17
Phthalic acid, phthalic anhydride	18
Suberic acid	19
4-Oxopimelic acid	20
p-Hydroxybenzoic acid	21
Terephthalic acid	22
Isophthalic acid	23
Vanillic acid	24
Azelaic acid	25
3,4-Dimethoxybenzoic acid	26
Sebacic acid	27
Myristic acid	28
Pentadecanoic acid	30
Palmitic acid	33
Heptadecanoic acid	35
Linoleic acid	37
Oleic acid	38
Stearic acid	39
Nonadecanoic acid	41
Dehydroabietic acid	43
Eicosanoic acid	44
Heneicosanoic acid	45
Docosanoic acid	46
Tricosanoic acid	47
Tetracosanoic acid	48

[a] From Yokouchi and Ambe (1986).

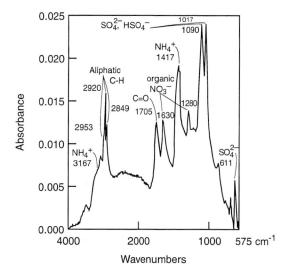

FIGURE 9.49 Infrared spectrum of submicron (0.050–0.075 μm) particles collected in Los Angeles (adapted from Mylonas et al., 1991).

remote region. Further rinsing with water removed most of the remaining peaks assigned to ammonium and sulfate.

Consistent with the earlier discussion of the contribution of crustal materials to larger particles, rinsing particles with diameters in the 1- to 2-μm range with water removed the peaks due to ammonium etc. but left peaks in the 1000- and 500-cm^{-1} regions, which are characteristic of minerals such as kaolinite and serpentine (Fig. 9.51).

Similar observations regarding the importance of polar organics in particles in rural and remote regions

Tennessee (Blando et al., 1998). Peaks very similar to those observed in the urban sample are seen. (Note, however, the different size ranges of the samples.) In addition, a peak at 877 cm^{-1}, characteristic of an S–O–C asymmetric stretch, suggests the presence of organosulfur compounds. Gentle rinsing with hexane to remove nonpolar organics gave no significant change in the spectrum, whereas rinsing with acetone removed a number of the larger peaks (e.g., those due to C=O and O–H as well as the organosulfur species), showing the predominance of polar organics in this relatively

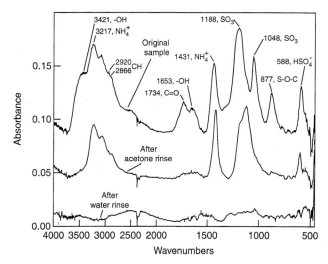

FIGURE 9.50 FTIR spectra of particles from the Smoky Mountains, Tennessee, with diameters between 0.5 and 1.0 μm (top spectrum), after rinsing with acetone (middle spectrum) and then with water (bottom spectrum) (adapted from Blando et al., 1998).

have been reported by a number of researchers. For example, Mazurek *et al.* (1997) analyzed particles collected in the Grand Canyon area of the United States and found that the concentration of organics in fine particles was about equal to that of sulfate. Of the organics, about 25–50% could be analyzed by GC-FID, and of this, about half were highly polar compounds. Figure 9.52 shows the particle fine mass concentrations of the elutable organics at Hopi Point, divided into those that are acid (determined using derivatization to the methyl esters) and those that are neutral, as a function of elution time/molecular size (Mazurek *et al.*, 1997). Clearly, a large portion of the organics are acidic. For comparison, similar distributions are shown for the urban area in West Los Angeles, upwind at San Nicholas Island in the Pacific Ocean, and downwind at Rubidoux, near Riverside. The differences in distributions between the Grand Canyon sample and the urban samples suggest that long-range transport from this urban area was not important and that the organic portion of the particles at Grand Canyon may reflect naturally emitted biogenic hydrocarbons and their oxidation products (see Chapter 6.J).

Figure 9.52 also illustrates the effects of transport and oxidation in an urban area. Thus, the total amount of elutable organics and the acid fraction are both much smaller at the upwind San Nicholas Island site. However, as the air mass travels downwind over an urban area, both the concentration of particulate organics and the concentrations of acids increase (e.g., compare West Los Angeles to Rubidoux).

Although a clear anthropogenic signature was not evident in the particulate organics from the Grand Canyon area in Fig. 9.52, it has been reported in other studies of this region. For example, Cui *et al.* (1997) analyzed particles from Meadview, Arizona. They reported that lower molecular weight fatty acids ($<C_{20}$) predominate over the larger acids expected from plant waxes (see Fig. 9.44) and for the $<C_9$ acids, there was no preference for the even C numbered species, indicating that they were likely due to anthropogenic sources. Phthalic acid was also observed and attributed to auto exhaust emissions.

In addition to inorganic nitrates, a variety of other forms of organic nitrogenous species also exist in particles. For example, Novakov and co-workers, using X-ray photoelectron spectroscopy (XPS), observed reduced nitrogen species tentatively identified as amines, amides, and possibly nitriles (Novakov *et al.*, 1972; Gundel *et al.*, 1979; Chang *et al.*, 1982). Similarly, Gundel and co-workers (1993) report the presence of nitro compounds, organic nitrates or nitrites, amines, and amides in the polar fraction of organics in particles collected in Elizabeth, New Jersey, and Kneip *et al.* (1983) report *N*-nitroso compounds in particles collected in New York City. These compounds, which they call N_x species, often comprise a major portion of the particulate nitrogen; for example, in particles collected in Berkeley, California, in November 1976, of the total nitrogen present, 50% was identified as the reduced N_x species. Much of this appears to be in the form of organic amides, which can be hydrolyzed to the acid

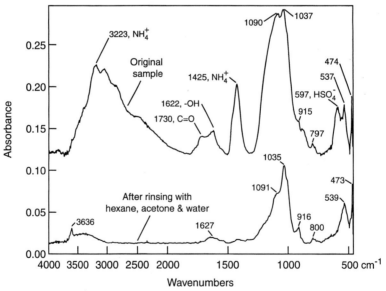

FIGURE 9.51 FTIR spectra of particles from the Smoky Mountains, Tennessee, with diameters from 1.0 to 2.0 μm (top spectrum) and after rinsing with hexane, acetone, and water (bottom spectrum) (adapted from Blando *et al.*, 1998).

C. CHEMICAL COMPOSITION OF TROPOSPHERIC AEROSOLS 401

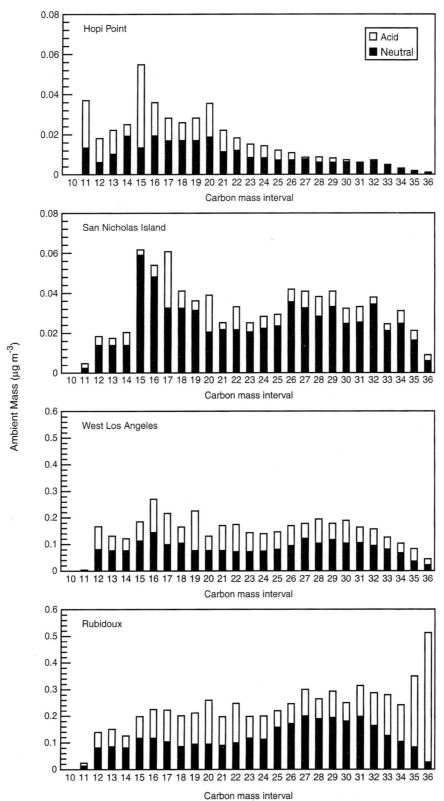

FIGURE 9.52 Mass concentrations of acidic and neutral elutable organics in fine particles (<2.1 μm) at Hopi Point (Grand Canyon region), at San Nicholas Island in the Pacific Ocean upwind of Los Angeles, in West Los Angeles, and downwind at Rubidoux. The bars represent all of the organics that elute between the n-alkanes C_n and C_{n+1}, where n is shown under each bar. Note the different scales for concentrations (adapted from Mazurek *et al.*, 1997).

and NH_4^+:

$$R-\underset{\underset{O}{\|}}{C}-NH_2 + H_2O \xrightarrow{H^+} R-\underset{\underset{O}{\|}}{C}-OH + NH_4^+$$

Thus, caution should be exercised in interpreting NH_4^+ concentrations obtained using extraction procedures. Thus in one set of samples, ~85% of the N_x species originally present were removed by water extraction while, simultaneously, the NH_4^+ concentration in the extract increased proportionally. Evidence for the presence of amides in ambient particulates has also been found using high-resolution mass spectroscopy (Cronn et al., 1977).

Other oxygenated organic components reported in air include dihydroxybenzene and phthalates (e.g., Appel et al., 1980). The latter may arise from direct emissions, since phthalates are used as plasticizers.

The nature and size of the multifunctional and highly oxidized organics shown in Tables 9.18 and 9.19 are consistent with their production from atmospheric reactions of unsaturated hydrocarbon precursors. For example, cyclic olefins and dialkenes can be oxidized to give stable species containing two substituted groups in the same molecule. The products therefore tend to be of high polarity and low volatility. For example, Table 9.20 shows the vapor pressures of likely oxidation products of some simple alkenes, cyclic alkenes, and a dialkene. Also shown in Table 9.20 are the minimum alkene concentrations needed to form the least volatile product at a concentration in excess of its saturation concentration, assuming complete conversion of the precursor to the product. The concentrations of even very long chain alkenes required to form condensable acids exceed those commonly observed in ambient air. Only the cyclic alkenes and the dialkene are present in some cases in the concentrations required to form condensed-phase secondary particles. It is noteworthy that many biogenic organics such as α-pinene are cyclic and readily form secondary organic aerosols (e.g., see Hatakeyama et al., 1989; Pandis et al., 1991; and Zhang et al., 1992).

It is important to note, however (see later), that it is not necessary that the gas-phase concentration exceed the saturation vapor pressure for an organic to exist in particles. Thus, partitioning into the organic phase of existing particles can occur at lower gas-phase concentrations (see Section D on gas–particle partioning of semivolatile compounds). However, even in this case it is necessary that the organic not be highly volatile for this to occur.

Care must be taken in extrapolating the results of laboratory studies to the lower concentrations and generally higher relative humidities (RH) found in ambient air. For example, Izumi et al. (1988) showed that the organic aerosol yield for the cyclohexene–O_3 reaction decreased in a nonlinear fashion as the initial reactant concentrations decreased from ~5 ppm; in addition, the concentration of condensation nuclei increased steeply with RH above ~30%. This may be at least in part due to the effects of gas–particle partitioning on the measured aerosol yields discussed in more detail below.

A variety of laboratory studies of the organic condensed-phase products of model hydrocarbon–NO_x–air systems have established that multifunctional compounds such as those observed in ambient air are

TABLE 9.20 Lowest Ambient Alkene Concentration Required to Form the Corresponding Condensable Species in Excess of Its Saturation Concentration[a]

Olefinic precursor	Least volatile photooxidation product	Product vapor pressure (Torr)	Minimum precursor concentration[b]
Propylene	Acetic acid	16	21,000 ppm
1-Butene	Propionic acid	4	5,200 ppm
1-Hexene	Pentanoic acid	0.25	327 ppm
1-Heptene	Hexanoic acid	0.02	26 ppm
1-Octene	Heptanoic acid	$\sim 9 \times 10^{-3}$	~12 ppm
1-Decene	Nonanoic acid	$\sim 6 \times 10^{-4}$	~0.8 ppm
1-Tridecene	Dodecanoic acid	10^{-5}	13 ppb
Cyclopentene	Glutaric acid	2×10^{-7}	~0.3 ppb
Cyclohexene	Adipic acid	6×10^{-8}	0.08 ppb
1,7-Octadiene	Adipic acid	6×10^{-8}	0.08 ppb
3-Methylcyclohexene	Methyladipic acid	$\sim 2 \times 10^{-8}$	~0.03 ppb

[a] From Grosjean and Friedlander (1980).
[b] Lower limits calculated assuming complete conversion of the precursor to the least volatile of the possible products.

indeed formed. Schwartz (1974) carried out one of the first detailed product analyses of the organic component of particles formed in the oxidation of ~10 ppm of cyclohexene, toluene, or α-pinene in the presence of ~2–5 ppm NO_x in air. The portion of the aerosol that was extractable into methylene chloride but that was insoluble in water was analyzed to focus on the aerosol products that had not undergone extensive secondary oxidations, in the hopes that their mechanisms of formation could be more easily related to the parent hydrocarbons. A variety of multifunctional products were tentatively identified in these studies, with structures analogous to those observed in ambient aerosols.

Since then, a number of studies of model systems have confirmed that dialkenes, cyclic alkenes, and aromatics form substituted monocarboxylic acids, dicarboxylic acids, and organic nitrates in the condensed phase (e.g., see O'Brien et al., 1975a; Grosjean and Friedlander, 1979; Dumdei and O'Brien, 1984; Izumi and Fukuyama, 1990; and Forstner et al., 1997a, 1997b). For example, Table 9.21 shows the products identified in particles formed in the 1-octene– and 1-decene–NO_x–ambient air systems. In both bases, only ~40% of the total particle mass could be identified, and the yields shown in Table 9.21 are those relative to the total identified compounds. That is, the absolute product yields are about factor of 2.5 larger. As expected from the known oxidation mechanisms (see Chapter 6.E), heptanal and heptanoic acid are the major condensed-phase oxidation products of 1-octene and nonanal and nonanoic acid from 1-decene (see Problem 4). The mechanism of formation of the furanones, which are formed in relatively high yields, is not known. Secondary oxidation of the aldehydes is one possibility:

$$C_3H_7(CH_2)_3CHO + OH \xrightarrow{O_2, NO} C_3H_7(CH_2)_3C(O)O$$
$$\xrightarrow{cyclization} \text{[furanone structure with } H_7C_3\text{]}$$

However, Forstner et al. (1997a) also point out that the formation of furanones from the cyclization of γ-hydroxycarboxylic acids (e.g., $C_3H_7CH(OH)CH_2CH_2COOH$) in the condensed phase is well known. If γ-hydroxycarboxylic acids are formed from OH reactions with the alkenes, the furanones can be formed from this cyclization in the particles. However, Forstner and co-workers also indicated that such a reaction could occur during sample workup and hence the true yields of the furanones could be significantly smaller.

Not surprisingly, based on their complex (and not yet well understood) oxidation mechanisms (see Chapter 6.G), aromatic hydrocarbons are efficient precursors of secondary organic aerosol particles, SOA. Indeed, Odum et al. (1997a) have shown, based on smog chamber studies, that the secondary organic aerosol formation from whole gasoline vapor can be due essentially totally to the aromatic content. For example, Table 9.22 lists some of the major condensed-phase products identified in particles formed in the VOC–NO_x oxidations in air of some simple aromatic hydrocarbons. Note that the yields are expressed as a percentage of the total identifiable mass. Only 15–30% of the extractable mass that was eluted through the GC could be identified in each case, so that the absolute

TABLE 9.21 Major Products Observed in the Particles Formed in the VOC–NO_x Oxidations in Air of 1-Octene and 1-Decene[a]

Reactant	Products	Structure	Percentages of total identifiable products
1-Octene	Heptanal	$CH_3(CH_2)_5CHO$	31
	Heptanoic acid	$CH_3(CH_2)_5COOH$	27
	Dihydro-5-propyl-2(3H)-furanone	H_7C_3–furanone	30
1-Decane[b]	Nonanal	$CH_3(CH_2)_7CHO$	43
	Nonanoic acid	$CH_3(CH_2)_7COOH$	26
	Dihydro-5-pentyl-2(3H)-furanone	$H_{11}C_5$–furanone	14

[a] From Forstner et al. (1997a).
[b] Results of three experiments that were reasonably consistent; fourth experiment had much higher yields of furanone (44%) and lower yields of nonanoic acid (79%).

TABLE 9.22 Major Products Identified in the Particles Formed from the VOC–NO_x Oxidations in Air of Some Aromatic Hydrocarbons[a]

Aromatic hydrocarbon precursor	Products	Structure	Percentage of total identifiable product
Toluene	3-Methyl-2,5-furandione		26
	Dihydro-2,5-furandione		22
	2-Methyl-4-nitrophenol		10
	2,5-Furandione		9.6
m-Xylene	3-Methyl-2,5-furandione		61
	m-Toluic acid		9.2
Ethylbenzene	Acetophenone		19
	3-Methyl-2,5-furandione		17
	2,5-Furandione		16
p-Ethyltoluene	3-Ethyl-2,5-furandione		35
	3-Methyl-2,5-furandione		16
	4'-Methylacetophenone		15

(continues)

TABLE 9.22 (continued)

Aromatic hydrocarbon precursor	Products	Structure	Percentage of total identifiable product
1,2,4-Trimethylbenzene	4-Methylphthalic acid	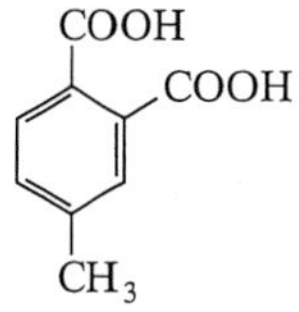 (structure)	28
	3-Methyl-2,5-furandione	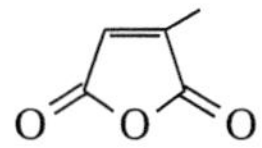 (structure)	27
	3,4-Dimethylbenzoic acid	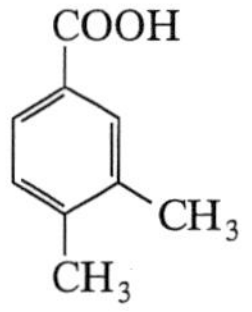 (structure)	12
	3-Methyl-2,5-hexanedione	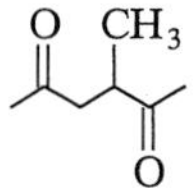 (structure)	9.9

[a] From Forstner et al. (1997b); only products whose yields based on the total identifiable mass in the particle phase and >10% are shown. A wide variety of additional minor products are reported by Forstner et al. (1997b), as well.

yields are much larger than shown in Table 9.22, by factors of 3–6.

Some of the products shown in Table 9.22 arise from abstraction reactions from the alkyl side groups (e.g., see Problem 6), although most arise from addition of OH to the aromatic ring followed by ring-opening reactions such as those discussed in Chapter 6.G. For example, as shown in Fig. 6.13, addition of OH to toluene followed by ring scission forms butenedial. As seen below, further reaction of butenedial gives 2,5-furandione, one of the organic particulate products (Table 9.22) of the toluene oxidation:

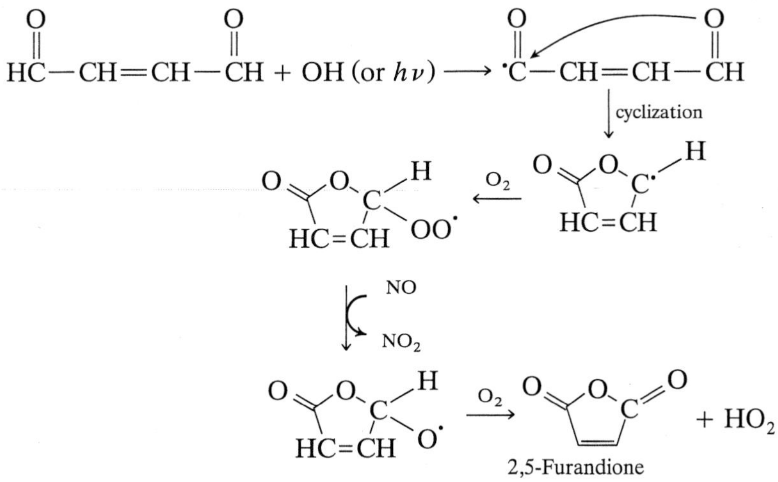

$$HC(=O)-CH=CH-CH(=O) + OH \text{ (or } h\nu\text{)} \longrightarrow {}^{\bullet}C(=O)-CH=CH-CH(=O)$$

$$\downarrow \text{cyclization}$$

(cyclic intermediate) $\xleftarrow{O_2}$ (cyclic intermediate)

$$\downarrow \begin{array}{c} NO \\ NO_2 \end{array}$$

(cyclic alkoxy intermediate) $\xrightarrow{O_2}$ 2,5-Furandione + HO_2

Similar chemistry leading to the formation of the products shown in Table 9.22 is discussed by Forstner *et al.* (1997b). Some of these products may accelerate the photodegradation of less reactive species in the condensed phase (e.g., McDow *et al.*, 1996).

The yields of secondary organic aerosols from a series of aromatic hydrocarbon–NO_x oxidations have been measured by Odum *et al.* (1997a, 1997b). They showed that the total secondary organic aerosol formed from a mixture of aromatic hydrocarbons can be approximated as the sum of the individual contributions. Based on their experiments, the yield of secondary organic aerosols expressed as the total organic particle mass concentrations formed, ΔM_0 (in μg m^{-3}), divided by the mass concentration of aromatic precursor reacted, Δ(aromatic), is given by

$$\frac{\Delta M_0}{\Delta(\text{aromatic})} = 4.2 \times 10^{-3}(\Delta M_0)^{0.6023}. \quad \text{(KK)}$$

The yield of secondary organic aerosol depends on the organic particle mass concentration because of the gas–particle partitioning of the semivolatile organic products (see later). Thus, Odum *et al.* (1996) showed that the yield of secondary organic aerosol, Y, is given by

$$Y = M_0 \Sigma \left[\frac{\alpha_i K_{\text{om},i}}{1 + M_0 K_{\text{om},i}} \right]. \quad \text{(LL)}$$

In Eq. (LL), M_0 is the concentration of the condensed-phase organic (in μg m^{-3}) available to absorb semivolatile organic products, α_i is a constant that relates the concentration of the ith secondary organic aerosol component formed, C_i, to the amount of parent precursor organic reacted [i.e., C_i (ng m^{-3}) = $1000\alpha_i\Delta$(parent organic in μg m^{-3})], and $K_{\text{om},i}$ is the gas–particle partioning coefficient for the ith component. As discussed in more detail in Section D, $K_{\text{om},i}$ is in effect an equilibrium constant between the condensed- and gas-phase concentrations.

Thus, if a particle secondary oxidation product does not get partitioned efficiently into the condensed phase (i.e., $K_{\text{om},i}$ is small) or the available organic condensed phase for uptake of the semivolatile product is small, Eq. (LL) reduces to $Y = M_0 \Sigma \alpha_i K_{\text{om},i}$ and the secondary organic aerosol yield is proportional to the amount of condensed phase available for uptake of the low-volatility gaseous products. On the other hand, if $K_{\text{om},i}$ and M_0 are large, Eq. (LL) becomes $Y = \Sigma \alpha_i$, independent of the amount of condensed phase available for product uptake.

Figure 9.53, for example, shows a plot of the yield of secondary organic aerosol from the VOC–NO_x oxidation in air of some aromatic compounds as a function

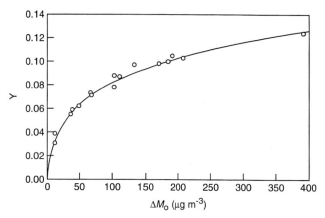

FIGURE 9.53 Yield (Y) of secondary organic aerosol as a function of the amount of aerosol generated, ΔM_0, during the VOC–NO_x oxidations in air of some aromatic hydrocarbons (adapted from Odum *et al.*, 1997b).

of ΔM_0 (Odum *et al.*, 1997b). The rapid initial increase in the yield is due to the increasing availability of condensed phase for uptake of the semivolatile products. However, as the amount of available condensed phase continues to increase, this becomes less limiting, and the yields approach the final, limiting yield determined by α_i.

In short, the same types of aerosol organic products have been identified both in model systems and in polluted urban ambient air and can generally be rationalized based on the oxidation of known constituents of air. The measured yields of organics in the particles can depend on the amount of particle phase available for uptake of the organic if it is semivolatile and partitions between the gas and condensed phases. This partitioning, and its dependence on the amount of condensed phase available, may be at least in part responsible for discrepancies in the yields of secondary organic aerosol reported in a number of studies.

Grosjean (1989, 1992) has used estimates of the emissions of gaseous secondary organic aerosol precursors to estimate the amounts and types of organic compounds expected in particles in the Los Angeles area under typical moderately polluted conditions. These calculations suggest that a variety of organics in particles should be formed by reactions of the precursors and that carbonyl compounds, organic acids, nitrated derivatives, and phenols should all be formed. Gaseous aromatics were predicted to predominate in the formation of organic aerosols, which has been confirmed in smog chamber experiments (Odum *et al.*, 1997a, 1997b).

The complexity of organics found in particles as well as the possibility of multiple sources, both primary and secondary, often makes identification of sources difficult. There are some organics that are "markers" for particular sources, e.g., diterpenoid acids and retene as a marker of wood smoke (Mazurek et al., 1991; Rogge et al., 1993d; Standley and Simoneit, 1994; Radzi bin Abas et al., 1995). Another approach is to use receptor models similar to those applied to inorganics (see Section C.1.c). For example, Greaves et al. (1987) applied factor analysis to particles collected at Boulder, Colorado, an urban site. They identified more than three dozen organic components and four factors: factor 1 was a photochemical factor that included oxygenated organics such as carboxylic acids, aldehydes, furans, lactones, ketones, and phthalates; factors 2 and 3 were different biogenic sources and included such compounds as terpenoids, bornyl acetate, camphor, phytone (6,10,14-trimethylpentadecan-2-one), and high molecular weight n-alkanes; and factor 4 was a vehicular source that included some saturated hydrocarbons and an unidentified branched-chain carboxylic acid.

Similarly, Rogge et al. (1993d) and Schauer et al. (1996) analyzed particles collected in southern California and identified and measured more than 80 organic compounds. Some (e.g., the $C_{29}-C_{33}$ n-alkanes) could be identified with biogenic sources, some with primary anthropogenic emissions (e.g., oleic acid from meat cooking), and some with secondary products from the oxidation of precursors (e.g., some aliphatic dicarboxylic acids). Figure 9.54, for example, shows the composition of particles as a percentage of the total mass at (a) the west end of the Los Angeles air basin and (b) downwind at Rubidoux. Evidence of secondary aerosol generation is found in the increased total particle mass at the downwind location, 42.1 compared to 24.5 μg m^{-3}, the increase in the relative amount of nitrate, and the increased aliphatic dicarboxylic acids.

The relative importance of secondary organic particle formation from reactions of gaseous precursors has been examined in a number of studies, particularly in the Los Angeles area. For example, Gray and co-workers (1986) reported that under conditions of high photochemical activity, as much as 27–38% of the organic particulate carbon may be due to secondary aerosol formation at downwind locations in this area. Similarly, Pandis et al. (1992) estimate that 15–22% of the particulate organic carbon is secondary in nature at downwind locations with aged air masses, compared to only 5–8% at upwind locations. Turpin and Huntzicker (1995) have shown that at Claremont, California, located east of Los Angeles, under many conditions primary organic emissions such as those discussed by Hildemann et al. (1994a) predominate in the organic portion of the particles. However, during periods of smog formation, about 40% was typically secondary in nature, with as much as 50% being secondary in Pasadena (Turpin et al., 1991). Similar conclusions have been reached by Hildemann et al. (1994b).

The fraction of total carbon and nitrogen in air that exists in the condensed phase is highly variable. For example, the percentage of organics in particles has been reported to be as high as 12.5% of the total carbon in the Los Angeles area, while as much as 25% of the nitrogen was found in particles (Fraser et al., 1996).

Most studies of the chemical composition of particles in the troposphere to date have used analysis of bulk samples, which are usually collected in the boundary layer close to the earth's surface. As discussed in Chapter 6. J.3, there is a great deal of interest in the chemistry of the upper troposphere. Much less is known about the chemical composition in this region, particularly of particles. However, it appears that organics are also important constituents of particles in this region as well. For example, Novakov et al. (1997) in studies of particles both onshore and offshore of the eastern United States found that the mass fraction of the particles due to carbon compounds increased as a function of altitude. In the boundary layer, the fraction was typically ~10–40%, increasing to ~50–90% at an altitude of 2–3 km.

Figure 9.55a shows the results of single-particle analysis (see Chapter 11.B.4a) of a typical particle in the upper troposphere (Murphy et al., 1998). In the negative ion spectra, a variety of fragments due to organics are observed, along with sulfates and some halogens. In other particles, soot and minerals were also common constituents. For comparison, Fig. 9.55b shows that a typical particle in the stratosphere is primarily sulfate (see Chapter 12.C.5).

(3) Surfactants in aerosol particles The presence of long-chain organics having one or more polar functional groups (e.g., the carboxylic acids and the nitrates) in the condensed phase suggests that these may act as surfactants in aqueous atmospheric aerosols, forming an organic coating over the surface of the aerosol. Molecules that have long-chain ($>C_5$), nonpolar groups attached to polar tails, such as those in Table 9.23, can form a surface film on droplets by lining up with the polar ends in the water and nonpolar, hydrophobic ends projecting into air as shown schematically in Fig. 9.56.

The degree of compression of the film would be expected to alter its effects on uptake and evaporation from the particle. Thus, as organic films of linear-chain surfactants are compressed, they go through different stages, from a disordered 2-D "gas-like" phase at low

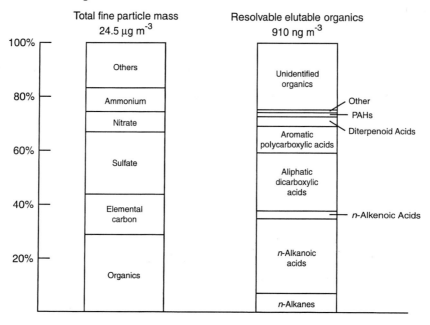

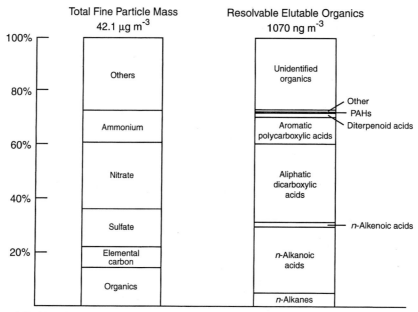

FIGURE 9.54 Composition of particles in Los Angeles, at west end of air basin, and in Rubidoux, at east end (adapted from Rogge *et al.*, 1993d).

degrees of compression to a liquid-like state to finally a highly ordered, solid condensed state (Gaines, 1966; MacRitchie, 1990). The compressed films would be expected to be less permeable. There is some evidence for this. For example, Rubel and Gentry (1985) measured the accommodation coefficient (see Chapter 5.E.1) for water as well as ammonia on acid droplets coated with hexadecanol. The water accommodation coefficient decreased from 8×10^{-3} as the alcohol coverage increased, i.e., as the degree of compression of the organic film increased, to 4×10^{-4}, with a sharp change at the point that the film underwent a phase transition from the liquid to the solid condensed state. Similarly, Däumer *et al.* (1992) showed that coating an H_2SO_4 aerosol with straight-chain organics retarded the rate of neutralization by ammonia, whereas

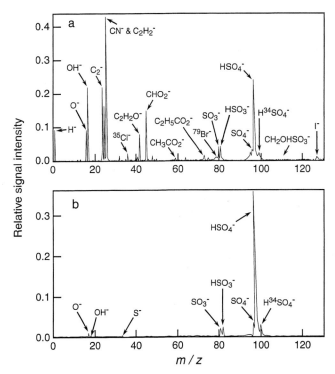

FIGURE 9.55 Typical negative ion mass spectra for a single particle in the (a) upper troposphere (14.6 km, 22°N) and (b) stratosphere (19 km, 31°N) (adapted from Murphy et al., 1998).

TABLE 9.23 Structures of Some Classes of Surface-Active Molecules Found in the Atmosphere[a]

Class of compound	Structure
Alcohol	R—CH$_2$OH
Acid	R—C(=O)OH
Aldehyde	R—C(=O)H
Ketone	R—C(=O)R′
Ester	R—C(=O)OR′
Amine	R—NH$_2$

[a] From Gill et al. (1983).

branched molecules did not, presumably because the permeability of the films was much larger.

The representation in Fig. 9.56 is simplified in that such an orderly arrangement applies to surfactants having saturated hydrophobic chains that can compress to an ordered solid condensed phase. However, this is not the case for all potential surface-active compounds in the atmosphere; for example, the presence of a double bond as in the case of oleic acid, CH$_3$(CH$_2$)$_7$CH=CH(CH$_2$)$_7$COOH, gives the molecule a "crooked" shape, which requires more surface area per molecule and which does not lead to a well-ordered solid condensed phase at the interface. Thus, Xiong et al. (1998) measured effects of organic films on the uptake of water into particles of H$_2$SO$_4$. The saturated straight-chain lauric and stearic acids significantly reduced the uptake of water when present at amounts equivalent to one monolayer, whereas oleic acid had no effect at this concentration.

The possibility of such organic films being formed on aerosol particles in the atmosphere as well as on fog, cloud, and rain droplets and snowflakes has been discussed in detail by Gill and co-workers (1983). As seen from our earlier discussions on the types of organics that have been observed in both urban and nonurban aerosols, there is no question that surface-active species

that can form organic films on water do exist in the atmosphere. However, Gill and co-workers (1983) have estimated, based on the limited data available, that only in aerosol particles does sufficient surface-active organic clearly exist to form a film around the particle. They suggest that the presence of surface-active films on cloud droplets, snowflakes, and raindrops is unlikely.

While the focus has been on long-chain surfactant-like organics, Donaldson and Anderson (1999) suggest, based on their measurements of the standard free energy of adsorption of gases onto water, that there may also be significant surface coverages of smaller organics under atmospheric conditions.

There is some field evidence for the existence of organic films on the surfaces of particles. For example, Fig. 9.57 shows the results of electron microscopy of haze aerosol collected in Los Angeles (Husar and Shu, 1975). The droplets are "wrinkled" in appearance, and they suggest this is due to "haze aerosol" droplets being coated with an organic layer that collapsed when the water in the particle evaporated under vacuum. Husar and Shu propose that the wrinkled appearance is due to a nonvolatile layer of organics that shrunk after water evaporated from the particle during analysis;

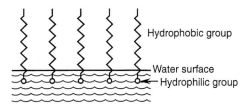

FIGURE 9.56 The orientation of surface-active organic molecules at the water surface (adapted from Gill et al., 1983).

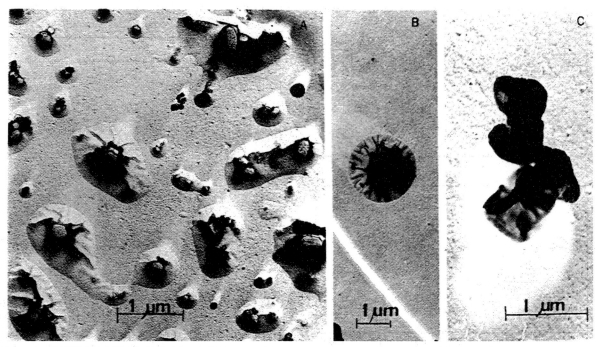

FIGURE 9.57 Electron micrographs of haze aerosol collected in Pasadena in 1973 (A) and of (B) unheated and (C) heated haze particle (adapted from Husar and Shu, 1975).

they described heated particles in Fig. 9.57C as appearing like an "evacuated, thick-walled, rubber ball."

Similarly, Pósfai *et al.* (1998) carried out AFM and TEM studies of aerosol particles collected over the North Atlantic Ocean. The size of the particles measured under the vacuum conditions of the TEM were smaller than those measured by AFM, due to evaporation of water from the particles. However, the TEM images also showed "halos" around the particles at the same diameters as the AFM measurements, suggesting that the particles under ambient conditions had an organic coating. This was supported by energy-dispersive X-ray spectrometry (EDS), which showed that these residues contained S, O, and C (although at least some of the C signal may have been due to the substrate).

Organic films may have some or all of the following effects: (1) reduction of the rate of evaporation of water from the droplets, (2) inhibition of the transport of stable molecules and of highly reactive free radicals such as OH and HO_2 from the gas phase into the droplet, and (3) reduction of the efficiency with which the particles are scavenged by larger cloud and rain droplets (Gill *et al.*, 1983). Thus, the presence of organic films may increase the lifetime of such particles in the atmosphere compared to those expected if the films were not present (Toossi and Novakov, 1985).

In addition, such films may impede the uptake of species other than water. For example, Jefferson *et al.* (1997) showed that a stearic acid coating inhibited the uptake of H_2SO_4 on $(NH_4)_2SO_4$ and NaCl particles; the mass accommodation coefficients for NaCl, for example, decreased from 0.79 to 0.19–0.31 for different coverages of stearic acid.

There is experimental evidence for a reduced rate of evaporation in the presence of organic films. For example, Shulman *et al.* (1997) showed that the presence of difunctional oxygenated organics such as oxalic or *cis*-pinonic acids decreased the rate of evaporation of water from particles. Chang and Hill (1980) exposed water drops of ~19-μm diameter to a stream of air containing the alkene decene (53–220 ppm) and O_3 (20 ppm); the rate of water evaporation was reduced by the presence of decene and O_3, and a major product of the reaction, nonanal, also reduced the evaporation rate when added separately. Larson and co-workers (Andrews and Larson, 1993; Wagner *et al.*, 1996) reported that coating NaCl with an organic surfactant reduces the uptake of water, although the form of the organic on the particle and how it exerts its effect on water uptake are complex (Hansson *et al.*, 1990; Wagner *et al.*, 1996). Surfactant on carbon black has been reported to increase the hygroscopicity of the particles. However, Hameri and co-workers (1992) reported that a high molecular weight alkane or carboxylic acids added to NaCl particles did not significantly alter the water absorption by the particles.

Uptake of water by aerosol particles is a complex function not only of the presence of surfactants but also of their overall composition, which makes discerning the presence and effects of a surfactant in ambient particles difficult. McMurry and Stolzenburg (1989) give a review of work in this area up to 1989. In their studies in a polluted urban atmosphere, they applied a tandem differential mobility analyzer (TDMA) (see Chapter 11.B.2). In essence, this is two differential mobility analyzers (DMA) in which particles of a given size are first selected in one DMA and then exposed to air of a known relative humidity (RH). The change in particle size is measured using the second DMA. The difference in particle volume, obtained from the size change, between a given RH and 0% RH gives the water content. They observed that some particles were nonhygroscopic (i.e., did not absorb water), while others were hygroscopic, and that particles in the 0.4- to 0.5-μm range absorbed more water than those in the 0.05- to 0.2-μm range.

Zhang et al. (1993) showed that particles that were more hygroscopic tended to consist largely of sulfates and nitrates, whereas the less hygroscopic species tended to be associated with carbon.

More recently, the uptake of water by tropospheric particles in relatively remote locations near the Grand Canyon and in a polluted urban area near Los Angeles was studied by Saxena et al. (1995) using a TDMA similar to the studies of McMurry and co-workers. Figure 9.58 shows the measured total water content of these particles in Claremont, California, east of Los Angeles, as a function of the water calculated to be associated with inorganics. As already discussed, a variety of inorganics such as NaCl, sulfates, and nitrates take up water, and their composition, including the water content, can be calculated from the composition, RH, and temperature. It is seen in Fig. 9.58 that, on average, about 25–35% less water is associated with these particles. Saxena et al. (1995) suggest this may be due to the fact that many of the organics in an urban area such as this are primary in nature, and hence may be hydrophobic. These organics may act as surfactants on the particle surface, inhibiting uptake of water, as suggested by Gill et al. (1983). An alternate explanation is that the calculated water contents are incorrect due to the equilibrium thermodynamics of inorganics being altered in these complex mixtures.

On the other hand, the opposite effect is observed for particles at the more remote site near the Grand Canyon. Figure 9.59 shows that these particles appear to contain more water than expected based on the inorganics. Figure 9.60 shows the relationship between the excess water in the particles and the organics in 0.1-μm particles. Clearly, water appears to be associated with the organics; i.e., the organics are behaving as if they are hydrophilic. Saxena and co-workers estimate that for RH of 80–88%, about 25–40% of the total water uptake is associated with organics and conclude that this behavior is consistent with the organics being primarily due to formation in the atmosphere and hence highly oxidized and hydrophilic in nature.

Clearly, this is a complex area that requires much more research.

(4) Elemental vs organic carbon Because the organics found in particles in urban areas are largely associated with human activities, and particularly combustion

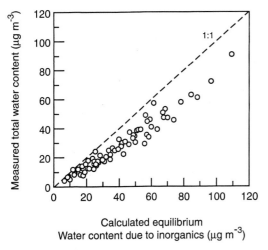

FIGURE 9.58 Measured and calculated water content of particles in Claremont, California, in 1987 at an RH of 80–85% (adapted from Saxena et al., 1995).

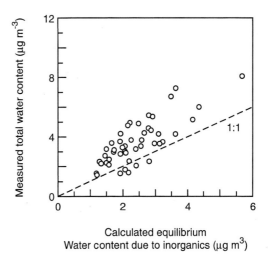

FIGURE 9.59 Measured and calculated water content of particles near the Grand Canyon in 1992 at an RH of 80–85% (adapted from Saxena et al., 1995).

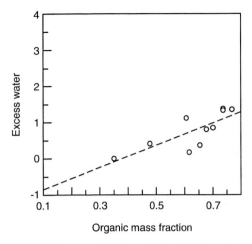

FIGURE 9.60 Excess water in particles with diameters 0.1 μm near the Grand Canyon in 1992 as a function of organic mass fraction (adapted from Saxena *et al.*, 1995).

processes, one would expect elemental carbon to also be a significant component of such particles and indeed, it is. Although it has been found in all of the size ranges to some extent, it is generally predominantly in the accumulation mode (e.g., see Berner *et al.*, 1996; and Smith and O'Dowd, 1996) as expected for a combustion product. Its concentrations are sensitive to the lcoation relative to sources, of course. For example, Cooke *et al.* (1997) report a geometric mean concentration of black carbon at Mace Head, Ireland, from 1989 to 1996 of 14.2 ng m^{-3} for clean air masses. For anthropogenically influenced air from Europe and the United Kingdom, the concentration was 216 ng m^{-3} compared to levels of $\sim 1 \mu$g m^{-3} in urban locations in that region.

The ratio of elemental (EC) to organic carbon (OC) has been measured in many studies and is commonly found to be less than one. For example, Shah *et al.* (1986) measured EC and OC in particles from both urban and rural areas across the United States and found the urban average concentrations to be 6.6 μg m^{-3} for OC and 3.8 μg m^{-3} for EC, i.e., a ratio of EC/OC of 0.6. Similarly, in the Los Angeles area, EC typically represents about a third of the carbonaceous component of particles (Rogge *et al.*, 1993d). For rural areas, the concentrations were about a third of those in the urban areas, but the ratio was about the same, at ~ 0.5, consistent with ratios of $\sim 0.4-0.5$ measured in the Ohio River Valley by Huntzicker *et al.* (1986). Novakov *et al.* (1997) measured carbonaceous aerosol off the east coast of the United States at altitudes from 0.2 to 3.0 km and report that the ratio of EC/OC is ~ 0.1.

Such ratios of EC/OC are typical of many areas but, as expected, depend heavily on the particular mix of sources. For example, in Rio de Janeiro, the ratio is higher, about 1, due to the contribution of EC from diesel-powered buses (Daisey *et al.*, 1987).

Finally, while elemental carbon is a common constituent of urban aerosols due to its combustion source, it is also found in particles in remote areas due to long-range transport (e.g., see O'Dowd *et al.*, 1993a; and Saxena *et al.*, 1995).

It is interesting that the surfaces of elemental carbon particles also hold significant amounts of water, likely due to the presence of polar functional groups such as carboxylic acids. For example, Smith and co-workers (Chughtai *et al.*, 1996) have shown that about half of the surface of fresh soot takes up water, but if oxidized by high concentrations of NO_2/N_2O_4, this increases to almost 100%. Reactions of soot with nitrogen and sulfur oxides, which can alter its hygroscopic properties are discussed in Chapters 7.B.3b, 7.D.2, and 8.C.4. The incorporation of trace metals into the soot also increased the uptake of water, as did oxidation by reaction with O_3 (Smith and Chughtai, 1996).

D. GAS–PARTICLE DISTRIBUTION OF SEMIVOLATILE ORGANICS

There has been increasing recognition that a number of organics in air have vapor pressures in an intermediate regime such that they exist in part in the gas phase and partially associated with particles. These species are commonly referred to as semivolatile organic compounds (SOC) and include a number of high molecular weight alkanes, polycyclic aromatic hydrocarbons (PAH; see Chapter 10), organochlorine compounds, phthalic acid esters, aldehydes and ketones, and aliphatic as well as organic acids (Cautreels and Van Cauwenberghe, 1978). The distribution between the two phases is of interest for determining the total concentrations in air and indeed for understanding measurements made in field studies. For example, Kadowaki (1994) has shown that differences between summer and winter in the concentrations of $C_{21}-C_{28}$ *n*-alkanes in particles collected in Nagoya, Japan, are not due to differences in emissions but rather to changes in the gas-to-particle partitioning occurring with changes in temperature. In addition, as discussed earlier, understanding the yields of secondary organic aerosols from the atmospheric oxidation of a compound or group of compounds requires understanding how the products partition between the gas and condensed phases. Finally, as treated in Chapter 10, some PAH have different reaction mechanisms and products for the gas phase compared to the condensed phase.

In short, understanding the gas–particle partitioning of SOC is important in a number of areas in atmospheric chemistry as well as related aspects of toxicology.

The earliest work in this area assumed that particles in the atmosphere were solid and that the uptake of SOC involved *adsorption* to a solid or solid-like surface. It was subsequently recognized that many atmospheric particles are liquid or have liquid-like outer layers, and hence the uptake of gases could be treated as *absorption* into a liquid. These approaches are summarized in the following. It should be noted that these treat the *equilibria* between the gas- and condensed-phase species; i.e., it is assumed that thermodynamics rather than kinetics controls the distribution between the phases. The implications of this assumption are discussed later.

Given these caveats, we first treat the case of *adsorption* to a solid surface, and then *absorption* into a liquid particle or liquid surface layer on a particle. As we shall see, the distribution of SOC between the gas and condensed phases can be used to infer the nature of the sorbent sites.

1. Adsorption on Solid Particles

A gas–particle partition coefficient, K_p, is commonly used to describe the distribution of a SOC between the gas and particle phase. It is defined as the ratio of the SOC in particles (in units such as ng μg^{-1}) to that in the gas phase (in units such as ng m^{-3}). In essence, it is the fraction of the mass of total suspended particulate matter (TSP) that is the SOC of interest divided by the SOC gas-phase concentration. This gas–particle partition coefficient, which has units of m^3 μg^{-1}, can be calculated from the following:

$$K_p = (F/\text{TSP})/A. \quad \text{(MM)}$$

In Eq. (MM), F (from *f*ilter-associated material) is the concentration of the SOC in air that is in the particle phase, in ng m^{-3}, A (from the use of an *a*dsorbent to collect the gas) is the gas-phase concentration in ng m^{-3}, and TSP is the concentration of *t*otal *s*uspended *p*articles, in μg m^{-3}. (Note that this is the definition in common usage today; however, some earlier papers (e.g., Yamasaki *et al.*, 1982; Pankow, 1987) defined K_p as the inverse, i.e., as equal to $A(\text{TSP})/F$.) Because measurements of F, TSP, and A are difficult to make in an artifact-free manner and because equilibrium may not always hold in the atmosphere (see later), the quantity $(F/\text{TSP})/A$ is often referred to as the *measured* partition coefficient, in contrast to the true, thermodynamic partitioning coefficient, K_p. As shown in Box 9.1, Eq. (MM) can be shown to be equivalent to assuming that adsorption of the gas on the particle surface follows a Langmuir isotherm (Yamasaki *et al.*, 1982; Pankow, 1987).

Pankow (1987) showed that this assumption of a linear Langmuir isotherm is consistent with the pioneering work of Junge (1977) on the adsorption of SOCs on particles. The fraction of the total SOC in the atmosphere adsorbed on aerosol particles, denoted ϕ, was hypothesized by Junge to be related to the surface area (S_T, cm^2 per cm^3 of air) of the TSP and the saturation vapor pressure of the SOC by

$$\phi = cS_T/(p_L^\circ + cS_T). \quad \text{(TT)}$$

In this case p_L° is the saturation vapor pressure of the adsorbing gas at that temperature and c is a constant characteristic of the compound and the temperature.

As discussed in detail by Pankow (1987) and Pankow and Bidleman (1992), the Junge approach can be reduced to an expression of the form

$$K_p = \{(N_s A_{\text{TSP}} T)/16 p_L^\circ\} \exp[(\Delta H_d - \Delta H_{\text{vap}})/RT], \quad \text{(UU)}$$

where N_s is the number of moles of surface adsorption sites per cm^2, A_{TSP} is the specific surface area (in cm^2 μg^{-1}) of the TSP, ΔH_d is the enthalpy of desorption for the SOC directly from the surface, ΔH_{vap} is the enthalpy of vaporization of the (subcooled) liquid, T is the temperature (in K), and R is the gas constant. In many cases, the compound is a solid at that temperature, rather than a liquid, in which case p_L° (Torr) is the vapor pressure of the subcooled liquid; the reason for choosing the pressure of the subcooled liquid is the assumption that the adsorbed (or absorbed; see later) states resemble disordered liquids more than highly ordered crystalline solids (Pankow, 1994b).

Thus the relationship between log K_p and log p_L° should be of the form

$$\log K_p = m_r \log p_L^\circ + b_r, \quad \text{(VV)}$$

where m_r and b_r are constants. Using a somewhat different approach, Mackay *et al.* (1986) predicted a similar relationship for partitioning of semivolatile compounds in the atmosphere. From Eq. (UU), m_r should be equal to -1 and b_r should be equal to $\log\{(N_s A_{\text{TSP}} T/16)\exp(\Delta H_d - \Delta H_{\text{vap}})/RT\}$.

Values of p_L° for a series of PAH determined experimentally at 25°C by Yamasaki *et al.* (1984), and subsequently temperature-corrected to 20°C by Pankow and Bidleman (1992), are given in Table 9.24, along with the values for the parameters c and d which describe

> # BOX 9.1
> # LANGMUIR ADSORPTION OF SOC ON PARTICLES
>
> Let S_0 equal the number of sites on the surface that are occupied. If the total number of sites is S, then $(S - S_0)$ is the number of sites on the surface that are not occupied and available for adsorption of SOC. At equilibrium, the rate of adsorption and that of evaporation from the surface must be equal:
>
> $$\text{SOC}_{(g)} \underset{k_{-1}}{\overset{k_1}{\rightleftarrows}} \text{SOC}_{(ads)}.$$
>
> If the gas-phase pressure of $\text{SOC}_{(g)}$ is P, then
>
> $$k_1 P(S - S_0) = k_{-1} S_0. \quad \text{(NN)}$$
>
> Rearranging, one obtains
>
> $$S_0/S = b_L P/(1 + b_L P) = \theta_L, \quad \text{(OO)}$$
>
> where $b_L = k_1/k_{-1}$ (the subscript L is used for Langmuir). That is, the fraction of the total surface sites that are occupied by adsorbed molecules is given by the expression in (OO) and is usually denoted θ_L.
>
> In the limit of low gas-phase pressures of the adsorbing compound, which is true for SOC in the atmosphere, $1 \gg b_L P$ and (OO) reduces to
>
> $$S_0/S = \Delta_L = b_L P. \quad \text{(PP)}$$
>
> However, as first developed by Yamasaki *et al.* (1982) and subsequently by Pankow (1987), the fraction of the total surface sites that hold adsorbed molecules, Δ_L, is just proportional to the concentration of adsorbed species on the particles, F (ng m^{-3}), divided by the total concentration of total suspended particles, TSP (ng m^{-3}), where the proportionality constant M contains the conversion constants to convert the TSP concentration into the total surface area and F into the area of the adsorbed molecules:
>
> $$\Delta_L = M(F/\text{TSP}). \quad \text{(QQ)}$$
>
> Similarly, the right-hand side of Eq. (PP) can be converted into a term involving the gas-phase concentration A in Eq. (MM) using the ideal gas law, $P = (n/V)RT$. The number of moles per m^3, (n/V), is converted to ng m^{-3} using the molecular weight (MW) of the SOC. Combining the MW, R, T, and other conversion factors into one constant N, $P = NA$, and the right side of Eq. (PP) becomes $b_L P = b_L NA$. Equation (PP) therefore becomes
>
> $$M(F/\text{TSP}) = b_L NA$$
>
> or
>
> $$b_L = (M/N)(F/\text{TSP})/A. \quad \text{(RR)}$$
>
> However, as seen in Eq. (OO), $b_L = k_1/k_{-1}$; i.e., b_L is in effect an equilibrium constant for the adsorption of $\text{SOC}_{(g)}$ on the solid and the reevaporation of $\text{SOC}_{(ads)}$ from the solid surface. Hence $b_L = k_1/k_{-1} = K'$ and Eq. (RR) becomes
>
> $$K' = (\text{Constant})(F/\text{TSP})/A. \quad \text{(SS)}$$
>
> In short, the form of the gas–particle partitioning defined in Eq. (MM) is consistent with Langmuir adsorption of the SOC on the surface of the TSP.

the temperature dependence of p_L° as $\log p_L^\circ$ (Torr) $= c/T + d$.

Figure 9.61 shows some typical plots to test Eq. (VV) for some SOC in the form of higher molecular weight alkanes and some PAH (see Chapter 10) at various relative humidities (Storey *et al.*, 1995). In this case, the gas–particle partitioning coefficient K_p has been normalized to take into account different surface areas by using a surface area normalized coefficient, $K_{p,s}$, defined as

$$K_{p,s} = K_p/(\text{Specific surface area of adsorbing substrate}). \quad \text{(WW)}$$

Figure 9.61 shows that $\log K_{p,s}$ is indeed linear with $\log p_L^\circ$, and the slopes are typically close to -1 as expected. Fig. 9.61 also illustrates the effects of relative humidity (RH) on the adsorption of these compounds on a quartz fiber filter (QFF); the values of $K_{p,s}$ decreased by about an order of magnitude as the RH increased from ~ 30 to 70% which was attributed to changes in the properties of the surface as water adsorbed onto it. Furthermore, the values for adsorption of these compounds on urban particulate matter (UPM) that had been observed in other studies (Yamasaki *et al.*, 1982; Foreman and Bidleman, 1990) are clearly much larger than on the QFF. This suggests that if mineral oxide particle surfaces in the atmosphere behave like the QFF, adsorption to such inorganic sur-

TABLE 9.24 Subcooled Liquid Vapor Pressures (p_L°) at 20°C for a Series of Polycyclic Aromatic Hydrocarbons[a] and Their Temperature Dependence[b]

PAH	log p_L° (20°C) (Torr)	p_L° (Torr)	c	d
Fluorene	−2.72	1.9×10^{-3}	−3632	9.68
Phenanthrene	−3.50	3.2×10^{-4}	−3982	10.09
Anthracene	−3.53	3.0×10^{-4}	−4004	10.14
Fluoranthene	−4.54	2.9×10^{-5}	−4464	10.70
Pyrene	−4.73	1.9×10^{-5}	−4529	10.73
Benzo[a]fluorene	−5.24	5.8×10^{-6}	−4792	11.11
Benzo[b]fluorene	−5.22	6.0×10^{-6}	−4814	11.21
Benz[a]anthracene	−6.02	9.6×10^{-7}	−5179	11.66
Chrysene	−6.06	8.7×10^{-7}	−5200	11.69
Triphenylene	−6.06	8.7×10^{-7}	−5208	11.71
Benzo[b]fluoranthene	−7.12	7.6×10^{-8}	−5711	12.37
Benzo[k]fluoranthene	−7.13	7.4×10^{-8}	−5711	12.36
Benzo[a]pyrene	−7.33	4.7×10^{-8}	−5777	12.39
Benzo[e]pyrene	−7.37	4.3×10^{-8}	−5799	12.42

[a] From Pankow and Bidleman (1992). Original data were for 25°C and published by Yamasaki et al. (1984). They were corrected to 20°C by Pankow and Bidleman (1992).
[b] Temperature dependence given by log p_L° (Torr) = $c/T + d$.

faces will not be significant in urban areas (but may be more important in remote areas).

It should be noted that a similar trend in K_p with relative humidity appears to apply to the gas-to-particle distribution of PAH in urban areas (Pankow et al., 1993), although in studies by Cotham and Bidleman (1992), RH did not appear to significantly affect the adsorption of organochlorine pesticides, at least over the range of RH of 30–95%.

While many sets of data appear to follow Eq. (VV) relatively well, with slopes of $m_r = -1$ as predicted, deviations in the values of m_r and b_r are often observed. There are a number of reasons for such deviations (e.g., see Pankow and Bidleman, 1992). For example, changes in temperature, concentrations of SOC, and relative humidity during sampling, nonattainment of equilibrium, and sampling artifacts can all lead to deviations from the predicted, equilibrium relationship. In addition, if $(\Delta H_d - \Delta H_{vap})$ in Eq. (UU) is not constant along the series, relationship (VV) will not hold because the value of b_r is changing.

However, it is also the case that slopes different from $m_r = -1$ can occur even under equilibrium conditions, due to the nature of the molecular interactions involved in adsorption on a surface. For example, Goss and Schwarzenbach (1998) propose a modified formulation of Eq. (VV) that more explicitly takes into account the interactions between the SOC and the surface on a molecular level. Thus, based on work by Goss (1997), they express the relationship between a gas–particle partition coefficient K_i^{ads}, defined as the adsorbed concentration on the surface (in mg m^{-2}) divided by the gas-phase concentration (in mg m^{-3}), and p_L° in the following form:

$$\ln K_i^{ads} = \left[-0.133(\gamma^{vdw})^{0.5}\right]\ln p_L^\circ + 2.09(\gamma^{vdw})^{0.5} + 2.08\beta_i(\gamma^+)^{0.5} + 1.37\alpha_i(\gamma^-)^{0.5} - 19.5. \quad (XX)$$

The first two terms represent van der Waals interactions between the adsorbed SOC and the surface, which would apply to all SOC. The second two terms represent Lewis acid–base interactions, which can be important for compounds containing O, N, or aromatic rings, for example, the adsorption of alkyl ethers on the polar surface of quartz. The γ coefficients (in mJ m^{-2}) describe the *surface* properties, where γ^{vdw} is associated with its van der Waals interactions with adsorbing gases, γ^+ describes its electron-acceptor interactions, and γ^- describes the electron-donor interactions of the surface. On the other hand, the properties of the *adsorbing species* are described by $\ln p_L^\circ$ for the van der Waals interactions and by the dimensionless parameters β_i and α_i, which relate to the electron-donor and electron-acceptor properties (if any), respectively, of the adsorbing molecule.

This formulation explicitly accounts for both the properties of the surface and those of the adsorbing SOC that determine on a molecular level the amount of adsorption of the gas on the surface, and hence the gas–particle partitioning. For example, Goss and Schwarzenbach (1998) describe the implications of three

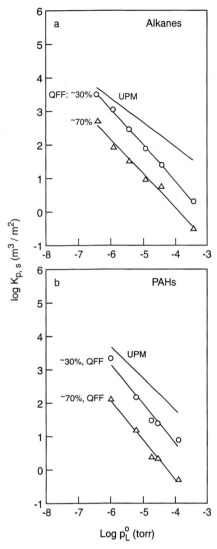

FIGURE 9.61 Surface area normalized gas–particle partition coefficients as a function of liquid vapor pressure for some alkanes and PAHs on urban particulate matter (UPM) and quartz fiber filters (QFF) at 30% RH and 70% RH (adapted from Storey *et al.*, 1995; data for UPM from Yamasaki *et al.* (1982) and Foreman and Bidleman (1990)).

different cases for the slopes of plots of ln K_i^{ads} against ln p_L^o: (1) where no Lewis acid–base interactions occur (e.g., $\beta_i = \alpha_i = 0$ and/or $\gamma^+ = \gamma^- = 0$), for example, the adsorption of alkanes on Teflon; (2) compounds with the same Lewis acid–base interactions with the surface but differing van der Waals components, e.g., alkylbenzenes on quartz; and (3) compounds with different functionalities having different Lewis acid–base interactions.

In the first case, slopes of ln K_i^{ads} against ln p_L^o should reflect the van der Waals interaction term $[-0.133(\gamma^{vdw})^{0.5}]$. In the second case, the slope is again given by $[-0.133(\gamma^{vdw})^{0.5}]$ but there is an extra term in the intercept, corresponding to the γ^+ or γ^- term in Eq. (XX); hence the line will parallel that for the first case but be shifted from it by a constant amount. For the third case, the terms γ^+ and γ^- vary with the compound, but often in a manner that follows ln p_L^o as well, because of the common molecular properties that determine both terms. For example, the aromatic rings of PAH act as electron donors, with the electron-donor parameter β_i a measure of the Lewis acid–base interaction. However, as this interaction changes for a series of PAH, there is a concomitant change in their vapor pressures, p_L^o. As a result, the term in Eq. (XX) expressing the Lewis acid-base interaction, $2.08\beta_i(\gamma^+)^{0.5}$, rises proportionally to ln p_L^o. The slope of a plot of ln K_i^{ads} against ln p_L^o is thus larger than for the first two cases.

In short, the slope of plots of the natural logarithm of the gas–particle partitioning coefficient against ln p_L^o for a series of adsorbing SOC and surfaces can help to elucidate on a molecular level the types of interactions between the two.

The partitioning of an SOC between the gas and particle would be expected to depend on temperature. Intuitively, one expects that an increase in temperature would result in less adsorption and a higher gas-phase concentration. Indeed, the temperature dependence of K_p in Eq. (MM) is usually (Pankow, 1987, 1991, 1992) expressed as

$$\log K_p = c_p/T + d_p. \quad (YY)$$

This expression is developed in detail by Pankow (1987), and its origin treated in Box 9.2.

Figure 9.62 shows some typical plots by Pankow (1991) of $\log[(F/TSP)/A]$ against $1/T$ for some PAH measured by Yamasaki *et al.* (1982). The plots are reasonably linear, as expected from Eq. (YY). Pankow (1991, 1992) shows that since d_p is expected to be similar for similar compounds [Eq. (BBB)], assuming a single value of d_p for such a group seems to be justified.

In summary, *adsorption* of semivolatile organic compounds (SOC) on solid particles in the atmosphere is expected to occur, leading to partitioning of such compounds between the gas and condensed phases. As expected, this partitioning is temperature dependent, with increasing amounts adsorbed on the particles as the temperature is lowered. The relationship between the logarithm of the measured gas–particle partitioning coefficient and the logarithm of the vapor pressure of the liquid SOC at that temperature (subcooled, if necessary) is expected to be linear, and a slope of −1 is common. However, this slope and deviations from −1

> **BOX 9.2**
>
> **TEMPERATURE DEPENDENCE OF K_p**
>
> The origin of the relationship between K_p and temperature in Eq. (YY) can be seen by reexamining equations (RR) and (SS), which show that K_p is directly proportional to b_L, i.e., to the ratio of the rate constants for adsorption and desorption, k_1/k_{-1}. The rate constant for desorption from the surface, k_{-1}, can be expressed as a function of the heat of desorption, ΔH_d (Adamson, 1982):
>
> $$k_{-1} = k_0 \exp(-\Delta H_d/RT), \qquad (ZZ)$$
>
> where k_0 is approximately 10^{12}–10^{13} s^{-1}; i.e., if the heat of desorption is zero, the molecule desorbs within the period of one vibration. Since b_L and hence K_p depend on $1/k_{-1}$, they should vary with $\exp(+\Delta H_d/RT)$; i.e., $\log K_p$ should vary with $1/T$, with the slope related to the heat of desorption of the SOC from the solid surface.
>
> The rate constant for adsorption, k_1, is also temperature dependent, but the dependence is small compared to that for k_{-1}. The value and temperature dependence of k_1 are determined by the rate of gas–solid collisions, which from kinetic molecular theory is given by
>
> Rate of collisions of gas with surface
>
> $$= [G](RT/2\pi M)^{1/2} \text{ (number per cm}^2\text{ s}^{-1}), \qquad (AAA)$$
>
> where [G] is the gas concentration (number cm^{-3}), T is the temperature, and M is the molecular weight of the gas. Replacing [G] by P/RT using the ideal gas law, the rate of collisions of the SOC with the surface, and hence k_1, varies with $T^{-1/2}$, which is usually small compared to the exponential dependence on temperature of k_{-1}.
>
> In short, a plot of $\log[(F/TSP)/A]$ against T^{-1} should, in principle, be linear over the relatively narrow range of temperatures commonly found in the troposphere, and hence be well-described by Eq. (YY). Thus such plots should give a straight line of slope c_p and intercept d_p. Pankow (1987, 1991) shows that c_p and d_p should be given by
>
> $$c_p = (\Delta H_d/2.303R) - T_{\text{amb}}/4.606 \qquad (BBB)$$
>
> and
>
> $$d_p = \log\{A_{\text{TSP}}/275 k_0 (MW/T_{\text{amb}})^{1/2}\} + 1/4.606, \qquad (CCC)$$
>
> where T_{amb} (in K) is the middle of the temperature range over which K_p is studied, A_{TSP} is the specific surface area for the particles (cm^2 μg^{-1}), k_0 (s^{-1}) is as defined in Eq. (ZZ), and MW is the molecular weight of the SOC.

can provide useful insights into the nature of the interactions between the SOC and the solid surface (assuming that equilibrium between the gas and solid is really achieved).

2. Absorption into Liquids

As discussed by Bidleman (1988), there is a variety of evidence from both laboratory and field studies that the gas–particle partitioning in many cases in the atmosphere is consistent with partitioning to a liquid rather than adsorption on a solid; i.e., the gaseous SOC is *absorbed* into a liquid particle or a liquid on the surface of a solid. For example, tetrachlorodibenzo-*p*-dioxin (TCDD) is found in both the vapor and particle phases, whereas none would be found in the gas phase if the solid vapor pressure controlled the distribution. Similarly, the distribution of PCBs between the gas and particle phases is consistent with absorption into a liquid (Falconer and Bidleman, 1994). In this case, the gas-phase concentration of the organic need not exceed its saturation vapor pressure to partition into the particles since the process, in effect, involves dissolving the gas into a solution.

If an SOC is absorbed into a liquid organic layer on the particle, a relationship between K_p and p_L° that is similar to that developed for adsorption onto a solid can be derived (Pankow *et al.*, 1994a, 1994b). In this case, the gas–particle partitioning coefficient for the *i*th compound is defined as

$$K_p = (F_{i,\text{om}}/TSP)/A_i, \qquad (DDD)$$

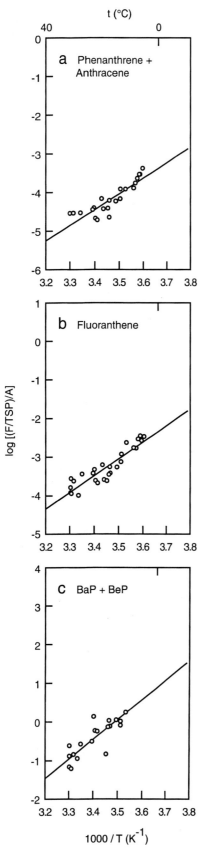

FIGURE 9.62 Plots of $\log[(F/TSP)/A]$ against T^{-1} for some PAH [from Pankow (1991) based on data of Yamasaki et al. (1982)].

where $F_{i,\text{om}}$ is the concentration of i in air (ng m^{-3}) that is particle associated and dissolved in the liquid organic material, "om," phase, TSP has units of μg m^{-3}, and A is the gas-phase concentration in units of ng m^{-3}. As for adsorbed SOC, Eq. (MM), K_p has units of m^3 per μg. The thermodynamics of absorption are described in Box 9.3.

The expression for the gas–particle partitioning coefficient K_p given by Eq. (DDD) in units of m^3 per μg is developed by Pankow (1994a, 1994b) as

$$K_p = \frac{F_{i,\text{om}}/\text{TSP}}{A_i} = \frac{760 RT f_{\text{om}}}{10^6 \text{MW}_{\text{om}} \gamma_{\text{om}} p_L^\circ}, \quad \text{(JJJ)}$$

where MW_{om} (g mol^{-1}) is the mean molecular weight of the species constituting the *om* (liquid organic matter) phase, f_{om} is the weight fraction of the total suspended particulate matter that is the absorbing *om* phase (in addition to organics, this includes any inorganics and water that may be present), and the other symbols have their usual meaning discussed earlier.

Taking the logarithm of (JJJ), one obtains Eq. (KKK):

$$\log K_p = -\log p_L^\circ - \log \gamma_{\text{om}} + \log \frac{760 RT f_{\text{om}}}{10^6 \text{MW}_{\text{om}}}. \quad \text{(KKK)}$$

Thus, Eq. (KKK) [and the analogous logarithmic form of Eq. (III) in Box 9.3] predicts that a plot of $\log K_p$ against $\log p_L^\circ$ for the partitioning of a series of compounds into liquid particles or into a liquid layer on particles should be a straight line with a slope of -1 *if* the activity coefficients in the liquid phase, γ_{om}, remain constant.

Figure 9.63, for example, plots $\log K_p$ against $\log p_L^\circ$ for the partitioning of a series of PAH (see Chapter 10) between the gas phase and particles of either dioctyl phthalate (DOP) or secondary organic aerosol (SOA) generated from the photooxidation of gasoline vapor (Liang et al., 1997). The slope of the plot for uptake into DOP is -1.09 and that for uptake into SOA is -1.05, in agreement with Eq. (KKK) if the activity coefficients are the same for these compounds.

However, as pointed out earlier, a slope of -1 is only expected when the activity coefficients do not change along the series of SOC. In fact, γ_i frequently does change. For example, Jang and Kamens (1998) reported that the activity coefficients for the partitioning of alkanes and PAH on wood soot particles increased with the relative humidity. Furthermore, changes in the activity coefficients are often correlated with the changes in p_L° (Goss and Schwarzenbach, 1998). In that case, since $\ln \gamma_i$ is proportional to $\ln p_L^\circ$,

> **BOX 9.3**
>
> **RELATIONSHIP BETWEEN ABSORPTION AND SUBCOOLED VAPOR PRESSURE OF SOC**
>
> Following the approach of Goss and Schwarzenbach (1998), for equilibrium between a species, i, in the gas phase at pressure P_i and dissolved in a liquid organic material at a mole fraction $X_{i,\text{om}}$, the chemical potentials (μ) in the two phases must be equal, i.e.,
>
> $$\mu_{i,g} = \mu_{i,\text{om}}. \quad \text{(EEE)}$$
>
> The chemical potential in the gas phase is given by
>
> $$\mu_{i,g} = \mu_i^\circ + RT \ln(P_i/P^0), \quad \text{(FFF)}$$
>
> where P° is a standard-state pressure and $\mu_{i,g}^\circ$ is the associated chemical potential of the standard state. Taking the standard state as the pure liquid, P° is the saturation vapor pressure over pure liquid (subcooled, where appropriate), i.e., $P^\circ = p_L$. Similarly, the chemical potential for the ith compound in the liquid organic layer is given by
>
> $$\mu_{i,\text{om}} = \mu_i^\circ + RT(\ln \gamma_i X_{i,\text{om}}), \quad \text{(GGG)}$$
>
> where the standard state is the pure liquid at 1 bar pressure. Combining Eqs. (EEE) through (GGG), one obtains
>
> $$P_i/p_L^\circ = \gamma_i X_{i,\text{om}}. \quad \text{(HHH)}$$
>
> Goss and Schwarzenbach (1998) define a unitless gas–particle partitioning coefficient, K_i, as the ratio of molar concentrations in the condensed, om, phase ($c_{i,\text{om}}$) to that in the gas phase ($c_{i,g}$). They show that using Eq. (HHH), K_i can be given by
>
> $$K_i = \frac{c_{i,\text{om}}}{c_{i,g}} = \frac{RT}{\gamma_i p_L^\circ V_{\text{om}}}, \quad \text{(III)}$$
>
> where V_{om} is the molar volume of the liquid organic layer.

the slope becomes $-(1 + s)$, where s reflects the simultaneous changes in $\ln \gamma_i$ when p_L° changes. They conclude that the slope should lie between about $+0.2$ and -1.0 when the possible changes in the activity coefficients are taken into account. For example, the slope of $\ln K_i$ defined by Eq. (III) as a function of $\ln p_L^\circ$ for the partitioning of a series of alkanes and alkylbenzenes between the gas phase and liquid octanol at 25°C is about -1; however, that for partitioning of chlorobenzenes into octanol is only -0.59. Goss and Schwarzenbach (1998) attribute this to the smaller attractive forces between the chlorobenzenes and octanol compared to those in the pure liquid chlorobenzenes.

Because the slopes of log K_p against log p_L° can be -1 for both *absorption* into a liquid layer and *adsorption* onto a solid, the slope alone cannot be used to differentiate the mechanism of gas–particle partitioning. However, a combination of the slope and the absolute values of K_p can be used to test the two mechanisms for cases where only van der Waals interactions occur, since K_p^{ads} can be calculated from Eq. (XX). For example, Goss and Schwarzenbach (1998) show that the calculated values of K_p^{ads} for a series of alkanes and PCBs partitioning into typical urban aerosols are much smaller than reported in the literature, indicating that *absorption* into a liquid phase must be important.

As is the case for adsorption, the trend in gas–solid partitioning coefficients for absorption can be used to probe for similarities and differences in interactions between the gaseous SOC and liquid on a molecular level. For example, Goss and Schwarzenbach (1998) point out that if two liquids and their interactions with SOC are the same, a plot of the gas–particle partitioning coefficients for a series of gases in one liquid, $K_{i,1}$, against the analogous coefficients in the second liquid, $K_{i,2}$, should be a straight line with a slope of 1.0. If the interactions are different, however, this correlation will not hold. For example, such a log–log plot of the gas–solid partitioning coefficients for the uptake of a series of alkanes and ethanol in n-heptane and dibutyl ether shows that the correlation does not hold for ethanol. This indicates that the two liquids are not chemically similar in terms of their interactions with the SOC, which is not surprising in this case because of the strong interactions between ethanol and the ether. If the liquids were ambient air particles of unknown surface composition, such a plot would show that the

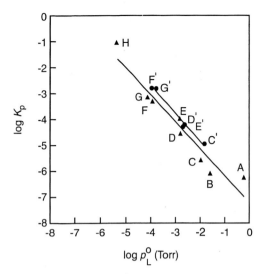

FIGURE 9.63 Plots of om-phase-normalized gas–particle partitioning constant log $K_{p,om}$ vs logarithm of the subcooled liquid vapor pressure, log p_L^o, for a series of semivolatile PAHs partitioning on (●) dioctyl phthalate (DOP) or (▲) secondary organic aerosol (SOA) from photooxidized gasoline vapor. PAHs are as follows: naphthalene, A; acenaphthalene, B; fluorene, C and C′; phenanthrene, D and D′; anthracene, E and E′; fluoranthene, F and F′; pyrene, G and G′; chrysene, H (adapted from Liang et al., 1997).

particles are quite different in their chemical properties and hence their effects on gas–particle partitioning.

In short, differentiating between adsorption on a solid and absorption into a liquid for partitioning of semivolatile compounds in the atmosphere is often difficult to do in an unambiguous manner. However, the use of a combination of approaches can help to differentiate these two mechanisms and, perhaps more important, give some insight into the mechanisms of interaction of the SOC with the condensed-phase material.

3. Octanol–Air Partitioning Coefficients

One problem with the use of p_L^o as a key parameter in both *adsorption* and *absorption* is the difficulty in obtaining accurate values for p_L^o for solid SOCs, since they are not experimentally accessible and must be estimated (e.g., see Finizio et al., 1997, and references therein). In addition, as discussed in the preceding section with respect to absorption into a liquid phase, slopes of -1 for plots of log K_p against log p_L^o are only expected if the activity coefficients, γ_i, do not change along a series of compounds.

An alternate approach has been proposed that avoids the use of p_L^o and introduces a ratio of activity coefficients rather than an absolute, single value. The advantage in terms of the activity coefficients is that the ratio is usually less sensitive to changes along a series of compounds than is a single value of the activity coefficient. In this approach, the parameter p_L^o is replaced by a different descriptor of an SOC's volatility, its *octanol–air partition coefficient*, K_{OA} (Finizio et al., 1997; Harner and Bidleman, 1998; Pankow, 1998). This is defined for a given SOC partitioned between liquid octanol and air as

$$K_{OA} = C_O/C_A, \quad \text{(LLL)}$$

where C_O and C_A are the concentrations of the SOC in octanol and in air, respectively, both in units of mol m^{-3}. A distinct advantage of the use of K_{OA} compared to p_L^o is that it can be directly measured (e.g., see Harner and Mackay, 1995; Harner and Bidleman, 1996) or can be estimated from knowledge of two well-established and widely used partition coefficients, K_{OW} for octanol–water and K_{AW} for air–water (e.g., see Mackay et al., 1992; and Baum, 1998).

The relationship between K_p and K_{OA} can be readily developed based on thermodynamic principles (e.g., see Finizio et al., 1997); see Box 9.4.

Thus, the relationship between K_p and K_{OA} by Finizio et al. (1997) leads to Eq. (PPP):

$$\log K_p = \log K_{OA} + \log\left[\frac{\gamma_O}{\gamma_{om}}\right] + \log\left[\frac{760 f_{om}(MW)_O}{10^9 MW_{om} \rho_O}\right] \quad \text{(PPP)}$$

To test the validity of using log K_{OA} rather than p_L^o as a direct descriptor of the volatility of SOCs, Finizio and co-workers (1997) calculated values of log K_p (normalized to 25°C) from published ambient air data on gas-to-particle partitioning of PAHs (as well as PCBs and organochlorine pesticides) at seven sampling sites ranging from urban to rural airsheds. K_{OA} values were obtained from the literature or calculated from the ratio K_{OW}/K_{AW}, where K_{OW} is the octanol–water partition coefficient (e.g., see Mackay et al., 1992; and Baum, 1998) and K_{AW} is the dimensionless air–water partition coefficient calculated from literature values of Henry's law constants. Figure 9.64 shows a plot of log K_p against log K_{OA} for the PAH data. A linear relationship is observed, as expected from Eq. (PPP). The slope of 0.79 is below the theoretical value of 1.0, reflecting the fact that the slopes plotted in the conventional manner, i.e., log K_p vs log p_L^o, were also significantly less than the theoretical value of $m_r = -1.00$. The authors cite several possible reasons, including nonequilibrium conditions as described by Kamens et al. (1995).

Figure 9.65a shows the percentages of the 3- and 4-ring PAHs, as well as some PCBs and polychlorinated naphthalenes (PCNs), found on aerosol particles in the

BOX 9.4
RELATIONSHIP BETWEEN K_p AND K_{OA}

First, using the ideal gas law, the partial pressure (which is equal to the fugacity) of the SOC in air is given by $C_A RT$. Over the octanol solution, the fugacity (f_i) of the SOC is given by $P_i/p_L^°$; i.e., its partial pressure over the octanol solution is given by $P_i = f_i p_L^°$. Using $f_i = X_{i,O} \gamma_O$, the partial pressure of the SOC over an octanol solution containing a mole fraction $X_{i,O}$ of the SOC is given by $P_i = (X_{i,O} \gamma_O p_L^°)$, where the symbols have their usual meaning of mole fraction, activity coefficient, and liquid vapor pressure, respectively, all in octanol. The mole fraction can be replaced in this expression, since $X_{i,O} = [C_O(MW_O)/10^3 \rho_O]$, where C_O is the concentration of the SOC in octanol in units of mol m^{-3}, MW_O is the molecular weight of octanol, and ρ_O is the density (kg m^{-3}) of octanol. Thus, the fugacity of the SOC above the octanol solution must be $P_i = [C_O(MW_O) \gamma_O p_L^°/10^3 \rho_O]$, and we know that $P_i = C_A RT$. Hence

$$C_A RT = \frac{C_O(MW_O)\gamma_O p_L^°}{10^3 \rho_O}. \quad \text{(MMM)}$$

Equation (LLL) then becomes

$$K_{OA} = \frac{C_O}{C_A} = \frac{10^3 \rho_O RT}{(MW_O)\gamma_O p_L^°}. \quad \text{(NNN)}$$

This, then, gives the relationship between $p_L^°$ and the octanol–air partitioning coefficient, K_{OA}. Rearranging Eq. (NNN) to obtain an expression for $p_L^°$ and substitution into Eq. (JJJ) give (Finizio et al., 1997)

$$K_p = \frac{760 RT f_{om}}{10^6 MW_{om} \gamma_{om}} \cdot \frac{(MW_O)\gamma_O K_{OA}}{10^3 \rho_O RT}$$

$$= K_{OA}\left(\frac{\gamma_O}{\gamma_{om}}\right)\left[\frac{760 f_{om}(MW)_O}{10^9 MW_{om} \rho_O}\right]. \quad \text{(OOO)}$$

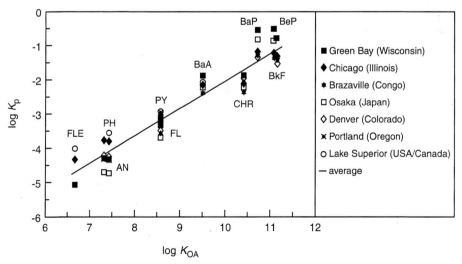

FIGURE 9.64 Plot of log K_p for 10 PAHs in ambient air samples collected at seven sites worldwide as a function of calculated values of log K_{OA}. FLE = fluorene, PH = phenanthrene, AN = anthracene, PY = pyrene, FL = fluoranthene, BaA = benz[a]anthracene, CHR = chrysene, BaP = benzo[a]pyrene, BeP = benzo[e]pyrene, BkF = benzo[k]fluoranthene (adapted from Finizio et al., 1997).

urban Chicago area in February and March (Harner and Bidleman, 1998). The solid line shows the predicted percentages calculated using the Junge–Pankow *adsorption* model, Eq. (TT). Figure 9.65b treats the same data set in terms of *absorption* into a liquid using the octanol–water partitioning coefficient assuming the fraction of organic matter, f_{om}, in the particles is either 10 or 20%. Both models are in reasonable agreement with the PAH data for winds from the southwest. With winds from the northeast, enrichment of the PAH in particles is observed compared to model predictions using the absorption model (Fig. 9.65b), which the authors suggest could be due to nonattainment of equilibrium or to the trapping of nonexchangeable PAHs in the particles. The *adsorption* model overpredicts the percentages of PCBs expected to be found in the particles, but the K_{OA} *absorption* model is effective in explaining the field partitioning data for PCBs and PCNs.

In short, data such as those in Figs. 9.64 and 9.65 support the use of the octanol–air partitioning coefficient as a useful parameter for characterizing gas–particle partitioning of SOC into liquid particles or liquid layers on particles in air.

As discussed elsewhere in this book, there is increasing evidence for reactions at the air–water *interface* in the atmosphere. Pankow (1997) has treated partitioning of gases to the interface as well and predicts that as for adsorption on a solid and absorption into a liquid, there should be a linear relationship between $\log K_p$ and $\ln p_L^\circ$ with a slope of approximately -1.

This discussion of gas–particle partitioning has focused on the idea that equilibrium between the two phases is attained in the atmosphere. However, it should be noted that equilibrium cannot always be assumed in the atmosphere. For example, Wania *et al.* (1998) examined the exchange of SOC between the air and the surface under two assumptions, one being equilibrium and one that treats the kinetics of the exchange of the SOC with the soil and with incoming and outgoing air. They show that some atmospheric observations can be explained by the kinetic effects. For example, the concentration of γ-hexachlorocyclohexane (HCH) in the gas phase has a significant temperature dependence while α-HCH does not. This is unexpected for two such similar molecules if equilibrium gas–particle partitioning controls the gaseous concentrations. Wania *et al.* (1998) suggest that α-HCH concentrations are determined largely by the kinetics of transport of air containing this compound to the measurement sites, since it is distributed globally and hence there are large reservoirs such as the oceans. γ-HCH, on the other hand, has lower global "background" levels but has been used extensively in industrialized countries, leading to higher concentrations downwind. Hence they attribute the γ-HCH levels, measured at sampling sites located downwind, to a shift in the equilibrium toward the gas phase as soil temperatures increased, leading to increased evaporation.

Because of such "real-world" nonequilibrium situations, some efforts, for example, by Turco and co-workers (Jacobson *et al.*, 1996), Kamens and co-workers (Odum *et al.*, 1994), and Seinfeld and co-workers (Bowman *et al.*, 1997), have focused on developing dynamic models to describe gas–particle distributions.

In a similar vein, the time scales to achieve equilibrium for inorganics have been examined by Meng and Seinfeld (1996), who show that small (submicron) particles can come to equilibrium with the gas phase in less than a few hours typically but that larger particles may not. The major factors determining the time needed to reach equilibrium are the aerosol size distribution,

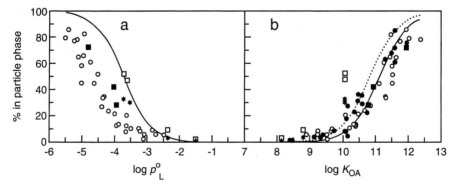

FIGURE 9.65 Observed percentages in the particle phase of PAHs (($\square$) N-E air; (*) S-W air), PCBs (($\blacksquare$) nonortho, ($\bigcirc$) monoortho, and multiortho PCBs, respectively), and ($\bullet$) PCNs in Chicago air compared to model predicted values. (a) Solid line is calculated with Junge–Pankow (J-P) *adsorption* model (Eq. (TT)). (b) Solid and dotted lines are calculated with *absorption* model for aerosols assumed to contain 10 and 20% organic matter (om), respectively (adapted from Harner and Bidleman, 1998).

temperature, and accommodation coefficient for uptake of the gas into the particle (see Chapter 5).

In short, traditional equilibrium theories of gas adsorption provide at least a qualitative framework for describing the partitioning of SOC between the gas and particle phases in the atmosphere, and more recent theories are providing further insight into these partitioning processes in the atmosphere.

E. PROBLEMS

1. Use the Kelvin equation (Chapter 14.C.2) to show that it is true that the vapor pressure of pure water at 25°C is only 0.1% greater over a 1-μm radius particle than over a flat surface, but 11% greater over a 0.01-μm radius particle. The surface tension of water is 72 dyn cm^{-1} at 25°C.

2. At what particle radius would the vapor pressure of water be twice that over a flat surface at 25°C? The surface tension of water is 72 dyn cm^{-1} at 25°C.

3. Use Eq. (GG) to predict the critical concentration of gaseous H_2SO_4 in ppb for nucleation at a relative humidity of 40% and a temperature of 25°C. How does this compare to the theoretical predictions and experimental observations shown in Fig. 9.30?

4. Write reaction mechanisms for the oxidation of 1-octene in the atmosphere that will give the major observed condensed-phase products heptanal and heptanoic acid.

5. In smog chamber studies of the oxidation of 1-decene, octanal was the condensed-phase product formed in the next highest yield after nonanal, nonanoic acid, and dihydro-5-pentyl-2(3H)furanone (Forstner et al., 1997a). Write a plausible mechanism for its formation.

6. Show how m-toluic acid could be formed from the abstraction reaction of OH with m-xylene (Table 9.22).

7. An important aspect of Eq. (LL) is that it shows that the yield of secondary organic aerosol depends on the mass of the condensed-phase organic available to take up semivolatile organic products. Show that Eq. (LL) can be derived for a single species. Use the relationship between C_i and Δ (organic) and $K_{i,\mathrm{om}} = F_{i,\mathrm{om}}/(A_i M_0)$. $F_{i,\mathrm{om}}$ is the concentration of the species (ng m^{-3}) in the condensed ("om") phase and M_0 is the mass of condensed-phase organic available for uptake of the semivolatile organic. You will also need to use the mass balance restraint $C_i = A_i + F_{i,\mathrm{om}}$ and the definition of the yield of secondary organic aerosol,

$$Y[10^3 V_c \Delta(\mathrm{organic})] = \sum F_i V_c,$$

where V_c is the volume of air in which the reaction takes place.

References

Adamson, A. W., *A Textbook of Physical Chemistry*, Academic Press, New York, 1973.

Adamson, A. W. *Physical Chemistry of Surfaces*, Wiley, New York, 1982.

Aitchison, J., and J. A. C. Brown, *The Lognormal Distribution*, Cambridge Univ. Press, London, 1957.

Allen, H. C., J. M. Laux, R. Vogt, B. J. Finlayson-Pitts, and J. C. Hemminger, "Water-Induced Reorganization of Ultrathin Nitrate Films on NaCl: Implications for the Tropospheric Chemistry of Sea Salt Particles," *J. Phys. Chem., 100*, 6371–6375 (1996).

Alonso, C. D., M. H. R. B. Martins, J. Romano, and R. Godinho, "São Paulo Aerosol Characterization Study," *J. Air Waste Manage. Assoc., 47*, 1297–1300 (1997).

Andreae, M. O., E. V. Browell, M. Garstang, G. L. Gregory, R. C. Harriss, G. F. Hill, D. J. Jacob, M. C. Pereira, G. W. Sachse, A. W. Setzer, P. L. Silva Dias, R. W. Talbot, A. L. Torres, and S. C. Wofsy, "Biomass-Burning Emissions and Associated Haze Layers over Amazonia," *J. Geophys. Res., 93*, 1509–1527 (1988).

Andrews, E., and S. M. Larson, "Effect of Surfactant Layers on the Size Changes of Aerosol Particles as a Function of Relative Humidity," *Environ. Sci. Technol., 27*, 857–865 (1993).

Appel, B. R., S. M. Wall, and R. L. Knights, "Characterization of Carbonaceous Materials in Atmospheric Aerosols by High Resolution Mass Spectrometric Thermal Analysis," *Adv. Environ. Sci. Technol., 10*, 353–365 (1980).

Appel, B. R., Y. Tokiwa, J. Hsu, E. L. Kothny, and E. Hahn, "Visibility as Related to Atmospheric Aerosol Constitutents," *Atmos. Environ., 19*, 1525–1534 (1985).

Arimoto, R., R. A. Duce, B. J. Ray, A. D. Hewitt, and J. Williams, "Trace Elements in the Atmosphere of American Samoa: Concentrations and Deposition to the Tropical South Pacific," *J. Geophys. Res., 92*, 8465–8479 (1987).

Artaxo, P., H. Storms, F. Bruynseels, R. Van Grieken, and W. Maenhaut, "Composition and Sources of Aerosols from the Amazon Basin," *J. Geophys. Res., 93*, 1605–1615 (1988).

Artaxo, P., F. Gerab, M. A. Yamasoe, and J. V. Martins, "Fine Mode Aerosol Composition at Three Long-Term Atmospheric Monitoring Sites in the Amazon Basin," *J. Geophys. Res., 99*, 22857–22868 (1994).

Artaxo, P., and H.-C. Hansson, "Size Distribution of Biogenic Aerosol Particles from the Amazon Basin," *Atmos. Environ., 29*, 393–402 (1995).

Avila, A., I. Queralt-Mitjans, and M. Alarcón, "Mineralogical Composition of African Dust Delivered by Red Rains over Northeastern Spain," *J. Geophys. Res., 102*, 21977–21996 (1997).

Baum, E. J., *Chemical Property Estimation*, Lewis Publishers, Boca Raton, FL, 1998.

Beichert, P., and B. J. Finlayson-Pitts, "Knudsen Cell Studies of the Uptake of Gaseous HNO_3 and Other Oxides of Nitrogen on Solid NaCl: The Role of Surface-Adsorbed Water," *J. Phys. Chem., 100*, 15218–15228 (1996).

Berko, H. N., P. C. McCaslin, and B. J. Finlayson-Pitts, "Formation of Gas-Phase Bromine Compounds by Reaction of Solid NaBr with Gaseous $ClONO_2$, Cl_2, and BrCl at 298 K," *J. Phys. Chem., 95*, 6951–6958 (1991).

Berner, A., S. Sidla, Z. Galambos, C. Kruisz, R. Hitzenberger, H. M. ten Brink, and G. P. A. Kos, "Modal Character of Atmospheric Black Carbon Size Distributions," *J. Geophys. Res., 101*, 19559–19565 (1996).

Bidleman, T. F., "Atmospheric Processes," *Environ. Sci. Technol., 22*, 361–367 (1988).

Blanchard, D. C., and A. H. Woodcock, "The Production, Concentration, and Vertical Distribution of the Sea-Salt Aerosol," *Ann. N.Y. Acad. Sci., 338*, 330–347 (1980).

Blanchard, D. C., "The Oceanic Production of Atmospheric Sea Salt," *J. Geophys. Res., 90,* 961–963 (1985).

Blando, J. D., R. J. Porcja, T.-H. Li, D. Bowman, P. J. Lioy, and B. J. Turpin, "Secondary Formation and the Smoky Mountain Organic Aerosol: An Examination of Aerosol Polarity and Functional Group Composition during SEAVS," *Environ. Sci. Technol., 32,* 604–613 (1998).

Bodhaine, B. A., J. M. Harris, J. A. Ogren, and D. J. Hofmann, "Aerosol Optical Properties at Mauna Loa Observatory: Long-Range Transport from Kuwait?" *Geophys. Res. Lett., 19,* 581–584 (1992).

Bodhaine, B. A., "Aerosol Absorption Measurements at Barrow, Mauna Loa, and the South Pole," *J. Geophys. Res., 100,* 8967–8975 (1995).

Bohren, C. F., and D. R. Huffman, *Absorption and Scattering by Small Particles,* Wiley, New York, 1983.

Boon, K. F., L. Kiefert, and G. H. McTainsh, "Organic Matter Content of Rural Dust in Australia," *Atmos. Environ., 32,* 2817–2823 (1998).

Bowman, F. M., J. R. Odum, J. H. Seinfeld, and S. N. Pandis, "Mathematical Model for Gas–Particle Partitioning of Secondary Organic Aerosols," *Atmos. Environ., 31,* 3921–3931 (1997).

Braaten, D. A., and T. A. Cahill, "Size and Composition of Asian Dust Transported to Hawaii," *Atmos. Environ., 20,* 1105–1109 (1986).

Brook, J. R., A. H. Wiebe, S. A. Woodhouse, C. V. Audette, T. F. Dann, S. Callaghan, M. Piechowski, E. Dabek-Zlotorzynska, and J. F. Dloughy, "Temporal and Spatial Relationships in Fine Particle Strong Acidity, Sulphate, PM_{10}, and $PM_{2.5}$ across Multiple Canadian Locations," *Atmos. Environ., 31,* 4223–4236 (1997).

Buseck, P. R., and M. Pósfai, "Airborne Minerals and Related Aerosol Particles: Effects on Climate and the Environment," *Proc. Natl. Acad. Sci. USA, 96,* 3372–3379 (1999).

Cadle, R. D., and R. C. Robbins, "Kinetics of Atmospheric Chemical Reactions Involving Aerosols," *Discuss. Faraday Soc., 30,* 155–161 (1960).

Cahill, T. A., P. H. Wakabayashi, and T. A. James, "Chemical States of Sulfate at Shenandoah National Park during Summer, 1991," *NIM B, 109 / 110,* 542–547 (1996).

Cantrell, W., G. Shaw, R. Benner, and D. Veazey, "Evidence for Sulfuric Acid Coated Particles in the Arctic Air Mass," *Geophys. Res. Lett., 24,* 3005–3008 (1997).

Caquineau, S., A. Gaudichet, L. Gomes, M.-C. Magonthier, and B. Chatenet, "Saharan Dust: Clay Ratio as a Relevant Tracer to Assess the Origin of Soil-Derived Aerosols," *Geophys. Res. Lett., 25,* 983–986 (1998).

Cautreels, W., and K. Van Cauwenberghe, "Experiments on the Distribution of Organic Pollutants between Airborne Particulate Matter and the Corresponding Gas Phase," *Atmos. Environ., 12,* 1133–1141 (1978).

Chang, D. P. Y., and R. C. Hill, "Retardation of Aqueous Droplet Evaporation by Air Pollutants," *Atmos. Environ., 14,* 803–807 (1980).

Chang, S. G., R. Brodzinsky, L. A. Gundel, and T. Novakov, "Chemical and Catalytic Properties of Elemental Carbon," in *Particulate Carbon: Atmospheric Life Cycle* (G. T. Wolff and R. L. Klimsch, Eds.), pp. 159–181, Plenum, New York, 1982.

Charlson, R. J., H. Horvath, and R. F. Pueschel, "The Direct Measurement of Atmospheric Light Scattering Coefficient for Studies of Visibility and Pollution," *Atmos. Environ., 1,* 469–478 (1967).

Charlson, R. J., A. H. Vanderpol, D. S. Covert, A. P. Waggoner, and N. C. Ahlquist, "$H_2SO_4/(NH_4)_2SO_4$ Background Aerosol: Optical Detection in the St. Louis Region," *Atmos. Environ., 8,* 1257–1267 (1974a).

Charlson, R. J., A. H. Vanderpol, D. S. Covert, A. P. Waggoner, and N. C. Ahlquist, "Sulfuric Acid–Ammonium Sulfate Aerosol: Optical Detection in the St. Louis Region," *Science, 184,* 156–158 (1974b).

Chen, X., and B. R. T. Simoneit, "Epicuticular Waxes from Vascular Plants and Particles in the Lower Troposphere: Analysis of Lipid Classes by Iatroscan Thin-Layer Chromatography with Flame Ionization Detection," *J. Atmos. Chem., 18,* 17–31 (1994).

Chiapello, I., G. Bergametti, B. Chatenet, P. Bousquet, F. Dulac, and E. Santos Soares, "Origins of African Dust Transported over the Northeastern Tropical Atlantic," *J. Geophys. Res., 102,* 13701–13709 (1997).

Chughtai, A. R., M. E. Brooks, and D. M. Smith, "Hydration of Black Carbon," *J. Geophys. Res., 101,* 19505–19514 (1996).

Cicerone, R. J., "Halogens in the Atmosphere," *Rev. Geophys. Space Phys., 19,* 123–139 (1981).

Clarke, A. D., and R. J. Charlson, "Radiative Properties of the Background Aerosol: Absorption Component of Extinction," *Science, 229,* 263–265 (1985).

Clarke, A. D., D. Davis, V. N. Kapustin, F. Eisele, G. Chen, I. Paluch, D. Lenschow, A. R. Bandy, D. Thornton, K. Moore, L. Mauldin, D. Tanner, M. Litchy, M. A. Carroll, J. Collins, and G. Albercook, "Particle Nucleation in the Tropical Boundary Layer and Its Coupling to Marine Sulfur Sources," *Science, 282,* 89–92 (1998).

Clegg, S. L., and P. Brimblecombe, "Equilibrium Partial Pressures and Mean Activity and Osmotic Coefficients of 0–100% Nitric Acid as a Function of Temperature," *J. Phys. Chem., 94,* 5369–5380 (1990), and references therein.

Cobourn, W. G., R. B. Husar, and J. D. Husar, "Continuous *In Situ* Monitoring of Ambient Particulate Sulfur Using Flame Photometry and Thermal Analysis," *Atmos. Environ., 12,* 89–98 (1978).

Cofer, W. R., III, J. S. Levine, D. I. Sebacher, E. L. Winstead, P. J. Riggin, J. A. Brass, and V. G. Ambrosia, "Particulate Emissions from a Mid-Latitude Prescribed Chaparral Fire," *J. Geophys. Res., 93,* 5207–5212 (1988).

Conner, W. D., R. L. Bennett, W. S. Weathers, and W. E. Wilson, "Particulate Characteristics and Visual Effects of the Atmosphere at Research Triangle Park," *J. Air Waste Manage. Assoc., 41,* 154–160 (1991).

Cooke, W. F., S. G. Jennings, and T. G. Spain, "Black Carbon Measurements at Mace Head, 1989–1996," *J. Geophys. Res., 102,* 25339–25346 (1997).

Cooper, J. E., and E. E. Bray, "A Postulated Role of Fatty Acids in Petroleum Formation," *Geochim. Cosmochim. Acta, 27,* 1113–1127 (1963).

Cotham, W. E., and T. F. Bidleman, "Laboratory Investigations of the Partitioning of Organochlorine Compounds between the Gas Phase and Atmospheric Aerosols on Glass Fiber Filters," *Environ. Sci. Technol., 26,* 469–478 (1992).

Covert, D. S., A. P. Waggoner, R. E. Weiss, N. C. Ahlquist, and R. J. Charlson, "Atmospheric Aerosols, Humidity, and Visibility," *Adv. Environ. Sci. Technol., 10,* 559–581 (1980).

Covert, D. S., "North Pacific Marine Background Aerosol: Average Ammonium to Sulfate Molar Ratio Equals 1," *J. Geophys. Res., 93,* 8455–8458 (1988).

Covert, D. S., V. N. Kapustin, P. K. Quinn, and T. S. Bates, "New Particle Formation in the Marine Boundary Layer," *J. Geophys. Res., 97,* 20581–20589 (1992).

Cronn, D. R., R. J. Charlson, R. L. Knights, A. L. Crittenden, and B. R. Appel, "A Survey of the Molecular Nature of Primary and Secondary Components of Particles in Urban Air by High Resolution Mass Spectrometry," *Atmos. Environ., 11,* 929–937 (1977).

Cui, W., J. Machir, L. Lewis, D. J. Eatough, and N. L. Eatough, "Fine Particulate Organic Material at Meadview during the Project MOHAVE Summer Intensive Study," *J. Air Waste Manage. Assoc., 47,* 357–369 (1997).

Cunningham, P. T., B. D. Holt, S. A. Johnson, D. L. Drapcho, and R. Kumar, "Acidic Aerosols: Oxygen-18 Studies of Formation and

Infrared Studies of Occurrence and Neutralization," in *Chemistry of Particles, Fogs, and Rain* (J. L. Durham, Ed.), Acid Precipitation Series, Vol. 2, pp. 53–130 (J. I. Teasley, Series Ed.), Butterworth, Stoneham, MA, 1984.

Daisey, J. M., A. H. Miguel, J. B. de Andrade, P. A. P. Pereira, and R. L. Tanner, "An Overview of the Rio de Janeiro Aerosol Characterization Study," *J. Air Pollut. Control Assoc.*, 37, 15–23 (1987).

Däumer, B., R. Niessner, and D. Klockow, "Laboratory Studies of the Influence of Thin Organic Films on the Neutralization Reaction of H_2SO_4 Aerosol with Ammonia," *J. Aerosol Sci.*, 23, 315–325 (1992).

Davies, C. N., Ed., *Aerosol Science*, Academic Press, London, 1966.

Day, D. E., W. C. Malm, and S. M. Kreidenweis, "Seasonal Variations in Aerosol Composition and Acidity at Shenandoah and Great Smoky Mountains National Parks," *J. Air Waste Manage. Assoc.*, 47, 411–418 (1997).

De Haan, D. O., T. Brauers, K. Oum, J. Stutz, T. Nordmeyer, and B. J. Finlayson-Pitts, "Heterogeneous Chemistry in the Troposphere: Experimental Approaches and Applications to the Chemistry of Sea Salt Particles," *Int. Rev. Phys. Chem.*, 18(3) (1999).

Dentener, F. J., G. R. Carmichael, and Y. Zhang, "The Role of Mineral Aerosol as a Reactive Surface in the Global Troposphere," *J. Geophys. Res.*, 101, 22869–22889 (1996).

Dixon, R. W., and H. Aasen, "Measurement of Hydroxymethanesulfonate in Atmospheric Aerosols," *Atmos. Environ.*, 33, 2023–2029 (1999).

Dodd, J. A., J. M. Ondov, G. Tuncel, T. G. Dzubay, and R. K. Stevens, "Multimodal Size Spectra of Submicrometer Particles Bearing Various Elements in Rural Air," *Environ. Sci. Technol.*, 25, 890–903 (1991).

Donaldson, D. J., and D. Anderson, "Adsorption of Atmospheric Gases at the Air–Water Interface. II: C_1–C_4 Alcohols, Acids, and Acetone," *J. Phys. Chem. A*, 103, 871–876 (1999).

Duce, R. A., W. H. Zoller, and J. L. Moyers, "Particulate and Gaseous Halogens in the Antarctic Atmosphere," *J. Geophys. Res.*, 78, 7802–7811 (1973).

Duce, R. A., C. K. Unni, B. J. Ray, J. M. Prospero, and J. T. Merrill, "Long-Range Atmospheric Transport of Soil Dust from Asia to the Tropical North Pacific: Temporal Variability," *Science*, 209, 1522–1524 (1980).

Duce, R. A., V. A. Mohnen, P. R. Zimmerman, D. Grosjean, W. Cautreels, R. Chatfield, R. Jaenicke, J. A. Ogren, E. D. Pellizzari, and G. T. Wallace, "Organic Material in the Global Troposphere," *Rev. Geophys. Space Phys.*, 21, 921–952 (1983).

Dumdei, B. E., and R. J. O'Brien, "Toluene's Degradation Products under Simulated Atmospheric Conditions," *Nature*, 311, 248–250 (1984).

Dunwoody, C. L., "Rapid Nitrate Loss from PM10 Filters," *J. Air Pollut. Control Assoc.*, 36, 817–818 (1986).

Dzubay, T. G., R. K. Stevens, C. W. Kewis, D. H. Hern, W. J. Courtney, J. W. Tesch, and M. A. Mason, "Visibility and Aerosol Composition in Houston, Texas," *Environ. Sci. Technol.*, 16, 514–525 (1982).

Dzubay, T. G., R. K. Stevens, G. E. Gordon, I. Olmez, A. E. Sheffield, and W. J. Courtney, "A Composite Receptor Method Applied to Philadelphia Aerosol," *Environ. Sci. Technol.*, 22, 46–52 (1988).

Eatough, D. J., D. A. Eatough, L. Lewis, and E. A. Lewis, "Fine Particulate Chemical Composition and Light Extinction at Canyonlands National Park Using Organic Particulate Material Concentrations Obtained with a Multisystem, Multichannel Diffusion Denuder Sampler," *J. Geophys. Res.*, 101, 19515–19531 (1996).

Eisele, F. L., and P. H. McMurry, "Recent Progress in Understanding Particle Nucleation and Growth," *Philos. Trans. R. Soc. London*, 352, 191–201 (1997).

Eldering, A., S. M. Larson, J. R. Hall, K. J. Hussey, and G. R. Cass, "Development of an Improved Image Processing Based Visibility Model," *Environ. Sci. Technol.*, 27, 626–635 (1993).

Eldering, A., and G. R. Cass, "Source-Oriented Model for Air Pollutant Effects on Visibility," *J. Geophys. Res.*, 101, 19343–19369 (1996).

Eldering, A., J. R. Hall, K. J. Hussey, and G. R. Cass, "Visibility Model Based on Satellite-Generated Landscape Data," *Environ. Sci. Technol.*, 30, 361–370 (1996).

Erickson, D. J., J. T. Merrill, and R. A. Duce, "Seasonal Estimates of Global Atmospheric Sea-Salt Distributions," *J. Geophys. Res.*, 91, 1067–1072 (1986).

Falconer, R. L., and T. F. Bidleman, "Vapor Pressures and Predicted Particle/Gas Distributions of Polychlorinated Biphenyl Congeners as Functions of Temperature and Ortho-Chlorine Substitution," *Atmos. Environ.*, 28, 547–554 (1994).

Ferman, M. A., G. T. Wolff, and N. A. Kelly, "The Nature and Sources of Haze in the Shenandoah Valley/Blue Ridge Mountains Area," *J. Air Pollut. Control Assoc.*, 31, 1074–1082 (1981).

Finizio, A., D. Mackay, T. Bidleman, and T. Harner, "Octanol–Air Partition Coefficient as a Predictor of Partitioning of Semi-Volatile Organic Chemicals to Aerosols," *Atmos. Environ.*, 31, 2289–2296 (1997).

Finlayson-Pitts, B. J., and S. N. Johnson, "The Reaction of NO_2 with NaBr: Possible Source of BrNO in Polluted Marine Atmospheres," *Atmos. Environ.*, 22, 1107–1112 (1988).

Finlayson-Pitts, B. J., F. E. Livingston, and H. N. Berko, "Ozone Destruction and Bromine Photochemistry at Ground Level in the Arctic Spring," *Nature*, 343, 622–625 (1990).

Fitzgerald, J. W., "Marine Aerosols—A Review," *Atmos. Environ.*, 25A, 533–545 (1991).

Flagan, R. C., S.-C. Wang, F. Yin, J. H. Seinfeld, G. Reischl, W. Winklmayr, and R. Karch, "Electrical Mobility Measurements of Fine-Particle Formation during Chamber Studies of Atmospheric Photochemical Reactions," *Environ. Sci. Technol.*, 25, 883–890 (1991).

Fogg, T. R., and R. A. Duce, "Boron in the Troposphere: Distribution and Fluxes," *J. Geophys. Res.*, 90, 3781–3796 (1985).

Foreman, W. T., and T. F. Bidleman, "Semivolatile Organic Compounds in Ambient Air in Denver, Colorado," *Atmos. Environ.*, 24A, 2405–2416 (1990).

Forstner, H. J. L., R. C. Flagan, and J. H. Seinfeld, "Molecular Speciation of Secondary Organic Aerosol from Photooxidation of the Higher Alkenes: 1-Octene and 1-Decene," *Atmos. Environ.*, 31, 1953–1964 (1997a).

Forstner, H. J. L., R. C. Flagan, and J. H. Seinfeld, "Secondary Organic Aerosol from the Photooxidation of Aromatic Hydrocarbons: Molecular Composition," *Environ. Sci. Technol.*, 31, 1345–1358 (1997b).

Fraser, M. P., D. Grosjean, E. Grosjean, R. A. Rasmussen, and G. R. Cass, "Air Quality Model Evaluation Data for Organics. 1. Bulk Chemical Composition and Gas/Particle Distribution Factors," *Environ. Sci. Technol.*, 30, 1731–1743 (1996).

Friedlander, S. K., *Smoke, Dust, and Haze: Fundamentals of Aerosol Behavior*, Wiley-Interscience, New York, 1977.

Friedlander, S. K., "A Note on New Particle Formation in the Presence of Aerosol," *J. Colloid Interface Sci.*, 67, 387–388 (1978).

Friedlander, S. K., "Future Aerosols of the Southwest: Implications for Fundamental Aerosol Research," *Ann. N.Y. Acad. Sci.*, 338, 588–598 (1980).

Friedlander, S. K., "Dynamics of Aerosol Formation by Chemical Reaction," *Ann. N.Y. Acad. Sci.*, 404, 354–364 (1983).

Fuchs, N. A., *The Mechanics of Aerosols*, Pergamon, Oxford, 1964.

Fuchs, N. A., and A. G. Sutugin, *Highly Dispersed Aerosols*, Ann Arbor Science Publishers, Ann Arbor, MI, 1970.

Fukuta, N., and P. E. Wagner, *Nucleation and Atmospheric Aerosols*, A. Deepak Publishing, Hampton, VA, 1992.

Gaines, G. L., *Insoluble Monolayers at Liquid–Gas Interfaces*, Wiley, New York, 1966.

Gard, E. E., M. J. Kleeman, D. S. Gross, L. S. Hughes, J. O. Allen, B. D. Morrical, D. P. Fergenson, T. Dienes, M. E. Gälli, R. J. Johnson, G. R. Cass, and K. A. Prather, "Direct Observation of Heterogeneous Chemistry in the Atmosphere," *Science, 279*, 1184–1187 (1998).

Gatz, D. F., and J. M. Prospero, "A Large Silicon–Aluminum Aerosol Plume in Central Illinois: North African Desert Dust?" *Atmos. Environ., 30*, 3789–3799 (1996).

Gill, P. S., T. E. Graedel, and C. J. Weschler, "Organic Films on Atmospheric Aerosol Particles, Fog Droplets, Cloud Droplets, Raindrops, and Snowflakes," *Rev. Geophys. Space Phys., 21*, 903–920 (1983).

Goldberg, R. N., "Evaluation Activity and Osmotic Coefficients for Aqueous Solutions: Thirty-Six Unibivalent Electrolytes," *J. Phys. Chem. Ref. Data, 10*, 671–764 (1981).

Gong, S. L., and L. A. Barrie, "Modeling Sea-Salt Aerosols in the Atmosphere. 1. Model Development," *J. Geophys. Res., 102*, 3805–3818 (1997).

Gong, S. L., L. A. Barrie, J. M. Prospero, D. L. Savoie, G. P. Ayers, J.-P. Blanchet, and L. Spacek, "Modeling Sea-Salt Aerosols in the Atmosphere. 2. Atmospheric Concentrations and Fluxes," *J. Geophys. Res., 102*, 3819–3830 (1997).

Gordon, G. E., "Receptor Models," *Environ. Sci. Technol., 22*, 1132–1142 (1988).

Gordon, R. J., N. J. Trivedi, B. P. Singh, and E. C. Ellis, "Characterization of Aerosol Organics by Diffuse Reflectance Fourier Transform Infrared Spectroscopy," *Environ. Sci. Technol., 22*, 672–677 (1988).

Goss, K.-U., "Conceptual Model for the Adsorption of Organic Compounds from the Gas Phase to Liquid and Solid Surfaces," *Environ. Sci. Technol., 31*, 3600–3605 (1997).

Goss, K.-U., and R. P. Schwarzenbach, "Gas/Solid and Gas/Liquid Partitioning of Organic Compounds: Critical Evaluation of the Interpretation of Equilibrium Constants," *Environ. Sci. Technol., 32*, 2025–2032 (1998).

Graedel, T. E., D. T. Hawkins, and L. D. Claxton, *Atmospheric Chemical Compounds—Sources, Occurrence, and Bioassay*, Academic Press, New York, 1986.

Gray, H. A., G. R. Cass, J. J. Huntzicker, E. K. Heyerdahl, and J. A. Rau, "Characteristics of Atmospheric Organic and Elemental Carbon Particle Concentrations in Los Angeles," *Environ. Sci. Technol., 20*, 580–589 (1986).

Gray, H. A., and G. R. Cass, "Source Contributions to Atmospheric Fine Carbon Particle Concentrations," *Atmos. Environ., 32*, 3805–3825 (1998).

Greaves, R. C., R. M. Barkely, R. E. Sievers, and R. R. Meglen, "Covariations in the Concentrations of Organic Compounds Associated with Springtime Atmospheric Aerosols," *Atmos. Environ., 21*, 2549–2561 (1987).

Groblicki, P. J., G. T. Wolff, and R. J. Countess, "Visibility Reducing Species in the Denver Brown Cloud. I," *Atmos. Environ., 15*, 2473–2484 (1981).

Grosjean, D., and S. K. Friedlander, "Formation of Organic Aerosols from Cyclic Olefins and Diolefins," *Adv. Environ. Sci. Technol., 10*, 435–473 (1980).

Grosjean, D., and J. H. Seinfeld, "Parameterization of the Formation Potential of Secondary Organic Aerosols," *Atmos. Environ., 23*, 1733–1747 (1989).

Grosjean, D., "*In Situ*, Organic Aerosol Formation during a Smog Episode: Estimated Production and Chemical Functionality," *Atmos. Environ., 26A*, 953–963 (1992).

Gundel, L. A., S. G. Chang, M. S. Clemenson, S. S. Markowitz, and T. Novakov, "Characterization of Particulate Amines," in *Nitrogenous Air Pollutants. Chemical and Biological Implications* (D. Grosjean, Ed.), pp. 211–220, Ann Arbor Science Publishers, Ann Arbor, MI, 1979.

Gundel, L. A., J. M. Daisey, L. R. F. de Carvalho, N. Y. Kado, and D. Schuetzle, "Polar Organic Matter in Airborne Particles: Chemical Characterization and Mutagenic Activity," *Environ. Sci. Technol., 27*, 2112–2119 (1993).

Hahn, J., "Organic Constituents of Natural Aerosols," *Ann. N.Y. Acad. Sci., 338*, 359–376 (1980).

Hameri, K., M. J. Rood, and H.-C. Hansson, "Hygroscopic Properties of NaCl Aerosol Coated with Organic Compounds," *J. Aerosol Sci., 23*, Suppl. 1, S437–S440 (1992).

Hänel, G., and B. Zankl, "Aerosol Size and Relative Humidity: Water Uptake by Mixtures of Salts," *Tellus, 31*, 478–486 (1979).

Hansson, H.-C., A. Wiedensohler, M. J. Rood, and D. S. Covert, "Experimental Determination of the Hygroscopic Properties of Organically Coated Aerosol Particles," *J. Aerosol Sci., 21*, S241–S244 (1990).

Harner, T., and D. Mackay, "Measurement of Octanol–Air Partition Coefficients for Chlorobenzenes, PCBs, and DDT," *Environ. Sci. Technol., 29*, 1599–1606 (1995).

Harner, T., and T. F. Bidleman, "Measurements of Octanol–Air Partition Coefficients for Polychlorinated Biphenyls," *J. Chem. Eng. Data, 41*, 895–899 (1996).

Harner, T., and T. F. Bidleman, "Octanol–Air Partition Coefficient for Describing Particle/Gas Partitioning of Aromatic Compounds in Urban Air," *Environ. Sci. Technol., 32*, 1494–1502 (1998).

Hatakeyama, S., K. Izumi, T. Fukuyama, and H. Akimoto, "Reactions of Ozone with α-Pinene and β-Pinene in Air: Yields of Gaseous and Particulate Products," *J. Geophys. Res., 94*, 13013–13024 (1989).

Hatch, T., and S. P. Choate, "Statistical Description of the Size Properties of Non-Uniform Particulate Substances," *J. Franklin Inst., 207*, 369–387 (1929).

Hawthorne, S. B., D. J. Miller, R. M. Barkley, and M. S. Krieger, "Identification of Methoxylated Phenols as Candidate Tracers for Atmospheric Wood Smoke Pollution," *Environ. Sci. Technol., 22*, 1191–1196 (1988).

Hegg, D., T. Larson, and P.-F. Yuen, "A Theoretical Study of the Effect of Relative Humidity on Light Scattering by Tropospheric Aerosols," *J. Geophys. Res., 98*, 18435–18439 (1993).

Hegg, D. A., P. V. Hobbs, R. J. Ferek, and A. P. Waggoner, "Measurements of Some Aerosol Properties Relevant to Radiative Forcing on the East Coast of the United States," *J. Appl. Meteorol., 34*, 2306–2315 (1995).

Hegg, D. A., D. S. Covert, M. J. Rood, and P. V. Hobbs, "Measurements of Aerosol Optical Properties in Marine Air," *J. Geophys. Res., 101*, 12893–12903 (1996).

Hegg, D. A., J. Livingston, P. V. Hobbs, T. Novakov, and P. Russell, "Chemical Apportionment of Aerosol Column Optical Depth Off the Mid-Atlantic Coast of the United States," *J. Geophys. Res., 102*, 25293–25303 (1997).

Heisler, S. L., and S. K. Friedlander, "Gas-to-Particle Conversion in Photochemical Smog: Aerosol Growth Laws and Mechanisms for Organics," *Atmos. Environ., 11*, 157–168 (1977).

Hering, S., A. Eldering, and J. H. Seinfeld, "Bimodal Character of Accumulation Mode Aerosol Mass Distributions in Southern California," *Atmos. Environ., 31*, 1–11 (1997).

Hering, S. V., and S. K. Friedlander, "Origins of Aerosol Sulfur Size Distributions in the Los Angeles Basin," *Atmos. Environ.,* 16, 2647–2656 (1982).

Hildemann, L. M., G. R. Markowski, M. C. Jones, and G. R. Cass, "Submicrometer Aerosol Mass Distributions of Emissions from Boilers, Fireplaces, Automobiles, Diesel Trucks, and Meat-Cooking Operations," *Aerosol Sci. Technol.,* 14, 138–152 (1991a).

Hildemann, L. M., M. A. Mazurek, G. R. Cass, and B. R. T. Simoneit, "Quantitative Characterization of Urban Sources of Organic Aerosol by High-Resolution Gas Chromatography," *Environ. Sci. Technol.,* 25, 1311–1325 (1991b).

Hildemann, L. M., D. B. Klinedinst, G. A. Klouda, L. A. Currie, and G. R. Cass, "Sources of Urban Contemporary Carbon Aerosol," *Environ. Sci. Technol.,* 28, 1565–1576 (1994a).

Hildemann, L. M., M. A. Mazurek, G. R. Cass, and B. R. T. Simoneit, "Seasonal Trends in Los Angeles Ambient Organic Aerosol Observed by High-Resolution Gas Chromatography," *Aerosol Sci. Technol.,* 20, 303–317 (1994b).

Hildemann, L. M., W. F. Rogge, G. R. Cass, M. A. Mazurek, and B. R. T. Simoneit, "Contribution of Primary Aerosol Emissions from Vegetation-Derived Sources to Fine Particle Concentrations in Los Angeles," *J. Geophys. Res.,* 101, 19541–19549 (1996).

Hinds, W. C., *Aerosol Technology,* Wiley, New York, 1982.

Hitzenberger, R., U. Dusek, and A. Berner, "Black Carbon Measurements Using an Integrating Sphere," *J. Geophys. Res.,* 101, 19601–19606 (1996).

Ho, W. W., G. M. Hidy, and R. M. Govan, "Microwave Measurements of the Liquid Water Content of Atmospheric Aerosols," *J. Appl. Meteorol.,* 13, 871–879 (1974).

Holmes, J., T. Samberg, L. McInnes, J. Zieman, W. Zoller, and J. Harris, "Long-Term Aerosol and Trace Acidic Gas Collection at Mauna Loa Observatory 1979–1991," *J. Geophys. Res.,* 102, 19007–19019 (1997).

Hopke, P. K., *Receptor Modeling in Environmental Chemistry,* Wiley, New York, 1985.

Hoppel, W. A., J. W. Fitzgerald, and R. E. Larson, "Aerosol Size Distributions in Air Masses Advecting Off the East Coast of the United States," *J. Geophys. Res.,* 90, 2365-2379 (1985).

Hoppel, W. A., G. M. Frick, and R. E. Larson, "Effect of Nonprecipitating Clouds on the Aerosol Size Distribution in the Marine Boundary Layer," *Geophys. Res. Lett.,* 13, 125–128 (1986).

Hoppel, W. A., G. M. Frick, J. W. Fitzgerald, and R. E. Larson, "Marine Boundary Layer Measurements of New Particle Formation and the Effects Nonprecipitating Clouds Have on Aerosol Size Distribution," *J. Geophys. Res.,* 99, 14443–14459 (1994).

Horvath, H., "Atmospheric Light Absorption—A Review," *Atmos. Environ.,* 27A, 293–317 (1993).

Horvath, H., "Size Segregated Light Absorption Coefficient of the Atmospheric Aerosol," *Atmos. Environ.,* 29, 875–883 (1995).

Horvath, H., L. Catalan, and A. Trier, "A Study of the Aerosol of Santiago de Chile. III: Light Absorption Measurements," *Atmos. Environ.,* 31, 3737–3744 (1997).

Howell, S. G., and B. J. Huebert, "Determining Marine Aerosol Scattering Characteristics at Ambient Humidity from Size-Resolved Chemical Composition," *J. Geophys. Res.,* 103, 1391–1404 (1998).

Huffman, D. R., and J. L. Stapp, "Optical Measurements on Solids of Possible Interstellar Importance," in *Interstellar Dust and Related Topics* (J. M. Greenberg and H. C. Van de Hulst, Eds.), Reidel, Boston, 1973.

Hughes, L. S., G. R. Cass, J. Gone, M. Ames, and I. Olmez, "Physical and Chemical Characterization of Atmospheric Ultrafine Particles in the Los Angeles Area," *Environ. Sci. Technol.,* 32, 1153–1161 (1998).

Huntzicker, J. J., E. K. Heyerdahl, S. R. McDow, J. A. Rau, W. H. Griest, and C. S. MacDougall, "Combustion as the Principal Source of Carbonaceous Aerosol in the Ohio River Valley," *J. Air Pollut. Control Assoc.,* 36, 705–709 (1986).

Husar, R. B., K. T. Whitby, and B. Y. H. Liu, "Physical Mechanisms Governing the Dynamics of Los Angeles Smog Aerosol," *J. Colloid Interface Sci.,* 39, 211–224 (1972).

Husar, R. B., and W. R. Shu, "Thermal Analysis of the Los Angeles Smog Aerosol," *J. Appl. Meteorol.,* 14, 1558–1564 (1975).

IAMAP, "Terminology and Units of Radiation Quantities and Measurements," Radiation Commission of the International Association of Meteorology and Atmospheric Physics, Boulder, CO, 1978.

Imre, D. G., J. Xu, I. N. Tang, and R. McGraw, "Ammonium Bisulfate/Water Equilibrium and Metastability Phase Diagrams," *J. Phys. Chem. A,* 101, 4191–4195 (1997).

Izumi, K., K. Murano, M. Mizuochi, and T. Fukuyama, "Aerosol Formation by the Photooxidation of Cyclohexene in the Presence of Nitrogen Oxides," *Environ. Sci. Technol.,* 22, 1207–1215 (1988).

Izumi, K., and T. Fukuyama, "Photochemical Aerosol Formation from Aromatic Hydrocarbons in the Presence of NO_x," *Atmos. Environ.,* 24A, 1433–1441 (1990).

Jacobson, M. Z., A. Tabazadeh, and R. P. Turco, "Simulating Equilibrium within Aerosols and Nonequilibrium between Gases and Aerosols," *J. Geophys. Res.,* 101, 9079–9091 (1996).

Jaecker-Voirol, A., and P. Mirabel, "Heteromolecular Nucleation in the Sulfuric Acid–Water System," *Atmos. Environ.,* 23, 2053–2057 (1989).

Jaenicke, R., "Vertical Distribution of Atmospheric Aerosols," in *Nucleation and Atmospheric Aerosols* (N. Fukuta and P. E. Wagner, Eds.), pp. 417–425, A. Deepak Publishing, Hampton, VA, 1992.

Jang, M., and R. M. Kamens, "A Thermodynamic Approach for Modeling Partitioning of Semi-Volatile Organic Compounds on Atmospheric Particulate Matter: Humidity Effects," *Environ. Sci. Technol.,* 32, 1237–1243 (1998).

Japar, S. M., W. W. Brachaczek, R. A. Gorse, Jr., J. M. Norbeck, and W. R. Pierson, "The Contribution of Elemental Carbon to the Optical Properties of Rural Atmospheric Aerosols," *Atmos. Environ.,* 20, 1281–1289 (1986).

Jefferson, A., F. L. Eisele, P. J. Ziemann, R. J. Weber, J. J. Marti, and P. H. McMurry, "Measurements of the H_2SO_4 Mass Accommodation Coefficient onto Polydisperse Aerosol," *J. Geophys. Res.,* 102, 19021–19028 (1997).

Jennings, S. G., T. G. Spain, B. G. Doddridge, H. Maring, B. P. Kelly, and A. D. A. Hansen, "Concurrent Measurements of Black Carbon Aerosol and Carbon Monoxide at Mace Head," *J. Geophys. Res.,* 101, 19447–19454 (1996).

John, W., S. M. Wall, J. L. Ondo, and W. Winklmayr, "Modes in the Size Distributions of Atmospheric Inorganic Aerosol," *Atmos. Environ.,* 24A, 2349–2359 (1990).

Johnson, S. A., and R. Kumar, "Composition and Spectral Characteristics of Ambient Aerosol at Mauna Loa Observatory," *J. Geophys. Res.,* 96, 5379–5386 (1991).

Junge, C. E., "Recent Investigations in Air Chemistry," *Tellus,* 18, 127–139 (1956).

Junge, C. E., *Air Chemistry and Radioactivity,* Academic Press, New York, 1963.

Junge, C. E., in *Fate of Pollutants in the Air and Water Environments* (I. H. Suffet, Ed.), Part I, pp. 7–26, Wiley, New York, 1977.

Kadowaki, S., "Characterization of Carbonaceous Aerosols in the Nagoya Urban Area. 2. Behavior and Origin of Particulate *n*-Alkanes," *Environ. Sci. Technol.,* 28, 129–135 (1994).

Kamens, R., J. Odum, and Z.-H. Fa, "Some Observations on Times to Equilibrium for Semi-Volatile Polycyclic Aromatic Hydrocarbons," *Environ. Sci. Technol.*, 29, 43–50 (1995).

Kao, A. S., and S. K. Friedlander, "Temporal Variations of Particulate Air Pollution: A Marker for Free Radical Dosage and Adverse Health Effects?" *Inhalation Toxicol.*, 7, 149–156 (1995).

Katrinak, K. A., J. R. Anderson, and P. R. Buseck, "Individual Particle Types in the Aerosol of Phoenix, Arizona," *Environ. Sci. Technol.*, 29, 321–329 (1995).

Kavouras, I. G., N. Stratigakis, and E. S. Stephanou, "*Iso-* and *Anteiso-*Alkanes: Specific Tracers of Environmental Tobacco Smoke in Indoor and Outdoor Particle-Size Distributed Urban Aerosols," *Environ. Sci. Technol.*, 32, 1369–1377 (1998).

Kawamura, K., and I. R. Kaplan, "Motor Exhaust Emissions as a Primary Source for Dicarboxylic Acids in Los Angeles Ambient Air," *Environ. Sci. Technol.*, 21, 105–110 (1987).

Kawamura, K., R. Seméré, Y. Imai, Y. Fujii, and M. Hayashi, "Water Soluble Dicarboxylic Acids and Related Compounds in Antarctic Aerosols," *J. Geophys. Res.*, 101, 18721–18728 (1996a).

Kawamura, K., A. Yanase, T. Eguchi, T. Mikami, and L. A. Barrie, "Enhanced Atmospheric Transport of Soil Derived Organic Matter in Spring over the High Arctic," *Geophys. Res. Lett.*, 23, 3735–3738 (1996b).

Kerker, M., *The Scattering of Light and Other Electromagnetic Radiation*, Academic Press, New York, 1969.

Kerminen, V.-M., and A. S. Wexler, "Growth Laws for Atmospheric Aerosol Particles: An Examination of the Bimodality of the Accumulation Mode," *Atmos. Environ.*, 29, 3263–3275 (1995).

Kerminen, V.-M., A. S. Wexler, and S. Potukuchi, "Growth of Freshly Nucleated Particles in the Troposphere: Roles of NH_3, H_2SO_4, and HCl," *J. Geophys. Res.*, 102, 3715–3724 (1997).

Kerminen, V.-M., R. E. Hillamo, T. Mäkelä, J.-L. Jaffrezo, and W. Maenhaut, "The Physicochemical Structure of the Greenland Summer Aerosol and Its Relation to Atmospheric Processes," *J. Geophys. Res.*, 103, 5661–5670 (1998).

Ketseridis, G., J. Hahn, R. Jaenicke, and C. Junge, "The Organic Constituents of Atmospheric Particulate Matter," *Atmos. Environ.*, 10, 603–610 (1976).

Kim, Y., H. Sievering, J. Boatman, D. Wellman, and A. Pszenny, "Aerosol Size Distribution and Aerosol Water Content Measurements during Atlantic Stratocumulus Transition Experiment/Marine Aerosol and Gas Exchange," *J. Geophys. Res.*, 100, 23027–23038 (1995).

Kitto, M. E., D. L. Anderson, G. E. Gordon, and I. Olmez, "Rare Earth Distributions in Catalysts and Airborne Particles," *Environ. Sci. Technol.*, 26, 1368–1375 (1992).

Kleeman, M. J., G. R. Cass, and A. Eldering, "Modeling the Airborne Particle Complex as a Source-Oriented External Mixture," *J. Geophys. Res.*, 102, 21355–21372 (1997).

Kleeman, M. J., and G. R. Cass, "Source Contributions to the Size and Composition Distribution of Urban Particulate Air Pollution," *Atmos. Environ.*, 32, 2803–2816 (1998).

Kneip, T. J., J. M. Daisey, J. J. Solomon, and R. J. Hershman, "*N*-Nitroso Compounds: Evidence for Their Presence in Airborne Particles," *Science*, 221, 1045–1046 (1983).

Koloutsou-Vakakis, S., and M. J. Rood, "The $(NH_4)_2SO_4$–Na_2SO_4–H_2O System: Comparison of Deliquescence Humidities Measured in the Field and Estimated from Laboratory Measurements and Thermodynamic Modeling," *Tellus*, 46B, 1–15 (1994).

Kreidenweis, S. M., and J. H. Seinfeld, "Nucleation of Sulfuric Acid–Water and Methanesulfonic Acid–Water Solution Particles: Implications for the Atmospheric Chemistry of Organosulfur Species," *Atmos. Environ.*, 22, 283–296 (1988a).

Kreidenweis, S. M., and J. H. Seinfeld, "Effect of Surface Tension of Aqueous Methanesulfonic Acid Solutions upon Nucleation and Growth of Aerosol," *Atmos. Environ.*, 22, 1499–1500 (1988b).

Kuhlbusch, T. A. J., A.-M. Hertlein, and L. W. Schütz, "Sources, Determination, Monitoring, and Transport of Carbonaceous Aerosols in Mainz, Germany," *Atmos. Environ.*, 32, 1097–1110 (1998).

Kulmala, M., and P. E. Wagner, *Nucleation and Atmospheric Aerosols*, 1996, Elsevier Science Ltd., Oxford, U.K., 1996.

Kulmala, M., A. Laaksonen, and L. Pirjola, "Parameterizations for Sulfuric Acid/Water Nucleation Rates," *J. Geophys. Res.*, 103, 8301–8307 (1998).

Lapple, C. E., "The Little Things in Life," *SRI J.*, 5, 95–102 (1961).

Larson, S. M., and G. R. Cass, "Characteristics of Summer Midday Low-Visibility Events in the Los Angeles Area," *Environ. Sci. Technol.*, 23, 281–289 (1989).

Larson, T. V., N. C. Ahlquist, R. E. Weiss, D. S. Covert, and A. P. Waggoner, "Chemical Speciation of H_2SO_4–$(NH_4)_2SO_4$ Particles Using Temperature and Humidity Controlled Nephelometry," *Atmos. Environ.*, 16, 1587–1590 (1982).

Laulainen, N., and E. Trexler, "Assessing the Relative Contribution of Biogenic and Fossil Processes to Visibility-Scattering Aerosols Found in Remote Areas: Near-Term Organic Research Program," *J. Air Waste Manage. Assoc.*, 47, 212–215 (1997).

Lawrence Berkeley Laboratory (LBL), *Instrumentation for Environmental Monitoring Air*, Vol. 1, Part 2, September 1979.

Lawrence, J., and P. Koutrakis, "Measurement and Speciation of Gas and Particle Phase Organic Acidity in an Urban Environment. 1. Analytical," *J. Geophys. Res.*, 101, 9159–9169 (1996a).

Lawrence, J., and P. Koutrakis, "Measurement and Speciation of Gas and Particle Phase Organic Acidity in an Urban Environment. 2. Analytical," *J. Geophys. Res.*, 101, 9171–9184 (1996b).

Leinen, M., J. M. Prospero, E. Arnold, and M. Blank, "Mineralogy of Aeolian Dust Reaching the North Pacific Ocean. 1. Sampling and Analysis," *J. Geophys. Res.*, 99, 21017–21023 (1994).

Li, S.-M., A. M. Macdonald, J. W. Strapp, Y.-N. Lee, and X.-L. Zhou, "Chemical and Physical Characterizations of Atmospheric Aerosols over Southern California," *J. Geophys. Res.*, 102, 21341–21353 (1997).

Liang, C., J. F. Pankow, J. R. Odum, and J. H. Seinfeld, "Gas/Particle Partitioning of Semi-Volatile Organic Compounds to Model Inorganic, Organic, and Ambient Smog Aerosols," *Environ. Sci. Technol.*, 31, 3086–3092 (1997).

Li-Jones, X., and J. M. Prospero, "Variations in the Size Distribution of Non-Sea-Salt Sulfate Aerosol in the Marine Boundary Layer at Barbados: Impact of African Dust," *J. Geophys. Res.*, 103, 16073–16084 (1998).

Lin, Z. C., J. M. Ondov, W. R. Kelly, P. J. Paulsen, and R. K. Stevens, "Tagging Diesel and Residential Oil Furnace Emissions in Roanoke, Virginia, with Enriched Isotopes of Samarium," *J. Air Waste Manage. Assoc.*, 42, 1057–1062 (1992).

Liousse, C., H. Cachier, and S. G. Jennings, "Optical and Thermal Measurements of Black Carbon Aerosol Content in Different Environments: Variation of the Specific Attenuation Cross Section, Sigma (σ)," *Atmos. Environ.*, 27A, 1203–1211 (1993).

Liu, B. Y. H., Ed., *Fine Particles: Aerosol Generation Measurement, Sampling, and Analysis*, Academic Press, New York, 1976.

Liu, L., and M. H. Smith, "Urban and Rural Aerosol Particle Optical Properties," *Atmos. Environ.*, 29, 3293–3301 (1995).

Lodge, J. P., Jr., "Final Comment to Discussions of Paper Entitled Non-Health Effects of Airborne Particulate Matter," *Atmos. Environ.*, 17, 899–909 (1983).

Lodge, J. P., Jr., "A Comparison of Urban and Rural Aerosol Composition Using Dichotomous Samplers," *Atmos. Environ.*, 19, 840 (1985).

Loranger, S., and J. Zayed, "Environmental Contamination and Human Exposure to Airborne Total and Respirable Manganese in Montreal," *J. Air Waste Manage. Assoc.*, 47, 983–989 (1997).

Lowenthal, D. H., and K. A. Rahn, "The Mn/V Ratio in Retrospect," *J. Air Pollut. Control Assoc.*, 37, 829–830 (1987).

Lowenthal, D. H., K. R. Wunschel, and K. A. Rahn, "Tests of Regional Elemental Tracers of Pollution Aerosols. 1. Distinctness of Regional Signatures, Stability during Transport, and Empirical Validation," *Environ. Sci. Technol.*, 22, 413–420 (1988).

Lowenthal, D. H., and K. A. Rahn, "Tests of Regional Elemental Tracers of Pollution Aerosols. 2. Sensitivity of Signatures and Apportionments to Variations in Operating Parameters," *Environ. Sci. Technol.*, 22, 420–426 (1988).

Mackay, D., S. Paterson, and W. H. Schroeder, "Model Describing the Rates of Transfer Processes of Organic Chemicals between Atmosphere and Water," *Environ. Sci. Technol.*, 20, 810–816 (1986).

Mackay, D., W. Y. Shiu, and K. C. Ma, *Illustrated Handbook of Physical–Chemical Properties and Environmental Fate for Organic Chemicals*, Vol. II, Polynuclear Aromatic Hydrocarbons, Polychlorinated Dioxins, and Dibenzofurans, Lewis Publishers, Chelsea, MI, 1992.

MacRitchie, F., *Chemistry at Interfaces*, Academic Press, San Diego, 1990.

Main, H. H., and S. K. Friedlander, "Dry Deposition of Atmospheric Aerosols by Dual Tracer Method—I. Area Source," *Atmos. Environ.*, 24A, 102–108 (1990).

Malm, W., J. Molenar, and L. L. Chan, "Photographic Simulation Techniques for Visualizing the Effect of Uniform Haze on a Scenic Resource," *J. Air Pollut. Control Assoc.*, 33, 126–129 (1983).

Malm, W. C., J. V. Molenar, R. A. Eldred, and J. F. Sisler, "Examining the Relationship among Atmospheric Aerosols and Light Scattering and Extinction in the Grand Canyon Area," *J. Geophys. Res.*, 101, 19251–19265 (1996).

Malm, W. C., and K. A. Gebhart, "Source Apportionment of Sulfur and Light Extinction Using Receptor Modeling Techniques," *J. Air Waste Manage. Assoc.*, 47, 250–268 (1997).

Malm, W. C., and S. M. Kreidenweis, "The Effects of Models of Aerosol Hygroscopicity on the Apportionment of Extinction," *Atmos. Environ.*, 31, 1965–1976 (1997).

Mamane, Y., and T. G. Dzubay, "Characteristics of Individual Particles at a Rural Site in the Eastern United States," *J. Air Pollut. Control Assoc.*, 36, 906–911 (1986).

Marti, J. J., A. Jefferson, X. P. Cai, C. Richert, P. H. McMurry, and F. Eisele, "H_2SO_4 Vapor Pressure of Sulfuric Acid and Ammonium Sulfate Solutions," *J. Geophys. Res.*, 102, 3725–3735 (1997a).

Marti, J. J., R. J. Weber, and P. H. McMurry, "New Particle Formation at a Remote Continental Site: Assessing the Contributions of SO_2 and Organic Precursors," *J. Geophys. Res.*, 102, 6331–6339 (1997b).

Mason, C. B., and B. Moore, *Principles of Geochemistry*, 4th ed., Wiley, New York, 1982.

Mateu, J., R. Forteza, M. Colom-Altes, and V. Cerda, "Background Levels and Long-Range Transport of Major Inorganic Components of Aerosols over the Balearic Islands," *J. Environ. Sci. Health.*, A29(2), 321–337 (1994).

Mazurek, M. A., G. R. Cass, and B. R. T. Simoneit, "Biological Input to Visibility-Reducing Aerosol Particles in the Remote Arid Southwestern United States," *Environ. Sci. Technol.*, 25, 684–694 (1991).

Mazurek, M., K. A. Hallock, M. Leach, M. C. Masonjones, H. D. Masonjones, L. G. Salmon, and G. R. Cass, "Visibility-Reducing Organic Aerosols in the Vicinity of Grand Canyon National Park: 1. Properties Observed by High Resolution Gas Chromatography," *J. Geophys. Res.*, 102, 3779–3793 (1997).

McDow, S. R., M. Jang, Y. Hong, and R. M. Kamens, "An Approach to Studying the Effect of Organic Composition on Atmospheric Aerosol Photochemistry," *J. Geophys. Res.*, 101, 19593–19600 (1996).

McInnes, L., M. Bergin, J. Ogren, and S. Schwartz, "Apportionment of Light Scattering and Hygroscopic Growth to Aerosol Composition," *Geophys. Res. Lett.*, 25, 513–516 (1998).

McMurry, P. H., and J. C. Wilson, "Growth Laws for the Formation of Secondary Ambient Aerosols: Implications for Chemical Conversion Mechanisms," *Atmos. Environ.*, 16, 121–134 (1982).

McMurry, P. H., and D. J. Rader, "Aerosol Wall Losses in Electrically Charged Chambers," *Aerosol Sci. Technol.*, 4, 249–268 (1985).

McMurry, P. H., and M. R. Stolzenburg, "On the Sensitivity of Particle Size to Relative Humidity for Los Angeles Aerosols," *Atmos. Environ.*, 23, 497–507 (1989).

McMurry, P. H., X. Zhang, and C.-T. Lee, "Issues in Aerosol Measurement for Optics Assessments," *J. Geophys. Res.*, 101, 19189–19197 (1996).

Meng, Z., and J. H. Seinfeld, "On the Source of the Submicrometer Droplet Mode of Urban and Regional Aerosols," *Aerosol Sci. Technol.*, 20, 253–265 (1994).

Meng, Z., and J. H. Seinfeld, "Time Scales to Achieve Atmospheric Gas–Aerosol Equilibrium for Volatile Species," *Atmos. Environ.*, 30, 2889–2900 (1996).

Mercer, T. T., *Aerosol Technology in Hazard Evaluation*, Academic Press, New York, 1973.

Merrill, J., E. Arnold, M. Leinen, and C. Weaver, "Mineralogy of Aeolian Dust Reaching the North Pacific Ocean. 2. Relationship of Mineral Assemblages to Atmospheric Transport Patterns," *J. Geophys. Res.*, 99, 21025–21032 (1994).

Middleton, W. E. K., *Vision through the Atmosphere*, Univ. of Toronto Press, Toronto, 1952.

Milford, J. B., and C. I. Davidson, "The Sizes of Particulate Trace Elements in the Atmosphere—A Review," *J. Air Pollut. Control Assoc.*, 35, 1249–1260 (1985).

Milford, J. B., and C. I. Davidson, "The Sizes of Particulate Sulfate and Nitrate in the Atmosphere—A Review," *J. Air Pollut. Control Assoc.*, 37, 125–134 (1987).

Millikan, R. A., "The General Law of a Small Spherical Body through a Gas, and Its Bearing upon the Nature of Molecular Reflection from Surfaces," *Phys. Rev.*, 22, 1–23 (1923).

Mishchenko, M. I., A. A. Lacis, B. E. Carlson, and L. D. Travis, "Nonsphericity of Dust-Like Tropospheric Aerosols: Implications for Aerosol Remote Sensing and Climate Modeling," *Geophys. Res. Lett.*, 22, 1077–1080 (1995).

Molenar, J. V., W. C. Malm, and C. E. Johnson, "Visual Air Quality Simulation Techniques," *Atmos. Environ.*, 28, 1055–1063 (1994).

Moosmüller, H., W. P. Arnott, C. F. Rogers, J. C. Chow, C. A. Frazier, L. E. Sherman, and D. L. Dietrich, "Photoacoustic and Filter Measurements Related to Aerosol Light Absorption during the Northern Front Range Air Quality Study (Colorado 1996/1997)," *J. Geophys. Res.*, 103, 28149–28157 (1998).

Mosher, B. W., and R. A. Duce, "A Global Atmospheric Selenium Budget," *J. Geophys. Res.*, 92, 13289–13298 (1987).

Moulin, C., C. E. Lambert, F. Dulac, and U. Dayan, "Control of Atmospheric Export of Dust from North Africa by the North Atlantic Oscillation," *Nature*, 387, 691–694 (1997).

Mouri, H., and K. Okada, "Shattering and Modification of Sea-Salt Particles in the Marine Atmosphere," *Geophys. Res. Lett.*, 20, 49–52 (1993).

Moyers, J. L., and R. A. Duce, "Gaseous and Particulate Iodine in the Marine Atmosphere," *J. Geophys. Res.*, 77, 5229–5238 (1972a).

Moyers, J. L., and R. A. Duce, "Gaseous and Particulate Bromine in the Marine Atmosphere," *J. Geophys. Res.*, 77, 5330–5338 (1972b).

Moyers, J. L., and G. Colovos, "Discussions—Lead and Bromine Particle Size Distribution in the San Francisco Bay Area," *Atmos. Environ.*, L8, 1339–1347 (1974).

Murphy, D. M., and D. S. Thomson, "Chemical Composition of Single Aerosol Particles at Idaho Hill: Negative Ion Measurements," *J. Geophys. Res.*, 102, 6353–6368 (1997).

Murphy, D. M., D. S. Thomson, and A. M. Middlebrook, "Bromine, Iodine, and Chlorine in Single Aerosol Particles at Cape Grim," *Geophys. Res. Lett.*, 24, 3197–3200 (1997).

Murphy, D. M., D. S. Thomson, and M. J. Mahoney, "*In Situ* Measurements of Organics, Meteoritic Material, Mercury, and Other Elements in Aerosols at 5 to 19 Kilometers," *Science*, 282, 1664–1669 (1998).

Mylonas, D. T., D. T. Allen, S. H. Ehrman, and S. E. Pratsinis, "The Sources and Size Distributions of Organonitrates in Los Angeles Aerosol," *Atmos. Environ.*, 25A, 2855–2861 (1991).

National Research Council, *Airborne Particles*, University Park Press, Baltimore, MD, 1979.

National Research Council, *Research Priorities for Airborne Particulate Matter—Immediate Priorities and a Long-Range Research Portfolio*, National Academy Press, Washington, DC, 1998.

Nielsen, T., A. H. Egelov, K. Granby, and H. Skov, "Observations on Particulate Organic Nitrates and Unidentified Components of NO_y," *Atmos. Environ.*, 29, 1757–1769 (1995).

Nielsen, T., J. Platz, K. Granby, A. B. Hansen, H. Skov, and A. H. Egelov, "Particulate Organic Nitrates: Sampling and Night/Day Variation," *Atmos. Environ.*, 32, 2601–2608 (1998).

Nilsson, E. D., and M. Kulmala, "The Potential for Atmospheric Mixing Processes to Enhance the Binary Nucleation Rate," *J. Geophys. Res.*, 103, 1381–1389 (1998).

Noble, C. A., and K. A. Prather, "Real-Time Size Measurement of Correlated Size and Composition Profiles of Individual Atmospheric Aerosol Particles," *Environ. Sci. Technol.*, 30, 2667–2680 (1996).

Novakov, T., P. K. Mueller, A. E. Alcocer, and J. W. Otvos, "Chemical Composition of Pasadena Aerosol by Particle Size and Time of Day. III. Chemical States of Nitrogen and Sulfur by Photoelectron Spectroscopy," *J. Colloid Interface Sci.*, 39, 225–234 (1972).

Novakov, T., D. A. Hegg, and P. V. Hobbs, "Airborne Measurements of Carbonaceous Aerosols on the East Coast of the United States," *J. Geophys. Res.*, 102, 30023–30030 (1997).

Nyeki, S., U. Baltensperger, I. Colbeck, D. T. Jost, E. Weingartner, and H. W. Gäggeler, "The Jungfraujoch High-Alpine Research Station (3454 m) as a Background Clean Continental Site for the Measurement of Aerosol Parameters," *J. Geophys. Res.*, 103, 6097–6107 (1998).

Oatis, S., D. Imre, R. McGraw, and J. Xu, "Heterogeneous Nucleation of a Common Atmospheric Aerosol: Ammonium Sulfate," *Geophys. Res. Lett.*, 25, 4469–4472 (1998).

O'Brien, R. J., J. R. Holmes, and A. H. Bockian, "Formation of Photochemical Aerosol from Hydrocarbons. Chemical Reactivity and Products," *Environ. Sci. Technol.*, 9, 568–576 (1975a).

O'Brien, R. J., J. H. Crabtree, J. R. Holmes, M. C. Hoggan, and A. H. Bockian, "Formation of Photochemical Aerosol from Hydrocarbons. Atmospheric Analysis," *Environ. Sci. Technol.*, 9, 577–582 (1975b).

O'Dowd, C. D., and M. H. Smith, "Submicron Particle, Radon, and Soot Carbon Characteristics over the Northeast Atlantic," *J. Geophys. Res.*, 98, 1123–1135 (1993a).

O'Dowd, C. D., and M. H. Smith, "Physicochemical Properties of Aerosols over the Northeast Atlantic: Evidence for Wind-Speed-Related Submicron Sea-Salt Aerosol Production," *J. Geophys. Res.*, 98, 1137–1149 (1993b).

O'Dowd, C. D., M. H. Smith, I. E. Consterdine, and J. A. Lowe, "Marine Aerosol Sea-Salt and the Marine Sulfur Cycle–A Short Review," *Atmos. Environ.*, 31, 73–80 (1997).

Odum, J. R., J. Yu, and R. M. Kamens, "Modeling the Mass Transfer of Semivolatile Organics in Combustion Aerosols," *Environ. Sci. Technol.*, 28, 2278–2285 (1994).

Odum, J. R., T. Hoffmann, F. Bowman, D. Collins, R. C. Flagan, and J. H. Seinfeld, "Gas/Particle Partitioning and Secondary Organic Aerosol Yields," *Environ. Sci. Technol.*, 30, 2580–2585 (1996).

Odum, J. R., T. P. W. Jungkamp, R. J. Griffin, R. C. Flagan, and J. H. Seinfeld, "The Atmospheric Aerosol-Forming Potential of Whole Gasoline Vapor," *Science*, 276, 96–99 (1997a).

Odum, J. R., T. P. W. Jungkamp, R. J. Griffin, H. J. L. Forstner, R. C. Flagan, and J. H. Seinfeld, "Aromatics, Reformulated Gasoline, and Atmospheric Organic Aerosol Formation," *Environ. Sci. Technol.*, 31, 1890–1897 (1997b).

Ohta, S., M. Hori, S. Yamagata, and N. Murao, "Chemical Characterization of Atmospheric Fine Particles in Sapporo with Determination of Water Content," *Atmos. Environ.*, 32, 1021–1025 (1998).

Olmez, I., and G. E. Gordon, "Rare Earths: Atmospheric Signatures for Oil-Fired Power Plants and Refineries," *Science*, 229, 966–968 (1985).

Olmez, I., A. E. Sheffield, G. E. Gordon, J. E. Houck, L. C. Pritchett, J. A. Cooper, T. G. Dzubay, and R. L. Bennett, "Compositions of Particles from Selected Sources in Philadelphia for Receptor Modeling Applications," *J. Air Pollut. Control Assoc.*, 38, 1392–1402 (1988).

Ondov, J. M., and W. R. Kelly, "Tracing Aerosol Pollutant with Rare Earth Isotopes," *Anal. Chem.*, 63, 691–697 (1991).

Ondov, J. M., W. R. Kelly, J. Z. Holland, Z. C. Lin, and S. A. Wight, "Tracing Fly Ash Emitted from a Coal-Fired Power Plant with Enriched Rare-Earth Isotopes: An Urban Scale Test," *Atmos. Environ.*, 26B, 453–462 (1992).

Ondov, J. M., and A. S. Wexler, "Where Do Particulate Toxins Reside? An Improved Paradigm for the Structure and Dynamics of the Urban Mid-Atlantic Aerosol," *Environ. Sci. Technol.*, 32, 2547–2555 (1998).

Ouimette, J. R., R. C. Flagan, and A. R. Kelso, "Chemical Species Contributions to Light Scattering by Aerosols at a Remote Arid Site," in *Atmospheric Aerosol: Source/Air Quality Relationships*, ACS Symposium Series, No. 167 (E. S. Macias and P. K. Hopke, Eds.), pp. 125–156, Am. Chem. Soc., Washington, DC, 1981.

Ouimette, J. R., and R. C. Flagan, "The Extinction Coefficient of Multicomponent Aerosols," *Atmos. Environ.*, 16, 2405–2419 (1982).

Ozkaynak, H., A. D. Schatz, G. D. Thurston, R. G. Isaacs, and R. B. Husar, "Relationships between Aerosol Extinction Coefficients Derived from Airport Visual Range Observations and Alternative Measures of Airborne Particle Mass," *J. Air Pollut. Control Assoc.*, 35, 1176–1185 (1985).

Pandis, S. N., S. E. Paulson, J. H. Seinfeld, and R. C. Flagan, "Aerosol Formation in the Photooxidation of Isoprene and α-Pinene," *Atmos. Environ., 25A,* 997–1008 (1991).

Pandis, S. N., R. A. Harley, G. R. Cass, and J. H. Seinfeld, "Secondary Organic Aerosol Formation and Transport," *Atmos. Environ., 26A,* 2269–2282 (1992).

Pandis, S. N., A. S. Wexler, and J. H. Seinfeld, "Dynamics of Tropospheric Aerosols," *J. Phys. Chem., 99,* 9646–9659 (1995).

Pankow, J. F., "Review and Comparative Analysis of the Theories on Partitioning between the Gas and Aerosol Particulate Phases in the Atmosphere," *Atmos. Environ., 21,* 2275–2283 (1987).

Pankow, J. F., "Common y-Intercept and Single Compound Regressions of Gas–Particle Partitioning Data vs $1/T$," *Atmos. Environ., 25A,* 2229–2239 (1991).

Pankow, J. F., "Application of Common y-Intercept Regression Parameters for Log K_p vs $1/T$ for Predicting Gas–Particle Partitioning in the Urban Environment," *Atmos. Environ., 26A,* 2489–2497 (1992).

Pankow, J. F., and T. F. Bidleman, "Interdependence of the Slopes and Intercepts from Log–Log Correlations of Measured Gas–Particle Partitioning and Vapor Pressure—I. Theory and Analysis of Available Data," *Atmos. Environ., 26A,* 1071–1080 (1992).

Pankow, J. F., J. M. E. Storey, and H. Yamasaki, "Effects of Relative Humidity on Gas/Particle Partitioning of Semivolatile Organic Compounds to Urban Particulate Matter," *Environ. Sci. Technol., 27,* 2220–2226 (1993).

Pankow, J. F., "An Absorption Model of Gas/Particle Partitioning of Organic Compounds in the Atmosphere," *Atmos. Environ., 28,* 185–188 (1994a).

Pankow, J. F., "An Absorption Model of the Gas/Aerosol Partitioning Involved in the Formation of Secondary Organic Aerosol," *Atmos. Environ., 28,* 189–193 (1994b).

Pankow, J. F., "Partitioning of Semi-Volatile Organic Compounds to the Air/Water Interface," *Atmos. Environ., 31,* 927–929 (1997).

Pankow, J. F., "Further Discussion of the Octanol/Air Partition Coefficient K_{oa} as a Correlating Parameter for Gas/Particle Partitioning Coefficients," *Atmos. Environ., 32,* 1493–1497 (1998).

Parekh, P. P., and L. Husain, "Fe/Mg Ratio: A Signature for Local Coal-Fired Power Plants," *Atmos. Environ., 21,* 1707–1712 (1987).

Patel, J. K., C. H. Kapedia, and D. B. Owen, *Handbook of Statistical Distributions*, Dekker, New York, 1976.

Penner, J. E., and T. Novakov, "Carbonaceous Particles in the Atmosphere: A Historical Perspective to the Fifth International Conference on Carbonaceous Particles in the Atmosphere," *J. Geophys. Res., 101,* 19373–19378 (1996).

Perry, K. D., T. A. Cahill, R. A. Eldred, D. D. Dutcher, and T. E. Gill, "Long-Range Transport of North African Dust to the Eastern United States," *J. Geophys. Res., 102,* 11225–11238 (1997).

Petzold, A., C. Kopp, and R. Niessner, "The Dependence of the Specific Attenuation Cross-Section on Black Carbon Mass Fraction and Particle Size," *Atmos. Environ., 31,* 661–672 (1997).

Phalen, R. F., *Inhalation Studies: Foundations and Techniques*, CRC Press, Boca Raton, FL, 1984.

Pickle, T., D. T. Allen, and S. E. Pratsinis, "The Sources and Size Distributions of Aliphatic and Carbonyl Carbon in Los Angeles Aerosol," *Atmos. Environ., 24A,* 2221–2228 (1990).

Pirjola, L., A. Laaksonen, P. Aalto, and M. Kulmala, "Sulfate Aerosol Formation in the Arctic Boundary Layer," *J. Geophys. Res., 103,* 8309–8321 (1998).

Pósfai, M., H. Xu, J. R. Anderson, and P. R. Buseck, "Wet and Dry Sizes of Atmospheric Aerosol Particles: An AFM–TEM Study," *Geophys. Res. Lett., 25,* 1907–1910 (1998).

Pratsinis, S., T. Novakov, E. C. Ellis, and S. K. Friedlander, "The Carbon Containing Component of the Los Angeles Aerosol: Source, Apportionment, and Contributions to the Visibility Budget," *J. Air Pollut. Control Assoc., 34,* 643–650 (1984).

Prospero, J. M., and R. T. Nees, "Dust Concentration in the Atmosphere of the Equatorial North Atlantic: Possible Relationship to the Sahelian Drought," *Science, 196,* 1196–1198 (1977).

Prospero, J. M., R. A. Glaccum, and R. T. Nees, "Atmospheric Transport of Soil Dust from Africa to South America," *Nature, 289,* 570–572 (1981).

Prospero, J. M., and R. T. Nees, "Impact of the North African Drought and El Niño on Mineral Dust in the Barbados Trade Winds," *Nature, 320,* 735–738 (1986).

Pruppacher, H. R., and J. D. Klett, *Microphysics of Clouds and Precipitation*, 2nd ed., Kluwer Academic, Dordrecht/Norwell, MA, 1997.

Pryor, S. C., R. Simpson, L. Guise-Bagley, R. Hoff, and S. Sakiyama, "Visibility and Aerosol Composition in the Fraser Valley during REVEAL," *J. Air Waste Manage. Assoc., 47,* 147–156 (1997).

Quinn, P. K., D. S. Covert, T. S. Bates, V. N. Kapustin, D. C. Ramsey-Bell, and L. M. McInnes, "Dimethylsulfide/Cloud Condensation Nuclei/Climate System: Relevant Size-Resolved Measurements of the Chemical and Physical Properties of Atmospheric Aerosol Particles," *J. Geophys. Res., 98,* 10411–10427 (1993).

Rabano, E. S., N. T. Castillo, K. J. Torre, and P. A. Solomon, "Speciation of Arsenic in Ambient Aerosols Collected in Los Angeles," *J. Air Pollut. Control Assoc., 39,* 76–80 (1989).

Radzi bin Abas, M., B. R. T. Simoneit, V. Elias, J. A. Cabral, and J. N. Cardoso, "Composition of Higher Molecular Weight Organic Matter in Smoke Aerosol from Biomass Combustion in Amazonia," *Chemosphere, 30,* 995–1015 (1995).

Rahn, K. A., "The Mn/V Ratio as a Tracer of Large-Scale Sources of Pollution Aerosol for the Arctic," *Atmos. Environ., 15,* 1457–1464 (1981).

Rahn, K. A., and D. H. Lowenthal, "Elemental Tracers of Distant Regional Pollution Aerosols," *Science, 223,* 132–139 (1984).

Rahn, K. A., and D. H. Lowenthal, "Reply to Comments on Thurston and Laird, 'Tracing Aerosol Pollution,'" *Science, 227,* 1407–1412 (1985).

Reid, J. S., R. G. Flocchini, T. A. Cahill, and R. S. Ruther, "Local Meteorological, Transport, and Source Aerosol Characteristics of Late Autumn Owens Lake (Dry) Dust Storms," *Atmos. Environ., 28,* 1699–1706 (1994).

Richards, L. W., "Sight Path Measurements for Visibility Monitoring and Research," *J. Air Pollut. Control Assoc., 38,* 784–791 (1988).

Richards, L. W., M. Stoelting, and R. G. M. Hammarstrand, "Photographic Method for Visibility Monitoring," *Environ. Sci. Technol., 23,* 182–186 (1989).

Rivera-Carpio, C. A., C. E. Corrigan, T. Novakov, J. E. Penner, C. F. Rogers, and J. C. Chow, "Derivation of Contributions of Sulfate and Carbonaceous Aerosols to Cloud Condensation Nuclei from Mass Size Distributions," *J. Geophys. Res., 101,* 19483–19493 (1996).

Robbins, R. C., R. D. Cadle, and D. L. Eckhardt, "The Conversion of Sodium Chloride to Hydrogen Chloride in the Atmosphere," *J. Meteorol., 16,* 53–56 (1959).

Rogge, W. F., L. M. Hildemann, M. A. Mazurek, G. R. Cass, and B. R. T. Simoneit, "Sources of Fine Organic Aerosol. 1. Charbroilers and Meat Cooking Operations," *Environ. Sci. Technol., 25,* 1112–1125 (1991).

Rogge, W. F., L. M. Hildemann, M. A. Mazurek, G. R. Cass, and B. R. T. Simoneit, "Sources of Fine Organic Aerosol. 2. Noncatalyst and Catalyst-Equipped Automobiles and Heavy-Duty Diesel Trucks," *Environ. Sci. Technol., 27*, 636–651 (1993a).

Rogge, W. F., L. M. Hildemann, M. A. Mazurek, G. R. Cass, and B. R. T. Simoneit, "Sources of Fine Organic Aerosol. 4. Particulate Abrasion Products from Leaf Surfaces of Urban Plants," *Environ. Sci. Technol., 27*, 2700–2711 (1993b).

Rogge, W. F., L. M. Hildemann, M. A. Mazurek, G. R. Cass, and B. R. T. Simoneit, "Sources of Fine Organic Aerosol. 5. Natural Gas Home Appliances," *Environ. Sci. Technol., 27*, 2736–2744 (1993c).

Rogge, W. F., M. A. Mazurek, L. M. Hildemann, G. R. Cass, and B. R. T. Simoneit, "Quantification of Urban Organic Aerosols at a Molecular Level: Identification, Abundance and Seasonal Variation," *Atmos. Environ., 27A*, 1309–1330 (1993d).

Rogge, W. F., L. M. Hildemann, M. A. Mazurek, G. R. Cass, and B. R. T. Simoneit, "Sources of Fine Organic Aerosol. 3. Road Dust, Tire Debris, and Organometallic Brake Lining Dust: Roads as Sources and Sinks," *Environ. Sci. Technol., 27*, 1892–1904 (1993e).

Rogge, W. F., L. M. Hildemann, M. A. Mazurek, G. R. Cass, and B. R. T. Simoneit, "Sources of Fine Organic Aerosol. 6. Cigarette Smoke in the Urban Atmosphere," *Environ. Sci. Technol., 28*, 1375–1388 (1994).

Rogge, W. F., L. M. Hildemann, M. A. Mazurek, G. R. Cass, and B. R. T. Simoneit, "Mathematical Modeling of Atmospheric Fine Particle-Associated Primary Organic Compound Concentrations," *J. Geophys. Res., 101*, 19379–19394 (1996).

Rogge, W. F., L. M. Hildemann, M. A. Mazurek, G. R. Cass, and B. R. T. Simoneit, "Sources of Fine Organic Aerosol. 7. Hot Asphalt Roofing Tar Pot Fumes," *Environ. Sci. Technol., 31*, 2726–2730 (1997a).

Rogge, W. F., L. M. Hildemann, M. A. Mazurek, G. R. Cass, and B. R. T. Simoneit, "Sources of Fine Organic Aerosol. 8. Boilers Burning No. 2 Distillate Fuel Oil," *Environ. Sci. Technol., 31*, 2731–2737 (1997b).

Rogge, W. F., L. M. Hildemann, M. A. Mazurek, G. R. Cass, and B. R. T. Simoneit, "Sources of Fine Organic Aerosol. 9. Pine, Oak, and Synthetic Log Combustion in Residential Fireplaces," *Environ. Sci. Technol., 32*, 13–22 (1998).

Rood, M. J., T. V. Larson, D. S. Covert, and N. C. Ahlquist, "Measurement of Laboratory and Ambient Aerosols with Temperature and Humidity Controlled Nephelometry," *Atmos. Environ., 19*, 1181–1190 (1985).

Rood, M. J., D. S. Covert, and T. V. Larson, "Temperature and Humidity Controlled Nephelometry: Improvements and Calibration," *Aerosol Sci. Technol., 7*, 57–65 (1987).

Rubel, G. O., and J. W. Gentry, "Measurement of Water and Ammonia Accommodation Coefficients at Surfaces with Adsorbed Monolayers of Hexadecanol," *J. Aerosol Sci., 16*, 571–574 (1985).

Ruby, M. G., "Visibility Measurement Methods: I. Integrating Nephelometer," *J. Air Pollut. Control Assoc., 35*, 244–248 (1985).

Sanders, C. C., F. T. Cross, G. E. Dagle, and J. A. Mahaffey, Eds., *Pulmonary Toxicology of Respirable Particles*, CONF-791002, National Technical Information Service, Springfield, VA, 1980.

Satsumabayashi, H., H. Kurita, Y. Yokouchi, and H. Ueda, "Mono- and Di-carboxylic Acids under Long-Range Transport of Air Pollution in Central Japan," *Tellus, 41B*, 219–229 (1989).

Satsumabayashi, H., H. Kurita, Y. Yokouchi, and H. Ueda, "Photochemical Formation of Particulate Dicarboxylic Acids under Long-Range Transport in Central Japan," *Atmos. Environ., 24A*, 1443–1450 (1990).

Savoie, D. L., J. M. Prospero, and E. S. Saltzman, "Non-Sea-Salt Sulfate and Nitrate in Trade Wind Aerosols at Barbados: Evidence for Long-Range Transport," *J. Geophys. Res., 94*, 5069–5080 (1989).

Saxena, P., L. M. Hildemann, P. H. McMurry, and J. H. Seinfeld, "Organics Alter Hygroscopic Behavior of Atmospheric Particles," *J. Geophys. Res., 100*, 18755–18770 (1995).

Saylor, R. D., "An Estimate of the Potential Significance of Heterogeneous Loss to Aerosols as an Additional Sink for Hydroperoxy Radicals in the Troposphere," *Atmos. Environ., 31*, 3653–3658 (1997).

Schauer, J. J., W. F. Rogge, L. M. Hildemann, M. A. Mazurek, and G. R. Cass, "Source Apportionment of Airborne Particulate Matter Using Organic Compounds as Tracers," *Atmos. Environ., 30*, 3837–3855 (1996).

Schmidt, D. S., R. A. Schmidt, and J. D. Dent, "Electrostatic Force on Saltating Sand," *J. Geophys. Res., 103*, 8997–9001 (1998).

Schneider, J. K., and R. B. Gagosian, "Particle Size Distribution of Lipids in Aerosols Off the Coast of Peru," *J. Geophys. Res., 90*, 7889–7898 (1985).

Schroeder, W. H., and P. Urone, "Formation of Nitrosyl Chloride from Salt Particles in Air," *Environ. Sci. Technol., 8*, 756–758 (1974).

Schroeder, W. H., M. Dobson, D. M. Kane, and N. D. Johnson, "Toxic Trace Elements Associated with Airborne Particulate Matter: A Review," *J. Air Pollut. Control Assoc., 37*, 1267–1285 (1987).

Schuetzle, D., D. Cronn, A. L. Crittenden, and R. J. Charlson, "Molecular Composition of Secondary Aerosol and Its Possible Origin," *Environ. Sci. Technol., 9*, 838–845 (1975).

Schwartz, W., *Chemical Characterization of Model Aerosols*, U.S. Environmental Protection Agency, Report No. EPA-650/3—74-011, August 1974.

Sempere, R., and K. Kawamura, "Low Molecular Weight Dicarboxylic Acids and Related Polar Compounds in the Remote Marine Rain Samples Collected from Western Pacific," *Atmos. Environ., 30*, 1609–1619 (1996).

Shah, J. J., R. L. Johnson, E. K. Heyerdahl, and J. J. Huntzicker, "Carbonaceous Aerosol at Urban and Rural Sites in the United States," *J. Air Pollut. Control Assoc., 36*, 254–257 (1986).

Shaw, G. E., "Transport of Asian Desert Aerosol to the Hawaiian Islands," *J. Appl. Meteorol., 19*, 1254–1259 (1980).

Sheridan, P. J., C. A. Brock, and J. C. Wilson, "Aerosol Particles in the Upper Troposphere and Lower Stratosphere: Elemental Composition and Morphology of Individual Particles in Northern Midlatitudes," *Geophys. Res. Lett., 21*, 2587–2590 (1994).

Shulman, M. L., R. J. Charlson, and E. J. Davis, "The Effects of Atmospheric Organics on Aqueous Droplet Evaporation," *J. Aerosol Sci., 28*, 737–752 (1997).

Sicre, M.-A., R. B. Gagosian, and E. T. Peltzer, "Evaluation of the Atmospheric Transport of Marine-Derived Particles Using Long-Chain Unsaturated Ketones," *J. Geophys. Res., 95*, 1789–1795 (1990a).

Sicre, M.-A., J.-C. Marty, and A. Saliot, "*n*-Alkanes, Fatty Acid Esters, and Fatty Acid Salts in Size Fractionated Aerosols Collected over the Mediterranean Sea," *J. Geophys. Res., 95*, 3649–3657 (1990b).

Simoneit, B. R. T., R. E. Cox, and L. J. Standley, "Organic Matter of the Troposphere—IV: Lipids in Harmattan Aerosols of Nigeria," *Atmos. Environ., 22*, 983–1004 (1988).

Simoneit, B. R. T., "Organic Matter of the Troposphere—V: Application of Molecular Marker Analysis to Biogenic Emissions into the Troposphere for Source Reconciliations," *J. Atmos. Chem., 8*, 251–275 (1989).

Simoneit, B. R. T., J. N. Cardoso, and N. Robinson, "An Assessment of the Origin and Composition of Higher Molecular Weight Organic Matter in Aerosols over Amazonia," *Chemosphere, 21,* 1285–1301 (1990).

Simoneit, B. R. T., G. Sheng, X. Chen, J. Fu, J. Zhang, and Y. Xu, "Molecular Marker Study of Extractable Organic Matter in Aerosols from Urban Areas of China," *Atmos. Environ., 25A,* 2111–2129 (1991a).

Simoneit, B. R. T., J. N. Cardoso, and N. Robinson, "An Assessment of Terrestrial Higher Molecular Weight Lipid Compounds in Aerosol Particulate Matter over the South Atlantic from about 30–70°S," *Chemosphere, 23,* 447–465 (1991b).

Simoneit, B. R. T., W. F. Rogge, M. A. Mazurek, L. J. Standley, L. M. Hildemann, and G. R. Cass, "Lignin Pyrolysis Products, Lignans, and Resin Acids as Specific Tracers of Plant Classes in Emissions from Biomass Combustion," *Environ. Sci. Technol., 27,* 2533–2541 (1993).

Sinclair, D., in *Handbook on Aerosols*, U.S. Atomic Energy Commission, Washington, DC, 1950.

Sisler, J. F., and W. C. Malm, "Characteristics of Winter and Summer Aerosol Mass and Light Extinction on the Colorado Plateau," *J. Air Waste Manage. Assoc., 47,* 317–330 (1997).

Slinn, W. G. N., "Predictions for Particle Deposition to Vegetative Canopies," *Atmos. Environ., 16,*, 1785–1794 (1982).

Slinn, W. G. N., in *Air–Sea Exchange of Gases and Particles* (P. S. Liss and W. G. N. Slinn, Eds.), NATO ASI Series 108, pp. 299–405, Reidel, Boston, 1983.

Slinn, W. G. N., "Aerosol Sinks," in *Department of Energy Atmospheric Chemistry Program Newsletter*, Vol. 4, No. 9, September 1993, pp. 5–10.

Sloane, C. S., "Effect of Composition on Aerosol Light Scattering Efficiencies," *Atmos. Environ., 20,* 1025–1037 (1986).

Smith, D. M., and A. R. Chughtai, "Reaction Kinetics of Ozone at Low Concentrations with *n*-Hexane Soot," *J. Geophys. Res., 101,* 19607–19620 (1996).

Smith, M. H., and C. D. O'Dowd, "Observations of Accumulation Mode Aerosol Composition and Soot Carbon Concentrations by Means of a High-Temperature Volatility Technique," *J. Geophys. Res., 101,* 19583–19591 (1996).

Solomon, P. A., and J. L. Moyers, "A Chemical Characterization of Wintertime Haze in Phoenix, Arizona," *Atmos. Environ., 20,* 207–213 (1986).

Standley, L. J., and B. R. T. Simoneit, "Characterization of Extractable Plant Wax, Resin, and Thermally Matured Components in Smoke Particles from Prescribed Burns," *Environ. Sci. Technol., 21,* 163–169 (1987).

Standley, L. J., and B. R. T. Simoneit, "Preliminary Correlation of Organic Molecular Tracers in Residential Wood Smoke with the Source of Fuel," *Atmos. Environ., 24B,* 67–73 (1990).

Standley, L. J., and B. R. T. Simoneit, "Resin Diterpenoids as Tracers for Biomass Combustion Aerosols," *J. Atmos. Chem., 18,* 1–15 (1994).

Stevens, R. K., T. G. Dzubay, C. W. Kewis, and A. P. Altshuller, "Discussions of Paper Entitled 'Non-Health Effects of Airborne Particulate Matter,'" *Atmos. Environ., 17,* 899–903 (1983).

Storey, J. M. E., W. Luo, L. M. Isabelle, and J. F. Pankow, "Gas/Solid Partitioning of Semivolatile Organic Compounds to Model Atmospheric Solid Surfaces as a Function of Relative Humidity. 1. Clean Quartz," *Environ. Sci. Technol., 29,* 2420–2428 (1995).

Sturges, W. T., and R. M. Harrison, "Bromine in Marine Aerosols and the Origin, Nature, and Quantity of Natural Atmospheric Bromine," *Atmos. Environ., 20,* 1485–1496 (1986).

Sturges, W. T., and R. M. Harrison, "Thermal Speciation of Atmospheric Nitrate and Chloride: A Critical Evaluation," *Environ. Sci. Technol., 22,* 1305–1311 (1988).

Svenningsson, I. B., H.-C. Hansson, A. Wiedensohler, J. A. Ogren, K. J. Noone, and A. Hallberg, "Hygroscopic Growth of Aerosol Particles in the Po Valley," *Tellus, 44B,* 556–569 (1992).

Sverdrup, G. M., and K. T. Whitby, "The Effect of Changing Relative Humidity on Aerosol Size Distribution Measurements," *Adv. Environ. Sci. Technol., 10,* 527–538 (1980a).

Sverdrup, G. M., and K. T. Whitby, "The Variation of the Aerosol Volume to Light-Scattering Coefficient," *Adv. Environ. Sci. Technol., 10,* 539–558 (1980b).

Swietlicki, E., S. Prui, H.-C. Hansson, and H. Edner, "Urban Air Pollution Source Apportionment Using a Combination of Aerosol and Gas Monitoring Techniques," *Atmos. Environ., 30,* 2795–2809 (1996).

Talbot, R. W., R. C. Harriss, E. V. Browell, G. L. Gregory, D. I. Sebacher, and S. M. Beck, "Distribution and Geochemistry of Aerosols in the Tropical North Atlantic Troposphere: Relationship to Saharan Dust," *J. Geophys. Res., 91,* 5173–5182 (1986).

Tang, I. N., "Phase Transformation and Growth of Aerosol Particles Composed of Mixed Salts," *J. Aerosol Sci., 7,* 361–371 (1976).

Tang, I. N., H. R. Munkelwitz, and J. G. Davis, "Aerosol Growth Studies. IV. Phase Transformation of Mixed Salt Aerosols in a Moist Atmosphere," *J. Aerosol Sci., 9,* 505–511 (1978).

Tang, I. N., "Deliquescence Properties and Particle Size Change of Hygroscopic Aerosols," in *Generation of Aerosols and Facilities for Exposure Experiments* (K. Willeke, Ed.), Chap. 7, pp. 153–167, Ann Arbor Science Publishers, Ann Arbor, MI, 1980.

Tang, I. N., and H. R. Munkelwitz, "Composition and Temperature Dependence of the Deliquescence Properties of Hygroscopic Aerosols," *Atmos. Environ., 27A,* 467–473 (1993).

Tang, I. N., and H. R. Munkelwitz, "Water Activities, Densities, and Refractive Indices of Aqueous Sulfates and Sodium Nitrate Droplets of Atmospheric Importance," *J. Geophys. Res., 99,* 18801–18808 (1994a).

Tang, I. N., and H. R. Munkelwitz, "Aerosol Phase Transformation and Growth in the Atmosphere," *J. Appl. Meteorol., 33,* 791–796 (1994b).

Tang, I. N., K. H. Fung, D. G. Imre, and H. R. Munkelwitz, "Phase Transformation and Metastability of Hygroscopic Microparticles," *Aerosol Sci. Technol., 23,* 443–453 (1995).

Tang, I. N., "Chemical and Size Effects of Hygroscopic Aerosols on Light Scattering Coefficients," *J. Geophys. Res., 101,* 19245–19250 (1996).

Ten Brink, H. M., J. P. Veefkind, A. Waijers-Ijpelaan, and J. C. Van der Hage, "Aerosol Light-Scattering in the Netherlands," *Atmos. Environ., 30,* 4251–4261 (1996).

Thomsen, J., "Enrichment of Na^+, Mn^{2+}, Fe^{3+} and Cu^{2+} at Water Surfaces Covered by Artificial Multilayer Films," *J. Phys. Chem., 87,* 4974–4978 (1983).

Thurston, G. D., and N. M. Laird, "Tracing Aerosol Pollution," *Science, 227,* 1406–1408 (1985).

Toossi, R., and T. Novakov, "The Lifetime of Aerosols in Ambient Air: Consideration of the Effects of Surfactants and Chemical Reactions," *Atmos. Environ., 19,* 127–133 (1985).

Turco, R. P., J.-X. Zhao, and F. Yu, "A New Source of Tropospheric Aerosols: Ion–Ion Recombination," *Geophys. Res. Lett., 25,* 635–638 (1998).

Turpin, B. J., J. J. Huntzicker, S. M. Larson, and G. R. Cass, "Los Angeles Summer Midday Particulate Carbon: Primary and Secondary Aerosol," *Environ. Sci. Technol., 25,* 1788–1793 (1991).

Turpin, B. J., and J. J. Huntzicker, "Identification of Secondary Organic Aerosol Episodes and Quantitation of Primary and Secondary Organic Aerosol Concentrations during SCAQS," *Atmos. Environ., 29,* 3527–3544 (1995).

Turpin, B. J., P. Saxena, G. Allen, P. Koutrakis, P. McMurry, and L. Hildemann, "Characterization of the Southwestern Desert Aerosol, Meadview, AZ," *J. Air Waste Manage. Assoc., 47,* 344–356 (1997).

Uematsu, M., R. A. Duce, and J. M. Prospero, "Deposition of Atmospheric Mineral Particles in the North Pacific Ocean," *J. Atmos. Chem., 3,* 123–138 (1985).

Van de Hulst, H. C., *Light Scattering by Small Particles,* Wiley, New York, 1957.

Vega, E., I. Garcia, D. Apam, M. E. Ruiz, and M. Barbiaux, "Application of a Chemical Mass Balance Receptor Model to Respirable Particulate Matter in Mexico City," *J. Air Waste Manage. Assoc., 47,* 524–529 (1997).

Veltkamp, P. R., K. J. Hansen, R. M. Barkley, and R. E. Sievers, "Principal Component Analysis of Summertime Organic Aerosols at Niwot Ridge, Colorado," *J. Geophys. Res., 101,* 19495–19504 (1996).

Vogt, R., and B. J. Finlayson-Pitts, "A Diffuse Reflectance Infrared Fourier Transform Spectroscopic (DRIFTS) Study of the Surface Reaction of NaCl with Gaseous NO_2 and HNO_3," *J. Phys. Chem., 98,* 3747–3755 (1994); *ibid., 99,* 13052 (1995).

Vogt, R., C. Elliott, H. C. Allen, J. M. Laux, J. C. Hemminger, and B. J. Finlayson-Pitts, "Some New Laboratory Approaches to Studying Tropospheric Heterogeneous Reactions," *Atmos. Environ., 30,* 1729–1737 (1996).

Vossier, T. L., and E. S. Macias, "Contribution of Fine Particle Sulfates to Light Scattering in St. Louis Summer Aerosol," *Environ. Sci. Technol., 20,* 1235–1243 (1986).

Waggoner, A. P., and R. J. Charlson, "Measurements of Aerosol Optical Parameters," in *Fine Particles: Aerosol Generation, Measurement, Sampling, and Analysis* (B. Y. H. Liu, Ed.), pp. 511–533, Academic Press, New York, 1976.

Waggoner, A. P., R. E. Weiss, N. C. Ahlquist, D. S. Covert, S. Will, and R. J. Charlson, "Optical Characteristics of Atmospheric Aerosols," *Atmos. Environ., 15,* 1891–1909 (1981).

Waggoner, A. P., "Reply to Discussions of Paper Entitled 'Non-Health Effects of Airborne Particulate Matter,'" *Atmos. Environ., 17,* 900–903 (1983).

Waggoner, A. P., R. E. Weiss, and T. V. Larson, "*In-Situ,* Rapid Response Measurement of H_2SO_4 Aerosols in Urban Houston: A Comparison with Rural Virginia," *Atmos. Environ., 17,* 1723–1731 (1983a).

Waggoner, A. P., R. E. Weiss, and N. C. Ahlquist, "The Color of Denver Haze," *Atmos. Environ., 17,* 2081–2086 (1983b).

Wagner, J., E. Andrews, and S. M. Larson, "Sorption of Vapor Phase Octanoic Acid onto Deliquescent Salt Particles," *J. Geophys. Res., 101,* 19533–19540 (1996).

Wallace, L., and T. Slonecker, "Ambient Air Concentrations of Fine ($PM_{2.5}$) Manganese in U.S. National Parks and in California and Canadian Cities: The Possible Impact of Adding MMT to Unleaded Gasoline," *J. Air Waste Manage. Assoc., 47,* 642–652 (1997).

Wania, F., J.-E. Haugen, Y. D. Lei, and D. Mackay, "Temperature Dependence of Atmospheric Concentrations of Semi-Volatile Organic Compounds," *Environ. Sci. Technol., 32,* 1013–1021 (1998).

Weber, R. J., P. H. McMurry, F. L. Eisele, and D. J. Tanner, "Measurement of Expected Nucleation Precursor Species and 3–500-nm Diameter Particles at Mauna Loa Observatory, Hawaii," *J. Atmos. Sci., 52,* 2242–2257 (1995).

Weber, R. J., J. J. Marti, P. H. McMurry, F. L. Eisele, D. J. Tanner, and A. Jefferson, "Measured Atmospheric New Particle Formation Rates: Implications for Nucleation Mechanisms," *Chem. Eng. Commun., 151,* 53–64 (1996).

Weber, R. J., J. J. Marti, P. H. McMurry, F. L. Eisele, D. J. Tanner, and A. Jefferson, "Measurements of New Particle Formation and Ultrafine Particle Growth Rates at a Clean Continental Site," *J. Geophys. Res., 102,* 4375–4385 (1997).

Weber, R. J., P. H. McMurry, L. Mauldin, D. J. Tanner, F. L. Eisele, F. J. Brechtel, S. M. Kreidenweis, G. L. Kok, R. D. Schillawski, and D. Baumgardner, "A Study of New Particle Formation and Growth Involving Biogenic and Trace Gas Species Measured during ACE 1," *J. Geophys. Res., 103,* 16385–16396 (1998).

Weiss, R. E., A. P. Waggoner, R. J. Charlson, and N. C. Ahlquist, "Sulfate Aerosol: Its Geographical Extent in the Midwestern and Southern United States," *Science, 195,* 979–981 (1977).

Weiss, R. E., T. V. Larson, and A. P. Waggoner, "*In Situ* Rapid-Response Measurement of $H_2SO_4/(NH_4)_2SO_4$ Aerosols in Rural Virginia," *Environ. Sci. Technol., 16,* 525–532 (1982).

Wesolowski, J. J., A. E. Alcocer, and B. R. Appel, "The Validation of the Lundgren Impactor," *Adv. Environ. Sci. Technol., 10,* 125–146 (1980).

Wexler, A. S., and J. H. Seinfeld, "Second-Generation Inorganic Aerosol Model," *Atmos. Environ., 25A,* 2731–2748 (1991).

Whitby, K. T., B. Y. H. Liu, R. B. Husar, and N. J. Barsic, "The Minnesota Aerosol Analyzing System Used in the Los Angeles Smog Project," *J. Colloid Interface Sci., 39,* 136–164 (1972a).

Whitby, K. T., R. B. Husar, and B. Y. H. Liu, "The Aerosol Size Distribution of Los Angeles Smog," *J. Colloid Interface Sci., 39,* 177–204 (1972b).

Whitby, K. T., and G. M. Sverdrup, "California Aerosols: Their Physical and Chemical Characteristics," *Adv. Environ. Sci. Technol., 8,* 477–525 (1980).

White, W. H., "On the Theoretical and Empirical Basis for Apportioning Extinction by Aerosols: A Critical Review," *Atmos. Environ., 20,* 1659–1672 (1986).

Wight, S. A., J. M. Ondov, and Z.-C. Lin, "Tagging In-Stack Suspended Particles from a Coal-Fired Power Plant with an Enriched Isotopic Tracer," *Aerosol Sci. Technol., 15,* 191–200 (1991).

Willeke, K., Ed., *Generation of Aerosols,* Ann Arbor Science Publishers, Ann Arbor, MI, 1980.

Williams, M. D., E. Treiman, and M. Wecksung, "Plume Blight Visibility Modeling with a Simulated Photograph Technique," *J. Air Pollut. Control Assoc., 30,* 131–134 (1980).

Witz, S., R. W. Eden, C. S. Liu, and M. W. Wadley, "Water Content of Collected Aerosols in the South Coast and Southeast Desert Air Basins," *J. Air Pollut. Control Assoc., 38,* 418–419 (1988).

Wolff, G. T., M. A. Ferman, N. A. Kelley, D. P. Stroup, and M. S. Ruthkosky, "The Relationships between the Chemical Composition of Fine Particles and Visibility in the Detroit Metropolitan Area," *J. Air Pollut. Control Assoc., 32,* 1216–1220 (1982a).

Wolff, G. T., P. J. Groblicki, S. H. Cadle, and R. J. Countess, "Particulate Carbon at Various Locations in the United States," in *Particulate Carbon: Atmospheric Life Cycle* (G. T. Wolff and R. L. Klimisch, Eds.), pp. 297–315, Plenum, New York, 1982b.

Wolff, G. T., M. S. Ruthkosky, D. P. Stroup, P. E. Korsog, M. A. Ferman, G. J. Wendel, and D. H. Stedman, "Measurements of SO_x, NO_x, and Aerosol Species on Bermuda," *Atmos. Environ., 20,* 1229–1239 (1986).

Wyslouzil, B. E., J. H. Seinfeld, R. C. Flagan, and K. Okuyama, "Binary Nucleation in Acid–Water Systems. II. Sulfuric Acid–Water and a Comparison with Methanesulfonic Acid–Water," *J. Chem. Phys., 94,* 6842–6850 (1991).

Xiong, J. Q., M. Zhong, C. Fang, L. C. Chen, and M. Lippmann, "Influence of Organic Films on the Hygroscopicity of Ultrafine Sulfuric Acid Aerosol," *Environ. Sci. Technol., 32,* 3536–3541 (1998).

Yamasaki, H., K. Kuwata, and H. Miyamoto, "Effects of Ambient Temperature on Aspects of Airborne Polycyclic Aromatic Hydrocarbons," *Environ. Sci. Technol., 16,* 189–194 (1982).

Yamasaki, H., K. Kuwata, and Y. Kuge, "Determination of Vapor Pressure of Polycyclic Aromatic Hydrocarbons in the Supercooled Liquid Phase and Their Adsorption on Airborne Particulate Matter," *Nippon Kagaku Kaishi, 8,* 1324–1329 (1984) (*Chem. Abstr., 101,* 156747p (1984)).

Yokouchi, Y., and Y. Ambe, "Characterization of Polar Organics in Airborne Particulate Matter," *Atmos. Environ., 20,* 1727–1734 (1986).

Yoshizumi, K., and K. Asakuno, "Characterization of Atmospheric Aerosols in Chichi of the Ogasawara (Bonin) Islands," *Atmos. Environ., 20,* 151–155 (1986).

Zhang, S.-H., M. Shaw, J. H. Seinfeld, and R. C. Flagan, "Photochemical Aerosol Formation from α-Pinene and β-Pinene," *J. Geophys. Res., 97,* 20717–20729 (1992).

Zhang, X. Q., P. H. McMurry, S. V. Hering, and G. S. Casuccio, "Mixing Characteristics and Water Content of Submicron Aerosols Measured in Los Angeles and at the Grand Canyon," *Atmos. Environ., 27A,* 1593–1607 (1993).

Zhang, X. Y., R. Arimoto, and Z. S. An, "Dust Emission from Chinese Desert Sources Linked to Variations in Atmospheric Circulation," *J. Geophys. Res., 102,* 28041–28047 (1997).

Zieman, J. J., J. L. Holmes, D. Connor, C. R. Jensen, and W. H. Zoller, "Atmospheric Aerosol Trace Element Chemistry at Mauna Loa Observatory. 1. 1979–1985," *J. Geophys. Res., 100,* 25979–25994 (1995).

Zweidinger, R. B., R. K. Stevens, C. W. Lewis, and H. Westburg, "Identification of Volatile Hydrocarbons as Mobile Source Tracers for Fine-Particulate Organics," *Environ. Sci. Technol., 24,* 538–542 (1990).

CHAPTER

10

Airborne Polycyclic Aromatic Hydrocarbons and Their Derivatives
Atmospheric Chemistry and Toxicological Implications

Polycyclic aromatic hydrocarbons (PAHs) are members of a unique class of air pollutants relevant to many scientific and societal issues having a variety of aspects: chemical, toxicological, engineering, technological, public health, economic and regulatory, and legislative. They are products of incomplete combustion formed during the burning or pyrolysis of organic matter and are released into ambient air as constituents of highly complex mixtures of polycyclic organic matter, POM. As defined in the U.S. Clean Air Act Amendments of 1990 (CAAA, 1990), POM "includes organic compounds with more than one benzene ring, and which have a boiling point greater than or equal to 212°F (100°C)." While we generally refer to specific PAHs and polycyclic aromatic compounds (PACs) throughout this chapter, it should be noted that the complex mixture represented by the term POM is the subject of many studies and regulatory designations.

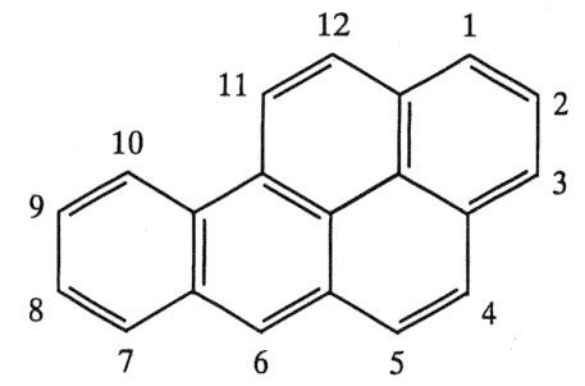

Atmospheric PAHs occur in the form of gases (e.g., 2-ring, highly volatile naphthalene, **III**), solids adsorbed/absorbed to the surfaces of fine respirable aerosol particles (e.g., 5-ring benzo[*a*]pyrene, BaP, **I**), and 3- and 4-ring semivolatile compounds that are distributed between the gas- and the particle-phases (e.g., the semivolatile 3-ring phenanthrene, **IV**, and 4-ring pyrene, **II**, and fluoranthene, **V**); for gas–particle partitioning, see Chapter 9D and Section A.4 in this chapter.

The ubiquitous nature of these airborne PAHs is evident from the fact that the 16 U.S. Environmental Protection Agency "Priority Polycyclic Aromatic Hydrocarbon Pollutants" shown in Table 10.1 (U.S. EPA, 1988) are found, as we shall see in this chapter, in urban airsheds throughout the world. Their widespread presence is due to their emissions from a wide range of combustion sources, including diesel and gasoline engines, biomass burning of agricultural and forest fuels (Jenkins *et al.*, 1996), and outdoor wood smoke (Watts *et al.*, 1988) (for summaries of the extensive literature

I
Benzo[*a*]pyrene

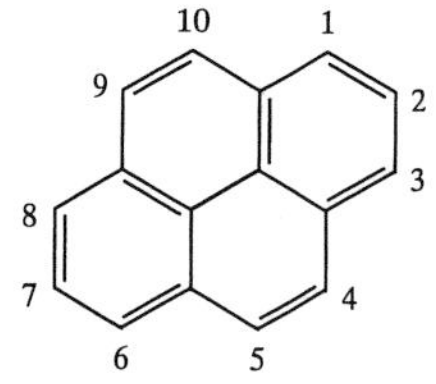

II
Pyrene

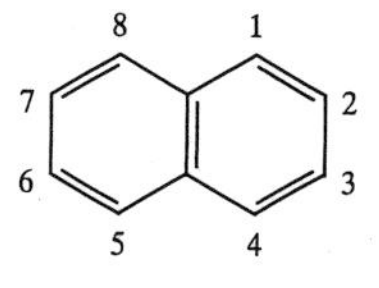

III
Naphthalene

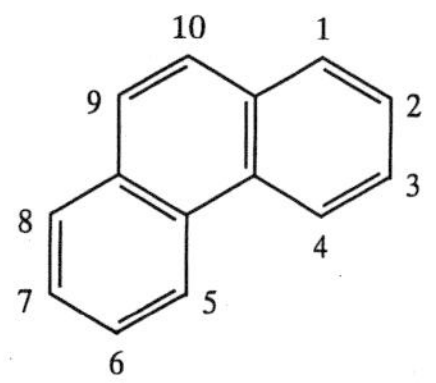

IV
Phenanthrene

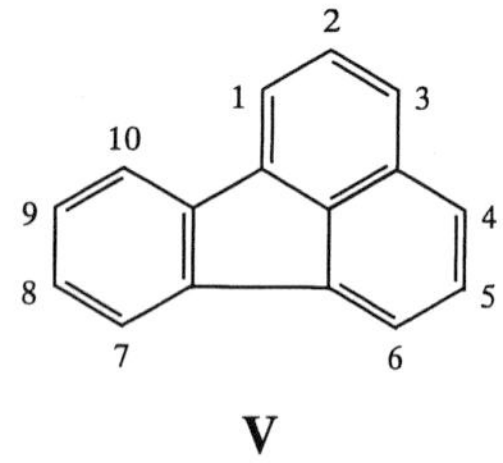

V
Fluoranthene

TABLE 10.1 Structures, Common Names, Empirical Formulas, Molecular Weights, Melting Points, Boiling Points, and CAS Numbers for the 16 U.S. EPA "Priority PAH Pollutants"[a,b]

Structure	Common name	Empirical formula	MW (g mol^{-1})	Mp (°C)	Bp (°C)	CAS number
	Naphthalene	C_8H_{10}	128.18	80.5	218	91-20-3
	Acenaphthylene	$C_{12}H_8$	152.20	92	265–275	208-96-8
	Acenaphthene	$C_{12}H_{10}$	154.20	96.2	277.5	83-32-9
	Fluorene	$C_{13}H_{10}$	166.23	116	295	86-73-7
	Phenanthrene*	$C_{14}H_{10}$	178.2	101	339	85-01-8
	Anthracene*	$C_{14}H_{10}$	178.2	216.2	340	120-12-7
	Pyrene	$C_{16}H_{10}$	202.3	156	360	129-00-0
	Fluoranthene	$C_{16}H_{10}$	202.3	111	375	206-44-0
	Benz[a]anthracene	$C_{18}H_{12}$	228.3	160	435	56-55-3
	Chrysene	$C_{18}H_{12}$	228.3	255	448	218-01-9

(continues)

TABLE 10.1 (continued)

Structure	Common name	Empirical formula	MW (g mol^{-1})	Mp (°C)	Bp (°C)	CAS number
	Benzo[a]pyrene	$C_{20}H_{12}$	252.3	175	495	50-32-8
	Benzo[b]fluoranthene	$C_{20}H_{12}$	252.32	168	481	205-99-2
	Benzo[k]fluoranthene	$C_{20}H_{12}$	252.32	217	481	207-08-9
	Benzo[ghi]perylene	$C_{22}H_{12}$	276.34	277	525	191-24-2
	Indeno[1,2,3-cd]pyrene	$C_{22}H_{12}$	276.34	163	—	193-39-5
	Dibenz[a,h]anthracene	$C_{22}H_{14}$	278.35	267	524	53-70-3

[a] Adapted from Mackay et al. (1992); data on indeno[1,2,3-cd]pyrene from Harvey (1997). Structures and their numbering are based on IUPAC recommendations as described by Loening et al. (1990).

[b] U.S. EPA (1998) designation. Exceptions are noted by asterisks.

on combustion sources, see Björseth (1983), Baek et al. (1991a), Venkataraman and Friedlander (1994c), Schauer et al. (1996), Harrison et al. (1996), Howsam and Jones (1998), and Simoneit (1998); for diesel and gasoline engines, see IARC (1989), Benner et al. (1989), Westerholm et al. (1991), Bagley et al. (1992), Johnson et al. (1994), Lowenthal et al. (1994), Hammerle et al. (1994), WHO (1996), Miguel et al. (1998), and Schauer et al. (1999); for coal fly ash, see Gohda et al. (1993); and for municipal incinerators, see Lee et al. (1993). PAHs are also common constituents of air indoors, arising from coal and wood combustion (Mumford et al., 1990), wood combustion (Alfheim and Ramdahl, 1984), and environmental tobacco smoke, ETS (Gundel et al., 1995b; and the California EPA, 1997).

The physical and chemical processes by which PAHs and PACs are formed in combustion are very complex and beyond the scope of this book; the reader is referred to articles in the literature such as those by Badger (1962), Haynes (1991), and Vander Wal and co-workers (1997). Only a relatively few PAHs (ca. 100) are stable enough to survive the combustion–pyrolysis process and enter our air environment as primary pollutants in complex combustion-generated mixtures in amounts sufficient to be of concern. The *toxicity* of such combustion-generated mixtures is reflected in the fact that POM is one of the 189 hazardous air pollutants (HAPs) cited in the 1990 U.S. Clean Air Act Amendments (Kelly *et al.*, 1994; Kao, 1994). Their *complexity* is evident from the fact that the term POM includes not only PAHs, the most abundant and intensively studied chemical class in POM emissions and ambient air, but also a wide range of N-, O-, and S-atom polycyclic aromatic compounds, including nitroarenes and azaarenes; PAH lactones, ketones, and quinones; and thioarenes. Structures and nomenclatures for selected PAHs and PACs relevant to atmospheric chemistry are discussed in Section A.2.

Some of these N-, O-, and S-atom-functionalized polycyclic aromatic compounds (PACs) are powerful bacterial mutagens and animal, and possible human, carcinogens, e.g., the exocyclic nitro-substituted PAH 1-nitropyrene, **VI**, which is a primary pollutant in diesel exhaust (IARC, 1989). The endocyclic heterocyclic PAC dibenz[*a,h*]acridine, **VII**, is also classified as a "possible human carcinogen" (IARC, 1987; see Sections B and C).

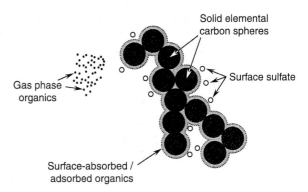

FIGURE 10.1 Schematic of a diesel soot particle consisting of an agglomeration of elemental carbon spheres (0.01- to 0.08-μm diameter). Its surface is covered with absorbed/adsorbed particle-phase organics, including 5-ring (e.g., BaP) and 6-ring PAHs. Gas-phase organics include all of the highly volatile 2-ring PAHs (e.g., naphthalene and methylnapthalenes). Semivolatile 3-ring (e.g., phenanthrene and anthracene) and 4-ring PAHs (e.g., pyrene (**II**) and fluoranthene (**V**)) are distributed between both phases. Sulfate is also associated with diesel particles. (Adapted with permission from Johnson *et al.*, 1994, SAE Paper 940233 © 940233 Society of Automotive Engineers, Inc.; see also Schauer *et al.*, 1999.)

VI 1-Nitropyrene

VII Dibenz[*a,h*]acridine

The physical and chemical complexity of primary combustion-generated POM is illustrated in Fig. 10.1 (Johnson *et al.*, 1994), a schematic diagram of a diesel exhaust particle and associated copollutants. The gas-phase regime contains volatile (2-ring) PAHs and a fraction of the semivolatile (3- and 4-ring) PAHs. The particle-phase contains the remainder of the semivolatile PAHs ("particle-associated") along with the 5- and 6-ring heavy PAHs adsorbed/absorbed to the surface of the elemental carbon spheres that constitute the "backbone" of the overall diesel soot particle. Also present is sulfate formed from oxidation of sulfur present in the diesel fuel and gas- and particle-phase PACs.

The elemental carbon core particles ("black carbon") range from ca. 0.01- to 0.08-μm aerodynamic diameter. They not only add significantly to the total mass of the aerosol particle but, as seen in Chapter 9, are also a significant cause of visibility degradation in polluted atmospheres (e.g., see Larson *et al.*, 1989).

Polycyclic aromatic hydrocarbons and their associated more polar (hence more water soluble) heteroatom derivatives (polycyclic aromatic compounds, PACs; see later) not only are present in air as gases and particles but also are present, for example, in urban "street dust" samples (Takada *et al.*, 1990) and other soil and water environments throughout the world due to wet and dry deposition (see, for example, Schwarzenbach *et al.* (1993), Wild and Jones (1995), and Neilson (1998)). Furthermore, through long-range transport of polluted air masses, PAHs can be found in ambient air at receptor sites far from their original sources. For example, PAHs from the European continent and Great Britain have been observed at a "background site" at Birkenes, Norway, and in Copenhagen, Denmark (Björseth *et al.*, 1979; Björseth and Olufsen, 1983; Nielsen *et al.*, 1999a, 1999b). PAHs from the former Soviet Union have been reported to reach the Norwegian Arctic (Pacgna and Oehme, 1988) and sources in Africa and in northern Europe (e.g., Germany, Belgium, and France) have been reported to impact Corsica, France (Masclet *et al.*, 1988).

Of particular interest in terms of atmospheric chemistry are reactions of certain PAHs in VOC–NO$_x$–air atmospheres to form biologically active polycyclic aromatic compounds, PACs. Thus, not only is the fundamental chemistry of the formation and fates of these secondary air pollutants of interest, but it can also have major toxicological implications. For example, in some airsheds certain PACs that are reaction products (e.g., nitro-PAH and nitro-PAH lactones) contribute significantly more to the overall, direct-acting bacterial mutagenicities of the gaseous and particle phases of ambient air than do the PACs in the *primary* emissions directly emitted by sources. Furthermore, certain of them, e.g., 2-nitrofluoranthene (**XXVII**), are human cell mutagens:

2-Nitrofluoranthene

XXVII

Nitro-PAHs, regardless of whether directly emitted or formed in air, are of concern because many of them are animal, and possible human, carcinogens (IARC, 1989; see also review of environmental carcinogens by Tokiwa *et al.* (1998) and references therein).

The goal of this chapter is to illustrate the atmospheric chemistry of PAHs and PACs. However, because of their unique biological properties, we also provide some perspective on their relevance to air pollution toxicology and the development of sound scientific health risk assessments for specific carcinogenic PAHs such as benzo[*a*]pyrene (California Air Resources Board, 1994) and for complex combustion-generated emissions such as diesel exhaust (e.g., IARC, 1989; WHO, 1996; and California Air Resources Board, 1998). For discussions of such broad topics as research on analytical methods for the determination of PAHs in environmental samples, see, for example, the review of chromatographic methods by Poster, Sander, and Wise (1998) and references therein. Similarly, for reviews of their environmental chemistry and related carcinogenicities, see Neilson (1998) and Harvey (1997). The book *Environmental Organic Chemistry* by Schwarzenbach, Gschwend, and Imboden (1993) provides a useful perspective on the subject and contains helpful specific examples relating to PAHs.

A. NOMENCLATURE AND SELECTED PHYSICAL AND SPECTROSCOPIC PROPERTIES OF POLYCYCLIC AROMATIC HYDROCARBONS (PAHs) AND POLYCYCLIC AROMATIC COMPOUNDS (PACs)

1. Combustion-Generated PAHs and PACs

Historical, chemical, and toxicological interest in PAHs and PACs goes back over two centuries when Sir Percival Pott proposed that the high rate of cancer of the scrotum incurred by London's chimney sweeps was due to the presence of certain chemicals in the fireplace soot (i.e., POM) to which they were heavily exposed (Pott, 1775). Some 150 years later, Passey (1925) reported that organic extracts of such "domestic soot" induced tumors in experimental animals.

In the early 1940s, Leiter *et al.* (1942) demonstrated that a similar phenomenon occurred with organic extracts of ambient air particles—that is, injection of "tars extracted from atmospheric dusts" collected at locations throughout New York City produced subcutaneous sarcomas in mice. Shortly thereafter, Leiter and Shear (1943) reported that "marginal doses of 3,4-benzpyrene" (known today as benzo[*a*]pyrene, BaP, **I**), the powerful carcinogen earlier isolated from coal tar and synthesized by Cook *et al.* (1933), also produced subcutaneous tumors in mice.

These observations, coupled with the discoveries of BaP in chimney soot (Goulden and Tipler, 1949) and in ambient air particles collected at 10 stations throughout Great Britain (Waller, 1952), and the tumorigenic/carcinogenic properties of extracts of ambient particles collected during episodes of Los Angeles photochemical smog (Kotin *et al.*, 1954) were key factors in establishing the atmospheric chemistry of PAHs and PACs as a new field in air pollution research.

Since then, many monographs, handbooks, symposia proceedings, and specialized chapters (in addition to thousands of research papers) dealing with the chemistry and biological and toxicological aspects of PAHs, POM, and PACs have appeared in the scientific, engineering, and medical sciences literature. Examples are cited in Box 10.1.

2. Structures and IUPAC Rules for Nomenclature

a. PAHs

Over the decades several significantly different PAH and PAC numbering/nomenclature systems have been proposed and widely used in the older literature, e.g., that of Clar (1964). Unfortunately, even today this can lead to confusion on the part of those unfamiliar with the history of different systems of nomenclature.

We follow the 1979 IUPAC recommendations summarized in *Polynuclear Aromatic Hydrocarbons: Nomenclature Guide* (Loening *et al.* 1990). The American Chemical Society also publishes the *Ring Systems Handbook*, which, ca. 1990, contained structural diagrams for over 70,000 unique ring systems (American Chemical Society, 1977 to present).

A detailed discussion of these rules and nomenclature is beyond the scope of this book. However, we

BOX 10.1
SELECTED REFERENCES TO THE CHEMISTRY AND TOXICOLOGY OF AIRBORNE PAHs AND PACs

Review/evaluations by government agencies: U.S. National Research Council/National Academy of Sciences (NRC, 1972, 1983, 1988); International Agency for Research on Cancer monographs (IARC, 1983, 1987, 1989); World Health Organization; Environmental Health Criteria (EHC) monographs (e.g., Diesel Fuel and Exhaust Emissions, EHC 171, WHO, 1996); U.S. Department of Health and Human Services, Public Health Services and National Toxicology Program, the *Reports on Carcinogens* (8th Report, U.S. DHHS, 1998); the California Air Resources Board (1994, 1997, 1998); the California EPA (1997); and the Danish Environmental Protection Agency (e.g., see Nielsen *et al.*, 1994, 1997, 1998).

Other publications include the following: handbooks and collections of chapters on PAHs and PACs (e.g., Bjørseth, 1983; Grimmer, 1983b; Bjørseth and Ramdahl, 1985; White, 1985; Howard *et al.*, 1990; Neilson, 1998); the series *Polynuclear Aromatic Hydrocarbons* published by Battelle Press (e.g., see Cooke *et al.*, 1982); monographs on polycyclic aromatic hydrocarbons (e.g., Mackay *et al.* 1992; and Harvey, 1997); chapters dealing with PAHs and PACs in books on atmospheric chemistry and environmental organic chemistry (e.g., in Finlayson-Pitts and Pitts (1986), Schwarzenbach *et al.* (1993), and Graedel, Hawkins, and Claxton (1986)); review articles on atmospheric reactions and fates of PAHs by Nielsen *et al.* (1983), Nielsen (1984), Nikolaou *et al.* (1984), Van Cauwenberghe (1985), Pitts (1987, 1993a, 1993b), Baek *et al.* (1991), Atkinson and Arey (1994), and Arey (1998a).

outline them in Box 10.2 and give examples in Tables 10.1 and 10.2 of how they are generally employed with respect to atmospherically relevant PAHs. The nomenclature for selected N-, O-, and S-atom-functionalized PACs is described briefly in Section A.2.b and examples are found in Tables 10.3–10.5.

The same molecule may appear with different orientations. For example, in the current literature, the structure for the 3-ring PAH phenanthrene is drawn as (**IVa**), (**IVb**), or (**IVc**). While one or more IUPAC rules may be broken, their ring numberings are internally consistent.

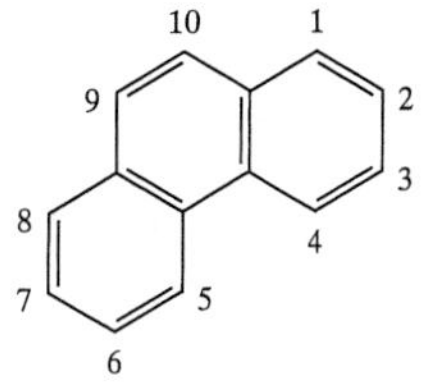

IVa

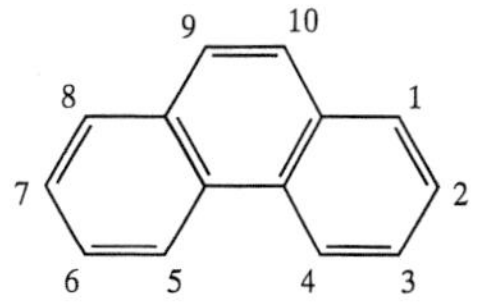

IVb

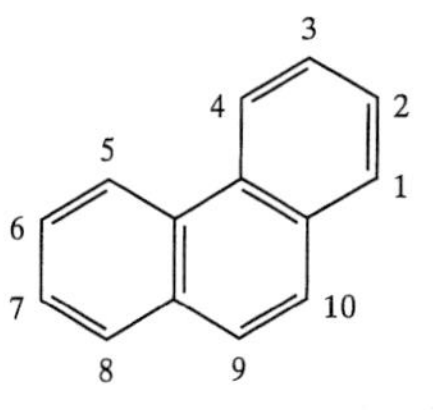

IVc

Some PAHs may have two hydrogens at a specific carbon position, the *indicated hydrogens*. Such an indicated hydrogen should be mentioned by carbon number, even when it is further oxidized to a carbonyl group; consider, for example, 7H-benz[*de*]anthracene (**XII**) and 7H-benz[*de*]anthracen-7-one (**XIII**, benzanthrone).

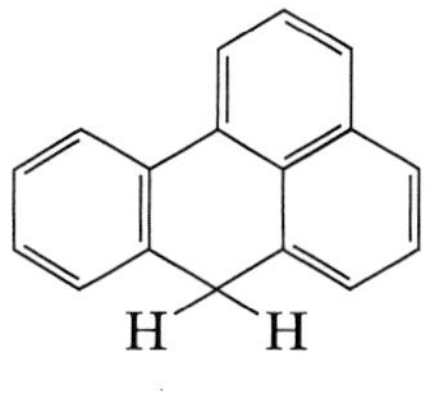

XII

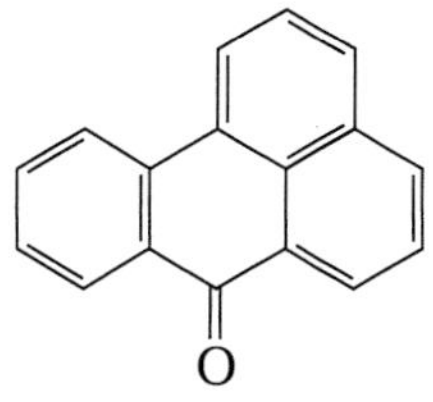

XIII

Similarly, the structure of a nitro-PAH lactone found in ambient aerosols and also formed in laboratory irradiations of phenathrene–NO$_x$–air mixtures, 2-nitro-6H-dibenzo[*b,d*]pyran-6-one, which is a powerful, direct-acting bacterial mutagen and potent human cell mutagen, is commonly shown as (**XI**) (Helmig *et al.*, 1992a; Arey *et al.*, 1992; Sasaki *et al.*, 1995, 1997b; Arey, personal communication).

Finally, a possible source of confusion when assigning structures to a given PAH is the "correct" number of the double bonds within the fused rings. The *Fries Rule* states that "the most stable form of a polynuclear

BOX 10.2
SELECTED IUPAC RULES FOR NAMING, ORIENTING, AND NUMBERING SELECTED PAHs

One starts with the IUPAC "preferred name" (sometimes referred to as the "trivial name") used for certain of the simple PAHs, e.g., anthracene, phenanthrene, pyrene, and fluoranthene. More complicated structures are then built up by the addition of, for example, benzo, dibenzo, or naphtho groups to the skeleton of the simple PAH.

The sequence of steps to follow when drawing a structure of a fused PAH ring system is the following:

1. Define the molecular structure for which a preferred name exists in the IUPAC list (1979).

2. Place the maximum number of rings in a horizontal row.

3. Place the greatest number of remaining rings above and to the right of a vertical axis drawn through the center of the molecule (i.e., in the upper right-hand quadrant). Structure I shown earlier for benzo[a]pyrene illustrates these points.

4. After properly orientating it, number the carbon atoms in the molecule in a clockwise direction starting with the carbon atom *not* common to another ring (i.e., not involved with fusion of two rings) that is in the most counter-clockwise position of the uppermost ring that is farthest to the right (see structures for PAHs I–V and Table 10.1). Carbon atoms *common* to two or more rings are *not* numbered.

5. Letter the faces of the ring in alphabetical order.

6. Starting with "a," letter the faces of the rings in alphabetical order beginning with that side between carbon atoms 1 and 2. Continue clockwise around the molecule, noting, however, that ring faces common to two rings are *not* lettered.

See fluoranthene, **V**, and benzo[k]fluoranthene, **XXVI**.

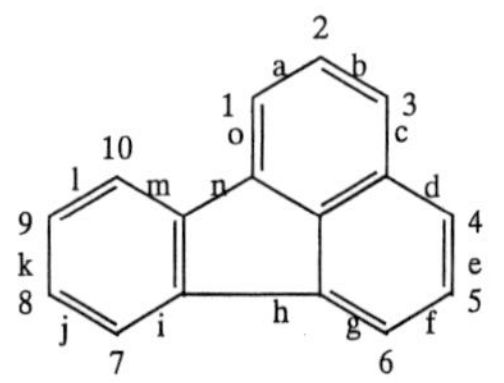

Since addition of another ring is defined by the bond(s) involved in the fusion, in this case the k face of fluoranthene, the name becomes benzo[k]fluoranthene.

The numbering in the final molecule can be different from the one used in the original compound, since the addition of more rings may change its orientation in the frame of reference. For example, consider anthracene (**VIII**), 9,10-dimethylanthracene (**IX**), and 7,12-dimethylbenz[a]anthracene (**X**):

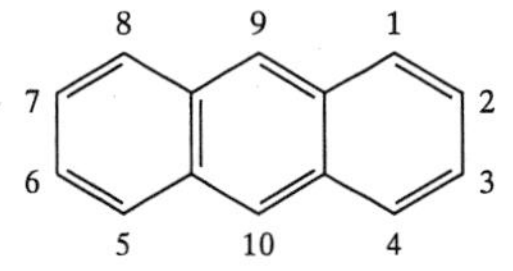

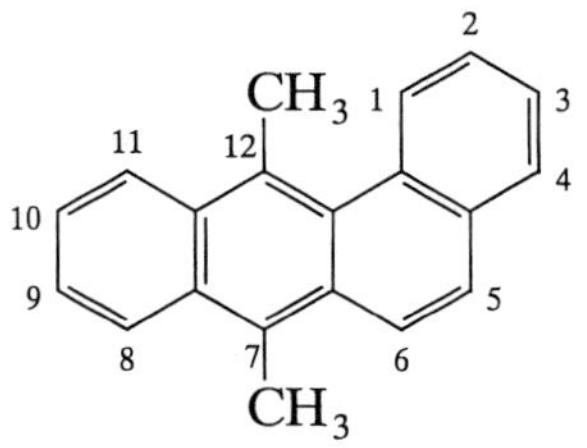

Note that the numbering of anthracene and 9,10-dimethylanthracene does not conform to the foregoing IUPAC rules. This "common" numbering is derived from anthraquinone, in which the middle, or *meso*, positions are numbered 9 and 10.

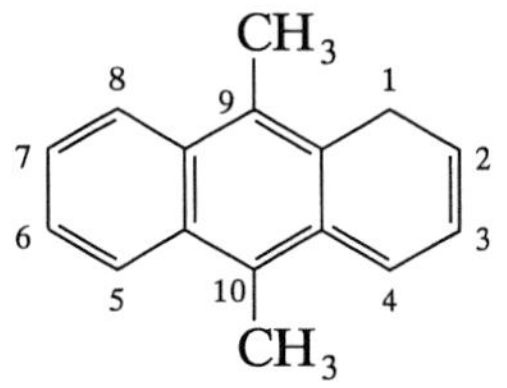

V
Fluoranthene

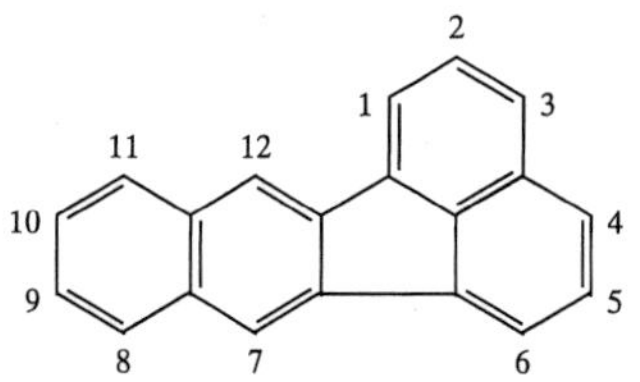

XXVI
Benzo[k]fluoranthene

TABLE 10.2 Structures, Common Names, Empirical Formulas, Molecular Weights, Melting Points, Boiling Points, and CAS Numbers for Selected PAHs of Atmospheric Interest[a,b]

Structure	Common name	Empirical formula	MW (g mol^{-1})	Mp (°C)	Bp (°C)	CAS number
	1-Methylnaphthalene	$C_{11}H_{10}$	142.2	−22	244.6	90-12-0
	2-Methylnaphthalene	$C_{11}H_{10}$	142.2	34.6	241.9	91-57-6
	Biphenyl	$C_{12}H_{10}$	154.21	71	256	92-52-4
	1-Methylphenanthrene	$C_{15}H_{12}$	192.26	123	359	832-69-6
	Cyclopenta[cd]pyrene	$C_{18}H_{10}$	226.28	175	–	27208-37-3
	Benzo[ghi]fluoranthene	$C_{18}H_{10}$	226.28	149	432[b]	203-12-3
	Triphenylene	$C_{18}H_{12}$	228.3	199	438	217-59-4
	Benzo[e]pyrene	$C_{20}H_{12}$	252.3	178	493[b]	192-97-2
	Benzo[j]fluoranthene	$C_{20}H_{12}$	252.3	166	~480[b]	205-82-3

(continues)

TABLE 10.2 (continued)

Structure	Common name	Empirical formula	MW (g mol^{-1})	Mp (°C)	Bp (°C)	CAS number
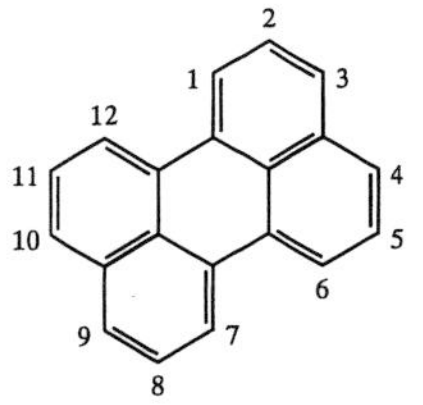 (perylene structure)	Perylene	$C_{20}H_{12}$	252.3	277	495	198-55-0
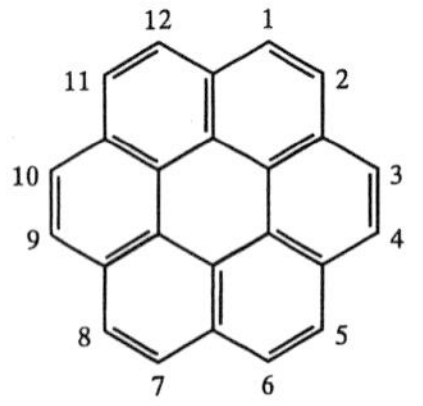 (coronene structure)	Coronene	$C_{24}H_{12}$	300.36	>350	525	191-07-1
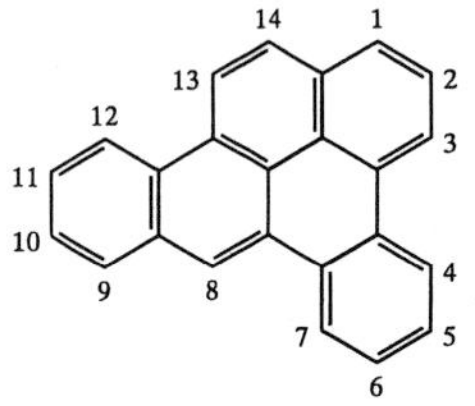 (dibenzo[a,e]pyrene structure)	Dibenzo[a,e]pyrene	$C_{24}H_{14}$	302.28	234	—	192-65-4

[a] Adapted from Mackay et al. (1992) and Mackay and Callcott (1998). Structures and numbering are based on IUPAC recommendations; see Loening et al. (1990).
[b] From Bjørseth (1983).

hydrocarbon is that in which the maximum number of rings have the benzenoid arrangement of three double bonds" (Fieser and Fieser, 1956). This statement is consistent with a subsequent proposal by Clar (1972) that for such polyarenes one should maximize the number of "aromatic sextets."

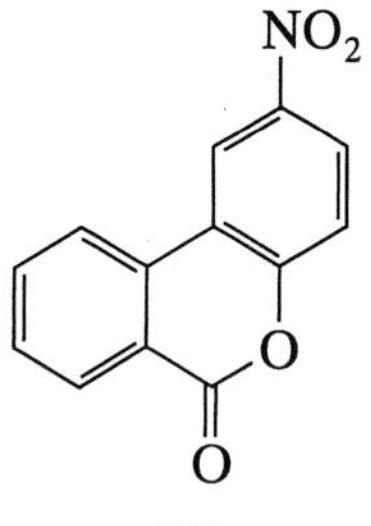

XI

2-Nitro-6H-dibenzo[b,d]pyran-6-one

For example, of the three possible electronic structures for naphthalene, the symmetrical structure (**a**) with two benzenoid rings (aromatic sextets) would be the more stable, and hence the preferable representation.

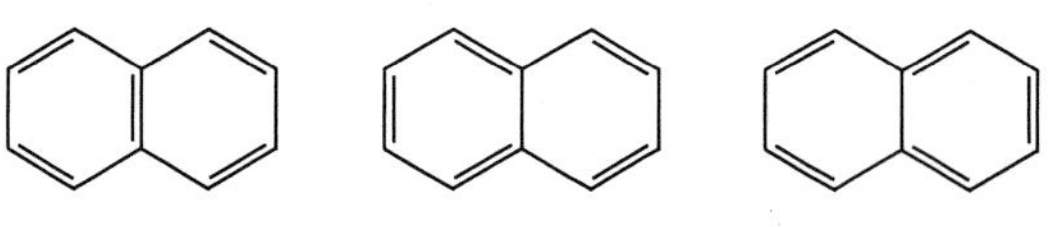

a **b** **c**

[Note that when the Clar convention is used for Kekulé structures, the benzenoid rings are often shown as circles (e.g., see Moyano and Paniagua, 1991).]

Finally, since our focus is on the atmospheric chemistry of unsubstituted (or methyl substituted) PAHs, we do not discuss major classes of PAHs from natural sources that are not found at significant levels in ambient air but may be important in soil and water environments, e.g., alkylated PAHs (see the review by Simoneit, 1998).

b. PACs

Despite the generally lower emission factors and ambient levels of PACs compared to PAHs (there are some exceptions), interest in certain airborne N-, O-, and S-atom PACs has increased significantly, in large part because of their mutagenic and carcinogenic properties. For example, in 1976, Cautreels and Van Cauwenberghe used GC–MS to identify more than 100 compounds in three different fractions (neutral, acidic,

TABLE 10.3 Structures, Common Names, Empirical Formulas, and Molecular Weights of Selected Nitrogen-Containing Mono- and Polycyclic Aromatic Compounds (N-PACs) Identified in Ambient Air

Structure	Common name	Empirical formula	MW (g mol^{-1})	Reference
	Pyridine	C_5H_5N	79	—
	Quinoline	C_9H_7N	129	a, b, c
	Isoquinoline	C_9H_7N	129	b, c
	Benzo[f]quinoline (5,6-Benzoquinoline)	$C_{13}H_9N$	179	a, b, c
	Benzo[h]quinoline (7,8-Benzoquinoline)	$C_{13}H_9N$	179	a, b, c
	Phenanthridine	$C_{13}H_9N$	179	a, b, c
	Acridine*	$C_{13}H_9N$	179	b, c, d
	Benz[a]acridine (1,2-Benzacridine)	$C_{17}H_{11}N$	229	d
	Benzo[c]acridine (3,4-Benzacridine)	$C_{17}H_{11}N$	229	a

(continues)

TABLE 10.3 (continued)

Structure	Common name	Empirical formula	MW (g mol^{-1})	Reference
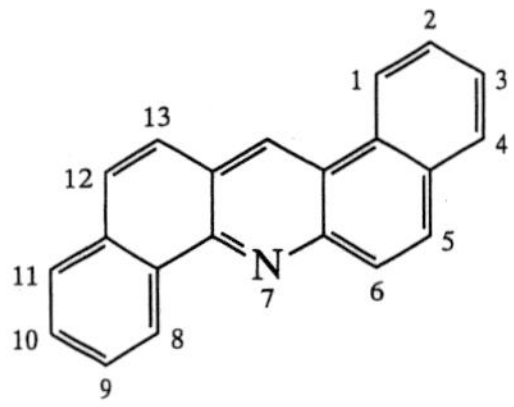	Dibenz[a,h]acridine	$C_{21}H_{13}N$	279	a, d
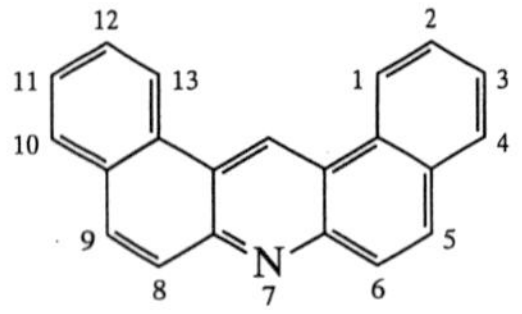	Dibenz[a,j]acridine	$C_{21}H_{13}N$	279	a
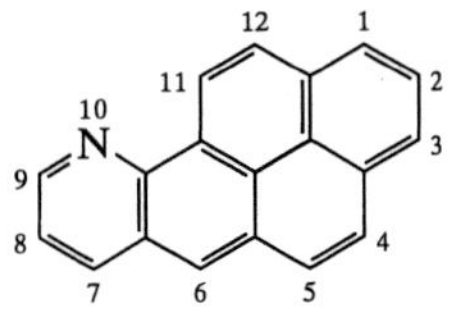	10-Azabenzo[a]pyrene	$C_{19}H_{11}N$	253	d
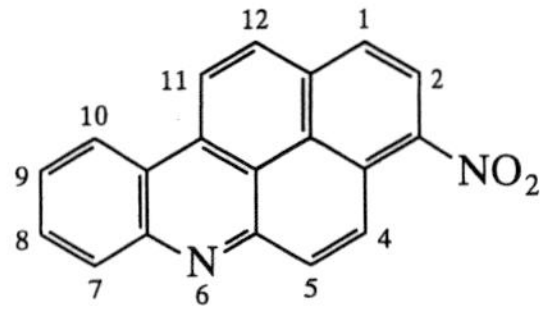	3-Nitro-6-azabenzo[a]pyrene	$C_{19}H_{10}O_2N_2$	298	e
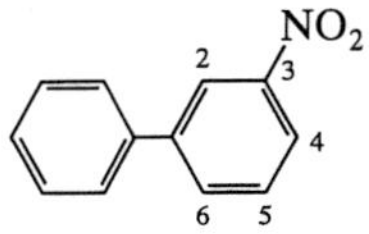	3-Nitrobiphenyl	$C_{12}H_9NO_2$	199	f, g, h, i, j, l
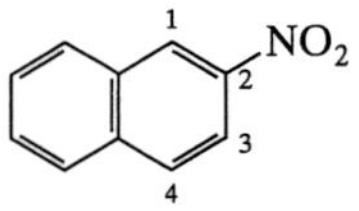	2-Nitronaphthalene	$C_{10}H_7NO_2$	173	f, g, h, i, j, l, m
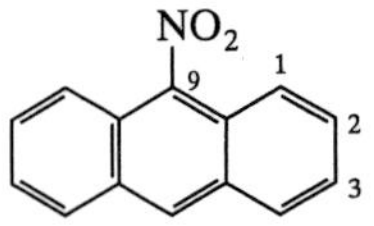	9-Nitroanthracene	$C_{14}H_9NO_2$	223	f, g, h, i, k, l, m
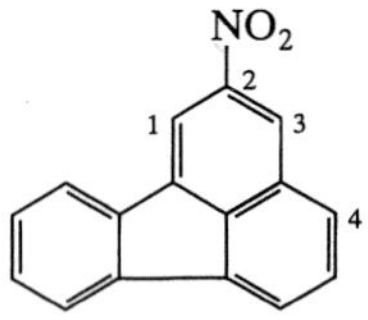	2-Nitrofluoranthene	$C_{16}H_9NO_2$	247	f, g, h, i, k, l, m
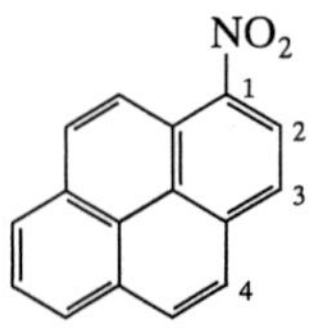	1-Nitropyrene	$C_{16}H_9NO_2$	247	f, g, h, i, k, l, m
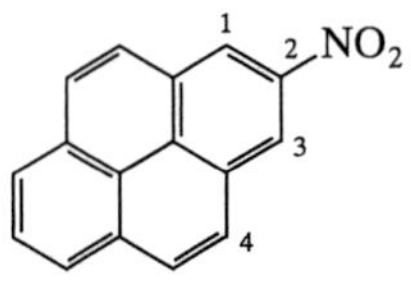	2-Nitropyrene	$C_{16}H_9NO_2$	247	g, l

(continues)

A. NOMENCLATURE AND SELECTED PHYSICAL AND SPECTROSCOPIC PROPERTIES OF PAHs AND PACs 447

TABLE 10.3 (continued)

Structure	Common name	Empirical formula	MW (g mol^{-1})	Reference
	1,6-Dinitropyrene	$C_{16}H_8N_2O_4$	292	h, m
	Pyrrole	C_4H_5N	63	—
	Indole	C_8H_7N	117	—
	Carbazole*	$C_{12}H_9N$	167	—

Note: Exceptions to IUPAC rules are designated with asterisks.
[a] Nielsen et al. (1986).
[b] Chen and Preston (1998).
[c] Dong et al. (1977).
[d] Yamauchi and Handa (1987).
[e] Sera et al. (1994).
[f] Wilson et al. (1995).
[g] Atkinson and Arey (1994) and references therein.
[h] Lewtas and Nishioka (1990) and references therein.
[i] Zielinska et al. (1990) and references therein.
[j] Fraser et al. (1998).
[k] Hannigan et al. (1998).
[l] Arey (1998a) and references therein.
[m] Tokiwa and Ohnishi (1986), Tokiwa et al. (1987), and references therein.

and basic) of extracts of ambient aerosols; the compounds included not only PAHs and aliphatics but also polar oxygenated substances, including several O-PACs and N-PACs. Similarly, Gundel and co-workers (1993) identified O-PACs and organic nitrates and nitrites as well as amides, amines, and nitro compounds in polar fractions of extracts of airborne particulate matter from particles ($D_{50} \leq 15$ μm) in ambient air in Elizabeth, New Jersey, and in the U.S. National Institute of Standards and Technology (NIST) Standard Reference Material SRM 1649, "Urban Dust/Organics" (see Box 10.3). Other investigators have identified S-PACs (e.g., dibenzothiophene) in the gas and particle phases of both direct emissions and ambient air.

Hence, some understanding of the nomenclature used for O-, N-, and S-PACs is also useful.

Given the presence of N-, O-, and S-heteroatoms as members of fused PAH ring systems (endocyclic), attached to such fused PAH rings (exocyclic), or both, the nomenclature of PACs is more complex than for PAHs. However, Chapter 2 of *Polynuclear Aromatic Hydrocarbons—Nomenclature Guide* by Loening et al. (1990) and *Systematic Nomenclature of the Nitrogen, Oxygen, and Sulfur Functional Polycyclic Aromatic Compounds* (Later et al., 1990) provide a useful synopsis of the IUPAC rules for monofunctional N-, O-, and S-PACs (IUPAC, 1979). Many of the IUPAC rules for the parent, nonsubstituted PAHs described in Box 10.2 are applicable to these heteroatom derivatives. These rules are summarized in Box 10.4 (N-PAC), Box 10.5 (O-PAC), and Box 10.6 (S-PAC), respectively.

As an example, both replacement (aza) and substitutive ($-NO_2$) nomenclature are used in naming the powerful bacterial mutagen 3-nitro-6-azabenzo-

TABLE 10.4 Structures, Common Names, Empirical Formulas, and Molecular Weights for Selected Oxygen-Containing Polycyclic Aromatic Compounds (O-PACs) Identified in Ambient Air

Structure	Common name	Empirical formula	MW (g mol^{-1})	Reference
Ketones				
	1-Acenaphthenone	$C_{12}H_8O$	168	a
	Fluoren-9-one (9H-Fluoren-9-one)	$C_{13}H_8O$	180	a, b, d, f, g
	Benzanthrone (7H-Benz[de]anthracen-7-one)	$C_{17}H_{10}O$	230	a, b, f, g
	11H-Benzo[b]fluoren-11-one	$C_{17}H_{10}O$	230	a, b, g
	6H-Benzo[cd]pyren-6-one	$C_{19}H_{10}O$	254	a, f, g
Quinones				
	9,10-Anthracenedione (Anthraquinone)	$C_{14}H_8O_2$	208	b, d, f
	9,10-Phenanthrenedione (Phenanthraquinone)	$C_{14}H_8O_2$	208	a
	7,12-Benz[a]anthracenedione (Benz[a]anthracene-7,12-quinone)	$C_{18}H_{10}O_2$	258	b, c, d, f, g
	5,12-Naphthacenedione (5,12-Naphthacenequinone)	$C_{18}H_{10}O_2$	258	a, b, g

(continues)

TABLE 10.4 (continued)

Structure	Common name	Empirical formula	MW (g mol^{-1})	Reference
Anhydrides				
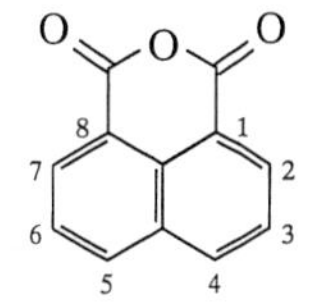	Naphthalene-1,8-dicarboxylic anhydride (1,8-Naphthalic anhydride)	$C_{12}H_6O_3$	198	a, b, c, f, g
Lactones				
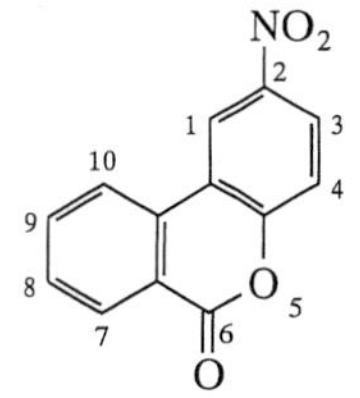	2-Nitro-6H-dibenzo[b,d]pyran-6-one	$C_{13}H_7NO_4$	241	e
Endocyclic O-PACs				
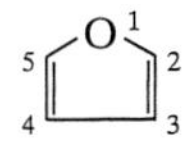	Furan	C_4H_4O	68	–
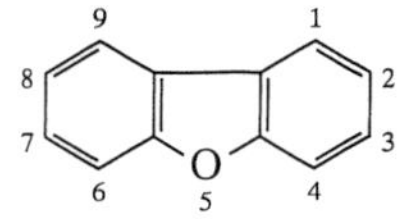	Dibenzofuran	$C_{12}H_8O$	168	d

[a] Allen et al. (1997).
[b] König et al. (1983).
[c] Wilson et al. (1995).
[d] Ligocki and Pankow (1989).
[e] Helmig et al. (1992a).
[f] Fraser et al. (1998).
[g] Hannigan et al. (1998).

[a]pyrene (**XXXI**) found in diesel exhaust and ambient air (e.g., see Section C and Sera et al. (1994)).

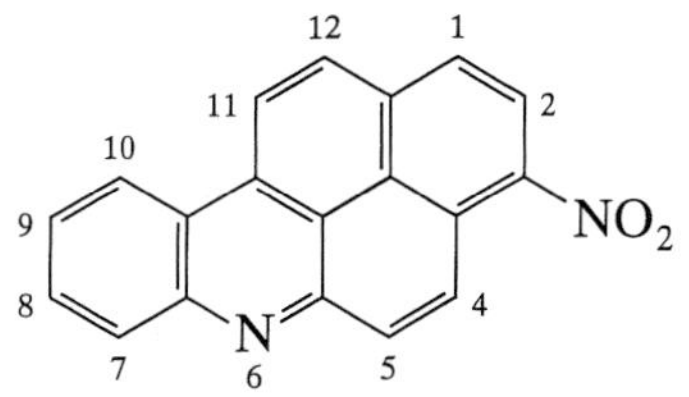

XXXI

3-Nitro-6-azabenzo[a]pyrene

Structures of several N-PACs identified in ambient air (and references) are shown in Table 10.3.

As an example, we saw earlier, replacing the CH_2 group in the parent PAH 7H-benz[de]anthracene, **XII**, by a keto group, C=O, gives 7H-benz[de]anthracen-7-one, **XIII** (benzanthrone). It is well known as a major O-PAC in combustion-generated emissions and ambient air (see Table 10.4) and the 3-nitro derivative, 3-nitrobenzanthrone (3-nitro-7H-benz[de]anthracen-7-one, **XXIV**) was recently identified by Enya et al. (1997) in both diesel exhaust and ambient air; it is an extremely powerful, direct bacterial mutagen:

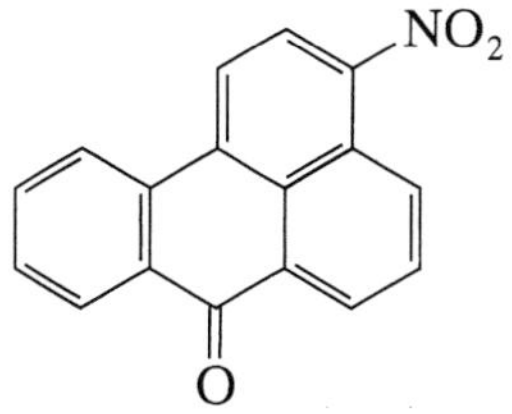

XXIV

3-Nitro-7H-benz[de]anthracen-7-one
or 3-Nitrobenzanthrone

Structures of compounds representing several classes of O-PACs identified in ambient air, e.g., PAH ketones, quinones, anhydrides, and lactones, are shown in Table 10.4.

TABLE 10.5 Structures, Common Names, Empirical Formulas, and Molecular Weights for Selected Sulfur-Containing Mono- and Polycyclic Aromatic Compounds (S-PACs) in Ambient Air[a,e,f] and Combustion-Generated Emissions,[c,d] Including Cigarette Smoke[b]

Structure	Common name	Empirical formula	MW (g mol^{-1})	Reference
	Thiophene	C_4H_4S	84	—
	Benzo[b]thiophene	C_8H_6S	134	b
	Dibenzothiophene	$C_{12}H_8S$	184	a, b, c, e, f
	Benzonaphthothiophene	$C_{16}H_{10}S$	234	c
	3-Nitrobenzothiophene	$C_8H_5SNO_2$	179	d

[a] Ligocki and Pankow (1989); ambient air.
[b] Schmid et al. (1985); cigarette smoke condensate.
[c] Tong and Karasek (1984); diesel particulate matter (DPM).
[d] Schuetzle and Daisey (1990) and references therein.
[e] Fraser et al. (1998); ambient air.
[f] Atkinson et al. (1988a); ambient air.

BOX 10.3
SOURCE OF NIST STANDARD REFERENCE MATERIAL SRM 1649, "AIR PARTICLES"

Throughout this chapter, we cite examples of the use of the NIST Standard Reference Material SRM 1649, which is referred to as "Air Particles" or "Urban Air Particulate Matter," (a) to validate analytical procedures for determination of PAHs and PACs in samples of complex mixtures of particulate matter in ambient air and (b) for laboratory intercomparisons of methodologies for bacterial bioassays and bioassay-directed fractionations of organic extracts of such mixtures (e.g., see Claxton et al., 1992a; Lewtas et al., 1990a, 1992; and May et al., 1992).

SRM 1649 was originally collected in a "bag house" over a period of one year in the late 1970s at a site near Washington, DC, screened through a 200-mesh sieve (cutoff point 125 μm), and since then apparently stored in bottles at room temperature (Lewtas et al., 1990a). Understandably, Claxton et al. (1992a) caution "While the air particles in SRM 1649 are similar to other air particulate samples, they do not represent a typical air particulate sample as collected by most researchers—they are intended as reference materials and not as samples for assessment of levels of airborne toxicants."

> **BOX 10.4**
>
> **SELECTED IUPAC NOMENCLATURE RULES FOR N-ATOM POLYCYCLIC AROMATIC COMPOUNDS**
>
> *Endocyclic PACs.* There are two approaches to naming endocyclic N-PACs, e.g., dibenz[*a,h*]acridine, **VII** (see Later, 1985; and Later *et al.*, 1990). In one style, the ending *ine* is used for endocyclic 6-membered rings, e.g., pyridine and quinoline. As with PAHs, if a benzo group is fused to a given side, the letter for that side is placed in brackets; e.g., addition of a benzo group to the *h* side of quinoline forms benzo[*h*]quinoline (note letters and numbers *within* square brackets are italicized, as with PAHs). Alternatively, it is named relative to the carbon atoms involved in the benzo-ring fusion (that constitute the ends of side *h*), i.e., 7,8-benzoquinoline:
>
> Benzo[*h*]quinoline
> 7,8-Benzoquinoline
>
> Similarly, benzo[*f*]quinoline ≡ 5,6-benzoquinoline.
>
> The second approach for endocyclic N-PACs is to use *replacement nomenclature*, in which the term *aza* is a prefix to the corresponding PAH. The position of the N-atom in the fused-ring system of the PAH precedes the term *aza*. Thus, for example, quinoline is 1-azanaphthalene. We generally use this convention for three or more ring N-PACs with endocyclic nitrogen heteroatoms.
>
> *Substitutive nomenclature* is employed for exocyclic N-PACs (and the O- and S-PACs). Functional groups such as amino ($-NH_2$), cyano ($-C{\equiv}N$), nitro ($-NO_2$), and hydroxyl ($-OH$) are attached as prefixes to the parent PAH, e.g., 1-nitropyrene, **VI**.

3. Solubilities and Vapor Pressures

PAHs have low solubilities in water (Table 10.6) as expected from their nonpolar character. These decrease dramatically in going from the 2- and 3-ring compounds (e.g., naphthalene, with a solubility of 31 mg L^{-1}) through the 3- and 4-ring semivolatile organics (e.g., fluoranthene, with a solubility of 0.26 mg L^{-1}) to 5-ring BaP, with a solubility of only 0.0038 mg L^{-1} (Mackay *et al.*, 1992).

However, reactions of PAHs in ambient air to form more polar species (e.g., nitro-PAHs, ketones, quinones, lactones, and dicarboxylic acids) greatly enhance their solubilities in aqueous systems. This has major implications when one considers the distribution of PAHs, and their atmospherically formed PAC derivatives, through the air, water, and soil environments. These increases in solubility upon reaction are important not only from an environmental chemistry perspective but also in terms of possible impacts on public health and ecosystems, e.g., in both the exposure and the health effect portions of risk assessments of PAHs (e.g., see Mackay *et al.*, 1992; Schwarzenbach *et al.*, 1993; Neilsen, 1998; Baum, 1998; the review by Mackay and Callcott, 1998; and references therein).

Values recommended by Mackay *et al.* (1992) and Mackay and Callcott (1998) for the vapor pressures of a number of PAHs of atmospheric interest and several properties used in theoretical calculations of their gas–particle partitioning in ambient air are given in Table 10.6: for examples of their use, see discussion of gas–particle partitioning theory in Chapter 9.D and references cited above.

Measured vapor densities of several volatile and semivolatile PAHs over a temperature range from 10 to 50°C have been used by Sonnefeld and co-workers (1983) to generate vapor pressures as a function of temperature:

$$\log P° = -A/T + B$$

Values for A and B are given in Table 10.7.

> # BOX 10.5
> ## SELECTED IUPAC NOMENCLATURE RULES FOR O-ATOM POLYCYCLIC AROMATIC COMPOUNDS
>
> The naming of endocyclic O-atom heterocycles is based on the same principles as for N-PACs; e.g., the structures of furan (**XVII**) and dibenzofuran (**XVIII**) are
>
> **XVII**
> Furan
>
> **XVIII**
> Dibenzofuran
>
> Similarly, with exocyclic O-PACs, substitutive nomenclature and common suffixes are used to describe the O-atom compounds formed when an H-atom on a PAH is replaced by another atom or functional group, e.g., 6-hydroxybenzo[a]pyrene, **XIX**, or 2-methoxychrysene, **XX**:
>
> **XIX**
> 6-Hydroxybenzo[a]pyrene
>
> **XX**
> 2-Methoxychrysene
>
> When PAHs are substituted with two exocyclic functional groups, e.g., the PAC **XXI**, one may employ either a prefix and a suffix, e.g., 1-nitropyren-2-ol, or two prefixes, 2-hydroxy-1-nitropyrene (we use the latter):
>
> **XXI**
> 2-Hydroxy-1-nitropyrene
>
> For ketones in which a CH_2 group is replaced by a $C=O$, one adds the suffix *one*. For example, 9*H*-fluorene (**XXII**) becomes 9*H*-fluoren-9-one or 9-fluorenone (**XXIII**).
>
> **XXII**
> 9*H*-Fluorene
> or Fluorene
>
> **XXIII**
> 9*H*-Fluoren-9-one
> or 9-Fluorenone

> **BOX 10.6**
>
> **SELECTED IUPAC NOMENCLATURE RULES FOR S-ATOM POLYCYCLIC AROMATIC COMPOUNDS**
>
> The IUPAC nomenclature of S-atom PACs is also similar to that for N-PACs. Thus the "trivial" name thiophene (**XIV**) is a basis for the common names of S-heterocycles containing one endocyclic S-atom in a 5-membered fused ring. Addition of two benzo groups gives dibenzothiophene (**XV**):
>
> **XIV**
> Thiophene
>
> **XV**
> Dibenzothiophene
>
> Names of S-heterocycles with two S-atoms in a 6-membered ring are based on the thianthrene structure, **XVI** (Later et al., 1990):
>
> **XVI**
> Thianthrene
>
> Examples of common names and structures of several S-PACs and nitro-S-PACs identified in combustion-generated emissions (including tobacco smoke condensate) and ambient air are given in Table 10.5.

4. Gas–Particle Partitioning, Sampling Techniques, and Ambient Levels of Selected PAHs and PACs

Whether or not a given PAH exists virtually entirely in the gas phase or in the particle phase, or is partitioned between them, is a critical factor in determining its physical and chemical fates in ambient air and in subsequent intra- and intermedia transport through our air/water/soil environments. This is true not only for physical processes such as wet and dry deposition but also for their chemical reactivity, lifetimes, and fates in VOC–NO_x systems characteristic of polluted airsheds. For example, the homogeneous gas-phase reactions of pyrene and fluoranthene differ dramatically from the rates, mechanisms, and products of their particle-associated heterogeneous reactions (Sections E and F).

Similarly, the toxicological effects may depend on whether the compound is inhaled as a gas, inhaled as a particle, or adsorbed/absorbed on the surfaces of coexisting particles. The distribution between these forms varies with the particular compound as well as with a number of parameters such as temperature. For example, in hot weather in ambient air in Claremont and Riverside, California, up to 25% of the PAC 2-nitro-6H-dibenzo[b,d]pyran-6-one (structure **XI**; vide supra) was in the gas phase, rather than being adsorbed/absorbed on surfaces (Arey et al., 1994). This nitrophenanthrene lactone, formed in a gas-phase atmospheric reaction of phenanthrene, is widely distributed in urban ambient air and is a powerful bacterial mutagen (Helmig et al., 1992a, 1992b; Arey et al., 1992), as well as being mutagenic in the MCL-5 human cell assay (Sasaki et al., 1997b).

Gas–particle partitioning of semivolatile organics, including PAHs, is discussed in more detail in Chapter 9.D. The reader is also referred to books by Mackay et al. (1992), Schwarzenbach et al. (1993), and Baum (1998) and articles by Pupp et al. (1974), Junge (1977), Cautreels and Van Cauwenberghe (1978), Yamasaki et al. (1982), Pankow (1987, 1991, 1994a, 1994b, 1998), Pankow et al. (1993, 1994), Bidleman and Foreman (1987), Bidleman (1988), Ligocki and Pankow (1989), Foreman and Bidleman (1990), Pankow and Bidleman (1992), Storey et al. (1995), Wilson et al. (1995), Liang and Pankow (1996), Jang et al. (1997b), Finizio et al. (1997a, 1997b), Liang et al. (1997), Chen and Preston (1997), Goss (1997), Goss and Schwarzenbach (1998), Harner and Bidleman (1998), Jang and Kamens (1998), Mackay and Callcott (1998), and Feilberg et al. (1999a).

We now consider sampling techniques and typical ambient concentrations of PAHs and PACs.

TABLE 10.6 Vapor Pressures, Subcooled Liquid Vapor Pressures (p_L^o), Solubilities in Water (S, C^S), Logarithm of Octanol/Water Partition Coefficient (log K_{OW}), Calculated Henry's Law Constant (H), and Calculated Logarithm of the Octanol-Air Partition Coefficient (log K_{OA}) at 25°C for Selected PAHs of Atmospheric Interest[a,e]

PAH[b]	Vapor pressure				Solubility		log K_{OW}	Calculated H^d (Pa m³ mol⁻¹)	Calculated log K_{OA}[c]
	P		p_L^o		S	C^S			
	Pa	Torr	Pa	Torr	(mg/L)	(mmol/m³)			
Naphthalene	10.4	7.8×10^{-2}	36.8	2.76×10^{-1}	31	241	3.37	43	5.13[e]
1-Methylnaphthalene	8.84	6.63×10^{-2}	8.8	6.6×10^{-2}	28	197	3.87	45	5.61[e]
2-Methylnaphthalene	9	7×10^{-2}	11.2	8.4×10^{-2}	25	176	3.86	51	—
Biphenyl	1.3	9.8×10^{-3}	3.7	2.8×10^{-2}	7	45	3.90	29	—
Acenaphthene	3×10^{-1}	2×10^{-3}	1.5	1.1×10^{-3}	3.80	24.6	4.00[e]	12.17[e]	6.23[e]
Acenaphthylene	9×10^{-1}	6.8×10^{-3}	4.1	3.1×10^{-2}	16.1	107	4.00	8.4	—
Fluorene	9×10^{-2}	7×10^{-4}	7.2×10^{-1}	5.4×10^{-3}	1.90	11.4	4.18	7.87[e]	6.68
Phenanthrene	2×10^{-2}	1.5×10^{-4}	1.1×10^{-1}	8×10^{-4}	1.10	6.17	4.57	3.24[e]	7.45
Anthracene	1×10^{-3}	8×10^{-6}	7.78×10^{-2}	5.84×10^{-4}	0.045	0.253	4.54	3.96[e]	7.34
Pyrene	6.0×10^{-4}	4.5×10^{-6}	1.19×10^{-2}	8.93×10^{-5}	0.132	0.652	5.18	0.92	8.43[e]
Fluoranthene	1.23×10^{-3}	9.2×10^{-6}	8.72×10^{-3}	6.55×10^{-5}	0.26	1.19	5.22	1.04	8.60
Chrysene	5.7×10^{-7}	4.3×10^{-9}	1.07×10^{-4}	8.03×10^{-7}	2×10^{-3e}	—	5.75[e]	0.0122[e]	10.44
Triphenylene	2.3×10^{-6}	1.7×10^{-8}	1.21×10^{-4}	9.1×10^{-7}	0.043	0.188	5.49	0.012	—
B[a]A	2.80×10^{-5}	2.1×0^{-7}	6.06×10^{-4}	4.55×10^{-6}	0.011	0.048	5.91	0.581[e]	10.80[e]
Benzo[a]pyrene	7.0×10^{-7}	5.3×10^{-9}	2.13×10^{-5}	1.60×10^{-7}	0.0038	0.0151	6.04	0.046	10.71[e]
Benzo[e]pyrene	7.4×10^{-7}	5.6×10^{-9}	2.41×10^{-5}	1.81×10^{-7}	0.004	0.016	—	0.020	11.13
Perylene	1.4×10^{-8}	1.1×10^{-10}	—	—	0.0004	0.0016	6.25	0.003	11.70[e]
B[b]F	—	—	—	—	0.0015	0.0060	5.80	—	—
B[k]F	5.2×10^{-8}	3.9×10^{-10}	4.12×10^{-6}	3.09×10^{-8}	0.0008	0.0032	6.00	0.016	11.19
B[ghi]P	—	—	2.25×10^{-5}	1.69×10^{-7}	0.00026	0.00097	6.50	0.075	—
DB[a,h]A	3.7×10^{-10}	2.8×10^{-12}	9.2×10^{-8}	6.9×10^{-10}	0.0006	0.0022	6.75	0.00017[e]	13.91[e]
Coronene	2.0×10^{-10}	1.5×10^{-12}	—	—	0.00014	0.00047	6.75	—	—

[a] Adapted from Mackay et al. (1992). There are their "selected" (i.e., "best") values given in their Summary Table 2.2 (some were rounded off). See also Sonnefeld et al. (1983) for "conventional" vapor pressures.

[b] Abbreviations: B[a]A, benz[a]anthracene; B[b]F, benzo[b]fluoranthene; B[k]F, benzo[k]fluoranthene; B[ghi]P, benzo[ghi]perylene; DB[a,h]A, dibenz[a,h]anthracene.

[c] Source: Finizio et al. (1997).

[d] This H is for Henry's law expressed as $H = P/[PAH]$, where P is the gas-phase concentration in pascals and [PAH] is the liquid-phase concentration in moles per cubic meter. Traditionally in atmospheric chemistry, Henry's law is expressed as $H = [X]/P_X$.

[e] Source: Table 1, Mackay and Callcott (1998).

a. PAHs

Sampling quantitatively each compound in the spectrum of relevant PAHs present in urban ambient air is challenging because their conventional vapor pressures cover a range of some 10^{10} Torr, e.g., $\sim 8 \times 10^{-2}$ for naphthalene vs 1.5×10^{-12} for coronene (see Table 10.6).

Furthermore, the concentrations of the gas phase 2- and 3-ring PAHs are generally far higher than those of the 5- and 6-ring particle-phase species. Thus, as seen in Fig. 10.2, average concentrations in urban southern California air at four sites (shown in Figure 10.23) were ~ 6000, 30, and 50 ng m⁻³ for gas phase naphthalene, fluorene, and phenanthrene, compared to ~ 0.14, 0.29, and 0.77 ng m⁻³ for BaP, indeno[1,2,3-cd]pyrene and benzo[ghi]perylene, respectively in the particle phase (Fraser et al., 1998). Although the temperatures were quite high during this period (27°C day and 22°C night), similar relative concentrations of gas-phase vs particle-phase PAHs have also been seen in other studies at lower temperatures. For example, Fig. 10.3 shows measured PAHs in Chicago, Illinois, during February/March 1995 when the mean day/night temperatures were ~ 1°C (Harner and Bidleman, 1998). As seen in Fig. 10.3, $\sim 96\%$ of the fluorene and phenanthrene were in the gas phase (naphthalene was not sampled) and accounted for $\sim 56\%$ of the total mass of the measured PAHs.

Table 10.8 summarizes some measured concentrations of various PAHs in a variety of locations as well as the percentage of each found in the particle phase. Consistent with the data in Figs. 10.2 and 10.3, the percentage found in the particle phase increases with the size of the PAH. For example, phenanthrene con-

TABLE 10.7 Vapor Pressures (Pascals)[a] as a Function of Temperature between 283.15 and 323.15 K for Several Polycyclic Aromatic Hydrocarbons[b]

Compound	$\log P° = -A/T + B$	
	A	B
Naphthalene	3960.03 ± 58.9	14.299 ± 0.200
Naphthalene-d_8	3689.52 ± 27.6	13.392 ± 0.09
Acenaphthylene	3821.55 ± 23.7	12.768 ± 0.079
Acenaphthene	4535.39 ± 47.3	14.669 ± 0.159
Fluorene	4616.07 ± 30.3	14.385 ± 0.101
Phenanthrene	4962.77 ± 32.8	14.852 ± 0.109
Phenanthrene-d_{10}	4704.13 ± 31.5	14.060 ± 0.11
Anthracene	4791.87 ± 50.3	12.977 ± 0.170
Fluoranthene	4415.56 ± 46.2	11.901 ± 0.155
Benz[a]anthracene	4246.51 ± 132.2	9.684 ± 0.431
Pyrene	4760.73 ± 26.2	12.748 ± 0.087

[a] 1 Pa = 7.5×10^{-3} mm.
[b] Adapted from Sonnefeld et al. (1983).

centrations range from 5.1 to 78 ng m^{-3}, with 0–4.5% in the particle phase, whereas for fluoranthene, they range from ~3.5 to 12.4 ng m^{-3}, with 0.7–23% reported in the particle phase.

As discussed in Chapter 9.D, another important factor determining the gas–particle partitioning is the mass of total suspended particulate matter (TSP) in the air parcel being sampled, and the size and chemical and physical properties of their surfaces. For example, Kamens and co-workers (1995) calculated the effects (Table 10.9) of PAH vapor pressure and the size and concentrations of coexisting aerosols on the gas–particle partitioning of phenanthrene (3 rings), pyrene (4 rings), and benzo[a]pyrene (5-rings). They concluded, for example, that the percentage of pyrene in the particle phase would increase from ~3 to 77% in going from an air mass at 25°C containing 10 μg m^{-3} of 0.5-μg-size aerosols to one with 500 μg m^{-3} of 0.25-μm-size aerosols.

There is an important caveat. These, and other experimental values cited for the gas- and particle-phase concentrations of semivolatile PAHs, do not necessarily represent their actual equilibrium concentrations in the ambient air being sampled. Thus, there can be both positive and negative artifacts associated with the sampling methodologies. Furthermore, sorption equilibrium may not be achieved in the polluted ambient air.

Let us now briefly consider experimental methods for sampling, beginning with the gas phase. Naphthalene, the methylnaphthalenes, and other abundant and highly volatile 2-ring PAHs are quantitatively trapped with Tenax-GC or Tenax-TA solid absorbents (e.g., see Arey et al., 1989a; Baek et al., 1992; and Zielinska et

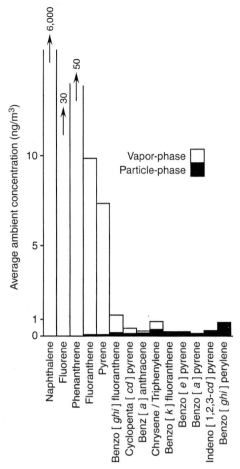

FIGURE 10.2 Average ambient concentrations (ng m^{-3}) of gas- and particle-phase PAHs collected at four urban sampling sites across the Los Angeles Air Basin for eight 4-h sampling periods over two hot days, September 8–9, 1993, during a heavy smog episode (peak ozone 290 ppb at Claremont, California). For concentration data and site locations, see Table 10.8 and Fig. 10.23. Naphthalene was collected in a VOC stainless steel canister; 3- to 6-ring PAHs were sampled independently using a quartz fiber filter/PUF plug, Hi-Vol dichotomous virtual impactor system similar to that shown in Fig. 10.4 (adapted from Fraser et al., 1998).

al., 1996). For example, Arey and co-workers (1989a) have used Tenax-GC in "low-flow" (~1 L min^{-1}) and "high-flow" cartridges (~10 L min^{-1}). Naphthalene was sampled with the low-flow cartridge because of its great abundance (e.g., the highest naphthalene concentration in a 12-h sample was 6100 ng m^{-3} = 1.1 ppb), and the remaining seven PAHs with a high-flow cartridge.

Table 10.10 gives the concentrations of volatile PAHs averaged over a 9-day event in August 1986, with nine 12-h daytime and nine 12-h nighttime sample collection periods. The sampling site was less than 1 km from a major freeway so that these data are characteristic of air parcels heavily impacted by motor vehicles. The nighttime average concentrations for naphthalene were

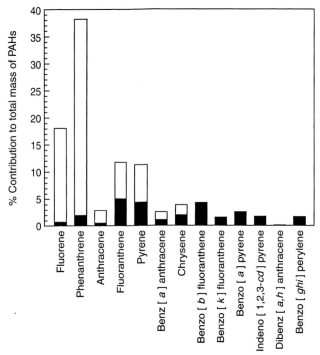

FIGURE 10.3 Average mass percent contributions (gas + particle phase) of 13 PAHs collected in urban Chicago air during 15 sampling events in February/March 1995. Open bars, approximate percentage in gas phase; solid bars, approximate percentage in particle phase. 2-Ring PAHs not determined. Average percent of total mass in the particle phase was 4.4 and 42.9 for 3- and 4-ring PAHs, respectively. Averages: ΣPAHs = 58 ng m^{-3}; TSP = 66 μg m^{-3}; temperature 1°C (range 0 to −4°C). Sampling apparatus, double glass fiber filters and two PUF plugs (adapted from Harner and Bidleman, 1998).

collect naphthalene and the 1- and 2-methylnaphthalenes and several monoaromatics, e.g., benzene and toluene. Semivolatile 3- and 4-ring PAHs were sampled with a glass fiber filter (GFF)/PUF system (see later).

Naphthalene concentrations averaged 6000 ng m^{-3} and ranged from 0.0 to 22,600 ng m^{-3}, consistent with the measurements of Arey *et al.* (1989a) discussed earlier (see Table 10.10).

However, Fraser *et al.* (1998) noted that this canister sampling technique may underestimate the methylnaphthalenes. Thus, their methylnaphthalenes/naphthalene ratios were lower than those obtained by Arey *et al.* (1989a) using Tenax-GC solid adsorbent. This may be due to significant adsorption of the methyl isomers to the canister (Arey, personal communication). Zielinska and co-workers (1996) evaluated measurement methods for VOCs up to C$_{20}$ emitted from motor vehicles and reported that C$_8$–C$_{12}$ hydrocarbons were more stable on the Tenax cartridge than in canisters. Similar problems with canister sampling for organics are discussed in Chapter 11.A.4e.

A variety of filters have been used to collect 5- and 6-ring particulate PAHs. These include glass and quartz fiber, GFF and QFF (e.g., Ligocki *et al.*, 1989; Fraser *et al.*, 1998), Teflon membrane, TM (e.g., Ligocki *et al.*, 1989; Smith and Harrison, 1996), and Teflon-impregnated glass fiber filters, TIGF (e.g., Arey *et al.*, 1987; Feilberg *et al.*, 1999a).

A common type of apparatus for determining total PAH and PAC concentrations and their gas–particle partitioning that has found widespread use is shown in Fig. 10.4 and is illustrated by studies such as those by Cautreels and Van Cauwenberghe (1978), Thrane and Mikalsen (1981), Yamasaki *et al.* (1982), Atkinson *et al.* (1988a), Arey *et al.* (1989a, 1992), Harger *et al.* (1992), Sasaki *et al.* (1995), Gupta *et al.* (1996), and Fraser *et al.* (1998). A Teflon-impregnated glass fiber filter that collects particle-associated PAHs is followed by three glass cylinders packed with polyurethane foam ("PUF plugs") that trap the gas-phase PAHs that pass through the particle filter. The filter and the PUF plugs are extracted by organic solvents (e.g., CH$_2$Cl$_2$) and the particle-phase and gas-phase PAHs (and PACs) identified and quantified by various techniques. Generally, a conventional Hi-Vol sampler draws the polluted ambient air through the apparatus; however, given the availability of highly sensitive analytical techniques and microbiological assays (e.g., the "microsuspension modification" to the Ames *Salmonella typhimurium* bacterial assay; see Section C.2.c), medium- and low-volume samplers are often used as alternatives.

Two to four PUF plugs are typically used in the sampling train, depending on sampling conditions (e.g., ambient temperature) and the volatility of the PAHs

consistently higher than for the daytime periods, ranging from 3000 to 6100 ng m^{-3} (overall average 4300) at night compared to 2000–4300 (overall average 3100 during the day). This was also true for the methylnaphthalenes and the six other volatile PAHs. Concentrations of the 2-methyl isomer were always significantly higher than those of the 1-methyl isomer, averaging 240 vs 130 ng m^{-3} daytime and 710 vs 370 ng m^{-3} nighttime.

Concentrations of the gaseous nitroarenes, 1- and 2-nitronaphthalene, and 3-nitrobiphenyl in these 18 samples were determined concurrently with the PAHs, in this case sampling with a Teflon-impregnated glass fiber filter (TIGF)/polyurethane foam (PUF) apparatus (Fig. 10.4). The PUF plugs quantitatively trapped the gas-phase nitroaromatics, but not naphthalene and biphenyl and several 3-ring PAHs.

Cass and Fraser (private communication) and Fraser *et al.* (1998) have used another technique, collection in an internally electropolished stainless steel canister, to

TABLE 10.8 Sum of the Particle-Associated (Filter) and Gas-Phase (Solid Adsorbent, PUF Plugs[f] or Tenax Cartridges[g]) Concentrations of EPA Priority PAH Pollutants and the Percentage of Each PAH in the Particle Phase

Polycyclic aromatic hydrocarbon, formula, and molecular weight	Total concentration (filter + solid adsorbent, ng m^{-3}) and percentage (in parentheses) of specific PAHs in particle phase				
	Birmingham University, U.K.[a]	London, U.K.[b]	Los Angeles Air Basin[c]	Torrance, CA[d]	Portland, OR[e]
Naphthalene, C_8H_{10}, 128	–	–	6000[b] (0)	3300[g] (0)	–
Acenaphthylene, $C_{12}H_8$, 152	15.4 (4.0)	–	15.6 (0)	–	32.0 (0.1)
Acenaphthene, $C_{12}H_{10}$, 154	13.5 (11.9)	–	–	–	–
Fluorene, $C_{13}H_{10}$, 166	13.7 (7.7)	–	29.8 (0)	–	11.1 (0.6)
Phenanthrene, $C_{14}H_{10}$, 178	24.1 (4.5)	5.12 (2.1)	50.3 (0)	78[k] (0.4)	26.3 (1.0)
Anthracene, $C_{14}H_{10}$, 178	4.49 (8.7)	2.84 (6.3)	3.04 (0)	6.1 (0.5)	3.44 (0.9)
Fluoranthene, $C_{16}H_{10}$, 202	12.4 (9.4)	3.46 (23)	9.85 (0.7)	8.0 (5.8)	8.42 (4.9)
Pyrene, $C_{16}H_{10}$, 202	38.0 (6.2)	3.79 (21)	7.30 (1.0)	8.0 (7.5)	7.33 (7.1)
Cyclopenta[cd]pyrene,[l] $C_{18}H_{10}$, 226	–	4.08 (64)	0.40 (35)	–	–
Benz[a]anthracene, $C_{18}H_{12}$, 228	5.59 (26)	1.41 (56)	0.25 (60)	–	1.57 (73)
Chrysene, $C_{18}H_{12}$, 228	6.49 (34)	1.62 (75)	0.78[i] (44)	–	2.06 (65)
Benzo[b]fluoranthene, $C_{20}H_{12}$, 252	2.15 (87)	1.78 (90)	–	–	3.71[j] (96)
Benzo[k]fluoranthene, $C_{20}H_{12}$, 252	1.20 (93)	0.75 (91)	0.22 (100)	–	–
Benzo[e]pyrene,[l] $C_{20}H_{12}$, 252	–	2.31 (87)	0.22 (100)	2.1 (100)	–
Benzo[a]pyrene, $C_{20}H_{12}$, 252	0.81 (90)	1.63 (88)	0.14 (100)	0.59 (98)	–
Benzo[ghi]perylene, $C_{22}H_{12}$, 276	1.97 (97)	3.31 (100)	0.77 (100)	–	–
Indeno[1,2,3-cd]pyrene, $C_{22}H_{12}$, 276	1.96 (100)	1.57 (100)	0.29 (100)	–	–
Dibenz[a,h]anthracene, $C_{22}H_{14}$, 278	0.83 (94)	0.12 (100)	–	–	–

[a] From Smith and Harrison (1996). Sampler elevation 15 m, situated ~300 m from a high traffic "spine roadway." Particle phase collected on Teflon membrane filter (TMF); gas-phase species trapped downstream on PUF plugs. Daily 24-h samples collected during February 1992; mean TSP during that period 60 μg m^{-3}.

[b] From Baek et al. (1992). Annual average concentrations derived from 48 week-long curbside samples collected throughout 1987 at an urban, traffic-dominated site in south Kensington, London. Sampler elevation 5 m; particle-associated PAHs collected on a PTFE filter (Millipore Ltd.) and gaseous PAHs on a cartridge containing Tenax-TA adsorbent.

[c] From Fraser et al. (1998). Average concentrations of eight 4-h samples collected day and night at four sites in southern California (see Fig. 10.2) during an intense two-day photochemical smog episode, September 8–9, 1993 (1-h average O_3 peaked at 0.29 ppm at Claremont, California). Volatile PAHs collected in stainless steel canisters, with the remainder collected with a quartz fiber filter/PUF solid adsorbent system. Average daytime and nighttime temperatures (°C): Long Beach, 22.2 (19.6); downtown Los Angeles, 25.8 (20.6); Azusa, 28.5 (22.9); Claremont, 30.4 (23.1) (Cass and Fraser, personal communication, 1999).

[d] From Arey et al. (1987). Site, a roof (9-m elevation) at El Camino Community College in Torrance, California (20 km south of central Los Angeles). Daytime 12-h sampling period, 0600 to 1800 hours, February 25, 1986; maximum temperature 35°C at 1100 hours. Sampling systems: Tenax-GC cartridge for naphthalene and phenanthrene; Teflon-impregnated glass fiber filter (TIGF) and PUF plugs for phenanthrene and heavier PAHs (see diagram, Fig. 10.4).

[e] From Ligocki and Pankow (1989). One sampling system had a glass fiber filter (GFF) and backup PUF plugs. Others employed either two GFFs or a single Teflon membrane filter (TMF); both had downstream PUF plugs. See original article for experimental details. Samples collected during February and April 1984 and February and April 1985; overall mean sampling temperatures were 8 and 5°C, respectively. Sampling periods 5–30 h. For gas- and particle-phase concentrations of those PACs that were copollutants with these PAHs, see Table 10.11.

[f] Gas-phase PAHs collected on PUF plugs unless otherwise noted (e.g., on Tenax-GC or Tenax-TA adsorbents).

[g] Collected on a Tenax-filled cartridge.

[h] Collected in a canister sampler.

[i] Chrysene/triphenylene.

[j] Benzo[b + j + k]fluoranthene.

[k] Same value for collection on Tenax-GC cartridge and the sum of the concentrations of phenanthrene trapped on three PUF plugs in series.

[l] Not EPA priority pollutants.

TABLE 10.9 Calculated Effect of Vapor Pressure and Aerosol Concentration on the Percentage in the Aerosol Phase of Phenanthrene, Pyrene, and Benzo[a]pyrene[a]

Aerosol concentration[c] ($\mu g\ m^{-3}$)	Aerosol size (μm)	Percentage of PAH in particle phase		
		Phenanthrene (5.3×10^{-4})[b]	Pyrene (3.3×10^{-5})[b]	Benzo[a]pyrene (9×10^{-8})[b]
10	0.5	0.2	3	93
100	0.35	3.0	32	99
500	0.25	17	77	100

[a] Adapted from Kamens et al. (1995).
[b] Subcooled liquid vapor pressures in Torr at 25°C calculated by Kamens et al. (1995) from Hawthorne et al. (1989).
[c] Concentration of aerosol, $\mu g\ m^{-3}$; particle density 1.25.

and PACs being sampled. However, Arey and co-workers (1989a) found that even with four PUF plugs, fluorene (and certainly not naphthalene and the methylnaphthalenes) was not collected quantitatively and some breakthrough of phenanthrene also occurred, especially during daytime sampling periods. Thus seven of the volatile 2- and 3-ring PAHs were collected on Tenax-GC high-flow cartridges, and naphthalene, because of its very high concentrations (up to 6100 ng m^{-3}), on a low-flow cartridge. Tenax-TA solid adsorbent also effectively traps these volatile PAHs (Zielinska et al., 1996).

Because of these sampling difficulties, the reported gas-phase concentrations of the semivolatile 3- and 4-ring PAHs are "operationally defined"; i.e., they are the quantity of a given PAH extracted from the solid adsorbent PUF plugs (or other solid sorbents such as Tenax cartridges). Similarly, the 5- and 6-ring PAHs along with that fraction of the semivolatile PAHs collected on the "upstream" filter are "operationally" defined as being in the particle phase.

We emphasize the term "operational" when describing the gas- and particle-phase concentrations of PAHs sampled using this popular filter/sorbent technique, because it has been recognized for some time that this and other methods are subject to both positive and negative artifacts (e.g., see Van Vaeck and Van Cauwenberghe, 1984; Fitz et al., 1984; Coutant et al., 1988; Baek et al., 1991b; Kaupp and Umlauf, 1992; Hart and Pankow, 1994; Kamens and Coe, 1997; and Feilberg et al., 1999a). These artifacts may lead to significant deviations of the measured gas and particle concentrations of semivolatile organics (SOCs) from their "true" equilibrium concentrations used in theoretical treatments of the gas-particle partitioning of PAHs (see Chapter 9.D and references cited above).

As examples, the term "blow off" refers to a phenomenon in which the pressure drops across a filter can cause particle-associated SOCs (e.g., 3- and 4-ring PAHs) to be stripped from the collected particulate matter and trapped downstream on the PUF sorbent, leading to an underestimate of their true particle-phase concentrations and an overestimate of their gas-phase levels. However, several studies (e.g., McDow and Huntziker, 1990; Turpin et al., 1994) suggest that this is less of a problem than adsorption of organics; see following discussion).

Conversely, gaseous semivolatile organics can adsorb to filter surfaces, resulting in artificially overestimating their particle-phase concentrations and underestimating their gas-phase levels. For example, Hart and Pankow (1994) conducted a study using two filter/PUF sorbent samplers similar to the apparatus in Fig. 10.4 but each incorporating a "backup" filter immediately downstream from the regular filters used to collect particle-phase PAHs. They found that for a typical sampling event in ambient air, gas adsorption to a single quartz fiber filter in a conventional Hi-Vol sampler could shift the apparent gas-particle partitioning parameter in favor of the particle phase, leading to overestimations by factors ranging from ~1.2 to 1.6. However, a second sampler operating in parallel and using a single Teflon membrane filter gave gas-particle distributions approximately the same as the corrected quartz fiber filter values.

Gas-particle partitioning is also impacted by changes in ambient temperature and aerosol concentrations during sampling, resulting in artifacts in the relative particle- and gas-phase concentrations (see Hart and Pankow, 1994).

A sampling system designed to reduce artifact problems was employed by Liang and co-workers (1997) in an experimental chamber study of the gas-particle partitioning of several PAHs (and n-alkanes) on three types of model aerosols and ambient urban particulate matter. The system consisted of a dual glass fiber filter system as described by Hart and Pankow (1994), followed by two parallel sampling trains for trapping gas-phase species. One had two sequential PUF plugs and the other had two Tenax-GC cartridges in sequence to trap the more volatile species.

An alternative approach is that using denuder-based samplers for determining gas–particle partitioning of semivolatile PAHs and other organics (see Feilberg et al., 1999a; Chapter 11.A.3b has a description of denuders.) In principle, gas-phase PAHs in ambient air are first trapped on a denuder surface coated with a sorbent material (e.g., a resin); particles pass through and are collected on a follow-up filter. In practice, experimental problems exist, for example, in extracting quantitatively species from the sorbent coating on the denuder walls. Gundel and co-workers (1995a) addressed

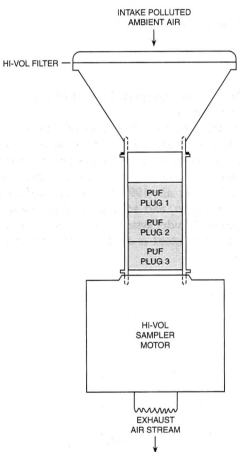

FIGURE 10.4 Diagram of a sampling apparatus used to collect 3- to 6-ring PAHs at a variety of sites in California; similar systems have been used by other researchers throughout the world since circa 1980. Particle-phase PAHs are collected on a Teflon-impregnated glass fiber filter (TIGF) labeled as "Hi-Vol Filter," and gas-phase PAHs are trapped on three polyurethane foam (PUF) plugs (in sequence) downstream from the TIGF. A conventional Hi-Vol sampler pulls ambient air through the TIGF/PUF plug apparatus (adapted from Atkinson et al., 1988a; see, e.g., Arey et al. (1989a), Harger et al. (1992), and Sasaki et al. (1995) for examples of its use in laboratory as well as field studies).

TABLE 10.10 Average Daytime (0800 to 2000 hours PDT) and Nighttime (2000 to 0800 hours PDT) Concentrations of Volatile 2- and 3-Ring PAHs Collected over Eighteen 12-h Sampling Periods (9 Daytime, 9 Nighttime) on Tenax-GC Solid Adsorbent at Glendora, California, August 12–21, 1986[a]

Day/night	Average concentration (ng m^{-3})							
	Naphthalene	1-Methylnaphthalene	2-Methylnaphthalene	Acenaphthylene	Biphenyl	Acenaphthlene	Fluorene	Phenanthrene
Daytime	3100	130	240	3	55	5	28	23
Nighttime	4300	370	710	12	100	25	44	48

[a] Adapted from Arey et al. (1989a). Sampling conditions were hot weather and moderate photochemical air pollution.

TABLE 10.11 Average Gas- and Particle-Phase Concentrations of Several O-Atom and One S-Atom PACs and the Percentage of a Specific PAC in the Particle Phase in Ambient Air in Portland, Oregon, February and April 1984 and 1985,[a,d,e] and the Los Angeles Air Basin, September 1993[b,d,e]

Polycyclic aromatic compound	Gas phase (ng m^{-3})		Particle phase (ng m^{-3})		Σ(Gas + particle) (ng m^{-3})		Percentage in particle phase	
	Portland, OR	L.A. air basin	Portland, OR	L.A. air basin	Portland, OR	L.A. air basin	Portland, OR	L.A. air basin
Dibenzofuran	19	20.0	0.10	ND[c]	19.1	20.0	0.5	0.0
9-Fluorenone	7.0	4.10	0.14	0.29	7.14	4.39	2.0	6.6
Dibenzothiophene	1.8	6.34	0.039	ND	1.84	6.34	2.1	0.0
9,10-Anthracenedione	2.5	2.52	0.59	0.36	3.09	2.88	19	13
7H-Benz[de]anthracenone	0.067	ND	1.7	0.20	1.77	0.20	96	100

[a] From Ligocki and Pankow (1989).
[b] From Fraser et al. (1998).
[c] ND = not detected.
[d] For sampling conditions, see Table 10.8.
[e] For sampling conditions, concentrations, and gas–particle distributions of copollutant PAHs, see Table 10.8.

this problem in studies of PAHs in indoor air by depositing a thin film of ground particles of the adsorbent resin XAD-4 on sandblasted surfaces of annular glass denuders to strip the gas-phase PAHs from the incoming Low-Vol airstream. Particle-phase PAHs passed through the denuder and were collected on a TIGF filter. A "postfilter" denuder trapped any volatile PAHs blown off from this primary filter. Similarly, Kamens and co-workers (1995) employed such an apparatus in environmental chamber studies of "time to equilibrium" in gas–particle partitioning of several PAHs (see later).

As is the case for other types of sampling systems, artifacts can occur, e.g., evaporation of semivolatile PAHs from the particles as the gas phase is removed by the denuder surface (e.g., see Kamens and Coe, 1997; and Feilberg et al., 1999a). For information on the design, operation, and accuracy and precision of various types of denuder systems, see Chapter 11.A.3b and, for example, Coutant et al. (1988, 1989, 1992), Eatough

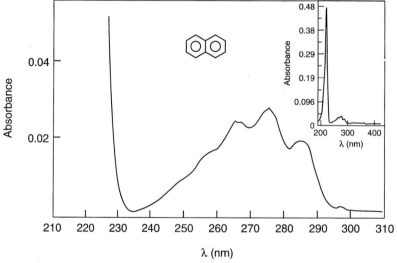

FIGURE 10.5 UV absorption spectrum of naphthalene ($\approx 8 \times 10^{-6}$ M in cyclohexane) (spectrum taken by Alisa Ezell, 1998).

et al. (1993), Turpin *et al.* (1993), Gundel *et al.* (1995a), Kamens and Coe (1997), and Feilberg *et al.* (1999a).

b. PACs

Compared to PAHs, much less has been published about the gas–particle distributions of PACs in ambient air. Table 10.11 gives average gas- and particle-phase concentrations, and their percentage in the particle phase, of several O-PAC and a widely distributed S-PAC, dibenzothiophene, determined at Portland, Oregon (Ligocki and Pankow, 1989), and in the Los Angeles area (Fraser *et al.*, 1998). Again, an increase in the percentage found in the particle phase is seen with the larger compounds.

In summary, the smallest PAHs and PACs are found primarily in the gas phase and the largest compounds in the particle phase. Those of intermediate sizes partition between the gas and particle phases, with the distribution determined by a number of factors such as the temperature and mass and size of available particles to absorb/adsorb the PAH or PAC (see Chapter 9.D).

5. Absorption and Emission Spectra of Selected PAHs and PACs

As seen in Figs. 10.5–10.11, polycyclic aromatic hydrocarbons absorb in the actinic UV, $\lambda > 290$ nm (Karcher *et al.*, 1985). Their $\pi \to \pi^*$ transitions are strong (much more intense than the corresponding $n \to \pi^*$ transitions in aromatic carbonyl compounds), so PAHs have relatively large molar extinction coefficients, ε (the exception is naphthalene, where the $\pi \to \pi^*$ transition is "forbidden").

PAHs also generally have well-structured emission spectra (see Figs. 10.6–10.10) and relatively large fluorescence quantum yields. For example, in degassed *n*-heptane at room temperature, the fluorescence quantum yields are as follows: fluoranthene, 0.35; benz[*a*]anthracene, 0.23; chrysene, 0.18; BaP, 0.60; BeP, 0.11; and benzo[*ghi*]perylene, 0.29 (Heinrich and Güsten,1980). Cyclopenta[*cd*]pyrene, however, does not fluoresce.

These large fluorescence quantum yields provide a sensitive method of analysis for PAHs. During the 1950s and 1960s, strongly emitting trace impurities were often a major source of experimental artifacts that could negate the advantage of much greater sensitivity (by factors of $\sim 10^2 - 10^3$) of fluorescence over UV–visible absorption spectroscopy for PAH analysis. Since then, major advances have been made in the separation procedures and in the spectroscopic detection, identification, and quantification of small amounts of individual PAHs. For example, Mahanama and co-workers (1994) used a combination of absorption and fluorescence spectroscopy to identify and quantify concentrations of key PAHs in simulated and real environmental tobacco smoke (ETS) as well as the NIST Standard Reference Material SRM 1649 "Air Particles" (see Box 10.3). Table 10.12 summarizes the programmed wavelengths selected for the excitation and fluorescence of the individual PAHs and the results of these studies for SRM 1649 (Mahanama *et al.*, 1994;

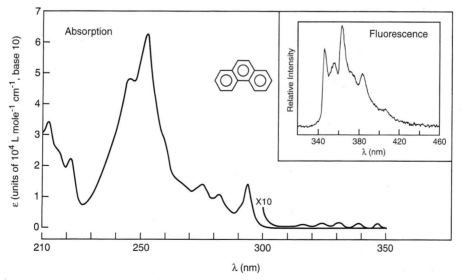

FIGURE 10.6 UV absorption and fluorescence spectra of phenanthrene in cyclohexane (adapted from Karcher *et al.*, 1985).

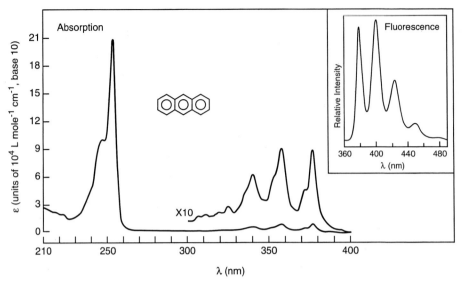

FIGURE 10.7 UV absorption and fluorescence spectra of anthracene in cyclohexane (adapted from Karcher et al., 1985).

Gundel et al., 1995b). Advantages of using this approach include high sensitivity (50 ng per gram of sample in these particular studies) and the ability to discriminate between compounds such as benzo[b]fluoranthene and benzo[k]fluoranthene, which is difficult by other techniques such as gas chromatography.

Solvents also affect the measured absorption spectra. Thus, the π,π^* bands shift to longer wavelengths (a "red" bathochromic shift) in polar solvents. For example, the long-wavelength band of anthracene shifts from ~375 nm in n-hexane to 381 nm in acetonitrile (Wehry, 1983).

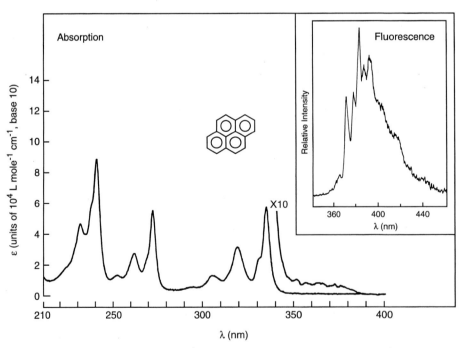

FIGURE 10.8 UV absorption and fluorescence spectra of pyrene in cyclohexane (adapted from Karcher et al., 1985).

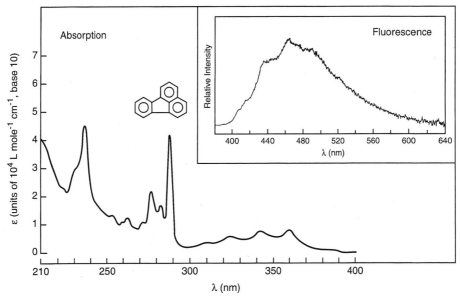

FIGURE 10.9 UV absorption and fluorescence spectra of fluoranthene in cyclohexane (adapted from Karcher et al., 1985).

Adding N-, O-, or S-atom functionalities to a PAH can cause major changes in its UV–visible absorption spectrum. For example, as seen in Fig. 10.12, addition of NO_2 groups to BaP to form the 1-, 3-, and 6-nitro isomers results in pronounced red shifts in their absorption spectra (Pitts et al., 1978). This enhanced ability to absorb solar radiation has significant implications with respect to the atmospheric reactions, lifetimes, and fates of PACs. Thus, as discussed in Section F, photolysis significantly exceeds OH radical attack as an "efficient" loss process for some gas-phase nitroarenes such as 1-nitronaphthalene (e.g., see Atkinson et al., 1989; and Feilberg et al., 1999a).

The physical state of a PAH also can have a dra-

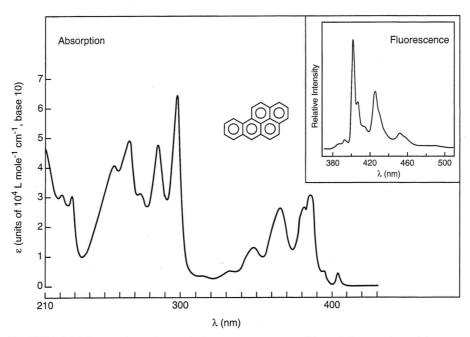

FIGURE 10.10 UV absorption and fluorescence spectra of benzo[a]pyrene in cyclohexane (adapted from Karcher et al., 1985).

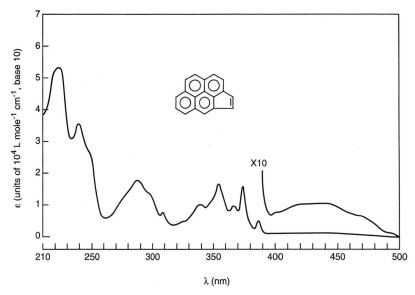

FIGURE 10.11 UV–visible absorption spectrum of cyclopenta[cd]pyrene in cyclohexane (adapted from Karcher et al., 1985). This PAH does not fluoresce.

TABLE 10.12 Concentrations of 10 PAHs in National Institute of Standards and Technology (NIST) Standard Reference Material SRM 1649 Air Particles Determined by Dual Programmable Fluorescence Detector Method Compared to Their NIST Reference Concentrations[a]

PAH	Wavelength change (at time in min)	Excitation wavelength (nm)	Emission wavelength (nm)	Concentration (µg/g)	
				Reference[b] (NIST)	Measured
Detector 1[c]					
Phenanthrene	0.0	250	370	4.5 ± 0.3	7.3 ± 0.6
Pyrene	11.5	235	380	6.3 ± 0.4	5.6 ± 1.0
Benz[a]anthracene	19.8	225	395	2.6 ± 0.3	2.8 ± 0.1
Benzo[a]pyrene	27.5	290	410	2.9 ± 0.5	2.8 ± 0.2
Indeno[1,2,3-cd]pyrene	30.3	245	480	3.3 ± 0.5	4.0 ± 0.1
Detector 2[c]					
Fluoranthene	0.0	230	450	7.1 ± 0.5	6.5 ± 0.7
Chrysene	19.4	260	370	3.5 ± 0.1	3.4 ± 0.1
Benzo[b]fluoranthene	22.9	230	430	6.2 ± 0.3	5.7 ± 0.3
Benzo[k]fluoranthene	22.9	230	430	2.0 ± 0.1	2.2 ± 0.1
Benzo[ghi]perylene	30.4	225	415	4.5 ± 1.1	3.4 ± 0.2

[a] Adapted from Mahanama et al. (1994).
[b] Certificate of analysis, Standard Reference Material 1649, Urban Dust/Organics (see Box 10.3).
[c] Detector 1 follows the column, detector 2 follows detector 1.

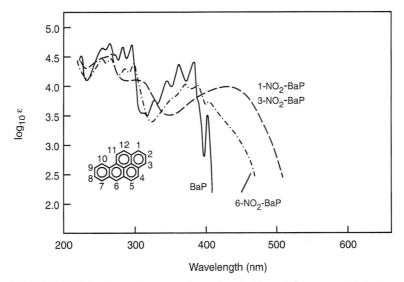

FIGURE 10.12 Absorption spectra in methanol of benzo[a]pyrene and its 1-, 3-, and 6-nitro derivatives (adapted from Pitts *et al.*, 1978).

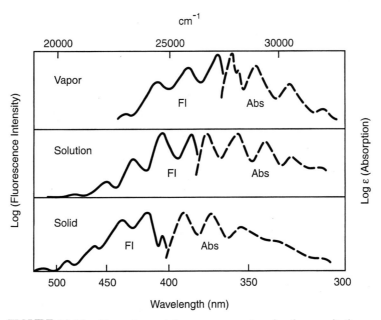

FIGURE 10.13 Absorption and fluorescence spectra of anthracene in three phases. Pronounced red shifts occur going from the vapor state to solution in dioxane to the solid state; separation of the 0–0 bands also increases (adapted from Bowen, 1946).

matic effect on its UV-visible absorption spectrum. Figure 10.13 shows the pronounced red shifts for anthracene for both absorption and fluorescence accompanying the phase changes from gas to solution to solid. This phenomenon, demonstrated by E. J. Bowen over a half century ago, is discussed in his classic monograph *Chemical Aspects of Light* (1946).

Infrared spectra of PAHs and PACs are also available, in part because of interest in PAHs in interstellar space (e.g., see Hudgins and Sandford, 1998a, 1998b, 1998c; and Langhoff et al., 1998). Their theoretical treatment and experimental infrared data may also prove useful in studies related to the troposphere.

For a detailed discussion of the fundamental aspects of the spectroscopy and photochemistry of PAHs and PACs, see monographs by Calvert and Pitts (1966), Turro (1978), Wayne (1988), and Gilbert and Baggott (1991).

B. BIOLOGICAL PROPERTIES OF PAHs AND PACs. I: CARCINOGENICITY

1. Historical Perspective: Benzo[a]pyrene, the "Classic Chemical Carcinogen"

In the latter part of the nineteenth century, workers in the paraffin refining, shale oil, and coal tar industries had high incidences of skin cancer. A possible cause emerged during the period 1915–1918 when Japanese scientists discovered that painting the ears of rabbits and mice with coal tar extracts produced tumors, some of which were malignant (Yamagiwa and Ichikawa, 1918).

The subsequent search for the "coal tar carcinogen" by chemists and medical researchers is classic (Phillips, 1983; Osborne and Crosby, 1987). Starting in 1922, research by a team of British chemists at the Institute of Cancer Research in London culminated in the synthesis in 1930 of the first pure chemical compounds to demonstrate carcinogenic activity, dibenz[a,h]anthracene (DBA), **XXV**, and its 3-methyl derivative (Kenneway, 1930; Kenneway, 1955, references therein). After distilling and fractionating *two tons* of pitch from a "Gas, Light, and Coke Company," they ultimately isolated several *grams* of two $C_{20}H_{12}$ polycyclic aromatic hydrocarbons. Three years later, Cook and co-workers (1933) synthesized for the first time benzo[a]pyrene and benzo[e]pyrene and proved them identical to these two "coal tar PAHs." Ultimate confirmation of the carcinogenicity of BaP came when all five survivors of a group of ten mice whose backs had been painted with synthetic BaP developed tumors; BeP was not carcinogenic (Cook et al., 1933). Subsequently, BaP was identified in carcinogenic extracts of ambient air particles (Leiter et al., 1942; Leiter and Shear, 1943; Waller, 1952; Kotin, 1954) and in chimney soot (Goulden and Tipper, 1949).

XXV

Dibenz[a,h]anthracene

Since then, long-term sampling data show that annual averages of BaP concentrations in major urban areas have dropped dramatically over a period of several decades. For example, at a roadside sampling site (1.5 m above ground level) located on Fleet Street, London, BaP fell from 39 ng m^{-3} in 1962–1963 to 10 ng m^{-3} in 1972–1973 (Commings and Hampton, 1976), similar to 2.0 ng m^{-3} in 1987 measured at a site 5 m above ground in Central London (Baek et al., 1992). (Note that concentrations in urban areas tend to decrease with sampling height, i.e., with distance from mobile sources.) Key reasons for the major decline in BaP concentrations in London and other urban airsheds throughout the world have been the enactment and enforcement of "clean air" legislation and the trend to cleaner fuels (see, for example, NRC, 1972, 1983; Hoffman and Wynder, 1977; Bjørseth, 1983; Grimmer, 1983a, 1983b; Holmberg and Ahlborg, 1983; Osborne and Crosby, 1987; Baek et al., 1992; California Air Resources Board, 1994, 1998; and references therein).

There is, however, a word of caution. Holmberg and Ahlborg point out in their 1983 *Consensus Report: Mutagenicity and Carcinogenicity of Car Exhaust and Coal Combustion Emissions*, "It should be stressed, however, that a reduction in the BaP level does not necessarily mean a reduction in the potential health hazards, since the spectrum of pollutants has also changed with time." For example, in their long-term study of BaP concentrations and the bacterial mutagenicity of ambient particles collected in Sapporo, Japan, from 1974 to 1992, Matsumoto and co-workers (1998) reported that BaP levels dropped 75–80%, but the level of overall bacterial mutagenicity remained "relatively unchanged." Similar concerns with focusing solely on BaP have been expressed by other researchers (e.g., see Lane and Katz, 1977; Pitts, 1983;

Rosenkranz and Mermelstein, 1985a; Tokiwa and Ohnishi, 1986; Lewtas, 1993a; Atkinson and Arey, 1994; Rosenkranz, 1996; Nielsen et al., 1996; Finlayson-Pitts and Pitts, 1997; and Collins et al., 1998).

However, BaP has often been used as a "marker" for POM to set air quality and emission standards (e.g., see Nielsen et al., 1995). For example, the Netherlands Environmental Programme 1988-1991 draft document gives an *"acceptable level"* for the annual average concentration of BaP in ambient air of 0.5 ng m^{-3} and a *"tolerable level"* of 5 ng m^{-3} (Montizaan et al., 1989). Based on "technical and economic feasibility" as well as concentrations in other western European cities, Germany has an "orienting value" of an annual average of BaP of 10 ng m^{-3} (Montizaan et al., 1989); a basis for this approach, as stated by The Umwelt Bundesamt (Federal Environmental Agency), is that "dose–effect relationships for man do not exist" (Montizaan et al., 1989; Nielsen et al., 1995, 1999a). Such relationships have, however, been established for PAHs in occupational exposures (e.g., see Mastrangelo et al., 1996).

Because certain PAHs are carcinogens, and thus there is no "safe level," the WHO (1987) does not recommend one. However, it has developed risk assessments based on BaP "as an indicator" (Nielsen et al., 1995).

2. Carcinogenicity of PAHs, Cancer Potencies, and Potency Equivalence Factors

There are a variety of sources of data on the carcinogenicity of environmental PAHs and PACs. A key source is a series of monographs published by the International Agency for Research on Cancer (IARC), *The Evaluation of Carcinogenic Risks to Humans*. This series includes *Polynuclear Aromatic Compounds*, Vol. 32, Part 1, *Chemical, Environmental, and Experimental Data* (1983); *Overall Evaluations of Carcinogenicity: An Updating of IARC Monographs*, Volumes 1–42 (1987); and *Diesel and Gasoline Engine Exhausts and Some Nitroarenes*, Vol. 46 (1989). Other useful evaluations of the carcinogenicity of specific PAHs and PACs include, for example, those of the U.S. Environmental Protection Agency (1986), the California Air Resources Board (1994), and the U.S. Department of Health and Human Services, Public Health Service, National Toxicology Program (U.S. DHHS, 1998).

Evaluations of the carcinogenicity of selected PAHs and PACs are summarized in Tables 10.13 and 10.14. Designations are defined in Box 10.7.

Unless otherwise noted, we use IARC definitions and symbols for the carcinogenicity of a given agent; for example, BaP and benzo[*b*]fluoranthene (BbF) are Class 2A and 2B animal carcinogens and "probable" and "possible" human carcinogens, respectively. A note of caution is appropriate, however, regarding differences in the classification schemes. For example, the EPA classifications for BaP and BbF are both B2, "sufficient evidence from animal studies."

Because it is not only BaP but also a variety of other PAHs and PACs that are of concern in terms of the possible inhalation cancer risk to humans of complex mixtures of combustion-generated POM, a number of approaches have been developed for evaluating the potencies of various compounds (e.g., Holmberg and Ahlborg, 1983; IARC, 1989; Nisbet and La Goy, 1992; Lewtas, 1985b, 1993a, 1993b; 1994; Heinrich et al., 1994; CARB, 1994; Mastrangelo et al., 1996; Nielsen et al., 1996; OEHHA, 1998; CARB, 1998; Collins et al., 1998; Tokiwa et al., 1998). One approach is to calculate the inhalation "unit risks" for excess lung cancer for BaP and each of its copollutant carcinogenic PAHs and PACs in polluted ambient air. The latter values are divided by the unit risk for BaP to obtain their individual potency equivalence factors, PEFs, based on BaP = 1.00, e.g., Nielsen et al., 1996, and Collins et al., 1998. These PEFs are listed in Table 10.13 for PAHs and Table 10.14 for PACs. The PEFs range from 0.01 for chrysene and 2-nitrofluorene to 10 for each of the three 6-ring dibenzopyrenes ($C_{24}H_{14}$, MW 302) and for the N-PACs 6-nitrochrysene and 1,6-dinitropyrene. Interestingly, Cavalieri et al. (1989, 1994) reported that the dibenzo[*a,l*]pyrene isomer is actually 100–200 times more tumorigenic than BaP, and they termed it "the most potent carcinogenic aromatic hydrocarbon."

To assess both the relative and absolute contributions of various PAHs and PACs to health impacts, the potencies must be combined with concentrations of the individual PAHs and PACs in air. The levels of compounds other than BaP can be quite substantial and hence contribute significantly to the overall carcinogenicity and mutagenicity. For example, Allen and co-workers (1998) reported that, while the individual concentrations of the biologically active 6-ring PAHs identified and quantified in urban Boston air are relatively small (e.g., the concentration of dibenzo[*a,e*]pyrene was 0.133 ng m^{-3}), their *total* concentration of ~1.5 ng m^{-3} is comparable to the BaP present as a copollutant in the sample.

Collins et al. (1998) applied the PEF data for PAHs and PACs (Tables 10.13 and 10.14) to daytime concentrations measured in ambient air in Riverside, California (~90 km east and downwind of Los Angeles; see Fig. 10.23), during the months of July and August 1994 (Atkinson and Arey, 1997; Krieger et al., 1997). As seen in Table 10.15, the "PEF adjusted concentration," 269 pg m^{-3}, is ~40% of the total mass of PAHs and PACs

TABLE 10.13 Carcinogenicities of Selected PAHs in Ambient Air As Evaluated by IARC, U.S. EPA, and U.S. DHHS, Cancer Potency Equivalence Factors Relative to BaP = 1.00 from Nielsen et al. (1996), CARB (1994), and Collins et al. (1998), and Human Cell Mutagenicities Relative to BaP = 1.00

Structure	Name	IARC Class[a]		U.S. EPA Class[b]	U.S. DHHS Class[c]	Potency equivalence factors		Relative human cell mutagenicity (BaP = 1.00)[h-j]
		Animals	Human			Nielsen[d,f]	CARB / OEHHA[e-g]	
	Benz[a]anthracene	S	2A	B2	S RAHC	0.005	0.1	0.022
	Chrysene	L	3	B2	—	0.03	0.01	0.017
	5-Methylchrysene	S	2B	NC	S RAHC	—	1.0	0.63
	Benzo[a]pyrene	S	2A	B2	S RAHC	1.0	1.0	1.00
	Benzo[b]fluoranthene	S	2B	B2	S RAHC	0.05	0.1	0.25
	Benzo[j]fluoranthene	S	2B	NC	S RAHC	0.05	0.1	0.26
	Benzo[k]fluoranthene	S	2B	B2	S RAHC	0.05	0.1	0.11

Benzo[ghi]perylene	I	–	–	–	0.02	–	0.19
Indeno[1,2,3-cd]pyrene	S	2B	B2	S RAHC	0.1	0.1	0.31
Dibenz[a,h]anthracene	S	2A	B2	S RAHC	–	0.4	0.29
Cyclopenta[cd]pyrene	L	3	–	–	0.02	–	6.9
Dibenzo[a,e]pyrene	S	2B	NC	S RAHC	–	1.0	2.9
Dibenzo[a,h]pyrene	S	2B	NC	S RAHC	–	10.0	–
Dibenzo[a,i]pyrene	S	2B	NC	S RAHC	–	10.0	–

(continues)

TABLE 10.13 (continued)

Structure	Name	IARC Class[a] Animals	IARC Class[a] Human	U.S. EPA Class[b]	U.S. DHHS Class[c]	Potency equivalence factors Nielsen[d,f]	Potency equivalence factors CARB/OEHHA[e–g]	Relative human cell mutagenicity (BaP = 1.00)[h–j]
	Dibenzo[a,l]pyrene	S	2B	NC	S RAHC	—	10.0	24

[a] From Supplement 7 (IARC, 1987), unless otherwise indicated: I, inadequate evidence; L, limited evidence; ND, no adequate data; S, sufficient evidence; 1, Group 1, the agent is carcinogenic to humans; 2A, Group 2A, the agent is probably carcinogenic to humans; 2B, Group 2B, the agent is possibly carcinogenic to humans; 3, Group 3, the agent is not classifiable as to its carcinogenicity to humans.

[b] Source: U.S. EPA *Carcinogen Classifications* (1986). Group A, human carcinogen; Group B, probable human carcinogen; Subgroup B1, "agents showing limited evidence of carcinogenicity from epidemiologic studies;" Subgroup B2, "agents for which there is 'sufficient' evidence from animal studies and for which there is is 'inadequate evidence' or 'no data' from epidemiologic studies;" Group C, possible human carcinogen; Group D, not classifiable as to human carcinogenicity; NC, not classified.

[c] U.S. DHHS/PHS/NTP (1998). Class S is sufficient evidence of carcinogenicity in animals. All PAHs in this table that have been evaluated are classified "reasonably anticipated to be human carcinogens (RAHC)."

[d] Nielsen et al. (1996) "City Air Pollution of Polycyclic Aromatic Hydrocarbons and Other Mutagens: Occurrence, Sources, and Health Effects."

[e] California Air Resources Board/Office of Environmental Health Hazard Assessment, *Benzo[a]pyrene as a Toxic Air Contaminant* (1994); Office of Environmental Health Hazard Assessment/California Environmental Protection Agency, *Air Toxics Hot Spot Program Risk Assessment Guideline, Part II: Technical Support Document for Describing Available Cancer Potency Factors* (1998); Collins et al. (1998).

[f] For the actual values and a discussion on various approaches by a range of scientific groups to calculating such values of *unit risks for cancer* for the PAHs and PACs in Tables 10.14 and 10.15, see Nielsen et al. (1995, 1996, 1999b), OEHHA/Cal EPA (1998), Collins et al. (1998), and references therein.

[g] Collins et al. (1998).

[h] Source: Durant et al. (1996).

[i] Recalculated from data in Durant et al. (1996).

[j] Several of these relative potencies differ from those given in Hannigan et al. (1998); see Table 10.25.

TABLE 10.14 Carcinogenicities of Selected PACs in Emissions and Ambient Air[a] As Evaluated by the IARC, U.S. EPA, and U.S. DHHS, Cancer Potency Equivalence Factors As Evaluated by CARB/OEHHA (BaP = 1.0), and Human Cell Mutagenicities Relative to BaP = 1.0

Structure	Name	IARC Class[b] Animals	IARC Class[b] Humans	U.S. EPA Class[c]	U.S. DHHS Class[d]	CARB/OEHHA Potency equivalence factor[e,f]	Relative human cell mutagenicity (BaP = 1.0[g])
	1,6-Dinitropyrene	S	2B	NC	RAHC	10.0	0.20
	1,8-Dinitropyrene	S	2B	NC	RAHC	1.0	1.2
	6-Nitrochrysene	S	2B	NC	–	10.0	–
	2-Nitrofluorene	S	2B	NC	–	0.01	0.05
	1-Nitropyrene	S	2B	NC	RAHC	0.1	0.025
	4-Nitropyrene	S	2B	NC	RAHC	0.1	–
	1-Nitronaphthalene	I	3	–	–	–	–
	2-Nitronaphthalene	I	3	–	–	–	–
	9-Nitroanthracene	N.D.	3	–	–	–	–

(continues)

TABLE 10.14 (continued)

Structure	Name	IARC Class[b]		U.S. EPA Class[c]	U.S. DHHS Class[d]	CARB / OEHHA potency equivalence factor[e,f]	Relative human cell mutagenicity (BaP = 1.0[g])
		Animals	Humans				
	6-Nitrobenzo[a]pyrene	L	3	—	—	—	—
	3-Nitrofluoranthene	I	3	—	—	—	0.0026
	Dibenz[a,j]acridine	S	2B	NC	RAHC	0.1	—
	Dibenz[a,h]acridine	S	2B	NC	RAHC	0.1	—
	7H-Dibenzo[c,g]carbazole	S	2B	NC	RAHC	1.0	—

[a] For definitions of the classifications by the various organizations, see Table 10.13.
[b] From IARC (Supplement 7, 1987; 1989).
[c] From U.S. EPA (1986). NC, not classified.
[d] From U.S. Department of Health and Human Services (1998); RAHC, reasonably anticipated to be a human carcinogen.
[e] From California Air Resources Board (1994) and Office of Environmental Health Hazard Assessment, California Environmental Protection Agency (Cal EPA) (1998).
[f] From Collins et al. (1998).
[g] From Durant et al. (1996).

(651 pg/m^3). That is, the "total inhalation risk" associated with these PAHs and PACs is ~40% of that predicted *if one assumes each compound has the same carcinogenic potency as BaP*. One important caveat was pointed out by Collins et al. (1998): since PAHs are "multipathway" carcinogens, their calculated *inhalation* risk may account for only 5–10% of the estimated total *multipath* exposure risk.

Nielsen and co-workers (1996) employed a similar "potency equivalence factor" methodology in their risk assessment evaluation of the health impacts of PAHs and PACs for typical ambient levels encountered at several sites in Copenhagen, Denmark. For other examples of comparative risk evaluations, see, for example, the calculation by Collins et al. (1998) of the equivalent inhalation risk of ambient aerosols in London, using the 1991 data of Halsall et al. (1994), and Lewtas (1993a, 1993b, 1994, and references therein).

Finally, while several volatile and semivolatile PAHs, e.g., naphthalene, the methylnaphthalenes, phenanthrene, pyrene, and fluoranthene, are not significant mutagens or carcinogens (hence not included in Table 10.13), they are *precursors* to powerful direct bacterial mutagens formed in gas-phase atmospheric reactions with hydroxyl during the day and nitrate radicals at night (see Section F). Furthermore, 2-nitrofluoranthene,

> **BOX 10.7**
> **IARC, U.S. EPA, AND U.S. DEPARTMENT OF HEALTH AND HUMAN SERVICES (DHHS) CLASSIFICATIONS FOR CARCINOGENICITY OF CHEMICAL AGENTS**
>
> IARC *Overall Evaluations of Carcinogenicity: An Updating of IARC Monographs, Volumes 1–42, Supplement 7 (1987)*:
>
> *I* Inadequate evidence
> *L* Limited evidence
> *ND* No adequate data
> *S* Sufficient evidence
> *1, Group 1* The agent is carcinogenic to humans
> *2A, Group 2A* The agent is probably carcinogenic to humans
> *2B, Group 2B* The agent is possibly carcinogenic to humans
> *3, Group 3* The agent is not classifiable as to its carcinogenicity to humans
>
> U.S. EPA *Carcinogen Classifications (1986)*:
>
> *Group A* Human carcinogen
> *Group B* Probably human carcinogen
> *Subgroup B1* Agents showing limited evidence of carcinogenicity from epidemiologic studies
> *Subgroup B2* Agents for which there is 'sufficient' evidence from animal studies and for which there is 'inadequate evidence' or 'no data' from epidemiologic studies
> *Group C* Possible human carcinogen
> *Group D* Not classifiable as to human carcinogenicity
>
> U.S. DHHS *Report on Carcinogens (1998)*:
>
> Class S is sufficient evidence of carcinogenicity in animals. RAHC is the abbreviation for their listing "Reasonably Anticipated to be Human Carcinogens"

an atmospheric reaction product of fluoranthene, is a potent human cell mutagen present in ambient urban/suburban air environments throughout the world (Ciccioli *et al.*, 1996; Arey, 1998a, and references therein).

Clearly, a sound evaluation of the *total* mutagenic/carcinogenic potencies of a complex mixture of POM emissions (e.g., diesel exhaust) should include not only the PEFs of the *primary* particle- and vapor-phase PAHs and PACs but also those of the mutagens formed in atmospheric reactions of precursor PAHs (see, for example, Arey *et al.* (1992), Lewtas (1993b), Atkinson and Arey (1994), Nielsen *et al.* (1996), Arey (1998a), and Section F). For examples of such formal scientific health risk assessments prepared by the State of California Air Resources Board and Office of Environmental Health Hazard Assessment, see "Benzo[*a*]pyrene as a Toxic Air Contaminant" (CARB, 1994) and "Identification of Diesel Exhaust as a Toxic Air Contaminant" (CARB, 1998).

3. Carcinogenicity of Nitroarenes and Other Nitro-PACs

In the mid-1970s with the advent of a relatively simple short-term bacterial assay, the "Ames Test" (Box 10.9), a new class of mutagenic compounds was discovered, the nitroarenes. Within a short time, they were identified as directly mutagenic copollutants present along with well-known promutagenic PAHs in combustion-generated emissions and distributed worldwide in urban ambient air (see Section D).

Since then, mono- and dinitro-PAHs have been the subject of intense research interest in atmospheric chemistry and toxicology. This is in large part because a number of them have been evaluated by IARC (1989) and other evaluating bodies, e.g., the U.S. Department of Health and Human Services (U.S. DHHS, 1998) and the State of California Air Resources Board and Office of Environmental Health Hazard Assessment (CARB, 1994; OEHHA, 1998), and classified as being "possible" or "reasonably anticipated to be" human carcinogens. These are shown in Table 10.14, along with two dibenzacridines and a dibenzocarbazole. Note that the PEFs of the 1,6- and 1,8-dinitropyrenes are 10 times that of BaP and equal to BaP, respectively; 6-nitrochrysene is also estimated as 10 times as potent as BaP.

We discuss the bacterial and human cell mutagenicities and related atmospheric chemistry of these biologically active N-PACs in subsequent sections. Box 10.8 contains examples of literature on this subject.

TABLE 10.15 Concentrations of Carcinogenic PAHs and PACs Measured in Ambient Air in Riverside, California, Their Cancer Potency Equivalence Factors (PEF), and Calculated PEF-Adjusted Concentrations[a]

PAH and PAC	Measured concentration (pg m^{-3})[b]	PEF	PEF-adjusted concentration (pg m^{-3})
Benzo[a]pyrene	36	1.0	36
Benz[a]anthracene	39	0.1	3.9
Benzo[b + j + k]fluoranthenes	360	0.1	36
Dibenzo[a,e]pyrene	1.7	1.0	1.7
Dibenzo[a,h]pyrene	<0.5[c]	10	—
Dibenzo[a,i]pyrene	<0.5[c]	10	—
Dibenzo[a,l]pyrene	18	10	180
Indeno[1,2,3-cd]pyrene	87	0.1	8.7
Chrysene	61	0.01	0.61
Dibenz[a,h + a,c]anthracene	8.2	0.1[d]	0.82
1-Nitropyrene	4.4	0.1	0.44
2-Nitrofluoranthene	36	0.01	0.36
4-Nitropyrene	≤1.6[c]	0.1	—
1,6-Dinitropyrene	0	10	0
1,8-Dinitropyrene	No data		
6-Nitrochrysene	≤5.4[c]	10	—
2-Nitrofluorene	0	0.01	0
Total carcinogenic PAHs	651		269

[a] Adapted from Collins et al. (1998).
[b] Daytime values during July and August; from Atkinson and Arey (1997) and Krieger et al. (1997).
[c] Not included in total.
[d] Assumes equal PEF (0.1) for both dibenzanthracenes.

BOX 10.8
EXAMPLES OF LITERATURE ON THE MUTAGENICITY AND CARCINOGENICITY OF NITROARENES

There is a large body of literature on the mutagenicity and carcinogenicity of PACs. Useful reviews include, for example Rosenkranz and Mermelstein (1985a, 1985b), Tokiwa and Oshini (1986), Tokiwa et al. (1986), Tokiwa et al. (1987), Tokiwa et al. (1998), and Fu (1990). For a detailed review of 2-nitrofluorene, see Beije and Möller (1988). Examples of nitro-PACs in diesel exhaust that are powerful direct mutagens include, for example, several nitroazabenzo[a]pyrene derivatives (Sera et al., 1994) (see structure **XXXI** and Table 10.3) and 3-nitrobenzanthrone (Enya et al., 1997) (structure **XXIV**). Monographs, symposia proceedings, and chapters therein include, for example, *Nitrated Polycyclic Aromatic Hydrocarbons* (White, 1985), *Carcinogenic and Mutagenic Responses to Aromatic Amines and Nitroarenes* (King et al., 1988), and *Nitroarenes: Occurrence, Metabolism, and Biological Impact* (Howard et al., 1990). Other evaluations of a range of mutagenic chemicals include papers by Shelby (1988), Shelby et al. (1993), and Morita et al. (1997). Examples of papers on the carcinogenicity and mutagenicity of nitroarenes include Ohgaki et al. (1984), El-Bayoumy and Hecht (1986), Wislocki et al. (1986), Rosenkranz (1987), Imaida et al. (1988), Hecht and El-Bayoumy (1990), Tokiwa et al. (1993), and Rosenkranz (1996). Evaluations and references to the original literature for 1-nitropyrene, 4-nitropyrene, 6-nitrochrysene, and 1,6- and 1,8-dinitropyrene are found in the IARC 1989 Report, the *8th Report on Carcinogens, Summary* (U.S. DHHS, 1998), and Collins et al. (1998).

C. BIOLOGICAL PROPERTIES OF PAHs AND PACs. II: MUTAGENICITY

1. Short-Term Tests for Genetic and Related Effects

A 1972 report from the U.S. National Academy of Sciences concluded that organic extracts of respirable particulate POM collected from both combustion sources and ambient air exhibited a carcinogenicity in animals significantly greater than could be accounted for by the amounts of known carcinogenic PAHs and PACs determined analytically to be present in the samples (NRC, 1972; see also Kotin et al., 1954; Hueper et al., 1962; Epstein et al., 1966; Grimmer, 1983b, and chapters therein). A crucial question for researchers in air pollution chemistry and the health sciences became, "What unknown compounds are responsible for this excess carcinogenicity?"

Concurrently, Gordon and co-workers (1973) reported another important phenomenon: benzene extracts of airborne particles collected in the Los Angeles Air Basin had 100–1000 times the cell transformation activity of that which could be attributed to the measured levels of known PAHs in the samples. Strikingly, the polar (methanol) fraction of these extracts, which amounted to only ~3% of the total mass in the sample of ambient particles, had an activity equal to the neutral benzene extract that contained the remaining 97% of the PAHs (including BaP).

Scientists were faced with formidable experimental challenges in trying to determine which compounds were responsible for this biological activity. Not only are they present in trace amounts and in chemically very complex mixtures of airborne POM, but *in vivo* animal assays for suspected new genotoxic and carcinogenic agents were then, and remain, time-consuming, labor intensive, and expensive.

Fortunately, a number of *in situ*, short-term bioassays to detect genotoxic and related effects have become available. These include a variety of measured "endpoints" such as aneuploids, chromosal aberrations, DNA damage, dominant lethal mutation, gene mutation, inhibition of intercellular communication, micronuclei, mitotic recombination and gene conversions, and sister chromatid exchange and cell transformation (IARC, 1989). A detailed discussion of these tests is beyond the scope of this book. However, such tests are important from our perspective as atmospheric chemists because, as we shall see, they can be used to detect biologically active compounds in very complex mixtures, and hence serve to focus chemical analysis efforts (IARC, 1989, p. 20). We emphasize in advance the IARC (1989) statement that, with respect to results from any *one* specific short-term assay, "The relative potency of agents in tests for mutagenicity and related effects is not a reliable indicator of carcinogenic potency."

For a critical appraisal of long-term and short-term assays, see, for example, WHO (1985a, 1985b) and Montesano et al. (1986). For discussions of their use in characterizing the cancer risk of POM, see, for example, in addition to the IARC Reports (1983, 1987, 1989) and the U.S. DHHS *8th Report on Carcinogens, Summary* (U.S. DHHS, 1998), articles by Lewtas (1993a, 1993b, 1994), Nielsen et al. (1996), and Collins et al. (1998) and references therein.

2. The Ames *Salmonella typhimurium* Reversion Assay

In 1973, the introduction of a novel *in vitro*, short-term bacterial assay for the detection of chemical mutagens by Ames and co-workers provided a major breakthrough in experimental approaches to meeting the analytical and toxicological challenges posed by complex environmental mixtures, including ambient POM (Ames et al., 1973, 1975; McCann et al., 1975; Maron and Ames, 1983). Since then, this relatively inexpensive, fast (3–4 days), and highly sensitive "reverse mutation" assay employing histidine-requiring (his^-) mutants of the *Salmonella typhimurium* bacterium has played a major role in elucidating not only the toxicology but also the associated atmospheric chemistry of PAHs and PACs. Indeed, throughout the remainder of this chapter, we illustrate how, in an "operational" sense, the Ames plate incorporation bacterial assay can serve as a highly sensitive "detector," especially when used with what is known as the "microsuspension modification" (Kado et al., 1983, 1986, 1991, 1992); Atkinson et al., 1991; Claxton et al., 1992a, 1992b; Bagley et al., 1992, and references therein). The application of this assay to determine the mutagenic activities of organic extracts of whole samples of gases and/or particles collected from primary emissions, from ambient air, and from products of laboratory chamber studies of simulated polluted atmospheres, as well as of fractions of the original "total" extracts of polycyclic organic matter, and finally, of the individual PAHs and PACs isolated and characterized from these fractions, has proven to be extremely useful, as we shall illustrate in the following.

a. Principle of Method: Direct vs Activatable Bacterial Mutagens

Normal *Salmonella* bacteria (his^+) do not require added histidine for growth since they generate their own. The *Ames test* employs mutant strains (his^-) that are histidine auxotrophs—that is, they lack the ability to produce their own histidine and thus can grow only in histidine-enriched media. However, attack of certain chemicals at appropriate sites on the bacterial DNA produces *reverse* mutations in which the bacteria are converted back to their wild forms (his^+). These "histidine-independent" phenotypes produce the histidine necessary for cell growth, and colonies develop on the plates. Such genotoxic chemical agents are referred to as *direct mutagens*.

A critical element of this assay is the provision for addition of a mammalian "microsomal activation" system (+S9) that provides for the detection of *promutagenic* chemicals such as BaP and other biologically active PAHs. Thus, in contrast to the direct mutagens cited, these PAHs must be converted to their mutagenic metabolites, i.e., "activated," by addition of this exogenous system before they can express their genotoxicity. Hence they are referred to as *activatable* mutagens or *promutagens*. The shorthand description in the literature is +S9 for assays with microsomal activation (detects promutagens) and −S9 for assays without microsomal activation (detects direct mutagens). The enzyme system (+S9) is obtained, for example, from the homogenates of livers of male Sprague–Dawley rats that have been injected intraperitoneally with the inducer Arochlor 1254.

This assay employs a number of different strains of *Salmonella typhimurium* mutants that are histidine auxotrophs. The first tester strains developed and tested on extracts of diesel and ambient POM in the mid to late 1970s included TA1535, which responds to base-pair substitution chemical mutagens (e.g., β-propiolactone), and TA1537 and TA1538, which respond to frameshift mutagens, e.g., BaP and other PAHs.

In 1983, Ames and colleagues reported several new strains with significantly greater sensitivities that remain in widespread use today by toxicologists and atmospheric chemists (Maron and Ames, 1983). Strain TA100 (TA1535 with the plasmid pKM101) responds both to base-pair substitution and to frameshift mutagens. Compared to TA1538, strain TA98 (TA1538 containing the same pKM101 plasmid) has a greatly enhanced response to frameshift mutagens, including both direct-acting mono- and dinitro-PAH and their promutagenic parent PAHs.

An important microbiological clue to the chemical nature of genotoxic species in primary emissions and ambient POM has been provided by strain TA98NR (Mermelstein *et al.*, 1981; Rosenkranz and Mermelstein, 1983, 1985a, 1985b). This isolate of TA98 is deficient in the "classical" bacterial nitroreductase that catalyzes the bioactivation of most mononitro-PAHs to

TABLE 10.16 Direct Mutagenicities of Selected Nitro-PAHs in the *Salmonella typhimurium* Plate Incorporation Assay (−S9) with Strains TA98, TA98NR, and TA98/1,8-DNP$_6$[a,b]

		Specific activity (rev μg^{-1}), −S9		
Chemical	MW	TA98	TA98NR	TA98/1,8-DNP$_6$
2-Nitrofluorene	211	420	49	59
2-Nitrofluoranthene	247	4,200	1,000	730
3-Nitrofluoranthene	247	(31,000)[b]	21,000	3,900
8-Nitrofluoranthene	247	(74,000)[b]	30,000	6,000
1-Nitropyrene	247	2,300	310	1,300
2-Nitropyrene	247	16,000	1,600	2,300
1,3-Dinitropyrene	292	130,000	76,000	2,200
1,6-Dinitropyrene	292	420,000	410,000	180,000
1,8-Dinitropyrene	292	990,000	1,100,000	21,000
1-Nitrobenzo[*a*]pyrene[c]	297	2,500	450	2,700
3-Nitrobenzo[*a*]pyrene[c]	297	5,300	940	4,400
6-Nitrobenzo[*a*]pyrene[c]	297	0	0	0

[a] Adapted from Winer and Atkinson (1987).
[b] From Arey *et al.* (1988b).
[c] Mutagenicities with microsomal activation (TA98, +S9) for BaP, 1-NO$_2$-BaP, 3-NO$_2$-BaP, and 6-NO$_2$-BaP are 390, 3,100, 2,900, and 1,500 rev μg^{-1} (Pitts *et al.* 1984).
[d] rev μg^{-1} = 10^3 × MW^{-1} × rev nmol^{-1}.

their ultimate mutagenic forms. Thus, if one conducts two separate assays (−S9) of a given POM extract, a significantly *lower* response on TA98NR relative to TA98 indicates the probable presence of mononitro-PAH in the sample. The data in Table 10.16 illustrate these differences (e.g., see 2-nitrofluoranthene, 1- and 2-nitropyrene, and 2-nitrofluorene).

Another useful strain of TA98 developed by Rosenkranz and colleagues is TA98/1,8-DNP$_6$. While possessing a functioning nitroreductase, it is deficient in a second enzyme that activates the very potent direct bacterial mutagens 1,8- and 1,6-dinitropyrene, and it is therefore insensitive to these compounds. Relative to TA98 and TA98NR, a significant decrease in response of this strain suggests the presence of dinitropyrenes in a given sample (see, for example, the data in Table 10.16).

Finally, four more *Salmonella typhimurium* tester strains that are derivatives of TA98 and TA100 and have significantly increased sensitivities to various classes of mutagenic chemicals are now available: YG1021, YG1026, YG1024, and YG1029 (Watanabe *et al.*, 1989, 1990; Einistö and co-workers, 1991). Strain YG1021 is especially suited for assays of nitroarenes; e.g., it is ~24 times more sensitive to 1-nitropyrene than TA98 (12,172 vs 514 rev/nmol; −S9). Strain

BOX 10.9
PROCEDURE FOR THE AMES *Salmonella typhimurium* REVERSION ASSAY

Figure 10.14 is a simplified diagram of the test procedure. A suitable organic solvent, e.g., dichloromethane, or solvent mixture (see Nielsen, 1992) is used to extract the POM from the environmental sample, the solvent is evaporated, and the residual POM is redissolved in dimethyl sulfoxide (DMSO). Serial dilutions of this DMSO solution are then added to a series of tubes, each containing agar and one of the several his^- test strains of bacteria, e.g., TA98, which is widely used for frameshift mutagens.

The resulting solution is then overlaid onto Petri dishes containing a minimal glucose agar medium (with a trace of histidine to allow cell division) and incubated in the dark for 48-72 h at 37°C. The plates are removed from the incubator and any colonies present are counted.

A significant increase in the number of his^+ colonies above the background count (the number of colonies on Control Plate C when *no* sample is added) on Test Plate A indicates that the POM sample contains a chemical(s) that is (are) a *direct* mutagen with the Ames strain employed. If no direct activity is observed on Test Plate A, it does not necessarily mean there are *no* bacterial mutagens in the sample. As noted earlier, carcinogenic PAHs such as BaP and benz[*a*]anthracene are *promutagens* and must first be metabolized to reactive intermediates (metabolites), which then attack the bacterial DNA and cause $his^- \rightarrow his^+$ reversions. Therefore, in practice, a small amount (e.g., 1-2 mg per plate) of an enzyme system is added to another portion of the original test sample, Test Plate B (+S9), and the assay conducted in parallel with Test Plate A (-S9).

After incubation, the spontaneous $his^- \rightarrow his^+$ reversions found on background Plate C are subtracted from the colony count on Test Plate B. If a statistically significant number of colonies remain (i.e., "net revertants"), the POM sample contains one or more chemicals that are promutagens. Sam-

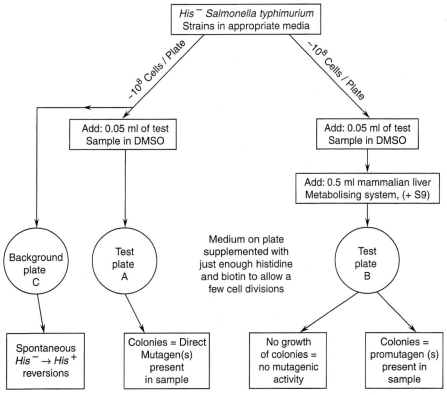

FIGURE 10.14 Diagram of procedure for the standard Ames *Salmonella typhimurium* reversion assay for chemical mutagens (Maron and Ames, 1983).

ples of POM collected from primary emissions and polluted ambient air contain both direct (e.g., nitroarenes and 1-nitropyrene) and activatable (e.g., BaP) bacterial mutagens.

Dose–response curves are generated from the two sets of data (+S9 and −S9) by running two series of plates prepared from a range of serial dilutions (duplicate or triplicate plates for each dilution) of the original sample. The slope of the ideally linear portion of this curve is the *specific activity* of a pure chemical and is expressed in units of revertants per microgram (rev μg^{-1}) or revertants per nanomole (rev $nmol^{-1}$).

The *mutagenic potency* of a POM extract is generally expressed in units of revertants per microgram (rev μg^{-1}) of the dried organic extract of the sample (sometimes, however, as rev mg^{-1} of the original particulate matter). The concentration of mutagens in ambient air (or synthetic laboratory atmospheres) is the mutagen density, generally given in units of revertants per cubic meter of ambient air (or chamber) sampled (rev m^{-3}).

It should be noted that in the *forward* mutation *Salmonella typhimurium* TM677 assay (see later discussion), mutagenic potency is generally expressed in units of (mutagen fraction $\times 10^5$)/μg of "equivalent organic carbon (EOC)," and mutagen density (i.e., mutagen concentration) in units of (mutant fraction $\times 10^5$)/m^3 of air. Assays are carried out with (for promutagens), and without (for direct mutagens), "postmitochondrial supernatant," +PMS and −PMS, respectively. Since this is equivalent to +S9 or −S9 in the Ames test, we use the latter abbreviation for both assays (see Hannigan *et al.*, 1996, and references therein). In this book, unless otherwise specified, the term "bacterial mutagenicity" refers to data obtained with the more commonly used Ames *Salmonella typhimurium reversion assay*.

YG1024 is ~30 times more sensitive than TA98 to 1,8-dinitropyrene (2,150,000 vs 72,200 rev/nmol; −S9). For example, Legzdins and co-workers (1994) used the enhanced response of strain YG1021 to determine the mutagenicities of 14 PAHs, and several nitroarenes, S-atom PACs, and O-atom PACs in the ambient air of Hamilton, Ontario, Canada.

b. Assay Procedure

A brief outline of the essential features of the standard plate incorporation assay (Maron and Ames, 1983) is given in Box 10.9. Protocols for pure chemicals and for complex mixtures of primary emissions and ambient aerosols are also described by Belser *et al.* (1981) and Alfheim *et al.* (1984b); intra- and interlaboratory comparisons are discussed later in this chapter.

Specific activities of individual airborne PAHs and PACs determined by the "standard" plate incorporation assay cover a huge range; e.g., the mono- and dinitro-PAHs in Table 10.16 have direct activities (−S9) ranging from several hundred to ~10^6 rev μg. This is also true for different isomers of the same compound. For example, the 1-, 3-, and 6-NO_2-BaP isomers formed in laboratory exposures of BaP particles to NO_2 (plus a trace of HNO_3) in air have direct activities (TA98, −S9) of 2500, 5300, and <1 rev μg^{-1}, respectively (Pitts *et al.*, 1984b).

c. The "Microsuspension Modification"

Given the very small amounts of material found in atmospheric samples, improvements in the Ames assay that use smaller amounts of sample and/or have improved sensitivity are clearly desirable. Yahagi *et al.* (1977) described one such approach. Subsequently, Kado and co-workers (1983) reported an experimentally simple, but highly effective, modification. which is now widely used. The term "microsuspension modification" comes from the use of small volumes, ~0.2 mL. Increased numbers (~10^9) of bacterial cells are added to the sample being tested, along with (+S9) or without (−S9) liver homogenate mix. After incubation for 90 min at 37°C, it is processed according to the standard Ames test protocol.

Use of the Kado microsuspension preincubation modification has become increasingly popular, in large part because it provides a major increase in sensitivity compared to the standard plate incorporation assay (Maron and Ames, 1983) and requires less sample. As seen in Tables 10.17 and 10.18, this is especially true for the volatile nitroarenes 2-nitrofluorene and 1- and 2-nitronaphthalene; less enhancement is achieved with the larger nitroarenes, e.g., 2-nitrofluoranthene (Arey, personal communication). The increased sensitivity is also illustrated by an intercomparison of the "modified" vs standard plate assay using the reference complex mixtures SRM 1649, "Air Particles" (Box 10.3), and SRM 1650, "Diesel Particles." The amount of material needed to detect and quantify mutagenicity was improved by a factor of 10–20 and the sensitivity was increased by a factor of ~30 (Claxton *et al.*, 1992a, 1992b; see also Bagley and co-workers (1992) for re-

TABLE 10.17 Comparison of Specific Mutagenic Activities of Several Standard Nitroarenes Determined on Strain TA98 (−S9) Using the *Salmonella typhimurium* Assay with and without the Kado Microsuspension Modification[a–c]

Compound	Mutagenic activity (rev nmol^{-1})	
	Microsuspension assay	Plate assay
2-Nitrofluorene	4100[a]	93[a]
2-Nitrofluoranthene	3500[a]	960[a]
1-Nitronaphthalene	30[a]	0.05,[b] 0.4[b]
2-Nitronaphthalene	680[a]	0.2,[b] 0.9[b]

[a] Adapted from Arey (1998a).
[b] Cited in reviews by Rosenkranz and Mermelstein (1985a, 1985b).
[c] Kado et al. (1983, 1986, 1991).

sults of a comparison study conducted in their laboratory).

This modification has been applied in a variety of studies, including the mutagenic activity of "fine" ambient air particles as a function of particle size (Kado et al., 1986), personal exposure to ETS (Kado et al., 1991), and vapor-phase mutagens (Kado et al., 1992). Similarly, it has been applied to 2-nitronaphthalene, which is the product of a nighttime atmospheric reaction of NO_3 and the daytime OH-initiated reaction of naphthalene. This nitronaphthalene is present at relatively high levels in ambient air and is a human cell as well as a direct bacterial mutagen (Arey, 1998a, and references therein; see Section F.2).

The large increase in the mutagenic activity of 2-nitronaphthalene determined earlier with the standard assay, 0.2–0.9 rev/nmol (Rosenkranz and Mermelstein, 1985a, 1985b) compared to 680 rev/nmol with the microsuspension assay (Arey, personal communication), is important in studies of ambient air. Other examples of the application of a microsuspension modification to the Ames assay include ambient air particles (Kado et al., 1986; Atkinson et al., 1991; Zinbo et al., 1992; Matsumoto et al., 1998), ambient vapor-phase and particulate mutagens (Kado et al., 1992; Harger et al., 1992; Watanabe et al., 1995), vapor-phase mutagens in ambient air (Gupta et al., 1996), laboratory studies of the photooxidations of 2- to 4-ring PAHs (Sasaki et al., 1995), vapor-phase mutagens in diesel exhaust (Hsieh et al., 1993), personal exposure to environmental tobacco smoke (Kado et al., 1991), and diesel particles (Bagley et al., 1992).

d. The Salmonella typhimurium Reversion Assay for Gas-Phase Mutagens

Until the early to mid-1980s, research on the mutagenicity of respirable POM focused almost exclusively on the particulate phase. Another aspect of tropospheric chemistry with significant health implications is the application of the Ames bacterial assay (with or without the microsuspension modification) to the detection and identification of mutagenic *vapor-phase* PAHs and PACs. For example, Harger and co-workers (1992) reported that the mutagenicities of concurrently collected samples of vapor-phase and particle-phase organics in southern California (Claremont, California) ambient air were comparable.

As discussed in Section A.4, there are a variety of approaches to separating the gas and particle phases that allow their contributions to the total mutagenic activity to be determined independently. These include, for example, a combination of a TIGF filter and three PUF plugs, discussed earlier and shown in Fig. 10.4 (Arey et al., 1989a), or a combination of a filter for particles and XAD-2 resin to trap the gas-phase species (e.g., see Alfheim et al., 1985; Pyysalo et al., 1987; and Tuominen et al., 1988). Such approaches have been applied to studying the gas-phase mutagenic PAHs and PACs in ambient air, in environmental chamber studies of 2- to 4-ring PAHs and of VOC–NO_x systems, and in

TABLE 10.18 Direct Mutagenic Activities, TA98 (−S9), in the *Salmonella typhimurium* Assay (Microsuspension Preincubation) of Standard Samples of 14 Methylnitronaphthalene Isomers[a] and Selected Nitroarenes[b]

Nitroarene[a]	MW	Mutagenic activity (rev nmol^{-1})[c]
2-Nitrofluorene	211	4100 (93)[b]
1-Nitronaphthalene	173	30
2-Nitronaphthalene	173	680
2-Nitrofluoranthene	247	3500 (960)[b]
2-Methyl-1-nitronaphthalene	187	0.8
1-Methyl-8-nitronaphthalene	187	5
2-Methyl-8-nitronaphthalene	187	3
2-Methyl-4-nitronaphthalene	187	60
1-Methyl-5-nitronaphthalene	187	340
2-Methyl-5-nitronaphthalene	187	60
1-Methyl-4-nitronaphthalene	187	210
2-Methyl-3-nitronaphthalene	187	None detected
1-Methyl-6-nitronaphthalene	187	710
1-Methyl-2-nitronaphthalene	187	60
1-Methyl-3-nitronaphthalene	187	1300
1-Methyl-7-nitronaphthalene	187	860
2-Methyl-7-nitronaphthalene	187	1400
2-Methyl-6-nitronaphthalene	187	1700

[a] From Arey (1998a). See also Gupta et al. (1996) for the mutagenicities of the nitronaphthalenes and methylnitronaphthalenes are essentially those found in Gupta et al. (1996). See this paper for experimental details.
[b] Values in parentheses obtained by Harger et al. (1992) using standard plate assay.
[c] rev nmol^{-1} = 10^{-3} × MW × rev μg^{-1}.

wood smoke and automobile exhaust (see, for example, the following: for ambient air, Claxton (1985), Wong et al. (1991), Arey et al. (1992), Harger et al. (1992), Kado et al. (1992), and Gupta et al. (1996); for environmental chamber photooxidations of 2- to 4-ring PAHs, Sasaki et al. (1995); for a variety of VOCs Shepson et al. (1985, 1986, 1987), Kleindienst et al. (1985a, 1985b, 1990), and Kleindienst (1994); for wood smoke and automobile exhaust, Kleindienst et al. (1986, 1992) and Hsieh et al. (1990, 1993)).

e. Accuracy and Precision

Both intra- and interlaboratory mutagenic potencies reported in the literature for a given chemical (e.g., a common reference standard, 2-nitrofluorene), or complex mixture (e.g., extracts of particulate POM), can vary widely. Given their widespread international use, knowing the accuracy and precision of the mutagenic potencies obtained from plate incorporation assays is essential if they are to be used to correlate mutagen levels in primary emissions and ambient air with changes in parameters such as fuel composition and operating conditions or with meteorological conditions, season of the year, gaseous pollutant levels (e.g., VOCs, NO_x, and O_3), and emission inventories.

Some of the experimental variability associated with this assay were addressed, for example, by Belser et al. (1981) and Maron and Ames (1983), who recommended procedures to improve the accuracy and precision of the assay. These included concurrent assays conducted with "positive controls," i.e., reference chemicals of known mutagenicities such as 2-nitrofluorene for strains TA98 and TA100 with and without activation; spontaneous reversion rate (SRR) controls; checks for toxicity; and attention to such parameters as uniformity of the soft agar thickness, cell density, and temperature uniformity during incubation. For examples of the effectiveness of these and other measures used to improve accuracy and precision, see, Fitz et al. (1983), Ball et al. (1984), Hsieh et al. (1990), Claxton et al. (1992a, 1992b), Lewtas et al. (1992), May et al. (1992), Greenberg et al. (1993), and Gupta et al. (1996).

International collaborative studies have also been carried out to investigate the accuracy and precision of such bioassays when applied to complex mixtures (e.g., Claxton et al., 1992a, 1992b; Lewtas et al., 1992; May et al., 1992). The conclusions from one of the major studies (Claxton et al., 1992b) are representative of such intercomparisons:

- The intralaboratory variance of the bioassay results ranged from 16 to 88%, depending on the Standard Reference Material used and the particular bioassay conditions, such as tester strain and metabolic activation; interlaboratory variance, i.e., reproducibility, ranged from 33 to 152%.
- About 55–95% of the total variation for the three environmental samples studied was due to between-laboratory variability.
- The variation in the mutagenic potency of the control compounds was similar to that of the environmental samples; the exception was 1-nitropyrene, for which the reproducibility ranged from 127 to 132%.

f. Some Mutagenic Potencies in the Salmonella typhimurium Assay

Table 10.16 contains the direct mutagenicities for a number of nitroarenes of atmospheric interest determined in one laboratory (strains TA98, TA98NR, and TA98/1,8-DNP_6; rev/μg, $-$S9) (Winer and Atkinson, 1987; Arey et al., 1988b) using the Salmonella typhimurium plate incorporation assay and the quality control measures described by Belser et al. (1981) and Maron and Ames (1983). Zielinska and co-workers (1988b) have also determined the mutagenic activities in these strains for a series of nitrofluoranthenes.

The genotoxicity, mutagenicity, and carcinogenicity of nitroarenes, as well as their occurrence and modes of formation, have been critically reviewed by Rosenkranz and Mermelstein (1985a, 1985b) and Tokiwa and Oshini (1986). Specific activities (in rev $nmol^{-1}$) of some 80 nitroarenes (plate incorporation assay) on strain TA98, without ($-$S9) and with ($+$S9) microsomal activation, are cited by Rosenkranz and Mermelstein (1985a), whereas Tokiwa and Oshini give the direct activities (rev $nmol^{-1}$, $-$S9) on strains TA98 and TA100 (plate incorporation assay) for 53 nitroarenes.

More recently, Einestö et al. (1991) determined the mutagenicities on strains YG1021 and YG1024 (vide supra) and strains TA98, TA98NR, and TA98/1,8-DNP_6 for 18 different chemicals, including 2-nitrofluorene, 1-nitropyrene, and 1,8-dinitropyrene (all $-$S9). Additionally, mutagenicities of 12 chemicals, including BaP ($+$S9) and 2-nitronaphthalene ($-$S9), were determined on strains YG1026 and YG1029 vs TA100, TA100NR, and TA100/1,8-DNP_6 (see the paper for a description of their assay procedures). As seen in Table 10.19, 2-nitrofluorene is significantly more sensitive on strains YG1021 and YG1024 than on TA98, 1,8-dinitropyrene is significantly more sensitive on strain YG1024 than TA98, 2-nitronaphthalene ($-$S9) is more sensitive on strains YG1026 and YG1029 than on TA100, and the promutagen BaP ($+$S9) is about the same on all strains.

Direct mutagenicities (rev/nmol; $-$S9) for both the plate incorporation and microsuspension assay are given in Table 10.20 for four isomeric nitrofluorenes and four nitrophenanthrene lactones (Arey et al., 1992; Atkinson

TABLE 10.19 Mutagenic Activities of Several Nitroarenes (−S9) and Benzo[a]pyrene (+S9) Assayed with Strains TA98 and TA100 and Several Derivative Strains[a−c]

Compound	MW	S9 mix	Mutagenic activity (rev nmol^{-1})		
			TA98	YG1021	YG1024
2-Nitrofluorene	211	−	293 (1.0)	2,050 (7.0)	3,200 (11)
1,8-Dinitropyrene	292	−	72,000 (1.0)	83,000 (1.1)	2,150,000 (30)
			TA100	YG1026	YG1029
2-Nitronaphthalene	173	−	25 (1.0)	83 (3.3)	69 (2.8)
Benzo[a]pyrene	212	+	134 (1.0)	89 (0.7)	159 (1.2)

[a] Adapted from Einistö et al. (1991).
[b] Mutagenic activities on strains TA98 and TA100 assigned a value of 1.0. Relative mutagenicities on other strains given in parentheses.
[c] For mutagenic activities on TA98NR, TA98/1,8-DNP$_6$, TA100NR, and TA100/1,8-DNP, see Einistö et al. (1991).

and Arey, 1993); Table 10.18 reports values for 14 methylnitronaphthalenes assayed with the microsuspension modification (Gupta et al., 1996; Arey, 1998a). Both studies include mutagenicities for 1- and 2-nitronaphthalene. The fact that the 2- and 3-nitrofluorene isomers (Table 10.20) are the most active (with the 2-isomer more active than the 3-isomer) is consistent with an earlier report and theoretical considerations (see Arey et al., 1992, and references therein).

The wide range of mutagenicities of the methylnitronaphthalene isomers, e.g., from "none detected" for 2-methyl-3-nitronaphthalene to 1700 rev/nmol for 2-methyl-6-nitronaphthalene, and the greater activity of 2-nitro- vs 1-nitronaphthalene (Table 10.18), is consistent with the proposal of Yu and co-workers (1992) that within a class of isomeric nitro-PAH, the one with the nitro group located at the longest molecular axis should be the most potent direct mutagen in *Salmonella typhimurium* strain TA98 (Gupta et al., 1996). Other studies dealing with the experimental and theoretical aspects of structure–mutagenicity relationships in nitro-PAHs include, for example, those by El-Bayoumy et al. (1981), Ball et al. (1984), Hirayama et al. (1986), and Jung et al. (1991).

In the early 1990s, several nitro-polycyclic aromatic compounds that are powerful direct mutagens were identified in ambient particulate matter, including the nitrophenanthrene lactones 2- and 4-nitro-6H-dibenzo[b,d]pyran-6-one, whose mutagenicities are given in Table 10.20. The 2-isomer (**XI**) is not only very

TABLE 10.20 Direct Mutagenicities in Strain TA98 (−S9) for 1- and 2-Nitronaphthalene and Several Nitrofluorenes and Nitrodibenzopyranones in the *Salmonella typhimurium* Plate Incorporation and Microsuspension Preincubation Assays[a]

Compound	MW	Microsuspension assay, TA98 − S9 (rev nmol^{-1})	Plate incorporation assay, TA98 − S9 (rev nmol^{-1})
1-Nitronaphthalene	173	48	0.05,[c] 0.4[c]
2-Nitronaphthalene	173	890	0.2,[c] 0.9[c]
1-Nitrofluorene	211	860	4.2
2-Nitrofluorene	211	4,100	93
3-Nitrofluorene	211	5,500	41
4-Nitrofluorene	211	34	0.40
2-Nitro-6H-dibenzo[b,d]-pyran-6-one	241	58,600	d
3-Nitro-6H-dibenzo[b,d]-pyran-6-one	241	10,000	d
4-Nitro-6H-dibenzo[b,d]-pyran-6-one[b]	241	480	d
8-Nitro-6H-dibenzo[b,d]pyran-6-one[b]	241	6,500	d

[a] From Arey et al. (1992).
[b] From Atkinson and Arey (1993).
[c] From Rosenkranz and Mermelstein (1983).
[d] No data available.

potent, it is present in relatively high levels in ambient particles; indeed, it may contribute up to ~20% of their direct bacterial mutagenicity (TA98 −S9) in samples from southern California. Furthermore, in hot weather a significant fraction of **XI** has been reported in the gas phase (Arey *et al*., 1994):

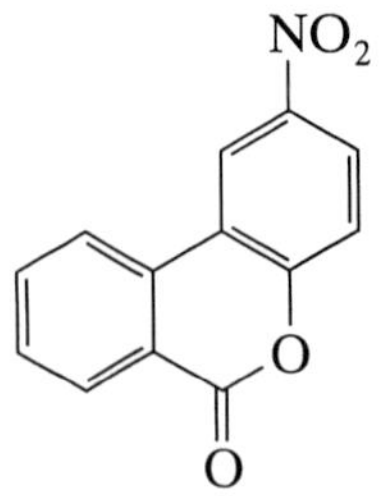

XI

Interestingly, these nitro-PAH lactones are *not* found in primary emissions of particulate POM; they are formed in atmospheric reactions (Atkinson and Arey, 1994; Arey, 1998a, and references therein; see Section F.2).

Recently, Enya and co-workers (1997) identified another very powerful type of direct mutagen, the nitro-PAH ketone 3-nitrobenzanthrone (3-nitro-7*H*-benz[*de*]anthracen-7-one, **XXIV**), in extracts of particles from ambient air and diesel exhaust:

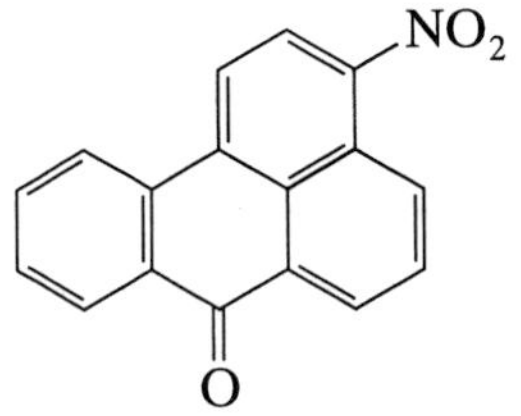

XXIV

While present only in relatively small amounts in ambient aerosols (e.g., 5.2–11.5 pg m^{-3}), its mutagenicity, 208,000 rev nmol^{-1} in TA98 (*Salmonella typhimurium*, plate incorporation assay with preincubation, −S9 mix) and 6,290,000 rev nmol^{-1} in strain YG1024, increases its importance in contributing to the total mutagenicity.

Members of another class of powerful direct mutagens, four nitroazabenzo[*a*]pyrene derivatives, were identified by Sera and co-workers (1994) in the basic fraction of extracts of diesel exhaust and in ambient POM. Structures and direct mutagenic activities on strains TA98 and YG1024 (−S9 mix) of the 1- and 3-nitro-6-azabenzo[*a*]pyrenes and the 1- and 3-nitro-6-azabenzo[*a*]pyrene-*N*-oxides are shown in Table 10.21 (Sera *et al*., 1992; Fukuhara *et al*., 1992). They have been measured in ambient air in Fukuoka, Japan, at concentrations of 1.1, 1.2, 0.8, and 0.3 ng/g, respectively, and in diesel emissions at concentrations of 4.9, 7.7, 2.2, and 3.8 ng/g, respectively (Sera *et al*., 1994).

Another powerful direct mutagen identified in ambient particulate matter collected in Santiago, Chile, is 3,6-dinitrobenzo[*a*]pyrene. Although its concentration is low (0.002 ng m^{-3} of air), its specific direct activities on strains TA98 and YG1024 (−S9 mix) are high, 400,000 and 4,800,00 rev μg^{-1}, respectively (Sera *et al*, 1991). For a discussion of polar forms of BaP in ambient particles, see Ismail *et al*. (1998).

g. Microbiology and Analytical Chemistry: Bioassay-Directed Fractionation and Chemical Analysis of Complex Mixtures of POM

Analysis of the complex mixtures of gaseous and/or particulate POM in primary emissions or ambient air is a daunting task, given the huge numbers of species and the small concentrations. The development of the *Salmonella typhimurium* assay has helped to direct such analysis through the technique of bioassay-directed chemical analysis.

With this approach, fractionation by a suitable HPLC technique separates the organic extract of a complex mixture of POM into subfractions of increasing polarities. These are then assayed for mutagenicity, e.g., using the microsuspension modification of the Ames assay. The resulting *mutagram* shows how the "total" mutagenic activity of the original POM extract is distributed among several chemically distinct classes of mutagenic subfractions. Conventional analytical techniques are then employed to determine the chemical structures and concentrations of the mutagenic species present in each subfraction.

Finally, one bioassays each of the newly identified agents and compares the sum of their mutagenic potencies with the total specific activity of the original complex POM mixture. These quantities may or may not be equal, depending on several factors, including the possibility of synergistic or antagonistic effects due to chemical interactions (e.g., see Donnelly *et al*., 1998). For examples of the use of bioassay-directed fractionations and chemical analysis, see Epler *et al*. (1979), Guerin *et al*. (1979), Huisingh *et al*. (1979), Pellizzari *et al*. (1979), Epler (1980), Pitts *et al*. (1980), Waters *et al*. (1979, 1981, 1983), Bjørseth (1983), Holmberg and Ahlborg (1983), Alfheim *et al*. (1983, 1984a, 1984b), Thilly *et al*. (1983), Schuetzle *et al*. (1985) and references therein, Nishioka *et al*. (1985, 1988), Salmeen *et al*. (1984, 1985), Schuetzle and Lewtas (1986), Lewtas *et al*. (1990a, 1990b), Arey *et al*. (1992), Harger *et al*. (1992), Zinbo *et al*. (1992), Greenberg *et al*. (1993), Sasaki *et al*. (1995), and Gupta *et al*. (1996).

TABLE 10.21 Direct Mutagenicities of Two Nitroazabenzo[a]pyrenes, Their N-Oxide Derivatives,[a] and 3,6-Dinitrobenzo[a]pyrene[c] Detected in Ambient and Diesel Particulate Matter[b] and, for Comparison, 1- and 3-Nitrobenzo[a]pyrene

Structure	Chemical	Mutagenicity (rev µg^{-1}), −S9[d]	
		TA98	YG1024
	1-Nitrobenzo[a]pyrene	2,200	26,000
	3-Nitrobenzo[a]pyrene	4,600	22,000
	1-Nitro-6-azabenzo[a]pyrene[b]	352,000	2,750,000
	3-Nitro-6-azabenzo[a]pyrene[b]	348,000	7,670,000
	1-Nitro-6-azabenzo[a]pyrene-N-oxide[b]	115,000	1,930,000
	3-Nitro-6-azabenzo[a]pyrene-N-oxide[b]	1,260,000	30,500,000
	3,6-Dinitrobenzo[a]pyrene[c]	400,000	4,800,000

[a] Adapted from Sera et al. (1992).
[b] Detected in "semivolatile phase" of ambient air and diesel emissions (Sera et al. 1994).
[c] From Sera et al. (1991).
[d] Plate incorporation assays on strains TA98 and YG1024, −S9 mix.

3. The *Salmonella* TM677 "Forward Mutation" Assay

In the late 1970s, a *forward* mutagen assay using a different strain of *Salmonella* was introduced by Skopek et al. (1978a). The "genetic marker" is resistance to 8-azaguanine. This is produced when the "normal" strain TM677—which cannot survive in the presence of this purine analog— is mutated (e.g., by an airborne mutagen) to forms that can survive (Skopek et al., 1978a, 1978b; Kaden et al., 1979; Hannigan et al., 1994, 1996, and references therein).

As with the *Salmonella* reversion assay, this short-term test is conducted both without (−PMS) and with metabolic activation produced by addition of "postmitochondrial supernatant" containing rat liver enzymes (+PMS). These terms are equivalent to −S9 and +S9 in the Ames reversion assay; we use the latter designation for both types of bacterial assays. A more sensitive micro-forward mutation bioassay using this TM677 strain to determine the mutagenicity of indoor air particles, including ETS and wood smoke, is described by Lewtas *et al.* (1987).

For a comparison of the two techniques, reverse vs forward mutation *Salmonella* assays, see Skopek *et al.* (1978b) and Lewtas *et al.* (1990b). Examples of its use in atmospheric chemistry/air pollution research include the following: application to the mutagenicity of soot and 70 PAHs (Kaden *et al.*, 1979), indoor air particles, using a modification that increased assay sensitivity (Lewtas *et al.*, 1987), urban aerosol sources compared to atmospheric samples (Hannigan *et al.*, 1994), mutagenicities of mono- and dinitropyrenes in *Salmonella typhimurium* strain TM677 (Busby *et al.*, 1994a), and seasonal and spatial trends in the mutagenicity of fine organic aerosols in southern California (Hannigan *et al.*, 1996) (vide infra).

4. Human Cell Mutagenicities of PAHs and PACs

The two short-term *Salmonella* bacterial assays have proven useful indeed from the perspectives of both atmospheric chemistry and toxicology. However, for evaluating the possible impacts on human health of biologically active PAHs and PACs, assay systems for determining *human cell* mutagenicities using human liver cells may be potentially more relevant. Nevertheless, Durant and co-workers (1998) caution that, although certain PAHs and PACs may be active in a given human cell line, "they may not necessarily pose the same risks to lymphoblastoid or other human cells *in vivo*."

Two human cell lines with good sensitivities have proven useful for laboratory and field studies; MCL-5 (Crespi *et al.*, 1991), and h1A1v2 (Penman *et al.*, 1994). Durant and co-workers (1996) report specific activities determined with the h1A1v2 assay for standard (reference) samples of 67 PAHs and PACs. Fifty-five have been identified or are "suspected to be present" in ambient urban aerosols. The human cell potencies (relative to BaP = 1.00) of selected PAHs and PACs, along with their cancer potency equivalence factors (PEFs), are given in Tables 10.13 and 10.14. Dibenzo[*a,l*]pyrene (**XXIX**), cyclopenta[*cd*]pyrene

(**XXVIII**), and dibenzo[*a,e*]pyrene (**XXX**) are 24, 6.9, and 2.9 times more mutagenic in this human cell assay than BaP (Durant *et al.*, 1996). Dibenzo[*a,l*]pyrene is nearly 50 times more powerful than BaP in the MCL-5 human cell assay (see Busby and co-workers (1995) for a discussion of its mutagenicity, bacterial and human cell, and animal carcinogenicity).

XXIX

Dibenzo[*a,l*]pyrene

Durant and co-workers (1998) analyzed an organic extract of the NIST reference complex mixture SRM 1649 (see Box 10.3) using the same human cell line (h1A1v2) and bioassay-directed fractionation. In the nonpolar fraction, cyclopenta[*cd*]pyrene, benzo[*a*]pyrene, and benzo[*b*]fluoranthene were responsible for ~7, ~4, and ~2%, respectively, of the total extract mutagenicity. Only one potent O-PAC mutagen was identified in the semipolar fraction, the ketone, 6*H*-benzo[*cd*]pyren-6-one (~0.5%) (see discussion in Section D.3).

Dibenzo[*a,l*]pyrene, a potent mutagen in the MCL-5 assay, was "not detected" in this analysis of SRM 1649. However, this 6-ring PAH **XXIX** was identified (but not quantified) by comparison to an authentic standard by Allen and co-workers (1998) in their analysis of size-segregated aerosols in urban Boston ambient air. Furthermore, in their August 1994 study of biologically active, particle-associated PAHs in ambient Riverside, California, Atkinson and Arey (1997) measured concentrations of ~18, 20, and 15 pg m^{-3} of this compound for one daytime and two nighttime sampling events, respectively. These concentrations were ~40–50% of the BaP concentrations in the same aerosol samples.

In a laboratory environmental chamber study of the gas-phase photooxidation of naphthalene and phenanthrene, Sasaki and co-workers (1997b) found two products, 2-nitronaphthalene and 2-nitrodibenzopyranone (**XI**), that displayed significant genotoxicity in the MCL-5 human cell assay. This finding emphasized the importance of atmospheric reactions in forming mutagens, since the concentrations of such compounds are relatively high in ambient air compared to those expected for nitroarenes directly emitted from primary combustion sources (see Section F).

> **BOX 10.10**
> **CYCLOPENTA[cd]PYRENE IN AIR**
>
> Cyclopenta[cd]pyrene is a significant contributor to the total mass of biologically active, particle-associated PAHs in emissions from light- and heavy-duty diesel engines, e.g., 869–1671 μg/g of extract vs 208–588 μg/g of extract for BaP (Tong and Karasek, 1984, in Table 6 in IARC, 1989; see also, e.g., Westerholm et al., 1991), and from gasoline engines without catalytic converters, e.g., 750–987 μg/L of combusted fuel vs 50–81 μg/L of BaP (see Table 10 in IARC, 1989; and Westerholm et al. (1988). For a comparison of cyclopenta[cd]pyrene emissions (and other PAHs) from California automobiles with catalytic converters (low emissions) and without catalysts (high emissions) and for heavy-duty diesel trucks, see Rogge et al. (1993a), Hannigan et al. (1994), and Schauer et al. (1996); for medium-duty diesel trucks, see Schauer et al. (1999); see also Westerholm and Egebäck (1994) for results from the Swedish Urban Air Project.
>
> Cyclopenta[cd]pyrene is also present in ambient aerosols collected in urban airsheds throughout the world at levels approximately equal to those of BaP and in some locations at significantly higher concentrations, e.g., Elverum, Norway (Ramdahl, 1983a); Kokkola, Finland (Pyysalo et al., 1987); La Spezia, Italy (Barale et al., 1994); London, England (Baek et al., 1992); Copenhagen, Denmark (Nielsen et al., 1996); in "an industrial city" in Germany (Grimmer and Misfeld, 1983); Taiwan (Lee et al., 1995); and at six sites at selected locations in California (Atkinson et al., 1988a). Furthermore, in their study of concentrations of 86 vapor-phase, semivolatile, and particle-phase aromatics in ambient Los Angeles smog (during two very hot days in September 1993), Fraser and co-workers (1998) reported cyclopenta[cd]pyrene was not only present in both the *gas* and *particle* phases but also that there was almost twice as much in the vapor phase, 0.26 (gas) vs 0.14 ng m^{-3} (particles).
>
> Given its demonstrated toxicology, relatively high levels in diesel exhaust and non-catalyst-equipped motor vehicles, widespread distribution in ambient air, and unique structure–reactivity aspects, further research on the fundamental and applied atmospheric chemistry and toxicology of cyclopenta[cd]pyrene would seem useful.

Cyclopenta[cd]pyrene (CPP, **XXVIII**) is a powerful human cell mutagen in strain h1A1v2, ~7 times more potent than BaP (Durant et al., 1996; Hannigan et al., 1998), and a strong promutagen in both the Ames *Salmonella typhimurium* reversion assay (Eisenstadt and Gold, 1978) and the *Salmonella typhimurium* strain TM677 forward mutation assay (Kaden et al., 1979):

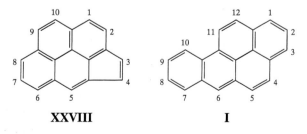

XXVIII
Cyclopenta[cd]pyrene

I
Benzo[a]pyrene

Not only is it interesting toxicologically, it has a unique structure with a highly localized double bond between the carbon atoms 3 and 4 in the cyclopenta ring.

Analogous to acenaphthylene (Tables 10.1 and 10.36 and Sections E and F), this provides a reactive site for attack by ozone and OH and NO$_3$ radicals (see Eisenstadt and Gold, 1978; Sangaiah and Gold, 1985; Goldring et al., 1988; and Zielinska et al., 1988a). This is consistent with its rapid decay during transport as observed, for example, by Nielsen (1988) in Denmark, by Greenberg (1989) at three urban sites in New Jersey, and by Fraser et al. (1998) at four sites in southern California. See discussion in Box 10.10.

The human cell mutagenicities of several mono- and dinitropyrenes were determined by Busby and co-workers (1994b) in a forward mutation assay using a metabolically competent line of MCL-5 cells. Minimum detectable mutagen concentrations (MMC, in nmol ml^{-1}; the smaller the detectable concentration, the greater the potency) were as follows: 1,6-dinitropyrene, 0.8; 1,8-dinitropyrene, 1.5; 4-nitropyrene, 3.1; 1-nitropyrene, 9.1. Other PACs tested, 2-nitropyrene, 1,3-dinitropyrene, and pyrene, were nonmutagens in this assay. There was an 11-fold range from the most (1,6-DNP) to the least (1-NP) potent mutagen.

In 1997, Busby and co-workers reported that 2-nitrofluoranthene, an important product of atmospheric transformations (vide infra) was inactive in MCL-5 cells but a potent mutagen in h1A1v2 cells; another important atmospheric reaction product, the nitrophenanthrene lactone 2-nitrodibenzopyranone (**XI**), was inactive in both h1A1v2 and MCL-5 cells. Furthermore, it was nonmutagenic in the forward mutation bacterial assay in the absence of rat liver postmitochondrial supernatant ($-$S9) but was mutagenic with the addition of S9 mix.

However, Sasaki and co-workers (1997b) reported that this nitro-PAH lactone is mutagenic in the MCL-5, and possibly the h1A1v2, human cell assays (Arey, personal communication), and it is a powerful direct-acting mutagen in the *Salmonella* reversion bacterial assay, e.g., 58,600 rev/nmol, TA98, $-$S9 (see Table 10.20, Arey *et al.* (1992), and Arey (1998a), and references therein). Additionally, DNA adducts have been found in the liver DNA from rats treated with this common air pollutant (Watanabe *et al.*, 1996). 2-Nitronaphthalene, a volatile atmospheric reaction product of naphthalene (Sasaki *et al.*, 1997a; Feilberg *et al.*, 1999a), present in relatively high concentrations in ambient air (Gupta *et al.*, 1996), is also genotoxic in the MCL-5 line of human cells (Sasaki *et al.*, 1997b).

Busby and co-workers (1995) compared the mutagenicities of BaP and five dibenzopyrenes in the *Salmonella typhimurium* TM677 forward mutation bacterial assay ($+$PMS) to three in the MCL-5 human cell assay. The powerful carcinogen dibenzo[*a,l*]pyrene was 50 times more potent than BaP in human cells (vide supra); however, it was only 1.7 times as potent as BaP in the bacterial assay. Interestingly, there was a 10,000-fold range between the most and least mutagenic PAH in MCL-5 human cells vs a range of only ∼4 in the bacterial cells.

Durant *et al.* (1996, 1998) reported that several nitroarenes, e.g., 2-nitrofluoranthene, were less mutagenic than BaP in the h1A1v2 cell line and concluded, "For these compounds to be important h1A1v2 cell mutagens, their concentrations would need to be in the order of ∼1 μg/g or higher." Although Durant and coworkers did not detect 2-nitrofluoranthene in SRM 1649, Arey and co-workers (1988b) reported levels of 2-nitrofluoranthene as high as 10 μg/g in samples of ambient air from Claremont, California, ∼60 km east of Los Angeles. Furthermore, it has been found in airsheds throughout the world (e.g., see Atkinson and Arey, 1994; Legzdins *et al.*, 1994; Ciccioli *et al.*, 1996; Hannigan *et al.*, 1997; Fraser *et al.*, 1998; and Arey, 1998a). Thus, 2-nitrofluoranthene may make a contribution to the h1A1v2 human cell mutagenicity of respirable particles in ambient air, not only in the Los Angeles area but also at many other sites as well.

The toxicology-based conclusion that the minimum concentration for 2-nitrofluoranthene to be an important human cell mutagen is 1 μg/g, coupled with air quality sampling data showing its concentrations in respirable particles sampled from ambient air can in fact reach 10 μg/g, provides a useful example of a productive symbiotic interaction between atmospheric chemists and toxicologists. Such interactions are essential for reliable risk assessments of air pollution and human health effects of complex combustion-generated mixtures of gases and particles.

D. BACTERIAL AND HUMAN CELL MUTAGENICITIES OF POLLUTED AMBIENT AIR

With the carcinogenic and mutagenic properties of certain key airborne PAHs and PACs as background, let us now examine the contributions of these genotoxic compounds to the bacterial and human cell mutagenicities of ambient air.

1. Bacterial Mutagenicity of Urban Air: A Worldwide Phenomenon

a. Background

Tokiwa and co-workers reported in 1976 that organic extracts of particles collected in Ohmuta and Fukuoka, Japan, were active in the Ames *Salmonella typhimurium* reversion assay with microsomal activation ($+$S9); that is, they contained promutagens (Tokiwa *et al.*, 1976). This was reasonable since such extracts were well known to contain many carcinogenic/mutagenic PAHs that are promutagens, e.g., BaP. However, shortly thereafter, a major new discovery was reported: organic extracts of ambient aerosols collected in Duisburg, Germany (Dehnen *et al.*, 1977); Ohmuta and Fukuota, Japan (Tokiwa *et al.*, 1977); Berkeley, California; and Buffalo, New York (Talcott and Wei, 1977); and the Los Angeles Air Basin (Pitts *et al.*, 1977) also showed strong, *direct* mutagenic activity ($-$S9). Similar results were reported for ambient particles in Chicago, Illinois (Commoner *et al.*, 1978); Kobe, Japan (Teranishi *et al.*, 1978); Stockholm, Sweden (Löfroth, 1981); and Contra Costa County, California (Wesolowski *et al.*, 1981). Furthermore, unknown "direct" chemical mutagens were associated with fine respirable particles having diameters less than 2.5 μm (Commoner *et al.*, 1979; Pitts *et al.*, 1979; Talcott and Harger, 1980; Tokiwa *et al.*, 1980; Löfroth, 1981).

The Ames assay provided significant microbiological "clues" as to the chemical nature of these airborne mutagens. Extracts of ambient air particles at all locations showed direct ($-S9$) frameshift activity with strain TA98 (or with earlier related, but less sensitive, strains TA1537 and TA1538), but little or no activity with strain TA1535 (base-pair substitution mutagens). They also displayed mutagenicity on strain TA100.

Concurrently, *Salmonella* assays of extracts of primary combustion-generated particulate *emissions* from various sources were shown to be mutagenic in Ames frameshift strains TA98 and TA100 *without* microsomal activation ($-S9$). Therefore, fly ash from coal-fired power plants (Chrisp et al., 1978), wood smoke (Löfroth, 1978; Ramdahl et al., 1982a), automobile exhaust (Wang et al., 1978), and diesel soot (Huisingh et al., 1979) must also contain direct mutagens. The question became, what were the specific compounds?

A breakthrough came in 1978, when nitroarenes were reported to be direct-acting, frameshift mutagens present both in ambient air particles collected in Prague (Jäger, 1978; 3-nitrofluoranthene and 6-nitro-BaP) and in extracts of particulate POM emissions in auto exhaust (Wang et al., 1978). These discoveries resulted in major research efforts to identify and quantify nitroarenes and other direct-acting mutagens in emissions and ambient POM; by 1983, some 100 mono- and dinitro-PAHs had been reported to be associated with diesel particles (e.g., Paputa-Peck et al., 1983; Schuetzle and Perez, 1983; MacCrehan et al., 1988).

In 1983, Tokiwa and co-workers reported dinitropyrenes (which are powerful bacterial mutagens; see Tables 10.16 and 10.19) in ambient particles from Santiago, Chile (Tokiwa et al., 1983), and Nielsen (1983) identified and quantified 1-nitropyrene and several other nitro-PAHs in complex POM mixtures. He and his co-workers also identified several nitro-PAHs in ambient particles collected in a rural area of Denmark (Nielsen et al., 1984). In the United States, nitroarenes were identified and quantified in extracts of ambient particles from St. Louis, Missouri (Ramdahl et al., 1982b); Detroit, Michigan (Gibson, 1982); and southern Ontario, Canada (Sweetman et al., 1982). For more detailed treatments and references, see, for example, articles by Alfheim et al. (1983, 1984b) and Rosenkranz (1996) and reviews by Rosenkranz and Mermelstein (1985a, 1985b), Tokiwa and Ohnishi (1986), and van Houdt (1990).

Reports that certain of the mono- and dinitroarenes found in the exhausts from diesel and gasoline engines, and in ambient air, were animal carcinogens and possible human carcinogens (e.g., see Table 10.14 and IARC, 1989) accelerated research in this new area. In subsequent years, the phenomenon of public exposure to both the direct and activatable bacterial mutagens associated with respirable ambient particles was documented for a variety of air environments throughout the world. These included cities in the San Francisco Bay area (Kado et al., 1986; Flessel et al., 1991), ten localities in the Genoa municipality, Italy (De Flora et al., 1989), four locations in Athens, Greece (Viras et al., 1990), a three-site comparative study in Rio de Janeiro, Brazil, Camden, New Jersey, and the Caldecott Tunnel, California (Miguel et al., 1990), Rome (Crebelli et al., 1991), and a number of locations in Italy (Nardini and Clonfero, 1992; Scarpato et al., 1993; Barale et al., 1994; Pagano et al., 1996), Hamilton, Ontario, Canada (Legzdins et al., 1994), Barcelona City, Spain (Bayona et al., 1994), five stations in metropolitan Mexico City (Villalobos-Pietrini et al., 1995), Copenhagen, Denmark (Nielsen et al., 1998a, 1999a, 1999b), and Sapporo, Japan (Matsumoto et al., 1998).

b. Particle Size Distribution of PAHs and Mutagenicity

An important aspect of inhalable PAHs is their distribution as a function of particle size in ambient aerosols since size is a key parameter in determining aerosol lung deposition efficiencies (see Chapter 2.A.5).

There have been a number of studies of the size distribution of PAHs and of the associated mutagenicity over the past three decades. For example, Pierce and Katz (1975) measured mass concentrations as a function of particle size for aerodynamic diameters >0.5 μm for BaP and several other PAHs and two O-PACs in aerosol samples collected at five sites in, and near, Toronto, Canada. The size distribution of the particles was approximately log-normal, and the majority of the mass of the associated PAHs and PACs was found in particles with diameters below 3 μm. Miguel and Friedlander (1978) used an eight-stage, low-pressure impactor with 50% cutoffs down to 0.05-μm aerodynamic diameter (Hering et al., 1978, 1979) to determine the mass distributions of BaP and coronene in aerosols collected in Pasadena, California. Approximately 75% of the BaP and 85% of the coronene were associated with particles with diameters less than 0.26 μm and ~50% of the mass fell in the narrow size range from 0.075 to 0.12 μm. Similar studies have been conducted at various locations around the world, including Belgium, The Netherlands, and the United Kingdom (e.g., see Van Vaeck and Van Cauwenberghe, 1978; and Baek et al., 1991).

To better understand the effects of atmospheric processes (reactions, gas-particle partitioning, etc.) on the size distributions of PAHs in ambient aerosols, Venkataraman and Friedlander (1994b) carried out measurements of gases and particles during winter and summer periods at several sites in the Los Angeles Air Basin. Concurrently, they determined the mass distributions of these PAHs in aerosol particles in vehicular emissions using the same instrumentation and analytical procedures (Venkataraman et al., 1994a). As seen in Fig. 10.15a, the distribution of fluoranthene in ambient aerosols collected during the winter is bimodal with a primary emissions mode at 0.05–0.5 μm (mode I) and an accumulation mode at 0.5–4.0 μm (mode II). A significant fraction of the fluoranthene present in the total sample is present in the accumulation mode, II. Three other 4-ring, semivolatile PAHs sampled (pyrene, chrysene, and benz[a]anthracene) had similar mass distributions.

In contrast, during the winter, the nonvolatile 5- and 6-ring PAHs BaP (Fig. 10.15b), benzo[b]fluoranthene, dibenzanthracene, benzo[ghi]perylene, benzo[k]fluoranthene, and indeno[cd]pyrene had ~63–82% of their masses in the 0.05- to 0.5-μm range, primary emission mode I. However, as seen in Fig. 10.15c, the pattern shifts significantly to larger sizes for aerosols sampled in the summer.

In a similar study, Allen and co-workers (1996) determined the particle size distribution for 15 PAHs with molecular weights ranging from 178 (e.g., phenanthrene) to 300 (coronene) and associated with urban aerosols in Boston, Massachusetts. As for BaP in the winter (Venkataraman and Friedlander, 1994b), PAHs with MW > 228 were primarily present in the fine aerosol fraction ($D_p < 2$ μm). A study of 6-ring, MW 302 PAH at the same site showed bimodal distributions, with most of the mass in the 0.3- to 1.0-μm particle size size range; a smaller fraction was in the "ultrafine" mode particles (0.09–0.14 μm) (Allen et al., 1998). For PAHs with MW 178—202, the compounds were approximately evenly distributed between the "fine" and "coarse" ($D_p > 2$ μm) fractions. Polycyclic aromatic hydrocarbons in size-segregated aerosols collected a month later at a rural site were present to a greater degree in the coarse fraction than those collected in urban Boston, consistent with other observations (e.g., Pierce and Katz, 1975; Van Vaeck and Van Cauwenberghe, 1985; Venkataraman and Friedlander, 1994b). These size distributions are consistent with the condensation of the large, nonvolatile PAHs on small particles during cooling of the exhaust. However, the smaller, semivolatile PAHs become distributed between the smaller and larger particles via continuing vaporization and condensation processes in the atmosphere.

Furthermore, a number of studies have shown that the direct mutagenicity of particles is primarily associated with particles having $D_p < 2.5$ μm. For example, Viras and co-workers (1990) report that at two sites in Athens, Greece, 81 and 92% of the total *direct* activities (TA98 −S9) were associated with particles having $D_p < 3.3$ μm. Furthermore, ~60 and 80%, respectively, were in submicron particles, $D_p < 1.0$ μm.

Similar results have been reported for the particle size distribution of promutagenic activity (TA98 +S9) of ambient particles collected by Pagano and co-workers (1996) near a busy road in Bologna, Italy. Figure 10.16 shows their data for fractions ranging from <0.4 to <3.3 μm for each of seven, week-long sampling events conducted from November 15, 1994, to March 31, 1995. Except for the first sampling period, the highest activity was in the <0.4-μm fraction.

c. Variables Influencing Mutagenicity Levels

A number of variables influence the overall levels of mutagenicity to which the general public is exposed through inhalation of ambient air particles. (Note, however, that total exposure also includes gas-phase mutagens; see later.) The factors include, for example, (a) the inherent mutagenic potencies (rev μg^{-1} of extract) of the fine particles emitted by each type of combustion

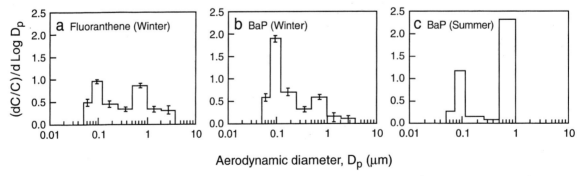

FIGURE 10.15 Mass distributions as a function of particle size of fluoranthene, winter ($C = 1.01$ ng/m^3), and benzo[a]pyrene, winter ($C = 0.060$ ng/m^3) and summer ($C = 0.21$ ng/m^3), in ambient aerosols sampled at an urban/suburban air monitoring station in Pico Rivera, near downtown Los Angeles (adapted from Venkataraman and Friedlander, 1994b).

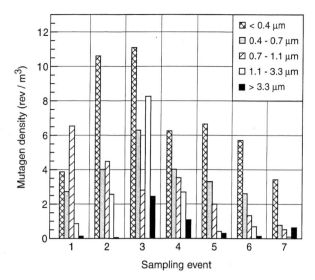

FIGURE 10.16 Particle size vs activatable mutagenicity (mutagen density, rev m^{-3}; TA98 + S9) for ambient particulate POM collected in seven week-long sampling events from November 15, 1994, to March 31, 1995. The site was "near a very busy road in Bologna, Italy" (adapted from Pagano et al., 1996).

source in that area (determined by the specific activities (rev μg^{-1}) and concentrations of the individual biologically active compounds associated with the particles), (b) the emission strengths of the particles (emission factors) from each source class, (c) the mutagenic potencies and concentrations of the secondary mutagenic PACs *formed* in atmospheric reactions of non-mutagenic PAHs (e.g., gas-phase reactions of naphthalene, phenanthrene, pyrene, and fluoranthene; see Section F), (d) the decay of reactive mutagenic/carcinogenic PAHs during transport (e.g., of BaP and cyclopenta[cd]pyrene; see Section E), and (e) the condensation of semivolatile mutagenic compounds on particles after their emission as gases.

Outdoor mutagenicity levels (rev m^{-3}) are, of course, also influenced by seasonal, spatial, meteorological, and air pollution variables and may also reflect the sampling methodology. For example, in the mid-1980s, Alfheim and co-workers reported possible artifactual formation or degradation of mutagenic species when comparing ambient aerosols sampled with different types of Hi-Vol filters and XAD-2 resin (Alfheim and Lindskog, 1984a; Alfheim et al., 1985). In similar tests of sampling with glass fiber filters (GFF) and Teflon-coated fiberglass filters, Daisey and co-workers (1986a) concluded "the filter medium can significantly influence the extractable mass, chemical composition, and mutagenic activity of the organic fractions of the ambient aerosol."

However, in a study involving four kinds of Hi-Vol filters (glass, quartz, Teflon, and Teflon-impregnated glass fiber filters), Fitz and co-workers (1984) reported no large differences due to filter artifacts. Subsequently, de Raat and colleagues (1990) found that differences in sampling for particle-phase mutagenicity using glass fiber filters vs Teflon or Teflon-coated filters were too small to support one type of filter over the other.

Sampling for different periods of time, e.g., 3 h, 24 h, monthly, seasonal, and annual averages, conveys different types of information, chemically and toxicologically. For example, mutagenic and chemical product evidence of the relative efficiencies of OH radical attack on PAH during daylight hours vs NO$_3$ radical attack at night can be deduced from 4- to 12-h averages of direct mutagenicity (e.g., see Arey et al., 1989a) but would be masked in 24-h, monthly, seasonal, or yearly averages (see, e.g., discussion by Masclet et al. (1986) on the impacts of atmospheric reactions, sampling parameters, and air pollution/meteorological variables).

The location of a sampling site relative to the sources, meteorology, topography, etc. characteristic of that area is also clearly important. Thus, in some cases the sampling site is close to a strong source(s) of emissions that dominates the local air quality. Thus, in urban areas worldwide, with very different climates, meteorology, and topography, emissions of PAHs and PACs from diesel- and gasoline-powered light- and heavy-duty motor vehicles (e.g., diesel trucks and non-catalyst-equipped automobiles) have been, and continue to be, major contributors to the particle-phase *promutagenicity* of respirable ambient particulate POM (e.g., see Benner et al., 1989, and the review by van Houdt, 1990). This has been shown to be true, for example, in urban air in Oslo, Norway, and Stockholm, Sweden (Alfheim et al., 1983), Copenhagen, Denmark (Nielsen, 1996; Nielsen et al., 1996, 1999a, 1999b), the San Francisco Bay Area (Flessel et al., 1985; Kado et al., 1986; Flessel et al., 1991), southern California cities (Pitts et al., 1982b, 1985a; Atkinson et al., 1988a; Venkataraman and Friedlander, 1994b, 1994c; Schauer et al., 1996; Hannigan et al., 1994, 1996, 1997), Helsinki and Lahati, Finland (Tuominen et al., 1988), Athens, Greece (Viras et al., 1990), Padova, Italy (Nardine and Clonfero, 1992), The Netherlands (De Raat et al., 1991), and Sapporo, Japan (Matsumoto et al., 1998).

The impact of close proximity to a major source on the mutagenicity levels of ambient air particles (rev m^{-3}) is seen in Fig. 10.17 for a site immediately adjacent to, and generally downwind from, a heavily traveled freeway in west Los Angeles (Pitts et al., 1985a). Peak levels were ~120 (+S9) and 100 (−S9) rev m^{-3} during the midmorning rush hour and ~110 (±S9)

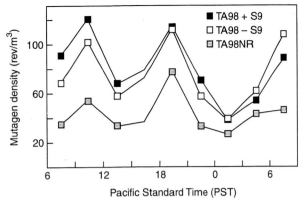

FIGURE 10.17 Diurnal variations in mutagen densities (rev m^{-3}, TA98 +S9, TA98 −S9, and TA98NR) of 3-h averaged samples of ambient particles collected on TIGF filters at a site downwind from, and adjacent to, a heavily traveled freeway in west Los Angeles (March 9–10, 1983). Midmorning and early evening peaks reflect the two rush hour traffic periods and demonstrate the utility of short-time, 3-h vs 24-h average sampling periods (adapted from Pitts *et al.*, 1985a).

during the early evening traffic "jam," compared to ~60 rev m^{-3} in the early afternoon and ~40 rev m^{-3} in the early morning hours. The results for TA98NR were significantly lower, indicating the presence of nitroarenes.

Alternatively, some sampling sites are at locations that are downwind from polluted urban centers, so that under common meteorological/weather conditions, one is sampling "aged" air parcels that have been subjected to the phenomena of transport and atmospheric reactions (e.g., Azusa and Riverside, California, east and usually downwind from central Los Angeles; see map, Fig. 10.23). Such transport can occur over long distances. For example, although the mutagenicity of atmospheric aerosols sampled at a "background site" in Birkenes, Norway, rarely exceeded 1 rev m^{-3}, Alfheim and Møeller (1979) showed that a substantial fraction of it arose from long-range transported air pollution.

T. Nielsen and co-workers investigated, at a site in downtown Copenhagen, the mutagenicity of polluted air masses transported from the European continent compared to that originating from local sources. The direct mutagenicity was higher by a factor of 5–7 compared to "average" local levels; arrival of the air mass undergoing long-range transport was also associated with an increase in the local SO$_2$ level by a factor of ~2.5 (Nielsen *et al.*, 1999a, 1999b).

An interesting aspect of the long-range transport of mutagenic compounds is a field study by De Pollok and co-workers (1997), who collected and measured ambient air particles at three different levels of a TV tower (surface, <1 m; mid, 240 m, and top 433 m) near Raleigh, North Carolina, for three periods prior to, during, and after Hurricane Gordon in November 1994. The surface samples were not mutagenic on strains YG1021 and YG1026 (vide supra), but the top- and mid-level samples showed significant mutagenicity for the "posthurricane, normal weather samples." The authors suggest this arises from long-range transport of mutagenic nitroarenes.

While mobile sources are generally very important, seasonal patterns of emissions from various stationary local sources combined with the particular meteorology (e.g., "tight" low-radiation inversions in the winter) can have a significant impact on the nature and extent of mutagenicity. For example, residential wood combustion plays a major wintertime role in many regions of the world such as Elverum, Norway, a small city in a heavily wooded region (Ramdahl *et al.*, 1984a), Albuquerque, New Mexico (Lewis *et al.*, 1988), Juneau, Alaska (Watts *et al.*, 1988), and Christchurch, New Zealand (Cretney *et al.*, 1985) where domestic wood and coal combustion contributes. Similarly, Pyysalo *et al.* (1987) reported seasonal variations (late spring vs early winter) in the genotoxicity of particulate and vapor phases of ambient air—and associated levels of PAHs and PACs—in a small industrial town in Finland impacted by emissions from domestic and industrial energy sources (coal, heavy oil, wood, waste wood materials, and peat).

Another seasonal combustion source with major local, and in some cases regional, impacts on particle loadings, PAH levels, and ambient mutagenicities in areas worldwide is large-scale, prescribed open burning of biomass to dispose of crop and forest residues, e.g., sugar cane, orchard prunings, wheat, barley, and rice straws, and Douglas fir and ponderosa pine slash (for emission factors, see Jenkins *et al.*, 1996). For example, Mast and co-workers (1984) reported that in the early 1980s, some 2–3 million tons (U.S.) of these straws were generated annually in California alone, with virtually all of the rice straw burned in the field during October and November and early spring. In one study, they collected rice straw particulate matter both in the field during such a "prescribed burn" and in laboratory experiments in an instrumented burning tower. Particle extracts from all samples were active on strain TA98 with microsomal activation (+S9). A wide range of PAHs and PACs were isolated and characterized (as well as other organics), including alkylated phenanthrenes, which they suggested contributed to the particle mutagenicity.

During wintertime, in most urban areas, mutagenicity levels are generally higher than during the summer. For example, Fig. 10.18 shows the results of Viras and co-workers (1990) for 24-h average mutagenicities (rev

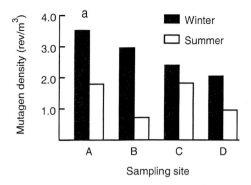

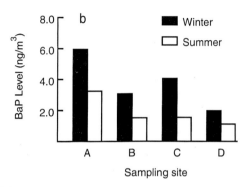

FIGURE 10.18 Seasonal variation in the mutagen density (rev m^{-3}, TA98, $-S9$) of extracts of ambient particles, and the associated BaP levels (ng m^{-3}), collected at four sites in Athens, Greece, February 1984–January 1985. The annual average was 1.9 rev m^{-3} (adapted from Viras et al., 1990).

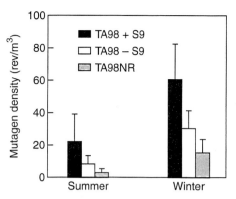

FIGURE 10.19 Mean (and standard deviations) of the mutagenicities of ambient particles (rev m^{-3}; TA98 $+S9$, TA98 $-S9$, and TA98NR) sampled for 3-h periods winter and summer at eight streets with different widths and traffic volumes in Padova, Italy, June 1990–February 1991. The decrease on strain TA98NR indicates the presence of nitroarenes (adapted from Nardini and Clonfero, 1992).

m^{-3}, TA98 $-S9$) and the concurrent BaP levels of ambient aerosol samples taken at four sites in Athens once or twice weekly over a period of one year (February 1984 to January 1985). The mean annual values of mutagenicity (1.5–2.5 rev m^{-3}) are low relative to other polluted urban areas.

Figure 10.19 shows results from a similar study conducted in Padova, Italy, by Nardini and Clonfero (1992). Sampling (TA98, $+S9/-S9$) was at street level for 3-h periods in the morning for 23 summer and 22 winter events between June 1990 and February 1991; stations were at eight streets with different widths and traffic volumes. Mean values on TA98NR for all samples ($+S9$ and $-S9$) were lower in both the summer (by ~38%) and the winter (by ~49%), indicating the presence of nitroarenes (see Section C.2). Similar examples of higher mutagenicities in winter have been reported by others (e.g., Tuominen et al., 1988; De Flora et al., 1989; Flessel et al., 1991; Bodzek et al., 1993; Scarpato et al., 1993; Nielsen et al., 1999a, 1999b; Matsumoto et al., 1998). Note, however, that winter averages are not always higher than summer averages. For example, Bayona and co-workers (1994) found the highest levels of direct-acting mutagenicity in Barcelona, Spain, were in the fall and spring, whereas the promutagenicity was highest in the summer.

Of course, short-term meteorological variables such as atmospheric stability, height and strength of inversion layers (e.g., strong wintertime, low-radiation inversions), wind speeds, ambient temperature, and precipitation can have major impacts on gas- and particle-phase ambient mutagenicity levels. In terms of air pollution parameters, lead in fine ambient particles sampled from major urban airsheds correlates strongly with mutagenicity and motor vehicle traffic. A number of investigators have also explored possible correlations of ambient levels of PAHs and PACs as well as O_3, SO_2, NO_x, and VOCs in a number of locations [e.g., Scandinavian cities (Møller et al., 1982); downtown Los Angeles and two downwind sites (Pitts et al., 1982b); Detroit, Michigan (Wolff et al., 1986); four San Francisco Bay Area cities (Kado et al., 1986); Wageningen (rural) and Terschelling (background), The Netherlands (van Houdt et al., 1987); two cities and a rural area in Finland (Tuominen et al., 1988); seven sites throughout California (Atkinson et al., 1988a); Athens, Greece (Viras et al., 1990); 17 Italian towns (Barale et al., 1994); Copenhagen, Denmark (Nielsen, 1996; Nielsen et al., 1999a, 1999b)].

2. Sources, Ambient Levels, Transport, and Transformation: Some Case Studies

a. Bacterial Mutagenicity

Let us now consider some case studies of PAHs and PACs in ambient air. A number of studies of bacterial mutagenicity of fine ambient aerosols (primary and secondary) have been conducted using different tests for genotoxicity, particularly the Ames *Salmonella typhimurium* reversion bacterial assay and the *Salmonella* TM677 forward mutation bacterial assay. Human cell assays are discussed in the following section.

Possible health effects on the general public of mutagenic/carcinogenic PAHs and PACs in polluted ambient air are generally viewed as being related to long-term exposures to these agents. In an important contribution to the area, Matsumoto and co-workers (1998) reported the results from an 18-year study (1974–1992) in Sapporo, Japan, of the impacts of seasons and sources on the mutagenicities of seasonal composites of ambient particulate matter and concurrent concentrations of BaP (and eight other PAHs) and its gaseous copollutants NO_2 and SO_2. Sapporo (population 1.7 million in 1992) has a great deal of commercial, but relatively little industrial, activity. Outdoor temperatures ranged from $-10°C$ in the winter to over $30°C$ in the summer. The two major sources of air pollution were combustion of fossil fuels for heating homes and offices (mainly coal) and exhaust gases and particles from gasoline- and diesel-fueled motor vehicles. The composites were assayed on strain TA98 (and TA100; see article) without S9 mix ($-S9$, direct activity) and with added S9 ($+S9$, indirect activity, promutagens) using a "preincubation procedure" (Yahagi *et al.*, 1977) of the Ames *Salmonella typhimurium* reversion assay (Ames *et al.*, 1975; Maron and Ames, 1983).

As seen in Figs. 10.20a and 10.20b with strain TA98, both the direct ($-S9$) and indirect ($+S9$) activities of the ambient particles showed pronounced seasonal variations, being highest in winter and lowest in summer. Furthermore, over the 18-year period, indirect activities for each season (Fig. 10.20b), and for the corresponding annual averages (not shown), declined to ~60–50% of their initial values. However, the *direct* activities of the seasonal and annual aerosol samples did not change significantly. This is reflected in the 3-year moving-average ratios of direct to indirect activity ($-S9/+S9$), which rose almost linearly from ~0.6 in 1976 to ~1.2 in 1990. Concurrently, the concentrations of BaP, Fig. 10.20c, showed matching seasonal influences, but the composites and annual averages dropped dramatically, e.g., from ~4.8 ng m^{-3} in 1975–1976 to ~0.7 ng m^{-3} in 1989–1991. Similar results were found for the sums of the eight other PAHs.

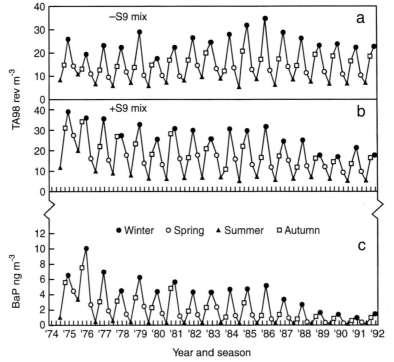

FIGURE 10.20 Seasonal concentrations of benzo[*a*]pyrene (ng m^{-3}) and direct ($-S9$ mix) and indirect ($+S9$ mix) mutagenicities (rev m^{-3}) on TA98, Ames *Salmonella typhimurium* assay with the "preincubation procedure" of Yahagi *et al.* (1977) of ambient aerosols collected annually during the winter, spring, summer, and autumn for 18 years (1974–1992) in Sapporo, Japan (adapted from Matsumoto *et al.*, 1998).

Matsumoto and co-workers rationalize their results for BaP levels and the increasing ratio of direct/indirect in terms of changes in source strengths. Thus, there was a large decrease in coal consumption in the Sapporo area, from ~375,000 tons in 1975 to only ~20 tons in 1992. Concurrently, there was a large increase in the number of motor vehicles in that area, from ~300,000 in 1975 to over one million in 1992, and the percentage of diesel-powered vehicles went from <10% in 1975 to 32% in 1992.

Additionally, starting in the mid-1970s, there were stricter regulations on the sulfur content of fuel oil. These, coupled with the large decrease in coal use, are reflected in ambient SO_2 levels, which, for example, in the wintertime dropped from ~30 ppb in 1975–1976 to ~7 ppb in 1991–1992.

They concluded that, given the fact that over 18 years the direct mutagenicity of ambient particles remained unchanged while the concentration of BaP dropped significantly, BaP is not a good indicator of the mutagenicity/carcinogenicity of extracts of ambient particles. Finally, in agreement with other studies (see Sections E and F), they suggested more attention be paid to nitro-PACs "emitted directly from diesel-powered vehicles" as well as those formed secondarily in atmospheric reactions involving NO_x.

A number of studies of bacterial mutagenicity and the PAHs and PACs responsible for it have been carried out in southern California. At the risk of seeming somewhat parochial, we describe briefly the results of some of them, since they are unique in the breadth of parameters and species measured concurrently. A key objective of these studies was to better understand the role played by atmospheric reactions in determining the composition and mutagenicity of fine particles in polluted air parcels at or near their point of origin (e.g., Long Beach and downtown Los Angeles) and subsequently during transport to cities ~40–110 km downwind. While specific results from this area are presented, the major conclusions should be generally applicable to major polluted airsheds throughout the world, e.g., Mexico City (Villalobos-Pietrini et al., 1995).

For example, Atkinson and co-workers (1988a) sampled at seven sites across California in 1986–1987, selected because they had different emission sources (see Table 10.22). The eighth site, San Nicholas Island (~130 km offshore in the Pacific Ocean, Fig. 10.21), was chosen as the "upwind" regional background site. A total of 118 sets of 12-h duration daytime and nighttime gas- and particle-phase samples were obtained and 35 PAHs, 9 nitroarenes, and dibenzothiophene were identified and quantified. Particle-phase mutagenicities of samples were assayed on strains TA98 +S9 and −S9 as well as strains TA98NR and TA98/1,8-DNP_6, all without activation (−S9).

As seen in Fig. 10.22, at the six mainland sites, there was a correlation between direct-acting POM mutagen densities (rev m^{-3}) and measured concentrations (pg m^{-3}) of the strong, direct mutagen 2-nitropyrene (16,000 rev μg^{-1}). This nitroarene is not present in primary combustion emissions (possible exceptions are rare); rather, it is formed in both ambient and simulated polluted atmospheres as a product of the homogeneous gas-phase reaction of OH radicals with pyrene (see Section F, articles by Nielsen et al. (1984), Pitts et al. (1985b), Nielsen and Ramdahl (1986), Sweetman et al. (1986), Winer and Atkinson (1987), and Zielinska et al. (1989a), and reviews by Pitts (1987), Zielinska et al. (1990), Atkinson and Arey (1994) and Arey (1998a)). Furthermore, *changes* in the ambient PAH burden that

TABLE 10.22 Average (+S9 and −S9) and Highest (−S9) Values of Mutagenicity[b] (rev m^{-3}) of Ambient Particles Collected[c] at Seven Cities/Sites in California[d] with Different Types of Emission Sources[a]

Sampling dates	Location	Source characterization	Average mutagenicity (rev m^{-3}), TA98		Highest value (−S9)
			+S9	−S9	
8/86	Glendora	Motor vehicle	33	35	61
10/86	Yuba City	Agricultural burning	24	30	95
12/86–1/87	Concord	Industrial	63	62	130
2–3/87	Mammoth Lakes	Wood burning	22	7	17 (84 with S9)
3–4/87	Oildale	Oil production	10	9	20
5–6/87	Reseda	Residential	19	22	50
7/87	Pt. Arguello	Rural	0.2	0.4	0.5

[a] Adapted from Atkinson et al. (1988a).
[b] Mutagen density (rev m^{-3}) on strain TA98, standard Ames plate incorporation *Salmonella typhimurium* reversion array.
[c] 12-h daytime or nighttime duration; Hi-Vol sampler with 10-μm cutoff.
[d] See Fig. 10.21.

FIGURE 10.21 Twenty-four-hour average values of mutagenicity levels of ambient particulate matter [rev m^{-3}, TA98 with (*) and without S9] collected at seven sites in California significantly impacted by different types of combustion-generated emission sources: Glendora (motor vehicle), Yuba City (agricultural burning), Concord (industrial), Mammoth Lakes (major wintertime ski area, ~2500-m elevation, wood-burning and motor vehicle emissions), Oildale (oil production), Reseda (residential), and Point Arguello (rural); also shown is San Nicholas Island (upwind regional background). Standard Ames plate incorporation *Salmonella typhimurium* reversion assay; 12-h sampling periods, day and night: 8/86, 10/86, 12/86–1/87, 2–3/87, 3–4/87, 5–6/87, and 7/87; San Francisco and Los Angeles are included for reference to the locations of the actual sampling sites (adapted from Atkinson *et al.*, 1988a).

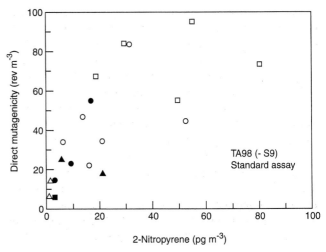

FIGURE 10.22 Direct mutagenicity of ambient particles (mutagen density, rev m^{-3}, TA98, −S9) as a function of ambient concentrations of 2-nitropyrene, a directly mutagenic product of a gas-phase atmospheric reaction initiated by OH radical attack on pyrene. Samples collected at six sites in California with different types of emissions: (●) Glendora; (○) Yuba City; (□) Concord; (■) Mammoth Lakes; (△) Oildale; (▲) Reseda (see Fig. 10.22) (adapted from Atkinson *et al.*, 1988a).

took place during transport of a given air parcel from its initial source to the sampler at each downwind site were "mirrored" by the measured direct-acting mutagenicities of the transported aerosols, suggesting that the direct mutagenicity is due to the atmospheric reaction products of the primary PAH emissions. Finally, based on the measured concentrations of directly mutagenic MW 247 nitroarenes (2-nitrofluoranthene, 8-nitrofluoranthene, 1-nitropyrene, and 2-nitropyrene) in the POM samples collected at the different locations and the specific activities of the compounds, these nitro-PAHs were estimated to contribute no more than ~10% (and generally ≤5%) to the total ambient direct mutagen densities at those sites; the remaining ~90% of the direct mutagenicity is at least partially due to more polar compounds formed in atmospheric reactions (Atkinson *et al.*, 1988a).

This conclusion was subsequently confirmed in further field and laboratory studies by the discovery of a class of powerful, direct-acting mutagens, the nitrodibenzopyranones (Table 10.20; see Arey *et al.*, 1992; Harger *et al.*, 1992; and Helmig *et al.*, 1992a, 1992b). These are present not only in ambient particulate POM but also as particle-phase products of gas-phase reactions of phenanthrene in simulated polluted atmospheres (e.g., Arey *et al.*, 1992, 1994; Sasaki *et al.*, 1995; Atkinson and Arey, 1994; Arey, 1998a and references therein; and Section F).

From a risk assessment/risk management perspective, such results (and those from other studies; see later) demonstrate that knowing the sources, mutagenic potencies, and associated emission levels of primary combustion-generated particles in a given air parcel originating near the source(s) gives one useful measure of the particle-phase mutagen density in the polluted ambient air at that location. However, under common air pollution scenarios (Los Angeles Air Basin, Mexico City, etc.) the situation becomes more complex. Thus there is the formation of direct mutagens as reaction products during transport simultaneously with the concurrent decay of primary promutagenic PAHs by various reactions discussed in Sections E and F. Indeed, under conditions of stable meteorology (e.g., "tight inversions") and severe photochemical smog, significant formation and decay of mutagenic species in ambient air also can take place close to the original site of the

TABLE 10.23 1993 Annual Averages of Bacterial Mutagenic Properties (*Salmonella typhimurium* TM677 Forward Mutation Assay)[b,d] of Fine Ambient Particles Collected at Upwind Background (San Nicolas Island), Urban (Long Beach and Central Los Angeles), and Downwind (Azusa and Rubidoux) Sites across Southern California[a,c,d]

Site	Class	Mutagenic potency (mutant fraction × 10^5 / μg of EOC)[b]		Mutagenic density (mutant fraction × 10^5 / m^3 of air)[d]	
		− S9	+ S9	− S9	+ S9
Long Beach	Urban	3.0 ± 0.15	0.69 ± 0.04	21.8 ± 1.4	4.8 ± 0.34
Central Los Angeles	Urban	2.5 ± 0.12	0.95 ± 0.05	22.6 ± 1.0	8.5 ± 0.41
Azusa	Downwind	2.0 ± 0.10	0.44 ± 0.04	17.3 ± 0.8	3.7 ± 0.34
Rubidoux	Downwind	1.1 ± 0.05	0.46 ± 0.03	12.1 ± 0.7	5.4 ± 0.39
San Nicolas Island	Background	0.52 ± 0.03	0.07 ± 0.03	0.7 ± 0.02	0.2 ± 0.02

[a] Adapted from Hannigan *et al.* (1996).
[b] For discussion of assay, see Busby *et al.* (1994a); EOC = equivalent organic carbon.
[c] See map of these sites in southern California, Fig. 10.23.
[d] These are not the values in the article by Hannigan *et al.* (1996). They are larger by a factor of 1.7 to "reflect recalibration of airflow rates." The mutagenic potency data are not affected (Cass, personal communication; see Hannigan *et al.*, 1998).

primary emissions (e.g., central Los Angeles) prior to transport downwind (see review by Arey (1998a) and references therein).

In an analogous set of studies at four cities across southern California (see Fig. 10.23), Hannigan and co-workers (1996) employed the *Salmonella typhimurium* TM677 forward mutation bacterial assay (Skopek *et al.*, 1978b; Busby *et al.*, 1994a) to determine the seasonal and spatial variation of the bacterial mutagenicities of fine ambient aerosols. The 1993 annual average bacterial mutagenic properties of the fine aerosols collected at the four urban and one upwind "background" site are summarized in Table 10.23. Consistent with the 1988a study of Atkinson and co-workers, the mutagen densities of ambient air at the four urban locations are more than an order of magnitude higher than at San Nicolas Island upwind. In this study, at most sites the monthly averages did not show a significant seasonal variation.

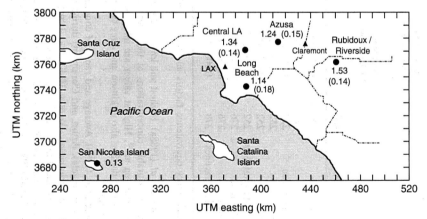

FIGURE 10.23 The 1993 annual averages for the h1A1v2 human cell mutagenic potencies shown in parentheses (induced mutant fraction × (10^6) per μg of equivalent organic carbon) and mutagen densities shown without parentheses (induced mutant fraction × (10^6) per m^3) of samples collected at four urban community air monitoring sites across southern California and at a "regional background" site, San Nicolas Island (adapted from Hannigan *et al.*, 1997). The grid is in universal coordinates. Los Angeles International Airport (LAX) and Claremont are shown for their locations relative to the four mainland sampling sites. Note that the values shown here for mutagen densities are larger by a factor of 1.7 than those in the Hannigan *et al.*, 1997, article to "reflect a recalibration of the air flow rate through the sample." The mutagenic *potency* data were not affected and remain unchanged (Cass, private communication; see also Hannigan *et al.*, 1998).

TABLE 10.24 Average Concentrations (ng m^{-3}) and Their Ranges[a] of Selected PAH which Are Animal Carcinogens and/or Human Cell Mutagens in Ambient Aerosols[a] Collected at Six Sites in California; Four Southern California Cities; Hamilton, Ontario, Canada; London, U.K.; and Copenhagen, Denmark

PAH	MW	Six sites in California[b] Range of highest and lowest values	Four southern California cities[c] Mean (range)	Hamilton, Ontario, Canada[d] Mean (range)	London, U.K.[e] Mean (range)	Copenhagen, Denmark[f] Mean City street	Copenhagen, Denmark[f] Mean City park
Cyclopenta[cd]pyrene	226	0.01–11.8	0.40 (0.00–2.74)[c]	—	0.80 (0.16–4.31)	6.1	1.5
Benz[a]anthracene	228	0.13–18.3	0.25 (0.00–2.38)[c]	2.2 (0.01–12.8)	1.48 (0.41–5.64)[g]	4.1	1.0
Chrysene + triphenylene	228	0.300–23.6	0.78 (0.00–3.32)	2.2 (0.01–13.6)	0.56 (ND–3.15)	7.9	2.3
Benzo[a]pyrene	252	0.130–12.5	0.14 (0.00–0.80)	2.0 (0.01–12.9)	ND	4.4	1.4
Benzo[e]pyrene	252	0.25–8.0	0.22 (0.02–1.00)	2.3 (0.02–16.4)	2.15 (0.54–7.77)[h]	4.4	1.3
Benzo[b + j + k]fluoranthenes	252	0.430–23.2	0.24 (0.00–1.17)[j]	5.1 (0.04–30)	ND	9.3	2.9
Indeno[1,2,3-cd]pyrene	276	0.110–12.3	0.29 (0.02–1.38)	3.0 (0.03–21.2)	4.40 (0.49–16.8)	4.5	1.1
Benzo[ghi]perylene	276	0.230–11.4	0.77 (0.03–4.23)	4.3 (0.07–25.4)	0.33 (ND–1.35)[i]	8.0	1.5
Dibenz[a,c + a,h]anthracenes	278	0.008–3.6	ND	0.8 (0.01–5.7)		ND	

[a] For cancer potency equivalence factors and human cell mutagenicities, see Table 10.13. See text for more information on sampling times and dates collected.
[b] From Atkinson et al. (1988a). Highest and lowest 12-h concentrations measured in 1986–1987 at six sites in California with different emissions (see Fig. 10.21). The only measurable concentrations at the "clean air" site, Point Arguello, were benz[a]anthracene, 0.007; chrysene + triphenylene, 0.050; and benzo[e]pyrene, 0.007 ng m^{-3}. Highest values are all from wintertime measurements in the Mammoth Lakes ski area; lowest values are generally at the residential site in Reseda. See also Atkinson and Arey (1997) and Table 10.15 for data collected in 1994 on concentrations of these carcinogenic PAHs in Riverside, California.
[c] From Fraser et al. (1998). Average and range of concentration (ng m^{-3}) of 4-h samples collected every 6 h at four sites in southern California (Long Beach, central Los Angeles, Azusa, and Claremont) during hot weather and severe smog episode (peak O$_3$ = 290 ppb, 1-h average), September 8–9, 1993. Concentrations of cyclopenta[cd]pyrene (0.40) and benz[a]anthracene (0.25) are the sums of their vapor-phase (0.26 and 0.10) and particle-phase concentrations (0.14 and 0.15) ng m^{-3}, respectively.
[d] From Legzdins et al. (1994); mean and range of 24-h samples.
[e] From Halsall et al. (1994). Values reported are the 1992 annual averages of the sums of the vapor- and particle-phase concentrations of PAHs in samples collected weekly. The ranges given in parentheses are those for week-long samples taken biweekly.
[f] From Nielsen et al. (1996). Values are mean concentrations for seventy-six 24-h samples and 12 "short-term" samples in the heavily traveled street and fifty-one 24-h samples in the city park.
[g] Only chrysene.
[h] Sum of benzo[k]fluoranthene and benzo[b]fluoranthene.
[i] Only dibenzo[a,c]anthracene.
[j] Sum of benzo[k]fluoranthene and benzo[j]fluoranthene.

b. Human Cell Mutagenicity

The first studies of *human cell* mutagenicity of fine particles across an air basin were carried out by Hannigan *et al.* (1997, 1998) using the h1A1v2 human cell assay of Penman *et al.* (1994). Figure 10.23 shows the 1993 annual average h1A1v2 human cell mutagenic potencies (values in parentheses expressed as induced mutant fraction IMF ($\times 10^6$) per μg of equivalent organic carbon) and mutagen densities (values, no parentheses, expressed as induced mutant fraction ($\times 10^6$) per m^3 of air). There were no systematic seasonal variations in the 1993 annual averages of the human cell mutagenic potencies of aerosols collected at any of the four urban sites, suggesting that no single, major seasonal emission source dominated the release of human cell mutagens into southern California air. Rather, they arise from a mixture of primary sources operating throughout the year, dominated by heavy motor vehicle emissions (e.g., see Schauer *et al.*, 1996). In addition, as was the case for bacterial mutagens, if a significant portion of the human cell mutagens in Los Angeles air are formed in atmospheric reactions (e.g., 2-nitrofluoranthene), such reactions must take place basinwide, and in wintertime, as well as in the fall periods of maximum photochemical smog.

Although aerosols collected at the farthest downwind site (Rubidoux) show approximately the same mutagenic potency as the two upwind sites at central Los Angeles and Azusa (~ 0.14 induced mutant fraction ($\times 10^6$) per μg of equivalent organic carbon), the downwind site has the highest annual loading of airborne fine particles and associated highest concentration of organic carbon. Hence the product of the mutagenic potency and the equivalent organic carbon loading results in the higher average mutagen density, 1.53 induced mutant fraction ($\times 10^6$) per m^3 of air at the downwind site. Not surprisingly, given the sources in urban areas, the mutagen density for the samples collected at the regional background site (not an annual average) was a factor of ~ 10 less than at any of the four onshore urban sites.

3. Bioassay-Directed Chemical Analysis for PAHs and PACs in Fine Ambient Aerosols Using a Human Cell Assay

In the studies of Hannigan *et al.* (1997) described in the previous subsection, the contributions of *specific* PAHs and PACs to the total human cell mutagenicity of fine ambient aerosols in southern California air were not identified and quantified. However, previous investigators had determined ambient levels of a number of carcinogenic PAHs (IARC classes 2A and 2B; see Table 10.13) present in the polluted Los Angeles Air Basin (e.g., see Atkinson *et al.*, 1988a; Atkinson and Arey, 1997; and Venkataraman and Friedlander, 1994b), many of which are mutagenic in the h1A1v2 cell line (Durant *et al.*, 1996). Furthermore, such PAH human cell mutagens/animal carcinogens are ubiquitous, as illustrated in Table 10.24 for several urban sites in Canada, the United States, and western Europe. It could be expected that at least some of the biologically active PAHs in Table 10.24 would contribute significantly to the "total" human cell mutagenicities of the 1993 fine aerosols assayed by Hannigan *et al.* (1997).

As discussed earlier, with respect to bacterial mutagens, one of the applications of mutagen assays is their use in focusing analytical studies directed to determining the chemical composition. This approach was also used by Hannigan and co-workers (1998), who carried out a bioassay-directed chemical analysis using the h1A1v2 human cell line of a composite sample made up of a portion of every 24-h filter sample.

The approach is illustrated in Fig. 10.24. Briefly, HPLC was first used to separate the whole fraction extract into four portions: *nonpolar 1* (PAHs and alkanes); *nonpolar 2* (high molecular weight PAHs and some smaller nitro-PACs, e.g., 9-nitroanthracene); *semipolar* (nitro-PAHs and other nitro-PACs and oxy-PACs, e.g., polycyclic ketones, quinones, aldehydes, amides, and aza-PACs); and *polar* (PAC anhydrides and acids, aliphatic aldehydes, and other organics). These four major fractions were then separated into a total of 19 subfractions. Those that were mutagenic have bold outlines; the exception is fraction 6a, which is double boxed and, along with fraction 7a, contains a significant amount of the moderately strong mutagen 6*H*-benzo[*cd*]pyren-6-one.

The sum of the absolute potencies of these mutagenic subfractions, 243 ± 35 induced mutant fraction ($\times 10^6$)/mg of equivalent organic carbon, is significantly larger ($\sim 160\%$) than the value for the unfractionated extract (150 ± 31), a phenomenon that has been observed in other environmental human cell mutagenicity studies (e.g., Durant *et al.*, 1994). Hannigan *et al.* (1998) proposed that this could be due to the presence of fewer interfering compounds in the less complex, fractionated material.

Hannigan *et al.* (1998) reported that most of the mutagenicity of the extract of the "whole," composite sample of respirable ambient aerosols was in the "whole" aromatics fraction, 83 vs 17% in the aliphatic portion. Table 10.25 summarizes the results of this bioassay-directed chemical analysis for the aromatics. As seen in Table 10.26, six PAHs, cyclopenta[*cd*]pyrene,

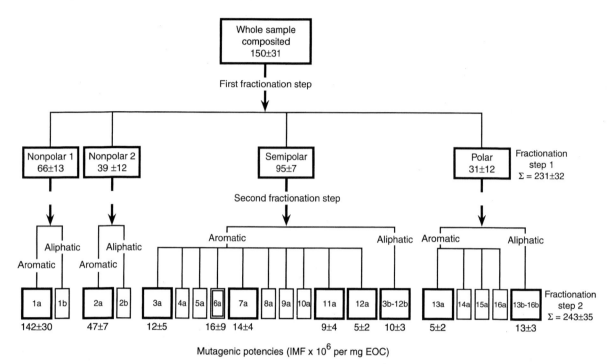

FIGURE 10.24 Flow chart for bioassay-directed chemical analysis of h1A1v2 human cell mutagens in composite sample of fine ambient aerosols collected in the year 1993 at four urban community monitoring stations in the Los Angeles Air Basin. Mutagenic potencies of the whole sample (150 ± 31), the individual subfractions (1a → 16b), and the sum of the subfractions (243 ± 35) are in units of induced mutant fraction, IMF, ($\times 10^6$) per mg of equivalent organic carbon, EOC. Standard deviations for subfractions are omitted for clarity. Fractions that were mutagenic have bold outlines or have a double box (adapted from Hannigan et al., 1998.)

BaP, benzo[*ghi*]perylene, benzo[*b*]fluoranthene, indeno[1,2,3-*cd*]pyrene, and benzo[*k*]fluoranthene, contribute the major portion of the identifiable mutagenicity of the extract of the whole unfractionated sample, accounting for 8.6, 2.5, 1.7, 1.4, 1.2, and 0.8%, respectively, of the total mutagenicity of the whole sample. Two semipolar mutagenic PACs were also present at significant levels: 2-nitrofluoranthene, a product of atmospheric reactions, and 6*H*-benzo[*cd*]pyren-6-one, a primary O-PAC pollutant present in exhaust emissions from diesel engines and non-catalyst-equipped cars (see Sections E and F). These account for an additional 0.8 and 1.6%, respectively, of the identified whole sample mutagenic potency (see Table 10.26).

The remaining PAHs and PACs in Table 10.25 are present at much smaller levels. Thus, overall only about 20% of the *total* whole composite sample mutagenicity is assignable to individual, "positively identified," quantified, and assayed PAHs and PACs. Furthermore, as seen from the mutagenicities of the extracts of the four samples produced in the first fractionation step [nonpolar 1 (66 ± 13), nonpolar 2 (39 ± 12), semipolar (95 ± 7), and polar (31 ± 12)], more than half of the total sample mutagenicity is found in the semipolar and polar fractions—in which only ~3% of the contributing PACs have been identified. The lack of closure may be due in part to the incomplete data base on the mutagenicity of individual compounds where pure samples have not been available for testing. For example, Hannigan et al. (1998) noted that all methyl isomers of PAHs tested to date in the h1A1v2 assay are more potent than the unsubstituted parent PAHs (e.g., 5-methylchrysene is 37 times more potent than chrysene itself, 0.63 vs 0.017, both relative to BaP = 1.00; Durant et al., 1996). This may be relevant because significant concentrations of four methylphenanthrene isomers (3-, 2-, 9-, and 1-) were present, but had not been tested.

However, there are several caveats that should be kept in mind before extrapolating these results to other parts of the world. First, as discussed in more detail in Chapter 16, since the mid-1960s reductions in emissions of volatile organic compounds and of respirable "soot" particles and the associated PAHs and PACs have occurred in California due to increasing controls on mobile, as well as stationary, sources. Standards for emissions from mobile sources are the most stringent in the world at the present time. Second, even over the time period from the mid-1980s to the mid-1990s when the bacterial and human cell mutagen studies discussed

TABLE 10.25 Mass Concentrations of Individual Genotoxic PAHs and PACs in the "Whole Sample" and in Subfractionated Extracts,[i] Their Absolute Mutagenic Potencies and Potencies Relative to Benzo[a]pyrene = 1.00, and Contributions of Each to the Overall Total Potency of a Composite Mixture of Ambient Aerosols Collected at Four Urban Sites in Southern California for One Year (1993) and Subjected to Bioassay-Directed Chemical Analysis Using the h1A1v2 Human Cell Mutagenicity Assay

Compound	Formula	Concentration (ng / mg of EOC)[b]		Mutagenic potency[c]	Potency relative to BaP = 1.00[h]	Contribution to potency of sample extracts IMF ($\times 10^6$) mg of EOC	
		Whole sample	Sub fraction			Whole sample	Sub fraction
Aromatic nonpolar subfraction 1a							
PAHs							
Benz[a]anthracene	$C_{18}H_{12}$	23.9	12.2	4.0×10^{-4}	3.6×10^{-3}	9.6×10^{-3}	4.8×10^{-3}
Cyclopenta[cd]pyrene	$C_{18}H_{10}$	20.3	8.3	0.64	5.8	12.9	5.3
Chrysene + triphenylene	$C_{18}H_{12}$	47.2	24.8	4.0×10^{-4}	3.6×10^{-3}	0.02	0.01
Benzo[k]fluoranthene	$C_{20}H_{12}$	76.7	17.0	0.015	0.14	1.17	0.26
Benzo[b]fluoranthene	$C_{20}H_{12}$	71.5	17.7	0.029	0.26	2.06	0.51
Benzo[j]fluoranthene	$C_{20}H_{12}$	15.3	3.9	0.013	0.12	0.20	0.05
Benzo[e]pyrene	$C_{20}H_{12}$	84.5	17.6[f]	2.1×10^{-4}	1.9×10^{-3}	1.8×10^{-2}	3.7×10^{-3}
Benzo[a]pyrene	$C_{20}H_{12}$	33.9	7.9	0.11	1.00	3.72	0.87
Indeno[1,2,3-cd]pyrene	$C_{22}H_{12}$	87.2	11.8[f]	0.020	0.18	1.73	0.24
Benzo[ghi]perylene	$C_{22}H_{12}$	190	27.9[f]	0.013	0.12	2.43	0.36
Dibenz[a,h]anthracene	$C_{22}H_{14}$	41.9	3.7	0.020	0.18	0.82	0.07
Dibenzo[a,l]pyrene	$C_{24}H_{14}$	ND	ND	1.73	15.7	0.00	0.00
Dibenzo[a,e]pyrene	$C_{24}H_{14}$	ND	ND	0.24	2.2	0.00	0.00
Naphtho[2,1-a]pyrene	$C_{24}H_{14}$	–	0.4[e]	0.024	0.22	–	0.01
Aromatic nonpolar subfraction 2a							
PAHs							
Indeno[1,2,3-cd]pyrene	$C_{22}H_{12}$	0.7	0.1[f]	0.020	0.18	0.01	0.00
Benzo[ghi]perylene	$C_{22}H_{12}$	3.4	0.5[f]	0.013	0.12	0.04	0.01
Dibenzo[a,e]pyrene	$C_{24}H_{14}$	–	0.6[e]	0.24	2.2	–	0.14
Nitro-PAHs							
9-Nitroanthracene	$C_{14}H_9NO_2$	–	3.8[e]	3.9×10^{-4}	3.5×10^{-3}	–	1.5×10^{-3}
Aromatic semipolar subfraction 3a							
Nitro-PAHs							
2-Nitrofluoranthene	$C_{16}H_9NO_2$	28.4	6.4	0.019	0.17	1.27	0.12
1-Nitropyrene	$C_{16}H_9NO_2$	–	0.2[e]	0.0013	0.012	–	2.6×10^{-4}
2-Nitropyrene[f]	$C_{16}H_9NO_2$	–	2.3[e]	0.00	0.00[g]	0.00	0.00
Aromatic semipolar subfraction 6a							
Polycyclic aromatic ketones							
6H-Benzo[cd]pyren-6-one	$C_{19}H_{10}O$	90.7	44.5	0.018	0.16	1.63	0.80
Aromatic semipolar subfraction 7a							
Polycyclic aromatic ketones							
6H-Benzo[cd]pyren-6-one	$C_{19}H_{10}O$	39.7	19.5	0.018	0.16	0.72	0.35

[a] Source: Adapted from Hannigan et al. (1998).
[b] Concentrations of each PAH/PAC quantified in extracts of "whole composite samples" and of extracts of various mutagenic subfractions. The latter values are generally lower because of losses during separations and extractions. The annual average airborne fine particle equivalent organic carbon (EOC) concentration for this Los Angeles basinwide 1993 composite sample was 8.89 $\mu g/m^3$ of EOC. The latter is defined as "organic carbon on the atmospheric filter sample as determined by thermal evolution and combustion analysis prior to sample extraction."
[c] Single-compound potencies in the standard 12-mL h1A1v2 assay determined using data of Durant et al. (1996). Units are induced mutant fraction $\times 10^6$/ng of PAH or PAC tested. ND, not detected but included for reference.
[d] All PAHs and PACs listed in this modified table were identified and quantified at the "positive" confidence level.
[e] Compound measured only in subfractions.
[f] Compound also measured in small amounts in another subfraction; for clarity, not included here.
[g] 2-Nitropyrene is not mutagenic in this human cell assay. Concentrations of this atmospheric reaction product are included for comparison to concentrations of 1-nitropyrene (are predominantly from primary emissions) in this composite ambient sample (see Section F).
[h] Single-compound potencies relative to BaP taken as 1.00. Note: Some of these relative human cell potency equivalence factors for certain PAHs and PACs seem to differ significantly from those reported by Durant et al. (1996); e.g., compare their data (e.g., see Table 10.13) with those reported in the Hannigan et al. (1998) article.
[i] See Table 10.24 for concentrations of these PAHs in various urban atmospheres.

TABLE 10.26 Concentrations, Potencies, and Percentage Contributions of Selected PAHs and PACs Tested in the h1A1v2 Human Cell Assay[b] and Positively Identified in the Bioassay-Directed Chemical Analysis of a Composite "Whole" Sample Made Up of Fine Ambient Aerosols Collected at Four Urban Sites in the Los Angeles Air Basin[a,c]

				Contributions to mutagenic potency of "whole" sample 150 ± 31 IMF ($\times 10^6$) mg of EOC	
Compound	Formula	Concentration[a] in "whole" sample	Mutagenic potency[b]	Concentration-weighted potency[d]	Percentage of "whole" sample potency
PAHs					
Cyclopenta[cd]pyrene	$C_{18}H_{10}$	20.3	0.64	12.9	8.6
Benzo[a]pyrene	$C_{20}H_{12}$	33.9	0.11	3.72	2.5
Benzo[ghi]perylene	$C_{22}H_{12}$	190	0.013	2.48	1.7
Benzo[b]fluoranthene	$C_{20}H_{12}$	71.5	0.029	2.06	1.4
Indeno[1,2,3-cd]pyrene	$C_{22}H_{12}$	87.2	0.020	1.73	1.2
Benzo[k]fluoranthene	$C_{20}H_{12}$	76.7	0.015	1.17	0.8
				$\Sigma = 24.0$	$\Sigma = 16.2$
Nitro-PAHs					
2-Nitrofluoranthene	$C_{16}H_9NO_2$	28.4	0.019	1.27	0.8
Polycyclic aromatic ketones					
6H-Benzo[cd]pyren-6-one	$C_{19}H_{10}O$	130.4	0.018	2.4	1.6
				Total $\Sigma = 27.7$	$\Sigma = 18.6$

[a] Adapted from Hannigan et al. (1998). Concentrations are expressed in ng/mg of EOC.
[b] Durant et al. (1996). Units are induced mutant fraction $\times 10^6$/ng of PAH or PAC tested. See footnote c of Table 10.25.
[c] Hannigan et al. (1998) and references therein; e.g., Durant et al. (1996). See Table 10.25 and Fig. 10.23.
[d] Units are IMF $\times 10^6$/mg of EOC.

earlier were carried out, there was a significant change in the mobile source emissions.

These changes are illustrated by the California Air Resources Board (1996) usage and emissions data in Tables 10.27 and 10.28 for the Los Angeles Air Basin in 1982 and 1993. For example, particle emissions from light-duty automobiles decreased by more than a factor of two, from 5.2 to 2.2 tons per day, from 1982 to 1993, while those from medium- to heavy-heavy-duty trucks increased from 14.8 to 18.5 tons per day. Most of the latter is due to heavy-heavy-duty trucks, which in total number of in-use vehicles is much smaller (∼70,000) than the number of in-use light-duty automobiles (∼6 million). This increase is important in that diesel engines are major sources of benzo[b]fluoranthene and benzo[k]fluoranthene, which are significant contributors to ambient air mutagenicity (e.g., see Table 10.26).

The data in Tables 10.27 and 10.28 also show that the percentage of in-use, gasoline-fueled noncatalyst light-duty automobiles dropped from 44% in 1982 to 8% in 1993 and the percentage of their vehicle miles traveled went from 33% to only 4% during that period. Furthermore, while diesel cars accounted for only 1.9 and 1.5% of the in-use light-duty automobiles in 1982 and 1993, respectively, they were responsible for 45 and 40% of the *total* auto exhaust particulate matter. Particulate emissions from noncatalyst cars dropped from 28 to 12% and those from catalyst cars increased from 27% to 48% of the total. These changes have important implications for potential changes in the contributions of various PAHs and PACs to ambient air mutagenicity. For example, three of the most important human cell mutagens identified to date in ambient southern California aerosols, cyclopenta[cd]pyrene, indeno[1,2,3-cd]pyrene, and benzo[ghi]perylene, have "historically been emitted largely from noncatalyst-equipped, gasoline-powered motor vehicles in Los Angeles" (Hannigan et al., 1998; see Schauer, 1996), whose emissions have decreased dramatically over this 11-year period (Table 10.27).

Another potential difference between the conditions of the studies discussed above and other areas is that the Los Angeles Air Basin still has relatively high ambient levels of nitrate radicals, ozone, and other photochemical oxidants. This is expected to result in higher ambient levels of certain secondary, semipolar, and polar PACs, including some bacterial and human cell mutagens, formed in atmospheric reactions during transport. An example is 2-nitrofluoranthene, which, as we have seen, is a potent human cell mutagen also found at significant levels in urban airsheds throughout

TABLE 10.27 Motor Vehicle Use and Emissions Inventory for California's South Coast Air Basin[b] for the Years 1982 and 1993[a,f]

	1982			1993		
	Number of in-use vehicles	Vehicle miles traveled	PM[e] in exhaust emissions (tons/day)	Number of in-use vehicles	Vehicle miles traveled	PM in exhaust emissions (tons/day[f])
Light-duty automobiles						
Gasoline						
Noncatalyst	2.20×10^6	44×10^6	1.46	0.50×10^6	8.19×10^6	0.27
Catalyst	2.69×10^6	85×10^6	1.41	5.69×10^6	180×10^6	1.06
Diesel	9.68×10^4	4.2×10^6	2.33	9.4×10^4	2.55×10^6	0.87
Total light-duty automobiles	4.98×10^6	133×10^6	5.20	6.28×10^6	190×10^6	2.20
Medium-heavy-duty trucks[c]						
Gasoline						
Noncatalyst	5.96×10^4	2.62×10^6	0.16	2.58×10^4	0.93×10^6	0.33
Catalyst	0	0	0	0.47×10^4	0.44×10^6	0.03
Diesel	3.23×10^4	0.81×10^6	2.17	4.93×10^4	3.40×10^6	4.42
Total medium-heavy-duty trucks	9.19×10^4	3.43×10^6	2.33	7.98×10^4	4.77×10^6	4.78
Heavy-heavy-duty trucks[d]	8.80×10^4	4.45×10^6	12.5	7.12×10^4	8.53×10^6	13.7
Total tons/day of particulate matter exhaust emissions			20.0			20.7

[a] From California Air Resources Board "Motor Vehicle Inventory MVE17G Corrected" (1996); in *Methodology for Estimating Emissions from On-Road Motor Vehicles*, Vol. 1, *Introduction and Overview* (1996).
[b] Also known as the Los Angeles Air Basin or southern California Air Basin.
[c] 14,000 to 33,000 pounds.
[d] > 33,000 pounds.
[e] PM = particulate matter.
[f] Two classes of on-road motor vehicles, light- and medium-duty trucks, that had relatively lower tons/day of exhaust emissions of PM are omitted for clarity. The three categories of light-duty trucks, noncatalyst, catalyst, and diesel, emitted a total of 1.43 and 0.99 tons/day of POM in 1982 and 1993, respectively. PM exhaust emissions totaled 0.04 and 0.05 tons/day in 1982 and 1993, respectively, for catalyst and noncatalyst, gasoline-fueled medium-duty trucks.

the world (Atkinson and Arey, 1994; Ciccioli *et al.*, 1996; Arey, 1998a; see Section F).

Finally, from the "exposure" aspects of a health risk assessment of ambient POM, recall that the data for concentrations of PAHs and PACs in the composite whole sample are basinwide annual averages for 1993. The actual short-term (and probably annual) averages of concentrations of human cell mutagens at each of the four urban sampling sites located many kilometers apart (Fig. 10.23) can be significantly different. For

TABLE 10.28 Usage and Exhaust Emissions of Particulate Matter (PM) for Light-Duty Automobiles in California's South Air Basin[b] in 1982 and 1993[a]

	Percentage of in-use light-duty automobiles		Percentage of vehicle miles traveled		Relative percentage of exhaust emissions of PM by light-duty automobiles		Emissions of PM by light-duty automobiles (tons/day)	
	1982	1993	1982	1993	1982	1993	1982	1993
Gasoline light-duty automobiles								
Noncatalyst	44.2	8.0	33.1	4.3	28	12	1.46	0.27
Catalyst	53.9	90.5	63.9	94.4	27	48	1.41	1.06
Diesel light-duty automobiles	1.9	1.5	3.0	1.3	45	40	2.33	0.87
							$\Sigma = 5.20$	$\Sigma = 2.20$

[a] Calculated from data in California Air Resources Board "Motor Vehicle Inventory MVE17G Corrected" (1996); see data in Table 10.27.
[b] Formal designation. For clarity and frame of reference, it is often referred to (e.g., in this chapter) as the Los Angeles Air Basin, or simply southern California Air Basin (see map Fig. 10.23).

example, as discussed in more detail shortly, Fraser et al. (1998) reported that the 48-h average concentrations of the reactive PAHs benzo[*a*]pyrene and cyclopenta[*cd*]pyrene in samples collected in September 1993 were much higher in central Los Angeles, 0.29 and 1.04 ng m^{-3}, respectively, than in downwind Claremont, 0.03 and 0.03 ng m^{-3}. Conversely, concentrations of 2-nitronaphthalene, a major atmospheric reaction product (see Section F.2) and mutagen in the MCL-5 human cell line (Sasaki et al., 1997b), averaged 3.81 ng m^{-3} in central Los Angeles but 8.11 ng m^{-3} at the downwind Claremont site.

4. Bioassay-Directed Chemical Analysis for Vapor-Phase and Particle-Phase PAHs and PACs in Ambient Air Using Bacterial Assays

Since the initial discovery of the mutagenicity of air samples in 1977, most of the research has focused on genotoxicity of organic extracts of the airborne particulate matter. However, this emphasis has been shifting and there is increasing interest in, and concern over, the vapor-phase mutagenicity of urban atmospheres and its toxicological implications. We discuss here examples of some of the studies in which bacterial assays have been used to help in the identification of both gas- and particle-phase PAHs and PACs in ambient air.

In one of the first studies of the vapor-phase mutagenicity of polluted urban air, Alfheim and co-workers (1985) collected both ambient particles and vapor-phase compounds and used the *Salmonella typhimurium* reversion assay. The direct activities (−S9) of the extracts generally exceeded the promutagenicities (+S9), and furthermore, the vapor-phase mutagenicity ranged from 0 to 88% of the total activity.

Subsequently, at several sites in Finland with different emission sources (using a similar sampling methodology) Pyysalo and co-workers (1987) found direct mutagenicity in both gas and particle phases with the *Salmonella*/microsome test and activity in the sister chromatoid exchange (SCE) assay employing eukaryotic Chinese hamster ovary cells. In urban air in Helsinki, Tuominen and co-workers (1988) found the total direct and indirect activities to be about the same. Furthermore, the majority of the activity in each phase was due to polar compounds. Interestingly, the total SCE-inducing activity of unfractionated vapor-phase extracts was consistently higher than that of particle-phase extracts.

A major advance in this area came in 1992 when Harger and co-workers (1992) reported the direct mutagenicity of vapor- and particle-phase organics separated using HPLC in samples collected during the summer of 1987 in Claremont, California (see map in Fig. 10.23). This HPLC procedure had several significant advantages over earlier studies. First, mutagenic compounds were separated from possibly interfering toxic compounds and hence were more likely to be detected. Second, a mutagen polarity profile (mutagram or "fingerprint") of the directly mutagenic *gas-phase* PAC was obtained that could be compared with the mutagram of the concurrently sampled particulate organics collected on the filter and with the chromatographic characteristics (e.g., polarity) of individual (or classes of) known or suspected direct mutagens.

Figure 10.25 shows these mutagrams for the vapor and particle phases, respectively. Interestingly, the total direct mutagenicity of the vapor phase, 210 rev m^{-3}, was actually greater than that of the particle phase, 160 rev m^{-3}; furthermore, its mutagenicity profile was substantially different. Thus, fraction 4 is the major peak for the vapor-phase sample whereas most of the particle-phase mutagenicity is in the more polar peaks 6 and 7. Similar enhancements in the contributions of more polar species were reported for bioassay-directed fractionation of SRM 1649 urban air particulate matter (Schuetzle and Lewtas, 1986; Nishioka et al., 1988;

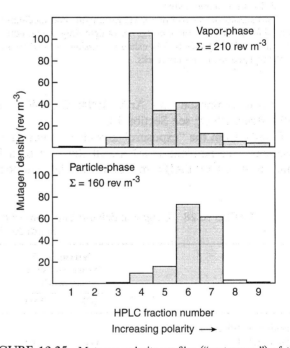

FIGURE 10.25 Mutagen polarity profiles ("mutagrams") of the direct-acting mutagenicities of HPLC-separated fractions of dichloromethane extracts of polyurethane foam solid adsorbent (PUF, "vapor phase") and TIGF filter ("particle phase") samples collected concurrently in ambient air in Claremont, California, August 1987. Mutagen densities (rev m^{-3}) on strain TA98, −S9, microsuspension modification of standard *Salmonella typhimurium* reversion assay (adapted from Harger et al., 1992).

Lewtas et al., 1990a). Similarly, Greenberg and co-workers (1990, 1993) found that the polar neutral compounds in inhalable particle samples at Newark, New Jersey, were the greatest contributors to ambient mutagenicities, regardless of the season of the year. The importance of more polar compounds in the direct-acting TA98 mutagenicity has also been seen in the emissions of POM from heavy-duty diesel trucks and wood smoke (e.g., Schuetzle and Lewtas, 1986). (Note, however, that the structures, concentrations, and mutagenicity polarity profiles of diesel exhaust emissions can be strongly impacted by such parameters as fuel composition, engine type, age, operating conditions, control devices, etc.)

A breakthrough in identifying possible particle-phase polar compounds responsible for the direct mutagenicity of fraction 6 in the Claremont, California, study of Harger et al. (Fig. 10.25) came with the isolation and identification by Helmig and co-workers (1992a, 1992b) of a new class of direct mutagens in particulate matter collected in Riverside, California, the nitrolactones of phenanthrene, specifically, the 2- and 4-nitro-6H-dibenzo[b,d]pyran-6-ones. The 2-nitro isomer (structure XI) is a very powerful, direct mutagen (240,000 rev μg^{-1}) and Harger et al. (1992) ascribe a major portion of the mutagenicity in their fraction 6 to this widely distributed product of the gas-phase OH radical attack on phenanthrene (see Atkinson and Arey (1994), Sasaki et al. (1995), Arey (1998a), and references therein; and Section F).

Based on results from concurrent environmental chamber studies of the photooxidations of naphthalene, fluorene, and phenanthrene by Arey and co-workers (1992), Harger and co-workers attributed a significant fraction of the vapor-phase direct activity of fraction 4 (Fig. 10.25) to 2-nitronaphthalene (2-NN) present in these 12-h daytime samples as a result of an OH-initiated homogeneous gas-phase reaction of naphthalene in ambient air. Although 1-nitronaphthalene (1-NN) is also present at significant concentrations in both ambient and laboratory systems, it is a much weaker, direct mutagen (see Table 10.20; Arey et al., 1992).

Harger and co-workers noted that methylnitronaphthalenes (MNN) were also abundant in southern California air, but their mutagenic activities in the microsuspension were unknown. Subsequently, Gupta and co-workers (1996) determined them for the 14 methylnitronaphthalenes (see Table 10.18) and used these data to quantify contributions of nitronaphthalenes and methylnitronaphthalenes to the vapor-phase mutagenicity of 12-h daytime and 12-h nighttime samples collected on PUF plugs at an air monitoring site in

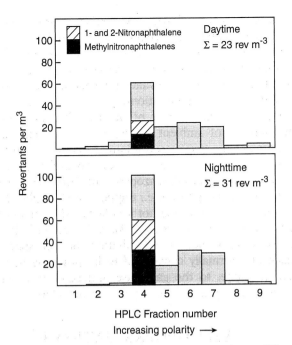

FIGURE 10.26 Mutagrams of direct mutagenicity on different HPLC fractions for composite 12-h daytime and nighttime vapor-phase samples collected on PUF plugs in Redlands, California, in October 1994. Mutagen densities (rev m^{-3}) on TA98, $-S9$, microsuspension modification (adapted from Gupta et al., 1996).

Redlands, California, a receptor site ~ 110 km downwind from central Los Angeles.

Bioassay-directed fractionations of the daytime and nighttime samples produced the mutagrams shown in Fig. 10.26. The ambient nitronaphthalenes and methylnitronaphthalenes in fraction 4 contribute $\sim 18\%$ of the total daytime mutagenic activity of 23 rev m^{-3}; they are formed by OH radical attack on the parent PAHs (see Sasaki et al., 1995, and Section F). The total nighttime activity is higher than the daytime activity, 31 rev m^{-3}, and was attributed to more efficient formation of the nitronaphthalenes and methylnitronaphthalenes in NO_3 radical initiated reactions (Atkinson and Arey, 1994).

In summary, Gupta and co-workers (1996) reported that ambient concentrations of the vapor-phase nitronaphthalenes and methylnitronaphthalenes are generally greater than those of other particle-phase nitro-PAHs present simultaneously (Arey et al., 1987; Wilson et al., 1995). This is particularly important in that, as noted earlier (Pitts et al., 1985c; Pitts, 1987), 2-nitronaphthalene has been reported on intraperitoneal injection to be converted to 2-hydroxyaminonaphthalene, the same metabolite that 2-aminonaphthalene (Johnson and Cornish, 1978), an animal and human carcinogen (IARC, 1987; U.S. DHHS, 1998), is converted to. Additionally, Tokiwa and Ornishi (1986) report that in feeding studies on animals, 2-nitronaphthalene has shown carcinogenic activity. This would seem to raise relevant atmospheric chemistry/inhalation toxicology questions.

Finally, in the context of the overall vapor-phase mutagenicity of ambient air, we note that significant fractions of the two powerful human cell and bacterial mutagens discussed earlier, cyclopenta[cd]pyrene (**XXVIII**) and 2-nitrodibenzopyranone (**XI**), have been found in the gas phase (i.e., trapped on PUF plugs) in samples collected during hot weather at sites in southern California (Fraser et al., 1998, and Arey et al., 1994, respectively). Hence the contributions of such species, which are normally considered to be primarily in the particle phase, to the gas-phase mutagenicity at high ambient temperatures should also be considered.

E. ATMOSPHERIC FATES OF PARTICLE-ASSOCIATED PAHs: HETEROGENEOUS REACTIONS

Atmospheric reactions of PAHs fall into two broad categories: (1) *heterogeneous* processes involving particle-associated compounds (including the semivolatile 4-ring and 5- and 6-ring PAHs), e.g., photolysis/photooxidation and gas–particle interactions, and (2) *homogeneous* gas-phase reactions of volatile 2- and 3-ring and semivolatile 4-ring PAHs initiated by OH (daytime) and NO_3 radicals (nighttime) and ozone (24 h).

As we have seen, a great deal is known about emission sources and strengths, ambient levels, and mutagenic/carcinogenic properties of the particle-phase PAHs in airborne POM. However, because of the tremendous physical and chemical complexity of the aerosol surfaces on which photolysis, photooxidations, and gas–particle interactions take place in "real" polluted ambient air, much less is known about the structures, yields, and absolute rates and mechanisms of formation of PAH and PAC reaction products, especially for the more polar PACs. This is one area in which there exists a major gap in our knowledge of their atmospheric chemistry and toxicology.

1. Background

In 1954, Kotin and co-workers reported the "carcinogenicity of atmospheric extracts" of Los Angeles air. Subsequently, in 1956, they reported the carcinogenic activity of oxidation products of aliphatic hydrocarbons and, in 1958, of ozonized gasoline. Concurrently, Falk and co-workers (1956) published results of their laboratory studies of photooxidations of PAHs. A decade later, Tebbens et al. (1966) and Thomas and co-workers (1968) reported that major decreases in the concentrations of BaP and perylene occurred in smoke that was irradiated while passing through flow chambers (e.g., a 60% decrease in the BaP content of soot).

Based in part upon this pioneering research, a National Academy of Sciences Committee suggested in their 1972 document *Particulate Polycyclic Organic Matter* that photolysis, photooxidations, and gas-particle interactions might lead to significant degradation of PAHs in ambient particles and formation of products more polar than the parent PAHs (NRC, 1972).

Shortly thereafter, laboratory studies of the photooxidation and ozonolysis of PAHs deposited on several substrates (e.g., glass surfaces and thin-layer chromatography plates) and exposed to sunlight and/or various levels of ozone in air found significant differences in the reactivities of individual PAHs. Half-lives of BaP and benz[a]anthracene were short (~1 h or less), benzo[e]pyrene (BeP) and pyrene intermediate, and benzofluoranthenes long (Lane and Katz, 1977; Katz et al., 1979). Among the reaction products were three BaP diones that had previously been identified in ambient aerosols collected in Toronto, Canada (Pierce and Katz, 1976), and subsequently at Duisburg, West Germany (König et al., 1983), and Elverum, Norway (Ramdahl, 1983b).

During the same time frame, extracts of ambient aerosols collected throughout the world were found to be directly mutagenic in the Ames bacterial assay. Concurrently, BaP deposited on various substrates (e.g., glass fiber filters) and exposed to ppm and sub-ppm levels of ozone was shown to react readily to form direct-acting frameshift mutagens (Pitts et al., 1978). Furthermore, exposures of particle-associated 4- and 5-ring PAHs to similar levels of NO_2 (and a required "trace of gaseous HNO_3") in air resulted in significant yields of nitro-PAHs; these products, which reflected a range of reactivities of the parent compounds, were also direct-acting frameshift mutagens (Jäger, 1978; Jäger and Hanüs, 1980; Pitts et al., 1978; Pitts, 1979; Hughes et al., 1980; Tokiwa et al., 1981; Nielsen et al., 1983; vide infra). For example, an 8-h exposure of BaP deposited on a glass fiber filter to 250 ppb of NO_2 (the California air quality standard) and traces of HNO_3 gave an ~20% yield of products, predominantly 6-NO_2-BaP with lesser amounts of the 1- and 3-isomers. Soon thereafter, mutagenicity studies by Wang and co-workers (1980) implied the presence of nitroarenes in ambient air. Subsequently, several nitroarenes were isolated and positively identified by Ramdahl and co-workers (1982b). We note, however, that these initial researchers cautioned that their rate, mechanism, product, and mutagenic data from laboratory exposures of PAHs deposited in several ways on a variety of substrates (evaporation of a BaP solution on a Hi-Vol filter, vapor deposition of PAHs on synthetic and real particle substrates, etc.) and exposed to relatively high levels of simulated sunlight and O_3 and NO_2 in clean air may not be truly representative of "real world" heterogeneous reaction systems.

We discuss in this section four key aspects of heterogeneous reactions: (1) theoretical and experimental structure and reactivity relationships; (2) field measurements of relative and absolute PAH decay rates in near-source ambient air and during downwind transport; (3) laboratory studies of the photolysis/photooxidation and gas–particle interactions with O_3 and NO_2 of key 4- and 6-ring PAHs adsorbed on model substrates or ambient aerosols; and (4) environmental chamber studies of the reactions of such PAHs associated with several physically and chemically different kinds of combustion-generated aerosols (e.g., diesel soot, wood smoke, and coal fly ash). Where such data are available, we also briefly consider some toxicological ramifications of these reactions.

Space considerations preclude extensive discussions of these broad topics; for details, see reviews and critiques by Nielsen and co-workers (1983), Van Cauwenberghe and Van Vaeck (1983), Pitts (1983), Van Cauwenberghe (1985), Bjørseth and Olufsen (1983), Valerio et al. (1984), Nikolaou et al. (1984), Finlayson-Pitts and Pitts (1986), Pitts (1987, 1993a, 1993b), Baek et al. (1991), Atkinson and Arey (1994) (gas-phase reactions only), and Arey (1998a).

2. Theoretical and Experimental Structure–Reactivity Relationships

In their thoughtful 1983 review, Nielsen and co-workers noted that particles of diesel soot or wood smoke can absorb significant amounts of water. Thus, they suggested that the most plausible mechanism(s) for nitration (and possibly other electrophilic reactions) of particle-associated PAHs in ambient air may involve reactions both in a liquid film and on solid surfaces and that fundamental laboratory studies of the rates, products, and mechanisms of PAHs in polar solvents would be atmospherically relevant for reactions in the liquid films. Based on this, they proposed a classification scheme for the reactivities of key PAHs in electrophilic reactions, which was subsequently described in detail (Nielsen, 1984).

Given the complexities of the system involved, it is interesting that this reactivity scale has proven to be a useful predictor not only for overall relative decay rates of PAHs associated with aerosols in ambient air but also for the relative rates of their electrophilic heterogeneous reactions in simulated atmospheres.

The original suggestion of Nielsen and co-workers (1983) that these heterogeneous processes involved particle-phase PAHs interacting with other compounds, organic and inorganic, present in "liquid" (or quasi-liquid) layers on the surfaces of carbonaceous particles (e.g., wood smoke) is consistent with results from subsequent studies. For example, Strommen and Kamens (1997) cite as evidence for this phenomenon the fact that a viscous, oily slick, with little particle integrity, results when particles of diesel or wood soot impact a metal foil (McDow et al., 1994). Furthermore, Kamens, McDow, and co-workers have demonstrated the strong influence of aerosol water content (humidity), solvent effects, and dissolved aerosol constituents on such processes as the photodegradation of benz[a]anthracene (vide infra; also see Odum et al., 1994a; McDow et al., 1994, 1995, 1996; Jang and McDow, 1997; and Jang and Kamens, 1998). The diffusion model of Odum and co-workers (1994b), which describes the gas–particle partitioning of semivolatile organics (including PAHs), incorporates the idea of diffusion of the SOCs into viscous, liquid-like, organic layers on particles of diesel soot. Subsequently, Strommen and Kamens (1997) extended this concept with a dual-impedance model that incorporated rapid diffusion into an outer layer of liquid-like organic material and slower diffusion into an inner core of discrete solids. We discuss in the following the implications of such liquid-like organic layers to the rates and mechanisms of the photochemical and gas–particle reactions of particle-associated PAHs and nitro-PAHs (e.g., see Fan et al., 1996a, 1996b; and Feilberg and Nielsen, 1999b).

In developing the reactivity scale, Nielsen first investigated the transformation rates of 25 PAHs and four derivatives of anthracene in water–methanol–dioxane solutions, taken as a model of "wet particles," and containing small amounts of dinitrogen tetroxide and nitric and nitrous acids. The measured half-lives and relative rates are shown in Table 10.29. The range of reactivities in solution for PAHs of different structures is remarkable, from 100,000 (arbitrarily set) for anthanthrene (**XXXII**) to <0.2 for the least reactive compounds:

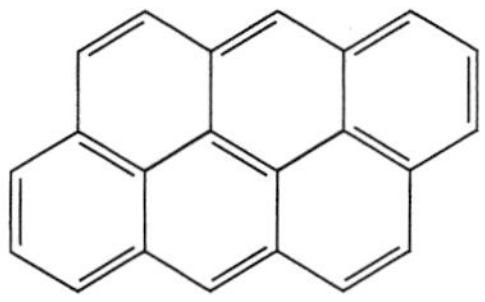

XXXII

Anthanthrene

The half-lives of the powerful carcinogens increase from 200 min for BaP to >300 days for benzo[b]fluoranthene and >800 days for indeno[1,2,3-cd]pyrene, with the relative reactivities decreasing from 1100 to <0.5 and <0.2, respectively, an overall factor of ~5500.

Even taken qualitatively, these reactivity data have important toxicological as well as chemical implications regarding the composition of PAHs and PACs in and on the surfaces of aerosols in polluted air parcels, both near-source and during transport (downwind). Thus, under certain conditions (e.g., daytime, summer season, and high oxidant levels) over a period of hours BaP concentrations in ambient air could be expected to decay dramatically as a result of reactions, while those of the benzofluoranthenes and indeno[1,2,3-cd]pyrene would be expected to remain relatively constant. Of course, their absolute concentrations also change as a result of dilution of the air parcel caused by increased mixing depth over time and transport. However, impacts of such physical processes are minimized if one considers ratios of concentrations of reactive to nonreactive PAHs, e.g., BaP/BeP (vide infra). Certainly, these reactivity data call into question the use of BaP concentrations *alone* as a reliable indicator of the overall inhalable, particle-phase carcinogenicity of polluted ambient air (e.g., see Lane and Katz (1977) and later text for other examples).

Electron-donating substituents (e.g., methyl) generally lead to increased relative transformation rates; for example, the relative reactivity of anthracene is 1200 compared to 21,000, 6500, and 1600 for 9,10-dimethyl-, 9-methyl-, and 2-methylanthracene, respectively (Table 10.29). Electron-withdrawing substituents, e.g., nitro, decreased rates. These effects are characteristic of electrophilic reactions.

Based on the laboratory data in Table 10.29, information on electrophilic nitrations by Dewar *et al.* (1956) and correlations with several spectroscopic constants, Nielsen proposed the structure–reactivity classification shown in Table 10.30 for electrophilic reactions "in the environment under dark or weak lighting conditions." Classifications of several PAHs in Table 10.30 differ somewhat from their ranking in the laboratory results

TABLE 10.29 Half-Lives, Relative Rates of Nitration,[c] and Reactivity Classifications of 25 PAHs in a Mixture of Polar Solvents[a,b]

Compound	Half-life	Relative reactivity	Classification[e]
Anthanthrene	2.2 min	100,000[d]	II
Perylene	~3 min	73,000	II
9,10-Dimethylanthracene	11 min	21,000	–
9-Methylanthracene	34 min	6,500	–
Picene	100 min	2,100	III
2-Methylanthracene	140 min	1,600	–
Anthracene	190 min	1,200	II
Benzo[a]pyrene	200 min	1,100	II
1-Methylanthracene	260 min	860	–
Benzo[ghi]perylene	5.5 days	28	III
Pyrene	5.7 days	27	III
Coronene	23 days	6.8	IV
Chrysene	24 days	6.3	IV
Dibenz[a,h]anthracene	29 days	5.3	IV
Benz[a]anthracene	34 days	4.5	III
Benzo[e]pyrene	>100 days	<1	IV
Benzo[b]fluorene	<200 days	<0.9	–
Benzo[ghi]fluoranthene	>200 days	<0.6	–
Benzo[b]fluoranthene	>300 days	<0.5	V
Benzo[k]fluoranthene	>300 days	<0.5	V
Triphenylene	>300 days	<0.5	V
Phenanthrene	>70 days	<2	V
Fluoranthene	>400 days	<0.4	V
Benzo[j]fluoranthene	>800 days	<0.2	V
Indeno[1,2,3-cd]pyrene	>800 days	<0.2	V

[a] Adapted from Nielsen (1984).
[b] Water–methanol–dioxane, 40:36:24.
[c] $[HNO_3]_{added}$ = 0.16 M, $[NaNO_2]_{added}$ = 0.016 M, plus trace of N_2O_4; 25°C.
[d] Relative reactivity of anthanthrene set at 100,000.
[e] See text and Table 10.30.

TABLE 10.30 Reactivity Scale[b] for the Electrophilic Reactions of PAHs (and Benzene)[e]
[Italics Are Particle-Phase, IARC Class 2A "Probable" or Class 2B "Possible" Human Carcinogens[a,c,d]]

I:	Benzo[*a*]tetracene, *dibenzo[a,h]pyrene*, pentacene, tetracene
II:	Anthanthrene, anthracene, *benzo[a]pyrene*, cyclopenta[*cd*]pyrene, *dibenzo[a,l]pyrene*, *dibenzo[a,i]pyrene*, dibenzo[*a,c*]tetracene, perylene
III:	*Benz[a]anthracene*, benzo[*g*]chrysene, benzo[*ghi*]perylene, *dibenzo[a,e]pyrene*, picene, pyrene
IV:	Benzo[*c*]chrysene, benzo[*c*]phenanthrene, benzo[*e*]pyrene, chrysene, coronene, *dibenzanthracene*, dibenzo[*e,l*]pyrene
V:	Acenaphthylene, *benzofluoranthenes*, fluoranthene, indeno[1,2,3-*cd*]fluoranthene, *indeno[1,2,3-cd]pyrene*, naphthalene, phenanthrene, triphenylene
VI:	Benzene[e]

[a] Adapted from Nielsen (1984).
[b] Reactivity *decreases* in the order I to VI.
[c] See Table 10.29.
[d] See Table 10.13.
[e] Vapor phase, IARC Class 1, "the agent is carcinogenic to humans."

shown in Table 10.29. Alkyl (methyl) substituted compounds generally have higher reactivities than the parent PAHs; these are placed in the same group or in one with a higher reactivity. This finding is consistent with results of a field study by Fraser *et al.* (1998) that established the higher reactivity of a series of methylphenanthrenes compared to phenanthrene (vide infra). Finally, Nielsen stressed that this scale is *not* valid for nonionic free radical reactions of PAH (e.g., homogeneous gas-phase, OH radical initiated reactions of the type discussed in Section F).

In the ensuing years, predictions of the relative reactivities of particle-associated PAHs based on this classification have proven to be generally consistent with results from field studies conducted in airsheds throughout the world and laboratory and environmental studies on various substrates of the type discussed in the following sections.

3. Field Studies of Atmospheric Reactions: Transport and Transformation

a. Decay of Particle-Associated PAHs in Ambient Air

In their study of the ambient concentrations of PAHs, biphenyl, and nitroarenes in a high-NO_x wintertime episode in Torrance, California (near Los Angeles Airport, Fig. 10.23), Arey and co-workers (1987) collected daytime (0600 to 1800 hours) and nighttime (1800 to 0600 hours) samples. Ratios of the concentrations (ng m^{-3}) of high-reactivity BaP and perylene (Group II, Table 10.30) to low-reactivity BeP (Group IV) were lower during the day than at night. Thus [BaP]/[BeP] = 0.6/2.1 = 0.29 during the day compared to 1.6/2.1 = 0.76 at night. Similarly, perylene/BeP ratios were 0.2/2.1 = 0.10 during the day compared to 0.5/2.1 = 0.24 at night. Arey *et al.* (1987) suggested that these results reflect daytime chemical loss processes (e.g., photodegradation and gas–particle reactions) for BaP and perylene and/or different nighttime sources of these PAHs.

In 1988, Nielsen reported further phenomenological evidence for rapid atmospheric decay of the high-reactivity (Group II, Table 10.30) particle-phase carcinogens and human cell mutagens BaP and cyclopenta[*cd*]pyrene during transport of combustion-generated particulate POM. Thus, in ambient aerosols collected at the Risø National Laboratory (near Copenhagen, Denmark), ratios of the concentrations of BaP to BeP (Class IV), [BaP]/[BeP], and of cyclopenta-[*cd*]pyrene to "low-reactivity" [chrysene (Class IV) + triphenylene (Class V)] were strongly dependent on wind direction and PAH pollution levels. In both cases, the ratios were significantly lower in air masses coming from a more distant, clean air region than in winds from the nearby city of Roskilde (population 40,000), which is the major source for PAHs at Risø. The [BaP]/[BeP] and [cyclopenta[*cd*]pyrene]/[chrysene + triphenylene] ratios were respectively 29 and 45% lower in air masses from the more distant region, indicating that BaP and cyclopenta[*cd*]pyrene (CPP) were degraded during transport. Given that this study was conducted in the winter with low ambient temperatures, solar intensities, and NO_2 levels, higher decay rate constants would be expected in the summer (Nielsen, 1988). In contrast to BaP and CPP, no evidence could be found for decay of the more stable compounds BeP, the benzo[*b + j + k*]fluoranthenes (Class V), chrysene + triphenylene, benzo[*ghi*]perylene (Class III), and indeno[1,2,3-*cd*]pyrene (Class V), consistent with the ranking of these PAHs on the reactivity scale discussed earlier (Tables 10.29 and 10.30; Nielsen, 1984).

Similarly, Greenberg (1989) determined the concentrations of six selected PAHs at three urban sites in

New Jersey over two summer periods and one winter period and ratioed them to the stable PAH benzo[b]fluoranthene, BbF (Class V, Table 10.30). The [BaP]/[BbF] ratios were 0.61 in the summer of 1982 compared to 0.92 in the winter of 1983; for [CPP]/[BbF] they were 0.17 during the summer compared to 0.57 in the winter. In contrast, winter to summer ratios for the carcinogenic Class V PAHs benzo[k]fluoranthene and indeno[1,2,3-cd]pyrene (Table 10.30) are equal and close to unity, indicative of the low reactivity predicted by Nielsen (1984, 1988).

Masclet and co-workers (1986) have also developed a relative PAH decay index. They used it, for example, to identify various major sources of urban pollution and developed a model for PAH concentrations at receptor sites. An interesting and relevant area that is beyond the scope of this chapter is the use of PAHs as organic tracers and incorporating their relative decay rates (reactivities) into such receptor-source, chemical mass balance models. Use of relative rates can significantly improve such model performances (e.g., see Daisey et al., 1986; Masclet et al., 1986; Pistikopoulos et al., 1990a, 1990b; Lee et al., 1993; Li and Kamens, 1993; and Venkataraman and Friedlander, 1994a, 1994b, 1994c).

b. Time–Concentration Profiles of Particle-Phase PAHs during Transport across an Air Basin

Fraser and co-workers (1998) measured the particle- and vapor-phase concentrations of 86 aromatic compounds, including PAHs ranging from naphthalene to benzo[ghi]perylene, at four sites in southern California (Long Beach, central Los Angeles, Azusa, and Claremont; see map in Fig. 10.23) over a two-day period September 8–9, 1993, when there was a photochemical smog episode (peak O_3 = 290 ppb at Claremont) (see also Fig. 10.2 and Table 10.8.)

Temporal and spatial distributions across the air basin of the particle-phase primary pollutants BaP and cyclopenta[cd]pyrene, as well as the vapor-phase secondary pollutant 3-nitrobiphenyl, are shown in Fig. 10.27 in order of increasing distance inland, i.e., in order of increasing transport time. The major ~6 to 10 a.m. peaks of BaP and cyclopenta[cd]pyrene at central Los Angeles and Azusa reflect the high density of motor vehicle emissions at this morning "rush" period.

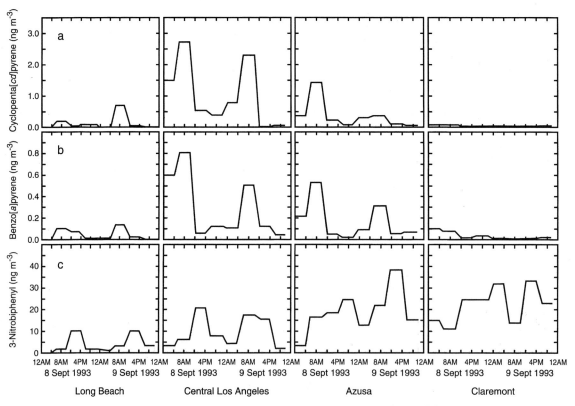

FIGURE 10.27 Spatial and temporal distributions of two reactive particle-associated PAHs, benzo[a]pyrene and cyclopenta[cd]pyrene, and the gas-phase atmospheric reaction product 3-nitrobiphenyl during transport of a heavily polluted air parcel across southern California, September 8–9, 1993 (adapted from Fraser et al., 1998).

TABLE 10.31 Ambient Concentrations (ng m^{-3}) at Four Sites of Two "Reactive" Particle-Associated PAHs^c and Three Nitro-PAHs Formed in Atmospheric Reactions During Transport of a Heavily Polluted Air Mass^b across Southern California^a

	Sampling site^d			
	Long Beach	Los Angeles	Azusa	Claremont
PAHs				
Cyclopenta[cd]pyrene	0.14	1.04	0.36	0.03
Benzo[a]pyrene	0.05	0.29	0.17	0.03
Nitro-PAHs				
1-Nitronaphthalene	4.69	10.7	19.9	16.4
2-Nitronaphthalene	1.34	3.81	8.13	8.11
3-Nitrobiphenyl	3.88	9.47	18.7	21.8

^{a} Adapted from Fraser et al. (1998); study carried out September 8–9, 1993. Samples aggregated and averaged over 48 h at each sampling site.
^{b} Peak ozone-290 ppb at Claremont.
^{c} Class II reactivity scale (Nielsen, 1984; see Table 10.30).
^{d} See map, Fig. 10.23.

TABLE 10.32 Ratios of Two-Day^c Average Ambient Concentrations of Two "Reactive" (Class II)^b and Two "Low-Reactivity" (Class V)^b PAHs in Fine Aerosols at Claremont and Central Los Angeles (Claremont/Central Los Angeles)^{a,d}

Compound	Ratio of concentrations Claremont/Central Los Angeles
Cyclopenta[cd]pyrene	0.03
Benzo[a]pyrene	0.10
Benzo[k]fluoranthene	0.21
Indeno[cd]pyrene	0.20

^{a} Adapted from Fraser et al. (1998).
^{b} Reactivity classification (Nielsen, 1984; see Table 10.30).
^{c} September 8–9 (48 h), 1993. Sampling at Claremont occurred during a major photochemical smog episode (peak O$_3$ = 293 ppb).
^{d} See map, Fig. 10.23.

The rapid depletion of these reactive PAHs during transport of this polluted air parcel is dramatic. Thus, as seen in Table 10.31, their concentrations (averaged over all times) drop from 1.04 and 0.29 ng m^{-3} for cyclopenta[cd]pyrene and BaP, respectively, at central Los Angeles to 0.03 ng m^{-3} for both PAHs at Claremont.

The physical effect of dilution of the aerosols due to increased mixing depths during transport of the air parcel inland is factored out by taking the ratios of the concentrations at Claremont to those at central Los Angeles. As seen in Table 10.32, the ratios are 0.21 and 0.20 for the low-reactivity PAHs benzo[k]fluoranthene and indeno[1,2,3-cd]pyrene (Class V, Table 10.30), reflecting dilution, whereas the ratios are 0.10 and 0.03 for reactive BaP and cyclopenta[cd]pyrene (Class II), respectively, indicative of reactions in addition to dilution.

Conversely, as seen in Fig. 10.27 and Table 10.31, the concentrations of the gas-phase atmospheric reaction product 3-nitrobiphenyl increase during transport across the basin, going from 9.5 ng m^{-3} in central Los Angeles to 21.8 ng m^{-3} in Claremont. Concentrations of 1- and 2-nitronaphthalene (recall the latter is a human cell mutagen in the MCL-5 cell line; Sasaki et al., 1997b), which are also formed in gas-phase reactions (see Section F), increase significantly during transit. Finally, some O-PAHs, e.g., 1,8-naphthalic anhydride and cyclopenta[cd]phenanthrone, were also observed to accumulate during downwind transport due to their formation in atmospheric reactions.

These spatial and temporal distributions of biologically active PAHs and PACs during transport across an air basin, e.g., the rapid decay of reactive mutagens/carcinogens, the long lifetimes of low-reactivity PAH carcinogens, and the significant rate of formation of human cell nitro-PAH mutagens, have significant exposure and public health implications, e.g., in risk assessments of the health effects of benzo[a]pyrene and diesel emissions (e.g., see California Air Resources Board reports, 1994 and 1998, respectively; WHO, 1996; and Collins et al., 1998).

From the perspective of atmospheric chemistry and toxicology, recall that the human cell mutagenic potencies of fine ambient aerosols were essentially the same in samples collected concurrently at four sites across southern California (see Fig. 10.23 and Sections D.2 and D.3). However, by the time a heavily polluted air parcel from central Los Angeles reaches the downwind site of Claremont, the ambient concentrations of two of the most powerful mutagens, BaP and cyclopenta[cd]pyrene, have dropped dramatically (Fig. 10.27). Key questions include: What are the chemical structures, ambient levels, and mechanisms of formation of the genotoxic PACs in fine ambient aerosols that make up this human cell "mutagenicity gap?" Are they biologically active O- and N-PAC reaction *products* of particle-phase PAHs, so that as the primary promutagens decay, biologically active reaction products are formed? Alternatively, are mutagens formed by reactions in air during transport of nonmutagenic PAH? Of course, both mechanisms may be occurring simultaneously.

4. Photochemical Reactions of Particle-Associated PAHs

The reviews "Toxicological Indications of the Organic Fraction of Aerosols: A Chemist's View" by Van Cauwenberghe and Van Vaeck (1983) and "Atmospheric Reactions of PAH" by Van Cauwenberghe (1985) provide critical assessments and extensive literature references of the status of research to 1985 in the complex area of heterogeneous photochemical reactions and of the interactions of PAHs on laboratory substrates, primary combustion particles, and ambient particulate matter with ozone and NO_2 in air. In the following sections, we briefly summarize results from this earlier era and address subsequent studies on heterogeneous atmospheric reactions of PAHs in simulated and real atmospheres.

It is now generally agreed that photodegradation is the major chemical pathway for the loss of 4- to 6-ring PAHs in ambient aerosols. Reaction rates, mechanisms, and products of this overall phenomenon depend not only on the structures and UV–visible absorption spectra (on and in aerosol surfaces) of specific PAHs but also on the physical and chemical natures of the surfaces of the particles on/in which they are adsorbed/absorbed (i.e., sorbed). For example, the critical role played by substrate surfaces is reflected in the photolytic half-lives of ~1000 h for benz[a]anthracene (BaA) adsorbed on "black" coal fly ash and irradiated with a laboratory photoreactor (Behymer and Hites, 1988) compared to ~25 min for BaA associated with wood soot exposed midday outdoors in a Teflon chamber under moderate conditions of temperature and relative humidity (Kamens et al., 1988). For this atmospheric aerosol (as well as diesel soot and secondary organic aerosol, SOA), as mentioned earlier, the surface can be viewed as an organic liquid-like layer containing water and a wide range of different organic chemicals (e.g., see McDow et al., 1996, and references therein).

In contrast to their photostability on a black coal fly ash substrate, reactive PAHs such as BaP and BaA photodecompose quite rapidly with half-lives in the range of ~1-4 h (depending on experimental conditions) at the solid–air interfaces of certain water-insoluble inorganic oxides such as SiO_2 and Al_2O_3 (e.g., Behymer and Hites, 1985; see also the reviews on PAH–SiO_2 systems by Sigman et al. (1997) and Dabestani (1997) and articles on specific PAHs discussed herein).

These photochemical reactions in liquids and on solid surfaces are discussed in the following sections.

a. Photooxidation in Solution and "Liquid-like" Surfaces of Organic Aerosols

In 1963, E. J. Bowen published his classic review "The Photochemistry of Aromatic Hydrocarbon Solutions," in which he described two major reaction pathways for PAHs irradiated in organic solvents: photodimerization and photooxidation mediated by the addition of singlet molecular oxygen, $O_2(^1\Delta)$ (or simply $^1O_2$), to a PAH (e.g., anthracene). For details on the spectroscopy and photochemistry of this lowest electronically excited singlet state of molecular oxygen, see Chapter 4.A, the monograph by Wayne (1988), and his review article (1994). For compilations of quantum yields of formation and of rate constants for the decay and reactions of $O_2(^1\Delta)$, see Wilkinson et al., 1993 and 1995, respectively.

Khan and co-workers suggested in 1967 that in polluted urban atmospheres $^1O_2$ might be generated by triplet energy transfer from "strongly absorbing polynuclear aromatic hydrocarbons to normal oxygen," i.e., that PAHs could act as photosensitizers. Furthermore, the absorbing molecule, e.g., a PAH, "could be in the solid, liquid, or gaseous state, or adsorbed on a solid" (Pitts et al., 1969). This suggestion was confirmed directly in experiments by Eisenberg and co-workers (1984) employing several model PAHs deposited on surfaces as sensitizers. Interestingly, they suggested that the $^1O_2$ so formed could then react with a second molecule of the same PAH (in its ground electronic state) to form oxygenated products, some of which are direct-acting mutagens in the Ames assay. The overall process, shown in the following reaction scheme, is referred to as autooxidation.

$$BaP(S_0) + h\nu \rightarrow BaP^*(S_1)$$

$$BaP^*(S_1) \xrightarrow[\text{crossing}]{\text{intersystem}} BaP^{**}(T_1)$$

$$BaP^{**}(T_1) + O_2(^3\Sigma_g^-) \rightarrow BaP(S_0) + O_2(^1\Delta_g)$$

The $O_2(^1\Delta_g)$ can then react with another BaP molecule (or a different PAH receptor molecule in a POM mixture), e.g.,

$$BaP(S_0) + O_2(^1\Delta_g) \rightarrow \text{Reaction products}$$

Note: The symbols S_0, S_1, and T_1 represent the ground electronic state and first excited singlet and triplet states, respectively, of BaP. $O_2(^1\Delta_g)$ is 22.5 kcal mol^{-1} above the ground state of O_2, while for BaP the S_1 and T_1 states are 71 and 42 kcal mol^{-1} above the S_0 ground state, respectively.

Fox and Olive (1979) described the photooxidation by $O_2(^1\Delta)$ of anthracene associated with ambient particulate POM. McCoy and Rosenkranz (1980) reacted $^1O_2$ with chrysene and 3-methylcholanthrene and reported products that are direct-acting mutagens. Furthermore, Seed and co-workers (1989) have shown that exposure to singlet molecular oxygen oxidizes several BaP metabolites, giving chemical products that are "mutagenic in the absence of mammalian metabolic action," i.e., that are direct bacterial mutagens in the *Salmonella typhimurium* forward mutation assay. Such phenomena are relevant to current interest in the possible formation during transport of directly mutagenic species in the heterogeneous degradation of certain PAHs associated with ambient aerosols.

The key role of $O_2(^1\Delta)$ in the direct photooxidation of benzo[a]pyrene dissolved in aerated benzene solutions (10^{-3} M) and irradiated using a 500-W visible light source with a Pyrex filter is described by Lee-Ruff and co-workers (1986). They also carried out a sensitized photooxidation of BaP using such singlet oxygen sensitizers as tetraphenylporphyrin and methylene blue and found the same distribution of products as with the direct photooxidation, thus establishing the key role of $O_2(^1\Delta)$ in both systems. Products included a mixture of 1,6-, 3,6-, and 6,12-benzo[a]pyrenedione, which had earlier been reported in ambient aerosols in Toronto, Canada (Pierce and Katz, 1976), Germany (König et al., 1983), and Scandinavia (Ramdahl, 1983b). Additionally, they isolated and identified a major product not previously reported, the 5a,6-secobenzo[a]pyrene derivative, the aldehyde **XXXIII**:

BaP $\xrightarrow{h\nu/O_2}$ **XXXIII**

This product suggests a singlet oxygen mediated mechanism for quinone formation (Lee-Ruff et al., 1986).

Subsequently, Lee-Ruff and Wang (1991) conducted a similar study of 6-methylbenzo[a]pyrene, a methyl isomer whose carcinogenicity is approximately equal to that of BaP. It photooxidized ~20 times faster than BaP and, in addition to quinones, formed as a major product a seco ketone analogous to **XXXIII** from BaP. Its formation (as with BaP) is ascribed to a $^1O_2$-mediated mechanism.

Since the mid-1980s, Kamens, McDow, and co-workers have carried out a number of studies on factors influencing the rates, mechanisms, and products of the photodegradation of PAHs. They employed both model laboratory systems in which solutions containing various organics found in ambient aerosols are irradiated in photoreactors and environmental chamber studies in which real organic combustion aerosols relevant to ambient atmospheres, e.g., diesel soot and wood smoke, are irradiated with natural or simulated sunlight under a variety of ambient conditions.

For example, Kamens and co-workers (1988) reported that daytime photodegradation rates of PAHs on wood smoke and gasoline combustion soot irradiated by natural sunlight in 25-m^3 outdoor Teflon film chambers were clearly influenced by such variables as solar intensity, relative humidity, and ambient temperature. Thus, under "moderate conditions" (e.g., T = 20°C, "moderate" humidity, and a solar intensity of 1 cal cm^{-2} min^{-1}), PAH half-lives for the most reactive PAHs (Class II, Table 10.30), cyclopenta[cd]pyrene and BaP, were 0.3 and 0.5 h, respectively, compared to 0.8 and 1.3 h for the low-reactivity (Class V) benzofluoranthenes (k and b). At very low values of sunlight intensity, water vapor concentrations, and ambient temperatures ($-10°C$), half-lives increased significantly; e.g., for cyclopenta[cd]pyrene and BaP, the half-lives increased to 2 and 6 h, respectively, and for the benzofluoranthenes to ~10 h. These PAH relative rates are generally consistent with the Nielsen reactivity scale (Nielsen, 1984; Table 10.30). Note that McDow and co-workers (1995) reported that the water content of combustion particles of wood smoke and gasoline soot increased with increasing relative humidity and that PAH photodegradation rates are likely to increase with increasing water content (BaP was an exception) (see also a recent modeling study of humidity effects by Jang and Kamens (1998)).

McDow, Kamens, and co-workers also conducted laboratory experiments on the effects of common organic constituents (e.g., methoxyphenols) on the rates, mechanisms, and products of the solution-phase photodegradation of PAHs associated with wood smoke and diesel soot (see, for example, Odum et al. (1994a), and McDow et al. (1994, 1995, 1996)). Figure 10.28, for example, shows the degradation of the reactive BaP (Class II reactivity) compared to BeP (Class V reactivity) in two solvents, hexadecane, taken as representative of aliphatics in diesel soot, or a mixture of 11 methoxyphenols found in particulate matter (McDow et al., 1994). As expected, BaP decays much more rapidly than BeP. In addition, the decay in the mixture of methoxyphenols is much faster than that in hexadecane.

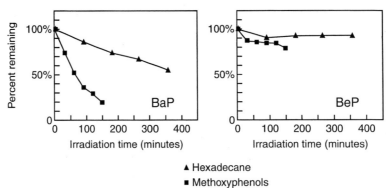

FIGURE 10.28 Rates of photodegradation of benzo[a]pyrene (Class II reactivity; see Table 10.30) and benzo[e]pyrene (Class V) in irradiated solutions of hexadecane or a mixture of methoxyphenols that are representative of important classes of organics present in particles of diesel soot and wood smoke, respectively (adapted from McDow et al., 1994).

Vanillin

Similarly, Jang and McDow (1995, 1997a) investigated the rates, products, and mechanisms of the photooxidation of BaA carried out in three solvents (toluene, benzene, and benzene-d_6) that contained three common constituents of ambient aerosols that are known accelerators of the photodegradation process: 9,10-anthraquinone, 9-xanthone, and the substituted methoxyphenol, vanillin. On the basis of this and previous studies, the researchers proposed "that there were at least two mechanisms" (see also Jang and McDow, 1995). One is singlet oxygen mediated and observed with BaA dissolved in benzene and benzene-d_6 solvents; it involves $^1O_2$ addition to an aromatic ring of BaA followed by ring opening and products such as phthalic acid. The second mechanism is most important in toluene; it features H-atom abstraction from the solvent by aromatic carbonyl compounds, leading to complex free radical initiated oxidations. Jang and McDow (1995) have also suggested as other possible mechanisms for the photooxidation of PAH triplet–triplet energy transfer to the PAH by a photosensitizer (e.g., polycyclic aromatic quinones or ketones and aromatic aldehydes) or free radical chain oxidations initiated by electronically excited PAH. In short, efficiencies of PAH photooxidation processes may depend in part on the H-atom-donating properties of some of the organic constituents (e.g., methoxyphenols) and on properties such as oxygen solubility and $^1O_2$ lifetimes in the liquid-like layer of organic combustion aerosols.

b. Photooxidations on Inorganic Solid–Air Surfaces

Just as with organic combustion aerosols, the chemical and physical nature of inorganic solid substrates can have a dramatic impact on the photoreactivity of adsorbed PAH. In 1980, Korfmacher and co-workers reported that BaP, pyrene, and anthracene all photolyzed efficiently in liquid solution but were resistant to photodegradation when adsorbed on coal fly ash. Subsequent studies confirmed this observation and revealed that the carbon content of the ash (and the associated darkening of color) is a key factor in establishing the photostability of these PAHs. Indeed, they were stabilized at relatively small percentages of carbon, e.g., 5% or less (Behymer and Hites, 1985, 1988; Yokley et al., 1986; Dunstan et al., 1989; Miller et al., 1990).

PAHs adsorbed on particles of carbon black were also photostabilized (Behymer and Hites, 1988). However, Barofsky and Baum (1976) demonstrated that BaP, anthracene, BaA, and pyrene deposited on carbon microneedle field desorption emitters and exposed to UV radiation were all photooxidized to carbonyl compounds. Similarly, PAHs can photodegrade efficiently in air when adsorbed to substrates of silica gel, alumina, or glass plates (e.g., see Lane and Katz, 1977; Kormacher et al., 1980; Behymer and Hites, 1985; Yokely et al., 1986).

The effect of the physical state of BaP and perylene adsorbed on fused-silica plates on their reaction rates with ozone in air was studied by Wu and co-workers (1984), who measured the fluorescence of the two PAHs as functions of concentration, reaction time, and O_3 concentration. Interestingly, their reaction rates in aggregated states (inferred from eximer formation) were much slower than those in the dispersed state (only monomer fluorescence). They concluded that the reaction of O_3 with aggregated PAH molecules was much slower than with monomers on the surface. Alebić-Juretić and co-workers (1990) reported a similar observation for the PAH–silica gel–O_3 reaction.

Such a "surface/bulk" reactivity phenomenon may in part be responsible for the low (or zero) reactivity reported for BaP deposited on, or present in, a variety of substrates and exposed to ambient levels of O_3 (e.g., 100 ppb) in air (see Grosjean et al., 1983; and Coutant et al., 1988). These observations can be rationalized by assuming that, while BaP in fact does react rapidly with O_3 in ambient particles, not all of it is at (or close enough to) the surface to be available for reaction (Atkinson et al., 1988a; Arey, 1998a).

In the early 1990s, a series of articles appeared that described the rates, mechanisms, and products of the photooxidations of several simple gas-phase PAHs adsorbed on the surfaces of SiO_2 and Al_2O_3 particles. These are models for some of the inorganic particles found in air (see Chapter 9). Studies of naphthalene and 1-methylnaphthalene (Barbas et al., 1993), acenaphthylene (Barbas et al., 1994), anthracene (Dabestani et al., 1995), phenanthrene (Barbas et al., 1996), and 1-methoxynaphthalene on SiO_2 (Sigman et al., 1996; see also below and the reviews by Dabestini (1997) and Sigman (1997)) in the presence and absence of air and using different photolytic wavelengths were carried out.

Sigman and co-workers (1997) and Dabestini (1997) reported studies of the photochemistry of 2- and 3-ring PAHs adsorbed on SiO_2. Photooxidations were shown to proceed by two distinct mechanisms. Type I involves electron transfer from the photoexcited singlet state (S_1) of the PAH to atmospheric oxygen, forming superoxide, $O_2^{\cdot -}$ (or transfer to "surface-active sites"). Indeed, Barbas and co-workers (1993) reported the room temperature EPR spectrum of $O_2^{\cdot -}$ when naphthalene adsorbed on the surface of SiO_2 is irradiated. The type II mechanism is a singlet oxygen mediated, photooxidation process such as that described earlier for BaP in liquids—that is, triplet–triplet energy transfer from the T_1 state of an electronically excited PAH molecule to the triplet ground state of an O_2 molecule ($^3\Sigma_g$), producing $O_2(^1\Delta)$, which then adds to the ground-state PAH, ultimately forming oxidized photoproducts.

PAHs photooxidized exclusively by the type I electron transfer–superoxide mechanism include naphthalene and 1-methylnaphthalene (Barbas et al., 1993), fluorene (Barbas et al., 1997), and acenaphthene (Reyes et al., 1998). Typical products formed in the type I photooxidation of naphthalene and 1-methylnaphthalene are shown in Fig. 10.29A.

PAHs photooxidized by the type II singlet oxygen mediated mechanism include acenaphthylene, whose oxidized products and yields are shown in Fig. 10.29B (Barbas et al., 1994), phenanthrene (Barbas et al., 1996), anthracene (Dabestini et al., 1995), and tetracene (Dabestini et al., 1996). An additional photochemical process, the formation of photodimers, is also observed for acenaphthylene, anthracene, and tetracene.

Interestingly, Barbas and co-workers (1996) noted that, once sorbed to the SiO_2 surface, phenanthrene did not desorb into the gas phase even under high vacuum. As in previous studies of the PAHs cited above, no dark reactions were observed at the SiO_2–phenanthrene interface. The authors suggested that the photostability of phenanthrene adsorbed on carbon black (half-life >1000 h) and fly ash (half-life ~49 h) reported by Behymer and Hites (1985) may be due to a number of factors, including competitive absorption of incident radiation, energy transfer from excited phenanthrene to other PAHs in carbon black and "more facile reactions of singlet oxygen with the substrate."

5. Gas–Particle Reactions

a. Ozonolysis

Lane and Katz reported in 1977 that the dark reaction of BaP deposited on the surface of glass Petri dishes with air containing 200 ppb of ozone was fast, with a half-life of ~38 min. Katz and co-workers (1979) exposed nine PAHs on thin-layer chromatography plates of cellulose in the dark to 200 ppb of O_3 in air and found pronounced differences in their reactivities, e.g., half-lives of 36 min for BaP, 2.9 h for BaA, 7.6 h for BeP, and 53 h for benzo[b]fluoranthene. Subsequently, in good agreement with Lane and Katz, a half-life of ~1 h was determined for BaP deposited on glass fiber filters and exposed (passively in a controlled atmosphere) to 200 ppb of O_3 in the dark (Pitts et al., 1980).

Pitts et al. (1986) exposed five individual PAHs, pyrene, fluoranthene, benz[a]anthracene, BeP, and BaP, deposited on glass fiber and Teflon-impregnated glass fiber filter (TIGF) substrates "passively" for 3 h in the dark in a 360-L Teflon environmental chamber to 50–300 ppb of O_3 in air at several relative humidities. These experimental conditions more nearly resemble the actual exposure of ambient particles to O_3 (in the dark) during transport than do exposures in Hi-Vol flow systems. Consistent with earlier studies, BaP, BaA,

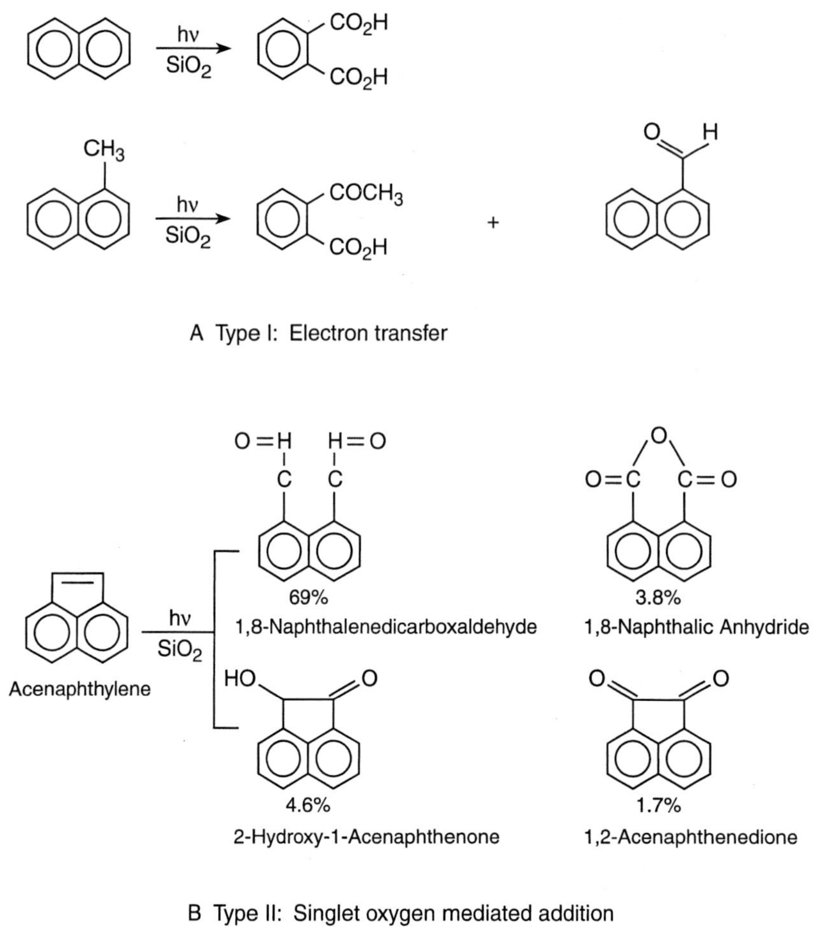

A Type I: Electron transfer

B Type II: Singlet oxygen mediated addition

FIGURE 10.29 Typical photoproducts observed (A) in the irradiation (λ = 300 nm) in air of naphthalene and 1-methylnaphthalene adsorbed on silica and formed by a Type I electron transfer (superoxide) mechanism (Barbas *et al.*, 1993) and (B) in the irradiation in air of acenaphthylene (λ = 350 nm) adsorbed on silica formed by a Type II singlet oxygen mechanism (Barbas *et al.*, 1994) (adapted from Dabestani, 1997).

and pyrene were the most reactive (~60–80% degraded at 1% RH) and, except for BaP, degradation was much less at 50% RH than at 1% RH.

Fresh ambient particulate POM sampled near a freeway was also exposed in this passive system for 3 h in the dark to 200 ppb of O_3 in air at 1% RH. Concentrations of specific PAHs determined in the ambient particles and their percent reacted were similar for samples collected on both kinds of filters (glass fiber and TIGF). Again, under passive exposure conditions to approximately ambient levels of O_3, BaP and BaA were found to be significantly more reactive than BeP.

Results of a study of the ozonolysis of primary combustion aerosols by Van Vaeck and Van Cauwenberghe (1984a) are illustrated in Fig. 10.30. Shown are percent conversion profiles as a function of time for the decay of several 5- and 6-ring PAHs in diesel exhaust particulate matter (D_p < 0.5 μm) collected on glass fiber filters and exposed in the dark to 1.5 ppm O_3 in air under Hi-Vol sampling conditions. Half-lives for degradation range from ~15–20 min for BaP to ~2–3 h for low-reactivity benzo[*k*]fluoranthene. The authors noted that experiments at ambient levels of O_3 of 100 ppb also "include significant conversion." Similarly, Lindskog and co-workers (1985) reported several PAHs in soot particles that were generated in a smoke gas generator and collected on a glass fiber filter were "transformed" when exposed to ozone. Overall, such results suggest that in ambient aerosols, the highly reactive PAHs, e.g., BaP and cyclopenta[*cd*]pyrene, could be susceptible to ozone degradation under certain atmospheric conditions, as well as during Hi-Vol collection of samples.

Van Cauwenberghe (1985) proposed two different mechanisms for the degradation of unsubstituted PAHs by ozone; one is a one-step electrophilic–nucleophilic attack on olefin bonds with high electron density. This forms an unstable primary ozonide that decomposes, with ring opening giving products with aldehyde and/or carboxylic acid groups. Peroxides or α-keto hydroperoxides may also be formed. The second mechanism is a two-step electrophilic attack involving initial formation of a σ complex and ultimately quinones.

Complex mixtures of oxidized products are produced in the ozonolysis of BaP on solid substrates. For example, BaP deposited on a glass fiber filter and exposed to 1 ppm O_3 in air primarily formed products characteristic of the ring-opening mechanism, e.g., dialdehydes and dicarboxylic and ketocarboxylic acids. However, some quinones and phenols, were also formed. For details on the mechanisms and O-PAH products of these BaP–O_3, gas–solid substrate reactions, see Van Cauwenberghe et al. (1979) and Van Cauwenberghe and Van Vaeck (1983 and references therein).

Mixtures of products formed in the reactions of BaP deposited on glass fiber filters with 100–200 ppb of O_3 in air are not only complex chemically, but their extracts are also directly mutagenic in the Ames assay (strain TA98, −S9). Using the method of bioassay-directed fractionation and chemical analysis, as illustrated in Figs. 10.31a to 10.31d, Pitts and co-workers (1980) isolated, identified, and obtained the specific activity, 1600 rev μg^{-1} (TA98, −S9), of the major stable contributor to the activity of such mixtures, benzo[a]pyrene-4,5-oxide (Fig. 10.31d). It is a well-known BaP metabolite, a powerful direct-acting frameshift mutagen, and a weak carcinogen on mouse skin (Levin et al., 1976). The value of 1600 rev μg^{-1} is in good agreement with those of other researchers (e.g., Wislocki et al., 1976, ~1370 rev μg^{-1}; Chang et al., 1979).

Extrapolation of results from the ozonolysis or photooxidation of BaP deposited on solid substrates (e.g., filters) to possible reaction rates and products for BaP sorbed on or in ambient aerosols in real-world polluted atmospheres should be done with caution. For example, the yield of BaP epoxides is low (0.5%) for exposures in the dark and even less when the BaP-coated filter is exposed concurrently to actinic light (0.07%). However, this reaction, forming a BaP metabolite, is interesting per se and may have significant atmospheric and toxicological relevance if it were to take place on the surfaces of fine respirable ambient aerosols (Pitts et al., 1980; Pitts, 1983). Van Cauwenberghe and Van Vaeck (1983) noted that because of its increased solubility in lung tissues, its pulmonary absorption may be enhanced over nonpolar BaP. Similar considerations might apply to as yet unidentified products of the heterogeneous reactions of ozonolysis and photooxidation of, for example, the highly reactive and powerful human cell mutagen cyclopenta[cd]pyrene (**XXVIII**).

b. Nitration

Certain 4- to 6-ring PAHs, deposited on a variety of substrates and exposed to a range of concentrations of gaseous NO_2 in the presence of trace amounts of HNO_3 near ambient levels, react to form mono- and dinitro-PAHs (Pitts et al., 1978; Pitts, 1979; Jäger and Hanus, 1980; Tokiwa et al., 1981).

In the first product study in a laboratory-simulated atmosphere, ~20% of the BaP deposited on a GF filter and exposed for 8 h to air containing 0.25 ppm of NO_2 and traces of gaseous HNO_3 was converted to 6-NO_2-BaP; smaller amounts of the 1- and 3-isomers were also formed (Pitts et al., 1978; Pitts, 1979). Interestingly, in this heterogeneous process, a promutagen was converted to a stronger promutagen, 6-NO_2-BaP, and two powerful *direct* mutagens, the 1- and 3-NO_2-BaP isomers (TA98, −S9; see Table 10.16), were formed.

The generality of this reaction of NO_2 with particle-associated PAHs was demonstrated when, under similar laboratory conditions, perylene (a weak promutagen) was converted to 3-nitroperylene, and pyrene (a nonmutagen), at a slower rate, to 1-nitropyrene; both nitro-PAHs are direct-acting frameshift mutagens. In a similar experiment, chrysene was not nitrated; this is consistent with its low reactivity in the Nielsen reactivity scale (Table 10.30).

Tokiwa et al. (1981) reported that direct-acting mutagenic derivatives were formed when pyrene, phenanthrene, fluorene, fluoranthene, and chrysene (and carbazole) were deposited on filter paper and exposed to

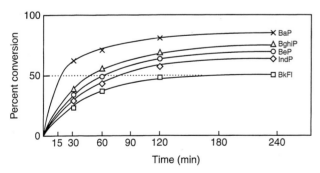

FIGURE 10.30 Percent conversion–time profiles for the decay of 5 PAHs in diesel exhaust particulate matter ($D_p = 0.5$ μm) collected on glass fiber filters and exposed to 1.5 ppm of ozone in air under Hi-Vol sampling conditions. Half-lives (dotted line) decrease in order of the Nielsen (1984) electrophilic reactivity scale (Table 10.30): BaP, benzo[a]pyrene; BghiP, benzo[ghi]perylene; BeP, benzo[e]pyrene; IndP, indeno[1,2,3-cd]pyrene; BkFl, benzo[k]fluoranthene (adapted from Van Vaeck and Van Cauwenberghe, 1984).

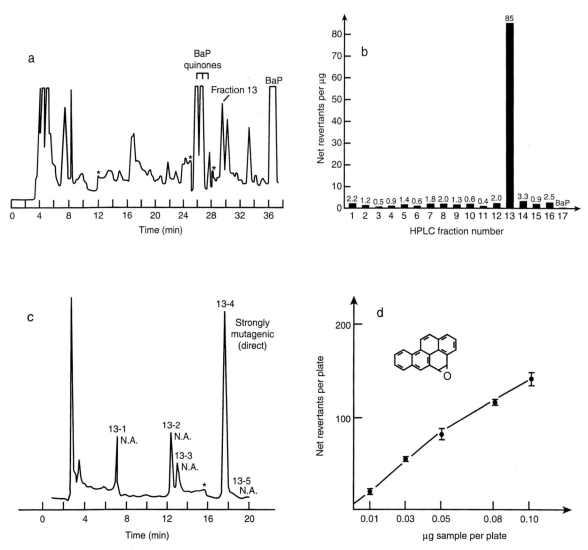

FIGURE 10.31 (a) Preparative HPLC separation (acetonitrile–water solvent) of products of the dark reaction of BaP deposited on a glass fiber filter with ~ 200 ppb O_3 in air (asterisk marks scale change). (b) Specific direct-acting mutagenic activity of HPLC fractions (rev μg^{-1}) on TA98, −S9. (c) Preparative HPLC separation of mutagenic fraction 13. (d) Dose–response curve for fraction 13-4 (TA98, −S9), BaP-4,5-oxide, whose specific activity is 1600 rev μg^{-1} (adapted from Pitts *et al.*, 1980). N.A., no activity.

10 ppm NO_2. The 24-h nitration yields for pyrene rose from 0.02% with 1 ppm NO_2 in air to 2.85% when traces of HNO_3 were present.

Jäger and Hanus (1980) found the order of reactivity for the reactions of PAHs adsorbed on several substrates with 1.3 ppm of NO_2 in air to be silica gel > fly ash > deactivated aluminum oxide > carbon. The qualitative composition of nitro-PAH products, however, was independent of the substrate.

Subsequently, Ramdahl and co-workers (1984b) adsorbed six PAHs on several substrates and exposed them to 0.5 ppm of NO_2 in air containing water vapor and traces of HNO_3. The highest reactivity was observed for PAHs adsorbed on silica. The yields of three nitro-PAHs detected on alumina were only 14–24% of those on silica. The relative PAH reactivity order was perylene > benzo[*a*]pyrene > pyrene > chrysene > fluoranthene = phenanthrene, similar to that found by other researchers in their solution-phase studies (e.g., see Dewar *et al.*, 1956; and Nielsen, 1984). A similar order for relative reactivity was observed by Butler and Crossley (1981) for the loss of PAH deposited on soot particles and exposed to NO_2 in air.

Guo and Kamens (1991) describe a system for studying gas–particle reactions on the surfaces of combustion aerosols in which they report a half-life of ~ 80 h for "high loadings" of particle-bound BaP in wood smoke particles reacting with ~ 200 ppb of NO_2 in air.

Relatively small amounts of gaseous nitric acid in the NO_2–air mixtures used to nitrate particle-phase

PAHs on various substrates in simulated ambient air or combustion systems appear to play an important role in the reactivity (e.g., see Pitts et al., 1978; Pitts, 1979; Hughes et al., 1980; Lindskog et al., 1985; and Yokley et al., 1985). Grosjean and co-workers (1983) found no reaction of nitric acid free NO_2 in air with BaP, perylene, and 1-nitropyrene deposited on several substrates. Although Wu and Niki (1985) did not report an important role for HNO_3 in their spectroscopic study of the reaction of NO_2 with pyrene deposited on a silicon surface, acidity could have been provided to some extent by the silica surface. Thus, once again, substrate complexities may be involved.

The nitration of PAHs by NO_2/HNO_3 also occurs under laboratory conditions approximating plume gases, that is, higher concentrations of gases and deposition on coal fly ash as a substrate. Thus, Hughes and co-workers (1980) reported that BaP and pyrene reacted with 100 ppm of NO_2; the presence of nitric acid (possibly on the surface of the fly ash) enhanced the rate of reaction. Reactions proceeded more rapidly on silica gel than fly ash substrates, and for pyrene, both mono and dinitro isomers were formed. At the 100 ppm plume gas level, neither NO nor SO_2 reacted with BaP or pyrene on the substrates studied; both PAHs reacted with SO_3, but products were not characterized.

In a study simulating stack gas sampling, Brorström-Lundén and Lindskog (1985) found that addition of NO_2 caused substantial degradation of the reactive PAHs present in soot generated from a propane flame. A strong enhancing effect was observed when gaseous HCl was added to the laboratory stack gases, for example, 90% loss of BaP in 1 h with added HCl compared to 20% in its absence. They suggested that, in addition to other processes that might occur in the hot effluent stream, under these acidic conditions, sampling artifacts may be a major problem in sampling stack gases, often in times as short as 15–30 min or less.

The possibility of artifactual formation of nitro-PAHs during the sampling of diesel exhaust was addressed soon after their discovery in diesel particles (e.g., Lee et al., 1980; Lee and Schuetzle, 1983). Schuetzle (1983) concluded that artifactual formation of nitro-PAHs "is a minor problem" (between 10 and 20% of the measured 1-nitropyrene) at short sampling times [e.g., 23 min, which is one federal test procedure (FTP) driving cycle], at low sampling temperatures (42°C), and in diluted exhaust containing NO_2.

The question of formation of nitroarenes during Hi-Vol sampling of ambient POM was considered in early studies (Pitts et al., 1978; Pitts, 1979) and addressed in several of the studies of PAH nitrations discussed above. In a definitive evaluation, Arey and co-workers (1988a) coated several perdeuterated PAHs (e.g., perdeuterofluoranthene, perdeuteropyrene, and perdeutero-BaP) onto Hi-Vol filters loaded with previously collected ambient POM and exposed them for 7–10 h to ambient air during a high-NO_x episode in southern California. Less than 3% of the total 1-nitropyrene collected during the episode was formed in the sampling process, and no formation of nitrofluoranthene was observed. Hence the authors concluded artifactual formation of nitroarenes during Hi-Vol sampling (e.g., the nitrofluorenes and nitropyrenes) is not significant (see Arey et al., 1988a, for references to other studies).

Overall, while the combinations of substrate effects, ambient NO_2 levels, and other gas–particle phenomena preclude a definitive answer, the formation of significant amounts of nitroarenes in heterogeneous particle-phase NO_2-PAH, atmospheric reactions seems unlikely, e.g., much slower than photooxidation or ozonolysis. This conclusion also applies to heterogeneous reactions of N_2O_5 with particle-bound PAHs on diesel and wood soot (Kamens and co-workers, 1990; see also Pitts et al., 1985c, 1985d, 1985e).

XXVIII

Cyclopenta[cd]pyrene   Acenaphthylene

An exception may be particle-associated cyclopenta[cd]pyrene (**XXVIII**), which, with its exocyclic double bond (analogous to acenaphthalene), may react at a significant rate with the gas-phase NO_3 radical (or N_2O_5) and even faster in heterogeneous photooxidations or ozonolysis. Certainly, by analogy with acenaphthylene, whose atmospheric lifetimes under typical atmospheric conditions are 6 min for NO_3 radicals, 43 min for O_3, and 1.6 h for OH radicals (see Atkinson and Arey, 1994), gas-phase cyclopenta[cd]pyrene would also be expected to decay rapidly by these reactions in nighttime and daytime polluted air environments. Recall that Fraser and co-workers (1998) found that over 50% of the cyclopenta[cd]pyrene was in the gas phase (see Figs. 10.2 and 10.27) during hot weather. As noted earlier, the rates of formation, mechanisms, chemical structures, and yields of products, as well as the mutagenic activities, bacterial and human cell, of such cyclopenta[cd]pyrene reactions are currently unknown but constitute an interesting and relevant research challenge to atmospheric chemists and toxicologists alike.

6. Atmospheric Fates of Particle-Associated Nitroarenes

As we have seen, key nitroarenes found in extracts of ambient particulate matter are 1-nitropyrene (1-NO_2-Py), predominant in primary combustion emissions, and 2-nitrofluoranthene and 2-nitropyrene, major products of gas-phase atmospheric reactions. Here we focus simply on their atmospheric fates as particle-bound species participating in heterogeneous decay processes. Formation of such nitro-PAHs in gas-phase reactions is addressed in Section F.

a. Gas–Particle Reactions

Heterogeneous reactions in sunlight of particle-bound nitro-PAHs with ambient O_3 or NO_2 are generally believed to be minor relative to photodegradation. For example, based on their experimental results and modeling study of the formation and decay of nitro-PAHs in diesel exhaust emissions aged in the light and dark in an outdoor smog chamber, Fan and co-workers (1995) concluded that "photodecomposition was the main loss pathway for nitro-PAH in the atmosphere."

Subsequently, Fan and co-workers (1996a) conducted an experimental and modeling study of the reactions of O_3 and NO_2 with nitro-PAHs on heterogeneous soot particles. They concluded that while photodegradation is the major daytime loss process for the nitro-PAH, during the night, particle oxidation by O_3 may be the most important decay pathway.

b. Photochemical Reactions

In 1966, Chapman and co-workers proposed a nitro–nitrite photorearrangement as an efficient primary photochemical process for nitroarenes in which the nitro group is out of the plane of the aromatic rings. This is followed by dissociation into NO and a phenoxy-type radical; ultimately quinones and other oxy products are formed (Chapman et al., 1966).

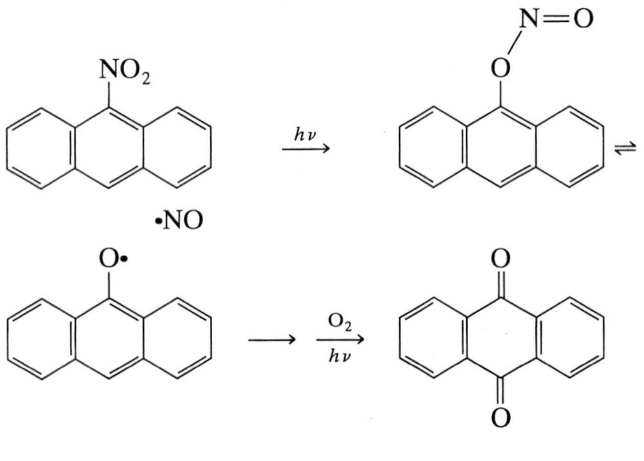

Subsequently, Ioki (1977) used ESR spectroscopy to confirm the production of the benzo[a]pyrene-6-oxyl radical in the nitro–nitrite photorearrangement of 6-NO_2-BaP irradiated in benzene solution:

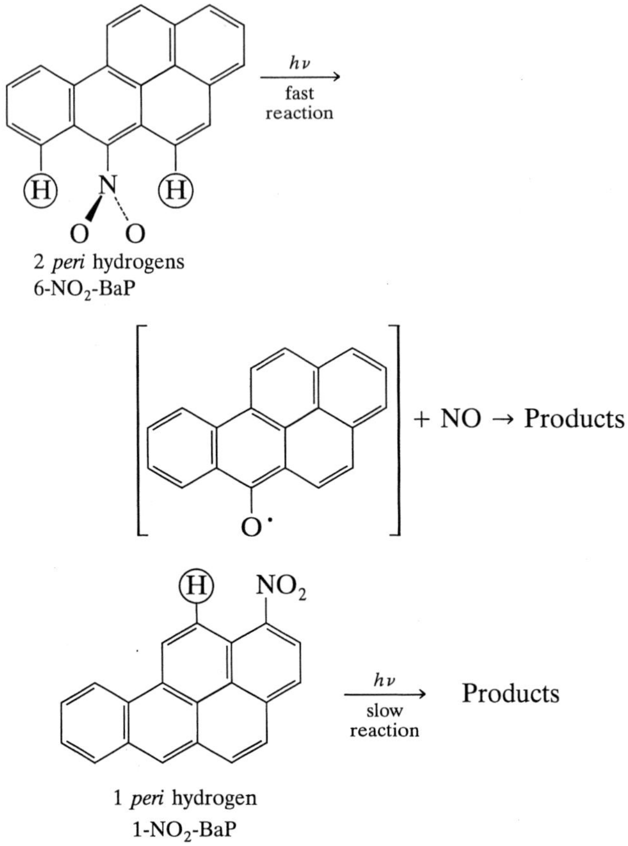

Based on the Chapman mechanism, Pitts (1983) proposed that 6-NO_2-BaP, with two *peri* hydrogens and the NO_2 "out of plane," should be less stable photochemically than the 1- and 3-NO_2 isomers with only one *peri* hydrogen. This proved to be the case. Thus, in solution-phase irradiations of these isomers, 6-NO_2-BaP decomposed rapidly whereas the 1- and 3-isomers were much more stable (Zielinska, 1985). A key question then was whether or not these results could be extrapolated to give their relative photodecomposition rates when irradiated as particle-bound species on the surfaces of primary combustion products and ambient aerosols (vide infra; see also Feilberg and Nielsen, 1999b).

Subsequently, Benson (1985) reported 1-nitropyrene deposited on glass photodecomposed in sunlight with a half-life of 14 h. The reaction was accompanied by loss of the nitro group, formation of a phenolic derivative and possibly quinones, and a significant reduction in mutagenicity, consistent with the Chapman mechanism and previous results on nitro-BaP isomers (Finlayson-Pitts and Pitts, 1986).

Stärk and co-workers (1985) reported that irradiation of a 0.1 mM solution of 1-nitropyrene in 2-propanol with light from 320 to 418 nm changes its absorption spectrum and concurrently results in almost total loss of its direct ($-$S9) or activatable ($+$S9) mutagen activity in the Ames *Salmonella* assay.

A similar concomitant loss of mutagenicity with loss of compound was observed when Holloway and co-workers (1987) irradiated ($\lambda > 310$ nm) 1-nitropyrene, 1,8-dinitropyrene, and 3-nitrofluoranthene coated onto silica or in a dimethyl sulfoxide solution. Half-lives for photodecomposition of 1-nitropyrene in solution compared to those on silica were 1.2 and 6 days, respectively; for 3-nitrofluoranthene the half-lives were 12.5 days in solution and >20 days on silica. Interestingly, 1,8-dinitropyrene photodecomposed with half-lives of 0.7 day in dimethyl sulfoxide compared to 5.7 days on silica; a major photodecomposition product was 1-nitropyren-8-ol.

Van den Braken-van Leersum and co-workers (1987) reported that on irradiation in methanol solutions ($\lambda > 300$ nm), 1-nitropyrene undergoes a rapid conversion via the nitro–nitrite rearrangement, forming 1-hydroxypyrene (88%) and 1-hydroxy-2-nitropyrene (7%). Under the same conditions, the 4-nitro isomer is more stable than the 1-nitro isomer and 2-nitropyrene is very stable; it does not react either with or without oxygen present.

In a study of substrate effects on the photodecomposition of several key nitroarenes, Fan and co-workers (1996b) added several key nitro-PAHs and their deuterated analogs, along with particles of diesel exhaust or wood smoke, to a 190-m^3 outdoor smog chamber. Rapid degradation was observed when they were aged in natural sunlight at temperatures from -19 to $+38$°C. For example, the half-lives on diesel soot particles at noon on June 15 were 0.8 h for both 1-nitropyrene and 2-nitropyrene and 1.2 h for 2-nitrofluoranthene. The half-life was 0.5 h for 1-nitropyrene-d_9 on particles of wood soot. Overall, the authors concluded that the photodecomposition rates of these nitro-PAHs are strongly influenced by the physical and chemical nature of the substrates as seen in the photooxidation of PAHs discussed earlier, and on the solar radiation.

This is consistent with a study by Feilberg and Nielsen (1999b), who investigated the influence of other aerosol components on the photodegradation rates of representative particle-associated nitro-PAHs in a model system consisting of the nitro-PAH dissolved in cyclohexane along with various known constituents of diesel exhaust and wood smoke particles. These "cosolutes" included PAHs, substituted phenols, hydroxy-PAHs, oxy-PAHs, and substituted benzaldehydes.

In the absence of cosolutes, the photodegradation rates depended on the orientation of the nitro group. Thus 1-nitropyrene decayed relatively fast by the nitro–nitrite primary intramolecular photorearrangement process, followed by secondary radical reactions. However, 2-nitropyrene and 2-nitrofluoranthene were stable toward photolysis, consistent with the NO$_2$ group being in the same plane as the aromatic rings.

However, when H-atom-donating cosolutes, e.g., certain phenols, were added, the photodegradation rates of both 1-nitropyrene and 3-nitrofluoranthene increased. In this case, the reaction occurred via H-atom abstraction from the phenol by the electronically excited nitro-PAHs. Feilberg and Nielsen concluded that the photodegradation of nitro-PAHs on both diesel particles and wood smoke proceeds primarily by radical formation. However, H-atom abstraction by the excited triplet states of 1-nitropyrene and 2-nitrofluoranthene may also contribute.

F. REACTIONS OF GAS-PHASE PAHs: ATMOSPHERIC FORMATION OF MUTAGENIC NITROARENES

From the late 1970s to the mid-1980s, the major source of nitro-PAHs in ambient air was thought to be combustion emissions of particulate POM. However, field and laboratory evidence subsequently pointed to a second major source of NO$_2$-PAHs, their *in situ* atmospheric formation through gas-phase reactions of 2- to 4-ring volatile and semivolatile PAHs. Although both are potential sources of nitro-PAHs in air, as we shall see, the distributions of various isomers and compounds provide compelling evidence for a significant contribution of atmospheric reactions to the direct mutagenicities of polluted airsheds throughout the world. We discuss first their direct emissions and then their formation in atmospheric reactions of PAHs.

1. Combustion-Generated Primary Emissions of Nitroarenes

Initially, major attention was focused on emissions of diesel soot from light- and heavy-duty motor vehicles, because the particle extracts contained such direct-acting bacterial mutagens and animal and possible human carcinogens (IARC Group 2B, 1989) as 1-nitropyrene (1-NO$_2$-PY) and the 1,6- and 1,8-dinitropyrenes (DNP), as well as a third isomer, 1,3-dinitropyrene, IARC Group III. For example, Paputa-Peck and co-workers (Paputa-Peck *et al.*, 1983; IARC, 1989) reported ~100 nitro-PAHs (17 positively identified) in an extract of a light-duty diesel particulate sample. Among these, as seen in Table 10.33, 1-nitropyrene was the major nitroarene at levels of 75 µg/g compared to 0.30, 0.40, and 0.53 µg/g for 1,3-, 1,6-, and 1,8-DNP, respectively. For a comparison as to how the absolute and relative emission strengths of 1-nitropyrene and the three dinitropyrenes can vary, see the study of Japanese diesel-powered vehicles by Hayakawa and co-workers (1994). Also, for comparison with the 1983 Paputa-Peck et al. study, see the report of Feilberg and co-workers (1999a), who recently identified and quantified the semivolatile 1-nitronaphthalene in diesel exhaust (vide infra).

Generally, 1-nitropyrene and 2-nitrofluorene are the most abundant nitro-PAHs in diesel exhaust (e.g., see also Schuetzle and Perez, 1983; and Beije and Möller, 1988). However, emission rates of these and other nitro-PAHs vary significantly with engine type, fuel, operating conditions, etc. (e.g., see Schuetzle and Frazier, 1986; IARC, 1989; and WHO, 1996). Thus generalizations should be viewed with caution (e.g., 2-nitrofluoranthene is greater than 2-nitrofluorene in Table 10.33). Furthermore, the distribution of diesel compared to gasoline engine powered vehicles varies markedly from country to country (indeed, state to state), with associated variations in exhaust emission impacts on ambient air quality. For example, in 1994, diesel-powered vehicles constituted 18% of all vehicles in Japan, with the proportion continuing to increase. Murahashi and co-workers (1995) reported total dinitropyrene concentrations [Σ(1,3-DNP + 1,6-DNP + 1,8-DNP)] of 3.6 pg m^{-3} in samples taken adjacent to a busy intersection in downtown Kanazawa, Japan. In ambient air, Berlin, Germany, concentrations of 5.2 ng m^{-3} of 2-nitrofluorene (also emitted in diesel exhaust) were reported by Beije and Möller (1988).

As seen in Table 10.34, the isomeric distribution of the nitro-PAHs in diesel exhaust particles is consistent with electrophilic nitrations of their parent PAHs. The dominance of 1-nitropyrene and the isomer distributions of the nitropyrenes and nitrofluoranthenes observed in diesel exhaust are generally consistent with the higher reactivity of pyrene on the electrophilic reactivity scale (Nielsen, 1984; see Table 10.30) and with the Ruehle et al. (1985) assignment of major electrophilic nitration products. [For evaluations of diesel exhaust, see IARC (1989), WHO (1996); and CARB (1998); reviews include Schuetzle (1983), White (1985), Tokiwa and Ohnishi (1986), and Schuetzle and Daisey (1990); articles include, e.g., Xu et al. (1982), Schuetzle et al. (1982), Gibson (1983), Schuetzle and Perez (1983), Paputa-Peck et al. (1983), Schuetzle and Frazier (1986), Robbat et al. (1986), Beije and Möller (1988), MacCrehan et al. (1988), Schneider et al. (1990), Westerholm et al. (1991), Rogge et al. (1993a), Hammerle et al. (1994), Hayakawa et al. (1994), Johnson et al. (1994), Lowenthal et al. (1994), Westerholm and Egebäck (1994), Murahashi et al. (1995), and Nielsen (1995, 1996).]

Other combustion sources of nitro-PAHs and nitro-PACs include, for example, the following: *gasoline exhaust*, Wang et al., 1978; Alsberg et al., 1985; IARC, 1989; *coal fly ash*, Fisher et al., 1979; Fisher, 1983; White, 1985; *wood smoke*, Alfheim et al., 1984b; *indoor air particulate matter*, e.g., emissions from a kerosene heater, Kinouchi et al., 1988; see also the review by van Houdt (1990) and references therein.

2. Atmospheric Formation of Nitro-PAHs and Nitro-PACs

In the mid-1980s, the discovery of significant amounts of 2-nitropyrene in ambient particulate matter collected in a rural region of Denmark by Nielsen and co-workers (1984) and 2-nitrofluoranthene in southern California by Pitts and colleagues (1985b) provided unique initial evidence for the possible formation of nitroarenes by reactions of precursor PAHs in ambient air. Thus, these compounds are not electrophilic nitration products of their parent PAHs and are generally not observed in combustion sources such as diesel soot; see, for example, Table 10.34 and the report of Ciccioli and co-workers on the detection of emissions of 2-nitrofluoranthene and 2-nitropyrene solely from a "very minor" Italian industrial source (see Ciccioli et al., 1993, 1995, 1996, and references therein).

Subsequently, researchers confirmed the presence of 2-nitropyrene and 2-nitrofluoranthene in extracts of samples of ambient POM collected at sites throughout the world and reported 2-nitrofluoranthene levels that

TABLE 10.33 Concentrations of Selected Nitroarenes in Extracts of Diesel Particulate POM[a,b]

Nitroarene	Concentration (µg/g)
1-Nitronaphthalene	0.95
2-Nitronaphthalene	0.35
2-Nitrofluorene	1.2
1-Nitropyrene	75
3-Nitrofluoranthene	3.5
8-Nitrofluoranthene	1.3
6-Nitrobenzo[a]pyrene	4.2
1,3-Dinitropyrene	0.30
1,6-Dinitropyrene	0.40
1,8-Dinitropyrene	0.53

[a] Adapted from Paputa-Peck et al. (1983) and IARC (1989).
[b] See also Table 10.35.

TABLE 10.34 Distribution and (Yields) of Nitro-PAH Isomers (a) Formed from the Gas-Phase Reactions of Selected Volatile and Semivolatile PAHs with OH Radicals and NO$_3$ Radicals (Each in the Presence of Oxides of Nitrogen), (b) Formed by Electrophilic Nitration, and (c) Present in Diesel Exhaust[a,b]

Parent PAH	Gas-phase OH radical initiated reaction (yield)[b]	Gas-phase NO$_3$ radical initiated reaction (yield)[b]	Electrophilic nitration[c]	NO$_2$-PAH in diesel exhaust[g]
Naphthalene	1-Nitronaphthalene (0.3%) 2-Nitronaphthalene (0.3%)	1-Nitronaphthalene (17%) 2-Nitronaphthalene (7%)	1-Nitronaphthalene > 2-Nitronaphthalene	1-Nitronaphthalene 2-Nitronaphthalene
1-Methylnaphthalene	1M5NN[d] > 1M4NN ≥ 1M6NN (total yield ~0.4%)[h]	1M3NN > 1M5NN ≥ 1M4NN (total yield ~30%)[h]	1M4NN > 1M2NN > 1M5NN[d]	None reported
2-Methylnaphthalene	2M5NN > 2M6NN ~ 2M7NN (total yield ~0.2%)[h]	2M4NN > 2M1NN ~ 2M5NN (total yield ~30%)[h]	2M1NN ≫ 2M8NN > 2M4NN[d]	2-Methyl-1-nitronaphthalene
Fluorene	3-Nitrofluorene (~1.4%) 1-Nitrofluorene (~0.6%)	No data	2-Nitrofluorene	2-Nitrofluorene
Anthracene[f]	1-Nitroanthracene (low yield) 2-Nitroanthracene (low yield)	1-Nitroanthracene (low yield) 2-Nitroanthracene (low yield)	9-Nitroanthracene	9-Nitroanthracene
Fluoranthene	2-Nitrofluoranthene (~3%) 7-Nitrofluoranthene (~1%) (~24%) 8-Nitrofluoranthene (~0.3%)	2-Nitrofluoranthene (~24%)	3-Nitrofluoranthene > 8-Nitrofluoranthene > 7-Nitrofluoranthene	3-Nitrofluoranthene > 8-Nitrofluoranthene
Pyrene	2-Nitropyrene (~0.5%) 4-Nitropyrene (~0.06%)	4-Nitropyrene (0.06%)	1-Nitropyrene	1-Nitropyrene
Biphenyl	3-Nitrobiphenyl (5%)	No reaction observed	2-Nitrobiphenyl 4-Nitrobiphenyl[e]	2-Nitrobiphenyl

[a] Adapted from Atkinson and Arey (1997) and Arey (1998a).
[b] Data from Atkinson and Arey (1994).
[c] From Ruehle et al. (1985).
[d] From Eaborn et al. (1968). Nomenclature: e.g., 1M5NN = 1-methyl-5-nitronaphthalene.
[e] From Zielinska et al. (1990).
[f] Although 9-nitroanthracene was observed in both OH and NO$_3$ reactions, it may not be a product as it is also found in exposure to NO$_2$/HNO$_3$ systems (Arey, 1998a).
[g] From Paputa-Peck et al. (1983); see Table 10.33.
[h] From Zielinska et al. (1989b).

generally exceeded those of 1-nitropyrene (e.g., Nielsen and Ramdahl, 1986; Ramdahl et al., 1986; Sweetman et al., 1986; Arey et al., 1987, 1988b; Atkinson et al., 1987a, 1988a; Nishioka et al., 1988; Zielinska et al., 1989a; Ciccioli et al., 1993, 1995, 1996; Legzdins et al., 1994; Atkinson and Arey, 1994; Wilson et al., 1995; Arey, 1998a).

For example, Legzdins and co-workers (1994) used the bioassay-directed fractionation and chemical analysis technique to isolate, identify, and quantify 2-nitrofluoranthene in extracts of ambient particles collected in Hamilton, Ontario, Canada. They found it accounted for ~70% of the total nonpolar direct bacterial mutagenicity (strain YG1021, standard reversion assay, Maron and Ames, 1983).

Additionally, in two different monitoring campaigns conducted in the center of Milan, Italy, Ciccioli and co-workers (1993) reported 2-nitrofluoranthene, 2-nitropyrene, and 1-nitropyrene were the only nitroarenes detected. Subsequently, in a comprehensive study of the atmospheric formation and transport of 2-nitrofluoranthene and 2-nitropyrene, they established their presence and levels in ambient particles collected at sites located in urban, suburban, forest, and remote areas in Europe, Asia, America, and Antarctica (Ciccioli et al., 1996, and references therein; see also Ciccioli et al., 1995).

As an example of typical experimental data, Fig. 10.32 is a GC–MS selected ion monitoring (SIM) profile (m/z 247) for the nitrofluoranthenes and nitropyrenes in an extract of ambient particles collected in southern California (Arey et al., 1988b). The 1-nitropyrene (1-NP) and 3-nitrofluoranthene (3-NF) presumably are from diesel emissions (Tables 10.33 and 10.34), but the dominance of 2-nitrofluoranthene and 2-nitropyrene reflects a second major source.

Shortly after discovery of the 2-nitro isomers of pyrene and fluoranthene in ambient air, an OH radical initiated mechanism analogous to that of Nielsen and co-workers (1984) for 2-nitropyrene was proposed and experimental laboratory evidence reported for the *in situ* formation of 2-nitrofluoranthene (Pitts et al., 1985b; Ramdahl et al., 1986; Nielsen and Ramdahl, 1986; Sweetman et al., 1986; Arey et al., 1986; for background, see the discussion by Atkinson et al., 1987a). As seen in Fig. 10.33, the major pathway involves (a) OH radical attack at the sites of highest electron densities (the 3-position is favored for fluoranthene and the 1-position for pyrene, (b) the addition of NO_2 in the *ortho* (2-) position to the hydroxycyclohexadienyl-type radical, and (c) loss of water and formation of 2-nitrofluoranthene (and by analogy 2-nitropyrene). Note that analogous to the reactions of OH with simple aromatics, O_2 can compete with NO_2 in adding to the

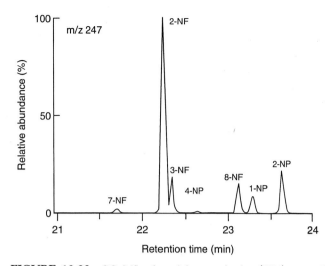

FIGURE 10.32 GC–MS selected ion monitoring (SIM) scan of the molecular ions (m/z 247) of nitrofluoranthene (NF) and nitropyrene (NP) isomers in extracts of ambient particles collected in Torrance, California, January 1986 (adapted from Arey et al., 1988b, and Atkinson and Arey, 1997).

OH adduct radical (e.g., see Atkinson, 1994, and Chapter 6.G). However, in the presence of sufficient NO_2, 2-nitrofluoranthene and 2-nitropyrene are formed and subsequently condense out on particle surfaces (see Pitts, 1987, Atkinson and Arey, 1994, Arey, 1998a, and references therein).

For this mechanism to be effective, several requirements must be met:

- Under ambient conditions, a substantial portion of the fluoranthene and pyrene must be in the gas phase, which is indeed the case (see Figs. 10.2 and 10.3).
- Attack by OH radicals on the gas-phase fluoranthene and pyrene must be fast. As seen in Table 10.35, again this is true. In Table 10.36, the calculated atmospheric lifetimes of selected gas-phase PAHs due to reaction with OH are shown, e.g., lifetimes of ~2.9 h for fluoranthene and pyrene.
- Products (and their mutagenicities) of the gas-phase reactions of these and other 2- to 4-ring PAHs (fluorene, naphthalene, etc.) carried out under simulated atmospheric conditions should be consistent with the nitroarenes and nitro-PACs that have been identified in ambient air. This criterion has been well established over the years in environmental chamber studies conducted in several laboratories (e.g., see reviews by Atkinson and Arey, 1994; and Arey, 1998a; articles by Kamens et al., 1994; Fan et al., 1995; Feilberg et al., 1999a; and references therein).

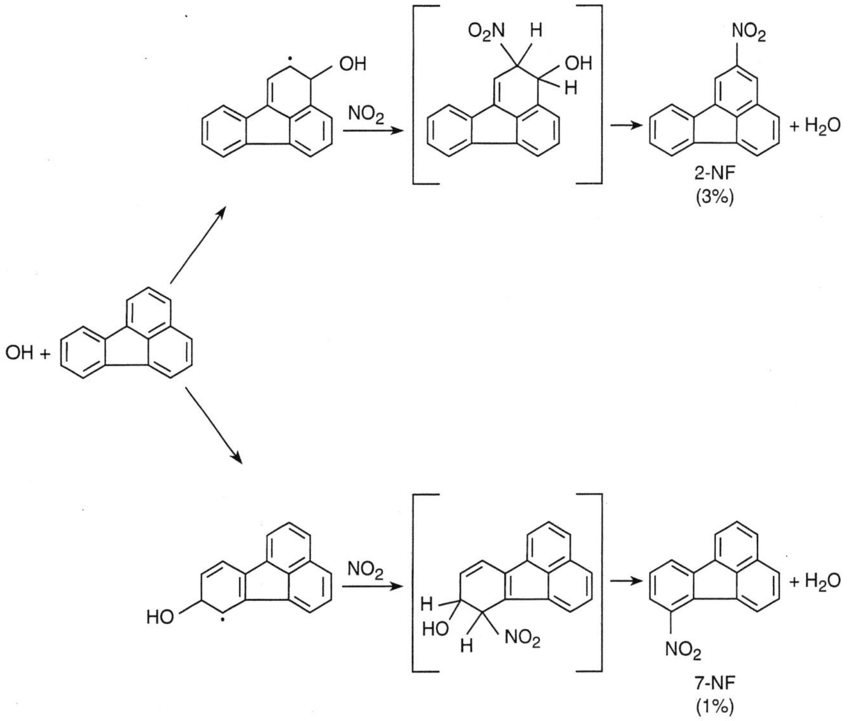

FIGURE 10.33 Mechanism of oxidation of fluoranthene by OH in air (adapted from Arey, 1998a).

For example, Fig. 10.34 shows the mutagram (TA98, −S9) of an extract of ambient particles collected in Claremont, California, August 1987 (Harger et al., 1992) together with mutagrams of extracts of environmental chamber reaction products for the simulated OH radical initiated reactions of phenanthrene, fluoranthene, and pyrene, respectively (Arey et al., 1992; Sasaki et al., 1995; Atkinson and Arey, 1994; Arey, 1998a). It can be seen that, in combination, the mutagrams of the photooxidation reaction products of phenanthrene, pyrene, and fluoranthene are very similar to the mutagen profile of the ambient air sample.

2-Nitrofluoranthene, found in fraction 4, dominated the fluoranthene reaction products, consistent with extracts of ambient POM. The more polar fraction 6 of the OH radical initiated phenanthrene reaction products contained the two nitrophenanthrene lactones previously identified in ambient air (Helmig et al., 1992a, 1992b), the potent direct mutagen 2-nitro-6H-dibenzo[b,d]pyran-6-one (**XI**) and its less mutagenic 4-NO_2 isomer (see Table 10.20). Recall that this 2-nitro-PAC **XI** has been estimated to account for up to 20% of the direct mutagenicity of an ambient aerosol extract (Helmig et al., 1992b). Two nitropyrene lactone isomers were tentatively identified in fraction 6 of the pyrene reaction products. El-Bayoumy and Hecht (1986) reported earlier that two nitropyrene lactones were strong mutagens in the standard Ames plate incorporation assay (Maron and Ames, 1983).

Figure 10.35 is the GC–MS m/z 247 profile of the nitrofluoranthenes and nitropyrenes in an extract of an ambient particle sample collected at night (Arey et al., 1988b). The high ratio of 2-nitrofluoranthene to 2-nitropyrene observed in this nighttime sample is indicative of nighttime gas-phase NO_3 radical reactions (for a review, see Kwok et al., 1994b). An NO_3 radical initiated mechanism for atmospheric formation of 2-nitrofluoranthene is shown in Fig. 10.36 (Atkinson and Arey, 1997; Arey, 1998a, and references therein). Analogous to the OH reaction, NO_3 is proposed to add to the ring to form a fluoranthene–NO_3 adduct, followed by *ortho* addition of NO_2 and subsequent loss of HNO_3. This reaction is noteworthy because of its selectivity; i.e., only 2-nitrofluoranthene is formed, and in high yield (24%) compared to the OH-initiated daytime reaction (3%).

Another example is the atmospheric formation levels, fates, and mutagenicities of 1- and 2-nitronaphthalenes (1-NN and 2-NN) and certain methylnitronaphthalene isomers (MNN). Naphthalene is the most abundant of the PAHs and its role, and that of the abundant MNNs, in the atmospheric chemistry of PAHs are being increasingly recognized as being important; e.g., see Pitts et al., 1985c; Atkinson et al., 1987b,

TABLE 10.35 Room Temperature Rate Constants, k, for the Gas-Phase Reactions of Selected PAHs and Nitro-PAHs with the Hydroxyl Radical, the Nitrate Radical, and Ozone (from Arey 1998a)

	k (cm^3 molecule^{-1} s^{-1}) for reaction with		
PAH or nitro-PAH	OH[a]	NO$_3$[b]	O$_3$[c]
Naphthalene	2.16×10^{-11}	3.6×10^{-28} [NO$_2$]	$<2 \times 10^{-19}$ [d]
1-Methylnaphthalene	5.3×10^{-11}	7.7×10^{-28} [NO$_2$]	$<1.3 \times 10^{-19}$
2-Methylnaphthalene	5.2×10^{-11}	1.08×10^{-27} [NO$_2$]	$<4 \times 10^{-19}$
Acenaphthylene	1.1×10^{-10}	5.5×10^{-12}	5.5×10^{-16}
Acenaphthene	1.0×10^{-10}	4.6×10^{-13} [e]	$<5 \times 10^{-19}$
Biphenyl	7.2×10^{-12}	$<5 \times 10^{-30}$ [NO$_2$]	$<2 \times 10^{-19}$ [d]
Fluorene[f]	1.6×10^{-11}	3.5×10^{-14}	$<2 \times 10^{-19}$
Phenanthrene[g]	1.3×10^{-11}	1.2×10^{-13} [h]	4.0×10^{-19}
Anthracene	1.7×10^{-11} [f]		
Fluoranthene	$\sim 5 \times 10^{-11}$ [i]	5.1×10^{-28} [NO$_2$]	
Pyrene	$\sim 5 \times 10^{-11}$ [i]	1.6×10^{-27} [NO$_2$][j]	
1-Nitronaphthalene	5.4×10^{-12}	3.0×10^{-29} [NO$_2$]	$<6 \times 10^{-19}$
2-Nitronaphthalene	5.6×10^{-12}	2.7×10^{-29} [NO$_2$]	$<6 \times 10^{-19}$

[a] Taken from Atkinson (1989) except as indicated.
[b] Taken from Atkinson (1991) except as indicated; [NO$_2$] = concentration of NO$_2$ in molecules cm^{-3} units.
[c] Taken from Atkinson (1994) except as indicated.
[d] Taken from Atkinson and Arey (1994).
[e] Full rate expression $k = 4.6 \times 10^{-13} + 1.7 \times 10^{-27}$ [NO$_2$], but with ambient NO$_2$ levels, only direct reaction is important.
[f] Taken from Kwok et al. (1997).
[g] Taken from Kwok et al. (1994a).
[h] Full rate expression $k = 1.2 \times 10^{-13} + 7.0 \times 10^{-28}$ [NO$_2$], but with ambient NO$_2$ levels, only direct reaction is important.
[i] Estimated as described by Arey et al. (1986); measured quantity was ($k \times$ yield) (from Atkinson et al., 1990).
[j] Measured in the presence of ppm levels of NO$_2$; additional direct reaction as observed for phenanthrene cannot be ruled out.

TABLE 10.36 Calculated Atmospheric Lifetimes of Selected PAHs and Nitro-PAHs Due to Gas-Phase Reactions with the OH Radical, the NO$_3$ Radical, and Ozone and from Photolysis (from Arey, 1998a)

	Lifetime due to reaction with			
PAH or nitro-PAH	OH[a]	NO$_3$[b]	O$_3$[c]	Photolysis[d]
Naphthalene	6.8 h	1.5 yr	>80 days	
1-Methylnaphthalene	2.8 h	250 days	>125 days	
2-Methylnaphthalene	2.8 h	180 days	>40 days	
Acenaphthylene	1.3 h	6 min	~43 min	
Acenaphthene	1.5 h	1.2 h	>30 days	
Biphenyl	1.7 days	>105 yr	>80 days	
Fluorene	9.1 h	1.3 days	>80 days	
Phenanthrene	11.2 h	4.6 h	41 days	
Anthracene	8.6 h			
Fluoranthene	~2.9 h[e]	~1 yr		
Pyrene	~2.9 h[e]	~120 days[f]		
1-Nitronaphthalene	2.3 days	18 yr	>28 days	1.7 h
2-Nitronaphthalene	2.2 days	20 yr	>28 days	2.2 h

[a] For a 12-h daytime OH radical concentration of 1.9×10^6 molecules cm^{-3}.
[b] For a 12-h nighttime NO$_3$ radical concentration of 5×10^8 molecules cm^{-3} and an NO$_2$ concentration of 2.4×10^{11} molecules cm^{-3}.
[c] For a 24-h O$_3$ concentration of 7×10^{11} molecules cm^{-3}.
[d] For average 12-h daytime NO$_2$ photolysis rate of $J_{NO_2} = 5.2 \times 10^{-3}$ s^{-1}; data on nitronaphthalenes from Atkinson et al. (1989). No evidence has been reported for the gas-phase photolysis of the 2- to 4-ring PAHs; Feilberg et al. (1999a) reported that 2-nitronaphthalene photolyzed at a significantly slower rate than the 1-nitro isomer (see text).
[e] Rate constants estimated.
[f] Does not include possible direct reaction with NO$_3$; see Table 10.35.

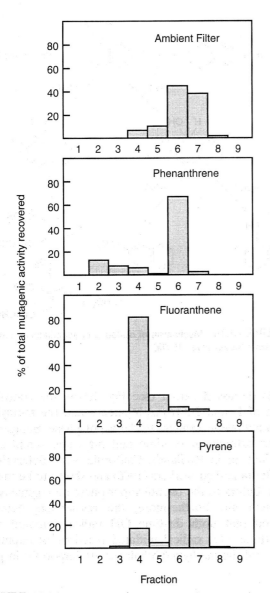

FIGURE 10.34 Mutagrams (Ames assay, microsuspension modifications, TA98, −S9) of extracts of ambient particulate matter collected on a filter in Claremont, California, August 28, 1987 (Arey et al., 1992; Harger et al., 1992) and the percentage of the total mutagenic activity from environmental chamber reaction products vs HPLC fraction number of the gas-phase atmospheric reactions of the semivolatile phenanthrene, fluoranthene, and pyrene, respectively (Sasaki et al., 1995). The mutagenicity of fraction 6 from the phenanthrene reaction is due to the product 2-nitro-6H-dibenzo[b,d]pyran-6-one (the nitrophenanthrene lactone XI); 2-nitrofluoranthene (XXVII) is the mutagenic compound in fraction 4 from the fluoranthene reaction. Two nitropyrene lactones were tentatively identified in fraction 6 from the pyrene chamber reaction (adapted from Atkinson and Arey, 1994, and Arey, 1998a).

1990a,b, 1994; Zielinska et al., 1989b; Arey et al., 1992; Lane and Tang, 1994; Bunce and Zhu, 1994; Bunce et al., 1997; Sasaki et al., 1995, 1997a; and Feilberg et al., 1999a; see also reviews by Zielinska et al., 1990; Atkinson and Arey, 1994; and Arey, 1998a.

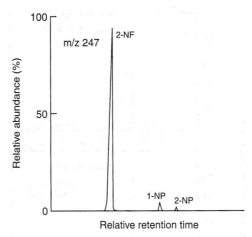

FIGURE 10.35 GC–MS selected ion monitoring (SIM) of the molecular ions (m/z 247) of nitrofluoranthene (NF) and nitropyrene (NP) isomers in an extract of particles collected at night in Claremont, California, 1800 to 2400 hours on September 14, 1985 (adapted from Arey et al., 1988b).

For example, Fig. 10.37 shows a mutagram of an extract of ambient gas-phase POM collected on polyurethane foam plugs in Claremont, California, along with mutagrams of the extracts of PUF samples collected from environmental chamber studies of the OH radical initiated reactions of fluorene and naphthalene, respectively (Arey et al., 1992; Harger et al., 1992; Atkinson and Arey, 1994). Clearly, the mutagrams from the OH-initiated reactions are very similar to that of ambient air.

Thus, most of the activity of the products of the photooxidation of these two PAHs is in fraction 4. In the naphthalene reaction, key products in fraction 4 are 1- and 2-nitronaphthalene (Arey et al., 1992), with

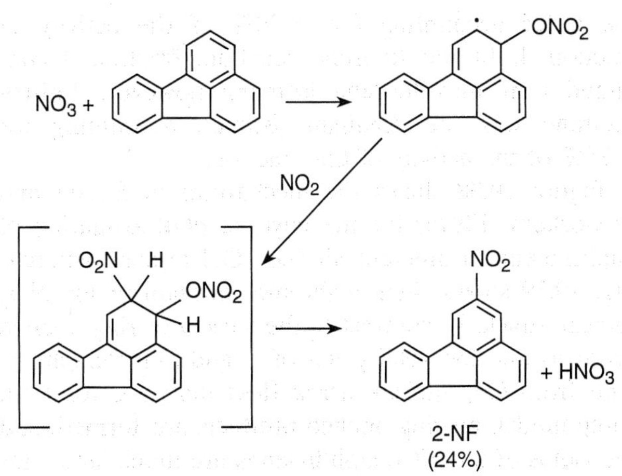

FIGURE 10.36 Mechanism of oxidation of fluoranthene by the nitrate radical (adapted from Atkinson and Arey, 1997).

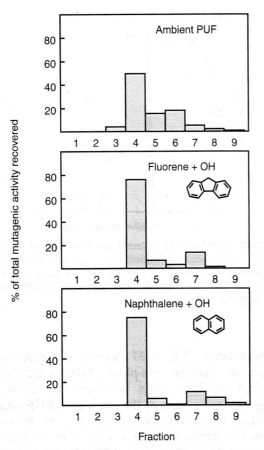

FIGURE 10.37 Mutagrams of extracts of an ambient sample of gas-phase direct mutagens collected on a PUF plug in Claremont, California, and, for comparison, extracts of the vapor-phase products from the OH radical initiated gas-phase reactions of naphthalene and fluorene in an environmental chamber. For naphthalene, the mutagenicity of fraction 4 is primarily due to the presence of 1- and 2-nitronaphthalene. Fraction 4 from the fluorene reaction contains four nitrofluorene isomers, with 3-nitrofluorene being dominant (adapted from Arey et al., 1992; Harger et al., 1992).

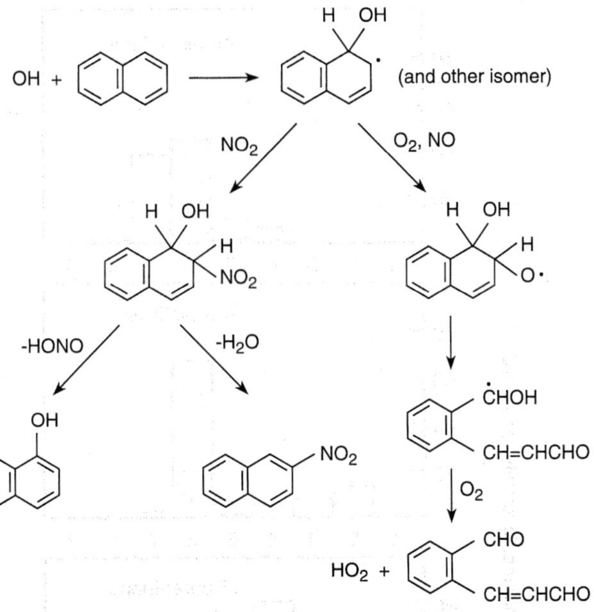

FIGURE 10.38 Mechanism of oxidation of naphthalene in air by OH (from Sasaki et al., 1997a).

As discussed earlier (see Fig. 10.26), the contributions of 1- and 2-nitronaphthalenes and the methylnitronaphthalene isomers to the vapor-phase mutagenicity of extracts of daytime and nighttime samples of ambient air in Redlands, California, were determined by Gupta and co-workers (1996) and shown to be major contributors to the overall vapor-phase mutagenicity of ambient air. Furthermore, the researchers demonstrated that both daytime OH radical initiated and nighttime NO_3 radical initiated reactions of naphthalene and the methylnaphthalenes are important in pol-

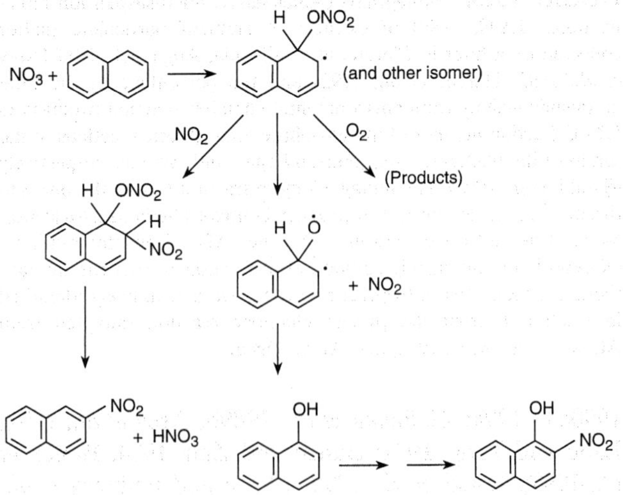

the latter accounting for ~90% of the activity of fraction 4. In the fluorene reaction, fraction 4 contained four nitrofluorene isomers; however, 3-nitrofluorene was the dominant isomer, accounting for ~75% of the activity of this fraction.

Figure 10.38 shows the mechanism of Sasaki and co-workers (1997a) for the daytime photooxidation of naphthalene in ambient air (i.e., OH attack), whereas Fig. 10.39 shows their nighttime mechanism for NO_3 radical attack. In contrast to the variety of ring-opened products and low total yields of 1- and 2-nitronaphthalene from OH radical attack (less than 1% for both compounds), no ring-opened products are formed, and the yields of the nitronaphthalenes are much larger for the NO_3 radical reaction, with the ratio of the 1- to 2-isomers being 2.4:1 (Sasaki et al., 1997a).

FIGURE 10.39 Mechanism of NO_3 radical reaction with naphthalene in air (from Sasaki et al., 1997a).

luted ambient air atmospheres. Thus the MNN isomer pattern found in the daytime ambient PUF samples was significantly different from the nighttime profile. Furthermore, there was excellent agreement between the laboratory MNN isomer product patterns observed for OH attack and the profile from field measurements during the day and between the laboratory NO$_3$ studies and the nighttime isomer profiles.

Recently, to test their model for the formation, decay, and gas–particle partitioning of nitronaphthalenes, Feilberg and co-workers (1999a) conducted outdoor environmental chamber studies of the reactions of naphthalene in the presence of diesel exhaust, propylene (to enhance OH radical formation), and additional NO. Experimentally, they determined that the gas-phase photolysis of 1-nitronaphthalene was its major degradation pathway; however, 2-nitronaphthalene photolyzed at a significantly slower rate so that other degradation processes may be important. These results differed from those of Atkinson *et al.* (1989), who found their photolysis rates comparable (see Table 10.36). Additionally, Feilberg *et al.* propose a somewhat modified mechanism of Atkinson *et al.* (1989) and Sasaki *et al.* (1997a) for the OH-initiated gas-phase reaction of naphthalene to form 1-nitronaphthalene.

For the kinetics and mechanisms of the gas-phase reactions of other 2- to 4-ring PAHs with OH radicals, NO$_3$ radicals, and O$_3$, see, e.g., phenanthrene, Kwok *et al.* (1994a), and indan, indene, fluorene, and 9,10-dihydroanthracene, Kwok *et al.* (1997); for dibenzothiophene, a volatile S-PAC that is ubiquitous in polluted ambient air environments and forms mutagenic photooxidation products, see Kwok *et al.* (1999); see also Kwok *et al.* (1994b) for a review of gas-phase NO$_3$ radical reactions with aromatics.

In summary, as Arey (1998a) points out, one can assign much of the direct-acting mutagenicity (TA98, $-$S9, microsuspension modification) of fractions 4 and 6 of particle-phase ambient samples to nitroarenes (fraction 4) and nitroarene lactones (fraction 6). Certain gas-phase nitroarenes are also major contributors to fraction 4. If one sums up the particle- and gas-phase contributions of these gas-phase reaction products of semivolatile and volatile PAHs, they could account for over 50% of the total vapor- and particle-phase direct mutagenicity of ambient air. Clearly this phenomenon has toxicological implications that should be addressed in risk assessments of the exposure and health effects of combustion-generated POM, e.g., diesel exhaust.

References

Aceves, M., and J. O. Grimalt, "Seasonally Dependent Size Distributions of Aliphatic and Polycyclic Aromatic Hydrocarbons in Urban Aerosols from Densely Populated Areas," *Environ. Sci. Technol., 27,* 2896–2908 (1993).

Adams, J., E. L. Atlas, and C.-S. Glam, "Ultratrace Determination of Vapor-Phase Nitrogen Heterocyclic Bases in Ambient Air," *Anal. Chem., 54,* 1515–1518 (1982).

Alebić-Juretić, A., T. Cvita, and L. Klasinc, "Heterogeneous Polycyclic Aromatic Hydrocarbon Degradation with Ozone on Silica Gel Carrier," *Environ. Sci. Technol., 24,* 62–66 (1990).

Alfheim, I., and M. Møller, "Mutagenicity of Long-Range Transported Atmospheric Aerosols," *Sci. Total Environ., 13,* 275–278 (1979).

Alfheim, I., G. Löfroth, and M. Møller, "Bioassay of Extracts of Ambient Particulate Matter," *Environ. Health Perspect., 47,* 227–238 (1983).

Alfheim, I., G. Becher, J. K. Hongslo, and T. Ramdahl, "Mutagenicity Testing of High Performance Liquid Chromatography Fractions from Wood Stove Emission Samples Using a Modified *Salmonella* Assay Requiring Smaller Sample Volumes," *Environ. Mutagen., 6,* 91–102 (1984a).

Alfheim, I., A. Bjørseth, and M. Moller, "Characterization of Microbial Mutagens in Complex Samples—Methodology and Application," *Crit. Rev. Environ. Control, 14,* 91–150 (1984b).

Alfheim, I., and A. Lindskog, "A Comparison between Different High Volume Sampling Systems for Collecting Ambient Airborne Particles for Mutagenicity Testing and for Analysis of Organic Compounds," *Sci. Total Environ., 34,* 203–222 (1984a).

Alfheim, I., and T. Ramdahl, "Contribution of Wood Combustion to Indoor Air Pollution As Measured by Mutagenicity in *Salmonella* and Polycyclic Aromatic Hydrocarbon Concentration," *Environ. Mutagen., 6,* 121–130 (1984b).

Alfheim, I., A. Jebens, and S. Johansen, "Sampling Ambient Air for Mutagenicity Testing by High-Volume Filtration on Glass-Fibre Filters and on XAD-2," *Environ. Int., 11,* 111–118 (1985).

Allen, J. O., N. M. Dookeran, K. A. Smith, A. F. Sarofim, K. Taghizadeh, A. L. Lafleur, "Measurement of Polycyclic Aromatic Hydrocarbons Associated with Size-Segregated Atmospheric Aerosols in Massachusetts," *Environ. Sci. Technol., 30,* 1023–1031 (1996).

Allen, J. O., N. M. Dookeran, K. Taghizadeh, A. L. Lafleur, K. A. Smith, and A. F. Sarofim, "Measurement of Oxygenated Polycyclic Aromatic Hydrocarbons with a Size-Segregated Urban Aerosol," *Environ. Sci. Technol., 31,* 2064–2070 (1997).

Allen, J. O., J. L. Durant, N. M. Dookeran, K. Taghizadeh, E. F. Plummer, A. L. Lafleur, A. F. Sarofim, and K. A. Smith, "Measurement of $C_{24}H_{14}$ Polycyclic Aromatic Hydrocarbons Associated with a Size-Segregated Urban Aerosol," *Environ. Sci. Technol., 32,* 1928–1932 (1998).

Alsberg, T., U. Stenberg, R. Westerholm, M. Strandell, U. Rannug, A. Sundvall, L. Romert, V. Bernson, B. Pettersson, R. Toftgård, B. Franzén, M. Jansson, J. Å. Gustafsson, K. E. Egebäck, and G. Tejle, "Chemical and Biological Characterization of Organic Material from Gasoline Exhaust Particles," *Environ. Sci. Technol., 19,* 43–50 (1985).

American Chemical Society, *Ring Systems Handbook*, Am. Chem. Soc., Washington, DC, 1977–present.

Ames, B. N., W. E. Durston, E. Yamasaki, and F. D. Lee, "Carcinogens Are Mutagens: A Simple Test System Combining Liver Homogenates for Activation and Bacteria for Detection," *Proc. Natl. Acad. Sci. U.S.A., 70,* 2281–2285 (1973).

Ames, B. N., J. McCann, and E. Yamasaki, "Methods for Detecting Carcinogens and Mutagens with the *Salmonella*/Mammalian-Microsome Mutagenicity Test," *Mutat. Res., 31,* 347–364 (1975).

Arey, J., "Atmospheric Reactions of PAHs Including Formation of Nitroarenes," in *The Handbook of Environmental Chemistry, PAHs and Related Compounds* (A. H. Neilson, Ed.), Vol. 3, Part I, pp. 347–385, Springer-Verlag, Berlin, 1998a.

Arey, J., B. Zielinska, R. Atkinson, A. M. Winer, T. Ramdahl, and J. N. Pitts, Jr., "The Formation of Nitro-PAH from the Gas-Phase Reactions of Fluoranthene and Pyrene with the OH Radical in the Presence of NO_x," *Atmos. Environ.*, 20, 2339–2345 (1986).

Arey, J., B. Zielinska, R. Atkinson, and A. M. Winer, "Polycyclic Aromatic Hydrocarbon and Nitroarene Concentrations in Ambient Air during a Wintertime High-NO_x Episode in the Los Angeles Basin," *Atmos. Environ.*, 21, 1437–1444 (1987).

Arey, J., B. Zielinska, R. Atkinson, and A. M. Winer, "Formation of Nitroarenes during Ambient High-Volume Sampling," *Environ. Sci. Technol.*, 22, 457–462 (1988a).

Arey, J., B. Zielinska, W. P. Harger, R. Atkinson, and A. M. Winer, "The Contribution of Nitrofluoranthenes and Nitropyrenes to the Mutagenic Activity of Ambient Particulate Organic Matter Collected in Southern California," *Mutat. Res.*, 207, 45–51 (1988b).

Arey, J., R. Atkinson, B. Zielinska, and P. A. McElroy, "Diurnal Concentrations of Volatile Polycyclic Aromatic Hydrocarbons and Nitroarenes during a Photochemical Air Pollution Episode in Glendora, California," *Environ. Sci. Technol.*, 23, 321–327 (1989a).

Arey, J., B. Zielinska, R. Atkinson, and S. M. Aschmann, "Nitroarene Products from the Gas-Phase Reactions of Volatile Polycyclic Aromatic Hydrocarbons with the OH Radical and N_2O_5," *Int. J. Chem. Kinet.*, 21, 775–799 (1989b).

Arey, J., R. Atkinson, S. M. Aschmann, and D. Schuetzle, "Experimental Investigation of the Atmospheric Chemistry of 2-Methyl-1-nitronaphthalene and a Comparison of Predicted Nitroarene Concentrations with Ambient Air Data," *Polycyclic Aromat. Compd.*, 1, 33–50 (1990).

Arey, J., W. P. Harger, D. Helmig, and R. Atkinson, "Bioassay-Directed Fractionation of Mutagenic PAH Atmospheric Photooxidation Products and Ambient Particulate Extracts," *Mutat. Res.*, 281, 67–76 (1992).

Arey, J., R. Atkinson, W. P. Harger, D. Helmig, and J. Sasaki, "Formation of Mutagens from the Atmospheric Photooxidants of PAH and Their Occurrence in Ambient Air," Final Report, Contract No. A132-075, California Air Resources Board, 1994.

Atkinson, R., "Kinetics and Mechanisms of the Gas-Phase Reactions of the Hydroxyl Radical with Organic Compounds," *J. Phys. Chem. Ref. Data Monogr.*, 1, 1–246 (1989).

Atkinson, R., "Kinetics and Mechanisms of the Gas-Phase Reactions of the NO_3 Radical with Organic Compounds," *J. Phys. Chem. Ref. Data*, 20, 459–507 (1991).

Atkinson, R., "Gas-Phase Tropospheric Chemistry of Organic Compounds," *J. Phys. Chem. Ref. Data Monogr.*, 2, 1–216 (1994).

Atkinson, R., S. M. Aschmann, and J. N. Pitts, Jr., "Kinetics of the Reaction of Naphthalene and Biphenyl with OH Radicals and with O_3 at 294 ± 1 K," *Environ. Sci. Technol.*, 18, 110–113 (1984).

Atkinson, R., and S. M. Aschmann, "Kinetics of the Reactions of Naphthalene, 2-Methylnaphthalene, and 2,3-Dimethylnaphthalene with OH Radicals and with O_3 at 295 ± 1 K," *Int. J. Chem. Kinet.*, 18, 569–573 (1986a).

Atkinson, R., A. M. Winer, and J. N. Pitts, Jr., "Estimation of Nighttime N_2O_5 Concentrations from Ambient NO_2 and NO_3 Radical Concentrations and the Role of N_2O_5 in Nighttime Chemistry," *Atmos. Environ.*, 20, 331–339 (1986b).

Atkinson, R., J. Arey, B. Zielinska, J. Pitts, Jr., and A. M. Winer, "Evidence for the Transformation of Polycyclic Organic Matter in the Atmosphere," *Atmos. Environ.*, 21, 2261–2264 (1987a).

Atkinson, R., J. Arey, B. Zielinska, and S. M. Aschmann, "Kinetics and Products of the Gas-Phase Reactions of OH Radicals and N_2O_5 with Naphthalene and Biphenyl," *Environ. Sci. Technol.*, 21, 1014–1022 (1987b).

Atkinson, R., J. Arey, A. M. Winer, B. Zielinska, T. Dinoff, W. Harger, and P. McElroy, "A Survey of Ambient Concentrations of Selected Polycyclic Aromatic Hydrocarbons (PAH) at Various Locations in California," Final Report to the California Air Resources Board, Contract No. A5-185-32, May 1988a.

Atkinson, R., and S. M. Aschmann, "Kinetics of the Reactions of Acenaphthene and Acenaphthylene and Structurally-Related Aromatic Compounds with OH and NO_3 Radicals, N_2O_5, and O_3 at 296 ± 2 K," *Int. J. Chem. Kinet.*, 20, 513–539 (1988b).

Atkinson, R., S. M. Aschmann, J. Arey, and B. Zielinska, "Gas-Phase Atmospheric Chemistry of 1- and 2-Nitronaphthalene and 1,4-Naphthoquinone," *Atmos. Environ.*, 23, 2679–2690 (1989).

Atkinson, R., J. Arey, B. Zielinska, and S. M. Aschmann, "Kinetics and Nitro-Products of the Gas-Phase OH and NO_3 Radical Initiated Reaction of Naphthalene-d_8, Fluoranthene-d_{10}, and Pyrene," *Int. J. Chem. Kinet.*, 22, 999–1014 (1990a).

Atkinson, R., E. C. Tuazon, and J. Arey, "Reactions of Naphthalene in N_2O_5–NO_3–NO_2–Air Mixtures," *Int. J. Chem. Kinet.*, 22, 1071–1082 (1990b).

Atkinson, R., J. Arey, W. P. Harger, D. Hemig, and P. A. McElroy, "Hydroxynitro-PAH and Other PAH Derivatives in California's Atmosphere and Their Contribution to Ambient Mutagenicity," Final Report to the California Air Resources Board, Contract No. A732-154, 1991.

Atkinson, R., and J. Arey, "Lifetimes and Fates of Toxic Air Contaminants in California's Atmosphere," Final Report to the California Air Resources Board, Contract No. A032-055, 1993.

Atkinson, R., and J. Arey, "Atmospheric Chemistry of Gas-Phase Polycyclic Aromatic Hydrocarbons," *Environ. Health Perspect.*, 102 (Suppl. 4), 117–126 (1994).

Atkinson, R., and S. M. Aschmann, "Products of the Gas-Phase Reactions of Aromatic Hydrocarbons: Effect of NO_2 Concentration," *Int. J. Chem. Kinet.*, 26, 929–944 (1994).

Atkinson, R., E. C. Tuazon, I. Bridier, and J. Arey, "Reactions of NO_3–Naphthalene Adducts with O_2 and NO_2," *Int. J. Chem. Kinet.*, 26, 605–614 (1994).

Atkinson, R., and J. Arey, "Lifetimes and Fates of Toxic Air Contaminants in California's Atmosphere," Final Report to the California Air Resources Board, Contract No. 93-307, March 1997.

Badger, G. M., "Mode of Formation of Carcinogens in Human Environment," *Natl. Cancer Inst. Monogr.*, 9, 1–16 (1962).

Baek, S. O., R. A. Field, M. E. Goldstone, P. W. Kirk, J. N. Lester, and R. Perry, "A Review of Atmospheric Polycyclic Aromatic Hydrocarbons: Sources, Fate, and Behavior," *Water, Air, Soil Pollut.*, 60, 279–300 (1991a).

Baek, S. O., M. E. Goldstone, P. W. W. Kirk, J. N. Lester, and R. Perry, "Phase Distribution and Particle Size Dependency of Polycyclic Aromatic Hydrocarbons in the Urban Atmosphere," *Chemosphere*, 22, 503–520 (1991b).

Baek, S. O., M. E. Goldstone, P. W. W. Kirk, J. N. Lester, and R. Perry, "Concentrations of Particulate and Gaseous Polycyclic Aromatic Hydrocarbons in London Air Following a Reduction in the Lead Content of Petrol in the United Kingdom," *Sci. Total Environ.*, 111, 169–199 (1992).

Bagley, S. T., S. L. Stoltz, D. M. Becker, and R. E. Keen, "Characterization of Organic Extracts from Standard Reference Materials 1649, 'Urban Dust/Organics,' and 1650, 'Diesel Particulate Matter,' Using a Microsuspension Assay. A WHO/IPCS/CSCM Study," *Mutat. Res.*, 276, 81–86 (1992).

Baker, J. E., and S. J. Eisenreich, "Concentrations and Fluxes of Polycyclic Aromatic Hydrocarbons and Polychlorinated Biphenyls across the Air–Water Interface of Lake Superior," *Environ. Sci. Technol., 24,* 342–352 (1990).

Ball, L. M., M. J. Kohan, L. D. Claxton, and J. Lewtas, "Mutagenicity of Derivatives and Metabolites of 1-Nitropyrene: Activation by Rat Liver S9 and Bacterial Enzymes," *Mutat. Res., 138,* 113–125 (1984).

Barale, R., L. Giromini, S. Del Ry, B. Barnini, M. Bulleri, I. Barrai, F. Valerio, M. Pala, J. He, "Chemical and Mutagenic Patterns of Airborne Particulate Matter Collected in 17 Italian Towns," *Environ. Health Perspect., 102,* 67–73 (1994).

Barbas, J. T., M. E. Sigman, A. C. Buchanan, and E. A. Chevis, "Photolysis of Substituted Naphthalenes on SiO_2 and Al_2O_3," *Photochem. Photobiol., 58,* 155–158 (1993).

Barbas, J. T., R. Dabestani, and M. E. Sigman, "A Mechanistic Study of Photodecomposition of Acenaphthylene on a Dry Silica Surface," *Photochem. Photobiol., 80,* 103–111 (1994).

Barbas, J. T., M. E. Sigman, and R. Dabestani, "Photochemical Oxidation of Phenanthrene Sorbed on Silica Gel," *Environ. Sci. Technol., 30,* 1776–1780 (1996).

Barbas, J. T., M. E. Sigman, R. Arce, and R. Dabestani, "Spectroscopy and Photochemistry of Fluorene at a Silica Gel/Air Interface," *J. Photochem. Photobiol. A: Chem., 109,* 229–236 (1997).

Bartle, K. D., M. L. Lee, and S. A. Wise, "Modern Analytical Methods for Environmental Polycyclic Aromatic Compounds," *Chem. Soc. Rev., 10,* 113–158 (1981).

Barofsky, D. F., and E. J. Baum, "Exploratory Field Description Mass Analysis of the Photoconversion of Adsorbed Polycyclic Aromatic Hydrocarbons," *J. Am. Chem. Soc., 98,* 8286–8287 (1976).

Baum, E. J., *Chemical Property Estimation*, Lewis Publishers, Boca Raton, FL, 1998.

Baumgard, K. J., and J. H. Johnson, "The Effect of Fuel and Engine Design on Diesel Exhaust Particle Size Distributions," SAE 960131, Special Publication 1140, p. 37, 1996.

Bayona, J. M., M. Casellas, P. Fernández, A. M. Solanas, and J. Albaigés, "Sources and Seasonal Variability of Mutagenic Agents in the Barcelona City Aerosol," *Chemosphere, 29,* 441–450 (1994).

Behymer, T. D., and R. A. Hites, "Photolysis of Polycyclic Aromatic Hydrocarbons Adsorbed on Simulated Atmospheric Particulates," *Environ. Sci. Technol., 19,* 1004–1006 (1985).

Behymer, T. D., and R. A. Hites, "Photolysis of Polycyclic Aromatic Hydrocarbons Adsorbed on Fly Ash," *Environ. Sci. Technol., 22,* 1311–1319 (1988).

Beije, B., and L. Möller, "2-Nitrofluorene and Related Compounds: Prevalence and Biological Effects," *Mutat. Res., 196,* 177–209 (1988).

Belser, W. L., Jr., S. D. Shaffer, R. D. Bliss, P. M. Hynds, L. Yamamoto, J. N. Pitts, Jr., and J. A. Winer, "A Standardized Procedure for Quantification of the Ames *Salmonella*/Mammalian Microsome Mutagenicity Test," *Environ. Mutagen., 3,* 123–139 (1981).

Benner, B. A., Jr., G. E. Gordon, and S. A. Wise, "Mobile Sources of Atmospheric Polycyclic Aromatic Hydrocarbons: A Roadway Tunnel Study," *Environ. Sci. Technol., 23,* 1269–1278 (1989).

Benson, J. M., A. L. Brooks, Y. S. Cheng, T. R. Henderson, and J. E. White, "Environmental Transformation of 1-Nitropyrene on Glass Surfaces," *Atmos. Environ., 19,* 1169–1174 (1985).

Bidleman, T. F., "Atmospheric Processes," *Environ. Sci. Technol., 2,* 361–367 (1988).

Bidleman, T. F., and W. T. Foreman, "Vapor–Particle Partitioning of Semi-Volatile Organic Compounds," in *Sources and Fates of Aquatic Pollutants* (R. A. Hites and S. J. Eisenreich, Eds.), pp. 29–56, Am. Chem. Soc., Washington, DC, 1987.

Biermann, H. W., H. MacLeod, R. Atkinson, A. M. Winer, and J. N. Pitts, Jr., "Kinetics of the Gas-Phase Reactions of the Hydroxyl Radical with Naphthalene, Phenanthrene, and Anthracene," *Environ. Sci. Technol., 19,* 244–248 (1985).

Bjørseth, A., Ed., *Handbook of Polycyclic Aromatic Hydrocarbons*, Dekker, New York, 1983.

Bjørseth, A., G. Lunde, and A. Lindskog, "Long-Range Transport of Polycyclic Aromatic Hydrocarbons," *Atmos. Environ., 13,* 45–53 (1979).

Bjørseth, A., and B. S. Olufsen, "Long-Range Transport of Polycyclic Aromatic Hydrocarbons," in *Handbook of Polycyclic Aromatic Hydrocarbons* (A. Bjørseth, Ed.), pp. 507–524, Dekker, New York, 1983.

Bjørseth, A., and T. Ramdahl, Eds., *Handbook of Polycyclic Aromatic Hydrocarbons, Emission Sources and Recent Advances in Analytical Chemistry*, Vol. 2, Dekker, New York, 1985.

Bodzek, D., K. Luks-Betlej, and L. Warzecha, "Determination of Particle-Associated Polycyclic Aromatic Hydrocarbons in Ambient Air Samples from the Upper Silesia Region of Poland," *Atmos. Environ., 27A,* 759–764 (1993).

Bowen, E. J., *Chemical Aspects of Light*, 2nd ed., Oxford Univ. Press, Oxford, 1946.

Bowen, E. J., "The Photochemistry of Aromatic Hydrocarbon Solutions," *Adv. Photochem., 1,* 23–42 (1963).

Brooks, A. L., A. P. Li, J. S. Dutcher, C. R. Clark, S. J. Rothenberg, R. Kiyoura, W. E. Bechtold, and R. O. McClellan, "A Comparison of Genotoxicity of Automotive Exhaust Particles from Laboratory and Environmental Sources," *Environ. Mutagen., 6,* 651–668 (1984).

Brorström-Lundén, E., and A. Lindskog, "Degradation of Polycyclic Aromatic Hydrocarbons during Simulated Stack Gas Sampling," *Environ. Sci. Technol., 25,* 313–316 (1985).

Bunce, N. J., and J. Zhu, "Products from Photochemical Reactions of Naphthalene in Air," *Polycyclic Aromat. Compd., 5,* 123–130 (1994).

Bunce, N. J., L. Liu, J. Zhu, and D. A. Lane, "Reaction of Naphthalene and Its Derivatives with Hydroxyl Radicals in the Gas Phase," *Environ. Sci. Technol., 31,* 2252–2259 (1997).

Busby, W. F., H. Smith, W. W. Bishop, and W. G. Thilly, "Mutagenicity of Mono- and Dinitropyrenes in the *Salmonella typhimurium* TM677 Forward Mutation Assay," *Mutat. Res., 322,* 221–232 (1994a).

Busby, W. F., Jr., B. W. Penman, and C. L. Crespi, "Human Cell Mutagenicity of Mono- and Dinitropyrenes in Metabolically Competent MCL-5 Cells," *Mutat. Res., 322,* 233–242 (1994b).

Busby, W. F., H. Smith, C. L. Crespi, and B. W. Penman, "Mutagenicity of Benzo[*a*]pyrene and Dibenzopyrenes in the *Salmonella typhimurium* TM677 and the MCL-5 Human Cell Forward Mutation Assays," *Mutat. Res., 342,* 9–16 (1995).

Busby, W. F., Jr., H. Smith, C. L. Crespi, B. W. Penman, and A. L. Lafleur, "Mutagenicity of the Atmospheric Transformation Products 2-Nitrofluoranthene and 2-Nitrodibenzopyranone in *Salmonella* and Human Cell Forward Mutation Assays," *Mutat. Res., 389,* 261–270 (1997).

Butler, J. D., and P. Crossley, "Reactivity of Polycyclic Aromatic Hydrocarbons Adsorbed on Soot Particles," *Atmos. Environ., 15,* 91–94 (1981).

Butler, J. P., T. J. Kneip, and J. M. Daisey, "An Investigation of Interurban Variations in the Chemical Composition and Mutagenic Activity of Airborne Particulate Organic Matter Using an Integrated Chemical Class/Bioassay System," *Atmos. Environ., 21,* 883–892 (1987).

California Air Resources Board (CARB), *Identification of Benzo[a]pyrene as a Toxic Air Contaminant*, 1994.

California Air Resources Board (CARB), *Methodology for Estimating Emissions from On-Road Motor Vehicles*, Vol. I: *Introduction and Overview*, L. Hrynchuk and B. Effa, Planning and Technical Support Division Mobile Source Emission Inventory Branch, 1996. (For information about documents or MVEI7G Model, contact Public Information Office or Transportation Activity Section, California Air Resources Board. We thank Mr. E. E. Yotter and Dr. John Holmes, Chief, Research Division, for providing this information (1999).)

California Air Resources Board (CARB), *Identification of Diesel Exhaust as a Toxic Air Contaminant*, 1998.

California Environmental Protection Agency, Office of Environmental Health Hazard Assessment, "Health Effects of Exposure to Environmental Tobacco Smoke," Final Report, September 1997.

Calvert, J. G., and J. N. Pitts, Jr., *Photochemistry*, Wiley, New York, 1966.

Cautreels, W., and K. Van Cauwenbergh, "Determination of Organic Compounds in Airborne Particulate Matter by Gas Chromatography–Mass Spectrometry," *Atmos. Environ., 10*, 447–457 (1976).

Cautreels, W., and K. Van Cauwenberghe, "Experiments on the Distribution of Organic Pollutants between Airborne Particulate Matter and the Corresponding Gas Phase," *Atmos. Environ., 12*, 1133–1141 (1978).

Cavalieri, E. L., E. G. Rogan, S. Higginbotham, P. Cremonesi, and S. Salmasi, "Tumor-Initiating Activity in Mouse Skin and Carcinogenicity in Rat Mammary Gland of Dibenzo[a]pyrenes: The Very Potent Environmental Carcinogen Dibenzo[a,l]pyrene," *J. Cancer Res. Clin. Oncol., 115*, 67–72 (1989).

Cavalieri, E. L., S. Higginbotham, and E. G. Rogan, "Dibenzo[a,l]pyrene: The Most Potent Carcinogenic Aromatic Hydrocarbon," *Polycyclic Aromat. Compd., 6*, 177–183 (1994).

Chang, R. L., A. W. Wood, W. Levin, H. D. Mah, D. R. Thakker, D. M. Jerina, and A. H. Conney, "Differences in Mutagenicity and Cytotoxicity of (+)- and (−)-Benzo[a]pyrene-4,5-oxide: A Synergistic Interaction of Enantiomers," *Proc. Natl. Acad. Sci. U.S.A., 76*, 4280–4284 (1979).

Chapman, O. L., C. Heckert, J. W. Reasoner, and S. P. Thackaberry, "Photochemical Studies on 9-Nitroanthracene," *J. Am. Chem. Soc., 88*, 5550–5554 (1966).

Chen, H.-Y., and M. R. Preston, "Gas/Particle Partitioning of Azaarenes in an Urban Atmosphere," *Environ. Pollut., 97*, 169–174 (1997).

Chen, H.-Y., and M. R. Preston, "Azaarenes in the Aerosol of an Urban Atmosphere," *Environ. Sci. Technol., 32*, 577–583 (1998).

Chen, S.-J., S.-H. Liao, W.-J. Jian, S.-C. Chiu, and G.-C. Fang, "Particle-Bound Composition of Polycyclic Aromatic Hydrocarbons and Aerosol Carbons in the Ambient Air," *J. Environ. Sci. Health, A32*, 585–604 (1997).

Chrisp, C. E., G. L. Fisher, and J. E. Lammert, "Mutagenicity of Filtrates from Respirable Coal Fly Ash," *Science, 199*, 73–75 (1978).

Chuang, J., M. Nishioka, and B. Petersen, "Bioassay-Directed Fractionation of the Organic Extract of SRM 1649 Urban Air Particulate Matter," *Int. J. Environ. Anal. Chem., 39*, 245–256 (1990).

Chuang, J. C., S. A. Wise, S. Cao, and J. L. Mumford, "Chemical Characterization of Mutagenic Fractions of Particles from Indoor Coal Combustion: A Study of Lung Cancer in Xuan Wei, China," *Environ. Sci. Technol., 26*, 999–1004 (1992).

Ciccioli, P., A. Cecinato, R. Cabella, and E. Brancaleoni, "The Contribution of Gas-Phase Reactions to the Nitroarene Fraction of Molecular Weight 247 Present in Carbon Particles Sampled in an Urban Area of Northern Italy," *Atmos. Environ., 27A*, 1261–1270 (1993).

Ciccioli, P., A. Cecinato, E. Brancaleoni, M. Frattoni, P. Zacchei, P. de Castro Vasconcellos, "The Ubiquitous Occurrence of Nitro-PAH of Photochemical Origin in Airborne Particles," *Ann. Chim., 85*, 455–469 (1995).

Ciccioli, P., A. Cecinato, E. Brancaleoni, M. Frattoni, and P. Zacchei, "Formation and Transport of 2-Nitrofluoranthene and 2-Nitropyrene of Photochemical Origin in the Troposphere," *J. Geophys. Res., 101*, 19567–19581 (1996).

Clar, E., *Polycyclic Hydrocarbons*, 2 Vols., Academic Press, New York, 1964.

Clar, E., *The Aromatic Sextet*, Wiley, New York, 1972.

Claxton, L. D., "Assessment of Bacterial Mutagenicity Methods for Volatile and Semi-Volatile Compounds and Mixtures," *Environ. Int. J., 11*, 375–382 (1985).

Claxton, L. D., G. Douglas, D. Krewski, J. Lewtas, H. Matshushita, and H. Rosenkranz, "Overview, Conclusions, and Recommendations of the IPCS Collaborative Study on Complex Mixtures," *Mutat. Res., 276*, 61–80 (1992a).

Claxton, L. D., J. Creason, B. Lerous, E. Agurell, S. Bagley, D. W. Bryant, Y. A. Courtois, G. Douglas, C. B. Clare, S. Goto, P. Quillardet, D. R. Jagannath, K. Kataoka, G. Mohn, P. A. Nielsen, T. Ong, T. C. Pederson, H. Shimizu, L. Nylund, H. Tokiwa, G. J. Vink, Y. Wang, and D. Warshawsky, "Results of the IPCS Collaborative Study on Complex Mixtures," *Mutat. Res., 276*, 23–32 (1992b).

Clean Air Act Amendments of 1990, Conference Report to Accompany S. 1630, Report No. 101-952, pp. 139–162, U.S. Government Printing Office, Washington, DC, 1990.

Collins, J. F., J. P. Brown, G. V. Alexeeff, and A. G. Salmon, "Potency Equivalency Factors for Some Polycyclic Aromatic Hydrocarbons and Polycyclic Aromatic Hydrocarbon Derivatives," *Regulat. Toxicol. Pharmacol., 28*, 45–54 (1998).

Commings, B. T., and L. Hampton, "Changing Pattern in Concentrations of Polycyclic Aromatic Hydrocarons in the Air of Central London," *Atmos. Environ., 10*, 561–562 (1976).

Commoner, B., P. Madyastha, A. Bronsdon, and A. J. Vithayathil, "Environmental Mutagens in Urban Air Particulates," *J. Toxicol. Environ. Health, 4*, 59–77 (1978).

Commoner, B., A. J. Vithayathil, and P. Dolara, "Mutagenic Analysis of Complex Samples of Aqueous Effluents, Air Particulates, and Foods," in *Application of Short-Term Bioassays in the Fractionation and Analysis of Complex Environmental Mixtures* (M. D. Waters, S. Nesnow, J. L. Huisingh, S. Sandhu, and L. Claxton, Eds.), pp. 529–570, Plenum, New York, 1979.

Cook, J. W., C. L. Hewett, and I. Hieger, "The Isolation of a Cancer-Producing Hydrocarbon from Coal Tar. Parts I, II, and II," *J. Chem. Soc.*, 395–405 (1933).

Cooke, M., A. J. Dennis, and G. L. Fisher, Eds., *Polynuclear Aromatic Hydrocarbons: Physical and Biological Chemistry*, Battelle Press, Columbus, OH, 1982.

Cotham, W. E., and T. F. Bidleman, "Polycyclic Aromatic Hydrocarbons and Polychlorinated Biphenyls in Air at an Urban and Rural Site near Lake Michigan," *Environ. Sci. Technol., 29*, 2782–2789 (1995).

Countess, R. J., G. T. Wolff, and S. H. Cadle, "The Denver Winter Aerosol: A Comprehensive Chemical Characterization," *J. Air Pollut. Control Assoc., 30*, 1194–2000 (1980).

Coutant, R. W., L. Brown, J. C. Chuang, R. M. Riggin, and R. G. Lewis, "Phase Distribution and Artifact Formation in Ambient Air Sampling for Polynuclear Aromatic Hydrocarbons," *Atmos. Environ., 22*, 403–409 (1988).

Coutant, R. W., P. J. Callahan, M. R. Kuhlman, and R. G. Lewis, "Design and Performance of a High-Volume Compound Annular Denuder," *Atmos. Environ., 23,* 2205–2211 (1989).

Coutant, R. W., P. J. Callahan, and J. C. Chuang, "Efficiency of Silicone-Grease-Coated Denuders for Collection of Polynuclear Aromatic Hydrocarbons," *Atmos. Environ., 26A,* 2831–2834 (1992).

Crebelli, R., S. Fuselli, G. Conti, L. Conti, and A. Carere, "Mutagenicity Spectra in Bacterial Strains of Airborne and Engine Exhaust Particulate Extracts," *Mutat. Res., 261,* 237–248 (1991).

Crespi, C. L., and W. G. Thilly, "Assay for Gene Mutation in a Human Lymphoblast Line, AHH-1, Competent for Xenobiotic Metabolism," *Mutat. Res., 128,* 221–230 (1984).

Crespi, C. L., F. J. Gonzalez, D. T. Steimel, T. R. Turner, H. V. Gelboin, B. W. Penman, and R. Langenbach, "A Metabolically Competent Human Cell Line Expressing Five cDNAs Encoding Procarcinogen-Activating Enzymes: Application to Mutagenicity Testing," *Chem. Res. Toxicol., 4,* 566–572 (1991).

Cretney, J. R., H. K. Lee, G. J. Wright, W. H. Swallow, and M. C. Taylor, "Analysis of Polycyclic Aromatic Hydrocarbons in Air Particulate Matter from a Lightly Industrialized Urban Area," *Environ. Sci. Technol., 19,* 397–404 (1985).

Dabestani, R., K. J. Ellis, and M. E. Sigman, "Photodecomposition of Anthracene on Dry Surfaces: Products and Mechanism," *J. Photochem. Photobiol. A, 86,* 231–239 (1995).

Dabestani, R., M. Nelson, and M. E. Sigman, "Photochemistry of Tetracene Adsorbed on Dry Silica—Products and Mechanism," *Photochem. Photobiol., 64,* 80–86 (1996).

Dabestani, R., "Photophysical and Photochemical Behavior of Polycyclic Aromatic Hydrocarbons on Silica Surfaces," *Inter-Am. Photochem. Soc. Newsl., 20,* 24–36 (1997).

Daisey, J. M., C. F. Allen, G. McGarrity, T. Atherholt, J. Louis, L. McGeorge, and P. J. Lioy, "Effects of Filter Type on the Organic Composition and Mutagenicity of Inhalable Particulate Matter," *Aerosol Sci. Technol., 5,* 69–80 (1986a).

Daisey, J. M., J. L. Cheney, and P. J. Lioy, "Profiles of Organic Particulate Emissions from Air Pollution Sources; Status and Needs for Receptor Source Apportionment Modeling," *J. Air Pollut. Control Assoc., 36,* 17–33 (1986b).

Davies, R. L., C. L. Crespi, K. Rudo, T. R. Turner, and R. Langenbach, "Development of a Human Cell Line by Selection and Drug-Metabolizing Gene Transfection with Increased Capacity to Activate Promutagens," *Carcinogen, 10,* 885–891 (1989).

De Flora, S., M. Bagnasco, A. Izzotti, F. d'Agostini, M. Pala, and F. Valerio, "Mutagenicity of Polycyclic Aromatic Hydrocarbon Fractions Extracted from Urban Air Particulates," *Mutat. Res., 224,* 305–318 (1989).

Dehnen, W., N. Pitz, and R. Tomingas, "The Mutagenicity of Airborne Particulate Pollutants," *Cancer Lett., 4,* 5–12 (1977).

De Pollok, F. S., V. P. Aneja, T. J. Hughes, and L. D. Claxton, "Chemical and Mutagenic Analysis of Volatile Organic Compounds in Raleigh Air Samples at Three Different Elevations before, during, and after Hurricane Gordon," *Chemosphere, 35,* 879–893 (1997).

De Raat, W. K., F. L. Schulting, E. Burghardt, and F. A. de Meijere, "Application of Polyurethane Foam for Sampling Volatile Mutagens from Ambient Air," *Sci. Total Environ., 63,* 175–189 (1987).

De Raat, W. K., G. L. Bakker, and F. A. de Meijere, "Comparison of Filter Materials Used for Sampling of Mutagens and Polycyclic Aromatic Hydrocarbons in Ambient Airborne Particles," *Atmos. Environ., 24A,* 2875–2887 (1990).

De Raat, W. K., and F. A. de Meijere, "Polycyclic Aromatic Hydrocarbon (PAH) Concentrations in Ambient Airborne Particles from Local Traffic and Distant Sources; Variation of the PAH Profile," *Sci. Total Environ., 103,* 1–17 (1991).

Dewar, M. J. S., T. Mole, and E. W. T. Warford, "Electrophilic Substitution. Part VI. The Nitration of Aromatic Hydrocarbons; Partial Rate Factors and Their Interpretation," *J. Chem. Soc., Part III,* 3581–3586 (1956).

Dong, M. W., D. C. Locke, and D. Hoffman, "Characterization of Aza-Arenes in Basic Organic Portion of Suspended Particulate Matter," *Environ. Sci. Technol., 11,* 612–618 (1977).

Donnely, K. C., L. D. Claxton, H. J. Huebner, and J. L. Capizzi, "Mutagenic Interactions of Model Chemical Mixtures," *Chemosphere, 37,* 1253–1261 (1998).

Dunstan, T. D. J., R. F. Mauldin, Z. Jinxlan, A. D. Hipps, E. L. Wehry, and G. Mamantov, "Adsorption and Photodegradation of Pyrene on Magnetic, Carbonaceous, and Mineral Subfractions of Coal Stack Ash," *Environ. Sci. Technol., 23,* 303–308 (1989).

Durant, J. L., W. G. Thilly, H. F. Hemond, and A. L. Lafleur, "Identification of the Principal Human Cell Mutagen in an Organic Extract of a Mutagenic Sediment," *Environ. Sci. Technol., 28,* 2033–2044 (1994).

Durant, J. L., W. F. Busby, Jr., A. L. Lafleur, B. W. Penman, and C. L. Crespi, "Human Cell Mutagenicity of Oxygenated, Nitrated, and Unsubstituted Polycyclic Aromatic Hydrocarbons Associated with Urban Aerosols," *Mutat. Res., 371,* 123–157 (1996).

Durant, J. L., A. L. Lafleur, E. F. Plummer, K. Taghizadeh, W. F. Busby, Jr., and W. G. Thilly, "Human Lymphoblast Mutagens in Urban Airborne Particles," *Environ. Sci. Technol., 32,* 1894–1906 (1998).

Eaborn, C., P. Golborn, R. E. Spillett, and R. Taylor, "Aromatic Reactivity. Part XXXVII: Detritiation of Substituted 1- and 2-Tritionaphthalenes," *J. Chem. Soc. B,* 1112–1123 (1968).

Eatough, D. J., A. Wadsworth, D. A. Eatough, J. W. Crawford, L. D. Hansen, and E. A. Lewis, "A Multiple-System, Multi-Channel Diffusion Denuder Sampler for the Determination of Fine-Particulate Organic Material in the Atmosphere," *Atmos. Environ., 27A,* 1213–1219 (1993).

Einistö, P., M. Watanabe, M. Ishidate, and T. Nohmi, "Mutagenicity of 30 Chemicals in *Salmonella typhimurium* Strains Possessing Different Nitroreductase or O-Acetyltransferase Activities," *Mutat. Res., 259,* 95–102 (1991).

Eisenberg, W. C., K. Taylor, and R. W. Murray, "Production of Singlet Delta Oxygen by Atmospheric Pollutants," *Carcinogenesis, 5,* 1095–1096 (1984).

Eisenstadt, E., and A. Gold, "Cyclopenta[*cd*]pyrene: A Highly Mutagenic Polycyclic Aromatic Hydrocarbon," *Proc. Natl. Acad. Sci. U.S.A., 75,* 1667–1669 (1978).

El-Bayoumy, K., E. J. Lavoie, S. S. Hecht, E. A. Fow, and D. Hoffmann, "The Influence of Methyl Substitution on the Mutagenicity of Nitronaphthalenes and Nitrobiphenyls," *Mutat. Res., 81,* 143–153 (1981).

El-Bayoumy, K., and S. S. Hecht, "Mutagenicity of K-Region Derivatives of 1-Nitropyrene; Remarkable Activity of 1- and 3-Nitro-5*H*-phenanthro[4,5-*bcd*]pyran-5-one," *Mutat. Res., 170,* 31–40 (1986).

Enya, T., H. Suzuki, T. Watanabe, T. Hirayama, and Y. Hisamatsu, "3-Nitrobenzanthrone, a Powerful Bacterial Mutagen and Suspected Human Carcinogen Found in Diesel Exhaust and Airborne Particulates," *Environ. Sci. Technol., 31,* 2772–2776 (1997).

Epler, J. L., B. R. Clark, C. Ho, M. R. Guerin, and T. K. Rao, "Short-Term Bioassay of Complex Organic Mixtures: Part II, Mutagenicity Testing," in *Application of Short-Term Bioassays in the Fractionation and Analysis of Complex Environmental Mixtures* (M. D. Waters, S. Nesnow, J. Huisingh, S. Sandhu, and L. Claxton, Eds.), pp. 269–289, Plenum, New York, 1979.

Epler, J. L., "The Use of Short-Term Tests in the Isolation and Identification of Chemical Mutagens in Complex Mixtures," in *Chemical Mutagens: Principles and Methods for Their Detection* (F. J. DeSerres and A. Hollander, Eds.), Vol. 6, pp. 239–270, Plenum, New York, 1980.

Epstein, S. S., S. Joshi, J. Andrea, N. Mantel, E. Sawicki, T. Stanley, and E. C. Tabor, "Carcinogenicity of Organic Particulate Pollutants in Urban Air after Administration of Trace Quantities to Neonatal Mice," *Nature, 212,* 1305–1307 (1966).

Falconer, L. R., and T. F. Bidleman, "Vapor Pressures and Predicted Particle/Gas Distributions of Polychlorinated Biphenyl Congeners as Functions of Temperature and Ortho-Chlorine Substitution," *Atmos. Environ., 28,* 547–554 (1994).

Falk, H. L., I. Markul, and P. Kotin, "Aromatic Hydrocarbons. IV. Their Fate Following Emission into the Atmosphere and Experimental Exposure to Washed Air and Synthetic Smog," *Arch. Ind. Health, 13,* 13–17 (1956).

Fan, Z., D. Chen, P. Birla, and R. M. Kamens, "Modeling of Nitro-Polycyclic Aromatic Hydrocarbon Formation and Decay in the Atmosphere," *Atmos. Environ., 29,* 1171–1181 (1995).

Fan, Z., R. M. Kamens, J. Zhang, and J. Hu, "Ozone–Nitrogen Dioxide–NPAH Heterogeneous Soot Particle Reactions and Modeling NPAH in the Atmosphere," *Environ. Sci. Technol., 30,* 2821–2827 (1996a).

Fan, Z., R. M. Kamens, J. Hu, J. Zhang, and S. McDow, "Photostability of Nitro-Polycyclic Aromatic Hydrocarbons on Combustion Soot Particles in Sunlight," *Environ. Sci. Technol., 30,* 1358–1364 (1996b).

Feilberg, A., and T. Nielsen, "Model Systems to Simulate Photochemistry in the Organic Liquid Phase of Combustion Aerosols," presented at the 6th FECS Conference on Chemistry and the Environment, Atmospheric Chemistry and Air Pollution, University of Copenhagen, Copenhagen, Denmark, August 26–28, 1998.

Feilberg, A., R. M. Kamens, M. R. Strommen, and T. Nielsen, "Modeling the Formation, Decay, and Partitioning of Semivolatile Nitro-Polycyclic Aromatic Hydrocarbons (Nitronaphthalenes) in the Atmosphere," *Atmos. Environ., 33,* 1231–1243 (1999a).

Feilberg, A., and T. Nielsen, "Effect of Aerosol Chemical Composition on the Photodegradation of Nitro-Polycyclic Aromatic Hydrocarbons," submitted to *Environ. Sci. Technol.* (1999b).

Fieser, L. F., and M. Fieser, *Organic Chemistry,* 3rd ed., Reinhold, New York, 1956.

Finizio, A., D. Mackay, T. Bidleman, and T. Harner, "Octanol–Air Partition Coefficient as a Predictor of Partitioning of Semi-Volatile Organic Chemicals to Aerosols," *Atmos. Environ., 31,* 2289–2296 (1997a).

Finizio, A., M. Vighi, and D. Sandroni, "Determination of n-Octanol/Water Partition Coefficients (K_{ow}) of Pesticides: Critical Review and Comparison of Methods," *Chemosphere, 34,* 131–161 (1997b).

Finlayson-Pitts, B. J., and J. N. Pitts, Jr., *Atmospheric Chemistry: Fundamentals and Experimental Techniques,* Wiley, New York, 1986.

Finlayson-Pitts, B. J., and J. N. Pitts, Jr., "Atmospheric Chemistry of Tropospheric Ozone Formation: Scientific and Regulatory Implications," *J. Air Waste Manage. Assoc., 43,* 1091–1100 (1993).

Finlayson-Pitts, B. J., and J. N. Pitts, Jr., "Volatile Organic Compounds: Ozone Formation, Alternative Fuels, and Toxics," *Chem. Ind.,* 796–800, October (1993).

Finlayson-Pitts, B. J., and J. N. Pitts, Jr., "Tropospheric Air Pollution: Ozone, Airborne Toxics, Polycyclic Aromatic Hydrocarbons, and Particles," *Science, 276,* 1045–1052 (1997).

Fisher, G. L., "Biomedically Relevant Chemical and Physical Properties of Coal Combustion Products," *Environ. Health Perspect., 47,* 189–199 (1983).

Fisher, G. L., C. E. Chrisp, and O. G. Raabe, "Physical Factors Affecting the Mutagenicity of Fly Ash from a Coal-Fired Power Plant," *Science, 204,* 879–881 (1979).

Fitz, D. R., G. J. Doyle, and J. N. Pitts, Jr., "An Ultra-High Volume Sampler for the Multiple Filter Collection of Respirable Particulate Matter," *J. Air Pollut. Control Assoc., 33,* 877–879 (1983).

Fitz, D. R., D. M. Lokensgard, and G. J. Doyle, "Investigation of Filtration Artifacts When Sampling Ambient Particulate Matter for Mutagen Assay," *Atmos. Environ., 18,* 205–213 (1984).

Flessel, C. P., G. N. Quirquis, J. C. Cheng, K.-I. Chang, E. S. Hahn, S. Twiss, and J. J. Wesolowski, "Sources of Mutagens in Contra Costa County Community Aerosols during Pollution Episodes: Diurnal Variations and Relations to Source Emissions Tracers," *Environ. Int. J., 11,* 293–301 (1985).

Flessel, P., Y. Y. Wang, K.-I. Chang, J. J. Wesolowski, G. N. Guirguls, I.-S. Kim, D. Levaggi, and W. Siu, "Seasonal Variations and Trends in Concentrations of Filter-Collected Polycyclic Aromatic Hydrocarbons (PAH) and Mutagenic Activity in the San Francisco Bay Area," *J. Air Waste Manage. Assoc., 41,* 276–281 (1991).

Foreman, W. T., and T. F. Bidleman, "Semi-volatile Organic Compounds in the Ambient Air of Denver, Colorado," *Atmos. Environ., 24A,* 2405–2416 (1990).

Fox, M. A., and S. Olive, "Photooxidation of Anthracene on Atmospheric Particulate Matter," *Science, 205,* 582–583 (1979).

Fraser, M. P., G. R. Cass, B. R. T. Simoneit, and R. A. Rasmussen, "Air Quality Model Evaluation Data for Organics. 5. C_6–C_{22} Nonpolar and Semipolar Aromatic Compounds," *Environ. Sci. Technol., 32,* 1760–1770 (1998).

Fu, P. P., "Metabolism of Nitro-Polycyclic Aromatic Hydrocarbons," *Drug Metab. Rev., 22,* 209–268 (1990).

Fukuhara, K., A. Hakura, N. Sera, H. Tokiwa, and N. Miyata, "1- and 3-Nitro-6-azabenzo[a]pyrenes and Their N-Oxides: Highly Mutagenic Nitrated Azaarenes," *Chem. Res. Toxicol., 5,* 149–153 (1992).

Gibson, T. L., "Nitro Derivatives of Polynuclear Aromatic Hydrocarbons in Airborne and Source Particulate Matter," *Atmos. Environ., 16,* 2037–2040 (1982).

Gibson, T. L., "Sources of Direct-Acting Nitroarene Mutagens in Airborne Particulate Matter," *Mutat. Res., 122,* 115–121 (1983).

Gilbert, A., and J. Baggott, *Essentials of Molecular Photochemistry,* CRC Press, Boca Raton, FL, 1991.

Gohda, H., H. Hatano, T. Hanai, I. Miyaji, N. Takahashi, Z. Sun, Z. Dong, H. Yu, T. Cao, I. Albrecht, K. Naikwadi, and F. W. Karasek, "GC and GC–MS Analysis of Polychlorinated Dioxins, Dibenzofurans, and Aromatic Hydrocarbons in Fly Ash from Coal-Burning Works," *Chemosphere, 27,* 9–15 (1993).

Goldring, J. M., L. M. Ball, R. Sangaiah, and A. Gold, "Synthesis and Biological Activity of Nitro-Substituted Cyclopenta-Fused PAH," in *Polynuclear Aromatic Hydrocarbons: A Decade of Progress* (M. Cooke and A. J. Dennis, Eds.), pp. 285–299, Battelle Press, Columbus, OH, 1988.

Gonzalez, F. J., C. L. Crespi, and H. V. Gelboin, "cDNA-Expressed Human Cytochrome P450s: A New Age of Molecular Toxicology and Human Risk Assessment," *Mutat. Res., 247,* 113–127 (1991).

Gordon, R. J., R. J. Bryan, J. S. Rhim, C. Demoise, R. G. Wolford, A. E. Freeman, and R. J. Huebner, "Transformation of Rat and Mouse Embryo Cells by a New Class of Carcinogenic Compounds Isolated from Particles in City Air," *Int. J. Cancer, 12,* 223–232 (1973).

Goss, K.-U., "Conceptual Model for the Adsorption of Organic Compounds from the Gas Phase to Liquid and Solid Surfaces," *Environ. Sci. Technol., 31,* 3600–3605 (1997).

Goss, K.-U., and R. P. Schwarzenbach, "Gas/Solid and Gas/Liquid Partitioning of Organic Compounds: Critical Evaluation of the Interpretation of Equilibrium Constants," *Environ. Sci. Technol., 32,* 2025–2032 (1998).

Goulden, F., and M. M. Tipler, "Experiments on the Identification of 3,4-Benzpyrene in Domestic Soot by Means of the Fluorescence Spectrum," *Br. J. Cancer, 3,* 157–160 (1949).

Graedel, T. E., D. T. Hawkins, and L. D. Claxton, *Atmospheric Chemical Compounds: Sources, Occurrence, and Bioassay,* Academic Press, Orlando, FL, 1986.

Grafe, A., I. E. Mattern, and M. Green, "A European Collaborative Study of the Ames Assay. I. Results and General Interpretation," *Mutat. Res., 85,* 391–410 (1981).

Greenberg, A., "Phenomenological Study of Benzo[*a*]pyrene and Cyclopenteno[*cd*]pyrene Decay in Ambient Air Using Winter/Summer Comparisons," *Atmos. Environ., 23,* 2797–2799 (1989).

Greenberg, A., Y. Wang, F. B. Darack, R. Harkov, J. B. Louis, and T. Atherholt, "Biologically-Directed Fractionation of Four Seasonal Extracts of Airborne Particulates and Related Chemical Studies," *Polycyclic Aromat. Compd., 1,* 233–250 (1990).

Greenberg, A., J.-H. Lwo, T. B. Atherholt, R. Rosen, T. Hartman, J. Butler, and J. Louis, "Bioassay-Directed Fractionation of Organic Compounds Associated with Airborne Particulate Matter: An Interseasonal Study," *Atmos. Environ., 27A,* 1609–1626 (1993).

Greibrokk, T., G. Löfroth, L. Nilsson, R. Toftgard, J. Carlstedt-Duke, and J. Gustafsson, "Nitroarenes: Mutagenicity in the Ames *Salmonella*/Microsome Assay and Affinity to the TCDD-Receptor Protein," in *Toxicity of Nitroaromatic Compounds* (D. E. Rickert, Ed.), pp. 166–183, Hemisphere Publishing, Washington, DC, 1985.

Grimmer, G., "Chemistry," in *Environmental Carcinogens: Polycyclic Aromatic Hydrocarbons* (G. Grimmer, Ed.), pp. 27–60, CRC Press, Boca Raton, FL, 1983a.

Grimmer, G., Ed., *Environmental Carcinogens: Polycyclic Aromatic Hydrocarbons,* CRC Press, Boca Raton, FL, 1983b.

Grimmer, G., and J. Misfeld, "Environmental Carcinogens: A Risk for Man? Concept and Strategy of the Identification of Carcinogens in the Environment," in *Environmental Carcinogens: Polycyclic Aromatic Hydrocarbons* (G. Grimmer, Ed.), pp. 1–26, CRC Press, Boca Raton, FL, 1983.

Grimmer, G., and F. Pott, "Occurrence of PAH," in *Environmental Carcinogens: Polycyclic Aromatic Hydrocarbons* (G. Grimmer, Ed.), pp. 61–129, CRC Press, Boca Raton, FL, 1983.

Grimmer, G., H. Brune, R. Deutsch-Wenzel, G. Dettbarn, J. Jacob, K.-W. Naujack, U. Mohr, and H. Ernst, "Contribution of Polycyclic Aromatic Hydrocarbons and Nitro-Derivatives to the Carcinogenic Impact of Diesel Engine Exhaust Condensate Evaluated by Implantation into the Lungs of Rats," *Cancer Lett., 37,* 173–180 (1987).

Grosjean, D., K. Fung, and J. Harrison, "Interactions of Polycyclic Aromatic Hydrocarbons with Atmospheric Pollutants," *Environ. Sci. Technol., 17,* 673–679 (1983).

Grover, P. L., and P. Sims, "Effect of Anti-inflammatory Compounds on Actomyosihadenosine Triphosphate Interaction," *Biochem. Pharmacol., 19,* 2251–2259 (1970).

Guerin, M. R., B. R. Clark, C. Ho, J. L. Epler, and T. K. Rao, "Short-Term Bioassays of Complex Organic Mixtures: Part 1, Chemistry," in *Application of Short-Term Bioassays in the Fractionation and Analysis of Complex Environmental Mixtures* (M. D. Waters, S. Nesnow, J. L. Huisingh, S. Sandhu, and L. Claxton, Eds.), pp. 249–268, Plenum, New York, 1979.

Gundel, L. A., J. M. Daisey, L. R. F. de Carvalho, N. Y. Kado, and D. Schuetzle, "Polar Organic Matter in Airborne Particles: Chemical Characterization and Mutagenic Activity," *Environ. Sci. Technol., 27,* 2112–2119 (1993).

Gundel, L. A., V. C. Lee, K. R. R. Mahanama, R. K. Stevens, and J. M. Daisey, "Direct Determination of the Phase Distributions of Semi-Volatile Polycyclic Aromatic Hydrocarbons Using Annular Denuders," *Atmos. Environ., 29,* 1719–1733 (1995a).

Gundel, L. A., K. R. R. Mahanama, and J. M. Daisey, "Semi-volatile and Particulate Polycyclic Aromatic Hydrocarbons in Environmental Tobacco Smoke: Cleanup, Speciation, and Emission Factors," *Environ. Sci. Technol., 29,* 1607–1614 (1995b).

Guo, Z., and R. M. Kamens, "An Experimental Technique for Studying Heterogeneous Reactions of Polyaromatic Hydrocarbons on Particle Surfaces," *J. Atmos. Chem., 12,* 137–151 (1991).

Gupta, P., W. P. Harger, and J. Arey, "The Contribution of Nitro- and Methylnitronaphthalenes to the Vapor-Phase Mutagenicity of Ambient Air Samples," *Atmos. Environ., 30,* 3157–3166 (1996).

Gustafson, K. E., and R. M. Dickhut, "Particle/Gas Concentrations and Distributions of PAHs in the Atmosphere of Southern Chesapeake Bay," *Environ. Sci. Technol., 31,* 140–147 (1997).

Gustafson, K. E., and R. M. Dickhut, "Comment on 'Particle/Gas Concentrations and Distributions of PAHs in the Atmosphere of Southern Chesapeake Bay,' " *Environ. Sci. Technol., 31,* 3738–3739 (1997).

Haagen-Smit, A. J., and M. M. Fox, "Ozone Formation in Photochemical Oxidation of Organic Substances," *Ind. Eng. Chem., 48,* 1484–1497 (1956).

Halsall, C. J., P. J. Coleman, B. J. Davis, V. Burnett, K. S. Waterhouse, P. Harding-Jones, and K. C. Jones, "Polycyclic Aromatic Hydrocarbons in U.K. Urban Air," *Environ. Sci. Technol., 28,* 2380–2386 (1994).

Hammerle, R., D. Schuetzle, and W. Adams, "A Perspective on the Potential Development of Environmentally Acceptable Light-Duty Diesel Vehicles," *Environ. Health Perspect., 102,* 25–30 (1994).

Hannigan, M. P., G. R. Cass, A. L. Lafleur, J. P. Longwell, and W. G. Thilly, "Bacterial Mutagenicity of Urban Organic Aerosol Sources in Comparison to Atmospheric Samples," *Environ. Sci. Technol., 28,* 2014–2024 (1994).

Hannigan, M. P., G. R. Cass, A. L. Lafleur, W. F. Busby, Jr., and W. G. Thilly, "Seasonal and Spatial Variation of the Bacterial Mutagenicity of Fine Organic Aerosol in Southern California," *Environ. Health Perspect., 104,* 428–436 (1996).

Hannigan, M. P., G. R. Cass, B. W. Penman, C. L. Crespi, A. L. Lafleur, W. F. Busby, Jr., and W. G. Thilly, "Human Cell Mutagens in Los Angeles Air," *Environ. Sci. Technol., 31,* 438–447 (1997).

Hannigan, M. P., G. R. Cass, B. W. Penman, C. L. Crespi, A. L. Lafleur, W. F. Busby, Jr., W. G. Thilly, and B. R. T. Simoneit, "Bioassay-Directed Chemical Analysis of Los Angeles Airborne Particulate Matter Using a Human Cell Mutagenicity Assay," *Environ. Sci. Technol., 32,* 3502–3514 (1998).

Harger, W. P., J. Arey, and R. Atkinson, "The Mutagenicity of HPLC-Separated Vapor-Phase and Particulate Organics in Ambient Air," *Atmos. Environ., 26A,* 2463–2466 (1992).

Harner, T., and D. Mackay, "Measurement of Octanol–Air Partition Coefficients for Chlorobenzenes, PCBs, and DDT," *Environ. Sci. Technol., 29,* 1599–1606 (1995).

Harner, T., and T. F. Bidleman, "Measurements of Octanol–Air Partition Coefficients for Polychlorinated Biphenyls," *J. Chem. Eng. Data, 41,* 895–899 (1996).

Harner, T., and T. F. Bidleman, "Octanol–Air Partition Coefficient for Describing Particle/Gas Partitioning of Aromatic Compounds in Urban Air," *Environ. Sci. Technol., 32,* 1494–1502 (1998).

Harrison, R. M., D. J. T. Smith, and L. Luhana, "Source Apportionment of Atmospheric Polycyclic Aromatic Hydrocarbons Collected from an Urban Location in Birmingham, U.K.," *Environ. Sci. Technol., 30,* 825–832 (1996).

Hart, K. M., and J. F. Pankow, "High-Volume Air Sampler for Particle and Gas Sampling. 2. Use of Backup Filters to Correct for the Adsorption of Gas-Phase Polycyclic Aromatic Hydrocarbons to the Front Filter," *Environ. Sci. Technol., 28,* 655–661 (1994).

Harvey, R. G., *Polycyclic Aromatic Hydrocarbons*, Wiley-VCH, New York, 1997.

Hawthorne, S. B., M. S. Krieger, D. J. Miller, and M. B. Mathiason, "Collection and Quantitation of Methoxylated Phenol Tracers for Atmospheric Pollution from Residential Wood Stoves," *Environ. Sci. Technol., 22,* 470–475 (1989).

Hayakawa, K., M. Butoh, Y. Hirabayashi, and M. Miyazaki, "Determination of 1,3-, 1,6-, 1,8-Dinitropyrenes, and 1-Nitropyrene in Vehicle Exhaust Particulates," *Jpn. J. Toxicol. Environ. Health, 40,* 20–25 (1994).

Haynes, B. S., "Soot and Hydrocarbons in Combustion," in *Fossil Fuel Combustion: A Source Book* (W. Bartock and A. F. Sarofim, Eds.), pp. 261–326, Wiley, New York, 1991.

Hecht, S. S., and K. El-Bayoumy, "The Possible Role of Nitroarenes in Human Cancer," in *Nitroarenes* (P. C. Howard *et al.*, Eds.), pp. 309–316, Plenum, New York, 1990.

Heinrich, G., and H. Güsten, "Fluorescence Spectroscopic Properties of Carcinogenic and Airborne Polynuclear Aromatic Hydrocarbons," in *Polynuclear Aromatic Hydrocarbons: Chemistry and Biological Effects* (A. Bjørseth and A. J. Dennis, Eds.), pp. 983–1003, Battelle Press, Columbus, OH, 1980.

Heinrich, U., M. Roller, and F. Pott, "Estimation of a Lifetime Unit Lung Cancer Risk for Benzo[a]pyrene Based on Tumour Rates in Rats Exposed to Coal Tar/Pitch Condensation Aerosol," *Toxicol. Lett., 72,* 155–161 (1994).

Helmig, D., J. Arey, W. P. Harger, R. Atkinson, and J. López-Cancio, "Formation of Mutagenic Nitrodibenzopyranones and Their Occurrence in Ambient Air," *Environ. Sci. Technol., 26,* 622–624 (1992a).

Helmig, D., J. López-Cancio, J. Arey, W. P. Harger, and R. Atkinson, "Quantification of Ambient Nitrodibenzopyranones: Further Evidence for Atmospheric Mutagen Formation," *Environ. Sci. Technol., 26,* 2207–2213 (1992b).

Helmig, D., J. Arey, R. Atkinson, W. P. Harger, and P. A. McElroy, "Products of the OH Radical-Initiated Gas-Phase Reaction of Fluorene in the Presence of NO_x," *Atmos. Environ., 26A,* 1735–1745 (1992c).

Hering, S. V., R. C. Flagan, and S. K. Friedlander, "Design and Evaluation of New Low-Pressure Impactor. 1," *Environ. Sci. Technol., 12,* 667–673 (1978).

Hering, S. V., S. K. Friedlander, J. J. Collins, and L. W. Richards, "Design and Evaluation of a New Low-Pressure Impactor. 2," *Environ. Sci. Technol., 13,* 184–188 (1979).

Herreno-Saenz, D., F. E. Evans, T. Heinze, J. Lewtas, and P. P. Fu, "*In Vitro* Metabolism and DNA Adduct Formation from the Mutagenic Environmental Contaminant 2-Nitrofluoranthene," *Chem. Res. Toxicol., 5,* 863–869 (1992).

Hieger, I., "The Spectra of Cancer-Producing Tars and Oils and of Related Substances," *Biochemistry, 24,* 505–511 (1930).

Hinckley, D. A., T. F. Bidleman, and W. T. Foreman, "Determination of Vapor Pressures for Nonpolar and Semipolar Organic Compounds from Gas Chromatographic Retention Data," *J. Chem. Eng. Data, 35,* 232–237 (1990).

Hirayama, T., H. Kusakabe, T. Watanabe, S. Ozasa, Y. Fujioka, and S. Fukui, "Relationship between Mutagenic Potency in *Salmonella typhimurium* Strains and the Chemical Structure of Nitro Biphenyls," *Mutat. Res., 163,* 101–107 (1986).

Hof, R. M., K. A. Brice, and C. J. Halsall, "Nonlinearity in the Slopes of Clausius–Clapeyron Plots for SVOCs," *Environ. Sci. Technol., 32,* 1793–1798 (1998).

Hoff, R. M., and K.-W. Chan, "Measurement of Polycyclic Aromatic Hydrocarbons in the Air along the Niagara River," *Environ. Sci. Technol., 21,* 556–561 (1987).

Hoffman, D., and E. L. Wynder, "Organic Particulate Pollutants—Chemical Analysis and Bioassays for Carcinogenicity, in *Air Pollution* (A. C. Stern, Ed.), Vol. II, 3rd ed., pp. 361–455, Academic Press, New York, 1977.

Holder, P. S., E. L. Wehry, and G. Mamantov, "Photochemical Transformation of 1-Nitropyrene Sorbed on Coal Fly Ash Fractions," *Polycyclic Aromat. Compd., 4,* 135–139 (1994).

Holloway, M. P., M. C. Biaglow, E. C. McCoy, M. Anders, H. S. Rosenkranz, and P. C. Howard, "Photochemical Instability of 1-Nitropyrene, 3-Nitrofluoranthene, 1,8-Dinitropyrene and Their Parent Polycyclic Aromatic Hydrocarbons," *Mutat. Res., 187,* 199–207 (1987).

Hollstein, M., J. McCann, F. A. Angelosanto, and W. W. Nichols, "Short-Term Tests for Carcinogens and Mutagens," *Mutat. Res., 65,* 133–226 (1979).

Holmberg, B., and U. Ahlborg, Eds., Symposium on Biological Tests in the Evaluation of Mutagenicity and Carcinogenicity of Air Pollutants with Special Reference to Motor Exhausts and Coal Combustion Products, Stockholm, Sweden, February 8–12, in *Environ. Health Perspect., 47,* 1–345 (1983).

Howard, P. C., S. S. Hecht, and F. A. Beland, Eds., *Nitroarenes: Occurrence, Metabolism, and Biological Impact, Environmental Science Research*, Vol. 40, Plenum, New York, 1990.

Howsam, M., and K. C. Jones, "Sources of PAHs in the Environment," in *The Handbook of Environmental Chemistry, PAHs and Related Compounds* (A. H. Neilson, Ed.), Vol. 3, Part I, Springer-Verlag, Berlin (1998).

Hsieh, P., N. Y. Kado, J. N. Seiber, and T. Shibamoto, "Methods Development for Assessment of Vapor-Phase Mutagens and Carcinogens," Final Report to the California Air Resources Board, Contract No. A6—174-32, June 1990.

Hsieh, D. P. H., N. Y. Kado, and R. Okamoto, "Measurement and Chemical Characterization of Vapor-Phase Mutagens in Diesel Exhaust," Final Report to the California Air Resources Board, Contract No. A032-095, March 1993.

Hudgins, D. M., and S. A. Sandford, "Infrared Spectroscopy of Matrix Isolated Polycyclic Aromatic Hydrocarbons. 1. PAHs Containing Two to Four Rings," *J. Phys. Chem. A, 102,* 329–343 (1998a).

Hudgins, D. M., and S. A. Sandford, "Infrared Spectroscopy of Matrix Isolated Polycyclic Aromatic Hydrocarbons. 2. PAHs Containing Five or More Rings," *J. Phys. Chem. A, 102,* 344–352 (1998b).

Hudgins, D. M., and S. A. Sandford, "Infrared Spectroscopy of Matrix Isolated Polycyclic Aromatic Hydrocarbons. 3. Fluoranthene and the Benzofluoranthenes," *J. Phys. Chem. A, 102,* 353–360 (1998c).

Hueper, W. C., P. Kotin, E. C. Tabor, W. W. Payne, H. L. Falk, and E. Sawicki, "Carcinogenic Bioassays on Air Pollutants," *Arch. Pathol. Lab. Med., 74,* 89–116 (1962).

Hughes, M. M., D. F. S. Natusch, D. R. Taylor, and M. V. Zeller, "Chemical Transformations of Particulate Polycyclic Organic Matter," in *Polynuclear Aromatic Hydrocarbons: Chemistry and Biological Effects* (A. Bjørseth and A. J. Dennis, Eds.), pp. 1–8, Battelle Press, Columbus, OH, 1980.

Hughes, T. J., D. M. Simmons, L. G. Monteith, and L. D. Claxton, "Vaporization Technique to Measure Mutagenic Activity of Volatile Organic Chemicals in the Ames/*Salmonella* Assay," *Environ. Mutagen., 9,* 421–441 (1987).

Huisingh, J., R. Bradow, R. Jungers, L. Claxton, R. Zweidinger, S. Tejada, J. Bumgarner, F. M. Waters, V. F. Simmon, C. Hare, C. Rodgriguez, and L. Snow, "Application of Bioassay to Characterization of Diesel Particle Emissions," in *Applications of Short-Term Bioassay in the Fractionation and Analysis of Complex Environmental Mixtures* (M. D. Waters, S. Nesnow, J. L. Huisingh, S. Sandhu, and L. Claxton, Eds.), pp. 383–418, Plenum, New York, 1979.

Imaida, K., L. K. Tay, M.-S. Lee, C. Y. Wang, N. Ito, and C. M. King, "Tumor Induction by Nitropyrenes in the Female CD Rat," in *Carcinogenic and Mutagenic Responses to Aromatic Amines and Nitroarenes,*" Proceedings of the Third International Conference on Carcinogenic and Mutagenic *N*-Substituted Aryl Compounds, April 25–28, 1987, Dearborn, Michigan (C. M. King, L. J. Romano, and D. Schuetzle, Eds.), pp. 187–197, Elsevier, New York, 1988.

International Agency for Research on Cancer (IARC), Monographs in the series *Evaluation of Carcinogenic Risk of Chemicals to Humans*: Vol. 32, *Polynuclear Aromatic Compounds, Part 1*, 1984a; Vol. 33, *Polynuclear Aromatic Compounds, Part 2, Carbon Blacks, Mineral Oils, and Some Nitroarenes*, 1984b; Vol. 34, *Polynuclear Aromatic Compounds, Part 3, Industrial Exposures in Aluminum Production, Coal Gasification, Coke Production, and Iron and Steel Foundry*, 1984c; Vol. 35, *Polynuclear Aromatic Compounds, Part 4, Bitumens, Coal-Tars and Derived Products, Shale-Oils, and Soots*, 1985; Supplement No. 7, *Overall Evaluations of Carcinogenicity: An Updating of IARC Monographs, Vols. 1–42*, 1987; Vol. 46, *Diesel and Gasoline Exhausts and Some Nitroarenes*, 1989.

International Union of Pure and Applied Chemistry, *Nomenclature of Organic Chemistry*, Sections A, B, C, D, E, F, and H, Pergamon, Oxford, 1979.

Ioki, Y., "Aryloxyl Radicals by Photorearrangement of Nitro-Compounds," *J. Chem. Soc., Perkin Trans. II,* 1240–1242 (1977).

Ismail, Y., A. L. Lafleur, and R. W. Giese, "Polar Forms of Benzo[*a*]pyrene May Be Ubiquitous in the Environment," *Environ. Sci. Technol., 32,* 2494–2497 (1998).

Izumi, K., and T. Fukuyama, "Photochemical Aerosol Formation from Aromatic Hydrocarbons in the Presence of NO_x," *Atmos. Environ., 24A,* 1433–1441 (1990).

Jäger, J., "Detection and Characterization of Nitro Derivatives of Some Polycyclic Aromatic Hydrocarbons by Fluorescence Quenching after Thin-Layer Chromatography: Application to Air Pollution Analysis," *J. Chromatogr., 152,* 575–578 (1978).

Jäger, J., and V. Hanus, "Reaction of Solid Carrier-Adsorbed Polycyclic Aromatic Hydrocarbons with Gaseous Low-Concentrated Nitrogen Dioxide," *J. Hyg. Epidemiol. Microbiol. Immunol., 24,* 1–12 (1980).

Jang, M., and S. R. McDow, "Benz[*a*]anthracene Photodegradation in the Presence of Known Organic Constituents of Atmospheric Aerosols," *Environ. Sci. Technol., 29,* 2654–2660 (1995).

Jang, M., and S. R. McDow, "Products of Benz[*a*]anthracene Photodegradation in the Presence of Known Organic Constituents of Atmospheric Aerosols," *Environ. Sci. Technol., 31,* 1046–1053 (1997a).

Jang, M., R. M. Kamens, K. B. Leach, and M. R. Strommen, "A Thermodynamic Approach Using Group Contribution Methods to Model the Partitioning of Semi-Volatile Organic Compounds on Atmospheric Particulate Matter," *Environ. Sci. Technol., 31,* 2805–2811 (1997b).

Jang, M., and R. M. Kamens, "A Thermodynamic Approach for Modeling Partitioning of Semi-Volatile Organic Compounds on Atmospheric Particulate Matter: Humidity Effects," *Environ. Sci. Technol., 32,* 1237–1243 (1998).

Jenkins, B. M., A. D. Jones, S. Q. Turn, and R. B. Williams, "Particle Concentrations, Gas–Particle Partitioning, and Species Intercorrelations for Polycyclic Aromatic Hydrocarbons (PAH) Emitted during Biomass Burning," *Atmos. Environ., 30,* 3825–3835 (1996).

Johnson, D. E., and H. H. Cornish, "Metabolic Conversion of 1- and 2-Nitronaphthalene to 1- and 2-Naphthylamine in the Rat," *Toxicol. Appl. Pharmacol., 46,* 549–553 (1978).

Johnson, J. H., S. T. Bagley, L. D. Gratz, and D. G. Leddy, "A Review of Diesel Particulate Control Technology and Emissions Effects—1992 Horning Memorial Award Lecture," *SAE, Spec. Publ. SP-1020,* 1–35 (1994).

Jung, H., A. U. Shaikh, R. H. Heflich, and P. P. Fu, "Nitro Group Orientation, Reduction Potential, and Direct-Acting Mutagenicity of Nitro-Polycyclic Aromatic Hydrocarbons," *Environ. Mol. Mutagen., 17,* 169–180 (1991).

Junge, C. E., "Basic Considerations about Trace Constituents in the Atmosphere as Related to the Fate of Global Pollutants," in *Fate of Pollutants in the Air and Water Environments* (I. H. Suffet, Ed.), pp. 7–26, Wiley, New York, 1977.

Kaden, D. A., R. A. Hites, and W. G. Thilly, "Mutagenicity of Soot and Associated Polycyclic Aromatic Hydrocarbons to *Salmonella typhimurium*," *Cancer Res., 39,* 4152–4159 (1979).

Kado, N. Y., D. Langley, and E. Eisenstadt, "A Simple Modification of the *Salmonella* Liquid Incubation Assay," *Mutat. Res., 121,* 25–32 (1983).

Kado, N. Y., G. N. Guirguis, C. P. Flessel, R. C. Chan, K.-I. Chang, and J. J. Wesolowski, "Mutagenicity of Fine (<2.5 μm) Airborne Particles: Diurnal Variation in Community Air Determined by a *Salmonella* Micro Preincubation (Microsuspension) Procedure," *Environ. Mutagen., 8,* 53–66 (1986).

Kado, N. Y., S. A. McCurdy, S. J. Tesluk, S. K. Hammond, D. P. H. Hsieh, J. Jones, and M. B. Schenker, "Measuring Personal Exposure to Airborne Mutagens and Nicotine in Environmental Smoke," *Mutat. Res., 261,* 75–82 (1991).

Kado, N. Y., J. M. Wing, P. A. Kuzmicky, J. E. Woodrow, H. Ning, J. N. Seiber, and D. P. H. Hsieh, "Quantitative Integration of the *Salmonella* Microsuspension Assay with Supercritical Fluid Extraction of Model Airborne Vapor-Phase Mutagens," *Mutat. Res., 271,* 253–260 (1992).

Kamens, R., D. Bell, A. Dietrich, J. Perry, R. Goodman, L. Claxton, and S. Tejada, "Mutagenic Transformation of Dilute Wood Smoke Systems in the Presence of Ozone and Nitrogen Dioxide. Analysis of Selected High-Pressure Liquid Chromatography Fractions from Wood Smoke Particle Extracts," *Environ. Sci. Technol., 19,* 63–69 (1985).

Kamens, R. M., J. N. Fulcher, and G. Zhishi, "Effects of Temperature on Wood Soot PAH Decay in Atmospheres with Sunlight and Low NO_x," *Atmos. Environ., 20,* 1579–1587 (1986).

Kamens, R., Z. Guo, J. Fulcher, and D. Bell, "Influence of Humidity, Sunlight, and Temperature on the Daytime Decay of Polyaromatic Hydrocarbons on Atmospheric Soot Particles," *Environ. Sci. Technol., 22,* 103–108 (1988).

Kamens, R. M., H. Karam, J. Guo, J. M. Perry, and L. Stockburger, "The Behaviour of Oxygenated Polycyclic Aromatic Hydrocarbons on Atmospheric Soot Particles," *Environ. Sci. Technol., 23,* 801–806 (1989).

Kamens, R. M., J. Guo, Z. Guo, and S. R. McDow, "Polynuclear Aromatic Hydrocarbon Degradation by Heterogeneous Reactions with N_2O_5 on Atmospheric Particles," *Atmos. Environ., 24A,* 1161–1173 (1990).

Kamens, R. M., F. Zhi-Hua, Y. Yao, D. Chen, S. Chen, and M. Vartiainen, "A Methodology for Modeling the Formation and

Decay of Nitro-PAH in the Atmosphere," *Chemosphere, 28,* 1623–1632 (1994).

Kamens, R., J. Odum, and Z.-H. Fa, "Some Observations on Times to Equilibrium for Semi-Volatile Polycyclic Aromatic Hydrocarbons," *Environ. Sci. Technol., 29,* 43–50 (1995).

Kamens, R. M., and D. L. Coe, "A Large Gas-Phase Stripping Device to Investigate Rates of PAH Evaporation for Airborne Diesel Particles," *Environ. Sci. Technol., 31,* 1830–1833 (1997).

Kao, A. S., "Formation and Removal Reactions of Hazardous Air Pollutants," *J. Air Waste Manage. Assoc., 44,* 683–696 (1994).

Karcher, W., R. J. Fordham, J. J. Dubois, P. G. J. M. Glaude, and J. A. M. Ligthart, Eds., *Spectral Atlas of Polycyclic Aromatic Compounds*, Reidel, Dordrecht, The Netherlands, 1985.

Karonis, D., E. Lois, S. Stournas, and F. Zannikos, "Correlations of Exhaust Emissions from Diesel Engine with Diesel Fuel Properties," *Energy Fuels, 12,* 230–238 (1998).

Katz, M., C. Chan, H. Tosine, and T. Sakuma, "Relative Rates of Photochemical and Biological Oxidation (*in Vitro*) of Polycyclic Aromatic Hydrocarbons, in *Polynuclear Aromatic Hydrocarbons* (P. W. Jones and P. Leber, Eds.), pp. 171–189, Ann Arbor Science Publishers, Ann Arbor, MI, 1979.

Kaupp, H., and G. Umlauf, "Atmospheric Gas–Particle Partitioning of Organic Compounds: Comparison of Sampling Methods," *Atmos. Environ., 26A,* 2259–2267 (1992).

Keller, C. D., and T. F. Bidleman, "Collection of Airborne Polycyclic Aromatic Hydrocarbons and Other Organics with a Glass Fiber Filter–Polyurethane Foam System," *Atmos. Environ., 18,* 837–845 (1984).

Kelly, T. J., R. Mukund, C. W. Spicer, and A. J. Pollack, "Concentrations and Transformations of Hazardous Air Pollutants," *Environ. Sci. Technol., 28,* 378A–387A (1994).

Kennaway, E., "Further Experiments on Cancer-Producing Substances," *Biochem. J., 24,* 497–504 (1930).

Kennaway, E. L., "Identification of Carcinogenic Compound in Coal-Tar," *Br. Med. J., 2,* 749–752 (1955).

Khan, A. U., J. N. Pitts, Jr., and E. B. Smith, "The Role of Singlet Molecular Oxygen in the Production of Photochemical Air Pollution," *Environ. Sci. Technol., 1,* 656–657 (1967).

King, C. M., L. J. Romano, and D. Schuetzle, Eds., *Carcinogenic and Mutagenic Responses to Aromatic Amines and Nitroarenes*, Elsevier, New York, 1988.

Kinouchi, T., K. Nishifuji, H. Tsutsui, S. L. Hoare, and Y. Ohnishi, "Mutagenicity and Nitropyrene Concentration of Indoor Air Particulates Exhausted from a Kerosene Heater," *Jpn. J. Cancer Res. (Gann), 79,* 32–41 (1988).

Kleindienst, T. E., P. B. Shepson, E. O. Edney, L. T. Cupitt, and L. D. Claxton, "The Mutagenic Activity of the Products of Propylene Photooxidation," *Environ. Sci. Technol., 19,* 620–627 (1985a).

Kleindienst, T. E., P. B. Shepson, E. O. Edney, and L. D. Claxton, "Peroxyacetyl Nitrate: Measurement of Its Mutagenic Activity Using the *Salmonella*/Mammalian Microsome Reversion Assay," *Mutat. Res., 157,* 123–128 (1985b).

Kleindienst, T. E., P. B. Shepson, E. O. Edney, L. T. Cupitt, and L. D. Claxton, "Wood Smoke: Measurement of the Mutagenic Activities of Its Gas- and Particle-Phase Photooxidation Products," *Environ. Sci. Technol., 20,* 493–501 (1986).

Kleindienst, T. E., P. B. Shepson, D. F. Smith, E. E. Hudgens, and C. M. Nero, "Comparison of Mutagenic Activities of Several Peroxyacyl Nitrates," *Environ. Mol. Mutagen., 16,* 70–80 (1990).

Kleindienst, T. E., D. F. Smith, E. E. Hudgens, R. F. Snow, E. Perry, L. D. Claxton, J. J. Bufalini, F. M. Black, and L. T. Cupitt, "The Photooxidation of Automobile Emissions: Measurements of the Transformation Products and Their Mutagenic Activity," *Atmos. Environ., 26A,* 3039–3053 (1992).

Kleindienst, T. E., "Recent Developments in the Chemistry and Biology of Peroxyacetyl Nitrate," *Res. Chem. Intermed., 20,* 335–384 (1994).

Kömp, P., and M. S. McLachlan, "Interspecies Variability of the Plant/Air Partitioning of Polychlorinated Biphenyls," *Environ. Sci. Technol., 31,* 2944–2948 (1997).

König, J., E. Balfanz, W. Funcke, and T. Romanowski, "Determination of Oxygenated Polycyclic Aromatic Hydrocarbons in Airborne Particulate Matter by Capillary Gas Chromatography and Gas Chromatography/Mass Spectrometry," *Anal. Chem., 55,* 599–603 (1983).

Korfmacher, W. A., E. L. Wehry, G. Mamantov, and D. F. S. Natusch, "Resistance to Photochemical Decomposition of Polycyclic Aromatic Hydrocarbons Vapor-Adsorbed on Coal Fly Ash," *Environ. Sci. Technol., 14,* 1094–1099 (1980).

Korfmacher, W. A., L. G. Rushing, J. Arey, B. Zielinska, and J. N. Pitts, Jr., "Identification of Mononitropyrenes and Mononitrofluoranthenes in Air Particulate Matter Via Fused Silica Gas Chromatography Combined with Negative Ion Atmospheric Pressure Ionization Mass Spectrometry," *J. High Resolut. Chromatogr. Chromatogr. Commun., 10,* 641–646 (1987).

Kotin, P., H. L. Falk, P. Mader, and M. Thomas, "Aromatic Hydrocarbons. I. Presence in the Los Angeles Atmosphere and the Carcinogenicity of Atmospheric Extracts," *Arch. Ind. Hyg., 9,* 153–163 (1954).

Kotin, P., H. L. Falk, and M. Thomas, "Production of Skin Tumors in Mice with Oxidation Products of Aliphatic Hydrocarbons," *Cancer, 9,* 905–909 (1956).

Kotin, P., H. L. Falk, and G. J. McCammon, "The Experimental Induction of Pulmonary Tumors and Changes in the Respiratory Epithelium in C57B1 Mice Following Their Exposure to an Atmosphere of Ozonized Gasoline," *Cancer, 11,* 473–481 (1958).

Krieger, R. K., J. N. Wright, and J. Arey, "Ambient Monitoring of Selected PAH in California," Paper 97-WP100A.06, Air & Waste Management Association 90th Annual Meeting & Exhibition, June 8–13, 1997, Toronto, Ontario, Canada. (Available on CD-ROM.)

Kwok, E. S. C., W. P. Harger, J. Arey, and R. Atkinson, "Reactions of Gas-Phase Phenanthrene under Simulated Atmospheric Conditions," *Environ. Sci. Technol., 28,* 521–527 (1994a).

Kwok, E. S. C., R. Atkinson, and J. Arey, "Kinetics and Mechanisms of the Gas-Phase Reactions of the NO_3 Radical with Aromatic Compounds," *Int. J. Chem. Kinet., 26,* 511–525 (1994b).

Kwok, E. S. C., R. Atkinson, and J. Arey, "Kinetics of the Gas-Phase Reactions of Indan, Indene, Fluorene, and 9,10-Dihydroanthracene with OH Radicals, NO_3 Radicals, and O_3," *Int. J. Chem. Kinet., 29,* 299–309 (1997).

Kwok, E. S. C., R. Atkinson, and J. Arey, "Kinetics of the Gas-Phase Reactions of Dibenzothiophene with OH Radicals, NO_3 Radicals, and O_3," *Polycyclic Aromat. Compd.,* in press (1999).

Lane, D. A., and H. Tang, "Photochemical Degradation of Polycyclic Aromatic Compounds. I. Naphthalene," *Polycyclic Aromat. Compd., 5,* 131–138 (1994).

Lane, D. A., and M. Katz, "The Photomodification of Benzo[*a*]pyrene, Benzo[*b*]fluoranthene, and Benzo[*k*]fluoranthene under Simulated Atmospheric Conditions," in *Fate of Pollutants in the Air and Water Environments, Part 2* (I. A. Suffet, Ed.), pp. 137–154, Wiley-Interscience, New York, 1977.

Langhoff, S. R., C. W. Bauschlicher, Jr., D. M. Hudgins, S. A. Sandford, and L. J. Allamandola, "Infrared Spectra of Substituted Polycyclic Aromatic Hydrocarbons," *J. Phys. Chem. A, 102,* 1632–1646 (1998).

Lao, R. C., and R. S. Thomas, "The Volatility of PAH and Possible Losses in Ambient Sampling," in *Polycyclic Aromatic Hydrocar-*

bons (A. Bjørseth and A. J. Dennis, Eds.), pp. 829–839, Battelle Press, Columbus, OH, 1980.

Larson, S. M., G. R. Cass, and H. A. Gray, "Atmospheric Carbon Particles and the Los Angeles Visibility Problem," *Aerosol Sci. Technol., 10,* 118–130 (1989).

Later, D. W., "Contribution of Nitropyrene to the Mutagenic Activity of Coal Fly Ash," *Environ. Mutagen., 6,* 131–144 (1984).

Later, D. W., "Nitrogen Polycyclic Aromatic Compounds in Coal-Derived Materials," in *Handbook of Polycyclic Aromatic Hydrocarbons* (A. Bjørseth and T. Ramdahl, Eds.), Vol. 2, Dekker, New York, 1985.

Later, D. W., C. W. Wright, K. L. Loening, and J. E. Merritt, "Systematic Nomenclature of the Nitrogen, Oxygen, and Sulfur Functional Polycyclic Aromatic Compounds," in *Polynuclear Aromatic Hydrocarbons: Nomenclature Guide* (K. Loening, J. Merritt, D. Later, and W. Wright, Eds.), pp. 27–47, Battelle Press, Columbus, OH, 1990.

Lee, F. S. C., W. R. Pierson, and J. Ezike, "The Problem of PAH Degradation during Filter Collection of Airborne Particulates—An Evaluation of Several Commonly Used Filter Media," in *Polynuclear Aromatic Hydrocarbons: Chemical and Biological Effects*
(A. Bjørseth and A. J. Dennis, Eds.), pp. 543–563, Battelle Press, Columbus, OH, 1980.

Lee, F., and D. Schuetzle, "Sampling, Extraction, and Analysis of Polycyclic Aromatic Hydrocarbons from Internal Combustion Engines," in *Handbook of Polycyclic Aromatic Hydrocarbons* (A. Bjørseth, Ed.), pp. 27–94, Dekker, New York, 1983.

Lee, M. L., M. V. Novotny, and K. D. Bartle, *Analytical Chemistry of Polycyclic Aromatic Compounds,* Academic Press, New York, 1981.

Lee, W.-J., Y.-F. Wang, T.-C. Lin, Y.-Y. Chen, W.-C. Lin, C.-C. Ku, and J.-T. Cheng, "PAH Characteristics in the Ambient Air of Traffic-Source," *Sci. Total Environ., 159,* 185–200 (1995).

Lee, W.-M., H.-C. Tong, and S.-Y. Yeh, "Partitioning Model of PAHs between Gaseous and Particulate Phases with Consideration of Reactivity of PAHs in an Urban Atmosphere," *J. Environ. Sci. Health, A28,* 563–583 (1993).

Lee, W.-M. G., Y.-S. Yuan, and J.-C. Chen, "Polycyclic Aromatic Hydrocarbons Adsorbed in Fly Ash from Municipal Incinerator," *J. Environ. Sci. Health, A28,* 1017–1035 (1993).

Lee-Ruff, E., H. Kazarians-Moghaddam, and M. Katz, "Controlled Oxidations of Benzo[a]pyrene," *Can. J. Chem., 64,* 1297–1303 (1986).

Lee-Ruff, E., and C. Wang, "Photooxidation of 6-Methylbenzo[a]pyrene and Related Compounds," in *Polycyclic Aromatic Compounds,* Proceedings of the Thirteenth International Symposium on Polynuclear Aromatic Hydrocarbons (P. Garrigues and M. Lamotte, Eds.), Bordeaux, France, October 1–4, 1991, pp. 93–101.

Legzdins, A. E., B. E. McCarry, and D. W. Bryant, "Polycyclic Aromatic Compounds in Hamilton Air: Their Mutagenicity, Ambient Concentrations, and Relationships with Atmospheric Pollutants," *Polycyclic Aromat. Compd., 5,* 157–165 (1994).

Leiter, J., M. B. Shimkin, and M. J. Shear, "Production of Subcutaneous Sarcomas in Mice with Tars Extracted from Atmospheric Dusts," *J. Natl. Cancer Inst., 3,* 155–165 (1942).

Leiter, J., and M. J. Shear, "Quantitative Experiments on the Production of Subcutaneous Tumors in Strain A. Mice with Marginal Doses of 3,4-Benzpyrene," *J. Natl. Cancer Inst., 3,* 455–477 (1943).

Levin, W., A. W. Wood, H. Yagi, P. M. Dansette, D. M. Jerina, A. H. Conney, "Carcinogenicity of Benzo[a]pyrene-4,5-, 7,8-, and 9,10-Oxides on Mouse Skin," *Proc. Natl. Acad. Sci. U.S.A., 73,* 243–247 (1976).

Lewis, C., W. Baumgardner, R. Stevens, L. Claxton, and J. Lewtas, "Contribution of Wood Smoke and Motor Vehicle Emissions to Ambient Aerosol Mutagenicity," *Environ. Sci. Technol., 22,* 968–971 (1988).

Lewtas, J., Ed., *Toxicological Effects of Emissions from Diesel Engines,* Elsevier, New York, 1982.

Lewtas, J., "Evaluation of the Mutagenicity and Carcinogenicity of Motor Vehicle Emissions in Short-Term Bioassays," *Environ. Health Perspect., 47,* 141–152 (1983).

Lewtas, J., "Combustion Emissions: Characterization and Comparison of Their Mutagenic and Carcinogenic Activity," in *Carcinogens and Mutagens in the Environment* (H. F. Stich, Ed.), Vol. V, *The Workplace: Sources of Carcinogens,* pp. 59–74, CRC Press, Boca Raton, FL, 1985a.

Lewtas, J., "Development of a Comparative Potency Method for Cancer Risk Assessment of Complex Mixtures Using Short-Term *in Vivo* and *in Vitro* Bioassays," *Toxicol. Indust. Health, 1,* 193–203 (1985b).

Lewtas, J., "Complex Mixtures of Air Pollutants: Characterizing the Cancer Risk of Polycyclic Organic Matter," *Environ. Health Perspect., 100,* 211–218 (1993a).

Lewtas, J., "Airborne Carcinogens," *Pharmacol. Toxicol., 72,* 55–63 (1993b).

Lewtas, J., "Human Exposure to Complex Mixtures of Air Pollutants," *Toxicol. Lett., 72,* 163–169 (1994).

Lewtas, J., S. Goto, K. Williams, J. C. Chuang, B. A. Peterson, and N. K. Wilson, "The Mutagenicity of Indoor Air Particles in a Residential Pilot Field Study: Application and Evaluation of New Methodologies," *Atmos. Environ., 21,* 443–449 (1987).

Lewtas, J., J. Chuang, M. Nishioka, and B. Petersen, "Bioassay-Directed Fractionation of the Organic Extract of SRM 1649 Urban Air Particulate Matter," *Int. J. Environ. Anal. Chem., 39,* 245–256 (1990a).

Lewtas, J., L. King, K. Williams, L. Ball, and D. DeMarini, "Bioassay-Directed Fractionation of 1-Nitropyrene Metabolites: Generation of Mutagrams by Coupling Reverse-Phase HPLC with Microsuspension Mutagenicity Assays," *Mutagenesis, 6,* 481–489 (1990b).

Lewtas, J., and M. G. Nishioka, "Nitroarenes: Their Detection, Mutagenicity, and Occurrence in the Environment," in *Nitroarenes: Occurrence, Metabolism, and Biological Impact* (P. C Howard, S. S. Hecht, and F. A. Beland, Eds.), pp. 61–72, Plenum, New York, 1990.

Lewtas, J., L. D. Claxton, H. S. Rosenkranz, D. Schuetzle, M. Shelby, H. Matsushita, F. E. Würgler, F. K. Zimmermann, G. Löfroth, W. E. May, D. Krewski, T. Matsushima, Y. Ohnishi, H. N. G. Gopalan, R. Sarin, and G. C. Becking, "Design and Implementation of a Collaborative Study of the Mutagenicity of Complex Mixtures in *Salmonella typhimurium,*" *Mutat. Res., 276,* 3–9 (1992).

Li, C. K., and R. M. Kamens, "The Use of Polycyclic Aromatic Hydrocarbons as Source Signatures in Receptor Modeling," *Atmos. Environ., 27A,* 523–532 (1993).

Liang, C., and J. F. Pankow, "Gas/Particle Partitioning of Organic Compounds to Environmental Tobacco Smoke: Partition Coefficient Measurements by Desorption and Comparison to Urban Particulate Material," *Environ. Sci. Technol., 30,* 2800–2805 (1996).

Liang, C., J. F. Pankow, J. R. Odum, and J. H. Seinfeld, "Gas/Particle Partitioning of Semi-Volatile Organic Compounds to Model Inorganic, Organic, and Ambient Smog Aerosols," *Environ. Sci. Technol., 31,* 3086–3092 (1997).

Ligocki, M. P., and J. F. Pankow, "Measurements of the Gas/Particle Distributions of Atmospheric Organic Compounds," *Environ. Sci. Technol., 23,* 75–83 (1989).

Lindskog, A., E. Brorström-Lundén, and Å. Sjödin, "Transformation of Reactive PAH on Particles by Exposure to Oxidized Nitrogen Compounds and Ozone," *Environ. Int.*, 11, 125–130 (1985).

Loening, K., and J. Merritt, "Some Aids for Naming Polycyclic Aromatic Hydrocarbons and Their Heterocyclic Analogs," in *Polynuclear Aromatic Hydrocarbons: Nomenclature Guide*, 1st ed., pp. 1–25, Battelle Press, Columbus, OH, (1990).

Loening, K., J. Merritt, D. Later, and W. Wright, *Polynuclear Aromatic Hydrocarbons: Nomenclature Guide*, 1st ed., Battelle Press, Columbus, OH, 1990.

Löfroth, G., "Mutagenicity Assay of Combustion Emissions," *Chemosphere*, 7, 791–798 (1978).

Löfroth, G., "Comparison of the Mutagenic Activity in Carbon Particulate Matter and in Diesel and Gasoline Engine Exhaust, in *Short-Term Bioassays in the Analysis of Complex Environmental Mixtures, II* (M. D. Waters, S. Sandhu, J. Lewtas-Huisingh, L. Claxton, and S. Nesnow, Eds.), pp. 319–336, Plenum, New York, 1981.

Lowenthal, D. H., B. Zielinska, J. C. Chow, J. G. Watson, M. Gautam, D. H. Ferguson, G. R. Neuroth, and K. D. Stevens, "Characterization of Heavy-Duty Diesel Vehicle Emissions," *Atmos. Environ.*, 28, 731–743 (1994).

Lunde, G., and A. Bjørseth, "Polycyclic Aromatic Hydrocarbons in Long-Range Transported Aerosols," *Nature*, 268, 518–519 (1977).

MacCrehan, W. A., W. E. May, S. D. Yang, and G. A. Benner, Jr., "Determination of Nitro Polynuclear Aromatic Hydrocarbons in Air and Diesel Particulate Matter Using Liquid Chromatography with Electrochemical and Fluorescence Detection," *Anal. Chem.*, 60, 194–199 (1988).

Mackay, D., and W. Y. Shiu, "Aqueous Solubility of Polynuclear Aromatic Hydrocarbons," *J. Chem. Eng. Data*, 22, 399–402 (1977).

Mackay, D., S. Paterson, and W. H. Schroeder, "Model Describing the Rates of Transfer Processes of Organic Chemicals between Atmosphere and Water," *Environ. Sci. Technol.*, 20, 810–816 (1986).

Mackay, D., W. Y. Shiu, and K. C. Ma, *Illustrated Handbook of Physical–Chemical Properties and Environmental Fate for Organic Chemicals*, Vol. II, Polynuclear Aromatic Hydrocarbons, Polychlorinated Dioxins and Dibenzofurans, Lewis Publishers, Chelsea, MI, 1992.

Mackay, D., and D. Callcott, "Partitioning and Physical Chemical Properties of PAHs," in *PAHs and Related Compounds, Chemistry* (A. Neilson, Ed.), pp. 325–346, Springer-Verlag, Berlin, 1998.

Mahanama, K. R. R., L. A. Gundel, and J. M. Daisey, "Selective Fluorescence Detection of Polycyclic Aromatic Hydrocarbons in Environmental Tobacco Smoke and Other Airborne Particles," *Int. J. Environ. Anal. Chem.*, 56, 289–309 (1994).

Maron, D. M., and B. N. Ames, "Revised Methods for the *Salmonella* Mutagenicity Test," *Mutat. Res.*, 113, 173–215 (1983).

Masclet, P., G. Mouvier, and K. Nikolaou, "Relative Decay Index and Sources of Polycyclic Aromatic Hydrocarbons," *Atmos. Environ.*, 20, 439–446 (1986).

Masclet, P., P. Pistikopoulos, S. Beyne, and G. Mouvier, "Long Range Transport and Gas/Particle Distribution of Polycyclic Aromatic Hydrocarbons at a Remote Site in the Mediterranean Sea," *Atmos. Environ.*, 22, 639–650 (1988).

Mast, T. J., D. P. H. Hsieh, and J. N. Seiber, "Mutagenicity and Chemical Characterization of Organic Constituents in Rice Straw Smoke Particulate Matter," *Environ. Sci. Technol.*, 18, 338–348 (1984).

Mastrangelo, G., E. Fadda, and V. Marzia, "Polycyclic Aromatic Hydrocarbons and Cancer in Man," *Environ. Health Perspect.*, 104, 1166–1170 (1996).

Matsumoto, Y., S. Sakai, T. Kato, T. Nakajima, and H. Satoh, "Long-Term Trends of Particulate Mutagenic Activity in the Atmosphere of Sapporo. 1. Determination of Mutagenic Activity by the Conventional Tester Strains TA98 and TA100 during an 18-Year Period (1974–1992)," *Environ. Sci. Technol.*, 32, 2665–2671 (1998).

May, W. E., B. A. Benner, Jr., S. A. Wise, D. Schuetzle, and J. Lewtas, "Standard Reference Materials for Chemical and Biological Studies of Complex Environmental Samples," *Mutat. Res.*, 276, 11–22 (1992).

McCann, J., E. Choi, E. Yamasaki, and B. N. Ames, "Detection of Carcinogens as Mutagens in the *Salmonella*/Microsome Test: Assay of 300 Chemicals," *Proc. Natl. Acad. Sci. U.S.A.*, 72, 5135–5139 (1975).

McCann, J., "*In Vitro* Testing for Cancer-Causing Chemicals," *Hosp. Prac.*, 73–85, September (1983).

McCoy, E. C., and H. S. Rosenkranz, "Activation of Polycyclic Aromatic Hydrocarbons to Mutagens by Singlet Oxygen: An Enhancing Effect of Atmospheric Pollutants," *Cancer Lett.*, 9, 35–42 (1980).

McCoy, E. C., E. J. Rosenkranz, H. S. Rosenkranz, and R. Mermelstein, "Nitrated Fluorene Derivatives Are Potent Frameshift Mutagens," *Mutat. Res.*, 90, 11–20 (1981).

McDow, S. R., and J. J. Huntzicker, "Vapor Adsorption Artifact in the Sampling of Organic Aerosol—Face Velocity Effects," *Atmos. Environ.*, 24A, 2563–2571 (1990).

McDow, S. R., Q. Sun, M. Vartiainen, Y. Hong, Y. Yao, T. Fister, R. Qao, and R. M. Kamens, "Effect of Composition and State of Organic Components on Polycyclic Aromatic Hydrocarbon Decay in Atmospheric Aerosols," *Environ. Sci. Technol.*, 28, 2147–2153 (1994).

McDow, S. R., M. Vartiainen, Q. Sun, Y. Hong, Y. Yao, and R. M. Kamens, "Combustion Aerosol Water Content and Its Effect on Polycyclic Aromatic Hydrocarbon Reactivity," *Atmos. Environ.*, 29, 791–797 (1995).

McDow, S. R., M. Jang, Y. Hong, and R. M. Kamens, "An Approach to Studying the Effect of Organic Composition on Atmospheric Aerosol Photochemistry," *J. Geophys. Res.*, 101, 19593–19600 (1996).

Menzie, C. A., B. B. Potocki, and J. Santodonato, "Exposure to Carcinogenic PAHs in the Environment," *Environ. Sci. Technol.*, 26, 1278–1284 (1992).

Mermelstein, R., D. K. Demosthenes, N. Butler, E. C. McCoy, and H. S. Rosenkranz, "The Extraordinary Mutagenicity of Nitropyrenes in Bacteria," *Mutat. Res.*, 89, 187–196 (1981).

Mermelstein, R., E. C. McCoy, and H. S. Rosenkranz, "The Mutagenic Properties of Nitroarenes: Structure–Activity Relationships," in *Toxicity of Nitroarenes* (D. E. Rickert, Ed.), Chap. 14, Hemisphere Publishing, Washington, DC, 1985.

Miguel, A. H., J. M. Daisey, and J. A. Sousa, "Comparative Study of the Mutagenic and Genotoxic Activity Associated with Inhalable Particulate Matter in Rio de Janeiro Air," *Environ. Mol. Mutagen.*, 15, 36–43 (1990).

Miguel, A. H., and S. K. Friedlander, "Distribution of Benzo[a]pyrene and Coronene with Respect to Particle Size in Pasadena Aerosols in the Submicron Range," *Atmos. Environ.*, 12, 2407–2413 (1978).

Miguel, A. H., T. W. Kirchstetter, R. A. Harley, and S. V. Hering, "On-Road Emissions of Particulate Polycyclic Aromatic Hydrocarbons and Black Carbon from Gasoline and Diesel Vehicles," *Environ. Sci. Technol.*, 32, 450–455 (1998).

Miller, V. R., E. L. Wehry, and G. Mamantov, "Photochemical Transformation of Pyrene Vapor Deposited on Eleven Subfractions of a High-Carbon Coal Stack Ash," *Environ. Toxicol. Chem.*, 9, 975–980 (1990).

Mohr, U., H. Reznik-Schuller, G. Reznik, G. Grimmer, and J. Misfeld, "Investigations on the Carcinogenic Burden by Air Pollution in Man. XIV. Effects of Automobile Exhaust Condensate on the Syrian Golden Hamster Lung," *Zentralbl. Bacteriol. Parasitenkd. Infectionshr. Hyg. Abt. Orig. Reihe. B, 163,* 425–432 (1976).

Møller, M., I. Alfheim, S. Larssen, and A. Mikalsen, "Muagenicity of Airborne Particles in Relation to Traffic and Air Pollution Parameters," *Environ. Sci. Technol., 16,* 221–225 (1982).

Montesano, R., H. Bartsch, H. Vainio, J. Wilbourn, and Y. Yamasaki, Eds., *Long-Term and Short-Term Assays for Carcinogenesis—A Critical Appraisal (IARC Scientific Publications No. 83)*, International Agency for Research on Cancer, Lyon, 1986.

Montizaan, G. K., P. G. N. Kramers, J. A. Janus, and R. Posthumus, "Integrated Criteria Document PAH: Addendum 758474011, National Institute of Public Health and Environmental Protection (RIVM), Bilthoven, The Netherlands (1989).

Morawska, L., N. D. Bofinger, L. Kocis, and A. Nwankwoala, "Submicrometer and Supermicrometer Particles from Diesel Vehicle Emissions," *Environ. Sci. Technol., 32,* 2033–2042 (1998).

Morita, T., N. Asano, T. Awogi, Y. F. Sasaki, S.-I. Sato, H. Shimada, S. Sutou, T. Suzuki, A. Wakata, T. Sofuni, and M. Hayashi, "Evaluation of the Rodent Micronucleus Assay in the Screening of IARC Carcinogens (Groups 1, 2A, and 2B). The Summary Report of the 6th Collaborative Study by CSGMT/JEMS-MMS," *Mutat. Res., 389,* 3–122 (1997).

Morrow, P. E., J. K. Haseman, C. H. Hobbs, K. E. Driscoll, V. Vu, and G. Oberdörster, "The Maximum Tolerated Dose for Inhalation Bioassays: Toxicity vs. Overload," *Fundam. Appl. Toxicol., 29,* 155–167 (1996).

Motykiewicz, G., J. Michalska, J. Szeliga, and B. Cimander, "Mutagenic and Clastogenic Activity of Direct-Acting Components from Air Pollutants of the Silesian Industrial Region," *Mutat. Res., 204,* 289–296 (1988).

Moyano, A., and J.-C. Paniagua, "A Simple Approach for the Evaluation of Local Aromaticities," *J. Org. Chem., 56,* 1858–1866 (1991).

Mumford, J. L., C. T. Helmes, X. Lee, J. Seidenberg, and S. Nesnow, "Mouse Skin Tumorigenicity Studies of Indoor Coal and Wood Combustion Emissions from Homes of Residents in Xuan Wei, China with High Lung Cancer Mortality," *Carcinogenesis, 11,* 397–403 (1990).

Murahashi, T., M. Miyazaki, R. Kakizawa, Y. Yamagishi, M. Kitamura, and K. Hayakawa, "Diurnal Concentrations of 1,3-, 1,6-, 1,8-Dinitropyrenes, 1-Nitropyrene, and Benzo[a]pyrene in Air in Downtown Kanazawa and the Contribution of Diesel-Engine Vehicles," *Jpn. J. Toxicol. Environ. Health, 41,* 328–333 (1995).

Murov, S. L., I. Carmichael, and G. Hug, *Handbook of Photochemistry*, 2nd ed., Dekker, New York, 1993.

Nardini, B., and E. Clonfero, "Mutagens in Urban Air Particulate," *Mutagenesis, 7,* 421–425 (1992).

National Research Council, *Particulate Polycyclic Organic Matter*, Committee on Biologic Effects of Atmospheric Pollutants, National Academy of Sciences, National Academy Press, Washington, DC, 1972.

National Research Council, *Polycyclic Aromatic Hydrocarbons: Evaluation of Sources and Effects*, Committee on Pyrene and Selected Analogues, National Academy of Sciences, National Academy Press, Washington, DC, 1983.

National Research Council, *Complex Mixtures*, National Academy of Sciences, National Academy Press, Washington, DC, 1988.

National Research Council, *Rethinking the Ozone Problem in Urban and Regional Air Pollution*, National Academy of Sciences, National Academy Press, Washington, DC, 1991.

Natusch, D. F. S., and B. A. Tomkins, "Theoretical Considerations of the Adsorption of Polynuclear Aromatic Hydrocarbon Vapor onto Fly Ash in a Coal-Fired Power Plant," in *Carcinogenesis*, Vol. 3, *Polynuclear Aromatic Hydrocarbons* (P. W. Jones and R. I. Freudenthal, Eds.), Raven Press, New York, 1978.

Neilson, A. H., *The Handbook of Environmental Chemistry*, Vol. 3, Part I, *PAHs and Related Compounds*, Springer-Verlag, Berlin, 1998.

Ngabe, B., and T. F. Bidleman, "Occurrence and Vapor Particle Partitioning of Heavy Organic Compounds in Ambient Air in Brazzaville, Congo," *Environ. Pollut., 76,* 147–156 (1992).

Nielsen, P. A., "Mutagenicity Studies on Complex Environmental Mixtures: Selection of Solvent System for Extraction," *Mutat. Res., 276,* 117–123 (1992).

Nielsen, T., "Isolation of Polycyclic Aromatic Hydrocarbons and Nitro-Derivatives in Complex Mixtures by Liquid Chromatography," *Anal. Chem., 55,* 286–290 (1983).

Nielsen, T., "Reactivity of Polycyclic Aromatic Hydrocarbons towards Nitrating Species," *Environ. Sci. Technol., 18,* 157–163 (1984).

Nielsen, T., "The Decay of Benzo[a]pyrene and Cyclopenteno[cd]pyrene in the Atmosphere," *Atmos. Environ., 22,* 2249–2254 (1988).

Nielsen, T., "Traffic PAH and Other Mutagens in Air in Denmark," Miljøprojekt No. 275, Danish Environmental Protection Agency, Ministry of the Environment and Energy, Denmark, 1995.

Nielsen, T., "Traffic Contribution of Polycyclic Aromatic Hydrocarbons in the Center of a Large City," *Atmos. Environ., 30,* 3481–3490 (1996).

Nielsen, T., "Isolation of Polycyclic Aromatic Hydrocarbons and Nitro Derivatives in Complex Samples by Means of High-Performance Liquid Chromatography," *Anal. Chem., 55,* 286–290 (1983).

Nielsen, T., T. Ramdahl, and A. Bjørseth, "The Fate of Airborne Polycyclic Organic Matter," *Environ. Health Perspect., 47,* 103–114 (1983).

Nielsen, T., B. Seitz, and T. Ramdahl, "Occurrence of Nitro-PAH in the Atmosphere in a Rural Area," *Atmos. Environ., 18,* 2159–2165 (1984).

Nielsen, T., P. Clausen, and F. P. Jensen, "Determination of Basic Azaarenes and Polynuclear Aromatic Hydrocarbons in Airborne Particulate Matter by Gas Chromatography," *Anal. Chim. Acta, 187,* 223–231 (1986).

Nielsen, T., and T. Ramdahl, "Determination of 2-Nitrofluoranthene and 2-Nitropyrene in Ambient Particulate Matter: Evidence for Atmospheric Reactions," *Atmos. Environ., 20,* 1507–1509 (1986).

Nielsen, T., H. E. Jørgensen, J. C. Larsen, and M. Poulsen, "City Air Pollution of PAH and Other Mutagens: Occurrence, Sources, and Health Effects," *Sci. Total Environ., 190,* 41–49 (1996).

Nielsen, T., A. Feilberg, M. L. Binderup, and J. Tønnesen, "The Impact of Regulations on Traffic Emissions on Street Levels of PAH, Other PAC, and Mutagens in Air in Copenhagen," Miljøprojekt No. xxx, Danish Environmental Protection Agency, Ministry of the Environment and Energy; Denmark, 1998a.

Nielsen, T., A. Feilberg, and M.-L. Binderup, "The Variation of Street Air Levels of PAH and Other Mutagenic PAC in Relation to the Regulations of Traffic Emissions and the Impact of Atmospheric Processes," *Environ. Sci. Pollut. Res., 6,* 126–130 (1999a).

Nielsen, T., A. Feilberg, M.-L. Binderup, and J. Tønnesen, "Impact of Regulations of Traffic Emissions on PAH Level in Air," Miljøprojekt No. 447, National Agency of Environmental Protection, Copenhagen, Denmark, 1999b.

Nikolaou, K., P. Masclet, and G. Mouvier, "Sources and Chemical Reactivity of Polynuclear Aromatic Hydrocarbons in the Atmosphere—A Critical Review," *Sci. Total Environ., 32,* 103–132 (1984).

Nikula, K. J., M. B. Snipes, E. B. Barr, W. C. Griffith, R. F. Henderson, and J. L. Mauderly, "Comparative Pulmonary Toxicities and Carcinogenicities of Chronically Inhaled Diesel Exhaust and Carbon Black in F344 Rats," *Fundam. Appl. Toxicol.*, 25, 80–94 (1995).

Nisbet, I. C. T., and P. K. LaGoy, "Toxic Equivalency Factors (TEFs) for Polycyclic Aromatic Hydrocarbons (PAHs)," *Regulat. Toxicol. Pharmacol.*, 16, 290–300 (1992).

Nishioka, M. G., C. C. Chuang, and B. A. Petersen, "Development and Quantitative Evaluation of a Compound Class Fractionation Scheme for Bioassay-Directed Characterization of Ambient Air Particulate Matter," *Environ. Int.*, 11, 137–146 (1985).

Nishioka, M. G., C. C. Howard, D. A. Contos, L. M. Ball, and J. Lewtas, "Detection of Hydroxylated Nitro Aromatic and Hydroxylated Nitro Polycyclic Aromatic Compounds in an Ambient Air Particulate Extract Using Bioassay-Directed Fractionation," *Environ. Sci. Technol.*, 22, 908–915 (1988).

Odum, J. R., S. R. McDow, and R. M. Kamens, "Mechanistic and Kinetic Studies of the Photodegradation of Benz[*a*]anthracene in the Presence of Methoxyphenols," *Environ. Sci. Technol.*, 28, 1285–1290 (1994a).

Odum, J. R., J. Yu, and R. M. Kamens, "Modeling the Mass Transfer of Semi-Volatile Organics in Combustion Aerosols," *Environ. Sci. Technol.*, 28, 2278–2285 (1994b).

Office of Environmental Health Hazard Assessment, California Environmental Protection Agency, *Air Toxics Hot Spot Program Risk Assessment Guideline: Part II: Technical Support Document for Describing Available Cancer Potency Factors*, 1998.

Ohgaki, H., C. Negishi, K. Wakabayashi, K. Kusama, S. Sato, and T. Sugimura, "Introduction of Sarcomas in Rats by Subcutaneous Injection of Dinitropyrenes," *Carcinogen*, 5, 583–585 (1984).

Osborne, M. R., and N. T. Crosby, *Benzopyrenes*, Cambridge Univ. Press, Cambridge, UK, 1987.

Pacyna, J. M., and M. Oehme, "Long-Range Transport of Some Organic Compounds to the Norwegian Arctic," *Atmos. Environ.*, 22, 243–257 (1988).

Pagano, P., T. De Zaiacomo, E. Scarcella, S. Bruni, and M. Calamosca, "Mutagenic Activity of Total and Particle-Sized Fractions of Urban Particulate Matter," *Environ. Sci. Technol.*, 30, 3512–3516 (1996).

Pankow, J. F., "Review and Comparative Analysis of the Theories on Partitioning between the Gas and Aerosol Particulate Phases in the Atmosphere," *Atmos. Environ.*, 21, 2275–2283 (1987).

Pankow, J. F., "Common *y*-Intercept and Single Compound Regressions of Gas-Partitioning Data vs 1/T," *Atmos. Environ.*, 25A, 2229–2239 (1991).

Pankow, J. F., and T. F. Bidleman, "Interdependence of the Slopes and Intercepts from Log–Log Correlations of Measured Gas–Particle Partitioning and Vapor Pressure—I. Theory and Analysis of Available Data," *Atmos. Environ.*, 26A, 1071–1080 (1992).

Pankow, J. F., J. M. E. Storey, and H. Yamasaki, "Effects of Relative Humidity on Gas/Particle Partitioning of Semi-Volatile Organic Compounds to Urban Particulate Matter," *Environ. Sci. Technol.*, 27, 2220–2226 (1993).

Pankow, J. F., "An Absorption Model of Gas/Particle Partitioning of Organic Compounds in the Atmosphere," *Atmos. Environ.*, 28, 185–188 (1994a).

Pankow, J. F., "An Absorption Model of the Gas/Aerosol Partitioning Involved in the Formation of Secondary Organic Aerosol," *Atmos. Environ.*, 28, 189–193 (1994b).

Pankow, J. F., L. M. Isabelle, D. A. Buchholz, W. Luo, and B. D. Reeves, "Gas/Particle Partitioning of Polycyclic Aromatic Hydrocarbons and Alkanes to Environmental Tobacco Smoke," *Environ. Sci. Technol.*, 28, 363–365 (1994).

Pankow, J. F., "Further Discussion of the Octanol/Air Partition Coefficient K_{oa} as a Correlating Parameter for Gas/Particle Partitioning Coefficients," *Atmos. Environ.*, 32, 1493–1497 (1998).

Paputa-Peck, M. C., R. S. Marano, D. Schuetzle, T. L. Riley, C. V. Hampton, T. J. Prater, L. M. Skewes, T. E. Jensen, P. H. Ruehle, L. C. Bosch, and W. P. Duncan, "Determination of Nitrated Polynuclear Aromatic Hydrocarbons in Particulate Extracts by Capillary Column Gas Chromatography with Nitrogen Selective Detection," *Anal. Chem.*, 55, 1946–1954 (1983).

Passey, R. D., and J. Carter-Braine, "Experimental Soot Cancer," *J. Pathol. Bacteriol.*, 28, 133–144 (1925).

Pellizzari, E. D., L. W. Little, C. Sparacino, and T. J. Hughes, "Integrating Microbiological and Chemical Testing into the Screening of Air Samples for Potential Mutagenicity," in *Application of Short-Term Bioassays in the Fractionation and Analysis of Complex Environmental Mixtures* (M. D. Waters, S. Nesnow, J. L. Huisingh, S. Sandhu, and L. Claxton, Eds.), pp. 331–351, Plenum, New York, 1979.

Penman, B. W., L. Chen, H. V. Gelboin, F. J. Gonzalez, and C. L. Crespi, "Development of a Human Lymphoblastoid Cell Line Constitutively Expressing Human CYP1A1 cDNA: Substrate Specificity with Model Substrates and Promutagens," *Carcinogenesis*, 15, 1931–1937 (1994).

Peters, J., and B. Seifert, "Losses of Benzo[*a*]pyrene under the Conditions of High-Volume Sampling," *Atmos. Environ.*, 14, 117–119 (1980).

Phillips, D. H., "Fifty Years of Benzo[*a*]pyrene," *Nature*, 303, 468–472 (1983).

Pierce, R., and M. Katz, "Dependence of Polynuclear Aromatic Hydrocarbon Content on Size Distribution of Atmospheric Aerosols," *Environ. Sci. Technol.*, 9, 347–353 (1975).

Pierce, R., and M. Katz, "Chromatographic Isolation and Spectral Analysis of Polycyclic Quinones. Application to Air Pollution Analysis," *Environ. Sci. Technol.*, 10, 45–51 (1976).

Pistikopoulos, P., P. Masclet, and G. Mouvier, "A Receptor Model Adapted to Reactive Species: Polycyclic Aromatic Hydrocarbons; Evaluation of Source Contributions in an Open Urban Site—I. Particle Compounds," *Atmos. Environ.*, 24A, 1189–1197 (1990a).

Pistikopoulos, P., H. M. Wortham, L. Gomes, S. Masclet-Beyne, E. B. Nguyen, P. A. Masclet, and G. Mouvier, "Mechanisms of Formation of Particulate Polycyclic Aromatic Hydrocarbons in Relation to the Particle Size Distribution; Effects on Meso-Scale Transport," *Atmos. Environ.*, 24A, 2573–2584 (1990b).

Pitts, J. N., Jr., "Photochemical and Biological Implications of the Atmospheric Reactions of Amines and Benzo[*a*]pyrene," *Philos. Trans. R. Soc. London*, A290, 551–576 (1979).

Pitts, J. N., Jr., "Formation and Fate of Gaseous and Particulate Mutagens and Carcinogens in Real and Simulated Atmospheres," *Environ. Health Perspect.*, 47, 115–140 (1983).

Pitts, J. N., Jr., "Nitration of Gaseous Polycyclic Aromatic Hydrocarbons in Simulated and Ambient Urban Atmospheres: A Source of Mutagenic Nitroarenes," *Atmos. Environ.*, 21, 2531–2547 (1987).

Pitts, J. N., Jr., "Anthropogenic Ozone, Acids, and Mutagens: Half a Century of Pandora's NO_x," *Res. Chem. Intermed.*, 19, 251–298 (1993a).

Pitts, J. N., Jr., in *Occupational Medicine: State of the Art Reviews* (D. J. Shusterman and J. E. Peterson, Eds.), pp. 621–662, Hanley & Belfus, Inc., Philadelphia, PA, 1993b.

Pitts, J. N., Jr., A. Khan, E. B. Smith, and R. P. Wayne, "Singlet Oxygen in the Environmental Sciences: Singlet Molecular Oxygen and Photochemical Air Pollution," *Environ. Sci. Technol.*, 3, 241–247 (1969).

Pitts, J. N., Jr., D. Grosjean, T. M. Mischke, V. F. Simmon, and D. Poole, "Mutagenic Activity of Airborne Particulate Organic Pollutants," *Toxicol. Lett.*, 1, 65–70 (1977).

Pitts, J. N., Jr., K. A. Van Cauwenberghe, D. Grosjean, J. P. Schmid, D. R. Fitz, W. L. Belser, Jr., G. B. Knudson, and P. M. Hynds, "Atmospheric Reactions of Polycyclic Aromatic Hydrocarbons: Facile Formation of Mutagenic Nitro Derivatives," *Science, 202*, 515–519 (1978).

Pitts, J. N., Jr., D. Grosjean, T. M. Mischke, V. F. Simmon, and D. Poole, "Mutagenic Activity of Airborne Particulate Organic Pollutants," in *Biological Effects of Environmental Pollutants* (S. D. Lee and J. B. Mudd, Eds.), pp. 219–235, Ann Arbor Science Publishers, Ann Arbor, MI, 1979.

Pitts, J. N., Jr., D. M. Lokensgard, P. S. Ripley, K. A. Van Cauwenberghe, L. van Vaeck, S. D. Shaffer, A. J. Thill, and W. L. Belser, Jr., "'Atmospheric' Epoxidation of Benzo[a]pyrene by Ozone: Formation of the Metabolite Benzo[a]pyrene-4,5-oxide," *Science, 210*, 1347–1349 (1980).

Pitts, J. N., Jr., D. M. Lokensgard, W. Harger, T. S. Fisher, V. Mejia, J. Schuler, G. M. Scorziell, and Y. A. Katzenstein, "Mutagens in Diesel Exhaust: Identification and Direct Activities of 6-Nitrobenzo[a]pyrene, 9-Nitroanthracene, 1-Nitropyrene, and 5H-Phenanthro[4,5-bcd]pyran-5-one," *Mutat. Res., 103*, 241–249 (1982a).

Pitts, J. N., Jr., W. Harger, D. M. Lokensgard, D. R. Fitz, G. M. Scorziell, and V. Mejia, "Diurnal Variations in the Mutagenicity of Airborne Particulate Oganic Matter in California's South Coast Air Basin," *Mutat. Res., 104*, 35–41 (1982b).

Pitts, J. N., Jr., A. M. Winer, and J. A. Sweetman, "Particulate and Gas Phase Mutagens in Ambient and Simulated Atmospheres," Final Report, Contract No. A3-049-32, California Air Resources Board, 1984a.

Pitts, J. N., Jr., B. Zielinska, and W. P. Harger, "Isomeric Mononitrobenzo[a]pyrenes: Synthesis, Identification, and Mutagenic Activities," *Mutat. Res., 140*, 81–85 (1984b).

Pitts, J. N., Jr., A. M. Winer, D. M. Lokensgard, and J. A. Sweetman, "Identification of Particulate Mutagens in Southern California's Atmosphere," Final Report, Contract No. A1-155-32, California Air Resources Board, 1984c.

Pitts, J. N., Jr., J. Arey, W. Harger, D. Fitz, H. R. Pauer, and A. M. Winer, "Diurnal Mutagenicity of Airborne Particulate Organic Matter Adjacent to a Heavily Travelled West Los Angeles Freeway," *J. Air Pollut. Control Assoc., 35*, 638–643 (1985a).

Pitts, J. N., Jr., J. A. Sweetman, B. Zielinska, A. M. Winer, and R. Atkinson, "Determination of 2-Nitrofluoranthene and 2-Nitropyrene in Ambient Particulate Organic Matter: Evidence for Atmospheric Reactions," *Atmos. Environ., 19*, 1601–1608 (1985b).

Pitts, J. N., Jr., R. Atkinson, J. A. Sweetman, and B. Zielinska, "The Gas-Phase Reaction of Naphthalene with N_2O_5 to Form Nitronaphthalenes," *Atmos. Environ., 19*, 701–705 (1985c).

Pitts, J. N., Jr., B. Zielinska, J. A. Sweetman, R. Atkinson, and A. M. Winer, "Reactions of Adsorbed Pyrene and Perylene with Gaseous N_2O_5 under Simulated Atmospheric Conditions," *Atmos. Environ., 19*, 911–915 (1985d).

Pitts, J. N., Jr., J. A. Sweetman, B. Zielinska, A. M. Winer, R. Atkinson, and W. P. Harger, "Formation of Nitroarenes from the Reaction of Polycyclic Aromatic Hydrocarbons with Dinitrogen Pentoxide," *Environ. Sci. Technol., 19*, 1115–1121 (1985e).

Pitts, J. N., Jr., H.-R. Paur, B. Zielinska, J. Arey, A. M. Winer, T. Ramdahl, and V. Mejia, "Factors Influencing the Reactivity of Polycyclic Aromatic Hydrocarbons Adsorbed on Filters and Ambient POM with Ozone," *Chemosphere, 15*, 675–685 (1986).

Poster, D. L., L. C. Sander, and S. A. Wise, "Chromatographic Methods for Analysis for the Detection of PAHs in Environmental Samples," in *The Handbook of Environmental Chemistry, PAHs and Related Compounds* (A. H. Neilsen, Ed.), Vol. 3, Part I, pp. 77–135, Springer-Verlag, Berlin, 1998.

Pott, F., and G. Oberdörster, "Intake and Distribution of PAH," in *Environmental Carcinogens: Polycyclic Aromatic Hydrocarbons* (G. Grimmer, Ed.), pp. 129–156, CRC Press, Boca Raton, FL, 1983.

Pott, P., *Chirurgical Observations Relative to the Cataract, the Polypus of the Nose, the Cancer of the Scrotum, the Different Kinds of Ruptures, and the Mortification of the Toes and Feet* (L. Hawes, W. Clarke, and R. Collins, Eds.), London, 1775.

Pupp, C., R. C. Lao, J. J. Murray, and R. F. Pottie, "Equilibrium Vapor Concentrations of Some Polycyclic Aromatic Hydrocarbons, As_4O_6 and SeO_2, and the Collection Efficiencies of These Air Pollutants," *Atmos. Environ., 8*, 915–925 (1974).

Pyysalo, H., J. Tuominen, K. Wickström, E. Skyttä, L. Tikkanen, S. Salomaa, M. Sorsa, T. Nurmela, T. Mattila, and V. Pohjola, "Polycyclic Organic Material (POM) in Urban Air. Fractionation, Chemical Analysis, and Genotoxicity of Particulate and Vapor Phases in an Industrial Town in Finland," *Atmos. Environ, 21*, 1167–1180 (1987).

Ramdahl, T., I. Alfheim, S. Rustad, and T. Olsen, "Chemical and Biological Characterization of Emissions from Small Residential Stoves Burning Wood and Charcoal," *Chemosphere, 11*, 601–611 (1982a).

Ramdahl, T., G. Becher, and A. Bjørseth, "Nitrated Polycyclic Aromatic Hydrocarbons in Urban Air Particles," *Environ. Sci. Technol., 16*, 861–865 (1982b).

Ramdahl, T., "Retene—A Molecular Marker of Wood Combustion in Ambient Air," *Nature, 306*, 580–582 (1983a).

Ramdahl, T., "Polycyclic Aromatic Ketones in Environmental Samples," *Environ. Sci. Technol., 17*, 666–670 (1983b).

Ramdahl, T., J. Schjoldager, L. A. Currie, J. E. Hanssen, M. Møller, G. A. Klouda, and I. Alfheim, "Ambient Impact of Residential Wood Combustion in Elverum, Norway," *Sci. Total Environ., 36*, 81–90 (1984a).

Ramdahl, T., A. Bjørseth, D. Lokensgard, and J. N. Pitts, Jr., "Nitration of Polycyclic Aromatic Hydrocarbons Adsorbed to Different Carriers in a Fluidized Bed Reactor," *Chemosphere, 13*, 527–534 (1984b).

Ramdahl, T., J. A. Sweetman, B. Zielinska, W. P. Harger, A. M. Winer, and R. Atkinson, "Determination of Nitrofluoranthenes and Nitropyrenes in Ambient Air and Their Contribution to Direct Mutagenicity," presented at the Tenth Anniversary of the International Symposium on Polynuclear Aromatic Hydrocarbons, Columbus, OH, October 21–23, 1985b.

Ramdahl, T., J. A. Sweetman, B. Zielinska, R. Atkinson, A. M. Winer, and J. N. Pitts, Jr., "Analysis of Mononitro Isomers of Fluoranthene and Pyrene by High Resolution Capillary Gas Chromatography/Mass Spectrometry," *J. High Resolut. Chromatogr. Chromatogr. Commun., 8*, 849–852 (1985).

Ramdahl, T., B. Zielinska, J. Arey, R. Atkinson, A. M. Winer, and J. N. Pitts, Jr., "Ubiquitous Occurrence of 2-Nitrofluoranthene and 2-Nitropyrene in Air," *Nature, 321*, 425–427 (1986).

Reyes, C., M. E. Sigman, R. Arce, J. T. Barbas, and R. Dabestani, "Photochemistry of Acenaphthene at a Silica Gel/Air Interface," *J. Photochem. Photobiol. A: Chem., 112*, 277–283 (1998).

Robbat, A., Jr., N. P. Corso, P. J. Doherty, and M. H. Wolf, "Gas Chromatographic Chemiluminescent Detection and Evaluation of Predictive Models for Identifying Nitrated Polycyclic Aromatic Hydrocarbons in a Diesel Fuel Particulate Extract," *Anal. Chem., 58*, 2078–2084 (1986).

Rogge, W. F., L. M. Hildemann, M. A. Mazurek, G. R. Cass, and B. R. T. Simoneit, "Sources of Fine Organic Aerosol. 2. Noncatalyst and Catalyst-Equipped Automobiles and Heavy-Duty Diesel Trucks," *Environ. Sci. Technol., 27*, 636–651 (1993a).

Rosenkranz, H. S., "Predicting the Carcinogenic Potential of Environmental Nitropyrenes," *Environ. Mol. Mutagen.*, 10, 149–156 (1987).

Rosenkranz, H. S., "Mutagenic Nitroarenes, Diesel Emissions, Particulate-Induced Mutations and Cancer: An Essay on Cancer-Causation by a Moving Target," *Mutat. Res.*, 367, 65–72 (1996).

Rosenkranz, H. S., and R. Mermelstein, "Mutagenicity and Genotoxicity of Nitroarenes: All Nitro-Containing Chemicals Were Not Created Equal," *Mutat. Res.*, 114, 217–267 (1983).

Rosenkranz, H. S., and R. Mermelstein, "The Genotoxicity, Metabolism, and Carcinogenicity of Nitrated Polycyclic Aromatic Hydrocarbons," *J. Environ. Sci. Health, C3*, 221–272 (1985a).

Rosenkranz, H. S., and R. Mermelstein, "The Mutagenic and Carcinogenic Properties of Nitrated Polycyclic Aromatic Hydrocarbons," in *Nitrated Polycyclic Aromatic Hydrocarbons* (C. White, Ed.), pp. 267–297, Hüthig, Heidelberg, 1985b.

Ruehle, P. H., L. C. Bosch, and W. P. Duncan, "Synthesis of Nitrated Polycyclic Aromatic Hydrocarbons," in *Nitrated Polycyclic Aromatic Hydrocarbons* (C. M. White, Ed.), pp. 169–235, Hüthig, Heidelberg, 1985.

Salmeen, I. T., A. M. Pero, R. Zator, D. Schuetzle, and T. L. Riley, "Ames Assay Chromatograms and the Identification of Mutagens in Diesel Particle Extracts," *Environ. Sci. Technol.*, 18, 375–382 (1984).

Salmeen, I. T., R. A. Gorse, Jr., and W. R. Pierson, "Ames Assay Chromatograms of Extracts of Diesel Exhaust Particles from Heavy-Duty Trucks on the Road and from Passenger Cars on a Dynamometer," *Environ. Sci. Technol.*, 19, 270–273 (1985).

Sangaiah, R., and A. Gold, "A Short and Convenient Synthesis of Cyclopenta[*cd*]pyrene and Its Oxygenated Derivatives," in *Polynuclear Aromatic Hydrocarbons: Mechanisms, Methods and Metabolism* (M. Cooke and A. J. Dennis, Eds.), Eighth International Symposium on Polynuclear Aromatic Hydrocarbons, Battelle Press, Columbus, OH, 1985.

Sasaki, J., J. Arey, and W. P. Harger, "Formation of Mutagens from the Photooxidations of 2–4-Ring PAH," *Environ. Sci. Technol.*, 29, 1324–1335 (1995).

Sasaki, J., S. M. Aschmann, E. S. C. Kwok, R. Atkinson, and J. Arey, "Products of the Gas-Phase OH and NO_3 Radical-Initiated Reactions of Naphthalene," *Environ. Sci. Technol.*, 31, 3173–3179 (1997a).

Sasaki, J. C., J. Arey, D. A. Eastmond, K. K. Parks, and A. J. Grosovsky, "Genotoxicity Induced in Human Lymphoblasts by Atmospheric Reaction Products of Naphthalene and Phenanthrene," *Mutat. Res.*, 393, 23–35 (1997b).

Savard, S., R. Otson, G. R. Douglas, "Mutagenicity and Chemical Analysis of Sequential Organic Extracts of Airborne Particulates," *Mutat. Res.*, 276, 101–115 (1992).

Sawicki, E., "Airborne Carcinogens and Allied Compounds," *Arch. Environ. Health*, 14, 46–53 (1967).

Sawicki, E., S. P. McPherson, T. W. Stanley, J. Meeker, and W. C. Elbert, "Quantitative Composition of the Urban Atmosphere in Terms of Polynuclear Aza-Heterocyclic Compounds and Aliphatic and Polynuclear Aromatic Hydrocarbons," *Int. J. Air Pollut.*, 9, 515–524 (1965).

Scarpato, R., F. Di Marino, A. Strano, A. Curti, R. Campagna, N. Loprieno, I. Barrai, and R. Barale, "Two Years' Air Mutagenesis Monitoring in a Northwestern Rural Area of Italy with an Industrial Plant," *Mutat. Res.*, 319, 293–301 (1993).

Schauer, J. J., F. R. Wolfgang, L. M. Hildemann, M. A. Mazurek, and G. R. Cass, "Source Apportionment of Airborne Particulate Matter Using Organic Compounds as Tracers," *Atmos. Environ.*, 30, 3837–3855 (1996).

Schauer, J. J., M. J. Kleeman, G. R. Cass, and B. R. T. Simoneit, "Measurement of Emissions from Air Pollution Sources. 2. C_1 through C_{30} Organic Compounds from Medium Duty Diesel Trucks," *Environ. Sci. Technol.*, 33, 1578–1587 (1999).

Schmähl, D., R. P. Deutsch-Wenzel, H. Brune, P. Schneider, U. Mohr, M. Habs, F. Pott, and D. Steinhoff, "Biological Activity," in *Environmental Carcinogens: Polycyclic Aromatic Hydrocarbons* (G. Grimmer, Ed.), pp. 157–220, CRC Press, Boca Raton, FL, 1983.

Schmid, E. R., G. Bachlechner, K. Varmuza, and H. Klus, "Determination of Polycyclic Aromatic Hydrocarbons, Polycyclic Aromatic Sulfur, and Oxygen Heterocycles in Cigarette Smoke Condensate," *Fresenius' Z. Anal. Chem.*, 322, 213–219 (1985).

Schneider, E., P. Krenmayr, and K. Varmuza, "A Routine Method for the Analysis of Mononitro-PAH in Immission and Emission Samples," *Monatsh. Chem.*, 121, 393–401 (1990).

Schuetzle, D., D. Cronn, A L. Crittenden, and R. J. Charlson, "Molecular Composition of Secondary Aerosol and Its Possible Origin," *Environ. Sci. Technol.*, 9, 838–845 (1975).

Schuetzle, D., F. S.-C. Lee, T. J. Prater, and S. B. Tejada, "The Identification of Polynuclear Aromatic Hydrocarbon (PAH) Derivatives in Mutagenic Fractions of Diesel Particulate Extracts," *Int. J. Environ. Anal. Chem.*, 9, 93–144 (1981).

Schuetzle, D., T. L. Riley, T. J. Prater, T. M. Harvey, and D. F. Hunt, "Analysis of Nitrated Polycyclic Aromatic Hydrocarbons in Diesel Particulate," *Anal. Chem.*, 54, 265–271 (1982).

Schuetzle, D., "Sampling of Vehicle Emissions for Chemical Analysis and Biological Testing," *Environ. Health Perspect.*, 47, 65–80 (1983).

Schuetzle, D., and J. M. Perez, "Factors Influencing the Emissions of Nitrated-Polynuclear Aromatic Hydrocarbons (Nitro-PAH) from Diesel Engines," *J. Air Pollut. Control Assoc.*, 33, 751–755 (1983).

Schuetzle, D., and T. E. Jensen, "Analysis of Nitrated Polycyclic Aromatic Hydrocarbons (Nitro-PAH) by Mass Spectrometry," in *Nitrated Polycyclic Aromatic Hydrocarbons* (C. White, Ed.), pp. 121–167, A. Hüthig Verlag, Heidelberg, 1985.

Schuetzle, D., T. E. Jensen, and J. C. Ball, "Polar Polynuclear Aromatic Hydrocarbon Derivatives in Extracts of Particulates: Biological Characterization and Techniques for Chemical Analysis," *Environ. Int.*, 11, 169–181 (1985).

Schuetzle, D., and J. A. Frazier, "Factors Influencing the Emission of Vapor and Particulate Phase Components from Diesel Engines," in *Carcinogenic and Mutagenic Effects of Diesel Engine Exhaust* (N. Ishinishi, A. Koizumi, R. O. McClellan, and W. Stöber, Eds.), pp. 41–63, Elsevier, Amsterdam/New York, 1986.

Schuetzle, D., and J. Lewtas, "Bioassay-Directed Chemical Analysis in Environmental Research," *Anal. Chem.*, 58, 1060A–1075A (1986).

Schuetzle, D., and J. M. Daisey, "Identification of Genotoxic Agents in Complex Mixtures of Air Pollutants," in *Genetic Toxicology of Complex Mixtures* (M. D. Waters, F. B. Daniel, J. Lewtas, M. M. Moore, and S. Nesnow, Eds.), pp. 11–22, Plenum, New York, 1990.

Schwarzenbach, R. P., P. M. Gschwend, and D. M. Imboden, *Environmental Organic Chemistry*, Wiley, New York, 1993.

Seed, J. L., K. G. Specht, T. A. Dahl, and W. R. Midden, "Singlet Oxygen Induced Mutagenesis of Benzo[*a*]pyrene Derivatives," *Photochem. Photobiol.*, 50, 625–632 (1989).

Sera, N., M. Kai, K. Horikawa, K. Fukuhara, N. Miyata, and H. Tokiwa, "Detection of 3,6-Dinitrobenzo[*a*]pyrene in Airborne Particulates," *Mutat. Res.*, 263, 27–32 (1991).

Sera, N., K. Fukuhara, N. Miyata, K. Horikawa, and H. Tokiwa, "Mutagenicity of Nitro-azabenzo[*a*]pyrene and Its Related Compounds," *Mutat. Res.*, 280, 81–85 (1992).

Sera, N., K. Fukuhara, N. Miata, and H. Tokiwa, "Detection of Nitro-azabenzo[a]pyrene Derivatives in the Semi-volatile Phase Originating from Airborne Particulate Matter, Diesel and Gasoline Vehicles, *Mutagenesis, 9,* 47–52 (1994).

Shelby, M. D., "The Genetic Toxicity of Human Carcinogens and Its Implications," *Mutat. Res., 204,* 3–15 (1988).

Shelby, M. D., G. L. Erexson, G. J. Hook, and R. R. Tice, "Evaluation of a Three-Exposure Mouse Bone Marrow Micronucleus Protocol: Results with 49 Chemicals," *Environ. Mol. Mutagen., 21,* 160–179 (1993).

Shepson, P. B., T. E. Kleindienst, E. O. Edney, G. R. Namie, J. H. Pittman, L. T. Cupitt, and L. D. Claxton, "The Mutagenic Activity of Irradiated Toluene/NO_x/H_2O/Air Mixtures," *Environ. Sci. Technol., 19,* 249–255 (1985).

Shepson, P. B., T. E. Kleindienst, E. O. Edney, C. M. Nero, L. T. Cupitt, and L. D. Claxton, "Acetaldehyde: The Mutagenic Activity of Its Photooxidation Products," *Environ. Sci. Technol., 20,* 1008–1012 (1986).

Shepson, P. B., T. E. Kleindienst, C. M. Nero, D. N. Hodges, L. T. Cupitt, and L. D. Claxton, "Allyl Chloride: The Mutagenic Activity of Its Photooxidation Products," *Environ. Sci. Technol., 21,* 568–573 (1987).

Sheu, H.-L., W.-J. Lee, J.-H. Tsai, Y.-C. Fan, C.-C. Su, and H.-R. Chao, "Particle Size Distribution of Polycyclic Aromatic Hydrocarbons in the Ambient Air of a Traffic Intersection," *J. Environ. Sci. Health, A31,* 1293–1316 (1996).

Sicre, M. A., J. C. Marty, A. Saliot, X. Aparicio, J. Grimalt, and J. Albaiges, "Aliphatic and Aromatic Hydrocarbons in Different Sized Aerosols over the Mediterranean Sea: Occurrence and Origin," *Atmos. Environ., 21,* 2247–2259 (1987).

Sigman, M. E., J. T. Barbas, E. A. Chevis, and R. Dabestani, "Spectroscopy and Photochemistry of 1-Methoxynaphthalene on SiO_2," *New J. Chem., 20,* 243–248 (1996).

Sigman, M. E., R. Arce, C. Rayes, J. T. Barbas, and R. Dabestani, "Environmental Organic Photochemistry: PAH Photolysis at Solid–Air Interfaces," *Div. Environ. Chem., Prepr. Ext. Abstr., 37,* 313–315 (1997).

Simcik, M. F., H. Zhang, S. J. Eisenreich, and T. P. Franz, "Urban Contamination of the Chicago/Coastal Lake Michigan Atmosphere by PCBs and PAHs during AEOLOS," *Environ. Sci. Technol., 31,* 2141–2147 (1997).

Simcik, M. F., T. P. Franz, H. Zhang, and S. J. Eisenreich, "Gas–Particle Partitioning of PCBs and PAHs in the Chicago Urban and Adjacent Coastal Atmosphere: States of Equilibrium," *Environ. Sci. Technol., 32,* 251–257 (1998).

Simoneit, B. R. T., "Biomarker PAHs in the Environment," in *PAHs and Related Compounds, Chemistry* (A. H. Neilson, Ed.), pp. 175–221, Springer-Verlag, Berlin, 1998.

Skopek, T. R., H. L. Liber, D. A. Kaden, and W. G. Thilly, "Relative Sensitivities of Forward and Reverse Mutation Assays in *Salmonella typhimurium*," *Proc. Natl. Acad. Sci. U.S.A., 75,* 4465–4469 (1978a).

Skopek, T. R., H. L. Liber, J. J. Krolewski, and W. G. Thilly, "Quantitative Forward Mutation Assay in *Salmonella typhimurium* Using 8-Azaguanine Resistance as a Genetic Marker," *Proc. Natl. Acad. Sci. U.S.A., 75,* 410–414 (1978b).

Skopek, T. R., and W. G. Thilly, "Rate of Induced Forward Mutations at 3 Genetic Loci in *Salmonella typhimurium, Mutat. Res., 108,* 45–52 (1983).

Smith, D. J. T., and R. M. Harrison, "Concentrations, Trends, and Vehicle Source Profile of Polynuclear Aromatic Hydrocarbons in the U.K. Atmosphere," *Atmos. Environ., 30,* 2513–2525 (1996).

Smith, D. J. T., R. M. Harrison, L. Luhana, C. A. Pio, L. M. Castro, M. N. Tariq, S. Hayat, and T. Quraishi, "Concentrations of Particulate Airborne Polycyclic Aromatic Hydrocarbons and Metals Collected in Lahore, Pakistan," *Atmos. Environ., 30,* 4031–4040 (1996).

Solomon, P. A., J. L. Moyers, and R. A. Fletcher, "High-Volume Dichotomous Virtual Impactor for the Fractionation and Collection of Particles According to Aerodynamic Size," *Aerosol Sci. Technol., 2,* 455–464 (1983).

Sonnefeld, W. J., W. H. Zoller, and W. E. May, "Dynamic Coupled Column Liquid Chromatographic Determination of Ambient Temperature Vapor Pressures of Polynuclear Aromatic Hydrocarbons," *Anal. Chem., 55,* 275–280 (1983).

Stärk, G., J. Stauff, H. G. Miltenburger, and I. Stumm-Fischer, "Photodecomposition of 1-Nitropyrene and Other Direct-Acting Mutagens Extracted from Diesel-Exhaust Particulates," *Mutat. Res., 155,* 27–33 (1985).

Storey, J. M. E., W. Luo, L. M. Isabelle, and J. F. Pankow, "Gas/Solid Partitioning of Semi-Volatile Organic Compounds to Model Atmospheric Solid Surfaces as a Function of Relative Humidity. 1. Clean Quartz," *Environ. Sci. Technol., 29,* 2420–2428 (1995).

Strommen, M. R., and R. M. Kamens, "Development and Application of a Dual-Impedance Radial Diffusion Model to Simulate the Partitioning of Semi-Volatile Organic Compounds in Combustion Aerosols," *Environ. Sci. Technol., 31,* 2983–2990 (1997).

Sweetman, J. A., F. W. Karasek, and D. Schuetzle, "Decomposition of Nitropyrene during Gas Chromatographic–Mass Spectrometric Analysis of Air Particulate and Fly-Ash Samples," *J. Chromatogr., 247,* 245–254 (1982).

Sweetman, J. A., B. Zielinska, R. Atkinson, T. Ramdahl, A. M. Winer, and J. N. Pitts, Jr., "A Possible Formation Pathway for the 2-Nitrofluoranthene Observed in Ambient Particulate Organic Matter," *Atmos. Environ., 20,* 235–238 (1986).

Takada, H., T. Onda, and N. Ogura, "Determination of Polycyclic Aromatic Hydrocarbons in Urban Street Dusts and Their Source Materials by Capillary Gas Chromatography," *Environ. Sci. Technol., 24,* 1179–1186 (1990).

Talcott, R., and E. Wei, "Airborne Mutagens Bioassayed in *Salmonella typhimurium*," *J. Natl. Cancer Inst., 58,* 449–451 (1977).

Talcott, R., and W. Harger, "Airborne Mutagens Extracted from Particles of Respirable Size," *Mutat. Res., 79,* 177–180 (1980).

Tebbens, B. D., J. F. Thomas, and M. Mukai, "Fate of Arenes Incorporated with Airborne Soot," *Am. Ind. Hyg. Assoc. J., 27,* 415–422 (1966).

Tebbens, B. D., M. Mukai, and J. F. Thomas, "Fate of Arenes Incorporated with Airborne Soot: Effects of Irradiation," *Am. Ind. Hyg. Assoc. J., 32,* 365–372 (1971).

Teranishi, K., K. Hamada, and H. Watanabe, "Mutagenicity in *Salmonella typhimurium* Mutants of the Benzene-Soluble Organic Matter Derived from Airborne Particulate Matter and Its Five Fractions," *Mutat. Res., 56,* 273–280 (1978).

Thakkar, S., and M. Manes, "Adsorptive Displacement Analysis of Many-Component Priority Pollutants on Activated Carbon. 2. Extension of Low Parts per Million (Based on Carbon)," *Environ. Sci. Technol., 22,* 470–472 (1988).

Thilly, W. G., J. Longwell, and B. Andon, "General Approach to the Biological Analysis of Complex Mixtures," *Environ. Health Perspect., 48,* 129–136 (1983).

Thomas, J. F., M. Mukai, and B. D. Tebbens, "Fate of Airborne Benzo[a]pyrene," *Environ. Sci. Technol., 2,* 33–39 (1968).

Thrane, K. E., and A. Mikalsen, "High-Volume Sampling of Airborne Polycyclic Aromatic Hydrocarbons Using Glass Fiber Filters and Polyurethane Foam," *Atmos. Environ., 15,* 909–918 (1981).

Tokiwa, H., H. Takeyoshi, K. Morita, K. Takahashi, N. Saruta, and Y. Ohnishi, "Detection of Mutagenic Activity in Urban Air Pollutants," *Mutat. Res., 38,* 351 (1976).

Tokiwa, H., K. Morita, H. Takeyoshi, K. Takahashi, and Y. Ohnishi, "Detection of Mutagenic Activity in Particulate Air Pollutants," *Mutat. Res., 48,* 237–248 (1977).

Tokiwa, H., S. Kitamori, K. Takahashi, and Y. Ohnishi, "Mutagenic and Chemical Assay of Extracts of Airborne Particulates," *Mutat. Res., 77,* 99–108 (1980).

Tokiwa, H., R. Nakagawa, K. Morita, and Y. Ohnishi, "Mutagenicity of Nitro Derivatives Induced by Exposure of Aromatic Compounds to Nitrogen Dioxide," *Mutat. Res., 85,* 195–205 (1981).

Tokiwa, H., S. Kitamori, R. Nakagawa, K. Horikawa, and L. Matamala, "Demonstration of a Powerful Mutagenic Dinitropyrene in Airborne Particulate Matter," *Mutat. Res., 121,* 107–116 (1983).

Tokiwa, H., and Y. Ohnishi, "Mutagenicity and Carcinogenicity of Nitroarenes and Their Sources in the Environment," *CRC Crit. Rev. Toxicol., 17,* 23–60 (1986).

Tokiwa, H., T. Otofuji, R. Nakagawa, K. Horikawa, T. Maeda, N. Sano, I. Izumi, and H. Otsuka, "Dinitro Derivatives of Pyrene and Fluoranthene in Diesel Emission Particulates and Their Tumorigenicity in Mice and Rats," in *Carcinogenic and Mutagenic Effects of Diesel Engine Exhaust* (N. Ishinishi, A. Koizumi, R. O. McClellan, and W. Stöber, Eds.), pp. 253–270, Elsevier, Amsterdam/New York, 1986.

Tokiwa, H., R. Nakagawa, K. Horikawa, and A. Ohkubo, "The Nature of the Mutagenicity and Carcinogenicity of Nitrated, Aromatic Compounds in the Environment," *Environ. Health Perspect., 73,* 191–199 (1987).

Tokiwa, H., N. Sera, K. Horikawa, Y. Nakanishi, and N. Shigematu, "The Presence of Mutagens/Carcinogens in the Excised Lung and Analysis of Lung Cancer Induction," *Carcinogenesis, 14,* 1933–1938 (1993).

Tokiwa, H., Y. Nakanishi, N. Sera, N. Hara, and S. Inuzuka, "Analysis of Environmental Carcinogens Associated with the Incidence of Lung Cancer," *Toxicol. Lett., 99,* 33–41 (1998).

Tong, H. Y., and F. W. Karasek, "Quantitation of Polycyclic Aromatic Hydrocarbons in Diesel Exhaust Particulate Matter by High-Performance Liquid Chromatography Fractionation and High-Resolution Gas Chromatography," *Anal. Chem., 56,* 2129–2134 (1984).

Tuazon, E. C., R. Atkinson, A. M. Winer, and J. N. Pitts, Jr., "A Study of the Atmospheric Reactions of 1,3-Dichloropropene and Other Selected Organochlorine Compounds," *Arch. Environ. Contam. Toxicol., 13,* 691–700 (1984).

Tuominen, J., S. Salomaa, H. Pyysalo, E. Skyttä, L. Tikkanen, T. Nurmela, M. Sorsa, V. Pohjola, M. Sauri, and K. Himberg, "Polynuclear Aromatic Compounds and Genotoxicity in Particulate and Vapor Phases of Ambient Air: Effect of Traffic, Season, and Meteorological Conditions," *Environ. Sci. Technol., 22,* 1228–1234 (1988).

Turpin, B. J., S.-P. Liu, K. S. Podolske, M. S. P. Gomes, S. J. Elsenreich, and P. H. McMurry, "Design and Evaluation of a Novel Diffusion Separator for Measuring Gas/Particle Distributions of Semi-Volatile Organic Compounds," *Environ. Sci. Technol., 27,* 2441–2449 (1993).

Turpin, B. J., J. J. Huntzicker, and S. V. Hering, "Investigation of Organic Aerosol Sampling Artifacts in the Los Angeles Basin," *Atmos. Environ., 28,* 3061–3071 (1994).

Turro, N. J., *Modern Molecular Photochemistry,* Benjamin/Cummings, Menlo Park, CA, 1978.

U.S. Department of Health and Human Services, Public Health Services, National Toxicology Program, *Report on Carcinogens,* 8th ed., 1998, Summary.

U.S. Environmental Protection Agency, "Sampling and Analysis Procedures for Screening of Industrial Effluents for Priority Pollutants. Method 610, Polynuclear Aromatic Hydrocarbons," Environmental Monitoring and Support Laboratories, Cincinnati, OH, 1977.

U.S. Environmental Protection Agency, *Carcinogen Classifications,* National Center for Environmental Assessment, Office of Research and Development, Washington, DC, EPA-600/R-93-089, 1986.

U.S. Environmental Protection Agency, *Second Supplement to Compendium of Methods for the Determination of Toxic Organic Compounds in Ambient Air,* Atmospheric Research and Exposure Assessment Laboratory, Research Triangle Park, NC, EPA-600/4-89-018, pp. TO-13 to TO-97, 1988.

U.S. Environmental Protection Agency, *Provisional Guidance for Quantitative Risk Assessment of Polycyclic Aromatic Hydrocarbons,* Office of Research and Development, Washington, DC, EPA-600/R-93-089, July 1993.

U.S. Environmental Protection Agency, *Locating and Estimating Air Emissions from Sources of Polycyclic Organic Matter,* Office of Air Quality Planning and Standards, Research Triangle Park, NC, EPA-454/R-98-014, 1998.

Valerio, F., P. Bottino, D. Ugolini, M. R. Cimberle, G. Tozzi, and A. Frigerio, "Chemical and Photochemical Degradation of Polycyclic Aromatic Hydrocarbons in the Atmosphere," *Sci. Total Environ., 40,* 169–188 (1984).

Van Cauwenberghe, K., L. Van Vaeck, and J. N. Pitts, Jr., "Chemical Transformations of Organic Pollutants during Aerosol Sampling," *Adv. Mass Spectrom., 8B,* 1499–1507 (1979).

Van Cauwenberghe, K. A., and L. Van Vaeck, "Toxicological Implications of the Organic Fraction of Aerosols: A Chemist's View," *Mutat. Res., 116,* 1–20 (1983).

Van Cauwenberghe, K. A., "Atmospheric Reactions of PAH," in *Handbook of Polycyclic Aromatic Hydrocarbons* (A. Bjørseth and T. Ramdahl, Eds.), pp. 351–384, Dekker, New York, 1985.

Van den Braken-van Leersum, A. M., C. Tintel, M. van't Zelfde, J. Cornelisse, and J. Lugtenburg, "Spectroscopic and Photochemical Properties of Mononitropyrenes," *Recl. Trav. Chim. Pays-Bas, 106,* 120–128 (1987).

Van Houdt, J. J., G. M. Alink, and J. S. M. Boleij, "Mutagenicity of Airborne Particles Related to Meteorological and Air Pollution Parameters," *Sci. Total Environ., 61,* 23–36 (1987).

Van Houdt, J. J., "Mutagenic Activity of Airborne Particulate Matter in Indoor and Outdoor Environments," *Atmos. Environ., 24B,* 207–220 (1990).

Vander Wal, R. L., K. A. Jensen, and M. Y. Choi, "Simultaneous Laser-Induced Emission of Soot and Polycyclic Aromatic Hydrocarbons within a Gas-Jet Diffusion Flame," *Combust. Flame, 109,* 399–414 (1997).

Van Vaeck, L., and K. Van Cauwenberghe, "Cascade Impactor Measurements of the Size Distribution of the Major Classes of Organic Pollutants in Atmospheric Particulate Matter," *Atmos. Environ., 12,* 2229–2239 (1978).

Van Vaeck, L., and K. Van Cauwenberghe, "Conversion of Polycyclic Aromatic Hydrocarbons on Diesel Particulate Matter upon Exposure to PPM Levels of Ozone," *Atmos. Environ., 18,* 323–328 (1984).

Van Vaeck, L., K. Van Cauwenberghe, and J. Janssens, "The Gas–Particle Distribution of Organic Aerosol Constituents: Measurement of the Volatilization Artifact in Hi-Vol Cascade Impactor Sampling," *Atmos. Environ., 18,* 417–430 (1984).

Van Vaeck, L., and K. A. Van Cauwenberghe, "Characteristic Parameters of Particle Size Distributions of Primary Organic Constituents of Ambient Aerosols," *Environ. Sci. Technol., 19,* 707–716 (1985).

Venkataraman, C., J. M. Lyons, and S. K. Friedlander, "Size Distributions of Polycyclic Aromatic Hydrocarbons and Elemental Carbon. 1. Sampling, Measurement Methods, and Source Characterization," *Environ. Sci. Technol.*, 28, 555–562 (1994a).

Venkataraman, C., and S. K. Friedlander, "Size Distributions of Polycyclic Aromatic Hydrocarbons and Elemental Carbon. 2. Ambient Measurements and Effects of Atmospheric Processes," *Environ. Sci. Technol.*, 28, 563–572 (1994b).

Venkataraman, C., and S. K. Friedlander, "Source Resolution of Fine Particulate Polycyclic Aromatic Hydrocarbons Using a Receptor Model Modified for Reactivity," *J. Air Waste Manage. Assoc.*, 44, 1103–1108 (1994c).

Villalobos-Pietrini, R., S. Blanco, and S. Gomez-Arroyo, "Mutagenicity Assessment of Airborne Particles in Mexico City," *Atmos. Environ.*, 29, 517–524 (1995).

Viras, L. G., K. Athanasiou, and P. A. Siskos, "Determination of Mutagenic Activity of Airborne Particulates and of the Benzo[a]pyrene Concentrations in Athens Atmosphere," *Atmos. Environ.*, 24B, 267–274 (1990).

Waller, R. E., "The Benzopyrene Content of Town Air," *Br. J. Cancer*, 6, 8–21 (1952).

Wang, C. Y., M.-S. Lee, C. M. King, and P. O. Warner, "Evidence for Nitroaromatics as Direct-Acting Mutagens of Airborne Particles," *Chemosphere*, 9, 83–87 (1980).

Wang, Y. Y., S. M. Rappaport, R. F. Sawyer, R. E. Talcott, and E. T. Wei, "Direct-Acting Mutagens in Automobile Exhaust," *Cancer Lett.*, 5, 39–47 (1978).

Watanabe, M., M. Ishidate, Jr., and T. Nohmi, "A Sensitive Method for the Detection of Mutagenic Nitroarenes: Construction of Nitroreductase-Overproducing Derivatives of *Salmonella typhimurium* Strains TA98 and TA100," *Mutat. Res.*, 216, 211–220 (1989).

Watanabe, M., M. Ishidate, Jr., and T. Nohmi, "Sensitive Method for the Detection of Mutagenic Nitroarenes and Aromatic Amines: New Derivatives of *Salmonella typhimurium* Tester Strains Possessing Elevated *O*-Acetyltransferase Levels," *Mutat. Res.*, 234, 337–348 (1990).

Watanabe, T., M. J. Kohan, D. Walsh, L. M. Ball, D. M. DeMarini, and J. Lewtas, "Mutagenicity of Nitrodibenzopyranones in the *Salmonella* Plate-Incorporation and Microsuspension Assays," *Mutat. Res.*, 345, 1–9 (1995).

Watanabe, T., T. Hirayama, and J. Lewtas, "Mutagenicity and DNA Adduct Formation of Air Pollutants: Nitrodibenzopyranone Isomers and Nitrodibenzofuran Isomers," *Jpn. J. Toxicol. Environ. Health*, 42, 4 (1996).

Waters, M. D., S. Nesnow, J. L. Huisingh, S. Sandhu, and L. Claxton, Eds., *Application of Short-Term Bioassays in the Fractionation and Analysis of Complex Environmental Mixtures*, Plenum, New York, 1979.

Waters, M., S. Sandhu, J. Lewtas-Huisingh, L. Claxton, and S. Nesnow, Eds., *Short-Term Bioassays in the Analysis of Complex Environmental Mixtures, II*, Plenum, New York, 1981.

Waters, M. D., S. S. Sandhu, J. Lewtas, L. Claxton, N. Chernoff, and S. Nesnow, Eds., *Short-Term Bioassays in the Analysis of Complex Environmental Mixtures, III*, Plenum, New York, 1983.

Watts, R. R., R. J. Drago, R. G. Merrill, R. W. Williams, E. Perry, and J. Lewtas, "Wood Smoke Impacted Air: Mutagenicity and Chemical Analysis of Ambient Air in a Residential Area of Juneau, Alaska," *JAPCA*, 38, 652–660 (1988).

Wayne, R. P., *Principles and Applications of Photochemistry*, Oxford Univ. Press, Oxford, 1988.

Wayne, R. P., "Singlet Oxygen in the Environmental Sciences," *Res. Chem. Intermed.*, 20, 395–422 (1994).

Wehry, E. L., "Optical Spectrometric Techniques for Determination of Polycyclic Aromatic Hydrocarbons," in *Handbook of Polycyclic Aromatic Hydrocarbons* (A. Bjørseth, Ed.), pp. 323–396, Dekker, New York, 1983.

Wesolowski, J., P. Flessel, S. Twiss, J. Cheng, R. Chan, L. Garcia, J. Ondo, A. Fong, and S. Lum, "The Chemical and Biological Characterization of Particulate Matter as Part of an Epidemiological Cancer Study," *J. Aerosol Sci.*, 12, 208–212 (1981).

Westerholm, R. N., T. A. Alsberg, A. B. Frommelin, and M. E. Strandell, "Effect of Fuel Polycyclic Aromatic Hydrocarbon Content on the Emissions of Polycyclic Aromatic Hydrocarbons and Other Mutagenic Substances from a Gasoline-Fueled Automobile," *Environ. Sci. Technol.*, 22, 925–930 (1988).

Westerholm, R. N., J. Almén, H. Li, J. Rannung, K. Egebäck, and K. Grägg, "Chemical and Biological Characterization of Particulate-, Semi-volatile-, and Gas-Phase-Associated Compounds in Diluted Heavy-Duty Diesel Exhausts: A Comparison of Three Different Semi-volatile-Phase Samplers," *Environ. Sci. Technol.*, 25, 332–338 (1991).

Westerholm, R., and K. Egebäck, "Exhaust Emissions from Light- and Heavy-Duty Vehicles; Chemical Composition, Impact of Exhaust after Treatment and Fuel Parameters," *Environ. Health Perspect.*, 102(Suppl. 4), 13–23 (1994).

White, C. M., Ed., *Nitrated Polycyclic Aromatic Hydrocarbons*, Hüthig, Heidelberg, 1985.

Wild, S. R., and K. C. Jones, "Polynuclear Aromatic Hydrocarbons in the United Kingdom Environment: A Preliminary Source Inventory and Budget," *Environ. Pollut.*, 88, 91–108 (1995).

Wilkinson, F., W. P. Helman, and A. B. Ross, "Quantum Yields for the Photosensitized Formation of the Lowest Electronically Excited Singlet State of Molecular Oxygen in Solution," *J. Phys. Chem. Ref. Data*, 22, 113–262 (1993).

Wilkinson, F., W. P. Helman, and A. B. Ross, "Rate Constants for the Decay and Reactions of the Lowest Electronically Excited Singlet State of Molecular Oxygen in Solution. An Expanded and Revised Compilation," *J. Phys. Chem. Ref. Data*, 24, 663–1021 (1995).

Williams, R., C. Sparacino, B. Petersen, J. Bumgarner, R. H. Jungers, and J. Lewtas, "Comparative Characterization of Organic Emissions from Diesel Particles, Coke Oven Mains, Roofing Tar Vapors, and Cigarette Smoke Condensate," *Int. J. Environ. Anal. Chem.*, 26, 27–49 (1986).

Wilson, N. K., T. R. McCurdy, and J. C. Chuang, "Concentrations and Phase Distributions of Nitrated and Oxygenated Polycyclic Aromatic Hydrocarbons in Ambient Air," *Atmos. Environ.*, 29, 2575–2584 (1995).

Winer, A. M., and R. Atkinson, "The Role of Nitrogenous Pollutants in the Formation of Atmospheric Mutagens and Acid Deposition," Final Report, California Air Resources Board, Contract No. A4-081-32, 1987.

Wise, S. A., S. N. Chesler, L. R. Hilpert, W. E. May, R. E. Rebbert, C. R. Vogt, M. G. Nishioka, A. Austin, and J. Lewtas, "Quantification of Polycyclic Aromatic Hydrocarbons and Nitro-Substituted Polycyclic Aromatic Hydrocarbons and Mutagenicity Testing for the Characterization of Ambient Air Particulate Matter," *Environ. Int.*, 11, 147–160 (1985).

Wise, S. A., B. A. Benner, S. N. Chesler, L. R. Hilpert, C. R. Vogt, and W. E. May, "Characterization of the Polycyclic Aromatic Hydrocarbons from Two Standard Reference Material Air Particulate Samples," *Anal. Chem.*, 58, 3067–3077 (1986).

Wise, S. A., A. Deissler, and L. C. Sander, "Liquid Chromatographic Determination of Polycyclic Aromatic Hydrocarbon Isomers of Molecular Weight 278 and 302 in Environmental Standard Reference Materials," *Polycyclic Aromat. Compd.*, 3, 169–184 (1993).

Wislocki, P., A. Wood, R. Chang, W. Levin, H. Yagi, O. Hernandez, P. Dansette, D. Jerina, and A. Conney, "Mutagenicity and Cytotoxicity of Benzo[a]pyrene, Arene Oxides, Phenols, Quinones, and Dihydrodiols in Bacterial and Mammalian Cells," *Cancer Res., 36,* 3350–3357 (1976).

Wislocki, P. G., E. S. Bagan, A. Y. H. Lu, K. L. Dooley, P. P. Fu, H. Han-Hsu, F. A. Beland, and F. F. Kadlubar, "Tumorigenicity of Nitrated Derivatives of Pyrene, Benz[a]anthracene, Chrysene, and Benzo[a]pyrene in the Newborn Mouse Assay," *Carcinogenesis, 7,* 1317–1322 (1986).

Wolff, G. T., J.-S. Siak, T. L. Chan, and P. E. Korsog, "Multivariate Statistical Analyses of Air Quality Data and Bacterial Mutagenicity Data from Ambient Aerosols," *Atmos. Environ., 20,* 2231–2241 (1986).

Wong, J. M., N. Y. Kado, P. A. Kuzmicky, H.-S. Ning, J. E. Woodrow, D. P. H. Hsieh, and J. N. Seiber, "Determination of Volatile and Semi-volatile Mutagens in Air Using Solid Adsorbents and Supercritical Fluid Extraction," *Anal. Chem., 63,* 1644–1650 (1991).

World Health Organization (WHO), Environmental Health Criteria 47, "Carcinogens, Summary Report on the Evaluation of Short-Term *in Vitro* Tests," World Health Organization, Geneva, 1985a.

World Health Organization (WHO), Environmental Health Criteria 51, "Guide to Short-Term Tests for Detecting Mutagenic and Carcinogenic Chemicals," World Health Organization, Geneva, 1985b.

World Health Organization (WHO), "Polynuclear Aromatic Hydrocarbons (PAH), in *Air Quality Guidelines for Europe,* WHO Regional Publications, European Series No. 23, World Health Organization, Regional Office for Europe, Copenhagen, pp. 105–117, 1987.

World Health Organization (WHO), Environmental Health Criteria 109, "Carcinogens, Summary Report on the Evaluation of Short-Term *in Vivo* Tests," World Health Organization, Geneva, 1990.

World Health Organization, International Program on Chemical Safety (IPCS), Environmental Health Criteria 171, "Diesel Fuel and Exhaust Emissions," World Health Organization, Geneva, 1996.

Wu, C.-H., I. Salmeen, and H. Niki, "Fluorescence Spectroscopic Study of Reactions between Gaseous Ozone and Surface-Adsorbed Polycyclic Aromatic Hydrocarbons," *Environ. Sci. Technol., 18,* 603–607 (1984).

Wu, C.-H., and H. Niki, "Fluorescence Spectroscopic Study of Kinetics of Gas–Surface Reactions between Nitrogen Dioxide and Adsorbed Pyrene," *Environ. Sci. Technol., 19,* 1089–1094 (1985).

Xu, X. B., J. P. Nachtman, Z. L. Jin, E. T. Wei, S. M. Rappaport, and A. L. Burlingame, "Isolation and Identification of Mutagenic Nitro-PAH (Polycyclic Aromatic Hydrocarbon) in Diesel Exhaust Particulates," *Anal. Chim. Acta, 136,* 163–174 (1982).

Yahagi, T., M. Nagao, Y. Seino, T. Matshushima, and T. Sugimura, "Mutagenicities of *N*-Nitrosamines on *Salmonella,*" *Mutat. Res., 48,* 121–130 (1977).

Yamagiwa, K., and K. Ichikawa, "Experimental Study of the Pathogenesis of Carcinoma," *J. Cancer Res., 3,* 1–29 (1918).

Yamasaki, H., K. Kuwata, and H. Miyamoto, "Effects of Temperature on Aspects of Airborne Polycyclic Aromatic Hydrocarbons," *Environ. Sci. Technol., 16,* 189–194 (1982).

Yamasaki, H., K. Kuwata, and Y. Kuge, "Determination of Vapor Pressure of Polycyclic Aromatic Hydrocarbons in the Supercooled Liquid Phase and Their Adsorption on Airborne Particulate Matter," *Nippon Kagaku Kaishi, 8,* 1324–1329 (*Chem. Abstr., 101,* 156747p (1984).

Yamauchi, T., and T. Handa, "Characterization of Aza Heterocyclic Hydrocarbons in Urban Atmospheric Particulate Matter," *Environ. Sci. Technol., 21,* 1177–1181 (1987).

Yokely, R. A., A. A. Garrison, G. Mamantov, and E. L. Wehry, "The Effect of Nitrogen Dixoide on the Photochemical and Nonphotochemical Degradation of Pyrene and Benzo[a]pyrene Adsorbed on Coal Fly Ash," *Chemosphere, 14,* 1771–1778 (1985).

Yokley, R. A., A. A. Garrison, E. L. Wehry, and G. Mamantov, "Photochemical Transformation of Pyrene and Benzo[a]pyrene Vapor-Deposited on Eight Coal Stack Ashes," *Environ. Sci. Technol., 20,* 86–90 (1986).

Yu, S., D. Herreno-Saenz, D. W. Miller, F. F. Kadlubar, and P. P. Fu, "Mutagenicity of Nitro-Polycyclic Aromatic Hydrocarbons with the Nitro Substituent Situated at the Longest Molecular Axis," *Mutat. Res., 283,* 45–52 (1992).

Zielinska, B., University of California, Riverside, Statewide Air Pollution Research Center, unpublished results (1985).

Zielinska, B., J. Arey, R. Atkinson, and P. A. McElroy, "Nitration of Acephenanthrylene under Simulated Atmospheric Conditions and in Solution and the Presence of Nitroacephenanthrylene(s) in Ambient Particles," *Environ. Sci. Technol., 22,* 1044–1048 (1988a).

Zielinska, B., J. Arey, W. P. Harger, and R. W. K. Lee, "Mutagenic Activities of Selected Nitrofluoranthene Derivatives in *Salmonella typhimurium* Strains TA98, TA98NR, and TA98/1,8-DNP$_6$," *Mutat. Res., 206,* 131–140 (1988b).

Zielinska, B., J. Arey, R. Atkinson, and A. M. Winer, "The Nitroarenes of Molecular Weight 247 in Ambient Particulate Samples Collected in Southern California," *Atmos. Environ., 23,* 223–229 (1989a).

Zielinska, B., J. Arey, R. Atkinson, and P. A. McElroy, "Formation of Methylnitronaphthalene from the Gas-Phase Reactions of 1- and 2-Methylnaphthalene with OH Radicals and N_2O_5 and Their Occurrence in Ambient Air," *Environ. Sci. Technol., 23,* 723–729 (1989b).

Zielinska, B., J. Arey, and R. Atkinson, "The Atmospheric Formation of Nitroarenes and Their Occurrence in Ambient Air," in *Nitroarenes: Occurrence, Metabolism, and Biological Impact* (P. C. Howard, S. S. Hecht, and F. A. Beland, Eds.), pp. 73–84, Plenum, New York, 1990.

Zielinska, B., J. C. Sagebiel, G. Harshfield, A. W. Gertler, and W. R. Pierson, "Volatile Organic Compounds up to C_{20} Emitted from Motor Vehicles: Measurement Methods," *Atmos. Environ., 30,* 2269–2286 (1996).

Zinbo, M., D. Schuetzle, D. P. H. Hsieh, N. Y. Kado, J. M. Dasiey, and L. A. Gundel, "An Improved Fractionation Procedure for the Bioassay-Directed Chemical Analysis of Ambient Air Particulate Extracts," *Anal. Sci., 8,* 461–468 (1992).

CHAPTER 11

Analytical Methods and Typical Atmospheric Concentrations for Gases and Particles

As seen throughout this book, the chemistry of the troposphere and stratosphere is sufficiently complex that an interplay between laboratory, field, and modeling studies is critical to elucidate fundamental processes and their relationships to each other. Accurate data on the geographical and temporal distribution of pollutants and trace species are essential for testing of models that incorporate the results of laboratory studies. Clearly, the predictions of such models can only be as good as the data against which they are tested. Hence, methods of measurement that are sensitive, specific, and accurate and that have good time resolution (generally of the order of seconds or better) are an essential element of understanding atmospheric chemistry.

Historically, atmospheric compounds were measured using wet chemical techniques. For example, ozone was measured by bubbling air through a solution containing iodide, and the I_2 formed was measured using wet chemical techniques. Such methods were used as early as the mid-1800s to measure ozone in a number of locations worldwide, providing data on the increase in its concentrations since then, discussed in Chapter 14.B.2d.

However, not surprisingly, wet chemical methods are frequently subject to a number of potential interferences, both positive and negative. In the case of ozone measured using iodide oxidation, for example, SO_2 gives a 1:1 negative interference, whereas NO_2 gives a positive interference with a response that is equivalent to about 5–10% that of ozone on a molecular basis. In addition, calibrations can be sensitive to the exact procedure used. For a discussion of many of the factors affecting O_3 measurements made using this technique, and the air quality implications, see Finlayson-Pitts and Pitts (1986) and references therein. Because this technique is not specific for O_3, the measured values are often reported as "oxidant," rather than O_3, although the latter is, under most circumstances, the major contributor.

As a result, while such methods have been very useful in the past and continue to be applied for initial surveys of air quality in areas in which measurements have not been made in the past, they have generally been abandoned in favor of instrumental methods of analysis. As a result, this chapter focuses on the most commonly used instrumental, often spectroscopic, methods for measuring air pollutants, trace gases, and particles in air (e.g., see Roscoe and Clemitshaw, 1997). The focus is on tropospheric measurements, although, in most cases, the same techniques are used in the stratosphere.

In many countries around the world, air quality standards for specific gases are set to protect public health (see Chapter 2.D). Standards are often set for particles in terms of mass m^{-3} less than a certain size, e.g., 10 μm. With the increasing focus on the health effects of fine particles, there is also great interest in composition as a function of size. The gases for which air quality standards are set are generally referred to as "criteria pollutants" and include NO_2, O_3, CO, and SO_2 in the United States. Reference and equivalent methods, summarized in Table 11.1, have been established by the U.S. Environmental Protection Agency for the measurement of these compounds. Since nonmethane hydrocarbons and organics (for which a number of different names and acronyms are used; see later) are precursors to O_3 and a number of other species of atmospheric interest, a variety of methods for measuring the total organics or individual compounds have been developed for these as well.

TABLE 11.1 Reference and Equivalent Methods Designated by the U.S. Environmental Protection Agency for Monitoring Criteria Gaseous Air Pollutants

Gas	Reference or equivalent method
NO_2	Ozone chemiluminescence
	Differential optical absorption spectrometry
	Sodium arsenite
O_3	UV absorption
	Chemiluminescence
	Differential optical absorption spectrometry
CO	Nondispersive infrared
SO_2	UV fluorescence
	Differential optical absorption spectrometry
	Pararosaniline

The principles behind these and other techniques used to measure a variety of trace gases in the atmosphere, including the criteria pollutants and free radicals such as NO_3, OH, HO_2, and RO_2, are described in the following sections. In addition, typical tropospheric concentrations in regions from remote to urban areas are given.

In the second section of this chapter, techniques for measuring and characterizing particles are described.

A. GASES

1. Optical Spectroscopic Techniques

a. Chemiluminescence

As described in Chapter 3, the products of some chemical reactions are initially produced in electronically excited states. If the excited state has a sufficiently short radiative lifetime, it will emit light faster than collisional quenching by air molecules can occur (see Problem 1). The effective concentration of the emitting species (and hence emitted light intensity) is proportional to the concentrations of the reactants. As a result, the chemiluminescence intensity can be used to monitor one of the reactants if the second reactant is kept at a constant (excess) concentration.

For example, ozone reacts with nitric oxide to form electronically excited NO_2:

$$O_3 + NO \rightarrow NO_2^* + O_2, \quad (1)$$

$$NO_2^* \rightarrow NO_2 + h\nu. \quad (2)$$

The light emission from electronically excited NO_2 extends from 590 nm out to 2800 nm, a region that is relatively easily and sensitively monitored using conventional photomultipliers. This reaction is a standard technique for monitoring NO.

To measure NO using chemiluminescence, the air is mixed with a stream containing excess O_3 and the chemiluminescence intensity monitored. In principle, the reverse procedure can be used; i.e., excess NO can be added and the chemiluminescence intensity used to measure O_3. However, this requires a source of NO such as a gas cylinder, which is often not convenient for field studies.

Another chemiluminescence method for monitoring ozone involves the production of electronically excited formaldehyde in the O_3 reaction with ethene:

$$O_3 + C_2H_4 \rightarrow HCHO^* + \text{Other products}, \quad (3)$$

$$HCHO^* \rightarrow HCHO + h\nu. \quad (4)$$

The light emission is again in a region that can be followed quite sensitively, from ~300 to 550 nm. A disadvantage of this method is again the need to have a gas cylinder of ethene available.

A chemiluminescence method used for atmospheric measurements of NO_2 involves its reaction with luminol (5-amino-2,3-dihydro-1,4-phthalazinedione) in alkaline solution or on a wick wetted with such a solution (Maeda et al., 1980; Wendel et al., 1983; Schiff et al., 1986). Although this technique can be quite sensitive, down to ~5 ppt, there can be interferences from species such as PAN and O_3. Ozone can be selectively removed from the airstream prior to measurement, although simultaneous removal of some of the NO_2 by the scrubber has been problematic in some studies (Fehsenfeld et al., 1990). Alternatively, the contributions of interfering gases to the measured signal can be taken into account (assuming they are measured separately), which has been shown to be effective at NO_2 concentrations greater than 0.3 ppb (e.g., see Fehsenfeld et al., 1990). Nonlinearity at low concentrations has also been reported for this method (Kelly et al., 1990), requiring careful correction. In addition, NO and CO_2 suppress the NO_2 signal if they are present at high concentrations relative to NO_2 (Spicer et al., 1994a).

The application of chemiluminescence to atmospheric measurements is reviewed by Navas et al. (1997).

b. Fluorescence

Fluorescence is the basis of a number of measurement methods for atmospheric gases. In the case of SO_2, for example, a Zn (213.8 nm) or Cd (228.8 nm) lamp is used to excite the SO_2 and the fluorescence in the 200- to 400-nm region is monitored (Okabe et al., 1973; Schwarz et al., 1974). Again, this technique works for those species whose excited states are sufficiently short-lived that quenching does not totally dominate (see Problem 2). As discussed in more detail later, this technique is also used to measure NO as well as the OH free radical in air.

c. Infrared Spectroscopy (IR)

Infrared spectroscopy has been applied to ambient air measurements since the mid-1950s (Stephens, 1958). Indeed, PAN was first identified in laboratory systems by its infrared absorptions and dubbed "compound X" because its identity was not known (Stephens et al., 1956a, 1956b). It was subsequently measured in ambient air (Scott et al., 1957). Since then, IR has been applied in many areas and has provided unequivocal and artifact-free measurements of a number of compounds. Because of its specificity, it has often been used as a "standard" for intercomparison studies (e.g., for HNO_3; see later).

Other infrared absorption techniques are also used in ambient air measurements, including tunable diode laser spectroscopy (TDLS), nondispersive infrared (NDIR) spectroscopy, and matrix isolation spectroscopy. These are discussed in more detail later.

A major advantage of infrared absorption spectroscopy derives from the characteristic "fingerprints" associated with infrared-active molecules. On the other hand, interferences from common atmospheric components such as CO_2 and H_2O are significant, so that the sensitivity and detection limits that can be obtained are useful primarily for polluted urban air situations. For atmospheric work, long optical path lengths are needed. To obtain these, multiple-pass cells are commonly used. Such cells are also often used in UV–visible spectroscopic measurements in air, discussed in Section A.1.d.

(1) Multipass cells There are several different configurations of multipass cells in use. The most common approach is a three-mirror multiple-pass cell (Fig. 11.1a) known as a *White cell* after the individual who first put forth the basic design (White, 1942). The light is first focused on the entrance to the cell. The beam diverges and falls on spherical mirror M1, which reflects the image and refocuses it onto mirror M2, known as the field mirror. The diverging beam from M2 is reflected to spherical mirror M3, which, like M1, reflects and refocuses the image at the opposite end of the cell. If the mirrors are adjusted so that this image is at the exit aperture of the cell, the light beam leaves and strikes a detector. A total of four passes of the cell has therefore been made and the effective path length for absorption is $L = 4a$, where a is the length of the cell. However, the mirrors may be adjusted so that the reflected image from M3 falls on mirror M2 and is again reflected to M1 at a small angle to the original input light beam, leading to another set of four passes along the length of the cell.

For example, in the spot pattern shown in Fig. 11.1b, the beam enters the cell at the gap marked "0" in the field mirror and, after multiple reflections, exits at the gap in the field mirror on the opposite side, marked "28." A total of 28 passes (or $2n + 2$, where n is the

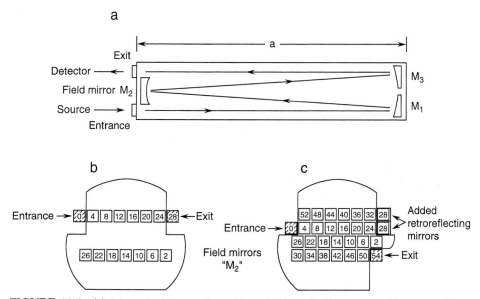

FIGURE 11.1 (a) Schematic diagram of a multipass White cell, (b) sequence of images on filled mirror for White cell design, and (c) sequence of images on field mirror for Horn and Pimentel design (1971). (Adapted from Finlayson-Pitts and Pitts, 1986; and Hanst and Hanst, 1994.)

number of spots on the field mirror) has been made in this case.

The advantage of such a White cell is that the source is reimaged on the field mirror M2 after each double traversal of the cell. This keeps the energy that enters the cell within the mirror system so that energy losses occur mainly through light absorption by the mirrors and, of course, by the gases in the cell. In practice, the loss of light energy through absorption by the mirrors imposes a major limitation on the number of passes that can be used. The fraction of the energy lost after n reflections from a mirror whose reflectivity is R is given by $(1 - R^n)$. Thus, if a mirror reflects 98% of the incident light and absorbs 2%, only 36% of the incident intensity will remain after 50 reflections from the mirror. After 100 reflections, only 13% of the incident intensity is left. While the path length and hence absorbance have increased, the energy loss may be so severe that such a large number of reflections becomes impractical.

The number of reflections is also limited by the size of the image striking the entrance and the size of the mirror M2. As seen in Fig. 11.1b, the images that are refocused from M1 and M3 onto the field mirror M2 are "stacked" beside each other. The width of M2 therefore determines how many of these images can be accommodated (i.e., how many reflections are possible).

A practical problem arising when the images are too closely spaced (i.e., at long path length) is one of adjustment; temperature changes, for example, can cause very small changes in the mirror adjustments which result in moving the exit beam away from the exit aperture.

Variations of the White cell are also in use. For example, Horn and Pimentel (1971) added a corner mirror assembly to redirect the beam that would normally exit the cell back into it. This doubles the number of passes, giving four rows of spots on the field mirror. The image pattern for such a design is shown in Fig. 11.1c (Hanst and Hanst, 1994).

More complex multiple-reflection systems that give a much greater number of traversals have also been developed. For example, Tuazon *et al.* (1980) describe a system using four collecting mirrors that focus the light onto four field mirrors. The advantages and disadvantages of such multiple-mirror cells are discussed by Hanst (1971) and Hanst and Hanst (1994).

An alternate design for folded optics was described by White in 1976. In this design, the light beam is folded back on itself, giving larger path lengths and greater optical stability. The effects of vibration, thermal expansion, and astigmatism are reduced and alignment errors are minimized with this design.

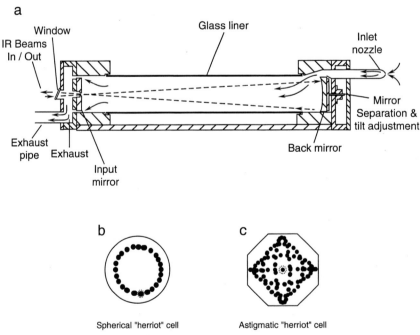

FIGURE 11.2 (a) Schematic diagram of multipass cell for infrared spectroscopy using astigmatic Herriott configuration (adapted from McManus *et al.*, 1995), (b) spot configurations for normal Herriott multipass cell, and (c) spot configurations for astigmatic configuration (adapted from Zahniser *et al.*, 1997).

A second multipass cell configuration is the Herriott cell (Herriott *et al.*, 1964; Herriott and Schulte, 1965). This is particularly useful for coherent light sources such as lasers used in tunable diode laser spectroscopy but has also been used with incoherent light sources using optical fibers cemented to a ball lens at the entrance to the cell (Zahniser, personal communication). Two spherical mirrors are separated by a distance close to their radius of curvature, and the light beam enters through a hole in one of them, directed in an off-axis direction. After multiple reflections between the two mirrors, the light beam exits through the same hole as it entered, but at a different angle (Fig. 11.2a). The beam remains collimated throughout, in contrast to the White cell system, and gives the spot pattern shown in Fig. 11.2b. The path length is changed by changing the distance between the mirrors; in practice this means that this design is most useful for fixed path length systems.

An astigmatic variant of the Herriott cell designed for use in ambient air studies is shown in Fig. 11.2a and described by McManus *et al.* (1995) and Zahniser *et al.* (1997). In this design, the two mirrors have different radii of curvature, giving the spot patterns shown in Fig. 11.2c. The spots more evenly fill the mirror, so that for a given number of passes, the spots are more widely spaced, or conversely, more passes can be obtained without problems of beam overlap (McManus *et al.*, 1995).

Major advantages of such cells are that they are relatively easy to align and folded optical paths can be obtained in small volumes. This is important when small amounts of sample are available, for example, in laboratory studies or when a fast response is needed; cells of smaller volume can be pumped out faster, giving shorter residence times in the cell.

(2) FTIR Fourier transform infrared spectroscopy has been used for many years to measure atmospheric gases. Because FTIR has become such a common analytical method, we do not describe the technique itself here but rather refer the reader to several excellent books and articles on the subject (e.g., see Griffiths and de Haseth, 1986; Wayne, 1987). For reviews of some atmospheric applications, see Tuazon *et al.* (1978, 1980), Marshall *et al.* (1994), and Hanst and Hanst (1994).

A problem in the application of FTIR to ambient air is that water vapor, CO_2, and CH_4 are all present in significant concentrations and absorb strongly in certain regions of the spectrum. As a result, the spectral regions that are useful for ambient air measurements are 760- to 1300-cm^{-1}, 2000- to 2230-cm^{-1}, and 2390- to 3000-cm^{-1}.

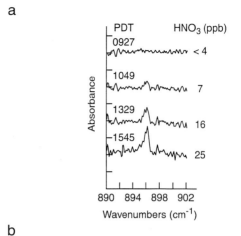

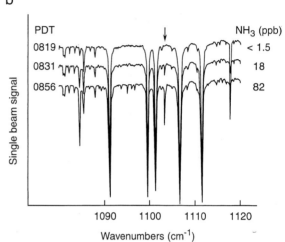

FIGURE 11.3 Typical FT-IR spectra in ambient air as a function of time in (a) the HNO_3 region and (b) the NH_3 region on September 14 and 16, 1985, respectively, in Claremont, California. NH_3 peak is marked by the arrow. Concentrations of each are shown on the right-hand side (adapted from Biermann *et al.*, 1988).

Figure 11.3 shows typical ambient air spectra in two regions in which HNO_3 (Fig. 11.3a) and NH_3 (Fig. 11.3b), respectively, have characteristic absorption bands (Biermann *et al.*, 1988). Figure 11.4 shows, for comparison, some typical reference spectra for HNO_3 and NH_3 taken at much higher concentrations in a 25-cm-long cell (see Problem 6). It can be seen that the absorption bands in air even in a polluted urban area are relatively weak. However, FTIR has also proven particularly useful as a standard for intercomparison studies in polluted urban atmospheres (e.g., see Hering *et al.*, 1988).

Table 11.2 summarizes the detection limits for FTIR measurements in the atmosphere for some gases of interest. Typical concentrations of each in remote to polluted atmospheres are discussed below with respect to the individual species; however, in general, it can be stated that FTIR is most suitable for measuring

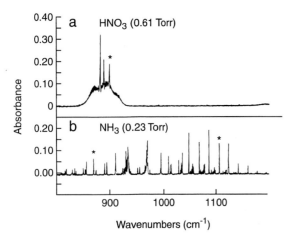

FIGURE 11.4 Reference spectra of gaseous HNO_3 and NH_3, respectively, at $L = 25$ cm and $P_{TOT} = 740$ Torr in N_2. Asterisks denote peaks used in analysis of ambient air (adapted from Biermann et al., 1988).

atmospheric trace gases in polluted urban areas or close to sources where they are found at the highest concentrations.

For example, Yokelson et al. (1996, 1997a, 1997b) have used FTIR to measure species emitted from combustion processes; this has permitted the simultaneous measurement of such species as HCHO, CH_3OH, CH_3OOH, C_2H_4, C_2H_2, C_3H_6, C_6H_5OH, CS_2, HCN, and NH_3 produced in fires, in addition to CO, CO_2, and CH_4. Indeed, such FTIR studies permitted the first identification of 2-hydroxyethanal ($HOCH_2CHO$) in smoke from fires (Yokelson et al., 1997a). Although such compounds could be detected and measured using chromatographic methods (see later), losses during sampling would likely be problematical, unlike open-pass FTIR, where the measurement can be made without direct sampling.

(3) Tunable diode laser spectroscopy (TDLS) A second technique based on infrared absorption spectrometry is tunable diode laser spectrometry, TDLS. The practice and application of TDLS in atmospheric measurements have been reviewed by Schiff et al. (1994a, 1994b) and Brassington (1995) and in the symposium proceedings edited by Grisar et al. (1992). As in the case of FTIR, this technique relies on measuring the absorbance at specific wavelengths due to the absorption of IR radiation by various pollutants. However, rather than using a continuous-wavelength light source and scanning the entire infrared spectrum, tunable diode laser spectroscopy employs a laser light source of very narrow linewidth that is tunable over a smaller (e.g., 100–200 cm^{-1}) wavelength range.

TABLE 11.2 Detection Limits for Some Trace Gases in Air by FTIR,[a] TDLS,[b] and Matrix Isolation IR[e]

Gas	FTIR detection wavenumber (cm^{-1})	FTIR detection limit[c] (ppb at $L = 1$ km)	TDLS detection wavenumber (cm^{-1})	TDLS detection limit[d] (ppb at $L = 150$ m)	Matrix isolation detection limit[e] (ppb)
SO_2	1133	25[g]	1360.7	0.5	0.01
NH_3	931	4	1065	0.025	
	967.5	3			
	993	4			
HCHO	2779, 2781.5	6	2781	0.05	0.03
HCOOH	1105	2	1107	1.0	0.02
HNO_3	896	6	1720	0.1	0.01
N_2O_5	740, 1248	4			0.02[f]
HONO	791 (*trans*)	10			0.01
	853 (*cis*)				
PAN	1162	3			0.05
H_2O_2	1251	40	1285.7	0.1	

[a] From Tuazon et al. (1980).
[b] From Schiff et al. (1994b).
[c] Resolution 0.5 cm^{-1}.
[d] 150 m, integration time 3–5 min.
[e] From Griffith and Schuster (1987); for a 15-L air sample.
[f] Based on laboratory spectra only.
[g] From E. Tuazon, personal communication, 1998.

The advantages of TDLS over FTIR are increased resolution and sensitivity. The widths of the laser lines are less than 10^{-4} cm^{-1}. This can be compared to typical pressure-broadened half-widths of infrared absorption bands of species of atmospheric interest, which are of the order of 0.05 cm^{-1} at atmospheric pressure; at low pressures (e.g., <1 Torr), where the linewidth is limited by Doppler broadening, typical half-widths are 0.0005–0.005 cm^{-1}. Thus the TDL output is usually sufficiently narrow to scan rotational absorption lines even at low pressures where Doppler broadening is the limiting factor on lineshape. This narrow laser linewidth allows one to measure weak absorptions between the ambient H_2O and CO_2 lines. Thus one can measure accurately small absorbances due to specific rotational lines in a vibration–rotation spectrum with high selectivity. However, for many molecules of interest, the presence of such rotational fine structure requires lowering the total pressure of the sample to ~10–30 Torr to minimize pressure broadening of the absorption lines. (For larger molecules, the absorption spectrum appears as a continuum even at these lowered pressures.)

A disadvantage of TDLS is that scanning the entire IR spectrum quickly is not possible since each diode normally covers a limited wavelength range and even the use of several diodes in one instrument does not provide the wide range of FTIR. Thus TDLS is more useful for following specific pollutants known to be present than for searching for previously unidentified species. In addition, the high-resolution capability is not of use for very large molecules with many overlapping bands. While reducing the pressure of the sample helps in reducing the absorbing linewidth, it also results in a loss of sensitivity through reductions in concentration and the possibility of interactions with the walls of the cell.

Commonly used tunable diode lasers are made of lead salt compounds such as $PbS_{1-x}Se_x$, $Pb_{1-x}Sn_xTe$, $Pb_{1-x}Ge_xTe$, $Pb_{1-x}Sn_xSe$, and $Pb_{1-x}Cd_xS$. Diodes made from Group III (Ga, Al, and In) and Group V (P, As, and Sb) elements are not in widespread use for atmospheric applications because they emit at wavelengths beyond 2 μm (5000 cm^{-1}) where the molecular absorptions are much weaker overtone and combinaton bands, limiting the detection sensitivity (Schiff et al., 1994a, 1994b; Brassington, 1995). A p–n junction is formed in the crystal and the diode is mounted onto a support such as copper that serves as a temperature controller during operation. When an electrical current is applied, the diode emits light spontaneously at a wavelength corresponding to the energy band gap in the semiconductor. Laser action results from reflections from the end faces of the crystal. This gap depends on the chemical composition of the laser and hence different wavelengths from 3 to 30 μm (3300–330 cm^{-1}) can be produced by altering the diode composition. The actual structure of these devices is more complex than a simple p–n junction, typically involving double heterostructures (e.g., see Brassington, 1995).

Tuning of the emitted wavelength can be accomplished, in principle, through variation of one of three possible parameters: applied magnetic field strength, diode temperature, and hydrostatic pressure. In practice, temperature, which can be controlled by changing the current through the diode, is used. Typical variations of output with temperature are about 3 cm^{-1} per K (Brassington, 1995). Figure 11.5, for example, shows the output of laser frequency as a function of temperature from a lead salt diode laser (Werle et al., 1992). The output at a given current is a series of longitudinal modes whose separation, typically about 2 cm^{-1}, is determined by $(2\eta L)^{-1}$, where η is the index of refraction of the salt (usually 4.5–7) and L is the length of the laser cavity, i.e., separation of the end faces of the crystal (typically 300–400 μm). Tuning of such semiconductor lasers over ~100–200 cm^{-1} can typically be carried out, which is sometimes sufficient to measure more than one pollutant with a single laser. Alternatively, several different diode lasers are included in the same apparatus.

A number of different modulation techniques can be used to increase the signal-to-noise ratio (e.g., see Schiff et al., 1994a, 1994b; and Brassington, 1995). For example, the laser beam can be mechanically chopped and detected using phase-sensitive detection with a lock-in amplifier. A more commonly used method for accurately measuring small absorbances is to modulate

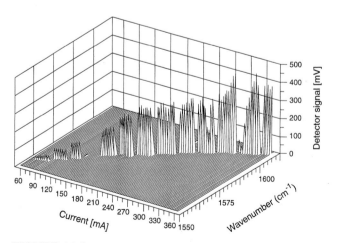

FIGURE 11.5 Variation of laser frequency and signal with current for a typical lead salt diode laser (adapted from Werle et al., 1992).

the frequency output of the laser by modulating the current and thus the temperature of the diode (Reid et al., 1978). Absorbances down to ~10^{-5} to $3-10^{-6}$ can be measured using multipass cells, corresponding to ppb to sub-ppb concentrations for many pollutants of atmospheric interest; the limits for open-reflection systems are not as good ($10^{-3}-10^{-4}$) due to interference from atmospheric turbulence (Schiff et al., 1994a, 1994b).

Figure 11.6 shows the major elements of a typical TDLS apparatus used for aircraft measurements (Hastie et al., 1983). Two diode lasers can be mounted on the dewar cold finger used for temperature/wavelength tuning; one is chosen for use by moving it into the appropriate position. A series of flat and off-axis parabolic mirrors are used to direct the laser beam into a White cell through which the air is pumped and back out to the sample detector. The He–Ne laser is used for alignment. A reference cell containing a high concentration of the species of interest can be inserted into the light path for calibration.

Figure 11.7 shows a typical $2f$ spectrum for the 1597-cm^{-1} line of NO_2 obtained using this apparatus, compared to a calibration obtained using 1.4 ppb NO_2. Fitting the ambient air spectrum to the reference gives an ambient air concentration of 72 ppt (Schiff et al., 1990).

TDLS is particularly useful for species such as H_2O_2 that are present at small concentrations and while very important in atmospheric chemistry, are difficult to measure. Figure 11.8 compares ambient H_2O_2 concentrations measured using TDLS and a wet scrubbing with enzyme fluorescence technique (Kleindienst et al., 1988a; Schiff et al., 1994a, 1994b). The two are generally in agreement to within about 30%.

Table 11.2 gives reported detection limits for some gases that have also been measured in the atmosphere by FTIR. As expected, the sensitivity of TDLS is significantly better than that of FTIR. For most species of atmospheric interest, detection limits are ~0.1 ppb for measurement times of 1 min in a 200-m White cell (G. Mackay, personal communication, 1998).

(4) Nondispersive infrared spectroscopy (NDIR) Figure 11.9 is a schematic diagram of the major components of an NDIR device (Skoog et al., 1998). As the name implies, it measures infrared-absorbing gases without dispersing the radiation or using FT techniques to derive wavelength-dependent signals. This method is also referred to as gas filter correlation. Infrared radiation is directed into two cells, one of which (the reference cell) is filled with a non-infrared-absorbing gas and the second of which (the sample cell) holds the sample (in a flow mode). The IR beams passing through the two cells then individually strike the compartments of the sensor cell, which are filled with the gas of interest and are separated by a thin, flexible metal diaphragm. When IR reaches this sensor cell, it is absorbed, causing heating and hence changes in pressure.

If the concentration of the absorbing gas is zero in the sample cell, the radiation striking both compartments is the same, and hence the heating is the same and there is no movement of the diaphragm separating

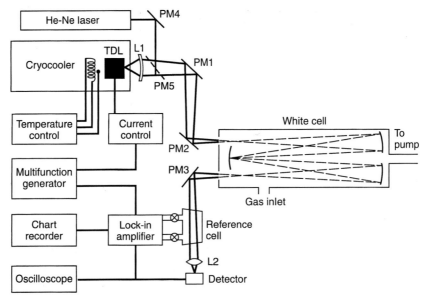

FIGURE 11.6 Schematic diagram of a TDLS apparatus (adapted from Hastie et al., 1983).

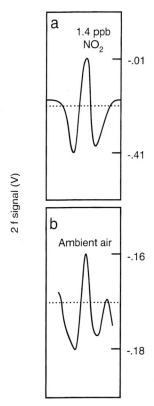

FIGURE 11.7 Typical TDLS spectra of NO_2 in the 1597-cm^{-1} region: (a) calibration spectra of 1.4 ppb NO_2; (b) ambient air, corresponding to 72 ppt NO_2 (adapted from Schiff *et al.*, 1990).

the two compartments of the sensor cell. However, if the gas of interest is present in the sample cell, it absorbs some of the IR, and less reaches that compartment of the sensor cell. This results in uneven heating of the two compartments of the sensor and higher pressures on the reference side. This moves the flexible metal diaphragm to the right, and the movement is measured by a change in capacitance between the diaphragm and a fixed capacitance plate. This method is used for CO, for example (Table 11.1). Atmospheric systems and applications of NDIR are described by Hanst and Hanst (1994).

(5) Matrix isolation spectroscopy (MI) Matrix isolation was first used in laboratory studies about four decades ago by Pimentel and co-workers (Whittle *et al.*, 1954). The method involves condensing the sample along with an inert "matrix" substance onto a cold infrared-transmitting or infrared-reflecting surface. At low temperatures in a matrix, rotation is essentially stopped for all but a very few small species. As a result, the infrared absorption is due solely to the vibrational transition, giving a single strong band instead of a series of rotational lines around the (0,0) vibrational

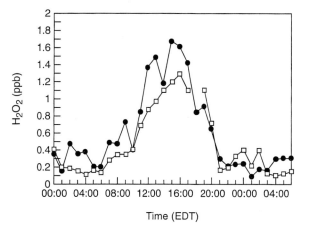

FIGURE 11.8 Ambient H_2O_2 concentrations measured by TDLS (●) and the continuous scrubbing enzyme fluorometric technique (□) during the period June 24–26 at Research Triangle Park, North Carolina (adapted from Schiff *et al.*, 1994a, 1994b).

transition. The low temperature of the matrix, lack of diffusion in the matrix, and the isolation of trapped molecules also help to minimize decomposition and other reactions of labile species.

This approach has also been used in the analysis of air by infrared spectroscopy (e.g., see Griffith, 1994). As discussed later, matrix isolation has also been used in conjunction with electron spin resonance (ESR) to measure free radical species, including NO_2, NO_3, HO_2, and RO_2.

In the matrix-FTIR studies, about 10–80 L of air is typically trapped using either a liquid nitrogen or liquid argon trap. At these temperatures, N_2, O_2, H_2, CH_4, and CO are not trapped, but CO_2 and trace gases are. The CO_2 in air acts as the inert matrix material when the sample is condensed on the infrared sample stage

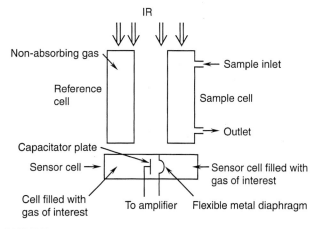

FIGURE 11.9 Schematic diagram of nondispersive infrared device (adapted from Skoog *et al.*, 1998).

in the second step. Water must be removed either before or after collection of the sample to minimize its contribution to absorption and scattering of IR. The cryogenically trapped air sample is then transferred to a low-temperature window for infrared analysis, usually by reflection–absorption spectroscopy. The CO_2 matrix is used as an internal standard, and because its concentration in air is well known (see Chapter 14), the concentrations of the trapped gases can be obtained from the strength of their infrared absorptions compared to those of CO_2.

Table 11.2 also shows the detection limits for some atmospheric gases using MI infrared spectoscopy and a 15-L air sample (Griffith and Schuster, 1987). Clearly, this technique can measure quite small concentrations, typically in the ppt range. The disadvantage is that in the configuration used to date, samples must be collected and brought back to the laboratory for analysis. As a result, it is not a "real-time" measurement, as is the case for FTIR and TDLS. In addition, the possibility of reactions during sampling and transfer onto the analysis window must be considered.

d. DOAS (UV–Visible Absorption Spectroscopy)

(1) Basis of technique Because of the relatively large absorption cross sections in the UV and visible for many gases of atmospheric interest, use of absorption spectroscopy in this region presents an obvious analytical approach. In the case of laboratory studies, measurement of the light intensity in the absence (I_0) and presence (I) of the species of interest is readily applied to obtain concentrations using the Beer–Lambert law (see Chapter 3.B):

$$A = \ln(I_0/I) = \sigma NL, \quad (A)$$

where σ is the absorption cross section (cm^2 molecule^{-1}), N is the concentration (molecules cm^{-3}), and L is the path length.

However, the fact that so many species in air absorb in this region presents a limitation in that one must be able to distinguish various species from each other as well as from background broad absorption and Rayleigh and Mie scattering of light by gases and particles. Because of these factors, UV–visible spectroscopy is, in practice, applied in air only to those species with banded structures, i.e., "fingerprints," of width ~5 nm or less. The technique used to do this is differential optical absorption spectrometry (DOAS). For reviews of DOAS, see Platt (1994) and Plane and Smith (1995).

Figure 11.10 illustrates the basis of this technique for a species that has narrow absorption bands at wavelengths λ_A and λ_B, superimposed on a slowly varying background. Because of Rayleigh and Mie scat-

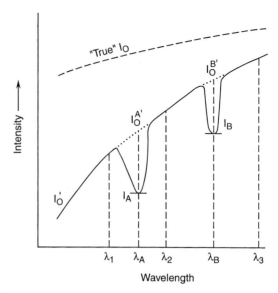

FIGURE 11.10 Light intensities relevant to DOAS spectrometry.

tering, the "true" I_0 shown by the upper dashed line, i.e., the intensity in the absence of air, cannot be measured. Scans of this spectral region do allow the broad background I_0', however, to be interpolated from the measurements of $I(\lambda)$. Thus, rather than measuring (I_0/I), the ratios $(I_0^{A'}/I_A)$ and $(I_0^{B'}/I_B)$ are measured and used to obtain the concentration of the absorbing species. That is, one is measuring the *differential* optical absorption (D) rather than the true optical absorption (A). However, this can be used for measuring concentrations as well since the differential optical absorption also follows a Beer–Lambert relationship:

$$D = \ln(I_0^{A'}/I_A) = \sigma' NL. \quad (B)$$

In this case, σ' is the differential optical absorption cross section for the absorption band. In practice, of course, there are many different absorbers, i, present at different concentrations N_i and absorbing at different wavelengths over the path length L.

Returning to Fig. 11.10, the relationship between I and the "true" I_0 can be expressed as

$$I(\lambda) = I_0(\lambda)A(\lambda)e^{\{-L[\Sigma \sigma_i(\lambda)N_i + \varepsilon_R(\lambda) + \varepsilon_M(\lambda)]\}}. \quad (C)$$

In Eq. (C), $A(\lambda)$ is an attenuation factor characteristic of the measurement system, ε_R and ε_M are the equivalent extinction coefficients due to Rayleigh and Mie scattering of gases and particles, and σ_i are the total absorption cross sections of the absorbing gases, all of which are wavelength dependent. Although the Rayleigh and Mie scattering contributions are not absorption processes, their contributions to the reduction in light intensity can be treated for DOAS measurements as if they were. The value of $\varepsilon_R(\lambda)$ is 1.3×10^{-6} cm^{-1} at 300 nm for STP conditions, reducing the light intensity by about 12% in each kilometer. The value of $\varepsilon_M(\lambda)$ strongly varies with aerosol loading. Typical values at 300 nm range from 1×10^{-6} cm^{-1} for clean maritime air (without sea spray) to $\sim 10^{-5}$ cm^{-1} for rural continental air. However, fog or heavy pollution can limit the application of DOAS because of the associated high values of the extinction.

The total absorption cross sections (σ_i) of a single trace gas i can be broken down into a contribution from the structured portion, σ_i', and one from the broadband portion that varies only slowly with wavelength, σ_i^B:

$$\sigma_i = \sigma_i' + \sigma_i^B. \qquad (D)$$

Substituting into Eq. (C), one obtains

$$I(\lambda) = \{I_0(\lambda)A(\lambda)e^{-L[\Sigma \sigma_i^B(\lambda)N_i + \varepsilon_R(\lambda) + \varepsilon_M(\lambda)]}\}e^{-L[\Sigma \sigma_i'(\lambda)N_i]}$$
$$= I_0'(\lambda)e^{-L[\Sigma \sigma_i' N_i]}. \qquad (E)$$

Taking natural logarithms, the differential optical absorbance (D') is given by

$$D'(\lambda) = \ln[I_0'(\lambda)/I(\lambda)] = L[\Sigma \sigma_i'(\lambda)N_i]. \qquad (F)$$

A major advantage of DOAS is its high sensitivity for species that meet the requirement of having narrow absorption bands in the UV–visible. Furthermore, because the differential optical absorption coefficients are fundamental spectroscopic properties of the molecule, the measurements need not be calibrated in the field.

(2) Analysis of spectra Different approaches to spectral analysis are described by Platt (1994) and Plane and Smith (1995). Calibration spectra of the absorbing species must be available for fitting the DOAS spectra. These spectra are usually obtained using the same instrument and settings. However, literature spectra of the same or higher resolution can be used if they are converted to the same resolution as used in the measurements.

To quantify the measured spectra, a combination of linear and nonlinear least-squares fitting routines are used, in which the measured intensities are fit to those of scaled reference spectra while minimizing the residual absorbance. Taking the natural logarithm of Eq. (E), one obtains

$$\ln I(\lambda) = \ln I_0'(\lambda) - L[\Sigma \sigma_i'(\lambda)N_i]. \qquad (G)$$

This is of the form

$$F(\lambda) = P(\lambda) + \Sigma a_j S_j(\lambda), \qquad (H)$$

where a_j are scaling factors for each species j chosen to give the best fit to the total spectrum and S_j are the known reference absorption spectra of each of the species. It has been observed that the term $P(\lambda)$ in Eq. (H), which contains the components that vary slowly with wavelength, i.e., $I_0(\lambda)$, $A(\lambda)$, $\varepsilon_R(\lambda)$, $\varepsilon_M(\lambda)$, and $\sigma_i^B(\lambda)$, can be approximated by a polynomial function of the form $P(\lambda) = \Sigma a_n \lambda^n$, where n is typically ~ 5. Thus, $\ln I(\lambda)$ is fit using least-squares analysis with combination of a polynomial and the second term to obtain the scaling factors a_j. From these scaling factors and the known path length, L, the concentration of the absorber j can be calculated. Care must be taken to ensure that the wavelengths are properly calibrated (e.g., using a low-pressure Hg lamp) and that small drifts in the spectra due to thermal drift (typically ~ 0.1 pixel K^{-1}) are taken into account. In addition, changes in air pressure can cause shifts, ~ 0.2 pixels in going from 1000 to 750 mbar. Such problems and the details of analysis of DOAS spectra, including methods of error estimation, are discussed by Stutz and Platt (1996).

(3) Typical apparatus Figure 11.11 is a schematic diagram of the components of a typical DOAS system. A broadband light source is needed, which, for example, can be a high-pressure Xe or incandescent quartz–iodine lamp, a broadband laser, or the sun or moon. The light traverses the air sample, either in a single-pass system or in a multipass system using an open White cell. The light strikes the entrance slit of a spectrograph which disperses the radiation. Detection as a function of wavelength of the dispersed light is carried out using a slotted-disk mechanism or, more commonly, a photodiode array (PDA) or charge-coupled device (CCD).

The use of the sun or moon as the light source allows one to measure the total column abundance, i.e., the concentration integrated through a column in the atmosphere. This approach has been used for a number of years (e.g., see Noxon (1975) for NO$_2$ measurements) and provided the first measurements of the nitrate radical in the atmosphere (Noxon *et al.*, 1978). As discussed later in this chapter, such measurements made as a function of solar zenith angle also provide information on the vertical distributions of absorbing species. Cloud-free conditions are usually used for such measurements; as discussed by Erle *et al.* (1995), the presence of tropospheric clouds can dramatically increase the effective path length (by an order of

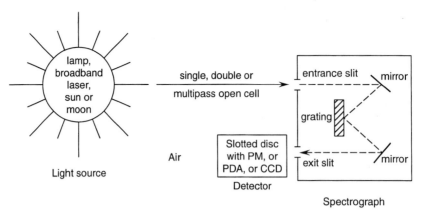

FIGURE 11.11 Schematic diagram of components of a DOAS system.

magnitude) through the atmosphere because of Mie scattering by the cloud droplets.

Surface-based instruments have also been developed for the application of DOAS to measure the integrated absorptions either over long direct path lengths or over folded light paths that give large total path lengths and hence high sensitivity but more closely approximate point measurements. There are two common approaches that have been used. In the earlier systems, a slotted-disk arrangement with a photomultiplier was used. These have been largely supplanted by the use of photodiode arrays.

In conventional spectroscopy, the grating of the spectrograph disperses the light so that the spectrum spreads out across the exit plane. The exit slit is stationary and wavelength scanning is achieved by slowly rotating the grating so that a series of wavelengths strike the exit sequentially and are detected by the photomultiplier. However, this is not suitable for ambient air studies where atmospheric turbulences with frequencies of <10 Hz make it desirable that spectra be scanned at rates >100 Hz. The use of the slotted disk, developed by Platt, Perner, and co-workers, allows one to attain the high scan rates needed. In this technique, the conventional exit slit is replaced by a mask that allows a 6- to 40-nm segment of the dispersed spectrum to fall on a rotating wheel, with the central wavelength set by the spectrograph wavelength setting. The rotating wheel contains a number of narrow slits (typically 50) around its perimeter. As seen in Fig. 11.12, as the wheel rotates, the slits "scan" the portion of the spectum dispersed across the monochromator exit slit. The slits in the rotating wheel are sufficiently well spaced that only one rotating slit is in the aperture at one time and also sufficiently narrow that only the light from a small portion of the dispersed spectrum passes through the rotating slit to the detector.

The signal, detected using a photomultiplier, is measured at several hundred different locations of the rotating slit across the exit aperture (i.e., at several hundred different wavelength intervals), and these signals are stored in different channels of a computer for subsequent data analysis. The light barrier on the edge of the mask shown in Fig. 11.12 triggers the computer so that as a rotating slit enters the mask aperture, data accumulation is started. As each rotating slit crosses the exit plane of the monochromator and performs one scan, the signals are added to the appropriate channels in the computer, resulting in many scans being superimposed; this signal averaging increases the signal-to-noise ratio.

As described in standard analytical chemistry books (e.g., Skoog *et al.*, 1998), photodiode arrays consist of a series (typically 1024) of side-by-side semiconductor rectangular detectors, or pixels. In this second type of DOAS instrument, the exit slit of the spectrograph is replaced by the photodiode array detector (PDA). Light striking the spectrograph grating is dispersed onto the PDA. The particular range of wavelengths striking the PDA is determined by the rotation of the grating, and

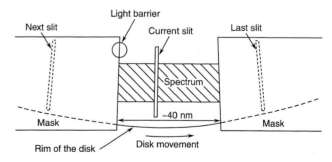

FIGURE 11.12 Schematic diagram of slotted-disk rapid-scanning mechanism used in DOAS studies (adapted from Platt, 1994).

the resolution, i.e., nanometers per pixel, by the entrance slit width. For example, a typical spectral range covered in one scan or set of scans is 40 nm, and with a 1024-element PDA, the resolution is then 40 nm/1024 pixels = 0.04 nm per pixel.

The advantage of using a PDA is that it records all wavelengths simultaneously, the so-called "multiplex" advantage. As a result, total photons detected are about 100–500 times greater in a given time period than for the slotted-disk arrangement, resulting in at least an order of magnitude increase in signal-to-noise (Stutz and Platt, 1997). However, there are some complications with using PDA that must be taken into account. First, the response of each of the pixels is not identical, which must be taken into account, for example, using multichannel scanning techniques described by Brauers et al. (1995). Second, under atmospheric conditions, different angles of incidence of the light on the PDA can give rise to "residual structures" in the spectrum that remain after all of the true absorptions have been removed; these can be quite large, of the order of 10^{-2} absorbance units, thus limiting the sensitivity to an order of magnitude less than the slotted-disk instruments. The use of a quartz fiber mode mixer overcomes this problem by acting as a diffuser, providing even illumination of the PDA with relatively small losses ($\sim 20\%$) in the intensity (Stutz and Platt, 1997).

(4) Typical DOAS spectra and detection limits Table 11.3 shows detection limits for some gases of atmospheric interest at a path length of 5 km for the slotted-disk and PDA techniques, respectively, and for the PDA at a path length of 15 km (Stutz and Platt, 1997). Also shown are detection limits for a 5-km path length estimated by Plane and Smith (1995). With the improvements in the PDA method described by Stutz and Platt (1997), the sensitivity is as good as, or better than, that using the slotted-disk approach. Detection limits for 15 km using the PDA vary from sub-ppt levels for NO_3 to about 100 ppt for HCHO.

Figure 11.13 shows a typical DOAS spectrum measured in air after correcting for atmospheric background light and an electronic offset (Stutz and Platt, 1997). Below the spectrum are shown reference spectra for the gases that contribute to the atmospheric spectrum, scaled by the a_j factors determined using Eq. (H). In this case, O_3, NO_2, SO_2, and HCHO all contribute, leaving a residual spectrum with a peak-to-peak absorbance of 6×10^{-4}.

DOAS has proven particularly useful for NO_3, for which other widely used methods are not available, and for HONO. In the latter case, denuder techniques have been applied, but a great deal of care must be exercised to recognize and, if possible, avoid artifacts (see later). Figure 11.14 shows the application of DOAS to the measurement of the nitrate radical during the night in Riverside, California. Since NO_3 photolyzes rapidly, it is only detectable at night. Bands at 623 and 662 nm can be seen growing in, peaking in this case at ~ 290 ppt around 8 p.m. local time (Platt et al., 1980b). As discussed in Chapter 7.D, the diurnal profile and time of the peak are quite variable, depending not only on its rate of formation but also on the scavenging processes.

DOAS has also been used for the measurement of the OH (see later) as well as BrO, ClO, and IO free radicals in the atmosphere (Platt and Hausmann, 1994; Platt and Janssen, 1995; Tuckermann et al., 1997; Hebestreit et al., 1999; Alicke et al., 1999), all of which have absorption bands in the UV (see Chapter 4 and DeMore et al. (1997)). For example, Fig. 11.15 shows OH concentrations measured as a function of time using DOAS (Dorn et al., 1996). The OH bands clearly

TABLE 11.3 Detection Limits for DOAS Measurements of Some Gases of Atmospheric Interest Using the Slotted-Disk or Photodiode Array (PDA) Techniques[a,b]

Gas	Technique	Path length	Detection limit Platt et al.[a]	Detection limit Plane et al.[b]
O_3	Slotted disk	5	2 ppb	2 ppb
	PDA	5	0.17–1.4 ppb	
	PDA	15	0.2–0.45 ppb	
SO_2	Slotted disk	5	100 ppt	10 ppt
	PDA	5	50–100 ppt	
	PDA	15	16–33 ppt	
NO_2	Slotted disk	5	200 ppt	50 ppt
	PDA	5	100–200 ppt	
	PDA	15	33–66 ppt	
HCHO	Slotted disk	5	500 ppt	50 ppt
	PDA	5	200–500 ppt	
	PDA	15	66–166 ppt	
HONO	Slotted disk	5	60 ppt	30 ppt
	PDA	5	30–60 ppt	
	PDA	15	10–20 ppt	
NO_3	Slotted disk	5	2 ppt	0.4 ppt
	PDA	5	1–3 ppt	
	PDA	15	0.33–1 ppt	
OH	PDA	5	1.5×10^6 cm^{-3} (0.06 ppt)	3×10^6 cm^{-3} (0.12 ppt)
ClO	PDA	5	1 ppt	20 ppt
BrO	PDA	5	0.5 ppt	30 ppt
IO	PDA	5	0.4 ppt	10 ppt

[a] Adapted from Stutz and Platt (1997) and Platt and Hausmann (1994).

[b] From Plane and Smith (1995) for $L = 5$ km and a minimum optical density of 10^{-4}.

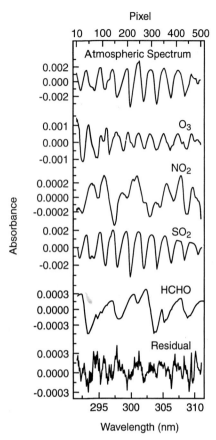

FIGURE 11.13 Typical spectrum measured using DOAS and its component contributions (see text) (adapted from Stutz and Platt, 1997).

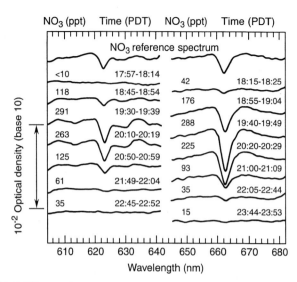

FIGURE 11.14 Measurement of NO_3 using DOAS in Riverside, California, on the evening of September 12, 1979 (adapted from Platt et al., 1980b).

environment in Ireland at concentrations up to 6 ppt (Alicke et al., 1999).

Table 11.3 gives detection limits reported for DOAS measurements of OH and the halogen oxide radicals at a path length of 5 km and assuming a detectable absorbance of 10^{-4}. This method provides ppt to sub-ppt sensitivities for these radicals.

grow in as the sun comes up, initiating photolysis which forms OH (see Chapter 1.B).

As discussed in Chapter 6.J.4, there is a halogen-catalyzed destruction of surface-level O_3 at polar sunrise in the Arctic and bromine atoms are believed to be the major reactant destroying O_3:

$$Br + O_3 \rightarrow BrO + O_2. \qquad (5)$$

In this case, BrO should be generated, and indeed, it has been observed by DOAS under these conditions at concentrations up to ~30 ppt (Tuckermann et al., 1997). Figure 11.16 shows a DOAS spectrum taken at polar sunrise at Alert in April 1992 and a reference spectrum of BrO (instrument features are included in this); clearly, BrO is present, in this case at a concentration of 17 ppt (Platt and Hausmann, 1994). BrO has also been detected at the Dead Sea, Israel, and attributed to heterogeneous reactions of the sea salt. ClO has also been detected at concentrations up to ~40 ppt under these conditions using DOAS (Tuckermann et al., 1997), and IO in a midlatitude coastal marine

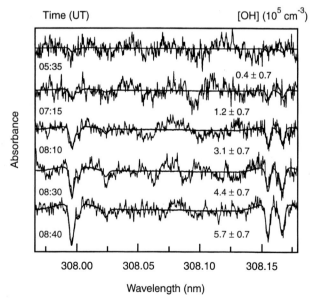

FIGURE 11.15 Measurement of OH using DOAS as a function of time (UT) after subtraction of the contributions of other known absorbers. The solid lines through the data are OH reference spectra (adapted from Dorn et al., 1996).

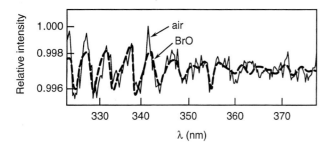

FIGURE 11.16 DOAS spectrum taken at Alert, N.W.T., on April 20, 1992, overlaid by fitted BrO reference spectrum (---) (adapted from Platt and Hausmann, 1994).

2. Mass Spectrometry

Mass spectrometry has the potential for being a very powerful analytical technique for atmospheric measurements, and indeed, it has been used for a number of decades in upper atmosphere measurements of ions and neutrals. Viggiano (1993) has reviewed ion chemistry and the application of mass spectrometry to tropospheric and stratospheric measurements through 1993. The first mass spectrometric measurements were made in the upper atmosphere from 64 to 112 km in 1963 (Narcisi and Bailey, 1965), followed by stratospheric measurements in 1977 (Arnold *et al.*, 1977) and, finally, tropospheric measurements in 1983 (Eisele, 1983; Heitmann and Arnold, 1983). They have also been extended to measurements in jet aircraft exhaust (e.g., Arnold *et al.*, 1998).

Table 11.4 summarizes measurements of various species in the stratosphere and troposphere by mass spectrometry through the early 1990s (Viggiano, 1993, and references therein). The altitude at which they were measured and the concentration ranges are shown, as well as whether they were detected using positive or negative ions (see later discussion).

Mass spectrometric measurements require four components: (1) an inlet to introduce the sample; (2) a means of ionizing the species of interest; (3) mass filtering/separation; (4) detection of the ions. Accomplishing this under atmospheric conditions is difficult due to the high sample pressure, which is incompatible with the high voltages used in the ion acceleration region and mass analyzers, and to the complexity of the mixtures found in air. Special considerations imposed by atmospheric conditions are discussed briefly next.

a. Sample Introduction

Because mass separation techniques use high voltages and hence require high vacuum, a means of transmitting the sample from the relatively high pressures found in the atmosphere into the low pressures in the analyzer is required. This typically involves differential pumping stages or the use of pulsed nozzles, which results in lowering of the sample pressure but also a proportionate loss of sensitivity.

TABLE 11.4 Some Species Measured by Mass Spectrometry in the Atmosphere up to about 1990[a]

Species	Altitude (km)	Detect by +/− Ion	Concentration range
C_5H_5N, pyridine	0–4	+	1–10 pptv
CH_3COCH_3, acetone	6–14	+	1–100 pptv
CH_3CN, acetonitrile	10–45	+	0.1–10 pptv
NH_3, ammonia	0–10	+	0.1–10^3 pptv
HOCl, hypochlorous acid	35–39	−	0.1–1 ppbv
SO_2, sulfur dioxide	0–11	−	1 ppbv
HNO_3, nitric acid	4–50	−	1–10^4 pptv
H_2SO_4, sulfuric acid	0–45	−	0.01–10 pptv
H_2O	38–40	+	1–10 ppmv
NO, nitric oxide	10	−	1 ppmv
NO_2, nitrogen dioxide	10	−	0.1 ppmv
HONO, nitrous acid	10	−	1 ppbv
CH_3SO_3H, methanesulfonic acid	0	−	≪ 1 pptv
C_6H_7N, picoline	0	+	1 pptv
C_7H_9N, lutidine	0	+	1 pptv
$C_3H_4O_4$, malonic acid	0	−	0.1–10 pptv
CH_3SCH_3, dimethyl sulfide	0	+	1–100 pptv
$C_{15}H_{24}$, β-caryophyllene	0	+	1–10 pptv
CH_3SOCH_3, dimethyl sulfoxide	0	+	1 pptv

[a] Adapted from Viggiano (1993); see references therein for original literature.

b. Ionization

Electron impact ionization is often used in conventional mass spectrometry. However, because of the often complex fragmentation patterns that result, it is not ideal for the direct analysis of complex mixtures such as air. In addition, the hot filaments used to generate the electron beam burn out at high pressures in air. If, in order to avoid this, the sample pressure is reduced prior to the ionization region, there is a loss of sensitivity that is sufficiently severe as to preclude the measurement of trace gases. As a result, alternate ionization processes must be used that either have higher sensitivity (and more selectivity) than electron impact so that the reduction of sample pressure in the ionization region does not present a sensitivity problem or that can ionize the trace species under conditions of atmospheric pressure. Chemical ionization using ion–molecule reactions and laser photoionization have both been used.

(1) Chemical ionization Ion chemistry in air is reasonably well understood (see Viggiano, 1993, and references therein). Ion–molecule reactions in the troposphere and stratosphere quickly ($\sim 10^{-3}$ s) give positively charged protonated water clusters, $H_3O^+(H_2O)_n$, where n is typically 3–5, and negatively charged O_2^- and its clusters with water. In the presence of CO_2, O_2^- and its clusters are rapidly ($\sim 10^{-3}$ s) converted into $CO_3^-(H_2O)_n$ ions. Subsequent ion–molecule reactions with trace gases (T) generate ionized products of T that are then detected.

For example, if a compound (T) has proton affinity > 170 kcal mol^{-1}, transfer of a proton from $H_3O^+(H_2O)_n$ occurs:

$$H_3O^+(H_2O)_n + T \rightarrow HT^+(H_2O)_n, \quad (6)$$

and if the water is stripped from the cluster, the ion HT^+ remains and can be separated and detected by MS. That is, an $(M + 1)$ peak results, where M is the molecular weight of the trace gas. Carbonyl compounds such as acetone have been detected with high sensitivity using this approach. For example, Fig. 11.17 is a schematic diagram of a mass spectrometer flown in an aircraft that was used to detect acetone in the lower stratosphere (Arnold and Hauck, 1985). The air is drawn in and the acetone reacts with $H_3O^+(H_2O)_n$ formed using a high-frequency glow discharge ion source. The $(M + 1)$ peak corresponding to $[H \cdot CH_3C(O)CH_3]^+$ is sampled downstream through an inlet orifice into the quadrupole mass spectrometer.

Near the earth's surface, there is sufficient ammonia in air that it undergoes a reaction with $H_3O^+(H_2O)_n$ to form $NH_4^+(H_2O)_n$, and this can also act as an ionizing agent. However, the proton affinity of the trace gas must be > 204 kcal mol^{-1} for proton transfer to occur from $NH_4^+(H_2O)_n$ to the trace gas.

In the negative ion mode, species with electron affinities greater than 44 kcal mol^{-1} can accept an electron from $O_2^-(H_2O)_n$ ions, forming a T^- ion with mass equal to the molecular weight, M:

$$O_2^-(H_2O)_n + T \rightarrow T^- + O_2 + nH_2O. \quad (7)$$

For example, tropospheric Cl_2 has been uniquely identified for the first time using its ionization in air to Cl_2^- followed by tandem MS detection (Spicer *et al.*, 1998).

If a molecule has a high gas-phase acidity, it can react with CO_3^- clusters as is the case for nitric acid, for example (e.g., Knop and Arnold, 1987a):

$$CO_3^-(H_2O)_n + HNO_3$$
$$\rightarrow NO_3^-(H_2O)_n + HCO_3,$$
$$\rightarrow NO_3^-(HCO_3)(H_2O)_{n-1} + H_2O.$$

Nitrate clusters with H_2O and/or HNO_3 such as $NO_3^-(HNO_3)_n$ are common in the atmosphere (Perkins and Eisele, 1984). Proton transfer to such clusters can occur, but clearly, the trace gas must be more acidic than HNO_3. This limits the number of trace gases that can be ionized through this mechanism but includes the important atmospheric species H_2SO_4 and methanesulfonic acid, CH_3SO_3H (Tanner and Eisele, 1991; Viggiano, 1993).

In short, positive and negative ions in air containing the trace gases of interest can be formed through discharge techniques and ions of the trace gases of interest generated via ion–molecule reactions. As discussed in more detail later, this approach has been used quite successfully to measure a number of species in air, including formic acid, acetic acid, dimethyl sulfide, and Cl_2 (Spicer *et al.*, 1994a, 1998). An alternate method is to add another compound to the mass spectrometer inlet, ionize this added species, and use its ion–molecule reactions to form ions and/or ion adducts of the species of interest. This has been used to measure HONO, for example, in air where a chloride ion adduct of HONO is formed when $CHCl_3$ is added in the corona discharge region (Spicer *et al.*, 1993a). Other examples include the measurement of HNO_3. For example, as described in Section A.4a(5), radioactive ionization of added SF_6 generates daughter ions that react with SiF_4 to give SiF_5^-. The SiF_5^- forms an adduct with HNO_3 and this adduct can be used to measure HNO_3 in air (Huey *et al.*, 1998).

(2) Laser photoionization Another ionization method with great potential for ambient air applications is

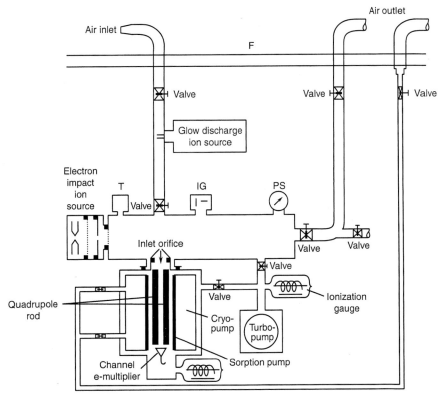

FIGURE 11.17 Schematic of mass spectrometer used for stratospheric measurements (IG = ion getter pump, PS = pressure sensor) (adapted from Arnold and Hauck, 1985).

laser photoionization (see Letokhov (1987) and Pfab (1995) for reviews). Trace gases can be ionized if sufficient energy in the form of light is pumped in; for example, polycyclic aromatic hydrocarbons (PAH; see Chapter 10) in combustion mixtures have been measured by two-photon ionization at 248 nm (e.g., Castaldi and Senkan, 1998).

In practice, for application to ambient air, efficient photoionization requires the use of pulsed lasers and multiphoton absorption methods. The terms "multiphoton ionization," or MPI, and "resonance-enhanced multiphoton ionization," or REMPI, are used to describe these processes.

Figure 11.18 illustrates the principles of application of REMPI to NO (discussed in more detail later). The electronically excited states of NO are shown in Fig. 11.18a and some potential ionization schemes in Fig. 11.18b (Pfab, 1995). Pulsed tunable lasers with wavelengths from ~190 to 1000 nm and spectral resolutions of 0.1 cm^{-1} are readily available. To ionize NO, the absorption of two, three, or four photons is needed. The first photon excites the NO into an intermediate state from which it is ionized using a second or, in some cases, two more photons. The transitions are described as an $(n + m)$ transition, where n is the number of photons that need to be absorbed simultaneously to reach the intermediate state and m is the number of photons to ionize the molecule from that state. The wavelength/energies of the photons involved in the various steps may be the same, which is referred to as a "one-color" process, or different, a "two-color" process. In the two-color case, the second photon is primed to indicate it is a different wavelength than the first photon. For example, in Fig. 11.18b, ionization via the A state can occur either by a $(1 + 1)$ process using 226 nm or by a $(1 + 1')$ process, where the A state is reached using 226 nm and ionization from this state occurs using 308 nm (Pfab, 1995; Hippler and Pfab, 1995). (The dashed arrows show transitions used for detecting NO by laser-induced fluorescence; see Section A.4a(1).)

The high spectral resolution of laser radiation provides selectivity. For example, Figure 11.19 shows the REMPI spectrum of the NO X (0,0) → A band using a $(1 + 1')$ process with 226- and 308-nm light to photoionize NO (e.g., see Pfab, 1995; and Lee et al., 1997). As the laser is tuned into resonance with a particular rotational transition in this band, ions are generated and detected using a conventional electron multiplier. Clearly, high selectivity is possible by tuning on and off

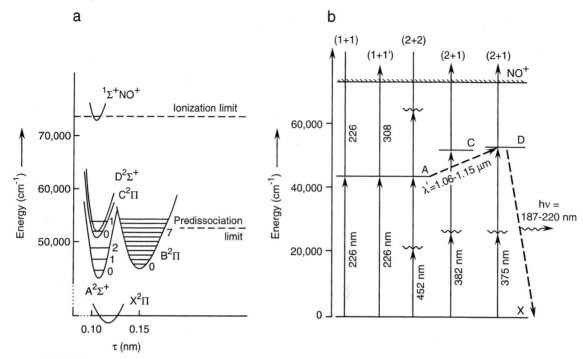

FIGURE 11.18 (a) Potential energy diagram and (b) REMPI schemes for the excitation and ionization of NO (adapted from Pfab, 1995). The ⌇⌇⌇ indicates a virtual state. The dashed arrows show other transitions used to detect NO in the atmosphere by laser-induced fluorescence (see Section A.4.a(1)).

the rotational transitions. (Note, however, as discussed later, interferences may result if NO is generated by photolysis of other species such as NO_2 in the laser beam.)

Pumping low-lying Rydberg states as the resonant intermediate is generally preferred over excitation of higher valence states because the latter often predissociate, undergo rearrangement, or intersystem cross before the subsequent absorption of photons can form ions. The Rydberg region can be reached by the simultaneous absorption of two or three photons, but with a loss of sensitivity (the absorption cross sections for two-photon absorption are of the order of 10^{-50} cm^4 compared to 10^{-18} cm^2 for one-photon absorption). On the other hand, pulse energies from conventional excimer or Nd:YAG pumped tunable laser systems are one to two orders of magnitude higher, since no second harmonic generation stages are required.

Other interferences from the use of two- or three-photon resonant excitation compared to $(1 + n)$ processes are photolysis of the analyte at the energy of the first (or second) photon and broadening of the absorption spectra due to the higher photon fluxes employed.

Table 11.5 summarizes some potential REMPI processes for the measurement of species of atmospheric interest (Pfab, 1995). This ionization technique clearly has a great deal of potential, although to date it has not been applied extensively to measurements in ambient air.

c. Mass Filters

Two types of mass analyzers have been used extensively in atmospheric applications: quadrupole mass filters and time-of-flight (TOF) instruments. The use of ion traps is also being increasingly explored for this application. For the fundamental principles of mass

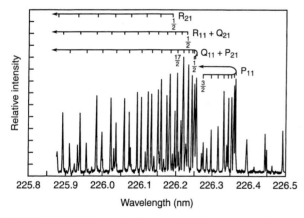

FIGURE 11.19 Two-color $(1 + 1')$ REMPI spectrum of the NO $(X^2\Pi, v' = 0 \rightarrow A^2\Sigma^+, v' = 0)$ band at 300 K (adapted from Pfab, 1995).

TABLE 11.5 Some Potential REMPI Transitions for Ionization of Molecules of Atmospheric Interest[a]

	Type of process	Resonant transition	Excitation wavelength (nm)
NO	1 + 1	X → A	226
	1 + 1'	X → A	226 + 308[b]
	1 + 1'	X → B, C, D	191 + 355 or 226
	2 + 1	X → C	380
CO	1 + 1'	X → B	115 + 345
	2 + 1	X → B	229–231
	2 + 1	X → E	215
NO_2	1 + 1' + 1	3p Rydberg +	484 + 248 + 484[c]
	1 + 1' + 1''	3p Rydberg +	e.g., 482 + 275 + 460[d]
N_2O	2 + 1	3p Rydberg +	230–250
NH_3	2 + 1	$\tilde{X} \to \tilde{B}$	340–355
		$\tilde{X} \to \tilde{D}$	286–289
CH_3CHO	2 + 1	n → 3s	346–365
CH_3COCH_3	2 + 2	n → 3s	372–392
	2 + 1	n → 3p	322–342[e]
C_6H_6	1 + 1	$S_0 \to S_1$	259[f]
Toluene	1 + 1	$S_0 \to S_1$	267
Naphthalene	1 + 1	$S_0 \to S_1$	270–310

[a] Adapted from Pfab (1995).
[b] Hippler and Pfab (1995).
[c] Benter et al. (1995).
[d] Campos et al. (1990).
[e] McDiarmid and Sabljić (1988).
[f] Boesl et al. (1980).

separation using these techniques, see standard analytical chemistry texts such as Skoog et al. (1998).

A potential limitation in the application of MS to near-surface measurements is the tremendous number of compounds in the atmosphere, particularly organics, and hence the increased complexity of interpretation of the single mass spectrum. In the MS ion source, the use of particular ion–molecule reactions to form the ions of interest or the ionization of one selected compound through resonant multiphoton absorption discussed earlier provides one means of specificity. A second method applied in the analyzer region is tandem mass spectrometry.

Figure 11.20, for example, is a schematic diagram of a tandem MS used for both surface and airborne measurements in the troposphere (Spicer et al., 1994a). Air is drawn into the sample inlet and ions are formed by a corona discharge generated by high voltage be-

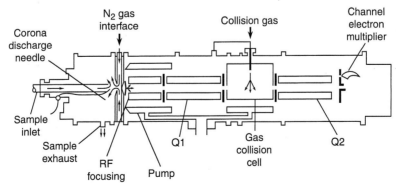

FIGURE 11.20 Schematic diagram of tandem mass spectrometer (adapted from Spicer et al., 1994a).

tween a needle and plate. Ions such as $H_3O^+(H_2O)_n$ are generated and undergo the ion-molecule reactions with trace gases (T) as described earlier. The ions then enter the interface, where water is stripped from the cluster by a stream of dry N_2, leaving the ion TH^+. Negative ions are generated and sampled by reversing the voltages on the needle and plate.

After being focused, the ions enter the first quadrupole (Q1), which can be used as a single mass spectrometer. However, the peaks observed using Q1 may not be parent ions. While the degree of fragmentation of ions formed using chemical ionization is generally much less than that using electron impact, it does occur. Hence observation of a particular peak corresponding to TH^+ in the positive ion mode, for example, does not guarantee that the trace gas T is responsible for the signal at this mass rather than a fragment from a larger molecule.

Tandem MS provides a powerful approach to this problem. In this mode, an ion exiting Q1 enters a cell containing a low pressure of a gas such as Ar or N_2 where it is collisionally dissociated. The fragments are then detected using the second quadrupole (Q2). The fragments are characteristic of the ion selected using Q1 and provide confirmation of the identity of the parent ion.

Figure 11.21, for example, shows the MS-MS of the peaks at m/e 70, 72, and 74, respectively, when a calibration sample of Cl_2 is sampled into the instrument shown in Fig. 11.20. The peak at m/e 70 fragments only to 35 amu, that at m/e 72 to both 35 and 37 amu, and that at m/e 74 only to 37 amu. Clearly, such fragmentation is consistent with the peaks in the Q1 scan being attributable to Cl_2 with isotopes ^{35}Cl and ^{37}Cl.

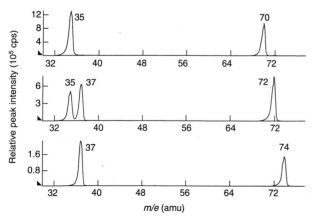

FIGURE 11.21 MS-MS of peaks at m/e 70, 72, and 74 due to Cl_2 (spectra taken by K. Oum).

The instrument can be run in various combinations of fixed or scanning modes for Q1 and Q2 (e.g., see Johnson and Yost, 1985). Particularly useful is the continuous mode, where particular peaks in the Q1 scan and certain fragments in the Q2 mass spectrometer are followed, rather than scanning one (or both) of the quadrupoles. Indeed, this method has been used to measure Cl_2 specifically in the marine boundary layer (Spicer et al., 1998). In these studies, Cl_2^- was generated as described earlier and the mass combinations (Q1/Q2) for 70/35, 72/35, 72/37, and 74/37 were followed. The combination of MS-MS and the isotope ratios provided unique confirmation that the species being measured was indeed Cl_2. Concentrations down to 15 ppt Cl_2 could be measured using this approach, with slightly better sensitivity for Br_2.

Table 11.6 shows some compounds that can be measured using this technique and estimated detection limits (Spicer et al., 1994a).

d. Detectors

The detectors used in mass spectrometers for atmospheric applications are essentially the same as for other MS applications and are commonly electron multipliers, either channeltrons or multichannel plate

TABLE 11.6 Estimated Detection Limits for Some Species of Atmospheric Interest by Atmospheric Pressure Ionization Mass Spectrometry[a]

Class	Example	Estimated detection limit (ppt)
Alcohols	Methanol	500
Aldehydes	Benzaldehyde	10
Alkaloids	Nictone	0.3
Amides	Dimethylformamide	20
Amines	Pyridine	2
	Ammonia	1
Carboxylic acids	Acetic acid	20
Esters	Ethyl butyrate	300
Ethers	Diethyl ether	50
Inorganic acids	Sulfuric acid	10
	Hydrogen chloride	20
Ketones	Acetone	40
Nitriles	Benzonitrile	4
Nitrosamines	Dimethylnitrosamine	10
Organometallics	Trimethylarsine	1
Pesticides	Sulfotep	100
Sulfur compounds (organic)	Dimethyl sulfoxide	10
	Dimethyl sulfide	2
Sulfur compounds (inorganic)	Sulfur dioxide	10
Terpenes (terpene-like compounds)	Linalool	500

[a] Adapted from Spicer et al. (1994a).

(MCP) detectors, respectively. The latter are preferred in TOF instruments since the area of detection is much larger and space charge distortions within the separated ion packets can be minimized.

In short, mass spectrometry is a powerful analytical tool that has been used successfully for a number of years at high altitudes and is now seeing increasing use in the troposphere, including at the earth's surface. A number of different approaches have been developed, including systems that are designed to measure species such as OH, NO, and HNO_3. They are described in more detail in the sections on measurement techniques for the individual species.

3. Filters, Denuders, Transition Flow Reactors, Mist Chambers, and Scrubbers

A variety of methods have been used to collect gases for subsequent quantification by techniques such as ion chromatography or colorimetric or other (e.g., electrochemical) analyses. These include filter methods, denuders, transition flow reactors, and scrubbers. Sampling must be carried out for sufficient periods of time to collect measurable amounts of the species of interest. From the total volume of air sampled and the amount of the analyte measured, the average concentration of the species in air over the collection period can be calculated. These techniques do not provide real-time analyses, although collection periods as short as ~0.5 h provide sufficient sample for analysis in some cases.

a. Filters

Air is drawn through a filter that consists of, or is coated with, a substance that takes up the species of interest. The filter is then extracted and the ions of interest are measured. For most species of interest, care must be taken to avoid interference from particles. For example, as discussed later, filter collection of HNO_3 will have a contribution from particulate nitrate if particles are not removed first. Figure 11.22 shows a typical filter pack used to measure gaseous HNO_3 and NH_3 (Anlauf et al., 1988).

As discussed below with respect to measurement techniques for individual compounds, the filter material is optimized for the compound of interest. For example, nylon has been found to be effective for the adsorption of gaseous HNO_3 (Spicer, 1977). However, care must be taken that the filter does not also remove other gases simultaneously that are measured downstream of the filter. For example, SO_2 has been shown to be taken up by nylon filters, forming sulfate, with the extent of conversion being quite variable (e.g., Chan et

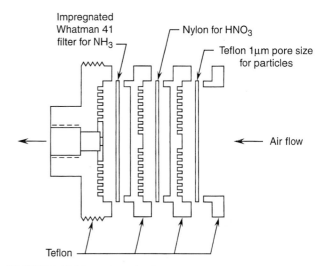

FIGURE 11.22 Schematic diagram of a typical filter pack used to measure gaseous HNO_3, particulate matter, and gaseous NH_3 (adapted from Anlauf et al., 1988).

al., 1986). In addition, changes in the characteristics of what is ostensibly the same filter material with time have been reported in the literature; for example, Cadle and Mulawa (1987) reported that nylon manufactured prior to May 1984 did not retain SO_2 to a significant extent (< 4%), whereas that manufactured at later times retained as much as 70% of gaseous SO_2.

In short, the performance of filter media must be carefully assessed prior to field deployment.

b. Denuders

Denuders (also known as diffusion denuders) are based on differences in the diffusion properties of gases compared to particles. The principle is illustrated in Fig. 11.23. A laminar flow of air is pulled through a tube. The inertia of the particles carries them through the tube, while the relatively high diffusivity of gases means that they will strike the walls of the tube a number of times while passing through it. If the walls are coated with a substance that will take up a particular compound, or group of compounds, from the gas phase, then these substances are removed from the gas stream. As discussed in detail by Durham et al. (1987), the depletion of the gas as a function of distance, x, as it travels along the tube is given by the Gormley–Kennedy equation:

$$\frac{C_x}{C_0} = 0.819 e^{(-11.49 Dx/Q)}. \quad (I)$$

This holds for values of C_x/C_0 less than 0.819, where the gas is collected with 100% efficiency. C_0 and C_x are the mass concentrations of the gas at the entrance to

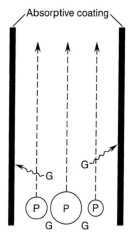

FIGURE 11.23 Schematic of principle of operation of denuders. G = gas, P = particles.

the tube and at distance x, respectively, D is the diffusion coefficient of the gas in air, and Q is the volumetric flow rate.

There are a variety of denuder designs, for example, ones incorporating a number of separate tubes in parallel or annular denuders in which the air is pulled through the annular space between two concentric tubes (e.g., Hering *et al.*, 1988; Krieger and Hites, 1992; Koutrakis *et al.*, 1993; Eatough *et al.*, 1993). In a variant of this method, the "coating" is a stream of water that continuously flows along the walls of the denuder and is collected for analysis (Buhr *et al.*, 1995), or, alternatively, a parallel plate with NaOH as the absorbing agent can be used (Simon and Dasgupta, 1995). The opposite approach is used in a diffusion separator developed for semivolatile organic compounds. In this case, the air containing the aerosol and gases flows along the outer walls of a tube, in the center of which is a core flow of clean air; only gases diffuse sufficiently rapidly to penetrate into the central core of clean air, which is sampled at the end of the tube onto a solid sorbent (Turpin *et al.*, 1993). However, despite differences in design in each case, the fundamental principle of using rapid gas diffusion to separate gases and particles is common to all methods.

Denuders have been used in several different ways. One is to extract the walls of the denuder and measure the adsorbed gas directly by ion chromatography. Denuders have also been used as "difference denuders." For example, in nitric acid measurements, the combination of gas-phase HNO_3 and particle nitrate has been measured using a nylon filter or Teflon–nylon filter combination in one sampling train. In a parallel sampling train, particulate nitrate alone is measured by first passing the airstream through a denuder to remove gaseous HNO_3. The difference between the two gives gaseous HNO_3.

As discussed with respect to the measurement of individual compounds, different coatings are used for the collection and measurement of different compounds. The criteria used to choose these coatings and interferences that can occur in the application of denuders to ambient air measurements are discussed by Perrino *et al.* (1990).

c. Transition Flow Reactors (TFRs)

These operate in a manner similar to that of denuders except that the gas flow is in the transition regime rather than being laminar flow and only a fraction (F) of the gas of interest is trapped at the walls. As described by Durham *et al.* (1986), TFRs can be treated as if there is a stagnant film of air adjacent to the wall and a core of turbulent air passing through the center of the tube. Uptake of the gas can be thought of as molecular diffusion through the stagnant air film. The fraction of the gas taken up is then given by

$$F = 1 - e^{(-2\pi r D x / Q \lambda)}, \quad (J)$$

where x, D, and Q are as defined in Eq. (I), r is the radius of the tube, and λ is the thickness of the stagnant air film at the wall. F is typically about 10% and in practice is determined by independent calibrations. The advantages of this sampler are that it has a high gas transfer coefficient and samples a greater volume of air (Durham *et al.*, 1986).

However, in at least one intercomparison study using diffusion denuders and transition flow reactors, different results were obtained for some important atmospheric gases such as SO_2, HNO_3, and H^+, where the TFR values were about 30, 80, and 85% higher, respectively, than those from the denuder system (Sickles *et al.*, 1989); the researchers attributed these differences to biases in the TFR measurements.

d. Mist Chambers and Scrubbers

Air is passed through a chamber where a mist of water or other aqueous solution is used to scrub out species of interest. The solution is then analyzed for the corresponding ions. As discussed shortly, this method has been used for several atmospheric gases, including HNO_3, carboxylic acids, and carbonyl compounds.

It has also been applied to measure inorganic chlorine gases and to differentiate HCl from other inorganics such as Cl_2 and HOCl (Keene *et al.*, 1993; Pszenny *et al.*, 1993). In this case, the first chamber has an acidic solution that scrubs out HCl, some Cl_2 and HOCl, and other chlorine-containing species such as ClNO,

ClNO$_2$, and ClONO$_2$. The air then passes into a second chamber with an alkaline scrubbing solution, which absorbs most of the Cl$_2$ and some HOCl. The two solutions are analyzed for chloride ion by ion chromatography. Differences in the chloride ion concentrations in the acid compared to the alkaline solutions provide a measure of chlorine-containing inorganics other than HCl.

4. Methods for, and Tropospheric Levels of, Specific Gases

a. NO, NO$_2$, NO$_x$, and NO$_y$

As we have seen in earlier chapters, NO is the major form of nitrogen oxides emitted from combustion processes, but in the atmosphere it is oxidized to NO$_2$ and other oxides of nitrogen. The term NO$_x$ is used for the sum of (NO + NO$_2$). The term NO$_y$ denotes the sum of NO, NO$_2$ (i.e., NO$_x$), plus all other oxides of nitrogen where the nitrogen is in an oxidation state of +2 or greater:

$$NO_y = NO + NO_2 + HNO_3 + NO_3 + 2N_2O_5$$
$$+ HONO + PAN + \text{higher peroxynitrates}$$
$$+ \text{alkyl nitrates} + \text{particulate nitrate} \ldots \quad (K)$$

The term NO$_z$ is also occasionally used in the literature. In these cases, it is defined by

$$NO_z = NO_y - NO_x. \quad (L)$$

Operationally, NO$_y$ is defined by the measurement method used to measure it, as discussed in more detail in Section A.4.a(2). Since NO, NO$_y$, and NO$_x$ are commonly measured simultaneously using variants of the same techniques, these are discussed together in the following sections, and in that order, for reasons that will become apparent.

(1) NO Nitric oxide is most commonly measured using the chemiluminescence from its reaction with O$_3$ described earlier. One such instrument designed for high-sensitivity (1- to 2-ppt detection limit in 10 s) is described by Ridley and Grahek (1990).

A second method is a two-photon laser-induced fluoresence (TP-LIF) technique (Bradshaw et al., 1985; Sandholm et al., 1990, 1997). Figure 11.18b illustrates the basis of this method. Ground-state NO (X$^2\Pi$) is pumped at 226 nm using a Nd:YAG pumped dye laser into the A$^2\Sigma$ state. This molecule is further excited by a second photon, λ', in the 1.06- to 1.15-μm range into the D$^2\Sigma$ electronically excited state, from which it fluoresces, returning to the ground state. Because the fluorescence occurs at higher energies and shorter wavelengths (187–220 nm) than the two pumping steps, interference from the excitation lasers is minimal. While the simplest approach is to carry out the second step using a fixed (1.1 μm) wavelength (Sandholm et al., 1990), there are advantages to being able to tune the IR laser, such as increasing the selectivity of the measurements and optimizing the pumping efficiency from the A$^2\Sigma$ state to the D$^2\Sigma$ state (Bradshaw et al., 1985). The sensitivity of this method is ~20 ppt for a 1-s integration time and 0.4 ppt for 100-s integration time at a signal-to-noise of 2:1 (Sandholm et al., 1990, 1997).

Intercomparison studies of these two measurement methods for NO generally show good agreement for levels of 25 ppt and greater (e.g., Hoell et al., 1987a; Gregory et al., 1990; Crosley, 1996). For example, Fig. 11.24 shows the results of one aircraft study in which the chemiluminescence method and the TP-LIF method were compared (Crosley, 1996). The slope of the plot in Fig. 11.24 was 0.94, with an intercept of -0.1 ± 0.8 ppt and $r^2 = 0.90$. For the data <25 ppt, although the slope was 0.989, the correlation was poorer, $r^2 = 0.66$.

A technique that has been used in laboratory studies for oxides of nitrogen and shows promise for field measurements is resonance-enhanced multiphoton ionization (REMPI) (Guizard et al., 1989; Lemire et al., 1993; Simeonsson et al., 1994). For example, Akimoto and co-workers (Lee et al., 1997) have reported a REMPI system in which a (1 + 1) two-photon absorption of light at 226 nm by NO results in ionization (vide supra). They report a detection limit of ~16 ppt in their laboratory studies. Other oxides of nitrogen such as NO$_2$ and HNO$_3$ can also photodissociate in the

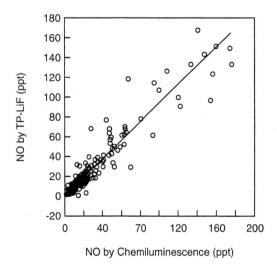

FIGURE 11.24 Measured NO concentrations using TP-LIF and chemiluminescence during one series of aircraft flights (adapted from Crosley, 1996).

laser beam to generate NO, causing interference in the NO measurement. However, since the ionization of NO is a two-photon process, the signal is expected to vary with the square of the laser power (P). On the other hand, since the production of NO from other compounds such as NO_2 requires three photons to generate and then photoionize NO, the dependence of the signal on the laser power is steeper. For example, Lee et al. (1997) report that the signal varies with $P^{1.75}$ for NO but $P^{2.4}$ for NO_2.

Figure 11.25 is a schematic diagram of one such REMPI system currently under development for ambient air analysis (Schmidt et al., 1999). The ions are generated in a two-photon process and then separated using time-of-flight mass spectrometry (TOF-MS), which provides an additional means of selectivity. For NO in laboratory air, the current detection limit using this system is 1 ppt. It has also been applied to the measurement of CO and CH_3CHO in laboratory systems using a $(2 + 1')$ two-color ionization process, with detection limits in synthetic air in laboratory studies of 10 and 1 ppt, respectively.

(2) NO_y NO_y is measured by passing the airstream containing NO and the other oxides of nitrogen over a catalyst to convert all of the other oxides of nitrogen into NO, which is then measured by one of the techniques just discussed. The resulting measurement is taken as the total oxides of nitrogen present.

The most common catalysts used are MoO at 375–400°C or Au at 300°C with added CO or H_2. The mechanism of reduction at the surfaces is not clear. Reaction of the various oxides of nitrogen on the metal surfaces may leave a surface oxide, which is then removed by reaction with the CO, forming CO_2, or with the H_2, forming H_2O (e.g., Kliner et al., 1997, and references therein).

This method of measurement of total oxides of nitrogen means that NO_y is defined operationally in terms of compounds that can be reduced to NO over these catalysts. It had been generally accepted that under typical operating conditions, species such as HCN, CH_3CN, N_2O, NH_3, and amines are not significantly reduced and hence did not contribute to NO_y (e.g., see Crosley, 1996). However, Kliner et al. (1997) showed that HCN, CH_3CN, and NH_3 can be converted to NO with high efficiencies under some conditions. For example, 85% of the HCN was converted using H_2 and 100% using CO with an Au catalyst at 300°C. Weinheimer et al. (1998) measured conversion efficiencies for HCN using three "outwardly identical" gold converters at 300°C with added CO. The conversion efficiency was 5–7% for ambient air sampled during aircraft flights with or without added water for two of the converters, with the efficiency doubled when synthetic air was sampled on the ground. The third converter had efficiencies for HCN of ~30% under all conditions. Bradshaw et al. (1998) reported conversion efficiencies ranging from 6 to 100% for HCN in gold converters. High conversion efficiencies were also found for organic nitrates, with the efficiencies being larger for the smaller nitrates such as nitroethane; differences were also noted between pure gold and gold-plated

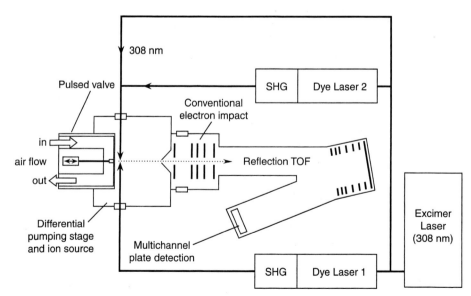

FIGURE 11.25 Schematic of REMPI-TOF. The conventional electron impact ionization source is just used for ion beam focusing (adapted from Schmidt et al., 1999).

converters, and in the latter case, depended on previous cleaning of the converter.

In short, it is clear that the conversion efficiencies have to be tested for each converter under conditions in which the field measurements are made.

With HCN concentrations of ~170 ppt in the stratosphere and upper troposphere (Coffey et al., 1981; Cicerone and Zellner, 1983; Zander et al., 1988; Schneider et al., 1997), and up to ~900 ppt at times (Rinsland et al., 1998), HCN could contribute significantly to NO_y, depending on the conversion efficiency. The same is true of acetonitrile, CH_3CN, whose concentrations are less well known; it has been measured over Europe at concentrations in the range of 150–200 ppt (e.g., Hamm et al., 1989) and in the lower stratosphere at concentrations of 110–160 ppt (Schneider et al., 1997). However, much smaller concentrations, of the order of a few tens of ppt, have also been reported in the atmosphere (Knop and Arnold, 1987a, 1987b). High concentrations of NH_3 are quite common in the troposphere, particularly near sources such as cattle feedlots (vide infra).

In addition to the potentially varying contributions of compounds such as HCN and NH_3 to NO_y, there are a number of other variables that can impact the measured values. One of the most important is the effect of the sampling lines, which can adsorb and desorb various gases. Nitric acid in particular is well known to be "sticky," readily adsorbing to various surfaces in a manner that is not reproducible and depends on such factors as the amount of water present on the surface. It is therefore not surprising that the agreement between various methods of measuring NO_y is not as good as for NO. Figure 11.26, for example, shows the NO_y measurements made during the flights for which the NO data are shown in Fig. 11.24 (Crosley, 1996). The slope of the regression line is 1.18 ± 0.04, with an r^2 value of only 0.37 for the scattered data.

In principle, if each of the compounds contributing to NO_y is individually measured, their concentrations should sum up to the measured NO_y. While this is sometimes the case, in many field studies the sum has been found to be less than the measured NO_y (e.g., Fahey et al., 1986; Ridley et al., 1990a; Atlas et al., 1992; Parrish et al., 1993; Sandholm et al., 1994; Nielsen et al., 1995; Williams et al., 1997). Cases where the sum of the individual components is significantly less than the NO_y measured simultaneously are often referred to as "missing NO_y."

Table 11.7, for example, summarizes measurements of the components of NO_y made at Niwot Ridge, Colorado, in mid-1987 (Ridley et al., 1990a). The sum of NO_x (NO + NO_2), HNO_3, particulate nitrate, PAN,

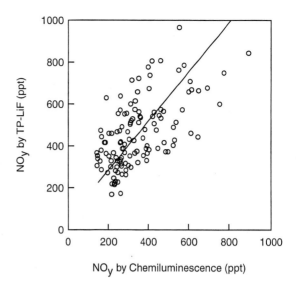

FIGURE 11.26 Measured NO_y concentrations using catalytic converters and TP-LIF or chemiluminescence to measure the NO produced in the same flights as NO data in Fig. 11.24 (adapted from Crosley, 1996).

PPN (peroxypropionyl nitrate), methyl nitrate, n-propyl nitrate, and 2-butyl nitrate was, on average, $76 \pm 13\%$ of the measured NO_y.

Figure 11.27 summarizes the ratio of the sum of the individual components of NO_y to the total NO_y measured using conversion to NO (Parrish et al., 1993). These data summarize measurements of NO, NO_2, PAN, HNO_3, and particulate nitrate as well as total NO_y at Whitetop Mountain (Tennessee), Bondville (Illinois), Scotia (Pennsylvania), and Egbert (Ontario, Canada). The median value (which is less influenced by extremes) of the percentage NO_y that can be accounted for ranges from 75 to 94%, with all but the Bondville site being within experimental error of 100%.

TABLE 11.7 Measured Components of NO_y at Niwot Ridge, Colorado, in Mid-1987[a]

Compound	Percentage of NO_y
NO_x + HNO_3 + NO_3^- + PAN	74
NO_x	32
PAN	24
PPN[b]	1.2
(NPN + 2BN)[c]	0.5
MN[d]	<0.2
Sum	76 ± 13

[a] Adapted from Ridley et al. (1990a).
[b] PPN = peroxypropionyl nitrate.
[c] NPN = n-propyl nitrate; 2BN = 2-butyl nitrate.
[d] MN = methyl nitrate.

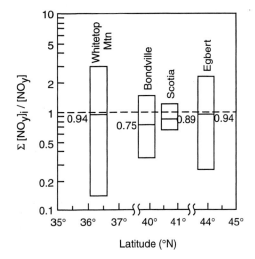

FIGURE 11.27 Ratio of the sum of (NO + NO$_2$ + PAN + HNO$_3$ + particulate nitrate) measured individually to total measured NO$_y$ at three sites in the United States (Whitetop Mountain, Tennessee; Bondville, Illinois; and Scotia, Pennsylvania) and one in Canada (Egbert, Ontario). The bars represent the range of results and the mid-range lines the median values (adapted from Parrish *et al.*, 1993).

However, note the wide range of total NO$_y$ that could be accounted for in Fig. 11.27. As discussed by Parrish *et al.* (1993), values above 100% must be due to systematic errors whereas those below 100% may reflect either systematic errors or true "missing NO$_y$."

Figure 11.28 shows similar data for measurements made at Idaho Hill, Colorado, in the fall of 1993 (Williams *et al.*, 1997). Measurements were made of NO, NO$_2$, PAN, PPN, HNO$_3$, and particulate nitrate, as well as total NO$_y$. Two sets of meteorological conditions were encountered, one where the wind was downslope and from the west where there were few sources nearby, and one where the wind was upslope, carrying pollutants from urban areas to the east. Figure 11.28a shows that for upslope air masses from the east with relatively fresh emissions, the sum of the measured compounds accounts, within experimental error, for the total NO$_y$. The average ratio of ΣNO$_y$/total NO$_y$ was 1.06 ± 0.15. On the other hand, during periods with cleaner, downslope air from the west (which has also had more time to react), the sum of the individual compounds frequently does not add up to the total measured NO$_y$ (Fig. 11.28b). The deficit ranges from 0 to 50% of the measured total NO$_y$.

The average contributions of the various oxides of nitrogen to NO$_y$ for the two conditions are shown in Fig. 11.29 (Williams *et al.*, 1997). The mean total NO$_y$ measured under the more polluted conditions was 4.3 ± 2.4 ppb, with essentially all of it accounted for by the measured individual compounds. Under the cleaner

As discussed by Parrish *et al.* (1993), the Bondville HNO$_3$ data may be artificially low, so that the apparent "missing NO$_y$" of 25% may be an overestimate. Given the difficulty in measuring individual components of NO$_y$ such as HNO$_3$ (vide infra) at the very low levels found in the atmosphere, these data suggest that extent of the "missing NO$_y$" is relatively small on average.

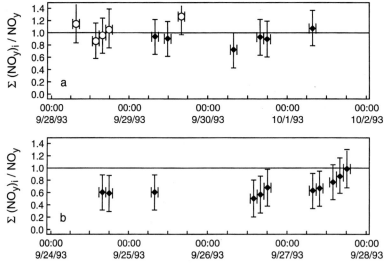

FIGURE 11.28 Ratio of sum of individual compounds (NO + NO$_2$ + PAN + PPN + HNO$_3$ + particulate nitrate) to total measured NO$_y$ under two types of overall meteorological conditions: (a) episodes with winds from the south and east with fresh emissions and (b) winds primarily from the west with cleaner but more aged air (adapted from Williams *et al.*, 1997).

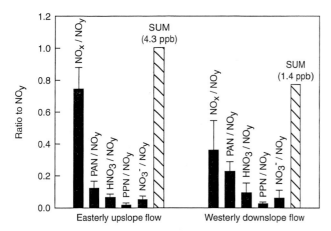

FIGURE 11.29 Ratio of measured individual compounds of NO_y to total NO_y at Idaho Hill, Colorado, with easterly winds (more polluted air) and with westerly winds (cleaner but more aged air), respectively (data from Williams *et al.*, 1997).

conditions of westerly flow (but where the air was more aged), the total NO_y was smaller, 1.4 ± 0.4 ppb, but only 77% was accounted for by the individual compounds. Interestingly, the deficit appeared to correlate with O_3, suggesting the compounds responsible are photochemically generated. A similar observation has been made in Denmark by Nielsen *et al.* (1995, 1998), who also report that the fraction of total NO_y that is in the form of particulate nitrate is small (0.17–0.28%). The deficit in accounting for NO_y at Idaho Hill also decreased as the air temperature decreased, which may reflect a correlation of temperature with the age of the air mass and/or that the species responsible for the missing NO_y are thermally unstable (Williams *et al.*, 1997).

As seen from the $VOC-NO_x$ chemistry in Chapter 6, organic nitrates are among the expected products of the oxidation of hydrocarbons in air containing NO_x. Williams *et al.* (1997) have considered the possible contribution of simple alkyl nitrates but, based on other measurements of these species, indicate that it is unlikely they are responsible for a significant portion of the "missing NO_y."

Multifunctional organics are also possible contributors. Nielsen *et al.* (1998) have examined the possible contribution of multifunctional compounds to "missing NO_y" in both the gas and particle phases. As discussed in Chapter 9, compounds with sufficiently high vapor pressures ($> 2 \times 10^{-5}$ Torr) exist essentially completely in the gas phase, those with low vapor pressures ($< 2 \times 10^{-9}$ Torr) in the condensed phase (i.e., on or in particles), and those in between the two extremes as both gases and particles. Nielsen and co-workers have developed a relationship between the expected vapor pressure of a multifunctional compound and its structure:

$$\log P = -(0.4069 \pm 0.0057)(\text{no. of C})$$
$$- (2.144 \pm 0.070)(\text{no. of nitrate groups})$$
$$- (1.961 \pm 0.057)(\text{no. of OH groups})$$
$$- (1.130 \pm 0.071)(\text{no. of carbonyl groups})$$
$$+ (4.466 \pm 0.077). \quad (M)$$

This relationship is based on data for 183 compounds, including C_7-C_{29} hydrocarbons, C_1-C_{18} alcohols, C_2-C_{10} diols, C_5-C_{18} carbonyls, C_1-C_{20} alkyl nitrates, and C_2-C_3 hydroxynitrates and dinitrates. Based on this analysis, Nielsen *et al.* (1998) suggest that the organic nitrates found in particles are probably bi- and multifunctional compounds and that they may also contribute to gas-phase NO_y and NO_z.

Such multifunctional compounds, however, are very difficult to collect, identify, and quantify and, in fact, need to be specifically targeted if they are of interest for a particular study. As a result, such compounds usually go undetected but may be responsible for some of the "missing NO_y."

In addition, given that the efficiency of conversion of compounds such as HCN and NH_3 over the catalysts may be higher than thought under some conditions (e.g., Kliner *et al.*, 1997; Weinheimer *et al.*, 1998; Bradshaw *et al.*, 1998), these compounds may also be responsible for a substantial portion of the "missing NO_y." However, Williams *et al.* (1998) argue that such interferences, if they exist in their measurements, are too small to account for the magnitude of the "missing NO_y" in their studies.

Because of the sensitivity of NO_y measurements to the particular catalyst used, its recent exposure, cleaning, etc., agreement between various measurements of NO_y and between NO_y and the sum of individual compounds would not necessarily be expected, especially in aged air masses and/or other types or air masses where compounds other than NO and NO_2 contribute significantly to NO_y. Indeed, this is the case (e.g., see discussion by Bradshaw *et al.* (1998) and Williams *et al.* (1998)). Agreement is generally reasonably good at higher concentrations and when NO_x is a major portion of NO_y, e.g., in urban and suburban areas (Williams *et al.*, 1998).

(3) NO_x and NO_2 NO_x is defined as the sum of (NO + NO_2). NO can be measured by the techniques described earlier. NO_2 is one of the compounds contributing to NO_y and in a relatively "young" air mass is often the primary contributor. However, separating out its contribution from other compounds contributing to NO_y obviously requires a different approach.

One approach that has been used is to photolyze the NO_2 at wavelengths below 400 nm to form NO and then measure the NO using chemiluminescence or TP-LIF as discussed earlier (Kley and McFarland, 1980; Ridley et al., 1988; Gao et al., 1994). The reactions are as follows:

$$NO_2 + h\nu \xrightarrow{k_p} NO + O, \quad (8)$$

$$NO + O_3 \xrightarrow{k_{9a}} NO_2^* + O_2 \text{ (in gas phase)}, \quad (9a)$$

$$\xrightarrow{k_{9b}} NO_2 + O_2 \text{ (on cell wall)}. \quad (9b)$$

From the differential equations for the change in NO_2 and NO with time, i.e., $d[NO_2]/dt$ and $d[NO]/dt$, based on reactions (8), (9a), and (9b), it can be shown that the fractional conversion of NO_2 to NO is given by Eq. (N) (Kley and McFarland, 1980; Gao et al., 1994):

$$\text{Fractional conversion} = \frac{k_p(1 - e^{-(k_p + k_9[O_3])\tau})}{k_p + k_9[O_3]}. \quad (N)$$

k_p is the photolysis rate constant for NO_2, reaction (8), $k_9 = k_{9a} + k_{9b}$, and τ is the residence time of the air in the photolysis cell. Fractional conversions of up to ~0.65 have been observed ((Kley and McFarland, 1980; Ridley et al., 1988; Gao et al., 1994). Photolysis of NO_2 at 353 nm using a XeF excimer laser has also been used (Sandholm et al., 1990). Measurement precision and detection limits are determined by a number of factors, including an artifact due to desorption of NO_x from the walls of the reaction vessel during irradiation. Gao et al. (1994) report the latter is equivalent to ~20-40 ppt using synthetic air in the laboratory, but in ambient air, may limit measurements of concentrations below 100 ppt.

As discussed earlier, TDLS can be used to measure NO_2. The detection limit cited for a path length of 33.5 m in a ground-based study is ~150 ppt (Mackay et al., 1988) and 25 ppt in an aircraft study (Schiff et al., 1990). The detection limit for DOAS with a path length of 800 m is ~4 ppb (Biermann et al., 1988).

Finally, matrix isolation combined with electron spin resonance has been used for NO_2 as well as for other free radicals such as HO_2, RO_2, and NO_3 (Mihelcic et al., 1985, 1990, 1993; Zenker et al., 1998). Trace gases in a sample of air (typically about 8 L) are trapped in a D_2O matrix at 77 K and the ESR spectrum obtained. Any paramagnetic species present has a characteristic ESR spectrum that can be used to identify it and, using reference spectra, obtain its concentration. Since NO_2 is the paramagnetic species present in the largest concentration, it is easily detected and measured.

Several intercomparison studies for NO_2 have been carried out (e.g., Fehsenfeld et al., 1990). At concentrations of NO_2 above 400 ppt, measurements using the photolysis of NO_2 and chemiluminescence for the NO generated by photolysis were in reasonably good agreement with TDLS measurements. At levels above about 300 ppt, the photolysis and luminol method corrected for ozone and PAN agreed reasonably well, with the slope of the corrected luminol versus photolysis data being 1.09 (Fehsenfeld et al., 1990).

An airborne intercomparison study (Gregory et al., 1990a) was also carried out using two photolysis methods (the 353-nm laser photolysis with TP-LIF detection of NO and a Xe arc lamp photolysis with chemiluminescence detection of NO) as well as TDLS. Overall, for NO_2 up to 200 ppt, the techniques agreed with the average values of all three by 20% or better and with each other to within 30%. However, below 50 ppt, there was very little correlation between the various measurement techniques (Gregory et al., 1990a).

An informal intercomparison study of NO_2 measurements was carried out in a remote atmosphere at Izaña, Tenerife (Zenker et al., 1998). Three techniques were used: TDLS, photolysis with a chemiluminescence detector, and matrix isolation–ESR. Agreement between the three methods was good, with plots of data from one technique against the others having slopes within experimental error of unity. For example, TDLS and the photolysis technique plotted against the matrix isolation measurements had slopes of 0.90 ± 0.47 and 1.04 ± 0.34, respectively, over a range of NO_2 concentrations from ~100 to 600 ppt.

In summary, there are a variety of methods of measuring NO_2 that are reasonably accurate for higher concentrations, particularly those found in polluted areas. However, at smaller concentrations found in the remote troposphere, there are significant discrepancies between the various methods.

In addition to these techniques, there are passive samplers for NO_2 that have been used for unique situations such as indoor measurements. For example, in the Palmes Tube, NO_2 diffuses through to a surface coated with triethanolamine and is trapped in the form of NO_2^-. The nitrite is subsequently measured colorimetrically (e.g., see Boleij et al., 1986; Miller, 1988; and Krochmal and Górski, 1991). As with most, if not all, such wet chemical methods, interferences can arise, for example, from PAN (Hisham and Grosjean, 1990) and HONO (Spicer et al., 1993b).

(4) Typical levels of NO, NO_2, and NO_y Figure 11.30 shows a summary of measurements of surface concentrations of NO, NO_x, and NO_y made at a variety of remote to rural sites in North America and Europe (Emmons et al., 1997). The bars encompass the central 90% of the values and the medians and means are

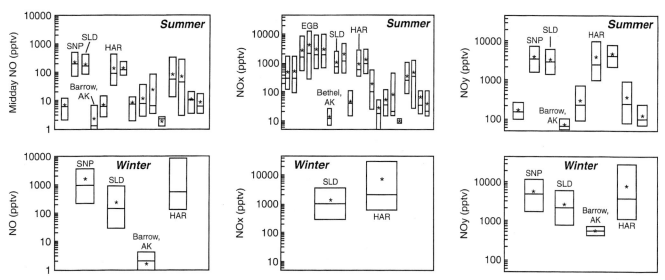

FIGURE 11.30 Range of surface concentrations of NO, NO_x, and NO_y at a variety of Northern Hemisphere locations. The asterisk is the mean and the horizontal line the median. The bar represents the central 90% of measured values. SNP, Shenandoah National Park, Virginia; SLD, Schauinsland, Germany; EGB, Egbert, Ontario, Canada; and HAR, Harvard Forest, Massachusetts. (Adapted from Emmons et al., 1997.)

shown by the horizontal lines and asterisks, respectively. The identity of some of the specific sites at which measurements were made are indicated. As expected, remote regions such as Barrow, Alaska, have the smallest concentrations of NO, typically less than 10 ppt. The most polluted (but still rural) areas have concentrations that in the winter are in the ppb range.

Similar conclusions hold for NO_x and NO_y. NO_x concentrations range from median values of 25 ppt in Alaska (remote) to 2.3 ppb at Egbert, a rural area in Ontario, Canada, in the summer. NO_y ranges from 69 ppt at Barrow, Alaska, to 5.0 ppb at Shenandoah National Park in Virginia (Emmons et al., 1997). Data from a number of studies are archived electronically and can be accessed as described by Carroll and Emmons (1996).

In urban areas, the concentrations can of course be much greater. For example, in Paris in late September 1997, NO_2 concentrations exceeded 210 ppb (*Environmental Science & Technology*, 1997). In metropolitan Toronto, Canada, peak 30-min average concentrations of ~40 ppb have been reported (Schiff et al., 1986). In the Los Angeles area, maximum 1-h concentrations of about 200 ppb are encountered (Air Quality Management District summary data; see Appendix IV).

Measurements of NO, NO_x, NO_y, PAN, HNO_3, and O_3 in the free troposphere obtained from 1985 to 1995 are summarized by Thakur et al. (1999) and can be obtained electronically as described in that paper.

(5) HNO_3 Analysis of HNO_3 at the low levels typically found in the atmosphere is difficult, in large part due to its tendency to adsorb very readily to surfaces. As a result, sampling HNO_3 in an artifact-free manner is often the limiting aspect in making accurate measurements.

Several different methods exist for measuring HNO_3, most commonly FTIR and TDLS, which were described earlier. Other techniques commonly used include filters, denuders, transition flow reactors, and scrubbers, followed by analysis of the collected material for nitrate, e.g., by ion chromatography. A modification of the luminol method has also been used. Finally, mass spectrometric methods look very promising as a sensitive and specific method of detection and measurement. A brief description of each of these methods that have not yet been treated follows.

Filters. HNO_3 is efficiently trapped out on nylon filters. Typically, two or more filters are connected in series. A schematic of such a filter pack was shown in Fig. 11.22 (Anlauf et al., 1988). A Teflon filter first removes particles from the airstream and a nylon filter then removes gaseous HNO_3. In this particular system, a third filter (Whatman 41 impregnated with an aqueous solution of glycerol and citric acid) was used to trap NH_3. After sample collection, each of the filters is extracted separately and nitrate, ammonium, and additional particle components collected on the Teflon filter are measured by ion chromatography. The sensitivity of this method for nitric acid and the other species is determined in part by filter blank values (i.e., nitrate on unexposed filters) and by the total amount collected and hence the sampling time used. Times of

4–6 h are often used, but can be as short as 0.5–2 h (Fehsenfeld *et al.*, 1998).

Other filters have also been used to collect nitric acid, such as Whatman 41 filters impregnated with NaCl (e.g., Anlauf *et al.*, 1986).

Potential interferences in the measurement of nitric acid using this method include removal of gaseous HNO_3 on the Teflon particle filter and/or volatilization of particle nitrate collected on this Teflon filter. As discussed in Chapter 7, NH_4NO_3 is a common particle component, but exists in equilibrium with gas-phase NH_3 and HNO_3:

$$NH_4NO_{3(s,aq)} \leftrightarrow NH_{3(g)} + HNO_{3(g)}. \quad (10, -10)$$

Shifts to the right, e.g., due to a temperature increase, release HNO_3 and NH_3, which are then collected on the nylon filter and Whatman impregnated filters, respectively, and measured as gas-phase nitric acid and ammonia. This was hypothesized to be responsible for higher filter pack values compared to those measured by mass spectrometry under some conditions, particularly at colder air temperatures (<15°C), where the equilibrium (10, −10) favors relatively larger amounts of ammonium nitrate in air (Fehsenfeld *et al.*, 1998). Talbot *et al.* (1990) observed higher HNO_3 concentrations by the nylon filter technique compared to a mist chamber (vide infra) and hypothesized that unknown (perhaps organic) nitrogenous compounds were also being collected on the nylon, forming nitrate. They also showed that O_3 at typical concentrations found in the troposphere could react with some unknown substance(s) on the nylon filter to generate a positive artifact. This artifact was significantly reduced by prewashing the nylon to remove water-soluble adsorbed species.

Denuders. A variety of denuder wall coatings have been used to collect HNO_3 (Perrino *et al.*, 1990). These include nylon fiber mats (e.g., Durham *et al.*, 1987), MgO (e.g., Solomon *et al.*, 1988, 1992), Na_2CO_3/glycerol (e.g., Ferm, 1986; Koutrakis *et al.*, 1988), and tungsten oxide (WO_3) (e.g., see Fox *et al.*, 1988). A variant of this is the tungstic acid technique. Air containing the HNO_3 is passed through tubes coated with tungstic acid. When the tube is subsequently heated, the HNO_3 decomposes and desorbs as NO or NO_2.

Transition flow reactors. TFRs have also been used to measure gaseous HNO_3 (e.g., Hering *et al.*, 1988). When operated in the configuration used by Durham *et al.* (1986), where the TFR is upstream of the Teflon particle filter, the problems of adsorbing gaseous HNO_3 on the particle filter or of evaporation of HNO_3 from the collected particles onto the gaseous HNO_3 sampler are avoided.

Scrubbers. Mist chamber scrubbers have also been used for HNO_3. The airstream passes through a Teflon filter to remove particles and then encounters a mist of water that scrubs the HNO_3 out of the air. The nitrate concentration is measured in the aqueous scrubbing solution using ion chromatography (Talbot *et al.*, 1990).

Luminol method. As described earlier, NO_2 undergoes a chemiluminescent reaction with luminol, and this has been used to measure NO_2. This has also been used to measure HNO_3 by difference. One airstream passes through a Teflon filter to remove particles, while another passes through a Teflon–nylon filter combination to remove both particles and gaseous HNO_3. The air is then passed over hot glass beads to convert NO and HNO_3 to NO_2, which is measured using the luminol method. The difference in signal between the two filtering methods then gives gaseous HNO_3 (Hering *et al.*, 1988).

All of these methods have potential interferences, some of which have been described in conjunction with the individual methods. For example, whenever a Teflon prefilter for particles is used, gaseous HNO_3 can be taken up on this filter, particularly when it has a high particle loading, leading to a negative artifact. Volatilization of ammonium nitrate particles collected on the Teflon filter can also occur, giving a positive artifact. Uptake and conversion of other nitrogen-containing compounds to nitrate gives a positive artifact (e.g., see Koutrakis *et al.*, 1988). In addition, as mentioned at the beginning, HNO_3 is notoriously "sticky," and the use of short (or ideally, no) sampling lines made of materials such as Teflon that minimize its adsorption are necessary to avoid negative artifacts when it is adsorbed and positive artifacts when it later desorbs. Finally, they are not "real-time" methods in that sampling for periods of time of the order of hours is typically required.

Given these considerations, it is perhaps not surprising that intercomparison studies have shown significant disagreements between the various techniques. Figure 11.31, for example, shows some data from one such intercomparison (the so-called "nitric acid shootout"), in which filter pack (FP), difference denuder (DD), annular denuder (AD), TDLS, and FTIR measurements were all taken simultaneously. It is seen that the filter pack measurements were consistently higher, and the annular denuder and TDLS consistently lower, than the FTIR measurements, which were used as the standard for comparison. While most measurements in this period were within the wide error bars for the FTIR measurements, note that the range of concentrations from the different methods spans more than a factor of two! In addition, there were significant differences in some of the methods from day to night.

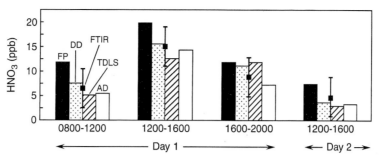

FIGURE 11.31 Comparison of gaseous HNO$_3$ measurements made simultaneously in Claremont, California, on two different days using FT-IR shown as ■; with error bars and by a filter pack (FP), a difference denuder (DD), an annular denuder (AD), and tunable diode laser spectroscopy (TDLS) (adapted from Hering et al., 1988).

These studies were carried out in a polluted urban atmosphere where the concentrations are relatively high; it might be expected that the agreement (or lack thereof) would certainly not improve at the much smaller concentrations found in rural and remote regions. Similar disagreements have been observed in other intercomparison studies (e.g., Anlauf et al., 1985; Gregory et al., 1990b; Huebert et al., 1990), even in synthetic atmospheres (e.g., Fox et al., 1988).

Because of such problems, spectroscopic methods, where available, are generally preferred. However, the sensitivity of long-path FTIR (4 ppb at a 1.15-km path length) is insufficient except for the most polluted atmospheres. TDLS measurements have better sensitivity (75 ppt) but it is still not sufficient for measurements in remote atmospheres. As a result, there continues to be a need for accurate, specific, and very sensitive methods of measurement of HNO$_3$ that also have good time resolution (seconds or better). Recent efforts using mass spectrometry look promising in this regard.

Mass spectrometry. Several instruments based on chemical ionization mass spectrometry (CIMS) have been developed and applied to ambient air (Huey et al., 1998; Mauldin et al., 1998). Figure 11.32 shows two such instruments that have undergone an informal intercomparison study between themselves and a filter pack (nylon) method (Fehsenfeld et al., 1998).

The first CIMS method (Fig. 11.32a) uses the equilibrium HNO$_3$ adduct with HSO$_4^-$ to measure gaseous HNO$_3$ (Mauldin et al., 1998):

$$HNO_3 + HSO_4^- \underset{}{\overset{K_{11,-11}}{\longleftrightarrow}} HSO_4^- \cdot HNO_3. \quad (11, -11)$$

From equilibrium considerations,

$$[HNO_3] = [HSO_4^- \cdot HNO_3]/K_{11,-11}[HSO_4^-], \quad (O)$$

and the ratio of the adduct and reagent ions is measured to obtain the concentration of HNO$_3$. Sampling and ionization of H$_2$SO$_4$ to generate the reagent HSO$_4^-$ ion using a radioactive ^{241}Am source occur at atmospheric pressure. The ions are transported into the mass spectrometer through a free jet expansion of dry N$_2$, which helps to strip water clustered to the ions, as well as through a collisional dissociation chamber where the collisional energies have been adjusted so that HSO$_4^-$(H$_2$O)$_n$ clusters are dissociated but the adduct HSO$_4^- \cdot$HNO$_3$ is not. Figure 11.33a shows a typical mass spectrum of air using this approach. The strong reagent HSO$_4^-$ ion and the adduct are clearly visible. Detection limits for a 1-s integration time are reported to be 1–3 ppt in clean air but higher in polluted air due to increased background signals (Mauldin et al., 1998).

In the second CIMS instrument, shown in Fig. 11.32b, air is drawn through a small orifice into a flow tube at ~20 Torr pressure where it is mixed with SiF$_5^-$ reagent ions (Huey et al., 1998). These ions are generated by alpha particle ionization where electrons attach to small amounts (~30 ppb) of SF$_6$ in the N$_2$ carrier gas, forming SF$_6^-$, SF$_5^-$, and F$^-$. These species react with SiF$_4$, present at ~300 ppb, forming SiF$_5^-$. The SiF$_5^-$ reagent ions react with gaseous HNO$_3$, setting up an equilibrium with the adduct (Huey and Lovejoy, 1996):

$$SiF_5^- + HNO_3 + M \underset{}{\overset{K_{12,-12}}{\longleftrightarrow}} SiF_5^- \cdot HNO_3 + M. \quad (12, -12)$$

A typical mass spectrum taken in air is shown in Fig. 11.33b. The reagent ion and the adduct are clearly visible, as are adducts with other species, including formic acid. As discussed by Huey et al. (1998), the latter should permit the same method to be used to measure HCOOH, but with a sensitivity that is likely to be about two orders of magnitude less than that for

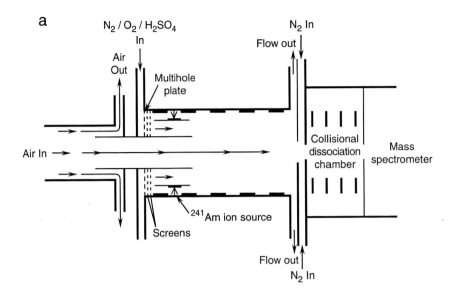

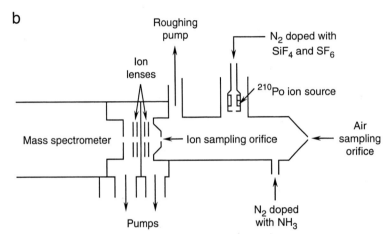

FIGURE 11.32 Schematic diagrams of two chemical ionization mass spectrometers used to measure HNO_3 (adapted from Mauldin *et al.*, 1998; and Huey *et al.*, 1998).

HNO_3 because of differences in the equilibrium constants for adduct formation. Since reaction (12, −12) is at equilibrium, the HNO_3 concentration is given by

$$[HNO_3] = [SiF_5^- \cdot HNO_3]/K_{12,-12}[SiF_5^-] \quad (P)$$

and measuring HNO_3 requires measuring the ratio of the adduct and reagent ions. Background signals occur from desorption of HNO_3 from the flow tube walls and from formation of radicals in the ionization region that generate HNO_3. The addition of small amounts of NH_3 was found to decrease the first signal but not react with HNO_3 in the air under the flow tube conditions. Detection limits were 5–50 ppt (depending on the background, which was larger at higher HNO_3 concentrations) for a 1-s measurement time.

An informal intercomparison of these two CIMS methods with a filter pack method shows generally excellent agreement between the mass spectrometric approaches and often, but not uniformly, good agreement with the filter pack method (Fehsenfeld *et al.*, 1998). The latter was often high, which was attributed to interference from decomposition of ammonium nitrate to HNO_3 + NH_3 on the Teflon particle prefilter, followed by absorption of the HNO_3 by the nylon filter.

In short, these chemical ionization mass spectrometry methods appear to be quite promising for the measurement of HNO_3, especially at the low levels found in the remote trosposphere.

Typical tropospheric concentrations of HNO_3. Given the difficulties in measuring atmospheric nitric acid,

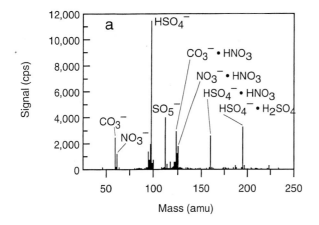

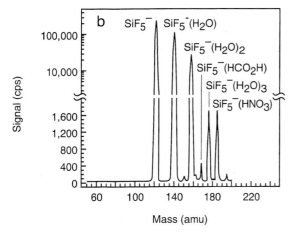

FIGURE 11.33 Typical chemical ionization mass spectra taken using (a) HSO_4^- as the chemical ionization reagent or (b) SiF_5^- as the CI reagent. Note the change to a logarithmic scale in (b) above ~ 2000 counts per second (CPS) (adapted from Mauldin et al., 1998; and Huey et al., 1998).

there is not a large data base of measurements in a variety of locations and types of air masses that is believed to accurately portray its concentrations, particularly in remote atmospheres. However, the intercomparison studies described by Hering et al. (1988) suggest that concentrations in polluted urban areas are as high as 25 ppb for 4- to 6-h averages, levels that have also been measured over much shorter time periods using FTIR (Biermann et al., 1988). Similar ppb levels (up to ~5 ppb) have been measured near Vienna, Austria, in the summer (Piringer et al., 1997).

A brief informal intercomparison study of the mass spectrometry methods with a nylon filter pack method near Boulder, Colorado, gave average levels of 0.38–1.6 ppb when the wind carried air from the direction of the greater metropolitan Denver urban area and 0.14–0.56 ppb when the wind was downslope and westerly, where there are fewer emissions sources; previous filter pack measurements at this site gave concentrations ranging from a few ppt, characteristic of remote regions, to several ppb, characteristic of polluted urban areas (Fehsenfeld et al., 1998).

Measurements of HNO_3 in the marine boundary layer are typically of the order of tens to hundreds of ppt. For example, Heikes et al. (1996) reported average concentrations of 160 ppt, with a range from 30 to 280 ppt. In the middle and upper troposphere, concentrations of ~100–400 ppt have been reported (e.g., Singh et al., 1998).

(6) NO_3 As discussed earlier, the nitrate radical can be measured using visible spectroscopy and its absorption bands, particularly the one at 662 nm. As a result, visible absorption spectroscopy has been the method of measurement used most extensively for NO_3. As discussed shortly, a matrix isolation technique has also been applied with success in some studies.

Noxon et al. (1978) were the first to report the detection of NO_3 and to estimate its column abundance in the atmosphere, using its absorption at 662 nm and the moon as the light source. Their initial hypothesis was that most of the NO_3 was in the stratosphere. However, Noxon et al. (1980) subsequently showed using the moon as the light source, or alternatively a surface-based lamp with a 10-km path length at the Fritz Peak Observatory in Colorado, that NO_3 was also present in the troposphere at concentrations up to a few hundred ppt. About the same time, NO_3 was also detected and measured in the polluted troposphere by Platt et al. (1980b). Since then, there have been a number of measurements of its column abundance and concentrations at specific locations in the troposphere (e.g., see Platt, 1994; and Plane and Smith, 1995), all of which are at night or at sunset or sunrise due to the rapid photolysis of NO_3 during the day.

Vertical profiles, and in particular the amounts of tropospheric NO_3, have been extracted from measurements of the column abundance as a function of solar zenith angle at sunrise using either the moon or scattered sky light as the light source (e.g., Smith and Solomon, 1990; Smith et al., 1993; Weaver et al., 1996; Aliwell and Jones, 1996a, 1996b, 1998). As the sun rises, the column abundance of NO_3 decreases due to photolysis. During the night, the sun is sufficiently below the horizon that the atmosphere is in darkness throughout the stratosphere and troposphere. As it rises to a solar zenith angle of ~97°, altitudes down to 40 km are exposed to direct sunlight, and by the time the solar zenith angle is 93°, only the region below 10 km is not exposed to direct sunlight. Because the photolysis of NO_3 is so fast, under these conditions any signal remaining must be attributable to tropospheric

NO$_3$ (e.g., Smith and Solomon, 1990; Weaver et al., 1996; Aliwell and Jones, 1998).

Using this approach in various locations, it has been shown that the relative contributions of stratospheric and tropospheric NO$_3$ vary considerably. For example, in the Antarctic in spring, essentially all of the NO$_3$ was in the stratosphere (Smith et al., 1993), whereas at Fritz Peak, Colorado, in the summer, about equal amounts were in the troposphere and stratosphere (Weaver et al., 1996). Assuming that this tropospheric NO$_3$ was in a 1-km-thick bounday layer, the average NO$_3$ radical concentration in this layer at sunrise was about 20 ppt. Aliwell and Jones (1998) using a similar approach at Cambridge, England, suggest the average concentration could be as high as 89 ppt.

Surface measurements of NO$_3$ have been made using folded light paths and a light source such as a Xe arc lamp (Platt, 1994; Plane and Smith, 1995). Concentrations as high as ~350 ppt have been observed in polluted urban areas (Platt et al., 1980b, 1981; Platt and Janssen, 1995), although many times even in polluted areas, the concentrations are below 20 ppt (Biermann et al., 1988). This likely reflects the balance between sources and sinks. For example, since NO$_3$ reacts rapidly with NO, significant concentrations of NO$_3$ will not be observed close to NO emission sources.

A second technique, matrix isolation–electron spin resonance (ESR), described earlier for NO$_2$ measurements, has also been used to measure NO$_3$ in the atmosphere (Mihelcic et al., 1985, 1990, 1993). Because there is a large concentration of NO$_2$ in air compared to other paramagnetic species, this dominates the spectra. However, the contribution of NO$_2$ can be subtracted using a reference spectrum and the residual then matched using the simultaneous fit of other contributing species (Mihelcic et al., 1990) to derive the contributions of species such as NO$_3$, HO$_2$, and RO$_2$. A sample spectrum containing contributions from all of these species is shown later (Fig. 11.53) in the discussion of HO$_2$ and RO$_2$ measurements. The detection limit of this method for NO$_3$ is 3 ppt (Mihelcic et al., 1993). The disadvantage is that it is not a "real-time" method, and as with any sampling of free radicals, care must be taken not to destroy the radicals before they are trapped and/or measured.

Typical tropospheric concentrations. Studies carried out in a remote region, at Izaña de Tenerife in the Canary Islands, showed average nighttime NO$_3$ concentrations in clean air from the mid-Atlantic to be ~8 ppt, with a maximum of ~20 ppt (Carslaw et al., 1997b), and the concentrations are often below the detection limits (e.g., <3 ppt at Loop Head, Ireland; Platt and Janssen, 1995).

In rural–suburban areas, concentrations between these two have been observed. For example, in central California peak concentrations were typically about 30 ppt, with a maximum value of ~80 ppt (Smith et al., 1995), and at Kap Arkona in the Baltic Sea, the average nighttime concentrations were 8 ppt (Heintz et al., 1996). However, even in such rural–suburban areas, high concentrations can occur, e.g., 280 ppt at Deuselbach, Germany (Platt et al., 1981; Platt and Janssen, 1995).

In short, the concentration of NO$_3$ in the troposphere can vary from very small, low-ppt concentrations to several hundred ppt, depending on the particular air mass. As discussed in Chapters 7 and 10, at typical tropospheric levels, it is believed to play a major role in nighttime chemistry, in some cases rivaling daytime OH for the net oxidation of certain organics, particularly alkenes (e.g., see Aliwell and Jones, 1998) as well as certain gaseous PAH.

(7) HONO Because of its importance as an OH source by photolysis at dawn, particularly in polluted areas, there have been a number of measurements reported for HONO. The two methods used most commonly have been DOAS and denuder methods. In addition, diffusion scrubber and photofragmentation—laser-induced fluorescence methods have been developed, although they have not seen widespread use.

DOAS. As discussed earlier, HONO has been measured at the earth's surface at a number of sites around the world by DOAS using its characteristic absorption bands in the 340- to 380-nm region. Because this is a spectroscopic method, it has high selectivity as well as sensitivity. The disadvantage is that it requires a unique set of equipment and, more importantly, skilled experimentalists to accurately extract the HONO signal from others present in air, particularly that of NO$_2$, which is essentially always present simultaneously.

Denuders. The principle behind denuders has been discussed earlier with respect to HNO$_3$. Several types of wall coatings have been used for trapping HONO, including Na$_2$CO$_3$ (Ferm and Sjödin, 1985) and a triple denuder coated first with tetrachloromercurate to remove SO$_2$ and HNO$_3$, followed by two Na$_2$CO$_3$-coated tubes (Febo et al., 1993; Febo and Perrino, 1995; Febo et al., 1996). A wet diffusion denuder using NaOH as the absorbing agent has also been described (Simon and Dasgupta, 1995). The nitrite ion is then extracted and measured, most commonly using ion chromatography but, in some cases, using colorimetric methods (e.g., Ferm and Sjödin, 1985).

Figure 11.34 shows one set of measurements of HONO made in Milan, Italy, using DOAS and a denuder method, respectively (Febo et al., 1996). In this

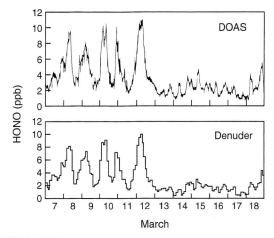

FIGURE 11.34 HONO measurements at Milan, Italy, using DOAS or denuders, respectively, in March 1994 (adapted from Febo et al., 1996).

particular study, the DOAS and denuder methods are in reasonably good agreement, although this is not always the case.

As is the case for denuder methods in general, artifacts are possible and this is particularly the case for nitrite, which can be oxidized relatively easily to nitrate. Positive artifacts due to high concentrations of NO_2 and PAN can occur, and SO_2 can also interfere (Perrino et al., 1990). For example, HONO has been reported during the day using denuders at levels much higher than would be expected given its rapid photolysis, suggesting the presence of significant positive artifacts (e.g., see Lammel and Cape, 1996).

Diffusion scrubber. In one such approach, air containing the HONO flows past a microporous membrane through which a flow of water is pumped. HONO (and other gases) diffuse through the membrane and are "scrubbed" into the water and measured by ion chromatography (Večeřa and Dasgupta, 1991). A second approach involves reducing the absorbed nitrite ion to NO using ascorbic acid and then measuring the NO by chemiluminescence (Kanda and Taira, 1990; Harrison et al., 1996). Since there is often significant NO present, HONO must be determined using difference methods.

Photofragmentation–laser-induced fluorescence. In an approach similar to that which has been applied to HNO_3, HONO can be photolyzed to generate OH and the OH measured using laser-induced fluorescence (Rodgers and Davis, 1989):

$$HONO + h\nu(355 \text{ nm}) \rightarrow OH(X^2\Pi)_{v''=0} + NO, \quad (13)$$

$$OH(X^2\Pi)_{v''=0} + h\nu(282 \text{ nm}) \rightarrow OH(A^2\Sigma)_{v'=1}$$
$$\xrightarrow{M} OH(A^2\Sigma)_{v'=0}, \quad (14)$$

$$OH(A^2\Sigma)_{v'=0} \rightarrow OH(X^2\Pi) + h\nu(309 \text{ nm}). \quad (15)$$

The detection of OH and the uncertainties associated with it are discussed in detail shortly. The detection limit is a few tens of ppt for a measurement time of 15 min (Rodgers and Davis, 1989).

Atmospheric pressure ionization mass spectrometry has also been used to meaure HONO in indoor air enviroments (Spicer et al., 1993a) and outdoors from a research aircraft (Berkowitz et al., 1998). In this case, HONO does not have a parent peak in either positive or negative ion modes, so that adduct formation must be used to form a parent ion. In this case, CCl_4 was introduced into the corona discharge region, forming Cl^-, which forms an adduct with HONO. Peaks at m/e 82 and 84 corresponding to the $^{35}Cl^-$ and $^{37}Cl^-$ adducts are observed in the negative ion mode and their collisionally induced fragmentation to m/e 46 followed.

Typical surface concentrations of HONO. Because it is so readily photolyzed, HONO builds up at night and its concentrations rapidly drop at dawn. Figure 11.35 summarizes the range of measured tropospheric HONO concentrations and their average values in remote to polluted urban regions (Lammel and Cape, 1996). The highest concentrations are generally found in polluted areas where there are higher concentrations of precursor NO_2 (see Chapter 7). Concentrations up to 10 ppb have been observed in Milan, Italy (Febo et al., 1996), and the Los Angeles area (Platt et al., 1980a; Harris et al., 1982; Atkinson et al., 1986; Winer and Biermann, 1994). Peak values of several ppb have been observed in many other urban locations, including, for example, Ispra, Italy (Andrés-Hernández et al., 1996), and Göteborg, Sweden (Sjödin, 1988).

(8) NH_3 A number of measurement techniques have been used for ammonia, including a spectroscopic method, denuder methods, and filter packs.

Photofragmentation–laser-induced fluorescence (PD-LIF). This spectroscopic method is based on the photofragmentation of NH_3 in a two-photon process using 193-nm radiation, followed by laser-induced fluorescence of the NH fragment (Schendel et al., 1990). The processes are as follows:

$$NH_3 + 2h\nu(193 \text{ nm}) \rightarrow NH(b^1\Sigma^+)_{v''=0} + H_2 \text{ or } 2H, \quad (16)$$

$$NH(b^1\Sigma^+)_{v''=0} + h\nu(450 \text{ nm}) \rightarrow NH(c^1\Pi)_{v'=0}, \quad (17)$$

$$NH(c^1\Pi)_{v'=0} \rightarrow NH(a^1\Delta)_{v''=0} + h\nu(325 \text{ nm}). \quad (18)$$

Selectivity is obtained by tuning the NH excitation laser to a specific rotational transition and following specific fluorescence transitions. Detection limits of ~5 ppt for

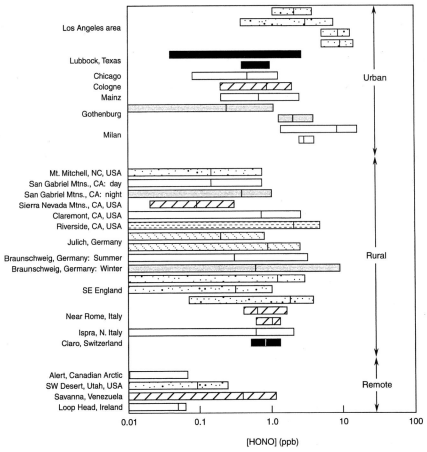

FIGURE 11.35 Range of tropospheric HONO concentrations observed in remote to polluted areas. The averages are shown by the vertical bars. Note the scale is logarithmic (adapted from Lammel and Cape, 1996).

an integration time of 5 min have been reported (Williams et al., 1992).

Denuder methods. As described earlier for HNO_3, diffusion denuders are also used for NH_3. For example, tungstic acid coated denuders take up both HNO_3 and NH_3. The ammonia is then thermally desorbed (as NH_3), oxidized to NO, and measured using chemiluminescence (Braman et al., 1982; Hering et al., 1988). In practice, since HNO_3 desorbs in the form of NO and NO_2, the NH_3 from the thermal desorption is usually first readsorbed on a second WO_3-coated tube to separate it from the HNO_3 signal, and a second thermal desorption followed by conversion to NO used for the measurement of NH_3.

Other denuder coatings for NH_3 include molybdenum oxide, from which ammonia is thermally desorbed and converted to NO for measurement as in the tungstic acid method; with MoO_3 as the coating, some of the NH_3 thermally desorbs directly in the form of NO (Langford et al., 1989). In addition, oxalic acid has been used as a coating and the NH_3 collected as ammonium ion is obtained by extracting the tube and measuring NH_4^+ by ion chromatography (Ferm, 1979).

Filter packs. As shown in Fig. 11.22, NH_3 can be collected on impregnated filters in filter packs designed to collect particles and gas-phase nitric acid. Oxalic acid or citric acid on Whatman filters is often used to absorb the gaseous ammonia, which is then measured by extraction into aqueous solution and ion chromatography or by a colorimetric method (e.g., see Anlauf et al., 1988; and Williams et al., 1992).

Williams et al. (1992) have carried out an intercomparison of PD-LIF with three denuder methods and one filter pack technique using both laboratory-prepared samples and ambient air. All methods agreed to within 10% when measuring an ammonia standard. When a measured amount of NH_3 was introduced as a spike into filtered air, the agreement was not as good. The PF-LIF and citric acid denuders gave 87 and 93% response, respectively, but the recoveries using the tungstic acid denuder, the molybdenum oxide annular denuder, and the filter pack were only 68, 37, and 22%, respectively. In the ambient air measurements, the PF-LIF and citric acid denuders were in good agreement. The tungstic acid technique gave values that were well correlated with the PF-LIF values but had an intercept that varied with the concentration range. The filter pack gave concentrations having a response about two-thirds of those methods, possibly due to uptake of NH_3 on the Teflon particle prefilter at the lower temperatures of these studies. The molybdenum oxide annular denuder measurements were about 64% of the PF-LIF values above 1 ppb but tended to be higher below this concentration.

Although other fluorescence (e.g., Rapsomanikis et al., 1988; Genfa et al., 1989), chemiluminescence (Maeda and Takenaka, 1992), and photoacoustic or photothermal methods (e.g., De Vries et al., 1995) have been proposed, they have not found widespread use.

Typical tropospheric concentrations of NH_3. In remote areas, NH_3 concentrations can be quite low, <50 ppt (e.g., Lewin et al., 1986; Alkezweeny et al., 1986), whereas close to sources such as agricultural areas and cattle feedlots, they can be about three orders of magnitude larger. For example, measurements of ammonia made at Boulder, Colorado, and at Niwot Ridge, west of Boulder, are typically ~5.5 ppb and ~200 ppt, respectively (Langford and Fehsenfeld, 1992). The higher value in Boulder is due to the proximity of agricultural areas and cattle feedlots. Similarly, Biermann et al. (1988) observed concentrations of NH_3 at Claremont, California, over a range from 57 ppb when the winds were blowing from the direction of an agricultural area with a high density of cattle feedlots and poultry and dairy farms to undetectable levels (<1–2 ppb) when the wind was from other directions.

b. O_3

Detection techniques. Detection techniques for surface-based measurements of ozone include (1) UV absorption at 254 nm; (2) chemiluminescence on reaction with NO (or ethene); (3) DOAS; (4) TDLS; and (5) wet chemical methods, mainly those involving the oxidation of I^- to I_2 and measurement of the I_2 colorimetrically or coulometrically. The wet chemical method and the principles behind DOAS and TDLS were discussed earlier and are not treated further here.

UV absorption relies on the strong peak absorption of O_3 at 254 nm (see Chapter 4.B), which coincidentally overlaps the strong 254-nm emission from low-pressure mercury lamps. Commercial instruments based on this absorption are in widespread use. Typically, air is drawn into a cell and the absorption of the 254-nm line from a Hg lamp is measured. The airstream is then switched to pass through a catalyst that destroys O_3 and the absorption is again measured to provide I_0.

Alternatively, two parallel cells are used, one of which has air from which O_3 has been scrubbed and the other the air containing ozone. These instruments appear to be relatively artifact-free, although interference from high concentrations of photooxidation products in laboratory studies and from contamination of the cell windows has been reported (Meyer et al., 1991; Kleindienst et al., 1993).

The principles behind the chemiluminescence methods using NO and ethene were discussed earlier. The instruments using NO also are widely used, and results obtained are in good agreement with those using UV absorption at 254 nm. However, there appears to be a water interference in the NO chemiluminescence method that gives a positive artifact of ~3.7% per percent water (e.g., see Kleindienst et al., 1993). This could be significant under some conditions. For example, Kleindienst et al. (1993) estimate that an error of 13 ppb O_3 could result at high relative humidities and temperatures (30°C and 60% RH).

Another approach is to obtain tropospheric ozone levels using satellite data as described in Chapter 13.C (e.g., see Fishman et al., 1990; and Munro et al., 1998). Vertical tropospheric ozone profiles can be extracted using satellite measurements of backscattered solar radiation at wavelengths where ozone has strong absorption bands.

Finally, passive samplers have also been developed for ozone, primarily for use in epidemiological studies. For example, Brauer and Brook (1995) describe the application of a passive sampler in which air containing ozone diffuses through a Teflon membrane and reacts with nitrite. The sampler is then extracted and the nitrate product measured using ion chromatography.

Typical ambient levels. As discussed in Chapter 14.B.2d, levels of ozone worldwide before the industrial revolution appear to have been ~10–15 ppb. However, at the present time, levels of 30–40 ppb are found in even the most remote regions. This increase has been attributed to increased anthropogenic emissions of oxides of nitrogen, since photolysis of NO_2 is the sole known significant source of anthropogenically derived ozone.

Peak levels in rural–suburban areas are typically in the 80- to 150-ppb range, reaching as high as 500 ppb or more in the most highly polluted urban areas that have few controls on emission sources.

c. CO

Detection techniques. Detection techniques commonly used for CO include two infrared techniques, TDLS and NDIR (also known as gas filter correlation, GFC), and gas chromatography with various detectors. The principles behind TDLS and NDIR have been discussed earlier. Sachse *et al.* (1987), for example, applied TDLS to measure CO using the P(5) line at 4.7 μm (2128 cm^{-1}). Measurements could be made in 1 s with an accuracy of ± 1.4 ppb.

The application of a commercial NDIR instrument to ambient CO measurements is described by Parrish *et al.* (1994); precision (1σ) of ~2 ppb with 1-h averaging times could be obtained. A similar detection principle has been used to measure middle-tropospheric CO from the space shuttle (Reichle *et al.*, 1990).

Finally, gas chromatography can be used to separate CO from the other constituents in air. Various detection methods have been used, including conversion of CO to CH$_4$ and measurement of CH$_4$ by flame ionization detection (e.g., Porter and Volman, 1962). A unique method is also used for CO in which it reacts with hot HgO, releasing Hg vapor, which is measured by atomic absorption of light from a mercury lamp, known as the GC–HgO method (e.g., see Greenberg *et al.*, 1996). Intercomparisons of chromatographic measurements and TDLS have shown that the two approaches are in good agreement (e.g., see Hoell *et al.*, 1985, 1987b). Intercomparisons of the GC–HgO methods with NDIR have also been carried out using a round-robin approach on prepared CO standards (Novelli *et al.*, 1998a); while agreement was good in many cases, the uncertainties associated with the NDIR method were larger by a factor of about 5 at low CO levels, ~50 ppb. Differences in the accuracy between laboratories were also noted, even among those using the same method. These were traced in some cases to inaccurate calibrations, but other factors such as nonlinearity in the GC–HgO detectors over the full range of atmospheric concentrations were also suggested as possible contributing factors.

Typical ambient levels. Typical levels of CO range from ~50–150 ppb in remote areas (e.g., see Parrish *et al.*, 1991, 1994; Novelli *et al.*, 1992, 1998a, 1998b; and Derwent *et al.*, 1998) to ~1000 ppb in rural–suburban areas up to ~10 ppm in very polluted areas such as Mexico City (e.g., Riveros *et al.*, 1998). It is interesting that the values that appear to be representative of clean, remote areas are about the same as those found using ice cores for the preindustrial era; for example, Haan *et al.* (1996) report preindustrial values of about 92 ppb for a Greenland ice core and ~55–60 ppb for an Antarctic ice core.

Figure 11.36, for example, shows the zonally averaged CO concentrations in the Northern and Southern Hemispheres, respectively, from 1990 to 1995 (Novelli *et al.*, 1998b). Concentrations are higher in the Northern than in the Southern Hemisphere, but both show a decreasing trend with time, -2.6 ± 0.3 ppb yr^{-1} in the Northern Hemisphere and -1.9 ± 0.1 ppb yr^{-1} in the

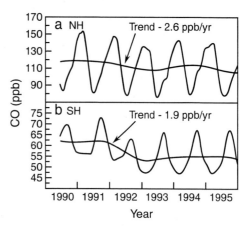

FIGURE 11.36 Zonal average concentrations of CO in (a) the Northern Hemisphere and (b) the Southern Hemisphere (adapted from Novelli *et al.*, 1998b).

Southern Hemisphere. The seasonal trends observed in both hemispheres reflect an anticorrelation with OH that removes CO from the atmosphere. CO in the NH, for example, peaks in March/April and is a minimum in July.

d. SO_2

Detection techniques. As shown in Table 11.1, common techniques for measuring SO$_2$ include UV fluorescence, DOAS, and a wet chemical (pararosaniline) technique in which SO$_2$ undergoes the Schiff reaction with pararosaniline, HCHO, and HCl to form a red-violet product whose absorbance at 560–580 nm is measured (West and Gaeke, 1956; Dasgupta and Gupta, 1986). Filter packs (e.g., Ferek and Hegg, 1993), diffusion scrubbers (e.g., Lindgren and Dasgupta, 1989), and denuders (e.g., see Pui *et al.*, 1990) have also been used for SO$_2$.

Infrared techniques, including TDLS (Schiff *et al.*, 1994b) and matrix isolation FTIR (Griffith and Schuster, 1987), have been applied to SO$_2$, with detection limits of 500 and 10 ppt, respectively.

Several chemiluminescence methods have also been developed for SO$_2$. For example, SO$_2$ has been shown to enhance the chemiluminescence signal from the luminol–NO$_2$ reaction so that the enhancement can be used as a measure of SO$_2$ at a fixed NO$_2$ concentration (Zhang *et al.*, 1985). Another method involves the formation of SO in a hydrogen flame followed by reaction with O$_3$ to generate electronically excited SO$_2$ whose emission is followed at 340 nm (e.g., see Benner and Stedman, 1990).

Gas chromatography with various detectors such as flame photometry have been applied in a number of studies; flame photometry involves the generation of electronically excited S_2 in a hydrogen-rich flame, and the emission at 394 nm is followed. The signal is proportional to the square of the concentration of the sulfur compound. Since all compounds containing sulfur will form S_2 in the flame, a separation technique such as GC must be used to differentiate SO_2 from other sulfur compounds.

Another GC method, isotope dilution GC–MS, involves the addition of an isotopomer of the analyte of interest to the sampling manifold (e.g., see Bandy et al., 1993; and Blomquist et al., 1993). In the case of SO_2, where the ambient SO_2 consists mainly of ^{16}O and ^{32}S, SO_2 containing the ^{34}S isotope is used. This labeled SO_2 at mass 66 is used as internal standard and has a number of additional advantages such as minimizing the loss of the analyte in the sampling system (Bandy et al., 1993). The air with the added isotopomer is trapped cryogenically and then sampled into a GC–MS for analysis.

The results of intercomparison studies suggest that measuring SO_2, particularly at low concentrations found in remote areas, is difficult. For example, Gregory et al. (1993) evaluated five techniques for measuring SO_2 in air: GC with flame photometric or mass spectrometric detection, a chemiluminescence method using reaction with $KMnO_4$, and a filter method with either the $KMnO_4$ chemiluminescence or ion chromatographic detection. Above ~200 ppt, agreement between pairs of techniques varied from about 30% to several orders of magnitude. At concentrations below this, there was no correlation between the measurements made using these five methods! Kok et al. (1990) compared a filter pack, diffusion denuder, and a commercial pulsed fluorescence analyzer over the range of 0.1–1 ppb. The results for this intercomparison were also quite scattered, although they did show some correlation with each other.

Typical ambient levels. Concentrations of SO_2 in remote areas are quite low, ~10–50 ppt (e.g., Bandy et al., 1993), since the only source is oxidation of biogenically produced organic sulfur compounds such as dimethyl sulfide (see Chapter 8.E). In rural–suburban areas, concentrations of ~1–20 ppb (e.g., Luria et al., 1987; Boatman et al., 1988) are common and in polluted urban areas, levels up to several hundred ppb are observed (e.g., Bennett et al., 1986).

e. NMHC and VOC

As discussed in detail in Chapters 6 and 16, organic compounds play a key role in the formation of ozone, particles, and other species of interest. While some of the individual species are of concern from the point of view of health effects (e.g., the carcinogens benzene and 1,3-butadiene), for most VOC it is because of their central role in the formation of ozone and associated pollutants that their emissions are regulated.

As discussed in Chapter 16.B, methane reacts relatively slowly with OH ($k^{298K} = 6.3 \times 10^{-15}$ cm^3 molecule^{-1} s^{-1}), with a lifetime with respect to this reaction of about 5 years at an OH concentration of 1×10^6 cm^{-3}. It is because of this long lifetime in the troposphere that methane is the only hydrocarbon to survive long enough to reach the stratosphere and participate in chemistry there. However, in terms of generating RO_2 radicals that convert NO to NO_2 on time scales of a few hours to days, CH_4 is not important compared to larger hydrocarbons (see Chapter 16.B for a discussion of reactivity and data in Table 16.8). This is the case even though its concentrations in air are much higher than those of other organics, 1.8 ppm versus a few ppb or less. Because of this, the regulatory focus has traditionally been on "non-methane hydrocarbons" (NMHC). (See Chapter 14 for a discussion of the role of CH_4 in global climate.)

However, it is not only hydrocarbons, but a wide variety of organics, that participate in the chemistry. Indeed, as discussed in Chapter 16, some of these such as the carbonyl compounds can be much more reactive than the parent hydrocarbon. As a result, the term "volatile organic compounds" (VOC) is often used to encompass all of the individual reactive organics in the troposphere, and we shall use this term here. Other terms used synonymously are ROG ("reactive organic gases") and NMOG ("non-methane organic gases").

There are two approaches to measuring VOC: measuring the total without segregating into individual species, or measuring the individual organics that contribute to VOC. The former is commonly used for routine monitoring purposes whereas the latter is clearly of interest from the point of view of understanding the details of atmospheric reactions.

(1) Sampling and measurement techniques for VOC Total NMOC. Total non-methane organics (NMOC) as a group (i.e., nonspeciated) are commonly measured by cryotrapping the air sample, e.g., in liquid argon, which does not trap CH_4. The contents of the trap can then be thermally desorbed directly into a flame ionization detector (FID) (McElroy et al., 1986).

In a second approach, CH_4 is separated and measured independently and NMOC obtained by difference. This can be accomplished using gas chromatography to separate the CH_4 or, alternatively, oxidizing all of the organics except CH_4 to $CO_2 + H_2O$ and measuring the CH_4 remaining by FID, which does not respond to CO_2 and water vapor. [Note, however, that water affects the GC baseline so that care must be taken in integrating peaks on top of this baseline (McElroy *et al.*, 1986).]

For urban data, total NMOC or NMHC is commonly reported in terms of ppm or ppb *carbon*, ppmC or ppbC, respectively. This is defined as follows:

ppbC = ppb of the organic

× number of carbon atoms per molecule.

For example, 70 ppb of *n*-butane would be reported as 280 ppbC. (For nonurban data, mixing ratios (ppb) of the individual compounds are often used instead, so the reader is cautioned to determine how such data are presented.)

The relative response of FIDs per carbon atom is quite constant across a spectrum of hydrocarbons typically found in ambient air, with some exceptions such as acetylene (e.g., see Apel *et al.*, 1994, 1995). Hence, calibrations are carried out using a known concentration of a reference hydrocarbon such as propane in air, and the results expressed as ppbC referenced to propane.

Individual VOC. The term *VOC* is commonly used to describe speciated measurements of individual organics. The almost universal approach to the identification and measurement of individual VOC is GC with either FID or mass spectrometry (MS). GC–MS is used to establish the identity of a particular compound through the combination of retention times and mass spectra and, of course, can also be used for quantification. However, for a given type of air mass, GC–FID is commonly used for more extensive quantitative measurements after the individual peaks have been identified. For reviews of various aspects of sampling and measurement of VOC in air, see Westberg and Zimmerman (1993), Apel *et al.* (1994), Klemp *et al.* (1994), Sacks and Akard (1994), and Dewulf and Van Langenhove (1997).

Because most organics are present at ppb–ppt concentrations in air, sample preconcentration is required. This is typically accomplished by direct cryotrapping of air, trapping on a solid sorbent, or sampling into an evacuated canister followed by cryotrapping the canister contents prior to injection onto the GC column. Combinations of these, such as cryotrapping followed by transfer to a solid sorbent, have also been used (e.g., Shepson *et al.*, 1987).

Cryotrapping involves pumping air through a cooled loop, which usually contains an inert material such as glass beads to improve the trapping efficiency. Liquid argon (bp $-186°C$) or nitrogen (bp $-196°C$) are commonly used and will trap all but CH_4, which is too volatile even at these temperatures (vapor pressure is about 9 Torr even in liquid nitrogen, ($-196°C$)) to be trapped. The trap is then warmed and the mixture of organics injected onto the head of a GC column. A scheme for carrying this out without the use of valves to minimize the formation of artifacts is described by Stephens (1989).

A common problem in trapping VOC is the presence of significant quantities of water vapor. Trapping this water can plug loops used in direct cryotrapping or in the cryotrapping of samples collected in canisters prior to injection into the GC. In the case of solid sorbents, adsorption of water can reduce the efficiency of trapping of the organics; as might be expected, the amount of water trapped depends on the particular sorbent used, with Tenax, for example, adsorbing relatively small amounts (less than 3 mg of water per gram of adsorbent) and Carboxen 569 adsorbing as much as 400 mg of water per gram of adsorbent (Helmig and Vierling, 1995). Water can also cause shifting baselines and changing retention times, as well as problems with various detectors (McElroy *et al.*, 1986; Helmig and Vierling, 1995). As a result, removing water, or at least decreasing its concentration before it is trapped, is important.

As a result, water traps are usually placed ahead of the cryotrap (e.g., see Goldan *et al.*, 1995). Alternatively, a two-trap system is used in which water and the organics are trapped in the first trap, which is then slowly warmed to desorb organics, but not most of the water, into a second trap (e.g., Greenberg *et al.*, 1996).

A variety of systems have been used to remove water prior to trapping the organics. For example, passing the airstream through a semipermeable membrane that allows water to diffuse through where it is removed by a flow of inert gas is common. However, depending on the preconditioning of the membrane material, some loss and rearrangement of some organics can occur. For example, Gong and Demerjian (1995) reported that Nafion that was preconditioned by heating caused depletion of some $C_4–C_6$ alkenes such as isoprene and rearrangements of others. Alternate approaches include the use of multisorbent tubes, purging with a dry gas such as He or N_2 to remove some water (e.g., see McClenny *et al.*, 1995), and using a combination of cryotrapping and adsorption (Shepson *et al.*, 1987).

Trapping of organics by passing the air through solid sorbents is an alternative to cryotrapping. The organics are then recovered either by thermal desorption or by solvent extraction of the sorbent. Typical sorbents used include carbon-based sorbents such as activated charcoal, porous carbon, Carbosieves, Carbotraps, and Carbopacks and polymers such as Tenax, Chromosorbs, polyurethane foams (see Chapter 10), and XAD resins. A description of solid sorbents and their application to ambient air measurements are discussed in detail by

Camel and Caude (1995) and Matisová and Škrabáková (1995).

As might be expected, these sorbents can have different efficiencies for the uptake of individual organics, so that results from trapping on one sorbent may not be the same as that on another one. For example, a comparison of Tenax-TA with a combination of Carbotraps B and C on a test atmosphere containing a number of compounds of atmospheric interest showed good agreement for most compounds; however, ethanethiol was not well retained by the Carbotrap, whereas these carbon sorbents were more effective for chlorodifluoromethane and 1,2-dichlorotetrafluoroethane (McCaffrey et al., 1994). A combination of adsorbents in one trap is often used to overcome such problems (e.g., see Helmig and Greenberg, 1994). The sorbents are arranged so that lightest organics pass through to the last layer and are trapped there, so that the most highly absorbing sorbent is the last one exposed to the airstream. The direction of the flow is reversed during thermal desorption (Matisová and Škrabáková, 1995).

Even for compounds that are efficiently trapped by a particular sorbent, care must be taken not to pass so much sample through the adsorbent that it becomes overloaded and hence "breakthrough" of the organic occurs (e.g., see Brown and Purnell (1979) for data on a variety of organics using Tenax-GC). Care must also be taken to investigate and, where possible, avoid artifacts when such sorbents are used. These can arise from desorption of organics such as benzene, either from the sorbent itself or from uptake of organics on it during storage (e.g., see Cao and Hewitt, 1994). To overcome such problems, the sorbents are cleaned prior to use, which is commonly done either by heating under a flow of gas or by extracting the solid with a solvent or combination of solvents followed by storage in a cold, organic-free container (Helmig, 1996).

Artifacts can also arise from reaction of the sorbent with components of the airstream, such as O_3 and NO_2. For example, Clausen and Wolkoff (1997) observed a number of products from the reactions of O_3 with Tenax, such as acetophenone, decanal, and benozic acid. Interestingly, 2,6-diphenyl-p-hydroquinone was generated when limonene was also present and was suggested to arise from the reaction with the Tenax of the radicals or the Criegee biradical generated in the ozone–limonene reaction. They also observed reactions of NO_2 with the Tenax sorbent.

Reaction of adsorbed organics with other components of the airstream during sampling, particularly ozone, is well known, leading to artificially low levels of the organic and, in some cases, the appearance of new species that are products of these artifact reactions (see Helmig (1997) for a review of these interferences). This is particularly the case for reactive compounds such as the unsaturated biogenics. It should be noted that such interferences due to ozone also occur with cryotrapping (Helmig, 1997). In this case, reactions between ozone and the organics can occur as the trap is warmed to desorb the organics onto the GC column. For example, Goldan et al. (1995) examined the effects of 100 ppb O_3 on a synthetic mixture of n-pentane, isoprene, 2,2-dimethylbutane, benzene, m-xylene, and α-pinene. Figure 11.37 shows the chromatograms of this mixture in the absence of O_3, in the presence of 100 ppb O_3 with no ozone trap, and finally, with a Na_2SO_3 scrubber for ozone upstream of the cryotrap. The presence of 100 ppb O_3 essentially completely removes α-pinene and most of the isoprene, and new peaks corresponding to oxidation products (methacrolein and methyl vinyl ketone) appear. However, this is not a problem if an ozone trap is used (Fig. 11.37c). Similar results have been reported for a series of terpenes, where O_3 at 8 ppb had no effect but at 61–125 ppb reduced the measured terpene concentrations and decreased the precision of the measurements (Larsen et al., 1997).

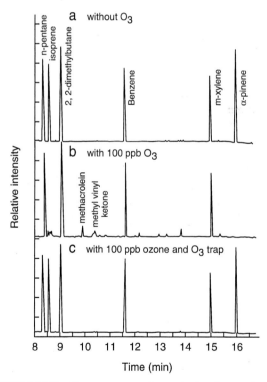

FIGURE 11.37 Chromatograms of a synthetic mixture of n-pentane, isoprene, 2,2-dimethylbutene, benzene, m-xylene, and α-pinene in synthetic air which are cryotrapped (a) in the absence of O_3, (b) in the presence of 100 ppb O_3, and (c) same as (b) but with an ozone scrubber upstream (from Goldan et al., 1995).

The severity of these problems is quite sensitive to the nature of the particular organic, as expected, and likely to the particular sampling configuration and conditions as well. For example, Calogirou *et al.* (1997) showed that saturated oxygenated terpenes adsorbed on Tenax were not affected by the presence of ~100 ppb O_3 whereas as much as 80–90% of the most reactive, unsaturated compounds reacted. Indeed, no α-terpinene was observed when ozone was present. On the other hand, Koppmann *et al.* (1995) report no significant interference problems with O_3 for the C_2–C_4 hydrocarbons which were sampled using a heated stainless steel inlet line, which destroys as much as half of the initial ozone, and then cryotrapped.

Because of these problems, removal of ozone prior to trapping the organics is highly desirable. A variety of approaches have been used, including the use of heated inlet lines as just described (Koppmann *et al.*, 1995), annular denuders coated with inorganics that react rapidly with O_3 such as KI (e.g., Williams and Grosjean, 1990), scrubbers containing O_3-reactive substances such as KI, crystalline Na_2SO_3, Na_2CO_3, or MnO_2, or polymeric sulfur scrubbers such as polyphenylene sulfide in which O_3 oxidizes the –S– group to =S(O)– without the formation of gas-phase products that could interfere with the analysis (e.g., Calogirou *et al.*, 1997). Titration of the O_3 with NO (e.g., Sirju and Shepson, 1995) has also been used. These and other techniques are reviewed in detail by Helmig (1997).

Collection of air samples in stainless steel canisters whose surfaces have been passivated is another common collection technique for VOCs. (Aluminum has also been used but the stability of polar organics in them is poor; Gholson *et al.*, 1990.) Indeed, this method is used not only for sampling air but in medical applications as well, where they have been used to sample organics in a single breath (Pleil and Lindstrom, 1995). Passivation of the canisters is often carried out using a process called SUMMA and hence referred to as "SUMMA canisters." The canisters also have to be thoroughly cleaned before use; an example of one such procedure is described by Blake *et al.* (1994). The sample is then typically preconcentrated by transfer to a cold trap prior to injection onto the GC column (e.g., see Blake *et al.*, 1994).

Loss of some organics to the walls of the canisters can occur, and these can subsequently desorb. Such negative changes depend on the nature, past use, and pretreatment of the canister surface, the nature of the compound, the canister pressure, the storage temperature, and interestingly, the water vapor present. The loss of organics to the walls is generally less in the presence of water, which has been attributed to water occupying active sites on the metal surface that would otherwise adsorb organics (e.g., Pate *et al.*, 1992; Apel *et al.*, 1994). The nature of the compound is particularly important, as might be expected, since highly reactive organics can react with other species such as ozone in the canister itself (Apel *et al.*, 1994). In addition, characteristics of the compound such as its vapor pressure, solubility in water, and polarity determine how readily it is taken up onto the canister surfaces or into a thin film of water on the surface. In fact, positive changes in some VOCs have been observed in canisters and attributed to uptake into a film of water on the canister surface. As samples are removed from the canister and the pressure reduced, the amount of surface water decreases, causing a release of dissolved organic into the gas phase. This appears then as a positive change with time.

Zielinska *et al.* (1996) and Kelly and Holdren (1995) have summarized the stability in canisters of organics, some of which are U.S. EPA designated HAPs (hazardous air pollutants). Kelly and Holdren propose that for compounds whose stability in canisters is not known, estimates can be made based on species of similar physical and chemical characteristics. These characteristics include their vapor pressure, polarizability, water solubility, Henry's law coefficient in water, and estimated lifetimes with respect to reactions in air and in the aqueous phase.

While these methods are most commonly used for sampling and analysis of ambient air, others have been applied as well, particularly in laboratory studies or source sampling. For example, flexible chambers made from thin films of Teflon or Tedlar are often used to store organics prior to sampling. The advantage is that as sample is removed, they collapse so that their pressure remains at atmospheric while the volume decreases. However, again some losses of organics can occur in these chambers, either on the walls or on the attached sampling hardware (e.g., see Wang *et al.*, 1996).

Passive samplers are used for specific applications such as for indoor air environments or as passive dosimeters. In this approach, the air containing the organic diffuses to and adsorbs on a solid sorbent without active pumping. The organics are subsequently thermally desorbed or extracted from the sorbent using a solvent (e.g., see Shields and Weschler, 1987).

(2) Intercomparison studies Because of the importance of accurate measurement of individual organics in air, there have been several intercomparison studies carried out in which a number of different laboratories have analyzed common samples, either synthetic or ambient air. In one such study (Apel *et al.*, 1994), a

16-component synthetic mixture was distributed to about three dozen laboratories for analysis. Of the 28 laboratories completing this analysis, 12 identified all 16 components correctly. Figure 11.38 shows the percentage difference for each component from the gravimetric values used in the mixture preparation averaged over all participating laboratories as well as the percent standard deviation for all of the measurements. Clearly, the accurate identification and analysis of individual hydrocarbons found in ambient air remains a difficult problem.

Figure 11.39 shows the results of another such intercomparison (Bernardo-Bricker et al., 1995). Some of the problems encountered and the approaches taken to solve them are also discussed by Bricker and co-workers. As seen in Fig. 11.39, the precision of the measurement of compounds present at concentrations less than 5 ppbC is much worse than for species present in higher concentrations, as might be expected. However, some of these compounds, such as limonene, are notoriously difficult to measure accurately even at higher concentrations, so that it is likely a combination of the small concentrations and nature of the compound that contributes to the poor precision seen in some cases.

(3) Typical atmospheric concentrations A large number of field studies have been carried out in which organics as well as a number of associated species such as NO_x, NO_y, and O_3 have been measured simultaneously, both at the earth's surface and at various altitudes. For the sake of brevity, we focus here primarily on surface measurements and, of necessity, cite only a few of the many studies that have been carried out. The accuracy of the absolute values in any case should be interpreted in light of the results of intercomparison studies such as those already described. However, it is clear that there are major differences in the concentrations, depending on where the sample is taken. Table 11.8 summarizes a few typical ranges of concentrations measured in regions that can be characterized as covering the range from remote to highly polluted urban areas.

f. Aldehydes, Ketones, Alcohols, and Carboxylic Acids

As discussed in Chapter 6, carbonyl compounds are particularly important in atmospheric chemistry because of their absorption of light in the 290- to 400-nm region that generates free radicals. With the increasing use of alcohol fuels (see Chapter 16) as well as the recognition that there are biogenic emissions of alcohols (see Chapter 6.J.1), the measurement of these compounds has also become of interest. Finally, carboxylic acids are now known to contribute significantly to the total acid burden and hence acid deposition in the atmosphere (Chapter 8.D). In this section, techniques for measuring these oxygenated compounds and some typical levels found in the troposphere are discussed.

(1) Aldehydes and ketones Spectroscopic techniques have proven particularly useful for the smaller aldehydes, which have distinct infrared and UV–visible absorption bands. As seen in Table 11.2 and discussed earlier, HCHO has been measured by FTIR in polluted urban areas as well as by TDLS and matrix isolation spectroscopy. In addition, as seen in Table 11.3, DOAS has high sensitivity for HCHO due to its strongly banded absorption in the 300- to 400-nm region (see Chapter 4.M).

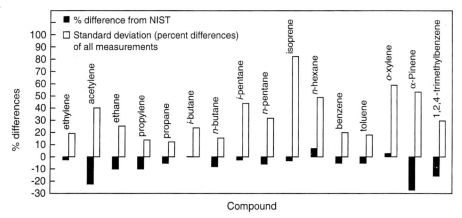

FIGURE 11.38 Average percent differences from the NIST standards for all participating laboratories (■) and the percent standard deviation of all measurements (□) (adapted from Apel et al., 1994).

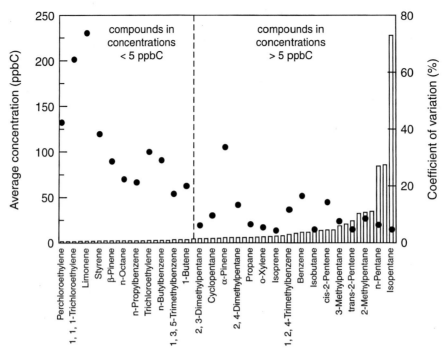

FIGURE 11.39 Results of intercomparison study of a mixture of organics found in ambient air. The concentrations (in ppbC) are shown as unfilled bars, along with the coefficient of variation (●) for these measurements (adapted from Bernardo-Bricker et al., 1995).

In addition to the spectroscopic methods, there are a number of *derivatization methods*, in which a derivative of the carbonyl compound that can be easily separated and measured is formed. The most common of these is the use of 2,4-dinitrophenylhydrazine (DNPH), which reacts to form the hydrazone:

The mixture of hydrazones formed from the reactions of the set of carbonyl compounds commonly found in air can then be separated using HPLC and detected using light absorption at 254 or ~365 nm (e.g., see Kuntz et al., 1980; Lipari and Swarin, 1982; and Kuwata et al., 1983).

The aldehydes and ketones can be collected in several different ways. One of the most common is the use of solid sorbents such as silica gel, Florisil, or C_{18} cartridges that are coated with DNPH (e.g., see Lipari and Swarin, 1985; Arnts and Tejada, 1989; Zhou and Mopper, 1990; Sirju and Shepson, 1995; and Kleindienst et al., 1998). Some care must be taken to avoid artifacts due to reactions of O_3, which has been shown to give positive interferences for C_{18} cartridges but negative interferences when silica gel is used as the sorbent (e.g., see Arnts and Tejada, 1989; Rodler et al., 1993; Sirju and Shepson, 1995; Kleindienst et al., 1998; Gilpin et al., 1997; and Apel et al., 1998b). The DNPH technique has also been used in passive samplers for HCHO (e.g., see Levin et al., 1989; and Grosjean and Williams, 1992). However, with care, concentrations in the 30- to 500-ppt range (depending on the particular carbonyl) can be measured using this technique and a 156-L air sample (Grosjean et al., 1996a).

Other approaches to collection of the carbonyl compounds include the use of impingers containing the DNPH in a solvent such as acetonitrile (e.g., Grosjean and Fung, 1982; Lipari and Swarin, 1982), scrubbers (e.g., Dasgupta et al., 1988, 1990; Lee and Zhou, 1993), mist chambers (e.g., Cofer and Edahl, 1986; Munger et al., 1995; Khare et al., 1997), and condensation collectors (Dawson and Farmer, 1988).

Alternate derivatization techniques have also been used. These include dansylhydrazine with fluorescence or chemiluminescence detection of the hydrazone (e.g., Nondek et al., 1992; Rodler et al., 1993), 3-methyl-2-

TABLE 11.8 Some Typical Concentration Ranges Measured for Some Small Hydrocarbons from Remote to Urban Areas (ppb)

Compound	Type of air mass		
	Urban	Rural–suburban	Remote
	Brazil,[a] Denmark,[b] India,[c] Japan,[d] Greece,[e] U.K.,[f] U.S.,[i,k,q] Canada,[l,o] Sweden,[m,n] France,[w] Italy,[x] Mexico[y]	Brazil,[a] Denmark,[b] U.S.,[i,k,q] Canada[l,o] Sweden,[m,n] France[w]	U.S.,[i,k,q] Canada,[l,o] France,[w] Antarctic,[p,s] North Atlantic[t]
Ethane	0.6–29	0.4–4	0.4–2
Ethene	0.7–168	0.1–3	0.07–0.4
Acetylene	0.7–44	0.1–3	0.01–0.7
Propane	0.4–221	0.2–2	0.04–0.9
Propene	0.1–39	0.02–2	0.02–0.2
n-Butane	0.02–96	0.1–1	0.1–0.75
Isobutane	0.3–45	0.04–0.6	0.003–0.2
1-Butene	0.2–7	0.01–0.04	0.005–0.014[w]
2-Methyl propene	3–18	0.04–0.16	—
2-Butene (cis and trans)	0.10–4	0.008–0.06	0.006–0.024[w]
n-Pentane	0.7–67	0.07–0.4	0.007–0.19
Isopentane	2–90	0.1–0.6	0.008–0.2
Benzene	0.9–26	0.1–0.6	0.008–0.2
Toluene	2–39	0.05–0.8	0.01–0.25
o-Xylene	0.4–6	0.02–0.2	0.001–0.03
m- and p-Xylene	0.3–30	0.02–0.12	0.002–0.008
1,3-Butadiene	0.1–6	0.03–0.6	0.02–0.13
Isoprene	0.1–2	0.005–5.5	

[a] Grosjean et al. (1998a, 1998b); Port Alegre, Brazil.
[b] Hansen et al. (1998); Copenhagen.
[c] Rao et al. (1997); measurements made close to industrial regions in Bombay.
[d] Morikawa et al. (1998); monthly means for Osaka.
[e] Moschonas and Glavas (1996); Athens in early morning.
[f] Dollard et al. (1995); range of annual means for eight sites in seven cities (Birmingham, Cardiff, Edinburgh, Bristol, Middlesbrough, Belfast, and two sites in London).
[g] Evans et al. (1992); 24-h medians in five cities (Boston, Chicago, Houston, Seattle, Tacoma).
[h] Edgerton et al. (1989); daily medians in nine cities (Phoenix, Chicago, Denver, Houston, Los Angeles, Philadelphia, Pittsburgh, San Jose, St. Louis).
[i] Zielinska and Fujita (1994); range for two sets of samples collected from 6:00 to 8:00 a.m. and 10:00 a.m. to noon, respectively, in downtown Los Angeles in 1991.
[j] Fraser et al. (1998); Los Angeles during severe smog episode in 1993.
[k] Gong and Demerjian (1997); Whiteface Mountain, New York.
[l] Bottenheim and Shepherd (1995); monthly means at four locations (Saturna Island, British Columbia; Egbert, Ontario; Lac la Flamme, Quebec; Kejimkujik National Park, Nova Scotia).
[m] Lindskog and Moldanova (1994); annual average based on monthly means at Rörvik on the west coast of Sweden for all but 1,3-butadiene and isoprene; for the latter, the ranges during the period April 23–27, 1990, at Rörvik from Lindskog et al. (1992) are reported.
[n] Hewitt et al. (1995); typical concentrations from forested sites in the United States, Europe, and Brazil.
[o] Blake et al. (1994); measured in northern Quebec and Ontario.
[p] Rudolph et al. (1989); either yearly average or range is given.
[q] Singh et al. (1988); average concentrations measured in lower troposphere over the land and ocean in Colorado and California.
[s] Clarkson et al. (1997); annual averages at Baring Head, New Zealand, and Scott Base, Antarctica, between 1991 and 1996.
[t] Penkett et al. (1993); range represents summer minimum to winter maximum over the North Atlantic Ocean with air mainly of polar marine origins.
[u] Mowrer and Lindskog (1991); mean concentrations measured at Rörvik from February 21 to April 8, 1989.
[v] Rappenglück et al. (1998); means of 20-min samples in two locations in Athens.
[w] Boudries et al. (1994); monthly average concentrations at Porspoder on the west coast, Brittany. Remote air masses were of oceanic origin. Those under "rural–suburban" were identified as being of continental origin.
[x] Possanzini et al. (1996); range measured in Rome from 0600 to 0900 hours.
[y] D. R. Blake and F. S. Rowland, personal communication, 1998; concentrations measured at various locations throughout the day in Mexico City in 1993.

benzothiazolinone hydrazone, which has been used for post-HPLC column derivatization (Igawa *et al.*, 1989), and *O*-(2,3,4,5,6-pentafluorobenzyl)hydroxylamine hydrochloride derivatization, which has been used in combination with GC–MS (Lacheur *et al.*, 1993; Yu *et al.*, 1995, 1998).

There are some methods that are specific to HCHO. For example, the Hantzsch reaction of HCHO, collected with a diffusion scrubber, with ammonium acetate, acetic acid, and acetylacetone to form diacetyldihydrolutidine, which is measured using its fluorescence at 470 nm, has been applied to air measurements (Dasgupta *et al.*, 1988, 1990; Kleindienst *et al.*, 1988a,b; Lawson *et al.*, 1990; Khare *et al.*, 1997). Reaction with 1,3-cyclohexanedione and ammonium acetate to form a dihydropyridine derivative that is measured by fluorescence has been used in conjunction with a diffusion scrubber (Fan and Dasgupta, 1994). Enzymatic methods have been used in which formaldehyde dehydrogenase catalyzes the oxidation of HCHO to HCOOH in the presence of β-nicotinamide adenine dinucleotide, NAD^+, which is reduced to NADH. The latter is measured by fluorescence at 450 nm (Lazrus *et al.*, 1988; Ho and Richards, 1990).

Finally, cryotrapping followed by GC–MS analysis has been used for a variety of carbonyl compounds, with the exception of HCHO, which is removed on surfaces (e.g., see Goldan *et al.*, 1995; and Leibrock and Slemr, 1997).

These spectroscopic and derivatization techniques have largely supplanted earlier wet chemical techniques such as that employing chromotropic acid (Altshuller and McPherson, 1963).

A number of intercomparison studies of the various methods of measurement of HCHO have been carried out. As might be expected given the specificity of spectroscopic methods, the results of FTIR, DOAS, and TDLS are generally in good agreement, within 15% of their mean value in one study in a polluted atmosphere (Lawson *et al.*, 1990). During the same study, the diacetyldihydrolutidine derivative method and the DNPH method were lower by ~15–25% than the spectroscopic mean, whereas the enzymatic method was higher by ~25%.

In another intercomparison using TDLS, several DNPH methods, the 1,3-cyclohexanedione diffusion scrubber, and the enzymatic method were compared using both spiked samples and ambient air. The TDLS was used as a standard for comparison. For ambient air measurements, results obtained with the 1,3-cyclohexanedione diffusion scrubber were about 30% higher than those obtained with TDLS, whereas results for the enzymatic method were about 35% lower. The DNPH cartridge measurements were quite variable, which may reflect in part the problem with interference by ozone reactions mentioned earlier. However, in an earlier study, the slope of a plot of TDLS measurements against DNPH on silica gel cartridges was within the relatively large scatter unity (0.95 ± 0.36), whereas that for the enzymatic method against TDLS was about 30% low, with a slope of 0.71 ± 0.09 (Kleindienst *et al.*, 1988b). Similarly, Sirju and Shepson (1995) showed that if O_3 was removed to avoid interference, TDLS and DNPH cartridge measurements were in agreement. Benning and Wahner (1998) compared DNPH cartridge measurements with O_3 removal with DOAS measurements of HCHO; there was reasonable overall agreement between the two, although there was significant scatter in individual measurements on a plot of DOAS concentrations against those measured using DNPH.

TDLS and DOAS measurements of HCHO were intercompared in a rural region in Colorado (Harder *et al.*, 1997b). The DOAS measurements were made over a 10.3-km path and the TDLS essentially at a point at one end of the DOAS path length (Fried *et al.*, 1997). When anthropogenic influences were thought to be small, the two techniques were in excellent agreement, to within 5%. At other times, higher concentrations were obtained using DOAS due to anthropogenic influences that did not affect the TDLS measurements.

In short, the spectroscopic methods appear to be reliable and specific for HCHO. The derivatization methods are generally in reasonable overall agreement with the spectroscopic methods where intercomparisons have been carried out, but there can be very large discrepancies in individual measurements. Part of the reason for these discrepancies may be related to the fact that some of the spectroscopic methods average over long distances whereas the derivatization methods sample at a point. On the other hand, the latter methods involve collecting the sample over a period of time, usually several hours, whereas the spectroscopic methods are real-time measurements. Finally, variations in collection efficiencies and possible interferences must be taken into account for the derivatization methods.

Typical tropospheric concentrations. Table 11.9 shows some typical concentrations of the major aldehydes and ketones measured near the earth's surface. Formaldehyde is usually present in the highest concentration, followed by acetaldehyde and then, at significantly smaller concentrations, higher aldehydes and ketones. Figure 11.40, for example, shows the percentages of the total carbonyl compounds due to each of the simple straight-chain aldehydes measured in one study in the Los Angeles area in which 23 different carbonyl compounds were identified and measured (Grosjean *et al.*, 1996a). These percentages are in terms of ppb of each

TABLE 11.9 Some Typical Concentrations of Aldehydes and Ketones in Ambient Air (ppb)

	Type of air mass		
	Urban	Rural–suburban	Remote
Compound	Denver,[a] Los Angeles,[b] Albuquerque,[c] various U.S. Cities,[d] Rome,[e] Copenhagen,[f] Paris,[g] Mexico City[h]	Germany,[j,k] Denmark,[l] U.S.,[i,o,p,s,u,v] Canada,[m] Venezuela[n]	Canada,[q] Caribbean Sea[r]
HCHO	1–60	0.1–10	0.3–2
CH_3CHO	1–18	0.1–4	0.1–1
CH_3CH_2CHO	0.1–3	0.004–0.2	0–0.2
$CH_3CH_2CH_2CHO$	0.2–1.4	0.1–0.3	—
$CH_3C(O)CH_3$	0.2–9	0.2–8	0–1
$CH_3C(O)CH_2CH_3$	0.3–8	0.1–0.5	0–0.18
Benzaldehyde	0.1–1	—	< 0.01–0.5
Acrolein	<0.04–1	~0.6[t]	—
Crotonaldehyde	0.1–0.5[o]	0.2–0.3[o]	~0.09[r]
Methacrolein	<0.7[c]	0.02–1.7	—
Methyl vinyl ketone	—	0.2–1.5	

[a] Anderson et al. (1994, 1996).
[b] Grosjean et al. (1996a).
[c] Gaffney et al. (1997).
[d] Salas and Singh (1986).
[e] Possanzini et al. (1996).
[f] Granby et al. (1997a).
[g] Kalabokas et al. (1988).
[h] Báez et al. (1989).
[i] Lee et al. (1995).
[j] Benning and Wahner (1998).
[k] Slemr et al. (1996).
[l] Granby et al. (1997).
[m] Shepson et al. (1991).
[n] Trapp and Serves (1995).
[o] Grosjean et al. (1996).
[p] Harder et al. (1997a,b); Fried et al. (1997).
[q] Tanner et al. (1996).
[r] Zhou and Mopper (1990, 1993).
[s] Goldan et al. (1995).
[t] Leibrock and Slemr (1997).
[u] Apel et al. (1998b).
[v] Riemer et al. (1998).

compound (not as ppbC). There were also 19 carbonyls whose identity could not be confirmed but likely included hydroxycarbonyl aliphatic and aromatic compounds. Nondek et al. (1992), for example, have tentatively identified biogenic emissions of p-hydroxybenzaldehyde in a forest.

Dicarbonyl compounds such as glyoxal (CHOCHO) and methylglyoxal ($CH_3C(O)CHO$) have also been measured in tropospheric air, in rural areas at small concentrations, and in polluted urban areas. For example, Munger et al. (1995) report an average concentration of 44 ppt glyoxal in central Virginia, and Lee et al. (1995) measured average glyoxal concentrations of 18–83 ppt and average methylglyoxal concentrations of 31–88 ppt at a rural site in Georgia. Higher concentrations are found in urban air, e.g., 0.78 ± 0.85 ppb glyoxal and 1.0 ± 0.6 ppb methylglyoxal in the Los Angeles area (Grosjean et al., 1996a).

(2) Alcohols There are far fewer measurements of alcohols than there are of the aldehydes and ketones. They are typically measured using GC–FID or GC–MS (e.g., see Goldan et al., 1995; Leibrock and Slemr, 1997; and Grosjean et al., 1998a,b,c). Measuring alcohols is difficult because of their polar nature and tendency to stick to surfaces. For example, Apel et al. (1998a) reported that GC measurements of methanol, ethanol, and n-butanol in calibration gas mixtures had higher variability than aldehydes and ketones.

Methanol and ethanol are the primary simple alcohols that have been identified in air, with concentrations of methanol in the ~1- to 20-ppb range and

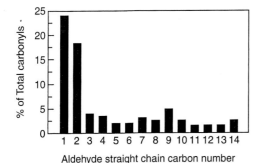

FIGURE 11.40 Percentage of total carbonyl compounds due to each of the straight-chain aldehydes in the Los Angeles area (adapted from Grosjean et al., 1996a).

ethanol in the ~0.1- to 1-ppb range in rural areas (Goldan *et al.*, 1995; Leibrock and Slemr, 1997; Riemer *et al.*, 1998). In areas where ethanol is used as a fuel, the concentrations can be much higher. For example, in Porto Alegre, Brazil, ethanol concentrations up to 68 ppb have been measured (Grosjean *et al.*, 1998b). Higher alcohols such as *n*-butanol have also been reported to be present in air in smaller concentrations; for example, Riemer *et al.* (1998) reported concentrations of ~55 ppt at a rural site in the southeastern United States.

As discussed in Chapter 6.J.1, there are also biogenic emissions of multifunctional alcohols, which are treated in that section.

(3) Carboxylic acids The smallest carboxylic acid, formic acid, can be measured using infrared spectroscopy (Table 11.2), since it has characteristic absorption bands. As discussed earlier and seen in Fig. 11.33b, mass spectrometry with chemical ionization using SiF_5^- also revealed HCOOH in an indoor environment (Huey *et al.*, 1998). However, since the sensitivity in these initial studies was about two orders of magnitude less than that for HNO_3, the detection limit may be about the same as that for FTIR and TDLS. Formic and acetic acids have been monitored continuously from aircraft (Chapman *et al.*, 1995) and their surface flux determined by eddy correlation (Shaw *et al.*, 1998) using atmospheric pressure ionization mass spectrometry. Detection limits are about 30 ppt.

Gas-phase carboxylic acids have been sampled using mist chambers (e.g., Andreae *et al.*, 1987; Talbot *et al.*, 1988), condensates (Dawson and Farmer, 1988), filters coated with alkaline compounds such as KOH, NaOH, K_2CO_3, and Na_2CO_3 (e.g., Grosjean *et al.*, 1990; Nolte *et al.*, 1997), and denuders coated with NaOH (Keene *et al.*, 1989). The acid anions are then separated and detected using ion chromatography. It should be noted that interferences have been reported for some of these methods. For example, the conversion of formaldehyde to formic acid and PAN to acetate on alkaline filters has been observed (Andreae *et al.*, 1987; Keene *et al.*, 1989; Grosjean and Parmar, 1990), and with some ion chromatography columns, coelution of several anions can be a problem (e.g., see Jaffrezo *et al.*, 1998). The results of one intercomparison study (Keene *et al.*, 1989) suggest that artifacts in these measurement methods occur episodically and that care should be taken in their application.

A promising method involves derivatization by reaction with pentafluorobenzyl bromide (Chien *et al.*, 1998). Carboxylic acids (RC(O)OH) react to form the esters, $RC(O)OCH_2C_6F_5$, which can be measured by GC–MS. This method has the advantage of increased sensitivity and selectivity.

Formic and acetic acids are found primarily (>98%) in the gas phase (e.g., Andreae *et al.*, 1987; Talbot *et al.*, 1988). Concentrations of gas-phase HCOOH and CH_3COOH in rural areas are typically ~0.3–3 and ~0.5–2 ppb, respectively (Andreae *et al.*, 1987; Talbot *et al.*, 1988; Dawson and Farmer, 1988; Sanhueza *et al.*, 1996; Nolte *et al.*, 1997; Granby *et al.*, 1997a,b), although higher concentrations, up to 32 ppb for CH_3COOH, have been observed in wood-burning areas (Gaffney *et al.*, 1997). In urban areas, HCOOH and CH_3COOH concentrations are about the same, typically in the range of ~1–10 ppb (e.g., see Tuazon *et al.*, 1981; Dawson and Farmer, 1988; Grosjean, 1990; Grosjean *et al.*, 1990; Khare *et al.*, 1997; Granby *et al.*, 1997a; Nolte *et al.*, 1997; and Gaffney *et al.*, 1997).

Multifunctional acids containing a carbonyl group such as pyruvic acid [$CH_3C(O)COOH$] are typically measured using the derivatization techniques used for aldehydes and ketones, such as the DNPH method (e.g., see Lee *et al.*, 1995).

g. PAN, Other Peroxynitrates, and Alkyl Nitrates

The formation and fate of peroxyacyl nitrates, $RC(O)OONO_2$, were discussed in Chapter 6.I. These compounds are almost universally measured using gas chromatography with electron capture detection (GC-ECD), although a luminol chemiluminescence detector has also been used in which PAN is thermally decomposed to NO_2 at the end of the column and the NO_2 measured (Burkhardt *et al.*, 1988; Blanchard *et al.*, 1990; Gaffney *et al.*, 1998). In polluted atmospheres where the concentrations are higher, FTIR has also been used (Table 11.2). For a summary of methods, see reviews by Gaffney *et al.* (1989) and Kleindienst (1994).

Of the peroxyacyl nitrates, the most prevalent compound is peroxyacetyl nitrate, $R = CH_3$, with peroxypropionyl nitrate (PPN, $R = C_2H_5$) typically being present at the next highest concentration. Because they are formed in the VOC–NO_x photochemical cycles, the highest levels of PAN are often seen downwind of urban areas rather than in the center. For example, in the Los Angeles area, some of the highest concentrations have been measured at a mountain site about 35 km northeast of Los Angeles (Grosjean *et al.*, 1993a, 1996b).

Peak concentrations of PAN in or downwind of major urban areas during periods of high photochemical activity can reach levels as high as ~35 ppb (e.g., see Tuazon *et al.*, 1981; Tanner *et al.*, 1988; Grosjean *et al.*, 1993a, 1996b; Altshuller, 1993; Williams *et al.*, 1993; Kleindienst, 1994; Suppan *et al.*, 1998; and Gaffney *et al.*, 1998, 1999). In rural areas, peak concentrations up to about a ppb are typical (e.g., see Corkum et al., 1986; Andersson-Sköld et al., 1993; Ridley et al., 1990a; Gaffney et al., 1993, 1997; Hastie et al., 1996; Nouaime et al., 1998; and Roberts et al., 1998). However, in remote areas where NO and NO_x levels are small (vide supra), a few tens of ppt to ca. several hundred ppt are common (e.g., see Rudolph et al., 1987; Singh et al., 1990; Ridley et al., 1990b, 1998; Perros, 1994; Talbot et al., 1994; Heikes et al., 1996; Beine et al., 1996, 1997; Solberg et al., 1997; and Singh et al., 1998).

As might be expected from the chemistry common to the formation of O_3 and PAN, the two are often highly correlated both temporally and geographically. Figure 11.41, for example, shows the median values for the diurnal variation of ozone and PAN at one site in Athens, Greece, during meteorological conditions conducive to photochemical smog formation (Suppan et al., 1998). Both increase about 9 a.m. and continue at relatively high levels until late in the afternoon.

Smaller concentrations of higher members of the series have also been observed. In and downwind of polluted urban areas, peak concentrations of PPN of ~0.4-4 ppb have been reported (e.g., see Grosjean et al., 1993a, 1996b; and Williams et al., 1993), whereas the concentrations in rural areas are about an order of magnitude smaller (e.g., Ridley et al., 1990a). For example, typical average PPN levels in the southeastern United States have been reported to be ~50 ppt, with the PPN/PAN ratio being about 0.15 in air masses impacted by anthropogenic emissions (Nouaime et al., 1998; Roberts et al., 1998). Peroxy-n-butyryl nitrate (n-$C_3H_7C(O)OONO_2$) and peroxymethacroyl nitrate (MPAN, $CH_2=C(CH_3)C(O)OONO_2$) have been measured at concentrations up to ~1-2 ppb (Williams et al., 1993; Grosjean et al., 1993a, 1993b; Gaffney et al., 1999). Since MPAN is an oxidation product of isoprene

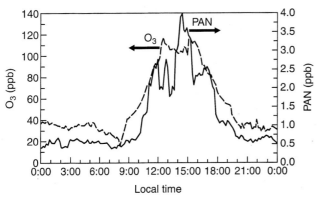

FIGURE 11.41 Diurnal variation of median concentrations of PAN and O_3 at a monitoring site in Athens, Greece, during periods of high photochemical activity (adapted from Suppan et al., 1998).

(see Chapter 6), it is found primarily in forested areas having significant isoprene emissions. For example, in rural areas near Nashville, Tennessee, the average concentration was about 30 ppt (Nouaime et al., 1998), with a typical MPAN/PAN ratio of 0.10-0.17 (Roberts et al., 1998).

Simple alkyl nitrates are also commonly measured using GC-ECD, usually with preconcentration either by cryotrapping or using a solid sorbent (e.g., Atlas and Schauffler, 1991; Ridley et al., 1997). Another approach is GC with an NO_y detector as described earlier (e.g., Flocke et al., 1998). In this approach, the compounds are converted to NO over a catalyst as they emerge from the GC column, and the NO measured by its chemiluminescence reaction with O_3.

A number of alkyl nitrates have been observed in the troposphere, including methyl nitrate and ethyl nitrate, as well as all of the isomers of the higher alkyl nitrates up to C_5 (e.g., see Buhr et al., 1990; Ridley et al., 1990a; O'Brien et al., 1995; and Flocke et al., 1998). Although the specific isomers were not identified, the C_6-C_8 alkyl nitrates have also been measured (O'Brien et al., 1995; Flocke et al., 1998). A summary of the measurements through about 1998 is found in Flocke et al. (1998).

Of the simple alkyl nitrates, methyl nitrate is present in the highest concentration. For example, in measurements made in Schauinsland, a rural area in Germany, concentrations of CH_3ONO_2 up to 216 ppt were measured. The median value, however, was only 19 ppt (Flocke et al., 1998). In the same studies, the median concentrations for ethyl nitrate, n-propyl nitrate, 2-propyl nitrate, and 1-butyl nitrate were 9, 3, 12, and 2 ppt, respectively. The sum of the C_1-C_8 alkyl nitrates averaged 120 ppt, which is only ~3% of the NO_y. Similarly, in rural Ontario, Canada, 17 different organic nitrates were identified in air, but their sum was only 0.5-3% of NO_y (O'Brien et al., 1995). In aircraft measurements over the Pacific Ocean near Hawaii, average values for methyl nitrate near the surface were ~6 ppt and the sum of C_1-C_5 alkyl nitrates was <5% of the total NO_y (Ridley et al., 1997).

In short, a variety of alkyl nitrates are present in air, but at relatively small concentrations compared to the peroxyacyl nitrates and to NO_y.

h. H_2O_2 and Organic Peroxides

There are a variety of methods for collecting and measuring H_2O_2 and organic peroxides in air. H_2O_2 is especially water soluble and hence partitions between the gas phase and clouds and fogs (e.g., Macdonald et al., 1995). While the collection techniques for air versus clouds and fogs are different, the analytical techniques are the same.

Collection of air for peroxide analysis has been accomplished using a number of approaches, including mist chamber sampling, diffusion scrubbers (e.g., see Dasgupta et al., 1988), impingers, and cryogenic trapping (e.g., see Sakugawa and Kaplan, 1987). Artifacts have been observed with many of the sampling systems. For example, Sakugawa and Kaplan (1987) reported that H_2O_2 collected by impingers was higher than by cryotrapping and attributed this to generation of H_2O_2 by aqueous-phase reactions in the bubbler. Indeed, the generation of H_2O_2 in water when O_3 is bubbled into it has been observed (e.g., Zika and Saltzman, 1982; Heikes, 1984). On the other hand, artifact formation of H_2O_2 and hydroxymethyl hydroperoxide during cryogenic trapping of air was reported by Staffelbach et al. (1995, 1996) and attributed to reactions of alkenes with O_3 in the traps. In addition, H_2O_2 is a sufficiently "sticky" molecule that it is readily lost to surfaces in sampling systems, so that such surfaces prior to scrubbing into solution must be minimized (e.g., Lee et al., 1991, 1993). Differences in collection efficiencies for different hydroperoxides must also be taken into account (e.g., de Serves and Ross, 1993).

Various techniques for measuring peroxides in air are reviewed by Gunz and Hoffmann (1990) and Sakugawa et al. (1990). H_2O_2 can be measured spectroscopically by FTIR and by TDLS (e.g., Slemr et al., 1986), although the FTIR detection limit is too high to be of value except for relatively rare, extremely large concentrations (see Table 11.2). More common are derivatization methods, and of these, one using p-hydroxyphenylacetic acid (POPHA) has been applied extensively to ambient air.

The POPHA method is based on the oxidation of horseradish peroxidase in the +3 state to its +5 state (Lazrus et al., 1985, 1986; Kok et al., 1986). This oxidized form is then reduced by electron transfer from POPHA, generating the POPHA free radical. The POPHA free radicals self-react to produce the dimer, which, upon excitation at 320 nm, fluoresces at 405 nm. The overall reaction is

Since both H_2O_2 and organic peroxides carry out this reaction, this method measures total peroxides. However, the contribution of H_2O_2 can be separated by adding catalase, which decomposes H_2O_2 but not the organic peroxides. The instruments therefore usually have two channels, one in which catalase is added to give the organic peroxides, and one in which it is not, giving total peroxides. The difference between the two channels gives H_2O_2. Another approach involves using the different solubilities of H_2O_2 and organic peroxides (Staffelbach et al., 1996).

The organic peroxides can be separated by HPLC prior to detection (Kok et al., 1995). Sauer et al. (1996, 1997), for example, used HPLC with a POPHA detector to measure peroxides in air and in rainwater in Germany and at a marine coastal site in France. Although no organic peroxides were found in air, several were identified in some rainwater samples, including hydroxymethyl hydroperoxide ($HOCH_2OOH$) and 1-hydroxyethyl hydroperoxide ($CH_3CH(OH)OOH$). Hydroxymethyl hydroperoxide is expected to be formed in the atmosphere from the reaction of the one-carbon Criegee biradical ($\cdot CH_2OO\cdot$) with water, i.e., from the reaction of ethene and terminal alkenes with O_3 in air (see Chapter 6.E.2) and perhaps from the reaction of $HOCH_2OO\cdot$ radical with HO_2 (see Chapter 6.E.2).

Fels and Junkermann (1994) reported both hydroxymethyl hydroperoxide and CH_3OOH in air in a rural area in Germany, with these two compounds comprising more than 90% of the total organic hydroperoxides. Ethyl hydroperoxide and peroxyacetic acid were also detected in some samples. Methyl hydroperoxide is expected in low-NO_x environments from the reaction of CH_3O_2 with HO_2 (see Chapter 6). The organic hydroperoxides were about 10–40% of the H_2O_2 concentrations measured simultaneously. This is similar to the observations of Tremmel et al. (1994), who found that organic hydroperoxides in air over the northeastern United States were typically about half that of H_2O_2.

Other hydroperoxides have also been detected at small concentrations in air. For example, Hewitt and Kok (1991) reported the presence of 1-hydroxyethyl hydroperoxide as well as an unidentified compound, perhaps hydroxybutyl hydroperoxide, in air in rural Colorado.

In remote areas, CH_3OOH is generally the major, and often the sole detectable, organic hydroperoxide present (e.g., see Staffelbach et al., 1996). This is not surprising, since CH_4 is often the major organic in such regions, and hence the $CH_3O_2 + HO_2$ reaction is important.

A three-channel approach was developed by Lee et al. (1993) to distinguish H_2O_2 from hydroxymethyl hydroperoxide and total peroxides. In this approach, one channel is used to scrub the air sample into a POPHA solution to obtain total peroxides. In a second channel, the air sample is scrubbed into Fenton reagent solution at a pH of 3. This converts the H_2O_2 into OH radicals:

$$Fe(II) + H_2O_2 \rightarrow Fe(III) + OH^- + OH.$$

The OH radicals are trapped by reaction with benzoic acid, forming hydroxybenzoic acid, which is measured by fluorescence. Organic peroxides ROOH form RO + OH^-, but the derivatives of benzoic acid formed by the reaction of the RO radicals do not fluoresce under the conditions chosen to measure H_2O_2. Thus, in principle, this second channel measures H_2O_2. However, in practice, it was found to give about a 30% response to hydroxymethyl hydroperoxide as well, so that the results from this channel must be corrected for this contribution (Lee et al., 1993).

In the third channel of this instrument, air is scrubbed into a solution containing Fenton reagent at a pH of 9. At this high pH, hydroxymethyl hydroperoxide is rapidly hydrolyzed to give H_2O_2. Thus the third channel gives the sum of H_2O_2 and hydroxymethyl hydroperoxide and the difference between this and the second channel (corrected) gives hydroxymethyl hydroperoxide.

An intercomparison of TDLS, POPHA, and a luminol chemiluminescence method for H_2O_2 was carried out using zero air, irradiated VOC–NO_x mixtures, and ambient air (Kleindienst et al., 1988a). The TDLS and two POPHA methods using different sampling approaches (continuous scrubbing and a diffusion scrubber, respectively) were in reasonably good agreement. However, the luminol method exhibited positive and negative interferences under different conditions and hence has not since been applied extensively to ambient air measurements.

Another intercomparison was carried out at the Mauna Loa Observatory in 1991 and 1992 (Staffelbach et al., 1996). TDLS was used to measure H_2O_2. In addition, the POPHA method with catalase was used to distinguish between H_2O_2 and organic peroxides, the POPHA method with aqueous solubility differences was employed to discriminate between these compounds, and HPLC was used to separate and detect different hydroperoxides. For H_2O_2, while the measurements using the wet chemical methods and TDLS showed similar trends, there was a significant amount of scatter in individual measurements. For example, the correlation coefficients for plots of the TDLS versus the two POPHA techniques varied from 0.20 to 0.60. HPLC showed that CH_3OOH was the only organic hydroperoxide present. However, individual measurements of CH_3OOH made using this method compared to those using POPHA with catalase only had a correlation coefficient of 0.14 in wet air and 0.49 in dry air, whereas the corresponding correlation coefficients for the two POPHA measurements of organic hydroperoxides were 0.33 and 0.48, respectively.

In short, while wet chemical techniques are valuable for measurement of H_2O_2 and organic hydroperoxides, the absolute accuracy and precision remain a subject of concern and research.

Typical tropospheric concentrations. Measurements of H_2O_2 in air up to approximately 1990 are summarized by Gunz and Hoffmann (1990) and Sakugawa et al. (1990). Concentrations of H_2O_2 observed in air near the earth's surface are typically about 1–5 ppb (e.g., Daum et al., 1990; Van Valin et al., 1990; Claiborn and Aneja, 1991; Tremmel et al., 1993; Lee et al., 1993; Das and Aneja, 1994; Tanner and Schorran, 1995; Macdonald et al., 1995; Staffelbach et al., 1996; Sanhueza et al., 1996; Ridley et al., 1997; Martin et al., 1997; Balasubramanian and Husain, 1997; Weinstein-Lloyd et al., 1998). Although it may initially appear surprising, concentrations in remote and rural areas are not tremendously different from those in more polluted urban areas. For example, Heikes et al. (1996) reported levels of 0.3–5 ppb in the marine boundary layer, and Weinstein-Lloyd et al. (1998) measured concentrations of 1–4 ppb in the continental boundary layer midday in a rural area in the southeastern United States. The reason is that although there is a great deal more photochemical activity in the polluted regions, which might be expected to lead to H_2O_2, there is also more NO. Since H_2O_2 is formed by the $HO_2 + HO_2$ self-reaction and since HO_2 also reacts rapidly with NO, higher NO levels tend to inhibit the formation of the peroxide.

In addition to H_2O_2, methyl hydroperoxide has also been measured. For example, in air over Hawaii, concentrations of ~0.1–0.5 ppb were typical, although concentrations as high as 1.6 ppb have been observed in remote areas (Staffelbach et al., 1996; Heikes et al., 1996; Sanhueza et al., 1996; Ridley et al., 1997). Weinstein-Lloyd et al. (1998) measured concentrations up to ~2.5 ppb in the rural continental boundary layer. There are insufficient studies to firmly establish the relative contribution of CH_3OOH and perhaps other organic hydroperoxides to the total atmospheric levels, but the data available indicate that H_2O_2 is the major hydroperoxide present in air. For example, Tanner and Schorran (1995) found that H_2O_2 typically comprised about 90% of the total peroxides in the Grand Canyon area of the United States and Ayers et al. (1996) found in a limited set of measurements at Cape Grim, Tasmania, that H_2O_2 was ~60% of the total. Similarly, the median concentration of CH_3OOH measured by Weinstein-Lloyd et al. (1998) in the rural continental boundary layer was 1.7 ppb, representing about a third of the total measured hydroperoxides.

Hydroxymethyl hydroperoxide has also been identified and measured in rural continental areas as well as in the marine boundary layer. For example, Lee et al. (1993) report concentrations as high as 5 ppb in rural Georgia, and Heikes et al. (1996) measured concentrations from 0.6 to 1.6 ppb in the marine boundary layer over the south Atlantic Ocean. The median value in the rural continental boundary layer in the southeast United States was reported to be 0.97 ppb, with individual measurements ranging from ~0.2 to 3 ppb (Weinstein-Lloyd et al., 1998).

In summary, H_2O_2 is ubiquitous in air throughout the troposphere. CH_3OOH and $HOCH_2OOH$ have also been observed, generally at smaller concentrations than H_2O_2. There is at present little evidence for significant contributions of larger organic peroxides.

i. HO_x Free Radicals

As seen throughout this book, OH and HO_2 are central to the chemistry of both the lower and upper atmosphere. As a result, accurate measurement of their concentrations is critical to a quantitative understanding of atmospheric chemistry.

(1) OH Estimates of globally averaged OH concentrations have been obtained by applying a mass balance type of approach to certain compounds whose major removal from the atmosphere is believed to be reaction with OH. For example, the emissions of methylchloroform, CH_3CCl_3, are well known and its concentrations have been measured at a number of locations around the world. Using 3-D models, one can calculate the concentrations of OH and their geographical distribution that remove CH_3CCl_3 at appropriate rates to generate the measured concentrations (e.g., see Altshuller review, 1989; Spivakovsky et al., 1990; Prinn et al., 1992, 1995; and Krol et al., 1998). A similar approach has been taken using ^{14}CO (Brenninkmeijer et al., 1992).

In particular air masses, estimates of OH concentrations have also been derived from the relative rates of decay of a series of hydrocarbons in the air mass whose rate constants for reactions with OH are well known (e.g., Blake et al., 1993). Alternatively, organics can be added as tracers; criteria for the choice of suitable compounds are discussed by Davenport and Singh (1987). However, such approaches can be complicated by the effects of transport and mixing of the air mass with ones of different composition (e.g., McKeen et al., 1990) and by the possible contribution of oxidants other than OH to the decay of the organic (e.g., Blake et al., 1993). In addition, average OH concentrations rather than point or local measurements are derived from such data, and both this and the mass balance approach are indirect. However, Ehhalt et al. (1998) have proposed an alternate approach to using hydrocarbon concentrations to determine OH, which minimizes the assumptions inherent in this method.

Clearly, direct techniques for measuring OH are needed that provide concentrations either at a point or over relatively restricted spatial scales. Two (absorption and laser-induced fluorescence) are direct, spectroscopic methods and two others (mass spectrometry and a radiocarbon method) rely on conversion of OH to another species that is measured. Each of these approaches and some of the intercomparisons that have been carried out are discussed briefly in the following sections. A good overview of these methods is found in a review by Eisele and Bradshaw (1993) and articles by Crosley (1994, 1995a, 1995b) and papers in a special issue of the *Journal of the Atmospheric Sciences* [52 (19), October 1, 1995].

Absorption spectroscopy. OH undergoes an allowed transition between its $X^2\Pi$ ground state and the first electronically excited $A^2\Sigma^+$ state. Because it is a small species, absorption lines due to the individual vibrational and rotational transitions can be resolved experimentally. As a result, it has a very characteristic banded absorption structure around 308 nm whose features make it an ideal candidate for DOAS measurements.

Atmospheric OH has been measured for a number of years using its absorption. For example, vertical column abundances of OH have been measured in a number of studies using the sun as a light source (e.g., Burnett and Burnett, 1981, 1996; Burnett et al., 1988; Burnett and Minschwaner, 1998). Over the past several decades, beginning with the measurements of Perner et al. (1976), absorption spectroscopy has been used to make measurements of OH over much shorter paths in the troposphere. The fundamental principles behind this technique have been described earlier in the discussion of DOAS spectrometry. Here we briefly discuss some of the aspects of the measurements that are unique to OH, as well as some typical applications.

Figure 11.42a shows an energy level diagram and some of the allowed lines in the $v'' = 0$ level of $X^2\Pi$ to the $v' = 0$ level of $A^2\Sigma^+$ (Mount, 1992). Figure 11.42b shows the absorption spectrum of OH obtained using a butane flame as the source in this case. The emission of a frequency-doubled dye laser whose full width at half-maximum is 0.41 nm is also shown (Dorn et al., 1995a). The laser emission is sufficiently broad

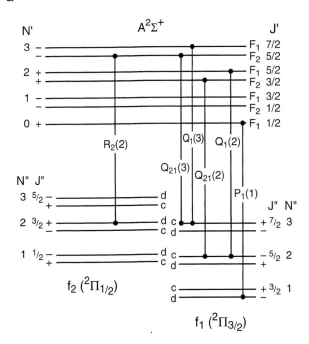

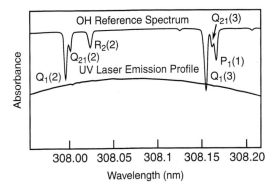

FIGURE 11.42 (a) Energy levels and some allowed transitions for the OH($X^2\Pi \to A^2\Sigma^+$) absorption, (b) a typical broadband laser emission line profile, and an OH reference spectrum with absorption lines in this region. (Adapted from Mount, 1992; and Dorn et al., 1995a.)

compared to the OH linewidth that six different OH lines can be measured simultaneously using a high-resolution spectrograph and linear photodiode array. In an open multipass White cell with base path of 38.5 m and a total path length of 1.85 km, a detection limit for OH of 8.7×10^5 radicals cm^{-3} can be obtained (Brandenburger et al., 1998).

Open-beam double-pass instruments using a retroflector array to return the light beam back to the spectrograph are also in use (e.g., Mount, 1992; Mount and Harder, 1995; Harder et al., 1997a). Another variation is the use of a fast-scanning technique with picosecond pulse lengths and a photomultiplier detector, rather than the use of a photodiode array. For example, Armerding and co-workers (Armerding et al., 1994, 1995, 1996, 1997) tune the laser wavelength rapidly over the region of interest, covering up to ~ 13 cm^{-1} in spectral width in 0.1 ms. Multiple scans, taken with a repetition rate of 1.3 kHz, are averaged to improve the signal-to-noise ratio.

A typical example of such measurements was shown in Fig. 11.15.

A major advantage of this approach is that the fundamental spectroscopic parameters for OH, including the absorption cross sections for various transitions, are well known (e.g., see Mount, 1992; Dorn et al., 1995b), so that absolute concentrations of OH can be calculated based solely on the absorption spectra. Another major advantage is that the laser beams are expanded so that generation of OH in the beam itself by photolysis of ozone is not the problem that it has been in LIF measurements (vide infra).

A disadvantage is that the use of long paths gives average concentrations over the whole distance, over which there could be considerable variability. Folded paths using White cells provide a more restricted measurement distance, but reflection losses on the mirrors preclude increasing the total path length and hence detection sensitivity beyond a certain point. Whether such a system can be made sufficiently stable for aircraft use is also not clear. Finally, the sensitivity of such systems is in the 10^5 to low 10^6 radical cm^{-3} range. While this is sufficient for daytime measurements when there are significant photolytic sources of OH, it is not adequate for nighttime measurements where much lower concentrations (likely of the order of 10^4 cm^{-3} or less) are generated by such processes as PAN decomposition.

Laser-induced fluorescence (LIF). Laser-induced fluorescence measurements have been applied to the atmosphere since the suggestion of Baardsen and Terhune in 1972 that this method should be feasible. Figure 11.43 shows the energy levels and transitions involved in LIF measurements. OH is excited from its ground $X^2\Pi$ state into the first electronically excited $A^2\Sigma$ state. The $v'' = 0$ to $v' = 0$ transition is around 308 nm and the $v'' = 0$ to $v' = 1$ at 282 nm. Two schemes have been used: excitation using 282 nm into $v' = 1$ of the upper electronic state, or excitation using 308 nm into $v' = 0$ of the upper state. Collisional quenching deactivates some of the $v' = 1$ into $v' = 0$ in competition with fluorescence, mainly in the (1,1) band of the electronic transition (that is, from $v' = 1$ of the upper state into $v'' = 1$ of the lower state). Collisional deactivation of $v' = 0$ then occurs in competition with fluorescence in the (0,0) band at 308 nm

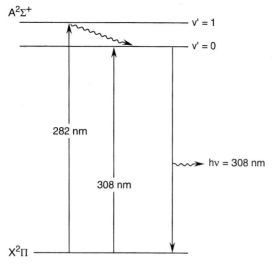

FIGURE 11.43 Schematic diagram of OH energy levels used in LIF measurements.

(e.g., see Copeland *et al.*, 1985; Copeland and Crosley, 1986; Crosley, 1989).

The major problem with LIF measurements in the past has been what might be called the "atmospheric uncertainty principle;" i.e., in the act of carrying out the measurement, the system is perturbed and artifact formation of OH can occur (e.g., see Smith and Crosley, 1990; and Hard *et al.*, 1992b). This is primarily due to the photolysis of O_3 to generate $O(^1D)$, which in the presence of water vapor forms OH:

$$O_3 + h\nu(\lambda < 350 \text{ nm}) \rightarrow O(^1D) + O_2,$$

$$O(^1D) + H_2O \rightarrow 2OH,$$

$$+ M \rightarrow O(^3P) + M.$$

This artifact formation of OH is less severe when the excitation is at 308 nm, rather than 282 nm, since both the absorption cross sections and quantum yields for ozone photolysis decrease rapidly with wavelength in this region (see Chapter 4.B). Another advantage is that the absorption cross section for the (0,0) transition is about a factor of four larger than for the (0,1) transition, increasing the amount of excited OH. As a result, most LIF systems now use 308-nm excitation (e.g., Chan *et al.*, 1990; Stevens *et al.*, 1994; Hard *et al.*, 1995; Holland *et al.*, 1995). The disadvantage is a larger background signal at the fluorescence wavelength due to scattered laser light.

A second approach to minimizing the artifactual formation of OH in these measurements has been to sample the air through a nozzle into a low-pressure region operated at ~4 Torr. This was pioneered by O'Brien, Hard, and co-workers (Hard *et al.*, 1984; Chan *et al.*, 1990; Hard *et al.*, 1995) and is known as the FAGE technique (*f*luorescence *a*ssay with *g*as *e*xpansion). The advantage is that the rate of generation of OH from the $O(^1D) + H_2O$ reaction is smaller, providing less *in situ* generation of OH in the laser beam. While the OH concentration in air is reduced proportionately with the pressure, collisional quenching of the electronically excited OH is as well; the result is that the OH LIF signal does not change substantially on reducing the pressure.

The "zero" signal in such instruments is usually established by adding an organic such as isobutane (e.g., Hard *et al.*, 1992b) or C_6F_6 (Stevens *et al.*, 1994; Dubey *et al.*, 1996) that reacts rapidly with the OH. The difference in signal when the compound is added compared to when it is not is then a measure of the OH present. Another approach is to tune the laser on and off resonance with the OH absorption, permitting measurement of the background signal, which can be subtracted (e.g., Hofzumahaus *et al.*, 1996).

Figure 11.44 is a schematic diagram of one LIF instrument (Stevens *et al.*, 1994; Brune *et al.*, 1998). An air-cooled copper-vapor laser pumps a dye laser whose output at 616 nm is doubled to generate the 308-nm exciting radiation. An OH reference cell in which OH is generated from the thermal dissociation of water

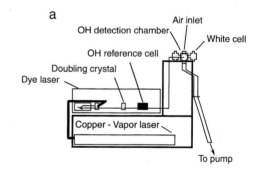

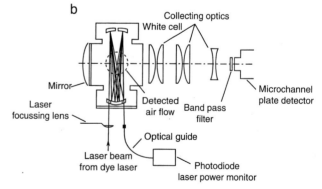

FIGURE 11.44 (a) Overall schematic diagram of an LIF instrument used for OH and HO_2 and (b) sample chamber in this instrument. (Adapted from Stevens *et al.*, 1994.)

provides the reference for tuning the dye laser into resonance with the OH absorption. The beam is directed into a multipass cell with White cell optics as shown in Fig. 11.44b.

One disadvantage of LIF compared to absorption measurements is the need for field calibration. It is a nontrivial issue to generate known concentrations of OH under ambient conditions for this purpose. A variety of approaches are used. These include photolysis of water vapor at 185 nm where the H_2O absorption cross section as well as that of O_2 are needed (e.g., Holland et al., 1998). However, there has been considerable uncertainty associated with these absorption cross sections (e.g., see Lazendorf et al., 1997; and Hofzumahaus et al., 1997, 1998). Stevens et al. (1994) used an internal calibration by titration of known concentrations of NO_2 with an excess of H atoms which generates OH via $H + NO_2 \rightarrow OH + NO$ combined with external calibration using water vapor photolysis to account for transmission of OH through the sampling system. Sampling from a sample chamber in which a VOC–NO_x mixture is irradiated and the rate of decay of the organics used to obtain the OH concentration has also been used (Hard et al., 1984; Chan et al., 1990).

As discussed later, LIF has also been used to measure HO_2 by conversion to OH by reaction with NO.

Mass spectrometry. Reaction of OH to form an ion, HSO_4^-, which can be measured by mass spectrometry was first demonstrated by Eisele and Tanner (1991). Figure 11.45 is a schematic diagram of this approach (Tanner et al., 1997). Air is sampled through an inlet system described in detail by Eisele et al. (1997) and mixed with isotopically labeled $^{34}SO_2$, forming $H_2^{34}SO_4$ via reactions discussed in Chapter 8.C.2:

$$OH + {}^{34}SO_2 + M \rightarrow H{}^{34}SO_3 + M,$$

$$H{}^{34}SO_3 + O_2 \rightarrow HO_2 + {}^{34}SO_3,$$

$$^{34}SO_3 + H_2O + M \rightarrow H_2{}^{34}SO_4 + M.$$

Sufficient $^{34}SO_2$ is added to convert more than 99% of the OH in air to the acid. The use of isotopically labeled SO_2 forms labeled H_2SO_4 which is not present in measurable quantities in air. Thus, labeled H_2SO_4 is equal to the initial OH and allows $H_2^{32}SO_4$ present in air to be measured simultaneously. Periodically during the measurements, propane is added simultaneously with the $^{34}SO_2$ at concentrations that will remove most of the OH, providing a background signal.

As discussed shortly, HO_2 and RO_2 react in the presence of NO to regenerate OH, which will lead to an overestimate of the OH concentration. To minimize this, propane is added downstream of the $^{34}SO_2$ injec-

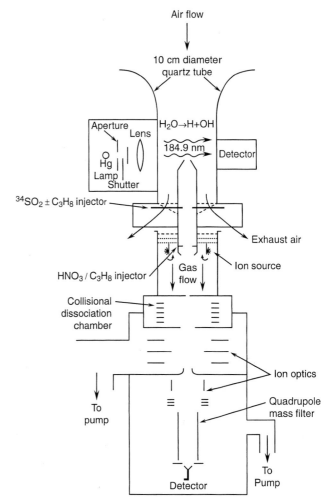

FIGURE 11.45 Schematic diagram of mass spectrometer used for OH measurements using derivatization approach (adapted from Tanner et al., 1997).

tor to remove any of this regenerated OH. However, as discussed by Tanner et al. (1997), at high NO_x concentrations, some regeneration does occur and the measurements must be corrected for that.

At this downstream port, HNO_3 is also added at concentrations such that the $NO_3^-(HNO_3)$ ion adduct is the major nitrate ion (see discussion of mass spectrometry in Section A.2). Since H_2SO_4 is a stronger acid, it proton transfers to the cluster:

$$H_2^{34}SO_4 + NO_3^-(HNO_3) \rightarrow H{}^{34}SO_4^-(HNO_3) + HNO_3.$$

Figure 11.46a shows a typical mass spectrum. In addition to the $NO_3^-(HNO_3)$ ionizing agent, smaller amounts of NO_3^- and $NO_3^-(HNO_3)_2$ are present. The HNO_3 adducts of both the naturally occurring ^{32}S and the added ^{34}S isotopes of HSO_4^- are seen as well as the corresponding HSO_4^- ions. These ions then enter a

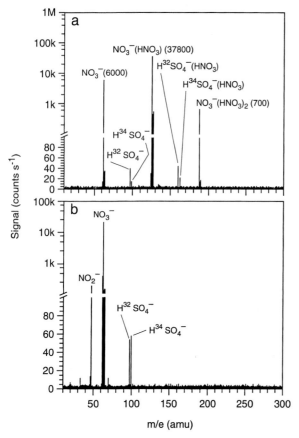

FIGURE 11.46 Typical mass spectra of ambient air (a) without collisional dissociation of the adducts and (b) with collisional dissociation. The counts for NO_3^- and its adducts with HNO_3 are shown in parentheses in (a). Note changes in scale from linear to logarithmic (adapted from Tanner *et al.*, 1997).

collisional dissociation chamber where the HNO_3 adducts of HSO_4^- are fragmented. Figure 11.46b shows a typical mass spectrum (not for the same conditions as Fig. 11.46a, however) after collisional dissociation. The HSO_4^- ions are now the sole form of these ions and can be cleanly measured using mass spectrometry.

As is the case for LIF, calibration to obtain absolute concentrations is a challenge. In the instrument shown in Fig. 11.45, a calibration source based on the photolysis of water at 185 nm is installed in the inlet. From the absorption cross section of H_2O gas at 185 nm, its concentration, the light intensity, and the sample flow rate, the concentration of OH generated by the photolysis can be calculated. However, not only is there significant uncertainty in the absorption cross section for H_2O at 185 nm (e.g., see Lazendorf *et al.*, 1997; Hofzumahaus *et al.*, 1997, 1998; and Tanner *et al.*, 1997), but the measured calibration factor was highly variable from day to day, by as much as a factor of two (Tanner *et al.*, 1997).

In summary, the measurement of OH by reaction to form isotopically labeled $H_2^{34}SO_4$ and measurement of the latter by chemical ionization mass spectrometry is promising. As discussed later, measurements made using this technique compare well with those using UV absorption, and the sensitivity is good, $\sim 2 \times 10^5$ radicals cm^{-3}. With sufficiently long integration times, even much smaller nighttime concentrations in the range of low 10^4 radicals cm^{-3} have been measured (Tanner and Eisele, 1995). Figure 11.47, for example, shows OH measurements made throughout the night at the Mauna Loa Observatory in Hawaii in May 1992. Concentrations of OH just after sunset are in the low 10^5 cm^{-3} range and fall off during the night to levels indistinguishable from zero in this instrument.

The disadvantages are the need to correct for regeneration of OH in the instrument from HO_2 and RO_2 reactions at high ambient NO concentrations and, as with LIF, the uncertainty in the absolute calibration.

Radiocarbon technique. Campbell and Sheppard (Campbell *et al.*, 1979) developed a method for OH based on its oxidation of isotopically labeled ^{14}CO that is added to the air sample. Assuming that CO is oxidized to CO_2 by the reaction with OH, the rate of formation of CO_2 is given by

$$d[CO_2]/dt = k[CO][OH],$$

where k is the rate constant for the OH + CO reaction under the appropriate conditions. The added ^{14}CO and reaction time are such that the ^{14}CO concentration remains essentially constant. Integration of this rate expression results in the following expression:

$$[OH] = \frac{[^{14}CO_2]}{[^{14}CO]} \frac{1}{kt}. \tag{Q}$$

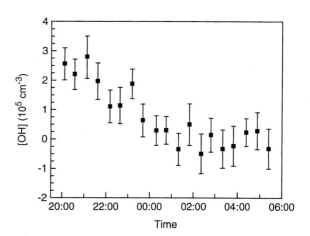

FIGURE 11.47 Nighttime measurements of OH at Mauna Loa Observatory, Hawaii, in May 1992 made by the mass spectrometry derivatization technique (adapted from Tanner and Eisele, 1995).

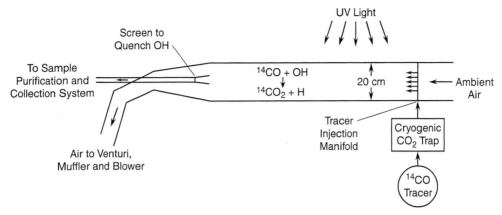

FIGURE 11.48 Schematic diagram of radiochemical OH measurement apparatus (adapted from Felton et al., 1990).

Figure 11.48 is a schematic diagram of this apparatus (Felton et al., 1990). Air is introduced into a sampling manifold consisting of a quartz tube where it is mixed with the ^{14}CO. The air is collected downstream after a measured reaction time and analyzed for $^{14}CO_2$.

There are several assumptions inherent in this method (e.g., see Felton et al., 1990, 1992). For example, the concentration of $^{14}CO_2$ in ambient air must be negligible compared to that formed in the reaction and the OH concentration in air is assumed to be unperturbed either by the addition of ^{14}CO or by the sampling system itself, e.g., by loss on the walls. While straightforward in principle, as discussed by Felton et al. (1990, 1992), it is experimentally challenging. For example, accurately measuring the small concentrations of $^{14}CO_2$ formed is difficult, imposing stringent requirements on the purity of the ^{14}CO tracer and on the purification techniques used for the product $^{14}CO_2$.

Intercomparisons. A number of intercomparison studies have been carried out for the different OH measurement techniques (e.g., see Beck et al., 1987; Mount and Eisele, 1992; Eisele et al., 1994; Campbell et al., 1995; Brauers et al., 1996; Mount et al., 1997a, 1997b; and Hofzumahaus et al., 1998). Overall, given the extreme difficulty in sampling and measuring this highly reactive free radical at the sub-ppt concentrations found in air, the agreement is generally quite good.

Figure 11.49, for example, shows measurements of the diurnal variation of OH made using LIF and UV absorption, respectively, on two different days in a rural area in Germany (Hofzumahaus et al., 1998). The agreement is, in most cases, excellent. These data also illustrate a typical diurnal variation of OH, being below the detection limits of the instruments at night (5×10^5 radicals cm^{-3} for LIF and 1.5×10^6 radicals cm^{-3} for DOAS) and rising to a peak of $\sim 10^7$ radicals cm^{-3} at noon when photolysis of its precursors peaks. Similar diurnal behavior has been observed in remote areas such as the Mauna Loa Observatory (e.g., Eisele et al., 1996) and in more polluted areas as well (e.g., Felton et al., 1990; Hard et al., 1995; Mount et al., 1997b). Typical peak OH concentrations are usually in the range of $\sim (2-10) \times 10^6$ radicals cm^{-3}.

Figure 11.50 shows for this particular intercomparison study a plot of OH measured by DOAS against those obtained simultaneously by LIF. The correlation coefficient is $r = 0.85$. Disagreement was greatest when the wind was from a particular direction, which gave higher DOAS readings. The reason for this is not clear, but Hofzumahaus and co-workers propose that it may

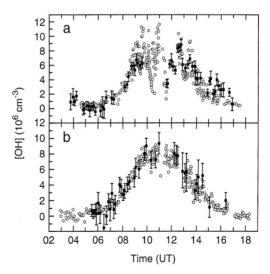

FIGURE 11.49 Diurnal variation of OH measured using LIF (○) and DOAS (●) in a rural area in Germany on the (a) 16th and (b) 17th of August 1994. (Adapted from Hofzumahaus et al., 1998.)

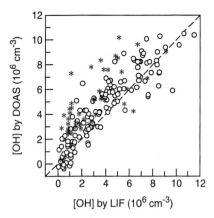

FIGURE 11.50 Correlation between OH measurements made by DOAS and by LIF in a rural area in Germany in August 1994. The data indicated by asterisks were measurements made when the wind was from a particular direction suggesting it might contain unrecognized OH sources affecting the long-path DOAS measurements (adapted from Hofzumahaus *et al.*, 1998).

be due to unrecognized OH sources that affected the long-path DOAS measurements more than the point measurements made by LIF. Exclusion of those data points improves the correlation ($r = 0.90$) and the slope of the line is 1.09 ± 0.04 with an intercept within experimental error of zero.

Similarly, agreement between UV absorption and the mass spectrometer technique is quite good. Figure 11.51 shows a plot of OH concentrations measured using the mass spectrometry derivatization technique compared to those measured using long-path UV absorption in a rural area in Colorado for clear days with NO_x below 450 ppt measured by an *in situ* technique and NO_2 below 500 ppt averaged over the long path

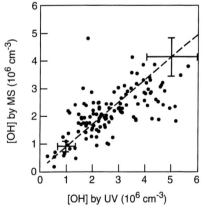

FIGURE 11.51 Correlation between measurements made by the mass spectrometry–derivatization technique and long-path UV absorption in rural Colorado for lower NO_x conditions (adapted from Mount *et al.*, 1997a).

(Mount *et al.*, 1997a). The slope of the line is 0.82 ± 0.06; i.e., the mass spectrometer point measurements were about 20% lower than the UV measurements over a path length of 10.3 km. About 25% of the data were different by amounts outside the experimental errors. Such discrepancies may be due to comparing distance-averaged to point values and/or to calibration inaccuracies.

An intercomparison of the mass spectrometer method with an LIF instrument, however, was not as good. While the slope of the plot of LIF versus the MS measurements was 0.73, the r value was only 0.26, in part due to poor laser performance in the LIF instrument during the studies (Mather *et al.*, 1997).

Extensive intercomparisons using the radiocarbon technique have not been carried out. Campbell *et al.* (1995) compared measurements using the radiocarbon technique to those from an LIF instrument (Chan *et al.*, 1990). The values obtained were frequently near the detection limits of the instruments, but despite that, were reasonably well correlated ($r^2 = 0.74$). However, the slope of a plot of the radiocarbon versus LIF absolute concentrations was 2.9, i.e., there was a difference of about a factor of three.

In short, given the challenges associated with measuring OH, the disagreement between the various methods is not surprising and the discrepancies appear to be improving as the methods are developed further.

(2) HO_2 and RO_2 There are three approaches that are used to measure HO_2 and/or RO_2: (1) conversion of HO_2 and/or RO_2 to OH and measurement of the latter using techniques already described, (2) a chemical amplifier method, and (3) matrix isolation ESR.

Conversion of HO_2 to OH. HO_2 can be measured by conversion into OH by its fast reaction with NO,

$$HO_2 + NO \rightarrow OH + NO_2,$$

followed by measurement of the OH by one of the methods described in the preceding section. For example, LIF detection of OH generated by reacting HO_2 with NO has been used to measure HO_2 at both remote and urban sites (Hard *et al.*, 1984, 1992a).

Another approach combines the mass spectrometric derivatization approach with chemical amplification (Reiner *et al.*, 1997, 1998). In this instrument, HO_2 and RO_2 are converted to OH through the reactions in the chemical amplifier approach discussed below, and the OH is then converted to H_2SO_4 by reaction with SO_2 and measured by chemical ionization mass spectrometry using $NO_3^-(HNO_3)$ clusters as described earlier. In this case, the use of isotopically labeled SO_2 is not necessary, since the ambient H_2SO_4 concentration is much smaller than that of the peroxy radicals.

Because HO_2 radical concentrations in the troposphere are typically about two orders of magnitude larger than those of OH, the contribution of ambient OH to the signal does not present a problem.

Chemical amplifier method. Another approach, known as the chemical amplifier method, pioneered by Cantrell and Stedman (Cantrell and Stedman, 1982; Cantrell *et al.*, 1984) has been used extensively to measure the combination of HO_2 and RO_2 (although the latter is not necessarily with 100% efficiency; vide infra). This method involves the conversion of HO_2 to OH in a chain reaction with a length of ~100–200. Figure 11.52 is a schematic diagram of one such instrument (Cantrell *et al.*, 1993). Air containing HO_2, RO_2, OH, and other species is sampled into the instrument, where it is mixed with NO, typically at ~3 ppm, and CO, at about 7–10% of the total flow. HO_2 reacts with NO as above to generate OH. In the presence of large concentrations of CO, HO_2 is regenerated:

$$OH + CO \xrightarrow{O_2} HO_2 + CO.$$

Thus, a chain reaction is set up in which HO_2 converts NO to NO_2 and is subsequently regenerated by the OH + CO reaction. The NO_2 is measured using techniques such as those described earlier; in the case of the system in Fig. 11.52, the luminol chemiluminescence technique is used. Termination of the chain occurs via reactions such as

$$OH + NO + M \rightarrow HONO + M,$$
$$HO_2 + NO_2 + M \rightarrow HO_2NO_2 + M,$$
$$HO_2 \rightarrow \text{Wall loss}.$$

The HO_2 concentration is given by

$$[HO_2] = \Delta NO_2/\text{chain length},$$

where the chain length is defined as the number of NO_2 molecules formed per initial HO_2 radical.

In addition to HO_2, organic peroxy free radicals are also measured, although not necessarily with 100% efficiency. For example, if CH_3O_2 is also present, the following reactions occur:

$$CH_3O_2 + NO \rightarrow CH_3O + NO_2,$$
$$CH_3O + O_2 \rightarrow HO_2 + HCHO.$$

The HO_2 then reacts as above in a chain reaction. While CH_3O_2 forms HO_2 in a straightforward series of reactions, larger RO_2 radicals may not. For example, as discussed in Chapter 6, a significant fraction of the reactions of larger RO_2 radicals with NO generates stable organic nitrates, $RONO_2$, rather than RO + NO_2. In addition, larger alkoxy radicals may not solely undergo reaction with O_2 to generate HO_2; indeed, as seen in Chapter 6, this is a minor path for some organic peroxy radicals, where decomposition and/or isomerization may predominate. As a result, the chemical amplifier measures HO_2 and some weighted fraction of RO_2 radicals.

For example, Cantrell and co-workers (1993) estimate the efficiency of conversion of simple alkyl peroxy radicals to vary from 0.93 for $CH_3CH_2O_2$ to 0.47 for $(CH_3)_2CO_2$, and it may be even less for larger alkyl peroxy radicals. This may be the reason that in some intercomparison studies, the matrix isolation–ESR technique (vide infra), which measures the sum of RO_2, gives some higher concentrations for some individual measurements than the chemical amplifier method (e.g., Zenker *et al.*, 1998).

Calibration has been carried out using known HO_2/RO_2 sources such as the thermal decomposition of PAN or H_2O_2 (e.g., Cantrell *et al.*, 1993), photolysis of H_2O_2 or water vapor (e.g., Schultz *et al.*, 1995), and the photolysis of CH_3I in the presence of O_2 (e.g., Clemitshaw *et al.*, 1997). This in effect allows the chain length to be determined so that peroxy radical concentrations can be derived from the increase in NO_2 as given above. However, there appear to be some factors affecting the sensitivity that are not well understood. For example, the chain length has been shown to be sensitive to the concentration of water vapor in air in at least one instrument, for reasons that are not clear (Mihele and Hastie, 1998).

Matrix isolation–electron spin resonance. A third method used to measure HO_2 and RO_2 is matrix isolation with ESR (see earlier description of matrix isolation). Because HO_2 and RO_2 have distinct ESR signals, they can be differentiated (Mihelcic *et al.*, 1985, 1990, 1993). For example, Fig. 11.53, part A, shows the ESR spectrum obtained when approximately

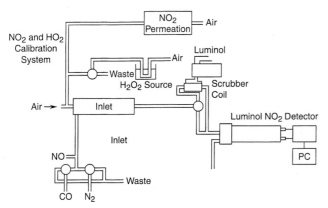

FIGURE 11.52 Schematic diagram of chemical amplifier apparatus for measurement of HO_2 and RO_2 (adapted from Cantrell *et al.*, 1993).

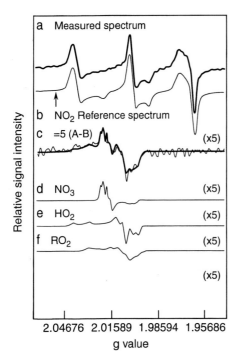

FIGURE 11.53 Matrix isolation–ESR measurement of NO$_2$ (680 ppt), NO$_3$ (5.2 ppt), HO$_2$ (10 ppt), and ΣRO$_2$ (5 ppt) in Schavinsland, Germany, in August 1990 (adapted from Mihelcic et al., 1993).

8 L of air in rural Germany was trapped in a polycrystalline matrix of D$_2$O at 77 K (Mihelcic et al., 1993). Spectrum b shows the ESR spectrum of NO$_2$; it can be seen that most of the observed ESR signals are due to NO$_2$, calculated from reference spectra to be present at a concentration of 0.68 ppb in this sample. Spectrum c is the difference between spectra a and b, magnified by a factor of five. Spectra d, e, and f are those of NO$_3$, HO$_2$, and RO$_2$, respectively, and their sum is shown by the heavy line through spectrum c. Clearly, the signals in spectrum c reflect contributions from these three radicals, at concentrations of 5.2 ppt NO$_3$, 10 ppt HO$_2$, and 5 ppt RO$_2$ in this particular sample. Detection limits for this method are 5 ppt for HO$_2$ and RO$_2$, respectively (Mihelcic et al., 1993).

Fewer intercomparison studies have been carried out for peroxy radicals than for OH. Two chemical amplification methods were compared during a measurement campaign in Brittany, France (Cantrell et al., 1996). Although the measurements tended to track one another, there is more scatter than might be expected, given the similar nature of the instruments. For example, a plot of the data from one instrument against those from the second had a slope of 0.71 but a correlation coefficient of only $r = 0.36$. In another study (Zenker et al., 1998), comparison of three chemical amplifier techniques to matrix isolation–ESR gave agreement to within 25% for two of the chemical amplifier methods and the ESR approach. The third chemical amplifier technique gave on average values that were about 65% of the matrix isolation–ESR values.

Measurements using the chemical amplifier technique were also carried out at the same time as the mass spectrometer derivatization method was used, with titration of the HO$_2$ to OH (Cantrell et al., 1997a). The chemical amplifier values were a factor of 2–3 times higher than those measured using the mass spectrometer approach, possibly because the latter measured HO$_2$ whereas the former measured HO$_2$ and some weighted fraction of RO$_2$. Finally, comparison of chemical amplifier measurements to those using matrix isolation–ESR (Volz-Thomas et al., 1995; cited by Cantrell et al., 1997b) shows agreement within about 40% for clean or moderately polluted air masses. For more heavily polluted air, the chemical amplifier was systematically lower, suggesting that there were significant concentrations of larger RO$_2$ radicals to which the chemical amplifier was less sensitive.

Typical tropospheric concentrations. Figure 11.54 shows the diurnal variation of average typical peroxy radical concentrations made using the chemical amplifier technique in Cape Grim, Tasmania, and Mace Head, Ireland (Carpenter et al., 1997). As is the case for OH, HO$_2$ and RO$_2$ typically peak around noon, when photolysis is maximum, and are much smaller at night, particularly in low-NO$_x$ environments where there is little nighttime NO$_3$ (e.g., Monks et al., 1996; Carslaw et al., 1997a; Stevens et al., 1997). Peak concentrations are in the 10^8–10^9 cm^{-3} range in remote areas (e.g., Carpenter et al., 1997; Fischer et al., 1998), with higher concentrations in polluted areas. For example, in downtown Denver, peak concentrations of 3×10^9 radicals cm^{-3} have been measured (Hu and Stedman, 1995).

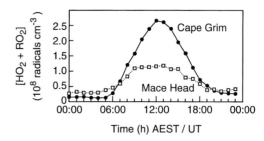

FIGURE 11.54 Diurnal profile of average (HO$_2$ + RO$_2$) concentrations measured at Cape Grim, Tasmania (●), and at Mace Head, Ireland (■), under clean air conditions using a chemical amplification technique. (Adapted from Carpenter et al., 1997.)

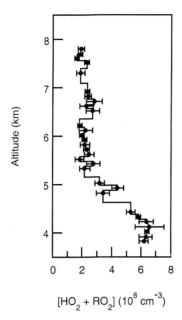

FIGURE 11.55 Altitude profiles for $HO_2 + RO_2$ in the free troposphere over southern Germany determined by conversion to OH and measuring OH by the mass spectrometric derivatization technique (adapted from Reiner et al., 1998).

Figure 11.55 shows an altitude profile for peroxy radicals measured above the boundary layer over southern Germany using chemical amplification with the mass spectrometric derivatization measurement of OH (Reiner et al., 1998). Concentrations are again seen to be in the range of $10^8–10^9$ cm^{-3}.

In summary, HO_2 and RO_2 radical concentrations are substantially greater than those of OH, typically by several orders of magnitude. There are several different approaches to measuring these peroxy radicals, and the results from these are in overall agreement as to the magnitude of the concentrations and their diurnal variation. However, there have not been a significant number of intercomparison studies of these methods, so evaluation of the absolute accuracies will require further work.

5. Generation of Standard Gas Mixtures

As seen throughout this discussion of the measurement of gases in the atmosphere, a critical component is the accurate calibration of the technique for the gas(es) of interest. This clearly requires sources of such calibration gases, which however, vary depending on the particular gas.

In the simplest case, the gas of interest can be purchased in a gas cylinder with known concentration provided by the supplier. In the United States, NIST has some mixtures relevant to atmospheric measurements. This approach has been used, for example, for simple hydrocarbons that are readily available and relatively stable. Preparation of standards in cylinders can also be carried out by the individual laboratory (e.g., see Apel et al., 1998a). Such standards are frequently at higher concentrations than those to be measured in air. In this case, dynamic dilution systems are used to dilute the cylinder mixtures to the desired concentration range.

Caution must be exercised in using cylinder gases in some cases, however. For example, NO_2 in air from cylinders commonly contains a few percent HNO_3 as an impurity, and nickel carbonyls are present in CO stored in cylinders.

In other cases, the species cannot be preprepared as a mixture in air and hence flow systems must be used to generate them. For example, HNO_3 strongly adsorbs to surfaces and hence it is not possible to make a stable calibration mixture that can be stored. Some larger organics also do not have long-term stability when stored in gas cylinders.

In such cases, permeation tubes or diffusion cells are commonly used to generate the species in a flow of air, which can then be introduced into the measuring device. Permeation tubes are permeable tubes whose ends are sealed off and which contain the species of interest as a liquid in equilibrium with its vapor. The vapor permeates through the walls of the device at a rate that depends on temperature. The rate of permeation at a given temperature is normally supplied by the manufacturer and can be determined independently by weighing the permeation tube before and after use. From a knowledge of the flow rate of the gas passing over the tube, which entrains the vapor, the concentration of the species of interest in the air flow can be calculated. This approach is commonly used for species such as HNO_3, Cl_2, and HCl.

A similar approach is the use of diffusion cells. In this case, the liquid is held in a container that has a capillary of fixed length and diameter through which the vapor over the liquid diffuses. The vapor exiting the capillary is swept into a flow of gas to provide the gas mixture; this approach has been used to prepare mixtures of terpenes in air, for example (Larsen et al., 1997). The concentration of the gas can be varied by using capillaries of varying internal diameter and length.

In some cases, the compound itself is sufficiently unstable that it cannot be purchased and must be synthesized. This is the case for compounds such as O_3 and HONO. Ozone at ppb to ppm concentrations in air is generated either by photolyzing O_2, e.g., using a low-pressure mercury lamp, or by a discharge in O_2;

when discharges are used, care must be taken to exclude air from the discharge region to avoid the simultaneous formation of oxides of nitrogen. In the case of HONO, a flow of gaseous HCl over $NaNO_2$ salt is often used to generate this compound in a flow system (Febo et al., 1995). For other more "exotic" species such as $ClONO_2$ and $ClNO_2$, synthesis of the compounds is more involved and the literature should be consulted for methods of synthesis.

B. PARTICLES

With the increasing epidemiological evidence for significant health impacts of particles (see Chapter 2.A.5), measurement of particle characteristics has taken on new urgency. With particles, both the chemical composition and size distribution of each component are important, and a wide range of sizes from ultrafine particles to coarse particles must be analyzed. While there is no fixed definition of "ultrafine" particles, those with diameters <10 nm are often referred to as ultrafine (although in some cases, up to 100 nm has been included in this description). In addition, the chemical components encompass almost the entire periodic table and include inorganic and organic as well as elemental and complex molecular species. Hence the area of particle characterization is a very challenging one.

Traditionally, particles have been collected and then analyzed for the distribution of mass and chemical composition. Various size ranges, or "bins," have been used, ranging from simple cutoffs at 10 μm, for example, to multibin analyses in which particles in six or more size ranges are collected and analyzed individually. Such approaches have produced the vast majority of the data in the literature, and the techniques used are summarized briefly in the following sections.

However, one might clearly expect significant variations in chemical composition between particles even within one range, and hence analysis of individual particles by size and composition is important. In addition, measuring such size-resolved properties in real time is desirable to elucidate sources, the atmospheric chemistry of particles, and the processes involved in their formation and fate. While techniques are now becoming available that address these concerns, this area of real-time and single-particle measurement could be considered to be in its infancy. Some of the instrumental techniques that have been successfully applied to ambient air are described in the following sections, along with some promising new approaches.

1. Sampling and Collection of Particles

The first steps in traditional analysis of the physical and chemical properties of atmospheric particulate matter are sampling, that is, obtaining a representative sample over the desired size range, and collection, that is, separating the particles from air. During sampling and collection, such parameters as humidity, temperature, and particle concentration must be controlled to maintain the sample integrity.

Sampling of particles presents some different considerations compared to sampling of gases. The larger mass of particles results in a much greater inertia, so that when the gas flow curves sharply, the particles tend to go straight ahead. High or low inlet velocities as well as bends in tubing used to sample for particles can thus lead to significant particle size bias and should be avoided. In addition, the sampling lines should be as short as possible to minimize particle loss by gravitational settling and turbulent deposition. Losses can also occur on the sampling surfaces if an electrostatic charge is allowed to build up.

Because it is particles in the smaller size range, $\lesssim 2.5$ μm ($PM_{2.5}$), that are of greatest interest with respect to health effects, inlet systems are normally used that exclude larger particles. These size exclusion inlets are usually based on filters, cyclone collectors, or impactors, the principle behind which is discussed shortly. Inlet cutoff diameters from 2.2 to 15 μm are achieved in commercial instruments using these techniques (Chow, 1995).

Collection of particles is based on filtration, gravitational and centrifugal sedimentation, inertial impaction and impingement, diffusion, interception, or electrostatic or thermal precipitation (e.g., see Spurny, 1986, Chapter 3). The choice of method depends on a number of parameters such as the composition and size of the particles, the purpose of the sample, and acceptable sampling rates. Table 11.10 summarizes some of the commonly used methods and the size ranges over which they are effective.

a. Filters

Filters collect liquid and solid particles by mechanisms including diffusion, impaction, interception, electrostatic attraction, and sedimentation onto the filter while allowing the gas to pass through. The types commonly used in atmospheric particulate collection are membranes, fibrous mats, or porous sheets. Different filter materials are used depending on the particular type of measurement being carried out, including Teflon, quartz fiber, nylon, silver, cellulose filters, glass fibers, and polycarbonate. The characteristics of each are summarized by Chow (1995).

TABLE 11.10 Some Commonly Used Methods of Collecting Atmospheric Particles

Method	Approximate range of diameters[a] (μm)
Filters	>0.03
Sedimentation collectors	
Gravitational	≥10
Centrifugal	0.1–10
Impactors	
Atmospheric pressure	≥0.5
Low pressure	≥0.05
Precipitators	
Electrostatic	0.05–5
Thermal	0.005–5

[a] The upper size ranges are usually related to inlet losses that prevent large particles from reaching the sampling surface.

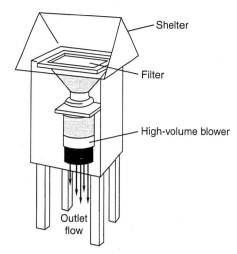

FIGURE 11.56 Schematic of Hi-Vol particulate sampler (adapted from Lawrence Berkeley Laboratory, 1979).

Different filters have unique characteristics, which include the collection efficiency as a function of particle size, the pressure drop at a given flow velocity, and types of reactions that occur on the filter surfaces. Perhaps somewhat surprisingly, sieving action is not the only filtration mechanism. The major filtering action is due to forces that bring the particles into contact with the filter surface where they may stick; these include impaction, interception, diffusion, sedimentation, and electrostatic attraction. At larger particle diameters and high flows, impaction is efficient, whereas at smaller diameters and flows, collection by diffusion to the surface is important; this increased efficiency at large and small diameters results in a minimum at ~0.3 μm in the curve of collection efficiency against particle diameter at the usual sampling rate per unit surface.

Fibrous mat type filters include the frequently used paper (cellulose) fiber filter, for example, the Whatman paper filter, and glass fiber filters. A common fibrous mat filter used for sample collection and air cleaning is known as the HEPA filter (*h*igh *e*fficiency *p*articulate *a*ir filter) and is made of a combination of cellulose and mineral fibers. A widely used type of fibrous mat filter is the *high-volume* filter, commonly referred to as *Hi-Vol*, shown in Fig. 11.56. A modification of the Hi-Vol filter to increase the total air flow allows the collection of sufficient particulate matter in relatively short time periods (e.g., 2 h) to carry out chemical analysis; this is important for studies of the diurnal variation of various chemical components of the aerosol as well as for minimizing sampling artifacts (Fitz *et al.*, 1983).

Porous materials and membranes used as filters have a number of small, often tortuous, pores. This type of filter includes sintered glass filters, organic membrane filters, and silver membrane filters. Two types of membrane filters are Nucleopore and Millipore filters, named after their principal manufacturers. Nucleopore filters are thin films with smooth surfaces and straight, uniform cylindrical pores made by irradiating a thin polycarbonate plastic sheet in contact with a uranium sheet with slow neutrons. The neutrons cause fission of ^{235}U and the resulting fragments produce ionization tracks through the plastic; these tracks are then chemically etched to a desired and uniform size using a sodium hydroxide solution.

Millipore filters have twisted, interconnecting pores that are much more complex than those in Nucleopore filters. They are available in different materials such as Teflon, polycarbonate, quartz, silver, and cellulose acetate.

Membrane filters are particularly useful when surface analytical techniques, such as optical and electron microscopy and X-ray fluorescence analysis, are to be used subsequent to collection, because most of the particles remain on the surface of the filter.

Filter sampling is also accompanied by potential reactions of pollutant gases with the particles on the filter or with the filter medium (including binders that are used in some filters) during sampling and the absorption of water from humid air. In the first case, conversion of gaseous SO_2 and HNO_3 to particulate sulfate and nitrate, respectively, has been observed on some filters. Some filters, especially paper filters, are hygroscopic and thus tend to adsorb water vapor from humid air. Glass fiber filters are relatively (but not entirely) insensitive to humidity, which is a major reason they have been used in the Hi-Vol reference method. However, even here the particulate matter collected on the filter may be hygroscopic and adsorb or desorb water. To minimize this problem, Hi-Vol filters are equilibrated at temperatures between 15 and 35°C and in air with a relative humidity ≤50% for 24 h prior to weighing before and after sampling.

Other problems with collection using filters, such as interference of impurities contained in the filter itself with chemical analysis of the collected particles, are discussed by Chow (1995).

b. Impactors

Impactors are based on the principle that particles in an airstream will tend to continue in a straight line due to their inertia when the flow of air bends sharply; if a surface to which they can adhere is present, they will impact on it and may stick. In practice, a collection plate is placed in the flow of air, causing the gas flow to stream around the obstacle; particles, however, may strike the plate and stick. Obviously, the larger the particle, the greater its inertia and the greater the impaction on the plate.

The impaction efficiency (η) for particles depends directly on the particle diameter (D), the flow velocity of the air (V), and the particle density (ρ); it varies inversely with the gas viscosity (μ) and with a parameter (D_b) that is representative of the impactor's physical dimensions (e.g., the inlet nozzle diameter) and that is related to the curvature of the airstream.

$$\eta = D^2 V \rho / 18 \mu D_b. \qquad (R)$$

Thus, the impaction efficiency should be greatest for larger, denser particles and higher flow velocities. The factors involved in particle impaction on surfaces are discussed in detail by John (1995).

There are two overall types of impactors in widespread use: cascade and virtual impactors.

Cascade impactors. Impactors have been used to obtain different size fractions of ambient particles in the range of diameters ~0.5–30 μm. The range can be extended down to 0.05 μm by operating some of the later stages at reduced pressures (Hering *et al.*, 1978, 1979). The cascade impactor, as its name implies, is a series of impactor plates connected in series or in parallel (Fig. 11.57). The diameters of the nozzles or slits above each impactor plate become increasingly smaller as the air moves through the impactor so that the air moves increasingly faster through these orifices and smaller and smaller particles impact on the plates [see Eq. (R)].

Impactors with various designs as well as different types of impaction surfaces are in use (e.g., see Chow, 1995). Examples include the Lundgren impactor, the Anderson sampler, the Mercer impactor, and the Uni-

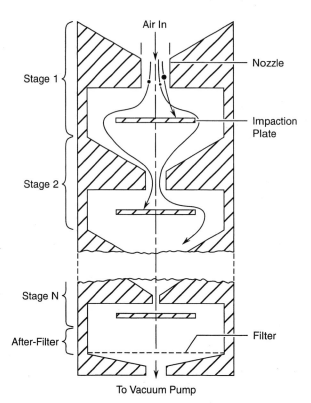

FIGURE 11.57 Principle of operation of cascade impactor (adapted from Marple and Willeke, 1979).

versity of Washington Mark III impactor. An impactor that is in wide use is the MOUDI (*M*icro*o*rifice *U*niform *D*eposit *I*mpactor) (Marple *et al.*, 1991). This device collects particles down to 0.056 μm in aerodynamic diameter and, as the name implies, gives a uniform particle deposit on the plates. This uniform deposit helps in carrying out chemical analysis by such techniques as X-ray fluorescence. The uniformity in deposition is obtained by using multiple nozzles located at specific distances from the center of the impactor plates and rotating the plates beneath the nozzles.

Two problems with particle collection by impactors are bounce-off and reentrainment (John, 1995). Reentrainment is the resuspension of a previously collected particle from the surface into the gas flow due either to the motion of the air over the surface or to impact of an incoming particle. When a particle strikes a surface, if it does not stick, it can bounce off back into the gas steam, break into fragments, or cause a previously adsorbed particle to be knocked off into the gas stream; in all three cases the collection efficiency is lowered and the net effect is referred to as bounce-off. To minimize these problems, the surface of the impactor is often coated with a soft, energy-absorbing substance such as oil, water, grease, resin, or paraffin, which helps to absorb the kinetic energy of the striking particle; a summary of the types of agents used to minimize bounce-off and reentrainment is given by Marple and Willeke (1979), Cahill (1979), and Turner and Hering (1987).

While the use of soft surfaces would seem to be mandated by the foregoing discussion of bounce-off problems, there are a number of disadvantages to coating the impactor surfaces with a substance such as grease. For example, it makes accurate mass determinations difficult and can introduce such a large background of certain chemicals that the chemical analysis of these elements in the particles becomes difficult. In addition, with such surfaces one cannot use chemical analytical techniques that only probe the upper surface layer because the coating surrounds some of the collected particles.

Virtual impactors. The virtual impactor is a modified type of impactor, an example of which is shown in Fig. 11.58; one commonly used type of virtual impactor is known as the dichotomous sampler. The basis of virtual impactors is that the airstream impacts against a mass of relatively still air rather than against a plate. The inertia of the particles carries them into the still air mass, which is slowly withdrawn through a filter to collect the particles. This type of impactor avoids the problem of particle reentrainment from the impaction surface caused by air motion over the collected particles or by dislodging due to collisions of incoming particles with the impactor surface. It also avoids the problem of bounce-off or of using greases that may interfere with subsequent chemical analysis.

c. Electrostatic Precipitators

Electrostatic precipitators operate on the principle of the attraction of a charged particle for an oppositely charged collector. They have been used for both collecting particles for further analysis and for controlling particulate emissions from sources. In one common design, the particles in air can be charged if introduced into a cylindrical chamber containing a wire down the axis of the cylinder that is at a high negative voltage (e.g., 5–50 kV) relative to the walls of the chamber. A corona discharge is set up around the wire and this produces ions; the negatively charged ions are attracted to the positively charged outer walls. These ions collide with the particles in the air, charging them and causing them to move to the outer walls to be captured there. In place of the corona discharge, ions may also be generated using radioactive bombardment of the particles.

While electrostatic precipitators have relatively high collection efficiencies (99–100%) over a wide range of particle sizes (~ 0.05–5 μm), there are a number of disadvantages. These include the lack of size information, particle reentrainment due to sparking, and practical problems such as high cost and shock hazards. As a result, they have not been widely used in ambient air studies.

An example of a study in which this approach was applied involved the use of a transmission electron microscopy (TEM) grid as the collector plate in the electrostatic precipitator (Witkowski *et al.*, 1988). After sample collection, analysis by TEM (vide infra) could then be carried out.

A related area is that of single-particle levitation, which has been used in a number of studies to isolate a single particle and study its properties (e.g., see papers by Tang and co-workers in Chapter 9). A review of this area is given by Davis (1997).

d. Sedimentation Collectors

These collectors are used primarily for large particles (≥ 2.5 μm), that is, those in the coarse particle range. They include collection by gravitational sedimentation (e.g., dustfall jars) as well as by centrifugal

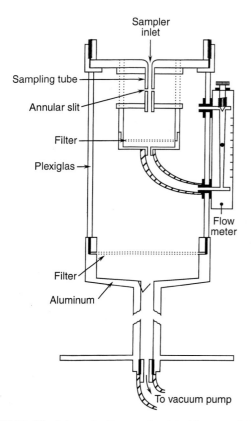

FIGURE 11.58 Schematic diagram of a virtual impactor (adapted from Conner, 1966).

sedimentation, which allows collection in the submicrometer range (e.g., centrifuges and cyclone collectors).

Gravitational sedimentation only collects the large particles that settle out of the atmosphere fairly quickly. This *dustfall* generally consists of particles that are relatively large and, as such, are not particularly relevant to the focus of this book. Thus dustfall collectors will not be discussed further.

The principle of centrifugal collection is, of course, well known. Collection of particles using centrifugation involves passing the aerosol at a controlled rate through a rapidly spinning air mass. Collection of particles in ranges as small as ~0.1–1 μm has been reported using this technique. The cyclone collector, a modification of the centrifuge technique, is based on bringing the air samples into a stationary cylindrical vessel at high velocity; a vortex is formed by the entry of the air tangential to the length of the vessel and particles in this vortex are subjected to a centrifugal force that depends on their size (Fig. 11.59). As a result, particles of different sizes are deposited at different locations along the length of the cyclone separator. Although cyclone collectors have been applied to size distribution measurements by using a series of cyclones in parallel, each having a different cut size, they are most commonly used as precollectors to remove larger particles (~3- to 30-μm diameter) before the air sample enters a device such as an impactor designed for the measurement of particles in smaller size ranges.

2. Measurement of Physical Characteristics: Mass and Size

a. Mass

The total mass of particles per unit volume of air is one of the major parameters used to characterize particles in air and, along with size, is the basis of air quality standards for particulate matter (see Chapter 2). Methods of mass measurement include gravimetric methods, β-ray attenuation, piezoelectric devices, and the oscillating microbalance.

(1) Gravimetric methods The most straightforward method of determining the particle loading of the atmosphere is to weigh a collection substrate such as a filter before and after sampling. However, care must be taken to be sure that both temperature and relative humidity are carefully controlled when weighing both the loaded and clean substrate. As discussed earlier, some filters and/or the collected particles are hygroscopic and unless care is taken to equilibrate them at a fixed temperature and relative humidity, the change in water content may completely mask the change in mass due to the particles. In addition, problems such as forces due to static electricity on the filter that interfere with accurate weight measurements must be controlled. Finally, particulate loading can change the sampling air flow rate and lead to large errors in determining the actual volume of air sampled.

(2) β-Ray attenuation β-Particle beams (electrons) emitted from a radioactive source are attenuated when they pass through a filter on which particulate matter has been collected. (β-particle beams rather than α-particle beams or γ-rays are used because α particles do not penetrate typical thicknesses of filter well and γ-rays are too penetrating and hence would require large sample thicknesses.) Figure 11.60 shows a

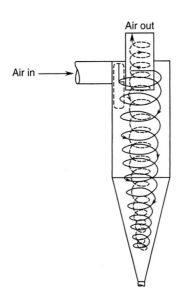

FIGURE 11.59 Schematic diagram of one type of cyclone collector (adapted from Ayer and Hochstrasser, 1979).

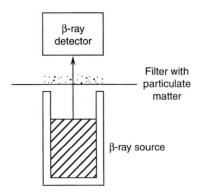

FIGURE 11.60 Schematic diagram of a typical β-ray attenuation device for measuring particulate mass.

schematic of a β-ray attenuation device, which consists essentially of a β source such as ^{14}C, a β detector, and a means of positioning the filter paper containing the particulate matter between the two. The ratio of the transmission of β-rays through a clean and loaded portion of the filter, respectively, is related to the particle loading via a Beer–Lambert type of relationship:

$$I/I_0 = e^{-\mu X}. \quad (S)$$

I_0 and I are the intensities of the β-rays that have passed through the clean and loaded portions of the filter, respectively, X is the thickness of the deposit, and μ is an attenuation constant that is approximately proportional to the density (ρ) of the material deposited. The mass per unit area deposited on the filter, given by ρX, is the parameter desired in this measurement. Rearranging Eq. (S), one obtains

$$\ln(I_0/I) = (\mu/\rho)\rho X. \quad (T)$$

The parameter μ/ρ is a constant known as the *mass absorption coefficient*; with the assumption that this is independent of the type of absorbing particles (an assumption that generally holds well enough to cause ≲10% uncertainty), the value of $\ln(I_0/I)$ is directly related to the parameter of interest, ρX = mass per unit area.

Such measurements can be carried out on filters with different cutoff sizes to obtain size resolution as well (e.g., see Spagnolo and Paoletti, 1994).

(3) Piezoelectric microbalance The piezoelectric microbalance is a resonant frequency device. The piezoelectric effect is the development of a charge on some crystals such as quartz when a stress is applied; the stress may be mechanical (e.g., added weight) or electrical. Such crystals may be used as part of a resonance circuit to provide very stable, narrow-band frequencies; the quartz crystal is plated on two sides with a thin conducting layer and leads are connected to the resonance circuit so the crystal replaces an *LC* network. The obtained frequency of vibration (v_o) depends on a number of parameters of the crystal but is usually ~5–10 MHz. However, if a mass (Δm) becomes attached to one side of the crystal, it changes the resonant frequency by an amount Δv_o such that

$$\Delta v_0/v_0 = \Delta m/m \quad (U)$$

as long as the increase in mass Δm is much smaller than the mass (m) of the active part of the crystal.

Particulate matter from ambient air can be deposited on the crystal in various ways, for example, by using it as an impaction device. The mass of the collected particles can then be determined by following the change in the frequency. Alternatively, a reference crystal held at the same temperature and pressure as the crystal on which the particles are collected can be used, and the difference in frequencies between the two crystals can be determined.

The piezoelectric microbalance is very sensitive, capable of detecting ~10^{-8}–10^{-9} g. The particles collected on the crystal surface can be chemically analyzed after collection using surface-sensitive techniques. One limitation is possible overloading of the crystal; thus when the collected mass reaches ~0.5–1% of the mass per unit of the crystal, the surface must be cleaned.

(4) Oscillating microbalance The tapered-element oscillating microbalance is based on a similar principle to the piezoelectric microbalance. A hollow glass piece is mounted with the wider end fixed and a filter attached to the narrower end. The tip oscillates at a particular frequency in an applied magnetic field. As particles are collected on the filter, the resonant frequency of the glass tends to decrease. A feedback is used to maintain the oscillation frequency and provides a measurement of the collected mass (e.g., see Patashnick and Rupprecht, 1991). Good agreement has generally been observed between measurements made using the tapered-element oscillating microbalance and Hi-Vol filter methods (e.g., Eldering and Glasgow, 1998).

b. Size

There are several different approaches that are commonly used to determine particle size distributions in air. One of them, impaction, has been discussed earlier. Multistage impactors with different cut points are used extensively to obtain both mass and chemical composition data as a function of size for particles with diameters ≳0.2 μm. Others, including methods based on optical properties, electrical or aerodynamic mobility, and diffusion speeds, are described briefly in the following section. The condensation particle counter (CPC) is used as a detector in combination with some of these size-sorting methods.

The reader is cautioned to keep in mind that atmospheric particles are not all spherical nor even necessarily simple in shape. Thus, as discussed in Chapter 9.A, the term *size* cannot be uniquely defined for atmospheric particles. As a result, a measurement of the distribution of sizes using an impactor that is based on inertial characteristics, for example, may not give the same results as a size measurement based on optical techniques that use light scattering. With this caveat in mind, let us examine the most commonly used methods of determining the size distribution of atmospheric particles.

(1) Optical methods Optical counters, optical microscopy, and electron microscopy fall under this heading. A review of optical methods is given by Baron *et al.* (1993).

Single-particle optical counters. These instruments are used to measure particles in the ~0.1- to 10-μm range by measuring the amount of light scattered by a single particle (Martens and Keller, 1968). As discussed in Chapter 9.A.4, the amount of this Mie scattering depends not only on the refractive index but also on the radius of the particle; hence the intensity of scattered light is a measure of the particle size. Assuming that the particles are spherical, smooth, and of known refractive index, one can calculate, using Mie theory, the intensity of scattered light of wavelength λ at various angles (θ) to the incident beam for a particle of a given size. Integrating over all scattering angles and wavelengths (since "white" incandescent sources are normally used in these instruments), one obtains the theoretical response of the single-particle counter, that is, the curve of scattered light intensity as a function of the particle diameter. Typical theoretical response curves are shown in Fig. 11.61 (Cooke and Kerker, 1975).

Calibration of these single-particle counters is usually carried out using monodisperse polystyrene latex or polyvinyl latex spheres, which are available in sizes from ~0.1 to 3 μm and have a refractive index of 1.6; alternatively, aerosols with lower refractive indices may be generated from liquids such as dioctyl phthalate ($m = 1.49$). Whitby and Willeke (1979) discuss the importance of instrument calibration using standardized aerosols with an index of refraction as close as possible to the sample being measured; since the refractive index of atmospheric particles varies from 1.33 for water to 1.7 for minerals, they recommend using a calibration aerosol with $m \cong 1.5$. Because light scattering is very dependent on the particle shape, when measuring irregularly shaped particles such as coal dust, one should calibrate the instrument with aerosols generated from the same material. Figure 11.62, for example, shows the instrument response as a function of particle diameter for an ideal calibration aerosol of dioctyl phthalate and for coal dust particles.

Potential problems with using single-particle counters in ambient measurements and ways to minimize these are discussed in detail by Whitby and Willeke (1979).

Optical counters allow relatively rapid measurements of the size distribution and, unlike some of the other methods of size fractionation, include volatile particles in the measurement. However, some care must be taken in interpreting the detailed shape of the size distribution spectrum because of some anomalies that have been observed; for example, around the 1-μm region, interference from light that is reflected or refracted from the front and back of the particle gives a "knee" in many calibration curves of number of particles versus their diameter (LBL, 1979).

Electron and optical microscopies. Counting the particles and measuring their sizes can be done by optical or electron microscopy, the former for particles with diameters from ~0.4 μm to several hundred microns,

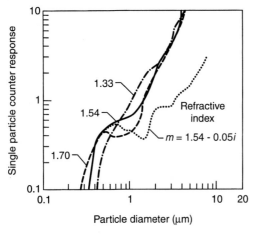

FIGURE 11.61 Theoretical response of a typical single-particle counter (adapted from Whitby and Willeke, 1979; data from Cooke and Kerker, 1975).

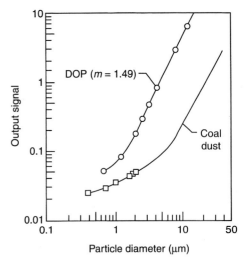

FIGURE 11.62 Experimental calibration curves for a commercial single-particle counter and two types of calibration aerosols: dioctyl phthalate (DOP) and coal dust (adapted from Whitby and Willeke, 1979).

and the latter for particles from ~0.001 μm and larger. The particles must be on the surface of the substrate and form less than a monolayer to minimize overlap of the particles. For electron microscopy, the sample and substrate must also be able to be subjected to high vacuum, heat, and electron bombardment without degradation over a period of time sufficient to make the measurement.

Because of the instrumental requirements, these are usually not routine monitoring techniques. However, unlike other methods, they give detailed information on particle shapes. In addition, chemical composition information can be obtained using transmission electron microscopy (TEM) or scanning electron microscopy (SEM) combined with energy-dispersive spectrometry (EDS). The electron beam causes the sample to emit fluorescent X-rays that have energies characteristic of the elements in the sample. Thus a map showing the distribution of elements in the sample can be produced as the electron beam scans the sample.

For example, Fig. 11.63 shows the TEM image of an NaCl crystal (Fig. 11.63A) and of the same crystal after exposure to gaseous HNO_3 (Fig. 11.63B) and then small amounts of water vapor (Fig. 11.63C) (Allen et al., 1996). After the crystal is dried, the formation of new microcrystallites attached to the NaCl is observed. These can be shown using EDS to be crystals of $NaNO_3$. Thus Fig. 11.64b shows the EDS spectra obtained from the larger, original NaCl crystal (but after exposure to HNO_3 and water vapor) and of small microcrystallites attached to it (Fig. 11.64a). Only Na and Cl are seen in the first case, but Na, N, and O in the second, and in the correct ratio for $NaNO_3$. The HNO_3 has reacted with the NaCl surface to generate metastable surface nitrate ions. Exposure to water vapor generates a mobile quasi-liquid layer on the surface that contains Na^+, Cl^-, and NO_3^- ions. On drying, the ions segregate to form separate microcrystallites of $NaNO_3$ and a fresh NaCl crystal (Allen et al., 1996).

The combination of SEM with EDS has also been applied to atmospheric particles (e.g., Pósfai et al., 1995; Anderson et al., 1996; McMurry et al., 1996; Ganor et al., 1998). For example, individual sea salt particles were analyzed using TEM combined with EDS as well as selected-area electron diffraction (SAED) by Pósfai et al. (1995) and Anderson et al. (1996). The crystal shapes correlated well with the chemical composition determined using EDS and SAED. For example, cubic crystals of NaCl were observed. Sulfate occurred in either rod-shaped crystals, which had significant concentrations of (Mg + K + Ca) compared to Na, or tubular crystals, with much smaller concentrations of these three metals. In the latter case, the EDS showed

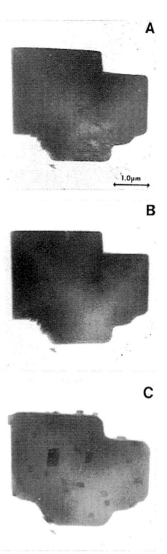

FIGURE 11.63 TEM images of an NaCl crystal (A) before reaction and (B) after reaction with gaseous HNO_3 (1.2×10^{15} cm^{-3} for 15 min) and then (C) exposure to water vapor (<15 Torr) followed by drying (adapted from Allen et al., 1996).

spectra with large contributions from Na, O, and S, and the SAED patterns were similar to that of Na_2SO_4.

It should be noted that as with all analytical techniques that involve subjecting the sample to vacuum conditions before and/or during the analysis, separation of components via selective crystallization is expected (e.g., Ge et al., 1998a). Hence these particles may not have actually existed in these crystalline forms at relative humidities above their deliquescence points in the atmosphere, although the various constituents observed were clearly present.

(2) Atomic force microscopy (AFM) and scanning tunneling microscopy (STM) These methodologies for probing the morphological details of a surface down to

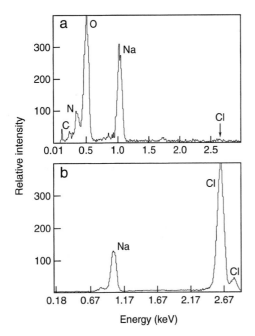

FIGURE 11.64 Energy-dispersive spectroscopy light-element spectra acquired from (a) one of the small microcrystallites attached to NaCl after exposure to gaseous HNO_3 (see Fig. 11.63C) and (b) an adjacent area of the NaCl crystal (adapted from Allen *et al.*, 1996).

the atomic scale have been used extensively in many laboratory studies to map atomic and molecular structures. In STM, which is used with electrically conducting materials, a probe with a very fine metal tip scans over the surface of the sample, which is held at a potential relative to the tip. When the tip is very close to the surface, there is a tunneling current between the tip and the sample, whose magnitude depends exponentially on the distance between the tip and the sample. The tip is kept at a constant distance from the sample using a feedback circuit to measure and maintain the tunneling current at a constant value. The tip thus moves up and down with the surface topography. For insulating surfaces, AFM accomplishes similar topographical mapping. In this case, the force acting between the surface and a fine tip attached to a cantilever is kept constant and the deflections of the cantilever required to do this are monitored, for example using optical means.

AFM has been used in only a few studies to explore the sizes and morphology of airborne particles (e.g., Friedbacher *et al.*, 1995; Pósfai *et al.*, 1998). In this case, atomic scale resolution is not used, but rather much lower resolution that provides information on particle sizes and shapes in the micron and submicron size range under ambient conditions. This has the advantage that effects due to the application of vacuum to the particles do not occur, as is the case for TEM (vide supra). AFM combined with TEM has been applied by Pósfai *et al.* (1998), for example, to explore the loss of water from particles upon exposure to vacuum conditions.

(3) Electrical mobility analyzers Several types of instruments for measuring particle sizes in the atmosphere depend on the mobility of charged particles in an electric field (e.g., see Yeh (1993) and Flagan (1998) for a review and history of the development of this field). The electrical mobility analyzer developed by Whitby and co-workers at the University of Minnesota, in particular, has been used extensively to measure particles in the range ~0.003 to ~1 μm (Whitby and Clark, 1966; Eisele and McMurry, 1997).

Figure 11.65a illustrates the principles of the electrical aerosol analyzer. The essential components are the aerosol charger, the mobility analyzer, and the detector (shown in Fig. 11.65a as the current-collecting filter). The air containing the particulate matter is first introduced into the aerosol charger, where a corona discharge generates positive ions for particle charging. The positively charged particles are introduced as a thin layer around the outside of the tubular mobility analyzer. Clean air flows down the central portion of the tube between the layer of ambient aerosol at the walls and the charged collection rod in the center of the tube. A negative voltage is applied to the collection rod, causing the positively charged particles to move from the outer wall through the clean air to the collection rod.

The particles with the highest mobilities reach the collection rod first and are removed from the gas stream; those that do not reach the rod before the flow passes out of the region of the electric field pass through to a detector and are measured. Increasing the voltage on the collection rod increases the number of charged particles that reach it before passing out of the field and hence decreases the number reaching the detector. The relationship of the particle count at the detector to the voltage in the analyzer is thus dependent on the particle mobility in the analyzer, which depends on particle size. Thus size distributions can be obtained by studying the detector output as a function of collection rod voltage. The detector may be a current-sensing device, as in Fig. 11.65a, or other type such as a condensation nuclei counter (vide infra).

Details of the calibration, use, performance, and artifactual problems are given in a proceedings entitled *Aerosol Measurement* (Lundgren *et al.*, 1979); this also shows data for the mobility distribution for monodisperse aerosols.

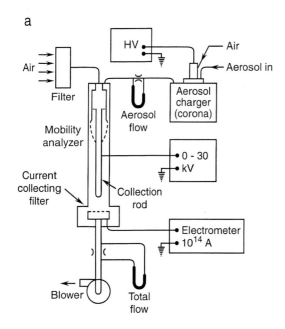

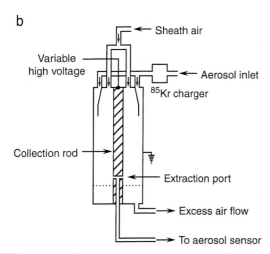

FIGURE 11.65 (a) Electrical aerosol analyzer (adapted from Whitby and Clark, 1966). (b) Schematic diagram of differential mobility analyzer (adapted from Yeh, 1993).

A widely used form of the electrical mobility analyzer now in use is called a differential mobility analyzer (DMA), which operates on the same principle. Figure 11.65b is a schematic diagram of a DMA. Charging is carried out using a radioactive source such as ^{85}Kr, ^{90}Sr, ^{210}Po, or ^{241}Am to produce ions of both signs that become attached to the particles. In this case, an extraction port is located in the center rod. Smaller particles of opposite electrical charge to the center rod and having higher electrical mobilities strike the central rod before this port and are removed; larger particles are carried beyond the extraction port and out with the major flow. Only a narrow range of particles will strike the extraction port, providing a narrow range of particle sizes at the aerosol outlet. This monodisperse aerosol exiting the mobility analyzer is then directed to a measuring device such as a condensation particle counter (CPC; see later) or a Faraday cup electrometer (e.g., Winklmayr et al., 1991). A modified version that extends the range down to 1 nm, approaching molecular ions, has also been developed (Rosell-Llompart et al., 1996).

Results of particle size distributions in air using different methods tend to be in reasonably good agreement when the different sampling times are taken into account. For example, Hoff et al. (1996) made particle size measurements in a rural area in Ontario, Canada, using a DMA, an eight-stage impactor and light scattering instruments. The number size distributions obtained using each technique were in excellent agreement, assuming for the impactor samples that the composition of the particles had a density of 2 g cm^{-3}.

(4) Diffusion separators As discussed in Chapter 9.A.3, small particles with diameters $\lesssim 0.05$ μm undergo diffusion via Brownian motion sufficiently rapidly that this can be used to separate particles. Thus the aerosol can be passed through a tube in which the smaller particles diffuse more rapidly to the walls and are removed there, leaving the larger, more slowly diffusing particles to pass through. Variation of residence time in the tubes by varying the flow rates and tube lengths leads to different size cutoffs (but not high resolution). Hence size fractionation of small particles can be achieved using such diffusion separators. The particles exiting the tube can be measured using techniques such as the CPC. The design and testing of a typical diffusion battery-CPC apparatus are described by Raes and Reineking (1985).

(5) Aerodynamic particle size This technique is based on measuring the velocity lag of particles in accelerating air flows (Wilson and Liu, 1980; Baron et al., 1993). A laser beam is split into two coherent beams using a beam splitter, and these two beams are then focused onto a point, forming an interference pattern. When a particle passes through this interference pattern, it scatters light, with the scattered light intensity oscillating as the particle passes through the interference fringes. The frequency of the oscillation of the scattered light multiplied by the spacing of the fringes gives the velocity of the particle perpendicular to the fringes. From the particle velocity, the size can be obtained.

(6) Condensation particle counter (CPC) Very small particles in the Aitken range act as condensation nuclei for the formation of larger particles in a supersaturated

vapor. If these very small particles are injected into air that is supersaturated with water or another vapor such as an alcohol, the vapor condenses on them to form droplets. In the condensation particle counter (CPC), supersaturation of the air containing these particles is achieved by passing them through a higher temperature region saturated with the vapor and then into a lower temperature region in which the alcohol vapor condenses on the nuclei to form larger droplets. These can be counted as is done in absolute nuclei counters, for example, by measuring the pulses of scattered light by a single droplet as it passes through the viewing volume. Alternatively, the particles can be measured using techniques such as total light extinction or scattering. In this case (sometimes called photoelectric nuclei counters), calibration against some other reference is required. CPCs are applicable in the size range from ~3 to 1000 nm. Note that a 10-nm particle contains ~10^4 molecules, whereas a 2.7-nm particle contains only ~10^2 molecules (Eisele and McMurry, 1997). Hence these very small ultrafine particles are approaching molecular clusters.

Measurement of ultrafine particles, those ≤ 10 nm in diameter, has become of increasing interest due to their importance in acting as cloud condensation nuclei (see Chapter 14) and in elucidating rates and mechanisms of homogeneous nucleation. Methods based on the foregoing principles have been developed and applied for these very small particles (e.g., see Stolzenburg and McMurry, 1991; McDermott et al., 1991; Wiedensohler et al., 1993, 1994; Saros et al., 1996; Marti et al., 1996; and Weber et al., 1998). For example, CPCs have been used in conjunction with pulse height analysis to measure particles with diameters in the 3- to 15-nm range. Higher supersaturations of the alcohol vapor are required to grow the smaller particles into the light-scattering range, and hence these particles travel further in the measuring system before activation occurs. As a result, for particles less than 15 nm, the final size of the light-scattering particles formed from smaller particles is smaller as well, whereas above 15 nm, the final particle size is relatively independent of the initial particle size (Saros et al., 1996). For particles in this smaller size range, the height of the detector pulses produced by the scattered light decreases with particle size. Thus, for ultrafine particles, pulse height analysis provides a means of determining particle size (e.g., Stolzenburg and McMurry, 1991; Saros et al., 1996; Marti et al., 1996; Weber et al., 1998).

(7) *Summary* In summary, no one technique is capable of measuring the size distribution of atmospheric

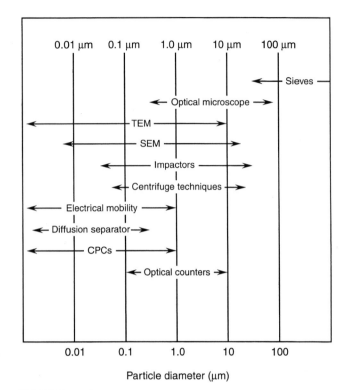

FIGURE 11.66 Summary of size ranges covered by various analytical techniques for atmospheric aerosols. TEM, transmission electron microscopy; SEM, scanning electron microscope (adapted from Hinds, 1982).

aerosols from the smallest to the largest diameters of interest, a range covering approximately five orders of magnitude. However, a combination of methods can be used to provide information over this range. Figure 11.66 summarizes the size ranges covered by the various techniques.

c. Typical Particle Concentrations in the Atmosphere

Given the complexity of particle size distributions in the atmosphere (see Chapter 9.A), as well as the large number of chemical components (Chapter 9.C) that are not distributed equally throughout the various sizes, characterizing a "typical" collection of particles in the atmosphere is not possible. However, some indication of particle levels in the atmosphere is provided by mass measurements of PM_{10} (i.e., total mass less than 10 μm in diameter), for which extensive measurements have been made for regulatory purposes.

Concentrations of PM_{10} in urban areas around the world as high as ~300 μg m^{-3} have been reported (e.g., Reponen et al., 1996; Kasparian et al., 1998; Morawaska et al., 1998; Lam et al., 1999). In extreme situations such as the fires in Indonesia, 24-h average concentrations of PM_{10} as high as ~900 μg m^{-3} were measured (Brauer and Hisham-Hashim, 1998). More typical mass concentrations in cities are $\sim 20-100$ μg m^{-3} (e.g., Berico *et al.*, 1997; Brook *et al.*, 1997; Harrison *et al.*, 1999). In suburban-rural areas, peak concentrations are about an order of magnitude lower (e.g., Reponen *et al.*, 1996; Eldering and Glasgow, 1998). For example, studies of PM$_{10}$ at 28 sites in 10 countries in Europe reported median concentrations in rural Scandinavia of 11 μg m^{-3} (Hoek *et al.*, 1997).

With the new regulations on PM$_{2.5}$, there are also increasing data on this smaller particle fraction as well. In rural areas, maximum 24-h average concentrations typically range from ~ 1 to 50 μg m^{-3} (e.g., Hoff *et al.*, 1996; Sweet and Gatz, 1998). Because a significant fraction of the smaller particles is formed by reactions in air rather than by direct emissions, and as we have seen such reactions occur during transport, differences in concentrations between rural and urban areas may not be extremely large (Sweet and Gatz, 1998).

Thus, typical concentrations of PM$_{2.5}$ in urban areas are $\sim 50-80$ μg m^{-3} (Sweet and Gatz, 1998). For example, Lam *et al.* (1999) report median PM$_{2.5}$ levels of $\sim 70-150$ μg m^{-3} in Hong Kong, and levels up to ~ 200 μg m^{-3} have been observed in Mexico City (Vega *et al.*, 1997). 1-h average concentrations of PM$_3$ measured in Pocatello, Idaho, during January 1996 were from 0 to about 45 μg m^{-3} (Eldering and Glasgow, 1998). The mean concentration of PM$_{2.5}$ in 19 locations in Canada was reported to be 13.9 μg m^{-3}, with a standard deviation of 9.5 μg m^{-3} (Brook *et al.*, 1997); in these studies, the ratio of PM$_{2.5}$ to PM$_{10}$ was quite variable, ranging from 0.36 to 0.65, with ratios similar to the high end of this range reported for Birmingham, U.K., by Harrison *et al.* (1999).

Concentrations in remote areas are, of course, generally much smaller, as indicated by the much higher visual range in these areas (see Chapter 9).

3. Measurement of Chemical Composition

As discussed in Chapter 9.C, ambient particulate matter contains inorganic elements and ions, including trace metals, as well as graphitic (elemental) carbon and a wide variety of organic compounds and water. Techniques in common use to measure these species are discussed very briefly here. For further details of the principles behind these techniques, the reader should consult instrumental analysis texts (e.g., Skoog *et al.*, 1998). Specific applications of various methods to particles in the atmosphere are described in the book edited by Spurny (1986) as well as the references at the end of this chapter.

a. Inorganic Elements

Table 11.11 summarizes the major methods used to measure the inorganic elements in atmospheric particulate matter.

Colorimetry. Colorimetric methods, that is, wet chemical methods in which reagents are added to generate a light-absorbing species whose absorbance can

TABLE 11.11 Some Common Methods Used to Measure Inorganic Elements in Atmospheric Particles

Element	Analytical methods
Aluminum	XRF,[a] PIXE,[b] AA,[d] NA,[e] ICP[j]
Antimony	Col,[c] AA, XRF, NA, ASV,[f] ICP
Arsenic	Col, AA, XRF, NA, ASV, MS,[g] PIXE, ICP
Beryllium	Col, AA, ES,[h] ICP
Bismuth	AA, ASV, ICP
Bromine	XRF, NA, MS, PIXE
Cadmium	Col, AA, XRF, NA, ASV, MS, ES, ICP
Calcium	XRF, PIXE, NA, AA, ICP
Chlorine	XRF, NA, MS, PIXE
Chromium	AA, XRF, NA, MS, ES, PIXE, ICP
Cobalt	AA, NA, XRF, ICP, ES
Copper	AA, XRF, NA, ASV, MS, ES, PIXE, ICP
Gallium	XRF, NA, AA, PIXE, ICP
Iodine	NA, MS
Iron	Col, AA, XRF, NA, PIXE, ASV, MS, ES, ICP
Lead	Col, AA, XRF, ESCA,[i] ASV, MS, ES, PIXE, ICP
Magnesium	NA, PIXE, AA, ICP
Manganese	Col, AA, XRF, NA, MS, ES, PIXE, ICP
Mercury	Col, AA, XRF, NA, ES, ICP
Molybdenum	Col, AA, ES, MS, XRF, PIXE, ICP
Nickel	AA, XRF, NA, MS, ES, PIXE, ICP
Nitrogen	ESCA
Potassium	XRF, PIXE, NA, AA
Rubidium	XRF, NA, PIXE
Selenium	Col, AA, NA, ASV, ES, XRF, PIXE, MS, ICP
Silicon	XRF, PIXE, AA, ICP
Sodium	NA, AA, PIXE, ICP
Strontium	XRF, NA, PIXE, AA, ICP
Sulfur	NA, XRF, ESCA, MS, PIXE, ICP
Tin	AA, XRF, MS, ES, ICP
Titanium	AA, XRF, NA, ES, PIXE, ICP
Vanadium	AA, XRF, NA, MS, ES, PIXE, ICP
Zinc	AA, XRF, NA, ASV, MS, ES, PIXE, ICP

Source: Adapted from LBL (1979) and Chow (1995).
[a] XRF = X-ray fluorescence analysis.
[b] PIXE = particle-induced X-ray emission.
[c] Col = colorimetry.
[d] AA = atomic absorption spectrometry.
[e] NA = neutron activation analysis.
[f] ASV = anodic stripping voltammetry.
[g] MS = mass spectrometry.
[h] ES = emission spectrometry.
[i] ESCA = electron spectroscopy for chemical analysis; also known as XPS = X-ray photoelectron spectroscopy.
[j] ICP = inductively coupled plasma spectroscopy.

be quantified using conventional absorption spectroscopy, have been used rather extensively in the past. An example is the measurement of Cl^- in aerosols from remote regions by the mercury thiocyanate method (Huebert and Lazrus, 1980). In this technique, chloride ions react with $Hg(SCN)_2$ in a dioxane–ethanol solution to form $HgCl_2$, $HgCl_4^{2-}$, and SCN^-. Upon addition of Fe^{3+} in a nitric acid solution, an orange solution due to $FeSCN^{2+}$ results, whose absorbance can be measured at its 460-nm peak (Iwaski et al., 1956).

While colorimetric methods have the advantages of being relatively inexpensive, simple to carry out, and applicable to a large number of elements, they are increasingly being replaced by other physical techniques. The major reason for this is that, as discussed earlier in this chapter, wet chemical methods are more likely to suffer from unrecognized interferences, particularly in complex environmental samples. However, when the aerosol composition is sufficiently well known that one can be confident of the absence of interfering species, colorimetric methods are useful.

X-ray fluorescence (XRF). The sample is irradiated with monochromatic X-rays that eject electrons from the inner shells of the elements. When an electron from an outer shell of the ion drops into the vacancy, it emits characteristic X-rays whose wavelength is used to identify the element and whose intensity is related to the amount present. XRF is used primarily for elements heavier than magnesium because of the weak fluorescence of lighter elements and absorption of the X-rays within the particles. The combination of transmission or scanning electron microscopy (TEM/SEM) with X-ray fluorescence, also known as energy-dispersive spectrometry (EDS), was discussed in Section B.2b.

Particle-induced X-ray emission (PIXE). Elements heavier than sodium can be analyzed using PIXE. In this method, the sample is bombarded with a beam of particles, usually protons, that excites the elements in the sample in a manner similar to that for XRF, causing them to emit X-rays at wavelengths characteristic of the elements (Johansson et al., 1975). Closely related methods of analysis use other ions such as α particles to bombard the sample and induce the X-ray emission. As in the case of XRF, the lighter elements (hydrogen through fluorine) cannot be easily measured with this technique; however, backscattering of α particles used to bombard the sample can be measured, and the energy lost in the nuclear recoil can be used to identify the scattering element for these lighter species. These ion-excited X-ray analytical techniques (IXA) are reviewed by Cahill (1980, 1981a, 1981b) and by Traxel and Wätjen (1986). An example of the application to PM_{10} and $PM_{2.5}$ in Brisbane, Australia, is discussed by Chan et al. (1997).

Atomic absorption spectrometry (AA). This is a standard laboratory analytical tool for metals. The metal is extracted into a solution and then vaporized in a flame. A light beam with a wavelength absorbed by the metal of interest passes through the vaporized sample; for example, to measure zinc, a zinc resonance lamp can be used so that the emission and absorbing wavelengths are perfectly matched. The absorption of the light by the sample is measured and Beer's law is applied to quantify the amount present.

Emission spectrometry (ES). Emission spectrometry is based on the excitation of an element to an upper electronically excited state, from which it returns to the ground state by the emission of radiation. As discussed in Chapter 3, the wavelength emitted is characteristic of the emitted species, and, under the approximate conditions, the emission intensity is proportional to its concentration. Means of excitation include arcs and sparks, plasma jets (see ICP), and lasers.

Inductively coupled plasma spectroscopy (ICP). ICP has become a well-established analytical technique for a variety of trace metals. The sample is introduced into a plasma formed by a rf discharge in a gas such as argon. Ions and electrons generated in the plasma are induced to travel in annular paths by interaction with a fluctuating magnetic field generated by the rf induction coil. Elements in the plasma are excited and emit at their characteristic wavelengths. The particular elements can thus be identified from the emission wavelengths and the amounts of each from the emission intensity.

ICP can also be coupled with mass spectrometry (ICP–MS) for very high sensitivity and is finding increasing use for elemental analysis (e.g., Skoog et al., 1998).

Neutron activation (NA). The sample is bombarded with neutrons and the radioactivity induced in the sample is then measured. Both β and γ radiation can be monitored, but γ radiation is more frequently used because of the discrete wavelengths associated with emission that can be used to identify the emitter.

Anodic stripping voltammetry (ASV). This is an electrochemical technique in which the element to be analyzed is first deposited on an electrode and then redissolved, that is, "stripped," from the electrode to form a more concentrated solution. For example, a drop of mercury hanging from a platinum electrode in a solution containing the species to be measured has been used as the deposition electrode. A potential slightly more negative than the half-wave potential for the ion of interest is applied to deposit the element on the electrode. After deposition of the metal for a given time period, stirring of the solution is stopped and the voltage decreased at a constant rate toward the anodic potential while the anodic current is measured. The peak anodic current, corrected for the residual current, is proportional to the elemental concentration under controlled conditions, for example, fixed deposition time.

Mass spectrometry (MS). Mass spectrometry is a common method for detecting and measuring organics (vide infra), but it has also been used for certain inorganic elements and ions as well. For example, Schuetzle et al. (1973) volatilized ambient particulate matter into the source region of a high-resolution mass spectrometer by heating the sample continuously from 20 to 400°C. The elements sulfur, cadmium, and iodine were identified and measured using their masses, ion intensities, and vaporization temperatures. In combination with ICP, MS provides a powerful analytical technique for trace metals.

Secondary ion mass spectrometry (SIMS) and secondary neutral mass spectrometry (SNMS) have also been used for the surface analysis of atmospheric particles. In the SIMS approach, the sample is collected and bombarded with high-energy (keV) atoms or molecules, typically Ar^+, causing ejection of material from the surface of the particles into the gas phase. The emitted species include positive and negative ions that are then measured by mass spectrometry. In the SNMS method, the sample is located behind an orifice that contains an rf plasma, for example in argon. The sample holder is held at negative potential, which extracts Ar^+ from the plasma and accelerates them toward the sample where they eject surface materials. Neutral species ejected from the surface become partially ionized as they travel back through the plasma, and these are then detected by mass spectrometry (SNMS). The depth analyzed is typically a few monolayers. The application of these techniques to atmospheric particles is described by Klaus (1986) and, as discussed in more detail below, in a series of papers by Goschnick and co-workers (Goschnick et al., 1994a,b; Bentz et al., 1995a, 1995b; Faude and Goschnick, 1997).

Mass spectrometry has also been shown to be a promising method for differentiating the oxidation states of some metals. This is important because the oxidation state in some cases determines the toxicity of the element. For example, Cr(VI) is a carcinogen whereas Cr(III) is not. Laser ionization mass spectrometry studies of the oxides of chromium and arsenic suggest that some cluster ions are characteristic of the oxidation state. For example, Neubauer et al. (1995) have shown using single-particle laser ionization mass spectrometry (vide infra) that the ratios of ions such as $Cr_3O_9^-/Cr_2O_6^-$, $HCr_2O_7^-/Cr_2O_6^-$, and $Cr_2O_5^-/Cr_2O_6^-$ can be used to determine the relative amounts of Cr(III) and Cr(VI). However, the relative signal intensities also depended on a number of other parameters such as particle size, water content, laser irradiance, and counterions. As a result, at present this approach is applicable only to well-controlled situations such as process analysis rather than ambient air.

Similarly, Allen et al. (1996) have used laser ionization mass spectrometry to differentiate the oxidation states of arsenic. In this case, a bulk sample was collected on a sampling stage and inserted into the instrument (rather than using single particles). As_2O_3, i.e., As(III), was shown to give a characteristic $As_3O_5^-$ ion whereas As_2O_5, i.e., As(V), gave an $As_3O_8^-$ ion. This approach has also been shown to be promising for some organics. For example, it has been used to screen for the presence of nitro-PAHs (see Chapter 10) in diesel exhaust particles (Bezabeh et al., 1997). Quantification was not possible due to such factors as matrix effects.

While these techniques are promising for ambient air analysis, this is clearly going to be complex due to the many different species present in air and the many parameters that affect the ionization process.

Electron spectroscopy for chemical analysis (ESCA / X-ray photoelectron spectroscopy (XPS). The sample is irradiated with X-rays of a fixed frequency, causing ejection of electrons whose kinetic energy is measured. Conservation of energy dictates that the kinetic energy of the electron plus its binding energy must equal the energy of the exciting photon; since the latter is known and the kinetic energy of the electron is measured, the binding energy can be calculated. Since the binding energies are characteristic of each element, this can be used for elemental analysis. In addition, the binding energies of inner-shell electrons are influenced to some extent by the bonding electrons that determine the oxidation state of the element. For example, ESCA was used by Novakov and co-workers (1972) to elucidate the forms of nitrogen and sulfur in atmospheric particulate matter. Similarly, Faude and Goschnick (1997) used XPS to identify a variety of components of aerosol particles in the upper Rhine Valley in Germany, including sulfate and chlorine. Nitrogen in the form of ammonium and organo-nitrogen compounds (but, interestingly, not nitrate) was observed and carbon in organic or elemental as well as in oxidized forms attached to oxygen was noted. Other atmospheric applications are discussed in the review of Cox and Linton (1986).

Intercomparison studies. A number of intercomparison studies have been carried out to determine the accuracy and precision of measurements of various elements found in particles. For example, Nejedly et al.

(1998) compared the analysis of ambient air particles using ion chromatography, PIXE, and X-ray fluorescence. Two samplers operated side-by-side with PIXE analysis of the filters to assess precision were generally in excellent agreement, to within ~10% for a series of elements including Na, Mg, Ti, Cu, Al, Si, Mn, Fe, Ca, Zn, and S as well as for mass of $PM_{2.5}$ and the light absorption coefficient.

In another portion of the study, filters were analyzed both by X-ray fluorescence and by PIXE for samples collected in a remote area and in an urban area. The normalized percentage difference compared to the mean of the two was ≤ 10% for the elements S, K, Ca, Mn, Fe, and Zn. However, at small concentrations of sulfur, the X-ray fluorescence data were about 20–30% higher than the PIXE analyses for unknown reasons. The agreement for Si was poorer, perhaps due to the greater absorption of X-rays by Si and/or matrix effects.

In the third portion of the study, the results using five different sampler and analytical method combinations were compared. When obvious outliers were excluded from the data, the normalized percentage differences compared to the mean value for sulfur varied from −21 to +23%. Pairwise comparisons for other elements showed similar variability. The agreement overall for X-ray fluorescence compared to PIXE was good, although there was scatter in the individual measurements, perhaps due to differences in sampling (Nejedly et al., 1998).

Similarly, the concentrations of 17 elements in particles sampled using a variety of methods at Mace Head, Ireland, were compared (Francois et al., 1995). Sampling was carried out using a Hi-Vol sampler, a stacked filter unit, Nucleopore filters, and cascade impactors. Analytical techniques included AA, ICP, NA, PIXE, and X-ray fluorescence. Concentrations obtained using the Hi-Vol sampler were higher, which was attributed to differences in collection efficiencies, particularly of larger particles. Ratios of the concentrations of elements determined using cascade impactors compared to stacked filter units ranged from 0.48 ± 0.12 for Na to 1.31 ± 0.19 for Ti, but for most elements were in good agreement. A comparison of two different cascade impactors with PIXE analysis gave relatively good agreement for the total elemental concentrations, although the size distributions in the smaller size range differed for S and Pb. The latter may reflect differences in cutoff diameters for the states in the two impactors and/or bounce and particle reentrainment problems (Francois et al., 1995).

In summary, there is relatively good agreement overall between different methods of elemental analysis for atmospheric particles, with many of the observed discrepancies due to differences in sampling rather than analysis.

b. Inorganic Ions

Inorganic ions such as NH_4^+, SO_4^{2-}, and NO_3^- are major components of ambient particulate matter and a wide variety of methods have been used to measure their concentrations. A few of the methods most commonly used are summarized in Table 11.12 and discussed briefly in the following sections.

Colorimetry. A variety of colorimetric techniques have been used to measure ions such as NH_4^+, SO_4^{2-}, and NO_3^- in ambient particles. For example, nitrate can be measured by reduction to nitrite using hydrazine in the presence of a copper catalyst, followed by its conversion to a colored azo dye, which can be measured by its absorbance at 524 nm (Mullin and Riley, 1955). Sulfate has been determined using an exchange reaction between sulfate and a barium–nitrosulfonazo(III) chelate in aqueous acetonitrile; the chelate has an absorbance peak at 642 nm and hence the decrease in this peak can be followed as a measure of the amount of sulfate present that has exchanged with the chelate (Hoffer *et al.*, 1979). Similarly, NH_4^+ can be measured by the indophenol blue method (Weatherburn, 1967).

Ion chromatography (IC). Ion chromatography has become one of the most widely used methods for the determination of ion concentrations in ambient particles. As the name implies, ions are separated using ion exchange chromatography and are detected usually using electrical conductivity. For example, sulfate and nitrate can be separated on a column containing a strong basic resin using a carbonate solution as the eluant. To overcome the high conductivity of the eluant, which would mask the signal due to the sulfate and

TABLE 11.12 Some Common Methods of Measuring the Major Inorganic Ions in Atmospheric Particles

Ion	Analytical methods
NH_4^+	Col,[a] ESCA,[b] IC,[c] SIE,[d] IR[e]
SO_4^{2-}	Col, ESCA, IC, IR
NO_3^-	Col, ESCA, SIE, IC, CC,[f] IR

Source: Adapted from LBL (1979).
[a] Col = colorimetry.
[b] ESCA = electron spectroscopy for chemical analysis.
[c] IC = ion chromatography.
[d] SIE = selective ion electrodes.
[e] IR = infrared spectroscopy.
[f] Chemical conversion followed by detection of the product of the NO_3^- reaction.

nitrate, the solutions then pass into a suppression column that contains a strong acid resin; this converts the carbonate into $CO_2 + H_2O$, which has a low conductivity, and the sulfate and nitrate into their acids, which have high conductivities and hence can be easily detected against the suppressed eluant background (Mulik et al., 1976). This *eluant suppression* was the key to the development of IC to measure sulfate and nitrate. Since this first application of IC in ambient aerosols, a variety of anions and cations in ambient aerosols have been separated and measured using this technique. An example of its application to the measurement of nitrate, sulfate, chloride, and ammonium in PM_{10} in Taiwan is discussed by Tsai and Perng (1998).

Selective ion electrodes (SIE). Selective ion electrodes are essentially variants of the well-known pH meter. They are membrane indicator types of electrodes in which a potential is developed across a membrane in the presence of the ion; the size of the potential is related to the concentration and hence can be used to quantitatively detect and measure the species. However, instead of a glass membrane, as in the pH meter, the membranes consist of organics that are immersible in water. For example, anion-sensitive electrodes use a solution of an anion exchange resin in an organic solvent; the liquid can be held in the form of a gel, for example, in polyvinyl chloride. The ion reacts with the organic membrane, setting up an equilibrium between the free ion in solution and the ion bound to the membrane, generating a potential difference, which is measured.

Membrane electrodes used to measure species such as NH_4^+ that are in equilibrium with the gaseous form (i.e., NH_3) in solution are known as gas-sensing electrodes. In this case, the solution to be analyzed is separated from the analyzing solution by a gas-permeable membrane. The gas in the solution to be analyzed diffuses through the membrane and changes the pH of the internal solution, which is monitored using a standard glass electrode.

Infrared and Raman spectroscopy. Stephens and Price (1970, 1972) used infrared spectroscopy to examine both ambient and laboratory-generated aerosols. They identified sulfate, nitrate, and ammonium ion absorption bands in ambient particles as well as bands indicating the presence of organics in diesel exhaust (C–H) and oxidized organics in irradiated hydrocarbon–NO_x mixtures. Since then, many studies using IR have been carried out and a variety of species identified, including CO_3^{2-}, PO_4^{3-}, and SiO_4^{4-}. See Chapter 9.C.2 and Figs. 9.49, 9.50, and 9.51 for some typical FTIR spectra of atmospheric particles.

A variety of infrared approaches have been used, including transmission IR, photoacoustic IR, diffuse reflectance IR, and attenuated total reflectance. The principles behind these methods and their application to atmospheric aerosols have been reviewed by Allen and Palen (1989).

Raman spectroscopy (reviewed by Schrader, 1986) has also been applied to single particles in laboratory systems. For example, Fung and Tang (1991, 1992a, 1992b) and Fung et al. (1994) have applied resonance Raman spectroscopy to particles containing nitrate and sulfate, both very common constituents of atmospheric particles. The detection limits for nitrate and sulfate in aqueous droplets of 15-μm diameter were reported to be about 0.0025 M (Fung et al., 1994), suggesting that this method might prove applicable to ambient particles as well.

A variant of Raman spectroscopy that has been used to probe interfaces in large aqueous particles (e.g., of the order of several hundred microns) in laboratory studies is nonlinear morphology-dependent stimulated Raman scattering (e.g., Zhang and Aker, 1993; Aker and Zhang, 1994). In this method, light generated inside the particle in effect undergoes internal reflections at the interface; when the wavelength of the light is an integral factor times the circumference of the particle, it gets "trapped," in effect increasing the optical path length and hence the net absorption by species dissolved in the particle. As with resonance Raman, this technique has not yet been applied to particles in ambient air.

Mass spectrometry. Laser microprobe mass spectrometry (LMMS) has also been applied to atmospheric particles to measure primarily inorganic elements and ions (e.g., see Bruynseels et al., 1985; Kaufmann, 1986; Wieser and Wurster, 1986; Dierck et al., 1992; and Hara et al., 1996). The particles are collected using techniques such as impactors described earlier and subsequently analyzed. A laser pulse, e.g., at 266 nm, is used to volatilize a selected particle or a group of particles and the gaseous fragments produced are analyzed by mass spectrometry. Both positive and negative ions can be analyzed. The mass spectra of ambient particles can be quite complex and include many fragments as well as clusters. It is because of the extensive fragmentation that specific organic compounds cannot be identifed, although clusters of carbon atoms from soot, for example, can be seen. As discussed shortly, this method has been applied more recently to single particles suspended in air, and typical positive and negative ion spectra are shown in Section B.4a.

c. Total Carbon: Organic versus Graphitic (Elemental)

The separate determination of organic and elemental carbon in atmospheric particles has been addressed in a number of ways by many workers over a period of years; despite this, there is still no accepted accurate and reliable standard method of sampling and analysis for these important aerosol species.

Four major methods have been used to separate the organic elemental carbon: thermal methods, digestion, extraction, and optical techniques. These are discussed in detail in the volume on particulate carbon edited by Wolff and Klimisch (1982) and in the article by Cadle *et al.* (1983).

In the thermal methods, the sample is heated to increasingly higher temperatures, with most steps being carried out in the presence of O_2. The basis of this method is that volatile organics will vaporize first and then other organic compounds will be oxidized. Only at the highest temperatures will graphitic carbon oxidize. The carbon thus ejected into the vapor phase at various temperatures is detected in the form of CO_2 or, alternatively, after catalytic reduction, as CH_4.

For example, in one thermal method, shown in Fig. 11.67, the sample is oxidized and volatilized with an O_2–He mixture at 350°C; the volatilized carbon is oxidized to CO_2 in an MnO_2 bed and reduced to CH_4 so it can be measured using the sensitive technique of flame ionization detection (FID) (e.g., Huntzicker *et al.*, 1982; Japar *et al.*, 1984; Huffman, 1996). The purge gas is then replaced by pure He and the temperature is raised to 600°C; in this step the remaining organic carbon is volatilized, oxidized to CO_2 by the MnO_2 catalyst, and reduced to CH_4 for measurement. Finally, elemental carbon is determined by heating in an O_2–He mixture from 400 to 600°C. In this particular apparatus, a light pipe, He–Ne laser, and photocell are used to monitor the reflectance of the filter as an indication of the changes in graphitic carbon on the filter (vide infra).

Although relatively fast and simple, such thermal methods can suffer from the possibility of carbonization of organics during heating in an inert atmosphere; thus elemental carbon can be formed from organic carbon during the analysis, leading to significant errors. Corrections for this can be applied by following the sample reflectance during the heating (e.g., Huffman, 1996). Fung (1990) suggests that this error can be minimized by using another approach in which the sample is oxidized by MnO_2 during rapid heating to a maximum of 525°C, during which organic carbon is oxidized but elemental carbon is not. Heating to 850°C then leads to oxidation of elemental carbon by MnO_2.

A second approach to analyzing organic and elemental carbon has been to digest the sample in a strongly oxidizing solution (e.g., nitric acid) to remove the organics. The remaining carbon on the filter is then measured using standard methods with the assumption that only graphitic carbon remains on the filter after digestion. Organic carbon is then the difference between the total carbon on the filter before and after digestion, respectively. However, it has been shown that during digestion, some elemental carbon is removed, in addition to organic carbon (Cadle *et al.*, 1983). Thus digestion has no clear advantages over thermal methods.

Extraction of the organics from filters using various solvents has also been used. Total carbon analysis of portions of the filter after extraction gives graphitic carbon directly, and organic carbon is obtained by the difference between this and total carbon before extraction (e.g., Japar *et al.*, 1984). As with thermal and digestion techniques, there are problems in establishing that the organics and elemental carbon are clearly and accurately separated.

There are a variety of optical methods used to measure graphitic carbon alone, the most widely used being visible light absorption or reflectance techniques. Visible light absorption is the basis of what is known as the integrating plate method (IPM) (Lin *et al.*, 1973), shown schematically in Fig. 11.68. Particles are collected on a Nucleopore filter and inserted between the light source and the detector; the light transmitted through the filter is compared to that transmitted through a clean filter, that is, one not containing particles. Opal glass is placed between the filter and the detector to transmit an isotropic light flux from scattered and transmitted light through the filter. Scattering of light by the particles, which would interfere with an absorption measurement, is minimized by using a filter with a refractive index approximately equal to that of the particles.

As might be expected for a measurement based on simple light absorption in a complex sample, there are

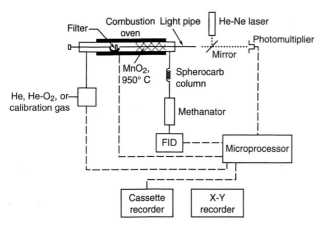

FIGURE 11.67 Schematic diagram of one type of thermal analyzer for organic and graphite carbon (adapted from Huntzicker *et al.*, 1982).

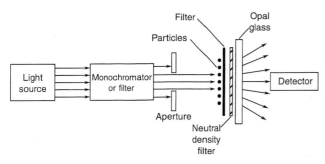

FIGURE 11.68 Schematic of integrating plate method (IPM) for measuring graphite carbon (adapted from Weiss and Waggoner, 1982).

a number of potential problems. For example, there may be other light-absorbing organics or other species present in the sample (e.g., Huffman, 1996); in addition, it is not clear what value should be used for the absorptivity of combustion-derived carbon particles. Thus, Horvath (1997) showed that the light absorption coefficient of carbon measured using the integrating plate method was systematically high and that this is expected theoretically for measurements made using a combination of transmission and integration of the scattered light.

Reflectance techniques, like the IPM, are based on the absorption of visible light by graphitic carbon. However, rather than measuring the decrease in light transmitted through a filter due to absorption, the decrease in light reflected from the carbon-containing surface is measured; the higher the elemental carbon, the more light will be absorbed and the less reflected. Thus $\log(R_0/R)$, where R_0 is the reflectance in the absence of carbon and R the reflectance in its presence, has been shown to be linearly related to the elemental carbon concentrations (Delumyea *et al.*, 1980). Because this is a light absorption/reflectance measurement, it suffers from the same types of problems as the IPM. However, it has an advantage in terms of its simplicity. In addition, in some urban areas there are historical records of filter sample reflectances that can be calibrated against more recent methods to examine historical trends in graphitic carbon (e.g., see Cass *et al.*, 1983).

Although carbonate has been observed in some ambient samples (e.g., see Cunningham *et al.*, 1984), it is generally believed to be present at insignificant concentrations compared to organic and graphitic carbon.

Although it has been generally assumed that elemental carbon is the only component of particles that absorbs visible light, as discussed in Chapter 9, this may not be the case. Instrumentation for measuring total light absorption by all particle components based on the heating of the surrounding gas caused by the absorbed energy is discussed by Moosmüller *et al.* (1997).

d. Speciation of Organics

As seen in Chapter 9.C.2, a very wide variety of organics are found in particles in ambient air and in laboratory model systems. The most common means of identification and measurement of these species is mass spectrometry (MS), combined with either thermal separation or solvent extraction and gas chromatographic separation combined with mass spectrometry and/or flame ionization detection. For larger, low-volatility organics, high-performance liquid chromatography (HPLC) is used, combined with various detectors such as absorption, fluorescence, and mass spectrometry. For applications of HPLC to the separation, detection, and measurement of polycyclic aromatic hydrocarbons, see Wingen *et al.* (1998) and references therein.

Thermal desorption was described earlier with respect to differentiating organic and elemental carbon. Once organics have been desorbed by heating, they can be identified and measured individually using chromatographic techniques. While this technique works well for a number of organics, as discussed shortly, some compounds are thermally unstable and decompose during the desorption process. In addition, it may not completely vaporize high molecular weight compounds of low volatility.

Solvent extraction of the sample is also frequently used in the analysis of particulate matter. Through the appropriate choice of solvents, the organics can be separated into acid, base, and neutral fractions, polar and nonpolar fractions, and so on. This grouping of compounds according to their chemical properties using extraction techniques simplifies the subsequent analysis. Each fraction can then be analyzed by GC–MS, with the GC retention time and the mass spectrum used for identification and measurement.

A more recent extraction technique involves the use of supercritical fluids such as CO_2. This has a number of advantages (e.g., see Skoog *et al.*, 1998) in that it avoids the use of large quantities of solvent and the need to concentrate the extract during which losses of the analyte may occur. Because the extracting fluid under atmospheric conditions is a gas, separating the fluid from the analyte only requires lowering the total pressure to release the "solvent" as a gas. In addition, it is quite fast since the rate of extraction depends on the rate of diffusion in the supercritical fluid and its viscosity, both of which are faster than for liquid extractions (e.g., minutes to hours versus hours to days). Low temperatures can be used for many supercritical fluids such as CO_2, minimizing thermal decomposition and/or reactions that may occur using thermal desorption techniques.

Supercritical fluid extraction (SFE) has been applied to ambient air particles with some success. For example, Hansen *et al.* (1995) developed a technique in which particles collected on a filter were extracted online using CO_2 and simultaneously transferred to the cooled head of a GC column, a process that took only 20 min. Recovery of long-chain alcohols and carboxylic acids spiked onto filters was much better for SFE than for thermal desorption but about the same or worse for some compounds such as nicotine. For a sample of urban aerosol particles, SFE detected some compounds not seen by thermal desorption such as larger alcohols. This was attributed to a lack of volatilization of these compounds with the temperatures used in the thermal desorption method and/or to decomposition during heating. On the other hand, some compounds such as benzaldehyde and the alkene 1-nonacosene were observed using thermal desorption but not using SFE. The alkene was thought not to be present in the aerosol, but rather was produced by thermal decomposition of some other compound and hence would not be generated during SFE. The amount of benzaldehyde may have been below the detection limit of the SFE–GC system.

Two types of ionization sources are in widespread use—electron impact and chemical ionization. The traditional means of ionization by electron impact often causes extensive fragmentation of molecules so that only peaks corresponding to the fragments are seen in the mass spectrum. Particularly in a complex environmental sample, this may preclude positive compound identification. Chemical ionization complements electron impact mass spectra and is particularly useful for establishing the molecular weight of the compound. In chemical ionization sources, an electron beam is used to ionize a reagent gas such as CH_4. The sample is then ionized by collisions with the ionized fragments from CH_4. This often results in relatively strong peaks at masses one greater or one less than the parent peak, MH, through reactions such as the following:

$$CH_5^+ + MH \rightarrow MH_2^+ + CH_4,$$
$$C_2H_5^+ + MH \rightarrow M^+ + C_2H_6.$$

Other types of mass spectrometry have also been used to examine ambient particulate samples. One such technique is secondary ion mass spectrometry (SIMS) in which the surface of the sample is bombarded with a beam of ions or neutral atoms that cause ejection of fragments from the surface. The fragments may be neutral atoms or molecules, positively or negatively charged species, electrons, or photons. The charged species, that is, the secondary ions, can be analyzed using MS, generating a SIMS spectrum. Elements such as potassium and sodium as well as functional groups such as COOH, sulfates, and nitrates can be detected by SIMS.

e. Artifacts

It is evident from the earlier discussion of sampling and collection of bulk samples of atmospheric particles that there is ample opportunity for the formation of artifacts, which can be either positive or negative, depending on the particular species measured and the techniques used. These can arise from a variety of processes such as the following: reactions of collected particles with other particle components or with gases (e.g., of NaCl with gaseous HNO_3), volatilization of compounds from collected particles (e.g., NH_4NO_3), adsorption and/or reactions of gases on filters or on the particles previously collected [e.g., SO_2 uptake and oxidation on particles (Eatough *et al.*, 1995) and adsorption of gas-phase organics on quartz fiber filters (e.g., Appel *et al.*, 1989; McDow and Huntzicker, 1990; Turpin *et al.*, 1994)], and reactions of gases with the filter medium, as discussed earlier for SO_2 on nylon filters, for example (e.g., Chan *et al.*, 1986; Cadle and Mulawa, 1987). In addition, the particle composition may change during collection due to shifts in gas-particle equilibria from changes in temperature, pressure drop across the collecting medium, or composition of the sampled air (e.g., Zhang and McMurry, 1987, 1991; Kaupp and Umlauf, 1992).

In addition to these chemical artifacts, physical artifacts can also occur. For example, the problems of particle bounce (e.g., see Wedding *et al.*, 1986) and reentrainment in impactors were discussed earlier. In addition, air turbulence is known to have a significant effect on the overall sampling efficiency of particle inlets (e.g., Wiener *et al.*, 1988; Francois *et al.*, 1995).

In short, care must be taken in sampling and analysis of airborne particles, as well as in the data interpretation, to minimize or at least recognize potential artifact problems. Such problems, along with a need to understand not only the bulk composition of a collection of airborne particles but also that of individual particles, have contributed to the development of real-time and single-particle analysis techniques discussed in the following section.

4. Real-Time Monitoring Techniques for Particles

Efforts to develop and apply real-time monitoring techniques for particles have been underway for more than two decades. Early approaches typically involved the impaction of particles on a hot filament, surface, or oven that volatilized and ionized the species in the particles, with the ions detected by mass spectrometry (e.g., Myers and Fite, 1975; Davis, 1977a, 1997b; Allen and Gould, 1981; Stoffels, 1981a,b; Stoffels and Lagergren, 1981; Sinha *et al.*, 1982, 1985; Sinha and Friedlander, 1986; Stoffels and Allen, 1986). Laser ionization was also investigated (Sinha, 1984). The ions produced were measured by mass spectrometry. Since many species volatilized but did not ionize, neutrals were detected in some studies by electron impact ionization subsequent to volatilization (e.g., Allen and Gould, 1981; Sinha *et al.*, 1982). More recent developments of this approach (e.g., Tobias *et al.*, 1999; Jayne *et al.*, 1999) are discussed later in this section.

While the early studies paved the way for further development of real-time and single-particle monitoring techniques, their application to ambient air was limited by a number of factors. These included the fact that the burst of ions produced was very short, <10 ms, so that scanning mass spectrometric methods such as quadrupoles could not scan sufficiently rapidly to capture the large mass range of interest. As a result, only a single mass or very limited number of masses could be recorded for each scan. Other limitations included extensive fragmentation of organics and a dependence of the efficiency of ionization on the composition of the particles. The use of time-of-flight mass spectrometry proposed by Allen and Gould in 1981 has helped to overcome the first problem. As we shall see, the fragmentation and variable ionization efficiencies continue to present challenges.

a. Single-Particle Laser Ionization Techniques

Since the first use of laser ionization by Sinha in 1984 to detect single particles, there has been a great deal of activity and development of this method for application to the atmosphere (see Johnston and Wexler (1995) for a review). In the early work, Sinha (1984) developed a method for simultaneous sizing of particles in which the particle first scattered light from one He–Ne laser, followed by a second He–Ne laser. The time interval between the two was used to obtain the speed of the particle and hence its size. The Nd:YAG ionizing laser was collinear with the second He–Ne laser and was fired at a set delay time after the particle was detected by the first He–Ne laser. Only particles of a given size whose speed is such that they reach the ionizing laser as it fired were detected. Particles of different sizes could be detected by varying the delay time between the scattering and ionizing lasers.

Subsequently, Marijnissen *et al.* (1988) proposed a single-particle system in which the amount of scattered light could be used along with the index of refraction of the particle to calculate its size. Shortly thereafter, McKeown *et al.* (1991) demonstrated the analysis of single particles using this approach, combined with time-of-flight mass spectrometry.

Since then, there has been a substantial development of such instruments. For example, Hinz *et al.* (1994) reported the first real-time monitoring of ambient particles in laboratory air using this technique of laser ionization combined with time-of-flight mass spectrometry. Carbon peaks from soot, metals attributed to abrasion of laboratory devices, and nicotine after "enriching the ambient air with tobacco smoke" were observed. Prather *et al.* (1994) and Mansoori *et al.* (1994) reported the application to inorganic and organic particles prepared in the laboratory, and Dale *et al.* (1994) and Yang *et al.* (1995a) reported analysis of organics adsorbed to the surface of particles of silicon carbide generated in the laboratory, using laser desorption combined with ion trap mass spectrometry. Reilly *et al.* (1998) have used laser ablation with an ion trap MS to identify polycyclic aromatic hydrocarbons in the particles from diesel engines.

Subsequently, Carson *et al.* (1995) and Neubauer *et al.* (1996, 1997) reported the ability to provide speciation of some aerosol components such as ammonium sulfate, ammonium sulfite, and methanesulfonic acid through control of the ionizing laser pulse energy, and Reents *et al.* (1995) showed that parent peaks could be obtained even for components such as SiO_2 that are difficult to ionize. Hinz *et al.* (1996) reported the simultaneous detection of both positive and negative ions produced by laser ionization of a single particle using a dual TOF system.

Prather and co-workers (Prather *et al.*, 1994; Noble *et al.*, 1994; Nordmeyer and Prather, 1994) introduced a significant improvement in particle size measurement and determining which particles are ionized. Light from the first laser is scattered by the particle and detected by a photomultiplier. It then travels a known distance where it encounters and scatters light from a second laser, which is also detected. The delay time between the two scattered light pulses is determined by the speed, i.e., the size, of the particle. This delay time is used to trigger the ionizing laser located further downstream at exactly the time that the particle should be in its optical line of sight. This use of three lasers allows both the determination of particle size and synchronization between detection of the particle and its ionization.

As illustrated below, the mass spectra of particles in ambient air can be (not surprisingly) quite complex. The use of tandem mass spectrometry would therefore be quite valuable, and indeed, such an instrument has

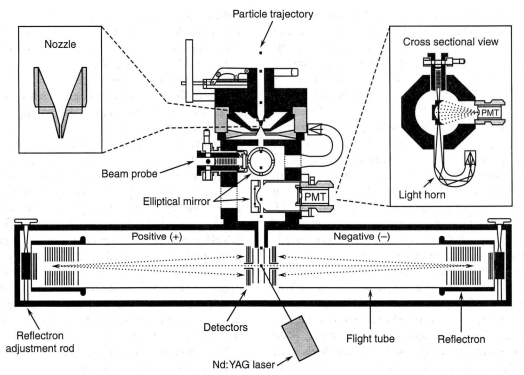

FIGURE 11.69 Schematic diagram of single-particle laser ionization mass spectrometer (adapted from Gard et al., 1997).

been developed using an ion trap mass spectrometer (March, 1992). Its application to aerosols generated in the laboratory has been explored for relatively simple systems (Yang et al., 1995a, 1996) and looks promising, although application to ambient air awaits further investigation.

Figure 11.69 is a schematic diagram of a single-particle laser ionization mass spectrometer with the particle sizing and ionization synchronization scheme of Prather and co-workers (Gard et al., 1997). Ionization is produced using light at 266 nm using a Nd:YAG laser and both positive and negative ions from the single particle are detected using a dual-ion coaxial set of TOF mass spectrometers. Figure 11.70 shows both the positive and negative mode mass spectra acquired from a single particle that was generated in the laboratory from wood burning. Hydrocarbon fragments are seen in both positive and negative ion modes, with potassium also present in the positive ion spectrum and HSO_4^- in the negative ion spectrum.

Figure 11.71 shows some single-particle mass spectra obtained in the positive ion mode in a rural area in Colorado using a laser ionization single-particle mass spectrometer (Murphy and Thomson, 1995, 1997a,b). Figure 11.71a is an example of a mass spectrum of a particle containing organics with fragments occurring up to higher amu; indeed, there are peaks appearing at most masses, suggesting a complex mixture. On the other hand, the spectrum in Fig. 11.71b shows mainly peaks due to C_n, which has been assigned to elemental carbon (soot particles). Figure 11.71c shows evidence for ammonium ions, perhaps due in part to ammonium nitrate as seen from the peaks in this mass spectrum.

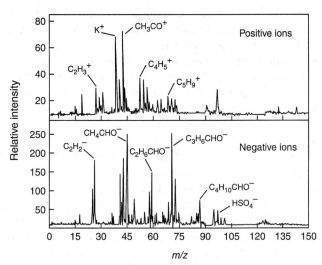

FIGURE 11.70 Positive and negative ions detected in a single particle from wood smoke generated in the laboratory (adapted from Gard et al., 1997).

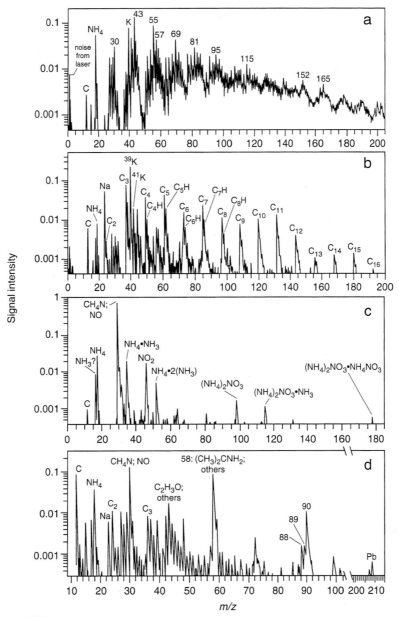

FIGURE 11.71 Typical laser ionization positive ion mass spectra of single particles in rural Colorado (adapted from Murphy and Thomson, 1997a,b).

Clearly, laser ionization–TOF mass spectrometry is a promising tool for real-time single-particle analysis. However, there are some important aspects of particle characterization on which data are not provided by these techniques at the present time. For example, while the identification and grouping of elements in single particles in ambient air using principal-component analysis provide insight into their sources (e.g., see Noble and Prather, 1996; Liu et al., 1997; Murphy and Thomson, 1997a, 1997b; Middlebrook et al., 1997; and Wood and Prather, 1998), independent quantification has not yet been achieved. Indeed, it may prove to be elusive due to the sensitivity of ion formation to the particular conditions, including the wavelength of the ionizing laser, the laser fluence, the chemical composition of the particle, etc. (e.g., see Thomson et al., 1997; and Ge et al., 1998b).

For example, Neubauer et al. (1998) have shown that the spectra can be very sensitive to the amount of water present and whether the particle is aqueous, i.e., above the deliquescence point, or solid (but holding adsorbed water on the surface). Figure 11.72 shows the

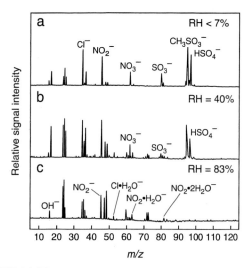

FIGURE 11.72 Negative ion laser ionization mass spectrometry of particles generated in the laboratory that contain equimolar amounts of NaCl, NH_4NO_3, $(NH_4)_2SO_4$, and CH_3SO_3H at (a) 7%, (b) 40%, and (c) 83% relative humidity (adapted from Neubauer et al., 1998).

negative ion mass spectra of particles containing equimolar amounts of NaCl, NH_4NO_3, $(NH_4)_2SO_4$, and methanesulfonic acid at relative humidities of <7, 40, and 83%, respectively. At the lowest relative humidity where the particle is a solid, peaks due to all of the salts appear, and the peak due to the anion of methanesulfonic acid is clear. At 83% RH, the latter has almost completely disappeared, as has that due to the sulfate.

However, relative peak intensities can be useful if the nature of the particles has not changed significantly. For example, Gard et al. (1998) followed peaks characteristic of chloride and nitrate in a coastal region in southern California and showed that they had an inverse correlation in single particles. Such behavior is expected from the well-known reaction of gaseous HNO_3 (and perhaps other gaseous oxides of nitrogen) with NaCl to give gaseous HCl and solid $NaNO_3$ (e.g., see De Haan et al., 1999, and references therein).

A second issue is that while organics can be identified from C_nH_m peaks, speciation is generally not possible. For example, although negative ions corresponding to organic acid anions were observed in particles in rural Colorado, Murphy and Thomson (1997b) indicate that these could be due to fragmentation from larger organics. Fragmentation patterns of specific organics can, however, provide clues to the presence of certain classes of compounds. For example, Fig. 11.71d shows the positive ion mass spectrum of a particle that Murphy and Thomson (1997a) attribute to the presence of amines or amides in the particle.

A third issue is illustrated by the mass spectrum in Fig. 11.71a, which, as described by Murphy and Thomson (1997a), appears to have a peak at almost every mass. Thus, in many instances in the atmosphere, particularly in polluted urban areas, the spectra may be so complex that only major classes of compounds may be discernible.

Finally, the approaches to sizing rely to date on light scattering. As discussed in Chapter 9.A.4, visible light scattering peaks in the 0.1- to 1-μm range so that particles smaller than this cannot be detected and hence measured using this approach. Although particles down to ~10 nm in size can be detected by free firing of the laser (e.g., Reents et al., 1995; Carson et al., 1997b), this clearly gives a rather random selection of particles detected and may not work at low particle concentrations typical of the atmosphere. New approaches for detecting smaller particles in such systems are needed.

b. Alternate Potential Mass Spectrometric Methods for Sizing and Chemical Composition

Figure 11.73 is a schematic diagram of another approach to single-particle size and composition measurements that is applicable to volatile and semivolatile species over the size range from ~0.05 to 1 μm (Jayne et al., 1999). This approach is similar in principle to that described by Allen and Gould (1981) and Sinha et al. (1982). Particles are sampled through an aerodynamic inlet that provides a narrow beam of particles with near unit efficiency (Liu et al., 1995a, 1995b; Schreiner et al., 1998). As the air containing the particle beam expands into the vacuum at the end of the inlet, the particles are accelerated, with smaller particles attaining higher speeds and vice versa. The beam of particles entering the sizing chamber is chopped to provide a time-of-flight

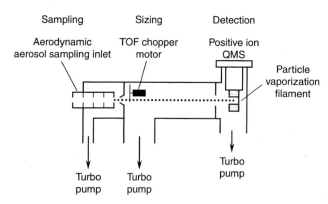

FIGURE 11.73 Schematic diagram of aerosol mass spectrometer for volatile and semivolatile particle sizing and composition measurement (graciously provided by J. Jayne, D. Worsnop, and C. Kolb, 1999).

measurement of the time for a particle to reach the detector, from which the particle size can be determined. In the third chamber, the particle collides with a heated surface that flash vaporizes volatile and semivolatile components. The vaporized species are ionized by electron impact and a quadrupole mass spectrometer is used to obtain the mass spectrum. The detection process is sufficiently sensitive to be able to detect single particles larger than 50 nm, so that the number of particles can be determined as a function of particle size.

At present, the full mass range cannot be scanned for one particle so that single-particle analysis is currently not possible with this approach; i.e., a complete mass spectrum cannot be obtained for one particle. Nonvolatile species are also not detected, so that important particle components in soil dust and soot cannot be measured. The advantages compared to single-particle laser ionization techniques are that some of the important volatile and semivolatile components such as ammonium sulfate and nitrate can be quantified. There is also much less fragmentation of organics. (However, given the complexity of the organic composition of particles in ambient air (see Chapter 9.C.2), it is not clear that specific organics in particles will be able to be identified.) Finally, the size and complexity of the instrument are significantly reduced without the need to incorporate a laser.

Figure 11.74 shows a similar approach to the measurement of a continuous beam of volatile and semivolatile particles with diameters as small as ~0.02 μm (Tobias et al., 1999). Upon exiting the aerodynamic lensing system, the particles enter a small (~0.1 cm^3) cell whose temperature can be regulated and is typically in the range of ~100–200°C. The particles vaporize in this cell and the vapors are sampled into a quadrupole mass spectrometer with electron impact ionization. The temperature of the vaporization cell can be optimized for particular compounds of interest, so that fragmentation of organic compounds, for example, can be minimized. Sizing of particles in this system can be provided by using a differential mobility analyzer prior to the aerodynamic lenses.

Both of these approaches are hence useful for a continuous stream of particles of similar composition where the mass spectrum can be continuously scanned, for example in laboratory environmental chamber studies.

In summary, the use of mass spectrometric methods, combined with various approaches to vaporizing and ionizing the particles, is gaining increasing popularity and interest for the analysis of continuous sources of particles or single particles. The problem of quantification of the components seen by single-particle laser ionization techniques remains to be solved. On the other hand, the vaporization approaches can provide quantitative data on some volatile and semivolatile components but cannot measure the nonvolatile species and, at present, do not provide a full mass spectrum for a single particle.

c. Depth Profiling of Particle Composition

As seen in the preceding sections, the technology associated with particle measurement is developing from bulk analyses of chemical composition to that of single particles. Further "splitting hairs," one would like to be able to depth profile particle composition. For example, as discussed in Chapter 9, it is believed that some particles have an organic coating on them, which can affect the particle properties such as the uptake and evaporation of species between the particle and the gas phase. Another example is the surface reaction of NaCl in sea salt particles with gaseous oxides of nitrogen, generating NaNO$_3$. Whether the surface is coated with a layer of nitrate, preventing further oxidation of the underlying NaCl, or whether the nitrate recrystallizes into separate microcrystallites as suggested by laboratory studies (see Figs. 11.63 and 11.64, Vogt and Finlayson-Pitts (1994), and Allen et al. (1996)), leaving a fresh surface of NaCl that can react further, is of interest.

Depth profiling of solids is commonly carried out in surface science studies, for example, using sputtering of the surface with argon ions. Such methods have also been applied to bulk samples of collected atmospheric particles (e.g., Jach and Powell, 1984; Bentz et al., 1994, 1995a, 1995b; Goschnick et al., 1994a, 1994b; Faude and Goschnick, 1997). For example, Faude and

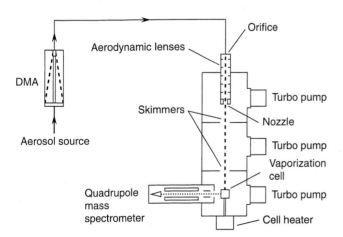

FIGURE 11.74 Schematic diagram of aerosol particle mass spectrometer for measurement of composition of continuous beams of volatile and semivolatile particles (graciously provided by P. Ziemann, 1998).

Goschnick (1997) applied a combination of X-ray photoelectron spectroscopy (XPS), secondary neutral mass spectrometry (SNMS), and secondary ion mass spectrometry (SIMS) to particles collected on a five-stage impactor in the upper Rhine Valley in Germany and found that the surface was enriched in organic compounds as well as ammonium sulfate and chloride. In a suburban area near Karlsruhe, Germany, SNMS and SIMS studies showed that in submicron particles, the cores consisted primarily of carbon (both elemental and organic), whereas the surface was enriched in ammonium sulfate; on the other hand, larger particles with diameters greater than 2 μm had cores of aluminosilicates surrounded by organics, carbonates, and nitrates and then a surface layer of ammonium sulfate and hydrogen-enriched organics (Bentz et al., 1995a,b).

Depth profiling of single airborne particles has been reported by Carson et al. (1995, 1997a), who showed that the use of variable laser fluences in single-particle laser ionization mass spectrometry can be used to probe thin films on particles in laboratory systems. At low laser intensities, only the surface layer is volatilized and ionized, whereas the entire particle can be vaporized and detected at higher intensities.

Similarly, Ge et al. (1996, 1998a) have studied the composition of single particles formed from the crystallization of solutions of mixtures of salts such as KCl/NaCl and $(NH_4)_2SO_4/NH_4NO_3$ using single-particle laser ionization mass spectrometry. As the relative humidity decreases, the least soluble salt crystallizes out. With further decreases in the relative humidity, the other salt reaches its saturation and the two salts precipitate out together. The composition of the aqueous phase at this point is referred to as the eutonic composition. Thus, a solid core of the least soluble species surrounded by a mixture of the salts is expected. Ge et al. (1996) analyzed the relative peak heights due to individual components in single particles crystallized from solutions containing mixtures of salts as a function of the solution composition and laser fluence. They were able to establish that for a KCl/NaCl mixture, for example, the surface layer was enriched in the minor component relative to the original solution, as expected.

Emission of electrons from the particle surface has also been used in laboratory studies to probe surface composition. Electron emission has been induced by UV irradiation, for example, by Burtscher and Schmidt-Ott (1986) to probe perylene on the surface of carbon particles. In a series of laboratory studies, Ziemann et al. (1995, 1997, 1998) have demonstrated the potential utility of secondary electron yield measurements as a technique for probing particle surface composition. In this method, particles are bombarded with

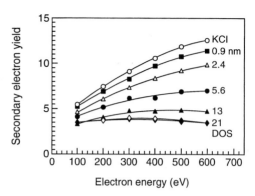

FIGURE 11.75 Measured secondary electron yields from the electron bombardment of 89-nm KCl particles, 128-nm particles of dioctyl subacate (DOS), and KCl coated with DOS of film thickness (in nm) shown (adapted from Ziemann and McMurry, 1998).

an electron beam and the change in the charge of the particles is measured using a Faraday cup. The number of secondary electrons emitted per primary electron incident on the particle, defined as the secondary electron yield, is characteristic of the chemical composition and of particle size. Figure 11.75, for example, shows the measured secondary electron yields for an 89-nm KCl particle, a 128-nm particle of the ester dioctyl sebacate (DOS), and KCl coated with DOS of various film thicknesses as a function of the energy of the bombarding electron beam (Ziemann and McMurray, 1998). Clearly, the yields are quite different for the organic and KCl and are a sensitive function of the thickness of the DOS film on the KCl.

Ziemann and McMurray (1998) also measured secondary electron yields and transport efficiencies of NaCl particles coated with octacosane upon exposure to relative humidities above the NaCl deliquescence point and then after drying. These data suggested that upon drying, the original organic film becomes localized on one side of the particle, giving it an irregular shape, in much the same manner as observed for the effect of water on surface nitrate (Figs. 11.63 and 11.64). Similarly, the secondary electron yields indicated that mixtures of NaCl and $NaNO_3$ crystallized heterogeneously whereas those of NaCl and NH_4Cl were homogeneously mixed (Ziemann and McMurray, 1997).

5. Generation of Calibration Aerosols

Calibration of the instruments to measure the size distribution and chemical composition requires methods of generating aerosols of well-defined sizes and known composition. Generating aerosols of known characteristics, both physical and chemical, is also necessary for carrying out studies of aerosol effects (e.g., on health or visibility) under well-defined experimental conditions.

A monodisperse aerosol is one with a narrow size distribution, which, for log-normal-distributed particles, usually means a geometric standard deviation of about 1.2 or smaller. Monodisperse particles are expected to have simple shapes and uniform composition with respect to size. A polydisperse aerosol, on the other hand, is one containing a wide range of particle sizes, but which may otherwise be homogeneous in terms of the basic physical and chemical properties that are not related to size. The term heterodisperse is also used occasionally; this describes aerosols varying widely in physical and chemical characteristics, as well as size.

As discussed in detail by Raabe (1976), an investigator's use of the terms monodisperse and polydisperse aerosols may depend on the particular properties of importance in the study; thus an aerosol may consist of particles of the same size, that is, be monodisperse with respect to size, but may vary in settling speed due to variations in density, that is, be polydisperse with respect to settling speed.

There are a number of techniques for generating aerosols, and these are discussed in detail in the LBL report (1979) and in volumes edited by Willeke (1980) and Liu et al. (1984). We briefly review here the major methods currently in use; these include atomizers and nebulizers, vibrating orifices, spinning disks, the electrical mobility analyzer discussed earlier, dry powder dispersion, tube furnaces, and condensation of vapors from the gas phase.

a. Atomizers and Nebulizers

Aerosols may be produced by atomizing liquids or suspensions of solids in liquids. Nebulizers are a type of atomizer in which both large and small particles are initially produced but in which the large particles are removed by impaction within the nebulizer. As a result, only particles with diameters ≤ 10 μm exit most nebulizers.

There are two basic means of generating particles from liquids in nebulizers: compressed air or ultrasonic vibration. Figure 11.76 shows one relatively simple type of compressed air nebulizer. The compressed air shoots out of a small orifice at high velocity, creating a reduced pressure in the region of the orifice; a feed tube connected to the liquid through a small opening is subjected to this region of lowered pressure (the Venturi effect) and hence liquid is drawn up from the reservoir and exits as a thin stream. The flow of high-velocity air striking the liquid stream breaks it up into small droplets and carries this aerosol toward the exit.

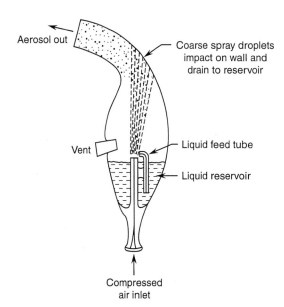

FIGURE 11.76 Schematic diagram of compressed air nebulizer (from Hinds, 1982).

The larger droplets are removed by impaction on the curved wall, and the smaller particles exit the device. Detailed descriptions of other types of compressed air nebulizers that differ somewhat in design are found in Raabe (1976) and Hinds (1982).

These compressed air nebulizers produce polydisperse aerosols. After the aerosol is produced, the size distribution may change due to evaporation of liquid from the droplets. In addition, the particles may be electrically charged due to an ion imbalance in the droplets as they form; if such charges become further concentrated due to evaporation, the particle may break up into smaller particles. Thus electrical neutralization of the aerosol, for example, by exposure to a radioactive source, is usually necessary to prevent electrostatic effects from dominating the particle motion, coagulation, and other behavior.

Nebulization may be used to produce suspensions of liquid droplets in air by using the pure liquid as the fluid or by using liquids with low vapor pressures dissolved in volatile solvents that then evaporate off the particle. Suspensions of solid particles in air may also be generated using the nebulization of suspensions of insoluble materials (insoluble plastic particles suspended in organic solvents, aqueous colloidal suspensions, e.g., of ferric hydroxide, etc.) or of soluble materials dissolved in water (e.g., salts in water). Drying the aerosol after its generation is an important factor in the final aerosol produced since this may alter both the physical and chemical nature of the particles; for example, rapid drying may produce low-density particles that are basically hollow shells formed by crystallization on the surface of the drying droplet.

Although these devices produce polydisperse aerosols, monodisperse aerosols can be generated by following the nebulizer with a size fractionating device such as an aerosol centrifuge or a differential mobility analyzer. Alternatively, nebulizers can be used to produce monodisperse aerosols of solid particles if suspensions of particles of one size are used as the generating fluid; for example, polystyrene and polyvinyltoluene latex beads of uniform size from 0.1 to 3.5 μm are commercially available in water suspensions. However, care must be taken to ensure that the vast majority of droplets formed contain only one sphere; otherwise, when the liquid evaporates, clusters of spheres will be formed. In addition, care must be taken to be sure they do not carry significant electrical charge (an anionic surfactant is added to the suspension to inhibit coagulation of the spheres). Finally, under some conditions, significant concentrations of other small particles may be simultaneously produced due to drying of empty droplets that contain impurities in the suspending liquid.

b. Vibrating-Orifice Generator

Another common method of generating particles with diameters 0.5–50 μm is the Berglund–Liu (1973) vibrating-orifice generator shown schematically in Fig. 11.77. The solution to be aerosolized is pumped through a small orifice 5–20 μm in diameter. The orifice is oscillated by a piezoelectric crystal so that the liquid stream is broken on each oscillation, forming a small liquid particle that is carried away in a stream of air.

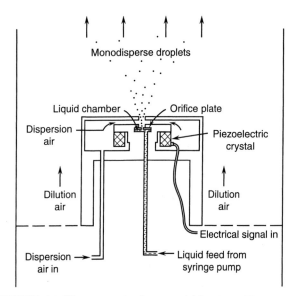

FIGURE 11.77 Schematic diagram of vibrating-orifice aerosol generator (from Hinds, 1982).

These droplets are then mixed with more air to dry the particles. The rate of droplet formation is equal to the oscillation frequency (f) of the piezoelectric crystal. From the volumetric flow rate of the liquid and the frequency f, the volume of the individual drops and particle diameter can be calculated.

c. Spinning-Disk Generator

A third means of producing aerosols is the spinning-disk aerosol generator. The liquid is fed to the center of the spinning disk and then moves by centrifugal force to the outer edge. It accumulates at the edge until there is sufficient liquid that the centrifugal force exceeds the surface tension forces and a droplet of liquid is thrown off. Some smaller "satellite droplets" are also produced, but these are separated from the primary drops, which are thrown out further by using a flow of air. Monodisperse aerosols with diameters ≥ 0.5 μm, and more typically in the range ~ 20–30 μm, are produced, the size being determined by the radius of the disk and the speed of the rotation (Hinds, 1982).

d. Dry Powder Dispersion

When an aerosol consisting of solid particles is required, they are usually generated by the dispersion of a dry powder. One of the most common devices that is particularly effective for dry, hard materials such as silica is known as the Wright dust feed (Wright, 1950). The dust is packed into a plug and a sharp blade is used to scrape dust off the surface of the plug; the particles are then swept away by a stream of air along the outer edge of the scraper blade and exit through a jet onto an impactor to break up particle clusters. The aerosol produced has particle diameters ≤ 10 μm. Variations on this type of aerosol generator are described in the LBL report (1979).

Fluidized beds have also been used for generating suspensions of solid particles with diameters in the range of ~ 0.5–40 μm. Air flows through the fluidized bed, which contains beads kept suspended by the motion of the air; dust injected into the bed is broken up into small particles and carried out with the air flow (Raabe, 1976).

A device such as an impactor or cyclone is frequently used at the exit of these dry powder dispersion devices to eliminate the large particles. A charge neutralizer is usually used to reduce the electrostatic charges on the dispersed particles.

e. Tube Furnaces

Monodisperse particles of salts, metals, metal oxides, and carbon have also been generated using electrically heated tube furnaces (e.g., Scheibel and

Porstendörfer, 1983; Ramamurthi and Leong, 1987). Salt particles, for example, can be generated by inserting the salt in a "boat" inside the tube furnace with a flow of N_2 passing over the salt. The salt vaporizes at the high temperatures and is carried downstream where the gas stream is cooled and the salt undergoes condensation into small particles.

Particles consisting of metals and metal derivatives such as the oxides can be generated in a two-step process. Primary aerosol particles are first generated using a vibrating-orifice generator and then dried by passing through a heated tube. This dried aerosol then passes into a tube furnace where the secondary aerosol is generated. For example, particles of nickel or nickel oxide can be generated from a primary aerosol of nickel formate using a tube furnace with N_2 or air as the carrier gas, respectively. Similarly, particles of ferric oxide or iron can be obtained from the thermal decomposition of ferrous sulfate heptahydrate in N_2 or in a mixture of H_2/N_2, respectively (Ramamurthi and Leong, 1987).

f. Condensation

When nuclei are present under conditions of supersaturation, condensation occurs on the nuclei, causing them to grow. If the particle radius is much less than the average distance between the particles, the growth rate of the particle with time is uniform, and since the nuclei are so much smaller than the final particles formed, nuclei of different initial sizes all give rise ultimately to particles of the same size, that is, a monodisperse aerosol, with diameters typically in the range 0.003–1.0 μm.

In practice then, a liquid having a low vapor pressure such as oleic acid or lubricating oils is carried by an inert gas to a heater where it vaporizes. As it leaves the heated area, it condenses to form the aerosol. Different generator designs based on condensation of a supersaturated vapor are discussed in the LBL report (1979).

However, care must be taken to ensure that seed nuclei are present on which condensation can occur. If a very clean system is used in which nuclei are not present, spontaneous nucleation may occur; this process is such that nuclei do not appear uniformly either in space or in time, and the initial particle growth rate depends on the degree of supersaturation. As a result, a polydisperse aerosol is produced under these conditions.

While condensation from a supersaturated vapor can be used to produce liquid particles, it is not as easily applied to the generation of solid particles except those that can be liquefied at modest elevated temperatures. However, these may be generated using different techniques. Methods used in the past have included the vaporization of wires or of salts fused onto wires, the use of "exploding" wires, and the use of electric arcs. For example, heating metal wires in an inert gas such as N_2 to a sufficiently high temperature produces small particles in the condensation nuclei and Aitken nuclei range. Thus tungsten wire at ~1000–1200°C produces tungsten metal particles, whereas nichrome wire gives chromium particles.

Solid salt particles may be produced by fusing the salt onto the wire by immersing the wire in a saturated solution of the salt and passing an electrical current through it. When the wire is heated, salt particles are produced. Theoretical and practical considerations for such generators are described by Fuchs and Sutugin (1970).

The exploding wire method involves putting a large amount of energy into a wire suddenly, causing it to "explode." If O_2 is present, a metal oxide aerosol is produced, whereas particles of pure metal are formed in an inert atmosphere such as helium. Exploding wire generators and their size distribution characteristics have been discussed by Phalen (1972).

Finally, electric arcs have also been used in some cases to produce solid particles of the electrode material, such as graphite, as have argon plasmas at high temperatures which produce a finely divided aerosol on rapid cooling.

All these methods generally give (≤ 1 μm) polydisperse aerosols of the solid particles and, unless rapid air dilution is provided, coagulation leads to large agglomerates of the small primary particles.

Gas-phase reactions can also be used to produce products of low volatility that condense to give an aerosol. The reaction of gaseous NH_3 with HCl to form particles of solid ammonium chloride and the reaction of gaseous SO_3 with water vapor to form H_2SO_4 are typical examples. Such methods tend to give submicron particles.

C. PROBLEMS

1. Chemiluminescence is useful as a monitoring technique in some cases. One of the requirements is that the excited state emit sufficiently fast that quenching does not completely overwhelm radiative processes.

   a. Calculate the number of collisions per second of NO_2 in an excited state with air molecules at 1 atm pressure and 298 K. Assume the diameters of air and the excited state can both be taken as 0.40 nm.

   b. Estimate the lifetime of the excited state if 10% of the excited molecules emit light rather than being quenched by air and if the quenching occurs with a probability of one (i.e., on every collision).

c. What would your answer be for (b) if the quenching occurs with an efficiency of 1%?

2. Laser-induced fluorescence (LIF) is a technique commonly used to monitor OH in air. One approach discussed in this chapter to improving the signal-to-noise ratio is to lower the total pressure to reduce quenching of the excited OH. The lifetime of the emitting state of OH ($A^2\Sigma^+$) is about 700 ns and typical values (they depend on rotational level) of the quenching rate constants (in units of 10^{-10} cm^3 molecules^{-1} s^{-1}) have been measured around room temperature to be 0.31 for N_2, 1.41 for O_2, and 4.1 for CO_2 (Copeland et al., 1985; Copeland and Crosley, 1986; Crosley, 1989). What percentage of the excited OH emits rather than being quenched at 1 atm and 298 K? By what factor does this increase if the total pressure is 10^{-3} atm? Take all molecular diameters to be 0.40 nm.

3. (a) For the mixture of organics reported in Table 11.8 for urban atmospheres, what would you expect a total VOC analyzer to report for the organic concentration if the lowest values reported are used? What would it be if the highest values were used? (b) Repeat part (a) for the data given for remote areas and compare to your answer in (a).

4. Develop Eq. (N) starting from the differential equations for NO_2 and NO in the reaction system consisting of reactions (8) and (9).

5. As discussed in the chapter, UV–visible spectroscopy provides sensitive detection for many atmospheric gases. However, a criterion for applying DOAS is that the absorption have a banded structure. Why do so few molecules of interest have a detailed, banded structure in the UV–visible compared, for example, to their infrared spectra?

6. Reference infrared spectra are shown for HNO_3 and NH_3 in a 25-cm path cell in Fig. 11.4. By what factor are the concentrations increased over even the peak concentrations of these species observed in the ambient air spectra of Fig. 11.3?

7. Derive from first principles the expression for [OH] as a function of the measured ^{14}CO and ^{14}CO$_2$ concentrations, OH + CO rate constant, and reaction time shown in Eq. (Q).

References

Aker, P. M., and J.-X. Zhang, "Morphology-Dependent Stimulated Raman Scattering (MDSRS)," *J. Photochem. Photobiol. A: Chem.*, 80, 381–388 (1994).

Alicke, B., K. Hebestreit, J. Stutz, and U. Platt, "Iodine Oxide in the Marine Boundary Layer," *Nature*, 397, 572–573 (1999).

Aliwell, S. R., and R. L. Jones, "Measurement of Atmospheric NO$_3$. 1. Improved Removal of Water Vapour Absorption Features in the Analysis for NO$_3$," *Geophys. Res. Lett.*, 23, 2585–2588 (1996a).

Aliwell, S. R., and R. L. Jones, "Measurement of Atmospheric NO$_3$. 2. Diurnal Variation of Stratospheric NO$_3$ at Midlatitude," *Geophys. Res. Lett.*, 23, 2589–2592 (1996b).

Aliwell, S. R., and R. L. Jones, "Measurements of Tropospheric NO$_3$ at Midlatitude," *J. Geophys. Res.*, 103, 5719–5727 (1998).

Alkezweeny, A. J., G. L. Laws, and W. Jones, "Aircraft and Ground Measurements of Ammonia in Kentucky," *Atmos. Environ.*, 20, 357–360 (1986).

Allen, D. T., and E. Palen, "Recent Advances in Aerosol Analysis by Infrared Spectroscopy," *J. Aerosol Sci.*, 20, 441–455 (1989).

Allen, J., and R. K. Gould, "Mass Spectrometric Analyzer for Individual Aerosol Particles," *Rev. Sci. Instrum.*, 52, 804–809 (1981).

Allen, T. M., D. Z. Bezabeh, C. H. Smith, E. M. McCauley, A. D. Jones, D. P. Y. Chang, I. M. Kennedy, and P. B. Kelly, "Speciation of Arsenic Oxides Using Laser Desorption/Ionization Time-of-Flight Mass Spectrometry," *Anal. Chem. A*, 68, 4052–4059 (1996).

Altshuller, A. P., and S. P. McPherson, "Spectrophotometric Analysis of Aldehydes in the Los Angeles Atmosphere," *J. Air Pollut. Control Assoc.*, 13, 109–111 (1963).

Altshuller, A. P., "Ambient Air Hydroxyl Radical Concentrations: Measurements and Model Predictions," *JAPCA*, 39, 704–708 (1989).

Altshuller, A. P., "PANs in the Atmosphere," *J. Air Waste Manage. Assoc.*, 43, 1221–1230 (1993).

Anderson, J. R., P. R. Buseck, and T. L. Patterson, "Characterization of the Bermuda Tropospheric Aerosol by Combined Individual-Particle and Bulk-Aerosol Analysis," *Atmos. Environ.*, 30, 319–338 (1996).

Anderson, L. G., J. A. Lanning, and P. Wolfe, "Acetone in the Urban Atmosphere: A Study in Denver, Colorado," *Isr. J. Chem.*, 34, 341–353 (1994).

Anderson, L. G., J. A. Lanning, R. Barrell, J. Miyagishima, R. H. Jones, and P. Wolfe, "Sources and Sinks of Formaldehyde and Acetaldehyde: An Analysis of Denver's Ambient Concentration Data," *Atmos. Environ.*, 30, 2113–2123 (1996).

Andersson-Sköld, Y., J. Moldanova, and A. Lindskog, "Comparison of Simulated and Measured Concentrations of Ozone, PAN, and Organic Species—Influence of Chemical Activity and Emission Pattern," in *The Proceedings of EUROTRAC Symposium '92* (P. M. Borrel et al., Eds.), pp. 433–436, SPB Academic Publishing BV, The Hague, The Netherlands, 1993.

Andreae, M. O., R. W. Talbot, and S.-M. Li, "Atmospheric Measurements of Pyruvic and Formic Acid," *J. Geophys. Res.*, 92, 6635–6641 (1987).

Andrés-Hernández, M. D., J. Notholt, J. Hjorth, and O. Schrems, "A DOAS Study on the Origin of Nitrous Acid at Urban and Nonurban Sites," *Atmos. Environ.*, 30, 175–180 (1996).

Anlauf, K. G., P. Fellin, H. A. Wiebe, H. I. Schiff, G. I. Mackay, R. S. Braman, and R. Gilbert, "A Comparison of Three Methods for Measurement of Atmospheric Nitric Acid and Aerosol Nitrate and Ammonium," *Atmos. Environ.*, 19, 325–333 (1985).

Anlauf, K. G., H. A. Wiebe, and P. Fellin, "Characterization of Several Integrative Sampling Methods for Nitric Acid, Sulphur Dioxide, and Atmospheric Particles," *JAPCA*, 36, 715–723 (1986).

Anlauf, K. G., D. C. MacTavish, H. A. Wiebe, H. I. Schiff, and G. I. Mackay, "Measurement of Atmospheric Nitric Acid by the Filter Method and Comparisons with the Tuneable Diode Laser and Other Methods," *Atmos. Environ.*, 22, 1579–1586 (1988).

Apel, E. C., J. G. Calvert, and F. C. Fehsenfeld, "The Nonmethane Hydrocarbon Intercomparison Experiment (NOMHICE): Tasks 1 and 2," *J. Geophys. Res.*, 99, 16651–16664 (1994).

Apel, E. C., J. G. Calvert, R. Zika, M. O. Rodgers, V. P. Aneja, J. F. Meagher, and W. A. Lonneman, "Hydrocarbon Measurements during the 1992 Southern Oxidants Study Atlanta Intensive: Protocol and Quality Assurance," *J. Air Waste Manage. Assoc., 45,* 521–528 (1995).

Apel, E. C., J. G. Calvert, J. P. Greenberg, D. Riemer, R. Zika, T. E. Kleindienst, W. A. Lonneman, K. Fung, and E. Fujita, "Generation and Validation of Oxygenated Volatile Organic Carbon Standards for the 1995 Southern Oxidants Study Nashville Intensive," *J. Geophys. Res., 103,* 22281–22294 (1998a).

Apel, E. C., J. G. Calvert, D. Riemer, W. Pos, R. Zika, T. E. Kleindienst, W. A. Lonneman, K. Fung, E. Fujita, P. B. Shepson, T. K. Starn, and P. T. Roberts, "Measurements Comparison of Oxygenated Volatile Organic Compounds at a Rural Site during the 1995 SOS Nashville Intensive," *J. Geophys. Res., 103,* 22295–22316 (1998b).

Appel, B. R., W. Cheng, and F. Salaymeh, "Sampling of Carbonaceous Particles in the Atmosphere—II," *Atmos. Environ., 23,* 2167–2175 (1989).

Armerding, W., M. Spiekermann, and F. J. Comes, "OH Multipass Absorption: Absolute and *In Situ* Method for Local Monitoring of Tropospheric Hydroxyl Radicals," *J. Geophys. Res., 99,* 1225–1239 (1994).

Armerding, W., M. Spiekermann, J. Walter, and F. J. Comes, "MOAS: An Absorption Laser Spectrometer for Sensitive and Local Monitoring of Tropospheric OH and Other Trace Gases," *J. Atmos. Sci., 52,* 3381–3392 (1995).

Armerding, W., M. Spiekermann, J. Walter, and F. J. Comes, "Multipass Optical Absorption Spectroscopy: A Fast Scanning Laser Spectrometer for the *In-Situ* Determination of Atmospheric Trace Gas Components, in Particular OH," *Appl. Opt., 35,* 4206–4219 (1996).

Armerding, W., F. J. Comes, H. J. Crawack, O. Forberich, G. Gold, R. Ruger, M. Spiekermann, J. Walter, E. Cuevas, A. Redondas, R. Schmitt, and P. Matuska, "Testing the Daytime Oxidizing Capacity of the Troposphere: 1994 OH Field Campaign at the Izana Observatory, Tenerife," *J. Geophys. Res., 102,* 10603–10611 (1997).

Arnold, F., D. Krankowsky, and K. H. Marien, "First Mass Spectrometric Measurements of Positive Ions in the Stratosphere," *Nature, 267,* 30–31 (1977).

Arnold, F., and G. Hauck, "Lower Stratosphere Trace Gas Detection Using Aircraft-Borne Active Chemical Ionization Mass Spectrometry," *Nature, 315,* 307–309 (1985).

Arnold, F., K.-H. Wohlfrom, M. W. Klemm, J. Schneider, K. Gollinger, U. Schumann, and R. Busen, "First Gaseous Ion Composition Measurements in the Exhaust Plume of a Jet Aircraft in Flight: Implications for Gaseous Sulfuric Acid, Aerosols, and Chemiions," *Geophys. Res. Lett., 25,* 2137–2140 (1998).

Arnts, R. R., and S. B. Tejada, "2,4-Dinitrophenylhydrazine-Coated Silica Gel Cartridge Method for Determination of Formaldehyde in Air: Identification of an Ozone Interference," *Environ. Sci. Technol., 23,* 1428–1430 (1989).

Atkinson, R., A. M. Winer, and J. N. Pitts, Jr., "Estimation of Nighttime N_2O_5 Concentrations from Ambient NO_2 and NO_3 Radical Concentrations and the Role of N_2O_5 in Nighttime Chemistry," *Atmos. Environ., 20,* 331–339 (1986).

Atlas, E., and S. Schauffler, "Analysis of Alkyl Nitrates and Selected Halocarbons in the Ambient Atmosphere Using a Charcoal Preconcentration Technique," *Environ. Sci. Technol., 25,* 61–67 (1991).

Atlas, E. L., B. A. Ridley, G. Hubler, J. G. Walega, M. A. Carroll, D. D. Montzka, B. J. Huebert, R. B. Norton, F. E. Grahek, and S. Schauffler, "Partitioning and Budget of NO_y Species during the Mauna Loa Observatory Photochemistry Experiment," *J. Geophys. Res., 97,* 10449–10462 (1992).

Ayer, H. E., and J. M. Hochstrasser, "Cyclone Discussion," in *Aerosol Measurement* (D. A. Lundgren, F. S. Harris, Jr., W. H. Marlow, M. Lippmann, W. E. Clark, and M. D. Durham, Eds.), pp. 70–79, University Presses of Florida, Gainesville, FL, 1979.

Ayers, G. P., S. A. Penkett, R. W. Gillett, B. Bandy, I. E. Galbally, C. P. Meyer, C. M. Elsworth, S. T. Bentley, and B. W. Forgan, "The Annual Cycle of Peroxides and Ozone in Marine Air at Cape Grim, Tasmania," *J. Atmos. Chem., 23,* 221–252 (1996).

Baardsen, E. L., and R. W. Terhune, "Detection of OH in the Atmosphere Using a Dye Laser," *Appl. Phys. Lett., 21,* 209–211 (1972).

Baez, A. P., R. D. Belmont, O. G. Gonzalez, and I. P. Rosas, "Formaldehyde Levels in Air and Wet Precipitation at Mexico City, Mexico," *Environ. Pollut., 62,* 153–169 (1989).

Balasubramanian, R., and L. Husain, "Observations of Gas-Phase Hydrogen Peroxide at an Elevated Rural Site in New York," *J. Geophys. Res., 102,* 21209–21220 (1997).

Bandy, A. R., D. C. Thornton, and A. R. Driedger III, "Airborne Measurements of Sulfur Dioxide, Dimethyl Sulfide, Carbon Disulfide, and Carbonyl Sulfide by Isotope Dilution Gas Chromatography/Mass Spectrometry," *J. Geophys. Res., 98,* 23423–23433 (1993).

Baron, P. A., M. K. Mazumder, and Y. S. Cheng, "Direct-Reading Techniques Using Optical Particle Detection," in *Aerosol Measurement: Principles, Techniques, and Applications* (K. Willeke and P. A. Baron, Eds.), pp. 381–409, Van Nostrand Reinhold, New York, 1993.

Beck, S. M., R. J. Bendura, D. S. McDougal, J. M. Hoell, Jr., G. L. Gregory, H. J. Curfman, Jr., D. D. Davis, J. Bradshaw, M. O. Rodgers, C. C. Wang, L. I. Davis, M. J. Campbell, A. L. Torres, M. A. Carroll, B. A. Ridley, G. W. Sachse, G. F. Hill, E. P. Condon, and R. A. Rasmussen, "Operational Overview of NASA GTE/CITE 1 Airborne Instrument Intercomparisons: Carbon Monoxide, Nitric Oxide, and Hydroxyl Instrumentation," *J. Geophys. Res., 92,* 1977–1985 (1987).

Beine, H. J., D. A. Jaffe, D. R. Blake, E. Atlas, and J. Harris, "Measurements of PAN, Alkyl Nitrates, Ozone, and Hydrocarbons during Spring in Interior Alaska," *J. Geophys. Res., 101,* 12613–12619 (1996).

Beine, H. J., D. A. Jaffe, J. A. Herring, J. A. Kelley, T. Krognes, and F. Stordal, "High-Latitude Springtime Photochemistry. Part I: NO_x, PAN, and Ozone Relationships," *J. Atmos. Chem., 27,* 127–153 (1997).

Benner, R. L., and D. H. Stedman, "Field Evaluation of the Sulfur Chemiluminescence Detector," *Environ. Sci. Technol., 24,* 1592–1596 (1990).

Bennett, M., C. Rogers, and S. Sutton, "Mobile Measurements of Winter SO_2 Levels in London, 1983–84," *Atmos. Environ., 20,* 461–470 (1986).

Benning, L., and A. Wahner, "Measurements of Atmospheric Formaldehyde (HCHO) and Acetaldehyde (CH_3CHO) during POPCORN 1994 Using 2,4-DNPH Coated Silica Cartridges," *J. Atmos. Chem., 31,* 105–117 (1998).

Benter, T., M. Liesner, V. Sauerland, and R. N. Schindler, "Mass Spectrometric *In-Situ* Determination of NO_2 in Gas Mixtures by Resonance Enhanced Multiphoton Ionization," *Fresenius' J. Anal. Chem., 351,* 489–492 (1995).

Bentz, J. W. G., M. Fichtner, J. Goschnick, and H.-J. Ache, "Depth-Resolved Speciation of Nitrogen Compounds in Environmental Solids," *Fresenius' J. Anal. Chem., 349,* 205–207 (1994).

Bentz, J. W. G., J. Goschnick, J. Schuricht, and H. J. Ache, "Depth-Resolved Investigation of the Element and Compound Inventory of Aerosol Particles from Outdoor Air," *Fresenius J. Anal. Chem., 353,* 559–564 (1995a).

Bentz, J. W. G., J. Goschnick, J. Schuricht, H. J. Ache, J. Zehnpfennig, and A. Benninghoven, "Analysis and Classification of Individual Outdoor Aerosol Particles with SIMS Time-of-Flight Mass Spectrometry," *Fresenius' J. Anal. Chem., 353,* 603–608 (1995b).

Berglund, R. N., and B. Y. H. Liu, "Generation of Monodisperse Aerosol Standards," *Environ. Sci. Technol., 7,* 147–153 (1973).

Berico, M., A. Luciani, and M. Formignani, "Atmospheric Aerosol in an Urban Area—Measurements of TSP and PM10 Standards and Pulmonary Deposition Assessments," *Atmos. Environ., 31,* 3659–3665 (1997).

Berkowitz, C. M., J. D. Fast, S. Springston, J. M. Hubbe, R. A. Plastridge, R. Larson, C. W. Spicer, and P. Doskey, "Formation Mechanisms and Chemical Characteristics of Elevated Photochemical Layers over the Northeast United States," *J. Geophys. Res., 103,* 10631–10647 (1998).

Bernardo-Bricker, A., C. Farmer, P. Milne, D. Riemer, R. Zika, and C. Stoneking, "Validation of Speciated Nonmethane Hydrocarbon Compound Data Collected during the 1992 Atlanta Intensive as Part of the Southern Oxidants Study (SOS)," *J. Air Waste Manage. Assoc., 45,* 591–603 (1995).

Bezabeh, D. Z., T. M. Allen, E. M. McCauley, P. B. Kelly, and A. D. Jones, "Negative Ion Laser Desorption Ionization Time-of-Flight Mass Spectrometry of Nitrated Polycyclic Aromatic Hydrocarbons," *J. Am. Soc. Mass Spectrom., 8,* 630–636 (1997).

Biermann, H. W., E. C. Tuazon, A. M. Winer, T. J. Wallington, and J. N. Pitts, Jr., "Simultaneous Absolute Measurements of Gaseous Nitrogen Species in Urban Ambient Air by Long Pathlength Infrared and Ultraviolet–Visible Spectroscopy," *Atmos. Environ., 22,* 1545–1554 (1988).

Blake, D. R., T. W. Smith, Jr., T.-Y. Chen, W. J. Whipple, and F. S. Rowland, "Effects of Biomass Burning on Summertime Nonmethane Hydrocarbon Concentrations in the Canadian Wetlands," *J. Geophys. Res., 99,* 1699–1719 (1994).

Blake, N. J., S. A. Penkett, K. C. Clemitshaw, P. Anwyl, P. Lightman, A. R. W. Marsh, and G. Butcher, "Estimates of Atmospheric Hydroxyl Radical Concentrations from the Observed Decay of Many Reactive Hydrocarbons in Well-Defined Urban Plumes," *J. Geophys. Res., 98,* 2851–2864 (1993).

Blanchard, P., P. B. Shepson, K. W. So, H. I. Schiff, J. W. Bottenheim, A. J. Gallant, J. W. Drummond, and P. Wong, "A Comparison of Calibration and Measurement Techniques for Gas Chromatographic Determination of Atmospheric Peroxyacetyl Nitrate (PAN)," *Atmos. Environ., 24A,* 2839–2846 (1990).

Blomquist, B. W., A. R. Bandy, D. C. Thornton, and S. Chen, "Grab Sampling for the Determination of Sulfur Dioxide and Dimethyl Sulfide in Air by Isotope Dilution Gas Chromatography/Mass Spectrometry," *J. Atmos. Chem., 16,* 23–30 (1993).

Boatman, J. F., M. Luria, C. C. Van Valin, and D. L. Wellman, "Continuous Atmospheric Sulfur Gas Measurements Aboard an Aircraft: A Comparison between the Flame Photometric and Fluorescence Methods," *Atmos. Environ., 22,* 1949–1955 (1988).

Boesl, U., H. J. Neusser, and E. W. Schlag, "Visible and UV Multiphoton Ionization and Fragmentation of Polyatomic Molecules," *J. Chem. Phys., 72,* 4327–4333 (1980).

Boleij, J. S. M., E. Lebret, F. Hoek, D. Noy, and B. Brunekreef, "The Use of Palmes Diffusion Tubes for Measuring NO_2 in Homes," *Atmos. Environ., 20,* 597–600 (1986).

Bottenheim, J. W., and M. F. Shepherd, "C_2–C_6 Hydrocarbon Measurements at Four Rural Locations across Canada," *Atmos. Environ., 29,* 647–664 (1995).

Boudries, H., G. Toupance, and A. L. Dutot, "Seasonal Variation of Atmospheric Nonmethane Hydrocarbons on the Western Coast of Brittany, France," *Atmos. Environ., 28,* 1095–1112 (1994).

Bradshaw, J. D., M. O. Rodgers, S. T. Sandholm, S. KeSheng, and D. D. Davis, "A Two-Photon Laser-Induced Fluorescence Field Instrument for Ground-Based and Airborne Measurements of Atmospheric NO," *J. Geophys. Res., 90,* 12861–12873 (1985).

Bradshaw, J., S. Sandholm, and R. Talbot, "An Update on Reactive Odd-Nitrogen Measurements Made during Recent NASA Global Tropospheric Experiment Programs," *J. Geophys. Res., 103,* 19129–19148 (1998).

Braman, R. S., T. J. Shelley, and W. A. McClenny, "Tungstic Acid for Preconcentration and Determination of Gaseous and Particulate Ammonia and Nitric Acid in Ambient Air," *Anal. Chem., 54,* 358–364 (1982).

Brandenburger, U., T. Brauers, H.-P. Dorn, M. Hausmann, and D. H. Ehhalt, "*In-Situ* Measurements of Tropospheric Hydroxyl Radicals by Folded Long-Path Laser Absorption during the Field Campaign POPCORN," *J. Atmos. Chem., 31,* 181–204 (1998).

Brassington, D. J., "Tunable Diode Laser Absorption Spectroscopy for the Measurement of Atmospheric Species," in *Spectroscopy in Environmental Science* (R. J. H. Clark and R. E. Hester, Eds.), pp. 85–148, Wiley, New York, 1995.

Brauer, M., and J. R. Brook, "Personal and Fixed-Site Ozone Measurements with a Passive Sampler," *Air Waste Manage. Assoc., 45,* 529–537 (1995).

Brauer, M., and J. Hisham-Hashim, "Fires in Indonesia: Crisis and Reaction," *Environ. Sci. Technol.,* 404A–407A, September 1 (1998).

Brauers, T., H.-P. Dorn, and U. Platt, in *Physico-Chemical Behaviour of Atmospheric Pollutants, Proceedings of the 5th European Symp.,* Varese, Italia (G. Angeletti, Ed.), pp. 237–242, Kluwer Academic, Dordrecht, 1990.

Brauers, T., M. Hausmann, U. Brandenburger, and H.-P. Dorn, "Improvement of Differential Optical Absorption Spectroscopy with a Multichannel Scanning Technique," *Appl. Opt., 34,* 4472–4479 (1995).

Brauers, T., U. Aschmutat, U. Brandenburger, H.-P. Dorn, M. Hausmann, M. Beßling, A. Hofzumahaus, F. Holland, C. Plass-Dülmer, and D. H. Ehhalt, "Intercomparison of Tropospheric OH Radical Measurements by Multiple Folded Long-Path Laser Absorption and Laser Induced Fluorescence," *Geophys. Res. Lett., 23,* 2545–2548 (1996).

Brenninkmeijer, C. A. M., M. R. Manning, D. C. Lowe, G. Wallace, R. J. Sparks, and A. Volz-Thomas, "Interhemispheric Asymmetry in OH Abundance Inferred from Measurements of Atmospheric ^{14}CO," *Nature, 356,* 50–52 (1992).

Brook, J. R., T. F. Dann, and R. T. Burnett, "The Relationship among TSP, PM_{10}, $PM_{2.5}$, and Inorganic Constituents of Atmospheric Particulate Matter at Multiple Canadian Locations," *J. Air Waste Manage. Assoc., 47,* 2–19 (1997).

Brown, R. H., and C. J. Purnell, "Collection and Analysis of Trace Organic Vapor Pollutants in Ambient Atmospheres. The Performance of a Tenax-GC Absorbent Tube," *J. Chromatogr., 178,* 79–90 (1979).

Brune, W. H., I. C. Faloona, D. Tan, A. J. Weinheimer, T. Campos, B. A. Ridley, S. A. Vay, J. E. Collins, G. W. Sachse, L. Jaeglé, and D. J. Jacob, "Airborne *In-Situ* OH and HO_2 Observations in the Cloud-Free Troposphere and Lower Stratosphere during SUCCESS," *Geophys. Res. Lett., 25,* 1701–1704 (1998).

Bruynseels, F., H. Storms, T. Tavares, and R. Van Grieken, "Characterization of Individual Particle Types in Coastal Air by Laser Microprobe Mass Analysis," *Int. J. Environ. Anal. Chem., 23,* 1–14 (1985).

Buhr, M. P., D. D. Parrish, R. B. Norton, F. C. Fehsenfeld, R. E. Sievers, and J. M. Roberts, "Contribution of Organic Nitrates to the Total Reactive Nitrogen Budget at a Rural Eastern U.S. Site," *J. Geophys. Res., 95,* 9809–9816 (1990).

Buhr, S. M., M. P. Buhr, F. C. Fehsenfeld, J. S. Holloway, U. Karst, R. B. Norton, D. D. Parrish, and R. E. Sievers, "Development of a Semi-Continuous Method for the Measurement of Nitric Acid Vapor and Particulate Nitrate and Suflate," *Atmos. Environ., 29,* 2609–2624 (1995).

Burkhardt, M. R., N. I. Maniga, D. H. Stedman, and R. J. Paur, "Gas Chromatographic Method for Measuring Nitrogen Dioxide and Peroxyacetyl Nitrate in Air without Compressed Gas Cylinders," *Anal. Chem., 60,* 816–819 (1988).

Burnett, C. R., and E. B. Burnett, "Spectroscopic Measurements of the Vertical Column Abundance of Hydroxyl (OH) in the Earth's Atmosphere," *J. Geophys. Res., 86,* 5185–5202 (1981).

Burnett, C. R., K. R. Minschwaner, and E. B. Burnett, "Vertical Column Abundance Measurements of Atmospheric Hydroxyl from 26°, 40°, and 65°N," *J. Geophys. Res., 93,* 5241–5253 (1988).

Burnett, C. R., and E. B. Burnett, "The Regime of Decreased OH Vertical Column Abundances at Fritz Peak Observatory, CO: 1991–1995," *Geophys. Res. Lett., 23,* 1925–1927 (1996).

Burnett, C. R., and K. Minschwaner, "Continuing Development in the Regime of Decreased Atmospheric Column OH at Fritz Peak, Colorado," *Geophys. Res. Lett., 25,* 1313–1316 (1998).

Burtscher, H., and A. Schmidt-Ott, "*In Situ* Measurement of Adsorption and Condensation of a Polyaromatic Hydrocarbon on Ultrafine C Particles by Means of Photoemission," *J. Aerosol Sci., 17,* 699–703 (1986).

Cadle, S. H., P. J. Groblicki, and P. A. Mulawa, "Problems in the Sampling and Analysis of Carbon Particulate," *Atmos. Environ., 17,* 593–600 (1983).

Cadle, S. H., and P. A. Mulawa, "The Retention of SO_2 by Nylon Filters," *Atmos. Environ., 21,* 599–603 (1987).

Cahill, T. A., "Comments on Surface Coatings for Lundgren-Type Impactors," in *Aerosol Measurement* (D. A. Lundgren, F. S. Harris, Jr., W. H. Marlow, M. Lippmann, W. E. Clark, and M. D. Dunham, Eds.), pp. 131–134, University Presses of Florida, Gainesville, FL, 1979.

Cahill, T. A., "Proton Microprobes and Particle-Induced X-Ray Analytical Systems," *Annu. Rev. Nucl. Part. Sci., 30,* 211–252 (1980).

Cahill, T. A., "Innovative Aerosol Sampling Devices Based Upon PIXE Capabilities," *Nucl. Instrum. Methods, 181,* 473–480 (1981a).

Cahill, T. A., "Ion Beam Analysis of Environmental Samples," *Adv. Chem. Ser., 197,* 511–522 (1981b).

Calogirou, A., M. Duane, D. Kotzias, M. Lahaniati, and B. R. Larsen, "Polyphenylenesulfide, Noxon, an Ozone Scavenger for the Analysis of Oxygenated Terpenes in Air," *Atmos. Environ., 31,* 2741–2751 (1997).

Camel, V., and M. Caude, "Trace Enrichment Methods for the Determination of Organic Pollutants in Ambient Air," *J. Chromatogr. A, 710,* 3–19 (1995).

Campbell, M. J., J. C. Sheppard, and B. F. Au, "Measurement of Hydroxyl Concentration in Boundary Layer Air by Monitoring CO Oxidation," *Geophys. Res. Lett., 6,* 175–178 (1979).

Campbell, M. J., B. D. Hall, J. C. Sheppard, P. L. Utley, R. J. O'Brien, T. M. Hard, and L. A. George, "Intercomparison of Local Hydroxyl Measurements by Radiocarbon and FAGE Techniques," *J. Atmos. Sci., 52,* 3421–3427 (1995).

Campos, F. X., Y. Jiang, and E. R. Grant, "Triple-Resonance Spectroscopy of the Higher Excited States of NO_2: Rovibronic Interactions, Autoionization, and *l*-Uncoupling in the (100) Manifold," *J. Chem. Phys., 93,* 2308–2327 (1990).

Cantrell, C. A., and D. H. Stedman, "A Possible Technique for the Measurement of Atmospheric Peroxy Radicals," *Geophys. Res. Lett., 9,* 846–849 (1982).

Cantrell, C. A., D. H. Stedman, and G. J. Wendel, "Measurement of Atmospheric Peroxy Radicals by Chemical Amplification," *Anal. Chem., 56,* 1496–1502 (1984).

Cantrell, C. A., R. E. Shetter, J. A. Lind, A. H. McDaniel, and J. G. Calvert, "An Improved Chemical Amplifier Technique for Peroxy Radical Measurements," *J. Geophys. Res., 98,* 2897–2909 (1993).

Cantrell, C. A., R. E. Shetter, and J. Calvert, "Peroxy Radical Chemistry during FIELDVOC 1993 in Brittany, France," *Atmos. Environ., 30,* 3947–3957 (1996).

Cantrell, C. A., R. E. Shetter, J. G. Calvert, F. L. Eisele, and D. J. Tanner, "Some Considerations of the Origin of Nighttime Peroxy Radicals Observed in MLOPEX 2c," *J. Geophys. Res., 102,* 15899–15913 (1997a).

Cantrell, C. A., R. E. Shetter, J. G. Calvert, F. L. Eisele, E. Williams, K. Baumann, W. H. Brune, P. S. Stevens, and J. H. Mather, "Peroxy Radicals from Photostationary State Deviations and Steady State Calculations during the Tropospheric OH Photochemistry Experiment at Idaho Hill, Colorado, 1993," *J. Geophys. Res., 102,* 6369–6378 (1997b).

Cao, X.-L., and C. N. Hewitt, "Build-up of Artifacts on Adsorbents during Storage and Its Effect on Passive Sampling and Gas Chromatography-Flame Ionization Detection of Low Concentrations of Volatile Organic Compounds in Air," *J. Chromatogr. A, 688,* 368–374 (1994).

Carpenter, L. J., P. S. Monks, B. J. Bandy, S. A. Penkett, I. E. Galbally, and C. P. Meyer, "A Study of Peroxy Radicals and Ozone Photochemistry at Coastal Sites in the Northern and Southern Hemispheres," *J. Geophys. Res., 102,* 25417–25427 (1997).

Carroll, M. A., and L. Emmons, "Access NO_x and NO_y Measurements On-Line," *EOS, 77,* 34 (1996).

Carslaw, N., L. J. Carpenter, J. M. C. Plane, B. J. Allan, R. A. Burgess, K. C. Clemitshaw, H. Coe, and S. A. Penkett, "Simultaneous Observations of Nitrate and Peroxy Radicals in the Marine Boundary Layer," *J. Geophys. Res., 102,* 18917–18933 (1997a).

Carslaw, N., J. M. C. Plane, H. Coe, and E. Cuevas, "Observations of the Nitrate Radical in the Free Troposphere at Izana de Tenerife," *J. Geophys. Res., 102,* 10613–10622 (1997b).

Carson, P. G., K. R. Neubauer, M. V. Johnston, and A. S. Wexler, "On-Line Chemical Analysis of Aerosols by Rapid Single-Particle Mass Spectrometry," *J. Aerosol Sci., 26,* 535–545 (1995).

Carson, P. G., M. V. Johnston, and A. S. Wexler, "Real-Time Monitoring of the Surface and Total Composition of Aerosol Particles," *Aerosol Sci. Technol., 26,* 291–300 (1997a).

Carson, P. G., M. V. Johnston, and A. S. Wexler, "Laser Desorption/Ionization of Ultrafine Aerosol Particles," *Rapid Commun. Mass Spectrosc., 11,* 993–996 (1997b).

Cass, G. R., M. H. Conklin, J. J. Shah, J. J. Huntzicker, and E. S. Macias, "Elemental Carbon Concentrations: Estimation of and Historical Data Base," *Atmos. Environ., 18,* 153–162 (1983).

Castaldi, M. J., and S. M. Senkan, "Real-Time, Ultrasensitive Monitoring of Air Toxics by Laser Photoionization Time-of-Flight Mass Spectrometry," *J. Air Waste Manage. Assoc., 48,* 77–81 (1998).

Chan, C. Y., T. M. Hard, A. A. Mehrabzadeh, L. A. George, and R. J. O'Brien, "Third-Generation FAGE Instrument for Tropospheric Hydroxyl Radical Measurement," *J. Geophys. Res., 95,* 18569–18576 (1990).

Chan, W. H., D. B. Orr, and D. H. S. Chung, "An Evaluation of Artifact SO_4^{2-} Formation on Nylon Filters under Field Conditions," *Atmos. Environ., 20,* 2397–2401 (1986).

Chan, Y. C., R. W. Simpson, G. H. McTainsh, and P. D. Vowles, "Characterisation of Chemical Species in PM$_{2.5}$ and PM$_{10}$ Aerosols in Brisbane, Australia," *Atmos. Environ., 31*, 3773–3785 (1997).

Chapman, E., K. Busness, J. Thorp, D. V. Kenny, and C. W. Spicer, "Continuous Airborne Measurements of Gaseous Formic and Acetic Acids over the Western North Atlantic," *Geophys. Res. Lett., 22*, 405–408 (1995).

Chien, C.-J., M. J. Charles, K. G. Sexton, and H. E. Jeffries, "Analysis of Airborne Carboxylic Acids and Phenols as Their Pentafluorobenzyl Derivatives: Gas Chromatography/Ion Trap Mass Spectrometry with a Novel Chemical Ionization Reagent, PFBOH," *Environ. Sci. Technol., 32*, 299–309 (1998).

Chow, J. C., "Measurement of Methods to Determine Compliance with Ambient Air Quality Standards for Suspended Particles," *J. Air Waste Manage. Assoc., 45*, 320–382 (1995).

Cicerone, R. J., and R. Zellner, "The Atmospheric Chemistry of Hydrogen Cyanide (HCN)," *J. Geophys. Res., 88*, 10689–10696 (1983).

Claiborn, C. S., and V. P. Aneja, "Measurements of Atmospheric Hydrogen Peroxide in the Gas Phase and in Cloud Water at Mt. Mitchell, North Carolina," *J. Geophys. Res., 96*, 18771–18787 (1991).

Clarkson, T. S., R. J. Martin, and J. Rudolph, "Ethane and Propane in the Southern Marine Troposphere," *Atmos. Environ., 31*, 3763–3771 (1997).

Clausen, P. A., and P. Wolkoff, "Degradation Products of Tenax TA Formed during Sampling and Thermal Desorption Analysis: Indicators of Reactive Species Indoors," *Atmos. Environ., 31*, 715–725 (1997).

Clemitshaw, K. C., L. J. Carpenter, S. A. Penkett, and M. E. Jenkin, "A Calibrated Peroxy Radical Chemical Amplifier for Ground-Based Tropospheric Measurements," *J. Geophys. Res., 102*, 25405–25416 (1997).

Cofer, W. R., III, and R. A. Edahl, Jr., "A New Technique for Collection, Concentration, and Determination of Gaseous Tropospheric Formaldehyde," *Atmos. Environ., 20*, 979–984 (1986).

Coffey, M. T., W. G. Mankin, and R. J. Cicerone, "Spectroscopic Detection of Stratospheric Hydrogen Cyanide," *Science, 214*, 333–335 (1981).

Conner, W. D., "An Inertial-Type Particle Separator for Collecting Large Samples," *J. Air Pollut. Control Assoc., 16*, 35–38 (1966).

Cooke, D. D., and M. Kerker, "Response Calculations for Light Scattering Aerosol Particle Counters," *Appl. Opt., 14*, 734–739 (1975).

Copeland, R. A., M. J. Dyer, and D. R. Crosley, "Rotational-Level-Dependent Quenching of $A^2\Sigma^+$ OH and OD," *J. Chem. Phys., 82*, 4022–4032 (1985).

Copeland, R. A., and D. R. Crosley, "Temperature Dependent Electronic Quenching of OH($A^2\Sigma^+$, $v' = 0$) between 230 and 310 K," *J. Chem. Phys., 84*, 3099–3105 (1986).

Corkum, R., W. W. Biesbrecht, T. Bardsley, and E. A. Cherniak, "Peroxyacetyl Nitrate (PAN) in the Atmosphere at Simcoe, Canada," *Atmos. Environ., 20*, 1241–1248 (1986).

Cox, X. B., and R. W. Linton, "Particle Analysis by X-Ray Photoelectron Spectroscopy," in *Physical and Chemical Characterization of Individual Airborne Particles* (K. R. Spurny, Ed.), Chap. 18, pp. 341–357, Ellis Horwood, Chichester, 1986.

Crosley, D. R., "Rotational and Translational Effects in Collisions of Electronically Excited Diatomic Hydrides," *J. Phys. Chem., 93*, 6273–6282 (1989).

Crosley, D. R., "Local Measurement of Tropospheric HO$_x$," in *NASA Conference Publication 3245*, Summary of Workshop Held at SRI International, Menlo Park, California, 1994.

Crosley, D. R., "The Measurement of OH and HO$_2$ in the Atmosphere," *J. Atmos. Sci., 52*, 3299–3314 (1995a).

Crosley, D. R., "Laser Fluorescence Detection of Atmospheric Hydroxyl Radicals," in *Problems and Progress in Atmospheric Chemistry* (J. R. Barker, Ed.), *Advanced Series in Physical Chemistry* (C.-Y. Ng, Ed.), Vol. 3, World Scientific, Singapore, 1995b.

Crosley, D. R., "NO$_y$ Blue Ribbon Panel," *J. Geophys. Res., 101*, 2049–2052 (1996).

Cunningham, P. T., B. D. Holt, S. A. Johnson, D. L. Drapcho, and R. Kumar, "Acidic Aerosols: Oxygen-18 Studies of Formation and Infrared Studies of Occurrence and Neutralization," in *Chemistry of Particles, Fogs, and Rain* (J. L. Durham, Ed.), Acid Precipitation Series (J. I. Teasley, Series Ed.), pp. 53–130, Butterworth, Stoneham, MA, 1984.

Dale, J. M., M. Yang, W. B. Whitten, and J. M. Ramsey, "Chemical Characterization of Single Particles by Laser Ablation/Desorption in a Quadrupole Ion Trap Mass Spectrometer," *Anal. Chem., 66*, 3431–3435 (1994).

Damrauer, L., "Luminol-Based Nitrogen Dioxide Detector," *Anal. Chem., 55*, 937–940 (1983).

Das, M., and V. P. Aneja, "Analysis of Gaseous Hydrogen Peroxide Concentrations in Raleigh, North Carolina," *J. Air Waste Manage. Assoc., 44*, 176–180 (1994).

Dasgupta, P. K., and V. K. Gupta, "Membrane-Based Flow Injection System for Determination of Sulfur(IV) in Atmospheric Water," *Environ. Sci. Technol., 20*, 524–526 (1986).

Dasgupta, P. K., S. Dong, H. Hwang, H.-C. Yang, and Z. Genfa, "Continuous Liquid-Phase Fluorometry Coupled to a Diffusion Scrubber for the Real-Time Determination of Atmospheric Formaldehyde, Hydrogen Peroxide, and Sulfur Dioxide," *Atmos. Environ., 22*, 949–964 (1988).

Dasgupta, P. K., S. Dong, and H. Hwang, "Diffusion Scrubber-Based Field Measurements of Atmospheric Formaldehyde and Hydrogen Peroxide," *Aerosol Sci. Technol., 12*, 98–104 (1990).

Daum, P. H., L. I. Kleinman, A. J. Hills, A. L. Lazrus, A. C. D. Leslie, K. Busness, and J. Boatman, "Measurement and Interpretation of Concentrations of H$_2$O$_2$ and Related Species in the Upper Midwest during Summer," *J. Geophys. Res., 95*, 9857–9871 (1990).

Davenport, J. E., and H. B. Singh, "Systematic Development of Reactive Tracer Technology to Determine Hydroxyl Radical Concentrations in the Troposphere," *Atmos. Environ., 21*, 1969–1981 (1987).

Davis, E. J., "A History of Single Aerosol Particle Levitation," *Aerosol Sci. Technol., 26*, 212–254 (1997).

Davis, W. D., "Continuous Mass Spectrometric Analysis of Particulates by Use of Surface Ionization," *Environ. Sci. Technol., 11*, 587–592 (1977a).

Davis, W. D., "Continuous Mass Spectrometric Determination of Concentration of Particulate Impurities in Air by Use of Surface Ionization," *Environ. Sci. Technol., 11*, 593–596 (1977b).

Dawson, G. A., and J. C. Farmer, "Soluble Atmospheric Trace Gases in the Southwestern United States. 2. Organic Species HCHO, HCOOH, CH$_3$COOH," *J. Geophys. Res., 93*, 5200–5206 (1988).

Delumyea, R. G., L.-C. Chu, and E. S. Macias, "Determination of Elemental Carbon Component of Soot in Ambient Aerosol Samples," *Atmos. Environ., 14*, 647–652 (1980).

De Haan, D., T. Brauers, K. Oum, J. Stutz, T. Nordmeyer, and B. J. Finlayson-Pitts, "Heterogeneous Chemistry in the Troposphere: Experimental Approaches and Applications to the Chemistry of Sea Salt Particles," *Int. Rev. Phys. Chem., 18*, (1999).

DeMore, W. B., S. P. Sander, D. M. Golden, R. F. Hampson, M. J. Kurylo, C. J. Howard, A. R. Ravishankara, C. E. Kolb, and M. J. Molina, "Chemical Kinetics and Photochemical Data for Use in

Stratospheric Modeling," in JPL Publication 97-4, Jet Propulsion Laboratory, Pasadena, CA, January 15, 1997.

Derwent, R. G., P. G. Simmonds, S. Seuring, and C. Dimmer, "Observation and Interpretation of the Seasonal Cycles in the Surface Concentrations of Ozone and Carbon Monoxide at Mace Head, Ireland from 1990 to 1994," *Atmos. Environ., 32,* 145–157 (1998).

de Serves, C., and H. B. Ross, "Comparison of Collection Devices for Atmospheric Peroxides," *Environ. Sci. Technol., 27,* 2712–2718 (1993).

De Vries, H. S. M., F. J. M. Harren, G. P. Wyers, R. P. Otjes, J. Slanina, and J. Reuss, "Non-intrusive, Fast, and Sensitive Ammonia Detection by Laser Photothermal Deflection," *Atmos. Environ., 29,* 1069–1074 (1995).

Dewulf, J., and H. Van Langenhove, "Analytical Techniques for the Determination and Measurement Data of 7 Chlorinated C_1- and C_2-Hydrocarbons and 6 Monocyclic Aromatic Hydrocarbons in Remote Air Masses: An Overview," *Atmos. Environ., 31,* 3291–3307 (1997).

Dierck, I., D. Michaud, L. Wouters, and R. Van Grieken, "Laser Microprobe Mass Analysis of Individual North Sea Aerosol Particles," *Environ. Sci. Technol., 26,* 802–808 (1992).

Dollard, G. J., T. J. Davies, B. M. R. Jones, P. D. Nason, J. Chandler, P. Dumitrean, M. Delaney, D. Watkins, and R. A. Field, "The UK Hydrocarbon Monitoring Network," in *Volatile Organic Compounds in the Atmosphere, Issues in Environmental Science and Technology* (R. E. Hester and R. M. Harrison, Eds.), pp. 37–50, The Royal Society of Chemistry, Cambridge, U.K., 1995.

Dorn, H.-P., U. Brandenburger, T. Brauers, and M. Hausmann, "A New *In-Situ* Laser Long-Path Absorption Instrument for the Measurement of Tropospheric OH Radicals," *J. Atmos. Sci., 52,* 3373–3380 (1995a).

Dorn, H.-P., R. Neuroth, and A. Hofzumahaus, "Investigation of OH Absorption Cross Sections of Rotational Transitions in the $A^2\Sigma^+$, $v' = 0 \leftarrow X^2\Pi$, $v'' = 0$ Band under Atmospheric Conditions: Implications for Tropospheric Long-Path Absorption Measurements," *J. Geophys. Res., 100,* 7397–7409 (1995b).

Dorn, H.-P., U. Brandenburger, T. Brauers, M. Hausmann, and D. H. Ehhalt, "*In-Situ* Detection of Tropospheric OH Radicals by Folded Long-Path Laser Absorption. Results from the POPCORN Field Campaign in August 1994," *Geophys. Res. Lett., 23,* 2537–2540 (1996).

Dubey, M. K., T. F. Hanisco, P. O. Wennberg, and J. G. Anderson, "Monitoring Potential Photochemical Interference in Laser-Induced Fluorescence Measurements of Atmospheric OH," *Geophys. Res. Lett., 23,* 3215–3218 (1996).

Durham, J. L., T. G. Ellestad, L. Stockburger, K. T. Knapp, and L. L. Spiller, "A Transition-Flow Reactor Tube for Measuring Trace Gas Concentrations," *JAPCA, 36,* 1228–1232 (1986).

Durham, J. L., L. L. Spiller, and T. G. Ellestad, "Nitric Acid–Nitrate Aerosol Measurements by a Diffusion Denuder: A Performance Evaluation," *Atmos. Environ., 21,* 589–598 (1987).

Eatough, D. J., A. Wadsworth, D. A. Eatough, J. W. Crawford, L. D. Hansen, and E. A. Lewis, "A Multiple-System, Multi-Channel Diffusion Denuder Sampler for the Determination of Fine-Particulate Organic Material in the Atmosphere," *Atmos. Environ., 27A,* 1213–1219 (1993).

Eatough, D. J., L. J. Kewis, M. Eatough, and E. A. Lewis, "Sampling Artifacts in the Determination of Particulate Sulfate and $SO_2(g)$ in the Desert Southwest Using Filter Pack Samplers," *Environ. Sci. Technol., 29,* 787–791 (1995).

Edgerton, S. A., M. W. Holdren, D. L. Smith, and J. J. Shah, "Inter-urban Comparison of Ambient Volatile Organic Compound Concentrations in U.S. Cities," *JAPCA, 39,* 729–732 (1989).

Ehhalt, D. H., F. Rohrer, A. Wahner, M. J. Prather, and D. R. Blake, "On the Use of Hydrocarbons for the Determination of Tropospheric OH Concentrations," *J. Geophys. Res., 103,* 18981–18997 (1998).

Eisele, F. L., "Direct Tropospheric Ion Sampling and Mass Identification," *Int. J. Mass Spectrom. Ion Processes, 54,* 119–126 (1983).

Eisele, F. L., and D. J. Tanner, "Ion-Assisted Tropospheric OH Measurements," *J. Geophys. Res., 96,* 9295–9308 (1991).

Eisele, F. L., and J. D. Bradshaw, "The Elusive Hydroxyl Radical. Measuring OH in the Atmosphere," *Anal. Chem., 65,* 927A–939A (1993).

Eisele, F. L., G. H. Mount, F. C. Fehsenfeld, J. Hardner, E. Marovich, D. D. Parrish, J. Roberts, M. Trainer, and D. Tanner, "Intercomparison of Tropospheric OH and Ancillary Trace Gas Measurements at Fritz Peak Observatory, Colorado," *J. Geophys. Res., 99,* 18605–18626 (1994).

Eisele, F. L., D. J. Tanner, C. A. Cantrell, and J. G. Calvert, "Measurements and Steady State Calculations of OH Concentrations at Mauna Loa Observatory," *J. Geophys. Res., 101,* 14665–14679 (1996).

Eisele, F. L., R. L. Mauldin III, D. J. Tanner, J. R. Fox, T. Mouch, and T. Scully, "An Inlet/Sampling Duct for Airborne OH and Sulfuric Acid Measurements," *J. Geophys. Res., 102,* 27993–28001 (1997).

Eisele, F. L., and P. H. McMurry, "Recent Progress in Understanding Particle Nucleation and Growth," *Philos. Trans. R. Soc. London B, 352,* 191–201 (1997).

Eldering, A., and R. M. Glasgow, "Short-Term Particulate Matter Mass and Aerosol-Size Distribution Measurements: Transient Pollution Episodes and Bimodal Aerosol-Mass Distributions," *Atmos. Environ., 32,* 2017–2024 (1998).

Emmons, L. K., M. A. Carroll, D. A. Hauglustaine, G. P. Brasseur, C. Atherton, J. Penner, S. Sillman, H. Levy II, F. Rohrer, W. M. F. Wauben, P. F. J. Van Velthoven, Y. Wang, D. Jacob, P. Bakwin, R. Dickerson, B. Doddridge, C. Gerbig, R. Honrath, G. Hübler, D. Jafe, Y. Kondo, J. W. Munger, A. Torres, and A. Volz-Thomas, "Climatologies of NO_x and NO_y: A Comparison of Data and Models," *Atmos. Environ., 31,* 1851–1904 (1997).

Environmental Science & Technology/News, "European News," *Environ. Sci. Technol., 31,* 557A (1997).

Erle, F., K. Pfeilsticker, and U. Platt, "On the Influence of Tropospheric Clouds on Zenith-Scattered-Light Measurements of Stratospheric Species," *Geophys. Res. Lett., 22,* 2725–2728 (1995).

Evans, G. F., T. A. Lumpkin, D. L. Smith, and M. C. Somerville, "Measurements of VOCs from the TAMS Network," *J. Air Waste Manage. Assoc., 42,* 1319–1323 (1992).

Fahey, D. W., G. Hübler, D. D. Parrish, E. J. Williams, R. B. Norton, B. A. Ridley, H. B. Singh, S. C. Liu, and F. C. Fehsenfeld, "Reactive Nitrogen Species in the Troposphere: Measurements of NO, NO_2, HNO_3, Particulate Nitrate, Peroxyacetyl Nitrate (PAN), O_3, and Total Reactive Odd Nitrogen (NO_y) at Niwot Ridge, Colorado," *J. Geophys. Res., 91,* 9781–9793 (1986).

Fan, Q., and P. K. Dasgupta, "Continuous Automated Determination of Atmospheric Formaldehyde at the Parts per Trillion Level," *Anal. Chem., 66,* 551–556 (1994).

Faude, F., and J. Goschnick, "XPS, SIMS, and SNMS Applied to a Combined Analysis of Aerosol Particles from a Region of Considerable Air Pollution in the Upper Rhine Valley," *Fresenius' J. Anal. Chem., 358,* 67–72 (1997).

Febo, R., C. Perrino, and M. Cortiello, "A Denuder Technique for the Measurement of Nitrous Acid in Urban Atmospheres," *Atmos. Environ., 27A,* 1721–1728 (1993).

Febo, A., C. Perrino, M. Gherardi, and R. Sparapani, "Evaluation of a High-Purity and High-Stability Continuous Generation System for Nitrous Acid," *Environ. Sci. Technol., 29,* 2390–2395 (1995).

Febo, R., and C. Perrino, "Measurement of High Concentrations of Nitrous Acid inside Automobiles," *Atmos. Environ., 29*, 345–351 (1995).

Febo, A., C. Perrino, and I. Allegrini, "Measurement of Nitrous Acid in Milan, Italy, by DOAS and Diffusion Denuders," *Atmos. Environ., 30*, 3599–3609 (1996).

Fehsenfeld, F. C., R. R. Dickerson, G. Hübler, W. T. Luke, L. J. Nunnermacker, E. J. Williams, J. M. Roberts, J. G. Calvert, C. M. Curran, A. C. Delany, C. S. Eubank, D. W. Fahey, A. Fried, B. W. Gandrud, A. O. Langford, P. C. Murphy, R. B. Norton, K. E. Pickering, and B. A. Ridley, "A Ground-Based Intercomparison of NO, NO_x, and NO_y Measurement Techniques," *J. Geophys. Res., 92*, 14710–14722 (1987).

Fehsenfeld, F. C., J. W. Drummond, U. K. Roychowdhury, P. J. Galvin, E. J. Williams, M. P. Buhr, D. D. Parrish, G. Hübler, A. O. Langford, J. G. Calvert, B. A. Ridley, F. Grahek, B. G. Heikes, G. L. Kok, J. D. Shetter, J. G. Walega, C. M. Elsworth, R. B. Norton, D. W. Fahey, P. C. Murphy, C. Hovermale, V. A. Mohnen, K. L. Demerjian, G. I. Mackay, and H. I. Schiff, "Intercomparison of NO_2 Measurement Techniques," *J. Geophys. Res., 95*, 3579–3597 (1990).

Fehsenfeld, F. C., L. G. Huey, D. T. Sueper, R. B. Norton, E. J. Williams, F. L. Eisele, R. L. Mauldin III, and D. J. Tanner, "Ground-Based Intercomparison of Nitric Acid Measurement Techniques," *J. Geophys. Res., 103*, 3343–3353 (1998).

Fels, M., and W. Junkermann, "The Occurrence of Organic Peroxides in Air at a Mountain Site," *Geophys. Res. Lett., 21*, 341–344 (1994).

Felton, C. C., J. C. Sheppard, and M. J. Campbell, "The Radiochemical Hydroxyl Radical Measurement Method," *Environ. Sci. Technol., 24*, 1841–1847 (1990).

Felton, C. C., J. C. Sheppard, and M. J. Campbell, "Precision of the Radiochemical OH Measurement Method," *Atmos. Environ., 26*, 2105–2109 (1992).

Ferek, R. J., and D. A. Hegg, "Measurements of Dimethyl Sulfide and SO_2 during GTE/CITE 3," *J. Geophys. Res., 98*, 23435–23442 (1993).

Ferm, M., "Method for Determination of Atmospheric Ammonia," *Atmos. Environ., 13*, 1385–1393 (1979).

Ferm, M., and A. Sjödin, "A Sodium Carbonate Denuder for Determination of Nitrous Acid in the Atmosphere," *Atmos. Environ., 19*, 979–983 (1985).

Ferm, M., "A Na_2CO_3-Coated Denuder and Filter for Determination of Gaseous HNO_3 and Particulate NO_3^- in the Atmosphere," *Atmos. Environ., 20*, 1193–1201 (1986).

Finlayson-Pitts, B. J., and J. N. Pitts, Jr., *Atmospheric Chemistry: Fundamentals and Experimental Techniques*, Wiley, New York, 1986.

Fischer, H., C. Nikitas, U. Parchatka, T. Zenker, G. W. Harris, P. Matuska, R. Schmitt, D. Mihelcic, P. Muesgen, H.-W. Paetz, M. Schultz, and A. Volz-Thomas, "Trace Gas Measurements during the Oxidizing Capacity of the Tropospheric Atmosphere Campaign 1993 at Izana," *J. Geophys. Res., 103*, 13505–13518 (1998).

Fishman, J., C. E. Watson, J. C. Larsen, and J. A. Logan, "Distribution of Tropospheric Ozone Determined from Satellite Data," *J. Geophys. Res., 95*, 3599–3617 (1990).

Fitz, D. R., G. J. Doyle, and J. N. Pitts, Jr., "An Ultrahigh Volume Sampler for the Multiple Filter Collection of Respirable Particulate Matter," *J. Air Pollut. Control Assoc., 33*, 877–879 (1983).

Flagan, R. C., "History of Electrical Aerosol Measurements," *Aerosol Sci. Technol., 28*, 301–380 (1998).

Flocke, F., A. Volz-Thomas, H.-J. Buers, W. Pätz, H.-J. Garthe, and D. Kley, "Long-Term Measurements of Alkyl Nitrates in Southern Germany. 1. General Behavior and Seasonal and Diurnal Variation," *J. Geophys. Res., 103*, 5729–5746 (1998).

Fox, D. L., L. Stockburger, W. Weathers, C. W. Spicer, G. I. Mackay, H. I. Schiff, D. J. Eatough, F. Mortensen, L. D. Hansen, P. B. Shepson, T. E. Kleindienst, and E. O. Edney, "Intercomparison of Nitric Acid Diffusion Denuder Methods with Tunable Diode Laser Absorption Spectroscopy," *Atmos. Environ., 22*, 575–585 (1988).

Francois, F., W. Maenhaut, J.-L. Colin, R. Losno, M. Schulz, T. Stahlschmidt, L. Spokes, and T. Jickells, "Intercomparison of Elemental Concentrations in Total and Size-Fractionated Aerosol Samples Collected during the Mace Head Experiment, April 1991," *Atmos. Environ., 29*, 837–849 (1995).

Fraser, M. P., G. R. Cass, B. R. T. Simoneit, and R. A. Rasmussen, "Air Quality Model Evaluation Data for Organics. 5. C_6–C_{22} Nonpolar and Semipolar Aromatic Compounds," *Environ. Sci. Technol., 32*, 1760–1770 (1998).

Fried, A., S. Sewell, B. Henry, B. P. Wert, T. Gilpin, and J. R. Drummond, "Tunable Diode Laser Absorption Spectrometer for Ground-Based Measurements of Formaldehyde," *J. Geophys. Res., 102*, 6253–6266 (1997).

Friedbacher, G., M. Grasserbauer, Y. Meslmani, N. Klaus, and M. J. Higatsberger, "Investigation of Environmental Aerosol by Atomic Force Microscopy," *Anal. Chem., 67*, 1749–1754 (1995).

Fuchs, N. A., and A. G. Sutugin, *Highly Dispersed Aerosols*, Ann Arbor Science Publishers, Ann Arbor, MI, 1970.

Fung, K., "Particulate Carbon Speciation by MnO_2 Oxidation," *Aerosol Sci. Technol., 12*, 122–127 (1990).

Fung, K. H., and I. N. Tang, "Relative Raman Scattering Cross-Section Measurements with Suspended Particles," *Appl. Spectrosc., 45*, 734–737 (1991).

Fung, K. H., and I. N. Tang, "Analysis of Aerosol Particles by Resonance Raman Scattering Technique," *Appl. Spectrosc., 46*, 159–162 (1992a).

Fung, K. H., and I. N. Tang, "Aerosol Particle Analysis by Resonance Raman Spectroscopy," *J. Aerosol Sci., 23*, 301–307 (1992b).

Fung, K. H., D. G. Imre, and I. N. Tang, "Detection Limits for Sulfates and Nitrates in Aerosol Particles by Raman Spectroscopy," *J. Aerosol Sci., 25*, 479–485 (1994).

Gaffney, J. S., N. A. Marley, and E. W. Prestbo, "Peroxyacyl Nitrates (PANs): Their Physical and Chemical Properties," in *The Handbook of Environmental Chemistry* (O. Hutzinger, Ed.), Vol. 4, Part B, pp. 4–38, Springer-Verlag, Berlin, 1989.

Gaffney, J. S., N. A. Marley, and E. W. Prestbo, "Measurements of Peroxyacetyl Nitrate at a Remote Site in the Southwestern United States: Tropospheric Implications," *Environ. Sci. Technol., 27*, 1905–1910 (1993).

Gaffney, J. S., N. A. Marley, R. S. Martin, R. W. Dixon, L. G. Reyes, and C. J. Popp, "Potential Air Quality Effects of Using Ethanol-Gasoline Fuel Blends: A Field Study in Albuquerque, New Mexico," *Environ. Sci. Technol., 31*, 3053–3061 (1997).

Gaffney, J. S., R. M. Bornick, Y.-H. Chen, and N. A. Marley, "Capillary Gas Chromatographic Analysis of Nitrogen Dioxide and PANs with Luminol Chemiluminescent Detection," *Atmos. Environ., 32*, 1445–1454 (1998).

Gaffney, J. S., N. A. Marley, and P. V. Doskey, "Peroxyacetyl Nitrate and Hydrocarbon Measurements in Mexico City," Invited Paper, Special Session on Mexico City Air Quality, Spring Meeting of the American Geophysical Union (AGU), Boston, Massachusetts, May 25–28, 1998, *Atmos. Environ.*, in press (1999).

Ganor, E., Z. Levin, and R. Van Griekens, "Composition of Individual Aerosol Particles above the Israelian Mediterranean Coast during the Summer Time," *Atmos. Environ., 32*, 1631–1642 (1998).

Gao, R. S., E. R. Keim, E. L. Woodbridge, S. J. Ciciora, M. H. Proffitt, T. L. Thompson, R. J. Mclaughlin, and D. W. Fahey, "New Photolysis System for NO_2 Measurements in the Lower Stratosphere," *J. Geophys. Res., 99,* 20673–20681 (1994).

Gard, E., M. J. Kleeman, D. S. Gross, L. S. Hughes, J. O. Allen, B. D. Morrical, D. P. Fergenson, T. Dienes, M. E. Gälli, R. J. Johnson, G. R. Cass, and K. A. Prather, "Direct Observation of Heterogeneous Chemistry in the Atmosphere," *Science, 279,* 1184–1187 (1998).

Gard, E., J. E. Mayer, B. D. Morrical, T. Dienes, D. P. Fergenson, and K. A. Prather, "Real-Time Analysis of Individual Atmospheric Aerosol Particles: Design and Performance of a Portable ATOFMS," *Anal. Chem., 69,* 4083–4091 (1997).

Ge, Z., A. S. Wexler, and M. V. Johnston, "Multicomponent Aerosol Crystallization," *J. Colloid Interface Sci., 183,* 68–77 (1996).

Ge, Z., A. S. Wexler, and M. V. Johnston, "Deliquescence Behavior of Multicomponent Aerosols," *J. Phys. Chem. A, 102,* 173–180 (1998a).

Ge, Z., A. S. Wexler, and M. V. Johnston, "Laser Desorption/Ionization of Single Ultrafine Multicomponent Aerosols," *Environ. Sci. Technol., 32,* 3218–3223 (1998b).

Genfa, Z., P. K. Dasgupta, and S. Dong, "Measurement of Atmospheric Ammonia," *Environ. Sci. Technol., 23,* 1467–1474 (1989).

Gholson, A. R., R. K. M. Jayanty, and J. F. Storm, "Evaluation of Aluminum Canisters for the Collection and Storage of Air Toxics," *Anal. Chem., 62,* 1899–1902 (1990).

Gilpin, T., E. Apel, A. Fried, B. Wert, J. Calvert, Z. Genfa, P. Dasgupta, J. W. Harder, B. Heikes, B. Hopkins, H. Westberg, T. Kleindienst, Y.-N. Lee, X. Zhou, W. Lonneman, and S. Sewell, "Intercomparison of Six Ambient $[CH_2O]$ Measurement Techniques," *J. Geophys. Res., 102,* 21161–21188 (1997).

Goldan, P. D., W. C. Kuster, F. C. Fehsenfeld, and S. A. Montzka, "Hydrocarbon Measurements in the Southeastern United States: The Rural Oxidants in the Southern Environment (ROSE) Program 1990," *J. Geophys. Res., 100,* 25945–25963 (1995).

Gong, Q., and K. L. Demerjian, "Hydrocarbon Losses on a Regenerated Nafion Dryer," *J. Air Waste Manage. Assoc., 45,* 490–493 (1995).

Gong, Q., and K. L. Demerjian, "Measurement and Analysis of C_2–C_{10} Hydrocarbons at Whiteface Mountain, New York," *J. Geophys. Res., 102,* 28059–28069 (1997).

Goschnick, J., J. Schuricht, and H. J. Ache, "Calibration of Depth Profiles of Microparticles Measured with Plasma-Based Secondary Neutral Mass Spectrometry," *Fresenius' J. Anal. Chem., 349,* 203–205 (1994a).

Goschnick, J., J. Schuricht, and H. J. Ache, "Depth-Structure of Airborne Microparticles Sampled Downwind from the City of Karlsruhe in the River Rhine Valley," *Fresenius' J. Anal. Chem., 350,* 426–430 (1994b).

Granby, K., C. S. Christensen, and C. Lohse, "Urban and Semi-Rural Observations of Carboxylic Acids and Carbonyls," *Atmos. Environ., 31,* 1403–1415 (1997a).

Granby, K., A. H. Egelov, T. Nielsen, and C. Lohse, "Carboxylic Acids: Seasonal Variation and Relation to Chemical and Meteorological Parameters," *J. Atmos. Chem., 28,* 195–207 (1997b).

Greenberg, J. P., D. Helmig, and P. R. Zimmerman, "Seasonal Measurements of Nonmethane Hydrocarbons and Carbon Monoxide at the Mauna Loa Observatory during the Mauna Loa Observatory Photochemistry Experiment 2," *J. Geophys. Res., 101,* 14581–14598 (1996).

Gregory, G. L., J. M. Hoell, Jr., M. A. Carroll, B. A. Ridley, D. D. Davis, J. Bradshaw, M. O. Rodgers, S. T. Sandholm, H. I. Schiff, D. R. Hastie, D. R. Karecki, G. I. Mackay, G. W. Harris, A. L. Torres, and A. Fried, "An Intercomparison of Airborne Nitrogen Dioxide Instruments," *J. Geophys. Res., 95,* 10103–10127 (1990a).

Gregory, G. L., J. M. Hoell, Jr., B. J. Huebert, S. E. Van Bramer, P. J. LeBel, S. A. Vay, R. M. Marinaro, H. I. Schiff, D. R. Hastie, G. I. Mackay, and D. R. Karecki, "An Intercomparison of Airborne Nitric Acid Measurements," *J. Geophys. Res., 95,* 10089–10102 (1990b).

Gregory, G. L., J. M. Hoell, Jr., A. L. Torres, M. A. Carroll, B. A. Ridley, M. O. Rodgers, J. Bradshaw, S. Sandholm, and D. D. Davis, "An Intercomparison of Airborne Nitric Oxide Measurements: A Second Opportunity," *J. Geophys. Res., 95,* 10129–10138 (1990c).

Gregory, G. L., D. D. Davis, N. Beltz, A. R. Bandy, R. J. Ferek, and D. C. Thornton, "An Intercomparison of Aircraft Instrumentation for Tropospheric Measurements of Sulfur Dioxide," *J. Geophys. Res., 98,* 23325–23352 (1993).

Griffith, D. W. T., and G. Schuster, "Atmospheric Trace Gas Analysis Using Matrix Isolation-Fourier Transform Infrared Spectroscopy," *J. Atmos. Chem., 5,* 59–81 (1987).

Griffith, D. W. T., "Matrix Isolation Spectroscopy in Atmospheric Chemistry," in *Air Monitoring by Spectroscopic Techniques* (M. W. Sigrist, Ed.), *Chemical Analysis Series*, Vol. 127, Chap. 7, pp. 471–514, Wiley, New York, 1994.

Griffiths, P. R., and J. A. de Haseth, *Fourier Transform Infrared Spectrometry*, Wiley, New York, 1986.

Grisar, R., H. Bottner, M. Tacke, and G. Restelli, Eds., *Monitoring of Gaseous Pollutants by Tunable Diode Lasers*, Proceedings of the International Symposium, Freiburg, Germany, October 17–18, 1991, Kluwer Academic, Dordrecht/Norwell, MA, 1992.

Grosjean, D., and K. Fung, "Collection Efficiencies of Cartridges and Microimpingers for Sampling of Aldehydes in Air as 2,4- Dinitrophenylhydrazones," *Anal. Chem., 54,* 1221–1224 (1982).

Grosjean, D., "Liquid Chromatography Analysis of Chloride and Nitrate with 'Negative' Ultraviolet Detection: Ambient Levels and Relative Abundance of Gas-Phase Inorganic and Organic Acids in Southern California," *Environ. Sci. Technol., 24,* 77–81 (1990).

Grosjean, D., and S. S. Parmar, "Interferences from Aldehydes and Peroxyacetyl Nitrate When Sampling Urban Air Organic Acids on Alkaline Traps," *Environ. Sci. Technol., 24,* 1021–1026 (1990).

Grosjean, D., E. C. Tuazon, and E. Fujita, "Ambient Formic Acid in Southern California Air: A Comparison of Two Methods, Fourier Transform Infrared Spectroscopy and Alkaline Trap-Liquid Chromatography with UV Detection," *Environ. Sci. Technol., 24,* 144–146 (1990).

Grosjean, D., and E. L. Williams, II, "A Passive Sampler for Airborne Formaldehyde," *Atmos. Environ., 26A,* 2923–2928 (1992).

Grosjean, D., E. L. Williams, II, and E. Grosjean, "Peroxyacyl Nitrates at Southern California Mountain Forest Locations," *Environ. Sci. Technol., 27,* 110–121 (1993a).

Grosjean, D., E. L. Williams, II, and E. Grosjean, "Ambient Levels of Peroxy-*n*-butyryl Nitrate at a Southern California Mountain Forest Smog Receptor Location," *Environ. Sci. Technol., 27,* 326–331 (1993b).

Grosjean, E., D. Grosjean, M. P. Fraser, and G. R. Cass, "Air Quality Model Evaluation Data for Organics. 2. C_1–C_{14} Carbonyls in Los Angeles Air," *Environ. Sci. Technol., 30,* 2687–2703 (1996a).

Grosjean, E., D. Grosjean, M. P. Fraser, and G. R. Cass, "Air Quality Model Evaluation Data for Organics. 3. Peroxyacetyl Nitrate and Peroxypropionyl Nitrate in Los Angeles Air," *Environ. Sci. Technol., 30,* 2704–2714 (1996b).

Grosjean, E., D. Grosjean, R. Gunawardena, and R. A. Rasmussen, "Ambient Concentrations of Ethanol and Methyl *tert*-Butyl Ether in Porto Alegre, Brazil, March 1996–April 1997," *Environ. Sci. Technol., 32,* 736–742 (1998a).

Grosjean, E., D. Grosjean, and R. A. Rasmussen, "Ambient Concentrations, Sources, Emission Rates, and Photochemical Reactivity of C_2-C_{10} Hydrocarbons in Porto Alegre, Brazil," *Environ. Sci. Technol.*, 32, 2061–2069 (1998b).

Grosjean, E., R. A. Rasmussen, and D. Grosjean, "Ambient Levels of Gas Phase Pollutants in Porto Alegre, Brazil," *Atmos. Environ.*, 32, 3371–3379 (1998c).

Guizard, S., D. Chapoulard, M. Horani, and D. Gauyacq, "Detection of NO Traces Using Resonantly Enhanced Multiphoton Ionization: A Method for Monitoring Atmospheric Pollutants," *Appl Phys. B*, 48, 471–477 (1989).

Gunz, D. W., and M. R. Hoffmann, "Atmospheric Chemistry of Peroxides: A Review," *Atmos. Environ.*, 24A, 1601–1633 (1990).

Haan, D., P. Martinerie, and D. Raynaud, "Ice Core Data of Atmospheric Carbon Monoxide over Antarctica and Greenland during the Last 200 Years," *Geophys. Res. Lett.*, 23, 2235–2238 (1996).

Hamm, S., G. Helas, and P. Warneck, "Acetonitrile in the Air over Europe," *Geophys. Res. Lett.*, 16, 483–486 (1989).

Hansen, A. B., H. Skov, and T. Nielsen, "Comparison of Volatile Non-Methane Hydrocarbon Concentrations and Profiles at Different Locations: Open Land, Urban Background, and Street," personal communication (1998).

Hansen, K. J., B. N. Hansen, E. Cravens, and R. E. Sievers, "Supercritical Fluid Extraction—Gas Chromatographic Analysis of Organic Compounds in Atmospheric Aerosols," *Anal. Chem.*, 67, 3541–3549 (1995).

Hanst, P. L., "Spectroscopic Methods for Air Pollution Measurement," *Adv. Environ. Sci. Technol.*, 2, 91–213 (1971).

Hanst, P. L., and S. T. Hanst, "Gas Measurement in the Fundamental Infrared Region," in *Air Monitoring by Spectroscopic Techniques* (M. W. Sigrist, Ed.), *Chemical Analysis Series*, Vol. 127, Chap. 6, pp. 335–470, Wiley, New York, 1994.

Hara, K., T. Kikuchi, K. Furuya, M. Hayashi, and Y. Fujii, "Characterization of Antarctic Aerosol Particles Using Laser Microprobe Mass Spectrometry," *Environ. Sci. Technol.*, 30, 385–391 (1996).

Hard, T. M., R. J. O'Brien, C. Y. Chan, and A. A. Mehrabzadeh, "Tropospheric Free Radical Determination by FAGE," *Environ. Sci. Technol.*, 18, 768–777 (1984).

Hard, T. M., C. Y. Chan, A. A. Mehrabzadeh, and R. J. O'Brien, "Diurnal HO_2 Cycles at Clean Air and Urban Sites in the Troposphere," *J. Geophys. Res.*, 97, 9785–9794 (1992a).

Hard, T. M., A. A. Mehrabzadeh, C. Y. Chan, and R. J. O'Brien, "FAGE Measurements of Tropospheric HO with Measurements and Model of Interferences," *J. Geophys. Res.*, 97, 9795–9817 (1992b).

Hard, T. M., L. A. George, and R. J. O'Brien, "FAGE Determination of Tropospheric HO and HO_2," *J. Atmos. Sci.*, 52, 3354–3372 (1995).

Harder, J. W., R. O. Jakoubek, and G. H. Mount, "Measurement of Tropospheric Trace Gases by Long-Path Differential Absorption Spectroscopy during the 1993 OH Photochemistry Experiment," *J. Geophys. Res.*, 102, 6215–6226 (1997a).

Harder, J. W., A. Fried, S. Sewell, and B. Henry, "Comparison of Tunable Diode Laser and Long-Path Ultraviolet-Visible Spectroscopic Measurements of Ambient Formaldehyde Concentrations during the 1993 OH Photochemistry Experiment," *J. Geophys. Res.*, 102, 6267–6282 (1997b).

Harris, G. W., W. P. L. Carter, A. M. Winer, J. N. Pitts, Jr., U. Platt, and D. Perner, "Observations of Nitrous Acid in the Los Angeles Atmosphere and Implications for Predictions of Ozone-Precursor Relationships," *Environ. Sci. Technol.*, 16, 414–419 (1982).

Harrison, R. M., J. D. Peak, and G. M. Collins, "Tropospheric Cycle of Nitrous Acid," *J. Geophys. Res.*, 101, 14429–14439 (1996).

Harrison, R. M., M. Jones, and G. Collins, "Measurements of the Physical Properties of Particles in the Urban Atmosphere," *Atmos. Environ.*, 33, 309–321 (1999).

Hartsell, B. E., V. P. Aneja, and W. A. Lonneman, "Relationships between Peroxyacetyl Nitrate, O_3, and NO_y at the Rural Southern Oxidants Study Site in Central Piedmont, North Carolina, Site SONIA," *J. Geophys. Res.*, 99, 21033–21041 (1994).

Hastie, D. R., G. I. Mackay, T. Iguchi, B. A. Ridley, and H. I. Schiff, "Tunable Diode Laser Systems for Measuring Trace Gases in Tropospheric Air," *Environ. Sci. Technol.*, 17, 352A–364A (1983).

Hastie, D. R., P. B. Shepson, N. Reid, P. B. Roussel, and O. T. Melo, "Summertime NO_x, NO_y, and Ozone at a Site in Rural Ontario," *Atmos. Environ.*, 30, 2157–2165 (1996).

Hebestreit, K., J. Stutz, D. Rosen, V. Matveiv, M. Peleg, M. Luria, and U. Platt, "DOAS Measurements of Tropospheric Bromine Oxide in Mid-Latitudes," *Science*, 283, 55–57 (1999).

Heikes, B. G., "Aqueous H_2O_2 Production from O_3 in Glass Impingers," *Atmos. Environ.*, 18, 1433–1445 (1984).

Heikes, B., M. Lee, D. Jacob, R. Talbot, J. Bradshaw, H. Singh, D. Blake, B. Anderson, H. Fuelberg, and A. M. Thompson, "Ozone, Hydroperoxides, Oxides of Nitrogen, and Hydrocarbon Budgets in the Marine Boundary Layer over the South Atlantic," *J. Geophys. Res.*, 101, 24221–24234 (1996).

Heintz, F., U. Platt, H. Flentje, and R. Dubois, "Long-Term Observation of Nitrate Radicals at the Tor Station, Kap Arkona (Rügen)," *J. Geophys. Res.*, 101, 22891–22910 (1996).

Heitmann, H., and F. Arnold, "Composition Measurements of Tropospheric Ions," *Nature*, 306, 747–751 (1983).

Helmig, D., and J. P. Greenberg, "Automated *in situ* Gas Chromatographic–Mass Spectrometric Analysis of ppt Level Volatile Organic Trace Gases Using Multistage Solid-Adsorbent Trapping," *J. Chromatogr. A*, 677, 123–132 (1994).

Helmig, D., and L. Vierling, "Water Adsorption Capacity of the Solid Adsorbents Tenax TA, Tenax GR, Carbotrap, Carbotrap C, Carbosieve SIII, and Carboxen 569 and Water Management Techniques for the Atmospheric Sampling of Volatile Organic Trace Gases," *Anal. Chem.*, 67, 4380–4386 (1995).

Helmig, D., "Artifact-Free Preparation, Storage, and Analysis of Solid Adsorbent Sampling Cartridges Used in the Analysis of Volatile Organic Compounds in Air," *J. Chromatogr. A*, 732, 414–417 (1996).

Helmig, D., "Ozone Removal Techniques in the Sampling of Atmospheric Volatile Organic Trace Gases," *Atmos. Environ.*, 31, 3635–3651 (1997).

Hering, S. V., R. C. Flagan, and S. K. Friedlander, "Design and Evaluation of New Low-Pressure Impactor. I.," *Environ. Sci. Technol.*, 12, 667–673 (1978).

Hering, S. V., S. K. Friedlander, J. J. Collins, and L. W. Richards, "Design and Evaluation of a New Low-Pressure Impactor. 2.," *Environ. Sci. Technol.*, 13, 184–188 (1979).

Hering, S. V., D. R. Lawson, I. Allegrini, A. Febo, C. Perrino, M. Possanzini, J. E. Sickles, II, K. G. Anlauf, A. Wiebe, B. R. Appel, W. John, J. Ondo, S. Wall, R. S. Braman, R. Sutton, G. R. Cass, P. A. Solomon, D. J. Eatough, N. L. Eatough, E. C. Ellis, D. Grosjean, B. B. Hicks, J. D. Womack, J. Horrocks, K. T. Knapp, T. G. Ellestad, R. J. Paur, W. J. Mitchell, M. Pleasant, E. Peake, A. MacLean, W. R. Pierson, W. Brachaczek, H. I. Schiff, G. I. Mackay, C. W. Spicer, D. H. Stedman, A. M. Winer, H. W. Biermann, and E. C. Tuazon, "The Nitric Acid Shootout: Field Comparison of Measurement Methods," *Atmos. Environ.*, 22, 1519–1539 (1988).

Herriott, D., H. Kogelnik, and R. Kompfner, "Off-Axis Paths in Spherical Mirror Interferometers," *Appl. Opt.*, 3, 523–526 (1964).

Herriott, D. R., and H. J. Schulte, "Folded Optical Delay Lines," *Appl. Opt*, 4, 883–889 (1965).

Herzberg, G., *Molecular Spectra and Molecular Structure*, 2nd ed., Van Nostrand–Reinhold, New York, 1950.

Hewitt, C. N., and G. L. Kok, "Formation and Occurrence of Organic Hydroperoxides in the Troposphere: Laboratory and Field Observations," *J. Atmos. Chem., 12,* 181–194 (1991).

Hewitt, C. N., X.-L. Cao, C. Boissard, and S. C. Duckham, "Atmospheric VOCs from Natural Sources," in *Volatile Organic Compounds in the Atmosphere, Issues in Environmental Science and Technology* (R. E. Hester and R. M. Harrison, Eds.), pp. 17–36, The Royal Society of Chemistry, Cambridge, UK, 1995.

Hinds, W. C., *Aerosol Technology,* Wiley, New York, 1982.

Hinz, K.-P., R. Kaufmann, and B. Spengler, "Laser-Induced Mass Analysis of Single Particles in the Airborne State," *Anal. Chem., 66,* 2071–2076 (1994).

Hinz, K.-P., R. Kaufmann, and B. Spengler, "Simultaneous Detection of Positive and Negative Ions from Single Airborne Particles by Real-Time Laser Mass Spectrometry," *Aerosol Sci. Technol., 24,* 233–242 (1996).

Hippler, M., and J. Pfab, "Detection and Probing of Nitric Oxide (NO) by Two-Colour Laser Photoionisation (REMPI) Spectroscopy on the A ← X Transition," *Chem. Phys. Lett., 243,* 500–505 (1995).

Hisham, M. W. M., and D. Grosjean, "Sampling of Atmospheric Nitrogen Dioxide Using Triethanolamine: Interference from Peroxyacetyl Nitrate," *Atmos. Environ., 24,* 2523–2525 (1990).

Ho, M. H., and R. A. Richards, "Enzymatic Method for the Determination of Formaldehyde," *Environ. Sci. Technol., 24,* 201–204 (1990).

Hoek, G., B. Forsberg, M. Borowska, S. Hlawiczka, E. Vaskövi, H. Welinder, M. Branis, I. Benes, F. Kotesovec, L. O. Hagen, J. Cyrys, M. Jantunen, W. Roemer, and B. Brunekreef, "Wintertime PM10 and Black Smoke Concentrations across Europe: Results from the Peace Study," *Atmos. Environ., 31,* 3609–3622 (1997).

Hoell, J. M., Jr., G. L. Gregory, D. S. McDougal, G. W. Sachse, and G. F. Hill, "An Intercomparison of Carbon Monoxide Measurement Techniques," *J. Geophys. Res., 90,* 12881–12889 (1985).

Hoell, J. M., Jr., G. L. Gregory, D. S. McDougal, A. L. Torres, D. D. Davis, J. Bradshaw, M. O. Rodgers, B. A. Ridley, and M. A. Carroll, "Airborne Intercomparison of Nitric Oxide Measurement Techniques," *J. Geophys. Res., 92,* 1995–2008 (1987a).

Hoell, J. M., Jr., G. L. Gregory, D. S. McDougal, G. W. Sachse, G. F. Hill, E. P. Condon, and R. A. Rasmussen, "Airborne Intercomparison of Carbon Monoxide Measurement Techniques," *J. Geophys. Res., 92,* 2009–2019 (1987b).

Hoff, R. M., L. Guise-Bagley, R. M. Staebler, H. A. Wiebe, J. Brook, B. Georgi, and T. Düsterdiek, "Lidar, Nephelometer, and *in Situ* Aerosol Experiments in Southern Ontario," *J. Geophys. Res., 101,* 19199–19209 (1996).

Hoffer, E. M., E. L. Kothny, and B. R. Appel, "Simple Method for Microgram Amounts of Sulfate in Atmospheric Particulates," *Atmos. Environ., 13,* 303–306 (1979).

Hofzumahaus, A., U. Aschmutat, M. Heßling, F. Holland, and D. H. Ehhalt, "The Measurement of Tropospheric OH Radicals by Laser-Induced Fluorescence Spectroscopy during the POPCORN Field Campaign," *Geophys. Res. Lett., 23,* 2541–2544 (1996).

Hofzumahaus, A., T. Brauers, U. Aschmutat, U. Brandenburger, H.-P. Dorn, M. Hausmann, M. Heßling, F. Holland, C. Plass-Dülmer, M. Sedlacek, M. Weber, and D. H. Ehhalt, "Reply" to paper by Brauers *et al.*, 1996, *Geophys. Res. Lett., 24,* 3039–3040 (1997).

Hofzumahaus, A., U. Aschmutat, U. Brandenburger, T. Brauers, H.-P. Dorn, M. Hausmann, M. Heßling, F. Holland, C. Plass-Dülmer, and D. H. Ehhalt, "Intercomparison of Tropospheric OH Measurements by Different Laser Techniques during the POPCORN Campaign 1994," *J. Atmos. Chem., 31,* 227–246 (1998).

Holland, F., M. Hessling, and A. Hofzumahaus, "*In-Situ* Measurement of Tropospheric OH Radicals by Laser-Induced Fluorescence—A Description of the KFA Instrument," *J. Atmos. Sci., 52,* 3393–3401 (1995).

Holland, F., U. Aschmutat, M. Heßling, A. Hofzumahaus, and D. H. Ehhalt, "Highly Time Resolved Measurements of OH during POPCORN Using Laser-Induced Fluorescence Spectroscopy," in *Atmospheric Measurements during POPCORN—Characterization of the Photochemistry over a Rural Area,* pp. 205–225, Kluwer Academic, Dordrecht/Norwell, MA, 1998.

Horn, D., and G. C. Pimentel, "2.5-km Low-Temperature Multiple-Reflection Cell," *Appl. Opt., 10,* 1892–1898 (1971).

Horvath, H., "Experimental Calibration for Aerosol Light Absorption Measurements Using the Integrating Plate Method—Summary of the Data," *Aerosol Sci., 28,* 1149–1161 (1997).

Hu, J., and D. H. Stedman, "Atmospheric RO_x Radicals at an Urban Site: Comparison to a Simple Theoretical Model," *Environ. Sci. Technol., 29,* 1655–1659 (1995).

Hübler, G., D. Perner, U. Platt, A. Toennissen, and D. H. Ehhalt, "Ground-Level OH Radical Concentration: New Measurements by Optical Absorption," *J. Geophys. Res., 89,* 1309–1319 (1984).

Huebert, B. J., and A. L. Lazrus, "Tropospheric Gas-Phase and Particulate Nitrate Measurements," *J. Geophys. Res., 85,* 7322–7328 (1980).

Huebert, B. J., S. E. Vanbramer, P. J. Lebel, S. A. Vay, A. L. Torres, H. I. Schiff, D. Hastie, G. Hubler, J. D. Bradshaw, M. A. Carroll, D. D. Davis, B. A. Ridley, M. O. Rodgers, S. T. Sandholm, and S. Dorris, "Measurements of Nitric Acid to NO_x Ratio in the Troposphere," *J. Geophys. Res., 95,* 10193–10198 (1990).

Huey, L. G., and E. R. Lovejoy, "Reactions of SiF_5^- with Atmospheric Trace Gases: Ion Chemistry for Chemical Ionization Detection of HNO_3 in the Troposphere," *Int. J. Mass Spectrom. Ion Processes, 155,* 133–140 (1996).

Huey, L. G., E. J. Dunlea, E. R. Lovejoy, D. R. Hanson, R. B. Norton, F. C. Fehsenfeld, and C. J. Howard, "Fast Time Response Measurements of HNO_3 in Air with a Chemical Ionization Mass Spectrometer," *J. Geophys. Res., 103,* 3355–3360 (1998).

Huffman, H. D., "Comparison of the Light Absorption Coefficient and Carbon Measures for Remote Aerosols: An Independent Analysis of Data from the Improve Network—I," *Atmos. Environ., 30,* 73–83 (1996).

Huntzicker, J. J., R. L. Johnson, J. J. Shah, and R. A. Cary, "Analysis of Organic and Elemental Carbon in Ambient Aerosols by a Thermal-Optical Method," in *Particulate Carbon: Atmospheric Life Cycles* (G. T. Wolff and R. L. Klimisch, Eds.), pp. 79–88, Plenum, New York, 1982.

Igawa, M., J. W. Munger, and M. R. Hoffmann, "Analysis of Aldehydes in Cloud- and Fogwater Samples by HPLC with a Postcolumn Reaction Detector," *Environ. Sci. Technol., 23,* 556–561 (1989).

Iwaski, I., S. Utsumi, K. Hagino, and T. Ozawa, "A New Spectrophotometric Method for the Determination of Small Amounts of Chloride Using the Mercury Thiocyanate Method," *Bull. Chem. Soc. Jpn., 29,* 860–864 (1956).

Jach, T., and C. J. Powell, "X-Ray Photoemission Spectroscopy of Environmental Particles," *Environ. Sci. Technol., 18,* 58–61 (1984).

Jaffrezo, J. L., N. Calas, and M. Bouchet, "Carboxylic Acids Measurements with Ionic Chromatography," *Atmos. Environ., 32,* 2705–2708 (1998).

Japar, S. M., A. C. Szkariat, R. A. Gorse, Jr., E. K. Heyerdahl, R. I. Johnson, J. A. Rau, and J. J. Huntzicker, "Comparison of Solvent

Extraction and Thermal-Optical Carbon Analysis Methods: Application to Diesel Vehicle Exhaust Aerosol," *Environ. Sci. Technol., 18,* 231–234 (1984).

Jayne, J. T., D. C. Leard, X. Zhang, P. Davidovits, K. A. Smith, C. E. Kolb, and D. R. Worsnop, "Development of an Aerosol Mass Spectrometer for Size and Composition Analysis of Submicron Particles," *Aerosol Sci. Technol.,* in press (1999).

Johansson, T. B., R. E. Van Grieken, J. W. Nelson, and J. W. Winchester, "Elemental Trace Analysis of Small Samples by Proton Induced X-Ray Emission," *Anal. Chem., 47,* 855–860 (1975).

John, W., "Particle-Surface Interactions: Charge Transfer, Energy Loss, Resuspension, and Deagglomeration," *Aerosol Sci. Technol., 23,* 2–24 (1995).

Johnson, J., and R. Yost, "Tandem Mass Spectrometry for Trace Analysis," *Anal. Chem., 57,* 758–768 (1985).

Johnston, M. V., and A. S. Wexler, "MS of Individual Aerosol Particles," *Anal. Chem., 67,* 721A–726A (1995).

Kalabokas, P., P. Carlier, P. Fresnet, G. Mouvier, and G. Toupance, "Field Studies of Aldehyde Chemistry in the Paris Area," *Atmos. Environ., 22,* 147–155 (1988).

Kanda, Y., and M. Taira, "Chemiluminescent Method for Continuous Monitoring of Nitrous Acid in Ambient Air," *Anal. Chem., 62,* 2084–2087 (1990).

Kasparian, J., E. Frejafon, P. Rambaldi, J. Yu, B. Vezin, J. P. Wolf, P. Ritter, and P. Viscardi, "Characterization of Urban Aerosols Using SEM-Microscopy, X-Ray, and Lidar Measurements," *Atmos. Environ., 32,* 2957–2967 (1998).

Kaufmann, R. L., "Laser Microprobe Mass Spectroscopy (LAMMA) of Particulates," in *Physical and Chemical Characterization of Individual Airborne Particles* (K. R. Spurny, Ed.), Chap. 13, pp. 227–250, Ellis Horwood, Chichester, 1986.

Kaupp, H., and G. Umlauf, "Atmospheric Gas–Particle Partitioning of Organic Compounds: Comparison of Sampling Methods," *Atmos. Environ., 26A,* 2259–2267 (1992).

Keene, W. C., R. W. Talbot, M. O. Andreae, K. Beecher, H. Berresheim, M. Castro, J. C. Farmer, J. N. Galloway, M. R. Hoffmann, S.-M. Li, J. R. Maben, J. W. Munger, R. B. Norton, A. A. P. Pszenny, H. Puxbaum, H. Westberg, and W. Winiwarter, "An Intercomparison of Measurement Systems for Vapor and Particulate Phase Concentrations of Formic and Acetic Acids," *J. Geophys. Res., 94,* 6457–6471 (1989).

Keene, W. C., J. R. Maben, A. A. P. Pszenny, and J. N. Galloway, "Measurement Technique for Inorganic Chlorine Gases in the Marine Boundary Layer," *Environ. Sci. Technol., 27,* 866–874 (1993).

Kelly, T. J., C. W. Spicer, and G. F. Ward, "An Assessment of the Luminol Chemiluminescence Technique for Measurement of NO_2 in Ambient Air," *Atmos. Environ., 24A,* 2397–2403 (1990).

Kelly, T. J., and M. W. Holdren, "Applicability of Canisters for Sample Storage in the Determination of Hazardous Air Pollutants," *Atmos. Environ., 29,* 2595–2608 (1995).

Khare, P., G. S. Satsangi, N. Kimar, K. M. Kumari, and S. S. Srivastava, "Surface Measurements of Formaldehyde and Formic and Acetic Acids at a Subtropical Semiarid Site in India," *J. Geophys. Res., 102,* 18997–19005 (1997).

Klaus, N., "Aerosol Analysis by Secondary-Ion Mass-Spectrometry," in *Physical and Chemical Characterization of Individual Airborne Particles* (K. R. Spurny, Ed.), Chap. 17, pp. 331–340, Ellis Horwood, 1986.

Kleindienst, T. E., P. B. Shepson, D. N. Hodges, C. M. Nero, R. R. Arnts, P. K. Dasgupta, H. Hwang, G. L. Kok, J. A. Lind, A. L. Lazrus, G. I. Mackay, L. K. Mayne, and H. I. Schiff, "Comparison of Techniques for Measurement of Ambient Levels of Hydrogen Peroxide," *Environ. Sci. Technol., 22,* 53–61 (1988a).

Kleindienst, T. E., P. B. Shepson, C. M. Nero, R. R. Arnts, S. B. Tejada, G. I. Mackay, L. K. Mayne, H. I. Schiff, J. A. Lind, G. L. Kok, A. L. Lazrus, P. K. Dasgupta, and S. Dong, "An Intercomparison of Formaldehyde Measurement Techniques at Ambient Concentrations," *Atmos. Environ., 22,* 1931–1939 (1988b).

Kleindienst, T. E., E. E. Hudgens, D. F. Smith, F. F. McElroy, and J. J. Bufalini, "Comparison of Chemiluminescence and Ultraviolet Ozone Monitor Responses in the Presence of Humidity and Photochemical Pollutants," *J. Air Waste Manage. Assoc., 43,* 213–222 (1993).

Kleindienst, T. E., "Recent Developments in the Chemistry and Biology of Peroxyacetyl Nitrate," *Res. Chem. Intermed., 20,* 335–384 (1994).

Kleindienst, T. E., E. W. Corse, F. T. Blanchard, and W. A. Lonneman, "Evaluation of the Performance of DNPH-Coated Silica Gel and C_{18} Cartridges in the Measurement of Formaldehyde in the Presence and Absence of Ozone," *Environ. Sci. Technol., 32,* 124–130 (1998).

Klemp, M., A. Peters, and R. Sacks, "High-Speed GC Analysis of VOCs: Sample Collection and Inlet Systems," *Environ. Sci. Technol., 28,* 369A–376A (1994).

Kley, D., and M. McFarland, "Chemiluminescence Detector for NO and NO_2," *Atmos. Technol., 12,* 63–69 (1980).

Kliner, D. A. V., B. C. Daube, J. D. Burley, and S. C. Wofsy, "Laboratory Investigation of the Catalytic Reduction Technique for Measurement of Atmospheric NO_y," *J. Geophys. Res., 102,* 10759–10776 (1997).

Knop, G., and F. Arnold, "Stratospheric Trace Gas Detection Using a New Balloon-Borne ACIMS Method: Acetonitrile, Acetone, and Nitric Acid," *Geophys. Res. Lett., 14,* 1262–1265 (1987a).

Knop, G., and F. Arnold, "Atmospheric Acetonitrile Measurements in the Tropopause Region Using Aircraft-Borne Active Chemical Ionization Mass Spectrometry," *Planet. Space Sci., 35,* 259–266 (1987b).

Kok, G. L., K. Thompson, and A. L. Lazrus, "Derivatization Technique for the Determination of Peroxides in Precipitation," *Anal. Chem., 58,* 1192–1194 (1986).

Kok, G. L., A. J. Schanot, P. F. Lindgren, P. K. Dasgupta, D. A. Hegg, P. V. Hobbs, and J. F. Boatman, "An Airborne Test of Three Sulfur Dioxide Measurement Techniques," *Atmos. Environ., 24A,* 1903–1908 (1990).

Kok, G. L., S. E. McLaren, and T. A. Staffelbach, "HPLC Determination of Atmospheric Organic Hydroperoxides," *J. Atmos. Oceanic Technol., 12,* 282–289 (1995).

Koppmann, R., F. J. Johnen, A. Khedim, J. Rudolph, A. Wedel, and B. Wiards, "The Influence of Ozone on Light Nonmethane Hydrocarbons during Cryogenic Preconcentration," *J. Geophys. Res., 100,* 11383–11391 (1995).

Koutrakis, P., J. M. Wolfson, J. L. Slater, M. Brauer, J. D. Spengler, R. K. Stevens, and C. L. Stone, "Evaluation of an Annular Denuder/Filter Pack System to Collect Acidic Aerosols and Gases," *Environ. Sci. Technol., 22,* 1463–1468 (1988).

Koutrakis, P., C. Sioutas, S. T. Ferguson, J. M. Wolfson, J. D. Mulik, and R. M. Burton, "Development and Evaluation of a Glass Honeycomb Denuder/Filter Pack System to Collect Atmospheric Gases and Particles," *Environ. Sci. Technol., 27,* 2497–2501 (1993).

Krieger, M. S., and R. A. Hites, "Diffusion Denuder for the Collection of Semivolatile Organic Compounds," *Environ. Sci. Technol., 26,* 1551–1555 (1992).

Krochmal, D., and L. Górski, "Determination of Nitrogen Dioxide in Ambient Air by Use of a Passive Sampling Technique and Triethanolamine as Absorbent," *Environ. Sci. Technol., 25,* 531–535 (1991).

Krol, M., P. J. van Leeuwen, and J. Lelieveld, "Global OH Trend Inferred from Methylchloroform Measurements," *J. Geophys. Res.*, 103, 10697–10711 (1998).

Kuntz, R., W. Lonneman, G. Namie, and L. A. Hull, "Rapid Determination of Aldehydes in Air Analyses," *Anal. Lett.*, 13, 1409–1415 (1980).

Kuwata, K., M. Uebori, H. Yamasaki, Y. Kuge, and Y. Kiso, "Determination of Aliphatic Aldehydes in Air by Liquid Chromatography," *Anal. Chem.*, 55, 2013–2016 (1983).

Lacheur, R. M., L. B. Sonnenberg, P. C. Singer, R. F. Christman, and M. J. Charles, "Identification of Carbonyl Compounds in Environmental Samples," *Environ. Sci. Technol.*, 27, 2745–2753 (1993).

Lam, G. C. K., D. Y. C. Leung, M. Niewiadomski, S. W. Pang, A. W. F. Lee, and P. K. K. Louie, "Street-Level Concentrations of Nitrogen Dioxide and Suspended Particulate Matter in Hong Kong," *Atmos. Environ.*, 33, 1–11 (1999).

Lammel, G., and J. N. Cape, "Nitrous Acid and Nitrite in the Atmosphere," *Chem. Soc. Rev.*, 25, 361–369 (1996).

Langford, A. O., P. D. Goldan, and F. C. Fehsenfeld, "A Molybdenum Oxide Annular Denuder System for Gas Phase Ambient Ammonia Measurements," *J. Atmos. Chem.*, 8, 359–376 (1989).

Langford, A. O., and F. C. Fehsenfeld, "Natural Vegetation as a Source or Sink for Atmospheric Ammonia: A Case Study," *Science*, 255, 581–583 (1992).

Larsen, B., T. Bomboi-Mingarro, E. Brancaleoni, A. Calogirou, A. Cecinato, C. Coeur, I. Chatzianestis, M. Duane, M. Frattoni, J.-L. Fugit, U. Hansen, V. Jacob, N. Mimikos, T. Hoffmann, S. Owen, R. Perez-Pastor, A. Reichmann, G. Seufert, M. Staudt, and R. Steinbrecher, "Sampling and Analysis of Terpenes in Air. An Interlaboratory Comparison," *Atmos. Environ.*, 31, 35–49 (1997).

Lawrence Berkeley Laboratory (LBL), *Instrumentation for Environmental Monitoring, Air*, Vol. 1, Part 2, September 1979, pp. 301–378.

Lawson, D. R., H. W. Biermann, E. C. Tuazon, A. M. Winer, G. I. Mackay, H. I. Schiff, G. L. Kok, P. K. Dasgupta, and K. Fung, "Formaldehyde Measurement Methods Evaluation and Ambient Concentrations during the Carbonaceous Species Methods Comparison Study," *Aerosol Sci. Technol.*, 12, 64–76 (1990).

Lazendorf, E. J., T. F. Hanisco, N. M. Donahue, and P. O. Wennberg, "Comment on: 'The Measurement of Tropospheric OH Radicals by Laser-Induced Fluorescence Spectroscopy during the POPCORN Field Campaign,' by Hofzumahaus et al. and 'Intercomparison of Tropospheric OH Radical Measurements by Multiple Folded Long-Path Laser Absorption and Laser Induced Fluorescence,' by Brauers et al.," *Geophys. Res. Lett.*, 24, 3037–3038 (1997).

Lazrus, A. L., G. L. Kok, S. N. Gitlin, and J. A. Lind, "Automated Fluorometric Method for Hydrogen Peroxide in Atmospheric Precipitation," *Anal. Chem.*, 57, 917–922 (1985).

Lazrus, A. L., G. L. Kok, J. A. Lind, S. N. Gitlin, B. G. Heikes, and R. E. Shetter, "Automated Fluorometric Method for Hydrogen Peroxide in Air," *Anal. Chem.*, 58, 594–597 (1986).

Lazrus, A. L., K. L. Fong, and J. A. Lind, "Automated Fluorometric Determination of Formaldehyde in Air," *Anal. Chem.*, 60, 1074–1078 (1988).

Le Lacheur, R. M., L. B. Sonnenberg, P. C. Singer, R. F. Christman, and M. J. Charles, "Identification of Carbonyl Compounds in Environmental Samples," *Environ. Sci. Technol.*, 27, 2745–2753 (1993).

Lee, J. H., Y. Chen, and I. N. Tang, "Heterogeneous Loss of Gaseous H_2O_2 in an Atmospheric Air Sampling System," *Environ. Sci. Technol.*, 25, 339–342 (1991).

Lee, J. H., D. F. Leahy, I. N. Tang, and L. Newman, "Measurement and Speciation of Gas Phase Peroxides in the Atmosphere," *J. Geophys. Res.*, 98, 2911–2915 (1993).

Lee, S.-H., J. Hirokawa, Y. Kajii, and H. Akimoto, "New Method for Measuring Low NO Concentrations Using Laser Induced Two Photon Ionization," *Rev. Sci. Instrum.*, 68, 2891–2897 (1997).

Lee, Y.-N., and X. Zhou, "Method for the Determination of Some Soluble Atmospheric Carbonyl Compounds," *Environ. Sci. Technol.*, 27, 749–756 (1993).

Lee, Y.-N., X. Zhou, and K. Hallock, "Atmospheric Carbonyl Compounds at a Rural Southeastern United States Site," *J. Geophys. Res.*, 100, 25933–25944 (1995).

Leibrock, E., and J. Slemr, "Method for Measurement of Volatile Oxygenated Hydrocarbons in Ambient Air," *Atmos. Environ.*, 31, 3329–3339 (1997).

Lemire, G. W., J. B. Simeonsson, and R. C. Sausa, "Monitoring of Vapor-Phase Nitro Compounds Using 226-nm Radiation: Fragmentation with Subsequent No Resonance-Enhanced Multiphoton Ionization Detection," *Anal. Chem.*, 65, 529–533 (1993).

Letokhov, V. S., *Laser Photoionization Spectroscopy*, Academic Press, San Diego, 1987.

Levin, J.-O., R. Lindahl, and K. Andersson, "Monitoring of Parts-Per-Billion Levels of Formaldehyde Using a Diffusive Sampler," *JAPCA*, 39, 44–47 (1989).

Lewin, E. E., R. G. de Pena, and J. P. Shimshock, "Atmospheric Gas and Particle Measurements at a Rural Northeastern U.S. Site," *Atmos. Environ.*, 20, 59–70 (1986).

Lin, C.-I., M. Baker, and R. J. Charlson, "Absorption Coefficient of Atmospheric Aerosol: A Method for Measurement," *Appl. Opt.*, 12, 1356–1363 (1973).

Lindgren, P. F., and P. K. Dasgupta, "Measurement of Atmospheric Sulfur Dioxide by Diffusion Scrubber Coupled Ion Chromatography," *Anal. Chem.*, 61, 19–24 (1989).

Lindskog, A., Y. Andersson-Sköld, P. Grennfelt, and J. Mowrer, "Concentration Profiles of Hydrocarbons during Episodes in Relation to Emission Pattern, Model Calculations, and Oxidants," *J. Atmos. Chem.*, 14, 425–438 (1992).

Lindskog, A., and J. Moldanova, "The Influence of the Origin, Season, and Time of the Day on the Distribution of Individual NMHC Measured at Rörvik, Sweden," *Atmos. Environ.*, 18, 2383–2398 (1994).

Lipari, F., and S. J. Swarin, "Determination of Formaldehyde and Other Aldehydes in Automobile Exhaust with an Improved 2,4-Dinitrophenylhydrazine Method," *J. Chromatogr.*, 247, 297–306 (1982).

Lipari, F., and S. J. Swarin, "2,4-Dinitrophenylhydrazine-Coated Florisil Sampling Cartridges for the Determination of Formaldehyde in Air," *Environ. Sci. Technol.*, 19, 70–74 (1985).

Liu, B. Y. H., D. Y. H. Pui, and H. J. Fissan, *Aerosols, Science, Technology, and Industrial Applications of Airborne Particles*, Elsevier, New York, 1984.

Liu, D.-Y., D. Rutherford, M. Kinsey, and K. A. Prather, "Real-Time Monitoring of Pyrotechnically Derived Aerosol Particles in the Troposphere," *Anal. Chem.*, 69, 1808–1814 (1997).

Liu, P., P. J. Ziemann, D. B. Kittelson, and P. H. McMurry, "Generating Particle Beams of Controlled Dimensions and Divergence: I. Theory of Particle Motion in Aerodynamic Lenses and Nozzle Expansions," *Aerosol Sci. Technol.*, 22, 293–313 (1995a).

Liu, P., P. J. Ziemann, D. B. Kittelson, and P. H. McMurry, "Generating Particle Beams of Controlled Dimensions and Divergence: II. Experimental Evaluation of Particle Motion in Aerodynamic Lenses and Nozzle Expansions," *Aerosol Sci. Technol.*, 22, 314–324 (1995b).

Lundgren, D. A., F. S. Harris, Jr., W. H. Marlow, M. Lippmann, W. E. Clark, and M. D. Durham, Eds., *Aerosol Measurement*, University Press of Florida, Gainesville, FL, 1979.

Luria, M., C. C. Van Valin, J. F. Boatman, D. L. Wellman, and R. F. Pueschel, "Sulfur Dioxide Flux Measurements over the Western Atlantic Ocean," *Atmos. Environ., 21,* 1631–1636 (1987).

Macdonald, A. M., K. G. Anlauf, C. M. Banic, W. R. Leaitch, and H. A. Wiebe, "Airborne Measurements of Aqueous and Gaseous Hydrogen Peroxide during Spring and Summer in Ontario, Canada," *J. Geophys. Res., 100,* 7253–7262 (1995).

Mackay, G. I., H. I. Schiff, A. Wiebe, and K. Anlauf, "Measurements of NO_2, H_2CO, and HNO_3 by Tunable Diode Laser Absorption Spectroscopy during the 1985 Claremont Intercomparison Study," *Atmos. Environ., 22,* 1555–1564 (1988).

Maeda, Y., K. Aoki, and M. Munemori, "Chemiluminescence Method for the Determination of Nitrogen Dioxide," *Anal. Chem., 52,* 307–311 (1980).

Maeda, Y., and N. Takenaka, "Chemiluminescence Determination of Trace Amounts of Ammonia and Halogen Species in the Environment," in *Optical Methods in Atmospheric Chemistry,* SPIE, Vol. 1715, pp. 185–193, Bellingham, WA, 1992.

Mansoori, B. A., M. V. Johnston, and A. S. Wexler, "Quantitation of Ionic Species in Single Microdroplets by On-Line Laser Desorption/Ionization," *Anal. Chem., 66,* 3681–3687 (1994).

March, R. E., "Ion Trap Mass Spectrometry," *Int. J. Mass Spectrom. Ion Processes, 118/119,* 71–135 (1992).

Marijnissen, J., B. Scarlett, and P. Verheijen, "Proposed On-Line Aerosol Analysis Combining Size Determination, Laser-Induced Fragmentation, and Time-of-Flight Mass Spectroscopy," *J. Aerosol Sci., 19,* 1307–1310 (1988).

Marple, V. A., and K. Willeke, "Inertial Impactors," in *Aerosol Measurement* (D. A. Lundgren, F. S. Harris, Jr., W. H. Marlow, M. Lippmann, W. E. Clark, and M. D. Durham, Eds.), pp. 90–107, University Press of Florida, Gainesville, FL, 1979.

Marple, V. A., K. L. Rubow, and S. M. Behm, "A Microorifice Uniform Deposit Impactor (MOUDI): Description, Calibration, and Use," *Aerosol Sci. Technol., 14,* 434–446 (1991).

Marshall, T. L., C. T. Chaffin, R. M. Hammaker, and W. G. Fateley, "An Introduction to Open-Path FT-IR Atmospheric Monitoring," *Environ. Sci. Technol., 28,* 224–232 (1994).

Martens, A. E., and J. D. Keller, "An Instrument for Sizing and Counting Airborne Particles," *Am. Ind. Hyg. Assoc. J., 29,* 257–267 (1968).

Marti, J. J., R. J. Weber, M. T. Saros, J. G. Vasiliou, and P. H. McMurry, "Modification of the TSI 3025 Condensation Particle Counter for Pulse Height Analysis," *Aerosol Sci. Technol., 25,* 214–218 (1996).

Martin, D., M. Tsivou, B. Bonsang, C. Abonnel, T. Carsey, M. Springer-Young, A. Pszenny, and K. Suhre, "Hydrogen Peroxide in the Marine Atmospheric Boundary Layer during the Atlantic Stratocumulus Transition Experiment/Marine Aerosol and Gas Exchange Experiment in the Eastern Subtropical North Atlantic," *J. Geophys. Res., 10,* 6003–6015 (1997).

Mather, J. H., P. S. Stevens, and W. H. Brune, "OH and HO_2 Measurements Using Laser-Induced Fluorescence," *J. Geophys. Res., 102,* 6427–6436 (1997).

Matisová, E., and S. Škrabáková, "Carbon Sorbents and Their Utilization for the Preconcentration of Organic Pollutants in Environmental Samples," *J. Chromatogr. A, 707,* 145–179 (1995).

Mauldin, R. L., III, D. J. Tanner, and F. L. Eisele, "A New Chemical Ionization Mass Spectrometer Technique for the Fast Measurement of Gas Phase Nitric Acid in the Atmosphere," *J. Geophys. Res., 103,* 3361–3367 (1998).

McCaffrey, C. A., J. MacLachlan, and B. I. Brookes, "Adsorbent Tube Evaluation for the Preconcentration of Volatile Organic Compounds in Air for Analysis by Gas Chromatography–Mass Spectrometry," *Analyst, 119,* 897–902 (1994).

McClenny, W. A., K. D. Oliver, and E. H. Daughtrey, Jr., "Analysis of VOCs in Ambient Air Using Multisorbent Packings for VOC Accumulation and Sample Drying," *J. Air Waste Manage. Assoc., 45,* 792–800 (1995).

McDermott, W. T., R. C. Ockovic, and M. R. Stolzenburg, "Counting Efficiency of an Improved 30-A Condensation Nucleus Counter," *Aerosol Sci. Technol., 14,* 278–287 (1991).

McDiarmid, R., and A. SabljiX, "Experimental Assignments of the $3p$ Rydberg States of Acetone," *J. Chem. Phys., 89,* 6086–6095 (1988).

McDow, S. R., and J. J. Huntzicker, "Vapor Adsorption Artifact in the Sampling of Organic Aerosol: Face Velocity Effects," *Atmos. Environ., 24A,* 2563–2571 (1990).

McElroy, F. F., V. L. Thompson, D. M. Holland, W. A. Lonneman, and R. L. Sella, "Cryogenic Preconcentration-Direct FID Method for Measurement of Ambient NMOC: Refinement and Comparison with GC Speciation," *JAPCA, 36,* 710–714 (1986).

McKeen, S. A., M. Trainer, E. Y. Hsie, R. K. Tallamraju, and S. C. Liu, "On the Indirect Determination of Atmospheric OH Radical Concentrations from Reactive Hydrocarbon Measurements," *J. Geophys. Res., 95,* 7493–7500 (1990).

McKeown, P. J., M. V. Johnston, and D. M. Murphy, "On-Line Single-Particle Analysis by Laser Desorption Mass Spectrometry," *Anal. Chem., 63,* 2069–2073 (1991).

McManus, J. B., P. L. Kebabian, and M. S. Zahniser, "Astigmatic Mirror Multiple Pass Absorption Cells for Long Pathlength Spectroscopy," *Appl. Opt., 34,* 3336–3348 (1995).

McMurry, P. H., M. Litchy, P.-F. Huang, X. Cai, B. J. Turpin, W. D. Dick, and A. Hanson, "Elemental Composition and Morphology of Individual Particles Separated by Size and Hygroscopicity with the TDMA," *Atmos. Environ., 30,* 101–108 (1996).

Meyer, C. P., C. M. Elsworth, and I. E. Galbally, "Water Vapor Interference in the Measurement of Ozone in Ambient Air by Ultraviolet Absorption," *Rev. Sci. Instrum., 62,* 223–228 (1991).

Middlebrook, A. M., D. S. Thomson, and D. M. Murphy, "On the Purity of Laboratory-Generated Sulfuric Acid Droplets and Ambient Particles Studied by Laser Mass Spectrometry," *Aerosol Sci. Technol., 27,* 293–307 (1997).

Mihelcic, D., P. Müsgen, and D. H. Ehhalt, "An Improved Method of Measuring Tropospheric NO_2 and RO_2 by Matrix Isolation and Electron Spin Resonance," *J. Atmos. Chem., 3,* 341–361 (1985).

Mihelcic, D., A. Volz-Thomas, H. W. Pätz, D. Kley, and M. Mihelcic, "Numerical Analysis of ESR Spectra from Atmospheric Samples," *J. Atmos. Chem., 11,* 271–297 (1990).

Mihelcic, D., D. Klemp, P. Müsgen, H. W. Pätz, and A. Volz-Thomas, "Simultaneous Measurements of Peroxy and Nitrate Radicals at Schauinsland," *J. Atmos. Chem., 16,* 313–335 (1993).

Mihele, C. M., and D. R. Hastie, "The Sensitivity of the Radical Amplifier to Ambient Water Vapour," *Geophys. Res. Lett., 25,* 1911–1913, 3167 (1998).

Miller, D. P., "Low-Level Determination of Nitrogen Dioxide in Ambient Air Using the Palmes Tube," *Atmos. Environ., 22,* 945–947 (1988).

Monks, P. S., L. J. Carpenter, S. A. Penkett, and G. P. Ayers, "Nighttime Peroxy Radical Chemistry in the Remote Marine Boundary Layer over the Southern Ocean," *Geophys. Res. Lett., 23,* 535–538 (1996).

Moosmüller, H., W. P. Arnott, and C. F. Rogers, "Methods for Real-Time, *In Situ* Measurement of Aerosol Light Absorption," *J. Air Waste Manage. Assoc., 47,* 157–166 (1997).

Morawaska, L., S. Thomas, N. Bofinger, D. Wainwright, and D. Neale, "Comprehensive Characterization of Aerosols in a Subtropical Urban Atmosphere: Particle Size Distribution and Correlation with Gaseous Pollutants," *Atmos. Environ., 32,* 2467–2478 (1998).

Morikawa, T., S. Wakamatsu, M. Tanaka, I. Uno, T. Kamiura, and T. Maeda, "C_2–C_5 Hydrocarbon Concentrations in Central Osaka," *Atmos. Environ., 32,* 2007–2016 (1998).

Moschonas, N., and S. Glavas, "C_3–C_{10} Hydrocarbons in the Atmosphere of Athens, Greece," *Atmos. Environ., 30,* 2769–2772 (1996).

Mount, G. H., "The Measurement of Tropospheric OH by Long Path Absorption. 1. Instrumentation," *J. Geophys. Res., 97,* 2427–2444 (1992).

Mount, G. H., and F. L. Eisele, "An Intercomparison of Tropospheric OH Measurements at Fritz Peak Observatory, Colorado," *Science, 256,* 1187–1190 (1992).

Mount, G. H., and J. W. Harder, "The Measurement of Tropospheric Trace Gases at Fritz Peak Observatory, Colorado, by Long-Path Absorption: OH and Ancillary Gases," *J. Atmos. Sci., 52,* 3342–3353 (1995).

Mount, G. H., F. L. Eisele, D. J. Tanner, J. W. Brault, P. V. Johnston, J. W. Harder, E. J. Williams, A. Fried, and R. Shetter, "An Intercomparison of Spectroscopic Laser Long-Path and Ion-Assisted *in-Situ* Measurements of Hydroxyl Concentrations during the Tropospheric OH Photochemistry Experiment, Fall 1993," *J. Geophys. Res., 102,* 6437–6455 (1997a).

Mount, G. H., J. W. Brault, P. V. Johnston, E. Marovich, R. O. Jakoubek, C. J. Volpe, J. Harder, and J. Olson, "Measurement of Tropospheric OH by Long-Path Laser Absorption at Fritz Peak Observatory, Colorado, during the OH Photochemistry Experiment, Fall 1993," *J. Geophys. Res., 102,* 6393–6413 (1997b).

Mowrer, J., and A. Lindskog, "Automatic Unattended Sampling and Analysis of Background Levels of C_2–C_5 Hydrocarbons," *Atmos. Environ., 25A,* 1971–1979 (1991).

Mulik, J., R. Puckett, D. Williams, and E. Sawicki, "Ion Chromatographic Analysis of Sulfate and Nitrate in Ambient Aerosols," *Anal. Lett., 9,* 653–663 (1976).

Mullin, J. B., and J. P. Riley, "The Spectrophotometric Determination of Nitrate in Natural Waters with Particulate Reference to Sea Water," *Anal. Chim. Acta, 12,* 464–480 (1955).

Munger, J. W., D. J. Jacob, B. C. Daube, L. W. Horowitz, W. C. Keene, and B. G. Heikes, "Formaldehyde, Glyoxal, and Methylglyoxal in Air and Cloudwater at a Rural Mountain Site in Central Virginia," *J. Geophys. Res., 100,* 9325–9333 (1995).

Munro, R., R. Siddans, W. J. Reburn, and B. J. Kerridge, "Direct Measurement of Tropospheric Ozone Distributions from Space," *Nature, 392,* 168–171 (1998).

Murphy, D. M., and D. S. Thomson, "Laser Ionization Mass Spectroscopy of Single Aerosol Particles," *Aerosol Sci. Technol., 22,* 237–249 (1995).

Murphy, D. M., and D. S. Thomson, "Chemical Composition of Single Aerosol Particles at Idaho Hill: Positive Ion Measurements," *J. Geophys. Res., 102,* 6341–6352 (1997a).

Murphy, D. M., and D. S. Thomson, "Chemical Composition of Single Aerosol Particles at Idaho Hill: Negative Ion Measurements," *J. Geophys. Res., 102,* 6353–6368 (1997b).

Myers, R. L., and W. L. Fite, "Electrical Detection of Airborne Particulates Using Surface Ionization Techniques," *Environ. Sci. Technol., 9,* 334–336 (1975).

Narcisi, R. S., and A. D. Bailey, "Mass Spectrometric Measurements of Positive Ions at Altitudes from 64 to 112 Kilometers," *J. Geophys. Res., 70,* 3687–3700 (1965).

Navas, M. J., A. M. Jiménez, and G. Galán, "Air Analysis: Determination of Nitrogen Compounds by Chemiluminescence," *Atmos. Environ., 31,* 3603–3608 (1997).

Nejedly, Z., J. L. Campbell, W. J. Teesdale, J. F. Dlouhy, T. F. Dann, R. M. Hoff, J. R. Brook, and H. A. Wiebe, "Inter-Laboratory Comparison of Air Particulate Monitoring Data," *J. Air Waste Manage. Assoc., 48,* 386–397 (1998).

Neubauer, K. R., M. V. Johnston, and A. S. Wexler, "Chromium Speciation in Aerosols by Rapid Single-Particle Mass Spectrometry," *Int. J. Mass Spectrom. Ion Processes, 151,* 77–87 (1995).

Neubauer, K. R., S. T. Sum, M. V. Johnston, and A. S. Wexler, "Sulfur Speciation in Individual Aerosol Particles," *J. Geophys. Res., 101,* 18701–18707 (1996).

Neubauer, K. R., M. V. Johnston, and A. S. Wexler, "On-Line Analysis of Aqueous Aerosols by Laser Desorption Ionization," *Int. J. Mass Spectrom. Ion Processes, 163,* 29–37 (1997).

Neubauer, K. R., M. V. Johnston, and A. S. Wexler, "Humidity Effects on the Mass Spectra of Single Aerosol Particles," *Atmos. Environ., 32,* 2521–2529 (1998).

Nielsen, T., A. H. Egeløv, K. Granby, and H. Skov, "Observations on Particulate Organic Nitrates and Unidentified Components of NO_y," *Atmos. Environ., 29,* 1757–1769 (1995).

Nielsen, T., J. Platz, K. Granby, A. B. Hansen, H. Skov, and A. H. Egeløv, "Particulate Organic Nitrates: Sampling and Night/Day Variation," *Atmos. Environ., 32,* 2601–2608 (1998).

Noble, C. A., T. Nordmeyer, K. Salt, B. Morrical, and K. A. Prather, "Aerosol Characterization Using Mass Spectrometry," *Trends Anal. Chem., 13,* 218–222 (1994).

Noble, C. A., and K. A. Prather, "Real-Time Measurement of Correlated Size and Composition Profiles of Individual Atmospheric Aerosol Particles," *Environ. Sci. Technol., 30,* 2667–2680 (1996).

Nolte, C. G., P. A. Solomon, T. Fall, L. G. Salmon, and G. R. Cass, "Seasonal and Spatial Characteristics of Formic and Acetic Acids Concentrations in the Southern California Atmosphere," *Environ. Sci. Technol., 31,* 2547–2553 (1997).

Nondek, L., D. R. Rodler, and J. W. Birks, "Measurement of Sub-ppbv Concentrations of Aldehydes in a Forest Atmosphere Using a New HPLC Technique," *Environ. Sci. Technol., 26,* 1174–1178 (1992).

Nordmeyer, T., and K. A. Prather, "Real-Time Measurement Capabilities Using Aerosol Time-of-Flight Mass Spectrometry," *Anal. Chem., 66,* 3540–3542 (1994).

Nouaime, G., S. B. Bertman, C. Seaver, D. Elyea, H. Huang, P. B. Shepson, T. K. Starn, D. D. Riemer, R. G. Zika, and K. Olszyna, "Sequential Oxidation Products from Tropospheric Isoprene Chemistry: MACR and MPAN at a NO_x-Rich Forest Environment in the Southeastern United States," *J. Geophys. Res., 103,* 22463–22471 (1998).

Novakov, T., P. K. Mueller, A. E. Alcocer, and J. W. Otvos, "Chemical Composition of Pasadena Aerosol by Particle Size and Time of Day. III. Chemical States of Nitrogen and Sulfur by Photoelectron Spectroscopy," *J. Colloid. Interface Sci., 39,* 225–234 (1972).

Novelli, P. C., L. P. Steele, and P. P. Tans, "Mixing Ratios of Carbon Monoxide in the Troposphere," *J. Geophys. Res., 97,* 20731–20750 (1992).

Novelli, P. C., V. S. Connors, H. G. Reichle, Jr., B. E. Anderson, C. A. M. Brenninkmeijer, E. G. Brunke, B. G. Doddridge, V. W. J. H. Kirchhoff, K. S. Lam, K. A. Masarie, T. Matsuo, D. D. Parrish, H. E. Scheel, and L. P. Steele, "An Internally Consistent Set of Globally Distributed Atmospheric Carbon Monoxide Mixing Ratios Developed Using Results from an Intercomparison of Measurements," *J. Geophys. Res., 103,* 19285–19293 (1998a).

Novelli, P. C., K. A. Masarie, and P. M. Lang, "Distributions and Recent Changes of Carbon Monoxide in the Lower Troposphere," *J. Geophys. Res., 103,* 19015–19033 (1998b).

Noxon, J. F., "Nitrogen Dioxide in the Stratosphere and Troposphere Measured by Ground-Based Absorption Spectroscopy," *Science, 189,* 547–549 (1975).

Noxon, J. F., R. B. Norton, and W. R. Henderson, "Observation of Atmospheric NO_3," *Geophys. Res. Lett., 5,* 675–678 (1978).

Noxon, J. F., R. B. Norton, and E. Marovich, "NO$_3$ in the Troposphere," *Geophys. Res. Lett., 7,* 125–128 (1980).

O'Brien, J. M., P. B. Shepson, K. Muthuramu, C. Hao, H. Niki, D. R. Hastie, R. Taylor, and P. B. Roussel, "Measurements of Alkyl and Multifunctional Organic Nitrates at a Rural Site in Ontario," *J. Geophys. Res., 100,* 22795–22804 (1995).

Okabe, H., P. L. Splitstone, and J. J. Ball, "Ambient and Source SO$_2$ Detector Based on a Fluorescence Method," *J. Air Pollut. Control Assoc., 23,* 514–516 (1973).

Parrish, D. D., M. Trainer, M. P. Buhr, B. A. Watkins, and F. C. Fehsenfeld, "Carbon Monoxide Concentrations and Their Relation to Concentrations of Total Reactive Oxidized Nitrogen at Two Rural U.S. Sites," *J. Geophys. Res., 96,* 9309–9320 (1991).

Parrish, D. D., M. P. Buhr, M. Trainer, R. B. Norton, J. P. Shimshock, F. C. Fehsenfeld, K. G. Anlauf, J. W. Bottenheim, Y. Z. Tang, H. A. Wiebe, J. M. Roberts, R. L. Tanner, L. Newman, V. C. Bowersox, K. J. Olszyna, E. M. Bailey, M. O. Rodgers, T. Wang, H. Berresheim, U. K. Roychowdhury, and K. L. Demerjian, "The Total Reactive Oxidized Nitrogen Levels and the Partitioning between the Individual Species at Six Rural Sites in Eastern North America," *J. Geophys. Res., 98,* 2927–2939 (1993).

Parrish, D. D., J. S. Holloway, and F. C. Fehsenfeld, "Routine, Continuous Measurement of Carbon Monoxide with Parts per Billion Precision," *Environ. Sci. Technol., 28,* 1615–1618 (1994).

Patashnick, H., and E. G. Rupprecht, "Continuous PM$_{10}$ Measurements Using the Tapered Element Oscillating Microbalance," *J. Air Waste Manage. Assoc., 41,* 1079–1083 (1991).

Pate, B., R. K. M. Jayanty, M. R. Peterson, and G. F. Evans, "Temporal Stability of Polar Organic Compounds in Stainless Steel Canisters," *J. Air Waste Manage. Assoc., 42,* 460–462 (1992).

Penkett, S. A., N. J. Blake, P. Lightman, A. R. W. Marsh, P. Anwyl, and G. Butcher, "The Seasonal Variation of Nonmethane Hydrocarbons in the Free Troposphere over the North Atlantic Ocean: Possible Evidence for Extensive Reaction of Hydrocarbons with the Nitrate Radical," *J. Geophys. Res., 98,* 2865–2885 (1993).

Perkins, M. D., and F. L. Eisele, "First Mass Spectrometric Measurements of Atmospheric Ions at Ground Level," *J. Geophys. Res., 89,* 9649–9657 (1984).

Perner, D., D. H. Ehhalt, H. W. Pätz, U. Platt, E. P. Röth, and A. Volz, "OH-Radicals in the Lower Troposphere," *Geophys. Res. Lett., 3,* 466–468 (1976).

Perner, D., and U. Platt, "Detection of Nitrous Acid in the Atmosphere by Differential Optical Absorption," *Geophys. Res. Lett., 6,* 917–920 (1979).

Perrino, C., F. De Santis, and A. Febo, "Criteria for the Choice of a Denuder Sampling Technique Devoted to the Measurement of Atmospheric Nitrous and Nitric Acids," *Atmos. Environ., 24A,* 617–626 (1990).

Perros, P. E., "Large-Scale Distribution of Peroxyacetylnitrate from Aircraft Measurements during the TROPOZ II Experiment," *J. Geophys. Res., 99,* 8269–8279 (1994).

Pfab, J., "Laser-Induced Fluorescence and Ionization Spectroscopy of Gas Phase Species," in *Spectroscopy in Environmental Science* (R. J. H. Clark and R. E. Hester, Eds.), Chap. 4, pp. 149–222, Wiley, New York, 1995.

Phalen, R. F., "Evaluation of an Exploded-Wire Aerosol Generator for Use in Inhalation Studies," *Aerosol Sci., 3,* 395–409 (1972).

Piringer, M., E. Ober, H. Puxbaum, and H. Kromp-Kolb, "Occurrence of Nitric Acid and Related Compounds in the Northern Vienna Basin during Summertime Anticyclonic Conditions," *Atmos. Environ., 31,* 1049–1057 (1997).

Plane, J. M. C., and N. Smith, "Atmospheric Monitoring by Differential Optical Absorption Spectroscopy," in *Spectroscopy in Environmental Science* (R. J. H. Clark and R. E Hester, Eds.), pp. 223–262, Wiley, New York, 1995.

Platt, U., D. Perner, G. W. Harris, A. M. Winer, and J. N. Pitts, Jr., "Observations of Nitrous Acid in an Urban Atmosphere by Differential Optical Absorption," *Nature, 285,* 312–314 (1980a).

Platt, U., D. Perner, A. M. Winer, G. W. Harris, and J. N. Pitts, Jr., "Detection of NO$_3$ in the Polluted Troposphere by Differential Optical Absorption," *Geophys. Res. Lett., 7,* 89–92 (1980b).

Platt, U., D. Perner, J. Schröder, C. Kessler, and A. Toennissen, "The Diurnal Variation of NO$_3$," *J. Geophys. Res., 86,* 11965–11970 (1981).

Platt, U., "Differential Optical Absorption Spectroscopy (DOAS)," in *Air Monitoring by Spectroscopic Techniques* (M. W. Sigrist, Ed.), Chemical Analysis Series, Vol. 127, pp. 27–84, Wiley, New York, 1994.

Platt, U., and M. Hausmann, "Spectroscopic Measurement of the Free Radicals NO$_3$, BrO, IO, and OH in the Troposphere," *Res. Chem. Intermed., 20,* 557–578 (1994).

Platt, U., and C. Janssen, "Observation and Role of the Free Radicals NO$_3$, ClO, BrO, and IO in the Troposphere," *Faraday Discuss., 100,* 175–198 (1995).

Pleil, J. D., and A. B. Lindstrom, "Collection of a Single Alveolar Exhaled Breath for Volatile Organic Compounds Analysis," *Am. J. Indust. Med., 28,* 109–121 (1995).

Porter, K., and D. H. Volman, "Flame Ionization Detection of Carbon Monoxide for Gas Chromatographic Analysis," *Anal. Chem., 34,* 748–749 (1962).

Pósfai, M., J. R. Anderson, P. R. Buseck, and H. Sievering, "Compositional Variations of Sea-Salt-Mode Aerosol Particles from the North Atlantic," *J. Geophys. Res., 100,* 23063–23074 (1995).

Pósfai, M., H. Xu, J. R. Anderson, and P. R. Buseck, "Wet and Dry Sizes of Atmospheric Aerosol Particles: An AFM–TEM Study," *Geophys. Res. Lett., 25,* 1907–1910 (1998).

Possanzini, M., V. Di Palo, M. Petricca, R. Fratarcangeli, and D. Brocco, "Measurements of Lower Carbonyls in Rome Ambient Air," *Atmos. Environ., 30,* 3757–3764 (1996).

Prather, K. A., T. Nordmeyer, and K. Salt, "Real-Time Characterization of Individual Aerosol Particles Using Time-of-Flight Mass Spectrometry," *Anal. Chem., 66,* 1403–1407 (1994).

Prinn, R., D. Cunnold, P. Simmonds, F. Alyea, R. Boldi, A. Crawford, P. Fraser, D. Gutzler, D. Hartley, R. Rosen, and R. Rasmussen, "Global Average Concentration and Trend for Hydroxyl Radicals Deduced from ALE/GAGE Trichloroethane (Methyl Chloroform) Data for 1978–1990," *J. Geophys. Res., 97,* 2445–2461 (1992).

Prinn, R. G., R. F. Weiss, B. R. Miller, J. Huang, F. N. Alyea, D. M. Cunnold, P. J. Fraser, D. E. Hartley, and P. G. Simmonds, "Atmospheric Trends and Lifetime of CH$_3$CCl$_3$ and Global OH Concentrations," *Science, 269,* 187–192 (1995).

Pszenny, A. A., W. C. Keene, D. J. Jacob, S. Fan, J. R. Maben, M. P. Zetwo, M. Springer-Young, and J. N. Galloway, "Evidence of Inorganic Chlorine Gases Other Than Hydrogen Chloride in Marine Surface Air," *Geophys. Res. Lett., 20,* 699–702 (1993).

Pui, D. Y. H., C. W. Lewis, C.-J. Tsai, and B. Y. H. Liu, "A Compact Coiled Denuder for Atmospheric Sampling," *Environ. Sci. Technol., 24,* 307–312 (1990).

Raabe, O. G., "The Generation of Aerosols of Fine Particles," in *Fine Particles: Aerosol Generation, Measurement, Sampling, and Analysis* (B. Y. L. Liu, Ed.), pp. 57–110, Academic Press, New York, 1976.

Raes, F., and A. Reineking, "A New Diffusion Battery Design for the Measurement of Sub-20 nm Aerosol Particles: The Diffusion Carrousel," *Atmos. Environ., 19,* 385–388 (1985).

Ramamurthi, M., and K. H. Leong, "Generation of Monodisperse Metallic, Metal Oxide, and Carbon Aerosols," *J. Aerosol Sci., 18,* 175–191 (1987).

Rao, A. M. M., G. G. Pandit, P. Sain, S. Sharma, T. M. Krishnamoorthy, and K. S. V. Nambi, "Non-Methane Hydrocarbons in Industrial Locations of Bombay," *Atmos. Environ., 31,* 1077–1085 (1997).

Rappenglück, P. Fabian, P. Kalabokas, L. G. Viras, and I. C. Ziomas, "Quasi-Continuous Measurements of Non-Methane Hydrocarbons (NMHC) in the Greater Athens Area during Medcaphot-Trace," *Atmos. Environ., 32,* 2103–2121 (1998).

Rapsomanikis, S., M. Wake, A.-M. N. Kitto, and R. M. Harrison, "Analysis of Atmospheric Ammonia and Particulate Ammonium by a Sensitive Fluorescence Method," *Environ. Sci. Technol., 22,* 948–952 (1988).

Reents, W. D., Jr., S. W. Downey, A. B. Emerson, A. M. Mujsce, A. J. Muller, D. J. Siconolfi, J. D. Sinclair, and A. G. Swanson, "Single Particle Characterization by Time-of-Flight Mass Spectrometry," *Aerosol Sci. Technol., 23,* 263–270 (1995).

Reichle, H. G., Jr., V. S. Connors, J. A. Holland, R. T. Sherrill, H. A. Wallio, J. C. Casas, E. P. Condon, B. B. Gormsen, and W. Seiler, "The Distribution of Middle Tropospheric Carbon Monoxide during Early October 1984," *J. Geophys. Res., 95,* 9845–9856 (1990).

Reid, J., J. Shewchun, B. K. Garside, and E. A. Ballik, "High Sensitivity Pollution Detection Employing Tunable Diode Lasers," *Appl. Opt., 17,* 300–307 (1978).

Reilly, P. T. A., R. A. Gieray, W. B. Whitten, and J. M. Ramsey, "Real-Time Characterization of the Organic Composition and Size of Individual Diesel Engine Smoke Particles," *Environ. Sci. Technol., 32,* 2672–2679 (1998).

Reiner, T., M. Hanke, and F. Arnold, "Atmospheric Peroxy Radical Measurements by Ion Molecule Reaction-Mass Spectrometry: A Novel Analytical Method Using Amplifying Chemical Conversion to Sulfuric Acid," *J. Geophys. Res., 102,* 1311–1326 (1997).

Reiner, T., M. Hanke, and F. Arnold, "Aircraft-Borne Measurements of Peroxy Radicals in the Middle Troposphere," *Geophys. Res. Lett., 25,* 47–50 (1998).

Reponen, A., J. Ruuskanen, A. Mirme, E. Pärjälä, G. Hoek, W. Roemer, J. Hosiokangas, J. Pekkanen, and M. Jantunen, "Comparison of Five Methods for Measuring Particulate Matter Concentrations in Cold Winter Climate," *Atmos. Environ., 30,* 3873–3879 (1996).

Ridley, B. A., M. A. Carroll, G. L. Gregory, and G. W. Sachse, "NO and NO_2 in the Troposphere: Technique and Measurements in Regions of a Folded Tropopause," *J. Geophys. Res., 93,* 15813–15830 (1988).

Ridley, B. A., and F. E. Grahek, "A Small, Low Flow, High Sensitivity Reaction Vessel for NO Chemiluminescence Detectors," *J. Atmos. Oceanic Technol., 1,* 307–311 (1990).

Ridley, B. A., J. D. Shetter, J. G. Walega, S. Madronich, C. M. Elsworth, F. E. Grahek, F. C. Fehsenfeld, R. B. Norton, D. D. Parrish, G. Hübler, M. Buhr, E. J. Williams, E. J. Allwine, and H. H. Westberg, "The Behavior of Some Organic Nitrates at Boulder and Niwot Ridge, Colorado," *J. Geophys. Res., 95,* 13949–13961 (1990a).

Ridley, B. A., J. D. Shetter, B. W. Gandrud, L. J. Salas, H. B. Singh, M. A. Carroll, G. Hübler, D. L. Albritton, D. R. Hastie, H. I. Schiff, G. I. Mackay, D. R. Karechi, D. D. Davis, J. D. Bradshaw, M. O. Rodgers, S. T. Sandholm, A. L. Torres, E. P. Condon, G. L. Gregory, and S. M. Beck, "Ratios of Peroxyacetyl Nitrate to Active Nitrogen Observed during Aircraft Flights over the Eastern Pacific Ocean and Continental United States," *J. Geophys. Res., 95,* 10179–10192 (1990b).

Ridley, B. A., E. L. Atlas, J. G. Walega, G. L. Kok, T. A. Staffelbach, J. P. Greenberg, F. E. Grahek, P. G. Hess, and D. D. Montzka, "Aircraft Measurements Made during the Spring Maximum of Ozone over Hawaii: Peroxides, CO, O_3, NO_y, Condensation Nuclei, Selected Hydrocarbons, Halocarbons, and Alkyl Nitrates between 0.5 and 9 km Altitude," *J. Geophys. Res., 102,* 18935–18961 (1997).

Ridley, B., J. Walega, G. Hübler, D. Montzka, E. Atlas, D. Hauglustaine, F. Grahek, J. Lind, T. Campos, R. Norton, J. Greenberg, S. Schauffler, S. Oltmans, and S. Whittlestone, "Measurements of NO_x and PAN and Estimates of O_3 Production over the Seasons during Mauna Loa Observatory Photochemistry Experiment 2," *J. Geophys. Res., 103,* 8323–8339 (1998).

Riemer, D., W. Pos, P. Milne, C. Farmer, R. Zika, E. Apel, K. Olszyna, T. Kleindienst, W. Lonneman, S. Bertman, P. Shepson, and T. Starn, "Observations of Nonmethane Hydrocarbons and Oxygenated Volatile Organic Compounds at a Rural Site in the Southeastern United States," *J. Geophys. Res., 103,* 28111–28128 (1998).

Rinsland, C. P., M. R. Gunson, P.-H. Wang, R. F. Arduini, B. A. Baum, P. Minnis, A. Goldman, M. C. Abrams, R. Zander, E. Mahieu, R. J. Salawitch, H. A. Michelsen, F. W. Irion, and M. J. Newchurch, "ATMOS/ATLAS 3 Infrared Profile Measurements of Trace Gases in the November, 1994 Tropical and Subtropical Upper Troposphere," *J. Quant. Spectrosc. Radiat. Transfer, 60,* 891–901 (1998).

Riveros, H. G., A. Alba, P. Ovalle, B. Silva, and E. Sandoval, "Carbon Monoxide Trend, Meteorology, and Three-Way Catalysts in Mexico City," *J. Air Waste Manage. Assoc., 48,* 459–462 (1998).

Roberts, J. M., J. Williams, K. Baumann, M. P. Buhr, P. D. Goldan, J. Holloway, G. Hübler, W. C. Kuster, S. A. McKeen, T. B. Ryerson, M. Trainer, E. J. Williams, F. C. Fehsenfeld, S. B. Bertman, G. Nouaime, C. Seaver, G. Grodzinsky, M. Rodgers, and V. L. Young, "Measurements of PAN, PPN, and MPAN Made during the 1994 and 1995 Nashville Intensives of the Southern Oxidant Study: Implications for Regional Ozone Production from Biogenic Hydrocarbons," *J. Geophys. Res., 103,* 22473–22490 (1998).

Rodgers, M. O., and D. D. Davis, "A UV-Photofragmentation/Laser-Induced Fluorescence Sensor for the Atmospheric Detection of HONO," *Environ. Sci. Technol., 23,* 1106–1112 (1989).

Rodler, D. R., L. Nondek, and J. W. Birks, "Evaluation of Ozone and Water Vapor Interferences in the Derivatization of Atmospheric Aldehydes with Dansylhydrazine," *Environ. Sci. Technol., 27,* 2814–2820 (1993).

Roscoe, H. K., and K. C. Clemitshaw, "Measurement Techniques in Gas-Phase Tropospheric Chemistry: A Selective View of the Past, Present, and Future," *Science, 276,* 1065–1072 (1997).

Rosell-Llompart, J., I. G. Loscertales, D. Bingham, and J. F. de la Mora, "Sizing Nanoparticles and Ions with a Short Differential Mobility Analyzer," *J. Aerosol Sci., 27,* 695–719 (1996).

Rudolph, J., B. Vierkorn-Rudolph, and F. X. Meixner, "Large-Scale Distribution of Peroxyacetylnitrate Results from the STRATOZ III Flights," *J. Geophys. Res., 92,* 6653–6661 (1987).

Rudolph, J., A. Khedim, and D. Wagenbach, "The Seasonal Variation of Light Nonmethane Hydrocarbons in the Antarctic Troposphere," *J. Geophys. Res., 94,* 13039–13044 (1989).

Sachse, G. W., G. F. Hill, L. O. Wade, and M. G. Perry, "Fast-Response, High-Precision Carbon Monoxide Sensor Using a Tunable Diode Laser Absorption Technique," *J. Geophys. Res., 92,* 2071–2081 (1987).

Sacks, R., and M. Akard, "High-Speed GC Analysis of VOCs: Tunable Selectivity and Column Selection," *Environ. Sci. Technol., 28,* 428A–433A (1994).

Sakugawa, H., and I. R. Kaplan, "Atmospheric H_2O_2 Measurement: Comparison of Cold Trap Method with Impinger Bubbling Method," *Atmos. Environ., 21,* 1791–1798 (1987).

Sakugawa, H., I. R. Kaplan, W. Tsai, and Y. Cohen, "Atmospheric Hydrogen Peroxide," *Environ. Sci. Technol., 24,* 1452–1461 (1990).

Salas, L. J., and H. B. Singh, "Measurements of Formaldehyde and Acetaldehyde in the Urban Ambient Air," *Atmos. Environ., 20,* 1301–1304 (1986).

Sandholm, S. T., J. D. Bradshaw, K. S. Dorris, M. O. Rodgers, and D. D. Davis, "An Airborne Compatible Photofragmentation Two-Photon Laser-Induced Fluorescence Instrument for Measuring Background Tropospheric Levels of NO, NO_x, and NO_2," *J. Geophys. Res., 95,* 10155–10161 (1990).

Sandholm, S., J. Olson, J. Bradshaw, R. Talbot, H. Singh, G. Gregory, D. Blake, B. Anderson, G. Sachse, J. Barrick, J. Collins, K. Klemm, B. Lefer, O. Klemm, K. Gorzelska, D. Herlth, and D. O'Hara, "Summertime Partitioning and Budget of NO_y Compounds in the Troposphere over Alaska and Canada: ABLE 3B," *J. Geophys. Res., 99,* 1837–1861 (1994).

Sandholm, S., S. Smyth, R. Bai, and J. Bradshaw, "Recent and Future Improvements in Two-Photon Laser-Induced Fluorescence NO Measurement Capabilities," *J. Geophys. Res., 102,* 28651–28661 (1997).

Sanhueza, E., M. Santana, D. Trapp, C. de Serves, L. Figueroa, R. Romero, A. Rondón, and L. Donoso, "Field Measurement Evidence for an Atmospheric Chemical Source of Formic and Acetic Acids in the Tropic," *Geophys. Res. Lett., 23,* 1045–1048 (1996).

Saros, M. T., R. J. Weber, J. J. Marti, and P. H. McMurry, "Ultrafine Aerosol Measurement Using a Condensation Nucleus Counter with Pulse Height Analysis," *Aerosol Sci. Technol, 25,* 200–213 (1996).

Sauer, F., G. Schuster, C. Schäfer, and G. K. Moortgat, "Determination of H_2O_2 and Organic Peroxides in Cloud- and Rain-Water on the Kleiner Feldberg during FELDEX," *Geophys. Res. Lett., 23,* 2605–2608 (1996).

Sauer, F., S. Limbach, and G. K. Moortgat, "Measurements of Hydrogen Peroxide and Individual Organic Peroxides in the Marine Troposphere," *Atmos. Environ., 31,* 1173–1184 (1997).

Scheibel, H. G., and J. Porstendörfer, "Generation of Monodisperse Ag- and NaCl-Aerosols with Particle Diameters between 2 and 300 nm," *J. Aerosol Sci., 14,* 113–126 (1983).

Schendel, J. S., R. E. Stickel, C. A. van Dijk, S. T. Sandholm, D. D. Davis, and J. D. Bradshaw, "Atmospheric Ammonia Measurement Using a VUV/Photofragmentation Laser-Induced Fluorescence Technique," *Appl. Opt., 29,* 4924–4937 (1990).

Schiff, H. I., D. R. Hastie, G. I. Mackay, T. Iguchi, and B. A. Ridley, "Tunable Diode Laser Systems for Measuring Trace Gases in Tropospheric Air," *Environ. Sci. Technol., 17,* 352A–364A (1983).

Schiff, H. I., G. I. Mackay, C. Castledine, G. W. Harris, and Q. Tran, "Atmospheric Measurements of Nitrogen Dioxide with a Sensitive Luminol Instrument," *Water, Air Soil Pollution, 30,* 105–114 (1986).

Schiff, H. I., D. R. Karecki, G. W. Harris, D. R. Hastie, and G. I. Mackay, "A Tunable Diode Laser System for Aircraft Measurements of Trace Gases," *J. Geophys. Res., 95,* 10147–10153 (1990).

Schiff, H. I., G. I. Mackay, and J. Bechara, "The Use of Tunable Diode Laser Absorption Spectroscopy for Atmospheric Measurements," *Res. Chem. Intermed., 20,* 525–556 (1994a).

Schiff, H. I., G. I. Mackay, and J. Bechara, "The Use of Tunable Diode Laser Absorption Spectroscopy for Atmospheric Measurements," in *Air Monitoring by Spectroscopic Techniques* (M. W. Sigrist, Ed.), *Chemical Analysis Series,* Vol. 127, pp. 239–333, Wiley, New York, 1994b.

Schmidt, S., M. F. Appel, R. M. Garnica, R. N. Schindler, and Th. Benter, "Atmospheric Pressure Laser Ionization (APLI). A New Analytical Technique for Highly Selective Detection of Ultra-Low Concentrations in the Gas Phase," *Anal Chem.,* in press (1999).

Schneider, J., V. Bürger, and F. Arnold, "Methyl Cyanide and Hydrogen Cyanide Measurements in the Lower Stratosphere: Implications for Methyl Cyanide Sources and Sinks," *J. Geophys. Res., 102,* 25501–25506 (1997).

Schrader, B., "Micro Raman, Fluorescence, and Scattering Spectroscopy of Single Particles," in *Physical and Chemical Characterization of Individual Airborne Particles* (K. R. Spurny, Ed.), Chap. 19, pp. 358–379, Ellis Horwood, Chichester, 1986.

Schreiner, J., C. Voigt, K. Mauersberger, P. McMurry, and P. Ziemann, "Aerodynamic Lens System for Producing Particle Beams at Stratospheric Pressures," *Aerosol Sci. Technol., 29,* 50–56 (1998).

Schuetzle, D., A. L. Crittenden, and R. J. Charlson, "Application of Computer Controlled High Resolution Mass Spectrometry to the Analysis of Air Pollutants," *J. Air Pollut. Control Assoc., 23,* 704–709 (1973).

Schultz, M., M. Heitlinger, D. Mihelcic, and A. Volz-Thomas, "Calibration Source for Peroxy Radicals with Built-In Actinometry Using H_2O and O_2 Photolysis at 185 nm," *J. Geophys. Res., 100,* 18811–18816 (1995).

Schwarz, F. P., H. Okabe, and J. K. Whittaker, "Fluorescence Detection of Sulfur Dioxide in Air at the Parts Per Billion Level," *Anal. Chem., 46,* 1024–1028 (1974).

Scott, W. E., E. R. Stephens, P. L. Hanst, and R. C. Doerr, "Further Developments in the Chemistry of the Atmosphere," Paper presented at the 22nd Midyear Meeting of the American Petroleum Institute's Division of Refining, Philadelphia, PA, May 14, 1957.

Shaw, W. J., C. W. Spicer, and D. V. Kenny, "Eddy Correlation Fluxes of Trace Gases Using a Tandem Mass Spectrometer," *Atmos. Environ., 32,* 2887–2898 (1998).

Shepson, P. B., T. E. Kleindienst, and H. B. McElhoe, "A Cryogenic Trap/Porous Polymer Sampling Technique for the Quantitative Determination of Ambient Volatile Organic Compound Concentrations," *Atmos. Environ., 21,* 579–587 (1987).

Shepson, P. B., D. R. Hastie, H. I. Schiff, M. Polizzi, J. W. Bottenheim, K. Anlauf, G. I. Mackay, and D. R. Karecki, "Atmospheric Concentrations and Temporal Variations of C_1–C_3 Carbonyl Compounds at Two Rural Sites in Central Ontario," *Atmos. Environ., 25A,* 2001–2015 (1991).

Shields, H. C., and C. J. Weschler, "Analysis of Ambient Concentrations of Organic Vapors with a Passive Sampler," *JAPCA, 37,* 1039–1045 (1987).

Sickles, J. E., II, L. L. Hodson, E. E. Rickman, Jr., M. L. Saeger, D. L. Hardison, A. R. Turner, C. K. Sokol, E. D. Estes, and R. J. Paur, "Comparison of the Annular Denuder System and the Transition Flow Reactor for Measurements of Selected Dry Deposition Species," *JAPCA, 39,* 1218–1224 (1989).

Simeonsson, J. B., G. W. Lemire, and R. C. Sausa, "Laser-Induced Photofragmentation/Photoionization Spectrometry: A Method for Detecting Ambient Oxides of Nitrogen," *Anal. Chem., 66,* 2272–2278 (1994).

Simon, P. K., and P. K. Dasgupta, "Continuous Automated Measurement of Gaseous Nitrous and Nitric Acids and Particulate Nitrite and Nitrate," *Environ. Sci. Technol., 29,* 1534–1541 (1995).

Singh, H. B., W. Viezee, and L. J. Salas, "Measurements of Selected C_2–C_5 Hydrocarbons in the Troposphere: Latitudinal, Vertical, and Temporal Variations," *J. Geophys. Res., 93,* 15861–15878 (1988).

Singh, H. B., E. Condon, J. Vedeer, D. O'Hara, B. A. Ridley, B. W. Gandrud, J. D. Shetter, L. J. Salas, B. Huebert, G. Hübler, M. A.

Carroll, D. L. Albritton, D. D. Davis, J. D. Bradshaw, S. T. Sandholm, M. O. Rodgers, S. M. Beck, G. L. Gregory, and P. J. LeBel, "Peroxyacetyl Nitrate Measurements during CITE 2: Atmospheric Distribution and Precursor Relationships," *J. Geophys. Res., 95,* 10163–10178 (1990).

Singh, H. B., W. Viezee, Y. Chen, A. N. Thakur, Y. Kondo, R. W. Talbot, G. L. Gregory, G. W. Sachse, D. R. Blake, J. D. Bradshaw, Y. Wang, and D. J. Jacob, "Latitudinal Distribution of Reactive Nitrogen in the Free Troposphere over the Pacific Ocean in Late Winter/Early Spring," *J. Geophys. Res., 103,* 28237–28246 (1998).

Sinha, M. P., C. E. Giffin, D. D. Norris, T. J. Estes, V. L. Vilker, and S. K. Friedlander, "Particle Analysis by Mass Spectrometry," *J. Colloid Interface Sci., 87,* 140–153 (1982).

Sinha, M. P., "Laser-Induced Volatilization and Ionization of Microparticles," *Rev. Sci. Instrum., 55,* 886–891 (1984).

Sinha, M. P., and S. K. Friedlander, "Real-Time Measurement of Sodium Chloride in Individual Aerosol Particles by Mass Spectrometry," *Anal. Chem., 57,* 1880–1883 (1985).

Sinha, M. P., and S. K. Friedlander, "Mass Distribution of Chemical Species in a Polydisperse Aerosol: Measurement of Sodium Chloride in Particles by Mass Spectrometry," *J. Colloid Inteface Sci., 112,* 573–582 (1986).

Sirju, A.-P., and P. B. Shepson, "Laboratory and Field Investigation of the DNPH Cartridge Technique for the Measurement of Atmospheric Carbonyl Compounds," *Environ. Sci. Technol., 29,* 384–392 (1995).

Sjödin, A., "Studies of the Diurnal Variation of Nitrous Acid in Urban Air," *Environ. Sci. Technol., 22,* 1086–1089 (1988).

Skoog, D. A., F. J. Holler, and T. A. Nieman, *Principles of Instrumental Analysis,* 5th ed., Harcourt Brace & Company, Philadelphia, PA, 1998.

Slemr, F., G. W. Harris, D. R. Hastie, G. I. Mackay, and H. I. Schiff, "Measurement of Gas Phase Hydrogen Peroxide in Air by Tunable Diode Laser Absorption Spectroscopy," *J. Geophys. Res., 91,* 5371–5378 (1986).

Slemr, J., W. Junkermann, and A. Volz-Thomas, "Temporal Variations in Formaldehyde, Acetaldehyde, and Acetone and Budget of Formaldehyde at a Rural Site in Southern Germany," *Atmos. Environ., 30,* 3667–3676 (1996).

Smith, G. P., and D. R. Crosley, "A Photochemical Model of Ozone Interference Effects in Laser Detection of Tropospheric OH," *J. Geophys. Res., 95,* 16427–16442 (1990).

Smith, J. P., and S. Solomon, "Atmospheric NO_3. 3. Sunrise Disappearance and the Stratospheric Profile," *J. Geophys. Res., 95,* 13819–13827 (1990).

Smith, J. P., S. Solomon, R. W. Sanders, H. L. Miller, L. M. Perliski, J. G. Keys, and A. L. Schmeltekopf, "Atmospheric NO_3. 4. Vertical Profiles at Middle and Polar Latitudes at Sunrise," *J. Geophys. Res., 98,* 8983–8989 (1993).

Smith, N., J. M. C. Plane, C.-F. Nien, and P. A. Solomon, "Nighttime Radical Chemistry in the San Joaquin Valley," *Atmos. Environ., 29,* 2887–2897 (1995).

Solberg, S., T. Krognes, F. Stordal, Ø. Hov, H. J. Beine, D. A. Jaffe, K. C. Clemitshaw, and S. A. Penkett, "Reactive Nitrogen Compounds at Spitsbergen in the Norwegian Arctic," *J. Atmos. Chem., 28,* 209–225 (1997).

Solomon, P. A., S. M. Larson, T. Fall, and G. R. Cass, "Basinwide Nitric Acid and Related Species Concentrations Observed during the Claremont Nitrogen Species Comparison Study," *Atmos. Environ., 22,* 1587–1594 (1988).

Solomon, P. A., L. G. Salmon, T. Fall, and G. R. Cass, "Spatial and Temporal Distribution of Atmospheric Nitric Acid and Particulate Nitrate Concentrations in the Los Angeles Area," *Environ. Sci. Technol., 26,* 1594–1601 (1992).

Spagnolo, G. S., and D. Paoletti, "Automatic System for Three Fractions Sampling of Aerosol Particles in Outdoor Environments," *J. Air Waste Manage. Assoc., 44,* 702–706 (1994).

Spicer, C. W., "Photochemical Atmospheric Pollutants Derived from Nitrogen Oxides," *Atmos. Environ., 11,* 1089–1095 (1977).

Spicer, C. W., D. V. Kenny, G. F. Ward, and I. H. Billick, "Transformations, Lifetimes, and Sources of NO_2, HONO, and HNO_3 in Indoor Environments," *J. Air Waste Manage. Assoc., 43,* 1479–1485 (1993a).

Spicer, C. W., Y. Yanagisawa, J. D. Mulik, and I. H. Billick, "The Prevalence of Nitrous Acid in Indoor Air and Its Impact on NO_2 Measurement Made by Passive Samplers," *Indoor Air '93, 3,* 277–282 (1993b).

Spicer, C. W., D. V. Kenny, W. J. Shaw, K. M. Busness, and E. G. Chapman, "A Laboratory in the Sky—New Frontiers in Measurements Aloft," *Environ. Sci. Technol., 28,* 412A–420A (1994a).

Spicer, C. W., D. V. Kenny, G. F. Ward, I. H. Billick, and N. P. Leslie, "Evaluation of NO_2 Measurement Methods for Indoor Air Quality Applications," *J. Air Waste Manage Assoc., 44,* 163–168 (1994b).

Spicer, C. W., E. G. Chapman, B. J. Finlayson-Pitts, R. A. Plastridge, J. M. Hubbe, J. M. Fast, and C. M. Berkowitz, "Unexpectedly High Concentrations of Molecular Chlorine in Coastal Air," *Nature, 394,* 353–356 (1998).

Spivakovsky, C. M., R. Yevich, J. A. Logan, S. C. Wofsy, M. B. McElroy, and M. J. Prather, "Tropospheric OH in a Three-Dimensional Chemical Tracer Model: An Assessment Based on Observations of CH_3CCl_3," *J. Geophys. Res., 95,* 18441–18471 (1990).

Spurny, K. R., "Sampling Methods and Sampling Preparation," in *Physical and Chemical Characterization of Individual Airborne Particles* (K. R. Spurny, Ed.), Chap. 3, pp. 40–71, Ellis Horwood, Chichester, 1986.

Staffelbach, T., A. Neftel, and P. K. Dasgupta, "Artifact Peroxides Produced during Cryogenic Sampling of Ambient Air," *Geophys. Res. Lett., 22,* 2605–2608 (1995).

Staffelbach, T. A., G. L. Kok, B. G. Heikes, B. McCully, G. I. Mackay, D. R. Karecki, and H. I. Schiff, "Comparison of Hydroperoxide Measurements Made during the Mauna Loa Observatory Photochemistry Experiment 2," *J. Geophys. Res., 101,* 14729–14739 (1996).

Stephens, E. R., P. L. Hanst, R. C. Doerr, and W. E. Scott, "Reactions of Nitrogen Dioxide and Organic Compounds in Air," *Ind. Eng. Chem., 48,* 1498–1504 (1956a).

Stephens, E. R., W. E. Scott, P. L. Hanst, and R. C. Doerr, "Recent Developments in the Study of the Organic Chemistry of the Atmosphere," *J. Air Pollut. Control Assoc., 6,* 159–165 (1956b).

Stephens, E. R., "Long-Path Infrared Spectroscopy for Air Pollution Research," *Soc. Appl. Spectrosc., 12,* 80–84 (1958).

Stephens, E. R., and M. A. Price, "Smog Aerosol: Infrared Spectra," *Science, 168,* 1584–1586 (1970).

Stephens, E. R., and M. A. Price, "Comparison of Synthetic and Smog Aerosols," *J. Colloid Interface Sci., 39,* 272–286 (1972).

Stephens, E. R., "Valveless Sampling of Ambient Air for Analysis by Capillary Gas Chromatography," *JAPCA, 39,* 1202–1205 (1989).

Stevens, P. S., J. H. Mather, and W. H. Brune, "Measurement of Tropospheric OH and HO_2 by Laser-Induced Fluorescence at Low Pressure," *J. Geophys. Res., 99,* 3543–3557 (1994).

Stevens, P. S., J. H. Mather, W. H. Brune, F. Eisele, D. Tanner, A. Jefferson, C. Cantrell, R. Shetter, S. Sewall, A. Fried, B. Henry, E. Williams, K. Baumann, P. Goldan, and W. Kuster, "HO_2/OH and RO_2/HO_2 Ratios during the Tropospheric OH

Photochemistry Experiment: Measurement and Theory," *J. Geophys. Res., 102,* 6379–6391 (1997).

Stoffels, J. J., "A Direct-Inlet Mass Spectrometer for Real-Time Analysis of Airborne Particles," *Int. J. Mass Spectrom. Ion Phys., 40,* 217–222 (1981a).

Stoffels, J. J., "A Direct Inlet for Surface-Ionization Mass Spectrometry of Airborne Particles," *Int. J. Mass Spectrom. Ion Phys., 40,* 223–234 (1981b).

Stoffels, J. J., and C. R. Lagergren, "On the Real-Time Measurement of Particles in Air by Direct-Inlet Surface-Ionization Mass Spectrometry," *Int. J. Mass Spectrom. Ion Phys., 40,* 243–254 (1981).

Stoffels, J. J., and J. Allen, "Mass Spectrometry of Single Particles *In Situ*," in *Physical and Chemical Characterization of Individual Airborne Particles* (K. R. Spurny, Ed.), Chap. 20, pp. 380–399, Ellis Horwood, Chichester, 1986.

Stolzenburg, M. R., and P. H. McMurry, "An Ultrafine Aerosol Condensation Nucleus Counter," *Aerosol Sci. Technol., 14,* 48–65 (1991).

Stutz, J., and U. Platt, "Numerical Analysis and Estimation of the Statistical Error of Differential Optical Absorption Spectroscopy Measurements with Least-Squares Methods," *Appl. Opt., 35,* 6041–6053 (1996).

Stutz, J., and U. Platt, "Improving Long-Path Differential Optical Absorption Spectroscopy with a Quartz-Fiber Mode Mixer," *Appl. Opt., 36,* 1105–1115 (1997).

Suppan, P., P. Fabian, L. Vyras, and S. E. Gryning, "The Behaviour of Ozone and Peroxyacetyl Nitrate Concentrations for Different Wind Regimes during the MEDCAPHOT-TRACE Campaign in the Greater Area of Athens, Greece," *Atmos. Environ., 32,* 2089–2102 (1998).

Sweet, C. W., and D. F. Gatz, "Summary and Analysis of Available $PM_{2.5}$ Measurements in Illinois," *Atmos. Environ., 32,* 1129–1133 (1998).

Talbot, R. W., K. M. Beecher, R. C. Harriss, and W. R. Cofer III, "Atmospheric Geochemistry of Formic and Acetic Acids at a Mid-Latitude Temperate Site," *J. Geophys. Res., 93,* 1638–1652 (1988).

Talbot, R. W., A. S. Vijgen, and R. C. Harriss, "Measuring Tropospheric HNO_3: Problems and Prospects for Nylon Filter and Mist Chamber Techniques," *J. Geophys. Res., 95,* 7553–7561 (1990).

Talbot, R. W., J. D. Bradshaw, S. T. Sandholm, H. B. Singh, G. W. Sachse, J. Collins, G. L. Gregory, B. Anderson, D. Blake, J. Barrick, E. V. Browell, K. I. Klemm, B. L. Lefer, O. Klemm, K. Gorzelska, J. Olson, D. Herlth, and D. O'Hara, "Summertime Distribution and Relations of Reactive Odd Nitrogen Species and NO_y in the Troposphere over Canada," *J. Geophys. Res., 99,* 1863–1885 (1994).

Tanner, D. J., and F. L. Eisele, "Ions in Oceanic and Continental Air Masses," *J. Geophys. Res., 96,* 1023–1031 (1991).

Tanner, D. J., and F. L. Eisele, "Present OH Measurement Limits and Associated Uncertainties," *J. Geophys. Res., 100,* 2883–2892 (1995).

Tanner, D. J., A. Jefferson, and F. L. Eisele, "Selected Ion Chemical Ionization Mass Spectrometric Measurement of OH," *J. Geophys. Res., 102,* 6415–6425 (1997).

Tanner, R. L., A. H. Miguel, J. B. de Andrade, J. S. Gaffney, and G. E. Streit, "Atmospheric Chemistry of Aldehydes: Enhanced Peroxyacetyl Nitrate Formation from Ethanol-Fueled Vehicular Emissions," *Environ. Sci. Technol., 22,* 1026–1034 (1988).

Tanner, R. L., and D. E. Schorran, "Measurements of Gaseous Peroxides near the Grand Canyon—Implication for Summertime Visibility Impairment from Aqueous-Phase Secondary Sulfate Formation," *Atmos. Environ., 29,* 1113–1122 (1995).

Tanner, R. L., B. Zielinska, E. Uberna, G. Harshfield, and A. P. McNichol, "Concentrations of Carbonyl Compounds and the Carbon Isotopy of Formaldehyde at a Coastal Site in Nova Scotia during the NARE Summer Intensive," *J. Geophys. Res., 101,* 28961–28970 (1996).

Thakur, A. N., H. B. Singh, P. Mariani, Y. Chen, Y. Wang, D. J. Jacob, G. Brasseur, J.-F. Müller, and M. Lawrence, "Distribution of Reactive Nitrogen Species in the Remote Free Troposphere: Data and Model Comparisons," *Atmos. Environ., 33,* 1403–1422 (1999).

Thomson, D. S., A. M. Middlebrook, and D. M. Murphy, "Thresholds for Laser-Induced Ion Formation from Aerosols in a Vacuum Using Ultraviolet and Vacuum-Ultraviolet Laser Wavelengths," *Aerosol Sci. Technol., 26,* 544–559 (1997).

Tobias, H. J., P. M. Kooiman, K. S. Docherty, and P. J. Ziemann, "Real-Time Chemical Analysis of Organic Aerosols Using a Thermal Desorption Particle Beam Mass Spectrometer," *Aerosol Sci. Technol.,* in press (1999).

Trapp, D., and C. De Serves, "Intercomparison of Formaldehyde Measurements in the Tropical Atmosphere," *Atmos. Environ., 29,* 3239–3243 (1995).

Traxel, K., and U. Wätjen, "Particle-Induced X-Ray Emission Analysis (PIXE) of Aerosols," in *Physical and Chemical Characterization of Individual Airborne Particles* (K. R. Spurny, Ed.), Chap. 16, pp. 298–330, Ellis Horwood, Chichester, 1986.

Tremmel, H. G., W. Junkermann, and F. Slemr, "On the Distribution of Hydrogen Peroxide in the Lower Troposphere over the Northeastern United States during Late Summer 1988," *J. Geophys. Res., 98,* 1083–1099 (1993).

Tremmel, H. G., W. Junkermann, and F. Slemr, "Distribution of Organic Hydroperoxides during Aircraft Measurements over the Northeastern United States," *J. Geophys. Res., 99,* 5295–5307 (1994).

Tsai, C.-J., and S.-N. Perng, "Artifacts of Ionic Species for Hi-Vol PM_{10} and PM_{10} Dichotomous Samplers," *Atmos. Environ., 32,* 1605–1613 (1998).

Tuazon, E. C., R. A. Graham, A. M. Winer, R. R. Easton, J. N. Pitts, Jr., and P. L. Hanst, "A Kilometer Pathlength Fourier-Transform Infrared System for the Study of Trace Pollutants in Ambient and Synthetic Atmospheres," *Atmos. Environ., 12,* 865–875 (1978).

Tuazon, E. C., A. M. Winer, R. A. Graham, and J. N. Pitts, Jr., "Atmospheric Measurements of Trace Pollutants by Kilometer-Pathlength FT-IR Spectroscopy," *Adv. Environ. Sci. Technol., 10,* 259–300 (1980).

Tuazon, E. C., A. M. Winer, and J. N. Pitts, Jr., "Trace Pollutant Concentrations in a Multiday Smog Episode in the California South Coast Air Basin by Long Path Length Fourier Transform Infrared Spectroscopy," *Environ. Sci. Technol., 15,* 1232–1237 (1981).

Tuckermann, M., R. Ackermann, C. Gölz, H. Lorenzen-Schmidt, T. Senne, J. Stutz, B. Trost, W. Unold, and U. Platt, "DOAS-Observation of Halogen Radical-Catalysed Arctic Boundary Layer Ozone Destruction during the ARCTOC-Campaigns 1995 and 1996 in Ny-Ålesund, Spitsbergen," *Tellus, 49B,* 533–555 (1997).

Turner, J. R., and S. V. Hering, "Greased and Oiled Substrates as Bounce-Free Impaction Surfaces," *J. Aerosol Sci., 18,* 215–224 (1987).

Turpin, B. J., S.-P. Liu, K. S. Podolske, M. S. P. Gomes, S. J. Eisenreich, and P. H. McMurry, "Design and Evaluation of a Novel Diffusion Separator for Measuring Gas/Particle Distributions of Semivolatile Organic Compounds," *Environ. Sci. Technol., 27,* 2441–2449 (1993).

Turpin, B. J., J. J. Huntzicker, and S. V. Hering, "Investigation of Organic Aerosol Sampling Artifacts in the Los Angeles Basin," *Atmos. Environ., 28,* 3061–3071 (1994).

Van Valin, C. C., M. Luria, J. D. Ray, and J. F. Boatman, "Hydrogen Peroxide and Ozone over the Northeastern United States in June 1987," *J. Geophys. Res., 95*, 5689–5695 (1990).

Večeřa, Z., and P. K. Dasgupta, "Measurement of Ambient Nitrous Acid and a Reliable Calibration Source for Gaseous Nitrous Acid," *Environ. Sci. Technol., 25*, 255–260 (1991).

Vega, E., I. García, D. Apam, M. E. Ruíz, and M. Barbiaux, "Application of a Chemical Mass Balance Receptor Model to Respirable Particulate Matter in Mexico City," *J. Air Waste Manage. Assoc., 47*, 524–529 (1997).

Viggiano, A. A., "*In situ* Mass Spectrometry and Ion Chemistry in the Stratosphere and Troposphere," *Mass Spectrom. Rev., 12*, 115–137 (1993).

Vogt, R., and B. J. Finlayson-Pitts, "A Diffuse Reflectance Infrared Fourier Transform Spectroscopic (DRIFTS) Study of the Surface Reaction of NaCl with Gaseous NO_2 and HNO_3," *J. Phys. Chem., 98*, 3747–3755 (1994); correction, *ibid, 95*, 13052 (1995).

Volz-Thomas, A., *et al.*, "The PRICE (Peroxy Radical Intercomparison Exercise): An Introduction," *Proc. XXI IUGG*, 13263 (1995).

Wang, Y., T. S. Raihala, A. P. Jackman, and R. St. John, "Use of Tedlar Bags in VOC Testing and Storage: Evidence of Significant VOC Losses," *Environ. Sci. Technol., 30*, 3115–3117 (1996).

Wayne, R. P., "Fourier Transformed," *Chem. Br., 23*, 440–445 (1987).

Weatherburn, M. W., "Phenol–Hypochlorite Reaction for Determination of Ammonia," *Anal. Chem., 39*, 971–974 (1967).

Weaver, A., S. Solomon, R. W. Sanders, K. Arpag, and H. L. Miller, Jr., "Atmospheric NO_3. 5. Off-Axis Measurements at Sunrise: Estimates of Tropospheric NO_3 at 40°N," *J. Geophys. Res., 101*, 18605–18612 (1996).

Weber, R. J., M. R. Stolzenburg, S. N. Pandis, and P. H. McMurry, "Inversion of Ultrafine Condensation Nucleus Counter Pulse Height Distributions to Obtain Nanoparticle (~3–10 nm) Size Distributions," *J. Aerosol Sci., 29*, 601–615 (1998).

Wedding, J. B., Y. J. Kim, and J. P. Lodge, Jr., "Interpretation of Selected EPA Field Data on Particulate Matter Samplers: Rubidoux and Phoenix II," *JAPCA, 36*, 164–170 (1986).

Weinheimer, A. J., T. L. Campos, and B. A. Ridley, "The In-Flight Sensitivity of Gold-Tube NO_y Converters to HCN," *Geophys. Res. Lett., 25*, 3943–3946 (1998).

Weinstein-Lloyd, J. B., J. H. Lee, P. H. Daum, L. I. Kleinman, L. J. Nunnermacker, S. R. Springston, and L. Newman, "Measurements of Peroxides and Related Species during the 1995 Summer Intensive of the Southern Oxidants Study in Nashville, Tennessee," *J. Geophys. Res., 103*, 22361–22373 (1998).

Weiss, R. E., and A. P. Waggoner, "Optical Measurements of Airborne Soot in Urban, Rural and Remote Locations," in *Particulate Carbon: Atmospheric Life Cycle* (G. T. Wolff and R. L. Klimisch, Eds.), pp. 317–325, Plenum, New York, 1982.

Wendel, G. J., D. H. Stedman, C. A. Cantrell, and L. Damrauer, "Luminol-Based Nitrogen Dioxide Detector," *Anal. Chem., 55*, 937–940 (1983).

Werle, P., R. Muecke, and F. Slemr, "Development of a Prototype IR-FM Absorption Spectrometer: Design Criteria and System Performance," in *Monitoring of Gaseous Pollutants by Tunable Diode Lasers* (R. Grisar, H. Boettner, M. Tacke and G. Restelli, Eds.), pp. 169–182, Kluwer Academic, Dordrecht, The Netherlands, 1992.

West, P. W., and G. C. Gaeke, "Fixation of Sulfur Dioxide as Disulfitomercurate(II) and Subsequent Colorimetric Estimation," *Anal. Chem., 28*, 1816–1819 (1956).

Westberg, H., and P. Zimmerman, "Analytical Methods Used to Identify Nonmethane Organic Compounds in Ambient Atmospheres," *Adv. Chem. Ser., 232*, 275–290 (1993).

Whitby, K. T., and W. E. Clark, "Electric Aerosol Particle Counting and Size Distribution for the 0.015 to 1 μm Size Range," *Tellus, 18*, 573–586 (1966).

Whitby, K. T., and K. Willeke, "Single Particle Optical Counters: Principles and Field Use," in *Aerosol Measurement* (D. A. Lundgren, F. S. Harris, Jr., W. H. Marlow, M. Lippman, W. E. Clark, and M. D. Durham, Eds.), pp. 145–182, University Press of Florida, Gainesville, FL, 1979.

White, J. U., "Long Optical Paths of Large Aperture," *J. Opt. Soc. Am., 32*, 285–288 (1942).

White, J. U., "Very Long Optical Paths in Air," *J. Opt. Soc. Am., 66*, 411–416 (1976).

Whittle, E., D. A. Dows, and G. C. Pimentel, "Matrix Isolation Method for the Experimental Study of Unstable Species," *J. Chem. Phys., 22*, 1943 (1954).

Wiedensohler, A., P. Aalto, D. Covert, J. Heintzenberg, and P. McMurry, "Intercomparison of Three Methods to Determine Size Distributions of Ultrafine Aerosols with Low Number Concentrations," *J. Aerosol Sci., 24*, 551–554 (1993).

Wiedensohler, A., P. Aalto, D. Covert, J. Heintzenberg, and P. H. McMurry, "Intercomparison of Four Methods to Determine Size Distributions of Low-Concentration (~100 cm^{-3}), Ultrafine Aerosols ($3 < D_p < 10$ nm) with Illustrative Data from the Arctic," *Aerosol Sci. Technol., 21*, 95–109 (1994).

Wiener, R. W., K. Okazaki, and K. Willeke, "Influence of Turbulence on Aerosol Sampling Efficiency," *Atmos. Environ., 22*, 917–928 (1988).

Wieser, P., and R. Wurster, "Application of Laser-Microprobe Mass Analysis to Particle Collections," in *Physical and Chemical Characterization of Individual Airborne Particles* (K. R. Spurny, Ed.), Chap. 14, pp. 251–270, Ellis Horwood, 1986.

Willeke, K., Ed., *Generation of Aerosols*, Ann Arbor Science Publishers, Ann Arbor, MI, 1980.

Williams, E. J., S. T. Sandholm, J. D. Bradshaw, J. S. Schendel, A. O. Langford, P. K. Quinn, P. J. LeBel, S. A. Vay, P. D. Roberts, R. B. Norton, B. A. Watkins, M. P. Buhr, D. D. Parrish, J. G. Calvert, and F. C. Fehsenfeld, "An Intercomparison of Five Ammonia Measurement Techniques," *J. Geophys. Res., 97*, 11591–11611 (1992).

Williams, E. J., J. M. Roberts, K. Baumann, S. B. Bertman, S. Buhr, R. B. Norton, and F. C. Fehsenfeld, "Variations in NO_y Composition at Idaho Hill, Colorado," *J. Geophys. Res., 102*, 6297–6314 (1997).

Williams, E. J., K. Baumann, J. M. Roberts, S. B. Bertman, R. B. Norton, F. C. Fehsenfeld, S. R. Springston, L. J. Nunnermacker, L. Newman, K. Olszyna, J. Meagher, B. Hartsell, E. Edgerton, J. R. Pearson, and M. O. Rodgers, "Intercomparison of Ground-Based NO_y Measurement Techniques," *J. Geophys. Res., 103*, 22261–22280 (1998).

Williams, E. L., II, and D. Grosjean, "Removal of Atmospheric Oxidants with Annular Denuders," *Environ. Sci. Technol., 24*, 811–814 (1990).

Williams, E. L., II, E. Grosjean, and D. Grosjean, "Ambient Levels of the Peroxyacyl Nitrates PAN, PPN, and MPAN in Atlanta, Georgia," *J. Air Waste Manage. Assoc., 43*, 873–879 (1993).

Wilson, J. C., and B. Y. H. Liu, "Aerodynamic Particle Size Measurement by Laser-Doppler Velocimetry," *J. Aerosol Sci., 11*, 139–150 (1980).

Winer, A. M., and H. W. Biermann, "Long Pathlength Differential Optical Absorption Spectroscopy (DOAS) Measurements of Gaseous HONO, NO_2, and HCHO in the California South Coast Air Basin," *Res. Chem. Intermed., 20*, 423–445 (1994).

Wingen, L. M., J. C. Low, and B. J. Finlayson-Pitts, "Chromatography, Absorption, and Fluorescence: A New Instrumental Analysis

Experiment on the Measurement of Polycyclic Aromatic Hydrocarbons in Cigarette Smoke," *J. Chem. Educ.*, 75, 1599–1603 (1998).

Winklmayr, W., G. P. Reischl, A. O. Lindner, and A. Berner, "A New Electromobility Spectrometer for the Measurement of Aerosol Size Distributions in the Size Range from 1 to 100 nm," *J. Aerosol Sci.*, 22, 289–296 (1991).

Witkowski, R. E., W. A. Cassidy, and G. W. Penney, "The Design and Deployment of an ESP Atmospheric Particle Collector at the South Pole Clean Air Facility," *JAPCA*, 38, 1168–1171 (1988).

Wolff, G. T., and R. L. Klimisch, Eds., *Particulate Carbon: Atmospheric Life Cycle*, Plenum, New York, 1982.

Wood, S. H., and K. A. Prather, "Time-of-Flight Mass Spectrometry Methods for Real Time Analysis of Individual Aerosol Particles," *Trends Anal. Chem.*, 17, 346–356 (1998).

Wright, B. M., "A New Dust-Feed Mechanism," *J. Sci. Instrum.*, 27, 12–15 (1950).

Yang, M., J. M. Dale, W. B. Whitten, and J. M. Ramsey, "Laser Desorption Mass Spectrometry of a Levitated Single Microparticle in a Quadrupole Ion Trap," *Anal. Chem.*, 67, 1021–1025 (1995a).

Yang, M., J. M. Dale, W. B. Whitten, and J. M. Ramsey, "Laser Desorption Tandem Mass Spectrometry of Individual Microparticles in an Ion Trap Mass Spectrometer," *Anal. Chem.*, 67, 4330–4334 (1995b).

Yang, M., P. T. A. Reilly, K. B. Boraas, W. B. Whitten, and J. M. Ramsey, "Real-Time Chemical Analysis of Aerosol Particles Using an Ion Trap Mass Spectrometer," *Rapid Commun. Mass Spectrom.*, 10, 347–351 (1996).

Yeh, H.-C., "Electrical Techniques," in *Aerosol Measurement: Principles, Techniques, and Applications* (K. Willeke and P. A. Baron, Eds.), pp. 410–426, Van Nostrand-Reinhold, New York, 1993.

Yokelson, R. J., D. W. T. Griffith, and D. E. Ward, "Open-Path Fourier Transform Infrared Studies of Large-Scale Laboratory Biomass Fires," *J. Geophys. Res.*, 101, 21067–21080 (1996).

Yokelson, R. J., D. E. Ward, R. A. Susott, J. Reardon, and D. W. T. Griffith, "Emissions from Smoldering Combustion of Biomass Measured by Open-Path Fourier Transform Infrared Spectroscopy," *J. Geophys. Res.*, 102, 18865–18877 (1997a).

Yokelson, R. J., J. G. Goode, R. A. Susott, R. E. Babbitt, D. E. Ward, S. P. Baker, W. M. Hao, and D. W. T. Griffith, "Smoke Chemistry Measurements by Airborne Fourier Transform Infrared Spectroscopy (AFTIR)," IGAC International Symposium on Atmospheric Chemistry and Future Global Environment, Nagoya, Japan, November 11–13, 1997b.

Yu, J., H. E. Jeffries, and R. M. Le Lacheur, "Identifying Airborne Carbonyl Compounds in Isoprene Atmospheric Photooxidation Products by Their PFBHA Oximes Using Gas Chromatography/Ion Trap Mass Spectrometry," *Environ. Sci. Technol.*, 29, 1923–1932 (1995).

Yu, J., R. C. Flagan, and J. H. Seinfeld, "Identification of Products Containing –COOH, –OH, and –C=O in Atmospheric Oxidation of Hydrocarbons," *Environ. Sci. Technol.*, 32, 2357–2370 (1998).

Zahniser, M. S., D. D. Nelson, J. B. McManus, J. H. Shorter, J. Wormhoudt, and C. E. Kolb, "Tunable Infrared Laser Spectroscopy for Atmospheric Trace Gas Detection," Presented at the National Meeting of the American Chemical Society, Division of Environmental Chemistry, Preprints of Extended Abstracts, 37, 273–276 (1997).

Zander, R., C. P. Rinsland, C. B. Farmer, J. Namkung, R. H. Norton, and J. M. Russell III, "Concentrations of Carbonyl Sulfide and Hydrogen Cyanide in the Free Upper Troposphere and Lower Stratosphere Deduced from ATMOS/Spacelab 3 Infrared Solar Occultation Spectra," *J. Geophys. Res.*, 93, 1669–1678 (1988).

Zenker, T., H. Fischer, C. Nikitas, U. Parchatka, G. W. Harris, D. Mihelcic, P. Müsgen, H. W. Pätz, M. Schultz, A. Volz-Thomas, R. Schmitt, T. Behmann, M. Weißenmayer, and J. P. Burrows, "Intercomparison of NO, NO_2, NO_y, O_3, and RO_x Measurements during the Oxidizing Capacity of the Tropospheric Atmosphere (OCTA) Campaign 1993 at Izana," *J. Geophys. Res.*, 103, 13615–13634 (1998).

Zhang, D., Y. Maeda, and M. Munemori, "Chemiluminescence Method for Direct Determination of Sulfur Dioxide in Ambient Air," *Anal. Chem.*, 57, 2552–2555 (1985).

Zhang, J.-X., and P. M. Aker, "Spectroscopic Probing of Aerosol Particle Interfaces," *J. Chem. Phys.*, 99, 9366–9375 (1993).

Zhang, X., and P. H. McMurry, "Theoretical Analysis of Evaporative Losses of Adsorbed or Absorbed Species during Atmospheric Aerosol Sampling," *Environ. Sci. Technol.*, 25, 456–459 (1991).

Zhang, X. Q., and P. H. McMurry, "Theoretical Analysis of Evaporative Losses from Impactor and Filter Deposits," *Atmos. Environ.*, 21, 1779–1789 (1987).

Zhou, X., and K. Mopper, "Measurement of Sub-Parts-per-Billion Levels of Carbonyl Compounds in Marine Air by a Simple Cartridge Trapping Procedure Followed by Liquid Chromatography," *Environ. Sci. Technol.*, 24, 1482–1485 (1990).

Zhou, X., and K. Mopper, "Carbonyl Compounds in the Lower Marine Troposphere over the Caribbean Sea and Bahamas," *J. Geophys. Res.*, 98, 2385–2392 (1993).

Zielinska, B., and E. Fujita, "The Composition and Concentration of Hydrocarbons in the Range of C2 to C18 in Downtown Los Angeles, CA," *Res. Chem. Intermed.*, 20, 321–334 (1994).

Zielinska, B., J. C. Sagebiel, G. Harshfield, A. W. Gertler, and W. R. Pierson, "Volatile Organic Compounds up to C_{20} Emitted from Motor Vehicles: Measurement Methods," *Atmos. Environ.*, 30, 2269–2286 (1996).

Ziemann, P. J., P. Liu, D. B. Kittelson, and P. H. McMurry, "Electron Impact Charging Properties of Size-Selected, Submicrometer Organic Particles," *J. Phys. Chem.*, 99, 5126–5138 (1995).

Ziemann, P. J., and P. H. McMurry, "Spatial Distribution of Chemical Components in Aerosol Particles As Determined from Secondary Electron Yield Measurements: Implications for Mechanisms of Multicomponent Aerosol Crystallization," *J. Colloid Interface Sci.*, 193, 250–258 (1997).

Ziemann, P. J., and P. H. McMurry, "Secondary Electron Yield Measurements as a Means for Probing Organic Films on Aerosol Particles," *Aerosol Sci. Technol.*, 28, 77–90 (1998).

Zika, R. G., and E. S. Saltzman, "Interaction of Ozone and Hydrogen Peroxide in Water: Implications for Analysis of H_2O_2 in Air," *Geophys. Res. Lett.*, 9, 231–234 (1982).

CHAPTER

12

Homogeneous and Heterogeneous Chemistry in the Stratosphere

Although widespread interest in the chemistry of the stratosphere followed that of the troposphere, there has been an explosion of research in this area over the past three decades. This was prompted by the recognition that oxides of nitrogen (Crutzen, 1971) emitted directly into the stratosphere by the proposed fleet of supersonic transports, SSTs (Johnston, 1971; Crutzen, 1972, 1974; McElroy et al., 1974), and chlorine (Stolarski and Cicerone, 1974) from chlorofluorocarbons (CFCs) emitted at the earth's surface (Molina and Rowland, 1974; Cicerone et al., 1974; Rowland and Molina, 1975; Stolarski and Rundel, 1975) could lead to depletion of ozone in the stratosphere. While the large fleets and extensive usage of SSTs did not materialize as originally anticipated, similar issues have again been discussed with respect to a potential next generation of aircraft, the high-speed civil transport (HSCT). Meanwhile, the discovery of the Antarctic "ozone hole" in 1985 was a dramatic confirmation of the role of CFCs in ozone depletion and also highlighted the importance of heterogeneous chemistry, that is, reactions of gases on and/or in particles.

We describe in this chapter the homogeneous and heterogeneous chemistry of the stratosphere and the effects of anthropogenic perturbations on it. The next chapter describes the trends in stratospheric ozone and the control measures that have been instituted to abate its depletion. As we shall see, much of the chemistry is qualitatively the same as that in the troposphere. However, there are major differences due to the atmospheric conditions in the stratosphere. Thus, as seen in Chapter 3.C, ultraviolet light down to about 180 nm penetrates into the stratosphere and is absorbed by O_2 and O_3. Light absorption by O_3, combined with the energy released from the recombination of O with O_2 to form O_3, is responsible for the increase in temperature with altitude characterizing the stratosphere. Upon light absorption, the excess energy that does not go into bond-breaking is released as heat (see Chapter 4.B

for a description of ozone photochemistry). Total pressures are also lower, in the ~1- to 100-Torr range, as is the temperature, ~185–275 K, and it is much drier, typically 2–6 ppm water vapor. As we shall see in this chapter, these differences provide a unique environment for some very interesting and complex chemistry.

A. CHEMISTRY OF THE UNPERTURBED STRATOSPHERE

Figure 12.1 shows typical average vertical ozone profiles measured at Edmonton, Alberta, Canada, in the January–April period in 1980–1982 and in 1993, respectively. The high concentrations of ozone in the stratosphere are evident, demonstrating why this region is often referred to colloquially as the "ozone layer." (The significant decrease in stratospheric ozone from the period 1980–1982 to 1993 will be discussed below. While a small decrease in tropospheric ozone is seen in these particular data, ozone levels have generally increased in the troposphere; see Chapter 14.)

The total ozone integrated through a column in the atmosphere from the earth's surface is often used as a measure of stratospheric ozone, since as seen in Fig. 12.1, about 85–90% of the total ozone is found in this region. *Dobson units* are used to express the amount of total column ozone. One Dobson unit (DU) is the height of the layer of pure gaseous ozone in units of 10^{-5} m that one would have if one separated all of the atmospheric O_3 and compressed it into a layer at 1 atm and 273 K. That is, 100 DU is equivalent to a layer of pure ozone of thickness of 1 mm.

The importance of this high concentration of ozone in the stratosphere resides in its effects on radiation and associated feedbacks. Figure 12.2 shows the ozone absorption spectrum, which was discussed in detail

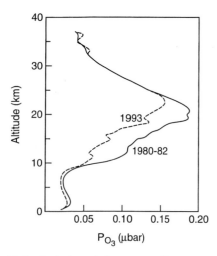

FIGURE 12.1 Typical vertical ozone profiles at Edmonton, Alberta, Canada, in the January–April period for 1980–1982 (solid line) and 1993 (dashed line) (adapted from Kerr et al., 1993).

along with its photochemistry in Chapter 4.B. The strong absorption below 300 nm acts as a filter for the UV solar radiation. As discussed in Chapter 3, this absorption determines the actinic cutoff at the earth's surface of 290 nm. Less total ozone in the stratosphere translates into more UV radiation at the earth's surface, with concomitant increases in skin cancer, for example (Brasseur et al., 1995). Changes in the radiation balance may also occur due to complex feedbacks. For example, more UV might be expected to lead to more ozone formation in the troposphere under conditions where there is sufficient NO_x, but less under remote conditions or in the upper troposphere (see, for example, Liu and Trainer, 1988; Schnell et al., 1991; Thompson, 1991; and Taalas et al., 1997). As discussed in Chapters 14 and 16, not only is ozone a toxic gas for which air quality standards have been set, but it is also a greenhouse gas. In addition, increased radiation reaching the troposphere will increase OH production due to O_3 photolysis to $O(^1D)$, followed by its reaction with gaseous H_2O. This could have major impacts on essentially all species whose concentrations are controlled directly or indirectly by OH. For example, Toumi et al. (1994) suggest that increased oxidation of SO_2 to aerosol sulfate due to this increased OH caused by stratospheric ozone destruction could have climate effects (see Chapter 14.D). Finally, effects on various ecosystems, which are currently not well understood, are anticipated.

Before we discuss the chemistry of the stratosphere in detail, let us first briefly consider how chemicals emitted at the earth's surface are transported into the stratosphere and, conversely, how stratospheric species are transported into the troposphere. This issue is important as it determines which species end up in the stratosphere; as we shall see, it is only those that survive for a sufficiently long time in the troposphere that are transported into the stratosphere. Thus, reactive organics (larger than CH_4) from the troposphere do not survive to reach the stratosphere in appreciable quantities. Similarly, transport from the stratosphere to the troposphere is the major mechanism for removal of the products of the reactions of anthropogenic species that occur in the upper atmosphere. In addition, some O_3 is injected into the troposphere, which is a natural source upon which anthropogenic production of O_3 is imposed (see Chapter 16).

1. Stratosphere–Troposphere Exchange (STE)

As discussed in Chapter 2, the troposphere is characterized by decreasing temperature with altitude, whereas the opposite is true for the stratosphere. In the troposphere, vertical mixing occurs on a time scale

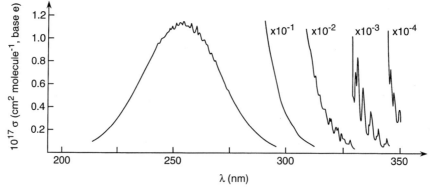

FIGURE 12.2 UV absorption of O_3 at room temperature in the Hartley (200–300 nm) and Huggins (300–360 nm) bands (adapted from Daumont et al., 1992).

of hours to days, whereas the time scale for similar mixing in the stratosphere is months to years.

Figure 12.3 schematically depicts large, global-scale motions that transport tropospheric air into the stratosphere in the tropics. This has been described recently in terms of a "global-scale fluid dynamical suction pump" (Holton et al., 1995), which slowly moves air from the troposphere into the stratosphere in the tropics. This transport is what largely controls the tropospheric lifetimes of species such as the CFCs and N_2O that have no significant tropospheric chemical sinks. As a result of this motion, the mixing ratios of such species in the stratosphere peak in the tropical regions and are approximately the same as their tropospheric mixing ratios. This large-scale upward motion in the tropics and downward motion toward the poles is what is known as *Hadley circulation*.

As air is transported rapidly upward, for example in a convective system, cooling occurs (see Chapter 2), leading to the condensation of water as ice crystals. Because of this removal of water as moist tropospheric air rises, air in the stratosphere is very dry, of the order of a few ppm. Some water is also produced directly in the stratosphere from the oxidation of CH_4 and H_2. The so-called extratropical pump then moves the air poleward and downward at higher latitudes (Path I), warming the air as it descends.

Using the terminology of Hoskins (1991), Holton and co-workers (1995) distinguish between the "overworld," the "lowermost stratosphere," and the "underworld." The overworld and underworld are the regions of the stratosphere and troposphere, respectively, in which air moving along surfaces of constant potential temperature, i.e., isentropic surfaces (see Chapter 2), remains in those regions. The lowermost stratosphere is characterized by the fact that isentropic surfaces cross into the troposphere, as shown in Fig. 12.3. In these regions, the instantaneous tropopause can be strongly deformed, and diabatic heating or cooling and turbulent mixing can occur, providing another mechanism (Path II) for stratosphere–troposphere exchange of air (STE). This horizontal mixing along the isentropic surfaces in the lower stratosphere and the troposphere can be faster than mixing between the lowermost stratosphere and the overworld, due to the slow vertical mixing in the stratosphere. This mechanism for mixing between the lowermost stratosphere and the troposphere, which has been observed from SAGE II measurements of O_3 during the spring (Wang et al., 1998), can be particularly important for species found in, or emitted predominantly into, the lowermost stratosphere, e.g., emissions from supersonic aircraft (vide infra).

There appears to be a subtropical eddy transport barrier (of variable strength) to horizontal mixing of air between the tropical overworld stratosphere and the extratropical regions. This has been described by Plumb (1996) as a "tropical pipe," shown schematically in Fig. 12.4; in this model, meridional circulation from the tropics poleward by large-scale "pumping" described earlier occurs along with isentropic mixing within the Northern and Southern Hemispheres, respectively. The region of limited horizontal mixing around the equator is shown as the tropical pipe. Evidence for this barrier to horizontal mixing in the tropical overworld has been found in different correlations between the concentrations of trace gases in the tropical troposphere compared to the midlatitudes or Antarctica (Plumb, 1996) and in a "tape recorder effect" (see following) on the vertical distribution of water vapor in the tropics (Mote

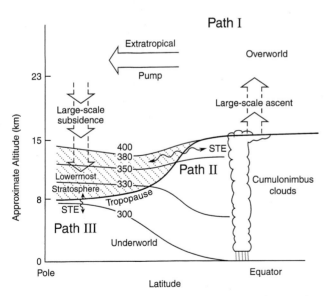

FIGURE 12.3 Schematic of wave-driven extratropical pump that drives global-scale transport from the tropical troposphere to the stratosphere and then poleward. Three possible paths for stratosphere–troposphere exchange (STE) are shown. (Adapted from Holton et al., 1995.)

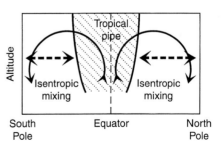

FIGURE 12.4 Schematic of "tropical pipe" model (adapted from Plumb, 1996).

et al., 1995, 1996; Holton et al., 1995). In addition, reduced horizontal transport of aerosols in the tropics in the altitude range of 21–28 km has been observed (Trepte and Hitchman, 1992).

The so-called tape recorder effect is based on the fact that the amount of water vapor entering the stratosphere in the tropics should vary with the tropopause temperature, which has an annual cycle, because water freezes into ice, which then sediments. As a result, water vapor entering the stratosphere is essentially determined by the vapor pressure of ice at the tropopause temperature (e.g., Dessler, 1998). If horizontal mixing is weak, layers with distinct water vapor concentrations result as each layer of air with its characteristic water signature is moved upward from the tropopause. This barrier to horizontal mixing appears to be relatively weak near the bottom of the overworld, stronger a few kilometers up, and then weaker again in the middle stratosphere (Boering et al., 1995; Mote et al., 1995, 1996).

Water vapor concentrations have also been used to show that stratospheric air in the midlatitudes cannot all have originated via the tropical pump. i.e., path I in Fig. 12.3. For example, Dessler et al. (1995b) have shown that water vapor concentrations in the lowermost stratosphere at 37.4°N, 122.1°W are higher than expected for an air mass that has passed through the cold tropical tropopause. Their data are consistent with path II, although as they point out, these measurements do not exclude path III, which represents convective transport from the troposphere to the stratosphere at mid and high latitudes. Lelieveld et al. (1997) report aircraft measurements of CO, O_3, and HNO_3 over western Europe that suggest that tropospheric air can be mixed into the lower stratosphere.

Water vapor, N_2O, CO_2, and O_3 measured in the stratosphere have been used to explore STE. CO_2 concentrations are higher in the midlatitude upper troposphere compared to regions near the equator and hence can be used to distinguish air that has reached the stratosphere via path III compared to path I or II. Similarly, Hintsa and co-workers (1998b) showed that high water vapor mixing ratios (9–20 ppm), which are occasionally encountered in the stratosphere, are consistent with mixing of tropospheric air through paths II and III.

Volk et al. (1996) and Minschwaner et al. (1996) have used measurements of trace gases having a wide range of lifetimes to follow the entrainment of midlatitude air in the lower stratosphere into the tropical stratosphere. They show that while the tropical stratosphere is relatively isolated from the more rapidly horizontally mixed midlatitude region, significant entrainment of midlatitude air into the tropical stratosphere can still occur because the time scale for the ascending air mass is similar to that for entrainment. For example, they conclude that almost half of the tropical stratospheric air at 21 km originated in midlatitudes during the approximately 8 months it took to ascend from the tropopause to this altitude. This implies that emissions in the midlatitudes from supersonic aircraft can be transported to the middle and upper stratosphere, in agreement with the interpretation of NO_y and O_3 measurements by Avallone and Prather (1996).

Mixing of tropospheric air masses between the Northern and Southern Hemispheres is relatively slow. For example, a molecule released in one of the hemispheres typically takes about 15 months to reach the other hemisphere, with mixing occurring primarily through the mid to upper troposphere. The region in the tropics in which the large-scale upward motion of tropospheric air occurs is known as the *intertropical convergence zone* (ITCZ). The ITCZ varies with season, lying north of the equator in July and south of the equator in January. Exchange of air between the hemispheres occurs during this seasonal displacement of the ITCZ. Such interhemispheric exchange also occurs due to turbulent mixing in the upper troposphere near the equator.

In short, exchange of air between the Northern and Southern Hemispheres is slow, as is that between the troposphere and stratosphere, both being on the time scale of about a year (Warneck, 1988). The mechanisms of stratosphere–troposphere exchange are complex but a detailed understanding of these is critical to the assessment of the atmospheric fates of many species, particularly those emitted in the lowermost stratosphere. For reviews of these processes, see Holton et al. (1995), Salby and Garcia (1990), and Mahlman (1997) and for some relevant studies, Langford et al. (1996) and Folkins and Appenzeller (1996).

2. Chapman Cycle and NO_x Chemistry

Chapman (1930) first proposed the fundamental ozone-forming and destruction reactions that lead to a steady-state concentration of O_3 in the stratosphere. These reactions are now known as the *Chapman cycle*:

$$O_2 + h\nu(\lambda < 242 \text{ nm}) \rightarrow 2O, \qquad (1)$$

$$O + O_2 \xrightarrow{M} O_3, \qquad (2)$$

$$O + O_3 \rightarrow 2O_2, \qquad (3)$$

$$O_3 + h\nu(\lambda \leq 336 \text{ nm}) \rightarrow O(^1D) + O_2. \qquad (4)$$

Both O_2 and O_3 absorb light below 242 nm (see Chapter 4.A and 4.B). When stratospheric O_3 is removed by reactions described in detail in this chapter, there is more light available for photolysis of O_2 and the formation of more ozone via reactions (1) and (2). This so called self-healing effect in part mitigates the destruction of ozone by other species such as CFCs, halons, and NO_x.

In the upper stratosphere and mesosphere, a number of models predict ozone concentrations that are smaller by ~10–25% in the 40- to 45-km region compared to the observed concentrations, with the magnitude of the discrepancy increasing with altitude (Allen and Delitsky, 1991; Eluszkiewicz and Allen, 1993). This has been dubbed the "ozone deficit" problem (see Slanger (1994), for a brief review). Part of the discrepancy in the upper stratosphere can be explained from the production of highly vibrationally excited O_2 ($v \geq 26$) produced in the ozone photolysis, reaction (4), followed by its reaction with O_2 to generate O_3 (e.g., see Miller et al. (1994) and references therein). As discussed in more detail later, another possible explanation of this discrepancy lies in a previously unrecognized minor channel in the OH + ClO reaction which generates HCl + $O_2(^1\Delta_g)$, rather than HO_2 + Cl (Lipson et al., 1997; Dubey et al., 1998).

Although there has been some controversy over whether there is indeed a true ozone deficit problem (e.g., Crutzen et al., 1995), a combination of measured concentrations of OH, HO_2, and ClO with photochemical modeling seems to indicate that it may, indeed, exist (Osterman et al., 1997; Crtuzen, 1997), although the source of the discrepancy remains unclear. Measurements of ClO in the upper stratosphere have found concentrations that are much smaller (by a factor of ~2) than predicted by the models (e.g., Dessler et al., 1996; Michelsen et al., 1996). Because of the chlorine chemistry discussed later, model overestimates of ClO will also result in larger predicted losses of O_3 and hence smaller concentrations.

A number of studies have measured the isotopic distribution in atmospheric ozone. There are three naturally occurring isotopes of oxygen, ^{16}O, ^{17}O, and ^{18}O, which might initially be expected to be represented statistically in atmospheric ozone. However, both stratospheric and tropospheric ozone have been measured to be enriched in the heavier isotopes over what one would expect statistically (e.g., see Mauersberger, 1981; Mauersberger et al., 1993; Krankowsky et al., 1995). A variety of explanations of this fractionation have been put forth, including nuclear symmetry restrictions on the O_2 + O reaction that forms O_3 (Gellene, 1996), the preferential dissociation of heavy ozone to form vibrationally excited O_2 ($v \geq 26$) that then reacts with O_2 as discussed to form O_3 (Houston et al., 1996), and contribution of metastable states, effects of mass, symmetry, and third-body reaction efficiencies in the O + O_2 reaction (Sehested et al., 1995, 1998). For a discussion of this area, see Thiemens et al. (1995), Anderson et al. (1995), and Krankowsky and Mauersberger (1996).

During the 1960s as the kinetics of reactions (1)–(4) became well established (Schiff, 1969), it became clear that the cycle represented by reactions (1)–(4) was inconsistent with the observed levels of stratospheric O_3 and that another loss mechanism for O_3 must be operative (Crutzen, 1969). Alternatives involving HO_x species such as

$$O(^1D) + H_2O \rightarrow 2OH, \qquad (5)$$

$$O_3 + OH \rightarrow O_2 + HO_2, \qquad (6)$$

$$O + HO_2 \rightarrow O_2 + OH, \qquad (7)$$

$$HO_2 + O_3 \rightarrow OH + 2O_2, \qquad (8)$$

were also insufficient to explain the observed O_3 in the 30- to 35-km range.

Bates and Hays (1967) showed that the major atmospheric fate of nitrous oxide, N_2O, produced at the earth's surface by biological processes, was transport to the stratosphere followed by photodissociation:

$$N_2O + h\nu \rightarrow N_2 + O(^1D), \qquad (9a)$$
$$\rightarrow NO + N(^4S). \qquad (9b)$$

At that time, the quantum yield of (9b) was believed to be about 0.2, and Bates and Hays estimated concentrations of (NO + NO_2) that could result from this path. Crutzen (1970, 1971) then used these projected concentrations to show that the reactions of oxides of nitrogen, formed from naturally produced N_2O, with O_3 were the key "missing link" in determining stratospheric ozone concentrations:

$$NO + O_3 \rightarrow NO_2 + O_2, \qquad (10)$$

$$NO_2 + O \rightarrow NO + O_2, \qquad (11)$$

$$NO_2 + h\nu \rightarrow NO + O. \qquad (12)$$

Reactions (10) and (11) form a chain for destruction of O_3. This work was cited in the award of the 1995 Nobel Prize in Chemistry to Crutzen, jointly with Rowland and Molina for their work on CFCs (vide infra).

It is now known that the quantum yield of (9b) is less than 1% (Greenblatt and Ravishankara, 1990). However, N_2O is still the major source of oxides of nitrogen in the stratosphere (McElroy and McConnell, 1971) via its minor (compared to loss by photolysis) reaction with electronically excited oxygen atoms,

O(^{1}D), of which ~60% proceeds via (13b) (Cantrell et al., 1994):

$$N_2O + O(^1D) \to N_2 + O_2, \quad (13a)$$

$$k_{13a}^{298K} = 4.9 \times 10^{-11} \text{ cm}^3 \text{ molecule}^{-1} \text{ s}^{-1}$$

(DeMore et al., 1997),

$$\to 2NO, \quad (13b)$$

$$k_{13b}^{298K} = 6.7 \times 10^{-11} \text{ cm}^3 \text{ molecule}^{-1} \text{ s}^{-1}$$

(DeMore et al., 1997).

As discussed in detail in Chapter 14.B.2c, chemical sources of N_2O have also been proposed, in addition to the biological sources, to explain discrepancies in the measured isotope distribution of N_2O in the atmosphere and potential imbalances in the sources and sinks of N_2O (e.g., Prather, 1998). For example, Zipf and Prasad (1998b) observed N_2O formation when O_2 was dissociated to generate oxygen atoms in the presence of N_2. They proposed that highly vibrationally excited O_3 formed in the $O + O_2$ recombination reacted with N_2 to generate N_2O.

Figure 12.5 shows the results of one model (the Lawrence Livermore National Laboratory, LLNL, model) calculation of the vertical ozone profile predicted using only the Chapman cycle reactions of O_x and that for a reference atmosphere containing levels of NO_x, ClO_x, and HO_x believed to be typical of 1960. As expected from the foregoing discussion, the inclusion of the additional species and their chemistry reduces the predicted peak O_3 concentration. These calculations assumed a stratospheric chlorine level of 1.1 ppb; however, at this concentration the predicted effect on the profile is relatively small so that the model predictions do not change significantly if ClO_x chemistry is removed from the model (Kinnison et al., 1988). The inclusion of NO_x and HO_x chemistry is seen in Fig. 12.5 to modify the Chapman cycle predictions significantly, giving predicted profiles that are in much better agreement with measurements.

The production of NO_2, with NO as a possible precursor to NO_2, has been observed when synthetic air or O_2/N_2 mixtures are photolyzed using a deuterium lamp, an argon flash lamp, or a 185-nm mercury line (Zipf and Prasad, 1998a). They proposed that this occurs from the reaction of electronically excited $O_2(B^3\Sigma_u)$ with N_2, or photodissociation of $O_2 \cdot N_2$ dimer, and that the rate of NO_x production from this process could be comparable to that from reaction (13b) (Zipf and Prasad, 1998a; Prasad, 1998). If this proves to be the case, there must be some unidentified NO_x sinks to be consistent with the measured NO_x concentrations in the upper atmosphere.

In 1971, Johnston suggested that anthropogenic emissions of NO_x from a proposed fleet of supersonic transports (SSTs) could cause a reduction in ozone due to the set of chain reactions (10) and (11). At the time, a fleet of 500 SSTs flying seven hours a day in the stratosphere by 1985 was projected, and based on that, Johnston (1971) showed that the emissions would be expected to lead to significant ozone depletion. This was never realized because of the much smaller use of SSTs than projected. However, a subsequent proposal for the development of a high-speed civil transport (HSCT) raised some of the same issues, as discussed in the following section.

B. HIGH-SPEED CIVIL TRANSPORT (HSCT), ROCKETS, AND THE SPACE SHUTTLE

One recent area of concern has been emissions from potential future supersonic aircraft and from existing rocket and space shuttle launches. The emissions from these sources and their impacts on stratospheric chemistry are the focus of this section. Although continuing development of the HSCT has been halted, the chemistry is relevant and hence is discussed here.

1. HSCT

The basis for concern is that the HSCTs are designed to fly at much higher altitudes than the existing commercial jets and hence increase direct emissions into the stratosphere. The HSCT was proposed to carry about 250–300 passengers and to fly from Los Angeles to Tokyo, for example, in about 4 h, compared to more than 10 h for subsonic aircraft. These aircraft emit NO_x, CO, CO_2, H_2O, SO_2, and soot. Our understanding of the chemistry of the stratosphere is now much more detailed. The effects of emissions of NO_x as well

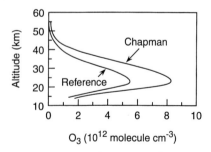

FIGURE 12.5 Model-calculated ozone vertical profiles for a Chapman or O_x model, with only O_2, O, and O_3 as reactive species and the reference atmosphere chosen to be typical of 1960 conditions (adapted from Kinnison et al., 1988).

as other pollutants are now known to be sensitive not only to the altitude and latitude of the emissions but also to the coexistence of other species such as the halogens (vide infra).

At the present time, subsonic aircraft fly in the upper troposphere and, to some extent, the lower stratosphere. For example, Hoinka and Reinhardt (1993) showed that about 44% of the cruising time of aircraft above the North Atlantic was spent in the stratosphere in the 1989–1991 period. This already provides a significant source of NO_x in these regions. On the basis of 3-D model studies, Brasseur and co-workers (Brasseur et al., 1996; Lamarque et al., 1996) estimate that in the upper troposphere at midlatitudes, 15–30% of the NO_x is due to aircraft exhaust, 15–25% from surface sources, and 15–60% from lightning. These estimates are reasonably consistent with earlier suggestions of Pommereau et al. (1989), Ehhalt et al. (1992), and Beck et al. (1992). Model predictions suggest that NO_x emissions from subsonic aircraft lead to O_3 increases in the upper troposphere, although the episodic nature and magnitude would make it difficult to detect (Strand and Hov, 1996; Flatøy and Hov, 1996).

Figure 12.6 shows the anticipated change in the altitude at which fuel is consumed by aircraft if 500 HSCTs (Mach 2.4) are included in the fleet in the year 2015 compared to a completely subsonic aircraft fleet. This distribution of fuel consumption is a measure of the projected change in emissions as a function of altitude. It is seen that a significant increase in fuel consumption with the associated emissions is projected at altitudes around 20 km, which is the optimum cruise altitude with respect to frictional drag for the HSCT flying at Mach 2.4. The residence time of compounds emitted at this altitude is quite long, of the order of months to years, so that emissions can have substantial effects on ozone and other stratospheric species (Stolarski et al., 1995).

Table 12.1 summarizes the potential effects of HSCT emissions on stratospheric ozone and on global climate through changes in radiative forcing (Stolarski et al., 1995) as discussed in Chapter 14. Since the effect of NO_x on O_3 is anticipated to be the most important stratospheric issue, we focus on this here. The other issues are treated briefly where appropriate.

Figure 12.7 shows the model-predicted changes in ozone as a function of altitude attributable to a fleet of 500 HSCTs with a so-called emission index of 5. The emission index (EI_{NO_x}) is the number of grams of NO_2 emitted per kilogram of fuel burned (although the emissions are primarily in the form of NO, they are expressed as if they were NO_2). An EI_{NO_x} of 5 or less is the goal of HSCT engine design. (It should be noted that at very high speeds, NO may produced not only by the engine but also in a thin layer surrounding the aircraft in which the air heats up due to viscous effects at the high speeds. For example, Brooks et al. (1993) show that at speeds of Mach 8 and above, this source could become important and, in fact, at Mach 16, may be about equal to that from combustion in the engine.) While there are clearly significant quantitative differences between the models (due in part to differences in the treatment of transport processes), they all predict ozone loss at altitudes above 24 km. Most models predict ozone increases from 14 to 20 km due to HSCT emissions for reasons to be discussed (Stolarksi et al., 1995).

The chemistry involving NO_x is closely intertwined with that of the halogens (ClO_x and BrO_x) and of HO_x, so that the predicted effects of a given set of emissions from the HSCT depend on these species as well. Because halogen chemistry is treated in more detail in later sections, we shall focus here primarily on the reasons for the different effects of NO_x emissions at different altitudes. How closely these chemistries are intertwined will be apparent in the treatment below of destruction of stratospheric ozone by chlorofluorocarbons (CFCs) and brominated compounds.

The effect of NO_x emissions on stratospheric ozone depends to a great extent on the competition between (11) and (12) for the fate of NO_2 once it has been formed in the NO + O_3 reaction (10):

$$NO_2 + O \rightarrow NO + O_2, \tag{11}$$

$$NO_2 + h\nu \rightarrow NO + O. \tag{12}$$

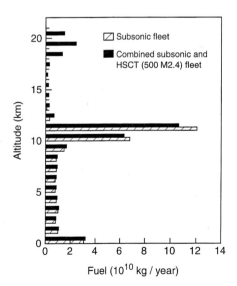

FIGURE 12.6 Total estimated fuel usage at various altitudes for an all-subsonic fleet and for a 2015 fleet that includes a modified subsonic fleet plus 500 Mach 2.4 HSCTs in the year 2015 (adapted from Stolarski et al., 1995).

TABLE 12.1 Some Potential Effects of a Fleet of 500 HSCTs[a]

Pollutant	Potential perturbation	Potential atmospheric interactions
		On Ozone
NO_x	Peak 50–100% increase for an index of 5 g of NO_2/kg of fuel	Ozone depletion by NO_x catalysis at higher altitudes; interference with ClO_x, HO_x, and BrO_x catalysis at lower altitudes
H_2O	Peak 10–20% increase	Increased HO_x formation and hence ozone depletion; interference with NO_x catalysis, enhanced ClO_x catalysis
Sulfur	10–200% increase[b] in surface area in sulfate particles	Increased aerosol surface area, enhanced ozone depletion by ClO_x, decreased ozone depletion by NO_x
Soot	Highly uncertain: 0–300% increase	Additional nucleation sites for aerosols and surface for catalyzed SO_2 oxidation to H_2SO_4
Hydrocarbons	~0.1% increase compared to CH_4	Source of CO, HO_x, and H_2O
CO	5–20% increase	Modification of catalysis by HO_x and NO_x
		On Radiative Forcing
CO_2	Current subsonic, ~3% of CO_2 from fossil fuel; HSCT, ~1%	Direct change in IR radiative forcing
H_2O	Peak 10–20% increase	Direct change in IR radiative forcing; NAT/ice condensation, cirrus cloud formation, change in radiative forcing
Sulfur	10–200% increase[b] in surface area of sulfate particles	Increased aerosol mass loading, change in radiative forcing
Soot	Highly uncertain: 0–300% increase	Additional nucleation sites for aerosols, increased surface area change in radiative forcing
NO_x	Peak 50–100% increase	Ozone depletion, change in radiative forcing

[a] From Stolarski *et al.* (1995).
[b] Depends on assumptions concerning gas-to-particle conversion in the plume.

Reaction (11) destroys an oxygen atom that could otherwise add to O_2 to form O_3, reaction (2), and, in addition, generates another NO that will also react with O_3 directly. The net cycle is then

$$O + O_3 \to 2O_2,$$

i.e., ozone loss. This cycle is the major source of NO_x destruction of ozone in the middle and upper stratosphere. The reaction of NO_2 with O_3 to form the nitrate radical, followed by its photolysis, can also contribute:

$$NO + O_3 \to NO_2 + O_2 \quad (10)$$
$$NO_2 + O_3 \to NO_3 + O_2 \quad (14)$$
$$\underline{NO_3 + h\nu \to NO + O_2} \quad (15)$$
$$\text{Net: } 2O_3 \to 3O_2$$

On the other hand, if the photolysis of NO_2, reaction (12), predominates, there is a cycle formed by reactions (10), (12), and (2):

$$NO + O_3 \to NO_2 + O_2, \quad (10)$$
$$NO_2 + h\nu \to NO + O, \quad (12)$$
$$O + O_2 \overset{M}{\to} O_3. \quad (2)$$

This cycle alone leads to no net change in O_3.

However, an *increase* in O_3 at lower altitudes can result due to interactions with the ClO_x, BrO_x, and HO_x cycles. The reason for this is that at the lower altitudes, these species play a much larger role in determining O_3 concentrations than the NO_x family does (e.g., Garcia and Solomon, 1994; WMO, 1995). An example of the loss rates for O_3 by NO_x, (ClO_x + BrO_x), and HO_x from 17 to 21 km calculated for 38°N in May 1993, based on measurements of OH, HO_2, ClO, NO, and O_3, is shown in Fig. 12.8a (Wennberg *et al.*, 1994). Below ~25 km, the major determinants of O_3 loss are HO_x and halogen chemistry, with NO_x playing a much smaller role; as discussed earlier, the major source of O_3 is photolysis of O_2, reactions (1) and (2) in the Chapman cycle.

The HO_x destruction of O_3 shown in Fig. 12.8a is due to reactions (6) and (8):

$$OH + O_3 \to HO_2 + O_2 \quad (6)$$
$$\underline{HO_2 + O_3 \to OH + 2O_2} \quad (8)$$
$$\text{Net: } 2O_3 \to 3O_2$$

For the conditions under which the measurements in Fig. 12.8a were made, this cycle accounts for about 30–50% of the total O_3 removal rate. Increasing water concentrations have been observed in the stratosphere, so that the contribution of the HO_x cycle in the upper

B. HIGH-SPEED CIVIL TRANSPORT (HSCT), ROCKETS, AND THE SPACE SHUTTLE

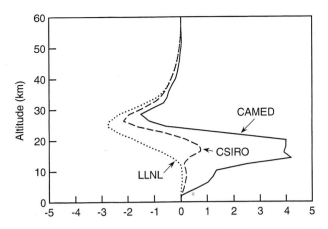

FIGURE 12.7 Typical calculated changes in ozone concentrations as a function of altitude for a fleet of 500 HSCTs flying at Mach 2.4. These profiles are for 45°N latitude for the month of March using three different models (CAMED, CSIRO, and LLNL). Emissions of NO_x are assumed to be equivalent to 5 g of NO_2/kg of fuel burned (from Stolarski et al., 1995).

stratosphere particularly may be increasing (Evans et al., 1998).

Chlorine atoms produced by the photolysis of CFCs, and bromine atoms from halons, also destroy O_3 by cycles that are discussed later. As seen in Fig. 12.8a, these cycles are similar in magnitude to the HO_x cycle in terms of O_3 removal.

Because NO_x plays such a small role in the removal of O_3 in the lower stratosphere, its effects in this region arise primarily because of its interactions with the halogen and HO_x cycles that do control O_3 loss here. Thus NO_2 interferes with the ClO_x and BrO_x cycles by forming chlorine and bromine nitrate, $ClONO_2$ and $BrONO_2$, respectively:

$$ClO + NO_2 \overset{M}{\rightarrow} ClONO_2, \quad (16)$$

$$BrO + NO_2 \overset{M}{\rightarrow} BrONO_2. \quad (17)$$

Chlorine and bromine nitrate serve as temporary reservoirs for chlorine and bromine, taking them out of their ozone destruction cycles. (While theoretical studies suggest that other forms of bromine such as O_2BrONO_2 could in principle also act as reservoirs (Lee et al., 1999a, 1999b), there is no evidence at the present time that these are important under atmospheric conditions.)

NO_x also interacts with the HO_x cycle via the reaction of HO_2 with NO:

$$HO_2 + NO \rightarrow OH + NO_2. \quad (18)$$

This reduces the HO_2 concentration, and hence its contribution to O_3 destruction via reaction (8) of HO_2 with O_3. In addition, the rates of reaction of HO_2 with ClO and BrO decrease with less HO_2, which lowers the rate of O_3 destruction through cycles outlined later.

NO_2 reacts with OH to form nitric acid:

$$OH + NO_2 \overset{M}{\rightarrow} HNO_3. \quad (19)$$

(See Chapter 7.E for a discussion of the kinetics and Brown et al. (1999) for a recent study under low-temperature and low-pressure conditions representative of the stratosphere.) This reaction removes NO_x from the ozone destruction cycle since HNO_3 does not readily regenerate active forms of NO_x and is removed by transport to the troposphere followed by rainout and washout. Photolysis of HNO_3 to OH + NO_2 is

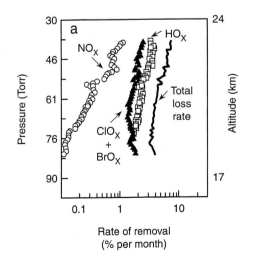 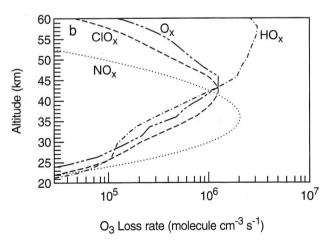

FIGURE 12.8 (a) Rates of removal of O_3 at 38°N in May 1993 due to NO_x, ($ClO_x + BrO_x$), and HO_x chemistry, respectively, as a function of altitude in the stratosphere (adapted from Wennberg et al., 1994); (b) 24-h average rates of removal of O_3 as a function of altitude (adapted from Osterman et al., 1997).

relatively slow, although some model calculations find some sensitivity of the predicted ozone destruction to the HNO_3 photolysis rate (Jones *et al.*, 1993).

As a result, increasing NO_x emissions does not have a significant direct effect at lower altitudes as it does at higher ones but rather has indirect effects on the halogen and HO_x cycles, which reduce the ozone destruction due to these species. The net result, then, is interference in these other ozone-destroying cycles, leading to an increase in ozone at these altitudes as seen in the model predictions in Fig. 12.7. (In the very low stratosphere, NO_x can also produce O_3 through the VOC–NO_x chemistry discussed in Chapter 6.)

Figure 12.8b shows the contribution of the various cycles to O_3 loss over a larger range of altitudes, deduced from a combination of measurements of OH, HO_2, ClO, and NO_2 and calculations using these measured concentrations (Osterman *et al.*, 1997). It can be seen that the NO_x cycles dominate at altitudes from ~25 to 38 km, with HO_x and, to a lesser extent, O_x (O + O_3) and ClO_x being important above that. The NO_x catalytic ozone destruction cycle has been proposed to be responsible for the low O_3 concentrations observed at high latitudes in the Northern Hemisphere from 20 to 31 km during the summer (Brühl *et al.*, 1998). The importance of various cycles at altitudes from 20 to 65 km and the importance of the chain length in determining these are discussed in detail by Lary (1997).

HSCTs also emit particles and SO_2, with the latter being oxidized to H_2SO_4 and sulfate particles. Measurements of particle concentrations in the plume of the Concorde SST showed much larger particle concentrations than anticipated (Fahey *et al.*, 1995a). Furthermore, a much larger portion of the SO_2 in the exhaust was oxidized to H_2SO_4 particles than expected based on the OH levels measured in the exhaust plume (Hanisco *et al.*, 1997), suggesting that there are some as yet unknown mechanisms of SO_2 oxidation in the plume.

Kärcher *et al.* (1996) suggest that this additional oxidation occurs on the soot particles that have been observed in the stratosphere and attributed to aircraft emissions (Pueschel *et al.*, 1992a; Blake and Kato, 1995). The oxidation of SO_2 on soot particles is known to occur in the troposphere as well (see Chapter 8.C.4). If the same is true of the exhaust from HSCTs, their emissions could lead to significant increases in both the number of particles in the lower stratosphere and as their associated surface area.

For example, modeling studies by Weisenstein *et al.* (1996) predict that the surface area of stratospheric particles could increase by as much as 75%, if 10% of the SO_2 is rapidly converted to H_2SO_4. Yu and Turco (1997) propose that the presence of ions in the exhaust may promote the nucleation of sulfuric acid aerosols if ~20–30% of the sulfur has been oxidized to H_2SO_4.

Another contributing factor may be that more highly oxidized sulfur-containing compounds than SO_2 are formed in the aircraft engines and emitted in significant amounts. For example, modeling studies (R. C. Brown *et al.*, 1996a, 1996b; Kärcher and Fahey, 1997) suggest that if a significant portion (~35%) of the total sulfur emissions are in the form of SO_3, it would form H_2SO_4 through the rapid reaction with gaseous water (see Chapter 8). Condensation of H_2SO_4 to form new particles or on existing soot particles could act to further catalyze SO_2 oxidation.

Such catalysis on particles is potentially important since heterogeneous reactions on such particles are now recognized as playing a key role in the chemistry of the stratosphere. A number of nitrogen-containing species such as N_2O_5, $ClONO_2$, $BrONO_2$, and HO_2NO_2 are now known to react on particles to form nitric acid, which ties up the NO_x and ultimately removes it from the stratosphere. Hence particles act to remove oxides of nitrogen from the stratosphere, lowering the predicted effects of HSCT NO_x emissions compared to the gas-phase only case (e.g., see Pitari *et al.*, 1993; Weisenstein *et al.*, 1993; Bekki and Pyle, 1993; and Considine *et al.*, 1995). However, the same chemistry leads to increased O_3 destruction by halogens and HO_x. Randeniya *et al.* (1996a, 1996b), for example, suggest based on modeling calculations that $BrONO_2$ hydrolysis may be particularly important since it occurs during the day. In contrast, N_2O_5 formation occurs at night (because of the rapid photolysis of its precursor NO_3 during the day) and its removal by hydrolysis can be limited ultimately by its rate of formation, which involves the relatively slow O_3–NO_2 reaction. As discussed in detail in Section D, $BrONO_2$ hydrolysis generates HOBr, which rapidly photolyzes to OH + Br, thus increasing HO_x while tying up NO_x as HNO_3 in the particles.

The direct destruction of O_3 by its reaction on soot particles generated by aircraft at midlatitudes has also been proposed, but given the large uncertainties in the mechanism and kinetics of this reaction, it is not clear that this will prove to be significant (Bekki, 1997; Lary *et al.*, 1997).

HSCT emissions may also interact with polar stratospheric clouds, PSCs, in much the same way as with particles (Pitari *et al.*, 1993). That is, reaction of a number of nitrogenous species on PSCs leads to the formation of HNO_3, which can remain adsorbed on or in the PSC. The larger cloud particles sediment to lower altitudes in the stratosphere, redistributing NO_y, or into the troposphere, permanently removing NO_x

from the stratosphere (so-called denitrification; Fahey *et al.*, 1990). An additional issue with respect to PSCs is the addition of water vapor from the HSCT exhaust, as well as NO_x. Peter *et al.* (1991) estimate that the emissions of NO_x and H_2O from HSCTs could lead to a doubling in the occurrence of Type I PSCs and as much as an order of magnitude increase in the occurrence of Type II PSCs (see later for a description of Type I versus Type II PSCs).

There are several important points with respect to the effects of any future HSCT emissions. First, ozone concentrations at a particular location and time depend not only on the local chemistry but on transport processes as well. In the lower stratosphere, transport processes occur on time scales comparable to the rates of ozone formation and loss so that taking into account such transport is particularly important. However, in the middle and upper stratosphere, production and removal of O_3 are much faster than transport so that a steady state exists between these two processes.

Second, as already discussed, recent advances in understanding stratosphere–troposphere exchange suggest that even though HSCT emissions would occur primarily in the 20-km region, there are mechanisms for transporting these into the middle and upper stratosphere in the tropics, leading to more ozone destruction than otherwise might have been anticipated. For example, one 3-D model simulation predicts that about 15–25% of the exhaust released in the region from 30°N toward the pole is transported into the tropics (Weaver *et al.*, 1996). However, field measurements of NO_y and O_3 and their ratio in the lower stratosphere show a steep gradient near the tropics. This is expected if the "tropical pipe" model discussed earlier applies, where there is a barrier that reduces the exchange between the tropics and midlatitudes (Fahey *et al.*, 1996).

Third, there are uncertainties in the actual emissions estimates. While the few in-flight measurements of NO_x emissions that have been carried out on subsonic and supersonic aircraft to date show NO_x emissions in good agreement with those predicted from ground-based tests (e.g., see Zheng *et al.*, 1994; Schulte and Schlager, 1996; and Fahey *et al.*, 1995a, 1995b), there are significant uncertainties in other emissions, particularly the particles as discussed earlier.

Figure 12.9 shows some model-calculated percent changes in total column ozone due to a HSCT fleet that was projected in 2015 assuming the emission goal of $EI_{NO_x} = 5$ g of NO_2/kg of fuel was met (Stolarski *et al.*, 1995). These calculations compare the change in O_3 due to this fleet compared to a completely subsonic fleet in that year using the three different models for which predicted altitude changes were shown in Fig.

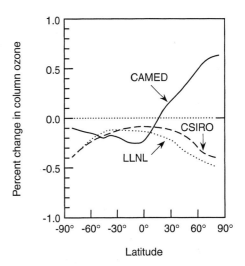

FIGURE 12.9 Calculated percent change in total column ozone during March as a function of latitude due to a Mach 2.4 HSCT fleet from the three models for which results were shown in Fig. 12.7, assuming NO_x emission radiation of 5 g of NO_2/kg of fuel. These are the predicted changes due to the projected HSCT fleet compared to a projected solely subsonic fleet (adapted from Stolarski *et al.*, 1995).

12.7. Small changes are predicted by all models in the tropics, but larger changes poleward, up to ~0.5%. Much larger changes, approaching 1.7%, are predicted if the EI_{NO_x} is 15.

An assessment of the effects of HSCTs on stratospheric ozone is given by Stolarski *et al.* (1995), and the interactions between NO_x and ClO_x cycles at various concentrations are treated by Kinnison *et al.* (1988), Johnston *et al.* (1989), and Considine *et al.* (1995). A discussion of some of the general issues involved in the development and possible future use of the HSCT is found in Zurer (1995).

2. Space Shuttle and Solid Rocket Motors

The launch of the space shuttle and other vehicles such as the Titan launch vehicles results in emissions directly into the troposphere and the stratosophere. Exhaust emissions include Al_2O_3 (30% by weight), CO (24%), HCl (21%), H_2O (10%), N_2 (9%), CO_2 (4%), and H_2 (2%) (Danilin, 1993).

The major focus on the effects of exhaust emissions has been on the HCl component and its role in ozone depletion and on the Al_2O_3 particles, which could provide a surface for the heterogeneous conversion of HCl to active forms of chlorine. It has been proposed that if the HCl were converted to photochemically active forms relatively rapidly, a mini "ozone hole" could form in the flight path of the vehicle (Aftergood, 1991; McPeters *et al.*, 1991; Karol *et al.*, 1992).

Modeling studies have concluded that the longer term, global-scale, perturbation on the stratosphere due to the HCl emissions is likely to be small (Prather *et al.*, 1990; Denison *et al.*, 1994; Jackman *et al.*, 1996, 1998). For example, Jackman *et al.* (1998) predicted that the annually averaged loss of global total ozone by 1997 due to the previous launches of the space shuttle and Titan III and IV rockets was only 0.025%; two-thirds of this was attributed to HCl reactions and one-third to reactions on alumina (vide infra). Even on the local scale along the flight path, the perturbation is predicted to be quite small if most of the chlorine is in the form of HCl. The conversion of HCl to atomic chlorine by reaction with OH, and to a lesser extent by reaction with O or direct photolysis, is relatively slow. However, if even a small percentage of the HCl is converted in the exhaust into more active forms such as Cl_2, or if Cl_2 is emitted directly, the predicted ozone depletion could be significant in the plume itself, via chemistry similar to that occurring over Antarctica in winter (see Sections C.4 and C.5).

For example, modeling studies in which 1% of the HCl is assumed to be in the form of Cl_2 in the exhaust predict that ozone depletions of 2–70% can occur inside the exhaust plume, increasing from 2% at about 10-km altitude to 70% at ~40 km (Danilin, 1993). Observing and measuring significant depletion of ozone in the exhaust plume using satellites such as TOMS (*T*otal *O*zone *M*apping *S*pectrometer) are not straightforward since the spatial resolution covers a much larger area than that over which ozone loss is anticipated (Syage and Ross, 1996; Ross, 1996). However, in one study in which stratospheric ozone was followed in the wake of two Titan IV rockets, O_3 concentrations were observed to fall to near zero within a half hour of the launch and then to recover in the next half hour (Ross *et al.*, 1997). Similar effects have been observed in the plume of a Delta rocket (D. W. Toohey, personal communication). This occurred over a fairly limited geographical region, ~4–8 km wide. Thus, such launches are not believed to be important in ozone destruction on a global scale.

The impact on the generation of active forms of chlorine by heterogeneous reactions on the Al_2O_3 particles is somewhat controversial (Prather *et al.*, 1990; Danilin, 1993; Denison *et al.*, 1994; Jackman *et al.*, 1996). For example, the reaction of $ClONO_2$ with HCl to generate Cl_2, for example, proceeds relatively rapidly (reaction probability of ~0.02) at 208–223 K on hydroxylated α-alumina (Molina *et al.*, 1997). However, it appears that this is not unique to the Al_2O_3 surface but rather is due to water layers adsorbed on the solid.

A potential direct effect is the destruction of O_3 directly on the particles, which appears, based on laboratory studies, not to be very important (Hanning-Lee *et al.*, 1996). It has also been proposed that these Al_2O_3 particles could act as seed particles for nucleation of other aerosol particles and clouds (Turco *et al.*, 1982).

Robinson and co-workers (Dai *et al.*, 1997; Robinson *et al.*, 1997) have observed that γ-Al_2O_3 that has been preheated to 1000 K under vacuum to dehydroxylate the surface and generate active sites for reaction can dissociate halomethanes such as CF_3Cl, CF_2Cl_2, and CCl_4. In these experiments, the gases were adsorbed on the preheated solid at low temperatures (100 K) and subsequently warmed. HCl and CF_xCl_y fragments desorbed at ~150 K, and CO_2 between 240 and 320 K. Infrared as well as X-ray photoelectron spectroscopy (XPS) and temperature-programmed reaction (TPR) studies showed the formation of carbonate, bicarbonate, and/or formate and inorganic as well as organic forms of fluorine on the surface at 150–200 K. However, while such reactions may have some impact in the rocket plume itself, the low mass accommodation coefficient ($\sim 10^{-4}$–10^{-5}) combined with surface saturation effects will preclude significant effects on a global scale.

Injection of species into the stratosphere associated with these launches includes emissions not only from the rocket exhaust but also from ablation of the solid rocket motors, the paint on the outer hulls, and hardware from satellites and discarded portions of rockets in the atmosphere (Zolensky *et al.*, 1989). The increase in launches of such vehicles has led to a significant increase in particles associated with solid rocket use. Figure 12.10, for example, shows the concentration of large (>1-μm diameter) solid stratospheric particles in the 17- to 19-km altitude region from 1976 to 1984,

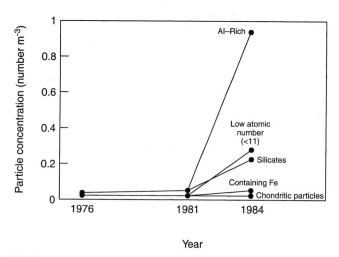

FIGURE 12.10 Concentration of stratospheric particles with >1 μm at altitudes of 17–19 km from 1976 to 1984 (adapted from Zolensky *et al.*, 1989).

classified by their major chemical composition. The largest increase was in aluminum-rich particles, thought to be from the exhaust of solid rockets using fuels containing an aluminum additive and perhaps from ablation of the spacecraft paint. Silicate particles also increased, which may be due to ablation of the solid rocket motors and the spacecraft paint. Iron is found in exhaust and in alloys used to make the solid rocket motors. Noteworthy in Fig. 12.10 is the lack of significant increase in chondritic particles from extraterrestrial sources. The fact that they did not also increase over this period of time suggests that the measured increase in the other particles is a real effect rather than a sampling artifact, which has been of concern (Zolensky et al., 1989).

In short, the launch of space vehicles is an additional source of emissions to the stratosphere, along with the proposed fleet of HSCTs.

C. CHLOROFLUOROCARBONS

1. Types, Nomenclature, and Uses

Chlorofluorocarbons, that is, compounds containing chlorine, fluorine, carbon, and possibly hydrogen, have been used extensively in the industrialized nations in the past decades primarily as propellants in aerosol spray cans, as refrigerants, and as blowing agents, for example, for producing polyurethane foam. Their chemical characteristics have made them ideally suited for such uses in that they are generally nontoxic and chemically inert. Thus, they can be used around open flames, and leaks in refrigeration units do not present a health hazard as older units operated on coolants such as SO_2 once did.

The principal chlorofluorocarbons which have been used are CCl_3F, CCl_2F_2, and $CHClF_2$. These are often referred to as CFC-11, CFC-12, and CFC-22, respectively; alternatively, the abbreviations F-11, F-12, and F-22 are used, after the Dupont trade name Freon. The numbers are based on a system developed by chemists at DuPont. The first number is the number of hydrogen atoms, plus one, the second gives the number of fluorine atoms, and the remainder are chlorine atoms.

For chlorofluorocarbons containing two or more carbon atoms, a three-digit numbering system is used; the first digit gives the number of carbons minus one, and as for the one-carbon compounds, the second is the number of hydrogens plus one and the third is the number of fluorine atoms. Thus, CCl_2FCClF_2 is F-113, or CFC-113 (whose use for cleaning electronic components increased in the mid to late 1970s), and $CClF_2CClF_2$ is F-114, or CFC-114. When more than one isomer is possible, the most symmetrical one has only a number; the letters a and b are added to distinguish the less symmetrical isomers.

To provide some historical perspective, Table 12.2 gives the distribution of sales by use for CFC-11 and CFC-12, broken down into aerosol propellants, blowing agents, refrigerants, and other uses, from 1976 to 1992. On a global basis, the use as aerosol propellants accounted for more than half of the sales of CFC-11 and CFC-12 in 1976. Essentially all of this as well as most of that used in other applications has been released into the atmosphere.

Figure 12.11 shows the history of the estimated annual global release rates of CFC-11 and CFC-12 from 1952 to 1980. Over this period of about three decades, the release rate into the atmosphere increased dramatically until about 1974, as these compounds found increasing use in our industrialized society. Figure 12.12 shows similar data from 1972 to 1992. The decrease in emissions as the Montreal Protocol took effect can be clearly seen (see Chapter 13.A). Data from 1986 to 1996 from the European Union show similar trends (McCulloch and Midgley, 1998).

Figure 12.13 shows the trend in surface CFC-11 and CFC-12 concentrations in air measured at latitudes

TABLE 12.2 Total CFC Sales by Use from 1976 to 1992 (in Thousands of Metric Tons)[a]

Compound	Year	Use				
		Aerosol propellant	Blowing agent	Refrigeration	Other	Total
CFC-11	1976	195.0	97.0	26.9	21.0	339.9
	1984	97.5	173.9	23.9	17.0	312.3
	1992	8.2	155.5	17.8	5.0	186.5
CFC-12	1976	237.3	15.0	127.8	30.7	410.8
	1984	121.3	49.2	187.5	24.1	382.1
	1992	14.7	14.3	177.4	9.8	216.2

[a] From AFEAS (1993).

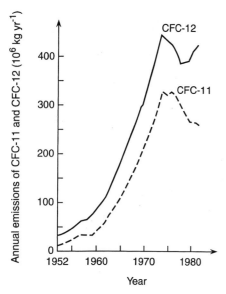

FIGURE 12.11 Estimated annual worldwide releases of CFC-11 and CFC-12 from 1952 to 1980. Data from Chemical Manufacturers' Association (adapted from National Research Council, 1984).

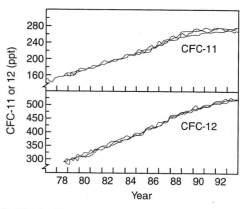

FIGURE 12.13 Concentrations of CFC-11 and CFC-12 in air in the 30°N to 90°N region as a function of time. The different curves represent measurements made at various locations (adapted from WMO, 1995).

from 30°N to 90°N at a variety of locations, including the United States, Canada, and Ireland. The increases seen at the surface in the 1985 to 1994 period have also been observed by remote sensing techniques on the space shuttle (e.g., Zander *et al.*, 1996). The rate of growth has now started to decrease due to the decreased emissions. For example, CFC-11 was growing at about 9–11 ppt/year around 1980; this fell to ~7 ppt/year in 1990, 3 ppt/year by 1993 (WMO, 1995), and, more recently, to slightly negative values (Montzka *et al.*, 1996; Cunnold *et al.*, 1997; Derwent *et al.*, 1998).

2. Lifetimes and Atmospheric Fates of CFCs and Halons

In considering the atmospheric fate of CFCs and halons, it is useful to examine the total atmospheric lifetime of a compound X, τ_X. This is in effect the time required for a pulse emitted into the atmosphere to decay to $1/e$ of its initial value (see Chapter 5.A.1c). It can be calculated from

$$\tau_X = \frac{\Sigma \text{Burden}_{atm}}{\Sigma \text{Loss Rate}},$$

where $\Sigma \text{Burden}_{atm}$ represents the total amount of X in the atmosphere and $\Sigma \text{Loss Rate}$ the globally integrated loss rate due to all processes, including reactions, uptake into oceans, wet and dry deposition, etc. The relationship between the total atmospheric lifetime τ_X and the lifetimes with respect to the individual processes that contribute to the removal of the compound X is given by

$$\tau_X^{-1} = \tau_{trop}^{-1} + \tau_{strat}^{-1} + \tau_{ocean}^{-1} + \tau_{dep}^{-1} + \tau_{other}^{-1},$$

where the individual terms on the right side of the equation represent the lifetimes with respect to tropospheric reactions, stratospheric chemistry, removal by the oceans, deposition, and any other processes that might contribute, respectively. For a compound that reacts only with OH in the troposphere, for example, $\tau_{trop} = \tau_{OH} = 1/k_{OH}[OH]$, where k_{OH} is the rate constant for the X–OH reaction. Similarly, if there are no known tropospheric sinks, the atmospheric lifetime is the same as the stratospheric lifetime. (Note that this approach addresses removal from the entire atmosphere and does not take into account a number of factors that may be important for individual species.

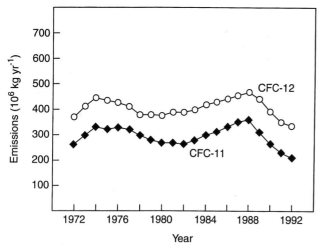

FIGURE 12.12 Estimated global annual emissions of CFC-11 and CFC-12 (adapted from World Meteorological Organization, 1995).

For example, some may have short lifetimes and/or nonuniform distributions and hence be sensitive to such factors as the location of the emissions, sunlight intensity, season, etc.)

The lifetime of CFCs in the atmosphere can also be estimated using a mass balance approach. Knowing the atmospheric concentrations of CFCs, one can calculate the total amount in the atmosphere. This amount must be the result of a balance between emissions into, and loss from, the atmosphere. If the emission rates are known, the loss rate required to give the observed atmospheric concentrations can be calculated, and from this, a lifetime obtained. Such calculations may be based on either the absolute atmospheric concentrations of CFCs or, alternatively, the observed relative rates of change in the concentrations.

The chlorofluorocarbons (CFCs) have very long lifetimes in the troposphere. This is a consequence of the fact that they do not absorb light of wavelengths above 290 nm and do not react at significant rates with O_3, OH, or NO_3. In addition to the lack of chemical sinks, there do not appear to be substantial physical sinks; thus they are not very soluble in water and hence are not removed rapidly by rainout. While laboratory studies have shown that some of the CFCs decompose on exposure to visible and near-UV present in the troposphere when the compounds are adsorbed on siliceous materials such as sand (Ausloos et al., 1977; Gäb et al., 1977, 1978), the lifetimes for CFC-11 and CFC-12 with respect to these processes have been estimated to be ~540 and 1800 years, respectively (National Research Council, 1979). Similarly, an observed thermal decomposition when adsorbed on sand appears to be an insignificant loss process under atmospheric conditions.

As a result, CFCs reside in the troposphere for years and are slowly transported up across the tropopause into the stratosphere, primarily in the tropics as discussed earlier. For example, the estimated lifetime of CFC-11 in the global atmosphere, that is, the time to diffuse to the stratosphere and undergo photolysis, is approximately 40–80 years, with that for CFC-12 being about twice as long (WMO, 1995). Tropospheric losses are negligible, so that this is determined by the time to reach the stratosphere and then to dissociate. The stratospheric lifetimes of relevant compounds have been estimated based on their measured concentrations (e.g., Volk et al., 1997; Avallone and Prather, 1997). For example, using $\tau_{CFC-11} = 45 \pm 7$ years, Volk et al. (1997) obtained the following lifetime estimates: $\tau_{N_2O} = 122 \pm 24$ years, $\tau_{CH_4} = 93 \pm 18$ years, $\tau_{CFC-12} = 87 \pm 17$ years, $\tau_{CFC-113} = 100 \pm 32$ years, $\tau_{CCl_4} = 32 \pm 6$ years, $\tau_{CH_3CCl_3} = 34 \pm 7$ years, and $\tau_{Halon-1211} = 24 \pm 6$ years.

While there are a variety of other chlorinated organics such as methylchloroform (CH_3CCl_3) that are emitted, these have relatively short tropospheric lifetimes because they have an abstractable hydrogen atom (e.g., see WMO, 1995). For example, while the stratospheric lifetime of methylchloroform is estimated to be 34 ± 7 years (Volk et al., 1997), its overall atmospheric lifetime is only 5–6 years, primarily due to the removal by OH in the troposphere ($\tau_{OH} \sim 6.6$ years), with a much smaller contribution from uptake by the ocean ($\tau_{ocean} \sim 85$ years) (WMO, 1995).

Since the wavelength distribution of solar radiation shifts to shorter wavelengths with increasing altitude (see Chapter 3.C), the CFCs eventually become exposed to wavelengths of light that they can absorb. Figure 12.14 shows the absorption cross sections of some halogenated methanes from 160 to 280 nm (see also Chapter 4.V). The absorptions become very weak beyond ~240 nm in the case of CFC-11, and 220 nm in the case of CFC-12. Recall from Chapter 3 that both O_3 and O_2 absorb radiation in the ultraviolet. Figure 12.15 shows these absorption cross sections for O_2 and O_3 from 120 to 360 nm; there is a *window* in the overlapping O_2 and O_3 absorptions from ~185 to 210 nm, that is, a region where the total light absorption is in a shallow minimum. This is a region in which the CFCs also absorb light (Fig. 12.14).

The C–Cl bond dissociation energy in CF_2Cl_2 is 76 kcal mol^{-1}, whereas that for the strong C–F bond is 110 kcal mol^{-1}. As a result, the weaker C–Cl bond can break at longer wavelengths:

$$CF_2Cl_2 + h\nu(\lambda < 240 \text{ nm}) \rightarrow CF_2Cl + Cl. \quad (20)$$

Subsequent reactions of the CF_2Cl radical also release the chlorine atoms tied up in this fragment, so that all of the chlorine in the original molecule becomes available for ozone destruction.

Fluorine chemistry in the stratosphere was also considered and it was concluded that ozone depletion by chlorine was $>10^4$ more efficient than that by fluorine (Rowland and Molina, 1975; Stolarksi and Rundel, 1975). Since then, the kinetics of reaction of F atoms with O_2 to form the FO_2 radical and its thermal decomposition have been measured (e.g., see Pagsberg et al., 1987; Lyman and Holland, 1988; Ellerman et al., 1994; and review in DeMore et al., 1997). The equilibrium constant for the $F-FO_2$ system

$$F + O_2 + M \leftrightarrow FO_2 + M, \quad (21)$$

is given by $K_{eq} = 3.2 \times 10^{-25}\exp(6100/T)$ (DeMore et al., 1997), so that under stratospheric conditions it lies far to the right, with $[FO_2]/[F] \sim 10^4$. Potential reactions of FO_2 that could lead to the destruction of

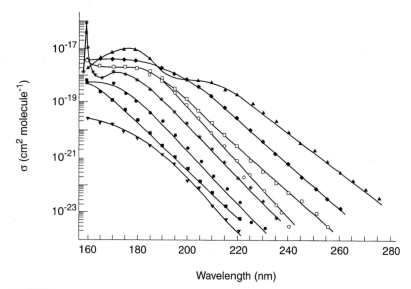

FIGURE 12.14 Semilogarithmic plot of the absorption cross sections of the halogenated methanes at 298K: ∗, $CHCl_3$; ■, $CHClF_2$; □, CH_2Cl_2; ●, CH_2ClF; ▲, CCl_4; ◆, CCl_3F (CFC-11); ○, CCl_2F_2 (CFC-12); ▼, $CClF_3$ (adapted from Hubrich and Stuhl, 1980).

ozone through such cycles as

$$FO_2 + O_3 \rightarrow FO + 2O_2, \quad (22)$$

$$k_{22}^{298K} < 3.4 \times 10^{-16} \text{ cm}^3 \text{ molecule}^{-1} \text{ s}^{-1}$$

(DeMore et al., 1997),

$$FO + O_3 \rightarrow FO_2 + O_2, \quad (23)$$

$$k_{23}^{298K} < 1 \times 10^{-14} \text{ cm}^3 \text{ molecule}^{-1} \text{ s}^{-1}$$

(DeMore et al., 1997),

have been examined by determining the kinetics of these reactions as well as those of FO_2 with NO, NO_2, and the organics CH_4 and C_2H_6 (Sehested et al., 1994;

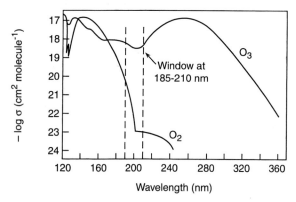

FIGURE 12.15 Absorption cross sections for O_2 and O_3 from 120 to 360 nm, showing the window from ~185 to 210 nm (adapted from Rowland and Molina, 1975).

Li et al., 1995b). The FO_2 reaction with O_3 is slow (Li et al., 1995b; DeMore et al., 1997), as are the reactions with the organics (Sehested et al., 1994; Li et al., 1995b). While fluorine is cycled between F, FO_2, and FO, fluorine atoms are removed efficiently by their fast reactions with CH_4 and H_2O to form HF,

$$F + CH_4 \rightarrow HF + CH_3, \quad (24)$$

$$k_{24}^{298K} = 6.7 \times 10^{-11} \text{ cm}^3 \text{ molecule}^{-1} \text{ s}^{-1}$$

(DeMore et al., 1997),

$$F + H_2O \rightarrow HF + OH, \quad (25)$$

$$k_{25}^{298K} = 1.4 \times 10^{-11} \text{ cm}^3 \text{ molecule}^{-1} \text{ s}^{-1}$$

(DeMore et al., 1997),

which, unlike HCl, does not itself react with OH (indeed, reaction (25) is exothermic). The HF formed in these reactions also appears to be unreactive on ice surfaces typical of polar stratospheric clouds (Hanson and Ravishankara, 1992a).

While the FO_2 reactions with NO and NO_2 are moderately fast, with room temperature rate constants of the order of 10^{-12} and 10^{-13} cm^3 molecule^{-1} s^{-1}, respectively (Sehested et al., 1994; Li et al., 1995b), the concentrations of NO and NO_2 are sufficiently small that they do not represent major atmospheric loss processes for FO_2. It is interesting, however, that the FO_2 + NO reaction proceeds by transfer of the F atom to form FNO (which photolyzes) rather than by transfer of an oxygen atom, which is more common for peroxy radicals. Li *et al.* (1995b) attribute this to the antibonding interactions between the $3p_z$ orbitals on fluorine and oxygen.

In short, the net effect of fluorine atom chemistry on ozone destruction is very small, 10^3–10^4 times smaller than the effect of chlorine on a per-atom basis (Sehested *et al.*, 1994; Ravishankara *et al.*, 1994; Li *et al.*, 1995b; Lary, 1997).

More recently, the possible contribution of organofluorine free radicals to ozone destruction has been considered with respect to the introduction of CFC transitional alternates and longer term replacements; however, as discussed in Chapter 13.D, this also does not appear to be important in terms of ozone destruction.

3. Gas-Phase Chemistry in the Stratosphere

The Cl atom released by photolysis of the CFCs reacts in a catalytic chain reaction that leads to the destruction of O_3:

$$Cl + O_3 \rightarrow ClO + O_2 \tag{26}$$

$$\underline{ClO + O \rightarrow Cl + O_2} \tag{27}$$

$$\text{Net: } O_3 + O \rightarrow 2O_2$$

Because the concentration of oxygen atoms increases with altitude, the reaction cycle represented by (26) and (27) is important primarily in the middle and upper stratosphere (e.g., Garcia and Solomon, 1994; WMO, 1995). For the lower stratosphere, however, it is only responsible for about 5% of the portion of the total ozone loss that is due to halogens at 15 km and ~25% at 21 km (see Fig. 12.8; Wennberg *et al.*, 1994). Most of the O_3 loss associated with ClO_x and BrO_x at the relatively low altitudes in Fig. 12.8 is due to the following cycle (Solomon *et al.*, 1986; Crutzen and Arnold, 1986):

$$Cl + O_3 \rightarrow ClO + O_2 \tag{26}$$

$$ClO + HO_2 \rightarrow HOCl + O_2 \tag{28}$$

$$HOCl + h\nu \rightarrow Cl + OH \tag{29}$$

$$\underline{OH + O_3 \rightarrow HO_2 + O_2} \tag{6}$$

$$\text{Net: } 2O_3 \rightarrow 3O_2$$

This cycle accounts for ~30% of the ozone loss due to halogens in the lower stratosphere, and the corresponding cycle for bromine for ~20–30% (Wennberg *et al.*, 1994). Reaction of ClO with HO_2, reaction (28), produces $HOCl + O_2$ with a yield $\geq 95\%$ at temperatures from 210 to 300 K; however, at the lowest end of this temperature range, there is evidence for the production of $HCl + O_3$, with a yield of <5%, which may be significant in the net ozone destruction both in the polar regions and in midlatitudes (Finkbeiner *et al.*, 1995).

The reaction of OH with ClO,

$$OH + ClO \rightarrow HO_2 + Cl, \tag{30a}$$

$$\rightarrow HCl + O_2(^1\Delta_g), \tag{30b}$$

$$k_{30} = k_{30a} + k_{30b} = (5.5 \pm 1.6) \times 10^{-12} e^{(292 \pm 72)/T}$$

(Lipson *et al.*, 1997),

may also play an important role. While it produces mainly $HO_2 + Cl$, even a small contribution from the second channel (30b) could be important because it generates the reservoir species HCl rather than the reactive chlorine atom. Experimental studies report a yield of $5 \pm 2\%$ (Lipson *et al.*, 1997). Dubey *et al.* (1998) show that even a small branching ratio for (30b), which likely proceeds through a four-center transition state, in the 7–10% range gives larger predicted O_3 concentrations in the upper stratosphere due to smaller production of ozone-destroying atomic chlorine in (30a). This brings the model predictions and measurements into better agreement. Dubey *et al.* (1998) also show that the predicted branching ratio for the minor channel is very sensitive to the reaction dynamics, especially to the energy difference between the transition state for the formation of the HOOCl* adduct and the formation of the four-center transition state.

In addition, the ClO_x and BrO_x cycles are interconnected by the reaction of ClO and BrO (Yung *et al.*, 1980; Prather *et al.*, 1984; McElroy *et al.*, 1986; Wayne *et al.*, 1995):

$$ClO + BrO \rightarrow Br + OClO, \tag{31a}$$

$$\rightarrow Br + ClOO, \tag{31b}$$

$$\rightarrow BrCl + O_2. \tag{31c}$$

The recommended rate constants at 298 K for reactions (31a), (31b), and (31c) are 6.8×10^{-12}, 6.1×10^{-12}, and 1.0×10^{-12} cm^3 molecule^{-1} s^{-1}, respectively (DeMore *et al.*, 1997). Reaction (31) is responsible for much of the uncertainty (~21%) in model predictions of ozone loss in the Arctic (Fish and Burton, 1997).

While more than 90% of reaction (31) at 298 K produces bromine atoms directly, the minor channel producing BrCl is important in the atmosphere under certain conditions (e.g., McKinney *et al.*, 1997). (Measurements of OClO formed in reaction (31a) in the Antarctic and Arctic stratosphere are discussed later.) For example, McKinney *et al.* (1997) report large fractions (50–95%) of total bromine in the form of BrO in the chemically perturbed region of the Arctic vortex from 17 to 23 km at sunrise, along with high concentrations of ClO. Other studies have also reported higher measured concentrations of BrO than expected (e.g., Wahner and Schiller, 1992; Avallone et al., 1995). Comparison of the formation of BrO as a function of solar zenith angle to model predictions suggests that bromine atoms (and then BrO from the reaction with O_3) must be produced by the rapid photolysis of a precursor, which McKinney et al. suggest to be BrCl. $BrONO_2$, usually considered to be the bromine reservoir, does not photolyze sufficiently rapidly to be consistent with their observations. In addition, the enhanced ClO levels in the chemically perturbed region suggest NO_2, and hence $BrONO_2$, must be low. That is, BrCl from reaction (31c) is acting as a nighttime reservoir for bromine, rapidly releasing it by photolysis at dawn.

Chlorine atoms are formed subsequently from the thermal decomposition of ClOO formed in reaction (31b) to $Cl + O_2$ (vide infra). Photolysis of OClO in the gas phase formed in reaction (31a) gives $O + ClO$ with a quantum yield of unity (DeMore et al., 1997). However, this does not lead to net ozone loss since O_3 is regenerated from the reaction of O with O_2. The alternate photolysis path giving $Cl + O_2$ has a quantum yield of $<5 \times 10^{-4}$ in the gas phase and hence is not important (Lawrence et al., 1990). However, the photochemistry is very sensitive to the environment. For example, the photolysis of OClO in solution or adsorbed in or on an ice matrix at 80 K gives ClOO (e.g., see Vaida and Simon, 1995; Dunn et al., 1995; and Pursell et al., 1995). If this were to occur at stratospheric temperatures, chlorine atoms would be regenerated and lead to net ozone destruction. On the other hand, photolysis of isolated OClO at 150 K and either 360 or 367 nm generates a $ClClO_2$ species, proposed to be formed by the photochemical reactions of aggregates of OClO (Graham et al., 1996b; Pursell et al., 1996). However, the results of laboratory studies suggest that less than 10^{-6} monolayers of OClO will exist on the ice surface at equilibrium at typical Antarctic springtime stratospheric temperatures and pressures (L. A. Brown et al., 1996; Graham et al., 1996a). In this case, such condensed-phase photochemistry is likely not important in stratospheric ozone depletion.

Since BrCl, produced in the minor channel (31c), absorbs strongly in the UV and visible (see Table 4.30), it also ultimately generates atomic bromine. Bromine atoms then react with O_3 as well,

$$Br + O_3 \rightarrow BrO + O_2, \qquad (32)$$

leading to a net loss of O_3. This ClO_x–BrO_x cycle is believed to be responsible for about 20–25% of the loss due to halogen chemistry in the 16- to 20-km region at midlatitudes (Wennberg et al., 1994; Lary, 1997) and is a significant cycle in the polar lower (<14 km) stratosphere (Lary, 1997).

The destruction of O_3 by chlorine and bromine can be "short-circuited" by removing either Cl and Br or, alternatively, ClO and BrO. For chlorine atoms, this occurs by reaction with methane that has been transported from the troposphere:

$$Cl + CH_4 \rightarrow HCl + CH_3. \qquad (33)$$

ClO forms chlorine nitrate by reaction with NO_2:

$$ClO + NO_2 \xrightarrow{M} ClONO_2. \qquad (16)$$

As a result, concentrations of ClO and NO_2 would be expected to be inversely correlated in the atmosphere, as has been observed in field studies (e.g., Stimpfle et al., 1994).

Both HCl and $ClONO_2$ are known as "temporary reservoirs" for chlorine, since active chlorine can be regenerated via the reactions

$$HCl + OH \rightarrow Cl + H_2O, \qquad (34)$$

$$ClONO_2 + h\nu \rightarrow Cl + NO_3. \qquad (35)$$

(There is also a second channel in reaction (35) with a similar branching ratio, but producing $ClO + NO_2$; see Chapter 4.N). As discussed in Section D, bromine is particularly effective in destroying ozone because unlike chlorine atoms, atomic bromine does not react rapidly with CH_4, the only organic that is sufficiently long-lived to reach the stratosphere. As a result, the analogous reaction to (33) does not occur for Br. However, Br does react with HO_2 to form HBr, although the reaction is slower than the corresponding chlorine atom reaction:

$$Br + HO_2 \rightarrow HBr + O_2. \qquad (36)$$

Analogous to chlorine chemistry, the formation of bromine nitrate represents the major "short circuit" in its ozone destruction cycle:

$$BrO + NO_2 \xrightarrow{M} BrONO_2. \qquad (17)$$

As discussed in Section D, $BrONO_2$ undergoes even more rapid heterogeneous reactions than $ClONO_2$, forming HOBr, and the photolysis of both $BrONO_2$ and HOBr is relatively fast. Thus bromine spends more time in its catalytically active form, making it very effective in ozone destruction.

Crutzen et al. (1992) suggested that the methylperoxy radical formed in the CH_4 oxidation may also play a role in the gas-phase halogen chemistry. (The role of halogens in the oxidation of methane in the lower stratosphere and upper troposphere and the impact on other cycles are discussed by Lary and Toumi (1997).) This reaction has at least two channels that are important, although the contribution of additional channels cannot be ruled out with certainty:

$$CH_3O_2 + ClO \rightarrow CH_3O + ClOO, \quad (37a)$$
$$\rightarrow CH_3OCl + O_2, \quad (37b)$$
$$k_{37}^{298K} = k_{37a} + k_{37b} = 2.2 \times 10^{-12} \text{ cm}^3 \text{ molecule}^{-1} \text{ s}^{-1}$$

(DeMore et al., 1997).

The relative contributions of (37a) and (37b) are controversial, with the branching ratio for (37a) reported to be 0.3–0.85 (DeMore et al., 1997 and references therein; Kukui et al., 1994; Helleis et al., 1994; Biggs et al., 1995; Daële and Poulet, 1996). If the reaction proceeds via (37a), the ClOO will thermally decompose as discussed earlier, regenerating a chlorine atom. Chlorine atoms can also be regenerated if the reaction proceeds via (37b) to form methyl hypochlorite, CH_3OCl, which then photolyzes to $CH_3O + Cl$. Alternatively, Biggs et al. (1995) suggest that CH_3OCl could undergo a heterogeneous reaction with HCl, similar to those discussed later for $ClONO_2$, $BrONO_2$, and N_2O_5:

$$CH_3OCl + HCl \xrightarrow{\text{ice/particles}} CH_3OH + Cl_2. \quad (38)$$

Crowley et al. (1994) have measured the absorption cross sections of CH_3OCl and calculate a lifetime with respect to photolysis under stratospheric conditions of 4 h at a solar zenith angle of 80°. The rate of the heterogeneous reaction (38) is not known.

In short, the chemistry of the halogens, NO_x, and HO_x is intimately connected. As we saw earlier with respect to the HSCT, effects on one of these can affect the other cycles significantly as well, and indeed, the overall effects on stratospheric ozone may be due mainly to these "secondary" interactions involving other families of compounds.

4. Antarctic "Ozone Hole"

In 1985, Farman et al. reported that the total column ozone at Halley Bay in the Antarctic had decreased substantially at polar sunrise each year for about 5–10 years. Figure 12.16 shows the Farman et al. data supplemented by measurements taken since then (Jones and Shanklin, 1995). Clearly a major drop in column ozone has been occurring since the mid to late 1970s. The extent of this change, and the rapidity with which it occurred, were unprecedented and focused the atmospheric chemistry community's attention on the reasons for this massive destruction of stratospheric ozone in the Antarctic spring.

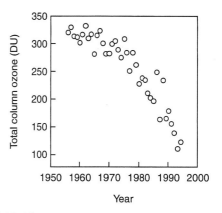

FIGURE 12.16 Average total column ozone measured in October at Halley Bay, Antarctica, from 1957 to 1994 [DU = Dobson units (see text)] (adapted from Jones and Shanklin, 1995).

Figure 12.17 shows the ozone profiles over the U.S. Amundsen-Scott Station at the South Pole in 1993 on August 23 prior to formation of the ozone hole and on October 12 after the ozone hole had developed. The total column ozone decreased from 276 DU on August 23 to only 91 DU on October 12, and, in addition, there was essentially no ozone in the region from 14 to 19 km (Hofmann et al., 1994a). During the same period at the McMurdo Station in Antarctica, the total column ozone decreased from 275 to 130 DU (B. J. Johnson et al., 1995). While similar profiles have been observed since the discovery of the ozone hole, these data show some of the most extensive ozone destruction ever observed, although 1994 and 1995 showed almost as much O_3

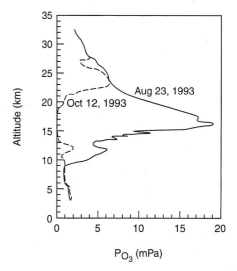

FIGURE 12.17 Vertical O_3 profile before (August 23) and after (October 12) development of the ozone hole at the U.S. Amundsen–Scott Station, South Pole, in 1993 (adapted from Hofmann et al., 1994a).

loss (Nardi *et al.*, 1997). It is noteworthy that loss of O_3 in the isolated interior regions of the polar vortex may have been maximized even by the mid-1980s (Jiang *et al.*, 1996).

There are several reasons for the dramatic ozone destruction (see Fig. 2.17): low temperatures may have prolonged the presence of polar stratospheric clouds, which play a key role in the ozone destruction, the polar vortex was very stable, there were increased sulfate aerosols from the 1991 Mount Pinatubo volcanic eruption, which also contribute to heterogeneous chemistry, and chlorine levels had continued to increase. These issues are treated in more detail shortly.

Soon after the discovery of the hole, a number of different theories to explain this remarkable observation were advanced (Solomon, 1988). These included atmospheric dynamics involving vertical advection, which introduces upper tropospheric air with its lower ozone levels (e.g., see Tung *et al.*, 1986; Mahlman and Fels, 1986; Shiotani and Gille, 1987; Tung and Yang, 1988; and Rosenfield *et al.*, 1988), and solar proton events that produce more oxides of nitrogen and destroy O_3 through the cycles discussed earlier for the HSCT (Callis and Natarajan, 1986; Stephenson and Scourfield, 1991; Shumilov *et al.*, 1992). However, if the first explanation were correct, other long-lived trace gases produced in the troposphere such as N_2O should be enhanced; this was found not to be the case. In fact, stratospheric N_2O mixing ratios decline rapidly as the South Pole is approached (e.g., see Schoeberl and Hartmann, 1991; and Tuck, 1989).

Increased production of oxides of nitrogen through solar proton events associated with the 11-year cycle in solar activity would be expected to be most important in the upper stratosphere, above the region where the majority of the ozone depletion was observed; in addition, lower, rather than higher, concentrations of gas-phase oxides of nitrogen appear to be associated with the ozone depletion (e.g., see Noxon, 1978; McKenzie and Johnston, 1984; Thomas *et al.*, 1988; Keys and Gardiner, 1991; and Solomon and Keys, 1992). Hence both of these explanations are not consistent with atmospheric observations.

Through a variety of studies, it is now generally accepted that the observed losses are associated with chlorine derived from CFCs and that heterogeneous chemistry on polar stratospheric clouds plays a major role. The chemistry in this region is the result of the unique meteorology. As described in detail by Schoeberl and Hartmann (1991) and Schoeberl *et al.* (1992), a polar vortex develops in the stratosphere during the winter over Antarctica. The air in this vortex remains relatively isolated from the rest of the stratosphere, allowing photochemically active products to build up during the polar winter and setting the stage for the rapid destruction of ozone when the sun comes up and the polar vortex dissipates, apparently from the top down (Bevilacqua *et al.*, 1995). There is, however, some exchange of the air mass, especially that in the vortex edges, with regions outside the vortex (e.g., see Wauben *et al.*, 1997a, 1997b). Whether it is better described as a "flow reactor" rather than a "containment vessel" remains somewhat controversial (McIntyre, 1995; Tuck and Proffitt, 1997; Wauben *et al.*, 1997a, 1997b).

A second critical component of the meteorology in this region is that the stratospheric temperatures during the winter can be very low. As the sunlight decreases in the fall, radiative cooling of air in the upper stratosphere occurs, which causes sinking of this polar air mass. Adiabatic heating occurs as the air sinks, which partially offsets the temperature decrease due to radiative cooling. Radiative equilibrium is reached at altitudes below about 30 km, decreasing the descent of the air mass. The cooling takes the air temperature to values as low as ~ 185 K. The temperature difference between the polar region and midlatitudes leads to a vortex with wind speeds at its edge of about 100 m s^{-1}—hence the designation "polar vortex."

At these low temperatures, even the relatively small amount of water in the stratosphere (about 5-6 ppm at the beginning of winter, dropping to 2-3 ppm when dehydration occurs during July and August) forms ice crystals. In addition, at slightly higher temperatures, crystalline nitric acid trihydrate (NAT) also forms, which was initially thought to represent one type of PSC. As discussed in more detail shortly, the formation of polar stratospheric clouds (PSCs) is quite complex and is now believed to involve ternary solutions of HNO_3, H_2SO_4, and water as well. PSCs can have a quite remarkable appearance, with various colors depending on their altitude and the presence of clouds (Sarkissian *et al.*, 1991). As discussed in detail later, they play a critical role in ozone depletion by providing surfaces for heterogeneous chemistry.

There are a number of factors that determine the amount of ozone destruction over Antarctica each year. Clearly, the concentrations of chlorine and bromine are major determinants, and these have increased from about 1.1 ppb in the 1960s to 2.4 to 3.2 ppb over the 1986-1995 decade (WMO, 1995). Lower temperatures lead to more polar stratospheric clouds and more heterogeneous chemistry. As discussed later, it has also been increasingly recognized that lower temperatures have a significant effect on heterogeneous chemistry by increasing the solubility of HCl in liquid aerosol particles. Aerosol concentrations are important because they serve as hosts for heterogeneous chemistry as well as assisting in the formation of polar stratospheric clouds (e.g., see Hofmann and Oltmans, 1993; Deshler *et al.*, 1996; Portmann *et al.*, 1996; and Molina *et al.*, 1996). Finally, there appears to be an association between the depth of the Antarctic ozone hole and quasi-biennial oscillations, QBO (e.g., see Bojkov, 1986; Garcia and Solomon, 1987; and Angell, 1993b), which may be associated with temperature changes due to enhanced transport from the tropics to the poles in the months preceding development of the ozone hole.

As discussed earlier, the destruction of O_3 by chlorine in midlatitudes is controlled by tying up chlorine atoms as HCl via its reaction (33) with CH_4 or by tying up ClO in the form of chlorine nitrate, $ClONO_2$, via its reaction (16) with NO_2. These temporary chlorine reservoirs only slowly regenerate atomic chlorine so they are important in determining how much ozone destruction occurs. The reaction of $ClONO_2$ with HCl in the gas phase is slow, with a rate constant of less than $\sim 1 \times 10^{-20}$ cm^3 $molecule^{-1}$ s^{-1} (DeMore *et al.*, 1997). However, as first shown in the mid-1980s (Molina *et al.*, 1987; Tolbert *et al.*, 1987, 1988a; Leu, 1988a), it proceeds quite rapidly on the surfaces of ice found in the stratosphere, generating Cl_2 and HNO_3:

$$HCl_{(ads)} + ClONO_2 \xrightarrow[\text{fast}]{\text{ice/particle}} Cl_2 + HNO_{3(ads)}. \quad (39)$$
(gas, slow)

(Of course, this reaction does not involve the simultaneous collision of HCl and $ClONO_2$ at the particle surface; HCl is taken up by the particle and $ClONO_2$ subsequently collides and reacts with it.) Because of the very "sticky" nature of nitric acid, it stays on the ice or in the solution. This has the added effect of removing oxides of nitrogen from the gas phase, which then frees up additional ClO that might otherwise be tied up in the form of chlorine nitrate. As discussed in more detail shortly, it is now known that stratospheric particles may not only be solids, but under some conditions liquid solutions containing mixtures of H_2SO_4 and water or ternary solutions of HNO_3 with H_2SO_4 and water. However, reactions such as (39) can occur not only at the surfaces of these liquid particles but in the bulk as well. Indeed, as we shall see, some of these reactions are much faster in and on liquid particles than on ice.

Farman and co-workers (1985) suggested that the reaction between HCl and $ClONO_2$ may play a key role if it were fast enough, which at the time did not seem to be the case for the gas-phase reaction. Subsequently, Solomon *et al.* (1986) proposed that enhancement of this reaction on the ice surfaces of polar stratospheric clouds could explain the development of the ozone hole via the production of Cl_2 during the winter. It is also consistent with sequestering oxides of nitrogen in the form of HNO_3 on the ice surface. It has been suggested that similar chemistry may occur on cirrus clouds near the tropopause (Borrmann *et al.*, 1996, 1997a; Solomon *et al.*, 1997), where a significant amount of HCl can be taken up by ice particles at equilibrium (Thibert and Dominé, 1997). Whether sufficient cloud surface area and inorganic chlorine compounds coexist in the same region to cause this chemistry is not clear.

Similarly, the reaction of HCl with N_2O_5 is slow in the gas phase, but was shown in the late 1980s to occur rapidly on ice surfaces or in the solutions found in stratospheric particles (Tolbert *et al.*, 1988b; Leu, 1988b):

$$HCl_{(ads)} + N_2O_5 \xrightarrow[\text{fast}]{\text{ice/particle}} ClNO_2 + HNO_{3(ads)}. \quad (40)$$
(gas, slow)

As a result of the enhancement of reactions (39) and (40) in or on surfaces provided by PSCs, HCl and $ClONO_2$, which normally act as reservoirs, are converted over the winter into the photochemically active Cl_2 and nitryl chloride, $ClNO_2$. When the sun comes up in the spring, these species are rapidly photolyzed to generate chlorine atoms, setting off the chain destruction of ozone. However, in the case of $ClNO_2$ from reaction (40), NO_2 is generated simultaneously; this gaseous NO_2 can then sequester chlorine in the form of $ClONO_2$. Hence the heterogeneous reaction (39) is much more important. The nature of PSCs and heterogeneous reactions on PSCs and aerosol surfaces are discussed in more detail in the following section.

While reaction (39) shows the overall reaction that occurs between the two chlorine reservoirs, it has been proposed that it may actually occur in several steps (e.g., Hanson and Ravishankara, 1991, 1993a; Abbatt *et al.*, 1992). Thus $ClONO_2$ has been shown to hydrolyze on the surfaces of solid and liquid particles, generating HOCl:

$$H_2O + ClONO_2 \xrightarrow{\text{particle}} HOCl + HNO_3. \quad (41)$$

HOCl can then react with HCl on or in the particle, generating Cl_2 (Prather, 1992):

$$HOCl + HCl \xrightarrow{\text{particle}} Cl_2 + H_2O. \quad (42)$$

The net effect of these two reactions is that shown as reaction (39).

However, Oppliger *et al.* (1997) suggest that while reaction (39) of $ClONO_2$ with HCl on ice proceeds through a direct mechanism, the hydrolysis reaction (41) does not. For example, while $ClONO_2$ is rapidly taken up by ice at 180 K, the formation of HOCl is delayed (see also Hanson and Ravishankara, 1992b). Rossi and co-workers (Oppliger et al., 1997) propose that $ClONO_2$ forms an intermediate ($H_2OCl^+ \cdots NO_3^-$) that subsequently releases HOCl to the gas phase, while generating hydrated HNO_3. Infrared studies by Sodeau and co-workers (Sodeau et al., 1995; Koch et al., 1997; Horn et al., 1998) support the formation of $[H_2OCl]^+$ under conditions of low water availability, with this intermediate ultimately reacting with water to generate HOCl. *Ab initio* calculations, however, suggest that the hydrolysis occurs by a concerted nucleophilic attack of an oxygen from a water (ice) molecule on the chloride of $ClONO_2$, simultaneously with a proton transfer from the attacking water to the ice (Bianco and Hynes, 1998).

There are also important differences in the gas-phase chemistry of the Antarctic ozone hole compared to the chemistry at midlatitudes. One is the formation and photolysis of the ClO dimer. In the Antarctic spring, recycling of ClO back to chlorine atoms via reaction (27) with oxygen atoms does not play a major role because of the relatively small oxygen atom concentrations at the low UV levels at that time. Molina and Molina (1987) proposed that the formation of a dimer of ClO could, however, lead to regeneration of atomic chlorine through the following reactions:

$$ClO + ClO \xrightarrow{M} (ClO)_2 \quad (43a)$$

$$(ClO)_2 + h\nu \to Cl + ClOO \quad (44)$$

$$ClOO \xrightarrow{M} Cl + O_2 \quad (45)$$

Photolysis of the dimer, reaction (44), proceeds primarily via generation of Cl + ClOO (Cox and Hayman, 1988; Molina et al., 1990). For example, Molina et al. (1990) reported the quantum yield for this channel at 308 nm to be unity, with an uncertainty of ~30%. Okumura and co-workers (Moore et al., 1999) and Schindler and co-workers (Schmidt et al., 1998) have reported that the quantum yield is less than 1.0. For example, Schmidt et al. (1998) used resonance-enhanced multiphoton ionization (REMPI) with time-of-flight (TOF) mass spectrometry to follow the production of oxygen and chlorine atoms as well as ClO in vibrational levels up to $v'' = 5$ in the photolysis of the dimer. At a photolysis wavelength of 250 nm, the quantum yield for chlorine atom production was measured to be 0.65 ± 0.15, but ClO was not observed. Assuming that all of the excited dimer dissociates, this suggests that the production of ClO in vibrational levels $v'' > 5$ accounted for about a third of the reaction. Moore et al. (1999) used TOF-MS to follow the products of photolysis (including Cl atoms, ClO, and O_2) of the dimer at 248 and 308 nm, respectively. At 248 nm, chlorine atom production from three primary processes was observed and attributed to reaction (44), reaction (45), and a concerted process producing 2Cl + O_2 directly. They also observed the production of ClO at both 248 and 308 nm. At 248 nm, the Cl:ClO product yields were in the ratio 0.88:0.12 and at 308 nm, 0.90:0.10. Their studies suggest that under stratospheric conditions, the quantum yield for the channel producing chlorine atoms is 0.90 ± 0.10 and that for producing ClO is 0.10 ± 0.10 with an upper limit of 0.31, which is not inconsistent with the work of Molina et al. (1990) and Schmidt et al. (1998).

Atmospheric measurements of ClO, BrO, O_3, and N_2O confirm the importance of reactions (43a)–(45) in the destruction of O_3. For example, Anderson et al. (1989) showed that this cycle is the largest contributor to ozone loss in the Antarctic vortex from 14–18 km.

In summary, reactions (43a)–(45) have generally been taken to represent the chemistry occurring in the ozone hole. However, reduced efficiency of chlorine atom production in the photolysis of $(ClO)_2$, reaction (44), and hence ozone destruction, needs to be modeled and tested against the atmospheric observations.

The self-reaction of ClO has both a termolecular component shown as (43a) and a bimolecular component with three possible sets of products (Hayman et al., 1986; Nickolaisen et al., 1994; Horowitz et al., 1994):

$$ClO + ClO \to Cl_2 + O_2, \quad (43b)$$
$$k_{43b} = 1.01 \times 10^{-12} e^{-1590/T} \text{ cm}^3 \text{ molecule}^{-1} \text{ s}^{-1},$$
$$\to Cl + ClOO, \quad (43c)$$
$$k_{43c} = 2.98 \times 10^{-11} e^{-2450/T} \text{ cm}^3 \text{ molecule}^{-1} \text{ s}^{-1},$$
$$\to Cl + OClO, \quad (43d)$$
$$k_{43d} = 3.5 \times 10^{-13} e^{-1370/T} \text{ cm}^3 \text{ molecule}^{-1} \text{ s}^{-1}.$$

Figure 12.18 shows the energetics involved in the self-reaction. As is often the case for termolecular reactions (see Chapter 5), the formation of a dimer from ClO + ClO has a very small activation energy. However, the effective bimolecular channels do have significant activation energies as seen in the rate constants of Nickolaisen et al. (1994) given in reactions (43b)–(43d). The branching ratios at room temperature for the bimolecular reactions, i.e., (43x)/[(43b) + (43c) + (43d)], where x = b, c, or d, measured by Nickolaisen and co-workers, 0.29 for (43b), 0.50 for (43c), and 0.21 for (43d), are in good agreement with values of 0.29, 0.41, and 0.20, respectively, reported by Horowitz et al. (1994).

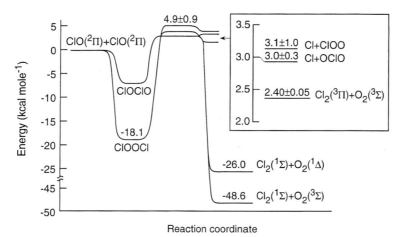

FIGURE 12.18 Energetics of the ClO + ClO reaction (adapted from Nickolaisen *et al.*, 1994).

However, the formation of the dimer in the termolecular reaction is sufficiently fast under stratospheric conditions that the bimolecular reactions are not important. For example, using the recommended termolecular values (DeMore *et al.*, 1997) for the low-pressure-limiting rate constant of $k_0^{300} = 2.2 \times 10^{-32}$ cm^6 molecule^{-2} s^{-1} and the high-pressure-limiting rate constant of $k_\infty^{300} = 3.5 \times 10^{-12}$ cm^3 molecule^{-1} s^{-1} with temperature-dependent coefficients $n = 3.1$ and $m = 1.0$ (see Chapter 5), the effective rate constant at 25 Torr pressure and 300 K is 1.6×10^{-14} cm^3 molecule^{-1} s^{-1}, equal to the sum of the bimolecular channels (Nickolaisen *et al.*, 1994). At a more typical stratospheric temperature of 220 K and only 1 Torr pressure, the effective second-order rate constant for the termolecular reaction already exceeds that for the sum of the bimolecular channels, 2.4×10^{-15} versus 1.9×10^{-15} cm^3 molecule^{-1} s^{-1}.

In short, under stratospheric conditions the self-reaction of ClO to form the dimer (ClO)$_2$ is the most important channel and has been shown to be consistent with observations of ozone destruction in the Antarctic (Molina and Molina, 1987; Sander *et al.*, 1989; Trolier *et al.*, 1990; Nickolaisen *et al.*, 1994).

This cycle is believed to be responsible for approximately 75% of the ozone destruction in the 13- to 19-km region in the Antarctic ozone hole (Anderson *et al.*, 1991). The cycles involving ClO + O, reaction (27), and ClO + HO$_2$, reactions (28) and (29), each account for approximately 5% of the ozone loss and the remainder is due to the ClO + BrO interaction, reactions (31) and (32) (Anderson *et al.*, 1991).

It should be noted that around the edges of the vortex where exchange with the surrounding air occurs, there is less extensive denitrification in this "collar" region (e.g., Toon *et al.*, 1989; Ricaud *et al.*, 1998). As a result, ClO was trapped as ClONO$_2$ and there was less ozone destruction.

Evidence for the contribution of the ClO + BrO interaction is found in the detection and measurement of OClO that is formed as a major product of this reaction, reaction (31a). This species has a very characteristic banded absorption structure in the UV and visible regions, which makes it an ideal candidate for measurement using differential optical absorption spectrometry (see Chapter 11). With this technique, enhanced levels of OClO have been measured in both the Antarctic and the Arctic (e.g., Solomon *et al.*, 1987, 1988; Wahner and Schiller, 1992; Sanders *et al.*, 1993). From such measurements, it was estimated that about 20–30% of the total ozone loss observed at McMurdo during September 1987 and 1991 was due to the ClO + BrO cycle, with the remainder primarily due to the formation and photolysis of the ClO dimer (Sanders *et al.*, 1993). The formation of OClO from the ClO + BrO reaction has also been observed outside the polar vortex and attributed to enhanced contributions from bromine chemistry due to the heterogeneous activation of BrONO$_2$ on aerosol particles (e.g., Erle *et al.*, 1998).

It is interesting that enhanced OClO levels were observed at McMurdo as early as late June, which is not expected since light is not available at that time to generate Cl and Br and hence the ClO and BrO precursors (Sanders *et al.*, 1993). It appears that portions of the polar vortex can be exposed to sunlight even during the polar winter due to the size of the vortex and some displacement of the vortex edge into sunlit regions. This leads to the generation of enhanced ClO, BrO, and their product OClO as well as reduced NO$_3$ and increased NO$_2$ (e.g., see Tuck, 1989; Solomon *et al.*, 1993; Jiang *et al.*, 1996). This effect has also been proposed to have increased the wintertime loss of O$_3$ so that the threshold for development of the ozone hole is lowered (Jiang *et al.*, 1996). [It should be noted that another mechanism for producing OClO is reaction (43d) of ClO + ClO; for example, this has been invoked to explain the observed levels of ClO in the Arctic vortex under warmer (225 K) conditions (Pierson *et al.*, 1999).]

Evidence for the role of chlorine is seen in Fig. 12.19. This shows ClO and O$_3$ measured on August 23, 1987, prior to development of the ozone hole, and on September 16, after the hole had formed, as the sampling aircraft flew south into the polar vortex (Anderson *et al.*, 1991). The rapid drop in O$_3$ and accompanying increase in ClO due to the chemistry discussed above is clearly seen on September 16. On August 23, ClO is observed because some of the air in the polar vortex has been exposed previously to sunlight around the edges of the vortex, forming ClO. This leads to small amounts of ozone destruction via reactions (44), (45), and (26), with O$_3$ losses of ~10–20% prior to development of the hole. However, the total available light at this time is not sufficient to drive substantial ozone depletion, which is in effect determined by the total integrated solar exposure available to cause the chemistry. As the solar exposure increases through September, the chain destruction of O$_3$ above is greatly enhanced, as seen in Fig. 12.19. Typical ozone loss

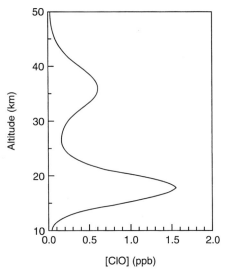

FIGURE 12.20 Vertical profile of ClO on September 19–20, 1992, at McMurdo Station, Antarctica (adapted from Emmons *et al.*, 1995).

rates during development of the ozone hole are ~3–4% per day when light is available all day for photolysis (Rosen *et al.*, 1993).

Figure 12.20 shows measurements of the vertical profile of ClO at McMurdo Station, Antarctica, on September 19–20, 1992, as the ozone hole was developing (Emmons *et al.*, 1995). As expected based on the foregoing chemistry, the concentration peaks in the 15- to 20-km range, in the same region as the greatest depletion of O$_3$ is observed. (The peak at higher altitudes is that normally observed globally in the stratosphere due to gas-phase chemistry.)

In short, the overall features of the chemistry involved with the massive destruction of ozone and formation of the ozone hole are now reasonably well understood and include as a key component heterogeneous reactions on the surfaces of polar stratospheric clouds and aerosols. However, there remain a number of questions relating to the details of the chemistry, including the microphysics of dehydration and denitrification, the kinetics and photochemistry of some of the ClO$_x$ and BrO$_x$ species, and the nature of PSCs under various conditions. PSCs and aerosols, and their role in halogen and NO$_x$ chemistry, are discussed in more detail in the following section.

5. Polar Stratospheric Clouds (PSCs) and Aerosols

a. Nature of Aerosols and PSCs

Sulfate aerosol particles with diameters typically in the 0.1- to 0.3-μm range are well known to be formed

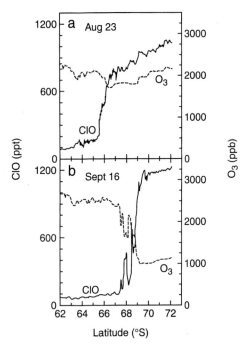

FIGURE 12.19 Aircraft measurements of ClO and O$_3$ on (a) August 23 and (b) September 16, 1987, as the aircraft flew south (adapted from Anderson *et al.*, 1991).

in the stratosphere from a number of sources, forming what is known as the Junge layer. Carbonyl sulfide, COS, is produced in the troposphere by both natural and anthropogenic processes (Chin and Davis, 1993, 1995). It reacts with OH, but the reaction is quite slow, with a room temperature rate constant of 1.9×10^{-15} cm^3 molecule^{-1} s^{-1} (DeMore et al., 1997); this corresponds to a calculated tropospheric lifetime with respect to this one reaction of approximately 17 years at an OH concentration of 1×10^6 radicals cm^{-3}. Based on measured atmospheric concentrations of COS and estimated source strengths, Chin and Davis (1995) estimate a global atmospheric lifetime of about 4 years. As a result of this lifetime, significant quantities of COS reach the stratosphere, where it is ultimately oxidized to sulfuric acid (Crutzen, 1976; Kourtidis et al., 1995; Zhao et al., 1995). In the absence of volcanic injections of SO_2, this is the major source of stratospheric sulfate aerosols (SSA) (Crutzen, 1976).

However, the eruption of large volcanoes also injects large quantities of SO_2 into the stratosphere, increasing the concentration of SSA significantly. For example, typical number concentrations of SSA are about 1–10 particles cm^{-3} under nonvolcanically perturbed conditions; the number density increases by 1–2 orders of magnitude after major volcanic eruptions (e.g., see Russell et al., 1996).

These aerosols play a major role in stratospheric chemistry by directly providing surfaces for heterogeneous chemistry (discussed in more detail later) as well as serving as nuclei for polar stratospheric cloud formation. Figure 12.21 schematically shows the processes believed to be involved in PSC formation. The thermodynamic stability of the various possible forms of PSCs at stratospherically relevant temperatures and the transitions between them are discussed in detail by Koop et al. (1997a).

The concentration of sulfuric acid in SSA is typically 50–80 wt% under mid- and low-latitude stratosphere conditions. However, as the temperature drops, these particles take up increasing amounts of water, which dilutes the particles to as low as 30 wt% H_2SO_4. Gaseous nitric acid is also absorbed by these solutions, forming ternary H_2SO_4–H_2O–HNO_3 solutions with as much as 30 wt% in each acid.

As more and more HNO_3 and H_2O are taken up into solution from the gas phase, the relative amount of H_2SO_4 diminishes until the particle is primarily an HNO_3–H_2O mixture. Continued reduction in the temperature results in nitric acid and sulfuric acid hydrates freezing out. Based on laboratory studies, it has been proposed that nitric acid trihydrate (NAT) freezes out of solution first (Hanson and Mauersberger, 1988a, 1988b; Molina et al., 1993; Iraci et al., 1994, 1995; Beyer et al., 1994), although some studies suggest that much lower temperatures, ~170 K, would be required for this to be sufficiently fast in the stratosphere (Bertram and Sloan, 1998b). It has also been proposed that other hydrates such as nitric acid dihydrate (NAD), which nucleates rapidly at stratospheric temperatures from 2:1 H_2O:HNO_3 solutions (Tisdale et al., 1997), are formed. Nitric acid pentahydrate (NAP), ternary hydrates such as $H_2SO_4 \cdot HNO_3 \cdot 5H_2O$, or higher hydrates in the form of a water-rich metastable solid phase (vide infra) may also be formed as intermediates prior to the formation of the more stable NAT (Tolbert and Middlebrook, 1990; Marti and Mauersberger, 1993; 1994; Worsnop et al., 1993; Fox et al., 1995; Tabazadeh and Toon, 1996).

Sulfuric acid tetrahydrate (SAT) also ultimately freezes out of these ternary solutions (Molina et al., 1993; Iraci et al., 1995). At higher temperatures found at higher altitudes in the middle and low latitudes, sulfuric acid monohydrate (SAM) may also be stable (Zhang et al., 1995).

In a number of laboratory studies (Molina et al., 1993; Iraci et al., 1994, 1995; Beyer et al., 1994; Kolb et al., 1995), these crystallizations had been observed to form the solid nitric acid and sulfuric acid hydrates a few degrees above the ice frost point, defined as the temperature at which the air is saturated with respect to the formation of a plane surface of ice. [See Marti and Mauersberger (1993) for the vapor pressure of ice at stratospherically relevant temperatures.] However, it appears that this does not occur to a significant extent in the atmosphere. Thus, Carslaw et al. (1994) and Koop et al. (1995, 1997b) report studies showing that ternary HNO_3–H_2SO_4–H_2O solutions do not freeze above the frost point; they suggest that these solutions remain liquid in the stratosphere until the temperature falls below the frost point, where the ice crystals formed act as nuclei for the crystallization of nitric and sulfuric acid hydrates. Similarly, Anthony et al. (1997) followed aerosols composed of solutions of sulfuric and nitric acids and water as a function of time in a low-temperature chamber using FTIR and observed that they remained as supercooled liquids for the duration of the experiments, up to 3 h.

The nitric acid concentration may be a major determinant of the extent of supercooling that occurs for these ternary mixtures (Molina et al., 1993; Song, 1994). For example, Molina et al. (1993) observed that HNO_3 did not affect the supercooling of H_2SO_4–H_2O mixtures at temperatures above 196 K, but below this temperature, the presence of HNO_3 rapidly promoted freezing. In addition, the availability of seed crystals to promote crystallization appears to be a critical issue. As discussed in detail by MacKenzie et al. (1995), a variety of potential seed crystals and/or surfaces that

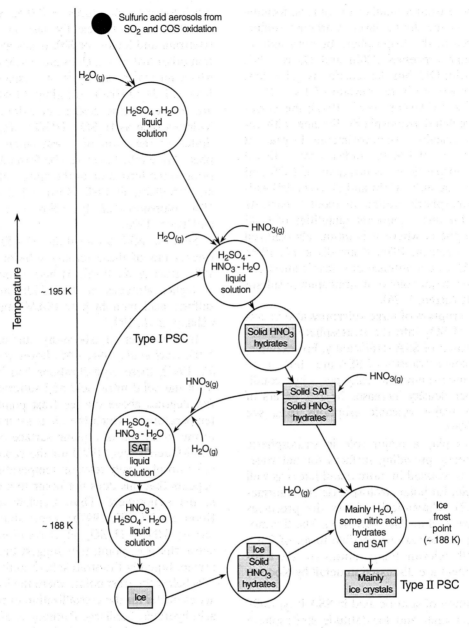

FIGURE 12.21 Schematic of polar stratospheric cloud (PSC) formation.

can assist in crystallization are available in typical laboratory studies, which may not be the case for solution droplets in the stratosphere; even under laboratory conditions, whether NAT or SAT first crystallize from the ternary solution may depend on the particular seed crystal/nucleation mechanism.

At any rate, these PSCs, which can be either liquid or solid, are known as Type I PSCs and form at temperatures about 2–5 K above the ice frost point. They are believed to contain large amounts of nitric acid and water (Fahey *et al.*, 1989; Pueschel *et al.*, 1989, 1990) due to the uptake of these species by the initial SSA as the temperature falls. Typical particle radii lie in the range from 0.1 to ~5 μm, and number concentrations in Antarctica are about 1–10 particles cm^{-3} (Kinne *et al.*, 1989; Hamill and Toon, 1991; Hofmann and Deshler, 1991).

This understanding of the mechanism of formation of Type I PSCs is consistent with atmospheric measurements. For example, Massie *et al.* (1997) showed that gas-phase HNO$_3$ over Scandinavia in January 1992 decreased as the temperature fell while the volume of the PSC particles increased simultaneously. Beyerle et al. (1997) also showed using multiwavelength LIDAR over Scandinavia that the volume density of particles increased as the temperature approached the frost point in a manner consistent with the formation of liquid ternary H_2SO_4–HNO_3–H_2O solutions. Accompanying this was a depletion of gas-phase HNO_3, as it was taken up by the PSCs.

The exact composition and even the phase of these Type I PSCs under specific conditions is not well established (Tolbert, 1994, 1996). Particles collected by Pueschel et al. (1989) by impaction on gold wires had a low collection efficiency for nitric acid, which they suggested was indicative of nitric acid being in the solid state. However, infrared spectra of Type I PSCs in Antarctica in September 1987 indicate that under these particular conditions, the Type I PSCs were likely liquid ternary solutions of nitric and sulfuric acids and water (Toon and Tolbert, 1995). A similar conclusion was reached by Dye et al. (1996) during studies carried out in 1994, based on relationships between the temperature, PSC particle volume, and NO_y, and by Stefanutti et al. (1995) using ground-based LIDAR measurements. Tabazadeh et al. (1995) report evidence for both liquid and solid Type I PSCs under different sampling conditions. Perhaps the most definitive evidence comes from direct mass spectrometric measurements of PSCs in the Arctic by Schreiner et al. (1999), who found the molar ratio of H_2O to HNO_3 to be greater than 10 at temperatures from 189 to 192 K, consistent with them being supercooled ternary solutions.

Aerosols and films consisting of binary solutions of H_2SO_4 and water readily undergo supercooling (Anthony et al., 1995; Iraci et al., 1995; Bertram et al., 1996; Carleton et al., 1997; Koop et al., 1998). However, once they do freeze, the tetrahydrate of sulfuric acid (SAT) is thought to be the stable form under stratospheric conditions (Peter et al., 1992; Luo et al., 1992; Middlebrook et al., 1993; Zhang et al., 1993). (Although the octahydrate $H_2SO_4 \cdot 8H_2O$ has also been observed in laboratory experiments under conditions different from those found in the stratosphere (Imre et al., 1997), it appears unlikely to be important in the stratosphere.) SAT surfaces take up about a monolayer of HNO_3 from the gas phase under conditions typical of the stratosphere; NAT formation on a SAT surface only occurs at high concentrations of gaseous HNO_3 and water vapor. Interestingly, the SAT surface can be "preactivated" by prior formation and evaporation of NAT from the surface; that is, the gaseous concentrations of HNO_3 and H_2O required to form a NAT film on SAT are much lower if NAT is previously formed on the surface and then removed (Zhang et al., 1996).

Zhang et al. (1996) suggest that this preactivation changes the crystal structure of SAT at the interface to more closely match the NAT lattice. This nucleation and growth of NAT on solid SAT was proposed as one potential mechanism for Type I PSC formation (e.g., see Tolbert, 1994), although this process now appears to be less important because of the low probability for binary nucleation of nitric acid and water on SAT (MacKenzie et al., 1995).

Browell et al. (1990) used polarized laser light to probe Type I PSCs using LIDAR techniques and observed that in some cases, the reflected light had undergone substantial depolarization, whereas in other cases it had not. Particles that caused significant depolarization have been dubbed Type 1a particles and the others Type 1b. Toon et al. (1990a) have shown that Type 1a PSCs are not spherical and are quite large, with radii equivalent to >1 μm *if* they were treated as being spherical. These particles may be crystalline NAT or NAD (Rosen et al., 1993; Meilinger et al., 1995; Tabazadeh et al., 1996; Tabazadeh and Toon, 1996; Larsen et al., 1997) which can nucleate from liquid solutions as the temperature falls (e.g., Bertram and Sloan, 1998a,b; Prenni et al., 1998).

Type 1b particles are spherical or nearly so and have typical radii of ~ 0.5 μm, in agreement with the observations of Stefanutti et al. (1991, 1995). These particles are thought to be ternary HNO_3–H_2SO_4–H_2O solutions (Carslaw et al., 1994; Tabazadeh et al., 1994a, 1994b; Hamill et al., 1996; Larsen et al., 1997). The probability of their occurrence in the Arctic has been shown to increase significantly at temperatures at which these ternary particles are expected to grow (Rosen et al., 1997). Like binary H_2SO_4–H_2O solutions, the ternary solutions have been shown in laboratory studies to undergo supercooling (Anthony et al., 1997), in agreement with the atmospheric observations.

Some field measurements of HNO_3 suggest that the formation of liquid or solid Type I PSCs depends on the initial background sulfate aerosols on which the PSCs form. If they are liquid, then liquid ternary solution PSCs tend to form first as the temperature drops below 192 K, whereas if the sulfate particles are initially solids, solid Type 1c PSCs may be generated (Santee et al., 1998).

As the temperature falls below the ice frost point, water condenses out as ice, forming large particles (Fig. 12.21). These are known as Type II PSCs. They are formed at lower temperatures corresponding to the frost point of water (~ 188 K for stratospheric conditions), or possibly 2–3 K below that (Tabazadeh et al., 1997). They are much larger than Type I PSCs, of the order of 5–50 μm in diameter, and consist mainly of ice; their number concentrations are also smaller, typically in the range of ~10^{-2}–10^{-3} cm^{-3} (Kinne et al., 1989; Hamill and Toon, 1991; Hofmann and Deshler, 1991). Because of their size, their rate of gravitational settling is relatively large, and settling rates of about 1 km per day can occur (Hamill and Toon, 1991). This acts to permanently remove cocondensing nitric acid, i.e., denitrifies the stratosphere, and also to dehydrate it (e.g., see Vömel et al., 1997).

For example, there is a loss of stratospheric gas-phase nitric acid over the South Pole in June and July as PSC formation occurs (e.g., de Zafra et al., 1997; Santee et al., 1998). However, as temperatures increased in the spring, an increase in gaseous HNO$_3$ is not observed, consistent with the prior removal of HNO$_3$ by settling out of PSCs holding nitrate. By removing water, this gravitational settling also dehydrates the stratosphere (Gandrud et al., 1990). While these Type II PSCs are mainly water, it has been suggested that under some circumstances they may have a coating of NAT that inhibits the evaporation of water from the particles (Tolbert and Middlebrook, 1990; Peter et al., 1994; Middlebrook et al., 1996; Biermann et al., 1998).

Fourier transform infrared reflection–absorption spectroscopy studies (FTIR-RAS) by Tolbert and co-workers (Zondlo et al., 1998) of the uptake of HNO$_3$ on ice at 185 K have shown that a supercooled liquid forms on the surface; upon evaporation of water, the ice film becomes more concentrated in HNO$_3$ and at stoichiometries of 3:1 and 2:1 H$_2$O:HNO$_3$, respectively, NAT and NAD crystallize out. The reactions of ClONO$_2$ and N$_2$O$_5$ with the ice also led to the formation of supercooled H$_2$O–HNO$_3$ liquid layers on the ice surface.

Toon and Tolbert (1995) suggest that if Type I PSCs are primarily ternary solutions rather than crystalline NAT, the higher vapor pressure of HNO$_3$ over the solution would in effect "distill" nitric acid from Type I to Type II PSCs, assisting in denitrification of the stratosphere. This overcomes the problem that if Type II PSCs have nitric acid only by virtue of the initial core onto which the water vapor condenses, the amount of HNO$_3$ they could remove may not be very large. The supercooled H$_2$O–HNO$_3$ liquid layer observed by Zondlo et al. (1998) clearly may also play an important role in terms of the amount of HNO$_3$ that can exist on the surface of these PSCs.

LIDAR measurements of stratospheric aerosols (Browell et al., 1990) show that above the frost point, PSCs can be solids, perhaps solid SAT. Pure SAT, which does not form PSCs very efficiently, does not melt until quite high temperatures, about 210–215 K (Middlebrook et al., 1993; Iraci et al., 1995). However,

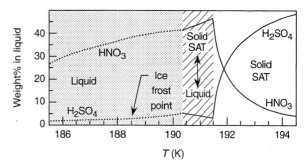

FIGURE 12.22 Composition of liquid in equilibrium with sulfuric acid tetrahydrate (SAT) as the temperature is lowered and SAT deliquesces in the presence of 5 ppm water vapor and 10 ppb HNO$_3$ at 50 m bar altitude (adapted from Koop and Carslaw, 1996).

Koop and Carslaw (1996) have shown that as solid SAT is cooled in the presence of gaseous water and nitric acid, the SAT deliquesces. That is, SAT takes up water as well as nitric acid and forms an equilibrium with water–nitric acid–sulfuric acid solutions (Fig. 12.21).

Figure 12.22 shows the composition in terms of the weight percent HNO$_3$ and H$_2$SO$_4$ as a function of temperature as solid SAT is cooled from 194 K under conditions corresponding to a pressure of 50 mbar in an atmosphere containing 5 ppm H$_2$O and an HNO$_3$ concentration of 10 ppb (Koop and Carslaw, 1996). Under these particular conditions, as the temperature falls below 192 K, the SAT is in equilibrium with a liquid film on the particle containing both HNO$_3$ and H$_2$O. The particular temperature at which SAT deliquesces is a function of the water vapor and gaseous nitric acid concentrations as shown in Fig. 12.23. As the temperature falls further and more HNO$_3$ and H$_2$O are taken up into the liquid, the solid SAT dissolves completely, forming a ternary solution of the two acids and water. This solution can then act again to nucleate PSCs.

However, at lower HNO$_3$ concentrations than assumed, e.g., in a denitrified atmosphere, the formation of the liquid is shifted to temperatures about 3 K lower than shown in Fig. 12.22 (Martin et al., 1998). In addition, Martin et al. (1998) predict that under these conditions, SAT will not deliquesce to a liquid solution at temperatures above the frost point as shown in Fig. 12.22. Their experiments also suggest that the formation of the liquid, although thermodynamically favored, may be too slow to be important under stratospheric conditions.

Actually measuring the composition and phase of PSCs and aerosols in the stratosphere is extremely difficult. Some direct measurements have been made by collecting aerosol samples and subsequently analyzing them using techniques such as X-ray energy-disper-

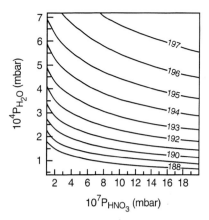

FIGURE 12.23 Temperatures (K) at which SAT deliquesces at different gas-phase pressures of H_2O and HNO_3 typical of the stratosphere (from Koop and Carslaw, 1996).

sive analysis. For example, Pueschel *et al.* (1994) collected and analyzed aerosols after the Mount Pinatubo eruptions and identifed ash particles at lower altitudes that contained Si, Al, Mg, and Na that were coated with H_2SO_4. Particles at higher altitudes were shown to be predominantly $H_2SO_4-H_2O$ mixtures.

Another method of probing sulfuric acid aerosols is to heat the sample intake sufficiently to vaporize sulfuric acid–water aerosols but not other particles such as those containing ash minerals; the difference between the measured particles with and without intake heating provides a measure of the contribution of sulfuric acid–water. Using this technique, Deshler *et al.* (1992), for example, have shown that more than 90% of the stratospheric particles above Laramie, Wyoming, after the Mount Pinatubo eruption were composed of sulfuric acid–water mixtures.

Light scattering and absorption techniques have also been used, for example to obtain the index of refraction of the particles and then to compare these atmospheric measurements to laboratory measurements of NAT, NAD, etc. determined in laboratory studies. Adriani and co-workers (1995), for example, using light scattering in the visible, report four types of particles over McMurdo Station, Antarctica, in 1992: ice, volcanic aerosol particles, and two types of nitric acid–water PSCs. The characteristics of these four types are shown in Table 12.3. The index of refraction measured at 532 nm for particles identified as ice, 1.32 ± 0.01, is consistent with the known index of refraction for ice. Two types of nitric acid containing particles were observed, which had indices of refraction in the range of 1.39–1.42; one substantially depolarized the light beam and one did not. These are consistent with laboratory measurements of the refractive indices of thin films of amorphous, or possibly crystalline, nitric acid trihydrate or with higher amorphous hydrates of HNO_3 (Berland *et al.*, 1994; Middlebrook *et al.*, 1994; Niedziela *et al.*, 1998). The value of 1.43 ± 0.04 observed for the volcanic aerosol in this and earlier studies (e.g., Grainger *et al.*, 1993; Santer *et al.*, 1988) is consistent with sulfuric acid–water solutions (Russell *et al.*, 1996; Luo *et al*, 1996). Refractive indices of ternary solutions of $HNO_3-H_2SO_4-H_2O$ are now available for use in identifying particle compositions in such studies (Luo *et al.*, 1996).

Similar measurements of particles and PSCs have been made in the Arctic region (e.g., see Pueschel *et al.*, 1992b; Dye *et al.*, 1992; Wilson *et al.*, 1992; Deshler *et al.*, 1994; and Brogniez *et al.*, 1997).

Baumgardner *et al.* (1996) measured the refractive index of particles at altitudes from 4 to 20 km at locations from 43.5°S to 37.5°N. While the measured refractive indices were consistent with sulfuric acid–water mixtures at lower temperatures (around 195 K), at higher temperatures the values were smaller than predicted. Furthermore, a significant decrease in the average refractive index of the particles to 1.34 was observed at lower altitudes, from approximately 4 to 9 km, which Baumgardner and co-workers suggest is due to to the presence of nonspherical or light-absorbing particles.

Other light scattering techniques, e.g., using several different scattering wavelengths, have also been used to probe PSCs and aerosols (e.g., Larsen *et al.*, 1995;

TABLE 12.3 Characteristics of Particles over McMurdo Station, Antarctica, in 1992[a]

	Nitric acid–water PSC		Ice	Volcanic aerosol
	Nondepolarizing	Depolarizing		
Concentration (cm^{-3})	9.4–23	5.7–31	6.7–12	6.7–91
Surface area ($\mu m^2\ cm^{-3}$)	19–50	1–32	4.7–126	3.8–44
Volume ($\mu m^3\ cm^{-3}$)	3.3–15	0.2–5.4	0.5–86	0.3–8.6
Index of refraction at 532 nm	1.39 ± 0.03	1.42 ± 0.04	1.32 ± 0.01	1.43 ± 0.04

[a] From Adriani *et al.* (1995).

Gobbi *et al.*, 1998). Although the relevant optical properties in the infrared have been investigated in laboratory studies (e.g., Toon *et al.*, 1994; Richwine *et al.*, 1995), they have not been probed for atmospheric particles. However, such studies clearly also have the potential to shed light on the composition of PSCs.

Knowing whether the PSCs are solid or liquid solutions is important because, as we shall see in the following section, the reaction probabilities for various reactions on PSCs can differ, depending on the nature of the particle. As a result, chlorine activation and ozone destruction are sensitive to this as well. For example, N_2O_5 hydrolysis is much faster in a liquid ternary solution than on NAT (see Table 12.5), and the $ClONO_2 + H_2O$ and $HOCl + HCl$ reactions are also faster. Chlorine activation is therefore faster on ternary solution PSCs (Ravishankara and Hanson, 1996; Borrmann *et al.*, 1997b; Del Negro *et al.*, 1997).

For a review of PSCs, see the article by Peter (1997).

b. Uptake of HCl into PSCs

Given the preceding discussion of the nature of PSCs and stratospheric aerosols, it is clear that the uptake and subsequent reactions of HCl with $ClONO_2$, HOCl, and N_2O_5 both on solid surfaces and into liquid solutions consisting of various combinations of HNO_3, H_2O, and H_2SO_4 must be considered. We first discuss the uptake of HCl onto ice surfaces, which is relevant to Type II PSCs, and then uptake into solutions that are thought to be representative of aerosols and Type I PSCs.

For the heterogeneous reactions of HCl on PSCs and aerosols to be important, there must be mechanisms to continuously provide HCl to the surface. This could occur, for example, if HCl is sufficiently soluble in ice and if it diffuses at a sufficient rate from the bulk to the surface. However, the solubility and diffusion rates have been shown to be sufficiently small that these processes are not expected to be important under stratospheric conditions (see Wolff and Mulvaney, 1991; Dominé *et al.*, 1994; and Thibert and Dominé, 1997).

However, HCl has been shown in a number of studies to be taken up by ice and NAT surfaces, with the amount depending on a number of factors including temperature and the partial pressure of HCl in the gas phase (e.g., see summary in DeMore *et al.*, 1997). The amount of HCl that can be taken up has been shown to correspond to a significant fraction of a monolayer. While the formation of hydrates such as $HCl \cdot 6H_2O$ has been observed in laboratory systems (e.g., see Koehler *et al.*, 1993; Chu *et al.*, 1993; Graham and Roberts, 1994, 1995; and Banham *et al.*, 1996), consideration of the phase equilibria under stratospheric conditions suggests that these will not be important at the low HCl partial pressures and higher temperatures of the stratosphere (Wooldridge *et al.*, 1995).

One of the interesting chemical aspects of the heterogeneous chemistry of HCl is why its reactions on ice surfaces are so much more efficient than in the gas phase. A compelling explanation is that HCl ionizes on the solid surfaces, so that the reaction does not involve covalently bound HCl, but rather, the chloride ion. This is consistent with the fact that chloride ions react very rapidly in the gas phase with the relevant species such as $ClONO_2$ (Haas *et al.*, 1994) and with the observation that chloride ions from NaCl undergo analogous reactions at room temperature with $ClONO_2$ and N_2O_5 (Finlayson-Pitts *et al.*, 1989; Livingston and Finlayson-Pitts, 1991; Finlayson-Pitts, 1993).

There is infrared evidence for the ionization of HCl on ice (Horn *et al.*, 1992; Delzeit *et al.*, 1993; Banham *et al.*, 1996; Koch *et al.*, 1997) and molecular dynamics simulations also support this view (Robertson and Clary, 1995; Gertner and Hynes, 1996). In the simulations, HCl becomes incorporated into the ice via hydrogen bonding between the chlorine of HCl and a hydrogen of a surface water or between the hydrogen of HCl and the oxygen of a surface water as depicted in Fig. 12.24. George and co-workers (Haynes *et al.*, 1992) have shown that under stratospheric conditions, the ice surface is very dynamic, with continuous, rapid evaporation of water molecules from the surface and recondensation. At temperatures of 180–210 K, the rate of water condensation and evaporation corresponds to $10-10^3$ monolayers per second. Thus as HCl is taken

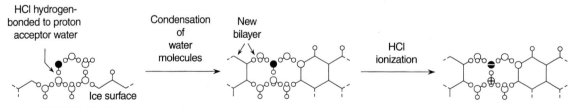

FIGURE 12.24 Schematic of the incorporation of HCl from the gas phase onto the surface of ice via hydrogen bonding, followed by condensation of water and ionization of the HCl (adapted from Gertner and Hynes, 1996).

up at the surface and ionizes, it can also be, in effect, "buried" as surface water molecules evaporate and recondense on top of it (Fig. 12.24).

It is noteworthy that there is some laboratory evidence that HBr, in contrast to HCl, may form a hydrate, $HBr \cdot 3H_2O$, under polar stratospheric cloud formation conditions (Chu and Heron, 1995).

Sodeau and co-workers (Sodeau et al., 1995; Koch et al., 1997) have infrared evidence that chlorine nitrate also ionizes on ice at 180 K, forming an intermediate identified as $[H_2OCl]^+$ through the initial solvation. Hence heterogeneous reactions on ice may be rapid not only because of the ionization of HCl but also because of the ionization or partial ionization of $ClONO_2$ (Horn et al., 1998). A similar mechanism has been proposed for N_2O_5 hydrolysis on surfaces (Koch et al., 1997). It should be noted, however, as discussed shortly, that Bianco and Hynes (1998) propose, based on *ab initio* calculations, that the intermediate observed is not $[H_2OCl]^+$ but rather solvated HNO_3.

Molina and co-workers have proposed that the surface layer can be thought of as a "quasi-liquid layer" with significant mobility of the species, particularly in the presence of higher partial pressures of HCl (Abbatt et al., 1992). Thus the uptake of HCl can be treated as uptake and solvation in this quasi-liquid layer. The nature of this surface is not well understood, however. Although the existence of a quasi-liquid layer on ice surfaces near the freezing point has been recognized for more than a century, the nature of the ice surface under various conditions even in the absence of other species such as HCl continues to be the subject of debate (e.g., see Hobbs, 1973; Conklin and Bales, 1993; Knight, 1996a, 1996b; Baker and Dash, 1996; Pruppacher and Klett, 1997; and papers in "Physics and Chemistry of Ice," Petrenko et al., 1997).

There is again an analogy to NaCl surfaces at room temperature. Thus when solid NaCl having even small amounts of surface nitrate (formed by reaction with HNO_3 or NO_2) is exposed to low pressures of gaseous water, well below the deliquescence points of bulk NaCl and $NaNO_3$, a very mobile surface layer is formed; when the water is pumped off, the ions in this mobile liquid layer selectively recrystallize into separate microcrystallites of $NaNO_3$ and NaCl (Vogt and Finlayson-Pitts, 1994; Vogt et al., 1996; Allen et al., 1996; Laux et al., 1996).

Because Type I PSCs may consist of NAT under some conditions, uptake of HCl onto crystalline NAT as well as ice surfaces is of interest. As reviewed by DeMore et al. (1997), the mass accommodation coefficient for HCl on both ice and NAT at stratospheric temperatures is very large, approaching unity.

HCl is efficiently absorbed into H_2SO_4–H_2O and into HNO_3–H_2SO_4–H_2O solutions, which as discussed earlier, are found in the stratosphere in the form of aerosol particles and Type I PSCs under some conditions (Wolff and Mulvaney, 1991). The solubility of HCl in these liquid solutions can be expressed in terms of the usual Henry's law constant (Elrod et al., 1995; Abbatt, 1995; Luo et al., 1995; Hanson, 1998). Table 12.4 shows some typical measurements of the Henry's law constants for HCl in several typical binary and ternary solutions, respectively. Hanson (1998) has shown that the solubility data for HCl in binary mixtures of H_2SO_4 and water in these and other studies can be fit by the form

$$H^*_{HCl} = [e_0 + e_1 x + e_2 x^2] \exp[c_0 + (d_0 + d_1 x)/T],$$

where x is the mole fraction of H_2SO_4, $d_0 = +6922$, $d_1 = -9800$, and the fit parameters c_0, e_0, e_1, and e_2 are given by $c_0 = -9.021$, $e_0 = +0.363$, $e_1 = -2.616$, and $e_2 = +4.995$. The Henry's law constants in sulfuric acid–water solutions increase as the temperature decreases and as the dilution of the solution increases. This increase in HCl solubility as the temperature falls

TABLE 12.4 Some Measured Values of Henry's Law Constant for HCl in H_2SO_4–H_2O or H_2SO_4–HNO_3–H_2O Solutions at Stratospherically Relevant Temperatures[a]

Solution (wt% with H_2O)	T(K)	H^* (mol L^{-1} atm^{-1}) Elrod et al. (1995)[a]	Hanson (1998)[b]
50% H_2SO_4	225	4.0×10^5	
	222.3		3.23×10^5
	219	7.2×10^5	
	216	9.8×10^5	
	209.8		1.36×10^7
	208	2.3×10^6	
	205.1		2.44×10^6
43–45% H_2SO_4	226	2.5×10^6	
	225		1.11×10^6
	218	6.8×10^6	
	214		3.75×10^6
	208	1.8×10^7	
	204.8		1.3×10^7
48% H_2SO_4 + 3.5% HNO_3	231	1.1×10^5	
	226	1.7×10^5	
	216	5.4×10^5	
36.2% H_2SO_4 + 12.5% HNO_3	228	5.5×10^5	
	218	1.7×10^6	
	208	4.7×10^6	

[a] 43% H_2SO_4.
[b] 45% H_2SO_4.

is a major factor in maintaining the high efficiency with which the temporary chlorine reservoirs are converted into photochemically active forms.

c. Heterogeneous Chemistry on PSCs and Aerosols

The reactions of $ClONO_2$, N_2O_5, and HOCl with HCl and H_2O on solid and liquid surfaces relevant to PSCs have been the subject of numerous laboratory studies. The measured reaction probabilities depend on the nature (i.e., solid or liquid) and composition of the surface, the temperature and the relative humidity, and the concentrations of the gases. The dependence on the latter arises because of surface "saturation" effects that quickly arise at high reactant concentrations, as well as other effects such as surface "melting" and preactivation, which are less well understood.

Figure 12.25 summarizes some results of laboratory studies of the reaction probabilities for the reaction of $ClONO_2$ with HCl and H_2O and of HOCl with HCl on various surfaces that are believed to be present in the stratosphere under various conditions. Table 12.5 summarizes typical reaction probabilities for these heterogeneous reactions and for the reaction of N_2O_5 with HCl and H_2O. It should be noted that in a number of studies of these reactions where the PSC is liquid (either a binary H_2SO_4–H_2O or ternary HNO_2–H_2SO_4–H_2O mixture), the laboratory data are better fit by a model that includes two reactions, one in the bulk and one at the surface (e.g., see Hanson, 1998).

A number of models of these heterogeneous reactions have been developed that are consistent with the laboratory observations. The reader is referred to papers by Elliott *et al.* (1991), Burley and Johnston (1992b), Mozurkewich (1993), Tabazadeh and Turco (1993), Henson *et al.* (1996), and Koch *et al.* (1997) for some illustrative approaches.

The hydrolysis of N_2O_5 on surfaces is important in that it provides a significant path for "denoxification" of the stratosphere, i.e., for the conversion of NO and NO_2 to other oxides of nitrogen such as HNO_3. This does not permanently remove NO_x from the stratosphere (as is the case with denitrification), since HNO_3 can ultimately be photolyzed in the gas phase back to OH + NO_2. However, since this is relatively slow, denoxification at least helps to tie up NO_x temporarily so that the chain destruction of O_3 via reaction with Cl can proceed more readily since less ClO is tied up in the form of $ClONO_2$.

The reactions tend to be fast on ice as well as on liquid solutions characteristic of the stratosphere. This indicates that they should occur on Type II PSCs as well as on H_2SO_4–H_2O mixtures characteristic of SSA and on HNO_3–H_2SO_4–H_2O ternary solutions which

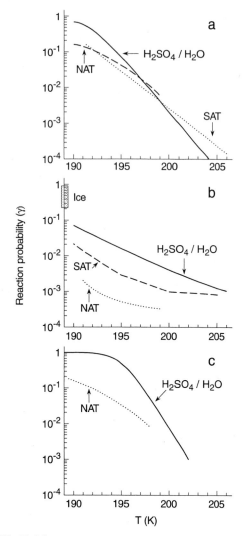

FIGURE 12.25 Typical measured reaction probabilities for (a) $ClONO_2$ + HCl, (b) $ClONO_2$ + H_2O, and (c) HOCl + HCl for different surfaces that can be present and promote heterogeneous chemistry under typical stratospheric conditions (adapted from Ravishankara and Hanson, 1996, and references therein).

may represent Type I PSCs under some conditions. While the reactions of $ClONO_2$ and HOCl with HCl are also fast on crystalline NAT, the hydrolyses of $ClONO_2$ and N_2O_5, as well as the N_2O_5–HCl reactions on NAT, are relatively slow. In addition, Zondlo *et al.* (1998) have shown that the hydrolyses of $ClONO_2$ and N_2O_5 on the supercooled H_2O–HNO_3 liquid layer formed by their uptake on ice also are slow, comparable to those on NAT. Hence the contributions of the latter three reactions to ozone depletion may depend critically on the composition of Type I PSCs.

Recent aircraft studies in the Southern Hemisphere are generally consistent with the laboratory kinetics. Thus, Kawa *et al.* (1997) showed a sharp increase in

TABLE 12.5 Some Values of Reaction Probabilities (γ) for the Heterogeneous Chlorine Activation Reactions under Typical Stratospheric Conditions[a]

Reaction	Solid or solution[b]				
	Ice	NAT	SAT	Liquid $H_2SO_4-H_2O$	Liquid $H_2SO_4-HNO_3-H_2O$
$ClONO_2 + HCl \rightarrow Cl_2 + HNO_3$	0.2	0.1	$\sim 10^{-3}-10^{-1\,c}$	$0.01-0.5^d$	$0.02-0.2^d$
$N_2O_5 + HCl \rightarrow ClNO_2 + HNO_3$	0.03	3×10^{-3}		—	—
$HOCl + HCl \rightarrow Cl_2 + H_2O$	0.3	0.1		$\geq 0.1^e$	$\geq 0.1^e$
$ClONO_2 + H_2O \rightarrow HOCl + HNO_3$	≥ 0.1	$\sim 10^{-4}-10^{-2\,i}$	$\sim 10^{-2}-10^{-3\,e}$	$\sim 1 \times 10^{-4}-0.1^f$	
$N_2O_5 + H_2O \rightarrow HOCl + HNO_3$	0.02	3×10^{-4}	$\sim 10^{-2}$	0.1^g	$0.06-0.095^g$
$HO_2NO_2 + HCl \rightarrow HOCl + HNO_3$				$<1 \times 10^{-4\,h}$	

[a] From DeMore et al. (1997) and Ravishankara and Hanson (1996) and references therein.

[b] NAT = nitric acid trihydrate (solid); SAT = sulfuric acid tetrahydrate (solid).

[c] Hanson and Ravishankara, 1993b; Zhang et al. (1994a), see Fig. 12.25a.

[d] Zhang et al. (1994b), Elrod et al. (1995) and Hanson (1998). γ increases as temperature falls primarily due to increased solubility of HCl. γ decreases as percentage of H_2SO_4 increases, and Hanson (1998) reports that it also decreases with increased HNO_3; see Fig. 12.25a.

[e] Zhang et al. (1994b); bimolecular rate constant in the liquid phase is $\sim 1.4 \times 10^6$ L mol^{-1} s^{-1} for 60 wt% H_2SO_4 at 251 K (Hanson and Lovejoy, 1996).

[f] Hanson (1998) and Ball et al. (1998) and references therein. γ decreases from ~ 0.1 at 35 wt% H_2SO_4 to $\sim 10^{-4}$ at 75 wt% H_2SO_4. It also has a small temperature dependence, especially at 75 wt% H_2SO_4, where the reaction probability increases with temperature.

[g] Hanson (1997); HNO_3 from 0.8 to 15%.

[h] Zhang et al. (1997).

[i] Depends on the amount of surface water; see Barone et al. (1997) and references therein.

ClO and decrease in HCl as the temperature fell, even though PSCs were not present (although it is possible that they were present at some earlier time). These observations were shown to be consistent with heterogeneous reactions on liquid binary and ternary solutions, with the temperature dependence reflecting the increased reaction probability for HCl + $ClONO_2$ due to increased solubility of HCl at the lower temperatures.

It should be noted that laboratory-measured reaction probabilities are often not *directly* applicable to the calculation of uptake and reaction in atmospheric particles due to the need to take diffusion and reaction into the droplets into account as well (Hanson et al., 1994). As discussed in detail in Chapter 5, under most laboratory conditions where uptake into a relatively thick film is measured, large concentration gradients are often formed in the liquid, leading to large measured uptake coefficients. However, for the case of small liquid particles in the atmosphere, the concentration gradient will be smaller. Figure 12.26 shows the calculated correction factors for the experimental uptake coefficient measured on plane thick films, γ_m, as a function of l/a, where a is the droplet radius and l is the "diffusoreactive length" (Hanson et al., 1994). The latter is a measure of the distance within the droplet over which the reaction is taking place. At small values of l/a, approximately <0.1, the use of the measured value of γ_m is appropriate because there is a large concentration gradient to drive the uptake. However, at larger values of l/a, the correction is significant; for example, at $l/a = 1$, the effective value of the uptake coefficient is about a third of the measured value because there is not a large concentration gradient under these conditions to drive the uptake. In this case, the droplet concentrations are essentially in Henry's law equilibrium with the gas phase and the reaction

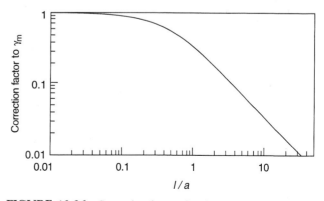

FIGURE 12.26 Correction factors for the measured uptake coefficient, γ_m, as a function of the ratio of the diffusoreactive length (l) to the droplet radius (a) (adapted from Hanson et al., 1994).

kinetics can be described by the usual aqueous-phase expressions.

There may be additional mechanisms that contribute to the generation of active chlorine in the stratosphere as well. For example, Burley and Johnston (1992a) have proposed that nitrosyl sulfuric acid ($NO^+HSO_4^-$) may be formed in the stratosphere and react with HCl to give nitrosyl chloride, ClNO. De Haan et al. (1997) also suggest that the heterogeneous reaction of ClOOCl with HCl may be important, depending on the fate of the product HOOCl. This would provide a means of removing the excess HCl remaining after the $ClONO_2$ + HCl reaction.

Most of the research to date has focused on aerosols and PSCs containing inorganic species such as nitric and sulfuric acids. While CH_4 is the only hydrocarbon that is sufficiently unreactive in the troposphere to reach the stratosphere, it is oxidized to compounds such as HCHO that can be taken up into sulfuric acid particles (Tolbert et al., 1993). The effects of such uptake and subsequent chemistry are not well established.

In short, the heterogeneous chemistry that drives the Antarctic ozone hole can occur not only on solid surfaces but also in and on liquid solutions containing combinations of HNO_3, H_2SO_4, and H_2O. As discussed in the following section, it is believed that this is why volcanic eruptions have such marked effects on stratospheric ozone on a global basis.

6. Effects of Volcanic Eruptions

The finding that the heterogeneous chemistry that occurs on polar stratospheric clouds also occurs in and on liquid solutions in the form of liquid aerosol particles and droplets in the atmosphere provided a key link in understanding the effects of volcanic eruptions on stratospheric ozone in both the polar regions and midlatitudes. As discussed herein, the liquid particles formed from volcanic emissions are typically 60–80 wt% H_2SO_4–H_2O, and hence the chemistry discussed in the previous section can also occur in these particles (Hofmann and Solomon, 1989). We discuss briefly in this section the contribution of volcanic emissions to the chemistry of the stratosphere and to ozone depletion on a global scale. For a brief review of this area, see McCormick et al. (1995).

Volcanic eruptions can be sufficiently energetic that they inject large quantities of gases and particles directly into the stratosphere (rather than by diffusion from the upper troposphere). The gases include SO_2, HCl, HF, and SiF_4 (Mankin and Coffey, 1984; Symonds et al., 1988; Bekki, 1995; Francis et al., 1995). The particles can include inorganic mineral particles such as silicates, halide salts, and sulfates (e.g., see Woods et al., 1985; Snetsinger et al., 1987; and Pueschel et al., 1994).

Additional sulfates continue to form after the eruption as gaseous SO_2 is oxidized to sulfuric acid and sulfates. While we shall focus here on the effects of these sulfate particles on the heterogeneous chemistry of the stratosphere, there may be other important effects on the homogeneous chemistry as well. For example, model calculations by Bekki (1995) indicate that this oxidation of SO_2 by OH leads to reduced OH levels, which alters its associated chemistry.

Finally, there are a variety of effects associated particularly with the increased particle loading that we shall not discuss in detail. These include, for example, the potential for a change in actinic flux at the earth's surface [which could be either negative or even positive due to "trapping" of photons and multiple scattering between the aerosol layer and the surface (Michelangeli et al., 1992; Minnis et al., 1993)], warming of the stratosphere due to absorption of solar and terrestrial radiation by aerosols (Angell, 1993a; Chandra, 1993), cooling of the troposphere (Dutton and Christy, 1992; see also Chapter 14), changes in stratospheric circulation (Tie et al., 1994), effects on cloud formation (Minnis et al., 1993), and altered rates of photolysis of O_2 and O_3 (Huang and Massie, 1997). For a more detailed discussion, see Fiocco et al. (1996).

The eruption of Mount Pinatubo in the Philipines in mid-June of 1991 is believed to be the largest volcanic perturbation on the stratosphere in this century, injecting 14–20 Mt of gaseous SO_2 and 21–40 Mt of sulfuric acid/sulfate aerosol particles (Russell et al., 1996, and references therein). At many locations around the world, stratospheric particle concentrations were measured to increase by 1–2 orders of magnitude after this eruption (e.g., see Shibata et al., 1994; Ansmann et al., 1996; Anderson and Saxena, 1996; and Deshler et al., 1996). Figure 12.27, for example, shows one set of measurements of the aerosol optical depth during 1991, 1992, and part of 1993 at 19.5°N (Russell et al., 1996). The increase in light scattering due to the injection of volcanic aerosols is dramatic.

Figure 12.28 shows the particle surface area size distribution before the Mount Pinatubo eruption (Fig. 12.28a), inside the main aerosol layer several months after the eruption (Fig. 12.28b), and almost two years after the eruption (Fig. 12.28c). (See Chapter 9.A.2 for a description of how particle size distributions are normally characterized.) Prior to the eruption, the surface area distribution is unimodal, with typical radii of 0.05–0.09 μm and a number concentration of ~1–20 particles cm^{-3}. In the main stratospheric aerosol layer formed after the eruption, the distribution is bimodal

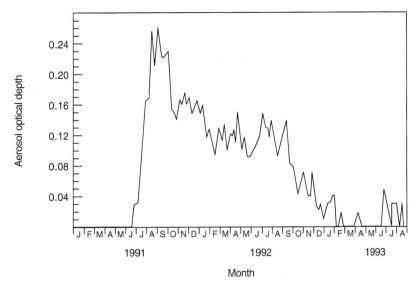

FIGURE 12.27 Zonally averaged optical depth at 19.5°N and derived from the satellite-based Advanced Very High Resolution Radiometer (AVHRR) (adapted from Russell et al., 1996).

(Fig. 12.28b), and both small and larger particles with radii <0.2 and >0.6 μm, respectively, increase by 1–2 orders of magnitude (e.g., see Deshler et al., 1993; Stone et al., 1993; Goodman et al., 1994; and Russell et al., 1996). The smaller particles are likely newly formed sulfuric acid–water particles, whereas the larger ones are the inorganic minerals (Deshler et al., 1992; Grainger et al., 1993; Wilson et al., 1993; Pueschel et al., 1994). Almost two years later, the bimodal distribution is still evident, but the number of smaller particles has decreased by about an order of magnitude while the larger particles have not changed significantly. This change in the particle size distribution reflects growth of the larger particles at the expense of the smaller ones via condensation and coagulation. As reviewed by Russell et al. (1996), a variety of measurements indicated that these particles are primarily concentrated (~65–80 wt%) sulfuric acid–water mixtures, as expected.

Figure 12.29 shows the ratio of the particle surface area at an altitude of 20 km and 45°N latitude to that in 1978–1979 for the period from 1979 to 1995 based on satellite measurements (Solomon et al., 1996). The increases due to volcanic eruptions are evident, particularly the Mount Pinatubo eruption.

Although the increase in stratospheric sulfate aerosols after volcanic eruptions is dramatic, there is some evidence that these events may be superimposed on a longer term trend to increased stratospheric sulfate concentrations (Hofmann, 1990). Whether this is due to increased anthropogenic or natural sources such as COS or to an increased residual volcanic layer, i.e.,

we have not returned to prevolcanic conditions (Thomason et al., 1997), is not clear. A database on stratospheric aerosols from 1850 to 1990 has been described by Sato et al. (1993) and is available on the Internet.

As discussed earlier, it is now well established that a variety of heterogeneous reactions can occur on such aerosols as well as on PSCs. These include the following:

$$N_2O_5 + H_2O \xrightarrow{aerosol} 2HNO_3, \quad (46)$$

$$ClONO_2 + HCl \xrightarrow{aerosol} Cl_2 + HNO_3, \quad (39)$$

$$ClONO_2 + H_2O \xrightarrow{aerosol} HOCl + HNO_3, \quad (41)$$

$$HCl + HOCl \xrightarrow{aerosol} Cl_2 + H_2O, \quad (42)$$

$$BrONO_2 + H_2O \xrightarrow{aerosol} HOBr + HNO_3, \quad (47)$$

$$HCl + HOBr \xrightarrow{aerosol} BrCl + H_2O. \quad (48)$$

Laboratory studies of the uptake of ClO into sulfuric acid (Martin et al., 1979, 1980), taken in light of a deficit in the inorganic chlorine budget at ~17 km after the Mount Pinatubo eruption, led Jaeglé et al. (1996) to propose that a heterogeneous reaction of ClO on sulfuric acid aerosols to form perchloric acid, $HClO_4$, may also occur.

The kinetics of these reactions in liquid solutions characteristic of the stratosphere, such as concentrated $H_2SO_4-H_2O$ or ternary solutions with HNO_3, depend on temperature as expected and in some cases at least, on acidity as well. For example, Donaldson et al. (1997) have shown that the second-order rate constant for the

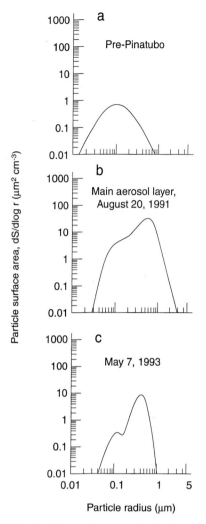

FIGURE 12.28 Particle surface area distributions in the stratosphere (a) before Mount Pinatubo eruption, (b) August 20, 1991, over California, and (c) May 7, 1993, over California (adapted from Russell et al. (1996) and Goodman et al. (1994)).

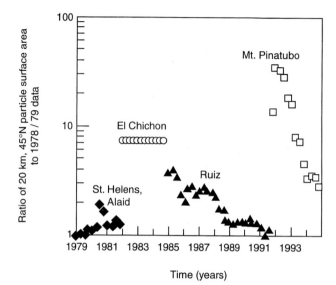

FIGURE 12.29 Ratio of particle surface areas to those in the winter of 1978–1979, from 1979 to 1995 at 45°N and 20-km altitude. Data are not available for 1982–1984. (Adapted from Solomon et al., 1996.)

reaction (42) of HOCl with HCl increases by more than an order of magnitude from 2.8×10^5 to 4.9×10^6 L mol^{-1} s^{-1} at 250 K as the weight percent sulfuric acid in a mixture with water increased from 49 to 67%. They propose that this is due to increasing protonation of the HOCl to H_2OCl^+ in highly acidic solutions, followed by reaction of the protonated form with Cl$^-$.

Of these heterogeneous reactions, the hydrolysis of N_2O_5 is particularly important in midlatitudes. For example, Fig. 12.30 shows the measured NO_x ($NO_x = NO + NO_2$) to NO_y ($NO_y = NO_x + HNO_3 + N_2O_5 + \cdots$) ratio at different latitudes compared to the predicted ratio using a gas-phase model as well as to a model that incorporates the N_2O_5 hydrolysis on aerosol particles (Fahey et al., 1993). Clearly, the inclusion of this reaction is necessary to bring the measurements and models into better agreement (e.g., Rodriguez et al., 1991; Fahey et al., 1993; Kondo et al., 1997; Sen et al., 1998). This reaction (46) on aerosol particles of course occurs not only in midlatitudes but at the poles as well (Keys et al., 1993).

As discussed by Morris et al. (1997), the photolysis of HNO_3 to $OH + NO_2$ and its reaction with OH to generate NO_3 are the major sources of NO_x during the day. With large aerosol concentrations, the loss of NO_x at night is largely through the formation of N_2O_5 ($NO_3 + NO_2 \leftrightarrow N_2O_5$), followed by the hydrolysis of N_2O_5. These reactions shift the NO_x/NO_y ratio to smaller values in the presence of high particle concentrations.

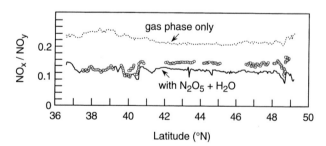

FIGURE 12.30 Comparison of predicted ratio NO_x/NO_y as a function of latitude at 19.5–20.5 km using only gas-phase chemistry (dotted line) or with the N_2O_5 hydrolysis on aerosol particles (solid line) compared to measured values shown as unfilled circles (from Fahey et al., 1993).

There are a number of measurements documenting changes in NO_x and NO_y in the stratosphere after the Mount Pinatubo eruption and which have been attributed to the removal of oxides of nitrogen due to reactions on aerosol particles. For example, a decrease in stratospheric NO_2 after the eruption followed by a return to normal levels has been reported (e.g., see Van Roozendael et al., 1997; and De Mazière et al., 1998). Similarly, NO_x decreases of up to 70% were reported, as well as increases in gaseous HNO_3 (much of that produced on the sulfate particles is released to the gas phase) (e.g., see Coffey and Mankin, 1993; Koike et al., 1993, 1994; David et al., 1994; Webster et al., 1994; and Rinsland et al., 1994).

Not only does this heterogeneous hydrolysis alter the NO_x reactions that can lead to ozone destruction or formation, but it also changes the halogen cycles because less NO_2 is available to trap ClO as the nitrate. In addition, HO_x levels are increased. Thus there is relatively more ClO and the ClO_x cycles leading to ozone destruction become more effective (e.g., McElroy et al., 1992; Avallone et al., 1993a, 1993b; Schoeberl et al., 1993b). Heterogeneous chemistry would be expected to shift the partitioning of chlorine away from HCl toward more active forms. While increasing ratios of HCl/Cl_y subsequent to the Mount Pinatubo eruption have been reported in some studies, suggesting decreasing contributions from heterogeneous chemistry after the eruption (e.g., an increase in HCl/Cl_y from 0.40 in late 1991 to 0.70 in 1996 based on in situ measurements; Webster et al., 1998), satellite data suggest smaller changes in HCl/Cl_y (16 ± 9% from 1992/1993 to 1995/1996) (e.g., Dessler et al., 1997).

This effect can be seen in the midlatitude stratospheric measurements of Keim et al. (1996) shown in Fig. 12.31. In the tropopause region (shown by the dotted line), the aerosol surface area increases. A significant increase in ClO and decrease in NO is seen at the same time, while NO_y increases. This was attributed to the heterogeneous reaction of $ClONO_2$ with HCl to form HNO_3 on the aerosol particles. The Cl_2 product generates Cl atoms, which react with O_3 to give enhanced ClO. Both $ClONO_2$ and particle HNO_3 are measured as NO_y so that conversion of one to the other should not lead to enhanced NO_y. The latter was attributed to highly efficient sampling of sulfate aerosol particles containing nitrate. Similar observations of enhanced ClO and suppressed NO have been reported in other studies as well (e.g., Fahey et al., 1993; Toohey et al., 1993). Dessler et al. (1993) invoked heterogeneous reaction to explain measured ClO concentrations that were larger than expected for gas-phase chlorine chemistry.

The hydrolysis reaction (46) of N_2O_5 under many conditions in the atmosphere becomes limited by the rate of N_2O_5 formation, which only occurs at a significant rate at night (because of the rapid photolysis of the NO_3 precursor during the day). Hence under these conditions, reactions (39), (41), and (42) followed by photolysis of the chlorine-containing products become primarily responsible for the removal of gas-phase NO_y and increase in ClO (Keim et al., 1996).

Figure 12.32 shows the results of model calculations of the effects of the increased aerosols for October 1986 at 43.5°N (Solomon et al., 1996). The calculated change in the odd-oxygen loss rate when the measured aerosol particle surface area is incorporated into the

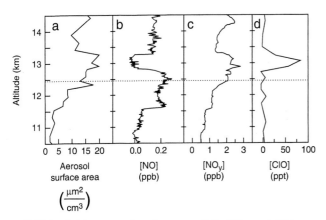

FIGURE 12.31 Aerosol surface area, NO, NO_y, and ClO as a function of altitude (adapted from Keim et al., 1996).

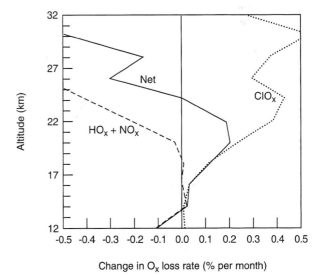

FIGURE 12.32 Calculated change in loss rate for odd oxygen as a function of altitude in October 1986 at 43.5°N compared to the predictions for a constant particle concentration typical of winter 1978–1979 levels (adapted from Solomon et al., 1996).

model compared to the assumption of a constant particle concentration at the winter 1978/1979 levels is shown. Also shown are the contributions from the ClO_x and $HO_x + NO_x$ cycles to the total. The effects of the $HO_x + NO_x$ cycles on O_3 destruction below about 20 km due to increased volcanic aerosols are negligible. At higher altitudes, removing oxides of nitrogen results in less ozone destruction because in this region NO_x chemistry dominates (Fig. 12.8b). For example, Mickley et al. (1997a, 1997b) show that as oxides of nitrogen recovered (i.e., increased) at an altitude of ~28 km in the years after the Mount Pinatubo eruption, O_3 decreased simultaneously due to the increased contribution of NO_x to its loss. However, in the lower stratosphere where the ClO_x cycles are important, the contribution of the chlorine cycles to ozone destruction is enhanced by the removal of oxides of nitrogen on aerosol particles. In short, the effects of volcanic aerosols on stratospheric ozone depend on altitude and, critically, on the halogen concentrations.

Indeed, modeling calculations (Tie and Brasseur, 1995) indicate that at the lower stratospheric halogen concentrations before 1980, the overall effect of increased aerosols due to volcanic eruptions would have been to *increase* stratospheric ozone due to the removal of N_2O_5 on aerosols. Interestingly, the predicted effect on ozone is not very sensitive to the amount of aerosol injected under these pre-CFC conditions because the rate of the N_2O_5 hydrolysis becomes limited by the rate of its formation in the $NO_2 + NO_3$ reaction (e.g., see Fahey et al., 1993; Tie et al., 1994; and Tie and Brasseur, 1996). However, after 1980, the halogen concentrations had increased to the point that the net effect was a decrease in total column ozone. The halogen effect on ozone is predicted to depend on the amount of volcanic aerosol injected because unlike N_2O_5, the hydrolysis of $ClONO_2$ on the particles is not limited by its rate of formation.

Based on modeling studies, it has been suggested that the depth of the Antarctic ozone hole may also be impacted in part by the presence of volcanic aerosols in addition to PSCs. For example, calculations by Portmann et al. (1996) have shown that the combination of increased halogens and volcanic aerosols may have been combined to give the dramatic reduction in O_3 that was first reported by Farman and co-workers in 1985. They propose that there are four critical cycles in the formation of the ozone hole: (1) development of the concentrations of the important species, e.g., HCl and $ClONO_2$, before winter; (2) conversion of these into active halogen forms during winter and denitrification and dehydration of the stratosphere; (3) continued conversion into active forms in the spring while ozone depletion is occurring, the so-called maintenance period; and finally, (4) termination of the ozone destruction cycles.

Model-predicted effects of continued activation of chlorine during ozone depletion and the effects of the extent of conversion of chlorine species into active forms on total column ozone at 75°S in 1990 are shown in Fig. 12.33 (Portmann et al., 1996). The calculated total column ozone is shown for the cases where HCl/Cl_y = 1.0, 0.4, and 0.0 at day 180 and for the case where this ratio is 0 but the heterogeneous chemistry that converts $ClONO_2$ to active forms ceases at day 220. A ratio of 0 for HCl/Cl_y corresponds to total conversion of HCl into active forms, with a ratio of 1 corresponding to no such conversion. Figure 12.33 shows that the onset of ozone depletion would be significantly delayed if there has been no heterogeneous conversion of HCl to active forms during the winter, but large ozone losses are still expected due to activation during the spring. If the conversion is assumed to be complete on day 180, but the heterogeneous chemistry ceases at day 220, the depth of the hole on October 1 (day 270) is seen to be much reduced. Hence the presence of aerosols in the absence of PSCs could provide a vehicle for continued heterogeneous chemistry during this "maintenance period." Temperature is another important factor due to its role not only in the formation of PSCs but also in determining the kinetics of both the heterogeneous and homogeneous reactions. Smaller effects on total column ozone are calculated due to denitrification and dehydration of the polar stratosphere (Portmann et al., 1996).

Support for the importance of aerosols in maintaining chlorine in an active form during the maintenance period is found in Fig. 12.34. This shows the satellite-derived average total O_3 in the vortex as a function of

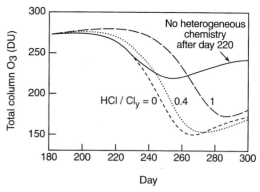

FIGURE 12.33 Calculated change in total column O_3 in the Antarctic springtime for various assumptions described in the text (adapted from Portmann et al., 1996).

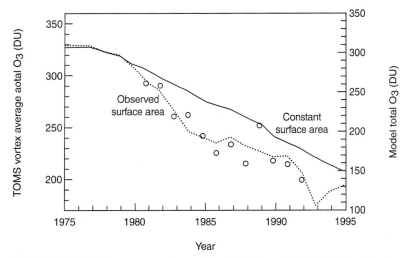

FIGURE 12.34 Vortex-averaged total O$_3$ (DU) from TOMS satellite data for October at 75°S compared to model predictions using the assumption of a constant aerosol surface area or the measured surface areas (adapted from Portmann et al., 1996).

year compared to model calculations with the assumption of either a constant aerosol or the observed aerosol surface areas (Portmann et al., 1996). Clearly, the data are better fit by the model that includes variable aerosols due to the volcanic eruptions, which contributes significantly to continued ozone destruction during the spring by maintaining enhanced levels of active chlorine over a longer period of time. A similar conclusion has been reached by Shindell and de Zafra (1997). It should be noted that including the effects of not only the particle surface area but also temperature is important. Thus, Solomon et al. (1998) show that including temperature fluctuations in the model for midlatitudes improves the match of the model predictions and the observations, since lower temperatures enhance the net chlorine activation through the heterogeneous chemistry described in this chapter.

In short, it is predicted that the combination of high aerosol concentrations and low temperatures will lead to the development of particularly strong ozone holes.

Although decreases in total column ozone are anticipated due to this heterogeneous chemistry, there are also indirect effects of volcanic injections on stratospheric ozone and indeed, these may predominate under some conditions. For example, the volcanic aerosol can absorb both long-wavelength radiation emitted by the earth's surface and direct solar radiation, both of which lead to local heating in the stratosphere (although the former is the larger effect; see Chapter 14). This can cause upwelling of the aerosol layer, bringing air with lower ozone concentrations to higher altitudes that normally have higher O$_3$ concentrations (e.g., see Schoeberl et al., 1993a).

On the other hand, a reduction in stratospheric ozone leads to less heating of the stratosphere; Zhao et al. (1996) suggest that this feedback could lead in the mid- and high-latitude lower stratosphere to sufficient net cooling that the breakup of the Arctic polar vortex (see Section C.7) could be delayed. Another indirect effect is the reduction in light intensity below the aerosol layer due to absorption and backscattering of sunlight, accompanied by an increase in actinic flux above the layer. The net effect is an increase in the photolysis rate of O$_3$ above the layer, which is not countered by an increased rate of production via O$_2$ photolysis. These effects are thought to be most important in the first 6 months to a year after an eruption, with the effects of heterogeneous chemistry predominating subsequently after the aerosol has been dispersed globally (e.g., see Kinne et al., 1989; Michelangeli et al., 1992; Pitari et al., 1993; and Tie et al., 1994).

In addition to these indirect effects of volcanic emissions, there are a variety of nonvolcanic parameters that, of course, can change O$_3$ as well, and these must be taken into account in assessing the role of the volcanic emissions alone. For example, there is a natural solar variability, part of which cycles on a time scale of about 11 years and part of which is on a much longer time scale (Lean, 1991; Lean et al., 1995a, 1995b; Labitzke and van Loon, 1996). In addition, stratospheric ozone levels vary with the quasi-biennial oscillation (QBO), which is associated with a periodic variation in the zonal winds at the equator between 20 and

35 km (Garcia and Solomon, 1987; Chipperfield et al., 1994; WMO, 1995). The mean period for the QBO is about 27 months, but it can vary from 23 to 34 months (Zawodny and McCormick, 1991; WMO, 1995). For example, Chandra and Stolarksi (1991) point out that while a decrease in total ozone of 5–6% occurred in the winter following the El Chichon eruption, much of this could be due to the QBO and at most 2–4% could be attributed to the El Chichon emissions.

Long-term trends due to CFCs must also be removed from the data to examine the effects of volcanic emissions. Finally, one must take into account the possible contributions of air that has been processed through the polar vortices and of meteorological influences that are unique to certain locations (e.g., see Ansmann et al., 1996).

Despite these difficulties in quantifying the effects of volcanic emissions on stratospheric ozone and the uncertainties in the relative importance of direct versus indirect effects, there are ample data to support a decrease in stratospheric ozone due to volcanic emissions. Figure 12.35, for example, shows the ozone above Brazzaville, Congo, measured using electrochemical (ECC) sondes in the 16- to 28-km altitude range from 1990 to 1992 (Grant et al., 1992, 1994). Also shown is the expected ozone based on satellite measurements (SAGE II) from October 1984 to June 1991, corrected for normal cyclical variations and long-term trends. This "ozone climatology" can be compared to the measured ozone before the eruption of Mount Pinatubo as well as afterward. It is seen that before the eruption, the two are in reasonably good agreement, but afterward, the measured values lie some 15–33 Dobson units (DU) below the values expected based on the climatology, which is outside two standard deviations associated with the satellite data. The maximum decreases correspond to the loss of about 12% of the average total ozone. How much of this decrease is due to heterogeneous chemistry and how much is due to changes in heating, radiation, etc. are not known.

While the data in Fig. 12.35 are for a tropical site, similar data have been gathered at mid- and high-latitude locations as well (e.g., Gleason et al., 1993; McGee et al., 1994; Rodriguez et al., 1994; Hofmann et al., 1994b). For example, decreases of approximately 10% of the total column ozone over the Observatoire de Haute Provence in southern France were observed in July and August 1992, with decreased ozone observed at altitudes that overlapped those having increased volcanic aerosols (McGee et al., 1994).

A similar relationship was observed in Germany. Figure 12.36, for example, shows the deviation of the monthly mean ozone concentration after corrections for seasonal variations, long-term trends, the QBO and vortex effects, and the associated particle surface area concentration from 1991 to 1994 (Ansmann et al., 1996). The increase in the particle surface area due to Mount Pinatubo is clear; associated with this increase in aerosol particles are negative monthly mean deviations in ozone that persist until fall 1993, when the surface area approaches the preeruption values. Similarly, the decrease in the total column ozone from 1980–1982 to 1993 observed at Edmonton, Alberta, Canada, and shown at the beginning of this chapter in Fig. 12.1 has been attributed to the effects of the Mount Pinatubo eruption (Kerr et al., 1993).

7. Ozone Depletion in the Arctic

Given the dramatic decrease in stratospheric ozone in the Antarctic during spring, a similar phenomenon might be expected in the Arctic as well. However, it is now clear that while ozone depletion occurs in the

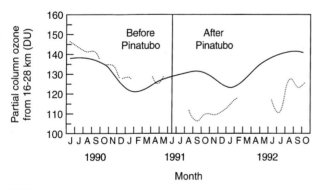

FIGURE 12.35 Monthly integrated ozone partial column in the (16- to 28-km) measured at Brazzaville (dotted line) and expected from SAGE II climatology (solid line) (adapted from Grant et al., 1994).

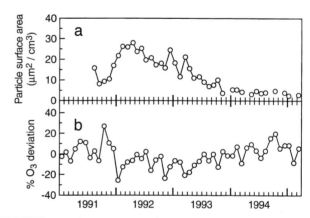

FIGURE 12.36 Deviation of (a) particle surface area at 16–20 km and (b) monthly mean ozone at 15.3–19.8 km measured in Germany from 1991 to 1994 (adapted from Ansmann et al., 1996).

Arctic stratosphere due to chemistry that is qualitatively similar to that in the Antarctic, an analogous "ozone hole" is not formed. The major reason for this difference is the different meteorology and dynamics (Schoeberl *et al.*, 1992; Manney and Zurek, 1993; Manney *et al.*, 1996).

First, the temperatures found in the Arctic stratosphere are warmer by about 10 K compared to those in the Antarctic. The Arctic stratospheric temperatures do not drop below 195 K as frequently (e.g., see Pawson *et al.*, 1995), so that PSCs, particularly Type II, which consists primarily of ice and requires temperatures of ~188 K, do not form as readily nor persist for the lengths of time that they do in the Antarctic polar vortex. In addition, as discussed in Section C.5b, the uptake of HCl, a key species in the heterogeneous chemistry, into liquid solutions found in the stratosphere is highly temperature dependent, with the Henry's law constant increasing as the temperature decreases (Table 12.4). The heterogeneous reaction probabilities also depend on temperature. However, it appears that mountain-induced gravity waves cause local reductions of up to 10–15 K in the temperature of the stratosphere, which can increase PSC formation and hence increased halogen activation in the Arctic stratosphere (Carslaw *et al.*, 1998a).

Second, the northern polar vortex is much less stable and hence less isolated from mixing with external air masses compared to the Antarctic case; events in January and February in which there was substantial mixing of air from midlatitudes into the vortex have been reported (e.g., see Browell *et al.*, 1993; Plumb *et al.*, 1994). This makes it particularly important to make both measurements and model predictions with sufficient resolution (Edouard *et al.*, 1996). In addition, the Arctic polar vortex tends to break up earlier than the Southern Hemisphere polar vortex; since ozone destruction is determined to a large degree by the extent of exposure to sunlight, the earlier breakup and mixing with air external to the vortex cuts the ozone loss short.

Finally, the dynamics are quite different, with ozone concentrations in the Arctic stratosphere usually increasing in December and into the early spring due to the normal large-scale transport of air containing higher ozone concentrations from the tropics at higher altitudes, followed by downward transport (see Section A.1). Any decreases due to the chemical destruction processes already described are superimposed on these normal increases. Hence the chemically induced losses of total column ozone can be at least in part masked by natural variations (e.g., see Proffitt *et al.*, 1990, 1993; Waters *et al.*, 1993; Manney *et al.*, 1994a, 1994b; Henriksen *et al.*, 1994; Santee *et al.*, 1995; Solomon *et al.*, 1996; and Zhao *et al.*, 1996).

For example, while the vortex-averaged O_3 concentration at one altitude in the Arctic in the spring of 1994 was measured to decrease by ~10%, the net chemical loss was estimated at ~20% but this was partially compensated by an increase due to transport of air containing higher ozone concentrations from higher altitudes (Manney *et al.*, 1995). Similar amounts of chemical ozone loss in the Arctic polar vortex have been calculated based on measurements of ClO, BrO, and O_3 (e.g., Brune *et al.*, 1991; Salawitch *et al.*, 1993).

Despite these differences, it is clear that ozone destruction due to CFCs and halons also occurs in the the lower stratosphere in the Arctic. For example, total O_3 losses of the order of 50–100 DU have been deduced in the Arctic polar vortex during the 1991–1995 winters (e.g., see Larsen *et al.*, 1994; and Müller *et al.*, 1996). Again, the increasing importance of heterogeneous chemistry at lower temperatures is evident. For example, Fig. 12.37 shows the measured concentrations of ClO as a function of the minimum temperatures experienced by the air masses obtained using back trajectories; also shown is the deficit in HCl, defined as the difference between the measured HCl concentrations and those expected in these air masses based on the concentrations of N_2O, which can be used as a tracer (Toohey *et al.*, 1993; Webster *et al.*, 1993a). Clearly, at the lower temperatures, below 196 ± 4 K, where heterogenous chemistry is expected to convert HCl and $ClONO_2$ to active forms of chlorine, there is a greater HCl deficit than otherwise expected and much higher levels of ClO. Furthermore, these studies showed that the amounts of active chlorine in the form of ClO and its dimer, Cl_2O_2, were equivalent to twice the observed HCl deficit, consistent with the heterogeneous reaction of HCl with $ClONO_2$ (Webster *et al.*, 1993a).

Similarly, Fig. 12.38 shows some typical measurements of ClO and the HCl deficit at latitudes both outside and inside the Arctic polar vortex (Webster *et al.*, 1993b). As expected based on the known chemistry, there is a significant HCl deficit inside the vortex, accompanied by increased ClO concentrations (the more gradual increase in ClO at the edge of the vortex is attributed to air that has undergone PSC chemistry in the past but is now partially recovered; Webster *et al.*, 1993b). Indeed, almost complete conversion to active forms of chlorine has been measured in the winter Arctic vortex. Figure 12.39, for example, shows one estimate of the partitioning of total inorganic chlorine, Cl_y (Cl_y = Cl + ClO + $2Cl_2O_2$ + HCl + $ClONO_2$ + HOCl; OClO, BrCl, and $2Cl_2$ are also included in this if present), both outside and inside the Arctic polar vortex in January and February 1989 based on measurements of NO, NO_y, ClO, N_2O, and total organic

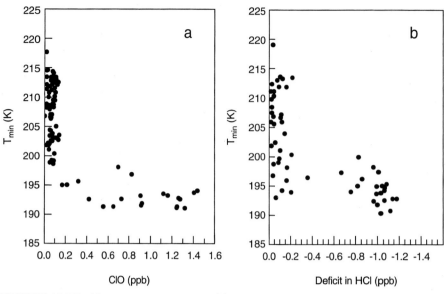

FIGURE 12.37 (a) ClO concentrations and (b) HCl deficit at various minimum temperatures experienced by the air masses in the Arctic stratosphere during October 1991–February 1992 (adapted from Toohey et al. (1993) and Webster et al. (1993a)).

chlorine (CCl$_y$) (Kawa et al., 1992b). Conversion of HCl and ClONO$_2$, the major forms of Cl$_y$ outside the vortex, which has been confirmed by atmospheric measurements (e.g., see Dessler et al., 1995a), to ClO and its dimer inside the vortex is dramatic.

As is the case for the Antarctic polar vortex, the extent of ozone depletion is governed to a large extent by the number of hours of sunlight available to drive the chemistry (Lefèvre et al., 1994; von der Gathen et al., 1995; Rex et al., 1998). Figure 12.40, for example, shows the depletion of ozone in the Arctic stratosphere in the winter of 1991–1992 as a function of the number of hours to which the air mass had been exposed to sunlight (von der Gathen et al., 1995). Depletion rates up to 10 ppb per hour in sunlight were observed (Rex et al., 1998). During this period, temperatures were sufficiently low for PSC formation during December and the first half of January. As seen in Fig. 12.41, mean ozone depletion rates of up to ~1.5% per day were observed during this period, a little less but similar in magnitude to the rates measured in the Antarctic polar vortex (von der Gathen et al., 1995).

Both denitrification and dehydration are very common in the Antarctic polar vortex, but they do not appear to be as common in the Arctic regions (e.g., see Ramaswamy, 1988; Fahey et al., 1990; Toon et al, 1990b; Arnold et al., 1992; Kawa et al., 1992a; Tuck et al., 1994; Santee et al., 1995; Van Allen et al., 1995; and Sugita et al., 1998). For example, Santee and co-workers (1995) have shown, using satellite-based data, that in the Antarctic polar vortex in 1992, gas-phase concentrations of HNO$_3$ and H$_2$O were both very small in mid-August, at the ClO peak. As the temperature rose above that where evaporation of PSCs should have occurred, their concentrations remained small, suggesting that the atmosphere was both denitrified and dehydrated; circumstantial support for denitrification of the Antarctic stratosphere is also found in nitrate peaks

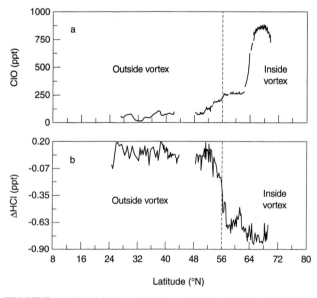

FIGURE 12.38 (a) Concentrations of ClO and (b) the HCl deficit in February and March 1992 outside and inside the Arctic vortex (adapted from Webster et al., 1993b).

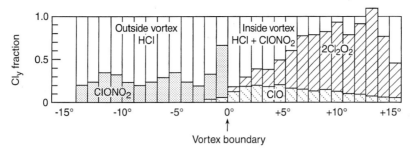

FIGURE 12.39 Estimated partitioning of the Cl_y reservoir outside and inside the Arctic polar vortex at a potential temperature of 420 K (adapted from Kawa et al., 1992b).

found in Antarctic firn cores corresponding to spring and early summer (Mulvaney and Wolff, 1993).

In the Arctic, however, observations of dehydration and denitrification are relatively few. For example, Toon et al. (1992) measured a variety of gases in the stratosphere during the 1989 Arctic winter using infrared spectrometry. They measured higher concentrations of both gaseous HNO_3 and H_2O inside the vortex than outside in most cases, indicating that neither dehydration nor denitrification had occurred. However, in one case where the stratospheric temperatures were <190 K, HNO_3 decreased significantly but increased again as the temperature rose. This suggested that while nitric acid had been temporarily frozen out on particles at the colder temperatures, it evaporated back into the gas phase as the temperature rose; i.e., denitrification did not take place, despite this brief condensation onto particles.

A similar phenomenon has been reported by Santee et al. (1996) during the 1995–1996 winter. Figure 12.42 shows their measurements of gas-phase HNO_3 interpolated to the isentropic surface at 465 K as well as the true stratospheric temperature. A strong correlation between the temperature and HNO_3 concentrations is evident, which they attribute to condensation of gaseous HNO_3 onto particles as the temperature falls, followed by reevaporation into the gas phase at higher temperatures.

While there is a variety of evidence from these and other measurements (e.g., see Notholt, 1994) that denitrification is more episodic and less widespread in the Arctic compared to the Antarctic, it does not mean that it does not occur. Indeed, there is good evidence for denitrification of the Arctic polar vortex under some conditions (e.g., see Schlager et al., 1990; Kondo et al., 1992; Wilson et al., 1992; Dye et al., 1992; Pueschel et al., 1992b; Tuck et al., 1994; and Höpfner et al., 1996). For example, direct measurements of stratospheric water and NO_y at the edge of the Arctic polar vortex in one study showed that in that case, both dehydration and denitrification had occurred (Hintsa et al., 1998a).

When the Arctic polar vortex is not denitrified, more gas-phase HNO_3 is available as the sunlight intensity increases, and this photolyzes, regenerating NO_2:

$$HNO_3 + h\nu \rightarrow OH + NO_2. \qquad (49)$$

The production of NO_2 then assists in terminating the

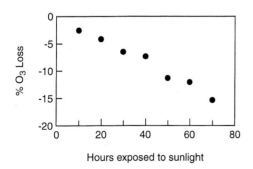

FIGURE 12.40 Loss of O_3 in Arctic polar vortex as a function of hours of exposure of the air mass to sunlight (adapted from von der Gathen et al., 1995).

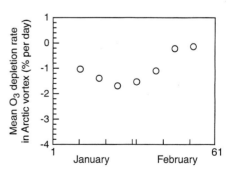

FIGURE 12.41 Mean depletion rate (% day^{-1}) in Arctic polar vortex in January and February 1992 (adapted from von der Gathen et al., 1995).

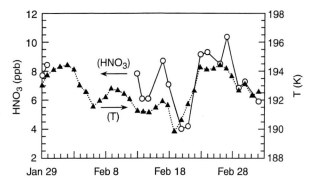

FIGURE 12.42 Gas-phase HNO$_3$ concentrations (○) on the 465 K isentropic surface and stratospheric temperatures (▲) in the Arctic in winter 1996 (adapted from Santee *et al.*, 1996).

ozone depletion cycles via the formation of chlorine nitrate and bromine nitrate:

$$ClO + NO_2 \xrightarrow{M} ClONO_2, \quad (16)$$

$$BrO + NO_2 \xrightarrow{M} BrONO_2. \quad (17)$$

In addition, the reaction of OH with HNO$_3$,

$$OH + HNO_3 \rightarrow H_2O + NO_3, \quad (50)$$

is important at high solar zenith angles. Thus, while conversion of HCl and ClONO$_2$ to active forms of chlorine also occurs in the Arctic winter, the resulting ozone depletion due to chemical processes is not as dramatic as that in the Antarctic (e.g., see Salawitch *et al.*, 1993; Webster *et al.*, 1993a; and Notholt *et al.*, 1995). Combined with the different meteorology and dynamics discussed earlier, the net decreases in column O$_3$ observed in the Arctic polar vortex have been significantly smaller than those in the Antarctic. For example, total column ozone losses of 10–30% were observed during the 1997 winter/spring (Manney *et al.*, 1997; Newman *et al.*, 1997). A series of papers on Arctic statospheric ozone during the 1996–1997 season is found in the November 15, 1997, issue of *Geophysical Research Letters*.

8. Ozone Destruction in the Midlatitudes

The chemistry in the midlatitude stratosphere follows that discussed throughout this chapter. As seen in the previous sections, the heterogeneous chemistry that was once thought to be unique to PSCs also occurs in and on the liquid solutions characteristic of sulfate particles distributed globally, with their relative importance being determined by the temperature, composition, and phase of the condensed phase.

In brief, HO$_x$ chemistry [reactions (8) and (6) of HO$_2$ and OH with O$_3$], as well as ClO$_x$–HO$_x$ [reactions (6) and (26)–(29)] and ClO$_x$–BrO$_x$ [reaction (31)] interactions, primarily control the chemistry of the lower stratosphere; NO$_x$ chemistry becomes important at altitudes above ~25 km (e.g., McElroy *et al.*, 1992; Avallone *et al.*, 1993a; Wennberg *et al.*, 1994). Removal of NO$_2$ from the gas phase (where it "quenches" the halogen chemistry by tying up ClO and BrO as their respective nitrates) occurs via its reaction to form N$_2$O$_5$, followed by hydrolysis on stratospheric sulfate aerosols, which increase dramatically after volcanic eruptions. Other heterogeneous reactions such as the hydrolysis of ClONO$_2$ become more important at lower temperatures (e.g., see Fig. 12.25).

Figure 12.43 summarizes the model-predicted relative importance of these cycles for conditions at ~20-km altitude and 43.5°N, assuming total chlorine and bromine levels found in 1990 (Solomon *et al.*, 1996). The importance of the HO$_x$ cycles, removal of NO$_x$ by the hydrolysis of N$_2$O$_5$, and the increased importance of halogen chemistry as the particle surface area available for heterogeneous reactions increase, are all illustrated.

Of course, it is not just the chemistry but also the dynamics that determine the net effect on total column ozone in midlatitudes. Transport of air from the tropics to midlatitudes was discussed earlier in Section A.1. There is also evidence for the influence of high-latitude air on ozone at midlatitudes. It has been proposed, for example, that the Arctic polar vortex acts more like a "flowing processor" than an isolated air mass. In this scenario, air flows through the polar vortex and as it does, undergoes the chemistry described earlier for the

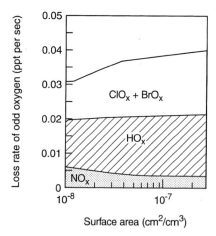

FIGURE 12.43 Model-predicted removal rates of odd oxygen as a function of particle surface area for 1990 levels of total chlorine and bromine at 20-km altitude and 43.5°N (adapted from Solomon *et al.*, 1996).

Antarctic ozone hole formation. Outflow to lower latitudes then provides a source of air that has been processed by the polar vortex and PSCs (e.g., see Proffitt et al., 1990, 1993; Randel and Wu, 1995).

A second mechanism for the polar vortex to influence midlatitudes is during the breakup of the polar vortex in the spring when the vortex air becomes mixed with air at lower latitudes (e.g., see Manney et al., 1994a or 1994b). Similarly, Atkinson and Plumb (1997) suggest that on some occasions, ozone depletion observed at midlatitudes may be due to transport of ozone-depleted air from the Antarctic ozone hole. Finally, chemistry occurring on PSCs and aerosols in the region outside the edges of the vortex followed by transport to lower latitudes may influence midlatitude chemistry.

Evidence for each of these mechanisms of influence of the polar regions on midlatitudes is discussed in Chapter 4 of the WMO (1995, 1999) document.

D. CONTRIBUTION OF BROMINATED ORGANICS

In Section C.3, we saw that gas-phase chlorine chemistry in the stratosphere is inextricably intertwined with bromine chemistry. Because of this close interrelationship, altering the concentrations of only one of the halogens (e.g., through controls) may not have the proportional quantitative result that might be initially expected. We explore in this section in more detail the role of brominated organics in stratospheric ozone destruction and the interrelationship with chlorine chemistry.

1. Sources and Sinks of Brominated Organics

Brominated organics are emitted into the atmosphere by a variety of natural and anthropogenic sources. Methyl bromide, CH_3Br, is the most abundant organobromine compound emitted into the atmosphere, although others such as dibromomethane and chlorobromomethane may also contribute to a significant extent (e.g., see Kourtidis et al., 1996). Halons, especially CF_2ClBr (Halon-1211) and CF_3Br (Halon-1301), are used as fire suppressants in situations where the use of water is not appropriate (e.g., around electronic equipment and on civilian aircraft; Freemantle, 1995). The numbering system for halons is in the following order: #C, #F, #Cl, and #Br, where # indicates the number of each kind of atom in the molecule; when there is no chlorine in the compound, a zero is used in the third position. Thus $C_2F_4Br_2$, for example, is Halon-2402 (O'Sullivan, 1989).

TABLE 12.6 Estimated Annual Global Emissions of Halon-1211 (CF_2ClBr) and Halon-1301 (CF_3Br) from 1963 to 1990[a] (in 10^6 kg/year)

Year	Halon-1211	Halon-1301
1963	0.033	0.004
1970	0.832	0.072
1975	2.5	0.73
1980	4.6	1.8
1985	8.9	5.1
1990	9.5	3.0

[a] From McCulloch (1992).

Table 12.6 shows one estimate of the emissions of Halon-1211 and Halon-1301 into the atmosphere from 1963 to 1990 (McCulloch, 1992). Table 12.7 gives the 1995 WMO estimate of the methyl bromide sources, both natural and anthropogenic (which are about equal in magnitude). Despite the fact that compounds such as methyl bromide and chlorobromomethane can be removed by reaction with OH in the troposphere (e.g., Orkin et al., 1997), whereas the halons cannot, the emissions of methyl bromide in particular are of a sufficient magnitude that some CH_3Br does reach the stratosphere. One set of measurements of the vertical profiles of CH_3Br, CH_2Br_2, Halon-1211, and Halon-1301 suggests that CH_3Br is responsible for ~55–70% of the bromine carried into the stratosphere by these compounds (Kourtidis et al., 1998).

Methyl bromide is used as a fumigant for soils (the agricultural use shown in Table 12.7) and shipments of fruits and vegetables as well as for buildings for termite control (shown as "structural purposes" in Table 12.7). Large amounts are released during biomass burning (e.g., see Manö and Andreae, 1994; Cicerone, 1994)

TABLE 12.7 Estimated Annual Emissions of CH_3Br[a]

Source	Best estimates		
	Range (10^6 kg/year)	Anthropogenic (10^6 kg/year)	Natural (10^6 kg/year)
Ocean	60–160	0	90
Agriculture	20–60	35	0
Biomass burning	10–50	25	5
Gasoline additives	0.5–22	1–15	0
Structural purposes	4	4	0
Industrial emissions	2	2	0
Totals	97–298	67–81	95

[a] From World Meteorological Organization (1995); see 1999 document for update.

and small amounts from the combustion of leaded gasoline containing bromine additives (e.g., Thomas *et al.*, 1997; Chen *et al.*, 1999). Emissions from terrestrial higher plants have also been reported (Gan *et al.*, 1998).

As indicated by the ranges in Table 12.7, there are significant uncertainties associated with these estimates (WMO, 1995; Butler, 1995). For example, the bromine from gasoline additives arises from the use of ethylene dibromide in leaded gasoline to prevent the accumulation of lead deposits in the engine. A significant percentage of this bromine is emitted in the form of CH_3Br as well as in the form of particles. With the introduction of catalytic converters on automobiles, and the associated phase-out of lead, which poisons the catalysts, the use of ethylene dibromide has also decreased. However, emissions still arise from the continued use of leaded gasolines in many regions of the world. Thomas *et al.* (1997) have examined the likely decrease in CH_3Br emissions from combustion of gasoline between 1984 and 1992 and the increase in that due to its use as an agricultural fumigant. They conclude that within a large uncertainty, it is possible that much of the increased agricultural use could have been counterbalanced by declining emissions associated with gasoline combustion.

A large uncertainty is associated with the ocean source, which, given the potential size of its contribution, is a major key to understanding its role in determining methyl bromide concentrations in the atmosphere. Biological processes that are not well understood produce CH_3Br. However, the ocean also acts as a sink for CH_3Br, which hydrolyzes and also reacts via a nucleophilic displacement with Cl^- (Elliott and Rowland, 1993; Butler, 1994; Lobert *et al.*, 1995; Jeffers and Wolfe, 1996; Yvon and Butler, 1996; Yvon-Lewis and Butler, 1997). As a result of the balancing of these two effects, i.e., production and uptake, reductions of anthropogenic emissions may not be linearly reflected in a corresponding change in the atmospheric concentrations (Butler, 1994).

Whether production or destruction predominates depends on a number of factors, including temperature and the rate of biological production of CH_3Br. As a result, the ocean can serve as either a net source or a net sink, depending on the conditions (e.g., see Anbar *et al.*, 1996; and Pilinis *et al.*, 1996).

The amount of CH_3Br that is applied to soils as a fumigant and that escapes to the atmosphere is also uncertain. For example, Cicerone and co-workers (Yagi *et al.*, 1993, 1995) measured a range from 34 to 87% of the methyl bromide applied to a field that escaped to the atmosphere. Since approximately 80% of synthetic CH_3Br use is due to soil fumigation (Shorter *et al.*, 1995), such variations are significant in accurately assessing the methyl bromide budget. It has been suggested based on field experiments that the emissions from soil fumigation could be reduced to very low levels by covering the soil with a film that is impermeable, or nearly so, to CH_3Br (Yates *et al.*, 1998).

Bacterial processes in soils are known to act as a sink for CH_3Br (Oremland *et al.*, 1994a, 1994b; Shorter *et al.*, 1995; Serça *et al.*, 1998; Varner *et al.*, 1999). However, since the magnitude depends on a variety of factors, including the type of soil, moisture content, temperature, etc., accurate extrapolation to a global scale is not easily done, with estimates from $\sim 42 \pm 32$ Gg/year (Shorter *et al.*, 1995) to 94 ± 54 Gg/year (Serça *et al.*, 1998). Uptake by the foliage of plants has also been observed and suggested to be of the same order of magnitude as that by soils (Jeffers *et al.*, 1998).

Probably the best known portion of the CH_3Br budget is the removal by reaction with OH (e.g., Wingenter *et al.*, 1998). The rate constant for reaction (51),

$$OH + CH_3Br \rightarrow CH_2Br + H_2O, \qquad (51)$$

at 298 K is $k_{51} = 2.9 \times 10^{-14}$ cm^3 molecule^{-1} s^{-1}, which gives a calculated lifetime with respect to reaction with OH at a concentration of 5×10^5 radicals cm^{-3} of 2.2 years. However, since it is also taken up by soils and ocean hydrolysis, the atmospheric lifetime is less than this, 0.8 ± 0.1 years (Colman *et al.*, 1998; Ko *et al.*, 1998). It should be noted, however, that the response of the bromine already in the lower stratosphere is slower, so that the recovery of ozone in response to surface reductions takes much longer than this (Prather, 1997).

2. Bromine Chemistry in the Stratosphere

Figure 12.44 shows the major organobromine compounds measured at the earth's surface from 1988 to 1996 (Wamsley *et al.*, 1998). These compounds have also been measured at the tropopause in the tropics, i.e., at the point at which the air is believed to enter the stratosphere. Methyl bromide and the halons are the major species observed. For example, in one set of measurements, of 17.4 ppt organic bromine, 55% was from CH_3Br, 38% from the halons, 6% from CH_2Br_2, and 0.8% from the combination of CH_2BrCl and $CHBrCl_2$ (Schauffler *et al.*, 1998).

Figure 12.45 summarizes the most important chemistry of bromine in the stratosphere, both gas phase and heterogeneous (shown as the darker lines). Once the bromine-containing organics reach the stratosphere, they absorb light and photolyze in a manner analogous to the chlorofluorocarbons. As seen in

D. CONTRIBUTION OF BROMINATED ORGANICS

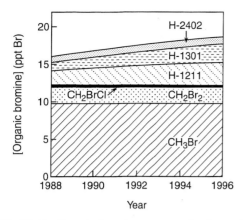

FIGURE 12.44 Organic bromine compounds observed at the earth's surface from 1988 to 1996 (adapted from Wamsley et al., 1998).

Chapter 4, the absorption cross sections of the major sources of bromine to the stratosphere, CH_3Br, $CBrClF_2$ (Halon-1211), and CBr_3F (Halon-1301), are substantial well out into the 250-nm region. The C–Br bond is even weaker than the C–Cl bond, as is generally the case for bromine compared to chlorine bonds (e.g., Lee et al., 1999a). For example, the C–Br bond strength is ~ 70 kcal mol^{-1} versus 85 kcal mol^{-1} for C–Cl and ~ 110 kcal mol^{-1} for C–F; the C–Br bond breaks first:

$$CBr_3F + h\nu \rightarrow Br + CBr_2F. \quad (52)$$

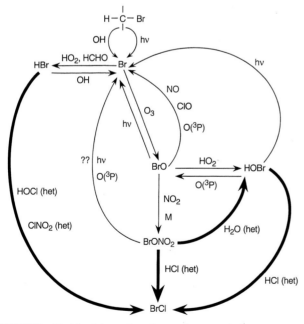

FIGURE 12.45 Schematic of gas-phase and heterogeneous bromine chemistry in the stratosphere. The heavier dark lines show the heterogeneous (het) chemistry.

In the case of hydrogen-containing compounds such as CH_3Br, reaction (51) with OH also generates a bromine-containing organic radical. As is the case for the chlorine compounds, subsequent reactions of these radicals are expected ultimately to release the remaining bromine atoms as well. For example, the CH_2Br radical formed from the CH_3Br + OH reaction has been shown (Chen et al., 1995; Orlando et al., 1996) in laboratory studies in 1 atm air or O_2/N_2 mixtures, and at temperatures from 228 to 298 K, to undergo the following reactions, as might be anticipated from the mechanisms discussed in Chapter 6:

$$CH_2Br + O_2 \xrightarrow{M} CH_2BrOO, \quad (53)$$

$$CH_2BrOO \xrightarrow{NO} CH_2BrO, \quad (54)$$

$$CH_2BrO \xrightarrow{decomposition} HCHO + Br. \quad (55)$$

The reaction of the alkoxy radical with O_2 to give formyl bromide, HC(O)Br + HO_2, was shown to be negligible compared to its decomposition. In short, the oxidation of brominated organics by OH and other species also ultimately generates atomic bromine. Oxidation of the organic bromine compounds to generate inorganic bromine is sufficiently fast that after a few years, inorganic forms comprise almost all of the stratospheric bromine. For example, Wamsley et al. (1998) showed that 97.5% of stratospheric bromine was in the form of inorganic species in an air parcel that had been in the stratosphere for about 6 years.

The atmospheric chemistry of other bromo-chloromethanes that are present in smaller amounts in the troposphere (Fig. 12.44) is discussed by Orkin et al. (1997) and Bilde et al. (1998).

Bromine is more effective, per atom, in destroying O_3 than is chlorine for several reasons. First, unlike chlorine atoms, atomic bromine does not react rapidly with organics. For example, the reaction of Br with CH_4 is too slow to represent a significant loss process, unlike the analogous chlorine atom reaction. Atomic bromine does react with HCHO at a reasonable rate ($k^{298K} = 1.1 \times 10^{-12}$ cm^3 molecule^{-1} s^{-1}) as well as with higher aldehydes (whose concentrations in the stratosphere are not large enough for this to be important, however). As a result, Br is converted into its reservoir species, HBr, primarily through its reactions with HO_2 and HCHO:

$$Br + HO_2 \rightarrow HBr + O_2, \quad (36)$$

$$k_{36} = 1.5 \times 10^{-11} e^{-600/T} \text{ cm}^3 \text{ molecule}^{-1} \text{ s}^{-1}$$

(DeMore et al., 1997),

$$Br + HCHO \rightarrow HBr + CHO, \quad (56)$$

$$k_{56} = 1.7 \times 10^{-11} e^{-800/T} \text{ cm}^3 \text{ molecule}^{-1} \text{ s}^{-1}$$

(DeMore *et al.*, 1997),

These sinks for Br compete with its reaction (32) with O_3:

$$Br + O_3 \rightarrow BrO + O_3, \quad (32)$$

$$k_{32} = 1.7 \times 10^{-11} e^{-800/T} \text{ cm}^3 \text{ molecule}^{-1} \text{ s}^{-1}$$

(DeMore *et al.*, 1997).

Like HCl, HBr is relatively slowly converted back to Br via reaction with OH:

$$HBr + OH \rightarrow Br + H_2O, \quad (57)$$

$$k_{57} = 1.1 \times 10^{-11} e^{-0/T} \text{ cm}^3 \text{ molecule}^{-1} \text{ s}^{-1}$$

(DeMore *et al.*, 1997).

Measurements of the solubility of HBr in sulfuric acid at 220 K gave Henry's law constants from 8.5×10^3 M atm^{-1} for 72 wt% H_2SO_4 to 1.5×10^7 M atm^{-1} for 54 wt% H_2SO_4 (Williams *et al.*, 1995; Abbatt, 1995). Application of these values to stratospheric aerosol particles typical of midlatitude conditions gives very small equilibrium concentrations of dissolved HBr; i.e., most of the HBr will remain in the gas phase.

Another difference between bromine and chlorine compounds is their photochemistry. Thus, absorption cross sections of bromine compounds tend to be larger than those of the corresponding chlorine compounds, especially at longer wavelengths where the intensity of solar radiation available for photolysis is larger. As a result, conversion of bromine compounds into active forms is generally faster. For example, BrO has a peak absorption cross section of 1.5×10^{-17} cm^2 molecule^{-1} at 298 K, with the absorption occurring out beyond 360 nm; ClO, on the other hand, has a peak absorption cross section of 6×10^{-18} cm^2 molecule^{-1}, and absorption approaches zero at 310 nm. As a result, conversion of BrO to Br + O through photolysis is important for BrO but the analogous reaction for ClO is not. Indeed, the lifetime of BrO with respect to photolysis in the stratosphere is of the order of seconds (Lary, 1996).

BrO can also be converted back to atomic bromine via its reaction (31) with ClO discussed earlier, its reaction with NO,

$$BrO + NO \rightarrow Br + NO_2, \quad (58)$$

$$k_{58} = 8.8 \times 10^{-12} e^{+260/T} \text{ cm}^3 \text{ molecule}^{-1} \text{ s}^{-1}$$

(DeMore *et al.*, 1997),

and, in the upper stratosphere, where there are higher ground-state oxygen atom concentrations, its reaction with O(^{3}P). Despite these rapid sink processes for BrO, it still represents the major form of bromine in the lower stratosphere (Lary, 1996).

HOBr is formed in the stratosphere by the reaction of BrO with HO_2:

$$BrO + HO_2 \rightarrow HOBr + O_2, \quad (59)$$

$$k_{59} = 3.4 \times 10^{-12} e^{540/T} \text{ cm}^3 \text{ molecule}^{-1} \text{ s}^{-1}$$

(DeMore *et al.*, 1997).

One issue of interest with respect to reaction (59) is whether there is a second channel that produces the reservoir species HBr, rather than the reactive HOBr, which would have a major effect on the partitioning of reactive bromine in the stratosphere (Lary, 1996). Larichev *et al.* (1995) place an upper limit of 1.5% on this potential path over the temperature range 233–298 K, which is consistent with indirect estimates of an upper limit of 0.01% (Mellouki *et al.*, 1994) as well as with measurements of HBr and upper limits for the HOBr concentration in the stratosphere (D. G. Johnson *et al.*, 1995). Cronkhite *et al.* (1998) observe no dependence of the overall rate constant on pressure between 12 and 25 Torr, indicating that a third possible channel producing an $HO_2 \cdot BrO$ adduct is not important.

The best match of observed HBr concentrations at midlatitudes to modeled values is obtained with a branching ratio of ~1% (Chartrand and McConnell, 1998). This assumes that there is no HBr production from a minor channel in the OH + BrO reaction, analogous to the OH + ClO reaction discussed earlier. (The major path is OH + BrO → Br + HO_2.) The atmospheric observations are also consistent with no production of HBr in reaction (59), but rather, a branching ratio for the production of HBr + O_2 in the OH + BrO reaction in the range of 1–3% (Chipperfield *et al.*, 1997; Chartrand and McConnell, 1998). In this case, the OH + BrO reaction would be the major source of HBr in the 20- to 35-km region (Chipperfield *et al.*, 1997).

Another possible source of HOBr is the heterogeneous hydrolysis of $BrONO_2$. Once formed, HOBr photolyzes rapidly, with a lifetime of ~15 min at noon at the equinox at midlatitudes in the lower stratosphere (Lary, 1996). Above about 25 km, the reaction with O(^{3}P) becomes important:

$$HOBr + O(^3P) \rightarrow OH + BrO, \quad (60)$$

$$k_{60} = 1.4 \times 10^{-10} e^{-430/T} \text{ cm}^3 \text{ molecule}^{-1} \text{ s}^{-1}$$

(Nesbitt *et al.*, 1995).

Similarly, the absorption cross sections for the reservoir species $BrONO_2$ are much larger than those of

ClONO$_2$ toward 400 nm, so that its lifetime with respect to photolysis at noon at the equinox in midlatitudes is only a few minutes (Lary, 1996). As discussed in Chapter 4.N, the photolysis products are not well established but appear to include Br + NO$_3$ as well as BrO + NO$_2$, with possibly a small contribution from channels producing oxygen atoms. If the products are BrO + NO$_2$, then the formation of BrONO$_2$ becomes a "do-nothing" reaction and will not contribute to ozone depletion. The same is true of the channel producing Br + NO$_3$, since NO$_3$ photolyzes primarily to O + NO$_2$ (see Chapter 4.G).

As discussed in Section C.3, ClO$_x$ chemistry and BrO$_x$ chemistry are closely interrelated via the gas-phase reactions of ClO with BrO:

$$ClO + BrO \rightarrow Br + OClO, \quad (31a)$$
$$\rightarrow Br + ClOO, \quad (31b)$$
$$\rightarrow BrCl + O_2. \quad (31c)$$

Indeed, these reactions play an important role in the Antarctic ozone hole and they have important implications for control strategies, particularly of the brominated compounds. For example, Danilin *et al.* (1996) examined the effects of ClO$_x$–BrO$_x$ coupling on the cumulative loss of O$_3$ in the Antarctic ozone hole from August 1 until the time of maximum ozone depletion. Increased bromine increased the rate of ozone loss under the denitrified conditions assumed in the calculations by converting ClO to Cl, primarily via reactions (31b) and (31c) (followed by photolysis of BrCl). Danilin *et al.* (1996) estimate that the efficiency of ozone destruction per bromine atom (α) is 33–55 times that per chlorine atom (the "bromine enhancement factor") under these conditions in the center of the Antarctic polar vortex, $\alpha \sim 60$ calculated as a "global average" over all latitudes, seasons, and altitudes (WMO, 1999).

However, although as much as 50% of the loss of O$_3$ could be attributed to bromine chemistry at a Br$_y$ concentration of 25 ppt, a reduction in bromine did not give a proportional change in the total destruction of O$_3$. Figure 12.46 shows the predicted cumulative O$_3$ loss as a function of the Br$_y$ concentration; as bromine decreases, the contribution of the BrO–ClO cycle decreases as well. However, the net effect on total ozone loss is quite small because near-total ozone destruction occurs even without a significant contribution from BrO.

It should be noted that in the Arctic, the integrated concentrations of ClO are generally smaller than in the Antarctic; thus, the BrO + ClO cycle is relatively more important than the ClO + ClO cycle (Salawitch *et al.*, 1993; WMO, 1995). In addition, only ~30–50% of the

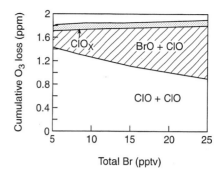

FIGURE 12.46 Model-calculated cumulative loss of ozone from August 1 to the day of maximum ozone depletion as a function of stratospheric bromine ([Cl$_y$] = 12.5 ppbv, [NO$_y$] = 2 ppb, 70°S at an altitude corresponding to 50 mbar total pressure in these calculations) (adapted from Danilin *et al.*, 1996).

O$_3$ in a given air mass is destroyed over the Arctic, compared to near-total ozone destruction over Antarctica. As a result, bromine does make a significant contribution to the total O$_3$ destruction in the Arctic, and as a result, control of brominated organics is expected to have a greater effect on minimizing ozone destruction in this region.

The BrO + BrO self-reaction occurs in a manner analogous to the ClO + ClO self-reaction:

$$BrO + BrO \rightarrow 2Br + O_2, \quad (61a)$$
$$\rightarrow Br_2 + O_2, \quad (61b)$$
$$\xrightarrow{M} Br_2O_2. \quad (61c)$$

The rate constant for paths a and b have been determined to be $k_{61a} = 5.3 \times 10^{-12} e^{-211/T}$ and $k_{61b} = 1.1 \times 10^{-14} e^{983/T}$ cm^3 molecule^{-1} s^{-1}, respectively (Harwood *et al.*, 1998). At temperatures below 250 K, the termolecular reaction forming the dimer becomes relatively more important (Harwood *et al.*, 1998). However, this self-reaction is not important because of the relatively small concentrations of BrO in the stratosphere. Reactions with ClO and HO$_2$ are much more important.

Although there are fewer studies of the heterogeneous chemisty of BrONO$_2$ and HOBr than of the corresponding chlorine compounds, it is clear from the laboratory studies that have been done that analogous chemistry occurs, and at least as fast as for the chlorine compounds. Table 12.8 shows some of the most important reactions and typical values of the reaction probabilities. On ice, the hydrolyses of ClONO$_2$ and BrONO$_2$ proceed at comparable rates (Tables 12.5 and 12.8). However, toward midlatitudes the particles are largely concentrated sulfuric acid–water mixtures, and on this surface the ClONO$_2$ hydrolysis reaction probability

TABLE 12.8 Some Typical Reaction Probabilities under Stratospheric Conditions for Some Heterogeneous Reactions Involving Bromine

Reaction	Reaction probability (γ)	
	Ice	Liquid H_2SO_4–H_2O
$BrONO_2 + H_2O \xrightarrow{het} HOBr + HNO_3$	$\geq 0.3^b$	$\simeq 0.8^a$
$BrONO_2 + HCl \xrightarrow{het} BrCl + HNO_3$	$\geq 0.3^b$	$\geq 0.9^d$
$HOBr + HCl \xrightarrow{het} BrCl + H_2O$	0.25^c	$\sim 0.2^d$
$HOBr + HBr \xrightarrow{het} Br_2 + H_2O$	0.1^c	0.25^c
$HBr + ClONO_2 \xrightarrow{het} BrCl + HNO_3$	$> 0.3^e$	—

[a] From Hanson et al. (1996).
[b] From Hanson and Ravishankara (1993) and Allanic et al. (1997).
[c] From Abbatt (1994); γ for $HOBr + HBr \rightarrow Br_2 + H_2O$ on 69 wt% H_2SO_4–H_2O at 228 K corresponds to a liquid-phase rate constant of $> 5 \times 10^4$ L mol^{-1} s^{-1} (Allanic et al., 1997).
[d] From Hanson and Ravishankara (1995); bimolecular rate constant in 70 wt% H_2SO_4 is $\sim 1.4 \times 10^5$ L mol^{-1} s^{-1} (Abbatt, 1995).
[e] From Hanson and Ravishankara (1992).

decreases. As a result, chlorine activation becomes less important at midlatitudes while bromine activation continues. Because of this, HOBr becomes an important intermediate in bromine chemistry. For example, model calculations by Randeniya et al. (1997) suggest that the increased OH produced upon photolysis of HOBr increases the loss of O_3 in the 12- to 20-km range under conditions of large aerosol particle concentrations such as those after the Mount Pinatubo eruption. Another factor may be the strong interaction of ice and HOBr, which increases its residence time on the particles and hence the opportunity to carry out heterogeneous chemistry (Allanic et al., 1997).

HOBr also serves to couple bromine and chlorine chemistry in an indirect manner. Thus, photolysis of HOBr generates increased OH concentrations, which then cause a faster recycling of HCl back into chlorine atoms (Lary et al., 1996; Randeniya et al., 1996a,b; Tie and Brasseur, 1996). Lary et al. (1996) estimated that the lifetime of HCl can be reduced by as much as a factor of three through this effect and suggest that the unexplained rapid rise in OH reported by Salawitch et al. (1994) at dawn may be due to the photolysis of HOBr formed overnight rather than of a nitrogen species such as HONO.

Another coupling of these halogen chemistries takes place when HOBr reacts with HCl in the condensed phase (Abbatt, 1994; Lary et al., 1996), forming BrCl, which is then released to the gas phase. Like Cl_2 and Br_2, BrCl rapidly photolyzes so that active chlorine is again generated from what was HCl in the particle. Reactions of $BrONO_2$ with HCl and of HBr with HOCl or $ClONO_2$ give the same net result.

Thus, the effect of heterogeneous bromine chemistry is primarily to amplify the chlorine-catalyzed destruction of ozone through the more rapid conversion of the reservoir species HCl back into active forms of chlorine (Lary et al., 1996; Tie and Brasseur, 1996). This becomes particularly important under conditions of enhanced aerosol particles, e.g., after major volcanic eruptions.

In summary, there are a variety of paths by which bromine can contribute to stratospheric chemistry. Excellent reviews of the gas-phase chemistry and of the heterogeneous chemistry of bromine relevant to the stratosphere are found in Lary (1996) and Lary et al. (1996), and of the thermochemistry of bromine oxides in Chase (1996).

E. CONTRIBUTION OF IODINE-CONTAINING ORGANICS

In addition to chlorinated and brominated organics, iodine-containing organics are also emitted into the troposphere, primarily by biological processes in the oceans. Methyl iodide is believed to be the major species emitted, but others such as $ClCH_2I$ and CH_2IBr may also be generated (e.g., see Cicerone, 1981; Klick and Abrahamsson, 1992; Moore and Tokarczyk, 1992; Schall and Heumann, 1993; Gribble, 1994; Happell and Wallace, 1996; and Carpenter et al., 1999), and ethyl iodide has also been measured recently (Yokouchi et al., 1997).

There is a substantial difference in their tropospheric chemistry from that of the chlorine and bromine compounds, however (e.g., see Huie and Laszlo, 1995). The carbon–halogen bond is very weak, 57 kcal mol^{-1} in CH_3–I compared to 70 kcal mol^{-1} for CH_3–Br, 85 kcal mol^{-1} for CH_3–Cl, and 108 kcal mol^{-1} for CH_3–F. In addition, the absorption spectra are red-shifted for the iodine compounds, so that their absorption spectra better overlap with increasing solar intensity. As a result, organoiodine compounds photolyze readily in the troposphere to generate iodine atoms (Calvert and Pitts, 1966). Combined with other fates such as reaction with OH and NO_3, their tropospheric lifetimes are sufficiently short that they are not expected to reach the stratosphere in sufficient quantities to contribute to ozone destruction.

Because of these rapid removal processes in the troposphere, the contribution of iodine to stratospheric photochemistry has not received much attention. However, Solomon et al. (1994) suggested that rapid transport from the lower troposphere into the upper troposphere and lower stratosphere via convective clouds could provide a mechanism for injecting such compounds into the stratosphere. While the relevant chemistry of iodine is not well known, it would be expected to interact with the ClO_x cycles in much the same way as BrO, e.g.,

$$I + O_3 \rightarrow IO + O_2 \quad (62)$$
$$ClO + IO \rightarrow I + OClO \quad (63a)$$
$$\rightarrow I + ClOO \quad (63b)$$
$$\rightarrow ICl + O_2 \quad (63c)$$
$$\rightarrow I + Cl + O_2 \quad (63d)$$
$$\rightarrow OIO + Cl \quad (63e)$$
$$\rightarrow IOO + Cl \quad (63f)$$
$$\xrightarrow{M} IO_2Cl \quad (63g)$$

The overall rate constant for the ClO + IO reaction has been measured to be $k_{63} = 5.1 \times 10^{-12} e^{280/T}$ cm^3 molecule^{-1} s^{-1}, with a branching ratio of 0.14 ± 0.04 for all channels not producing I atoms at 298 K. (Turnipseed et al., 1997). This is in agreement with branching ratios for (63a) of 0.55 ± 0.03, (63c) of 0.20 ± 0.02, and (63d) of 0.25 ± 0.02 reported by Bedjanian et al. (1997a).

In addition, BrO–IO cross interactions would be expected; the major channel in this reaction appears to generate Br + OIO, with a branching ratio of ~1 within an uncertainty of ~35% (Bedjanian et al., 1997b, 1998; Laszlo et al., 1997; Gilles et al., 1997). Reaction of IO with HO_2, O, and NO and photolysis will also occur (DeMore et al., 1997):

$$IO + HO_2 \rightarrow HOI + O_2 \quad (64)$$
$$IO + O \rightarrow O_2 + I, \quad (65)$$
$$IO + NO \rightarrow I + NO_2, \quad (66)$$
$$IO + h\nu \rightarrow I + O. \quad (67)$$

Solomon et al. (1994) proposed that below ~20 km, iodine could make a major contribution to O_3 destruction if there were 1 ppt of total iodine in the stratosphere. Episodic transport of iodine compounds to the upper troposphere clearly happens on some occasions, as evidenced by the observation of concentrations of CH_3I as high as ~1 ppt at 10–12 km when a typhoon provided strong vertical upward motion (Davis et al., 1996). However, it may be that this is the exception rather than the rule.

For example, Wennberg et al. (1997) used high-resolution spectra taken from the Kitt Peak National Solar Observatory to search for evidence of IO. Combined with simulations using assumed IO chemistry, they conclude that the total stratospheric iodine is ~0.2 ppt, with an upper limit of ~0.3 ppt. Similarly, Pundt et al. (1998) conclude there must be <0.2 ppt iodine at altitudes <20 km, based on solar spectra obrained using balloon platforms. If these small concentrations based on a few measurements are typical, iodine will not be responsible for significant ozone destruction.

In short, it appears likely that sufficient iodine does not reach the stratosphere to make a significant contribution to ozone destruction.

F. SUMMARY

In summary, the chemistry of the stratosphere and the effects of anthropogenic perturbations on it have a rich history, with new chemistry that continues to unfold. For reviews of various aspects of the chemistry and history, see Cicerone (1981, 1987), Rowland (1989, 1992, 1993), Molina (1991), Rowland and Molina (1994), Toohey (1995), Brasseur et al. (1995), chapters by Li et al. (1995a), Anderson, and Sander et al. in the book edited by Barker (1995), chapters by Brune, Middlebrook and Tolbert, Wilson, and Brasseur et al. in the book edited by Macalady (1998), and the World Meteorological Organization (WMO) 1995 and 1999 reports "Scientific Assessment of Ozone Depletion."

G. PROBLEMS

1. If the bond dissociation energies for C–F, C–Cl, C–Br, and C–I are 113, 83, 70, and 57 kcal mol^{-1},

respectively, what are the minimum wavelengths needed to break these bonds? Comment on the relevance with respect to why (1) organoiodine compounds have such short lifetimes in the troposphere and (2) the lifetimes of organoiodine compounds are shorter than those of the analogous organobromine compounds.

2. CH_3Br reacts with OH with a room temperature rate constant of 2.9×10^{-14} cm^3 molecule^{-1} s^{-1}.

(a) What is the lifetime of methyl bromide with respect to reaction with OH at a global mean concentration of 5×10^5 radicals cm^{-3}?

(b) Colman et al. (1998) report an atmospheric lifetime of 0.8 ± 0.1 years for methyl bromide. Since it is also taken up by soils and oceans, what must the lifetime with respect to uptake by soils and oceans be?

(c) If the global mean OH concentration is really 1×10^6 radicals cm^{-3}, what must the lifetime be with respect to uptake by soils and oceans?

3. Keim et al. (1996) have shown that there is a significant reduction in NO and increase in NO_y and ClO in a layer above the tropopause that has increased aerosol surface areas (Fig. 12.31). They attribute this to increased heterogeneous reactions of $ClONO_2$ on particles to form HNO_3 and active chlorine.

(a) Estimate the ratio of the rates of $ClONO_2$ loss and HNO_3 production in going from an altitude of 11 km below the aerosol layer where the particle surface area is ~ 4 μm^2 cm^{-3} to 13 km where the surface area is ~ 20 μm^2 cm^{-3}. Assume the temperature and $ClONO_2$ concentrations remain approximately constant over this altitude range.

(b) Estimate the ratio of the lifetimes for $ClONO_2$ with respect to heterogeneous hydrolysis in going from 14 km above the aerosol layer to 13 km, at the top of the layer. Assume the reaction probability is 10^{-3} for the particle composition at 14 km but increases to 0.1 at 13 km due to the increased water content of these particles in the aerosol layer. Use Fig. 12.31 to estimate the change in particle surface area from 14 to 13 km and assume the temperature at both altitudes is approximately 210 K. Does this support the conclusion of Keim et al. (1996) that activation on particles is a reasonable explanation for their observations?

4. McKinney et al. (1997) report measurements of ClO and BrO in a chemically perturbed region of the stratosphere over Sweden. They observe that the production of BrO with decreasing solar zenith angle is so rapid that it implies the bromine atom source is very photolabile, likely BrCl, rather than $BrONO_2$. Using data from Chapters 3 and 4, calculate the lifetimes of BrCl and $BrONO_2$ at a solar zenith angle of 86° to estimate how much faster BrCl photolysis would be compared to $BrONO_2$ at 25 km. Assume the earth–sun distance correction factor is 1.0.

5. If the aerosol surface area in the McKinney et al. studies in Problem 4 was 5 μm^2 cm^{-3}, what would the lifetime be for $BrONO_2$ with respect to hydrolysis on particles? Take a temperature of 200 K and assume a reaction probability of 0.8 (Table 12.8) and that the reaction is not diffusion-limited. Compare this to the lifetime with respect to photolysis calculated in Problem 4.

6. The hydrolysis of $BrONO_2$ does not generate active bromine directly, but rather HOBr, which must then photolyze to OH + Br. Calculate the lifetime of HOBr with respect to photolysis at a solar zenith angle of 86° and an altitude of 25 km. Assume the earth–sun distance correction factor is 1.0. Is the hydrolysis of $BrONO_2$ or the photolysis of HOBr the rate-determining step?

References

Abbatt, J. P. D., K. D. Beyer, A. F. Fucaloro, J. R. McMahon, P. J. Wooldridge, R. Zhang, and M. J. Molina, "Interaction of HCl Vapor with Water-ice: Implications for the Stratosphere," *J. Geophys. Res., 97*, 15819–15826 (1992).

Abbatt, J. P. D., "Heterogeneous Reaction of HOBr with HBr and HCl on Ice Surfaces at 228 K," *Geophys. Res. Lett., 21*, 665–668 (1994).

Abbatt, J. P. D., "Interactions of HBr, HCl, and HOBr with Supercooled Sulfuric Acid Solutions of Stratospheric Composition," *J. Geophys. Res., 100*, 14009–14017 (1995).

Adriani, A., T. Deshler, G. Di Donfrancesco, and G. P. Gobbi, "Polar Stratospheric Clouds and Volcanic Aerosol during Spring 1992 over McMurdo Station, Antarctica: Lidar and Particle Counter Comparisons," *J. Geophys. Res., 100*, 25877–25897 (1995).

"AFEAS, Alternative Fluorocarbons Environmental Acceptability Study," Washington, DC, 1993; see web site in Appendix IV.

Aftergood, S., "The Space Shuttle's Impact on the Stratosphere—Comment," *J. Geophys. Res., 96*, 17377 (1991).

Allanic, A., R. Oppliger, and M. J. Rossi, "Real-Time Kinetics of the Uptake of HOBr and $BrONO_2$ on Ice and in the Presence of HCl in the Temperature Range 190–200 K," *J. Geophys. Res., 102*, 23529–23541 (1997).

Allen, H. C., J. M. Laux, R. Vogt, B. J. Finlayson-Pitts, and J. C. Hemminger, "Water-Induced Reorganization of Ultrathin Nitrate Films on NaCl: Implications for the Tropospheric Chemistry of Sea Salt Particles," *J. Phys. Chem., 100*, 6371–6375 (1996).

Allen, M., and M. L. Delitsky, "A Test of Odd-Oxygen Photochemistry Using Spacelab 3 Atmospheric Trace Molecule Spectroscopy Observations," *J. Geophys. Res., 96*, 12883–12891 (1991).

Anbar, A. D., Y. L. Yung, and F. P. Chavez, "Methyl Bromide: Ocean Sources, Ocean Sinks, and Climate Sensitivity," *Global Biogeochem. Cycles, 10*, 175–190 (1996).

Anderson, J., and V. K. Saxena, "Temporal Changes of Mount Pinatubo Aerosol Characteristics over Northern Midlatitudes Derived from SAGE II Extinction Measurements," *J. Geophys. Res., 101*, 19455–19463 (1996).

Anderson, J. G., W. H. Brune, S. A. Lloyd, D. W. Toohey, S. P. Sander, W. L. Starr, M. Loewenstein, and J. R. Podolske, "Kinetics of O_3 Destruction by ClO and BrO within the Antarctic Vortex: An Analysis Based on *in Situ* ER-2 Data," *J. Geophys. Res., 94*, 11480–11520 (1989).

Anderson, J. G., D. W. Toohey, and W. H. Brune, "Free Radicals within the Antarctic Vortex: The Role of CFC's in Antarctic Ozone Loss," *Science, 251,* 39–46 (1991).

Anderson, J. G., "Polar Processes in Ozone Depletion," in *Progress and Problems in Atmospheric Chemistry, Advanced Series in Physical Chemistry* (J. R. Barker, Ed.), Vol. 3, pp. 744–770, World Scientific, Singapore, 1995.

Anderson, S. M., K. Mauersberger, and J. Morton, "The Ozone Molecule: Isotope Effects and Electronic Structure," in *Progress and Problems in Atmospheric Chemistry, Advanced Series in Physical Chemistry* (J. R. Barker, Ed.), Vol. 3, pp. 473–499, World Scientific, Singapore, 1995.

Angell, J. K., "Comparison of Stratospheric Warming Following Agung, El Chichon, and Pinatubo Volcanic Eruptions," *Geophys. Res. Lett., 20,* 715–718 (1993a).

Angell, J. K., "Reexamination of the Relation between Depth of the Antarctic Ozone Hole, and Equatorial QBO and SST, 1962–1992," *Geophys. Res. Lett., 20,* 1559–1562 (1993b).

Ansmann, A., F. Wagner, U. Wandinger, and I. Mattis, "Pinatubo Aerosol and Stratospheric Ozone Reduction: Observations over Central Europe," *J. Geophys. Res., 101,* 18775–18785 (1996).

Anthony, S. E., R. T. Tisdale, R. Disselkamp, M. A. Tolbert, and J. C. Wilson, "FTIR Studies of Low Temperature Sulfuric Acid Aerosols," *Geophys. Res. Lett., 22,* 1105–1108 (1995).

Anthony, S. E., T. B. Onasch, R. T. Tisdale, R. S. Disselkamp, and M. A. Tolbert, "Laboratory Studies of Ternary $H_2SO_4/HNO_3/H_2O$ Particles: Implications for Polar Stratospheric Cloud Formation," *J. Geophys. Res, 102,* 10777–10784 (1997).

Arnold, F., K. Petzoldt, and E. Reimer, "On the Formation and Sedimentation of Stratospheric Nitric Acid Aerosols: Implications for Polar Ozone Destruction," *Geophys. Res. Lett., 19,* 677–680 (1992).

Atkinson, R. J., and R. A. Plumb, "Three-Dimensional Ozone Transport during the Ozone Hole Breakup in December 1987," *J. Geophys. Res., 102,* 1451–1466 (1997).

Ausloos, P., R. E. Rebbert, and L. Glasgow, "Photodecomposition of Chloromethanes Adsorbed on Silica Surfaces," *J. Res. Natl. Bur. Stand. U.S., 82,* 1–8 (1977).

Avallone, L. M., D. W. Toohey, W. H. Brune, R. J. Salawitch, A. E. Dessler, and J. G. Anderson, "Balloon-Borne *in Situ* Measurements of ClO and Ozone: Implications for Heterogeneous Chemistry and Mid-Latitude Ozone Loss," *Geophys. Res. Lett., 20,* 1795–1798 (1993a).

Avallone, L. M., D. W. Toohey, M. H. Proffitt, J. J. Margitan, K. R. Chan, and J. G. Anderson, "*In Situ* Measurements of ClO at Mid-latitudes: Is There an Effect from Mt. Pinatubo?" *Geophys. Res. Lett., 20,* 2519–2522 (1993b).

Avallone, L. M., D. W. Toohey, S. M. Schauffler, W. H. Pollock, L. E. Heidt, E. L. Atlas, and K. R. Chan, "*In Situ* Measurements of BrO during AASE II," *Geophys. Res. Lett., 22,* 831–834 (1995).

Avallone, L. M., and M. J. Prather, "Photochemical Evolution of Ozone in the Lower Tropical Stratosphere," *J. Geophys. Res., 101,* 1457–1461 (1996).

Avallone, L. M., and M. J. Prather, "Tracer–Tracer Correlations: Three-Dimensional Model Simulations and Comparisons to Observations," *J. Geophys. Res., 102,* 19233–19246 (1997).

Baker, M. B., and J. G. Dash, "Comment on 'Surface Layers on Ice' by C. A. Knight," *J. Geophys. Res., 101,* 12929–12931 (1996).

Ball, S. M., A. Fried, B. E. Henry, and M. Mozurkewich, "The Hydrolysis of $ClONO_2$ on Submicron Liquid Sulfuric Acid Aerosol," *Geophys. Res. Lett., 25,* 3339–3342 (1998).

Banham, S. F., J. R. Sodeau, A. B. Horn, M. R. S. McCoustra, and M. A. Chesters, "Adsorption and Ionization of HCl on an Ice Surface," *J. Vac. Sci. Technol. A, 14,* 1620–1626 (1996).

Barker, J. R., Ed., *Progress and Problems in Atmospheric Chemistry,* in Advanced Series in Physical Chemistry, Vol. 3, World Scientific, Singapore, 1995.

Barone, S. B., M. A. Zondlo, and M. A. Tolbert, "A Kinetic and Product Study of the Hydrolysis of $ClONO_2$ on Type Ia Polar Stratospheric Cloud Materials at 285 K," *J. Phys. Chem. A, 101,* 8643–8652 (1997).

Bates, D. R., and P. B. Hays, "Atmospheric Nitrous Oxide," *Planet. Space Sci., 15,* 189–197 (1967).

Baumgardner, D., J. Dye, B. Gandrud, K. Barr, K. Kelly, and K. R. Chan, "Refractive Indices of Aerosols in the Upper Troposphere and Lower Stratosphere," *Geophys. Res. Lett., 23,* 749–752 (1996).

Beck, J. P., C. E. Reeves, F. A. A. M. De Leeuw, and S. A. Penkett, "The Effect of Aircraft Emissions on Tropospheric Ozone in the Northern Hemisphere," *Atmos. Environ., Part A, 26,* 17–29 (1992).

Bedjanian, Y., G. Le Bras, and G. Poulet, "Kinetics and Mechanism of the IO + ClO Reaction," *J. Phys. Chem. A, 101,* 4088–4096 (1997a).

Bedjanian, Y., G. Le Bras, and G. Poulet, "Kinetic Study of the Br + IO, I + BrO, and Br + I_2 Reactions. Heat of Formation of the BrO Radical," *Chem. Phys. Lett., 266,* 233–238 (1997b).

Bedjanian, Y., G. Le Bras, and G. Poulet, "Kinetics and Mechanism of the IO + BrO Reaction," *J. Phys. Chem. A, 102,* 10501–10511 (1998).

Bekki, S., "On the Possible Role of Aircraft-Generated Soot in the Middle Latitude Ozone Depletion," *J. Geophys. Res., 102,* 10751–10758 (1997).

Bekki, S., and J. A. Pyle, "Potential Impact of Combined NO_x and SO_x Emissions from Future High Speed Civil Transport Aircraft on Stratospheric Aerosols and Ozone," *Geophys. Res. Lett., 20,* 723–726 (1993).

Bekki, S., "Oxidation of Volcanic SO_2: A Sink for Stratospheric OH and H_2O," *Geophys. Res. Lett., 22,* 913–916 (1995).

Berland, B. S., D. R. Haynes, K. L. Foster, M. A. Tolbert, S. M. George, and O. B. Toon, "Refractive Indices of Amorphous and Crystalline HNO_3/H_2O Films Representative of Polar Stratospheric Clouds," *J. Phys. Chem., 98,* 4358–4364 (1994).

Bertram, A. K., D. D. Patterson, and J. J. Sloan, "Mechanisms and Temperatures for the Freezing of Sulfuric Acid Aerosols Measured by FTIR Extinction Spectroscopy," *J. Phys. Chem., 100,* 2376–2383 (1996).

Bertram, A. K., and J. J. Sloan, "Temperature-Dependent Nucleation Rate Constants and Freezing Behavior of Submicron Nitric Acid Dihydrate Aerosol Particles under Stratospheric Conditions," *J. Geophys. Res., 103,* 3553–3561 (1998a).

Bertram, A. K., and J. J. Sloan, "The Nucleation Rate Constants and Freezing Mechanism of Nitric Acid Trihydrate Aerosol under Stratospheric Conditions," *J. Geophys. Res., 103,* 13261–13265 (1998b).

Bevilacqua, R. M., K. W. Hoppel, J. S. Hornstein, R. L. Lucke, E. P. Shettle, T. L. Ainsworth, D. Debrestian, M. D. Fromm, S. S. Krigman, J. Lumpe, W. Glaccum, J. J. Olivero, R. T. Clancy, C. E. Randall, D. W. Rusch, E. Chasséfere, C. Daladier, C. Deniel, C. Brogniez, and J. Lenoble, "First Results from POAM II: The Dissipation of the 1993 Antarctic Ozone Hole," *Geophys. Res. Lett., 22,* 909–912 (1995).

Beyer, K. D., S. W. Seago, H. Y. Chang, and M. J. Molina, "Composition and Freezing of Aqueous H_2SO_4/HNO_3 Solutions Under polar Stratospheric Conditions," *Geophys. Res. Lett., 21,* 871–874 (1994).

Beyerle, G., B. Luo, R. Neuber, T. Peter, and I. S. McDermid, "Temperature Dependence of Ternary Solution Particle Volumes As Observed by Lidar in the Arctic Stratosphere during Winter 1992/1993," *J. Geophys. Res., 102,* 3603–3609 (1997).

Bianco, R., and J. T. Hynes, "Ab Initio Model Study of the Mechanism of Chlorine Nitrate Hydrolysis on Ice," *J. Phys. Chem. A*, 102, 309–314 (1998).

Biermann, U. M., J. N. Crowley, T. Huthwelker, G. K. Moortgat, P. J. Crutzen, and Th. Peter, "FTIR Studies on Lifetime Prolongation of Stratospheric Ice Particles Due to NAT Coating," *Geophys. Res. Lett.*, 25, 3939–3942 (1998).

Biggs, P., C. E. Canosa-Mas, J.-M. Fracheboud, D. E. Shallcross, and R. P. Wayne, "Efficiency of Formation of CH_3O in the Reaction of CH_3O_2 and ClO," *Geophys. Res. Lett.*, 22, 1221–1224 (1995).

Bilde, M., T. J. Wallington, C. Ferronato, J. J. Orlando, G. S. Tyndall, E. Estupiñan, and S. Haberkorn, "Atmospheric Chemistry of CH_2BrCl, $CHBrCl_2$, $CF_3CHBrCl$ and CBr_2Cl_2," *J. Phys. Chem. A*, 102, 1976–1986 (1998).

Blake, D. F., and K. Kato, "Latitudinal Distribution of Black Carbon Soot in the Upper Troposphere and Lower Stratosphere," *J. Geophys. Res.*, 100, 7195–7202 (1995).

Boering, K. A., E. J. Hintsa, S. C. Wofsy, J. G. Anderson, B. C. Daube, Jr., A. E. Dessler, M. Loewenstein, M. P. McCormick, J. R. Podolske, E. M. Weinstock, and G. K. Yue, "Measurements of Stratospheric Carbon Dioxide and Water Vapor at Northern Midlatitudes: Implications for Troposphere-to-Stratosphere Transport," *Geophys. Res. Lett.*, 22, 2737–2740 (1995).

Bojkov, R. D., "The 1979–1985 Ozone Decline in the Antarctic As Reflected in Ground Based Observations," *Geophys. Res. Lett.*, 13, 1236–1239 (1986).

Borrmann, S., S. Solomon, J. E. Dye, and B. Luo, "The Potential of Cirrus Clouds for Heterogeneous Chlorine Activation," *Geophys Res. Lett.*, 23, 2133–2136 (1996).

Borrmann, S., S. Solomon, L. Avallone, D. Toohey, and D. Baumgardner, "On the Occurrence of ClO in Cirrus Clouds and Volcanic Aerosol in the Tropopause Region," *Geophys. Res. Lett.*, 24, 2011–2014 (1997a).

Borrmann, S., S. Solomon, J. E. Dye, D. Baumgardner, K. K. Kelly, and K. R. Chan, "Heterogeneous Reactions on Stratospheric Background Aerosols, Volcanic Sulfuric Acid Droplets, and Type I Polar Stratospheric Clouds: Effects of Temperature Fluctuations and Differences in Particle Phase," *J. Geophys. Res.*, 102, 3639–3648 (1997b).

Brasseur, G. P., J. C. Gille, and S. Madronich, "Ozone Depletion," in *Future Climates of the World: A Modelling Perspective* (A. Henderson-Sellers, Ed.), Chapter 11, pp. 399–432, Elsevier, Amsterdam, 1995.

Brasseur, G. P., J.-F. Müller, and C. Granier, "Atmospheric Impact of NO_x Emissions by Subsonic Aircraft: A Three-Dimensional Model Study," *J. Geophys. Res.*, 101, 1423–1428 (1996).

Brasseur, G. P., X. Tie, P. J. Rasch, and F. Lefèvre, "A Three-Dimensional Simulation of the Antarctic Ozone Hole: Impact of Anthropogenic Chlorine on the Lower Stratosphere and Upper Troposphere," *J. Geophys. Res.*, 102, 8909–8930 (1997).

Brasseur, G. P., F. Lefèvre, and A. K. Smith, "Chemical-Transport Models of the Atmosphere," in *Perspectives in Environmental Chemistry* (D. L. Macalady, Ed.), pp. 369–399, Oxford Univ. Press, New York, 1998.

Brogniez, C., J. Lenoble, R. Ramananahérisoa, K. H. Fricke, E. P. Shettle, K. W. Hoppel, R. M. Bevilacqua, J. S. Hornstein, J. Lumpe, M. D. Fromm, and S. S. Krigman, "Second European Stratospheric Arctic and Midlatitude Experiment Campaign: Correlative Measurements of Aerosol in the Northern Polar Atmosphere," *J. Geophys. Res.*, 102, 1489–1494 (1997).

Brooks, S. B., M. J. Lewis, and R. R. Dickerson, "Nitric Oxide Emissions from the High-Temperature Viscous Boundary Layers of Hypersonic Aircraft within the Stratosphere," *J. Geophys. Res.*, 98, 16755–16760 (1993).

Browell, E. V., C. F. Butler, S. Ismail, P. A. Robinette, A. F. Carter, N. S. Higdon, O. B. Toon, M. R. Schoeberl, and A. F. Tuck, "Airborne Lidar Observations in the Wintertime Arctic Stratosphere: Polar Stratospheric Clouds," *Geophys. Res. Lett.*, 17, 385–388 (1990).

Browell, E. V., C. F. Butler, M. A. Fenn, W. B. Grant, S. Ismail, M. R. Schoeberl, O. B. Toon, M. Loewenstein, and J. R. Podolske, "Ozone and Aerosol Changes during the 1991–1992 Airborne Arctic Stratospheric Expedition," *Nature*, 261, 1155–1158 (1993).

Brown, L. A., V. Vaida, D. R. Hanson, J. D. Graham, and J. T. Roberts, "Uptake of Chlorine Dioxide by Model PSCs under Stratospheric Conditions," *J. Phys. Chem.*, 100, 3121–3125 (1996).

Brown, R. C., M. R. Anderson, R. C. Miake-Lye, C. E. Kolb, A. A. Sorokin, and Y. Y. Buriko, "Aircraft Exhaust Sulfur Emissions," *Geophys. Res. Lett.*, 23, 3603–3606 (1996a).

Brown, R. C., R. C. Miake-Lye, M. R. Anderson, and C. E. Kolb, "Effect of Aircraft Exhaust Sulfur Emissions on Near Field Plume Aerosols," *Geophys. Res. Lett.*, 23, 3607–3610 (1996b).

Brown, S. S., R. K. Talukdar, and A. R. Ravishankara, "Reconsideration of the Rate Constant for the Reaction of Hydroxyl Radicals with Nitric Acid," *J. Phys. Chem. A*, 103, 3031–3037 (1999).

Brühl, C., P. J. Crutzen, and J.-U. Grooß, "High-Latitude, Summertime NO_x Activation and Seasonal Ozone Decline in the Lower Stratosphere: Model Calculations Based on Observations by HALOE on UARS," *J. Geophys. Res.*, 103, 3587–3597 (1998).

Brune, W. H., J. G. Anderson, D. W. Toohey, D. W. Fahey, S. R. Kawa, R. L. Jones, D. S. McKenna, and L. R. Poole, "The Potential for Ozone Depletion in the Arctic Polar Stratosphere," *Science*, 252, 1260–1266 (1991).

Brune, W. H., "Stratospheric Chemistry—Perspectives in Environmental Chemistry," in *Perspectives in Environmental Chemistry* (D. L. Macalady, Ed.), pp. 292–324, Oxford Univ. Press, New York, 1998.

Burley, J. D., and H. S. Johnston, "Nitrosyl Sulfuric Acid and Stratospheric Aerosols," *Geophys. Res. Lett.*, 19, 1363–1366 (1992a).

Burley, J. D., and H. S. Johnston, "Ionic Mechanisms for Heterogeneous Stratospheric Reactions and Ultraviolet Photoabsorption Cross Sections for NO_2^+, HNO_3, and NO_3^- in Sulfuric Acid," *Geophys. Res. Lett.*, 19, 1359–1362 (1992b).

Butler, J. H., "The Potential Role of the Ocean in Regulating Atmospheric CH_3Br," *Geophys. Res. Lett.*, 21, 185–188 (1994).

Butler, J. H., "Methyl Bromide under Scrutiny," *Nature*, 376, 469–470 (1995).

Callis, L. B., and M. Natarajan, "The Antarctic Ozone Minimum: Relationship to Odd Nitrogen, Odd Chlorine, the Final Warming, and the 11-Year Solar Cycle," *J. Geophys. Res.*, 91, 10771–10796 (1986).

Caloz, F., F. F. Fenter, K. D. Tabor, and M. J. Rossi, "I: Design and Construction of a Knudsen-Cell Reactor for the Study of Heterogeneous Reactions over the Temperature Range 130–750 K, Performances and Limitations," *Rev. Sci. Instrum.*, 68, 3172–3179 (1997).

Calvert, J. G., and Pitts, J. N., Jr., *Photochemistry*, Wiley, New York, 1966.

Cantrell, C. A., R. E. Shetter, and J. G. Calvert, "Branching Ratios for the $O(^1D) + N_2O$ Reaction," *J. Geophys. Res.*, 99, 3739–3743 (1994).

Carleton, K. L., D. M. Sonnenfroh, W. T. Rawlins, B. E. Wyslouzil, and S. Arnold, "Freezing Behavior of Single Sulfuric Acid Aerosols Suspended in a Quadrupole Trap," *J. Geophys. Res.*, 102, 6025–6033 (1997).

Carpenter, L. J., W. T. Sturges, S. A. Penkett, P. S. Liss, B. Alicke, K. Hebestreit, and U. Platt, "Short-Lived Alkyl Iodides and

Bromides at Mace Head, Ireland: Links to Biogenic Sources and Halogen Oxide Production," *J. Geophys. Res., 104,* 1679–1689 (1999).

Carslaw, K. S., B. P. Luo, S. L. Clegg, Th. Peter, P. Brimblecombe, and P. J. Crutzen, "Stratospheric Aerosol Growth and HNO_3 Gas Phase Depletion from Coupled HNO_3 and Water Uptake by Liquid Particles," *Geophys. Res. Lett., 21,* 2479–2482 (1994).

Carslaw, K. S., M. Wirth, A. Tsias, B. P. Luo, A. Dörnbrack, M. Leutbecher, H. Volkert, W. Renger, J. T. Bacmeister, E. Reimers, and Th. Peter, "Increased Stratospheric Ozone Depletion Due to Mountain-Induced Atmospheric Waves," *Nature, 391,* 675–678 (1998a).

Carslaw, K. S., M. Wirth, A. Tsias, B. P. Luo, A. Dörnbrack, M. Leutbecher, H. Volkert, W. Renger, J. T. Bacmeister, and T. Peter, "Particle Microphysics and Chemistry in Remotely Observed Mountain Polar Stratospheric Clouds," *J. Geophys. Res., 103,* 5785–5796 (1998b).

Chandra, S., "Changes in Stratospheric Ozone and Temperature Due to the Eruptions of Mt. Pinatubo," *Geophys. Res. Lett., 20,* 33–36 (1993).

Chandra, S., and R. S. Stolarski, "Recent Trends in Stratospheric Total Ozone: Implications of Dynamical and El Chichon Perturbations," *Geophys. Res. Lett., 18,* 2277–2280 (1991).

Chapman, S., "A Theory of Upper-Atmospheric Ozone," in *Memoirs of the Royal Meteorological Society,* Vol. III, No. 26 (1930).

Chartrand, D. J., and J. C. McConnell, "Evidence for HBr Production Due to Minor Channel Branching at Mid-Latitudes," *Geophys. Res. Lett., 25,* 55–58 (1998).

Chase, M. W., "NIST-JANAF Thermochemical Tables for the Bromine Oxides," *J. Phys. Chem. Ref. Data, 25,* 1069–1111 (1996).

Chen, J., V. Catoire, and H. Niki, "Mechanistic Study of the $BrCH_2O$ Radical Degradation in 700 Torr Air," *Chem. Phys. Lett., 245,* 519–528 (1995).

Chen, T.-Y., D. R. Blake, J. P. Lopez, and F. S. Rowland, "Estimation of Global Vehicular Methyl Bromide Emissions: Extrapolation from a Case Study in Santiago, Chile," *Geophys. Res. Lett., 26,* 283–286 (1999).

Chin, M., and D. D. Davis, "Global Sources and Sinks of OCS and CS_2 and Their Distributions," *Global Biogeochem. Cycles, 7,* 321–337 (1993).

Chin, M., and D. D. Davis, "A Reanalysis of Carbonyl Sulfide as a Source of Stratospheric Background Sulfur Aerosol," *J. Geophys. Res., 100,* 8993–9005 (1995).

Chipperfield, M. P., L. J. Gray, J. S. Kinnersley, and J. Zawodny, "A Two-Dimensional Model Study of the QBO Signal in SAGE II NO_2 and O_3," *Geophys. Res. Lett., 21,* 589–592 (1994).

Chipperfield, M. P., D. E. Shallcross, and D. J. Lary, "A Model Study of the Potential Role of the Reaction BrO + OH in the Production of Stratospheric HBr," *Geophys. Res. Lett., 24,* 3025–3028 (1997).

Chu, L. T., M.-T. Leu, and L. F. Keyser, "Uptake of HCl in Water Ice and Nitric Acid Ice Films," *J. Phys. Chem., 97,* 7779–7785 (1993).

Chu, L. T., and J. W. Heron, "Uptake of HBr on Ice at Polar Atmospheric Conditions," *Geophys. Res. Lett., 22,* 3211–3214 (1995).

Cicerone, R. J., R. S. Stolarski, and S. Walters, "Stratospheric Ozone Destruction by Man-made Chlorofluoromethanes," *Science, 185,* 1165–1167 (1974).

Cicerone, R. J., "Halogens in the Atmosphere," *Rev. Geophys. Space Phys., 19,* 123–139 (1981).

Cicerone, R. J., "Changes in Stratospheric Ozone," *Science, 237,* 35–42 (1987).

Cicerone, R. J., "Fires, Atmospheric Chemistry, and the Ozone Layer," *Science, 263,* 1243–1244 (1994).

Coffey, M. T., and W. G. Mankin, "Observations of the Loss of Stratospheric NO_2 Following Volcanic Eruptions," *Geophys. Res. Lett., 20,* 2873–2876 (1993).

Colman, J. J., D. R. Blake, and F. S. Rowland, "Atmospheric Residence Time of CH_3Br Estimated from the Junge Spatial Variability Relation," *Science, 281,* 392–396 (1998).

Conklin, M. H., and R. C. Bales, "SO_2 Uptake on Ice Spheres: Liquid Nature of the Ice–Air Interface," *J. Geophys. Res., 98,* 16851–16855 (1993).

Considine, D. B., A. R. Douglass, and C. H. Jackman, "Sensitivity of Two-Dimensional Model Predictions of Ozone Response to Stratospheric Aircraft: An Update," *J. Geophys. Res., 100,* 3075–3090 (1995).

Cox, R. A., and G. D. Hayman, "The Stability and Photochemistry of Dimers of the ClO Radical and Implications for Antarctic Ozone Depletion," *Nature, 332,* 796–800 (1988).

Cronkhite, J. M., R. E. Stickel, J. M. Nicovich, and P. H. Wine, "Laser Flash Photolysis Studies of Radical–Radical Reaction Kinetics: The HO_2 + BrO Reaction," *J. Phys. Chem. A, 102,* 6651–6658 (1998).

Crowley, J. N., F. Helleis, R. Müller, G. K. Moortgat, P. J. Crutzen, and J. J. Orlando, "CH_3OCl: UV/Visible Absorption Cross Sections, J Values and Atmospheric Significance," *J. Geophys. Res., 99,* 20683–20688 (1994).

Crutzen, P. J., "Determination of Parameters Appearing in the 'Dry' and 'Wet' Photochemical Theories for Ozone in the Stratosphere," *Tellus, 21,* 368–388 (1969).

Crutzen, P. J., "The Influence of Nitrogen Oxides on the Atmospheric Ozone Content," *Q. J. Roy. Meteorol. Soc., 96,* 320–325 (1970).

Crutzen, P. J., "Ozone Production Rates in an Oxygen–Hydrogen–Nitrogen Oxide Atmosphere," *J. Geophys. Res., 76,* 7311–7327 (1971).

Crutzen, P. J., "A Threat to the Earth's Ozone Shield," *Ambio, 1,* 41–51 (1972).

Crutzen, P. J., "A Review of Upper Atmospheric Photochemistry," *Can. J. Chem., 52,* 1569–1581 (1974).

Crutzen, P. J., "The Possible Importance of CSO for the Sulfate Layer of the Stratosphere," *Geophys. Res. Lett., 3,* 73–76 (1976).

Crutzen, P. J., and F. Arnold, "Nitric Acid Cloud Formation in the Cold Antarctic Stratosphere: A Major Cause for the Springtime 'Ozone Hole,'" *Nature, 324,* 651–655 (1986).

Crutzen, P. J., R. Müller, C. Brühl, and Th. Peter, "On the Potential Importance of the Gas Phase Reaction CH_3O_2 + ClO → ClOO + CH_3O and the Heterogeneous Reaction HOCl + HCl → H_2O + Cl_2 in 'Ozone Hole' Chemistry," *Geophys. Res. Lett., 19,* 1113–1116 (1992).

Crutzen, P. J., J.-U. Grooß, C. Brühl, R. Müller, and J. M. Russell III, "A Reevaluation of the Ozone Budget with HALOE UARS Data: No Evidence for the Ozone Deficit," *Science, 268,* 705–708 (1995).

Crutzen, P. J., "Mesospheric Mysteries," *Science, 277,* 1951–1952 (1997).

Cunnold, D. M., R. F. Weiss, R. G. Prinn, D. Hartley, P. G. Simmonds, P. J. Fraser, B. Miller, F. N. Alyea, and L. Porter, "GAGE/AGAGE Measurements Indicating Reductions in Global Emissions of CCl_3F and CCl_2F_2 in 1992–1994," *J. Geophys. Res., 102,* 1259–1269 (1997).

Daële, V., and G. Poulet, "Kinetics and Products of the Reactions of CH_3O_2 with Cl and ClO," *J. Chim. Phys. Phys.-Chim. Biol., 93,* 1081–1099 (1996).

Dai, Q., G. N. Robinson, and A. Freedman, "Reactions of Halomethanes with γ-Alumina Surfaces. 1. An Infrared Spectroscopic Study," *J. Phys. Chem. B, 101,* 4940–4946 (1997).

Danilin, M. Yu., "Local Stratospheric Effects of Solid-Fueled Rocket Emissions," *Ann. Geophys., 11,* 828–836 (1993).

Danilin, M. Y., N.-D. Sze, M. K. W. Ko, J. M. Rodriguez, and M. J. Prather, "Bromine–Chlorine Coupling in the Antarctic Ozone Hole," *Geophys. Res. Lett., 23,* 153–156 (1996).

Daumont, D., J. Brion, J. Charbonnier, and J. Malicet, "Ozone UV Spectroscopy. I: Absorption Cross-Sections at Room Temperature," *J. Atmos. Chem., 15,* 145–155 (1992).

David, S. J., F. J. Murcray, A. Goldman, C. P. Rinsland, and D. G. Murcray, "The Effect of the Mt. Pinatubo Aerosol on the HNO_3 Column over Mauna Loa, Hawaii," *Geophys. Res. Lett., 21,* 1003–1006 (1994).

Davis, D., J. Crawford, S. Liu, S. McKeen, A. Bandy, D. Thornton, F. Rowland, and D. Blake, "Potential Impact of Iodine on Tropospheric Levels of Ozone and Other Critical Oxidants," *J. Geophys. Res., 101,* 2135–2147 (1996).

De Haan, D. O., I. Fløisand, and F. Stordal, "Modeling Studies of the Effects of the Heterogeneous Reaction $ClOOCl + HCl \rightarrow Cl_2 + HOOCl$ on Stratospheric Chlorine Activation and Ozone Depletion," *J. Geophys. Res., 102,* 1251–1258 (1997).

De Mazière, M., M. Van Roozendael, C. Hermans, P. C. Simon, P. Demoulin, G. Roland, and R. Zander, "Quantitative Evaluation of the Post-Mount Pinatubo NO_2 Reduction and Recovery, Based on 10 Years of Fourier Transform Infrared and UV-Visible Spectroscopic Measurements at Jungfraujoch," *J. Geophys. Res., 103,* 10849–10858 (1998).

Del Negro, L. A., D. W. Fahey, S. G. Donnelly, R. S. Gao, E. R. Keim, R. C. Wamsley, E. L. Woodbridge, J. E. Dye, D. Baumgardner, B. W. Gandrud, J. C. Wilson, H. H. Jonsson, M. Loewenstein, J. R. Podolske, C. R. Webster, R. D. May, D. R. Worsnop, A. Tabazadeh, M. A. Tolbert, K. K. Kelly, and K. R. Chan, "Evlauating the Role of NAT, NAD, and Liquid $H_2SO_4/H_2O/HNO_3$ Solutions in Antarctic Polar Stratospheric Cloud Aerosol: Observations and Implications," *J. Geophys. Res., 102,* 13255–13282 (1997).

Delzeit, L., B. Rowland, and J. P. Devlin, "Infrared Spectra of HCl Complexed/Ionized in Amorphous Hydrates and at Ice Surfaces in the 15–90 K Range," *J. Phys Chem., 97,* 10312–10318 (1993).

DeMore, W. B., S. P. Sander, D. M. Golden, R. F. Hampson, M. J. Kurylo, C. J. Howard, A. R. Ravishankara, C. E. Kolb, and M. J. Molina, "Chemical Kinetics and Photochemical Data for Use in Stratospheric Modeling," in JPL Publication 97-4, Jet Propulsion Laboratory, Pasadena, CA, January 15, 1997.

Denison, M. R., J. J. Lamb, W. D. Bjorndahl, E. Y. Wong, and P. D. Lohn, "Solid Rocket Exhaust in the Stratosphere: Plume Diffusion and Chemical Reactions," *J. Spacecraft Rockets, 31,* 435–442 (1994).

Derwent, R. G., P. G. Simmonds, S. O'Doherty, and D. B. Ryall, "The Impact of the Montreal Protocol on Halocarbon Concentrations in Northern Hemisphere Baseline and European Air Masses at Mace Head, Ireland over a Ten Year Period from 1987–1996," *Atmos. Environ., 32,* 3689–3702 (1998).

Deshler, T., D. J. Hofmann, B. J. Johnson, and W. R. Rozier, "Balloonborne Measurements of the Pinatubo Aerosol Size Distribution and Volatility at Laramie, Wyoming during the Summer of 1991," *Geophys. Res. Lett., 19,* 199–202 (1992).

Deshler, T., B. J. Johnson, and W. R. Rozier, "Balloonborne Measurements of Pinatubo Aerosol during 1991 and 1992 at 41°N: Vertical Profiles, Size Distribution, and Volatility," *Geophys. Res. Lett., 20,* 1435–1438 (1993).

Deshler, T., Th. Peter, R. Müller, and P. Crutzen, "The Lifetime of Leewave-Induced Ice Particles in the Arctic Stratosphere: I. Balloonborne Observations," *Geophys. Res. Lett., 21,* 1327–1330 (1994).

Deshler, T., B. J. Johnson, D. J. Hofmann, and B. Nardi, "Correlations between Ozone Loss and Volcanic Aerosol at Altitudes below 14 km over McMurdo Station, Antarctica," *Geophys. Res. Lett., 23,* 2931–2934 (1996).

Dessler, A. E., R. M. Stimpfle, B. C. Daube, R. J. Salawitch, E. M. Weinstock, D. M. Judah, J. D. Burley, J. W. Munger, S. C. Wofsy, J. G. Anderson, M. P. McCormick, and W. P. Chu, "Balloon-Borne Measurements of ClO, NO, and O_3 in a Volcanic Cloud: An Analysis of Heterogeneous Chemistry between 20 and 30 km," *Geophys. Res. Lett., 20,* 2527–2530 (1993).

Dessler, A. E., D. B. Considine, G. A. Morris, M. R. Schoeberl, J. M. Russell, III, A. E. Roche, J. B. Kumer, J. L. Mergenthale, J. W. Waters, J. C. Gille, and G. K. Yue, "Correlated Observations of HCl and $ClONO_2$ from UARS and Implications for Stratospheric Chlorine Partitioning," *Geophys. Res. Lett., 22,* 1721–1724 (1995a).

Dessler, A. E., E. J. Hintsa, E. M. Weinstock, J. G. Anderson, and K. R. Chan, "Mechanisms Controlling Water Vapor in the Lower Stratosphere: 'A Tale of Two Stratospheres,'" *J. Geophys. Res., 100,* 23167–23172 (1995b).

Dessler, A. E., S. R. Kawa, D. B. Considine, J. W. Waters, L. Froidevaux, and J. B. Kumer, "UARS Measurements of ClO and NO_2 at 40 and 46 km and Implications for the Model 'Ozone Deficit,'" *Geophys. Res. Lett., 23,* 339–342 (1996).

Dessler, A. E., D. B. Considine, J. E. Rosenfield, S. R. Kawa, A. R. Douglass, and J. M. Russell, III, "Lower Stratospheric Chlorine Partitioning during the Decay of the Mt. Pinatubo Aerosol Cloud," *Geophys. Res. Lett., 24,* 1623–1626 (1997).

Dessler, A. E., "A Re-examination of the 'Stratospheric Fountain' Hypothesis," *Geophys. Res. Lett., 25,* 4165–4168 (1998).

De Zafra, R. L., V. Chan, S. Crewell, C. Trimble, and J. M. Reeves, "Millimeter Wave Spectroscopic Measurements over the South Pole. 3. The Behavior of Stratospheric Nitric Acid through Polar Fall, Winter, and Spring," *J. Geophys. Res., 102,* 1399–1410 (1997).

Dominé, F., E. Thibert, F. Van Landeghem, E. Silvente, and P. Wagnon, "Diffusion and Solubility of HCl in Ice: Preliminary Results," *Geophys. Res. Lett., 21,* 601–604 (1994).

Donaldson, D. J., A. R. Ravishankara, and D. R. Hanson, "Detailed Study of $HOCl + HCl \rightarrow Cl_2 + H_2O$ in Sulfuric Acid," *J. Phys. Chem. A, 101,* 4717–4725 (1997).

Dubey, M. K., M. P. McGrath, G. P. Smith, and F. S. Rowland, "HCl Yield from OH + ClO: Stratospheric Model Sensitivities and Elementary Rate Theory Calculations," *J. Phys. Chem. A, 102,* 3127–3133 (1998).

Dunn, R. C., B. N. Flanders, and J. D. Simon, "Solvent Effects on the Spectroscopy and Ultrafast Photochemistry of Chlorine Dioxide," *J. Phys. Chem., 99,* 7360–7370 (1995).

Dutton, E. G., and J. R. Christy, "Solar Radiative Forcing at Selected Locations and Evidence for Global Lower Tropospheric Cooling Following the Eruptions of El Chichon and Pinatubo," *Geophys. Res. Lett., 19,* 2313–2316 (1992).

Dye, J. E., D. Baumgardner, B. W. Gandrud, S. R. Kawa, K. K. Kelly, M. Loewenstein, G. V. Ferry, K. R. Chan, and B. L. Gary, "Particle Size Distributions in Arctic Polar Stratospheric Clouds, Growth, and Freezing of Sulfuric Acid Droplets, and Implications for Cloud Formation," *J. Geophys. Res., 97,* 8015–8034 (1992).

Dye, J. E., D. Baumgardner, B. W. Gandrud, K. Drdla, K. Barr, D. W. Fahey, L. A. Del Negro, A. Tabazadeh, H. H. Jonsson, J. C. Wilson, M. Loewenstein, J. R. Podolske, and K. R. Chan, "*In-situ* Observations of an Antarctic Polar Stratospheric Cloud: Similarities with Arctic Observations," *Geophys. Res. Lett., 23,* 1913–1916 (1996).

Edouard, S., B. Legras, F. Lefèvre, and R. Eymard, "The Effect of Small-Scale Inhomogeneities on Ozone Depletion in the Arctic," *Nature, 384,* 444–447 (1996).

Ehhalt, D. H., F. Rohrer, and A. Wahner, "Sources and Distribution of NO_x in the Upper Troposphere at Northern Midlatitudes," *J. Geophys. Res., 97,* 3725–3738 (1992).

Ellerman, T., J. Sehested, O. J. Nielsen, P. Pagsberg, and T. J. Wallington, "Kinetics of the Reaction of F Atoms with O_2 and UV Spectrum of FO_2 Radicals in the Gas Phase at 295 K," *Chem. Phys. Lett., 218,* 287–294 (1994).

Elliott, S., R. P. Turco, O. B. Toon, and P. Hamill, "Application of Physical Adsorption Thermodynamics to Heterogeneous Chemistry on Polar Stratospheric Clouds," *J. Atmos. Chem., 13,* 211–224 (1991).

Elliott, S. M., and F. S. Rowland, "Nucleophilic Substitution Rates and Solubilities for Methyl Halides in Seawater," *Geophys. Res. Lett., 20,* 1043–1046 (1993).

Elrod, M. J., R. E. Koch, J. E. Kim, and M. J. Molina, "HCl Vapour Pressures and Reaction Probabilities for $ClONO_2$ + HCl on Liquid H_2SO_4–HNO_3–HCl–H_2O Solutions," *Faraday Discuss., 100,* 269–278 (1995).

Eluszkiewicz, J., and M. Allen, "A Global Analysis of the Ozone Deficit in the Upper Stratosphere and Lower Mesosphere," *J. Geophys. Res., 98,* 1069–1082 (1993).

Emmons, L. K., D. T. Shindell, J. M. Reeves, and R. L. de Zafra, "Stratospheric ClO Profiles from McMurdo Station, Antarctica, Spring 1992," *J. Geophys. Res., 100,* 3049–3055 (1995).

Erle, F., A. Grendel, D. Perner, U. Platt, and K. Pfeilsticker, "Evidence of Heterogeneous Bromine Chemistry on Cold Stratospheric Sulphate Aerosols," *Geophys. Res. Lett., 25,* 4329–4332 (1998).

Etheridge, D. M., L. P. Steele, R. J. Francey, and R. L. Langenfelds, "Atmospheric Methane between 1000 A.D. and Present: Evidence of Anthropogenic Emissions and Climatic Variability," *J. Geophys. Res., 103,* 15970–15993 (1998).

Evans, S. J., R. Toumi, J. E. Harries, M. P. Chipperfield, and J. M. Russell III, "Trends in Stratospheric Humidity and the Sensitivity of Ozone to These Trends," *J. Geophys. Res., 103,* 8715–8725 (1998).

Fahey, D. W., K. K. Kelly, G. V. Ferry, L. R. Poole, J. C. Wilson, D. M. Murphy, M. Loewenstein, and K. R. Chan, "*In Situ* Measurements of Total Reactive Nitrogen, Total Water, and Aerosol in a Polar Stratospheric Cloud in the Antarctic," *J. Geophys. Res., 94,* 11299–11315 (1989).

Fahey, D. W., K. K. Kelly, S. R. Kawa, A. F. Tuck, M. Loewenstein, K. R. Chan, and L. E. Heidt, "Observations of Denitrification and Dehydration in the Winter Polar Stratospheres," *Nature, 344,* 321–324 (1990).

Fahey, D. W., S. R. Kawa, E. L. Woodbridge, P. Tin, J. C. Wilson, H. H. Jonsson, J. E. Dye, D. Baumgardner, S. Borrmann, D. W. Toohey, L. M. Avallone, M. H. Proffitt, J. Margitan, M. Loewenstein, J. R. Podolske, R. J. Salawitch, S. C. Wofsy, M. K. W. Ko, D. E. Anderson, M. R. Schoeberl, and K. R. Chan, "*In Situ* Measurements Constraining the Role of Sulphate Aerosols in Mid-latitude Ozone Depletion," *Nature, 363,* 509–514 (1993).

Fahey, D. W., E. R. Keim, K. A. Boering, C. A. Brock, J. C. Wilson, S. Anthony, T. F. Hanisco, P. O. Wennberg, R. C. Miake-Lye, R. J. Salawitch, N. Louisnard, E. L. Woodbridge, R. S. Gao, S. G. Donnelly, R. C. Wamsley, L. A. Del Negro, B. C. Daube, S. C. Wofsy, C. R. Webster, R. D. May, K. K. Kelly, M. Loewenstein, J. R. Podolske, and K. R. Chan, "Emission Measurements of the Concorde Supersonic Aircraft in the Lower Stratosphere," *Science, 270,* 70–74 (1995a).

Fahey, D. W., E. R. Keim, E. L. Woodbridge, R. S. Gao, K. A. Boering, B. C. Daube, S. C. Wofsy, R. P. Lohmann, E. J. Hintsa, A. E. Dessler, C. R. Webster, R. D. May, C. A. Brock, J. C. Wilson, R. C. Miake-Lye, R. C. Brown, J. M. Rodriguez, M. Loewenstein, M. H. Proffitt, R. M. Stimpfle, S. W. Bowen, and K. R. Chan, "*In Situ* Observations in Aircraft Exhaust Plumes in the Lower Stratosphere at Midlatitudes," *J. Geophys. Res., 100,* 3065–3074 (1995b).

Fahey, D. W., S. G. Donnelly, E. R. Keim, R. S. Gao, R. C. Wamsley, L. A. Del Negro, E. L. Woodbridge, M. H. Proffitt, K. H. Rosenlof, M. K. W. Ko, D. K. Weisenstein, C. J. Scott, C. Nevison, S. Solomon, and K. R. Chan, "*In Situ* Observations of NO_y, O_3, and the NO_y/O_3 Ratio in the Lower Stratosphere," *Geophys. Res. Lett., 23,* 1653–1656 (1996).

Farman, J. C., B. G. Gardiner, and J. D. Shanklin, "Large Losses of Total Ozone in Antarctica Reveal Seasonal ClO_x/NO_x Interaction," *Nature, 315,* 207–210 (1985).

Finkbeiner, M., J. N. Crowley, O. Horie, R. Müller, G. K. Moortgat, and P. J. Crutzen, "Reaction between HO_2 and ClO: Product Formation between 210 and 300 K," *J. Phys. Chem., 99,* 16264–16275 (1995).

Finlayson-Pitts, B. J., M. J. Ezell, and J. N. Pitts, Jr., "Formation of Chemically Active Chlorine Compounds by Reactions of Atmospheric NaCl Particles with Gaseous N_2O_5 and $ClONO_2$," *Nature, 337,* 241–244 (1989).

Finlayson-Pitts, B. J., "Chlorine Atoms as a Potential Tropospheric Oxidant in the Marine Boundary Layer," *Res. Chem. Intermed., 19,* 235–249 (1993).

Fiocco, G., D. Fua, and G. Visconti, Eds., *The Effects of Mt. Pinatubo Eruption on the Atmosphere and Climate,* NATO ASI Series Volume 42, Subseries I, Global Environmental Change, Springer-Verlag, Berlin/New York, 1996.

Fish, D. J., and M. R. Burton, "The Effect of Uncertainties in Kinetic and Photochemical Data on Model Predictions of Stratospheric Ozone Depletion," *J. Geophys. Res., 102,* 25537–25542 (1997).

Flatøy, F., and Ø. Hov, "Three-Dimensional Model Studies of the Effect of NO_x Emissions from Aircraft on Ozone in the Upper Troposphere over Europe and the North Atlantic," *J. Geophys. Res., 101,* 1401–1422 (1996).

Folkins, I., and C. Appenzeller, "Ozone and Potential Vorticity at the Subtropical Tropopause Break," *J. Geophys. Res., 101,* 18787–18792 (1996).

Fox, L. E., D. R. Worsnop, M. S. Zahniser, and S. C. Wofsy, "Metastable Phases in Polar Stratospheric Aerosols," *Science, 267,* 351–355 (1995).

Francis, P., A. Maciejewski, C. Oppenheimer, C. Chaffin, and T. Caltabiano, "SO_2:HCl Ratios in the Plumes from Mt. Etna and Volcano Determined by Fourier Transform Spectoscopy," *Geophys. Res. Lett., 22,* 1717–1720 (1995).

Freemantle, M., "Search for Halon Replacements Stymied by Complexities of Fires," *Chem. Eng. News,* January 30, 1995, pp. 25–31.

Gäb, S., J. Schmitzer, H. W. Thamm, H. Parlar, and F. Korte, "Photomineralization Rate of Organic Compounds Adsorbed on Particulate Matter," *Nature, 270,* 331–333 (1977).

Gäb, S., J. Schmitzer, H. W. Thamm, and F. Korte, "Mineralization of Chlorofluorocarbons in the Sunlight of the Troposphere," *Angew. Chem., Int. Ed. Engl., 17,* 366–367 (1978).

Gan, J., S. R. Yates, H. D. Ohr, and J. J. Sims, "Production of Methyl Bromide by Terrestrial Higher Plants," *Geophys. Res. Lett., 25,* 3595–3598 (1998).

Gandrud, B. W., J. E. Dye, D. Baumgardner, G. V. Ferry, M. Loewenstein, K. R. Chan, L. Sanford, B. Gary, and K. Kelly, "The January 30, 1989 Arctic Polar Stratospheric Clouds (PSC) Event: Evidence for a Mechanism of Dehydration," *Geophys. Res. Lett., 17,* 457–460 (1990).

Garcia, R. R., and S. Solomon, "A Possible Relationship between Interannual Variability in Antarctic Ozone and the Quasi-biennial Oscillation," *Geophys. Res. Lett., 14,* 848–851 (1987).

Garcia, R. R., and S. Solomon, "A New Numerical Model of the Middle Atmosphere. 2. Ozone and Related Species," *J. Geophys. Res., 99,* 12937–12951 (1994).

Gellene, G. I., "An Explanation for Symmetry-Induced Isotopic Fractionation in Ozone," *Science, 274,* 1344–1346 (1996).

Gertner, B. J., and J. T. Hynes, "Molecular Dynamics Simulation of Hydrochloric Acid Ionization at the Surface of Stratospheric Ice," *Science, 271,* 1563–1566 (1996).

Gilles, M. K., A. A. Turnipseed, J. B. Burkholder, A. R. Ravishankara, and S. Solomon, "Kinetics of the IO Radical. 2. Reaction of IO with BrO," *J. Phys. Chem. A, 101,* 5526–5534 (1997).

Gleason, J. F., P. K. Bhartia, J. R. Herman, R. McPeters, P. Newman, R. S. Stolarski, L. Flynn, G. Labow, D. Larko, C. Seftor, C. Wellemeyer, W. D. Komhyr, A. J. Miller, and W. Planet, "Record Low Global Ozone in 1992," *Science, 260,* 523–526 (1993).

Gobbi, G. P., D. Di Donfrancesco, and A. Adriani, "Physical Properties of Stratospheric Clouds during the Antarctic Winter of 1995," *J. Geophys. Res., 103,* 10859–10873 (1998).

Goodman, J., K. G. Snetsinger, R. F. Pueschel, G. V. Ferry, and S. Verma, "Evolution of Pinatubo Aerosol near 19 km Altitude over Western North America," *Geophys. Res. Lett., 21,* 1129–1132 (1994).

Graham, J. D., and J. T. Roberts, "Interaction of Hydrogen Chloride with an Ultrathin Ice Film: Observation of Adsorbed and Absorbed States," *J. Phys. Chem., 98,* 5974–5983 (1994).

Graham, J. D., and J. T. Roberts, "Interaction of HCl with Crystalline and Amorphous Ice: Implications for the Mechanisms of Ice-Catalyzed Reactions," *Geophys. Res. Lett., 22,* 251–254 (1995).

Graham, J. D., J. T. Roberts, L. A. Brown, and V. Vaida, "Uptake of Chlorine Dioxide by Model Polar Stratospheric Cloud Surfaces: Ultrahigh-Vacuum Studies," *J. Phys. Chem., 100,* 3115–3120 (1996a).

Graham, J. D., J. T. Roberts, L. D. Anderson, and V. H. Grassian, "The 367 nm Photochemistry of OClO Thin Films and OClO Adsorbed on Ice," *J. Phys. Chem., 100,* 19551–19558 (1996b).

Grainger, R. G., A. Lambert, F. W. Taylor, J. J. Remedios, C. D. Rodgers, M. Corney, and B. J. Kerridge, "Infrared Absorption by Volcanic Stratospheric Aerosols Observed by IAMS," *Geophys. Res. Lett., 20,* 1283–1286 (1993).

Grant, W. B., J. Fishman, E. V. Browell, V. G. Brackett, D. Nganga, A. Minga, B. Cros, R. E. Veiga, C. F. Butler, M. A. Fenn, and G. D. Nowicki, "Observations of Reduced Ozone Concentrations in the Tropical Stratosphere after the Eruption of Mt. Pinatubo," *Geophys. Res. Lett., 19,* 1109–1112 (1992).

Grant, W. B., E. V. Browell, J. Fishman, V. G. Brackett, R. E. Veiga, D. Nganga, A. Minga, B. Cros, C. F. Butler, M. A. Fenn, C. S. Long, and L. L. Stowe, "Aerosol-Associated Changes in Tropical Stratospheric Ozone Following the Eruption of Mount Pinatubo," *J. Geophys. Res., 99,* 8197–8211 (1994).

Greenblatt, G. D., and A. R. Ravishankara, "Laboratory Studies on the Stratospheric NO_x Production Rate," *J. Geophys. Res., 95,* 3539–3547 (1990).

Gribble, G. W., "The Natural Production of Chlorinated Compounds," *Environ. Sci. Technol., 28,* 310A–319A (1994).

Haas, B.-M., K. C. Crellin, K. T. Kuwata, and M. Okumura, "Reaction of Chloride Ions with Chlorine Nitrate and Its Implications for Stratospheric Chemistry," *J. Phys. Chem., 98,* 6740–6745 (1994).

Hamill, P., and O. B. Toon, "Polar Stratospheric Clouds and the Ozone Hole," *Physics Today, 44,* 34–42 (1991).

Hamill, P., A. Tabazadeh, S. Kinne, O. B. Toon, and R. P. Turco, "On the Growth of Ternary System $HNO_3/H_2SO_4/H_2O$ Aerosol Particles in the Stratosphere," *Geophys. Res. Lett., 23,* 753–756 (1996).

Hanisco, T. F., P. O. Wennberg, R. C. Cohen, J. G. Anderson, D. W. Fahey, E. R. Keim, R. S. Gao, R. C. Wamsley, S. G. Donnelly, L. A. Del Negro, R. J. Salawitch, K. K. Kelly, and M. H. Proffitt, "The Role of HO_x in Super- and Subsonic Aircraft Exhaust Plumes," *Geophys. Res. Lett., 24,* 65–68 (1997).

Hanning-Lee, M. A., B. B. Brady, L. R. Martin, and J. A. Syage, "Ozone Decomposition on Alumina: Implications for Solid Rocket Motor Exhaust," *Geophys. Res. Lett., 23,* 1961–1964 (1996).

Hanson, D., and K. Mauersberger, "Laboratory Studies of the Nitric Acid Trihydrate: Implications for the South Polar Stratosphere," *Geophys. Res. Lett., 15,* 855–858 (1988a).

Hanson, D., and K. Mauersberger, "Vapor Pressures of HNO_3/H_2O Solutions at Low Temperatures," *J. Phys. Chem., 92,* 6167–6170 (1988b).

Hanson, D. R., and A. R. Ravishankara, "The Reaction Probabilities of $ClONO_2$ and N_2O_5 on 40% to 75% Sulfuric Acid Solutions," *J. Geophys. Res., 96,* 17307–17314 (1991).

Hanson, D. R., and A. R. Ravishankara, "Heterogeneous Chemistry of HBr and HF," *J. Phys. Chem., 96,* 9441–9446 (1992a).

Hanson, D. R., and A. R. Ravishankara, "Investigation of the Reactive and Nonreactive Processes Involving $ClONO_2$ and HCl on Water and Nitric Acid Doped Ice," *J. Phys. Chem., 96,* 2682–2691 (1992b).

Hanson, D. R., and A. R. Ravishankara, "Reactions of Halogen Species on Ice Surfaces," in *The Tropospheric Chemistry of Ozone in the Polar Regions* (H. Niki and K. H. Becker, Eds.), NATO ASI Series, Vol. 17, Springer-Verlag, Berlin/New York, 1993a.

Hanson, D. R., and A. R. Ravishankara, "Reaction of $ClONO_2$ with HCl on NAT, NAD and Frozen Sulfuric Acid and Hydrolysis of N_2O_5 and $ClONO_2$ on Frozen Sulfuric Acid," *J. Geophys. Res., 98,* 22931–22936 (1993b).

Hanson, D. R., A. R. Ravishankara, and S. Solomon, "Heterogeneous Reactions in Sulfuric Acid Aerosols: A Framework for Model Calculations," *J. Geophys. Res., 99,* 3615–3629 (1994).

Hanson, D. R., and A. R. Ravishankara, "Heterogeneous Chemistry of Bromine Species in Sulfuric Acid under Stratospheric Conditions," *Geophys. Res. Lett., 22,* 385–388 (1995).

Hanson, D. R., and E. R. Lovejoy, "Heterogeneous Reactions in Liquid Sulfuric Acid: HOCl + HCl as a Model System," *J. Phys. Chem., 100,* 6397–6405 (1996).

Hanson, D. R., A. R. Ravishankara, and E. R. Lovejoy, "Reaction of $BrONO_2$ with H_2O on Submicron Sulfuric Acid Aerosol and the Implications for the Lower Stratosphere," *J. Geophys. Res., 101,* 9063–9069 (1996).

Hanson, D. R., "Reaction of N_2O_5 with H_2O on Bulk Liquids and on Particles and the Effect of Dissolved HNO_3," *Geophys. Res. Lett., 24,* 1087–1090 (1997).

Hanson, D. R., "Reaction of $ClONO_2$ with H_2O and HCl in Sulfuric Acid and $HNO_3/H_2SO_4/H_2O$ Mixtures," *J. Phys. Chem. A, 102,* 4794–4807 (1998).

Happell, J. D., and D. W. R. Wallace, "Methyl Iodide in the Greenland/Norwegian Seas and the Tropical Atlantic Ocean: Evidence for Photochemical Production," *Geophys. Res. Lett., 23,* 2105–2108 (1996).

Harwood, M. H., D. M. Rowley, R. A. Cox, and R. L. Jones, "Kinetics and Mechanism of the BrO Self-Reaction: Temperature- and Pressure-Dependent Studies," *J. Phys. Chem. A, 102*, 1790–1802 (1998).

Hayman, G. D., J. M. Davies, and R. A. Cox, "Kinetics of the Reaction ClO + ClO → Products and Its Potential Relevance to Antarctic Ozone," *Geophys. Res. Lett., 13*, 1347–1350 (1986).

Haynes, D. R., N. J. Tro, and S. M. George, "Condensation and Evaporation of H_2O on Ice Surfaces," *J. Phys. Chem., 96*, 8502–8509 (1992).

Helleis, F., H. B. Crowley, and G. K. Moortgat, "Temperature Dependent CH_3COCl Formation in the Reaction between CH_3O_2 and ClO," *Geophys. Res. Lett., 21*, 1795–1798 (1994).

Henriksen, K., S. H. H. Larsen, O. I. Shumilov, and B. Thorkelsson, "Ozone Variations in the Scandinavian Sector of the Arctic during the AASE Campaign and 1989," *Geophys. Res. Lett., 21*, 1775–1778 (1994).

Henson, B. F., K. R. Wilson, and J. M. Robinson, "A Physical Adsorption Model of the Dependence of $ClONO_2$ Heterogeneous Reactions on Relative Humidity," *Geophys. Res. Lett., 23*, 1021–1024 (1996).

Hintsa, E. J., P. A. Newman, H. H. Jonsson, C. R. Webster, R. D. May, R. L. Herman, L. R. Lait, M. R. Schoeberl, J. W. Elkins, P. R. Wamsley, G. S. Dutton, T. P. Bui, D. W. Kohn, and J. G. Anderson, "Dehydration and Denitrification in the Arctic Polar Vortex during the 1995–1996 Winter," *Geophys. Res. Lett., 25*, 501–504 (1998a).

Hintsa, E. J., K. A. Boering, E. M. Weinstock, J. G. Anderson, B. L. Gary, L. Pfister, B. C. Daube, S. C. Wofsy, M. Loewenstein, J. R. Podolske, J. J. Margitan, and T. P. Bui, "Troposphere-to-Stratosphere Transport in the Lowermost Stratosphere from Measurements of H_2O, CO_2, N_2O, and O_3," *Geophys. Res. Lett., 25*, 2655–2658 (1998b).

Hobbs, P. V., "Ice Physics," Clarendon, Oxford, 1973.

Hofmann, D. J., and S. Solomon, "Ozone Destruction through Heterogeneous Chemistry Following the Eruption of El Chichon," *J. Geophys. Res., 94*, 5029–5041 (1989).

Hofmann, D. J., "Increase in the Stratospheric Background Sulfuric Acid Aerosol Mass in the Past 10 Years," *Science, 248*, 996–1000 (1990).

Hofmann, D. J., and T. Deshler, "Stratospheric Cloud Observations during Formation of the Antarctic Ozone Hole in 1989," *J. Geophys. Res., 96*, 2897–2912 (1991).

Hofmann, D. J., and S. J. Oltmans, "Anomalous Antarctic Ozone during 1992: Evidence for Pinatubo Volcanic Aerosol Effects," *J. Geophys. Res., 98*, 18555–18561 (1993).

Hofmann, D. J., S. J. Oltmans, J. A. Lathrop, J. M. Harris, and H. Vömel, "Record Low Ozone at the South Pole in the Spring of 1993," *Geophys. Res. Lett., 21*, 421–424 (1994a).

Hofmann, D. J., S. J. Oltmans, W. D. Komhyr, J. M. Harris, J. A. Lathrop, A. O. Langford, T. Deshler, B. J. Johnson, A. Torres, and W. A. Matthews, "Ozone Loss in the Lower Stratosphere over the United States in 1992–1993: Evidence for Heterogeneous Chemistry on the Pinatubo Aerosol," *Geophys. Res. Lett., 21*, 65–68 (1994b).

Hoinka, K. P., and M. E. Reinhardt, "North Atlantic Air Traffic within the Lower Stratosphere: Cruising Times and Corresponding Emissions," *J. Geophys. Res., 98*, 23113–23131 (1993).

Holton, J. R., P. H. Haynes, M. E. McIntyre, A. R. Douglass, R. B. Rood, and L. Pfister, "Stratosphere–Troposphere Exchange," *Rev. Geophys., 33*, 403–439 (1995).

Höpfner, M., C. E. Blom, T. Blumenstock, H. Fischer, and T. Gulde, "Evidence for the Removal of Gaseous HNO_3 inside the Arctic Polar Vortex in January 1992," *Geophys. Res. Lett., 23*, 149–152 (1996).

Horn, A. B., M. A. Chesters, M. R. S. McCoustra, and J. R. Sodeau, "Adsorption of Stratospherically Important Molecules on Thin D_2O Ice Films Using Reflection Absorption Infrared Spectroscopy," *J. Chem. Soc., Faraday Trans., 88*, 1077–1078 (1992).

Horn, A. B., J. R. Sodeau, T. B. Roddis, and N. A. Williams, "Mechanism of the Heterogeneous Reaction of Hydrogen Chloride with Chlorine Nitrate and Hypochlorous Acid on Water Ice," *J. Phys. Chem. A, 102*, 6107–6120 (1998).

Horowitz, A., J. N. Crowley, and G. K. Moortgat, "Temperature Dependence of the Product Branching Ratios of the ClO Self-Reaction in Oxygen," *J. Phys. Chem., 98*, 11924–11930 (1994).

Hoskins, B. J., "Towards a PV–θ View of the General Circulation," *Tellus, Ser. A, 43*, 27–35 (1991).

Houston, P. L., A. G. Suits, and R. Toumi, "Isotopic Enrichment of Heavy Ozone in the Stratosphere," *J. Geophys. Res., 101*, 18829–18834 (1996).

Huang, T. Y. W., and S. T. Massie, "Effect of Volcanic Particles on the O_2 and O_3 Photolysis Rates and Their Impact on Ozone in the Tropical Stratosphere," *J. Geophys. Res., 102*, 1239–1249 (1997).

Hubrich, C., and F. Stuhl, "The Ultraviolet Absorption of Some Halogenated Methanes and Ethanes of Atmospheric Interest," *J. Photochem., 12*, 93–107 (1980).

Huie, R. E., and B. Laszlo, "The Atmospheric Chemistry of Iodine Compounds," in *Halon Replacements—Technology and Science, ACS Symposium Series 611* (A. W. Miziolek and W. Tsang, Eds.), Chap. 4, Am. Chem. Soc., Washington, DC, 1995.

Imre, D. G., J. Xu, and A. C. Tridico, "Phase Transformations in Sulfuric Acid Aerosols: Implications for Stratospheric Ozone Depletion," *Geophys. Res. Lett., 24*, 69–72 (1997).

Iraci, L. T., A. M. Middlebrook, M. A. Wilson, and M. A. Tolbert, "Growth of Nitric Acid Hydrates on Thin Sulfuric Acid Films," *Geophys. Res. Lett., 21*, 867–870 (1994).

Iraci, L. T., A. M. Middlebrook, and M. A. Tolbert, "Laboratory Studies of the Formation of Polar Stratospheric Clouds: Nitric Acid Condensation on Thin Sulfuric Acid Films," *J. Geophys. Res., 100*, 20969–20977 (1995).

Jackman, C. H., D. B. Considine, and E. L. Fleming, "Space Shuttle's Impact on the Stratosphere: An Update," *J. Geophys. Res., 101*, 12523–12529 (1996).

Jackman, C. H., D. B. Considine, and E. L. Fleming, "A Global Modeling Study of Solid Rocket Aluminum Oxide Emission Effects on Stratospheric Ozone," *Geophys. Res. Lett., 25*, 907–910 (1998).

Jaeglé, L., Y. L. Yung, G. C. Toon, B. Sen, and J.-F. Blavier, "Balloon Observations of Organic and Inorganic Chlorine in the Stratosphere: The Role of $HClO_4$ Production on Sulfate Aerosols," *Geophys. Res. Lett., 23*, 1749–1752 (1996).

Jeffers, P. M., and N. L. Wolfe, "On the Degradation of Methyl Bromide in Sea Water," *Geophys. Res. Lett., 23*, 1773–1776 (1996).

Jeffers, P. M., N. L. Wolfe, and V. Nzengung, "Green Plants: A Terrestrial Sink for Atmospheric CH_3Br," *Geophys. Res. Lett., 25*, 43–46 (1998).

Jiang, Y., Y. L. Yung, and R. W. Zurek, "Decadal Evolution of the Antarctic Ozone Hole," *J. Geophys. Res., 101*, 8985–8999 (1996).

Johnson, B. J., T. Deshler, and R. Zhao, "Ozone Profiles at McMurdo Station, Antarctica during the Spring of 1993; Record Low Ozone Season," *Geophys. Res. Lett., 22*, 183–186 (1995).

Johnson, D. G., W. A. Traub, K. V. Chance, and K. W. Jucks, "Detection of HBr and Upper Limit for HOBr: Bromine Partitioning in the Stratosphere," *Geophys. Res. Lett., 22*, 1373–1376 (1995).

Johnston, H., "Reduction of Stratospheric Ozone by Nitrogen Oxide Catalysts from Supersonic Transport Exhaust," *Science, 173*, 517–522 (1971).

Johnston, H. S., D. E. Kinnison, and D. J. Wuebbles, "Nitrogen Oxides from High-Altitude Aircraft: An Update of Potential Effects on Ozone," *J. Geophys. Res., 94*, 16351–16363 (1989).

Jones, A. E., S. Bekki, and J. A. Pyle, "Sensitivity of Supersonic Aircraft Modelling Studies to HNO_3 Photolysis Rate," *Geophys. Res. Lett., 20*, 2231–2234 (1993).

Jones, A. E., and J. D. Shanklin, "Continued Decline of Total Ozone over Halley, Antarctica, Since 1985," *Nature, 376*, 409–411 (1995).

Kärcher, B., M. M. Hirschberg, and P. Fabian, "Small-Scale Chemical Evolution of Aircraft Exhaust Species at Cruising Altitudes," *J. Geophys. Res., 101*, 15169–15190 (1996).

Kärcher, B., and D. W. Fahey, "The Role of Sulfur Emission in Volatile Particle Formation in Jet Aircraft Exhaust Plumes," *Geophys. Res. Lett., 24*, 389–392 (1997).

Karol, I. L., Y. E. Ozolin, and E. V. Rozanov, "Effect of Space Rocket Launches on Ozone," *Ann. Geophys., 10*, 810–814 (1992).

Kawa, S. R., D. W. Fahey, K. K. Kelly, J. E. Dye, D. Baumgardner, B. W. Gandrud, M. Loewenstein, G. V. Ferry, and K. R. Chan, "The Arctic Polar Stratospheric Cloud Aerosol: Aircraft Measurements of Reactive Nitrogen, Total Water, and Particles," *J. Geophys. Res., 97*, 7925–7938 (1992a).

Kawa, S. R., D. W. Fahey, L. E. Heidt, W. H. Pollock, S. Solomon, D. E. Anderson, M. Loewenstein, M. H. Proffitt, J. J. Margitan, and K. R. Chan, "Photochemical Partitioning of the Reactive Nitrogen and Chlorine Reservoirs in the High-Latitude Stratosphere," *J. Geophys. Res., 97*, 7905–7923 (1992b).

Kawa, S. R., P. A. Newman, L. R. Lait, M. R. Schoeberl, R. M. Stimpfle, J. G. Anderson, D. W. Kohn, C. R. Webster, R. D. May, D. Baumgardner, J. E. Dye, J. C. Wilson, K. R. Chan, and M. Loewenstein, "Activation of Chlorine in Sulfate Aerosol as Inferred from Aircraft Observations," *J. Geophys. Res., 102*, 3921–3933 (1997).

Keim, E. R., D. W. Fahey, L. A. Del Negro, E. L. Woodbridge, R. S. Gao, P. O. Wennberg, R. C. Cohen, R. M. Stimpfle, K. K. Kelly, E. J. Hintsa, J. C. Wilson, H. H. Jonsson, J. E. Dye, D. Baumgardner, S. R. Kawa, R. J. Salawitch, M. H. Proffitt, M. Loewenstein, J. R. Podolske, and K. R. Chan, "Observations of Large Reductions in the NO/NO_y Ratio near the Mid-latitude Tropopause and the Role of Heterogeneous Chemistry," *Geophys. Res. Lett., 23*, 3223–3226 (1996).

Kerr, J. B., D. I. Wardle, and D. W. Tarasick, "Record Low Ozone Values over Canada in Early 1993," *Geophys. Res. Lett., 20*, 1979–1982 (1993).

Keys, J. G., and B. G. Gardiner, "NO_2 Overnight Decay and Layer Height at Halley Bay, Antarctica," *Geophys. Res. Lett., 18*, 665–668 (1991).

Keys, J. G., P. V. Johnston, R. D. Blatherwick, and F. J. Murcray, "Evidence for Heterogeneous Reactions in the Antarctic Autumn Stratosphere," *Nature, 361*, 49–51 (1993).

Kinne, S., O. B. Toon, G. C. Toon, C. B. Farmer, E. V. Browell, and M. P. McCormick, "Measurements of Size and Composition of Particles in Polar Stratospheric Clouds from Infrared Solar Absorption Spectra," *J. Geophys. Res., 94*, 16481–16491 (1989).

Kinnison, D., H. Johnston, and D. Wuebbles, "Ozone Calculations with Large Nitrous Oxide and Chlorine Changes," *J. Geophys. Res., 93*, 14165–14175 (1988).

Klick, S., and K. Abrahamsson, "Biogenic Volatile Iodated Hydrocarbons in the Ocean," *J. Geophys. Res., 97*, 12683–12687 (1992).

Knight, C. A., "Surface Layers on Ice," *J. Geophys. Res., 101*, 12921–12928 (1996a); "Reply," *ibid., 101*, 12933–12936 (1996b).

Ko, M. K. W., N. D. Sze, C. Scott, J. M. Rodriguez, and D. K. Weisenstein, "Ozone Depletion Potential of CH_3Br," *J. Geophys. Res., 103*, 28187–28195 (1998).

Koch, T. G., S. F. Banham, J. R. Sodeau, A. B. Horn, M. R. S. McCoustra, and M. A. Chesters, "Mechanisms for the Heterogeneous Hydrolysis of Hydrogen Chloride, Chlorine Nitrate and Dinitrogen Pentoxide on Water-Rich Atmospheric Particle Surfaces," *J. Geophys. Res., 102*, 1513–1522 (1997).

Koehler, B. G., L. S. McNeill, A. M. Middlebrook, and M. A. Tolbert, "Fourier Transform Infrared Studies of the Interaction of HCl with Model Polar Stratospheric Cloud Films," *J. Geophys. Res., 98*, 10563–10571 (1993).

Koike, M., Y. Kondo, W. A. Matthews, P. V. Johnston, and K. Yamazaki, "Decrease of Stratospheric NO_2 at 44°N Caused by Pinatubo Volcanic Aerosols," *Geophys. Res. Lett., 20*, 1975–1978 (1993).

Koike, M., N. B. Jones, W. A. Matthews, P. V. Johnston, R. L. McKenzie, D. Kinnison, and J. Rodriguez, "Impact of Pinatubo Aerosols on the Partitioning between NO_2 and HNO_3," *Geophys. Res. Lett., 21*, 597–600 (1994).

Kolb, C. E., D. R. Worsnop, M. S. Zahniser, P. Davidovits, L. F. Keyser, M.-T. Leu, M. J. Molina, D. R. Hanson, A. R. Ravishankara, L. R. Williams, and M. A. Tolbert, "Laboratory Studies of Atmospheric Heterogeneous Chemistry," in *Current Problems in Atmospheric Chemistry* (J. R. Barker, Ed.), Advances in Physical Chemistry Series, World Scientific, Singapore, 1995.

Kondo, Y., P. Aimedieu, M. Koike, Y. Iwasaka, P. A. Newman, U. Schmidt, W. A. Matthews, M. Hayashi, and W. R. Sheldon, "Reactive Nitrogen, Ozone, and Nitrate Aerosols Observed in the Arctic Stratosphere in January 1990," *J. Geophys. Res., 97*, 13025–13038 (1992).

Kondo, Y., T. Sugita, R. J. Salawitch, M. Koike, and T. Deshler, "Effect of Pinatubo Aerosols on Stratospheric NO," *J. Geophys. Res., 102*, 1205–1213 (1997).

Koop, T., U. M. Biermann, W. Raber, B. P. Luo, P. J. Crutzen, and Th. Peter, "Do Stratospheric Aerosol Droplets Freeze above the Ice Frost Point?" *Geophys. Res. Lett., 22*, 917–920 (1995).

Koop, T., and K. S. Carslaw, "Melting of $H_2SO \cdot 4H_2O$ Particles upon Cooling: Implications for Polar Stratospheric Clouds," *Science, 272*, 1638–1641 (1996).

Koop, T., K. S. Carslaw, and Th. Peter, "Thermodynamic Stability and Phase Transitions of PSC Particles," *Geophys. Res. Lett., 24*, 2199–2202 (1997a).

Koop, T., B. Luo, U. M. Biermann, P. J. Crutzen, and T. Peter, "Freezing of $HNO_3/H_2SO_4/H_2O$ Solutions at Stratospheric Temperatures: Nucleation Statistics and Experiments," *J. Phys. Chem. A, 101*, 1117–1133 (1997b).

Koop, T., H. P. Ng, L. T. Molina, and M. J. Molina, "A New Optical Technique to Study Aerosol Phase Transitions: The Nucleation of Ice from H_2SO_4 Aerosols," *J. Phys. Chem. A, 102*, 8924–8931 (1998).

Kourtidis, K. A., R. Borchers, P. Fabian, and J. Harnisch, "Carbonyl Sulfide (COS) Measurements in the Arctic Polar Vortex," *Geophys. Res. Lett., 22*, 393–396 (1995).

Kourtidis, K., R. Borchers, and P. Fabian, "Dibromomethane (CH_2Br_2) Measurements at the Upper Troposphere and Lower Stratosphere," *Geophys. Res. Lett., 23*, 2581–2583 (1996).

Kourtidis, K., R. Borchers, and P. Fabian, "Vertical Distribution of Methyl Bromide in the Stratosphere," *Geophys. Res. Lett., 25*, 505–508 (1998).

Krankowsky, D., F. Bartecki, G. G. Klees, K. Mauersberger, and K. Schellenbach, "Measurement of Heavy Isotope Enrichment in Tropospheric Ozone," *Geophys. Res. Lett., 22*, 1713–1716 (1995).

Krankowsky, D., and K. Mauersberger, "Heavy Ozone—A Difficult Puzzle to Solve," *Science, 274,* 1324–1325 (1996).

Kukui, A. S., T. P. W. Jungkamp, and R. N. Schindler, "Determination of the Rate Constant and of Product Branching Ratios in the Reaction of CH_3O_2 with OCl between 233 and 300 K," *Ber. Bunsenges. Phys. Chem., 98,* 1298–1302 (1994).

Labitzke, K., and H. van Loon, "The Stratospheric Decadal Oscillation: Is It Associated with the 11-Year Sunspot Cycle?" *Atmos. Environ., 30,* xv–xvii (1996).

Lamarque, J.-F., G. P. Brasseur, P. G. Hess, and J.-F. Müller, "Three-Dimensional Study of the Relative Contributions of the Different Nitrogen Sources in the Troposphere," *J. Geophys. Res., 101,* 22955–22968 (1996).

Langford, A. O., C. D. Masters, M. H. Proffitt, E.-Y. Hsie, and A. F. Tuck, "Ozone Measurements in a Tropopause Fold Associated with a Cut-off Low System," *Geophys. Res. Lett., 23,* 2501–2504 (1996).

Larichev, M., F. Maguin, G. Le Bras, and G. Poulet, "Kinetics and Mechanism of the $BrO + HO_2$ Reaction," *J. Phys. Chem., 99,* 15911–15918 (1995).

Larsen, N., B. Knudsen, I. S. Mikkelsen, T. S. Jørgensen, and P. Eriksen, "Ozone Depletion in the Arctic Stratosphere in Early 1993," *Geophys. Res. Lett., 21,* 1611–1614 (1994).

Larsen, N., J. M. Rosen, N. T. Kjome, and B. Knudsen, "Deliquescence and Freezing of Stratospheric Aerosol Observed by Balloonborne Backscattersondes," *Geophys. Res. Lett., 22,* 1233–1236 (1995).

Larsen, N., B. M. Knudsen, J. M. Rosen, N. T. Kjome, R. Neuber, and E. Kyrö, "Temperature Histories in Liquid and Solid Polar Stratospheric Cloud Formation," *J. Geophys. Res., 102,* 23505–23517 (1997).

Lary, D. J., "Gas Phase Atmospheric Bromine Photochemistry," *J. Geophys. Res., 101,* 1505–1516 (1996).

Lary, D. J., M. P. Chipperfield, R. Toumi, and T. Lenton, "Heterogeneous Atmospheric Bromine Chemistry," *J. Geophys. Res., 101,* 1489–1504 (1996).

Lary, D. J., "Catalytic Destruction of Stratospheric Ozone," *J. Geophys. Res., 102,* 21515–21526 (1997).

Lary, D. J., A. M. Lee, R. Toumi, M. J. Newchurch, M. Pirre, and J. B. Renard, "Carbon Aerosols and Atmospheric Photochemistry," *J. Geophys. Res., 102,* 3671–3682 (1997).

Lary, D. J., and R. Toumi, "Halogen-Catalyzed Methane Oxidation," *J. Geophys. Res., 102,* 23421–23428 (1997).

Laszlo, B., R. E. Huie, M. J. Kurylo, and A. W. Miziolek, "Kinetic Studies of the Reactions of BrO and IO Radicals," *J. Geophys. Res., 102,* 1523–1532 (1997).

Laux, J. M., T. F. Fister, B. J. Finlayson-Pitts, and J. C. Hemminger, "X-ray Photoelectron Spectroscopy Studies of the Effects of Water Vapor on Ultra-thin Nitrate Layers on NaCl," *J. Phys. Chem., 100,* 19891–19897 (1996).

Lawrence, W. G., K. C. Clemitshaw, and V. A. Apkarian, "On the Relevance of OClO Photodissociation to the Destruction of Stratospheric Ozone," *J. Geophys. Res., 95,* 18591–18595 (1990).

Lean, J., "Variations in the Sun's Radiative Output," *Rev. Geophys., 29,* 505–535 (1991).

Lean, J., J. Beer, and R. Bradley, "Reconstruction of Solar Irradiance Since 1610: Implications for Climate Change," *Geophys. Res. Lett., 22,* 3195–3198 (1995a).

Lean, J. L., O. R. White, and A. Skumanich, "On the Solar Ultraviolet Spectral Irradiance during the Maunder Minimum," *Global Biogeochem. Cycles, 9,* 171–182 (1995b).

Lee, T. J., C. E. Dateo, and J. E. Rice, "An Analysis of Chlorine and Bromine Oxygen Bonding and Its Implications for Stratospheric Chemistry," *Mol. Phys., 96,* 633–643 (1999a).

Lee, T. J., S. Parthiban, and M. Head-Gordon, "Accurate Calculations on Excited States: New Theories Applied to the -X, -XO, and $-XO_2$ (X = Cl and Br) Chromophores and Implications for Stratospheric Bromine Chemistry," *Spectrochim. Acta A, 55,* 561–574 (1999b).

Lefèvre, F., G. P. Brasseur, I. Folkins, A. K. Smith, and P. Simon, "Chemistry of the 1991–1992 Stratospheric Winter: Three-Dimensional Model Simulations," *J. Geophys. Res., 99,* 8183–8195 (1994).

Lelieveld, J., B. Bregman, F. Arnold, V. Bürger, P. J. Crutzen, H. Fischer, A. Waibel, P. Siegmund, P. F. J. van Velthoven, "Chemical Perturbation of the Lowermost Stratosphere through Exchange with the Troposphere," *Geophys. Res. Lett., 24,* 603–606 (1997).

Leu, M.-T., "Laboratory Studies of Sticking Coefficients and Heterogeneous Reactions Important in the Antarctic Stratosphere," *Geophys. Res. Lett., 15,* 17–20 (1988a).

Leu, M.-T., "Heterogeneous Reactions of N_2O_5 with H_2O and HCl on Ice Surfaces: Implications for Antarctic Ozone Depletion," *Geophys. Res. Lett., 15,* 851–854 (1988b).

Li, Z., T. S. Dibble, and J. S. Francisco, "Experimental and Theoretical Progress in Understanding the Role of CX_3 Radicals in Atmospheric Chemical Processes," in *Progress and Problems in Atmospheric Chemistry, Advanced Series in Physical Chemistry* (J. R. Barker, Ed.), Vol. 3, pp. 686–743, World Scientific, Singapore, 1995a.

Li, Z., R. R. Friedl, and S. P. Sander, "Kinetics of FO_2 with NO, NO_2, O_3, CH_4, and C_2H_6," *J. Phys. Chem., 99,* 13445–13451 (1995b).

Lipson, J. B., M. J. Elrod, T. W. Beiderhase, L. T. Molina, and M. J. Molina, "Temperature Dependence of the Rate Constant and Branching Ratio for the OH + ClO Reaction," *J. Chem. Soc., Faraday Trans., 93,* 2665–2673 (1997).

Liu, S. C., and M. Trainer, "Responses of the Tropospheric Ozone and Odd Hydrogen Radicals to Column Ozone Change," *J. Atmos. Chem., 6,* 221–233 (1988).

Livingston, F. E., and B. J. Finlayson-Pitts, "The Reaction of Gaseous N_2O_5 with Solid NaCl at 298 K: Estimated Lower Limit to the Reaction Probability and Its Potential Role in Tropospheric and Stratospheric Chemistry," *Geophys. Res. Lett., 18,* 17–20 (1991).

Lobert, J. M., J. H. Butler, S. A. Montzka, L. S. Geller, R. C. Myers, and J. W. Elkins, "A Net Sink for Atmospheric CH_3Br in the East Pacific Ocean," *Science, 267,* 1002–1005 (1995).

Luo, B., K. S. Carslaw, T. Peter, and S. L. Clegg, "Vapour Pressures of $H_2SO_4/HNO_3/HCl/HBr/H_2O$ Solutions to Low Stratospheric Temperatures," *Geophys. Res. Lett., 22,* 247–250 (1995).

Luo, B., U. K. Krieger, and Th. Peter, "Densities and Refractive Indices of $H_2SO_4/HNO_3/H_2O$ Solutions to Stratospheric Temperatures," *Geophys. Res. Lett., 23,* 3707–3710 (1996).

Luo, B. P., Th. Peter, and P. J. Crutzen, "Homogeneous Freezing of Sulfuric Acid Droplets: I. Formation of $H_2SO_4 \cdot 4H_2O$ (SAT)," in *Nucleation and Atmospheric Aerosols* (N. Fukuta and P. E. Wagner, Eds.), pp. 225–228, A. Deepak, Hampton, VA, 1992.

Lyman, J. L., and R. Holland, "Oxygen Fluoride Chemical Kinetics," *J. Phys. Chem., 92,* 7232–7241 (1988).

Macalady, D. L., Ed., *Perspectives in Environmental Chemistry,* Oxford Univ. Press, New York, 1998.

MacKenzie, A. R., M. Kulmala, A. Laaksonen, and T. Vesala, "On the Theories of Type 1 Polar Stratospheric Cloud Formation," *J. Geophys. Res., 100,* 11275–11288 (1995).

Mahlman, J. D., and S. B. Fels, "Antarctic Ozone Decreases: A Dynamical Cause," *Geophys. Res. Lett., 13,* 1316–1319 (1986).

Mahlman, J. D., "Dynamics of Transport Processes in the Upper Troposphere," *Science, 276,* 1079–1083 (1997).

Mankin, W. G., and M. T. Coffey, "Increased Stratospheric Hydrogen Chloride in the El Chichon Cloud," *Science, 226,* 170–172 (1984).

Manney, G. L., and R. W. Zurek, "Interhemispheric Comparison of the Development of the Stratospheric Polar Vortex during Fall: A 3-Dimensional Perspective for 1991–1992," *Geophys. Res. Lett., 20,* 1275–1278 (1993).

Manney, G. L., L. Froidevaux, J. W. Waters, R. W. Zurek, W. G. Read, L. S. Elson, J. B. Kumer, J. L. Mergenthaler, A. E. Roche, A. O'Neill, R. S. Harwood, I. MacKenzie, and R. Swinbank, "Chemical Depletion of Ozone in the Arctic Lower Stratosphere during Winter 1992–93," *Nature, 370,* 429–434 (1994a).

Manney, G. L., R. W. Zurek, M. E. Gelman, A. J. Miller, and R. Nagatani, "The Anomalous Arctic Lower Stratospheric Polar Vortex of 1992–1993," *Geophys. Res. Lett., 21,* 2405–2408 (1994b).

Manney, G. L., R. W. Zurek, L. Froidevaux, and J. W. Waters, "Evidence for Arctic Ozone Depletion in Late February and Early March 1994," *Geophys. Res. Lett., 22,* 2941–2944 (1995).

Manney, G. L., M. L. Santee, L. Froidevaux, J. W. Waters, and R. W. Zurek, "Polar Vortex Conditions during the 1995–96 Arctic Winter: Meteorology and MLS Ozone," *Geophys. Res. Lett., 23,* 3203–3206 (1996).

Manney, G. L., L. Froidevaux, M. L. Santee, R. W. Zurek, and J. W. Waters, "MLS Observations of Arctic Ozone Loss in 1996–97," *Geophys. Res. Lett., 24,* 2697–2700 (1997).

Manö, S., and M. O. Andreae, "Emission of Methyl Bromide from Biomass Burning," *Science, 263,* 1255–1257 (1994).

Marti, J., and K. Mauersberger, "Laboratory Simulations of PSC Particle Formation," *Geophys. Res. Lett., 20,* 359–362 (1993).

Marti, J., and K. Mauersberger, "A Survey and New Measurements of Ice Vapor Pressure at Temperatures between 170 and 250 K," *Geophys. Res. Lett., 20,* 363–366 (1993).

Marti, J. J., and K. Mauersberger, "Evidence for Nitric Acid Pentahydrate Formed under Stratospheric Conditions," *J. Phys. Chem., 98,* 6897–6899 (1994).

Martin, L. R., A. G. Wren, and M. Wun, "Chlorine Atom and ClO Wall Reaction Products," *Int. J. Chem. Kinet., 11,* 543–547 (1979).

Martin, L. R., H. S. Judeikis, and M. Wun, "Heterogeneous Reactions of Cl and ClO in the Stratosphere," *J. Geophys. Res., 85,* 5511–5518 (1980).

Martin, S. T., D. Salcedo, L. T. Molina, and M. J. Molina, "Deliquescence of Sulfuric Acid Tetrahydrate Following Volcanic Eruptions or Denitrification," *Geophys. Res. Lett., 25,* 31–34 (1998).

Massie, S. T., J. E. Dye, D. Baumgardner, W. J. Randel, F. Wu, X. Tie, L. Pan, F. Figarol, G. P. Brasseur, M. L. Santee, W. G. Read, R. G. Grainger, A. Lambert, J. L. Mergenthaler, and A. Tabazadeh, "Simultaneous Observations of Polar Stratospheric Clouds and HNO_3 over Scandinavia in January, 1992," *Geophys. Res. Lett., 24,* 595–598 (1997).

Mauersberger, K., "Measurement of Heavy Ozone in the Stratosphere," *Geophys. Res. Lett., 8,* 935–937 (1981).

Mauersberger, K., J. Morton, B. Schueler, J. Stehr, and S. M. Anderson, "Multi-Isotope Study of Ozone: Implications for the Heavy Ozone Anomaly," *Geophys. Res. Lett., 20,* 1031–1034 (1993).

McCormick, M. P., L. W. Thomason, and C. R. Trepte, "Atmospheric Effects of the Mt. Pinatubo Eruption," *Nature, 373,* 399–404 (1995).

McCulloch, A., "Global Production and Emissions of Bromochlorodifluoromethane and Bromotrifluoromethane (Halons 1211 and 1301)," *Atmos. Environ., 26A,* 1325–1329 (1992).

McCulloch, A., and P. M. Midgley, "Estimated Historic Emissions of Fluorocarbons from the European Union," *Atmos. Environ., 32,* 1571–1580 (1998).

McElroy, M. B., and J. C. McConnell, "Nitrous Oxide: A Natural Source of Stratospheric NO," *J. Atmos. Sci., 28,* 1095–1098 (1971).

McElroy, M. B., S. C. Wofsy, J. E. Penner, and J. C. McConnell, "Atmospheric Ozone: Possible Impact of Stratospheric Aviation," *J. Atmos. Sci., 31,* 287–303 (1974).

McElroy, M. B., R. J. Salawitch, S. C. Wofsy, and J. A. Logan, "Reductions of Antarctic Ozone Due to Synergistic Interactions of Chlorine and Bromine," *Nature, 321,* 759–762 (1986).

McElroy, M. B., and R. J. Salawitch, "Changing Composition of the Global Stratosphere," *Science, 243,* 763–770 (1989).

McElroy, M. B., R. J. Salawitch, and K. Minschwaner, "The Changing Stratosphere," *Planet Space Sci., 40,* 373–401 (1992).

McGee, T. J., P. Newman, M. Gross, U. Singh, S. Godin, A.-M. Lacoste, and G. Megie, "Correlation of Ozone Loss with the Presence of Volcanic Aerosols," *Geophys. Res. Lett., 21,* 2801–2804 (1994).

McIntyre, M. E., "The Stratospheric Polar Vortex and Sub-Vortex: Fluid Dynamics and Midlatitude Ozone Loss," *Philos. Trans. R. Soc. London, A, 352,* 227–240 (1995).

McKenzie, R. L., and P. V. Johnston, "Springtime Stratospheric NO_2 in Antarctica," *Geophys. Res. Lett., 11,* 73–75 (1984).

McKinney, K. A., J. M. Pierson, and D. W. Toohey, "A Wintertime *in Situ* Profile of BrO between 17 and 27 km in the Arctic Vortex," *Geophys. Res. Lett., 24,* 853–856 (1997).

McPeters, R., M. Prather, and S. Doiron, "Reply—S. Aftergood Comment on 'The Space Shuttle's Impact on the Stratosphere,'" *J. Geophys. Res., 96,* 17379–17381 (1991).

Meilinger, S. K., T. Koop, B. P. Luo, T. Huthwelker, K. S. Carslaw, U. Krieger, P. J. Crutzen, and Th. Peter, "Size-Dependent Stratospheric Droplet Composition in Lee Wave Temperature Fluctuations and Their Potential Role in PSC Freezing," *Geophys. Res. Lett., 22,* 3031–3034 (1995).

Mellouki, A., R. K. Talukdar, and C. J. Howard, "Kinetics of the Reactions of HBr with O_3 and HO_2: The Yield of HBr from HO_2 + BrO," *J. Geophys. Res., 99,* 22949–22954 (1994).

Michelangeli, D. V., M. Allen, Y. L. Yung, R.-L. Shia, D. Crisp, and J. Eluszkiewicz, "Enhancement of Atmospheric Radiation by an Aerosol Layer," *J. Geophys. Res., 97,* 865–874 (1992).

Michelsen, H. A., R. J. Salawitch, M. R. Gunson, C. Aellig, N. Kämpfer, M. M. Abbas, M. C. Abrams, T. L. Brown, A. Y. Chang, A. Goldman, F. W. Irion, M. J. Newchurch, C. P. Rinsland, G. P. Stiller, and R. Zander, "Stratospheric Chlorine Partitioning: Constraints from Shuttle-Borne Measurements of [HCl], [$ClNO_3$], and [ClO]," *Geophys. Res. Lett., 23,* 2361–2364 (1996).

Mickley, L. J., J. P. D. Abbatt, J. E. Frederick, and J. M. Russell III, "Response of Summertime Odd Nitrogen and Ozone at 17 mbar to Mount Pinatubo Aerosol over the Southern Midlatitudes: Observations from the Halogen Occultation Experiment," *J. Geophys. Res., 102,* 23573–23582 (1997a).

Mickley, L. J., J. P. D. Abbatt, J. E. Frederick, and J. M. Russell III, "Evolution of Chlorine and Nitrogen Species in the Lower Stratosphere during Antarctic Spring: Use of Tracers to Determine Chemical Change," *J. Geophys. Res., 102,* 21479–21491 (1997b).

Middlebrook, A. M., L. T. Iraci, L. S. McNeill, B. G. Koehler, M. A. Wilson, O. W. Saastad, and M. A. Tolbert, "Fourier Transform-Infrared Studies of Thin H_2SO_4/H_2O Films: Formation, Water Uptake, and Solid–Liquid Phase Changes," *J. Geophys. Res., 98,* 20473–20481 (1993).

Middlebrook, A. M., B. S. Berland, S. M. George, and M. A. Tolbert, "Real Refractive Indices of Infrared-Characterized Nitric-Acid/Ice Films: Implications for Optical Measurements of Polar Stratospheric Clouds," *J. Geophys. Res., 99,* 25655–25666 (1994).

Middlebrook, A. M., M. A. Tolbert, and K. Drdla, "Evaporation Studies of Model Polar Stratospheric Cloud Films," *Geophys. Res. Lett., 23,* 2145–2148 (1996).

Middlebrook, A. M., and M. A. Tolbert, "Laboratory Studies of Heterogeneous Chemistry in the Stratosphere," in *Perspectives in Environmental Chemistry* (D. L. Macalady, Ed.), pp. 325–343, Oxford Univ. Press, New York, 1998.

Miller, R. L., A. G. Suits, P. L. Houston, R. Toumi, J. A. Mack, and A. M. Wodtke, "The 'Ozone Deficit' Problem: $O_2(X, v > 26) + O(^3P)$ from 226-nm Ozone Photodissociation," *Science, 265*, 1831–1838 (1994).

Minnis, P., E. F. Harrison, L. L. Stowe, G. G. Gibson, P. M. Denn, D. R. Doelling, and W. L. Smith, Jr., "Radiative Climate Forcing by Mount Pinatubo Eruption," *Science, 259*, 1411–1415 (1993).

Minschwaner, K., A. E. Dessler, J. W. Elkins, C. M. Volk, D. W. Fahey, M. Loewenstein, J. R. Podolske, A. E. Roche, and K. R. Chan, "Bulk Properties of Isentropic Mixing into the Tropics in the Lower Stratosphere," *J. Geophys. Res., 101*, 9433–9439 (1996).

Molina, M. J., and F. S. Rowland, "Stratospheric Sink for Chlorofluoromethanes: Chlorine Atom-Catalysed Destruction of Ozone," *Nature, 249*, 810–812 (1974).

Molina, L. T., and M. J. Molina, "Production of Cl_2O_2 from the Self-Reaction of the ClO Radical," *J. Phys. Chem., 91*, 433–436 (1987).

Molina, M. J., T.-L. Tso, L. T. Molina, and F. C.-Y. Wang, "Antarctic Stratospheric Chemistry of Chlorine Nitrate, Hydrogen Chloride, and Ice: Release of Active Chlorine," *Science, 238*, 1253–1257 (1987).

Molina, M. J., A. J. Colussi, L. T. Molina, R. N. Schindler, and T.-L. Tso, "Quantum Yield of Chlorine-Atom Formation in the Photodissociation of Chlorine Peroxide (ClOOCl) at 308 nm," *Chem. Phys. Lett., 173*, 310–315 (1990).

Molina, M. J., "Heterogeneous Chemistry on Polar Stratospheric Clouds," *Atmos. Environ., 25A*, 2535–2537 (1991).

Molina, M. J., R. Zhang, P. J. Wooldridge, J. R. McMahon, J. E. Kim, H. Y. Chang, and K. D. Beyer, "Physical Chemistry of the $H_2SO_4/HNO_3/H_2O$ System: Implications for Polar Stratospheric Clouds," *Science, 261*, 1418–1423 (1993).

Molina, M. J., L. T. Molina, and D. M. Golden, "Environmental Chemistry (Gas and Gas–Solid Interactions): The Role of Physical Chemistry," *J. Phys. Chem., 100*, 12888–12896 (1996).

Molina, M. J., L. T. Molina, R. Zhang, R. F. Meads, and D. D. Spencer, "The Reaction of $ClONO_2$ with HCl on Aluminum Oxide," *Geophys. Res. Lett., 24*, 1619–1622 (1997).

Montzka, S. A., J. H. Butler, R. C. Myers, T. M. Thompson, T. H. Swanson, A. D. Clarke, L. T. Lock, and J. W. Elkins, "Decline in the Tropospheric Abundance of Halogen from Halocarbons: Implications for Stratospheric Ozone Depletion," *Science, 272*, 1318–1322 (1996).

Moore, R. M., and R. Tokarczyk, "Chloro-Iodomethane in N. Atlantic Waters: A Potentially Significant Source of Atmospheric Iodine," *Geophys. Res. Lett., 19*, 1779–1782 (1992).

Moore, T. A., M. Okumura, J. W. Seale, and T. K. Minton, "UV Photolysis of ClOOCl", *J. Phys. Chem. A, 103*, 1691–1695 (1999).

Morris, G. A., D. B. Considine, A. E. Dessler, S. R. Kawa, J. Kumer, J. Mergenthaler, A. Roche, and J. M. Russell III, "Nitrogen Partitioning in the Middle Stratosphere As Observed by the Upper Atmosphere Research Satellite," *J. Geophys. Res., 102*, 8955–8965 (1997).

Mote, P. W., K. H. Rosenlof, J. R. Holton, R. S. Harwood, and J. W. Waters, "Seasonal Variations of Water Vapor in the Tropical Lower Stratosphere," *Geophys. Res. Lett., 22*, 1093–1096 (1995).

Mote, P. W., K. H. Rosenlof, M. E. McIntyre, E. S. Carr, J. C. Gille, J. R. Holton, J. S. Kinnersley, H. C. Plumphrey, J. M. Russell III, and J. W. Waters, "An Atmospheric Tape Recorder: The Imprint of Tropical Tropopause Temperatures on Stratospheric Water Vapor," *J. Geophys. Res., 101*, 3989–4006 (1996).

Mozurkewich, M., "Effect of Competitive Adsorption on Polar Stratospheric Cloud Reactions," *Geophys. Res. Lett., 20*, 355–358 (1993).

Müller, R., P. J. Crutzen, J.-U. Grooß, C. Brühl, J. M. Russell III, and A. F. Tuck, "Chlorine Activation and Ozone Depletion in the Arctic Vortex: Observations by the Halogen Occultation Experiment on the Upper Atmosphere Research Satellite," *J. Geophys. Res., 101*, 12531–12554 (1996).

Mulvaney, R., and E. W. Wolff, "Evidence of Winter/Spring Denitrification of the Stratosphere in the Nitrate Record of Antarctic Firn Cores," *J. Geophys. Res., 98*, 5213–5220 (1993).

Nardi, B., T. Deshler, M. E. Hervig, and L. D. Oolman, "Ozone Measurements over McMurdo Station, Antarctica during Spring 1994 and 1995," *Geophys. Res. Lett., 24*, 285–288 (1997).

National Research Council, *Stratospheric Ozone Depletion by Halocarbons: Chemistry and Transport*, Panel on Chemistry and Transport, Committee on Impacts of Stratospheric Change, Assembly of Mathematical and Physical Sciences, National Academy of Sciences, Washington, DC, 1979.

National Research Council, *Causes and Effects of Changes in Stratospheric Ozone: Update, 1983*, Committee on Causes and Effects of Changes in Stratospheric Ozone, Update, 1983, Environmental Studies Board, Commission on Physical Sciences, Mathematics, and Resources, National Academy Press, Washington, DC, 1984.

Nesbitt, F. L., P. S. Monks, W. A. Payne, and L. J. Stief, "The Reaction $O(^3P)$ + HOBr: Temperature Dependence of the Rate Constant and Importance of the Reaction as an HOBr Stratospheric Loss Process," *Geophys. Res. Lett., 22*, 827–830 (1995).

Newman, P. A., J. F. Gleason, R. D. McPeters, and R. S. Stolarski, "Anomalously Low Ozone over the Arctic," *Geophys. Res. Lett., 24*, 2689–2692 (1997).

Nickolaisen, S. L., R. R. Friedl, and S. P. Sander, "Kinetics and Mechanism of the ClO + ClO Reaction: Pressure and Temperature Dependences of the Bimolecular and Termolecular Channels and Thermal Decomposition of Chlorine Peroxide," *J. Phys. Chem., 98*, 155–169 (1994).

Niedziela, R. F., R. E. Miller, and D. R. Worsnop, "Temperature- and Frequency-Dependent Optical Constants for Nitric Acid Dihydrate from Aerosol Spectroscopy," *J. Phys. Chem. A, 102*, 6477–6484 (1998).

Notholt, J., "The Moon as a Light Source for FTIR Measurements of Stratospheric Trace Gases during the Polar Night: Application for HNO_3 in the Arctic," *J. Geophys. Res., 99*, 3607–3614 (1994).

Notholt, J., P. von der Gathen, and S. Peil, "Heterogeneous Conversion of HCl and $ClONO_2$ during the Arctic Winter 1992–1993 Initiating Ozone Depletion," *J. Geophys. Res., 100*, 11269–11274 (1995).

Noxon, J. F., "Stratospheric NO_2 in the Antarctic Winter," *Geophys. Res. Lett., 5*, 1021–1022 (1978).

Oppliger, R., A. Allanic, and M. J. Rossi, "Real-Time Kinetics of the Uptake of $ClONO_2$ on Ice and in the Presence of HCl in the Temperature Range 160 K $\leq T \leq$ 200 K," *J. Phys. Chem. A, 101*, 1903–1911 (1997).

Oremland, R. S., L. G. Miller, C. W. Culbertson, T. L. Connell, and L. Jahnke, "Degradation of Methyl Bromide by Methanotrophic Bacteria in Cell Suspensions and Soils," *Appl. Environ. Microbiol., 60*, 3640–3646 (1994a).

Oremland, R. S., L. G. Miller, and F. R. Strohmaier, "Degradation of Methyl Bromide in Anaerobic Sediments," *Environ. Sci. Technol., 28*, 514–520 (1994b).

Orkin, V. L., V. G. Khamaganov, A. G. Guschin, R. E. Huie, and M. J. Kurylo, "Atmospheric Fate of Chlorobromomethane: Rate Constant for the Reaction with OH, UV Spectrum, and Water Solubility," *J. Phys. Chem. A, 101*, 174–178 (1997).

Orlando, J. J., G. S. Tyndall, and T. J. Wallington, "Atmospheric Oxidation of CH_3Br: Chemistry of the CH_2BrO Radical," *J. Phys. Chem.*, 100, 7026–7033 (1996).

Osterman, G. B., R. J. Salawitch, B. Sen, G. C. Toon, R. A. Stachnik, H. M. Pickett, J. J. Margitan, J.-F. Blavier, and D. B. Peterson, "Balloon-Borne Measurements of Stratospheric Radicals and Their Precursors: Implications for the Production and Loss of Ozone," *Geophys. Res. Lett.*, 24, 1107–1110 (1997).

O'Sullivan, D. A., "International Gathering Plans Ways to Safeguard Atmospheric Ozone," *Chem. Eng. News*, 33–36, June 26 (1989).

Pagsberg, P., E. Ratajczak, A. Sillesen, and J. T. Jodkowski, "Spectrokinetic Studies of the Gas-Phase Equilibrium $F + O_2 \leftrightarrow FO_2$ between 295 and 359 K," *Chem. Phys. Lett.*, 141, 88–94 (1987).

Pawson, S., B. Naujokat, and K. Labitzke, "On the Polar Stratospheric Cloud Formation Potential of the Northern Stratosphere," *J. Geophys. Res.*, 100, 23215–23225 (1995).

Peter, Th., C. Brühl, and P. J. Crutzen, "Increase in the PSC-Formation Probability Caused by High-Flying Aircraft," *Geophys. Res. Lett.*, 18, 1465–1468 (1991).

Peter, Th., B. P. Luo, and P. J. Crutzen, "Homogeneous Freezing of Sulfuric Acid Droplets: II. Implications for PSCs and Heterogeneous Chemistry," in *Nucleation and Atmospheric Aerosols* (N. Fukuta and P. E. Wagner, Eds.), pp. 229–232, A. Deepak, Hampton, VA, 1992.

Peter, T., R. Müller, P. J. Crutzen, and T. Deshler, "The Lifetime of Leewave-Induced Ice Particles in the Arctic Stratosphere: II. Stabilization Due to NAT-Coating," *Geophys. Res. Lett.*, 21, 1331–1334 (1994).

Peter, T., "Microphysics and Heterogeneous Chemistry of Polar Stratospheric Clouds," *Annu. Rev. Phys. Chem.*, 48, 785–822 (1997).

Petrenko, V. C., S. Colbeck, I. Baker, E. M. Schulson, N. Khusnatdinov, and G. Ashton, Organizing Committee, "Physics and Chemistry of Ice, 1996," *J. Phys. Chem. B*, 101, 6079–6312 (1997).

Pierson, J. M., K. A. McKinney, D. W. Toohey, J. Margitan, U. Schmidt, A. Engel, and P. A. Newman, "An Investigation of ClO Photochemistry in the Chemically Perturbed Arctic Vortex," *J. Atmos. Chem.*, 32, 61–81 (1999).

Pilinis, C., D. B. King, and E. S. Saltzman, "The Oceans: A Source or a Sink of Methyl Bromide?" *Geophys. Res. Lett.*, 23, 817–820 (1996).

Pitari, G., V. Rizi, L. Ricciardulli, and G. Visconti, "High-Speed Civil Transport Impact: Role of Sulfate, Nitric Acid Trihydrate, and Ice Aerosols Studied with a Two-Dimensional Model Including Aerosol Physics," *J. Geophys. Res.*, 98, 23141–23164 (1993).

Plumb, R. A., D. W. Waugh, R. J. Atkinson, P. A. Newman, L. R. Lait, M. R. Schoeberl, E. V. Browell, A. J. Simmons, and M. Loewenstein, "Intrusions into the Lower Stratospheric Arctic Vortex During the Winter of 1991–1992," *J. Geophys. Res.*, 99, 1089–1105 (1994).

Plumb, R. A., "A 'Tropical Pipe' Model of Stratospheric Transport," *J. Geophys. Res.*, 101, 3957–3972 (1996).

Pommereau, J. P., F. Goutail, Y. Kondo, W. A. Matthews, and M. Helten, "An NO_x Source in the Upper Troposphere?" in *Ozone in the Atmosphere* (R. D. Bojkov and P. Fabian, Eds.), pp. 328–331, A. Deepak, Hampton, VA, 1989.

Portmann, R. W., S. Solomon, R. R. Garcia, L. W. Thomason, L. R. Poole, and M. P. McCormick, "Role of Aerosol Variations in Anthropogenic Ozone Depletion in the Polar Regions," *J. Geophys. Res.*, 101, 22991–23006 (1996).

Prasad, S. S., "Potential New Atmospheric Sources and Sinks of Odd Nitrogen: Sources Involving the Excited O_2 and the $N_2O \cdot O_3$ Species," *Geophys. Res. Lett.*, 25, 2173–2176 (1998).

Prather, M. J., M. B. McElroy, and S. C. Wofsy, "Reductions in Ozone at High Concentrations of Stratospheric Halogens," *Nature*, 312, 227–231 (1984).

Prather, M. J., M. M. García, A. R. Douglass, C. H. Jackman, M. K. W. Ko, and N. D. Sze, "The Space Shuttle's Impact on the Stratosphere," *J. Geophys. Res.*, 95, 18583–18590 (1990).

Prather, M. J., "More Rapid Polar Ozone Depletion through the Reaction of HOCl with HCl on Polar Stratospheric Clouds," *Nature*, 355, 534–537 (1992).

Prather, M. J., "Timescales in Atmospheric Chemistry: CH_3Br, the Ocean, and Ozone Depletion Potentials," *Global Biogeochem. Cycles*, 11, 393–400 (1997).

Prather, M. J., "Time Scales in Atmospheric Chemistry: Coupled Perturbations to N_2O, NO_y, and O_3," *Science*, 279, 1339–1341 (1998).

Prenni, A. J., T. B. Onasch, R. T. Tisdale, R. L. Siefert, and M. A. Tolbert, "Composition-Dependent Freezing Nucleation Rates for HNO_3–H_2O Aerosols Resembling Gravity-Wave-Perturbed Stratospheric Particles," *J. Geophys. Res.*, 103, 28439–28450 (1998).

Proffitt, M. H., J. J. Margitan, K. K. Kelly, M. Loewenstein, J. R. Podolske, and K. R. Chan, "Ozone Loss in the Arctic Polar Vortex Inferred from High-Altitude Aircraft Measurements," *Nature*, 347, 31–36 (1990).

Proffitt, M. H., K. Aikin, J. J. Margitan, M. Loewenstein, J. R. Podolske, A. Weaver, K. R. Chan, H. Fast, and J. W. Elkins, "Ozone Loss inside the Northern Polar Vortex during the 1991–1992 Winter," *Science*, 261, 1150–1154 (1993).

Pruppacher, H. R., and J. D. Klett, in *Microphysics of Clouds and Precipitation*, 2nd ed., pp. 229–230, Reidel, Dordrecht, 1997.

Pueschel, R. F., K. G. Snetsinger, J. K. Goodman, O. B. Toon, G. V. Ferry, V. R. Oerbeck, J. M. Livingston, S. Verma, W. Fong, W. L. Starr, and K. R. Chan, "Condensed Nitrate, Sulfate, and Chloride in Antarctic Stratospheric Aerosols," *J. Geophys. Res.*, 94, 11271–11284 (1989).

Pueschel, R. F., K. G. Snetsinger, P. Hamill, J. K. Goodman, and M. P. McCormick, "Nitric Acid in Polar Stratospheric Clouds: Similar Temperature of Nitric Acid Condensation and Cloud Formation," *Geophys. Res. Lett.*, 17, 429–432 (1990).

Pueschel, R. F., D. F. Blake, K. G. Snetsinger, A. D. A. Hansen, S. Verma, and K. Kato, "Black Carbon (Soot) Aerosol in the Lower Stratosphere and Upper Troposphere," *Geophys. Res. Lett.*, 19, 1659–1662 (1992a).

Pueschel, R. F., G. V. Ferry, K. G. Snetsinger, J. Goodman, J. E. Dye, D. Baumgardner, and B. W. Gandrud, "A Case of Type I Polar Stratospheric Cloud Formation by Heterogeneous Nucleation," *J. Geophys. Res.*, 97, 8105–8114 (1992b).

Pueschel, R. F., P. B. Russell, D. A. Allen, G. V. Ferry, K. G. Snetsinger, J. M. Livingston, and S. Verma, "Physical and Optical Properties of the Pinatubo Volcanic Aerosol: Aircraft Observations with Impactors and a Sun-Tracking Photometer," *J. Geophys. Res.*, 99, 12915–12922 (1994).

Pundt, I., J.-P. Pommereau, C. Phillips, and E. Lateltin, "Upper Limit of Iodine Oxide in the Lower Stratosphere," *J. Atmos. Chem.*, 30, 173–185 (1998).

Pursell, C. J., J. Conyers, P. Alapat, and R. Parveen, "Photochemistry of Chlorine Dioxide in Ice," *J. Phys. Chem.*, 99, 10433–10437 (1995).

Pursell, C. J., J. Conyers, and C. Denison, "Photochemistry of Chlorine Dioxide in Polycrystalline Ice ($T = 140$–185 K): Production of Chloryl Chloride, Cl–(OClO)," *J. Phys. Chem.*, 100, 15450–15453 (1996).

Ramaswamy, V., "Dehydration Mechanism in the Antarctic Stratosphere during Winter," *Geophys. Res. Lett.*, 15, 863–866 (1988).

Randel, W. J., and F. Wu, "TOMS Total Ozone Trends in Potential Vorticity Coordinates," *Geophys. Res. Lett., 22,* 683–686 (1995).

Randeniya, L. K., P. F. Vohralik, I. C. Plumb, K. R. Ryan, and S. L. Baughcum, "Impact of the Heterogeneous Hydrolysis of $BrONO_2$ on Calculated Ozone Changes Due to HSCT Aircraft and Increased Sulphate Aerosol Levels," *Geophys. Res. Lett., 23,* 343–346 (1996a).

Randeniya, L. K., P. F. Vohralik, I. C. Plumb, K. R. Ryan, and S. L. Baughcum, "Impact of Heterogeneous $BrONO_2$ Hydrolysis on Ozone Trends and Transient Ozone Loss during Volcanic Periods," *Geophys. Res. Lett., 23,* 1633–1636 (1996b).

Randeniya, L. K., P. F. Vohralik, I. C. Plumb, and K. R. Ryan, "Heterogeneous $BrONO_2$ Hydrolysis: Effect on NO_2 Columns and Ozone at High Latitudes in Summer," *J. Geophys. Res., 102,* 23543–23557 (1997).

Ravishankara, A. R., A. A. Turnipseed, N. R. Jensen, S. Barone, M. Mills, C. J. Howard, and S. Solomon, "Do Hydrofluorocarbons Destroy Stratospheric Ozone?" *Science, 263,* 71–75 (1994).

Ravishankara, A. R., and D. R. Hanson, "Differences in the Reactivity of Type I Polar Stratospheric Clouds Depending on Their Phase," *J. Geophys. Res., 101,* 3885–3890 (1996).

Rex, M., P. von der Gathen, N. R. P. Harris, D. Lucic, B. M. Knudsen, G. O. Braathen, S. J. Reid, H. De Backer, H. Claude, R. Fabian, H. Fast, M. Gil, E. Kyrö, I. S. Mikkelsen, M. Rummukainen, H. G. Smit, J. Stähelin, C. Varotsos, and I. Zaitcev, "*In Situ* Measurements of Stratospheric Ozone Depletion Rates in the Arctic Winter 1991/1992: A Lagrangian Approach," *J. Geophys. Res., 103,* 5843–5853 (1998).

Ricaud, P., E. Monnier, F. Goutail, J.-P. Pommereau, C. David, S. Godin, L. Froidevaux, J. W. Waters, J. Mergenthaler, A. E. Roche, H. Pumphrey, and M. P. Chipperfield, "Stratosphere over Dumont d'Urville, Antarctica, in Winter 1992," *J. Geophys. Res., 103,* 13267–13284 (1998).

Richwine, L. J., M. L. Clapp, R. E. Miller, and D. R. Worsnop, "Complex Refractive Indices in the Infrared of Nitric Acid Trihydrate Aerosols," *Geophys. Res. Lett., 22,* 2625–2628 (1995).

Rinsland, C. P., M. R. Gunson, M. C. Abrams, L. L. Lowes, R. Zander, E. Mahieu, A. Goldman, M. K. W. Ko, J. M. Rodriguez, and N. D. Sze, "Heterogeneous Conversion of N_2O_5 to HNO_3 in the Post-Mount Pinatubo Eruption Stratosphere," *J. Geophys. Res., 99,* 8213–8219 (1994).

Robertson, S. H., and D. C. Clary, "Solvation of Hydrogen Halides on the Surface of Ice," *Faraday Discuss., 100,* 309–320 (1995).

Robinson, G. N., Q. Dai, and A. Freedman, "Reactions of Halomethanes with γ-Alumina Surfaces. 2. X-ray Photoelectron and Temperature-Programmed Reaction Spectroscopic Studies," *J. Phys. Chem. B, 101,* 4947–4953 (1997).

Rodriguez, J. M., M. K. W. Ko, and N. D. Sze, "Role of Heterogeneous Conversion of N_2O_5 on Sulphate Aerosols in Global Ozone Losses," *Nature, 352,* 134–137 (1991).

Rodriguez, J. M., M. K. W. Ko, N. D. Sze, C. W. Heisey, "Ozone Response to Enhanced Heterogeneous Processing after the Eruption of Mt. Pinatubo," *Geophys. Res. Lett., 21,* 209–212 (1994).

Rosen, J. M., N. T. Kjome, and S. J. Oltmans, "Simultaneous Ozone and Polar Stratospheric Cloud Observations at South Pole Station during Winter and Spring 1991," *J. Geophys. Res., 98,* 12741–12751 (1993).

Rosen, J. M., N. T. Kjome, N. Larsen, B. M. Knudsen, E. Kyrö, R. Kivi, J. Karhu, R. Neuber, and I. Beninga, "Polar Stratospheric Cloud Threshold Temperatures in the 1995–1996 Arctic Vortex," *J. Geophys. Res., 102,* 28195–28202 (1997).

Rosenfield, J. E., M. R. Schoeberl, and P. A. Newman, "Antarctic Springtime Ozone Depletion Computed from Temperature Observations," *J. Geophys. Res., 93,* 3833–3849 (1988).

Ross, M., "Local Effects of Solid Rocket Motor Exhaust on Stratospheric Ozone," *J. Spacecraft Rockets, 33,* 144–153 (1996).

Ross, M. N., J. R. Benbrook, W. R. Sheldon, P. F. Zittel, and D. L. McKenzie, "Observation of Stratospheric Ozone Depletion in Rocket Exhaust Plumes," *Nature, 390,* 62–64 (1997).

Rowland, F. S., and M. J. Molina, "Chlorofluoromethanes in the Environment," *Rev. Geophys. Space Phys., 13,* 1–35 (1975).

Rowland, F. S., "Chlorofluorocarbons and the Depletion of Stratospheric Ozone," *Am. Sci., 77,* 36–45 (1989).

Rowland, F. S., "Stratospheric Ozone Depletion," in *Annual Review of Physical Chemistry* (H. L. Strauss, G. T. Babcock, and S. R. Leone, Eds.), Vol. 42, pp. 731–768, Annual Reviews, Palo Alto, CA, 1991.

Rowland, F. S., "The CFC Controversy: Issues and Answers," *ASHRAE J.,* December, 20–27 (1992).

Rowland, F. S., President's Lecture: The Need for Scientific Communication with the Public," *Science, 290,* 1571–1576 (1993).

Rowland, F. S., and M. J. Molina, "Ozone Depletion: 20 Years after the Alarm," *Chem. Eng. News,* August 15, 8–13 (1994).

Russell, P. B., J. M. Livingston, R. F. Pueschel, J. J. Bauman, J. B. Pollack, S. L. Brooks, P. Hamill, L. W. Thomason, L. L. Stowe, T. Deshler, E. G. Dutton, and R. W. Bergstrom, "Global to Microscale Evolution of the Pinatubo Volcanic Aerosol Derived from Diverse Measurements and Analyses," *J. Geophys. Res., 101,* 18745–18763 (1996).

Salawitch, R. J., S. C. Wofsy, E. W. Gottlieb, L. R. Lait, P. A. Newman, M. R. Schoeberl, M. Loewenstein, J. R. Podolske, S. E. Strahan, M. H. Proffitt, C. R. Webster, R. D. May, D. W. Fahey, D. Baumgardner, J. E. Dye, J. C. Wilson, K. K. Kelly, J. W. Elkins, K. R. Chan, and J. G. Anderson, "Chemical Loss of Ozone in the Arctic Polar Vortex in the Winter of 1991–1992," *Science, 261,* 1146–1149 (1993).

Salawitch, R. J., S. C. Wofsy, P. O. Wennberg, R. C. Cohen, J. G. Anderson, D. W. Fahey, R. S. Gao, E. R. Keim, E. L. Woodbridge, R. M. Stimpfle, J. P. Koplow, D. W. Kohn, C. R. Webster, R. D. May, L. Pfister, E. W. Gottlieb, H. A. Michelsen, G. K. Yue, M. J. Prather, J. C. Wilson, C. A. Brock, H. H. Jonsson, J. E. Dye, D. Baumgardner, M. H. Proffitt, M. Loewenstein, J. R. Podolske, J. W. Elkins, G. S. Dutton, E. J. Hintsa, A. E. Dessler, E. M. Weinstock, K. K. Kelly, K. A. Boering, B. C. Daube, K. R. Chan, and S. W. Bowen, "The Diurnal Variation of Hydrogen, Nitrogen, and Chlorine Radicals: Implications for the Heterogeneous Production of HNO_2," *Geophys. Res. Lett., 21,* 2551–2554 (1994).

Salby, M. L., and R. R. Garcia, "Dynamical Perturbations to the Ozone Layer," *Phys. Today,* March, 38–46 (1990).

Sander, S. P., R. R. Friedl, and Y. L. Yung, "Rate of Formation of the ClO Dimer in the Polar Stratosphere: Implications for Ozone Loss," *Science, 245,* 1095–1098 (1989).

Sander, S. P., R. R. Friedl, and J. S. Francisco, "Experimental and Theoretical Studies of Atmospheric Inorganic Chlorine Chemistry," in *Progress and Problems in Atmospheric Chemistry, Advanced Series in Physical Chemistry* (J. R. Barker, Ed.), Vol. 3, pp. 876–921, World Scientific, Singapore, 1995.

Sanders, R. W., S. Solomon, J. P. Smith, L. Perliski, H. L. Miller, G. H. Mount, J. G. Keys, and A. L. Schmeltekopf, "Visible and Near-Ultraviolet Spectroscopy at McMurdo Station, Antarctica. 9. Observations of OClO from April to October 1991," *J. Geophys. Res., 98,* 7219–7228 (1993).

Santee, M. L., W. G. Read, J. W. Waters, L. Froidevaux, G. L. Manney, D. A. Flower, R. F. Jarnot, R. S. Harwood, and G. E. Peckham, "Interhemispheric Differences in Polar Stratospheric HNO_3, H_2O, ClO, and O_3," *Science, 267* 849–852 (1995).

Santee, M. L., G. L. Manney, W. G. Read, L. Froidevaux, and J. W. Waters, "Polar Vortex Conditions during the 1995–96 Arctic Winter: MLS ClO, and HNO$_3$," *Geophys. Res. Lett., 23,* 3207–3210 (1996).

Santee, M. L., A. Tabazadeh, G. L. Manney, R. J. Salawitch, L. Froidevaux, W. G. Read, and J. W. Waters, "UARS Microwave Limb Sounder HNO$_3$ Observations: Implications for Antarctic Polar Stratospheric Clouds," *J. Geophys. Res., 103,* 13285–13313 (1998).

Santer, R., M. Herman, D. Tanré, and J. Lenoble, "Characterization of Stratospheric Aerosol from Polarization Measurements," *J. Geophys. Res., 93,* 14209–14221 (1988).

Sarkissian, A., J. P. Pommereau, and F. Goutail, "Identification of Polar Stratospheric Clouds from the Ground by Visible Spectrometry," *Geophys. Res. Lett., 18,* 779–782 (1991).

Sato, M., J. E. Hansen, M. P. McCormick, and J. B. Pollack, "Stratospheric Aerosol Optical Depths, 1850–1990," *J. Geophys. Res., 98,* 22987–22994 (1993).

Schall, C., and K. G. Heumann, "GC Determination of Volatile Organoiodine and Organobromine Compounds in Arctic Seawater and Air Samples," *Fresenius' J. Anal. Chem., 346,* 717–722 (1993).

Schauffler, S. M., E. L. Atlas, F. Flocke, R. A. Lueb, V. Stroud, and W. Travnicek, "Measurements of Bromine Containing Organic Compounds at the Tropical Tropopause," *Geophys. Res. Lett., 25,* 317–320 (1998).

Schiff, H. I., "Neutral Reactions Involving Oxygen and Nitrogen," *Can. J. Chem., 47,* 1903–1916 (1969).

Schlager, H., F. Arnold, D. Hofmann, and T. Deshler, "Balloon Observations of Nitric Acid Aerosol Formation in the Arctic Stratosphere: I. Gaseous Nitric Acid," *Geophys. Res. Lett., 17,* 1275–1278 (1990).

Schmidt, S., R. N. Schindler, and T. Benter, "Photodissociation Dynamics of ClO and ClOOCl: Branching Ratios, Kinetic Energy and Quantum Yield of Primary Photoproducts," Presented at the XXIII Informal Conference on Photochemistry, May 10–15, 1998, Pasadena, CA.

Schnell, R. C., S. C. Liu, S. J. Oltmans, R. S. Stone, D. J. Hofmann, E. G. Dutton, T. Deshler, W. T. Sturges, J. W. Harder, S. D. Sewell, M. Trainer, and J. M. Harris, "Decrease of Summer Tropospheric Ozone Concentrations in Antarctica," *Nature, 351,* 726–729 (1991).

Schoeberl, M. R., and D. L. Hartmann, "The Dynamics of the Stratospheric Polar Vortex and Its Relation to Springtime Ozone Depletions," *Science, 251,* 46–52 (1991).

Schoeberl, M. R., L. R. Lait, P. A. Newman, and J. E. Rosenfield, "The Structure of the Polar Vortex," *J. Geophys. Res., 97,* 7859–7882 (1992).

Schoeberl, M. R., P. K. Bhartia, and E. Hilsenrath, "Tropical Ozone Loss Following the Eruption of Mt. Pinatubo," *Geophys. Res. Lett., 20,* 29–32 (1993a).

Schoeberl, M. R., A. R. Douglass, R. S. Stolarski, P. A. Newman, L. R. Lait, D. Toohey, L. Avallone, J. G. Anderson, W. Brune, D. W. Fahey, and K. Kelly, "The Evolution of ClO and NO along Air Parcel Trajectories," *Geophys. Res. Lett., 20,* 2511–2514 (1993b).

Schreiner, J., C. Voigt, A. Kohlmann, F. Arnold, K. Mauersberger, and N. Larsen, "Chemical Analysis of Polar Stratospheric Cloud Particles," *Science, 283,* 968–970 (1999).

Schulte, P., and H. Schlager, "In-Flight Measurements of Cruise Altitude Nitric Oxide Emission Indices of Commercial Jet Aircraft," *Geophys. Res. Lett., 23,* 165–168 (1996).

Sehested, J., K. Sehested, O. J. Nielsen, and T. J. Wallington, "Atmospheric Chemistry of FO$_2$ Radicals: Reaction with CH$_4$, O$_3$, NO, NO$_2$, and CO at 295 K," *J. Phys. Chem., 98,* 6731–6739 (1994).

Sehested, J., O. J. Nielsen, H. Egsgaard, N. W. Larsen, T. Pedersen, L. K. Christensen, and M. Wiegell, "First Direct Kinetic Study of Isotopic Enrichment of Ozone," *J. Geophys. Res., 100,* 20979–20982 (1995).

Sehested, J., O. J. Nielsen, H. Egsgaard, N. W. Larsen, T. S. Andersen, and T. Pedersen, "Kinetic Study of the Formation of Isotopically Substituted Ozone in Argon," *J. Geophys. Res., 103,* 3545–3552 (1998).

Sen, B., G. C. Toon, G. B. Osterman, J.-F. Blavier, J. J. Margitan, R. J. Salawitch, and G. K. Yue, "Measurements of Reactive Nitrogen in the Stratosphere," *J. Geophys. Res., 103,* 3571–3585 (1998).

Serça, D., A. Guenther, L. Klinger, D. Helmig, D. Hereid, and P. Zimmerman, "Methyl Bromide Deposition to Soils," *Atmos. Environ., 32,* 1581–1586 (1998).

Shibata, T., T. Itabe, K. Mizutani, and K. Asai, "Pinatubo Volcanic Aerosols Observed by Lidar at Wakkanai, Japan," *Geophys. Res. Lett., 21,* 197–200 (1994).

Shindell, D. T., and R. L. de Zafra, "Limits on Heterogeneous Processing in the Antarctic Spring Vortex from a Comparison of Measured and Modeled Chlorine," *J. Geophys. Res., 102,* 1441–1449 (1997).

Shiotani, M., and J. C. Gille, "Dynamical Factors Affecting Ozone Mixing Ratios in the Antarctic Lower Stratosphere," *J. Geophys. Res., 92,* 9811–9824 (1987).

Shorter, J. H., C. E. Kolb, P. M. Crill, R. A. Kerwin, R. W. Talbot, M. E. Hines, and R. C. Harriss, "Rapid Degradation of Atmospheric Methyl Bromide in Soils," *Nature, 377,* 717–719 (1995).

Shumilov, O. I., K. Henriksen, O. M. Raspopov, and E. A. Kasatkina, "Arctic Ozone Abundance and Solar Proton Events," *Geophys. Res. Lett., 19,* 1647–1650 (1992).

Slanger, T. G., "Energetic Molecular Oxygen in the Atmosphere," *Science, 265,* 1817–1818 (1994).

Snetsinger, K. G., G. V. Ferry, P. B. Russell, R. F. Pueschel, V. R. Oberbeck, D. M. Hayes, and W. Fong, "Effects of El Chichon on Stratospheric Aerosols Late 1982 to Early 1984," *J. Geophys. Res., 92,* 14761–14771 (1987).

Sodeau, J. R., A. B. Horn, S. F. Banham, and T. G. Koch, "Ionization of Chlorine Nitrate on Ice at 180 K," *J. Phys. Chem., 99,* 6258–6262 (1995).

Solomon, S., R. R. Garcia, F. S. Rowland, and D. J. Wuebbles, "On the Depletion of Antarctic Ozone," *Nature, 321,* 755–758 (1986).

Solomon, S., G. H. Mount, R. W. Sanders, and A. L. Schmeltekopf, "Visible Spectroscopy at McMurdo Station, Antarctica. 2. Observations of OClO," *J. Geophys. Res., 92,* 8329–8338 (1987).

Solomon, S., "The Mystery of the Antarctic Ozone 'Hole'," *Rev. Geophys., 26,* 131–148 (1988).

Solomon, S., G. H. Mount, R. W. Sanders, R. O. Jakoubek, and A. L. Schmeltekopf, "Observations of the Nighttime Abundance of OClO in the Winter Stratosphere above Thule, Greenland," *Science, 242,* 550–555 (1988).

Solomon, S., and J. G. Keys, "Seasonal Variations in Antarctic NO$_x$ Chemistry," *J. Geophys. Res., 97,* 7971–7978 (1992).

Solomon, S., J. P. Smith, R. W. Sanders, L. Perliski, H. L. Miller, G. H. Mount, J. G. Keys, and A. L. Schmeltekopf, "Visible and Near-Ultraviolet Spectroscopy at McMurdo Station, Antarctica. 8. Observations of Nighttime NO$_2$ and NO$_3$ from April to October 1991," *J. Geophys. Res., 98,* 993–1000 (1993).

Solomon, S., R. R. Garcia, and A. R. Ravishankara, "On the Role of Iodine in Ozone Depletion," *J. Geophys. Res., 99,* 20491–20499 (1994).

Solomon, S., R. W. Portmann, R. R. Garcia, L. W. Thomason, L. R. Poole, and M. P. McCormick, "The Role of Aerosol Variations in Anthropogenic Ozone Depletion at Northern Midlatitudes," *J. Geophys. Res., 101,* 6713–6727 (1996).

Solomon, S., S. Borrmann, R. R. Garcia, R. Portmann, L. Thomason, L. R. Poole, D. Winker, and M. P. McCormick, "Heterogeneous Chlorine Chemistry in the Tropopause Region," *J. Geophys. Res., 102,* 21411–21429 (1997).

Solomon, S., R. W. Portmann, R. R. Garcia, W. Randel, F. Wu, R. Nagatani, J. Gleason, L. Thomason, L. R. Poole, and M. P. McCormick, "Ozone Depletion at Mid-Latitudes: Coupling of Volcanic Aerosols and Temperature Variability to Anthropogenic Chlorine," *Geophys. Res. Lett., 25,* 1871–1874 (1998).

Song, N., "Freezing Temperatures of $H_2SO_4/HNO_3/H_2O$ Mixtures—Implications for Polar Stratospheric Clouds," *Geophys. Res. Lett., 21,* 2709–2712 (1994).

Stefanutti, L., M. Morandi, M. Del Guasta, S. Godin, G. Megie, J. Brechet, and J. Piquard, "Polar Stratospheric Cloud Observations over the Antarctic Continent at Dumont D'urville," *J. Geophys. Res., 96,* 12975–12987 (1991).

Stefanutti, L., M. Morandi, M. Del Guasta, S. Godin, and C. David, "Unusual PSCs Observed by LIDAR in Antarctica," *Geophys. Res. Lett., 22,* 2377–2380 (1995).

Stephenson, J. A. E., and M. W. J. Scourfield, "Importance of Energetic Solar Protons in Ozone Depletion," *Nature, 352,* 137–139 (1991).

Stimpfle, R. M., J. P. Koplow, R. C. Cohen, D. W. Kohn, P. O. Wennberg, D. M. Judah, D. W. Toohey, L. M. Avallone, J. G. Anderson, R. J. Salawitch, E. L. Woodbridge, C. R. Webster, R. D. May, M. H. Proffitt, K. Aiken, J. Margitan, M. Loewenstein, J. R. Podolske, L. Pfister, and K. R. Chan, "The Response of ClO Radical Concentrations in the Lower Stratosphere," *Geophys. Res. Lett., 21,* 2543–2546 (1994).

Stolarski, R. S., and R. J. Cicerone, "Stratospheric Chlorine: A Possible Sink for Ozone," *Can. J. Chem., 52,* 1610–1615 (1974).

Stolarski, R. S., and R. D. Rundel, "Fluorine Photochemistry in the Stratosphere," *Geophys. Res. Lett., 2,* 443–445 (1975).

Stolarski, R. S., S. L. Baughcum, W. H. Brune, A. R. Douglass, D. W. Fahey, R. R. Friedl, S. C. Liu, R. A. Plumb, L. R. Poole, H. L. Wesoky, and D. R. Worsnop, "1995 Scientific Assessment of the Atmospheric Effects of Stratospheric Aircraft," NASA Reference Publication 1381, November 1995.

Stone, R. S., J. R. Key, and E. G. Duton, "Properties and Decay of Stratospheric Aerosols in the Arctic Following the 1991 Eruptions of Mount Pinatubo," *Geophys. Res. Lett., 20,* 2359–2362 (1993).

Strand, A., and Ø. Hov, "The Impact of Man-Made and Natural NO_x Emissions on Upper Tropospheric Ozone: A Two-Dimensional Model Study," *Atmos. Environ., 30,* 1291–1303 (1996).

Sugita, T., Y. Kondo, H. Nakajima, U. Schmidt, A. Engel, H. Oelhaf, G. Wetzel, M. Koike, and P. A. Newman, "Denitrification Observed inside the Arctic Vortex in February 1995," *J. Geophys. Res., 103,* 16221–16233 (1998).

Syage, J. A., and M. N. Ross, "An Assessment of the Total Ozone Mapping Spectrometer for Measuring Ozone Levels in a Solid Rocket Plume," *Geophys. Res. Lett., 23,* 3227–3230 (1996).

Symonds, R. B., W. I. Rose, and M. H. Reed, "Contribution of Cl- and F-Bearing Gases to the Atmosphere by Volcanoes," *Nature, 334,* 415–418 (1988).

Taalas, P., J. Damski, and E. Kyrö, "Effect of Stratospheric Ozone Variations on UV Radiation and on Tropospheric Ozone at High Latitudes," *J. Geophys. Res., 102,* 1533–1539 (1997).

Tabazadeh, A., and R. P. Turco, "A Model for Heterogeneous Chemical Processes on the Surfaces of Ice and Nitric Acid Trihydrate Particles," *J. Geophys. Res., 98,* 12727–12740 (1993).

Tabazadeh, A., R. P. Turco, K. Drdla, and M. Z. Jacobson, "A Study of Type I Polar Stratospheric Cloud Formation," *Geophys. Res. Lett., 21,* 1619–1622 (1994a).

Tabazadeh, A., R. P. Turco, K. Drdla, M. Z. Jacobson, and O. B. Toon, "Correction to 'A Study of Type I Polar Stratospheric Cloud Formation,'" *Geophys. Res. Lett., 21,* 2869 (1994b).

Tabazadeh, A., O. B. Toon, and P. Hamill, "Freezing Behavior of Stratospheric Sulfate Aerosols Inferred from Trajectory Studies," *Geophys. Res. Lett., 22,* 1725–1728 (1995).

Tabazadeh, A., and O. B. Toon, "The Presence of Metastable HNO_3/N_2O Solid Phases in the Stratosphere Inferred from ER-2 Data," *J. Geophys. Res., 101,* 9071–9078 (1996).

Tabazadeh, A., O. B. Toon, B. L. Gary, J. T. Bacmeister, and M. R. Schoeberl, "Observational Constraints on the Formation of Type Ia Polar Stratospheric Clouds," *Geophys. Res. Lett., 23,* 2109–2112 (1996).

Tabazadeh, A., O. B. Toon, and E. J. Jensen, "Formation and Implications of Ice Particle Nucleation in the Stratosphere," *Geophys. Res. Lett., 24,* 2007–2010 (1997).

Thibert, E., and F. Dominé, "Thermodynamics and Kinetics of the Solid Solution of HCl In Ice," *J. Phys. Chem. B, 101,* 3554–3565 (1997).

Thiemens, M. H., T. Jackson, E. C. Zipf, P. W. Erdman, and C. van Egmond, "Carbon Dioxide and Oxygen Isotope Anomalies in the Mesosphere and Stratosphere," *Science, 270,* 969–972 (1995).

Thomas, R. J., K. H. Rosenlof, R. T. Clancy, and J. M. Zawodny, "Stratospheric NO_2 over Antarctica As Measured by the Solar Mesosphere Explorer during Austral Spring, 1986," *J. Geophys. Res., 93,* 12561–12568 (1988).

Thomas, V. M., J. A. Bedford, and R. J. Cicerone, "Bromine Emissions from Leaded Gasoline," *Geophys. Res. Lett., 24,* 1371–1374 (1997).

Thomason, L. W., G. S. Kent, C. R. Trepte, and L. R. Poole, "A Comparison of the Stratospheric Aerosol Background Periods of 1979 and 1989–1991," *J. Geophys. Res., 102,* 3611–3616 (1997).

Thompson, A. M., "New Ozone Hole Phenomenon," *Nature, 352,* 282–283 (1991).

Tie, X., G. P. Brasseur, B. Briegleb, and C. Granier, "Two-Dimensional Simulation of Pinatubo Aerosol and Its Effect on Stratospheric Ozone," *J. Geophys. Res., 99,* 20545–20562 (1994).

Tie, X., and G. Brasseur, "The Response of Stratospheric Ozone to Volcanic Eruptions: Sensitivity to Atmospheric Chlorine Loading," *Geophys. Res. Lett., 22,* 3035–3038 (1995).

Tie, X., and G. Brasseur, "The Importance of Heterogeneous Bromine Chemistry in the Lower Stratosphere," *Geophys. Res. Lett., 23,* 2505–2508 (1996).

Tisdale, R. T., A. M. Middlebrook, A. J. Prenni, and M. A. Tolbert, "Crystallization Kinetics of HNO_3/H_2O Films Representative of Polar Stratospheric Clouds," *J. Phys. Chem. A, 101,* 2112–2119 (1997).

Tolbert, M. A., M. J. Rossi, R. Malhotra, and D. M. Golden, "Reaction of Chlorine Nitrate with Hydrogen Chloride and Water at Antarctic Stratospheric Temperatures," *Science, 238,* 1258–1260 (1987).

Tolbert, M. A., M. J. Rossi, and D. M. Golden, "Heterogeneous Interactions of Chlorine Nitrate, Hydrogen Chloride, and Nitric Acid with Sulfuric Acid Surfaces at Stratospheric Temperatures," *Geophys. Res. Lett., 15,* 847–850 (1988a).

Tolbert, M. A., M. J. Rossi, and D. M. Golden, "Antarctic Ozone Depletion Chemistry: Reactions of N_2O_5 with H_2O and HCl on Ice Surfaces," *Science, 240,* 1018–1021 (1988b).

Tolbert, M. A., and A. M. Middlebrook, "Fourier Transform Infrared Studies of Model Polar Stratospheric Cloud Surfaces: Growth and

Evaporation of Ice and Nitric Acid/Ice," *J. Geophys. Res., 95,* 22423–22431 (1990).

Tolbert, M. A., J. Pfaff, I. Jayaweera, and M. J. Prather, "Uptake of Formaldehyde by Sulfuric Acid Solutions: Impact on Stratospheric Ozone," *J. Geophys. Res., 98,* 2957–2962 (1993).

Tolbert, M. A., "Sulfate Aerosols and Polar Stratospheric Cloud Formation," *Science, 264,* 527–528 (1994).

Tolbert, M. A., "Polar Clouds and Sulfate Aerosols," *Science, 272,* 1597–1598 (1996).

Toohey, D. W., L. M. Avallone, L. R. Lait, P. A. Newman, M. R. Schoeberl, D. W. Fahey, E. L. Woodbridge, and J. G. Anderson, "The Seasonal Evolution of Reactive Chlorine in the Northern Hemisphere Stratosphere," *Science, 261,* 1134–1136 (1993).

Toohey, D. W., "A Critical Review of Stratospheric Chemistry Research in the U.S.: 1991–1994," Paper 95RG00400, *Rev. Geophys. Suppl.,* 759–773 (1995).

Toon, G. C., C. B. Farmer, L. L. Lowes, P. W. Schaper, J.-F. Blavier, and R. H. Norton, "Infrared Aircraft Measurements of Stratospheric Composition over Antarctica during September 1987," *J. Geophys. Res., 94,* 16571–16596 (1989).

Toon, G. C., C. B. Farmer, P. W. Schaper, L. L. Lowes, and R. H. Norton, "Composition Measurements of the 1989 Arctic Winter Stratosphere by Airborne Infrared Solar Absorption Spectroscopy," *J. Geophys. Res., 97,* 7939–7961 (1992).

Toon, O. B., E. V. Browell, S. Kinne, and J. Jordan, "An Analysis of Lidar Observations of Polar Stratospheric Clouds," *Geophys. Res. Lett., 17,* 393–396 (1990a).

Toon, O. B., R. P. Turco, and P. Hamill, "Denitrification Mechanisms in the Polar Stratospheres," *Geophys. Res. Lett., 17,* 445–448 (1990b).

Toon, O. B., M. A. Tolbert, B. G. Koehler, A. M. Middlebrook, and J. Jordan, "The Infrared Optical Constants of H_2O-Ice, Amorphous Nitric Acid Solutions, and Nitric Acid Hydrates," *J. Geophys. Res., 99,* 25631–25654 (1994).

Toon, O. B., and M. A. Tolbert, "Spectroscopic Evidence against Nitric Acid Trihydrate in Polar Stratospheric Clouds," *Nature, 375,* 218–221 (1995).

Toumi, R., S. Bekki, and K. S. Law, "Indirect Influence of Ozone Depletion on Climate Forcing by Clouds," *Nature, 372,* 348–351 (1994).

Trepte, C. R., and M. H. Hitchman, "Tropical Stratospheric Circulation Deduced from Satellite Aerosol Data," *Nature, 355,* 626–628 (1992).

Trolier, M., R. L. Mauldin III, and A. R. Ravishankara, "Rate Coefficient for the Termolecular Channel of the Self-Reaction of ClO," *J. Phys. Chem., 94,* 4896–4907 (1990).

Tuck, A. F., "Synoptic and Chemical Evolution of the Antarctic Vortex in Late Winter and Early Spring, 1987," *J. Geophys. Res., 94,* 11687–11737 (1989).

Tuck, A. F., D. W. Fahey, M. Loewenstein, J. R. Podolske, K. K. Kelly, S. J. Hovde, D. M. Murphy, and J. W. Elkins, "Spread of Denitrification from 1987 Antarctic and 1988–1989 Arctic Stratospheric Vortices," *J. Geophys. Res., 99,* 20573–20583 (1994).

Tuck, A. F., and M. H. Proffitt, "Comment on 'On the Magnitude of Transport out of the Antarctic Polar Vortex,' by W. M. F. Wauben et al.," *J. Geophys. Res., 102,* 28215–28218 (1997).

Tung, K. K., M. K. W. Ko, J. M. Rodriguez, and N. D. Sze, "Are Antarctic Ozone Variations a Manifestation of Dynamic or Chemistry?" *Nature, 333,* 811–814 (1986).

Tung, K. K., and H. Yang, "Dynamical Component of Seasonal and Year-to-Year Changes in Antarctic and Global Ozone," *J. Geophys. Res., 93,* 12537–12559 (1988).

Turco, R. P., O. B. Toon, R. C. Whitten, and R. J. Cicerone, "Space Shuttle Ice Nuclei," *Nature, 298,* 830–832 (1982).

Turnipseed, A. A., M. K. Gilles, J. B. Burkholder, and A. R. Ravishankara, "Kinetics of the IO Radical. 1. Reaction of IO with ClO," *J. Phys. Chem. A, 101,* 5517–5525 (1997).

Vaida, V., and J. D. Simon, "The Photoreactivity of Chlorine Dioxide," *Science, 268,* 1443–1448 (1995).

Van Allen, R., X. Liu, and F. J. Murcray, "Seasonal Variation of Atmospheric Nitric Acid over the South Pole in 1992," *Geophys. Res. Lett., 22,* 49–52 (1995).

Van Roozendael, M., M. De Mazière, C. Hermans, P. C. Simon, J.-P. Pommereau, F. Goutail, X. X. Tie, G. Brasseur, and C. Granier, "Ground-Based Observations of Stratospheric NO_2 at High and Midlatitudes in Europe after the Mount Pinatubo Eruption," *J. Geophys. Res., 102,* 19171–19176 (1997).

Varner, R. K., P. M. Crill, R. W. Talbot, and J. H. Shorter, "An Estimate of the Uptake of Atmospheric Methyl Bromide by Agricultural Soils," *Geophys. Res. Lett., 26,* 727–730 (1999).

Vogt, R., and B. J. Finlayson-Pitts, "A Diffuse Reflectance Infrared Fourier Transform Spectroscopic (DRIFTS) Study of the Surface Reaction of NaCl with Gaseous NO_2 and HNO_3," *J. Phys. Chem., 98,* 3747–3755 (1994); *ibid., 99,* 13052.

Vogt, R., C. Elliott, H. C. Allen, J. M. Laux, J. C. Hemminger, and B. J. Finlayson-Pitts, "Some New Laboratory Approaches to Studying Tropospheric Heterogeneous Reactions," *Atmos. Environ., 30,* 1729–1737 (1996).

Volk, C. M., J. W. Elkins, D. W. Fahey, R. J. Salawitch, G. S. Dutton, J. M. Gilligan, M. H. Proffitt, M. Loewenstein, J. R. Podolske, K. Minschwaner, J. J. Margitan, and K. R. Chan, "Quantifying Transport between the Tropical and Mid-Latitude Lower Stratosphere," *Science, 272,* 1763–1768 (1996).

Volk, C. M., J. W. Elkins, D. W. Fahey, G. S. Dutton, J. M. Gilligan, M. Loewenstein, J. R. Podolske, K. R. Chan, and M. R. Gunson, "Evaluation of Source Gas Lifetimes from Stratospheric Observations," *J. Geophys. Res., 102,* 25543–25564 (1997).

Vömel, H., M. Rummukainen, R. Kivi, J. Karhu, T. Turunen, E. Kyrö, J. Rosen, N. Kjome, and S. Oltmans, "Dehydration and Sedimentation of Ice Particles in the Arctic Stratospheric Vortex," *Geophys. Res. Lett., 24,* 795–798 (1997).

von der Gathen, P., M. Rex, N. R. P. Harris, D. Lucic, B. M. Knudsen, G. O. Braathen, H. De Backer, R. Fabian, H. Fast, M. Gil, E. Kyrö, I. S. Mikkelsen, M. Rummukainen, J. Stähelin, and C. Varotsos, "Observational Evidence for Chemical Ozone Depletion over the Arctic in Winter 1991-92," *Nature, 375,* 131–134 (1995).

Wahner, A., and C. Schiller, "Twilight Variation of Vertical Column Abundances of OClO and BrO in the North Polar Region," *J. Geophys. Res., 97,* 8047–8055 (1992).

Wamsley, P. R., J. W. Elkins, D. W. Fahey, G. S. Dutton, C. M. Volk, R. C. Myers, S. A. Montzka, J. H. Butler, A. D. Clarke, P. J. Fraser, L. P. Steele, M. P. Lucarelli, E. L. Atlas, S. M. Schauffler, D. R. Blake, F. S. Rowland, W. T. Sturges, J. M. Lee, S. A. Penkett, A. Engel, R. M. Stimpfle, K. R. Chan, D. K. Weisenstein, M. K. W. Ko, and R. J. Salawitch, "Distribution of Halon-1211 in the Upper Troposphere and Lower Stratosphere and the 1994 Total Bromine Budget," *J. Geophys. Res., 103,* 1513–1526 (1998).

Wang, P.-H., D. M. Cunnold, J. M. Zawodny, R. B. Pierce, J. R. Olson, G. S. Kent, and K. M. Skeens, "Seasonal Ozone Variations in the Isentropic Layer between 330 and 380 K As Observed by SAGE II: Implications of Extratropical Cross-Tropopause Transport," *J. Geophys. Res., 103,* 28647–28659 (1998).

Warneck, P., "Chemistry of the Natural Atmosphere," in *International Geophysics Series* (R. Dmowska and J. R. Holton, Eds.), Vol. 41, Academic Press, San Diego, 1988.

Waters, J. W., L. Froidevaux, W. G. Read, G. L. Manney, L. S. Elson, D. A. Flower, R. F. Jarnot, and R. S. Harwood, "Stratospheric

ClO and Ozone from the Microwave Limb Sounder on the Upper Atmosphere Research Satellite," *Nature, 362,* 597–603 (1993).

Wauben, W. M. F., R. Bintanja, P. F. J. van Velthoven, and H. Kelder, "On the Magnitude of Transport out of the Antarctic Polar Vortex," *J. Geophys. Res., 102,* 1229–1238 (1997a).

Wauben, W. M. F., P. F. J. van Velthoven, H. Kelder, and R. Bintanja, "Reply," *J. Geophys. Res., 102,* 28219–28221 (1997b).

Wayne, R. P., G. Poulet, P. Biggs, J. P. Burrows, R. A. Cox, P. J. Crutzen, G. D. Hayman, M. E. Jenkin, G. Le Bras, G. K. Moortgat, U. Platt, and R. N. Schindler, "Halogen Oxides—Radicals, Sources and Reservoirs in the Laboratory and the Atmosphere," *Atmos. Environ., 29,* 2677–2881 (1995).

Weaver, C. J., A. R. Douglass, and D. B. Considine, "A 5-Year Simulation of Supersonic Aircraft Emission Transport Using a Three-Dimensional Model," *J. Geophys. Res., 101,* 20975–20984 (1996).

Webster, C. R., R. D. May, D. W. Toohey, L. M. Avallone, J. G. Anderson, P. Newman, L. Lait, M. R. Schoeberl, J. W. Elkins, and K. R. Chan, "Chlorine Chemistry on Polar Stratospheric Cloud Particles in the Arctic Winter," *Science, 261,* 1130–1134 (1993a).

Webster, C. R., R. D. May, D. W. Toohey, L. M. Avallone, J. G. Anderson, and S. Solomon, "*In Situ* Measurements of the ClO/HCl Ratio: Heterogeneous Processing on Sulfate Aerosols and Polar Stratospheric Clouds," *Geophys. Res. Lett., 20,* 2523–2526 (1993b).

Webster, C. R., R. D. May, M. Allen, L. Jaegle, and M. P. McCormick, "Balloon Profiles of Stratospheric NO_2 and HNO_3 for Testing the Heterogeneous Hydrolysis of N_2O_5 on Sulfate Aerosols," *Geophys. Res. Lett., 21,* 53–56 (1994).

Webster, C. R., R. D. May, H. A. Michelsen, D. C. Scott, J. C. Wilson, H. H. Jonsson, C. A. Brock, J. E. Dye, D. Baumgardner, R. M. Stimpfle, J. P. Koplow, J. J. Margitan, M. H. Proffitt, L. Jaeglé, R. L. Herman, H. Hu, G. J. Flesch, and M. Loewenstein, "Evolution of HCl Concentrations in the Lower Stratosphere from 1991 to 1996 Following the Eruption of Mt. Pinatubo," *Geophys. Res. Lett., 25,* 995–998 (1998).

Weisenstein, D. K., M. K. W. Ko, J. M. Rodriguez, and N.-D. Sze, "Effects on Stratospheric Ozone from High-Speed Civil Transport: Sensitivity to Stratospheric Aerosol Loading," *J. Geophys. Res., 98,* 23133–23140 (1993).

Weisenstein, D. K., M. K. W. Ko, N.-D. Sze, and J. M. Rodriguez, "Potential Impact of SO_2 Emissions from Stratospheric Aircraft on Ozone," *Geophys. Res. Lett., 23,* 161–164 (1996).

Wennberg, P. O., R. C. Cohen, R. M. Stimpfle, J. P. Koplow, J. G. Anderson, R. J. Salawitch, D. W. Fahey, E. L. Woodbridge, E. R. Keim, R. S. Gao, C. R. Webster, R. D. May, D. W. Toohey, L. M. Avallone, M. H. Proffitt, M. Loewenstein, J. R. Podolske, K. R. Chan, and S. C. Wofsy, "Removal of Stratospheric O_3 by Radicals: *In Situ* Measurements of OH, HO_2, NO, NO_2, ClO, and BrO," *Science, 266,* 398–404 (1994).

Wennberg, P. O., J. W. Brault, T. F. Hanisco, R. J. Salawitch, and G. H. Mount, "The Atmospheric Column Abundance of IO: Implications for Stratospheric Ozone," *J. Geophys. Res., 102,* 8887–8898 (1997).

Williams, L. R., D. M. Golden, and D. L. Huestis, "Solubility of HBr in Sulfuric Acid at Stratospheric Temperatures," *J. Geophys. Res., 100,* 7329–7335 (1995).

Wilson, J. C., M. R. Stolzenburg, W. E. Clark, M. Loewenstein, G. V. Ferry, K. R. Chan, and K. K. Kelly, "Stratospheric Sulfate Aerosol in and near the Northern Hemisphere Polar Vortex: The Morphology of the Sulfate Layer, Multimodal Size Distributions, and the Effect of Denitrification," *J. Geophys. Res., 97,* 7997–8013 (1992).

Wilson, J. C., H. H. Jonsson, C. A. Brock, D. W. Toohey, L. M. Avallone, D. Baumgardner, J. E. Dye, L. R. Poole, D. C. Woods, R. J. DeCoursey, M. Osborn, M. C. Pitts, K. K. Kelly, K. R. Chan, G. V. Ferry, M. Loewenstein, J. R. Podolske, and A. Weaver, "*In Situ* Observations of Aerosol and Chlorine Monoxide after the 1991 Eruption of Mount Pinatubo: Effect of Reactions on Sulfate Aerosol," *Science, 261,* 1140–1143 (1993).

Wilson, J. C., "The Stratospheric Aerosol and Its Impact on Stratospheric Chemistry, in *Perspectives in Environmental Chemistry* (D. L. Macalady, Ed.), pp. 344–368, Oxford Univ. Press, New York, 1998.

Wingenter, O. W., C. J.-L. Wang, D. R. Blake, and F. S. Rowland, "Seasonal Variation of Tropospheric Methyl Bromide Concentrations: Constraints on Anthropogenic Input," *Geophys. Res. Lett., 25,* 2797–2800 (1998).

Wolff, E. W., and R. Mulvaney, "Reactions on Sulphuric Acid Aerosol and on Polar Stratospheric Clouds in the Antarctic Stratosphere," *Geophys. Res. Lett., 18,* 1007–1010 (1991).

Woods, D. C., R. L. Chuan, and W. I. Rose, "Halite Particles Injected into the Stratosphere by the 1982 El Chichon Eruption," *Science, 230,* 170–172 (1985).

Wooldridge, P. J., R. Zhang, and M. J. Molina, "Phase Equilibria of H_2SO_4, HNO_3, and HCl Hydrates and the Composition of Polar Stratospheric Clouds," *J. Geophys. Res., 100,* 1389–1396 (1995).

World Meteorological Organization (WMO), "Scientific Assessment of Ozone Depletion: 1994," Global Ozone Research and Monitoring Project, Report No. 37, published February 1995; 1998 update, Report No. 44, published February, 1999.

Worsnop, D. R., L. E. Fox, M. S. Zahniser, and S. C. Wofsy, "Vapor Pressures of Solid Hydrates of Nitric Acid: Implications for Polar Stratospheric Clouds," *Science, 259,* 71–74 (1993).

Yagi, I., J. Williams, N.-Y. Wang, and R. J. Cicerone, "Agricultural Soil Fumigations as a Source of Atmospheric Methyl Bromide," *Proc. Natl. Acad. Sci. U.S.A., 90,* 8420–8423 (1993).

Yagi, K., J. Williams, N.-Y. Wang, and R. J. Cicerone, "Atmospheric Methyl Bromide (CH_3Br) from Agricultural Soil Fumigations," *Science, 267,* 1979–1981 (1995).

Yates, S. R., D. Wang, J. Gan, F. F. Ernst, and W. A. Jury, "Minimizing Methyl Bromide Emissions from Soil Fumigation," *Geophys. Res. Lett., 25,* 1633–1636 (1998).

Yokouchi, Y., H. Mukai, H. Yamamoto, A. Otsuki, C. Saitoh, and Y. Nojiri, "Distribution of Methyl Iodide, Ethyl Iodide, Bromoform and Dibromomethane over the Ocean (East and Southeast Asian Seas and the Western Pacific)," *J. Geophys. Res., 102,* 8805–8809 (1997).

Yu, F., and R. P. Turco, "The Role of Ions in the Formation and Evolution of Particles in Aircraft Plumes," *Geophys. Res. Lett., 24,* 1927–1930 (1997).

Yung, Y. L., J. P. Pinto, R. T. Watson, and S. P. Sander, "Atmospheric Bromine and Ozone Perturbations in the Lower Stratosphere," *J. Atmos. Sci., 37,* 339–353 (1980).

Yvon, S. A., and J. H. Butler, "An Improved Estimate of the Oceanic Lifetime of Atmospheric CH_3Br," *Geophys. Res. Lett., 23,* 53–56 (1996).

Yvon-Lewis, S. A., and J. H. Butler, "The Potential Effect of Oceanic Biological Degradation on the Lifetime of Atmospheric CH_3Br," *Geophys. Res. Lett., 24,* 1227–1230 (1997).

Zander, R., E. Mahieu, M. R. Gunson, M. C. Abrams, A. Y. Chang, M. Abbas, C. Aellig, A. Engel, A. Goldman, F. W. Irion, N. Kämpfer, H. A. Michelsen, M. J. Newchurch, C. P. Rinsland, R. J. Salawitch, G. P. Stiller, and G. C. Toon, "The 1994 Northern Midlatitude Budget of Stratospheric Chlorine Derived from ATMOS/ATLAS-3 Observations," *Geophys. Res. Lett., 23,* 2357–2360 (1996).

Zawodny, J. M., and M. P. McCormick, "Stratospheric Aerosol and Gas Experiment II Measurements of the Quasi-Biennial Oscillations in Ozone and Nitrogen Dioxide," *J. Geophys. Res., 96,* 9371–9377 (1991).

Zhang, R., P. J. Wooldridge, J. P. D. Abbatt, and M. J. Molina, "Physical Chemistry of the H_2SO_4/H_2O Binary System at Low Temperatures: Stratospheric Implications," *J. Phys. Chem., 97,* 7351–7358 (1993).

Zhang, R., J. T. Jayne, and M. J. Molina, "Heterogeneous Interactions of $ClONO_2$ and HCl with Sulfuric Acid Tetrahydrate: Implications for the Stratosphere," *J. Phys. Chem., 98,* 867–874 (1994a).

Zhang, R., M.-T. Leu, and L. F. Keyser, "Heterogeneous Reactions of $ClONO_2$, HCl, and HOCl on Liquid Sulfuric Acid Surfaces," *J. Phys. Chem., 98,* 13563–13574 (1994b).

Zhang, R., M.-T. Leu, and L. F. Keyser, "Sulfuric Acid Monohydrate: Formation and Heterogeneous Chemistry in the Stratosphere," *J. Geophys. Res., 100,* 18845–18854 (1995).

Zhang, R., M.-T. Leu, and M. J. Molina, "Formation of Polar Stratospheric Clouds on Preactivated Background Aerosols," *Geophys. Res. Lett., 23,* 1669–1672 (1996).

Zhang, R., M.-T. Leu, and L. F. Keyser, "Heterogeneous Chemistry of HO_2NO_2 in Liquid Sulfuric Acid," *J. Phys. Chem. A, 101,* 3324–3330 (1997).

Zhao, X., R. P. Turco, C.-Y. J. Kao, and S. Elliott, "Numerical Simulation of the Dynamical Response of the Arctic Vortex to Aerosol-Associated Chemical Perturbations in the Lower Stratosphere," *Geophys. Res. Lett., 23,* 1525–1528 (1996).

Zhao, Z., R. E. Stickel, and P. H. Wine, "Quantum Yield for Carbon Monoxide Production in the 248 nm Photodissociation of Carbonyl Sulfide (OCS)," *Geophys. Res. Lett., 22,* 615–618 (1995).

Zheng, J., A. J. Weinheimer, B. A. Ridley, S. C. Liu, G. W. Sachse, B. E. Anderson, and J. E. Collins, Jr., "An Analysis of Aircraft Exhaust Plumes from Accidental Encounters," *Geophys. Res. Lett., 21,* 2579–2582 (1994).

Zipf, E. C., and S. S. Prasad, "Evidence for New Sources of NO_x in the Lower Atmosphere," *Science, 279,* 211–213 (1998a).

Zipf, E. C., and S. S. Prasad, "Experimental Evidence That Excited Ozone Is a Source of Nitrous Oxide," *Geophys. Res. Lett., 25,* 4333–4336 (1998b).

Zolensky, M. E., D. S. McKay, and L. A. Kaczor, "A Tenfold Increase in the Abundance of Large Solid Particles in the Stratosphere, As Measured over the Period 1976–1984," *J. Geophys. Res., 94,* 1047–1056 (1989).

Zondlo, M. A., S. B. Barone, and M. A. Tolbert, "Condensed-Phase Products in Heterogeneous Reactions: N_2O_5, $ClONO_2$ and HNO_3 Reacting on Ice Films at 185 K," *J. Phys. Chem. A, 102,* 5735–5748 (1998).

Zurer, P. S., "NASA Cultivating Basic Technology for Supersonic Passenger Aircraft," *Chem. Eng. News,* 10–16, April 24 (1995).

CHAPTER 13

Scientific Basis for Control of Halogenated Organics

Despite basic understanding of the fundamental gas-phase chemistry discussed in Chapter 12 and the expectation based on this that CFCs would cause stratospheric ozone destruction, there was no clear and incontrovertible evidence that this was occurring until about the mid-1980s. The reason was the difficulty of reliably extracting small changes in stratospheric ozone superimposed on top of natural variations arising from a number of sources. However, with the discovery of the Antarctic "ozone hole" and ruling out natural causes for this dramatic ozone loss, it became firmly established that CFCs were indeed leading to stratospheric ozone destruction. Subsequent studies in the Arctic and at midlatitudes (discussed herein) have shown that ozone destruction has occurred globally.

This area illustrates an interesting policy and regulatory dilemma that was faced in the late 1970s, because of the long delay times between release of the CFCs at the earth's surface and their impact on stratospheric ozone. Thus, although the science indicated the urgent need for controls starting with the seminal Molina and Rowland paper in 1974, there was at that time no concrete evidence for stratospheric ozone destruction from CFCs. However, had no policy decisions or regulatory actions been taken until such evidence was available and universally accepted, the magnitude and time horizon for ozone destruction would have been much more severe.

In this chapter, we discuss the international agreements for phasing out halogenated organics and the resulting trends in the atmospheric concentrations of CFCs. The assessment of the relative importance of various compounds in terms of their ozone depletion potential (ODP) is summarized. Trends in stratospheric ozone and associated changes in surface UV radiation are treated, and finally, the chemistry of alternatives to CFCs that are finding increasing use as the phase-outs of CFCs take place.

A. INTERNATIONAL AGREEMENTS ON PHASEOUT OF HALOGENATED ORGANICS

Because of the recognition in the mid-1970s that CFCs were a potential threat to stratospheric ozone, their use in aerosol sprays was banned in the United States in 1978. On an international level, the Vienna Convention for the Protection of the Ozone Layer agreed in March 1985 on the need to control emissions of CFCs and other chlorine-containing substances. This was followed by the Montreal Protocol in September 1987, which limited the production and import internationally of CFCs as well as halons. Specifically, the consumption of CFC-11, -12, -113, -114, and -115 was to be reduced to their 1986 levels by 1990, to 80% of these levels by 1994, and to 50% by 1999. The Halons-1211, -1301, and -2402 were specified to return to the 1986 levels by 1992. For developing countries, the target for phaseout was the year 2010 (see *International Legal Materials*, 1987a,b, for details).

However, about this time, a variety of research indicated that even with full implementation of the Montreal Protocol, the atmospheric abundance of chlorine could reach as much as 6–9 ppb between the years 2050 and 2075. This delay is due to the relatively long time between emission of these compounds into the troposphere and when they reach the stratosphere and photolyze to produce an active chlorine atom. Figure 13.1, for example, compares the estimated equivalent effective stratospheric chlorine from 1960 until the year 2100 with no controls and a 3% increase per year in CFC and methylchloroform emissions to those with the controls agreed to in the Montreal Protocol. Equivalent effective stratospheric chlorine loading depends on emissions as well as removal processes, which determine what fraction of the CFCs emitted at the earth's

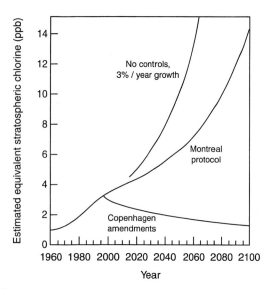

FIGURE 13.1 Estimated equivalent effective stratospheric chlorine for a continued 3% growth per year, for controls contained in the Montreal Protocol, and for those in the Copenhagen amendments (adapted from World Meteorological Organization, 1995).

surface reach the stratosphere. In addition, the effectiveness of compounds in destroying stratospheric ozone depends on how readily they dissociate to form chlorine or bromine. As discussed in Chapter 12.D, bromine contributes significantly to ozone destruction as well and, indeed, is more efficient than chlorine under the current atmospheric conditions by about a factor of 60 (World Meteorological Organization, 1999). The contribution of bromine can be translated into equivalent chlorine by taking this increased efficiency into account. The net effect of chlorine and bromine is then expressed as "equivalent effective stratospheric chlorine" and it is this quantity that is shown in Fig. 13.1.

While the growth in stratospheric chlorine should clearly be slowed by the Montreal Protocol agreements, it was still substantial and expected to lead to quite large losses of ozone. This recognition, bolstered by the dramatic appearance of the Antarctic ozone hole, led to further major amendments to the Montreal Protocol.

The Montreal Protocol was first modified in June 1990 in what is known as the London Amendments. These amendments speeded up the phaseout of CFCs and halons to 50% by 1995, 85% by 1997, and 100% by 2000, or 2010 for the developing countries (e.g., see Chatterjee, 1995; and Parson and Greene, 1995). Carbon tetrachloride was to be fully phased out by 2000 and methylchloroform (CH_3CCl_3) by 2005. Included in the controls were a variety of additional CFCs, including CFC-13, -111, -112, -211, -212, -213, -214, -215, -216, and -217, not covered in the original protocol. A non-binding portion of the agreement included phasing out the interim replacement compounds for CFCs, the HCFCs (hydrochlorofluorocarbons), with a target date of 2020, but not later than 2040. (See *International Legal Materials*, 1991, for details.)

A further amendment was subsequently signed in November 1992 in Copenhagen. This speeded up the complete phaseout of CFCs, CCl_4, and CH_3CCl_3 to 1996 and the halons from 2000 to 1994. The phaseout target for developing nations continued to be the year 2010. The HCFCs were to be reduced by 35% by 2004, 65% by 2010, 90% by 2015, and 99.5% by 2020, with full phaseout by 2030. Starting in 1995, the production of methyl bromide in developed countries was held at 1991 levels. In mid-July 1994, most countries had agreed to the restrictions; approximately 40 countries, with ~6% of the world's population, had not agreed to the controls (Holmes and Ellis, 1996).

Figure 13.1 also shows the estimated equivalent effective stratospheric chlorine content as a function of year under the Copenhagen Amendments. Chlorine is expected to peak around the year 2000 and then decrease. As discussed by Holmes and Ellis (1996), noncompliance and allowed exemptions could lead to a much slower decline than projected in Fig. 13.1.

In November 1995, it was agreed that CH_3Br usage would be eliminated by 2010 in developed countries; in developing countries, its use is frozen in the year 2002 at the average 1995–1998 levels. Furthermore, HCFCs would be phased out in developed countries by 2020, with some production allowed until 2030 for application in existing refrigeration and air conditioning units. In developing countries, the use of HCFCs will be frozen at 2015 levels in the year 2016, with complete elimination by 2040 (Zurer, 1995).

Figure 13.2 summarizes these international agreements for developed countries through ~1995 (Parson and Greene, 1995). In September 1997, it was agreed that the use of methyl bromide by developed countries would end earlier, in 2005, and in developing countries by 2015 (Spurgeon, 1997a). In addition, steps to try to reduce smuggling of CFCs, which had become a problem (Spurgeon, 1997b), were taken.

As discussed in more detail in this chapter, detecting trends in stratospheric ozone and deconvoluting the causes are complex, particularly outside the polar regions. However, it is estimated that for the Antarctic, where the most dramatic loss of ozone has been observed, recovery may be experimentally observable by the year 2008 if the Montreal Protocol and associated amendments are followed (e.g., Hofmann *et al.*, 1994).

When first developed as refrigerants in 1928 by the DuPont chemists Midgley and Henne (1930), CFCs appeared to have ideal properties for this application

A. INTERNATIONAL AGREEMENTS ON PHASEOUT OF HALOGENATED ORGANICS

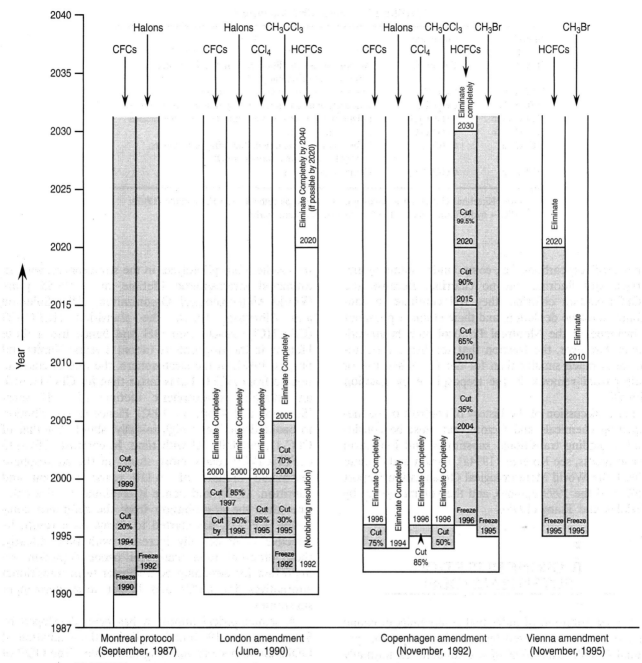

FIGURE 13.2 Summary of Montreal Protocol and subsequent amendments (adapted from Parson and Greene, 1995).

and, subsequently, for use as aerosol propellants and blowing agents. They were nontoxic, nonflammable, and highly stable. Of course, as described in the previous chapter, it is now known that it is these very properties that render them virtually nonreactive in the troposphere and hence give them sufficiently long lifetimes in the lower atmosphere to reach the stratosphere.

Table 13.1 shows the most common CFC substitutes currently in use. The physical and chemical properties that are needed for substitute compounds are described in an article by Manzer (1990) and the book edited by Miziolek and Tsang (1995). Overall, the approach in the short term has been to make these compounds more reactive in the troposphere by adding one or more abstractable hydrogens. The term "HCFC" is used for hydrochlorofluorocarbons, i.e., compounds containing hydrogen, chlorine, and fluorine, and "HFC"

TABLE 13.1 Some CFC Alternates[a]

Name[b]	Compound	Use
HFC-134a	CH_2FCF_3	Automobile air conditioners, commercial chillers, residential refrigerators
HFC-125	CHF_2CF_3	Industrial and commercial refrigeration
HFC-152a	CH_3CHF_2	Industrial and commercial refrigeration, aerosol sprays
HCFC-141b	CH_3CCl_2F	Polyurethane foams, solvent and cleaning applications
HCFC-142b	CH_3CF_2Cl	Polyurethane foams
HCFC-22	$CHClF_2$	Industrial refrigerant, manufacturing of polymers, polystyrene foams, aerosol sprays
HCFC-123	$CHCl_2CF_3$	Commercial chillers

[a] From Kirschner (1994); for a discussion of halon replacements, see Freemantle (1995).
[b] HFC = hydrofluorocarbon; HCFC = hydrochlorofluorocarbon.

for hydrofluorocarbons, i.e., compounds containing hydrogen and fluorine but no chlorine. Because the HCFCs contain chlorine, they will continue to contribute to ozone depletion, and their ultimate phaseout is included in the Montreal Protocol and its amendments. However, the fraction that reaches the stratosphere is much smaller than for the CFCs because of their partial removal in the troposphere by reaction with OH.

For a discussion of the history of controls on ozone-depleting chemicals and factors that must be considered in finding transitional substitutes and long-term replacements, see Ko et al. (1994a), Parson and Greene (1995), the World Meteorological Organization Report (1995 and the 1999 update), and the volume edited by Miziolek and Tsang (1995).

B. OZONE DEPLETION POTENTIALS (ODP)

The contribution of individual compounds to ozone depletion is characterized by the ozone depletion potential (ODP). The ODP of a compound as normally defined is the ratio of the global loss of ozone (i.e., integrated over latitude, altitude, and time) from that compound at steady state per unit mass emitted relative to the loss of ozone due to emission of unit mass of a reference compound, usually taken as CFC-11 ($CFCl_3$) (Wuebbles, 1983; Fisher et al., 1990; Solomon et al., 1992). The ODP thus provides a relative measure of the *overall* impact of a compound on ozone destruction over the long term. Values for ODPs have been derived using a variety of models (e.g., see World Meteorological Organization, 1995, 1999).

However, the time scale for contribution to ozone depletion is not the same for all compounds (Fisher et al., 1990). For example, CFC-11 does not react in the troposphere but photolyzes in the stratosphere, with an estimated stratospheric lifetime of ~50–55 years (World Meteorological Organization, 1995; Solomon and Albritton, 1992). The transition HCFC-123 (CF_3CHCl_2) reacts with OH and hence has a finite lifetime in the troposphere (about 2 years; Nimitz and Skaggs, 1992). In the stratosphere, the rate of chlorine release from HCFC-123 is faster than for CFC-11, with an estimated stratospheric lifetime of ~47 years (Solomon and Albritton, 1992). Hence its contribution to ozone depletion is high initially, about 50% that of CFC-11, but falls off with time. In contrast, CFC-113 ($CFCl_2CF_2Cl$) reacts more slowly in the stratosphere (estimated lifetime of ~110 years; Solomon and Albritton, 1992), and hence it continues to give chlorine after the contribution from the reference compound, CFC-11, has started to decay. As a result, its effective ODP actually increases with time. Clearly, these different time scales for ozone depletion are important for designing both shorter term transitional alternatives for CFCs and longer term permanent substitutes.

A semiempirical approach has been developed by Solomon et al. (1992) and applied to the estimation of ODPs on both short and long time scales. The ODP of compound X is calculated from Eq. (A) (Solomon and Albritton, 1992):

$$\text{ODP}(t) = \frac{F_X}{F_{\text{CFC-11}}} \frac{M_{\text{CFC-11}}}{M_X} \frac{n_X}{3} \alpha \frac{\int_{t_s}^{t} e^{-(t-t_s)/\tau_X} dt}{\int_{t_s}^{t} e^{-(t-t_s)/\tau_{\text{CFC-11}}} dt}.$$

(A)

The ratio $F_X/F_{\text{CFC-11}}$ is the fraction of the compound X that has been dissociated in the stratosphere compared to that of CFC-11, determined from atmospheric measurements. This term is most important at short times, i.e., right after release of the compound, in terms of determining the ODP. M denotes the molecular weight of each species, to convert the ODP to mass emissions. n_X is the number of chlorine, bromine, or iodine atoms in X (with the 3 in the denominator representing the three chlorine atoms in CFC-11), and α is an enhancement factor that reflects the relative efficiency of ozone destruction by bromine (or iodine, should that prove ultimately to be important). Values of α for bromine are ~40–100 (World Meteorological Organization, 1999; Ko et al., 1998), and are larger for iodine, up to ~92,000 (Solomon et al., 1994a,b). (Note, however, that as discussed in Chapter 12.E, current evidence suggests that there is insufficient iodine in the stratosphere to contribute significantly to ozone depletion.) The time t_s is the time for a molecule to be transported from the surface to the region of the stratosphere being examined, and t is the total time; i.e., the $(t - t_s)$ term represents the total time in the stratosphere available for conversion into active forms of chlorine. The total atmospheric lifetimes of X and CFC-11 are denoted by τ_X and τ_{CFC-11}, respectively.

The total atmospheric lifetime of a compound is the time required for a pulse emitted into the atmosphere to decay to $1/e$ of its initial value (see Chapters 5.A.1c and 12.C). It can be calculated from

$$\tau_X = \frac{\Sigma \text{Burden}_{atm}}{\Sigma \text{Loss rate}}, \quad (B)$$

where $\Sigma\text{Burden}_{atm}$ represents the total amount of X in the atmosphere and Σ(Loss rate) the globally integrated loss rate due to all processes, including reactions, deposition, etc. τ_X can also be calculated from the global source strengths and the atmospheric concentrations.

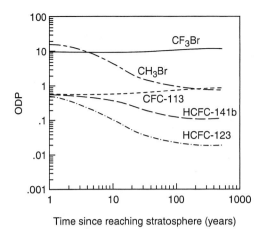

FIGURE 13.3 Time dependence of ODPs (adapted from World Meteorological Organization, 1995).

The integrals in Eq. (A) are of the form e^{-kt}, where $k_X = 1/\tau_X$ and t is time in that region of the stratosphere. The ratio of these integrals in the last term in Eq. (A) can be thought of as the ratio of the time-dependent concentrations of X and CFC-11 as they decay. As $t \to \infty$, the ODP calculated from Eq. (A) approaches the overall, time-integrated ODP calculated from models. This last term is the part of Eq. (A) that determines the time dependence of the ODP, and it generally dominates the ODP at longer times.

The semiempirical nature of this calculation arises from basing the ratio F_X/F_{CFC-11} on measurements of the compound of interest and CFC-11, combined with calculated atmospheric lifetimes.

Figure 13.3 and Table 13.2 demonstrate the effects of different time scales for ozone depletion. The data

TABLE 13.2 Time Dependence of Ozone Depletion Potentials for Some Compounds Using the Semiempirical Approach[a]

Compound	Time Horizon (years)					
	5	15	25	40	100	500
CFC-113	0.55	0.58	0.60	0.64	0.78	1.09
CH$_3$CCl$_3$	1.03	0.57	0.38	0.26	0.15	0.12
HCFC-142b	0.17	0.15	0.13	0.12	0.08	0.07
HCFC-22	0.19	0.15	0.13	0.10	0.07	0.05
HCFC-141b	0.54	0.38	0.30	0.22	0.13	0.11
HCFC-123	0.51	0.11	0.07	0.04	0.03	0.02
CH$_3$Br[b]	15.3	3.1	1.8	1.2	0.69	0.57[b]
H-1121	11.3	9.7	8.5	7.1	4.9	4.1
H-1301	10.3	10.5	10.6	10.8	11.5	12.5
H-2402	12.8	11.6	10.6	9.4	7.0	5.9

[a] Using Eq. (A); from Solomon and Albritton (1992), for conditions of the Arctic lower stratosphere.
[b] Ko et al. (1998) estimated an ODP of 0.39 based on kinetic and photochemical data from DeMore et al. (1997).

in Table 13.2 were derived using measurements of F_X/F_{CFC-11} characteristic of polar regions. However, as discussed by Solomon and Albritton (1992), these are not expected to differ by large amounts from the globally averaged values of the ODPs. As expected from the earlier discussion, the ODP for HCFC-123 is initially high and then falls off rapidly, approaching a long-time, steady-state value of ~ 0.02. This rapid decline reflects its short atmospheric lifetime of ~ 1.7 years (Solomon and Albritton, 1992) due largely to its removal by reaction with OH in the troposphere (e.g., see Nimitz and Skaggs (1992) for a discussion of tropospheric lifetimes and ODPs). However, the ODP of CFC-113, with no known chemistry in the troposphere and a longer stratospheric lifetime than CFC-11, increases with time since it continues to generate atomic chlorine while CFC-11 decays.

These data also demonstrate the impact of bromine chemistry on the stratosphere (see Chapter 12.D). The initial ODP for methyl bromide is 15, due primarily to the large α factor associated with bromine chemistry. However, since it is removed by reaction with OH in the troposphere as well as by other processes such as hydrolysis in the oceans and uptake by soils and foliage (see Chapter 12.D), it has a short atmospheric lifetime of ~ 1.3 years and hence the ODP decreases rapidly with time, toward a long-term steady-state value.

The ODP for CH_3Br also illustrates the sensitivity of these estimates to the input data. For example, Ko et al. (1998) have summarized changes in the estimated efficiency of bromine in stratospheric ozone destruction from $\alpha = 36$ in 1990 to $\alpha = 58$ in 1997 due to improved understanding of the kinetics and photochemistry of bromine compounds that are important in the stratosphere. They also show changes in the estimated atmospheric lifetimes of CH_3Br over this period of time, from 2.1 to 0.7 years. The net result of these changes is a variation in the estimated ODP from 0.73 in 1990 up to 0.99 using 1992 data and then to their estimate of 0.39 based on 1997 kinetic and photochemical data.

Table 13.3 summarizes some of the values of globally and time-averaged ODPs (corresponding to the long times in Fig. 13.2) obtained using either models or the semiempirical method based on Eq. (A) at long times (Solomon et al., 1992). Also shown are the total atmospheric lifetimes for these gases.

Note that the HFCs in Table 13.3 have such small ODPs that only upper limits can be placed on them. That is, their ODPs are essentially zero for all practical purposes. The reason is the obvious one, that they do not contain chlorine and, as discussed in Chapter 12.C, fluorine does not participate in ozone destruction cycles (Molina and Rowland, 1974; Rowland and Molina, 1975; Stolarkski and Rundel, 1975; Cicerone, 1979; Ravishankara et al., 1993; Sehested et al., 1994; Li et al., 1995).

In summary, a combination of the time-dependent

TABLE 13.3 Atmospheric Lifetimes and Steady-State Ozone Depletion Potentials (ODP) Predicted Using Either a Two-Dimensional Model or a Semiempirical Method[a,b]

Potential trace gas	Atmospheric lifetime (years)	Steady-state ozone depletion	
		Model	Semiempirical[a]
CFC-11	50	1.0	1.0
CFC-12	102	0.82	0.9
CFC-113	85	0.90	0.9
CH_3CCl_3	5.4	0.12	0.12
HCFC-22	13.3	0.04	0.05
HCFC-123	1.4	0.014	0.02
HCFC-141b	9.4	0.10	0.1
HCFC-142b	19.5	0.05	0.066
HFC-134a	14	$<1.5 \times 10^{-5}$	$<5 \times 10^{-4}$
HFC-125	36	$<3 \times 10^{-5}$	
CH_3Br	1.3 (0.7)[d]	0.64	0.57 (0.39)[d]
H-1301		12	13
H-1211		5.1	5
CH_2ClBr	0.23–0.36[c]	0.098–0.15[c]	
$CH_2BrCH_2CH_3$	0.029[c]	0.026[c]	

[a] Based on Eq. (A) as $t \to \infty$.

[b] From World Meteorological Organization (1995).

[c] From Wuebbles et al. (1998, 1999); the lifetimes and ODPs given for CH_2ClBr are the range with and without an ocean sink.

[d] Values from Ko et al. (1998) estimated using the more recent kinetic and photochemical data from DeMore et al. (1997).

ODPs and the final, steady-state values can be used to assess the relative importance of various compounds to stratospheric ozone depletion on both short and long time scales. For a more detailed discussion of ODPs, see Ko et al. (1994a) and the World Meteorological Organization Report (1995, 1999).

C. TRENDS IN CFCs, THEIR REPLACEMENTS, STRATOSPHERIC O_3, AND SURFACE UV

1. Trends in CFCs and Their Replacements

On a historical note, it is interesting that tropospheric observations of CFCs and chlorinated organics date back to the early 1970s when Lovelock (1972) first reported such measurements. Such observations by Lovelock (1972, 1974, 1975), Lovelock et al. (1973), and others (e.g., Su and Goldberg, 1973; Wilkniss et al., 1975) first raised the issue of increased tropospheric concentrations of CFCs due to their industrial use and also highlighted the contributions of natural emissions of halogenated organics. These early papers on CFCs pointed out their lack of reactivity in the troposphere, which ultimately led to the classic Rowland and Molina papers proposing their transport into the stratosphere and subsequent contribution to ozone depletion. Since that time, there have been numerous measurements of these compounds in both the stratosphere and troposphere, which demonstrate increased trends in their concentrations that follow their increased usage (e.g., see World Meteorological Organization, 1995, 1999).

Clearly, the first effect of the imposition of the controls outlined in the Montreal Protocol and its subsequent amendments should be a reduction in the rate of increase of CFCs in the troposphere. This indeed has been observed to be the case (see, for example, Butler et al., 1992; Elkins et al., 1993; Khalil and Rasmussen, 1993; Fraser et al., 1994; Simmonds et al., 1996; Montzka et al., 1996a; Cunnold et al., 1997; and Derwent et al., 1998). Figure 13.4 shows the measured monthly mean tropospheric concentrations of the major CFCs used in the past, CFC-11, -12, and -113, as well as CCl_4 and methylchloroform from 1991 to 1996. As expected based on the much higher rates of consumption in the industrialized countries of the Northern Hemisphere, concentrations of these compounds are higher in this region. However, in both the Northern and Southern Hemispheres, the rates of increase in the troposphere of CFC-11, CFC-12, and CFC-113 have fallen substantially since 1991 (Montzka et al., 1996a; Cunnold et al., 1997; Derwent et al., 1998), and indeed, for CFC-11 and CFC-113 now appear to be decreasing.

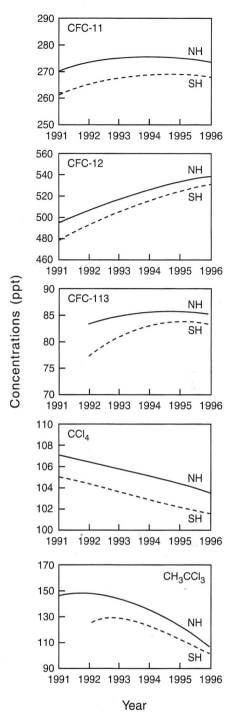

FIGURE 13.4 Fits to monthly mean tropospheric concentrations of some CFCs and other chlorinated organics falling under the Montreal Protocol and its amendments in the Northern Hemisphere (NH) (———) and Southern Hemisphere (SH) (– – –), respectively (adapted from Montzka et al., 1996a).

For example, before the Montreal Protocol, the annual growth rates of CFC-11, CFC-12, and CFC-113 were approximately 10, 18, and 6 ppt per year, respectively; however, the growth rates in mid-1995 were estimated to be reduced to (0 to −0.6), 6–7, and −0.7 ppt per year, respectively (Montzka *et al.*, 1996a; Cunnold *et al.*, 1997). Derwent *et al.* (1998) report that between 1995 and 1996, CFC-11 decreased by −0.8 ± 0.2 ppt per year, and CFC-113 by about −0.5 ppt per year. That is, the tropospheric concentrations of CFC-11 and CFC-113 have started to decrease in response to the controls. (For a review of the history and atmospheric trends in CFC-11 until 1992, see Khalil and Rasmussen (1993).) The same is seen in Fig. 13.4 to be true for methylchloroform and CCl_4. The latter is used in the production of CFCs. CCl_4 peaked at 104 ± 3 ppt in 1989–1990 and has been decreasing at 0.7 ± 0.1 ppt per year since then (Simmonds *et al.*, 1998a; Derwent *et al.*, 1998). Part of the reason for the rapid response of CH_3CCl_3 concentrations to changes in emissions (Fig. 13.4), with a decrease of about −18 ± 2 ppt per year between mid-1995 and mid-1996 (Derwent *et al.*, 1998), is its short tropospheric lifetime (~5 years) due to the presence of abstractable hydrogen atoms.

From the discussion of the chemistry of CFCs in the stratosphere in the preceding chapter, it might be expected that HCl and HF from the reactions of the chlorine and fluorine atoms produced on photolysis of the CFCs would also have increased in the stratosphere. This has also been observed, and the magnitude and timing of the increases are clearly consistent with the increase in CFCs (Rinsland *et al.*, 1991; Russell *et al.*, 1996). For example, Wallace *et al.* (1997) have followed stratospheric HCl using ground-based infrared spectroscopy with the sun as the source from 1971 through April of 1997. Over this period, HCl increased by a factor of 3–4. There is an indication in the data that the rate of increase has slowed since about 1990, as expected given the trend in CFCs.

Figure 13.5 shows the measured global mean tropospheric concentrations of the halons H-1301, H-1211, and H-2402 (Butler *et al.*, 1998). At the end of 1996, the mixing ratio of H-1301 was 2.3 ± 0.1 ppt, with a growth rate of 0.044 ± 0.011 ppt yr^{-1}. As seen in Fig. 13.5 and in other analyses of the trend (Montzka *et al.*, 1996a), the growth rate appears to be slowing. H-1211 was present at a mixing ratio of 3.5 ± 0.1 ppt at the end of 1996, with a growth rate of 0.16 ± 0.016 ppt yr^{-1}. Fewer data are available for H-2402, which was present at a concentration of 0.45 ± 0.03 ppt at the end of 1996 and had a growth rate of 0.009 ± 0.001 ppt yr^{-1}.

At the same time that the production and usage of the compounds shown in Figs. 13.4 and 13.5 were decreasing due to the international agreements, the use of alternates (Table 13.1) was increasing (e.g., see Midgley and McCulloch, 1997; and McCulloch and

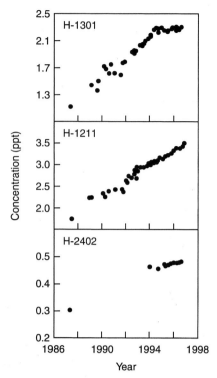

FIGURE 13.5 Global mean mixing ratios for the halons H-1301, H-1211, and H-2402 from 1987 to 1997 (adapted from Butler *et al.*, 1998).

Midgley, 1998). It might be expected that the tropospheric concentrations of these CFC alternatives would therefore increase, and this is indeed the case (e.g., see Montzka *et al.*, 1993, 1994, 1996a, 1996b; Irion *et al.*, 1994; Schauffler *et al.*, 1995; Simmonds *et al.*, 1998b; and Miller *et al.*, 1998).

Figure 13.6 shows the average tropospheric concentrations of HCFC-22 ($CHClF_2$), -141b (CH_3CCl_2F), and -142b (CH_3CF_2Cl) as well as of HFC-134a (CH_2FCF_3), all of which have increased substantially over the past few years (Montzka *et al.*, 1996a, 1996b). In the case of HFC-134a, for example, the concentrations were below the detection limit of 0.01 ppt at Cape Grim, Australia, prior to 1990, but the concentrations have increased since, with an exponential growth of ~200% per year (Oram *et al.*, 1996). At Mace Head, Ireland, the rate of increase of HFC-134a from October 1994 to March 1997 was 2.05 ± 0.02 ppt per year, which is very large given that the absolute concentration midway through these measurements was 3.67 ppt (Simmonds *et al.*, 1998b). On a global scale, increases of ~100% per year in HFC-134a have been reported (Montzka *et al.*, 1996b).

Sharp increases in the concentrations of HCFC-141b and -142b since 1989 have also been noted at Cape

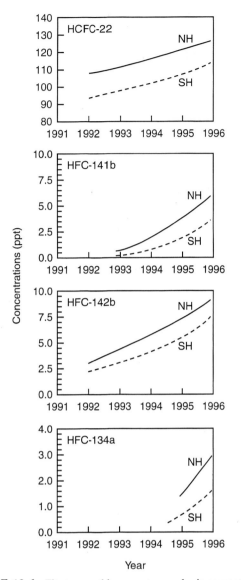

FIGURE 13.6 Fits to monthly mean tropospheric concentrations of some CFC alternates in the Northern (———) and Southern (- - -) Hemispheres (adapted from Montzka et al., 1996a).

rate are similar to those shown in Fig. 13.6 and measured from 1992 to 1997 at La Jolla, California, where the growth rate was 5.5 ppt yr^{-1}.

Figure 13.7 shows the effective total tropospheric concentration of chlorine from halocarbons from 1992 to 1996 (Montzka et al., 1996a). The concentration peaked in 1994 at ~3.0 ppb, but when methyl chloride (CH_3Cl) and other chlorinated organics are taken into account, the peak was likely ~3.7 ppb. The total tropospheric chlorine concentration in mid-1995 decreased at a rate of approximately 25 ppt per year, in contrast to *increases* of 110 ppt per year in 1989 (Montzka et al., 1996a; Cunnold et al., 1997). Bromine compounds show the same trend. As a result, the stratospheric levels of chlorine and bromine are expected to peak around the year 2000 (Montzka et al., 1996a; World Meteorological Organization, 1995, 1999).

Note, however, that synergistic interactions with the effects of greenhouse gases have the potential to alter the timetable for recovery of stratospheric ozone. For example, while CO_2 causes warming in the troposphere, it produces cooling in the stratosphere by emitting infrared radiation out to space (see Chapter 14). This cooling may increase the formation of polar stratospheric clouds (PSCs), especially in the Arctic, where temperatures are often not sufficiently low to generate PSCs for extended periods of time. As a result, as discussed in Chapter 14.B.2d, this cooling may delay recovery of stratospheric ozone.

In addition to CFCs and their replacements, there are some fully fluorinated compounds that are emitted to the atmosphere during various industrial processes, including the manufacture of HCFCs and HFCs. Because of the strong C–F bonds, these compounds have long atmospheric lifetimes (e.g., see Cicerone, 1979; and Ravishankara et al., 1993) and hence have been used as tracers to determine the age of stratospheric

Grim (Oram et al., 1995) and at Mace Head. In the latter case, HCFC-141b increased at a rate of 2.49 ± 0.03 ppt yr^{-1}, compared to a midpoint concentration of 7.38 ppt, and HCFC-142b increased at a rate of 1.24 ± 0.02 ppt yr^{-1}, with a midpoint concentration over the October 1994 to March 1997 time period of 8.78 ppt (Simmonds et al., 1998b). HCFC-141b and -142b have also been detected in the stratosphere (e.g., see Lee et al., 1995).

HCFC-22 ($CHClF_2$) has been measured continuously from 1978 to 1996 (Miller et al., 1998). Its growth rate over this period was 6.0 ppt yr^{-1}, with a 1996 concentration of 117 ppt. This concentration and growth

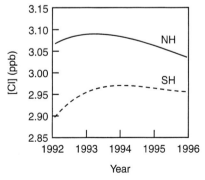

FIGURE 13.7 Effective total tropospheric concentration of chlorine from halocarbons from 1992 to 1996 in the Northern (———) and Southern (- - -) Hemispheres (adapted from Montzka et al., 1996a).

air parcels (e.g., see Patra *et al.*, 1997; and Harnisch *et al.*, 1998). For example, CF_4 and C_2F_6 are generated during aluminum production, and SF_6 during its use in electrical insulation and switching and as an inert gas for handling reactive molten metals such as aluminum. The tropospheric concentrations of all of these compounds have been increasing rather dramatically over the past several decades (Harnisch *et al.*, 1996a; Maiss and Levin, 1994; Maiss *et al.*, 1996; Maiss and Brenninkmeijer, 1998). For example, the growth rate for SF_6 in early 1996 was measured to be $6.9 \pm 0.2\%$ (0.24 ± 0.01 ppt) per year on top of a global average mixing ratio of 3.4 ppt (Geller *et al.*, 1997). Similarly, Maiss and Brenninkmeijer (1998) showed that SF_6 increased from 0.24 ppt in 1970 to 3.8 ppt in 1996; furthermore, the observed increase in its atmospheric concentrations was best fit by a function with a squared dependence on time.

For CF_4, there is some evidence (Harnisch *et al.*, 1996b) that the observed increase has occurred on top of a "background" due to natural emissions whose sources are as yet not well understood. Both CF_4 and SF_6 have been found in natural fluorites (CaF_2) and hence degassing of these compounds from the earth's surface may be at least one natural source of these compounds, particularly CF_4 (Harnisch *et al.*, 1998).

Although not fully fluorinated, HFC-23 (CHF_3) also has a long lifetime (~ 250 years; WMO, 1995) because its reaction with OH is slow, $k^{298} = 2.8 \times 10^{-16}$ cm^3 molecule^{-1} s^{-1} (DeMore *et al.*, 1997). A major source is the production of HCFC-22 ($CHClF_2$), where HFC-23 is a byproduct. As the use of HCFC-22 and its atmospheric levels have increased (Fig. 13.6), the levels of HFC-23 would be expected to increase as well, and indeed, this is the case. Oram *et al.* (1998) reported the first atmospheric measurements of this compound at Cape Grim, Australia. HFC-23 increased from ~ 2 ppt in 1978 to 11 ppt in 1995, with a growth rate of 5% per year in 1995.

In short, the trends in the tropospheric concentrations of CFCs, halons, and their substitutes follow trends in their emissions. The effects of the controls imposed by the Montreal Protocol and its subsequent amendments are evident in the trends and have been used to show that the associated impact on ozone destruction is expected to begin about the turn of the century. The following section briefly describes the observed trends in stratospheric ozone.

2. Trends in Stratospheric O_3

As discussed in Chapter 12, trends in stratospheric ozone in the Antarctic spring during formation of the "ozone hole" are clear. However, as treated in detail in that chapter, there are unique meteorological and chemical circumstances that are responsible for the dramatic loss of ozone in the Antarctic spring. It is perhaps of even greater interest as to whether there is evidence for a trend in midlatitude stratospheric ozone, since this is where the bulk of the population resides.

Detecting and quantifying ozone trends in midlatitudes from anthropogenic perturbations is complex due to the effects of natural variations in stratospheric ozone and to the interactions between various effects (e.g., Krzyscin, 1994; Brasseur *et al.*, 1995; Callis *et al.*, 1997; Zerefos *et al.*, 1997; Hood, 1997; Callis *et al.*, 1997). Thus, long-term trends in ozone must be extracted from variability due to the solar cycle, which has an 11-year period associated with it, as well as the quasi-biennial cycle (QBO), which is an oscillation of zonal winds in the stratosphere around the equator and which has a 26–30 month cycle (e.g., see Kane *et al.*, 1998). For example, Bjarnason *et al.* (1993) examined column O_3 measurements made at Reykjavík from 1957 to 1990 using a Dobson spectrometer and applied a stratospheric model that included variations due to seasons, the solar cycle, the QBO, and a linear trend. The combination of the data and model showed a variation of $3.5 \pm 0.8\%$ in column O_3 over a solar cycle and $2.1 \pm 0.6\%$ over a QBO, on top of a linear trend of decreasing O_3.

In addition, there is an observed correlation between total column ozone and the El Niño Southern Oscillation (ENSO) in the tropical troposphere, with decreases in total ozone in middle and sometimes polar latitudes following the ENSO by several months; the period associated with the ENSO is ~ 43 months (Zerefos *et al.*, 1992). While the association between the ENSO and ozone is not well understood, it has been proposed that the warming of the troposphere in the tropics over the Pacific Ocean causes increases in the upper troposphere air temperatures and tropopause height and an upwelling in the lower stratosphere. If sufficiently large, this could have more widespread impact than just in the tropics (e.g., see Zerefos *et al.*, 1992; and Kalicharran *et al.*, 1993).

There is also a significant correlation between temperature fluctuations in the lower stratosphere and fluctuations in total ozone. There are two sources of this correlation, radiative and dynamical (McCormack and Hood, 1994). Thus, increased ozone leads to increased absorption of solar radiation and increased heating. In addition, dynamical effects associated with vertical and meridional air motions also give a positive correlation between ozone and stratospheric temperature. For example, Randel and Cobb (1994) analyzed total column O_3 and temperatures in the lower stratosphere from 1979 to 1992. Correlations between O_3

and temperature associated with the solar cycle, the QBO, and the ENSO were identified. The variations were ~13 DU per K for the solar cycle effect, 14–16 DU per K for the tropical and subtropical QBO, and ~6–9 DU per K for the ENSO.

Heterogeneous chemistry leading to ozone destruction can also lead to ozone–temperature correlations since many of the important reactions forming active chlorine are faster at lower temperatures (see Chapter 12). In addition, there is more PSC formation at the poles and hence more ozone destruction in these regions associated with lower temperatures (see Fig. 13.14 and associated discussion below).

Some trend analyses have also observed correlation between total column O_3 and the temperature of the free troposphere, which also has dynamical origins. For example, Staehelin et al. (1998b) report that at Arosa, Switzerland, the change in O_3 per degree change in tropospheric temperature was $-(4.34 \pm 0.20)$ DU per K, attributed to the cold advection of polar air having high total O_3 or warm subtropical air containing low levels of O_3.

A further complication in such trend analyses is potential systematic errors associated with various measurement techniques used to determine ozone. Several different techniques have been in use for sufficient lengths of time that they can be used to assess longer term trends in ozone; ground-based techniques include Dobson measurements, Umkehr techniques and ozonesondes, and satellite approaches include TOMS (*T*otal *O*zone *M*apping *S*pectrometer), SBUV (*S*olar *B*ackscatter *U*ltra*V*iolet), and SAGE (*S*tratospheric *A*erosol and *G*as *E*xperiment satellite measurement) (Kaye, 1995).

Dobson instruments are based on the measurement of solar radiation transmitted through the atmosphere at pairs of wavelengths around 300 nm, one of which ozone absorbs more strongly. Such ground-based measurements have been made since 1926 at Arosa, Switzerland. Given that this is the longest data record for column O_3, there has been considerable effort expended to provide "quality control" on the data as different instruments, detection devices, monitoring wavelengths, and calibration procedures were substituted over the years (Staehelin et al., 1998a). More than 30 such instruments have been operating worldwide since 1957. The Umkehr technique uses a Dobson spectrometer to make measurements at dawn and dusk, i.e., at high solar zenith angles, to obtain the vertical distribution of ozone from the effects on light absorption of changing the total path through the atmosphere.

Ozonesondes are *in situ* measurements launched from the ground to altitudes of ~30 km to measure the vertical distribution of ozone. They typically use chemical analytical techniques such as oxidation of I^- to I_2; as discussed in Chapter 11, although such instruments respond to O_3, they also respond positively to some extent to other compounds such as NO_2 and negatively to others such as SO_2.

The SBUV instrument is based on satellite measurements of backscattered light at 12 wavelengths between 255 and 340 nm; from comparisons of wavelengths where ozone absorbs to those where Rayleigh light scattering occurs, total column ozone and vertical profiles with low resolution (5–15 km, depending on the altitude) can be obtained. One such instrument operated from 1978 to 1990 and another from 1989 to 1994, with two more operating in 1996. Some of the difficulties in extracting the vertical O_3 distribution using SBUV data are described by Ziemke and Chandra (1998).

TOMS is also a satellite-based method based on a similar approach to that of SBUV, in which the earth's albedo is measured at several wavelengths around 300 nm. Unlike SBUV, vertical distributions of O_3 are not derived using TOMS, but it provides better horizontal resolution (Ziemke et al., 1998). TOMS has also been used to measure tropospheric aerosols using the wavelength dependence of UV reflectivity at wavelengths that are not absorbed by O_3 (e.g., Hsu et al., 1996; Herman et al., 1997; Torres et al., 1998).

SAGE instruments have been operating since 1978, although with a gap between 1981 and 1984. Figure 13.8 illustrates the geometry associated with these measurements (Rusch et al., 1998). These instruments are solar occultation instruments that measure the light intensity at seven wavelengths at satellite sunrise and sunset, giving about a dozen such measurements in one 24-h day. Three of the wavelengths are used to derive O_3 (0.6 μm), NO_2 (0.448 μm), and water vapor (0.94 μm), whereas the other four (1.02, 0.525, 0.453, and

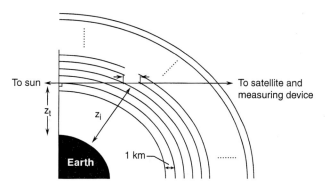

FIGURE 13.8 Schematic of geometry associated with SAGE measurements (adapted from Rusch et al., 1998).

0.385 μm) are used to extract aerosol particle concentrations. Light transmission data are generated by comparison to the unattenuated solar flux and, along with ancillary measurements of other parameters, can be used to obtain extinction profiles for O_3 and for other species, including aerosol particles, water vapor, and NO_2 (Rusch et al., 1998).

However, extracting ozone concentrations and their vertical profiles from SAGE data (and indeed other long-path, remote techniques) is not straightforward. Typically, the contribution from Rayleigh scattering is first removed and then the contribution of each gas to the total extinction is derived as a function of the tangent altitude (z_t in Fig. 13.8) using the wavelength chosen to be characteristic of light absorption by gas (the so-called species inversion). Finally, the contribution of each gas as a function of altitude (z_i) is extracted (the "geometric" or "altitude" inversion) (e.g., see Cunnold et al., 1989). Rusch et al. (1998) have shown that the order of applying the inversions can make a difference; for example, carrying out the altitude inversion first and then the species inversion improves agreements for O_3 below 22 km with concentrations derived using ozonesondes.

An important factor in deriving O_3 concentrations is the presence of aerosol particles, which also scatter light at 0.6 μm. Thus, correction for their contribution to extinction at this wavelength must by applied to derive the ozone concentrations. This requires some assumptions regarding aerosol particle properties such as the size distribution, which is not known. It is also commonly assumed that the optical properties of particles do not change with altitude. Such problems introduce uncertainties into the calculation of the particle contribution (e.g., Steele and Turco, 1997a, 1997b; Thomason et al., 1997; Fussen, 1998) and hence into the ozone concentrations extracted from such data.

For example, Fig. 13.9a shows the trends in column O_3 as a function of latitude in the stratosphere (at altitudes above pressures corresponding to 82.5 mbar) for the period from November 1984 to May 1991 derived using SAGE II measurements (Cunnold et al., 1996). Also shown are the trends derived from TOMS data made for the same locations; the latter include changes in tropospheric ozone as well. Figure 13.9b shows similar data for a smaller period of time, from January 1988 to May 1991, during which time stratospheric aerosol concentrations were quite stable and small. In the 1984–1988 period included in Fig. 13.9a, the aerosol particle concentrations were larger and were decreasing with time. The agreement between the TOMS and SAGE II data is much better for the period with small particle concentrations, suggesting that they were responsible for much of the discrepancy seen in

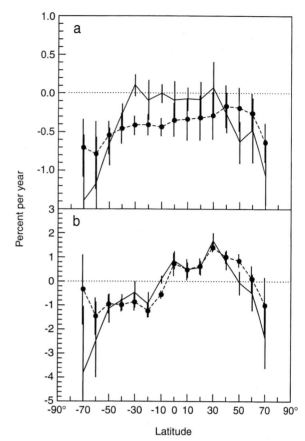

FIGURE 13.9 Trend in column O_3 (a) from November 1984 to May 1991 based on SAGE II measurements in the stratosphere (at pressures above 82.5 mbar, ● and dashed lines) and TOMS measurements (solid line) and (b) from January 1988 to May 1991, which was a period of relatively small and stable aerosol particle concentrations. Note change in vertical scale. (Adapted from Cunnold et al., 1996.)

Fig. 13.9a. Cunnold et al. (1996) indicate that these errors in accurately extracting the aerosol contribution were responsible for about half of the discrepancy and increases in tropospheric ozone the other half. Similar discrepancies between SAGE measurements and the results from other techniques such as ozonesondes have also been reported and attributed largely to the problem of accurately removing the aerosol particle contributions (e.g., Veiga et al., 1995; Steele and Turco, 1997b).

All of these approaches, as well as others such as the MLS (*M*icrowave *L*imb *S*ounder) that have not been used for sufficient lengths of time to give long-term trends, have some uncertainties associated with them, as is the case for all analytical techniques (e.g., see Harris et al., 1997). For example, regular calibration is particularly difficult with satellite-based measurements and care must be taken to ensure that a continuous degradation in the instrument sensitivity, for example, is not interpreted as a trend in total ozone.

As a result of such complexities, the magnitude of trends in ozone and even the direction of the changes do not always agree, particularly when segregated by altitude and latitude (e.g., see Rusch *et al.*, 1994; McPeters *et al.*, 1994; Reinsel *et al.*, 1994a,b; and Miller *et al.*, 1995). For example, Rusch *et al.* (1994) report that in the high-latitude upper stratosphere, SBUV data give a negative trend in O_3 from 1979 to 1991, in the range -0.4 to -1.0% per year; however, the trend inferred from SAGE data is zero to slightly positive, 0 to 0.5% per year. They also showed that while data from the TOMS and SBUV instruments were in good agreement, those from the SAGE were not.

Similarly, Miller *et al.* (1995) compared ozone trends derived from Umkehr and balloon ozonesondes. Figure 13.10 shows that while both predict the same trend qualitatively in the lower stratosphere, they do not agree quantitatively. In addition, from ~25 to 35 km, even the direction of the changes is not in agreement. Harris *et al.* (1997) report that at altitudes of 16–17 km in midlatitudes, the ozone trend measured using SAGE is as much as $-20 \pm 8\%$ per decade, whereas ozonesonde measurements in the Northern Hemisphere give an average of $-7 \pm 3\%$ per decade.

In addition, as seen in Fig. 13.9, changes in tropospheric composition can also impact the results. For example, the trend in total ozone over 20 years measured in Belgium was $-1.38 \pm 0.50\%$ per decade but after correcting for changes in tropospheric SO_2, no significant trend in ozone could be discerned (De Muer and De Backer, 1992). There is also a concern regarding the effects of trends in tropospheric aerosols on the derived trends in O_3.

Because of such uncertainties, trend analysis is an active area of current research. However, it is clear that many studies have shown qualitatively a decreasing trend in total column ozone at midlatitudes, and it is generally accepted that there is indeed a decrease associated with CFCs and halons. Given the complexities in such trend analyses, it is not surprising that there remains considerable uncertainty in the *quantitative* values at particular locations and times.

Figure 13.11 is one example of variations in total column ozone in Europe, Eastern Siberia and the Far East, and Western Siberia (Bojkov *et al.*, 1994). These data have been smoothed using a 12-month running mean, but variations due to the QBO etc. have not been removed. Figure 13.11c shows for the Western Siberia data the deviations specifically attributed to the QBO; while this contributes to a significant extent to the observed trends, it clearly does not account for the entire downward trend in total ozone with time.

Similarly, Fig. 13.12 shows the percentage deviation in regionally averaged stratospheric ozone for North America, Europe, and the Far East after variations due to the solar cycle, seasonal variations, the QBO, and atmospheric nuclear tests were subtracted out. Negative deviations are consistently seen in recent years, suggesting a long-term trend on top of the natural variability (Stolarski *et al.*, 1992).

Such ozone trends generally tend to be more negative in the winter and spring than in the summer and fall and larger at higher latitudes compared to equatorial regions (e.g., see Kerr, 1991; Kundu and Jain, 1993; Reinsel *et al.*, 1994b; Hollandsworth *et al.*, 1995; Harris *et al.*, 1997; Bojkov and Fioletov, 1997; Taalas *et al.*, 1997; Komhyr *et al.*, 1997; and Staehelin *et al.*, 1998b).

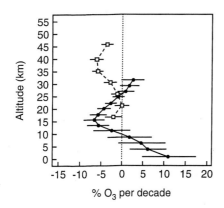

FIGURE 13.10 Trends in O_3 based on ozonesonde measurements (●) and Umkehr measurements (□) (adapted from Miller *et al.*, 1995).

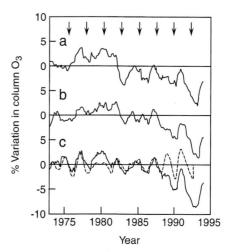

FIGURE 13.11 Percentage variations in total column ozone smoothed using a 12-month running mean for a network of stations in (a) Europe, (b) Eastern Siberia and the Far East, and (c) Western Siberia from 1973 to March 1994. The arrows show the expected QBO. In (c) the dashed line shows the component that has a periodicity expected for the QBO (adapted from Bojkov *et al.*, 1994).

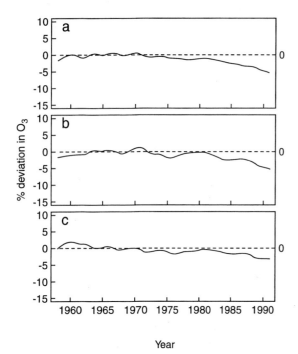

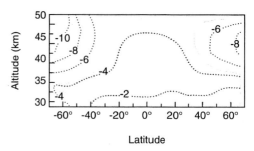

FIGURE 13.13 Trends in annual average ozone (% per decade) as a function of latitude and altitude derived from SBUV data (adapted from Hollandsworth *et al.*, 1995).

FIGURE 13.12 Fits to percentage deviation in regionally averaged ozone in (a) North America, (b) Europe, and (c) the Far East after subtracting components due to the solar cycle, seasonal variations, the QBO, and atmospheric nuclear tests (adapted from Stolarksi *et al.*, 1992).

For example, analysis of a combined set of data from ground-based measurements using Dobson spectrometers and related approaches as well as ozonesondes and SBUV from 1979 to 1994 indicates that at midlatitudes in the Northern Hemisphere, the trend in total ozone is as much as −7% per decade in the winter and spring compared to −3% per decade in the summer and fall; in the Southern Hemisphere at midlatitudes, there is less seasonal variation but negative trends in total ozone in the range −3 to −6% per decade are observed (Harris *et al.*, 1997). Trends in the tropics were not clearly statistically significant.

At Arosa, Switzerland, where there are records back to 1926 (the longest data record available), the trend in annual mean O_3 has been determined to be $-(2.3 \pm 0.6)\%$ per decade. When contributions due to the solar cycle, temperature, and stratospheric aerosol concentrations are taken into account, the trend is $-(1.9 \pm 0.6)\%$ per decade. However, the total measured column O_3 includes both stratospheric and a smaller tropospheric contribution, and the latter has been increasing (see Chapters 14 and 16). This would tend to mask part of a decrease in stratospheric O_3. Applying an estimate of the increase in tropospheric ozone gives a trend in stratospheric O_3 of $-(3.0 \pm 0.6)\%$ per decade at Arosa (Staehelin *et al.*, 1998b).

Figure 13.13 shows SBUV-derived trends in annual average ozone as a function of latitude and altitude; the magnitude of the trend is generally small in the equatorial regions and larger at higher latitudes and altitudes (Hollandsworth *et al.*, 1995; Hilsenrath *et al.*, 1992; Claude *et al.*, 1994; Chandra *et al.*, 1995, 1996; Harris *et al.*, 1997; Bojkov and Fioletov, 1997).

Superimposed on long-term trends are "spikes," some of which may be related to the chemistry discussed in Chapter 12. For example, total column ozone was at a record low in 1992–1993, likely due to heterogeneous chemistry on aerosol particles from Mount Pinatubo (e.g., Gleason *et al.*, 1993; Kerr *et al.*, 1993; Bojkov *et al.*, 1993; Hofmann *et al.*, 1994; Komhyr *et al.*, 1994). In January–February 1995, low ozone values were observed in the Northern Hemisphere and were as much as 25–35% below the long-term average over Siberia. This may be related to the displacement of the polar vortex to this location, accompanied by low stratospheric temperatures conducive to the formation of polar stratospheric clouds (Bojkov *et al.*, 1995a). Figure 13.14, for example, shows the measured deviation of ozone from the long-term mean and the tem-

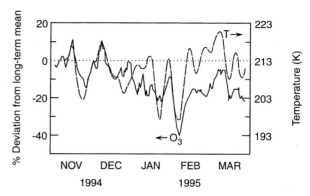

FIGURE 13.14 Percentage deviation of total ozone from long-term mean and stratospheric temperature at an altitude corresponding to 50 hPa near Siberia in late 1994 and early 1995 (adapted from Bojkov *et al.*, 1995a).

perature in the stratosphere at a pressure of 50 hPa; the correlation between the two is striking. Both measurements (e.g., Solomon et al., 1998) and modeling studies (e.g., Tie et al., 1997) support this relationship between colder temperatures and increased O_3 depletion due to increased chlorine activation through heterogeneous reactions.

A similar observation of record low total column ozone over Lauder, New Zealand, down to 222 DU compared to the 1985–1996 average of ~340 DU was reported by Brinksma et al. (1998). They attributed the low ozone in part to a portion of the Antarctic polar vortex passing over this location at altitudes of ~25–35 km and in part to injection at lower altitudes (~22 km) of ozone-poor subtropical air.

In short, there is strong evidence for negative trends in total column ozone in mid and high latitudes, particularly during the winter and spring, superimposed on natural variations. However, the absolute magnitudes at a particular location and time have a high degree of uncertainty associated with them and large fluctuations from year to year are evident. Clearly, this is an area that warrants further attention.

3. Trends in Surface Ultraviolet Radiation

Because of the strong absorption of ultraviolet (UV) radiation starting at ~320 nm by O_3, one of the major impacts of decreased stratospheric ozone is expected to be increased UV at the earth's surface, with associated effects such as increases in skin cancer and cataracts and damage to plants and other ecosystem components. It has therefore been of great interest to determine whether such a relationship can be detected and, if so, what the magnitude of the effect is. The latter is commonly expressed as an "amplification factor" (AF) or radiation amplification factor (RAF), defined as the fractional change in radiation (R) per fractional change in total column ozone (O_3):

$$\text{RAF} = \text{AF} = \{dR/R\}/\{d(O_3)/O_3\}. \quad \text{(C)}$$

Because it is the biological effects that are of interest, R may be weighted using the so-called "action spectrum," which is the wavelength dependence of the particular effect of concern (Brasseur et al., 1995). Figure 13.15, for example, shows the action spectra for damage to DNA, to plants, and for the induction of erythema, or reddening of the skin (Madronich, 1992).

The ultraviolet region is commonly divided into the UV-A region from 315 to 400 nm and the UV-B region from 280 to 315 nm. It is the UV-B region that is of greatest concern in terms of the impacts of UV radiation, and because of the strong absorption of light by

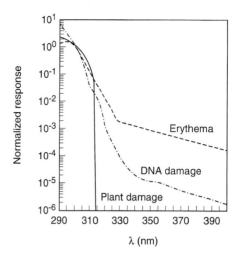

FIGURE 13.15 Normalized action spectra taking response = 1.0 at 300 nm (adapted from Madronich, 1992).

ozone here, it is also the region most impacted by decreases in stratospheric ozone.

As is the case for assessing trends in total column O_3, there are a number of complexities in determining if there are associated trends in UV at the earth's surface (e.g., see Frederick, 1990; Crutzen, 1992; Correll et al., 1992; Seckmeyer et al., 1994; and Madronich et al., 1998). For example, changes in cloudiness and increased aerosol particle concentrations can alter the radiation at the earth's surface (see Chapter 14). This is particularly true in highly polluted urban areas such as Mexico City (Galindo et al., 1995). Light scattering by aerosol particles may have decreased UV in the 280- to 320-nm range by 2–18% since the beginning of the industrial revolution, offsetting at least in part increased UV from ozone depletion in the Northern Hemisphere, where most of the industrial emissions have occurred (e.g., Liu et al., 1991; Sabziparvar et al., 1998). Similarly, Wenny et al. (1998) measured UV-B radiation at a mountain top and in a nearby valley in North Carolina and correlated decreasing UV-B intensities with increased aerosol optical depth in the layer of air between the two. However, increased UV at large solar zenith angles has also been predicted (see, for example, Davies, 1993; and Tsitas and Yung, 1996).

Toumi et al. (1994) also suggested there is a feedback between reduced stratospheric ozone and particles in that the increased UV due to ozone depletion may increase sulfate particle formation by increasing the concentrations of tropospheric OH.

In short, changes in aerosol concentrations over industrialized regions can complicate the interpretation of UV trends (e.g., see Justus and Murphey, 1994). The same is true of clouds, which play a major role in the

UV reaching the earth's surface. Studies of the interannual cloud variability suggest that by about the year 2000, trends in UV should be sufficiently large compared to this variability that it should be detectable (Lubin and Jensen, 1995; see also Madronich, 1995).

Whether or not simultaneous changes in aerosol particles and clouds have indeed altered UV at the earth's surface is not clear. For example, Herman *et al.* (1996) have analyzed satellite data for ozone and reflectivity from 1979 to 1992 and conclude that over this period of time there was no significant trend in the UV albedo due to changes in clouds or aerosols.

UV at the earth's surface can also be affected by concentrations of tropospheric gases that absorb in the same region as O_3. These include, of course, tropospheric ozone, as well as SO_2. Brühl and Crutzen (1989), for example, have shown that increased tropospheric ozone can lead to decreases of UV at the earth's surface. Sabziparvar *et al.* (1998) calculated that at low latitudes, increases in tropospheric O_3 occurring since preindustrial times (see Chapter 14.B.2d) may have decreased surface UV in the 280- to 320-nm range by up to 5% and the erythemally weighted UV by up to 9%. Similarly, increased emissions of SO_2, which absorbs light in the 300-nm region as well (see Chapter 4.K), can counteract UV increases due to stratospheric ozone depletion (De Muer and De Backer, 1992).

Typical values of the amplification factor are of the order of 1–2. For example, that for erythema at the Mauna Loa Observatory was measured to be approximately 1.4 at solar zenith angles ≤60°, decreasing to 0.6 at 85° (Bodhaine *et al.*, 1996, 1997).

Figure 13.16 shows an example of some of these factors calculated for the most extreme conditions that might be encountered during a year at Thessaloniki, Greece (Bais *et al.*, 1993). The amplification factors (AF) for the fractional change in UV per fractional change in total column O_3 and SO_2, respectively, as well as change in cloudiness, are shown as a function of wavelength. The values of AF are less than one for SO_2 and clouds, as expected, and greater than one for depletion of total column ozone. As expected, the amplification factor for O_3 increases at shorter wavelengths in a nonlinear manner (Bodhaine *et al.*, 1997). Increases in UV of about 12% per year at 305 nm and 3% per year at 325 nm have been reported at Thessaloniki from 1990 to 1993 (Zerefos *et al.*, 1995).

Clearly, given the magnitude of ozone depletion in Antarctica in the spring, as well as its large geographical extent, an increase in UV might be expected to be most clearly observed there and indeed, this is the case, with UV increases as much as a factor of two over that

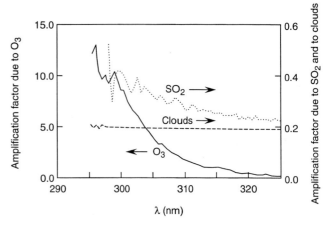

FIGURE 13.16 Calculated UV amplification factors attributable to changes in total column O_3, SO_2, and clouds, respectively, at Thessaloniki, Greece (adapted from Bais *et al.*, 1993).

normally expected (e.g., see Lubin *et al.*, 1989a, 1989b, 1992; Stamnes *et al.*, 1990, 1992; and Beaglehole and Carter, 1992). Figure 13.17, for example, shows the relationship between the radiation amplification factor for the DNA-effective region (Fig. 13.15) and the total column ozone measured at Palmer Station, Antarctica (Lubin *et al.*, 1992). A clear (but nonlinear) relationship is evident. Increased surface UV has also been measured in southern Argentina (e.g., see Frederick *et al.*, 1993; and Bojkov *et al.*, 1995b) and New Zealand (McKenzie *et al.*, 1991; Seckmeyer and McKenzie, 1992), possibly due in part to ozone-depleted air that was processed in the polar vortex.

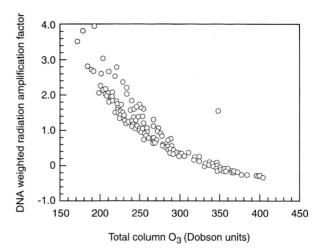

FIGURE 13.17 Radiation amplification factor weighted by DNA action spectrum as a function of total column O_3 at Palmer Station, Antarctica, during Austral springs 1988 and 1990 (adapted from Lubin *et al.*, 1992).

Calculations using satellite measurements of total column ozone indicate that there should be a trend in surface UV at midlatitudes as well, with increases in UV being higher for the higher latitudes. For example, Sabziparvar et al. (1998) have calculated that the daily integrated radiation in the 280- to 320-nm range may have increased by as much as 60% in October and 20% in April at high latitudes since preindustrial times. Figure 13.18 shows calculated monthly changes in the effective UV dose weighted for DNA damage (see Fig. 13.15) as a function of latitude and month. Clearly, substantial increases in UV are expected, based on the ozone depletion data.

In midlatitudes, the actual measured trends in UV at the earth's surface are not so clear-cut (e.g., see Madronich et al., 1998). Some studies report no trend or even a decreasing trend in UV at the earth's surface, whereas others report an increase. For example, in one of the first studies of this issue, no increase was detected over the period 1974 to 1985 using data from the Robertson–Berger network (RB) at National Weather Stations in the United States (Scotto et al., 1988a), although the effects of urban air pollution were suggested to possibly have offset any increases in UV due to ozone depletion (Grant, 1988; Scotto et al., 1988b). Weatherhead and co-workers (1997) have carried out an in-depth analysis of the RB data and concluded that instrumental problems such as calibration difficulties render this data set unsuitable for the detection of long-term trends (see also Basher et al., 1994; and Krzyscin, 1996).

The first convincing evidence of an upward trend in UV was reported by Kerr and McElroy (1993). They monitored radiation at wavelength intervals of 0.5 nm between 290 and 325 nm with a resolution of approximately 0.5 nm in Toronto, Canada. Between 1989 and 1993, UV at 300 nm increased by about 35% in the winter and 7% in the summer, but there was no such increase in the 320- to 325-nm range where light absorption by O_3 is weak. Figure 13.19 shows their reported trends in UV-B radiation over this period of time as a function of wavelength; the dependence is what one would expect due to decreased column ozone, with larger decreases at shorter wavelengths. However, some caution is prudent in extrapolating this to much longer time scales (e.g., see Michaels et al., 1994; and Kerr and McElroy, 1994). It has been suggested that the relationship between UV and total O_3 can be used to estimate total O_3 at sites where UV is measured but column O_3 is not (Fioletov et al., 1997).

Because one of the major impacts of increased UV at the earth's surface is expected to be an increase in skin cancer, several countries now include UV forecasts as part of their weather reports (e.g., see Kerr, 1994), and estimates of skin cancer increases due to ozone depletion have been made. Figure 13.20, for example, shows the estimated number of excess cases of skin cancers for the United States and for northwestern Europe if no controls had been imposed on CFCs and halons and those expected with the Copenhagen Amendments (Slaper et al., 1996). Clearly, there is expected to be a major impact of the control strategies on the incidence of skin cancer, although the number of excess cases will still be about 33,000 per year in the United States and 14,000 per year in northwestern Europe at the projected worst-case year of 2050.

Seckmeyer et al. (1997) showed that even though the monthly mean spectral UV over Germany in 1995 was not significantly different from that in previous years, periods of very high and very low irradiance levels

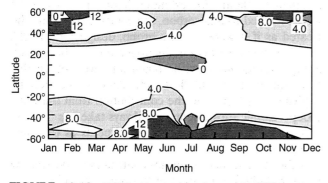

FIGURE 13.18 Calculated monthly changes in UV exposure weighted by the DNA action spectrum over the period 1979–1992 (adapted from Herman et al., 1996).

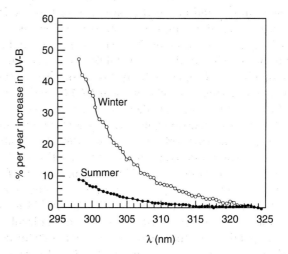

FIGURE 13.19 Measured trends in UV-B radiation from 1989 to 1993 at Toronto, Canada, as a function of wavelength (adapted from Kerr and McElroy, 1993).

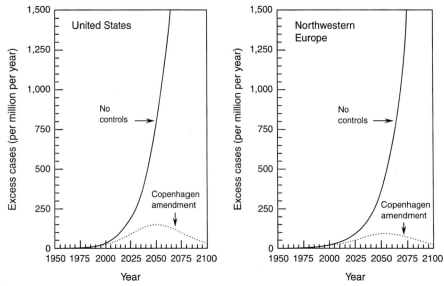

FIGURE 13.20 Estimated excess cases of skin cancer in the United States and northwestern Europe if no controls had been imposed on CFCs and halons and those under the Copenhagen Amendment (adapted from Slaper *et al.*, 1996).

occurred. They suggest that in terms of biological effects, such extremes may have impacts that are not reflected in monthly averages.

D. TROPOSPHERIC CHEMISTRY OF ALTERNATE CFCs

1. Kinetics of OH Reactions

The replacements and alternates for the CFCs (Table 13.1) are characterized by having abstractable hydrogen atoms, and hence they are removed to varying extents by reaction with OH in the troposphere before reaching the stratosphere. The HFCs do not contain chlorine at all, so that their ODPs are very small, essentially zero (Table 13.3). In this section we discuss briefly the tropospheric chemistry of HCFCs and HFCs.

Because many of the alternates and replacements for CFCs have an abstractable hydrogen atom, reaction with OH in the troposphere dominates their loss. Table 13.4 gives some rate constants for the reaction of OH with these compounds; the kinetics summary of DeMore *et al.* (1997) should be consulted for other compounds. It is seen that the rate constants at 298 K are typically in the range of 10^{-13}–10^{-15} cm^3 molecule^{-1} s^{-1}, depending on the degree of halogen substitution and the nature of the halogen, e.g., F, Cl, or Br. Typical A factors are of the order of 1×10^{-12} cm^3 molecule^{-1} s^{-1} per H atom (DeMore, 1996).

As discussed in Chapter 6, group additivity approaches have been developed for estimating the rate constants for the reactions of a variety of species such as OH with organics. DeMore (1996) has applied a similar approach for estimating the rate constants for the reaction of OH with halogenated alkanes. The rate constant per hydrogen atom at 298 K is calculated using Eq. (D):

$$\log_{10} k = -14.79 + \Sigma G_i. \qquad (D)$$

The first term, -14.79, is the logarithm of the rate constant per H atom for CH$_4$, and the factors G_i are corrections for the effects of other groups attached to the carbon atom from which the hydrogen is being abstracted. When two other groups are attached, their contributions are treated separately. In the unique case that the two groups are both atoms, a single factor is assigned to them. For example, if one F and one Cl were attached to this carbon, the contribution would be treated as if it were a single group, designated "F,Cl".

When there is only one hydrogen on a carbon atom, group interactions must be taken into account. In this case, the best overall fit to the data is obtained by applying a multiplier to the contribution from the third group. Atomic substituents are always taken as the first and second groups, and larger substituents as the third group.

Table 13.5 shows the group contributions derived by DeMore (1996) to be used in this approach. For example, let us use this method to estimate the rate constant at 298 K for the reaction of OH with HCFC-122a,

TABLE 13.4 Rate Constants for the Reactions of OH with Some Alternates and Replacements for CFCs[a-c]

Name	Compound	$k(298\ K)$ (cm^3 molecule^{-1} s^{-1})	A(cm^3 molecule^{-1} s^{-1})	E_a/R (K)
HFC-134a	CH_2FCF_3	4.2×10^{-15}	1.5×10^{-12}	1750
HFC-125	CHF_2CF_3	1.9×10^{-15}	5.6×10^{-13}	1700
HFC-143a	CH_3CF_3	1.2×10^{-15}	1.8×10^{-12}	2170[b]
HFC-152a	CH_3CHF_2	3.5×10^{-14}	2.4×10^{-12}	1260
HFC-161	CH_3CH_2F	1.7×10^{-13}	7.0×10^{-12}	1100
HFC-245fa	$CHF_2CH_2CF_3$	7.0×10^{-15}	0.61×10^{-12}	1331[b]
HFC-245cb	$CH_3CF_2CF_3$	1.5×10^{-15}	4.41×10^{-13}	1690[c]
HCFC-141b	CH_3CCl_2F	5.7×10^{-15}	1.7×10^{-12}	1700
HCFC-142b	CH_3CF_2Cl	3.1×10^{-15}	1.3×10^{-12}	1800
HCFC-22	$CHClF_2$	4.7×10^{-15}	1.0×10^{-12}	1600
HCFC-123	$CHCl_2CF_3$	3.4×10^{-14}	7.0×10^{-13}	900
[d]	$CH_3OCF_2CF_2H$	2.0×10^{-13}		
[d]	$CH_3CH_2OCF_2CF_2H$	4.3×10^{-13}		
[d]	CH_3OCF_2CFClH	1.7×10^{-13}		
[d]	$CH_2=CHCH_2OCF_2CF_2H$	1.9×10^{-11}		
[e]	$CH_3OC_4F_9$	7.2×10^{-15}		
[e]	$CHF_2OCF_2OCHF_2$	2.4×10^{-15}		
[e]	$CHF_2O(CF_2)_2OCHF_2$	4.7×10^{-15}		
[e]	$CHF_2O(CF_2)_2OCF_2OCHF_2$	4.6×10^{-15}		

[a] From DeMore *et al.* (1997).
[b] From Orkin *et al.* (1996).
[c] From Orkin *et al.* (1997).
[d] From Heathfield *et al.* (1998).
[e] From Cavalli *et al.* (1998).

CHFClCFCl$_2$. The groups attached to the same carbon as the abstractable hydrogen are F, Cl, and $-CCl_2F$. The F and Cl are taken as the first and second substituents and treated as a single F,Cl group. The $-CFCl_2$ group is then the third group to which the multiplication factor 0.37 is applied.

$$\log_{10} k = -14.79 + G(F,Cl) + 0.37G(CCl_2F),$$
$$= -14.79 + 0.98 + 0.37(0.08) = -13.78,$$

and $k(298\ K) = 1.7 \times 10^{-14}$ cm^3 molecule^{-1} s^{-1}, within 15% of the experimentally measured value of 1.5×10^{-14} cm^3 molecule^{-1} s^{-1} (Hsu and DeMore, 1995).

DeMore (1996) has also applied this to larger halocarbons. For example, for the reaction of HCFC-236ea, $CF_3CHFCHF_2$, the rate constant is the sum of those for abstraction of the two hydrogens. For the first hydrogen, the F atom is taken as the first substituent, but it is not clear which of the two remaining substituents, $-CF_3$ or $-CHF_2$, should be taken as the second and third, respectively, with the multiplier of 0.37 being applied to the third group only. In this case, DeMore recommends calculating the rate constant both ways and taking an average. Thus for the first hydrogen, two different values are calculated depending on which group is taken as the third group. If $-CF_2H$ is taken as the third group, then for this hydrogen

$$\log_{10} k = -14.79 + G(F) + G(CF_3) + 0.37G(CF_2H),$$
$$= -14.79 + 0.56 - 0.50 + 0.37(-0.08)$$
$$= -14.76.$$

On the other hand, if $-CF_3$ is taken as the third group,

$$\log_{10} k = -14.79 + G(F) + G(CF_2H) + 0.37G(CF_3),$$
$$= -14.79 + 0.56 - 0.08 + 0.37(-0.50)$$
$$= -14.50.$$

The average of these two values is -14.63, and $k = 2.3 \times 10^{-15}$ cm^3 molecule^{-1} s^{-1} for this hydrogen. For the second hydrogen, the calculation is straightforward:

$$\log_{10} k = -14.79 + G(F,F) + 0.37G(CHFCF_3),$$
$$= -14.79 + 0.35 + 0.37(-0.08) = -14.47,$$

and $k = 3.4 \times 10^{-15}$ cm^3 molecule^{-1} s^{-1} for this hydrogen, where the G factor for $-CHF_2$ is used as a surrogate for $-CHFCF_3$. The total rate constant for abstraction of the two hydrogens is thus (2.3 + 3.4) or 5.7×10^{-15} cm^3 molecule^{-1} s^{-1}, again within 15% of the experimentally determined value of 5.3×10^{-15} cm^3 molecule^{-1} s^{-1} (DeMore *et al.*, 1997).

In short, a number of the rate constants for the reactions of OH with HFCs, HCFCs, and other potential CFC replacements such as partially fluorinated

TABLE 13.5 Estimated Group Contributions for Estimating the Rate Constants for OH + Halohydrocarbons[a]

Group	G
F	0.56
F,F	0.35
CF_3	−0.50
CH_2F	0.44
CHF_2	−0.08
CH_3	1.21
Cl	0.73
2Cl	1.46
CH_2Cl	0.77
$CHCl_2$	0.60
CHFCl	0.05
CF_2Cl	−0.19
F,Cl	0.98
CCl_3	0.29
$CFCl_2$	0.08
Br	0.83
Br,Br	1.54
Br,Cl	1.43
F,Br	1.1
Third group multiplier	0.37

[a] From DeMore (1996).

ethers have been measured, and group additivity methods similar to those used for simple hydrocarbons (Chapter 6) are also available to permit estimating their values when experimental data are not available. Other approaches for estimating reaction kinetics include, for example, correlations with the HOMO (highest occupied molecular orbital) of the molecule, which has been applied to fluorinated compounds (e.g., Dhanya and Saini, 1997).

2. Tropospheric Chemistry

As examples of the tropospheric chemistry, let us take HFC-134a (CH_2FCF_3), HFC-125 (CHF_2CF_3), HCFC-123 ($CHCl_2CF_3$), and HCFC-141b (CH_3CCl_2F). The chemistry in many respects parallels that of alkanes discussed in detail in Chapter 6. We treat here these selected compounds as typifying the types of chemistry that occur with these CFC substitutes. For a more comprehensive treatment of the chemistry of a number of compounds, see Wallington *et al.* (1994a) and Midgley (1995).

a. Chemistry of HFC-134a (CH_2FCF_3)

The initial reaction of OH with HFC-134a, CH_2FCF_3, produces a halogenated alkyl radical that quickly adds O_2 to form a peroxy radical:

$$CF_3CH_2F + OH \rightarrow H_2O + CF_3CHF, \quad (1)$$

$$CF_3CHF + O_2 \xrightarrow{M} CF_3CHFOO. \quad (2)$$

With a value for k_1 of 4.2×10^{-15} cm^3 molecule^{-1} s^{-1}, the lifetime with respect to reaction with OH at a concentration of 1×10^6 radicals cm^{-3} is $\tau = (k_1[OH])^{-1}$ ~8 years. Reaction with O_2 is sufficiently fast that reaction (2) represents the sole fate of any significance of the alkyl radicals in air.

As is the case for unsubstituted alkylperoxy radicals, when sufficient NO is present, these radicals react rapidly to form an alkoxy radical:

$$CF_3CHFOO + NO \rightarrow CF_3CHFO + NO_2. \quad (3a)$$

Wallington *et al.* (1996a) and Møgelberg *et al.* (1997b) report that about 64% of the CF_3CHFO radicals formed in this reaction are sufficiently excited to decompose to CF_3 + HC(O)F at a pressure of 800 Torr. They attribute this to the generation of vibrationally excited (CF_3CHFO) during the decomposition of the initially formed adduct ($CF_3CHFOO \cdots NO$)*. Thus a second channel must be included:

$$CF_3CHFOO + NO \rightarrow NO_2 + (CF_3CHFO)^{\neq}$$
$$\rightarrow CF_3 + HC(O)F. \quad (3b)$$

However, $(CF_3CHFO)^{\neq}$ formed by the reaction of CF_3CHFOO with itself at low NO levels does not contain sufficient energy for decomposition to be significant, so that the alkoxy radical CF_3CHFO is the major product (Wallington *et al.*, 1996a). The CF_3 formed in (3b) adds O_2 to form CF_3OO.

With an overall room temperature rate constant of $k_3 = 1.3 \times 10^{-11}$ cm^3 molecule^{-1} s^{-1} (Wallington and Nielsen, 1991), the lifetime of CF_3CHFOO is only about half a minute at 0.1 ppb NO. At very small concentrations of NO, reaction with HO_2 (or other RO_2) can occur, e.g.,

$$CF_3CHFOO + HO_2 \rightarrow CF_3CHFOOH + O_2. \quad (4)$$

With a rate constant at 298 K of $k_4 = 4 \times 10^{-12}$ cm^3 molecule^{-1} s^{-1} (Maricq *et al.*, 1994b; Hayman and Battin-Leclerc, 1995), the lifetime of the CF_3CHFOO radical with respect to HO_2 at 5×10^8 radicals cm^{-3} is about 8 min. That is, at this HO_2 concentration, NO would have to fall to 6 ppt for reactions (3) and (4) to occur at equal rates.

The alkoxy radical CF_3CHFO has, in principle, two fates: reaction with O_2 and decomposition. For this particular radical, both are important under tropospheric conditions (Wallington et al., 1992, 1996a; Edney and Driscoll, 1992; Tuazon and Atkinson, 1993a; Rattigan et al., 1994; Bednarek et al., 1996; Hasson et al., 1998):

$$CF_3CHFO \xrightarrow{decomposition} CF_3 + HC(O)F, \quad (5)$$

$$CF_3CHFO + O_2 \rightarrow CF_3C(O)F + HO_2. \quad (6)$$

The absolute kinetics of the decomposition, reaction (5), compared to reaction (6) with O_2 have been somewhat uncertain (e.g., Wallington et al., 1992; Tuazon and Atkinson, 1993a; Rattigan et al., 1994). Hasson et al. (1998) studied the oxidation of HFC-134a by FTIR from 298 to 357 K at various concentrations of O_2 and report that the ratio of rate constants is given by $k_6/k_5 = (3.8 \pm 1.6) \times 10^{-24} \, e^{(2400 \pm 500)/T}$ cm^3 molecule^{-1}. Because of the strong temperature dependence, the relative importance of reaction (5) decreases significantly as one moves from the earth's surface to the much colder tropopause. Thus, about 20–25% of the loss of CF_3CHFO is estimated to proceed via (6) at the surface, increasing to ~80% at the tropopause (Tuazon and Atkinson, 1993a; Rattigan et al., 1994).

As discussed shortly, $CF_3C(O)F$ generated in reaction (6) ultimately hydrolyzes to trifluoroacetic acid (CF_3COOH) in the atmosphere.

The CF_3 radical adds O_2 to form CF_3O_2, which then forms the alkoxy radical CF_3O:

$$CF_3 + O_2 \xrightarrow{M} CF_3O_2, \quad (7)$$

$$CF_3O_2 + NO \rightarrow CF_3O + NO_2. \quad (8)$$

The values of k_0 and k_∞ for reaction (7) are 3.0×10^{-29} cm^6 molecule^2 s^{-1} and 4.0×10^{-12} cm^3 molecule^{-1} s^{-1}, respectively, at 300 K (DeMore et al., 1997), giving an effective bimolecular rate constant of 3.7×10^{-12} cm^3 molecule^{-1} s^{-1} at 1 atm pressure at 300 K. The lifetime of CF_3 with respect to reaction with O_2 is thus only ~50 ns under these conditions, and other reactions such as that with NO_2 (e.g., see Sehested et al., 1996; and Vakhtin, 1997) cannot compete.

Reaction (8), reaction of CF_3O_2 with NO, is also sufficiently fast that this is a major fate of CF_3O_2 in the troposphere. Thus the rate constant at 298 K for reaction (8) is 1.6×10^{-11} cm^3 molecule^{-1} s^{-1} (e.g., see Sehested and Nielsen, 1993; Bevilacqua et al., 1993; Turnipseed et al., 1994; and Bhatnagar and Carr, 1994), which gives a lifetime for CF_3O_2 of 4 min at 10 ppt NO. At small NO concentrations, reaction with HO_2 (or RO_2) can also occur:

$$CF_3OO + HO_2 \rightarrow CF_3OOH + O_2. \quad (9)$$

The rate constant for this reaction is $k_9 = (4.0 \pm 2.0) \times 10^{-12}$ cm^3 molecule^{-1} s^{-1} (Sehested et al., 1997).

At an HO_2 concentration of 1×10^9 cm^{-3}, the lifetime of CF_3O_2 with respect to reaction (9) with HO_2 is also ~4 min. Thus, under these assumed concentrations of NO and HO_2, the reactions are equal in magnitude.

CF_3O_2 (and similar halogenated peroxy radicals) can also react with NO_2 to form trifluoromethyl peroxynitrate, CF_3OONO_2, but the relatively fast thermal decomposition of peroxynitrates makes this only a temporary reservoir under the conditions at the earth's surface, as is the case for unsubstituted alkylperoxy radicals discussed in Chapter 6 (Chen et al., 1993a; Niki et al., 1994; Møgelberg et al., 1994b). For example, the lifetime of CF_3OONO_2 is estimated to be only ~1 min at room temperature and 1 atm total pressure; however, under typical conditions at the tropopause, the lifetime with respect to thermal decomposition is much longer, about 1 year (Mayer-Figge et al., 1996). Similar considerations apply to halogenated PAN analogs. For example, the lifetime of $CF_3C(O)OONO_2$ is calculated to be only ~3 h at room temperature but to increase by about seven orders of magnitude under the conditions at 12 km! (Zabel et al., 1994; Wallington et al., 1994e).

It should be noted that while the formation of hydroperoxides as shown in reaction (9) is common for simple hydrocarbon alkylperoxy radicals (see Chapter 6), recent work on chlorine-containing RO_2 radicals indicates that this is not always the case. Catoire et al. (1996) studied the reactions of $CHCl_2O_2$ and CCl_3O_2 with HO_2. In the latter reaction, the yield of phosgene, $COCl_2$, was ~100%. The reaction was interpreted to proceed via formation of the tetraoxide intermediate CCl_3OOOOH, which decomposes via a cyclic transition state to $COCl_2 + HOCl + O_2$. The tetraoxide intermediate from the $CHCl_2O_2$ reaction was proposed to have two analogous decomposition paths available, one giving $COCl_2 + H_2O + O_2$ (29% at 296 K and 700 Torr total pressure) and the other $HC(O)Cl + HOCl + O_2$ (71% at 296 K and 700 Torr). In neither case was the formation of a stable hydroperoxide observed, in contrast to the reaction of CH_3O_2 with HO_2, for example (see Chapter 6). In the case of CH_2FO_2 radicals, both pathways occur, with 29% of the reaction with HO_2 giving the CH_2FOOH and 71% $HC(O)F$ as products (Wallington et al., 1994d).

Other reactions of CF_3O_2 such as that with O_3 are slow and cannot compete (e.g., see Nielsen and Sehested, 1993; Turnipseed et al., 1994; and Meller and Moortgat, 1995).

There are no abstractable hydrogen atoms on CF_3O, so that reaction with O_2 is negligible ($k < 4 \times 10^{-17}$ cm^3 molecule^{-1} s^{-1} at 373 K; Turnipseed et al., 1994; Niki et al., 1994). In addition, the C–F bond is too strong for decomposition to give a fluorine atom and CF_2O to take place. However, CF_3O reacts with NO, possibly with H$_2$O and with organics (e.g., Niki et al., 1994):

$$CF_3O + NO \rightarrow CF_2O + FNO, \quad (10)$$

$$CF_3O + H_2O \rightarrow CF_3OH + OH, \quad (11)$$

$$CF_3O + RH \rightarrow CF_3OH + R. \quad (12)$$

The reaction (10) with NO is fast, $k_{10} = (5-6) \times 10^{-11}$ cm^3 molecule^{-1} s^{-1} at 298 K (e.g., see Bevilacqua et al., 1993; Sehested and Nielsen, 1993; Bhatnagar and Carr, 1994; Jensen et al., 1994; Turnipseed et al., 1994; and Dibble et al., 1995), and has been shown to give CF$_2$O and FNO as products (e.g., Chen et al., 1992a; Niki et al., 1994; Dibble et al., 1995). FNO reacts only slowly with atmospheric reactive species such as O$_3$, O(^{3}P), HO$_2$, and HCl so that photolysis appears to be its major atmospheric fate (Wallington et al., 1995). CF$_3$O has also been observed to react with NO$_2$ to form CF$_3$ONO$_2$ (Niki et al., 1994), but the kinetics were not reported.

The importance of the reaction of CF$_3$O with H$_2$O is less certain; an upper limit to the rate constant k_{11} at 298 K is 1×10^{-16} cm^3 molecule^{-1} s^{-1} (Wallington et al., 1993b; Turnipseed et al., 1995). However, because of the relatively large amounts of water vapor in the troposphere, application of this upper limit gives a lifetime for CF$_3$O with respect to reaction with water vapor of ≥ 30 ms at 50% RH. This can be compared to a lifetime with respect to reaction (10) with 100 ppt NO of 7 s.

CF$_3$O abstracts a hydrogen atom from alkanes in much the same way as OH, forming CF$_3$OH (Chen et al., 1992b; Saathoff and Zellner, 1993; Bevilacqua et al., 1993; Sehested and Wallington, 1993; Kelly et al., 1993; Niki et al., 1994; Barone et al., 1994; Jensen et al., 1994; Bednarek et al., 1995; Wallington and Ball, 1995). It also reacts rapidly with alkenes and aromatics, in a manner suggesting addition as the initial reaction (e.g., see Chen et al., 1993b; Niki et al., 1994; Kelly et al., 1994; and Wallington and Ball, 1995). Table 13.6, for example, gives room temperature rate constants for the reaction of CF$_3$O with some organics. Even the reaction with CH$_4$ is sufficiently fast that the lifetime of CF$_3$O with respect to 1.7 ppm CH$_4$ is only 1 s. Hence reaction with organics is expected to be a major fate of CF$_3$O in the troposphere. In the case of reaction with alkanes, the final product, CF$_3$OH, is the same as the reaction with H$_2$O.

CF$_3$O has also been observed to react with CO with kinetics that are consistent with two channels, a pressure-independent channel and a pressure-dependent one. The existence of the pressure-dependent channel suggests CF$_3$O adds to CO to form an adduct, analogous to the OH + CO reaction; this reaction may play

TABLE 13.6 Rate Constants at Room Temperature for the Reaction of CF$_3$O with Some Organics

Organic	k(298 K) (cm^3 molecule^{-1} s^{-1})	Reference
CH$_4$	2.2×10^{-14}	Wallington and Ball, 1995
C$_2$H$_6$	1.4×10^{-12}	Wallington and Ball, 1995
C$_3$H$_8$	4.7×10^{-12}	Kelly et al., 1993
i-C$_4$H$_8$	6.1×10^{-12}	Wallington and Ball, 1995
C$_2$H$_4$	3.0×10^{-11}	Kelly et al., 1993
i-C$_4$H$_{10}$	7.2×10^{-12}	Kelly et al., 1993
C$_6$H$_6$	3.6×10^{-11}	Kelly et al., 1993, 1994
CH$_3$OH	2.5×10^{12}	Wallington and Ball, 1995

a minor role in the tropospheric removal of CF$_3$O as well (Turnipseed et al., 1995; Wallington and Ball, 1995; Meller and Moortgat, 1997).

While OH removes a significant amount of the CFC alternatives in the troposphere, some fraction does reach the stratosphere. One possible chain that has been considered in the stratosphere involves reactions of CF$_3$O and CF$_3$O$_2$ with O$_3$ (e.g., see Ko et al., 1994b). However, the rate constants for these reactions were measured subsequently and shown to be too slow relative to other removal processes such as CF$_3$O + NO to be significant (Nielsen and Sehested, 1993; Maricq and Szente, 1993; Wallington et al., 1993a, 1994c; Maricq et al., 1994b; Ravishankara et al., 1994; Fockenberg et al., 1994; Meller and Moortgat, 1995; Mörs et al., 1995).

The atmospheric fate of CF$_3$OH appears to be primarily uptake into clouds in the troposphere and possibly into sulfate aerosols in the stratosphere. Lovejoy et al. (1995), for example, measured uptake coefficients for CF$_3$OH on sulfuric acid–water mixtures from $\sim 10^{-3}$ to $\sim 10^{-1}$, the larger values occurring at the highest water concentrations. From these data, they suggest an uptake coefficient of about 0.1 for clouds, which gives a lifetime with respect to such uptake of the order of seconds under typical cloud conditions. As a result, uptake into tropospheric clouds would be controlled by the frequency of cloud formation. On the other hand, in the stratosphere, where the particle concentration is relatively low and the acid concentration high, the lifetime could be of the order of several years (Lovejoy et al., 1995). Other fates such as decomposition, photolysis, or reaction with species such as OH and Cl are believed to be negligible (Wallington and Schneider, 1994; Schneider et al., 1995; Huey et al., 1995).

Figure 13.21 summarizes the chemistry of HFC-134a.

b. Chemistry of HCFC-125 (CHF$_2$CF$_3$)

With sufficient NO present, HCFC-125 (CHF$_2$CF$_3$) reacts in an analogous manner to generate an alkoxy

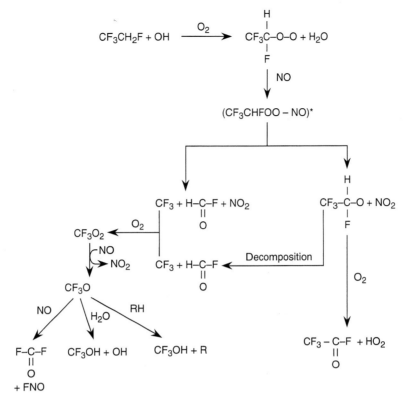

FIGURE 13.21 Summary of major reaction paths in the tropospheric chemistry of HFC-134a.

radical:

$$CF_3CHF_2 + OH \rightarrow H_2O + CF_3CF_2, \quad (13)$$

$$CF_3CF_2 + O_2 \xrightarrow{M} CF_3CF_2OO, \quad (14)$$

$$CF_3CF_2OO + NO \rightarrow CF_3CF_2O + NO_2. \quad (15)$$

With no abstractable hydrogen atoms on the alkoxy radical, reaction with O_2 cannot occur and it decomposes by scission of the C–C bond (Edney and Driscoll, 1992; Tuazon and Atkinson, 1993b; Sehested et al., 1993):

$$CF_3CF_2O \rightarrow COF_2 + CF_3. \quad (16)$$

The CF_3 radical then reacts as described earlier, forming COF_2, FNO, and CF_3OH.

Figure 13.22 summarizes the chemistry of HCFC-125.

c. Chemistry of HCFC-123 ($CHCl_2CF_3$)

The initial OH chemistry of HCFC-123 ($CHCl_2CF_3$) produces a fully halogenated alkoxy radical where further hydrogen abstraction by O_2 is not an option:

$$CF_3CHCl_2 + OH \rightarrow H_2O + CF_3CCl_2, \quad (17)$$

$$CF_3CCl_2 + O_2 \xrightarrow{M} CF_3CCl_2OO, \quad (18)$$

$$CF_3CCl_2OO + NO \rightarrow CF_3CCl_2O + NO_2. \quad (19)$$

However, recall that the C–Cl bond is much weaker than a C–F bond. As a result, one additional potential route of decomposition in this case is elimination of a chlorine atom:

$$CF_3CCl_2O \rightarrow Cl + CF_3C(O)Cl. \quad (20)$$

In fact, this predominates over the competing C–C bond scission,

$$CF_3CCl_2O \rightarrow COCl_2 + CF_3, \quad (21)$$

with yields of $CF_3C(O)Cl$ approaching 100% (Edney et al., 1991; Tuazon and Atkinson, 1993b; Hayman et al., 1994). Similar chlorine atom elimination has been observed for other chlorine-containing alkoxy radicals (e.g., see Wu and Carr, 1992; and Bhatnagar and Carr, 1995). The $CF_3C(O)Cl$ hydrolyzes in clouds and fogs to form CF_3COOH, trifluoroacetic acid.

It should be noted that, although it is not important in this particular case, there is an additional path available for some alkoxy radicals formed in the oxidation of certain halogenated organics, that is, the intramolecular elimination of HCl from α-monochloroalkoxy radicals. For example, the alkoxy radical $CH_2ClCHClO$ both reacts with O_2 and eliminates HCl,

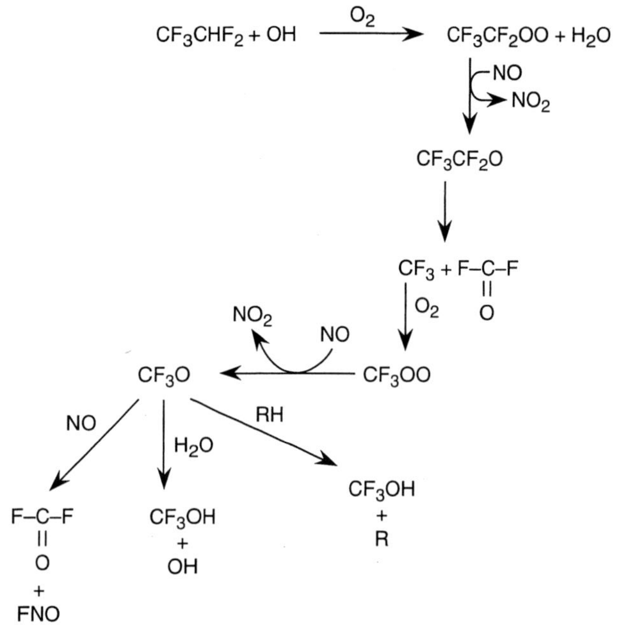

FIGURE 13.22 Summary of major tropospheric reactions of HCFC-125.

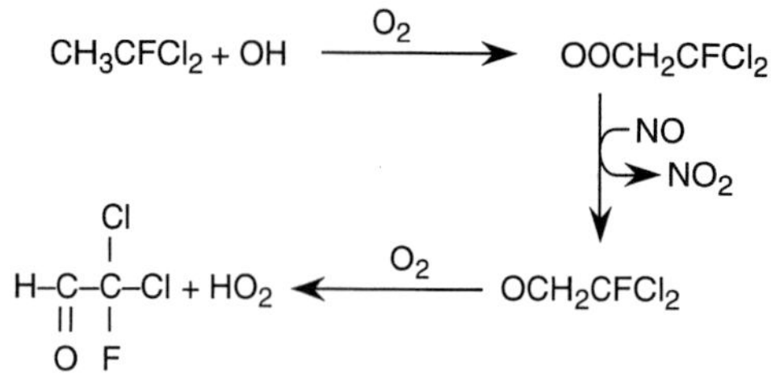

FIGURE 13.24 Summary of major tropospheric reactions of HCFC-141b.

dominates over decomposition by C–C bond scission (Edney *et al.*, 1991; Tuazon and Atkinson, 1993b, 1994):

$$OCH_2CFCl_2 \rightarrow HCHO + CFCl_2. \quad (26)$$

Further reaction of $CFCl_2$ with O_2 and subsequently with NO generates $ClC(O)F$.

Figure 13.24 summarizes the chemistry of HCFC-141b.

In short, the tropospheric chemistry of the CFC replacement compounds containing abstractable hydrogen atoms is very similar to the VOC chemistry discussed in Chapter 6. The major differences are found in the relative importance of the various potential fates of the alkoxy radicals formed in these reactions, which depends on the structure of the parent compound. Table 13.7 summarizes the major and minor products formed by the tropospheric oxidation of the four particular typical CFC replacement compounds treated here. For the chemistry of other CFC replacements, see papers by Chen *et al.* (1997) and a series of papers by Wallington, Nielsen, and co-workers (e.g., Wallington *et al.*, 1994a; Wallington and Nielsen, 1995; Møgelberg *et al.*, 1995a, 1995b, 1995c, 1996, 1997; Giessing *et al.*, 1996). The atmospheric chemistry of a series of hydrofluoroethers (HFE), such as $C_4F_9OCH_3$ (HFE-7100), $C_4F_9O_4C_2H_5$ (HFE-7200), and $CF_3CH_2OCH_2CF_3$, used as CFC alternates is discussed by Wallington *et al.* (1997, 1998), Christensen *et al.* (1998), and Cavalli *et al.* (1998)

with the latter favored at 1 atm in air (Wallington *et al.*, 1996b).

Figure 13.23 summarizes the chemistry of HCFC-123.

d. Chemistry of HCFC-141b (CH_3CFCl_2)

Hydrogen atom abstraction from HCFC-141b (CH_3CFCl_2) produces the alkoxy radical OCH_2CFCl_2:

$$CH_3CFCl_2 + OH \rightarrow H_2O + CH_2CFCl_2, \quad (22)$$

$$CH_2CFCl_2 + O_2 \xrightarrow{M} OOCH_2CFCl_2, \quad (23)$$

$$OOCH_2CFCl_2 + NO \rightarrow NO_2 + OCH_2CFCl_2. \quad (24)$$

In this case, reaction of the alkoxy radical with O_2,

$$OCH_2CFCl_2 + O_2 \rightarrow HO_2 + HC(O)CFCl_2, \quad (25)$$

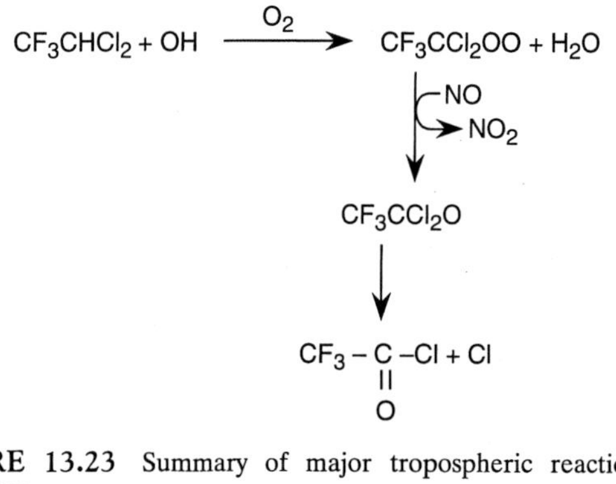

FIGURE 13.23 Summary of major tropospheric reactions of HCFC-123.

e. Tropospheric Fates of Halogenated Products of HCFC Oxidation

As seen in Table 13.7, oxidation of HCFCs by OH generates a variety of halogenated aldehydes and ketones as well as phosgene ($COCl_2$), its fluorine analog (COF_2), $ClC(O)F$ and $HC(O)F$, and the alcohol CF_3OH. The ultimate atmospheric fate of these products depends on their structures, of course, which determines their absorption cross sections as well as reactivity with OH, and their solubility in aqueous solutions such as clouds, rainwater, and the oceans.

TABLE 13.7 Typical Organic Products of the Tropospheric Oxidation of Some of the CFC Replacement Compounds

Replacement	Compound	Major fates of alkoxy radicals	Major products	Minor products
HFC-134a	CH_2FCF_3	Decomposition and reaction with O_2	$HC(O)F, COF_2, CF_3OH, CF_3C(O)F$	
HFC-125	CHF_2CF_3	Decomposition by C–C bond scission	COF_2, CF_3OH	
HCFC-123	$CHCl_2CF_3$	Cleavage of C–Cl bond, small fraction by C–C bond scission	$CF_3C(O)Cl$	$COF_2, CF_3OH, COCl_2$
HCFC-141b	CH_3CCl_2F	Reaction with O_2 primarily	$HC(O)CFCl_2$	$ClC(O)F$

The fate of CF_3OH described earlier is believed to be primarily uptake into clouds. The same is true of $COCl_2$, COF_2, $HC(O)F$, and $CF_3C(O)F$. For all of these species, photolysis at the wavelengths found in the troposphere is negligible, as is reaction with OH (Nölle et al., 1992; Rattigan et al., 1993; Wallington et al., 1994a; Zachariah et al., 1995; World Meteorological Organization, 1995). Table 13.8 summarizes estimates of the lifetimes of these halogenated product species with respect to uptake by the oceans, clouds, and rainwater. Uptake into clouds followed by hydrolysis is the major removal mechanism from the atmosphere for these compounds. Hydrolysis forms HCl and HF in the case of $COCl_2$, COF_2, and $HC(O)F$ as well as formic acid in the latter case. $CF_3C(O)F$ hydrolyzes to HF and trifluoroacetic acid, $CF_3C(O)OH$.

The effects on human health of trifluoroacetic acid (TFA) from the oxidation of HFC-134a, halothane (1,1,1-trifluoro-2-bromo-2-chloroethane, used as an anesthetic), and some of the other CFC replacements, such as HCFC-123 and HCFC-124, have been of some concern. The toxicology of this compound is reviewed by Ball and Wallington (1993) and its effects on plants are discussed by Tromp et al. (1995). The reaction of TFA in the gas phase with OH is relatively slow ($k \sim 1.7 \times 10^{-13}$ cm^3 molecule^{-1} s^{-1} at 296 K) and is estimated to account for only ~10–20% of the loss of TFA, with the remainder being removed by rainout (Møgelberg et al., 1994a). Calculated maximum concentrations of TFA in rainwater in the future have been suggested to be in the range of 1–80 nmol L^{-1} (Ball and Wallington, 1993; Tromp et al., 1995), but a 3-D modeling study suggests that global annually averaged rainwater concentrations would be ~120 ng L^{-1}, and in northern midlatitudes, monthly averaged concentrations could be as large as 450 ng L^{-1} in the summer (Kotamarthi et al., 1998). These higher estimated values are consistent with measurements of 40–1200 ng L^{-1} of TFA in rainwater in California and Nevada (Wujcik et al., 1997). Frank and co-workers (1996) report that in 1995, concentrations of TFA in Europe in rainwater were 0.26–2.1 nmol L^{-1}, with up to 5.5 nmol L^{-1} in rivers; in the Dead Sea, where the TFA may have been concentrated through evaporation processes, the concentration was 56 nmol L^{-1}.

$CF_3C(O)Cl$ from the oxidation of HCFC-123 is rapidly taken up into cloudwater (Table 13.8). However, it also photolyzes (Rattigan et al., 1993; Wallington, 1994a; WMO, 1995), with an estimated tropospheric lifetime of ~33 days assuming a quantum yield for dissociation of unity (Rattigan et al., 1993):

$$CF_3C(O)Cl + h\nu \rightarrow CF_3 + CO + Cl. \quad (27)$$

The subsequent reactions of CF_3 are as discussed earlier. In the troposphere, the most likely fate of Cl is reaction with organics (see Chapter 6).

The aldehyde $HC(O)CFCl_2$ formed in the oxidation of HCFC-141b is expected to photolyze in the troposphere, forming in part $CFHCl_2$ (Wallington et al., 1994a), which is itself oxidized in the troposphere by reaction with OH. By analogy to the photolysis of

TABLE 13.8 Estimated Lifetimes of Halogenated Carbonyl Compounds in the Aqueous Phase[a]

Compound	Ocean	Clouds	Rainout
$HC(O)F$	3 months	4 days	180 years
COF_2[b]	4 months	4 days	72 years
$ClC(O)F$	2 years	6 days	265 years
$CF_3C(O)F$[c]	2 years	6 days	675 years
$CF_3C(O)Cl$[d]	3 years	6 days	900 years

[a] From Kanakidou et al. (1995).

[b] Tropospheric lifetime estimated to be ~0.5–3 days by De Bruyn et al. (1992, 1995) based on measurements of uptake by water surfaces; see also George et al. (1994a,b).

[c] Tropospheric lifetime estimated to be ~0.5–3 days by De Bruyn et al. (1992, 1995) based on measurements of uptake by water surfaces; see also George et al. (1994a,b).

[d] Tropospheric lifetime estimated to be ~2–10 days by De Bruyn et al. (1992, 1995) based on measurements of uptake by water surfaces; see also George et al. (1994a,b).

nonhalogenated aldehydes, formation of HCO + $CFCl_2$ is also expected.

Because $HC(O)CFCl_2$ has an abstractable hydrogen atom, it reacts with OH in the troposphere:

$$HC(O)CFCl_2 + OH \rightarrow C(O)CFCl_2 + H_2O. \quad (28)$$

With a rate constant at 298 K of 1.2×10^{-12} cm^3 molecule^{-1} s^{-1}, the lifetime of this aldehyde with respect to OH at 1×10^6 radicals cm^{-3} is about 10 days (Scollard et al., 1993). Either decomposition of $C(O)CFCl_2$ to CO + $CFCl_2$ or its reaction with O_2 can occur (Tuazon and Atkinson, 1994):

$$C(O)CFCl_2 \rightarrow CO + CFCl_2, \quad (29)$$

$$C(O)CFCl_2 + O_2 \xrightarrow{M} OOC(O)CFCl_2. \quad (30)$$

Tuazon and Atkinson (1994) showed that $79 \pm 7\%$ of the reaction proceeds via decomposition (29) and $21 \pm 5\%$ by reaction (30) with O_2. However, the relative importance of these two possible fates depends on the particular radical. For example, for CF_3CO, more than 99% proceeds by reaction with O_2 (Wallington et al., 1994b), whereas for $C(O)CF_2Cl$, 39% decomposes and 61% reacts with O_2 (Tuazon and Atkinson, 1994).

Reaction of the alkylperoxy radical generated in (30) with NO followed by decomposition will generate the $CFCl_2$ radical:

$$OOC(O)CFCl_2 + NO \rightarrow OC(O)CFCl_2 + NO_2, \quad (31)$$

$$OC(O)CFCl_2 \rightarrow CO_2 + CFCl_2. \quad (32)$$

Reaction with NO_2 to form the peroxyacyl nitrate $CFCl_2C(O)OONO_2$ can also occur.

Kanakidou et al. (1995) have carried out three-dimensional modeling studies of the global tropospheric fates of HFC-134a and other HCFCs using projected emissions and the chemistry described earlier. Table 13.9 shows their calculated lifetimes for some of the CFC replacements with respect to oxidation by OH, loss in the stratosphere, and the overall lifetime from both processes. Calculated lifetimes are in the range of 1–20 years. The model predicts that the oxidation products shown in Table 13.7 will not accumulate significantly in the troposphere, with concentrations typically 1% or less of their parent compound.

The reason that the ODPs of these CFC replacements are much smaller than those of the original CFCs is the presence of an abstractable hydrogen with which OH can react. However, this also means that they can also contribute to ozone formation in the troposphere. Hayman and Derwent (1997) have used their photochemical trajectory model to calculate tropospheric ozone-forming potentials of some of these CFC replacements. Table 13.10 summarizes these relative ozone-forming potentials, expressed taking that for ethene as 100. Clearly, although they react in the troposphere, their contribution to tropospheric ozone formation is expected to be very small.

While we have focused here on CFC replacements, similar chemistry applies to replacements for the bromine-containing halons. For example, CF_2BrH is a potential halon substitute that will react with OH in the troposphere (DeMore et al., 1997). Through the subsequent reaction with O_2 and then NO, the alkoxy radical CF_2BrO is formed. This decomposes via scission of the weak C–Br bond to form COF_2 (Bilde et al., 1996).

Similarly, CF_3I is a potential halon substitute. However, as for CH_3I, it photolyzes rapidly to generate an iodine atom with an estimated lifetime of less than 2 days (Solomon et al., 1994a). While iodine in the stratosphere is expected to be very effective in ozone de-

TABLE 13.9 Estimated Tropospheric, Stratospheric, and Overall Lifetimes of Some CFC Replacements[a]

Compound	Lifetime for oxidation by OH (years)	Lifetime for stratospheric loss (years)	Overall lifetime (years)
HCFC-22	13.7	214	12.9
HFC-134a	13.9	357	13.4
HCFC-123	1.2	62	1.2
HCFC-124	6.0	87	5.6
HCFC-141b	9.5	90	8.6
HCFC-142b	20.3	389	19.3

[a] From Kanakidou et al. (1995).

TABLE 13.10 Calculated Ozone Formation Potentials for Some CFC Replacements[a]

Compound	Formula	Ozone formation potential[b]
HCFC-22	CHF_2Cl	0.1
HCFC-123	CF_3CHCl_2	0.3
HCFC-124	CF_3CHFCl	0.1
HCFC-141b	CH_3CFCl_2	0.1
HCFC-142b	CH_3CF_2Cl	0.1
HCFC-225ca	$CF_3CF_2CHCl_2$	0.2
HFC-23	CHF_3	0.0
HFC-32	CH_2F_2	0.2
HFC-125	CF_3CHF_2	0.0
HFC-134a	CF_3CH_2F	0.1
HFC-143a	CH_3CF_3	0.0
HFC-152a	CH_3CHF_2	1.0
HFC-227ea	CF_3CHFCF_3	0.0

[a] From Hayman and Derwent (1997).
[b] Relative to 100 for ethene, C_2H_4.

struction if present in sufficient quantities (Solomon et al., 1994a, 1994b), as discussed in Chapter 12.E, it appears unlikely that sufficient amounts reach the stratosphere to contribute significantly to ozone destruction (e.g., Wennberg et al., 1997; Pundt et al.,1998). The ODP is estimated to be <0.008 (Solomon et al., 1994a).

E. SUMMARY

It is clear from the data presented in this chapter that the effects of control strategies developed for CFCs and halons are already measurable. Although loss of stratospheric ozone with accompanying increases in ultraviolet radiation in some locations have clearly occurred, the tropospheric concentrations of CFCs are not increasing nearly as fast as in the past. Indeed, the concentrations of CFC-11 and CFC-113 appear to have peaked and have started to decline. The equivalent effective stratospheric chlorine concentrations are predicted to have peaked about 1997 and to return to levels found around 1980 at about the year 2050 (World Meteorological Organization, 1995). The significance of the 1980 level is that these levels resulted in detectable Antarctic ozone depletion.

While the tropospheric concentrations of CFCs and halons are not increasing as rapidly in the past due to controls outlined in the Montreal Protocol and subsequent amendments, those of the CFC replacements are increasing. However, due to their different structures and reactivities, the ozone depletion potentials associated with these compounds are significantly less that those of the compounds they replace. This truly represents a success story in terms of application of atmospheric chemistry to the development of effective control strategies.

F. PROBLEMS

1. CF_3O undergoes some reactions in the atmosphere that the analogous CH_3O does not. Calculate the reaction enthalpies at 298 K for reactions (10)–(12) of CF_3O and compare to the analogous reactions of CH_3O, and for reaction (12), compare also to the OH reaction. Take RH in (12) to be ethane, C_2H_6. Comment on why these reactions are important for CF_3O but not CH_3O.

2. One of the approaches to the development of alternate CFCs is to use compounds with one or more abstractable hydrogen atoms so that their tropospheric lifetimes are reduced and less reaches the stratosphere.

(a) Calculate the tropospheric lifetimes with respect to reaction with OH of CCl_4, $CHCl_3$, CH_2Cl_2, and CH_3Cl. Assume for the sake of these calculations a temperature of 298 K and an OH concentration of 1×10^6 radicals cm^{-3}. The rate constants are $<5 \times 10^{-16}$, 1.0×10^{-13}, 1.1×10^{-13}, and 3.6×10^{-14} cm^3 $molecule^{-1}$ s^{-1}, respectively.

(b) Are there any potential negative implications of their reactivity with OH in terms of tropospheric air quality?

3. Organoiodine compounds are somewhat unusual in that their absorption cross sections are sufficiently red-shifted that they absorb in the actinic region ($\lambda > 290$ nm) and hence photolyze in the troposphere. Use the data for CH_3I in Table 4.42, Chapter 4, to estimate the lifetime of this compound with respect to photolysis at the earth's surface at a solar zenith angle of 50° on May 1. How does this compare to its lifetime with respect to reaction with OH at 1×10^6 radicals cm^{-3} if the rate constant is 7.2×10^{-14} cm^3 $molecule^{-1}$ s^{-1}?

4. Using the heats of formation given in Appendix I, calculate $\Delta H(298$ K$)$ for the reactions of CH_4 with Cl and Br, respectively. As discussed in Chapter 12, the Cl + CH_4 reaction plays an important role in the stratosphere whereas the analogous reaction of bromine atoms does not. Comment on whether this difference is due to enthalpy.

References

Bais, A. F., C. S. Zerefos, C. Meleti, I. C. Ziomas, and K. Tourpali, "Spectral Measurements of Solar UVB Radiation and Its Relations to Total Ozone, SO_2, and Clouds," *J. Geophys. Res., 98*, 5199–5204 (1993).

Ball, J. C., and T. J. Wallington, "Formation of Trifluoroacetic Acid from the Atmospheric Degradation of Hydrofluorocarbon 134a: A Human Health Concern?" *Air Waste, 43,*, 1260–1262 (1993).

Barone, S. B., A. A. Turnipseed, and A. R. Ravishankara, "Kinetics of the Reactions of the CF_3O Radical with Alkanes," *J. Phys. Chem., 98*, 4602–4608 (1994).

Basher, R. E., X. Zheng, and S. Nichol, "Ozone-Related Trends in Solar UV-B Series," *Geophys. Res. Lett., 21*, 2713–2716 (1994).

Beaglehole, D., and G. G. Carter, "Antarctic Skies. 1. Diurnal Variations of the Sky Irradiance, and UV Effects of the Ozone Hole, Spring 1990," *J. Geophys. Res., 97*, 2589–2596 (1992).

Bednarek, G., J. P. Kohlmann, H. Saathoff, and R. Zellner, "Temperature Dependence and Product Distribution for the Reaction of CF_3O Radicals with Methane," *Int. J. Res. Phys. Chem. Chem. Phys., 188*, 1–15 (1995).

Bednarek, G., M. Breil, A. Hoffmann, J. P. Kohlmann, V. Moers, and R. Zellner, "Rate and Mechanism of the Atmospheric Degradation of 1,1,1,2-Tetrafluoroethane (HFC-134a)," *Ber. Bunsenges. Phys. Chem. Chem. Phys., 100*, 528–539 (1996).

Bevilacqua, T. J., D. R. Hanson, and C. J. Howard, "Chemical Ionization Mass Spectrometric Studies of the Gas-Phase Reactions CF_3O_2 + NO, CF_3O + NO, and CF_3O + RH," *J. Phys. Chem., 97*, 3750–3757 (1993).

Bhatnagar, A., and R. W. Carr, "Flash Photolysis Time-Resolved Mass Spectrometric Investigations of the Reactions of CF_3O_2 and CF_3O Radicals with NO," *Chem. Phys. Lett., 231,* 454–459 (1994).

Bhatnagar, A., and R. W. Carr, "Temperature Dependence of the Reaction of CF_3CFClO_2 Radicals with NO and the Unimolecular Decomposition of the CF_3CFClO Radical," *J. Phys. Chem., 99,* 17573–17577 (1995).

Bilde, M., J. Sehested, T. E. Møgelberg, T. J. Wallington, and O. J. Nielsen, "Atmospheric Chemistry of CF_2BrO Radicals," *J. Phys. Chem., 100,* 7050–7059 (1996).

Bjarnason, G. G., O. E. Rögnvaldsson, T. I. Sigfússon, T. Jakobsson, and B. Thorkelsson, "Total Ozone Variations at Reykjavík Since 1957," *J. Geophys. Res., 98,* 23059–23077 (1993).

Bodhaine, B. A., R. L. McKenzie, P. V. Johnston, D. J. Hofmann, E. G. Dutton, R. C. Schnell, J. E. Barnes, S. C. Ryan, and M. Kotkamp, "New Ultraviolet Spectroradiometer Measurements at Mauna Loa Observatory," *Geophys. Res. Lett., 23,* 2121–2124 (1996).

Bodhaine, B. A., E. G. Dutton, D. J. Hofmann, R. L. McKenzie, and P. V. Johnston, "UV Measurements at Mauna Loa: July 1995 to July 1996," *J. Geophys. Res., 102,* 19265–19273 (1997).

Bojkov, R. D., C. S. Zerefos, D. S. Balis, I. C. Ziomas, and A. F. Bais, "Record Low Total Ozone during Northern Winters of 1992 and 1993," *Geophys. Res. Lett., 20,* 1351–1354 (1993).

Bojkov, R. D., V. E. Fioletov, and A. M. Shalamjansky, "Total Ozone Changes over Eurasia Since 1973 Based on Reevaluated Filter Ozonometer Data," *J. Geophys. Res., 99,* 22985–22999 (1994).

Bojkov, R. D., V. E. Fioletov, D. S. Balis, C. S. Zerefos, T. V. Kadygrova, and A. M. Shalamjansky, "Further Ozone Decline during the Northern Hemisphere Winter–Spring of 1994–1995 and the New Record Low Ozone over Siberia," *Geophys. Res. Lett., 22,* 2729–2732 (1995a).

Bojkov, R. D., V. E. Fioletov, and S. B. Diaz, "The Relationship Between Solar UV Irradiance and Total Ozone from Observations Over Southern Argentina," *Geophys. Res. Lett., 22,* 1249–1252 (1995b).

Bojkov, R. D., and V. E. Fioletov, "Changes of the Lower Stratospheric Ozone over Europe and Canada," *J. Geophys. Res., 102,* 1337–1347 (1997).

Brasseur, G. P., J. C. Gille, and S. Madronich, "Ozone Depletion," in *Future Climates of the World: A Modelling Perspective*, World Survey of Climatology, Vol. 16 (A. Henderson-Sellers, Ed.), Chap. 11, pp. 399–431, Elsevier, Amsterdam, 1995.

Brinksma, E. J., Y. J. Meijer, B. J. Connor, G. L. Manney, J. B. Bergwerff, G. E. Bodeker, I. S. Boyd, J. B. Liley, W. Hogervorst, J. W. Hovenier, N. J. Livesey, and D. P. J. Swart, "Analysis of Record-Low Ozone Values during the 1997 Winter over Lauder, New Zealand," *Geophys. Res. Lett., 25,* 2785–2788 (1998).

Brühl, C., and P. J. Crutzen, "On the Disproportionate Role of Tropospheric Ozone as a Filter against Solar UV-B Radiation," *Geophys. Res. Lett., 16,* 703–706 (1989).

Butler, A. D., J. M. Lobert, and J. W. Elkins, "Growth and Distribution of Halons in the Atmosphere," *J. Geophys. Res., 103,* 1503–1511 (1998).

Butler, J. H., J. W. Elkins, B. D. Hall, S. O. Cummings, and S. A. Montzka, "A Decrease in the Growth Rates of Atmospheric Halon Concentrations," *Nature, 359,* 403–405 (1992).

Callis, L. B., M. Natarajan, J. D. Lambeth, and R. E. Boughner, "On the Origin of Midlatitude Ozone Changes: Data Analysis and Simulations for 1979–1993," *J. Geophys. Res., 102,* 1215–1228 (1997).

Catoire, V., R. Lesclaux, W. F. Schneider, and T. J. Wallington, "Kinetics and Mechanisms of the Self-Reactions of CCl_3O_2 and $CHCl_2O_2$ Radicals and Their Reactions with HO_2," *J. Phys. Chem., 100,* 14356–14371 (1996).

Cavalli, F., M. Glasius, J. Hjorth, B. Rindone, and N. R. Jensen, "Atmospheric Lifetimes, Infrared Spectra, and Degradation Products of a Series of Hydrofluoroethers," *Atmos. Environ., 32,* 3767–3773 (1998).

Chandra, S., C. H. Jackman, and E. L. Fleming, "Recent Trends in Ozone in the Upper Stratosphere: Implications for Chlorine Chemistry," *Geophys. Res. Lett., 22,* 843–846 (1995).

Chandra, S., L. Froidevaux, J. W. Waters, O. R. White, G. J. Rottman, D. K. Prinz, and G. E. Brueckner, "Ozone Variability in the Upper Stratosphere during the Declining Phase of the Solar Cycle 22," *Geophys. Res. Lett., 23,* 2935–2938 (1996).

Chatterjee, K., "Implications of Montreal Protocol: With Particular Reference to India and Other Developing Countries," *Atmos. Environ., 29,* 1883–1903 (1995).

Chen, J., T. Zhu, and H. Niki, "FTIR Spectroscopic Study of the Reactions of CF_3O with NO: Evidence for $CF_3O + NO \rightarrow CF_2O + FNO$," *J. Phys. Chem., 96,* 6115–6117 (1992a).

Chen, J., T. Zhu, H. Niki, and G. J. Mains, "Long Path FT-IR Spectroscopic Study of the Reactions of CF_3O Radicals with Ethane and Propane," *Geophys. Res. Lett., 19,* 2215–2218 (1992b).

Chen, J., V. Young, T. Zhu, and H. Niki, "Long Path Fourier Transform Infrared Spectroscopic Study of the Reactions of CF_3OO and CF_3O Radicals with NO_2," *J. Phys. Chem., 97,* 11696–11698 (1993a).

Chen, J., T. Zhu, V. Young, and H. Niki, "Long Path FTIR Spectroscopic Study of the Reactions of CF_3O Radicals with Alkenes," *J. Phys. Chem., 97,* 7174–7176 (1993b).

Chen, J., V. Young, H. Niki, and H. Magid, "Kinetic and Mechanistic Studies for Reactions of $CF_3CH_2CHF_2$ (HFC-245fa) Initiated by H-Atom Abstraction Using Atomic Chlorine," *J. Phys. Chem. A, 101,* 2648–2653 (1997).

Christensen, L. K., J. Sehested, O. J. Nielsen, M. Bilde, T. J. Wallington, A. Guschin, L. T. Molina, and M. J. Molina, "Atmospheric Chemistry of HFE-7200 ($C_4F_9OC_2H_5$): Reaction with OH Radicals and Fate of $C_4F_9OCH_2CH_2O(\cdot)$ and $C_4F_9OCHO(\cdot)CH_3$ Radicals," *J. Phys. Chem. A, 102,* 4839–4845 (1998).

Cicerone, R. J., "Atmospheric Carbon Tetrafluoride: A Nearly Inert Gas," *Science, 206,* 59–61 (1979).

Claude, H., F. Schönenborn, W. Steinbrecht, and W. Vandersee, "New Evidence for Ozone Depletion in the Upper Stratosphere," *Geophys. Res. Lett., 21,* 2409–2412 (1994).

Correll, D. L., C. O. Clark, B. Goldberg, V. R. Goodrich, D. R. Hayes, Jr., W. H. Klein, and W. D. Schecher, "Spectral Ultraviolet-B Radiation Fluxes at the Earth's Surface: Long-Term Variations at 39°N, 77°W," *J. Geophys. Res., 97,* 7579–7591 (1992).

Crutzen, P. J., "Ultraviolet on the Increase," *Nature, 356,* 104–105 (1992).

Cunnold, D. M., W. P. Chu, R. A. Barnes, M. P. McCormick, and R. E. Veiga, "Validation of SAGE II Ozone Measurements," *J. Geophys. Res., 94,* 8447–8460 (1989).

Cunnold, D. M., H. Wang, W. P. Chu, and L. Froidevaux, "Comparisons between Stratospheric Aerosol and Gas Experiment II and Microwave Limb Sounder Ozone Measurements and Aliasing of SAGE II Ozone Trends in the Lower Stratosphere," *J. Geophys. Res., 101,* 10061–10075 (1996).

Cunnold, D. M., R. F. Weiss, R. G. Prinn, D. Hartley, P. G. Simmonds, P. J. Fraser, B. Miller, F. N. Alyea, and L. Porter, "GAGE/AGAGE Measurements Indicating Reductions in Global Emissions of CCl_3F and CCl_2F_2 in 1992–1994," *J. Geophys. Res., 102,* 1259–1269 (1997).

Davies, R., "Increased Transmission of Ultraviolet Radiation to the Surface Due to Stratospheric Scattering," *J. Geophys. Res., 98,* 7251–7253 (1993).

De Bruyn, W. J., S. X. Duan, X. Q. Shi, P. Davidovits, D. R. Worsnop, M. S. Zahniser, and C. E. Kolb, "Tropospheric Heterogeneous Chemistry of Haloacetyl and Carbonyl Halides," *Geophys. Res. Lett., 19,* 1939–1942 (1992).

De Bruyn, W. J., J. A. Shorter, P. Davidovits, D. R. Worsnop, M. S. Zahniser, and C. E. Kolb, "Uptake of Haloacetyl and Carbonyl Halides by Water Surfaces," *Environ. Sci. Technol., 29,* 1179–1185 (1995).

DeMore, W. B., "Experimental and Estimated Rate Constants for the Reactions of Hydroxyl Radicals with Several Halocarbons," *J. Phys. Chem., 100,* 5813–5820 (1996).

DeMore, W. B., S. P. Sander, D. M. Golden, R. F. Hampson, M. J. Kurylo, C. J. Howard, A. R. Ravishankara, C. E. Kolb, and M. J. Molina, "Chemical Kinetics and Photochemical Data for Use in Stratospheric Modeling," in JPL Publication 97-4, Jet Propulsion Laboratory, Pasadena, California, January 15, 1997.

De Muer, D., and H. De Backer, "Revision of 20 Years of Dobson Total Ozone Data at Uccle (Belgium): Fictitious Dobson Total Ozone Trends Induced by Sulfur Dioxide Trends," *J. Geophys. Res., 97,* 5921–5937 (1992).

Derwent, R. G., P. G. Simmonds, S. O'Doherty, and D. B. Ryall, "The Impact of the Montreal Protocol on Halocarbon Concentrations in Northern Hemisphere Baseline and European Air Masses at Mace Head, Ireland over a Ten Year Period from 1987–1996," *Atmos. Environ., 32,* 3689–3702 (1998).

Dhanya, S., and R. D. Saini, "Rate Constants of OH Radical Reactions in Gas Phase with Some Fluorinated Compounds: A Correlation with Molecular Parameters," *Int. J. Chem. Kinet., 29,* 187–194 (1997).

Dibble, T. S., M. M. Maricq, J. J. Szente, and J. S. Francisco, "Kinetics of the Reaction of CF_3O and NO," *J. Phys. Chem., 99,* 17394–17402 (1995).

Edney, E. O., B. W. Gay, and D. J. Driscoll, "Chlorine Initiated Oxidation Studies of Hydrochlorofluorocarbons—Results for HCFC-123 (CF_3CHCl_2) and HCFC-141b ($CFCl_2CH_3$)," *J. Atmos. Chem., 12,* 105–120 (1991).

Edney, E. O., and D. J. Driscoll, "Chlorine Initiated Photooxidation Studies of Hydrochlorofluorocarbons (HCFCs) and Hydrofluorocarbons (HFCs)—Results for HCFC-22 ($CHClF_2$); HFC-41 (CH_3F); HCFC-124 ($CClFHCF_3$); HFC-125 (CF_3CHF_2); HFC-134a (CF_3CH_2F); HCFC-142b ($CClF_2CH_3$); and HFC-152a (CHF_2CH_3)," *Int. J. Chem. Kinet., 24,* 1067–1081 (1992).

Elkins, J. W., T. M. Thompson, T. H. Swanson, J. H. Butler, B. D. Hall, S. O. Cummings, D. A. Fisher, and A. G. Raffo, "Decrease in the Growth Rates of Atmospheric Chlorofluorocarbons 11 and 12," *Nature, 364,* 780–783 (1993).

Fioletov, V. E., J. B. Kerr, and D. I. Wardle, "The Relationship between Total Ozone and Spectral UV Irradiance from Brewer Observations and Its Use for Derivation of Total Ozone from UV Measurements," *Geophys. Res. Lett., 24,* 2997–3000 (1997).

Fisher, D. A., C. H. Hales, D. L. Filkin, M. K. W. Ko, N. D. Sze, P. S. Connell, D. J. Wuebbles, I. S. A. Isaksen, and F. Stordal, "Model Calculations of the Relative Effects of CFCs and Their Replacements on Stratospheric Ozone," *Nature, 344,* 508–512 (1990).

Fockenberg, Ch., H. Saathoff, and R. Zellner, "A Laser Photolysis/LIF Study of the Rate Constant for the Reaction $CF_3O + O_3 \rightarrow$ Products," *Chem. Phys. Lett., 218,* 21–28 (1994).

Frank, H., A. Klein, and D. Renschen, "Environmental Trifluoroacetate," *Nature, 382,* 34 (1996).

Fraser, P., D. Fisher, P. Bloomfield, S. P. Sander, and M. K. W. Ko, "Report on Concentrations, Lifetimes, and Trends of CFCs, Halons, and Related Species," *NASA Reference Publication 1339,* January 1994.

Frederick, J. E., "Trends in Atmospheric Ozone and Ultraviolet Radiation: Mechanisms and Observations for the Northern Hemisphere," *Photochem. Photobiol., 51,* 757–763 (1990).

Frederick, J. E., P. F. Soulen, S. B. Diaz, I. Smolskaia, C. R. Booth, T. Lucas, and D. Neuschuler, "Solar Ultraviolet Irradiance Observed from Southern Argentina: September 1990 to March 1991," *J. Geophys. Res., 98,* 8891–8897 (1993).

Freemantle, M., "Search for Halon Replacements Stymied by Complexities of Fires," *Chem. Eng. News,* Jan. 30, 25–31 (1995).

Fussen, D., "A Critical Analysis of the Stratospheric Aerosol and Gas Experiment II Spectral Inversion Algorithm," *J. Geophys. Res., 103,* 8455–8464 (1998).

Galindo, I., S. Frenk, and H. Bravo, "Ultraviolet Irradiance over Mexico City," *J. Air Waste Manage. Assoc., 45,* 886–892 (1995).

Geller, L. S., J. W. Elkins, J. M. Lobert, A. D. Clarke, D. F. Hurst, J. H. Butler, and R. C. Myers, "Tropospheric SF_6: Observed Latitudinal Distribution and Trends, Derived Emissions, and Interhemispheric Exchange Time," *Geophys. Res. Lett., 24,* 675–678 (1997).

George, C., J. LaGrange, P. LaGrange, P. Mirabel, C. Pallares, and J. L. Ponche, "Heterogeneous Chemistry of Trichloroacetyl Chloride in the Atmosphere," *J. Geophys. Res.-Atmos., 99,* 1255–1262 (1994a).

George, C., J. Y. Saison, J. L. Ponche, and P. Mirabel, "Kinetics of Mass Transfer of Carbonyl Fluoride, Trifluoroacetyl Fluoride, and Trifluoroacetyl Chloride at the Air/Water Interface," *J. Phys. Chem., 98,* 10857–10862 (1994b).

Giessing, A. M. B., A. Feilberg, T. E. Møgelberg, J. Sehested, M. Bilde, T. J. Wallington, and O. J. Nielsen, "Atmospheric Chemistry of HFC-227ca: Spectrokinetic Investigation of the $CF_3CF_2CF_2O_2$ Radical, Its Reactions with NO and NO_2, and the Atmospheric Fate of the $CF_3CF_2CF_2O$ Radical," *J. Phys. Chem., 100,* 6572–6579 (1996).

Gleason, J. F., P. K. Bhartia, J. R. Herman, R. McPeters, P. Newman, R. S. Stolarski, L. Flynn, G. Labow, D. Larko, C. Seftor, C. Wellemeyer, W. D. Komhyr, A. J. Miller, and W. Planet, "Record Low Global Ozone in 1992," *Science, 260,* 523–526 (1993).

Grant, W. B., "Global Stratospheric Ozone and UVB Radiation," *Science, 242* 1111 (1988).

Harnisch, J., R. Borchers, P. Fabian, and M. Maiss, "Tropospheric Trends for CF_4 and C_2F_6 Since 1982 Derived from SF_6 Dated Stratospheric Air," *Geophys. Res. Lett., 23,* 1099–1102 (1996a).

Harnisch, J., R. Borchers, P. Fabian, H. W. Gäggeler, and U. Schotterer, "Effect of Natural Tetrafluoromethane," *Nature, 384,* 32 (1996b).

Harnisch, J., and A. Eisenhauer, "Natural CF_4 and SF_6 Earth," *Geophys. Res. Lett., 25,* 2401–2404 (1998).

Harris, N. R. P., G. Ancellet, L. Bishop, D. J. Hofmann, J. B. Kerr, R. D. McPeters, M. Prendez, W. J. Randel, J. Staehelin, B. H. Subbaraya, A. Volz-Thomas, J. Zawodny, and C. S. Zerefos, "Trends in Stratospheric and Free Tropospheric Ozone," *J. Geophys. Res., 102,* 1571–1590 (1997).

Hasson, A. S., C. M. Moore, and I. W. M. Smith, "The Fluorine Atom Initiated Oxidation of CF_3CFH_2(HFC-134a) Studied by FTIR Spectroscopy," *Int. J. Chem. Kinet., 30,* 541–554 (1998).

Hayman, G. D., M. E. Jenkin, T. P. Murrells, and C. E. Johnson, "Tropospheric Degradation Chemistry of HCFC-123 (CF_3CHCl_2) —A Proposed Replacement Chlorofluorocarbon," *Atmos. Environ., 28,* 421–437 (1994).

Hayman, G. D., and F. Battin-Leclerc, "Kinetics of the Reactions of the HO_2 Radical with Peroxy Radicals Derived from Hydrochlorofluorocarbons and Hydrofluorocarbons," *J. Chem. Soc., Faraday Trans., 91,* 1313–1323 (1995).

Hayman, G. D., and R. G. Derwent, "Atmospheric Chemical Reactivity and Ozone-Forming Potentials of Potential CFC Replacements," *Environ. Sci. Technol., 31*, 327–336 (1997).

Heathfield, A. E., C. Anastasi, P. Pagsberg, and A. McCulloch, "Atmospheric Lifetimes of Selected Fluorinated Ether Compounds," *Atmos. Environ., 32*, 711–717 (1998).

Herman, J. R., P. K. Bhartia, J. Ziemke, Z. Ahmad, and D. Larko, "UV-B Increases (1979–1992) from Decreases in Total Ozone," *Geophys. Res. Lett., 23*, 2117–2120 (1996).

Herman, J. R., P. K. Bhartia, O. Torres, C. Hsu, C. Seftor, and E. Celarier, "Global Distribution of UV-Absorbing Aerosols from Nimbus-7/TOMS Data," *J. Geophys. Res., 102*, 16911–16922 (1997).

Hilsenrath, E., R. P. Cebula, and C. H. Jackman, "Ozone Depletion in the Upper Stratosphere Estimated from Satellite and Space Shuttle Data," *Nature, 358*, 131–133 (1992).

Hofmann, D. J., S. J. Oltmans, J. M. Harris, J. A. Lathrop, G. L. Koenig, W. D. Komhyr, R. D. Evans, D. M. Quincy, T. Deshler, and B. J. Johnson, "Recovery of Stratospheric Ozone over the United States in the Winter of 1993–1994," *Geophys. Res. Lett., 21*, 1779–1782 (1994).

Hollandsworth, S. M., R. D. McPeters, L. E. Flynn, W. Planet, A. J. Miller, and S. Chandra, "Ozone Trends Deduced from Combined Nimbus 7 SBUV and NOAA 11 SBUV/2 Data," *Geophys. Res. Lett., 22*, 905–908 (1995).

Holmes, K. J., and J. H. Ellis, "Potential Environmental Impacts of Future Halocarbon Emissions," *Environ. Sci. Technol., 30*, 348A–355A (1996).

Hood, L. L., "The Solar Cycle Variation of Total Ozone: Dynamical Forcing in the Lower Stratosphere," *J. Geophys. Res., 102*, 1355–1370 (1997).

Hsu, K. J., and W. B. DeMore, "Rate Constants and Temperature Dependences for the Reactions of Hydroxyl Radical with Several Halogenated Methanes, Ethanes, and Propanes by Relative Rate Measurements," *J. Phys. Chem., 99*, 1235–1244 (1995).

Hsu, N. C., J. R. Herman, P. K. Bhartia, C. J. Seftor, O. Torres, A. M. Thompson, J. F. Gleason, T. F. Eck, and B. N. Holben, "Detection of Biomass Burning of Smoke from TOMS Measurements," *Geophys. Res. Lett., 23*, 745–748 (1996).

Huey, L. G., D. R. Hanson, and E. R. Lovejoy, "Atmospheric Fate of CF_3OH. 1: Gas Phase Thermal Decomposition," *J. Geophys. Res., 100*, 18771–18774 (1995).

International Legal Materials, 26, 1541–1561 (1987a).

International Legal Materials, 26, 874–887 (1987b).

International Legal Materials, 30, 537–554 (1991).

Irion, F. W., M. Brown, G. C. Toon, and M. R. Gunson, "Increase in Atmospheric CHF_2Cl (HCFC-22) over Southern California from 1985 to 1990," *Geophys. Res. Lett., 21*, 1723–1726 (1994).

Jensen, N. R., D. R. Hanson, and C. J. Howard, "Temperature Dependence of the Gas Phase Reactions of CF_3O with CH_4 and NO," *J. Phys. Chem., 98*, 8574–8579 (1994).

Justus, C. G., and B. B. Murphey, "Temporal Trends in Surface Irradiance at Ultraviolet Wavelengths," *J. Geophys. Res., 99*, 1389–1394 (1994).

Kalicharran, S., R. D. Diab, and F. Sokolic, "Trends in Total Ozone over Southern African Stations between 1979 and 1991," *Geophys. Res. Lett., 20*, 2877–2880 (1993).

Kanakidou, M., F. J. Dentener, and P. J. Crutzen, "A Global Three-Dimensional Study of the Fate of HCFCs and HFC-134a in the Troposphere," *J. Geophys. Res., 100*, 18781–18801 (1995).

Kane, R. P., Y. Sahai, and C. Casiccia, "Latitude Dependence of the Quasi-Biennial Oscillation and Quasi-Triennial Oscillation Characteristics of Total Ozone Measured by TOMS," *J. Geophys. Res., 103*, 8477–8490 (1998).

Kaye, J. A., "Space-Based Data in Atmospheric Chemistry," in *Progress and Problems in Atmospheric Chemistry*, Advanced Series in Physical Chemistry (J. R. Barker, Ed.), Vol. 3, pp. 569–615, World Scientific, Singapore, 1995.

Kelly, C., J. Treacy, H. W. Sidebottom, and O. J. Nielsen, "Rate Constants for the Reaction of CF_3O Radicals with Hydrocarbons at 298 K," *Chem. Phys. Lett., 207*, 498–503 (1993).

Kelly, C., H. W. Sidebottom, J. Treacy, and O. J. Nielsen, "Reactions of CF_3O Radicals with Selected Alkenes and Aromatics under Atmospheric Conditions," *Chem. Phys. Lett., 218*, 29–33 (1994).

Kerr, J. B., "Trends in Total Ozone at Toronto between 1960 and 1991," *J. Geophys. Res., 96*, 20703–20709 (1991).

Kerr, J. B., and C. T. McElroy, "Evidence for Large Upward Trends of Ultraviolet-B Radiation Linked to Ozone Depletion," *Science, 262*, 1032–1034 (1993).

Kerr, J. B., D. I. Wardle, and D. W. Tarasick, "Record Low Ozone Values over Canada in Early 1993," *Geophys. Res. Lett., 20*, 1979–1982 (1993).

Kerr, J. B., "Decreasing Ozone Causes Health Concern," *Environ. Sci. Technol., 28*, 514A–518A (1994).

Kerr, J. B., and C. T. McElroy, "Analyzing Ultraviolet-B Radiation: Is There a Trend?" *Science, 264*, 1341–1343 (1994).

Khalil, M. A. K., and R. A. Rasmussen, "The Environmental History and Probable Future of Fluorocarbon 11," *J. Geophys. Res., 98*, 23091–23106 (1993).

Kirschner, E., "Producers of CFC Alternatives Gear Up for 1996 Phaseout," *Chem. Eng. News*, July 4, 12–14 (1994).

Ko, M. K. W., N.-D. Sze, and M. J. Prather, "Better Protection of the Ozone Layer," *Nature, 367*, 505–508 (1994a).

Ko, M. K. W., N.-D Sze, J. M. Rodríguez, D. K. Weisenstein, C. W. Heisey, R. P. Wayne, P. Biggs, C. E. Canosa-Mas, H. W. Sidebottom, and J. Treacy, "CF_3 Chemistry: Potential Implications for Stratospheric Ozone," *Geophys. Res. Lett., 21*, 101–104 (1994b).

Ko, M. K. W., N.-D. Sze, C. Scott, J. M. Rodríguez, and D. K. Weisenstein, "Ozone Depletion Potential of CH_3Br," *J. Geophys. Res., 103*, 28187–28195 (1998).

Komhyr, W. D., R. D. Grass, R. D. Evans, R. K. Leonard, D. M. Quincy, D. J. Hofmann, and G. L. Koenig, "Unprecedented 1993 Ozone Decrease over the United States from Dobson Spectrophotometer Observations," *Geophys. Res. Lett., 21*, 201–204 (1994).

Komhyr, W. D., G. C. Reinsel, R. D. Evans, D. M. Quincy, R. D. Grass, and R. K. Leonard, "Total Ozone Trends at Sixteen NOAA/CMDL and Cooperative Dobson Spectrophotometer Observatories during 1979–1996," *Geophys. Res. Lett., 24*, 3225–3228 (1997).

Kotamarthi, V. R., J. M. Rodriguez, M. K. W. Ko, T. K. Tromp, N. D. Sze, and M. J. Prather, "Trifluoroacetic Acid from Degradation of HCFCs and HFCs: A Three-Dimensional Modeling Study," *J. Geophys. Res., 103*, 5747–5758 (1998).

Krzyscin, J. W., "On the Interannual Oscillations in the Northern Temperate Total Ozone," *J. Geophys. Res., 99*, 14527–14534 (1994).

Krzyscin, J. W., "UV Controlling Factors and Trends Derived from the Ground-Based Measurements Taken at Belsk, Poland, 1976–1994," *J. Geophys. Res., 101*, 16797–16805 (1996).

Kundu, N., and M. Jain, "Total Ozone Trends over Low Latitude Indian Stations," *Geophys. Res. Lett., 20*, 2881–2883 (1993).

Lee, J. M., W. T. Sturges, S. A. Penkett, D. E. Oram, U. Schmidt, A. Engel, and R. Bauer, "Observed Stratospheric Profiles and Stratospheric Lifetimes of HCFC-141b and HCFC-142b," *Geophys. Res. Lett., 22*, 1369–1372 (1995).

Li, Z., R. R. Friedl, and S. P. Sander, "Kinetics of FO_2 with NO, NO_2, O_3, CH_4, and C_2H_6," *J. Phys. Chem., 99*, 13445–13451 (1995).

Liu, S. C., S. A. McKeen, and S. Madronich, "Effect of Anthropogenic Aerosols on Biologically Active Ultraviolet Radiation," *Geophys. Res. Lett., 18,* 2265–2268 (1991).

Lovejoy, E. R., L. G. Huey, and D. R. Hanson, "Atmospheric Fate of CF_3OH. 2: Heterogeneous Reaction," *J. Geophys. Res., 100,* 18775–18780 (1995).

Lovelock, J. E., "Atmospheric Turbidity and CCl_3F Concentrations in Rural Southern England and Southern Ireland," *Atmos. Environ., 6,* 917–925 (1972).

Lovelock, J. E., R. J. Maggs, and R. J. Wade, "Variations in Hydrocarbon Gas Concentration during Supertanker Cleaning Operations," *Nature, 241,* 194–196 (1973).

Lovelock, J. E., "Atmospheric Halocarbons and Stratospheric Ozone," *Nature, 252,* 292–294 (1974).

Lovelock, J. E., "Natural Halocarbons in the Air and in the Sea," *Nature, 256,* 193–194 (1975).

Lubin, D., J. E. Frederick, C. R. Booth, T. Lucas, and D. Neuschuler, "Measurements of Enhanced Springtime Ultraviolet Radiation at Palmer Station, Antarctica," *Geophys. Res. Lett., 16,* 783–785 (1989a).

Lubin, D., J. E. Frederick, and A. J. Krueger, "The Ultraviolet Radiation Environment of Antarctica: McMurdo Station during September–October 1987," *J. Geophys. Res., 94,* 8491–8496 (1989b).

Lubin, D., B. G. Mitchell, J. E. Frederick, A. D. Alberts, C. R. Botoh, T. Lucas, and D. Neuschuler, "A Contribution toward Understanding the Biospherical Significance of Antarctic Ozone Depletion," *J. Geophys. Res., 97,* 7817–7828 (1992).

Lubin, D., and E. H. Jensen, "Effects of Clouds and Stratospheric Ozone Depletion on Ultraviolet Radiation Trends," *Nature, 377,* 710–713 (1995).

Madronich, S., "Implications of Recent Total Atmospheric Ozone Measurements for Biologically Active Ultraviolet Radiation Reaching the Earth's Surface," *Geophys. Res. Lett., 19,* 37–40 (1992).

Madronich, S., "The Radiation Equation," *Nature, 377,* 682–683 (1995).

Madronich, S., R. L. McKenzie, L. O. Björn, and M. M. Caldwell, "Changes in Biologically Active Ultraviolet Radiation Reaching the Earth's Surface," *J. Photochem. Photobiol. B: Biol., 46,* 5–19 (1998).

Maiss, M., and I. Levin, "Global Increase of SF_6 Observed in the Atmosphere," *Geophys. Res. Lett., 21,* 569–572 (1994).

Maiss, M., L. P. Steele, R. J. Francey, P. J. Fraser, R. L. Langenfelds, N. B. A. Trivett, and I. Levin, "Sulfur Hexafluoride—A Powerful New Atmospheric Tracer," *Atmos. Environ., 30,* 1621–1629 (1996).

Maiss, M., and C. A. M. Brenninkmeijer, "Atmospheric SF_6: Trends, Sources, and Prospects," *Environ. Sci. Technol., 32,* 3077–3086 (1998).

Manzer, L. E., "The CFC–Ozone Issue: Progress on the Development of Alternatives to CFCs," *Science, 249,* 31–35 (1990).

Maricq, M. M., and J. J. Szente, "Upper Limits for the Rate Constants of the Reactions of $CF_3O + O_3 \rightarrow CF_3O_2 + O_2$ and $CF_3O_2 + O_3 \rightarrow CF_3O + 2O_2$," *Chem. Phys. Lett., 213,* 449–456 (1993).

Maricq, M. M., J. J. Szente, T. S. Dibble, and J. S. Francisco, "Atmospheric Chemical Kinetics of FC(O)O," *J. Phys. Chem., 98,* 12294–12309 (1994a).

Maricq, M. M., J. J. Szente, M. D. Hurley, and T. J. Wallington, "Atmospheric Chemistry of HFC-134a: Kinetic and Mechanistic Study of the $CF_3CFHO_2 + HO_2$ Reaction," *J. Phys. Chem., 98,* 8962–8970 (1994b).

Mayer-Figge, A., F. Zabel, and K. H. Becker, "Thermal Decomposition of $CF_3O_2NO_2$," *J. Phys. Chem., 100,* 6587–6593 (1996).

McCormack, J. P., and L. L. Hood, "Relationship between Ozone and Temperature Trends in the Lower Stratosphere: Latitude and Seasonal Dependences," *Geophys. Res. Lett., 21,* 1615–1618 (1994).

McCulloch, A., and P. M. Midgley, "Estimated Historic Emissions of Fluorocarbons from the European Union," *Atmos. Environ., 32,* 1571–1580 (1998).

McKenzie, R. L., W. A. Matthews, and P. V. Johnston, "The Relationship between Erythemal UV and Ozone, Derived from Spectral Irradiance Measurements," *Geophys. Res. Lett., 18,* 2269–2272 (1991).

McPeters, R. D., T. Miles, L. E. Flynn, C. G. Wellemeyer, and J. M. Zawodny, "Comparison of SBUV and SAGE II Ozone Profiles: Implications for Ozone Trends," *J. Geophys. Res., 99,* 20513–20524 (1994).

Meller, R., and G. K. Moortgat, "Photolysis of $CF_3O_2CF_3$ in the Presence of O_3 in Oxygen: Kinetic Study of the Reactions of CF_3O and CF_3O_2 Radicals with O_3," *J. Photochem. Photobiol., 86,* 15–25 (1995).

Meller, R., and G. K. Moortgat, "The Reaction of Photochemically Generated CF_3O Radicals with CO," *Int. J. Chem. Kinet., 29,* 579–587 (1997).

Michaels, P. J., S. F. Singer, and P. C. Knappenberger, "Analyzing Ultraviolet-B Radiation—Is There a Trend?" *Science, 264,* 1341–1342 (1994).

Midgley, P. M., "Alternatives to CFCs and Their Behaviour in the Atmosphere," in *Volatile Organic Compounds in the Atmosphere, Issues in Environmental Science and Technology* (R. E. Hester and R. M. Harrison, Eds.), Vol. 4, Royal Chem. Soc., Cambridge, UK, 1995.

Midgley, P. M., and A. McCulloch, "Estimated National Releases to the Atmosphere of Chlorodifluoromethane (HCFC-22) during 1990," *Atmos. Environ., 31,* 809–811 (1997).

Midgley, T., and A. L. Henne, "Organic Fluorides as Refrigerants", *Ind. Eng. Chem., 22,* 542–545 (1930).

Miller, A. J., G. C. Tiao, G. C. Reinsel, D. Wuebbles, L. Bishop, J. Kerr, R. M. Nagatani, J. J. DeLuisi, and C. L. Mateer, "Comparisons of Observed Ozone Trends in the Stratosphere through Examination of Umkehr and Balloon Ozonesonde Data," *J. Geophys. Res., 100,* 11209–11217 (1995).

Miller, B. R., J. Huang, R. F. Weiss, R. G. Prinn, and P. J. Fraser, "Atmospheric Trend and Lifetime of Chlorodifluoromethane (HCFC-22) and the Global Tropospheric OH Concentration," *J. Geophys. Res., 103,* 13237–13248 (1998).

Miziolek, A. W., and W. Tsang, Eds., *Halon Replacements—Technology and Science,* ACS Symposium Series 611, Am. Chem. Soc., Washington, DC, 1995.

Møgelberg, T. E., O. J. Nielsen, J. Sehested, T. J. Wallington, and M. D. Hurley, "Atmospheric Chemistry of CF_3COOH. Kinetics of the Reaction with OH Radicals," *Chem. Phys. Lett., 226,* 171–177 (1994a).

Møgelberg, T., O. J. Nielsen, J. Sehested, T. J. Wallington, M. D. Hurley, and W. F. Schneider, "Atmospheric Chemistry of HFC-134a. Kinetic and Mechanistic Study of the $CF_3CFHO_2 + NO_2$ Reaction," *Chem. Phys. Lett., 225,* 375–380 (1994b).

Møgelberg, T. E., O. J. Nielsen, J. Sehested, T. J. Wallington, and M. D. Hurley, "Atmospheric Chemistry of HFC-272ca: Spectrokinetic Investigation of the $CH_3CF_2CH_2O_2$ Radical, Its Reactions with NO and NO_2, and the Fate of the $CH_3CF_2CH_2O$ Radical," *J. Phys. Chem., 99,* 1995–2001 (1995a).

Møgelberg, T. E., J. Platz, O. J. Nielsen, J. Sehested, and T. J. Wallington, "Atmospheric Chemistry of HFC-236fa: Spectrokinetic Investigation of the $CF_3CHO_2 \cdot CF_3$ Radical, Its Reaction with NO, and the Fate of the $CF_3CHO \cdot CF_3$ Radical," *J. Phys. Chem., 99,* 5373–5378 (1995b).

Møgelberg, T. E., A. Feilberg, A. M. B. Giessing, J. Sehested, and M. Bilde, "Atmospheric Chemistry of HFC-236cb: Spectrokinetic Investigation of the $CF_3CF_2CFHO_2$ Radical, Its Reaction with NO and NO_2, and the Fate of the CF_3CF_2CFHO Radical," *J. Phys. Chem., 99,* 17386–17393 (1995c).

Møgelberg, T. E., J. Sehested, M. Bilde, T. J. Wallington, and O. J. Nielsen, "Atmospheric Chemistry of $CF_3CFO \cdot CF_3$ Radical, Its Reactions with NO and NO_2, and Fate of the $CF_3CFO \cdot CF_3$ Radical," *J. Phys. Chem., 100,* 8882–8889 (1996).

Møgelberg, T. E., J. Sehested, G. S. Tyndall, J. J. Orlando, J.-M. Fracheboud, and T. J. Wallington, "Atmospheric Chemistry of HFC-236cb: Fate of the Alkoxy Radical CF_3CF_2CFHO," *J. Phys. Chem. A, 101,* 2828–2832 (1997a).

Møgelberg, T. E., J. Sehested, T. J. Wallington, and O. J. Nielsen, "Atmospheric Chemistry of HFC-134a: Kinetics of the Decomposition of the Alkoxy Radical of CF_3CFHO," *Int. J. Chem. Kinet., 29,* 209–217 (1997b).

Molina, M. J., and F. S. Rowland, "Stratospheric Sink for Chlorofluoromethanes: Chlorine Atom-Catalysed Destruction of Ozone," *Nature, 249,* 810–812 (1974).

Montzka, S. A., R. C. Myers, J. H. Butler, J. W. Elkins, and S. O. Cummings, "Global Tropospheric Distribution and Calibration Scale of HCFC-22," *Geophys. Res. Lett., 20,* 703–706 (1993).

Montzka, S. A., R. C. Myers, J. H. Butler, and J. W. Elkins, "Early Trends in the Global Tropospheric Abundance of Hydrochlorofluorocarbon-141b and 142b," *Geophys. Res. Lett., 21,* 2483–2486 (1994).

Montzka, S. A., J. H. Butler, R. C. Myers, T. M. Thompson, T. H. Swanson, A. D. Clarke, L. T. Lock, and J. W. Elkins, "Decline in the Tropospheric Abundance of Halogen from Halocarbons: Implications for Stratospheric Ozone Depletion," *Science, 272,* 1318–1322 (1996a).

Montzka, S. A., R. C. Myers, J. H. Butler, J. W. Elkins, L. T. Lock, A. D. Clarke, and A. H. Goldstein, "Observations of HFC-134a in the Remote Troposphere," *Geophys. Res. Lett., 23,* 169–172 (1996b).

Mörs, V., G. A. Argüello, A. Hoffmann, W. Malms, E. P. Röth, and R. Zellner, "Kinetics of the Reactions of FC(O)O Radicals with NO, NO_2, O_3, $O(^3P)$, CH_4, and C_2H_6," *J. Phys. Chem., 99,* 15899–15910 (1995).

Nielsen, O. J., and J. Sehested, "Upper Limits for the Rate Constants of the Reactions of CF_3O_2 and CF_3O Radicals with Ozone at 295 K," *Chem. Phys. Lett., 213,* 433–441 (1993).

Niki, H., J. Chen, and V. Young, "Long Path-FTIR Studies of Some Atmospheric Reactions Involving CF_3OO and CF_3O Radicals," *Res. Chem. Intermed., 20,* 277–301 (1994).

Nimitz, J. S., and S. R. Skaggs, "Estimating Tropospheric Lifetimes and Ozone-Depletion Potentials of One- and Two-Carbon Hydrofluorocarbons and Hydrochlorofluorocarbons," *Environ. Sci. Technol., 26,* 739–744 (1992).

Nölle, A., H. Heydtmann, R. Meller, W. Schneider, and G. K. Moortgat, "UV Absorption Spectrum and Absorption Cross Sections of COF_2 at 296 K in the Range 200–230 nm," *Geophys. Res. Lett., 19,* 281–284 (1992).

Oram, D. E., C. E. Reeves, S. A. Penkett, and P. J. Fraser, "Measurements of HCFC-142b and HCFC-141b in the Cape Grim Air Archive: 1978–1993," *Geophys. Res. Lett., 22,* 2741–2744 (1995).

Oram, D. E., C. E. Reeves, W. T. Sturges, S. A. Penkett, P. J. Fraser, and R. L. Langenfelds, "Recent Tropospheric Growth Rate and Distribution of HFC-134a (CF_3CH_2F)," *Geophys. Res. Lett., 23,* 1949–1952 (1996).

Oram, D. E., W. T. Sturges, S. A. Penkett, A. McCulloch, and P. J. Fraser, "Growth of Fluoroform (CHF_3, HFC-23) in the Background Atmosphere," *Geophys. Res. Lett., 25,* 35–38 (1998).

Orkin, V. L., R. E. Huie, and M. J. Kurylo, "Atmospheric Lifetimes of HFC-143a and HFC-245fa: Flash Photolysis Resonance Fluorescence Measurements of the OH Reaction Rate Constants," *J. Phys. Chem., 100,* 8907–8912 (1996).

Orkin, V. L., R. E. Huie, and M. J. Kurylo, "Rate Constants for the Reactions of OH with HFC-245cb ($CH_3CF_2CF_3$) and Some Fluoroalkenes (CH_2CHCF_3, CH_2CFCF_3, CF_2CFCF_3, and CF_2CF_2)," *J. Phys. Chem. A, 101,* 9118–9124 (1997).

Parson, E. A., and O. Greene, "The Complex Chemistry of the International Ozone Agreements," *Environment, 37,* 17–43 (1995).

Patra, P. K., S. Lal, B. H. Subbaraya, C. H. Jackman, and P. Rajaratnam, "Observed Vertical Profile of Sulphur Hexafluoride (SF_6) and Its Atmospheric Applications," *J. Geophys. Res., 102,* 8855–8859 (1997).

Pundt, I., J.-P. Pommereau, C. Phillips, and E. Lateltin, "Upper Limit of Iodine Oxide in the Lower Stratosphere," *J. Atmos. Chem., 30,* 173–185 (1998).

Randel, W. J., and J. B. Cobb, "Coherent Variations of Monthly Mean Total Ozone and Lower Stratospheric Temperature," *J. Geophys. Res., 99,* 5433–5447 (1994).

Rattigan, O. V., O. Wild, R. L. Jones, and R. A. Cox, "Temperature-Dependent Absorption Cross-Sections of CF_3COCl, CF_3COF, CH_3COF, CCl_2CHO, and CF_3COOH," *J. Photochem. Photobiol. A: Chem., 73,* 1–9 (1993).

Rattigan, O. V., D. M. Rowley, O. Wild, R. L. Jones, and R. A. Cox, "Mechanism of Atmospheric Oxidation of 1,1,1,2-Tetrafluoroethane (HFC-134a)," *J. Chem. Soc., Faraday Trans., 90,* 1819–1829 (1994).

Ravishankara, A. R., S. Solomon, A. A. Turnipseed, and R. F. Warren, "Atmospheric Lifetimes of Long-Lived Halogenated Species," *Science, 259,* 194–201 (1993).

Ravishankara, A. R., A. A. Turnipseed, N. R. Jensen, S. Barone, M. Mills, C. J. Howard, and S. Solomon, "Do Hydrofluorocarbons Destroy Stratospheric Ozone?" *Science, 263,* 71–75 (1994).

Reinsel, G. C., W.-K. Tam, and L. H. Ying, "Comparison of Trend Analyses for Umkehr Data Using New and Previous Inversion Algorithms," *Geophys. Res. Lett., 21,* 1007–1010 (1994a).

Reinsel, G. C., G. C. Tiao, D. J. Wuebbles, J. B. Kerr, A. J. Miller, R. M. Nagatani, L. Bishop, and L. H. Ying, "Seasonal Trend Analysis of Published Ground-Based and TOMS Total Ozone Data through 1991," *J. Geophys. Res., 99,* 5449–5464 (1994b).

Rinsland, C. P., J. S. Levine, A. Goldman, N. D. Sze, M. K. W. Ko, and D. W. Johnson, "Infrared Measurements of HF and HCl Total Column Abundances above Kitt Peak, 1977–1990: Seasonal Cycles, Long-Term Increases, and Comparisons with Model Calculations," *J. Geophys. Res., 96,* 15523–15540 (1991).

Rowland, F. S., and M. J. Molina, "Chlorofluoromethanes in the Environment," *Rev. Geophys. Space Phys., 13,* 1–35 (1975).

Rusch, D. W., R. T. Clancy, and P. K. Bhartia, "Comparison of Satellite Measurements of Ozone and Ozone Trends," *J. Geophys. Res., 99,* 20501–20511 (1994).

Rusch, D. W., C. E. Randall, M. T. Callan, M. Horanyi, R. T. Clancy, S. C. Solomon, S. J. Oltmans, B. J. Johnson, U. Koehler, H. Claude, and D. De Muer, "A New Inversion for Stratospheric Aerosol and Gas Experiment II Data," *J. Geophys. Res., 103,* 8465–8475 (1998).

Russell, J. M., III, M. Luo, R. J. Cicerone, and L. E. Deaver, "Satellite Confirmation of the Dominance of Chlorofluorocarbons in the Global Stratospheric Chlorine Budget," *Nature, 379,* 526–529 (1996).

Saathoff, H., and R. Zellner, "LIF Detection of the CF_3O Radical and Kinetics of Its Reactions with CH_4 and C_2H_6," *Chem. Phys. Lett., 206,* 349–354 (1993).

Sabziparvar, A. A., P. M. de F. Forster, and K. P. Shine, "Changes in Ultraviolet Radiation Due to Stratospheric and Tropospheric Ozone Changes Since Preindustrial Times," *J. Geophys. Res., 103,* 26107–26113 (1998).

Schauffler, S. M., W. H. Pollock, E. L. Atlas, L. E. Heidt, and J. S. Daniel, "Atmospheric Distributions of HCFC 142b," *Geophys. Res. Lett., 22819–22822* (1995).

Schneider, W. F., T. J. Wallington, K. Minschwaner, and E. A. Stahlberg, "Atmospheric Chemistry of CF_3OH: Is Photolysis Important?" *Environ. Sci. Technol., 29,* 247–250 (1995).

Scollard, D. J., J. J. Treacy, H. W. Sidebottom, C. Balestra-Garcia, G. Laverdet, G. Le Bras, H. MacLeod, and S. Téton, "Rate Constants for the Reactions of Hydroxyl Radicals and Chlorine Atoms with Halogenated Aldehydes," *J. Phys. Chem., 97,* 4683–4688 (1993).

Scotto, J., G. Cotton, F. Urbach, D. Berger, and T. Fears, "Biologically Effective Ultraviolet Radiation: Surface Measurements in the United States, 1974 to 1985," *Science, 239,* 762–763 (1988a).

Scotto, J., G. Cotton, F. Urbach, D. Berger, and T. Fears, "Response to 'Global Stratospheric Ozone and UVB Radiation,'" *Science, 242,* 1111–1112 (1988b).

Seckmeyer, G., and R. L. McKenzie, "Increased Ultraviolet Radiation in New Zealand (45°S) Relative to Germany (48°N)," *Nature, 359,* 135–137 (1992).

Seckmeyer, G., B. Mayer, R. Erb, G. Bernhard, "UV-B in Germany Higher in 1993 Than in 1992," *Geophys. Res. Lett., 21,* 577–580 (1994).

Seckmeyer, G., B. Mayer, G. Bernhard, R. Erb, A. Albold, H. Jäger, and W. R. Stockwell, "New Maximum UV Irradiance Levels Observed in Central Europe," *Atmos. Environ., 1,* 2971–2976 (1997).

Sehested, J., T. Ellerman, O. J. Nielsen, T. J. Wallington, and M. D. Hurley, "UV Absorption Spectrum, and Kinetics and Mechanism of the Self-Reaction of $CF_3CF_2O_2$ Radicals in the Gas Phase at 295 K," *Int. J. Chem. Kinet., 25,* 701–717 (1993).

Sehested, J., and O. J. Nielsen, "Absolute Rate Constants for the Reaction of CF_3O_2 and CF_3O Radicals with NO at 295 K," *Chem. Phys. Lett., 206,* 369–375 (1993).

Sehested, J., and T. J. Wallington, "Atmospheric Chemistry of Hydrofluorocarbon 134a. Fate of the Alkoxy Radical CF_3O," *Environ. Sci. Technol., 27,* 146–152 (1993).

Sehested, J., K. Sehested, O. J. Nielsen, and T. J. Wallington, "Atmospheric Chemistry of FO_2 Radicals: Reaction with CH_4, O_3, NO, NO_2 and CO at 295 K," *J. Phys. Chem. A, 98,* 6731–6739 (1994).

Sehested, J., O. J. Nielsen, C. A. Rinaldi, S. I. Lane, and J. C. Ferrero, "Kinetics and Mechanism of the Reaction of CF_3 Radicals with NO_2," *Int. J. Chem. Kinet., 28,* 579–588 (1996).

Sehested, J., T. Møgelberg, K. Fagerström, G. Mahmoud, and T. J. Wallington, "Absolute Rate Constants for the Self Reactions of HO_2, CF_3CFHO_2, and CF_3O_2 Radicals and the Cross Reactions of HO_2 with FO_2, HO_2 with CF_3CFHO_2, and HO_2 with CF_3O_2 at 295 K," *Int. J. Chem. Kinet., 29,* 673–682 (1997).

Simmonds, P. G., R. G. Derwent, A. McCulloch, S. O'Doherty, and A. Gaudry, "Long-Term Trends in Concentrations of Halocarbons and Radiatively Active Trace Gases in Atlantic and European Air Masses Monitored at Mace Head, Ireland from 1987–1994," *Atmos. Environ., 30,* 4041–4063 (1996).

Simmonds, P. G., D. M. Cunnold, R. F. Weiss, R. G. Prinn, P. J. Fraser, A. McCulloch, F. N. Alyea, and S. O'Doherty, "Global Trends and Emission Estimates of CCl_4 from *in Situ* Background Observations from July 1978 to June 1996," *J. Geophys. Res., 103,* 16017–16027 (1998a).

Simmonds, P. G., S. O'Doherty, J. Huang, R. Prinn, R. G. Derwent, D. Ryall, G. Nickless, and D. Cunnold, "Calculated Trends and the Atmospheric Abundance of 1,1,1,2-Tetrafluoroethane, 1,1-Dichloro-1-fluoroethane, and 1-Chloro-1,1-difluoroethane Using Automated *in Situ* Gas Chromatography–Mass Spectrometry Measurements Recorded at Mace Head, Ireland, from October 1994 to March 1997," *J. Geophys. Res., 103,* 16029–16037 (1998b).

Slaper, H., G. J. M. Velders, J. S. Daniel, F. R. de Gruijl, and J. C. van der Leun, "Estimates of Ozone Depletion and Skin Cancer Incidence to Examine the Vienna Convention Achievements," *Nature, 384,* 256–258 (1996).

Solomon, S., and D. L. Albritton, "Time-Dependent Ozone Depletion Potentials for Short- and Long-Term Forecasts," *Nature, 357,* 33–37 (1992).

Solomon, S., M. Mills, L. E. Heidt, W. H. Pollock, and A. F. Tuck, "On the Evaluation of Ozone Depletion Potentials," *J. Geophys. Res., 97,* 825–842 (1992).

Solomon, S., J. B. Burkholder, A. R. Ravishankara, and R. R. Garcia, "Ozone Depletion and Global Warming Potentials of CF_3I," *J. Geophys. Res., 99,* 20929–20935 (1994a).

Solomon, S., R. R. Garcia, and A. R. Ravishankara, "On the Role of Iodine in Ozone Depletion," *J. Geophys. Res., 99,* 20491–20499 (1994b).

Solomon, S., R. W. Portmann, R. R. Garcia, W. Randel, F. Wu, R. Nagatani, J. Gleason, L. Thomason, L. R. Poole, and M. P. McCormick, "Ozone Depletion at Mid-latitudes: Coupling of Volcanic Aerosols and Temperature Variability to Anthropogenic Chlorine," *Geophys. Res. Lett., 25,* 1871–1874 (1998).

Spurgeon, D., "End Agreed for Ozone-Destroying Pesticide," *Nature, 389,* 319 (1997a).

Spurgeon, D., "Ozone Treaty 'Must Tackle CFC Smuggling,'" *Nature, 389,* 219 (1997b).

Staehelin, J., A. Renaud, J. Bader, R. McPeters, P. Viatte, B. Hoegger, V. Bugnion, M. Giroud, and H. Schill, "Total Ozone Series at Orosa (Switzerland): Homogenization and Data Comparison," *J. Geophys. Res., 103,* 5827–5841 (1998a).

Staehelin, J., R. Kegel, and N. R. P. Harris, "Trend Analysis of the Homogenized Total Ozone Series of Arosa (Switzerland), 1926–1996," *J. Geophys. Res., 103,* 8389–8399 (1998b).

Stamnes, K., J. Slusser, and M. Bowen, "Biologically Effective Ultraviolet Radiation, Total Ozone Abundance, and Cloud Optical Depth at McMurdo Station, Antarctica, September 15, 1988 through April 15, 1989," *Geophys. Res. Lett., 17,* 2181–2184 (1990).

Stamnes, K., Z. Jin, and J. Slusser, "Several-Fold Enhancement of Biologically Effective Ultraviolet Radiation Levels at McMurdo Station, Antarctica during the 1990 Ozone 'Hole,'" *Geophys. Res. Lett., 19,* 1013–1016 (1992).

Steele, H. M., and R. P. Turco, "Retrieval of Aerosol Size Distributions from Satellite Extinction Spectra Using Constrained Linear Inversion," *J. Geophys. Res., 102,* 16737–16747 (1997a).

Steele, H. M., and R. P. Turco, "Separation of Aerosol and Gas Components in the Halogen Occultation Experiment and the Stratospheric Aerosol and Gas Experiment (SAGE) II Extinction Measurements: Implications for SAGE II Ozone Concentrations and Trends," *J. Geophys. Res., 102,* 19665–19681 (1997b).

Stolarski, R. S., and R. D. Rundel, "Fluorine Photochemistry in the Stratosphere," *Geophys. Res. Lett., 2,* 443–444 (1975).

Stolarski, R., R. Bojkov, L. Bishop, C. Zerefos, J. Staehelin, and J. Zawodny, "Measured Trends in Stratospheric Ozone," *Science, 256,* 342–349 (1992).

Su, C.-W., and E. D. Goldberg, "Chlorofluorocarbons in the Atmosphere," *Nature, 245,* 27 (1973).

Taalas, P., J. Damski, E. Kyrö, M. Ginzburg, and G. Talamoni, "Effect of Stratospheric Ozone Variations on UV Radiation and on Tropospheric Ozone at High Latitudes," *J. Geophys. Res., 102*, 1533–1539 (1997).

Thomason, L. W., L. R. Poole, and T. Deshler, "A Global Climatology of Stratospheric Aerosol Surface Area Density Deduced from Stratospheric Aerosol and Gas Experiment II Measurements: 1984–1994," *J. Geophys. Res., 102*, 8967–8976 (1997).

Tie, X.-X., C. Granier, W. Randel, and G. P. Brasseur, "Effects of Interannual Variation of Temperature on Heterogeneous Reactions and Stratospheric Ozone," *J. Geophys. Res., 102*, 23519–23527 (1997).

Torres, O., P. K. Bhartia, J. R. Herman, Z. Ahmad, and J. Gleason, "Derivation of Aerosol Properties from Satellite Measurements of Backscattered Ultraviolet Radiation: Theoretical Basis," *J. Geophys. Res., 103*, 17099–17110 (1998).

Toumi, R., S. Bekki, and K. S. Law, "Indirect Influence of Ozone Depletion on Climate Forcing by Clouds," *Nature, 372*, 348–351 (1994).

Tromp, T. K., M. K. W. Ko, J. M. Rodríguez, and N. D. Sze, "Potential Accumulation of a CFC-Replacement Degradation Product in Seasonal Wetlands," *Nature, 376*, 327–330 (1995).

Tsitas, S. R., and Y. L. Yung, "The Effect of Volcanic Aerosols on Ultraviolet Radiation in Antarctica," *Geophys. Res. Lett., 23*, 157–160 (1996).

Tuazon, E. C., and R. Atkinson, "Tropospheric Degradation Products of CH_2FCF_3 (HFC-134a)," *J. Atmos. Chem., 16*, 301–312 (1993a).

Tuazon, E. C., and R. Atkinson, "Tropospheric Transformation Products of a Series of Hydrofluorocarbons and Hydrochlorofluorocarbons," *J. Atmos. Chem., 17*, 179–199 (1993b).

Tuazon, E. C., and R. Atkinson, "Tropospheric Reaction Products and Mechanisms of the Hydrochlorofluorocarbon-141b, Hydrochlorofluorocarbon-142b, Hydrochlorofluorocarbon-225ca, and Hydrochlorofluorocarbon-225cb," *Environ. Sci. Technol., 28*, 2306–2313 (1994).

Turnipseed, A. A., S. B. Barone, and A. R. Ravishankara, "Kinetics of the Reactions of CF_3O_x Radicals with NO, O_3, and O_2," *J. Phys. Chem., 98*, 4594–4601 (1994).

Turnipseed, A. A., S. B. Barone, N. R. Jensen, D. R. Hanson, C. J. Howard, and A. R. Ravishankara, "Kinetics of the Reactions of CF_3O Radicals with CO and H_2O," *J. Phys. Chem., 99*, 6000–6009 (1995).

Vakhtin, A. B., "Kinetics and Mechanism of the $CF_3 + NO_2$ Reaction at $T = 298$ K," *Int. J. Chem. Kinet., 29*, 203–208 (1997).

Veiga, R. E., D. M. Cunnold, W. P. Chu, and M. P. McCormick, "Stratospheric Aerosol and Gas Experiments I and II Comparisons with Ozonesondes," *J. Geophys. Res., 100*, 9073–9090 (1995).

Wallace, L., W. Livingston, and D. N. B. Hall, "A Twenty-Five Year Record of Stratospheric Hydrogen Chloride," *Geophys. Res. Lett., 24*, 2363–2366 (1997).

Wallington, T. J., and O. J. Nielsen, "Pulse Radiolysis Study of CF_3CFHO_2 Radicals in the Gas Phase at 298 K," *Chem. Phys. Lett., 187*, 33–39 (1991).

Wallington, T. J., M. D. Hurley, J. C. Ball, and E. W. Kaiser, "Atmospheric Chemistry of Hydrofluorocarbon 134a—Fate of the Alkoxy Radical CF_3CFHO," *Environ. Sci. Technol., 26*, 1318–1324 (1992).

Wallington, T. J., M. D. Hurley, and W. F. Schneider, "Kinetic Study of the Reaction $CF_3O + O_3 \rightarrow CF_3O_2 + O_2$," *Chem. Phys. Lett., 213*, 442–448 (1993a).

Wallington, T. J., M. D. Hurley, W. F. Schneider, J. Sehested, and O. J. Nielsen, "Atmospheric Chemistry of CF_3O Radicals: Reaction with H_2O," *J. Phys. Chem., 97*, 7606–7611 (1993b).

Wallington, T. J., and W. F. Schneider, "The Stratospheric Fate of CF_3OH," *Environ. Sci. Technol., 28*, 1198–1200 (1994).

Wallington, T. J., D. R. Worsnop, O. J. Nielsen, J. Sehested, W. J. Debruyn, and J. A. Shorter, "The Environmental Impact of CFC Replacements—HFCs and HCFCs," *Environ. Sci. Technol., 28*, 320A–326A (1994a).

Wallington, T. J., M. D. Hurley, O. J. Nielsen, and J. Sehested, "Atmospheric Chemistry of C_3CO_x Radicals: Fate of CF_3CO Radicals, the UV Absorption Spectrum of $CF_3C(O)O_2$ Radicals, and Kinetics of the Reaction $CF_3C(O)O_2 + NO \rightarrow CF_3C(O)O + NO_2$," *J. Phys. Chem., 98*, 5686–5694 (1994b).

Wallington, T. J., T. Ellermann, O. J. Nielsen, and J. Sehested, "Atmospheric Chemistry of FCO_x Radicals: UV Spectra and Self-Reaction Kinetics of FCO and $FC(O)O_2$ and Kinetics of Some Reactions of FCO_x with O_2, O_3, and NO at 296 K," *J. Phys. Chem., 98*, 2346–2356 (1994c).

Wallington, T. J., M. D. Hurley, W. F. Schneider, J. Sehested, and O. J. Nielsen, "Mechanistic Study of the Gas-Phase Reaction of CH_2FO_2 Radicals with HO_2," *Chem. Phys. Lett., 218*, 34–42 (1994d).

Wallington, T. J., J. Sehested, and O. J. Nielsen, "Atmospheric Chemistry of $CF_3C(O)O_2$ Radicals. Kinetics of Their Reaction with NO_2 and Kinetics of the Thermal Decomposition of the Product $CF_3C(O)O_2NO_2$," *Chem. Phys. Lett., 226*, 563–569 (1994e).

Wallington, T. J., and J. C. Ball, "Atmospheric Chemistry of CF_3O Radicals: Reaction with CH_4, CD_4, CH_3F, CF_3H, ^{13}CO, C_2H_5F, C_2D_6, C_2H_6, CH_3OH, i-C_4H_8, and C_2H_2," *J. Phys. Chem., 99*, 3201–3205 (1995).

Wallington, T. J., and O. J. Nielsen, "Atmospheric Chemistry of Hydrofluorocarbons," in *Progress and Problems in Atmospheric Chemistry,* Advanced Series in Physical Chemistry (J. R. Barker, Ed.), Vol. 3, pp. 616–685, World Scientific, Singapore, 1995.

Wallington, T. J., W. F. Schneider, J. J. Szente, M. M. Maricq, O. J. Nielsen, and J. Sehested, "Atmospheric Chemistry of FNO and FNO_2: Reactions of FNO with O_3, $O(^3P)$, HO_2, and HCl and the Reaction of FNO_2 with O_3," *J. Phys. Chem., 99*, 984–989 (1995).

Wallington, T. J., M. D. Hurley, J. M. Fracheboud, J. J. Orlando, G. S. Tyndall, J. Sehested, T. E. Møgelberg, and O. J. Nielsen, "Role of Excited CF_3CFHO Radicals in the Atmospheric Chemistry of HFC-134a," *J. Phys. Chem., 100*, 18116–18122 (1996a).

Wallington, T. J., M. Bilde, T. E. Møgelberg, J. Sehested, and O. J. Nielsen, "Atmospheric Chemistry of 1,2-Dichloroethane: UV Spectra of $CH_2ClCHCl$ and $CH_2ClCHClO_2$ Radicals, Kinetics of the Reactions of $CH_2ClCHCl$ Radicals with O_2 and $CH_2ClCHClO_2$ Radicals with NO and NO_2, and Fate of the Alkoxy Radical $CH_2ClCHClO$," *J. Phys. Chem., 100*, 5751–5760 (1996b).

Wallington, T. J., W. F. Schneider, J. Sehested, M. Bilde, J. Platz, O. J. Nielsen, L. K. Christensen, M. J. Molina, L. T. Molina, and P. W. Wooldridge, "Atmospheric Chemistry of HFE-7100 ($C_4F_9OCH_3$): Reaction with OH Radicals, UV Spectra and Kinetic Data for $C_4F_9OCH_2$ and $C_4F_9OCH_2O_2$ Radicals, and the Atmospheric Fate of $C_4F_9OCH_2O\cdot$ Radicals," *J. Phys. Chem. A, 101*, 8264–8274 (1997).

Wallington, T. J., A. Guschin, T. N. N. Stein, J. Platz, J. Sehested, L. K. Christensen, and O. J. Nielsen, "Atmospheric Chemistry of $CF_3CH_2OCH_2CF_3$: UV Spectra and Kinetic Data for $CF_3CH(\cdot)OCH_2CF_3$ and $CF_3CH(OO\cdot)OCH_2CF_3$ Radicals and Atmospheric Fate of $CF_3CH(O\cdot)OCH_2CF_3$ Radicals," *J. Phys. Chem. A, 102*, 1152–1161 (1998).

Weatherhead, E. C., G. C. Tiao, G. C. Reinsel, J. E. Frederick, J. J. DeLuisi, D. Choi, and W.-K. Tam, "Analysis of Long-Term Behavior of Ultraviolet Radiation Measured by Robertson–Berger Meters at 14 Sites in the United States," *J. Geophys. Res. 102*, 8737–8754 (1997).

Wennberg, P. O., J. W. Brault, T. F. Hanisco, R. J. Salawitch, and G. H. Mount, "The Atmospheric Column Abundance of IO: Implications for Stratospheric Ozone," *J. Geophys. Res., 102*, 8887–8898 (1997).

Wenny, B. N., J. S. Schafer, J. J. DeLuisi, V. K. Saxena, W. F. Barnard, I. V. Petropavlovskikh, and A. J. Vergamini, "A Study of Regional Aerosol Radiative Properties and Effects on Ultraviolet-B Radiation," *J. Geophys. Res., 103*, 17083–17097 (1998).

Wilkniss, P. E., J. W. Swinnerton, D. J. Bressan, R. A. Lamontagne, and R. E. Larson, "CO, CCl_4, Freon-11, CH_4 and Rn-222 Concentrations at Low Altitude over the Arctic Ocean in January, 1974," *J. Atmos. Sci., 32*, 158–162 (1975).

World Meteorological Organization (WMO), "Scientific Assessment of Ozone Depletion: 1994," Global Ozone Research and Monitoring Project, Report No. 37, published February 1995; update, Report No. 44, February 1999.

Wu, F., and R. W. Carr, "Time-Resolved Observation of the Formation of CF_2O and CFClO in the $CF_2Cl + O_2$ and $CFCl_2 + O_2$ Reactions. The Unimolecular Elimination of Cl Atoms from CF_2ClO and $CFCl_2O$ Radicals," *J. Phys. Chem., 96*, 1743–1748 (1992).

Wuebbles, D. J., "Chlorocarbon Emission Scenarios: Potential Impact on Stratospheric Ozone," *J. Geophys. Res., 88*, 1433–1443 (1983).

Wuebbles, D. J., A. K. Jain, K. O. Patten, and P. S. Connell, "Evaluation of Ozone Depletion Potentials for Chlorobromomethane (CH_2ClBr) and 1-Bromopropane ($CH_2BrCH_2CH_3$)," *Atmos. Environ., 32*, 107–113 (1998).

Wuebbles, D. J., R. Kotamarthi, and K. O. Patten, "Updated Evaluation of Ozone Depletion Potentials for Chlorobromomethane (CH_2ClBr) and 1-Bromopropane ($CH_2BrCH_2CH_3$)," *Atmos. Environ., 33*, 1641–1643 (1999).

Wujcik, C. E., D. Zehavi, and J. N. Seiber, "Trifluoroacetate Levels in 1995–1996 Fog, Rain, Snow, and Surface Waters from California and Nevada," National Meeting of the American Chemical Society, April 1997, Division of Environmental Chemistry Preprints of Extended Abstracts, Vol. 37, pp. 31–33, 1997.

Zabel, F., F. Kirchner, and K. H. Becker, "Thermal Decomposition of $CF_3C(O)O_2NO_2$, $CClF_2C(O)O_2NO_2$, $CCl_2FC(O)O_2NO_2$, and $CCl_3C(O)O_2NO_2$," *Int. J. Chem. Kinet., 26*, 827–845 (1994).

Zachariah, M. R., W. Tsang, P. R. Westmoreland, and D. R. F. Burgess, Jr., "Theoretical Prediction of the Thermochemistry and Kinetics of Reactions of CF_2O with Hydrogen Atom and Water," *J. Phys. Chem., 99*, 12512–12519 (1995).

Zerefos, C. S., A. F. Bais, I. C. Ziomas, and R. D. Bojkov, "On the Relative Importance of Quasi-Biennial Oscillation and El Nino/Southern Oscillation in the Revised Dobson Total Ozone Records," *J. Geophys. Res., 97*, 10135–10144 (1992).

Zerefos, C. S., A. F. Bais, C. Meleti, and I. C. Ziomas, "A Note on the Recent Increase of Solar UV-B Radiation over Northern Middle Latitudes," *Geophys. Res. Lett., 22*, 1245–1247 (1995).

Zerefos, C. S., K. Tourpali, B. R. Bojkov, D. S. Balis, B. Rognerund, and I. S. A. Isaksen, "Solar Activity–Total Column Ozone Relationships: Observations and Model Studies with Heterogeneous Chemistry," *J. Geophys. Res., 102*, 1561–1569 (1997).

Ziemke, J. R., and S. Chandra, "Comment on 'Tropospheric Ozone Derived from TOMS/SBUV Measurements during TRACE A,'" by J. Fishman *et al., J. Geophys. Res., 103*, 13903–13906 (1998).

Ziemke, J. R., J. R. Herman, J. L. Stanford, and P. K. Bhartia, "Total Ozone/UVB Monitoring and Forecasting: Impact of Clouds and the Horizontal Resolution of Satellite Retrievals," *J. Geophys. Res., 103*, 3865–3871 (1998).

Zurer, P., "Controls Tightened on Methyl Bromide, HCFCs," *Chem. Eng. News*, December 18, 1995, p. 8.

CHAPTER 14

Global Tropospheric Chemistry and Climate Change

Over the past several decades, there has been increasing recognition in a number of areas of the environmental impacts, both realized and potential, of human activities not only on local and regional scales but also globally. This is particularly true of changes to the composition and chemistry of the atmosphere caused by such anthropogenic activities. One example, for which there is irrefutable evidence, is stratospheric ozone depletion by chlorofluorocarbons, discussed in detail in Chapters 12 and 13.

Another area of global-scale dimensions that is commanding increased attention is the potential impact of atmospheric trace gases and aerosol particles on climate, the subject of this chapter. *Climate* is the long-term statistical characterization of parameters describing what we commonly term "weather," such as surface temperature. For example, the mean surface temperature with its associated variability over some time period, typically taken as 30 years, is one measure of climate. Thus, climate is distinguished from short-term, e.g., day-to-day, variations, which are typically referred to as "weather."

The recognition that atmospheric gases play a central role in determining the earth's climate goes back more than a century to Joseph Fourier, who proposed in 1827 that heat is trapped by the atmosphere. In 1861, John Tyndall showed that O_2, N_2, and H_2 do not absorb infrared radiation but that CO_2 and N_2O (as well as a number of organic compounds) do. Subsequently, Arrhenius (1896) considered the role of changes in atmospheric CO_2 on the earth's temperature due to the absorption of infrared radiation. He estimated, for example, that the temperature in the Arctic would rise by 8–9°C from an increase in atmospheric CO_2 by a factor of 2.5–3 over that present in 1896. Beginning in the late 1800s, Chamberlin explored the relationship between climate and atmospheric composition. A brief history of this area is given by Weart (1997) and by Fleming (1992).

As we shall see, the interrelationships between atmospheric composition, chemistry, and climate are very complex. For example, as discussed in more detail herein, it is clear that CO_2 emissions, primarily from fossil fuel combustion, have increased dramatically over the past century, leading to substantial increases in its atmospheric concentrations. The concentrations of a number of other greenhouse gases have been increasing as well (Ramanathan *et al.*, 1985). In the simplest approach, these increases are expected to lead to a significant increase in the surface temperature, and indeed, there is general agreement that an increase of about 0.3–0.6°C over the past century has occurred (IPCC, 1996).

Thus, there is a sound scientific basis for anticipating that chemical changes in the atmosphere will impact climate. However, the interplay between all of the contributing factors and hence the ultimate *quantitative* impacts are, at present, not well understood and the subject of intense research activity. For example, the increase in surface temperatures over the past century has not been continuous, occurring primarily from approximately 1910 to 1940 and from 1975 to the present, with recent years being some of the warmest since extensive record keeping began about 1860. Ice core studies have documented changes over the past approximately 100,000 years in the earth's climate, prior to extensive fossil fuel use, with some of the changes occurring quite rapidly (over time scales of a decade or less). Accompanying these have been changes in the concentrations of a number of atmospheric trace gases such as CH_4, indicative of complex feedbacks occurring between the atmosphere, land masses, and oceans. Such natural variability complicates assessment of the anthropogenic influences on climate, particularly when the time scales for the effects of emissions due to human activity to be manifested can be a century or more for some gases.

Given the breadth and complexity of the scientific issues involved in global climate, we shall focus in this chapter primarily on the current state of understanding of the role of atmospheric composition and chemistry in determining the radiation balance of the atmosphere. However, some of the variables affecting global climate and potential feedback mechanisms with associated implications for climate change as well as the global climate record are briefly discussed as well; the reader should consult the references cited in those sections for further details. An excellent summary of research in this expanding and dynamic area through 1995 is found in *Climate Change 1995: The Science of Climate Change* by the Intergovernmental Panel on Climate Change (IPCC, 1996). Radiative transfer in the atmosphere is not treated here in detail. The reader should consult books by Liou (1980), Goody and Yung (1989), and Lenoble (1993) for excellent detailed treatments of this subject.

A. RADIATION BALANCE OF THE ATMOSPHERE: THE GREENHOUSE EFFECT

1. Global Absorption and Emission of Radiation

In Chapter 3 we examined the interaction of incoming solar radiation in the UV and visible regions of the spectrum with atmospheric gases, which drives atmospheric photochemistry. This incoming solar radiation also determines the temperature of the earth's surface through its absorption and reradiation. Although, as chemists, we tend to think of the absorption and emission of radiation in molecular terms, the greenhouse effect is best thought of in terms of the energy balance of the earth–atmosphere system taken as a whole (e.g., see Ramanathan *et al.*, 1987; Ramanathan, 1976, 1988a,b; and Wang *et al.*, 1995).

Figure 14.1 shows the solar flux outside the atmosphere, which is approximated by blackbody emission at 6000 K, and at sea level, respectively. Absorption of incoming solar radiation by O_3, O_2, H_2O, and CO_2 as the light passes through the atmosphere to the earth's surface is evident. Recall in Chapter 3 that the average total incoming light intensity per unit area normal to the direction of propagation outside the earth's atmosphere, i.e., the solar constant, is 1368 W m^{-2}. As seen in Fig. 14.2a, this is the energy density that would strike a planar disk of area πr^2 (where r is the radius of the earth) centered along the earth's axis. However, this incoming solar energy is spread over the entire $4\pi r^2$ surface area of the earth. The effective incoming solar radiation per unit area of the earth's surface is therefore $1368/4 = 342$ W m^{-2}. [It should be noted that, as discussed in Chapter 4.A, collision complexes of O_2 with a second O_2 molecule or with N_2 also may contribute an additional small amount to the absorption of incoming solar radiation; for example, $O_2 \cdot O_2$ and $O_2 \cdot N_2$ may contribute an additional 0.57–3 W m^{-2} to this total (Pfeilsticker *et al.*, 1997; Solomon *et al.*, 1998; Mlawer *et al.*, 1998).]

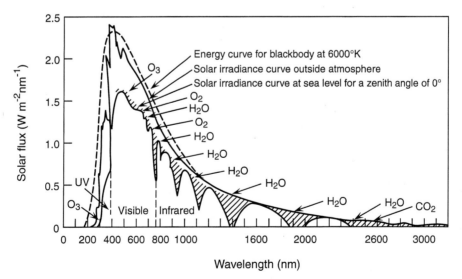

FIGURE 14.1 Solar flux outside the atmosphere and at sea level, respectively. The emission of a blackbody at 6000 K is also shown for comparison. The species responsible for light absorption in the various regions (O_3, H_2O, etc.) are also shown (adapted from Howard *et al.*, 1960).

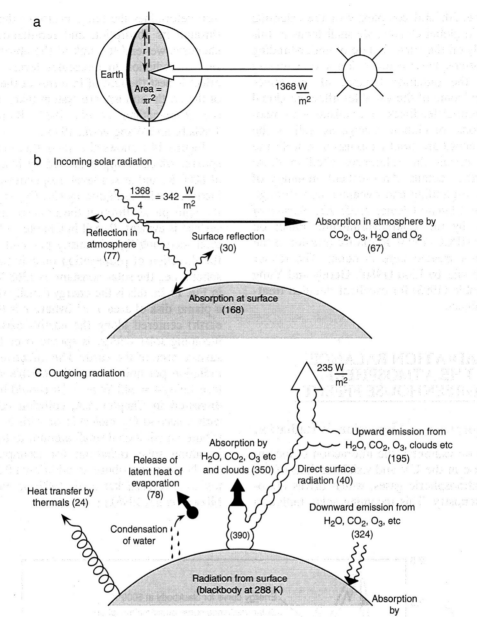

FIGURE 14.2 Global average mean radiation and energy balance per unit area of earth's surface. The numbers in parentheses are the energy in units of W m^{-2} typically involved in each path (adapted with permission from IPCC (1996), with numbers from Kiehl and Trenberth (1997)).

Figure 14.2b summarizes typical fates of this radiation. Of the incoming solar radiation, about 31% is reflected back to space either at the surface (30 W m^{-2}) or by the atmosphere itself (77 W m^{-2}). The remaining 235 W m^{-2} is absorbed, about 168 W m^{-2} by the earth's surface and 67 W m^{-2} by O_3, CO_2, H_2O, and O_2 (see Fig. 14.1) and by particles and clouds in the atmosphere.

The net absorption of 235 W m^{-2} by the earth's surface and atmosphere leads to heating and hence to the thermal emission of radiation. The Stefan–Boltzmann law can be applied to the combination of surface and atmosphere as a system (Ramanathan et al., 1987) approximated by a blackbody. Recall that this law says that the energy radiated by a blackbody at temperature T per unit time is given by $E = \sigma T^4$, where σ is the Stefan–Boltzmann constant, equal to 5.67×10^{-8} W m^{-2} K^{-4}. If this absorbed solar energy is radiated in accordance with the Stefan–Boltzmann law, the effective temperature, T_e, of the surface–atmosphere system can be estimated using $E = 235 \text{ W m}^{-2} = (5.67 \times 10^{-8} \text{ W m}^{-2} \text{ K}^{-4})T_e^4$, giving $T_e = 254$ K, or $-19°C$. This assumes no interaction with the atmosphere, which, we shall see, is certainly not the case.

Figure 14.3 is a schematic illustration of the wavelength distribution of the direct, incoming solar radiation and the outgoing, lower-energy, terrestrial radiation emitted by the earth's surface for a temperature of ~254 K. The global climate issues discussed in this chapter focus on the interactions of both the longer wavelength, terrestrial radiation and the shortwave, solar radiation with atmospheric gases and particles.

Clearly, 254 K is much colder than the typical temperatures around 288 K (15°C) found at the earth's surface. This difference between the calculated effective temperature and the true surface temperature is dramatically illustrated in Fig. 14.4, which shows the spectra of infrared radiation from earth measured from the Nimbus 4 satellite in three different locations, North Africa, Greenland, and Antarctica (Hanel *et al.*, 1972). Also shown by the dotted lines are the calculated emissions from blackbodies at various temperatures. Over North Africa (Fig. 14.3a), in the window between 850 and 950 cm^{-1}, where CO_2, O_3, H_2O, and other gases are not absorbing significantly, the temperature corresponds to blackbody emission at 320 K due to the infrared emissions from hot soil and vegetation.

The negative peaks that appear to be absorption bands superimposed on the continuous emission curve from the earth's surface are actually due to a combination of two processes involving the atmospheric greenhouse gases: (1) absorption of outgoing terrestrial infrared radiation by the gases, causing vibration–rotation transitions (and in the case of H_2O, pure rotational transitions), and (2) emission of infrared radiation by the greenhouse gases due to excited states populated by collisions. The population of these excited states is determined by the Boltzmann distribution (see Section A.2b) and hence the emission intensity is char-

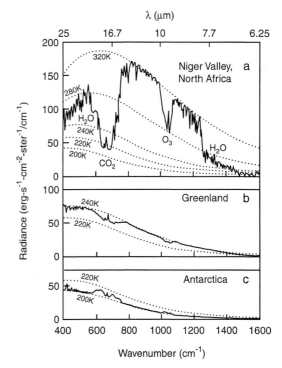

FIGURE 14.4 Infrared emission from earth measured from the Nimbus 4 satellite (a) over the Niger Valley, North Africa (14.8°N, 4.7°W) at 12:00 GMT; (b) over Greenland (72.9°LN, 41.1°W) at 12:18 GMT, and (c) over Antarctica (74.6°S, 44.4°E) at 11:32 GMT. Emissions from blackbodies at various temperatures are shown by the dotted lines for comparison (adapted from Hanel *et al.*, 1972).

acteristic of the particular temperature of the atmosphere where the emitting molecule is located. Contributions from CO_2 (600–750 cm^{-1}) and O_3 (1000–1070 cm^{-1}) as well as from H_2O [rotation bands below ~625 cm^{-1} (e.g., see Clough *et al.*, 1992) and vibration–rotation bands in the region from ~1200 to 2000 cm^{-1}] are evident. There are also smaller contributions from other greenhouse gases such as CH_4 and N_2O (vide infra). Figures 14.4b and 14.4c show that the surface temperatures of Greenland and Antarctica are much colder, ~240 and 200 K, respectively, at the times these spectra were recorded. The emissions from other greenhouse gases such as CFC-11 and CFC-12 have also been observed from the earth's surface in downward radiation (e.g., see Walden *et al.*, 1998).

The fact that both absorption and reemission contribute to the spectral features assigned to CO_2 and O_3 is illustrated by the spectra in Fig. 14.4 as well. Over North Africa (Fig. 14.4a), the air temperature is less than that at the surface, so that the emission intensity around 1040 cm^{-1} due to atmospheric O_3 is less than its absorption, leading to a distinct "negative" band superimposed on the continuous emission curve. The

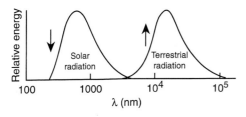

FIGURE 14.3 Schematic of wavelength dependence of energy emitted by the sun and hence entering the earth's atmosphere and the energy emitted by the earth's surface at a temperature of ~254 K. The absorptions of various atmospheric constituents have been omitted for clarity.

1040-cm^{-1} ozone band is very weak in the Greenland spectrum (Fig. 14.4b) not because O_3 is absent, but because the average temperature at which the atmospheric ozone is emitting at this location is about the same as the surface temperature, so that absorption and emission balance out. In Antarctica (Fig. 14.4c), the atmosphere is warmer than the surface so that infrared emissions due to CO_2 and O_3 more than counterbalance the absorption of terrestrial radiation and their bands actually appear as "positive" peaks on top of the colder surface emissions.

While there are obviously extreme variations in surface temperature, the measured emission profile in the "window" between the strong gas absorptions corresponds to a temperature of 288 K as an average over the earth's surface. From the Stefan–Boltzmann relationship, the measured average temperature of 288 K corresponds to an energy of emission of $E = \sigma T^4 = 390$ W m^{-2}. This temperature and the corresponding energy are clearly much greater than the effective temperature of 254 K calculated earlier assuming the earth is a blackbody emitting the absorbed solar radiation of 235 W m^{-2} with no interactions with the atmosphere above it.

Figure 14.2c schematically summarizes the transfer of thermal radiation and heat in the troposphere. Since there is a net absorption of 235 W m^{-2} of incoming solar radiation (Fig. 14.2b), a net 235 W m^{-2} in outgoing radiation is needed to balance this. Of the 390 W m^{-2} emitted as thermal infrared radiation by the earth (corresponding to the satellite-derived temperature of 288 K), approximately 40 W m^{-2} is radiated directly to space in the atmospheric "window" region from 7 to 13 μm where absorptions by CO_2, H_2O, and O_3 are relatively weak. It is this radiation that is detected as the "background" in Fig. 14.4 upon which the greenhouse gas bands are superimposed. The remaining 350 W m^{-2} is absorbed by the greenhouse gases and clouds. Water vapor is by far the most important greenhouse gas (e.g., Wang et al., 1976). For example, Kiehl and Trenberth (1997) calculate that in a standard atmosphere containing 353 ppm CO_2, 1.72 ppm CH_4, and 0.31 ppm N_2O as well as ozone and water vapor, water vapor contributes ~60% of the total radiative forcing (defined later). CO_2 is the next larger contributor, at ~26%, followed by O_3, at ~8%. Water vapor in the stratosphere, although present in small concentrations (see Chapter 12), is particularly important (Wang et al., 1976).

While we shall focus on the global view in this chapter, it is important to recognize that the processes shown in Fig. 14.2 are not homogeneous on a global scale. Thus, Fig. 14.5 shows the absorbed short-wavelength energy and the emitted long-wavelength energy

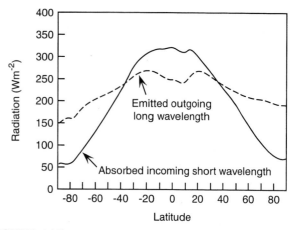

FIGURE 14.5 Annual mean radiation measured by satellite in 1988 at the top of the atmosphere as a function of latitude for incoming absorbed short-wavelength radiation and emitted outgoing long-wavelength radiation (adapted from Trenberth and Solomon, 1994).

at the top of the atmosphere, derived from satellite measurements in 1988, as a function of latitude (Trenberth and Solomon, 1994). Around the equator, a great deal more incoming solar energy is absorbed than is emitted at the longer wavelengths, whereas the opposite is true at high latitudes. As a result, low latitudes are warmed and high latitudes cooled, causing heat transport from the equator toward the poles by the atmosphere and oceans.

2. Radiative Transfer Processes in the Atmosphere

a. Macroscopic View

The spectral distribution of the radiation emitted by the earth's surface is determined by its temperature; i.e., the emission is that of a blackbody at an average temperature of 288 K. Thermal emission at the earth's surface produces a net upward flux of energy. Let us denote upward energy fluxes by F^+ and downward fluxes by F^-. The transfer of radiant energy in the troposphere can be thought of as occurring between vertical layers in the atmosphere as shown in Fig. 14.6a. The total energy flux (F^{net}) crossing a plane at a given altitude z is the difference between the upward and downward fluxes; i.e., $F^{net} = F^+ - F^-$. As shown in Fig. 14.6b, the flux of light energy through a volume of air is determined by a combination of transmission, absorption, scattering both in and out of the "beam," and thermal emission from the gas molecules (shown on a molecular level in Fig. 14.6c and described in more detail later).

A. RADIATION BALANCE OF THE ATMOSPHERE: THE GREENHOUSE EFFECT

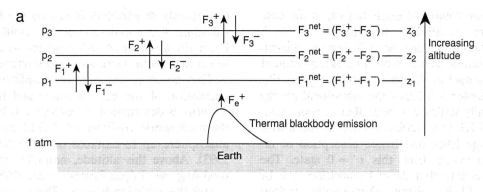

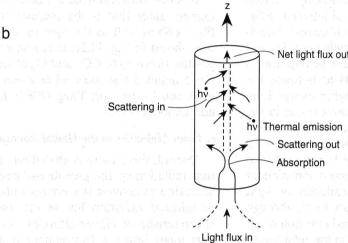

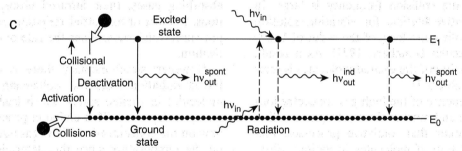

FIGURE 14.6 Schematic diagram of radiative transfer in the atmosphere: (a) between vertical layers in the atmosphere; (b) within a volume element; (c) on a molecular level. The filled circles on the energy levels portray qualitatively the much larger populations in the ground state.

In a particular layer of air in the atmosphere, the net change in energy of the layer is a result of the total fluxes crossing its boundaries from above and below. For example, in Fig. 14.6a for the layer between altitudes z_1 and z_2, if the net flux into the layer F_1^{net} is larger in magnitude than the net flux out, F_2^{net}, the difference $(F_1^{net} - F_2^{net})$ must go into heating the layer.

There are a variety of radiative transfer models for the atmosphere that incorporate all of the emission, absorption, and scattering processes as a function of altitude. These are used to predict the total radiance as a function of wavelength as well as the effect of changes in atmospheric gases, aerosol particles, and clouds on it. For details, see the books by Liou (1980), Goody and Yung (1989), and Lenoble (1993) as well as the article by Clough et al. (1992) for a typical line-by-line radiative transer model.

b. Molecular-Level View

On a molecular level, absorption of terrestrial infrared radiation of the appropriate wavelength corresponding to the energy-level splittings in the molecule causes vibrational–rotational excitation or, in the case of H_2O, pure rotational transitions as well. Vibration–rotation transitions occur if there is an oscillating dipole moment in the molecule, a requirement not met by homonuclear diatomics such as N_2 and O_2. Recall from Chapter 3.A.1 that the vibrational energy spacing is typically sufficiently large that at room temperature (~298 K), most molecules are in the lowest vibrational energy level, and hence absorption of infrared radiation occurs from this $v' = 0$ state. The selection rules dictate that $\Delta v = 1$ transitions are by far the most likely, giving vibrationally excited molecules in $v'' = 1$ upon absorption of infrared radiation emitted by the earth's surface. (Of course, associated rotational transitions occur simultaneously, and overtone and combination bands can also be important.)

Take the simplified case of transitions between the ground state (energy E_0) and one higher energy level (E_1) shown in Fig. 14.6c. Most molecules reside in the ground vibrational state at atmospheric temperatures and can therefore absorb energy corresponding to $E_1 - E_0$. Once a molecule has made the transition to the upper state, it can undergo spontaneous emission of light (shown as $h\nu_{out}^{spont}$), induced emission of light (shown as $h\nu_{out}^{ind}$), or deactivation back to E_0 through collisions with other molecules. Induced emission is not important in the atmosphere due to the low light levels. In the troposphere, the total gas densities are sufficiently high that the collision frequency is large. In addition, the radiative lifetime for vibration–rotation transitions is typically quite long, of the order of 1–100 ms for most molecules (Lambert, 1977). As a result, collisional deactivation of the excited molecules is their major fate (see Problem 1).

Another consequence of the high gas concentrations in the troposphere and most of the stratosphere is that it is collisions, rather than radiative processes, that control the population of molecules in various vibrational and rotational energy levels. As a result, excited molecules in E_1 are formed primarily by collisions (shown as "collisional activation" in Fig. 14.6c), not by absorption of radiation. Under these conditions, in the simplest (hypothetical) case of a molecule with ground-state energy E_0 of degeneracy g_0, and one in excited-state energy E_1 of degeneracy g_1, the ratio of the number of molecules in E_1 to that in the ground (E_0) state is given by the Boltzmann distribution:

$$\frac{N_1}{N_0} = \frac{g_1}{g_0} e^{-(E_1 - E_0)/kT}. \qquad (A)$$

Thus, gas collisions lead to a small equilibrium population of excited states. A small fraction of molecules in the excited states emit radiation rather than being collisionally deactivated. Based on the Boltzmann distribution, the population of the emitting states and hence the intensity of such radiation would be expected to decrease with decreasing temperature.

This situation with thermal equilibrium, where the population of the excited states and hence emission intensity is determined by collisions, is known as "local thermodynamic equilibrium" (LTE) and holds in the atmosphere up to altitudes of ~50–60 km (Lenoble, 1993). Above this altitude, non-LTE models must be used (e.g., see López-Puertas *et al.*, 1998a, 1998b).

It is this emission from the Boltzmann population of excited states that is the thermal emission shown in Fig. 14.6b as well as the upward and downward emission shown in Fig. 14.2c. It is also responsible for the positive bands to to CO_2 and O_3 observed in Fig. 14.4c. For a detailed discussion of emission (the *source function*), see Goody and Yung (1989), Liou (1980, 1992), and Lenoble (1993).

c. From Molecules to the Global Atmosphere

Overall, then, there is absorption of infrared terrestrial radiation by the greenhouse gases, collisional deactivation to convert this energy to heat, and emission of infrared radiation but at the lower temperatures characteristic of higher altitudes. As a result, the energy input into the troposphere is increased. This is clearly going to be a function of the concentrations of absorbing gases, their infrared absorption cross sections, the flux of terrestrial radiation, and the total gas pressure, which determines the rate of collisional deactivation.

However, simultaneously there is emission of infrared radiation from the Boltzmann distribution of molecules in excited states, which leads to a negative energy component. This emission process depends not only on the concentration of the gas but very sensitively on the temperature since this determines the population of the excited states that emit (Eq. (A)). As we shall see in some specific cases below, it is the balance between these two at any given altitude that determines the changes in fluxes and the ultimate impact of a change in a greenhouse gas concentration.

Because of efficient trapping of specific wavelengths of infrared radiation at lower altitudes by gases such as H_2O, CO_2, and O_3, radiation emitted to space by such infrared-active species generally originates from molecules at higher altitudes, where the temperature is lower. Because of the Boltzmann population temperature dependence, the relative proportion of excited states that are the source of the emission is lower. This leads to smaller total energy emission out to space compared to what would be the case for a higher temperature.

Recall that 235 W m^{-2} must be emitted to space to balance the net energy absorbed from incoming solar radiation (Fig. 14.2). While a small part of this (40 W m^{-2}) comes from direct emissions from the earth's surface in the atmospheric window where strong absorptions do not occur, the larger portion (195 W m^{-2}) comes from the lower temperature emissions from the greenhouse gases and the tops of clouds (see Fig. 14.2c). Since by the Stefan–Boltzmann law, 235 W m^{-2} corresponds to a temperature of 254 K, the emission of infrared radiation to space can then be thought of as occurring from an altitude where the temperature is 254 K, which is approximately 5.5 km above the earth's surface.

In effect, then, infrared emission out to space by the greenhouse gases and clouds occurs at lower temperatures than the corresponding absorptions. The 235 W m^{-2} of incoming solar radiation that is absorbed at the surface and in the atmosphere (Fig. 14.2b) is ultimately balanced by the outgoing radiation from the upward emission (at lower temperatures) of approximately 195 W m^{-2} from atmospheric constituents, including the greenhouse gases and clouds, and of 40 W m^{-2} from the surface that occurs in the atmospheric window between the strong absorptions due to CO_2, H_2O, and O_3.

The difference of 155 W m^{-2} between the 390 W m^{-2} emitted by the earth's surface (Fig. 14.2c) and the 235 W m^{-2} escaping from the atmosphere represents the amount of "trapped" radiation, the "greenhouse effect."

As seen in Fig. 14.2c, in addition to the 350 W m^{-2} absorbed in the atmosphere by the greenhouse gases and clouds, energy is also deposited in the atmosphere by convective, vertical mixing of surface heat through thermals (24 W m^{-2}), through the release of the latent heat of evaporation of water when it condenses into liquid water (78 W m^{-2}), and by direct absorption of light (Fig. 14.2b, 67 W m^{-2}). This total of 519 W m^{-2} (350 + 24 + 78 + 67 = 519 W m^{-2}) is balanced by emission of infrared radiation by gases and the tops of clouds upward into space (195 W m^{-2}) as well as downward (324 W m^{-2}), where it is absorbed by the surface and heats it.

In summary, in a hypothetical world unperturbed by anthropogenic emissions, the presence of H_2O, CO_2, and, to a lesser extent, O_3, CH_4, and N_2O in the atmosphere leads to a natural greenhouse effect that results in an average surface temperature of about 288 K, rather than 254 K, which is expected in the absence of these gases.

It is important to emphasize that because the greenhouse effect originates in radiative transfer processes in the earth–atmosphere system, the net effect of a

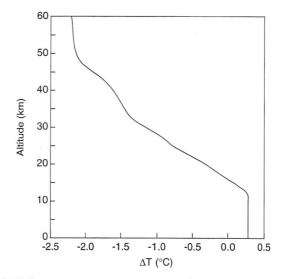

FIGURE 14.7 Model-calculated atmospheric temperature changes as a function of altitude due to an increase in CO_2 from 315 ppm in 1960 to 370 ppm projected for 2000 (no feedbacks taken into account) (adapted from Rind and Lacis, 1993).

greenhouse gas such as CO_2 on the temperature of the atmosphere depends on the altitude and temperature. For example, once CO_2 reaches the stratosphere, its density is too small to trap radiation to a significant extent. In addition, the temperature is increasing with altitude in the stratosphere. Thus, from the Boltzmann relationship (Eq. (A)), the relative concentrations of the excited emitting states are increasing, resulting in a greater net emission of energy to space. The result is that although CO_2 in the troposphere leads to warming, in the stratosphere it leads to cooling (e.g., see Roble and Dickinson, 1989; Cicerone, 1990; Rind et al., 1990; and Rind and Lacis, 1993).

Figure 14.7 shows one model calculation of the atmospheric temperature change due to increasing the CO_2 concentration from 315 to 370 ppm, corresponding to the change over the years from 1960 to 2000 (Rind and Lacis, 1993). Heating in the troposphere and cooling in the stratosphere are clearly evident. It is interesting that this cooling of the stratosphere due to CO_2 may have some interesting side effects. For example, Yung et al. (1997) estimate using model calculations that a doubling of CO_2 would increase the erythema-weighted UV by ~1% at the earth's surface due to the temperature effect on the UV absorption cross sections for O_3.

3. Dependence of Net Infrared Absorption on Atmospheric Concentrations

Net infrared absorption is determined by the intrinsic strength of the absorption for that particular molecule and transition (i.e., the absorption cross section), the effective path length, and the concentration of the absorbing gas. CO_2, H_2O, and to a lesser extent O_3, all absorb infrared radiation strongly in the atmosphere. As discussed in more detail shortly, other infrared-absorbing trace gases, particularly those that have strong absorptions in the relatively clean atmospheric window from 7 to 13 μm where CO_2, H_2O, and O_3 do not absorb strongly, also contribute to the net absorption of this radiation. However, as discussed in detail by Shine (1991), even a gas that absorbs in the same regions as the major greenhouse gases can contribute to trapping of infrared radiation; the contribution due to an increase in a particular trace gas depends on the combination of absorption region, initial concentration of the gas, and its absorption coefficients.

The dependence of absorption on concentration is linear only for weak absorption lines in the atmospheric window; this is the case, for example, for the chlorofluorocarbons. For stronger absorptions such as those due to O_3, CH_4, and N_2O, absorption at the peak of the absorption bands approaches saturation; in these case, the net absorption varies with the square root of the absorber concentration. For very strongly absorbing peaks such as those due to CO_2 and H_2O, absorption only occurs at the fringes of the band and the net absorption varies with the logarithm of the absorber concentration (Dickinson and Cicerone, 1986; Mitchell, 1989). The reasons for this are discussed in books devoted to the subject of atmospheric radiation, which should be consulted for details (e.g., see Liou, 1980; Goody and Yung, 1989; and Lenoble, 1993). A brief account is given in Box 14.1. The absorption cross sections of a variety of gases of atmospheric relevance that are needed for these calculations are available in the literature. See, for example, the HITRAN database (Rothman et al., 1992).

B. CONTRIBUTION OF TRACE GASES TO THE GREENHOUSE EFFECT

Increased atmospheric concentrations of CO_2, O_3, and other greenhouse gases over the past century have now been well documented (vide infra). All other things being equal, increasing the tropospheric concentrations of infrared-absorbing greenhouse gases will increase the difference between the amount of long-wavelength radiation absorbed and that emitted to space, leading to a net increase in the energy deposited in, and hence temperature of, the troposphere.

One way of thinking of this (Mitchell, 1989) is shown schematically in Fig. 14.9. Recall from Section A.2c that the emission of radiation to space can be thought of as occurring from a temperature of 254 K, corresponding to an altitude of ~5.5 km. An increased tropospheric concentration of a greenhouse gas means that trapping of the infrared radiation from the earth's surface will continue to occur to higher altitudes. The molecules that are emitting infrared radiation to space will therefore be at higher altitudes where the tropospheric temperatures are lower, T_1 in Fig. 14.9. At these lower temperatures, the populations of the emitting excited states will be lower (Boltzmann Eq. (A)) and hence the net emission of infrared energy smaller than that needed to balance the net absorbed incoming solar radiation of 235 W m^{-2}. To restore the energy balance between the incoming absorbed solar radiation and the outgoing infrared radiation, the atmosphere at the effective infrared emitting altitude must increase back to ~254 K. If the lapse rate remains the same (which, as discussed later, may not be the case), temperatures then must increase throughout the troposphere.

1. Infrared Absorption by Trace Gases

As seen in Figs. 14.1 and 14.2, CO_2 and H_2O, and to a lesser extent O_3, are major absorbers of infrared radiation in the troposphere. These gases, along with contributions from CH_4 and N_2O, are responsible for a greenhouse effect that exists without any emissions from anthropogenic activities. However, as seen earlier, any infrared-active atmospheric species can also act as a greenhouse gas. Thus, anthropogenic emissions of new infrared-absorbing species, or increased emissions of the traditional greenhouse gases, provide additional trapping of infrared radiation to higher altitudes, altering the energy balance of the troposphere.

For a new compound to be potentially important as a greenhouse gas in the atmosphere, it must have a sufficiently large infrared absorption cross section and be present in large enough concentrations to lead to significant absorption of infrared radiation. In addition, it will be most effective if it absorbs in the infrared window from approximately 7 to 13 μm between the CO_2 and H_2O absorptions. Finally, the atmospheric lifetime of the gas is important, in that a long-lived species such as the chlorofluorocarbons can make a larger contribution when integrated over time than a short-lived species (see Section B.3).

BOX 14.1
MOLECULAR BASIS FOR DIFFERING RELATIONSHIPS BETWEEN INFRARED ABSORPTION AND CONCENTRATION

The reason for the different dependencies of absorption on the concentration can be seen starting with the Beer–Lambert Law. If L_ν° is the radiance of frequency ν incident on an absorber with absorption coefficient σ present at concentration N and the effective path length is l, the transmitted radiance L_ν is given by

$$\frac{L_\nu}{L_\nu^\circ} = e^{-Nl\sigma}. \quad \text{(B)}$$

(For a definition of radiance, see Chapter 3.C.2.) The absorbed radiance is then given by

$$L_\nu^{abs} = L_\nu^\circ (1 - e^{-Nl\sigma}). \quad \text{(C)}$$

Since absorption lines have a finite width and since measurements of absorption in practice are carried out over a small but finite range of frequencies, the absorbed radiance over some interval $\Delta \nu$ is the parameter of interest:

$$L_{\Delta\nu}^{abs} = \int L_{\Delta\nu}^\circ (1 - e^{-Nl\sigma}) d\nu. \quad \text{(D)}$$

The incident radiation L_ν° usually does not change significantly over one line so that $L_{\Delta\nu}^\circ$ can be taken as a constant. The remainder of Eq. (D) is defined as the equivalent line width, W,

$$W = \int (1 - e^{-Nl\sigma}) d\nu, \quad \text{(E)}$$

where the integration is carried out over the range of frequencies that encompass the absorption line. The equivalent linewidth can be thought of as the width of a hypothetical absorption line of rectangular shape whose total absorption is equal to that of the real line (see Fig. 14.8a). Equation (C) can then be expressed as

$$L_{\Delta\nu}^{abs} = L_{\Delta\nu}^\circ W. \quad \text{(F)}$$

The absorption cross section σ depends on the frequency even for a single absorption line due to various line-broadening processes that impart a finite width and particular shape to the absorption line. Figure 14.8b, for example, shows a typical lineshape due to collisional broadening, the Lorentz

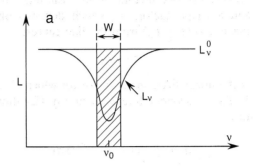

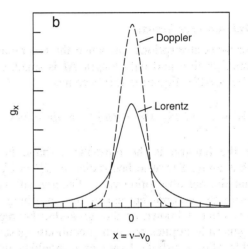

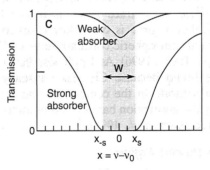

FIGURE 14.8 (a) Meaning of equivalent width, W; (b) Doppler and Lorentzian line-shapes for equivalent half-widths; (c) transmission curves for an absorption line for a weak and strong absorber, respectively (adapted from Lenoble, 1993).

lineshape, and a typical shape due to Doppler broadening. Under atmospheric conditions near the earth's surface, the linewidth is determined primarily by collisions; i.e., the Lorentz half-width is much larger than the natural linewidth or that due to

Doppler broadening. Collisional broadening becomes less important at the lower pressures found at higher altitudes, so that the Lorentzian half-width and the Doppler half-width become comparable at altitudes of approximately 30–40 km.

The absorption cross section can be expressed as the product of two factors, an intrinsic line strength S and a shape factor, g_x, which depends on the distance $x = (v - v_0)$ from the line center:

$$\sigma = S g_x. \qquad (G)$$

g_x is the normalized shape factor for which $\int_{-\infty}^{\infty} g_x \, dx = 1$. The equivalent linewidth in Eq. (E) thus becomes

$$W = \int_{-\infty}^{\infty} (1 - e^{-NlSg_x}) \, dx. \qquad (H)$$

a. Weak-Absorber Regime

For weak absorptions, i.e., when the combination of concentration and path length Nl is small, $(1 - e^{-NlSg}) \sim NlSg$. Equation (H) becomes

$$W = \int_{-\infty}^{\infty} (NlSg_x) \, dx = NlS \int_{-\infty}^{\infty} g_x \, dx = NlS, \qquad (I)$$

since the integral is the normalized shape factor, which is unity. The absorbed radiance $L_{\Delta\nu}^{\text{abs}} = L_{\Delta\nu}^\circ W$, so that the net absorption varies linearly with (Nl), i.e., with the column burden of the absorbing gas. This is what is known as the *weak-absorber regime* and generally applies to such greenhouse gases as the chlorofluorocarbons. However, it should be noted that even for some trace gases such as CF_4, the weak-absorber regime is only obeyed up to ~ 0.1 ppb (current atmospheric concentrations of CF_4 are ~ 0.07 ppb; IPCC, 1996). At 1 ppb, significant deviations are found because CF_4 has unusually sharp lines that saturate in the center and that are overlapped by the absorption bands of other atmospheric gases (Freckleton et al., 1996).

b. Strong-Absorber Regime

However, when the combination of concentration, N, and path length, l, is not small, the weak-absorber approximation is not valid, and the line-shape needs to be taken into account. The reason for this is that in the limit when absorption of light is already saturated at the center of the line, absorption due to added gas occurs only in the wings of the absorption line, which is sensitive to the lineshape. In the case of saturation at the center of the absorption line as shown in Fig. 14.8c, the absorption can be thought of as occurring in two regions, one from x_s to x_{-s} around the line center v_0, where the absorption is saturated, and one in the wings, where the absorption is weaker:

$$W = W_{\text{center}} + W_{\text{wings}}. \qquad (J)$$

In the center region from x_s to x_{-s} where the absorption is strong, $e^{-NlSg_x} \to 0$ and

$$W_{\text{cent}} = \int_{-x_s}^{x_s} (1 - e^{-NlSg_x}) \, dx \approx 2x_s. \qquad (K)$$

In the wings,

$$W_{\text{wings}} = 2 \int_{x_s}^{\infty} (1 - e^{-NlSg_x}) \, dx, \qquad (L)$$

where the factor of 2 takes into account symmetrical absorption in the wings both on the low- and high-frequency sides of the band center. Thus for strong absorption,

$$W = 2x_s + 2 \int_{x_s}^{\infty} (1 - e^{-NlSg_x}) \, dx, \qquad (M)$$

and the lineshape in the center of the line is not important, whereas that in the wings at frequencies beyond x_s is.

Let us return to the definition of equivalent linewidth in Eq. (H). Since lines at most pressures of interest here can generally be described as Lorentzian in shape, the shape factor in Eq. (H) is given by

$$g_x^L = \frac{\gamma_L}{\pi(x^2 + \gamma_L^2)}, \qquad (N)$$

where γ_L is the Lorentzian line half-width (i.e., peak width at half-maximum) and $x = (v - v_0)$ as before. This can be substituted into Eq. (H) and solved as described in detail elsewhere to get an expression for the equivalent width (e.g., see Liou, 1980; Goody and Yung, 1989; and Lenoble, 1993). The result for the limit of a strong absorber where absorption in the wings is important can be obtained readily for frequencies in the wings such that $x \gg \gamma_L$. In this case, the lineshape factor reduces to

$$g_x^L = \frac{\gamma_L}{\pi x^2}. \qquad (O)$$

In this region, then, the equivalent linewidth becomes

$$W_L = 2 \int (1 - e^{(-NlS\gamma_L/\pi x^2)}) \, dx = 2(NlS\gamma_L)^{1/2}. \qquad (P)$$

Since the absorbed radiance $L_{\Delta\nu}^{abs} = L_{\Delta\nu}^{\circ}W$, the net absorption in the *strong-absorber regime* varies as the square root of (Nl). This is the case for O_3, CH_4, and N_2O in the atmosphere.

c. Strong Absorptions by CO_2 and H_2O

The results of a number of laboratory studies [see Liou (1980); and Goody and Yung (1989) for descriptions of these] have shown that for strong absorptions of CO_2 and H_2O under conditions similar to those in the atmosphere, the total absorption band absorption area, A, can be described as the sum of three terms:

$$A = C + D \ln(Nl) + K \ln(p). \quad (Q)$$

The three parameters C, D, and K can be obtained by empirical fits to the data and p is the partial pressure of nonabsorbing gases present. Since this total absorption band area is directly related to the equivalent width and hence to the absorbed irradiance, there is a logarithmic dependence of the net absorption on (Nl), which is the case for the strong absorption bands of both water vapor and carbon dioxide in the atmosphere. As discussed by Goody and Yung (1989), the empirically observed logarithmic dependence of absorption on concentration can be shown to be consistent with theoretical expectations based on reasonable assumptions of bandshape and line intensities.

It should be noted that the foregoing considerations apply to the major absorption bands. In some cases, weaker absorption bands of the major greenhouse gases can be sufficiently weak to fall in the linear region. This is the case, for example, for light absorption by O_3 in the Chappius band, even if the strong Hartley–Huggins band (see Chapter 4.B) is saturated (e.g., see Lacis et al., 1990).

These weaker bands can have significant effects on the calculated outgoing infrared radiation. For example, Ho et al. (1998) show that much of the reported discrepancy between modeled outgoing long-wavelength radiation and satellite measurements can be attributed to not including weaker absorption bands due to CO_2 at 4.3 μm and O_3 at 14 μm and the weaker O_3 lines located far from the center of the strong 9.6-μm band.

For gases that satisfy these conditions, the effects can be proportionately quite large. For example, addition of one molecule of the chlorofluorocarbons (CFCs) CFC-11 and CFC-12 is equivalent to the addition of $\sim 10^4$ additional molecules of CO_2 due to the stronger absorption cross sections of the CFCs that occur in the atmospheric window and to the dependence of absorption on concentration for the CFCs but on the logarithm of concentration for CO_2 (Ramanathan et al., 1987).

Figure 14.10 shows the absorption bands and approximate absorption band strengths for a number of molecules found in the troposphere (Ramanathan et al., 1987; Ramanathan, 1988a, 1988b). There are many gases that, on the basis of intrinsic absorption strengths in the atmospheric window, can, in principle, contribute to tropospheric heating. However, the third requirement is that they be present in sufficient concentration to lead to significant infrared absorption. Of the molecules shown in Fig. 14.10, the ones that meet all of these requirements are CH_4, N_2O, the chlorofluorocarbons (CFCs) and other halocarbons such as methylchloroform, and some perfluorinated compounds such as SF_6 (see Chapters 12 and 13).

As we shall see in the next section, the concentrations of all of these "trace" greenhouse gases, as well as CO_2 and O_3, have been increasing over the past century or more.

2. Trends in Trace Gas Concentrations

a. CO_2

Carbon is, of course, extensively recycled through the earth system, including both the terrestrial biosphere and the oceans. Figure 14.11 summarizes this cycling and where the reservoirs of carbon are found. Anthropogenic activities contribute to atmospheric carbon mainly in the form of CO_2 emissions from fossil fuel combustion and, to a lesser extent, cement production, which total 5.5 Gt of C per year (where 1 Gt of C = 10^9 metric tons = 10^{15} g of carbon). The amount of carbon in hydrocarbons, including CH_4, and CO is less than 1% of the total atmospheric carbon (IPCC, 1996). Changes in land use, including biomass burning, also contribute to changing the balance, although the net quantitative contribution is less certain. Land use changes in the tropics during the decade from 1980 to 1990 are estimated to have contributed approximately 1.6 Gt of C per year (IPCC, 1996), but this does not

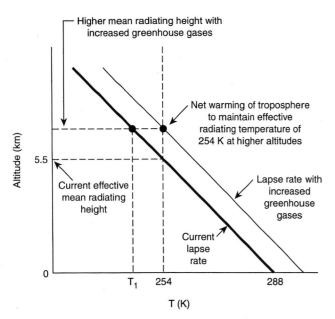

FIGURE 14.9 Warming of troposphere by increased greenhouse gas concentrations (adapted from Mitchell, 1989).

the Arctic, which they attribute to increased uptake of CO_2 by land vegetation during periods of warmer temperatures. [It is interesting that coverage by snow does not appear to terminate the exchange of CO_2 (Sommerfeld et al., 1993).]

There are different time scales associated with the various emissions and uptake processes. Two terms that are frequently used are *turnover time* and *response (or adjustment) time*. The turnover time is defined as the ratio of the mass of the gas in the atmosphere to its total rate of removal from the atmosphere. The response or adjustment time, on the other hand, is the decay time for a compound emitted into the atmosphere as an instantaneous pulse. If the removal can be described as a first-order process, i.e., the rate of removal is proportional to the concentration and the constant of proportionality remains the same, the turnover and the response times are approximately equal. However, this is not the case if the parameter relating the removal rate and the concentration is not constant. They are also not equal if the gas exchanges between several different reservoirs, as is the case for CO_2. For example, the turnover time for CO_2 in the atmosphere is about 4 years because of the rapid uptake by the oceans and terrestrial biosphere, but the response time is about 100 years because of the time it takes for CO_2 in the ocean surface layer to be taken up into the deep ocean. A pulse of CO_2 emitted into the atmosphere is expected to decay more rapidly over the first decade or so and then more gradually over the next century.

Figure 14.12 shows what has become classic data illustrating the increase in CO_2 concentrations at Mauna Loa, Hawaii, where continuous measurements have been made since 1958 (Keeling et al., 1995). The concentrations have risen from approximately 315 ppm in the late 1950s to 358 ppm in 1994. The cyclical pattern superimposed on the continuous increase reflects decreased CO_2 concentrations during summer and increased CO_2 during winter in response to seasonal differences in uptake during plant growth. The amplitude of the cyclical pattern in the Northern Hemisphere is largest at the most northerly locations, decreasing from ~15–20 to ~3 ppm near the equator, where plant growth is less dependent on season (Keeling et al., 1995).

Superimposed on the CO_2 concentration measurements in Fig. 14.12 are the concentrations expected if 55.9% of the cumulative CO_2 emissions from fossil fuel combustion and cement production remained in the atmosphere (Keeling et al., 1995). This percentage was chosen to match the atmospheric observations for the 20-year period between January 1, 1959, and January 1, 1979; the match between the two curves shows that

include a number of potential sinks due to increased growth of forests, which may be associated with increased atmospheric CO_2 and nitrogen deposition (see Chapter 7). For example, C. P. Keeling et al. (1996) have measured increases in the annual amplitude of the seasonal CO_2 cycle (vide infra) in Hawaii and in

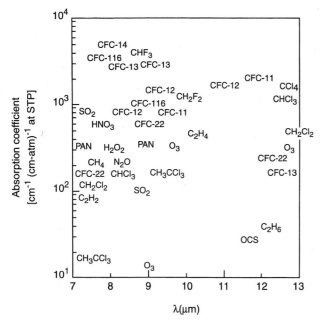

FIGURE 14.10 Intrinsic infrared absorption band strengths of some potential greenhouse gases in the atmospheric "window" (from Ramanathan, 1988a, 1988b).

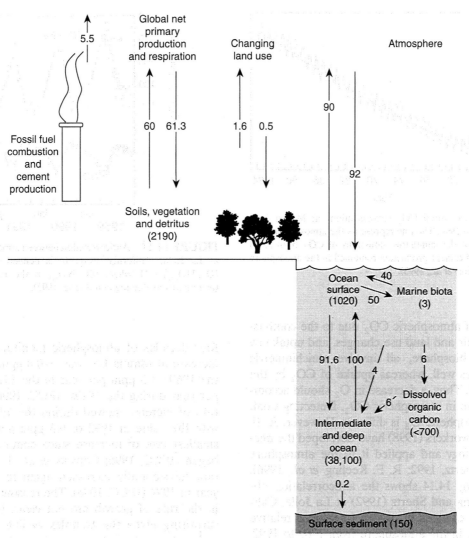

FIGURE 14.11 Summary of global carbon cycle. Amount (in gigatons of $C = 10^9$ metric tons $= 10^{15}$ g of C). Reservoirs are shown in parentheses, and fluxes (gigatons of C per year) are indicated by arrows. Note that the time scales associated with the various processes vary (adapted from IPCC, 1996).

slightly more than half of the CO_2 that has been emitted to date remains in the atmosphere. (A detailed analysis of the data in Fig. 14.12 shows a slight anomaly in the 1980s in that the CO_2 concentrations were higher than expected based on industrial emissions; Keeling and co-workers (1995) suggest this is due to changes in terrestrial and ocean sinks associated with changes in global temperatures.) Similar increases in CO_2 have been documented at locations around the world, including the South Pole, where measurements have been made since 1957. Most of the remaining emissions of CO_2 that have been removed from the atmosphere have been taken up by land ecosystems, with a small contribution from the oceans (Tans and White, 1998).

Figure 14.13 shows CO_2 concentrations measured in ice cores at the Byrd Station in Antartica from 5000 years before the present (bp) to 40,000 years bp (Anklin et al., 1997). The use of ice core data for elucidating atmospheric composition is discussed by Delmas (1992) and in more detail in Section E.1. As seen in Fig. 14.13, atmospheric CO_2 concentrations about 5000 years ago were only ~280 ppm. (Note that interpretation of such ice core data must be carried out with care since there is evidence that in some cases, CO_2 can be produced in the ice from decomposition of carbonate; e.g., see Smith et al., 1997.)

In short, it appears that CO_2 concentrations prior to the industrial age were even smaller than those measured starting in the late 1950s.

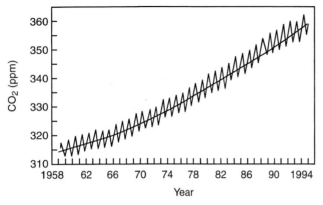

FIGURE 14.12 Measured CO_2 concentrations at Mauna Loa, Hawaii, from 1958 to 1994. The line represents the atmospheric CO_2 expected if 55.9% of the cumulative emissions of CO_2 from fossil fuel combustion and cement production remained in the atmosphere (adapted from Keeling et al., 1995).

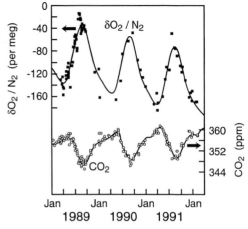

FIGURE 14.14 Anticorrelation between atmospheric O_2 and CO_2 at La Jolla, California. $\delta O_2/N_2$ is defined as $10^6\{[(O_2/N_2)_{air}/(O_2/N_2)_{ref}] - 1\}$, where $(O_2/N_2)_{ref}$ is the ratio for a reference (adapted from Keeling and Shertz, 1992).

Production of atmospheric CO_2 due to the combustion of fossil fuels and land use changes, and uptake by the terrestrial biosphere, all involve stoichiometric changes in O_2 as well, whereas uptake of CO_2 by the oceans does not. Thus, a decrease in O_2 should accompany the increase in atmospheric CO_2. Detecting small changes in atmospheric O_2 is difficult. However, R. F. Keeling and co-workers (1998) have developed the necessary methodology and applied it to the atmosphere (Keeling and Shertz, 1992; R. F. Keeling et al., 1996). For example, Fig. 14.14 shows the anticorrelation observed by Keeling and Shertz (1992) at La Jolla, California, between changes in oxygen (measured relative to N_2) and CO_2 in the atmosphere from 1989 to 1992. Similarly, Bender et al. (1994) measured changes in O_2 relative to N_2 in air bubbles trapped in ice cores at Vlostok, Antarctica, and report an increase in trapped O_2 and a decrease in CO_2 with increasing depth, i.e., as a function of time before the present.

As seen in Fig. 14.15, the rate of atmospheric increase of CO_2 has been quite variable over the past

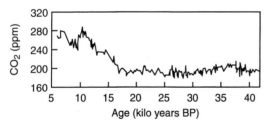

FIGURE 14.13 Concentrations of atmospheric CO_2 measured using gases trapped in ice cores from Byrd Station, Antarctica, from 5000 to 40,000 years before the present (bp) (adapted from Anklin et al., 1997).

four decades of atmospheric monitoring. The rate of increase at Mauna Loa was ~0.8 ppm per year during the 1960s, 1.3 ppm per year in the 1970s, and 1.5 ppm per year during the 1980s (IPCC, 1996). However, the rate of increase slowed during the 1989–1993 period, with the value in 1992 of 0.6 ppm per year being the smallest rate of increase since continuous monitoring began (IPCC, 1996; Conway et al., 1994). The growth rate subsequently increased again to over 2 ppm per year in 1994 (IPCC, 1996). The reasons for the changes in the rate of growth are not clear, but probably not surprising given the complex cycling mechanisms for carbon (Fig. 14.11). For example, exchange between the atmosphere and the terrestrial biosphere and the oceans is believed to have substantial year-to-year variability, which can be affected by such events as the El Niño Southern Oscillation (ENSO). Possible reasons for the variability are discussed in detail in the IPCC document (1996).

Current data on atmospheric CO_2 concentrations and temporal trends as well as those of other trace gases are available from the U.S. Department of Energy's Carbon Dioxide Information Analysis Center (CDIAC) (World Wide Web page is http://cdiac.esd.ornl.gov/cdiac). CDIAC also has available a number of additional data packages on global change issues such as trends in temperature and precipitation.

As discussed earlier, although CO_2 warms the troposphere, it cools the stratosphere since it efficiently radiates infrared out to space. This effect can contribute to changes in the temperature profile in the stratosphere and potentially have a signficant impact

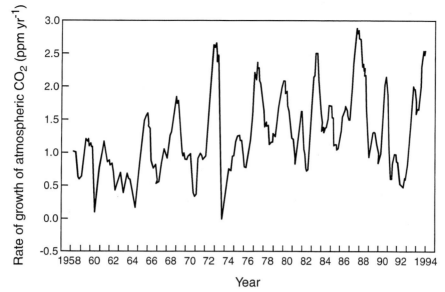

FIGURE 14.15 Rate of growth (ppm per year) of atmospheric CO_2 at Mauna Loa, Hawaii, from 1958 to 1994 (from Keeling and Worf as reported in IPCC, 1996).

on atmospheric circulation processes (e.g., see Rind *et al.*, 1990; and Rind and Lacis, 1993).

b. CH_4

Like CO_2, methane is emitted by both natural and anthropogenic processes. While the major sources are thought to have been identified, there is some uncertainty in the absolute magnitudes of their contributions as well as the factors that affect these (Cicerone and Oremland, 1988; Fung *et al.*, 1991). Table 14.1 shows one estimate of methane sources during the mid-1980s, expressed in units of teragrams (10^{12} g) of carbon per year (Crutzen, 1995). Approximately $60 \pm 10\%$ of the total emissions (370 out of a total of 630 Tg per year) is estimated to be associated with human activities. Of the 370 Tg of C per year, approximately 35% is due to losses during natural gas and oil production and distribution and coal mining. This estimate is reasonably consistent with measurements of the ^{14}C content of atmospheric methane, since fossil fuel derived methane is depleted in ^{14}C due to the long time frame for the fuel formation (Lowe *et al.*, 1988; Wahlen *et al.*, 1989).

The next largest source, approximately 30% of the 370 Tg of C per year, is due to emissions from domesticated ruminant livestock (e.g., see Johnson *et al.*, 1994) and from the decay of animal wastes. Emissions from rice fields (e.g., see Cicerone and Shetter, 1981; Cicerone *et al.*, 1983, 1992; and Tyler *et al.*, 1994) appear to comprise about 20%. The remainder of the 370 Tg of C per year is believed to be due about equally to emissions from sanitary landfills (e.g., see Bogner and Spokas, 1993) and from biomass burning (e.g., see Hao and Ward, 1993). It should be noted that the emissions in some cases, for example rice fields and landfills, represent the net flux of emission and microbially mediated oxidation processes so that both need to be understood in assessing the methane budget (e.g., Reeburgh *et al.*, 1993; Bogner and Spokas, 1993).

There were relatively few measurements of atmospheric methane concentrations prior to about 1980, except for a set from 1963 to 1970 by Stephens and co-workers (Stephens and Burleson, 1969; Stephens, 1985), which were in the 1.37–1.57 ppm range. The current global mean concentration of methane is 1.72 ppm, with higher concentrations in the Northern than

TABLE 14.1 Estimated Methane Sources during the Mid-1980s[a]

Source	Emissions (Tg of C/year)[b]
Natural	260 ± 30
Anthropogenic	370 ± 40
Gas leakage and oil production	85–105
Coal mining	25–45
Rice fields	20–150
Ruminants	65–100
Biomass burning	20–60
Animal wastes	20–40
Sanitary landfills	20–60

[a] From Crutzen (1995).
[b] Tg of C = 10^{12} g of C.

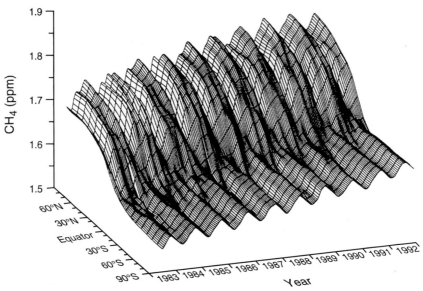

FIGURE 14.16 Averaged 3-D methane concentrations in the marine boundary layer. Lines are guides for the eye (adapted from Dlugokencky et al., 1994a).

in the Southern Hemisphere. Figure 14.16 shows the latitudinal distribution as a function of time from 1983 to 1992 (Dlugokencky et al., 1994a). The hemispheric distribution is clearly seen, as is the seasonal cycle due to changes in oxidation by OH and in methane emission sources. Even in this relatively short time span, the increase in the atmospheric concentrations is evident. Unlike CO_2, there does not appear to be a discernible trend in the amplitude of the seasonal cycles (Dlugokencky et al., 1997).

Ice core data show that the concentration prior to about the year 1750 was ~700 ppb, less than half of the current global average. The increase appears to have begun in the 1750–1800 period (e.g., see Khalil and Rasmussen, 1987, 1994b; Blunier et al., 1993; and Etheridge et al., 1998). Figure 14.17, for example, shows the concentrations of atmospheric methane for the past approximately 1000 years (Etheridge et al., 1998). The increase in concentration to the present value is well outside the variations of ~70 ppb observed prior to 1750. It is interesting that the increase appears to have begun prior to significant industrial activity, but parallels the increase in population growth in China; Blunier and co-workers (1993) suggest that this may be due to associated changes in emissions from rice fields.

The rate of increase of atmospheric CH_4 has been variable. From 1978 to 1987, the average growth rate was approximately 16 ppb per year (Blake and Rowland, 1988). However, the global growth rate slowed during the latter part of the 1980s (Steele et al., 1992; Dlugokencky et al., 1994a, 1994b, 1998). Figure 14.18

shows globally averaged methane concentrations (Fig. 14.18a) as well as the growth rate for CH_4 for latitudes from 82°N to 90°S from 1984 to 1996 (Fig. 14.18b) (Dlugokencky et al., 1998). Dlugokencky and co-workers suggest that the apparent leveling off of CH_4 may not reflect a change in sources or sinks but rather reflect an approach to steady state. The growth rates in the late 1980s were lower, but there was a sharp rise in 1991 immediately after the Mount Pinatubo volcanic eruption, followed by a sharp decrease to temporarily

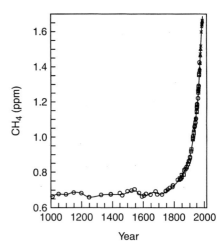

FIGURE 14.17 Atmospheric methane concentrations over the past 1000 years. Different symbols represent data from ice cores in Antarctica and Greenland and the Antarctic firn layer. Line from 1978 includes air measurements at Cape Grim, Tasmania (adapted from Etheridge et al., 1998).

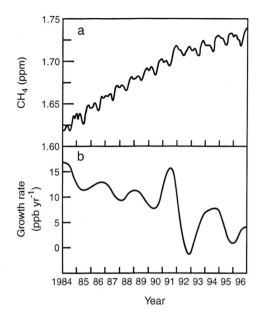

FIGURE 14.18 Globally averaged (a) CH_4 concentrations and (b) growth rates from 1984 to 1996 from 82°N to 90°S latitude (adapted from Dlugokencky et al., 1998).

negative values. The increase after the eruption is attributed by Dlugokencky and co-workers (1996, 1998) to reduced removal rates for CH_4 by its reaction with OH; light scattering by the volcanic aerosol particles and UV absorption by SO_2 decreased UV, and hence decreased OH would be expected in the troposphere. Reasons for the sharp decrease during late 1992 and 1993 are not clear but may involve such factors as smaller natural emissions from wetlands due to lower temperatures following the eruption (Hogan and Harriss, 1994; Dlugokencky et al., 1998) and/or changes in anthropogenically associated source strengths such as decreased emissions from the former U.S.S.R. and decreased biomass burning (e.g., see Dlugokencky et al., 1994b, 1998; Rudolph, 1994; and Crutzen, 1995). Increased removal by OH due to increased UV associated with stratospheric ozone destruction may also have contributed to these trends in CH_4 (Bekki et al., 1994).

While quantifying the sources and sinks of CH_4 has been difficult, isotopic measurements of $^{13}CH_4$ and CH_xD_{4-x} are promising. Various sources have characteristic isotopic signatures; e.g., as mentioned previously, fossil fuel derived CH_4 is depleted in ^{14}C (Lowe et al., 1988, 1994; Wahlen et al., 1989). The sinks of CH_4, e.g., reaction with OH, reaction with Cl, and uptake by soil bacteria, also exhibit kinetic isotope effects and these have been used to probe the causes of the observed recent changes in CH_4 growth rates (e.g., see Gupta et al., 1996, 1997). Measurements of isotopic fractionation of both ^{13}C and D have been used to estimate the fraction of CH_4 oxidized during transport through the coversoils in landfills, for example (e.g., see Bergamaschi et al., 1998a; and Liptay et al., 1998), and to identify sources of CH_4 in the troposphere (e.g., Bergamaschi et al., 1998b; Moriizumi et al., 1998). Recent studies of the isotopic composition in the upper troposphere show that the methane is enriched in ^{13}C in a manner that is not consistent with known kinetic isotope effects for CH_4 reactions (Tyler et al., 1998), again demonstrating the complexity of quantitatively defining the sources and sinks of methane in the atmosphere.

c. N_2O

Nitrous oxide is important not only as a greenhouse gas but, as discussed in Chapter 12, as the major natural source of NO_x in the stratosphere, where it is transported due to its long tropospheric lifetime (Crutzen, 1970). The major sources of N_2O are nitrification and denitrification in soils and aquatic systems, with smaller amounts directly from anthropogenic processes such as sewage treatment and fossil fuel combustion (e.g., see Delwiche, 1981; Khalil and Rasmussen, 1992; Williams et al., 1992; Nevison et al., 1995, 1996; Prasad, 1994, 1997; Bouwman and Taylor, 1996; and Prasad et al., 1997). The use of fertilizers increases N_2O emissions. For pastures at least, soil water content at the time of fertilization appears to be an important factor in determining emissions of N_2O (and NO) (Veldkamp et al., 1998).

Table 14.2 shows one estimate of the contribution of various sources to the N_2O budget (Bouwman and Taylor, 1996). While the major source of N_2O is known to be biological, there are several observations that

TABLE 14.2 Estimated Annual N_2O Budget[a]

Source	Emissions Tg of N/year)
Soils under natural vegetation and grasslands	5.7
Arable lands	1.0
Nitrogen fertilizer use	1.0
Animal wastes	1.0
Biomass burning	0.1
Agricultural waste burning	0.1
Postclearing enhanced soil flux	0.4
Fossil fuel combustion and traffic	0.3
Biofuel combustion	0.1
Industry	0.5
Oceans	3.6
Total sources	13.8

[a] From Bouwman and Taylor (1996).

indicate that the sources are not yet entirely characterized. For example, N$_2$O in the lower stratosphere has been shown to be isotopically enriched in both ^{18}O and ^{15}N relative to tropospheric N$_2$O (Kim and Craig, 1993). N$_2$O emissions from tropical rain forest soils, fertilized soils, and a wastewater treatment facility are lighter in both of these heavy isotopes than tropospheric N$_2$O (Yoshinari and Wahlen, 1985; Wahlen and Yoshinari, 1985; Kim and Craig, 1990, 1993). Either there is a source of N$_2$O in the stratosphere that selectively produces heavy N$_2$O or one that in the stratosphere selectively destroys light N$_2$O (Johnson *et al.*, 1995). Despite a number of studies, the source of this discrepancy is not yet clear (McElroy and Jones, 1996; Wingen and Finlayson-Pitts, 1998; Cliff and Thiemens, 1997; Rahn and Wahlen, 1997), although such processes as enhanced photolysis of the ^{14}N^{14}N^{16}O isotopomer (Yung and Miller, 1997; Rahn *et al.*, 1998) and/or formation of N$_2$O by reaction of highly vibrationally excited O$_3$ with N$_2$ (Zipf and Prasad, 1998) have been proposed.

One interesting potential source of N$_2$O is the heterogeneous oxidation of HONO on surfaces (Wiesen *et al.*, 1995; Pires and Rossi, 1997), which has been observed to form N$_2$O. This is likely responsible for the observation of significant amounts of N$_2$O in automobile exhaust, which was shown to be an artifact of sampling (Munzio and Kramlich, 1988). However, it may also occur on aerosol particles in the atmosphere (Clemens *et al.*, 1997), an area that warrants further investigation.

Like CO$_2$ and CH$_4$, the concentrations of N$_2$O in the atmosphere have also been increasing, from ~275 ppb in the preindustrial era to ~312 ppb in 1994 (IPCC, 1996). Figure 14.19, for example, shows one set of measurements of N$_2$O over the past 250 years obtained using ice core samples from Antarctica (Machida *et al.*, 1995). The rate of growth has been variable, averaging 0.8 ± 0.2 ppb per year from 1977 to 1988 (Khalil and Rasmussen, 1992), but was only 0.5 ppb per year in 1993 (IPCC, 1996).

d. O$_3$

As discussed in other chapters of this book and summarized in Chapter 16, the formation of tropospheric ozone from photochemical reactions of volatile organic compounds (VOC) and oxides of nitrogen (NO$_x$) involves many reactions. Concentrations are therefore quite variable geographically, temporally, and altitudinally. Additional complications come from the fact that there are episodic injections of stratospheric O$_3$ into the troposphere as well as a number of sinks for its removal. Because O$_3$ decomposes thermally, particularly on surfaces, it is not preserved in ice cores. All of these factors make the development of a global climatology for O$_3$ in a manner similar to that for N$_2$O and CH$_4$, for example, much more difficult. In addition, the complexity of the chemistry leading to O$_3$ formation from VOC and NO$_x$ is such that model-predicted ozone concentrations can vary from model to model (e.g., see Olson *et al.*, 1997).

Shortly after the discovery of ozone by Schönbein in 1839, measurements of this newly discovered atmospheric gas were initiated in a number of locations around the world, including Europe, South America, and North America (e.g., see Bojkov, 1986; Volz and Kley, 1988; McKeen *et al.*, 1989; Anfossi *et al.*, 1991; Sandroni *et al.*, 1992; and Marenco *et al.*, 1994). Although these very early measurements used wet chemical techniques (see Chapter 11), potential interferences can be estimated to make some approximate corrections to the data. Typical annual variations in tropospheric O$_3$ observed at Moncalieri in Italy from 1868 to 1893, Montsouris, France, from 1876 to 1886, and Zagreb, Croatia, in 1900 are compared in Fig. 14.20 to more recent data (1983) at Aikona, on an island believed to be relatively remote, in the Baltic Sea (Sandroni *et al.*, 1992). All of the early measurements peak around 10 ppb, whereas 30–40 ppb is a typical tropospheric O$_3$ concentration found essentially everywhere in the world today. Global increases in ozone have also been documented over more recent times, although the geographic distribution and temporal changes are complex (e.g., see Volz *et al.*, 1989; Janach, 1989; Lefohn *et al.*, 1992; Low *et al.*, 1990; Logan, 1994; Jiang and Yung, 1996; and Stockwell *et al.*, 1997). Berntsen *et al.* (1997) carried out model calculations that showed that significant radiative effects must have resulted from changes in O$_3$ since preindustrial times in the upper troposphere due to *in situ* formation from

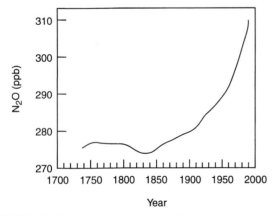

FIGURE 14.19 Fitted curve for atmospheric N$_2$O concentrations from 1735 to 1991 obtained from ice cores in Antarctica (adapted from Machida *et al.*, 1995).

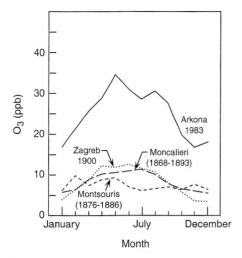

FIGURE 14.20 Monthly average O_3 concentrations at Moncalieri, Italy; Montsouris, France; and Zagreb, Croatia in the 1868–1900 period and Arkona (Baltic Sea) in 1983 (adapted from Sandroni et al., 1992).

the reactions of precursors transported from the boundary layer.

In short, the concentrations of tropospheric ozone, which is also a greenhouse gas, have also increased over the past century, an increase attributed to increased oxides of nitrogen emissions associated with fossil fuel combustion (e.g., see Volz and Kley, 1988; and Janach, 1989).

The potential effects of ozone on climate are particularly complex in that the net effect is very sensitive to the vertical distribution profile, with changes at the tropopause having the largest impact (Wang et al., 1980; Lacis et al., 1990). The reason for this is that O_3 near the ground is at temperatures close to those of the earth's surface. As a result, emission and absorption are occurring at essentially the same temperature, resulting in no contribution to the greenhouse effect. However, because the temperature falls with altitude up to the tropopause, the Boltzmann distribution (Eq. (A)) shifts to smaller relative populations in the excited states. Thus, as discussed earlier, the net emission from O_3 becomes smaller relative to absorption. While the same is true for other greenhouse gases such as CO_2 and CH_4, their sources and sinks are such they are relatively well mixed in the atmosphere and their vertical distributions are not subject to the variability associated with O_3. Another important difference is that O_3 absorbs strongly in the UV as well, which leads to heating in the stratosphere, in contrast to CO_2, which cools it. Thus, changes in the concentrations of ozone and its vertical distribution affect not only infrared but also solar UV radiation, with associated effects on climate (see also Section B.3b).

These phenomena are illustrated in Fig. 14.21a, which shows a model calculation of the change in global surface temperature when 10 Dobson units (DU) of O_3 (10 DU = total column O_3 equivalent to a layer of thickness 0.1 mm at 273 K and a pressure of 1 atm; see Chapter 12.A) are added one at a time to each of 33 vertical layers of the atmosphere (assuming no feedbacks). An increase in the global surface temperature is predicted when the ozone is added in layers up to

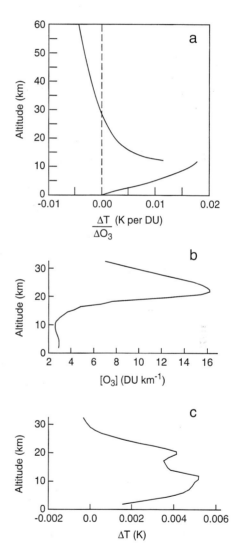

FIGURE 14.21 (a) Model-calculated change in global surface temperature when 10 Dobson units (DU) of O_3 is added to each of 33 vertical layers of the atmosphere, assuming no feedbacks (adapted from Lacis et al., 1990). (b) Global mean vertical O_3 profile (adapted from Forster and Shine, 1997). (c) Model-calculated change in global surface temperature due to increasing O_3 concentrations sequentially by 10% in each 1-km layer of the atmosphere (adapted from Forster and Shine, 1997).

~30 km. (The gap around 10 km is an artifact of the calculations.) However, increased O_3 above ~30 km causes cooling due to increased thermal emission to space and to increased absorption of solar radiation before it can reach the earth's surface (e.g., see Ramanathan *et al.*, 1985).

Addition of a constant, absolute amount of O_3 may be unrealistic, however, in that 10 DU is a large percentage change in the existing O_3 concentration at lower altitudes but a smaller percentage at the altitudes at which the ozone concentration peaks. Figure 14.21b, for example, shows one global mean vertical profile for O_3 (in Dobson units). Adding 10 DU at the tropopause corresponds to an increase in O_3 of ~400%, whereas at 24 km, adding 10 DU only increases O_3 by ~60% (Forster and Shine, 1997). Figure 14.21c shows a calculated change in surface temperature due to a systematic change in O_3 of 10% in each layer. While O_3 at the tropopause is still important, that at lower and higher altitudes is relatively more important compared to the case in Fig. 14.21a where an absolute increase in O_3 in each layer is assumed.

A number of model studies have explored the climate implications of changes in the vertical distribution of ozone (e.g., see Schwarzkopf and Ramaswamy, 1993; Wang *et al.*, 1993; Molnar *et al.*, 1994; and Chalita *et al.*, 1996). For example, Fig. 14.22a shows a model-calculated percentage change in tropospheric O_3 in July as a function of altitude and latitude from preindustrial times to the present (Chalita *et al.*, 1996). The largest increase in concentration is predicted in the boundary layer, with smaller increases at higher altitudes. However, as seen in Fig. 14.22b, the major contribution to radiative forcing (see Section B.3) comes from the relatively small ozone increase predicted for the 6- to 12-km region.

Changes in *stratospheric* ozone also impact atmospheric temperatures through three effects. As discussed in Chapter 3, UV absorption by stratospheric O_3 and the energy released in the $O + O_2$ reaction warm the stratosphere. As a result, destruction of stratospheric ozone due to chlorofluorocarbons results in cooling of the stratosphere. From the Boltzmann relationship (Eq. (A)), there is then less downward radiation across the tropopause from ozone at these lower temperatures. In addition, there is the direct effect of a smaller ozone concentration to emit infrared radiation, part of which is in a downward direction and which normally contributes to heating of the troposphere. These two effects of changes in stratospheric ozone lead to cooling of the troposphere. Counterbalancing these effects is the increased solar UV reaching the troposphere due to less absorption by stratospheric O_3, which is expected to cause surface heating (e.g., see Ramaswamy *et al.*, 1992, 1996; Zhong *et al.*, 1996; and Shine *et al.*, 1998).

It should be noted that while changes in stratospheric ozone can impact tropospheric heating and cooling, greenhouse gases may also impact stratospheric ozone destruction. As described earlier in this chapter, while CO_2 causes warming in the troposphere, it causes cooling in the stratosphere through the efficient emission of infrared to space. Some model calculations suggest that additional stratospheric cooling in the polar regions due to increased greenhouse gases may increase the formation of polar stratospheric clouds. Since these play such a key role in ozone destruction in those regions (see Chapter 12.C.5), increased ozone destruction is predicted, particularly in the Arctic, where temperatures are not as routinely cold as in the Antarctic (Austin *et al.*, 1992, 1994; Shindell *et al.*, 1998). (It should be noted, however, that warming of the troposphere by trapping of outgoing terrestrial radiation by PSCs during winter has also been proposed as being important historically at times of high methane concentrations that oxidized to form water; Sloan and Pollard, 1998.) In addition, it has been suggested that the recovery of these regions as the

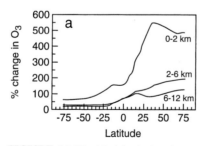 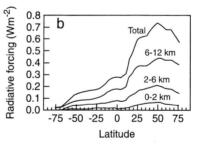

FIGURE 14.22 Model-calculated percentage (a) increase in O_3 (zonal average) in July and (b) the corresponding contributions to instantaneous radiative forcing calculated as a function of latitude and altitude from preindustrial times to the present (adapted from Chalita *et al.*, 1996).

emissions of ozone-destroying chlorine and bromine compounds decline may be delayed by about a decade due to this effect of greenhouse gases (Shindell et al., 1998).

There are a number of factors that affect the ultimate climate response to changes in tropospheric and stratospheric ozone. These include the altitude dependence of the forcing previously discussed, its role in absorbing solar UV in both the stratosphere and troposphere, its depletion through chain reactions of CFCs in the stratosphere, and finally, the large variability in its concentrations geographically, vertically, and temporally. Because of these complexities, the net effect is expected to also vary from one location to another, as well as temporally. The spatial and temporal effects due to ozone formed from emissions from biomass burning over large areas of the tropics are one example (Portmann et al., 1997). A number of studies have addressed the net effect of changes in O_3 on climate, and the reader is referred to them for more detailed information (e.g., see Hauglustaine et al., 1994; Marenco et al., 1994; Lelieveld and van Dorland, 1995; Forster et al., 1996; Chalita et al., 1996; Portmann et al., 1997; Berntsen et al., 1997; van Dorland et al., 1997; Graf et al., 1998; Haywood et al., 1998c; Brasseur et al., 1998; Stevenson et al., 1998; and Wang and Jacob, 1998).

e. CFCs, HCFCs, and HFCs

The atmospheric concentrations of chlorofluorocarbons (CFCs), hydrochlorofluorocarbons (HCFCs), and hydrofluorocarbons (HFCs) and the trends in these concentrations are discussed in detail in Chapter 13. In brief, the atmospheric concentrations of CFCs increased as they came into increasing use in the 1950s. As the phase-outs specified in the Montreal Protocol and its amendments have come into play (Figs. 12.13 and 13.4), the growth rates have fallen dramatically. For example, that for CFC-12 was ~18 ppt per year prior to the Montreal Protocol but in mid-1993 had fallen to 6–7 ppt per year (e.g., see Cunnold et al., 1997) and has since become slightly negative (e.g., Derwent et al., 1998a). However, the concentrations of their alternatives, the HCFCs and HFCs, are increasing as expected (see Fig. 13.6). Although they do not contribute significantly to radiative forcing at present, they could do so if their emissions approach those of the compounds they are replacing (Derwent et al., 1998b).

f. Other Gases

Other anthropogenically emitted gases such as CO have also been suggested to contribute to the greenhouse effect (e.g., see Evans and Puckrin, 1995). CO concentrations also increased during the 1980s but then decreased from 1988 to 1992 (e.g., see Khalil and Rasmussen, 1984, 1994a; Novelli et al., 1994; and Yurganov et al., 1997). CO is not believed to directly contribute significantly to the greenhouse effect (IPCC, 1996). However, increasing CO emissions may decrease the OH concentration, which would then increase the concentrations of other greenhouse gases that react with OH, such as CH_4. For example, Daniel and Solomon (1998) estimate that this indirect effect associated with anthropogenic emissions may be as or more significant over the next 15 years than that due to anthropogenic emissions of N_2O.

3. Radiative Forcing by Greenhouse Gases and Global Warming Potentials

In the simplest of worlds, the greenhouse gases would exert their influences independent of each other and of other factors such as aerosols and clouds, feedback mechanisms, and ozone depletion. This, of course, is not the case. However, it is useful before examining these "real-world" considerations to consider the direct effects on the radiation balance of the atmosphere of the greenhouse gases. These are commonly expressed in terms of the *radiative forcing*. Another tool used for examining the relative effects of various gases on the radiation balance of the atmosphere is the *global warming potential*.

a. Instantaneous and Adjusted Radiative Forcing

As discussed at the beginning of this chapter, changes in the radiation balance of the atmosphere can occur due to changes either in incoming solar radiation or in the outgoing infrared radiation. *Radiative forcing* is defined as a change in the average net radiation at the tropopause due to a particular perturbation of interest. This change (usually expressed in W m^{-2}) could be in either the incoming or outgoing radiation.

Using the flux at the tropopause to define radiative forcing is believed to be appropriate because of the rapid vertical mixing by convection and large-scale processes within the troposphere which closely couples the troposphere and the earth's surface (Wang et al., 1995). As a result, energy absorbed in the troposphere is assumed to be effective in warming the earth's surface (Lacis et al., 1990) and the change in the flux at the tropopause can be used to calculate the change in the surface temperature (Ramanathan, 1976).

Two approaches to calculating radiative forcing due to greenhouse gases have been taken. In the first, the immediate forcing due to increases in the greenhouse gas is calculated without allowing for a change in the stratospheric temperature. This is what is known as the *instantaneous radiative forcing*.

The second approach is to calculate the radiative forcing after allowing stratospheric temperatures to readjust to radiative equilibrium, but with the temperatures of the earth's surface and troposphere, as well as atmospheric moisture, fixed. This is known as the *adjusted radiative forcing*. The reason for allowing a stratospheric readjustment is that an increase in a greenhouse gas increases the net radiation absorbed in the troposphere. As a result, there is less upwelling radiation reaching the stratosphere (Fig. 14.2c). This causes cooling of the stratosphere, which decreases the net downward radiative flux from the stratosphere at the tropopause, contributing a negative component to the net radiative forcing attributable to an increase in a greenhouse gas (Wang *et al.*, 1995). The time for the stratosphere to adjust is of the order of months (Manabe and Strickler, 1964), so that for the longer term perturbations of interest, this adjustment will occur and decrease the net radiative forcing from the instantaneous value.

This is illustrated by the data in Table 14.3, which shows the calculated instantaneous and adjusted radiative forcing attributed to the increase in tropospheric O_3 from preindustrial times to the present (Berntsen *et al.*, 1997). Table 14.3 also illustrates the larger relative importance of absorption of long-wavelength IR by ozone compared to short-wavelength UV. [Note that these calculations represent the contribution due to changes in tropospheric ozone only; as discussed earlier and in the following text, the destruction of stratospheric ozone leads to a negative forcing, i.e., cooling. Indeed, it appears that this cooling effect is likely dominant at the present time (Hansen *et al.*, 1997b).]

Radiative forcing can be calculated for greenhouse gases in a fairly straightforward manner, particularly in the simplest case where there are no feedbacks or indirect effects on the chemistry of the atmosphere. However, translating these radiative forcings into real temperature changes at the earth's surface is much more uncertain due to the complex feedbacks involving chemistry, physics, and atmospheric dynamics. Empirical relationships of the form of Eq. (R),

$$\Delta T_s \, (°C) = \delta F_a, \quad (R)$$

between the change in temperature at the earth's surface, ΔT_s, and the adjusted radiative forcing F_a (in W m^{-2}) are often used (Hansen *et al.*, 1997b). The value of δ is model dependent, with typical values of 0.3–1.1 depending on the model and whether it includes the effects of feedbacks (Hansen *et al.*, 1997b, 1997d). This range corresponds to changes in predicted global temperatures from 1.5 to 4.5°C for a doubling of CO_2 (or the equivalent contributions from other greenhouse species).

As discussed in more detail in the following sections, some anthropogenic emissions are expected to cause positive radiative forcings whereas others are negative. Even though the two may be equal in magnitude, giving a net radiative forcing of zero, this does not mean that there will be no effects on climate. For example, negative radiative forcing caused by aerosol particles is expected to occur primarily over continents whereas the positive radiative forcing due to many greenhouse gases occurs globally. In addition, there are temporal differences, with the lifetime of particles typically being of the order of a week while those of many greenhouse gases are of the order of centuries. These differences can lead to impacts on climate, despite a net radiative forcing of zero. As a result of such considerations, radiative forcing is commonly used primarily for comparing the relative potential importance of various gases and particles on climate.

b. Absolute and Relative Global Warming Potentials

As was the case for ozone depletion potentials (see Chapter 13.B), the effects of greenhouse gases depend not only on the emissions but also on their lifetimes in the atmosphere (Ko *et al.*, 1993). *Global warming potentials* (GWP) express the time-integrated radiative forcing due to the instantaneous emission of a fixed amount (usually 1 kg) of the gas of interest. Thus, a time-scale horizon (TH) that will be considered in assessing the radiative effects of the gas must be specified. Both absolute and relative GWPs have been put forth, where these are defined as follows:

Absolute

$$\text{Absolute GWP} = \int_0^{TH} a_{gas}[gas]_t \, dt \quad (S)$$

TABLE 14.3 Radiative Forcing[a] Due To Changes in Tropospheric Ozone from Preindustrial Times to the Present Time for Clear Skies Calculated Using the Oslo Model[b]

Radiative forcing	Net	Long-wavelength thermal infrared	Short-wavelength solar UV
Instantaneous	0.48	0.42	0.06
Adjusted	0.39	0.33	0.06

[a] Global mean and annual mean radiative forcing in W m^{-2}.
[b] Adapted from Bernsten *et al.* (1997).

Relative

$$\text{Relative GWP} = \frac{\int_0^{TH} a_{\text{gas}}[\text{gas}]_t \, dt}{\int_0^{TH} a_{\text{ref}}[\text{ref}]_t \, dt} \quad (T)$$

In Eqs. (S) and (T), [gas] and [ref] represent the time-dependent concentrations of the gas of interest and the reference gas, respectively, which are assumed to decay with characteristic lifetimes or response times after the instantaneous injection of the pulse, and a_x (units of W m^{-2} per ppb or ppm) is the radiative forcing of the gas or reference per unit increase in their atmospheric concentrations. The value of a_x is assumed to be time-independent.

The reference gas often used for relative GWPs is CO_2 because it is the major greenhouse gas. As we have seen earlier, the cycling of CO_2 throughout the earth system is complex, occurs with different response times, and is not thoroughly understood in a quantitative manner at present. As a result, uncertainties in how it decays will be translated into uncertainties in the relative GWPs of other greenhouse gases.

There may, however, be some "cancellation of errors." For example, the concentration of atmospheric CO_2 ([ref]$_t$ in Eq. (T)) depends in a nonlinear fashion on the amount of total dissolved inorganic carbon in the ocean surface layer because of the equilibria with water (see Chapter 8.B) so that relatively less atmospheric CO_2 can be taken up by the oceans as its atmospheric concentrations increase. This would leave relatively more CO_2 in the atmosphere, increasing its greenhouse effect. On the other hand, since the strongest infrared absorption bands of CO_2 are already saturated (vide supra), the radiative forcing (a_{CO_2} in Eq. (T)) decreases as its concentrations increase.

Caldeira and Kasting (1993) show that these two factors largely cancel each other so that using CO_2 as the reference gas is still useful.

For the alternatives and proposed replacements for CFCs, CFC-11 has been used in some cases as the reference compound. In interpreting the GWPs, the reader should take note of which compound has been used as the reference.

Another index has been proposed as well, a *forcing equivalent index* (*FEI*) (Wigley, 1998), defined as

$$\text{FEI}_{\text{gas}}(t) = \frac{\Delta E_{CO_2}(t)}{\Delta E_{\text{gas}}(t)},$$

where $\Delta E_{\text{gas}}(t)$ is the emissions reduction in the gas of interest calculated year-by-year that is needed to give the same change in radiative forcing as changes in CO_2 emissions, $\Delta E_{CO_2}(t)$.

Table 14.4 summarizes the estimated total direct radiative forcing calculated for the period from preindustrial times to 1992 for CO_2, CH_4, N_2O, and O_3 (IPCC, 1996). The estimate for CH_4 includes the effects due to its impacts on tropospheric ozone levels or on stratospheric water vapor, both of which are generated during the oxidation of methane. That shown for O_3 is based on the assumption that its concentration increased from 25 to 50 ppb over the Northern Hemisphere. The total radiative forcing due to the increase in these four gases from preindustrial times to the present is estimated to be 2.57 W m^{-2}.

Also shown are the relative global warming potentials, using CO_2 as the reference and for the two time horizons of 20 and 100 years, respectively (IPCC, 1996). The apparently disproportionate effects of CH_4, N_2O, and O_3 relative to CO_2 are due to the fact that CO_2 was present from natural processes in large concentrations even in preindustrial times and is such a strong

TABLE 14.4 Direct Radiative Forcings and Global Warming Potential for the Major Greenhouse Gases[a] Relative to CO_2

Gas	Concentration		Total radiative forcing[c] (W m^{-2})	Relative global warming potential time horizon[d]	
	Preindustrial	1992		20 years	100 years
CO_2	278 ppm	356 ppm	1.56	1	1
CH_4	0.7 ppm	1.71 ppm	0.47[f]	56[e]	21[e]
N_2O	275 ppb	311 ppb	0.14	280	310
O_3	10–20 ppb	30–50 ppb	0.4[b]	—	—
Total			2.57		

[a] Adapted from IPCC (1996).
[b] Assuming an increase from 25 to 50 ppb throughout the Northern Hemisphere; see discussion for range and regional effects.
[c] Change from preindustrial times to 1992.
[d] Referenced to CO_2 and assuming CO_2 concentration is constant at mid-1990s levels.
[e] Includes estimated effects of CH_4 on production of tropospheric O_3 and stratospheric H_2O.
[f] Does not include indirect effects on tropospheric O_3 and stratospheric H_2O.

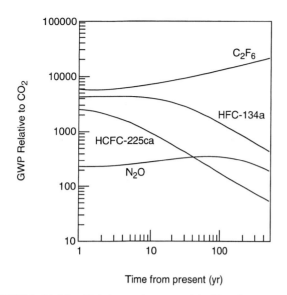

FIGURE 14.23 Global warming potentials (GWP) relative to CO_2 for N_2O, C_2F_6, HFC-134a, and HCFC-225ca as a function of time (adapted from IPCC, 1996).

absorber of infrared radiation that its absorption is already saturated at the peak of the absorption bands. As discussed earlier, the absorption due to CO_2 thus depends on the logarithm of its concentration, whereas it depends on the square root of the concentrations of the other three gases.

The importance of the time horizon over which the radiative forcing is considered is further illustrated by the data in Fig. 14.23, which shows the GWPs relative to CO_2 for N_2O, C_2F_6, HFC-134a (CH_2FCF_3), and HCFC-225ca ($CF_3CF_2CHCl_2$). The lifetime of HCFC-225ca at ~2.5 years is much shorter than for the other gases, and as a result, the GWP decreases relatively rapidly while that of N_2O, with a lifetime of ~120 years, increases over the same time frame. Thus, while the GWP of N_2O is initially an order of magnitude less than that for HCFC-225ca, it exceeds it after about four decades.

Table 14.5 shows the direct radiative forcing calculated for some CFCs, HCFC-22, and some other chlorine-containing gases due to the increase in their atmospheric concentrations from preindustrial times (when the concentrations of most of them were zero) to 1992. The direct radiative forcing due to the two most commonly used CFCs in the past, CFC-11 and CFC-12 (see Chapter 12.C.1), is approximately 8% of the total radiative forcing due to CO_2, CH_4, N_2O, and O_3 (Table 14.4).

However, the chlorinated and brominated gases exert not only *direct* radiative effects through their absorption of infrared radiation but also, as discussed in Section B.2, *indirect* effects through their destruction of stratospheric ozone (see Chapter 12). Decreases in stratospheric ozone have three effects: (1) less downward radiation across the tropopause from the lower O_3 concentrations; (2) lowered stratospheric temperatures, which also decrease the downward radiation through a decrease in the population of the emitting excited states (see Eq. (A)); and (3) an increase in the solar UV radiation reaching the earth's surface (e.g., see Ramaswamy *et al.*, 1996). The net effect is a decrease in the downward radiative flux at the tropopause, i.e., a negative radiative forcing (e.g., see Ramaswamy *et al.*, 1992). For example, Fig. 14.24 shows one estimate of the changes in the adjusted radiative forcing due to combined changes in total stratospheric and tropospheric O_3 compared to those to the other major greenhouse gases for the period from 1979 to 1994 (Hansen *et al.*, 1997b). The negative radiative forcing associated with the destruction of stratospheric ozone

TABLE 14.5 Direct Radiative Forcings for CFCs, HCFC-22, and Other Chlorine-Containing Gases[a]

Gas	Concentration (ppt)		Direct radiative forcing (W m^{-2})
	Preindustrial	1992	
CFC-11 (CCl$_3$F)	0	268	0.06[b]
CFC-12 (CCl$_2$F$_2$)	0	503	0.14
CFC-113 (CCl$_2$FCClF$_2$)	0	82	0.02
CFC-114 (CClF$_2$CClF$_2$)	0	20	0.007
CFC-115 (CF$_3$CClF$_2$)	0	<10	<0.003
HCFC-22 (CHClF$_2$)	0	100	0.02
CH$_3$Cl	~600	~600	0
CCl$_4$	0	132	0.01
CH$_3$CCl$_3$	0	135	0.007

[a] Adapted from IPCC (1996).
[b] Christidis *et al.* (1997) suggest this is too small by ~30% based on their new infrared absorption cross section measurements.

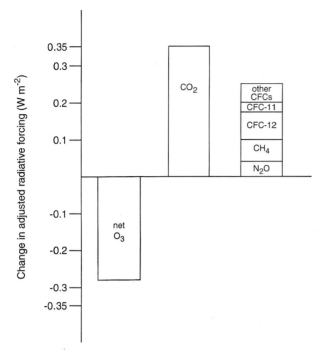

FIGURE 14.24 Calculated changes in adjusted radiative forcings from 1979 to 1994 due to changes in O_3 (both stratospheric and tropospheric based on SAGE data) and other greenhouse gases (adapted from Hansen et al., 1997b).

profile, which, of course, are related (e.g., see Zhong et al., 1996; and Shine et al., 1998). For example, using the changes in O_3 derived using a different data set (SAGE/TOMS data) gives a predicted net change in radiative forcing due to O_3 of -0.2 W m^{-2} compared to -0.28 W m^{-2} shown in Fig. 14.24. In addition, as seen later when all known anthropogenic perturbations are considered together, other estimates show the positive forcing by tropospheric ozone being larger than the negative forcing due to stratospheric ozone.

The chlorinated and brominated compounds, therefore, are expected to exert not only a *direct* positive radiative forcing (leading to warming) by their interaction with terrestrial infrared radiation in the usual manner of a greenhouse gas but also an *indirect* negative radiative forcing (leading to cooling) due to changes in stratospheric ozone which changes the radiative flux at the tropopause. The net effect is thus expected to be less than that predicted from the direct radiative forcings in Table 14.5, which only take into account the direct greenhouse gas effect.

Daniel et al. (1995), for example, have estimated the global warming potentials for a variety of chlorine- and bromine-containing gases involved in stratospheric ozone destruction, taking into account the indirect negative forcing as well as the direct positive forcing. Table 14.6 summarizes both the direct GWPs and the net values where the indirect, negative forcing due to ozone destruction has been taken into account. There are a number of simplifying assumptions that have been made in such calculations, but they are illustrative of the importance of considering the indirect as well as

counterbalances about half of the total positive forcing due to the other greenhouse gases. It should be noted that this estimated change in forcing due to the change in O_3 is quite sensitive both to the change in the vertical profile and to the change in the temperature

TABLE 14.6 Direct and Net Global Warming Potentials for Some Chlorine- and Bromine-Containing Gases and Halons Relative to CO_2[a]

	Time horizon			
	2010		2090	
Gas	Direct	Net	Direct	Net
CFC-11 (CCl_3F)	4360	1900	3170	1120
CFC-12 (CCl_2F_2)	6930	5720	6750	5540
CFC-113 (CCl_2FCClF_2)	4390	2940	4020	2630
HCFC-22 ($CHClF_2$)	3740	3430	1330	1200
CH_3Cl	26.6[b]	—	8.1[b]	—
CCl_4	1730	−1290	+1150	−1210
CH_3CCl_3	+320	−670	+90	−200
Halon-1301 (CF_3Br)	+5420	−19,740	+4460	−18,140
Halon-1211 (CF_2ClBr)	—	−18,960	—	−11,050
CH_3Br	12.8[b]	−4280	3.9[b]	−1190

[a] Except where noted, adapted from Daniel et al. (1995). Calculations based on a bromine enhancement factor of $\alpha = 40$ globally (see Chapter 12.D for discussion of α) and assuming phase-out of emissions as scheduled in the Copenhagen amendments to the Montreal Protocol (see Chapter 13.A for a description of these); note that WMO (1999) recommends $\alpha = 60$.

[b] From Grossman et al. (1997).

the direct effects of anthropogenic emissions. For example, as discussed in Chapter 12.D, atom for atom, bromine is much more effective in destroying stratospheric ozone than is chlorine. This leads to large changes in the predicted GWPs when the indirect effects of ozone destruction are taken into account. For example, the relative GWP for Halon-1301 (CF_3Br) changes from net heating for the direct effect to net cooling when the indirect effect of ozone destruction is taken into account.

Figure 14.25 illustrates the estimated relative contributions to the direct heating effect and to the indirect cooling effect in 1990 and 2040, respectively (Daniel et al., 1995). The enhanced impact of the halons on cooling is particularly apparent. The net contribution of halocarbons through about 2080 is still estimated by Daniel et al. (1995) to be significant, however, in the range of 0.15–0.25 W m^{-2}.

Such calculations also illustrate that the net effects of ozone-depleting gases on climate should change as a function of time due to changes in the chemistry. For example, Daniel et al. (1995) assume in their calculations that the stratospheric ozone loss prior to 1980 was negligible. As a result, the contribution of CFCs to radiative forcing until 1980 was estimated to be positive, in the range +0.05 to +0.10 W m^{-2} per decade. This rate of increase in radiative forcing should then have fallen as the indirect negative radiative forcing from ozone destruction came into play. Figure 14.26

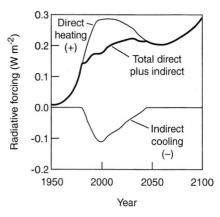

FIGURE 14.26 Estimated direct tropospheric heating (+) effects of chlorine- and bromine-containing gases and indirect cooling (−) effects due to their destruction of stratospheric ozone, assuming a bromine enhancement factor of $\alpha = 40$ (from Daniel et al., 1995).

illustrates their calculated temporal changes due to these two effects. As stratospheric ozone depletion becomes less severe in the future, the indirect, negative radiative forcing that has been in part counterbalancing the positive forcing will decrease; as a result, the net change in radiative forcing in the coming decades due to halocarbons is expected to be more steep than if their only effect was through direct positive radiative forcing, leading to heating (Solomon and Daniel, 1996; Hansen et al., 1997b).

As discussed in Chapter 13, the short- and long-term replacements for CFCs and halons in use or contemplated for use at the present time are typically compounds containing hydrogen (to shorten their tropospheric lifetimes, hence decreasing the amounts that reach the stratosphere) and/or fluorine (which does not participate in stratospheric ozone destruction to a significant extent). One concern with respect to these substitutes is their potential effect on climate since they are also greenhouse gases. Table 14.7 summarizes direct radiative forcings calculated for some of these compounds, as well as relative GWPs using either CFC-11 or CO_2 as the reference compound. Clearly, their contributions to radiative forcing may be significant and hence are taken into account in assessing the overall impacts of CFC replacements.

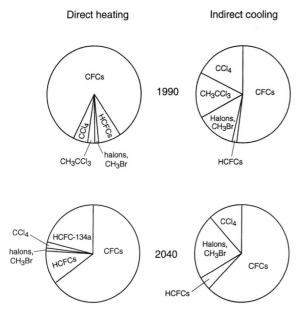

FIGURE 14.25 Partitioning of estimated direct tropospheric heating effect of stratospheric ozone-depleting gases in the years 1990 and 2040 and of the indirect cooling effect due to stratospheric ozone depletion (from Daniel et al., 1995).

C. AEROSOL PARTICLES, ATMOSPHERIC RADIATION, AND CLIMATE CHANGE

Aerosol particles have been thought for many years to play a role in global climate. For example, as discussed in Section E.1, while the mean global surface

TABLE 14.7 Direct Radiative Forcings per Unit Concentration for Some Fluorinated Compounds[a] and Relative Global Warming Potentials Taking GWP(CFC-11) = 1.0[b,c] or GWP (CO_2) = 1.0

Gas	Radiative forcing per ppb (W m^{-2} ppb^{-1})	Time horizon			
		GWP relative to CFC-11		GWP relative to CO_2	
		20 years	100 years	20 years	100 years
CFC-11 ($CFCl_3$)	d	(1.0)	(1.0)	(1.0)	(1.0)
HFC-134a (CH_2FCF_3)	0.17	0.58	0.29	3400	1300
HFC-134 (CF_2HCF_2H)	0.18	0.50[c]	0.22[c]	2900	1000
HFC-125 (CHF_2CF_3)	0.20	0.99	0.83	4600	2800
HFC-152a (CH_3CHF_2)	0.11	0.093	0.036	460	140
HFC-23 (CHF_3)	0.18	2.0	3.2	9100	11700
HFC-32 (CH_2F_2)	0.11	0.47	0.18	2100	650
HFC-41 (CH_3F)	0.02	0.10	0.038	490	150
HFC-227ea (C_3HF_7)	0.26	0.81	0.73	4300	2900
HFC-245ca ($C_3H_3F_5$)	0.20			1800	560
HFC-236fa ($CF_3CH_2CF_3$)	0.24			5610[e]	6940[e]
HFC-236ea ($CF_3CHFCHF_2$)				2200[e]	710[e]
CF_4	0.10[f]	0.88[f]	1.70[f]	3660[f]	4690[f]
C_2F_6	0.22[f]	1.26[f]	2.37[f]	5210[f]	6650[f]
C_3F_8	0.23[f]	0.97[f]	1.88[f]	4040[g]	5150[f]
SF_6	0.64			16300	23900

[a] Adapted from IPCC (1996). For recent computations that are in reasonable agreement with these values, see Papasavva *et al.* (1997).
[b] GWPs relative to CFC-11 = 1.0 from Pinock *et al.* (1995) except where indicated. GWPs relative to CO_2 from IPCC (1996) assuming current CO_2 concentrations.
[c] From Christidis *et al.* (1997) relative to CFC-11 = 1.0.
[d] Reported as 0.22 W m^{-2} in IPCC (1996). Christidis *et al.* (1997) report a revised value of 0.285 W m^{-2} ppb^{-1}.
[e] From Gierczak *et al.* (1996).
[f] From Roehl *et al.* (1995).

temperature has increased by ~0.3–0.6°C over the past century, the period from 1940 to the mid-1970s showed no such trend (IPCC, 1996). Indeed, this period was characterized by cooler than normal temperatures, which has often been qualitatively attributed to the scattering of incoming solar radiation by pollution-derived aerosol particles. In addition, as discussed in Chapter 9.A.4, elemental carbon in the particles, and perhaps some organics as well, also absorb solar radiation. A third effect discussed herein is the absorption and emission of long-wavelength infrared radiation by the inorganics found in mineral dust. These are the so-called *direct effects* of aerosol particles on global climate.

In addition, aerosol particles have *indirect effects*. The most important of these is their effect on cloud properties, since clouds obviously also have major effects on climate. In addition, since heterogeneous chemistry can occur on aerosol particles (see Chapter 5), it is possible that such chemistry can alter the concentrations of other contributors to the climate system, such as the greenhouse gases. One example is the formation of N_2O from reactions of HONO on the surface of aerosol particles (see Chapter 7.C).

As will be evident from the following, this is an area in which there are a number of uncertainties and which at the time of writing is a very active area of research. For reviews, see Penner *et al.* (1994), Schwartz (1994), Andreae (1995), Charlson and Heintzenberg (1995), IPCC (1996), National Research Council (1996), and Andreae and Crutzen (1997).

1. Direct Effects

a. Scattering of Solar Radiation

One of the important properties of aerosol particles is their ability to scatter light. The diameter of many airborne particles is approximately that of the wavelength of visible light, so that Mie scattering occurs. (As the particles become smaller, light scattering approaches the Rayleigh limit.) As discussed in Chapter 9.A.4, the intensity of Mie light scattering is a complex function of the wavelength of the incident light, the size and composition of the particle, and the scattering angle. However, for spherical particles of known composition and size, Mie theory can be used to predict the fraction of incident light that is scattered in various directions. One can therefore estimate how much light is scattered back in the upward direction into space, which is the important parameter for aerosol particle direct scattering effects on global climate. This cooling effect due to direct scattering of light back out to space has been dubbed the "whitehouse effect" (Schwartz, 1996).

The net effect of such scattering depends on the underlying surface (Andreae et al., 1995; Haywood and Shine, 1997). If the underlying surface is dark (e.g., oceans), backscattering by aerosol particles increases the planetary albedo, leading to cooling. However, if the surface is already highly reflecting, with an albedo greater than about 0.5, e.g., snow, scattering by particles can lead to a decrease in net reflection, especially at small solar zenith angles. The reason for this is that part of the light reflected from the surface is backscattered by the aerosol particles. This light then travels back through part of the atmosphere, undergoing enhanced absorption by particles (as well as gases), before being reflected as well as partly absorbed again at the surface. This increased opportunity for absorption due to multiple scattering results in a net decrease in the reflectance and therefore a positive value for the radiative forcing (Haywood and Shine, 1997). However, given what is currently known about aerosols and typical surface albedos, backscattering is believed to predominate on a global basis (Charlson et al., 1992b; Haywood and Shine, 1997).

As seen in Chapter 9, aerosol particles can, indeed most often do, have many chemical components and a wide range of particle sizes. However, sulfate is one of the most common species found in particles worldwide, being generated not only from the oxidation of anthropogenically derived SO_2 but also from the oxidation of sulfur compounds produced in biological processes (see Chapter 8). However, in terms of global climate *change* caused by human influences, estimation of the effects of sulfate particles generated from the SO_2 emissions, 90% of which occur in the industrialized Northern Hemisphere, has been of greatest interest. As we shall see, however, recent studies suggest that carbonaceous aerosol particles and soil dust may also be quite important (e.g., Tegen et al., 1997).

Over the past several decades, there have been a number of estimates of scattering of light by tropospheric sulfate particles, with the aim of quantifying the magnitude of the expected cooling effect due to scattering of solar radiation compared to warming due to the greenhouse gases discussed earlier (e.g., see Rasool and Schneider, 1971). One of the simplest approaches first taken was a box model in which the change in the short-wavelength radiative forcing averaged over the globe, ΔF_R, is treated in terms of a combination of effects/terms (Charlson et al., 1991, 1992a; Penner et al., 1994):

$$\Delta F_R = -\left(\frac{1}{4}F_T\right)(1 - A_c) \times \left[2T^2(1 - R_s)^2 \beta\right] \alpha^{RH} f^{RH} B_{SO_4^{2-}}. \quad (U)$$

The negative sign means that the direct effect of aerosols is expected to result in cooling of the troposphere, rather than heating as do the greenhouse gases discussed earlier. The terms in Eq. (U) are summarized in the following; also shown are the percentages of the total radiative forcing attributed to each factor by Charlson et al. (1992a) and Penner et al. (1994) and the uncertainty factor they assigned to each. (Subsequent estimates using global climate models will be discussed shortly.)

- The incoming solar intensity: $1/4 F_T$, where F_T is the average total incoming solar light intensity per unit area normal to the direction of propagation outside the earth's atmosphere, 1368 W m^{-2}, and the 1/4 factor takes into account that the solar energy is spread over the entire surface of the earth (see Fig. 14.2a).
- The fraction of the earth that is not covered by clouds, which is where increased scattering by aerosol particles is expected to be most important (see later, however, for a brief discussion of this assumption): $(1 - A_c) = 0.39$ with an uncertainty of a factor of 1.1, where A_c is the fraction covered by clouds.
- The fraction of incident radiation that is scattered upward (β), the fraction (T) of light transmitted through the atmosphere above the aerosol layer, and the surface albedo (R_s, which is the fraction of incident light that is reflected at the earth's surface), including multiple reflections between the surface and the aerosol layer: $2T^2(1 - R_s)^{2\beta}$. Charlson et al. (1992a) and Penner et al. (1994) estimated $T^2 = 0.58$ with an uncertainty factor of 1.4, $(1 - R_s)^2 = 0.72$ with an uncertainty factor of 1.2, and $\beta = 0.3$ with an uncertainty factor of 1.3. A model intercomparison study by Boucher et al. (1998) shows that the linearity in $(1 - R_s)^2$ holds up to $R_s \sim 0.2$–0.4; above this, model predictions for the dependence on this term diverge significantly.
- The light scattering efficiency of sulfate at a particular relative humidity (RH) chosen as a reference: α^{RH}; this was estimated as 4.8×10^2 m^2 mol^{-1} with an uncertainty of 1.4.
- The increase in scattering due to an increase in the RH above the reference value: f^{RH}; this was estimated as 1.7 with an uncertainty factor of 1.2.
- The amount of atmospheric aerosol sulfate, expressed as the column burden: $B_{SO_4^{2-}}$. This term can

be expressed as a combination of the source strength of SO_2 emitted and confined to an area A, Q_{SO_2} in moles of S per year, the fraction that is oxidized to sulfate particles, $Y_{sulfate}$, and the lifetime for sulfate particles in the atmosphere, $\tau_{sulfate}$:

$$B_{SO_4^{2-}} = Q_{SO_2} Y_{sulfate} \tau_{sulfate} / A. \qquad (V)$$

In the Charlson et al. (1992a) and Penner et al. (1994) estimates, $B_{SO_4^{2-}} = 3.3 \times 10^5$ mol m^{-2}, with uncertainties in Q_{SO_2}, $Y_{sulfate}$, and $\tau_{sulfate}$ of 1.15, 1.5, and 1.5, respectively. (Note, however, that some subsequent studies have suggested that direct radiative forcing may vary nonlinearly with the sulfate concentration due to chemical interactions with other particle constituents; e.g., see West et al., 1998.)

The net results of such calculations vary depending on the input assumptions. For example, use of the inputs cited above for each of the terms gives $\Delta F_R = -0.9$ W m^{-2}, with a possible range from -0.4 to -2 W m^{-2}. However, application of different models gives different absolute values for the radiative forcing. For example, Penner et al. (1998) calculate global average values from -0.55 to -0.81 W m^{-2}, depending on the assumptions regarding the effects of relative humidity.

Boucher et al. (1998) compared the results for direct radiative forcing by sulfate particles predicted by 12 different groups using 15 different models. Overall, the predicted forcings for various input assumptions of particle radii, surface albedo, and solar zenith angle were similar. The largest differences were for high surface albedos; the range of predicted globally averaged direct forcing by particles with 0.17-μm radius over a surface with albedo 0.05–0.15 was 27% (expressed as a percentage of the median value).

The largest uncertainties are associated with the aerosol radiative properties β, α^{RH}, and f^{RH}, which is particularly sensitive to the treament of the dependence on relative humidity (vide infra), and with the amount of sulfate (i.e., in $B_{SO_4^{2-}}$) available to scatter light. Other sensitivity analyses also suggest that uncertainties in $Y_{sulfate}$ and $\tau_{sulfate}$, and hence in $B_{SO_4^{2-}}$, as well as in the relative humidity factor are large sources of uncertainty (e.g., see Pan et al., 1997). Not included in Eq. (U) is the important fact that sulfate is not distributed evenly over the globe but is concentrated in urban/industrialized areas of the Northern Hemisphere and in downwind regions. As a result, there is a great deal of spatial as well as temporal variability in the direct forcing expected from sulfate aerosol (see later discussion). In addition, it should be noted that there are correlations between some of the terms in Eq. (U), so they do not all vary independently.

Recall from Chapter 9.A.4 that Mie scattering of light favors the forward direction and depends on particle size and composition and the wavelength of light. The fraction of light that is scattered in an upward direction relative to the earth's surface is therefore expected to be a strong function of the direction of the incident light, i.e., the solar zenith angle. This is illustrated in Fig. 14.27, which shows calculated values of β as a function of the solar zenith angle and particle radius r for light at 550 nm. As expected, the fraction of light scattered upward is smallest at a small zenith angle (i.e., overhead sun) and highest ($\sim 50\%$) for the sun on the horizon. It is also high for the smallest particles. It should also be noted that most calculations to date of the backscattered fraction (β) assume spherical aerosol particles, which may not be the case, particularly at low RH, where dry particles have shapes determined by their crystal structures. For example, Pilinis and Li (1998) have shown that this assumption can cause significant underestimation (as much as a factor of three) of the calculated direct aerosol forcing, particularly at small solar zenith angles.

Because the index of refraction enters into Mie scattering calculations, one would expect that the light scattering efficiency at a particular RH, α^{RH}, would depend on the particle composition as well as radius. As discussed in Chapter 9.A.4, this has indeed been shown to be the case. However, given that tropospheric particles are a complex mixture of inorganic and organic compounds, it is not clear that the linear and independent relationship between light scattering and the mass scattering coefficient for each species described by Eq. (EE) in Chapter 9 holds (Sloane, 1986; White, 1986; Hegg et al., 1993, 1994).

Water uptake affects light scattering by particles due to changing the particle size and the index of refraction. This has been discussed in detail in Chapter 9.B.4 and the reader is referred to that section for more

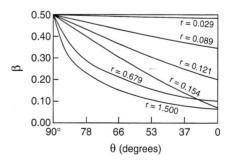

FIGURE 14.27 Calculated fraction of 550-nm light scattered upward (β) as a function of solar zenith angle (θ) and particle radius (μm). The refractive index of the particles is 1.4 (adapted from Nemesure et al., 1995).

details. The important point is that the combination ($\alpha^{RH} f^{RH}$) is a very sensitive function of RH, and indeed, this appears to be the most important aerosol parameter in direct forcing (Nemesure et al., 1995; Pilinis et al., 1995). Figure 14.28 shows one calculation of $\alpha^{RH} f^{RH}$ as a function of RH for $(NH_4)_2SO_4$, which is a common form of sulfate in tropospheric aerosols (Nemesure et al., 1995). The effect of relative humidity is large, increasing from ~ 1.5 m^2 per g of sulfate for dry particles to ~ 50 m^2 per g at 97% RH at a wavelength of 600 nm. Finally, as expected from Mie theory and demonstrated in Fig. 14.28, the light scattering efficiency varies widely over the wavelength region of interest in the troposphere.

The total emissions of SO_2 due to anthropogenic processes are believed to be relatively well known (see Chapter 2.A.4). A great deal is also known about the processes oxidizing tropospheric SO_2 to sulfate (see Chapter 8). In brief, while OH oxidizes SO_2 in the gas phase, oxidation in the liquid phase of fogs and clouds is generally more important. In the latter case, O_3 is a major oxidant at high pH values, but as the droplet becomes acidified, this slows down. Over most of the pH range characteristic of particles in the troposphere, H_2O_2 is an effective oxidant and hence is believed to be the major contributor to sulfate formation. S(IV) may also be oxidized in the marine boundary layer by HOCl and HOBr (Vogt et al., 1996). This means that the fraction of SO_2 forming sulfate will depend on such factors as the availability of liquid phase in the form of fogs and clouds, as well as the availability of oxidants such as H_2O_2, which is itself a complex function of the VOC–NO_x chemistry discussed throughout this book. In addition, a large contributor to the removal of sulfate is wet deposition, which, due to its spatial and temporal variability, results in a corresponding variability in the tropospheric lifetime of sulfate. Because of all of these factors, there is significant uncertainty associated with both $Y_{sulfate}$ and $\tau_{sulfate}$, and hence in the column burden of sulfate used in such calculations.

Finally, the term $(1 - A_c)$ in Eq. (U), where A_c is the fraction of the earth's surface covered by clouds, contains the implicit assumption that the direct scattering by aerosol particles is only significant in cloud-free regions. It is not clear, however, that this is the case. For example, while Haywood et al. (1997a) and Haywood and Shine (1997) report only a small contribution (5%) to aerosol particle direct forcing in cloudy regions, Boucher and Anderson (1995) report a much larger effect, with 22% of the total arising from cloudy areas. Other studies fall in between these values (e.g., see Haywood and Ramaswamy, 1998). For example, one calculation using a 1-D model suggests that cirrus clouds above a layer of particles of $(NH_4)SO_4$ can enhance the direct forcing at a solar zenith angle of 0° from -2.0 to -3.3 W m^{-2} for their assumed set of conditions (Liao and Seinfeld, 1998). This is due to the fact that the overlying cloud scatters the incoming beam so that the light incident on the aerosol layer is effectively at larger solar zenith angles than 0°, leading to increased upward scattering (see the dependence of the upscattered fraction, β, on solar zenith angle due to enhanced forward Mie scattering by particles, Fig. 14.27). A similar enhancement is predicted for a cloud layer of 100-m thickness assumed to be at an altitude of 900 m and embedded in a layer of aerosol taken to be uniformly distributed from 0 to 5 km. However, for much thicker clouds (e.g., 1000 m), the direct aerosol forcing due to sulfate may be decreased due to partial blocking of the incoming solar light intensity by the cloud (Liao and Seinfeld, 1998).

In short, the approach summarized in Eq. (U) provided a useful first approach to establishing that direct scattering of light by sulfate particles could be important in counterbalancing the expected warming due to the increased greenhouse gases. However, given the large number of variables that enter into each term in this equation and their spatial and temporal variability, the development and application of more sophisticated models are clearly needed. For some examples of such models, see Kiehl and Briegleb (1993), Taylor and Penner (1994), Cox et al. (1995), Mitchell et al. (1995), Boucher and Anderson (1995), Meehl et al. (1996), C. C. Chuang et al. (1997), Haywood and Shine (1997), Haywood et al. (1997a, 1997b), van Dorland et al. (1997), Tegen et al. (1997), Schult et al. (1997), Haywood and Ramaswamy (1998), and Penner et al. (1998).

For example, Fig. 14.29 shows one calculation of direct radiative forcing using a global climate model, GCM (Penner et al., 1998). Due to the preponderance of anthropogenic SO_2 emissions in the Northern Hemisphere, the direct radiative forcing due to sulfate

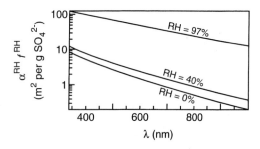

FIGURE 14.28 Calculated mass scattering efficiency term ($\alpha^{RH} f^{RH}$) as a function of wavelength for 0, 40, and 97% RH. Particle dry radius taken to be 0.096-μm $(NH_4)_2SO_4$ (adapted from Nemesure et al., 1995).

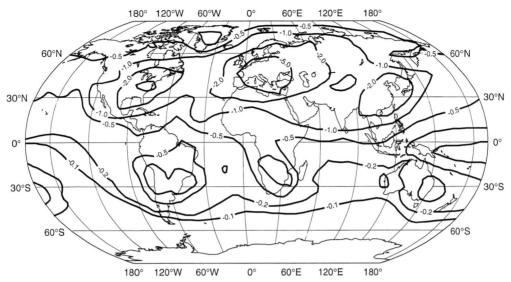

FIGURE 14.29 Calculated direct radiative forcing due to sulfate aerosol particles (adapted from Penner *et al.*, 1998).

aerosol is also predicted to occur in the Northern Hemisphere, particularly over the eastern United States, central and eastern Europe, and southeastern Asia. This model predicts a global average direct sulfate aerosol forcing of -1.2 W m^{-2} in the Northern Hemisphere compared to only -0.26 W m^{-2} in the Southern Hemisphere, where less than 10% of the anthropogenic sulfur emissions occur. Haywood and Ramaswamy (1998) predict very similar average values for the direct radiative forcing by sulfate, -1.4 W m^{-2} in the Northern Hemisphere and -0.24 W m^{-2} in the Southern Hemisphere, although there are some quantitative differences in the specific geographical dependencies.

It should be noted that the magnitude of the predicted forcing is quite sensitive to treatment of relative humidity (RH) in the model because of the effects on particle size and optical properties (e.g., Haywood and Shine, 1995; Haywood and Ramaswamy, 1998; Ghan and Easter, 1998; Haywood *et al.*, 1998a; Penner *et al.*, 1998). For example, in the calculations by Penner *et al.* (1998), when the particle properties were held fixed at the values for 90% RH for 90–99% RH, the predicted direct radiative forcing for sulfate particles decreased from -1.18 W m^{-2} to -0.88 W m^{-2} for the Northern Hemisphere and from -0.81 to -0.55 W m^{-2} globally.

Volcanic eruptions provide one test of the relationship between light scattering by sulfate particles and the resulting change in temperature, since they generate large concentrations of sulfate aerosol in the lower stratosphere and upper troposphere. These aerosol particles cause tropospheric cooling by backscattering solar radiation out to space. In principle, they can also cause tropospheric and stratospheric warming by absorbing and reemitting the long-wavength terrestrial infrared (McCormick *et al.*, 1995; Robock and Mao, 1995); calculations show that the warming increases substantially with particle size and should be sufficiently large to counterbalance the cooling from light scattering at area-weighted mean radii larger than ~2 μm (Lacis *et al.*, 1992). However, since the sulfate particles formed by gas-to-particle conversion of SO$_2$ are sub-μm (e.g., see Fig. 12.28), the backscattering effect leading to cooling generally predominates in the troposphere. Model calculations predicted that surface cooling of ~0.5°C should have occurred after the Mount Pinatubo eruption due to the scattering of solar radiation back out to space, and indeed, this is what was observed (e.g., Hansen *et al.*, 1992; Dutton and Christy, 1992; Minnis *et al.*, 1993; Lacis and Mishchenko, 1995; Robock and Mao, 1995; McCormick *et al.*, 1995; Parker *et al.*, 1996; Saxena and Yu, 1998).

Figure 14.30, for example, shows the predicted and measured monthly surface temperatures obtained prior to and following the eruption, demonstrating excellent agreement between observations and the model; the observed stratospheric warming is also in excellent agreement with model calculations (e.g., Lacis and Mishchenko, 1995; Angell, 1997). Temperature trends on a smaller, regional scale have also been attributed to the effects of such volcanic eruptions, although obtaining statistically robust results is difficult in these cases due to natural variability (e.g., Saxena *et al.*,

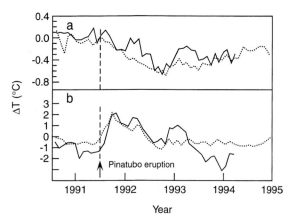

FIGURE 14.30 Measured (—) and model-predicted (···) change in monthly mean temperatures (a) at the earth's surface and (b) in the stratosphere (observations at 30 mbar and 10°S, model results for the 10- to 70-mbar layer from 8°S to 16°S) (adapted from Lacis and Mishchenko, 1995).

1997). Additional discussion of the effects of volcanic eruptions on surface temperatures is found in Section D.3.

Finally, while sulfate is a major component of aerosol particles, particularly in the Northern Hemisphere where fossil fuel burning is extensive, it is certainly not the sole particle component. As discussed in detail in Chapter 9, nitrate, organics and elemental carbon, inorganic soil elements, and sea salt components are all found in tropospheric particles to varying extents, depending on the region. Some regional sources such as biomass burning may also have quite widespread and global impacts (e.g., Kaufman *et al.*, 1991; Penner *et al.*, 1992). These other sources will also scatter light and contribute to direct aerosol forcing (e.g., see Andreae *et al.*, 1995). While less attention has been paid to assessing the contribution of these other components (Penner *et al.*, 1994), there are increasing indications that these may prove to be quite important in direct forcing by aerosol particles (e.g., see Andreae *et al.*, 1995; Penner *et al.*, 1998).

For example, Fig. 14.31 shows one set of calculations of direct radiative forcing by sulfate particles as well as those from biomass burning and from fossil fuel combustion in the Northern Hemisphere, Southern Hemisphere, and the global average, respectively (Penner *et al.*, 1998). In this case, the fossil fuel particles were assumed to contain both black carbon, which absorbs radiation and hence has a positive radiative forcing (see following section), and organic carbon, which scatters light (negative radiative forcing). The biomass particles were assumed to take up water as if they contained 30% $(NH_4)_2SO_4$ by mass and to scatter light. Figure 14.31 illustrates that the direct radiative forcing

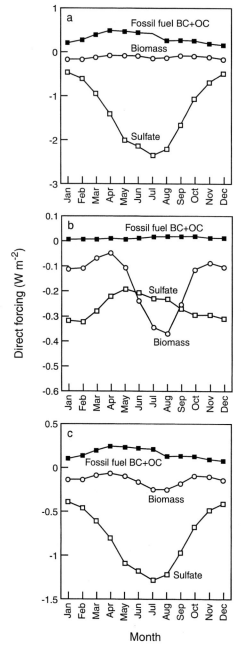

FIGURE 14.31 Calculated direct radiative forcing by sulfate, biomass, and fossil fuel black carbon (BC) + organic carbon (OC) particles in the (a) Northern Hemisphere, (b) Southern Hemisphere, and (c) global average (adapted from Penner *et al.*, 1998).

by biomass particles is predicted to be large in the Southern Hemisphere and negative because of their contribution to light scattering. However, over some regions, e.g., the Sahara desert, such particles are predicted to lead to positive radiative forcing because they are over an already highly reflecting surface (see the earlier discussion). (Note that the size distribution assumed for the particles affects the absolute value predicted for radiative forcing; for example, input of a different size distribution than was used for Fig. 14.31 for the biomass particles gave a minimum direct radiative forcing in the Southern Hemisphere of -0.52 W m^{-2}, significantly more negative than shown in Fig. 14.31b.) The contribution of fossil fuel black and organic carbon leads to a positive radiative forcing because of the absorption of solar radiation by black carbon (vide infra).

Organic constituents of particles are also likely to make a contribution to light scattering. For example, Li et al. (1998) studied aerosol particles over the east coast of Canada and found that unidentified species, likely organics, account for a large fraction ($\sim 2/3$) of the mass in the 0.1- to 1-μm range. In air masses with origins over the continental United States, these unidentified compounds were calculated to be responsible for 45–80% of the direct backscatter coefficient. There is also evidence of new particle formation likely involving organics from biological processes in coastal regions. For example, nucleation of new particles in the ultrafine size range (1.5–5 nm) has been observed in coastal regions at Mace Head, Ireland, and the Outer Hebrides and was correlated to solar radiation and low tide (O'Dowd et al., 1998). This suggests that photochemical processes may lead to new organic particle formation and that the precursors may be biogenics from the shore regions.

The contribution of carbonaceous components to tropospheric aerosols off the east coast of the United States has been reported to vary from $\sim 10\%$ at low altitudes to $>90\%$ of the total aerosol mass at altitudes of ~ 3 km (Novakov et al., 1997). These carbonaceous components include both organics and elemental carbon, the latter estimated to be $\sim 10\%$ of the total carbonaceous mass. The larger fraction at higher altitudes may reflect more rapid removal of the inorganic components such as sulfate and nitrate through wet deposition at the surface. These carbonaceous materials contributed 66 ± 16% of the light scattering coefficient (Hegg et al., 1997), i.e., were major contributors to the direct effect of aerosol particles. Figure 14.32 shows some typical contributions to the light scattering measured during these studies (Hegg et al., 1997). While the total scattering, shown here in terms of the aerosol optical depth, varied from one set of measurements to another, it can be seen that liquid water in the particles, carbonaceous compounds, and sulfate were consistently the largest contributors, and in that order. Light absorption was a minor contributor to the aerosol optical depth in these studies. Since these measurements were made close to the earth's surface, the relative importance of organics when averaged over the troposphere may, however, be somewhat less.

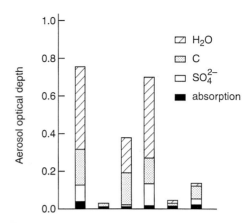

FIGURE 14.32 Typical contributions of aerosol liquid water, carbonaceous compounds, and sulfate in the lower troposphere of the east coast of the United States to the total aerosol optical depth. The contribution of light absorption is also shown. The different bars represent different sets of measurements during different flights (adapted from Hegg et al., 1997).

It should be noted that while the focus of most of these studies has been on the contributions of species that are believed to have changed over time, e.g., those associated with human activities, there are also contributions due to natural processes. The latter are presumably not changing with time, or at least not at a significant rate compared to those due to anthropogenic activities (see discussion in Section E, however). For example, the carbonaceous aerosols measured by Novakov et al. (1997b) and Hegg et al. (1997) over the east coast of the United States may have been largely natural in origin (see Chapter 9.C.2) and hence not expected to show secular changes associated with anthropogenic activities. Similarly, Murphy et al. (1998) have measured the size-dependent composition of aerosol particles in the marine boundary layer in the Southern Ocean and shown that the contribution of sea salt particles to backscattering dwarfs that due to non-sea salt sulfate (nss). In addition, these particles comprised more than 50% of the cloud condensation nuclei, CCN (vide infra). (Note, however, that since the concentration of sea salt particles is highest in the marine boundary layer, this contribution to backscattering will not be characteristic of the global troposphere.) Sea salt particles are generated by natural processes, so that this contribution to backscattering has undoubtedly always been present and will not contribute to a secular trend. However, understanding its contribution, as well as those from other natural processes, in scattering of solar radiation is important to place the contribution of anthropogenically derived aerosol particles in perspective. In addition, as discussed later with respect to indirect effects of sea salt particles, the effects of such natural aerosols often act in a synergistic manner with anthropogenic aerosol particles.

In short, the direct effects of aerosol particles in terms of backscattering solar radiation out to space and hence leading to cooling are reasonably well understood qualitatively and provided the aerosol composition, concentrations, and size distribution are known, their contribution can be treated quantitatively as well. However, major uncertainties exist in our knowledge of the physical and chemical properties, as well as the geographical and temporal variations, of aerosol particles and it is these uncertainties that primarily limit the ability to accurately quantify the direct effects at present.

b. Absorption of Solar Radiation

Depending on their chemical composition, aerosol particles can not only scatter incoming solar radiation but, in some cases, absorb it as well. This absorbed energy is converted to heat, which can contribute to warming of the troposphere. In addition, since energy absorbed by such particles does not reach the surface but heats the atmosphere directly, changes in the lapse rate may result as well, and this can contribute to global change by altering atmospheric circulation patterns (e.g., see Penner *et al.*, 1994; and Tegen *et al.*, 1997).

Although sulfate aerosols do not appreciably absorb incoming solar radiation, elemental or black (graphitic) carbon particles do. In addition, mineral dust particles absorb in the visible, primarily due to the presence of iron compounds such as hematite (Fe_2O_3) (Patterson, 1981; Sokolik *et al.*, 1993). As discussed in Chapter 9.A.4, some recent studies suggest there may be a previously unrecognized, but substantial, contribution of complex organics to light absorption (Malm *et al.*, 1996). However, since relatively little is known about the nature and concentrations of these complex organics, and the dust contribution to solar radiation absorption is generally assumed to be small compared to that of elemental carbon (which may not be justified in some regions; e.g., see Patterson, 1981), we focus here on the contribution of elemental carbon.

As described in Chapter 9 (see also Penner and Novakov, 1996), tropospheric particles containing carbon are often referred to as "carbonaceous aerosols." The form of the carbon may be organic or elemental, the latter often being referred to as graphitic or black carbon due to its strong absorption of visible light. The expression given in Eq. (U) for direct radiative forcing by aerosol particles can be modified to include contributions due to absorption (Chýlek and Wong, 1995):

$$\Delta F_R = -\left(\frac{1}{4}F_T\right)(1 - A_c)T^2$$
$$\times \left[(1 - R_s)^2 2\beta\tau_{scat} - 4R_s\tau_{abs}\right], \quad (W)$$

where $\tau_{scat} = \alpha^{RH}f^{RH}B_{SO_4^{2-}}$ is the effective aerosol optical scattering depth and τ_{abs} is the optical depth due to absorption. For no absorption, $\tau_{abs} = 0$ and Eq. (W) reduces to Eq. (U) as expected. Note that for absorption when scattering is small, ΔF_R becomes positive, i.e., warming results. In short, carbonaceous aerosols can both scatter solar radiation, causing negative radiative forcing, and absorb light, leading to positive forcing. The net effect depends on the composition of the aerosols as well as their particle size and vertical distribution.

Several groups have carried out detailed global 3-D model studies of carbonaceous aerosol particles and their effects on radiation balance (e.g., Liousse *et al.*, 1996; Haywood and Ramaswamy, 1998; Penner *et al.*, 1998). For example, Figure 14.33 shows the model estimates by Penner *et al.* (1998) for direct radiative forcing due to aerosol particles from fossil fuel combustion. These were assumed to include both black carbon, which absorbs solar radiation, and organic carbon, which scatters it. The black carbon over Europe, Asia, and, to a lesser extent, the eastern United States leads to positive radiative forcing in these areas due to direct absorption of solar radiation. Similar results are predicted by other models, although the absolute values of the radiative forcing may differ somewhat, depending on the details of the particular model used and the emission inventory chosen for the black carbon (e.g., see Schult *et al.*, 1997; Haywood and Ramaswamy, 1998; and Penner *et al.*, 1998).

Figure 14.34 shows one model prediction for the net radiative forcing due to a combination of sulfate, biomass, and fossil fuel particles containing both black and organic carbonaceous compounds (Penner *et al.*, 1998). The net result is predicted negative radiative forcing over the industrialized areas of the Northern Hemisphere due to aerosol particles (remember this does not include the positive forcing due to greenhouse gases). Positive radiative forcing is predicted over the highly reflecting ice-covered surfaces at high latitudes and over a small portion of Asia. The predicted average values of radiative forcing were -0.65 to -1.07 W m^{-2} for the Northern Hemisphere, -0.33 to -0.51 W m^{-2} for the Southern Hemisphere, and -0.51 to -0.88 W m^{-2} for the global average.

Jayaraman *et al.* (1998) measured the aerosol optical depth, aerosol size distribution, and the solar flux close to the coast of India, over the Arabian Sea, and then

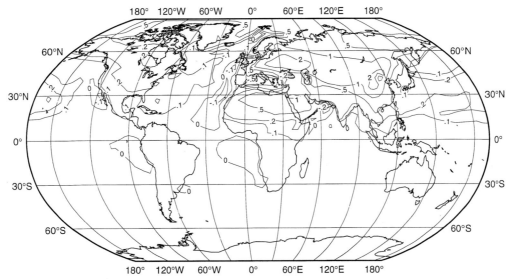

FIGURE 14.33 Calculated direct radiative forcing for the combination of black carbon and organic carbon from fossil fuel combustion. The numbers shown are the lower limits of the ranges included in each area. The boundaries of the regions are ± 10, ± 5, ± 2, ± 1, ± 0.5, ± 0.2, ± 0.1, and 0. Thus -5 represents the -5 to -10 W m^{-2} region, -2 represents the -2 to -5 W m^{-2} region, etc. (adapted from Penner et al., 1998).

over the more remote Indian Ocean. The aerosol optical depth increased from 0.1 or less in the remote region to 0.2–0.4 over the Arabian Sea to values up to 0.5 close to the Indian coast. This paralleled the trend in aerosol mass concentrations, which varied from a few μg m^{-3} over the Indian Ocean to ~ 80 μg m^{-3} near the coast. The data indicated that the contribution of light absorption by aerosols near the coast was larger than that over the remote ocean, suggesting a contribution from black carbon and perhaps other organics associated with anthropogenic activities.

The presence of clouds can also affect the net light absorption by black carbon, indeed even more than for sulfate (Haywood and Shine, 1997; Haywood and Ramaswamy, 1998; Liao and Seinfeld, 1998). For example, Liao and Seinfeld (1998) calculate that the net heating

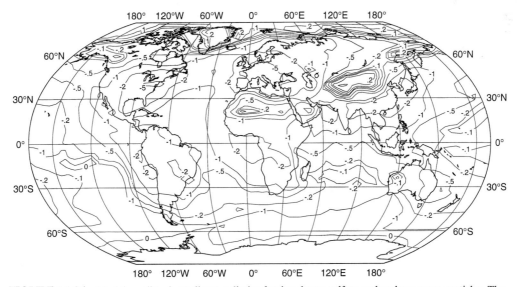

FIGURE 14.34 Model-predicted net direct radiative forcing due to sulfate and carbonaceous particles. The numbers shown are the lower limits of the ranges included in each area. The boundaries of the regions are ± 10, ± 5, ± 2, ± 1, ± 0.5, ± 0.2, ± 0.1, and 0. Thus, -5 represents the -5 to -10 W m^{-2} region, -2 represents the -2 to -5 W m^{-2} region, etc. (adapted from Penner et al., 1998).

effect can be enhanced by as much as a factor of three in the presence of a low, thick stratus cloud below the particles, due to the enhanced scattering of solar radiation back to the absorbing particles (see Chapter 3.C.2g). On the other hand, a thick cirrus cloud above the black carbon reduces its heating effect for an overhead sun because of reduced transmission of the direct solar radiation. As a result of this sensitivity to the location of clouds relative to the carbon particles, knowing the vertical distribution of black carbon is important (e.g., Haywood and Shine, 1997).

There is another interaction of absorbing aerosol particles with clouds that has been proposed. Hansen *et al.* (1997b, 1997d) suggest there is a "semi-direct" effect in that warming caused by light absorption may reduce cloud cover and result in net warming. Their calculations suggest that at values of the single-scatter albedo (defined as the fraction of extinction that appears as scattered radiation) below about 0.85, this semi-direct effect predominates and heating, rather than cooling, results.

In summary, aerosol particles containing compounds that can absorb solar radiation, such as elemental (or black) carbon and possibly some organic compounds as well, can also contribute to direct radiative forcing. This absorption of solar radiation generally results in positive radiative forcing. This effect occurs simultaneously with scattering, which results in negative radiative forcing. It should be noted that such particles also affect the amount of solar radiation reaching the low troposphere that is available for photochemistry (e.g., see Haywood and Shine, 1997). See Chapter 3.C.2f for a more detailed discussion of this issue.

c. Absorption of Long-Wavelength Infrared

Species such as sulfate and black carbon can absorb the long-wavelength thermal infrared emitted by the earth's surface, leading to positive radiative forcing. In principle, the same is true of other infrared-absorbing particle components such as nitrate, ammonium, formate, acetate, and oxalate for the bands that are not in the region of the spectrum that is already saturated (Marley *et al.*, 1993). It has been proposed that if these species and/or sulfate are present at sufficiently high concentrations in particles, for example in or downwind of urban areas, they can contribute to radiative forcing by direct absorption of infrared on local to regional scales (Marley *et al.*, 1993; Gaffney and Marley, 1998). On a global basis, Haywood and Shine (1997) and Haywood *et al.* (1997a) estimate that the contribution of sulfate and black carbon to long-wavelength direct forcing is at least an order of magnitude less than that due to the scattering, and in the case of black carbon, absorption of solar radiation.

However, this is not the case for airborne particles composed of crustal materials formed by erosion processes. As discussed in Chapter 9.C, mineral dust consists primarily of such crustal materials. Despite the fact that soil dust particles tend to be quite large, of the order of a micron and larger, they can be carried large distances. These particles not only scatter and absorb solar radiation but also absorb long-wavelength infrared emitted by the earth's surface.

Figure 14.35, for example, shows the real (n) and imaginary (k) parts of the index of refraction ($\eta = n - ki$) of three samples of dust collected in the Barbados, but thought to be transported from the Sahara Desert (Volz, 1973), from Afghanistan (Sokolik *et al.*, 1993), and from Whitehill, Texas, in the southwestern United States (Patterson, 1981), respectively (Sokolik *et al.*, 1998). Regions of absorption in the infrared due to some common dust components are also shown: the asymmetric C–O stretch of carbonate in calcite near 7 μm seen in the Texas dust, the asymmetric Si–O–Si

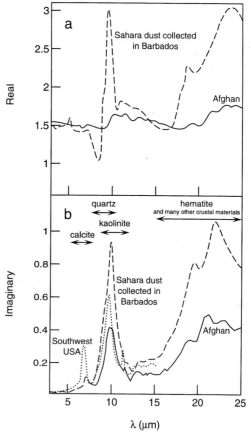

FIGURE 14.35 (a) Real (n) and (b) imaginary (k) parts of the index of refraction ($\eta = n - ki$) of some atmospheric dust samples from the Sahara collected in Barbados, Afghanistan, and Whitehill, Texas. Regions of strong absorption of some known common dust components are also shown (adapted from Sokolik *et al.*, 1998).

stretch in quartz near 9 μm, and absorptions due to kaolinite [$Al_2Si_2O_5(OH)_4$] in the 8.5- to 12-μm region (Salisbury et al., 1992). Many of these crustal materials also absorb in the 15- to 25-μm region; the major infrared absorption of hematite (α-Fe_2O_3) is also in this region.

This absorption also leads to positive radiative forcing. Figure 14.36, for example, shows a model estimate of the contribution to radiative forcing by dust particles due to scattering of solar radiation and absorption of infrared radiation (Tegen et al., 1996). As expected, the effects are calculated to be the largest in areas having the highest dust, around the Arabian Sea and over the Atlantic Ocean off the coast of Africa. Figure 14.36 shows that the calculated effect of infrared absorption by mineral dust particles can be equal to or greater than that of scattering of solar radiation. Alpert et al. (1998) have proposed that the previously unrecognized contribution of dust particles to heating of the atmosphere was responsible for inaccuracies in weather prediction models over the eastern tropical North Atlantic Ocean.

While dust particles are a natural component of the atmosphere, the amount of airborne dust is believed to have increased due to anthropogenic surface land modifications such as deforestation, cultivation, and shifts in vegetation (e.g., see Tegen and Fung, 1994, 1995; Tegen and Lacis, 1996; Li et al., 1996; Tegen et al., 1996; and Tegen et al., 1997). These activities may be responsible for as much as half of the total airborne dust (Tegen and Fung, 1995).

Based on their measurements of North African dust transported to the Barbados, Li et al. (1996) estimate that over a 10-year period, dust contributed about 56% of the total light scattering. Similarly, Tegen et al. (1997) estimate using a global transport model that scattering and absorption of light by submicron soil dust were about as important as that due to carbonaceous aerosols and scattering by sulfate aerosols. Sokolik and Toon (1996) point out that on a regional basis, direct radiative forcing due to dust aerosols can significantly exceed that of sulfate. Light scattering by mineral dust has been shown to be relatively insensitive to the relative humidity (Li-Jones et al., 1998).

Because scattering and absorption contribute to radiative forcing in opposite directions (Fig. 14.36), the positive radiative forcing and negative radiative forcing largely cancel at the top of the atmosphere. For example, for the model study shown in Fig. 14.36, the global mean net radiative forcing at the top of the atmosphere due to dust associated with disturbances from anthropogenic processes was only +0.09 W m^{-2}. However, the net radiative effect of this dust at the surface is calculated to be negative, -1 W m^{-2}, since both scattering and absorption reduce the sun intensity reaching the ground (Tegen et al., 1996). Finally, the absorption of infrared radiation by mineral dusts leads to direct heating of the atmosphere, which may alter atmospheric circulation processes (Tegen et al., 1996).

In short, the combination of absorption and scattering of light by mineral dusts, combined with an increase in these due to anthropogenic activities, has the potential to contribute to climate change. However, many uncertainties need to be removed before these effects can be confidently quantified. For example, the infrared absorption depends on the composition of the dust and as seen in Fig. 14.35, this can be quite variable from location to location and even as a function of time from one source. This one effect alone can lead to a large variability in the predicted effects on radiative forcing (Sokolik et al., 1998).

2. Indirect Effects of Aerosol Particles

a. Clouds

In addition to the direct effects on radiative forcing due to scattering and absorption of light, aerosol particles also have indirect effects, which may, in many instances, be more important than the direct radiative forcing. These indirect effects are based on the ability of some (but as we shall see, not all) aerosol particles to act as cloud condensation nuclei, CCN. This changes the number concentration of droplets in clouds and their size distribution, which can alter the precipitation rate. In addition, such changes in the cloud characteristics are believed to alter the lifetime and extent of the cloud (e.g., see Cess et al., 1997; and Lohmann and Feichter, 1997). As discussed in more detail shortly, clouds decrease the incoming solar radiation by reflecting a significant amount back out to space (the

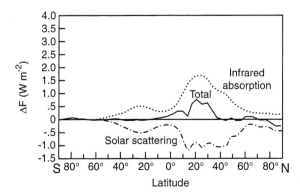

FIGURE 14.36 Predicted mean radiative forcing at the top of the atmosphere due to mineral dust from scattering, absorption, and their total (adapted from Tegen et al., 1996).

predominant effect), but high clouds can also lead to tropospheric warming through interaction with the longwave terrestrial thermal radiation. In addition, there are some data, which are presently controversial, suggesting that clouds absorb solar radiation directly to a larger extent than expected. If proven true, this can lead to thermal heating and effects on atmospheric circulation processes that are greater than have been understood to the present.

Twomey suggested in 1974 that anthropogenic emissions could affect cloud properties and albedo, i.e., have an *indirect effect* on global climate. Attention was further drawn to such indirect effects in 1987 when Charlson, Lovelock, Andreae, and Warren proposed a series of feedbacks involving dimethyl sulfide emitted by phytoplankton in seawater, CCN, and clouds. Dimethyl sulfide (DMS) is known to be oxidized in part to sulfate (see Chapter 8.E.1), which acts as a source of CCN and hence affects cloud properties, including albedo. Thus, DMS and its oxidation products such as methanesulfonate have been shown in a number of studies to correlate with CCN (Durkee *et al.*, 1991; Hegg *et al.*, 1991a,b; Ayers and Gras, 1991; Berresheim *et al.*, 1993; Quinn *et al.*, 1993; Putaud *et al.*, 1993; Lawrence, 1993; Andreae *et al.*, 1995; Ayers *et al.*, 1995). The relationship between DMS and CCN may not be linear, however. For example, the model of Pandis *et al.* (1994) predicts that small DMS emission fluxes (<1.3 μmol m^{-2} day^{-1}) do not lead to new CCN since much of the oxidation of SO$_2$ from DMS occurs in existing sea salt particles (e.g., see Sievering *et al.*, 1992; Chameides and Stelso, 1992). However, at fluxes > 2.3 μmol m^{-2} day^{-1}, new CCN are predicted by the model to be formed in an approximately linear relationship. Average DMS fluxes measured over the North Atlantic have been observed that span this range, from 1.2 to 12 μmol m^{-2} day^{-1}, with peaks up to 39 μmol m^{-2} day^{-1} (Tarrasón *et al.*, 1995). Over the Southern Ocean, a range from 0.2 to 5 μmol m^{-2} day^{-1} has been measured (Ayers *et al.*, 1995). This suggests that the effect of DMS on CCN may vary geographically and seasonally.

The seasonal cycle of CCN has also been shown to be correlated with that of cloud optical depth in one remote marine area (Boers *et al.*, 1994), and the isotope composition of non-sea salt sulfate over remote regions of the southern Pacific Ocean has been shown to be consistent with a DMS source (Calhoun *et al.*, 1991).

Based on such correlations, it is reasonable to assume that the Twomey proposal is applicable, i.e., that anthropogenic emissions of SO$_2$ and other species that form particles in the atmosphere may contribute to CCN and hence have indirect effects on climate.

Charlson *et al.* (1987) drew attention to the potential importance of feedback loops in which increased DMS emissions lead to increased sulfate CCN, increased clouds, and cloud albedo, followed by changes in the temperature and solar radiation at the surface. Changes in temperature and solar radiation might then alter DMS production, although whether in terms of increased or decreased emissions is uncertain. Despite many studies, the details and importance of such feedbacks remain to be elucidated (e.g., see Schwartz, 1988; Baker and Charlson, 1990; Lin *et al.*, 1992; Hegg, 1990, 1993a, 1993b; Chameides and Stelson, 1992; Raes and Van Dingenen, 1992, 1995; Sievering *et al.*, 1992; Lin *et al.*, 1993a, 1993b; Russell *et al.*, 1994; Pandis *et al.*, 1994, 1995; Raes, 1995; Capaldo and Pandis, 1997; and Andreae and Crutzen, 1997). For example, Bates and Quinn (1997) report that the DMS concentration in the surface seawater of the equatorial Pacific was relatively insensitive to changes in the properties of the atmosphere (e.g., cloud cover and precipitation) and oceans (e.g., sea surface temperature, mixed layer depths, and upwelling rates) associated with El Niño–Southern Oscillation events.

The following sections focus on the potential indirect effects of aerosol particles due to anthropogenic contributions, which, unlike the natural emissions, are expected to provide a contribution that changes with time.

Effect of aerosol particles on cloud drop number concentrations and size distributions Clouds and fogs are characterized by their droplet size distribution as well as their liquid water content. Fog droplets typically have radii in the range from a few μm to ~ 30–40 μm and liquid water contents in the range of 0.05–0.1 g m^{-3}. Clouds generally have droplet radii from 5 μm up to ~ 100 μm, with typical liquid water contents of ~ 0.05–2.5 g m^{-3} (e.g., see Stephens, 1978, 1979). For a description of cloud types, mechanisms of formation, and characteristics, see Wallace and Hobbs (1977), Pruppacher (1986), Cotton and Anthes (1989), Heymsfield (1993), and Pruppacher and Klett (1997).

There are several basic physical–chemical principles involved in the ability of aerosol particles to act as CCN and hence lead to cloud formation. These are the Kelvin effect (increased vapor pressure over a curved surface) and the lowering of vapor pressure of a solvent by a nonvolatile solute (one of the colligative properties). In Box 14.2, we briefly review these and then apply them to the development of the well-known Köhler curves that determine which particles will grow into cloud droplets by condensation of water vapor and which will not.

BOX 14.2
KELVIN EFFECT, VAPOR PRESSURE LOWERING, AND THE KÖHLER CURVES

1. *Kelvin effect.* Recall that the change in free energy of a gas due to a change in pressure (at constant T) from P_1 to P_2 is given by

$$\Delta G = nRT \ln(P_2/P_1). \qquad (X)$$

Consider the situation in Fig. 14.37 in which a number of moles dn is transferred from a bulk liquid with vapor pressure P_0 to a droplet of the same liquid having radius r and over which the vapor pressure is P. Assuming each system is at equilibrium, the free energies of the liquid and gas in each case must be equal. The change in free energy, dG, for transferring dn moles from the bulk liquid to the droplet is therefore the same as the accompanying free energy change for the gas. From Eq. (X), this is given by

$$dG = dn\,RT \ln(P/P_0). \qquad (Y)$$

However, there is also a free energy change due to the increase in surface area, dA, of the droplet caused by the transfer of the dn moles. The surface tension of the liquid, γ, is the work required per unit change in surface area to expand a surface against the intermolecular forces that tend to minimize the surface. The area of the initial droplet is $4\pi r^2$ and hence the change due to a small change dr caused by transfer of dn moles is $8\pi r\,dr$. The free energy change in the droplet due to an increase in its surface area is therefore

$$dG = \gamma\,dA = 8\pi r\gamma\,dr. \qquad (Z)$$

This transfer of dn moles causes a volume change $dV = 4\pi r^2\,dr$. If the molecular weight of the compound is MW and liquid density is ρ, then $\rho = (\text{MW})dn/(4\pi r^2\,dr)$. This expression can be used to replace dr in Eq. (Z):

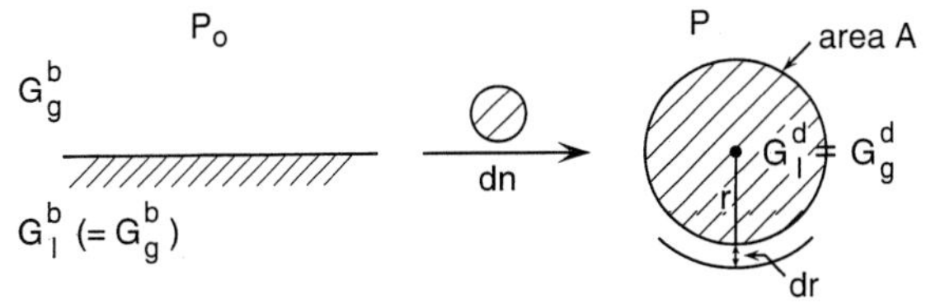

$$dG = 8\pi r\gamma(\text{MW})dn/4\pi r^2\rho = dn[2\gamma(\text{MW})/r\rho]. \qquad (AA)$$

Combining Eqs. (Y) and (AA) gives the Kelvin equation:

$$\ln(P/P_0) = [2\gamma(\text{MW})/r\rho RT] \qquad (BB)$$

or

$$P/P_0 = \exp[2\gamma(\text{MW})/r\rho RT].$$

This Kelvin equation says that the vapor pressure over a droplet depends exponentially on the inverse of the droplet radius. Thus, as the radius decreases, the vapor pressure over the droplet increases compared to that over the bulk liquid. This equation also holds for water coating an insoluble sphere (Twomey, 1977).

This has important implications for nucleation in the atmosphere. Condensation of a vapor such as water to form a liquid starts when a small number of water molecules form a cluster upon which other gaseous molecules can condense. However, the size of this initial cluster is very small, and from the Kelvin equation, the vapor pressure over the cluster would be so large that it would essentially immediately evaporate at the relatively small supersaturations found in the atmosphere, up to $\sim 2\%$ (Pruppacher and Klett, 1997). As a result, clouds and fogs would not form unless there was a preexisting particle upon which the water could initially condense. Such particles are known as *cloud condensation nuclei*, or CCN.

While water is a major component of tropospheric particles, and hence largely determines the surface tension (γ), organics found in particles may act as surfactants (see Chapter 9.C.2). In this case, their segregation at the air–water interface could potentially lead to a substantial surface tension lowering of such particles, which would lead to a lower equilibrium water vapor pressure over the droplet (Eq. (BB)) and hence activation at smaller supersaturations. This possibility is discussed in more detail in the next section.

2. *Vapor pressure lowering.* Raoult's law says that the vapor pressure of a solution component, A, whose pure vapor pressure is P_A° is proportional to

FIGURE 14.37 Basis of Kelvin effect for increased vapor pressure over small liquid droplets.

its mole fraction in solution, x_A, i.e., in a two-component solution of A and B, to $x_A = (1 - x_B)$:

$$P_A = x_A P_A^\circ = (1 - x_B) P_A^\circ \quad \text{i.e.,} \quad P_A/P_A^\circ = 1 - x_B$$

or alternatively

$$P_A^\circ - P_A = \Delta P = x_B P_A^\circ. \quad \text{(CC)}$$

Thus, if a nonvolatile solute is dissolved in water, the vapor pressure of water is lowered by an amount proportional to the mole fraction of dissolved solute, taking into account any dissociation that occurs (vide infra). It should be noted that this assumes ideal solution behavior.

As we have seen in Chapter 9, there are a variety of dissolved solutes in atmospheric particles, which will lower the vapor pressure of droplets compared to that of pure water. As a result, there is great interest in the nature and fraction of water-soluble material in atmospheric particles and their size distribution (e.g., Eichel *et al.*, 1996; Novakov and Corrigan, 1996; Hoffmann *et al.*, 1997). This vapor pressure lowering effect, then, works in the opposite direction to the Kelvin effect, which increases the vapor pressure over the droplet. The two effects are combined in what are known as the Köhler curves, which describe whether an aerosol particle in the atmosphere will grow into a cloud droplet or not under various conditions.

3. *Köhler curves.* Calculation of the mole fraction of dissolved solute, x_B, in a water droplet requires knowing the number of moles of water and of dissolved solute. Take a two-component solution such as NaCl in water, where the solute dissociates into i ions ($i = 2$ for NaCl). Assume m_B grams of salt of molecular weight MW_B are dissolved in water to form a solution of density ρ_s. The number of moles of dissolved ions is im_B/MW_B. The number of moles of water for a drop of volume $V = (4/3)\pi r^3$ is $[\rho_s V - m_B]/MW_A$, where MW_A is the molecular weight of water. The mole fraction of dissolved solute ions (B) is then given by Eq. (DD):

$$x_B = \frac{im_B/MW_B}{\left(\tfrac{4}{3}\pi r^3 \rho_s - m_B\right)/MW_A + im_B/MW_B}. \quad \text{(DD)}$$

Using Eq. (CC), the vapor pressure lowering due to m_B grams of dissolved salt that forms i ions per dissolved molecule is therefore given by

$$\frac{P_A}{P_A^\circ} = 1 - \frac{im_B/MW_B}{\left(\tfrac{4}{3}\pi r^3 \rho_s - m_B\right)/MW_A + im_B/MW_B}. \quad \text{(EE)}$$

For dilute solutions, where m_B is small, this reduces to

$$\frac{P_A}{P_A^\circ} = 1 - \frac{im_B/MW_B}{\tfrac{4}{3}\pi r^3 \rho_s/MW_A} = 1 - \left(\frac{im_B MW_A}{\tfrac{4}{3}\pi \rho_s MW_B}\right)\frac{1}{r^3}$$

$$= 1 - \frac{b}{r^3}, \quad \text{(FF)}$$

where $b = [im_B MW_A/((4/3)\pi \rho_s MW_B)]$.

This vapor pressure lowering by the solute acts simultaneously with, and counteracts, the vapor pressure increase due to the Kelvin effect [Eq. (BB)]. Multiplying the two, the net result for the vapor pressure above a solution containing a dissolved solute is given by

$$\frac{P_A}{P_A^\circ} = \left[1 - \frac{b}{r^3}\right]\exp\left[\frac{2\gamma MW_A}{r\rho_s RT}\right]. \quad \text{(GG)}$$

Applying the approximation $e^x = 1 + x + x^2/2! + \cdots$ and using only the first two terms, Eq. (GG) becomes

$$\frac{P_A}{P_A^\circ} = \left[1 - \frac{b}{r^3}\right]\left[1 + \frac{a}{r}\right] = 1 + \frac{a}{r} - \frac{b}{r^3}, \quad \text{(HH)}$$

where $a = 2\gamma MW_A/\rho_s RT$ and the r^{-4} term has been omitted since it is small compared to the other three terms for radii of atmospheric interest.

The term *supersaturation*, S, defined as $(P_A/P_A^\circ - 1)$ is often expressed in the form of percent supersaturation, i.e., as $100(P_A/P_A^\circ - 1)$, where P_A and P_A° are defined in Box 14.2. The relationship between the equilibrium vapor pressure over the droplet and that over the bulk liquid [Eq. (HH)] is often expressed in a simplified form using the supersaturation:

$$S = \frac{a}{r} - \frac{b}{r^3}. \quad \text{(II)}$$

Plots of S against radius are known as *Köhler curves*. Figure 14.38a shows a schematic diagram of such a curve. A more detailed thermodynamic treatment of Köhler curves is given by Reiss and Koper (1995).

Typical values of supersaturation found in clouds are between about 0.2 and up to 2%. For fogs, the values are lower by about an order of magnitude, typically between about 0.02 and 0.2% (Pruppacher and Klett, 1997).

C. AEROSOL PARTICLES, ATMOSPHERIC RADIATION, AND CLIMATE CHANGE

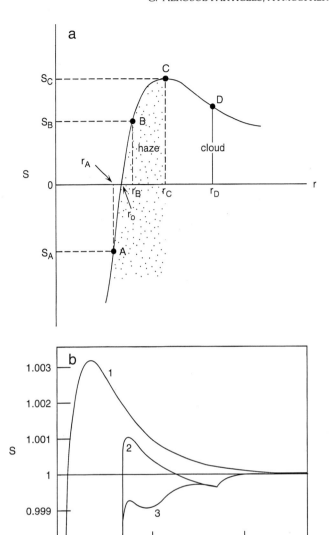

FIGURE 14.38 (a) Schematic diagram of traditional Köhler curve, where $S = P_A/P_A^\circ - 1$ is the supersaturation and r is the radius of the droplet. (b) Köhler curves for 30-nm dry particle of $(NH_4)_2SO_4$: (1) traditional curve; (2) for a 500-nm $CaSO_4$ particle (slightly soluble) and $(NH_4)_2SO_4$ as for curve 1; (3) as for curve 2 but in the presence of 1 ppb HNO_3 which is taken up by the particle (adapted from Kulmala et al., 1997).

The dependence of S on the radius r is such that at small radii, the second term due to the vapor pressure lowering dominates and S is negative. In this region, air with RH below 100% is in equilibrium with the particle; such diagrams are often plotted as a function of RH in this region, instead of S. At large radii, the first term due to the Kelvin effect dominates, and S becomes positive and ultimately reaches a maximum before decreasing again. The region to the left of the peak is known as the haze region for reasons that will become apparent shortly, whereas that to the right is known as the cloud droplet region.

Take as an example, a small dry particle of NaCl of a given mass (m_B) that is introduced into air at a water vapor pressure corresponding to S_A in Fig. 14.38a. Assuming that the RH is above the deliquescence point of NaCl, ~75% at 25°C, the particle will take up water, dissolve, and form a stable droplet of radius r_A. Similarly, if the air saturation ratio increases to S_B, the particle will, under equilibrium conditions, take up water and grow to radius r_B.

These particles are then in stable equilibrium with water in the air. Say the particle at point B loses some water molecules and starts to shrink. The equilibrium supersaturation for the smaller particle is lower than for the original particle. However, the supersaturation of the surrounding air remains higher, so that water will condense back out on the particle to bring it back to its original size. Similarly, if the particle at B gains some water molecules and the radius starts to increase, the value of S required to maintain this new size would be larger than that of the surrounding air and water would evaporate to restore the equilibrium size.

In short, particles to the left of the peak in Fig. 14.38a do not tend to shrink or grow. Because they are generally in the 0.1- to 1-μm size range which scatters light efficiently (see Chapter 9.A.4), these particles are known as *haze* particles or droplets. These often occur at relative humidities below 100%.

Consider, however, a particle at point D in Fig. 14.38a. If it gains some water molecules and the radius starts to increase, the surrounding air will have a larger supersaturation than the required equilibrium value of S for this larger particle. As a result, water will condense out on the droplet, causing it to grow further. Particles that lie to the right of the peak are thus in an unstable equilibrium (e.g., see Reiss and Koper, 1995) and can therefore activate into cloud droplets from the condensation of water.

There are two questions with respect to potential indirect effects of aerosol particles on properties of clouds: (1) What are the sources of new particles? (2) How do these new particles grow to sufficient size (>50 nm) to act as CCN?

The first issue is that of formation of new particles. As discussed in Chapter 9.B, nucleation of gases to form new particles in the atmosphere is not well understood. The observed rates of nucleation of H_2SO_4, for example, greatly exceed the calculated rates. An important contributor to the formation of new particles in the boundary layer (BL) under some conditions appears to be exchange between the BL and the free troposphere (e.g., Davison et al., 1996; Raes et al., 1997; Clarke et al., 1997). For example, some of the DMS from the oceans can be carried to the free troposphere, where it is oxidized to sulfate, generating new CCN. Mixing of the air mass back into the BL may then quench the formation of new CCN in that region by scavenging the low-volatility species such as H_2SO_4 before they can nucleate to form new particles (e.g., see Slinn, 1992; and Raes, 1995).

The second issue is how these new particles grow into a sufficient size that they can act as CCN. The size to which the particles must grow to act as CCN is determined under equilibrium conditions by the Köhler curves. The peak values of S and r on the Köhler curves (Fig. 14.38) are known as the critical values, S_c and r_c (see Problem 6). Whenever the supersaturation of the air mass is greater than S_c, condensation occurs to form cloud and fog droplets. However, if it is less than S_c, particles with radii less than the critical radius r_c will maintain an equilibrium size (e.g., point B in Fig. 14.38a) and not form clouds or fogs. Of course, in the real atmosphere, there are a variety of initial particle sizes containing varying amounts of dissolved solutes, and the supersaturation of the air mass also changes with time. However, the Köhler curves provide a basis for understanding which particles can grow into clouds and fogs and which will not.

It is important to note that the Köhler relationship assumes equilibrium. However, under real atmospheric conditions, the system may not be at equilibrium. P. Y. Chuang et al. (1997) and Hallberg et al. (1998) point out that if the time scale for growth of cloud droplets is larger than that for the particle to reach equilibrium, the growth of CCN into cloud droplets may be controlled by kinetics, rather than equilibrium. They suggest that ignoring this potential kinetic limitation may lead to overestimating the number of cloud droplets that will form from a given number of CCN. In addition, measurement techniques for CCN using cloud chambers may not give an accurate assessment of CCN under ambient conditions since the range of time scales encounted in air is much larger than that used in the measurements. At present, it is not clear how much of a problem such kinetic limitations present in the atmosphere. [Some parameterizations in use, however, do take into account the kinetic limitations (e.g., C. C. Chuang et al., 1997).]

In addition, these traditional Köhler curves do not take into account the effects of slightly soluble solutes or gases that can dissolve in the particles. Figure 14.38b compares the traditional Köhler curves for a 30-nm dry particle of $(NH_4)_2SO_4$ at 298 K with that for the same particle but containing a 500-nm core of slightly soluble $CaSO_4$ (Kulmala et al., 1997). The increased particle size reduces the Kelvin effect contribution and a minimum in the curve reflects the point at which all of the $CaSO_4$ dissolves. Also shown is a Köhler curve for the case where 1 ppb of gaseous HNO_3 is present and is taken up by the particles. It can be seen that the equilibrium supersaturation is less than 1 for sizes up to ~10 μm; i.e., for this more complex (but realistic) case, cloud droplets can form at RH below 100%.

Possible mechanisms of growth for small particles into a sufficient size that they are on the right side of the Köhler curve include uptake of small particles into existing cloud droplets where in-cloud oxidation of gaseous species such as SO_2 to sulfate occurs. Evaporation of the cloud then leaves a larger particle containing the additional oxidation products (e.g., see Hoppel and Frick, 1990; Hegg, 1990; Van Dingenen et al., 1995; and Hoppel et al., 1996). Other possible mechanisms of particle growth include coagulation of fine particles or the growth of existing particles by condensation of low-volatility products (e.g., see Lin et al., 1992, 1993a, 1993b; Hegg, 1990, 1993). However, coagulation is not expected to be important in remote areas since the number concentration of particles is much smaller than in polluted areas where coagulation can be important (Lin et al., 1992).

As we have already seen, the critical supersaturation S_c corresponding to the peak of the Köhler curve depends on a number of parameters unique to the aerosol particle. Thus, at a given supersaturation some particles will form cloud droplets and some will not. As a result, the total number of CCN will vary with the supersaturation used in the CCN measurement. This is illustrated in Fig. 14.39, which shows the concentration of CCN measured in Antarctica as a function of the percentage supersaturation for CCN that grow into droplets larger than 0.3 and 0.5 μm, respectively (Saxena, 1996). This particular set of measurements

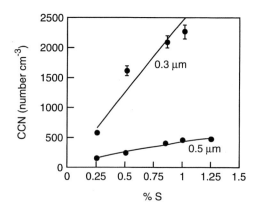

FIGURE 14.39 CCN concentrations measured during a cloud event at Palmer Station, Antarctica, as a function of percent supersaturation (%S). The two lines represent particles that grew to sizes greater than 0.3 and 0.5 μm, respectively (from Saxena, 1996).

was carried out during a burst of CCN production due to cloud processing (vide infra).

The relationship between the number concentration (N) of CCN and the supersaturation (S) is often expressed in the form $N = CS^k$, where C and k are empirical coefficients characteristic of the particular air mass (Pruppacher and Klett, 1997). However, alternate forms such as

$$N = N_0[1 - e^{-(BS^k)}],$$

where B and k are empirical coefficients and N_0 is the number concentration of CCN at infinite supersaturation, have been suggested to match data from laboratory studies where the aerosol composition is relatively simple (Ji and Shaw, 1998).

As discussed earlier, if organics congregate at the air–water interface of particles and act as surfactants, they can lead to a reduced Kelvin effect and hence activation at lower supersaturations. In addition, if they dissolve, they will also contribute to Raoult's law effects of the dissolved species. As discussed by Shulman et al. (1996), the dissolution of organics in particles may lead to modifed Köhler curves having two maxima instead of one. If the second maximum is at a higher supersaturation than the first, droplets could become partially activated. They also suggest that with the distribution of chemical compositions and particle sizes in the atmosphere, the region beyond the maxima could be relatively flat in shape, rather than having a steep negative slope. This would result in the formation of cloud droplets with radii corresponding to some characteristic metastable size.

The effects of surfactants on particles is discussed in detail in Chapter 9.C.2b. In some cases, the interaction of organic surfactants with particles has been observed to enhance water uptake and hence the cloud nucleating properties of the particle, whereas in others, it inhibits water uptake. An example of the former case is a study by Cruz and Pandis (1998), who coated particles of $(NH_4)_2SO_4$ with the C_5 dicarboxylic acid, glutaric acid. A coating of glutaric acid on ammonium sulfate increased the size of the particle and increased its cloud nucleating properties, but in a manner consistent with that expected from Köhler curves for a two-component solution assuming the organic does not alter the surface tension. On the other hand, coating with the water-insoluble dioctyl phthalate did not alter the activation of the inorganic salt. Similarly, Kotzick et al. (1997) and Weingartner et al. (1997) reported that oxidation of carbon or diesel soot particles by O_3 increased their CCN activation properties due to the formation of polar surface groups.

As discussed in detail in Chapter 9, Saxena et al. (1995) have measured the hygroscopic behavior of particles in the Los Angeles area (i.e., urban) and in the Grand Canyon, Arizona (i.e., nonurban). They found that organics in the urban particles inhibit water uptake, i.e., are hydrophobic in nature, whereas those in the nonurban particles are hydrophilic, i.e., increase water uptake. It may be that these differences are due to the formation of smaller, more oxidized, water-soluble organics during long-range transport to the nonurban site. Saxena et al. (1995) suggest that hydrophobic organics in the urban aerosol may form a surfactant film on the particles that inhibits water uptake. For example, difunctional acids such as oxalic and adipic acids have been shown to slow the rate of evaporation of water from droplets once they are sufficiently concentrated by the initial evaporation of water (Shulman et al., 1997). At any rate, as might be expected, the effects of organics on water uptake for real atmospheric particles clearly can be negative or positive, depending on their particular composition.

As expected from the earlier discussion of the Köhler curves, not all particles act as CCN. For example, only about 15–20% of the Aiken nuclei (see Chapter 9.A.2) in a marine air mass off the coast of Washington state acted as CCN at 1% supersaturation (Hegg et al., 1991b). Similarly, in a marine air mass in Puerto Rico, between 24 and 70% of the particles measured at 0.5% supersaturation before cloud formation led to cloud droplet formation (Novakov et al., 1994).

Gillani, Leaitch, and co-workers (1995) carried out a detailed study of the fraction of accumulation mode particles (diameters from 0.17 to 2.07 μm) that led to cloud droplet formation in continental stratiform clouds near Syracuse, New York. When the air mass was relatively clean, essentially all of the particles were activated to form cloud droplets in the cloud interior and the number concentration of cloud droplets increased linearly with the particle concentration. However, when the air mass was more polluted, the fraction of particles that were activated in the cloud interior was significantly smaller than one. This is illustrated by Fig. 14.40, which shows the variation of this fraction (F) as a function of the total particle concentration, N_{tot}. In the most polluted air masses (as measured by large values of N_{tot}), the fraction of particles activated was 0.28 ± 0.08, whereas in the least polluted, it was as high as 0.96 ± 0.05. The reason for this is likely that in the more polluted air masses, the higher number of particles provided a larger sink for water vapor, decreasing the extent of supersaturation.

In short, while anthropogenically produced particles can act as cloud condensation nuclei, only a fraction of them actually do so. This fraction can be close to one

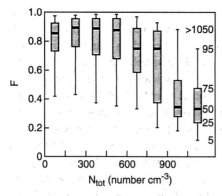

FIGURE 14.40 Fraction (F) of aerosol particles that are activated to form cloud droplets as a function of the total number of particles (N_{tot}). The horizontal line represents the 50th percentile for 10 sets of aircraft measurements. The 5th, 25th, 75th, and 95th percentiles are also shown (adapted from Gillani et al., 1995).

or as much as an order of magnitude smaller, depending on a number of factors, the most important of which are the particle sizes, the total particle concentration, and the local cooling rate in the cloud (Gillani et al., 1995).

An increase in aerosol particles that can act as CCN can increase the number of cloud droplets and their size distribution, both of which can affect the light scattering properties of clouds and hence climate. We first briefly discuss the effects of clouds on climate and then the potential impacts of anthropogenic aerosols on the formation and properties of clouds.

Clouds and global climate. Clouds in the troposphere interact with both solar and terrestrial radiation in complex ways, and either warming or cooling can result (e.g., see Ramanathan et al., 1989; Ramaswamy and Ramanathan, 1989; Fouquart et al., 1990; Harrison et al., 1990; Liou, 1992; King, 1993; Hartmann, 1993; Rossow and Zhang, 1995; Ramanathan, 1995; Crutzen and Ramanathan, 1996; and Baker, 1997). Thus, marine stratiform clouds found in the boundary layer backscatter solar radiation, leading to negative radiative forcing and a net cooling. Although such clouds also absorb terrestrial infrared radiation, they emit at about the same temperature as the earth's surface. As a result, as discussed for the greenhouse gases, there is little net effect (see Section A.2). However, cirrus clouds and deep convective cloud anvils found near the tropopause emit long-wavelength infrared out to space (Fig. 14.2c) at the colder temperatures characteristic of this region of the atmosphere. Because this energy emission is at lower temperatures, the net infrared emission out to space when they are present is smaller, leading to a positive radiative forcing, i.e., to warming (Twomey, 1991). As a result, the net radiative forcing due to low clouds over the oceans is generally negative, whereas it is positive over some continental regions with high clouds. On a global basis, the mean net effect is negative, ~ -20 W m^{-2} (Baker, 1997). For reviews of the relationship between clouds and climate, and anthropogenic effects on them, see Hobbs (1993a, 1993b), King (1993), Hartmann (1993), Andreae (1995), Schwartz and Slingo (1996), Schwartz (1996), and Baker (1997).

Since some aerosol particles, which may be solids or liquids, e.g., H_2SO_4, can serve as cloud condensation nuclei (CCN), increased particle emissions from anthropogenic processes have the potential for increasing the number of CCN. The concentration of droplets in a cloud is therefore expected to increase, although not necessarily in a linear fashion, with the increased concentrations of aerosol particles. The formation of a larger number of droplets for a given liquid water content will lead to each droplet being smaller, i.e., shift the size distribution to smaller droplets. This effect increases the cloud albedo and hence can contribute to global climate change (Twomey, 1974, 1977a,b, 1991; Twomey et al., 1984). Evidence for these effects is discussed shortly. Finally, the shift to smaller drop sizes may decrease the precipitation rate from clouds, increasing their lifetimes and hence the average amount of cloud cover (Albrecht, 1989; Lohmann and Feichter, 1997). This is also expected to have a significant effect on global climate.

It should be noted that while aerosol particles affect clouds by serving as CCN for cloud formation, the reverse is also true, i.e., clouds also affect the formation and size distribution of aerosol particles. For example, the oxidation of SO_2 to sulfate in clouds generates larger particles whose light scattering cross sections are larger than for smaller particles formed by gas-phase oxidation processes (e.g., see Lelieveld and Heintzenberg, 1992). Aerosol particles incorporated into a cloud droplet will reappear as particles when the cloud evaporates. However, if new aerosol constituents are formed by in-cloud oxidation (e.g., of SO_2 to sulfate), the size of the resulting particle will be larger than the original particle. As discussed earlier, this is a potentially important process for growing aerosol particles, which are too small to serve as CCN at low values of S_{max} (e.g., for marine stratus), into sufficiently large sizes (i.e., to the right of the peak in Fig. 14.38) that they can act as CCN under the appropriate conditions of supersaturation (e.g., see Hegg, 1990; and Kaufman and Tanré, 1994).

For example, as seen in Fig. 14.41, aerosol particle number size distributions in the clean marine boundary layer outside of clouds are often observed to have a bimodal distribution. The larger mode above 0.1 μm

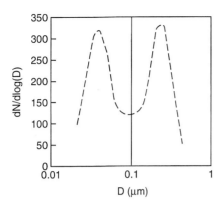

FIGURE 14.41 Typical particle size number distributions for marine aerosols outside of clouds where the total aerosol number concentration was < 500 cm^{-3} (adapted from Anderson et al., 1994).

has been attributed to aerosol particles that have been "cloud processed" (Hoppel et al., 1986; Hoppel and Frick, 1990; Anderson et al., 1994). That is, these particles served as CCN upon which clouds formed, followed by aqueous-phase reactions and evaporation of the cloud droplets to leave larger particles. Indeed, in some studies (e.g., Van Dingenen et al., 1995), it is assumed that particles in this mode can be taken as a measure of the CCN that were available for cloud formation in the prior cloud event. As discussed in Chapter 9.A.2, such aqueous-phase processes in the atmosphere are also believed to lead to two peaks in accumulation mode particles in urban areas.

The albedo (R) of a thick, boundary layer cloud that does not absorb solar radiation over a surface with zero albedo can be approximated (Twomey, 1991; Schwartz and Slingo, 1996; Baker, 1997) by

$$\text{Albedo} = R \approx \frac{\tau}{\left[\left(\frac{a}{1-g}\right) + \tau\right]} \tag{JJ}$$

The value of the factor a is usually taken as 1 or 2 and τ is the optical depth of the cloud, defined by $I/I_0 = e^{-\tau}$, where I and I_0 refer to the transmittance of direct solar radiation in the presence and absence of the cloud, respectively. The optical depth can be approximated by $\tau \cong 2\pi r_{\text{eff}}^2 Nh$, where r_{eff} is an effective average droplet radius for scattering of solar radiation, h is the thickness of the cloud, N is the number of cloud droplets per unit volume, and g is the asymmetry parameter for single scattering, defined as the average of the cosine of the scattering angle. A typical value of g is ~0.85 for cloud droplets. (Note that the cloud height h and N are not independent, since the number of cloud droplets affects the precipitation rate, and this alters the cloud height; e.g., see Pincus and Baker,

1994.) Typical cloud albedos for thick clouds in the boundary layer are ~0.5 over the ocean in midlatitudes; i.e., half of the incoming solar radiation is scattered back out to space (Baker, 1997). This approximation, Eq. (JJ), illustrates why a change in the number of cloud droplets and their size affects the cloud albedo and hence the radiative forcing (see Problem 9).

An important aspect of quantifying the indirect effects of anthropogenic emissions is the recognition that the changes generated in the cloud albedo are not constant for all clouds but rather depend on the particular cloud properties. For example, the effects of an absolute increase in CCN on a cloud with a low droplet number concentration will be larger than for one with a larger droplet number concentration, even if the two have the same albedo (e.g., see Platnick and Twomey, 1994; and Taylor and McHaffie, 1994). This "susceptibility" has been expressed in different ways, for example as $dR/d(\ln N)$. Platnick and Twomey (1994) have derived an expression applicable to nonabsorbing clouds for the sensitivity (dR/dN) of cloud albedo, R, to changes in the number of cloud droplets, N, which they term cloud susceptibility:

$$\frac{dR}{dN} = \tau \frac{\delta R}{\delta \tau} r_v^3 \frac{4\pi \rho_w}{9W}. \tag{KK}$$

In Eq. (KK), $\tau \cong 2\pi r_{\text{eff}}^2 Nh$ is the optical thickness defined above, r_v is the volume-weighted moment of the cloud droplet size distribution, which can be approximated by r_{eff}, ρ_w is the density of liquid water, and W is the liquid water content of the cloud. Using Eq. (JJ), the term $\tau(\delta R/\delta \tau)$ can be shown to be equal to $R(1-R)$ (see Problem 10).

Platnick and Twomey (1994) have applied Eq. (KK) to marine clouds off the coast of California and southern Africa, to fogs in central California, and to ship tracks. Figure 14.42 shows a typical range of susceptibilities as a function of cloud droplet size. The measured susceptibilities in these studies covered three orders of magnitude, from 5×10^{-5} cm^3 for fogs to 0.8×10^{-3} cm^3 for marine clouds off south Africa and 2×10^{-2} cm^3 for thin stratus clouds off the California coast. Similarly, Taylor and McHaffie (1994) report cloud susceptibilities in the range from 10^{-4} to $>8 \times 10^{-3}$ at various locations around the world. The highest susceptibilities were those with the smallest aerosol particle concentrations below the cloud base. As the particle concentration increased beyond ~500 cm^3, the susceptibility was relatively constant at ~5×10^{-4} cm^3. This means that the addition of new particles to a relatively clean air mass is far more effective than for a polluted one in terms of the effect on clouds. In short,

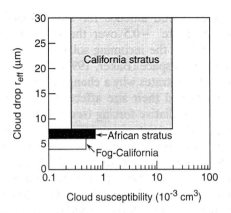

FIGURE 14.42 Cloud susceptibilities (logarithmic scale) and cloud droplet radii for stratus clouds off Africa and California and from fogs in California (adapted from Platnick and Twomey, 1994).

it is important in quantifying the indirect effects of anthropogenic emissions that the individual cloud properties be taken into account.

It should be noted that while most studies have focused on the effects of sulfate on stratocumulus clouds, model studies suggest that sulfate may also exert significant effects on convective clouds in tropical systems as well (Andronache et al., 1998).

The first major link between the indirect effects of aerosol particles and climate is whether there has been an increase in particles and in CCN due to anthropogenic activities. As discussed in Chapter 2, anthropogenic emissions of particles and of gas-phase precursors to particles such as SO_2 have clearly increased since preindustrial times, and it is reasonable that CCN have also increased. Ice core data provide a record of some of the species that can act as CCN. Not surprisingly, sulfate and nitrate in the ice cores have increased substantially over the past century (Mayewski et al., 1986, 1990; Laj et al., 1992; Fischer et al., 1998). For example, Figure 14.43 shows the increases in sulfate and nitrate since preindustrial times in an ice core in central Greenland (Laj et al., 1992). Sulfate has increased by ~300% and nitrate by ~200%. This suggests that sulfate and nitrate CCN also increased, although not necessarily in direct proportion to the concentrations in the ice core measurements.

Modeling studies by Langner et al. (1992) suggest that at most 6% of the anthropogenic SO_2 emissions can form new particles, since removal of SO_2 by direct deposition is large (~50%) and the portion that is oxidized in clouds does not lead to new particles. Taking these factors into account, Langner et al. (1992) estimate that new sulfate particles may have doubled since preindustrial times.

There is other evidence that supports increased CCN due to increased anthropogenic emissions. Thus, typical

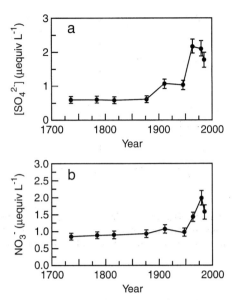

FIGURE 14.43 (a) Sulfate and (b) nitrate (in μequiv L^{-1}) in central Greenland ice cores from ~1750 to 1985 (adapted from Laj et al., 1992).

concentrations of CCN over industrialized continental areas are much larger than those in more remote regions. Over remote oceans, CCN concentrations measured using typical supersaturations of 0.7–1.25% are typically <100 cm^{-3} (e.g., Hegg et al., 1991a,b; Hudson and Li, 1995; Saxena, 1996) whereas over industrialized continents, number concentrations as high as 5000 cm^{-3} have been observed (e.g., Hudson, 1991; Pruppacher and Klett, 1997).

Another piece of evidence for anthropogenic emissions leading to increased CCN and hence effects on cloud properties such as albedo and extent is found in "ship tracks." These are lines of clouds that trace ship movements, either in initially cloud-free regions (Conover, 1966; Platnick and Twomey et al., 1994) or superimposed on preexisting clouds (Coakley et al., 1987). Emissions associated with the ship exhausts serve as CCN. This allows clouds to form where the background CCN concentration is too small for cloud formation. Alternatively, the CCN can modify existing cloud properties in the exhaust plume by changing the number and size distribution of the cloud droplets as well as the liquid water content (e.g., Ferek et al., 1998).

For example, Fig. 14.44 shows the cloud number concentration (N), effective radius (r_{eff}), and liquid water content measured simultaneously during an aircraft flight through two ship tracks (King et al., 1993). In addition, the upwelling and downwelling radiation at 744 nm and at 2.2 μm, respectively, are shown. Consistent with the foregoing discussion, the upwelling radiation at 744 nm increased in the ship tracks while the downwelling radiation decreased. However, both the upward and downward infrared radiation at 2.2 μm decreased due to increased absorption of this wavelength in the cloud. It is noteworthy that the cloud susceptibility, i.e., change in reflectance per change in cloud droplet number, in ship tracks has been measured to be smaller than outside of the tracks, as expected from the earlier discussion (Platnick and Twomey, 1994).

Another piece of evidence supporting the relationship between anthropogenic emissions and CCN is the observation by Hudson and Li (1995) of higher CCN concentrations associated with higher O_3 levels and with higher particle concentrations in air masses near the Azores. Falkowski *et al.* (1992) and Kim and Cess (1993) also report enhanced cloud albedos near continental coastal regions having higher sulfate concentrations.

A number of field studies have quantitatively examined the relationship between the CCN number concentration and/or cloud droplet concentration and the mass concentration of non-sea salt sulfate (nss). Figure 14.45 shows one summary of some of these studies in the form of a log–log plot (Van Dingenen *et al.*, 1995). Given the variety of measured parameters and wide range of conditions encompassed by these data (CCN

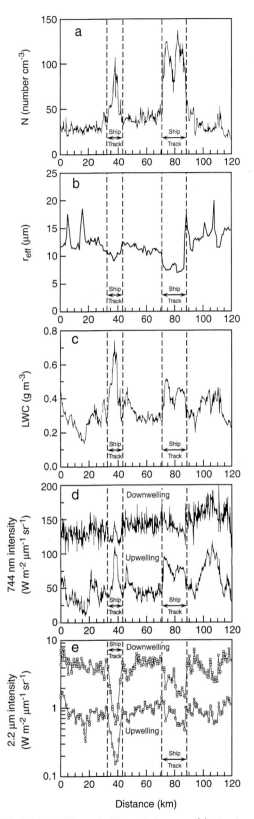

FIGURE 14.44 Effect of ship emissions on (a) cloud number concentration, N, (b) effective cloud droplet radius, r_{eff}, (c) cloud liquid water content, LWC, and (d, e) down- and upwelling radiation at (d) 744 nm and (e) 2.2 μm (adapted from King *et al.*, 1993).

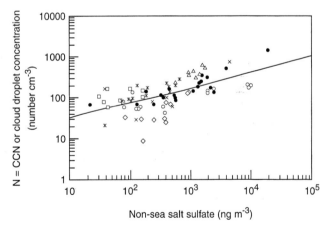

FIGURE 14.45 Log–log plot of CCN or cloud droplet concentration measured in a number of field studies as a function of non-sea salt sulfate (nss). Data are from (□) Cape Grim, Tasmania, CCN measured at 0.46% supersaturation (Gras, 1989; Ayers and Gillett, 1989); (△) Puerto Rico, CCN measured at 0.5% supersaturation (Novakov and Penner, 1993); (◇) western Washington, CNN measured at 0.3% supersaturation (Berresheim *et al.*, 1993); (○) the Azores (Hegg *et al.*, 1993); (×) Whiteface Mountain, New York, cloud droplets measured (Pueschel *et al.*, 1986); (∗) central Ontario, Canada, cloud droplets measured (Leaitch *et al.*, 1992); and (●) North Atlantic, cloud-processed accumulation mode particles measured (Van Dingenen *et al.*, 1995) (adapted from Van Dingenen *et al.*, 1995).

at different supersaturations versus cloud droplet concentrations, the large range in sulfate concentrations, etc.), such correlations ($r^2 = 0.42$) suggest that it is possible to relate particle mass to CCN and cloud droplet number concentration, and ultimately to changes in cloud albedo, albeit within a large uncertainty.

However, some caution is needed in applying such correlations, as might be expected, particularly given our current relatively rudimentary understanding of the indirect effects of aerosol particles on clouds. For example, the data in Fig. 14.45 include both CCN and cloud droplet number concentrations, which both appear to be correlated to nss. However, this is not always the case. While CCN and non-sea salt mass concentrations were observed to be highly correlated at a marine site in Puerto Rico, cumulus cloud droplet number concentrations were not, and stratocumulus cloud droplet number concentrations showed a very low sensitivity to nss (Novakov *et al.*, 1994). Similarly, Anderson and co-workers (1994) did not observe a convincing relationship between the cloud droplet number concentration and the aerosol number or volume at a coastal mountain site in the state of Washington. Entrainment and mixing processes in the clouds may have played major roles in these (and, of course, many other) studies. Finally, some studies suggest that particularly at very small sulfate concentrations, a wide range of CCN can be observed (e.g., Hegg, 1994), which may be related to non-sulfate species acting as CCN.

Indeed, while the initial focus on the indirect effects of anthropogenic aerosols has been on sulfate, there is increasing evidence that other species may also not only contribute significantly to CCN but actually dominate it under many circumstances. For example, although CCN at 1% supersaturation were correlated with sulfate in air masses over the northeast Pacific and the northeast Atlantic, the slope of the curve relating the two was much higher for the relatively clean northeast Pacific (Hegg *et al.*, 1993). The authors suggest that this is consistent with DMS as the major source of sulfate over the Pacific. Over the Atlantic, however, the slope of CCN versus sulfate was smaller and there was a significant intercept, suggesting that much of the CCN was not formed from sulfate. Observations at a coastal site in the state of Washington also suggested that components other than sulfate may be important in CCN formation at 0.9% supersaturation (Berresheim *et al.*, 1993).

Novakov and Penner (1993) measured the mass size distributions of sulfur, organic carbon, and chlorine (characteristic of sea salt) as well as the CCN concentration (at 0.5% supersaturation), nss, and Aitken nuclei concentrations at a mountain peak in Puerto Rico. They concluded that about 63% of the CCN at this site was due to organic aerosol particles, possibly due to some unspecified anthropogenic sources. Similar measurements at Point Reyes, California, gave variable contributions of organics and sulfate to CCN, ranging from 4 to 78% for organic particles, from 19 to 64% for sulfate, and from 9 to 31% for NaCl in sea salt particles (Rivera-Carpio *et al.*, 1996). While Andrews *et al.* (1997) suggest that the organic aerosol particles in the Puerto Rican studies may have originated from the rain forest below the mountain peak sampling site, subsequent studies at the mountain site and at a site in the Atlantic Ocean suggest that a large fraction of this organic aerosol may originate from the ocean (Novakov *et al.*, 1997a). A similar oceanic source of CCN measured in Antarctica was suggested by Saxena (1996). It is particularly interesting that most of the organic aerosol particles in the Puerto Rican study were water soluble; in addition, their average mass concentration (390 ng m^{-3}) was larger than that of sulfate (270 ng m^{-3}). This combination of water solubility and relatively high mass fraction suggests that the organic particles may be particularly effective as CCN (Novakov *et al.*, 1997a).

Recent studies have provided additional evidence for the contribution of organics to CCN. For example, Matsumoto *et al.* (1998) measured CCN at 0.5 and 1% supersaturations, along with the aerosol particle composition and size distribution at the Ogasawara Islands in the northwest Pacific Ocean. In agreement with earlier studies, air masses affected by continental emissions had CCN concentrations of $\sim 150-1000$ cm^{-3} (at 1% supersaturation) compared to 30–150 cm^{-3} for clean air masses. Sulfate, nitrate, and ammonium in the particles were correlated with ^{222}Rn, which is a tracer of continental air masses. Oxalate was also found in the particles, primarily in the accumulation mode (<1.1 μm), and was highly correlated with ^{222}Rn, indicating an anthropogenic source. However, formate and acetate were not well correlated with ^{222}Rn, suggesting marine biogenic sources for these species. A major contributor to the aerosol mass (~ 80% of the total mass) was unspecified water-soluble organics.

As discussed in Section C.1a, sea salt particles in the marine boundary layer have been shown to likely play a major role in backscattering of solar radiation (Murphy *et al.*, 1998), i.e., to the direct effect of aerosol particles. However, they also contribute to the indirect effect involving cloud formation, since they can also act as CCN. Since such particles are a natural component of the marine atmosphere, their contribution will not play a role in climate *change*, unless their concentration were somehow to be changed by anthropogenic activities, e.g., through changes in wind speed over the oceans, which largely determines the concentration of sea salt particles (Gong *et al.*, 1997a, 1997b). However, the presence of sea salt particles can still have an impact on the effects of *anthropogenically* derived species. Thus, activation of sea salt particles results in lowering of the peak supersaturation in the cloud. From the Köhler curves (Fig. 14.38), this means that the size of other particles such as non-sea salt sulfate (nss) must be larger in order to activate into cloud droplets. If fewer anthropogenic particles grow into this larger size, the number of cloud drops formed is reduced and the anthropogenic contribution to cloud droplet formation is proportionately smaller (O'Dowd *et al.*, 1997a, 1997b).

Sea salt particles also provide an aqueous medium for the oxidation of SO_2 to sulfate (e.g., see Sievering *et al.*, 1992; Chameides and Stelson, 1992; and O'Dowd *et al.*, 1997a, 1997b) and hence play a role in both the direct and indirect forcing by sulfate particles. As discussed in Chapter 8.C.3, such aqueous-phase processes, which usually dominate the overall conversion of SO_2 to sulfate, are pH dependent. This is in part due to the decreasing concentrations of dissolved S(IV) in the aqueous phase as the pH falls and in part due to the dependence of the reaction kinetics on pH. Seawater is basic (pH ~8) so that newly formed sea salt particles are likely basic when initially formed. Under these conditions, oxidation by O_3 is important (see Chapter 8.C.3d). This oxidation dominates until the alkalinity of the droplet has been consumed by the acid formed. As the pH of the droplet falls, oxidation by H_2O_2 and the gas-phase oxidation by OH become relatively more important. While the pH of atmospheric sea salt particles has not been well established, experimental studies in Bermuda under moderately polluted conditions suggest that it can be in the range 3.5–4.5 (Keene and Savoie, 1998, 1999). Oxidation by HOCl, believed to be an important intermediate in halogen chemistry in the marine boundary layer, has been proposed to be important as well (e.g., see Vogt *et al.*, 1996; Keene *et al.*, 1998; and Chapter 8.C.3).

The significance of this oxidation of S(IV) in sea salt particles is that if it occurs in existing aerosol particles, sulfate formation will not result in new particles and hence potentially new CCN, but rather contribute to the mass of existing particles (e.g., O'Dowd *et al.*, 1997b). A significant fraction of all particulate nss is believed to be generated by this oxidation in existing sea salt particles.

It has also been proposed that the uptake of gases such as HNO_3 and HCl onto particles may alter their ability to act as CCN (e.g., see Kulmala *et al.*, 1993, 1995, 1998; and Laaksonen *et al.*, 1997). Clearly, these are areas that need much further investigation.

In short, it is becoming clear that although the focus to date has been mainly on sulfate, the effects of other components, including both natural and anthropogenic species, need to be taken into account in both the direct and indirect effects of particles on global climate.

Given the evidence for a relationship between anthropogenic emissions and CCN, the next link to global climate is the assumption that increased CCN lead to increased cloud droplet concentrations (N). As seen in Eqs. (JJ) and (KK), increased concentrations affect both cloud albedo and its sensitivity to changes in the cloud droplet number. There is a great deal of evidence gathered over decades for a relationship between increased CCN and increased concentration of droplets in clouds. For example, some 30 years ago Warner and Twomey (1967) measured cloud droplet number concentrations and condensation nuclei at 0.5% supersaturation below the base of clouds upwind (over the ocean) and downwind of a region in which sugar cane was burning. The average concentration of CCN was 280 cm^{-3} over the ocean but 750 cm^{-3} downwind of the fire; the cloud droplet number concentration similarly increased from 300 to 920 cm^{-3}. The relationship between the average cloud droplet number and that expected from transport of the below-cloud CCN into the cloud was approximately linear in that particular case, as well as in other sets of measurements carried out at other locations under more normal conditions (e.g., Twomey and Warner, 1967).

Similarly, Martin and co-workers (1994) measured aerosol particles in the size range from 0.05 to 1.5 μm below the base of stratocumulus clouds, along with cloud droplet number concentrations in maritime and in continental air masses. Figure 14.46 shows the relationship between cloud droplet number concentration and the aerosol particle concentration for a set of flights carried out in the vicinity of the British Isles and in the South Atlantic (Martin *et al.*, 1994). There is an almost linear relationship between the two for maritime air masses. Given that the cutoff for particle measurements was 0.05 μm, these concentrations may have been underestimated, so that the slope of the line for maritime air masses can be taken as unity. That is, essentially all of the maritime particles at the cloud base could act as CCN under the range of supersaturations in these studies.

However, this relationship did not hold true for continental air masses. The fraction of aerosol particles that lead to cloud droplet formation is clearly less than one, in agreement with the studies of Gillani *et al.* (1995) discussed earlier. In addition, the relationship is much more scattered, indicating that the chemical

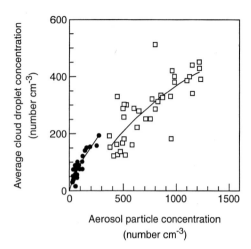

FIGURE 14.46 Average cloud droplet number concentration as a function of subcloud aerosol particle concentration (0.05–1.5 μm) in marine (●) and continental (□) air masses (adapted from Martin et al., 1994).

composition and hence ability to act as CCN are much more variable over the continents.

Number concentrations of ice crystals in cirrus clouds have also been observed to increase with aerosol particle concentrations (with diameters >0.018 μm) and, in particular, with the concentration of light-absorbing materials in the ice crystals (Ström and Ohlsson, 1998).

Not only do CCN affect the number of cloud droplets formed, but they also affect the size distribution of these droplets. This also affects cloud albedo and its sensitivity to changes in the number concentration (see Eqs. (JJ) and (KK)). Figure 14.47, for example, shows the size distribution for cloud droplets measured in urban and nonurban air around Denver, Colorado (Alkezweeny et al., 1993). The median volume diameter was 14 μm for the urban air cloud, and this was only ~50% of that of the much larger droplets in the nonurban air cloud. As expected, the cloud number concentration in the urban air cloud was also larger, 226 cm^{-3} compared to 22 cm^{-3} for the nonurban cloud. Analogous results have been reported in other studies (e.g., Pueschel et al., 1986). For example, based on satellite data, the radii of cloud drops over the oceans were observed to be 2–3 μm larger than over continental regions; in addition, marine cloud droplet radii in the Southern Hemisphere are ~1 μm larger than in the Northern Hemisphere (Han et al., 1994).

Hudson and Li (1995) measured aerosol particle concentrations below clouds, as well as various cloud parameters, in polluted as well as clean air masses in aircraft measurements around the Azores. Table 14.8 summarizes some of these data. Consistent with the studies discussed above, the polluted air mass had increased aerosol particle concentrations, increased CCN, and larger numbers of, but smaller sized, cloud droplets.

Albrecht (1989) suggested that increasing CCN concentrations would lead to decreasing cloud drop size and decreased drizzle production in marine stratocumulus and fair-weather cumulus clouds, leading to an increase in the geographical extent of clouds as well as their lifetime. Modeling studies suggest that this could be a significant effect (Lohmann and Feichter, 1997). Clearly, this too could play a role in global climate change. The studies by Hudson and Li (1995) also reported evidence for this effect in that the number of "drizzle drops" with diameters >50 μm was smaller (by an order of magnitude) in the cloud in the polluted air mass (Table 14.8). Related to this is the observation by Parungo et al. (1994) that there has been an increase in total oceanic clouds from 1930 to 1981, with the change in the Northern Hemisphere (2.3%) being about double that for the Southern Hemisphere (1.2%).

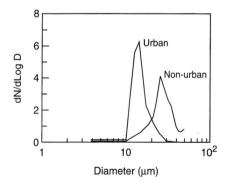

FIGURE 14.47 Cloud droplet size distributions for stratiform clouds in the Denver area for urban and nonurban air masses (adapted from Alkezweeny et al., 1993).

TABLE 14.8 Some Cloud and Subcloud Aerosol Properties Measured in Polluted and Clean Air Masses near the Azores[a]

Property	Clean air	Polluted air
Aerosol particle concentration	176 cm^{-3}	806 cm^{-3}
CCN concentration		
at $S = 0.7\%$	116 cm^{-3}	668 cm^{-3}
at $S = 0.04\%$	2 cm^{-3}	119 cm^{-3}
Cloud drop concentration[b]	10–100 cm^{-3}	220–370 cm^{-3}
Mean cloud drop diameter[b]	18 μm	7–9 μm
Drizzle drop concentration[c]	800 L^{-1}	80 L^{-1}

[a] Adapted from Hudson and Li (1995).
[b] Diameters 3–66 μm; upper end overlaps lower end of drizzle drop range.
[c] Diameters 50–600 μm defined as drizzle drops.

Parungo and co-workers suggest this may be due to the indirect effects of increasing SO_2 emissions.

The indirect effect of aerosols on climate, which at present contributes a major uncertainty in understanding anthropogenic perturbations on climate, is a very active area of research. For some typical model treatments of this indirect effect and how it interacts with those due to other, simultaneous, perturbations, see, for example, Jones et al. (1994), Hansen et al. (1997a–d), C. C. Chuang et al. (1997), Lohmann and Feichter (1997), and Pan et al. (1998).

Figure 14.48 shows one assessment (Hansen et al., 1997d) of the contributions of anthropogenic emissions to the average global radiative forcing from preindustrial times to the present as well as that due to changes in solar intensity over the past 200 years (see Section D.3). The contributions due to an increase in tropospheric O_3 from preindustrial times to 1980 and that due to stratospheric ozone destruction from 1979 to 1995 are predicted to essentially cancel out. Three contributions due to changes in tropospheric aerosol particles are included. Desert aerosols give a positive radiative forcing because of their absorption of light discussed earlier, whereas sulfate and biomass particles scatter light, leading to a negative radiative forcing. The indirect effect of particles on clouds has very large uncertainties associated with it and is shown as -1 W m^{-2} in Fig. 14.48. Finally, changes in vegetation are estimated to have contributed -0.2 W m^{-2}, due the reduction in the area of forests, which are dark.

As discussed in IPCC (1996), the confidence level associated with these values ranges from high for the greenhouse gases to very low for tropospheric aerosols, and in particular for the indirect effects. For example, the calculations of Penner et al. (1998) suggest a larger direct radiative forcing due to sulfate aerosol particles (-0.81 W m^{-2}) than that shown in Fig. 14.48 and a global average contribution of $+0.16$ W m^{-2} for fossil fuel black and organic carbon particles. The contribution of changes in the solar flux and uncertainties in this are discussed in Section D.3.

The uncertainties in the indirect effects on clouds are very large. As discussed earlier, increased CCN can alter the properties of clouds in several ways that can impact climate. Thus, they can lead to changes in cloud albedo and, in addition, alter the size distribution of cloud droplets, changing the precipitation rate and hence cloud lifetime. For example, Lohmann and Feichter (1997) carried out model studies of the indirect effects of sulfate on clouds and predicted increases in shortwave cloud forcing ranging from -1.4 to -4.8 W m^{-2}. A significant portion of the effects was due to changes in cloud lifetime; for example, for the -1.4 W m^{-2} case, about 40% was attributed to changes in the cloud lifetime and 60% to changes in cloud albedo.

It is important to note that such globally and annually averaged estimates of contributions to radiative forcing are not expected to be the sole measures of effects on climate. The inference may be mistakenly drawn that negative radiative forcing, e.g., through

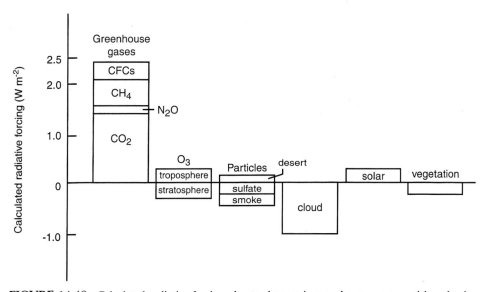

FIGURE 14.48 Calculated radiative forcings due to changes in greenhouse gases, particles, clouds, solar radiation, and vegetation from preindustrial times to 1995. That due to changes in stratospheric ozone is for the 1979–1995 period (adapted from Hansen et al., 1997d).

tropospheric aerosol particles, may largely counterbalance the positive radiative forcing due to the greenhouse gases and hence there will be no net change in climate. This is not expected to be the case since the effects operate on different geographical and temporal scales. Thus, many of the greenhouse gases (e.g., CO_2, CH_4, N_2O, and the CFCs) are sufficiently long-lived to be globally distributed. Their contributions to radiative forcing vary geographically, from about 3 W m^{-2} over hot regions such as the Sahara to ~0.6 W m^{-2} over the South Pole (National Research Council, 1996). Shorter-lived greenhouse gases such as O_3 have much more spatial and temporal variability, with associated differences in their contributions to radiative forcing. However, the contribution of all greenhouse gases to radiative forcing operates both day and night since it involves their interaction with terrestrial radiation.

On the other hand, aerosol particles from anthropogenic activities tend to be concentrated over or near industrial regions in the continents. Because both the direct and indirect effects of particles are predominantly in terms of scattering solar radiation, their effects are expected primarily during the day.

For example, model studies by Sinha and Harries (1997) have explored a hypothetical case in which CO_2 is doubled to 710 ppm and the amount of tropospheric aerosol is increased about a factor of four, giving no net change in the predicted equilibrium surface temperature. However, even with a predicted net surface temperature change of zero, significant effects on climate are still predicted. The solar radiation at the surface at mid and low latitudes is predicted for this hypothetical case to decrease by as much as -6 W m^{-2} in January. Similarly, the vertical distribution of the rate of total radiative heating is predicted to change by more than 4% at some altitudes, which would be expected to lead to changes in the lapse rate, potentially affecting atmospheric circulation processes.

In addition to the differences in geographical distribution of the greenhouse gases compared to the aerosol particles and the day–night differences, there are also differences in their temporal behavior. As discussed earlier, typical residence times for sulfate particles are about a week, whereas that of CO_2 is about 100 years. As a result, the impacts of sulfate aerosols are almost immediately manifested, whereas those due to CO_2 occur over decades to centuries (Schwartz, 1993).

Hansen and co-workers have carried out modeling studies that examine the effects of various perturbations, both anthropogenic and natural, on climate (Hansen et al., 1997a–c). The altitude and geographical location of the forcings are shown to be important determinants of the effects on climate, rather than simply the magnitude of the forcing. For example, the addition or removal of heat in the upper troposphere is partially compensated by changes in radiation to space, which does not occur close to the earth's surface (Hansen et al., 1997b).

In short, while net radiative forcing is a convenient means for examining the potential importance of various anthropogenic perturbations for climate, it cannot be used in an additive manner for gases and aerosol particles to predict the ultimate impacts.

b. Heterogeneous Chemistry Involving Climate Species

Another potential contribution of aerosol particles to global climate is that of heterogeneous chemistry. For example, particle surfaces could in principle destroy greenhouse gases such as ozone that are surface sensitive. Another example is the formation of greenhouse gases such as N_2O on surfaces. Thus, nitrous acid (HONO), which is itself formed by heterogeneous reactions on surfaces, has been shown to react on acid surfaces to generate N_2O by a mechanism that is not well understood (e.g., see Wiesen et al., 1995; Pires et al., 1996; and Pires and Rossi, 1997). Given the present lack of understanding of the reaction mechanism, it is not possible to assess the importance of such heterogeneous chemistry for N_2O formation in the atmosphere. However, it does illustrate the potential for heterogeneous chemistry on aerosol particles to impact global climate through the effect on gas-phase species.

D. SOME OTHER FACTORS AFFECTING GLOBAL CLIMATE

As discussed at the beginning of this chapter, the focus here is on the relationship between atmospheric chemistry and global climate change, rather than on the magnitude of this change and its causes. However, to place the role of atmospheric chemistry in context, we briefly treat in this section some other important factors known to be involved, or thought to be potentially involved, in climate change. For a more detailed discussion of these and other related issues, see IPCC (1996).

1. Absorption of Solar Radiation by Clouds

Although clouds form on existing particles, the "solutions" formed are quite dilute and hence are not expected to absorb solar radiation significantly. As a result, it has been commonly accepted that clouds will predominantly reflect solar radiation and that absorption will not be significant. However, as early as 1951, it was suggested that clouds appeared to be absorbing

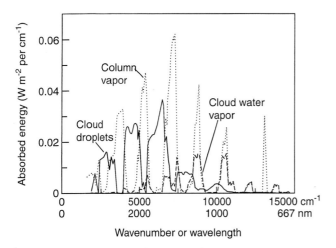

FIGURE 14.49 Absorption of light from an overhead sun by water associated with a 1-km stratus cloud with its top at an altitude of 2 km. The solid line is the absorption due to liquid water, the dashed line water vapor inside the cloud, and the dotted line water vapor in a column in the atmosphere (adapted from Davies *et al.*, 1984).

significantly as well (Fritz, 1951). This has been dubbed the "cloud absorption anomaly" [see reviews by Stephens and Tsay (1990), Liou (1992), and Ramanathan and Vogelmann (1997)]. Although this area might not be considered to fall in the realm of "atmosphere chemistry" per se, it is clearly potentially very important in the relationship between clouds and global climate.

Figure 14.49 shows the absorption of light from an overhead sun by liquid cloud droplets, water vapor inside the cloud, and water vapor in a column in the atmosphere for a 1-km stratus cloud whose top is 2 km above the ground (Davies *et al.*, 1984; see also Goldstein and Penner, 1964). There are small amounts of absorption in the tail end of the red region of the visible attributed to water vapor in and outside the cloud. The absorption increases into the near-IR (the region from ~780 to 2500 nm or 12,800–4000 cm^{-1}) and mid-IR (2.5–50 μm or 4000–200 cm^{-1}) where liquid water in the cloud absorbs (e.g., see Evans and Puckrin, 1996).

Several different approaches have been taken to investigate whether there is more absorption of visible light by clouds than expected based on current models of radiative transfer in the atmosphere. Some of these approaches and results are discussed in Box 14.3.

While there thus appears to be evidence for apparent excess absorption of solar radiation by clouds, there is substantial controversy over whether this is indeed true absorption or whether there is some other explanation for the discrepancies (e.g., see Stephens and Tsay, 1990; Imre *et al.*, 1996; Stephens, 1996; Cess and Zhang, 1996; Pilewskie and Valero, 1996; and Ramanathan and Vogelmann, 1997). For example, Li *et al.* (1995) also analyzed solar flux surface and satellite data over a 4-year period to obtain values of the ratio of shortwave cloud forcing at the surface to that at the top of the atmosphere. These varied from about 1.4 in the tropics, in agreement with Cess *et al.* (1995), to values less than 1 in polar regions. They concluded that, within the uncertainties, their analysis does not provide support for excess cloud absorption of solar radiation (Li *et al.*, 1995; Li and Moreau, 1996), although Zhang *et al.* (1997) suggest there may have been some unrecognized complexities in the analysis of the satellite data.

Similarly, Chou *et al.* (1998) used measurements of surface radiative fluxes and satellite radiance data in the Pacific warm pool region to conclude that the effect of clouds was similar to that expected, i.e., that the excess absorption, if it exists, is small.

Based on aircraft measurements, Francis *et al.* (1997) suggested that the excess cloud absorption was insignificant. In comparing satellite and ground-based observations of solar flux to model predictions, Arking (1996) also found no evidence for significant cloud excess absorption, although he reported a discrepancy between models and measurements that is not found over the central equatorial Pacific Ocean under clear skies (Conant *et al.*, 1997). Imre *et al.* (1996) used collocated satellite and surface observations of short-wavelength fluxes at a site in Oklahoma to probe for the contribution of enhanced absorption by clouds, but found none. They suggested that uncertainties and biases in the analyses, particularly in the clear-sky references used, can give rise to apparent excess cloud absorption that is an artifact of the analysis.

In short, whether excess absorption of solar radiation by clouds even occurs and why there are discrepancies between models and measurements remain controversial. For example, absorption into the visible region can be enhanced by the presence of strongly absorbing species such as soot, either in the cloud droplets themselves or as aerosol particles suspended between the cloud droplets, i.e., as interstitial aerosol particles (e.g., see Stephens and Tsay, 1990; Chýlek and Hallett, 1992; and Mel'nikova and Mikhaylov, 1994). However, as discussed shortly, the enhanced absorption of solar radiation by clouds has been reported in many locations globally, including in remote regions, and hence appears less likely to be so directly associated with anthropogenic emissions. It has also been suggested that measurements of cloud drop size distributions have missed the presence of larger "drizzle drops"

BOX 14.3
SOME INVESTIGATIONS OF THE CLOUD ABSORPTION ANOMALY

One approach to the cloud absorption anomaly has been to examine the energy budget of the so-called "warm pool" area in the western Pacific Ocean, which is the region from approximately 140°E to 170°E and 10°N to 10°S (Ramanathan et al., 1995). This area has relatively high annual mean sea-surface temperatures (SST), up to 302.5 K, resulting in a humid and cloudy atmosphere with which is associated frequent deep convection. The annual mean heat transport (D) both horizontally and vertically out of the ocean surface mixed layer is known to be small, $D \leq 20$ W m^{-2}. To give a constant temperature of the mixed layer, this small heat transport, D, out of the layer must be balanced by the net energy flux (H) at the ocean surface. As shown in Fig. 14.50, H is a balance between positive contributions from incoming solar radiation, S, and negative contributions from net outgoing thermal radiation, F (i.e., up minus down), and evaporative (E) and turbulent sensible (h) heat fluxes, i.e., $H = S - F - (E + h)$. Using available data to estimate values for all of the energy terms except S leads to a calculated value for the solar input of 175 W m^{-2}. However, the solar radiation input under clear skies is well known to be 275 W m^{-2}. If the total solar input is treated as the sum of incoming solar radiation and a contribution due to clouds, then the latter must be -100 W m^{-2}.

The outgoing solar radiation at the top of the atmosphere in this location based on 5 years of data from the Earth Radiation Budget Experiment (ERBE) on the Earth Radiation Budget Satellite (ERBS) is estimated to be -66 W m^{-2} (the negative sign indicating cooling by reflection of solar radiation by clouds). The ratio of the shortwave cloud energy flux at the surface to that at the top of the atmosphere must be $100/66 = 1.5$, representing a long-term average for this effect.

A similar conclusion is reached using direct measurements of solar fluxes at the top of the atmosphere (TOA) and at surface sites under clear compared to cloudy conditions (e.g., Cess et al., 1995, 1996b; Evans et al., 1995). Figure 14.51a shows the absorptance, defined as the fraction of the down-

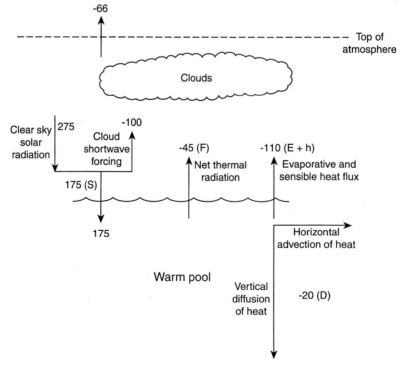

FIGURE 14.50 Schematic of energy balance in "warm pool" in western Pacific Ocean used to deduce the net effect of clouds on solar radiation. All numbers are given in W m^{-2} (adapted from Ramanathan et al., 1995).

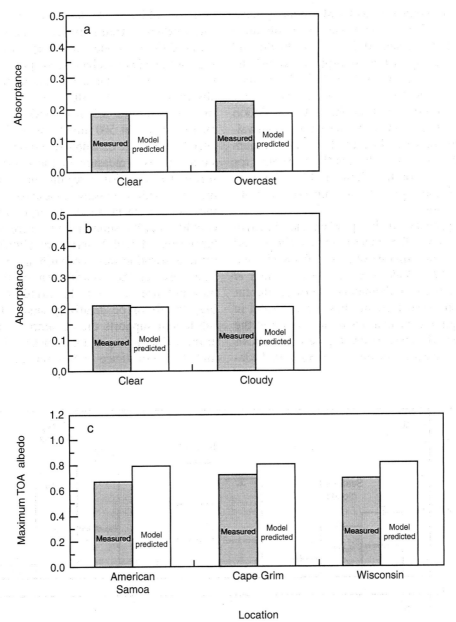

FIGURE 14.51 (a) Fraction of downward solar radiation at the top of the atmosphere that is absorbed, measured at Boulder, Colorado, under clear or cloudy skies compared to CCM2 model predictions; (b) same as (a) but in warm pool area over the central Pacific compared to the CCM3 model; (c) measured maximum albedo at the top of the atmosphere (TOA) at three locations compared to CCM2 model predictions (adapted from Cess et al., 1995, 1996b; and Valero et al., 1997a).

ward solar radiation at the top of the atmosphere that is absorbed, at Boulder, Colorado, under clear and cloudy skies (Cess et al., 1995, 1996b). Also shown are the values predicted using one climate model (Community Climate Model version 2, CCM2), which does not include enhanced cloud absorption of solar radiation. The agreement between observations and model predictions is excellent for clear skies. However, for cloudy skies the observed absorption is, on average, ~30% larger than the model predictions. Similar conclusions are reached using data over the warm pool (Fig. 14.51b). Such data give values for the ratio of the shortwave cloud forcing at the surface to that at the top of the atmosphere of 1.5, the same as that obtained using the energy balance approach over the warm pool.

Related to this are the maximum albedos measured at the top of the atmosphere at three locations, which are compared to CCM2 model predictions in Fig. 14.51c. These values were obtained from plots of broadband (0.2–5 μm) albedo (α) measured at the top of the atmosphere against the transmittance of the atmosphere (T), determined from the ratio of downward solar flux at the surface to that at the top of the atmosphere. Extrapolation to zero transmittance gives the maximum mean values of albedo shown in Fig. 14.51c. The maximum measured value is ~0.7, whereas the maximum for model predictions for this limit of thick clouds is ~0.8, perhaps indicative of some unaccounted absorption by clouds.

Another approach to this problem has been to make simultaneous flux measurements above and below clouds using aircraft (e.g., see Pilewskie and Valero, 1995, 1996; Valero *et al.*, 1997a; Zender *et al.*, 1997). The use of radiometers covering different spectral regions in such studies has proven useful in deconvoluting the contributions of absorption in the visible and near-IR, respectively. Figure 14.52a shows absorptances measured in one such study in Oklahoma, and Fig. 14.52b shows the corresponding measured values of transmittance (Valero *et al.*, 1997a). Three different values of absorptance are shown in Fig. 14.52a: (1) broadband absorptance from 224 nm to 3.91 μm; (2) the contribution of near-infrared absorptances, in this study defined as the region from 680 nm to 3.3 μm; and (3) spectral band measurements at 500 nm (10-nm width). The difference between the broadband and near-infrared measurements is a measure of the contribution of absorption in the visible region, and it is seen that it increases with increasing cloudiness. Consistent with the evidence for increased absorption with increased cloudiness is the simultaneous decrease in transmittance (Fig. 14.52b). Valero *et al.* (1997a) suggest that if the absorption were primarily in the near-IR, one would expect the contribution of the near-IR to transmittance seen for the cloudiest conditions in Fig. 14.51b to be relatively small; that it is still substantial supports the contention that absorption in the visible is important. However, it is interesting that the absorptance at 500 nm (Fig. 14.51a) shows

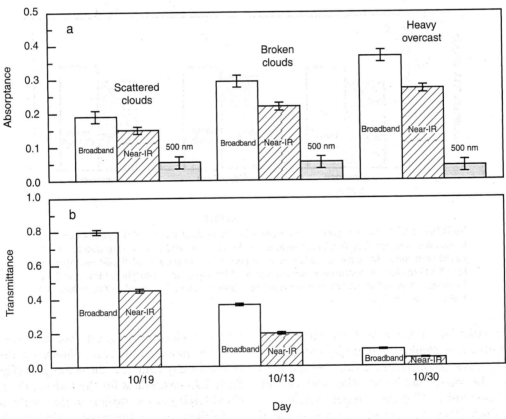

FIGURE 14.52 (a) Absorptance and (b) transmittance measured on days with varying degrees of cloudiness using aircraft colocated above and below the clouds where broadband (224 nm to 3.91 μm), near-IR (680 nm to 3.30 μm), and spectral band (10-nm width centered at 500 nm) measurements were made (adapted from Valero *et al.*, 1997a).

no evidence of increased absorption with increasing cloudiness; whatever the absorbing species is (if indeed there is one; see below) it does not appear to absorb light at 500 nm.

The difference between the broadband measured absorption (224 nm to 3.91 μm) and that from the near-IR radiometer (680 nm to 3.3 μm) was used as a measure of the visible light absorption (Zender *et al.*, 1997). On the cloudiest day during these studies, about 25% of the shortwave absorption was attributable to visible light, compared to 10% predicted by a model (Zender *et al.*, 1997). These studies also report values for the ratio of the shortwave cloud forcing at the surface to that at the top of the atmosphere in the range of 1.36–1.65, in agreement with the warm pool energy balance studies, but again significantly larger than model-predicted values of 1.12–1.14. Zender *et al.* (1997) also show that several other measures of cloud absorption, such as the slope of plots of albedo versus transmittance, are consistent with excess absorption of solar radiation that is not included in the models. Measurements of the single-scattering albedo (the fraction of incident energy that appears as scattered radiation) inside a marine stratocumulus cloud implied a small contribution from enhanced absorption as well (King *et al.*, 1990).

(Wiscombe and Welch, 1986). Cloud absorption is related to droplet size in a complex way (Stephens and Tsay, 1990) so that errors in droplet size measurements can alter the model-predicted absorption by a cloud. The treatment of absorption due to water vapor is another possibility. As discussed by Crisp (1997), the treatment of water vapor in models is simplified and may not properly reflect, for example, continuum absorptions between major bands in the near-IR. Model calculations suggest that the presence of a thin, saturated layer of water vapor above the clouds, for example, leads to increased absorption by 2–6% (Davies *et al.*, 1984; Podgorny *et al.*, 1998). However, the Crisp calculations indicate that this cannot account for all of the observed excess absorption.

In addition, it has been suggested that the excess absorption is really due to radiation escaping from the sides of clouds, which would not be observed in measurements carried out above and below an isolated cloud (Ackerman and Cox, 1981). However, Valero *et al.* (1997a) argue that this should not give significant errors for continuous measurements made over clouds and analyzed with sufficiently long averaging times. Finally, whether the models properly capture the radiative effects, e.g., due to the heterogeneity of drop sizes in clouds and the cloud shape, is not clear (e.g., see Chou *et al.*, 1995; Lubin *et al.*, 1996; Byrne *et al.*, 1996; and Loeb and Davies, 1996). For example, while many models assume plane parallel cloud geometry, the absorption for other shapes such as wavy, broken clouds can be different by as much as 10–15%, depending on the solar zenith angle (Podgorny *et al.*, 1998).

One important aspect of radiation and clouds that may ultimately prove to be important in this issue of excess cloud absorption is the very long effective path lengths for light inside clouds due to multiple scattering processes. For example, Pfeilsticker *et al.* (1997) measured the absorption of solar radiation by the oxygen collision complex $(O_2)_2$ under clear and cloudy sky conditions. The absorption is sufficiently weak that it is in the linear, "weak-absorber" regime (see Section A.3). Based on the Beer–Lambert law, the effective path length can be calculated from the known absorption cross section and atmospheric concentrations, combined with measurements of its atmospheric absorbance. For one cloud, for example, this approach gave an extra effective path length of 135 km, about an order of magnitude larger than the clear-sky geometrical path length! The presence of even a relatively weak absorption that is not accounted for in the models could have a disproportionate effect with such large effective path lengths (e.g., see Kondrat'ev *et al.*, 1996a,b).

If this "excess absorption" by clouds is ultimately shown to be a real phenomenon, then an increased cloud formation and extent due to anthropogenic emissions may alter the radiative balance of the atmosphere not only through increased reflectance but also through increased absorption of solar radiation. Such an effect could impact atmospheric temperatures, their vertical distribution, and circulation, as well as surface wind speeds and the surface latent heat flux (Kiehl *et al.*, 1995). Hence establishing if this is truly excess absorption, and if so, its origins, is a critical issue that remains to be resolved.

2. Feedbacks: Water Vapor, Clouds, and the "Supergreenhouse Effect"

Given the complexity of the ocean–atmosphere–biosphere system, it is not surprising that there are a number of feedbacks that greatly complicate the accurate prediction of the effects of anthropogenic or natural emissions. For example, warming is expected to lead to decreased amounts of clouds and hence to changes in cloud contributions to radiative forcing [see a comparison of model predictions for cloud feedbacks in Cess *et al.* (1996b)]. The DMS–CCN–cloud formation/reflectance discussed earlier is another such example.

Another important example involves water vapor in the atmosphere. Water vapor is the most important greenhouse gas, and its concentration in the atmosphere is a function of temperature as given by the Clausius–Clapeyron equation:

$$\ln \frac{P_2}{P_1} = -\frac{\Delta H_{vap}}{R}\left[\frac{1}{T_2} - \frac{1}{T_1}\right]. \quad \text{(LL)}$$

In Eq. (LL), ΔH_{vap} is the heat of vaporization of water and R is the gas constant. Thus the vapor pressure of water has an exponential dependence on temperature. This suggests that there may be a water vapor feedback associated with global climate change. If the atmosphere warms, for example due to increased greenhouse gases such as CO_2, increased concentrations of gaseous water are expected in accordance with Eq. (LL). The increased water vapor traps more thermal infrared radiation, warming the atmosphere further (e.g., Raval and Ramanathan, 1989; Stenchikov and Robock, 1995).

However, there may be additional feedbacks that limit what would otherwise be a runaway system. For example, Ramanathan and Collins (1991) proposed that there is a natural "thermostat" mechanism over the warm pool in the Pacific Ocean that limits the sea surface temperature (SST) from rising above 305 K. This mechanism consists of triggering deep convection when the SST exceeds ~300 K, resulting in the formation of thick anvil clouds. Reflection of solar radiation back to space then acts to cool this region, providing a negative feedback and acting like a thermostat. Consistent with this thermostat mechanism, Waliser *et al.* (1993) report that for SST between 299 and 303 K in the western Pacific, the frequency of highly reflective clouds and decreases in outgoing longwave radiation are correlated with the SST. Interestingly, between 303 and 305 K, these relationships reversed; i.e., the frequency of highly reflective clouds decreased and outgoing longwave radiation increased with SST.

This hypothesis has been somewhat controversial. For example, evaporative cooling of the ocean surface (Fu *et al.*, 1992) and large-scale dynamical feedbacks (Wallace, 1992) have also been suggested as being important in the feedback, suggesting that changes in cloud cover reflect changes in atmospheric circulation rather than changes in SST. In addition, Chou *et al.* (1998) found that the regions of maximum sea surface temperature did not necessarily coincide with those of maximum cloudiness, as expected if such feedbacks were operative. Some of these uncertainties are summarized by Stephens and Slingo (1992) and further discussion is found in Ramanathan and Collins (1992, 1993), Lau *et al.* (1994), Ramanathan *et al.* (1994), and Fu *et al.* (1993).

However, there are additional data supporting the relationship between the greenhouse effect and SST. As discussed by Valero *et al.* (1997b), the greenhouse effect (G) can be expressed as

$$G = \sigma(\text{SST})^4 - F^+, \quad \text{(MM)}$$

where σ is the Stefan–Boltzmann constant (5.67×10^{-8} W m^{-2} K^{-4}), $\sigma(\text{SST})^4$ is the thermal emission from the surface (see Section A.1), and F^+ is the outgoing radiation flux at the top of the atmosphere. The rate of change of thermal emission by the ocean surface with SST is given by

$$\frac{d[\sigma(\text{SST})^4]}{d(\text{SST})} = 4\sigma(\text{SST})^3, \quad \text{(NN)}$$

which for an ocean temperature of 300 K, typical of the tropics, is 6.1 W m^{-2} K^{-1}. This is what one would expect for the increase in thermal emission from the ocean surface as the SST increases. If the outgoing flux at the top of the atmosphere remains constant, the rate of change of the greenhouse effect with SST should also be about 6.1 W m^{-2} K^{-1}. However, measured values in the tropics under clear skies exceed this value by a factor of about two, which has been dubbed the "supergreenhouse effect."

For example, Fig. 14.53 shows G as a function of SST for SST > 300 K measured using airborne infrared

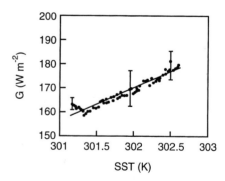

FIGURE 14.53 Measured values of the clear-sky greenhouse effect G [see Eq. (MM)] using measured upwelling infrared irradiance at an altitude corresponding to 191 mbar as a function of sea surface temperature (SST) over the central equatorial Pacific for SST > 300 K (adapted from Valero *et al.*, 1997b).

radiometers over the central equatorial Pacific to obtain F^+ and a combination of satellite and *in situ* data to obtain SST (Valero *et al.*, 1997b). It is clear that the greenhouse effect increases approximately linearly with SST. The slope of such plots gave values of $dG/d(SST)$ from 13.5 to 15.3 W m^{-2} K^{-1}, more than double the 6.1 W m^{-2} K^{-1} expected for the increase in the ocean surface thermal emission [Eq. (NN)]. Furthermore, the extent of the area over which this occurred was quite large, about half of the tropical ocean between 20°N and 20°S. The absorbed energy is radiated back to the surface to further contribute to surface warming (Valero *et al.*, 1997b).

In addition to these feedbacks involving water vapor and clouds, others are expected involving ice and snow. These surfaces are highly reflecting so that if warming leads to increased exposure of the underlying, darker surfaces, further warming will occur to give a positive feedback.

3. Solar Variability

Because it is the sun that drives the earth's energy balance (Fig. 14.2), even small variations in its output can significantly alter the earth's climate. Orbital variations of the earth relative to the sun which resulted in changes in the geographical distribution of solar radiation and, to a lesser extent, small changes (<1%) in the annual and global average solar intensity are believed to have affected global climate over the approximately past million years. This is often referred to as the Milankovitch mechanism (e.g., see Imbrie and Imbrie, 1979; Crowley and North, 1991; Lindzen, 1994; and Bryant, 1997). These solar variations have characteristic periodicities of ~20,000, 40,000, 100,000, and 400,000 years, respectively. These changes in solar insolation, and particularly in its geographical distribution, are expected to have affected climate by altering circulation patterns and heat transport in the atmosphere (Lindzen, 1994).

On a much shorter time scale, the radiant energy from the sun, the "solar constant," currently averages 1368 W m^{-2}. However, there is natural variability around this mean due to bright solar faculae and dark sunspots. In particular, there is a solar cycle approximately 11 years in length that occurs with an amplitude for total irradiance changes of about 0.1%; it is sometimes treated in terms of a "Hale cycle" of ~22 years in length (e.g., see Wilson, 1998). The variation in intensity during the solar or Hale cycles is not constant across the solar spectrum, but is larger in the UV (e.g., see Lean *et al.*, 1995b, 1997). In addition, changes in solar output are modified by the atmosphere before reaching the earth's surface, so that the magnitude of

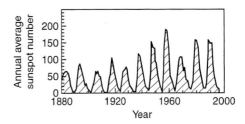

FIGURE 14.54 Annual average number of sunspots from 1880 to 2000, showing the 11-year cycle (adapted from Cliver *et al.*, 1998).

changes due to the solar cycle depends on wavelength, latitude, and altitude (Haigh, 1994). This may have indirect effects on the troposphere by altering stratospheric chemistry (see discussion by Robock (1996) and references therein).

Figure 14.54, for example, shows the annual average number of sunspots from 1880 to the present, which clearly shows this cycle (Cliver *et al.*, 1998). Both the sunspot number and the aa geomagnetic index have been used as proxies for the solar cycle. For the relatively short time period covered by available instrumental temperature records, both the sunspot number and the aa geomagnetic index are correlated to surface temperature (e.g., see Cliver *et al.*, 1998; and Wilson, 1998).

However, there is increasing evidence that longer term solar variations are measurable over the past few centuries as well, and understanding these is very important for discerning anthropogenic effects on global climate (e.g., Lean *et al.*, 1995a; Willson, 1997). Figure 14.55, for example, shows a reconstruction of total solar irradiance from 1610 to the present, in which the 11-year cycle and a component having much longer term variability are both included (Lean *et al.*, 1995a). This two-component model is consistent with the Maunder Minimum that occurred during the years from 1645 to 1715 (Eddy, 1976). During this period, the 11-year cycle did not occur for a number of decades, which was also the coldest period in the "Little Ice

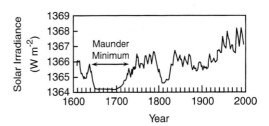

FIGURE 14.55 Reconstructed total solar irradiance from 1610 to 1995 using an 11-year solar cycle plus a longer term component of variability (adapted from Lean *et al.*, 1995a).

Age" from 1450 to 1850. The long-term component has been scaled to agree with the estimate of an overall increase in total irradiance from the Maunder Minimum to the present of 0.24% (Lean et al., 1992, 1995b).

Lean et al. (1995a) have used this reconstruction to estimate how much of the increase in surface temperatures in the Northern Hemisphere can be explained due to solar variability. They concluded that about half of the observed Northern Hemisphere surface temperature increase of 0.55°C since 1860 is due to solar variability but that it only accounts for about a third of the 0.36°C increase since 1970. Similar conclusions have been reached by a number of researchers (e.g., see Kelly and Wigley, 1992; Schlesinger and Ramankutty, 1992; Scuderi, 1993; Crowley and Kim, 1996; Solanki and Fligge, 1998; Cliver et al., 1998; and Wilson, 1998). However, Fröhlich and Lean (1998) have reexamined the solar irradiance record since 1978; they concluded that the irradiances in 1986 and 1996 were similar and that changes in the solar flux during this period could not have contributed significantly to the observed changes in global mean surface temperature.

In short, it is clear that variations in solar output have played a major role in determining the earth's climate in the past, and understanding and quantifying this variability are critical for understanding anthropogenic influences on global climate. The observed temperature increases over approximately the past three decades are larger than expected from solar variability and have been interpreted by many researchers in this field to be the first signs of anthropogenic perturbations on climate.

4. Volcanic Eruptions

As discussed in Section C.1a, major volcanic eruptions have been observed to alter the earth's climate through injection of large amounts of SO_2 into the stratosphere. There it is oxidized to sulfate particles that scatter incoming solar radiation, leading to cooling at the earth's surface. These particles also absorb long-wavelength terrestrial infrared radiation, warming the stratosphere (Fig. 14.30). While this absorption of infrared increases the downward emission of infrared from the stratosphere into the troposphere, i.e., causes a positive radiative forcing, the effect is much smaller than the direct scattering of solar radiation. As a result, the major overall net effect of volcanic eruptions is cooling (Robock and Mao, 1995).

However, it should be noted that the effect is somewhat geographically and temporally variable. For example, Robock and Mao (1995) have examined climate records since about 1850 and correlated them to volcanic eruptions both before and after removal of the effects of the El Niño–Southern Oscillation (ENSO) signal. While the effect of volcanic eruptions on the global mean surface temperature is cooling, there are circumstances where the effect is not only smaller than the mean, but warming was observed. For example, in the first winter following a number of different volcanic eruptions, the Northern Hemisphere and Eurasia on average warmed, in contrast to northern Africa and southwestern Asia, which cooled. Robock and Mao propose that the warming is due to changes in the winter circulation pattern, associated with an enhanced polar vortex, which lowers the extent of normal winter cooling.

Figure 14.56 demonstrates the overall cooling effect of volcanic eruptions in the Northern Hemisphere over the past six centuries, reconstructed using the effects of temperature on tree ring densities (Briffa et al., 1998). The relationship between the average summer monthly mean land and marine temperatures in the Northern Hemisphere and tree ring density was determined for the period from 1881 to 1960. This was then applied to measured tree ring densities to obtain the temperature anomalies for the entire 600 years compared to the 1881–1960 period. Some of the major volcanic eruptions are also marked on the diagram, clearly demonstrating their association with significant cooling. Similar conclusions have been reached using ice core data (e.g., White et al., 1997; Clausen et al., 1997; Taylor et al., 1997; Zielinski et al., 1997).

Volcanic eruptions provide an opportunity for testing not only our current understanding of the direct effects of aerosol particles due to backscattering but also the sensitivity of the climate system to such perturbations. Thus, after the initial short-term effects on temperature, the coupled atmosphere–ocean–land system responds on a longer time scale through a complex set of feedback mechanisms. As discussed by Lindzen and Giannitsis (1998), the effects of multiple volcanic eruptions should provide a better test of our understanding of such feedbacks than is provided by a single eruption.

5. Oceans

Oceans have an enormous effect on climate through many different mechanisms that are beyond the scope of this book. Globally, oceans absorb heat and greenhouse gases such as CO_2 from the atmosphere (Fig. 14.11), both moderating such changes (e.g., Schneider et al., 1997; Bush and Philander, 1998) and providing a time lag in the response to atmospheric perturbations (e.g., Wigley, 1995). Other phenomena such as the El Niño–Southern Oscillation (ENSO) and the North Atlantic Oscillation (NAO) clearly also have substantial

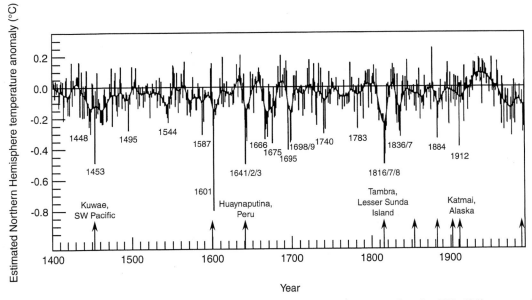

FIGURE 14.56 Temperature anomalies in the Northern Hemisphere compared to the 1881–1960 mean, calculated based on measured changes in tree ring densities. The 95% confidence limit is ~ ±0.3°C. The line shows bidecadal smoothed levels. Arrows on the lower axis mark some of the major volcanic eruptions (adapted from Briffa et al., 1998).

large-scale impacts on climate (e.g., see Enfield and Mayer, 1997; and Hurrell and Van Loon, 1997).

While the atmosphere and oceans are closely linked, how changes in one impact changes in the other is not as clear (e.g., Broecker and Denton, 1990). The oceans have well-documented circulation systems that play a major role in determining climate. For example, the Atlantic thermohaline circulation, often referred to as the "conveyor," consists of a complex combination of ocean currents that result in the transport of warmer surface waters from the North Pacific into the Indian Ocean, around the African continent, and into the northern Atlantic (Broecker, 1997). This provides a source of heat to air masses moving east in the winter, resulting in much warmer winters in Europe than would otherwise be the case. As discussed by Broecker (1997), this conveyor appears in the past to have jumped from one mode of operation to another, initiating substantial and rapid global climate changes. Furthermore, he suggests that it is possible that substantial increases in greenhouse gases such as CO_2 may also initiate such changes in the thermohaline circulation, with associated, and perhaps surprising, effects on global climate.

For example, increased water vapor at high latitudes and the associated increased precipitation, combined with melting glaciers, due to global warming could provide a layer of less dense surface water in the northern Atlantic. Since the conveyor is driven by high-density salt water, this could shut down this global ocean circulation system. Such a shutdown is expected to lead to cooling in the Northern Hemisphere but warming in the Southern Hemisphere since the heat transport associated with the conveyor no longer occurs (Kerr, 1998). That is, anthropogenic emissions that one normally associates with greenhouse warming may trip the ocean–atmosphere system in such a way that cooling could result in the Northern Hemisphere and warming in the Southern Hemisphere. Such feedbacks, hypothesized to have been triggered by closing of the Panamanian Isthmus, have been postulated to explain the Northern Hemisphere glaciation that occurred about 3 million years ago (Driscoll and Haug, 1998). Clearly, this is an area that needs to be explored further.

E. OBSERVATIONS OF CLIMATE CHANGES

1. Observed Temperature Trends

a. Trends over the Past Century

One of the obvious manifestations of anthropogenic emissions is expected to be an increase in the temperature of the air and sea surface (SST). As a result, there have been many analyses of such temperatures, for which there are substantial records based on instrumental measurements made in a number of locations back to approximately 1860 and in at least one location, Armagh Observatory, North Ireland, to 1795 (Wilson, 1998). A review of these data, as well as more limited temperature data at higher altitudes in the troposphere and stratosphere, is found in Bradley and Jones (1993) and IPCC (1996).

Figure 14.57 shows the globally averaged temperature anomalies for land and sea surface measurements from 1861 to 1994, relative to the 1961–1990 period (IPCC, 1996; Jones *et al.*, 1994). Such data indicate there has been an increase in near-surface temperatures of ~0.3–0.6°C over this period, with an uncertainty of about 0.15°C. Measurements of underground temperatures from 358 boreholes in central Europe, southern Africa, Australia, and eastern North America show a similar temperature trend (Pollack *et al.*, 1998). However, the increase has not been continuous, with substantial increases in temperature occurring between about 1920 and 1940, followed by a decrease and then an increase to the present time. The increase in the global average temperature over the past 40 years has been about 0.2–0.3°C. Similarly, the changes in temperature vary geographically and seasonally. For example, warming has occurred in the Northern Hemisphere over the continents, while cooling has occurred over the midlatitude North Pacific and over the northwestern Atlantic. The geographical and seasonal dependencies are summarized in IPCC (1996).

While the surface temperatures have clearly been increasing, some satellite measurements have suggested that the air temperatures in the troposphere have been cooling at altitudes where this was not expected. However, this is controversial (e.g., Pielke *et al.*, 1998a, 1998b). For example, Wentz and Schabel (1998) have shown that the loss of satellite altitude with time can introduce an artifact into the data, which, if not corrected, leads to artifact cooling trends.

The five warmest years for which there are surface temperature records have all been since 1990 (Jones *et al.*, 1998), with the most recent year for which there are data (at the time of writing), 1997, being the warmest in the past century (see Kerr, 1998, and references therein). Mann *et al.* (1998) have used a variety of indirect indicators for temperature (e.g., ice core data; see later) over the past 600 years in the Northern Hemisphere and report that mean annual temperatures for three of the eight years up to and including 1995 are higher than any since 1400 A.D.

An interesting aspect of the surface temperature changes is that in many locations, particularly continental regions, the minimum daily temperature has increased more than the maximum daily temperature (e.g., Hansen *et al.*, 1997d). As a result, the daily temperature *range* has decreased in these regions. Figure 14.58, for example, shows globally averaged maximum and minimum temperatures, as well as the diurnal temperature range, from 1950 to 1993 based on approximately 4100 nonurban stations (Easterling *et al.*, 1997). These are expressed as deviations from the mean for all stations in 5° × 5° latitude–longitude grid boxes during the period from 1961 to 1985. The trend in the maximum temperature is 0.82°C per century, but that in the minimum is larger, 1.79°C per century. As a result, the diurnal temperature range decreases, with a slope of −0.79°C per century.

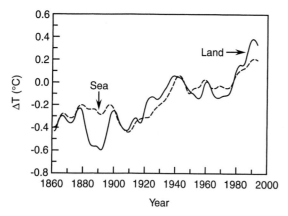

FIGURE 14.57 Global average temperature anomaly (ΔT) for land and sea surface measurements relative to the period from 1961 to 1990 (adapted from IPCC, 1996).

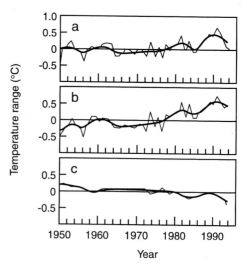

FIGURE 14.58 Global nonurban annual average temperature anomalies for the (a) maximum temperature, (b) minimum temperature, and (c) diurnal range of temperatures from 1950 to 1993 for ~4100 stations in both the Northern and Southern Hemispheres (adapted from Easterling *et al.*, 1997).

These trends vary, depending on location. For example, the diurnal temperature range did not decrease over mid-Canada or parts of southwest Asia, southern Africa, the interior of Australia, the western tropical Pacific Islands, and Europe (Easterling *et al.*, 1997). Similarly, in some European mountain locations both the minimum and maximum of daily temperatures have been observed to increase (Weber *et al.*, 1994) while in India, the maximum increased but there was no trend in the minimum (Kumar *et al.*, 1994).

There are many possible reasons for the decrease in the diurnal temperature range, where it occurs (Kukla and Karl, 1993). These include the urban heat island effect, which is strongest at night. However, the data in Fig. 14.58 excluded population centers of 50,000 or more; including data from urban areas leads to a slight increase in the slopes of maximum, minimum, and diurnal range of temperatures. Another potential contributor is an increase in cloudiness, which during the day scatters incoming solar radiation and leads to cooling. At night, the ground cools by thermal emission of infrared, which is counterbalanced in part by downward thermal infrared emission from atmospheric constituents. The downward emission is greatest when clouds and higher water vapor concentrations are present, leading to less net cooling of the surface at night and an increase in the minimum temperature. Soil moisture, which is affected by irrigation, drying of wetlands, deforestation, etc., is another factor. Evaporative cooling in moist soils occurs in the afternoon, but moist soils are warmer at night, leading to a decreased diurnal temperature range.

Finally, anthropogenic aerosol particles and greenhouse gases may also affect the diurnal temperature range. As discussed in detail earlier, aerosol particles cool during the day by scattering incoming solar radiation. At night, they can contribute to warming by absorbing terrestrial infrared radiation, but this effect is small relative to the daytime cooling effect. Greenhouse gases such as CO_2, of course, absorb the terrestrial infrared, leading to warming, an effect that does not vary as strongly with time of day as do those involving solar radiation. As a result, increased CO_2 leads to increased heating without a strong diurnal variation, whereas aerosol particles lead primarily to cooling during the day. These two effects can lead to a reduced diurnal temperature range, although if these were the only two effects operating, one would expect the reduced temperature range to be due more to changes in daytime temperatures. This is the opposite of what has been observed in many locations (Fig. 14.58).

Stenchikov and Robock (1995) suggest that feedbacks may actually be more important than these direct effects in reducing the diurnal temperature range. For example, in a warmer climate, more evaporation of water occurs, leading to increased gas-phase water vapor concentrations and possibly increased clouds. Stenchikov and Robock (1995) suggest that the major effect of such feedbacks in reducing the diurnal temperature range is not through the usual greenhouse effect but rather through increased absorption of solar radiation in the near-IR by the increased atmospheric water.

Hansen *et al.* (1995, 1997d) have modeled various contributions to changes in the diurnal temperature range and the increase in global temperatures and concluded that the observed changes are only consistent with a combination of factors. These include a contribution from direct forcing by anthropogenic aerosols and an increase in cloud cover (the indirect effect of aerosols discussed earlier) primarily over continental regions, which are about 50% of the magnitude of anthropogenic greenhouse gas global forcings, but in the opposite direction.

As discussed throughout this chapter, there are a variety of anthropogenic emissions that are expected to lead to warming (e.g., the greenhouse gases) or to cooling (e.g., increased aerosol particle concentrations). To quantify such contributions to changes in surface temperatures, it is necessary first to understand and account for changes due to natural processes such as the solar variability discussed earlier. How to do so in an accurate and reliable manner remains a very complex and controversial area (e.g., Lindzen, 1994; Jones, 1995; Mahasenan *et al.*, 1997; Legates and Davis, 1997; Wigley *et al.*, 1997). However, the weight of evidence at the present time suggests that the observed recent increase in global mean surface temperature is in part due to anthropogenic influences (e.g., see IPCC, 1996; Santer *et al.*, 1996; Overpeck *et al.*, 1997; Kaufmann and Stern, 1997; and Mann *et al.*, 1998).

b. *Temperatures and Other Proxies for Climate Change over the Past $\sim 10^5$ Years*

One approach to elucidating the contribution of natural variability to recent temperature trends is to examine markers for temperature over much longer time scales, prior to the industrial revolution. A major source of such data is ice cores (see also Section B.2a). These ice cores provide a record of climate and atmospheric composition for at least 110,000 years, for which there is agreement among various studies. Data are available for 250,000 years before the present (bp), but there is some uncertainty in the dating of the layers corresponding to these older ice core depths (Chappellaz *et al.*, 1997).

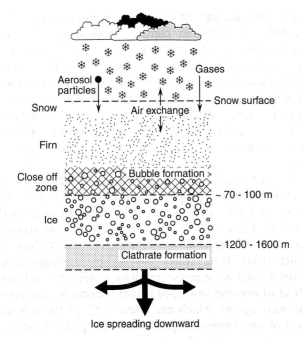

FIGURE 14.59 Schematic diagram of uptake and incorporation of gases and aerosol particles into ice cores (adapted from Delmas, 1992).

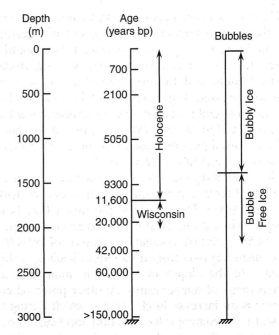

FIGURE 14.60 Relationship of ice core depth to years before the present (bp) and to the type of ice for the Central Greenland Ice Sheet. The Holocene and Wisconsin periods are also marked (adapted from Gow et al., 1997).

Figure 14.59 shows schematically the processes by which gases and particles are trapped from the atmosphere into snow and ice at high latitudes (Delmas, 1992). As snow is deposited, the surface is initially quite porous. As more snow accumulates, it compacts the underlying layers, forming a porous structure known as firn. Atmospheric gases continue to penetrate the porous firn. At some depth (typically ~100 m), there is a close-off zone in which recrystallization starts to seal off the pores, trapping the atmospheric constituents in bubbles in the ice. At larger depths, the bubbles are completely sealed off and the trapped gas is preserved. As snow accumulation continues, the ice below is further compacted and the ice sheet spreads down and out. At depths larger than ~1200 m, the hydrostatic pressure is sufficiently large that the air is forced into the ice to form clathrates so that distinct bubbles are no longer evident (Miller, 1969). Clathrates are solid ice lattices that incorporate another molecule such as CO_2 or CH_4 into their crystal lattice (e.g., see Kvenvolden, 1993). Once an ice core is drilled, various depths can be dated and the air trapped in the ice as either bubbles or clathrates is recovered for analysis, for example, by crushing the ice sample (Wilson and Long, 1997).

The bubbles found at larger depths therefore correspond to older atmospheres. Figure 14.60 shows the relationship between the ice core depth and age (in number of years before the present time, bp) as well as the characteristics of the ice for samples from the Greenland Ice Sheet Project 2, GISP2 (Gow et al., 1997). Also shown are the periods corresponding to the Holocene (the past 10,000 years) and the Wisconsin ice age, for which some data are shown below. Ice at depths of ~260, 1000, 2430, and 2759 m corresponds to ages of 1000, 5100, 50,000, and 103,000 years bp, respectively (Grootes and Stuiver, 1997).

Because the firn is ventilated by atmospheric air while the bubbles are forming over a period of time and ice depths, the air eventually trapped in the bubbles is a time-integrated sample that is younger than the snow deposit itself. For example, in one recent study (Smith et al., 1997), the air bubbles were, on average, 220–700 years younger than the ice in which they were embedded, but the difference can be as much as several thousand years (e.g., see Rommelaere et al., 1997). These exchange processes with the atmosphere, gas diffusion, and the porosity and tortuosity of the ice pores have to be taken into account in relating the depth of the core to the age of the trapped air.

As seen earlier (Section B.2), air trapped in these ice cores can be recovered and analyzed to provide a snapshot of the composition of the atmosphere tens or even hundreds of thousands of years ago (but note cautions with respect to potential artifacts, e.g., *in situ* formation of CO_2 from carbonate in the bubbles). In addition, ice core composition can be used to infer the local temperature of the atmosphere when the snow/ice was deposited, so-called paleothermometry (Delmas, 1992). The isotopic composition, particularly the $^{18}O/^{16}O$ and D/H ratios, of the ice is related to the temperature at the level of the precipitating cloud that generated the snow/ice. Isotopic fractionation occurs during the natural water cycle and this leads to a relationship between the isotopic composition and the precipitation temperature (Dansgaard, 1964). Once this relationship is established, the isotopic composition of water in the ice core can be used to estimate the corresponding atmospheric temperature. Such relationships, using ^{18}O as an example, are usually expressed in the form

$$T = a\delta^{18}O + b, \qquad \text{(OO)}$$

where $\delta^{18}O$ in per mil (‰) is defined as $1000(R - R_0)/R_0$, R is the isotope ratio of the sample, and R_0 is the ratio of a standard sample. In the case of oxygen, the standard is usually standard mean ocean water (SMOW). The values of the slope and intercept (a and b, respectively) vary from location to location and likely with time as well (Cuffey and Clow, 1997; Jouzel et al., 1997). However, it appears that such relationships are still useful for inferring historic temperatures (Jouzel et al., 1997; Salamatin et al., 1998).

Ice core data have provided evidence that quite rapid and large oscillations in climate have occurred over the period of record. Figure 14.61, for example, shows the temperature changes in central Greenland for the past ~110,000 years. Recent millenia are characterized by relatively small rates of temperature change. Indeed, in summarizing the results of the Greenland Summit Ice Core Projects (GISP2 and GRIP) published in a special issue of the *Journal of Geophysical Research* (Vol. 102 (C12), pp. 26315–26886, November 30, 1997), Hammer, Mayewski, Peel, and Stuiver state

> The ice-core records tell a clear story: humans have come of age agriculturally and industrially in the most stable climate regime of the last 110,000 years. However, even this relatively stable period is marked by change... [which is] more characteristic of the Earth's climate than is stasis.

Figure 14.61 illustrates the much larger changes, more than 20°C from one extreme to the other, that have occurred historically when viewed over these long time periods. Associated with decreases in temperature are decreased methane concentrations, increased dust loadings, and decreased snow accumulations (e.g., see Fig. 14.62). (It should be noted that the amplitude of the temperature changes can vary from site to site; for example, Dahl-Jensen et al. (1998) showed that the

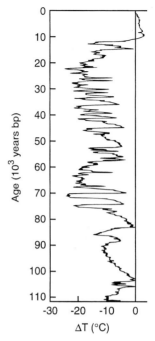

FIGURE 14.61 Calculated temperature changes over the past 100,000 years from the European Greenland Ice Core Program (GRIP) (adapted from Jouzel et al., 1997).

amplitude of the temperature record from the Dye 3 site in Greenland 865 km south of the GRIP site and 730 m lower in elevation was 50% larger than that at the GRIP site.)

Some of the indicators of climate change have shown very rapid changes, over decades or less, during particular time periods in the past. For example, Fig. 14.62 shows ice core measurements of (a) electrical conductivity, (b) snow accumulation rate, and (c) concentration of calcium at the start of the Holocene period some 11,600 years ago (Taylor et al., 1997). Electrical conductivity is a measure of the acidity of the ice, since H^+ is the major charge carrier; the direct current is measured between two electrodes that have a voltage difference of several thousand volts. As seen in Fig. 14.62a, such measurements provide excellent spatial and hence time resolution. The transition from the Wisconsin to the Holocene period is seen to be characterized by an increase in electrical conductivity of the ice core and a decrease in calcium. The two are inversely related since high concentrations of $CaCO_3$ neutralize strong acids, decreasing the conductivity. The rate of accumulation of snow (Fig. 14.62b) also increases. These changes occurred in less than about two decades. Taylor and co-workers also point out that the data in Fig. 14.62 suggest there is a "flicker" just prior to the rapid transition to the alternate climate state.

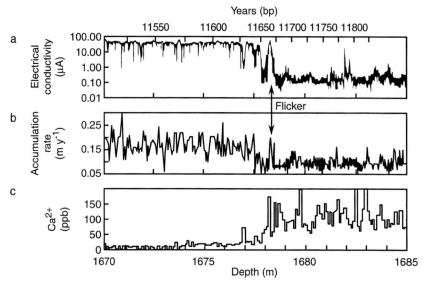

FIGURE 14.62 Evidence of rapid climate changes at the start of the Holocene period ~11,600 years before the present (bp): (a) electrical conductivity; (b) rate of accumulation of snow; (c) calcium concentration. The depth of the ice cores is shown on the bottom axis and the corresponding age in years before the present on the top axis (adapted from Taylor et al., 1997).

In short, ice core and other long-term records show that there have been dramatic climate changes in the past, some of them within or shorter than a typical human life span. Separating out such natural variability from anthropogenic perturbations remains a major challenge, particularly when it is possible that the anthropogenic emissions may act to hasten or "jolt" the climate system into a relatively rapid transition from one state to another.

2. Other Climate Changes

There are a variety of other climate changes that might be expected to occur simultaneously with changes in temperature. These include changes in precipitation, an increase in the mean sea level, and more variability in the climate. As discussed in detail in IPCC (1996), changes in precipitation patterns and cloudiness have been noted over the past approximately four decades and there is evidence that the sea level has risen by ~10–25 cm. The IPCC document should be consulted for detailed evidence for these effects and their possible relationship to anthropogenic perturbations.

F. THE FUTURE

As seen throughout this chapter, the parameters controlling climate are extremely varied and complex, with multiple feedbacks between them. Some possible future scenarios based on the state of the science as of about 1995 are described in the IPCC (1996) document. Figure 14.63 shows one model estimate for temperature changes due only to direct radiative forcing by CO_2 from 1990 to 2100 based on three scenarios (Wigley, 1998). While as we have seen, many gases contribute to radiative forcing, the calculations in Fig. 14.63 are based on expressing these changes in terms of equivalent CO_2 reductions. Note that this does not take into account other contributing factors such as aerosol particles which may contribute in the opposite

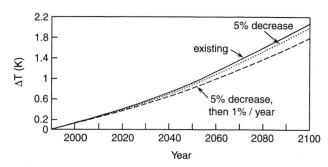

FIGURE 14.63 Calculated temperature changes relative to 1990 for existing policies (—), for a 5% decrease in equivalent CO_2 as required by Kyoto protocol from 1990 to 2010 followed by no further emissions reductions (···), and for further reductions of 1% per year (compounded) from 2010 to 2100 (- - -) assuming a climate sensitivity of 2.5 K for a doubling of CO_2 (adapted from Wigley, 1998).

direction. Hence, Fig. 14.63 should be taken as illustrative of the effects on direct radiative forcing by gases and not the net result of all contributing factors and feedbacks.

Some of complexities of climate are reflected in the natural variability, observed through such "proxies" as the ice core and tree-ring records. However, there are reasons, based on our current state of scientific understanding, to believe that anthropogenic activities that have resulted in changes in atmospheric composition may affect climate locally, regionally, and globally. Despite the uncertainties in the deconvolution of anthropogenic and natural contributions to the relatively recent observed global climate changes, the weight of scientific opinion at the end of the twentieth century is reflected in the IPCC (1996) summary:

> Nevertheless, the balance of evidence suggests that there is a discernible human influence on global climate.

Elucidating this influence remains a challenge for the twenty-first century.

G. PROBLEMS

1. Using gas kinetic molecular theory, show that under typical atmospheric conditions of pressure and temperature corresponding to an altitude of 5 km (see Appendix V) collisional deactivation of a CO_2 molecule will be much faster than reemission of the absorbed radiation. Take the collision diameter to be 0.456 nm and the radiative lifetime of the 15-μm band of CO_2 to be 0.74 s (Goody and Yung, 1989).

2. Assume that a typical greenhouse gas absorbs infrared radiation at 10 μm resulting in a vibrational energy change in the molecule. If there were a temperature change in the upper troposphere (where the temperature is typically 220 K) of +5°C, estimate using the Boltzmann equation the factor (i.e., ratio) by which the emission of energy out to space would change.

3. Calculate for liquid water the factor by which the vapor pressure increases over droplets of the following sizes compared to that over the bulk liquid at 298 K: (a) 1, (b) 0.1, (c) 0.01, and (d) 0.001 μm. If typical supersaturations of water vapor in the atmosphere are of the order of 0.1%, which of these could be stable in the atmosphere? The surface tension of water at room temperature is 72 dyn cm^{-1}.

4. Calculate for sulfuric acid ($\gamma = 55.1$ dyn cm^{-1}, $\rho = 1.84$) the percentage increase in vapor pressure compared to the bulk over droplets of the following sizes at 298 K: (a) 1, (b) 0.1, (c) 0.01, and (d) 0.001 μm. The vapor pressure of sulfuric acid is sufficiently low that it exists in the atmosphere primarily as a liquid, although there will be very small concentrations in the gas phase. By what percentage will the gas-phase concentration increase over that of the bulk liquid if H_2SO_4 exists primarily in 10-nm particles?

5. Derive an expression for r_0 in Fig. 14.38, corresponding to an RH of 100% or a supersaturation of zero, in terms of a and b in Eq. (II).

6. Derive expressions for the critical radius r_c and the critical supersaturation S_c at the peak of the Köhler curve in Fig. 14.38 in terms of the a and b parameters in Eq. (II). What is the relationship between r_c and r_0?

7. Calculate the critical radius r_c and critical supersaturation S_c for activation into a cloud droplet of a 10^{-15}-g NaCl particle. Assume the surface tension is 72 dyn cm^{-1} and the liquid density is that of water.

8. Repeat Problem 7 for a 10^{-16}-g $(NH_4)SO_4$ particle.

9. (a) Using Eq. (JJ), calculate the albedo of a 250-m-thick cloud with 5×10^7 droplets per m^3 and an effective mean radius for light scattering of 10 μm. Take $g = 0.8$.

(b) Now assume that the number of CCN have decreased sufficiently that the number of cloud droplets is 5×10^4 droplets m^{-3} but the liquid water content is the same. Calculate the new effective radius and the cloud albedo.

10. As seen in Eq. (KK), cloud susceptibility depends on $\tau(\delta R/\delta \tau)$. Using Eq. (JJ), show that $\tau(\delta R/\delta \tau) = R(1 - R)$.

References

Ackerman, S. A., and S. K. Cox, "Aircraft Observations of Shortwave Fractional Absorptance of Non-Homogeneous Clouds," *J. Appl. Meteorol., 20,* 1510–1515 (1981).

Albrecht, B. A., "Aerosols, Cloud Microphysics, and Fractional Cloudiness," *Science, 245,* 1227–1230 (1989).

Alkezweeny, A. J., D. A. Burrows, and C. A. Grainger, "Measurements of Cloud-Droplet-Size Distributions in Polluted and Unpolluted Stratiform Clouds," *J. Appl. Meteorol., 32,* 106–115 (1993).

Alpert, P., Y. J. Kaufman, Y. Shay-El, D. Tanre, A. da Silva, S. Schubert, and J. H. Joseph, "Quantification of Dust-Forced Heating of the Lower Troposphere," *Nature, 395,* 367–370 (1998).

Anderson, T. L., D. S. Covert, and R. J. Charlson, "Cloud Droplet Number Studies with a Counterflow Virtual Impactor," *J. Geophys. Res., 99,* 8249–8256 (1994).

Andreae, M. O., W. Elbert, and S. J. de Mora, "Biogenic Sulfur Emissions and Aerosols over the Tropical South Atlantic. 3. Atmospheric Dimethylsulfide, Aerosols, and Cloud Condensation Nuclei," *J. Geophys. Res., 100,* 11335–11356 (1995).

Andreae, M. O., and P. J. Crutzen, "Atmospheric Aerosols: Biogeochemical Sources and Role in Atmospheric Chemistry," *Science, 276,* 1052–1058 (1997).

Andrews, E., S. M. Kreidenweis, J. E. Penner, and S. M. Larson, "Potential Origin of Organic Cloud Condensation Nuclei Observed at Marine Site," *J. Geophys. Res., 102,* 21997–22012 (1997).

Andronache, C., L. J. Donner, V. Ramaswamy, C. J. Seman, and R. S. Hemler, "The Effects of Atmospheric Sulfur on the Radiative Properties of Convective Clouds: A Limited Area Modeling Study," *Geophys. Res. Lett.*, 25, 1423–1426 (1998).

Anfossi, D., S. Sandroni, and S. Viarengo, "Tropospheric Ozone in the Nineteenth Century: The Moncalieri Series," *J. Geophys. Res.*, 96, 17349–17352 (1991).

Angell, J. K., "Stratospheric Warming Due to Agung, El Chichón, and Pinatubo Taking into Account the Quasi-Biennial Oscillation," *J. Geophys. Res.*, 102, 9479–9485 (1997).

Anklin, M., J. Schwander, B. Stauffer, J. Tschumi, and A. Fuchs, "CO_2 Record between 40 and 8 kyr B.P. from the Greenland Ice Core Project Ice Core," *J. Geophys. Res.*, 102, 26539–26545 (1997).

Arking, A., "Absorption of Solar Energy in the Atmosphere: Discrepancy between Model and Observations," *Science*, 273, 779–782 (1996).

Arrhenius, S., "On the Influence of Carbonic Acid in the Air upon the Temperature of the Ground," *Philos. Mag.*, 41, 237–276 (1896).

Austin, J., and N. Butchart, "The Influence of Climate Change and the Timing of Stratospheric Warmings on Arctic Ozone Depletion," *J. Geophys. Res.*, 99, 1127–1145 (1994).

Austin, J., N. Butchart, and K. P. Shine, "Possibility of an Arctic Ozone Hole in a Doubled-CO_2 Climate," *Nature*, 360, 221–225 (1992).

Ayers, G., and R. Gillett, "Data Summaries. Quadrupod Aerosol Sampler," in *Baseline Atmospheric Program 89* (S. R. Wilson and J. L. Gras, Eds.), pp. 57–59, Bureau of Meteorology in association with CSIRO Division of Atmospheric Research, Melbourne, 1989.

Ayers, G. P., and J. L. Gras, "Seasonal Relationship between Cloud Condensation Nuceli and Aerosol Methanesulphonate in Marine Air," *Nature*, 353, 834–835 (1991).

Ayers, G. P., S. T. Bentley, J. P. Ivey, and B. W. Forgan, "Dimethylsulfide in Marine Air at Cape Grim, 41°S," *J. Geophys. Res.*, 100, 21013–21021 (1995).

Baker, M. B., and R. J. Charlson, "Bistability of CCN Concentrations and Thermodynamics in the Cloud-Topped Boundary Layer," *Nature*, 345, 142–145 (1990).

Baker, M. B., "Cloud Microphysics and Climate," *Science*, 276, 1072–1078 (1997).

Bates, T. S., and P. K. Quinn, "Dimethylsulfide (DMS) in the Equatorial Pacific Ocean (1982 to 1996): Evidence of a Climate Feedback?" *Geophys. Res. Lett.*, 24, 861–864 (1997).

Bekki, S., K. S. Law, and J. A. Pyle, "Effect of Ozone Depletion on Atmospheric CH_4 and CO Concentrations," *Nature*, 371, 595–597 (1994).

Bender, M. L., T. Sowers, J.-M. Barnola, and J. Chappellaz, "Changes in the O_2/N_2 Ratio of the Atmosphere during Recent Decades Reflected in the Composition of Air in the Firn at Vostok Station, Antarctica," *Geophys. Res. Lett.*, 21, 189–192 (1994).

Bergamaschi, P., C. Lubina, R. Königstedt, and H. Fischer, "Stable Isotopic Signatures ($\delta^{13}C$, δD) of Methane from European Landfill Sites," *J. Geophys. Res.*, 103, 8251–8265 (1998a).

Bergamaschi, P., C. A. M. Brenninkmeijer, M. Hahn, T. Röckmann, D. H. Scharfee, P. J. Crutzen, N. F. Elansky, I. B. Belikov, N. B. A. Trivett, and D. E. J. Worthy, "Isotope Analysis Based Source Identification for Atmospheric CH_4 and CO Sampled across Russia Using the Trans-Siberian Railroad," *J. Geophys. Res.*, 103, 8227–8235 (1998b).

Berntsen, T. K., I. S. A. Isaksen, G. Myhre, J. S. Fuglestvedt, F. Stordal, T. Alsvik Larsen, R. S. Freckleton, and K. P. Shine, "Effects of Anthropogenic Emissions on Tropospheric Ozone and Its Radiative Forcing," *J. Geophys. Res.*, 102, 28101–28126 (1997).

Berresheim, H., F. L. Eisele, D. J. Tanner, L. M. McInnes, D. C. Ramsey-Bell, and D. S. Covert, "Atmospheric Sulfur Chemistry and Cloud Condensation Nuclei (CCN) Concentrations over the Northeastern Pacific Coast," *J. Geophys. Res.*, 98, 12701–12711 (1993).

Bigg, E. K., "Discrepancy between Observation and Prediction of Concentrations of Cloud Condensation Nuclei," *Atmos. Res.*, 20, 82–86 (1986).

Blake, D. R., and F. S. Rowland, "Continuing Worldwide Increase in Tropospheric Methane, 1978 to 1987," *Science*, 239, 1129–1131 (1988).

Blunier, T., J. A. Chappellaz, J. Schwander, J.-M. Barnola, T. Desperts, B. Stauffer, and D. Raynaud, "Atmospheric Methane, Record from a Greenland Ice Core over the Last 1000 Years," *Geophys. Res. Lett.*, 20, 2219–2222 (1993).

Boers, R., G. P. Ayers, and J. L. Gras, "Coherence between Seasonal Variation in Satellite-Derived Cloud Optical Depth and Boundary Layer CCN Concentrations at a Mid-Latitude Southern Hemisphere Station," *Tellus*, 46B, 123–131 (1994).

Bogner, J. E., and K. A. Spokas, "Landfill Methane: Rates, Fates, and Role in Global Carbon Cycle," *Chemosphere*, 26, 1–4 (1993).

Bojkov, R. D., "Surface Ozone during the Second Half of the Nineteenth Century," *J. Clim. Appl. Meteorol.*, 25, 343–352 (1986).

Boucher, O., and T. L. Anderson, "General Circulation Model Assessment of the Sensitivity of Direct Climate Forcing by Anthropogenic Sulfate Aerosols to Aerosol Size and Chemistry," *J. Geophys. Res.*, 100, 26117–26134 (1995).

Boucher, O., S. E. Schwartz, T. P. Ackerman, T. L. Anderson, B. Bergstrom, B. Bonnel, P. Chýlek, A. Dahlback, Y. Fouquart, Q. Fu, R. N. Halthore, J. M. Haywood, T. Iversen, S. Kato, S. Kinne, A. Kirkevåg, K. R. Knapp, A. Lacis, I. Laszlo, M. I. Mishchenko, S. Nemesure, V. Ramaswamy, D. L. Roberts, P. Russell, M. E. Schlesinger, G. L. Stephens, R. Wagener, M. Wang, J. Wong, and F. Yang, "Intercomparison of Models Representing Direct Shortwave Radiative Forcing by Sulfate Aerosols," *J. Geophys. Res.*, 103, 16979–16998 (1998).

Bouwman, A. F., and J. A. Taylor, "Testing High-Resolution Nitrous Oxide Emission Estimates against Observations Using an Atmospheric Transport Model," *Global Biogeochem. Cycles*, 10, 307–318 (1996).

Bradley, R. S., and P. D. Jones, "'Little Ice Age' Summer Temperature Variations: Their Nature and Relevance to Recent Global Warming Trends," *The Holocene*, 3, 367–376 (1993).

Brasseur, G. P., J. T. Kiehl, J.-F. Müller, T. Schneider, C. Granier, X. Tie, and D. Hauglustaine, "Past and Future Changes in Global Tropospheric Ozone: Impact on Radiative Forcing," *Geophys. Res. Lett.*, 25, 3807–3810 (1998).

Briffa, K. R., P. D. Jones, F. H. Schweingruber, and T. J. Osborn, "Influence of Volcanic Eruptions on Northern Hemisphere Summer Temperature over the Past 600 Years," *Nature*, 393, 450–455 (1998).

Broecker, W. S., and G. H. Denton, "The Role of Ocean–Atmosphere Reorganizations in Glacial Cycles," *Quaternary Sci. Rev.*, 9, 305–341 (1990).

Broecker, W. S., and J. P. Severinghaus, "Diminishing Oxygen," *Nature*, 358, 710–711 (1992).

Broecker, W. S., "Thermohaline Circulation, the Achilles Heel of Our Climate System: Will Man-Made CO_2 Upset the Current Balance?" *Science*, 278, 1582–1588 (1997).

Bryant, E., *Climate Process & Change*, Cambridge Univ. Press, Cambridge, UK, 1997.

Bush, A. B. G., and S. G. H. Philander, "The Role of Ocean–Atmosphere Interactions in Tropical Cooling during the Last Glacial Maximum," *Science*, 279, 1341–1344 (1998).

Byrne, R. N., R. C. J. Somerville, and B. Subasilar, "Broken-Cloud Enhancement of Solar Radiation Absorption," *J. Atmos. Sci., 53,* 878–886 (1996).

Caldeira, K., and J. F. Kasting, "Insensitivity of Global Warming Potentials to Carbon Dioxide Emission Scenarios," *Nature, 366,* 251–253 (1993).

Calhoun, J. A., T. S. Bates, and R. J. Charlson, "Sulfur Isotope Measurements of Submicrometer Sulfate Aerosol Particles over the Pacific Ocean," *Geophys. Res. Lett., 18,* 1877–1880 (1991).

Capaldo, K. P., and S. N. Pandis, "Dimethylsulfide Chemistry in the Remote Marine Atmosphere: Evaluation and Sensitivity Analysis of Available Mechanisms," *J. Geophys. Res., 102,* 23251–23267 (1997).

Cess, R. D., M. H. Zhang, P. Minnis, L. Corsetti, E. G. Dutton, B. W. Forgan, D. P. Garber, W. L. Gates, J. J. Hack, E. F. Harrison, X. Jing, J. T. Kiehl, C. N. Long, J.-J. Morcrette, G. L. Potter, V. Ramanathan, B. Subasilar, C. H. Whitlock, D. F. Young, and Y. Zhou, "Absorption of Solar Radiation by Clouds: Observations versus Models," *Science, 267,* 496–503 (1995).

Cess, R. D., and M. H. Zhang, "Technical Comments on 'How Much Solar Radiation Do Clouds Absorb?'" *Science, 271,* 1133–1134 (1996).

Cess, R. D., M. H. Zhang, W. J. Ingram, G. L. Potter, V. Alekseev, H. W. Barker, E. Cohen-Solal, R. A. Colman, D. A. Dazlich, A. D. Del Genio, M. R. Dix, V. Dymnikov, M. Esch, L. D. Fowler, J. R. Fraser, V. Galin, W. L. Gates, J. J. Hack, J. T. Kiehl, H. Le Treut, K. I.-W. Lo, B. J. McAvaney, V. P. Meleshko, J.-J. Morcrette, D. A. Randall, E. Roeckner, J.-F. Royer, M. E. Schlesinger, P. V. Sporyshev, B. Timbal, E. M. Volodin, K. E. Taylor, W. Wang, and R. T. Wetherald, "Cloud Feedback in Atmospheric General Circulation Models: An Update," *J. Geophys. Res., 101,* 12791–12794 (1996a).

Cess, R. D., M. H. Zhang, Y. Zhou, X. Jing, and V. Dvortsov, "Absorption of Solar Radiation by Clouds: Interpretations of Satellite, Surface, and Aircraft Measurements," *J. Geophys. Res., 101,* 23299–23309 (1996b).

Cess, R. D., M. H. Zhang, G. L. Potter, V. Alekseev, H. W. Barker, S. Bony, R. A. Colman, D. A. Dazlich, A. D. Del Genio, M. Déqué, M. R. Dix, V. Dymnikov, M. Esch, L. D. Fowler, J. R. Fraser, V. Galin, W. L. Gates, J. J. Hack, W. J. Ingram, J. T. Kiehl, Y. Kim, H. Le Treut, X.-Z. Liang, B. J. McAvaney, V. P. Meleshko, J. J. Morcrette, D. A. Randall, E. Roeckner, M. E. Schlesinger, P. V. Sporyshev, K. E. Taylor, B. Timbal, E. M. Volodin, W. Wang, W. C. Wang, and R. T. Wetherald, "Comparison of the Seasonl Change in Cloud-Radiative Forcing from Atmospheric General Circulation Models and Satellite Observations," *J. Geophys. Res., 102,* 16593–16603 (1997).

Chalita, S., D. A. Hauglustaine, H. Le Treut, and J.-F. Müller, "Radiative Forcing Due to Increased Tropospheric Ozone Concentrations," *Atmos. Environ., 30,* 1641–1646 (1996).

Chameides, W. L., and A. W. Stelson, "Aqueous Phase Chemical Processes in Deliquescent Sea-Salt Aerosols: A Mechanism that Couples the Atmospheric Cycles of S and Sea Salt," *J. Geophys. Res., 97,* 20565–20580 (1992).

Chappellaz, J., E. Brook, T. Blunier, and B. Malaizé, "CH_4 and $\delta^{18}O$ of O_2 Records from Antarctic and Greenland Ice: A Clue for Stratigraphic Disturbance in the Bottom Part of the Greenland Ice Core Project and the Greenland Ice Sheet Project 2 Ice Cores," *J. Geophys. Res., 102,* 26547–26557 (1997).

Charlson, R. J., J. E. Lovelock, M. O. Andreae, and S. G. Warren, "Oceanic Phytoplankton, Atmospheric Sulphur, Cloud Albedo, and Climate," *Nature, 326,* 655–661 (1987).

Charlson, R. J., J. Langer, H. Rodhe, C. B. Leovy, and S. G. Warren, "Perturbation of the Northern Hemisphere Radiative Balance by Backscattering from Anthropogenic Sulfate Aerosols," *Tellus, 43AB,* 152–163 (1991).

Charlson, R. J., S. E. Schwartz, J. M. Hales, R. D. Cess, J. A. Coakley, Jr., J. E. Hansen, and D. J. Hofmann, "Climate Forcing by Anthropogenic Aerosols," *Science, 255,* 423–430 (1992a).

Charlson, R. J., S. E. Schwartz, J. M. Hales, R. D. Cess, J. A. Coakley, Jr., J. E. Hansen, and D. J. Hofmann, "Response to 'Aerosols and Global Warming,'" *Science, 256,* 598–599 (1992b).

Charlson, R. J., and J. Heintzenberg, Eds., *Aerosol Forcing of Climate,* Wiley, New York, 1995.

Chou, M. D., A. Arking, J. Otterman, and W. L. Ridgway, "The Effect of Clouds on Atmospheric Absorption of Solar Radiation," *Geophys. Res. Lett., 22,* 1885–1888 (1995).

Chou, M. D., and W. Zhou, "Estimation and Model Validation of the Surface Solar Radiation and Cloud Radiative Forcing Using TOGA COARE Measurements," *J. Clim., 10,* 610–620 (1997).

Chou, M.-D., W. Zhao, and S.-H. Chou, "Radiation Budgets and Cloud Radiative Forcing in the Pacific Warm Pool during TOGA COARE," *J. Geophys. Res., 103,* 16967–16977 (1998).

Christidis, N., M. D. Hurley, S. Pinnock, K. P. Shine, and T. J. Wallington, "Radiative Forcing of Climate Change by CFC-11 and Possible CFC Replacements," *J. Geophys. Res., 102,* 19597–19609 (1997).

Chuang, C. C., J. E. Penner, K. E. Taylor, A. S. Grossman, and J. J. Walton, "An Assessment of the Radiative Effect of Anthropogenic Sulfate," *J. Geophys. Res., 102,* 3761–3778 (1997).

Chuang, P. Y., R. J. Charlson, and J. H. Seinfeld, "Kinetic Limitations on Droplet Formation in Clouds," *Nature, 390,* 594–596 (1997).

Chýlek, P., and J. Hallett, "Enhanced Absorption of Solar Radiation by Cloud Droplets Containing Soot Particles in Their Surface," *Q. J. R. Meteorol. Soc., 118,* 167–172 (1992).

Chýlek, P., and J. Wong, "Effect of Absorbing Aerosols on Global Radiation Budget," *Geophys. Res. Lett., 22,* 929–931 (1995).

Cicerone, R. J., and J. D. Shetter, "Sources of Atmospheric Methane: Measurements in Rice Paddies and a Discussion," *J. Geophys. Res., 86,* 7203–7209 (1981).

Cicerone, R. J., J. D. Shetter, and C. C. Delwiche, "Seasonal Variation of Methane Flux from a California Rice Paddy," *J. Geophys. Res., 88,* 11022–11024 (1983).

Cicerone, R. J., and R. S. Oremland, "Biogeochemical Aspects of Atmospheric Methane," *Global Biogeochem. Cycles, 2,* 299–327 (1988).

Cicerone, R. J., "Greenhouse Cooling Up High," *Nature, 344,* 104–105 (1990).

Cicerone, R. J., C. C. Delwiche, S. C. Tyler, and P. R. Zimmerman, "Methane Emissions from California Rice Paddies with Varied Treatments," *Global Biogeochem. Cycles, 6,* 233–248 (1992).

Clarke, A. D., T. Uehara, and J. N. Porter, "Atmospheric Nuclei and Related Aerosol Fields over the Atlantic: Clean Subsiding Air and Continental Pollution during ASTEX," *J. Geophys. Res., 102,* 25281–25292 (1997).

Clausen, H. B., C. U. Hammer, C. S. Hvidberg, D. Dahl-Jensen, J. P. Steffensen, J. Kipfstuhl, and M. Legrand, "A Comparison of the Volcanic Records over the Past 4000 Years from the Greenland Ice Core Project and Dye 3 Greenland Ice Cores," *J. Geophys. Res., 102,* 26707–26723 (1997).

Clemens, J., J. Burkhardt, and H. Goldbach, "Abiogenic Nitrous Formation on Aerosols," *Atmos. Environ., 31,* 2961–2964 (1997).

Cliff, S. S., and M. H. Thiemens, "The $^{18}O/^{16}O$ and $^{17}O/^{16}O$ Ratios in Atmospheric Nitrous Oxide: A Mass-Independent Anomaly," *Science, 278,* 1774–1776 (1997).

Cliver, E. W., V. Boriakoff, and J. Feynman, "Solar Variability and Climate Change: Geomagnetic aa Index and Global Surface Temperature," *Geophys. Res. Lett., 25,* 1035–1038 (1998).

Clough, S. A., M. J. Iacono, and J. L. Moncet, "Line-by-Line Calculations of Atmospheric Fluxes and Cooling Rates: Application to Water Vapor," *J. Geophys. Res., 97,* 15761–15785 (1992).

Coakley, J. A., Jr., R. L. Bernstein, and P. A. Durkee, "Effect of Ship-Stack Effluents on Cloud Reflectivity," *Science, 237,* 1020–1022 (1987).

Conant, W. C., V. Ramanathan, F. P. J. Valero, and J. Meywerk, "An Examination of the Clear-Sky Solar Absorption over the Central Equatorial Pacific: Observations versus Models," *J. Clim., 10,* 1874–1884 (1997).

Conover, J. H., "Anomalous Cloud Lines," *J. Atmos. Sci., 23,* 778–785 (1966).

Conway, T. J., P. P. Tans, L. S. Waterman, K. W. Thoning, D. R. Kitzis, K. A. Maserie, and N. Zhang, "Evidence for Interannual Variability of the Carbon Cycle from the NOAA/CMDL Global Air Sampling Network," *J. Geophys. Res., 99,* 22831–22855 (1994).

Cotton, W. R., and R. A. Anthes, *Storm and Cloud Dynamics*, Academic Press, San Diego, 1989.

Cox, S. J., W.-C. Wang, and S. E. Schwartz, "Climate Response to Radiative Forcings by Sulfate Aerosols and Greenhouse Gases," *Geophys. Res. Lett., 22,* 2509–2512 (1995).

Crisp, D., "Absorption of Sunlight by Water Vapor in Cloudy Conditions: A Partial Explanation for the Cloud Absorption Anomaly," *Geophys. Res. Lett., 24,* 571–574 (1997).

Crowley, T. J., and G. R. North, *Paleoclimatology*, Oxford Univ. Press, New York, 1991.

Crowley, T. J., and K.-Y. Kim, "Comparison of Proxy Records of Climate Change and Solar Forcing," *Geophys. Res. Lett., 23,* 359–362 (1996).

Crutzen, P. J., "The Influence of Nitrogen Oxides on the Atmospheric Ozone Content," *Q. J. R. Meteorol. Soc., 96,* 320–325 (1970).

Crutzen, P. J., "On the Role of CH_4 in Atmospheric Chemistry; Sources, Sinks, and Possible Reductions in Anthropogenic Sources," *Ambio, 24,* 52–55 (1995).

Crutzen, P. J., and V. Ramanathan, *Clouds, Chemistry and Climate*, NATO ASI Series, Series I, Vol. 35, Springer-Verlag, Berlin, 1996.

Cruz, C. N., and S. N. Pandis, "The Effect of Organic Coating on the Cloud Condensation Nuclei Activation of Inorganic Atmospheric Aerosol," *J. Geophys. Res., 103,* 13111–13123 (1998).

Cuffey, K. M., and G. D. Clow, "Temperature, Accumulation, and Ice Sheet Elevation in Central Greenland through the Last Deglacial Transition," *J. Geophys. Res., 102,* 26383–26396 (1997).

Cunnold, D. M., R. F. Weiss, R. G. Prinn, D. Hartley, P. G. Simmonds, P. J. Fraser, B. Miller, F. N. Alyea, and L. Porter, "GAGE/AGAGE Measurements Indicating Reductions in Global Emissions of CCl_3F and CCl_2F_2 in 1992–1994," *J. Geophys. Res., 102,* 1259–1269 (1997).

Dahl-Jensen, D., K. Mosegaard, N. Gundestrup, G. D. Clow, S. J. Johnsen, A. W. Hansen, and N. Balling, "Past Temperatures Directly from the Greenland Ice Sheet," *Science, 282,* 268–271 (1998).

Daniel, J. S., and S. Solomon, "On the Climate Forcing of Carbon Monoxide," *J. Geophys. Res., 103,* 13249–13260 (1998).

Daniel, J. S., S. Solomon, and D. L. Albritton, "On the Evaluation of Halocarbon Radiative Forcing and Global Warming Potentials," *J. Geophys. Res., 100,* 1271–1285 (1995).

Dansgaard, W., "Stable Isotopes in Precipitation," *Tellus, 16,* 436–468 (1964).

Davies, R., W. L. Ridgway, and K.-F. Kim, "Spectral Absorption of Solar Radiation in Cloudy Atmospheres: A 20 cm^{-1} Model," *J. Atmos. Sci., 41,* 2126–2138 (1984).

Davison, B., C. N. Hewitt, C. D. O'Dowd, J. A. Lowe, M. H. Smith, M. Schwikowski, U. Baltensperger, and R. M. Harrison, "Dimethyl Sulfide, Methane Sulfonic Acid, and Physicochemical Aerosol Properties in Atlantic Air from the United Kingdom to Halley Bay," *J. Geophys. Res., 101,* 22855–22867 (1996).

Delmas, R. J., "Environmental Information from Ice Cores," *Rev. Geophys., 30,* 1–21 (1992).

Delwiche, C. E., Ed., *Denitrification, Nitrification, and Atmospheric Nitrous Oxide*, Wiley, New York, 1981.

Derwent, R. G., P. G. Simmonds, S. O'Doherty, and D. B. Ryall, "The Impact of the Montreal Protocol on Halocarbon Concentrations in Northern Hemisphere Baseline and European Air Masses at Mace Head, Ireland over a Ten Year Period from 1987–1996," *Atmos. Environ., 32,* 3689–3702 (1998a).

Derwent, R. G., P. G. Simmonds, S. O'Doherty, P. Ciais, and D. B. Ryall, "European Source Strengths and Northern Hemisphere Baseline Concentrations of Radiatively Active Trace Gases at Mace Head, Ireland," *Atmos. Environ., 32,* 3703–3715 (1998b).

Dickinson, R. E., and R. J. Cicerone, "Future Global Warming from Atmospheric Trace Gases," *Nature, 319,* 109–115 (1986).

Dlugokencky, E. J., L. P. Steele, P. M. Lang, and K. A. Masarie, "The Growth Rate and Distribution of Atmospheric Methane," *J. Geophys. Res., 99,* 17021–17043 (1994a).

Dlugokencky, E. J., K. A. Masarie, P. M. Lang, P. P. Tans, L. P. Steele, and E. G. Nisbet, "A Dramatic Decrease in the Growth Rate of Atmospheric Methane in the Northern Hemisphere during 1992," *Geophys. Res. Lett., 21,* 45–48 (1994b).

Dlugokencky, E. J., E. G. Dutton, P. C. Novelli, P. P. Tans, K. A. Masarie, K. O. Lantz, and S. Madronich, "Changes in CH_4 and CO Growth Rates after the Eruption of Mt. Pinatubo and Their Link with Changes in Tropical Tropospheric UV Flux," *Geophys. Res. Lett., 23,* 2761–2764 (1996).

Dlugokencky, E. J., K. A. Masarie, P. P. Tans, T. J. Conway, and X. Xiong, "Is the Amplitude of the Methane Seasonal Cycle Changing?" *Atmos. Environ., 31,* 21–26 (1997).

Dlugokencky, E. J., K. A. Masarie, P. M. Lang, and P. P. Tans, "Continuing Decline in the Growth Rate of the Atmospheric Methane Burden," *Nature, 393,* 447–450 (1998).

Driscoll, N. W., and G. H. Haug, "A Short Circuit in Thermohaline Circulation: A Cause for Northern Hemisphere Glaciation?" *Science, 282,* 436–438 (1998).

Durkee, P. A., F. Pfeil, E. Frost, and R. Shema, "Global Analysis of Aerosol Particle Characteristics," *Atmos. Environ., 25A,* 2457–2471 (1991).

Dutton, E. G., and J. R. Christy, "Solar Radiative Forcing at Selected Locations and Evidence for Global Lower Tropospheric Cooling Following the Eruptions of El Chichón and Pinatubo," *Geophys. Res. Lett., 19,* 2313–2316 (1992).

Easterling, D. R., B. Horton, P. D. Jones, T. C. Peterson, T. R. Karl, D. E. Parker, M. J. Salinger, V. Razuvayev, N. Plummer, P. Jamason, and C. K. Folland, "Maximum and Minimum Temperature Trends for the Globe," *Science, 277,* 364–367 (1997).

Eddy, J. A., "The Maunder Minimum," *Science, 192,* 1189–1202 (1976).

Eichel, C., M. Krämer, L. Schütz, and S. Wurzler, "The Water-Soluble Fraction of Atmospheric Aerosol Particles and Its Influence on Cloud Microphysics," *J. Geophys. Res., 101,* 29499–29510 (1996).

Enfield, D. B., and D. A. Mayer, "Tropical Atlantic Sea Surface Temperature Variability and Its Relation to El Nino–Southern Oscillation," *J. Geophys. Res., 102,* 929–945 (1997).

Etheridge, D. M., L. P. Steele, R. J. Francey, and R. L. Langenfelds, "Atmospheric Methane Between 1000 A.D. and Present: Evidence of Anthropogenic Emissions and Climate Variability," *J. Geophys. Res., 103,* 15979–15993 (1998).

Evans, W. F. J., and E. Puckrin, "An Observation of the Greenhouse Radiation Associated with Carbon Monoxide," *Geophys. Res. Lett., 22,* 925–928 (1995).

Evans, W. F. J., C. Reinhart, and E. Puckrin, "A Ground Based Measurement of the Anomalous Cloud Absorption Effect," *Geophys. Res. Lett., 22,* 2135–2138 (1995).

Evans, W. F. J., and E. Puckrin, "Near-Infrared Spectral Measurements of Liquid Water Absorption by Clouds," *Geophys. Res. Lett., 23,* 1941–1944 (1996).

Falkowski, P. G., Y. Kim, Z. Kolber, C. Wilson, C. Wirick, and R. Cess, "Natural versus Anthropogenic Factors Affecting Low-Level Cloud Albedo over the North Atlantic," *Science, 256,* 1311–1313 (1992).

Ferek, R. J., D. A. Hegg, P. V. Hobbs, P. Durkee, and K. Nielsen, "Measurements of Ship-Induced Tracks in Clouds off the Washington Coast," *J. Geophys. Res., 103,* 23199–23206 (1998).

Fischer, H., D. Wagenbach, and J. Kipfstuhl, "Sulfate and Nitrate Firn Concentrations on the Greenland Ice Sheet 2. Temporal Anthropogenic Deposition Changes," *J. Geophys. Res., 103,* 21935–21942 (1998).

Fleming, J. R., "T. C. Chamberlin and H_2O Climate Feedbacks: A Voice from the Past," *EOS, 73,* November 24, 1992.

Forster, P. M. de F., C. E. Johnson, K. S. Law, J. A. Pyle, and K. P. Shine, "Further Estimates of Radiative Forcing Due to Tropospheric Ozone Changes," *Geophys. Res. Lett., 23,* 3321–3324 (1996).

Forster, P. M. de F., and K. P. Shine, "Radiative Forcing and Temperature Trends from Stratospheric Ozone Changes," *J. Geophys. Res., 102,* 10841–10855 (1997).

Fouquart, Y., J. C. Buriez, M. Herman, and R. S. Kandel, "The Influence of Clouds on Radiation: A Climate-Modeling Perspective," *Rev. Geophys., 28,* 145–166 (1990).

Francis, P. N., J. P. Taylor, P. Hignett, and A. Slingo, "On the Question of Enhanced Absorption of Solar Radiation by Clouds," *Q. J. R. Meteorol. Soc., 123,* 419–434 (1997).

Freckleton, R. S., S. Pinnock, and K. P. Shine, "Radiative Forcing of Halocarbons: A Comparison of Line-by-Line and Narrow-Band Models Using CF_4 as an Example," *J. Quant. Spectrosc. Radiat. Transfer, 55,* 763–769 (1996).

Fritz, S., "Solar Radiant Energy," in *Compendium of Meteorology* (T. F. Malone, Ed.), pp. 14–29, Wiley, New York, 1951.

Fröhlich, C., and J. Lean, "The Sun's Total Irradiance: Cycles, Trends, and Related Climate Change Uncertainties Since 1976," *Geophys. Res. Lett., 25,* 4377–4380 (1998).

Fu, R., A. D. Del Genio, W. B. Rossow, and W. T. Liu, "Cirrus-Cloud Thermostat for Tropical Sea Surface Temperatures Tested Using Satellite Data," *Nature, 358,* 394–397 (1992).

Fu, R., W. T. Liu, A. D. Del Genio, and W. B. Rossow, "A Thermostat in the Tropics—Reply," *Nature, 361,* 412 (1993).

Fung, I., J. John, J. Lerner, E. Matthews, M. Prather, L. P. Steele, and P. J. Fraser, "Three-Dimensional Model Synthesis of the Global Methane Cycle," *J. Geophys. Res., 96,* 13033–13065 (1991).

Gaffney, J. S., and N. A. Marley, "Uncertainties of Aerosol Effects in Global Climate Models," *Atmos. Environ., 32,* 2873–2874 (1998).

Ghan, S. J., and R. C. Easter, "Comments on 'A Limited-Area-Model Case Study of the Effects of Sub-Grid Scale Variations in Relative Humidity and Cloud upon the Direct Radiative Forcing of Sulfate Aerosol,'" *Geophys. Res. Lett., 25,* 1039–1040 (1998).

Gierczak, T., R. K. Talukdar, J. B. Burkholder, R. W. Portmann, J. S. Daniel, S. Solomon, and A. R. Ravishankara, "Atmospheric Fate and Greenhouse Warming Potentials of HFC 236fa and HFC 236ea," *J. Geophys. Res., 101,* 12905–12911 (1996).

Gillani, N. V., S. E. Schwartz, W. R. Leaitch, J. W. Strapp, and G. A. Isaac, "Field Observations in Continental Stratiform Clouds: Partitioning of Cloud Particles between Droplets and Unactivated Interstitial Aerosols," *J. Geophys. Res., 100,* 18687–18706 (1995).

Goldstein, R., and S. S. Penner, "Transmission of Infrared Radiation through Liquid Water and through Water Vapor near Saturation," *J. Quant. Spectrosc. Radiat. Transfer, 4,* 359–361 (1964).

Gong, S. L., L. A. Barrie, and J. P. Blandet, "Modeling Sea-Salt Aerosols in the Atmosphere. 1. Model Development," *J. Geophys. Res., 102,* 3805–3818 (1997a).

Gong, S. L., L. A. Barrie, J. M. Prospero, D. L. Savoie, G. P. Ayers, J.-P. Blanchet, and L. Spacek, "Modeling Sea-Salt Aerosols in the Atmosphere. 2. Atmospheric Concentrations and Fluxes," *J. Geophys. Res., 102,* 3819–3830 (1997b).

Goody, R. M., and Y. L. Yung, *Atmospheric Radiation—Theoretical Basis,* 2nd ed., Oxford Univ. Press, New York, 1989.

Gow, A. J., D. A. Meese, R. B. Alley, J. J. Fitzpatrick, S. Anandakrishnan, G. A. Woods, and B. C. Elder, "Physical and Structural Properties of the Greenland Ice Sheet Project 2 Ice Core: A Review," *J. Geophsy. Res., 102,* 26559–26575 (1997).

Graf, H.-F., I. Kirchner, and J. Perlwitz, "Changing Lower Stratospheric Circulation: The Role of Ozone and Greenhouse Gases," *J. Geophys. Res., 103,* 11251–11261 (1998).

Gras, J. L., "Data Summaries, Particles," in *Baseline Atmospheric Program 89* (S. R. Wilson and J. L. Gras, Eds.), Bureau of Meteorology in association with the CSIRO Division of Atmospheric Research, Melbourne, 1989.

Grootes, P. M., and M. Stuiver, "Oxygen 18/16 Variability in Greenland Snow and Ice with 10^{-3}- to 10^5-Year Time Resolution," *J. Geophys. Res., 102,* 26455–26470 (1997).

Grossman, A. S., K. E. Grant, W. E. Blass, and D. J. Wuebbles, "Radiative Forcing Calculations for CH_3Cl and CH_3Br," *J. Geophys. Res., 102,* 13651–13656 (1997).

Gupta, M., S. Tyler, and R. Cicerone, "Modeling $\delta^{13}CH_4$ and the Causes of Recent Changes in Atmospheric CH_4 Amounts," *J. Geophys. Res., 101,* 22923–22932 (1996).

Gupta, M. L., M. P. McGrath, R. J. Cicerone, F. S. Rowland, and M. Wolfsberg, "$^{12}C/^{13}C$ Kinetic Isotope Effects in the Reactions of CH_4 with OH and Cl," *Geophys. Res. Lett., 24,* 2761–2764 (1997).

Haigh, J. D., "The Role of Stratospheric Ozone in Modulating the Solar Radiative Forcing of Climate," *Nature, 370,* 544–546 (1994).

Hallberg, A., K. J. Noone, and J. A. Ogren, "Aerosol Particles and Clouds: Which Particles Form Cloud Droplets?" *Tellus, 50B,* 59–75 (1998).

Hammer, C., P. A. Mayewski, D. Peel, and M. Stuiver, "Preface to Issue on Greenland Summit Ice Cores," *J. Geophys. Res., 102,* 26315–26316 (1997).

Han, Q., W. B. Rossow, and A. A. Lacis, "Near-Global Survey of Effective Droplet Radii in Liquid Water Clouds Using ISCCP Data," *J. Clim., 7,* 465–497 (1994).

Hanel, R. A., B. J. Conrath, V. G. Kunde, C. Prabhakara, I. Revah, V. V. Salomonson, and G. Wolford, "The Nimbus 4 Infrared Spectroscopy Experiment 1. Calibrated Thermal Emission Spectra," *J. Geophys. Res., 77,* 2629–2641 (1972).

Hansen, J., A. Lacis, R. Ruedy, and M. Sato, "Potential Climate Impact of Mount Pinatubo Eruption," *Geophys. Res. Lett., 19,* 215–218 (1992).

Hansen, J., M. Sato, and R. Ruedy, "Long-Term Changes of the Diurnal Temperature Cycle: Implications about Mechanisms of Global Climate Change," *Atmos. Res., 37,* 175–209 (1995).

Hansen, J., R. Ruedy, A. Lacis, G. Russell, M. Sato, J. Lerner, D. Rind, and P. Stone, "Wonderland Climate Model," *J. Geophys. Res., 102*, 6823–6830 (1997a).

Hansen, J., M. Sato, and R. Ruedy, "Radiative Forcing and Climate Response," *J. Geophys. Res., 102*, 6831–6864 (1997b).

Hansen, J., M. Sato, R. Ruedy, A. Lacis, K. Asamoah, K. Beckford, S. Borenstein, E. Brown, B. Cairns, B. Carlson, B. Curran, S. de Castro, L. Druyan, P. Etwarrow, T. Ferede, M. Fox, D. Gaffen, J. Glascoe, H. Gordon, S. Hollandsworth, X. Jiang, C. Johnson, N. Lawrence, J. Lean, J. Lerner, K. Lo, J. Logan, A. Luckett, M. P. McCormick, R. McPeters, R. Miller, P. Minnis, I. Ramberran, G. Russell, P. Russell, P. Stone, I. Tegen, S. Thomas, L. Thomason, A. Thompson, J. Wilder, R. Willson, and J. Zawodny, "Forcings and Chaos in Interannual to Decadal Climate Change," *J. Geophys. Res., 102*, 25679–25720 (1997c).

Harsen, J., M. Sato, A. Lacis, and R. Ruedy, "The Missing Climate Forcing," *Philos. Trans. R. Soc. London B, 352*, 231–240 (1997d).

Hao, W. M., and D. E. Ward, "Methane Production from Global Biomass Burning," *J. Geophys. Res., 98*, 20657–20661 (1993).

Harrison, E. F., P. Minnis, B. R. Barkstrom, V. Ramanathan, R. D. Cess, and G. G. Gibson, "Seasonal Variation of Cloud Radiative Forcing Derived from the Earth Radiation Budget Experiment," *J. Geophys. Res., 95*, 18687–18703 (1990).

Hartmann, D. L., "Radiative Effects of Clouds on Earth's Climate", in *Aerosol–Cloud–Climate Interactions* (P. V. Hobbs, Ed.), pp. 151–173, Academic Press, San Diego, 1993.

Hauglustaine, D. A., C. Granier, G. P. Brasseur, and G. Mégie, "The Importance of Atmospheric Chemistry in the Calculation of Radiative Forcing on the Climate System," *J. Geophys. Res., 99*, 1173–1186 (1994).

Haywood, J. M., and K. P. Shine, "The Effect of Anthropogenic Sulfate and Soot Aerosol on the Clear Sky Planetary Radiation Budget," *Geophys. Res. Lett., 22*, 603–606 (1995).

Haywood, J. M., and K. P. Shine, "Multi-Spectral Calculations of the Direct Radiative Forcing of Tropospheric Sulphate and Soot Aerosols Using a Column Model," *Q. J. R. Meteorol. Soc., 123*, 1907–1930 (1997).

Haywood, J. M., D. L. Roberts, A. Slingo, J. M. Edwards, and K. P. Shine, "General Circulation Model Calculations of the Direct Radiative Forcing by Anthropogenic Sulphate and Fossil-Fuel Soot Aerosol," *J. Clim., 10*, 1562–1577 (1997a).

Haywood, J. M., R. J. Stouffer, R. T. Wetherald, S. Manabe, and V. Ramaswamy, "Transient Response of a Coupled Model to Estimated Changes in Greenhouse Gas and Sulfate Concentrations," *Geophys. Res. Lett., 24*, 1335–1338 (1997b).

Haywood, J. M., V. Ramaswamy, and L. J. Donner, "Reply," *Geophys. Res. Lett., 25*, 1041 (1998a).

Haywood, J. M., and V. Ramaswamy, "Global Sensitivity Studies of the Direct Radiative Forcing Due to Anthropogenic Sulfate and Black Carbon Aerosols," *J. Geophys. Res., 103*, 6043–6058 (1998b).

Haywood, J. M., M. D. Schwarzkopf, and V. Ramaswamy, "Estimates of Radiative Forcing Due to Modeled Increases in Tropospheric Ozone," *J. Geophys. Res., 103*, 16999–17007 (1998c).

Hegg, D. A., "Heterogeneous Production of Cloud Condensation Nuclei in the Marine Atmosphere," *Geophys. Res. Lett., 17*, 2165–2168 (1990).

Hegg, D. A., R. J. Ferek, P. V. Hobbs, and L. F. Radke, "Dimethyl Sulfide and Cloud Condensation Nucleus Correlations in the Northeast Pacific Ocean," *J. Geophys. Res., 96*, 13189–13191 (1991a).

Hegg, D. A., L. F. Radke, and P. V. Hobbs, "Measurements of Aitken Nuclei and Cloud Condensation Nuclei in the Marine Atmosphere and Their Relation to the DMS–Cloud–Climate Hypothesis," *J. Geophys. Res., 96*, 18727–18733 (1991b).

Hegg, D. A., "Comment on 'A Model Study of the Formation of Cloud Condensation Nuclei in Remote Marine Areas,' by Lin *et al.*," *J. Geophys. Res., 98*, 7127–7128 (1993).

Hegg, D. A., R. J. Ferek, and P. V. Hobbs, "Light Scattering and Cloud Condensation Nucleus Activity of Sulfate Aerosol Measured over the Northeast Atlantic Ocean," *J. Geophys. Res., 98*, 14887–14894 (1993).

Hegg, D. A., "Cloud Condensation Nucleus–Sulfate Mass Relationship and Cloud Albedo," *J. Geophys. Res., 99*, 25903–25907 (1994).

Hegg, D. A., R. J. Ferek, and P. V. Hobbs, "Reply to Comment by T. L. Anderson *et al.*," *J. Geophys. Res., 99*, 25951–25954 (1994).

Hegg, D. A., J. L. Livingston, P. V. Hobbs, T. Novakov, and P. Russell, "Chemical Apportionment and Aerosol Column Optical Depth off the Mid-Atlantic Coast of the United States," *J. Geophys. Res., 102*, 25293–25303 (1997).

Heimann, M., "Dynamics of the Carbon Cycle," *Nature, 375*, 629–630 (1995).

Henderson-Sellers, A., Ed., *Future Climates of the World: A Modelling Perspective*, Vol. 16, Elsevier, Amsterdam/New York, 1995.

Heymsfield, A. J., "Microphysical Structures of Stratiform and Cirrus Clouds," in *Aerosol–Cloud–Climate Interactions* (P. V. Hobbs, Ed.), pp. 97–121, Academic Press, San Diego, 1993.

Ho, C.-H., M. D. Chou, M. Suarez, K.-M. Lau, and M. M.-H. Yan, "Comparison of Model-Calculated and ERBE-Retrieved Clear-Sky Outgoing Longwave Radiation," *J. Geophys. Res., 103*, 11529–11536 (1998).

Hobbs, P. V., Ed., *Aerosol–Cloud–Climate Interactions*, Academic Press, San Diego, 1993a.

Hobbs, P. V., "Aerosol–Cloud Interactions," in *Aerosol–Cloud–Climate Interactions* (P. V. Hobbs, Ed.), pp. 33–73, Academic Press, San Diego, 1993b.

Hoffmann, P., A. N. Dedik, F. Deutsch, T. Sinner, S. Weber, R. Eichler, S. Sterkel, C. S. Sastri, and H. M. Ortner, "Solubility of Single Chemical Compounds from an Atmospheric Aerosol in Pure Water," *Atmos. Environ., 31*, 2777–2785 (1997).

Hogan, K. B., and R. C. Harriss, "Comment on 'A Dramatic Decrease in the Growth Rate of Atmospheric Methane in the Northern Hemisphere during 1992' by E. J. Dlugokencky *et al.*," *Geophys. Res. Lett., 21*, 2445–2446 (1994).

Hoppel, W. A., G. M. Frick, and R. E. Larson, "Effect of Nonprecipitating Clouds on the Aerosol Size Distribution in the Marine Boundary Layer," *Geophys. Res. Lett., 13*, 125–128 (1986).

Hoppel, W. A., and G. M. Frick, "Submicron Aerosol Size Distributions Measured over the Tropical and South Pacific," *Atmos. Environ., 24A*, 645–659 (1990).

Hoppel, W. A., G. M. Frick, and J. W. Fitzgerald, "Deducing Droplet Concentration and Supersaturation in Marine Boundary Layer Clouds from Surface Aerosol Measurements," *J. Geophys. Res., 101*, 26553–26565 (1996).

Howard, J. N., J. I. F. King, and P. R. Gast, "Thermal Radiation," in *Handbook of Geophysics*, Ch. 16, Macmillan, New York, 1960.

Hudson, J. G., "Observations of Anthropogenic Cloud Condensation Nuclei," *Atmos. Environ., 25A*, 2449–2455 (1991).

Hudson, J. G., and H. Li, "Microphysical Contrasts in Atlantic Stratus," *J. Atmos. Sci., 52*, 3031–3040 (1995).

Hurrell, J. W., and H. Van Loon, "Decadal Variations in Climate Associated with the North Atlantic Oscillation," *Clim. Change, 36*, 301–326 (1997).

Imbrie, J., and K. P. Imbrie, *Ice Ages—Solving the Mystery*, Enslow Publishers, Short Hills, NJ, 1979.

Imre, D. G., E. H. Abramson, and P. H. Daum, "Quantifying Cloud-Induced Shortwave Absorption: An Examination of Uncertainties and of Recent Arguments for Large Excess Absorption," *J. Appl. Meteorol., 35*, 1991–2010 (1996).

IPCC, Intergovernmental Panel on Climate Change, Contribution of Working Group I to the Second Assessment Report (J. T. Houghton, L. G. Meira Filho, B. A. Callander, N. Harris, A. Kattenberg, and K. Maskell, Eds.), *Climate Change 1995: The Science of Climate Change*, Cambridge Univ. Press, Cambridge, UK, 1996.

Janach, W. E., "Surface Ozone: Trend Details, Seasonal Variations, and Interpretation," *J. Geophys. Res., 94,* 18289–18295 (1989).

Jayaraman, A., D. Lubin, S. Ramachandran, V. Ramanathan, E. Woodbridge, W. D. Collins, and K. S. Zalpuri, "Direct Observations of Aerosol Radiative Forcing over the Tropical Indian Ocean during the January–February 1996 Pre-INDOEX Cruise," *J. Geophys. Res., 103,* 13827–13836 (1998).

Ji, Q., and G. E. Shaw, "On Supersaturation Spectrum and Size Distributions of Cloud Condensation Nuclei," *Geophys. Res. Lett., 25,* 1903–1906 (1998).

Jiang, Y., and Y. L. Yung, "Concentrations of Tropospheric Ozone from 1979 to 1992 over Tropical Pacific South America from TOMS Data," *Science, 272,* 713–716 (1996).

Johnson, J. C., S. S. Cliff, and M. H. Thiemens, "Measurement of Multioxygen Isotopic ($\delta^{18}O$ and $\delta^{17}O$) Fractionation Factors in the Stratospheric Sink Reactions of Nitrous Oxide," *J. Geophys. Res., 100,* 16801–16804 (1995).

Johnson, K., M. Huyler, H. Westberg, B. Lamb, and P. Zimmerman, "Measurement of Methane Emissions from Ruminant Livestock Using a SF_6 Tracer Technique," *Environ. Sci. Technol., 28,* 359–362 (1994).

Jones, A., D. L. Roberts, and A. Slingo, "A Climate Model Study of Indirect Radiative Forcing by Anthropogenic Sulphate Aerosols," *Nature, 370,* 450–453 (1994).

Jones, P. D., "Observations from the Surface: Projection from Traditional Meteorological Observations," in *Future Climates of the World: A Modelling Perspective*, Chapter 5, pp. 151–189, Elsevier, Amsterdam/New York, 1995.

Jones, P. D., D. E. Parker, T. J. Osborn, and K. R. Briffa, "Global and Hemispheric Temperature Anomalies–Land and Marine Instrumental Records, 1856–1997," in *Trends: A Compendium of Data on Global Change*, Carbon Dioxide Information Analysis Center, Oak Ridge National Laboratory, Oak Ridge, TN, 1998; see Web site: http://cdiac.esd.ornl.gov/ trends/ temp/ jonescru/jones.html.

Jouzel, J., R. B. Alley, K. M. Cuffey, W. Dansgaard, P. Grootes, G. Hoffmann, S. J. Johnsen, R. D. Koster, D. Peel, C. A. Shuman, M. Stievenard, M. Stuiver, and J. White, "Validity of the Temperature Reconstruction from Water Isotopes in Ice Cores," *J. Geophys. Res., 102,* 26471–26487 (1997).

Kaufman, Y. J., R. S. Fraser, and R. L. Mahoney, "Fossil Fuel and Biomass Burning Effect on Climate—Heating or Cooling?" *J. Clim., 4,* 578–588 (1991).

Kaufman, Y. J., and D. Tanré, "Effect of Variations in Super-Saturation on the Formation of Cloud Condensation Nuclei," *Nature, 369,* 45–48 (1994).

Kaufmann, R. K., and D. I. Stern, "Evidence for Human Influence on Climate from Hemispheric Temperature Relations," *Nature, 388,* 39–44 (1997).

Keeling, C. D., and S. R. Shertz, "Seasonal and Interannual Variations in Atmospheric Oxygen and Implications for the Global Carbon Cycle," *Nature, 358,* 723–727 (1992).

Keeling, C. D., T. P. Worf, M. Wahlen, and J. van der Pilcht, "Interannual Extremes in the Rate of Rise of Atmospheric Carbon Dioxide Since 1980," *Nature, 375,* 666–670 (1995).

Keeling, C. D., J. F. S. Chin, and T. P. Worf, "Increased Activity of Northern Vegetation Inferred from Atmospheric CO_2 Measurements," *Nature, 382,* 146–149 (1996).

Keeling, R. F., S. C. Piper, and M. Heimann, "Global and Hemispheric CO_2 Sinks Deduced from Changes in Atmospheric O_2 Concentration," *Nature, 381,* 218–221 (1996).

Keeling, R. F., A. C. Manning, E. M. McEvoy, and S. R. Shertz, "Methods of Measuring Changes in Atmospheric O_2 Concentration and Their Application in Southern Hemisphere Air," *J. Geophys. Res., 103,* 3381–3397 (1998).

Keene, W. C., R. Sander, A. A. P. Pszenny, R. Vogt, P. J. Crutzen, and J. N. Galloway, "Aerosol pH in the Marine Boundary Layer: A Review and Model Evaluation," *J. Aerosol. Sci., 29,* 339–356 (1998).

Keene, W. C., and D. L. Savoie, "The pH of Deliquesced Sea-Salt Aerosol in Polluted Marine Air," *Geophys. Res. Lett., 25,* 2181–2184 (1998); *26,* 1315 (1999).

Kellog, W. W., "Aerosols and Global Warming," *Science, 256,* 598 (1992).

Kelly, P. M., and T. M. L. Wigley, "Solar Cycle Length, Greenhouse Forcing and Global Climate," *Nature, 360,* 328–330 (1992).

Kerr, R. A., "The Hottest Year, By a Hair," *Science, 279,* 315–316 (1998).

Khalil, M. A. K., and R. A. Rasmussen, "Carbon Monoxide in the Earth's Atmosphere: Increasing Trend," *Science, 224,* 54–56 (1984).

Khalil, M. A. K., and R. A. Rasmussen, "Atmospheric Methane: Trends over the Last 10,000 Years," *Atmos. Environ., 21,* 2445–2452 (1987).

Khalil, M. A. K., and R. A. Rasmussen, "The Global Sources of Nitrous Oxide," *J. Geophys. Res., 97,* 14651–14660 (1992).

Khalil, M. A. K., and R. A. Rasmussen, "Global Decrease in Atmospheric Carbon Monoxide Concentration," *Nature, 370,* 639–641 (1994a).

Khalil, M. A. K., and R. A. Rasmussen, "Global Emissions of Methane during the Last Several Centuries," *Chemosphere, 29,* 833–842 (1994b).

Kiehl, J. T., and B. P. Briegleb, "The Relative Roles of Sulfate Aerosols and Greenhouse Gases in Climate Forcing," *Science, 260,* 311–314 (1993).

Kiehl, J. T., J. J. Hack, M. H. Zhang, and R. D. Cess, "Sensitivity of a GCM Climate to Enhanced Shortwave Cloud Absorption," *J. Clim., 8,* 2200–2212 (1995).

Kiehl, J. T., and K. E. Trenberth, "Earth's Annual Global Mean Energy Budget," *Bull. Am. Meteorol. Soc., 78,* 197–208 (1997).

Kim, K.-R., and H. Craig, "Two-Isotope Characterization of N_2O in the Pacific Ocean and Constraints on Its Origin in Deep Water," *Nature, 347,* 58–61 (1990).

Kim, K.-R., and H. Craig, "Nitrogen-15 and Oxygen-18 Characteristics of Nitrous Oxide: A Global Perspective," *Science, 262,* 1855–1857 (1993).

Kim, Y., and R. D. Cess, "Effect of Anthropogenic Sulfate Aerosols on Low-Level Cloud Albedo over Oceans," *J. Geophys. Res., 98,* 14883–14885 (1993).

King, M. D., L. F. Radke, and P. V. Hobbs, "Determination of the Spectral Absorption of Solar Radiation by Marine Stratocumulus Clouds from Airborne Measurements within Clouds," *J. Atmos. Sci., 47,* 894–907 (1990).

King, M. D., "Radiative Properties of Clouds," in *Aerosol-Cloud-Climate Interactions* (P. V. Hobbs, Ed.), pp. 123–149, Academic Press, San Diego, 1993.

King, M. D., L. F. Radke, and P. V. Hobbs, "Optical Properties of Marine Stratocumulus Clouds Modified by Ships," *J. Geophys. Res., 98,* 2729–2739 (1993).

Ko, M. K. W., N. D. Sze, and G. Molnar, "Global Warming from Chlorofluorocarbons and Their Alternatives: Time Scales of Chemistry and Climate," *Atmos. Environ., 27A,* 581–587 (1993).

Kondrat'ev, K. Y., V. I. Binenko, and I. N. Melnikova, "Absorption of Solar Radiation by Cloudy and Cloudless Atmosphere," *Russ. Meteorol Hydrol.*, No. 2, 8–15 (1996a).

Kondrat'ev, K. Y., V. I. Binenko, and I. N. Melnikova, "Solar Radiation Absorption by Cloudy and Cloudless Atmospheres," *Meteorol. Gidrol.*, No. 2, 14–23 (1996b).

Kotzick, R., U. Panne, and R. Niessner, "Changes in Condensation Properties of Ultrafine Carbon Particles Subjected to Oxidation by Ozone," *J. Aerosol Sci.*, 28, 725–735 (1997).

Kreidenweis, S. M., and J. H. Seinfeld, "Nucleation of Sulfuric Acid–Water and Methanesulfonic Acid–Water Solution Particles. Implications for the Atmospheric Chemistry of Organosulfur Species," *Atmos. Environ.*, 22, 283–296 (1988).

Kukla, G., and T. R. Karl, "Nighttime Warming and the Greenhouse Effect," *Environ. Sci. Technol.*, 27, 1468–1474 (1993).

Kulmala, M., A. Laaksonen, P. Korhonen, T. Vesala, T. Ahonen, and J. C. Barrett, "The Effect of Atmospheric Nitric Acid Vapor on Cloud Condensation Nucleus Activation," *J. Geophys. Res.*, 98, 22949–22958 (1993).

Kulmala, M., P. Korhonen, A. Laaksonen, and T. Vesala, "Changes in Cloud Properties Due to NO_x Emissions," *Geophys. Res. Lett.*, 22, 239–242 (1995).

Kulmala, M., A. Laaksonen, R. J. Charlson, and P. Korhonen, "Clouds without Supersaturation," *Nature*, 388, 336–337 (1997).

Kulmala, M., A. Toivonen, T. Mattila, and P. Korhonen, "Variations of Cloud Droplet Concentrations and the Optical Properties of Clouds Due to Changing Hygroscopicity: A Model Study," *J. Geophys. Res.*, 103, 16183–16195 (1998).

Kumar, K. R., K. K. Kumar, and G. B. Pant, "Diurnal Asymmetry of Surface Temperature Trends over India," *Geophys. Res. Lett.*, 21, 677–680 (1994).

Kvenvolden, K. A., "Gas Hydrates—Geological Perspective and Global Change," *Rev. Geophys.*, 31, 173–187 (1993).

Laaksonen, A., J. Hienola, M. Kulmala, and F. Arnold, "Supercooled Cirrus Cloud Formation Modified by Nitric Acid Pollution of the Upper Troposphere," *Geophys. Res. Lett.*, 24, 3009–3012 (1997).

Lacis, A., J. Hansen, and M. Sato, "Climate Forcing by Stratospheric Aerosols," *Geophys. Res. Lett.*, 19, 1607–1610 (1992).

Lacis, A. A., D. J. Wuebbles, and J. A. Logan, "Radiative Forcing of Climate by Changes in the Vertical Distribution of Ozone," *J. Geophys. Res.*, 95, 9971–9981 (1990).

Lacis, A. A., and M. I. Mishchenko, "Climate Forcing, Climate Sensitivity, and Climate Response: A Radiative Modeling Perspective on Atmospheric Aerosols," in *Aerosol Forcing of Climate* (R. J. Charlson and J. Heintzenberg, Eds.), Wiley, New York, 1995.

Laj, P., J. M. Palais, and H. Sigurdsson, "Changing Sources of Impurities to the Greenland Ice Sheet over the Last 250 Years," *Atmos. Environ.*, 26A, 2627–2640 (1992).

Lambert, J. D., *Vibrational and Rotational Relaxation in Gases*, Clarendon, Oxford, 1977.

Langner, J., H. Rodhe, P. J. Crutzen, and P. Zimmermann, "Anthropogenic Influence on the Distribution of Tropospheric Sulphate Aerosol," *Nature*, 359, 712–716 (1992).

Lau, K. M., C. H. Sui, M. D. Chou, and W. K. Tau, "An Inquiry into the Cirrus-Cloud Thermostat Effect for Tropical Sea Surface Temperature," *Geophys. Res. Lett.*, 21, 1157–1160 (1994a).

Lau, K. M., C. H. Sui, M. D. Chou, and W. K. Tau, "Comment on the Paper 'An Inquiry into the Cirrus-Cloud Thermostat Effect for Tropical Sea Surface Temperature,'" *Geophys. Res. Lett.*, 21, 1185–1186 (1994b).

Lawrence, M. G., "An Empirical Analysis of the Strength of the Phytoplankton–Dimethylsulfide–Cloud–Climate Feedback Cycle," *J. Geophys. Res.*, 98, 20663–20673 (1993).

Leaitch, W. R., G. A. Isaac, J. W. Strapp, C. M. Banic, and H. A. Wiebe, "The Relationship between Cloud Droplet Number Concentrations and Anthropogenic Pollution: Observations and Climatic Implications," *J. Geophys. Res.*, 97, 2463–2474 (1992).

Lean, J., A. Skumanich, and O. White, "Estimating the Sun's Radiative Output during the Maunder Minimum," *Geophys. Res. Lett.*, 19, 1591–1594 (1992).

Lean, J., J. Beer, and R. Bradley, "Reconstruction of Solar Irradiance Since 1610: Implications for Climate Change," *Geophys. Res. Lett.*, 22, 3195–3198 (1995a).

Lean, J. L., O. R. White, and A. Skumanich, "On the Solar Ultraviolet Spectral Irradiance during the Maunder Minimum," *Global Biogeochem. Cycles*, 9, 171–182 (1995b).

Lean, J. L., G. J. Rottman, H. Lee Kyle, T. N. Woods, J. R. Hickey, and L. C. Puga, "Detection and Parameterization of Variations in Solar Mid- and Near-Ultraviolet Radiation (200–400 nm)," *J. Geophys. Res.*, 102, 29939–29956 (1997).

Lefohn, A. S., D. S. Shadwick, U. Feister, and V. A. Mohnen, "Surface-Level Ozone: Climate Change and Evidence for Trends," *J. Air Waste Manage. Assoc.*, 42, 136–144 (1992).

Legates, D. R., and R. E. Davis, "The Continuing Search for an Anthropogenic Climate Change Signal: Limitations of Correlation-Based Approaches," *Geophys. Res. Lett.*, 24, 2319–2322 (1997).

Lelieveld, J., and J. Heintzenberg, "Sulfate Cooling Effect on Climate through In-Cloud Oxidation of Anthropogenic SO_2," *Science*, 258, 117–120 (1992).

Lelieveld, J., and R. van Dorland, "Ozone Chemistry Changes in the Troposphere and Consequent Radiative Forcing of Climate," in *NATO ASI Series*, 132, (1995).

Lenoble, J., *Atmospheric Radiative Transfer*, A. Deepak Publishing, Hampton, VA, 1993.

Li, S.-M., K. B. Strawbridge, W. R. Leaitch, and A. M. Macdonald, "Aerosol Backscattering Determined from Chemical and Physical Properties and Lidar Observations over the East Coast of Canada," *Geophys. Res. Lett.*, 25, 1653–1656 (1998).

Li, X., H. Maring, D. Savole, K. Voss, and J. M. Prospero, "Dominance of Mineral Dust in Aerosol Light-Scattering in the North Atlantic Trade Winds," *Nature*, 380, 416–419 (1996).

Li, Z., H. W. Barker, and L. Moreau, "The Variable Effect of Clouds on Atmospheric Absorption of Solar Radiation," *Nature*, 376, 486–490 (1995).

Li, Z., and L. Moreau, "Alteration of Atmospheric Solar Absorption by Clouds: Simulation and Observation," *J. Appl. Meteorol.*, 35, 653–670 (1996).

Liao, H., and J. H. Seinfeld, "Effect of Clouds on Direct Aerosol Radiative Forcing of Climate," *J. Geophys. Res.*, 103, 3781–3788 (1998).

Li-Jones, X., H. B. Maring, and J. M. Prospero, "Effect of Relative Humidity on Light Scattering by Mineral Dust Aerosol As Measured in the Marine Boundary Layer over the Tropical Atlantic Ocean," *J. Geophys. Res.*, 103, 31113–31121 (1998).

Lin, X., W. L. Chameides, C. S. Kiang, A. W. Stelson, and H. Berresheim, "A Model Study of the Formation of Cloud Condensation Nuclei in Remote Marine Areas," *J. Geophys. Res.*, 97, 18161–18171 (1992).

Lin, X., W. L. Chameides, C. S. Kiang, A. W. Stelson, and H. Berresheim, Reply, *J. Geophys. Res.*, 98, 10815–10817 (1993a).

Lin, X., W. L. Chameides, C. S. Kiang, A. W. Stelson, and H. Berresheim, "Comment on 'A Model Study of the Formation of Cloud Condensation Nuclei in Remote Marine Areas,' Reply," *J. Geophys. Res.*, 98, 20815–20816 (1993b).

Lindzen, R. S., "Climate Dynamics and Global Change," *Annu. Rev. Fluid Mech.*, 26, 353–378 (1994).

Lindzen, R. S., and C. Giannitsis, "On the Climatic Implications of Volcanic Cooling," *J. Geophys. Res., 103*, 5929–5941 (1998).

Liou, K.-N., *An Introduction to Atmospheric Radiation*, Academic Press, New York, 1980.

Liou, K.-N, *Radiation and Cloud Processes in the Atmosphere: Theory, Observation, and Modeling*, Oxford Univ. Press, New York, 1992.

Liousse, C., J. E. Penner, C. Chuang, J. J. Walton, H. Eddleman, and H. Cachier, "A Global Three-Dimensional Model Study of Carbonaceous Aerosols," *J. Geophys. Res., 101*, 19411–19432 (1996).

Liptay, K., J. Chanton, P. Czepiel, and B. Mosher, "Use of Stable Isotopes to Determine Methane Oxidation in Landfill Cover Soils," *J. Geophys. Res., 103*, 8243–8250 (1998).

Loeb, N. G., and R. Davies, "Observational Evidence of Plane Parallel Model Biases: Apparent Dependence of Cloud Optical Depth on Solar Zenith Angle," *J. Geophys. Res., 101*, 1621–1634 (1996).

Logan, J. A., "Trends in the Vertical Distribution of Ozone: An Analysis of Ozonesonde Data," *J. Geophys. Res., 99*, 25553–25585 (1994).

Lohmann, U., and J. Feichter, "Impact of Sulfate Aerosols on Albedo and Lifetime of Clouds: A Sensitivity Study with the ECHAM4 GCM," *J. Geophys. Res., 102*, 13685–13700 (1997).

López-Puertas, M., G. Zaragoza, M. Á. López-Valverde, and F. W. Taylor, "Nonlocal Thermodynamic Equilibrium (LTE) Atmospheric Limb Emission at 4.6 μm. 1. An Update of the CO_2 Non-LTE Radiative Transfer Model," *J. Geophys. Res., 103*, 8499–8513 (1998a).

López-Puertas, M., G. Zaragoza, M. Á. López-Valverde, and F. W. Taylor, "Nonlocal Thermodynamic Equilibrium (LTE) Atmospheric Limb Emission at 4.6 μm. 2. An Analysis of the Daytime Wideband Radiances As Measured by UARS Improved Stratospheric and Mesospheric Sounder," *J. Geophys. Res., 103*, 8515–8530 (1998b).

Low, P. S., T. D. Davies, P. M. Kelly, and G. Farmer, "Trends in Surface Ozone at Hohenpeissenberg and Arkona," *J. Geophys. Res., 95*, 22441–22453 (1990).

Lowe, D. C., C. A. M. Brenninkmeijer, M. R. Manning, R. Sparks, and G. Wallace, "Radiocarbon Determination of Atmospheric Methane at Baring Head, New Zealand," *Nature, 332*, 522–525 (1988).

Lowe, D. C., C. A. M. Brenninkmeijer, G. W. Brailsford, K. R. Lassey, A. J. Gomez, and E. G. Nisbet, "Concentration and ^{13}C Records of Atmospheric Methane in New Zealand and Antarctica: Evidence for Changes in Methane Sources," *J. Geophys. Res., 99*, 16913–16925 (1994).

Lubin, D., J.-P. Chen, P. Pilewskie, V. Ramanathan, and F. P. J. Valero, "Microphysical Examination of Excess Cloud Absorption in the Tropical Atmosphere," *J. Geophys. Res., 101*, 16961–16972 (1996).

Machida, T., T. Nakazawa, Y. Fujii, S. Aoki, and O. Watanabe, "Increase in the Atmospheric Nitrous Oxide Concentration during the Last 250 Years," *Geophys. Res. Lett., 22*, 2921–2924 (1995).

Madronich, S., and C. Granier, "Impact of Recent Total Ozone Changes on Tropospheric Ozone Photodissociation, Hydroxyl Radicals, and Methane Trends," *Geophys. Res. Lett., 19*, 465–467 (1992).

Mahasenan, N., R. G. Watts, and H. Dowlatabadi, "Low-Frequency Oscillations in Temperature-Proxy Records and Implications for Recent Climate Change," *Geophys. Res. Lett., 24*, 563–566 (1997).

Malm, W. C., J. V. Molenar, R. A. Eldred, and J. F. Sisler, "Examining the Relationship among Atmospheric Aerosols and Light Scattering and Extinction in the Grand Canyon Area," *J. Geophys. Res., 101*, 19251–19265 (1996).

Manabe, S., and R. F. Strickler, "Thermal Equilibrium of the Atmosphere with a Convective Adjustment," *J. Atmos. Sci., 21*, 361–385 (1964).

Mann, M. E., R. S. Bradley, and M. K. Hughes, "Global-Scale Temperature Patterns and Climate Forcing over the Past Six Centuries," *Nature, 392*, 779–787 (1998).

Marenco, A., H. Gouget, P. Nédélec, J.-P Pagés, and F. Karcher, "Evidence of a Long-Term Increase in Tropospheric Ozone from Pic du Midi Data Series, Consequences: Positive Radiative Forcing," *J. Geophys. Res., 99*, 16617–16632 (1994).

Marley, N. A., J. S. Gaffney, and M. M. Cunningham, "Aqueous Greenhouse Species in Clouds, Fogs, and Aerosols," *Environ. Sci. Technol., 27*, 2864–2869 (1993).

Martin, G. M., D. W. Johnson, and A. Spice, "The Measurement and Parameterization of Effective Radius of Droplets in Warm Stratocumulus Clouds," *J. Atmos. Sci., 51*, 1823–1842 (1994).

Matsumoto, K., H. Tanaka, I. Nagao, and Y. Ishizaka, "Contribution of Particulate Sulfate and Organic Carbon to Cloud Condensation Nuclei in the Marine Atmosphere," *Geophys. Res. Lett., 24*, 655–658 (1997).

Matsumoto, K., I. Nagao, H. Tanaka, H. Miyaji, T. Iida, and Y. Ikebe, "Seasonal Characteristics of Organic and Inorganic Species and Their Size Distributions in Atmospheric Aerosols over the Northwest Pacific Ocean," *Atmos. Environ., 32*, 1931–1946 (1998).

Mayewski, P. A., W. B. Lyons, M. J. Spencer, M. Twickler, W. Dansgaard, B. Koci, C. I. Davidson, and R. E. Honrath, "Sulfate and Nitrate Concentrations from a South Greenland Ice Core," *Science, 232*, 975–977 (1986).

Mayewski, P. A., W. B. Lyons, M. J. Spencer, M. S. Twickler, C. F. Buck, and S. Whitlow, "An Ice Core Record of Atmospheric Response to Anthropogenic Sulphate and Nitrate," *Nature, 346*, 554–556 (1990).

McCormick, M. P., L. W. Thomason, and C. R. Trepte, "Atmospheric Effects of the Mt. Pinatubo Eruption," *Nature, 373*, 399–404 (1995).

McElroy, M. B., and D. B. A. Jones, "Evidence for an Additional Source of Atmospheric N_2O," *Global Biogeochem. Cycles, 10*, 651–659 (1996).

McKeen, S., D. Kley, and A. Volz, "The Historical Trend of Tropospheric Ozone over Western Europe: A Model Perspective," in *Ozone in the Atmosphere* (R. D. Bojkov and P. Fabian, Eds.), A. Deepak Publishing, Hampton, VA, 1989.

Meehl, G. A., W. M. Washington, D. J. Erickson III, B. P. Briegleb, and P. J. Jaumann, "Climate Change from Increased CO_2 and Direct and Indirect Effects of Sulfate Aerosols," *Geophys. Res. Lett., 23*, 3755–3758 (1996).

Mel'nikova, I. N., and V. V. Mikhaylov, "Spectral Scattering and Absorption Coefficients in Strati Derived from Aircraft Measurements," *J. Atmos. Sci., 51*, 925–931 (1994).

Miller, S. L., "Clathrate Hydrates of Air in Antarctic Ice," *Science, 165*, 489–490 (1969).

Minnis, P., E. F. Harrison, L. L. Stowe, G. G. Gibson, F. M. Denn, D. R. Doelling, and W. L. Smith, Jr., "Radiative Climate Forcing by the Mount Pinatubo Eruption," *Science, 259*, 1411–1415 (1993).

Mitchell, J. F. B., "The 'Greenhouse' Effect and Climate Change," *Rev. Geophys., 27*, 115–139 (1989).

Mitchell, J. F. B., T. C. Johns, J. M. Gregory, and S. F. B. Tett, "Climate Response to Increasing Levels of Greenhouse Gases and Sulphate Aerosols," *Nature, 376*, 501–504 (1995).

Mlawer, E. J., S. A. Clough, P. D. Brown, T. M. Stephen, J. C. Landry, A. Goldman, and F. J. Murcray, "Observed Atmospheric Collision-Induced Absorption in Near-Infrared Oxygen Bands," *J. Geophys. Res., 103*, 3859–3863 (1998).

Molnar, G. I., M. K. W. Ko, S. Zhou, and N. D. Sze, "Climatic Consequences of Observed Ozone Loss in the 1980s: Relevance to the Greenhouse Problem," *J. Geophys. Res., 99,* 25755–25760 (1994).

Moriizumi, J., K. Nagamine, T. Iida, and Y. Ikebe, "Carbon Isotopic Analysis of Atmospheric Methane in Urban and Suburban Areas: Fossil and Non-Fossil Methane from Local Sources," *Atmos. Environ., 32,* 2947–2955 (1998).

Munzio, L. J., and J. C. Kramlich, "An Artifact in the Measurements of N_2O from Combustion Sources," *Geophys. Res. Lett., 15,* 1369–1372 (1988).

Murphy, D. M., J. R. Anderson, P. K. Quinn, L. M. McInnes, F. J. Brechtel, S. M. Kreidenweis, A. M. Middlebrook, M. Pósfai, D. S. Thomson, and P. R. Buseck, "Influence of Sea-Salt on Aerosol Radiative Properties in the Southern Ocean Marine Boundary Layer," *Nature, 392,* 62–65 (1998).

National Research Council, "Aerosol Radiative Forcing and Climate Change," National Academy Press, Washington, DC, 1996.

Nemesure, S., R. Wagener, and S. E. Schwartz, "Direct Shortwave Forcing of Climate by the Anthropogenic Sulfate Aerosol: Sensitivity to Particle Size, Composition, and Relative Humidity," *J. Geophys. Res., 100,* 26105–26116 (1995).

Nevison, C. D., R. F. Weiss, and D. J. Erickson III, "Global Oceanic Emissions of Nitrous Oxide," *J. Geophys. Res., 100,* 15809–15820 (1995).

Nevison, C. D., G. Esser, and E. A. Holland, "A Global Model of Changing N_2O Emissions from Natural and Perturbed Soils," *Clim. Change, 32,* 327–378 (1996).

Novakov, T., and J. E. Penner, "Large Contribution of Organic Aerosols to Cloud-Condensation-Nuclei Concentrations," *Nature, 365,* 823–826 (1993).

Novakov, T., C. Rivera-Carpio, J. E. Penner, and C. F. Rogers, "The Effect of Anthropogenic Sulfate Aerosols on Marine Cloud Droplet Concentrations," *Tellus, 46B,* 132–141 (1994).

Novakov, T., and C. E. Corrigan, "Cloud Condensation Nucleus Activity of the Organic Components of Biomass Smoke Particles," *Geophys. Res. Lett., 23,* 2141–2144 (1996).

Novakov, T., C. E. Corrigan, J. E. Penner, C. C. Chuang, O. Rosario, and O. L. M. Bracero, "Organic Aerosols in the Caribbean Trade Winds: A Natural Source?" *J. Geophys. Res., 102,* 21307–21313 (1997a).

Novakov, T., D. A. Hegg, and P. V. Hobbs, "Airborne Measurements of Carbonaceous Aerosols on the East Coast of the United States," *J. Geophys. Res., 102,* 30023–30030 (1997b).

Novelli, P. C., K. A. Masarie, P. P. Tans, and P. M. Lang, "Recent Changes in Atmospheric Carbon Monoxide," *Science, 263,* 1587–1590 (1994).

O'Dowd, C. D., J. A. Lowe, and M. H. Smith, "Biogenic Sulphur Emissions and Inferred Non-Sea-Salt-Sulphate Cloud Condensation Nuclei in and around Antarctica," *J. Geophys. Res., 102,* 12839–12854 (1997a).

O'Dowd, C. D., M. H. Smith, I. E. Consterdine, and J. A. Lowe, "Marine Aerosol, Sea-Salt, and the Marine Sulphur Cycle: A Short Review," *Atmos. Environ., 31,* 73–80 (1997b).

O'Dowd, C. D., M. Geever, M. K. Hill, M. H. Smith, and S. G. Jennings, "New Particle Formation: Nucleation Rates and Spatial Scales in the Clean Marine Coastal Environment," *Geophys. Res. Lett., 25,* 1661–1664 (1998).

Olson, J., M. Prather, T. Berntsen, G. Carmichael, R. Chatfield, P. Connell, R. Derwent, L. Horowitz, S. Jin, M. Kanakidou, P. Kasibhatla, R. Kotamarthi, M. Kuhn, K. Law, J. Penner, L. Perliski, S. Sillman, F. Stordal, A. Thompson, and O. Wild, "Results from the Intergovernmental Panel on Climatic Change Photochemical Model Intercomparison (PhotoComp)," *J. Geophys. Res., 102,* 5979–5991 (1997).

Overpeck, J., K. Hughen, D. Hardy, R. Bradley, R. Case, M. Douglas, B. Finney, K. Gajewski, G. Jacoby, A. Jennings, S. Lamoureux, A. Lasca, G. MacDonald, J. Moore, M. Retelle, S. Smith, A. Wolfe, and G. Zielinski, "Arctic Environmental Change of the Last Four Centuries," *Science, 278,* 1251–1256 (1997).

Pan, W., M. A. Tatang, G. J. McRae, and R. G. Prinn, "Uncertainty Analysis of Direct Radiative Forcing by Anthropogenic Sulfate Aerosols," *J. Geophys. Res., 102,* 21915–21924 (1997).

Pan, W., M. A. Tatang, G. J. McRae, and R. G. Prinn, "Uncertainty Analysis of Indirect Radiative Forcing by Anthropogenic Sulfate Aerosols," *J. Geophys. Res., 103,* 3815–3823 (1998).

Pandis, S. N., L. M. Russell, and J. H. Seinfeld, "The Relationship between DMS Flux and CCN Concentration in Remote Marine Regions," *J. Geophys. Res., 99,* 16945–16957 (1994).

Pandis, S. N., L. M. Russell, and J. H. Seinfeld, "Reply," *J. Geophys. Res., 100,* 14357–14358 (1995).

Papasavva, S., S. Tai, K. H. Illinger, and J. E. Kenny, "Infrared Radiative Forcing of CFC Substitutes and Their Atmospheric Reaction Products," *J. Geophys. Res., 102,* 13643–13650 (1997).

Parker, D. E., H. Wilson, P. D. Jones, J. R. Christy, and C. K. Folland, "The Impact of Mount Pinatubo on Worldwide Temperatures," *Int. J. Climatol., 16,* 487–497 (1996).

Parungo, F., J. F. Boatman, H. Sievering, S. W. Wilkison, and B. B. Hicks, "Trends in Global Marine Cloudiness and Anthropogenic Sulfur," *J. Clim., 7,* 434–440 (1994).

Patterson, E. M., "Optical Properties of the Crustal Aerosol: Relation to Chemical and Physical Characteristics," *J. Geophys. Res., 86,* 3236–3246 (1981).

Pearman, G. I., and P. J. Fraser, "Sources of Increased Methane," *Nature, 332,* 489–490 (1988).

Penner, J. E., "Cloud Albedo, Greenhouse Effects, Atmospheric Chemistry, and Climate Change," *J. Air Waste Manage. Assoc., 40,* 456–461 (1990).

Penner, J. E., R. E. Dickinson, and C. A. O'Neill, "Effects of Aerosol from Biomass Burning on the Global Radiation Budget," *Science, 256,* 1432–1433 (1992).

Penner, J. E., R. J. Charlson, J. M. Hales, N. S. Laulainen, R. Leifer, T. Novakov, J. Ogren, L. F. Radke, S. E. Schwartz, and L. Travis, "Quantifying and Minimizing Uncertainty of Climate Forcing by Anthropogenic Aerosols," *Bull. Am. Meteorol. Soc., 75,* 375–400 (1994).

Penner, J. E., and T. Novakov, "Carbonaceous Particles in the Atmosphere: A Historical Perspective to the Fifth International Conference on Carbonaceous Particles in the Atmosphere," *J. Geophys. Res., 101,* 19373–19378 (1996).

Penner, J. E., C. C. Chuang, and K. Grant, "Climate Forcing by Carbonaceous and Sulfate Aerosols," *Clim., Dyn., 14,* 839–851 (1998).

Pfeilsticker, K., F. Erle, and U. Platt, "Notes and Correspondence: Absorption of Solar Radiation by Atmospheric O_4," *J. Atmos. Sci., 54,* 933–939 (1997).

Pielke, R. A., Sr., J. Eastman, T. N. Chase, J. Knaf, and T. G. F. Kittel, "1973–1996 Trends in Depth-Averaged Tropospheric Temperature," *J. Geophys. Res., 103,* 16927–16933 (1998a).

Pielke, R. A., Sr., J. Eastman, T. N. Chase, J. Knaf, and T. G. F. Kittel, "Correction to '1973–1996 Trends in Depth-Averaged Tropospheric Temperature,'" *J. Geophys. Res., 103,* 28909–28911 (1998b).

Pilewskie, P., and F. P. J. Valero, "Direct Observations of Excess Solar Absorption by Clouds," *Science, 267,* 1626–1629 (1995).

Pilewskie, P., and F. P. J. Valero, "Technical Comments on 'How Much Solar Radiation Do Clouds Absorb?'" *Science, 271,* 1134–1136 (1996).

Pilinis, C., S. N. Pandis, and J. H. Seinfeld, "Sensitivity of Direct Climate Forcing by Atmospheric Aerosols to Aerosol Size and Composition," *J. Geophys. Res., 100,* 18739–18754 (1995).

Pilinis, C., and X. Li, "Particle Shape and Internal Inhomogeneity Effects on the Optical Properties of Tropospheric Aerosols of Relevance to Climate Forcing," *J. Geophys. Res., 103,* 3789–3800 (1998).

Pincus, R., and M. B. Baker, "Effect of Precipitation on the Albedo Susceptibility of Clouds in the Marine Boundary Layer," *Nature, 372,* 250–252 (1994).

Pinnock, S., M. D. Hurley, K. P. Shine, T. J. Wallington, and T. J. Smyth, "Radiative Forcing of Climate by Hydrochlorofluorocarbons and Hydrofluorocarbons," *J. Geophys. Res., 100,* 23227–23238 (1995).

Pires, M., H. van den Bergh, and M. J. Rossi, "The Heterogeneous Formation of N_2O over Bulk Condensed Phases in the Presence of SO_2 at High Humidities," *J. Atmos. Chem., 25,* 229–250 (1996).

Pires, M., and M. J. Rossi, "The Heterogeneous Formation of N_2O in the Presence of Acidic Solutions: Experiments and Modeling," *Int. J. Chem. Kinet., 29,* 869–891 (1997).

Platnick, S., and S. Twomey, "Determining the Susceptibility of Cloud Albedo to Changes in Droplet Concentration with the Advanced Very High Resolution Radiometer," *J. Appl. Meteorol., 33,* 334–347 (1994).

Podgorny, I. A., A. M. Vogelmann, and V. Ramanathan, "Effects of Cloud Shape and Water Vapor Distribution on Solar Absorption in the Near Infrared," *Geophys. Res. Lett., 25,* 1899–1902 (1998).

Pollack, H. N., S. Huang, P.-Y. Shen, "Climate Change Record in Subsurface Temperatures: A Global Perspective," *Science, 282,* 279–281 (1998).

Portmann, R. W., S. Solomon, J. Fishman, J. R. Olson, J. T. Kiehl, and B. Briegleb, "Radiative Forcing of the Earth's Climate System Due to Tropical Tropospheric Ozone Production," *J. Geophys. Res., 102,* 9409–9417 (1997).

Prasad, S. S., "Natural Atmospheric Sources and Sinks of Nitrous Oxide. 1. An Evaluation Based on 10 Laboratory Experiments," *J. Geophys. Res., 99,* 5285–5294 (1994).

Prasad, S. S., "Potential Atmospheric Sources and Sinks of Nitrous Oxide. 2. Possibilities from Excited O_2, 'Embryonic' O_3, and Optically Pumped Excited O_3," *J. Geophys. Res., 102,* 21527–21536 (1997).

Prasad, S. S., E. C. Zipf, and X. Zhao, "Potential Atmospheric Sources and Sinks of Nitrous Oxide. 3. Consistency with the Observed Distributions of the Mixing Ratios," *J. Geophys. Res., 162,* 21537–21541 (1997).

Pruppacher, H. R., "The Role of Cloud Physics in Atmospheric Multiphase Systems: Ten Basic Statements," in *Chemistry of Multiphase Atmospheric Systems* (W. Jaeschke, Ed.), NATO ASI Series, Vol. G6, Springer-Verlag, Berlin/New York, 1986.

Pruppacher, H. R., and J. D. Klett, in *Microphysics of Clouds and Precipitation,* 2nd ed., pp. 229–230, Reidel, Dordrecht, 1997.

Pueschel, R. F., C. C. Van Valin, R. C. Castillo, J. A. Kadlecek, and E. Ganor, "Aerosols in Polluted versus Nonpolluted Air Masses: Long-Range Transport and Effects of Clouds," *J. Clim. Appl. Meteorol., 25,* 1908–1917 (1986).

Putaud, J.-P., S. Belviso, B. C. Nguyen, and N. Mihalopoulos, "Dimethylsulfide, Aerosols, and Condensation Nuclei over the Tropical Northeastern Atlantic Ocean," *J. Geophys. Res., 98,* 14863–14871 (1993).

Quinn, P. K., D. S. Covert, T. S. Bates, V. N. Kapustin, D. C. Ramsey-Bell, and L. M. McInnes, "Dimethylsulfide/Cloud Condensation Nuclei/Climate System: Relevant Size-Resolved Measurements of the Chemical and Physical Properties of Atmospheric Aerosol Particles," *J. Geophys. Res., 98,* 10411–10427 (1993).

Raes, F., and R. Van Dingenen, "Simulations of Condensation and Cloud Condensation Nuclei from Biogenic SO_2 in the Remote Marine Boundary Layer," *J. Geophys. Res., 97,* 12901–12912 (1992).

Raes, F., "Entrainment of Free Tropospheric Aerosols as a Regulating Mechanism for Cloud Condensation Nuclei in the Remote Marine Boundary Layer," *J. Geophys. Res., 100,* 2893–2903 (1995).

Raes, F., and R. Van Dingenen, "Comment on 'The Relationship between DMS Flux and CCN Concentration in Remote Marine Regions,'" *J. Geophys. Res., 100,* 14355–14356 (1995).

Raes, F., R. Van Dingenen, E. Cuevas, P. F. J. Van Velthoven, and J. M. Prospero, "Observations of Aerosols in the Free Troposphere and Marine Boundary Layer of the Subtropical Northeast Atlantic: Discussion of Processes Determining Their Size Distribution," *J. Geophys. Res., 102,* 21315–21328 (1997).

Rahn, T., and M. Wahlen, "Stable Isotope Enrichment in Stratospheric Nitrous Oxide," *Science, 278,* 1776–1778 (1997).

Rahn, T., H. Zhang, M. Wahlen, and G. A. Blake, "Stable Isotope Fractionation during Ultraviolet Photolysis of N_2O," *Geophys. Res. Lett., 25,* 4489–4492 (1998).

Ramanathan, V., "Radiative Transfer within the Earth's Troposphere and Stratosphere: A Simplified Radiative–Convective Model," *J. Atmos. Sci., 33,* 1330–1346 (1976).

Ramanathan, V., R. J. Cicerone, H. B. Singh, and J. T. Kiehl, "Trace Gas Trends and Their Potential Role in Climate Change," *J. Geophys. Res., 90,* 5547–5566 (1985).

Ramanathan, V., L. Callis, R. Cess, J. Hansen, I. Isaksen, W. Kihn, A. Lacis, F. Luther, J. Mahlman, R. Reck, and M. Schlesinger, "Climate–Chemical Interactions and Effects of Changing Atmospheric Trace Gases," *Rev. Geophys., 25,* 1441–1482 (1987).

Ramanathan, V., "The Greenhouse Theory of Climate Change: A Test by an Inadvertent Global Experiment," *Science, 240,* 293–299 (1988a).

Ramanathan, V., "The Radiative and Climatic Consequences of the Changing Atmospheric Composition of Trace Gases," in *The Changing Atmosphere* (F. S. Rowland and I. S. A. Isaksen, Eds.), pp. 160–186, Wiley, New York, 1988b.

Ramanathan, V., R. D. Cess, E. F. Harrison, P. Minnis, B. R. Barkstrom, E. Ahmad, and D. Hartmann, "Cloud-Radiative Forcing and Climate: Results from the Earth Radiation Budget Experiment," *Science, 243,* 57–63 (1989).

Ramanathan, V., and W. Collins, "Thermodynamic Regulation of Ocean Warming by Cirrus Clouds Deduced from Observations of 1987 El Nino," *Nature, 351,* 27–32 (1991).

Ramanathan, V., and W. Collins, "Thermostat and Global Warming," *Nature, 357,* 649 (1992).

Ramanathan, V., and W. Collins, "A Thermostat in the Tropics?" *Nature, 361,* 410–412 (1993).

Ramanathan, V., "Clouds and Climate," *Proc. Kon. Ned. Akad. Wetensch., 98,* 361–383 (1995).

Ramanathan, V., B. Subasilar, G. J. Zhang, W. Conant, R. D. Cess, J. T. Kiehl, H. Grassl, and L. Shi, "Warm Pool Heat Budget and Shortwave Cloud Forcing: A Missing Physics?" *Science, 267,* 499–503 (1995).

Ramanathan, V., and A. M. Vogelmann, "Atmospheric Greenhouse Effect, Excess Solar Absorption and the Radiation Budget: From the Arrhenius/Langley Era to the 1990s," *Ambio, 26,* 38–46 (1997).

Ramaswamy, V., and V. Ramanathan, "Solar Absorption by Cirrus Clouds and the Maintenance of the Tropical Upper Troposphere Thermal Structure," *J. Atmos. Sci., 46,* 2293–2310 (1989).

Ramaswamy, V., M. D. Schwarzkopf, and K. P. Shine, "Radiative Forcing of Climate from Halocarbon-Induced Global Stratospheric Ozone Loss," *Nature, 355,* 810–812 (1992).

Ramaswamy, V., and M. M. Bowen, "Effect of Changes in Radiatively Active Species upon the Lower Stratospheric Temperatures," *J. Geophys. Res., 99,* 18909–18921 (1994).

Ramaswamy, V., M. D. Schwarzkopf, and W. J. Randel, "Fingerpirnt of Ozone Depletion in the Spatial and Temporal Pattern of Recent Lower-Stratospheric Cooling," *Nature, 382,* 616–618 (1996).

Rasool, S. I., and S. H. Schneider, "Atmospheric Carbon Dioxide and Aerosols: Effects of Large Increases on Global Climate," *Science, 173,* 138–141 (1971).

Raval, A., and V. Ramanathan, "Observational Determination of the Greenhouse Effect," *Nature, 342,* 758–761 (1989).

Reeburgh, W. S., S. C. Whalen, and M. J. Alperin, "The Role of Methylotrophy in the Global CH_4 Budget," in *Microbial Growth on C-1 Compounds* (J. C. Murrell and D. P. Kelly, Eds.), pp. 1–14, Intercept, Andover, UK, 1993.

Reiss, H., and G. J. M. Koper, "The Kelvin Relation: Stability, Fluctuation, and Factors Involved in Measurement," *J. Phys. Chem., 99,* 7837–7844 (1995).

Rind, D., R. Suozzo, N. K. Balachandran, and M. J. Prather, "Climate Change and the Middle Atmosphere. Part I: The Doubled CO_2 Climate," *J. Atmos. Sci., 47,* 475–494 (1990).

Rind, D., and A. Lacis, "The Role of the Stratosphere in Climate Change," *Surv. Geophys., 14,* 133–165 (1993).

Rivera-Carpio, C. A., C. E. Corrigan, T. Novakov, J. E. Penner, C. F. Rogers, and J. C. Chow, "Derivation of Contributions of Sulfate and Carbonaceous Aerosols to Cloud Condensation Nuclei from Mass Size Distributions," *J. Geophys. Res., 101,* 19483–19493 (1996).

Roble, R. G., and R. E. Dickinson, "How Will Changes in Carbon Dioxide and Methane Modify the Mean Structure of the Mesosphere and Thermosphere?" *Geophys. Res. Lett., 16,* 1441–1444 (1989).

Robock, A., and J. Mao, "The Volcanic Signal in Surface Temperature Observations," *J. Clim., 8,* 1086–1103 (1995).

Robock, A., "Stratospheric Control of Climate," *Science, 272,* 972–973 (1996).

Roehl, C. M., D. Boglu, C. Bruhl, and G. K. Moortgat, "Infrared Band Intensities and Global Warming Potentials of CF_4, C_2F_6, C_3F_8, C_4F_{10}, C_5F_{12}, and C_6F_{14}," *Geophys. Res. Lett., 22,* 815–818 (1995).

Rommelaere, V., L. Arnaud, and J.-M. Barnola, "Reconstructing Recent Atmospheric Trace Gas Concentrations from Polar Firn and Bubbly Ice Data by Inverse Methods," *J. Geophys. Res., 102,* 30069–30083 (1997).

Rossow, W. B., and Y.-C. Zhang, "Calculation of Surface and Top of Atmosphere Radiative Fluxes from Physical Quantities Based on ISCCP Data Sets. 2. Validation and First Results," *J. Geophys. Res., 100,* 1167–1197 (1995).

Rothman, L. S., R. R. Gamache, R. H. Tipping, C. P. Rinsland, M. A. H. Smith, D. C. Benner, V. M. Devi, J.-M. Flaud, C. Camy-Peyret, A. Perrin, A. Goldman, S. T. Massie, L. R. Brown, and R. A. Toth, "The Hitran Molecular Database: Editions of 1991 and 1992," *J. Quant. Spectrosc. Radiat. Transfer, 48,* 469–507 (1992).

Rudolph, J., "Anomalous Methane," *Nature, 368,* 19–20 (1994).

Russell, L. M., S. N. Pandis, and J. H. Seinfeld, "Aerosol Production and Growth in the Marine Boundary Layer," *J. Geophys. Res., 99,* 20989–21003 (1994).

Salamatin, A. N., V. Y. Lipenkov, N. I. Barkov, J. Jouzel, J. R. Petit, and D. Raynaud, "Ice Core Age Dating and Paleothermometer Calibration Based on Isotope and Temperature Profiles from Deep Boreholes at Vostok Station (East Antarctica)," *J. Geophys. Res., 103,* 8963–8977 (1998).

Salisbury, J. W., L. S. Water, N. Vergo, and D. M. Dana, *Infrared (2.1–25 μm) Spectra of Minerals,* Johns Hopkins Press, Baltimore, MD, 1992.

Sandroni, S., D. Anfossi, and S. Viarengo, "Surface Ozone Levels at the End of the Nineteenth Century in South America," *J. Geophys. Res., 97,* 2535–2539 (1992).

Santer, B. D., K. E. Taylor, T. M. L. Wigley, J. E. Penner, P. D. Jones, and U. Cubasch, "Towards the Detection and Attribution of an Anthropogenic Effect on Climate," *Clim., Dyn., 12,* 77–100 (1995).

Santer, B. D., K. E. Taylor, T. M. L. Wigley, T. C. Johns, P. D. Jones, D. J. Karoly, J. F. B. Mitchell, A. H. Oort, J. E. Penner, V. Ramaswamy, M. D. Schwarzkopf, R. J. Stouffer, and S. Tett, "A Search for Human Influences on the Thermal Structure of the Atmosphere," *Nature, 382* 39–46 (1996).

Saxena, P., L. Hildemann, P. McMurry, and J. Seinfeld, "Organics Alter Hygroscopic Behavior of Atmospheric Particles," *J. Geophys. Res., 100,* 18755–18770 (1995).

Saxena, V. K., "Bursts of Cloud Condensation Nuclei (CCN) by Dissipating Clouds at Palmer Station, Antarctica," *Geophys. Res. Lett., 23,* 69–72 (1996).

Saxena, V. K., S. Yu, and J. Anderson, "Impact of Stratospheric Volcanic Aerosols on Climate: Evidence for Aerosol Shortwave and Longwave Forcing in the Southeastern U.S.," *Atmos. Environ., 31,* 4211–4221 (1997).

Saxena, V. K., and S. Yu, "Searching for a Regional Fingerprint of Aerosol Radiative Forcing in the Southeastern U.S.," *Geophys. Res. Lett., 25,* 2833–2836 (1998).

Schlesinger, M. E., and N. Ramankutty, "Implications for Global Warming of Intercycle Solar Irradiance Variations," *Nature, 360,* 330–333 (1992).

Schneider, E. K., R. S. Lindzen, and B. P. Kirtman, "A Tropical Influence on Global Climate," *J. Atmos. Sci., 54,* 1349–1358 (1997).

Schult, I., J. Feichter, and W. F. Cooke, "Effect of Black Carbon and Sulfate Aerosols on the Global Radiation Budget," *J. Geophys. Res., 102,* 30107–30117 (1997).

Schwartz, S. E., "Are Global Cloud Albedo and Climate Controlled by Marine Phytoplankton?" *Nature, 336,* 441–445 (1988).

Schwartz, S. E., "Does Fossil Fuel Combustion Lead to Global Warming?" *Energy, 18,* 1229–1248 (1993).

Schwartz, S. E., "Climate Forcing by Gases and Aerosols," *Pure Appl. Chem., 66,* 178–187 (1994).

Schwartz, S. E., "The Whitehouse Effect—Shortwave Radiative Forcing of Climate by Anthropogenic Aerosols: An Overview," *J. Aerosol Sci., 27,* 359–382 (1996).

Schwartz, S. E., and A. Slingo, "Enhanced Shortwave Cloud Radiative Forcing Due to Anthropogenic Aerosols," in *NATO ASI Series,* Vol. I35, *Clouds, Chemistry and Climate* (P. J. Crutzen and V. Ramanathan, Eds.), Springer-Verlag, Berlin/New York, 1996.

Schwarzkopf, M. D., and V. Ramaswamy, "Radiative Forcing Due to Ozone in the 1980s: Dependence on Altitude of Ozone Change," *Geophys. Res. Lett., 20,* 205–208 (1993).

Scuderi, L. A., "A 2000-Year Tree Ring Record of Annual Temperatures in the Sierra Nevada Mountains," *Science, 259,* 1433–1436 (1993).

Shindell, D. T., D. Rind, and P. Lonergan, "Increased Polar Stratospheric Ozone Losses and Delayed Eventual Recovery Owing to Increasing Greenhouse-Gas Concentrations," *Nature, 392,* 589–592 (1998).

Shine, K. P., "On the Cause of the Relative Greenhouse Strength of Gases Such As the Halocarbons," *J. Atmos. Sci., 48,* 1513–1518 (1991).

Shine, K. P., R. S. Freckleton, and P. M. de F. Forster, "Comment on 'Climate Forcing by Stratospheric Ozone Depletion Calculated from Observed Temperature Trends,' by Zhong et al.," *Geophys. Res. Lett., 25,* 663–664 (1998).

Shulman, M., M. Jacobson, R. Charlson, R. Synovec, and T. Young, "Dissolution Behavior and Surface Tension Effects of Organic Compounds in Nucleating Cloud Droplets," *Geophys. Res. Lett., 23,* 277–280 (1996).

Shulman, M. L., R. J. Charlson, and E. J. Davis, "The Effects of Atmospheric Organics on Aqueous Droplet Evaporation," *J. Aerosol Sci., 28,* 737–752 (1997).

Sievering, H., J. Boatman, E. Gorman, Y. Kim, L. Anderson, G. Ennis, M. Luria, and S. Pandis, "Removal of Sulphur from the Marine Boundary Layer by Ozone Oxidation in Sea-Salt Aerosols," *Nature, 360,* 571–573 (1992).

Sinha, A., and J. E. Harries, "Possible Change in Climate Parameters with Zero Net Radiative Forcing," *Geophys. Res. Lett., 24,* 2355–2358 (1997).

Slinn, W. G. N., "Structure of Continental Clouds before the Industrial Era: A Mystery To Be Solved," *Atmos. Environ., 26A,* 2471–2473 (1992).

Sloan, L. C., and D. Pollard, "Polar Stratospheric Clouds: A High Latitude Warming Mechanism in an Ancient Greenhouse World," *Geophys. Res. Lett., 25,* 3517–3520 (1998).

Sloane, C. S., "Effect of Composition on Aerosol Light Scattering Efficiencies," *Atmos. Environ., 20,* 1025–1037 (1986).

Smith, H. J., M. Wahlen, D. Mastroianni, K. Taylor, and P. Mayewski, "The CO_2 Concentration of Air Trapped in Greenland Ice Sheet Project 2 Ice Formed during Periods of Rapid Climate Change," *J. Geophys. Res., 102,* 26577–26582 (1997).

Sokolik, I. N., A. V. Andronova, and T. C. Johnson, "Complex Refractive Index of Atmospheric Dust Aerosols," *Atmos. Environ., 27A,* 2495–2502 (1993).

Sokolik, I. N., and O. B. Toon, "Direct Radiative Forcing by Anthropogenic Airborne Mineral Aerosols," *Nature, 381,* 681–683 (1996).

Sokolik, I. N., O. B. Toon, and R. W. Bergstrom, "Modeling the Radiative Characteristics of Airborne Mineral Aerosols at Infrared Wavelengths," *J. Geophys. Res., 103,* 8813–8826 (1998).

Solanki, S. K., and M. Fligge, "Solar Irradiance Since 1874 Revisited," *Geophys. Res. Lett., 25,* 341–344 (1998).

Solomon, S., and J. S. Daniel, "Impact of the Montreal Protocol and Its Amendments on the Rate of Change of Global Radiative Forcing," *Clim. Change, 32,* 7–17 (1996).

Solomon, S., R. W. Portmann, R. W. Sanders, and J. S. Daniel, "Absorption of Solar Radiation by Water Vapor, Oxygen, and Related Collision Pairs in the Earth's Atmosphere," *J. Geophys. Res., 103,* 3847–3858 (1998).

Sommerfeld, R. A., A. R. Mosier, and R. C. Musselman, "CO_2, CH_4, and N_2O Flux through a Wyoming Snowpack and Implications for Global Budgets," *Nature, 361,* 140–142 (1993).

Steele, L. P., E. J. Dlugokencky, P. M. Lang, P. P. Trans, R. C. Martin, and K. A. Masarie, "Slowing Down of the Global Accumulation of Atmospheric Methane during the 1980s," *Nature, 358,* 313–316 (1992).

Stenchikov, G. L., and A. Robock, "Diurnal Asymmetry of Climatic Response to Increased CO_2 and Aerosols: Forcings and Feedbacks," *J. Geophys. Res., 100,* 26211–26227 (1995).

Stephens, E. R., and F. R. Burleson, "Distribution of Light Hydrocarbons in Ambient Air," *J. Air Pollut. Control Assoc., 19,* 929–936 (1969).

Stephens, E. R., "Tropospheric Methane: Concentrations between 1963 and 1970," *J. Geophys. Res., 90,* 13076–13080 (1985).

Stephens, G., and T. Slingo, "An Air-Conditioned Greenhouse," *Nature, 358,* 369–370 (1992).

Stephens, G. L., "Radiation Profiles in Extended Water Clouds. I: Theory," *J. Atmos. Sci., 35,* 2111–2122 (1978).

Stephens, G. L., "Optical Properties of Eight Water Cloud Types," *CSIRO Aust. Div. Atmos. Phys. Tech. Pap., No. 36,* 1–35 (1979).

Stephens, G. L., and S.-C. Tsay, "On the Cloud Absorption Anomaly," *Q. J. R. Meteorol. Soc., 116,* 671–704 (1990).

Stephens, G. L., "How Much Solar Radiation Do Clouds Absorb?" *Science, 271,* 1131–1133 (1996).

Stevenson, D. S., C. E. Johnson, W. J. Collins, R. G. Derwent, K. P. Shine, and J. M. Edwards, "Evolution of Tropospheric Ozone Radiative Forcing," *Geophys. Res. Lett., 25,* 3819–3822 (1998).

Stockwell, W. R., G. Kramm, H.-E. Scheel, V. A. Mohnen, and W. Weiler, "1. Ozone Formation, Destruction, and Exposure in Europe and the United States," *Ecol. Stud., 127,* 1–39 (1997).

Ström, J., and S. Ohlsson, "*In Situ* Measurements of Enhanced Crystal Number Densities in Cirrus Clouds Caused by Aircraft Exhaust," *J. Geophys. Res., 103,* 11355–11361 (1998).

Tans, P. P., and J. W. C. White, "In Balance, with a Little Help from the Plants," *Science, 281,* 183–184 (1998).

Tarrasón, L., S. Turner, and I. Fløisand, "Estimation of Seasonal Dimethyl Sulphide Fluxes over the North Atlantic Ocean and Their Contribution to European Pollution Levels," *J. Geophys. Res., 100,* 11623–11639 (1995).

Taylor, J. P., and A. McHaffie, "Measurements of Cloud Susceptibility," *J. Atmos. Sci., 51,* 1298–1306 (1994).

Taylor, K. C., R. B. Alley, G. W. Lamorey, and P. Mayewski, "Electrical Measurements on the Greenland Ice Sheet Project 2 Core," *J. Geophys. Res., 102,* 26511–26517 (1997).

Taylor, K. E., and J. E. Penner, "Response of the Climate System to Atmospheric Aerosols and Greenhouse Gases," *Nature, 369,* 734–737 (1994).

Tegen, I., and I. Fung, "Modeling of Mineral Dust in the Atmosphere: Sources, Transport, and Optical Thickness," *J. Geophys. Res., 99,* 22897–22914 (1994).

Tegen, I., and I. Fung, "Contribution to the Mineral Aerosol Load from Land Surface Modification," *J. Geophys. Res., 100,* 18707–18726 (1995).

Tegen, I., and A. A. Lacis, "Modeling of Particle Size Distribution and Its Influence on the Radiative Properties of Mineral Dust Aerosol," *J. Geophys. Res., 101,* 19237–19244 (1996).

Tegen, I., A. A. Lacis, and I. Fung, "The Influence on Climate Forcing of Mineral Aerosols from Disturbed Soils," *Nature, 380,* 419–422 (1996).

Tegen, I., P. Hollrig, M. Chin, I. Fung, D. Jacob, and J. Penner, "Contribution of Different Aerosol Species to the Global Aerosol Extinction Optical Thickness: Estimates from Model Results," *J. Geophys. Res., 102,* 23895–23915 (1997).

Trenberth, K. E., and A. Solomon, "The Global Heat Balance: Heat Transports in the Atmosphere and Ocean," *Clim. Dyn., 10,* 107–134 (1994).

Twomey, S., and J. Warner, "Comparison of Measurements of Cloud Droplets and Cloud Nuclei," *J. Atmos. Sci., 24,* 702–703 (1967).

Twomey, S., "Pollution and the Planetary Albedo," *Atmos. Environ., 8,* 1251–1256 (1974).

Twomey, S., *Atmospheric Aerosols*, Elsevier, Amsterdam/New York, 1977a.

Twomey, S., "The Influence of Pollution on the Shortwave Albedo of Clouds," *J. Atmos. Sci., 34,* 1149–1152 (1977b).

Twomey, S. A., M. Piepgrass, and T. L. Wolfe, "An Assessment of the Impact of Pollution on Global Cloud Albedo," *Tellus, 36B,* 356–366 (1984).

Twomey, S., "Aerosols, Clouds, and Radiation," *Atmos. Environ., 25A*, 2435–2442 (1991).

Tyler, S. C., G. W. Brailsford, K. Yagi, K. Minami, and R. J. Cicerone, "Seasonal Variations in Methane $\delta^{13}CH_4$ Values for Rice Paddies in Japan and Their Implications," *Global Biogeochem. Cycles, 8,* 1–12 (1994).

Tyler, S. C., H. O. Ajie, M. L. Gupta, R. J. Cicerone, D. R. Blake, and E. J. Dlugokencky, "Stable Carbon Isotopic Composition of Atmospheric Methane: A Comparison of Surface Level and Upper Tropospheric Air," *J. Geophys. Res., 104,* 13895–13910 (1999).

Tyndall, J., "On the Absorption and Radiation of Heat by Gases and Vapours, and on the Physical Connexion of Radiation, Absorption, and Conduction—The Bakerian Lecture," *Philos. Mag. S., 22,* 273–285 (1861).

Uppenbrink, J., "Arrhenius and Global Warming," *Science, 272,* 1122 (1996).

Valero, F. P., R. D. Cess, M. Zhang, S. K. Pope, A. Bucholtz, B. Bush, and J. Vitko, Jr., "Absorption of Solar Radiation by the Cloudy Atmosphere: Interpretations of Collocated Aircraft Measurements," *J. Geophys. Res., 102,* 29917–29927 (1997a).

Valero, F. P. J., W. D. Collins, P. Pilewskie, A. Bucholtz, and P. J. Flatau, "Direct Radiometric Observations of the Water Vapor Greenhouse Effect over the Equatorial Pacific Ocean," *Science, 275,* 1773–1776 (1997b).

Van Dingenen, R., F. Raes, and N. R. Jensen, "Evidence for Anthropogenic Impact on Number Concentration and Sulfate Content of Cloud-Processed Aerosol Particles over the North Atlantic," *J. Geophys. Res., 100,* 21057–21067 (1995).

Van Dorland, R, F. J. Dentener, and J. Lelieveld, "Radiative Forcing Due to Tropospheric Ozone and Sulfate Aerosols," *J. Geophys. Res., 102,* 28079–28100 (1997).

Veldkamp, E., M. Keller, and M. Nunez, "Effects of Pasture Management on N_2O and NO Emissions from Soils in the Humid Tropics of Costa Rica," *Global Biogeochem. Cycles, 12,* 71–79 (1998).

Vogt, R., P. J. Crutzen, and R. Sander, "A Mechanism for Halogen Release from Sea-Salt Aerosol in the Remote Marine Boundary Layer," *Nature, 383,* 327–330 (1996).

Volz, A., and D. Kley, "Evaluation of the Montsouris Series of Ozone Measurements Made in the Nineteenth Century," *Nature, 332,* 240–242 (1988).

Volz, A., H. Geiss, S. McKeen, and D. Kley, "Correlation of Ozone and Solar Radiation at Montsouris and Hohenpoeissenberg: Indications for Photochemical Influence," in *Ozone in the Atmosphere* (R. D. Bojkov and P. Fabian, Eds.), A. Deepak Publishing, Hampton, VA, 1989.

Volz, F. E., "Infrared Optical Constants of Ammonium Sulfate, Sahara Dust, Volcanic Pumice, and Flyash," *Appl. Opt., 12,* 564–568 (1973).

Wahlen, M., and T. Yoshinari, "Oxygen Isotope Ratios in N_2O from Different Environments," *Nature, 313,* 780–782 (1985).

Wahlen, M., N. Tanaka, R. Henry, B. Deck, J. Zeglen, J. S. Vogel, J. Southon, A. Shemesh, R. Fairbanks, and W. Broecker, "Carbon-14 in Methane Sources and in Atmospheric Methane: The Contribution from Fossil Carbon," *Science, 245,* 286–290 (1989).

Walden, V. P., S. G. Warren, and F. J. Murcray, "Measurements of the Downward Longwave Radiation Spectrum over the Antarctic Plateau and Comparisons with a Line-by-Line Radiative Transfer Model for Clear Skies," *J. Geophys. Res., 103,* 3825–3846 (1998).

Waliser, D. E., N. E. Graham, and C. J. Gautier, "Comparison of the Highly Reflective Cloud and Outgoing Longwave Radiation Datasets for Use in Estimating Tropical Deep Convection," *J. Clim., 6,* 331–353 (1993).

Wallace, J. M., and P. V. Hobbs, *Atmospheric Science—An Introductory Survey*, Academic Press, New York, 1977.

Wallace, J. M., "Effect of Deep Convection on the Regulation of Tropical Sea Surface Temperature," *Nature, 357,* 230–231 (1992).

Wang, W. C., Y. L. Yung, A. A. Lacis, T. Mo, and J. E. Hansen, "Greenhouse Effects Due to Man-Made Perturbations of Trace Gases," *Science, 194,* 685–690 (1976).

Wang, W. C., J. P. Pinto, and Y. L. Yung, "Climatic Effects Due to Halogenated Compounds in the Earth's Atmosphere," *J. Atmos. Sci., 37,* 333–338 (1980).

Wang, W.-C., Y.-C. Zhuang, and R. D. Bojkov, "Climate Implications of Observed Changes in Ozone Vertical Distributions at Middle and High Latitudes of the Northern Hemisphere," *Geophys. Res. Lett., 20,* 1567–1570 (1993).

Wang, W.-C., M. P. Dudek, and X.-Z. Liang, "The Greenhouse Effect of Trace Gases," in *Future Climates of the World: A Modelling Perspective* (A. Henderson-Sellers, Ed.), Vol. 16, Elsevier, Amsterdam/New York, 1995.

Wang, Y., and D. J. Jacob, "Anthropogenic Forcing on Tropospheric Ozone and OH Since Preindustrial Times," *J. Geophys. Res., 103,* 31123–31135 (1998).

Warner, J., and S. Twomey, "The Production of Cloud Nuclei by Cane Fires and the Effect on Cloud Droplet Concentration," *J. Atmos. Sci., 24,* 704–706 (1967).

Warner, J., and W. G. Warne, "The Effect of Surface Films in Retarding the Growth by Condensation of Cloud Nuclei and Their Use in Fog Suppression," *J. Appl. Meteorol., 9,* 639–650 (1970).

Weart, S. R., "The Discovery of the Risk of Global Warming," *Phys. Today, 50,* 34–40 (1997).

Weber, R. O., P. Talkner, and G. Stefanicki, "Asymmetric Diurnal Temperature Change in the Alpine Region," *Geophys. Res. Lett., 21,* 673–676 (1994).

Weingartner, E., H. Burtscher, and U. Baltensperger, "Hygroscopic Properties of Carbon and Diesel Soot Particles," *Atmos. Environ., 31,* 2311–2327 (1997).

Wentz, F. J., and M. Schabel, "Effects of Orbital Decay on Satellite-Derived Lower-Tropospheric Temperature Trends," *Nature, 394,* 661–664 (1998).

West, J. J., C. Pilinis, A. Nenes, and S. N. Pandis, "Marginal Direct Climate Forcing by Atmospheric Aerosols," *Atmos. Environ., 32,* 2531–2542 (1998).

White, D. E., J. W. C. White, E. J. Steig, and L. K. Barlow, "Reconstructing Annual and Seasonal Climatic Responses from Volcanic Events Since A.D. 1270 As Recorded in the Deuterium Signal from the Greenland Ice Sheet Project 2 Ice Core," *J. Geophys. Res., 102,* 19683–19694 (1997).

White, W. H., "On the Theoretical and Empirical Basis for Apportioning Extinction by Aerosols: A Critical Review," *Atmos. Environ., 20,* 1659–1672 (1986).

Wiesen, P., J. Kleffmann, R. Kurtenbach, and K. H. Becker, "Mechanistic Study of the Heterogeneous Conversions of NO_2 into HONO and N_2O on Acid Surfaces," *Faraday Discuss., 100,* 121–127 (1995).

Wigley, T. M. L., "Global-Mean Temperature and Sea Level Consequences of Greenhouse Gas Concentration Stabilization," *Geophys. Res. Lett., 22,* 45–48 (1995).

Wigley, T. M. L., P. D. Jones, and S. C. B. Raper, "The Observed Global Warming Record: What Does It Tell Us?" *Proc. Natl. Acad. Sci. U.S.A., 94,* 8314–8320 (1997).

Wigley, T. M. L., "The Kyoto Protocol: CO_2, CH_4, and Climate Implications," *Geophys. Res. Lett., 25,* 2285–2288 (1998).

Williams, E. J., G. L. Hutchinson, and F. C. Fehsenfeld, "NO_x and N_2O Emissions from Soil," *Global Biogeochem. Cycles, 6,* 351–388 (1992).

Willson, R. C., "Total Solar Irradiance Trend during Solar Cycles 21 and 22," *Science, 277,* 1963–1965 (1997).

Wilson, A. T., and A. Long, "New Approaches to CO_2 Analysis in Polar Ice Cores," *J. Geophys. Res., 102*, 26601–26606 (1997).

Wilson, R. M., "Evidence for Solar-Cycle Forcing and Secular Variation in the Armagh Observatory Temperature Record (1844–1992)," *J. Geophys. Res., 103*, 11159–11171 (1998).

Wingen, L. M., and B. J. Finlayson-Pitts, "An Upper Limit on the Production of N_2O from the Reaction of $O(^1D)$ with CO_2 in the Presence of N_2," *Geophys. Res. Lett., 25*, 517–520 (1998).

Wiscombe, W. J., and R. M. Welch, "Reply: Comments on 'The Effects of Very Large Drops on Cloud Absorption, Part I: Parcel Models,'" *J. Atmos. Sci., 43*, 401–407 (1986).

World Meteorological Organization (WMO), "Scientific Assessment of Global Depletion," Report No. 44, published February, 1999.

Wuebbles, D. J., K. E. Grant, P. S. Connell, and J. E. Penner, "The Role of Atmospheric Chemistry in Climate Change," *JAPCA, 39*, 22–28 (1989).

Wurzler, S., A. I. Flossmann, H. R Pruppacher, and S. E. Schwartz, "The Scavenging of Nitrate by Clouds and Precipitation. I. A Theoretical Study of the Uptake and Redistribution of $NaNO_3$ Particles and HNO_3 Gas by Growing Cloud Drops Using an Entraining Air Parcel Model," *J. Atmos. Chem., 20*, 259–280 (1995).

Yoshinari, T., and M. Wahlen, "Oxygen Isotope Ratios in N_2O from Nitrification at a Wastewater Treatment Facility," *Nature, 317*, 349–350 (1985).

Yung, Y. L., Y. Jiang, H. Liao, and M. F. Gerstell, "Enhanced UV Penetration Due to Ozone Cross-Section Changes Induced by CO_2 Doubling," *Geophys. Res. Lett., 24*, 3229–3231 (1997).

Yung, Y. L., and C. E. Miller, "Isotopic Fractionation of Stratospheric Nitrous Oxide," *Science, 278*, 1778–1780 (1997).

Yurganov, L. N., E. I. Grechko, and A. V. Dzhola, "Variations of Carbon Monoxide Density in the Total Atmospheric Column over Russia between 1970 and 1995: Upward Trend and Disturbances, Attributed to the Influence of Volcanic Aerosols and Forest Fires," *Geophys. Res. Lett., 24*, 1231–1234 (1997).

Zender, C. S., B. Bush, S. K. Pope, A. Bucholtz, W. D. Collins, J. T. Kiehl, F. P. J. Valero, and J. Vitko, Jr., "Atmospheric Absorption during the Atmospheric Radiation Measurement (ARM) Enhanced Shortwave Experiment (ARESE)," *J. Geophys. Res., 102*, 29901–29915 (1997).

Zhang, M. H., R. D. Cess, and X. Jing, "Concerning the Interpretation of Enhanced Cloud Shortwave Absorption Using Monthly-Mean Earth Radiation Budget Experiment/Global Energy Balance Archive Measurements," *J. Geophys. Res., 102*, 25899–25905 (1997).

Zhong, W., R. Toumi, and J. D. Haigh, "Climate Forcing by Stratospheric Ozone Depletion Calculated from Observed Temperature Trends," *Geophys. Res. Lett., 23*, 3183–3186 (1996).

Zielinski, G. A., P. A. Mayewski, L. D. Meeker, K. Grönvold, M. S. Germani, S. Whitlow, M. S. Twickler, and K. Taylor, "Volcanic Aerosol Records and Tephrochronology of the Summit, Greenland, Ice Cores," *J. Geophys. Res., 102*, 26625–26640 (1997).

Zipf, E. C., and S. S. Prasad, "Experimental Evidence That Excited Ozone Is a Source of Nitrous Oxide," *Geophys. Res. Lett., 25*, 4333–4336 (1998).

CHAPTER

15

Indoor Air Pollution
Sources, Levels, Chemistry, and Fates

At the beginning of this book, we presented some discussion of health-based air quality standards. In the final chapter, which follows this one, the scientific bases of control measures for various pollutants are discussed. In between, the complex chemistry that occurs in both polluted and remote atmospheres, and that converts the primary pollutants into a host of secondary species, has been detailed. To provide further perspective on airborne gases and particles and human exposure levels, we briefly treat indoor air pollution in this chapter. As we shall see, for many species it is simply a question of emissions leading to elevated levels indoors. However, there is some chemistry that occurs in indoor atmospheres as well, and it is of interest to compare this to that occurring outdoors.

From the point of view of health impacts, it is the combination (not necessarily linear) of concentrations and duration of exposure to a pollutant or combination of pollutants that is important. In this regard, it is noteworthy that most individuals spend the majority of their time indoors, even in relatively moderate climates. For example, in California, people spend 87% of their time indoors on average, 7% in enclosed transit systems, and only 6% outdoors (Jenkins *et al.*, 1992). Similarly, Quackenboss *et al.* (1986) report that only 15% of the average day in Portage, Wisconsin, was spent outdoors in summer and less than 5% in winter. As a result, elevated concentrations of air pollutants indoors can have a significant impact on human health and can lead to enhanced chemical sensitivities (Hileman, 1991) as well as other health impacts such as cancer. Establishing the contribution of indoor air pollution to carcinogenicity and assessing the relative risks (e.g., Tancrède *et al.*, 1987) are complex and difficult. An interesting approach suggested recently is pet epidemiology, discussed in detail by Bukowski and Wartenberg (1997).

The term "indoors" is used in the literature to refer to a variety of environments, including homes, workplaces, and buildings used as offices or for recreational activities. In addition, a number of studies have been carried out to measure various compounds inside vehicles during commutes. As we shall see, and consistent with expectations, levels measured indoors are characteristic both of the particular sources present and, to a significant extent, of the outdoor concentrations of the species. We shall not, in general, distinguish in this chapter between the various types of indoor environments but rather focus on the sources of various compounds and their indoor chemistry.

Table 15.1 summarizes the major species of concern for indoor air pollution and some of their sources (Su, 1996). We focus in this chapter primarily on those species common to indoor and outdoor air environments, including oxides of nitrogen, volatile organic compounds (VOC), CO, ozone, the OH radical, SO_2, and particles. In addition, a brief discussion of radon is included since this has been one of the major foci of concern in the past with respect to indoor air pollution.

A. RADON

Radon (^{222}Rn) is formed by the radioactive decay of uranium, ^{238}U (Fig. 15.1a). As a result, the highest concentrations tend to be associated with soils derived from rocks with a high uranium content (Nazaroff and Nero, 1988; Boyle, 1988; Nero, 1989; Mose and Mushrush, 1997). Because radon is a gas that diffuses out of the soil, it can enter homes through cracks in the foundation, around loose-fitting pipes and wall joints, and through floor drains (e.g., Nero, 1989). The concentrations found in a home depend on the type of soil (including the moisture content) on which it sits and the extent of Rn penetration into the house. They also depend on the house ventilation rate and the particular location in the house in which the measurement is

TABLE 15.1 Some Indoor Air Pollutants and Their Sources[a]

Species	Sources
Radon (^{222}Rn)	Soil and some masonry building materials
Oxides of nitrogen	Combustion
Volatile organic compounds (including HCHO)	Building materials, carpets, solvents, paints, household and personal care products, air fresheners, pesticides, mothballs, humans, treated water
CO	Combustion
O_3	Outside air, photocopying machines, electrostatic air cleaners
SO_2	Combustion
Particles	Combustion
Asbestos	Building materials, handheld hair dryers
Microorganisms	Air conditioners, cool-mist humidifiers

[a] Adapted from Su (1996) and Nero (1988).

made, with the basement typically having the highest concentrations (e.g., Liu et al., 1991). Interestingly, because homes are often warmer than their surroundings, a "chimney effect" occurs that draws gases, including radon, into the house from the surroundings (e.g., Osborne, 1987; Nero, 1989; Turk et al., 1990; Hintenlang and Al-Ahmady, 1992).

Other sources include building materials such as concrete that are made from the earth's crustal materials and hence can contain significant amounts of uranium and radium (Nazaroff and Nero, 1988). Radon dissolves in water, and hence degassing from household water can also be a source. For example, Osborne (1987) reported that the radon concentration in a bathroom increased by more than two orders of magnitude during a 15-min period that a shower was running.

The health concerns associated with ^{222}Rn are primarily associated with its radon daughters. As a noble gas, radon is unreactive in air and is both readily inhaled and exhaled. However, a significant portion of its daughters are positively charged ions that are expected to attract water vapor and become hydrated; the formation of clusters with other ions is also likely (e.g., Castleman, 1991). Uptake on existing aerosol particles also occurs readily, and such particles can then be deposited in the respiratory tract, providing a source of radioactive emissions directly to the lung. Effects such as lung cancer may then ensue.

The health effects associated with radon, as well as sources and mitigation measures, are discussed in detail in several National Research Council reports (1988, 1991), in the book edited by Nazaroff and Nero (1988), and in the International Commission on Radiological Protection Report (1994). Initial risk assessments were based on data from underground miners who were exposed to relatively high levels of radon and its progeny. However, there has been considerable controversy over the extrapolation to lower levels in homes [e.g., see summaries by Nazaroff and Teichman (1990) and Peto and Darby (1994)].

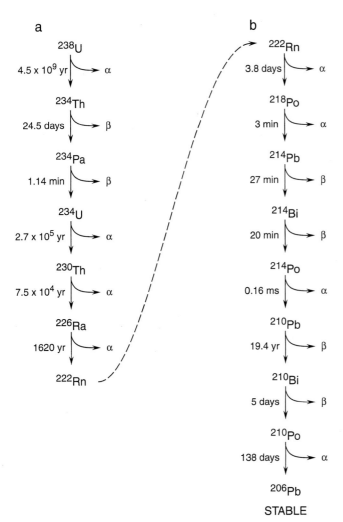

FIGURE 15.1 The major radioactive decay paths in (a) the uranium series leading to the formation of ^{222}Rn and (b) of ^{222}Rn to form ^{206}Pb. Half-lives are shown to the left of each process.

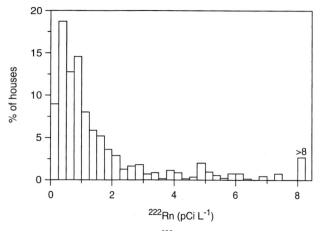

FIGURE 15.2 Distribution of ^{222}Rn measured in homes in the United States (adapted from Nero et al., 1986).

A number of studies have been carried out to determine the concentrations of radon in homes (e.g., see Nero et al., 1986; Alter and Oswald, 1987; Turk et al., 1990; Liu et al., 1991; and Mose and Mushrush, 1997). Units used to express the concentration of ^{222}Rn are picocuries per liter (pCi L^{-1}), with 1 pCi being the amount of substance that gives 2.2 radioactive decays per minute, or becquerels per cubic meter (Bq m^{-3}), where 1 pCi L^{-1} = 37 Bq m^{-3} (e.g., see Nazaroff and Teichman, 1990).

Figure 15.2 shows the distribution of levels measured in homes in the United States (Nero et al., 1986). As might be expected given the variables that affect ^{222}Rn concentrations in homes, the concentrations vary widely, from <0.1 to >8 pCi L^{-1}. Indeed, Alter and Oswald (1987) report a few single measurements of up to 4000 pCi L^{-1}. For comparison, the U.S. EPA recommends levels below 4 pCi L^{-1} (150 Bq m^{-3}), which is exceeded by about 7% of U.S. homes (Nero et al., 1986), and the International Commission on Radiological Protection (1993) recommends 5.4–16 pCi L^{-1} (200–600 Bq m^{-3}). Typical outdoor concentrations in continental areas are 0.1–0.4 pCi L^{-1} (Nazaroff and Nero, 1988).

B. OXIDES OF NITROGEN

1. Levels of NO$_x$

A large number of studies of NO and NO$_2$ have been carried out in many different indoor air environments. Because of air exchange, indoor levels are generally higher when outdoor levels increase (e.g., Hoek et al., 1989; Rowe et al., 1991; Hisham and Grosjean, 1991a; Spengler et al., 1994; Weschler et al., 1994; Baek et al., 1997). However, enhanced indoor levels are found when combustion sources are present. These include gas stoves, kerosene heaters, water heaters, and cigarette smoke (e.g., Wade et al., 1975; Marbury et al., 1988; Ryan et al., 1988; Petreas et al., 1988; Hoek et al., 1989; Pitts et al., 1989; Spengler et al., 1994; Levy et al., 1998). While combustion generates primarily NO, the focus indoors has been on NO$_2$ because of its health impacts (see Chapter 2.).

For example, Spengler et al. (1994) report that personal exposures to NO$_2$ were 10 ppb higher in homes in the Los Angeles area having gas ranges with pilot lights compared to those having electric ranges and 5 ppb higher if the gas range did not have a pilot light. Levy et al. (1998) report NO$_2$ concentrations indoors and outdoors, as well as personal exposures, in 18 different cities in 15 countries in the Northern Hemisphere over a two-day period in February and March when the use of combustion was expected to be higher. The ratio of indoor to outdoor concentrations ranged from 0.3 ± 0.2 in Berlin, Germany, to 2.8 ± 2.8 in Tokushima, Japan. Average NO$_2$ concentrations over the two-day period indoors ranged from 5.5 ppb in Kuopio, Finland, to 63 ppb in Mexico City. Outdoors, the range of concentrations was from 12 ppb in Geneva, Switzerland, to 52 ppb in Seoul, Korea.

Again, the use of gas stoves was highly correlated with indoor NO$_2$, with an indoor/outdoor concentration ratio of 1.19 for homes with a gas range compared to 0.69 for those without a gas stove. The ratio was even higher for homes with a kerosene space heater, 2.3 compared to 0.85 without such a heater (Levy et al., 1998). Both the indoor and outdoor concentrations of NO$_2$ were higher in cities where at least 75% of the homes had gas stoves; for example, the mean outdoor NO$_2$ concentration in such gas-intensive cities was 38 ± 20 ppb, compared to 14 ± 6 ppb in cities where fewer than 25% of the households had gas ranges.

High concentrations of NO$_2$ have also been measured in indoor skating rinks where the use of ice resurfacing machines powered by propane, gasoline, or diesel fuel results in significant emissions (e.g., Brauer and Spengler, 1994; Brauer et al., 1997; Pennanen et al., 1997). Mean concentrations of NO$_2$ of ~200 ppb have been reported, with some rinks having concentrations up to 3 ppm! The indoor-to-outdoor ratios of the arithmetic mean concentrations varied from about 1 to 41, with an overall mean of 20. Figure 15.3 shows the effect of the particular fuel used on the indoor NO$_2$ concentrations (Brauer et al., 1997). Propane gave the highest values, followed by gasoline and diesel. The NO$_2$ concentrations when the resurfacer was electric were similar to those outdoors.

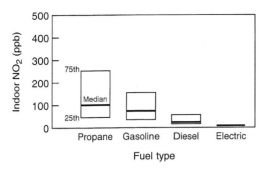

FIGURE 15.3 Indoor concentrations of NO_2 measured in ice skating rinks using different types of power for the ice resurfacers. The median and the 25th and 75th percentiles are shown in each case (adapted from Brauer *et al.*, 1997).

In the absence of such sources of NO_x, indoor and outdoor concentrations are quite similar (e.g., Weschler *et al.*, 1994), since removal of NO and NO_2 indoors, e.g., on surfaces, is relatively slow. However, as discussed shortly, although the surface reaction of NO_2 is relatively slow, it is still of interest since it generates nitrous acid (HONO). Different surfaces found inside homes have been found to have different removal rates for NO_2. Figure 15.4, for example, shows measured rates of removal of NO_2 by a number of common household materials (Spicer *et al.*, 1989). Large variations in removal rate (and hence the formation of products such as NO and HONO; see later) are evident, varying from negligible for plastic storm windows to quite large for wallboard.

In short, there is a variety of evidence that there are higher levels of NO_2 indoors when combustion sources are present and that the concentrations generated indoors can be quite substantial in some circumstances. One word of caution is in order, however, particularly with regard to earlier measurements of NO_2. As discussed in the following section, significant concentrations of HONO are generated both by a heterogeneous reaction of NO_2 on surfaces and by direct emissions from combustion sources. In some measurement methods used for NO_2, HONO is also detected and hence reported as NO_2. This is particularly true for the O_3 chemiluminescence method and for electrochemical sensors (e.g., Spicer *et al.*, 1994), so that NO_2 reported using these techniques should be regarded as upper limits to its concentrations. This problem can be circumvented through the use of denuders to remove HONO prior to sampling into the instrument.

2. HONO and HNO_3

As discussed in Chapter 7.C.1, HONO is formed by the reaction of NO_2 with water on surfaces. The reac-

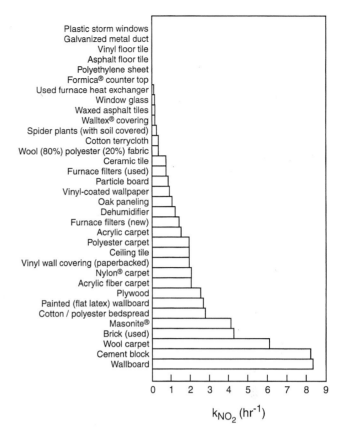

FIGURE 15.4 Rate constants for NO_2 removal on 3.3-m^2 samples of different materials commonly found in homes (adapted from Spicer *et al.*, 1989).

tion is usually represented as

$$2NO_2 + H_2O \xrightarrow{\text{surface}} HONO + HNO_3, \quad (1)$$

although the detailed mechanism is not known; gaseous HNO_3 is not generated in equivalent amounts, which has been attributed to its remaining adsorbed on the surface. This overall reaction occurs on a variety of surfaces in the laboratory and hence might be expected to also occur on surfaces in other environments, such as homes. This, indeed, is the case.

Pitts *et al.* (1985) first used differential optical absorption spectrometry (DOAS) to establish unequivocally that NO_2 injected into a mobile home forms HONO. Interestingly, the dependence of the rate of HONO generation on the NO_2 concentration was similar to that measured in laboratory systems, consistent with production in, or on, a thin film of water adsorbed on surfaces. A number of studies have confirmed that the behavior is similar to that in laboratory systems; i.e., the rate of production of HONO increases with NO_2 and with relative humidity. Indoor levels of HONO as high as 8 ppb as a 24-h average and 40 ppb as a 6-h average have been reported in normal, in-use buildings and homes (Febo and Perrino, 1991; Spengler et al., 1993; Weschler et al., 1994).

The ratio of HONO to NO_2 indoors can be quite large, up to ~ 0.15 (e.g., Febo and Perrino, 1991; Brauer et al., 1990, 1993; Spengler et al., 1993). This can be compared to typical values of a few percent outdoors (see Chapter 11). High levels of HONO (up to ~ 30 ppb) have also been measurd in automobiles in use in polluted urban areas, and again, the ratio of HONO to NO_2 was quite large, ~ 0.4, compared to 0.02–0.03 measured outdoors in the same study (Febo and Perrino, 1995).

Figure 15.5, for example, shows one set of measurements of NO_2, HONO, and NO after injection of NO_2 into a home used for research purposes (Spicer et al., 1993). As NO_2 decays, HONO is formed. Small amounts of NO are also generated, as has been observed in laboratory studies (e.g., Spicer et al., 1989). Delayed release of HONO was observed, suggesting that HONO was adsorbed in part on the surfaces and was subsequently released to the gas phase. For example, when the house was purged with outside air, the HONO levels dropped; however, on closing up the house again, the levels increased immediately to a few ppb. Degassing of HONO from the interior surfaces of automobiles has also been reported (Febo and Perrino, 1995). Interestingly, in measurements made in a commercial office building, indoor HONO concentrations were observed not to be enhanced under conditions of high indoor O_3, suggesting that dissolved HONO in a surface film of water is readily oxidized to nitrate by O_3 (Weschler et al., 1994).

The generation of NO in Fig. 15.5 was attributed by Spicer and co-workers to a reaction of gaseous NO_2 with adsorbed HONO:

$$NO_{2(g)} + HONO_{(ad)} \rightarrow H^+ + NO_3^- + NO_{(g)}. \quad (2)$$

The same process was hypothesized to explain some time periods in a commercial office building when indoor NO actually exceeded outdoor NO (Weschler et al., 1994).

As is the case in laboratory systems, equivalent amounts of HNO_3 are not observed as might be expected from the stoichiometry of reaction (1), likely due to HNO_3 remaining on the surface after formation and/or being taken up by surfaces. For example, Spicer et al. (1993) used atmospheric pressure ionization mass spectrometry (see Chapter 11.A.2), which could measure HNO_3 with a sensitivity of 50 ppt, during the NO_2 decay and HONO formation in the research home. HNO_3 was only observed, and in small concentrations (<1 ppb), when an unvented space heater was operating, and it decayed rapidly in the absence of sources. Similarly, Salmon et al. (1990) measured HNO_3 indoors and outdoors at five museums in the Los Angeles area and found that indoor concentrations were less than $\sim 40\%$ of those outdoors, and typically about 10%. The accumulation of nitrate on indoor surfaces in a commercial building has been reported by Weschler and Shields (1996a) and attributed to the formation and uptake of HNO_3 via reactions of NO_3 (see later) and/or oxidation of nitrite (i.e., adsorbed HONO) in an aqueous surface film.

Subsequently, it was shown that HONO is also directly emitted by gas stoves (Pitts et al., 1989). For example, Fig. 15.6 shows the concentrations of NO_2 and HONO measured using DOAS when two top burners of a new, residential gas kitchen stove were turned on in a mobile home with both the central ventilation and air conditioning running. NO_2 from the gas stove emissions reached almost 300 ppb, and HONO about

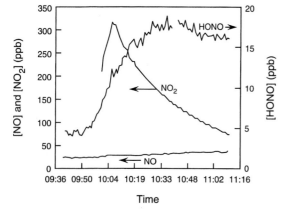

FIGURE 15.5 Concentrations of NO_2, HONO, and NO after injection of NO_2 into a research house (adapted from Spicer et al., 1993).

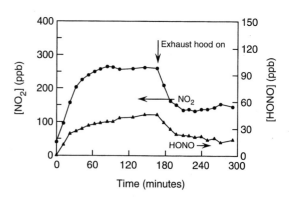

FIGURE 15.6 NO_2 and HONO measured in a mobile home with two burners of a kitchen stove on; the above-range exhaust was turned on at the time shown (adapted from Pitts et al., 1989).

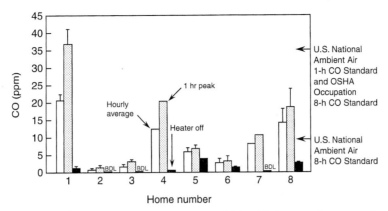

FIGURE 15.7 Measured concentrations of CO in eight homes using unvented kerosene heaters: dark shading, hourly average concentrations with heater off; no shading, hourly average concentrations with heater on; light shading, peak 1-h concentration (adapted from Mumford *et al.*, 1991).

45 ppb. When the exhaust hood (which was externally vented) above the range was turned on, the NO_2 and HONO decreased substantially. Similar production of HONO from kerosene and propane space heaters has been observed (e.g., see Pitts *et al.*, 1989; Brauer *et al.*, 1990; Febo and Perrino, 1991; and Vecera and Dasgupta, 1994). While the mechanism generating such substantial concentrations is not known, it may involve the recombination of OH with NO as the combustion gases cool.

In a house used for investigating indoor air pollution that had natural gas fueled appliances (a convective heater, a radiant heater, and a range with four burners), both the surface reaction of NO_2 and the direct combustion emissions contributed significantly to the measured indoor HONO. When an appliance was operational, the contribution of direct emissions was the more important source (Spicer *et al.*, 1993).

In short, the "dark reaction" of NO_2 with water on surfaces is ubiquitous and occurs not only in laboratory systems but also indoors. The combination of this heterogeneous reaction with combustion sources of HONO can produce significant concentrations of HONO indoors. As a result, there is a concern regarding the health impacts of nitrous acid, not only because it is an inhalable nitrite but also because it is likely the airborne acid present in the highest concentrations indoors.

C. CO AND SO_2

As for NO_x, combustion sources such as gas stoves and kerosene heaters can be significant sources of indoor CO. Figure 15.7, for example, shows measured CO concentrations in eight mobile homes with unvented kerosene heaters either off or on (Mumford *et al.*, 1991). Both the hourly average and peak 1-h concentrations are shown for the situation with the heater on, whereas only the 1-h average for the heater off is shown. Also shown are the United States 1- and 8-h standards and the Occupational Safety and Health Adminstration (OSHA) 8-h standard. In three of the homes, the average CO concentrations exceeded the 8-h standard and in one home, the 1-h ambient air and OSHA standard was exceeded.

The ratio of indoor to outdoor concentrations of CO in homes using gas stoves has been measured to be 1.2–3.8 (Wade *et al.*, 1975), with the highest ratios found close to the source. Similarly, higher CO levels indoors compared to outdoors have been reported for restaurants in Korea, with those using charcoal burners as well as gas giving much higher concentrations (Baek *et al.*, 1997). Figure 15.8, for example, shows the mean indoor-to-outdoor concentrations of CO and some other air pollutants measured in restaurants in Korea using either gas only or a combination of gas and

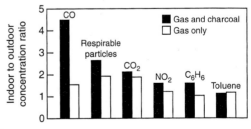

FIGURE 15.8 Median ratios of indoor-to-outdoor concentrations of CO and some other air pollutants measured in restaurants in Korea where either gas alone (□) or gas with charcoal (■) was used for cooking (adapted from Baek *et al.*, 1997).

charcoal. The use of charcoal increases this ratio to more than four; in one such restaurant, an indoor concentration of 90 ppm CO was measured.

In buildings where motor vehicle exhaust can be entrained from outdoors or attached parking garages, elevated indoor CO levels may also result (e.g., Hodgson et al., 1991).

On the other hand, in homes and offices where there was no direct indoor source of CO, the indoor-to-outdoor ratio was about one, and sometimes less. For example, in Riyadh, Saudi Arabia, CO concentrations were measured indoors and outdoors; the indoor-to-outdoor ratio varied from 0 to 2, but was typically below one (Rowe et al., 1989).

There have been a number of measurements of CO in the "indoor environment" of automobiles. Given that cars are major CO sources in urban areas, one might expect higher concentrations of CO during commutes and this is indeed the case. Typical CO concentrations of ~9–56 ppm have been measured inside automobiles during commutes in major urban areas (Flachsbart et al., 1987; Koushki et al., 1992; Ott et al., 1994; Dor et al., 1995; Fernandez-Bremauntz and Ashmore, 1995). This can be compared to peak outdoor levels of ~10 ppm in highly polluted urban areas (see Chapter 11.A.4c). Thus, a significant enhancement of CO inside automobiles during commutes is common. For example, Chan et al. (1991b) report a ratio of the in-vehicle CO concentration to that outdoors of ~4.5 in Raleigh, North Carolina.

As is the case for CO, SO_2 levels indoors and outdoors tend to be similar if there are no combustion sources indoors. For example, Hisham and Grosjean (1991b) report that the ratio of indoor-to-outdoor SO_2 concentrations averaged 0.89 for three museums in southern California, with a range from 0.36 to 1.92. On the other hand, quite high SO_2 concentrations can result when there are indoor combustion sources. For example, in China, where unvented stoves using coal are used extensively for cooking and heating, average SO_2 levels indoors are typically 250 μg m^{-3} (96 ppb) during the summer and 750 μg m^{-3} (287 ppb) during the winter, compared to average annual outdoor concentrations of 72–94 μg m^{-3} (28–36 ppb) (Florig, 1997; Ando et al., 1996).

D. VOLATILE ORGANIC COMPOUNDS

Volatile organic compounds (VOC) are ubiquitous components not only of ambient air but also of indoor air environments, including offices, commercial and retail buildings, and homes (Shah and Singh, 1988). There are three sources/categories for VOC: (1) entrainment of air from outside the building, (2) emissions from building materials, and (3) human activities inside buildings.

As might be expected given the nature of the sources, a very large variety of organic compounds have been identified and measured indoors (e.g., Brown et al., 1994; Crump, 1995; Kostiainen, 1995). These number in the hundreds of different compounds, with the particular species and their concentrations depending on the particular sources present as well as the air exchange rates. Table 15.2 summarizes some of the types of organics that have been measured in indoor air and typical sources (Tichenor and Mason, 1988; Crump, 1995). Because of the VOC sources present indoors, the indoor-to-outdoor concentration ratios are quite large for many compounds.

For example, Table 15.3 shows some typical ratios of indoor-to-outdoor concentrations for specific compounds found in each of the classes shown in Table 15.2, which are frequently present indoors (Brown et al., 1994). These data are based on a review of the literature and include data from a number of different countries. The ratio is for all but one compound substantially greater than one. Also shown in Table 15.3 is a typical range of concentrations expressed as the overall weighted average of the geometric mean, where the weighting was done using the number of available measurements. Some of the compounds associated with the three sources—entrainment from outdoors, emissions from building materials, and anthropogenic activities—are now briefly reviewed.

Entrainment of air from outdoor sources. Entrainment of outdoor air through ventilation systems brings with it the species found in ambient air, which have been discussed throughout this book. Some of them, such as HNO_3 and to a lesser extent O_3, can be removed on surfaces such as those in air conditioning systems, and hence the indoor concentrations tend to be lower than those outdoors. Others such as NO tend to have similar concentrations indoors and outdoors if there are no significant combustion sources indoors (e.g., Weschler et al., 1994). In the case of hydrocarbons, the concentrations of compounds that do not have significant indoor sources tend to be about the same as the outdoor concentrations. For example, Lewis and Zweidinger (1992) measured VOC in 10 homes in winter and showed that the concentrations of ethene, benzene, 2-methylpentane, methylcyclopentane, 2,2,4-trimethylpentane, and 2,3-dimethylbutane indoors were within experimental error of those outdoors.

TABLE 15.2 Some VOCs Measured Indoors and Their Sources[a]

Class of Compounds	Species	Typical Sources
Hydrocarbons	Aliphatic hydrocarbons	Paints, adhesives, gasoline, combustion products, floor waxes
	Aromatic hydrocarbons (toluene, xylenes, ethyl benzene, trimethylbenzenes, styrene, benzene)	Insulation, textiles, disinfectants, plastics, paints, smoking
	Terpenes (limonene, α-pinene)	Scented deodorizers, polishes, fabrics, fabric softeners, cigarettes, food, beverages
	PAHs	Combustion products (smoking, wood burning, kerosene heaters)
Oxygenated organics	Acrylic acid esters, epichlorohydrin	Monomers may escape from polymers
	Alcohols	Aerosols, window cleaners, paints, paint thinning, cosmetics, adhesives
	Ketones	Lacquers, varnishes, polish removers, adhesives
	Ethers	Resins, paints, varnishes, lacquers, dyes, soaps, cosmetics
	Esters	Plastics, resins, plasticizers, lacquer solvents, flavors, perfumes
	Ethylene oxide	Sterilizers (hospitals)
Other organics	Toluene diisocyanate	Polyurethane foam aerosols
	Phthalic acid anhydride	Epoxy resins
	Sodium dodecyl sulfate	Carpet shampoo
Chlorinated organics	Benzyl chloride	Vinyl tiles
	Tetrachloroethylene	Dry-cleaned clothes
	Chloroform	Chlorinated water
	1,1,1-Trichloroethane	Dry-cleaned clothes, aerosol sprays, fabric protectors
	Carbon tetrachloride	Industrial-strength cleaners
	p-Dichlorobenzene	Moth crystals, room deodorants

[a] Adapted from Crump (1995) and Tichenor and Mason (1988).

There are, however, some specific outdoor sources that can lead to higher concentrations of certain VOCs indoors than in the general outdoor air environment. For example, gases generated in landfills or from petroleum contamination can migrate through the soil and groundwater to adjacent buildings and homes to give larger indoor concentrations, particularly in basements and crawl spaces, than otherwise expected (e.g., Moseley and Meyer, 1992; Hodgson et al., 1992; Fischer et al., 1996). In one such case, the total hydrocarbon concentration was measured to be 120 ppm in a crawl space beneath the floor of a school where petroleum contamination was present from adjacent sources, compared to <80 ppb outdoors (Moseley and Meyer, 1992). Although concentrations in various rooms were lower, they were still elevated compared to outdoors, ranging from 0.13 to 3.4 ppm.

The use of pesticides *outside* buildings can also lead to enhanced concentrations of these compounds indoors. For example, Anderson and Hites (1988) measured the concentrations of chlorinated pesticides indoors and found elevated levels inside, e.g., a factor of 7 times higher for γ-chlordane compared to outdoor levels. One home that had the highest indoor concentrations had been treated with chlordane about a decade earlier, presumably by subsurface injection from which the pesticide migrated into the house through cracks in the basement walls. Enhanced levels of chlorpyrifos were observed indoors in homes where soil surrounding the home had been treated on a regular

TABLE 15.3 Ratio of Indoor-to-Outdoor Concentrations for Some VOCs and Typical Ranges of Concentrations[a]

Compound	Indoor / outdoor ratio (range of concentration ($\mu g\, m^{-3}$))	Compound	Indoor / outdoor ratio (range of concentration ($\mu g\, m^{-3}$))
n-Alkane		Alcohols	
n-Pentane	3	2-Propanol	>73
n-Hexane	9	n-Butanol	5 (<1)
n-Heptane	4 (1–5)		
n-Octane	7 (1–5)	Aldehydes	
n-Nonane	14 (1–5)	Acetaldehyde	5
n-Decane	19 (5–10)	Butanal	2 (1–5)
n-Undecane	20 (1–5)	Hexanal	>5
n-Dodecane	20 (1–5)	Nonanal	5 (5–10)
n-Tridecane	>6		
n-Tetradecane	16 (1–5)	Ketones	
n-Pentadecane	>5 (1–5)	Acetone	12 (20–50)
		Methyl ethyl ketone	4 (1–5)
Branched cycloalkanes			
2-Methylpentane	2	Esters	
2-Methylhexane	2	Ethyl acetate	15 (5–10)
3-Methylhexane	3		
Cyclohexane	4 (1–5)	Aromatic hydrocarbons	
		Styrene	10 (1–5)
Halogenated compounds		Benzene	3 (5–10)
Trichlorofluoromethane	10	Toluene	6 (20–50)
1,2-Dichloroethane	12 (<1)	Ethylbenzene	6 (5–10)
Dichloromethane	6 (10–20)	m- and p-xylene	6 (10–20)
Chloroform	5 (1–5)	o-Xylene	6 (5–10)
Carbon tetrachloride	2 (1–5)	n-Propylbenzene	4
1,1,1-Trichloroethane	9 (20–50)	1,3,5-Trimethylbenzene	4 (1–5)
1,1-Dichloroethene	13 (1–5)	1,2,4-Trimethylbenzene	15 (5–10)
Trichloroethylene	6 (1–5)	(1-Methylethenyl)benzene	5
Tetrachloroethylene	5 (5–10)	Naphthalene	4 (<1)
p-Dichlorobenzene	5 (5–10)		
m-Dichlorobenzene	0.4 (<1)	Terpenes	
		Camphene	20 (10–20)
		α-Pinene	23 (1–5)
		Limonene	80 (20–50)
		Total VOC	7

[a] Adapted from Brown et al. (1994) and references therein.

basis. Similarly, aldrin and dieldrin (used as termiticides in the United States from the 1950s to the 1970s) were shown to be present indoors in a home in which the surrounding soil and/or concrete blocks used for the basement walls had been treated during construction; as might be expected given the source, the highest levels were in the basement (Wallace et al., 1996).

Another source of VOC is motor vehicle emissions, which can be drawn into buildings from outdoors or parking garages (e.g., Perry and Gee, 1994; Daisey et al., 1994). For example, motor vehicles were major sources (responsible for ≥75%) of 12 of 39 individual compounds measured in a dozen buildings by Daisey et al. (1994). Of the 12 compounds, 5 were alkanes and 7 were aromatics. Similarly, Baek et al. (1997) report that vehicle emissions are important VOC sources indoors in Korea during the summer in homes and offices, as has been reported in the United States (e.g., Hodgson et al., 1991; Daisey et al., 1994).

Building materials. Emissions associated with building materials are major contributors to indoor levels of VOC. Table 15.4 summarizes some individual VOCs that have been associated with some building materials (Tichenor and Mason, 1988; Crump, 1995). The overall weighted average geometric mean concentrations of some individual organics measured indoors in buildings that were more than 3 months old are shown in Table 15.5 (Brown et al., 1994). New buildings often have higher concentrations of certain compounds compared to older buildings. For example, enhanced levels of

D. VOLATILE ORGANIC COMPOUNDS

TABLE 15.4 Some VOCs from Building Materials[a]

Source	Compounds
Carpet adhesive	Toluene
Floor adhesive (water based)	Nonane, decane, undecane, dimethyloctane, 2-methylnonane, dimethylbenzene
Particleboard	Formaldehyde, acetone, hexanal, propanol, butanone, benzaldehyde, benzene
Moth crystals	p-Dichlorobenzene
Floor wax	Nonane, decane, undecane, dimethyloctane, trimethylcyclohexane, ethylmethylbenzene
Wood stain	Nonane, decane, undecane, methyloctane, dimethylnonane, trimethylbenzene
Latex paint	2-Propanol, butanone, ethylbenzene, propylbenzene, 1,1'-oxybis[butane], butylpropionate, toluene, formic and acetic acids
Water-based acrylic wall paint	1,2-Propanediol, isomers of 2,4,4-trimethyl-1,3-pentanediol monoisobutyrate
Furniture polish	Trimethylpentane, dimethylhexane, trimethylhexane, trimethylheptane, ethylbenzene, limonene
Polyurethane floor finish	Nonane, decane, undecane, butanone, ethylbenzene, dimethylbenzene
Room freshener	Nonane, decane, undecane, ethylheptane, limonene, substituted aromatics (fragrances)
Particleboard	Formaldehyde, acetone, hexanol
Vinyl flooring	Alkyl aromatics, dodecane, 2,2,4-trimethyl-1,3-pentanediol diisobutylate, 2-ethyl-1-hexanol, phenol, cresol, ethyl hexyl acetate, ammonia
Floor varnish	Butyl acetate, N-methylpyrrolidone
Laminated cork floor tile	Phenol
Carpets	4-Phenylcyclohexene, styrene, 4-ethenylcyclohexene, 2-ethyl-1-hexanol, nonanol, heptanol
Silicone caulk	Methyl ethyl ketone, butyl propionate, 2-butoxyethanol, butanol, benzene, toluene
Paint	Dibutyl phthalate
Acrylic sealant	Hexane, dimethyloctanols
Creosote-impregnated timber	Naphthalene, methylnaphthalenes

[a] Adapted from Tichenor and Mason (1988), Crump (1995), Reiss et al. (1995b), and Wolkoff (1998).

n-dodecane, n-decane, and n-undecane, the xylenes, and 2-propanol have been measured in new buildings, and the total VOC concentration is generally larger (by factors of 4–23) compared to established buildings (Brown et al., 1994).

Kostiainen (1995) identified more than 200 individual VOCs indoors in 26 houses. In addition, they compared the VOC concentrations in normal houses to those where complaints of odors or illness had been registered. A number of different VOCs were present at increased concentrations in the houses with complaints compared to the normal houses; these included a variety of aromatic hydrocarbons, methylcyclohexane, n-propylcyclohexane, terpenes, and chlorinated compounds such as 1,1,1-trichloroethane and tetrachlorethene.

Carpets are a major source of VOCs in homes. For example, Sollinger et al. (1993, 1994) have identified 99 different VOCs emitted from a group of 10 carpet samples, and Schaeffer et al. (1996) identifed more than 100 different VOCs emitted from the carpet cushion alone. Emissions come not only from the carpet fibers but also from the backing materials and the adhesives used to bind the carpet to the backing. As a result, the individual compounds emitted by carpets can vary substantially, depending on the carpet construction. For example, Table 15.6 shows some of the compounds emitted by three different types of carpets: (1) with a styrene–butadiene rubber (SBR) latex adhesive to bind the primary and secondary backings; (2) with a polyvinyl chloride (PVC) secondary backing and made in the form of tiles; and (3) with a polyurethane foam secondary backing. With the SBR adhesive, the major VOCs emitted are 4-phenylcyclohexene (responsible for the "new carpet" smell) and styrene. With the PVC backing, the major compounds are vinyl ac-

TABLE 15.5 Overall Weighted Average Geometric Mean Concentrations of Some Individual VOCs Indoors[a]

Compound	Type of building	Number of measurements	Concentration ($\mu g\ m^{-3}$)
Hydrocarbons			
n-Hexane	Dwelling	656	5
	Office	26	12
n-Nonane	Dwelling	592	5
n-Decane	Dwelling	1085	5
Camphene	Mobile home	44	14
Limonene	Dwelling	584	21
Benzene	Dwelling	2171	8
Toluene	Dwelling	792	37
Ethylbenzene	Dwelling	1867	5
o-Xylene	Dwelling	1518	6
m- and p-xylene	Dwelling	1587	18
1,2,4-Trimethylbenzene	Dwelling	619	6
1,2,3-Trimethylbenzene	Office	152	9
m-Methylethylbenzene	Office	168	8
Oxygenated organics			
Acetone	Dwelling	86	32
Methanol	School	11	29
Ethanol	Dwelling	39	120
Acetic acid	School	5	12
Butyric acid	School	5	25
Methyl ethyl ketone	Dwelling	316	4–21
Diethyl ketone	School	12	6
Phenol	School	5	9
Nonanal	Dwelling	15	7
Ethyl acetate	Dwelling	302	8
	School	12	10
Chlorinated organics			
Dichloromethane	Dwelling	101	17
Chloroform	Office	20	10
1,2-Dichloroethene	Dwelling	35	11
1,1,1-Trichloroethane	Dwelling	1580	24
Tetrachloroethene	Dwelling	1919	7
p-Dichlorobenzene	Dwelling	1881	8

[a] Adapted from Brown et al. (1994).

etate (used as a copolymer in the production of PVC), 1,2-propanediol, and 2,2,4-trimethylpentane, whereas the polyurethane backing emitted primarily 2,6-di-*tert*-butyl-4-methylphenol (BHT), a compound often used as an antioxidant. Many of the compounds emitted are known to be used in the manufacturing processes (e.g., ε-caprolactam is used in Nylon-6 production) and/or are common solvents. For example, such species as 1,2-propanediol, 2,2,4-trimethylpentane, 1,1,1-trichloroethane, toluene, 1-butanol, and the dipropylene glycol methyl ethers may be used during the production of various carpet components.

Emissions of VOC from carpets tend to decrease with time and increase with temperature. For example, Fig. 15.9 shows the concentrations in a test chamber of 4-phenylcyclohexene, styrene, and 4-ethenylcyclohexene, another compound associated with SBR adhesives (Weschler et al., 1992b), from one carpet as a function of time (Hodgson et al., 1993). While styrene and 4-ethenylcyclohexene decrease rapidly with time, the decay of 4-phenylcyclohexene is much slower. In a house used for field studies, the concentration of 4-phenylcyclohexene remained above 2 ppb almost two months after installation, well above the odor threshold of 0.5 ppb.

Figure 15.10 shows the temperature dependence of the concentrations of two VOCs emitted by carpets, styrene and benzothiazole, into a test chamber (Sol-

TABLE 15.6 Some Organics Emitted from Carpets[a]

Compound	SBR[b] latex backing adhesive	PVC[b] backing	Polyurethane backing
4-Phenylcyclohexene	D	–	–
Styrene	D	–	–
Vinyl acetate	–	D	–
1,2-Propanediol	–	D	–
2,2,-4-Trimethylpentane	–	D	–
Formaldehyde	–	+	–
2-Ethyl-1-hexanol	–	+	–
2,6-Di-*tert*-butyl-4-methylphenol (BHT)	–	–	D
1-Butanol	–	–	+
Hexamethylcyclotrisiloxane	–	–	+
Dipropylene glycol methyl ethers	–	–	+
2-Methyl-1-propene	+	–	+
Acetaldehyde	–	+	–
Acetone	+	–	+
1- and 2-propanol	–	–	+
Trimethylsilanol	–	–	+
1,1,1-Trichloroethane	–	–	+
Acetic acid	–	+	–
Alkane HCs	–	+	–
2,2,5-Trimethylhexane	–	–	+
4-Ethenylcyclohexane	+	–	–
ε-Caprolactam	–	–	+

[a] Adapted from Hodgson *et al.* (1993); D = dominant emission, + = observed, – = not observed.
[b] SBR = styrene–butadiene rubber latex adhesive; PVC = polyvinyl chloride.

linger *et al.*, 1994). Both increase with temperature, but styrene at a much slower rate. Given the temperature control in many homes, the change in emissions over most temperature ranges encountered in dwellings may not be large. However, in homes without temperature control or in automobiles, for example, the temperature dependence could be significant.

Emission rates from other building materials such as flooring, paints, varnishes, and sealants also tend to increase, not surprisingly, with temperature (e.g.,

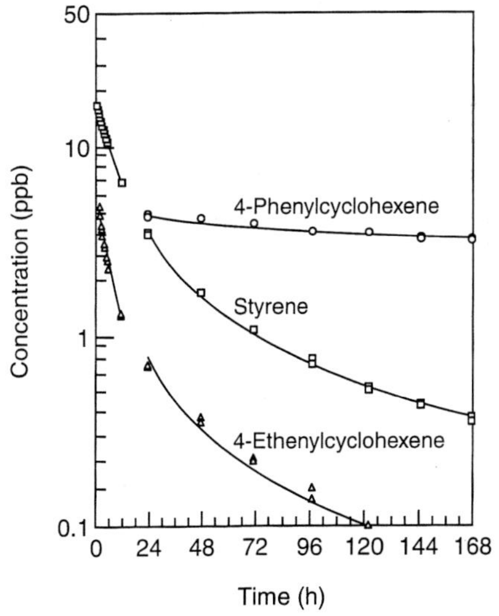

FIGURE 15.9 Concentrations of three VOCs emitted from a carpet sample made using SBR adhesives as a function of time in a test chamber (adapted from Hodgson *et al.*, 1993).

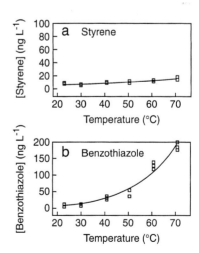

FIGURE 15.10 Temperature dependence of styrene and benzothiazole emitted from carpets into a test chamber (adapted from Sollinger *et al.*, 1994).

TABLE 15.7 Effect of Temperature, Relative Humidity, and Replacement of Air by N_2 on VOC Emissions from Some Common Building Materials[a,b]

Building product	Temperature (°C) 35	Temperature (°C) 60	RH (%)	Air replaced by N_2
Carpet				
2-Ethylhexanol	+	+	+	+
4-Phenylcyclohexene	+	+	NC	NC
PVC flooring				
2-Ethylhexanol	NC	+	NC	+
Phenol	NC	+	NC	+
Sealant				
Hexane	+	+	NC	NC
Dimethyloctanols	+	+	+	+
Varnish				
Butyl acetate	+	+	NC	NC
N-Methylpyrrolidone	+	+	NC	NC
Wall Paint				
1,2-Propanediol	NC	+	+	NC
Texanol[c]	NC	+	NC	NC

[a] Adapted from Wolkoff (1998).
[b] A "+" sign means an increase of 20% or more in concentration. NC means less than 20% change.
[c] Texanol = isomers of 2,4,4-trimethyl-1,3-pentanediol monoisobutyrate.

Wolkoff, 1998; Haghighat and de Bellis, 1998). Table 15.7, for example, shows the effects of temperature and relative humidity on the emissions of particular compounds associated with carpet, PVC flooring, sealants, varnish, and wall paint (Wolkoff, 1998). Interestingly, exposure of these samples to N_2 rather than air also increased the emissions in some cases. However, using increased temperatures to "bake-out" buildings and hence lower the concentrations of indoor VOCs does not appear to be particularly effective. For example, Bayer (1991) reports that the total VOC concentrations from particleboard are about the same after as before a 5-day bake-out at 88°C. Similarly, significant levels of HCHO have been observed in a mobile home even after 20 years of use in a hot ambient air environment (Pitts et al., 1989; see later).

The dependence of VOC emissions from building materials on relative humidity is more complex, with some emissions increasing with relative humidity, but others not. For example, Sollinger et al. (1994) report that the VOC emissions from carpets did not change with relative humidity over the range from 0 to 45% RH. On the other hand, the emissions of formic and acetic acids from latex paints have been reported to increase dramatically with relative humidity; for example, for one paint sample the emission rate for acetic acid almost tripled when the relative humidity was changed from 4–5% to 5–23% (Reiss et al., 1995b).

A number of different aldehydes have been measured indoors (e.g., see Crump and Gardiner, 1989; Lewis and Zweidinger, 1992; Zhang et al., 1994a; Daisey et al., 1994; and Reiss et al., 1995a), some of which are directly emitted and some of which are formed by chemical reactions indoors of VOCs such as styrene (see Section F). Of these, there is an enormous amount of evidence for direct emissions of HCHO from building materials. Interest in formaldehyde emissions and levels in homes and other buildings stems from its well-known health effects, which include possible human carcinogenicity and eye, skin, and respiratory tract irritation (Feinman, 1988). Formaldehyde is emitted from urea–formaldehyde foam insulation as well as from resins used in reconstituted wood products such as particleboard and plywood (Meyer and Reinhardt, 1986); urea–formaldehyde resins comprise about 6–8% of the weight of particleboard and 8–10% of medium-density fiberboard (Meyer and Hermanns, 1986). Other sources include permanent press fabrics (such as draperies and clothing), floor finishing materials, furniture, wallpaper, latex paint, varnishes, some cosmetics such as fingernail hardener and nail polish, and paper products (Kelly, 1996; Howard et al., 1998).

Many measurements of HCHO have been made in indoor air environments. In conventional homes, average concentrations are typically about 10–50 ppb (e.g., Stock, 1987; Zhang et al., 1994a; Reiss et al., 1995a). Sexton et al. (1989) measured concentrations of HCHO in 470 mobile homes in California and found geometric mean concentrations of 60–90 ppb, although maximum values of over 300 ppb were recorded in some cases. In a similar study in Wisconsin, levels up to 2.8 ppm were measured (Hanrahan et al., 1985). Higher levels are typically found in mobile homes because of the reconstituted wood products (e.g., particleboard and plywood) used in their construction. Interestingly, HCHO does not appear to be a significant product of natural gas combustion, as levels in dwellings with and without gas stoves turned on are not significantly different (e.g., Pitts et al., 1989; Zhang et al., 1994a).

Temperature is again an important determinant of HCHO levels. Figure 15.11, for example, shows the concentrations of HCHO as well as of formic acid and methanol measured using FTIR in a research mobile home as function of time as the temperature increased. At 70°F, the average HCHO concentration was 27 ppb but increased to 105 ppb at 100°F (Pitts et al., 1989).

D. VOLATILE ORGANIC COMPOUNDS 857

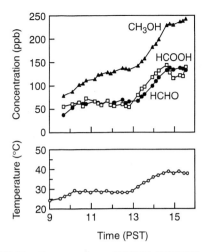

FIGURE 15.11 Concentrations of HCHO, HCOOH, and CH_3OH measured using long-path FTIR in a mobile home as a function of temperature. Although the gas stove burners were on, these were shown in separate experiments not to be the source of these organics (adapted from Pitts et al., 1989.)

HCOOH has also been observed from outgassing in a mobile home trailer by chemical ionization mass spectrometry during measurements of HNO_3 (Huey et al., 1998) and at lower concentrations (mean of 10 ppb) in conventional homes (Reiss et al., 1995a).

Figure 15.12 summarizes the ratio of indoor-to-outdoor concentrations of HCHO and higher aldehydes as well as formic and acetic acids measured in some conventional homes. Concentrations of all of these compounds, except possibly propionaldehyde, are significantly higher indoors, suggesting that not only HCHO but higher aldehydes and ketones as well as acids have significant indoor sources (Lewis and Zweidinger, 1992; Zhang et al., 1994a, 1994b; Zhang and Lioy, 1994; Reiss et al., 1995a). As discussed later, reactions of hydrocarbons with ozone indoors is a potential source, in addition to direct emissions.

It should be noted that while building materials are sources of a variety of VOCs, they can also adsorb organics as well (e.g., Van Loy et al., 1997). As a result, building surfaces and contents may act as reservoirs of organics, slowly releasing compounds over a period of time.

Anthropogenic activities. There are many sources of VOCs associated with human activities in buildings. For example, mixtures of C_{10} and C_{11} isoparaffinic hydrocarbons, which are characteristic of liquid process copiers and plotters, have been identified in office buildings in which these instruments were in use (Hodgson et al., 1991). Emissions of a number of hydrocarbons and aldehydes and ketones have been observed during operation of dry-process copiers; these include significant emissions of ethylbenzene, o-, m-, and p-xylenes, styrene, 2-ethyl-1-hexanol, acetone, n-nonanal, and benzaldehyde (Leovic et al., 1996). Enhanced levels of acetaldehyde in an office building in Brazil were attributed to the oxidation of ethanol used as a cleaning agent (Brickus et al., 1998), although levels outdoors were also enhanced due to the use of ethanol as a fuel (see Chapter 16.D.4). Pyrocatechol has been measured in an occupational environment where meteorological charts are mapped on paper impregnated with this compound (Ekinja et al., 1995), and

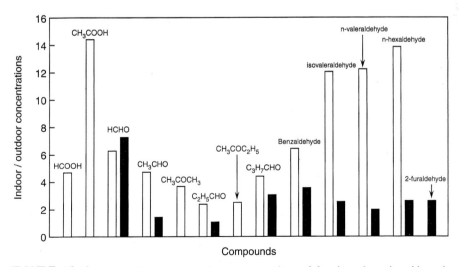

FIGURE 15.12 Ratio of indoor-to-outdoor concentrations of formic and acetic acids and some carbonyl compounds in some homes: unshaded, data from Reiss et al., 1995a, for summer data; shaded, data from Zhang et al., 1994a.

p-dichlorobenzene is observed when mothballs containing this compound are in use (e.g., Tichenor *et al.*, 1990; Chang and Krebs, 1992).

Elevated concentrations of the n-C_{13} to n-C_{18} alkanes and branched-chain and cyclic analogs were measured in a building having a history of air quality complaints; the source was found to be volatilization from hydraulic fluids used in the building elevators (Weschler *et al.*, 1990).

Enhanced levels of chlorinated compounds have been observed indoors due to human activity as well. For example, increased levels of perchloroethylene have been observed from unvented dry-cleaning units (e.g., Moschandreas and O'Dea, 1995) and volatilization of chlorinated organics such as chloroform from treated tap water can occur (e.g., McKone, 1987). Other sources include the use of household products. For example, chloroform emissions have been observed from washing machines when bleach containing hypochlorite was used (Shepherd *et al.*, 1996). It is interesting that emissions of organics associated with the use of washing machines are decreased when the machine is operated with clothes inside (Howard and Corsi, 1998).

Of course, activities such as smoking result in enhanced levels not only of nicotine (e.g., Thompson *et al.*, 1989) but also of a variety of other gases associated with cigarette smoke (e.g., California Environmental Protection Agency, 1997; Nelson *et al.*, 1998). For example, using 3-ethenylpyridine as a marker for cigarette smoke, Heavner *et al.* (1992) estimated that 0.2–39% of the benzene and 2–49% of the styrene measured in the homes of smokers were from cigarette smoke.

Humans emit a variety of VOCs such as pentane and isoprene (e.g., Gelmont *et al.*, 1981; Mendis *et al.*, 1994; Phillips *et al.*, 1994; Jones *et al.*, 1995; Foster *et al.*, 1996). In addition, emissions from personal care products have been observed. Decamethylcyclopentasiloxane (D5), a cyclic dimethylsiloxane with five Si–O units in the ring, and the smaller D4 analog, octamethylcyclotetrasiloxane, are used in such products as underarm deodorant and antiperspirants at concentrations up to 40–60% by weight (Shields and Weschler, 1992; Shields *et al.*, 1996). Increased concentrations of D5 have been measured in offices and are correlated to human activity, as expected if personal care products were the major source (Shields and Weschler, 1992). In some cases, increased concentrations attributable to emissions from silicone-based caulking materials were also observed (Shields *et al.*, 1996).

The use of pesticides indoors can lead to very large concentrations not only of the pesticide but of the additional VOCs used as a matrix for the pesticide, which represent most (>95%) of the mass of the material as purchased. For example, Bukowski and Meyer (1995) predict that VOC concentrations immediately after the application of a fogger could reach levels of more than 300 mg m^{-3}!

However, lower levels of pesticides themselves are common after use inside homes. For example, Lewis *et al.* (1988) reported the presence of 24 pesticides in homes, ranging in concentration from 0.002 to 15 μg m^{-3} (the latter for chlorpyrifos), while outdoor levels measured simultaneously were much lower, from <0.001 to 0.4 μg m^{-3}. The compounds present indoors at the highest concentrations were those used recently at the home. Similarly, Whitmore *et al.* (1994) reported 22 pesticides in homes in Jacksonville, Florida, and Springfield, Massachusetts, at indoor levels up to 0.5 μg m^{-3} indoors, but only 0.04 μg m^{-3} outdoors. The pesticides not only were present as gases but also adsorbed to dust particles in the home, particularly for the less volatile compounds. Indeed, higher concentrations of some pesticides have been found in dust than in air (e.g., Roinestad *et al.*, 1993).

There are a few data that suggest that pesticides can undergo reactions indoors. For example, Wallace *et al.* (1996) observed that the aldrin levels inside a home decreased with time, whereas those of dieldrin did not. Dieldrin had been applied with aldrin but is also an oxidation product of aldrin. One of the reasons for the lack of change in dieldrin may be that it was being formed as the aldrin decayed; however, this could not be differentiated from the effects of a lower vapor pressure of dieldrin, which could lead to lower overall removal rates. In the same study, pentachloroanisole was also measured inside the home and attributed to formation by degradation of pentachlorophenol, which is used as a wood preservative and termiticide.

Not surprisingly, some indoor organics are readily taken up on building surfaces, such as carpets and wallboard, and are subsequently released into the room. For example, Chang *et al.* (1998) showed that some of the alcohols found in indoor air environments are taken up by carpets and gypsum board and could be desorbed back into the gas phase later. However, revolatilization was observed to be slow, and Chang and co-workers estimated that it would take more than a year to remove the adsorbed organics. Similarly, Van Loy *et al.* (1998) showed that nicotine from environmental tobacco smoke can be readily adsorbed and then desorbed and that surfaces can hold significant amounts of nicotine. As a result of this reversible adsorption–desorption process, measurable levels of organics can be maintained indoors after the initial exposure by slow degassing from surfaces.

Not surprisingly, the concentrations of VOCs from automobile exhaust are higher in the "indoor environment" of automobiles during commutes. For example,

Duffy and Nelson (1997) report during commutes in Sydney, Australia, that the benzene concentrations inside vehicles were 10–25 times those in ambient air and that the concentrations of 1,3-butadiene were more than 55–115 times greater. The source appeared to be primarily from the exhausts of surrounding vehicles. Similar enhancements of benzene and other VOCs such as toluene, ethylbenzene, and the xylenes in automobiles and buses have been reported in many countries, including Korea (Jo and Choi, 1996), Taiwan (Chan et al., 1993), and the United States (e.g., Chan et al., 1991a,b; Lawryk and Weisel, 1996).

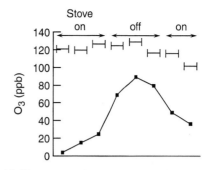

FIGURE 15.13 Measured concentrations of outdoor O_3 and indoor O_3 with and without a gas stove on: horizontal bars, outdoor O_3; solid squares, indoor O_3 (adapted from Zhang and Lioy, 1994).

E. OZONE

Because O_3 decomposes on surfaces, indoor levels are usually lower than those outdoors due to the decomposition that occurs as the air passes through air conditioning systems and impacts building surfaces (e.g., Reiss et al., 1994). The measured ratio of indoor-to-outdoor concentrations of ozone vary from 0.1 to 1, but are typically around 0.3–0.5 (e.g., Druzik et al., 1990; Hisham and Grosjean, 1991b; Liu et al., 1993; Weschler et al., 1989, 1994; Gold et al., 1996; Jakobi and Fabian, 1997; Avol et al., 1998; Drakou et al., 1998; Romieu et al., 1998). Buildings with low air exchange with outside air tend to have lower ratios, ~0.1–0.3 (Druzik et al., 1990; Weschler et al., 1994; Romieu et al., 1998). For example, Gold et al. (1996) estimate that at outdoor ozone concentrations of 170 ppb in Mexico City, the indoor-to-outdoor ratio of O_3 at a school was 0.71 ± 0.03 with the windows and doors open, which maximized the exchange with outside air, 0.18 ± 0.02 with the windows and doors closed and the air cleaner off, and 0.15 ± 0.02 with the windows and doors closed and the air cleaner on.

There are some additional sources of O_3 indoors. These include dry-process photocopying machines, laser printers, and electrostatic precipitators (e.g., Leovic et al., 1996; Wolkoff, 1999). Indeed, it is not unusual to detect O_3 by its odor during operation of some copy machines and laser printers.

In the "indoor environment" in cars, ozone levels tend to be significantly less than in the surrounding area. For example, Chan et al. (1991b) report that in-vehicle O_3 concentrations during commutes in Raleigh, North Carolina, were only about 20% of those measured in the local area at a fixed station. There are several contributing factors to these low concentrations. One is that NO concentrations are higher near roadways, so that O_3 is titrated to NO_2 by its rapid reaction with NO. A second is that O_3 can decompose on the surfaces of the automobile air conditioning system.

A similar titration effect has been observed inside homes where there are combustion sources of NO. Figure 15.13 shows measured outdoor and indoor levels of O_3 in one study with a gas stove turned on or off. Although the outdoor concentration remained relatively constant, the indoor levels were much lower when the gas stove was on compared to when it was off. Although NO_2 was not reported in this study, its levels presumably rose when the stove was on as the O_3 reacted with NO to form NO_2.

F. INDOOR VOC–NO_x–O_3 CHEMISTRY

As discussed in detail throughout this book, there is rich and complex chemistry involving volatile organic compounds (VOCs), oxides of nitrogen, and ozone in ambient air. One might therefore anticipate similar chemistry in indoor air environments, and although there are far fewer studies, this does indeed appear to be the case. Weschler and Shields (1997b) and Wolkoff et al. (1997, 1999) review VOC–NO_x chemistry that could potentially be important in indoor air enviroments and the implications for human exposures.

Although the chemistry occurring indoors is fundamentally the same as that occurring outdoors and discussed throughout this book, there are some important differences as well. For example, the time available for the chemistry to occur indoors is determined by the ventilation rate of the building. In addition, the light intensity and spectral distributions are quite different from those outdoors, decreasing the relative importance of photochemical reactions and increasing that of "dark reactions." For example, Pitts et al. (1985) followed the loss of methyl nitrite (CH_3ONO) in brightly lit rooms indoors in a mobile home used for the HONO studies; no loss was observed over a 5-h period, whereas this compound photolyzes rapidly outdoors.

Finally, the concentrations of the reactants, both absolute and relative, are different from those in most outdoor environments. As we have already seen, the concentrations of VOCs tend to be larger due to the large number of sources indoors, whereas that of O_3 tends to be smaller due to its removal on surfaces, e.g., of air conditioning systems. In the absence of combustion sources, concentrations of oxides of nitrogen tend to be similar to those outdoors.

A number of models that incorporate chemical reactions have been developed and applied to indoor air environments. For example, Özkaynak et al. (1982) developed a model that incorporated simple NO_x chemistry. A more comprehensive model that included VOC–NO_x chemistry as well was developed by Nazaroff and Cass (1986) for indoor air environments (it included ventilation, emissions, and also removal by wall loss) and applied to museum environments. Measurements of NO, NO_2, and O_3 both indoors and outdoors were also made for testing and refining the model predictions. As expected based on the discussion of ozone in Section E, O_3 concentrations indoors, both measured and predicted, were lower than those outdoors due to removal at surfaces; however, removal by reaction with NO, particularly in the morning when NO concentrations were high, was also a significant contributor. The model tended to consistently underpredict NO indoors, which they attribute to its generation in the surface reaction of NO_2 discussed earlier, which was not included in the model chemistry. Although other species were not measured, the results of the model suggest that NO_3 and N_2O_5 levels indoors are greater than those outdoors and that under some circumstances, the same may be true for HNO_3 and H_2O_2.

Weschler and co-workers (Weschler et al., 1992a, 1994) suggested that the formation of NO_3 indoors,

$$O_3 + NO_2 \rightarrow NO_3 + O_2, \quad (3)$$

followed by its well-known secondary chemistry to form HNO_3,

$$NO_3 + NO_2 \leftrightarrow N_2O_5, \quad (4)$$

$$N_2O_5 + H_2O \xrightarrow{\text{surface}} 2HNO_3, \quad (5)$$

$$NO_3 + RH \rightarrow HNO_3 + R, \quad (6)$$

may be an important indoor source of HNO_3. Indirect evidence for an indoor source is the measurement of the indoor-to-outdoor ratio (I/O) of HNO_3 and SO_2 in residences in several locations on the east coast of the United States during the summer months. The I/O ratio for HNO_3 was about the same as that for SO_2 at one location and larger at a second site. This is not expected, since HNO_3 is much more rapidly lost to surfaces than is SO_2. In a third location, the I/O ratio for HNO_3 was smaller than that of SO_2, but not by as much as would be expected from its rapid wall loss. Such data are indicative of an indoor source of HNO_3, and Weschler et al. (1992a) suggest reactions (3)–(6).

Evidence for such indoor chemistry has also been obtained by measuring NO, NO_y, and O_3 indoors and outdoors in an office building (Weschler et al., 1994). For example, as O_3 rises and NO decreases during the morning hours, the O_3 reaction with NO indoors leads to faster decay of NO than otherwise expected and a slower rate of increase of O_3. On the other hand, when NO is rising and O_3 falling in the early evening, this reaction speeds up the decay of O_3 and slows the increase in NO.

Weschler and Shields (1997b) suggest that with the higher concentrations of VOCs indoors, their reactions with O_3, NO_3, and OH may be important. There is some experimental evidence that this is indeed the case. For example, carpet exposed to O_3 in a chamber generated HCHO, benzaldehyde, benzoic acid, and acetophenone, all expected products from the reaction of O_3 with styrene (Zhang et al., 1994b) emitted from the latex adhesive used to bind the backing to the carpet; styrene decreased simultaneously (Weschler et al., 1992b). The formation of a series of C_5–C_{10} aldehydes was observed, which appeared to be from the reaction of O_3 with nonvolatile organics associated with the carpet fibers. Concentrations of HCHO increased by up to a factor of 3 and CH_3CHO by up to a factor of 20 in the presence of O_3. Interestingly, no additional effect was observed when NO_2 was also present, suggesting that the nitrate radical was not a significant contributor to the formation of these aldehydes and ketones.

Salthammer et al. (1999) examined emissions from commonly available coatings used on furniture and identified numerous oxidation products. These were observed without the addition of oxidants such as ozone, indicating that oxidation in air (perhaps including photodecomposition for some compounds) under typical conditions is sufficient to generate such products. For example, emissions of 2-ethylhexanol were identified from di-2-ethylhexyl phthalate, used as a plasticizer in many coatings.

Reiss et al. (1995b) exposed latex paint to O_3 and observed the production of HCHO as well as CH_3CHO and CH_3COCH_3 for some paint samples. They proposed that these were formed by the reactions of O_3 with some remaining double bonds that were not fully reacted during the process in which the $CH_2=CHR$

was polymerized to form the latex paint. Similarly, Chang and Guo (1998) and Fortmann *et al.* (1998) report emissions of hexanal during the drying of an alkyd paint; since hexanal was not a component of the paint itself, they proposed that it was formed by the oxidation of unsaturated fatty acid esters in the alkyd resin.

A correlation between indoor ozone and the concentrations of carbonyl compounds and organic acids in homes has been reported in several studies and attributed at least in part to indoor O_3 reactions (e.g., Reiss *et al.*, 1995a; Zhang and Lioy, 1994; Zhang *et al.*, 1994a–c). Not only HCHO and CH_3CHO but also larger aldehydes have been measured indoors, with indoor concentrations for all but possibly propionaldehyde being much larger than those outdoors (Fig. 15.12). The same is true for formic and acetic acids, which can be formed by ozone reactions from reaction of the Criegee biradicals with water vapor (see Chapter 6.E.2). Zhang *et al.* (1994c) report that indoor formic acid concentrations increased with the indoor concentration of O_3 and with relative humidity, as expected if the reaction of the HCHOO Criegee biradical with water vapor was a significant indoor source.

However, as discussed by Reiss *et al.* (1995a), separating the contribution of ozone reactions from other factors such as temperature and relative humidity, which also affect direct emissions, is difficult. For example, while the production rate of oxygenated organics is correlated with the ozone removal rate, the latter is also correlated with temperature. As a result, both reaction and increased direct emission rates due to higher temperatures may be contributing to these enhanced indoor levels.

Ozone can also react with components found in air ducts. For example, Morrison *et al.* (1998) reported that the sealant and neoprene gaskets used in the ducts emitted VOCs into the airstream, but at relatively low levels compared to the typical concentrations found indoors. However, reaction with O_3 led to increased emissions of aldehydes, particularly the C_5-C_{10} aldehydes.

The mechanism of the reaction of O_3 with alkenes was discussed in detail in Chapter 6.E.2. It was seen there that these reactions serve as an indoor source of OH through decomposition of the Criegee biradical. Weschler and Shields (1996b) proposed that the indoor reaction of O_3 with alkenes could serve as a source of OH and calculated that at 20 ppb O_3 and average indoor alkene concentrations, a steady-state OH concentration of about 2×10^5 molecules cm^{-3} might be expected.

Subsequently, they measured OH concentrations in an office building using the rate of decay of 1,3,5-trimethylbenzene, which reacts with OH but not O_3 (Weschler and Shields, 1997a). Although O_3 and *d*-limonene were injected during the experiment, their concentrations were chosen to be similar to that measured under normal operating conditions. An average OH concentration of 7×10^5 molecules cm^{-3} was obtained, which is about an order of magnitude smaller than daytime peak OH concentrations outdoors but also more than an order of magnitude larger than those outdoors at night (see Chapter 11.A.4i).

As discussed in Chapter 9.C.2, some of the larger alkenes such as terpenes form particles containing low-volatility organics on oxidation with ozone. Hence particle formation might be expected indoors in the presence of such compounds, and indeed this has been observed (Weschler and Shields, 1999).

In short, much of the chemistry that has been observed outdoors also occurs indoors. However, the relative importance of various reactions may be somewhat different due to the different absolute and relative concentrations of the reactants, the lower photolysis rates, the exchange of air with outdoor air, and the presence of relatively large surface areas, which can both remove various species and act as substrates for heterogeneous reactions.

G. PARTICLES

With the epidemiological studies suggesting increased mortality associated with particles (see Chapter 2), there has been increasing interest in indoor particle concentrations compared to outdoor levels. A number of studies have examined this over the years and are summarized in a review by Wallace (1996).

In general, if there are no indoor sources of particles, the levels indoors tend to reflect those outdoors. For example, application of a mass balance model to measurements of indoor and outdoor particle concentrations in Riverside, California, indicated that 75% of $PM_{2.5}$ and 65% of PM_{10} in a typical home were from outdoors (Wallace, 1996). Similar conclusions were reached by Koutrakis *et al.* (1991, 1992) for homes in two counties in New York. For example, they report that 60% of the mass of particles in homes is due to outdoor sources. However, the contribution to various individual elements in the particles varies from 22% for copper to 100% for cadmium.

There are some differences in indoor levels of particulate matter in areas with low outdoor compared to high outdoor levels. In the case of high outdoor levels, the indoor concentrations tend to be somewhat lower than those outdoors; for example, Colome *et al.* (1992)

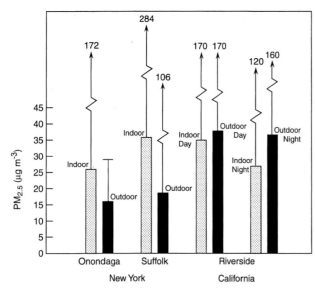

FIGURE 15.14 Geometric mean $PM_{2.5}$ concentrations indoors and outdoors in Onondaga and Suffolk Counties in New York state and in Riverside, California. The Riverside data are weighted means. The error bars represent the maximum values for the New York sites and the 98th percentile for the Riverside studies (data from Wallace, 1996; based on Sheldon *et al.*, 1989, and Pellizzari *et al.*, 1993).

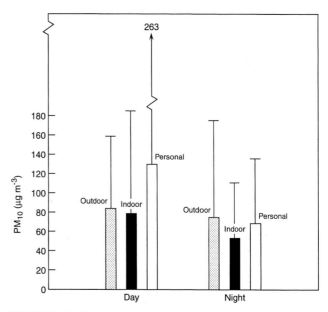

FIGURE 15.15 Geometric mean PM_{10} concentrations outdoors, indoors, and by personal exposure monitors during the day and at night in Riverside, California, weighted to provide estimates of concentrations for household-days or person-days. The error bars are 90th percentile (data from Clayton *et al.*, 1993).

report that the ratio of indoor-to-outdoor median concentrations of PM_{10} is 0.7 in residences in southern California. On the other hand, when outdoor levels are low, indoor levels tend to be higher. This is illustrated by the data in Fig. 15.14, which shows indoor and outdoor levels of $PM_{2.5}$ measured in two counties in New York state (Sheldon *et al.*, 1989) and in Riverside, California (Pellizzari *et al.*, 1993), as summarized by Wallace (1996).

Nighttime mass concentrations indoors tend to be smaller than those during the day, probably because of the decreased activity. Interestingly, when individuals wear personal exposure monitors to measure their actual exposure to particles, the measured mass concentrations tend to be higher than those measured with fixed monitors located indoors. Both of these are illustrated by the data in Fig. 15.15, which shows the geometric mean concentrations of PM_{10} in Riverside, California, measured outdoors, indoors, and with personal exposure monitors (Clayton *et al.*, 1993). Because of the study design, the measurements were weighted to provide estimates of the concentrations in terms of household-days or person-days, respectively. Indoor levels at night are about 50–70% of those during the day. However, the personal exposure concentrations are 165% of the measured indoor concentrations during the day and 128% at night. This has been dubbed the "body cloud" or "personal cloud" (Rodes, 1991). The reasons for the higher concentrations are not clear. While skin flakes and clothing fibers contribute to this, their levels do not appear to be sufficiently high to explain the observed levels (Özkaynak *et al.*, 1996). In addition, the elemental composition of the personal exposure particles in some studies has been observed to be very similar to that of indoor aerosol samples, indicating that different sources are not likely to be responsible (Özkaynak *et al.*, 1996; Wallace, 1996). It may be that the activity of the individual is responsible, at least in part. For example, carpets have been shown to be a reservoir of dust particles that can be suspended when walking, vacuuming, etc. (Thatcher and Layton, 1995; Leese *et al.*, 1997). Such activity leads to resuspension of the larger particles preferentially. For example, cleaning activities increased the airborne concentrations of 10- to 25-μm particles by a factor of 30, of 5- to 10-μm particles by a factor of 11, and of 0.5- to 1-μm particles insignificantly (Thatcher and Layton, 1995). Enhanced levels of particulate sulfate and H^+ in personal exposure monitors compared to indoor values (ratios of 1.25 and 3.1, respectively) have been reported (Suh *et al.*, 1992), although Özkaynak *et al.* (1996) report that sulfur in particles in the "personal cloud" was the only element not enhanced compared to indoor levels.

A major source of increased particles indoors is cigarette smoking. (e.g., Spengler *et al.*, 1981; Quackenboss *et al.*, 1989; Neas *et al.*, 1994). Figure 15.16 shows

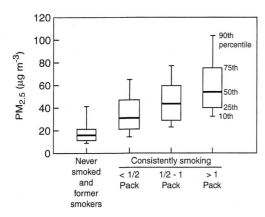

FIGURE 15.16 Annual average concentrations of $PM_{2.5}$ indoors as a function of smoking in the home. The data are shown in percentiles as marked. (Adapted from Neas *et al.*, 1994.)

the annual average concentrations of $PM_{2.5}$ measured indoors in homes with nonsmokers and for homes occupied by smokers as a function of the number of packs of cigarettes smoked per day (Neas *et al.*, 1994). There is a steady increase in the indoor levels of $PM_{2.5}$ as a function of the number of cigarettes smoked, with the increase in $PM_{2.5}$ estimated to be in the range of 25–45 $\mu g\ m^{-3}$ (Wallace, 1996).

In addition to the contribution to the *mass* concentrations of indoor particles, cigarette smoke is of concern because of the mutagens, carcinogens, and toxic air contaminants that are emitted (e.g., see Löfroth *et al.*, 1991; Chuang *et al.*, 1991; California Environmental Protection Agency, 1997; and Nelson *et al.*, 1998). Thus, a variety of both gaseous and particulate polycyclic aromatic hydrocarbons (PAH) and compounds (PAC) have been identified in buildings with cigarette smoke (e.g., Offermann *et al.*, 1991; Mitra and Ray, 1995). Indeed, in the homes of smokers, almost 90% of the total PAH was from tobacco smoke (Mitra and Ray, 1995). Higher levels of mutagenic particles have also been shown to be associated with indoor air containing cigarette smoke (e.g., Lewtas *et al.*, 1987; Löfroth *et al.*, 1988, 1991; Georghiou *et al.*, 1991).

Other significant sources identified in a number of studies are cooking, the use of kerosene heaters, wood burning, and humidifiers. For example, a study carried out under the auspices of the U.S. Environmental Protection Agency, the TEAM study (*T*otal *E*xposure *A*ssessment *M*ethodology), indicated that an increase in PM_{10} of $\sim 10-20\ \mu g\ m^{-3}$ could be attributed to cooking (Wallace, 1996). This source will obviously depend on the amount of cooking, the types of cooking, and the ventilation. For example, Löfroth *et al.* (1991) measured emissions of particles ranging from 0.07 to 3.5 mg per gram of food cooked, depending on the particular food. Baek *et al.* (1997) measured indoor and outdoor concentrations of particles in homes, offices, and restaurants in Korea and report ratios of 1.3, 1.3, and 2.4, respectively. The higher value in restaurants, even those using only gas and not charcoal, suggests a significant contribution from cooking.

Kerosene heaters can be significant sources of particles under some circumstances. For example, kerosene heaters were reported to contribute to indoor $PM_{2.5}$ in homes in Suffolk County, New York, but not Onondaga County; wood stoves and fireplaces and gas stoves did not contribute in either case (Koutrakis *et al.*, 1992; Wallace, 1996). A similar conclusion was reached in a study of eight mobile homes in North Carolina (Mumford *et al.*, 1991).

However, it should be noted that even where kerosene heaters do not contribute significantly to particle *mass* concentrations, they may still be important in terms of health effects. This is because of the composition of the particles emitted, which include polycyclic aromatic compounds and other mutagenic species, as well as sulfate (Leaderer *et al.*, 1990; Traynor *et al.*, 1990). For example, Traynor *et al.* (1990) studied the emissions from unvented kerosene space heaters and identified a number of PAHs (naphthalene, phenanthrene, fluoranthene, anthracene, chrysene, and indeno[*c,d*]pyrene) and nitro-PAHs (1-nitronaphthalene, 9-nitroanthracene, 3-nitrofluoranthene, and 1-nitropyrene), in addition to a host of other gaseous species. Baek *et al.* (1997) also reported increased levels of a number of gases indoors in homes and offices in Korea due to the use of kerosene heaters.

In studies of indoor air in eight mobile homes, Mumford *et al.* (1991) identified the PAHs and nitro-PAHs measured in emissions from kerosene heaters by Traynor *et al.* (1990), as well as a number of compounds that may be animal carcinogens, such as cyclopenta[*c,d*]pyrene, benz[*a*]anthracene, benzofluoranthenes, benzo-[*a*]pyrene, and benzo[*ghi*]perylene. While the mass concentrations of PM_{10} did not increase with the kerosene heater on in six of the eight homes studied, the particles in five of the homes had increased mutagenicity using TA98 with or without S9 added (see Chapter 10.C.4).

In short, not only the mass emissions but also the nature of the compounds emitted must be taken into account in assessing the health effects of indoor particles.

Humidifiers can be a significant source of airborne particles if tap water is used because as the water evaporates from the aerosol, the solids that were in water are left as particles. For example, Highsmith *et al.* (1988, 1992) showed that the airborne particle concentration increased linearly with the total dissolved solids in the water used in the humidifier. When an ultrasonic humidifier was operated in a closed single room using tap water with total dissolved solids of 303 mg L^{-1}, concentrations of fine and coarse particles of 6307 and 771 μg m^{-3}, respectively, were generated. When distributed over the whole house, the corresponding levels were 593 and 65 μg m^{-3}, respectively. The whole-house values fell to 41 and 13 μg m^{-3} when bottled water with total dissolved solids of 24 mg L^{-1} was used (Highsmith et al., 1992).

On the other hand, the use of evaporative ("swamp") coolers appears to decrease particulate matter indoors. For example, Quackenboss et al. (1989) report levels of PM$_{2.5}$ and PM$_{10}$ in homes having such coolers (for both smokers and nonsmokers) that are about half that of homes without them.

Where indoor heating and cooking involves the use of coal or biomass, indoor particle concentrations can be extremely large. For example, Florig (1997) and Ando et al. (1996) report that in China typical indoor total suspended particle (TSP) concentrations can be in the range from 250 to 900 μg m^{-3} in homes using coal and 950–3500 μg m^{-3} in those using biomass fuels. These levels can be compared to annual average outdoor concentrations of 250–410 μg m^{-3}. The high concentrations associated with coal burning combined with the mutagenic nature of the emissions have been suggested to be responsible for enhanced lung cancer in China (Mumford et al., 1987). Similarly, Davidson et al. (1986) measured TSP concentrations of 2900–42,000 μg m^{-3} in homes in Nepal that used biomass fuels, compared to outdoor levels of 280 μg m^{-3}. For particles with diameters less than 4 μm, the levels ranged from 870 to 14,000 μg m^{-3}.

Similar conclusions regarding the relative indoor and outdoor concentrations have been reached in studies of office and commercial buildings. For example, Ligocki et al. (1993) measured indoor and outdoor concentrations of particles and their components at five museums in southern California. The indoor-to-outdoor ratios of particle mass varied over a wide range, depending to a large extent on the ventilation and filtration systems in use. Ratios varied from 0.16 to 0.96 for particles with diameters less than 2.1 μm and from 0.06 to 0.3 for coarse particles with diameters greater than this.

The chemical composition of particles collected in the museums was also compared to that outdoors using a mass balance model. The results indicated that there were significant indoor sources of fine particle organics and that this source(s) was a significant fraction of the total indoor fine particle organic concentration (Ligocki et al., 1993). A similar conclusion was reached by Naik et al. (1991) in measurements made in a telephone switching office. The levels of the n-C$_{27}$ through n-C$_{33}$ alkanes in the fine particle fraction were all elevated, suggesting indoor sources such as waxes, polishes, and lubricants. Enhanced levels of dibutyl phthalate and di(2-ethylhexyl) phthalate were also observed in fine particles and attributed to plasticizers used in floor polishes and vinyl products.

Turk et al. (1989) measured particle concentrations in 38 buildings that had both smoking and nonsmoking areas. The average mass concentration of respirable particles in smoking areas was 70 μg m^{-3} compared to 19 μg m^{-3} in nonsmoking areas, whereas the outdoor concentrations were essentially the same, 19 μg m^{-3}.

In another study, Ott et al. (1996) measured particle concentrations inside and outside a tavern before and after a smoking ban was instated. Average respirable suspended particle mass concentrations were 57 μg m^{-3} above the outdoor concentrations prior to the ban, compared to 6–13 μg m^{-3} afterward. Cooking and resuspended dust also contributed to the indoor particle mass concentration, but at concentrations about 20–25% of that due to cigarette smoke.

In short, the indoor concentrations of particles depend on the outdoor levels, the ventilation system and exchange rates, and the presence of indoor sources such as cigarette smoke.

In many nonresidential buildings, deposition is of particular interest because of the potential for damage to materials in museums, offices, cultural objects, and industrial sites (e.g., Sinclair et al., 1988, 1990a,b; Nazaroff et al., 1990a; Salmon et al., 1994, 1995). Deposition onto both horizontal and vertical surfaces is of concern, but these can show different behaviors as a function of particle size. For example, deposition of particles to vertical and horizontal surfaces was measured inside five museums. Horizontal deposition velocities increased from $\sim 10^{-6}$ to 10^{-3} m s^{-1} as particle size increased from about 0.1 to 30 μm as expected for gravitational settling (see Chapter 9.A.3). However, the dependence on particle size of the uptake onto vertical surfaces, which is influenced by thermal and air flow fields (Nazaroff et al., 1990b), was variable, increasing with particle size in some cases, decreasing in others, and, in some, showing no dependence on particle size (Ligocki et al., 1990).

A number of models have been developed for particles indoors (e.g., Nazaroff and Cass, 1989a; Sinclair et al., 1990b; Nazaroff et al., 1990a; Weschler et al., 1996; Wallace et al., 1996, and references therein). This is a complex problem, given the number of potential sources, different deposition velocities for particles of different sizes (e.g., see Chapter 9.A.3 and Nazaroff and Cass (1989b)), the different particle compositions, and the effects of outdoor concentrations and ventilation rates. However, reasonably good agreement has been obtained in many cases between modeled and observed concentrations.

H. PROBLEMS

1. As discussed in this chapter, a relatively new area in indoor air pollution is that of hydroxyl radical chemistry. However, the importance of indoor OH chemistry (as well as O_3 and NO_3 chemistry) is determined by the rates of the reactions compared to the rate of air exchange. An OH concentration of 7×10^5 cm^{-3} has been reported in an indoor air environment by Weschler and Shields (1997a). Assess the importance of the OH reaction for the removal of limonene indoors compared to its removal by reaction with O_3. The O_3–limonene rate constant is 2×10^{-16} and the OH–limonene rate constant is 1.7×10^{-10} cm^3 molecule^{-1} s^{-1}. Take the O_3 concentration to be 20 ppb. How do these compare with a typical air exchange rate of 0.75 h^{-1} used in these experiments?

References

Alter, H. W., and R. A. Oswald, "Nationwide Distribution of Indoor Radon Measurements: A Preliminary Data Base," *JAPCA, 37*, 227–231 (1987).

Anderson, D. J., and R. A. Hites, "Chlorinated Pesticides in Indoor Air," *Environ. Sci. Technol., 22*, 717–720 (1988).

Ando, M., K. Katagiri, K. Tamura, S. Yamamoto, M. Matsumoto, Y. F. Li, S. R. Cao, R. D. Ji, and C. K. Liang, "Indoor and Outdoor Air Pollution in Tokyo and Beijing Supercities," *Atmos. Environ., 30*, 695–702 (1996).

Avol, E. L., W. C. Navidi, and S. D. Colome, "Modeling Ozone Levels in and around Southern California Homes," *Environ. Sci. Technol., 32*, 463–468 (1998).

Baek, S.-O., Y.-S. Kim, and R. Perry, "Indoor Air Quality in Homes, Offices, and Restaurants in Korean Urban Areas—Indoor/Outdoor Relationships," *Atmos. Environ., 31*, 529–544 (1997).

Bayer, C. M., "The Effect of 'Building Bake-Out' Conditions on Volatile Organic Compound Emissions," in *Indoor Air Pollution—Radon, Bioaerosols, & VOC's* (J. G. Kay, G. E. Keller, and J. F. Miller, Eds.), Chap. 9, Lewis Publishers, Chelsea, MI, 1991.

Boyle, M., "Radon Testing of Soils," *Environ. Sci. Technol., 22*, 1397–1399 (1988).

Brauer, M., P. B. Ryan, H. H. Suh, P. Koutrakis, J. D. Spengler, N. P. Leslie, and I. H. Billick, "Measurements of Nitrous Acid inside Two Research Houses," *Environ. Sci. Technol., 24*, 1521–1527 (1990).

Brauer, M., T. R. Rasmussen, S. K. Kjaergaard, and J. D. Spengler, "Nitrous Acid Formation in an Experimental Exposure Chamber," *Indoor Air, 3*, 94–105 (1993).

Brauer, M., and J. D. Spengler, "Nitrogen Dioxide Exposures inside Ice Skating Rinks," *Am. J. Public Health, 84*, 429–433 (1994).

Brauer, M., K. Lee, J. D. Spengler, R. O. Salonen, A. Pennanen, O. A. Braathen, E. Mihalikova, P. Miskovic, A. Nozaki, T. Tsuzuki, S. Rui-Jin, X. Xu, Z. Quing-Xiang, H. Drahonovska, and S. Kjaergaard, "Nitrogen Dioxide in Indoor Ice Skating Facilities: An International Survey," *J. Air Waste Manage. Assoc., 47*, 1095–1102 (1997).

Brickus, L. S. R., J. N. Cardoso, and F. R. De Quino Neto, "Distributions of Indoor and Outdoor Air Pollutants in Rio de Janeiro, Brazil: Implications to Indoor Air Quality in Bayside Offices," *Environ. Sci. Technol., 32*, 3485–3490 (1998).

Brown, S. K., M. R. Sim, M. J. Abramson, and C. N. Gray, "Concentrations of Volatile Organic Compounds in Indoor Air—A Review," *Indoor Air, 4*, 123–134 (1994).

Bukowski, J. A., and L. W. Meyer, "Simulated Air Levels of Volatile Organic Compounds Following Different Methods of Indoor Insecticide Application," *Environ. Sci. Technol., 29*, 673–676 (1995).

Bukowski, J. A., and D. Wartenberg, "An Alternative Approach for Investigating the Carcinogenicity of Indoor Air Pollution: Pets as Sentinels of Environmental Cancer Risk," *Environ. Health Perspect., 105*, 1312–1319 (1997).

California Environmental Protection Agency, Air Resources Board, Research Division, "Determination of Formaldehyde and Toluene Diisocyanate Emissions from Indoor Residential Sources," Contract No. 93-315, Final Report, November 1996.

California Environmental Protection Agency, Office of Environmental Health Hazard Assessment, "Health Effects of Exposure to Environmental Tobacco Smoke," Final Report, September 1997.

Castleman, A. W., Jr., "Consideration of the Chemistry of Radon Progeny," *Environ. Sci. Technol., 25*, 730–735 (1991).

Chan, C.-C., J. D. Spengler, H. Özkaynak, and M. Lefkopoulou, "Commuter Exposures to VOCs in Boston, Massachusetts," *J. Air Waste Manage. Assoc., 41*, 1594–1600 (1991a).

Chan, C.-C., H. Özkaynak, J. D. Spengler, and L. Sheldon, "Driver Exposure to Volatile Organic Compounds, CO, Ozone, and NO_2 under Different Driving Conditions," *Environ. Sci. Technol., 25*, 964–972 (1991b).

Chan, C.-C., S.-H. Lin, and G.-R. Her, "Student's Exposure to Volatile Organic Compounds While Commuting by Motorcycle and Bus in Taipei City," *J. Air Waste Manage. Assoc., 43*, 1231–1238 (1993).

Chang, J. C. S., and K. A. Krebs, "Evaluation of Para-Dichlorobenzene Emissions from Solid Moth Repellant as a Source of Indoor Air Pollution," *J. Air Waste Manage. Assoc., 42*, 1214–1217 (1992).

Chang, J. C. S., and Z. Guo, "Emissions of Odorous Aldehydes from Alkyd Paint," *Atmos. Environ., 32*, 3581–3586 (1998).

Chang, J. C. S., L. E. Sparks, and Z. Guo, "Evaluation of Sink Effects on VOCs from a Latex Paint," *J. Air Waste Manage. Assoc., 48*, 953–958 (1998).

Chuang, J. C., G. A. Mack, M. R. Kuhlman, and N. K. Wilson, "Polycyclic Aromatic Hydrocarbons and Their Derivatives in Indoor and Outdoor Air in an Eight-Home Study," *Atmos. Environ., 25B*, 369–380 (1991).

Clayton, C. A., R. I. Perritt, E. D. Pellizzari, K. W. Thomas, R. W. Whitmore, L. A. Wallace, H. Özkaynak, and J. D. Spengler, "Particle Total Exposure Assessment Methodology (PTEAM) Study: Distributions of Aerosol and Elemental Concentrations in Personal, Indoor, and Outdoor Air Samples in a Southern California Community," *J. Exposure Anal. Environ. Epidemiol., 3*, 227–250 (1993).

Colome, S. D., N. Y. Kado, P. Jaques, and M. Kleinman, "Indoor–Outdoor Air Pollution Relations: Particulate Matter Less Than 10 μm in Aerodynamic Diameter (PM 10) in Homes of Asthmatics," *Atmos. Environ., 26A*, 2173–2178 (1992).

Crump, D. R., and D. Gardiner, "Sources and Concentrations of Aldehydes and Ketones in Indoor Environments in the UK," *Environ. Int., 15*, 455–462 (1989).

Crump, D. R., "Volatile Organic Compounds in Indoor Air," in *Issues in Environmental Science and Technology* (R. E. Hester and R. M. Harrison, Eds.), Chap. 4, pp. 109–124, Royal Chem. Soc., Letchworth, UK, 1995.

Daisey, J. M., A. T. Hodgson, W. J. Fisk, M. J. Mendell, and J. Ten Brinke, "Volatile Organic Compounds in Twelve California Office Buildings: Classes, Concentrations, and Sources," *Atmos. Environ., 28,* 3557–3562 (1994).

Davidson, C. I., S.-F. Lin, J. F. Osborn, M. R. Pandey, R. A. Rasmussen, and M. A. K. Khalil, "Indoor and Outdoor Air Pollution in the Himalayas," *Environ. Sci. Technol., 20,* 561–567 (1986).

Dor, F., Y. Le Moullec, and B. Festy, "Exposure of City Residents to Carbon Monoxide and Monocyclic Aromatic Hydrocarbons during Commuting Trips in the Paris Metropolitan Area," *J. Air Waste Manage. Assoc., 45,* 103–110 (1995).

Drakou, G., C. Zerefos, I. Ziomas, and M. Voyatzaki, "Measurements and Numerical Simulations of Indoor O_3 and NO_x in Two Different Cases," *Atmos. Environ., 32,* 595–610 (1998).

Druzik, J. R., M. S. Adams, C. Tiller, and G. R. Cass, "The Measurement and Model Predictions of Indoor Ozone Concentrations in Museums," *Atmos. Environ., 24A,* 1813–1823 (1990).

Drye, E. E., H. Özkaynak, B. Burbank, I. H. Billick, J. D. Spengler, P. B. Ryan, P. E. Baker, and S. D. Colome, "Development of Models for Predicting the Distribution of Indoor Nitrogen Dioxide Concentrations," *J. Air Waste Manage. Assoc., 39,* 1169–1177 (1989).

Duffy, B. L., and P. F. Nelson, "Exposure to Emissions of 1,3-Butadiene and Benzene in the Cabins of Moving Motor Vehicles and Buses in Sydney, Australia," *Atmos. Environ., 31,* 3877–3885 (1997).

Ekinja, I., Z. Grabaric, and B. S. Grabaric, "Monitoring of Pyrocatechol Indoor Air Pollution," *Atmos. Environ., 29,* 1165–1170 (1995).

Febo, A., and C. Perrino, "Prediction and Experimental Evidence for High Air Concentration of Nitrous Acid in Indoor Environments," *Atmos. Environ., 25A,* 1055–1061 (1991).

Febo, A., and C. Perrino, "Measurement of High Concentration of Nitrous Acid inside Automobiles," *Atmos. Environ., 29,* 345–351 (1995).

Feinman, S. E., *Formaldehyde Sensitivity and Toxicity,* CRC Press, Boca Raton, FL, 1988.

Fernandez-Bremauntz, A. A., and M. R. Ashmore, "Exposure of Commuters to Carbon Monoxide in Mexico City—I. Measurement of In-Vehicle Concentrations," *Atmos. Environ., 29,* 525–532 (1995).

Fischer, M. L., A. J. Bentley, K. A. Dunkin, A. T. Hodgson, W. W. Nazaroff, R. G. Sextro, and J. M. Daisey, "Factors Affecting Indoor Air Concentrations of Volatile Organic Compounds at a Site of Subsurface Gasoline Contamination," *Environ. Sci. Technol., 30,* 2948–2957 (1996).

Flachsbart, P. G., G. A. Mack, J. E. Howes, and C. E. Rodes, "Carbon Monoxide Exposures of Washington Commuters," *JAPCA, 37,* 135–142 (1987).

Florig, H. K., "China's Air Pollution Risks," *Environ. Sci. Technol., 31,* 274A–279A (1997).

Fortmann, R., N. Roache, J. C. S. Chang, and Z. Guo, "Characterization of Emissions of Volatile Organic Compounds from Interior Alkyd Paint," *J. Air Waste Manage. Assoc., 48,* 931–940 (1998).

Foster, W. M., L. Jiang, P. T. Stetkiewicz, and T. H. Risby, "Breath Isoprene—Temporal Changes in Respiratory Output after Exposure to Ozone," *J. Appl. Physiol., 80,* 706–710 (1996).

Gelmont, D., R. A. Stein, and J. F. Mead, "Isoprene—The Main Hydrocarbon in Human Breath," *Biochem. Biophys. Res. Commun., 99,* 1456–1460 (1981).

Georghiou, P. E., P. Blagden, D. A. Snow, L. Winsor, and D. T. Williams, "Mutagenicity of Indoor Air Containing Environmental Tobacco Smoke: Evaluation of a Portable PM-10 Impactor Sampler," *Environ. Sci. Technol., 25,* 1496–1500 (1991).

Gold, D. R., G. Allen, A. Damokosh, P. Serrano, C. Hayes, and M. Castillejos, "Comparison of Outdoor and Classroom Ozone Exposures for School Children in Mexico City," *J. Air Waste Manage. Assoc., 46,* 335–342 (1996).

Haghighat, F., and L. de Bellis, "Material Emission Rates: Literature Review, and the Impact of Indoor Air Temperature and Relative Humidity," *Bldg. Environ., 33,* 261–277 (1998).

Hanrahan, L. P., H. A. Anderson, K. A. Daily, A. D. Eckmann, and M. S. Kanarek, "Formaldehyde Concentrations in Wisconsin Mobile Homes," *JAPCA, 35,* 1164–1167 (1985).

Heavner, D. L., M. W. Ogden, and P. R. Nelson, "Multisorbent Thermal Desorption/Gas Chromatography/Mass Selective Detection Method for the Determination of Target Volatile Organic Compounds in Indoor Air," *Environ. Sci. Technol., 26,* 1737–1746 (1992).

Highsmith, V. R., C. E. Rodes, and R. J. Hardy, "Indoor Particle Concentrations Associated with Use of Tap Water in Portable Humidifiers," *Environ. Sci. Technol., 22,* 1109–1112 (1988).

Highsmith, V. R., R. J. Hardy, D. L. Costa, and M. S. Germani, "Physical and Chemical Characterization of Indoor Aerosols Resulting from the Use of Tap Water in Portable Home Humidifiers," *Environ. Sci. Technol., 26,* 673–680 (1992).

Hileman, B., "Multiple Chemical Sensitivity," *Chem. Eng. News,* 26–42, July 22 (1991).

Hintenlang, D. E., and K. K. Al-Ahmady, "Pressure Differentials for Radon Entry Coupled to Periodic Atmospheric Pressure Variations," *Indoor Air, 2,* 208–215 (1992).

Hisham, M. W. M., and D. Grosjean, "Air Pollution in Southern California Museums: Indoor and Outdoor Levels of Nitrogen Dioxide, Peroxyacetyl Nitrate, Nitric Acid, and Chlorinated Hydrocarbons," *Environ. Sci. Technol., 25,* 857–862 (1991a).

Hisham, M. W. M., and D. Grosjean, "Sulfur Dioxide, Hydrogen Sulfide, Total Reduced Sulfur, Chlorinated Hydrocarbons, and Photochemical Oxidants in Southern California Museums," *Atmos. Environ., 25A,* 1497–1505 (1991b).

Hodgson, A. T., J. M. Daisey, and R. A. Grot, "Sources and Source Strengths of Volatile Organic Compounds in a New Office Building," *J. Air Waste Manage. Assoc., 41,* 1461–1468 (1991).

Hodgson, A. T., K. Garbesi, R. G. Sextro, and J. M. Daisey, "Soil-Gas Contamination and Entry of Volatile Organic Compounds into a House near a Landfill," *J. Air Waste Manage. Assoc., 42,* 277–283 (1992).

Hodgson, A. T., J. D. Wooley, and J. M. Daisey, "Emissions of Volatile Organic Compounds from New Carpets Measured in a Large-Scale Environmental Chamber," *J. Air Waste Manage. Assoc., 43,* 316–324 (1993).

Hoek, G., B. Brunekreef, and P. Hofschreuder, "Indoor Exposure to Airborne Particles and Nitrogen Dioxide during an Air Pollution Episode," *JAPCA, 39,* 1348–1349 (1989).

Howard, C., and R. L. Corsi, "Volatilization of Chemicals from Drinking Water to Indoor Air: The Role of Residential Washing Machines," *J. Air Waste Manage. Assoc., 48,* 907–914 (1998).

Howard, E. M., R. C. McCrillis, K. A. Krebs, R. Fortman, H. C. Lao, and Z. Guo, "Indoor Emissions from Conversion Varnishes," *J. Air Waste Manage. Assoc., 48,* 924–930 (1998).

Huey, L. G., E. J. Dunlea, E. R. Lovejoy, D. R. Hanson, R. B. Norton, F. C. Fehsenfeld, and C. J. Howard, "Fast Time Response Measurements of HNO_3 in Air with a Chemical Ionization Mass Spectrometer," *J. Geophys. Res., 103,* 3355–3360 (1998).

International Commission on Radiological Protection, *Ann. ICRP*, 23, No. 2 (1993).

International Commission on Radiological Protection, "Protection against Radon-222 at Home and at Work," *Ann. ICRP*, No. 65 (1994).

Jakobi, G., and P. Fabian, "Indoor/Outdoor Concentrations of Ozone and Peroxyacetyl Nitrate (PAN)," *Int. J. Biometeorol.*, 40, 162–165 (1997).

Jenkins, P. L., T. J. Phillips, E. J. Mulberg, and S. P. Hui, "Activity Patterns of Californians: Use of and Proximity to Indoor Pollutant Sources," *Atmos. Environ.*, 26A, 2141–2148 (1992).

Jo, W.-K., and S.-J. Choi, "Vehicle Occupants' Exposure to Aromatic Volatile Organic Compounds While Commuting on an Urban–Suburban Route in Korea," *J. Air Waste Manage. Assoc.*, 46, 749–754 (1996).

Jones, A. W., V. Lagesson, and C. Tagesson, "Origins of Breath Isoprene," *J. Clin. Pathol.*, 48, 979–980 (1995).

Kay, J. G., G. E. Keller, and J. F. Miller, *Indoor Air Pollution—Radon, Bioaerosols, & VOC's*, Lewis Publishers, Chelsea, MI, 1991.

Kelly, T. J., "Determination of Formaldehyde and Toluene Diisocyanate Emissions from Indoor Residential Sources," California Air Resources Board, Research Division, 2020 L Street, Sacramento, CA 95814, Final Report, Contract Number 93-315, November, 1996.

Kostiainen, R., "Volatile Organic Compounds in the Indoor Air of Normal and Sick Houses," *Atmos. Environ.*, 29, 693–702 (1995).

Koushki, P. A., K. H. Al-Dhowalia, and S. A. Niaizi, "Vehicle Occupant Exposure to Carbon Monoxide," *J. Air Waste Manage. Assoc.*, 42, 1603–1608 (1992).

Koutrakis, P., M. Brauer, S. L. K. Briggs, and B. P. Leaderer, "Indoor Exposures to Fine Aerosols and Acid Gases," *Environ. Health Perspect.*, 95, 23–28 (1991).

Koutrakis, P., S. L. K. Briggs, and B. P. Leaderer, "Source Apportionment of Indoor Aerosols in Suffolk and Onondaga Counties, New York," *Environ. Sci. Technol.*, 26, 521–527 (1992).

Lawryk, N. J., and C. P. Weisel, "Concentrations of Volatile Organic Compounds in the Passenger Compartments of Automobiles," *Environ. Sci. Technol.*, 30, 810–816 (1996).

Leaderer, B. P., P. M. Boone, and S. K. Hammond, "Total Particle, Sulfate, and Acidic Aerosol Emissions from Kerosene Space Heaters," *Environ. Sci. Technol.*, 24, 908–912 (1990).

Leese, K., "Indoor Air Measurements," *Environ. Sci. Technol. / News*, 31, 493A (1997).

Leese, K. E., E. C. Cole, R. M. Hall, and M. A. Berry, "Measurement of Airborne and Floor Dusts in a Nonproblem Building," *AIHA J.*, 58, 432–438 (1997).

Leovic, K. W., L. S. Sheldon, D. A. Whitaker, R. G. Hetes, J. A. Calcagni, and J. N. Baskir, "Measurement of Indoor Air Emissions from Dry-Process Photocopy Machines," *J. Air Waste Manage. Assoc.*, 46, 821–829 (1996).

Levy, J. I., K. Lee, J. D. Spengler, and Y. Yanagisawa, "Impact of Residential Nitrogen Dioxide Exposure on Personal Exposure: An International Study," *J. Air Waste Manage. Assoc.*, 48, 553–560 (1998).

Lewis, C. W., and R. B. Zweidinger, "Apportionment of Residential Indoor Aerosol VOC and Aldehyde Species to Indoor and Outdoor Sources, and Their Source Strengths," *Atmos. Environ.*, 26A, 2179–2184 (1992).

Lewis, R. G., A. E. Bond, D. E. Johnson, and J. P. Hsu, "Measurement of Atmospheric Concentrations of Common Household Pesticides: A Pilot Study," *Environ. Monitor. Assess.*, 10, 59–73 (1988).

Lewtas, J., S. Goto, K. Williams, J. C. Chuang, B. A. Petersen, and N. K. Wilson, "The Mutagenicity of Indoor Air Particles in a Residential Pilot Field Study: Application and Evaluation of New Methodologies," *Atmos. Environ.*, 21, 443–449 (1987).

Ligocki, M. P., H. I. H. Liu, G. R. Cass, and W. John, "Measurements of Particle Deposition Rates inside Southern California Museums," *Aerosol Sci. Technol.*, 13, 85–101 (1990).

Ligocki, M. P., L. G. Salmon, T. Fall, M. C. Jones, W. W. Nazaroff, and G. R. Cass, "Characteristics of Airborne Particles inside Southern California Museums," *Atmos. Environ.*, 27A, 697–711 (1993).

Liu, K.-S., S. B. Hayward, J. R. Girman, B. A. Moed, and F.-Y. Huang, "Annual Average Radon Concentrations in California Residences," *J. Air Waste Manage. Assoc.*, 41, 1207–1212 (1991).

Liu, L.-J. S., P. Koutrakis, H. H. Suh, J. D. Mulik, and R. M. Burton, "Use of Personal Measurements for Ozone Exposure Assessment: A Pilot Study," *Environ. Health Perspect.*, 101, 318–324 (1993).

Löfroth, G., P. I. Ling, and E. Agurell, "Public Exposure to Environmental Tobacco Smoke," *Mutat. Res.*, 202, 103–110 (1988).

Löfroth, G., C. Stensman, and M. Brandhorst-Satzkorn, "Indoor Sources of Mutagenic Aerosol Particulate Matter: Smoking, Cooking, and Incense Burning," *Mutat. Res.*, 261, 21–28 (1991).

Marbury, M. C., D. P. Harlos, J. M. Samet, and J. D. Spengler, "Indoor Residential NO_2 Concentrations in Albuquerque, New Mexico," *JAPCA*, 38, 392–398 (1988).

McKone, T. E., "Human Exposure to Volatile Organic Compounds in Household Tap Water: The Indoor Inhalation Pathway," *Environ. Sci. Technol.*, 21, 1194–1201 (1987).

Mendis, S., P. A. Sobotka, and D. E. Evler, "Pentane and Isoprene in Expired Air from Humans—Gas Chromatographic Analysis of a Single Breath," *Clin. Chem.*, 40, 1485–1488 (1994).

Meyer, B., and K. Hermanns, "Formaldehyde Release from Wood Products. An Overview," in *Formaldehyde Release from Wood Products*, ACS Symposium Series 316, Chapter 1, pp. 1–16, Am. Chem. Soc., Washington, D.C., 1986.

Meyer, B., B. A. Kottes Andrews, and R. M. Reinhardt, Eds., "Formaldehyde Release from Wood Products," in *ACS Symposium Series 316*, Am. Chem. Soc., Washington, DC, 1986.

Mitra, S., and B. Ray, "Patterns and Sources of Polycyclic Aromatic Hydrocarbons and Their Derivatives in Indoor Air," *Atmos. Environ.*, 29, 3345–3356 (1995).

Morrison, G. C., W. W. Nazaroff, J. Alejandro Cano-Ruiz, A. T. Hodgson, and M. P. Modera, "Indoor Air Quality Impacts of Ventilation Ducts: Ozone Removal and Emissions of Volatile Organic Compounds," *J. Air Waste Manage. Assoc.*, 48, 941–952 (1998).

Moschandreas, D. J., and D. S. O'Dea, "Measurement of Perchloroethylene Indoor Air Levels Caused by Fugitive Emissions from Unvented Dry-to-Dry Dry Cleaning Units," *J. Air Waste Manage. Assoc.*, 45, 111–115 (1995).

Mose, D. G., and G. W. Mushrush, "Variable Spacial and Seasonal Hazards of Airborne Radon," *Atmos. Environ.*, 31, 3523–3530 (1997).

Moseley, C. L., and M. R. Meyer, "Petroleum Contamination of an Elementary School: A Case History Involving Air, Soil-Gas, and Groundwater Monitoring," *Environ. Sci. Technol.*, 26, 185–192 (1992).

Mumford, J. L., X. Z. He, R. S. Chapman, S. R. Cao, D. B. Harris, X. M. Li, Y. L. Xian, W. Z. Jiang, C. W. Xu, J. C. Chuang, W. E. Wilson, and M. Cooke, "Lung Cancer and Indoor Air Pollution in Xuan Wei, China," *Science*, 235, 217–220 (1987).

Mumford, J. L., R. W. Williams, D. B. Walsh, R. M. Burton, D. J. Svendsgaard, J. C. Chuang, V. S. Houk, and J. Lewtas, "Indoor Air Pollutants from Unvented Kerosene Heater Emissions in

Mobile Homes: Studies on Particles, Semivolatile Organics, Carbon Monoxide, and Mutagenicity," *Environ. Sci. Technol., 25,* 1732–1738 (1991).

Naik, D. V., C. J. Weschler, and H. C. Shields, "Indoor and Outdoor Concentrations of Organic Compounds Associated with Airborne Particles: Results Using a Novel Solvent System," in *Indoor Air Pollution—Radon, Bioaerosols, & VOC's* (J. G. Kay, G. E. Keller, and J. F. Miller, Eds.), Chap. 6, pp. 59–70, Lewis Publishers, Chelsea, MI, 1991.

National Research Council, "Health Risks of Radon and Other Internally Deposited Alpha-Emitters," Committee on the Biological Effects of Ionizing Radiations, Commission on Life Sciences, National Academy Press, Washington, D.C., 1988.

National Research Council, *Comparative Dosimetry of Radon in Mines and Homes,* Panel on Dosimetric Assumption Affecting the Applications of Radon Risk Estimates, Board on Radiation Effects Research, Commission on Life Sciences, National Academy Press, Washington, D.C., 1991.

Nazaroff, W. W., and G. R. Cass, "Mathematical Modeling of Chemically Reactive Pollutants in Indoor Air," *Environ. Sci. Technol., 20,* 924–934 (1986).

Nazaroff, W. W., and A. V. Nero, Jr., *Radon and Its Decay Products in Indoor Air,* Wiley-Interscience, New York, 1988.

Nazaroff, W. W., and G. R. Cass, "Mathematical Modeling of Indoor Aerosol Dynamics," *Environ. Sci. Technol., 23,* 157–166 (1989a).

Nazaroff, W. W., and G. R. Cass, "Mass-Transport Aspects of Pollutant Removal at Indoor Surfaces," *Environ. Int., 15,* 567–584 (1989b).

Nazaroff, W. W., L. G. Salmon, and G. R. Cass, "Concentration and Fate of Airborne Particles in Museums," *Environ. Sci. Technol., 24,* 66–77 (1990a).

Nazaroff, W. W., M. P. Ligocki, T. Ma, and G. R. Cass, "Particle Deposition in Museums: Comparison of Modeling and Measurement Results," *Aerosol Sci. Technol., 13,* 332–348 (1990b).

Nazaroff, W. W., and K. Teichman, "Indoor Radon," *Environ. Sci. Technol., 24,* 774–782 (1990).

Neas, L. M., D. W. Dockery, J. H. Ware, J. D. Spengler, B. G. Ferris, Jr., and F. E. Speizer, "Concentration of Indoor Particulate Matter as a Determinant of Respiratory Health in Children," *Am. J. Epidemiol., 139,* 1088–1099 (1994).

Nelson, P. R., S. P. Kelly, and F. W. Conrad, "Studies of Environmental Tobacco Smoke Generated by Different Cigarettes," *J. Air Waste Manage. Assoc., 47,* 336–344 (1998).

Nero, A. V., M. B. Schwehr, W. W. Nazaroff, and K. L. Revzan, "Distribution of Airborne Radon-222 Concentrations in U.S. Homes," *Science, 234,* 992–997 (1986).

Nero, A., "Earth, Air, Radon, and Home," *Phys. Today,* 32–39, April (1989).

Nero, A. V., Jr., "Controlling Indoor Air Pollution," *Sci. Am., 258,* 42–48 (1988).

Offermann, F. J., S. A. Loiselle, A. T. Hodgson, L. A. Gundel, and J. M. Daisey, "A Pilot Study to Measure Indoor Concentrations and Emission Rates of Polycyclic Aromatic Hydrocarbons," *Indoor Air, 4,* 497–512 (1991).

Osborne, M. C., "Four Common Diagnostic Problems That Inhibit Radon Mitigation," *JAPCA, 37,* 604–606 (1987).

Ott, W., P. Switzer, and N. Willits, "Carbon Monoxide Exposures inside an Automobile Traveling on an Urban Arterial Highway," *J. Air Waste Manage. Assoc., 44,* 1010–1018 (1994).

Ott, W., P. Switzer, and J. Robinson, "Particle Concentrations inside a Tavern before and after Prohibition of Smoking: Evaluating the Performance of an Indoor Air Quality Model," *J. Air Waste Manage. Assoc., 46,* 1120–1134 (1996).

Özkaynak, H., P. B. Ryan, G. A. Allen, and W. A. Turner, "Indoor Air Quality Modeling: Compartmental Approach with Reactive Chemistry," *Environ. Int., 8,* 461–471 (1982).

Özkaynak, H., J. Xue, J. D. Spengler, L. A. Wallace, E. D. Pellizzari, and P. Jenkins, "Personal Exposure to Airborne Particles and Metals: Results from the Particle TEAM Study in Riverside, CA," *J. Exposure Anal. Environ. Epidemiol., 6,* 57–78 (1996).

Pellizzari, E. D., K. W. Thomas, C. A. Clayton, R. W. Whitmore, R. C. Shores, H. S. Zelon, and R. L. Perritt, "Particle Total Exposure Assessment Methodolgy (PTEAM): Riverside, California Pilot Study," Vol. 1, NTIS #PB 93-166957/AS, National Technical Information Service, Springfield, VA, 1993.

Pennanen, A. S., R. O. Salonen, S. Alm, M. J. Jantunen, and P. Pasanen, "Characterization of Air Quality Problems in Five Finnish Indoor Ice Arenas," *J. Air Waste Manage. Assoc., 47,* 1079–1086 (1997).

Perry, R., and I. L. Gee, "Vehicle Emissions and Effects on Air Quality: Indoors and Outdoors," *Indoor Environ., 3,* 224–236 (1994).

Peto, J., and S. Darby, "Radon Risk Reassessed," *Nature, 368,* 97–98 (1994).

Petreas, M., K.-S. Liu, B.-H. Chang, S. B. Hayward, and K. Sexton, "A Survey of Nitrogen Dioxide Levels Measured inside Mobile Homes," *JAPCA, 38,* 647–651 (1988).

Phillips, M., J. Greenberg, and J. Awad, "Metabolic and Environmental Origins of Volatile Organic Compounds in Breath," *J. Clin. Pathol., 47,* 1052–1053 (1994).

Pitts, J. N., Jr., T. J. Wallington, H. W. Biermann, and A. M. Winer, "Identification and Measurement of Nitrous Acid in an Indoor Environment," *Atmos. Environ., 19,* 763–767 (1985).

Pitts, J. N., Jr., H. W. Biermann, E. C. Tuazon, M. Green, W. D. Long, and A. M. Winer, "Time-Resolved Identification and Measurement of Indoor Air Pollutants by Spectroscopic Techniques: Gaseous Nitrous Acid, Methanol, Formaldehyde, and Formic Acid," *JAPCA, 39,* 1344–1347 (1989).

Quackenboss, J. J., J. D. Spengler, M. S. Kanarek, R. Letz, and C. P. Duffy, "Personal Exposure to Nitrogen Dioxide: Relationship to Indoor/Outdoor Air Quality and Activity Patterns," *Environ. Sci. Technol., 20,* 775–783 (1986).

Quackenboss, J. J., M. D. Lebowitz, and C. D. Crutchfield, "Indoor–Outdoor Relationships for Particulate Matter: Exposure Classifications and Health Effects," *Environ. Int., 15,* 353–360 (1989).

Reiss, R., P. B. Ryan, and P. Koutrakis, "Modeling Ozone Deposition onto Indoor Residential Surfaces," *Environ. Sci. Technol., 28,* 504–513 (1994).

Reiss, R., P. B. Ryan, S. J. Tibbetts, and P. Koutrakis, "Measurement of Organic Acids, Aldehydes, and Ketones in Residential Environments and Their Relation to Ozone," *J. Air Waste Manage. Assoc., 45,* 811–822 (1995a).

Reiss, R., P. B. Ryan, P. Koutrakis, and S. J. Tibbets, "Ozone Reactive Chemistry on Interior Latex Paint," *Environ. Sci. Technol., 29,* 1906–1912 (1995b).

Rodes, C. E., R. M. Kamens, and R. W. Wiener, "The Significance and Characteristics of the Personal Activity Cloud on Exposure Assessment Measurements for Indoor Contaminants," *Indoor Air, 2,* 123–145 (1991).

Roinestad, K. S., J. B. Louis, and J. D. Rosen, "Determination of Pesticides in Indoor Air and Dust," *J. AOAC Int., 76,* 1121–1126 (1993).

Romieu, I., M. C. Lugo, S. Colome, A. M. Garcia, M. H. Avila, A. Geyh, S. R. Velasco, and E. P. Rendon, "Evaluation of Indoor Ozone Concentration and Predictors of Indoor–Outdoor Ratio in Mexico City," *J. Air Waste Manage. Assoc., 48,* 327–335 (1998).

Rowe, D. R., K. H. Al-Dhowalia, and M. E. Mansour, "Indoor–Outdoor Carbon Monoxide Concentrations at Four Sites in Riyadh, Saudi Arabia," *J. Air Waste Manage. Assoc., 39,* 1100–1102 (1989).

Rowe, D. R., K. H. Al-Dhowalia, and M. E. Mansour, "Indoor–Outdoor Nitric Oxide and Nitrogen Dioxide Concentrations at Three Sites in Riyadh, Saudi Arabia," *J. Air Waste Manage. Assoc., 41,* 973–976 (1991).

Ryan, P. B., M. L. Soczek, J. D. Spengler, and I. H. Billick, "The Boston Residential NO_2 Characterization Study: I. Preliminary Evaluation of the Survey Methodology," *JAPCA, 38,* 22–27 (1988).

Salmon, L. G., W. W. Nazaroff, M. P. Ligocki, M. C. Jones, and G. R. Cass, "Nitric Acid Concentrations in Southern California Museums," *Environ. Sci. Technol., 24,* 1004–1012 (1990).

Salmon, L. G., C. S. Christoforou, and G. R. Cass, "Airborne Pollutants in the Buddhist Cave Temples at the Yungang Grottoes, China," *Environ. Sci. Technol., 28,* 805–811 (1994).

Salmon, L. G., C. S. Christoforou, T. J. Gerk, G. R. Cass, G. S. Casuccio, G. A. Cooke, M. Leger, and I. Olmez, "Source Contributions to Airborne Particle Deposition at the Yungang Grottoes, China," *Sci. Total Environ., 167,* 33–47 (1995).

Salthammer, T., A. Schwarz, and F. Fuhrmann, "Emission of Reactive Compounds and Secondary Products from Wood-Based Furniture Coating," *Atmos. Environ., 33,* 75–84 (1999).

Schaeffer, V. H., B. Bhooshan, S.-B. Chen, J. S. Sonenthal, and A. T. Hodgson, "Characterization of Volatile Organic Chemical Emissions from Carpet Cushions," *J. Air Waste Manage. Assoc., 46,* 813–820 (1996).

Sexton, K., M. X. Petreas, and K.-S. Liu, "Formaldehyde Exposures inside Mobile Homes," *Environ. Sci. Technol., 23,* 985–988 (1989).

Shah, J. J., and H. B. Singh, "Distribution of Volatile Organic Chemicals in Outdoor and Indoor Air," *Environ. Sci. Technol., 22,* 1381–1388 (1988).

Sheldon, L. S., T. D. Hartwell, B. G. Cox, J. E. Sickles II, E. D. Pellizzari, M. L. Smith, R. L. Perritt, and S. M. Jones, "An Investigation of Infiltration and Indoor Air Quality," Final Report, NY State ERDA Contract No. 736-CON-BCS-85, New York State Energy Research and Development Authority, Albany, NY (1989).

Shepherd, J. L., R. L. Corsi, and J. Kemp, "Chloroform in Indoor Air and Wastewater: The Role of Residential Washing Machines," *J. Air Waste Manage. Assoc., 46,* 631–642 (1996).

Shields, H. C., and C. J. Weschler, "Volatile Organic Compounds Measured at a Telephone Switching Center from 5/30/85–12/6/88: A Detailed Case Study," *J. Air Waste Manage. Assoc., 42,* 792–804 (1992).

Shields, H. C., D. M. Fleischer, and C. J. Weschler, "Comparisons among VOCs Measured in Three Types of U.S. Commercial Buildings with Different Occupant Densities," *Indoor Air, 6,* 2–17 (1996).

Sinclair, J. D., L. A. Psota-Kelty, and C. J. Weschler, "Indoor/Outdoor Ratios and Indoor Surface Accumulations of Ionic Substances at Newark, New Jersey," *Atmos. Environ., 22,* 461–469 (1988).

Sinclair, J. D., L. A. Psota-Kelty, C. J. Weschler, and H. C. Shields, "Deposition of Airborne Sulfate, Nitrate, and Chloride Salts as It Relates to Corrosion of Electronics," *J. Electrochem. Soc., 137,* 1200–1206 (1990a).

Sinclair, J. D., L. A. Psota-Kelty, C. J. Weschler, and H. C. Shields, "Measurement and Modeling of Airborne Concentrations and Indoor Surface Accumulation Rates of Ionic Substances at Neenah, Wisconsin," *Atmos. Environ., 24A,* 627–638 (1990b).

Sollinger, S., K. Levsen, and G. Wünsch, "Indoor Air Pollution by Organic Emissions from Textile Floor Coverings. Climate Chamber Studies under Dynamic Conditions," *Atmos. Environ., 27B,* 183–192 (1993).

Sollinger, S., K. Levsen, and G. Wünsch, "Indoor Pollution by Organic Emissions from Textile Floor Coverings: Climate Test Chamber Studies under Static Conditions," *Atmos. Environ., 28,* 2369–2378 (1994).

Spengler, J. D., D. W. Dockery, W. A. Turner, J. M. Wolfson, and B. G. Ferris, Jr., "Long-Term Measurements of Respirable Sulfates and Particles inside and outside Homes," *Atmos. Environ., 15,* 23–30 (1981).

Spengler, J. D., M. Brauer, J. M. Samet, and W. E. Lambert, "Nitrous Acid in Albuquerque, New Mexico, Homes," *Environ. Sci. Technol., 27,* 841–845 (1993).

Spengler, J., M. Schwab, P. B. Ryan, S. Colome, A. L. Wilson, I. Billick, and E. Becker, "Personal Exposure to Nitrogen Dioxide in the Los Angeles Basin," *J. Air Waste Manage. Assoc., 44,* 39–47 (1994).

Spicer, C. W., R. W. Coutant, G. F. Ward, D. W. Joseph, A. J. Gaynor, and I. H. Billick, "Rates and Mechanisms of NO_2 Removal from Indoor Air Residential Materials," *Environ. Int., 15,* 643–654 (1989).

Spicer, C. W., D. V. Kenny, G. F. Ward, and I. H. Billick, "Transformations, Lifetimes, and Sources of NO_2, HONO, and HNO_3 in Indoor Environments," *J. Air Waste Manage. Assoc., 43,* 1479–1485 (1993).

Spicer, C. W., D. V. Kenny, G. F. Ward, I. H. Billick, and N. P. Leslie, "Evaluation of NO_2 Measurement Methods for Indoor Air Quality Applications," *J. Air Waste Manage. Assoc., 44,* 163–168 (1994).

Stock, T. H., "Formaldehyde Concentrations inside Conventional Housing," *JAPCA, 37,* 913–918 (1987).

Su, W.-H., "Indoor Air Pollution," *Resour., Conserv. Recycl., 16,* 77–91 (1996).

Suh, H. H., J. D. Spengler, and P. Koutrakis, "Personal Exposures to Acid Aerosols and Ammonia," *Environ. Sci. Technol., 26,* 2507–2517 (1992).

Tancrède, M., R. Wilson, L. Zeise, and E. A. C. Crouch, "The Carcinogenic Risk of Some Organic Vapors Indoors: A Theoretical Survey," *Atmos. Environ., 21,* 2187–2205 (1987).

Thatcher, T. L., and D. W. Layton, "Deposition, Resuspension, and Penetration of Particles within a Residence," *Atmos. Environ., 29,* 1487–1497 (1995).

Thompson, C. V., R. A. Jenkins, and C. E. Higgins, "A Thermal Desorption Method for the Determination of Nicotine in Indoor Environments," *Environ. Sci. Technol., 23,* 429–435 (1989).

Tichenor, B. A., and M. A. Mason, "Organic Emissions from Consumer Products and Building Materials to the Indoor Environment," *JAPCA, 38,* 264–268 (1988).

Tichenor, B. A., L. A. Sparks, J. B. White, and M. D. Jackson, "Evaluating Sources of Indoor Air Pollution," *J. Air Waste Manage. Assoc., 40,* 487–492 (1990).

Traynor, G. W., M. G. Apte, H. A. Sokil, J. C. Chuang, W. G. Tucker, and J. L. Mumford, "Selected Organic Pollutant Emissions from Unvented Kerosene Space Heaters," *Environ. Sci. Technol., 24,* 1265–1270 (1990).

Turk, B. H., D. T. Grimsrud, J. T. Brown, K. L. Geisling-Sobotka, J. Harrison, and R. J. Prill, "Commercial Building Ventilation Rates and Particle Concentrations," *ASHRAE Trans., 95,* 422–433 (1989).

Turk, B. H., R. J. Prill, D. T. Grimsrud, B. A. Moed, and R. G. Sextro, "Characterizing the Occurrence, Sources, and Variability of Radon in Pacific Northwest Homes," *J. Air Waste Manage. Assoc., 40,* 498–506 (1990).

Van Loy, M. D., V. C. Lee, L. A. Gundel, J. M. Daisey, R. G. Sextro, and W. W. Nazaroff, "Dynamic Behavior of Semivolatile Organic Compounds in Indoor Air. 1. Nicotine in a Stainless Steel Chamber," *Environ. Sci. Technol., 31,* 2554–2561 (1997).

Van Loy, M. D., W. W. Nazaroff, and J. M. Daisey, "Nicotine as a Marker for Environmental Tobacco Smoke: Implications of Sorption on Indoor Surface Materials," *J. Air Waste Manage. Assoc., 48,* 959–968 (1998).

Vecera, Z., and P. K. Dasgupta, "Indoor Nitrous Acid Levels. Production of Nitrous Acid from Open-Flame Sources," *Int. J. Environ. Anal. Chem., 56,* 311–316 (1994).

Wade, W. A., III, W. A. Cote, and J. E. Yocom, "A Study of Indoor Air Quality," *JAPCA, 25,* 933–939 (1975).

Wallace, J. C., L. P. Brzuzy, S. L. Simonich, S. M. Visscher, and R. A. Hites, "Case Study of Organochlorine Pesticides in the Indoor Air of a Home," *Environ. Sci. Technol., 30,* 2715–2718 (1996).

Wallace, L., "Indoor Particles: A Review," *J. Air Waste Manage. Assoc., 46,* 98–126 (1996).

Weschler, C. J., H. C. Shields, and D. V. Naik, "Indoor Ozone Exposures," *JAPCA, 39,* 1562–1568 (1989).

Weschler, C. J., H. C. Shields, and D. Rainer, "Concentrations of Volatile Organic Compounds at a Building with Health and Comfort Complaints," *Am. Ind. Hyg. Assoc. J., 51,* 261–268 (1990).

Weschler, C. J., M. Brauer, and P. Koutrakis, "Indoor Ozone and Nitrogen Dioxide: A Potential Pathway to the Generation of Nitrate Radicals, Dinitrogen Pentaoxide, and Nitric Acid Indoors," *Environ. Sci. Technol., 26,* 179–184 (1992a).

Weschler, C. J., A. T. Hodgson, and J. D. Wooley, "Indoor Chemistry: Ozone, Volatile Organic Compounds, and Carpets," *Environ. Sci. Technol., 26,* 2371–2377 (1992b).

Weschler, C. J., H. C. Shields, and D. V. Naik, "Indoor Chemistry Involving O_3, NO, and NO_2 as Evidence by 14 Months of Measurements at a Site in Southern California," *Environ. Sci. Technol., 28,* 2120–2132 (1994).

Weschler, C. J., and H. C. Shields, "Chemical Transformations of Indoor Air Pollutants," in *Indoor Air '96* (S. Yoshizawa, K. Kimura, K. Ikeda, S. Tanabe, and T. Iwata, Eds.), Organizing Committee of the 7th International Conference of Indoor Air Quality and Climate, Vol. 1, pp. 919–924, Nagoya, Japan, 1996a.

Weschler, C. J., and H. C. Shields, "Production of the Hydroxyl Radical in Indoor Air," *Environ. Sci. Technol., 30,* 3250–3258 (1996b).

Weschler, C. J., H. C. Shields, and B. M. Shah, "Understanding and Reducing the Indoor Concentration of Submicron Particles at a Commercial Building in Southern California," *J. Air Waste Manage. Assoc., 46,* 291–299 (1996).

Weschler, C. J., and H. C. Shields, "Measurements of the Hydroxyl Radical in a Manipulated but Realistic Indoor Environment," *Environ. Sci. Technol., 31,* 3719–3722 (1997a).

Weschler, C. J., and H. C. Shields, "Potential Reactions among Indoor Pollutants," *Atmos. Environ., 31,* 3487–3495 (1997b).

Weschler, C. J., and H. C. Shields, "Indoor Ozone/Terpene Reactions as a Source of Indoor Particles," *Atmos. Environ., 33,* 2307–2318 (1999).

Whitmore, R. W., F. W. Immerman, D. E. Camann, A. E. Bond, R. G. Lewis, and J. L. Schaum, "Non-Occupational Exposures to Pesticides for Residents of Two U.S. Cities," *Arch. Environ. Contam. Toxicol., 26,* 47–59 (1994).

Wolkoff, P., P. A. Clausen, B. Jensen, G. D. Nielsen, and C. I. Wilkins, "Are We Measuring the Relevant Indoor Pollutants?" *Indoor Air, 7,* 92–106 (1997).

Wolkoff, P., "Impact of Air Velocity, Temperature, Humidity, and Air on Long-Term VOC Emissions from Building Products," *Atmos. Environ., 32,* 2659–2668 (1998).

Wolkoff, P., "Photocopiers and Indoor Air Pollution," *Atmos. Environ., 33,* 2129–2130 (1999).

Zhang, J., and P. J. Lioy, "Ozone in Residential Air: Concentrations, I/O Ratios, Indoor Chemistry, and Exposures," *Indoor Air, 4,* 95–105 (1994).

Zhang, J., Q. He, and P. J. Lioy, "Characteristics of Aldehydes: Concentrations, Sources, and Exposures for Indoor and Outdoor Residential Microenvironments," *Environ. Sci. Technol., 28,* 146–152 (1994a).

Zhang, J., W. E. Wilson, and P. J. Lioy, "Indoor Air Chemistry: Formation of Organic Acids and Aldehydes," *Environ. Sci. Technol., 28,* 1975–1982 (1994b).

Zhang, J., W. E. Wilson, and P. J. Lioy, "Sources of Organic Acids in Indoor Air: A Field Study," *J. Exposure Anal. Environ. Epidemiol., 4,* 25–47 (1994c).

CHAPTER 16

Applications of Atmospheric Chemistry
Air Pollution Control Strategies and Risk Assessments for Tropospheric Ozone and Associated Photochemical Oxidants, Acids, Particles, and Hazardous Air Pollutants

Understanding the chemical and physical processes discussed throughout this book is key to the development of cost-effective and health-protective air pollution control strategies. Application of atmospheric chemistry to reducing stratospheric ozone depletion was discussed in Chapter 13. Here we focus on its key role in strategies for controlling tropospheric pollutants, including ozone, acids, particles, and hazardous air pollutants.

A number of species have been designated "hazardous air pollutants" (HAPs) or toxic air contaminants (TACs). Most are directly emitted into the air, but some also have significant secondary sources, i.e., are formed by chemical reactions in air. Furthermore, the ultimate health impacts are determined not only by the emissions and formation of such compounds in air but also by their atmospheric fates. In short, some pollutants react in air to form less toxic species, whereas others form more toxic compounds. Thus, scientific risk assessments of these pollutants require an accurate and complete understanding of their atmospheric chemistry. Some specific examples are discussed in this chapter.

As we have seen, ozone and other photochemical oxidants, acid deposition, particles, and hazardous air pollutants are all intimately linked. Hence, we first treat VOC–NO_x systems in some detail and then the application to control of acids and particles. Finally, we discuss the application to risk assessments of hazardous air pollutants.

A. TROPOSPHERIC OZONE AND ASSOCIATED PHOTOCHEMICAL OXIDANTS

Since Haagen-Smit and co-workers established in the late 1950s that the key ingredients in the formation of tropospheric ozone are organics, NO_x, and sunlight, there has been a great deal of research devoted to developing and applying effective control strategies for ozone and associated photochemical oxidants. Their chemistry is very complex, sufficiently so that accurately predicting the impacts of controlling the precursor VOC and NO_x on ozone and other photochemically derived species at a particular location and time is difficult. However, as we shall see in this chapter, research over the past four decades has led to the successful implementation of a number of effective control strategies.

Developing control strategies for ozone is very different than for relatively unreactive species such as CO. In the latter case, the concentrations in air are a direct result of the emissions, and all things being equal, a reduction in emissions is expected to bring about an approximately proportional reduction in concentrations in ambient air. However, because O_3 is formed by chemical reactions in air, it does not necessarily respond in a proportional manner to reductions in the precursor emissions. Indeed, as we shall see, one can predict, using urban airshed or simple box models, that under some conditions, ozone levels *at a particular location* may actually increase when NO_x is decreased! This complexity has resulted in many years of discussions about the best control strategies and the issues involved in developing them. For a detailed treatment, the reader is referred to the National Academy of Sciences 1991 document "Rethinking the Ozone Problem in Urban and Regional Air Pollution" and to the U.S. Environmental Protection Agency 1993 document "The Role of Ozone Precursors in Tropospheric Ozone Formation and Control." In addition, there is a major cooperative research effort, NARSTO (*N*orth *A*merican *R*esearch *S*trategies for *T*ropospheric *O*zone), by the United States, Canada, and Mexico to address the scientific understanding of ozone formation and transport in the troposphere and the implications for control strategy development in these three countries. Critical review papers prepared as part of this assessment will be published in *Atmospheric Environment* in 1999; the NARSTO Web site is listed in Appendix IV.

A number of different approaches have been taken to understanding the VOC–NO_x chemistry and its application to control strategy developments. These include the use of environmental chambers, models ranging from simple linear rollback to complex Eulerian models, and field studies.

1. Environmental Chambers

Perhaps the most direct experimental means of examining the relationship between emissions and air quality is to simulate atmospheric conditions using large chambers. Measured concentrations of the primary pollutants are injected into these environmental (or *smog*) chambers, as they are called. These are then irradiated with sunlight or lamps used to mimic the sun, and the time–concentration profiles of the primary pollutants as well as the resulting secondary pollutants are measured. The primary pollutant concentrations as well as temperature, relative humidity, and so on can be systematically varied to establish the relationship between emissions and air quality, free from the complexities of continuously injected pollutant emissions and meteorology, both of which complicate the interpretation of ambient air data.

The results from such chamber studies are frequently used to test the chemical portion of various computer models for photochemical air pollution in order to provide a scientific basis for control strategies. While the interpretation of the results of smog chamber studies and their extrapolation to atmospheric conditions also have some limitations (vide infra), such studies do provide a highly useful means of initially examining the emissions–air quality relationship under controlled conditions.

In addition, smog chambers are useful as large reaction vessels to generate kinetic and mechanistic data on individual reactions believed to be important in the atmosphere. They have also been used extensively in the past as exposure chambers to study the effects of air pollution on people, animals, and plants.

a. Types of Chambers

Design criteria for these chambers are based on reproducing as faithfully as possible conditions in ambient air, excluding meteorology and the uncontrolled addition of pollutants. Although the general aims of all chamber studies are similar (i.e., to simulate reactions in ambient air under controlled conditions), the chamber designs and capabilities used to meet this goal vary widely. Thus chambers can differ in any or all of the following characteristics: (1) size and shape, (2) surface materials to which the pollutants are exposed, (3) range of pressures and temperatures that can be attained, (4) methods of preparation of reactants, including "clean air," (5) conditions (i.e., static or flow) under which experiments can be carried out, (6) analytical capabilities, and (7) spectral characteristics of the light source.

In discussing these chambers, we need to keep in mind one major caveat. The use of chambers by necessity involves the presence of surfaces in the form of the chamber walls, and this is the single largest uncertainty in using them as a surrogate for ambient air studies. Contributing to this uncertainty are possible unknown heterogeneous reactions occurring on both fresh (i.e., "clean") and conditioned chamber surfaces. Additionally, the outgassing of uncharacterized reactive vapors either deposited there during previous experiments or released from the plastic films used to make the chambers (e.g., nonpolymerized organics) can have pronounced effects in certain reaction systems, especially kinetic studies of low-reactivity organics (e.g., see Lonneman *et al.*, 1981; and Joshi *et al.*, 1982). Another, less severe, problem is reproducing the actinic radiation to which pollutants are exposed in ambient atmospheres. Finally, variations in the kinetics in the initial reaction stages due to inhomogeneities in the reactant concentrations during the mixing should be taken into account (Ibrahim *et al.*, 1987). The design of environmental chamber facilities attempts insofar as possible to minimize these variations from "real" air masses.

A general discussion of the nature and importance of these chamber characteristics, including "wall effects," follows. For detailed descriptions of various types of smog chamber facilities and their operation, one should consult the original literature, including, for example, indoor studies utilizing borosilicate glass cylinders (Joshi *et al.*, 1982; Behnke *et al.*, 1988), chambers made from Teflon (FEP) film with volumes up to

~250 m³ (Fitz *et al.*, 1981; Spicer, 1983; Evans *et al.*, 1986; Mentel *et al.*, 1996), and two similar 6-m³ evacuable and thermostated chambers (Darnall *et al.*, 1976; Winer *et al.*, 1980; Akimoto *et al.*, 1979). Fluorocarbon-film chambers range in size from 0.015–0.04 m³ (Lonneman *et al.*, 1981) to 200–2000 L (Kelly, 1982; Kelly *et al.*, 1985; Evans *et al.*, 1986) to large outdoor chambers with capacities of 25–200 m³ (Jeffries *et al.*, 1976; Kamens *et al.*, 1988; Leone *et al.*, 1985; Wängberg *et al.*, 1997).

(1) Glass reactors Many studies have been carried out in borosilicate glass reactors similar to those used in typical laboratory studies of gas-phase reactions. These are usually relatively small, a few liters up to approximately 100 L (0.1 m³). However, Doussin *et al.* (1997) have developed a borosilicate glass chamber with a volume of 977 L by using four cylinders held together by flanges.

While glass reactors are convenient, inexpensive, and readily available, there are some problems associated with their use. For example, Pyrex glass absorbs light at wavelengths ≤350 nm (the cutoff depends on the thickness of the glass, Fig. 16.10), which we have seen is a critical region for atmospheric photochemistry. Thus unless separate windows are included, species such as O_3 and HCHO which produce the free radicals OH and HO_2 are exposed to less short wavelength (e.g., 290–330 nm) radiation than is present under atmospheric conditions and hence undergo less photolysis.

In addition, such small vessels have high surface-to-volume (S/V) ratios, which may increase the relative contributions of reactions that occur on the surface. One such heterogeneous reaction is the decomposition on surfaces of O_3 to form O_2; obviously, the faster this decomposition, the lower the concentrations of O_3 that will be observed during the chamber run.

A more important reaction that may occur on surfaces in smog chambers is one generating HONO:

$$2NO_2 + H_2O \xrightarrow{\text{surface}} HONO + HNO_3.$$

As discussed in Sections B.3 and C of Chapter 7, this reaction has been shown to be too slow in aqueous solution to be significant in the atmosphere; it is faster on surfaces and has been proposed as a source of HONO in smog chambers (e.g., see Sakamaki *et al.*, 1983; Pitts *et al.*, 1984; Leone *et al.*, 1985; and Chapter 7.C). Since HONO is a major OH source in the early stages of irradiation in smog chambers, it is important to understand the mechanism of its formation and to quantify its rate of production under various experimental conditions. Thus, if this reaction only occurs at a significant rate on the surfaces typically encountered in chambers, the effects it has on the overall reactions should be removed when the results of chamber studies are extrapolated to ambient air. In the case of HONO generation, studies carried out in a mobile laboratory where the surfaces were very different than in environmental chambers also showed HONO formation (Pitts *et al.*, 1985), suggesting that this reaction may not be unique to smog chamber surfaces. Furthermore, as discussed in Chapter 7, there is evidence from field studies that HONO is formed from NO_x heterogeneous reactions on other surfaces as well (e.g., see review by Lammel and Cape, 1996).

The belief generally has been that the smaller the S/V ratio (i.e., the larger the smog chamber), the less important such surface reactions will be, and hence the more representative of the ambient atmosphere the results. While there is doubtless some justification for this approach, it must also be kept in mind that there are a variety of surfaces present in real atmospheres as well. These include not only the surfaces of the earth, buildings, and so on but also the surfaces of particulate matter suspended in air (Chapter 9). If the heterogeneous formation of HONO occurs not only on chamber surfaces but also on those found in urban atmospheres as well, then it is important to include it in extrapolating the chamber results to ambient air. In this case, the effects on the kinetics due to the different types and available amounts of surfaces in air compared to chambers must, of course, be taken into account.

(2) Collapsible reaction chambers As a result of these problems, larger smog chambers with surfaces thought to be relatively inert have found increasing use. Thus conditioned FEP Teflon films, for example, have been shown to have relatively low rates of surface destruction of a variety of reactive species. In addition, they typically transmit solar radiation in the 290- to 800-nm region (Fig. 16.1) and have low rates of hydrocarbon offgassing.

Collapsible smog chambers are easily constructed using flexible thin films of this material. In addition to the low rates of destruction of reactive species (e.g., see McMurry and Grosjean, 1985; Grosjean, 1985; and Kuster and Goldan, 1987) and their transparency to actinic UV, they have the advantage that the size of the chamber can be easily varied. An additional advantage of such chambers is that they may be easily divided into two sections simply by putting a heavy divider (e.g., bar) across the middle of the bag. One can then use one side of the bag as a "control" and the other to study the effects of varying one parameter such as the injection of additional pollutants.

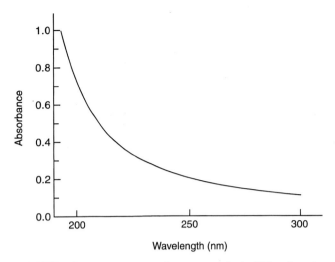

FIGURE 16.1 Typical absorption spectrum in the UV region of a 2-mil-thick Teflon film used to construct environmental chambers (spectrum taken by A. A. Ezell).

Figure 16.2 shows a schematic diagram of such a chamber (Fitz et al., 1981). Ports are included for the introduction of the primary pollutants and for sampling for product analysis. Because such bags do not have a rigid shape, they operate at atmospheric pressure. The volume of the chamber may be maintained during a run by introducing clean air at the same rate as sampling removes air from the chamber, thus diluting the mixture. Alternatively, these soft chambers can be allowed to collapse as air is removed for analysis; this maintains the pressure at 1 atm but results in an increasing S/V ratio during a run.

A potential problem with the use of these collapsible reaction chambers is contamination by outgassing of organics from the material used to fabricate the chambers. For example, the release of significant amounts of low molecular weight fluorocarbons from FEP Teflon has been reported (Lonneman et al., 1981). While this problem does not seem to occur with all samples (Kelly et al., 1985), investigators should clearly exercise caution with regard to possible contamination from this source.

Another possible problem is that the surfaces of Teflon films may be quite electrostatic. Transport to, and adsorption on, such surfaces can deplete the concentrations of particles in such chambers. For example, McMurry and Rader (1985) have shown that particles in the ~0.05- to 1.0-μm size range are removed in Teflon chambers primarily by electrostatic attraction to the Teflon surfaces. This leads to a large portion (~35–70%) of particles formed during smog chamber runs being deposited on the walls before the end of the experiment (McMurry and Grosjean, 1985). This problem can be minimized by the use of very large chambers, for example, the 256-m³ chamber at Jülich, Germany (Mentel et al., 1996), or the dual-chamber EUPHORE chambers at Valencia, Spain, each having a volume of ~200 m³ (Wängberg et al., 1997).

In any case, experience suggests that Teflon chambers first be conditioned by filling them with air containing O_3 and leaving them in the dark for several hours and/or filling them with "clean" air containing added NO_x and irradiating them for a period of time (Kelly, 1982). Such chamber "conditioning" and characterization studies appear to be an essential initial step in detailed chamber studies of the kinetics and mechanisms of atmospheric reactions.

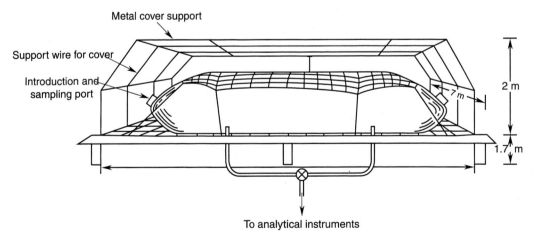

FIGURE 16.2 Schematic diagram of typical outdoor 40-m³ collapsible bag environmental chamber (adapted from Fitz et al., 1981).

(3) Evacuable chambers Ideally, one would like to be able to vary the pressure and temperature during environmental chamber runs in order to simulate various geographical locations, seasons, and meteorology and to establish the pressure and temperature dependencies of reactions. Varying the pressure and temperature also allows one to simulate the upper atmosphere (e.g., to study stratospheric and mesospheric chemistry).

Temperature control has an additional advantage with respect to the problem of chamber contamination. After a smog chamber has been used, some hydrocarbons and nitrogen compounds may remain adsorbed on the chamber walls. These may desorb in subsequent runs and, in some cases (e.g., HCHO), act as free radical sources to accelerate the photooxidation processes. The ability to "bake out" smog chambers while pumping to low pressures is therefore useful in reducing chamber contamination effects.

While glass reactors can be easily designed to include pressure and temperature control, they suffer from other limitations discussed earlier. In addition, the use of very large glass evacuable chambers at low pressures presents a potential safety problem. On the other hand, pressure and temperature are not easily controlled using collapsible reaction chambers.

As a result, some evacuable smog chambers have been designed and constructed to control both temperature and pressure. The one shown schematically in Fig. 16.3 is constructed with an aluminum alloy and the walls are coated with FEP Teflon. The end windows through which the mixture is irradiated are ultraviolet-grade quartz to allow transmission of the actinic UV. The radiation cutoff and spectral distribution in the 290- to 350-nm region can be varied using filters between the irradiation source and the chamber windows. The pumping system used to evacuate the chamber is hydrocarbon free; back-diffusion of organics from the use of conventional pump fluids can produce VOC concentrations that by themselves exceed the VOC air quality standard! In the particular case shown in Fig. 16.3, the pumping system consists of a liquid ring roughing pump (the fluid is water) with an air injection pump, cryosorption roughing pumps, a diffusion pump, and a mechanical pump. (Both the diffusion and mechanical pumps use unreactive perfluorinated oils rather than hydrocarbons as the working fluid.) Ports are included in the chamber for air and pollutant injection, for sampling (e.g., for GC or GC–MS analysis), and for *in situ* analysis using optical absorption spectroscopy (Darnall *et al.*, 1976; Winer *et al.*, 1980).

This type of chamber satisfies most design criteria in that both pressure and temperature can be varied, the intensity and spectrum of the irradiation can be altered, and the surface can be coated with a relatively inert material to minimize heterogeneous reactions and pollutant adsorption and offgassing. In addition, ports for both *in situ* spectroscopic product analysis and sampling can be easily included. The disadvantages are that they are relatively expensive, varying the S/V ratio through changing the dimensions of the chamber is not practical, and changing the nature of the chamber surface is difficult and time-consuming. Three such

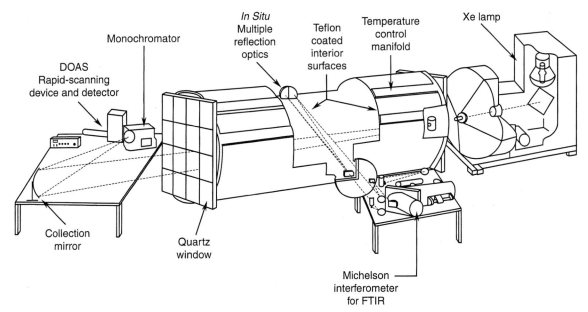

FIGURE 16.3 Schematic diagram of the evacuable chamber at the Air Pollution Research Center, University of California, Riverside.

chambers are described by Akimoto *et al.* (1979), Winer *et al.* (1980), and Doussin *et al.* (1997).

b. Preparation of Reactants, Including "Clean Air"

Different laboratories often use different sources and/or methods of preparation of the reactants NO_x, VOC, and "clean air." Frequently, the desired compounds can be purchased commercially. However, care must be taken when using commercially supplied materials to be sure that trace impurities, which can alter the experiment, are not present or are removed by purification before use. For example, gaseous HNO_3 at concentrations up to several percent is commonly present in commercially produced NO_2 that is stored in a gas cylinder. Similarly, small concentrations of alkenes are frequently present in commercial cylinders of the alkanes.

By far the largest component in smog chamber studies is air. It is especially imperative therefore that the air used to dilute the NO_x and VOC is "clean." Ambient air and commercially supplied air generally contain sufficient organic and NO_x impurities that extensive purification is needed to reduce these contaminants to low-ppb levels or below. One such purification system is described by Doyle *et al.* (1977); others are described in the papers on chamber facilities cited earlier.

c. Light Sources

(1) The sun At first glance, it might appear that the sun would be the ideal irradiation source for environmental chambers, and, indeed, it has been used successfully in outdoor chamber facilities (e.g., see Jeffries *et al.*, 1976; Leone *et al.*, 1985; and Kamens *et al.*, 1988). However, there are a number of practical problems. First, it requires either that the chamber be built outdoors or that the building housing the chamber have a suitable opening to admit the sunlight. Under these conditions, independent temperature control of the chamber becomes difficult. Second, the intensity of the sunlight can be altered by passing clouds in a manner that is difficult to measure and describe accurately. Third, experiments are limited to days of appropriate meteorology (e.g., rainy days are generally excluded). As an alternative, three types of lamps have been used to mimic irradiation from the sun. These are black fluorescent lamps, sunlamps, and xenon lamps.

(2) Black lamps A black lamp is a low-pressure mercury lamp whose envelope is covered with a phosphor such as strontium fluoroborate or barium disilicate. The type of phosphor determines the spectral distribution of the lamp output (Forbes *et al.*, 1976). Figure 16.4 shows a typical spectral distribution for $\lambda \leq 500$ nm from a black lamp as well as the solar

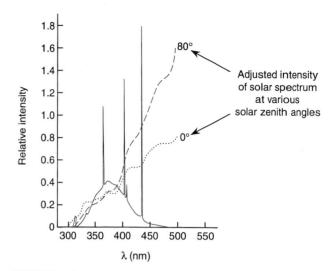

FIGURE 16.4 Relative spectral distributions for a typical black lamp and the solar spectrum at zenith angles of 0 and 80° normalized to the same NO_2 photolysis rate constant (adapted from Carter *et al.*, 1984).

spectrum at two zenith angles, where all curves have been normalized to the same NO_2 photolysis rate constant k_1 (Carter *et al.*, 1984). (See later for a discussion of k_1, which is a measure of the total light intensity.) Superimposed on the broad emission continuum from the phosphor are the low-pressure mercury lines at 313, 334, 365–366, 405–408, and 436 nm.

While such lamps provide good light intensity in the 340- to 400-nm region (and at the 313-nm mercury line) where important atmospheric photochemistry occurs, their spectral distribution is very different from that of the sun. Specifically, much of the intensity resides in the sharp mercury lines, and the output is poor in the critically important actinic UV region from 290 to 340 nm. In addition, the intensity falls off at $\lambda > 375$ nm, whereas the intensity of solar radiation is increasing significantly in this region. Such differences can significantly alter the photochemistry of important species such as O_3, NO_3, and HCHO, even if the lamp output is normalized to give the same NO_2 photolysis rate constant (k_1) as the sun.

(3) Sunlamps A sunlamp is similar to a black lamp, except that a different type of phosphor is used and the lamp envelope transmits UV. Figure 16.5 shows a typical spectral distribution from a commercial sunlamp. The wavelength corresponding to maximum power is shifted to lower wavelengths (~310 nm), compared to black lamps, and there is significant intensity down to ~270 nm. However, the intensity decreases rapidly above ~330 nm. The mercury lines can again be seen superimposed on the phosphor fluorescence.

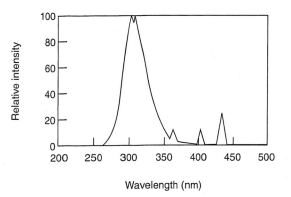

FIGURE 16.5 Typical spectral distribution from a sunlamp (reprinted with permission of North American Philips Lighting Corporation).

(4) Xenon lamps High-pressure xenon lamps provide the most faithful artificial reproduction of the solar energy distribution at the earth's surface in the wavelength region 290–700 nm. Figure 16.6 compares the output of an unfiltered xenon lamp to the zero air mass solar spectral irradiance (Winer *et al.*, 1979). Unlike black lamps, xenon lamps have substantial intensity in the critical region around 300 nm; the region ≤290 nm can be filtered out to match the solar energy distribution at the earth's surface using Pyrex of varying thickness (Fig. 16.10). In contrast to black lamps, the xenon lamp has maximum intensity at wavelengths above 400 nm.

The Xe lamp also has a series of peaks in the 800- to 1000-nm region that do not appear in sunlight. This relatively low-energy radiation does not cause significant photochemistry in the troposphere. However, if desired, the intensity of these peaks can be decreased with the use of appropriate filters.

(5) Measurement of light intensity In chamber studies, the spectral distribution of the irradiation source must be measured periodically (e.g., using a calibrated monochromator–photomultiplier combination) because the lamp and the windows in the chamber "age" (i.e., change with time). In addition to the spectral energy distribution of the lamp, the total absolute light intensity also must be measured. In particular, the intensity of the region below 430 nm where the most important photochemistry (e.g., of NO_2, O_3, and HCHO) occurs is of greatest interest. Both of these calibrations are tedious and must be carried out with care. However, they are sufficiently critical to data interpretation that they are carried out frequently. For example, measurement of the total absolute light intensity is typically carried out after every four or five runs, and in some cases where knowledge of the light intensity is essential, after every run.

The photolysis rate for NO_2 might be expected to be a good (although nonspecific) indicator of the intensity in the region <430 nm since it absorbs strongly (see Chapter 4.C) and is also one of the major photochemically active species in VOC–NO_x systems. Thus a standard procedure in smog chamber studies is to measure the rate of photolysis of NO_2 (k_1) as a relative measure of the total light intensity:

$$NO_2 + h\nu \xrightarrow{k_1} O(^3P) + NO. \qquad (1)$$

Determining k_1 is not as simple as measuring the loss of NO_2, however, since secondary reactions of the O and NO produced in (1) lead to nonexponential decays of NO_2. Thus plots of $\ln[NO_2]$ against irradiation time are observed to be curved. In early smog chamber studies, a parameter known as k_d was reported as a measure of the light intensity, where k_d was defined by

$$k_d = \left(\frac{-d\ln[NO_2]}{dt}\right)_{\text{limit } t \to 0}.$$

k_d was thus obtained experimentally by extrapolating the NO_2 concentration–time profile back to the beginning of the irradiation, $t = 0$.

Since k_1 is the fundamental parameter of interest, however, there has been emphasis on measuring and reporting k_1 rather than k_d in smog chamber studies. A procedure for determining k_1 from measured rates

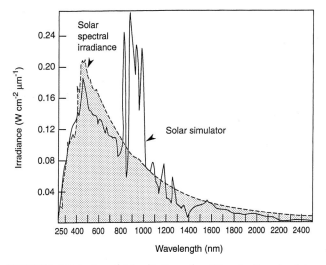

FIGURE 16.6 Spectral irradiation of unfiltered solar simulator compared to zero air mass solar spectral irradiance (adapted from Winer *et al.*, 1979).

of photolysis of NO_2 is described in detail by Holmes and co-workers (1973). In this procedure, NO_2 is photolyzed in the smog chamber. When O_2 is absent (i.e., in 1 atm of N_2), the reactions of interest are (1) and (2)–(7):

$$O + NO_2 \xrightarrow{k_2} NO + O_2, \quad (2)$$

$$O + NO_2 + M \xrightarrow{k_3} NO_3 + M, \quad (3)$$

$$O + NO + M \xrightarrow{k_4} NO_2 + M, \quad (4)$$

$$NO_3 + NO \rightarrow 2NO_2, \quad (5)$$

$$NO_3 + NO_2 \overset{M}{\leftrightarrow} N_2O_5, \quad (6,-6)$$

$$NO_3 + NO_2 \rightarrow NO + NO_2 + O_2. \quad (7)$$

With appropriate steady-state assumptions for O, NO_3, and N_2O_5, the kinetic expressions for these reactions can be solved to yield the following equation for k_1:

$$k_1 = \frac{1}{2t}\left[(1 + R_1 - R_2)\ln\frac{[NO_2]_0}{[NO_2]} + R_2\left(\frac{[NO_2]_0}{[NO_2]} - 1\right)\right],$$

where

$$R_1 = \frac{k_3[M]}{k_2} \quad \text{and} \quad R_2 = \frac{k_4[M]}{k_2}.$$

Thus, knowing the total pressure of N_2 and the rate constants k_2, k_3, and k_4, one can determine k_1 from the initial concentration of NO_2 and its loss with time.

The photolysis of NO_2 can also be carried out in the presence of O_2. However, in this case additional reactions (e.g., $O + O_2 + M \rightarrow O_3 + M$) must be considered and the kinetic expression for k_1 is more complex.

Of course, k_1 is not an absolute light intensity measurement per se but merely an indication of the intensity in one of the wavelength regions of interest for atmospheric chemistry. It has the advantage of being simple, convenient, and inexpensive, since only monitoring instruments for NO_2 are needed and these are generally standard components of the analytical apparatus. The disadvantage is that a number of photochemically active species (e.g., O_3 and HCHO) have absorption coefficients and wavelength dependencies different from NO_2 and, for these, k_1 will not necessarily be a good measurement of the rates of their photochemical reactions, depending on the spectral distribution of the light source.

Direct measurements of light intensity using radiometers as described in Chapter 3.2 can also be made.

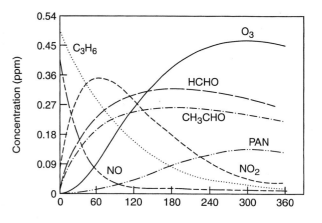

FIGURE 16.7 Typical primary and secondary pollutant profiles in a propene–NO_x irradiation in a smog chamber (adapted from Pitts et al., 1975).

d. Typical Time–Concentration Profiles of Irradiated VOC–NO_x–Air Mixtures

Figure 16.7 shows some typical concentration–time profiles for irradiation of a propene–NO mixture in the evacuable chamber of Fig. 16.3. The loss of the reactants, and the formation of the most commonly monitored secondary pollutants O_3, PAN, and the oxygenates HCHO and CH_3CHO are shown (Pitts et al., 1975).

With *in situ* spectroscopic techniques, critical data on the formation of such species as HONO, HNO_3, and NO_3, which are essential to understanding the chemistry of these systems, can also be obtained. Figure 16.8, for example, shows one portion of an FTIR spectrum obtained in a chamber run for a propene–NO_x

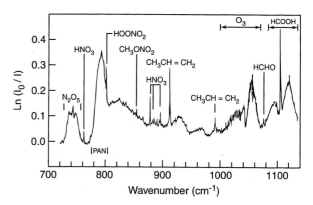

FIGURE 16.8 Infrared spectrum in the 700- to 1100-cm^{-1} region of a hydrocarbon–NO_x mixture irradiated for 139 min. Initial conditions were 10 ppm propene, 1 ppm *n*-butane, 1 ppm neopentane, and 5.4 ppm NO_x at 48°F. Path length, 85 m; resolution, 0.125 cm^{-1} (adapted from E. C. Tuazon, unpublished data, and Pitts et al., 1977).

mixture carried out at relatively high reactant concentrations and 48°F (Pitts *et al.*, 1977). A variety of species that are difficult to monitor with other techniques, such as HNO_3, N_2O_5, and peroxynitric acid (HO_2NO_2), are easily identified and measured by FTIR if their concentrations are sufficiently high.

Clearly, such experiments can be used to examine the relationship between primary emissions and the formation of a host of secondary pollutants. For example, runs can be carried out at varying initial concentrations of hydrocarbon and NO_x, and the effects on the formation of secondary pollutants such as O_3 studied. The reactivity of various hydrocarbons can be examined by studying them singly or in combination. In addition, such parameters as temperature, relative humidity and total pressure, presence of copollutants, and spectral distribution of the light source can be systematically varied.

One example of the use of chambers to study the effects of addition of copollutants is seen in Fig. 16.9 (Pitts *et al.*, 1976). As discussed earlier, one source of HO_2 free radicals in ambient air is the photolysis of formaldehyde:

$$HCHO + h\nu \rightarrow H + HCO,$$

$$H + O_2 \xrightarrow{M} HO_2,$$

$$HCO + O_2 \rightarrow HO_2 + CO.$$

One might anticipate that the addition of HCHO prior to irradiation would increase the rate of conversion of the reactants to secondary pollutants such as O_3 by providing an immediate source of the free radicals needed for the chain oxidations. As seen in Fig. 16.9, this is precisely what is observed.

Chambers also allow such parameters as the spectral distribution of the light source to be varied systematically. While this is impossible in ambient air studies, it is an important variable in order to simulate atmospheric chemistry at various altitudes. Since Pyrex glass absorbs radiation with $\lambda < 350$ nm (Calvert and Pitts, 1966), different thicknesses of Pyrex can be used to provide different amounts of filtering in this portion of the actinic UV. Figure 16.10, for example, shows the percentage transmission of two different thicknesses of glass (3/32 and 1/4 in.) in the 250- to 400-nm region (Winer *et al.*, 1979); as expected, the thicker the glass, the less transmitting it is below 350 nm. Since much of the important photochemistry (e.g., that of O_3 and HCHO) occurs in this region, one might expect the rates of free radical formation and hence the overall rates of product formation to be less for the most filtered case.

Figure 16.11 shows the formation of O_3 in a chamber as a function of time for a propene–*n*-butane–NO_x mixture irradiated using a xenon lamp (vide infra) with two glass filters of different thicknesses (Winer *et al.*, 1979). As expected, O_3 is formed more slowly and the peak concentration is lower for the most highly filtered light source.

In summary, chamber studies are a highly valuable experimental technique for studying atmospheric chemistry and the effects of varying parameters under controlled conditions.

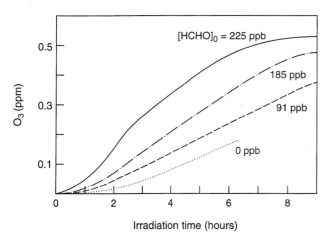

FIGURE 16.9 Effect of added HCHO on ozone formation in irradiations of a hydrocarbon–NO_x mixture. Average VOC was 2450 ppbC; average NO_x was 0.33 ppm (adapted from Pitts *et al.*, 1976).

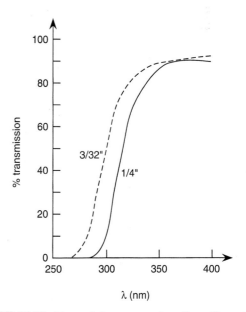

FIGURE 16.10 Transmission spectra of two Pyrex filters of thicknesses $\frac{3}{32}$ and $\frac{1}{4}$ in., respectively (adapted from Winer *et al.*, 1979).

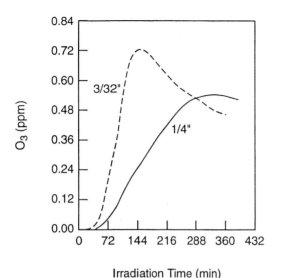

FIGURE 16.11 Ozone–time profiles from the irradiation of a hydrocarbon–NO_x mixture (initial concentrations were 0.4 ppm propene, 2.0 ppm n-butane, 0.4 ppm NO, and 0.1 ppm NO_2) in a smog chamber where the light source has been filtered with Pyrex of thickness $\frac{3}{32}$ or $\frac{1}{4}$ in., respectively (adapted from Winer et al., 1979).

e. Advantages and Limitations of Environmental Chambers

Clearly, environmental chamber studies are very useful tools in examining the chemical relationships between emissions and air quality and for carrying out related (e.g., exposure) studies. Use of these chambers has permitted the systematic variation of individual parameters under controlled conditions, unlike ambient air studies, where the continuous injection of pollutants and the effects of meteorology are often difficult to assess and to quantitatively incorporate into the data analysis. Chamber studies have also provided the basis for the validation of computer kinetic models. Finally, they have provided important kinetic and mechanistic information on some of the individual reactions occurring during photochemical smog formation.

However, as is true of any experimental system, there are limitations to chamber studies. One major problem has proven to be chamber contamination (e.g., Bufalini et al., 1977; Lonneman et al., 1981). For example, when clean air is irradiated in a smog chamber that has been in use, many of the characteristics of photochemical air pollution are observed. For example, nitrogen-containing compounds, and sometimes organics, are observed, and O_3 is formed. This is attributed to the adsorption of nitrogen compounds and organics on the walls of the smog chamber during a run, followed by desorption into the gas phase at a later time.

Indeed, offgassing of nitrogenous inorganics has been measured from the Teflon wall of the evacuable chamber shown in Fig. 16.3 (Carter, unpublished results, 1981). After a typical series of runs, the chamber was filled with pure air at ~5% relative humidity and air samples then removed for analysis. Offgassing of NO and a second nitrogen-containing species, probably HNO_3, was observed; HONO has also been observed. The rate of offgassing increased with temperature but fell substantially after an evacuated bake-out. Interestingly, in this particular case, no release of organics was observed, although it has been reported in other chambers.

Clearly, such adsorption–desorption processes on the surfaces of chambers potentially can have substantial effects on the observed levels of O_3 and other trace pollutants and on their rates of formation. While such effects can be minimized using bake-out while pumping if the chamber is evacuable, relatively few smog chambers have such capabilities at present. Even for evacuable chambers, contamination from adsorption on, and desorption from, the walls occurs. How to correct the results for this and reliably extrapolate the data to "real" atmospheres remains problematical.

A second concern has been indirect evidence pointing to the occurrence of chemical reactions on the walls of the chamber, followed by desorption of some of the products into the gas phase. These reactions manifest themselves by changing the observed gas-phase concentrations of one or more species in a manner that cannot be explained solely on the basis of known homogeneous gas-phase reactions.

For example, Fig. 16.12 shows the concentrations of

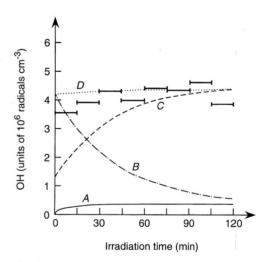

FIGURE 16.12 Concentrations of OH as a function of irradiation time during irradiation of a VOC–NO_x mixture in the evacuable smog chamber of Fig. 16.3. The initial conditions were as follows: $[NO]_0 = 0.5$ ppm, $[NO_2]_0 = 0.12$ ppm, $[C_3H_8]_0 = 0.013$ ppm, $[C_3H_6]_0 = 0.010$ ppm, $[HCHO]_0 = 0.02$ ppm, $T = 303$ K, RH = 50%, $k_1(NO_2$ photolysis$) = 0.49$ min^{-1}. Bars are experimental data; see text for description of curves A–D (adapted from Carter et al., 1982).

OH as a function of time when a VOC–NO_x mixture was irradiated in the evacuable smog chamber shown in Fig. 16.3 (Carter et al., 1981, 1982). The rates of decay of propane and propene were used to estimate the concentration of OH shown by the horizontal bars in Fig. 16.12 as a function of reaction time.

In this system the concentrations of propane, propene, and formaldehyde were kept sufficiently small that the predominant reactions were those of the NO_x species. By keeping the NO concentrations high, O_3 concentrations could be kept low because reaction (8) is fast:

$$NO + O_3 \rightarrow NO_2 + O_2. \quad (8)$$

Under these conditions, the foregoing reactions (1)–(8) together with reactions (9)–(13) are the major reactions:

$$O + O_2 + M \rightarrow O_3 + M, \quad (9)$$
$$2NO + O_2 \rightarrow 2NO_2, \quad (10)$$
$$OH + NO + M \rightarrow HONO + M, \quad (11)$$
$$HONO + h\nu \rightarrow OH + NO, \quad (12)$$
$$OH + NO_2 + M \rightarrow HNO_3 + M. \quad (13)$$

The rate constants for all these reactions are reasonably well established, so that the concentrations of OH expected can be predicted with some degree of reliability. (This would not be true if the concentration of organics was high enough to perturb the NO_x chemistry because of the many uncertainties in the mechanisms of the organic reactions.)

Line A in Fig. 16.12 shows the OH concentrations predicted using only the homogeneous gas-phase chemistry of reactions (1)–(13). Clearly, much larger concentrations of OH are observed than can be rationalized on the basis of gas-phase chemistry alone.

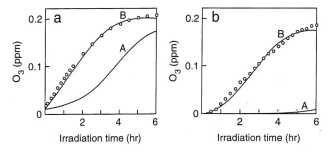

FIGURE 16.13 Comparison of observed O_3 concentration–time profiles (○) in two different evacuable chambers to predicted profiles if the heterogeneous production of HONO (reaction (14)) does not occur (curve A) and if reaction (14) with photoenhancement does occur (curve B). Results in (a) are results from the chamber of Akimoto et al. (1985) and those in (b) are from the evacuable chamber in Fig. 16.3 (adapted from Sakamaki and Akimoto, 1988).

As discussed earlier, one source of OH is the photolysis of HONO formed on surfaces by reaction (14),

$$2NO_2 + H_2O \xrightarrow{\text{surfaces}} HONO + HNO_3. \quad (14)$$

If one assumes that 10 ppb of HONO was present initially, the OH radical profile would follow curve B. This is in agreement with the initial OH concentrations but not with the larger concentrations at later times. The shape of curve B is expected since any initial HONO present will rapidly photolyze, producing a "pulse" of OH; however, after the HONO has photolyzed, lower concentrations of OH follow.

The shape of the observed OH profile suggests that some unrecognized source of OH must be present. Curve C is the predicted OH profile if it is assumed that a constant OH source exists during the run that produces OH at a rate of 0.245 ppb min^{-1}. In contrast to curve B, this leads to agreement with the observed OH levels and curve shape at long reaction times, but not at short times. (The lower OH concentrations predicted at short reaction times are due to the net reaction of OH with NO while HONO is building up to its equilibrium concentration.)

Combining curves B and C (i.e., using the assumption of both an initial HONO concentration and a constant radical source), one obtains curve D, which matches the observations quite well.

While a constant flux of OH was assumed in generating curves C and D of Fig. 16.12, this is obviously only a simplified construct taken to represent some species that forms OH in subsequent reactions. The flux of this unknown species appears to depend on the particular chamber and chamber history and is roughly proportional to the light intensity. It was also found that it increases significantly with temperature, relative humidity, and NO_2 concentrations but is independent of total pressure and the NO concentration (Carter et al., 1982).

Glasson and Dunker (1989) used the oxidation of CO by OH to remove the chamber radical source. When an excess of CO is used, the OH is converted to HO_2 and then reacted with NO to give NO_2:

$$OH + CO \xrightarrow{O_2} HO_2 + CO_2, \quad (15)$$
$$HO_2 + NO \rightarrow OH + NO_2. \quad (16)$$

The OH concentration is then calculated from the rate of loss of NO, the CO concentration, and the known rate constants. In contrast to the observations of Carter et al. (1982), the chamber OH radical source was found to depend on light intensity and temperature, but not on NO_2.

The identity and mechanism of formation of the precursor(s) to OH are unknown and subject to some controversy (Killus and Whitten, 1981; Besemer and Nieboer, 1985; Carter *et al.*, 1985; Leone *et al.*, 1985; Sakamaki and Akimoto, 1988), although HONO appears to be the most likely candidate. The dependence of its flux on the particular chamber characteristics suggests the involvement of unrecognized heterogeneous reactions on the chamber surfaces.

Sakamaki and Akimoto (1988) have used a combination of computer kinetic modeling and the results of environmental chamber experiments to show that the ozone concentration–time profiles are consistent with known chemistry *if* the photoenhancement of reaction (14) (Akimoto *et al.*, 1987) is taken into account.

Figure 16.13, for example, shows the concentration–time profiles for a run in the evacuable chamber shown in Fig. 16.3 and for one in the evacuable chamber of Akimoto *et al.* (1985). The calculation, which assumes no radical source, curve A, clearly underpredicts O_3 by a large margin. However, inclusion of a photoenhanced production of HONO via reaction (14), curve B, matches the observations quite well (Sakamaki and Akimoto, 1988).

However, without knowledge of the source of the increased OH flux, extrapolation of the concentration–time profiles of both the primary and secondary pollutants observed in such smog chamber studies to real atmospheres becomes less certain. For example, the reactions leading to the unknown precursor(s) to OH may occur only in smog chambers. Extrapolation to ambient air would thus require subtracting out this radical source. On the other hand, the same reactions may occur in ambient air where surfaces are available in the form of particulate matter, buildings, the earth, and so on; if this is true, then the rates would be expected to depend on the nature and types of surfaces available and may thus differ quantitatively from the smog chamber observations.

While chamber contamination and the presence of unknown surface reactions are probably the most important problems in extrapolating smog chamber data to atmospheric conditions, other minor problems exist as well. These include the need to measure carefully and frequently a number of chamber-specific parameters such as the decay rate of O_3 on the chamber walls and the initial formation of HONO. Such chamber-specific parameters raise the question again of how best to modify these parameters to describe ambient air.

However, despite these complications, smog chambers have proven extremely useful in studying the chemistry of photochemical air pollution under controlled conditions in which emissions and meteorology are not complicating factors. While there are some uncertainties and limitations in quantitatively extrapolating the results to ambient air, it may be that what appear to be chamber-specific complications may, in fact, apply in ambient air as well.

2. Isopleths for Ozone and Other Photochemically Derived Species

As we have seen, the chemistry connecting the primary emissions, VOC and NO_x, to the concentrations of secondary pollutants such as O_3, HNO_3, and particles is very complex, even when such effects as meteorology and new emissions into the air mass of interest are ignored. A result of this complex chemistry is that the concentrations of secondary pollutants are related to those of the precursors in ways that are, under many conditions, highly nonlinear. In principle, environmental chambers can be used to systematically examine these relationships. A first approach, dating back to the 1950s (e.g., see Haagen-Smit and Fox, 1954), is to examine the peak 1-h O_3 formed when mixtures of known initial concentrations of VOC and NO_x are irradiated in a laboratory chamber. The results are often displayed in the form of 2-dimensional isopleths such as those shown in Fig. 16.14a (Dodge, 1977a; Finlayson-Pitts and Pitts, 1993).

In practice, such isopleths are usually generated using computer models such as the EKMA (Empirical Kinetic Modeling Approach) model (vide infra) where the results of the model have been tested against environmental chamber data. Figure 16.14a shows the peak O_3 formed from the irradiation of mixtures of VOC and NO_x at the initial concentrations shown on the axes. Figure 16.14b reflects the same data in three dimensions (Finlayson-Pitts and Pitts, 1993). The overall shape of the "ozone hill" in Fig. 16.14b is useful in examining whether VOC or NO_x control, or both, would be most effective in controlling O_3. Thus, at high VOC/NO_x corresponding to point A, decreasing VOC alone at constant NO_x along the AB line gives only slowly decreasing O_3. However, decreasing NO_x at constant VOC, i.e., along the AC line, is very effective in rolling down the ozone hill. Thus in this case, the chemistry of the polluted air masses is NO_x-limited and NO_x control is most effective. This region of high VOC/NO_x is typical of suburban, rural, and downwind areas.

Low VOC/NO_x ratios, e.g., point D, have been found to be typical of polluted air masses found in many major urban centers, e.g., downtown Los Angeles (DTLA). Here, reducing VOC at constant NO_x along the line DE results in rolling down the ozone hill. However, reducing NO_x at constant VOC, along the

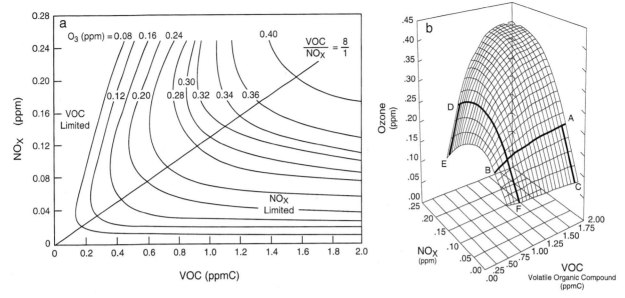

FIGURE 16.14 Typical peak ozone isopleths generated from initial mixtures of VOC and NO_x in air. (a) Two-dimensional depiction generated from the EKMA model (Dodge, 1977a); (b) three-dimensional depiction prepared by B. Dickerson. The VOC-limited region (e.g., at point D) is found in some highly polluted urban centers while the NO_x (e.g., at point A) is typical of downwind suburban and rural areas (adapted from Finlayson-Pitts and Pitts, 1993).

line DF, actually leads to an increase in O_3 initially until the ridgeline is reached. This behavior has been a major factor in the ozone control strategy controversy and used to argue (incorrectly, we believe—see following discussion) against NO_x controls.

This complex behavior is due to several effects. First, the reaction of O_3 with NO is rapid, so that O_3 is effectively titrated by the high NO concentrations. For example, measurements made in Nashville, Tennessee, show that during the early to mid-morning hours, CO and NO_y increased due to emissions, while O_3 decreased to almost zero due to the O_3–NO reaction (Nunnermacker et al., 1998). Decreased O_3 concentrations in the Vancouver, Canada, region have also been attributed to this reaction (Pryor, 1998).

A second reason for the response of O_3 to NO_x at low VOC/NO_x ratios is that at these high NO_x concentrations, NO_2 competes with VOC for the OH radical by forming HNO_3 (reaction (13)). This terminates the chain oxidation of VOC and removes NO_2 from the system without forming O_3. This chemistry has been confirmed by direct OH radical measurements; thus, Eisele et al. (1997) report that OH concentrations increase with the NO concentration up to ~1-2 ppb but decrease thereafter due to the OH + NO_2 reaction. This is consistent with model calculations of isopleths of OH concentrations as a function of the VOC/NO_x ratio, similar to the ozone isopleths in Fig. 16.14 (e.g., Kley, 1997); in the high NO_x–low VOC regime, predicted OH concentrations decrease as NO_x increases.

This behavior in the VOC-limited regime has been suggested to explain the apparent "weekend effect" on ozone concentrations (e.g., see Cleveland et al., 1974; Lebron, 1975; Elkus and Wilson, 1977; Graedel et al., 1977; Altshuler et al., 1995; and Brönnimann and Neu, 1997). Thus, in urban areas such as San Francisco it appears that ozone concentrations are actually slightly higher on weekends compared to weekdays. The same effect has been observed in other locations around the world, including Switzerland, where this occurs primarily under conditions when the meteorology is not conducive to ozone formation.

Altshuler et al. (1995) suggest that these higher ozone concentrations on weekends are due to a larger decrease in NO_x compared to VOC on weekends, effectively moving along the DF direction in Fig. 16.14b. Dreher and Harley (1998) have proposed that part of this decrease in NO_x emissions is due to reductions in diesel truck use on weekends. They also suggest that associated with lower diesel traffic there will be lower emissions on weekends of the strongly absorbing carbon particles from diesels (see Chapter 10). As discussed in Section F, this is expected to lead to increased surface light intensities and hence photolysis rates, resulting in increased O_3 formation.

In short, the effectiveness of VOC versus NO_x controls depends critically on the VOC/NO_x ratio.

While the VOC/NO_x ratio is easily defined in environmental chamber studies, one cannot as readily determine a unique ratio characteristic of various locations in an actual air basin. As discussed elsewhere (Altshuller, 1989; National Research Council, 1991; Wolff and Korsog, 1992), this ratio depends on a number of factors and varies spatially as well as temporally within an air basin and at one surface location can be highly variable from day to day. As Wolff and Korsog (1992) point out, limited surface measurements of VOC/NO_x at a single site cannot be taken as representative of the entire region. In addition, since air masses are transported during the day, the VOC/NO_x ratio at one location would not necessarily be expected to be directly related to ozone at that location, but rather at some location downwind. Even this correlation is not straightforward, however, since additional VOC and NO_x are emitted into the air mass as it travels downwind, and dilution, entrainment of air aloft (e.g., see MacGregor and Westberg, 1990), can occur.

However, the concept of using a VOC/NO_x ratio to explore qualitatively the implications of various control strategy options is very useful, based on our understanding of the chemistry leading to ozone formation, if one takes these factors into account. Figure 16.15, for example, shows model-predicted changes in VOC and NO_x concentrations over an 8-h period for several different concentrations but all with an initial VOC/NO_x ratio of 6:1, superimposed on isopleths showing the average rates of net O_3 production (Milford et al., 1994). As the air parcels age, they move toward higher VOC/NO_x ratios, i.e., from the VOC-limited regime to the NO_x-limited regime. The reasons for this are that oxides of nitrogen are removed more rapidly from the atmosphere and, at the same time, biogenic emissions contribute to the VOC concentrations. (Biogenic emissions of NO_x are small compared to anthropogenic emissions in urban areas; for example, Aneja et al. (1997) estimate that less than 1% of the total NO_x in Raleigh, North Carolina, is due to natural emissions.) As a result, the relative effectiveness of VOC versus NO_x controls is expected to change as the air parcel moves downwind and ages.

Because of this increase in the VOC/NO_x ratio as the air mass moves downwind, isopleths appropriate for one portion of an air basin, e.g., the upwind portion, will not necessarily be applicable to other regions, such as the downwind areas. The development of sophisticated grid-based models and their application to entire air basins have allowed modelers to include continuing emissions, transport, and transformation processes (see Section A.3).

For example, Fig. 16.16 shows the results of modeling by Milford et al. (1989) to examine the effects of control of VOC and NO_x on ozone under conditions typical of the base case, August 30–31, 1982, at several locations in southern California. The results for two of the sites modeled, downtown Los Angeles (DTLA) and Rubidoux, about 100 km east and downwind, are presented in the form of two-dimensional isopleths analogous to those in Fig. 16.14a as well as three dimensions analogous to Fig. 16.14b. In this case, points A and B represent the situation at these two locations in 1982, and the isopleths give the predicted changes in ozone resulting from various percentage reductions in basin-wide emissions of NO_x and VOC from the base case.

In the Los Angeles area during the day, the winds are predominantly from the west, carrying the pollutants inland. At night the direction is reversed and "aged" air masses are carried out over the Pacific Ocean, only to return the following morning carrying enhanced levels of OH sources such as nitrous acid (HONO) and formaldehyde. As a consequence, under multiday stagnant meteorological conditions, VOC oxidation and the NO to NO_2 conversion are enhanced and increasingly high peak O_3 levels are reached on successive episode days.

Figure 16.16 shows that in DTLA, which is most heavily impacted by direct mobile source emissions and has relatively low VOC/NO_x ratios, VOC control is predicted to be most effective. Thus, starting at the current emissions marked as point A, controlling VOC is clearly most effective in rolling down the ozone hill. Reducing NO_x in the absence of VOC control from

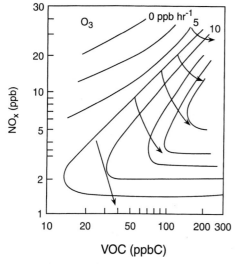

FIGURE 16.15 Isopleths of the calculated daytime average rates of net ozone production (ppb h^{-1}) as a function of initial VOC and NO_x. The model-predicted changes in VOC and NO_x concentrations over an 8-h time period as the air mass is transported and reacts are shown by the arrows (adapted from Milford et al., 1994).

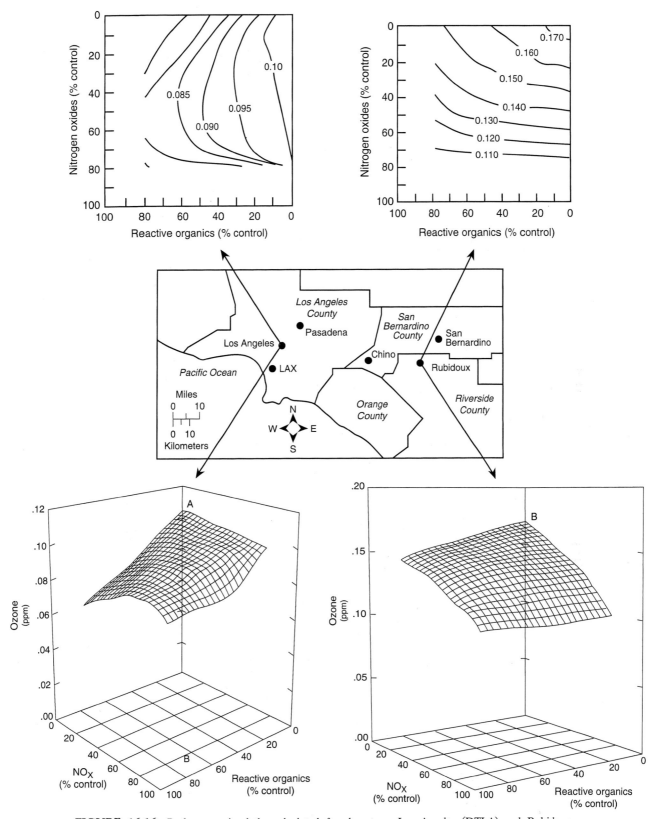

FIGURE 16.16 Peak ozone isopleths calculated for downtown Los Angeles (DTLA) and Rubidoux, approximately 100 km east and downwind of DTLA under typical meteorological conditions. Spatially uniform reductions of VOC and NO_x were employed in an airshed model by Milford et al. (1989). The top shows isopleths in two dimensions as presented by Milford et al. (1989), and the bottom shows these data extrapolated to three dimensions (from Finlayson-Pitts and Pitts, 1993).

point A does not produce a significant change in O_3. It is important, however, to note that DTLA is not where the O_3 peaks for the airshed as a whole are experienced.

In contrast, at downwind locations where the airshed ozone peaks are more typically encountered, reduction in VOC starting at point B without concurrent NO_x control does not lead to the rapid decrease in O_3 observed for DTLA; indeed, even an 80% reduction in VOC alone is not predicted to reach the U.S. air quality standard of 0.12 ppm O_3. In this case, control of NO_x reduces O_3 more rapidly than comparable control of VOC.

While the results in Fig. 16.16 were developed for the Los Angeles area, the same general features have been found in a number of other studies as well (e.g., see National Research Council, 1991; and Tesche and McNally, 1991). The approach and issues are thus qualitatively applicable to urban/suburban areas worldwide where pollutants are transported downwind to less densely populated areas and the VOC/NO_x ratio changes as the air mass is transported.

There are, however, some important changes in our understanding that have come about since the predictions in Fig. 16.16. First, as discussed in more detail shortly, the emissions inventory for VOC was underestimated at the time of the model calculations by a factor of two or more, at least for mobile source emissions. This tends to introduce a bias in favor of VOC control. Furthermore, this study only addressed one set of meteorological conditions characteristic of the particular episode modeled. For example, in a more recent study of a different air pollution episode in the same air basin where the meteorology was not as conducive to formation of high ozone levels, ozone was shown to be sensitive to increasing VOC from vehicle hot exhaust as far east as San Bernardino (Harley *et al.*, 1993a). This sensitivity to the particular meteorological conditions illustrates that the effectiveness of a particular control strategy may depend on whether one chooses to focus on the most severe episodes or, alternatively, on more typical conditions.

While recognizing these limitations, such isopleths are useful, however, in examining the differing responses of various locations within an air basin to control of VOC and NO_x. Since the air mass that starts upwind in the morning traverses the air basin during the day, a multidimensional approach to ozone control is clearly needed. Isopleths characteristic of different locations in an air basin such as those in Fig. 16.16 demonstrate that a combination of VOC and NO_x control is essential if air quality throughout a major air basin is to improve consistently (although not necessarily by equal amounts) in all locations.

3. Models

Ultimately, it is not possible to mimic experimentally all conditions of potential interest for control strategies. In addition, even if one assumed that one could carry out environmental chamber studies covering all of these conditions, as discussed earlier, there still remain a number of uncertainties in how to extrapolate these to the "real world" with its different surfaces and potential for heterogeneous reactions, complex meteorology, deposition processes, and varying new emissions. As a result, the development and application of mathematical models describing the chemistry, meteorology, and deposition are critical elements in the development and assessment of effective control strategies.

Of course, in developing these models, it is necessary to test them extensively against experimental data to ensure that their predictions are the result of appropriately simulating the chemical and physical processes involved, and not to a fortuitous cancellation of errors. For example, the chemical submodels are tested against environmental chamber data, and the final model results against ambient air measurements.

There are a variety of mathematical models used to describe the relationship between the precursors and the secondary pollutants they form upon reaction. In addition, there are some simple, often empirical, models that have been developed for application in particular areas. An example of these is also discussed in the following section.

a. Simple Models

(1) Linear rollback As the name implies, linear rollback is based on the assumption that pollutant concentrations will decrease proportionally to a decrease in the precursor emissions. For a pollutant such as CO, for example, the percentage reduction in emissions required to meet air quality standard for CO (A_{CO}) in a region that currently has observed concentrations as high as C_{CO} is given by linear rollback as:

$$\begin{aligned}\text{Percentage reduction in emissions} &= \frac{(C_{CO} - B_{CO}) - (A_{CO} - B_{CO})}{C_{CO} - B_{CO}} \times 100 \\ &= \frac{C_{CO} - A_{CO}}{C_{CO} - B_{CO}} \times 100\end{aligned}$$

Here B_{CO} is the background (i.e., clean air) concentration of CO, which must be taken into account.

For a *nonreactive* pollutant such as CO, linear rollback provides a reasonable control strategy, providing the location, temporal distribution, and relative strengths of the sources do not change. However, for a secondary pollutant such as O_3 that is formed through chemical reactions of primary pollutants, the complexity of the chemistry makes this approach inappropriate. In addition, since both VOC and NO_x are involved in oxidant formation, a decision must be made as to which pollutant—or both—linear rollback should be applied.

(2) Simple empirical models Simple empirical models have been developed for application to specific areas, primarily for forecasting purposes on a short-term basis. For example, in areas such as the Los Angeles region, daily air quality forecasts are made that are based primarily on past correlations of meteorological parameters with peak ozone concentrations (e.g., Zeldin and Thomas, 1975). Such models have proven useful for adjusting observed trends in measured pollutant concentrations to similar meteorological conditions, which is critical in assessing whether control strategies are having the intended effect.

Because such empirical models are based on historical relationships for a particular region, they may not be valid if significant changes occur, e.g., if the spatial or temporal distribution of the sources, or the chemical composition, changes. In addition, they are specific to the airshed for which they were developed.

b. Mathematical Models

(1) Components Clearly, the best approach to understanding, and ultimately predicting, the relationship between emissions of primary pollutants and the concentrations of secondary pollutants formed is to develop mathematical models that describe all of the inputs and factors that affect this relationship. The major components are (1) emissions of the primary pollutants, including their specific composition and spatial and temporal variations; (2) the meteorological and topographical features of the region, including such parameters as temperature, relative humidity, wind speed and direction in three dimensions, atmospheric stability, inversion height, and surface elevation and other terrain features; and (3) the chemistry, including both the kinetics and mechanisms of the reaction converting primary pollutants into secondary pollutants.

Figure 16.17a schematically shows the individual subcomponents of these three major modules (McRae *et al.*, 1982a), and Fig. 16.17b shows a typical chemical submodel in more detail (Dabdub and Seinfeld, 1996). In the following sections, we shall briefly refer to some of the individual subcomponents of these three modules, particularly with respect to effects on predicted concentrations of secondary species. However, consistent with the focus of this book on atmospheric chemistry, some important aspects of the chemistry module are discussed in more detail.

(2) Chemical component of models As we have seen from the examination of kinetics and mechanisms of atmospheric reactions thus far, the chemistry of even relatively simple organics can be quite complex. This chemistry has been described in terms of *explicit* chemical mechanisms, that is, a listing of the individual chemical reactions. The oxidation of even one organic in air includes hundreds of reactions.

In a VOC–NO_x mixture containing many different organics, the number of reactions becomes unmanageable for application in models used to describe an air basin or region. Thus the amount of computer time required for numerical integration of the rate equations associated with the thousands of individual species found in ambient air is prohibitive. Furthermore, even as computing power increases, in practice, the kinetics and mechanisms required as input are not all known.

As a result, these explicit chemical mechanisms are generally not used in such models, except to describe the inorganic NO_x chemistry, which by comparison is relatively straightforward (see Chapter 7). Rather, the mechanisms are condensed in various ways to reduce the number of organic reactions substantially. In these *lumped* mechanisms, the chemistry of the organics is treated by grouping or "lumping" together a number of reactions and/or chemical species. The overall rate constant and products of the lumped reactions are chosen to be representative of that group of reactions or reactants. Examples of how this can be accomplished for a group of non-aromatic VOCs are discussed in detail by Jenkin *et al.* (1997) and, for a group of alkanes, by Wang *et al.* (1998).

Two approaches have been used in developing condensed mechanisms for organics. The first groups organics for the most part by their traditional classifications, e.g., alkanes, alkenes, and aromatics. There are some exceptions, usually for the first member of a class, which often shows unique properties. For example, methane is not lumped together with other alkanes because of its low reactivity (vide infra). Two major chemical submodels that use this approach are those of Lurmann *et al.* (1986), often referred to as the LCC mechanism (named for the developers, *L*urmann, *C*arter, and *C*oyner), and the RADM mechanism (*R*egional *A*cid *D*eposition *M*odel) (Stockwell *et al.*, 1990). The latter has been updated by Stockwell *et al.* (1997) in a version known as RACM (*R*egional *A*tmospheric *C*hemistry *M*echanism). Table 16.1 shows a typical classification of organics used in the RADM and RACM submodels for gas-phase chemistry (Stockwell *et al.*,

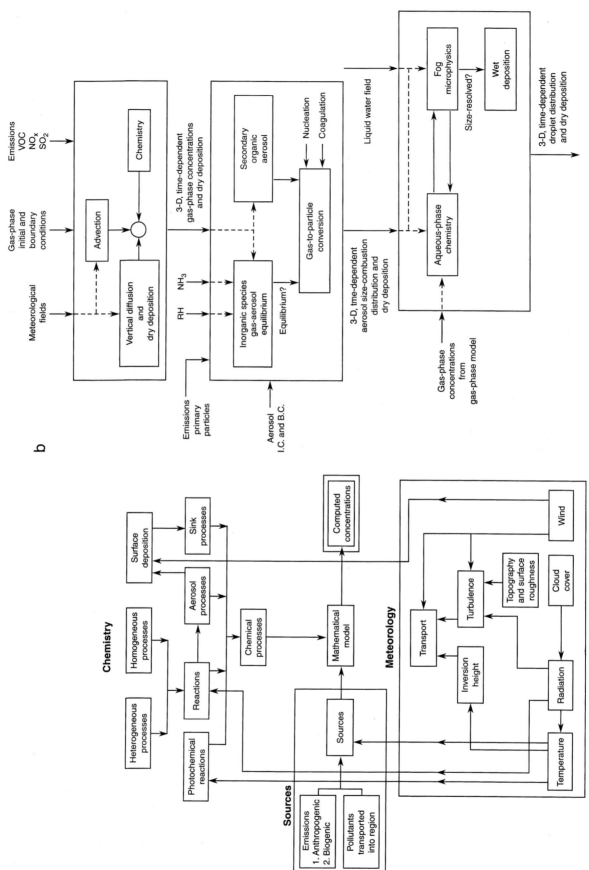

FIGURE 16.17 (a) Elements of a typical airshed model (from McRae et al., 1982a). (b) Elements of the chemical module (from D. Dabdub and J. Seinfeld, personal communication, 1996).

TABLE 16.1 Classification of Organics in the RCAM and RADM2 Mechanism[a]

Species	Acronym	Species	Acronym
Alkanes		Organic acids	
Methane	CH_4	Formic acid	ORA1
Ethane	ETH	Acetic acid and higher acids	ORA2
Alkanes, alcohols, esters, and alkynes with HO rate constants (298 K, 1 atm) between 2.7×10^{-13} and 3.4×10^{-12}	HC3	Organic short-lived intermediates	
		Peroxy radicals from alkanes	
		CH_3O_2	MO2
Alkanes, alcohols, esters, and alkynes with HO rate constants (298 K, 1 atm) between 3.4×10^{-12} and 6.8×10^{-12}	HC5	Peroxy radical formed from ethane, ETH	ETHP
		Peroxy radical formed from alkane, HC3	HC3P
		Peroxy radical formed from alkane, HC5	HC5P
Alkanes, alcohols, esters, and alkynes with HO rate constants (298 K, 1 atm) greater than 6.8×10^{-12}	HC8	Peroxy radical formed from alkane, HC8	HC8P
Alkenes		Peroxy radicals from alkenes	
Ethene	ETE (OL2)[b]	Peroxy radicals formed from ethene, ETE	ETEP (OL2P)[b]
Terminal alkenes	OLT	Peroxy radicals formed from alkene, OLT	OLTP
Internal alkenes	OLI	Peroxy radicals formed from alkene, OLI	OLIP
Butadiene and other anthropogenic dienes	DIEN	Peroxy radicals formed from ISO and DIEN	ISOP
Isoprene	ISO	Peroxy radicals formed from API	APIP
α-Pinene and other cyclic terpenes with one double bond	API	Peroxy radicals formed from LiM	LIMP
d-Limonene and other cyclic diene terpenes	LiM	Peroxy radicals from aromatics	
		Peroxy radical formed from aromatic, TOL	TOLP
Aromatics		Peroxy radical formed from aromatic, XYL	XYLP
Toluene and less reactive aromatics	TOL	Peroxy radicals formed from CSL	CSLP
Cresol and other hydroxy-substituted aromatics	CSL	Phenoxy and similar radicals	PHO
Xylene and more reactive aromatics	XYL	Aromatic-OH adduct from TOL	ADDT
		Aromatic-OH adduct from XYL	ADDX
Carbonyls		Aromatic-OH adduct from CSL	ADDC
Formaldehyde	HCHO		
Acetaldehyde and higher aldehydes	ALD	Peroxy radicals with carbonyl groups	
Ketones	KET	Acetylperoxy and higher saturated acylperoxy radical	ACO_3
Glyoxal	GLY		
Methylglyoxal and other α-carbonyl aldehydes	MGLY	Peroxy radicals formed from ketone, KET	KETP
Unsaturated dicarbonyls	DCB	Unsaturated acyl peroxy radicals	TCO_3
Methacrolein and other unsaturated monoaldehydes	MACR		
Unsaturated dihydroxy dicarbonyl	UDD	Peroxy radicals involving nitrogen	
Hydroxy ketone	HKET	NO_3–alkene adducts reacting to form carbonitrates + HO_2	OLNN (OLN)[b]
Organic nitrogen			
Peroxyacetyl nitrate and higher saturated PANs	PAN	Accounts for additional organic nitrate formation affected by the lumped organic species	XO_2 (XNO_2)[b]
Unsaturated PANs	TPAN		
Organic nitrate	ONIT	NO_3–alkene adducts reacing via decomposition	OLND
Organic peroxides			
Methyl hydrogen peroxide	OP1		
Higher organic peroxides	OP2		
Peroxyacetic acid and higher analogs	PAA		

[a] From Stockwell *et al.* (1997).
[b] Terminology used in Stockwell *et al.* (1990).

1990, 1997). The RADM mechanism is found in examples of the OZIPR model whose applications are used in this book (see problems at end of chapter).

The second approach groups organics in terms of bonding. This "carbon bond mechanism," referred to in the literature generally as CB(*X*), where *X* gives the version, divides organics into classes based on their chemical bonding. Table 16.2 gives a typical CB classification of organics (Gery *et al.*, 1989). This mechanism is also included in examples in the OZIPR model.

Regardless of the approach to lumping organics together, there are always uncertainties since there is a

TABLE 16.2 Classification of Organics in the Carbon Bond IV Mechanism, Expanded Version[a]

Species	Acronym	Species	Acronym
Formaldehyde (HCHO)	FORM	Toluene–hydroxyl radical adduct	TO2
Hydroxymethylperoxy radical (HOCH$_2$OO·)	FROX	Benzaldehyde	BZA
Organic peroxide (ROOH)	PROX	Peroxybenzoyl radical	BZO2
Formic acid (HCOOH)	FACD	Peroxybenzoyl nitrate	PBZN
High-molecular-weight aldehydes (RCHO, R > H)	ALD2	Phenylperoxy radical	PHO2
Peroxyacyl radical (CH$_3$C(O)OO·)	C2O3	Phenoxy radical	PHO
Peroxyacyl nitrate (CH$_3$C(O)OONO$_2$)	PAN	Nitrophenol	NPHN
Methylperoxy radical (CH$_3$OO·)	MEO2	Methylphenylperoxy radical	CRO2
Methylperoxy nitric acid (CH$_2$O$_2$NO$_2$)	MPNA	Methylphenoxy radical	CRO
Methoxy radical (CH$_3$O)	MEO	Nitrocresol	NCRE
Methyl nitrite (CH$_3$ONO)	MNIT	High-molecular-weight aromatic oxidation ring fragment	OPEN
Methyl nitrate (CH$_3$ONO$_2$)	MEN3		
Methanol (CH$_3$OH)	MEOH	Aromatic ring fragment acid	ACID
Acetone (CH$_3$C(O)CH$_3$)	AONE	Xylene (C$_6$H$_4$(CH$_3$)$_2$)	XYL
Acetylmethylperoxy radical (CH$_3$C(O)CH$_2$OO)	ANO2	Methylbenzylperoxy radical	XLO2
Paraffin carbon bond (C–C)	PAR	Xylene–hydroxyl radical adduct	XINT
Primary organic peroxy radical	RO2	Methylglyoxal (CH$_3$C(O)C(O)H)	MGLY
Secondary organic peroxy radical	RO2R	Peroxide radical of MGLY (CH$_3$C(O)C(O)OO)	MGPX
Organic nitrate	NTR	Peroxide radical of OPEN	OPPX
Secondary organic oxy radical	ROR	Isoprene	ISOP
Ketone carbonyl group (–C(O)–)	KET	Isoprene epoxide product	EPOX
Dimethyl secondary organic peroxide radical	AO2	Isoprene O-adduct	ISO1
Olefinic carbon bond (C=C)	OLE	Isoprene O-adduct	ISO2
Criegee biradical (H$_2$ĊOO·)	CRIG	Isoprene OH-adduct	ISO3
Methyl Criegee biradical (CH$_3$(H)ĊOO·)	MCRG	Dinitrate of isoprene	DISN
"Excited" formic acid	HOTA	Methylvinyl ketone OH-adduct	MV1
"Excited" acetic acid	HTMA	Methylvinyl ketone OH-adduct	MV2
Nitrated organic peroxy radical (–CH(ONO$_2$)–CH(OO)–)	PNO2	Methacrolein OH-adduct	MAC1
		Methacrolein OH-adduct	MAC2
C$_2$ dinitrate group	DNIT	Methylvinyl ketone nitrate	MVNT
Ethene (CH$_2$=CH$_2$)	ETH	Isoprene OH-adduct	ISO4
Ethanol peroxide radical (CH$_2$OH–CH$_2$OO)	ETO2	Methacrolein	MACR
Ozonide and further products	OZD	Methylvinyl ketone	MVK
Acetic acid (CH$_3$COOH)	ACAC	Nitrate of isoprene	ISNT
Toluene (C$_6$H$_5$CH$_3$)	TOL	Nitrate of isoprene	ISN
Benzylperoxy radical	BO2	Paraffin loss operator (–PAR)	X
Cresol and higher molecular weight phenols	CRES	Paraffine-to-peroxy radical operator	D

Total 87

[a] From Gery et al. (1989).

great deal of flexibility and judgment involved in choosing kinetics and products that are representative of a whole group of organics. For example, the rate constants for O_3 reacting with alkenes change by a factor of 40 from propene to 2-methyl-2-butene (Table 6.9). The greater the concentrations of the more reactive species in the air mass, the larger the chosen rate constant for this reaction should be. One approach is to weight the rate constants by the relative number of moles of the individual compounds in this class (e.g., see McRae et al., 1982a, 1982b).

Of course, as the reaction proceeds and the hydrocarbons are consumed, the composition of the remaining hydrocarbons shifts. If the rate constants for the organics reflect the mix, then they should also change with time. The severity of this problem is reduced somewhat by the fact that there is fresh injection of reactants during the day.

A number of intercomparisons of chemical submodels have been carried out. Figure 16.18, for example, shows some of the results from one such intercomparison for conditions chosen to be representative of moderately polluted conditions (Kuhn et al., 1998). The average final O_3 predicted by the models was 148 ppb, but there are clearly significant differences between the models. Thus the highest, the EMEP model (*E*uropean

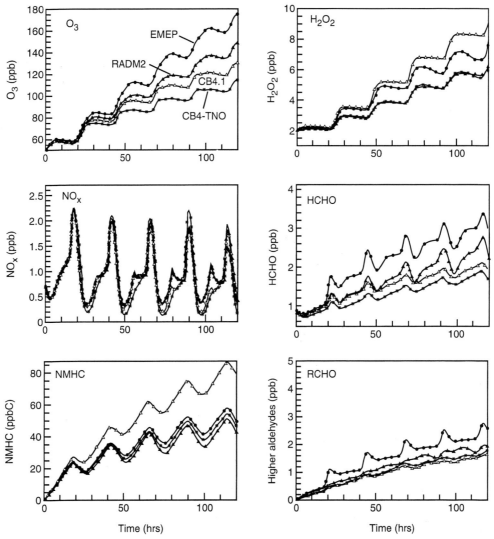

FIGURE 16.18 Concentration–time profiles for O_3, NO_x, NMHC, H_2O_2, HCHO, and higher aldehydes (RCHO) predicted using four different chemical submodels: two carbon bond four models (CB4.1 and CB4-TNO), a RADM model (RADM2), and the EMEP model (adapted from Kuhn *et al.*, 1998).

*M*onitoring and *E*valuation *P*rogram model; Simpson, 1995), predictions are 27% higher than the mean whereas the lowest, the CB4-TNO version of the carbon bond 4 mechanism, predicts ozone concentrations 35% below the mean. Other studies in which the carbon bond 4 mechanism was tested against environmental chamber data have also found that it underpredicts O_3 formation (e.g., Simonaitis *et al.*, 1997). The sensitivity of predicted O_3 by CB4 to the chemistry, particularly radical–radical reactions, has been discussed by Kasibhatla *et al.* (1997).

It is common in model intercomparisons that relatively good agreement is obtained for the major species NO_x and O_3 but the discrepancies can be larger for trace species such as HCHO and H_2O_2. This can be important since such species are radical sources that contribute substantially to continuing the chain oxidation of the organics. For example, for the model intercomparisons shown in Fig. 16.18, the range of predicted values expressed as the root mean square (rms) error for O_3 is 10%. However, the rms errors for H_2O_2, HCHO, and higher aldehydes are 22, 23, and 48%, respectively.

Similar results have been reported in an intercomparison study of models used to predict tropospheric ozone on a global scale (Olson *et al.*, 1997). Agreement for O_3 and NO_x was reasonably good for relatively clean atmospheres, with a larger spread for predicted H_2O_2. However, introduction of VOC chemistry increased the range of model predictions substantially, with the rms error for O_3 doubling and that for NO_x more than doubling, from 15 to 40%.

Given the numerous potential uncertainties due solely to uncertainties in the chemistry, particularly the VOC chemistry and how it is incorporated into models, it is clearly important to understand which are likely to have the most important net effects on predicted concentrations. A number of studies have been carried out to address this issue (e.g., see Hough and Reeves, 1988; Hough, 1988; Dodge, 1989, 1990; Chock *et al.*, 1995; and Olson *et al.*, 1997). Sensitivity analyses have been performed for a number of models. For typical approaches, see Derwent and Hov (1988), Milford *et al.* (1992), and Gao *et al.* (1995, 1996).

Once a chemical submodel has been developed, it must be tested extensively prior to its application in comprehensive computer models of an air basin or region. This is done by testing the chemical submodel predictions against the results of environmental chamber experiments. While agreement with the chamber experiments is necessary to have some confidence in the model, such agreement is not sufficient to confirm that the chemistry is indeed correct and applicable to real-world air masses. Some of the uncertainties include those introduced by condensing the organic reactions, uncertainties in kinetics and mechanisms of key reactions (e.g., of aromatics), and how to take into account chamber-specific effects such as the unknown radical source.

These chemical submodels are then incorporated into more comprehensive models of the type shown schematically in Fig. 16.17. These can vary from relatively simple box models to large-scale regional models, briefly described in the following text.

c. Simple Mathematical Models

(1) Box models (including EKMA) One type of simple model that has been applied to predict pollutant concentrations is known as the box model (Fig. 16.19) (e.g., Schere and Demerjian, 1978). The air mass over a region is treated as a box into which pollutants are emitted and undergo chemical reactions. Transport into and out of the box by meteorological processes and dilution is taken into account.

The box model is closely related to the more complex airshed models described below in that it is based on the conservation of mass equation and includes chemical submodels that represent the chemistry more accurately than many plume models, for example. However, it is less complex and hence requires less computation time. It has the additional advantage that it does not require the detailed emissions, meteorological, and air quality data needed for input and validation of the airshed models. However, the resulting predictions are

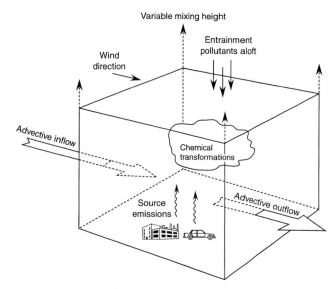

FIGURE 16.19 Schematic diagram showing basic elements of a simple box model (adapted from Schere and Demerjian, 1978).

correspondingly crude, especially in terms of spatial and temporal resolution.

A well-known box model that was developed initially for regulatory purposes is the *EKMA* model (*E*mpirical *K*inetic *M*odeling *A*pproach) (Dodge, 1977a, 1977b; Dimitriades and Dodge, 1983). It is based on a box of air containing VOC and NO_x at specified initial concentrations. Dilution occurs during the simulation, in which chemical reactions are converting the VOC and NO_x to O_3 and other secondary pollutants. Injection of pollutants during the day is included, along with a time-varying inversion height and entrainment of pollutants from aloft as the inversion height increases. Diurnal variation of photolysis rates, temperature, and relative humidity are also incorporated. Chemical mechanisms using both types of approach to "lumping" of the organics, i.e., the traditional organic classifications or the carbon bond approach, can be used with this box model. The model output can then be used to generate a series of isopleths for ozone or other secondary species via the ozone isopleth plotting package, or OZIPP. Applications of a research version of this model, called OZIPR (the R stands for research version), are provided with this book (Appendix III).

The "kinetic modeling" nomenclature arises from the incorporation of chemical kinetic submodels in EKMA. The "empirical" term comes from the use of *observed* O_3 peaks in combination with the model-predicted ozone isopleths to develop control strategy options. Thus, the approach historically was to use the model to develop a series of ozone isopleths using conditions specific for that area. The second highest hourly *observed* O_3 concentration and the measured

6–9 a.m. VOC/NO_x ratio were located on these isopleths. The isopleths were then used to determine how much reduction in VOC or NO_x would be required to reduce the peak O_3 concentration to the air quality standard.

While the EKMA model has been very useful as a first approach to incorporating the complex chemistry that links primary and secondary pollutants and to including some meteorological variables, it clearly is an oversimplification for large urban areas and for downwind regions. For example, it does not include complex meteorology such as mixing between the surface layer and higher altitudes or the effects of long-range transport, which are known to be important, nor is it useful for simulating multiday episodes, which generally produce the highest pollutant levels.

(2) Lagrangian models The next step in model development was the use of Lagrangian models depicted in Fig. 16.20 (Wayne *et al.*, 1973). These models consider a column of air containing certain initial pollutant concentrations and follow its trajectory as it moves along a trajectory. In effect, it is an expansion of a simple box model to a series of adjacent, interconnected boxes.

With the increase in computing power, both simple box models and Lagrangian models have been largely supplanted with grid-based models described in the following section. The idea of a Lagrangian approach, however, is still useful in field studies where the motion of an air parcel and changes in its chemical composition can be tracked as it moves downwind in a fashion similar to that depicted in Fig. 16.20. Such studies are often referred to as Langrangian experiments.

d. Grid Models: Urban to Regional Scales

Models currently in use for developing control strategy options are grid-based, or Eulerian, models, the principle of which is illustrated in Fig. 16.21 (Ames *et al.*, 1978). The area to be modeled is divided into grids, or boxes, in both the horizontal and vertical directions. Pollutant concentrations are calculated at fixed geographical locations at specified times based on their initial concentrations, new emissions, transport in and out of the box, dilution, and chemical reactions. By carrying out such calculations for each box, one can develop a 3-D map of pollutant concentrations as a function of time.

While this approach was first directed primarily to urban regions, regional models have been developed subsequently. The need for regional-scale models became apparent in situations such as the northeast of the United States, where long-range transport plays a major role in determining pollutant concentrations at various locations. With the combination of long-range transport, fresh emissions, and complex chemistry occurring along relatively large distances over highly populated centers, application of individual urban-scale models is not appropriate—hence the need for larger, regional models.

There are a number of urban-scale grid-based models. Examples include the UAM (*U*rban *A*irshed *M*odel; e.g., UAM-4 is version 4) (e.g., see Reynolds *et al.* (1973) and Tesche and McNally (1991)), the CIT model (*C*alifornia *I*nstitute of *T*echnology model) (see McRae *et al.* (1982b) and Russell *et al.* (1988)), and the SMOG model (*S*urface *M*eteorology and *O*zone *G*eneration model) (see Lu *et al.*, 1997a, 1997b). In such models, the horizontal size of each grid is of the order of a few kilometers, e.g., 4 or 5 km square grids with the vertical height split into 5–20 layers of increasing thickness beginning at ground level. For regional-scale models such as RADM (*R*egional *A*cid *D*eposition *M*odel; see Chang *et al.*, 1987), ADOM (*A*cid *D*eposition and *O*xidant *M*odel; Venkatram *et al.*, 1988), STEM II (*S*ulfur *T*ransport *E*ulerian *M*odel; Carmichael *et al.*, 1991), RTM-III (*R*egional *T*ransport *M*odel III; Liu *et al.*, 1984), LIRAQ (*L*ivermore *R*egional *A*ir *Q*uality Model; MacCracken *et al.*, 1978), CALGRO (Yamartino *et al.*, 1992), the model of McKeen *et al.* (1991a)), and ROM (*R*egional *O*xidant *M*odel; Lamb, 1983), the scale is of the order of 15–130 km. The number and size of the vertical layers can vary from 6 to 30.

Recently, models that incorporate urban to regional scales have also been developed. These models use the approach of nested grids in which small grids for the urban scale are "nested" within larger grids used for regional scales. [Although this sounds straightforward, it is not—see, for example, Sillman *et al.* (1990) and Kumar *et al.* (1994).] Such approaches, combined with

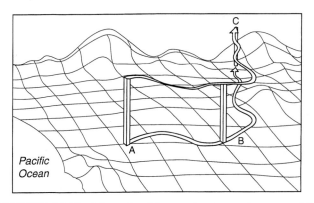

FIGURE 16.20 Schematic of Lagrangian-type trajectory model (adapted from Wayne *et al.*, 1973).

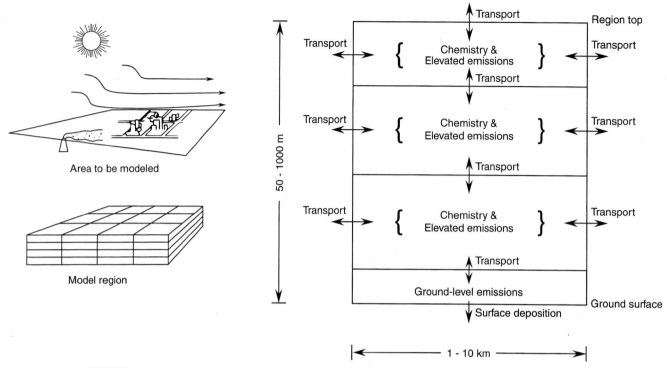

FIGURE 16.21 Schematic illustration of the grid used and treatment of atmospheric processes in one Eulerian airshed model (adapted from Ames *et al.*, 1978).

the development of a modular structure and "user friendly" software, are also under development for direct use in the regulatory arena [see, for example, the Models-3 description in Dennis *et al.* (1996) and Appleton (1996)]. The widespread application of such models will require a significant increase in computing power, such as the use of massively parallel computers (e.g, see Dabdub and Seinfeld, 1994a, 1994b, 1996).

While such models are becoming increasingly sophisticated and capable of incorporating more detailed emissions, meteorology, and chemistry, there remain a number of important issues and uncertainties in their inputs and, hence, predictions. A few of the important parameters are discussed briefly in the following.

(1) Meteorology Clearly, meteorology plays a determining role in pollutant concentrations at various locations. Simulating complex meteorology and topography is difficult, and in addition, the detailed 3-D data needed to test the meteorology submodels are not available for many regions.

For example, Kumar and Russell (1996) examined the effect on predicted ozone levels in the Los Angeles area of two different approaches now used for incorporating meteorology into a grid-based Eulerian model. The *diagnostic* approach is based on field measurements of the needed meteorological variables and in-terpolation and extrapolation of these data as necessary. The *prognostic* approach uses models to predict the evolution of the atmospheric system with time by integration of the equations of conservation of mass, motion, heat, and water in space and time. Figure 16.22 shows the ozone concentrations (dots) observed at three different locations during an air quality study in the Los Angeles area during August 27–29, 1987, in which detailed meteorological and chemical measurements were made. Also shown are model predictions using the diagnostic approach (solid line) and the prognostic approach (dashed line), respectively; in general, the prognostic approach leads to lower predicted O_3 concentrations for reasons that are discussed in detail by Kumar and Russell (1996).

Another example of the importance of accurately specifying meteorological parameters is described by Sistla *et al.* (1996). They investigated the effect on model predictions for ozone in the New York Metropolitan area of using either a spatially invariant mixing height or one that varied spatially. The latter was shown to predict O_3 concentrations that were in better agreement with the observations. Figure 16.23, for example, shows the observed and modeled O_3 concentrations for Westpoint, New York, where the spatially varying mixing height assumption was in reasonable agreement with the observed values but the

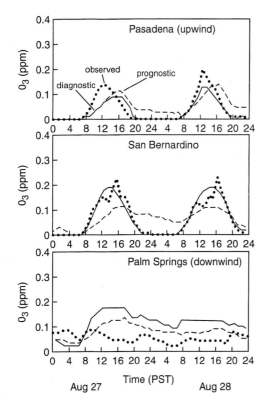

FIGURE 16.22 Observed (●) and model-predicted concentration–time profiles for O$_3$ at three locations in southern California using two different meteorological inputs to the model: (———) diagnostic approach; (- - -) prognostic approach (from Kumar and Russell, 1996).

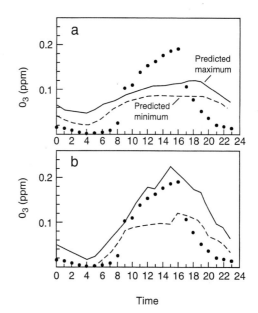

FIGURE 16.23 Observed (●) and model-predicted O$_3$ concentrations at West Point, New York, July 7, 1988, assuming (a) a spatially invariant mixing height or (b) a spatially varying mixing height. The solid line is the predicted maximum and the dashed line the predicted minimum O$_3$ concentration (from Sistla et al., 1996).

spatially invariant mixing height assumption was not (Sistla et al., 1996). Perhaps more important, the peak O$_3$ concentrations predicted using these two assumptions for various controls of VOC and NO$_x$ were different by as much as 30–50%.

Similar predictions have been made by other modeling studies. For example, Duncan and Chameides (1998) predict lower peak ozone concentrations under meteorological conditions typical of northern midlatitude summertime conditions than under conditions chosen to represent stagnant conditions favoring ozone formation where the wind speeds and mixing height are lower and the temperature is higher. More interesting is that the meteorology also affects the predicted sensitivity of ozone to reductions in VOC and NO$_x$. Thus for the stagnant conditions in a low-VOC/NO$_x$ regime, reduction of organics is predicted to have the largest effect on O$_3$ during the first 12 h, as expected from the ozone isopleths (Figs. 16.14 and 16.15). However, under the more typical meteorological conditions, the transition to increased sensitivity to NO$_x$ is predicted to occur after only 7 h because of the increased dilution and hence more rapid lowering of NO$_x$ levels.

It is important to note that it is not just the surface meteorological parameters that must be specified accurately, but those aloft as well. For example, many decades ago it was argued by some that high ozone levels downwind in Southern California could not result from emissions in the metropolitan Los Angeles area because the surface wind speeds were too slow to transport the pollutants in the observed time frame. However, tracer studies established that wind speeds aloft were, in fact, higher and that transport at higher elevations and mixing to the surface were, in fact, responsible for the higher observed concentrations. Accurate measurement of the vertical distribution of meteorological parameters is particularly important when there are significant emissions sources aloft such as power plants (e.g., see Al-Wali and Samson, 1996).

Figure 16.24 illustrates the vertical transport processes and complex interactions that can occur between urban areas, downwind regions, and the free troposphere (Fast, 1998). Emissions from urban regions into the surface layer can be transported into the mixed layer as well as into the free troposphere by several mechanisms, including venting up mountain slopes due to solar heating of the surface, which creates a chimney effect. As shown in Fig. 16.24, clouds also play a role in these vertical transport processes.

In many field studies, elevated concentrations of O$_3$, particles, and other air pollutants have been found in

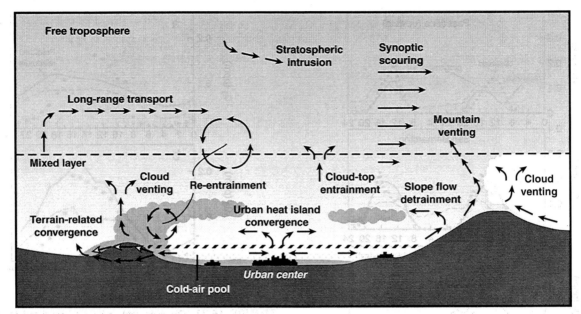

FIGURE 16.24 Schematic diagram of vertical mixing processes for pollutants (the authors thank Jerome Fast of the Pacific Northwest National Laboratory for graciously providing this diagram).

layers at higher altitudes; indeed, as seen earlier in Fig. 2.19, it has been known for a number of years that the concentrations aloft in the Los Angeles area often exceed the concentrations found simultaneously at the surface (Edinger, 1973). These observations are consistent with the results of model studies of this region (Lu and Turco, 1996). For example, Fig. 16.25 shows model-predicted ozone concentrations for one day (August 27, 1987) across this air basin and as a function of altitude and time. Figure 16.25a, representing noon PST, shows the ozone at the eastern end of the air basin held primarily in the surface layer. However, some venting up the mountain slopes is also predicted. By 4:00 p.m., transport of precursors while they react to form O_3 into the eastern region has occurred, generating higher surface concentrations. Increased venting along the mountain slopes injects these high O_3 levels into the free troposphere. By 8:00 p.m. (Fig. 16.25c), a stable boundary layer has formed (see Fig. 2.20), and the aged air containing high concentrations of O_3 and other photochemically formed species, including particles, are found in the region of the temperature inversion; the inversion itself is enhanced by the movement of cooler marine air inland from the west.

This phenomenon of elevated layers of pollutants aloft and subsequent mixing to the surface is not restricted to areas such as Los Angeles that have large meteorological effects caused by the mountains that ring this region. For example, Berkowitz and co-workers report similar observations near Nashville, Tennessee, over the eastern United States, and over the western region of the North Atlantic Ocean (Berkowitz and Shaw, 1997; Berkowitz et al., 1995, 1998).

These layers containing higher concentrations of pollutants provide an important mechanism for transport of ozone, particles, and their precursors to the free troposphere. In addition, in the morning when solar heating causes turbulent mixing (Fig. 2.20), these pollutants are mixed down to the surface. This not only increases the surface concentrations but also provides species that can initiate the VOC–NO_x chemistry that leads to more ozone formation. As a result, there is a carryover from one day to the next, leading to smog episodes in which the pollutant concentrations increase from day to day.

Most Eulerian models assume that emissions are immediately mixed within the grid box into which they are emitted. However, depending on the size of the grid and the particular meteorological conditions, this may not be a good assumption. Stockwell (1995) has examined the effects of this assumption on model predictions and suggested how such considerations might be used in developing appropriate grid structures in models.

Finally, temperature is an important meterological parameter that affects the formation of ozone and associated species. It is known from environmental chamber studies (e.g., Hatakeyama et al., 1991), modeling studies, and air quality data (e.g., Cardelino and Chameides, 1990; Sillman and Samson, 1995) that higher temperatures are more conducive to ozone for-

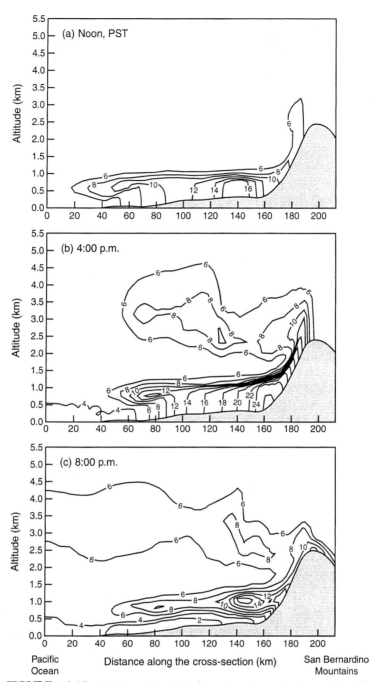

FIGURE 16.25 Model-predicted O_3 (in pphm, where 1 pphm = 10 ppb) west to east across the Los Angeles area South Coast air basin and as a function of altitude and time for August 27, 1987 (adapted from Lu and Turco, 1996).

mation. There are a number of reasons for this, but increased thermal decomposition of PAN and HO_2NO_2 and increased biogenic emissions are believed to be significant contributors.

Clearly, the meteorological inputs have a critical impact on model predictions and effects in uncertainties in these inputs need to be considered in assessing model predictions.

(2) Initial and boundary conditions As the name implies, initial conditions are the initial concentrations of all of the species considered in the model. Since

transport from one cell to the next is also an important determinant of subsequent concentrations, boundary conditions, that is, the concentrations of species entering the region being modeled, also need to be specified.

In practice, models are usually run to simulate a multiday episode. In this case, the effect of the initial concentrations is reduced in the later days of the simulation due to the chemistry and transport that takes place, but their effect can be detected on day 2 and sometimes on day 3 of the simulation. The assumed boundary conditions, however, can be more important since they represent direct transport into the region. When one models the effects of future control strategies, there are a number of possible assumptions one can use for boundary conditions. For example, one could decrease the emissions in the modeling region but assume that the boundary conditions remain the same, or alternatively, the boundary conditions could be proportionately decreased if it is assumed that the emissions reductions in the region modeled also apply to sources outside the region that are transported in.

Figure 16.26, for example, shows ozone isopleths predicted for the Los Angeles area using the CIT Eulerian model under two different assumptions for the boundary and initial conditions (Winner *et al.*, 1995). In Fig. 16.26a, the pollutant concentrations measured in the past were used to set the boundary and initial conditions. In Fig. 16.26b, concentrations measured in clean air over the Pacific Ocean upwind of the modeled area were used instead. The axes are given in terms of the percentage of VOC or NO_x emissions compared to the baseline level on August 28, 1987, rather than in terms of absolute concentrations as for some of the isopleths seen earlier. Not only are the predicted ozone concentrations for a given percentage reduction in the precursors different, but the shapes of the isopleths also change. Under the assumption of clean boundary and initial conditions, the isopleth shape is similar to the traditional isopleths derived from box models and environmental chamber studies (e.g., see Fig. 16.14). However, they become more L-shaped when the historical boundary and initial conditions are used. This is because the highest O_3 concentrations occur downwind due to the delayed production of O_3 at low VOC/NO_x ratios, so that the isopleths are similar to those shown in Fig. 16.16 for the eastern end of the air basin.

In a similar vein, Kuklin and Seinfeld (1995) have shown that the predicted reductions in VOC needed to attain the federal air quality standard of 0.12 ppm for O_3 in the Los Angeles area for a given level of NO_x emissions can differ by almost a factor of two, depending on the initial boundary conditions assumed.

The critical point is that the optimal control strategy derived from such models will be clearly impacted considerably by which set of isopleths, i.e., which set of boundary and initial conditions, is chosen.

(3) Emissions If the meteorology, chemistry, and emissions submodels are all reasonably correct, one should be able to predict the concentrations of various species, particularly the organics that are directly emitted. An alternative that removes some of the uncertainties associated with the meteorology is to compare measured with predicted *ratios* of compounds.

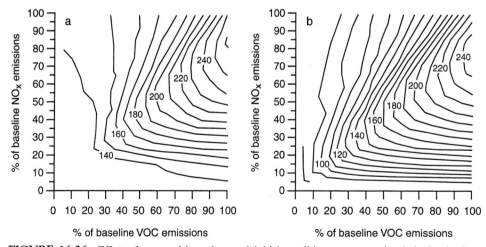

FIGURE 16.26 Effect of assumed boundary and initial conditions on ozone isopleths in the Los Angeles area for reductions in VOC and NO_x from August 28, 1987 levels: (a) using historical values of measured concentrations; (b) clean air values for boundary and initial conditions (adapted from Winner *et al.*, 1995).

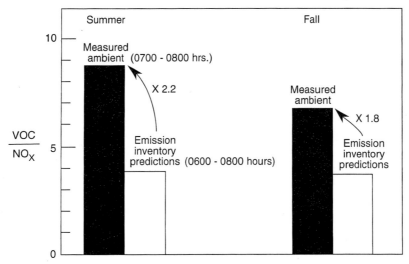

FIGURE 16.27 Measured ratio of VOC to NO_x in the Los Angeles area during the 1987 Southern California Air Quality Study (SCAQS) to that calculated from emission inventories. (Data from Fujita *et al.*, 1992.)

Figure 16.27 shows the results of one such comparison for the Los Angeles area during a comprehensive field campaign in 1987 when extensive chemical, meteorological, and emissions measurements were made (Fujita *et al.*, 1992). The VOC/NO_x ratio measured in both the summer and fall was approximately a factor of two larger than predicted from the emissions inventories for this region. A similar discrepancy was observed for the CO/NO_x ratio. These discrepancies are not unique to Los Angeles. For example, Fig. 16.28 shows a similar comparison between measured values of this ratio and that calculated from the emissions inventories for a number of urban areas in the United States. Again, the observed VOC/NO_x ratio was much higher than expected.

Clearly, such results suggest that either VOC emissions are larger than the estimates, NO_x emissions are lower, or both. It is now known that it is primarily increased VOC emissions that are responsible for this discrepancy and that it is to a large extent the mobile source emissions that had been underestimated. This conclusion comes primarily from tunnel studies in which measurements of VOC, CO, and NO_x concentrations are made in highway tunnels and compared to the calculated concentrations (or ratios of concentrations) predicted by mobile source emissions models. The latter have been developed at both the federal and state levels in the United States using emissions measurements on cars made under specified combinations of operating conditions and loads designed to represent an "average" driving cycle.

Figure 16.29 shows one such comparison using the 7C version of the California Air Resources Board mobile source emissions model known as EMFAC, which was in use at the time the ambient air measurements in Fig. 16.27 were made. The measured VOC/NO_x ratios in the tunnel were a factor of 4.0 ± 1.8 times larger than was predicted, and the CO/NO_x a factor of 2.7 ± 0.8 times larger. (Ratios are more accurately measured in such studies than absolute emissions, due to the need to correct for aerodynamic effects; e.g., see Rogak *et al.*, 1998a.) As discussed by Pierson *et al.* (1990) and the National Research Council report (1991), the same sort of discrepancy was also observed in other tunnel studies as well as in measurements made by the side of the road that have been used to estimate emission rates. It was also shown in such tunnel studies that the NO_x levels varied from 60 to 140% of those predicted by the emissions models. This suggests that it is primarily the VOC and CO that are being underestimated.

Since then, mobile source emissions models have been revised to take these increased emissions into account. As expected, the agreement between the observed and predicted concentration ratios is improved in most cases (e.g., see Gertler and Pierson, 1994; Gertler *et al.*, 1996; Robinson *et al.*, 1996; Pierson *et al.*, 1996; Kirchstetter *et al.*, 1996; McLaren *et al.*, 1996b; and Pollack *et al.*, 1998).

Similarly, increasing the on-road exhaust vehicle emissions by a factor of three in a Eulerian model was found to improve the agreement between predicted and observed concentrations of O_3 and organics in the

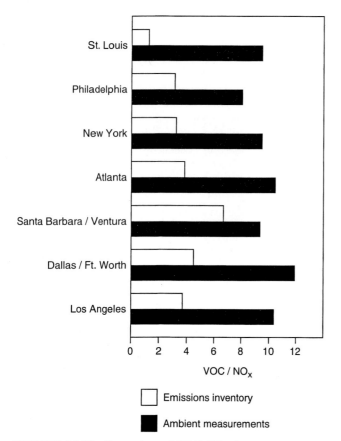

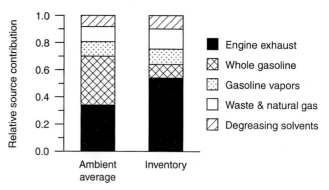

FIGURE 16.30 Relative source contributions to VOC from a chemical mass balance model and from an emissions inventory (adapted from Harley *et al.*, 1992). Note that this does not include all sources of VOC in this area.

FIGURE 16.28 Comparison of VOC/NO_x from measurements in ambient air to those calculated from emission inventories for some cities in the United States (adapted from National Research Council, 1991, based on data in Morris, 1990).

Los Angeles area (Harley *et al.*, 1993a, 1993b; Harley and Cass, 1995). This is not surprising in this case since emissions associated with automobile use are estimated to be a large fraction of the total organic emissions. Figure 16.30, for example, shows the relative contribu-

tions of engine exhaust, whole gasoline, gasoline vapors, waste and natural gas, and degreasing solvents to the total VOC estimated using a chemical mass balance model (Harley *et al.*, 1992), compared to that in the basinwide emissions inventory; it should be noted that *not* included in these classifications are contributions from biogenic organics (which may be quite substantial in this area; e.g., see Harley and Cass, 1995), domestic solvent use, or surface-coating activities. However, it is clear from Fig. 16.30 that for the five particular source categories shown, the combination of organics from engine exhaust, whole gasoline, and gasoline vapors associated with the use of motor vehicles forms a very large portion, about 80%, with exhaust alone accounting for approximately a third of these five categories. Hence it is not surprising that increasing the vehicle emissions brings the models into better agreement with observations.

There are a variety of potential sources of this historical underestimation of VOC and CO mobile

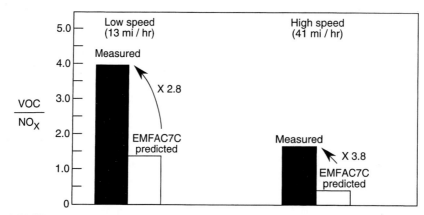

FIGURE 16.29 Ratio of VOC to NO_x measured in the 1987 Southern California Air Quality Study (SCAQS) in a tunnel compared to predictions using the California mobile source model, EMFAC7C (see Pierson *et al.*, 1990; and Ingalls *et al.*, 1989).

source emissions (e.g., see Ross et al., 1998). These include (1) inadequate consideration of evaporative emissions that occur while the car is running and the fuel warms up, as well as while the car is parked, either due to residual engine heat or to changes in the external temperature; (2) actual driving habits, e.g., accelerations, decelerations, and idling, being poorly represented by the test cycles used to develop the mobile source emissions models; (3) tampering with emission control devices and engine parameters; (4) overestimated effectiveness of catalysts, especially as they age; and (5) overestimated effectiveness of vehicle inspection and maintenance programs.

There is evidence for contributions from all of these factors. For example, the ambient air data summarized in Fig. 16.30 suggest that emissions of whole gasoline are significant (Harley et al., 1992). The source is not clear but likely includes unburned fuel emitted in the exhaust as well as fuel spillage etc. Similar observations have been made in other locations such as Chicago (Doskey et al., 1992), Atlanta (Conner et al., 1995), Mexico City (Riveros et al., 1995), and Toronto, Canada, where unburned gasoline appeared to account for up to 37% of total gasoline-related organics in the summer (McLaren et al., 1996a). Estimates of the contribution of whole gasoline to exhaust emissions based on tunnel studies cover a wide range. This contribution of whole gasoline to VOC in tunnels has been estimated to be from <5 to 25%, depending on the vehicle exhaust emission profile used to analyze the data (Fraser et al., 1998), and as high as 63% when the data are analyzed using the composition of gasoline as a surrogate for exhaust emissions (McLaren et al., 1996b).

It is also clear that on-road driving habits may not be well represented by the test cycles used to develop the emissions models. For example, Kelly and Groblicki (1993) point out that lower acceleration rates have been historically used in the test cycles in order to avoid slipping of the belt-driven dynamometers than are encountered in real driving. Figure 16.31 shows the results of on-board tailpipe emissions measurements during a portion of a real driving cycle in which the driver was accelerating up a hill (Kelly and Groblicki, 1993). At point A, the throttle changed from 20 to 50% open, resulting in a rich air/fuel ratio and increase in the CO emission from <0.1 to more than 3.5 g s^{-1}. At point B, the throttle was released slightly, and at point C, the throttle was opened again until point D. During this particular, 29-s period, a total of 165 g of CO was emitted, equivalent to the emissions from driving almost 50 miles at the U.S. federal standard of 3.4 g of CO per mile! Organic emissions are also increased during such accelerations, but not by as large a factor as the CO (Kelly and Groblicki, 1993). For example, over a similar acceleration period, the hydrocarbon emission rate would be equivalent to driving about 17 miles at the U.S. standard of 0.41 g of hydrocarbon per mile (see Table 16.3).

Similar results have been obtained by other researchers. For example, De Vlieger (1997) measured emissions while driving in a small town, in a rural area, and on a highway, respectively, in Belgium. In the urban driving, "aggressive" driving (also noted as "sporty") led to emissions of CO and hydrocarbons that were a factor of 2–3 times those during normal driving while those of NO_x were about 50% higher; differences during highway driving were much smaller, ~5–20%. Similarly, Cicero-Fernández et al. (1997) measured emissions from a vehicle during various driving cycles in the Los Angeles area and found significant increases in emissions while traveling up grades. The increases in emissions for each 1% increase in the grade were 3.0 g per mile for CO and 0.04 g per mile for hydrocarbons.

Since the first tunnel and roadway studies suggested that organic and CO mobile source emissions were being underestimated, remote-sensing techniques that can be applied to individual automobiles as they travel past a sensor have been developed and applied. The technique has been used not only as cars pass by the sensor on freeway on-ramps, etc. (e.g., see Bishop and Stedman, 1990, 1996; Stephens and Cadle, 1991; Rueff, 1992; Bishop et al., 1992, 1993, 1996, 1997; Lawson et al., 1990; and Cadle and Stephens, 1994), but also in tunnels (Bishop et al., 1994). The device is based on using an infrared beam to measure absorbances due to CO, hydrocarbons, and CO_2; from the ratios of CO and hydrocarbons to CO_2, the percentage of these two species in the exhaust and the grams emitted per gallon of fuel used can be derived. TDLS systems have also been developed for such measurements (Nelson et al., 1998). NO emissions have also been measured (Zhang et al., 1996b; Jiménez et al., 1999). Simultaneously, an image of the license plate is recorded so that its participation in mandated emission control programs can be traced. High CO emitters identified by remote sensors have been shown to correlate well with high emitters

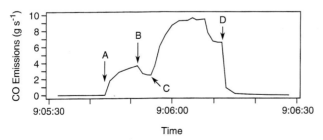

FIGURE 16.31 Increased CO emissions due to rapid acceleration uphill (adapted from Kelly and Groblicki, 1993).

identified by the traditional test cycle approach (Stephens *et al.*, 1996a), although a fraction of automobiles appear to have sufficiently variable emissions that they may escape identification by this method (Bishop *et al.*, 1996). The measurement for hydrocarbons appears not to be as consistent. For example, Stephens *et al.* (1996b) compared the results of remote sensors to other approaches, including gas chromatography, flame ionization, FTIR, and nondispersive IR. Individual VOC were measured in vehicle exhaust generated using dynamometers. For the exhaust samples, the VOC/CO_2 ratios measured using remote sensing were smaller than measured at the dynamometer bench, suggesting that remote sensing may not as accurately reflect total organics in vehicle exhaust.

Remote sensor studies as well as tunnel studies using conventional sampling (e.g., Rogak *et al.*, 1998b) have provided substantial evidence for increased emissions due to poor maintenance, tampering with emission control devices, and overestimating the effectiveness of catalysts, other emission control systems, and inspection and maintenance (I/M) programs (e.g., see Tiao *et al.*, 1989; Bishop and Stedman, 1990; Lawson *et al.*, 1990; Lawson, 1993; Zhang *et al.*, 1993, 1995, 1996a; Bishop *et al.*, 1993, 1996; Calvert *et al.*, 1993; Sjödin, 1994; Stephens, 1994; Harrington and McConnell, 1994; Beaton *et al.*, 1995; Gabele, 1995; Pierson, 1996; and Stedman *et al.*, 1997, 1998).

Remote sensor studies have also been effective in identifying the problems of "superemitters," that is, cars that have extremely high emissions. It has now been shown in a variety of studies that a small fraction of automobiles account for a large fraction of the total mobile source emissions of both CO and hydrocarbons. For example, during studies in Utah and in California, approximately 10% of the vehicles were found to be responsible for 50% of the CO emissions as well as 50% of the on-road hydrocarbon emissions; some (~5%) were super-emitters for both pollutants (Bishop *et al.*, 1993; Stephens, 1994; Beaton *et al.*, 1995). Similarly, remote-sensing studies in Mexico City showed that 12% of the fleet was responsible for half of the hydrocarbon emissions and 24% of the fleet produced half of the CO emissions (Beaton *et al.*, 1992).

Figure 16.32, for example, shows the average hydrocarbon exhaust emissions measured in four cities worldwide as a function of model year and divided into five groups by percentages, i.e., quintiles (Zhang *et al.*, 1995). Note that the scales are quite different, with the vehicles in Leicester emitting far higher concentrations than those in other cities. Zhang *et al.* (1995) propose that, given the similarity in vehicles and fuels in the two European fleets, the lower emissions in Gothenburg are due to better maintenance. However, in all cases, it is seen that the automobiles with the highest 20%

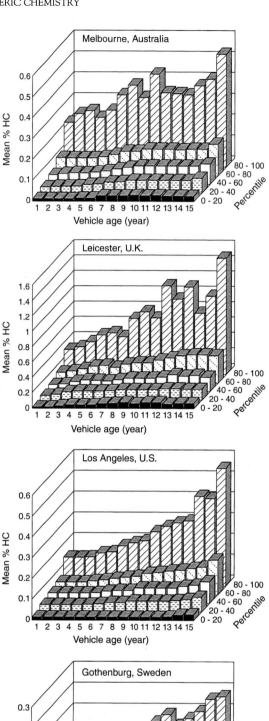

FIGURE 16.32 Average percentage of hydrocarbon in exhaust emissions by model year and grouped into five groups from lowest emissions (front) to highest (back) in four cities. Note the different scales (adapted from Zhang *et al.*, 1995).

hydrocarbon emissions are emitting hydrocarbons at concentrations that are a factor of two or more, on average, than the next quintile. Similar distributions were measured for CO emissions. In addition, it is not exclusively the older vehicles that fall into this superemitter class; even some relatively new automobiles had much higher emissions than expected, indeed higher than many of the well-performing older vehicles. However, the fraction of cars that are super-emitters declines for newer vehicles (Stephens, 1994).

Similar conclusions have been reached with respect to NO emissions; in an initial application of this remote-sensing technology, the highest 10% of automobile emitters were responsible for almost half of the NO emissions (Zhang et al., 1996b; Jiménez et al., 1999).

Because of large variations in emissions from car to car and from one set of driving habits to another, it has been suggested that a mobile source emissions inventory based on total fuel consumption rather than on predetermined test cycles may be more appropriate in the future (e.g., see Singer and Harley, 1996). This is supported by tunnel studies in which emissions of organics in grams per gallon of fuel used were relatively independent of whether the vehicle was going uphill or downhill, despite the fact that emissions on a grams per mile basis were much greater for the uphill vehicles (as was the fuel consumed) (Sagebiel et al., 1996).

However, while these unanticipated deviations from control programs are clearly important in devising future control strategies, it is important to recognize the tremendous achievements that have occurred in mobile source emission controls over the past two decades. Table 16.3 shows the history of light-duty motor vehicle emission standards for NO_x, CO, and hydrocarbons in the United States and the more stringent California standards. Figure 16.33 shows the distribution of exhaust emissions of CO and hydrocarbons measured using a remote-sensing technique for three different model years (Stephens, 1994). Clearly, a dramatic decline has occurred in these emissions as the emissions standards have been made more stringent. A similar

TABLE 16.3 Light-Duty Motor Vehicle Emission Standards in the United States and California[a]

	Federal					California			
Model year	HCs (g/mi)	CO (g/mi)	NO_x (g/mi)	Evap[b] (g/test)	Particulate[i] matter (g/mi)	HCs (g/mi)	CO (g/mi)	NO_x (g/mi)	Evap[b] (g/test)
Precontrol	10.60[c]	84.0	4.1	47		10.60[c]	84.0	4.1	47
1966						6.30	51.0	(6.0)	
1968	6.30	51.0	(6.0)[d]			6.30	51.0		
1970	4.10	34.0				4.10	34.0		6
1971	4.10	34.0				4.10	34.0	4.0	6
1972	3.00	28.0				2.90	34.0	3.0	2
1973	3.00	28.0	3.0			2.90	34.0	3.0	2
1974	3.00	28.0	3.0			2.90	34.0	2.0	2
1975	1.50	15.0	3.1[e]	2		0.90	9.0	2.0	2
1977	1.50	15.0	2.0	2		0.41	9.0	1.5	2
1978	1.50	15.0	2.0	6[e]		0.41	9.0	1.5	6[e]
1980	0.41	7.0	2.0	6		0.39[f]	9.0	1.0	2
1981	0.41	3.4	1.0	2		0.39[g]	7.0	0.7	2
1983	0.41	3.4	1.0	2		0.39	7.0	0.4[h]	2
1984	0.41	3.4	1.0	2	0.6	0.39	7.0	0.4	2
1985	0.41	3.4	1.0	2	0.4	0.39	7.0	0.4	2
1986	0.41	3.4	1.0	2	0.2	0.39	7.0	0.4	2
1987	0.41	3.4	1.0	2	0.08	0.39	7.0	0.4	2
1989	0.41	3.4	1.0	2	0.08	0.39	7.0	0.4	2
1990	0.41	3.4	1.0	2	0.08	0.39	7.0	0.4	2
1991	0.41	3.4	1.0	2	0.08	0.39	7.0	0.4	2
1992	0.41	3.4	1.0	2	0.08	0.39	7.0	0.4	2
1993	0.41	3.4	1.0	2	0.08	0.25	3.4	0.4	2

[a] From Calvert et al. (1993).
[b] Evaporative emissions.
[c] Crankcase emissions of 4.1 g/mi not included; fully controlled.
[d] Emissions of NO_x (no standard) increased with control of HCs and CO.
[e] Change in test procedure.
[f] Non-methane HC standard (or 0.41 g/mi for total HCs).
[g] Optional HC standard, 0.3 g/mi, requires 7-year or 75,000-mi limited recall authority.
[h] A 0.7 NO_x optional standard for 1983 and later, but requires limited recall authority for 7 years or 70,000 mi.
[i] Diesel passenger cars only.

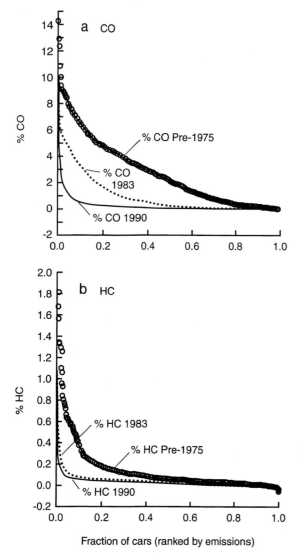

FIGURE 16.33 Distribution of exhaust CO and hydrocarbon (HC) vehicle exhaust emissions as a function of model year in the United States (adapted from Stephens, 1994).

was also lower for the catalyst-equipped vehicles, 2.1 g of O_3 per km driven compared to 8.2 g of O_3 per km, due to a reduction in the relative amounts of alkenes and substituted aromatics (Duffy et al., 1999).

In Sweden, three-way catalysts have been required on all cars since 1989, and tax incentives were offered to purchase such vehicles in the 1987 and 1988 model years. Figure 16.34 shows the CO and hydrocarbon exhaust emissions as a function of model year of gasoline-powered cars, measured using a remote-sensing technique (Sjödin, 1994). There is a large decrease in the emissions from 1987 to 1988 and 1989, supporting the effectiveness of these motor vehicle exhaust controls.

For a description of some of the issues involved in motor vehicle emissions control, see Calvert et al. (1993), for those involving diesel emissions, see Walsh (1995), and for motorcycle emisions compared to passenger cars, see Chan et al. (1995).

While there has been a great deal of work on emissions from motor vehicles, with emphasis on why the VOC and CO emissions have been historically underestimated, a similar problem appears to exist with respect to stationary source emissions, at least in some areas. For example, Henry et al. (1997) measured organic gases in an industrial area in Houston, Texas, and compared them to reported emissions inventories. Application of a multivariate receptor model revealed large inconsistencies between the measurements and expected concentrations.

Another group of sources that have been increasingly recognized as being potentially important are biogenic in nature (see Section A.2 in Chapter 2 and Sections J.1 and J.2 in Chapter 6). Organics such as isoprene are emitted by deciduous (hardwood) trees and to a smaller extent from other sources such as phytoplankton in the ocean (Graedel et al., 1986; Bonsang et al., 1992; Moore et al., 1994; Milne et al., 1995; McKay et al., 1996), and conifers (softwoods) are sources of organics such as α-pinene. It should be noted that recent studies indicate that some of the organics that are normally thought of as solely biogenic in origin, such as isoprene, may also be produced in automobile exhaust (e.g., see McLaren et al., 1996a). Methyl vinyl ketone and methacrolein, the major atmospheric oxidation products of isoprene, also appear to be generated in automobile exhaust (Biesenthal and Shepson, 1997). dl-Limonene has been reported to be formed in significant yields in the pyrolysis of used automobile tires (Pakdel et al., 1991).

Such organics are highly reactive with essentially all oxidants of tropospheric interest, including OH, O_3, NO_3, and chlorine atoms. For example, the lifetimes of

decline in CO has been observed in tunnel study measurements (Pierson, 1995).

A similar observation has been made in other countries where exhaust controls have been instituted. For example, vehicle emissions measurements made by remote sensing in Monterrey, Nuevo Léon, Mexico, showed that 1995 model years emitted 75% less CO, 70% less hydrocarbons, and 65% less NO compared to the pre-1991 vehicles without emission controls (Bishop et al., 1997). In Australia, emissions for pre-1986 vehicles were substantially larger than those from newer, catalyst-equipped cars, a factor of ~ 4 for hydrocarbons, ~ 2.5 for CO, and ~ 2 for NO_x; in addition, the reactivity (see Section 16.B) of the exhaust emissions

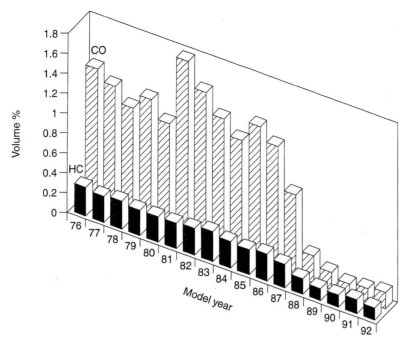

FIGURE 16.34 Average percentages of CO and HC in automobile exhaust by model year in Sweden (adapted from Sjödin, 1994).

isoprene with respect to OH at 1×10^6 cm^{-3}, O_3 at 30 ppb, NO_3 at 10 ppt, and Cl at 1×10^5 cm^{-3} are approximately 3, 30, 2, and 6 h, respectively. Because these compounds are so reactive, they can contribute significantly to the formation of O_3 and other secondary pollutants (vide infra) and hence are important to include in both urban- and regional-scale models.

For example, Chameides and co-workers (1988) suggested that the emissions of biogenic VOC in the southeastern United States may be sufficiently large that without concurrent NO_x controls, even 100% control of the anthropogenic emissions of VOC would not be sufficient to meet the air quality standard for O_3. When the high reactivity of biogenics is taken into account (vide infra), their contribution to O_3 formation is proportionately much larger than expected simply on the basis of concentrations (e.g., see Cardelino and Chameides, 1995). At some locations even in relatively polluted urban areas such as Los Angeles, the contribution may be more important than previously realized (e.g., see Harley and Cass, 1995). However, in the latter case, a majority of the emissions appears to occur in the mountains on the northern and eastern ends of the airshed where ozone is most sensitive to NO_x; hence their contribution to ozone formation may be less than would be implied by their total mass (e.g., Benjamin et al., 1997).

The relative importance of biogenic emissions depends on the nature of the air mass into which they are emitted, which thus includes where in a particular air basin they are emitted as well. For example, model calculations suggest that if there are large NO_x sources, the interaction with biogenic emissions can lead to significant O_3 formation. (On the other hand, if biogenic emissions are small compared to anthropogenic sources, this is not the case (Roselle et al., 1991).)

As expected, then, inclusion of biogenic emissions in models can have a significant effect under some conditions on the predicted effects of VOC versus NO_x control. For example, Pierce et al. (1998) show that when increased isoprene emissions are included in the RADM model, ozone formation in many regions of eastern North America is predicted to be more sensitive to reductions in NO_x rather than in VOC.

The relative contributions of biogenics clearly will be location dependent. For example, studies based on radiocarbon (^{14}C) abundances suggested that the biogenic contribution to the total VOC concentration in Atlanta, Georgia, was only 9–17%, with a level of uncertainty that it could be as low as 0% (Klouda et al., 1996). On the other hand, modeling studies of the northeast United States suggest that based on reactivity (vide infra), biogenic isoprene emissions could contribute up to 30% of the total organic reactivity at midday (Mathur et al., 1994).

Urban vegetation can impact ozone levels not just in terms of their chemical reactions, but also indirectly.

TABLE 16.4 Properties of Some Models That Incorporate Particles[a]

Property	Pilinis and Seinfeld (1988)	Lurmann et al. (1997)	Meng et al. (1998)	Binkowski and Shankar (1995)
Basic model[b]	CIT	UAM	CIT	RADM 2
Resolution (km)				
Horizontal	5	4–10	5	80
Vertical	5 layers to ~1.1 km	4 layers to ~1.5 km	5 layers to ~1.1 km	15 layers to ~15 km
Particle size resolution				
Number of bins	3	8	8	2
Range of D_p (μm)	0.05–10	0.04–10	0.04–10	0.01 and 0.07
Aerosol particles				
Organics	Yes	Yes	Yes	No
Elemental carbon	No	No	Yes	No
Inorganics	Yes	Yes	Yes	Yes
Nucleation	H_2SO_4/H_2O	H_2SO_4/H_2O	H_2SO_4/H_2O	H_2SO_4/H_2O
Coagulation	Yes	No	No	Yes
Condensation/evaporation	Volatile inorganics: G–P equilibrium[c] Sulfate and organics: transport	Volatile inorganics: G–P equilibrium[c] Sulfate and organics: transport	H_2O and CO_2: equilibrium Others: transport	Volatile inorganics: G–P equilibrium[c] Sulfate transport
Chemical reactions	No	No	No	No
Activation/interactions with fogs and clouds	No	No	No	No

[a] Adapted from Meng et al. (1998).
[b] CIT = California Institute of Technology/Carnegie Institute of Technology; UAM = Urban Airshed Model; RADM 2 = Regional Acid Deposition Model.
[c] G–P = gas–particle.

For example, Taha (1996) suggests that increased urban vegetation with low organic emission rates may lead to a net *decrease* in O_3 formation by lowering surface temperatures and biogenic emission rates as well as increasing dry deposition of pollutants.

In summary, at the present time there are large uncertainties in both the magnitude and speciation of emissions of organics in various areas, which introduces significant uncertainties into the model predictions. This is an area that clearly warrants continuing attention.

(4) Chemistry Clearly, the kinetics and mechanisms of reactions in the chemistry submodels play a key role in the predicted concentrations of O_3 and other secondary pollutants. Since key issues involved in incorporating this complex chemistry faithfully into models have been discussed above and indeed are the focus of this book, we shall not consider this further.

In summary, the development and application of urban and regional grid-scale models are critical to the development and implementation of cost-effective control strategies. However, these models are very complex, not just in their chemistry but also in a variety of other critical parameters, such as meteorology, emissions, and deposition. Discussion of these issues is found in Seinfeld (1988), in the 1991 National Research Council report "Rethinking the Ozone Problem in Urban and Regional Air Pollution," in a series of papers in "Regional Photochemical Specialty Measurement and Modeling Studies" (*J. Geophys. Res., 100* (D11), 1995), and in the proceedings of a conference on regional photochemical measurements and modeling (*J. Air Waste Manage. Assoc., 45*, 253, 1995; *Atmos. Environ., 30* (12), 1996).

e. Models Incorporating Particles

For a number of years, models included gas-phase chemistry only. However, increasing computational power and a larger data base on the physical and chemical properties of particles in the atmosphere and their interaction with gases have permitted interactions with aerosol particles to be included in more recent years. Table 16.4 summarizes some of the characteristics of several models that incorporate particles. These models treat the transport of the gases to the particles, their uptake, and ultimately equilibria between the gas and aqueous phases. Although they all include inorganics in the particles, they vary in whether they also include organics and elemental carbon in the condensed phase. The next generation of models will incorporate real-time chemistry as well, rather than assuming equilibrium as is commonly done for many species.

B. REACTIVITY OF VOC

It has been recognized since the early work of Haagen-Smit and co-workers (e.g., see Haagen-Smit *et al.*, 1953; Haagen-Smit and Fox, 1956) that different types of organic compounds react at different rates to form O_3 and other secondary pollutants in irradiated VOC–NO_x mixtures. Indeed, it was this recognition of differing reactivities that has led to the development of traditional control strategies for non-methane hydrocarbons, NMHC. Thus, methane reacts sufficiently slowly over the time scale of hours to a few days that it does not contribute significantly to photochemical smog formation. It reacts only with the OH radical at a significant rate (also with chlorine atoms in coastal regions), with a calculated lifetime of about 5 years at an OH concentration of 1×10^6 cm^{-3}. Thus although methane is important in the global troposphere and stratosphere (see Chapters 12 and 14), it is essentially unreactive in terms of photochemical oxidant production on urban and regional scales. On the other extreme are most alkenes and aldehydes, which are highly reactive (vide infra).

Because of this recognition of different reactivities of organics, various regulations have been promulgated over the years that attempt to take into account the fact that 2-butene, for example, will produce more O_3 than ethane. An early example is what was known as "Rule 66," implemented by the Los Angeles Air Pollution Control District in 1966 to limit solvent emissions based on their reactivity. For a discussion of some of the reactivity scales used prior to 1975, see Darnall *et al.* (1976), Bufalini *et al.* (1976), Bufalini and Dodge (1983), Dodge (1984), Hough and Derwent (1987), Derwent and Jenkin (1991), and the papers on applications to motor vehicle exhaust discussed below.

More recently, there has been a major focus on applying the same principle to the emissions from motor vehicles. We discuss here the scientific basis for this approach, with some recent examples of their application.

1. Typical Reactivity Scales

The concept of *reactivity scales* is based on ranking organics in terms of their potential for ozone production. A number of different parameters have been used to rank organics by their reactivity, including observed rates of reaction, product yields, and effects observed from irradiated VOC–NO_x mixtures. For example, the rates of O_3 or NO_2 formation or the hydrocarbon loss have been used to develop reactivity scales, as have the yields or dosages of products such as O_3 and PAN in

TABLE 16.5 OH Reactivity Scale for Hydrocarbons

Class	$k_{OH}{}^a$ (298 K)	$\tau_{1/2}^{OH\,b}$	Typical hydrocarbons
I	$\leq 8 \times 10^{-14}$	≥ 100 days	CH_4
II	$(8-80) \times 10^{-14}$	10–100 days	Acetylene, ethane, benzene
III	$(8-80) \times 10^{-13}$	1–10 days	Ethene, propane, toluene
IV	$(8-80) \times 10^{-12}$	2–24 h	Propene, o-, m-, and p-xylene, 1,2,4- and 1,2,5-trimethylbenzene
V	$\geq 8 \times 10^{-11}$	<2 h	2-Methyl-2-butene, d-limonene

Source: Adapted from Darnall *et al.* (1976).
[a] Units of cm^3 molecule^{-1} s^{-1}; note that k (OH + CH_4) has been revised downward (Chapter 6).
[b] Half-lives in atmosphere with respect to reaction with OH assuming [OH] = 1.5×10^6 radicals cm^{-3}.

smog chamber studies. Effects such as plant damage, eye irritation, or visibility reduction have also been shown to be useful parameters.

The importance of OH radicals in atmospheric chemistry is the basis of another reactivity scale for organics that do not photolyze in actinic radiation (Darnall *et al.*, 1976; Wu *et al.*, 1976). This scale is based on the fact that, for most hydrocarbons, attack by OH is responsible for the majority of the hydrocarbon consumption, and this process leads to the free radicals (HO_2, RO_2) that oxidize NO to NO_2, which then leads to O_3 formation. Even for alkenes, which react with O_3 at significant rates, consumption by OH still predominates in the early portion of the irradiation before O_3 has formed. It has therefore been suggested that the rate constant for reaction between OH and the hydrocarbon should reflect the overall reactivity of the hydrocarbon. The faster the hydrocarbon reacts with OH, the faster HO_2 and RO_2 are produced (and in higher concentrations) and the faster NO is oxidized to NO_2, and O_3 forms. The rate constant for initial OH attack should thus be related to other measures of reactivity of the hydrocarbon (e.g., the rates of NO_2 formation, hydrocarbon loss, and O_3 formation).

Table 16.5 shows the results of a historical grouping of hydrocarbons based on their rate constants for reaction with OH (Darnall *et al.*, 1976), and Table 16.6 is a summary of reactivity using a variety of different assessment parameters, including the yields of ozone/oxidant, PAN, HCHO, or aerosol particles, as well as eye irritation and plant damage (Altshuller, 1966). In general, regardless of the reactivity scale

TABLE 16.6 Summary of 1966 Organic Reactivities Using Various Reactivity Scales

Substance or subclass	Response on 0–10 scale						
	Ozone or oxidant	Peroxyacyl nitrate	Formaldehyde	Aerosol particles	Eye irritation	Plant damage	Averaged response
C_4–C_5 paraffins	0	0	0	0	0	0	0
Acetylene	0	0	0	0	0	0	0
Benzene	0	0	0	0	0	0	0
C_6^+ paraffins[a]	0–4	0[b]	0[b]	0	0[b]	0	1
Toluene (and other monoalkylbenzenes)	4	c	2	2	4	0–3	3
Ethene	6	0	6	1–2	5	+[d]	4
Terminal alkenes[e]	6–10	4–6	7–10	4–8	4–8	6–8	7
Diolefins	6–8	0–2	8–10	10	10	0[b]	6
Dialkyl- and trialkyl benzenes	6–10	5–10	2–4	+[d]	4–8	5–10	6
Internally double-bonded olefins[f]	5–10	8–10	4–6	6–10	4–8	10	8
Aliphatic aldehydes	5–10	+[d]	+[d]	ND[c]	+[d]	+[d]	–

Source: Altshuller (1966).
[a] Averaged over straight-chain and branched-chain paraffins.
[b] Very small yields or effects may occur after long irradiations.
[c] Not reported in this study.
[d] Effect noted experimentally, but data insufficient to quantitate.
[e] Includes measurements on propylene through 1-hexene, 2-ethyl-1-butene, and 2,4,4-trimethyl-1-pentene.
[f] Includes measurements on straight-chain butenes through heptenes with double bond in 2 and 3 position, 2-methyl-2-butene, 2,3-dimethyl-2-butene, and cyclohexene.

chosen, the reactivity of hydrocarbons tends to be in the order [alkenes with internal double bonds] > [di- and trialkyl aromatics, terminal alkenes] > ethylene > [monoalkyl aromatics] > [C_5 and larger alkanes] > [C_2–C_5 alkanes].

The reactivities derived using two different scales are generally in agreement with each other (and with subsequent scales; see below), but there are some notable exceptions. For example, non-methane alkanes are predicted to be relatively more important on the basis of OH reactivity than the other scales would imply. This arises because the OH scale ignores important mechanistic aspects of reactions in irradiated VOC–NO_x mixtures. Thus the initial rate of OH attack does not reflect whether that reaction ultimately leads to the generation of free radicals and hence to continued photooxidation; if the organic does not do so, that is, if it acts as an inhibitor of the chain photooxidations by removing OH, the initial rate of OH attack may assign too high a reactivity. In addition, the OH reactivity scale also does not take into account the nature of the products formed. For example, organics producing highly photolabile species that photolyze to form free radicals may be more reactive overall than indicated by the initial rate of OH attack. In addition, if products such as plant phytotoxicants or lachrymators are formed, the reactivity on the OH scale may be too slow. However, despite these potential deficiencies, the reactivity of organic mixtures in ambient air assessed using the OH reactivity scale has been found to give results that are generally consistent with the reactivity based on ozone formation (e.g., see Uno et al., 1985).

The concept of OH reactivity has been applied to give a "first-cut" assessment of the contribution of various individual organics and sources to photochemical oxidant formation in a number of situations. For example, Chameides et al. (1992) scaled the contribution of various VOC concentrations in a variety of atmospheres from remote to polluted urban areas using OH reactivity. They concluded that while NO_x concentrations decreased from polluted urban areas to rural to remote regions, the total VOC reactivity assessed in this manner was comparable at all continental areas from remote to polluted.

In short, while the OH reactivity scale has a number of caveats associated with its use, it has proven useful in providing at least an initial assessment of relative contributions of organics to photochemical smog formation.

2. Application to Control of Mobile Source Emissions

There has been extensive work on quantitatively assigning reactivities to individual organics in auto exhaust based on their ozone-forming potential. This has in part been catalyzed (no pun intended!) by the development in California of reactivity-based emission standards for organics from passenger cars, light-duty trucks, and medium-duty vehicles (Table 16.7) (Croes et al., 1992). Four classes of "low-emission" vehicles were established: Transition (TLEV), Low (LEV), Ultra Low (ULEV), and electricity-powered, zero-emission vehicles (ZEV).

The TLEV, LEV, and ULEV standards incorporate the concept of *reactivity-weighted mass emissions* of VOC, concurrent with increasingly strict NO_x control. The intent is to regulate based on *equal ozone-forming potentials* of the VOC emissions rather than simply on their total mass. That is, the emission standards for organics are set in terms of the amount of ozone formed in the atmosphere per mile traveled by a given vehicle/fuel combination rather than in terms of the simple total mass of VOC emitted per mile.

The central element to this new strategy is a quantity called the reactivity adjustment factor (RAF). The RAF for a given "new generation" fuel is defined as the ratio of the ozone formed from 1 g of VOC exhaust emissions from a vehicle operating on that test fuel to the ozone formed by 1 g of VOC exhaust emissions from that vehicle operating on current, conventional, industry-average, base gasoline. The number of grams of ozone per gram of VOC exhaust emissions is defined

TABLE 16.7 California's Low-Emission Vehicle Standards for Passenger Cars and Light-Duty Trucks: Introduction of Ozone Reactivity Adjustment Factors[a]

Vehicle category	Grams/mile by pollutant			
	NMOG[b]	NO_x	CO	HCHO
1993	0.250	0.4	3.4	0.015[c]
TLEV (transition)	0.125	0.4	3.4	0.015
LEV (low)	0.075	0.2	3.4	0.015
ULEV (ultralow)	0.040	0.2	1.7	0.008
ZEV[d] (zero)	0.000	0.0	0.0	0.000

[a] These are 50,000-mi exhaust emission standards; fleet average NMOG requirements begin at 0.25 g/mi in 1994 and are progressively reduced in subsequent years to a level of 0.062 g/mi. Any combination of TLEV, LEV, ULEV, ZEV, and 1993 conventional vehicles can be used.

[b] NMHC (non-methane hydrocarbons) for current and 1993 standards, NMOG (non-methane organic gases) with reactivity adjustments for others.

[c] Methanol-fueled vehicles only.

[d] Does not include power generation emissions, which are equivalent to 0.004, 0.02, and 0.017 g/mol for NMOG, NO_x, and CO, respectively, based on South Coast air basin power plant emissions associated with electric vehicles.

as *specific reactivity*. Hence the RAF is given by

$$\text{RAF} = \frac{\text{Specific reactivity of emissions with test fuel}}{\text{Specific reactivity of emissions with base fuel}}. \quad \text{(A)}$$

Determination of the specific reactivity of the exhaust emissions requires accurate knowledge of both the types and amounts of compounds emitted as well as how each contributes to O_3 formation. The latter factor, the ozone-forming potential, is treated in terms of its *incremental reactivity* (IR), which is defined as the number of molecules of ozone formed per VOC carbon atom added to an initial "surrogate" atmospheric reaction mixture of VOC and NO_x:

$$\text{IR} = \frac{\Delta[O_3]}{\Delta[\text{C atoms of VOC added}]}. \quad \text{(B)}$$

Table 16.8 shows incremental reactivities calculated for CO and some typical VOCs found in exhaust emissions (Carter and Atkinson, 1987, 1989; Carter, 1994). An important feature is that the IR of a VOC varies with the VOC/NO_x ratio of the air mass into which it is introduced. The IR values peak at a ratio of ~6 and drop off significantly at higher values, actually becoming negative for benzene and toluene. The reason for the latter is that during the oxidation of these organics, they act not only as free radical generators to convert NO to NO_2 but also as NO_x sinks. The IR of benzaldehyde is negative at all VOC/NO_x since its photooxidation does not result in radical generation, yet does remove NO_x.

The incremental reactivity of a VOC is the product of two fundamental factors, its *kinetic reactivity* and its *mechanistic reactivity*. The former reflects its rate of reaction, particularly with the OH radical, which, as we have seen, with some important exceptions (ozonolysis and photolysis of certain VOCs) initiates most atmospheric oxidations. Table 16.8, for example, also shows the rate constants for reaction of CO and the individual VOC with OH at 298 K. For many compounds, e.g., propene vs ethane, the faster the initial attack of OH on the VOC, the greater the IR. However, the second factor, reflecting the oxidation mechanism, can be determining in some cases as, for example, discussed earlier for benzaldehyde. For a detailed discussion of the factors affecting kinetic and mechanistic reactivities, based on environmental chamber measurements combined with modeling, see Carter *et al.* (1995) and Carter (1995).

The *peak* IR value of a VOC is known as its *maximum incremental reactivity* (MIR). The MIR of some VOCs are given in Table 16.9 and shown schematically in Fig. 16.35. (Note that the units of MIR used are grams of O_3 per gram of VOC added, rather than on a molecule per C atom basis as for the IRs in Table 16.8.) Note the very low reactivity for methane, as discussed earlier. These reactivities are in generally good agreement with experimental values measured

TABLE 16.8 Typical Calculated Incremental Reactivities and Maximum Ozone as a Function of the VOC/NO_x Ratio[a]

VOC/NO_x ratio	4	6	8	10	12	16	20	40
Base case max O_3 (ppb)[b]	72	160	214	215	209	194	180	139
Molecule (k^{OH})[c]	Incremental reactivity (Molecules of O_3/C atoms of VOC added)							
CO (2.4×10^{-13})[c]	0.011	0.025	0.022	0.018	0.016	0.012	0.010	0.005
Ethane (2.5×10^{-13})	0.024	0.054	0.041	0.031	0.026	0.018	0.015	0.007
n-Butane (2.4×10^{-12})	0.10	0.22	0.16	0.12	0.098	0.069	0.052	0.019
Ethene (8.5×10^{-12})	0.85	1.65	0.90	0.64	0.50	0.33	0.30	0.14
Propene (2.6×10^{-11})	1.28	2.04	1.03	0.61	0.51	0.39	0.25	0.14
trans-2-Butene (6.4×10^{-11})	1.42	2.02	0.97	0.62	0.48	0.31	0.23	0.054
Benzene (1.2×10^{-12})	0.038	0.082	0.033	0.011	0.003	−0.002	−0.004	−0.002
Toluene (6.0×10^{-12})	0.26	0.52	0.16	0.04	−0.021	−0.036	−0.058	−0.051
Formaldehyde (9.2×10^{-12})	2.42	3.28	1.20	0.77	0.48	0.32	0.24	0.051
Acetaldehyde (1.6×10^{-11})	1.34	1.83	0.83	0.55	0.42	0.29	0.24	0.098
Benzaldehyde (1.3×10^{-11})	−0.11	−0.15	−0.27	−0.34	−0.37	−0.41	−0.41	−0.40
Methanol (9.3×10^{-13})	0.12	0.27	0.17	0.12	0.091	0.066	0.055	0.029
Ethanol (3.2×10^{-12})	0.18	0.37	0.22	0.14	0.10	0.065	0.038	0.006

[a] From Carter and Atkinson (1989) calculated using the EKMA model with low dilution and an eight-component surrogate mixture of organics chosen to be representative of emissions into California's South Coast air basin. Note that the absolute values used may have changed for some components since the original publication due to further refinement of the model.

[b] Peak ozone predicted from photolysis of initial base case mixture with the specified VOC/NO_x ratio.

[c] Rate constants for reaction with OH at 298 K in units of cm^3 $molecule^{-1}$ s^{-1} taken from Atkinson (1989, 1994) and Atkinson *et al.* (1997a, 1997b).

TABLE 16.9 Maximum Incremental Reactivities (MIR) for Some VOCs

VOC	MIR[a] (grams of O_3 formed per gram of VOC added)
Carbon monoxide	0.054
Methane	0.015
Ethane	0.25
Propane	0.48
n-Butane	1.02
Ethene	7.4
Propene	9.4
1-Butene	8.9
2-Methylpropene (isobutene)	5.3
1,3-Butadiene	10.9
2-Methyl-1,3-butadiene (isoprene)	9.1
α-Pinene	3.3
β-Pinene	4.4
Ethyne (acetylene)	0.50
Benzene	0.42
Toluene	2.7
m-Xylene	8.2
1,3,5-Trimethylbenzene	10.1
Methanol	0.56
Ethanol	1.34
Formaldehyde	7.2
Acetaldehyde	5.5
Benzaldehyde	−0.57
Methyl tert-butyl ether	0.62
Ethyl tert-butyl ether	2.0
Acetone	0.56
C_4 ketones	1.18
Methyl nitrite	9.5

[a] From Carter (1994).

using environmental chambers (e.g., Kelly and Chang, 1999).

The RAF defined by Eq. (A) is calculated from the amounts of the individual compounds emitted and their MIR values as follows (Croes *et al.*, 1992; California Air Resources Board, 1992):

$$\text{RAF} = \sum F_{T,i}(\text{MIR}_i) / \sum F_{B,i}(\text{MIR}_i). \quad (C)$$

In Eq. (C) $F_{T,i}$ is the mass fraction of compound i in the exhaust from the test fuel, $F_{B,i}$ is the mass fraction of compound i in the exhaust from the base fuel, and MIR_i is the maximum incremental reactivity of VOC i. Thus, a specific vehicle/test fuel combination with an RAF of 1.0 is expected to have the same contribution to ozone formation (in terms of specific reactivity, grams of O_3 per gram of VOC exhaust emissions) as that vehicle operating on base gasoline. An RAF < 1.0 means that 1 g of the VOC exhaust emissions will form less ozone than when the vehicle is run on the base fuel. As might be expected, different fuel compositions have a significant effect on the RAF; the amount of aromatics and oxygenates such as methyl tert-butyl ether (MTBE) in the fuel, sulfur content, and distillation temperature are important determinants of the RAF (e.g., see Ho and Winer, 1998).

Table 16.10 shows the reactivity adjustment factors determined for light-duty motor vehicles operated on reformulated gasoline, M85 (85% methanol, 15% gasoline), compressed natural gas, and liquefied petroleum gas. There is no universally accepted definition of "reformulated gasoline"; that shown in Table 16.10 is the "Phase 2" gasoline mandated in California starting in 1996. It has a lower vapor pressure (which will lower evaporative emissions) as well as a lower sulfur content, the latter improving the efficiency of catalytic converters used for hydrocarbons, CO, and NO_x. Other changes include lower concentrations of toxics such as benzene and added oxygenated organics such as alcohols or ethers, which may reduce CO. For a detailed description of reformulated gasolines, see Calvert *et al.* (1993) and Ho and Winer (1998).

The allowed organic (NMOG) mass emissions (grams per mile) for TLEV, LEV, and ULEV are calculated as the appropriate standard given in Table 16.7 divided by the RAF. For example, for an RAF of 0.41, typical of TLEV fueled with M85, the allowed emissions are (0.125/0.41) = 0.30 g per mile (California Air Resources Board, 1993). Table 16.11 summarizes the allowable emissions for various potential fuels. The standards allow a larger total emission of organics as long as the ozone-forming potential of the emissions does not exceed those from conventional gasoline-fueled vehicles. For example, vehicles fueled on compressed natural gas can emit more than double the mass of those on reformulated gasoline because of the relatively low reactivity of the organics emitted. Indeed, the allowed emissions would be even larger in this case if only unburned methane was emitted; however, there are small concentrations of larger organics in the exhaust (and in the fuel itself) that increase the net reactivity of this fuel.

It should be noted that while the RAFs give the *relative* contribution of a test fuel to O_3 formation, they do not reflect changes in total mass emissions that may also occur with the change in fuel. The *net* impact of a fuel–vehicle combination can only be assessed in terms of the amount of O_3 formed per vehicle mile traveled. For example, in California in 1996 the only gasoline allowed was the reformulated Phase 2 gasoline. The RAF for TLEV and LEV operating on Phase 2 gasoline is unity. However, the *total* organic mass emissions

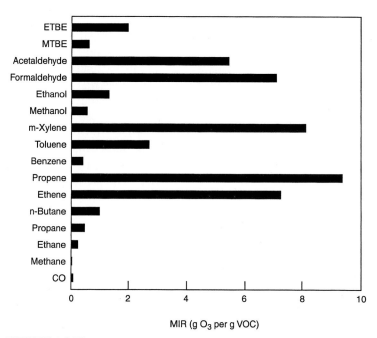

FIGURE 16.35 Maximum incremental reactivities of some organics (grams of O₃ produced per gram of VOC) (data graciously provided by B. Croes, personal communication).

are significantly lower, so that the contribution to net O₃ formation is expected to be substantially reduced.

Similarly, Black *et al.* (1998) measured emissions from a number of vehicle–fuel combinations, including a vehicle fueled on compressed natural gas. They calculate a value for the RAF of 0.87, about twice that reported in Table 16.10. However, the RAF values appear to decrease as the emission rates increase, due to increased contributions from the unreactive unburned fuel. The lower values are consistent with hydrocarbon emission rates of ~0.125 g per mile.

Other ways of expressing the reactivity of organics in terms of ozone formation and their interrelationships are discussed in detail by Carter (1994), and extension of the principles of this approach to other compounds such as nitric acid, PAN, and aerosol particles is discussed by Bowman and Seinfeld (1994), Bowman *et al.* (1995), and Derwent *et al.* (1998).

Because the values of IR and MIR shown in Tables 16.8 and 16.9 were developed using the EKMA box model tested against environmental chamber data (vide

TABLE 16.10 Reactivity Adjustment Factors for Light-Duty Vehicle – Fuel Combinations[a]

Fuel[b]	TLEV[c]	LEV[c]	ULEV[c]
Reformulated gasoline	0.98	0.94	0.94
M85	0.41	0.41	0.41
CNG	1.00	0.43	0.43
LPG	1.00	0.51	0.50

[a] RAFs are adopted in Section 13 of the "California Exhaust Emission Standards and Test Procedures for 1988 and Subsequent Model Passenger Cars, Light-Duty Trucks, and Medium-Duty Vehicles," and the process for establishing RAFs is in Appendix VIII of that document. The document is incorporated by reference in Title 13, California Code of Regulations, Section 1960.1(k); see also California Air Resources Board (1992a, 1993). The last amended date is March 19, 1998.

[b] M85 = mixture of 85% methanol, 15% gasoline; CNG = compressed natural gas; LPG = liquefied petroleum gas.

[c] TLEV = transition low-emission vehicle; LEV = low-emission vehicle; ULEV = ultralow-emission vehicle.

TABLE 16.11 Allowable NMOG Emissions (g/mi) for Various Potential Fuels for Light-Duty Motor Vehicles[a]

Fuel[b]	TLEV	LEV	ULEV
RF-A	0.125	0.075	0.040
RFG	0.128	0.080	0.043
M85	0.305	0.183	0.098
CNG	0.125	0.174	0.093
LPG	0.125	0.150	0.080

[a] See Title 13, California Code of Regulations, Section 1960.1(g)(1).

[b] RF-A is industry-average gasoline; RFG is California reformulated gasoline; M85 is a mixture of 85% methanol, 15% gasoline, CNG = compressed natural gas; LPG = liquefied petroleum gas.

supra), it is important to establish whether the same values for the various reactivity measures would be obtained using an airshed model. Russell and co-workers have applied the CIT airshed model to this problem and concluded that there is some difference, but it is small, ≤ ±15%, and is primarily due to temporal and spatial differences in the emission patterns (McNair et al., 1994; Russell et al., 1995). Chang and Rudy (1990a) have also shown that this approach should be valid even if the total VOC emissions are substantially altered. Other uncertainties in the reactivities due to uncertainties in the kinetics and mechanisms are discussed by Yang et al. (1995, 1996).

This approach has also been applied to sources other than automobiles to assess the relative importance of various organics and sources. For example, Blake and Rowland (1995) used the concept of maximum incremental reactivity to assess the relative importance of various organics in Mexico City. They concluded that liquefied petroleum gas was a major contributor to ozone formation and that relatively small fractions of highly reactive alkenes in the gas contributed disproportionately to ozone formation.

The application of reactivity approaches to consumer products is discussed in detail by Dimitriades (1996).

Analogous approaches to assessing VOC reactivities have been developed by Derwent and co-workers (e.g., see Hough and Derwent, 1987; Derwent and Jenkin, 1991; and Derwent et al., 1996) where a trajectory model is used to calculate the additional ozone production due to the addition of a particular VOC under conditions typical of air masses advected across northwest Europe toward the British Isles. The photochemical ozone creation potential (POCP) index thus calculated is a measure of the reactivity of the particular VOC in terms of O_3 formation. Andersson-Sköld and co-workers (1992) followed a similar approach for the summer conditions in southern Sweden.

Table 16.12 compares the POCP values derived by Derwent et al. (1996, 1998) and Andersson-Sköld et al. (1992) to the MIR approach of Carter (1994). While the general trends in reactivities predicted by each approach are qualitatively similar, there are quantitative differences. For example, the POPC values for the simple alkanes relative to ethene are larger than the MIR values. This reflects in part the details of the mechanisms used in the calculations and the time scale over which the reactions are followed as well as differences in the assumed pollutant mix into which the VOC is injected, such as the VOC/NO_x ratio.

In short, the application of the principles of varying reactivities of organics as one component of the development of cost-effective control strategies is increasingly being accepted as a sound approach to control of ozone and other photochemical oxidants. As discussed briefly below, this is also expected to impact the control of acids and particles as well.

TABLE 16.12 Some Measures of Reactivity for Various Organics[a]

HC or VOC	MIR[b]	POCP[c]	POCP[d]
Methane	0.2	0.6	
Ethane	3	12	13
Propane	6	18	50
n-Butane	14	35	47
Isobutane	16	31	41
n-Pentane	14	40	30
n-Hexane	13	48	45
Cyclohexane	17	29	
Ethylene	100	100	100
Isoprene	123	109	
Benzene	6	22	40
Toluene	36	64	47
o-Xylene	88	105	
m-Xylene	111	111	47
p-Xylene	89	101	47
Methanol	8	13	21
Ethanol	18	39	23
2-Propanol	7	14	20
n-Butanol	36	61	21
Formaldehyde	97	52	26
Acetaldehyde	74	64	19
Acetone	8	9.4	12
Butanone	16	37	18
Formic acid		3	
Acetic acid		10	
Dimethyl ether	10	17	
Methyl tert-butyl ether	8	15	

[a] Adapted from Derwent et al. (1996, 1998). Based on taking ethene as 100 for each scale.
[b] MIR = maximum incremental reactivity scale; from Carter (1994).
[c] POCP = photochemical ozone creation potential; from Derwent et al. (1998).
[d] POCP = photochemical ozone creation potential under typical NO_x conditions for Sweden; from Andersson-Sköld et al. (1992).

C. FIELD OBSERVATIONS OF VOC, NO_x, AND O_3

The most direct approach to assessing the effectiveness of a control strategy or strategies would at first glance appear to be an examination of the relationship between reductions in emissions and measured changes in air quality. As discussed in detail in "Rethinking the Ozone Problem in Urban and Regional Air Pollution" (National Research Council, 1991), this unfortunately has not been possible for a variety of reasons. For example, as discussed earlier, pollutant concentrations are very sensitive to the particular meteorological conditions. In comparing air quality statistics, it is then important to remove variations due to meteorological changes, or at least to reduce them to the same set of conditions. Unfortunately, there is no accepted methodology for doing so; for some relevant treatments, see Rao *et al.* (1992, 1994, 1995, 1996), Lefohn *et al.* (1993), Davidson (1993), and Zurbenko *et al.* (1995). In addition, there are relatively few comprehensive data sets of concentrations of the primary pollutants that can be used to assess whether, in fact, their concentrations have indeed decreased as expected from the mandated reductions in emissions.

It is important to note that there are also some sources of O_3 other than emissions in populated areas. For example, episodes of increased O_3 due to stratospheric intrusion do occur (e.g., see Logan, 1989). In addition, there has been a well-documented (see Chapter 14) increase in global tropospheric ozone levels on top of which the local and regional sources are superimposed.

In short, while a number of field studies and routine monitoring over the years have been directed to examining the relationship between emissions and air quality, with obvious implications for control strategy development, interpretation of the data is not straightforward or unambiguous. With these caveats in mind, we briefly discuss a few of the many studies that have been carried out.

One of the urban areas in which long-term monitoring data for the criteria pollutants are available, in addition to very detailed meteorological and chemical data collected during some intensive field campaigns, is the Los Angeles area. This area has also had stringent controls on both mobile and stationary source emissions for a number of years. Hence it provides a comprehensive case for examining the effects of control strategies. Figure 16.36 shows the measured number of days that O_3 exceeded 0.20 ppm for 1 h for the years 1976 to 1991, as well as the number adjusted for differences in meteorology using an empirical model (Davidson, 1993). An overall decrease of 64% in the number of days can be deduced from the weather-adjusted trends, in spite of a large increase in the number of vehicle miles traveled due to population increases during this time. The number of days (meteorology adjusted) above the federal air quality standard of 0.12 ppm for 1 h decreased 22% over the same period.

Figure 16.37 shows trends in NO_x and particle nitrate as well as SO_2 and particle sulfate in the Los Angeles region over a similar time period (Dolislager and Croes, 1996). These are shown as the mean of the top 30 annual concentrations of each species. While they have not been adjusted for changes in meteorology, effects of such changes are deemphasized by using a 3-year moving mean. It is seen that the trend in NO_x and the associated particle nitrate formed from it (see Chapter 7) is clearly downward, with a decrease of about 30% from 1977 to 1993, again despite the large increase in vehicle usage in this area (see Table 10.27).

It is interesting that over the same period in many other regions of the United States and in Europe and Japan, ozone levels did not appear to change as dramatically (National Research Council, 1991; Lindsay *et al.*, 1989; Rao *et al.*, 1992, 1994, 1995, 1996; Zurbenko *et al.*, 1995; Fiore *et al.*, 1998; Oltmans *et al.*, 1998). The major difference in control strategies in California compared to the U.S. federal approach has been an emphasis on both NO_x and VOC control, rather than primarily on VOC as has been the case at the federal level. For example, Table 16.3 shows the more stringent control of both NO_x and VOC from motor vehicles in California beginning in the mid-1970s. Since 1980, however, VOC emission standards in California have been comparable to the federal standards while the allowed NO_x emissions have been smaller by a factor of two or more.

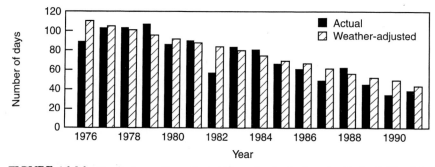

FIGURE 16.36 Trends in weather-adjusted Stage I episode days for O_3 in California's South Coast air basin during the months from May through October, 1976–1991. Stage I episodes are > 0.20 ppm O_3, 1-h average (adapted from Davidson, 1993).

C. FIELD OBSERVATIONS OF VOC, NO_x, AND O_3

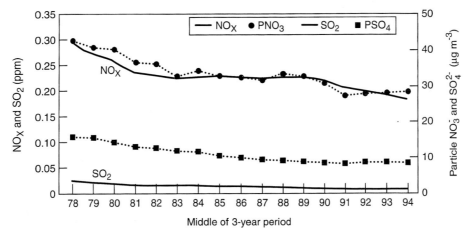

FIGURE 16.37 Trends in NO_x, particle nitrate (PNO_3), SO_2, and particle sulfate (PSO_4) as a function of year. The 3-year means of the annual top 30 daily means are shown. NO_x and SO_2 were measured at Long Beach (western end) and particle nitrate and sulfate at Riverside (the eastern end) (kindly provided by Leon Dolislager and B. Croes, California Air Resources Board).

Chameides and co-workers (1992) examined the observed concentrations of ozone and its precursors, NO_x and VOC, in a variety of tropospheric locations, from remote marine areas to polluted urban regions. Figure 16.38 shows ranges of observed NO_x and OH-reactivity adjusted VOC (expressed relative to propene) in four different types of atmospheres: urban–suburban, rural typical of the eastern United States, remote tropical Brazilian forests, and remote marine boundary layer. Also shown are isopleths for O_3 production during midday calculated using a box model.

As discussed by Chameides and co-workers, the data

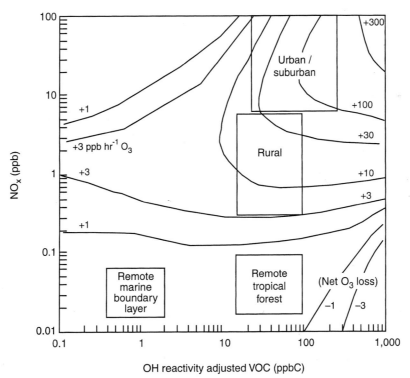

FIGURE 16.38 Observed NO_x and OH-reactivity-adjusted VOC (expressed as propene) in various regions of the troposphere. Isopleths shown are midday rates of O_3 production (ppb h^{-1}) calculated using a box model (adapted from Chameides et al., 1992).

in Fig. 16.38 indicate a strong relationship between O_3 and NO_x, but little with the reactivity-adjusted VOC concentrations in continental areas. For example, while the reactivity-adjusted VOC increases by almost two orders of magnitude from the remote marine region to the tropical forest sites due to biogenic organic emissions, O_3 (and NO_x) remain approximately constant. These trends are similar to the isopleths predicted using a simple photochemical box model and suggest that in remote and rural regions, O_3 is most sensitive to NO_x control, whereas in polluted urban–suburban areas, it can be sensitive to either NO_x or VOC control depending on the particular conditions, i.e., which side of the ridge line one is on in the top right corner of Fig. 16.14.

Given our knowledge of the detailed chemistry leading to O_3 formation from VOC and NO_x (see Chapters 6 and 7), one might expect that any NO_x that is oxidized will form either O_3 or other oxidized nitrogen compounds such as PAN, HNO_3, and N_2O_5. Trainer and co-workers (1993) showed that there was indeed a linear relationship between the amount of reacted oxides of nitrogen, expressed as $NO_z = (NO_y - NO_x)$ (see Chapter 7 for definitions of NO_z, NO_y, and NO_x), and O_3, with the slope giving 8.5 ppb of O_3 per ppb of NO_x oxidized. This slope, i.e., the number of molecules of O_3 generated per molecule of NO_x oxidized, is known as the *ozone production efficiency*.

Figure 16.39 shows a similar relationship measured at a site downwind of metropolitan Toronto, Canada (Roussel *et al.*, 1996). The slope of this line gives 13.8 ppb of O_3 formed per ppb of NO_x oxidized. The intercept in Fig. 16.39 is 26 ppb, which should represent the "background" O_3. (Note that as discussed in Chapter 14, this is not the level of O_3 in an atmosphere unperturbed by anthropogenic emissions but rather the current global level, which is a factor of 2–3 times higher than prior to the industrial revolution due to anthropogenic influences.) Both the slope and intercept are in good agreement with values of 6–12 reported for the slope and 24–42 reported for the intercept in other studies carried out in the eastern and southeastern United States (e.g., see Trainer *et al.*, 1993; Kleinman *et al.*, 1994; Olszyna *et al.*, 1994; Hastie *et al.*, 1996; and Ridley *et al.*, 1998). These intercept values are also consistent with measurements of current boundary layer levels of O_3 measured independently in a number of studies (e.g., Altshuller and Lefohn, 1996). Similar values for the ozone production efficiency have been reported in other areas around the world, including Athens (e.g., Peleg *et al.*, 1997) and Ll. Valby, Denmark (Skov *et al.*, 1997).

A slightly modified approach has also been used in a number of studies in which the sum of $(O_3 + NO_2)$ is plotted against NO_z (e.g., St. John *et al.*, 1998). This minimizes the effects of short-term variations in O_3 caused by its rapid reaction with NO. Thus, when O_3 is titrated by the NO reaction, the measured O_3 concentrations will be small; however, the NO_2 generated is a source of O_3 through its subsequent photolysis. Hence the sum of $(O_3 + NO_2)$ is sometimes chosen as a measure of the ultimate formation of ozone. The ozone production efficiency determined from slopes of plots of $(O_3 + NO_2)$ against NO_z in the Nashville, Tennessee, area was measured to be typically 5–6 if it was assumed that NO_y is not removed by other processes. The production efficiency appeared to be about the same for the general urban plume and for an air mass in which a plume from a power plant was also embedded. Including other losses for NO_y such as deposition lowers the estimated production efficiency by about a factor of two (St. John *et al.*, 1998; Nunnermacker *et al.*, 1998).

This relationship between O_3 and $NO_z = (NO_y - NO_x)$ in areas downwind of urban centers can be anticipated, based on the oxidation of NO to NO_2 by HO_2 and RO_2 radicals and the subsequent photolysis of NO_2 to form O_3. As discussed in Chapter 6.J, at lower NO_x concentrations, reactions of HO_2 and RO_2 with themselves and each other compete with their reactions with NO. However, the oxidation of NO to NO_2 leads to O_3 formation since photolysis of NO_2, generating O_3, is a major fate for NO_2. This then gives rise to the observed relationship between NO_z and O_3. Under these low-NO_x conditions, the formation of H_2O_2 and other peroxides is important and deter-

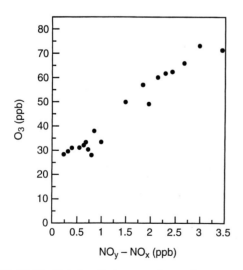

FIGURE 16.39 Relationship between observed concentrations of O_3 and reacted oxides of nitrogen expressed as $(NO_y - NO_x)$ at a site downwind of Toronto, Canada (adapted from Roussel *et al.*, 1996).

mined largely by the rate of formation of the precursor free radicals (e.g., Kleinman, 1991, 1994).

However, when the rate of NO_x emissions is larger than the rate of radical production, this relationship would not be expected to be as clear. In this region, reaction of NO_2 with OH to form HNO_3 becomes important, removing both the free radical OH and NO_2 without forming O_3, weakening the O_3–NO_z relationship. Under these high-NO_x conditions, the HO_2 + HO_2 or RO_2 reactions are also less important, leading to a decreased formation of peroxides.

Kleinman and co-workers (Kleinman, 1991, 1994; Kleinman et al., 1997) examined the utility of treating VOC–NO_x chemistry in terms of these two regimes defined in terms of the relative rates of free radical production compared to emissions of NO_x. Given that the rates of free radical formation vary rapidly and from location to location, transitions from one regime to another can occur diurnally and seasonally, as well as geographically. For example, such a transition can occur during the fall when photolysis of O_3 to form $O(^1D)$ and subsequently OH by its reaction with water decreases due to seasonal decreases in UV and relative humidity. Depending on the relative strength of the NO_x emissions, this can result in a transition from the low-NO_x to the high-NO_x regime, accompanied by decreases in H_2O_2 production and a weakening of the correlation between O_3 and NO_z.

This behavior is consistent with the observations of a number of field studies. For example, Jacob et al. (1995) report that in Shenandoah National Park in Virginia (U.S.) in early September, there was a good correlation between O_3 and NO_z, with a slope of 18 compared to the range of 8.5–14 observed in other studies. In the latter part of September, the correlation was weaker ($r^2 = 0.23$ vs 0.49 earlier) and the slope was only 7. This weakening of the relationship between O_3 and NO_z was accompanied by a decrease in concentrations of H_2O_2 from an average of 0.86 ppb to 0.13 ppb, as expected for a transition from the NO_x-limited to the VOC-limited regime.

A number of modeling studies, combined with field measurements, suggest that VOC control may be more effective than NO_x in controlling O_3 at some locations, primarily urban. This is consistent with both box and airshed model predictions in that if one is in effect in the low VOC/NO_x regime on the top of the ridge line in Fig. 16.14a, a decrease in NO_x could actually lead to an initial increase in O_3 before it decreases. It should be noted, however, that these highly polluted locations are generally not those at which the ozone peaks occur. As illustrated in Fig. 16.14, as the air parcel moves downwind from these low VOC/high NO_x regions, it generally moves into the NO_x-limited regime and it is in these downwind areas that O_3 generally peaks. This type of behavior, i.e., transition from the VOC- to the NO_x-limited regime, has been observed in many regions, including the Los Angeles area, Toronto (Fuentes and Dann, 1994), Munich (Fabian et al., 1994), and the eastern U.S. (McKeen et al., 1991b; Mathur et al., 1994). Modeling studies on the generation and transport of O_3 in urban areas are also consistent with this increased sensitivity to NO_x as the air mass moves downwind (e.g., Duncan and Chameides, 1998).

Based on the chemistry discussed above, the use of indicator species has been proposed to differentiate air masses in which ozone formation is more sensitive to NO_x than to VOC and vice versa (e.g., Milford et al., 1994; Sillman, 1995; Sillman et al., 1997; Lu and Chang, 1998; Sillman, 1999). For example, model calculations suggest that high values of the ratios $O_3/\{NO_y - NO_x\}$, $HCHO/NO_y$, and H_2O_2/HNO_3 reflect air masses in the NO_x-sensitive regime whereas low values reflect the VOC-sensitive regime (e.g., Sillman et al., 1997; Lu and Chang, 1998). The reasons for this are found in the complex chemistry discussed above, but some generalizations can be made. For example, high O_3 and low NO_y suggest relatively large free radical sources and smaller radical sinks such as the NO_2 + OH reaction. HCHO is a measure of the oxidation products of both anthropogenic and biogenic VOC; thus, the higher HCHO relative to NO_y, the further to the right of the isopleths (Fig. 16.14a) is the air mass, i.e., toward the NO_x-sensitive regime. Similarly, high concentrations of H_2O_2 relative to HNO_3 indicate that the air mass is in the high-VOC/NO_x regime where ozone is most sensitive to NO_x control. Some of the uncertainties in the application of such indicators in field studies are discussed by Sillman (1995) and Lu and Chang (1998).

Related to the use of indicator species is the use of "species age" in photochemical modeling studies (Venkatram et al., 1998). In this approach, the VOC and NO_x are calculated at a particular point of interest in an air basin assuming no chemical reactions, i.e., only transport occurs. The age of the VOC and NO_x since the time of emission is also calculated. The amount of O_3 formed is then estimated using the VOC–NO_x chemistry for that time period. This approach separates transport and chemistry in an explicit manner and allows the calculation of the effectiveness of various VOC and NO_x reductions at a particular location.

In short, a combined VOC–NO_x reduction strategy seems to be the optimum approach to controlling ozone and other secondary air pollutants, and there is evidence from the experience in southern California that this approach is effective. For differing viewpoints, however, see articles by scientists from General Motors

Research Laboratories (Chock *et al.*, 1981, 1983; Klimisch and Heuss, 1983; Kumar and Chock, 1984; Wolff, 1993).

D. ALTERNATE FUELS

Motor vehicles are major sources of air pollutants worldwide, and the number of vehicles is anticipated to continue growing (e.g., see Walsh, 1990). A major focus of control strategy development for mobile sources in recent years has been on the development of alternate fuels. These range from relatively minor changes in the traditional composition of gasoline, such as reformulated gasolines, to compressed natural gas (CNG), liquefied petroleum gas (LPG), alcohol fuels and their blends with gasoline, or hydrogen. There have also been significant developments in electric vehicles fueled either by batteries or fuel cells.

One might expect that vehicle emissions would be related to the composition of the fuel used, and a number of studies have confirmed this (e.g., see Schuetzle *et al.*, 1994; Siegl *et al.*, 1992; and Kaiser *et al.*, 1991, 1992, 1993). For example, emissions coming from a single-cylinder engine have shown that the mass emissions increase as the molecular weight of a single-component fuel increases and that benzene emissions decrease as the aromatic content of the fuel decreases (Schuetzle *et al.*, 1994).

Hence, a feasible control strategy should be the use of fuels with smaller mass emissions, reduced reactivity of the emissions, or both. We discuss briefly some of the chemical implications of the use of some of these alternate fuels. For a more comprehensive treatment of the advantages and disadvantages of alternate fuels and technologies, see the National Research Council report (1991), and for a discussion of a variety of issues associated with motor vehicle emissions, see Cadle *et al.* (1996, 1997a, 1997b) and Chang *et al.* (1991).

1. Reformulated Gasolines

As discussed by Calvert *et al.* (1993), there is no fixed, agreed-upon, definition of reformulated gasoline, although decreased VOC emissions and toxics are common goals. Table 16.13 shows the composition of what is known as Phase 2 reformulated gasoline compared to an average for gasolines sold in California in 1990. The vapor pressure at 100°F (known as the Reid vapor pressure, RVP) and the sulfur and benzene contents are lower, as are the aromatic and olefinic contents. The reduced vapor pressure reduces evaporative emissions (which can be as much as half of the total VOC emissions), while reducing the sulfur content improves

TABLE 16.13 Specifications of Reformulated "Phase 2" Gasoline[a]

Measure	Flat limits[b]	1990 California average
RVP (psi)[c]	7.0	8.5
Sulfur (ppmw)	40	150
Benzene (vol%)	1.0	2.0
Aromatic HCs (vol%)	25.0	32
Olefin (vol%)	6.0	9.9
Oxygen (wt%)	1.8–2.2	0
T_{90} (°F)	300	330
T_{50} (°F)	210	212

[a] From California Air Resources Board, R. Pasek and Bart Croes, personal communication.
[b] Flat limits met by each gallon; see Title 13, California Code of Regulations, Section 2262.
[c] Applies to summertime only; definition of summertime varies by location.

catalyst operation for the reduction of VOC, CO, and NO_x. The addition of oxyen-containing compounds such as methyl *tert*-butyl ether (MTBE), ethyl *tert*-butyl ether (ETBE), methanol, or ethanol is intended to counter the octane reduction due to reduced aromatic and olefinic contents and to decrease emissions of CO.

Tunnel studies of vehicle emissions carried out before and after there was a significant increase in the oxygen content of fuel sold in the San Francisco area (from 0.3 to 2.0%) showed CO emissions decreased by $21 \pm 7\%$, and VOC by $18 \pm 10\%$, respectively, with no change in NO_x (Kirchstetter *et al.*, 1996). Formaldehyde, however, which is an oxidation product of MTBE added to the fuel, increased by $13 \pm 6\%$ as did MTBE (see Problem 1).

A similar decrease of $18 \pm 11\%$ in CO emissions attributed to the use of oxygenated fuels was reported by Johnson *et al.* (1997). They used remote-sensing techniques to compare CO from automobiles leaving Las Vegas, where oxygenated fuels were in use at that time in winter (1991/92), to that from automobiles entering from California, where these fuels were not then in use.

It is not entirely clear from measurements of CO in ambient air, however, whether such substantial decreases in CO are a general phenomenon. For example, Mannino and Etzel (1996) have analyzed ambient CO concentrations measured before and after oxygenated fuels were used in a number of western states in the United States and compared these data to analogous measurements made in states where oxygenated fuels were not used. They report an only slightly greater decrease in CO in the areas using oxygenated fuels compared to those not using oxygenated fuels.

TABLE 16.14 Organic Composition of Exhaust and Evaporative Emissions from Gasoline and Some Alternate Fuels[a,b]

Organic	Gasoline[c]	Methanol		Ethanol	Liquefied petroleum gas	Compressed natural gas
		M85	M100			
Alkanes	0.632	0.224	0.023	0.077	0.797	0.170
Alkenes	0.040	0.007	0.001	0.002	0.062	0.031
Formaldehyde	0.021	0.067	0.050	0.010	0.041	0.023
Aldehydes	0.004	0.004	0.001	0.050	0.005	0.005
Ethene	0.031	0.005	0.001	0.034	0.082	0.017
Toluene	0.199	0.032	0.009	0.023	0.007	0.007
Aromatics	0.059	0.023	0.005	0.010	0.003	0.014
Methyl ethyl ketone	0.015	0.005	0.001	0.002	0.003	0.009
Methanol	0	0.633	0.911	0	0	0
Ethanol	0	0	0	0.791	0	0

[a] From California Air Resources Board (1989).
[b] Compositions given as fractions; the tests relied on a small number of vehicles, and the tests are not likely to be representative of the actual fleet of vehicles operating on each fuel.
[c] Indolene used as the reference gasoline.

One concern with the use of MTBE in fuels is its high solubility in water and hence contamination of groundwater and surface water from fuel leakage and spillage (e.g., Squillace et al., 1997; Reuter et al., 1998); because of this, a phase-out of MTBE is occurring in some places, e.g., California. Increased concentrations of this toxic, highly volatile compound in air are also of concern. In one study in Finland, for example, where the gas contained 11% MTBE, concentrations around the perimeter of the gas stations were measured to average 0.1–0.4 ppb; however, concentrations at the pumping islands were as high as 37 ppb (Vainiotalo et al., 1998).

It is interesting that the calculated reactivity of the speciated exhaust of oxygenated fuels containing MTBE did not change significantly because the lower reactivity of MTBE (Japar et al., 1991; Carter, 1994) is offset by the increased emissions of the highly reactive HCHO and isobutene. (It should be noted that these estimates of MTBE reactivity are for gas-phase reactions; heterogeneous oxidation on particles may also be important (e.g., see Idriss and Seebauer, 1996).) This is in agreement with the reactivity adjustment factors for reformulated gasoline in Table 16.10, which are all within 10% of that for the base gasoline. Gains in terms of ozone formation are then expected to be due primarily to changes in the mass of VOC emitted (although this is somewhat controversial—see Hoekman, 1992; Venturini, 1993; and Kirchstetter et al., 1996). This suggests that the major benefit of such reformulated gasolines and increased oxygen content may be reduced CO rather than reduced ozone formation, depending on the VOC/NO_x regime into which the exhaust is emitted. On the other hand, modeling studies suggest that the reduction in the vapor pressure and the sulfur and olefinic contents associated with the use of reformulated gasoline should contribute to reduction in ozone production (Dunker et al., 1996).

2. Compressed Natural Gas (CNG)

Because CNG is primarily methane, it is expected to have relatively low reactivity, with the small amounts of reactive "impurities" such as small olefins and alkanes being responsible for most of its reactivity (see Table 16.14). Emissions of CO are smaller than from gasoline-powered vehicles, while the effect on NO_x emissions is not clear (National Research Council, 1991). As seen in Tables 16.10 and 16.11, CNG shows the highest promise for low-reactivity exhaust emissions, and this appears to be the case for its use in "real" vehicles (Gabele, 1995). Figure 16.40, for example, shows the estimated ozone production per mile traveled for a vehicle fueled on CNG compared to vehicles fueled on reformulated gasoline (RFG) or the alcohol fuels M85 or E85 (vide infra). These measurements and estimates based on them include the contributions from both exhaust (including CO) and evaporative emissions (Black et al., 1998). Clearly, the reactivity of the CNG-powered vehicle emissions was substantially smaller than for the other vehicle–fuel combinations.

Additional advantages of CNG are that because of the nature of the fuel, evaporative emissions are of significantly lower reactivity than conventional gasoline and since the emissions coming directly from the engine are lower, the likelihood of such vehicles being "super-emitters" is much less.

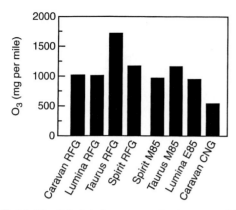

FIGURE 16.40 Calculated ozone production per vehicle mile traveled for various car–fuel combinations. RFG = reformulated gasoline; M85 = 85% methanol, 15% gasoline; E85 = 85% ethanol, 15% gasoline; CNG = compressed natural gas (adapted from Black et al., 1998).

3. Liquefied Petroleum Gas (LPG)

Liquefied petroleum gas is primarily propane but generally also contains significant amounts of olefins, which increase its reactivity substantially (Table 16.14). For example, LPG in the Los Angeles area contains about 2 mol% of alkenes whereas that in Mexico City contains almost 5% (Blake and Rowland, 1995). Hence while it is a significant improvement over gasoline in terms of reactivity (Tables 16.10 and 16.11), the exact amount of improvement is highly dependent on the nature and concentrations of these reactive impurities (Gabele, 1995).

4. Alcohol Fuels and Blends with Gasoline

Methanol and ethanol fuels are used both "neat" and as a blend with gasoline. The terminology "M85" signifies a blend of 85% methanol, 15% gasoline (by volume), and similarly for "E85," used to denote ethanol–gasoline blends. The advantages of using an alcohol–gasoline blend are that the flame is visible (a flame from pure methanol is not), the vapor pressure is higher, which aids in ignition, and there is flexibility in fueling, depending on the availability of gasoline or alcohol fuels.

Methanol is expected to oxidize to formaldehyde, both during combustion and after emission to the atmosphere. As discussed in Chapter 6.H, OH reacts with methanol primarily at the methyl group:

$$CH_3OH + OH \rightarrow CH_2OH + H_2O \quad (85\%), \quad (17a)$$
$$\rightarrow CH_3O + H_2O \quad (15\%). \quad (17b)$$

Both CH_2OH and CH_3O then react with O_2 to give HCHO (and HO_2). However, as discussed in detail in the National Research Council report (1991) and references cited therein, the results of modeling studies suggest that the amount of HCHO formed by oxidation of methanol in the atmosphere in many locations will not be significant compared to that formed by other processes. (Direct emissions of formaldehyde may be important, however, in some circumstances with low dilution rates such as underground parking garages or tunnels (e.g., see Chang and Rudy, 1990b).) Because methanol itself has low reactivity (Tables 16.8 and 16.9), some modest improvements in ozone may result from the use of methanol (e.g., see Dunker, 1990; National Research Council, 1991; Lloyd et al., 1989). It is interesting that the reactivity of the evaporative emissions from M85 is higher than that of the exhaust (Black et al., 1998).

There has been an emphasis on the addition of ethanol to gasoline, due to the availability from grain sources. Ethanol is somewhat more reactive than methanol (Tables 16.8 and 16.9) and forms acetaldehyde upon oxidation (see Problem 2). Further oxidation of acetaldehyde produces PAN (see Chapters 6 and 7). There is evidence for such an oxidation sequence in Brazil, where extensive use is made of ethanol as a fuel. Thus relatively high concentrations of PAN have been observed in this area compared to those found elsewhere (Tanner et al., 1988; Grosjean et al., 1990). Increases in PAN have also been reported in urban areas such as Albuquerque, New Mexico, using ethanol–gasoline blends for CO reduction (e.g., Gaffney et al., 1997, 1998; Whitten, 1998); concentrations of aldehydes, particularly CH_3CHO, were also higher.

Alcohol fuels have also been reported to have higher emission rates of VOC as well as toxics such as HCHO, CH_3CHO, 1,3-butadiene, and benzene (Gabele, 1995; Black et al., 1998). However, Stump et al. (1996) report decreased emissions of VOC, benzene, and 1,3-butadiene with a gasoline fuel containing 8.8% ethanol but, in agreement with the other studies, increased emissions of HCHO and CH_3CHO. Knapp et al. (1998) tested emissions from 11 vehicles at temperatures from -20 to $+75°F$ fueled on either gasoline or a blend with 10% ethanol and found the ethanol blend resulted in higher emissions of CH_3CHO, in some cases almost by almost an order of magnitude. The changes in the emissions of benzene, HCHO, and 1,3-butadiene were variable with respect to both amount and sign.

An additional problem with alcohol–gasoline blends is the increase in vapor pressure of gasoline in the mixture (e.g., see National Research Council, 1991; Calvert et al., 1993; and Timpe and Wu, 1995). This can contribute to much higher Reid vapor pressures, increasing the relative importance of evaporative emissions.

Table 16.14 shows the VOC composition of the combination of exhaust and evaporative emissions measured on a limited number of vehicles. Similar data have been reported by Gabele (1995). The increased aldehydes associated with the use of alcohol fuels is evident.

Overall, the use of ethanol blends is believed not to be effective in reducing ozone, but may actually increase it (Calvert *et al.*, 1993; Dunker *et al.*, 1996).

5. Hydrogen

Since the oxidation of H_2 produces water, it should be one of the cleanest possible fuels, and this indeed appears to be the case, with emissions consisting primarily of NO_x and very small amounts of VOC and CO from the combustion of some oil in the system. While it is being used in some buses, for example, it has not yet found widespread use due to technical and safety problems encountered on the passenger vehicle scale.

6. Electric Vehicles

Electric passenger vehicles powered by batteries have been reintroduced in the United States. Although there are no tailpipe emissions associated with electric vehicles, charging the batteries requires power with which there are associated emissions. However, emissions from large sources are relatively easier to control than from millions of individual automobiles. In addition, the emissions of VOCs and CO from power plants are generally low, while those of NO_x, SO_2, and particles depend on the type of fuel used. Natural gas-fired power plants generally have relatively low SO_2 and particle emissions, whereas coal-fired power plants have higher emissions of these pollutants and of NO_x. Studies by Austin described by Cadle *et al.* (1996) suggest that the use of electric vehicles would result in lower total emissions of NO_x in the western United States, where gas is used at power plants, but higher emissions in the east, where coal is used. Larger PM_{10} and SO_2 emissions were projected regardless of the power plant fuel source.

Another version of the electric vehicle that has been introduced into Japan and is scheduled to be introduced into the United States is a hybrid vehicle with both electric and gasoline engines. Such vehicles have the promise of lower emissions and much longer ranges than can currently be achieved with electric vehicles.

For a discussion of various aspects of electric vehicles, see Wilkinson (1997) and the May 1998 issue of *Environmental Manager*.

It should be noted that in addition to the change in *total* emissions, the change in the spatial and temporal nature of the emissions, i.e., replacing many individual sources that are dispersed in time and space by a single source, may also be very important in determining air quality at various locations in an air basin.

Fuel cells, in which a fuel such as methane, gasoline, or methanol is converted to electricity, to power vehicles also appear promising and indeed are currently being used on some buses. For a detailed discussion of this important area, see Lloyd (1992) and Lloyd *et al.* (1994).

E. CONTROL OF ACIDS

As discussed in Chapter 8, the major contributors to acid deposition are sulfuric and nitric acids, with a significant contribution being increasingly recognized from organic acids. The chemistry of formation of nitric and organic acids has been discussed in Chapters 7 and 8 and has been shown to be part of the complex VOC–NO_x chemistry that also leads to ozone formation. Hence control strategies applied for ozone will also impact the formation of these acids as well, although not necessary proportionately, as can be seen from the detailed chemistry.

For example, Meng *et al.* (1997) applied an Eulerian grid model that included both gas and aerosol chemistry to the Los Angeles area for conditions representative of a smog episode on August 27 and 28, 1987. Model predictions showed that the changes in the gas-phase HNO_3 due to reductions in VOC and NO_x were not proportionate, which is not surprising given the complex chemistry involved. For example, a 50% reduction in NO_x emissions alone gave a predicted 17% reduction in the maximum 1-h average concentration of HNO_3 at Riverside, in the eastern end of the air basin; a concurrent reduction of 50% in VOC gave a predicted reduction in peak HNO_3 of 39%. Interestingly, the peak HNO_3 was predicted to actually increase by 17% for VOC reductions from 35 to 50% without concurrent reductions in NO_x emissions. This increase in HNO_3 was associated with predicted simultaneous decreases in PAN of 58–76%. The HNO_3 increase was attributed to lowered concentrations of RCO_3 radicals, which resulted in less NO_2 being tied up in PAN and hence being available to react with OH to form HNO_3.

In short, the development of control strategies for HNO_3 is intimately tied with that of O_3. Although the control of organic acids has not been examined in detail, similar considerations are expected to apply there as well.

As discussed in Chapter 8, the formation of H_2SO_4 from SO_2 occurs largely in the aqueous phase. The major oxidants in fogs and clouds are H_2O_2 and O_3, so that control of photochemical oxidants is again expected to impact the rate of SO_2 oxidation and formation of H_2SO_4.

Given that the source of oxidants for SO_2 in both the gas and liquid phases is the VOC–NO_x chemistry discussed earlier and that a major contributor to acid deposition is nitric acid, it is clear that one cannot treat acid deposition and photochemical oxidant formation as separate phenomena. Rather, they are very closely intertwined and should be considered as a whole in developing cost-effective control strategies for both. For a representative description of this interaction, see the modeling study of Gao *et al.* (1996).

One of the key issues in developing effective control strategies for acid deposition has been what is known as "linearity." This term has been subject to a variety of interpretations and meanings and applied on microscopic, i.e., molecular, to macroscopic scales. A detailed treatment and discussion of linearity encompassing these scales is given by Hales and Renne (1992).

In its simplest form of interest for policy and regulatory purposes, linearity is often treated in terms of source–receptor relationships. That is, if the emissions of the precursor SO_2 are lowered by 50%, will the deposition of sulfate also decrease by 50% at all receptor sites?

A major factor involved in determining the relationship between SO_2 emissions and sulfate deposition is the chemistry. As discussed above, the oxidation of SO_2 by OH in the gas phase generates HO_2 and hence OH in the presence of NO. The regeneration of OH means that the oxidation will not be oxidant limited in the gas phase, and hence a reduction in SO_2 might be expected to be accompanied by a corresponding decrease in the formation of H_2SO_4.

However, the situation is not as clear-cut for the liquid-phase oxidation, which, we have seen, predominates in many (perhaps most) situations. In this case, a less than 1:1 relationship between the reduction in H_2SO_4 formed and SO_2 emitted may result for a number of reasons operating on the microscopic scale. For example, less H_2O_2 is available in many clouds than is needed to oxidize all of the S(IV) that is present, and hence the oxidation can be limited by the availability of oxidant (e.g., see Dutkiewicz *et al.*, 1995). Another important factor that comes into play is the interplay between the acidity of the aqueous phase, the reaction kinetics, and the solubility of S(IV). Thus, as seen in Chapter 8, the solubility of S(IV) decreases as the aqueous phase becomes more acidic, limiting the total sulfur available for oxidation. In addition, all oxidations in the aqueous phase except that by H_2O_2 are pH dependent; as a result, the contribution of various oxidants can vary significantly with time in an air mass as the oxidations take place and removal by rainout and washout occurs.

Examples of these complications have been reported in a number of field and modeling studies. For example, the amount of dissolved SO_2 in rain that has passed through a power plant plume has been found in some cases to be much less than might be expected, due to the acidification of the rain by other plume components that decreased the solubility of SO_2 (e.g., Dana *et al.*, 1975). Similarly, studies using the PLUVIUS model developed for application to gas-phase species and their interaction with clouds and precipitation (Hales, 1989) suggest that the deposition rate of sulfate from an urban source mixed in with contributions from the "background" may initially be nonlinear (Hales, 1991). However, at longer times this deposition rate becomes linear, and since most of the removal occurs at these longer times, the overall integrated deposition does indeed appear to be linear.

In short, as is the case for acids formed in VOC–NO_x chemistry, the chemical and physical processes associated with the formation and deposition of sulfuric acid are also quite complex.

There are a number of field measurements that have addressed this relationship between the mandated reductions in SO_2 emissions in the United States and the subsequent changes in sulfate deposition downwind. For example, one analysis of the trends in the atmospheric concentrations of sulfate in the northeastern United States suggests that from 1977 to 1989, the sulfate concentration decreased by about 22–28% during which the emissions of SO_2 were estimated to have decreased by 25% (Shreffler and Barnes, 1996).

Similarly, Husain *et al.* (1998) have reported trends in sulfate at two sites in New York state, from 1979 to 1996 at Whiteface Mountain in a remote area in the eastern part of the state and from 1983 to 1996 at Mayville, in the western part. The trends at both sites were highly correlated. Figure 16.41 shows the relationship between sulfate or total sulfur (defined as sulfate plus gas-phase SO_2) at Whiteface Mountain as a function of the estimated anthropogenic emissions of SO_2 upwind in the Midwest. The relationship is well described as linear, with a correlation coefficient of $r^2 = 0.81$. (However, a negative intercept suggests that this cannot be directly extrapolated down to very small SO_2 emissions.)

The results of a global 3-dimensional model simulation also suggest that in the boundary layer in the United States, a 50% reduction in anthropogenic SO_2 emissions in the United States will result in a similar (53%) annual reduction in the total (wet plus dry) deposition of sulfur (Chin and Jacob, 1996).

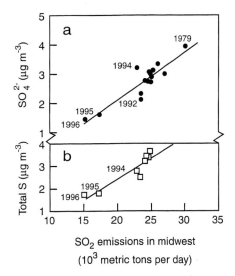

FIGURE 16.41 (a) Sulfate and (b) total sulfur (sulfate + $SO_{2(g)}$) at Whiteface Mountain, New York, as a function of estimated anthropogenic SO_2 emissions upwind in the Midwest from 1979 to 1996 (adapted from Husain et al., 1998).

In short, it appears that despite the potential complexities in the chemistry and meteorology involved in these long-range transport processes, at the SO_2 emission levels over the past decade, the deposition of sulfate has decreased approximately proportionately to the decrease in anthropogenic SO_2 emissions.

F. CONTROL OF PARTICLES

As discussed in Chapter 2, epidemiological studies have suggested that small particles may cause significant health effects, including increased mortality rates. Depending on location and time, such particles may be either primary in nature, i.e., directly emitted, or secondary, i.e., formed from reactions of gases in air.

Diesel engines are significant sources of particles, indeed the major source in some urban cores, and their use in many regions of the world is increasing (e.g., see Walsh (1997) and Chapter 10). As a result, there has been an emphasis on reducing these emissions, for example through engine redesigns. One concern has been the size distribution of the emissions. Thus, emission standards are written in terms of total mass, which reflects primarily the larger particles. In one laboratory study (Baumgard and Johnson, 1996), the total particulate mass emissions from a 1991 heavy-duty engine burning a 0.01 wt% S (low sulfur) fuel were much smaller than from a 1988 engine using the same fuel. However, in terms of the *number* of ultrafine particles (0.0075–0.046 μm), the newer engine emissions were about 30 times greater. The implications of such changes for the ultimate health impacts of primary emissions of particles from diesel engines are not clear.

The use of alternate fuels in heavy-duty vehicles has also been the subject of many studies, and it appears that a significant reduction in emissions of particles can be obtained using such fuels. For example, Wang et al. (1997) measured the emissions of particles, NO_x, CO, and hydrocarbons from more than 300 buses and heavy-duty trucks operating on either diesel fuel, natural gas, or alcohol fuels in 17 different states in the United States. Figure 16.42 shows their measurements of emissions of particulate matter and NO_x. Particle emissions using compressed natural gas were reduced by more than an order of magnitude compared to those using diesel fuels; improvements, although not as dramatic, were also seen with the alcohol fuels. On the other hand, only the alcohol fuels gave significant reductions in NO_x emissions.

As discussed in Chapter 9, in many locations secondary particles are more important contributors to atmospheric levels than are the primary emissions. The

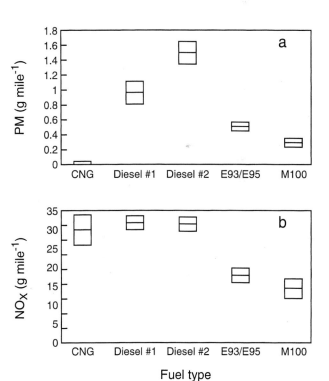

FIGURE 16.42 Measured emissions of (a) particles (PM) and (b) NO_x from more than 300 buses and heavy-duty trucks running on compressed natural gas (CNG), two diesel fuels, or alcohol fuels (E93, E95, and M100). The bars are the means and the boxes encompass the 95% confidence intervals. E93 = 93% ethanol, 5% methanol, 2% K-1 kerosene; E95 = 95% ethanol, 5% gasoline; M100 = 100% methanol (adapted from Wang et al., 1997).

chemical components of the secondary particles consist largely of sulfate, nitrate, and complex polar organics, all of which are formed by oxidation processes in the VOC–NO$_x$ system. This again suggests that control strategies applied to ozone should also impact the formation of secondary particles. For example, less NO$_x$ might be expected to give less particle nitrate in general, and less VOC should reduce the formation of secondary organics, including those in particles. The actual relationship will be more complicated than a one-to-one correlation, of course, in that the concentrations of reactive species such as OH needed to carry out the oxidations are also affected, but not in a linear manner, by changes in VOC and NO$_x$. A model for examining such relationships between individual organics and the production of secondary aerosol particles is discussed by Bowman *et al.* (1995).

Trends in particulate matter with diameters less than 10 μm, PM$_{10}$, in the Los Angeles air basin from 1985 to 1994 are shown in Fig. 16.43 (Prasad, personal communication, 1997). A downward trend in both the mean and the maximum levels is evident, occurring along with similar trends in O$_3$, NO$_x$, and particulate nitrate and sulfate in this region (see Figs. 16.36 and 16.37). This is qualitatively consistent with expectations based on the chemistry described in earlier chapters and again illustrates the strong interrelationships between various species in this complex atmospheric system.

Similar downward trends in PM$_{10}$ have been observed at surface measuring sites in the United States in urban, suburban, and rural areas. Figure 16.44, for example, shows the trends in the annual average PM$_{10}$ levels in these three types of air environments (Darlington *et al.*, 1997). Reductions of ~3–4% per year have been observed. Simultaneously, the annual average gas-phase concentrations of SO$_2$ and NO$_x$, precursors to sulfate and nitrate in particles, decreased. Downward trends of ~1.6–1.8% per year in the optically active aerosol over the United States has also been reported (e.g., Hofmann, 1993).

Until relatively recently, air quality models focused on gas-phase reactions. However, with increases in our understanding of the formation, interactions, and fates of suspended particles, as well as increase in computing power, significant progress is being made in incorporating aerosol chemistry and physics into local and regional models (e.g., see Wexler *et al.*, 1994; Eldering and Cass, 1996; and Meng *et al.*, 1997). For example, Meng *et al.* (1997) used an Eulerian grid model for the Los Angeles area to predict the relationships between the precursors VOC and NO$_x$ and their secondary oxidation products, including gaseous HNO$_3$ (discussed earlier), PM$_{2.5}$, and nitrate in PM$_{2.5}$. Because HNO$_3$ is predicted to increase with reductions in VOC when there are no simultaneous reductions in NO$_x$, both nitrate in PM$_{2.5}$ and total PM$_{2.5}$ also increase under this scenario. Reductions in NO$_x$ without concurrent reductions in VOC do lead to reductions in PM$_{2.5}$ and nitrate in PM$_{2.5}$, but not proportionately. For example, a 50% reduction in NO$_x$ is predicted to decrease both PM$_{2.5}$ and its nitrate component by ~20%.

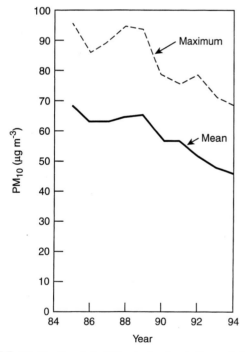

FIGURE 16.43 Trends in PM$_{10}$ in the South Coast air basin in California (Los Angeles area) from 1985 to 1994. The averages for six sites having complete data throughout this time period are shown as well as the maximum concentrations. The 1994 data are for January to October (kindly provided by Dr. Shankar Prasad, California Air Resources Board).

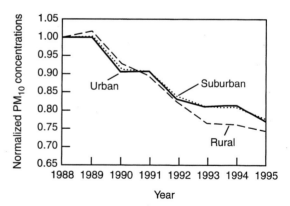

FIGURE 16.44 Concentrations of PM$_{10}$ normalized to 1988 as the base year in the United States for urban, suburban, and rural areas (adapted from Darlington *et al.*, 1997).

However, as expected from the chemistry discussed in Chapter 7.E, simultaneous control of NH_3 has a significant effect on particle nitrate formation, since the formation of ammonium nitrate is a major mechanism for conversion of gaseous HNO_3 to particulate nitrate. Thus, Meng *et al.* (1997) predict that a 50% reduction in both NO_x and NH_3 would give about the same reduction in particulate nitrate.

Not only does the chemistry generating O_3 affect the formation of particles, but the reverse is true; i.e., the presence of particles is thought to affect ozone formation. The major reason for this is the effect of aerosol particles on light scattering and hence the actinic flux and photolysis rates. For example, scattering of light by particles in effect increases the path length through the atmosphere (see Chapters 3.C.2f and 14.C), leading to increased rates of photolysis. On the other hand, absorption of light by particles decreases the effective actinic flux. For example, Dickerson *et al.* (1997) used the Urban Airshed Model to examine the effects of increased light scattering by aerosol particles in the boundary layer in the northeastern United States. Using a refractive index typical of particles in that region; they predict that the amount of ozone generated increased by about 30% due to the increased rates of photolysis, particularly NO_2. The opposite effect was predicted for absorbing aerosol particles. Similarly, Jacobson (1998) predict decreased O_3 formation (by 5–8%) in the Los Angeles area due to reduction in photolysis rates from UV-absorbing aerosol particles.

G. ATMOSPHERIC CHEMISTRY AND RISK ASSESSMENTS OF HAZARDOUS AIR POLLUTANTS

As discussed in Chapter 2.G, *risk assessment* and *risk management* are two separate elements of developing cost-effective control strategies for hazardous air pollutants, HAPs. The risk assessment portion ideally provides a complete understanding of the chemical and physical processes that apply to the HAP once it is emitted into the atmosphere. Thus, as we shall see in the examples that follow, some HAPs react in the atmosphere to form less toxic substances (sometimes referred to as "environmental deactivation"), whereas others form more toxic compounds ("environmental activation"). Understanding the atmospheric reactions of HAPs is clearly essential to developing appropriate health risk assessments upon which cost-effective control strategies can be based in the risk management phase.

In addition to the role of atmospheric reactions in the fate of airborne HAPs, atmospheric chemistry also plays a role in the formation of some of them, most notably formaldehyde and acetaldehyde. Thus, the potential formation of such compounds from the oxidation of precursors in the atmosphere must also be taken into account in their risk assessments.

In the United States, the Clean Air Act Amendments of 1990 defined a list of 189 compounds or mixtures of compounds as "hazardous air pollutants," shown in Table 16.15 (Kao, 1994; Kelly *et al.*, 1994). These include a range of chemicals, such as hydrocarbons, halogenated, oxygenated, and nitrogen- and sulfur-containing organics, pesticides, and inorganics. Although most are individual compounds, some such as "polycyclic organic matter" represent complex mixtures (see Chapter 10).

A detailed review of the sources, atmospheric chemistry, and fates of HAPs is beyond the scope of this book. Instead, we concentrate here on several organic HAPs to illustrate the role of atmospheric chemistry in risk assessments. For other species, reviews and summaries are available in the literature (e.g., see California Air Resources Board (1997) for summaries of 243 species and Lin and Pehkonen (1999) for a review of the atmospheric chemistry of mercury).

The California Air Resources Board has prepared risk assessments for a number of toxic airborne compounds and mixtures, designated as "toxic air contaminants," TACs (Table 16.15). For example, risk assessments for individual compounds such as benzene, benzo[*a*]pyrene (see Chapter 10), formaldehyde, and vinyl chloride have been carried out, in addition to complex mixtures such as diesel exhaust (California Air Resources Board, 1997a) and environmental tobacco smoke (California Environmental Protection Agency, 1997). These risk assessment documents form the basis for controls imposed as part of the risk management process (e.g., see Seiber, 1996).

Kelly *et al.* (1994) and Kao (1994) have reviewed what is known about ambient concentrations of the HAPs and their reactions in air. Not surprisingly, given the number of compounds involved, the concentrations and atmospheric reactions and fates of many of these species are not known, precluding the development of accurate risk assessments. Also complicating such risk assessments are the wide variety of sources and their temporal and spatial variations, especially in urban areas (e.g., see Spicer *et al.*, 1996; and Mukund *et al.*, 1996). Some of the atmospheric fates of selected VOCs are treated in a review by Pitts (1993), in a series of papers by Grosjean (Grosjean, 1990a–c; 1991a–d) and the 1997 California Air Resources Board report. We examine here selected HAPs and related compounds,

TABLE 16.15 Compounds or Mixtures Designated as Hazardous Air Pollutants (HAPs) in the United States[a] and Toxic Air Contaminants (TACs)[b] in the State of California[b,c]

I. Aliphatic and cyclic hydrocarbons

Saturated hydrocarbons
  Hexane
  2,2,4-Trimethylpentane

Unsaturated hydrocarbons
  *1,3-Butadiene

Saturated halogenated hydrocarbons
  Bromoform
  *Carbon tetrachloride
  *Chloroform
  1,2-Dibromo-3-chloropropane
  Ethyl chloride (chloroethane)
  *Ethylene dibromide (1,2-dibromoethane)
  *Ethylene dichloride (1,2-dichloroethane)
  Ethylidene dichloride (1,1-dichloroethane)
  Hexachloroethane
  Methyl bromide (bromomethane)
  Methyl chloride (chloromethane)
  Methylchloroform (1,1,1-trichloroethane)
  Methyl iodide (iodomethane)
  *Methylene chloride (dichloromethane)
  Propylene dichloride (1,2-dichloropropane)
  1,1,2,2-Tetrachloroethane
  1,1,2-Trichloroethane

Unsaturated halogenated hydrocarbons
  Allyl chloride
  Chloroprene
  1,3-Dichloropropene
  Hexachlorobutadiene
  Hexachlorocyclopentadiene
  *Tetrachloroethylene (perchloroethylene)
  *Trichloroethylene
  Vinyl bromide
  *Vinyl chloride
  Vinylidene chloride (1,1-dichloroethylene)

II. Aromatic compounds

Aromatic hydrocarbons
  *Benzene
  Biphenyl
  Catechol
  Coke oven emissions
  Cumene
  Ethylbenzene
  Naphthalene
  Polycyclic organic matter
  Styrene
  Toluene
  Xylenes (isomers and mixture)
  o-Xylene
  m-Xylene
  p-Xylene

Halogenated aromatic hydrocarbons
  Benzotrichloride
  Benzyl chloride
  Chlorobenzene
  1,4-Dichlorobenzene
  Hexachlorobenzene
  Polychlorinated biphenyls (aroclors)
  *2,3,7,8-Tetrachlorodibenzo-p-dioxin
  1,2,4-Trichlorobenzene

Phenolic compounds
  Cresols/cresylic acid (isomers and mixtures)
  o-Cresol
  m-Cresol
  p-Cresol
  Pentachlorophenol
  Phenol
  2,4,5-Trichlorophenol
  2,4,6-Trichlorophenol

Phthalates
  Bis(2-ethylhexyl) phthalate (DEHP)
  Dibutyl phthalate
  Dimethyl phthalate
  Phthalic anhydride

III. Nitrogenated organic compounds

Acetamide
Acetonitrile
2-(Acetylamino)fluorene
Acrylamide
Acrylonitrile
4-Aminobiphenyl
Aniline
o-Anisidine
Benzidine
Diazomethane
3,3-Dichlorobenzidene
Diethanolamine
N,N-Dimethylaniline
3,3-Dimethoxybenzidine
(Dimethylamino)azobenzene
3,3'-Dimethylbenzidine
Dimethylcarbamoyl chloride
Dimethylformamide
1,1-Dimethylhydrazine
4,6-Dinitro-o-cresol and salts
2,4-Dinitrophenol
2,4-Dinitrotoluene
1,2-Diphenylhydrazine
Ethyl carbamate (urethane)
Ethylene imine (aziridine)
Ethylenethiourea
Hexamethylene 1,6-diisocyanate
Hexamethylphosphoramide
Hydrazine
Methyl hydrazine
Methyl isocyanate
4,4'-Methylenebis(2-chloroaniline)
Methylene(diphenyl diisocyanate) (MDI)
4,4'-Methylenedianiline
Nitrobenzene
4-Nitrobiphenyl
4-Nitrophenol
2-Nitropropane
N-Nitroso-N-methylurea
N-Nitrosodimethylamine
N-Nitrosomorpholine
Pentachloronitrobenzene (quintobenzene)
p-Phenylenediamine
1,2-Propylene imine (2-methylaziridine)
Quinoline
2,4-Toluene diamine
2,4-Toluene diisocyanate
o-Toluidine
Triethylamine

(continues)

TABLE 16.15 *(continued)*

IV. Oxygenated organic compounds

Alcohols
  Methanol

Aldehydes
  *Acetaldehyde
  *Formaldehyde
  Propionaldehyde

α,β-Unsaturated carbonyls
  Acrolein

Carboxylic acids
  Acrylic acid
  Chloroacetic acid

Esters
  Ethyl acrylate
  Methyl methacrylate
  Vinyl acetate

Ethers
  Bis(chloromethyl) ether
  Chloromethyl methyl ether
  Dibenzofurans
  Dichloroethyl ether (bis(2-chloroethyl) ether)
  1,4-Dioxane (1,4-diethylene oxide)

  Glycol ethers
  Methyl *tert*-butyl ether (MTBE)

Ketones
  Methyl ethyl ketone (2-butanone)
  Methyl isobutyl ketone (hexone)

Oxides
  Epichlorohydrin (1-chloro-2,3-epoxypropane)
  1,2-Epoxybutane
  *Ethylene oxide
  Propylene oxide
  Styrene oxide

Other carbonyls and oxygenates
  Acetophenone
  Caprolactam
  2-Chloroacetophenone
  Ethylene glycol
  Hydroquinone
  Isophorone
  Maleic anhydride
  Phosgene
  1,3-Propane sultone
  β-Propiolactone
  Quinone

V. Pesticides and herbicides

Captan
Carbaryl
Chloramben
Chlordane
Chlorobenzilate
2,4-Dichlorophenoxyacetic acid
Dichlorodiphenyldichloroethylene (DDE)
Dichlorvos

Heptachlor
Lindane (all isomers)
Methoxychlor
Parathion[d]
Propoxur (baygon)
Toxaphene (chlorinated camphene)
Trifluralin

VI. Inorganic compounds

Antimony compounds
*Arsenic compounds (including arsine)
*Asbestos
Beryllium compounds
*Cadmium compounds
Calcium cyanamide
Carbon disulfide
Carbonyl sulfide
Chlorine
*Chromium compounds
Cobalt compounds
Cyanide compounds

Hydrochloric acid
Hydrofluoric acid
*Lead compounds
Manganese compounds
Mercury compounds
Mineral fibers (fine)
*Nickel compounds
Phosphine
Phosphorus
Radionuclides (including radon)
Selenium compounds
Titanium tetrachloride

VII. Sulfates

Diethyl sulfate

Dimethyl sulfate

---

[a] From Kao (1994) and Kelly *et al.* (1994).

[b] Asterisks indicate HAPs for which the State of California has prepared detailed risk assessments and identified them as Toxic Air Contaminants (TACs).

[c] For excellent summaries of general exposure and health effects data for 243 substances, see the California Air Resources Board report (1997b). Each summary describes the physical properties, sources, and concentrations both outdoors and indoors, atmospheric persistence, health effects, and other risk assessment information.

[d] For a risk assessment, see "Evaluation of Ethyl Parathion as a Toxic Air Contaminant," California Department of Food and Agriculture (1988).

and the role of atmospheric chemistry in their risk assessments.

Formaldehyde is a designated HAP/TAC. As we have seen on many occasions throughout this book, it not only is directly emitted by a number of sources, including by motor vehicles outdoors and by building materials indoors (see Chapter 15.D), but is also formed in air from the oxidation of both anthropogenic and biogenic organics.

These secondary sources of HCHO vary geographically, seasonally, and diurnally and include the oxidation of anthropogenically emitted VOC as well as biogenic organics such as isoprene. In California, for example, oxidation of VOC is estimated to generate most of the HCHO (about 150,000 tons per year, with a large uncertainty of 50%), whereas direct emissions account for only about 10% of the total, i.e., for ~18,000 tons per year (Cal EPA, 1992b; Harley and Cass, 1994). The importance of secondary formation of HCHO (and CH_3CHO) in the atmosphere is not unique to California (Kao, 1994). However, the relative amounts of direct emissions are undoubtedly much larger in urban areas where vehicles are responsible for ~80% of the direct emissions.

Formaldehyde reacts rapidly in air through photolysis (see Chapter 4.M) and through attack by OH, NO_3, and, in coastal areas, likely chlorine atoms as well:

$$HCHO + h\nu \rightarrow H + CHO, \quad (18a)$$
$$\rightarrow H_2 + CO, \quad (18b)$$
$$HCHO + X(X = OH, NO_3, Cl) \rightarrow XH + HCO. \quad (19)$$

Both H and HCO then react with O_2 to generate HO_2.

Formaldehyde is therefore an example of a HAP that has both primary and secondary sources and that is relatively rapidly "deactivated" in the atmosphere through photolysis and atmospheric reactions (see Problem 3).

A number of pesticides are listed as HAPs (see Table 16.16). These can be transported significant distances from their point of application, and during that time they undergo chemical transformations as well as deposition (Kurtz, 1990). Table 16.17 shows some pesticides and their transformation products in air (Seiber and Woodrow, 1995).

One example of a reactive HAP that is a pesticide is 1,3-dichloropropene, used as a soil fumigant for nematodes. Both the *cis* and *trans* forms react with OH and O_3 (Tuazon *et al.*, 1984), although the OH reaction is sufficiently fast that this reaction is expected to be the major atmospheric fate (see Problem 5). Thus, the rate constants for OH reaction with the *cis* and *trans* forms are 0.774×10^{-11} and 1.31×10^{-11} cm^3 molecule^{-1} s^{-1}, respectively, giving lifetimes with respect to OH at 2×10^6 radicals cm^{-3} of 18 and 11 h. The products of the OH reactions are formyl chloride (HC(O)Cl) and chloroacetaldehyde ($ClCH_2CHO$), with one molecule of each formed for each molecule of 1,3-dichloropropene reacted (Tuazon *et al.*, 1984) (see Problem 6).

A number of the pesticide HAPs are organophosphorus compounds. Seiber and co-workers have measured some organophosphorus pesticides, including diazinon, parathion, chlorpyrifos, and methidathion, in air and fogs and wet deposition in a variety of locations in California (e.g., Glotfelty *et al.*, 1990; Schomburg *et al.*, 1991; Zabik and Seiber, 1993; Seiber *et al.*, 1993; Baker *et al.*, 1996). Both the parent pesticides, all containing reactive P=S groups, and their oxidation products, the corresponding oxon (P=O), have been identified and quantified. The ratio of the oxons to the corresponding parent compounds increased with distance from the site of release and also increased after dawn, suggesting photochemical reactions were involved in the conversion.

These photochemical reactions likely involve OH chemistry. For example, Atkinson *et al.* (1989) reported that OH reacted with $(CH_3O)_3P=S$ to form $(CH_3O)_3P=O$ and, similarly, that $(CH_3O)_2P(S)CH_3$ formed $(CH_3O)_2P(O)CH_3$, in yields of 0.28 and 0.13, respectively. The reaction is believed to involve addition to the P=S bond, followed by secondary chemistry to generate the oxon (Goodman *et al.*, 1988).

Another example of this conversion of P=S found in pesticides to P=O is the oxidation of malathion in the atmosphere. Malathion itself is not a HAP and has relatively low acute mammalian toxicity because it is degraded by mammalian carboxylesterases. It is effective as a pesticide because in insects, it is activated to malaoxon, an acetylcholinesterase inhibitor. However, malathion itself typically contains impurities such as isomalathion whose mammalian toxicities are greater

TABLE 16.16 Some Pesticides and Pesticide Transformation Products Listed as Hazardous Air Pollutants[a]

1,3-Dichloropropene	Ethylene thiourea
2,3,7,7-Tetrachlorodibenzodioxin	Heptachlor
2,4,5-Trichlorophenol	Lindane
2,4-D salts and esters	Methoxychlor
4,6-Dinitro-*o*-cresol and salts	Methyl bromide
Acrolein	Parathion
Captan	Pentachloronitrobenzene
Carbaryl	Pentachlorophenol
Chloroamben	Phosgene
Chlordane	Phosphine
DDE	Propoxur
Dichlorvos	Toxaphene
Ethylene dibromide	Trifluralin
Ethylene oxide	

[a] From Seiber and Woodrow (1995); see Table 16.15.

G. ATMOSPHERIC CHEMISTRY AND RISK ASSESSMENTS OF HAZARDOUS AIR POLLUTANTS

TABLE 16.17 Some Pesticides and Their Transformation Products[a]

Pesticide	Structure	Reaction product
Merphos	$(RS)_3P$	$(RS)_3PO$ Def
Metam sodium	$H_3CN\overset{S}{\underset{}{\|\|}}C-S^- \, Na^+$	CH_3NCS MITC
Ethylenebis(dithiocarbamate) (EBDC)	$H_2C-NH-\overset{S}{\underset{}{\|\|}}C-S^-$ $H_2C-NH-\overset{S}{\underset{}{\|\|}}C-S^-$ M^{++}	ethylenethiourea ring structure Ethylenethiourea
Trifluralin	2,6-dinitro-N,N-dipropyl-4-(trifluoromethyl)aniline	N-propyl-2,6-dinitro-4-(trifluoromethyl)aniline
Ethyl parathion	$C_2H_5O-\overset{S}{\underset{OC_2H_5}{\|\|}}P-O-\text{C}_6\text{H}_4-NO_2$	$C_2H_5O-\overset{O}{\underset{OC_2H_5}{\|\|}}P-O-\text{C}_6\text{H}_4-NO_2$ Ethyl paraoxon

[a] Adapted from Sieber and Woodrow (1995).

than that of malathion itself (e.g., see Aldridge et al., 1979; and Ryan and Fukuto, 1985), as well as malaoxon.

Malathion in air readily undergoes oxidation to malaoxon:

Malathion → (atmospheric oxidation) → Malaoxon

Malathion has been used for aerial spraying over populated areas in southern California to control the Mediterranean fruit fly. Figure 16.45 shows one set of measurements of malathion and its oxidation product, malaoxon, made at three sites in this region before, during, and after spraying (Brown et al., 1993). Malaoxon is present initially as an impurity in the malathion. However, the effects of conversion of malathion to malaoxon by atmospheric reactions can be seen during the next 72 h, during which time the amount of the oxidation product is significantly increased relative to malathion. Brown et al. (1993) cite literature data that show that the oral toxicity of malaoxon in rats is much greater than that of the parent malathion. Thus, although malathion itself is not a HAP, it is a good example of a compound that is converted by atmospheric reactions into a more toxic species.

In addition to the toxicity issues associated with pesticide use, there are also reactivity concerns when they are used in urban areas. For example, chloropicrin (CCl_3NO_2) is used as a soil fumigant. It absorbs light in the actinic region (Moilanen et al., 1978; Carter et al., 1997), decomposing with a quantum yield of 0.87

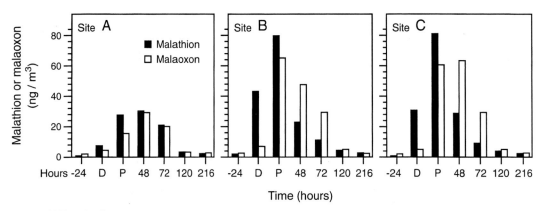

FIGURE 16.45 Concentrations of malathion and malaoxon measured in Garden Grove, southern California, in 1990 before aerial spraying (-24 h), during spraying (D) and immediately postspraying (P), and up to 216 h later (adapted from Brown *et al.*, 1993).

under simulated sunlight conditions (Carter *et al.*, 1997):

$$CCl_3NO_2 + h\nu \xrightarrow{O_2} Cl_2CO + ClNO. \quad (20)$$

Moilanen *et al.* (1978) observed that it was stable to irradiation when it was in a nitrogen carrier gas, indicating that molecular oxygen is involved in the photochemical decomposition; they suggest that O_2 may add to a biradical formed from the chloropicrin upon light absorption but that O_2 is regenerated in the decomposition of the adduct.

The reaction products, phosgene and nitrosyl chloride, photolyze to form chlorine atoms, which, as discussed in Chapter 6, are highly reactive with organics. Carter *et al.* (1997) used a combination of experiments and model calculations to show that this can lead to enhanced ozone formation, both through the initiation of the chain oxidation of VOC by atomic chlorine and by the addition of NO. They estimate that the reactivity of chloropicrin in terms of the O_3 generated per gram emitted is in the range of 0.5–1.5 times that of a typical VOC mixture in air, with the highest impact at the lowest NO_x levels.

Some HAPs impact not only the troposphere but also the stratosphere. The most obvious example is highly toxic methyl bromide, CH_3Br, used as a soil fumigant as well as for treatment of buildings for termites. As discussed in Chapter 12, this is a significant source of stratospheric bromine and hence contributes to stratospheric ozone depletion. Its continued use has been controversial and is being phased out (e.g., see Thomas, 1996; Ristaino and Thomas, 1997; and Duafala, 1996).

In summary, the atmospheric chemistry discussed throughout this book is an integral part of the risk assessment of many airborne chemicals. A firm understanding of such processes is critical to the development of cost-effective and health-protective control strategies.

H. PROBLEMS

1. Construct a mechanism that shows why methyl *tert*-butyl ether (MTBE), a common gasoline additive, oxidizes in air to give formaldehyde as well as acetone, *tert*-butyl formate ($HC(O)OC(CH_3)_3$), and methyl acetate ($CH_3C(O)OCH_3$).

2. High levels of acetaldehyde and PAN have been observed in Brazil, where ethanol is used extensively as a fuel. Show why this is expected.

3. Calculate the lifetime of HCHO with respect to reaction with OH at a concentration of 1×10^6 radicals cm^{-3}, NO_3 at 50 ppt, and Cl atoms at 5×10^4 atoms cm^{-3}. The rate constants are given in Chapter 6.

4. Use OZIPR (Appendix III) to quantitatively assess what effects the following parameters have on the peak levels of O_3 that are formed, as well as changes in the VOC and NO_x.

1. Effects of Time of Year, VOC and NO_x Levels, Reactivity, and HONO Formation (Use Example 1)

A. Time of Year

Run Example 1 (EX1.INP) for Los Angeles on June 21, 1990 (Latitude 34.058, Longitude 118.250, Time Zone 7). Consider this the base case for comparison for the remainder of Part I of this assignment.

Run the model again for December 31, 1990.

Plot O_3, VOC, and NO_x as a function of time of day for these two times of the year (you will need to use some other program to do this). You can put them all on one plot for June, and all three on a second for the December data, or if not too messy, all on one plot.

By what percentage does the predicted peak O_3 change simply due to changes in actinic flux?

For the remainder of Part 1, use Los Angeles, June 21, 1990.

B. VOC Reactivity

Run Example 1 for the case where there are no alkanes and all of the concentrations of alkanes in Example 1 have been put into internal olefins. *Note:* Be sure you know what species are used to represent the alkanes and the internal olefins in this model.

Plot VOC, NO_x, and O_3 for this case and compare to the base case (again use one plot). Give a brief description of why (chemically) or why not differences from the base case arise.

C. VOC and NO_x Levels

Run Example 1 for the cases where the initial levels of emitted species subject to control are increased by a factor of three in the following manner:

1. NO_x only
2. VOC only
3. CO only
4. NO_x, VOC, and CO are increased simultaneously

For each of these, plot O_3, VOC, and NO_x for the increased emissions and the base case for comparison. Give a brief description of why (chemically) or why not differences from the base case arise.

D. NO/NO_x Ratio

Run Example 1 for the case where the fraction of NO_x initially in the form of NO_2 is 0 rather than the default value of 0.25. Again, make plots to compare to the base case. Give a brief description of why (chemically) or why not differences from the base case arise.

E. Effects of Including HONO Formation

Run Example 1 with the mechanism (meccm.rad) modified to include HONO formation. While heterogeneous chemistry is not included in this mechanism, it can be crudely approximated by including it as a third-order reaction:

$$NO_2 + NO_2 + H_2O \rightarrow HONO + HNO_3$$

$$k = 1.0 \times 10^{-33} \text{ cm}^6 \text{ molecule}^{-2} \text{ s}^{-1}$$

Plot O_3, VOC, NO_2, HONO, and HNO_3 for this and the base case and comment on any differences.

2. Running the Isopleth Option (Example 8)

Note: This example generates isopleths of O_3 and other species. In order to print the isopleths, you will need a Hewlett Packard LaserJet or Epson RX or FX printer.

A. Changes in VOC Reactivity

Try changing the VOC reactivity from one extreme to another by first putting most of the VOC into what you think should be the least reactive form, and then into the most reactive form by modifying the file reac.cb4 appropriately. How much difference does this make to the O_3 isopleths and briefly, why?

Note: Don't put *all* of the VOC into OLE or just exchange PAR and OLE... the program crashes!

B. Changes in Reaction Mechanism

Think of some of the largest uncertainties in reaction mechanisms and choose one of them to study in more detail (e.g., the OH–aromatic reactions, the O_3–alkene reactions, etc.). Change the mechanism (file is meccm.cb4) to reflect an extreme version and see what difference this makes to the O_3 isopleth, if any. Be sure to describe clearly what you are changing and how. Comment on whether the isopleth changes and how.

Note: Remember that the output will be sensitive to the change in mechanism only to the extent that the organic involved is a major component of the VOC. You may also have to tinker with the concentrations to see an effect.

C. Adding HONO Formation

Repeat I.E above; i.e., add the possibility of HONO formation to the mechanism and examine the effects on all of the isopleths. Comment on the reason for differences, if any.

5. The rate constants for the reactions of *cis*- and *trans*-1,3-dichloropropene with O_3 have been measured to be 1.5×10^{-19} and 6.7×10^{-19} cm^3 molecule^{-1} s^{-1}, respectively (Tuazon *et al.*, 1984). Show that the loss of these compounds by reaction with O_3 is not important compared to their loss by reaction with OH under typical daytime tropospheric concentrations where O_3 is 80 ppb and OH is 5×10^6 cm^{-3}.

6. The OH reaction with 1,3-dichloropropene in air has been shown to give unit yields of both formyl chloride [HC(O)Cl] and chloroacetaldehyde [$CH_2ClCH(O)$]. Construct a mechanism consistent with this observation.

References

Akimoto, H., M. Hoshino, G. Inoue, F. Sakamaki, N. Washida, and M. Okuda, "Design and Characterization of the Evacuable and Bakable Photochemical Smog Chamber," *Environ. Sci. Technol., 13,* 471–475 (1979).

Akimoto, H., M. Hoshino, G. Inoue, F. Sakamaki, N. Washida, and M. Okuda, "Design and Characterization of the Evacuable and Bakable Photochemical Smog Chamber," *Environ. Sci. Technol., 19,* 507–512 (1985).

Akimoto, H., H. Takagi, and F. Sakamaki, "Photoenhancement of the Nitrous Acid Formation in the Surface Reaction of Nitrogen Dioxide and Water Vapor: Extra Radical Source in Smog Chamber Experiments," *Int. J. Chem. Kinet., 19,* 539–551 (1987).

Aldridge, W. N., J. W. Miles, D. L. Mount, and R. D. Verschoyle, "The Toxicological Properties of Impurities in Malathion," *Arch. Toxicol., 42,* 95–106 (1979).

Altshuler, S. L., T. D. Arcado, and D. R. Lawson, "Weekday vs. Weekend Ambient Ozone Concentrations: Discussion and Hypotheses with Focus on Northern California," *J. Air Waste Manage. Assoc., 45,* 967–972 (1995).

Altshuller, A. P., "An Evaluation of Techniques for the Determination of the Photochemical Reactivity of Organic Emissions," *J. Air Pollut. Control Assoc., 16,* 257–260 (1966).

Altshuller, A. P., "Nonmethane Organic Compound to Nitrogen Oxide Ratios and Organic Composition in Cities and Rural Areas," *J. Air Pollut. Control Assoc., 39,* 936–943 (1989).

Altshuller, A. P., and A. S. Lefohn, "Background Ozone in the Planetary Boundary Layer over the United States," *J. Air Waste Manage. Assoc., 46,* 134–141 (1996).

Al-Wali, K. I., and P. J. Samson, "Preliminary Sensitivity Analysis of Urban Airshed Model Simulations to Temporal and Spatial Availability of Boundary Layer Wind Measurements," *Atmos. Environ., 30,* 2027–2042 (1996).

Ames, J., T. C. Myers, L. E. Reid, D. C. Whitney, S. H. Golding, S. R. Hayes, and S. D. Reynolds, "The User's Manual for the SAI Airshed Model," Report to the EPA on Contract No. 68—02-2429 with the U.S. Environmental Protection Agency, August 18, 1978.

Andersson-Sköld, Y., P. Grennfelt, and K. Pleijel, "Photochemical Ozone Creation Potentials: A Study of Different Concepts," *J. Air Waste Manage. Assoc., 42,* 1152–1158 (1992).

Aneja, V. P., P. Roelle, and W. P. Robarge, "Contribution of Biogenic Nitric Oxide in Urban Ozone: Raleigh, NC, as a Case Study," *Atmos. Environ., 31,* 1531–1537 (1997).

Appleton, E. L., "Air Quality Modeling's Brave New World," *Environ. Sci. Technol., 30,* 200A–204A (1996).

Atkinson, R., "Kinetics and Mechanisms of the Gas-Phase Reactions of the Hydroxyl Radical with Organic Compounds," *J. Phys. Chem. Ref. Data Monogr. No. 1,* 1–246 (1989).

Atkinson, R., S. M. Aschmann, J. Arey, P. A. McElroy, and A. M. Winer, "Product Formation from the Gas-Phase Reactions of the OH Radical with $(CH_3O)_3PS$ and $(CH_3O)_2P(S)SCH_3$," *Environ. Sci. Technol., 23,* 243–244 (1989).

Atkinson, R., "Gas Phase Tropospheric Chemistry of Organic Compounds," *J. Phys. Chem. Ref. Data Monogr. No. 2,* 11–216 (1994).

Atkinson, R., D. L. Baulch, R. A. Cox, R. F. Hampson, Jr., J. A. Kerr, M. J. Rossi, and J. Troe, "Evaluated Kinetic and Photochemical Data for Atmospheric Chemistry: Supplement VI—IUPAC Subcommittee on Gas Kinetic Data Evaluation for Atmospheric Chemistry," *J. Phys. Chem. Ref. Data, 26,* 1329–1499 (1997a).

Atkinson, R., D. L. Baulch, R. A. Cox, R. F. Hampson, Jr., J. A. Kerr, M. J. Rossi, and J. Troe, "Evaluated Kinetic, Photochemical, and Heterogeneous Data for Atmospheric Chemistry. 5. IUPAC Subcommittee on Gas Kinetic Data Evaluation for Atmospheric Chemistry," *J. Phys. Chem. Ref. Data, 26,* 521–1011 (1997b).

Baker, L. T., D. L. Fitzell, J. N. Seiber, T. R. Parker, T. Shibamoto, M. W. Poore, K. E. Longley, R. P. Tomlin, R. Propper, and D. W. Duncan, "Ambient Air Concentrations of Pesticides in California," *Environ. Sci. Technol., 30,* 1365–1368 (1996).

Baumgard, K. J., and J. H. Johnson, "The Effect of Fuel and Engine Design on Diesel Exhaust Particle Size Distributions," *Soc. Automot. Eng. 960131* (Spec. Publ. 1140, p. 37), 1996.

Beaton, S. P., G. A. Bishop, and D. H. Stedman, "Emission Characteristics of Mexico City Vehicles," *J. Air Waste Manage. Assoc., 42,* 1424–1429 (1992).

Beaton, S. P., G. A. Bishop, Y. Zhang, L. L. Ashbaugh, D. R. Lawson, and D. H. Stedman, "On-Road Vehicle Emissions: Regulations, Costs, and Benefits," *Science, 268,* 991–993 (1995).

Behnke, W., W. Hollander, W. Koch, F. Nolting, and C. Zetzsch, "A Smog Chamber for Studies of the Photochemical Degradation of Chemicals in the Presence of Aerosols," *Atmos. Environ., 22,* 1113–1120 (1988).

Benjamin, M. T., M. Sudol, D. Vorsatz, and A. M. Winer, "A Spatially and Temporally Resolved Biogenic Hydrocarbon Emissions Inventory for the California South Coast Air Basin," *Atmos. Environ., 31,* 3087–3100 (1997).

Berkowitz, C. M., K. M. Busness, E. G. Chapman, J. M. Thorp, and R. D. Saylor, "Observations of Depleted Ozone within the Boundary Layer of the Western North Atlantic," *J. Geophys. Res., 100,* 11483–11496 (1995).

Berkowitz, C. M., and W. J. Shaw, "Airborne Measurements of Boundary Layer Chemistry during the Southern Oxidant Study: A Case Study," *J. Geophys. Res., 102,* 12795–12804 (1997).

Berkowitz, C. M., J. D. Fast, S. R. Springston, R. J. Larsen, C. W. Spicer, P. V. Doskey, J. M. Hubbe, and R. Plastridge, "Formation Mechanisms and Chemical Characteristics of Elevated Photochemical Layers over the Northeast United States," *J. Geophys. Res., 103,* 10631–10647 (1998).

Besemer, A. C., and H. Nieboer, "The Wall as a Source of Hydroxyl Radicals in Smog Chambers," *Atmos. Environ., 19,* 507–513 (1985).

Biesenthal, T. A., and P. B. Shepson, "Observations of Anthropogenic Inputs of the Isoprene Oxidation Products Methyl Vinyl Ketone and Methacrolein to the Atmosphere," *Geophys. Res. Lett., 24,* 1375–1378 (1997).

Bishop, G. A., and D. H. Stedman, "On-Road Carbon Monoxide Emission Measurement Comparisons for the 1988–1989 Colorado Oxy-Fuels Program," *Environ. Sci. Technol., 24,* 843–847 (1990).

Bishop, G. A., D. H. Stedman, and T. Jessop, "Infrared Emission and Remote Sensing," *J. Air Waste Manage. Assoc., 42,* 695–697 (1992).

Bishop, G. A., D. H. Stedman, J. E. Peterson, T. J. Hosick, and P. L. Guenther, "A Cost-Effectiveness Study of Carbon Monoxide Emissions Reduction Utilizing Remote Sensing," *J. Air Waste Manage. Assoc., 43,* 978–988 (1993).

Bishop, G. A., Y. Zhang, S. E. McLaren, P. L. Guenther, S. T. Beaton, J. E. Peterson, D. H. Stedman, W. R. Pierson, K. T. Knapp, R. B. Zweidinger, J. W. Duncan, A. Q. McArver, P. J. Groblicki, and J. F. Day, "Enhancements of Remote Sensing for Vehicle Emissions in Tunnels," *J. Air Waste Manage. Assoc., 44,* 169–175 (1994).

Bishop, G. A., and D. H. Stedman, "Measuring the Emissions of Passing Cars," *Acc. Chem. Res., 29,* 489–495 (1996).

Bishop, G. A., D. H. Stedman, and L. Ashbaugh, "Motor Vehicle Emissions Variability," *J. Air Waste Manage. Assoc., 46,* 667–675 (1996).

Bishop, G. A., D. H. Stedman, J. de La Garza Castro, and F. J. Davalos, "On-Road Remote Sensing of Vehicle Emissions in Mexico," *Environ. Sci. Technol., 31,* 3505–3510 (1997).

Black, F., S. Tejada, and M. Gurevich, "Alternative Fuel Motor Vehicle Tailpipe, and Evaporative Emissions Composition and Ozone Potential," *J. Air Waste Manage. Assoc., 48,* 578–591 (1998).

Blake, D. R., and F. S. Rowland, "Urban Leakage of Liquefied Petroleum Gas and Its Impact on Mexico City Air Quality," *Science, 269,* 953–956 (1995).

Bonsang, B., C. Polle, and G. Lambert, "Isoprene Production in Sea Water," *Geophys. Res. Lett., 19,* 1129–1132 (1992).

Bowman, F. M., and J. H. Seinfeld, "Ozone Productivity of Atmospheric Organics," *J. Geophys. Res., 99,* 5309–5324 (1994).

Bowman, F. M., C. Pilinis, and J. H. Seinfeld, "Ozone and Aerosol Productivity of Reactive Organics," *Atmos. Environ., 29,* 579–589 (1995).

Brönnimann, S., and U. Neu, "Weekend–Weekday Differences of Near-Surface Ozone Concentrations in Switzerland for Different Meteorological Conditions," *Atmos. Environ., 31,* 1127–1135 (1997).

Brown, M. A., M. X. Petreas, H. W. Okamoto, T. M. Mischke, and R. D. Stephens, "Monitoring of Malathion and Its Impurities and Environmental Transformation Products on Surfaces and in Air Following an Aerial Application," *Environ. Sci. Technol., 27,* 388–397 (1993).

Bufalini, J. J., T. A. Walter, and M. M. Bufalini, "Ozone Formation Potential of Organic Compounds," *Environ. Sci. Technol., 10,* 908–912 (1976).

Bufalini, J. J., T. A. Walter, and M. M. Bufalini, "Contamination Effects on Ozone Formation in Smog Chambers," *Environ. Sci. Technol., 11,* 1181–1185 (1977).

Bufalini, J. J., and M. C. Dodge, "Ozone-Forming Potential of Light Saturated Hydrocarbons," *Environ. Sci. Technol., 17,* 308–311 (1983).

Cadle, S. H., and R. D. Stephens, "Remote Sensing of Vehicle Exhaust Emissions," *Environ. Sci. Technol., 28,* 258A–264A (1994).

Cadle, S. H., B. K. Bailey, T. C. Belian, M. Carlock, R. A. Gorse, Jr., H. Haskew, K. T. Knapp, and D. R. Lawson, "Real World Vehicle Emissions: A Summary of the Fifth Coordinating Research Council On-Road Vehicle Emissions Workshop," *J. Air Waste Manage. Assoc., 46,* 355–369 (1996).

Cadle, S. H., R. A. Gorse, Jr., T. C. Belian, and D. R. Lawson, "Real-World Vehicle Emissions: A Summary of the Seventh Coordinating Research Council On-Road Vehicle Emissions Workshop," *J. Air Waste Manage. Assoc., 48,* 174–185 (1997a).

Cadle, S. H., R. A. Gorse, Jr., T. C. Belian, and D. R. Lawson, "Real-World Vehicle Emissions: A Summary of the Sixth Coordinating Research Council On-Road Vehicle Emissions Workshop," *J. Air Waste Manage. Assoc., 47,* 426–438 (1997b).

California Air Resources Board, "Definition of a Low-Emission Motor Vehicle in Compliance with the Mandates of Health and Safety Code Sections 39037.5 (Assembly Bill 234, Leonard, 1987). Available from the California Air Resources Board, Mobile Source Division, 9528 Telstar Ave., El Monte, CA, May 1989.

California Air Resources Board, "Initial Statement of Proposed Rulemaking, Amendments to the Low-Emission Vehicle Program," September 25, 1992, and Supplement, "Establishment of Reactivity Adjustment Factors" and "Speciated Vehicle Data and/or Airshed Modeling Results," November 13, 1992a. Available from the California Air Resources Board, 9528 Telstar Ave., El Monte, CA 91731.

California Air Resources Board, "Final Report on the Identification of Formaldehyde as a Toxic Air Contaminant," Sacramento, CA, 1992b.

California Air Resources Board, "Establishment of Corrections to Reactivity Adjustment Factors for Transitional Low-Emission Vehicles and Low-Emission Vehicles Operating on Phase 2 Reformulated Gasoline: Airshed Modeling Protocol and Updated Final Results," California Air Resources Board, Research Division, March 15, 1993.

California Air Resources Board, "Proposed Identification of Diesel Exhaust as a Toxic Air Contaminant, Public Comment and SRP Version," Sacramento, CA, May, 1997a.

California Air Resources Board, "Toxic Air Contaminant Identification List Summaries," September, 1997b.

California Environmental Protection Agency, "Health Effects of Exposure to Environmental Tobacco Smoke," Final Report, September 1997. Available from the Office of Environmental Health Hazard Assessment, 301 Capitol Mall, Second Floor, Sacramento, CA 95814; http://www.calepa.cahwnet.gov/oehha/.

Calkins, D. L., "Global Partnerships: A Collaborative Effort to Improve Air Quality in Developing Countries," *EM, J. Air Waste Manage. Assoc.,* October, 26–32 (1998).

Calvert, J. G., and J. N. Pitts, Jr., *Photochemistry,* Wiley, New York, 1966.

Calvert, J. G., J. B. Heywood, R. F. Sawyer, and J. H. Seinfeld, "Achieving Acceptable Air Quality: Some Reflections on Controlling Vehicle Emissions," *Science, 261,* 37–45 (1993).

Cardelino, C. A., and W. L. Chameides, "Natural Hydrocarbons, Urbanization, and Urban Ozone," *J. Geophys. Res., 95,* 13971–13979 (1990).

Cardelino, C. A., and W. L. Chameides, "An Observation-Based Model for Analyzing Ozone Precursor Relationships in the Urban Atmosphere," *J. Air Waste Manage. Assoc., 45,* 161–180 (1995).

Carmichael, G. R., L. K. Peters, and R. D. Saylor, "The STEM-II Regional Scale Acid Deposition and Photochemical Oxidant Model. I. An Overview of Model Development and Applications," *Atmos. Environ., 25,* 2077–2090 (1991).

Carter, W. P. L., R. Atkinson, A. M. Winer, and J. N. Pitts, Jr., "Evidence for Chamber-Dependent Radical Sources: Impact on Kinetic Computer Models for Air Pollution," *Int. J. Chem. Kinet., 13,* 735–740 (1981).

Carter, W. P. L., R. Atkinson, A. M. Winer, and J. N. Pitts, Jr., "Experimental Investigation of Chamber-Dependent Radical Sources," *Int. J. Chem. Kinet., 14,* 1071–1103 (1982).

Carter, W. P. L., R. Atkinson, A. M. Winer, and J. N. Pitts, Jr., "Experimental Protocol for Determining Photolysis Reaction Rate Constants," Report to the U.S. Environmental Protection Agency, Contract No. R80666-01, 1984.

Carter, W. P. L., R. Atkinson, A. M. Winer, and J. N. Pitts, Jr., "Comments on 'The Wall as a Source of Hydroxyl Radicals in Smog Chambers,'" *Atmos. Environ., 19,* 1977–1978 (1985).

Carter, W. P. L., and R. Atkinson, "An Experimental Study of Incremental Hydrocarbon Reactivity," *Environ. Sci. Technol., 21,* 670–679 (1987).

Carter, W. P. L., and R. Atkinson, "Computer Modeling Study of Incremental Hydrocarbon Reactivity," *Environ. Sci. Technol., 23,* 864–880 (1989).

Carter, W. P. L., "Development of Ozone Reactivity Scales for Volatile Organic Compounds," *J. Air Waste Manage. Assoc., 44,* 881–899 (1994).

Carter, W P. L., "Computer Modeling of Environmental Chamber Measurements of Maximum Incremental Reactivities of Volatile Organic Compounds," *Atmos. Environ., 29,* 2513–2527 (1995).

Carter, W. P. L., J. A. Pierce, D. Luo, and I. L. Malkina, "Environmental Chamber Study of Maximum Incremental Reactivities of Volatile Organic Compounds," *Atmos. Environ., 29,* 2499–2511 (1995).

Carter, W. P. L., D. Luo, and I. L. Malkina, "Investigation of the Atmospheric Reactions of Chloropicrin," *Atmos. Environ., 31,* 1425–1439 (1997).

Chameides, W. L., R. W. Lindsay, J. Richardson, and C. S. Kiang, "The Role of Biogenic Hydrocarbons in Urban Photochemical Smog: Atlanta as a Case Study," *Science, 241,* 1473–1475 (1988).

Chameides, W. L., F. Fehsenfeld, M. O. Rodgers, C. Cardelino, J. Martinez, D. Parrish, W. Lonneman, D. R. Lawson, R. A. Rasmussen, P. Zimmerman, J. Greenberg, P. Middleton, and T. Wang, "Ozone Precursor Relationships in the Ambient Atmosphere," *J. Geophys. Res., 97,* 6037–6055 (1992).

Chan, C.-C., C.-K. Nien, C.-Y. Tsai, and G.-R. Her, "Comparison of Tail-Pipe Emissions from Motorcycles and Passenger Cars," *J. Air Waste Manage. Assoc., 45,* 116–124 (1995).

Chang, J. S., R. A. Brost, I. S. A. Isaksen, S. Madronich, P. Middleton, W. R. Stockwell, and C. J. Walcek, "A Three-Dimensional Eulerian Acid Deposition Model: Physical Concepts and Formulations," *J. Geophys. Res., 92,* 14681–14700 (1987).

Chang, T. Y., and S. J. Rudy, "Ozone-Forming Potential of Organic Emissions from Alternative-Fueled Vehicles," *Atmos. Environ., 24A,* 2421–2430 (1990a).

Chang, T. Y., and S. J. Rudy, "Roadway Tunnel Air Quality Models," *Environ. Sci. Technol., 24,* 672–676 (1990b).

Chang, T. Y., R. H. Hammerle, S. M. Japar, and I. T. Salmeen, "Alternative Transportation Fuels and Air Quality," *Environ. Sci. Technol., 25,* 1190–1197 (1991).

Chin, M., and D. J. Jacob, "Anthropogenic and Natural Contributions to Tropospheric Sulfate: A Global Model Analysis," *J. Geophys. Res., 101,* 18691–18699 (1996).

Chock, D. P., A. M. Dunker, S. Kumar, and C. S. Sloane, "Effect of NO_x Emission Rates on Smog Formation in the California South Coast Air Basin," *Environ. Sci. Technol., 15,* 933–939 (1981).

Chock, D. P., A. M. Dunker, S. Kumar, and C. S. Sloane, "Letter to the Editor," *Environ. Sci. Technol., 17,* 58–62 (1983).

Chock, D. P., G. Yarwood, A. M. Dunker, R. E. Morris, A. K. Pollack, and C. H. Schleyer, "Sensitivity of Urban Airshed Model Results for Test Fuels to Uncertainties in Light-Duty Vehicle and Biogenic Emissions and Alternative Chemical Mechanisms. Auto/Oil Air Quality Improvement Research Program," *Atmos. Environ., 29,* 3067–3084 (1995).

Cicero-Fernández, P., J. R. Long, and A. M. Winer, "Effects of Grades and Other Loads on On-Road Emissions of Hydrocarbons and Carbon Monoxide," *J. Air Waste Manage. Assoc., 47,* 898–904 (1997).

Cleveland, W. S., T. E. Graedel, B. Kleiner, and J. L. Warner, "Sunday and Workday Variations in Photochemical Air Pollutants in New Jersey and New York," *Science, 186,* 1037–1038 (1974).

Conner, T. L., A. Lonneman, and R. L. Seila, "Transportation-Related Volatile Hydrocarbon Source Profiles Measured in Atlanta," *J. Air Waste Manage. Assoc., 45,* 383–394 (1995).

Croes, B. E., J. R. Holmes, and A. C. Lloyd, "Reactivity-Based Hydrocarbon Controls: Scientific Issues and Potential Regulatory Applications," *J. Air Waste Manage. Assoc., 42,* 657–661 (1992).

Dabdub, D., and J. H. Seinfeld, "Air Quality Modeling on Massively Parallel Computers," *Atmos. Environ., 28,* 1679–1687 (1994a).

Dabdub, D., and J. H. Seinfeld, "Numerical Advective Schemes Used in Air Quality Models—Sequential and Parallel Implementation," *Atmos. Environ., 28,* 3369–3385 (1994b).

Dabdub, D., and J. H. Seinfeld, "Parallel Computation in Atmospheric Chemical Modeling," *Parallel Comput., 22,* 111–130 (1996).

Dana, M. T., J. M. Hales, and M. A. Wolf, "Rain Scavenging of SO_2 and Sulfate from Power Plant Plumes," *J. Geophys. Res., 80,* 4119–4129 (1975).

Darlington, T. L., D. F. Kahlbaum, J. M. Heuss, and G. T. Wolff, "Analysis of PM_{10} Trends in the United States from 1988 through 1995," *J. Air Waste Manage. Assoc., 47,* 1070–1078 (1997).

Darnall, K. R., A. C. Lloyd, A. M. Winer, and J. N. Pitts, Jr., "Reactivity Scale for Atmospheric Hydrocarbons Based on Reaction with Hydroxyl Radical," *Environ. Sci. Technol., 10,* 692–696 (1976).

Davidson, A., "Update on Ozone Trends in California's South Coast Air Basin," *J. Air Waste Manage. Assoc., 43,* 226–227 (1993).

Dennis, R. L., D. W. Byun, J. H. Novak, K. J. Galluppi, C. J. Coats, and M. A. Vouk, "The Next Generation of Integrated Air Quality Modeling: EPA's Models-3," *Atmos. Environ., 30,* 1925–1938 (1996).

Derwent, R., and Ø. Hov, "Application of Sensitivity and Uncertainty Analysis Techniques to a Photochemical Ozone Model," *J. Geophys. Res., 93,* 5185–5199 (1988).

Derwent, R. G., and M. E. Jenkin, "Hydrocarbons and the Long-Range Transport of Ozone and PAN across Europe," *Atmos. Environ., 25A,* 1661–1678 (1991).

Derwent, R. G., M. E. Jenkin, and S. M. Saunders, "Photochemical Ozone Creation Potentials for a Large Number of Reactive Hydrocarbons under European Conditions," *Atmos. Environ., 30,* 181–199 (1996).

Derwent, R. G., M. E. Jenkin, S. M. Saunders, and M. J. Pilling, "Photochemical Ozone Creation Potentials for Organic Compounds in Northwest Europe Calculated with a Master Chemical Mechanism," *Atmos. Environ., 32,* 2429–2441 (1998).

De Vlieger, I., "On-Board Emission and Fuel Consumption Measurement Campaign on Petrol-Driven Passenger Cars," *Atmos. Environ., 31,* 3753–3761 (1997).

Dickerson, R. R., S. Kondragunta, G. Stechikov, K. L. Civerolo, B. G. Doddridge, and B. N. Holben, "The Impact of Aerosols on Solar Ultraviolet Radiation and Photochemical Smog," *Science, 278,* 827–830 (1997).

Dimitriades, B., and M. Dodge, Eds., *Proceedings of the Empirical Kinetic Modeling Approach (EKMA) Validation Workshop,* EPA Report No. EPA-600/9-83-014, August 1983.

Dimitriades, B., "Scientific Basis for the VOC Reactivity Issues Raised by Section 183(e) of the Clean Air Act Amendments of 1990," *J. Air Waste Manage. Assoc., 46,* 963–970 (1996).

Dodge, M. C., "Combined Use of Modeling Techniques and Smog Chamber Data to Derive Ozone-Precursor Relationships," in *Proceedings of the International Conference on Photochemical Oxidant Pollution and Its Control* (B. Dimitriades, Ed.), EPA-600/3-77-001b, Vol. II, pp. 881–889, 1977a.

Dodge, M. C., "Effect of Selected Parameters on Predictions of a Photochemical Model," U.S. Environmental Protection Agency Report No. EPA-60013-77-048, June 1977b.

Dodge, M. C., "Combined Effects of Organic Reactivity and $NMHC/NO_x$ Ratio on Photochemical Oxidant Formation—A Modeling Study," *Atmos. Environ., 18,* 1657–1665 (1984).

Dodge, M. C., "A Comparison of Three Photochemical Oxidant Mechanisms," *J. Geophys. Res., 94,* 5121–5136 (1989).

Dodge, M. C., "Formaldehyde Production in Photochemical Smog As Predicted by Three State-of-the-Science Chemical Oxidant Mechanisms," *J. Geophys. Res., 95,* 3635–3648 (1990).

Dolislager, L., and B. Croes, personal communication (1996).

Doskey, P. V., J. A. Porter, and P. A. Scheff, "Source Fingerprints for Volatile Non-Methane Hydrocarbons," *J. Air Waste Manage. Assoc., 42,* 1437–1445 (1992).

Doussin, J. F., D. Ritz, R. Durand-Jolibois, A. Monod, and P. Carlier, "Design of an Environmental Chamber for the Study of Atmospheric Chemistry: New Developments in the Analytical Device," *Analusis, 25,* 236–242 (1997).

Doyle, G. J., P. J. Bekowies, A. M. Winer, and J. N. Pitts, Jr., "Charcoal-Adsorption Air Purification System for Chamber Studies Investigating Atmospheric Photochemistry," *Environ. Sci. Technol., 11,* 45–51 (1977).

Dreher, D. B., and R. A. Harley, "A Fuel-Based Inventory for Heavy-Duty Diesel Truck Emissions," *J. Air Waste Manage. Assoc., 48,* 352–358 (1998).

Duafala, T., "Do We Need Further Controls of Agricultural Methyl Bromide?" *Atmos. Environ., 30,* R3–R4 (1996).

Duffy, B. L., P. F. Nelson, Y. Ye, and I. A. Weeks, "Speciated Hydrocarbon Profiles and Calculated Reactivities of Exhaust and Evaporative Emissions from 82 In-Use Light-Duty Australian Vehicles," *Atmos. Environ., 33,* 291–307 (1999).

Duncan, B. N., and W. L. Chameides, "Effects of Urban Emission Control Strategies on the Export of Ozone and Ozone Precursors from the Urban Atmosphere to the Troposphere," *J. Geophys. Res., 103,* 28159–28179 (1998).

Dunker, A. M., "Relative Reactivity of Emissions from Methanol-Fueled and Gasoline-Fueled Vehicles in Forming Ozone," *Environ. Sci. Technol., 24,* 853–862 (1990).

Dunker, A. M., R. E. Morris, A. K. Pollack, C. H. Schleyer, and G. Yarwood, "Photochemical Modeling of the Impact of Fuels and Vehicles on Urban Ozone Using Auto/Oil Program Data," *Environ. Sci. Technol., 30,* 787–801 (1996).

Dutkiewicz, V. A., E. G. Burkhard, and L. Husain, "Availability of H_2O_2 for Oxidation of SO_2 in Clouds in the Northeastern United States," *Atmos. Environ., 29,* 3281–3292 (1995).

Edinger, J. G., "Vertical Distribution of Photochemical Smog in Los Angeles Basin," *Environ. Sci. Technol., 7,* 247–252 (1973).

Eisele, F. L., G. H. Mount, D. Tanner, A. Jefferson, R. Shetter, J. W. Harder, and E. J. Williams, "Understanding the Production and Interconversion of the Hydroxyl Radical during the Tropospheric OH Photochemistry Experiment," *J. Geophys. Res., 102,* 6457–6465 (1997).

Eldering, A., and G. R. Cass, "Source-Oriented Model for Air Pollutant Effects on Visibility," *J. Geophys. Res., 101,* 19343–19369 (1996).

Elkus, B., and K. R. Wilson, "Photochemical Air Pollution: Weekend–Weekday Differences," *Atmos. Environ., 11,* 509–515 (1977).

Environmental Manager, published by the Air & Waste Management Association, May 1998.

Evans, L. F., I. A. Weeks, and A. J. Eccleston, "A Chamber Study of Photochemical Smog in Melbourne, Australia—Present and Future," *Atmos. Environ., 20,* 1355–1368 (1986).

Fabian, P., C. Haustein, G. Jakobi, B. Rappenglück, P. Suppan, and P. Steil, "Photochemical Smog in the Munich Metropolitan Area," *Beitr. Phys. Atmos., 67,* 39–56 (1994).

Fast, J., personal communication (1998).

Finlayson-Pitts, B. J., and J. N. Pitts, Jr., "Atmospheric Chemistry of Tropospheric Ozone Formation: Scientific and Regulatory Implications," *J. Air Waste Manage. Assoc., 43,* 1091–1100 (1991).

Finlayson-Pitts, B. J., and J. N. Pitts, Jr., "Volatile Organic Compounds: Ozone Formation, Alternative Fuels, and Toxics," *Chem. Ind.,* October 18, 796–800 (1993).

Fiore, A. M., D. J. Jacob, J. A. Logan, and J. H. Yin, "Long-Term Trends in Ground Level Ozone over the Contiguous United States, 1980–1995," *J. Geophys. Res., 103,* 1471–1480 (1998).

Fitz, D. R., M. C. Dodd, and A. M. Winer, "Photooxidation of α-Pinene at Near-Ambient Concentrations under Simulated Atmospheric Conditions," Paper No. 81-27.3, 74th Annual Meeting of the Air Pollution Control Association, Philadelphia, PA, June 21–26, 1981.

Forbes, P. D., R. E. Davies, L. C. D'Aloisio, and C. Cole, "Emission Spectrum Differences in Fluorescent Blacklight Lamps," *Photochem. Photobiol., 24,* 613–615 (1976).

Fraser, M. P., G. R. Cass, and B. R. T. Simoneit, "Gas-Phase and Particle-Phase Organic Compounds Emitted from Motor Vehicle Traffic in a Los Angeles Roadway Tunnel," *Environ. Sci. Technol., 32,* 2051–2060 (1998).

Fuentes, J. D., and T. F. Dann, "Ground-Level Ozone in Eastern Canada: Seasonal Variations, Trends, and Occurrences of High Concentrations," *J. Air Waste Manage. Assoc., 44,* 1019–1026 (1994).

Fujita, E. M., B. E. Croes, C. L. Bennett, D. R. Lawson, F. W. Lurmann, and H. H. Main, "Comparison of Emission Inventory and Ambient Concentration Ratios of CO, NMOG, and NO_x in California's South Coast Basin," *J. Air Waste Manage. Assoc., 42,* 264–276 (1992).

Gabele, P., "Exhaust Emissions from In-Use Alternative Fuel Vehicles," *J. Air Waste Manage. Assoc., 45,* 770–777 (1995).

Gaffney, J. S., N. A. Marley, R. S. Martin, R. W. Dixon, L. G. Reyes, and C. J. Popp, "Potential Air Quality Effects of Using Ethanol–Gasoline Fuel Blends: A Field Study in Albuquerque, New Mexico," *Environ. Sci. Technol., 31,* 3053–3061 (1997).

Gaffney, J. S., N. A. Marley, R. S. Martin, R. W. Dixon, L. G. Reyes, and C. J. Popp, "Response to Comment on 'Potential Air Quality Effects of Using Ethanol–Gasoline Fuel Blends: A Field Study in Albuquerque, New Mexico,'" *Environ. Sci. Technol., 32,* 3842–3844 (1998).

Gao, D., W. R. Stockwell, and J. B. Milford, "First-Order Sensitivity and Uncertainty Analysis for a Regional-Scale Gas-Phase Chemical Mechanism," *J. Geophys. Res., 100,* 23153–23166 (1995).

Gao, D., W. R. Stockwell, and J. B. Milford, "Global Uncertainty Analysis of a Regional-Scale Gas-Phase Chemical Mechanism," *J. Geophys. Res., 101,* 9107–9119 (1996).

Gertler, A. W., and W. R. Pierson, "Motor Vehicle Emissions Modeling Issues," *Sci. Total Environ., 146/147,* 333–338 (1994).

Gertler, A. W., E. M. Fujita, W. R. Pierson, and D. N. Wittorff, "Apportionment of NMHC Tailpipe vs. Non-tailpipe Emissions in the Fort McHenry and Tuscarora Mountain Tunnels," *Atmos. Environ., 30,* 2297–2305 (1996).

Gery, M. W., G. Z. Whitten, J. P. Killus, and M. C. Dodge, "A Photochemical Kinetics Mechanism for Urban and Regional Scale Computer Modeling," *J. Geophys. Res., 94,* 12925–12956 (1989).

Glasson, W. A., and A. M. Dunker, "Investigation of Background Radical Sources in a Teflon-Film Irradiation Chamber," *Environ. Sci. Technol., 23,* 970–978 (1989).

Glotfelty, D. E., M. S. Majewski, and J. N. Seiber, "Distribution of Several Organophosphorus Insecticides and Their Oxygen Analogues in a Foggy Atmosphere," *Environ. Sci. Technol., 24,* 353–357 (1990).

Goodman, M. A., S. M. Aschmann, R. Atkinson, and A. M. Winer, "Kinetics of the Atmospherically Important Gas-Phase Reactions of a Series of Trimethyl Phosphorothioates," *Arch. Environ. Contam. Toxicol., 17,* 281–288 (1988).

Graedel, T. E., L. A. Farrow, and T. A. Weber, "Photochemistry of the 'Sunday Effect,'" *Environ. Sci. Technol., 11,* 690–694 (1977).

Graedel, T. E., D. T. Hawkins, and L. D. Claxton, *Atmospheric Chemical Compounds— Sources, Occurrence, and Bioassay,* Academic Press, San Diego, 1986.

Grosjean, D., "Wall Loss of Gaseous Pollutants in Outdoor Teflon Chambers," *Environ. Sci. Technol., 19,* 1059–1065 (1985).

Grosjean, D., "Atmospheric Chemistry of Toxic Contaminants. 1. Reaction Rates and Atmospheric Persistence," *J. Air Waste Manage. Assoc., 40,* 1397–1402 (1990a).

Grosjean, D., "Atmospheric Chemistry of Toxic Contaminants. 2. Saturated Aliphatics: Acetaldehyde, Dioxane, Ethylene Glycol Ethers, Propylene Oxide," *J. Air Waste Manage. Assoc., 40,* 1522–1531 (1990b).

Grosjean, D., "Atmospheric Chemistry of Toxic Contaminants. 3. Unsaturated Aliphatics: Acrolein, Acrylonitrile, Maleic Anhydride," *J. Air Waste Manage. Assoc., 40,* 1664–1668 (1990c).

Grosjean, D., A. H. Miguel, and T. M. Tavares, "Urban Air Pollution in Brazil: Acetaldehyde and Other Carbonyls," *Atmos. Environ., 24B*, 1–6 (1990).

Grosjean, D., "Atmospheric Fate of Toxic Aromatic Compounds," *Sci. Total Environ., 100*, 367–414 (1991a).

Grosjean, D., "Atmospheric Chemistry of Toxic Contaminants. 4. Saturated Halogenated Aliphatics: Methyl Bromide, Epichlorhydrin, Phosgene," *J. Air Waste Manage. Assoc., 41*, 56–61 (1991b).

Grosjean, D., "Atmospheric Chemistry of Toxic Contaminants. 5. Unsaturated Halogenated Aliphatics: Allyl Chloride, Chloroprene, Hexachlorocyclopentadiene, Vinylidene Chloride," *J. Air Waste Manage. Assoc., 41*, 182–189 (1991c).

Grosjean, D., "Atmospheric Chemistry of Toxic Contaminants. 6. Nitrosamines: Dialkyl Nitrosamines and Nitrosomorpholine," *J. Air Waste Manage. Assoc., 41*, 306–311 (1991d).

Haagen-Smit, A. J., and M. M. Fox, "Photochemical Ozone Formation with Hydrocarbons and Automobile Exhaust," *J. Air Pollut. Control Assoc., 4*, 105–108 (1954).

Haagen-Smit, A. J., and M. M. Fox, "Ozone Formation in Photochemical Oxidation of Organic Substances," *Ind. Eng. Chem., 48*, 1484–1487 (1956).

Haagen-Smit, A. J., C. E. Bradley, and M. M. Fox, "Ozone Formation in Photochemical Oxidation of Organic Substances," *Ind. Eng. Chem., 45*, 2086–2089 (1953).

Hales, J. M., "A Generalized Multidimensional Model for Precipitation Scavenging and Atmospheric Chemistry," *Atmos. Environ., 23*, 2017–2031 (1989).

Hales, J. M., "A Modelling Investigation of Nonlinearities in the Wet Removal of SO_2 Emitted by Urban Sources," in *Atmospheric Chemistry: Models and Predictions for Climate and Air Quality* (C. S. Sloane and T. W. Tesche, Eds.), Chap. 8, pp. 117–130, Lewis Publishers, Chelsea, MI, 1991.

Hales, J. M., and D. S. Renne, "Source-Receptor Linearity: Definitions, Measurement, and Practical Implications," *Atmos. Environ., 26A*, 2111–2123 (1992).

Harley, R. A., M. P. Hannigan, and G. R. Cass, "Respeciation of Organic Gas Emissions and the Detection of Excess Unburned Gasoline in the Atmosphere," *Environ. Sci. Technol., 26*, 2395–2407 (1992).

Harley, R. A., A. G. Russell, G. J. McRae, G. R. Cass, and J. H. Seinfeld, "Photochemical Modeling of the Southern California Air Quality Study," *Environ. Sci. Technol., 27*, 378–388 (1993a).

Harley, R. A., A. G. Russell, and G. R. Cass, "Mathematical Modeling of the Concentrations of Volatile Organic Compounds: Model Performance Using a Lumped Chemical Mechanism," *Environ. Sci. Technol., 27*, 1638–1649 (1993b).

Harley, R. A., and G. R. Cass, "Modeling the Concentrations of Gas-Phase Toxic Organic Air Pollutants: Direct Emissions and Atmospheric Formation," *Environ. Sci. Technol., 28*, 88–98 (1994).

Harley, R. A., and G. R. Cass, "Modeling the Atmospheric Concentrations of Individual Volatile Organic Compounds," *Atmos. Environ., 29*, 905–922 (1995).

Harrington, W., and V. McConnell, "Modeling In-Use Vehicle Emissions and the Effects of Inspection and Maintenance Programs," *J. Air Waste Manage. Assoc., 44*, 791–799 (1994).

Hastie, D. R., P. B. Shepson, N. Reid, P. B. Roussel, and O. T. Melo, "Summertime NO_x, NO_y, and Ozone at a Site in Rural Ontario," *Atmos. Environ., 30*, 2157–2165 (1966).

Hatakeyama, S., H. Akimoto, and N. Washida, "Effect of Temperature on the Formation of Photochemical Ozone in a Propene–NO_x–Air–Irradiation System," *Environ. Sci. Technol., 25*, 1884–1890 (1991).

Henry, R. C., C. H. Spiegelman, J. F. Collins, and E. Park, "Reported Emissions of Organic Gases Are Not Consistent with Observations," *Proc. Natl. Acad. Sci. U.S.A., 94*, 6596–6599 (1997).

Ho, J., and A. M. Winer, "Effects of Fuel Type, Driving Cycle, and Emission Status on In-Use Vehicle Exhaust Reactivity," *J. Air Waste Manage. Assoc., 48*, 592–603 (1998).

Hoekman, S. K., "Speciated Measurements and Calculated Reactivities of Vehicle Exhaust Emissions from Conventional and Reformulated Gasolines," *Environ. Sci. Technol., 26*, 1206–1216 (1992).

Hofmann, D. J., "Twenty Years of Balloon-Borne Tropospheric Aerosol Measurements at Laramie, Wyoming," *J. Geophys. Res., 98*, 12753–12766 (1993).

Holmes, J. R., R. J. O'Brien, J. H. Crabtree, T. A. Hecht, and J. H. Seinfeld, "Measurement of Ultraviolet Radiation Intensity in Photochemical Smog Studies," *Environ. Sci. Technol., 7*, 519–523 (1973), and references therein.

Hough, A. M., and R. G. Derwent, "Computer Modeling Studies of the Distribution of Photochemical Ozone Production between Different Hydrocarbons," *Atmos. Environ., 21*, 2015–2033 (1987).

Hough, A. M., "An Intercomparison of Mechanisms for the Production of Photochemical Oxidants," *J. Geophys. Res., 93*, 3789–3812 (1988).

Hough, A. M., and C. Reeves, "Photochemical Oxidant Formation and the Effects of Vehicle Exhaust Emission Controls in the U.K. The Results from 20 Different Chemical Mechanisms," *Atmos. Environ., 22*, 1121–1135 (1988).

Husain, L., V. A. Dutkiewicz, and M. Das, "Evidence for Decrease in Atmospheric Sulfur Burden in the Eastern United States Caused by Reduction in SO_2 Emissions," *Geophys. Res. Lett., 25*, 967–970 (1998).

Ibrahim, S. S., R. W. Bilger, and N. R. Mudford, "Turbulence Effects on Chemical Reactions in Smog Chamber Flows," *Atmos. Environ., 21*, 2609–2621 (1987).

Idriss, H., and E. G. Seebauer, "Fast Photoreactions of Oxygenates on Tropospheric Fly Ash Particles," *J. Vac. Sci. Technol. A, 14*, 1627–1632 (1996).

Ingalls, M. N., L. R. Smith, and R. E. Kirksey, "Measurement of On-Road Vehicle Emission Factors in the California South Coast Air Basin—Vol. I: Regulated Emissions," Report No. SwRI-1604 from the Southwest Research Institute to the Coordinating Research Council, Atlanta, GA, June 1989; NTIS Document PB89220925.

Jacob, D. J., L. W. Horowitz, J. W. Munger, B. G. Heikes, R. R. Dickerson, R. S. Artz, and W. C. Keene, "Seasonal Transition from NO_x- to Hydrocarbon-Limited Conditions for Ozone Production over the Eastern United States in September," *J. Geophys. Res., 100*, 9315–9324 (1995).

Jacobson, M. Z., "Studying the Effects of Aerosols on Vertical Photolysis Rate Coefficient and Temperature Profiles over an Urban Airshed," *J. Geophys. Res., 103*, 10593–10604 (1998).

Japar, S. M., T. J. Wallington, S. J. Rudy, and T. Y. Chang, "Ozone-Forming Potential of a Series of Oxygenated Organic Compounds," *Environ. Sci. Technol., 25*, 415–420 (1991).

Jeffries, H. E., D. L. Fox, and R. Kamens, "Photochemical Conversion of NO to NO_2 by Hydrocarbons in an Outdoor Chamber," *J. Air Pollut. Control Assoc., 26*, 480–484 (1976).

Jenkin, M. E., S. M. Saunders, and M. J. Pilling, "The Tropospheric Degradation of Volatile Organic Compounds: A Protocol for Mechanism Development," *Atmos. Environ., 31*, 81–104 (1997).

Jiménez, J. L., M. D. Koplow, D. D. Nelson, M. S. Zahniser, and S. E. Schmidt, "Characterization of On-Road Vehicle NO Emissions by a TILDAS Remote Sensor," *J. Air Waste Manage. Assoc., 49*, 463–470 (1999).

Johnson, B. J., S. C. Huang, M. L. Pitchford, H. C. Ayoub, and M. H. Naylor, "Preliminary On-Road Measurement of the Effect of Oxygenated Fuel on CO Emissions near Las Vegas, Nevada," *J. Air Waste Manage. Assoc., 48*, 59–64 (1997).

Joshi, S. B., M. C. Dodge, and J. J. Bufalini, "Reactivities of Selected Organic Compounds and Contamination Effects," *Atmos. Environ., 16,* 1301–1310 (1982).

Kaiser, E. W., W. O. Siegl, Y. I. Henig, R. W. Anderson, and F. H. Trinker, "Effect of Fuel Structure on Emissions from a Spark-Ignited Engine," *Environ. Sci. Technol., 25,* 2005–2012 (1991); *erratum, 26,* 1672 (1992).

Kaiser, E. W., W. O. Siegl, D. F. Cotton, and R. W. Anderson, "Effect of Fuel Structure on Emissions from a Spark-Ignited Engine. 2. Napthalene and Aromatic Fuels," *Environ. Sci. Technol., 26,* 1581–1586 (1992).

Kaiser, E. W., W. O. Siegl, D. F. Cotton, and R. W. Anderson, "Effect of Fuel Structure on Emissions from a Spark-Ignited Engine. 3. Olefinic Fuels," *Environ. Sci. Technol., 27,* 1440–1447 (1993).

Kamens, R., Z. Guo, J. Fulcher, and D. Bell, "Influence of Humidity, Sunlight, and Temperature on the Daytime Decay of Polyaromatic Hydrocarbons on Atmospheric Soot Particles," *Environ. Sci. Technol., 22,* 103–108 (1988).

Kao, A. S., "Formation and Removal Reactions of Hazardous Air Pollutants," *J. Air Waste Manage. Assoc., 44,* 683–696 (1994).

Kasibhatla, P., W. L. Chameides, B. Duncan, M. Houyoux, C. Jang, R. Mathur, T. Odman, and A. Xiu, "Impact of Inert Organic Nitrate Formation on Ground-Level Ozone in a Regional Air Quality Model Using the Carbon Bond Mechanism 4," *Geophys. Res. Lett., 24,* 3205–3208 (1997).

Kelly, N. A., "Characterization of Fluorocarbon-Film Bags as Smog Chambers," *Environ. Sci. Technol., 16,* 763–770 (1982).

Kelly, N. A., K. L. Olson, and C. A. Wong, "Tests for Fluorocarbon and Other Organic Vapor Release by Fluorocarbon Film Bags," *Environ. Sci. Technol., 19,* 361–364 (1985).

Kelly, N. A., and P. J. Groblicki, "Real-World Emissions from a Modern Production Vehicle Driven in Los Angeles," *J. Air Waste Manage. Assoc., 43,* 1351–1357 (1993).

Kelly, N. A., and T. Y. Chang, "An Experimental Investigation of Incremental Reactivities of Volatile Organic Compounds," *Atmos. Environ., 33,* 2101–2110 (1999).

Kelly, T. J., R. Mukund, C. W. Spicer, and A. J. Pollack, "Concentrations and Transformations of Hazardous Air Pollutants," *Environ. Sci. Technol., 28,* 378–387 (1994).

Killus, J. P., and G. Z. Whitten, "Comments on 'A Smog Chamber and Modeling Study of the Gas Phase NO_x–Air Photooxidation of Toluene and the Cresols,'" *Int. J. Chem. Kinet., 13,* 1101–1103 (1981).

Kirchstetter, T. W., B. C. Singer, R. A. Harley, G. R. Kendall, and W. Chan, "Impact of Oxygenated Gasoline Use on California Light-Duty Vehicle Emissions," *Environ. Sci. Technol., 30,* 661–670 (1996).

Kleinman, L. I., "Seasonal Dependence of Boundary Layer Peroxide Concentration: The Low and High NO_x Regimes," *J. Geophys. Res., 96,* 20721–20733 (1991).

Kleinman, L. I., "Low and High NO_x Tropospheric Photochemistry," *J. Geophys. Res., 99,* 16831–16838 (1994).

Kleinman, L., Y.-N. Lee, S. R. Springston, L. Nunnermacker, X. Zhou, R. Brown, K. Hallock, P. Klotz, D. Leahy, J. H. Lee, and L. Newman, "Ozone Formation at a Rural Site in the Southeastern United States," *J. Geophys. Res., 99,* 3469–3482 (1994).

Kleinman, L. I., P. H. Daum, J. H. Lee, Y.-N. Lee, L. J. Nunnermacker, S. R. Springston, and L. Newman, "Dependence of Ozone Production on NO and Hydrocarbons in the Troposphere," *Geophys. Res. Lett., 24,* 2299–2302 (1997).

Kley, D., "Tropospheric Chemistry and Transport," *Science, 276,* 1043–1045 (1997).

Klimisch, R., and J. Heuss, "The Role of NO_x in Smog and Acid Rain: 25 Years of a Pollutant Looking for a Problem?" *Environ. Forum,* August, 28–32 (1983).

Klouda, G. A., C. W. Lewis, R. A. Rasmussen, G. C. Rhoderick, R. L. Sams, R. K. Stevens, L. A. Currie, D. J. Donahue, A. J. T. Jull, and R. L. Seila, "Radiocarbon Measurements of Atmospheric Volatile Organic Compounds: Quantifying the Biogenic Contribution," *Environ. Sci. Technol., 30,* 1098–1105 (1996).

Knapp, K. T., F. D. Stump, and S. B. Tejada, "The Effect of Ethanol Fuel on the Emissions of Vehicles over a Wide Range of Temperatures," *J. Air Waste Manage. Assoc., 48,* 646–653 (1998).

Kuhn, M., P. J. H. Builtjes, D. Poppe, D. Simpson, W. R. Stockwell, Y. Andersson-Sköld, A. Baart, M. Das, F. Fiedler, Ø. Hov, F. Kirchner, P. A. Makar, J. B. Milford, M. G. M. Roemer, R. Ruhnke, A. Strand, B. Vogel, and H. Vogel, "Intercomparison of the Gas-Phase Chemistry in Several Chemistry and Transport Models," *Atmos. Environ., 32,* 693–709 (1998).

Kuklin, A., and J. H. Seinfeld, "Emission Reductions Needed to Meet the Standard for Ozone in Southern California: Effect of Boundary Conditions," *J. Air Waste Manage. Assoc., 45,* 899–901 (1995).

Kumar, N., M. T. Odman, and A. G. Russell, "Multiscale Air Quality Modeling: Application to Southern California," *J. Geophys. Res., 99,* 5385–5397 (1994).

Kumar, N., and A. G. Russell, "Comparing Prognostic and Diagnostic Meteorological Fields and Their Impacts on Photochemical Air Quality Modeling," *Atmos. Environ., 30,* 1989–2010 (1996).

Kumar, S., and D. P. Chock, "An Update on Oxidant Trends in the South Coast Air Basin of California," *Atmos. Environ., 18,* 2131–2134 (1984).

Kurtz, D. A., *Long Range Transport of Pesticides,* Lewis Publishers, Chelsea, MI, 1990.

Kuster, W. C., and P. D. Goldan, "Quantitation of the Losses of Gaseous Sulfur Compounds to Enclosure Walls," *Environ. Sci. Technol., 21,* 810–815 (1987).

Lamb, R. G., "Regional Scale (1000 km) Model of Photochemical Air Pollution, Part 1. Theoretical Formulation," EPA/600/3-83-035, U.S. Environmental Protection Agency, Environmental Sciences Research Laboratories, Research Triangle Park, North Carolina, 1983.

Lammel, G., and J. N. Cape, "Nitrous Acid and Nitrite in the Atmosphere," *Chem. Soc. Rev., 25,* 361–369 (1996).

Lawson, D. R., P. J. Groblicki, D. H. Stedman, G. A. Bishop, and P. L. Guenther, "Emissions from In-Use Motor Vehicles in Los Angeles: A Pilot Study of Remote Sensing and the Inspection and Maintenance Program," *J. Air Waste Manage. Assoc., 40,* 1096–1105 (1990).

Lawson, D. R., "'Passing the Test'—Human Behavior and California's Smog Check Program," *J. Air Waste Manage. Assoc., 43,* 1567–1575 (1993).

Lebron, F., "A Comparison of Weekend-Weekday Ozone and Hydrocarbon Concentrations in the Baltimore-Washington Metropolitan Area," *Atmos. Environ., 9,* 861–863 (1975).

Lefohn, A. S., J. K. Foley, D. S. Shadwick, and B. E. Tilton, "Changes in Diurnal Patterns Related to Changes in Ozone Levels," *J. Air Waste Manage. Assoc., 43,* 1472–1478 (1993).

Leone, J. A., R. C. Flagan, D. A. Grosjean, and J. H. Seinfeld, "An Outdoor Smog Chamber and Modeling Study of Toluene-NO_x Photooxidation," *Int. J. Chem. Kinet., 17,* 177–216 (1985).

Lin, C.-J., and S. O. Pehkonen, "The Chemistry of Atmospheric Mercury: A Review," *Atmos. Environ., 33,* 2067–2079 (1999).

Lin, G.-Y., "Oxidant Prediction by Discriminant Analysis in the South Coast Air Basin, California," *Atmos. Environ., 16,* 135–143 (1982).

Lindsay, R. W., J. L. Richardson, and W. L. Chameides, "Ozone Trends in Atlanta, Georgia: Have Emission Controls Been Effective?" *J. Air Pollut. Control Assoc.*, 39, 40–43 (1989).

Liu, M. K., R. E. Morris, and J. P. Killus, "Development of a Regional Oxidant Model and Application to the Northeastern United States," *Atmos. Environ.*, 18, 1145–1161 (1984).

Lloyd, A. C., J. M. Lents, C. Green, and P. Nemeth, "Air Quality Management in Los Angeles: Perspectives on Past and Future Emission Control Strategies," *J. Air Waste Manage. Assoc.*, 39, 696–703 (1989).

Lloyd, A. C., "California Clean Air Initiatives—The Role of Fuel Cells," *J. Power Sources*, 37, 241–253 (1992).

Lloyd, A. C., J. H. Leonard, and R. George, "Fuel Cells and Air Quality—A California Perspective," *J. Power Sources*, 49, 209–223 (1994).

Logan, J. A., "Ozone in Rural Areas of the United States," *J. Geophys. Res.*, 94, 8511–8532 (1989).

Lonneman, W. A., J. J. Bufalini, R. L. Kuntz, and S. A. Meeks, "Contamination from Fluorocarbon Films," *Environ. Sci. Technol.*, 15, 99–103 (1981).

Lu, C.-H., and J. S. Chang, "On the Indicator-Based Approach to Assess Ozone Sensitivities and Emissions Features," *J. Geophys. Res.*, 103, 3453–3462 (1998).

Lu, R., and R. P. Turco, "Ozone Distributions over the Los Angeles Basin: Three-Dimensional Simulations with the Smog Model," *Atmos. Environ.*, 30, 4155–4176 (1996).

Lu, R., R. P. Turco, and M. Z. Jacobson, "An Integrated Air Pollution Modeling System for Urban and Regional Scales: 1. Structure and Performance," *J. Geophys. Res.*, 102, 6063–6079 (1997a).

Lu, R., R. P. Turco, and M. Z. Jacobson, "An Integrated Air Pollution Modeling System for Urban and Regional Scales: 2. Simulations for SCAQS 1987," *J. Geophys. Res.*, 102, 6081–6098 (1997b).

Lurmann, F. W., A. C. Lloyd, and R. Atkinson, "A Chemical Mechanism for Use in Long-Range Transport/Acid Deposition Computer Modeling," *J. Geophys. Res.*, 91, 10905–10936 (1986).

Lurmann, F. W., A. S. Wexler, S. N. Pandis, S. Musarra, N. Kumar, and J. H. Seinfeld, "Modelling Urban and Regional Aerosols—II. Application to California's South Coast Air Basin," *Atmos. Environ.*, 31, 2695–2715 (1997).

MacCracken, M. C., D. J. Wuebbles, J. J. Walton, W. H. Duewer, and K. E. Grant, "The Livermore Regional Air Quality Model. I. Concept and Development," *J. Appl. Meteorol.*, 17, 254–272 (1978).

MacGregor, L., and H. Westberg, "The Effect of NMOC and Ozone Aloft on Modeled Urban Ozone Production and Control Strategies," *J. Air Waste Manage. Assoc.*, 40, 1372–1377 (1990).

Mannino, D. M., and R. A. Etzel, "Are Oxygenated Fuels Effective? An Evaluation of Ambient Carbon Monoxide Concentrations in 11 Western States, 1986 to 1992," *J. Air Waste Manage. Assoc.*, 46, 20–24 (1996).

Mathur, R., K. L. Schere, and A. Nathan, "Dependencies and Sensitivity of Tropospheric Oxidants to Precursor Concentrations over the Northeast United States: A Model Study," *J. Geophys. Res.*, 99, 10535–10552 (1994).

McKay, W. A., M. F. Turner, B. M. R. Jones, and C. M. Halliwell, "Emissions of Hydrocarbons from Marine Phytoplankton—Some Results from Controlled Laboratory Experiments," *Atmos. Environ.*, 30, 2583–2593 (1996).

McKeen, S. A., E.-Y. Hsie, and S. C. Liu, "A Study of the Dependence of Rural Ozone on Ozone Precursors in the Eastern United States," *J. Geophys. Res.*, 96, 15377–15394 (1991a).

McKeen, S. A., E.-Y. Hsie, M. Trainer, R. Tallamraju, and S. C. Liu, "A Regional Model Study of the Ozone Budget in the Eastern United States," *J. Geophys. Res.*, 96, 10809–10845 (1991b).

McLaren, R., D. L. Singleton, J. Y. K. Lai, B. Khouw, E. Singer, Z. Wu, and H. Niki, "Analysis of Motor Vehicle Sources and Their Contribution to Ambient Hydrocarbon Distributions at Urban Sites in Toronto during the Southern Ontario Oxidants Study," *Atmos. Environ.*, 30, 2219–2232 (1996a).

McLaren, R., A. W. Gertler, D. N. Wittorff, W. Belzer, T. Dann, and D. L. Singleton, "Real-World Measurements of Exhaust and Evaporative Emissions in the Cassiar Tunnel Predicted by Chemical Mass Balance Modeling," *Environ. Sci. Technol.*, 30, 3001–3009 (1996b).

McMurry, P. H., and D. Grosjean, "Gas and Aerosol Wall Losses in Teflon Film Smog Chambers," *Environ. Sci. Technol.*, 19, 1176–1182 (1985).

McMurry, P. H., and D. J. Rader, "Aerosol Losses in Electrically Charged Chambers," *Aerosol Sci. Technol.*, 4, 249–268 (1985).

McNair, L. A., A. G. Russell, M. T. Odman, B. E. Croes, and L. Kao, "Airshed Model Evaluation of Reactivity Adjustment Factors Calculated with the Maximum Incremental Reactivity Scale for Transitional-Low Emission Vehicles," *J. Air Waste Manage. Assoc.*, 44, 900–907 (1994).

McRae, G. J., W. R. Goodin, and J. H. Seinfeld, "Mathematical Modeling of Photochemical Air Pollution," Final Report to the California Air Resources Board, Contract Nos. A5-046-87 and A7-187-30, April 27, 1982a.

McRae, G. J., W. R. Goodin, and J. H. Seinfeld, "Development of Second-Generation Mathematical Model for Urban Air Pollution—I. Model Formulation," *Atmos. Environ.*, 16, 679–696 (1982b).

Meng, Z., D. Dabdub, and J. H. Seinfeld, "Chemical Coupling between Atmospheric Ozone and Particulate Matter," *Science*, 277, 116–119 (1997).

Meng, Z., D. Dabdub, and J. H. Seinfeld, "Size-Resolved and Chemically Resolved Model of Atmospheric Aerosol Dynamics," *J. Geophys. Res.*, 103, 3419–3435 (1998).

Mentel, Th. F., D. Bleilebens, and A. Wahner, "A Study of Nighttime Nitrogen Oxide Oxidation in a Large Reaction Chamber—The Fate of NO_2, N_2O_5, and O_3 at Different Humidities," *Atmos. Environ.*, 30, 4007–4020 (1996).

Milford, J. B., A. G. Russell, and G. J. McRae, "A New Approach to Photochemical Pollution Control: Implications of Spatial Patterns in Pollutant Responses to Reductions in Nitrogen Oxides and Reactive Organic Gas Emissions," *Environ. Sci. Technol.*, 23, 1290–1301 (1989).

Milford, J. B., D. Gao, A. G. Russell, and G. J. McRae, "Use of Sensitivity Analysis to Compare Chemical Mechanisms for Air-Quality Modeling," *Environ. Sci. Technol.*, 26, 1179–1189 (1992).

Milford, J. B., D. Gao, S. Sillman, P. Blossey, and A. G. Russell, "Total Reactive Nitrogen (NO_y) as an Indicator of the Sensitivity of Ozone to Reductions in Hydrocarbon and NO_x Emissions," *J. Geophys. Res.*, 99, 3533–3542 (1994).

Milne, P. J., D. D. Riemer, R. G. Zika, and L. E. Brand, "Measurement of Vertical Distribution of Isoprene in Surface Seawater, Its Chemical Fate and Its Emission from Several Phytoplankton Monocultures," *Mar. Chem.*, 48, 237–244 (1995).

Moilanen, K. W., D. G. Crosby, J. R. Humphrey, and J. W. Giles, "Vapor-Phase Photodecomposition of Chloropicrin (Trichloronitromethane)," *Tetrahedron*, 34, 3345–3349 (1978).

Moore, R. M., D. E. Oram, and S. A. Penkett, "Production of Isoprene by Marine Phytoplankton Cultures," *Geophys. Res. Lett.*, 21, 2507–2510 (1994).

Morris, R., "Selection of Cities for Assessing the Effects of Test Fuels on Air Quality," Systems Applications International, Inc., San Rafael, CA, January 17, 1990.

Mukund, R., T. J. Kelly, and C. W. Spicer, "Source Attribution of Ambient Air Toxic and Other VOCs in Columbus, Ohio," *Atmos. Environ., 30,* 3457–3470 (1996).

National Research Council, "Rethinking the Ozone Problem in Urban and Regional Air Pollution," National Academy Press, Washington, DC, 1991.

Nelson, D. D., M. S. Zahniser, J. B. McManus, C. E. Kolb, and J. L. Jiménez, "A Tunable Diode Laser System for the Remote Sensing of On-Road Vehicle Emissions," *Appl. Phys. B, 67,* 433–441 (1998).

Nunnermacker, L. J., D. Imre, P. H. Daum, L. Kleinman, Y.-N. Lee, J. H. Lee, S. R. Springston, L. Newman, J. Weinstein-Lloyd, W. T. Luke, R. Banta, R. Alvarez, C. Senff, S. Sillman, M. Holdren, G. W. Keigley, and X. Zhou, "Characterization of the Nashville Urban Plume on July 3 and July 18, 1995," *J. Geophys. Res., 103,* 28129–28148 (1998).

Olson, J., M. Prather, T. Berntsen, G. Carmichael, R. Chatfield, P. Connell, R. Derwent, L. Horowitz, S. Jin, M. Kanakidou, P. Kasibhatla, R. Kotamarthi, M. Kuhn, K. Law, J. Penner, L. Perliski, S. Sillman, F. Stordal, A. Thompson, and O. Wild, "Results from the Intergovernmental Panel on Climatic Change Photochemical Model Intercomparison (PhotoComp)," *J. Geophys. Res., 102,* 5979–5991 (1997).

Olszyna, K. J., E. M. Bailey, R. Simonaitis, and J. F. Meagher, "O_3 and NO_y Relationships at a Rural Site," *J. Geophys. Res., 99,* 14557–14563 (1994).

Oltmans, S. J., A. S. Lefohn, H. E. Scheel, J. M. Harris, H. Levy II, I. E. Galbally, E.-G. Brunke, C. P. Meyer, J. A. Lathrop, B. J. Johnson, D. S. Shadwick, E. Cuevas, F. J. Schmidlin, D. W. Tarasick, H. Claude, J. B. Kerr, O. Uchino, and V. Mohnen, "Trends of Ozone in the Troposphere," *Geophys. Res. Lett., 25,* 139–142 (1998).

Pakdel, H., C. Roy, H. Aubin, G. Jean, and S. Coulombe, "Formation of *dl*-Limonene in Used Tire Vacuum Pyrolysis Oils," *Environ. Sci. Technol., 25,* 1646–1649 (1991).

Peleg, M., M. Luria, G. Sharf, A. Vanger, G. Kallos, V. Kotroni, K. Lagouvardos, and M. Varinou, "Observational Evidence of an Ozone Episode over the Greater Athens Area," *Atmos. Environ., 31,* 3969–3983 (1997).

Pierce, T., C. Geron, L. Bender, R. Dennis, G. Tonnesen, and A. Guenther, "Influence of Increased Isoprene Emissions on Regional Ozone Modeling," *J. Geophys. Res., 103,* 25611–25629 (1998).

Pierson, W. R., A. W. Gertler, and R. L. Bradow, "Comparison of the SCAQS Tunnel Study with Other On-Road Vehicle Emission Data," *J. Air Waste Manage. Assoc., 40,* 1495–1504 (1990).

Pierson, W. R., "Automotive CO Emission Trends Derived from Measurements in Highway Tunnels," *J. Air Waste Manage. Assoc., 45,* 831–832 (1995).

Pierson, W. R., "Motor Vehicle Inspection and Maintenance Programs—How Effective Are They?" *Atmos. Environ., 30,* i–iii (1996).

Pierson, W. R., A. W. Gertler, N. F. Robinson, J. C. Sagebiel, B. Zielinska, G. A. Bishop, D. H. Stedman, R. B. Zweidinger, and W. D. Ray, "Real-World Automotive Emissions—Summary of Studies in the Fort McHenry and Tuscarora Mountain Tunnels," *Atmos. Environ., 30,* 2233–2256 (1996).

Pitts, J. N., Jr., A. C. Lloyd, and J. L. Sprung, "Ecology, Energy, and Economics," *Chem. Br., 11,* 247–256 (1975).

Pitts, J. N., Jr., A. M. Winer, K. R. Darnall, G. J. Doyle, and J. M. McAfee, "Chemical Consequences of Air Quality Standards and of Control Implementation Programs," Final Report to the California Air Resources Board, Contract No. 4-214, May 1976.

Pitts, J. N., Jr., B. J. Finlayson-Pitts, and A. M. Winer, "Optical Systems Unravel Smog Chemistry," *Environ. Sci. Technol., 11,* 568–573 (1977).

Pitts, J. N., Jr., E. Sanhueza, R. Atkinson, W. P. L. Carter, A. M. Winer, G. W. Harris, and C. N. Plum, "An Investigation of the Dark Formation of Nitrous Acid in Environmental Chambers," *Int. J. Chem. Kinet., 16,* 919–939 (1984).

Pitts, J. N., Jr., T. J. Wallington, H. W. Biermann, and A. M. Winer, "Identification and Measurement of Nitrous Acid in an Indoor Environment," *Atmos. Environ., 19,* 763–767 (1985).

Pitts, J. N., Jr., "Atmospheric Formation and Fates of Toxic Ambient Air Pollutants," *Occup. Med.: State of the Art Rev., 8,* 621–662 (1993).

Pollack, A. K., A. M. Dunker, J. K. Fieber, J. G. Heiken, J. P. Cohen, S. B. Shepard, C. H. Schleyer, and G. Yarwood, "Revision of Light-Duty Vehicle Emission Inventories Using Real-World Measurements—Auto/Oil Program, Phase II," *J. Air Waste Manage. Assoc., 48,* 291–305 (1998).

Pryor, S. C., "A Case Study of Emission Changes and Ozone Responses," *Atmos. Environ., 32,* 123–131 (1998).

Rao, S. T., G. Sistla, and R. Henry, "Statistical Analysis of Trends in Urban Ozone Air Quality," *J. Air Waste Manage. Assoc., 42,* 1204–1211 (1992).

Rao, S. T., and I. G. Zurbenko, "Detecting and Tracking Changes in Ozone Air Quality," *J. Air Waste Manage. Assoc., 44,* 1089–1092 (1994).

Rao, S. T., E. Zalewsky, and I. G. Zurbenko, "Determining Temporal and Spatial Variations in Ozone Air Quality," *J. Air Waste Manage. Assoc., 45,* 57–61 (1995).

Rao, S. T., I. G. Zurbenko, P. S. Porter, J. Y. Ku, and R. F. Henry, "Dealing with the Ozone Non-Attainment Problem in the Eastern United States," *Environ. Manage., Air Waste Manage. Assoc.,* January, 17–31 (1996).

Reuter, J. E., B. C. Allen, R. C. Richards, J. F. Pankow, C. R. Goldman, R. L. Scholl, and J. S. Seyfried, "Concentrations, Sources, and Fate of the Gasoline Oxygenate Methyl *tert*-Butyl Ether (MTBE) in a Multiple-Use Lake," *Environ. Sci. Technol., 32,* 3666–3672 (1998).

Reynolds, S. D., P. M. Roth, and J. H. Seinfeld, "Mathematical Model of Photochemical Air Pollution—I. Formulation of the Model," *Atmos. Environ., 7,* 1033–1061 (1973).

Ridley, B. A., J. G. Walega, J.-F. Lamarque, F. E. Grahek, M. Trainer, G. Hübler, X. Lin, and F. C. Fehsenfeld, "Measurements of Reactive Nitrogen and Ozone to 5-km Altitude in June 1990 over the Southeastern United States," *J. Geophys. Res., 103,* 8369–8388 (1998).

Ristaino, J. B., and W. Thomas, "Agriculture, Methyl Bromide, and the Ozone Hole. Can We Fill the Gaps?" *Plant Dis.*, September, 964–977 (1997).

Riveros, H. G., J. Tejeda, L. Ortiz, A. Julián-Sánchez, and H. Riveros-Rosas, "Hydrocarbons and Carbon Monoxide in the Atmosphere of Mexico City," *J. Air Waste Manage. Assoc., 45,* 973–980 (1995).

Robinson, N. F., W. R. Pierson, A. W. Gertler, and J. C. Sagebiel, "Comparison of Mobile4.1 and Mobile5 Predictions with Measurements of Vehicle Emission Factors in Fort McHenry and Tuscarora Mountain Tunnels," *Atmos. Environ., 30,* 2257–2267 (1996).

Rogak, S. N., S. I. Green, and U. Pott, "Use of Tracer Gas for Direct Calibration of Emission-Factor Measurements in a Traffic Tunnel," *J. Air Waste Manage. Assoc., 48,* 545–552 (1998a).

Rogak, S. N., U. Pott, T. Dann, and D. Wang, "Gaseous Emissions from Vehicles in a Traffic Tunnel in Vancouver, British Columbia," *J. Air Waste Manage. Assoc., 48,* 604–615 (1998b).

Roselle, S. J., T. E. Pierce, and K. L. Schere, "The Sensitivity of Regional Ozone Modeling to Biogenic Hydrocarbons," *J. Geophys. Res., 96*, 7371–7394 (1991).

Ross, M., R. Goodwin, R. Watkins, T. Wenzel, and M. Q. Wang, "Real-World Emissions from Conventional Passenger Cars," *J. Air Waste Manage. Assoc., 48*, 502–515 (1998).

Roussel, P. B., X. Lin, F. Camacho, S. Laszlo, R. Taylor, O. T. Melo, P. B. Shepson, D. R. Hastie, and H. Niki, "Observations of Ozone and Precursor Levels at Two Sites around Toronto, Ontario, during Sontos 92," *Atmos. Environ., 30*, 2145–2155 (1996).

Rueff, R. M., "The Cost of Reducing Emissions from Late-Model High-Emitting Vehicles Detected via Remote Sensing," *J. Air Waste Manage. Assoc., 42*, 921–925 (1992).

Russell, A., J. Milford, M. S. Bergin, S. McBride, L. McNair, Y. Yang, W. R. Stockwell, and B. Croes, "Urban Ozone Control and Atmospheric Reactivity of Organic Gases," *Science, 269*, 491–495 (1995).

Russell, A. G., K. F. McCue, and G. R. Cass, "Mathematical Modeling of the Formation of Nitrogen-Containing Air Pollutants. 1. Evaluation of an Eulerian Photochemical Model," *Environ. Sci. Technol., 22*, 263–271 (1988).

Ryan, D. L., and T. R. Fukuto, "The Effect of Impurities on the Toxicokinetics of Malathion in Rats," *Pestic. Biochem. Physiol., 23*, 413–424 (1985).

Sagebiel, J. C., B. Zielinska, W. R. Pierson, and A. W. Gertler, "Real-World Emissions and Calculated Reactivities of Organic Species from Motor Vehicles," *Atmos. Environ., 30*, 2287–2296 (1996).

Sakamaki, F., S. Hatakeyama, and H. Akimoto, "Formation of Nitrous Acid and Nitric Oxide in the Heterogeneous Dark Reaction of Nitrogen Dioxide and Water Vapor in a Smog Chamber," *Int. J. Chem. Kinet., 15*, 1013–1043 (1983).

Sakamaki, F., and H. Akimoto, "HONO Formation as Unknown Radical Source in Photochemical Smog Chamber," *Int. J. Chem. Kinet., 20*, 111–116 (1988).

Schere, K. L., and K. L. Demerjian, "A Photochemical Box Model for Urban Air Quality," in *Proceedings of the 4th Joint Conference on Sensing of Environmental Pollutants*, pp. 427–433, Am. Chem. Soc., Washington, DC, 1978.

Schomburg, C. J., D. E. Glotfelty, and J. N. Seiber, "Pesticide Occurrence and Distribution in Fog Collected near Monterey, California," *Environ. Sci. Technol., 25*, 155–160 (1991).

Schuetzle, D., W. O. Siegl, T. E. Jensen, M. A. Dearth, E. W. Kaiser, R. Gorse, W. Kreucher, and E. Kulik, "The Relationship between Gasoline Composition and Vehicle Hydrocarbon Emissions—A Review of Current Studies and Future Research Needs," *Environ. Health Perspec., 102*, 3–12 (1994).

Seiber, J. N., B. W. Wilson, and M. M. McChesney, "Air and Fog Deposition Residues of Four Organophosphate Insecticides Used on Dormant Orchards in the San Joaquin Valley, California," *Environ. Sci. Technol., 27*, 2236–2243 (1993).

Seiber, J. N., and J. E. Woodrow, *ACS Conference Proceedings Series, Eighth International Congress of Pesticide Chemistry: Options 2000* (N. N. Ragsdale, P. C. Kearney, and J. R. Plimmer, Eds.), Am. Chem. Soc., Washington, DC, 1995.

Seiber, J. N., "Toxic Air Contaminants in Urban Atmospheres: Experience in California," *Atmos. Environ., 30*, 751–756 (1996).

Seinfeld, J. H., "Ozone Air Quality Models: A Critical Review," *J. Air Pollut. Control Assoc., 38*, 616–645 (1988).

Shreffler, J. H., and H. M. Barnes, "Estimation of Trends in Atmospheric Concentrations of Sulfate in the Northeastern United States," *J. Air Waste Manage. Assoc., 46*, 621–630 (1996).

Siegl, W. O., R. W. McCabe, W. Chun, E. W. Kaiser, J. Perry, Y. I. Henig, F. H. Trinker, and R. W. Anderson, "Speciated Hydrocarbon Emissions from the Combustion of Single Component Fuels. I. Effect of Fuel Structure," *J. Air Waste Manage. Assoc., 42*, 912–920 (1992).

Sillman, S., J. A. Logan, and S. C. Wofsy, "A Regional Scale Model for Ozone in the United States with Subgrid Representation of Urban and Power Plant Plumes," *J. Geophys. Res., 95*, 5731–5748 (1990).

Sillman, S., "The Use of NO_y, H_2O_2, and HNO_3 as Indicators for Ozone–NO_x–Hydrocarbon Sensitivity in Urban Locations," *J. Geophys. Res., 100*, 14175–14188 (1995).

Sillman, S., and P. J. Samson, "Impact of Temperature on Oxidant Photochemistry in Urban, Polluted Rural, and Remote Environments," *J. Geophys. Res., 100*, 11497–11508 (1995).

Sillman, S., D. He, C. Cardelino, and R. E. Imhoff, "The Use of Photochemical Indicators to Evaluate Ozone–NO_x–Hydrocarbon Sensitivity: Case Studies from Atlanta, New York, and Los Angeles," *J. Air Waste Manage. Assoc., 47*, 1030–1040 (1997).

Sillman, S., "The Relation between Ozone, NO_x and Hydrocarbons in Urban and Polluted Rural Environments," *Atmos. Environ., 33*, 1821–1845 (1999).

Simonaitis, R., J. F. Meagher, and E. M. Bailey, "Evaluation of the Condensed Carbon Bond (CB-IV) Mechanism against Smog Chamber Data at Low VOC and NO_x Concentrations," *Atmos. Environ., 31*, 27–43 (1997).

Simpson, D., "Biogenic Emissions in Europe. 2: Implications for Ozone Control Strategies," *J. Geophys. Res., 100*, 22891–22906 (1995).

Singer, B. C., and R. A. Harley, "A Fuel-Based Motor Vehicle Emission Inventory," *J. Air Waste Manage. Assoc., 46*, 581–593 (1996).

Sistla, G., N. Zhou, W. Hao, J.-Y. Ku, S. T. Rao, R. Bornstein, F. Freedman, and P. Thunis, "Effects of Uncertainties in Meteorological Inputs on Urban Airshed Model Predictions and Ozone Control Strategies," *Atmos. Environ., 30*, 2011–2025 (1996).

Sjödin, A., "On-Road Emission Performance of Late-Model TWC-Cars As Measured by Remote Sensing," *J. Air Waste Manage. Assoc., 44*, 397–404 (1994).

Skov, H., A. H. Egelov, K. Granby, and T. Nielsen, "Relationships between Ozone and Other Photochemical Products at Ll. Valby, Denmark," *Atmos. Environ., 31*, 685–691 (1997).

Spicer, C. W., "Smog Chamber Studies of NO_x Transformation Rate and Nitrate/Precursor Relationships," *Environ. Sci. Technol., 17*, 112–120 (1983).

Spicer, C. W., B. E. Buxton, M. W. Holdren, D. L. Smith, T. J. Kelly, S. W. Rust, A. D. Pate, G. M. Sverdrup, and J. C. Chuang, "Variability of Hazardous Air Pollutants in an Urban Area," *Atmos. Environ., 30*, 3443–3456 (1996).

Squillace, P. J., J. F. Pankow, N. E. Korte, and J. S. Zogorski, "Review of the Environmental Behaviour and Fate of Methyl *tert*-Butyl Ether," *Environ. Toxicol. Chem., 16*, 1836–1844 (1997).

St. John, J. C., W. L. Chameides, and R. Sayler, "Role of Anthropogenic NO_x and VOC as Ozone Precursors: A Case Study from the SOS Nashville/Middle Tennessee Ozone Study," *J. Geophys. Res., 103*, 22415–22423 (1998).

Stedman, D. H., G. A. Bishop, P. Aldrete, and R. S. Slott, "On-Road Evaluation of an Automobile Emission Test Program," *Environ. Sci. Technol., 31*, 927–931 (1997).

Stedman, D. H., G. A. Bishop, and R. S. Slott, "Repair Avoidance and Evaluating Inspection and Maintenance Programs," *Environ. Sci. Technol., 32*, 1544–1545 (1998).

Stephens, R. D., and S. H. Cadle, "Remote Sensing Measurements of Carbon Monoxide Emissions from On-Road Vehicles," *J. Air Waste Manage. Assoc., 41*, 39–46 (1991).

Stephens, R. D., "Remote Sensing Data and a Potential Model of Vehicle Exhaust Emissions," *J. Air Waste Manage. Assoc., 44*, 1284–1292 (1994).

Stephens, R. D., S. H. Cadle, and T. Z. Qian, "Analysis of Remote Sensing Errors of Omission and Commission under FTP Conditions," *J. Air Waste Manage. Assoc., 46*, 510–516 (1996a).

Stephens, R. D., P. A. Mulawa, M. T. Giles, K. G. Kennedy, P. J. Groblicki, S. H. Cadle, and K. T. Knapp, "An Experimental Evaluation of Remote Sensing-Based Hydrocarbon Measurements: A Comparison to FID Measurements," *J. Air Waste Manage. Assoc., 46*, 148–158 (1996b).

Stockwell, W. R., P. Middleton, and J. S. Chang, "The Second Generation Regional Acid Deposition Model Chemical Mechanism for Regional Air Quality Modeling," *J. Geophys. Res., 95*, 16343–16367 (1990).

Stockwell, W. R., "Effects of Turbulence on Gas-Phase Atmospheric Chemistry: Calculation of the Relationship between Time Scales for Diffusion and Chemical Reaction," *Meteorol. Atmos. Phys., 57*, 159–171 (1995).

Stockwell, W. R., F. Kirchner, M. Kuhn, and S. Seefeld, "A New Mechanism for Regional Atmospheric Chemistry Modeling," *J. Geophys. Res., 102*, 25847–25879 (1997).

Stump, F. D., K. T. Knapp, and W. D. Ray, "Influence of Ethanol-Blended Fuels on the Emissions from Three Pre-1985 Light-Duty Passenger Vehicles," *J. Air Waste Manage. Assoc., 46*, 1149–1161 (1996).

Taha, H., "Modeling Impacts of Increased Urban Vegetation on Ozone Air Quality in the South Coast Air Basin," *Atmos. Environ., 30*, 3423–3430 (1996).

Tanner, R. L., A. H. Miguel, J. B. de Andrade, J. S. Gaffney, and G. E. Streit, "Atmospheric Chemistry of Aldehydes: Enhanced Peroxyacetyl Nitrate Formation from Ethanol-Fueled Vehicular Emissions," *Environ. Sci. Technol., 22*, 1026–1034 (1988).

Tesche, T. W., and D. E. McNally, "Photochemical Modeling of Two 1984 SCCCAMP Ozone Episodes," *J. Appl. Meteorol., 30*, 745–763 (1991).

Thomas, W., "Methyl Bromide—Pesticide and Environmental Threat," *Atmos. Environ., 30*, i–iv (1996).

Tiao, G. C., L.-M. Liu, and G. B. Hudak, "A Statistical Assessment of the Effect of the Car Inspection/Maintenance Program on Ambient CO Air Quality in Phoenix, Arizona," *Environ. Sci. Technol., 23*, 806–814 (1989).

Timpe, R. C., and L. Wu, "Vapor Pressure Response to Denaturant and Water in E10 Blends," *J. Air Waste Manage. Assoc., 45*, 46–51 (1995).

Trainer, M., D. D. Parrish, M. P. Buhr, R. B. Norton, F. C. Fehsenfeld, K. G. Anlauf, J. W. Bottenheim, Y. Z. Tang, H. A. Wiebe, J. M. Roberts, R. L. Tanner, L. Newman, V. C. Bowersox, J. F. Meagher, K. J. Olszyna, M. O. Rodgers, T. Wang, H. Berresheim, K. L. Demerjian, and U. K. Roychowdhury, "Correlation of Ozone with NO_y in Photochemically Aged Air," *J. Geophys. Res., 98*, 2917–2925 (1993).

Tuazon, E. C., R. Atkinson, A. M. Winer, and J. N. Pitts, Jr., "A Study of the Atmospheric Reactions of 1,3-Dichloropropene and Other Selected Organochlorine Compounds," *Arch. Environ. Contam. Toxicol., 13*, 691–700 (1984).

United States Environmental Protection Agency, "The Role of Ozone Precursors in Tropospheric Ozone Formation and Control," EPA-454/R-93-024, July 1993.

Uno, I., S. Wakamatsu, R. A. Wadden, S. Konno, and H. Koshio, "Evaluation of Hydrocarbon Reactivity in Urban Air," *Atmos. Environ., 19*, 1283–1293 (1985).

Vainiotalo, S., Y. Peltonen, and P. Pfäffli, "MTBE Concentrations in Ambient Air in the Vicinity of Service Stations," *Atmos. Environ., 32*, 3503–3509 (1998).

Venkatram, A., P. K. Karamchandani, and P. K. Mistra, "Testing a Comprehensive Acid Deposition Model," *Atmos. Environ., 22*, 737–747 (1988).

Venkatram, A., S. Du, R. Hariharan, W. Carter, and R. Goldstein, "The Concept of Species in Photochemical Modeling," *Atmos. Environ., 32*, 3403–3413 (1998).

Venturini, P. D., "Comment on 'Speciated Measurements and Calculated Reactivities of Vehicle Exhaust Emissions from Conventional and Reformulated Gasolines,'" *Environ. Sci. Technol., 27*, 1453–1453 (1993).

Walsh, M., "Global Trends in Motor Vehicle Use and Emissions," *Am. Rev. Energy, 15*, 217–243 (1990).

Walsh, M., "Global Trends in Diesel Particulate Control—A 1995 Update," *SAE Tech. Pap. Ser., No. 950149*, February 27–March 2, 1995.

Walsh, M. P., "Global Trends in Diesel Emissions Control—A 1997 Update," *SAE Tech. Pap. Ser., No. 970179*, February 24–27, 1997.

Wang, S. W., P. G. Georgopoulos, G. Li, and H. Rabitz, "Condensing Complex Atmospheric Chemistry Mechanisms. 1. The Direct Constrained Approximate Lumping (DCAL) Method Applied to Alkane Photochemistry," *Environ. Sci. Technol., 32*, 2018–2024 (1998).

Wang, W. G., N. N. Clark, D. W. Lyons, R. M. Yang, M. Gautam, R. M. Bata, and J. L. Loth, "Emissions Comparisons from Alternative Fuel Buses and Diesel Buses with a Chassis Dynamometer Testing Facility," *Environ. Sci. Technol., 31*, 3132–3137 (1997).

Wängberg, I., T. Etzkorn, I. Barnes, U. Platt, and K. H. Becker, "Absolute Determination of the Temperature Behavior of the $NO_2 + NO_3 + (M) \Leftrightarrow N_2O_5 + (M)$ Equilibrium," *J. Phys. Chem. A, 101*, 9694–9698 (1997).

Wayne, L. G., A. Kokin, and M. I. Weisburd, "Controlled Evaluation of the Reactive Environmental Simulation Model (REM)," Vol. I, EPA Document No. R4-73-013a, NTIS Publication PB220456/8, February 1973.

Wexler, A. S., F. W. Lurmann, and J. H. Seinfeld, "Modelling Urban and Regional Aerosols—1. Model Development," *Atmos. Environ., 28*, 531–546 (1994).

Whitten, G. Z., "Comment on 'Potential Air Quality Effects of Using Ethanol-Gasoline Fuel Blends: A Field Study in Albuquerque, New Mexico,'" *Environ. Sci. Technol., 32*, 3840–3841 (1998).

Wilkinson, S. L., "Electric Vehicles Gear Up," *Chem. Eng. News*, October 13 18–24 (1997).

Winer, A. M., G. M. Breuer, W. P. L. Carter, K. R. Darnall, and J. N. Pitts, Jr., "Effects of Ultraviolet Spectral Distribution on the Photochemistry of Simulated Polluted Atmospheres," *Atmos. Environ., 13*, 989–998 (1979).

Winer, A. M., R. A. Graham, G. J. Doyle, P. J. Bekowies, J. M. McAfee, and J. N. Pitts, Jr., "An Evacuable Environmental Chamber and Solar Simulator Facility for the Study of Atmospheric Photochemistry," *Adv. Environ. Sci. Technol., 10*, 461–511 (1980).

Winner, D. A., G. R. Cass, and R. A. Harley, "Effect of Alternative Boundary Conditions on Predicted Ozone Control Strategy Performance: A Case Study in the Los Angeles Area," *Atmos. Environ., 29*, 3451–3464 (1995).

Wolff, G. T., and P. E. Korsog, "Ozone Control Strategies Based on the Ratio of Volatile Organic Compounds to Nitrogen Oxides," *J. Air Waste Manage. Assoc., 42*, 1173–1177 (1992).

Wolff, G. T., "On a NO_x-Focused Control Strategy to Reduce O_3," *Air Waste, 43*, 1593–1596 (1993).

Wolff, M. F., and J. N. Seiber, "Environmental Activation of Pesticides," *Occup. Med.: State of the Art Rev., 8*, 561–573 (1993).

Wu, C. H., S. M. Japar, and H. Niki, "Relative Reactivities of HO–Hydrocarbon Reactions from Smog Reaction Studies," *J. Environ. Sci. Health, Environ. Sci. Eng., A11,* 191–200 (1976).

Yamartino, R. J., J. S. Scire, G. R. Carmichael, and Y. S. Chang, "The CALGRID Mesoscale Photochemical Grid Model. I. Model Formulation," *Atmos. Environ., 26A,* 1493–1512 (1992).

Yang, Y.-J., W. R. Stockwell, and J. B. Milford, "Uncertainties in Incremental Reactivities of Volatile Organic Compounds," *Environ. Sci. Technol., 29,* 1336–1345 (1995).

Yang, Y.-J., W. R. Stockwell, and J. B. Milford, "Effect of Chemical Product Yield Uncertainties on Reactivities of VOCs and Emissions from Reformulated Gasolines and Methanol Fuels," *Environ. Sci. Technol., 30,* 1392–1397 (1996).

Zabik, J. M., and J. N. Seiber, "Atmospheric Transport of Organophosphate Pesticides from California's Central Valley to the Sierra Nevada Mountains," *J. Environ. Qual., 22,* 80–90 (1993).

Zeldin, M. D., and D. M. Thomas, "Ozone Trends in the Eastern Los Angeles Basin Corrected for Meteorological Variations," Paper presented at the International Conference on Environmental Sensing and Assessment, Las Vegas, NV, September 14–19, 1975.

Zhang, Y., D. H. Stedman, G. A. Bishop, P. L. Guenther, S. P. Beaton, and J. E. Peterson, "On-Road Hydrocarbon Remote Sensing in the Denver Area," *Environ. Sci. Technol., 27,* 1885–1891 (1993).

Zhang, Y., D. H. Stedman, G. A. Bishop, P. L. Guenther, and S. P. Beaton, "Worldwide On-Road Vehicle Exhaust Emissions Study by Remote Sensing," *Environ. Sci. Technol., 29,* 2286–2294 (1995).

Zhang, Y., D. H. Stedman, G. A. Bishop, S. P. Beaton, and P. L. Guenther, "On-Road Evaluation of Inspection/Maintenance Effectiveness," *Environ. Sci. Technol., 30,* 1445–1450 (1996a).

Zhang, Y., D. H. Stedman, G. A. Bishop, S. P. Beaton, P. L. Guenther, and I. F. McVey, "Enhancement of Remote Sensing for Mobile Source Nitric Oxide," *J. Air Waste Manage. Assoc., 46,* 25–29 (1996b).

Zurbenko, I. G., S. T. Rao, and R. F. Henry, "Mapping Ozone in the Eastern United States," *Environ. Manage., 1,* 24–39, (1995).

APPENDIX I

Enthalpies of Formation of Some Gaseous Molecules, Atoms, and Free Radicals at 298 K[a,c]

Species	$\Delta H°_{f298}$ (kJ mol^{-1})	Species	$\Delta H°_{f298}$ (kJ mol^{-1})
H	218.0	$CH_3O_2NO_2$	−44
H_2	0	CO_2	−393.5
$O(^3P)$	249.2	C_2H_2	228.0
$O(^1D)$	438.9	C_2H_4	52.2
O_2	0	C_2H_5	120.9
$O_2(^1\Delta)$	94.3	C_2H_6	−84.0
$O_2(^1\Sigma)$	156.9	CH_3CN	64.3
O_3	142.7	CH_2CO	−47.7
HO	39.3	CH_3CO	−10.0
HO_2	14.6	CH=CHOH	115
H_2O	−241.8	CH_3CHO	−165.8
H_2O_2	−136.3	C_2H_5O	−15.5
N	472.7	C_2H_5OH	−234.8
N_2	0	C_2H_4OH	−34
NH	352	$(CHO)_2$	−211.9
NH_2	168.7	CH_3CO_2	−207.5
NH_3	−45.9	CH_3CO_2H	−432.0
NO	90.3	$C_2H_5O_2$	−28.7
NO_2	33.2	CH_3OOCH_3	−125.7
NO_3	73.7	$CH_3C(O)O_2$	−172
N_2O	82.1	C_2H_5ONO	−103.8
N_2O_4	9.1	$C_2H_5ONO_2$	−154.0
N_2O_5	11.3	$C_2H_5OONO_2$	−63.2
HNO	112.9	C_3H_6	20.2
HNO_2	−79.5	n-C_3H_7	97.5
HNO_3	−135.1	i-C_3H_7	90.0
HO_2NO_2	−57	C_3H_8	−104.5
CH_3	146.4	CH_3COCH_2	−23.9
CH_4	−74.8	C_2H_5CHO	−187.4
CN	435	CH_3COCH_3	−217.2
HCN	135	C_3H_6OH	−74
HCO	41.8	n-C_3H_7O	−41.4
CH_2O	−108.6	i-C_3H_7O	−52.3
CH_3O	17.2	i-C_3H_7OH	−272.5
CH_2OH	−12.1[d]	CH_3COCHO	−271.1
CH_3OH	−201.6	$C_3H_5O_2$	87.9
CO	−110.5	i-$C_3H_7O_2$	−68.9
HCOOH	−378.8	n-$C_3H_7ONO_2$	−174.1
CH_3O_2	10.4	i-$C_3H_7ONO_2$	−190.8
CH_3OOH	−131	$CH_3C(O)O_2NO_2$	−258
$HOCH_2O_2$	−162.1	S	277.2
CH_3ONO	−65.3	HS	143.0
CH_3ONO_2	−119.7	H_2S	−20.6

(continues)

(continued)

Species	ΔH°_{f298} (kJ mol^{-1})	Species	ΔH°_{f298} (kJ mol^{-1})
HSO	−4	CHF$_2$Cl	−483.7
SO	5.0	CH$_3$CHFCl	−313.4
HSO$_2$	−222	CH$_3$CF$_2$Cl	−536.2
SO$_2$	−296.8	Cl$_2$	0
HOSO$_2$	−385	Cl$_2$O	77.2
SO$_3$	−395.7	Cl$_2$O$_2$	127.6
CH$_3$S	124.6	Cl$_2$O$_3$	153
CH$_3$SH	−22.9	CHCl$_2$	98.3
CH$_3$SCH$_2$	136.8	CHCl$_2$O$_2$	1.6
CH$_3$SCH$_3$	−37.24	COCl$_2$	−220.1
CS	272	CF$_2$Cl$_2$	−493.3
CH$_3$SO	−67	CH$_2$ClCF$_2$Cl	−543
CH$_3$SOO	75.7	CF$_3$CHCl$_2$	−740
OCS	−142	CF$_2$ClCHFCl	−724
CH$_3$SSCH$_3$	−24.3	CF$_2$ClCF$_2$Cl	−925.5
CS$_2$	117.2	CCl$_3$	71.1
HOCS$_2$	110.5	CCl$_3$O$_2$	−11.3
F	79.4	CCl$_3$O$_2$NO$_2$	−83.7
HF	−273.3	CHCl$_3$	−103.3
FO	109	CH$_3$CCl$_3$	−144.6
FO$_2$	25.4	CFCl$_3$	−284.9
FONO	67	CF$_2$ClCFCl$_2$	−726.8
FNO	−67[b]	CCl$_4$	−95.8
FNO$_2$	−108.8	C$_2$Cl$_4$	−12.4
FONO$_2$	10	Br	111.9
CH$_3$F	−232.6	HBr	−36.3
CH$_3$CH$_2$F	−263	HOBr	≥ −56.2
FCO	−171.5	BrO	125.8
F$_2$	0	BrNO	82.2
CH$_3$CHF$_2$	−501	BrONO	103
COF$_2$	−634.7	BrNO$_2$	63
CHF$_3$	−697.6	BrONO$_2$	47
CF$_3$	−467.4	CH$_2$Br	169.0
CH$_2$CF$_3$	−517.1	CH$_3$Br	−38.1
CH$_3$CF$_3$	−748.7	CF$_3$Br	−650
CH$_2$FCHF$_2$	−691	CF$_2$ClBr	−438
CH$_2$FCF$_3$	−895[b]	BrCl	14.6
CHF$_2$CF$_3$	−1105[b]	Br$_{2(g)}$	30.9
CF$_3$O	−655.6	Br$_2$O	107.1
CF$_3$OH	−923.4	CHBr$_2$	188
CF$_3$O$_2$	−614.0	CF$_2$Br$_2$	−379
CF$_4$	−933	CF$_2$BrCF$_2$Br	−789.9
Cl	121.3	CHBr$_3$	23.8
HCl	−92.3	I	106.8
HOCl	−78	HI	26.5
ClO	101.6	HOI	−90
ClOO	97.5	IO	126
OClO	95.6	INO	121.3
sym-ClO$_3$	232.6	INO$_2$	60.2
ClNO	51.7	CH$_2$I	230.1
ClNO$_2$	12.5	CH$_3$I	14.2
ClONO	56	CF$_3$I	−589
ClONO$_2$	22.9	I$_{2(g)}$	62.4
CH$_3$Cl	−82.0		

[a] Unless otherwise indicated, from D. R. Lide, *CRC Handbook of Chemistry and Physics*, 79th ed., CRC Press, Boca Raton, FL, 1998/1999, and Atkinson *et al.*, *J. Phys. Chem. Ref. Data*, **26**, 1329–1499 (1997).

[b] From DeMore *et al.*, JPL Publication No. 97-4, January 15, 1997.

[c] For estimating heats of formation of other stable species, use group additivity methods described by S. W. Benson, *Thermochemical Kinetics*, Wiley, New York, 1976, see review of N. Cohen and S. W. Benson for organic compounds ["Estimation of Heats of Formation of Organic Compounds by Additivity Methods," *Chem. Rev.*, **93**, 2419–2438 (1993)], and the paper by C.-J. Chen, D. Wong, and J. W. Bozzelli for chlorinated alkanes and alkenes ["Standard Chemical Thermodynamic Properties of Multichloro Alkanes and Alkenes: A Modified Group Additivity Scheme," *J. Phys. Chem. A*, **102**, 4551–4558 (1998)]. Application of group additivity to C$_2$–C$_6$ alkyl radicals is discussed by N. Cohen ["Thermochemistry of Alkyl for Radicals," *J. Phys. Chem.*, **96**, 9052–9058 (1992)].

[d] From J. Berkowitz, G. B. Ellison, and D. Gutman, "Three Methods to Measure RH Bond Energies," *J. Phys. Chem.*, **98**, 2744–2765 (1994).

APPENDIX II

Bond Dissociation Energies[a]

Bond broken	Bond dissociation energy (kJ mol^{-1})	Bond broken	Bond dissociation energy (kJ mol^{-1})
CH_3—H	438.9	C_2H_5O—OC_2H_5	158.6
C_2H_5—H	423.0	CH_3O—NO	174.9
n-C_3H_7—H	420.0	C_2H_5O—NO	175.7
i-C_3H_7—H	412.5	CH_3COO_2—NO_2	118.8
s-C_4H_9—H	410.9		
t-C_4H_9—H	403.8	$CH_3\overset{\overset{O}{\|}}{C}$—H	443[b]
Cyclopropyl—H	444.8		
Cyclobutyl—H	403.8		
Cyclopentyl—H	403.5	$C_2H_{53}\overset{\overset{O}{\|}}{C}$—H	445[b]
Cyclohexyl—H	399.6		
Cycloheptyl—H	387.0	$H\overset{\overset{O}{\|}}{C}$—H	368.5
$HOCH_2$—H	401.9		
$H_2C=C$—H	465.3		
$HC\equiv C$—H	556.1	$CH_3\overset{\overset{O}{\|}}{C}$—H	373.8
CH_3—CH_3	376.0		
CH_3—C_6H_5	317.1	H—$CH_2\overset{\overset{O}{\|}}{C}H$	394.6
t-C_4H_9—CH_3	425.9		
CH_3—CH_2CCH	318.0		
CH_3—$CH(CH_3)CCH$	305.4	$C_6H_5\overset{\overset{O}{\|}}{C}$—H	364[b]
CH_3—$CH_2C_6H_5$	332.2	CH_3—CHO	345[b]
C_2H_5—$CH_2C_6H_5$	294.1	C_6H_5—CHO	403[b]
$H_2C=CH_2$	733		
C_6H_5—H	465.3	$CH_3\overset{\overset{O}{\|}}{C}$—$CH_3$	340[b]
$C_6H_5CH_2$—H	370.3		
C_6H_5—CH_3	317	$CH_3\overset{\overset{O}{\|}}{C}CH_2$—H	411.3
H—CF_3	446.4		
H—CCl_3	392.5		
H—CBr_3	401.7	$CH_3\overset{\overset{O}{\|}}{C}$—$C_6H_5$	391
H—CH_2I	431		
CH_3—O_2	135.6	$CH_3\overset{\overset{O}{\|}}{C}$—$\overset{\overset{O}{\|}}{C}CH_3$	282.0
C_2H_5—O_2	147.2	HS—H	381.6
i-C_3H_7—O_2	157.9	CH_3S—H	365.3
$HOCH_2$—H	401.9	CH_3—SH	312.5
CH_3—OH	386[b]	H—CH_2SH	392.9
C_2H_5—OH	382[b]	t-C_4H_9—SH	286.2
$C_6H_5CH_2$—OH	340[b]	C_6H_5—SH	361.9
CH_3O—H	436.0	CH_3S—CH_3	307.9
C_2H_5O—H	437.7	CH_3SO_2—CH_3	279.5
C_6H_5O—H	361.9	CH_3—NH_2	354
CH_3O—CH_3	348[b]	C_2H_5—NH_2	354
CH_3O—C_2H_5	342[b]	C_6H_5—NH_2	427
C_6H_5O—CH_3	267[b]	C_6H_5NH—C_2H_5	298.7
CH_3O—OCH_3	157.3		

(continues)

(continued)

Bond broken	Bond dissociation energy (kJ mol^{-1})	Bond broken	Bond dissociation energy (kJ mol^{-1})
CH_3—CN	509.6	H—ONO_2	423.4
H—CH_2CN	392.9	HO—NO_2	206.7
H—CH_2NC	380.7	HO_2—NO_2	96
CH_3—F	472	O_2N—NO_2	56.9
CH_3—Cl	349	ON—NO_2	40.6
CH_3—Br	293		
CH_3—I	239	CS—S	430.5
CF_2Cl—F	490	CH_3S—SCH_3	272.8
CF_2Cl—Cl	346.0	OS—O	552
$CFCl_2$—F	462.3	HS—SH	276
$CFCl_2$—Cl	305	NH_2—H	452.7
		NH_2—NH_2	275.3
H—F	569.9	H—CN	527.6
H—Cl	431.6	CH_3—NO	167.4
H—Br	366.3	CH_3—NO_2	254.4
H—I	298.4	i-C_3H_7—NO	152.7
		i-C_3H_7—NO_2	246.9
OC=O	532.2		
H—H	436.0	HO—Cl	251
HO—H	498	O—ClO	247
HO—OH	213	HO—Br	234
HOO—H	369.0	Br—NO_2	82.0
CH_3OO—H	370.3	HO—I	234
H—ONO	327.6	I—NO	77.8
HO—NO	206.3	I—NO_2	76.6

[a] Except where otherwise indicated, from D. R. Lide, *CRC Handbook of Chemistry and Physics*, 79th ed., CRC Press, Boca Raton, FL, 1998/1999, and from J. Berkowitz, G. B. Ellison, and D. Gutman, "Three Methods to Measure RH Bond Energies," *J. Phys. Chem.*, 98, 2744–2765 (1994).

[b] From D. F. McMillen and D. M. Golden, *Annu. Rev. Phys. Chem.*, 33, 493–532 (1982).

APPENDIX III

Running the OZIPR Model

The EPA's OZIPR model (Ozone Isopleth Plotting Program, Research Version) was designed to be used primarily for calculating emissions reductions needed to reach the NAAQS for O_3. Both DOS-based and Windows versions are available and can be downloaded from the Web site: http://www.academicpress.com/pecs/download

OZIPR contains two comprehensive chemical mechanisms that use two different approaches to "lumping" organics. The two mechanisms used in these models, the RADM (*R*egional *A*cid *D*eposition *M*odel) and the Carbon Bond Mechanism (CBM), are discussed in Chapter 16.A.3b and in detail by Stockwell *et al.* [*J. Geophys. Res.*, 95, 16343 (1990) and *J. Geophys. Res.*, 102, 25847 (1997)] and by Gery *et al.* [*J. Geophys. Res.*, 94, 12925 (1989)].

EDITING *filename.inp* AND *meccm.???* FILES

The input file for Example 1, *ex1.inp*, has associated with it the RADM mechanism, called *meccm.rad* and a set of actinic fluxes as a function of solar zenith angles stored as *zen.rad*. A copy of the RADM chemical species list is found in Table 16.1.

The input file for Example 8 is *ex8.inp* and has the associated mechanism *meccm.cb4*, actinic fluxes as *zen.cb4* and *reac.cb4*. "CB4" (or CBIV) stands for "carbon bond mechanism, version 4." A species list is found in Table 16.2.

To run OZIPR, you need to do the following (what you will type in is in boldface):

DOS Version

1. Edit the input file *ex1.inp* as you would normally (e.g., using the DOS editor or a word-processing program) and then save it. You might want to change the title listed in the input file for each run while you are editing it. Be sure to leave the file saved as *.inp*, not as *.doc* etc. *Note:* Check the end of the file to make sure that there are no extra lines of "garbage," e.g., a line of open boxes. If there are, delete them or they will cause the program to crash.

2. Similarly, you can edit the *meccm.rad* file. Again be sure not to change the extension and check the end of the file for "garbage."

3. To do the computation after you have modified the inputs appropriately, type in **ozipr ex1**. Once you have done an OZIPR run using the *ex1.inp* input file, it generates an output file called *ex1.out*.

4. After you have run OZIPR for that particular application, you can print the output using **print ex1.out**

5. To run the isopleth program using Example 8, type in **ozipr ex8** (Running this takes much longer than running the first example.)

6. Once it has run, to view the isopleths type in **isoplot ex8.inf ex8.iso**. This will show the isopleths on the screen onc at a time. Press the Escape button to walk through them one at a time.

7. To plot the isopleths, use **isoplot ex8.inf / H ex8.iso** (Note: the /H is for an HP LaserJet printer; use /E for an Epson RX or FX printer.)

Windows Version

1. Edit the input and mechanism files as discussed above for the DOS version.

2. Under "Run Ozipr" in the toolbar, click once on the example you want to run. This runs the program and generates the output file. For example, if you have edited Example 1 inputs (either the *ex1.inp* file and/or the *meccm.rad* file), it will do the run and generate the *ex1.out* file.

3. The output file can be printed using the "File → Print" commands from the toolbar.

4. To run the options which generate the data for isopleths, e.g., Example 8, click once on the appropriate example under "Run Ozipr" on the toolbar. This generates the *ex8.out* file.

5. To view the isopleths on your screen, choose

"Run Isoplot" from the toolbar and then the example number. Pressing either the spacebar or the enter key sequentially steps you through the isopleths.

6. To plot the isopleths, choose "Example8(Plot)" under "Run Isoplot" from the toolbar. *Note:* This will only print to a Hewlett Packard LaserJet printer.

APPENDIX

IV

Some Relevant Web Sites

Note: The authors do not guarantee the operation of these Web sites nor the accuracy of the data in them. They are provided solely as examples of possibly useful information.

AIR QUALITY DATA AND STANDARDS

North America:

Environmental Protection Agency: "National Air Quality and Emissions Trends Report—1997"
http://www.epa.gov/oar/aqtrnd97/

Environmental Protection Agency: Real-time images of ground-level ozone levels throughout the Midwest and the Northeast
http://www.epa.gov/region01/oms/

U.S. Federal Air Quality Standards
http://www.epa.gov/oar/oaqps/
http://www.epa.gov/airs/criteria.html

North American Research Strategies for Tropospheric Ozone
http://www.cgenv.com/Narsto/

California

"The 1999 California Almanac of Emissions and Air Quality"
http://www.arb.ca.gov

CD-ROM, California Ambient Air Quality Data 1980–1996
http://www.arb.ca.gov/aqd/aqd.htm

The Air Quality Data Branch of the Air Resources Board now has available two CD-ROMs (CDs) with 17 years of air quality data.

The data on the first CD are stored in *Voyager* $*.voy$ files and are easily displayed using the Voyager data visualization software, also included on the CD. The CD also contains a startup program and 18 Voyager workbooks ($*.wkb$) to make it easier for users to locate and use data of interest. A tutorial on Voyager is also included. **Request CD #TSD-97-007-CD.**

The second CD contains the same basic data that are on the first CD but stores the data in ASCII files and other forms used by analysts that process their own data. **Request CD #TSD-97-008-CD.**

California Air Toxics Program
http://www.arb.ca.gov/toxics/toxics.html

Also see California Environmental Protection Agency, Office of Environmental Health Hazard Assessment
http://www.oehha.ca.gov

Los Angeles Area: South Coast Air Quality Management District—Currently, Internet customers can retrieve information such as smog levels, forecasts, and smog's health effects; meetings and workshops; governing board agenda, minutes, and committee agendas; and business assistance. AQMD's World Wide Web page address
http://www.aqmd.gov

EMISSIONS

Europe

1996 CORINAIR database
http://nfp-dk.eionet.eu.int
Use "test" as both username and password! Subject: "Air emissions" and subsubject: "Corinair-NAD."

Isoprene and Monoterpene Emitting Species

A database of isoprene and monoterpene emitting species is now available on the web. The database must first be downloaded and decoded, details of how to do this are shown on the page **http://www.es.lancs.ac.uk/es/people/pg/pas/download.html**

United States

Air Clearing House for Inventories and Emission Factors (*Air CHIEF*) http://www.epa.gov/ttnchie1/airchief.html

GLOBAL DATA

Data on atmospheric concentrations and trends in atmospheric CO_2 and a variety of other trace gases: U.S. Department of Energy Carbon Dioxide Information Analysis Center (CDIAC)
http://cdiac.esd.ornl.gov/cdiac

"Global and Hemispheric Temperature Anomalies—Land and Marine Instrumental Records, 1856–1997"
http://cdiac.esd.ornl.gov/trends/temp/jonescru/jones.html

Global Emissions Inventory Activity (GEIA)—GEIA inventories are available through anonymous file transfer protocol (FTP) from the Data Center at the National Center for Atmospheric Research via **ncardata.ucar.edu.**

International Global Atmospheric Chemistry Program (IGAC)
http://web.mit.edu/igac/www/

Reference Data for Atmosphere: Global Reference Atmospheric Models (GRAM)
http://elses1.msfc.nasa.gov/nee/nte.html

KINETICS AND PHOTOCHEMISTRY DATA BASES

Gas Phase

JPL Publication 97-4, *Chemical Kinetics and Photochemical Data for Use in Stratospheric Modeling, Evaluation Number 12*, by W. B. DeMore *et al.*, January 15, 1997. Files may be downloaded from **http://remus.jpl.nasa.gov/jpl97/** or may be copied via anonymous FTP from the Internet host **remus.jpl.nasa.gov** under **/pub/jpl97**.

Aqueous Phase

Radiation Laboratory at University of Notre Dame http://www.rcdc.nd.edu
Also you can go to the NIST site http://www.nist.gov/srd/nist40.htm

Henry's Law Constants (Kindly Provided by Dr. Rolf Sander)

http://www.mpch-mainz.mpg.de/~sander/res/henry.html

Absorption Spectra

"UV/Vis Spectra of Atmospheric Constituents," Version 1, Available on CD-ROM, ISBN 3-89100-030-8, A. Nölle, F. Pätzold, S. Pätzold, R. Meller, G. K. Moortgat, E. P. Röth, R. Ruhnke, and H. Keller-Rudek. Contact address: Deutsches Zentrum für Luft- und Raumfahrt e.V., ATMOS User Center DFD-AUC, D-82234 Wessling, Germany.
http://auc.dfd.dlr.de

STRATOSPHERE

Sources of Satellite Data on PSCs

SAGE/SAM
http://eosweb.larc.nasa.gov/

UARS
http://daac.gsfc.nasa.gov/DAAC_DOCS/gdaac_home.html

ILAS
http://www-ilas.nies.go.jp/

Results of the Tenth Meeting of the Parties to the Montreal Protocol, November 23–24, 1998, Cairo, Egypt
http://www.unep.org/ozone/10mop-results.htm

CFC PRODUCTION AND USES

Alternative Fluorocarbons Environmental Acceptability Study (AFEAS)
http://afeas.org/

APPENDIX V

Pressures and Temperatures for Standard Atmosphere

Altitude (km)	P mbar	P Torr	T (K)
0	1.01325×10^3	760.0	288.15
5	540.5	405.4	255.68
10	265.0	198.8	223.25
15	121.1	90.8	216.65
20	52.29	39.2	216.65
25	25.49	19.1	221.55
30	11.97	8.978	226.51
35	5.746	4.310	236.51
40	2.871	2.153	250.35
50	0.79779	0.5983	270.65

[a] From "U.S. Standard Atmosphere, 1976," NOAA, 1976.

APPENDIX VI

Answers to Selected Problems

CHAPTER 2

1. 1×10^6
2. 1 ppt = 4.09×10^{-5} MW μg m^{-3}, where MW = molecular weight
3. 13–35 ppm
4. 0.02 ppt
5. 301 K
6. 301 K
7. -9.8 K
8. 5.3×10^{18} kg

CHAPTER 3

1. For Los Angeles (a) 57, (b) 67, (c) 11, (d) 56, (e) 70
2. (a) 1.035, (b) 1.021, (c) 0.993, (d) 0.970, (e) 0.981, (f) 1.023
3. (a) 2.4×10^{-3} s^{-1}, (b) 2.2×10^{-3} s^{-1}, (c) 1.7×10^{-3} s^{-1}, (d) 8.3×10^{-4} s^{-1}, (e) 1.0×10^{-4} s^{-1}
4. SZA = 0° $k_p(Cl + NO_3) = 4.9 \times 10^{-5}$ s^{-1}
   $k_p(ClO + NO_2) = 4.3 \times 10^{-6}$ s^{-1}
   $k_p(total) = 5.3 \times 10^{-5}$ s^{-1}
   SZA = 40° $k_p(Cl + NO_3) = 4.0 \times 10^{-5}$ s^{-1}
   $k_p(ClO + NO_2) = 2.8 \times 10^{-6}$ s^{-1}
   $k_p(total) = 4.3 \times 10^{-5}$ s^{-1}
   SZA = 78° $k_p(Cl + NO_3) = 7.5 \times 10^{-6}$ s^{-1}
   $k_p(ClO + NO_2) = 1.5 \times 10^{-7}$ s^{-1}
   $k_p(total) = 7.7 \times 10^{-6}$ s^{-1}
5. (a) 9.7×10^{-7} s^{-1}, (b) 7.9×10^{-7} s^{-1}, (c) 4.7×10^{-7} s^{-1}, (d) 1.3×10^{-7} s^{-1}, (e) 7.3×10^{-9} s^{-1}
6. SZA = 40.3°; assume all QY = 1.0
   (a) $k_p(C_2H_5ONO_2) = 1.0 \times 10^{-6}$ s^{-1}
   (d) $k_p(ClONO_2) = 4.2 \times 10^{-5}$ s^{-1}
   (h) $k_p(H_2O_2) = 5.9 \times 10^{-6}$ s^{-1}
   (m) $k_p(CH_3I) = 4.9 \times 10^{-6}$ s^{-1}
7. SZA = 49.3°; assume all QY = 1.0
   (c) $k_p(PAN) = 5.0 \times 10^{-7}$ s^{-1}
   (e) $k_p(BrONO_2) = 1.1 \times 10^{-3}$ s^{-1}
   (i) $k_p(CH_3OOH) = 3.3 \times 10^{-6}$ s^{-1}
   (l) $k_p(HO_2NO_2) = 2.3 \times 10^{-6}$ s^{-1}

8.

Altitude (km)	Ratio 300–302 nm	400–405 nm	500–510 nm
15	4.3	1.4	1.1
25	12.5	1.4	1.1
40	32.5	1.4	1.1

Large effect at 300–302 nm is due to absorption by O_3 in this region.

9.

	Ratio (80%/"best")	
	SZA = 0°	SZA = 78°
298–300 nm	2.6	—
318–320 nm	3.0	2.5
400–405 nm	2.7	1.9

11. 2.4

CHAPTER 4

1. (a) 240 nm
   (b) 1 quantum of vibrational energy = 0.31×10^{-19} J and 240 nm = 8.28×10^{-19} J
   $\therefore$ Energy needed to dissociate = $(8.28 - 0.31) \times 10^{-19}$ J = 7.97×10^{-19} J $\equiv$ 249 nm
2. (a) $v' \cong 16$
   (b) $v' \cong 6$
   (c) $v' \cong 21$. This assumed O_2 behaves as a simple harmonic oscillator. However, at these high vibrational levels especially, this will not be the case and levels will become more closely spaced.
3. $k_p = 5.4 \times 10^{-6}$ s^{-1}; τ = 2 days
4. (a) BrCl is a factor of 4.7 greater than Cl_2 and Br_2 is a factor of 15 greater
   (b) BrCl is a factor of 6.6 greater than Cl_2 and Br_2 is a factor of 23 greater
5. (a) IO photolyzes a factor of 8.7 times faster than BrO at SZA = 20°

(b) IO photolyzes a factor of 15 times faster than BrO at SZA = 20°

6. 0 km: $k_p(SO_3) = 2.3 \times 10^{-7}$ s^{-1}
   $k[H_2O] \cong 4 \times 10^7$ s^{-1}

CHAPTER 5

1. (a) OH, 1.5 h; O_3, 7.8 h; NO_3, 16 h
   (b) 1.2 h
   (c) OH, 2.1 h; O_3, 11 h; NO_3, 24 h; overall, 1.7 h
2. (a) 9.5 years
   (b) 282 K
3. (a)

Reaction	k^{eff} (750 Torr, 300 K) (cm^3 molecule^{-1} s^{-1})
$O + O_2$	1.5×10^{-14}
$H + O_2$	1.2×10^{-12}
$OH + NO_2$	8.6×10^{-12}
$HO_2 + NO_2$	1.4×10^{-12}
$NO_2 + NO_3$	1.3×10^{-12}
$CH_3 + O_2$	1.1×10^{-12}
$C_2H_5 + O_2$	7.5×10^{-12}
$CH_3C(O)OO + NO_2$	8.6×10^{-12}
$ClO + NO_2$	2.3×10^{-12}
$BrO + NO_2$	2.8×10^{-12}
$OH + SO_2$	8.8×10^{-13}

3. (b) and (c)

Reaction	P (Torr)	k^{eff} (cm^3 molecule^{-1} s^{-1})
$O + O_2$	0	0
	150	3.0×10^{-15}
	300	5.8×10^{-15}
	450	8.8×10^{-15}
	600	1.2×10^{-14}
	630	1.2×10^{-14}
	750	1.5×10^{-14}
$CH_3 + O_2$	0	0
	150	0.598×10^{-12}
	300	0.817×10^{-12}
	450	0.958×10^{-12}
	600	1.05×10^{-12}
	630	1.07×10^{-12}
	750	1.12×10^{-12}
$CH_3C(O)OO + NO_2$	0	0
	150	8.0×10^{-12}
	300	8.3×10^{-12}
	450	8.5×10^{-12}
	600	8.5×10^{-12}
	630	8.6×10^{-12}
	750	8.6×10^{-12}

3. (d)

Reaction	$T_1 = -50°F = 227.6$ K	$T_2 = +120°F = 332.0$ K
$O + O_2$	$k_0 = 1.1 \times 10^{-33}$	$k_0 = 5.1 \times 10^{-34}$
	$k_\infty = $ —	$k_\infty = $ —
	$k^{eff} = 1.1 \times 10^{-33}$	$k^{eff} = 5.1 \times 10^{-34}$
$CH_3 + O_2$	$k_0 = 1.0 \times 10^{-30}$	$k_0 = 3.6 \times 10^{-31}$
	$k_\infty = 1.8 \times 10^{-12}$	$k_\infty = 1.8 \times 10^{-12}$
	$k^{eff} = 1.4 \times 10^{-12}$	$k^{eff} = 1.0 \times 10^{-12}$
$CH_3C(O)OO + NO_2$	$k_0 = 4.6 \times 10^{-28}$	$k_0 = 6.5 \times 10^{-29}$
	$k_\infty = 9.3 \times 10^{-12}$	$k_\infty = 9.3 \times 10^{-12}$
	$k^{eff} = 8.9 \times 10^{-12}$	$k^{eff} = 8.5 \times 10^{-12}$

4. (a) k(Cl + ethane) = 4.8×10^{-11} cm^3 molecule^{-1} s^{-1}
   (b) k(Cl + isobutane) = 1.7×10^{-10} cm^3 molecule^{-1} s^{-1}
9. $k(O_3 + I^-) = 2.8 \times 10^9$ L mol^{-1} s^{-1}
   $\alpha_{O_3} = 0.07$
11. $d_{hole} = 4.9$ mm
13. (a) $k_1 = 2.3 \times 10^{-12}$ cm^3 molecule^{-1} s^{-1}
    (b) $k_{-1} = 2.1 \times 10^{-3}$ s^{-1}
    (c) $\tau = 467$ s = 7.8 min
    (d) Rate of thermal decomposition is about two orders of magnitude faster than photolysis

CHAPTER 6

1. (a) $\tau = \{k_{OH}[OH] + k_{O_3}[O_3] + k_{NO_3}[NO_3]\}^{-1}$
   (b) $\tau = 14$ min
2. (a) 2.72×10^{-13} cm^3 molecule^{-1} s^{-1} (7% difference)
   (b) 2.2×10^{-12} cm^3 molecule^{-1} s^{-1} (10% difference)
   (c) 5.5×10^{-12} cm^3 molecule^{-1} s^{-1} (4% difference)
   (d) 3.2×10^{-12} cm^3 molecule^{-1} s^{-1} (6% difference)
   (e) 3.2×10^{-12} cm^3 molecule^{-1} s^{-1} (24% difference)
   (f) 9.7×10^{-12} cm^3 molecule^{-1} s^{-1} (3% difference)

(g) 1.1×10^{-11} cm^3 molecule^{-1} s^{-1} (0.9% difference)
3. (a) 4.3×10^{-17} cm^3 molecule^{-1} s^{-1} (7% difference)
   (b) 1.1×10^{-16} cm^3 molecule^{-1} s^{-1} (0% difference)
   (c) 1.8×10^{-16} cm^3 molecule^{-1} s^{-1} (6% difference)
   (d) 2.1×10^{-16} cm^3 molecule^{-1} s^{-1} (7% difference)
5. (b) $k = 8.2 \times 10^{-13}$ cm^3 molecule^{-1} s^{-1}; 9% difference from k_∞

T(°F)	τ
−30	1120 days
32	1.3 days
70	1.1 h
120	1.5 min

8. (b) 53, 71, and 253 min at [NO]/[NO$_2$] = 10, 1, and 0.1, respectively
9. τ_{NO_2} = 17 h compared to τ_{OH} = 11 min
   τ_{O_3} = 2.2 min and τ_{NO_3} = 11 s
13. (a) k^{eff}(HO$_2$ + HO$_2$) varies from 2.9×10^{-12} at an RH of 0% to 8.0×10^{-12} at an RH of 100%
    (b) Decreases by a factor of 0.77

CHAPTER 7

1. (a) 1.0×10^{-19} cm^3 molecule^{-1} s^{-1}
   (b) 47 days, 11 h, and 41 s at 0.1, 10, and 10,000 ppm, respectively
2. 1676 ppm
4. (a) 8.7×10^{-12} cm^3 molecule^{-1} s^{-1}
   (b) 3.7×10^{-13} cm^3 molecule^{-1} s^{-1}; changes by 96%
5. (a) τ = 17 days at 0 km, 5.9 h at 40 km. Photolysis back to NO$_2$ + OH is much more important in the stratosphere.
   (b) τ = 16 days at noon, 2 years at dawn/dusk. Photolysis at 0° solar zenith angle and reaction with OH at noon are comparable (but slow).
6. 26 ppt
7. (a) 7.9×10^{-12} cm^3 molecule^{-1} s^{-1}, 8.5% below JPL recommendation
   (b) 2.7×10^{-13} cm^3 molecule^{-1} s^{-1}, 27% below JPL recommendation
8. $k_{OH+HNO_3} = 3.8 \times 10^{-13}$ cm^3 molecule^{-1} s^{-1} for Eq. (D) compared to 5.9×10^{-13} for Eq. (E)
9. (a) [NO$_2$]/[HNO$_3$] = k_{53}/k_7 = 1.7×10^{-2} for JPL values and 1.9×10^{-2} for Brown *et al.* values (i.e., only 12% difference) at 300 K and 1 atm pressure
   (b) [NO$_2$]/[HNO$_3$] = 3.4×10^{-2} for JPL values compared to 5.9×10^{-2} for Brown *et al.* value, i.e., 75% difference at 220 K and 150 Torr. This would help resolve the discrepancy in the upper troposphere.

CHAPTER 8

2. SO$_{2(aq)}$: 2×10^{-13} mol L^{-1} s^{-1}; HSO$_3^-$ = 5×10^{-11} mol L^{-1} s^{-1}; SO$_3^{2-}$ = 1×10^{-11} mol L^{-1} s^{-1}; Total = 6×10^{-11} mol L^{-1} s^{-1}
3. (a) $H^*_{SO_2} = 1.65 \times 10^3$ mol L^{-1} atm^{-1}
   (b) [S(IV)] = 1.65×10^{-5} mol L^{-1}
4. (a) H^* = 2.9 mol L^{-1} atm^{-1}
   (b) [S(IV)] = 2.9×10^{-8} mol L^{-1}
   Concentrations in aqueous phase are much smaller due to shift in equilibria to left at lower pHs.
5. α_0 = fraction of SO$_{2(aq)}$ = 0.07
   α_1 = fraction of HSO$_3^-$ = 0.93
   α_2 = fraction of SO$_3^{2-}$ = 6×10^{-5}
   Total concentration of S(IV) in solution = 3.5×10^{-7} mol L^{-1}
6. $\alpha_0 = 1 \times 10^{-7}$; α_1 = 0.13; α_2 = 0.86
   S(IV) = 0.24 mol L^{-1}. Much higher concentration of dissolved S(IV) should increase relative importance of oxidation of S(IV) in sea salt particles.
7. 3% per h; i.e., only 10% of what is sometimes observed in troposphere
8. (a) 135% per h, (b) 1.4×10^3% per h
9. 0.7 ppt
10. 0.8, 0.08, and 2.2 ppb
11. τ(OH) = 4 h, τ(NO$_3$) = 1 h, τ(Cl) = 8 h
12. τ(OH) = 4.3×10^2 h compared to τ(Cl) = 8 h

CHAPTER 9

2. r = 1.5 nm
3. [H$_2$SO$_4$]$_{crit}$ = 0.08 ppb; about equal to theoretical but an order of magnitude less than experimental shown in Fig. 9.30.

CHAPTER 11

1. (a) 7.4×10^9 collisions s^{-1}
   (b) 1.4 ns
   (c) 0.14 μs
2. 0.1% emit at 1 atm; this increases to 52% at 10^{-3} atm
3. (a) 59–3462 ppbC urban
   (b) 1.9–19 ppbC remote
6. 3.2×10^4 for HNO$_3$ and 3.6×10^3 for NH$_3$

CHAPTER 12

1. C–F, 253 nm; C–Cl, 345 nm; C–Br, 408 nm; C–I, 502 nm

2. (a) 2.2 years, (b) 1.3 years, (c) 3.0 years
3. (a) Ratio of 13 km to 11 km = 5
   (b) Ratio of lifetime at 14 km to that at 13 km ≈ 200
4. $k_p(BrCl)/k_p(BrONO_2) = 8$
5. $\tau(BrONO_2$ hydrolysis$) = 1.6$ h compared to 13 min for photolysis
6. $\tau(HOBr) = 8.5$ min

CHAPTER 13

1. CF_3O reactions: $\Delta H_r^\circ = -136.3 + 13.3$ and -62.9 kJ mol^{-1}
   CH_3O reactions: $\Delta H_r^\circ = -103.2 + 62.3$ and -13.9 kJ mol^{-1}
   $OH-C_2H_6$ reaction: $\Delta H_r^\circ = -76.2$ kJ mol^{-1}
2. $\tau(CCl_4) > 63$ years; $\tau(CHCl_3) = 116$ days
   $\tau(CH_2Cl_2) = 105$ days; $\tau(CH_3Cl) = 322$ days
3. $\tau_{OH} = 160$ days; $\tau_{photolysis} = 3.3$ days
4. $\Delta H_r^\circ(CH_4 + Cl) = +7.6$ kJ mol^{-1}
   $\Delta H_r^\circ(CH_4 + Br) = +7.3$ kJ mol^{-1}
   Reactions are not enthalpy driven

CHAPTER 14

2. 1.2
3. (a) 1.001, (b) 1.01, (c) 1.11, (d) 2.85
4. (a) 1.002, (b) 1.02, (c) 1.27, (d) 10.7
5. $r = (b/a)^{1/2}$
6. $r_c = (3b/a)^{1/2}$
   $S_c = (4a^3/27b)^{1/2}$
   $r_c = \sqrt{3}r_0$
7. $r_c = 0.65$ μm, $S_c = 0.1\%$
8. $r_c = 0.17$ μm, $S_c = 0.4\%$
9. (a) $R = 0.4-0.6$
   (b) $R = 0.07-0.14$

CHAPTER 15

1. $\tau_{O_3} = 2.8$ h, $\tau_{OH} = 2.3$ h, $\tau_{air} = 1.3$ h

CHAPTER 16

1. $\tau_{OH} = 30$ h, $\tau_{NO_3} = 16$ days, $\tau_{Cl} = 3$ days
5. For cis form, $\tau_{O_3} = 39$ days while $\tau_{OH} = 7$ h
   For trans form, $\tau_{O_3} = 8.6$ days while $\tau_{OH} = 4$ h

Subject Index

A

α, mass accommodation coefficient for uptake into droplets or particles, 157
α, size parameter for particles, 157
Absolute rate constants
 kinetics analysis, 142
 laboratory techniques for determining, 141–149
 static techniques for, 148
Absorption coefficients, conversion factors, *see inside front cover*
Absorption spectra and cross sections
 of acetaldehyde, 108–111
 of acetone, 110–112
 of bromine monoxide, 113–114
 of bromine nitrate, 110–112
 of n-butyraldehyde, 109
 of i-butyraldehyde, 109
 of t-butyl nitrate, 109
 of chlorine monoxide, 113–114
 of chlorine nitrate, 110–112
 of ClOOCl, 114
 of OClO, 115
 of diethyl ketone, 109
 of dinitrogen pentoxide, 101
 of dinitrogen tetroxide, 96
 of ethyl nitrate, 105
 of formaldehyde, 108–109
 of halogenated methanes and ethanes, 117, 122–124
 of halogens F_2, Cl_2, Br_2 and I_2, 113–114
 of hydrogen bromide, 113
 of hydrogen chloride, 113
 of hydrogen peroxide, 107–108
 of hypohalous acids, HOCl, HOBr, HOI, 115
 of methyl ethyl ketone, 109
 of methyl nitrate, 103
 of nitrate radical, 100
 of nitric acid, 98
 of nitrogen dioxide, 95–96
 of nitrosyl chloride, 117, 120, 121
 of nitrous acid, 99
 of nitrous oxide, 101, 105
 of nitryl chloride, 117, 120, 121
 of organic nitrates, 103–106
 of oxygen, 86–89
 of ozone, 90–93
 of 2-pentanone, 109
 of peroxyacetyl nitrate, 102–107
 of peroxynitric acid, 100
 of propionaldehyde, 109
 of 2-propyl nitrate, 105
 of sulfur dioxide, 102
 of sulfur trioxide, 107
Absorption spectra, of polycyclic aromatic hydrocarbons, 461–466
Accommodation coefficient for gas uptake into liquids, 157–158, 307
Accumulation range of fine particles, 354–355
Acetaldehyde
 absorption spectrum, 108–111
 chlorine atom reaction, 214
 concentrations in troposphere, 592–593
 hydroxyl radical reaction, 214
 indoors, 857, 859
 measurement of, 565, 589–592
 nitrate radical reaction, 214
 photochemistry, quantum yields, 81–82, 109
Acetic acid
 concentrations in troposphere, 326
 contribution to acid deposition, 326–327
 indoors, 857, 861
 sources, 327
Acetone
 absorption spectra, 110–111
 concentrations in troposphere, 593
 measurement of, 565, 592–593
 photolysis, 110–112
 reaction in sulfuric acid particles, 241
Acetylene
 calculated troposphere lifetimes, 181
 hydroxyl radical reaction, 206–207
Acids
 base-equilibria in atmosphere, 303
 control of, 921–923
 dinitrogen pentoxide hydrolysis, 8, 279–280
 deposition, dry and wet, 30–33
 droplet size, effect on, 322–323
 free radical formation of, in aqueous phase, 315–322
 gas-phase formation of, 281, 298–301
 history, 8, 294
 hydrogen peroxide and organic peroxides, 313–314
 measurement of, 551–552, 566, 575–578
 nitrogen oxides in, 264, 266–272, 277–281, 314–315
 organic acids, 326–328, 594
 organic acids in particles, 394–407
 overview, 9
 and oxidation by O_2, catalyzed and uncatalyzed, 308–311
 and ozone oxidation, 311–313
 relative importance of various oxidants, 325–326
 sulfur dioxide oxidation rates, 301, 309
 surface reactions, 324–325
Acid fog
 chemistry, 323–324
 health effects, 324
Actinic flux
 calculated *vs.* measured, 75–76
 cloud effects on, 72–75
 derivation at earth's surface, 64–66
 earth-sun correction distance, 66
 effect of height above earth's surface, 69
 effects of latitude, 66–68
 season and time of day, 66–68
 estimates of, 65, 70, 71, 77–80
 sensitivity to surface albedo and particle and ozone concentrations, 69–72
 spherically integrated, 61
 surface elevation effect on, 68–69
 in stratosphere, 76–80
Aerodynamic diameter, 351
Aerosol chambers for studying heterogeneous reactions, 168–169
Aerosols, *see* Particles
Air mass at earth's surface, 58
Air pollution
 control of, 882–921
 history of, 3–9
 indoor, 13
 in London, 3, 4
 sulfurous *vs.* photochemical, 3–8
 system, 16

Air pollution system
  ambient concentrations of pollutants in, 33–37
  deposition, wet and dry, 30–33
  diagram of, 16
  economics and, 38
  emissions, 15–26
  meteorology, 26–30
  risk assessments, 38–39
  visibility loss, 37–38, 368–375
Air quality overview
  criteria and noncriteria pollutants, 35–37
  in major cities, 36
  primary and secondary, 36
  standards, 35
Airshed model, 888
Aitken nuclei, 354
Albedo, surface, 60
Alcohols
  concentrations in troposphere, 593–594
  reaction with hydroxyl radical, 216
  indoors, 851–858
Aldehydes
  in aqueous solution, 304–305
  absorption spectra, 107–110
  chlorine atom reactions, 213–214
  concentrations in ambient air, 592–593
  hydration constants, 304
  hydroxyl radical reaction, 213–214
  nitrate radical reactions, 213–214
  in particles, 395–409
  sulfur equilibria in aqueous solutions, 303
Aliphatic hydrocarbons
  concentrations in troposphere, 589–595
  measurement of, 566, 593–594
Alkaloids, measurement of, 566
Alkanes
  chlorine atom reactions, 184–185
  hydroxyl radical reaction, 182–184
  nitrate radical reaction, 184
Alkenes
  chlorine atom reactions, 205–206
  hydroxyl radical reactions, 191–196
  nitrate radical reaction, 201–205
  nitrogen dioxide reaction, 206, 272
  ozone reaction, 196–201
Alkoxy radicals
  atmospheric fates, relative importance of, 190–191
  decomposition of, 188–189
  hydroxyalkoxy, α and β, reactions of, 193–194
  isomerization of, 189
  reactions in ambient air, 188–191
  reaction with O_2, 188
Alkyl hydroperoxides, 186, 200–201
Alkyl nitrates
  from alkene-nitrate radical reaction, 202–205
  from RO_2 + NO reaction, 185–186, 220
  from RO + NO_2 reaction, 191
  hydroxyl radical reactions, 221
  photolysis of in troposphere, 102–103, 221
Alkyl nitrites, 221, 272
Alkyl radicals, reactions in ambient air, 185

Alkylperoxy radicals
  ambient air reactions, 185–188
  bimolecular self-reaction, rates and mechanisms, 186–187
  chemical amplification for detecting, 604–606
  concentrations in troposphere, 237–239
  hydroperoxyl radical reaction, 186
  measurement techniques, 604–606
  nitrate radical reactions, 187
  nitric oxide and nitrogen dioxide reactions, 185, 187–188
Alkyne reactions
  with hydroxyl radical, 206–207
Alternate fuels, 918–921
Ames test for bacterial mutagens, 475–480, 486–495; see also PAH, assay for bacterial mutagenicity
  accuracy and precision, 480
  activatable mutagens (promutagens, +S9), definition of, 476
  advantages of, 475–483
  for airborne particle-bound mutagens, 475–483, 486–495
  bioassay-directed fractionation and chemical analysis, 479, 482, 502–503, 515–516, 523–526
  chemical information from, 476–478
  direct mutagens (-S9), definition of, 476
  of diesel extracts, 487
  in gas-phase systems, 479–480
  interlaboratory comparison, 480
  microsuspension modification, 478–479
  motor vehicle exhaust
    diesel, 438, 439, 487, 492–493
    gasoline, 438, 492–493
  mutagenic potencies of nitroarenes and nitro-oxy PAC, 476, 479–483
  particle size and mutagenicity, 486
  principle and procedure, 475–478
  procedure diagram, 477
  strains used in, 476–478
  variability of, 480
Amines
  calculated lifetimes for OH attack, 221
  gas-phase reactions, 221–223
  hydroxyl radical reaction, 221–223
  measurement of, 566
  nitrosamines, see Nitrosamines
  nitrous acid reaction, 223
  sources, 221
Ammonia
  biological production, 286
  concentrations in troposphere, 583
  hydroxyl radical reaction, 286
  infrared spectrum, 551, 552
  interface species, 158
  measurement techniques, 551–552, 565, 581–583
  nitric acid reaction, 282–284, 286
Ammonium nitrate
  dissociation constant, 283–284
  formation from nitric acid reaction, 282
Anharmonic oscillator, 44–45

Aqueous phase reactions
  compilations of kinetics data, 173
  photochemistry, 316–317
  summary of, 315–322
Arctic
  bromine chemistry in, 242–244
  chlorine chemistry in, 241–244
  haze, 241
  ozone depletion, stratosphere, 696–700
  ozone depletion, surface level, 241
  typical pollutant concentrations in, 241
Aromatic hydrocarbons, see also Polycyclic aromatic hydrocarbons (PAH) and compounds (PAC)
  addition vs. abstraction by OH, 208
  ambient concentrations, 589–595
  chlorine atom reactions, 212
  hydroxyl radical reactions, 207–212
  measurement of, 565
  nitrate radical reaction, 212
  ring cleavage products, 209–212
Arrhenius equation
  and negative activation energies, 138–139
  for temperature-dependent rate constants, 138–139
Arsenic, in particles from dust storms, 393
Assay for bacterial mutagenicity, forward mutation, 483–484, 495–496
Assay for human cell mutagens, 484–486
Atmosphere, regions of, 2
Atomic force microscopy, of particles, 410
Atomizers, 633
Automobiles
  emissions of gases from, 899–907
  concentrations of pollutants inside, 858
  particles from, 923
  reactivity of emissions, 909–913
  standards for emissions, US and California, 903

B

$\bar{B}$, rotational constant characteristic of molecule, 45–46
b, total extinction coefficient, 365
Beer-Lambert absorption law, 53–55
Biogenic organics
  chemistry of, 231–234
  chlorine atom reactions, 205
  emissions of, 19, 225–231, 904
  hydroxyl radical reactions, 192
  nitrate radical reactions, 202
  oxygen-containing, 229–231
  ozone reactions, 196–199
  in particles, 393–396
  role in ozone formation, 904–907
Biomass burning
  composition of particles from, 393
  emissions from, 245–247
  HO_x production, 247
  nitric acid reactions on particles, 286
  phases of combustion, 245–246
  polycyclic aromatic hydrocarbons, 436
  remote sensing, 246
  uses, 244

Bisulfite ion, 301
Black lamps, 876
Boltzmann distribution, 46, 768
Bond dissociation energies, table of, see Appendix II
Broadening factor, of falloff curve for termolecular reactions, 136
Bromide ions, aqueous phase chemistry, 321–322
Bromine, Br_2
　absorption spectrum and photochemistry, 113–114
　compounds in troposphere, 242–244
Bromine chemistry
　in aqueous solution, 318–322
　in Arctic troposphere, 242–244
　in stratosphere, 701–706
Bromine monoxide
　absorption spectrum and photochemistry, 113–114
　dimethyl sulfide reaction, 333
　measurement of, 559
Bromine nitrate
　absorption spectrum, 110–112
　photochemistry, 113
Brownian diffusion of particles
　vs. gravitational settling, 364
　particle motion and, 363–364
Bubble apparatus for heterogeneous reactions, 168

C

C, Cunningham correction factor for particle settling velocity, 363
Cage effect in solution reactions, 152
Calibration aerosols, generation of, 632–635
Carbon
　black, elemental or graphitic, 367, 371, 373–375, 385, 623–625
　global cycle, 775
　light absorption by, 373–374, 789, 794–798
　light scattering by, 374–375
　measurement methods, 623–625
　structure, 374
　sulfur dioxide oxidation on, 324
Carbon bond mechanism, in chemical kinetic submodels, 887–892
Carbon dioxide
　aqueous phase equilibria, 295
　concentrations in troposphere, 773
　contribution to greenhouse effect, 763–766, 769–773
　cycling of, 775–776
　global warming potential, 785
　radiative forcing estimates, 785, 787
　sources of, 774–775
Carbon disulfide
　hydroxyl radical reaction, 335
　natural emissions, 335
Carbon monoxide
　alternate fuels and, 918–920
　from biogenic organic oxidations, 234
　concentrations in troposphere, 584
　emissions, discrepancies in, 899–907
　greenhouse effect and, 783
　hydroxyl radical reaction, 137–138
　linear rollback control method, 886
　measurement techniques, 565, 583–584
　from motor vehicles, 898–907
　sources, 20
Carbonate ion, chemistry, 295, 320
Carbonyl sulfide
　from dimethyl sulfide oxidation, 329
　hydroxyl radical reaction, 336
　sources, 335
Carboxylic acids
　concentrations in troposphere, 594
　hydroxyl radical reactions with, 216
　measurement of, 594
　in particles, 394–411
Cascade impactors, 610
Cavity ring down method for rate constant measurements, 147–148
Chapman cycle, in stratospheric chemistry, 2, 660–661
Chappuis bands in ozone absorption, 91
Chemiluminescence in measurement techniques, 548
CH_2OH, reaction with O_2, 194
Chloride ions, aqueous phase chemistry, 318–322
Chlorine atoms
　alkanes, reactions with, 184–185
　alkenes, reactions with, 205–206
　aromatic hydrocarbons, reactions with, 212–213
　concentrations, typical peak in troposphere, 328
　dimethyl sulfide reaction, 332–333
　oxygen-containing organics, reactions with, 214
　sources in troposphere, 180–181, 328
Chlorine chemistry
　in aqueous solutions, 318–322
　in Arctic troposphere, 241–244
Chlorine, molecular
　absorption spectrum and photochemistry, 113–114
　measurement of, 565–566
Chlorine monoxide
　absorption spectrum and photochemistry, 113–114
　dimer, 114, 678–680
　dimethyl sulfide reaction, 333
　measurement of, 559
Chlorine nitrate
　absorption spectrum, 110–112
　photochemistry, 112–113
　stratospheric chemistry, 669–680
Chlorofluorocarbons
　absorption cross sections of, 117, 122–124
　alternates for, 730, 744–753
　concentrations in troposphere, 733–736
　decomposition on surfaces, 671
　destruction in stratosphere, 671–673
　global warming potentials, 786–789
　greenhouse effect, 774, 783
　lifetimes, 670–671, 731–733
　numbering system for, 669
　ozone depletion potentials, 731–733
　phaseout protocols, 727–730
　radiative forcing by, 786–789
　sources and emissions, 669–670
Climate Change
　aerosol particles, direct effects on, 789–799
　aerosol particles, indirect effects on, 799–814
　carbon dioxide and, 763, 766–777
　carbon monoxide, 783
　chlorofluorocarbons and, 783
　clouds and, 799–814
　definition, 762
　feedbacks, 819–821
　global warming potentials of gases, 784–789
　hydrochlorofluorocarbons and, 783
　hydrofluorocarbons and, 783
　light absorption and, 794–799
　light scattering and, 790–799
　methane and, 766, 769–770, 777–779
　oceans and, 820, 822–823
　overview, 11–13
　ozone and, 763–766, 769–770, 774, 780–783
　radiation balance of earth, 763–766
　radiative forcing of gases, 783–789
　solar variability, 821–822
　sulfate particles, light scattering by, 790–796
　temperature trends, 823–828
　volcanic eruptions and, 822
　weather, compared to, 762
ClNO, see Nitrosyl chloride
ClOOCl
　absorption spectrum, 114, 118
　photochemistry, 114–115
　stratospheric chemistry, 678–680
Clouds
　absorption anomaly, 816–819
　absorption of infrared radiation by, 764, 815
　absorption of solar radiation by, 814–819
　actinic flux effects, 72–75
　albedo, 807
　cloud condensation nuclei and, 800–814
　drop size distributions, 808–813
　feedbacks in climate system, 819–821
　indirect effects of, 799–814
　Kelvin effect and, 801–802
　Köhler curves, 802–803
　number concentration of droplets, 808–813
　particle effects on, 800–814
　supersaturation in, 802
　susceptibility, 807–808
Coarse particles
　definition of, 354
　chemical composition of, 354–356, 381–385
Collision theory in gas-phase kinetics, 139–140
Complications of kinetics data, 172–173, see also Appendix IV

Concentrations
    conversion between units of, 33–34
Concentrations (*continued*)
    typical criteria pollutants in ambient air, 35–37
    typical noncriteria pollutants in ambient air, 35–37
Condensation methods, of aerosol generation, 635
Conversion, of units of
    concentration, *see* inside cover
    light absorption, coefficients, *see* inside cover
    rate constants, *see* inside cover
    energy, *see* inside cover
Cresols, nitrate radical reactions, 212
Criegee intermediate
    aldehydes, reaction with, 197
    in alkene-ozone reactions, 196
    carbon monoxide, reaction with, 201
    decomposition, 197–199
    hydroxyl radical production in, 197–199
    nitric oxide and nitrogen dioxide, reactions with, 210
    reaction rate constants, 201
    stabilization, 197–198
    sulfur dioxide, reaction with, 201, 299–300
    water, reaction with, 200–201
Cross sections
    collisional, 139
    reactive, 139
    light absorption, *see* absorption spectra and cross sections, *and individual compounds*
Cunningham correction factor, for gravitational settling of particles, 363

D

Deliquescence of salts in particles, 389–390
Denuders, 567–568
Deposition
    velocity of, 33
    wet *vs.* dry, 30–33
Diatomic molecules, energy transitions and spectroscopy, 43–49
Dicarboxylic acids in particles, 398–400
1,3-Dichloropropene, atmospheric lifetime and fate predictions, 928
Diesel exhaust, control technology and, 923–924; *see also* PAH and PAC
Differential optical absorption spectrometry (DOAS)
    analysis of spectra, 557
    basis of technique, 556–557
    detection limits, 559
    schematic diagram, 558
    typical spectra, 560–561
Diffuse reflectance infrared, Fourier transform spectroscopy, 171–172
Diffusion-controlled reactions
    in gas phase, 140, 152
    of uncharged nonpolar species in solution, 152–153

Diffusion denuder technique, 567–568
Diffusion separator, in particle size measurement, 617
Dimethyl disulfide
    hydroxyl radical reaction, 333–334
    natural emissions, 20–21
    nitrate radical reaction, 334
Dimethyl sulfide
    chorine atom reaction, 332–333
    global climate, role in, 328
    halogen oxides, reactions with, 333
    hydroxyl radical reaction, 328–332
    natural emissions, 20–21
    nitrate radical reaction, 332
    ozone reaction in aqueous solution, 333
Dimethyl sulfone, from hydroxyl radical–sulfur dioxide reaction, 329
Dimethyl sulfoxide
    from hydroxyl racial–sulfur dioxide reaction, 329
    hydroxyl radical, reaction with, 331
Dinitrogen pentoxide
    absorption spectra, 101
    alkenes, reactions with, 281
    equilibrium with NO_2 and NO_3, 267
    hydrolysis in troposphere, 279–280
    indoors, 860
    measurement of, 552
    photochemistry, 101
    reaction with water, 279–280
    sea salt, reations with, 280–281
    sodium halides, reactions with, 280–281
    thermal decomposition, 279
Dinitrogen tetroxide
    absorption spectra, 96
    equilibrium with NO_2, 95
    role in HONO formation, 270
Dinitrogen trioxide, as intermediate in nitrous acid formation, 272
Dissociation energies, bond, *see* Appendix II
Diterpenes, 226
DOAS, *see* Differential optical absorption spectrometry
Dobson units, 657
Droplet diameters, typical for fogs, clouds and rain, 308
Dry powder dispersion for particle generation, 634
Dust particles, 355–356, 381–382, 389, 391–392

E

Economics, 38
EKMA model, 892–893
Electronic transitions
    of diatomic molecules, 46–49
    of polyatomic molecules, 49–50
Electronically excited molecule
    energy level diagrams, 46–50
    intermolecular nonradiative processes, 50–51
    possible fates, 50–52

    primary and overall quantum yields, 51–52
    primary processes, 50–51
Electron microscopy of haze particles in Los Angeles area, 410
Electrophilic reactions, reactivity scale for reactions of polycyclic
    aromatic hydrocarbons, 506–507
Electrostatic precipitators, in particle collection, 611
Elementary reactions, definition, 130
Emissions
    anthropogenic, 15–26
    automobile emission standards in US and California, 903
    automobiles, discrepancies in emission inventories, 899–907
    biogenic organics, 225–231
    discrepancies in, 898–906
    factors, 16
    natural, 15–26
Energy
    levels and molecular absorption, 43–50
    transfer and collisional deactivation, 50–51
    typical regions, 53
    units commonly used, *see* inside cover
Enthalpies of formation, table of, *see* Appendix I,
Environmental chambers,
    advantages and limitations, 880–882
    "clean air" preparation, 876
    collapsible bags, 873–874
    contamination problems, 882
    effects of added formaldehyde, 879
    evacuable type, 875–876
    glass reactors, 873
    heterogeneous reactions in, 873
    hydroxyl radical concentrations in, 880–882
    light sources, 876–877
    nitrogen oxides in, 264–265, 274
    PAN in, 264–265
    particle growth in, 378
    reactant preparation, 876
    schematic diagrams of, 874–875
    solar simulators in, 877
    surface reactions in, 882
    types of, 872–876
    typical concentration-time profiles in, 878–880

F

F, broadening factor of falloff curve, 136
Factor analysis for source apportionment, 387
Falling droplet apparatus for heterogeneous reactions, 167–168
Falloff region, for termolecular rate constants, 135–136
Fast flow systems for rate constant measurements, 142–145
Fenton reaction, for hydroxyl production, 316

Filters
  ammonia monitoring using, 567
  mechanisms of action, 609
  nitric acid monitoring, 567
  nylon filters in nitric acid monitoring, 567
  in particle collection, 608–610
  sampling with, 567
  types and characteristics, 609
Fine particles
  chemical composition of, 356–357, 384–412
  definition of, 354–356
Flash photolysis systems for rate constant determinations, 145–146
Fluorescence
  definition of, 50
  in measurement techniques, 548
Fluorine
  chemistry in stratosphere, 671–673, 746–750
  fates of fluorinated products of hydrochlorofluorocarbon and hydrofluorocarbon oxidations, 750–753
  F_2, absorption spectrum and photochemistry, 113–114
  tetrafluoromethane, 736, 789
Fogs
  acid, 323–324
  health effects and, 324
  smog-fog-smog cycle, 323–324
Formaldehyde
  absorption spectra, 107–109
  in aqueous solution, 303–305
  chlorine atom reaction, 214
  concentrations in troposphere, 592–593
  as a hazardous air pollutant (HAP), 928
  Henry's law constant, 304
  hydration constant, 304
  hydroxyl radical reaction, 214
  hydroperoxy radical reaction, 216–217
  measurement techniques for, 552, 589–592
  nitrate radical reaction, 214
  nitric acid reaction, 241
  photochemistry of, 107–109
  sulfur dioxide reaction, aqueous phase, 303
  as a toxic air contaminant (TAC), 928
Formic acid
  in acid deposition, 326–327
  concentrations in troposphere, 326
  indoors, 857, 861
  measurement of, 552
  sources, 327
Fourier transform infrared spectroscopy in atmospheric studies, 551–556
"Fourth phase" in atmospheric chemistry, 164–165

G

γ, Activity coefficient, 154
Gaia hypothesis, 1
Gas
  compressed natural, 919–920
  liquefied petroleum, 920

Gasoline
  alcohol blends, 920–921
  reformulated, 918
Gas filter correlation spectroscopy, 554–555
Gas-particle partitioning, 406, 412–423
Gas phase kinetics, compilations of kinetics data, 172–173
Geometric standard deviation in log-normal distributions, definition of, 360
Geometric mean diameter of particles, 360–362
Global warming potentials
  absolute, 784–785
  definition, 784
  direct and indirect, 786–787
  gases, values for some, 785–789
  relative, 785
Glass UV light filters, 879
Gravitation settling of particles, 362–363
Greenhouse effect
  basis of, 763–766
  contribution of gases to, 766, 770–774
Growth of particles, mechanisms for, 375–380

H

Hadley circulation, 659
Half-lives
  calculating, 133–134
  definition, 132
Halogens
  measurement of, 565–566, 568
  tropospheric chemistry, 241–244, 321–322
Halogen oxides
  absorption spectra, 113–114
  dimethyl sulfide reaction, 333
Halons
  concentrations in troposphere, 734
  lifetimes, 731–733
  global warming potentials, 787–788
  nomenclature, 701
  ozone depletion potentials, 731–733
  phaseout, 727–730
  sources, 701–702
  stratospheric chemistry, 702–706
Hartley bands in ozone absorption, 90
Hatch-Choate equations, for fine particle, 361
Hazardous air pollutants (HAPs), 925–930
Haze, 410, 803
HBr, see Hydrogen bromide
HCl, see Hydrogen chloride
Heats of formation, see Appendix I
Henry's law
  for gas-liquid equilibrium, 151, 268, 295
  values for some gases, 152, 296
Heterogeneous reactions
  analysis, 158–165
  and climate change, 814
  compilations of kinetics data, 172–173
  in stratosphere, 677, 688–696, 705–706
High speed civil transport, 662–667
HNO_3, see Nitric acid
HO_2, see Hydroperoxyl radical

HO_x aqueous phase chemistry, 318–322
H_2O_2, see Hydrogen peroxide,
HO_2NO, see Peroxynitrous acid
HO_2NO_2, see Peroxynitric acid
Huggins bands in ozone absorption, 91
Humidograph, 390–391
Hydrazines
  hydroxyl radical reaction, 223
  nitric acid, reactions with, 225
  nitrogen dioxide reactions with, 225
  ozone reactions, 223–224
  sources, 223
Hydrazinium nitrate aerosols, 225
Hydrochlorofluorocarbons (HCFC)
  concentrations in troposphere, 734–735
  definition, 729
  global warming potentials, 786–789
  greenhouse effect, 783
  hydroxyl radical reaction, 744–746, 749–750
  lifetimes, 731–732, 752
  ozone depletion potentials in stratosphere, 731–732
  ozone formation potentials in troposphere, 752
  radiative forcing by, 786–789
  tropospheric chemistry of HCFC-125, -123, and -141b, 748–753
Hydrofluorocarbons (HFC)
  concentrations in troposphere, 734–735
  definition, 729–730
  global warming potentials, 786–789
  greenhouse effect, 783
  hydroxyl radical reaction, 744–746
  lifetimes, 731–732, 752
  ozone depletion potentials in stratosphere, 731–732
  ozone formation potentials in troposphere, 782
  radiative forcing by, 786–789
  tropospheric chemistry of HFC-134a, 746–748, 750–753
Hydrogen bromide, absorption spectrum, 113
Hydrogen chloride
  absorption spectrum, 113
  heterogeneous chemistry in stratosphere, 677, 688–696
  measurement of in troposphere, 568
  rotation-vibration infrared spectrum, 47
  uptake onto ice, 686–687
  uptake into sulfuric acid-water-nitric acid mixtures, 687–688
Hydrogen fuels, 921
Hydrogen peroxide
  absorption cross sections, 107–108
  formation of, 235
  concentrations in troposphere, 597–598
  measurement techniques for, 555, 595–597
  in natural troposphere, 234–239
  photochemistry, 107
  S(IV) reaction in solution, 313–314, 326
  tunable diode laser spectrometry and, 555
Hydrogen sulfide, hydroxyl radical reaction with, 335

Hydroperoxides, organic
  absorption spectra and photochemistry, 107–108
  formation in natural troposphere, 234–241
  Henry's law constants, 296
  oxidation of S(IV), 314
Hydroperoxyl radical
  aldehyde reactions, 216–217
  alkylperoxy radical reactions, 186, 234–241
  aqueous solution reactions, 320
  concentrations in troposphere, 238–239, 606–607
  fates in remote troposphere, 235–239
  formaldehyde reaction, 216–217
  Henry's law constant, 296
  methyl peroxy radical reaction, 235–239
  measurement techniques for, 604–606
  nitrate radical reaction, 235
  nitric oxide reaction, 235–239
  remote troposphere, rates of formation, 236–237
  self-reaction, 235
  sources in gas phase, 180
  sources in aqueous phase, 317, 320
  from sulfur dioxide oxidation, 298
Hydroxyalkoxy radicals
  α-, reactions of, 194
  β-, reactions of, 193–194
Hydroxyl radical
  alcohol reactions, 216
  aldehyde reactions, 213–214
  alkane reactions, 182–184
  alkene reactions, 191–196
  alkyl nitrate and nitrite reactions, 221
  alkyne reactions, 206–207
  amine reactions, 221–222
  aqueous phase chemistry, 278
  aromatic hydrocarbon reactions, 207–211
  biogenics, reactions with, 230–233
  carboxylic acid reactions, 216
  concentrations in troposphere, 560, 603–604
  dimethyl sulfide reaction, 328–332
  Henry's law constant, 296
  hydrazine reactions, 223
  measurement methods for, 559, 598–603
  from nitrate radical reaction with liquid water, 278
  nitric acid reaction, 281–282
  nitric oxide reaction, 274
  with nitrogen-containing organics, 220–225
  nitrogen dioxide reaction, 266–267
  with oxygen-containing organics, 213–217
  reactivity scale, 907–908
  remote troposphere, rates of formation in, 235–237
  sources in aqueous phase, 315–322
  sources in troposphere in gas phase, 179–180
  sulfur dioxide reaction, 298–299
Hydroxymethane sulfonate, 331–332
Hygroscopic growth factor for particles, 372–373
Hypobromous acid, HOBr
  absorption spectrum and photochemistry, 115, 119, 120

stratospheric chemistry and, 701–706
tropospheric chemistry and, 242–244
Hypochlorous acid, HOCl
  absorption spectrum and photochemistry, 115, 119
  stratospheric chemistry and, 677, 686
  tropospheric chemistry and, 242–244
Hypoiodous acid, absorption spectrum and photochemistry, 115, 120

I

Ice core data, 778, 780, 808, 825–828
Impactors, in particle size measurement, 610–611
Index of refraction of typical atmospheric compounds, 366–367
Indoor air pollution
  aldehydes, formation from VOC-NO_x chemistry, 860–861
  carbon monoxide, 849–850
  dinitrogen pentoxide, 860
  nitrate radical, 860
  nitric acid, 847–848, 860
  nitrous acid, 847–849
  ozone, 859–861
  PAH, 863
  particles, concentrations and sources, 861–865
  pesticides, 851–852, 858
  radon, 844–846
  sulfur dioxide, 849–850
  volatile organic compounds
    aldehydes, 856–857, 860, 861
    from carpets, 853–856, 858–859
    chlorinated compounds, 854, 858
    classes, observed, 851
    concentrations, 852
    formaldehyde, 856–857
    ketones, 857, 860–861
    organic acids, 857, 861
    pesticides, 851–852, 858
    sources, 852–858
    from smoking, 858–859
Infrared absorption
  dependence on concentrations in atmosphere, 769–773
  by gases, 763–773
  by particles, 798–799
  by clouds, 764, 815
Infrared spectroscopy, see also Fourier transform infrared spectrometry, tunable diode laser spectrometry, nondispersive infrared, and matrix isolation spectroscopy
  measurement of gases, 551–556
  of particles, 399–400
Interface reactions, 158, 164–165, 269, 307, 422
Intermolecular nonradiative processes, 51
Internal conversion, of electronically excited molecules, 50
Intersystem crossing, in electronically excited molecules, 50

Intertropical convergence zone, 660
Iodine, I_2, absorption spectrum and photochemistry, 113–114
Iodine
  in sea salt particles, 383
  in stratosphere, 706–707, 752–753
Iodine monoxide
  absorption spectrum and photochemistry, 113–114
  dimethyl sulfide reaction, 333
Ionic strength of solutions
  in atmosphere, 155
  definition, 154
Iron, aqueous phase chemistry and photochemistry, 309–311, 316, 325
Irradiance, 62
Isomerization, of alkoxy radicals, 189–191
Isopleths, of ozone concentrations, 882–886, 915
Isoprene
  chemical structure, 226
  emission rates, 226–227
  hydroxyl radical reaction, 194–196
  nitrate radical reaction, 203–205
  nitrogen dioxide reaction, 206
  sources and emissions, 225–227

J

J, quantum number of rotational state, 45
Jablonski diagram, 50

K

Kelvin effect and particles, 801–802
Ketones
  absorption spectra, 107–110
  concentrations in troposphere, 592–593
  measurement of, 565–566, 589–592
  photochemistry, 107–110
  reactions with hydroxyl and nitrate radicals and chlorine atoms, 214
Kinetic data, complications of, 172–173; see also Appendix IV
Kinetic techniques
  aqueous phase reactions, 155–156
  gas phase reactions, 142–149
  heterogeneous reactions, 156–172
Knudsen cell reactor
  gas–solid studies, 165–166
  schematic diagram, 165
Köhler curves, 802–805
Koschmieder equation, for visibility reduction, 369

L

Lagrangian approach
  for airshed models
  for source apportionment, 387–388
Langmuir adsorption of semivolatile organics, 414
Lapse rate, 26–28

SUBJECT INDEX

Lead, as anthropogenic emission, 25
Leighton relationship, for nitric oxide, nitrogen dioxide, and ozone reactions, 265–266
Lifetimes
  calculating, 133–134
  definition, 132
  and half-lives of pollutants, 134
Light absorption
  basic relationships, 52–53
  by particles, 365, 373–375, 789
  visibility reduction and, 368–373
Light intensity measurements, 61–64, 73–76, 877–878
Light sources
  in environmental chambers, 876–877
  intensity measurements, 61–64, 73–76, 877–878
Linear rollback model, of pollutant control, 886
Liquid jet apparatus for studying heterogeneous reactions, 169
Liquid water contents of fogs, clouds, and rain, 308–323, 808, 809
Log-normal distribution applied to ambient particles, 358–362
"London" (sulfurous) smog, overview, 3–4
Los Angeles (photochemical) smog, overview, 408

M

Malathion, 928–930
Mass accommodation coefficients
  definition, 157
  values for some atmospheric gases, 307
Mass measurement of particles in air,
  beta ray attenuation, 613
  gravimetric methods, 612
  oscillating microbalance, 613
  piezoelectric microbalance, 613
Mass spectrometry
  atmospheric pressure ionization (API-MS), 562
  chemical ionization, 562
  detectors, 566–567
  ionization methods, 562–564
  laser photoionization, 562–564
  mass filters, 564–566
  particle measurements using, 626–632
  resonance enhanced multiphoton ionization (REMPI), 563
  sample inlets, 561
Matrix isolation
  electron spin resonance, 605–606
  infrared spectroscopy, 555–556
Meinel band emission, in ozone-alkene reactions, 198
Measurement methods
  acetaldehyde, 565
  acetone, 565
  alkylperoxy radicals, 604–607
  ammonia, 552, 565
  benzene, 565
  bromine monoxide, 565
  calibration standards for gases, 607–608
  calibration standards for particles, 632–635
  carbon monoxide, 565
  chemiluminescence, 548
  dinitrogen pentoxide, 551–552
  fluorescence, 548
  formaldehyde, 552, 559
  formic acid, 552
  hydrogen peroxide, 552, 555
  hydroperoxyl radical, 604–607
  hydroxyl radical, 559
  infrared spectroscopy, 549–556
  iodine monoxide, 559
  naphthalene, 565
  nitrate radical, 559
  nitric acid, 551–552
  nitric oxide, 548, 563–565
  nitrogen dioxide, 555, 559, 565
  nitrous acid, 552, 559
  nitrous oxide, 565
  organics, 585–595
  ozone, 547, 548, 559, 736–739
  particle composition, 619–632
  particle size, 608–619
  peroxyacetyl nitrate, 552
  satellite methods, 737–739
  sulfur dioxide, 548, 552, 559
  toluene, 565
Mesopause, 2, 3
Mesophere, 2
Meterology
  in models, 888, 894–897
  and tropospheric chemistry, 26–30
Methane
  in troposphere, 777–779
  global warming potential, 785
  ice core data, 778
  isotopic composition, 779
  radiative forcing estimates, 785
  sinks of, 182–184, 779
  sources of, 777
  in stratosphere, 672, 674, 703
  tetrafluoro-,
    sources and concentrations, 736
Methanesulfonic acid
  in Antarctica, 331–332
  from dimethyl sulfide oxidation, 331–332
Methanol
  biogenic emissions of, 230
  hydroxyl radical reaction, 214–216
  nitrogen dioxide reaction with, 272
2-methyl-3-butene-2-ol
  emissions of, 229
  hydroxyl radical reaction, 234
  nitrate radical reaction, 234
  ozone reaction, 234
Methyl bromide
  sources, 701–702
  stratospheric reactions, 702–706
Methyl iodide
  atmospheric fate, 707, 752
  photolysis, 125
  sources, 706
3-Methylfuran, from isoprene oxidation, 194–196
Methyl mercaptan
  hydroxyl radical reaction with, 334–335
  natural emissions, 21
  nitrate radical reaction, 335
Methyl nitrate
  absorption spectra, 105
  concentration in ambient air, 595
Methyl nitrite, from methanol reaction, 272
Methylperoxy radical
  concentrations in troposphere, 238–239
  reaction with HO_2, 235–239
  remote troposphere, rates of formation in, 235–239
Mie light scattering, 365–368
  forward direction of, 366
  intensity vs. angle for water droplets, 367
  intensity vs. particle size, 368
Mist chambers, 568
Models
  airshed, elements of, 888
  boundary conditions, 898
  box models, 892
  carbon bond (CB) mechanism, 887–892
  chemical kinetic submodels, 887–892
  differences in predictions of, 891
  emissions in, 898–907
  EKMA, 892–893
  Eulerian, 893–894
  empirical, 887
  initial conditions, 897–898
  Lagrangian, 893
  linear rollback, 886
  meteorology, effects on predictions, 894–897
  OZIPR, 10, 889–892, see Appendix III
  particles in, 906–907
  regional acid deposition model (RADM), 887–892
Molecular oxygen, see Oxygen
Molozonide, 196
Monoterpenes
  atmospheric oxidation mechanisms, 231–234
  chemical structures, 226
  emissions of, 225–231
  temperature effects on emissions, 228
Morse potential energy curve, 43–45
MOUDI impactor, 610
Multiple linear regression for source apportionment, 387
Multiple pass cells
  differential optical absorption using, 558
  Herriott configuration, 550–551
  White cells, 549–551
Multiplicity, of molecular state, 47
Mutagens, see Ames test, PAH, and PAC

N

Natural lifetime, see Lifetime
Nebulizers, in aerosol generation, 633–634
Nephelometer
  extinction coefficients measured using, 368
  in humidograph technique, 390
  schematic diagram, 368

Nitrate ion
  in ice cores, 808
  photochemistry of, 317
  trends in, 808, 915
Nitrate radical
  absorption spectrum and cross sections, 100–101
  aldehyde reactions, 213–214
  alkane reactions, 8, 184
  alkene reactions, 201–205
  alkyperoxy radical reactions, 187
  aqueous phase reactions, 277–279
  aromatic hydrocarbon reactions, 212
  biogenic reactions, 202, 231–234
  bromide ion, reaction with, 278
  concentrations in troposphere, 560, 580
  cresol reactions, 212
  differential optical absorption spectrometry, 556–560, 578–580
  dimethyl sulfide reaction, 332
  equilibrium with NO_2 and N_2O_5, 267–268
  formation in nitrogen dioxide oxidation, 8, 267
  hydroperoxyl radical reaction, 235
  indoors, 860
  lifetime in ambient air, 277
  mass accommodation coefficient, 307
  measurement of, 559, 579–580
  nitric oxide reaction, 276
  nitrogen dioxide reaction, 8, 267–268, 276
  oxygen-containing organics, reactions with, 214
  α-phellandrene reaction, 234
  photochemistry, 101
  α-pinene reaction, 234
  pinonaldehyde, reaction with, 234
  sodium bromide, reaction with, 279
  sodium chloride, reaction with, 279
  relative humidity, effect on lifetime, 277
  from sulfate radical anion–nitrate reaction, 277–278
  sulfur (IV), reaction with, 278
  terpinene, reaction with, 234
  thermal decomposition, 276
  water, uptake into and reaction with, 277–278
Nitrates
  alkylperoxy, thermal decomposition of, 217–220
  in coarse particles, 384–385
  in fine particles, 385
  infrared spectroscopy, 398
  organic
    absorption spectra and photochemistry, 102–106
    in ambient air, 399, 594–595
    formation in nitrate radical–alkene reactions, 201–205
    in urban aerosols, 399
Nitric acid
  absorption spectra, 98
  in acid rain and fogs, 294
  ammonia reaction, 282–284
  ammonium formate, reaction with, 286
  calcium carbonate, reaction with, 286
  dry deposition, 33
  diffusion denuder measurement method, 567–568
  from dinitrogen pentoxide hydrolysis, 279–280
  formaldehyde reaction, 241
  Fourier transform infrared spectrometry and, 550–552
  hydroxyl radical reaction, 281–282
  infrared spectrometry and, 551–552
  intercomparison studies of, in ambient air, 576–577
  measurement techniques, 551–552, 575–578
  nitrogen dioxide oxidation to, 272
  nitrous acid reaction, 241
  photochemistry, 98
  in smog chambers, 264–265, 274
  sodium chloride reaction, 284–285, 379, 383
  soot, reactions with, 285–286
  tunable dioxide laser spectrometry and, 552
Nitric oxide
  alkylperoxy radical reaction, 185–186
  annual global emissions, 17–18
  aqueous phase reactions, 268–269
  chemiluminescence monitoring with, 548
  concentrations in troposphere, 574–575
  Criegee intermediate reactions, 201
  hydroperoxyl radical reaction, 235
  hydroxyl radical reaction, 274
  Leighton relationship, 265–266
  measurement techniques, 548, 563–565, 569–570
  nitrogen dioxide and water reaction, 271–272
  potential energy curves, 564
  REMPI spectrum, 564
  thermal oxidation by O_2, 265
Nitriles, measurement of, 566
Nitrite ion
  hydroxyl radical reaction, 275
  oxidation during freezing, 275
  photolysis of, in aqueous solution, 317
Nitrites, organic, 221
Nitrogen dioxide
  in aqueous solution, 268–269
  absorption spectra, 95–96
  alcohols, reactions with, 272
  alkene reaction, 206, 272
  alkoxy radical reactions, 191
  alkylperoxy radical reactions, 187–188
  aqueous phase reactions, 268–269
  concentrations in the troposphere, 574–575
  dimer, see dinitrogen tetroxide
  heterogeneous reaction, with water, 269–272
  hydrogen atom abstraction from organics, 272
  hydroxyl radical reaction, 266–267
  interface reaction, 158, 269
  Leighton relationship, 265–266
  light intensity measurements using, 75–76, 877–878
  measurement of, 555, 559–565
  mineral oxides, reactions with, 273
  nitrate radical reaction, 267–268
  nitrous acid production by, 269–272
  oxidation to nitric acid, 266–269
  ozone reaction, 267
  photochemistry, 96–97
  sea salt particles, reaction with, 272–273
  sodium bromide, reaction with, 272–273
  sodium chloride, reaction with, 272–273, 383
  sulfur dioxide, reaction in aqueous solution, 315
  water, reaction with, 268–272
Nitrogen trioxide, see Nitrate radical
Nitrosamines
  carcinogenic, 223
  measurement of, 566
  photolysis, 223
  sources, 223
N-Nitrosamines, see Nitrosamines
Nitrosyl bromide, BrNO, 273
Nitrosyl chloride, ClNO
  absorption spectra, 117, 120, 121
  from nitrous acid–hydrogen chloride reaction, 275–276
  photochemistry, 117
  from reaction with NaCl, 273, 383
Nitrous acid
  absorption spectra, 99
  and aerosol particles, 271
  from allylic hydrogen atom abstraction, 272
  amine reaction, 223
  in aqueous solution, 268–270
  concentrations in troposphere, 581
  differential optical absorption spectrometry, 559
  direct emissions from combustion systems, 274
  from formaldehyde–nitric acid reaction, 241
  hydrogen chloride, reaction with, 275–276
  as hydroxyl radical source, 7, 273
  hydroxyl radical reaction, 274–275
  ice, uptake onto, 275
  indoor formation, 273
  mass accommodation coefficient, 271
  measurement of, 552, 559, 580–581
  nitric acid reaction, 241
  from nitric oxide–hydroxyl radical reaction, 274
  from nitrogen dioxide reaction
    heterogeneous reaction on surfaces, 269–271
    with hydroperoxyl radical, 274
    with nitric oxide, 271–272
    on soot, 271
  nitrous oxide formation from, 271, 275
  from peroxynitric acid, 274
  photochemistry, 8, 99
  photolysis of, 99, 275
  sources, 8
  sulfur dioxide reaction with, 314–315
  sulfuric acid solutions, uptake into, 275
  surface sinks and sources, 271

Nitrous oxide
  absorption spectra, 101–102
  biological emissions, 17–18, 779–780
  concentrations in troposphere, 780
  global warming potential, 785
  ice core data, 780
  isotopic composition, 780
  measurement techniques for, 565
  from nitrous acid, 271, 814
  photochemistry, 101–102
  radiative forcing estimate, 785
  stratospheric role, 661–662
Nitryl bromide ($BrNO_2$) and iodide (INO_2)
  formation, 280–281
Nitryl chloride, $ClNO_2$
  absorption spectrum, 117, 120, 121
  photochemistry, 117
  formation, 280–281, 383
NMHC, Nonmethane hydrocarbons, see Organic compounds
NMOC, Nonmethane organic compounds, see Organic compounds
NO, see Nitric oxide
NO_2, see Nitrogen dioxide
NO_3, see Nitrate radical
NO_x
  concentrations in troposphere, 574–575, 915
  definition, 4, 264
  measurement techniques for, 573–574
  in smog chambers, 264–265
NO_y
  concentrations in troposphere, 574–575
  definition, 264, 286
  measurement techniques for, 570–573
  "missing," 205, 286
  in smog chambers, 264–265
NO_z, definition, 916
N_2O, see Nitrous oxide
N_2O_3, see Dinitrogen trioxide
N_2O_4, see Dinitrogen tetroxide
N_2O_5, see Dinitrogen pentoxide
Nondispersive infrared spectroscopy, 554–555
Nucleation of particles, 375–378

O

O_3, see Ozone
OClO
  absorption spectrum, 115, 118
  photochemistry, 115
Oceans, and climate change, 822–828
Octanol-air partitioning coefficients, 420–422
OH, see Hydroxyl radical
Oil well plumes, particles from burning of, 393
Olefins, see Alkenes
Optical counters, in particle size measurement, 614
Optical microscopy for particle measurements, 614–616
Order, reaction, 131

Organic acids
  in acid deposition, 326–327
  in aerosol particles, 393–409
  concentrations, typical in troposphere, 326
  measurement methods for, 594
Organic compounds
  acronyms, 18–19
  concentrations in troposphere, 589, 591–595
  emissions of, discrepancies in, 899–907
  as film on particles, 407–410
  gas-particle distribution of, 412–423
  incremental reactivity, 910
  maximum incremental reactivity, 910–913
  measurement methods for, 585–595
  nitrogen-containing, reactions in troposphere, 217–225
  oxygen-containing, reactions in troposphere, 213–217
  photochemical ozone creation potential (POCP), 913
  reactivity adjustment factors, 911–913
  reactivity scales, 907–913
  sources, 18–20
  tropospheric lifetimes of, 181
  typical reactivity scales, 907–909
  volatile organic, 18–19
Organic films on particles, 407–411
Organic nitrates
  absorption spectra, 102–103
  photochemistry, 103
  see also Nitrates, organic
Organic hydroperoxides
  absorption spectra, 107–108
  aqueous phase S(IV) oxidation, 313–314
  photochemistry, 107
Oxidants, definition of, 547
Oxides of nitrogen, see NO_x and NO_y
Oxidizing species for organics in troposphere, 179–181
Oxygen
  absorption spectrum, 86–89
  collision-induced absorptions, 89
  photochemistry, 89–90
  singlet molecular, 88–89
  van der Waals molecules, 89
Oxygen-containing compounds
  biogenic emissions, 229–231
  hydroxyl radical reaction, 213–217
  nitrate radical reaction, 213–217
OZIPR Model, 13, see Appendix III
Ozone
  absorption cross sections, 92–93
  absorption spectra, 90–91
  actinic flux sensitivity to, 69–71
  alkene reactions
    indoors, 860–861
    kinetics and mechanisms, 196–201
  amines, reactions with, 221–223
  attenuation of solar radiation by, 56
  biogenic organics, reactions with, 232–233
  concentrations
    in stratosphere, 657–658, 736–741, 781
    in troposphere, 583, 780–782, 914
  control strategies for, 882–921

  "deficit" problem, 661
  depletion potentials, 730–733
  global warming potential, 785
  hydrazine reactions, 223–225
  indicator species for, 916–917
  indoors, 859–861
  interference in hydroxyl radical measurement, 600
  isopleths, 882–886, 915
  isotopic composition in atmosphere, 661
  Leighton relationship, 265–266
  measurement methods for, 547, 548, 559, 583, 737–739
  nitrogen dioxide reaction, 267
  NO_x, NO_y, NO_z, relationship to, 916–918
  photochemistry, 91–95
  production efficiency, 916
  radiative forcing estimates, 784–788
  in remote troposphere, formation and loss processes, 235–237
  stratospheric injection into troposphere, 658–660
  sources in troposphere, 180
  trends in concentrations
    in stratosphere, 736–741
    in troposphere, 780–783
    in upper troposphere, 239–241
  vertical distribution, 657–658
Ozone-olefin reactions; see Ozone-alkene reactions
Ozonides
  primary (see also Molozonide), 196
  secondary, 197

P

PAN, see peroxyacetyl nitrate and peroxyacyl nitrates
Particles; see also Coarse particles, Fine particles
  absorption of light by, 365, 373–375, 789
  accumulation range, 354
  Aitken nuclei range, 354
  aerodynamic diameter, 351
  artifacts in measurement of, 626
  attenuation of solar radiation by, 393–396
  biogenic organics in, 393–396
  Brownian diffusion, 363–365
  carbon, elemental, 373–375, 411–412
  carbonate in, 625
  chemical composition of, 380–412, 619–632
  chemical mass balance and, 386–388
  climate change
    direct effects on, 789–799
    indirect effects on, 799–814
  coagulation of, 378–379
  coarse, definition, sources and sinks, 354–356
  concentrations in troposphere, 618–619, 924–926
  condensation mode, 356
  condensation on particles, 378
  control of, 923–925
  crustal elements in, 381–382

Particles (continued)
  deliquescence of, 389–390
  deposition rates, 364, 379
  depth profiling of, 631–632
  diameters
    aerodynamic, 351
    range of, 349
    Stokes, 351
  diesel, 439
  droplet mode, 356
  dust, transported, 355–356
  effluorescence of, 389
  electrostatic effects, 364–365
  elemental carbon, 367, 371, 373–375, 385, 623–625
  enrichment factors, crustal, 381
  external vs. internal mixtures, 357–358, 375, 388–391
  factor analysis and, 387
  fine particles, definition, sources and sinks, 354–358
  see also Fine particles,
  gas-particle partitioning, 406, 412–423
  gas reactions with, 379–380
  generation of, 632–636
  geometric mean diameters,
    types of, 359–360
    determination of, 361
  geometric standard deviation of size distribution, 360
  gravitational settling, 362–363
  growth of, 375–380
  Hatch-Choate equations for, 361
  health effects of, 22–23
  heterogeneous condensation, 378
  humidograph, 390
  hygroscopicity, 410–412
  hygroscopic growth factor, 372–373
  indoors, 861–865
  inorganic composition, 381–393
  inorganics, measurement of, 619–623
  ionic strength of, 155
  Lagrangian analysis and, 387–388
  light scattering and absorption by, 365–375, 789–799
  light scattering coefficients, 371–373
  mass spectrometry and, 626–632
  measurement methods for, 608–632
  mechanisms of formation, 375–380
  metals in, 358, 381–382, 385
  Mie light scattering by, 365–368
  in models, treatment of, 906–907
  modes, 351–358
  motion of, 362–365
  multimodal nature of, 351–358
  multiple linear regression and, 387
  mutagenicity of, see PAH and PAC
  nephelometer for measurements of, 368, 390
  nitrate in, 384–386
  nucleation and, 375–378
  number, mass, surface, and volume distributions, 351–358
  organic vs. elemental carbon, in particles, 411–412
  organic films on, 407–410
  organics in nonurban particles, 393–396
  organics in anthropogenically influenced and aged particles, 396–407
  primary and secondary, definition, 349
  real time monitoring techniques for, 626–632
  refractive indexes for components of, 367
  sea salt, 382–384
  secondary organic aerosols, 397–407
  semivolatile organics
    gas-particle partitioning, 412–423
    yields of, 406
  sensitivity of actinic flux to, 71–72
  settling velocities of, 362–365
  size distributions, 351–358
  size ranges, 351–358
  source apportionment, models of, 386–388
  sources of, 21–25, 354–358
  Stokes diameter, 351
  Stokes law, 362
  surfactants in, 407–410
  tracers of, 386
  ultrafine particles
    concentrations in polluted air masses, 358
    definition, 354
  visibility reduction and, 368–373
  water and light scattering, 372–373
  water uptake and evaporation, 410–412
Particle size
  calibration aerosols in measurement methods, 632–635
Particulate matter, see Particles
Partitioning coefficients, gas-particle
  into liquids, 417–420
  on solids, 413–417
Peroxides
  concentrations in troposphere, 597–598
  measurement methods for, 595–597
  organic, absorption spectra and photochemistry, 107–108
  oxidation of S(IV), 313–314, 326
Peroxyacetyl nitrate (PAN)
  absorption spectra, 103
  from acetaldehyde oxidation, 217
  concentrations in troposphere, 571, 594–595
  in biomass combustion, 217
  formation of, 217–218
  hydroxyl radical reaction, 220
  infrared spectrometry and, 217
  measurement techniques, 552, 594–595
  photochemistry, 103
  photolysis rate, 103, 220
  in smog chambers, 264–265
  thermal decomposition, 141, 218–220
Peroxyacetyl radical reaction with NO, 219
Peroxyacyl nitrates, PANs
  concentrations in troposphere, 594–595
  from isoprene oxidation, 218
  some common atmospheric species, 217–218
  thermal decomposition of, 219–220
  measurement methods for, 594–595

Peroxynitric acid
  absorption spectrum, 100
  photochemistry, 100
Pesticides
  atmospheric lifetimes and fate predictions, 928–930
  indoors, 851–852, 858
  measurement of, 566
Phosgene
  as product of tropospheric oxidations, 749–750
  tropospheric fate, 751
Phosphorescence, definition, 50
Photochemical air pollution, 4–8
Photochemical processes
  types of primary, 51
  primary vs. overall quantum yields, 51–52
Photochemical smog, history of, 4–8
Photochemistry, see also individual listings for compounds
  in aqueous solutions, 315–317
  basic relationships, 50–55
Photolysis
  calculation procedure for rates of, 61–64, 76–81
  rate constant, k_p, 61, 76–81
Photophysical processes
  of electronically excited molecules, 50
Photophysical transitions, Jablonski diagram, 50
Pinonaldehyde
  in air, 232
  nitrate radical reaction with, 234
  from α-pinene reactions, 232–234
Pinonic acids in air, 232
Polycyclic aromatic compounds, see PAH and PAC
  definitions of, 439
  chemical classes, 439
Polycyclic aromatic hydrocarbons (PAH) and polycyclic aromatic compounds (PAC)
  air quality and emissions standards, 466–467
  in airborne particulate matter, 436
  in ambient air
    PAH, 437–438, 443–444, 453–461, 493–494, 496, 497–502
    N-PAC, 445–447
    O-PAC, 448–449, 460–461
    S-PAC, 450, 461
  analysis of, spectroscopic, 461–462, 464
  artifacts in sampling and analysis, 458–460
  assay for bacterial mutagenicity, the Ames test, 475–480, 486–495
    accuracy and precision, 480
    activatible mutagens (promutagens), definition of, 476
    advantages, 475–483
    for airborne particle-bound mutagens, 475–483, 486–495
    bioassay-directed fractionation and chemical analysis, 479, 482
    chemical information from, 476–478
    direct mutagens (-S9), definition, 476
    for gas phase mutagens, 479–480

microsuspension modification, 478–479
mutagenicity of motor vehicle exhaust, diesel and gasoline, 487
mutagenic potencies of nitroarenes and nitro-oxy PACs, 476, 479–483
particle size and mutagenicity, 486
principles and procedures, 475–478
procedure diagram, 477
strains used in, 476–478
variability of, 480
assay for bacterial mutagenicity, forward mutation, 483–484, 495–496
assays for human cell mutagens, 484–486
atmospheric lifetimes (gas-phase), 527
benzo[a]pyrene, see also PAHs and PACs
air quality and emissions standards, 466–467
ambient levels, 455–461, 466
carcinogenicity, 466–475
historical, 440, 466–467
mutagenicity, bacterial, 481
mutagenicity, human cell, 484–485, 499–500
potency equivalence factors, 466–475
reactivity, 505–507
spectrum, UV/visible absorption and emission, 463, 465
structure, 436, 438
transport and transformation decay, 507–509
bioassay-directed fractionation and chemical analysis
bacterial assay, 479, 482, 484, 502–503
human cell assay, 497–500
cancer potencies and potency equivalence factors (PEF), 467–474
carcinogenicity, 466–475
PAH, 468–470
N-PAC, 471–474
concentrations in ambient air (gas and particle phases), 455–461, 467, 472, 473–474, 485
cyclopenta[c,d]pyrene, see also PAH and PAC
ambient levels, 455, 457, 469, 485, 497–500
carcinogenicity, 469
mutagenicity, human cell, 484–485, 499–500
potency equivalence factor, 469
reactivity, 505–508
spectrum, UV/visible absorption, 464
structure, 443
transport and transformations, decay, 507–509
emission controls and standards, impacts of, 500–501
emissions from
combustion sources, 436–439, 492
biomass burning, 436
coal fly ash, 438, 492–493
diesel engine exhaust, 438, 439, 492–493
environmental tobacco smoke, 438
fuel oil, 492–493
gasoline engine and exhaust, 438, 492–493
indoor air, 438
municipal incinerators, 438
wood smoke, 436, 438
fluorescence spectrometry, 461–462
formation of, 438
forward mutation bacterial assay, 483–484, 495–496
gas–particle partitioning, 415–423, 436, 453–461
gas phase reactions with OH, NO_3, and O_3 in simulated atmospheres, 519–527
gas phase reactions, rate constants for reactions with OH, NO_3, and O_3 and photolysis, 524
historical background, 466–467
human cell mutagens in ambient air, 484–486, 497–502, 509
indicated hydrogen, 441
long-range transport of, 439
molecular weights of PAH, 437–438, 443–444
molecular weights of N-, O-, S- PAC, 445–447, 448–449, 450
mutagenicity potencies of PAH and PAC, see Ames test
bacterial, 475–484
human cell, 484–486
mutagenicity of ambient air, 486–504
bacterial, 486–496
human cell, 497–500
effects of different emissions on, 492–494
factors influencing, 488–491
impacts of location, transport and atmospheric reactions, 493–494, 507–509
structure and reactivity, 505–507
nitration reactions in simulated atmospheres
particle phase NO_2/HNO_3, 515–517
gas phase, OH and NO_3 reactions, 520–527
N-PACs in ambient air, 445–447
nitroarenes
in ambient air, 446–447, 474, 503, 520–523
carcinogenicities and potency equivalency factors, 471–472, 474
emissions of, 519–520
photochemical decomposition, gas phase, 527
photochemical decomposition, particle phase, 518–519
nitro- and nitro-oxy derivatives, mutagenicity of, 476, 479–483
nomenclature, 437, 438, 440–450
O-PACs in ambient air, 448–449
ozonolysis in particle phase, 513–515
particle phase mutagenic nitro-PAH and nitro- O-PACs in ambient air, 503, 520–523
particle size distribution of PAH and mutagenicity, 487–488
photochemical reactions, particle associated PAH, 510–515
photoreactivity, structural effects, 511–513
photoreactivity, substrate effects, 510, 512–513
photodecomposition of nitro-derivatives, 518–519
photooxidation of, 510–514
physical properties of, 437–438, 443–444, 454, 455
potency equivalency factors, carcinogenicity, 467–474
reactions in simulated and ambient atmospheres
gas phase, 439, 509, 523–527
particle phase, 505–509, 513–518
reactivity scale and classification for electrophilic reactions, 506–507
risk assessments, 467–473, 501–502, 509
S-PACs in ambient air, 450
sampling methods, 454–461
singlet molecular oxygen reactions, 510–514
solid sorbents for, 456–460
solubilities in water of, 451, 454
sources, 436–438, 491
spectra
infrared, 466
UV/visible absorption and emission, 461–466
standard reference material, SRM 1649 air particles, 450, 464, 484
structures and nomenclature, 437–438, 440–453
transport and transformations
formation of bacterial mutagens, 493–495
formation of human cell mutagens, 497
decay of PAHs, 507–509
formation of nitroarenes and nitrolactones, 493–495, 508–509
US EPA priority pollutants, 436–438
vapor/particle phase partitioning of direct mutagens in ambient air, 502–503
vapor phase mutagens in ambient air, 502–504, 523, 525–527
vapor pressures, 451, 454, 455
Polycyclic organic matter, POM
definition of, 436
composition, 439
in diesel exhaust, 439
Pre-exponential factor, in rate constant, 138–139
Pressure dependence, of termolecular reaction rates, 135–136
Primary ozonide (Molozonide), 196
Primary pollutants, 15, 349
Pulse radiolysis
for studying gas-phase kinetics, 146–147
in solution phase reactions, 155–156
Pyrex filters, transmission spectra of, 879

Q

Quantum numbers
    molecular, 46–48
    rotation, 45
    vibration, 43–44
Quantum yields, 51–52

R

Radiance, 62
Radiation
    actinic, 55
    amplification factor, 741
    balance of incoming and outgoing, 763–766
    infrared absorption, dependence on gas concentrations, 769–773
    infrared emission by earth's surface and gases, 764–773
    thermal emission of earth, 764–766
    transfer processes in atmosphere, 766–769
    ultraviolet, at earth's surface, 741–766
Radiative forcing
    adjusted, 784
    clouds and, 799–814
    definition, 783
    gases, values for, 784–788, 813
    instantaneous, 783
    temperature change, relationship to, 784
    particles
        direct effects, 790–799, 813
        indirect effects, 799–814
Radiative transitions, of electronically excited molecules, 50
Radon, 844–846
Rates, reaction, definition of, 131
Rate constants
    compilations of, 171–172, and *see* Appendix IV
    conversion factors for, 132
    determination of, 141–156, 744–746
    first-, second-, and third-order, 131–132
    temperature dependence of, 138–141
    units of, 132
Rate laws, for elementary or overall reactions, 131
Rayleigh scattering of light, 58, 369
Reaction order, definition, 131
Reactions on surfaces, kinetics of, 158–172
Reactions in solution
    charged species, 153–155
    diffusion-controlled, 152
    experimental techniques, 155–156
    gas-phase species, comparison to, 152
    uncharged nonpolar species, 152–153
Reactivity
    adjustment factors for mobile source emissions, 910–913
    application to mobile sources, 909–913
    for electrophilic reactions of polycyclic aromatic hydrocarbons, 506–507
    incremental, 910
    maximum incremental, 910–913
    for organic compounds, 907–913
Relative rate constants, 149–150

Remote troposphere
    Arctic chemistry, 241–244
    biogenic organics
        emissions of, 225–231
        chemistry of, 231–234
    boundary layer chemistry, 234–239
    nitrogen oxides in, 9, 225
    overview, 9
    upper troposphere, chemistry of, 239–241
Resistance model, for heterogeneous reactions, 158–165
Respiratory tract, particulate matter and, 22–23
Response time for species emitted into atmosphere, 774
Risk assessments, 38–39, 467–473, 501–502, 509, 925

S

S(IV), definition, 301
S(VI), definition, 301
Salts, deliquescence points of inorganics, 389–390
Satellite measurement methods, 737–739
Schumann-Runge system, of O_2, 86–89
Scrubbers, use in measurements, 568
Sea salt
    and climate, 810–811
    cloud condensation nuclei and, 811
    coarse particle composition, 382–383
    gas-solid reactions, 284–285, 379, 383
    nitrogen dioxide, reaction with, 272–273, 383
Sea water composition, 383
Secondary pollutants, 15, 349
Sedimentation collectors, 611–612
Single particle optical counters, 614
Size measurement of particles
    aerodynamic particle sizer, 617
    atomic force microscopy, 615–616
    cascade impactors, 610
    condensation particle counter, 617–618
    diffusion separator, 617
    electrical mobility analyzer, 616–617
    generation of calibration aerosols, 632–636
    optical methods of, 614
    scanning electron microscopy, 614–615
    scanning tunneling microscopy, 615–616
    single particle optical counters, 614
Skin cancer, estimated increases from ozone depletion, 744
Smog-fog-smog cycles, 323–324
Smog chambers, *see* Environmental Chambers
Sodium chloride
    chlorine nitrate reaction, 686
    dinitrogen pentoxide reaction, 280–281
    nitric acid reaction, 284–285, 379, 383
    nitrogen dioxide reaction, 272–273, 383
    sulfuric acid reaction, 383
Solar radiation, *see also* Actinic flux
    actinic, 55
    attenuation by atmosphere, 55–60

Beer-Lambert law, 58
    at earth's surface, 65, 70–71
    intensity, measurement of, 61, 64, 73–76, 877–878
    striking a volume of air, 60
    outside atmosphere, 55–56
    variability, 821–822
Solar simulators, *see also* Light sources
    in environmental chambers, 877
    spectral distributions and intensities of, 877
Solar intensity variability, 821–822
Solar zenith angle
    definition, 56–57
    calculation of, 57
Solid rocket motors, effects of emission, 667–669
Solvent cage effect, in solution kinetics, 152
Soot
    nitric acid reaction, 286
    nitrogen dioxide reaction, 271
    nitrous acid formation from, 271
    sulfur dioxide oxidation on, 324–325
Source apportionment models, 386–388
Space shuttle, effects of emissions on stratosphere, 667–669
SSTs, effect on ozone in stratosphere, 662–667; *see also* High speed civil transport
Stefan Boltzmann Law, 764–765
Static techniques for determining rate constants, 148–149
Stokes diameter, 351
Stokes law, applied to particles, 362
Stratosphere, 2
Stratosphere-troposphere exchange, 658–660
Stratopause, 2, 3
Stratospheric chemistry
    bromine reactions in, 701–706
    bromine-chlorine interactions, 673–674, 705
    Chapman cycle, 660–661
    chlorofluorocarbons and, 669–680
    fluorine chemistry, 671–673
    heterogeneous chemistry and, 677, 686–700, 705–706
    high speed civil transport and, 662–667
    in Arctic, 696–700
    iodine in, 707–707
    mid-latitude chemistry, 700–701
    nitrous oxide in, 661–662
    ozone "hole," 675–680
    ozone injection into troposphere, 658–660
    particles in, 680–682, 690–696
    polar stratospheric clouds, 677, 680–690, 697–700
    protocols for protection of, 727–730
    solid rocket motors, 667–669
    space shuttle, 667–669
    volcanic eruptions, effects on, 690–700
Sulfates
    in coarse particles, 384–385
    in fine particles, 385–393
    in ice cores, 808
    trends in, 808, 915

Sulfur compounds, reduced
  emissions of, 20–21
  atmospheric reactions, 332–334
Sulfur dioxide
  absorption spectrum, 103–105
  aldehydes, formation of complexes with, 303–305
  aqueous phase reactions
    equilibria, 301–303
    concentrations, effect of pH on, 302
    kinetics, effects of pH on, 302–303
    size distribution in particles, effects on, 356–358, 380
    steps, chemical and physical in oxidation, 306–308
  characteristic times for diffusion and oxidation, 306
  Criegee biradical reaction, 201, 299–300
  concentrations of in troposphere, 585
  droplet size in fogs and clouds, effect on oxidation of, 322–323
  fluorescent measurement of, 548
  fog-smog-fog cycles and, 323–324
  formaldehyde reactions, 303
  free radical oxidation in aqueous systems, 315–322
  halogen compounds, oxidation by in aqueous solution, 318–322, 811
  Henry's law constant, 296
  hydrogen peroxide, oxidation by, 305, 313–314
  hydroxyl radical reaction
    in gas phase, 298–299
    in aqueous phase, 317–318
  interface reaction, 158, 269, 307
  mass accommodation coefficient, 307
  measurement of, 548, 552, 559, 584–585
  metal-catalyzed oxidation, 309–311, 325
  nitrate radical reaction, 277–278, 318
  nitrogen oxide reactions in aqueous solution, 314–315
  nitrous acid reaction in solution, 314–315
  organic peroxide oxidation, 314
  oxidation by O_2, catalyzed and uncatalyzed in solution, 308–311
  oxidation rates in troposphere, 296–298
  oxidation rates, calculating from kinetics, 300–301, 308–309
  ozone reaction in aqueous systems, 311–313
  photochemistry, 103–105
  relative importance of oxidation mechanisms, 325–326
  schematic of gas aqueous transfer, 306
  sea salt particles, oxidation in, 811
  soot, reaction on, 324
  sources, 20–21
  surface reactions in tropospheric oxidation, 324–325
  tracers to follow oxidation, 297
  trends, 915
  uncatalyzed oxidation in solution, 308–309
Sulfur hexafluoride
  concentrations in troposphere, 736
  global warming potential, 789
  radiative forcing by, 789
  sources, 736
Sulfuric acid
  control of, 922–923
  vs. nitric acid, 297
  sodium chloride reaction, 383
  vapor pressures, 297–298
Sulfur trioxide, absorption spectrum, 105, 107
Sun lamps, 876–877
Superoxide, in aqueous phase atmospheric chemistry, 315–322
Supersaturation, 802
Supersonic aircraft, see High speed civil transport
Surface albedo, 60
Surface reaction probability, 157
Surfactants, in aerosols, 407–410
Surface science techniques for studying heterogeneous reactions, 171–172

T

TAC, see Toxic air contaminants
Teflon film, ultraviolet transmission spectrum of, 873–874
Temperature
  effects on kinetics, 138–141
  from ice core data, 827
  inversions, 28–30
  potential, 28
  trends, 823–825
Termolecular reactions
  and pressure dependence of rates, 133–138
  temperature dependence of, 138–139
Ternary solutions in polar stratospheric clouds, 681–686
Terpenes, measurement of, 566
Tetraterpenes, 226
Thermal inversion, 28–30
Thermograph, for particle measurement, 391
Total carbon, measurement methods, 623–625
Toxic air contaminants (TAC), 925–930
Transient nuclei, see Aitken nuclei
Transition flow reactors, 568
Transition metal ions, in aqueous phase reactions, 312–322
Transition state theory, 140–141
Tropical pipe, 659
Tropopause, 2, 3
Troposphere, 2
Tube furnace for particle generation, 634
Tunable diode laser spectrometry
  application to atmospheric studies, 554, 901, 903
  basis of technique, 552–554
  detection limits, 552
Turnover time, for species emitted into atmosphere, 774
Type I polar stratospheric clouds, 682–690
Type II polar stratospheric clouds, 683–690

U

Upper troposphere
  hydroxyl radical in, 240
  oxides of nitrogen, 240–241
  sources of HO_x radicals, 240
Ultrasonic nebulizer, 633
Ultraviolet radiation
  trends in, at surface, 741–744
  UV-A, UV-B, UV-C, 55
US Clean Air Act Amendments of 1990, 436
US Environmental Protection Agency "Priority PAH," 436–438

V

V_g, deposition velocity, 33
Vapor pressure
  subcooled and gas-particle partition coefficient, 419–420
  subcooled of PAH, 415
Vibrating orifice aerosol generator, 634
Virtual impactors, 611
Visibility and air pollution, 37–38, 369
Visual range
  definition, 368
  light scattering and absorption, 368–369
  measurement of, 369
Volcanic eruptions
  climate change, 793–794, 822
  particles, 690–693
  stratospheric ozone, 690–696
  temperature changes, 794

W

Water content, fogs, clouds and rain, 308
Wet deposition, see Deposition
White cell, schematic diagram, 549
  see also Multiple pass cells

X

X-ray photoelectron spectroscopy, use in studying heterogeneous reactions, 171–173
Xenon lamps, as a light source in environmental chambers, 877

Z

Zenith angles, definition, 56–57